WORLD DIRECTORY OF CRYSTALLOGRAPHERS

AND OF OTHER SCIENTISTS

EMPLOYING CRYSTALLOGRAPHIC METHODS

EIGHTH EDITION

1990

GENERAL EDITOR

E. N. MASLEN

ASSOCIATE EDITOR

A. L. BEDNOWITZ

PUBLISHED FOR THE

INTERNATIONAL UNION OF CRYSTALLOGRAPHY

BY

SPRINGER-SCIENCE+BUSINESS MEDIA, B.V.

The text of this directory was formatted and prepared
on a computer controlled photocomposer printer

through the courtesy of

International Business Machines Corporation
Thomas J. Watson Research Center
Yorktown Heights, NY, U.S.A.

TABLE OF CONTENTS

Table of Contents

PREFACE TO THE EIGHTH EDITION

A brief historical account of the background leading to the publication of the first four editions of the *World Directory of Crystallographers* was presented by G. Boom in his preface to the Fourth Edition, published late in 1971. That edition was produced by traditional typesetting methods from compilations of biographical data prepared by national Sub-Editors. The major effort required to produce a directory by manual methods provided the impetus to use computer techniques for the Fifth Edition. The account of the production of the first computer assisted Directory was described by S.C. Abrahams in the preface of the Fifth Edition.

Computer composition, which required a machine readable data base, offered several major advantages. The choice of typeface and range of characters was flexible. Corrections and additions to the data base were rapid and, once established, it was hoped updating for future editions would be simple and inexpensive. The data base was put to other Union uses, such as preparation of mailing labels and formulation of lists of crystallographers with specified common fields of interest.

The Fifth Edition of the *World Directory of Crystallographers* was published in June 1977, the Sixth in May 1981 and the Seventh in May 1986. The Subject Indexes for the Fifth and Sixth Editions were printed in 1978 and 1981 respectively, both having a limited distribution. The Subject Index for the Seventh Edition was begun, but never completed.

Beginning in 1988, Sub-Editors of the Seventh edition were solicited for their assistance in producing the Eighth Edition. Detailed instructions were provided concerning submission of their entries. Sub-Editors were invited to commence immediately the process of contacting the members of their national crystallographic communities. Very nearly all the original Sub-Editors graciously agreed to undertake the task of updating their entries. In order to provide continuity for future Editions an attempt was made to obtain a Sub-Editor for every country listed.

In the Fourth quarter of 1988 the first updated entries began to arrive. Efforts were made to ensure that all countries were given the opportunity of crystallographic representation in the Eighth Edition. Two additions and one deletion were made to the list of included countries. In some cases there was a loss of contact with previous correspondent Sub-Editors. Where a contact could not be established, the entry from the Seventh Edition was used in order to maintain continuity of its representation in the Directory. Individual crystallographers inadvertently omitted from this edition are cordially invited to contact the General Editor for inclusion within the next edition.

The bulk of all biographical data was received in due course, although several countries did not provide their data until late in 1989. After all the entries were received and collected into manageable size files then were sent to Yorktown Heights for composition. Proofs were produced and examined by the Editors for correction. This process continued until July 1990 at which time the Name Index compilation was begun and the entries checked for duplications and other errors. The production of camera-ready prints commenced in October 1990.

The Name Index permits the country in which a listing appears to be easily identified. In cases where a listing is incomplete the information given reflects all that was available. In order to maintain the accuracy of each entry the Editors have faithfully tried to reproduce the information furnished to them. Changes were made, however, in the Fields of Interest section, in order to enhance the content of the planned Subject Index.

Errors in this edition are bound to be present. They may be eliminated from the permanent data base now established, with the cooperation of those who note them, if brought to the attention of the General or Associate Editor.

It is a great pleasure to thank the IUCr Executive Secretary Dr. J. N. King, General Secretary/Treasurer Prof. A. I. Hordvik and President Prof. A. Authier for their continued support and, with appreciation, to recognize all the Sub-Editors, without whose generous and excellent cooperation neither this nor any other edition of the *World Directory of Crystallographers* would be possible.

October 30, 1990

E. N. Maslen, General Editor
Crystallography Centre
University of Western Australia
Nedlands, Western Australia 6009, Australia

A. L. Bednowitz, Associate Editor
IBM T.J. Watson Research Center
Yorktown Heights, NY 10598-0218
U.S.A.

HOW IT WAS DONE

The text of this directory was prepared by a photocomposer printer at the IBM Thomas J. Watson Research Center in Yorktown Heights, New York. The Virtual Machine/Conversational Monitor System (VM/CMS), executive language (REX), and text-editing (XEDIT) facilities were used in combination to insert the appropriate (SCRIPT) control symbols for columnar format, headings and diacritical marks.

The addition of diacritical marks required an extensive control symbology. The procedure requested of the Sub-Editors, in order to indicate letters with diacritical marks, uses the SCRIPT Set Symbol facility (eg. "&ea." for the symbol "é"). with the IBM Document Composition Facility (GML - SCRIPT). The largest data set handled was the USA list with 21,456 lines of text.

Final proof readings were done by the General and Associate Editor before composing the finished product for offset printing. Although many typographical errors were encountered and corrected from the raw data it is expected that many are still embedded in the final text even though it has gone through several stages of proof reading. Even with the powerful assistance of a computerized editing facility there are still classes of error and omission which continue to be unavoidable in this kind of compilation. One serious problem encountered was with the automatic hyphenation of words at the end of a line. In some cases, especially in the handling of non-English words, oddities occurred.

EXPLANATORY NOTES

1. WHO IS INCLUDED?

In deciding who should be included in the Eighth Edition of the *World Directory of Crystallographers,* each Sub-Editor was invited to use his own judgement, aided by the general guidelines developed in previous Editions. Table 1 gives the number of entries included for each country in the last three editions. A country is defined in accordance with ICSU statute 8 from which we quote: "A national member adheres to the council either through its principal scientific academy, or its national research council, or any other institution or association of institutions. Such an institution effectively representing the independent scientific activity in a definite territory may be accepted as a national member, provided it can be listed under a name that will avoid any misunderstanding about the territory represented." In accordance with this the words "country" and "territory" are used synonymously in this directory.

Crystallography has been taken in its broadest sense as represented, for example, in the programs of the International Congresses of Crystallography. A crystallographer was recognized as a scientist with an active interest in crystallography, either for its own sake or for the contribution it could make to some other branch of science. It was clearly understood that the Seventh Edition would be of most value if it included all crystallographers. To help ensure none would inadvertently be omitted, it was hoped that Sub-Editors would enroll the aid of all who had previously been listed in their sections. Sub-Editors were provided with a listing of all entries appearing in the Seventh Edition for their section.

Generally, only crystallographers who returned completed Data Entry Forms are included in this edition. An exception was made by including all U.S.A. members of the American Crystallographic Association, as their current addresses were available to their Sub-Editor and their inclusion was regarded as useful. In addition, in those countries where updates were not readily available the entries from the Seventh Edition were included in an attempt to provide continuity.

The total number of entries has increased by 7 percent compared with the previous edition while the total number of represented countries has increased by one. As expected, changes have occurred in the crystallographic population of a number of countries, some increasing considerably with a few decreasing significantly.

2. FORMAT AND ARRANGEMENT OF THE INFORMATION

a. *Alphabetical order*

As in previous editions, the sections of this Eighth Edition are arranged in alphabetical order by countries, and by individuals within each country. Prefixes were handled differently depending on whether they are capitalized in spelling. For example, in **De Camp** with a capital "D", the prefix is given preceding the name which is hence placed in the alphabetical "D" group, whereas in **van der Meer** the uncapitalized prefixes follow the name as **Meer, van der** and this is placed in the alphabetical "M" group. Names that contain diacritical marks are generally handled as if there were no such marks, unless a Sub-Editor indicated a different practice for that country.

b. *Contents of the biographical entries*

Complete individual entries contain:

   (i)   family name, followed by title and given names
   (ii)   full institutional or correspondence address
   (iii)   year of birth, in parentheses
   (iv)   highest degree and field in which the degree was granted
   (v)   university or institution granting degree,
           country if different from that of address in (ii),
           and year degree was granted, all in parentheses
   (vi)   present position
   (vii)   telephone number, telex code, fax number and/or email address in parentheses
   (viii) major fields of scientific interest in the form of Keywords.

The name of each crystallographer listed is contained in the Name Index, together with the country in which the name is listed. Cross references are hence unnecessary, and have not been used.

Explanatory *Notes* precede most country listings, with information provided by the Sub-Editor. Variations in addresses for the same institution may be found: each is as provided by the individual, and is presumed to be acceptable by postal authorities.

The availability of direct dialing between many countries has led to provision of international telephone country codes in the *Notes* for most countries. In any given country, additional leading digits may be necessary: details for some countries are given in their *Notes*.

Table 1. *Number of Listings by Country*

Country	8th edition 1990	7th edition 1986	6th edition 1981	5th edition 1977	Country	8th edition 1990	7th edition 1986	6th edition 1981	5th edition 1977
Algeria	-	-	2	2	Libya	10	4	-	-
Argentina	49	44	28	29	Malaysia	30	30	20	15
Australia	192	197	200	170	Mexico	45	42	39	19
Austria	97	86	90	73	Netherlands	163	163	166	174
Bangladesh	22	22	25	23	New Zealand	46	42	49	65
Belgium	68	86	87	68	Nigeria	9	9	9	3
Bolivia	13	11	12	10	Norway	60	68	69	75
Brazil	86	81	108	109	Pakistan	64	60	31	29
Bulgaria	83	83	74	68	Panama	-	-	-	1
Burma	6	6	6	4	Peru	11	11	11	4
Canada	170	203	165	153	Philippines	14	13	8	4
Chile	43	44	43	42	Poland	175	150	153	136
China	245	223	10	-	Portugal	24	20	20	20
Colombia	30	30	28	22	Romania	31	31	32	30
Cuba	19	12	-	-	Saudi Arabia	16	14	10	9
Czechoslovakia	116	103	91	82	Singapore	12	10	6	4
Denmark	80	71	62	62	South Africa	74	98	80	72
Egypt, Arab. Rep.	71	71	77	77	Spain	167	113	113	87
Ethiopia	-	-	-	2	Sri Lanka	2	3	5	4
Finland	108	109	118	119	Sudan	2	2	2	1
France	180	312	348	320	Sweden	180	206	206	207
German Dem. Rep.	347	374	376	377	Switzerland	116	121	133	132
Germany, Fed. Rep.	650	589	455	428	Syrian Arab Rep.	2	2	-	1
Ghana	2	2	8	4	Taiwan	42	30	25	16
Greece	88	88	82	72	Tanzania	-	1	-	-
Hong Kong	6	7	6	4	Thailand	44	36	28	20
Hungary	94	74	71	63	Tunisia	37	32	18	4
Iceland	5	5	5	5	Turkey	79	84	60	27
India	350	370	340	318	Uganda	-	-	-	2
Indonesia	1	13	13	4	USSR	907	697	657	595
Iran	40	20	25	23	UK	636	732	651	598
Iraq	6	6	6	2	USA	2044	1621	1622	1642
Ireland	9	8	5	-	Uruguay	6	3	3	4
Israel	66	79	71	56	Venezuela	18	13	10	7
Italy	373	344	276	248	Vietnam	21	-	-	-
Ivory Coast	8	10	8	6	West Indies	-	-	-	2
Japan	595	565	516	446	Yugoslavia	154	138	128	118
Kenya	-	-	4	2	Zambia	3	-	-	-
Korea	26	17	15	11	Zimbabwe	-	-	2	5

Totals: 5th edition (1977): 71 countries, 7638 entries.
6th edition (1981): 68 countries, 8174 entries.
7th edition (1986): 69 countries, 8968 entries.
8th edition (1990): 70 countries, 9589 entries.

3. LANGUAGE

The language used throughout the bulk of this edition is English. However, addresses, degrees and positions are often given in the national language: in such cases, an explanation may generally be found in the *Notes*.

4. ABBREVIATIONS

Numerous abbreviations have been used in this edition, following the practice of previous editions. Common abbreviations are found in Table 2. Abbreviations peculiar to a given country are explained in their *Notes*. It should be noted that some of the above abbreviations are also valid in other languages, e.g. Ave, Blvd, Dept., Dr, Lab., Labs., Prof., Tel., and U.

Table 2. *English abbreviations*

AB, BA	Bachelor of Arts	Eng.	Engineer or Engineering
Acad. Sci.	Academy of Sciences	Esp.	Especially
AM	Master of Arts	Est.	Establishment
Alt. Addr.	Alternate Address	Ext.	Extension
Appl.	Applied	Fac.	Faculty
Apt.	Apartment	Fax	Facsimile
Assoc.	Associate	Grad.	Graduate
Asst.	Assistant	Inc.	Incorporated
Ave	Avenue	Inst.	Institute
BAgrSc	Bachelor of Agricultural Sciences	Instr.	Instructor
BASc	Bachelor of Applied Science	Jr.	Junior
BEE	Bachelor of Electrical Engineering	Lab.	Laboratory
Bldg.	Building	Labs.	Laboratories
Blvd.	Boulevard	Lect.	Lecturer
BMet	Bachelor of Metallurgy	Ltd.	Limited
BMetE	Bachelor of Metallurgical Engineering	MA	Master of Arts
BRD	Bundesrepublik Deutschland	Mbr.	Member
BSc	Bachelor of Science	MMet	Master of Metallurgy
C.	College	MD	Doctor of Medicine
Co.	Company	MSc, ScM	Master of Science
Coord.	Coordinator	Nat.	National
Corp.	Corporation	PhD	Doctor of Philosophy
CSSR	Czecho-Slovak Socialist Republic	P.O.	Post Office
Dept.	Department	Prof.	Professor
Dev.	Development	Rep.	Republic
DDR	Deutsche Demokratische Republik	Res.	Research or Researcher
DDS	Doctor of Dental Surgery	S., Sch.	School
DEng	Doctor of Engineering	Sci.	Science or Sciences
Dev.	Development	Scient.	Scientific or Scientist
Dir.	Director	Sr.	Senior
Div.	Division	St.	Saint or Street
DPharm	Doctor of Pharmacy	Techn.	Technical or Technology
DPhil	Doctor of Philosophy	Tel.	Telephone
Dr	Doctor	Th.	Theoretical
DSc, ScD	Doctor of Science	U.	University
Em.	Emeritus	UK	United Kingdom
Email	Electronic Mail	USA	United States of America
Emb.	Embankment	USSR	Union of Soviet Socialist Republics

ARGENTINA

Sub-Editor: **P.V. König di Perazzo**

Notes

1. International telephone country code - 54

2. Degrees confered by Argentine universities are *Doctor*(Dr) (aproximately equivalent to PhD), *Ingeniero*(Ing) (between PhD and MSc), and *Licenciado*(Lic) (aproximately equivalent to MSc).

3. In the list the following abbrevations have been used:
   CINDECA - Centro de investigación y Desarrollo en Procesos Catalíticos,
   CITEFA - Centro de Investigaciones Técnicas de las Fuerzas Armadas
   CONICET - Consejo Nacional de Investigaciones Científicas y Técnicas
   CNEA - Comisión Nacional de Energía Atómica
   LEMIT - Laboratorio de Ensayo de Materiales de Interés Tecnológico.
   PRINSO - Program of Research in Solid State Physics

   UNBA - Universidad Nacional de Buenos Aires
   UN Córdoba - Universidad Nacional de Córdoba
   UN Cuyo - Universidad Nacional de Cuyo
   UNLP - Universidad Nacional de La Plata
   UN del Sur - Universidad Nacional del Sur

Acuña, Dr Rodolfo José Gerencia de Investigaciones y Dessarrollo, Aluar, Aluminio Argentino S.A.I.C., Cangallo 525, Buenos Aires, Argentina. (1941) Dr, physics (UN Córdoba, 1971). Res. scient., crystallography lab. (tel. 01 + 493236). *Metal and alloy structures, X-ray diffraction.*

Alvarez, Dr Alberto Guillermo. CINDECA, 47 Nro 257, C.C. 544, La Plata, Buenos Aires 1900, Argentina. (1931) Dr, physics (UN Cuyo, 1960). Sr. scient. CONICET (tel. 021 + 21353 / 210711 / 240915). *X-ray crystallography, X-ray fluorescence.*

Baggio, Dr Ricardo. Dept. de Física, CNEA, Av. del Libertador 8250, Buenos Aires 1429, Argentina. (1946) Dr, physics (UNBA, 1975). Res. scient. (tel. 01 + 707711, ext. 337). *Inorganic and organic crystal structures, X-ray diffraction.*

Baggio, Dr Sergio. Gerencia de Investigaciones y Desarrollo, Aluar, Aluminio Argentino S.A.I.C. Cangallo 525, Buenos Aires, Argentina. (1940) Dr, chemistry (UNBA, 1964). Head, crystallography lab. (tel. 01 + 493236). *Crystal structure, powder methods, X-ray fluorescence.*

Benyacar, de, Lic María Angélica Rodríguez. Dept. de Física, CNEA, Av. del Libertador 8250, Buenos Aires 1429, Argentina. (1928) Lic, chemistry (UNBA, 1952). Head, crystallography div. (tel. 01 + 707711, ext. 337). *Crystal structure, physical properties - structure relationships.*

Bengochea, Dr Amado Leandro. Dept. de Geología, UN del Sur, Avda. Alem 1253, Bahía Blanca, Buenos Aires 8000, Argentina. (1945) Dr, geochemistry (UN del Sur, 1976). Asst. prof., res. scient., CONICET, (tel. 25196, ext. 354). *Geochemistry, X-ray powder diffraction, fluid inclusions, Lithogeochemistry.*

Bianchet, Dr Mario Antonio. Dept. de Física, UNLP, 115 y 49, La Plata, Buenos Aires 1900, Argentina. (1958) Dr, physics (UNLP, 1988). Ass. Prof. (tel. 021 + 39061, telex 31216 CESLA AR). *X-ray diffraction, solid state kinetics, inorganic and organic crystal structures.*

Bonetto, Miss Rita Dominga., CINDECA, Facultad de Ciencias Exactas, Dto. de Fisica, UNLP, 47 Nro 257, La Plata,Buenos Aires 1900, Argentina. (1951) Lic, physics (UN de Cordoba, 1976). Res. ass. CONICET (tel. 021 + 0711/1353/0915). *Powder diffraction, particle size, zeolites.*

Canepa, Dr Horacio Ricardo. PRINSO, CONICET - CITEFA, Zufriategui 4380, Villa Martelli, Buenos Aires 1603, Argentina. (1950) Dr, solid state physics (U. de Rennes, France, 1983). Res. scient. (tel. 01 + 7610031, ext. 212). *Crystal growth, polycrystals, semiconductors IV-VI. defects.*

Casanova, Lic Jorge Ramón. PRINSO, CONICET - CITEFA, Zufriategui 4380, Villa Martelli, Buenos Aires 1603, Argentina. (1944) Lic., physics (UNBA, 1979). Res. scient. (tel. 01 + 7610331, ext. 240). *Crystal growth, X-ray topography, X-ray diffraction, intercalation compounds.*

Cortelezzi, Dr César Rafael. Dept. de Geología, LEMIT, 52 121 y 122, La Plata, Buenos Aires 1900, Argentina. (1926) Dr, geology (UNLP, 1952). Head, res. dept. (tel. 021 + 31141). *Mineralogy, petrography, borate structures.*

Diodati, Dr Francisco Piero. Dept. de Física, Facultad de Ingeniería, UNBA, Paseo Colón 850, Buenos Aires 1063, Argentina. (1940) Dr, physics (UNLP, 1970). Res. scient. (tel. 01 + 346441, ext. 178). *Electron diffraction, gases, solid lasers.*

Dristas, Dr Jorge Anastasio. Dept. de Geología, UN del Sur, Avda. Alem 1253, Bahía Blanca, Buenos Aires 8000, Argentina. (1944) Dr, economic geology - mineralogy (UN del Sur, 1972). Assoc. prof., (tel. 362 + 091285). *X-ray powder diffraction, electron diffraction.*

Fernandez, Ing Juan Carlos. Dept. de Física, Facultad de Ingeniería, UNBA, Paseo Colón 850, Buenos Aires 1063, Argentina. (1951) Ing., electrical engineering (UNBA, 1974). Asst. prof. (tel. 01 + 346441, ext. 156). *Surface science, LEED, electron spectroscopy, Auger spectroscopy, surface crystallography, adsorbates, epitaxy.*

Figlas, Lic Norma Debora. Laboratorio de Biologia Molecular, Instituto de Investigaciones Bioquimicas de Bahia Blanca, CONICET, UN del Sur, C.C. 87,

Bahia Blanca 8000, Argentina. (1961) Lic, computer science (UN del Sur, 1988). Research fellow, (tel. 91 + 31342 ext. 230, telex 81712 DUJOR, fax 54-91-27276) *Protein crystallography, crystallographic computing.*

Gabelli, Lic Sandra Beatriz Laboratorio de Biologia Molecular, Instituto de Investigaciones Bioquimicas de Bahia Blanca, CONICET, UN del Sur, C.C. 87, Bahia Blanca 8000, Argentina. (1964) Lic, computer science (UN del Sur, 1988). Research fellow, (tel. 91 + 31342 ext. 230, telex 81712 DUJOR, fax 54-91-27276) *Protein crystallography, crystallographic computing.*

Gay, Dra Hebe Dina. Dept. de Geología, Facultad de Ciencias Exactas y Naturales, UN Córdoba., Velez Sarfield 299, Córdoba 5000, Argentina. (1927) Dr, geology (UN Córdoba, 1950). Full prof. (tel. 716131). *Crystallography, minerals.*

Guérin, Dr Diego Marcelo Alejandro. Laboratorio de Biologia Molecular, Instituto de Investigaciones Bioquimicas de Bahia Blanca, CONICET, UN del Sur, C.C. 857, Bahia Blanca 8000, Argentina. (1955) Dr, physics (UNLP, 1985). Res. scient. (tel. 91 + 31342/32741 ext. 230) *Protein crystallography, theory and methods in crystallography.*

Heredia, Lic Eduardo Armando. Dept. de Quimica, CITEFA, Zufriategui 4380, Buenos Aires 1661, Argentina. (1959) Lic., physics (UNBA, 1989). Res. scient. (tel. 01 + 7610231 ext. 231). *Crystal growth.*

Herren, Lic Gustavo Guillermo. Div. de Investigacion en Fisica del Solido, CITEFA, Zufriategui 4380, Villa Martelli, Buenos Aires 1603, Argentina. (1951) Lic., physics (UNBA, 1981). Res. scient. CONICET, (tel. 01 + 7610031/0081 ext. 158, telex 27057) *Inorganic crystals, Group theory.*

Hermida, Lic Jorge Daniel. Dept. de Materiales, CNEA, Av. del Libertador 8250, Buenos Aires 1429, Argentina. (1946) Lic., physics (UNBA, 1971). Res scient. (tel. 01 + 7550181, ext. 268, telex 18101 CAC AR). *X-ray diffraction, deformed structures.*

Iñiguez Rodríguez, Dr Adrián Mario. Centro de Investigacion Geologica, 1 Nro 644 La Plata, Buenos Aires 1900, Argentina. (1937) Dr, natural science (UNLP, 1962). Prof. *Mineralogy, clays and mixed layers, natural zeolites.*

Ipohorski Lenkiewicz, Dr Miguel. Dept. de Metalurgia, CNEA, Av. del Libertador 8250, Buenos Aires 1429, Argentina. (1939) Dr, physics (UN Cuyo, 1967). Res. scient. (tel. 01 + 7550181, ext. 282). *Electron microscopy, metals, metal physics.*

König di Perazzo, Lic Patricia Verónica. Dept. de Física, CNEA, Av. del Libertador 8250, Buenos Aires 1429, Argentina. (1941) Lic, physics (UNBA, 1965). Res. scient. (tel. 01 + 707711, ext. 337). *Inorganic crystal structures.*

Lovey, Dr Francisco Carlos. Div. Metales, CNEA, Centro Atomico Bariloche, San Carlos de Bariloche, Rio Negro 8400, Argentina. (1949) Dr, physics (UN Cuyo, 1981). *Phase transformations, metal and alloy structures, X-ray diffraction, electron microscopy.*

Levi, Dra Laura. Dept. de Física, CNEA, Av. del Libertador 8250, Buenos Aires 1429, Argentina. (1915) Dr, physics (U. Bologna, Italy, 1937). Res. scient. (tel. 01 + 707711, ext. 337). *Solid state, nucleation, cloud physics.*

Mahr von Staszewski, Dr Guillermo. PRINSO, CONICET - CITEFA, Zufriategui 4380, Villa Martelli, Buenos Aires 1603, Argentina. (1949) Dr, physics (UNBA, 1983). Co-Dir., PRINSO. (tel. 01 + 7610331, ext. 212). *Crystal growth.*

Maiza, Dr Pedro José. Dept. de Geología, UN del Sur, Avda. Alem 1253, Bahía Blanca, Buenos Aires 8000, Argentina. (1943) Dr, ore deposits (UN del Sur, 1972). Assoc. prof., res. scient., Argentine Res. Council. (tel. 25196, ext. 254). *Geochemistry, mineralogy, ore geology, X-ray powder diffraction.*

Manghi, Lic Estela Margarita. Dept. de Física, CNEA, Avda. del Libertador 8250, Buenos Aires 1429, Argentina. (1940) Lic, physics (UNBA, 1964). Res. scient.

(tel. 01 + 707711, ext. 337). *Solid state physics, X-ray topography, scanning electron microscopy.*

Marbec, Lic Ema Rosa. Dept. de Química, Instituto Nacional de Tecnología Industrial, CC 157, San Martin, Buenos Aires 1650, Argentina. (1939) Lic, chemistry (UNBA, 1966). Head, X-ray diffraction lab. (tel. 01 + 7556161, ext. 76). *X-ray fluorescence, metals, quantitative analysis, X-ray diffraction, iron oxides, uranyl compounds, organic acids.*

Mas, Dr Graciela Raquel. Dept. de Geología, UN del Sur, Avda. Alem 1253, Bahía Blanca, Buenos Aires 8000, Argentina. (1948) Dr, mineralogy (UN del Sur, 1976). Asst. prof., res. scient., CONICET,(tel. 91 + 25196, ext. 362). *Zeolite synthesis, X-ray powder diffraction, clay mineralogy.*

Nöllmann, Lic Ida. Dept. de Quimica, CITEFA, Zufriategui4380, Buenos Aires 1661, Argentina. (1941) Lic., physics (UNBA, 1966). Res. scient. (tel. 01 + 7610231, ext. 212). *Crystal growth.*

Padin, Lic Zulma Ernestina. Laboratorio de Biologia Molecular, Instituto de Investigaciones Bioquimicas de Bahia Blanca, CONICET, UN del Sur, C.C. 857, Bahia Blanca 8000, Argentina. (1944) Lic, computer science (UN del Sur, 1975) Professor. (tel. 91 + 31342/32741 ext. 230) *Protein crystallography.*

Piro, Dr Oscar Enrique. Dept. de Física, Facultad de Ciencias Exactas, UNLP, 49 y 115, C.C. 67 La Plata, Buenos Aires 1900, Argentina. (1944) Dr, physics (UNLP, 1977). Adjoint Prof. UNLP, Res. scient. CONICET. (tel. 021 + 39061, telex 31216 CESLA AR). *Crystal structure, electronic & vibrational properties, optical properties.*

Pochettino, Dr Alberto Antonio. Dept. de Metalurgia, CNEA, Av. del Libertador 8250, Buenos Aires 1429, Argentina. (1947) Dr, physics (UNLP, 1978). Res. scient. (tel. 01 + 7550181, ext. 282). *Electron microscopy, metals.*

Punte, Dr Graciela. Dept. de Física, UNLP, C.C. 67, La Plata, Buenos Aires 1900, Argentina. (1944) Dr, physics (UNLP, 1972). Sr. scient., CONICET (tel. 021 + 39061/241706, telex 31216 CESLA AR). *Inorganic and organic crystal structures, X-ray diffraction.*

Rigotti, Dr Graciela. Dept. de Física, UNLP, C.C. 67, La Plata, Buenos Aires 1900, Argentina. (1948) Dr, chemistry (UNLP, 1979). Sr. scient., CONICET, Buenos Aires. (tel. 021 + 39061). *Inorganic and organic crystal structures, X-ray diffraction, solid state kinetics.*

Rivero, Dr Blas Eduardo. Dept. de Física, UNLP, C.C. 67, La Plata, Buenos Aires 1900, Argentina. (1942) Dr, physics (UNLP, 1976). Sr. res. scient., CICPBA (tel. 021 + 39061/241706, telex 31216 CESLA AR). *Organic and inorganic crystal structures.*

Schmirgeld, Dr Lelia. Dept. de Física, CNEA, Av. del Libertador 8250, Buenos Aires 1429, Argentina. (1940) PhD (U. Warwick, UK, 1977). Res. scient. (tel. 01 + 707711, ext. 337). *Electron microscopy, physical properties.*

Silva, Dr Abelardo M. Laboratorio de Biologia Molecular, Instituto de Investigaciones Bioquimicas de Bahia Blanca, CONICET, UN del Sur, C.C. 857,

Bahia Blanca 8000, Argentina. (1948) PhD, physics (UNLP, 1978). Res. scient., CONICET. (tel. 91 + 31342/32741 ext. 230). *Protein crystallography, crystallographic theory and methods.*

Spinelli, Dr Silvia Haydeé Laboratorio de Biologia Molecular, Instituto de Investigaciones Bioquimicas de Bahia Blanca, CONICET, UN del Sur, C.C. 857, Bahia Blanca 8000, Argentina. (1954) Dr., physics (UNLP, 1985). Prof. (tel. 91 + 31342/32741 ext. 230). *Protein crystallography, crystallography theory and methods.*

Trigubo, Lic Alicia Beatriz. Div. de Investigacion en Fisica del Solido, CITEFA, Zufriategui 4380, Villa Martelli, Buenos Aires 1603, Argentina. (1944) Master, mat. science (USC, 1981). Res. scient. (tel. 01 + 7610131/0031, ext. 212) *Crystal growth, defect structures, X-ray diffraction, semiconductors, Electron microscopy, electron diffraction, crystallographic data, material science, microscopy.*

Vega, Lic Daniel Roberto. Dept. de Fisica, CNEA, Avda. del Libertador 8250, Buenos Aires 1429, Argentina. (1961) Lic, physics (UNBA, 1987). Res. ass. (tel. 01 + 707711, ext. 337). *Inorganic and organic crystal structures, physical properties, phase transformations.*

Versaci, Dr Raul Antonio. Dept. de Materiales, CNEA, Av. del Libertador 8250, Buenos Aires 1429, Argentina. (1945) Dr, physics (UNLP, 1979). Res. scient. (tel. 01 + 7550181, ext. 282, telex 18101 CAC AR). *Metals and alloys, structures, electron microscopy.*

Vidal, Lic Haydée Marta. Dept. de Química, Instituto Nacional de Tecnología Industrial, CC 157, San Martin, Buenos Aires 1650, Argentina. (1938) Lic, chemistry (UNBA, 1970). Res. scient. (tel.01 + 7556161, ext. 76). *Orientation, polymer fibers (natural and synthetic), powder methods.*

Viturro, Lic Hector Ruben. CINDECA, Dept. de Física, Facultad de Ciencias Exactas, UNLP, 47 Nro 257, La Plata, Buenos Aires 1900, Argentina. (1954) Lic, physics (UNLP, 1980). Ass. res. (tel. 021 + 39061). *Powder diffraction, molecular graphics, zeolites.*

Walsoe de Reca, Dr Elizabeth Noemí. PRINSO, CONICET - CITEFA, Zufriategui 4380, Villa Martelli, Buenos Aires 1603, Argentina. (1939) Dr, chemistry (UNBA, 1965). Dir., PRINSO. (tel. 01 + 7610081/7610031, ext. 158/145). *TEM, SEM, X-ray diffraction, X-ray topography.*

Zalba, Dr. Patricia Eugenia. Div. Geología, CETMIT, Cno Centenario 508 y 511, Gonnet, Argentina. (1944) Dr, earth science (UNLP, 1976). Head, mineralogy and petrography section. (tel 021 + 840247). *Quantitative analysis, X-ray techniques, minerals.*

Zimmerman, Lic Rosa. Dept. de Física del Sólido, CITEFA, Zufrategui y Varela, Villa Martelli, Buenos Aires 1603, Argentina. (1928) Lic, chemistry (UNBA, 1953). Res. scient. (tel. 01 + 7610131, ext. 156). *Thin film structure and properties, microelectronics.*

AUSTRALIA

Sub-Editor: A. W. Stevenson

Notes

1. International telephone country code - 61.

2. The abbreviation CSIRO is used for the Commonwealth Scientific and Industrial Research Organization.

Anstis, Dr Geoffrey Richard. Dept. of Appl. Physics, U. of Techn., Sydney, P.O. Box 123, Broadway, New South Wales 2007, Australia. (1949) PhD, mathematical physics (U. Adelaide, 1975). Lect. (tel. 02 + 218-9928, fax 02 + 281-2498, email granstis at utscsd.csd.uts.oz). *Electron diffraction theory, electron imaging theory.*

Avey, Dr Hugh Philip. Resource Materials Centre, Darling Downs Inst. of Advanced Education, P.O. Darling Heights, Queensland 4350, Australia. (1934) PhD, crystallography (U. London, UK, 1967). Head. (tel. 076 + 31-2462, telex 40010, fax 076 + 30-1182, email KEYLINK 07:ddn004). *Instrumentation, crystallographic methods, structure and function, biological molecules, information science.*

Bailey, Mr David Eric. School of Sci. and Techn., Nepean C. of Advanced Education, Second Ave., Kingswood, New South Wales 2750, Australia. MSc, physics (U. Western Australia, 1965). Sr. lect. (tel. 047 + 36-0429, fax 047 + 36-4186). *X-ray physics.*

Baker, Dr Anthony Thomas. Dept. of Chemistry, U. of Techn., Sydney, P.O. Box 123, Broadway, New South Wales 2007, Australia. (1954) PhD, chemistry (U. New South Wales, 1985). Lect. (tel. 02 + 218-9421, telex AA 75004, fax 02 + 281-2498). *Protein crystallography, isomorphous replacement, disordered structures.*

Bakshi, Dr Edward. Dept. of Chemistry, Monash U., Clayton, Victoria 3168, Australia. (1952) PhD, physics (Monash U., 1983). Res. fellow. (tel. 03 +

565-4526, telex AA 32691, fax 03 + 565-4597). *X-ray diffraction, neutron diffraction, magnetic materials, mathematical modelling, optics.*

Barnea, Dr Zwi. Sch. of Physics, U. of Melbourne, Parkville, Victoria 3052, Australia. (1932) PhD, physics (U. Melbourne, 1974). Reader. (tel. 03 + 344-7074, telex AA 35185, fax 03 + 347-4783). *Instrumentation, crystal physics, diffraction physics, anharmonicity.*

Beale, Dr John Phillip. Computing Services Dept., U. of New South Wales, P.O. Box 1, Kensington, New South Wales 2033, Australia. (1947) PhD, chemistry (U. New South Wales, 1970). Head, Systems Software Unit. (tel. 02 + 697-2919, telex AA 26054, fax 02 + 662-8665, email johnb at csdunixo.csd.unsw.oz). *Computing, protein crystallography, solid state chemistry.*

Beretka, Mr Julius. CSIRO Div. of Construction and Engineering, Graham Rd., Highett, Victoria 3190, Australia. (1930) MSc, chemistry (U. Adelaide, 1962). Principal res. scient. (tel. 03 + 556-2211, telex AA 33766, fax 03 + 553-2819). *Materials, reaction between solids, kinetics.*

Bevan, Prof David John Martin. Sch. of Physical Sci., The Flinders U. of South Australia, Sturt Rd., Bedford Park, South Australia 5042, Australia. (1926) PhD, chemistry (Imperial C., London, UK, 1957). Emeritus prof. (tel. 08 + 275-2190, fax 08 + 277-5523). *Structural inorganic chemistry, solid state chemistry.*

Blake, Mr Ronald George. Materials Div., Australian Nuclear Sci. and Techn. Organization, Private Mail Bag, Sutherland, New South Wales 2232, Australia. (1930) BSc, physics (U. New South Wales, 1967). Experimental officer. (tel. 02 + 543-3111). *Materials science, electron microscopy, electron diffraction.*

Boehm, Dr James M. Dept. of Radiation Oncology, Westmead Hospital, Westmead, New South Wales 2147, Australia. (1950) PhD, physics (U. Melbourne, 1978). Scientific officer. (tel. 02 + 633-6499, telex AA 120298, fax 02 + 633-4984). *X-ray diffraction, radiation protection, medical physics.*

Brown, Dr Roger Norman. Geological Services, AMDEL Ltd., 31 Flemington St., Frewville, South Australia 5063, Australia. (1931) PhD, crystallography (U. Adelaide, 1958). Sr. physicist. (tel. 08 + 372-2700, ext. 850, telex AMDEL AA 82520, fax 08 + 79-6623). *Powder diffraction, clay mineralogy.*

Browne, Mr Ian Bruce. Sietronics Pty. Ltd., P.O. Box 84, Hawker, Australian Capital Territory 2614, Australia. (1940) Assoc. Dipl., physics (Royal Melbourne Inst. of Techn., 1964). Director. (tel. 062 + 51-6611, telex AA 62754, fax 062 + 51-6659). *Powder diffraction.*

Bursill, Dr Leslie Arthur. Physics Dept., U. of Melbourne, Parkville, Victoria 3052, Australia. (1941) DSc, physics (U. Melbourne, 1981). Reader. (tel. 03 + 344-5431, telex AA 35185, UNIMEL, fax 03 + 347-4783). *High resolution electron microscopy, structural phase transitions, spiral crystallography, aperiodic solids.*

Calos, Mr Nicholas James. Dept. of Chemistry, U. of Queensland, St. Lucia, Brisbane, Queensland 4067, Australia. (1965) BSc(Hons), crystallography (U. Queensland, 1986). PhD student. (tel. 07 + 377-3296).

Cashion, Assoc. Prof. John Dixon. Dept. of Physics, Monash U., Wellington Rd., Clayton, Victoria 3168, Australia. (1942) DPhil, physics (Oxford U., UK, 1969). Assoc. Prof. (tel. 03 + 565-4000, ext. 3680, telex AA 32691, fax 03 + 565-3637). *Minerals, magnetic ordering, coals, soft lattice modes.*

Chang, Dr Wei-Jui. Dept. of Geology, U. of Western Australia, Nedlands, Western Australia 6009, Australia. (1948) PhD, solid state physics (Monash U., 1978). Scientific officer. (tel. 09 + 380-2694, telex AA 92992, fax 09 + 381-6427). *X-ray powder diffractometry, X-ray fluorescence spectrometry, material characterization, clay mineralogy.*

Cheary, Dr Robert Winston. School of Phys. Sci., U. of Techn., Sydney, P.O. Box 123, Broadway, Sydney, New South Wales 2007, Australia. (1946) PhD, X-ray diffraction (U. of Aston, Birmingham, UK, 1971). Sr. lect. in physics. (tel. 02 + 218-9517). *X-ray and neutron powder diffraction, X-ray line profile analysis.*

Cockayne, Dr David John Hugh. Electron Microscope Unit, U. of Sydney, Sydney, New South Wales 2006, Australia. (1942) DPhil, materials sci. (Oxford U., UK, 1970). Director. (tel. 02 + 692-2351, telex AA 26169 UNISYD, fax 02 + 692-3838). *Lattice defects, dynamical scattering.*

Colman, Dr Peter Malcolm. CSIRO Div. of Biotechnology, 343 Royal Parade, Parkville, Victoria 3052, Australia. (1944) PhD, physics (U. Adelaide, 1969). Chief res. scient. (tel. 03 + 342-4200, telex 33983, fax 03 + 347-5481). *Protein crystallography.*

Colmanet, Dr Silvano Francesco. Australian Radiation Lab., Lower Plenty Rd., Yallambie, Victoria 3085, Australia. (1955) PhD, chemistry (La Trobe U., 1989). Research scientist. (tel. 03 + 433-2211, ext. 263, telex AA 31726, fax 03 + 432-1835). *Crystallography, technetium, coordination chemistry, transition metals.*

Colvyas, Mr Kim. Technical Services Laboratories, BHP Steel, Rod and Bar Products Division, P.O.Box 196B, Newcastle, New South Wales 2300, Australia. (1950) MSc, chemistry (U. Newcastle, 1982) Principal chemist. (tel. 049 + 69-0620, telex AA 28006, fax 049 + 69-0605). *Powder diffraction, computing, X-ray fluorescence, analytical chemistry, statistics.*

Cook, Prof Alan Cecil. U. of Wollongong, P.O. Box 1144, Wollongong, New South Wales 2500, Australia. (1935) PhD, (Cantab, UK). Prof. of geology. (tel. 042 + 27-0841, telex 29022, fax 042 + 29-7768). *Optical properties, coals and cokes.*

Corbett, Dr Madeline. Dept. of Applied Chemistry, Swinburne Institute of Techn.,John St., Hawthorn, Victoria 3122, Australia. (1941) PhD, chemistry (crystallography) (U. Melbourne, 1970). Part-time lecturer. (tel. 03 + 819-8178). *Inorganic crystal structures.*

Coyle, Mr Richard Alan. Dept. of Physics, Monash U., Clayton, Victoria 3168, Australia. (1922) Fellow, physics (Melbourne Techn. C., 1948). Hon. Res. Assoc. (tel. 03 + 565-3601, telex AA32691, fax 03 + 565-3637). *Materials science, X-ray powder diffraction.*

Craig, Mr Donald Chadwick. School of Chemistry, U. of New South Wales, Kensington, New South Wales 2033, Australia. (1936) MSc, geology (U. New South Wales, 1964). Professional officer. (tel. 02 + 697-2222, ext. 4595, telex AA 26054, fax 02 + 662-2835). *Diffractometry, computing, small molecules, service crystallography.*

Creagh, Prof Dudley Cecil. Physics Dept., University C., U. of New South Wales, Northcott Drive, Campbell, Australian Capital Territory 2603, Australia. (1935) PhD, physics (U. New South Wales, 1975). Assoc. Prof. (tel. 062 + 68-8766, telex ADFADM AA 62030, fax 062 + 47-0702). *X-ray scattering, X-ray dispersion, characterization of materials (XRD, XRF, STEM/EDS, NMR), dynamical theory.*

Creek, Mr Russell Charles. Res. and New Techn., BHP Co. Pty. Ltd., 245 Wellington Rd., Mulgrave, Melbourne, Victoria 3170, Australia. (1958) MSc, physics (U. of Melbourne, 1983). Research officer. (tel. 03 + 566-7232, telex BHP AA 30408, fax 03 + 561-6709). *Electron diffraction, imaging theory, image processing.*

Cuff, Dr Christopher. Electron Microscope Unit and Dept. of Geology, James Cook U. of North Queensland, P.O. Douglas, Townsville, Queensland 4811, Australia. (1944) PhD, mineralogical crystallography (Imperial C., UK, 1971). Director, Electron Microscope Unit. (tel. 077 + 81-4496, telex AA 47009, fax 077 + 79-6371). *X-ray diffraction, neutron diffraction, mineral structures, silicate site occupancy problems, clay minerals, carbonates.*

Dance, Prof Ian Gordon. Sch. of Chemistry, U. of New South Wales, P.O. Box 1, Kensington, New South Wales 2033, Australia. (1940) PhD, inorganic chemistry (U. Manchester, UK, 1966). Prof. of inorganic chemistry (tel. 02 + 697-4703, telex AA 26054, fax 02 + 662-2835). *Crystal structure, structural inorganic chemistry, metals, sulphides, inclusion, solid state, synthesis, NMR, catalysis, clusters.*

Davis, Mr Paul Christopher. CSIRO Div. of Biotechnology, 343 Royal Parade, Parkville, Victoria 3052, Australia. (1948) Dip. Elec. Eng., electronics (Royal Melbourne Institute of Techn., 1972). (tel. 03 + 342-4200, ext. 326, fax 03 + 347-5481, email paul at dpca.dn.mu.oz). *Computing, protein crystallography.*

Davis, Dr Ronald Lindsay. Lucas Heights Research Laboratories, Australian Institute of Nuclear Science and Engineering, Private Mail Bag 1, Menai, New South Wales 2234, Australia. (1940) PhD, physics (Monash U., 1970). Leader, AINSE Neutron Scattering Group. (tel. 02 + 543-3111, ext. 3607, telex AA 24562, fax 02 + 543-5097). *Crystallography, magnetism, materials science, structure, neutron scattering, neutron scattering techniques.*

Dean, Dr Christopher. Qikdraw 3D, Qikdraw Systems, Innovation House West, Technology Park, The Levels, Adelaide, South Australia 5095, Australia. (1960) PhD, crystallography (U. Adelaide, 1986). Software Engineer. (tel. 08 + 260-8278, telex AA 88556, fax 08 + 260-8100). *Computer graphics, applied CAD, structural inorganic crystallography.*

Delaney, Dr William Timothy. 39 Davies Rd., Claremont, Western Australia 6010, Australia. (1944) PhD, quantum chemistry (U. Western Australia, 1972). Lect. (tel. 09 + 384-2247). *Charge density, structure analysis.*

Drennan, Dr John. CSIRO Div. of Materials Sci. and Techn., Locked Bag 33, Clayton, Melbourne, Victoria 3168, Australia. (1951) PhD, chemistry (Flinders U. South Australia, 1979). Sr. res. scient. (tel. 03 + 542-2777, ext. 2665, telex AA 32945 MATSCI, fax 03 + 544-1128). *Electron microscopy, diffraction, solid state chemistry.*

Duffy, Dr Douglas Neil. Dept. of Inorganic and Nuclear Chemistry, U. of New South Wales, P.O. Box 1, Kensington, Sydney, New South Wales 2033, Australia. (1955) D. Phil., organometallic chemistry (U. Waikato, New Zealand, 1981). Lect. (tel. 02 + 697-2222, ext. 4693, telex AA 26054, fax 02 + 662-2835). *Carbonyl, cluster, organometallic, transition metal.*

Eggleton, Dr Richard Anthony. Geology Dept. Australian Nat. U., G.P.O. Box 4, Canberra, Australian Capital Territory 2601, Australia. (1937) PhD, X-ray crystallography (Wisconsin, USA, 1965). Reader. (tel. 062 + 49-5111, ext. 2060, telex AA 62760 NATUNI, fax 062 + 48-9062). *Electron microscopy, X-ray diffraction, mineralogy.*

Elcombe, Dr Margaret Marion. Lucas Heights Res. Labs., Australian Nuclear Sci. and Technology Organization, Private Mail Bag 1, Menai, New South Wales 2234, Australia. (1942) PhD, lattice dynamics (Cambridge U., UK, 1966). Principal. res. scient. (tel. 02 + 543-3111, ext. 3611, telex AA 24562, fax 02 + 543-5097). *Lattice dynamics, neutron scattering.*

Etheridge, Ms Joanne. Dept. of Applied Physics, Royal Melbourne Inst. of Techn., G.P.O. Box 2476V, Melbourne, Victoria 3001, Australia. BSc(Hons), physics (U. Melbourne, 1987). Postgraduate student. (tel. 03 + 660-2600, telex AA 36406, fax 03 + 663-2764). *Electron microscopy, ceramic superconductors.*

Fallon, Dr Gary David. Dept. of Chemistry, Monash U., Wellington Rd., Clayton, Victoria 3168, Australia. (1948) PhD, chemical crystallography (Monash U., 1976). Professional officer. (tel. 03 + 565-4514). *Structure determination.*

Farrell, Miss Yvonne. Electonic Systems Division, Plessey Aust. Pty. Ltd., Faraday Park Railway Rd., Meadowbank, New South Wales 2114, Australia. (1963) B.App.Sc.(Hons), applied chemistry (U. Techn., 1987). Chemist. (tel. 02 + 807-0521, telex AA 121471, fax 02 + 808-2787). *Electron microscopy, ceramics, electronic materials.*

Field, Dr Donald William. Dept. of Physics, Queensland U. of Techn., G.P.O. Box 2434, Brisbane, Queensland 4001, Australia. (1947) PhD, physics (U. Adelaide, 1973). Lect. (tel. 07 + 223-2593, fax 07 + 229-6887) *X-ray diffraction, powder diffraction.*

Fields, Mr Barry Arthur. Dept. of Inorganic Chemistry, U. of Sydney, Sydney, New South Wales 2006, Australia. (1966) BSc(Hons), chemistry (U. of New South Wales, 1988). Postgrad. student and res. asst. (tel. 02 + 692-2830). *Protein crystallography.*

Figgis, Prof Brian Norman. School of Chemistry, U. of Western Australia, Nedlands, Western Australia 6009, Australia. (1930) DSc, chemistry (U. Western Australia, 1966). Prof. of inorganic chemistry. (tel. 09 + 380-3157, telex AA 92992, fax 09 + 388-1807, email bnf at xtal.uwa.oz (ACSNET)). *Chemical bonding, spin density, charge density, transition metal complexes, magnetochemistry.*

Finlayson, Dr Trevor Roy. Dept. of Physics, Monash U., Wellington Rd., Clayton, Victoria 3168, Australia. (1943) PhD, physics (Monash U., 1969). Sr. lect. (tel. 03 + 565-3683, telex AA 32691, fax 03 + 565-3637). *Powder diffraction, phase transformations, residual stress.*

Fletcher, Dr Neville Horner. CSIRO Div. of Radiophysics, c/o Australian Nat. U., R.S.Phys.S. - Maths., Canberra, Australian Capital Territory 2601, Australia. (1930) DSc, physics (U. Sydney, 1973). Chief res. scient. (tel. 062 + 49-4406, telex AA 62615, fax 062 + 49-1884). *Semiconductors, crystal growth, interfaces, ice physics, acoustics.*

Forwood, Dr Christopher Thomas. CSIRO Div. of Materials Sci. & Techn., Normanby Rd., Clayton, Victoria 3168, Australia. (1940) PhD, physics (U. Bristol, UK, 1965). Principal res. scient. (tel. 03 + 542-2865, telex AA 32945, fax 03 + 544-1128). *Defect structure of materials.*

Freeman, Prof Hans Charles. Dept. of Inorganic Chemistry, U. of Sydney, Sydney, New South Wales 2006, Australia. (1929) PhD, chemistry (U. Sydney, 1957). Prof. of inorganic chemistry. (tel. 02 + 692-2757, telex AA 26169 UNISYD, fax 02 + 692-3329). *Coordination complex structures, metalloprotein structures.*

Gable, Dr Robert William. Dept. of Chemistry, La Trobe U., Bundoora, Victoria 3083, Australia. (1954) PhD, inorganic chemistry (U. Melbourne, 1988). Grad. res. asst. *Structure analysis, coordination compounds, computer programming.*

Gatehouse, Dr Bryan Michael Kenneth Cummings. Dept. of Chemistry, Monash U., Clayton, Victoria 3168, Australia. (1932) DSc, inorganic chemistry (U. C., London, UK, 1977). Reader. (tel. 03 + 565-4561, fax 03 + 565-4597). *Single crystal structures, mixed metal oxides, inorganic structures.*

Gates, Dr Jeffrey Douglas. Dept. of Mining & Metallurgical Engineering, U. of Queensland, St. Lucia, Queensland 4067, Australia. (1958) PhD, materials engineering (Monash U., 1985). Res. fellow. (tel. 07 + 377-4183, telex UNIVQLD AA 40315, fax 07 + 371-5896). *Microstructure, alloy steels, localized corrosion.*

Goodman, Dr Peter. School of Physics, University of Melbourne, Parkville, Victoria 3052, Australia. (1928) DSc, electron diffraction (U. Melbourne, 1978). Sr. res. assoc. (tel. 03 + 344-7186, telex AA 35185, fax 03 + 347-4783). *Electron diffraction, electron microscopy, convergent-beam electron diffraction.*

Graham, Dr James. CSIRO Div. of Mineral Products, P.O. Wembley, Western Australia 6014, Australia. (1929) PhD, physical metallurgy (U. Birmingham, UK, 1956). Principal res. scient. (tel. 092 + 387-0371, telex AA 92178, fax 092 + 387-8642). *Minerals, sulphides, oxides.*

Grainger, Dr Colin Trevor. 9 Westward St., Kareela, New South Wales 2232, Australia. (1927). PhD, physics (U. New South Wales, 1967). Retired. (tel. 02 + 528-7681).

Grey, Dr Ian Edward. CSIRO Div. of Mineral Products, P.O. Box 124, Port Melbourne, Victoria 3207, Australia. (1944) PhD, inorganic chemistry (U. Tasmania, 1969). Sr. principal res. scient. (tel. 03 + 647-0268, fax 03 + 646-3223). *Structures, non-stoichiometric oxides, mineral structures, order-disorder, intergrowths.*

Guddat, Mr Luke William. St. Vincent's Institute of Medical Research, 9 Princes St., Fitzroy, Victoria 3065, Australia. (1962) BSc(Hons), chemistry (Monash U., 1983). Grad. student. (tel. 03 + 418-2373, email luke at svimra.dn.mu.oz). *Protein structure determination, crystal structures of mixed metal oxides.*

Gulbis, Ms Jacqueline Margot. Dept. of Chemistry, La Trobe U., Bundoora, Melbourne, Victoria 3083, Australia. (1961) BSc(Hons), chemistry (La Trobe U., 1984). PhD student. (tel. 03 + 479-2520, telex AA 33143, fax 03 + 478-5814). *X-ray structures, small biological molecules.*

Guss, Dr Jules Mitchell. School of Chemistry, U. of Sydney, Sydney, New South Wales 2006, Australia. (1946) PhD, inorganic chemistry (U. Sydney, 1970). Professional officer. (tel. 02 + 692-4302, telex AA 26169 UNISYD, fax 02 + 692-3329, email guss__m at summer.oz). *Biological macromolecules, structure determination and refinement, computer graphics.*

Hall, Prof Eric Ogilvie. 850 Sandy Bay Rd., Lower Sandy Bay, Tasmania 7005, Australia. (1925) PhD, solid state (Cambridge U., UK, 1952). Prof. *Metals, intermediate phase structures.*

Hall, Prof Sydney Reading. Crystallography Centre, U. of Western Australia, Nedlands, Western Australia 6009, Australia. (1939) PhD, physics (U. Western Australia, 1964). Assoc. prof. (tel. 09 + 380-2725, telex AA 92992, fax 09 + 388-1807, email hall%xtal.uwa.oz at munnari). *Computer applications and systems, phasing methodology.*

Hambley, Dr Trevor William. Sch. of Chemistry, U. of Sydney, Sydney, New South Wales 2006, Australia. (1955) PhD, inorganic chemistry (U. Adelaide, 1983). Lect. (tel. 02 + 692-2830, fax 02 + 692-3329, email hambley__t at summer.oz). *Metal complex structure, molecular mechanics, effects of steric stress and strain, drug design.*

Hammond, Mr Lloyd Charles. Dept. of Applied Physics, Curtin U. of Techn., G.P.O. Box U1987, Perth, Western Australia 6001, Australia. (1961) BSc, Grad. Dip., applied physics (Curtin U. of Techn., 1985). Postgraduate student. (tel. 09 + 350-7810, telex AA 92983, fax 09 + 458-4661). *Materials characterization, X-ray powder diffractometry/Rietveld analysis, X-ray fluorescence spectrometry, clay mineralogy.*

Hartley, Dr Richard H. Optoelectronic Division of the Surveillance Research Laboratory, P.O. Box 1650, Salisbury, South Australia 5108, Australia. (1939) PhD, thin film physics, electrical engineering (U. Western Australia, 1971). Principal res. scient. (tel. 08 + 259-6377, fax 08 + 259-5796, email Keylink 07 rhh001). *Solid state physics, thin films, MBE growth, mercury cadmium telluride.*

Hawkins, Ms Kathlein Darrelle. Electron Microscopy Group, Australian Nuclear Sci. and Techn. Organization, New Illawarra Rd., Menai, New South Wales 2234, Australia. (1963) BSc(Hons), metallurgy (U. New South Wales, 1985). Experimental officer grade 1. (tel. 02 + 543-3448, telex AA 24562, fax 02 + 543-5097). *Electron microscopy, ceramics.*

Hay, Dr David Gilbert. CSIRO Div. of Materials Sci. and Techn., Locked Bag 33, Clayton, Victoria 3168, Australia. (1948) PhD, chemistry (La Trobe U., 1982). Experimental scient. (tel. 03 + 542-2777). *X-ray powder diffraction, solid state chemistry.*

Healy, Dr Peter Conrad. Division of Sci. and Techn., Griffith U., Brisbane, Queensland 4111, Australia. (1947) PhD, chemistry (U. Western Australia, 1972). Sr. lect. (tel. 07 + 275-7442, telex AA 40362, fax 07 + 277-3759). *Structures, solid state NMR.*

Hicks, Dr Trevor John. Physics Dept., Monash U., Wellington Rd., Clayton, Victoria 3168, Australia. (1939) PhD, physics (Monash U., 1966). Reader (tel. 03 + 565-3681, telex AA 32691 MONASH, fax 03 + 565-3637). *Neutron diffraction, magnetic structure.*

Hill, Dr Roderick Jeffrey. CSIRO Division of Mineral Products, 339 Williamstown Rd., Port Melbourne, Victoria 3207, Australia. (1949) PhD, mineralogy, crystallography (U. Adelaide, 1976). Principal res. scient. (tel. 03 + 647-0211, telex AA 34349, fax 03 + 647-0395, email rodh!dmpmela at munnari.oz). *Powder diffraction, crystal chemistry, mineralogy, advanced ceramics.*

Hogan, Dr Leonard McNamara. Dept. of Mining and Metallurgical Eng., U. of Queensland, St Lucia, Brisbane, Queensland 4067, Australia. PhD, metallurgy (U. Queensland, 1965). Hon. res. consultant. (tel. 07 + 377-3737, telex UNIVQLD AA 40315, fax 07 + 371-5896). *Crystal growth, physical metallurgy, solidification theory.*

Horn, Dr Ernst. Dept. of Physical and Inorganic Chemistry, U. of Adelaide, G.P.O. Box 498, Adelaide, South Australia 5001, Australia. (1952) PhD, inorganic chemistry (U. Adelaide, 1983). Tutor/demonstrator. (tel. 08 + 223-4333, ext. 5712). *X-ray structure determination, inorgano-metallic complexes, organic compounds, metal carbonyl fluorides, optical resolution.*

Hoskins, Dr Bernard Foster. Dept. of Inorganic Chemistry, U. of Melbourne, Grattan St., Parkville, Victoria 3052, Australia. (1935) DSc, chemistry (U. Melbourne, 1976). Reader. (tel. 03 + 344-6471, telex AA 35185 UNIMELB, fax 03 + 347-5180). *Inorganic structures.*

Howard, Dr Christopher John. Lucas Heights Res. Labs., Australian Nuclear Sci. and Techn. Organization, Private Mail Bag 1, Menai, New South Wales 2234, Australia. (1942) PhD, nuclear magnetic resonance (U. Nottingham, UK, 1969). Principal res. scient. (tel. 02 + 543-3111, ext. 3609, telex AA 24562, fax 02 + 543-5097). *Neutron diffraction, high resolution neutron powder diffraction, accurate structure determination.*

Hugo, Mr Geoffrey Ronald. Dept. of Materials Engineering, Monash U., Wellington Rd., Clayton, Victoria 3168, Australia. (1963) BE(Hons), materials engineering (Monash U., 1987). Res. student. (tel. 03 + 565-4925, telex AA 32691, fax 03 + 565-3409). *Electron microscopy, phase transformations.*

Huq, Dr Fazlul. Dept. of Biological Sciences, Cumberland C. of Health Sci., East St., Lidcombe, NSW, Australia. (1945) PhD, X-ray Crystallography (U. London, UK, 1971) Lect. (tel. 02 + 646 6522, fax. 02 + 646 4853) *Membrane transport, nutrition for sports performance, biological structures, teaching, computing, structural chemistry.*

Hyde, Prof Bruce Godfrey. Res. Sch. of Chemistry, Australian Nat. U., P.O. Box 4, Canberra, Australian Capital Territory 2601, Australia. (1925) DSc, physical chemistry (U. Bristol, UK, 1974). Prof. in inorganic chemistry. (tel. 062 + 49-4401, fax 062 + 48-7817). *Electron microscopy, crystal chemistry.*

Jaeger, Mr Hans. CSIRO Div. of Materials Sci. and Techn., Locked Bag 33, Clayton, Victoria 3168, Australia. (1921) Dipl. Appl. Chem., chemistry (Eng. C., Essen, Germany, 1953). Principal res. scient. (retired). (tel. 03 + 542-2667, telex AA 32945, fax 03 + 544-1128). *Carbon fibres, supported metal catalysts, epitaxy.*

James, Dr Veronica Jean. Sch. of Physics, U. of New South Wales, P.O. Box 1, Kensington, New South Wales 2033, Australia. PhD, physics (U. New South Wales, 1970). Assoc. prof. (tel. 02 + 697-4631, telex AA 26054, fax 02 + 663-3420). *Collagen structure, diseased tissue.*

Johnson, Dr Andrew William Syme. Electron Microscopy Centre, U. of Western Australia, Nedlands, Western Australia 6009, Australia. PhD, physics (U. Western Australia, 1966). Director. (tel. 09 + 380-2764, telex AA 92992, fax 09 + 389-1053, email andy at earwax.uwa.oz). *Electron microscopy, electron diffraction, mineralogy.*

Jones, Dr John Brett. Dept. of Geology, U. of Adelaide, Adelaide, South Australia 5001, Australia. (1930) PhD, mineralogy (U. Wisconsin, USA, 1958). Reader. (tel. 08 + 228-5380, telex AA 89141, fax 08 + 224-0464). *Mineral structures.*

Jostsons, Dr Adam. Advanced Materials, Australian Nuclear Sci. and Techn. Organization, Private Mail Bag 1, Menai, New South Wales 2234, Australia. (1937) PhD, metallurgy (U. New South Wales, 1966). Director, advanced materials. (tel. 02 + 543-3265, telex AA 24562, fax 02 + 543-7179). *Materials science, electron microscopy, electron diffraction.*

Kalceff, Dr Walter. Dept. of Physics, U. of Techn., Sydney, P.O. Box 123, Broadway, Sydney, New South Wales 2007, Australia. (1950) PhD, rare earth magnetism (U. New South Wales, 1983). Lecturer. (tel. 02 + 2-0930, ext. 9552, fax 02 + 281-2498, email wkalceff at atom.oz (BITNET)). *Neutron scattering, magnetism.*

Kelly, Dr Patrick Manning. Deputy Director Research, Australian Nuclear Sci. and Techn. Organization, Private Mail Bag 1, Menai, New South Wales 2232, Australia. (1935) ScD, physical metallurgy (Cambridge U., UK, 1979). Deputy Dir. Res. (tel. 02 + 543-3315, telex AA 24562, fax 02 + 543-5097). *Materials science, electron microscopy, electron diffraction, martensitic transformations.*

Kennard, Dr Colin Harold Leslie. Dept. of Chemistry, U. of Queensland, St. Lucia, Queensland 4067, Australia. (1935) PhD, crystallography (U. New South Wales, 1961). Reader. (tel. 07 + 377-3296). *Neutron diffraction, structure analysis, pesticides, metal complexes, teaching methods.*

Kennon, Prof Noel Frederick. Dept. of Metallurgy and Materials Engineering, U. of Wollongong, Northfields Ave., Wollongong, New South Wales 2500, Australia. (1934) PhD, crystallography (U. New South Wales, 1967). Assoc. prof. (tel. 042 + 27-0555, ext. 3457). *Crystallography, phase transformation in steels, shape memory alloys, martensitic transformations, metallography.*

Kisi, Dr Erich Herold. Lucas Heights Res. Labs., Australian Nuclear Sci. and Techn. Organization, Private Mail Bag 1, Post Office Menai, New South Wales 2234, Australia. (1960) BMet(Hons), metallurgy (U. Newcastle, 1985). Scientific officer. (tel. 02 + 543-3111, ext. 3612, telex AA 24562, fax 02 + 543-5097). *Materials science, phase transformations, neutron powder diffraction, high temperature studies.*

Knott, Mr Robert Barry. Applied Physics Division, Australian Nuclear Sci. and Techn. Organization, Private Mail Bag, Menai, New South Wales 2234, Australia. (1946) MSc, physics (U. Newcastle, 1972). (tel. 02 + 543-3111, ext. 3741, telex AA 24562, fax 02 + 543-5097, email rbk at atom.oz). *Structural biology, neutron scattering, X-ray scattering.*

Kwietniak, Dr Mark Stefan. Applied Sci. Branch, Telecom Res. Labs., 770 Blackburn Rd., Clayton, Victoria 3168, Australia. (1945) PhD, solid state electronics (Polish Acad. of Sci., Poland, 1973). Senior researcher. (tel. 03 + 541-6637, fax 03 + 562-8364). *Materials science, structure properties of semiconductors.*

Kucharski, Dr Edward Stanislaw. Sch. of Chemistry, U. of Western Australia, Nedlands, Western Australia 6009, Australia. (1953) PhD, physical and inorganic chemistry (U. Western Australia, 1984). Res. officer. (tel. 09 + 380-3054, fax 09 + 388-1807). *Charge densities, transition metal complexes, structural studies, magnetically anomalous transition metal complexes, enzyme structures, neutron and polarized neutron diffraction studies.*

Lambert-Smith, Mr John Ernle Warwick. Callide 'B' Power Station, Queensland Electricity Commission, P.O. Box 392, Biloela, Queensland 4715, Australia. (1931) BSc, inorganic chemistry and X-ray crystallography (U. Sydney, 1957). Chemist - water management. (tel. 079 + 92-9389, telex QEGBCB AA 94842, fax 079 + 92-1715). *X-ray crystal structure analysis, crystallographic computer programming, coordination compounds of copper(II).*

Lambrianidis, Mr Louie Terry. A/Mat, D.S.T.O. Aero Research Laboratories, 506 Lorimer St., Port Melbourne, Victoria 3207, Australia. (1961) BSc(Hons), physics (U. Melbourne, 1984). Experimental officer. (tel. 03 + 647-7501, ext. 7509, fax 03 + 645-3869). *Electron microscopy, computing, metal alloys, diffraction physics, electronic instrumentation.*

Lancucki, Mr Christopher Joseph. CSIRO Div. of Building Construction and Engineering, Graham Rd., Highett, Victoria 3190, Australia. (1935) BSc, physics (U. Western Australia, 1959). Experimental scient. (tel. 03 + 556-2279). *Clay minerals, inorganic crystals, structures.*

Lawrence, Dr Michael Colin. CSIRO Div. of Biotechnology, 343 Royal Parade, Parkville, Melbourne, Victoria 3052, Australia. (1955) PhD, Theoretical physics (U. Cape Town, 1980). Sr. res. scientist. (tel. 03 + 342-4200, ext. 262, telex 33983, fax 03 + 347-5481, email mike at dpcb.dn.mu.oz). *Protein crystallography, image processing.*

Leverett, Dr Peter. Sch. of Sci. and Techn., Nepean C. of Advanced Education, Second Ave., Kingswood, New South Wales 2750, Australia. (1944) PhD, X-ray crystallography and oxide chemistry (Monash U., 1969). Dean. (tel. 047 + 36-0428, fax 047 + 36-4186). *X-ray structure determination, mixed metal oxides, synthesis, crystal chemistry.*

Lincoln, Dr Francis John. Dept. of Physical and Inorganic Chemistry, U. of Western Australia, Nedlands, Western Australia 6009, Australia. (1935) PhD, chemistry (U. Western Australia, 1967). Sr. lect. (tel. 09 + 380-3142, telex AA 92992 UNIWA, fax 09 + 388-1807). *Electron microscopy, minerals, solid state chemistry, inorganic materials, oxides, thermodynamic and structural examination, crystal chemistry.*

Lloyd, Dr Douglas James. Faculty of Education, Bendigo C.A.E., Edward's Rd., Bendigo, Victoria 3550, Australia. (1946) PhD, solid state chemistry (Monash U., 1972). Lect. (tel. 054 + 40-3222, ext. 300, telex BENCOL AA 38293, fax 054 + 40-3477). *Solid state chemistry, crystal structure determination.*

Lucas, Dr Brian William. Dept. of Physics, U. of Queensland, St. Lucia, Brisbane, Queensland 4067, Australia. PhD, physics (U. London, UK, 1965). Sr. lect. (tel. 07 + 377-3421). *X-ray diffraction, neutron diffraction, crystal physics.*

Lukaszewski, Dr George Michael. CSIRO Div. of Mineral Products, P.O. Box 124, Port Melbourne, Victoria 3207, Australia. (1935) PhD, inorganic chemistry (U. London, UK, 1960). Principal res. scient. (tel. 03 + 647-0211, ext. 210, telex 34349, fax 03 + 647-0395). *Mineral chemistry, magnesite reactivity, sulphide oxidation, thermoanalytical techniques.*

Lynch, Dr Denis Francis. CSIRO Div. of Materials Sci. and Techn., Locked Bag 33, Clayton, Victoria 3168, Australia. (1941) PhD, physics (U. New South Wales, 1965). Principal res. scient. (tel. 03 + 542-2927, telex AA 32945, fax 03 + 544-1128). *Electron diffraction, dynamical scattering, RHEED.*

Machin, Mr Ken James. St. Vincent's Institute of Medical Res., Victoria Parade, Fitzroy, Victoria 3065, Australia. (1956) BSc(Hons), biochemistry (Monash U., 1978). Res. officer. (tel. 03 + 418-2375, email ken at svimra.dn.mu.oz). *Protein crystallography.*

Mackay, Dr Maureen Florence. Dept. of Chemistry, La Trobe U., Bundoora, Victoria 3083, Australia. PhD, crystallography (U. Melbourne, 1968). Reader in chemistry. (tel. 03 + 479-2520, telex AA 33143, fax 03 + 478-5814). *Crystal structures, small biological molecules, natural products.*

Mackenzie, Dr James Kenneth. 15 Ronald St., Box Hill North, Victoria 3129, Australia. (1920). PhD, physics (U. Bristol, UK, 1949). Sr. principal res. scient. (retired). *Crystallography, phase transformations, crystal physics.*

Madsen, Mr Ian Charles. CSIRO Div. of Mineral Products, 339 Williamstown Rd., Port Melbourne, Victoria 3207, Australia. (1951) BApplSc, applied physics (South Australian Inst. of Techn., 1977). Experimental scient. (tel. 03 + 647-0211, ext. 366, telex AA 34349, fax 03 + 647-0395, email ianm at dmpmela.oz (ACSNET)). *X-ray powder diffractometry, single crystal diffractometry, Rietveld analysis, crystallographic computing.*

Malin, Dr Anthony Samuel. Comalco Research Centre, 15 Edgars Rd., Thomastown, Victoria 3074, Australia. (1932) PhD, metallurgy (U. New South Wales, 1978). Manager - rolling and corrosion group. (tel. 03 + 469-0777, telex AA 37437, fax 03 + 460-2718). *Electron microscopy, X-ray diffraction, recrystallization, strain deformation, metallography, corrosion.*

Maslen, Dr Edward Norman. Crystallography Centre, U. of Western Australia, Nedlands, Western Australia 6009, Australia. (1935) D.Phil, crystallography (Oxford U., UK, 1960). Director. (tel. 09 + 380-2738, fax 09 + 381-6427, email enm%xtal.uwa.oz). *Electron density.*

Mathieson, Prof Alexander McLeod. Chemistry Dept., La Trobe U., Bundoora, Victoria 3083, Australia. (1920) DSc, chemical crystallography (U. Melbourne, 1956). Prof. (Honorary). (tel. 03 + 479-2506, telex AA 33143, fax 03 + 478-5814). *X-ray apparatus, reflectivity measurement, extinction, two-dimensional intensity distributions.*

McCall, Dr Maxine June. CSIRO Div. of Biotechnology, Lab. for Molecular Biology, P.O. Box 184, North Ryde, New South Wales 2113, Australia. (1950) PhD, chemistry (Flinders U., 1980). Res. fellow. (tel. 02 + 886-4888, ext. 4936, telex AA 22970, fax 02 + 888-9271). *Nucleic acids, drug-DNA complexes, protein-DNA complexes.*

McKenzie, Dr David Robert. Sch. of Physics, U. of Sydney, Sydney, New South Wales 2006, Australia. (1946) PhD, physics (U. New South Wales, 1972). Reader. (tel. 02 + 692-3180, fax 02 + 660-2903). *Electron diffraction, lattice dynamics, amorphous and polycrystalline materials.*

McLaren, Dr Alexander Clark. Res. Sch. of Earth Sciences, Australian Nat. U., P.O. Box 4, Canberra; Australian Capital Territory 2601, Australia. (1928) ScD, physics (Cambridge U., UK, 1981). Professorial fellow. (tel. 062 + 49-2494, telex AA 62693, fax 062 + 48-9062). *Electron microscopy, minerals, defects in crystals and minerals, plastic deformation, rocks.*

McLaughlin, Mr George Millar. Res. Sch. of Chemistry, Australian Nat. U., P.O. Box 4, Canberra, Australian Capital Territory 2601, Australia. (1947) Grad. R.I.C. (Royal Inst. of Chemistry, UK, 1971). Laboratory manager. (tel. 062 + 49-2592, telex AA 62172, fax 062 + 48-7817). *Computing, biologically significant molecules, packing, conformation.*

Millar, Dr John Joseph. Physics Dept., Bendigo C. of Advanced Education, P.O. Box 199, Bendigo, Victoria 3550, Australia. (1946) PhD, physics (U. Melbourne, 1973). Head, Physics Dept. (tel. 054 + 40-3405, telex AA 38293, fax 054 + 40-3477). *Electron microscopy, energy dispersive analysis, materials analysis.*

Miller, Ms Sarah Ann. CSIRO Div. of Fuel Techn., Lucas Heights Res. Labs., Private Mail Bag 7, Menai, New South Wales 2234, Australia. (1961) BSc, chemistry (Flinders U. of South Australia, 1982). Experimental scient. (tel. 02 + 543-3111, ext. 3188, telex AA 73341). *Neutron diffraction, X-ray diffraction, powder diffraction, zeolites, industrial minerals.*

Mohyla, Dr Jury. South Australian C. of Advanced Education, Sturt Rd., Bedford Park, Adelaide, South Australia 5042, Australia. (1937) PhD, crystallography (Flinders U. of Sch. Austraia, 1979). Acad. computer co-ord. (tel. 08 + 275-5382, email j.mohyla at c.cflinders.oz (BITNET)). *Computing, solid state physics, education.*

Moodie, Dr Alexander Forbes. Dept. of Applied Physics, Royal Melbourne Inst. of Techn., 124 La Trobe St., Melbourne, Victoria 3000, Australia. (1923) DSc, physics (Royal Melbourne Inst. of Techn., 1982). Institute fellow. (tel. 03 + 660-2434, telex AA 36406, fax 03 + 663-2764). *Electron diffraction, electron microscopy, scattering theory.*

Moon, Assoc. Prof. Anthony Ronald. Dept. of Applied Physics, U. of Techn., Sydney, 17 Broadway, Sydney, New South Wales 2007, Australia. (1945) PhD, physics (U. Melbourne, 1970). Head, Dept. of Applied Physics. (tel. 02 + 2-0930, ext. 9468, fax 02 + 281-2498). *Electron diffraction, electron microscopy, optical spectroscopy.*

Moore, Dr Alan James William. 1 Hastings Rd., Hawthorn, Victoria 3123, Australia. (1920) PhD, physical chemistry (Cambridge U., UK, 1948). Sr. principal res. scient. (retired). (tel. 03 + 813-2075). *Surface structures, field ion microscopy.*

Morton, Dr Allan James. CSIRO Div. of Materials Sci. and Techn., Locked Bag 33, Clayton, Victoria 3168, Australia. (1939) PhD, metallurgy (U. New South Wales, 1964). Principal res. scient. (tel. 03 + 542-2777, telex AA 32945, fax 03 + 544-1128). *Electron microscopy, defects in metals, intermetallic alloy structures.*

Moss, Dr Barbara Kay. 40 Alto Avenue, Croydon, Victoria 3136, Australia. (1955) PhD, diffraction physics (U. Melbourne, 1983). Consultant. (tel. 03 + 723-1948). *Electron diffraction, polymers, anharmonic thermal motion.*

Muddle, Dr Barrington Charles. Dept. of Materials Eng., Monash U., Wellington Rd., Clayton, Victoria 3168, Australia. (1948) PhD, metallurgy (U. New South Wales, 1975). Sr. lect. (tel. 03 + 565-4908, telex AA 32691, fax 03 + 565-3409). *Electron microscopy, phase transformations.*

Nimmo, Dr John Kenneth. Physics Dept., U. of Queensland, St. Lucia, Queensland 4067, Australia. (1948) PhD, physics (U. Queensland, 1976). Sr. tutor. (tel. 07 + 377-2364). *Phase transitions, structure determination.*

Norman, Dr Peter David. Div. of Digital Techn., Chisholm Inst. of Techn., 900 Dandenong Rd., Caulfield East, Victoria 3145, Australia. (1932) PhD, physics (Monash U., 1987). Sr. lect. (tel. 03 + 573-2222, ext. 2354, telex AA 154087 CITVIC, fax 03 + 572-1298). *Electron diffraction, phase transformations, alloys, Fermiology, Kohn anomalies.*

O'Connor, Dr Brian Henry. Dept. of Appl. Physics, Curtin U. of Techn., G.P.O. Box U1987, Perth, Western Australia 6001, Australia. (1941) PhD, crystallography (U. Western Australia, 1964). Assoc. Prof. (tel. 09 + 350-7192, telex AA 92983, fax 09 + 458-4661). *Materials characterization, minerals, X-ray powder diffractometry, Rietveld analysis, X-ray fluorescence spectrometry, advanced materials.*

Parks, Dr Terrence Charles. CSIRO Div. of Mineral Products, Private Bag, P.O. Wembley, Western Australia 6014, Australia. (1940) PhD, physical chemistry (U. Western Australia, 1967). Sr. res. scient. (tel. 09 + 387-4233, ext. 388, telex AA 92178, fax 09 + 387-8642). *Defect structures, crystal chemistry, geochemical processes, mineral processing.*

Peng, Ms Ju Lin. School of Physics, U. of Melbourne, Parkville, Victoria 3052, Australia. (1939) solid state physics. Nat. res. fellow. (tel. 03 + 344-7072, telex AA 35185 UNIMEL). *HREM, crystalline solids, characterization, aperiodic solids, noncrystallographic symmetries, surface structure, atomic rearrangement studies by HREM.*

Phakey, Dr Prem P. Dept. of Physics, Monash U., Wellington Rd., Clayton, Victoria 3168, Australia. (1933) PhD, physics (Monash U., 1968). Sr. lect. (tel. 03 + 565-3613, telex AA 32691, fax 03 + 565-3637). *Defects in solids.*

Platts, Mr Simon Nicholas. Dept. of Inorganic Chemistry, Monash U., Wellington Rd., Clayton, Melbourne, Victoria 3168, Australia. (1964) BSc(Hons), science (Monash U., 1987). Research scholar. (tel. 03 + 565-4514). *Inorganic hydrates.*

Poppleton, Dr Bruce J. CSIRO Div. of Materials Sci. and Techn., Locked Bag 33, Clayton, Victoria 3168, Australia. (1936) PhD, chemistry (Monash U., 1973). Sr. res. scient. (tel. 03 + 542-2777, telex AA 32945, fax 03 + 544-1128). *Computing, small organic molecules.*

Prager, Dr Peter Robert. P.O. Box 17, Black Rock, Victoria 3193, Australia. (1944) PhD, physics (U. Melbourne, 1971). (tel. 03 + 589-2550). *Accurate structure analysis, crystal physics, polytypism.*

Pring, Dr Allan. Dept. of Mineralogy, South Australian Museum, North Terrace, Adelaide, South Australia 5000, Australia. (1956) PhD, high res. electron microscopy (Cambridge U., UK, 1983). Curator of minerals and meteorites. (tel. 08 + 223-8894). *Electron microscopy, mineral chemistry, sulphides, order-disorder, borates, magic angle spinning NMR.*

Rachinger, Prof William Albert. Dept. of Physics, Monash U., Wellington Rd., Clayton, Victoria 3168, Australia. (1927) PhD, metallurgy (U. Melbourne, 1952). Prof. of Exp. physics. (tel. 03 + 565-3631, telex AA 32691 MONASH, fax 03 + 565-3637). *Solid state physics, biological composite materials.*

Radoslovich, Dr Edward William. CSIRO Div. of Soils, G.P.O. Box 639, Canberra, Australian Capital Territory 2601, Australia. (1928) DSc, physics (U. Adelaide, 1968). Div. sec. (sci.). (tel. 062 + 46-5321, telex 61452, fax 062 + 47-5883). *Clay-organic complexes, mineral structures, research planning and management.*

Rae, Dr Alan David. Sch. of Chemistry, U. of New South Wales, Kensington, New South Wales 2033, Australia. (1941) PhD, chemistry (U. Auckland, New Zealand, 1964). Assoc. prof. (tel. 02 + 697-2222, ext. 4673, telex AA 26054, fax 02 + 662-2835). *Crystal structure refinement techniques, crystal structures that are twinned, disordered, modulated, incommensurate.*

Rao, Mr Zi He. St. Vincent's Institute of Medical Research, 41 Victoria Pde., Melbourne, Victoria 3065. Australia. (1950) MSc, protein crystallography (Academia Sinica, 1982). PhD student. (tel. 03 + 418-2374, email rao at svimra). *Protein crystallography, structure and function.*

Raston, Prof Colin Llewellyn. Div. of Sci. and Techn., Griffith U., Nathan, Queensland 4111, Australia. (1950) PhD, inorganic chemistry (U. Western Australia, 1976). Prof. (tel. 07 + 275-7564, telex AA 40362, fax 07 + 277-3759). *Coordination complexes, organo-metallic compounds, main group compounds.*

Raven, Mr Mark Derek. CSIRO Div. of Soils, Private Bag No. 2, Glen Osmond, South Australia 5064, Australia. (1960) BSc Grad. Dip., applied physics (Curtin U. of Techn., 1985). Technical officer. (tel. 08 + 274-9311, ext. 308, telex 82406, fax 08 + 338-1636). *Crystallography, X-ray powder diffractometry/Rietveld analysis, clay mineralogy.*

Reid, Dr Alan Forrest. CSIRO Inst. of Minerals- Energy and Construction, P.O. Box 93, North Ryde, New South Wales 2113, Australia. (1931) DSc, solid state chemistry (Australian Nat. U., 1969). Institute director. (tel. 02 + 887-8212, fax 02 + 887-8197). *Mineral structures.*

Reynolds, Dr Philip Andrew. Sch. of Chemistry, U. of Western Australia, Nedlands, Western Australia 6009, Australia. (1947) DPhil, chemistry (Oxford U., UK, 1973). Res. fellow. (tel. 09 + 293-4373). *Molecular crystals, spin density, disorder in crystals, transition metal complexes, charge densities.*

Robertson, Dr Glen Bradley. Res. Sch. of Chemistry, Australian Nat. U., G.P.O. Box 4, Canberra, Australian Capital Territory 2601, Australia. (1936) PhD, physics (U. Western Australia, 1966). Sr. fellow. (tel. 062 + 49-4380, fax 062 + 48-7817). *X-ray diffraction, neutron diffraction, organometallic complexes, low temperature analysis, charge density.*

Rossell, Dr Henry John. CSIRO Div. of Materials Sci. and Techn., Locked Bag 33, Clayton, Victoria 3168, Australia. (1936) PhD, physical chemistry (U. Western Australia, 1965). Principal res. scient. (tel. 03 + 542-2777, ext. 2702, telex AA 32945, fax 03 + 544-1128). *Oxide structures, refractory oxides.*

Rossouw, Dr Christopher John. CSIRO Div. of Materials Sci. and Techn., Locked Bag 33, Clayton, Victoria 3168, Australia. (1951) DPhil, physics (Oxford U., UK, 1977). Sr. res. scient. (tel. 03 + 542-2885, telex AA 32945, fax 03 + 544-1128). *Electron diffraction and channelling, inelastic scattering, HRTEM.*

Sabine, Prof Terence Murray. School of Physical Sciences, U. of Techn., Sydney, Broadway, Sydney, New South Wales 2007, Australia. (1930) DSc, physics (U. Melbourne, 1971). Prof. of physics. (tel. 02 + 218-9418, telex AA 75004, fax 02 + 281-2498, email tms at atomcom.oz). *Neutron diffraction, X-ray diffraction, synchrotron radiation, nuclear energy, engineering ceramics.*

Scott, Dr Henry Gordon. CSIRO Div. of Materials Sci. and Techn., Locked Bag 33, Clayton, Victoria 3168, Australia. (1933) PhD, physical chemistry (Cambridge U., UK, 1957). Program leader, solid state science. (tel. 03 + 542-2737, fax 03 + 544-1128). *X-ray powder diffraction, refractory oxide systems, crystal structures, solid state chemistry.*

Scudder, Dr Marcia Lorraine. Dept. of Inorganic Chemistry, U. of New South Wales, Anzac Parade, Kensington, New South Wales 2033, Australia. (1948) PhD, chemistry (U. Sydney, 1973). Professional officer. (tel. 02 + 697-2222, ext. 4595, telex AA 26054, fax 02 + 662-2835). *Complexes of metals with sulphur, ligands.*

Self, Dr Peter Geoffrey. CSIRO Div. of Soils, Private Bag 2, Glen Osmond, South Australia 5064, Australia. (1952) PhD, physics, electron microscopy (U. Melbourne, 1979). Res. scient. (tel. 08 + 274-9311, telex 82406, fax 08 + 338-1636). *Electron microscopy, high resolution electron microscopy, analytical electron microscopy.*

Sellar, Dr Jeffrey Ronald John. I.C.I. Australia Research Group, Newsom St., Ascot Vale, Victoria 3032, Australia. (1946) PhD, physics (Arizona State U., 1976). Sr. res. scient. (tel. 03 + 377-6418, fax 03 + 370-2250). *Electron diffraction, electron microscopy, solid state chemistry (esp. zirconia ceramics).*

Sinclair, Dr William John. Melbourne Research Laboratories, B.H.P., 245 Wellington Rd., Mulgrave, Melbourne, Victoria 3170, Australia. (1953) PhD, crystal chemistry (Australian Nat. U., 1982). Sr. res. officer. (tel. 03 + 560-7066, telex AA 30408, fax 03 + 561-6709). *Electron microscopy, materials science.*

Skelton, Dr Brian Warwick. Dept. of Physical and Inorganic Chemistry, U. of Western Australia, Nedlands, Western Australia 6009, Australia. (1948) PhD, chemical crystallography (U. Auckland, New Zealand, 1974). Res. officer. (tel. 09 + 380-2726). *Structures, inorganic and organic compounds.*

Slade, Dr Phillip Garland. CSIRO Div. of Soils, Waite Rd., Glen Osmond, Adelaide, South Australia 5064, Australia. (1941) PhD, chemical mineralogy (U. Adelaide, 1968). Principal res. scient. (tel. 08 + 274-9302, telex AA 82406, fax 08 + 338-1636). *Crystal chemistry, structures, clay minerals, micas, clay-organic intercalates.*

Smith, Dr Katherine Leah. Materials Division, Australian Nuclear Sci. and Techn. Organization, Private Mail Bag 1, Menai, New South Wales 2234, Australia. (1955) PhD, physics (Monash U., 1982). Research scient. (tel. 02 + 543-3111, ext. 3505, telex AA 24562, fax 02 + 543-5097). *Electron microscopy, mineralogy, ceramics.*

Smith, Dr Ross McDowall. Dept. of Metallurgy and Materials Engineering, U. of Wollongong, P.O. Box 1144, Wollongong, New South Wales 2500, Australia. (1954) PhD, metallurgy (U. Wollongong, 1988). Sr. res. metallurgist. (tel. 042 + 27-0009). *Precipitation hardening, HSLA steels, hot deformation, fatigue, titanium alloys.*

Smith, Prof Thomas Frederick. Physics Dept., Monash U., Wellington Rd., Clayton, Victoria 3168, Australia. (1939) PhD, physics (U. Sheffield, UK, 1963). Prof. of Exp. physics. (tel. 03 + 565-3630, telex AA 32691, MONASH, fax 03 + 565-3637). *Lattice dynamics, soft modes, lattice stability.*

Snow, Dr Michael Robert. Dept. of Physical and Inorganic Chemistry, U. of Adelaide, North Terrace, Adelaide, South Australia 5000, Australia. (1940)

PhD, inorganic chemistry (U. London, UK, 1966). Reader. (tel. 08 + 228-5559, telex UNIVAD AA 89141, fax 08 + 224-0464). *Coordination structures, organometallic structures.*

Spackman, Dr Mark Arthur. Dept. of Chemistry, U. of New England, Armidale, New South Wales 2351, Australia. (1954) PhD, theoretical chemistry (U. Western Australia, 1980). Lecturer. (tel. 067 + 73-2722, telex 166050, fax 067 + 73-3122, email mspackma%gara.une.oz at munnari.oz). *Electron density studies, electrostatic properties from diffraction data, intermolecular interactions, model studies of hydrogen bonding.*

Spadaccini, Dr Nicholas. Crystallography Centre, U. of Western Australia, Nedlands, Western Australia 6009, Australia. (1958) PhD, physics (U. Western Australia, 1988). Res. officer. (tel. 09 + 381-6426, email ns at xtal.uwa.oz). *Electron density studies.*

Spink, Mr John Arthur. CSIRO Div. of Materials Sci. and Techn., Locked Bag 33, Clayton, Victoria 3168, Australia. (1925) MSc, physical chemistry (U. Melbourne, 1955). Res. fellow (retired). (tel. 03 + 542-2777, telex AA 32945, fax 03 + 544-1128). *The history of diffraction techniques, structure analysis.*

Steffen, Dr William Lee. CSIRO Centre for Environmental Mechanics, G.P.O. Box 821, Canberra, Australian Capital Territory 2601, Australia. (1947) PhD, inorganic chemistry (U. Florida, USA, 1975). Scientific services officer. (tel. 062 + 46-5648, telex 62861). *Porous media physics, atmospheric physics, science editing and writing.*

Stephens, Dr Frederick Selwyn. Sch. of Chemistry, Macquarie U., North Ryde, New South Wales 2113, Australia. (1938) PhD, chemistry (U. New South Wales, 1963). Sr. lect. (tel. 02 + 805-7111, ext. 8278, telex MACUNI AA 122377, fax 02 + 887-4752). *Transition metal complexes.*

Stephenson, Prof Neville Charles. Sch. of Physical Sci., U. of Techn., Sydney, P.O. Box 123, Broadway, Sydney, New South Wales 2007, Australia. (1928) DSc, chemistry (U. New South Wales, 1970). Prof. of chemistry. (tel. 02 + 218-9472, fax 02 + 281-2498). *Solid state chemistry, structural systematics.*

Sterns, Dr Meta. Dept. of Chemistry, The Faculties, Australian Nat. U., G.P.O. Box 4, Canberra, Australian Capital Territory 2601, Australia. (1931) PhD, chemistry (U. Melbourne, 1965). Sr. tutor. (tel. 062 + 49-2025). *Structures, inorganic and organic compounds, X-ray powder diffractometry.*

Stevenson, Dr Andrew Wesley. CSIRO Div. of Materials Sci. and Techn., PO Box 160, Clayton, Victoria 3168, Australia. (1957) PhD, physics (U. Melbourne, 1984). Res. scient. (tel. 03 + 542-2917, telex AA 32945, fax 03 + 544-1128). *X-ray diffraction, diffractometry, materials science, instrumentation, thermal vibrations, semiconductors, X-ray optics, synchrotron radiation.*

Taylor, Dr John Charles. CSIRO Div. of Fuel Techn., Private Mail Bag 7, Menai, New South Wales 2234, Australia. (1935) DSc, chemistry (U. New South Wales, 1982). Sr. principal res. scient. (tel. 02 + 543-3013, fax 02 + 543-6774). *Powder diffraction, neutron and X-ray crystallography, actinide chemistry.*

Taylor, Dr Max Ronald. Sch. of Physical Sci., Flinders U. of South Australia, Sturt Rd., Bedford Park, South Australia 5042, Australia. (1936) PhD, chemistry (U. Sydney, 1964). Sr. lect. in chemistry. (tel. 08 + 275-2467, fax 08 + 277-5523). *Computing, biological metal complexes, charge density.*

Thompson, Dr John Gerard. Research School of Chemistry, Australian Nat. U., G.P.O. Box 4, Canberra, Australian Capital Territory 2601, Australia. (1954) PhD, geology (James Cook U., 1986). Res. fellow. (tel. 062 + 49-2620, telex AA 62172, fax 062 + 48-7817). *Solid state, crystal chemistry, clay mineralogy, superconductors.*

Tiekink, Dr Edward Richard Thomas. Dept. of Physical and Inorganic Chemistry, U. of Adelaide, Nth. Terrace, Adelaide, South Australia 5000, Australia. (1960) PhD, inorganic chemistry (U. Melbourne, 1986) Research assoc. (tel. 08 + 228-5943, telex UNIVAD AA 89141, fax 08 + 224-0464). *Structure analyses of small molecules, layer silicates, bioinorganic systems.*

Tong, Mr Hua. Dept. of Inorganic Chemistry, U. of Sydney, Sydney, New South Wales 2006, Australia. (1963) BSc, chemistry (Nankai U., 1982). Student. (tel. 02 + 692-2741). *Protein crystallography.*

Town, Dr Susan Lesley. School of Physics, U. of New South Wales, P.O. Box 1, Kensington, Sydney, New South Wales 2113, Australia. (1959) PhD, gamma-ray and neutron Kossel and Kikuchi effects (Monash U., 1987). Sr. res. assist. (tel. 02 + 697-4591, telex AA 26054, fax 02 + 663-3420). *X-ray diffraction, neutron diffraction, magnetism, texture, Heusler alloys, Kikuchi and Kossel effects, dynamical diffraction, high-temperature superconductors, gamma-ray scattering, solid state crystallography.*

Trefry, Mr Michael George. Physics Dept., U. of Western Australia, Hackett Drive, Crawley, Perth, Western Australia 6009, Australia. (1961) BSc(Hons), physics (U. Western Australia, 1983). Grad. student. (tel. 09 + 380-2738, email mgt at xtal.uwa.oz). *Electron density, chemical binding, solid state physics.*

Tulip, Mr William Richard. CSIRO Div. of Biotechnology, 343 Royal Parade, Parkville, Melbourne, Victoria 3052, Australia. (1964) BSc(Hons), chemistry (U. of Sydney, 1987). PhD student. (tel. 03 + 342-4200, ext. 270, telex 33983, fax 03 + 347-5481, email bill at dpca.dn.mu.oz). *Protein crystallography.*

Tulloch, Dr Peter Archibald. CSIRO Div. of Biotechnology, 343 Royal Parade, Parkville, Victoria 3052, Australia. (1942) PhD, physics (U. Melbourne, 1971). Principal res. scient. (tel. 03 + 342-4200, ext. 275, telex AA 33983, fax 03 + 347-5481, email fange at dpcb.dn.mu.oz). *Electron diffraction, electron microscopy, electron protein crystallography.*

Usher, Dr Brian Francis. Solid State Electronics Section, Telecom Australia Research Laboratories, 770 Blackburn Road, Clayton, Victoria 3168, Australia. (1949) PhD, physics (U. Western Australia, 1981). Principal scient. (tel. 03 + 541-6681, telex AA 33999, fax 03 + 543-4127). *X-ray diffraction, thin films and superlattices, optoelectronic materials.*

Vagg, Dr Robert Sylvester. Sch. of Chemistry, Macquarie U., New South Wales 2109, Australia. (1945) PhD, chemistry (Macquarie U., 1971). Sr. lect. in chemistry. (tel. 02 + 805-7111, ext. 8269, telex MACUNI AA 122377, fax 02 + 887-4752). *Transition metal complexes.*

Varghese, Dr Joseph Noozhumurry. CSIRO Div. of Biotechnology, 343 Royal Parade, Parkville, Victoria 3052, Australia. (1949) PhD, physics (U. Western Australia, 1974). Principal res. scient. (tel. 03 + 342-4200, ext. 277, telex 33983, fax 03 + 347-5481, email jose at dpca.dn.mu.oz (BITNET)). *Viruses, protein structure and function.*

Vilkins, Ms Louise Mary. School of Physical Sciences, The Flinders U. of South Australia, Sturt Road, Bedford Park, South Australia 5042, Australia. (1955) BSc(Hons), chemistry (The Flinders U. South Australia, 1986). Postgraduate student. (tel. 08 + 275-2462). *Charge density, small molecules.*

Wagenfeld, Dr Heinrich Karsten. Dept. of Appl. Physics, Royal Melbourne Inst. of Techn., 124 Latrobe St., Melbourne, Victoria 3000, Australia. (1928) Dr.Rer.Nat., physics (Freie U. Berlin, BRD, 1958). Dept. head. (tel. 03 + 660-2135, telex AA 36406). *Dynamical scattering, X-ray, neutron, electron scattering, inelastic scattering, ion implantation, secondary electron emission, Auger spectroscopy.*

Warminski, Dr Tadeusz Piotr. Telecom Techn. Branch, Telecom Research Laboratories, 770 Blackburn Rd., Clayton, Victoria 3168, Australia. (1940) Dr habil., physics (Polish Academy of Sciences, Poland, 1974). Principal scient. (tel. 03 + 541-6688, telex 33999 RLABA, fax 03 + 543-4127). *Electron microscopy, semiconductors, X-ray and electron probe analysis, heavy-metal fluoride glasses.*

Watson, Dr Kenneth John. 65 Circe Circle, Dalkeith, Western Australia 6009, Australia. (1935) PhD, physics (U. Western Australia, 1967). Consultant. (tel. 09 + 386-3330). *Chemical bonding, charge density, protein structure, function and engineering, exercise biochemistry and physiology.*

Watts, Mr John Andrew. CSIRO Div. of Mineral Products, 339 Williamstown Rd., Port Melbourne, Victoria 3207, Australia. (1933) BSc, chemistry (Ballarat Sch. of Mines, 1960). Experimental scient. (tel. 03 + 647-0211, ext. 272, telex AA 34349, fax 03 + 647-0395). *Structure and chemistry, sulphides, titanates and ferrites.*

Welberry, Dr Thomas Richard. Res. Sch. of Chemistry, Australian Nat. U., P.O. Box 4, Canberra, Australian Capital Territory 2601, Australia. (1945) PhD, chemical crystallography (London, UK, 1970). Sr. fellow (tel. 062 + 49-4122, telex AA 62172, fax 062 + 48-7817, email welberry at rsc0.anu.oz). *Disorder, phase transitions, statistical models, potential energy calculations.*

Westphalen, Dr John Arthur. South Australian Inst. of Techn., Levels Campus, Ingle Farm, South Australia 5098, Australia. (1929) PhD, X-ray crystallography (The Flinders U. of South Australia, 1984). Sr. lect. (tel. 085 + 260-2055, ext. 2065, telex 82565, fax 08 + 349-6939). *Structure analysis.*

Whillans, Dr Francis David. Dept. of Appl. Chemistry, Phillip Inst. of Techn., Plenty Rd., Bundoora, Victoria 3083, Australia. (1942) PhD, inorganic crystallography (U. Melbourne, 1971). Lect. (tel. 03 + 468-2479). *Chemical education, X-ray fluorescence analysis, thermal analysis.*

Whitaker. Miss Claire Rosemary. Physical and Inorganic Chemistry Dept., U. of Western Australia, Nedlands, Perth, Western Australia 6009, Australia. (1962) BSc(Hons), inorganic chemistry (U. Western Australia, 1984). PhD student. (tel. 09 + 380-3144, email cw at xtal.uwa.oz). *Lithium salt complexes, sodium salt complexes.*

White, Dr Allan Henry. Sch. of Chemistry, U. of Western Australia, Nedlands, Western Australia 6009, Australia. (1938) DSc, inorganic chemistry (U. Melbourne, 1981). Assoc. prof. (tel. 09 + 380-3144, telex AA 92992, fax 09 + 388-1005). *Structure, synthesis, inorganic and coordination chemistry.*

White, Prof John William. Research School of Chemistry, Australian Nat. U., G.P.O. Box 4, Canberra, Australian Capital Territory 2601, Australia. (1937) D. Phil., chemistry (Oxford U., U.K., 1962). Prof. and head of physical and theoretical chemistry. (tel. 062 + 49-3578, telex AA 62172, fax 062 + 48-7817, email heather!rsc0.anu at munnari.oz). *Neutron scattering, X-ray scattering, molecular electronics, polymers, surfaces, interfacial phenomena, simulation, catalysis.*

White, Dr Timothy John. School of Physics, U. of Melbourne, Parkville, Victoria 3052, Australia. (1958) PhD, chemistry (Australian Nat. U., 1982). Res. specialist. (tel. 03 + 344-5083, telex AA 35185 UNIMEL, fax 03 + 347-4783). *Electron microscopy, crystal chemistry, ceramics.*

Whitfield, Dr Harold John. CSIRO Div. of Materials Sci. and Techn., Locked Bag 33, Clayton, Victoria 3168, Australia. (1931) PhD, chemistry (Victoria U., Wellington, New Zealand, 1967). Principal res. scient. (tel. 03 + 542-2777, ext. 2924, telex AA 32945). *Electron diffraction, high resolution lattice imaging, superconductivity, inorganic structures.*

Wielunski, Dr Leszek Stanislaw. CSIRO Div. of Applied Physics, Private Mail Bag 7, Menai, New South Wales 2234, Australia. (1943) PhD, applied physics (Inst. Nucl. Res., Warsaw, Poland, 1972). Principal res. scient. (tel. 02 + 543-3302, telex AA 73341, fax 02 + 543-6774). *Materials science, semiconductors, semi-*

conductor structures, ion beam materials analysis, channelling, dechannelling on dislocations, crystal defects, interfaces, materials modification by ion implantation.

Wilkins, Dr Stephen William. CSIRO, Div. of Materials Sci. and Techn., Locked Bag 33, Clayton, Victoria 3168, Australia. (1946) PhD, physics (U. Melbourne, 1972). Principal res. scient. (tel. 03 + 542-2918, telex AA 32945, fax 03 + 544-1128). *X-ray optics and instrumentation, dynamical diffraction, diffraction physics, maximum-entropy methods, microstructural characterization.*

Williams, Mr Brian Edward. Aircraft Materials Div., Aeronautical Res. Laboratory, 506 Lorimer St., Port Melbourne, Victoria 3207, Australia. (1925) Fellowship Dipl., applied physics (Royal Melbourne Inst. of Techn., 1974). Experimental officer (retired). (tel. 03 + 647-7534). *Identification, powder methods.*

Williams, Dr Geoffrey Allan. Australian Radiation Lab., Lower Plenty Rd., Yallambie, Victoria 3085, Australia. (1950) PhD, chemistry (U. Melbourne, 1976). Sr. res. scient. (tel. 03 + 433-2211, ext. 398, telex 31726, fax 03 + 434-5654). *X-ray diffraction, spin density, charge density, Technetium chemistry, transition metal complexes.*

Wilson, Dr Alan Richard. Aircraft Materials Div., Aeronautical Res. Labs., Box 4331, G.P.O., Melbourne, Victoria 3001, Australia. (1953) PhD, physics,

HREM (U. Melbourne, 1982). Sr. res. scient. (tel. 03 + 647-7508, fax 03 + 645-3869). *Structure analysis, imaging, energy dispersive X-ray analysis, energy loss spectroscopy, weak-beam electron diffraction.*

Withers, Dr Raymond Leslie. Research School of Chemistry, Australian Nat. U., G.P.O. Box 4, Canberra, Australian Capital Territory 2601, Australia. (1954) PhD, diffraction physics (U. Melbourne, 1982). Res. fellow. (tel. 062 + 49-3714, telex AA 62172, fax 062 + 48-7817). *Electron diffraction, phase transitions, quasicrystals, modulated structures.*

Wood, Dr Graeme John. School of Physics, U. of Melbourne, Parkville, Victoria 3052, Australia. (1956) PhD, physics (U. Melbourne, 1981). Lecturer. (tel. 03 + 344-5438, fax 03 + 347-4783, email dwood at munda.mu.oz.au). *Materials science, electron microscopy, imaging theory, digital video processing.*

Wunderlich, Dr Jeffrey Alfred. CSIRO Div. of Mineral Products, P.O. Box 124, Port Melbourne, Victoria 3207, Australia. (1931) Dr. es. Sc., chemical crystallography (U. Paris, France, 1958). Principal res. scient. (tel. 03 + 647-0345, telex 34349, fax 03 + 647-0395). *Inorganic and mineral structures, XRD and XRF analysis.*

AUSTRIA

Sub-Editor: **A. Preisinger**

Notes

1. International telephone country code - 43.

2. Austrian universities grant various degrees. Those occurring below are, in ascending order, *Magister* (Mag.), conferred by a university and approximately equivalent to MA; *Diplom-Ingenieur* (Dipl.-Ing.), conferred by a technical university and approximately equivalent to MSc; *Doctor philosophiae* (Dr. phil.), conferred by a university, *Doctor technicae* (Dr. techn.), conferred by a technical university, and *Doctor rerum naturalium* (Dr. rer. nat.), conferred by both types of universities, all three doctor grades being approximately equivalent to PhD; and *Dozent* (Doz.), *venia legendi,* granted by both types of universities.

Aiginger, Prof. Dr Dipl.-Ing. Hannes. Atominst. der Österreichischen Universitäten, Schüttelstrasse 115, A-1020 Vienna, Austria. (1937) Doz., physics of electrons, X-rays and gamma-rays (Techn. U. Vienna, 1970). Full prof. (tel. 0222 + 21701, ext. 278). *Crystal orientation, X-ray metallography, textures, X-ray fluorescence analysis.*

Amthauer, Prof. Dr Georg. Inst. für Geowissenschaften, Abt. für Mineralogie, Universität Salzburg, Hellbrunnerstr. 34, A-5020 Salzburg, Austria. (1942) Doz., crystallography and mineralogy (U. Marburg/FRG, 1981). Full Prof. (tel. 0662 + 8044, ext. 5402). *Chemical bonding, high pressures, materials science, minerals, phase transitions, structural chemistry, X-ray diffraction.*

Badurek, Prof. Dr Dipl.-Ing. Gerald. Inst. für Kernphysik, Techn. U. Wien, Schüttelstrasse 115, A-1220 Vienna, Austria. (1948) Doz., nuclear solid state physics (Techn. U. Wien, 1983). Ass. prof. (tel. 0222 + 21701, ext. 229). *Diffractometry, instrumentation, neutron diffraction, small angle scattering.*

Bauer, Prof. Dr Günther Ernst. Inst. für Physik, Montanu. Leoben, Franz Josefstrasse 18, A-8700 Leoben, Austria. (1942) Doz., physics (Techn. U. Aachen, W. Germany, 1974). Full prof. (tel. 03842 + 42555, ext. 260; fax 46 010 40). *Crystal growth, IV-VI compounds, superlattices, ferroelectric phase transitions, FIR spectroscopy.*

Baumgartner, Prof. Dr Dipl.-Ing. Oswald. Inst. für Mineralogie, Kristallographie und Strukturchemie, Techn. U. Wien, Getreidemarkt 9, A-1060 Vienna, Austria. (1948) Dr. techn., chemistry (Techn. U. Vienna, 1977). Ass. Prof. (tel. 0222 + 58801, ext. 4743, email e171007 at awituw01). *Crystal structures, neutron diffraction.*

Becherer, Dr Karl. Inst. für Mineralogie und Kristallographie, U. Wien, Dr. Karl Lueger-Ring 1, A-1010 Vienna, Austria. (1926) Dr. phil., mineralogy-petrography (U. Vienna, 1961). Scient. officer. (tel. 0222 + 4300, ext. 2332). *Analytical chemistry, minerals, crystal optics, general crystallography.*

Beran, Prof. Dr Anton. Inst. für Mineralogie und Kristallographie, U. Wien, Dr. Karl Lueger-Ring 1, A-1010 Vienna, Austria. (1944) Doz., mineralogy (U. Vienna, 1980). Full Prof. (tel. 0222 + 4300, ext. 2320). *IR spectroscopy, microscopy, minerals.*

Betz, Prof. Dr Gerhard. Inst. für Allgemeine Physik, Techn. U. Wien, Wiedner Hauptstrasse 8-10, A-1040 Vienna, Austria. (1944) Doz., ionic physics (U. Vienna, 1981). Ass. Prof. (tel. 0222 + 58801, ext. 5591; telex 131000 tvfa wa). *Surface composition, chemical structure, sputtering, Auger electron spectroscopy.*

Blaschko, Dr Oskar. Inst. für Experimentalphysik, U. Wien, Strudlhofgasse 4, A-1090 Vienna, Austria. (1948) Dr. phil., solid state physics (U. Vienna, 1974). Asst. (tel. 0222 + 342630, ext. 226). *Phonons, phase transformations.*

Boller, Prof. Dr Herbert. Inst. für Chemie, Abt. für Allg. und Anorg. Chemie, U. Linz, Altenbergerstrasse 69, A-4040 Linz-Auhof, Austria. (1937) Doz., physical chemistry (U. Vienna, 1974). Full prof. (tel. 0732 + 2468, ext. 805; fax 0732

2468 10). *Structural chemistry, inorganic crystals, metals, materials science, magnetism.*

Brandstätter, Dr Franz. Mineralog.-petrograph. Abteilung, Naturhistorisches Museum, P.O. Box 417, Burgring 7, A-1014 Vienna, Austria. (1953) Dr. phil., mineralogy (U. Vienna, 1979). Res. asst. (tel. 0222 + 934145, ext. 265). *Minerals in meteorites.*

Breiter, Prof. Dr Manfred. Inst. für Technische Elektrochemie, Techn. U. Wien, Getreidemarkt 9, A-1060 Vienna, Austria. (1925) Doz., physical chemistry (Techn. U. Munich/FRG, 1957). Full Prof. and head. (tel. 0222 + 58801, ext. 4762; telex (61) 3222467 tuw) *Electron diffraction, inorganic crystals, surface structure, solid electrolytes, characterization, interfaces.*

Dirl, Prof. Dr Rainer. Inst. für Theoretische Physik, Techn. U. Wien, Wiedner Hauptstrasse 8-10, A-1040 Vienna, Austria. (1941) Doz., mathematical physics (Techn. U. Vienna, 1977). Ass. Prof. (tel. 0222 + 58801, ext. 5684). *Group theory.*

Eder, Prof. Dr Dipl.-Ing. Otto Josef. Physikinstitut, Österreichisches Forschungszentrum Seibersdorf, A-2244 Seibersdorf, Austria. (1936) Doz., experimental physics (U. Vienna, 1979). Head of physics department. (tel. 02254 + 80, ext. 3100; telex 014353 fzs a; fax 02254 80 2118). *Solid and liquid state physics, neutron scattering, neutron structure factors, statistical mechanics.*

Effenberger, Prof. Dr Herta. Inst. für Mineralogie und Kristallographie, U. Wien, Dr. Karl Lueger-Ring 1, A-1010 Vienna, Austria. (1954) Doz., mineralogy (U. Vienna, 1987). Ass. Prof. (tel. 0222 + 4300, ext. 2687). *Inorganic crystals, minerals, structural chemistry.*

Ellmeyer, Mag. Wolfgang. Inst. für Experimentalphysik, U. Wien, Strudlhofgasse 4, A-1090 Vienna, Austria. (1961) Mag., physics (U. Vienna, 1988). Scient. (tel. 0222 + 342630, ext. 225). *X-ray diffraction, symmetry, phase transitions, high pressures, crystallographic data.*

Ernst, Dr Gert. Physikinstitut, Österreichisches Forschungszentrum Seibersdorf, Lenaugasse 10, A-1082 Vienna, Austria. (1944) Dr. phil., physics (U. Vienna, 1971). Res. scient. (tel. 02254 + 80, ext. 3122; telex 014353). *Solid state physics, phase transformations, lattice dynamics, neutron spectrometry.*

Ettmayer, Prof. Dr Dipl.-Ing. Peter. Inst. für Chemische Technologie Anorganischer Stoffe, Techn. U. Wien, Getreidemarkt 9, A-1060 Vienna, Austria. (1934) Doz., chemical technology (Techn. U. Vienna, 1972). Full prof. (tel. 0222 + 58801, ext. 4799; fax 0222 5054800). *Diffractometry, materials science, phase determination.*

Fischer, Dr Richard. Inst. für Mineralogie und Kristallographie, U. Wien, Dr. Karl Lueger-Ring 1, A-1010 Vienna, Austria. (1933) Dr. phil., chemistry (U. Vienna, 1963). Asst. (tel. 0222 + 4300, ext. 2335). *Mineralogy, mineral deposits, inorganic crystal structures, lattice energies.*

Fuith, Dr Mag. Armin. Inst. für Experimentalphysik, U. Wien, Strudlhofgasse 4, A-1090 Vienna, Austria. (1949) Dr. phil., metal physics (U. Vienna, 1979). Asst. (tel. 0222 + 346515, ext. 225; telex 116222). *Crystal growth, defect structures, inorganic crystals, phase transitions.*

Glatter, Dr Otto. Inst. für Physikalische Chemie, U. Graz, Heinrichstrasse 28, A-8010 Graz, Austria. (1945) Dr. techn., physics (Techn. U. Graz, 1972). Asst. (tel. 0316 + 380, ext. 5433 or 5439; email glatter at edvz.uni-graz.ptt.at). *Small angle scattering, elastic light scattering, quasi-elastic light scattering, computer programming, inverse problems.*

Götzinger, Dr Michael Alois. Inst. für Mineralogie und Kristallographie, U. Wien, Dr. Karl Lueger-Ring 1, A-1010 Vienna, Austria. (1949) Dr. phil., mineralogy and petrography (U. Vienna, 1976). Asst. (tel. 0222 + 4300, ext. 2688). *Mineralogy, mineral deposites, X-ray diffraction phase analysis, inorganic crystal structures.*

Haditsch, Prof. Dr Johann Georg. Inst. für Geowissenschaften, Montanu. Leoben, A-8700 Leoben, Austria. (1934) Doz., mineralogy, economic geology (Montanu. Leoben, 1967). Full prof. (tel. 03842 + 42555, ext. 452). *X-ray diffraction phase analysis, ore microscopy, environmental geology.*

Halwax, Dr Dipl.-Ing. Erich Johann. Inst. für Mineralogie, Kristallographie und Strukturchemie, Techn. U. Wien, Getreidemarkt 9, A-1060 Vienna, Austria. (1951) Dr. techn., chemistry (Techn. U. Vienna, 1985). Asst. (tel. 0222 + 58801, ext. 4742 email e1710hc1 at awiiez11). *Inorganic crystals, structural chemistry, powder diffraction.*

Hausner, Dipl.-Ing. Robert. Forschungsinstitut, Veitscher Magnesitwerke A.G., Magnesitstrasse 2, A-8700 Leoben, Austria. (1941) Dipl.-Ing., physics (Techn. U. Vienna, 1966). Head of X-ray laboratory (tel. 03842 + 22581, ext. 361, telex 33323). *X-ray diffraction, inorganic crystal structures, carbon structures (pitch - coke - graphite), quantitative analysis, refractories, X-ray fluorescence analysis.*

Heritsch, Prof. em. Dr Haymo. Inst. für Mineralogie, Kristallographie und Petrologie, U. Graz, Universitätsplatz 2, A-8010 Graz, Austria. (1911) Doz., mineralogy-petrography (U. Graz, 1939). Prof. em. (tel. 0316 + 380, ext. 5540). *Inorganic crystal structures, crystal chemistry.*

Hiebl, Prof. Dr Kurt. Inst. für Physikalische Chemie, U. Wien, Währingerstrasse 42, A-1090 Vienna, Austria. (1948) Doz., physics (U. Vienna, 1986). Ass. Prof. (tel. 0222 + 343616, ext. 19). *Magnetism in solids, nuclear magnetic resonance in solids.*

Higatsberger, Prof. Dr Michael Josef. Inst. für Experimentalphysik, U. Wien, Boltzmanngasse 5, A-1090 Vienna, Austria. (1924) Doz., physics (U. Vienna, 1959). Full prof. and head. (tel. 0222 + 345232; telex 116222). *Solid state physics, instrumentation.*

Holub, Dr Fritz. Inst. für Werkstofftechnologie, Österreichisches Forschungszentrum Seibersdorf, A-2444 Seibersdorf, Austria. (1930) Dr. phil., chemistry (U. Vienna, 1961). Group leader (tel. 02254 + 80, ext. 3303; telex 014353 fzs). *Materials science, X-ray diffraction, metals, inorganic crystals, powder diffraction.*

Hörl, Prof. Dr Erwin M. Inst. für Werkstofftechnologie, Österreichisches Forschungszentrum Seibersdorf, A-2444 Seibersdorf, Austria. (1929) Doz., experimental solid state physics (Techn. U. Vienna, 1965). Head. (tel. 02254 + 80, ext. 3300). *Structures, solidified permanent gases, oxygen, stacking faults, computer programming, indexing Debye-Scherrer diagrams, powder patterns of low symmetry.*

Jánosi, Dr Dipl.-Ing. András. Inst. für Physikalische Chemie, U. Graz, Heinrichstrasse 28, A-8010 Graz, Austria. (1929) Dr. phil., chemistry (Katholische U. Löwen, 1959). Scient. (tel. 0316 + 380, ext. 5423). *Small angle scattering, wide angle X-ray scattering, synthetic and natural polymer systems.*

Kabelka, Dr Heinz I. Inst. für Experimentalphysik, U. Wien, Strudlhofgasse 4, A-1090 Vienna, Austria. (1949) Dr. phil., solid state physics (U. Vienna, 1976). Asst. (tel. 0222 + 342630, ext. 293; telex 116222). *Phase transitions in solids, ultrasonic phase velocity and attenuation, Raman scattering.*

Kahlert, Prof. Dr Hartmut. Inst. für Festkörperphysik, Techn. U. Graz, Petersgasse 16, A-8010 Graz, Austria. (1940) Doz., solid state physics (U. Vienna, 1976). Full prof. and head. (tel. 0316 + 7061, ext. 8460). *Crystal growth, quasi-one-dimensional solids, polymer preparation, polymer crystal structure, polyacethylene, polyparaphenylene, polysulphurnitride, high Tc-superconductors, layered perovskites.*

Kirchner, Prof. Dr Elisabeth Charlotte. Inst. für Geowissenschaften, U. Salzburg, Akademiestrasse 26, A-5020 Salzburg, Austria. (1935) Doz., mineralogy-petrography-crystallography (U. Salzburg, 1979). Full Prof. (tel. 0662 + 8044, ext. 5403). *Powder diffraction, crystal growth, inorganic crystals, structure-symmetry.*

Klepp, Prof. Dr Kurt Otto, Inst. für Chemie, Abt. für Allg. u. Anorg. Chemie, U. Linz, Altenbergerstrasse 69, A-4040 Linz, Austria. (1944) Doz., inorganic chemistry (U. Vienna, 1987). Ass. Prof. (tel. 0732 + 2468, ext. 800; fax 0723 2468 10; email k360870 at aearn). *Inorganic crystal structures, crystal chemistry.*

Kohlbeck, Prof. Dr Franz. Inst. für Geophysik, Techn. U. Wien, Gusshausstrasse 27-29, A-1040 Vienna, Austria. (1943) Doz., engineering geophysics (Techn. U. Vienna, 1981). Ass. Prof. (tel. 0222 + 58801, ext. 3803). *Computer programming, powder diffraction analysis.*

Komarek, Prof. Dr Kurt Ludwig. Inst. für Anorganische Chemie, U. Wien, Währingerstrasse 42, A-1090 Vienna, Austria. (1926) Dr. phil., chemistry (U. Vienna, 1949). Full prof. and head. (tel. 0222 + 345424, ext. 2). *Crystal growth, inorganic crystal structures, magnetism, materials science.*

Kratky, Dr Christoph. Inst. für Physikalische Chemie, U. Graz, Heinrichstrasse 28, A-8010 Graz, Austria. (1946) Dr. techn., chemistry (E.T.H. Zürich/Switzerland, 1976). Asst. (tel. 0316 + 380, ext. 5417). *Structural chemistry, proteins.*

Kratky, Prof. em. Dr Dipl.-Ing. Dr h.c. mult. Otto. Drosselweg 15, A-8010 Graz, Austria. (1902) Dr. techn., chemistry (Techn. U. Vienna, 1929). Prof. em. (tel. 0316 + 42073). *Small angle X-ray scattering.*

Krischner, Prof. Dr Dipl.-Ing. Harald. Inst. für Physikalische und Theoretische Chemie (Strukturforschung), Techn. U. Graz, Rechbauerstrasse 12, A-8010 Graz, Austria. (1930) Doz., physical chemistry (Techn. U. Graz, 1964). Full prof. (tel. 0316 + 7061, ext. 8223). *Inorganic crystals, structural chemistry.*

Kuchar, Doz. Dr Friedemar. Inst für Festkörperphysik, U. Wien, Strudlhofgasse 4, A-1090 Vienna, Austria. (1941) Doz., experimental solid state physics (U. Vienna, 1978). Asst. (tel. 0222 + 342630, ext. 264). *Impurities in semiconductors, IR properties, heterostructures, MOS structures.*

Kunsch, Dr Dipl.-Ing. Barnabas. Physikinstitut, Österreichisches Forschungszentrum Seibersdorf, A-2444 Seibersdorf, Austria. (1942) Dr. techn., physics (Techn. U. Vienna, 1970). Res. scient. (tel. 02254 + 80, ext. 3103; telex 014353). *Instrumentation, metals, neutron diffraction, powder diffraction.*

Kuzmany, Prof. Dr Hans. Inst. für Festkörperphysik, U. Wien, Strudlhofgasse 4, A-1090 Vienna, Austria. (1940) Doz., experimental physics (U. Vienna, 1976). Full Prof. (tel. 0222 + 342630, ext. 245; telex 116222; email a8231dad at awiuni11). *Materials science, polymers, vibrations, spectroscopy, high temperature superconductors.*

Laggner, Prof. Dr Peter. Inst. für Röntgenfeinstrukturforschung, Österreichische Akademie der Wissenschaften, Steyrergasse 17, A-8010 Graz, Austria. (1944) Doz., biochemistry-biophysics-physical chemistry (U. Graz, 1978). Managing director. (tel. 0316 + 812003; telex 311265; fax 77685; email laggner at stg.tu-graz.ptt.at). *Small angle X-ray scattering, biopolymers in solution, lipoproteins, membranes, amorphous materials, liquid crystals.*

Lengauer, Mr. Christian Leopold. Inst. für Geowissenschaften, Abt. für Mineralogie und Petrographie, U. Salzburg, Hellbrunnerstrasse 34/III, A-5020 Salzburg, Austria. (1959) Scient. (tel. 0662 + 8044, ext. 5445). *Computing, minerals, powder diffraction, phase determination.*

Lengauer, Dr Dipl.-Ing. Walter Oskar Franz. Inst. für Chemische Technologie Anorganischer Stoffe, Techn. U. Wien, Getreidemarkt 9, A-1060 Vienna, Austria. (1958) Dr. techn., chemistry (Techn. U. Vienna, 1986). Asst. (tel. 0222 + 58801, ext. 4812). *Materials science, phase transitions.*

Lihl, Prof. em. Dr Franz. Inst. für Angewandte und Technische Physik, Techn. U. Wien, Wiedner Hauptstrasse 8-10/137, A-1040 Vienna, Austria. (1906) Dr. phil., physics (U. Vienna, 1930). Prof. em. (tel. 0222 + 58801, ext. 3241). *Metal structures, magnetic structures, order-disorder and exsolution processes, low temperatures.*

Linke, Dr Walter. Wienerberger Baustoffe, Wienerbergstrasse 11, A-1100 Vienna, Austria. (1944) Dr. phil., mineralogy (U. Vienna, 1970). Scient. (tel. 0222 + 629241, ext. 370). *Crystal growth, crystal structures.*

Lottermoser, Dr Werner. Inst. für Geowissenschaften, U. Salzburg, Hellbrunnerstrasse 34, A-5020 Salzburg, Austria. (1955) Dr. phil. nat., solid state physics (U. Frankfurt/FRG, 1986). Asst. (tel. 0662 + 8044, ext. 5422). *Computing, diffractometry, magnetism, neutron diffraction, physics of minerals, spectroscopy, spin density.*

Ludwiczek, Dr Herbert. Inst. für Mineralogie, Kristallographie und Strukturchemie, Techn. U. Wien, Getreidemarkt 9, A-1060 Vienna, Austria. (1940) Dr. phil., physics (U. Vienna, 1973). Scient. (tel. 0222 + 58801, ext. 4749). *Crystal physics, computer programming.*

Mayer, Dr Dipl.-Ing. Helmut. Inst. für Mineralogie, Kristallographie und Strukturchemie, Techn. U. Wien, Getreidemarkt 9, A-1060 Vienna, Austria. (1939) Dr. techn., chemistry (Techn. U. Vienna, 1971). Scient. officer. (tel. 0222 + 58801, ext. 4742). *Inorganic crystal structures and minerals, thermal analysis.*

Mayr, Dr Dipl.-Ing. Michael. Labor für Strukturanalyse (RFP-3), Voest-Alpine, Postbox 3, A-4031 Linz, Austria. (1945) Dr. techn., physics (Techn. U. Vienna, 1975). Head of lab. (tel. 0732 + 585, ext. 2031; telex 2207461 vaa). *X-ray diffraction, textures, applications in steel making industry.*

Mereiter, Prof. Dr Kurt. Inst. für Mineralogie, Kristallographie und Strukturchemie, Techn. U. Wien, Getreidemarkt 9, A-1060 Vienna, Austria. (1945) Doz., mineralogy and crystallography (Techn. U. Vienna, 1987). Ass. Prof. (tel. 0222 + 58801, ext. 4747, email e171006 at awituw01). *Minerals, inorganic crystals, structural chemistry, morphology, crystal optics.*

Mikenda, Prof. Dr Werner. Inst. für Organische Chemie, U. Wien, Währingerstrasse 38, A-1090 Vienna, Austria. (1946) Doz., spectroscopy (U. Vienna, 1986). Ass. Prof. (tel. 0222 + 344630, ext. 74). *Chemical bonding, hydrogen bonding, structural chemistry, spectroscopy.*

Mikler, Dr Helga. Inst. für Anorganische Chemie, U. Wien, Währingerstrasse 42, A-1090 Vienna, Austria. (1932) Dr. phil., chemistry (U. Vienna,1958). Asst. (tel. 0222 + 345424, ext. 6). *Inorganic crystal structures.*

Müller, Doz. Dr Mag. Karl Werner. Inst. für Röntgenfeinstrukturforschung, Österreichische Akademie der Wissenschaften, Steyrergasse 17, A-8010 Graz, Austria. (1947) Doz., physical chemistry of biopolymers (U. Graz, 1985). Res.

Scient. (tel. 0316 + 812004, ext. 75). *Small angle X-ray scattering, polymers in solution, biopolymers, detergent micelles, gallstones.*

Neckel, Prof. Dr Adolf. Inst. für Physikalische Chemie, U. Wien, Währingerstrasse 42, A-1090 Vienna, Austria. (1926) Doz., physical chemistry (U. Vienna, 1965). Full prof. and head. (tel. 0222 + 343616). *Solid state chemistry, energy band structures and properties.*

Niedermayr, Dr Gerhard. Mineralogisch-Petrographische Abteilung, Naturhistorisches Museum, Burgring 7, A-1014 Vienna, Austria. (1941) Dr. phil., mineralogy and petrology (U. Vienna, 1965). Director of Staatliches Edelsteininstitut. (tel. 0222 + 934541, ext. 274). *Minerals, sedimentology, crystallographic data.*

Pertlik, Prof. Dr Franz. Inst. für Mineralogie und Kristallographie, U. Wien, Dr. Karl Lueger-Ring 1, A-1010 Vienna, Austria. (1943) Doz., mineralogy (U. Vienna, 1979). Ass. Prof. (tel. 0222 + 4300, ext. 2328). *Inorganic crystals, minerals, X-ray diffraction.*

Pilz, Prof. Dr Ingrid Edith. Inst. für Physikalische Chemie, U. Graz, Heinrichstrasse 28, A-8010 Graz, Austria. (1931) Doz., physical chemistry (U. Graz, 1970). Full Prof. (tel. 0316 + 380, ext. 5414) *Small angle scattering, biopolymer structure, immunoglobuline, protein conformational changes, repiratory proteins, hemoglobuline, cyanins.*

Pongratz, Dr Dipl.-Ing. Peter. Inst. für Angewandte Physik, Techn. U. Wien, Karlsplatz 13, A-1040 Vienna, Austria. (1947) Dr. techn., electron microscopy (Techn. U. Vienna, 1980). Asst. (tel. 0222 + 58801, ext. 5626). *Electron diffraction, electron microscopy, contrast theory, defects, semiconductors, ceramics, image processing, materials science.*

Preisinger, Prof. Dr Anton. Inst. für Mineralogie, Kristallographie und Strukturchemie, Techn. U. Wien, Getreidemarkt 9, A-1060 Vienna, Austria. (1925) Doz., mineralogy and crystal chemistry (U. Vienna, 1956). Full prof. and head. (tel. 0222 + 58801, ext. 4749; email e1710hc1 at awiiez11). *Crystal structures, crystal chemistry, biocrystallography, crystal surfaces, general crystallography.*

Reissner, Dipl.-Ing. Michael, Inst. für Angewandte und Technische Physik, Techn. U. Wien, Wiedner Hauptstrasse 8-10, A-1040 Vienna, Austria. (1956) Dipl.-Ing., technical physics (Techn. U. Vienna, 1981). Ass. (tel. 0222 + 58801, ext. 5635). *Magnetism, powder diffraction, X-ray diffraction, inorganic materials, Mössbauer spectroscopy.*

Rogl, Prof. Dr Peter Franz. Inst. für Physikalische Chemie, U. Wien, Währingerstrasse 42, A-1090 Vienna, Austria. (1945) Doz., physical chemistry (U. Vienna, 1980). Ass. Prof. (tel. 0222 + 343616, ext. 14). *Crystallography, structural chemistry, thermodynamics, phase diagrams, high melting systems, solid state chemistry, alloys, refractories, borides, carbides, magnetism.*

Sazedj-Khosrawan, Dr Feresteh. Gersthoferstrasse 150/1/8, A-1180 Vienna, Austria. (1954) Dr. phil., mineralogy and petrography (U. Vienna, 1983). Scient. (tel. 0222 + 4759753). *Mineralogy, crystal structures, crystal chemistry.*

Schattschneider, Prof. Dr Dipl.-Ing. Mag. Peter. Inst. für Angewandte und Technische Physik, Techn. U. Wien, Wiedner Hauptstrasse 8-10, A-1040 Vienna, Austria. (1950) Doz., electron physics (Techn. U. Vienna, 1987). Ass. Prof. (tel. 0222 + 58801, ext. 5626). *Electron energy loss spectroscopy, plasmons.*

Schranz, Dr Wilfried. Inst. für Experimentalphysik, U. Wien, Strudlhofgasse 4, A-1090 Vienna, Austria. (1960) Dr. rer. nat., physics (U. Vienna, 1988). Asst. (tel. 0222 + 342630, ext. 225). *Disorder, group theory, phase transitions, symmetry.*

Schroll, Prof. Dr Erich. Geotechnisches Institut, Bundesversuchs- und Forschungsanstalt Arsenal, Franz Grillstrasse 9, Objekt 214; postbox 8, A-1031 Vienna, Austria. (1923) Doz., mineralogy (U. Vienna, 1957). Head. (tel. 0222 + 782531, ext. 475; telex 1-36677). *X-ray fluorescence analysis, X-ray diffraction phase analysis.*

Schuster, Prof. Dr Julius Clemens. Inst. für Physikalische Chemie, U. Wien, Währingerstrasse 42, A-1090 Vienna, Austria. (1952) Dr. phil., chemistry (U. Vienna, 1977). Ass. Prof. (tel. 0222 + 343616). *Materials science, phase determination, structural chemistry.*

Schwarz, Prof. Dr Karlheinz. Inst. für Techn. Elektrochemie, Techn. U. Wien, Getreidemarkt 9, A-1060 Vienna, Austria. (1941) Doz., quantum chemistry (Techn. U. Vienna, 1975). Full Prof. (tel. 0222 + 58801, ext. 5097, email e1580sb1 at awiiez11). *Electronic structure of solids, energy band calculations, chemical bonding, electron densities, magnetism.*

Schwomma, Dr Otto. Entwicklungsabteilung, Siemens AG Österreich, Hainburgerstrasse 33, A-1030 Vienna, Austria. (1937) Dr. phil., physical chemistry (U. Vienna, 1964). Head. (tel. 0222 + 71711, ext. 5433). *Materials science.*

Seeger, Prof. Dr Karlheinz. Ludwig Boltzmanninst. für Festkörperphysik, Kopernikusgasse 15, A-1060 Vienna, Austria. (1927) Dr. rer. nat., physics (U. Heidelberg/FRG, 1955). Full prof. and head. (tel. 0222 + 563408, ext. 22). *Semiconductor physics and technology, piezoelectrics, quasi-one-dimensional conductors, intercalated graphite, FIR, Raman and microwave investigations.*

Seidl, Dr Erwin. Inst. für Neutronen und Festkörperphysik, Atominst. der Österreichischen Universitäten, Schüttelstrasse 115, A-1020 Vienna, Austria. (1939) Dr. phil., physics (U. Vienna, 1966). Asst. (tel. 0222 + 21701, ext. 259; fax 21 89 220). *Crystal growth, instrumentation, superconducting single crystal anisotropy, silicon crystals for neutron optics.*

Seifert, Dr Karl Josef. Inst. für Physikalische Chemie, U. Wien, Währingerstrasse 42, A-1090 Vienna, Austria. (1926) Dr. phil., chemistry (U. Vienna, 1962). Scient. officer. (tel. 0222 + 343616, ext. 36). *Inorganic crystal structures, documentation.*

Skalicky, Prof. Dr Peter. Inst. für Angewandte und Technische Physik, Techn. U. Wien, Wiedner Hauptstrasse 8-10/137, A-1040 Vienna, Austria. (1941) Doz., crystal physics (Techn. U. Vienna, 1973). Full prof. (tel. 0222 + 58801, ext. 5610). *Crystal physics, lattice defects, electron and X-ray diffraction, electron microscopy.*

Sobczak, Prof. Dr Rudolf Josef. Inst. für Physikalische Chemie, U. Linz, Altenbergerstrasse 69, A-4040 Linz, Austria. (1944) Doz., physical chemistry (U. Linz, 1980). Ass. Prof. (tel. 0732 + 2468, ext. 754). *Magnetism, instrumentation, polymers, teaching, X-ray diffraction, liquid crystals, materials science.*

Spindler, Mag. Peter. Inst. für Mineralogie und Kristallographie, U. Wien, Dr. Karl Lueger-Ring 1, A-1010 Vienna, Austria. (1959) Mag., mineralogy (U. Vienna, 1988). Scient. (tel. 0222 + 4300, ext. 2327). *Minerals.*

Stangler, Prof. Dr Ferdinand Karl Ludwig. Inst. für Festkörperphysik (Tieftemperaturphysik), U. Wien, Strudlhofgasse 4, A-1090 Vienna, Austria. (1928) Doz., experimental physics (U. Vienna, 1962). Full prof. and head. (tel. 0222 + 340673) *Crystal growth, electrical and magnetic properties, metal crystals, plasticity, crystal lattice defects, superconductivity, low temperature instruments.*

Steiner, Prof. Dr Walter. Inst. für Angewandte und Technische Physik, Techn. U. Wien, Wiedner Hauptstrasse 8-10/137, A-1040 Vienna, Austria. (1942) Doz., low temperature physics (Techn. U. Vienna, 1980). Full Prof. (tel. 0222 + 58801, ext. 5636). *Inorganic crystal structures, magnetism at low temperatures, Mössbauer spectroscopy.*

Stickler, Prof. Dr Roland. Inst. für Physikalische Chemie - Materialwissenschaften, U. Wien, Währingerstrasse 42, A-1090 Vienna, Austria. (1931) Dr. techn., metallurgy (Techn. U. Vienna, 1958). Full prof. (tel. 0222 + 343616, ext. 26). *Phase stability, phase analysis, defect structures, metals and alloys, semiconductor materials.*

Stumpfl, Prof. Dr Eugen Friedrich. Inst. für Mineralogie, Montanu. Leoben, Franz Josefstrasse 18, A-8700 Leoben, Austria. (1931) Dr. rer. nat., geological science (U. Heidelberg/FRG, 1956). Full prof. and head. (tel. 03842 + 42555, ext. 451). *Ore deposits, electron probe analysis, reflected light microscopy.*

Sturm, Prof. Dr Dipl.-Ing. Friedwin. Inst. für Physik (Angewandte Physik), Montanu. Leoben, Franz Josefstrasse 18, A-8700 Leoben, Austria. (1938) Doz., applied physics (Techn. U. Vienna, 1972). Full prof. (tel. 03842 + 42555, ext. 264). *X-ray metallography, computer programming, residual stresses.*

Uhl, Dr Eduard. Inst. für Pysikalische Chemie - Magnetochemie, U. Wien, Währingerstrasse 42, A-1090 Vienna, Austria. (1952) Dr. phil., chemistry (U. Vienna, 1980). Asst. (tel. 0222 + 343616) *Magnetic measurements, transition metal compounds, X-ray crystallography.*

Vana, Prof. Dr Dipl.-Ing. Norbert Johannes. Beschleunigerabteilung, Atominst. der Österreichischen Universitäten, Schüttelstrasse 115, A-1020 Vienna, Austria. (1940) Doz., optical and microwave spectroscopy (Techn. U. Vienna, 1975). Full Prof. (tel. 0222 + 21701, ext. 277). *Defect structures, radiation damage, archaeometry.*

Völlenkle, Prof. Dr Horst. Inst. für Mineralogie, Kristallographie und Strukturchemie, Techn. U. Wien, Getreidemarkt 9, A-1060 Vienna, Austria. (1938) Doz., structural inorganic chemistry (Techn. U. Vienna, 1980). Ass. Prof. (tel. 0222 + 58801, ext. 4742). *Inorganic and organic crystal structures, computer programming.*

Wagendristel, Prof. Dr Alfred Friedrich. Inst. für Angewandte Physik, Techn. U. Wien, Karlsplatz 13, A-1040 Vienna, Austria. (1941) Doz., thin films (Techn. U. Vienna, 1976). Full Prof. (tel. 0222 + 58801, ext. 5630). *Thin film structure, amorphous thin films, diffusion via lattice defects.*

Walitzi, Prof. Dr Eva Maria. Abt. für Mineralogie-Kristallographie, Inst. für Mineralogie-Kristallographie und Petrologie, U. Graz, Universitätsplatz 2, A-8010 Graz, Austria. (1930) Doz., mineralogy-crystallography (U. Graz, 1967). Full prof. (tel. 0316 + 380, ext. 5541). *Inorganic crystal structures, crystal chemistry.*

Walter, Dr Franz. Abt. für Mineralogie-Kristallographie, Inst. für Mineralogie-Kristallographie und Petrologie, U. Graz, Universitätsplatz 2, A-8010 Graz, Austria. (1952) Dr. phil., mineralogy-crystallography (U. Graz, 1980). Ass. (tel. 0316 + 380, ext. 5547). *Crystallographic data, minerals, crystal structures.*

Warhanek, Prof. Dr Hans. Inst. für Experimentalphysik, U. Wien, Strudlhofgasse 4, A-1090 Vienna, Austria. (1926) Dr. phil., physics (U. Vienna, 1953). Full Prof. (tel. 0222 + 342630, ext. 217; telex 116222). *Crystal growth, defect structures, inorganic crystals, microscopy, phase transitions.*

Weber, Prof. Dr Harald Wolfgang. Atominst. der Österreichischen Universitäten, Schüttelstrasse 115, A-1020 Vienna, Austria. (1944) Doz., low temperature physics (Techn. U. Vienna, 1975). Full Prof. (tel. 0222 + 21701, ext. 240; telex 3222467 tuw; fax 2189220; email weber at edvz.ati.ptt.at) *Superconductivity, crystallographic properties.*

Wildner, Mag. Manfred. Inst. für Mineralogie und Kristallographie, U. Wien, Dr. Karl Lueger-Ring 1, A-1010 Vienna, Austria. (1962) Mag., mineralogy (U. Vienna, 1989). Scient. (tel. 0222 + 4300, ext. 2327). *Inorganic crystals, crystallographic data, symmetry.*

Winter, Prof. Dr Dipl.-Ing. Hannspeter. Inst. für Allgemeine Physik, Techn. U. Wien, Wiedner Hauptstrasse 8-10, A-1040 Vienna, Austria. (1941) Dr. techn., physics (Techn. U. Vienna, 1970). Full prof. and head. (tel. 0222 + 58801, ext. 5710; telex 131000 tvfa wa). *Electron density, materials science, surface structure, plasma-surface interaction.*

Wobrauschek, Prof. Dr Dipl.-Ing. Peter. Abt. für Elektronen- und Röntgenphysik, Atominst. der Österreichischen Universitäten, Schüttelstrasse 115, A-1020 Vienna, Austria. (1939) Doz., X-ray physics (Techn. U. Vienna, 1983). Ass. Prof. (tel. 0222 + 21701, ext. 276; fax 2189220). *Crystal orientation, X-ray metallography, textures, X-ray fluorescence analysis.*

Zeilinger, Prof. Dr Anton Wolfgang. Abt. für Neutronen- und Festkörperphysik, Atominst. der Österreichischen Universitäten, Schüttelstrasse 115, A-1020 Vienna, Austria. (1945) Doz., neutron and solid state physics (Techn. U.

Vienna, 1979). Full Prof. (tel. 0222 + 21701, ext. 258). *Dynamical diffraction, perfect crystal neutron optics, neutron interferometry, quantum mechanics foundation.*

Zemann, Prof. Dr Josef. Inst. für Mineralogie und Kristallographie, U. Wien, Dr. Karl Lueger-Ring 1, A-1010 Vienna, Austria. (1923) Dr. phil., mineralogy (U. Vienna, 1946). Full prof. and head. (tel. 0222 + 4300, ext. 2333). *Minerals, inorganic crystals, structural chemistry, X-ray diffraction, neutron diffraction, hydrogen bonding, microscopy.*

Zipper, Prof. Dr Peter. Inst. für Physikalische Chemie, U. Graz, Heinrichstrasse 28, A-8010 Graz, Austria. (1941) Dr. phil., chemistry (U. Graz, 1970). Ass. Prof. (tel. 0316 + 380, ext. 5415). *Small angle scattering, diffuse scattering, polymers.*

Zobetz, Dr Erich. Inst. für Mineralogie, Kristallographie und Strukturchemie, Techn. U. Wien, Getreidemarkt 9, A-1060 Vienna, Austria. (1950) Dr. phil., mineralogy and petrography (U. Vienna, 1983). Asst. (tel. 0222 + 58801, ext. 4743). *Crystal structures, X-ray fluorescence analysis, Dirichlet domains.*

BANGLADESH

Sub-Editor: **Kh. A. I. F. M. Mannan**

Notes

1. International telephone country code - 880.

Ahmed, Dr A. H. Moinuddin. Dept. of Biochemistry, Dhaka U., Ramna, Dhaka-2, Bangladesh. (1938) PhD, inorganic chemistry and crystallography (Aberdeen U., UK, 1969). Assoc. prof. (tel. 245289). *Chemical crystallography, inorganic chemistry, bio-physical chemistry, bio-physics, cement chemistry, electron diffraction, electron microscopy.*

Ahmed, Dr Sultan. Physics Dept., Dhaka U., Curzon Hall, Dhaka-2, Bangladesh. (1936) PhD, solid state physics, (Southampton U., UK, 1971). Prof. (tel. Dhaka 50049). *Structural studies of crystalline and non-crystalline solids.*

Akhtar, Dr Farida. Dept. of Chemistry, Jahangirnagar U., Savar, Dhaka, Bangladesh. (1942) PhD, X-ray crystallography, (London U., UK, 1969). Assoc. prof. (tel. Savar 316071, ext. 19). *Structure, physico-chemical techniques, X-ray crystallography.*

Biswas, Dr Mohommad Alim. Glass and Ceramics Div., BCSIR Labs., Science Lab. Rd., Dhaka-2, Bangladesh. (1927) PhD, organic chemistry, (London U., UK, 1958). Head of division, (tel. 315563). *Inorganic crystal structures, clay minerals.*

Chawdhury, Prof. Sadruddin Ahmed. Physics Dept., Rajsahi U., Rajsahi, Bangladesh. (1934) PhD, crystallography (Manchester U., UK,). Prof. *Organic crystal structures, direct methods.*

Choudhury, Mrs Shamima. Physics Dept., Dhaka University, Bangladesh. (1951) M.S. Crystallography (U. of New South Wales, Australia) Asst. Prof. (tel. Dhaka 607913). *Organic and inorganic compounds, crystal structure.*

Chowdhury, Prof. Fazlul Halim. Senior Specialist, UNESCO House, 15 Jor Bagh, New Delhi-11003, India. (1930) PhD, physical chemistry (U. of Manchester, UK, 1956). Prof. (tel. Delhi-618092-93). *Large molecules, metal structures.*

Haider, Prof. Syed Zahir. Dept. of Chemistry, Dhaka U., Ramna, Dhaka-2, Bangladesh. (1927) PhD, inorganic chemistry (London U., UK, 1958). Prof. (tel. Dhaka 315991). *Metal phosphates, organoboron compounds, catalysts, coordination chemistry, building materials (low cost production).*

Husain, Dr Abul Hasanat Mohammad. Dept. of Physics, U. of Dhaka, Curzon Hall, Dhaka 2, Bangladesh. (1949) PhD, solid state physics (U. of Exeter, UK, 1977). Assoc. prof. (tel. 500449). *Ultrasonic applications, solar energy research, organic crystals.*

Ibrhim, Dr Muhammad. Physics Dept., Dhaka U., Curzon Hall, Dhaka-2, Bangladesh. (1945) PhD, surface physics (Southampton U., UK, 1972). Prof. *Surface structure, electrical properties, solids; thin films, adsorption phenomena.*

Islam, Prof. Aminul. Soil Science Dept., Dhaka U., 32-H, Isakhan Rd., Dhaka-2, Bangladesh. (1935) PhD, soil science (Michigan State U., USA, 1962). Prof. *Clay mineral structures.*

Islam, Mr Shafiqul. Physics Dept., U. of Rajshahi, Rajshahi, Bangladesh. (1951) MSc, crystallography (U. of Rajshahi, 1972). Lect. (tel. Rajshahi 2441, ext. 23). *Structure determination, organic molecules, direct methods.*

Khan, Dr Anwarur Rahman. Dept. of Applied Physics, U. of Dhaka, Ramna, Dhaka-2, Bangladesh. (1932) PhD, metal physics (U. of London, UK, 1967). Prof. (tel. 257859). *Electron microscopy, electron diffraction, thin films, structure and properties.*

Malik, Dr Khalifa Mohammad Abdul. Chemistry Dept., Dhaka University, Curzon Hall, Dhaka 1000, Bangladesh. (1946) PhD, crystallography (U. London, UK.,1974). Prof. *Synthetic and structural bio-inorganic chemistry, environmental chemistry.*

Mannan, Prof. Dr Kh. A. I. F. Mafizul. Physics Dept., Dhaka U., Ramna, Dhaka 1000, Bangladesh. (1939) DPhil, crystallography (Oxford U., UK, 1965). Prof. (tel. Dhaka 500449). *Organic and organo-metallic crystal structures, thin film structure and electrical properties, powder diffraction, amorphous films, jute fibre.*

Manzoor-I-Khuda, Dr Muhammad. T.R.C. Div., Jute Res. Inst., Tejgaon, Dhaka-15, Bangladesh. (1933) PhD, organic chemistry (U. of London, UK, 1957). Director. (tel. 310975). *Organic crystal structure, large molecules.*

Quader, Prof. Dr Mohammed Abdul. Physics Dept., Jahangirnagar U., Savar, Dhaka, Bangladesh. (1933) DPhil, physics (Calcutta U., 1962). Prof. (tel. Savar 316071). *Powder photography, phase diagrams, phase transformations, defects, faults and strains.*

Rahman, Dr Asadur. Physics Dept., Dhaka U., Curzon Hall, Dhaka 1000, Bangladesh. (1944) PhD, X-ray crystallography (U. of Dundee, UK, 1971). Prof. (tel. Dhaka 500449). *Organic and biological structures, nucleic acid structure and conformation, viruses, macromolecular structures, crystallographic computing.*

Rahman, Dr Sheikh Mohammed Mujibur. Physics Dept., U. of Dhaka, Dhaka 1000, Curzon Hall, Bangladesh. (1951) PhD, solid state physics (Dhaka U., 1976) PhD, solid state physics (Bristol U., UK, 1979). Assoc. prof. (tel. Dhaka 242935). *Electronic structure, Binary alloy structures, thermodynamic properties.*

Roy, Dr Ajoy Kumer. Dept. of Physics, U. of Dhaka, Ramna, Dhaka-2, Bangladesh. (1935) PhD, solid state physics (Leeds U., UK, 1966). Prof. (tel. Dhaka 500449). *Electron paramagnetic resonance spectroscopy, free radicals (oriented), paramagnetic centres.*

Syed, Dr A. Sattar. Industrial Physics Div., C.S.I.R. Sci. Lab. Rd., Dhaka-2, Bangladesh. (1935) PhD, solid state physics (U. of British Columbia, Canada, 1964). Sr. res. officer. (tel. Dhaka 315563, ext. 05). *Applied crystallography, manganese dioxide, thin films, crystal growth, organometallic complexes.*

Zaman, Dr (Mrs) Nazma, Dept. of Physics, Bangladesh U. of Eng. and Techn., Dhaka, Bangladesh. (1946) PhD, X-ray crystallography (U. of Manchester, UK, 1975). Asst. prof. (tel. Dhaka 252473). *Organic compounds, crystal structure.* lography (U. of Manchester, UK, 1975). Asst. prof. (tel. Dhaka 252473).

BELGIUM

Sub-Editor: **G.S.D. King**

Notes

1. International telephone country code - 32.

2. The academic degrees conferred in Belgium are : *Licencié en sciences* or *licentiaat in de wetenschappen* (Lic) (equivalent to MSc), *Docteur en sciences* or *doctor in de wetenschappen* (DSc) (equivalent to PhD), *Ingénieur civil* or *burgerlijk ingenieur* (Ir), *Agrégé de l'enseignement supérieur* (Agr. Ens. Sup.) or *geaggregeerde voor het hoger onderwijs* (Geaggr. H.O.)

3. The academic positions are : *Professeur ordinaire* or *gewoon hoogleraar*, *Professeur* or *hoogleraar* (both positions are referred to as Prof.), *Chargé de cours* or *docent*, *Professeur associé* or *geassocieerd hoogleraar*, *Chargé de cours associé* or *geassocieerd docent* (both positions are referred to as Assoc. Prof.), *Chef de travaux* or *werkleider*, *Premier assistant* or *eerstaanwezend assistent*, *Assistant* or *assistent* (both positions are referred to as Asst.).

4. All email addresses are on BITNET. The following nodes can be reached from other networks by replacing the node name by the appropriate domain addess below.

Node name	Domain address		Node name	Domain address
BANUIA51	uia82.vms.uia.ac.be		BLIULG11	vm1.earn-ulg.ac.be
BANUIA52	uia78.vms.uia.ac.be		BMLSCK11	sckmol.ac.be
BLEKUL11	cc1.kuleuven.ac.be		BNANDP11	scf.fundp.ac.be
BLEKUL21	cc2.kuleuven.ac.be		BUCLLN11	pythag.ucl.ac.be

Aernoudt, Prof. Dr Etienne. Dept. Metaalkunde en Toegepaste Materiaalkunde, Katholieke U. Leuven, De Croylaan 2, B-3030 Heverlee, Belgium. (1938) Dr Ir, metallurgy (T. H. Aachen, F. R. Germany, 1966). Prof. (tel. 016 + 220931, ext. 1302, fax 16 + 207995, email FHAAA17 at BLEKUL11). *Materials science, metals.*

Amelinckx, Prof. Dr Severin. Faculty of Sci., Universiteit Antwerpen, (RUCA), Groenenborgerlaan 171, B-2020 Antwerpen, Belgium. (1922) DSc, physics (U. Gent, 1948). Prof. (tel. 03 + 2180217). *Crystal physics, defect structures, electron diffraction, electron microscopy*

Baele, Miss Ingrid Albertina Frans Mariette. Fysica van de vaste stof, Universiteit Antwerpen, (RUCA), Groenenborgerlaan 171, B-2020 Antwerpen, Belgium. (1961) Lic, physics (U. Antwerp, 1984). Asst. (tel. 03 + 2180495). *Electron microscopy.*

Bauduin, Mrs Anne-Marie Ghislaine Gerardine. Facultés Universitaires de Namur, Groupe de Chimie-Physique, Rue De Bruxelles 61, B-5000 Namur, Belgium. (1961) Lic, chemistry (Fac. U. Namur, 1984). Res. student. (tel. 081 + 724111, ext. 2481, telex 59222, fax 081 + 230391, email BAUDUIN at BNANDP11). *Structural chemistry, structure-activity relationships.*

Bender, Dr Hugo J.M.R. Materials and packaging division, Interuniversity microelectronics center, Kapeldreef 75, B-3030 Leuven, Belgium. (1957) DSc, physics (U. Antwerp, 1984). Res. fellow. (tel. 016 + 281304, telex 26152, fax 016 + 229400). *Crystal physice, electron microscopy, semiconductors, X-ray diffraction.*

Blaton, Dr Norbert Louis. Lab. voor Analytische Chemie en Medicinale Fysicochemie, Inst. voor Farmaceutische Wetenschappen, Katholieke U. Leuven, Van Evenstraat 4, B-3000 Leuven, Belgium. (1945) DSc, chemistry (U. Leuven, 1974). Werkleider. (tel. 016 + 283419, email FKAAA02 at BLEKUL11). *Phase determination, powder diffraction, structural chemistry.*

Bracke, Mr Benedikt Rachel Frans. Chemistry dept., Universitaire Instelling Antwerpen, Universiteitsplein 1, B-2610 Wilrijk, Belgium. (1965) Lic, chemistry (U. Antwerpen, 1987). Asst. (tel. 03 + 8202365, fax 03 + 8202248, email ALSENOY at BANUIA52). *Electron density, phase determination.*

Brasseur, Prof. Dr Henri Alphonse Lambert. Crystallography Dept., U. of Liège, Inst. of Physics B5, Sart Tilman, B-4000 Liège, Belgium. (1905) Agr. Sc., physics (U. Liège, 1934). Retired. (tel. 041 + 525652). *Crystallography.*

Collin, Dr Sonia Bertha Josepha. Facultés Universitaires de Namur, Groupe de Chimie-Physique, Rue De Bruxelles 61, B-5000 Namur, Belgium. (1963) DSc, chemistry (Fac. U. Namur, 1984). Res. worker. (tel. 081 + 724111, ext. 2481, telex 59222, fax 081 + 230391, email COLLIN at BNANDP11). *Computation, structural chemistry, structure-activity relationships.*

De Bondt, Mr Hendrik Leon Augusta Jozef. Lab. voor Analytische Chemie en Medicinale Fysicochemie, Inst. voor Farmaceutische Wetenschappen, Katholieke U. Leuven, Van Evenstraat 4, B-3000 Leuven, Belgium. (1964) Pharmacist (U. Leuven, 1987). Res. Asst. NFWO. (tel. 016 + 283419, email FKAAA00 at BLEKUL11). *Electron density, structural chemistry, X-ray diffraction.*

Declercq, Prof. Jean Paul. Lab. Chimie-Phys. et Cristallographie, Université Catholique de Louvain, Pl. Pasteur 1, B-1348 Louvain-la-Neuve, Belgium. (1948) DSc, chemistry (U. Louvain, 1972). Chargé de cours. (tel. 010 + 472924, telex 59037, email DECLERCQ at BUCLLN11). *Computing, phase determination, proteins.*

De Gryse, Dr Roger Marc. Lab. voor Kristallografie en Studie van de Vaste Stof, Rijksuniversiteit Gent, Krijgslaan 281, B-9000 Gent, Belgium. (1941) Dr, applied science (U. Gent, 1973). Werkleider. (tel. 091 + 225715, ext. 2350). *Crystal physics.*

Dekeyser, Prof. Dr Willy Clement. Green Park, Pacificatielaan 63, B-9000 Gent, Belgium. (1910) DSc, physics (U. Gent, 1930). Prof. emeritus U. Gent. *Crystal physics, defect structures.*

Delaey, Prof. Dr Luc J. M. A. E. Dept. Metaalkunde en Toegepaste Materiaalkunde, Katholieke U. Leuven, De Croylaan 2, B-3030 Heverlee, Belgium. (1939) Dr. rer. nat., metallurgy (U. Stuttgart Germany, 1966). Prof. (tel. 016 + 220931, ext. 1272, fax 16 + 207995, email FHAAA17 at BLEKUL11). *Electron microscopy, metals.*

Deliens, Dr Michel. Section Minérologie et Pètrographie, Institut royal des Sciences naturelles de Belgique, Rue Vautier 29, B-1040 Bruxelles, Belgium. (1939) DSc, mineralogy (U. Louvain, 1972). Section head. (tel. 02 + 6480475 ext. 323). *Minerals, structural chemistry, X-ray diffraction.*

Deltour, Prof. Robert. Physique des solides CP 233, Université Libre de Bruxelles, Boulevard du Triomphe, B-1050 Bruxelles, Belgium. (1936) Ph.D physics (U. Bruxelles, 1964). Prof. (tel. 02 + 6400015 ext. 5752). *Crystal physics, semiconductors.*

De Ranter, Prof. Dr Camiel Joseph. Lab. voor Analytische Chemie en Medicinale Fysicochemie, Inst. voor Farmaceutische Wetenschappen, Katholieke U. Leuven, Van Evenstraat 4, B-3000 Leuven, Belgium. (1937) DSc, chemistry (U. Leuven, 1964). Prof. (tel. 016 + 283416, email FKAAA01 at BLEKUL11). *Structural chemistry, structure-activity relationships.*

Deruyttere, Prof. Dr André. Dept. Metaalkunde en Toegepaste Materiaalkunde, Katholieke U. Leuven, De Croylaan 2, B-3030 Heverlee, Belgium. (1925) PhD, metallurgy (U. Sheffield, UK, 1955). Prof. (tel. 016 + 220931, ext. 1271, fax 16 + 207995). *Materials science, metals.*

De Smedt, Mr Jozef Maria Adrianus Ludovicus. Chemistry dept., Universitaire Instelling Antwerpen, Universiteitsplein 1, B-2610 Wilrijk, Belgium. (1964) Lic, chemistry (U. Antwerpen, 1984). Asst. (tel. 03 + 8202365, telex 33646, fax 03 + 8202248, email NANCY at BANUIA52). *Electron density, phase determination.*

De Winter. Mr Hans Louis Jos Lab. voor Analytische Chemie en Medicinale Fysicochemie, Inst. voor Farmaceutische Wetenschappen, Katholieke U. Leuven, Van Evenstraat 4, B-3000 Leuven, Belgium. (1963) Pharmacist (U. Leuven, 1986). Asst. (tel. 016 + 283421, email KKAAA05 at BLEKUL21). *Structural chemistry, structure-activity relationships.*

Dideberg, Dr Otto. Crystallography Dept., U. of Liège, Inst. of Physics B5, Sart Tilman, B-4000 Liège, Belgium. (1942) DSc, physics (U. Liège, 1969). Assoc. Prof. (tel. 041 + 563762, telex 41397, fax 041 + 562355, email U210402 at BLIULG11). *Proteins, structural biology.*

Dupont, Dr Leon. Crystallography Dept., U. of Liège, Inst. of Physics B5, Sart Tilman, B-4000 Liège, Belgium. (1941) DSc, physics (U. Liège, 1969). Chef de travaux. (tel. 041 + 563762, telex 41397, fax 041 + 562355, email U210406 at BLIULG11). *Phase determination, structural chemistry.*

Durant, Prof. François Victor. Facultés Universitaires de Namur, Groupe de Chimie-Physique, Rue De Bruxelles 61, B-5000 Namur, Belgium. (1939) DSc, chemistry (U. Louvain, 1965). Prof. (tel. 081 + 724111, telex 59222, fax 081 + 230391, email DURANT at BNANDP11). *Computation, structural biology, structure-activity relationships.*

Evrard, Prof. Guy Henri. Facultés Universitaires de Namur, Groupe de Chimie-Physique, Rue De Bruxelles 61, B-5000 Namur, Belgium. (1943) DSc, chemistry (U. Louvain, 1969). Prof. (tel. 081 + 724111, telex 59222, fax 081 + 230391, email EVRARD at BNANDP11). *Phase determination, structural chemistry.*

Feneau-Dupont, Mrs Janine. Lab. Chimie-Phys. et Cristallographie, Université Catholique de Louvain, Pl. Pasteur 1, B-1348 Louvain-la-Neuve, Belgium. (1932) Lic, chemistry (U. Louvain, 1953). Collaboratrice scientifique. (tel. 010 + 472921). *Crystallography.*

Fiermans, Dr Lucien Victor August. Laboratorium voor Kristallografie en Studie van de Vaste Stof, Rijksuniversiteit Gent, Krijgslaan 281, B-9000 Gent, Belgium. (1937) Geaggr. H.O., physics (U. Gent, 1974). Werkleider. (tel. 091 + 225715). *Crystal physics, defect structures, semiconductors.*

Fontaine, Dr Frederic Desiré Albert. Exp. Physics Dept., U. of Liège, Inst. of Physics B5, Sart Tilman, B-4000 Liège, Belgium. (1932) DSc, chemistry (U. Liège, 1967). Chef de travaux. (tel. 041 + 563631, fax 041 + 562355, email U217304 at BLIULG11). *Polymers, small-angle scattering, structural chemistry.*

Fransolet, Dr André-Mathieu. Inst. for Mineralogy U. of Liège, B18, Sart-Tilman, B-4000 Liège, Belgium. (1947) DSc, geology and mineralogy (U. Liège, 1975). Res. assoc. FNRS, (tel. 041 + 562206). *Minerals.*

Gaspard, Prof. Jean-Pierre Exp. Physics Dept., U. of Liège, Inst. of Physics B5, Sart Tilman, B-4000 Liège, Belgium. (1945) DSc, physics (U. Paris-Sud, 1975). Chargé de cours. (tel. 041 + 563745, telex 41397, fax 041 + 562355, email U2150PL at BLIULG11). *Neutron diffraction, structural chemistry.*

Geise, Prof. Dr Herman Joseph Victor Heinrich. Chemistry dept., Universitaire Instelling Antwerpen, Universiteitsplein 1, B-2610 Wilrijk, Belgium. (1937) DSc, chemistry (U. Leiden, Netherlands,1964). Prof. (tel. 03 + 8202349, fax 03 + 8202249). *Electron diffraction, structural chemistry.*

Germain, Prof. Gabriel. Unité de Chimie physique moléculaire et de Cristallographie, Université Catholique de Louvain, Pl. Pasteur 1, B-1348 Louvain-la-Neuve, Belgium. (1933) DSc, chemistry (U. Louvain, 1958). Chargé de cours. (tel. 010 + 472833, email GERMAIN at BUCLLN11). *Computing, phase determination.*

Gibon, Dr Véronique Julie Jacques Laure. Facultés Universitaires de Namur, Groupe de Chimie-Physique, Rue De Bruxelles 61, B-5000 Namur, Belgium. (1958) DSc, chemistry (Fac. U. Namur, 1984). Asst. (tel. 081 + 724111, ext. 2481, telex 59222, fax 081 + 230391, email GIBON at BNANDP11). *Phase transitions, structural chemistry, structure-activity relationships.*

Kartheuser, Dr Edward Peter. Theoretical Physics Dept., U. of Liège, Sart Tilman, B-4000 Liège, Belgium. (1938) DSc, physics (U. Liège, 1968). Chargé de cours. (tel. 041 + 563639, fax 041 + 5623559). *Crystal physics, defect structures, semiconductors.*

King, Prof. Geoffrey Stephen Douglas. Laboratorium voor Kristallografie, Katholieke Universiteit Leuven, Celestijnenlaan 200C, B-3030 Leuven, Belgium. (1924) MSc, crystallography (U. London, UK, 1950). Prof. (tel. 016 + 201015, ext. 3582, telex 23674, fax 016 + 201368, email FGEEA01 at BLEKUL11). *Computing, crystallographic data, mineralogy, structural chemistry.*

Lamotte-Brasseur, Dr Josette Marie Louise. Crystallography Dept., U. of Liège, Inst. of Physics B5, Sart Tilman, B-4000 Liège, Belgium. (1943) DSc, physics (U. Liège, 1973). Chef de travaux. (tel. 041 + 563758, fax 041 + 562968, email U210405 at BLIULG11). *Computing, structural chemistry.*

Lenstra, Prof. Dr Albert Teun Hendrik. Chemistry dept., Universitaire Instelling Antwerpen, Universiteitsplein 1, B-2610 Wilrijk, Belgium. (1942) Dr, chemistry (U. Utrecht, Netherlands,1973). Prof. (tel. 03 + 8202365, fax 03 + 8202249, email LENSTRA at BANUIA52). *Computing, phase determination.*

Léonard, Dr André Jules Gérard. Groupe de Physico-Chimie Minérale et de Catalyse, Université Catholique de Louvain, Pl. Croix du Sud 1, B-1348 Louvain-la-Neuve, Belgium. (1936) DSc, crystallography (U. Louvain, 1959). Chef de travaux. (tel. 010 + 473588, fax 010 + 474745). *Crystal physics, minerals, X-ray diffraction.*

Maenhout - Van Der Vorst, Dr Mrs Wenefride Marguerite Romain. Lab. voor Kristallografie en Studie van de Vaste Stof, Rijksuniversiteit Gent, Krijgslaan 281, B-9000 Gent, Belgium. (1930) DSc, physics (U. Gent, 1957). Werkleider. (tel. 091 + 225715). *Crystal physics, surface structure.*

Marcoen, Dr Jean-Marie. Unité Sc. de la Terre Fac. Agronomie, B-5800 Gembloux, Belgium. (1948) Dr, agronomy (Gembloux, 1977). Maître de Conférences. (tel. 081 + 622209, telex 59482, fax 081 + 615965). *Mineralogy, X-ray diffraction.*

Matthys, Dr Paul Frederik André Edmond. Laboratorium voor Kristallografie en Studie van de Vaste Stof, Rijksuniversiteit Gent, Krijgslaan 281, B-9000 Gent, Belgium. (1949) Dr, physics (U. Gent, 1976). Werkleider. (tel. 091 + 225715, ext. 2365). *Crystal physics.*

Meunier - Piret, Dr (Mrs) Jacqueline. Lab. Chimie-Phys. et Cristallographie, U. Catholique de Louvain, Pl. Pasteur 1, B-1348 Louvain-la-Neuve, Belgium. (1934) DSc, chemistry (U. Louvain, 1961). Asst. (tel. 010 + 472922, email SCHEEFER at BUCLLN11). *Structural chemistry.*

Michel, Prof. Karl Heinrich Joseph Department of Physics, Universitaire Instelling Antwerpen, Universiteitsplein 1, B-2610 Wilrijk, Belgium. (1939) PhD, Statistical mechanics (U. Nijmegen, Netherlands,1965). Prof. (tel. 03 + 8202458, fax 03 + 8202249). *Crystal physics.*

Moreau, Prof. Jules Francois. Lab. de Minéralogie et de Géologie Appliquée, Université Catholique de Louvain, Batiment Mercator, Pl. L. Pasteur 3, B-1348 Louvain-la-Neuve, Belgium. (1931) Mining Eng. (U. Louvain, 1954). Prof. (tel. 010 + 472855). *Mineralogy.*

Naud, Dr Jean Marcel. Lab. de Minéralogie et de Géologie Appliquée, Université Catholique de Louvain, Batiment Mercator, Pl. L. Pasteur 3, B-1348 Louvain-la-Neuve, Belgium. (1942) DSc, chemistry (U. Louvain, 1968). Chef de travaux. (tel. 010 + 472851). *Mineralogy, powder diffraction.*

Peeters, Dr Oswald Maurice. Lab. voor Analytische Chemie en Medicinale Fysicochemie, Inst. voor Farmaceutische Wetenschappen, Katholieke U. Leuven, Van Evenstraat 4, B-3000 Leuven, Belgium. (1945) DSc, chemistry (U. Leuven, 1977). Werkleider. (tel. 016 + 283418, email FKAAA03 at BLEKUL11). *Phase determination, structural chemistry, structure-activity relationships.*

Piret, Prof Paul. Lab. Chimie-Phys. et Cristallographie, Université Catholique de Louvain, Pl. Pasteur 1, B-1348 Louvain-la-Neuve, Belgium. (1932) DSc, chemistry (U. Louvain, 1956). Prof. (tel. 010 + 472769, email SCHEEFER at BUCLLN11). *Mineralogy, structural chemistry.*

Point, Prof Jean-Jacques. Faculté des Sciences, Université de l'Etat at Mons, Av. Maistriau 21, B-7000 Mons, Belgium. (1928) PhD, applied science (U. Mons). Prof. (tel. 065 + 373353, telex 57764, fax 065 + 373054, email SPOINT at BMSUEM11). *Crystal physcis, polymers.*

Popelier, Mr Paul Lode Albert. Chemistry dept., Universitaire Instelling Antwerpen, Universiteitsplein 1, B-2610 Wilrijk, Belgium. (1964) Lic. chemistry (U. Antwerpen, 1986). Asst. (tel. 03 + 8202365, fax 03 + 8202249, email POPELIER at BANUIA52). *Electron density.*

Reynaers, Prof. Harry Louis. Dept. Scheikunde, Katholieke Universiteit Leuven, Celestijnenlaan 200F, B-3030 Heverlee, Belgium. (1938) DSc, chemistry (U. Leuven, 1964). Prof. (tel. 016 + 200656, ext. 3496, telex 23674, email KGCAB01 at BLEKUL21). *Polymers, small-angle scattering.*

Schryvers, Dr Dominique Maurits. Fysica van de vaste stof, Universiteit Antwerpen, (RUCA), Groenenborgerlaan 171, B-2020 Antwerpen, Belgium. (1959) DSc, physics (U. Antwerp, 1985). Res. asst. (tel. 03 + 2180495). *Electron microscopy, materials science.*

Sobry, Dr Roger. Exp. Physics Dept., U. of Liège, Inst. of Physics B5, Sart Tilman, B-4000 Liège, Belgium. (1946) DSc, physics (U. Liège, 1972). Chef de travaux. (tel. 041 + 563715, fax 041 + 562355, email U217302 at BLIULG11). *Small-angle scattering.*

Spirlet, Dr Marie-Rose. Exp. Physics Dept., U. of Liège, Inst. of Physics B5, Sart Tilman, B-4000 Liège, Belgium. (1946) DSc, chemistry (U. Liège, 1976). Chef de travaux. (tel. 041 + 563758, telex 41397, fax 041 + 562355, email U217303 at BLIULG11). *Structural chemistry.*

Tinant, Dr Bernard Guy André François. Lab. Chimie-Phys. et Cristallographie, Université Catholique de Louvain, Pl. Pasteur 1, B-1348 Louvain-la-Neuve, Belgium. (1951) DSc, chemistry (U. Louvain, 1978). Chercheur. (tel. 010 + 472924, email TINANT at BUCLLN11). *Phase determination, proteins.*

Tonnard, Dr Victor Edmond. Chaire Sc. de la Terre Fac. Agronomie, B-5800 Gembloux, Belgium. (1929) Dr, agronomy (Gembloux, 1957). Prof. (tel. 081 + 622269). *Minerals, X-ray diffraction.*

Toussaint, Prof. Jean. Crystallography Dept., U. of Liège, Inst. of Physics B5, Sart Tilman, B-4000 Liège, Belgium. (1916) DSc, physics (U. Liège, 1945). Retired. (tel. 041 + 334579). *Crystallography.*

Van Alsenoy, Dr Kris. Chemistry dept., Universitaire Instelling Antwerpen, Universiteitsplein 1, B-2610 Wilrijk, Belgium. (1948) DSc, chemistry (V.U. Brussels, 1977). Res. assoc. NFWO. (tel. 03 + 8202366, fax 03 + 8202248, email ALSENOY at BANUIA52). *Structural chemistry.*

Van den Bosch, Dr Adolf. Materials Sci. Dept., S.C.K.-C.E.N., B-2400 Mol, Belgium. (1928) DSc, chemistry (U. Gent, 1963). Res. physicist. (tel. 014 + 311801, ext. 2745, fax 014 + 315021). *Crystal physics.*

Van den Bossche, Dr Guy Ghislain Remy. Crystallography Dept., U. of Liège, Inst. of Physics B5, Sart Tilman, B-4000 Liège, Belgium. (1941) DSc, physics (U. Liège, 1973). Chef de travaux. (tel. 041 + 563763, telex 41397, fax 041 + 562355, email U210408 at BLIULG11). *Crystal physics.*

Vanhellemont, Mr Jan Hendrik. Materials and packaging division, Interuniversity microelectronics center, Kapeldreef 75, B-3030 Leuven, Belgium. (1953) Lic, physics (U. Antwerp, 1978). Res. asst. (tel. 016 + 281304, telex 26152, fax 016 + 229400). *Crystal physics, electron microscopy, semiconductors.*

Vanhouteghem, Mr Frankie Marie. Chemistry dept., Universitaire Instelling Antwerpen, Universiteitsplein 1, B-2610 Wilrijk, Belgium. (1961) Lic, chemistry (U. Antwerpen, 1983). Asst. (tel. 03 + 8202365, fax 03 + 8202248, email HOUTEGHE at BANUIA52). *Electron density, electron diffraction.*

Van Landuyt, Prof. Joseph Florent. Faculty of Sci., Universiteit Antwerpen, (RUCA), Groenenborgerlaan 171, B-2020 Antwerpen, Belgium. (1938) DSc, physics (U. Gent, 1965). Prof. (tel. 03 + 2180217). *Electron diffraction, electron microscopy.*

Van Meerssche, Prof. Maurice. Lab. Chimie-Phys. et Cristallographie, Université Catholique de Louvain, Pl. Pasteur 1, B-1348 Louvain-la-Neuve, Belgium. (1923) DSc, chemistry (U. Louvain, 1948). Prof. (tel. 010 + 472771). *Structural chemistry.*

Van Meervelt, Dr Luc. Lab. voor Kristallographie, Katholieke U. Leuven, Celestijnenlaan 200C, B-3030 Leuven, Belgium. (1958) Dr, chemistry (U. Leuven, 1986). Senior Res. Asst. NFWO. (tel. 016 + 201015, ext. 3584, telex

23674, fax 016 + 201368, email FGCEA01 at BLEKUL11). *Structural biology, structural chemistry.*

Van Tendeloo, Prof Gustaaf. Fysica van de vaste stof, Universiteit Antwerpen, (RUCA), Groenenborgerlaan 171, B-2020 Antwerpen, Belgium. (1950) DSc, physics (U. Antwerp, 1974). Asst. (tel. 03 + 2180262, fax 03 + 2180217). *Electron microscopy, metals, quasicrystals.*

Vennik, Prof. Ir Joost. Laboratorium voor Kristallografie en Studie van de Vaste Stof, Rijksuniversiteit Gent, Krijgslaan 281, B-9000 Gent, Belgium. (1927) Electrical Engineer, electronics (U. Gent, 1952). Prof. (tel. 091 + 225715, ext. 2325). *Crystal physics, materials science, semiconductors.*

Verbist, Prof. Jacques Jozef. Facultés Universitaires de Namur, Département de Chimie, Rue De Bruxelles 61, B-5000 Namur, Belgium. (1943) DSc, chemistry

(U. Louvain, 1969). Prof. (tel. 081 + 724111, ext. 2509, telex 59222, fax 081 + 230391). *Structural chemistry.*

Verlinde, Dr Christophe Louis-Marie Jos. Lab. voor Analytische Chemie en Medicinale Fysicochemie, Inst. voor Farmaceutische Wetenschappen, Katholieke U. Leuven, Van Evenstraat 4, B-3000 Leuven, Belgium. (1958) D.Pharm. (U. Leuven, 19881). Asst. (tel. 016 + 283418, email FKAAA07 at BLEKUL11). *Structural chemistry, structure - activity relationships.*

Wegener, Dr Wolter, Kernonderzoek, S.C.K.-C.E.N., B-2400 Mol, Belgium. (1941) Dr, physics (T.H.Aachen, 1970). Res. Physicist. (tel. 014 + 311801, ext. 2326, telex 31922, fax 014 + 315021, email NEUTRON2 at BMLSCK11). *Neutron diffraction, X-ray diffraction.*

BOLIVIA

Sub-Editor: A. Saavedra M.

Notes

1. International telephone country code - 591.

Alarcón, Mr Hugo. Inst. of Economic Geology, Fac. of Geological Sci., U. Mayor de San Andrés, La Paz, Bolivia. (1941) Lic, geology (U. Federal de Rio de Janeiro, Brazil, 1964). Prof., optical mineralogy (tel. 793392-359581). *Mineralogy, economic geology, ore microscopy and fluid inclusions.*

Arduz, Mr Marcelo. Fac. of Earth Sci., U. Mayor de San Andrés, La Paz, Bolivia. (1946) Lic, geological eng. (U. Nat. de la Plata, Argentina). Prof. optical mineralogy, petrography (tel. 793392-359581). *Optical mineralogy, petrography.*

Arellano, Mr J. Fac. of Earth Sci., U. Mayor de San Andrés, Casilla 5905, La Paz, Bolivia. (1947) Lic, geological eng. (U. Mayor de San Andrés, 1974). (tel. 793392-785262). *Petrography, X-ray crystallography.*

Avila-Salinas, Mr Waldo. Mineralogy and Petrography Div., Dept. of Labs., Geological Service of Bolivia, Fco. Zuazo 1673, Casilla 2729, La Paz, Bolivia. (1941) Lic, geological eng. (U. Mayor de San Andrés, 1965). Chief. (tel. 32692, personal tel.). *Inorganic X-ray crystallography.*

Portugal, Mr Remberto. Dept. of Physics, Fac. of Sci. and Techn., U. Mayor de San Simón, Casilla 2551, Cochabamba, Bolivia. (1947) MSc, physics (U. Estadual de Campinas, Brazil, 1979). Asst. prof. (tel. 042 + 25503, ext 318). *Small angle diffraction.*

Ricaldi, Mr Edgar. Dept. of Physics, U. Mayor de San Andrés, La Paz, Bolivia. (1947) Dipl, geophysics (Freiberg's Academy of Mines, DDR, 1973). Prof. (tel. 799299-792622). *Mineralogy, petrophysics.*

Saavedra, Mr Antonio. Fac. of Sci., U. Mayor de San Andrés, Casilla 604, La Paz, Bolivia. (1939) Lic, geological eng. (U. Mayor de San Andrés, 1964). Dean. (tel. 329701). *Mineralogy, petrography.*

Sanjinés, Mr Orlando. Fac. of Geological Sci., U. Mayor de San Andrés, La Paz, Bolivia. (1944) Lic, geological eng. (U. Mayor de San Andrés, 1968). Prof. (tel. 327429-793392-359581). *Mineralogy, economic geology.*

Santivañez, Mr Reynaldo. Fac. of Earth Sci., U. Mayor de San Andrés, Casilla 3698, La Paz, Bolivia. (1944) Lic, geological eng. (U. Mayor de San Andrés, 1978). Prof. (tel. 378587-793392-359581). *Petrography, petrogology, fluid inclusions.*

Vargas, Mr Edgar. Fac. of Geological Sci., U. Mayor de San Andrés, La Paz, Bolivia. casilla 6376, La Paz Bolivia. (1939) MSc Geomorphology (U. Sheffield, 1969) ITC Photogeologist (Delft, Holland) Prof. (tel. 790911-793392-359581). *Photogeology, geomorphology.*

Villalpondo, Mr Abelardo. Fac. of Geological Sci., U. Mayor de San Andrés, La Paz, Bolivia. casilla 3080, La Paz Bolivia. (1939) Dr rec nat (Freiberg's Academy of Mines, GDR, 1969) Prof. (tel. 794303-793392-359851). *Economic geology, mineralogy.*

Villegas, Dr Mario Oscar. Inst. de Investigaciones Fisicas, U. Mayor de San Andrés, La Paz, Bolivia. (1943) PhD, metallurgy (U. Nac. del Sur, Argentina, 1975). Res., prof. (tel. 799299-792622). *Phase identification, electron microscopy, X-ray techniques.*

Zelaya, Mr José Miguel. Inst. de Investigaciones Fisicas, U. Mayor de San Andrés, La Paz, Bolivia. (1947) MSc, mechanical eng. (U. Estadual de Campinas, Brazil, 1979). Res., prof. (tel. 799299-792622). *Crystal structure determination.*

BRAZIL

Sub-Editor: I.L. Torriani

Notes

1. International telephone country code - 55.

Abrão Pereira Mr Romeo. Centro Brasileiro de Pesquisas Físicas (CBPF) Rua Xavier Sigaud. 150 - Urca 22290 Rio de Janeiro, R.J., Brasil. (1947) MSc, physics (CBPF, 1988). Researcher (tel. 021+5410337, ext. 140). *Crystal physics, polymers, instrumentation.*

Almeida, Prof. Vasco Nogueira. Dept. de Física, U. Fed. de Goias, Goiania, Goias 74000, Brasil. (1923) MSc, physics (Fac. de Filos. Ciencias, Rio Claro, UNESP, 1966). Prof. adjunto. (tel. 062 + 2613088, ext. 168). *Crystal structures.*

Arguello, Dr Zoraide Primerano. Dept. de Estado Sólido, Inst. de Física, U. Est. de Campinas, Campinas, S.P. 13081, Brasil. (1938) PhD, physics (U. Est. de Campinas, 1972). Assoc. prof. (tel. 0192 + 391301, ext. 346). *Crystal growth.*

Arruda, Prof. Moacir Rabelo. Inst. de Geociencias, Dept. de Mineralogia e Petrografia, São Paulo, S. P. 01000, Brasil. (1925) PhD, geology (USP). Instr. (tel. 001 + 2122011). *Instrumentation, crystal optics, inorganic crystal structures.*

Baptista, Mr Augusto. Comissão Nac. de Energia Nuclear, Rio de Janeiro 20000, Brasil. (1930) BSc, chemistry (U. Fed. do Estado da Guanabara). Chemist. (tel. 021 + 2867002). *Organic and inorganic structures.*

Baptista, Mrs Neysa Rocha. Comissão Nac. de Energia Nuclear, Rio de Janeiro, Rio de Janeiro 20000, Brasil. (1930) BSc, chemistry (U. Fed. do Estado da Guanabara). Chemist. (tel. 021 + 2867002). *Inorganic crystal structures.*

Baran, Dr Zbigniew. Inst. de Física, U. Fed. da Bahia, Salvador, Bahia 40000, Brasil. (1930) PhD, Physics (U. Warsaw, Poland, 1970). Prof. (tel. 071 + 2472714). *Structural defects in crystalline solids.*

Barelli, Dr Nilso. Dept. de Tecnologia e Química de Aplicaçao, Inst. de Química, UNESP, Araraquara, São Paulo 14800, Brasil. (1944) PhD, mineralogy (Fac. de Filos. Ciencias e Letras de Araraquara, 1974). Asst. prof. (tel. 0162 + 320444, ext. 193). *Mineralogy, crystal growth, morphology, epitaxy.*

Beltrán Abrego, Dr José Ramón. Dept. de Física e Ciencia dos Materiais, Instituto de Física e Química , USP, São Carlos. C.P. 369 São Carlos, S.P. 13560, Brasil. (1952) PhD, physics (IFQSC-USP, 1987) Res. (tel. 0162 + 713365). *Small angle X-ray scattering, macromolecules in solution.*

Bezerra, Dr George Humberto. Dept. de Física Exp., Instituto de Física, USP. C.P. 20516, São Paulo, S.P. 01498, Brasil. (1956) PhD, physics (USP, 1988). Researcher (tel. 011 + 8155599) *Amorphous alloys, EXAFS.*

Bristoti, Dr Anildo. Dept. de Física, U. Fed. do Rio Grande do Sul, Porto Alegre, Rio Grande do Sul 90000, Brasil. (1936) PhD, metallurgical eng. (UCLA, USA, 1970). Assoc. prof. (tel. 0512 + 25-29-22). *X-ray crystallography, powder diffractometry, metallurgy, phase transitions, ferroelectrics.*

Bulhões, Mrs Iseli Angelica M. Dept. de Eng. de Materiais, U. Fed. de São Carlos, São Carlos, São Paulo 13560, Brasil. (1953) MSc, appl. physics (USP, 1979) Grad. student. (tel. 0162 + 718111, ext. 180). *Structures, X-ray diffraction.*

Campelo Farias, Prof. Carlinda. Dept. de Engenharia de Minas, U. Fed. de Pernambuco, Cidade U., Recife, Pernambuco 50000, Brasil. (1939) DSc, mineralogy (U. Fed. de Pernambuco, 1977) Assoc. prof. (tel. 081 + 2271208). *X-ray crystallography, powder diffractometry, minerals.*

Campos, Dr Cícero. Dept. de Estado Sólido, Inst. de Física, U. Est. de Campinas, Campinas, São Paulo 13100, Brasil. (1948) PhD, physics (U. Est. de Campinas, 1983). Asst. prof. (tel. 0192 + 391301, ext. 269). *Multiple X-ray scattering, dynamical theory.*

Carvalho da Silva, Prof. Jair. Dept. de Geologia, Escola Nac. de Minas e Metalurgia, Ouro Preto, Minas Gerais, Brasil. DSc, mineralogy (U. Brasil, 1957). Prof. *Minerals.*

Cassedane, Dr Jeannine. Inst. de Geociencias, U. Fed. do Rio de Janeiro, Ilha do Fundão, Rio de Janeiro 20000, Brasil. (1927) PhD, solid state (U. Strasbourg, France, 1969). Prof. (tel. 021 + 2305315). *X-ray crystallography, minerals.*

Castellano, Dr Eduardo Ernesto. Dept. de Fisica e Ciencia dos Materiais, Inst. de Física e Química de São Carlos USP, C.P. 369, São Carlos, São Paulo 13560, Brasil. (1941) PhD, physics (U. Nac. de La Plata, Argentina, 1968). Prof. (tel. 0162 + 713365). *Direct methods, macromolecules.*

Caticha Alfonso, Mr Ariel. Dept. do Estado Sólido, Inst. de Física, U. Est. de Campinas, Campinas, São Paulo 13100, Brasil. (1955) PhD, physics (Caltech, USA, 1985). Assist. prof. (tel. 0192 + 391301). *Scattering theory, dynamical theory.*

Caticha Ellis, Prof. Stephenson. Dept. do Estado Sólido, Inst. de Física U. Est. de Campinas, Campinas, São Paulo 13100, Brasil. (1927) Eng., engineering (U. de Montivídéo, Uruguay, 1954). Prof. (tel. 0192 + 391301, ext. 591). *Crystal physics, defects, instrumentation.*

Chang, Dr Shih- Lin. Dept. do Estado Sólido, Inst. de Física, U. Est. de Campinas, Departamento de Estado Sólido, C.P. 6165 Campinas, S.P. 13081, Brasil. (1946) PhD, physics (Polytechnic Inst. of New York, USA, 1975). Assoc. prof. (tel. 0192 + 391301, ext. 269). *X-ray scattering theory, dynamical theory, X-ray optics, Liquid phase epitaxy, thin films, instrumentation, X-ray interferometer, crystal physics.*

Correia Neves, Prof. José Marques. Inst. de Geociencias, U. Fed. de Minas Gerais, Belo Horizonte, Minas Gerais 30000, Brasil. (1929) DSc, mineralogy, geochemistry (U. Coimbra, Portugal, 1963). Prof. Titular UFMG. *Inorganic crystal structures, mineralogy.*

Costa Gouveia, Prof. Albany H. Dept. de Engenharia de Minas, U. Fed. de Pernambuco, Recife, Pernambuco 50000, Brasil. (1941) DSc, mineralogy (U. Fed. de Pernambuco, 1977). Assoc. prof. (tel. 081 + 2315205). *Optical crystallography, mineralogy.*

Costa Viana, Prof. Carlos Sergio da. Coppe, U. Fed. do Rio de Janeiro, Rio de Janeiro, Rio de Janeiro 20000, Brasil. (1942) PhD, philosophy (U. of Cambridge, UK, 1978). Assoc. prof. (tel. 021 + 2809322, ext. 242). *Texture, formability, anisotropy, mechanical properties.*

Craievich, Dr Aldo Felix. Laboratório Nacional de Luz Síncrotron (LNLS) C.P. 6192, Campinas, S.P. 13081, Brasil. (1939) PhD, Physics (U. Nac. de Cuyo, Argentina, 1969). Deputy Director (tel. 0192+512624). *Physical crystallography, small angle X-ray scattering, synchrotron radiation.*

Cusatis, Dr Cesar. Dept. de Física, U. Fed. do Paraná, C.P. 19091, Curitiba, Paraná 80000, Brasil. (1939) PhD, Physics (USP, 1973). Assoc. prof. (tel. 041 + 2669271). *X-ray optics, X-ray interferometry.*

Dias Rodrigues, Mrs Ana Maria Gonçalves. Dept. de Química e Física Molecular, Inst. de Física e Química de São Carlos, São Paulo 13560, Brasil. (1954) MSc, physical chemistry (USP, 1979). Asst. prof. (tel. 0162 + 721538). *Structure determination.*

Ferran, Dr Gustan. Coppe, U. Fed. do Rio de Janeiro, Rio de Janeiro, Rio de Janeiro 20000, Brasil. (1938) PhD, metalurgy (U. Madrid, Spain, 1966). Assoc. prof. (tel. 021 + 2609776). *Preferred orientation, texture, Kossel diffraction.*

Ferreira de Souza, Prof. Milton. Dept. de Física e Ciencias dos Materiais, Inst. de Física e Química de São Carlos, Av. Dr Carlos Botelho, 1465 São Carlos, São Paulo, Brasil. (1932) PhD, physics (USP, 1969). Prof. (tel. 0162 + 711016). *Solid state physics, defects in solids, crystal growth.*

Figueiredo Neto, Dr Antonio Martins. Dept. de Física Exp., Inst. de Física da USP, C.P. 20516, São Paulo S.P. 01498, Brasil. (1953) PhD, physics (USP, 1981). Assoc. prof. (tel. 0162 + 2114865). *Liquid crystals, small angle X-ray scattering.*

Folgueras Dominguez, Dr Sérvulo. Dept. de Química, U. Fed. de São Carlos, São Carlos, São Paulo 13560, Brasil. (1929) Dr, physics (U. Madrid, Spain, 1979). Assoc. Prof. (tel. 0162 + 718111, ext. 164). *Silicate structures, epitaxy.*

Formoso, Dr Milton Luiz. Inst. de Geociencias, U. Fed. do Rio Grande do Sul, Porto Alegre, Rio Grande do Sul 90000, Brasil. (1927) PhD, geology (USP, 1973). Prof. (tel. 0512 + 215422). *Geochemistry, mineralogy, geology.*

Francesconi, Dr Ricardo. Inst. de Geociencias, U. de São Paulo, São Paulo, S. P. 01000, Brasil. (1941) PhD, mineralogy (USP, 1966). Asst. prof. (tel. 011 + 2122011). *Mineralogy, mineral deposits.*

Francisco, Miss Regina Helena Porto. Dept. de Química e Física Molecular, Inst. de Física e Química de São Carlos USP, São Carlos, São Paulo 13560, Brasil. (1952) BSc, quimica (Fac. de Filos. Ciencias e Letras de Rib. Preto, 1973). Instr. (tel. 0162 + 712234, ext. 52). *Structure determination.*

Freire D'aguiar, Mr Manoel Marcos. Inst. Física, U. Fed. da Bahia, Salvador, Bahia 40000, Brasil. (1947) MSc, geophysics (U. Fed. da Bahia, 1974). Res. assoc. (tel. 071 + 2472714). *Crystal defects.*

Fujimore, Prof. Kenkichi. Inst. de Astronomia e Geofisica USP, Av. Miguel Estefano 4200 São Paulo, S.P. 01051, Brasil. (1929) PhD, physics (Tohoku U. Sendai, Japan). Prof. Mineralogy. *Inorganic crystal structures.*

Fulfaro, Dr Roberto. Inst. de Pesquisas Energéticas e Nucleares, C.P.11049, São Paulo, S.P. 05508, Brasil. (1938) PhD, neutron physics (U. Est. de Campinas, 1970). Head of center (COURP). (tel. 011 + 2116011, ext. 237). *Neutron scattering, lattice dynamics.*

Gomes, Mr Samuel Irati Novaes. Dept. de Materiais, Escola de Engenharia de São Carlos, São Carlos, São Paulo 13560, Brasil. (1939) BSc, physics (Fac. de Filos. Ciencias, Rio Claro). Instr. (tel. 0162 + 712234, ext. 95). *Metallurgy, defects, metals and alloys.*

Grundig, Prof. Werner. Inst. de Tecnologia, U. Fed. do Rio Grande do Sul, Porto Alegre, Rio Grande do Sul 90000, Brasil. MSc, engineer (U. Fed. do Rio Grande do Sul, 1937). Chief eng. *Powder diffraction.*

Herdade, Dr Silvio B. Inst. de Física, U. de São Paulo, C.P. 20516, São Paulo, S. P. 01000, Brasil. (1926) PhD, physics (U. de Campinas, 1969). Assoc. prof. (tel. 011 + 2116955). *Instrumentation, Solid state dielectric track detectors, track formation processes.*

Imakuma, Dr Kengo. Inst. de Pesquisas Energéticas e Nucleares, C.P. 11049, São Paulo, S.P. 05508, Brasil. (1943) PhD, physics (USP, 1973). Head, physics sect. (tel. 011 + 2116011, ext. 242). *Phase transitions, ceramics, alloys, radiation damage.*

Inglez, Mr Antonio Gabriel. Dept. de Mineralogia e Petrografia, Cidade U., São Paulo, São Paulo 010000, Brasil. (1943) BSc, geology. Instr. *Inorganic crystal structures, optical crystallography, computer programming.*

Kunrath, Mr José Irineu. Inst. de Física, U. Fed. do Rio Grande Do Sul, Porto Alegre, Rio Grande Do Sul 90000, Brasil. (1931) MSc, physics (U. Fed. Do Rio Grande Do Sul, 1960). Asst. prof. (tel. 0512 + 245817). *Solid state physics, powder diffraction.*

Labaki, Ms Lucila Chebel. Dept. de Estado Sólido, Inst. de Física, U. Est. de Campinas, C P. 6165, Campinas, São Paulo 13100, Brasil. (1943) MSc, biophysics (Sophia State U., Bulgaria, 1978) Grad. student. (tel. 0192 + 3913015, ext. 269). *Polymers, diffraction from chain molecules.*

Lechat, Dr Johannes Rudiger. Dept. de Química e Física Molecular, Inst. de Física e Química, USP, São Carlos, São Paulo 13560, Brasil. (1943) PhD, chemistry (USP, 1972). Asst. prof. (tel. 0162 + 712234, ext. 52). *Organic crystal structures.*

Leite, Dr Cirano Rocha. Dept. de Química Tecnológica e de Aplicação, Inst. de Química, UNESP, C.P. 174, Araraquara, São Paulo 14800, Brasil. (1941) PhD, mineralogy (USP, 1969). Prof. (tel.0162 + 320444, ext. 136). *Mineralogy, crystal growth, morphology, epitaxy.*

Madureira Filho, Prof. José Barbosa de. Inst. de Geociencias, Dept. de Mineralogia e Petrografia, U. de São Paulo, C.P. 20899, São Paulo, São Paulo 05508, Brasil. (1940) PhD, mineralogy and petrology (USP, 1983). Asst. Prof. (tel. 011 + 2122011). *Solid solutions, mineralogy.*

Martinez, Mr Luis Gallego. Instituto de Pesquisas Energéticas e Nucleares (IPEN) C.P. 11049, São Paulo, S.P. 05508, Brasil. (1957) MSc physics (IPEN/USP, 1988) Researcher (tel. 011 + 2116011, ext.242). *X-ray powder diffractometry, metals, ceramics, defects.*

Mascarenhas, Prof. Yvonne Primerano. Dept. de Física e Ciencia Dos Materiais, Inst. de Física e Química, USP, São Carlos, São Paulo 13560, Brasil. (1931) PhD, physics (USP). Prof. (tel. 0162 + 724496). *Crystal structures.*

Mascarenhas, Prof. Sergio. Dept. de Física e Ciencia dos Materiais, Inst. de Física e Química, Campus de São Carlos USP, C.P. 369, São Carlos, São Paulo 13560, Brasil. (1928) PhD, physics (USP, 1958). Prof. (tel. 0162 + 715381). *Biomolecular structure and function, bioelectrets, biophysics, bound water.*

Mazzaro, Mr Irineu. Dept. de Física, U. Fed. do Paraná, C.P. 19091, Curitiba, Paraná 80000, Brasil. (1953) MSc, physics (U. Est. de Campinas, 1979). Jr. lect. (tel. 041 + 2642855). *X-ray optics, multiple diffraction, crystal defects.*

Mazzocchi, Ms Vera. Instituto de Pesquisas Energéticas e Nucleares (IPEN) C.P. 11049 - Pinheiros, São Paulo, S.P. Brasil. (1955) MSc, physics (IPEN, 1984). Researcher (tel. 011 + 2116011, ext.141). *Neutron diffraction, neutron multiple diffraction, texture.*

Medeiros Rodrigues, Dr Maria Mabel. Dept. de Química e Física Molecular, Inst. de Física e Química de São Carlos USP, São Carlos, São Paulo 13560, Brasil. PhD, chemistry (USP, 1968). Asst. prof. (tel. 0162 + 712234, ext. 52). *Crystal structure determinations, organic compounds, biological interaction.*

Mestnik Filho, Dr José. Div. de Física Nuclear, Inst. de Pesquisas Ener. e Nucleares (IPEN) C.P. 11049-Pinheiros, São Paulo S.P. 01000, Brasil. (1949) PhD, neutron physics (IPEN, 1987). Researcher (tel. 011 + 2116011). *Neutron scattering, neutron diffraction, lattice dynamimcs, metal hydrides.*

Murta, Prof. Clecio. Companhia Brasileira de Tecnologia Nuclear, U. Fed. de Minas Gerais, Belo Horizonte, Minas Gerais 30000, Brasil. (1929) MSc, nuclear sci. (U. Fed. Minas Gerais, 1971). Head, instrumental analysis lab. (tel. 42-5422). *Diffractometry, X-ray fluorescence, electron microprobe, thermoanalysis, mineralogy.*

Oliva, Dr Glaucius. Dept. de Física e Cs. dos Materiais, Instituto de Física e Química de São Carlos, USP. C.P. 369, São Carlos, S.P. 13560, Brasil. (1959) PhD, physics (Birkbeck College, UK, 1988). Lecturer (tel. 0162 + 713365). *Protein crystallography, small molecules crystallography.*

Olivieri, Mr Johnny Rizzieri. Dept. de Física e Ciencia dos Materiais, Inst. de Física e Química de São Carlos USP, São Carlos, São Paulo 13560, Brasil. (1950) BSc, physics (U. de São Paulo, 1976). Grad. student. (tel. 0162 + 712234, ext. 33). *Small angle X-ray scattering, amorphous materials.*

Pavie Cardoso, Dr Lisandro. Dept. do Estado Sólido, Inst. de Física, U. Est. de Campinas, Campinas, São Paulo 13100, Brasil. (1950) PhD, physics (U. Est. de Campinas, 1983). Asst. prof. (tel. 0192 + 391301). *Crystal defects, multiple diffraction.*

Quaranta Cabral, Dr Ubirajara. U. Fed. do Rio de Janeiro, Rio de Janeiro, Rio de Janeiro 20000, Brasil. (1937) DSc, metallurgy (U. Paris, France). Asst. prof. (tel. 021 + 2609776). *Metal physics, structure.*

Queiroz do Amaral, Dr Lia. Inst. de Física, USP, C.P.20516, São Paulo, S. P.01498, Brasil. (1941) PhD, physics (USP, 1972). Assoc. prof. (tel. 011 + 8155599, ext 310). *Liquid crystals, small angle X-ray scattering, phase transitions, membranes, amorphous alloys.*

Ramos Parente, Dr Carlos Benedicto. Inst. de Pesquisas Energéticas e Nucleares, C.P.11049 Pinheiros, São Paulo 01000, Brasil. (1937) PhD, physics (USP, 1973). Res. (tel. 011 + 2116011, ext. 141). *Neutron diffraction,multiple neutron diffraction, texture.*

Regueira Teodósio, Prof. Joel. U. Fed. do Rio de Janeiro, Rio de Janeiro 20000, Brasil. (1943) MSc, metallurgy (U. Fed. do Rio de Janeiro, 1973). Asst. prof. (tel. 021 + 2609776). *X-ray powder diffraction, preferred orientation.*

Ribeiro Franco, Prof. Rui. Inst. de Pesquisas Energéticas e Nucleares, Div. de Física Nuclear, Cidade U., São Paulo, São Paulo 01000, Brasil. (1916) PhD, Petrology (USP, 1944). Head, teaching div. (tel. 011 + 2116011). *Crystal growth, optical crystalllography.*

Riella, Eng. Humberto Gracher. Metallurgy Div., Comissão Nac. de Energia Nuclear, Travessa R 400, São Paulo, São Paulo 05508, Brasil. (1953) Eng. (U. Fed. do Paraná, 1975). Res. (tel. 011 + 2116011, ext. 126). *Metals, diffusion in metals.*

Rodrigues da Silva, Dr Rilson. Dept. de Engenharia de Minas, Centro de Tecnologia, U. Fed. de Pernambuco, Recife, Pernambuco 50000, Brasil. (1932) DSc, mineralogy (U. Strasbourg, France,1969) Prof., crystallography and mineralogy (tel. 081 + 3612789). *X-ray diffraction, X-ray spectroscopy, mineralogy.*

Rodrigues, Dr Antonio Ricardo Dröher, Laboratório Nacional de Luz Sincrotron (LNLS) C. P. 6192, 13081 Campinas, S.P., Brasil. (1951) PhD, X-ray optics (King's C., London U., UK, 1979) Project Director (0192 + 512624) *X-ray optics, instrumentation.*

Rodrigues, Dr Edson. Dept. de Química e Física Molecular, Inst. de Física e Química de São Carlos USP, São Carlos, São Paulo 13560, Brasil. PhD, physics. Prof. (tel. 0162 + 711292). *Crystal physics, magnetic resonance.*

Rolim De Camargo, Prof. William Gerson. Inst. of Geosciences, U. of São Paulo, São Paulo, São Paulo 01000, Brasil. (1920) PhD, mineralogy, crystallography (USP, 1944). Prof., mineralogy. (tel. 011 + 2122011). *Mineralogy, uranium minerals, diamond, crystallography, accurate cell dimensions (methods).*

Santos, Dr Persio de Souza. Chemical Eng. Dept., U. de São Paulo, São Paulo, São Paulo 61548, Brasil. (1928) PhD, physical chemistry (USP, 1960). Head, Ind. chem. dept. (tel. 011 + 8159322). *Minerals (non-metallic), clays, zeolites.*

Santos, Mrs Regina Helena de Almeida. Dept. de Química e Física Molecular, Inst. de Física e Química de São Carlos USP, São Carlos, São Paulo 13560, Brasil. (1947) MSc, physical chemistry (USP, 1974). Asst. prof. (tel. 0162 + 712234, ext. 52). *Structural crystallography.*

Simone, Mr Carlos Alberto de. Dept. de Química, U. Fed. de Alagoas, Campus Universitário, Tabuleiro do Martins, Maceió, Alagoas 57000, Brasil. (1951) MSc, appl. physics (USP, 1983). Asst. Prof. (tel. 082 + 2421238). *Crystal structure, natural products, powder data refinement.*

Soledade Jr, Prof. Teomar. Inst. de Física, U. Fed. da Bahia, Salvador, Bahia 40000, Brasil. (1948) MSc, Physics (U. Est. de Campinas, 1976). Res. assoc. (tel. 071 + 2472714). *Crystal defects, divergent beam methods.*

Souza, Prof. Irineu Marques. Inst. de Geociencias USP, Dept. de Geologia Economica, Cidade U., São Paulo, São Paulo 01000, Brasil. (1940) BSc, geology. Instr. (tel. 011 + 2122011). *Mineralogy, crystal optics.*

Sousa, de, Mr José Carlos. Dept. de Física, Fundação U. Est. de Maringá, Av. Colombo 3690, Campus Univ., Maringá, Paraná 87100, Brasil. (1948) MSc, physics (U. Est. de Campinas, 1979). Grad. student. *Small angle X-ray diffraction, carbonaceous materials.*

Stojanoff, Dr Vivian. Instituto de Física, USP. C.P. 20516, São Paulo, S.P. 01498, Brasil. (1955) PhD, physics (USP, 1984). Assist. prof. (tel. 011 + 8155599). *Defects in semiconductors.*

Suzuki, Dr Carlos Kenichi. Faculdade de Engenharia de Campinas, Dept. de Engenharia de Materiais, C.P. 6122, Campinas, S.P. 13081, Brasil. (1945) PhD, applied physics engineering (U. Tokyo, Japan, 1980). Asst. prof. (tel. 0192 + 391301, ext.2952). *Dynamical X-ray theory, topography, mineralogy.*

Svisero, Dr Darcy Pedro. Dept. of Mineralogy and Petrology, Inst. of Geosciences U. of São Paulo, São Paulo, S.P. 01000, Brasil. (1940) PhD, geology (U. of São Paulo, 1971). Asst. prof. (tel. 2122011). *Mineralogy.*

Távora, Prof. Elysiario. Comissão Nac. de Energia Nuclear, Rio de Janeiro, Rio de Janeiro 20000, Brasil. DSc, natural sciences (U. Brasil, 1946). Prof., mineralogy, petrography. (tel. 021 + 2867002). *Theoretical and X-ray crystallography.*

Teixeira Mendes, Prof. Antonio Carlos. Escola Superior de Agricultura Luiz de Queiroz, Piracicaba, São Paulo, Brasil. Geology (Escola Superior de Agricultura Luiz de Queiroz). Prof. *Powder method, clay minerals.*

Tomita, Dr Koichi. Dept. de Fisico-Química, Inst. de Química, Araraquara, São Paulo 14800, Brasil. (1943) PhD, chemistry (Fac. de Filos. Ciencias e Letras de Araraquara, 1967). Asst. prof. (tel. 0162 + 320444, ext. 193). *Crystal structures.*

Torriani, Dr Iris Linares. Dept. de Estado Sólido, Inst. de Física, U. Est. de Campinas, C.P. 1170, Campinas, São Paulo 13100, Brasil. (1934) PhD, physics (U. Nac. de La Plata, Argentina, 1975). Assoc. prof. (tel. 0192 + 391301, ext. 269). *Small angle X-ray diffraction, biological applications, thin films and model membrane structures.*

Valarelli, Dr José Vicente. Inst. de Geociencias, Dept. de Mineralogia e Petrografia, C.P. 20899, São Paulo, São Paulo 01000, Brasil. (1939) PhD, mineralogy (USP, 1967) Assoc. prof. (tel. 011 + 2122011). *Applied mineralogy, mineral paragenesis, fluid inclusions.*

Varela, Mr José Arana. Dept de Ciencias Exatas, Fac. de Filos. Ciencias e Letras de Presidente Prudente, Presidente Prudente, São Paulo 19100, Brasil. (1944) MSc, Materials science ceramics (Inst. Tecnológico de Aeronáutica, 1975). Asst. prof. (tel. 0182 + 34116). *Materials science, ceramics, transformations.*

Villarroel, Prof. Hugo Sergio. Dept. Eng. Minas, U. Fed. de Pernambuco, Cidade U. Recife, Pernambuco 50000, Brasil. (1931) MSc, crystallography (U. Austral, Chile, 1970). Prof. (tel. 081 + 2271208). *X-ray diffraction, X-ray spectroscopy, minerals.*

Vinhas, Dr Laercio Antonio. Inst. de Pesquisas Energéticas e Nucleares, Cidade U., São Paulo, São Paulo 01000, Brasil. (1943) PhD, neutron physics (U. Est. de Campinas, 1970). Head res. (tel. 011 + 2116011, ext. 139). *Neutron scattering, neutron diffraction, molecular crystals, lattice dynamics.*

Willig, Prof. Cesar Dorneles. Escola de Geologia, U. Fed. do Rio Grande do Sul, Porto Alegre, Rio Grande do Sul 90000, Brasil. (1940) BSc, geology (U. Fed. do Rio Grande do Sul). Instr. *Inorganic structures, computer programming.*

Zukerman Schpector, Dr Julio, Laboratório de Cristalografia e Físico-quimica dos Materiais CCEN - Universidade Federal de Alagoas. 57000 Maceió, Alagoas, Brasil. (1948) PhD chemistry (U. de São Paulo, 1984) Assoc. Prof. (tel. 082 + 2421238) *Crystal structures, polymers.*

BULGARIA

Sub-Editor: **I. Bonev**

Notes

1. International telephone country code - 359, area code for Sofia - 02.

2. Degrees conferred in Bulgaria are *graduate* (Grad) (refers to a university diploma in physics, geochemistry, engineering, etc), *candidate of sciences* (CSc) and *doctor of sciences* (DSc). They are approximately equivalent to MSc, PhD and DSc, respectively. The highest honorary degrees conferred by the Bulgarian Academy of Sciences are *academician* (full member) and *corresponding member* of the Bulgarian Academy of Sciences.

3. The positions at the universities and the equivalent positions at the research institutes are, respectively: *assistant - research associate, docent* (associate professor) - *senior scientist* and *professor - research professor.*

Apostolov, Prof. Andrei. Chair of Solid State Physics, Sofia U., Blvd Anton Ivanov 5, Sofia 1126, Bulgaria. (1935) DSc, physics (Sofia U., 1977). Prof. and head. (tel. 623015). *Solid state physics, magnetism.*

Apostolov, Mr Anton. Central Inst. for Computing Technics, Acad G. Bonchev, Bl. 10, Sofia 1040, Bulgaria. (1951) Grad, physics (Sofia U., 1974). Res. assoc. (tel. 7131, ext. 2708). *Powder diffraction, high temperature X-ray experiments.*

Aslanian, Dr Selma. Geological Inst., Bulgarian Acad. Sci., Sofia 1113, Bulgaria. (1937) Dr.rer.nat., mineralogy (Leipzig U., DDR, 1969). Res assoc. (tel. 7131, ext. 3470). *Crystal growth, crystal chemistry.*

Atanassov, Dr Vassil. Chair of Mineralogy and Petrography, Higher Inst. of Mining and Geology, Sofia 1156, Bulgaria. (1933) CSc, mineralogy (Inst. of Geology and Geophysics, Novosibirsk, USSR, 1973). Docent. (tel. 62581, ext. 385). *Mineral morphology.*

Avramov, Dr Isak. Inst. of Physical Chemistry, Bulgarian Acad. Sci., Sofia 1040, Bulgaria. (1946) CSc, chemistry (Inst. of Physical Chemistry, 1980). Res. assoc. (tel. 7131, ext. 2566). *Crystal growth, glass transitions, thin films.*

Balkanov, Mr Ivan. Central Lab. for Electrochemical Power Sources, Bulgarian Acad. Sci., Sofia 1040, Bulgaria. (1949) Grad, geochemistry (Sofia U., 1974). Res. assoc. (tel. 7131, ext. 2714). *X-ray analysis.*

Bliznakov, Prof. Georgi. Inst. of General and Inorganic Chemistry, Bulgarian Acad. Sci., Sofia 1040, Bulgaria. (1920) Academician, chemistry (Bulgarian Acad. Sci., 1979). Prof., director; vice-pres. (tel. 877783). *Crystal growth, adsorption, catalysis, inorganic synthesis.*

Bonev, Dr Ivan. Geological Inst., Bulgarian Acad. Sci., Sofia 1113, Bulgaria. (1936) CSc, mineralogy (Geological Inst., 1972). Sr. scient. (tel. 7131, ext.2236). *Crystal structures, minerals, crystal growth, epitaxy, whiskers, X-ray crystallography.*

Bostanov, Mr Vesselin. Central Lab. for Electrochemical Power Sources, Bulgarian Acad. Sci., Sofia 1040, Bulgaria. (1933) Grad, metallurgy (Higher Inst. of Chemical Techn., 1956). Res. assoc. (tel. 7131, ext. 2761). *Electrocrystallization.*

Budevski, Prof. Evgeni. Central Lab. for Electrochemical Power Sources, Bulgarian Acad. Sci., Sofia 1040, Bulgaria. (1922) Corresp. member, electrochemistry (Bulgarian Acad. Sci., 1984). Director. (tel. 723454). *Electrocrystallization.*

Budurov, Prof. Stoyan. Chair of Inorganic Chemical Techn., Faculty of Chemistry, U. of Sofia, Blvd Anton Ivanov 1, Sofia 1126, Bulgaria. (1930) DSc, chemistry (Sofia U., 1976). Prof. (tel. 62561,ext. 337). *Crystallization, metals, solid state reactions.*

Delineshev, Dr Svetoslav. Inst. of General and Inorganic Chemistry, Bulgarian Acad. Sci., Sofia 1040, Bulgaria. (1940) CSc, chemistry (Inst. of General and Inorganic Chemistry, 1978). Sr. scient. (tel. 7131, ext. 2542). *Crystal growth, nucleation.*

Dimov, Mr Vergil. Inst. of Appl. Mineralogy, Bulgarian Acad. Sci., Moskowska 6, Sofia 1000, Bulgaria. (1946) Grad, physics (Sofia U., 1973). Res. assoc. (tel. 872450). *Electron microscopy.*

Djarova, Dr Maria. Faculty of Chemistry, Sofia U., Blvd Anton Ivanov 1, Sofia 1126, Bulgaria. (1945) CSc, chemistry (Sofia U., 1979). Res. assoc. (tel. 62561, ext. 479). *Growth and perfection of crystals.*

Dobrev, Dr Dobri. Inst. of Physical Chemistry, Bulgarian Acad. Sci., Sofia 1040, Bulgaria. (1935) CSc, chemistry (Inst. of Physical Chemistry, 1976). Sr. scient. (tel. 719307). *Thin films.*

Draganova, Dr Dragana. Faculty of Chemistry, Sofia U., Blvd Anton Ivanov 1, Sofia 1126, Bulgaria. (1936) CSc, chemistry (Sofia U., 1974). Asst. (tel. 62561, ext. 342). *Growth and perfection of crystals.*

Filizova, Dr Lyudmila. Inst. of Appl. Mineralogy, Bulgarian Acad. Sci., Moskowska 6, Sofia 1000, Bulgaria. (1937) CSc, mineralogy (Inst. of Geology and Geophysics, Novosibirsk, USSR). Res. assoc. (tel. 872450). *Synthesis, minerals, crystal growth.*

Grozdanov, Mr Lyudmil. Geological Inst., Bulgarian Acad. Sci., Sofia 1113, Bulgaria. (1934) Grad, geochemistry (Sofia U., 1957). Res. assoc. (tel. 7131, ext. 2260). *X-ray crystallography, amphiboles.*

Gutzow, Prof. Ivan. Inst. of Physical Chemistry, Bulgarian Acad. Sci., Sofia 1040, Bulgaria. (1933) DSc, chemistry (Inst. of Physical Chemistry, 1972). Res. prof. (tel. 719305). *Nucleation, crystal growth, crystallization in glass-forming systems.*

Iwanov, Mr Dantsho. Inst. of Physical Chemistry, Bulgarian Acad. Sci., Sofia 1040, Bulgaria. (1939) Grad, chemistry (Sofia U., 1967). Res. assoc. (tel. 7131, ext. 3584). *Crystal growth, thin films.*

Kaischew, Prof. Rostislaw. Inst. of Physical Chemistry, Bulgarian Acad. Sci., Sofia 1040, Bulgaria. (1908) Academician, physical chemistry (Bulgarian Acad. Sci., 1961). Prof., director. (tel. 727450). *Physical chemistry, crystal growth, nucleation.*

Kashchiev, Dr Dimcho. Inst. of Physical Chemistry, Bulgarian Acad. Sci., Sofia 1040, Bulgaria. (1942) CSc, chemistry (Inst. of Physical Chemistry, 1975). Sr. scient. (tel. 7131, ext. 2557). *Crystal growth, nucleation.*

Kirkova, Prof. Elena. Faculty of Chemistry, Sofia U., Blvd Anton Ivanov 1, Sofia 1126, Bulgaria. (1923) DSc, chemistry (Sofia U., 1983). Prof. (tel. 62561, ext. 215). *Crystal growth, adsorption.*

Kirov, Mr Georgi Kirilov. Inst. of Appl. Mineralogy, Bulgarian Acad. Sci., Moskowska 6, Sofia 1000, Bulgaria. (1932) Res. assoc. (tel. 885115, ext. 671). *Synthesis, minerals, crystal growth, morphology.*

Kirov, Doc Georgi Nikolov. Chair of Mineralogy and Crystallography, Sofia U., Blvd Russki 15, Sofia 1000, Bulgaria. (1930) Docent. (tel. 8581, ext. 256). *X-ray analysis, minerals.*

Konstantinov, Dr Ivan. Central Lab. of Photographic Processes, Bulgarian Acad. Sci., Sofia 1040, Bulgaria. (1943) CSc, chemistry (Inst. of Physical Chemistry, 1975). Res. assoc. (tel. 7131, ext. 3521). *Crystal growth, crystal physics.*

Kostov, Prof. Ivan. Nat. Natural History Museum, Bulgarian Acad. Sci., Blvd Russki 1, Sofia 1000, Bulgaria. (1913) Academician, mineralogy and crystallography (Bulgarian Acad. Sci., 1966). Prof., director. (tel. 882894). *Crystal growth, morphology, minerals, inorganic crystal structures.*

Kostov, Dr Ruslan. Inst. of Appl. Mineralogy, Bulgarian Acad. Sci., Moskowska 6, Sofia 1000, Bulgaria. (1956) CSc, mineralogy (IGEM, Moscow, USSR, 1984). Res. assoc. (tel. 872450). *Physics of minerals, crystal morphology.*

Kotsev, Dr Iosif. Faculty of Physics, Sofia U., Blvd Anton Ivanov 5, Sofia 1126, Bulgaria. (1942) CSc, physics (Moscow U., USSR, 1976). Sr. scient. (tel. 62561, ext. 458). *Symmetry.*

Kovachev, Dr Peter. Faculty of Chemistry, Sofia U., Blvd Anton Ivanov 1, Sofia 1126, Bulgaria. (1943) CSc, chemistry (Inst. of Physical Chemistry, 1975). Res. assoc. (tel. 62561, ext. 337). *Diffusion in metals.*

Krestev, Mr Venelin. Faculty of Physics, Sofia U., Blvd Anton Ivanov 5, Sofia 1126, Bulgaria. (1940) Grad, physics (Polytechn. Inst. Leningrad, USSR, 1966). Res. assoc. (tel. 62561, ext. 464). *Crystal imperfections, polymer physics.*

Kresteva, Dr Manya. Faculty of Physics, Sofia U., Blvd Anton Ivanov 5, Sofia 1126, Bulgaria. (1943) CSc, physics (Sofia U., 1982). Res. assoc. (tel. 62561, ext. 471). *Polymer physics.*

Maciček, Dr Josef. Inst. of Appl. Mineralogy, Bulgarian Acad. Sci., Moskowska 6, Sofia 1000, Bulgaria. (1953) CSc, chemistry (Moscow U., USSR, 1981). Res. assoc. (tel. 872450). *Crystal structure analysis.*

Maleev, Dr Michael. Inst. of Appl. Mineralogy, Bulgarian Acad. Sci., Moskowska 6, Sofia 1000, Bulgaria. (1941) CSc, mineralogy (Moscow U., USSR, 1968). Docent, director (tel. 872450). *Natural whiskers, electron microscopy.*

Malinowski, Prof. Yordan. Central Lab. of Photographic Processes, Bulgarian Acad. Sci., Sofia 1040, Bulgaria. (1923) Corresp. member, chemistry (Bulgarian Acad. Sci., 1979). Prof., director. (tel. 720073). *Crystal physics, crystal growth, lattice defects.*

Marinov, Dr Miko. Inst. of Physical Chemistry, Bulgarian Acad. Sci., Sofia 1040, Bulgaria. (1939) CSc, chemistry (Inst. of Physical Chemistry, 1981). Res. assoc. (tel. 7131, ext. 2533). *Crystal growth, electron microscopy.*

Markov, Dr Ivan. Inst. of Physical Chemistry, Bulgarian Acad. Sci., Sofia 1040, Bulgaria. (1941) CSc, chemistry (Inst. of Physical Chemistry, 1976). Res. assoc. (tel. 7131, ext. 2557). *Crystal growth and nucleation, epitaxy, thin films.*

Michailov, Mr Evgeni. Inst. of Physical Chemistry, Bulgarian Acad. Sci., Sofia 1040, Bulgaria. (1940) Grad, physics (Leningrad U., USSR, 1966). Res. assoc. (tel. 7131, ext. 2565). *Crystal growth, thin films, adsorption.*

Michailov, Mr Michail. Inst. of Solid State Physics, Bulgarian Acad. Sci., Blvd Lenin 72, Sofia 1184, Bulgaria. (1927) Res. assoc. (tel. 7341, ext. 440). *Electron microscopy, semiconductors, thin films.*

Milchev, Dr Alexander. Inst. of Physical Chemistry, Bulgarian Acad. Sci., Sofia 1040, Bulgaria. (1943) CSc, chemistry (Inst. of Physical Chemistry, 1982). Res. assoc. (tel. 7131, ext. 2558). *Crystal growth, electrocrystallization.*

Milchev, Dr Andrei. Inst. of Physical Chemistry, Bulgarian Acad. Sci., Sofia 1040, Bulgaria. (1946) Dr.rer.nat., physics (Leipzig U., DDR, 1979). Res. assoc. (tel. 7131, ext. 2566). *Phase transition, vitrification, two-dimensional systems.*

Miloshev, Prof. Georgi. Inst. of Geophysics, Bulgarian Acad. Sci., Sofia 1113, Bulgaria. (1933) DSc, physics (Inst. of Geophysics, 1974). Res. prof. (tel. 7131, ext. 3334). *Crystal growth.*

Mincheva-Stefanova, Prof. Yordanka. Mineralogy Section, Geological Inst., Bulgarian Acad. Sci., Sofia 1113, Bulgaria. (1923) Res. prof, section head. (tel. 7131, ext. 2282). *Morphology, minerals, crystal growth, sulphide minerals.*

Moldovanova, Prof. Maria. Chair of Semiconductor Physics, Sofia U., Blvd Anton Ivanov 1, Sofia 1126, Bulgaria. (1919) CSc, physics (Sofia U., 1952). Prof., head. (tel. 62561, ext. 322). *Semiconductor physics and technology.*

Nanev, Dr Christo. Inst. of Physical Chemistry, Bulgarian Acad. Sci., Sofia 1040, Bulgaria. (1938) CSc, chemistry (Inst. of Physical Chemistry, 1973). Sr. scient. (tel. 7131, ext. 2586). *Crystal growth, electrocrystallization, thin films, lattice defects.*

Nenow, Dr Dimiter. Inst. of Physical Chemistry, Bulgarian Acad. Sci., Sofia 1040, Bulgaria. (1931) CSc, chemistry (Inst. of Physical Chemistry, 1970). Sr. scient. (tel. 7131, ext. 2557). *Nucleation, crystal growth, crystal surfaces.*

Nikolaeva, Mrs Rumiana. Faculty of Chemistry, Sofia U., Blvd Anton Ivanov 1, Sofia 1126, Bulgaria. (1936) Grad, chemistry (Sofia U., 1959). Asst. (tel. 62561, ext. 342). *Crystal growth, defects.*

Pangarov, Prof. Nikola. Inst. of Physical Chemistry, Bulgarian Acad. Sci., Sofia 1040, Bulgaria. (1929) DSc, chemistry (Inst. of Physical Chemistry, 1968). Prof. (tel. 7131, ext. 2537). *Crystal growth, thin metal films.*

Pashov, Mr Nikolai. Inst. of Solid State Physics, Bulgarian Acad. Sci., Blvd Lenin 72, Sofia 1184, Bulgaria. (1929) Grad, physics (Sofia U., 1952). Sr. scient. (tel. 7341, ext. 641). *Crystal imperfections, electron microscopy, electron diffraction.*

Paunov, Dr Michael. Inst. of Physical Chemistry, Bulgarian Acad. Sci., Sofia 1040, Bulgaria. (1939) CSc, chemistry (Inst. of Physical Chemistry, 1969). Res. assoc. (tel. 7131, ext. 2586). *Crystal growth, heterogeneous nucleation, surfaces.*

Peneva, Dr Stefka. Faculty of Chemistry, Sofia U., Blvd Atnon Ivanov 1, Sofia 1126, Bulgaria. (1937) PhD, physics (U. of Delhi, India, 1970). Sr. scient. (tel. 62561, ext. 281). *Thin film growth, thin film perfection, thin film properties.*

Peshev, Prof Pavel. Inst. of General and Inorganic Chemistry, Bulgarian Acad. Sci., Sofia 1040, Bulgaria. (1933) DSc, chemistry (Inst. of General and Inorganic Chemistry, 1981). Res. prof. (tel. 7131, ext. 2573). *Inorganic synthesis, crystal growth.*

Petrov, Dr Kostadin. Inst. of General and Inorganic Chemistry, Bulgarian Acad. Sci., Sofia 1040, Bulgaria. (1940) CSc, chemistry (Inst. of General and Inorganic Chemistry, 1977). Sr. scient. (tel. 7131, ext. 2587). *X-ray crystallography.*

Petrov, Mr Ognyan. Inst. of Appl. Mineralogy, Bulgarian Acad. Sci., Moskowska 6, Sofia 1000, Bulgaria. (1952) Grad, geochemistry (Sofia U., 1980). Mineralogist. (tel. 872450). *Mineral structures.*

Petrov, Dr Srebri. Chair of Mineralogy and Crystallography, Sofia U., Blvd Russki 15, Sofia 1000, Bulgaria. (1942) CSc, mineralogy (Sofia U., 1984). Res. assoc. (tel. 8581, ext. 408). *X-ray analysis, minerals.*

Petrov, Dr Vasko. Inst. of General and Inorganic Chemistry, Bulgarian Acad. Sci., Sofia 1040, Bulgaria. (1941) CSc, techn. sciences (Moscow Inst. of Energetics, USSR, 1975). Res. assoc. (tel. 7131, ext. 2798). *Crystal growth.*

Philipov, Mr Alexander. Chair of Mineralogy and Crystallography, Sofia U., Blvd Russki 15, Sofia 1000, Bulgaria. (1944) Grad, mineralogy (Sofia U., 1972). Asst. (tel. 8581, ext. 256). *Crystal growth, minerals.*

Platikanova, Dr Vesselina. Central Lab. of Photographic Processes, Bulgarian Acad. Sci., Sofia 1040, Bulgaria. (1939) CSc, chemistry (Inst. of Physical Chemistry, 1971). Sr. scient. (tel. 723713). *Crystal physics, crystal growth.*

Popov, Dr Alexander. Central Lab. for Electrochemical Power Sources, Bulgarian Acad. Sci., Sofia 1040, Bulgaria. (1942) CSc, chemistry (Inst. of Physical Chemistry, 1980). Res. assoc. (tel. 7131, ext. 2758). *Electrocrystallization.*

Poulieff, Mr Christo. Central Lab. for Electrochemical Power Sources, Bulgarian Acad. Sci., Sofia 1040, Bulgaria. (1936) Grad, geochemistry (Sofia U., 1958).

Res. assoc. (tel. 7131, ext. 2707). *X-ray powder diffraction, infrared spectra, inorganic materials.*

Rachev, Mr Peter. Central Inst. for Computing Technics, Acad G. Bonchev, Bl. 10, Sofia 1040, Bulgaria. (1957) Grad, physics (Sofia U., 1979). Res. assoc. (tel. 7131, ext. 3991). *Electron microscopy, electron microanalysis.*

Rainov, Dr Nikola. Branch of Lab. Investigations, Geological Survey, G. A. Nassar 16, Sofia 1113, Bulgaria. (1939) CSc, mineralogy (Sofia U., 1975). Head of lab. (tel. 723651, ext. 032). *Feldspars, x-ray powder diffraction, infrared spectra, minerals, instrumentation.*

Rashkova, Dr Diana. Geological Inst., Bulgarian Acad. Sci., Sofia 1113, Bulgaria. (1952) CSc, mineralogy (Geological Inst., 1982). Res. assoc. (tel. 7131, ext. 2236). *Morphology, minerals.*

Russev, Dr Krassimir. Faculty of Chemistry, Sofia U., Blvd Anton Ivanov 1, Sofia 1126, Bulgaria. (1946) CSc, chemistry (Inst. of Physical Chemistry, 1975). Sr. scient. (tel. 62561, ext. 337). *Solid state reactions.*

Simov, Mr Stefan. Inst. of Solid State Physics, Bulgarian Acad, Sci., Blvd Lenin 72, Sofia 1184, Bulgaria. (1934) Grad, physics (Sofia U., 1960). Res. assoc. (tel. 7341, ext. 340). *Crystal growth, whiskers, thin films, electron microscopy.*

Staikov, Dr Georgy. Central Lab. for Electrochemical Power Sources, Bulgarian Acad. Sci., Sofia 1040, Bulgaria. (1943) CSc, chemistry (Inst. of Physical Chemistry, 1981). Res. assoc. (tel, 7131, ext. 2773). *Electrocrystallization.*

Stefanov, Mr Dechko Dimitrov. Geological Inst., Bulgarian Acad. Sci., Sofia 1113, Bulgaria. (1931) Grad, physics (Sofia U., 1954). Sr. scient. (tel. 7131, ext. 2207). *Clay minerals.*

Stefanov, Dr Stefan Rashkov. Inst. of Physical Chemistry, Bulgarian Acad. Sci., Sofia 1040, Bulgaria. (1935) DSc, chemistry (Inst. of Physical Chemistry, 1984). Sr. scient. (tel. 7131, ext. 3537). *Thin metal films, corrosion.*

Stoinova, Dr Margarita. Chair of Mineralogy and Petrography, Higher Inst. of Mining and Geology, Sofia 1156, Bulgaria. (1930) CSc, mineralogy (Sofia U., 1966). Docent. (tel. 62581, ext. 384). *Mineral morphology.*

Stoyanov, Dr Stoyan. Inst. of Physical Chemistry, Bulgarian Acad Sci., Sofia 1040, Bulgaria. (1941) CSc, physics (Inst. of Physical Chemistry, 1977). Sr. scient. (tel. 7131, ext. 2557). *Crystal growth.*

Stoychev, Mr Nikola. Inst. of Metal Sci. and Techn., Bulgarian Acad. Sci., Tchapaev 53, Sofia 1040, Bulgaria. (1939) CSc, physics (Inst. of Metallophysics, Kiev, USSR, 1977). Sr. scient. (tel. 71421, ext. 337). *Crystallization, metals.*

Tchehlarova, Mrs Irina. Branch of Lab. Investigations, Geological Survey, G. A. Nassar 16, Sofia 1113, Bulgaria. (1938) Grad, geochemistry (Sofia U., 1961). Geochemist. (tel. 723651, ext. 033) *X-ray powder diffraction, infrared spectra.*

Tchuneva, Mrs Vassilka. Branch of Lab. Investigations, Geological Survey, G. A. Nasser 16, Sofia 1113, Bulgaria. (1939) Grad, geochemistry (Sofia U., 1962). Geochemist. (tel. 723651, ext. 033). *X-ray powder methods.*

Tomov, Dr Ivan. Inst. of Physical Chemistry, Bulgarian Acad. Sci., Sofia 1040, Bulgaria. (1939) CSc, chemistry (Inst. of Physical Chemistry, 1981). Res. assoc. (tel. 7131, ext. 2533). *Textures, electrolytic layers.*

Topalova-Kalitzova, Mrs Maria. Inst. of Solid State Physics, Bulgarian Acad, Sci., Blvd Lenin 72, Sofia 1184, Bulgaria. (1940) Grad, physics (Sofia U., 1963). Res. assoc. (tel. 7341, ext. 340). *Electron microscopy, implantation, lattice defects.*

Toshev, Dr Alexander. Inst. of General and Inorganic Chemistry, Bulgarian Acad. Sci., Sofia 1040 Bulgaria. (1942). CSc, chemistry (Inst. of General and Inorganic Chemistry, 1978). Res. assoc. (tel. 7131, ext. 2563). *Crystal growth.*

Tsolovski, Dr Ilcho. Inst. of General and Inorganic Chemistry, Bulgarian Acad. Sci., Sofia 1040, Bulgaria. (1937) CSc, physics (Inst. of General and Inorganic Chemistry, 1976). Sr. scient. (tel. 7131, ext 2587). *X-ray crystallography.*

Vassilev, Dr Ivan. Inst. of Solid State Physics, Bulgarian Acad. Sci., Blvd Lenin 72, Sofia 1184, Bulgaria. (1932) CSc, physics (Inst. of Solid State Physics, 1976). Sr. scient. (tel. 7341, ext. 265). *X-ray diffraction, X-ray topography, lattice defects.*

Vesselinov, Mr Iliya. Inst. of Appl. Mineralogy, Bulgarian Acad. Sci., Moskowska 6, Sofia 1000, Bulgaria. (1942) Grad, geochemistry (Sofia U., 1967). Res. assoc. (tel. 885115, ext. 671). *Mineral synthesis, crystal growth, morphology.*

Vitanov, Dr Todor. Central Lab. for Electrochemical Power Sources, Bulgarian Acad. Sci., Sofia 1040, Bulgaria. (1926) CSc, chemistry (Inst. of Physical Chemistry, 1973). Sr. scient. (tel. 7131, ext. 2746). *Electrocrystallization, adsorption.*

Yaneva, Dr Svetlana. Inst. of Metal Sci. and Techn., Bulgarian Acad. Sci., Tchapaev 53, Sofia 1574, Bulgaria. (1942) CSc, chemistry (Sofia U., 1970). Res. assoc. (tel. 71421, ext. 337). *Eutectic crystallization.*

Zidarova, Mrs Bogdana. Inst. of Appl. Mineralogy, Bulgarian Acad. Sci., Moskovska 6, Sofia 1000, Bulgaria. (1943) Grad, geochemistry (Sofia U., 1971). Geochemist. (tel. 885115, ext. 671). *Synthesis, morphology, minerals.*

Zotov, Mr Nikolay, Geological Inst., Bulgarian Acad. Sci., Sofia 1113, Bulgaria. (1957) Grad, physics (Sofia U., 1982). Res. assoc. (tel. 7131, ext. 2232). *X-ray diffraction methods, computing, instrumentation.*

BURMA

Sub-Editor: **A.P. In**

Notes

1. International telephone country code - 95.

Htoon, Prof. Sein. Physics Dept., Moulmein U., University P.O., Moulmein, Burma. (1941) FInstP(DSc), physics (Inst. of Physics, UK, 1982). Prof. and Head. (tel. 032-22156). *Mathematical physics, crystallography.*

In, Mr Aung Paik. Physics Dept., Bassein Degree College, College P.O., Bassein, Burma. (1939) MSc, X-ray crystallography (Rangoon,1975). Lecturer. (tel. 042-21135, ext. Physics). *Theoretical physics, crystallography.*

Kyaw, Dr Htin. Physics Dept., Rangoon U., Rangoon, Burma. (1939) PhD, solid state electronics (U. of Salford, UK, 1971). Lecturer. *Field ion microscopy, crystallography, materials science.*

Mya Mya, Dr Khin. Physics Dept., Inst. of Medicine (I), Rangoon, Burma. (1948) PhD, X-ray crystallography (U. of York, UK, 1976). Asst. lecturer. *Direct methods, X-ray crystallography, Molecular biology (precursory), Astronomy.*

Tun, Mr Saw. Dept. of Analysis, Central Research Organisation, Kabaaye-Pagoda-Kanbe Roads, Rangoon, Yankin P.O., Burma. (1930) BSc, chemistry (U. of Rangoon, 1953). Sr. sci. (tel. 50544, ext. 24). *X-ray spectroscopy, powder diffraction, diffraction methods, analytical chemistry.*

Yin, Prof. Soe. Physics Dept., Bassein Degree College, College PO, Bassein, Burma. (1941) Dottore, theoretical solid state physics (Trieste U., Italy, 1980). Prof. and Head.(tel. 042-21135, ext. Physics). *Solid state theory, X-ray line broadening.*

CANADA

Sub-Editors: **M.E. Pippy and C.P. Huber**

Notes

1. International telephone country code - 001.

2. Abbreviations used for the Canadian provinces are as follows: Alta., Alberta; B.C., British Columbia; Man., Manitoba; N.B., New Brunswick; Nfld., Newfoundland; N.S., Nova Scotia; Ont., Ontario; P.E.I., Prince Edward Island; Qué., Québec; Sask., Saskatchewan.

Ahmed, Dr Farid Ramadan. Div. of Biological Sci., M-54, Nat. Res. Council of Canada, Ottawa, Ont. K1A 0R6, Canada. (1924) PhD, physical chemistry (U. Leeds, UK, 1953). Retired. (tel. 613 + 990-0858, fax 613 + 952-0583, email mep at nrcvm01). *Protein structures, area detectors, computer graphics.*

Bagchi, Prof. Subodh Nath. 5550 Bellerive, Brossard, Qué. J4Z 3C8, Canada. (1915) DSc, chemistry (U. Calcutta, India, 1946). Retired. *Amorphous structures, small angle scattering.*

Baranyi, Dr Anthony David. Glass & Ceramics Div., Ont. Res. Foundation, Sheridan Park, Mississauga, Ont. L5K 1B3, Canada. (1951) PhD, inorganic chemistry (McGill U., 1976). Assoc. res. sci. (tel. 416 + 822-4111, ext. 323). *Glass, glass - ceramics, ceramics, nuclear waste immobilization, solar thermal energy conversion, toxic materials release.*

Barrington-Leigh Dr John. Office of Dean of Medicine, Biomed. Technol. Liaison Offices, U. of Alberta, 2J2.25 WMC, Edmonton, Alta. T6G 2R7, Canada. (1943) PhD, natural science, physics (Cambridge U., UK, 1970). Hon. assoc. prof.; Adj. prof. (pharmacy) (tel. 403 + 432-6621, fax 403 + 432-7303). *Synchrotron X-ray Optics, Biological structures, fibre diffraction, energy transduction, structure prediction, immune regulation, cancer diagnostics.*

Bartlett, Dr Michael William. GBG Information Systems, IBM Canada Ltd., 101 Valleybrook Drive, Don Mills, Ont. M3B 3H1, Canada. (1943) PhD, X-ray crystallography (U. Waterloo, 1970). Programmer analyst. (tel. 416 + 443-3100).

Barton, Prof. Richard J. Dept. of Physics, U. of Regina, Regina, Sask. S4S 0A2, Canada. (1928) PhD, chemistry (Iowa State U., USA, 1956). Assoc. prof. (tel. 306 + 584-4653). *Coordination compounds, small molecules.*

Bassignana, Dr Isabella C. Advanced Technology Lab., Bell-Northern Research, Box 3511, Stn. C, Ottawa, Ont. K1Y 4H7, Canada. (1955) PhD, chemistry (U. California, Los Angeles, USA, 1982). Staff scientist. (tel. 613 + 763-3550, telex 0533175, fax 613 + 726-2626). *Double crystal X-ray diffraction, opto-electronic materials, topography, powder diffraction.*

Batchelor, Dr Raymond John. Chemistry Dept., Simon Fraser U., Burnaby, B.C. V5A 1S6, Canada. (1952) PhD (McMaster U., 1983). Res. Fellow. (tel. 604 + 291-4878, email userrayb at sfu.bitnet). *Structural inorganic chemistry.*

Bayliss, Prof. Peter. Dept. of Geology and Geophysics, U. of Calgary, Calgary, Alta. T2N 1N4, Canada. (1936) PhD, mineralogy (U. New South Wales, Australia, 1967). Prof. (tel. 403 + 284-5026, telex 03821545, fax 403 + 282-7298). *Mineralogy.*

Beauchamp, Prof. Dr André. Dépt. de Chimie, U. de Montréal, C.P. 6128, Succ. A, Montréal, Qué. H3C 3J7, Canada. (1940) PhD, chemistry (U. de Montréal, 1967). Prof. (tel. 514 + 343-7604, email 377 at umtlvr). *Structure, bio-inorganic interesting compounds, metal-metal bonding.*

Belanger-Gariepy, Mme Francine. Dépt. de Chimie, U. de Montréal, C.P. 6128, Succ. A, Montréal, Qué. H3C 3J7, Canada. (1955) MSc, chemistry (U. de Montréal, 1981). Res. asst. (tel. 514 + 343-7538). *Syntheses, characterization, crystal structure determination.*

Bensimon, Dr Corinne. Div. of Chemistry, Nat. Res. Council of Canada, Ottawa, Ont. K1A 0R6, Canada. (1962) PhD, inorganic chemistry (U. de Montréal, 1990). Res. assoc. (tel. 613 + 993-2527, email corinne at nrchem). *Cristallographie, inorganic chemistry.*

Besso, Ms Karyne. CANMET, Dept. of Energy, Mines and Resources, 555 Booth St., Ottawa, Ont. K1A 0G1, Canada. (1957) BSc, geology (Concordia U., 1977). Mineralogist, Process Mineralogy Section. (tel. 613 + 995-4752, fax 613 + 996-9673). *Powder X-ray diffraction, quantitative powder diffraction, automated powder diffractometry.*

Bird, Prof. Peter Hans. Dept. of Chemistry, Concordia U., 1455 de Maisonneuve Blvd. W., Blvd. W., Montréal, Qué. H3G 1M8, Canada. (1942) PhD, inorganic chemistry (U. Sheffield, UK, 1966). Prof. (tel. 514 + 848-3335). *Organometallic complexes of transition metals, small molecules.*

Birnbaum, Dr George I. Div. of Biological Sci., Nat. Res. Council of Canada, Ottawa, Ont. K1A 0R6, Canada. (1931) PhD, chemistry (Columbia U., USA, 1961). Sr. res. officer. (tel. 613 + 990-3245, telex 0533145, fax 613 + 952-0583, email gib at nrcvm01). *Proteins and other biological structures.*

Bluhm, Dr Terry Lee. Xerox Res. Centre of Canada, 2660 Speakman Dr., Mississauga, Ont. L5K 2L1, Canada. (1947) PhD, polymer chemistry (State U. of New York, USA, 1976). Scient. staff member. (tel. 416 + 823-7091, email bluhm.xrcc-ns at xerox.com). *Polymer crystallography, powder diffraction, small-angle X-ray scattering.*

Boorman, Dr Philip Michael. Chemistry Dept., U. of Calgary, Calgary, Alta. T2N 1N4, Canada. (1939) PhD, inorganic chemistry (U. Nottingham, UK, 1964). Prof. (tel. 403 + 220-5347, telex 03821545, fax 403 + 282-7298). *Tungsten complexes with sulfur ligands; coordinated sulfur-donor ligands, structure and reactions.*

Booth, Dr Andrew Donald. Autonetics Res. Assoc. Inc., Box 518, Sooke, B.C. V0S 1N0, Canada. (1918) DSc, crystallography (U. London, UK, 1951). Chairman of the board. (tel. 604 + 642-5352). *Computer applications to crystallography, numerical analysis, sound propagation in the ocean.*

Bottomley, Prof. Frank. Dept. of Chemistry, U. of New Brunswick, P.O. Box 4400, Fredericton, N.B. E3B 5A3, Canada. (1941) PhD, inorganic chemistry (U.

Toronto, 1968). Prof. (tel. 506 + 453-4774). *Inorganic chemistry, structural chemistry, catalysis.*

Brandon, Dr James Kenneth. Physics Dept., U. of Waterloo, Waterloo, Ont. N2L 3G1, Canada. (1940) PhD, physics (McMaster U., 1967). Assoc. prof. (tel. 519 + 885-1211, ext. 3494). *Inorganic structures, metal and alloy structures.*

Brayer, Prof. Gary David. Dept. of Biochemistry, U. of British Columbia, 2146 Health Sciences Mall, Vancouver, B.C. V6T 1W5, Canada. (1953) PhD, biochemistry (U. Alberta, 1979). Assoc. prof. (tel. 604 + 228-5216, fax 604 + 228-5227, email usergbbc at ubcmtsg.bitnet). *Macromolecular crystallography, enzyme structure and function, protein- nucleic acid interactions, complexation phenomena, electron transfer mechanisms.*

Brierley, Mr Cameron. Dept. of Chemistry, U. of Regina, Regina, Sask. S4S 0A2, Canada. (1953) BSc, chemistry (U. Saskatchewan, 1975). Grad. student. (tel. 306 + 584-4653). *Organic structures.*

Brisse, Dr François. Dépt. de Chimie, U. de Montréal, C.P. 6128, Succ. A, Montréal, Qué. H3C 3J7, Canada. (1935) PhD, chemistry (Dalhousie U., 1967). Prof. (tel. 514 + 343-7604). *Crystal structure determination, model molecules related to polymers, polymer structures, polyesters, polysaccharides, biologically interesting compounds, alkaloids, antibiotics, steroids, electron microscopy, electron diffraction.*

Brisson, Dr Josée. Dépt. de Chimie, U. de Montréal, C.P. 6128, Succ. A, Montréal, Qué. H3C 3J7, Canada. (1960) PhD, chimie (U. de Montréal, 1987). (tel. 514 + 343-7538). *Polyamide, polysaccharide, macromolecules, polymers, electron microscopy.*

Britten, Dr James Francis. Chemistry Dept., McGill U., 801 Sherbrooke St. W., Montréal, Qué. H3A 2K6, Canada. (1955) PhD, chemistry (McMaster U., 1984). Manager, McGill Chem. X-ray facility. (tel. 514 + 398-6728, telex 05268510, fax 514 + 398-3797, email cy81 at musica.mcgill.ca). *Inorganic and bioinorganic chemistry, solid state nmr, computer graphics.*

Brown, Dr Ian David. Inst. for Materials Res., McMaster U., Hamilton, Ont. L8S 4M1, Canada. (1932) PhD, crystallography (U. London, UK, 1959). Prof. of physics. (tel. 416 + 525-9140, ext. 4710, telex 0618347, fax 416 + 528-5030, email 1002332 at mcmaster). *Chemical bonding, inorganic solids, crystallographic data retrieval.*

Bushnell, Prof. Gordon William. Dept. of Chemistry, U. of Victoria, Victoria, B.C. V8W 2Y2, Canada. (1936) PhD, inorganic chemistry (U. West Indies, West Indies, 1966). Prof. (tel. 604 + 721-7163, email bushnell at uvvm). *Structural inorganic chemistry, single crystal X-ray diffraction, coordination compounds, proteins and nucleic acids.*

Buyers, Dr William James Leslie, FRSC. Neutron & Solid State Physics Div., Atomic Energy of Canada Ltd., Chalk River, Ont. K0J 1J0, Canada. (1937) PhD, physics (U. Aberdeen, UK, 1963). Manager, neutron & solid state physics. (tel. 613 + 584-3311, ext. 3974). *Neutron scattering, structural phase transitions, phonons, spin waves, disordered materials, liquids, crystalline electric field effects; defects.*

Camerman, Dr Norman. Dept. of Biochemistry, U. of Toronto, Toronto, Ont. M5S 1A8, Canada. (1939) PhD, chemistry (U. British Columbia, 1964). Prof. (tel. 416 + 978-7027). *Biological structures, biological structure - activity relation, molecular design.*

Cameron, Prof. Theodore Stanley. Dept. of Chemistry, Dalhousie U., Halifax, N.S. B3H 4J3, Canada. (1942) DPhil, chemistry (Oxford U., UK, 1969). Prof. (tel. 902 + 424-3305, telex 01921863, fax 902 + 424-8797, email cameron at ac.dal.ac). *Hydrogen bonding, molecular packing, graphical investigation, inorganic and organometallic structures, sulphur and phosphorus structures, residual electron density.*

Campbell, Dr Robert Laurence. Div. of Biological Sci., M-54, Nat. Res. Council of Canada, Ottawa, Ont. K1A 0R6, Canada. (1960) PhD, biological chemistry (Massachusetts Inst. Techn., USA, 1988). Res. assoc. (tel. 613 + 990-0889, email campbell at nrcvm01.nrc.ca). *Protein structure and function, enzyme mechanisms, molecular graphics, protein crystallography.*

Carty, Prof. Arthur John. Guelph-Waterloo Centre for Graduate Work in Chemistry, Waterloo Campus, U. of Waterloo, Waterloo, Ont. N2L 3G1, Canada. (1940) PhD, inorganic chemistry (U. Nottingham, UK, 1965). Prof. and Dir. (tel. 519 + 885-3544, ext. 3296). *Organometallic chemistry, phosphine and acetylene complexes, phosphinoacetylenes, mercury, cadmium, environmental pollution, structure determination by physical methods.*

Černý, Dr Petr. Dept. of Geological Sci., U. of Manitoba, Winnipeg, Man. R3T 2N2, Canada. (1934) PhD, mineralogy (Acad. of Sci. of CSSR, Czechoslovakia, 1966). Prof. (tel. 204 + 474-8765). *Morphology, structure, crystal chemistry, oxide minerals, silicate minerals.*

Chao, Dr George Y. Geology Dept., Carleton U., Ottawa, Ont. K1S 5B6, Canada. (1930) PhD, mineralogy & crystallography (U. Chicago, USA, 1958). Prof. of geology. (tel. 613 + 231-3885). *Structure, minerals.*

Charland, Dr Jean-Pierre. CANMET, BCC Bldg 3, Energy Res. Labs., 555 Booth St., Ottawa, Ont. K1A 0G1, Canada. (1957) PhD, chemistry (U. de Montréal, 1984). Res. sci. (tel. 613 + 995-5751). *Chemistry, inorganics, biomolecules, bioinorganic chemistry, surface science.*

Cheng, Dr Pei-Tak. Mount Sinai Hospital, Dept. of Labs, 600 University Ave., Toronto, Ont. M5G 1X5, Canada. (1940) PhD, crystallography (U. Toronto, 1972). Staff sci. (tel. 416 + 586-4468). *Physical chemistry, ultrastructure, crystalline deposit diseases, tissue crystallography.*

Chieh, Prof. Chung (Peter). Dept. of Chemistry, U. of Waterloo, Waterloo, Ont. N2L 3G1, Canada. (1939) PhD, chemistry (U. British Columbia, 1969). Prof. (tel. 519 + 888-4633, email chieh at watdcs). *Coordination, sulfur-containing ligands, theory of crystal structures.*

Clark, Dr Malcolm John Roy. Waste Management Branch, Ministry of Environment, 810 Blanchard St., Victoria, B.C. V8V 1X5, Canada. (1944) PhD, chemistry (U. New Brunswick, 1971). Environmental chemist. (tel. 604 + 387-9947). *Environmental chemistry, metals toxicity, computing.*

Cobbledick, Dr Roger Ernest. Bureau of Radiation & Medical Devices, Environmental Health Centre, Health & Welfare Canada, Ottawa, Ont. K1A 0L2, Canada. (1943) PhD, X-ray crystallography (U. Lancaster, UK, 1969). Clinical chemist. (tel. 613 + 954-0366). *Organic and organometallic structures, computing, clinical chemistry.*

Codding, Dr Penelope Wixson. Dept. of Chemistry, U. of Calgary, Calgary, Alta. T2N 1N4, Canada. (1946) PhD, chemistry (Michigan State U., USA, 1971). Assoc. prof. (tel. 403 + 220-7549, telex 03821545, fax 403 + 289-9488, email pcodding at uncamult.bitnet). *Structure - activity relationships, drugs, neurotoxins, peptides; DNA and protein crystallography.*

Cowie, Prof. Martin. Dept. of Chemistry, U. of Alberta, Edmonton, Alta. T6G 2G2, Canada. (1947) PhD, X-ray crystallography (U. Alberta, 1974). Prof. (tel. 403 + 432-5581, telex 0372979, fax 403 + 432-1664). *Small molecule activation and catalysis, transition metal complexes, metal-metal cooperativity, binuclear complexes, polynuclear complexes.*

Curzon, Prof. Albert Edward. Physics Dept., Simon Fraser U., Burnaby, B.C. V5A 1S6, Canada. (1934) PhD, physics (Imperial C., London, UK, 1959). Prof. (tel. 604 + 291-4181). *Electron microscopy, electron diffraction, thin film crystal structures.*

Cygler, Dr Mirek. Biotechnology Res. Inst., Nat. Res. Council of Canada, 6100 Royalmount Ave., Montréal, Qué. H4P 2R2, Canada. (1947) PhD, crystallography (U. Łódź, Poland, 1976). Assoc. res. officer. (tel. 514 + 496-6321, fax 514 + 496-6232, email cygler at nrcvm01). *Protein structure and function, protein-nucleic acid interactions, molecular replacement method.*

Deguire, Mrs Suzanne. Dépt. de Chimie, U. de Montréal, C.P. 6128, Succ. A, Montréal, Qué. H3C 3J7, Canada. (1962) MSc, chimie (U. de Montréal, 1985). Étudiante au doctorat. (tel. 514 + 343-7538). *Polyesters, polymers, microscopie electronique.*

Delbaere, Dr Louis Theophil Joseph. Dept. of Biochemistry, U. of Saskatchewan, Saskatoon, Sask. S7N 0W0, Canada. (1943) PhD, chemistry (U. Manitoba, 1970). Prof. (tel. 306 + 966-4373 or 966-4366, telex 0742659, fax 306 + 343-5914, email delbaere at sask). *Crystal structures, proteins, biologically important molecules; molecular recognition; structure-function relationship.*

Dichmann, Dr Klaus. Chemistry Dept., Vanier C., 821 Ste-Croix Blvd, Saint-Laurent, Qué. H4L 3X9, Canada. (1942) PhD, X-ray crystallography (U. Toronto, 1972). Prof. (tel. 514 + 488-2341, ext. 4052). *Crystal structure analysis, structural inorganic chemistry.*

Dion, Mrs Chantal. Dépt. de Chimie, U. de Montréal, C.P. 6128, Succ. A, Montréal, Qué. H3C 3J7, Canada. (1956) MSc, chemistry (U. du Québec à Montréal, 1983). Grad. student. (tel. 514 + 343-7538). *Platinum compounds, antitumoral activity.*

Dolling, Dr Gerald. Physics & Health Sci. Div., Atomic Energy of Canada Ltd., Chalk River Nuclear Labs, Chalk River, Ont. K0J 1J0, Canada. (1935) PhD, physics (Cambridge U., UK, 1961). Asst. to vice- president, physics. (tel. 613 + 584-3311, ext. 4011, telex 05334555, fax 613 + 589-2039, email 01735 at aeclcr.bitnet). *Neutron and X-ray diffraction; excitations in condensed matter, phonons, magnons; neutron inelastic scattering techniques.*

Donnay, Prof. Joseph Désiré Hubert. 516 Iberville, Mont St Hilaire, Qué. J3H 2V7, Canada. (1902) PhD, geology (Stanford U., USA, 1929). Emeritus, The Johns Hopkins U. (tel. 514 + 464-5652). *Crystal morphology, morphology-structure relationships, symmetry, crystalline aggregates.*

Drake, Prof. John E. Chemistry Dept., U. of Windsor, Windsor, Ont. N9B 3P4, Canada. (1936) DSc, inorganic chemistry (U. Southampton, UK, 1978). Prof. & dept. head. (tel. 519 + 253-4232, ext. 3521). *Spectroscopy, structure, organometallics and coordination compounds.*

Edwards, Dr William Donald. Communications Res. Centre, Dept. of Communications, P.O. Box 11490, Station H, Ottawa, Ont. K2H 8S2, Canada. (1926) PhD, physics (U. Birmingham, UK, 1949). Group leader. (tel. 613 + 596-9486). *Electronic crystalline materials, silicon, gallium arsenate, preparation and processing, structural properties, defects, X-ray topography, crystallography.*

Einstein, Prof. Frederick William Boldt. Dept. of Chemistry, Simon Fraser U., Burnaby, B.C. V5A 1S6, Canada. (1940) PhD, inorganic structural chemistry (U. Canterbury, New Zealand, 1965). Prof. (tel. 604 + 291-3594, email userfwbe at sfu.bitnet). *Inorganic structural chemistry, crystallographic computing, low temperature crystallography, molecular modelling, protein structure.*

Ercit, Dr Timothy Scott. Mineral Sci. Div., Nat. Museum of Natural Sciences, Ottawa, Ont. K1P 6P4, Canada. (1957) PhD, mineralogy (U. Manitoba, 1986). Res. sci. (tel. 613 + 952-3516). *Pegmatite mineralogy (oxides & phosphates), inorganic structures, paragenetic mineralogy.*

Falk, Dr Michael. Atlantic Res. Lab., Nat. Res. Council of Canada, 1411 Oxford St., Halifax, N.S. B3H 3Z1, Canada. (1931) DSc, physical chemistry (Laval U., 1958). Sr. res. officer. (tel. 902 + 426-8265, ext. 120). *Infrared spectra of crystals, crystalline hydrate structures.*

Fawcett, Dr John Keith. North Island C., 1480 Elm St., Campbell River, B.C. V9W 3A6, Canada. (1940) PhD, chemistry (U. British Columbia, 1965). Instructor. (tel. 604 + 286-8917, fax 604 + 286-8900). *Molecular structure, biological and organic compounds.*

Ferguson, Dr George. Chemistry Dept., U. of Guelph, Guelph, Ont. N1G 2W1, Canada. (1936) DSc, crystal structural analysis (U. Glasgow, UK, 1969). Prof. of chemistry. (tel. 519 + 824-4120, ext. 3548 and 3800, email chmferg at uoguelph). *Inorganic and organic structures.*

Ferguson, Prof. Robert Bury. Dept. of Geological Sci., U. of Manitoba, Winnipeg, Man. R3T 2N2, Canada. (1920) PhD, mineralogy (U. Toronto, 1948). Prof. Emeritus, mineralogy. (tel. 204 + 474-9786). *Crystallography, crystal structures, crystal chemistry, minerals, rock-forming silicates.*

Fleet, Dr Michael Edward. Dept. of Geology, U. of Western Ontario, London, Ont. N6A 5B7, Canada. (1938) PhD, geology (U. Manchester, UK, 1963). Prof. (tel. 519 + 661-3184). *Inorganic structures, structure related properties, crystal chemistry.*

Fortier, Dr Suzanne. Dept. of Chemistry, Queen's U., Kingston, Ont. K7L 3N6, Canada. (1949) PhD, crystallography (McGill U., 1976). Assoc. prof. (tel. 613 + 545-2654, fax 613 + 545-6300, email fortiers at qucdn). *Direct methods theory and applications.*

Franklin, Dr Kenneth James. Fuels and Materials Div., AECL, Chalk River, Ont. K0J 1J0, Canada. (1952) PhD, chemistry (McMaster U., 1983). Asst. res. officer. (tel. 613 + 584-3311, ext. 2560). *Structural inorganic chemistry, transition metal complexes, lanthanides, actinides.*

Franklin, Prof. Ursula Martius. Dept. of Metallurgy and Materials Sci., U. of Toronto, Toronto, Ont. M5S 1A4, Canada. (1921) PhD, applied physics (Techn. U., Berlin, BRD, 1948). Prof. (tel. 416 + 978-3012, telex 06-218915). *Characterisation, ancient materials, X-ray diffraction techniques, ternary alloy structures.*

Gabe, Dr Eric James. Div. of Chemistry, Nat. Res. Council of Canada, Ottawa, Ont. K1A 0R6, Canada. (1933) PhD, crystallography (U. Wales, UK, 1960). Sr. res. officer. (tel. 613 + 993-2527, email gabe at nrchem). *Automation, small computers, structures.*

Gait, Dr Robert Irwin. Dept. of Mineralogy & Geology, Royal Ontario Museum, 100 Queen's Park, Toronto, Ont. M5S 2C6, Canada. (1938) PhD, mineralogy (U. Manitoba, 1967). Curator of mineralogy. (tel. 416 + 586-5818, fax 416 + 586-5863). *General mineralogy; morphological crystallography; X-ray crystallography; descriptive mineralogy.*

Gaunt, Dr Paul. Dept. of Physics, U. of Manitoba, Winnipeg, Man. R3T 2N2, Canada. (1932) DPhil, metal physics (Oxford U., UK, 1958). Prof. (tel. 204 + 474-9589). *Lattice distortion, order-disorder phenomena, metals and alloys, magnetic properties related to structure, electron microscopy, electron diffraction.*

Gibbons, Dr Cyril Stephen. Dept. of Materials Chemistry, Ontario Res. Foundation, Sheridan park, Mississauga, Ont. L5K 1B3, Canada. (1945) PhD, structure analysis (U. British Columbia, 1971). Sr. res. sci. (tel. 416 + 822-4111, ext. 244).

Gopal, Dr Ramanathan. Industrial Batteries Div., Gould Manufacturing Canada Ltd., 275 Lewis St., Fort Erie, Ont. L2A 2R3, Canada. (1944) PhD, crystallography & physical chemistry (McMaster U., 1972). Sen. res. sci. (tel. 416 + 871-5600). *Inorganic molecules, small organic molecules, mineral crystallography; electrochemistry, electrocatalysts.*

Graham, Dr Albert Ronald. 515 St. George St. E., Fergus, Ont. N1M 1L1, Canada. (1917) PhD, mineralogy (U. Toronto, 1950). Retired. (tel. 519 + 843-1150). *Economic mineralogy.*

Grattan-Bellew, Dr Patrick Edward. Inst. for Res. in Construction, Nat. Res. Council of Canada, Ottawa, Ont. K1A 0R6, Canada. (1934) PhD, experimental mineralogy (Cambridge U., UK, 1969). Res. officer. (tel. 613 + 993-0096). *Cement chemistry, mineralogy in general.*

Grice, Dr Joel Denison. Mineral Sci. Div., Nat. Museum of Natural Sci., 1926 Merivale Road, Ottawa, Ont. K1P 6P4, Canada. (1946) PhD, mineralogy (U. Manitoba, 1973). Curator of minerals. (tel. 613 + 952-3513). *Crystal systematics, growth of minerals.*

Groat, Mr Lee Andrew. Dept. of Geological Sci., U. of Manitoba, Winnipeg, Man. R3T 2N2, Canada. (1959) BSc, geology (Queen's U., 1982). Grad. student. (tel. 204 + 474-8395). *Mineral structures, optical crystallography, electron microprobe analysis.*

Grundy, Dr Harry Douglas. Dept. of Geology, McMaster U., Hamilton, Ont. L8S 4M1, Canada. (1941) PhD, mineralogy (U. Manchester, UK, 1966). Prof. (tel. 416 + 525-9140, ext. 4516). *Silicate minerals, characterization, inorganic solids.*

Guay, Dr France. 3645 Collège, St. Hubert, Qué. J3Y 5R9, Canada. (1957) PhD, bioinorganic chemistry (U. de Montréal, 1984). (tel. 514 + 656-9584). *Inorganic chemistry, bioinorganic chemistry, heavy metals complexes.*

Hanson, Dr Alfred Wallace. Atlantic Res. Lab., Nat. Res. Council of Canada, 1411 Oxford St., Halifax, N.S. B3H 3Z1, Canada. (1923) PhD, crystallography (U. Manchester, UK, 1953). Retired. *X-ray crystallography; scanning electron microscopy; microprobe analysis.*

Hassan, Prof. Ishmael. Earth & Planetary Sciences, U. of Toronto, Erindale Campus, Mississauga, Ont. L5L 1C6, Canada. (1951) PhD, geochemistry (McMaster U., 1982). Asst. prof. (tel. 416 + 828-5419). *Geologically important materials, structure, chemistry.*

Hawthorne, Dr Frank Christopher. Dept. of Earth Sci., U. of Manitoba, Winnipeg, Man. R3T 2N2, Canada. (1946) PhD, geology (McMaster U., 1973). Assoc. prof. (tel. 204 + 474-9833, fax 204 + 269-6629, email hawthor at uofmcc). *Topological aspects of structure, minerals, graph theory, combinatorial theory, Rietveld method, spectroscopic methods.*

Hempel, Dr Andrew. Dept. of Biochemistry, U. of Toronto, University Ave., Toronto, Ont. M5S 1A8, Canada. (1944) PhD, organic chemistry (Techn. U. of Gdansk, Poland, 1975). Res. assoc. (tel. 416 + 978-3726, email andre at utoronto). *Biological structures, structure-activity relationships.*

Heyding, Dr Robert Donald. Dept. of Chemistry, Queen's U., Kingston, Ont. K7L 3N6, Canada. (1925) PhD, physical chemistry (McGill U., 1951). Prof. (tel. 613 + 545-2607). *Powder X-ray diffraction, phase transitions, inorganic structures.*

Huber, Dr Carol P. Div. of Biological Sci., M-54, Nat. Res. Council of Canada, Ottawa, Ont. K1A 0R6, Canada. (1937) DPhil, chemical crystallography (Oxford U., UK, 1963). Sr. res. officer. (tel. 613 + 990-0856, fax 613 + 952-0583, email cph at nrcvm01) *Protein crystallography, biologically interesting molecules.*

Hutcheon, Dr Wendy Lou (Brooks). Dept. of Biochemistry, U. of British Columbia, 2146 Health Sciences Mall, Vancouver, B.C. V6T 1W5, Canada. (1946) PhD, crystallography (U. Alberta, 1971). Res. assoc. (tel. 604 + 228-4509). *Protein crystallography, enzyme structure and function, protein-nucleic acid interactions.*

Hynes, Dr Rosemary Catherine. Div. of Chemistry, Nat. Res. Council of Canada, Ottawa, Ont. K1A 0R6, Canada. (1963) PhD, chemistry (U. Western Ontario, 1989). Res. assoc. (tel. 613 + 993-2527, email rosi at nrchem). *Inorganic chemistry, structures of small molecules.*

James, Prof. Michael N.G. Biochemistry Dept., U. of Alberta, Edmonton, Alta. T6G 2H7, Canada. (1940) DPhil, chemical crystallography (Oxford U., UK, 1966). Prof. (tel. 403 + 432-4550). *Protein structure, enzyme function, biological molecules, structure and function.*

Johnston, Mr Victor James. Chemistry Dept., Simon Fraser U., Burnaby, B.C. V5A 1S6, Canada. (1961) BSc, chemistry (Central Washington U., USA, 1985). Grad. student. (tel. 604 + 291-4878). *Organometallic chemistry, synthesis and structure.*

Jones, Dr Stephen John. Inst. for Marine Dynamics, Nat. Res. Council of Canada, Box 12093, Stn. A, St. Johns, Nfld. A1B 3T5, Canada. (1942) PhD, physics (U. Birmingham, 1967). Sr. res. officer. (tel. 709 + 772-5403, telex 0164784, fax 709 + 772-2462). *Ice physical properties, sea ice properties.*

Kayden, Dr Catherine Sheila. Div. of Biological Sci., M-54, Nat. Res. Council of Canada, Ottawa, Ont. K1A 0R6, Canada. (1960) PhD, chemistry (U. Victoria, 1989). Res. assoc. (tel. 613 + 990-0889, fax 613 + 952-0583, email kayden at nrcvm01). *Protein crystallography.*

Kerr, Dr Kathleen Ann. Dept. of Chemistry & Physics, U. of Calgary, Calgary, Alta. T2N 1N4, Canada. (1941) PhD, chemistry (U. Glasgow, UK, 1966). Assoc. prof. (tel. 403 + 284-5395). *Biologically active compounds, diffraction physics.*

King, Prof. Hubert Wylam. Dept. of Engineering Physics, Technical U. of Nova Scotia, P.O. Box 1000, Halifax, N.S. B3J 2X4, Canada. (1930) PhD, physical metallurgy (U. Birmingham, UK, 1956). Prof. and Dept. Head (tel. 902 + 429-8300, ext. 2205, telex (TUNS)019-21566, fax 902 + 422-7907). *High and low temperature phase transformations, ceramic oxides, thermal expansion, electrical and magnetic properties.*

Knop, Prof. Osvald. Dept. of Chemistry, Dalhousie U., Halifax, N.S. B3H 4J3, Canada. (1922) DSc, physical chemistry (Laval U., 1957). Harry Shirreff Prof. of Chemical Research. (tel. 902 + 424-3317). *Structural inorganic chemistry.*

Kocman, Dr Vladimir. DOMTAR Inc. Res. Centre, Box 300, Senneville, Qué. H9X 3L7, Canada. (1937) Doctorate, inorganic chemistry (U. J.E. Purkyne Brno, Czechoslovakia, 1968). Res. sci. (tel. 514 + 457-6810, ext. 284; telex 055-60625, fax 514 + 457-4527). *X-ray powder diffraction, X-ray fluorescence, differential thermal analysis, thermogravimetry, instrumental methods, analytical chemistry.*

Kodama, Dr Hideomi. Agriculture Canada, Land Resource Res. Centre, CEF, Carling Ave., Ottawa, Ont. K1A 0C6, Canada. (1931) PhD, mineralogy (Tokyo U. of Education, Japan, 1961). Sr. res. sci. (tel. 613 + 995-5011, ext. 7888; telex 053-3283, fax 613 + 995-6833). *Clay minerals, layer silicates, structural disorders, mineral component characterization.*

Le Page, Dr Yvon. Div. of Chemistry, Nat. Res. Council of Canada, Ottawa, Ont. K1A 0R6, Canada. (1943) PhD, physics engineering (U. Montréal, 1974). Sr. res. officer. (tel. 613 + 993-2527, fax 613 + 952-1275, email yvon at nrchem). *Accuracy in intensity measurements; symmetry; low temperatures; programming, small computers; crystal structures.*

Lea, Prof. Sydney George. Chemical and Metallurgical Dept., Ryerson Polytechnical Inst., 50 Gould St., Toronto, Ont. M5B 1E8, Canada. (1933) PhD, crystallography (U. London, UK, 1968). Prof. (tel. 416 + 595-5067). *Spectroscopy, inorganic structures, crystal physics.*

Lee, Mrs Florence Lan Fun. Div. of Chemistry, Nat. Res. Council of Canada, Ottawa, Ont. K1A 0R6, Canada. (1938) MEd, education (Ottawa U., 1966). Res. officer. (tel. 613 + 993-2527). *Structures, programming.*

Lee, Dr Xavier. Div. of Biological Sci., M-54, Nat. Res. Council of Canada, Ottawa, Ont. K1A 0R6, Canada. (1937) PhD, molecular biology (U. Grenoble, France, 1981). Res. assoc. (tel. 613 + 990-7387, email xlee at nrcvm01). *Tertiary structures of proteins, structure-function relation of biomacromolecules.*

Linden, Dr Anthony. Dept. of Chemistry, Dalhousie U., Halifax, N.S. B3H 4J3, Canada. (1957) PhD, inorganic chemistry (U. Melbourne, Australia, 1986). Postdoctoral fellow. (tel. 902 + 424-3318, email LINDEN at AC.DAL.CA). *Hydrogen bonding, residual electron density, main group elements, crystal packing, conformational analysis, molecular mechanics.*

Lock, Prof. Colin James Lyne. Labs for Inorganic Medicine, McMaster U., ABB-426, Hamilton, Ont. L8S 4M1, Canada. (1933) DSc, bioinorganic chemistry/crystallography (London U., UK, 1986). Prof. of chemistry and pathology. (tel. 416 + 525-9140, ext. 4760, telex 061-8347, fax 416 + 528-5030, email 1002525 at MCMASTER). *X-ray structural studies, metal complexes used in medicine.*

Loeb, Dr Stephen Joseph. Dept. of Chemistry, U. of Winnipeg, Winnipeg, Man. R3B 2E9, Canada. (1954) PhD, inorganic chemistry (U. Western Ontario, 1982). Assoc. prof. (tel. 204 + 786-9731, email UOWSJL at UOFMCC). *Inorganic, organometallic chemistry; catalysis.*

Lough, Dr Alan John. Dept. of Chem. & Biochemistry, U. of Guelph, Guelph, Ont. N1G 2W1, Canada. (1960) PhD, crystal structural analysis (Napier Polytechnic, Edinburgh, UK, 1988). Postdoc. res. fellow. (tel. 519 + 824-4120, ext. 3548, email chmajl at uoguelph). *Inorganic and organic structures.*

Luo, Mr Yao Guang. Chemistry Dept., U. of Regina, Regina, Sask. S4S 0A2, Canada. (1956) BSc, chemistry (Xiamen U., People's Rep. of China, 1982). Grad. student. (tel. 306 + 585-4653, email YLUO at UREGINA1). *X-ray crystallography, structure and function.*

Ma, Dr Lilian Yan Yan. 82 Haxby Private, Ottawa, Ont. K1T 3C2, Canada. (1946) PhD, chemistry (Simon Fraser U., 1971). (tel. 613 + 521-3442). *X-ray crystallography, biological activity - structural basis, molecular design.*

Mandarino, Dr Joseph Anthony. Dept. of Mineralogy, Royal Ontario Museum, 100 Queen's Park, Toronto, Ont. M5S 2C6, Canada. (1929) PhD, mineralogy (U. of Michigan, USA, 1958). Curator of mineralogy. (tel. 416 + 586-5819, fax 416 + 586-5863). *Descriptive mineralogy, X-ray diffraction, crystallography, crystal optics.*

Marchessault, Dr Robert Henry. Dept. of Chemistry, McGill U., 3420 University St., Montreal, Qué. H3A 2A7, Canada. (1928) PhD, physical chemistry (McGill U., 1954). NSERC/Xerox Prof. (tel. 514 + 398-6276, telex 05268510, fax 514 + 398-7249). *Polymer crystal structure, polymer morphology.*

Martin, Miss Lillian Ruth. Dept. of Chemistry, Simon Fraser U., Burnaby, B.C. V5A 1S6, Canada. (1947) BSc, physical and inorganic chemistry (Simon Fraser U.). Grad. student. (tel. 604 + 291-4408). *Organometallic, coordination and metal-cluster complexes.*

Matthews, Prof. Frederick White. Dalhousie U. (Retired, but teaching), 1168 Studley Ave., Halifax, N.S. B3H 3R7, Canada. (1915) PhD, physical chemistry (McGill U., 1941). (tel. 902 + 423-2155). *Powder data, data bases, information systems.*

Maxwell, Prof. George. Mathematics Dept., U. of British Columbia, #121 - 1984 Mathematics Rd., U. Campus, Vancouver, B.C. V6T 1Y4, Canada. (1946) PhD, mathematics (Queen's U., 1970). Assoc. prof. (tel. 604 + 228-6402). *Group theory; geometry; mathematical crystallography.*

Médicis, de, Dr Rinaldo M. Faculté de Médecine, U. de Sherbrooke, Chemin de Stooke, Sherbrooke, Qué. J1H 5M4, Canada. (1934) PhD, crystal chemistry (U. de Louvain, Belgium, 1967). Res. asst. (tel. 819 + 565-2144). *Crystal chemistry, X-ray and electron diffraction.*

Michel, Prof. André Gustave. Structural Chemistry Lab., U. of Sherbrooke, 2500 University Blvd., Sherbrooke, Qué. J1K 2R1, Canada. (1944) PhD, physical chemistry (Namur, Belgium, 1976). Prof. (tel. 819 + 565-3668). *X-ray crystal structure analysis, molecular modelling by computer.*

Middlemiss, Dr Nora E. Dept. of Glass & Ceramics, Ontario Res. Foundation, Sheridan Park, Mississauga, Ont. L5K 1B3, Canada. (1947) PhD, physical chemistry (McMaster U., 1978). Assoc. res. sci. (tel. 416 + 822-4111, ext. 247) *Ionic conductors, transition metal oxides, ceramics.*

Mihichuk, Prof. Lynn Michael. Chemistry Dept., U. of Regina, Regina, Sask. S4S 0A2, Canada. (1947) PhD, inorganic chemistry (U. British Columbia, 1975). Assoc. prof. (tel. 306 + 585-4793, email XRAY06 at UREGINAV). *Bioinorganic and organometallic structural chemistry.*

Mitchell, Dr Crighton Maurice. 8 Kingsford Cr., Kanata, Ont. K2K 1T3, Canada. (1917) PhD, physics (U. Toronto, 1952). Retired. (tel. 613 + 592-1186). *Deformation, residual stress, preferred orientation, crystal physics.*

Mitchell, Dr Keith A. R. Dept. of Chemistry, U. of British Columbia, 2036 Main Mall, Vancouver, B.C. V6T 1Y6, Canada. (1938) PhD, chemistry (U. London, UK, 1963). Prof. (tel. 604 + 228-5831). *Surface structure, LEED crystallography.*

Montgomery, Prof. (Em.) Henry. 502 - 2910 Cook St., Victoria, B.C. V8T 3S7, Canada. (1916) PhD, inorganic chemistry (U. Washington, USA, 1961). Retired. (tel. 604 + 383-8423). *X-ray structures, inorganics, metal organics.*

Muir, Dr Alastair Kerr. Biochemistry Dept., Faculty of Medicine, U. of Alberta, Edmonton, Alta. T6G 2H7, Canada. (1957) PhD, physical chemistry (U. Calgary, 1985). Postdoctoral Fellow. (tel. 403 + 432-4575, email userakmu at ualtamts). *DNA-binding proteins, immunoglobin structure, molecular interactions and motion, molecular dynamics, area detectors.*

Nachman, Dr Joseph. Biotech. Res. Inst., Nat. Res. Council of Canada, 6100 Royalmount Ave., Montréal, Qué. H4P 2R2, Canada. (1953) PhD, chemistry (U. Cambridge, England, 1986) Res. Assoc. (tel. 514 + 496-2619, fax 514 + 496-6232, email nachman at bri.nrc.ca; nachman at nrcvm01) *Macromolecular structures.*

Natarajan, Dr Mahadevan. 12 Court St., Antigonish, N.S. B2G 1Z6, Canada. (1940) PhD, phase transformations in solids (Indian Inst. of Techn., Kanpur, India, 1970). (tel. 902 + 863-6117). *Phase changes, electrical phenomena, IR, Raman, reflectance spectra, lattice dynamics.*

Nuffield, Prof. Edward Wilfrid. 1835 Morton Ave. Apt. 1603, Vancouver, B.C. V6G 1V3, Canada. (1914) PhD, mineralogy (U. of Toronto, 1944). Prof. Emeritus. (tel. 604 + 688-6062). *Crystal chemistry, ore minerals, X-ray diffraction methods.*

Olivier, M. Marc-J. Dépt. de Chimie, U. de Montréal, C.P. 6128, Succ. A, Montréal, Qué. H3C 3J7, Canada. (1951) MSc, chimie (U. de Montréal, 1980). Agent de recherches. (tel. 514 + 343-6747). *Heavy metal complexes with bases of DNA.*

Owen, Mr Charles Gordon. Technology Transfer Office, Dalhousie U., 6093 South St., Halifax, N.S. B3H 1T2, Canada. (1957) MSc, inorganic chemistry (Dalhousie U., 1982). Manager. (tel. 902 + 424-1648, telex 019-21863, fax 902 + 424-2319). *Technology transfer (medical, dental, physical sciences).*

Pandya, Mr Naresh. Chemistry Dept., Brock U., St. Catharines, Ont. L2S 3A1, Canada. (1960) BSc(Hons), chemical physics (U. Sussex, 1982). Grad. asst. (tel. 416 + 688-5550, ext.3406). *Structure correlation, modelling reaction transition states, crystallography.*

Payne, Dr Nicholas Charles. Chemistry Dept., U. of Western Ontario, London, Ont. N6A 5B7, Canada. (1942) PhD, inorganic chemistry (U. Sheffield, UK, 1967). Prof. (tel. 519 + 661-3793, fax 519 + 661-3022, email NOGGIN at UWOVAX.UWO.CA). *Preparation, single crystal X-ray, transition metal complexes, catalytically important complexes, asymmetric synthesis, absolute configuration, Bijvoet absorption edge technique.*

Pazdernik, Prof. LeRoy Joseph. Dépt. de Chimie-Biologie, U. du Québec à Trois-Rivières, C.P. 500, Trois-Rivières, Qué. G9A 5H7, Canada. (1942) PhD, inorganic chemistry (U. Iowa, USA, 1970). Prof. (tel. 819 + 376-5052). *Trace metal analysis, metal complexes, metals in environmental systems, metal coordination compounds.*

Perrault, Dr Guy. École Polytechnique, C.P. 6079, Montréal, Qué. H3C 3A7, Canada. (1927) PhD, mineralogy (U. Toronto, 1955). Prof. *Crystal structure, minerals, silicates, niobium minerals.*

Peterson, Dr Ronald Charles. Dept. of Geology, Queen's U., Kingston, Ont. K7L 3N6, Canada. (1953) PhD, geology (Virginia Polytechnic Inst., USA, 1981). Asst. prof. (tel. 613 + 545-2597, ext. 6180, email peterson at qucdn). *Silicate mineralogy, crystal chemistry.*

Ploc, Dr Robert Allen. Chemistry and Materials Div., Materials Science Branch, Atomic Energy of Canada Ltd., Chalk River Nuclear Labs., Stn. 82, Chalk River, Ont. K0J 1J0, Canada. (1939) PhD, metallurgy (Cambridge U., UK, 1965). Res. assoc. (tel. 613 + 584-4378, ext. 785). *Electron diffraction (computer analysis); oxide growth on Zr and Zr-based alloys; electron microscopy (transmission and scanning).*

Post, Dr Michael Leonard. Div. of Chemistry, Nat. Res. Council of Canada, Ottawa, Ont. K1A 0R6, Canada. (1945) PhD, chemistry (U. Surrey, UK, 1971). Sr. res. officer. (tel. 613 + 993-2101, fax 613 + 952-1275). *Metal alloy hydrides, ceramics, calorimetry of solid and gas reactions.*

Poulin, Mrs Suzie. Engineering Physics Dept., École Polytechnique, C.P. 6079, Succ. A, Montréal, Qué. H3C 3A7, Canada. (1954) MSc, chemistry (U. de Québec à Montréal, 1978). Res. assoc. (tel. 514 + 340-4308, telex 05-24146, fax 514 + 340-4440). *Surface science, nuclear physics, amorphous silicon devices.*

Powell, Dr Brian Mathieson. Neutron & Solid State Physics Div., Atomic Energy of Canada Ltd., Chalk River, Ont. K0J 1J0, Canada. (1938) PhD, physics (U. London, U.K, 1964). Sr. res. officer. (tel. 613 + 584-3311, ext. 3994, telex 053-34555, fax 613 + 589-2039, email 02124 at AECLCR). *Neutron diffraction, profile analysis, inelastic neutron scattering.*

Prasad, Dr Lata. Biochemistry Dept., U. of Saskatchewan, Saskatoon, Sask. S7N 0W0, Canada. (1945) PhD, crystallography (Flinders U. of S. Australia, 1972). Professional res. assoc. (tel. 306 + 477-1493, email prasad at sask). *Protein structures.*

Pringle, Mr Gordon James. Dept. of Energy, Mines and Resources, Geological Survey of Canada, 601 Booth St., Ottawa, Ont. K1A 0E8, Canada. (1944) MSc, mineralogy (U. New Brunswick, 1972). Electron microprobe analyst. (tel. 613 + 994-5023). *Electron microprobe analysis, feldspar optics, feldspar relations in basaltic rocks, powder diffraction techniques.*

Quail, Dr J. Wilson. Dept. of Chemistry, U. of Saskatchewan, Saskatoon, Sask. S7N 0W0, Canada. (1936) PhD, chemistry (McMaster U., 1963). Prof. (tel. 306 + 966-4663, fax 306 + 373-6088, email QUAIL at SASK). *Structure-function relationships of drugs and of proteins.*

Ramos, Mr Agustin Federico. Chemistry Dept., Simon Fraser U., Burnaby, B.C. V5A 1S6, Canada. (1960) MSc, chemistry (Simon Fraser U., 1987). Grad. student. (tel. 604 + 291-4878, email userramo at sfu.bitnet). *Theoretical inorganic chemistry, conformational analysis.*

Ranger, Dr Georges Joseph. R & D Div., Albright & Wilson Americas, 2 Gibbs Rd., Islington, Ont. M9B 1R1, Canada. (1955) PhD, inorganic chemistry (Wayne State U., USA, 1984). Res. chemist. (tel. 416 + 239-7111, ext. 301). *Inorganic and analytical chemistry, crystal structure determinations, heavy metal complexes.*

Raudsepp, Dr Mati. Dept. of Geological Sciences, U. of Manitoba, Winnipeg, Man. R3T 2N2, Canada. (1947) PhD, mineralogy (U. Manitoba, 1984). Res. assoc. (tel. 204 + 474-9833, email RAUDSEP at UOFMCC). *Experimental mineralogy, amphibole crystal chemistry, Rietveld structure analysis, infrared spectroscopy.*

Read, Dr Randy John. Medical Microbiology & Infectious Diseases, U. of Alberta, Edmonton, Alta. T6G 2H7, Canada. (1957) PhD, biochemistry (U. Alberta, 1986). Asst. prof. (tel. 403 + 492-4305, fax 403 + 492-7521, email USERRNDY at UALTAMTS). *Rational drug design, structure factor probabilities.*

Restivo, Dr Roderic John. Science Dept., Heritage College, C.P. 1757, Hull, Qué. J8X 3Y8, Canada. (1943) PhD, chemistry (U. Waterloo, 1969). Prof. (tel. 819 + 778-2270, ext. 243, fax 819 + 778-7364). *Bio-inorganic structural chemistry, organometallic chemistry.*

Rettig, Dr Steven John. U. of British Columbia, Chemistry, 2036 Main Mall, U. Campus, Vancouver, B.C. V6T 1Y6, Canada. (1948) PhD, chemistry (U. British Columbia, 1974). Res. assoc. (tel. 604 + 228-4865). *X-ray crystallography, organic and organometallic compounds; organoboron chemistry.*

Richard, Prof. Joseph Albert Pierre. Dépt. de Physique, U. du Québec à Montréal, C.P. 8888, Succ. A, Montréal, Qué. H3C 3P8, Canada. (1942) PhD, physics engineering (U. de Montréal, 1971). Prof. (tel. 514 + 282-7837). *Crystal structure, organometallics.*

Richardson, Dr Mary Frances. Chemistry Dept., Brock U., St. Catharines, Ont. L2S 3A1, Canada. (1941) PhD, inorganic chemistry (U. Kentucky, USA, 1967). Prof. (tel. 416 + 688-5550, ext. 3400, email crrichardson at brock.cdn). *X-ray structure determination, structure correlation, polymorphic transitions, solid state NMR.*

Roberts, Mr Andrew Clifford. Dept. of Energy, Mines and Resources, Geological Survey of Canada, 762 - 601 Booth St., Ottawa, Ont. K1A 0E8, Canada. (1950) MSc, mineralogy (Queen's U., 1976). X-ray diffraction mineralogist. (tel. 613 + 992-2802). *Single crystal studies, powder diffraction techniques.*

Robertson, Prof. Beverly Ellis. Dept. of Physics, U. of Regina, Regina, Sask. S4S 0A2, Canada. (1939) PhD, physics (McMaster U., 1967). Prof. (tel. 306 + 585-4264, fax 306 + 586-9862, email brob at uregina1). *Small molecule structures, statistics, charge densities.*

Rochon, Prof. Fernande D. Dépt. de Chimie, U. du Québec à Montréal, C.P. 8888, Succ. A, Montréal, Qué. H3C 3P8, Canada. PhD, inorganic chemistry (U. de Montréal, 1971). Prof. (tel. 514 + 282-4896 or 282-4119). *Inorganic chemistry, platinum complexes.*

Rose, Dr David Richard. Div. of Biological Sci., Nat. Res. Council of Canada, Ottawa, Ont. K1A 0R6, Canada. (1955) DPhil, molecular biophysics (Oxford U., UK, 1981). Assoc. res. officer. (tel. 613 + 990-0857, fax 613 + 952-0583, email rose at nrcbsp.nrc.ca). *Protein crystallography, antibodies, Fab fragments, molecular immunology.*

Santarsiero, Dr. Bernard I. Dominic M. Biochemistry, U. of Alberta, 425 Medical sciences Building, Edmonton, Alberta T6G 2H7 Canada. (1952) PhD physical chemistry (U. Washington,USA,1980) Res. Assoc. (tel. 403+492-2422). *Single crystal X-ray diffraction, instrumentation, macromolecular crystallography, variable temperature structure deformation.*

Sawyer, Dr Jeffrey Frederick. Lash Miller Chemical Labs, U. of Toronto, 80 St. George St., Toronto, Ont. M5S 1A1, Canada. (1954) PhD, chemistry (U. Warwick, UK, 1977). X-ray systems mgr. (tel. 416 + 978-6275). *Inorganic and organometallic structures, secondary bonding.*

Schrag, Dr Joseph D. Biotechnology Res. Inst., Nat. Res. Council of Canada, 6100 Royalmount Ave., Montréal, Qué. H4P 2R2, Canada. (1955) PhD, physiology (U. Illinois, USA, 1984). Res. assoc. (tel. 514 + 496-2557, fax 514 + 496-6232, email schrag at nrcbri). *Protein structure and function.*

Scott, Dr James Douglas. Kidd Creek Div., Falconbridge Ltd., Box 2002, Timmins, Ont. P4N 7K1, Canada. (1942) PhD, mineralogy (crystallography) (Queen's U., 1970). Sr. staff scient. (tel. 705 + 235-7573, telex 067-81648, fax 705 + 235-7580). *Sulfide structures, process mineralogy, uranium minerals, atomic substitution in minerals.*

Secco, Prof. Anthony Silvio. Dept. of Chemistry, U. of Manitoba, Winnipeg, Man. R3T 2N2, Canada. (1956) PhD, chemistry (U. British Columbia, 1982). Assoc. prof. (tel. 204 + 474-8379, telex 07-587721, fax 204 + 269-6629, email secco at uofmcc). *Biochemical crystallography, oligonucleotides, organic and inorganic structures.*

Sielecki, Dr Anita R. Biochemistry Dept., U. of Alberta, Edmonton, Alta. T6G 2H7, Canada. (1940) PhD, hydrodynamics (Hebrew U., Israel, 1969). Res. assoc. (tel. 403 + 432-2422, fax 403 + 432-7219, email usersiel at ualtamts.bitnet). *Protein crystallography.*

Simard, Mr Michel. Dépt. de Chimie, U. de Montréal, C.P. 6128, Succ. A, Montréal, Qué. H3C 3J7, Canada. (1958) MSc, chemistry (U. de Montréal, 1981). Grad. student. (tel. 514 + 343-7538). *Inorganic chemistry.*

Skowron, Mrs Anecita. Inst. for Materials Res., McMaster U., Hamilton, Ont. L8S 4M1, Canada. (1955) MSc, physics (U. Łódź, Poland, 1979). (tel. 416 + 525-9140, ext. 7092, telex 0618347, fax 416 + 528-5030). *X-ray crystallography, high resolution electron microscopy, sulphides.*

Smith, Prof. Vedene H., Jr. Chemistry Dept., Queen's U., Kingston, Ont. K7L 3N6, Canada. (1935) Fil. dr, quantum chemistry (U. Uppsala, Sweden, 1967). Prof. and dept. head. (tel. 613 + 545-2624, email smithvh at qucdn). *Electron density distributions (charge, spin and momentum).*

Stanley, Dr Eric. Physics Dept., U. of New Brunswick, P.O. Box 5050, Saint John, N.B. E2L 4L5, Canada. (1924) DSc, physics (U. Saskatchewan, 1972). Prof. (tel. 506 + 657-7310, ext. 242). *Direct methods, refinement, electron density distributions.*

Stephan, Dr D. W. Chemistry Dept., U. of Windsor, Windsor, Ont. N9B 3P4, Canada. (1953) PhD,(U. Western Ontario, 1980). (tel. 519 + 253-4232). *Organometallic and bioinorganic chemistry.*

Sundararajan, Dr Pudupadi R. Xerox Research Centre of Canada, 2660 Speakman Dr., Mississauga, Ont. L5K 2L1, Canada. (1943) DSc, physics (Madras U., India, 1982). Principal scient. (tel. 416 + 823-7091, ext. 219, email sundar.xrcc-ns at xerox.com). *Polymer crystallography, polymer morphology, polymer conformations in solid state and solution.*

Sunder, Dr Sham. Res. Chemistry Branch, AECL, Whiteshell Nuclear Res. Est., Pinawa, Man. R0E 1L0, Canada. (1942) PhD, chemistry (U. Alberta, 1972). Res. officer. (tel. 204 + 753-2311, ext. 2749, telex 07-57553, fax 204 + 753-2455). *Surface chemistry, vibrational spectroscopy, ESCA, XPS, nuclear fuel, X-ray diffraction, phase transitions, solid-liquid interface.*

Sutherland, Dr John Knox. Dept. of Minerals and Materials, Res. and Productivity Council, Box 6000, Fredericton, N.B. E3B 5H1, Canada. (1941) PhD, petrography (U. Manchester, UK, 1965). Mineralogist, chief of analytical services. (tel. 506 + 455-8994, ext. 247). *Mineralogy, petrology, metallurgy, mineral synthesis, analysis methods, rocks, ores, minerals.*

Sygusch, Prof. Jurgen. U. de Sherbrooke, Dept. of Biochemistry, Centre Hospitalier Universitaire, Sherbrooke, Qué. J1H 5N4, Canada. (1945) PhD, chemistry (U. Montréal, 1975). Prof. (tel. 819 + 564-5283, email s042 at udesvm). *Protein structure-function, protein crystallography, protein engineering.*

Szymański, Dr Jan Tomasz. CANMET, Dept. of Energy, Mines and Resources, 555 Booth St., Ottawa, Ont. K1A 0G1, Canada. (1938) PhD, inorganic chemistry (King's C., London U., UK, 1963) Res. sci., mineralogy section. (tel. 613 + 995-4077, fax 613 + 996-9673). *Mineral structures, sulphides, inorganic complexes.*

Taylor, Dr Peter. Res. Chemistry Branch, AECL, Whiteshell Nuclear Res. Est., Pinawa, Man. R0E 1L0, Canada. (1949) PhD, inorganic chemistry (U. Birmingham, UK, 1972). Res. chemist. (tel. 204 + 753-2311, ext. 3054). *Structure-solubility-stability relationships, inorganic oxide and salt systems.*

Theophanides, Prof. Theo. Dept. of Chemistry, U. of Montreal, Box 6128, Station A, Montreal, Qué. H3C 3J7, Canada. (1932) PhD, organometallic chemistry (U. Toronto, 1963). Assoc. prof. (tel. 514 + 343-6742). *Platinum coordination complexes, cancer chemotherapy synthesis, structure and bond properties, solubility-structure relationships.*

Traill, Dr Robert James. Dept. of Energy, Mines and Resources, Geological Survey of Canada, 601 Booth St., Ottawa, Ont. K1A 0E8, Canada. (1921) PhD, crystallography (Queen's U., 1956). Asst. dir., central lab. div. (tel. 613 + 994-5023). *Structures, minerals, X-ray diffraction, X-ray fluorescence analysis.*

Trotter, Prof. James. Dept. of Chemistry, U. of British Columbia, Vancouver, B.C. V6T 1Y6, Canada. (1933) DSc, chemistry (U. Glasgow, UK, 1963). Prof. (tel. 604 + 228-4527). *Organic and inorganic crystal structures.*

Tun, Dr Zin. Neutron and Solid State Physics, Chalk River Nuc. Lab., Chalk River, Ontario K0J 1J0, Canada. (1957) PhD,physics (McMaster U.,1985) Postdoctoral fellow. (tel. 613+584 3311, ext. 3995, telex 053 34555, fax 613+589 2039) *Disorder, high pressures, inorganic crystals, instrumentation, magnetism neutron diffraction, phase transitions, spin density, X-ray diffraction.*

Van Der Heijden, Dr Simon Petrus Nicolaas. Performance Engineering, SaskPower, 2025 Victoria Ave., Regina, Sask. S4P 0S1, Canada. (1943) PhD, chemistry-crystallography (U. Saskatchewan, 1974). Chief chemist. (tel. 306 + 566-3073, fax 306 + 566-2330). *Structural crystallography, crystal chemistry, coal gasification kinetics.*

Wang, Mr Hong. Chemistry Dept., U. of Regina, Regina, Sask. S4S 0A2, Canada. (1961) BSc, chemistry (Xiamen U., People's Rep. of China, 1981). Grad. student. (tel. 306 + 584-4902). *Inorganic structures, computing methods.*

Whitlow, Dr Simon Hugh. Environment Canada, 700 - 1200 W. 73 Ave., Vancouver, B.C. V6P 6H9, Canada. (1943) PhD, chemistry (U. British Columbia, 1969). Head, Data and Instrumentation. (tel. 604 + 666-7196). *Inorganic crystal structures, computer programming, data management.*

Wicks, Dr Frederick John. Dept. of Mineralogy, Royal Ontario Museum, 100 Queen's Park, Toronto, Ont. M5S 2C6, Canada. (1937) D. Phil., mineralogy (Oxford U., UK, 1969). Curator. (tel. 416 + 586-5820, fax 416 + 586-5863). *Serpentine minerals, crystal structures, crystal chemistry, X-ray diffraction, thermal analysis, asbestos.*

Wolbaum, Mr Keith Jonathon. Dept. of Physics, U. of Regina, Regina, Sask. S4S 0A2, Canada. (1953) BSc, physics (U. Regina, 1977). Res. asst. (tel. 306 + 584-4653). *Instrumentation.*

Wood, Dr Gordon H. CISTI, Nat. Res. Council of Canada, Ottawa, Ont. K1A 0S2, Canada. (1940) PhD, physics (U. British Columbia, 1969). CAN/SND manager. (tel. 613 + 993-3294, telex 053-3115, fax 613 + 952-7158, email num001 at nrcvm01). *Crystallographic databases.*

Yan, Mr Xiaoqian. Chemistry Dept., Simon Fraser U., Burnaby, B.C. V5A 1S6, Canada. (1956) MSc, inorganic chemistry (Nanjing U., China, 1983). Grad.

student. (tel. 604 + 291-4406). *Inorganic structural chemistry, organometallic compounds.*

Yang, Prof. Daniel Shun-Chung. Biochemistry, McMaster Universty, 1200 Main St. West, Hamilton, Ontario L8N3Z5, Canada. (1953) PhD, crystallography (U. Pittsburgh,1983) Ass. Prof. (tel. 416 + 5259140, ext. 2455, fax 416+5210048, email Danyang at Mcmaster.bitnet) *Protein structure and function, fourier methods in crystallography, ICE binding proteins.*

CHILE

Sub-Editor: O. Wittke

Notes

1. International telephone country code - 56.

2. The approximate equivalent of the Chilean degrees *Profesor de Matemáticas y Física* (Prof. de Mat. y Fís.), *Profesor de Física* (Prof. de Fís.), *Profesor de Química* (Prof. de Quím.), *Profesor de Biología y Química* (Prof. de Biol. y Quím.), *Licenciado en Física* (Lic. en Fís.), *Licenciado en Química* (Lic. en Quím.) and *Licenciado en Ciencias Naturales* (Lic. en C. Nat.) is between BSc and MSc in mathematics, physics or chemistry.

Aguilar, Mrs Adela. Lab. de Petrografía y Mineralogía, Langerfeldt y Aguilar Ltda. Casilla 4003, Santiago, Chile. (1934). Head. (tel. 2256986). *Minerals, microscopy.*

Almendras, Mrs Eliana. Dept. de Minas, Facultad de Ciencias Físicas y Matemáticas, U. de Chile, Casilla 2777, Santiago, Chile. (1940) Ingeniero de Minas (U. de Chile, 1962). Res. (tel. 6982071, ext. 477). *Minerals, microscopy.*

Barbagelata, Mr Franco. Dept. de Geología y Mineralogía Aplicada Centro de Investigación Minera y Metalúrgica (CIMM), Casilla 170, Santiago, Chile. (1942) Prof. de Quím. (U. de Chile, 1967). Head. (tel. 2289544). *Microscopy, minerals.*

Barrios, Mr Nelson. Dept. de Análisis Químico, Centro de Investigación Minera y Metalúrgica (CIMM), Casilla 170, Santiago 10, Chile. (1935) MSc, chemistry (U. Nacional Autónoma de México, México, 1971). Head. (tel. 2289544). *X-ray diffraction, emission spectrography.*

Besoaín, Dr Eduardo. Lab. de Fisicoquímica y Mineralogía, Inst. de Investigaciones Agropecuarias, Casilla 439)3, Santiago, Chile. (1929) Dr. landw., soil mineralogy (U. Bonn, BRD, 1969). Head. (tel. 5586061, ext. 236). *Minerals from soil, X-ray diffraction, X-ray spectrography, infrared spectrophotometry.*

Boys, Dr Daphne. Dept. de Física, Facultad de Ciencias Físicas y Matemáticas, U. de Chile, Casilla 487)3, Santiago, Chile. (1941) PhD, X-ray diffraction (Wales U., UK, 1972). Res. scient. (tel. 6960148, email dboys at uchcecvm.bitnet). *Coordination compounds, liquid crystals.*

Cid, Dr Hilda. Dept de Fisiología, Facultad de Ciencias Biológicas yRecursos Naturales, U. de Conceptción, Casilla 2407, Concepción, Chile. (1933) PhD, crystallography (M.I.T., USA, 1964). Prof., res. scient. (tel. 234985). *Proteins, macromolecules, structure and biological function.*

Chornik, Dr Boris. Dept. de Física, Facultad de Ciencias Físicas y Matemáticas, U. de Chile, Casilla 487)3, Santiago, Chile. (1941) PhD., physics (U. California., USA, 1970). Res. scient. (tel.6960148, email bchornik at uchcecvm.bitnet). *Surface structure, magnetism, materials science.*

Costamagna, Dr Juan Alberto. Dept. de Química, Fac. de Ciencia, U. de Santiago de Chile, Ecuador 3467, Santiago, Chile. (1940) Dr, chemistry (U. de Buenos Aires, Argentina, 1970). Res. scient. (tel. 95591). *Coordination compounds.*

Donoso, Mr Eduardo. Microscopía Electrónica, Dept. de Ciencias de los Materiales, Facultad de Ciencias Físicas y Matemáticas, U. de Chile, Casilla 1420, Santiago, Chile. (1947) Metalurgista (U. de Chile, 1972). Res. (tel. 6982071, ext. 316). *Electron microscopy.*

Escobar, Dr Carmen. Alejandro del Río 21877, Santiago, Chile. (1931) D. 3e cycle, crystallography (U. de Bordeaux, France, 1968). Res. scient. (tel. 2238203). *Coodination compounds, organic compounds.*

Garín, Dr Jorge. Dept. de Metalurgia, Facultad de Ingeniería, U. de Santiago de Chile, Casilla 10233, Santiago, Chile. (1943) PhD, metallurgy and materials science (U. Pennsylvania, USA, 1972). Head. (tel. 761045, email jgarin at usachvm1.bitnet). *Metals, inorganic crystals.*

Garland, Mrs María Teresa. Dept. de Física, Fac. de Ciencias Físicas y Matemáticas, U. de Chile, Casilla 487)3, Santiago, Chile. (1944) Lic. en Quím. (U. de Chile, 1969) Dr.Sc.(U. de Rennes I, France, 1986). Res. scient. (tel. 6960148, email mtgarlan at uchcecvm.bitnet). *Coordination compounds, organic compounds.*

González, Mrs Irma. Dept. de Geología, Fac. de Ciencias Físicas y Matemáticas, U. de Chile, Casilla 13518, Correo 21, Santiago, Chile. (1935) Prof. de Biol. y Quím. (U. de Chile, 1963). Res. (tel. 6982071, ext. 114). *Minerals.*

González, Mr Yanko. Dept. de Análisis Químico, Centro de Investigación Mineray Metalúrgica (CIMM), Casilla 170, Correo 10, Santiago, Chile. (1954) Lic. en Quí. (U. de Chille, 1980). Res. (tel. 2289544 ext. 282). *X-ray diffraction, X-ray fluorescence.*

Greene, Mr Fernando. Dept. de Microscopía. y Mineralogí Aplicada Centro de Investigación Minera y Metalúrgica (CIMM), Casilla 170, Santiago 10, Chile. (1943) Prof. de Quím. (U. de Chile, 1968). Assoc. res. (tel. 2289544 ext. 325). *Microscopy, minerals.*

Henríquez, Mr Fernando. Dept. de Minas, Facultad de Ingeniería, U. de Santiago de Chile, Casilla 10233, Santiago, Chile. (1942) MSc, geology (McGill U., Canada, 1972). Res. (tel. 95869). *Minerals.*

Hervé, Dr Francisco. Dept. de Geología y Geofísica. Facultad de Ciencias Físicas y Matemáticas, U. de Chile, Casilla 13518, Correo 21, Santiago, Chile. (1942) DSc, metamorphic petrology (Hokkaido U., Japan, 1974). Res. scient. (tel. 6982071, ext. 114). *Minerals, petrology.*

Infante, Dr Carlos. Dept. de Física, Facultad de Ciencias, U. de Chile, Casilla 653, Santiago, Chile. (1944) DPhil, inorganic chemistry (Oxford U., UK, 1975). Res. scient. (tel. 2712865 ext. 212). *Materials science,neutron diffraction.*

Joseph, Dr Günter. Sección Metales, Dept. de Ciencias de los Materiales, Facultad de Ciencias Físicas y Matemáticas, U. de Chile, Casilla 1420, Santiago, Chile. (1929) Dr rer. nat., physical chemistry of metals (U. Stuttgart,BRD, 1957). Prof., head. (tel. 6982071, ext. 140). *Materials science.*

Kittl, Mr Pablo. Microscopía Electrónica, Dept. de Ciencias de los Materiales, Facultad de Ciencias Físicas y Matemáticas, U. de Chile, Casilla 1420, Santiago, Chile. (1934) Lic. en Fís. (U. de Cuyo, Argentina, 1963). Res. (tel. 6982071, ext. 130). *Materials science.*

Kremer, Mr Germán. Dept. de Física, Facultad de Ciencias, U. de Chile, Casilla 653, Santiago, Chile. (1941) Prof. de Fís. (U. de Chile, 1966). Res. scient. (tel. 2712865 ext. 212). *Metals, thin films.*

Llanos, Dr Jaime. Dept. de Química, Facultad de Ciencias, U. del Norte, Casilla 1280, Antofagasta, Chile. (1952) Dr. rer. nat. (U. Stuttgart, BRD, 1984). Res. scient. (tel. 241148). *Inorganic crystals.*

Manríquez, Dr Víctor. Dept. de Química, Facultad de Ciencias, U. de Chile,, Casilla 653, Santiago, Chile. (1953) Dr. rer. nat. (U. Stuttgart, BRD, 1983). Res. scient. (tel. 2713888). *Inorganic crystals.*

Moraga, Mr Luís. Dept. de Física, Facultad de Ciencias, U. de Chile, Casilla 653, Santiago, Chile. (1942) Prof. de Fís. (U. de Chile, 1966). Res. scient. (tel. 2712865 ext. 212). *Metals, electron diffraction.*

Mujica, Dr Carlos. Dept. de Química, Facultad de Ciencias, U. del Norte, Casilla 1280, Antofagasta, Chile. (1954) Dr. rer. nat. (U. Stuttgart, BRD, 1984). Res. scient. (tel. 241148). *Inorganic crystals.*

Ossio, Miss Myriam. Sección Metales, Dept. de Ciencias de los Materiales, Facultad de Ciencias Físicas y Matemáticas, U. de Chile, Casilla 1420, Santiago, Chile. (1942) Lic. en Quím. (U. de Chile, 1967). Res. chemist. (tel. 6982071, ext. 140). *Metals.*

Peña, Miss Luzmila. Dept. de Física, Facultad de Ciencias Físicas y Matemáticas, U. de Chile, Casilla 487)3, Santiago, Chile. (1947) Tecnólogo médico (U. de Chile, 1969). Res. assistant. (tel. 6960148). *Powder diffraction, minerals.*

Pérez, Mrs Carmen. Inst. Investigaciones Tecnológicas (INTEC), Casilla 667, Santiago, Chile. (1945) Ingeniero de Metalurgia Extractiva (U. Técnica del Estado, 1968). Res. (tel. 2289066). *Minerals, microscopy.*

Perret, Mr Ramón. Sección Metales, Dept. de Ciencias de los Materiales, Facultad de Ciencias Físicas y Matemáticas, U. de Chile, Casilla 1420, Santiago, Chile. (1932) Lic. en Quím. (U. Católica de Chile, 1957). Res. scient. (tel. 6982071, ext. 140). *Metals.*

Rivera, Prof. Carlos. Palqui 2921, Santiago, Chile. (1925) Prof. de Fís. (U. de Chile, 1954). Res. scient. (tel. 2254625). *Microscopy.*

Schlein, Dr Werner. Centro de Investigación Minera y Metalúrgica (CIMM), Casilla 170, Correo 10, Santiago, Chile. (1936) PhD, chemistry (Brown U., USA, 1971). Director Ejecutivo. (tel. 2289544 ext.212). *Powder diffraction.*

Silva, Dr Elisa. Dept. de Física, Facultad de Ciencias Físicas y Matemáticas, U. de Chile, Casilla 487)3, Santiago, Chile. (1927) D. 3e cycle, sciences (U. de Paris, France, 1963). Res. scient. (tel. 6982071). *Electron microscopy.*

Souza, Mr Carlos, Dept. de Análisis Químico, Centro de Investigación Mineray Metalúrgica (CIMM), Casilla 170, Santiago 10, Chile. (1954) Lic. en Quím. (U. de Chile, 1980). Res. (tel. 2289544 ext.264). *X-ray diffraction, X-ray fluorescence.*

Spodine, Dr Evgenia, Dept. de Química Inorgánica y Analítica, Facultad de Ciencias Químicas y Farmacéuticas, U. de Chile, Casilla 233, Santiago, Chile. (1942) Dr, chemistry.(U. de Chile, 1987) Prof. res. scient. (tel. 371948). *Coordination compounds.*

Suwalsky, Dr Mario. Dept. de Química, Fac. de Ciencias, U. de Concepción, Casilla 3-C, Concepción, Chile. (1936) PhD, crystallography (Weizmann Inst., Israel, 1969). Prof., res. scient. (tel. 24985, ext. 2171). *Polymers, structural biology.*

Varschavsky, Mr Ari. Sección Metales, Dept. de Ciencias de los Materiales, Facultad de Ciencias Físicas y Matemáticas, U. de Chile, Casilla 1420, Santiago,

Chile. (1940) Ingeniero Eléctrico (U. de Chile, 1965). Res. scient. (tel. 6982071, ext. 150). *Metals, materials science.*

Vera, Mr Rafael. Dept. de Física, Facultad de Ciencias, U. de Concepción, Casilla 3-C, Concepción, Chile. (1927) Ingeniero Químico (U. de Concepción, 1951). Prof. (tel. 24985, ext. 2171). *X-ray diffraction.*

Vogel, Mrs Sonia. Lab. de Rayos X, Servicio Nacional de Geología y Minería, Casilla 10465, Santiago, Chile. (1942) Geólogo (U. de Chile, 1974). Head (tel. 375050 ext. 226). *Powder diffractometry.*

Ward, Mr José. Lab. de Análisis Químico, Centro de Estudios, Medicion y Certificacion de Calidad. (CESMEC), Casilla 14036, Santiago, Chile. (1934) Prof. de Mat. y Fís. (U. de Chile, 1962). Head. (tel. 2746088). *X-ray diffraction, X-ray fluorescence.*

Wittke, Prof. Oscar. Dept. de Física, Facultad de Ciencias Físicas y Matemáticas, U. de Chile, Casilla 5487, Santiago, Chile. (1929) Prof. de Mat. y Fís. (U. de Chile, 1961). Prof., head, X-ray diffraction lab. (tel. 6960148, email owittke at uchcecvm.bitnet). *X-ray diffraction.*

Zelada, Mr Gabriel. Lab. de Física, Comisión Chilena de Energía Nuclear, Casilla 188-D, Santiago, Chile. (1955) Lic. en ciencias mención fís. (U. de Chile, 1982). Res. scient. (tel. 2731827, ext. 841). *X-ray diffraction, neutron diffraction.*

Zlosilo, Mr Mario. Dept. de Análisis Instrumental, Centro de Investigación Minera y Metalúrgica (CIMM), Casilla 170, Santiago 10, Chile. (1943) Prof. de Quím. (U. de Chile, 1967). Assoc. res. (tel. 2289544 ext. 264). *Minerals, X-ray fluorescence.*

CHINA

Sub-Editor: Xu Xiaojie

Bai, Mr Chun-li. Lab. of Crystal Structure, Inst. of Chemistry, Academia Sinica, P.O. Box 2709, Beijing, People's Rep. of China. (1953) MS, structural chemistry (Graduate Sch., U. of Sci. and Techn. of China, 1981). Res. assoc. *Molecular mechanics, conformation, organic molecules, semiempirical MO calculations, EXAFS, spectroscopy.*

Bao, Prof. Guang-hong. Inst. of Basic Medical Sciences, Chinese Academy of Medical Sciences, 5 Dong Dan San Tiao, Beijing 100730, People's Rep. of China. (1933) physics. Assoc. Prof. *Natural products structure.*

Bi, Prof. Ru-chang. Dept. of Protein Crystallogrphy, Inst. of Biophysics, Academia Sinica, Zhong Guan Cun, Beijing 100080, People's Rep. of China. (1940) grad., molecular biophysics (Leningrad U., USSR, 1965). Assoc. prof., Dept. vice head. (tel. 28-1768). *X-ray structure analysis, protein crystallography, structure and function, biological macromolecules.*

Bi, Prof. Yu-run. Dept. of Geology, Peking U., Beijing, People's Rep. of China. (1933). Assoc. prof. (tel. 282471). *Crystal structure X-ray analysis.*

Cai, Prof. Jing-hua. Fuzhou Lab. of Structural Chemistry, Fujian Inst. of Res. on the Structure of Matter, Academia Sinica, P.O. Box 143, Fuzhou, Fujian, People's Rep. of China. (1935). Assoc. res. prof. *Transition metal complexes, X-ray crystallography.*

Cao, Prof. Ming-zhong. Dept. of Physics, Nankai U., 94 Weijin Road, Tianjin, People's Rep. of China. (1935) material sci. Assoc. prof. *Crystal structure analysis, noncrystalline materials.*

Cao, Prof. Zheng-min. Dept. of Geology, Peking U., Beijing, People's Rep. of China. (1933) geological prospecting. Assoc. prof. *Physical properties, mineral crystals, crystal morphology, skarn minerals.*

Chang, Mr Wen-rui. Dept. of Protein crystallography, Inst. of Biophys., Academia Sinica, Zhong Guan Cun, Beijing 100080, People's Rep. of China. (1940) chemistry. Assoc. prof. (tel. 28-1768). *Protein structure, insulin.*

Chen, Mrs Shi-zhi. Dept. of Protein Crystallography, Inst. of Biophys., Academia Sinica, Zhong Guan Cun, P.O. Box No. 349, Beijing 100080, People's Rep. of China. (1935) physical chemistry. Assoc. prof. (tel. 28-1768). *Protein crystallography, structural chemistry, X-ray crystallography.*

Chen, Mr Ben-ming. Inst. of Chemistry, Academia Sinica, Zhong Guan Cun, Beijing, People's Rep. of China. (1938). Res. assoc. *Small molecule structures, molecular mechanics.*

Chen, Prof. Dai-zhang. Mineralogy Div., China U. of Geosciences, Xue Yuan Road 29, Beijing 100083, People's Rep. of China. (1937) geology. Assoc. prof. *Mineralogy, wall-rock alteration, electron microprobe analysis.*

Chen, Mr Guo-ying. Dept. of Geological Sci., Lanzhou U., Lanzhou, Gansu, People's Rep. of China. (1937) geology. Instructor. *Mineralogy, oxidation zone, powder diffraction.*

Chen, Mr Jing-yi. Material Physics Div., Shanghai Inst. of Metallurgy, Academia Sinica, 865 Chang-ning Road, Shanghai 200050, People's Rep. of China. (1956) metal physics. Res. assoc. (tel. 511070, ext. 129). *X-ray double crystal diffraction, X-ray topography.*

Chen, Mr Jing-zhong. Test Center of Rocks and Minerals, China U. of Geosciences, Yu Jia Shan, Wuchang, Wuhan 430074, Hubei, People's Rep. of China. (1946) MS, mineralogy (Beijing Graduate Sch., Wuhan C. of Geology, 1981). Lect. (tel. 70481). *Crystal chemistry, minerals, electron microprobe analysis, mineralogy.*

Chen, Prof. Kuang-yuan. Genetic Mineralogy Div., China U. of Geosciences, Xue Yuan Road 29, Beijing 100083, People's Rep. of China. (1920) Fil. Lic., geology and mineralogy (Uppsala U., Swedan, 1951). Prof. (tel. 277460, ext. 418). *Oxide minerals, silicate minerals, genetic mineralogy, metallogenesis, ore prospecting.*

Chen, Prof. Li-quan. Solid State of Ionics Div., Inst. of Physics, Academia Sinica, Beijing, People's Rep. of China. (1940). Assoc. prof. *Solid state ionics, superionic conductor, crystal growth, crystal physics.*

Chen, Prof. Xian-qiu. Structure Res. Dept., Shanghai Inst. of Ceramics, Academia Sinica, 865 Chang-ning Road, Shanghai 200050, People's Rep. of China. (1925) BS, chemistry (National Zhong Shan U., 1949). Assoc. res. prof. (tel. 522470, telex 33309ASSIC CN). *Crystal growth, optical crystallography, optical mineralogy.*

Chen, Mr Yuan-zhu. Structural Chemistry Div., Fujian Inst. of Res. on the Structure of Matter, Fuzhou, Fujian 350002, People's Rep. of China. (1930) structural chemistry. Assoc. prof. *Metal complexes, natural substances, structure analysis.*

Chen, Mr Zhi-xue. Analyzing and Testing Center, Sichuan U., Wang Jia Road, Chengdu 64001, Sichuan, People's Rep. of China. (1935) solid state physics. Instructor. (tel. 316-24401, telex 2345). *Crystal growth, ceramic materials, powder diffraction.*

Chen, Dr Zhong-guo. Lab. for Structure of Matter, Inst. of Physical Chemistry, Peking U., Wei Xiu Yuan, Peking U., Beijing, People's Rep. of China. (1943) PhD, biochemistry (Munich Techn. U., BRD, 1982). Lect. (tel. 282471, ext. 3549). *Protein crystallography, kallikrein, birmann-birk inhibitor, trypsin, pancreatic trypsin inhibitor, superoxide dismutase.*

Cheng, Mr Min-chin. Chemistry Dept., Fudan U., 200 Handan Road, Shanghai 201903, People's Rep. of China. (1939). Lect. (tel. 480906, ext. 280). *Crystal structure analysis, inorganic and metal-organic compounds, drug action, powder diffraction.*

Chiang, Prof. Liang-jun. Geology Div., Central South Inst. of Mining and Metallurgy, Changsha, Hunan, People's Rep. of China. (1913). Prof. (tel. 82811, telex 4349). *Crystallography, spectroscopy, minerals.*

Cia, Mr Jin-hua. Lab. of Structural Chemistry, Fujian Inst. of Res. on the Structure of Matter, Academia Sinica, Fuzhou, Fujian, People's Rep. of China. (1934) Lect. *Transition metal complexes, metal cluster compounds, nitrogen fixation, X-ray crystallography.*

Cui, Prof. Wen-yuan. Dept. of Geplogy, Peking U., Beijing, People's Rep. of China. (1934). Assoc. prof. (tel. 282471, ext. 3227). *Genetic mineralogy.*

Dai, Mr Jin-bi. Dept. of Protein Crystallography, Inst. of Biophys., Academia Sinica, Zhong Guan Cun, Beijing 100080, People's Rep. of China. (1937) mathematics. Res. asst. (tel. 285529). *Direct methods, computing methods, protein crystallography.*

Dong, Mr Ji-he. Mineralogy Div., Qinghai Inst. of Salt Lake, Academia Sinica, No. 8 Xing Ning Street, Xining, Qinghai, People's Rep. of China. (1940). Res. asst. *Cell constants determination, quantitative analysis, powder diffraction.*

Dong, Prof. Yi-cheng. Dept. of Protein Crystallography, Inst. of Biophys., Academia Sinica, Zhong Guan Cun, P.O. Box No. 349, Beijing 100080, People's Rep. of China. (1939) physical chemistry. Assoc. prof. *Protein crystallography, X-ray crystal structure, structural chemistry.*

Dou, Mr Shi-qi. Dept. of Protein Crystallography, Inst. of Biophys., Academia Sinica, Zhong Guan Cun, P.O. Box No. 349, Beijing 100080, People's Rep. of China. (1940) spectroscopy. Res. assoc. (tel. 28-4483). *Protein crystallography, crystal structure determination, X-ray diffraction.*

Fan, Mr Guang-yu. Chemical Eng. Dept., Beijing Inst. of Techn., Bai Shi Qiao Road, Beijing, P.O. Box 327, People's Rep. of China. (1936) physical chemistry and polymers. Lect. (tel. 89-0321, ext. 605, telex 0055). *Physical chemistry, polymer diffraction, powder diffraction.*

Fan, Prof. Hai-fu. Lab. of X-ray Analysis, Inst. of Physics, Academia Sinica, P.O. Box 603, Zhong Guan Cun, Beijing 100080, People's Rep. of China. (1933) physical chemistry. Prof. *Methods in X-ray crystal structure analysis, image processing in high resolution electron microscopy, crystallographic computing.*

Fan, Mr Yu-guo. Inst. of Theoretical Chemistry, Jilin U., 79 Jie Fang Road, Changchun, Jilin, People's Rep. of China. (1940) MS, quantum chemistry (Jilin U., 1981). Lect. *Electron density.*

Fan, Mr Zhao-chang. The Fifth Dept., Shanghai Inst. of Organic Chemistry, 345 Lingling Road, Shanghai, People's Rep. of China. (1940) chemical kinetics. Res. assoc. (tel. 313300). *Structural chemistry, crystal structure, organometallic compounds, natural products.*

Fu, Prof. Heng. Lab. of Structural Chemistry, Inst. of Chemistry, Academia Sinica, Zhong Guan Cun, Beijing 100080, People's Rep. of China. (1929) chemistry. Assoc. res. prof. (tel. 284098). *Organic and organometallic structures, bonding in organic compounds.*

Fu, Prof. Ping-qiu. Mineralogy Div., Inst. of Geochemistry, Academia Sinica, Guan Shui Road, Guiyang, Guizhou, People's Rep. of China. (1933). Assoc. prof. (tel. 24757). *Crystal structure, new minerals.*

Fu, Mr Zheng-min. Inst. of Physics, Academia Sinica, P.O. Box 603, Beijing, People's Rep. of China. (1938) physics (Wuhan U.). Res. assoc. *Phase diagrams, phase transitions, crystal structure, DTA, X-ray powder method.*

Fu, Mr Zhu-ji. Lab. of Structural Chemistry, Fujian Inst. of Res. on the Structure of matter, Academia Sinica, Fuzhou, Fujian, People's Rep. of China. (1950) MS, structural chemistry (Fujian Inst. of Res. on the Structure of Matter, 1981). Lect. *X-ray crystallography, metal complexes, protein crystal structure.*

Gao, Mr Yi-guei. Dept. of Protein Crystallography, Inst. of Biophys., Academia Sinica, Zhong Guan Cun, Beijing 100080, People's Rep. of China. (1939) physical chemistry (Nan-Kai U., 1965). Res. assoc. (tel. 28-1768). *Protein crystallography, crystal structure determination.*

Gong, Prof. Xia-sheng. Mineralogy Div., Chengdu College of Geology, Chengdu, Sichuan 610059, People's Rep. of China. (1937) geology. Assoc. prof. *Clay mineralogy, crystal X-ray analysis.*

Gu, Prof. Xiao-cheng. Dept. of Biology, Peking U., Beijing, People's Rep. of China. (1930) botany. Assoc. prof. (tel. 282471, ext. 3240). *Protein crysatllography.*

Gu, Mr Yuan-xin. Lab. of X-ray Analysis, Inst. of Physics, Academia Sinica, Zhong Guan Cun, Beijing 100080, People's Rep. of China. (1938) physics. Res. assoc. (tel. 281866). *Direct methods, small molecules.*

Guan, Res. Ya-xian. Changchun Geological Inst., Changchun 130026, Jilin, People's Rep. of China. (1934) Miner. Dr, mineralogy and crystallography (Crystallography Inst., USSR, 1963). Assoc. prof. (tel. 375-24781). *Mineralogy, X-ray crystallography, geology, mineral crystal structure, crystal chemistry.*

Gui, Mrs Lu-lu. Dept. of Crystallography, Inst. of biophys., Academia Sinica, Zhong Guan Cun, Beijing 100080, People's Rep. of China. analytical chemistry. Assoc. prof. *Proteins, crystallography, growth, structure and function, biological macromolecules.*

Guo, Prof. Dong-yao. Inst. of Th. Chemistry, Jilin U., 79 Jiefang Road, Changchun, Jilin, People's Rep. of China. (1935). Assoc. prof. (tel. 23189, ext. 461). *Direct methods, algebraic research, symmetry groups, crystal structure determination.*

Guo, Mrs Fang. Dept. of Crystallography, Inst. of Chemistry, Academia Sinica, Zhong Guan Cun, Beijing, People's Rep. of China. (1939). Res. assoc. *Small molecule structures, computing software, methods, structure determination.*

Han, Mr Fu-son. Lab. of X-ray Analysis, Inst. of Physics, Academia Sinica, Zhong Guan Cun, Beijing 100080, People's Rep. of China. (1946) physics. Res. assoc. (tel. 281866). *Direct methods, small molecules, EXAFS.*

Han, Mr Shao-xu. The Res. Section, Crystal Structure and Crystal Chemistry of Minerals, China U. of Geosciences, Xue Yuan Road 29, Beijing 100083, People's Rep. of China. (1955) MS, crystal structure of minerals (Beijing Graduate Sch., Wuhan C. of Geology, 1982). Lect. (tel. 277461, ext.560). *Crystal structure, minerals, single crystal diffraction.*

Han, Mr Yu-zhen. Lab. for Structure of Matter, Inst. of Physical Chemistry, Peking U., Beijing, People's Rep. of China. (1940) computational mathematics. Lect. (tel. 282471, ext. 3549). *Crystallographic computing, computational chemistry, computational mathematics.*

He, Prof. Chong-fan. Crystal Lab., Shanghai Inst. of Ceramics, Academia Sinica, 865 Changning Road, Shanghai 200050, People's Rep. of China. (1926) chemistry. Assoc. res. prof. (tel. 522470, telex (761)33309 ASSIC CN). *Single crystal growth, synthetic mica properties, lead molybdate, bismuth germanate and silicate.*

He, Prof. Cun-heng. Lab. of X-ray Analysis, Inst. of Materia Medica, Chinese Academy of Medical Sciences, 1 Xian Nong Tan Street, Beijing 100050, People's Rep. of China. (1941) biophysics, chemistry. Assoc. Prof. *Direct methods, organic natural products, macromolecular structure.*

He, Mr Rei-ling. Geological Sci. Div., Lab. of Geology and Minerals Bureau of Shanxi Province, He Pin Road, Xi-an, Shanxi, People's Rep. of China. (1936) geology. Eng. (tel. 21024). *X-ray crystallography, powder diffraction, structure analysis.*

Hong, Mr Mao-chun. Fuzhou Lab. of Structural Chemistry, Fujian Inst. of Res. on the Structue of Matter, Academia Sinica, P.O. Box 143, Fuzhou, Fujian, People's Rep. of China. (1953) MS, structural chemistry (Fujian Inst. of Res. on the Structure of Matter, 1981). Res. assoc. *Transition metal complexes, metal cluster compounds, magnetochemistry, X-ray crystallography.*

Hou, Mr Yong-geng. Lab. of Crystal Structure, Inst. of Chemistry, Academia Sinica, P.O. Box 2709, Beijing, People's Rep. of China. (1935) structural chemistry. Res. assoc. *Direct method in crystallography.*

Hu, Mr Guo-zhi. Chengdu Center of Analysis and Test, Academia Sinica, Chengdu, Sichuang, People's Rep. of China. (1957). Asst. res. *Coordination chemistry, macromolecule structure, X-ray crystallography.*

Hu, Mr Heng-liang. Textile Chemistry Dept., Tianjin Textile C. of Techn., Chenglinzhang Road 89, Tianjin 300160, People's Rep. of China. (1942) mathematics. Lect. (tel. 43251). *Crystal structure, macromolecular materials, supermolecular structural parameters, polymers, WAXS and SAXS.*

Hu, Mr Sheng-zhi. Chemistry Dept., Xiamen U., Xiamen, Fuijan, People's Rep. of China. (1932) structure of matter. Assoc. prof. *Structure determination, chemical and biological molecules.*

Hua, Mr Zi-qian. Dept. of Biology, Peking U., Beijing, People's Rep. of China. (1933) biochemistry. Lect. (tel. 282471, ext. 3240). *Protein crystallography.*

Huang, Bao-quan. Structural Chemistry Div., Fujian Inst. of Res. on the Structure of Matter, Academia Sinica, Fuzhou, Fujian 350002, People's Rep. of China. (1957). Lect. *Transition metal complexes, metal cluster compounds, X-ray crystallography.*

Huang, Mr De-bin. Inst. of Structural Chemistry, Dept. of Chemistry, Fuzhou U., Fuzhou, Fujian, People's Rep. of China. (1951) MS, physical chemistry (Fuzhou U., 1981). Res. asst. *Crystal and molecular structure, transition metal complexes.*

Huang, Prof. Jin-ling. Fuzhou U., Fuzhou, Fujian 350002, People's Rep. of China. (1932) physical chemistry. Prof., U. president. (tel. 533218). *Crystal and molecular structure, transition metal complexes.*

Huang, Mr Jin-shun. Structural Chemistry Div., Fujian Inst. of Res. on the Structure of Matter, Academia Sinica, Fuzhou, Fujian, People's Rep. of China. (1939). Prof. *Transition metal complexes, metal cluster compounds, X-ray crystallography.*

Huang, Prof. Liang-ren. Fuzhou Lab. of Structural Chemistry, Fujian Inst. of Res. on the Structure of Matter, Academia Sinica, Fuzhou, Fujian, People's Rep. of China. (1940). Assoc. res. prof. *Transition metal complexes, metal cluster compounds, nitrogen fixation, X-ray crystallography.*

Huang, Ming-dong. Structural Chemistry Div., Fujian Inst. of Res. on the Structure of Matter, Academia Sinica, Fuzhou, Fujian 350002, People's Rep. of China. (1963). Lect. *Transition metal complexes, metal cluster compounds, X-ray crystallography.*

Huang, Mr Tai-shan. Chemistry Dept., Xiamen U., Xiamen, Fujian, People's Rep. of China. (1935) structural chemistry. Sr. lect. *X-ray crystal structure.*

Huang, Mr Zhi-ying. Fuzhou Lab. of Structural Chemistry, Fujian Inst. of Res. on the Structure of Matter, Fuzhou, Fujian, People's Rep. of China. (1941) MS, structural chemistry (Fujian Inst. of Res. on the Structure of Matter, 1981). Lect. *Small organic compound, transition metal complexes, electrochemistry.*

Ji, Prof. Shou-yuan. Dept. of Earth Sciences, Nanjing U., Hankou Road, Nanjing, Jinagsu 210008, People's Rep. of China. (1924) BS, geology (Central U., 1947). Prof. (tel. 34651, ext. 2448). *Crystal chemistry, silicate minerals.*

Jia, Prof. Shou-quan. Crystal Growth Div., Inst. of Physics, Academia Sinica, P.O. Box 603, Beijing 100080, People's Rep. of China. (1930). Assoc. prof., div. head. (tel. 281869). *Crystallography, crystal structure, crystal growth, high temperature and high pressure, growth from solution, hydrothermal reaction.*

Jiang, Mr An-bei. Laboratorial Center of Xiamen U., Si Ming Nan Lu, Xiamen, Fujian, People's Rep. of China. (1939) chemical physics. (tel. 25102-108). *Metal cluster compounds, structure, organometallic and organic compounds.*

Jiang, Mr Han-chen. Physics Dept., Jilin U., Je-fang Road, Chang-chun, Jilin, People's Rep. of China. (1936). Lect. *Electron theory - solids and molecules, structure, metals and alloys.*

Jiang, Mrs Shao-ying. State Bureau of Building Materials, Geological Inst., No. 11 Bei Shun Cheng Street, Xi Zhi Men, Beijing, People's Rep. of China. (1935).

Eng., Inst. deputy director. (tel. 551842). *Crystallography, mineralogy, mineral physics.*

Jiang, Prof Xiao-long. Material Physics Div., Shanghai Inst. of Metallurgy, Academia Sinica, 865 Chang Ning Road, Shanghai, People's Rep. of China. (1940) metal physics. Assoc. prof. (tel. 511070, ext. 129). *X-ray diffraction, X-ray topography, crystal growth, structure, metals and alloys, phase transition in solids.*

Jiang, Prof. Yan-dao. Crystal Growth Div., Inst. of Physics, Academia Sinica, Beijing, P.O. Box 603, People's Rep. of China. (1936) physics. Assoc. prof., div. head. *Metal growth, imperfection, transport, complex oxide crystals.*

Jin, Dr Wei-qing. Crystal Lab., Shanghai Inst. of Ceramics, Academia Sinica, 865 Chang-ning Road, Shanghai 200050, People's Rep. of China. (1941) Dr, Faculty of Sci. (Tokyo U., Japan, 1984). Res. asst. (tel. 522470, telex (761)33309 ASSIC CN). *Crystal growth mechanism, morphology, crystal growth from melt.*

Jin, Prof. Xiang-lin. Lab. for Structure of Matter, Inst. of Physical Chemistry, Peking U., Beijing, People's Rep. of China. (1940) physical chemistry. Assoc. prof. (tel. 282471, ext. 3549). *X-ray single crystal analysis, structural chemistry.*

Jin, Prof. Zhong-sheng. Changchun Inst. of Appl. Chemistry, Academia Sinica, Changchun 130022, People's Rep. of China. Assoc. prof. (tel. 882801, ext. 433).

Ke, Mr Heng-ming. Chemistry Dept., Zhongshan U., Guangzhou, Guangdong, People's Rep. of China. (1948) Polymer Chemistry and Physics. Assoc. lect. (tel. 46300, ext. 218). *Proteins, small molecules, structure, powder diffraction, crystallization kinetics, polymerization.*

Kong, Prof. You-hua. Center of Geological Res. and Analysis of Zhejiang Province, Ti Yu Chang Road 232, Hangzhou, Zhejiang 310007, People's Rep. of China. (1935). Assoc. prof. (tel. 24757). *Crystal structure, crystallochemistry, minerals, rock-forming minerals.*

Kuang, Mrs Bao. Dept. of Biology, Peking U., Beijing, People's Rep. of China. (1943) biophysics. Lect. (tel. 282471, ext. 3240). *Protein crystallography.*

Kuo, Prof. Ke-hsin. Beijing Electron Microscopy Lab., Academia Sinica, P.O. Box 2724, Beijing 100080, People's Rep. of China. (1923) dr tekn h.c., Physical Metallurgy (Royal Inst. of Techn., Sweden, 1980). Prof., foreign mbr. - Royal Swedish Academy of Eng. Sci. (tel. 2561422). *High resolution electron microscopy, surface crystallography.*

Li, Mr Da-ming. R&D Center in Advanced Inorganic Materials, Shanghai Inst. of Ceramics, Academia Sinica, 865 Chang-ning Road, Shanghai 200050, People's Rep. of China. (1933) geology. Head, synthetic diamond res. group. (tel. 522470, telex 33309 ASSIC CN). *Ultra-high pressure and high temperature technology, crystal growth.*

Li, Prof. De-yu. Lab. of Crystals, Shanghai Inst. of Ceramics, Academia Sinica, Shanghai 200050, People's Rep. of China. (1935) Cand. physicomathem (Inst. of Crystallogr. Acad. of Sci., USSR, 1966). Prof., head of lab. (tel. 512990). *Structure, characterization, powder diffraction methods, material science, X-ray crystal structure, crystal chemistry, phase diagrams, inorganic compound systems.*

Li, Mr Du. Chemistry Dept., Lanzhou U., Tian Shui Road, Lanzhou, Gansu, People's Rep. of China. (1939) physical chemistry. Lect. (tel. 22991). *Powder diffraction, structure determination, organic molecules, coordination compounds.*

Li, Prof. Run-shen. Material Physics Div., Shanghai Inst. of Metallurgy, Academia Sinica, 865 Chang-ning Road, Shanghai, People's Rep. of China. (1943) metal physics. Assoc. prof. (tel. 511070, ext. 129). *X-ray double crystal diffraction, X-ray topography.* **Li,** Mr Wan-mao. Dept. of Geological Sci., Lanzhou U., 418 Physics Building, Lanzhou, Gansu, People's Rep. of China. (1936) geology. Instructor. *Mineralogy, oxidation zone, powder diffraction.*

Liang, Prof. Dong-cai. Lab. of Protein Crystallography, Inst. of Biophys., Academia Sinica, Beijing 100080, People's Rep. of China. (1932) Cand. chemistry (Acad. of Sci., USSR, 1960). Prof., Academia Sinica member, fellow of Third World Academy of Sciences. (tel. 285529). *X-ray crystal structure, protein crystallography, biomolecular structure.*

Liang, Prof. Jing-kui. Lab. of Phase Transition and Diagram, Inst. of Physics, Academia Sinica, Zhong Guan Cun, P.O. Box 603, Beijing 100080, People's Rep. of China. (1931) Cand. Tech. Sci. (Acad. of Sci. USSR. 1960) Prof. (tel. 289131, ext. 284). *Polycrystal structure analysis, phase transition, phase diagram, inorganic compounds, alloys.*

Liang, Prof. Li. Dept. of Protein Crystallography, Inst. of Biophys., Academia Sinica, Beijing 100080, People's Rep. of China. (1941) physical chemistry. Assoc. prof. (tel. 28-1768). *Structure determination, macromolecules, structure and function, biological molecules, charge density, electrostatic potential.*

Liao, Prof. Bo-qang. Dept. of Applied Chemistry, Chongqin U., Sichuan, People's Rep. of China. (1929). Assoc. prof. *Powder diffraction.*

Lin, Mr Cheng-yi. Center of Material Analysis, Nanjing U., Hankou Road, Nanjing 210008, Jiangsu, People's Rep. of China. (1938) BS, mineralogy (Leningrad U., U. S. S. R., 1961) Assoc. prof. (tel. 34651, ext. 2523). *Crystal structure analysis, electron microprobe analysis.*

Lin, Prof. Chi-chang. Inst. of Structural Chemistry, Dept. of Chemistry, Fuzhou U., Fuzhou, Fujian, People's Rep. of China. (1933) physical chemistry. Assoc. prof. *Crystal and molecular structure, transition metal complexes.*

Lin, Mr Chuan. Inst. of Physics, Academia Sinica, Beijing, People's Rep. of China. (1940) physics. Inst. deputy director.

Lin, Mr Guang-da. Crystallography Div., Shanghai Inst. of Biochemistry, Academia Sinica, 320 Yue Yang Road, Shanghai 200031, People's Rep. of China. (1937) physics. Res. asst., head, crystallographic group (tel. 374430, telex 3933). *X-ray crystal structure analysis, biological macromolecules.*

Lin, Mr Xian-ti. Structural Chemistry Div., Fujian Inst. of Res. on the Structure of Matter, Academia Sinica, Fuzhou, Fujian 350002, People's Rep. of China. (1938). Assoc. Prof. *Transition metal complexes, metal cluster complexes, X-ray crystallography.*

Lin, Mr Xing-yuan. Inst. of Mineral Deposits, Chinese Academy of Geological Sciences, Beijing 100037, People's Rep. of China. (1938). Sr. res. *X-ray diffraction crystal structure.*

Lin, Prof. Yong-hua. Changchun Inst. of Applied Chemistry, Academia sinica, Changchun 130022, People's Rep. of China. BS, chemistry. Assoc. Prof. (tel. 882801-433). *Crystal structure.*

Lin, Mrs Yu-juan. Structural Chemistry Div., Fujian Inst. of Res. on the Structure of Matter, Academia Sinica, Fuzhou, Fujian, People's Rep. of China. (1939). Lect. *Transition metal complexes, X-ray crystallography, protein crystal structure.*

Lin, Prof. Zheng-jiong. Dept. of Protein Crystallography, Inst. of Biophys., Academia Sinica, Beijing 100080, People's Rep. of China. (1935) physical chemistry. Prof. (tel. 28-1768). *Protein crystallography, X-ray diffraction, biomocular structure.*

Liu, Mr Guang-zhao. R&D Center in Advanced Inorganic Materials, Shanghai Inst. of Ceramics, Academia Sinica, 865 Chang-ning Road, Shanghai 200050, People's Rep. of China. (1942) crystal growth. Group Leader, Large Single Diamand Crystal Res. (tel. 522470, telex 33309 ASSIC CN). *Crystal growth, Monte Carlo method, computer simulation.*

Liu, Prof. Han-qin. Fuzhou Lab. of structural chemistry, Fujian Inst. of Res. on the Structure of Matter, Fuzhou, Fujian, People's Rep. of China. (1937) PhD, chemical physics (U. of Chicago, USA, 1967). Res. prof. *Crystal and molecular structure, magnetic structure, metal cluster complexes, magnetic resonance.*

Liu, Ms Jing-ding. Inst. of Mineral Deposits, Chinese Academy of Geological Sciences, Baiwan Zhuang Road 26, Beijing 100037, People's Rep. of China. (1929). Assoc. Prof. *X-ray crystallography, crystallology, mineralogy.*

Liu, Mr Shi-xiong. Inst. of Structural Chemistry, Dept. of Chemistry, Fuzhou U., Fuzhou, Fujian, People's Rep. of China. (1943) physical chemistry. Lect. *Crystal and molecular structure, transition metal complexes.*

Liu, Mr Tian-liang. Inst. of Structural Chemistry, Dept. of Chemistry, Fuzhou U., Fuzhou, Fujian, People's Rep. of China. (1936) physical chemistry. Lect. *Crystal and molecular structure, transition metal complexes.*

Liu, Mr Wan. Changchun Geological Inst., Changchun, 130026, People's Rep. of China. (1934) X-ray crystallography. Assoc. Prof. (tel. 375-24781). *Minerology, X-ray crystallography, powder diffraction, crystal structure.*

Liu, Prof. Xue-lun. Zhengzhou Inst. of Multipurpose Utilization of Mineral Resources, Ministry of Geology and Mineral Resources, 26 Funin Road, Zhengzhou, Henan, People's Rep. of China. (1930) X-ray crystal structure analysis. Assoc. Prof. (tel. 48864). *X-ray crystallography, optical crystallography, powder diffraction.*

Liu, Prof. Yong-sheng. Changchun Inst. of Appl. Chemistry, Academia Sinica, Changchun 130022, People's Rep. of China. Assoc. prof. (tel. 882801, ext. 433).

Liu, Mr Zuo-cai. Dept. of Chemical Eng., Beijing Techn. Inst., Bai Shi Qiao Road, Beijing, P.O. Box 327, People's Rep. of China. (1946) MS, structural chemistry (Peking U., 1982). Lect. (tel. 89-0321, ext. 605). *Organic compounds, complexes, solid chemistry.*

Lou, Mrs Mei-zhen. Dept. of Protein Crystallography, Inst. of Biophys., Academia Sinica, Zhong Guan Cun, Beijing 100080, People's Rep. of China. (1935) organic chemistry. Res. assoc. (tel. 2563227). *Protein structure, insulin, crystal growth.*

Lu, Mrs Li Chengdu Center of Analysis and Test, Academia Sinica, Chengdu, Sichuan, People's Rep. of China. (1964). *Inorg. chemistry, crystal structure.*

Lu, Mrs Quang-ying. Dept. of Biology, Peking U., Beijing, People's Rep. of China. (1937) biochemistry. Lect. (tel. 28-2471, ext. 3240). *Protein crystallography.*

Lu, Prof. Jia-xi. Structural Chemistry Lab., Fujian Inst. of Res. on the Structure of Matter, Academia Sinica, P.O. Box No. 143, Fuzhou, Fujian 550002, People's Rep. of China. (1915) PhD, physical chemistry (U. of London, UK, 1939). Res. prof., Academia Sinica member, pres. (tel. 31835). *Transition metal cubanelike clusters, nitrogenase active center model compounds, crystal and molecular structure.*

Lu, Mr Kun-quan. X-ray Analysis Lab., Inst. of Physics, Academia Sinica, Zhong Guan Cun, Beijing, P.O. Box 603, People's Rep. of China. (1939) physics. Res. assoc. (tel. 281866). *EXAFS, X-ray diffraction, crystal growth.*

Lu, Prof. Qi. Test Center of Rocks and Minerals, China U. of Geosciences, Yu Jia Shan, Wuchang, Wuhan 430074, People's Rep. of China. (1942) MS, mineral crystal chemistry and structure (Beijing Graduate Sch., Wuhan C. of Geology, 1981). Lect. (tel. 70481). *Crystal chemistry, minerals, X-ray crystallography, clay minerology.*

Lu, Mr Shao-fang. Structural Chemistry Div., Fujian Inst. of Res. on the Structure of Matter, Academia Sinica, Fuzhou, Fujian, People's Rep. of China. (1940). Assoc. Prof. *Metal cluster compounds, transition metal complexes, new technical crystal, X-ray crystallography, structure-property relations.*

Lu, Prof. Yun-jin. Center of Material Analysis, Nanjing U., Hanlou Road, Nanjing, Jiangsu, People's Rep. of China. (1928) crystallography. Assoc. prof. (tel. 34651, ext. 2523). *Crystal structure, powder diffraction.*

Luo, Prof. Gu-feng. Dept. of Earth Sciences, Nanjing U., Hankou Road, Nanjing, Jiangsu 210008, People's Rep. of China. (1933) mineralogy and petrology. Prof. (tel. 634651, ext. 2675). *Crystal chemistry, rock-forming minerals, X-ray crystallography.*

Ma, Prof. Li-dun. Dept. of Chemistry, Fudan U., 200 Handan Road, Shanghai 201903, People's Rep. of China. (1935). Assoc. prof. (tel. 480906, ext. 353). *Structure, inorganic and metal-organic compounds, polycrystalline diffraction, EXAFS.*

Ma, Mr Xing-qi. Dept. of Protein Crystallography, Inst. of Biophys., Academia Sinica, Zhong Guan Cun, Beijing 100080, People's Rep. of China. (1936) physical chemistry. Res. assoc. *X-ray crystal structure, protein crystallography, structural chemistry.*

Ma, Prof. Zhe-sheng. Res. Section, Crystal Structure and Crystal Chemistry Minerals, China U. of Geosciences, Xue Yuan Road 29, Beijing 100083, People's Rep. of China. (1937) crystal structure, minerals. Assoc. prof. (tel. 277461, ext. 560). *Crystal structure, crystal chemistry, minerals, single crystal diffraction.*

Mai, Prof. Zhen-hong. Crystal Defects Div., Inst. of Physics, Academia Sinica, P.O. Box 603, People's Rep. of China. (1942). Assoc. prof., div. head. (tel. 281869). *Crystallography, crystal defects, X-ray diffraction, topography.*

Meng, Prof. Yi-min. Chemistry Dept., Lanzhou U., Tian Shui Road, Lanzhou, Gansu, People's Rep. of China. (1926) physical chemistry. Assoc. prof. (tel. 22991). *Crystal chemistry, single crystal diffraction.*

Miao, Mr Chun-sheng. Bureau of Geology and Mineral Resources of Liaoning Province, Exp. and Res. Center, Bei Ling Street, Shenyang, Liaoning, People's Rep. of China. (1940). Engineer (tel. 61768, telex 6012). *Powder diffraction, quantitative X-ray phase analysis, feldspar structure state, crystal lattice measurement.*

Miao, Prof. Fang-ming. Chemistry Dept., Tianjin Normal U., Ba Li Tai, Tianjin, People's Rep. of China. (1935) physical chemistry. Prof., dept. dean. (tel. 23489). *Crystal and molecular structure, small molecules.*

Mu, Mr Xiang-qi. Textile Chemistry Dept., Tianjin Textile C. of Techn., Cheng Lin Zhuang Road 89, Tianjin, People's Rep. of China. (1938) physical chemistry. Lect. (tel. 43251). *Polymers, structure, fiber diffraction methods.*

Ni, Prof. Chau-zhou. The Fifth Dept., Shanghai Inst. of Organic Chemistry, Abcademia Sinica, 345 Lingling Road, Shanghai, People's Rep. of China. (1937) physical chemistry. Assoc. Prof. (tel. 313300). *Small molecules, proteins, crystal structure, organometallic compounds, natural products.*

Pan, Prof. Dao-jun. The Res. Section Mineral physics, China U. of Geosciences, Xue Yuan Road 29, Beijing 100083, People's Rep. of China. (1931) physics. Prof. (tel. 2017461). *Mineral physics, solid physics.*

Pan, Prof. Ke-zhen. Dept. of Protein crystallography, Fujian Inst. of Res. on the Structure of Matter, Academia Sinica, Fuzhou, Fujian 350002, People's Rep. of China. (1933) structural chemistry. Res. prof., head. *Metal complexs, X-ray crystallography, proteins, crystal structure.*

Pan, Prof. Zhao-lu. Dept. of Geology, China U. of Geosciences,Yu Jia Shan, Wuhan 430074, Hubei, People's Rep. of China. (1925) BS, geology (Qinghua U., 1949). Prof. (tel. 70481). *Crystal chemistry, mineralogy.*

Peng, Mr Chang-qi. Mineralogy Div., Yichang Inst. of Geology and Mineral Resources, P.O. Box 502, Yichang, Hubei, People's Rep. of China. (1940). Asst. res. fellow. (tel. 21635). *Mineralogy, crystallochemistry, X-ray powder diffraction.*

Peng, Mrs Ming-shen. Geology Div., Central-south Inst. of Mining and savellurgy, Changsha, Hunan, People's Rep. of China. (1939). Lect. (tel. 82811, telex 4349). *Crystal field theory, spectroscopy, minerals, X-ray crystallography.*

Qi, Dr Zeng-du. R&D Center in Advanced Inorganic Materials, Shanghai Inst. of Ceramics, Academia Sinica, 865 Chang-ning Road, Shanghai 200050, People's Rep. of China. (1937) PhD, physics (Reading U., UK, 1981). Head, high pressure res. dept. (tel. 522470, telex 33309 ASSIC CN). *Crystal growth, high pressure physics.*

Qi, Prof. Zhi-ru. Teaching Res. Group of Petrology and Mineralogy, Xi-an Geological Inst., 6 Yan Ta Road, Xi-an Shanxi, People's Rep. of China. (1928). Assoc. prof. (tel. 52991). *Powder diffraction, crystal structure.*

Qian, Mrs Min-xie. Dept. of Crystallography, Inst. of Chemistry, Academia Sinica, Zhong Guan Cun, Beijing, People's Rep. of China. (1949) MS, structural chemistry (Peking U., 1981). Res. assoc. *Structure, small molecule, electron density, molecular mechanics.*

Ren, Prof. Lei-fu. Dept. of Geology, Peking U., Beijing, People's Rep. of China. (1930). Assoc. prof. (tel. 282471, ext. 3354). *Clay mineralogy.*

Shan, Mr Try-seo. Dept. of Chemical Eng., Beijing Inst. of Techn., Beijing, China. (1940) MS, structural chemistry (Peking U., 1981). Engineer. *X-ray crystallography and applications.*

Shang, Mr Mao-yu. Structural Chemistry Div., Fujian Inst. of Res. on the Structure of Matter, Academia Sinica, Fuzhou, Fujian, People's Rep. of China. (1944) MS, structural chemistry (Fujian Inst. of Res. on the Structure of Matter, 1981). Lect. *Transition metal complexes, metal cluster compounds, X-ray crystallography.*

Shao, Prof. Jie-lian. Dept. of Geology, China U. of Geosciences, Yu Jia Shan, Wuhan 430074, Hubei, People's Rep. of China. (1929) geology. Prof. (tel. 70481). *Crystallogeny, ore mineralogy.*

Shao, Prof. Mei-cheng. Lab. for Structure of Matter, Inst. of Physical Chemistry, Peking U., Zhong Guan Cun, Beijing, People's Rep. of China. (1931) structural chemistry. Prof., head, crystallographic structure res. group (tel. 282471, ext. 3549). *Transition metal complexes, X-ray crystallography.*

Shso, Prof. Wei. Genetic Mineralogy Div., China U. of Geosciences, Xue Yuan Road 29, Beijing 100083, People's Rep. of China. (1931) geology. Assoc. Prof. (tel. 2017461, ext. 418). *Crystallography, mineralogy.*

Shen, Miss Fu-ling. X-ray Crystal Structure Div., Inst. of Biophys., Academia Sinica, Zhong Guan Cun, Beijing 100080, People's Rep. of China. (1941). *X-ray crystal structure.*

Shen, Mr Cheng. Inst. of Th. Chemistry, Jilin U., 79 Jiefang Road, Changchun, Jilin, People's Rep. of China. (1946). Lect. *Crystal structure determination.*

Shen, Prof. Jin-chuan. Test Center of Rocks and Minerals, China U. of Geosciences, Yu Jia Shan, Wuchang, Wuhan 430074, Hubei, People's Rep. of China. (1936). Prof. (tel. 70481). *Crystal chemistry, minerals, rare earth compounds, X-ray crystallography.*

Shi, Mr Bi-de. Chemistry Dept., Xiamen U., Xiamen, Fujian, People's Rep. of China. (1943) structure of matter. Lect. *X-ray crystallography.*

Shi, Mr Da-shuang. Chemistry Dept., Xiamen U., Xiamen, People's Rep. of China. (1964) Ms, structural chemistry. Asst. *X-ray crystallography.*

Shi, Mr Ni-cheng. Res. Section, Crystal Structure and Crystal Chemistry of Minerals, China U. of Geosciences, Xue Yuan Road 29, Beijing 100083, People's Rep. of China. (1938) crystal structure of minerals. Assoc. prof. (tel. 2017461-616) *Crystal structure, minerals, crystallographic computing, single crystal diffraction.*

Shu, Mr Jin-fu. Test Center of Rocks and Minerals, China U. of Geosciences, Yu Jia Shan, Wuchang, Wuhan 430074, Hubei, People's Rep. of China. (1942) MS, mineral crystal chemistry and structure (Beijing Graduate Sch., Wuhan C. of Geology, 1981). Lect. (tel. 70481, telex 5378). *Crystal chemistry, minerals, X-ray crystallography, mineralogy, feldspar.*

Song, Mrs Shi-ying. Dept. of Protein Crystallography, Inst. of Biophys., Academia Sinica, Beijing 100080, People's Rep. of China. (1938) physical chemistry. Res. assoc. *Protein crystallography, X-ray diffraction, biomolecular structure.*

Sun, Mr Li-kun. Dept. of Applied Chemistry, Chongqi U., Sichuan, People's Rep. of China. (1964). *Direct methods, X-ray crystallography.*

Sun, Prof. Dai-sheng. Genetic Mineralogy Div., China U. of Geosciences, Xue Yuan Road 29, Beijing 100083, People's Rep. of China. (1935). Prof. *Crystallography, genetic mineralogy.*

Sun, Prof. Yi-jian. Petrology and Mineralogy Div., Nanjing Inst. of Geology and Mineral Resources, CAGS, 534 East Zhongshan Road, Nanjing, Jinagsu, People's Rep. of China. (1931) X-ray crystallography. Assoc. prof. (tel. 5206-43992). *Powder diffraction, rock-forming minerals, feldspar.*

Tan, Prof. Hao-ran. Crystal Lab., Shanghai Inst. of Ceramics, Academia Sinica, 865 Changning Road, Shanghai 200025, People's Rep. of China. (1927) BS, physics (Ling-nan U., 1949). Assoc. prof. (tel. 522740). *Powder diffraction, new crystal materials, acousto-optic properties of crystals.*

Tang, Prof. You-qi. Inst. of Physical Chemistry, Peking U., Beijing 100871, People's Rep. of China. (1920) PhD, chemistry (Caltech, 1950). Prof., Inst. director, Academia Sinica member. (tel. 282471). *X-ray structure analysis, symmetry, structural chemistry.*

Tang, Mr Zhi-kai. Central Lab., Geological Bureau of Sichuan, 101 Renmin Northern Road, Chengdu, Sichuan, People's Rep. of China. (1932) chemistry. Engineer (tel. 31097). *Mineral chemistry, mineral X-ray crystallography, powder diffraction, instrumentation.*

Wang, Prof. Da-cheng. Dept. of Protein Crystallography, Inst. of Biophys., Academia Sinica, Zhong Guan Cun, Beijing 100080, People's Rep. of China. (1940) biophysics. Prof., Inst. deputy director, Vice president biophys. society china. *Protein structure, hormones, insulin, enzyme, zymogen.*

Wang, Prof. Gen-yuan. Dept. of Geology, China U. of Geosciences, Yu Jia Shan, Wuhan 430074, Hubei, People's Rep. of China. (1933) geology. Assoc. prof. (tel. 70481). *History of crystallography- mineralogy and geology in China (especially the ancient period).*

Wang, Mr Guan-xin. X-ray Diffraction Lab., Inst. of Geochemistry, Academia Sinica, Guanshui Road, Guiyang, Guizhou, People's Rep. of China. (1936). Sr. engineer (tel. 24757). *Crystallography, mineralogy, systematic mineralogy, isomorphous substitution, minerals, X-ray diffraction, instrumentation, powder diffraction method.*

Wang, Prof. Jia-huai. Dept. of Protein Crystallography, Inst. of Biophys., Acaemia Sinica, Zhong Guan Cun, Beijing 100080, People's Rep. of China. (1941) biophysics. Prof., Dept. head. *Protein crystallography, X-ray crystal structure, structural chemistry.*

Wang, Prof. Kui-ren. Dept. of Earth and Space Sci., U. of Sci. and Techn. of China, 24 Jin Zhai Road, Hefei, Anhui, People's Rep. of China. (1934) X-ray Analysis of Mineral. Assoc. prof., dept. vicehead. *Minerals, genetical mineralogy, X-ray crystallography.*

Wang, Prof. Pu. Mineralogy Div., China U. of Geosciences, Xue Yuan Road 29, Beijing, People's Rep. of China. (1926) BS, geology (Tsin Hwa U., 1949). Prof. *Mineralogy, wall-rock alteration, crystal growth, structure, systematic mineralogy.*

Wang, Prof. Shun-jin. Dept. of Geology, China U. of Geosciences, Yu Jia Shan, Wuhan 430074, Hubei, People's Rep. of China. (1933) geology. Assoc. prof.

(tel. 70481). *Mineralogy of deposits, crystallography, crystallochemistry, minerals, magnetite, wolframite, scheelite, pyoxenes, amphiboles, plagioclase.*

Wang, Prof. Wen-kui. Dept. of Geology, China U. of Geosciences, Yu Jia Shan, Wuhan 430074, Hubei, People's Rep. of China. (1925) BS, geology (Peking U., 1948). Assoc. prof. (tel. 70481). *Morphology, surface microtopography of crystals.*

Wang, Mr Xing-xin. Lab. of Geology, Res. Inst. of Petroleum E&D of Daqing Oil Field, Daqing City, Heilongjiang, People's Rep. of China. (1936). Principle geologist. *Clay minerals, powder diffraction.*

Wang, Mrs Yao-ping. Dept. of Protein Crystallography, Inst. of Biophys., Academia Sinica, Zhong Guan Cun, Beijing 100080, People's Rep. of China. (1939) silicates. Res. assoc. *X-ray crystallography, protein crystallography, structural chemistry.*

Wang, Prof. Yu-ming. Dept. of Materials Science, Jilin U., Changchun 130023, People's Rep. of China. (1932) PhD (Moscow Inst. of Iron and Steel, 1961). Prof. (tel. 822333, ext. 704). *X-ray diffraction, crystal defects, metals and alloys, amorphous metals, structure, defects in thin films.*

Wang, Mr Zhao-zhou. Bureau of Geology and Mineral Resources of Liaoning Province, Exp. and Res. Center, Bei Ling Street, Shenyang, Liaoning, People's Rep. of China. (1937). Engineer (tel. 61768, telex 6012). *Powder diffraction, quantitative X-ray phase analysis, feldspar structure state, crystal lattice measurement.*

Wang, Prof. Zu-tao. Chemistry Dept., Tianjin Advanced Inst. of Sci. and Techn., No. 379 Heping Road, Tianjin, People's Rep. of China. (1924). Assoc. prof. (tel. 3-3467). *Crystal structure, organometallic compounds, drug structure - activity relationship, powder diffraction.*

Wei, Mr Ming-xiu. Dept. of Mineralogy and Petrology, Res. Inst. of Geology for Mineral Resources, San Li Dian, Guilin, Guangxi, People's Rep. of China. (1940) geochemistry. Commiteeman of Mineralogy and Crystallography Society of China. *X-ray diffraction, mineral structure, X-ray fluorescence analysis, mineralogy, petrology, geochemistry.*

Wei, Mr Xin-cheng. Dept. of Biology, Peking U., Beijing, People's Rep. of China. (1937) biophysics. Lect. (28-2471, ext. 3240). *Protein crystallography.*

Weng, Prof. ling-pao. Mineralogy Dept., China U. of Geosciences, Xue Yuan Road 29, Beijing, People's Rep. of China. (1932) earth science. Assoc. prof. *Mineralogy, wall-rock alteration, systematic mineralogy, rare element mineralogy.*

Wu, Mr Bo-mu. X-ray Crystal Structure Analysis Div., Inst. of Biophys., Academia Sinica, Zhong Guan Cun, Beijing 100080, People's Rep. of China. (1937) physical chemistry. Res. asst. *X-ray crystal structure analysis, X-ray data collection and treatment.*

Wu, Mr Ding-ming. Structural Chemistry Div., Fujian Inst. of Res. on the Structure of Matter, Academia Sinica, Fuzhou, Fujian, People's Rep. of China. (1943). Lect. *Transition metal complexes, metal cluster compounds, X-ray crystallography.*

Wu, Prof. Qian-zhang. Lab. of Crystallography, Inst. of Physics, Academia Sinica, P.O. Box 603, Beijing, People's Rep. of China. (1910) BS, physics (National Central U., 1933). Assoc. res. prof. (tel. 281869). *High temperature solution crystal growth, structure, phase transition, characterization and application.*

Wu, Mr Shen. Dept. of Protein Crystallography, Inst. of Biophys., Academia Sinica, Zhong Guan Cun, Beijing 100080, People's Rep. of China. (1940) physical chemistry. Res. assoc. *X-ray crystal structure, protein crystallography.*

Wu, Prof. Shou-yu. Dept. of Chemistry, Sichuan U., Wang Jia Road, Chengdu 64001, Sichuan, People's Rep. of China. (1930) chemistry. Assoc. prof. (tel. 24401-266). *Coordination chemistry, chemical bond, crystal structure, powder diffraction.*

Wu, Prof. Xin-tao. Fuzhou Lab. of Structural Chemistry, Fujian Inst. of Res. on the Structure of Matter, Academia Sinica, Fuzhou, Fujian, People's Rep. of China. (1939). Res. prof. *Metal cluster compounds, transition metal complexes, X-ray crystallography, structure-property relations, EXAFS.*

Wu, Mr Zi-wu. Lab. of Structural Chemistry, Fujian Inst. of Res. on the Structure of Matter, Fuzhou, Fujian, People's Rep. of China. (1938) structural chemistry. Lect. *Metal complexes, structural analysis, natural products, protein crystallography.*

Xia, Prof. Zong-xiang. The Fifth Dept., Shanghai Inst. of Organic Chemistry, 345Lingling Road, Shanghai, People's Rep. of China. (1942) physical chemistry. Assoc. prof. (tel. 313300). *X-ray crystal structure analysis, proteins, organometallic compounds, natural products.*

Xiao, Prof. Xu-gang. Dept. of Geology, Northeast U. of Techn., Wenhua Raod, Shenyang, Liaoning, People's Rep. of China. (1921) crystallography. Prof. (tel. 483081, telex. 80033 NEIT CN). *Crystal structure geometry, group theory, X-ray crystallography, electron microscopy, diffraction physics, defects, mineral and metal crystals, mineralogy.*

Xie, Dr Si-shen. Dept. A-10, Lab. of Phase Transition and Diagram, Inst. of Physics, Academia Sinica, P.O. Box 603, Beijing, People's Rep. of China. (1942) Dr, solid state physics (Inst. of Physics, Academia Sinica, 1983). Res. assoc. (tel. 28-2271). *Phase transformation, phase diagram, crystal structure, physical properties - structure relationship.*

Xie, Prof. Xian-de. Mineralogy Div., Inst. of Geochemistry, Academia Sinica, Guiyang, Guizhou, People's Rep. of China. (1934). Assoc. prof. *X-ray crystallography, lattice distortion, shocked minerals.*

XiMen, Mrs Lu-lu. Dept. of Material, Fudan U., 200 Fudan Road, Shanghai 201903, People's Rep. of China. (1934) geology. Assoc. prof. *Crystal structure, crystal chemistry, minerals, polycrystalline X-ray diffraction, single crystal analysis.*

Xu, Prof. Chang-fu. Inst. of Materia Medica, Chinese Academy of Medical Sciences, 1 Xian Nong Tan Street, Beijing 100050, People's Rep. of China. (1936) mathematics. Assoc. prof. *Natural products, structure, direct methods.*

Xu, Mr Ji-quan. Physical Chemistry Div., Inst. of Soil Sci., Academia Sinica, 71 Eastern Beijing Road, Nanjing, P.O. Box 821, Jiangsu, People's Rep. of China. (1928). Res. assoc. (tel. 33318). *Mixed layer minerals, layer silicates, powder diffraction.*

Xu, Prof. Jing-yang. Material Physics Div., Shanghai Inst. of Metallurgy, Academia Sinica, 865 Chang Ning Road, Shanghai 200050, People's Rep. of China. (1936) metal physics. Assoc. prof. (tel. 511070, ext. 129). *X-ray diffraction, X-ray topography, structure, metal and alloys, phase transition in solids.*

Xu, Mr Pei-cang. Mineralogy and Petrology Dept., Xi-an Inst. of Geology and Mineral Resources, Xi-an Inst. of Geol. and Res. Building, West End of You-yi Road, Xi-an, Shan-xi, People's Rep. of China. (1942) physics. Engineer (tel. 5-1266). *Crystal superstructure, minerals, X-ray analysis, spectrum analysis, solid physics, quantum mineralogy, ligand field theory.*

Xu, Prof. Shun-sheng. Materal Physics Div., Shanghai Inst. of Metallurgy, Academia Sinica, 865 Chang Ning Road, Shanghai 200050, People's Rep. of China. (1920) PhD, metal physics (U. of Notre Dame, USA, 1953). Res. prof. (tel. 511070, ext. 129). *X-ray diffraction, X-ray topography, crystal growth, structure, metals and alloys, electron microscopy and diffraction, phase transition in solids.*

Xu, Prof. Xiao-jie. Lab. for Structure of Matter, Inst. of Physical Chemistry, Peking U., Beijing, People's Rep. of China. (1937). Assoc. prof. (tel. 282471, ext. 3549). *Single crystal structure determination, molecular mechanics, molecular-dynamics.*

Xu, Prof. Zheng-yi. Optics and Electricity Lab., Inst. of Physics, Academia Sinica, Beijing, P.O. Box 603, People's Rep. of China. (1935) solid state theory. Assoc. prof. (tel. 28-1869). *Ionic conductors, ionic transport, physical properties, light scattering, defect - physical property relations, optical spectroscopy.*

Xue, Mrs Ji-yue. Dept. of Geology, Nanjing U., Hankou Road, Nanjing, Jiangsu 350002, People's Rep. of China. (1938) crystallography and mineralogy. Assoc. prof. (tel.34651, ext. 2448). *Crystal structure analysis, X-ray diffraction, minerals.*

Xue, Prof. Jun-zhi. Dept. of Geology, China U. of Geosciences, Yu Jia Shan, Wuhan 430074, Hubei, People's Rep. of China. (1935) mineralogy. Assoc. prof. (tel. 70481). *Ontogeny and phylogeny, genetic mineralogy, physical-chemistry, natural solid solutions.*

Xue, Prof. Zhi-lin. R&D Center in Advanced Inorganic Materials, Shanghai Inst. of Ceramics, Academia Sinica, 865 Chang-ning Road, Shanghai 200050, People's Rep. of China. (1929) physics. Assoc. res. prof., deputy head, R&D Center (tel. 522470, telex. 33309 ASSIC CN). *Crystal growth, physical properties, characterization, automation, microcomputers.*

Yan, Mr Qi-wei. Neutron Scattering Div., Inst. of Physics, Academia Sinica, P.O. Box 603, Beijing, People's Rep. of China. (1941) MS, neutron scattering (Inst. of Physics, Academia Sinica, 1981). (tel. 285047). *Neutron scattering, neutron inelastic scattering, neutron diffraction.*

Yan, Mr You-wei. Group of X-ray Crystallography, Shanghai Inst. of Biochemistry, Academia Sinica, 320 Yue Yang Road, Shanghai 200031, People's Rep. of China. (1946) MS, crystallography (Shanghai Inst. of Biochemistry, 1983). Res. asst. (tel. 374430, telex 3933). *X-ray protein crystallography.*

Yang, Mr Chuan-zheng. Material Physics Div., Shanghai Inst. of Metallurgy, Academia Sinica, 865 Chang-ning Road, Shanghai 200050, People's Rep. of China. (1939) metal physics. Assoc. prof. (tel. 511070, ext. 129). *X-ray diffraction, X-ray topography, crystal growth, structure, metals and alloys, electron microscopy and diffraction, phase transition in solids.*

Yang, Mr Guang-di. Inst. of Th. Chemistry, Jilin U., 79 Jiefang Road, Changchun, Jilin, People's Rep. of China. (1942). Lect. *Crystal structure determination.*

Yang, Ms Guang-ming. Test Center of Rocks and Minerals, China U. of Geosciences, Yu Jia Shan, Wuchang, Wuhan 430074, Hubei, People's Rep. of China. (1940) MS, crystal chemistry of minerals (Wuhan C. of Geology, 1981). Assoc. prof. *Crystal structure, crystallochemistry, minerals, rocks.*

Yang, Mr Hua-guang. Optics and Electricity Lab., Inst. of Physics, Academia Sinica, Beijing, P.O. Box 603, People's Rep. of China. (1936) crystal physics. Res. assoc. (tel. 281869). *Physical properties, phase transitions, light scattering in solids.*

Yang, Mr Hua-hui. Chemistry Dept., Xiamen U., Xiamen, Fujian, People's Rep. of China. (1934) Analytical Chemistry. Assoc. prof. *Powder diffraction, EXAFS, metal cluster compounds.*

Yang, Prof. Qi-bin. Inst. of Metal Res., Shenyang 110015, Liaoning, People's Rep. of China. (1938) metal physics. Assoc. prof. (tel. 483531, ext. 251). *Crystal structures, metals and alloys, electron diffraction and microscopy.*

Yang, Prof. Qing-chuan. Lab. for Structure of Matter, Inst. of Physical Chemistry, Peking U., Beijing, People's Rep. of China. (1940). Assoc. prof. (tel. 282471, ext. 3549). *Structural chemistry, accurate determination, molecular structure.*

Yang, Xian-jue. Rock-forming and Ore-forming Lab., Changchun Geological Inst., Changchun 130061, People's Rep. of China. (1922). Res. *Crystallochemistry, synthetic crystal and geometric crystallography.*

Yang, Mr Zuo-sheng. Marine Geology Dept., Shandong C. of Oceanology, 5 Yushan Road, Qindao, P.O. Box 90, Shandong, People's Rep. of China. (1937). Lect. *Powder diffraction, mineralogy.*

Yao, Mr Jia-xing. Lab. of X-ray Analysis, Inst. of Physics, Academia Sinica, Zhong Guan Cun, Beijing 100080, People's Rep. of China. (1941) radio. Res. assoc. (tel. 28-1489). *Direct methods, small molecules.*

Yao, Mr Xin-kan. Central Lab. Nankai U., 94 Weijin Road, Tianjin, People's Rep. of China. (1939). Assoc. prof. (tel. 334700, ext. 320). *Small molecules, powder diffraction, inorganic chemistry.*

Ye, Prof. Heng-qiang. Lab. of High-Resolution Electron Microscope, Inst. of Metal Res., Academia Sinica, Wenhua Road, Shenyang 110015, Liaoning, People's Rep. of China. (1940) metal physics. Assoc. prof. (tel. 483531, ext. 242). *Crystal structure, metals and alloys, crystal defects, electron diffraction.*

You, Mr Jun-ming. Dept. of Protein Crystallography, Inst. of Biophys., Academia Sinica, Zhong Guan Cun, Beijing 100080, People's Rep. of China. (1928) chemistry. Res. assoc. (tel. 285529). *Protein structure, insulin, crystal growth.*

You, Prof. Xiao-zeng. Inst. of Coordination Chemistry, Nanjing U., Nanjing, Jiangsu, People's Rep. of China. (1934) physical chemistry. Prof., Inst. Dir. (tel. 34651, telex 0909). *Crystal structure, coordination chemistry, structural chemistry.*

Yu, Mr Kai-bei. Chengdu Center of Analysis and Test, Academia Sinica, Chengdu, Sichuan, People's Rep. of China. (1945). Asst. res. *X-ray crystallography, twinning crystals.*

Yu, Prof. Rui-huang. Physics Dept., Jilin U., Chang-chun, Jilin, People's Rep. of China. (1906) PhD, physics (U. of Manchester, UK, 1938). Prof., Academia Sinica member. *Electron theory of solids and molecules, crystal and molecular structure, X-ray crystal structure analysis, syntheses of X-ray data.*

Yu, Prof. Wei-hai. Dept. of Physics, U. of Sci. and Techn. of China, Hefei, Anhui, People's Rep. of China. (1929). Assoc. prof., dept. vice-chairman. *Solid state physics.*

Yu, Prof. Xiu-fen. Inst. of Structural Chemistry, Dept. of Chemistry, Fuzhou U., Fuzhou, Fujian, People's Rep. of China. (1932) physical chemistry. Assoc. prof. *Crystal and molecular structure, transition metal complexes.*

Zhang, Mr Bu-sheng. Applied Physics Div., Geological Inst., Changchun, Jilin, People's Rep. of China. (1940) X-ray crystallography. Inst. (tel. 375-24781). *Crystallography, crystal physics, crystal construction, minerals, powder diffraction, computing.*

Zhang, Mr Ci-he. Inst. of Biophys., Academia Sinica, Zhong Guan Cun, Beijing 100080, People's Rep. of China. (1942) technical physics. *X-ray diffraction instrumentation, intellectual instruments.*

Zhang, Prof. Dao-biau. Crystal Lab., Shanghai Inst. of Ceramics, Academia Sinica, 865 Chang-ning Road, Shanghai 200050, People's Rep. Of China. (1931) chemical engineering. Assoc. res. prof. (tel. 522470). *Crystal growth from melt, crystal properties.*

Zhang, Mrs Gen-di. Dept. of Earth Sciences, Nanjing U., Hankou Road, Nanjing , Jiansu 210008, People's Rep. of China. (1939) mineralogy. Assoc. Scientist (tel. 634651, ext. 2393). *X-ray diffraction, minerals.*

Zhang, Prof. Guan-ying. Non-metallic Minerals Div., Wuhan Inst. of Building Materials, No. 8 Leshi Road, Wuchang, Wuhan, Hubei, People's Rep. of China. (1933) geology. Assoc. prof. *Non-metallic minerals, X-ray powder diffraction, X-ray fiber photography, chrysotiles, crystal symmetry.*

Zhang, Mr Guang-rong. State Bureau of Building Materials, Geological Inst., No. 11 Bei Shun Cheng Street, Xi Zhi Men, Beijing, People' s Rep. of China. (1933) Engineer (tel. 555309). *Crystallography, mineralogy.*

Zhang, Mr Han-hui. Inst. of Structural Chemistry, Dept. of Chemistry, Fuzhou U., Fuzhou, Fujian, People's Rep. of China. (1947) MS, physical chemistry (Fuzhou U., 1981). Res. asst. *Crystal and molecular structure, transition metal complexes.*

Zhang, Mr Han-qing. Inst. of Mineral Deposits, Chinese Academy of Geological Sciences, 26 Baiwan Zhuang Road, Beijing 100037, People's Rep. of China. (1936). Sr. res. (tel. 8311133, ext. 213). *X-ray crystallography, crystallology, mineralogy.*

Zhang, Prof. Jiang-hong. Res. Section, Crystal Structure and Crystal Chemistry of Minerals, China U. of Geosciences, Xue Yuan Road 29, Beijing 100083, People's Rep. of China. (1936) mineralogy, X-ray crystallography. Assoc. prof. (tel. 277461-560). *X-ray crystallography, minerology, crystal structure, crystal chemistry of minerals, new minerals.*

Zhang, Prof. Le-hui. Lab. of Crystallography, Inst. of Physics, Academia Sinica, P.O. Box 603, Beijing, People's Rep. of China. (1919) BS, physics (Tatung U., 1941). Assoc. res. prof. (tel. 28-1869). *Crystal growth, growth mechanism, high temperature oxide crystals.*

Zhang, Mr Li-xin. Dept. of Structure and Defect of Crystals, Inst. of Metal Res., Academia Sinica, 2-6 Wenhua Road, Shenyang, Liaoning, People's Rep. of China. (1935). Head, structure of surface layer res. group. (tel. 483531, ext. 242). *X-ray metallography, texture, surface layer structure.*

Zhang, Mrs Rong-ying. Group of X-ray Analysis of Mineral, Hubei Lab. of Geological Sci., 342 Jie Fang Da Dao, Wuhan, Hubei, People's Rep. of China. (1935) geology. Engineer (tel. 52847). *X-ray powder diffraction, rock and mineral identification.*

Zhang, Mr Rui-lin. Dept. of Physics, Ji-lin U., Je-fang Road, Chang-chun, Ji-lin, People's Rep. of China. (1934). Lect. *Crystal and molecular structure, electron theory of solids and molecules application.*

Zhang, Prof. Shao-hui. Chemistry Dept., Wuhan U., Wuhan, Hubei, People's Rep. of China. (1938). Assoc. prof. *Direct methods in crystallography.*

Zhang, Mr Shi-wei. Lab. for Structure of Matter, Inst. of Physical Chemistry, Peking U., Beijing, People's Rep. of China. (1945) MS, structural chemistry (Peking U., 1981). (tel. 282471, ext. 3549). *Structural chemistry.*

Zhang, Ms Shu-de. Dept. of Molecular Biology, Inst. of Biophys., Academia Sinica, Beijing 100080, People's Rep. of China. (1939). Res. assoc. (tel. 285529). *X-ray crystallography.*

Zhang, Mr Yong-mao. Structural Chemistry Div., Fujian Inst. of Res. on the Structure of Matter, Academia Sinica, Fuzhou, Fujian, People's Rep. of China. (1936) structural chemistry. Lect. *X-ray crystallography, metal complexes, protein crystal structure.*

Zhang, Prof. Yuan-long. Crystal Growth Dept., Shanghai Inst. of Ceramics, Academia Sinica, Chang Ning Road 865, Shanghai 200050, People's Rep. of China. (1915) BS, physics (Fu-Jen U., 1937). Res. prof. *Crystal growth theory and practice, phase transitions, structure-growth forms-physical properties correlation.*

Zhang, Mr Yue-ming. X-ray Diffraction Lab., Inst. of Geochemistry, Academia Sinica, Guan Shui Road, Guiyang, Guizhou, People's Rep. of China. (1937) powder diffraction of minerals. Engineer (tel. 24757). *X-ray powder diffraction, isomorphous substitution, minerals, high pressure minerals, programing, powder diffraction method.*

Zhang, Ms Ze-ying. Lab. for Structure of Matter, Inst. of Physical Chemitry, Peking U., Beijing, People's Rep. of China. (1940). Lect. (tel. 282471). *Structure analysis.*

Zhang, Prof. Zong. Div. of Mathematics and Physics, Academia Sinica, Beijing, People's Rep. of China. (1929) physics. Prof., div. director. (tel. 868361, ext. 448). *Crystallography, magnetism, neutron diffraction.*

Zhao, Prof. Qi-yuan. Marine Geology Dept., Shandong C. of Oceanology, 5 Yushan Road, Qindao, Shandong, P.O. Box 90, People's Rep. of China. (1929) crystallography and mineralogy. Assoc. prof. *Crystal synthesis, mineralogy, geochemistry.*

Zheng, Prof. Qi-tai. Inst. of Meteria Medica, Chinese Academy of Medical Sci., 1 Xian Nong Tan Street, Beijing 100050, People's Rep. of China. (1938) physics (Nan Kai U., Tian-Jing, 1961). Prof., crystallography. (tel. 33-8581-214). *Natural products, macromolecular structure, direct methods.*

Zheng, Prof. Pei-ju. Chemistry Dept., Fudan U., 200 Handan Road, Shanghai 201903, People's Rep. of China. (1933). Assoc. prof. *Crystal structure analysis, inorganic and metal-organic compounds, X-ray crystallography, powder diffraction, drug action.*

Zheng, Prof. Zhe. Dept. of Geology, Peking U., Beijing, People's Rep. of China. (1938). Assoc. prof. (tel. 282471). *Crystal symmetry, microstructures, high resolution electron microscopy, scanning tunnel microscopy.*

Zhong, Mr Na-tian. Inst. of Biophys., Academia Sinica, Zhong Guan Cun, Beijing 100080, People's Rep. of China. (1937) physical chemistry. *X-ray diffraction, instrumentation, computing.*

Zhong, Prof. Wei-zhuo. R&D Center in Advanced Inorganic Materials, Shanghai Inst. of Ceramics, Academia Sinica, 865 Chang-ning Road, Shanghai 200050, People's Rep. of China. (1932) crystallography. Assoc. res. prof. (tel. 522470, telex 33309ASSIC CN). *Crystal growth, morphology, defects.*

Zhou, Prof. Gong-du. Lab. for Structure of Matter, Inst. of Physical Chemistry, Peking U., Beijing, People's Rep. of China. (1931) physical chemistry. Assoc. prof. (tel. 282471, ext. 3549). *Structural chemistry.*

Zhou, Mr Gui-en. Central Lab. of Structural Analysis, U. of Sci. and Techn. of China, Hefei, Anhui, People's Rep. of China. (1937). Lect. (tel. 63300). *X-ray crystallography, powder diffraction, polymer structure.*

Zhou, Mr Kang-jing. Lab. of Structural Chemistry, Fujian Inst. of Res. on the Structure of Matter, Academia Sinica, Fuzhou, Fujian, People's Rep. of China. (1939) MS, structural chemistry (Fujian Inst. of Res. on the Structure of Matter, 1981). Lect. *X-ray crystallography, metal complexes, protein crystal structure.*

Zhou, Prof. Zhong-yuan. Chengdu Center of Analysis and Test, Academia Sinica, Chengdu, Sichuan, People's Rep. of China. (1938). Assoc. prof. *Coordination chemistry, chemical bond, X-ray crystallography.*

Zhou, Mr Zong-hua. Center of Analysis and Test, Sichuan U., Wangjiang Road, Chengdu, Sichuan, People's Rep. of China. (1935). Instructor. (tel. 316-24401). *Powder diffraction, X-ray single crystal structure analysis.*

Zhu, Mr He-bao. Dept. of Earth Sciences, Nanjing U., Hankou Road, Nanjing, Jiangsu 210008, Peolpe's Rep. of China. (1938) mineralogy. Res. Assoc. (tel. 634651, ext. 2448). *Crystal growth, physical properties, minerals.*

Zhu, Mr Nai-jue. Dept. of Crystallography, Inst. of Chemistry, Academia Sinica, Zhong Guan Cun, Beijing, People's Rep. of China. (1942). Res. assoc. *Structure, small molecules, charge density, neutron diffraction.*

Zhu, Mr Zhong-he. Center of Material Analysis, Nanjing U., Hankou Road, Nanjing Jiangsu, People's Rep. of China. (1939) coordination chemistry. Lect. (tel. 34651, ext. 2523). *Crystal structure.*

Zhuang, Mr Hong-hui. Structural Chemistry Div., Fujian Inst. of Res. on the Structure of Matter, Academia Sinica, Fuzhou, Fujian 350002, People's Rep. of China. (1938). Assoc. prof. *Physical chemistry, transition metal cluster compounds, X-ray crystallography.*

Zhuang, Mr Jian. Structural Chemistry Div., Fujian Inst. of Res. on the Structure of Matter, Academia Sinica, Fuzhou, Fujian, People's Rep. of China. (1942). Lect. *Metal cluster compounds, magnetochemistry, powder diffraction (X-ray and neutron), crystal growth.*

COLOMBIA

Sub-Editor: **E. Posada**

Notes

1. International telephone country code - 57.

2. The degrees *ingeniero* (Ing.), *físico, químico* and *geologo* are between BSc and MSc. The degrees *licenciado, magister* and *doctor* are equivalent to BSc, MSc and PhD, respectively.

3. For Bogotá D.E. the telephone numbers are preceded by number 2.

Acosta, Prof. Carlos Eduardo. Dept. de Geociencias, U. Nacional de Colombia, Ciudad Universitaria, Bogotá D.E. Colombia. (1919) Géologue, geology (Inst. Catholique de Paris, France, 1967). Prof. (tel. 442810). *Mineralogy.*

Alfonso U., Miss Ana Elena. Asesoría Industrial de la Gerencia, Banco de la República, Calle 16, 5-41, Bogotá D.E. Colombia. (1955) Químico, chemistry (U. Nacional de Colombia, 1977). Res. officer. (tel. 428977). *X-ray diffraction, metallurgy.*

Barriga Villalba, Prof. Antonio María. Museo de Numismática, Calle 11, 4-93, Bogotá D.E. Colombia. (1894) Dr, philosophy and letters, chemistry (Colegio Mayor de Nuestra Señora del Rosario, 1918). Honorary prof. and dir. (tel. 437200). *Inorganic crystal structures, metal physics, metals structures.*

Bernal de Ramírez, Prof. Inés. Dept. de Química, U. Nacional de Colombia, Ciudad Universitaria, Bogotá D.E. Colombia. (1940) Químico, chemistry (U. Nacional de Colombia, 1964). Assoc. prof. (tel. 699183). *Clay minerals.*

Brieva, Prof. Jorge Alfonso. Dept. de Geociencias, U. Nacional de Colombia, Ciudad Universitaria, Bogotá D.E. Colombia. (1935) MSc, sedimentary mineralogy, petrology (U. Tulsa, USA, 1963). Assoc. prof. (tel. 442810). *Sedimentary petrography, sedimentary petrology, clay minerals, computer programming.*

Calderón, Prof. Gómez Eduardo. Dept. de Química, U. Nacional de Colombia, Ciudad Universitaria, Bogotá D.E. Colombia. (1923) DSc, pharmacodynamics, pharmacology (U. Paris, France, 1956). Prof. (tel 699183). *Structure, organic compounds.*

Cortés, Dr Abdón. Subdirección Agrológica, Inst. Geográfico Agustín Codazzi, Carrera 30, 48-51, Bogotá D.E. Colombia. (1939) PhD, soil taxonomy (Purdue U., USA, 1971). Sub-dir. (tel. 442784). *Soil taxonomy, soil mineralogy, clays.*

Díaz Peraza, Prof. José Milcíades. Dept. de Física, U. Nacional de Colombia, Ciudad Universitaria, Bogotá D.E. Colombia. (1940) MSc, physics (U. South Carolina, USA, 1970). Asst. prof. (tel. 442874). *Paramagnetic resonance, X-ray diffraction, crystal growth.*

Erazo Plaza, Mr Antonio David. Colombia Cities Service Petroleoum Corp., Calle 37, 8-43 p11, Bogotá D.E. Colombia. Geólogo, geology (U. Nacional de Colombia, 1963). Geologist. (tel. 451560). *Crystal optics.*

Galvis, Mr Jaime. Integral S.A., Calle 29, 6-94 p6, Bogotá D.E. Colombia. (1941) Geólogo, geology (U. Nacional de Colombia, 1965). Res. officer. (tel. 324513). *Optical crystallography, mineralogy, petrography.*

Hernández, Prof. Luis C. Dept. de Física, U. Nacional de Colombia, Ciudad Universitaria, Bogotá D.E. Colombia. (1942) MSc solid state physics (U. Puerto Rico, USA, 1973). Assoc. prof. (tel. 442874). *X-ray diffraction, paramagnetic resonance.*

Jiménez Crespo, Prof. Augusto. Dept. de Física, U. Nacional de Colombia, Ciudad Universitaria, Bogotá D.E. Colombia. (1947) Físico, physics (U. Nacional de Colombia, 1971). Asst. prof. (tel. 730050). *X-ray diffraction, paramagnetic resonance.*

Lozano, Prof. José A. Dept. de Geociencias, U. Nacional de Colombia, Ciudad Universitaria, Bogotá D.E. Colombia. (1935) PhD, submarine geology (Columbia. U., USA, 1974). Assoc. prof. (tel. 442810). *Sea floor, rocks, sediments, geology.*

Luna, Dr Carlos Alfonso. Lab. de Suelos, Instituto Geográfico Agustín Codazzi, Carrera 30, 48-51, Bogotá D.E. Colombia. Dr, soil science (U. Utrecht, Netherlands, 1969). Chief. (tel. 447719). *Sand, clay minerals, soils.*

Llinás Rivera, Prof. Rubén Darío. Dept. de Geociencias, U. Nacional de Colombia, Ciudad Universitaria, Bogotá D.E. Colombia. (1940) Dipl.Geol., petrography, mineralogy (U. Stuttgart, BRD, 1971). Assoc. prof. (tel. 442810). *Optical crystallography, petrography.*

Macía Sanabria, Prof. Carlos A. Dept. de Geociencias, U. Nacional de Colombia, Ciudad Universitaria, Bogotá D.E. Colombia. (1947) Dr.rer.nat., geochemistry,

petrology (U. Stuttgart, BRD, 1978). Asst. prof. (tel. 442810). *Optical crystallography, petrology.*

Malagón Castro, Prof. Dimas. Facultad de Agronomía, U. Nacional de Colombia, Ciudad Universitaria, Bogotá D.E. Colombia. (1938) PhD, pedology (U. Nebraska, USA, 1973). Assoc. prof. (tel. 699111, ext 543). *Clay and sand minerals.*

Mejía Cifuentes, Prof. Leonidas. Centro Interamericano de Fotointerpretación, Carrera 30, 47A-57, Bogotá D.E. Colombia. (1933) MSc, soil taxonomy, mineralogy (U. North Carolina, USA, 1975). Prof. (tel 680300, ext. 64). *Clay mineralogy, sedimentary petrography.*

Merino de Matheus, Prof. Lucía Marina. Dept. de Química, U. Nacional de Colombia, Ciudad Universitaria, Bogotá D.E. Colombia. (1934) Dr, crystallography (U. São Paulo, Brasil, 1975). Asst. prof. (tel. 699183). *Inorganic crystal structures.*

Mora de González, Prof. Nery. Dept. de Química, U. Nacional de Colombia, Ciudad Universitaria, Bogotá D.E. Colombia. (1935) Químico, chemistry (U. Nacional de Colombia, 1956). Prof. (tel. 699183). *Clay minerals, X-ray diffraction.*

Ovalle de Bravo, Prof. Yolanda. Dept. de Química, U. Nacional de Colombia, Ciudad Universitaria, Bogotá D.E. Colombia. (1941) Químico, chemistry (U. Nacional de Colombia, 1965). Asst. prof. (tel. 699111, ext. 787). *Clay minerals.*

Posada G., Prof. Enrique. Instituto de Ciencias Naturales-Museo, U. Nacional de Colombia, Ciudad Universitaria, Bogotá D.E. Colombia. (1909) Dipl.Et.Sup., mineralogy, petrography (Inst. Catholique de Paris, France, 1935). Assoc. prof. (tel. 444403). *Optical crystallography, mineralogy.*

Quevedo, Dr Manuel M. Inst. Investigaciones Científicas y Técnicas, Calle 66A, 17-88 Apartado Aéreo 9403, Bogotá D.E. Colombia. (1915) Dr.rer.nat.habil., mathematics, physics, chemistry (U. Leipzig, DDR, 1969). Dir. (tel. 484321). *Inorganic crystal structures, large molecules, crystal physics, crystal optics.*

Rincón Saenz, Mr Luis Felipe. Museo Geológico, Inst. Nacional de Investigaciones Geológico Mineras, Diagonal 53, 34-35 Colombia. (1922) Chief (tel.443330). *Optical crystallography.*

Rodríguez S., Miss Gloria Inés. Sección de Petrografía y Mineralogía, Inst. Nacional de Investigaciones Geológico Mineras, Diagonal 53, 34-53 Bogotá D.E. Colombia. (1946) Geólogo, geology (U. Nacional de Colombia, 1972). Res. off. (tel. 443330, ext 40) *Mineralogy, optical crystallography.*

Rodríguez Lara, Prof. Jaime. Sección de Estado Sólido, Dept. de Física, U. Nacional de Colombia, Ciudad Universitaria, Bogotá D.E. Colombia. (1933) D. 3e cycle, solid state (U. Lille, France, 1966). Chief. (tel. 442874). *Paramagnetic resonance.*

Rubiano Lamouroux, Prof. Manuel. Dept. de Geociencias, U. Nacional de Colombia, Ciudad Universitaria, Bogotá D.E. Colombia. (1943) D.d'U., crystallography, mineralogy (U. Nancy, France, 1971). Assoc. prof. (tel. 442810). *Crystallography, mineralogy.*

Rubio de Cubides, Prof. Julia. Dept. de Química, U. Nacional de Colombia, Ciudad Universitaria, Bogotá D.E. Colombia. Químico, chemistry (U. Nacional de Colombia, 1956). Assoc. prof. (tel. 699183). *Clay minerals.*

Sanchez V., Mr Alfredo. Asesoría Industrial de la Gerencia, Banco de la República, Calle 16, 5-41, Bogotá D.E. Colombia. (1942) MSc, physical chemistry (U. Moscow, USSR, 1966). Chief. (tel. 428977). *X-ray diffraction, metallurgy.*

Varela Mora, Prof Juan de Dios. Dept. de Física, U. Nacional de Colombia, Ciudad Universitaria, Bogotá D.E. Colombia. (1942) Dr, crystallography, mineralogy (Leningrad State U., USSR, 1981). Assoc. prof. (tel. 440819). *Crystallography, X-ray diffraction, layer silicates.*

CUBA

Sub-Editor: C. Rodriquez Castellanos'

Calzadilla Amava, Mr Octavio. Dept. of General Physics, Fac. of Physics, Uni. of Havana, San Lazaro v L. Vedado, Havana, Cuba. (1949) M.Sc, Semiconductor structures (1976) *Crystallography.*

Cruz Gandrilla, Mr Francisco. Structural Analysis Lab., Inst. of Materials and Reagents for Electronics, Uni. of Havana, San Lazaro v L, Vedado, Havana, Cuba. (1951) M.Sc, electron microscopy (1979) (tel. 78956, telex 0512210,0511277) *Texture, X-ray diffraction, powder method, electron microscopy.*

Cruz Inclan, Dr.Sc. Carlos. Solid State Physics Dept., Centre of Appl. Studies to Nuclear Dev.(CEADEN), P.O. Box 6122, Havana, Cuba. (1950) Dr. physical sci. (Dresden, DDR, 1981). *Diffraction powder method, Mössbauer, ceramic materials.*

Curbelo Ramirez, Mr. Ciro. Structural Analysis Lab., Inst. of Materials and Reagents for Electronics, Uni. of Havana, San Lazaro v L, Vedado, Havana, Cuba. (1960) BSc, chemistry. (tel. 78956, telex 0512210.0511277). *X-ray diffraction, powder method.*

Daoo Morales, Dr.Sc. Anoel. X-ray Diffraction Lab., Nat. Centre of Scient. Res. (CNIC), Ave. 25 v 158, P.O. Box 6880, Cuba. (1949) Dr. chemical sci. *Crystal structure determination.*

Duoue Rodriguez, Mr. Julio. X-ray Diffraction Lab., Nat. Centre of Scient. Res. (CNIC), Ave. 25 v 158, P.O. Box 6880, Cuba. (1958) BSc, chemistry. *Crystal structure determination.*

Fajardo, Dr.Sc. Fabio, Inst. of Meteorology, P.O. Bix 17032, Habana 4, Cuba. (1948) Dr. mathematical & physical sci. *Crystal structure determination.*

Fuentes Betancourt, Prof.Dr.Sc. Juan E. Lab. of Purification and Growth, IMRE, U. of Havana, San Lazaro v L, Vedado, Havana, Cuba. (1940) Dr.Sc., Solid state physics. (tel. 78956, telex 0512210.0511277) *Semiconductors, piezo-electic materials, characterization.*

Fuentes Cobas, Dr.Sc. Luis E. Solid State Physics Dept., Centre of Appl. Studies to Nuclear Dev.(CEADEN), P.O. Box 6122, Havana, Cuba. (1945) Dr. physical sci. (Havana, 1982) *Neutron and X-ray diffraction, texture, phase analysis, imperfections.*

Guardiola Romero, Mr Rene. Physics-Mathematics Dept., Superior Inst. for Mining and Metallurgy of Moa, Las Coloradas, Moa, Holouin, Cuba. (1955) Geological Engineer. *Crystal structure determination.*

Herrera Palma, Mrs Victoria. Solid State Physics Dept., Centre of Appl. Studies to Nuclear Dev. (CEADEN), P.O. Box 6122, Havana, Cuba. (1953) M.Sc. solid state physics, semiconductors (USSR,1977) *X-ray diffraction, phase analysis, corrosion, residues.*

Infante de los Reves, Mr. Guillermo. X-ray Diffraction Lab., Centre of Res. for Mining and Metalurgy. Ave. 25 v 158, P.O. Box 6880, Cuba. (1950) BSc, chemistry. *X-ray diffraction, powder method.*

Lopez Guerra, Dr.Sc.Silio. X-ray Diffraction Lab., Dept. of Catalysis, Centre of Chemical Res.(CIQ), Washington 169, eso. Churruca, Cerro, C, Habana, Cuba. (1944) Dr. chemical sci. *X-ray diffraction, powder method.*

Pomés Hernandez, Prof.Dr.Sc. Ramón. Acad. Sci. of Cuba (ACC), Industria v San Jose, Habana, Cuba. (1947) Dr. mathematical & physical sci. (Humboldt U., GDR, 1982) Vice-pres. of ACC. *Crystal structure determination.*

Quinones Mena, Dr.Sc. Jose. Structural Analysis Lab., IMRE, U. of Havana, San Lazaro v L, Vedado, Havana, Cuba. (1944) Dr. physical sci. (tel. 78956, telex 0512210.0511277) *X-ray diffraction, powder method, electron microscope.*

Rodriguez Castellanos, Dr. Carlos. Dean, Fac. of Physics, U. Havana, San Lazaro y L, Ciudad de La Habana, Habana, 10400, Cuba. *Sub-editor.*

Serra Jones, Mr Alberto. Structural Analysis Lab., IMRE, U. of Havana, San Lazaro v L, Vedado, Havana, Cuba. (1943) M.Sc.N.M.R., metals. (tel. 78956, telex 0512210.0511277) *X-ray diffraction, powder method, double-crystal diffractometry, semiconductor devices.*

Tanus Alonso, Mrs. Mercedes. Structural Analysis Lab., IMRE, U. of Havana, San Lazaro v L, Vedado, Havana, Cuba. (1955) BSc,physics. (tel. 78956, telex 0512210.0511277) *X-ray diffraction, powder method, texture, electron microscopy.*

Tintorer Deloado Mr Oscar. Physics Dept. Aero-industrial Superior Inst. of Matanzas (ISAM), Km. 3 1/2 Carretera Varadero, Matanzas, Cuba. (1948) BSc, physics. *Electron microscopy, x-ray diffraction, powder method.*

CZECHOSLOVAKIA

Sub-Editor: M. Čerňanský

Notes

1. International telephone country code - 42.

2. Until 1953 the degree *rerum naturalium doctor* (RNDr., abbreviation Dr) was conferred by the faculties of sciences of the Czechoslovak universities and the degree *doctor of technical sciences* (Dr. techn., abbreviation Dr) by the technical universities. Since 1953 new degrees *candidatus scientiarum* (CSc) and *doctor scientiarum* (DrSc) are conferred by the faculties of sciences of the universities, by the technical universities, and by institutes of the Cz. Acad. Sci. and of the Slovak Acad. Sci. The lowest degree conferred by the faculties of sciences are *graduated physicist, graduated chemist, etc.;* the corresponding lowest degree conferred by technical universities is *inženýr* (Ing). In 1966 the degree RNDr. was re-established. CSc corresponds to PhD and DrSc to DSc in the British system of degrees.

3. At the universities (including the technical universities) persons with an academic training can be appointed in various positions: *profesor, docent* (associated professor; abbreviation doc.), *odb. asistent* (senior lecturer), *asistent* (lecturer).

4. In the list the following abbreviations are used: CSSR - Czechoslovak Socialistic Republic; Cz. Acad. Sci. - Czechoslovak Academy of Sciences; Slovak Acad. Sci. - Slovak Academy of Sciences; em. - (Emeritus) retired.

Baldrian, Dr Josef. Lab. of X-ray Polymer Structure Analysis, Inst. of Macromolecular Chemistry, Cz. Acad. Sci., Heyrovského nám. 2, 162 06 Praha 6, CSSR. (1938) CSc, physical chemistry (Inst. of Macromolecular Chemistry, Cz. Acad. Sci., 1965). Scient. (tel. 2 + 360341, ext. 388, telex 122019). *Polymers, small angle scattering.*

Barta, Ing Čestmír. Dept. of Material Res., Inst. of Physics, Cz. Acad. Sci., Na Slovance 2, 180 40 Praha 8, CSSR. (1926) CSc, crystal chemistry (Inst. of Chemical Techn., 1954). Head, crystal growth group. (tel. 2 + 225589, telex 122018). *Structural chemistry, materials science, cosmical technology.*

Bauer, Doc. Ing Jaroslav. Dept. of Crystallochemistry, Inst. of Chemical Technology, Suchbátarova 5, 166 28 Praha 6, CSSR. (1920) CSc, mineralogy (Inst. of Chemical Techn., 1962). Doc. of mineralogy, (tel. 2 + 3324083, telex 122744). *Minerals, structural chemistry, crystallography.*

Bouška, Doc. Dr Vladimír. Dept. of Mineralogy, Geochemistry and Crystallography, Faculty of Sci., Charles U., Albertov 6, 128 43 Praha 2, CSSR. (1933) DrSc, geochemistry (Charles U., 1979). Head of the Department of Geochemistry (tel. 2 + 297541, ext. 288). *Structural chemistry, crystal optics.*

Brádler, Ing Jaroslav. Dept. of Metal Physics, Inst. of Physics, Cz. Acad. Sci., Na Slovance 2, 180 40 Praha 8, CSSR. (1938) Ing., Materials science (Techn. U.,Plzeň, 1961) Scient. (tel. 2 + 8152608, telex 122018). *X-ray topography, defect structures.*

Broul, Ing Miroslav. Dept. of Chemical Eng., Res. Inst. of Inorganic Chemistry - Chemopetrol, Revoluční 86, 400 60 Ústí nad Labem, CSSR. (1945) CSc, physical chemistry (Inst. of Chemical Techn., 1979). Scient. (tel. 47 + 2182510, telex 184329). *Crystal growth, bulk crystallization from solutions.*

Březina, Ing Bohuslav. Dielectric Lab., Inst. of Physics, Cz. Acad. Sci., Na Slovance 2, 180 40 Praha 8, CSSR. (1928) CSc, chemistry (Inst. of Chemical Techn., 1956). Head of the group (tel. 2 + 8152641, telex 122018). *Crystal growth, defect structures, inorganic crystals, organic compounds, phase transitions, dielectrics.*

Bubáková, Dr Růžena. Dept. of Structures and Bonding, Inst. of Physics, Cz. Acad. Sci., Na Slovance 2, 180 40 Praha 8, CSSR. (1917) RNDr., physics (Charles U., 1953). Scient. (tel. 2 + 354241, ext. 87, telex 122018). *X-ray diffraction, perfect crystals.*

Chalupa, Ing Bohumil. Neutron Physics Dept., Nuclear Physics Inst., Cz. Acad. Sci., 250 68 Rež near Prague, CSSR. (1928) Ing., electrotechnics (Techn. U. of Prague, 1956). Head. (tel. 2 + 896231, ext. 2383, telex 122626). *Neutron diffraction, slow neutrons, nuclear physics solid state methods.*

Cuchý, Ing Zdeněk. Dept. of Crystal Growth, Monokrystaly, Leninova 175, 511 19 Turnov, CSSR. (1934) CSc, chemistry (Inst. of Chemical Techn., 1957). Head. (tel. 436 + 22751, telex 186333). *Crystal growth, structural chemistry, optical and spectroscopic properties.*

Čapková, Dr Pavla. Dept. of Physics of Semiconductors, Faculty of Mathematics and Physics, Charles U., Ke Karlovu 5, 121 16 Praha 2, CSSR. (1945) CSc, physics (Charles U., 1975). Sr. lect. (tel. 2 + 292141, ext. 389, telex 121673). *X-ray diffraction.*

Čech, Prof. Dr František. Dept. of Mineralogy, Geochemistry and Crystallography, Faculty of Sci., Charles U., Albertov 6, 128 43 Praha 2, CSSR. (1929) CSc, mineralogy (Charles U., 1961). Prof. of mineralogy (tel. 2 + 297541). *Minerals, crystallography.*

Čermák, Dr Jan. Dept. of Metal Physics, Inst. of Physics, Cz. Acad. Sci., Na Slovance 2, 180 40 Praha 8, CSSR. (1926) CSc, physics (Inst. of Solid State Physics, Cz. Acad. Sci., 1963). Scient. (tel. 2 + 8152898, telex 122018). *Metals, X-ray diffraction.*

Černohorský, Doc. Dr Martin. Dept. of General Physics, Faculty of Sci., J. E. Purkyně U., Kotlářská 2, 611 37 Brno, CSSR. (1923) CSc, physics (Inst. of Solid State Physics, Cz. Acad. Sci., 1963). Doc. of physics (tel. 5 + 51112). *Metals.*

Černý, Dr Radovan. Dept. of Physics of Semiconductors, Faculty of Mathematics and Physics, Charles U., Ke Karlovu 5, 121 16 Praha 2, CSSR. (1957) RNDr., solid state physics (Charles U., 1981). Res. asst. (tel. 2 + 292141, ext. 454, telex 121673). *X-ray diffraction, diffractometry.*

Čerňanský, Ing Marian. Dept. of Metal Physics, Inst. of Physics, Cz. Acad. Sci., Na Slovance 2, 180 40 Praha 8, CSSR. (1946) CSc, physics (Inst. of Physics, Cz. Acad. Sci., 1981) Scient. (tel. 2 + 8152898, telex 122018). *Metals, X-ray diffraction, surface structure, powder diffraction.*

Červeň, Doc. Dr Ivan. Dept. of Physics, Faculty of Electrotechnical Eng., Slovak Technical U., Mlynská dolina, 812 19 Bratislava, CSSR. (1933) CSc, physics (Slovak Techn. U., 1980). Doc. of physics. *Amorphous structures, symmetry.*

Červinka, Dr Ladislav. Dept. of Structures and Bonding, Inst. of Physics, Cz. Acad. Sci., Na Slovance 2, 180 40 Praha 8, CSSR. (1935) CSc, physics (Inst. of Solid State Physics, Cz. Acad. Sci., 1966). Scient. (tel. 2 + 354241, ext. 91, telex 122018). *Glasses, amorphous structures, defect structures.*

Číčel, Ing Blahoslav. Lab. of Crystal Chemistry and Reactivity of Hydrosilicates, Inst. of Inorganic Chemistry, Centre for Chem. Res., Slovak Acad. Sci., Dúbravská cesta, 842 36 Bratislava, CSSR. (1933) CSc, silicate techn. (Slovak Techn. U., 1963). Scient. (tel. 7 + 375488, ext. 2303). *Structural chemistry, minerals, silicate structures.*

Dlouhá, Ing Maja. Dept. of Solid State Physics Engineering, Faculty of Nuclear Science and Engineering, Czech Technical University, Břehová 7, 115 19 Praha 1, CSSR. (1938) CSc, physics (Czech Techn. U., 1984). Scient. (tel. 2 + 849951, ext. 410). *Neutron diffraction, texture.*

Dunaj-Jurčo, Doc. Ing Michal. Dept. of Inorganic Chemistry, Faculty of Chemical Techn., Slovak Technical U., Jánska 1, 880 37 Bratislava, CSSR. (1936) CSc, chemistry (Slovak Techn. U., 1967). Doc. of inorganic chemistry (tel. 7 + 56021, ext. 287, 434). *Structural chemistry, transition metal compounds.*

Ďurčanská, Dr Edita. Dept. of Inorganic Chemistry, Faculty of Chemical Techn., Slovak Technical U., Jánska 1, 880 37 Bratislava, CSSR. (1937) CSc, chemistry (Slovak Techn. U., 1970). Sr. lect. (tel. 7 + 56021, ext. 290). *Structural chemistry, transition metal compounds.*

Ďurovič, Ing Slavomil. Lab. of Diffraction Methods, Inst. of Inorganic Chemistry, Centre for Chem. Res., Slovak Acad. Sci., Dúbravská cesta, 842 36 Bratislava, CSSR. (1929) CSc, geology and mineralogy (Comenius U., 1962). Scient. (tel. 7 + 375488, ext. 2305). *Disorder, polytypes, silicate structures.*

Fiala, Dr. Jaroslav. Dept. of Metallurgy, Central Res. Inst. Škoda, Tylova 46, 316 00 Plzeň, CSSR. (1940) RNDr., solid state physics (Charles U., 1981). Head of the Lab. of X-Ray Diffraction (tel. 19 + 211,ext. 2302, telex 154247). *X-ray diffraction, phase determination.*

Fingerland, Dr Antonín. P.A.: Čermákova 1, 120 00 Praha 2, CSSR. (1923) CSc, physics (Inst. of Solid State Physics, Cz. Acad. Sci., 1967). Scient. (tel. 2 + 258903). *Scattering theory, symmetry, quantum chemistry.*

Ganev, Ing Nikolaj. Dept. of Solid State Physics Eng., Faculty of Nuclear Sci. and Physical Eng., Czech Technical University, Břehová 7, 115 19 Praha 1, CSSR. (1953) CSc, solid state physics (Czech Techn. U., 1986). Scient. (tel. 2 + 849951, ext. 413). *Metals, X-ray tensometry, solid state physics.*

Garaj, Prof. Dr Ján. Dept. of Analytical Chemistry, Slovak Techn. U., Jánska 1, 880 37 Bratislava, CSSR. (1934) DrSc, inorganic chemistry (Slovak Techn. U., 1976). Prof. of analytical chemistry (tel. 7 + 56043). *Analytical chemistry, structural chemistry, coordination compounds.*

Gosmanová, Dr Galina. Dept. of Solid State Physics Eng., Faculty of Nuclear Sci. and Physical Eng., Czech Technical University, Břehová 7, 115 19 Praha 1, CSSR. (1936) RNDr., physics (Charles U., 1976). Sr. lect. (tel. 2 + 849951, ext 415). *Metals, X-ray tensometry, solid state physics.*

Gruber, Dr Boris. Dept. of Mathematical Physics, Faculty of Mathematics and Physics, Charles U., Malostranské nám. 2/25, 118 00 Praha 1, CSSR. (1921) CSc, physics (Charles U., 1965). Scient. (tel. 2 + 532132, ext. 82). *Symmetry, computing.*

Gyepes, Doc. Dr Eduard. Dept. of Analytical Chemistry, Faculty of Sci., Comenius U., Mlynská dolina, Pavilón Chémie, 842 15 Bratislava, CSSR. (1935) CSc, chemistry (Comenius U., 1972). Doc. of analytical chemistry. (tel. 7 + 320003, ext. 386) *Structure chemistry, coordination compounds.*

Gyepesová, Dr Dalma. Lab. of Crystal Chemistry and Reactivity of Hydrosilicates, Inst. of Inorganic Chemistry, Centre for Chem. Res., Slovak Acad. Sci., Dúbravská cesta, 842 36 Bratislava, CSSR. (1936) CSc, chemistry (Comenius U., 1972). Head. (tel. 7 + 375 488, ext. 2164). *Silicate structures, structural chemistry.*

Handlovič, Ing Milan. Lab. of Diffraction Methods, Inst. of Inorganic Chemistry, Centre for Chem. Res., Slovak Acad. Sci., Dúbravská cesta, 842 36 Bratislava, CSSR. (1932) CSc, chemistry (Inst. of Inorganic Chemistry, Slovak Acad. Sci., 1968). Scient. (tel. 7 + 375488, ext. 2310). *Inorganic crystals.*

Hanic, Doc. Dr František. Lab. for Phase Analyses, Inst. of Inorganic Chemistry, Centre for Chem. Res., Slovak Acad. Sci., Dúbravská cesta, 842 36 Bratislava, CSSR. (1927) DrSc, inorganic chemistry (Inst. of Chemical Techn., 1966). Scient. (tel. 7 + 375 488, ext. 2317). *Structural chemistry, inorganic crystals, structure and properties.*

Hašek, Dr Jindřich. Lab. of X-Ray Polymer Structure Analysis, Inst. of Macromolecular Chemistry, Cz. Acad. Sci., Heyrovského nám. 2, 162 06 Praha 6, CSSR. (1945) CSc, physics (Charles U., 1972). Scient. (tel. 2 + 360341, ext. 387, telex 122019). *Crystal structure determination methods, molecular crystals, quant. chem. calculations, computing.*

Havlík, Ing Tomáš. Dept. of Non-ferous Metallurgy, Metallurgical Faculty, Techn. U. of Košice, Švermova 9, 043 85 Košice, CSSR. (1953) CSc, metallurgy (Techn. U. of Košice, 1983). Scient. (tel. 95 + 399063, ext 499). *Materials science, phase determination, solid state transformations.*

Hlavatá, Mrs Drahomíra. Lab. of X-Ray Polymer Structure Analysis, Inst. of Macromolecular Chemistry, Cz. Acad. Sci., Heyrovského nám. 2, 162 06 Praha 6, CSSR. (1942) CSc, physical chemistry (Inst. of Macromolecular Chemistry, Cz. Acad. Sci., 1972). Scient. (tel. 2 + 360341, ext. 207, telex 122019). *Organic compounds, structural chemistry, computing.*

Holý, Dr Václav. Dept. of Solid State Physics, Faculty of Sci., J. E. Purkyně U., Kotlářská 2, 611 37 Brno, CSSR. (1953) CSc, physics (J. E. Purkyně U., 1982). Sr. lect. (tel. 5 + 51112, ext. 46). *Scattering theory, dynamical theory, X-ray topography.*

Horváth, Ing Josef. Dept. of Metal Physics, Inst. of Physics, Cz. Acad. Sci., Na Slovance 2, 180 40 Praha 8, CSSR. (1950) CSc, physics (Inst. of Physics, Cz. Acad. Sci., 1982) Scient. (tel. 2 + 8152892, telex 122018). *Metals, scattering theory, surface structure.*

Hoschl, Dr Pavel. Inst. of Physics, Charles U., Ke Karlovu 5, 121 16 Praha 2, CSSR. (1938) CSc, physics (Charles U., 1967). Scient. (tel. 2 + 292141, ext. 266, telex telex 121673). *Crystal growth.*

Huml, Dr Karel. Lab. of X-Ray Polymer Structure Analysis, Inst. of Macromolecular Chemistry, Cz. Acad. Sci., Heyrovského nám. 2, 162 06 Praha 6, CSSR. (1934) DrSc, applied physics (Czech Techn. Univ., 1984). Head. (tel. 2 + 360341, ext. 390, telex 122019). *Organic compounds, computing, structural chemistry.*

Hybler, Dr Jiří. Dept. of Structures and Bonding, Inst. of Physics, Cz. Acad. Sci., Na Slovance 2, 180 40 Praha 8, CSSR. (1949) RNDr., crystallography (Charles U., 1973). Res. asst. (tel. 2 + 361337, telex 122018). *Orienting crystals, X-ray diffraction, minerals, inorganic crystals.*

Janovec, Dr Václav. Dept. of Dielectrics, Inst. of Physics, Cz. Acad. Sci., Na Slovance 2, 180 40 Praha 8, CSSR. (1930) CSc, physics (Inst. of Physics, Cz. Acad. Sci., 1958). Scient. (tel. 2 + 8152166, telex 122018). *Phase transitions, symmetry.*

Ječný, Ing Jiří. Lab. of X-Ray Polymer Structure Analysis, Inst. of Macromolecular Chemistry, Cz. Acad. Sci., Heyrovského nám. 2, 162 06 Praha 6, CSSR. (1938) Ing., instrumentation and automation (Charles U., 1962). Scient. (tel. 2 + 360341, ext. 372, telex 122019). *Organic compounds, structural chemistry, instrumentation.*

Jergel, Ing Matej. Dept. of Metal Physics, Inst. of Physics, Slovak Acad. Sci., Dúbravskú cesta 9, 842 28 Bratislava, CSSR. (1954) CSc, physics (Inst. of Physics, Slovak Acad. Sci., 1985). Scient. (tel. 7 + 375000, ext. 2481). *Amorphous structures, metals.*

Kabešová, Ing Mária. Dept. of Inorganic Chemistry, Faculty of Chemical Techn., Slovak Technical U., Jánska 1, 880 37 Bratislava, CSSR. (1932) CSc, chemistry (Slovak Techn. U., 1971). Sr. lect. (tel. 7 + 56021, ext. 290). *Structural chemistry, transition metal compounds.*

Karmazin, Dr Lubomír. Dept. of Phase Transformations, Inst. of Physical Metallurgy, Cz. Acad. Sci., Žižkova 22, 616 62 Brno, CSSR. (1938) CSc, physics

Kellö, Dr Eleonóra. Dept. of Analytical Chemistry, Faculty of Chemical Techn., Slovak Technical U., Jánska 1, 812 37 Bratislava, CSSR. (1949) CSc, chemistry (Slovak Techn. U., 1982). Res. asst. (tel. 7 + 56021, ext. 375). *Analytical chemistry, structural chemistry, coordination compounds.*

Kettmann, Ing Viktor. Dept. of Analytical Chemistry, Faculty of Pharmacy, Comenius U., Odbojárov 10, 832 32 Bratislava, CSSR. (1947) CSc, chemistry (Slovak Techn. U., 1979). Scient. (tel. 7 + 60451, ext. 223). *Structural chemistry, coordination compounds, conformation, drugs, pharmacological activity.*

Koman, Ing Marián. Dept. of Inorganic Chemistry, Faculty of Chemical Techn., Slovak Technical U., Jánska 1, 812 37 Bratislava, CSSR. (1953) CSc, chemistry (Slovak Techn. U., 1981). Res. asst. (tel. 7 + 56021, ext. 290). *Structural chemistry, coordination compounds, inorganic crystals.*

Komrska, Dr Jiří. Dept. of Electron Optics, Inst. of Instrument Techn., Cz. Acad. Sci., Královopolská 147, 612 64 Brno, CSSR. (1936) CSc, physics (J. E. Purkyně U., 1965). Scient. (tel. 5 + 54311, ext. 367, telex 62240). *Electron diffraction, optical diffraction techniques.*

Koreň, Mr Branislav. Dept. of Physics, Faculty of Chemical Techn., Slovak Technical U., Jánska 1, 812 37 Bratislava, CSSR. (1954) graduated physicist, physics (Comenius U., 1977). Sr. lect. (tel. 7 + 56021, ext. 343). *Inorganic crystals, structural chemistry, crystallographic statistics, coordination compounds.*

Kováčová, Dr Katarína. Dept. of Thermic Properties of Metals and Alloys, Inst. of Metal Materials, Slovak Acad. Sci., Februárového víťazstva 75, 801 00 Bratislava, CSSR. (1936) CSc, physical metallurgy (Res. Inst. of Welding, 1971). Scient. (tel. 7 + 673551, ext. 340). *Crystallization, crystal growth.*

Kožíšek, Ing Jozef. Dept. of Inorganic Chemistry, Faculty of Chemical Techn., Slovak Technical U., Jánska 1, 812 37 Bratislava, CSSR. (1952) CSc, chemistry (Slovak Techn. U., 1982). Res. asst. (tel. 7 + 56021, ext. 290). *Structural chemistry, coordination compounds, inorganic crystals.*

Kožíšková, Ing Zlatica. Dept. of Inorganic Chemistry, Faculty of Chemical Techn., Slovak Technical U., Jánska 1, 812 37 Bratislava, CSSR. (1953) Ing, chemistry (Slovak Techn. U., 1977). Res. asst. (tel. 7 + 56021, ext. 290). *Structural chemistry, coordination compounds, inorganic crystals.*

Králík, Dr František. Div. of Thermophysical Properties of Metals, Inst. of Metal Materials, Slovak Acad. Sci., Februárového víťazstva 75, 801 00 Bratislava, CSSR. (1919) RNDr., physics (Masaryk U., 1949). Head of Metal Res. Dept. (tel. 7 + 60088). *Metals.*

Králová, Dr Rudolfa. Dept. of Solid State Physics Eng., Faculty of Nuclear Sci. and Physical Eng., Czech Technical University, Břehová 7, 115 19 Praha 1, CSSR. (1941) CSc, physics (Czech Techn. U., 1984). Sr. lect. (tel. 2 + 849951, ext. 414). *Metals, X-ray tensometry, solid state physics.*

Kratochvil, Doc.Dr Bohumil. Dept. of Crystallochemistry, Inst. of Chemical Technology, Suchbátarova 5, 166 28 Praha 6, CSSR. (1949) CSc, inorganic chemistry (Charles U., 1977). Doc. of inorganic chemistry. (tel. 2 + 3113908, telex 122744). *Structural chemistry, X-ray diffraction.*

Kraus, Doc. Dr Ivo. Dept. of Solid State Physics Eng., Faculty of Nuclear Sci. and Physical Eng., Czech Technical University, Břehová 7, 115 19 Praha 1, CSSR. (1936) CSc, physics (Czech Techn. U., 1967). Doc. of experimental physics. (tel. 2 + 849951, ext 382). *Solid state physics, X-ray tensometry, metals.*

Křivý, Dr Ivan. Mathematics Department, College of Education, Dvořákova 7, 701 03 Ostrava 1, CSSR. (1940) CSc, physics (Charles U., 1973). Sr. lect. (tel. 69 + 233121, ext. 59). *Computing, statistical methods, thermal vibrations.*

Kuběna, Dr Josef. Dept. of Solid State Physics, Faculty of Sci., J. E. Purkyně U., Kotlářská 2, 611 37 Brno, CSSR. (1935) CSc, physics (J. E. Purkyně U., 1969). Scient. (tel. 5 + 51112, ext. 46). *Defect structures, X-ray topography.*

Kulda, Ing Jiří. Dept. of Neutron Physics, Nuclear Physics Institute, Cz. Acad. Sci., 250 68 Rež near Prague, CSSR. (1953) CSc, physics (Charles U., 1984). Scient. (tel. 2 + 896231, ext. 3125, telex 122626). *Small-angle scattering, dynamical theory, neutron diffraction.*

Kupka, Dr František. Dept. of Physical Chemistry, Inst. of Mineral Raw Materials, Vítězná 425, 284 03 Kutná Hora, CSSR. (1926) CSc, chemical technology of metals (Inst. of Chemical Techn., 1964). Head. (tel. 327 + 2577, ext. 76). *Structural chemistry, minerals, silicates.*

Kužel, Dr Radomír. Dept. of Physics of Semiconductors, Faculty of Mathematics and Physics, Charles U., Ke Karlovu 5, 121 16 Praha 2, CSSR. (1955) RNDr., solid state physics (Charles U., 1980). Res. asst. (tel. 2 + 292141, ext. 394, telex 121673). *X-ray diffraction, diffractometry.*

Kvapil, Ing Jiří. Physical Lab., Monokrystaly, Leninova 175, 511 19 Turnov, CSSR. (1937) CSc, inorganic technology (Inst. of Chemical Techn., 1967). Scient. (tel. 436 + 22751, telex 186333). *Crystal growth, laser crystals, colour centers.*

Lokaj, Ing Ján. Dept. of Microanalysis, Faculty of Chemical Techn., Slovak Technical U., Jánska 1, 812 37 Bratislava, CSSR. (1951) CSc, chemistry (Slovak Techn. U., 1982). Scient. (tel. 7 + 56021, ext. 206). *X-ray diffraction, structural chemistry, coordination compounds, Microanalysis.*

Loub, Dr Josef. Dept. of Inorganic Chemistry, Faculty of Sci., Charles U., Hlavova 8/2030, 128 40 Praha 2, CSSR. (1929) CSc, inorganic chemistry (Charles U., 1967). Scient. (tel. 2 + 292051, ext. 500). *Inorganic crystals, structural chemistry.*

Machajdík, Ing Daniel. Dept. of Superconductivity, Electrotechnical Institute, Slovak Acad. Sci., Dúbravská cesta, 809 32 Bratislava, CSSR. (1944) CSc, technical sciences (Inst. of Inorganic Chemistry, Slovak Acad. Sci., 1978). Scient. (tel. 7 + 45741, ext. 2311). *Superconducting materials, structural chemistry, phase determination.*

Maďar, Doc. Dr Ján. Dept. of Solid State Physics, Faculty of Mathematics and Physics, Comenius U., Mlynská dolina, Pavilón F2, 842 15 Bratislava, CSSR. (1926) CSc, physics (Comenius U., 1968). Doc. of exp. physics. (tel. 7 + 320003). *Structural chemistry, instrumentation.*

Maixner, Dr Jaroslav. X-ray laboratory, Inst. of Chemical Technology, Suchbátarova 5, 166 28 Praha 6, CSSR. (1961) RNDr., solid state physics (Charles U., 1986). Res. asst. (tel. 2 + 3324200, telex 122744). *Powder diffraction, diffractometry, phase determination, single crystals.*

Majling, Ing Ján. Dept. of Chemical Techn. of Silicates, Faculty of Chemical Techn., Slovak Techn. U., Jánska 1, 880 37 Bratislava, CSSR. (1942) CSc, silicate technology (Inst. of Inorganic Chemistry, Slovak Acad. Sci., 1971). Scient. (tel. 7 + 45341, ext. 2575). *Phase determination, silicates, phosphate systems.*

Malý, Mr Karel. Dept. of Structures and Bonding, Inst. of Physics, Cz. Acad. Sci., Na Slovance 2, 180 40 Praha 8, CSSR. (1952) CSc, physics (Inst. of Physics, Cz. Acad. Sci., 1986) Scient. (tel. 2 + 354241, ext. 96, telex 122018). *Structural chemistry, X-ray diffraction.*

Mánek, Ing Břetislav. Monokrystaly, Leninova 175, 511 19 Turnov, CSSR. (1933) Ing., inorganic chemistry (Inst. of Chemical Techn., 1956). Technical Director. (tel. 436 + 22751, telex 186333). *Crystal growth, oxide monocrystals, laser crystals, colour centers.*

Matherny, Prof. Dr-Ing Mikuláš. Dept. of Chemistry, Metallurgical Faculty, Technical U. in Košice, Švermova 9, 043 85 Košice, CSSR. (1930) DrSc, chemistry (Comenius U., 1980). Prof. of analytical chemistry (tel. 95 + 30298). *Analytical chemistry, experimental metallurgy, spectroscopy and spectrochemistry, X-ray diffraction.*

Melka, Dr Karel. X-Ray Lab., Geological Survey, Hradební 9, 110 00 Praha 1, CSSR. (1930) CSc, mineralogy (Charles U., 1964). Head of the X-Ray Lab. (tel. 2 + 65446). *Silicates, minerals.*

Michalec, Ing Rudolf. Dept. of Neutron Physics, Nuclear Physics Institute, Cz. Acad. Sci., 250 68 Rež near Prague, CSSR. (1937) DrSc, neutron physics (Charles U., 1981). Scient. (tel. 2 + 896231, ext. 3553, telex 122626). *Neutron diffraction, neutron spectrometry, small-angle scattering.*

Mikloš, Ing Dušan. Lab. of Diffraction Methods, Inst. of Inorganic Chemistry, Centre for Chem. Res., Slovak Acad. Sci., Dúbravská cesta, 842 36 Bratislava, CSSR. (1944) CSc, inorganic chemistry (Slovak Techn. U., 1976). Head. (tel. 7 + 375 488, ext. 2175). *Disorder, structural chemistry.*

Mikula, Dr Pavol. Dept. of Neutron Physics, Nuclear Physics Institute, Cz. Acad. Sci., 250 68 Rež near Prague, CSSR. (1947) CSc, physics, (Charles U., 1976). Scient. (tel. 2 + 896231, ext. 3125, telex 122626). *Neutron diffraction, small-angle scattering, instrumentation, Development of diffraction techniques.*

Moravec, Ing František. Dept. of Solid State Electronics, Inst. of Radio Engineering and Electronics, Cz. Acad. Sci., Lumumbova 1, 182 51 Praha 8, CSSR. (1931) CSc, crystallography (Inst. of Chemical Techn., 1967). Head of the Lab. (tel. 2 + 843741, ext. 317, telex 122646). *Crystal growth, single crystals.*

Moravcová, Dr Hana. X-Ray Laboratory, Geological Survey, Hradební 9, 110 15 Praha 1, CSSR. (1950) RNDr., physics (Charles U., 1975). Res. asst. (tel. 2 + 65446). *Phase determination, minerals, structural chemistry, silicates.*

Mrafko, Dr Peter. Dept. of Metal Physics, Inst. of Physics, Slovak Acad. Sci., Dúbravská cesta 9, 842 28 Bratislava, CSSR. (1940) CSc, physics (Inst. of Physics, Slovak Acad. Sci., 1975). Scient. (tel. 7 + 375000, ext. 2243). *Amorphous structures, metals, quasicrystals.*

Nehasil, Dr Miroslav. P.A.: Novorosijská 14, 100 00 Praha 10, CSSR. (1921) CSc, physics (Czech Techn. U., 1981). (tel. 2 + 7376203). *X-ray diffraction, X-ray tensometry.*

Novák, Ing Ctirad. Dept. of Structures and Bonding, Inst. of Physics, Cz. Acad. Sci., Na Slovance 2, 180 40 Praha 8, CSSR. (1921) CSc, mathematics (Inst. of Mathematics, Cz. Acad. Sci., 1961). Scient. (tel. 2 + 354241, ext. 96, telex 122018). *Structural chemistry, crystallography, computing, crystallographic data.*

Novotný, Ing Jiří. Dept. of Crystallochemistry, Inst. of Chemical Technology, Suchbátarova 5, 166 28 Praha 6, CSSR. (1962) Ing., solid state physics (Czech Techn. U., 1987). Res. asst. (tel. 2 + 3324200, telex 122744). *Structural chemistry, single crystals.*

Nývlt, Ing Jaroslav. Laboratory of Crystallization and Calorimetry, Inst. of Inorganic Chemistry, Cz. Acad. Sci., Majakovského 24, 160 00 Praha 6, CSSR. (1932) DrSc, physical chemistry (Inst. of Chemical Techn., 1967). Head. (tel. 2 + 256897). *Crystal growth, bulk crystallization from solutions.*

Ondráček, Ing Jan. Dept. of Crystallochemistry, Inst. of Chemical Technology, Suchbátarova 5, 166 28 Praha 6, CSSR. (1955) CSc, inorganic chemistry (Inst. of Chemical Techn., 1988). Scient. (tel. 2 + 3324285, telex 122744). *Structural chemistry, computing.*

Pavelčík, Ing František. Dept. of Analytical Chemistry, Faculty of Pharmacy, Comenius U., Odbojárov 10, 832 32 Bratislava, CSSR. (1945) CSc, inorganic chemistry (Slovak Techn. U., 1977). Scient. (tel. 7 + 60451, ext. 297). *Computing, coordination compounds, structural biology, molecular mechanics.*

Petříček, Dr Václav. Dept. of Structures and Bonding, Inst. of Physics, Cz. Acad. Sci., Na Slovance 2, 180 40 Praha 8, CSSR. (1948) CSc, physics (Inst. of Phys-

ics, Cz. Acad. Sci., 1981) Scient. (tel. 2 + 354241, ext. 96, telex 122018). *Structural chemistry, X-ray diffraction, computing.*

Pleštil, Ing Josef. Lab. of X-Ray Polymer Structure Analysis, Inst. of Macromolecular Chemistry, Cz. Acad. Sci., Heyrovského nám. 2, 162 06 Praha 6, CSSR. (1946) CSc, physics (Charles U., 1974). Scient. (tel. 2 + 360341, ext. 388, telex 122019). *Polymers, small angle scattering.*

Podbrdský, Dr Josef. Dept. of Electron Optics, Inst. of Scientific Instr., Cz. Acad. Sci., Královopolská 147, 612 64 Brno, CSSR. (1945) CSc, physics (J. E. Purkyně U., 1972). Scient. (tel. 5 + 54311, ext. 365, telex 62240). *Electron microscopy, electron diffraction.*

Podlahová, Dr Jana. Dept. of Inorganic Chemistry, Faculty of Sci., Charles U., Albertov 2030, 128 40 Praha 2, CSSR. (1937) CSc, inorganic chemistry (Charles U., 1964). Sr. lect. (tel. 2 + 292051, ext. 420). *Inorganic crystals, structural chemistry.*

Polcarová, Dr Milena. Dept. of Metal Physics, Inst. of Physics, Cz. Acad. Sci., Na Slovance 2, 180 40 Praha 8, CSSR. (1931) CSc, physics (Inst. of Physics, Cz. Acad. Sci., 1961). Scient. (tel. 2 + 8152608, telex 122018). *X-ray topography, defect structures.*

Pollert, Ing Emil. Dept. of Magnetism, Inst. of Physics, Cz. Acad. Sci., Na Slovance 2, 180 40 Praha 8, CSSR. (1938) CSc, chemistry (Inst. of Chemical Techn., 1972). Scient. (tel. 2 + 354241, ext. 67, telex 122018). *Material science, transition element oxides, phase determination.*

Rieder, Dr Milan. Inst. of Geological Sciences, Charles U., Albertov 6, 128 43 Praha 2, CSSR. (1940) PhD, geology (Johns Hopkins U., Baltimore, Md., USA, 1968). Scient. (tel. 2 + 297541, ext. 295). *Experimental petrology, minerals, crystallography.*

Rychlý, Ing Rudolf. State Glass Res. Inst., Škroupova 957, 501 92 Hradec Králové, CSSR. (1941) CSc, physical chemistry (Inst. of Chemical Techn., 1969). Scient. (tel. 49 + 32821, ext. 29, telex 194683). *Crystallization, glass, minerals.*

Schilder, Ing Jaroslav. Dept. of Electron Microscopy, Electrotechnical Inst., Slovak Acad. Sci., Dúbravská cesta, 809 32 Bratislava, CSSR. (1928) CSc, theoretical electrotechnics (Slovak Techn. U., 1960). Head of the laboratory (tel. 7 + 45139, 45651, ext. 2937). *Electron microscopy, thin films.*

Seidl, Ing Vlastimil. X-ray laboratory, Inst. of Chemical Technology, Suchbátarova 5, 166 28 Praha 6, CSSR. (1927) CSc, mineralogy (Inst. of Geochemistry and Mineral Resources, Cz. Acad. Sci., 1963). Head. (tel. 2 + 3324087, telex 122744). *Powder diffraction, zeolites.*

Simerská, Dr Marie. Dept. of Metal Physics, Inst. of Physics, Cz. Acad. Sci., Na Slovance 2, 180 40 Praha 8, CSSR. (1929) CSc, physics (Inst. of Solid State Physics, Cz. Acad. Sci., 1966). Scient. (tel. 2 + 8152778, telex 122018). *Metals, X-ray diffraction.*

Sivý, Dr Peter. Dept. of Physics, Faculty of Chemical Techn., Slovak Technical U., Jánska 1, 812 37 Bratislava, CSSR. (1951) RNDr., physics (Comenius U., 1981). Sr. lect. (tel. 7 + 56021, ext. 343). *Inorganic crystals, structural chemistry, crystallographic statistics, coordination compounds.*

Soldánová, Ing Jiřina. Dept. of Inorganic Chemistry, Faculty of Chemical Techn., Slovak Technical U., Jánska 1, 812 37 Bratislava, CSSR. (1946) CSc, chemistry (Slovak Techn. U., 1984). Scient. (tel. 7 + 56021, ext. 434). *Structural chemistry, coordination compounds.*

Steinhart, Dr Miloš. Lab. of X-ray Polymer Structure Analysis, Inst. of Macromolecular Chemistry, Cz. Acad. Sci., Heyrovského nám. 2, 162 06 Praha 6, CSSR. (1955) RNDr., physics (Charles U., 1979). Scient. (tel. 2 + 360341, ext. 388, telex 122019). *Small angle scattering, macromolecules.*

Symerský, Ing Jindřich. Dept. of Peptides, Institute of Organic Chemistry and Biochemistry, Cz. Acad. Sci., Flemingovo nám. 2, 166 10 Praha 6, CSSR. (1957) CSc, organic chemistry (Institute of Organic Chemistry and Biochemistry, Cz. Acad. Sci., 1988). Scient. (tel. 2 + 3312348). *X-ray diffraction, structural biology.*

Syneček, Doc. Dr Vladimír. Dept. of Metal Physics, Inst. of Physics, Cz. Acad. Sci., Na Slovance 2, 180 40 Praha 8, CSSR. (1929) CSc, physics (Inst. of Solid State Physics, Cz. Acad. Sci., 1956). Scient. (tel. 2 + 8152778, telex 122018). *Metals, X-ray diffraction.*

Šebo, Dr Pavel. Dept. of Structure Properties of Metals and Alloys, Inst. of Metal Materials, Slovak Acad. Sci., Februárového víťazstva 75, 801 00 Bratislava, CSSR. (1939) CSc, physics (Inst. of Solid State Physics, Cz. Acad. Sci., 1969). Scient. (tel. 7 + 673351, ext. 334). *Metals.*

Šedivý, Doc. Dr Josef. Dept. of Physics of Semiconductors, Faculty of Mathematics and Physics, Charles U., Ke Karlovu 5, 121 16 Praha 2, CSSR. (1919) RNDr., physics (Charles U., 1952). Doc. of experimental physics (tel. 2 + 292141, ext. 328, telex 121673). *X-ray diffraction, thermal vibrations.*

Šichová, Dr Hana. Dept. of Physics of Semiconductors, Faculty of Mathematics and Physics, Charles U., Ke Karlovu 5, 121 16 Praha 2, CSSR. (1931) CSc, physics (Charles U., 1965). Sr. lect. (tel. 2 + 292141, ext. 394, telex 121673). *X-ray diffraction, thermal vibrations.*

Šourek, Dr Zbyněk. Dept. of Structures and Bonding, Inst. of Physics, Cz. Acad. Sci., Na Slovance 2, 180 40 Praha 8, CSSR. (1948) CSc, physics (Inst. of Physics, Cz. Acad. Sci., 1979). Scient. (tel. 2 + 354241, ext. 92, telex 122018). *Perfect and real crystals, X-ray diffraction, x-ray topography.*

Šubrtová, Ing Věra. Dept. of Structures and Bonding, Inst. of Physics, Cz. Acad. Sci., Na Slovance 2, 180 40 Praha 8, CSSR. (1936) Ing., geology (Inst. of Mining and Metallurgy, 1959). Res. asst. (tel. 2 + 354241, ext. 96, telex 122018). *X-ray diffraction, structural chemistry.*

Ulický, Doc. Ing Ladislav. Dept. of Physical Chemistry, Slovak Techn. U., Jánska 1, 880 37 Bratislava, CSSR. (1931) CSc, physical chemistry (Comenius U., 1961). Doc. of physical chemistry (tel. 7 + 56021, ext. 427). *Small angle scattering, X-ray diffraction.*

Valach, Ing Fedor. Dept. of Physics, Faculty of Chemical Techn., Slovak Technical U., Jánska 1, 812 37 Bratislava, CSSR. (1946) CSc, chemistry (Slovak Techn. U., 1972). Sr. lect. (tel. 7 + 56021, ext. 343). *Inorganic crystals,structural chemistry, crystallographic statistics, coordination compounds.*

Valvoda, Dr Václav. Dept. of Physics of Semiconductors, Faculty of Mathematics and Physics, Charles U., Ke Karlovu 5, 121 16 Praha 2, CSSR. (1937) CSc, physics (Charles U., 1968). Sr. lect. (tel. 2 + 292141, ext. 395, telex 121673). *X-ray diffraction, thermal vibrations, texture, teaching.*

Vrábel, Ing Viktor. Dept. of Analytical Chemistry, Faculty of Chemical Techn., Slovak Technical U., Jánska 1, 812 37 Bratislava, CSSR. (1947) CSc, chemistry (Slovak Techn. U., 1980). Res. asst. (tel. 7 + 56021, ext. 375). *Structural chemistry, coordination compounds, analytical chemistry.*

Vratislav, Ing Stanislav. Dept. of Slid State Physics Engineering, Faculty of Nuclear Science and Physical Engineering, Czech Technical niversity, Břehová 7, 115 19 Praha 1, CSSR. (1939) Csc, physics (Czech Techn. U., 1974). Scient. (tel. 2 + 849951, ext. 411). *Neutron diffraction, instrumentation, texture.*

Weiss, Dr Zdeněk. X-Ray Lab., Dept. of Physics and Chemistry, Scientific Coal Res. Inst., Pikartská, 716 07 Ostrava-Radvanice, CSSR. (1942) RNDr., mineralogy and crystallography (Charles U., 1972). Head of X-Ray Lab. (tel. 69 + 215444, ext. 432, telex 52166). *Minerals, X-ray diffraction, computing, crystallography.*

Zikmund, Dr Zdeněk. Dept. of Dielectrics, Inst. of Physics, Cz. Acad. Sci., Na Slovance 2, 180 40 Praha 8, CSSR. (1939) CSc, physics, (Inst. of Solid State Physics, Cz. Acad. Sci., 1977). Scient. (tel. 2 + 8153010, telex 122018). *Symmetry, domain structures.*

Žák, Doc. Dr Lubor. Dept. of Mineralogy, Geochemistry and Crystallography, Faculty of Sci., Charles U., Albertov 6, 128 43 Praha 2, CSSR. (1925) CSc, mineralogy (Charles U., 1956). Doc. of mineralogy (tel. 2 + 297541, ext. 298). *Minerals.*

Žák, Dr Zdirad. Dept. of Inorganic Chemistry, Faculty of Sci., J. E. Purkyně U., Kotlářská 2, 611 37 Brno, CSSR. (1941) CSc, inorganic chemistry (J. E. Purkyně U.,1982). Sr. lect. (tel. 5 + 51112, ext. 69). *Inorganic crystals, structural chemistry, inorganic synthesis.*

DENMARK

Sub-Editor: N. Thorup

Notes

1. International telephone country code - 45 (to precede all telephone and fax numbers).

2. Degrees conferred by Danish Universities are *doctor philosophiae, doctor techniches,* etc. (Dr.phil., Dr.techn., etc.) (equivalent to DSc); *magister scientiarum, licentiatus pharmaciae, licentiatus technices,* etc. (Mag.scient., Lic.pharm., Lic.techn., etc.) (equivalent to PhD); a *candidatus magistrii, -pharmaciae, -polytechnices, -scientiarum* (Cand.mag., -pharm., -polyt., -scient.) (equivalent to MSc).

Andersen, Mr Erik Krogh. Dept. of Chemistry, U. of Odense, Campusvej 55, DK-5230 Odense M, Denmark. (1923) Dr.phil., chemistry (Odense U., 1971). Assoc. prof. (tel. 66158600, ext. 2538, fax: 66158780). *Chemistry.*

Andersen, Mrs Inger Grete Krogh. Dept. of Chemistry, U. of Odense, Campusvej 55, DK-5230 Odense M, Denmark. (1929) Cand.pharm. (Royal Danish Sch. of Pharmacy, 1954). Res. asst. (tel. 66158600, fax: 66158780). *Phase identifications.*

Andersen, Dr Niels Hessel. Physics and Metallurgy Dept., Risø National Lab., DK-4000 Roskilde, Denmark. (1945) Lic.scient., physics (Copenhagen U., 1976). Scient. (tel. 42371212, ext. 4700, 5721, telex 43116, fax: 42370115) *Ionic conductors, solid electrolytes, ceramic superconductors.*

Andersen, Dr Peter. Dept. of Inorg. Chemistry, U. of Copenhagen, Universitetsparken 5, DK-2100 Copenhagen, Denmark. (1938) Cand.polyt., chemistry (Techn. U. of Denmark, 1961). Assoc. prof. (tel. 31353133). *Coordination compounds.*

Andersen, Dr Stig Kjær. Inst. of Electronic Systems, Aalborg U., Strandvejen 19, DK-9000 Aalborg, Denmark. (1947) Lic.scient., experimental physics (Aarhus U., 1977). Sr. res. scient. (tel. 98138788). *Instrumentation, expert systems.*

Bang, Dr Eva Henriette. Dept. of Inorg. Chemistry, U. of Copenhagen, Universitetsparken 5, DK-2100 Copenhagen, Denmark. (1917) Cand.mag., chemistry (Copenhagen U., 1942). Assoc. prof. (tel. 31353133). *Coordination compounds.*

Bohr, Dr Jakob. Physics Dept., Risø National Lab., DK-4000 Roskilde, Denmark. (1957) Lic.techn., physics (Techn. U. of Denmark, 1984). Scient. (tel. 42371212, fax: 42370115) *Surface structure, synchrotron radiation, magnetic X-ray scattering.*

Brehm, Dr Lotte. Dept. of Chemistry BC, Royal Danish Sch. of Pharmacy, Universitetsparken 2, DK-2100 Copenhagen, Denmark. (1940) Lic.pharm., organic chemistry (Royal Danish Sch. of Pharmacy, 1969). Assoc. prof. (tel. 31370850, ext. 245, fax: 31375744). *Organic compounds, biological molecules.*

Broesby-Olsen, Mr Finn. Corporate Technology and Research, Danfoss A/S, L7 S38, DK-6430 Nordborg, Denmark. (1945) BSc, engineering. Scient. (tel. 74882562) *Powder diffraction, phase analysis, texture, strain measurements.*

Buchwald, Dr Vagn Fabritius. Inst. of Metallurgy, Techn. U. of Denmark, DTH 204, DK-2800 Lyngby, Denmark. (1929) Dr.scient., metallurgy (Copenhagen U., 1977). Assoc. prof. (tel. 42884022, ext. 3237). *Metals, solid state transformations, meteoritic minerals, history (metallurgy), archaeological metals.*

Buras, Prof. Bronislaw. Physics Dept., Risø National Lab., DK-4000 Roskilde, Denmark. (1915) MSc, physics (Warsaw U., 1949). Consultant. (tel. 42371212, ext. 4713, telex 43116, fax: 42370115) *Solid state physics, neutron and synchrotron radiation diffraction.*

Christensen, Dr Axel Nørlund. Dept. of Chemistry, U. of Aarhus, Langelandsgade 140, DK-8000 Aarhus, Denmark. (1934) Dr.phil., inorganic chemistry (Aarhus U., 1967). Assoc. prof. (tel. 86124633, fax 86139919, email anc at kemi.aau.dk). *Crystal growth, inorganic crystals.*

Clausen, Dr Kurt N. Physics Dept., Risø National Lab., DK-4000 Roskilde, Denmark. (1952) Lic. techn., physics (Techn U. of Denmark, 1981). Sr. scient. (tel. 42371212, ext. 4704, telex 43116, fax 42370115, email clausen at risoe.dk) *Neutron scattering.*

Cour, la, Dr Troels Frederik Marstrand. Dept. of Chemistry, U. of Aarhus, Langelandsgade 140, DK-8000 Aarhus, Denmark. (1944) Cand.scient., X-ray crystallography (Aarhus U., 1970). Assoc. prof. (tel. 86124633, ext. 217, fax 86139919). *Biological macromolecules.*

Danielsen, Dr Jacob. Dept. of Chemistry, U. of Aarhus, Langelandsgade 140, DK-8000 Aarhus, Denmark. (1937) Dr.phil., inorganic chemistry (Aarhus U., 1961). Assoc. prof. (tel. 86124633, ext. 230, fax 86139919, email jda at kemi.aau.dk) *Crystal growth, computing.*

Drenck, Prof. Kaj. Dept. of Physics, Royal Danish Sch. of Pharmacy, Universitetsparken 2, DK-2100 Copenhagen, Denmark. (1921) Dr.phil., crystallography (Copenhagen U., 1959). Prof. (tel. 31370850, ext. 270). *Instrumentation, powders, particle size, ferroelectrics.*

Fabius, Mrs Birgit. Dept. of Chemistry, U. of Aarhus, Langelandsgade 140, DK-8000 Aarhus, Denmark. (1959) Cand.scient., inorganic chemistry (Aarhus U., 1987). Res. fellow. (tel. 86124633, ext. 235, fax 86139919) *In situ methods, catalysis.*

Feidenhans'l, Dr Robert. Physics Dept., Risø National Lab., DK-4000 Roskilde, Denmark. (1958) Lic.scient., physics (Aarhus U., 1986). Scient. (tel. 42371212, ext. 4708, telex 43116, fax 42370115) *Surface structure.*

Gajhede, Dr Michael. Chemical Lab. IV, U. of Copenhagen, Universitetsparken 5, DK-2100 Copenhagen, Denmark. (1954) Lic.scient., chemistry (Copenhagen U., 1986). Asst. prof. (tel. 31353133, fax 31350609, email michael at kl4vax.nbi.dk) *Electrostatic potentials, crystal structures.*

Gerward, Dr Leif. Lab. of Applied Physics III, Techn. U. of Denmark, DTH 307, DK-2800 Lyngby, Denmark. (1939) Tekn.dr., physics (Chalmers U. of Techn., 1970). Assoc. prof. (tel. 42882488, ext. 2348, email a168001 at neumvs1). *Energy dispersive diffraction, high pressure crystal structures, X-ray interaction with matter.*

Grey, Dr Francois. Physics Dept., Risø National Lab., DK-4000 Roskilde, Denmark. (1963) Lic.scient., physics (Copenhagen U., 1988). Scient. (tel. 42371212, ext. 4707, telex 43116, fax 42370115, email grey at risoe.dk) *Surface structure.*

Grundvig, Dr Sidsel. Dept. of Geology, U. of Aarhus, C.F. Møllers Alle, DK-8000 Aarhus, Denmark. (1941) Lic.pharm., crystallography (Royal Danish Sch. of Pharmacy, 1967). Assoc. prof. (tel. 86128233). *Powder diffraction, X-ray fluorescence analysis, electron microprobe analysis.*

Grønlund, Dr Finn. Chemical Lab. IV, U. of Copenhagen, Universitetsparken 5, DK-2100 Copenhagen, Denmark. (1925) DSc, physical chemistry (U. de Paris, 1957). Assoc. prof. (tel. 31353133, ext. 242). *Surface structure, electron diffraction.*

Gråbæk, Mr Lars. Physics Dept., Risø National Lab., DK-4000 Roskilde, Denmark. (1958) Cand.scient., physics (Copenhagen U., 1987). Stud. lic. (tel. 42371212, ext. 4722, fax 42370115, email friis at risoe.dk) *Neutron and X-ray scattering, inclusions.*

Haagensen, Mr Carl Olaf. Dept. of Chemistry, U. of Aarhus, Langelandsgade 140, DK-8000 Aarhus, Denmark. (1929) Cand.polyt., chemistry (Techn. U. of Denmark, 1955). Assoc. prof. (tel. 86124633). *Crystal structures.*

Hazell, Dr Alan Charles. Dept. of Chemistry, U. of Aarhus, Langelandsgade 140, DK-8000 Aarhus, Denmark. (1935) PhD, inorganic and structural chemistry (Leeds U., 1962). Assoc. prof. (tel. 86124633, ext. 236, fax 86139919, email ach at kemi.aau.dk). *Crystal structures, platinum metal complexes, sulfur-nitrogen compounds, anomalous scattering.*

Hazell, Mrs Rita Grønbæk. Dept. of Chemistry, U. of Aarhus, Langelandsgade 140, DK-8000 Aarhus, Denmark. (1936) Cand.polyt., chemistry (Techn. U. of Denmark, 1961). Assoc. prof. (tel. 86124633, ext. 232, fax 86139919, email haz at kemi.aau.dk). *Crystal structures, methods of determination, refinement methods.*

Hjorth, Mr Michael. Chemistry Dept. B, Techn. U. of Denmark, DTH 301, DK-2800 Lyngby, Denmark. (1962) Cand.polyt., chemistry (Techn. U. of Denmark, 1987). Stud. lic. (tel. 42881777, fax 42882239, email kebnthor at neuvm1). *Structural chemistry, computing.*

Honoré, Dr Tage. A/S Ferrosan, Res. Div., Sydmarken 5, DK-2860 Soeborg, Denmark. (1951) Lic.pharm., organic chemistry (Royal Danish Sch. of Pharmacy, 1978). Head, biochemical. dept. (tel. 31692111, ext. 242). *Neurochemistry.*

Jensen, Mr Aage. Inst. of Mineralogy, U. of Copenhagen, Østervoldgade 10, DK-1350 Copenhagen, Denmark. (1930) Cand.mag., mineralogy (Copenhagen U., 1957). Assoc. prof. (tel. 33112232). *Oxides (Fe and Ti), gem minerals.*

Jensen, Dr Birthe. Dept. of Chemistry BC, Royal Danish Sch. of Pharmacy, Universitetsparken 2, DK-2100 Copenhagen, Denmark. (1938) Dr.pharm., organic chemistry (Royal Danish Sch. of Pharmacy, 1984). Assoc. prof. (tel. 31370850, ext. 246, fax 31375744) *Organic structures, biological structures, interatomic forces.*

Jensen, Mr Gunnar Bent. Dept. of Electrophysics, Techn. U. of Denmark, Anker Engelundsvej, DK-2800 Lyngby, Denmark. (1929) Cand.mag., theoretical physics (Copenhagen U., 1955). Assoc. prof. (tel. 42881188, ext. 2715). *Crystal physics.*

Jensen, Dr Stig Jorgo. Dept. of Technology, Royal Dental C., Vennelyst Blvd., DK-8000 Aarhus, Denmark. (1932) Dr.phil., chemistry (Aarhus U., 1969). Assoc. prof. (tel. 86132533, ext. 256). *Dental materials crystal structures.*

Jerslev Lund, Prof. Bodil. Dept. of Chemistry BC, Royal Danish Sch. of Pharmacy, Universitetsparken 2, DK-2100 Copenhagen, Denmark. (1919) Dr.phil., chemistry (Copenhagen U., 1958). Prof. (tel. 31370850, ext. 240, fax 31375744). *Organic structures, organic powder identification.*

Johnsen, Mr Ole. Mineral Collection, Geological Museum, U. of Copenhagen, Østervoldgade 5-7, DK-1350 Copenhagen, Denmark. (1940) Cand.scient., mineralogy (Copenhagen U., 1971). Assoc. prof. (tel. 33135001). *Silicate crystal structures, crystal chemistry, minerals.*

Juul Jensen, Mrs Dorte. Physics and Metallurgy Dept., Risø National Lab., DK-4000 Roskilde, Denmark. (1957) Lic.techn., metallurgy (Techn. U. of Denmark, 1983). Scient. (tel. 42371212, ext. 4756, telex 43116, fax 42351173). *Texture, neutron scattering, recrystallization, metals.*

Jørgensen, Dr Jens-Erik. Dept. of Chemistry, U. of Aarhus, Langelandsgade 140, DK-8000 Aarhus, Denmark. (1955) Lic.scient., inorganic chemistry (Aarhus U., 1984). Asst. prof. (tel. 86124633, ext. 295, fax 86139919, email jenserik at kemi.aau.dk). *Solid state chemistry, X-ray and neutron scattering.*

Jørgensen, Mr Ole. Inst. of Mineralogy, U. of Copenhagen, Østervoldgade 10, DK-1350 Copenhagen, Denmark. (1939) Mag.scient., geology (Copenhagen U., 1969). Assoc. prof. (tel. 33112232). *Silicate minerals, zeolites, crystal optics, crystal growth, microtopography.*

Kaas, Dr Karen. Chemistry Dept., Royal Veterinary and Agricultural U., Thorvaldsensvej 40, DK-1871 Copenhagen, Denmark. (1942) Lic.pharm., inorganic chemistry (Royal Danish Sch. of Pharmacy, 1973). Assoc. prof. (tel. 31351788, ext. 2404). *Coordination compounds.*

Karup-Møller, Dr Sven. Inst. of Mineral Industry, Techn. U. of Denmark, DTH 204, DK-2800 Lyngby, Denmark. (1936) Dr.scient., mineralogy (Copenhagen U., 1980). Assoc. prof. (tel. 42882222). *Ore deposit mineralogy.*

Kjeldgaard, Mr Morten. Dept. of Chemistry, U. of Aarhus, Langelandsgade 140, DK-8000 Aarhus, Denmark. (1953) Cand.scient., chemistry (Aarhus U., 1982). Res. fellow. (tel. 86124633, ext. 267, fax 86131160, email mok at kemi.aau.dk) *Protein crystallography, computer graphics.*

Kjær, Dr Kristian. Physics Dept., Risø National Lab., DK-4000 Roskilde, Denmark. (1955) Lic.techn., physics (Techn. U. of Denmark, 1981). Post-doctoral fellow.

(tel. 42371212, ext. 4709, 4716, telex 43116, fax 42370115, email kkjaer at risoe.dk) *Monolayers, surface structure.*

Kristensen, Mrs Bente Saustrup. Chemistry Dept. A, Techn. U. of Denmark, DTH 207, DK-2800 Lyngby, Denmark. (1931) Cand.polyt., chemistry (Techn. U. of Denmark, 1956). Assoc. prof. (tel. 42883111, ext. 3357). *Inorganic structures.*

Langer, Prof. Ebbe Wang. Dept. for Metallurgy, Techn. U. of Denmark, Lundtoftevej 100, DK-2800 Lyngby, Denmark. (1927) Dr.techn., metallurgy (Techn. U. of Denmark, 1967). Prof. in metallurgy. (tel. 42884022, ext. 3233). *Electron diffraction and electron microscopy, metals.*

Larsen, Mr Finn Krebs. Dept. of Chemistry, U. of Aarhus, Langelandsgade 140, DK-8000 Aarhus, Denmark. (1941) Cand.scient., inorganic chemistry (Aarhus U., 1966). Assoc. prof. (tel. 86124633, ext. 241, fax 86139919, email kre at kemi.aau.dk). *Accurate structure analysis, X-ray and neutron diffraction, electron density determination, low temperature crystallography, structural studies with synchrotron radiation.*

Larsen, Dr Ingrid Kjøller. Dept. of Chemistry BC, Royal Danish Sch. of Pharmacy, Universitetsparken 2, DK-2100 Copenhagen, Denmark. (1935) Lic.pharm., organic chemistry (Royal Danish Sch. of Pharmacy, 1965). Assoc. prof. (tel. 31370850, ext. 244, fax 31375744). *Organic structures, biological structures.*

Larsen, Mrs Sine. Chemical Lab. IV, U. of Copenhagen, Universitetsparken 5, DK-2100 Copenhagen, Denmark. (1943) Cand.scient., chemistry (Copenhagen U., 1968). Assoc. prof. (tel. 31353133, ext. 252, fax 31350609, email km4sine at dkuccc11 or sine at kl4vax.nbi.dk). *Coordination compounds, electron density determinations, intermolecular interactions.*

Lebech, Mrs Bente. Physics Dept., Risø National Lab., DK-4000 Roskilde, Denmark. (1937) Cand.polyt., physics (Techn. U. of Denmark, 1962). Sr. Scient. (tel. 42371212, ext. 4705, telex 43116, email lebech at risoe.dk). *Neutron and X-ray scattering, magnetic structures.*

Leffers, Dr Torben. Metallurgy Dept., Risø National Lab., DK-4000 Roskilde, Denmark. (1936) Dr.techn., physical metallurgy (Techn. U. of Denmark, 1975). Sr. scient. (tel. 42371212, ext. 5708). *Structure of metals, deformation of metals, irradiation of metals.*

Leonardsen, Mr Erik Sverre. Inst. of Mineralogy, U. of Copenhagen, Østervoldgade 10, DK-1350 Copenhagen, Denmark. (1934) Cand.real., inorganic chemistry (Oslo U., 1961). Assoc. prof. (tel. 33112232). *Minerals, mineral identification, computing.*

Lindegaard-Andersen, Prof. Asger. Lab. of Applied Physics III, Techn. U. of Denmark, DTH 307, DK-2800 Lyngby, Denmark. (1925) Dr.techn., physics (Techn. U. of Denmark, 1967). Prof. (tel. 42882488, ext. 2345, fax 42882239). *Crystal defects, crystal growth, X-ray topography, diffuse X-ray scattering, biomineralization.*

Lindgreen, Dr Holger. Geological Survey of Denmark, Thorasvej 31, DK-2400 Copenhagen NV, Denmark. (1947) Lic.agro., (Royal Veterinary and Agricultural U., 1974). Sr. scient. (tel. 31106600). *Clay minerals, clay diagnosis.*

Lorentzen, Mr Torben. Metallurgy Dept., Risø National Lab., DK-4000 Roskilde, Denmark. (1959) Cand.polyt., structural mechanics (Aalborg U., 1984). Stud. lic. (tel. 42371212, ext. 5729, fax 42370115) *Neutron scattering, strain measurements.*

Makovicky, Dr Emil. Inst. of Mineralogy, U. of Copenhagen, Østervoldgade 10, DK-1350 Copenhagen, Denmark. (1940) RNDr, mineralogy (Bratislava U., 1967). PhD, geology (Mc Gill U. Montreal, 1970). Assoc. prof. (tel. 33112232). *Sulfide and sulfosalt crystal structures, crystal chemistry, minerals, computing.*

Micheelsen, Prof. Harry. Inst. of Mineralogy, U. of Copenhagen, Østervoldgade 10, DK-1350 Copenhagen, Denmark. (1931) Dr.phil., mineralogy (Copenhagen U., 1966). Prof. (tel. 33112232). *Minerals ,crystal optics.*

Møller, Dr Christian Knakkergaard. Chemistry Dept., U. of Odense, Campusvej 55, DK-5230 Odense M, Denmark. (1920) Dr.phil., chemistry (Copenhagen U., 1961). Prof. (tel. 66158600, ext. 2534). *Crystal chemistry, solid state spectroscopy.*

Nielsen, Dr Anders. Res. and Dev. Div., Haldor Topsøe A/S, Nymøllevej 55, DK-2800 Lyngby, Denmark. (1919) Dr.techn., chemical engineering (Techn. U. of Denmark, 1950). Manager of res. and develop. div. (tel. 42878100). *Imperfect crystal structures, catalysts.*

Nielsen, Mr Kurt. Chemistry Dept. B, Techn. U. of Denmark, DTH 301, DK-2800 Lyngby, Denmark. (1943) Cand.scient., chemistry (Aarhus U., 1969). Assoc. prof. (tel. 42881777, ext. 2112, fax 42882239, email kebnthor at neuvm1). *Crystal structures, crystallographic methods.*

Norrestam, Prof. Rolf. Chemistry Dept. B, Techn. U. of Denmark, DTH 301, DK-2800 Lyngby, Denmark. (1937) Fil.dr., chemistry (Stockhlom U., 1972). Prof. (tel. 42881777, ext. 2180, telex 37529). *Crystal structures, crystallographic methods, computer graphics.*

Nyborg, Dr Jens. Dept. of Chemistry, U. of Aarhus, Langelandsgade 140, DK-8000 Aarhus, Denmark. (1942) Fil.dr., crystallography (Gothenburg U., 1971). Assoc. prof. (tel. 86124633, ext. 267, fax 86131160, email jnb at kemi.aau.dk or jnb at dkaauche). *Protein and nucleic acid crystallography, direct methods.*

Nørskov-Lauritsen, Dr Leif. Novo Research Institute, Novo Alle, DK-2880 Bagsværd, Denmark. (1953) Lic.scient., physical organic chemistry (Copenhagen U., 1984). Scient. (tel. 42982333, ext. 2834). *Proteins, molecular modelling.*

Olsen, Dr Janus Staun. Physics Lab. II, H.C. Ørsted Institute, Universitetsparken 5, DK-2100 Copenhagen, Denmark. (1937) Cand.mag. et mag.scient., physics (Copenhagen U., 1963). Assoc. prof. (tel. 31353133). *Energy dispersive diffraction, phase transformation, high pressures.*

Pauly, Prof. Hans. Inst. of Mineral Industry, Techn. U. of Denmark, DTH 204, DK-2800 Lyngby, Denmark. (1921) Dr.phil., mineralogy (Copenhagen U., 1958). Prof. (tel. 42884022, ext. 3216). *Ore-mineralogy, mineralogy, fluoride minerals.*

Pedersen, Dr Jan Skov. Physics Dept., Risø National Lab., DK-4000 Roskilde, Denmark. (1959) Lic.scient., physics (Copenhagen U., 1988). Scient. (tel. 42371212, ext. 4707, telex 43116, fax 42370115, email skov at risoe.dk) *Surface structure.*

Petersen, Dr Ole Valdemar. Mineralogical Div., Geological Museum, Østervoldgade 10, DK-1350 Copenhagen, Denmark. (1939) Lic.scient., mineralogy (Copenhagen U., 1970). Assoc. prof. (tel. 33112232). *Crystal optics, minerals.*

Plough-Sørensen, Mrs Gudrun. Dept. of Chemistry, U. of Odense, Campusvej 55, DK-5230 Odense M, Denmark. (1938) Cand.pharm. (Royal Danish Sch. of Pharmacy, 1962). Assoc. prof. (tel. 66158600). *Crystal structures.*

Rasmussen, Mrs Hanne. Dept. of Chemistry, U. of Aarhus, Langelandsgade 140, DK-8000 Aarhus, Denmark. (1961) Cand.pharm. (Royal Danish Sch. of Pharmacy, 1985). Scient. (tel. 86124633, ext. 267, fax 86131160, email hbr at kemi.aau.dk) *Protein crystallography, computer graphics.*

Rasmussen, Prof. Svend Erik. Dept. of Chemistry, U. of Aarhus, Langelandsgade 140, DK-8000 Aarhus, Denmark. (1925) Dr.phil., chemical crystallography (Copenhagen U., 1960). Prof. (tel. 86124633, ext. 229, fax 86139919, email ser at kemi.aau.dk). *Refractory compounds, crystal growth, powder diffraction.*

Rindorf, Mrs Grethe. Chemistry Dept. B, Techn. U. of Denmark, DTH 301, DK-2800 Lyngby, Denmark. (1919) Cand.polyt., chemistry (Techn. U. of Denmark, 1945). Assoc. prof. (tel. 42881777, ext. 2113, fax 42882239, email kebnthor at neuvm1). *Organic conductors.*

Rose-Hansen, Dr John. Inst. of Petrology, U. of Copenhagen, Østervoldgade 10, DK-1350 Copenhagen, Denmark. (1937) Mag.scient., petrology and mineralogy (Copenhagen U., 1963). Assoc. prof. (tel. 33112232). *Petrology, mineralogy, experimental petrology, fluid inclusions in minerals.*

Rønsbo, Mr Jørn. Inst. of Mineralogy, U. of Copenhagen, Østervoldgade 10, DK-1350 Copenhagen, Denmark. (1941) Cand.scient., geology (Copenhagen U., 1971). Assoc. prof. (tel. 33112232). *Mineral chemistry, electron microprobe analysis.*

Schmidt-Nielsen, Mr Søren. Sct. Knuds Gymnasium, Læ søegade, DK-5000 Odense, Denmark. (1942) Cand.scient., organic chemistry (Aarhus U., 1971). Lect. (tel. 65901614). *Small angle scattering.*

Simonsen, Mr Ole. Dept. of Chemistry, U. of Odense, Campusvej 55, DK-5230 Odense M, Denmark. (1937) Cand.scient., organic chemistry (Copenhagen U., 1966). Assoc. prof. (tel. 66158600, fax 66158780). *Organic structures.*

Sørensen, Dr Alex Mehlsen. Dept. of Chemistry BC, Royal Danish Sch. of Pharmacy, Universitetsparken 2, DK-2100 Copenhagen, Denmark. (1935) Lic.pharm., organic chemistry (Royal Danish Sch of Pharmacy, 1966). Assoc. prof. (tel. 31370850, ext. 226). *Organic structures, powder diffraction, polymorphism.*

Sørensen, Mr Ole. Res. Div., Physic. and Analyt. Dept., Haldor Topsøe A/S, Nymøllevej 55, DK-2800 Lyngby, Denmark. (1935) Cand.polyt., physics (Techn. U. of Denmark, 1960). Res. engineer. (tel. 42878100). *Electron microscopy, electron diffraction, electron microprobe analysis.*

Søtofte, Mrs Inger. Chemistry Dept. B, Techn. U. of Denmark, DTH 301, DK-2800 Lyngby, Denmark. (1940) Cand.mag., chemistry (Copenhagen U., 1965). Assoc. prof. (tel. 42881777, ext. 2141, fax 42882239, email kebnthor at neuvm1). *Crystal structures, electron diffraction.*

Thirup, Mr Søren. Dept. of Chemistry, U. of Aarhus, Langelandsgade 140, DK-8000 Aarhus, Denmark. (1955) Cand.scient., chemistry (Aarhus U., 1984). Res. fellow. (tel. 86124633, ext. 267, fax 86131160, email soren at kemi.aau.dk) *Protein crystallography, computer graphics.*

Thorup, Mr Niels. Chemistry Dept. B, Techn. U. of Denmark, DTH 301, DK-2800 Lyngby, Denmark. (1939) Cand.polyt., chemistry (Techn. U. of Denmark, 1963). Assoc. prof. (tel. 42881777, ext. 2140, telex 37529, fax 42882239, email kebnthor at neuvm1). *Structural chemistry, organic conductors, ceramic superconductors, databases.*

Villadsen, Mr Jørgen. Res. and Dev. Div., Haldor Topsøe A/S, Nymøllevej 55, DK-2800 Lyngby, Denmark. (1930) Cand.polyt., chemistry (Techn. U. of Denmark, 1953). Sr. sci., manager of phys. and anal. dept. (tel. 42878100, fax 42878494). *Inorganic crystal structures, crystal chemistry, phase analysis.*

Zachau-Christiansen, Dr Birgit. Fysisk-Kemisk Inst., Techn. U. of Denmark, DK-2800 Lyngby, Denmark. (1951) Lic.techn., chemistry (Techn. U. of Denmark, 1986). Scient. (tel. 42883111, ext. 3405). *Intercalation compounds.*

EGYPT, Arab Rep.

Sub-Editor: **M.M. Radwan**

Notes

1. International telephone country code - 20.

Abdel Aal, Mr Fawzi Amer. Geological Survey and Mineral Res. Organization, Salah Salem Street, Abbassia, Arab Rep. of Egypt. BSc, geology (Cairo U., 1945). Res. officer. (tel. 830782). *Economic geology.*

Abdel Hady, Prof. Seham. Physics Dept., Fac. of Sci., Helwan U., Helwan, Cairo, Arab Rep. of Egypt. PhD, crystallography (Cairo U., 1965). Dept. head. (tel. 802129). *Organic structures.*

Abdel Kader, Dr Abdel Aziz. X-ray Crystallography Unit, Nat. Res. Centre, Dokki-Cairo, Arab Rep. of Egypt. PhD, geology (London U., UK, 1964). Res. (tel. 802129). *Igneous petrology.*

Abdel Kader, Prof. (Miss) Naima. X-ray Crystallography Unit, Nat. Res. Centre, Dokki-Cairo, Arab Rep. of Egypt. PhD, crystallography (Cairo U., 1965). Res. (tel. 802129). *Organic structures.*

Abdel Kader, Mrs Zeinab Mohamed. Fac. of Sci., U. of Cairo, University Street, Giza-Cairo, Arab Rep. of Egypt. MSc, geology (Cairo U., 1969). Teaching asst. (tel. 841722, 841713). *Economic geology.*

Abdel Mohsen, Dr Hussein. Atomic Energy Commission, Inshas-Cairo, Arab Rep. of Egypt. DSc, geology Nancy U., France, 1959). Prof. (tel. 862013). *Radioactive mineralogy.*

Abdel Rehim, Dr Amin Mohamed. Dept. of Physics, U. of Tanta, Fac. of Sci., Tanta, Arab Rep. of Egypt. PhD, crystallography (London U., UK, 1955). Prof. *Crystal growth.*

Abdu, Prof. Fayez Madi. Fac. of Agriculture, Shoubra el Kheima, U. of Ain Shams, Abbassia-Cairo, Arab Rep. of Egypt. PhD, soils (Ain Shams U., 1957). Prof. (tel. 948716, 942711). *Soil chemistry, soil mineralogy.*

Abou-Saif, Dr Elhamy Aziz. Lab. of Electron Microscopy and Thin Films, Physics Dept., Nat. Res. Cntr., Altahrir Street, Dokki-Cairo, Arab Republic of Egypt. (1939)PhD, solid state physics (Ain-Shams U., 1972). Assoc. prof. *Electron diffraction.*

Ahmed, Dr Mohamed Saleh. Dept. of Physics, U. of Alexandria, Alexandria, Arab Rep. of Egypt. PhD, physics (London U., UK, 1950). Prof. *Diffuse scattering, crystal analysis.*

Akkad, Dr Mohamed Kamal. Dept. of Geology, U. of Tanta, Fac. of Sci., Tanta, Arab Rep. of Egypt. PhD, geology (London U., UK, 1952). Prof. *Petrology.*

Akkad, Mr Salah el Din. Geological Survey and Mineral Res. Organization, Salah Salem Street, Abbassia, Arab Rep. of Egypt. BSc, geology (Cairo U., 1942). Sub-director. (tel. 830782). *Petrology, ore deposits.*

Anwar, Dr Yehia. Fac. of Sci., Moharram Bey, U. of Alexandria, Alexandria, Arab Rep. of Egypt. PhD, crystallography (Durham U., UK, 1950). Prof. *Petrology.*

Arafa, Prof. Salah Arafa Mohamed. Sci. Dept., The American U. in Cairo, 113 Kasr El-Aini St., Cairo, Arab Rep. of Egypt. (1941) PhD, solid state (Cairo U., 1969). Prof. (tel. 22968 169, ext. 5301, telex 92224 AUCAI-UN) *Powder diffraction, crystal growth, optical characterization magnetic characterization, teaching.*

Ashry, Dr Mamdouh. U. of Assiut, Fac. of Sci., Assiut, Arab Rep. of Egypt. PhD, geology (Assiut U., 1967). Lect. *Geochemistry.*

Azer, Dr Nazmi. Fac. of Sci., U. of Cairo, University Street, Giza-Cairo, Arab Rep. of Egypt. Dr.phil, geology (Wien U., Austria, 1956). Asst. prof. (tel. 841722, 841713). *Economic geology.*

Barakat, Prof. Nayel. Dept. of Physics, Faculty of Sci., U. of Ain Shams, Abbassia-Cairo, Arab Rep. of Egypt. PhD, physics (London U., UK, 1952). Prof. (tel. 821096, 821633). *Spectrography.*

Badr, Dr Yehia Abd-El Hamid. Physics Dept., Fac. of Sci., U. of Cairo, Giza-Cairo, Egypt. (1946) PhD, molecular physics (Leningrad State U., USSR, 1974). Lect. *Phase transitions in ionic crystals.*

Basha, Dr Ahmed Fouad. Physics Dept., Fac. of Sci., U. of Cairo, Cairo, Egypt. (1941) PhD, solid state physics (Moscow State U., USSR, 1974). Lect. *Dielectric properties in molecular solids.*

Bassiouny, Dr Mohamed Khafagi. Fac. of Sci., U. of Ain Shams, Abbassia-Cairo, Arab Rep. of Egypt. Dr.phil., geology (Wien U., Austria, 1969). Lect. (tel. 821096, 821633). *Petrology.*

Elbadri, Dr H. Dept. of Mining, Fac. of Eng., U. of Cairo, University Street, Giza-Cairo, Arab Rep. of Egypt. PhD, geology (London U., UK, 1951). Prof. (tel. 89 69 26). *Crystal optics, radioactive minerals.*

El Demerdash, Dr Saad. Desert Inst., El Mataria, Cairo, Arab Rep. of Egypt. PhD, soils (Cairo U., 1970). Res. *Soil chemistry, soil mineralogy.*

El Gabi, Dr Sami. U. of Assiut, Fac. of Sci., Assiut, Arab Rep. of Egypt. Dr.rer.nat., geology (U. München, BRD, 1963). Asst. prof. *Igneous petrology.*

El Naggar, Dr Mohamed. U. of Assiut, Fac. of Sci., Assiut, Arab Rep. of Egypt. PhD, geology (London U., UK, 1969). Lect. *Carbonate geochemistry.*

El Ramly, Mr Mohamed Fawzi. Geological Survey and Mineral Res. Organization, Salah Salem Street, Abbassia, Arab Rep. of Egypt. BSc, geology (Cairo U., 1944). Res. officer. (tel. 830782). *Petrology.*

El Sayed, Prof. (Mrs) Karimat. Fac. of Sci., U. of Ain Shams, Abbassia-Cairo, Arab Rep. of Egypt. PhD, physics (London U., UK, 1965). Lect. (tel. 821096, 821633). *X-ray crystallography.*

El Shaabini, Prof. (Mrs) Aida Moustafa. X-ray Crystallography Unit, Nat. Res. Centre, Dokki-Cairo, Arab Rep. of Egypt. MSc, physics (Cairo U., 1969). Res. asst. (tel. 802129). *X-ray metallography.*

El Shanshury, Dr Ismail. Atomic Energy Commission, Inshas-Cairo, Arab Rep. of Egypt. PhD, physics (Rensselaer Polytechnic Inst., USA, 1963). Asst. prof. (tel. 862013). *Metals.*

El Sharkawi, Dr Mohamed Abdel Hamid. Fac. of Sci., U. of Cairo, University Street, Giza-Cairo, Arab Rep. of Egypt. PhD, geology (London U., UK, 1964). Lect. (tel. 841722, 841713). *Ore deposits.*

El Shazli, Dr El Shazli Mohamed. Atomic Energy Commission, Inshas-Cairo, Arab Rep. of Egypt. PhD, geology (London U., UK, 1950). Prof. (tel. 862013). *Ore deposits.*

Elwan, Dr Ahmed Abdel Salam. Desert Inst., El Mataria, Cairo, Arab Rep. of Egypt. PhD, soils (Ain Shams U., 1975). Res. *Soil chemistry, soil mineralogy.*

Fayed, Dr (Mrs) Leila. Fac. of Sci., U. of Cairo, University Street, Giza-Cairo, Arab Rep. of Egypt. PhD, geology (Sheffield U., UK, 1966). Lect. (tel. 841722, 841713). *Rock mechanics.*

Gad, Dr Gamal Mohamed. Refractories & Ceramic Lab., Nat. Res. Centre, Dokki-Cairo, Arab Rep. of Egypt. PhD, geochemistry (London U., UK, 1950). Prof. (tel. 802129). *Clay mineralogy.*

Ghebrial, Dr Mounir Guirgis. Geological Survey and Mineral Res. Organization, Salah Salem Street, Abbassia, Arab Rep. of Egypt. PhD, geology (London U., UK, 1953). Res. officer. (tel. 830782). *Structure petrology.*

Guindi, Dr Amin Riad. Fac. of Sci., U. of Alexandria, Alexandria, Arab Rep. of Egypt. PhD, geology (London U., UK, 1950). Prof. *Petrology.*

Gweifel, Dr Ismail. Fac. of Agriculture, Chatby, U. of Alexandria, Alexandria, Arab Rep. of Egypt. PhD, soils (Alexandria U., 1967). Lect. *Soil morphology, soil mineralogy.*

Hamdi, Prof. Hassan Mahmoud. Fac. of Agriculture, Shoubra el Kheima, U. of Ain Shams, Abbassia-Cairo, Arab Rep. of Egypt. DSc, (ETH Zürich, Switzerland, 1942). Prof. (tel. 948716, 942711). *Clay mineral structures.*

Harga, Dr Ahmed Amin. Desert Inst., El Mataria, Cairo, Arab Rep. of Egypt. PhD, soils (Ain Shams U., 1971). Res. *Soil chemistry, soil mineralogy.*

Hassan, Prof. Mohamed Youssef. Fac. of Sci., U. of Ain Shams, Abbassia-Cairo, Arab Rep. of Egypt. PhD, geology (Bristol U., UK, 1951). Prof. (tel. 821096, 821633). *Paleontology, stratigraphy.*

Helmi, Prof. Mohamed Ezzeldin. Fac. of Sci., U. of Ain Shams, Abbassia-Cairo, Arab Rep. of Egypt. PhD, geology (Michigan U., USA, 1951). Prof. (tel. 821096, 821633). *Mineralogy, crystal structure.*

Hinawi, Prof. Essam. Nat. Res. Centre, Dokki-Cairo, Arab Rep. of Egypt. PhD, geology (Cairo U., 1961). Asst. prof., res. (tel. 802129). *Economic geology.*

Kabish, Dr Lotfi. Nat. Res. Centre, Dokki-Cairo, Arab Rep. of Egypt. PhD, geology (Cairo U., 1948). Asst. prof., researcher. (tel. 802129). *Igneous petrology.*

Kamel, Prof. Dr Raafat Wasef. Physics Dept., Fac. of Sci., U. of Cairo, Giza-Cairo, Egypt. (1926) DSc, solid state physics (Cairo U., 1968). Prof. and head. (tel. 894095). *Defect structure in crystalline solids.*

Khadr, Dr Moustafa. Fac. of Agriculture, Chatby, U. of Alexandria, Alexandria, Arab Rep. of Egypt. PhD, soils (Alexandria U., 1956). Prof. *Soil morphology, soil mineralogy.*

Khalifa, Prof. (Mrs) Berlant. Fac. of Sci., U. of Ain Shams, Abbassia-Cairo, Arab Rep. of Egypt. PhD, physics (London U., UK, 1967). Lect. (tel. 821096, 821633). *Electron diffraction.*

Khalifa, Dr (Mrs) B. Abdel Meguid. Atomic Energy Commission, Inshas-Cairo, Arab Rep. of Egypt. PhD, chemistry (London U., UK, 1968). Res. (tel. 862013). *Inorganic structures.*

Khidr, Prof. (Mrs) Fatma Abdel Hakim. X-ray Crystallography Unit, Nat. Res. Centre, Dokki-Cairo, Arab Rep. of Egypt. MSc, physics (Cairo U., 1967). Res. asst. (tel. 802129). *X-ray metallography.*

Kholeif, Dr Mahmoud. Nat. Res. Centre, Dokki-Cairo, Arab Rep. of Egypt. Cand. geol.-min., geology (Moscow U., USSR, 1964). Res. (tel. 802129). *Ore deposits.*

Kishk, Dr Fawzi Mohamed. Fac. of Agriculture, Chatby, U. of Alexandria, Alexandria, Arab Rep. of Egypt. PhD, soils (U. California, Berkeley, USA, 1967). Asst. prof. *Soil chemistry, soil mineralogy.*

Labib, Dr (Mrs) Fawkia. X-ray Crystallography Unit, Nat. Res. Centre, Dokki-Cairo, Arab Rep. of Egypt. PhD, soils (Ain Shams U., 1970). Res. (tel. 802129). *Soil chemistry, soil mineralogy.*

Labib, Dr Tarik. Fac. of Agriculture, U. of Assiut, Assiut, Arab Rep. of Egypt. PhD, soils (U. California, Davis, USA, 1968). Lect. *Soil morphology, soil mineralogy.*

Lashin, Dr A. Mohamed. Fac. of Sci., Moharram Bey, U. of Alexandria, Alexandria, Arab Rep. of Egypt. PhD, crystallography (London U., UK, 1952). Prof. *Crystal structures.*

Lotfy, Dr Mohamed. Fac. of Sci., U. of Cairo, University Street, Giza-Cairo, Arab Rep. of Egypt. PhD, geology (Cairo U., 1957). Asst. prof. (tel. 841722, 841713). *Geology, petrology.*

Mansour, Mr Saber Moustapha. Geological Survey and Mineral Res. Organization, Salah Salem Street, Abbassia, Arab Rep. of Egypt. BSc, geology (Cairo U., 1944). Res. officer. (tel. 830782). *Petrology.*

Naga, Dr Mohamed Abdel Hamid. Fac. of Agriculture, U. of Cairo, Giza-Cairo, Arab Rep. of Egypt. PhD, soils (Cairo U., 1954). Prof. (tel. 89 65 86, 89 67 66). *Soil mineralogy.*

Nakhla, Dr Fakhry. Dept. of Mining, Fac. of Engineering, U. of Cairo, University Street, Giza-Cairo, Arab Rep. of Egypt. PhD, geology (London U., UK, 1951). Prof. (tel. 84 06 55). *Minerals.*

Philipp, Dr George. Fac. of Sci., U. of Cairo, University Street, Giza-Cairo, Arab Rep. of Egypt. PhD, geology (Cairo U., 1956). Asst. prof. (tel. 841722, 841713). *Sedimentary petrology.*

Rabie, Dr (Mrs) Farida Hamed. Fac. of Agriculture, Shoubra el Kheima, U. of Ain Shams, Abbassia-Cairo, Arab Rep. of Egypt. PhD, soils (U. California, Davis, USA, 1967). Asst. prof. (tel. 948716, 942711). *Soil morphology, soil mineralogy.*

Radwan, Dr Mostafa Mohsen Abdel-Razik. Physics Dept., Military Tech. C., Qubri-El-Quba, Cairo, Arab Rep. of Egypt. (1947) PhD, physics (Dundee U., UK, 1982). Lect. (tel. 853983). *Biomolecular structures, virus structures, amorphous structures.*

Ragab, Dr Abdel Ghani. Fac. of Sci., U. of Ain Shams, Abbassia-Cairo, Arab Rep. of Egypt. MSc, geology (Ain Shams U., Cairo, 1968). Teaching asst. (tel. 821096, 821633). *Petrology.*

Sadek, Dr Gamil. Nat. Res. Centre, Dokki-Cairo, Arab Rep. of Egypt. DSc, geology Lyon U., France, 1964). Res. (tel. 802129). *Ore deposits.*

Salem, Dr Safia Mahmoud. Physics Dept., Fac. of Sci., U. of Cairo, Egypt. PhD, X-ray crystallography (Cairo U., 1961). Asst. prof. (tel. 802129). *X-ray crystallography.*

Shoukri, Dr Nasri M. Fac. of Sci., U. of Cairo, University Street, Giza-Cairo, Arab Rep. of Egypt. PhD, geology (London U., UK, 1940). Prof. (tel. 841722, 841713). *Petrology.*

Soliman, Mr F. Abdel Aal. Geological Survey and Mineral Res. Organization, Salah Salem Street, Abbassia, Arab Rep. of Egypt. BSc, geology (Cairo U., 1944). Res. officer. (tel. 830782). *Petrology.*

Soliman, Dr Mohamed Soliman. Fac. of Sci., U. of Ain Shams, Abbassia-Cairo, Arab Rep. of Egypt. PhD, geology (Stanford U., USA, 1958). Asst. prof. (tel. 821096, 821633). *Sedimentation.*

Takla, Mr Maher Azmi. Fac. of Sci., U. of Cairo, University Street, Giza-Cairo, Arab Rep. of Egypt. MSc, geology (Cairo U., 1968). Teaching asst. (tel. 841722, 841713). *Economic geology.*

Thabet, Dr Atef. Geological Survey and Mineral Res. Organization, Salah Salem Street, Abbassia, Arab Rep. of Egypt. PhD, geology (Cairo U., 1963). Res. officer. (tel. 830782). *Petrology.*

Tousson, Dr Salama. Fac. of Sci., Moharram Bey, U. of Alexandria, Alexandria, Arab Rep. of Egypt. DSc, geology Nancy U., France, 1957). Asst. prof. *Mineral structures.*

Youssef, Dr I. Mourad. Fac. of Sci., U. of Ain Shams, Abbassia-Cairo, Arab Rep. of Egypt. PhD, geology (Alexandria U., 1950). Prof. (tel. 821096, 821633). *Structural geology.*

Zaghloul, Dr Mohamed Zaki. U. of Mansoura, Fac. of Sci., Mansoura, Arab Rep. of Egypt. PhD, geology (Bristol U., UK, 1958). Prof. *Crystal structures.*

Zatout, Mr Mohamed Abdel Meguid. General Egyptian Mining Organisation, Geheni Street, Dokki-Cairo, Arab Rep. of Egypt. *Petrology.*

FINLAND

Sub-Editor: **A. Vahvaselkä**

Notes

1. International telephone country code - 358.

Åberg, Prof. Teijo. Lab. of Physics, Helsinki U. of Techn., 02150 Espoo, Finland. (1937) PhD, physics (U. Helsinki, 1969). Assoc. prof. (tel. 0 460144, fax. 0 465077). *Inelastic X-ray scattering.*

Ahlgrén, Dr, Assoc. Prof. Markku Jouko. Dept. of Chemistry, U. of Joensuu, 80101 Joensuu, Finland. (1944) PhD, inorganic chemistry (U. Helsinki, 1979). Assoc. prof. (tel. 73 1513368). *Metal complexes, clusters.*

Ahtee, Dr Sisko-Maija. Dept. of Physics, U. of Helsinki, 00170 Helsinki, Finland. (1939) PhD, physics (U. Helsinki, 1971). Lect. (tel. 0 650211, telex 122229 nuphu sf, fax 0 174072, e-mail AHTEE at FINUHCB.BITNET). *X-ray and neutron crystallography, ferroelectric materials.*

Aikala, Dr Oiva Jaakko Mikael. Dept. of Physical Sciences, U. of Turku, 20500 Turku, Finland. (1940) PhD, theoretical physics (U. Turku, 1977). Docent. (tel. 21 645689). *Compton scattering, momentum and charge density theory, LCAO-methods.*

Aksela, Prof. Seppo Olavi. Dept. of Physics, U. of Oulu, Linnanmaa, 90570 Oulu, Finland. (1942) PhD, physics (U. Oulu, 1971). Assoc. prof. (tel. 81 345411). *Electron spectrometry.*

Blomberg, Miss Merja Kristiina. Dept. of Physics, U. of Helsinki, 00170 Helsinki, Finland. (1957) MSc, physics (U. Helsinki, 1986). Asst. (tel. 0 650211, ext. 23, telex 122229 nuphu sf, fax 0 174072, e-mail BLOMBERG at FINUHCB.BITNET). *X-ray crystallography, materials science.*

Carlson, Dr Sirkka Liisa. Dept. of Geology, U. of Helsinki, 00170 Helsinki, Finland. (1943) PhD, geology and mineralogy (1982). Asst. (tel. 0 1914427). *Crystallography, mineralogy.*

Friman, Dr Rauno Kalevi. Dept. of Physical Chemistry, Åbo Akademi, 20500 Åbo, Finland. (1946) PhD, physical chemistry (Åbo Akademi, 1983). Lect. (tel. 21 335133, ext. 255). *Small angle X-ray scattering, soap and detergent micelle structure, biological membranes.*

Haapala, Prof. Ilmari Johannes. Dept. of Geology, U. of Helsinki, 00170 Helsinki, Finland. (1939) PhD, geology and mineralogy (U. Helsinki, 1966). Prof. (tel. 0 1914426). *Crystallography, mineralogy.*

Hämäläinen, Mr Keijo Johannes. Dept. of Physics, U. of Helsinki, 00170 Helsinki, Finland. (1963) MSc, physics (U. Helsinki, 1987) Asst. (tel. 0 650211, telex 122229 nuphu sf, fax 0 174072, e-mail HAMALAINEN at FINUHCB.BITNET). *Inelastic X-ray scattering.*

Hämäläinen, Dr Reijo Pertti. Dept. of Chemistry, U. of Helsinki, 00100 Helsinki, Finland. (1940) PhD, inorganic chemistry (U. Helsinki, 1972). Docent. (tel. 0 650211, ext. 510). *Metal complexes.*

Harle, Mrs Säde Pirjo Anneli. Geological Survey of Finland, 02150 Espoo, Finland. (1962) MSc, geology and mineralogy (U. Helsinki, 1987). Geologist. (tel. 0 46931, telex 123185, fax 0 462205). *X-ray diffraction, mineralogy.*

Heleskivi, Dr Jouni Martti. Semiconductor Lab., Techn. Res. Centre of Finland, 02150 Espoo, Finland. (1938) DSc, electron physics (Helsinki U. of Techn., 1972). Director of Lab. (tel. 0 4566300, fax. o 4550115). *Semiconductor constants.*

Hiismäki, Prof. Pekka Eljas. Reactor Lab., Techn. Res. Centre of Finland, 02150 Espoo, Finland. (1939) DSc, physics (Helsinki U. of Techn., 1970). Director. (tel. 0 4566320). *Neutron diffraction, magnetic structures.*

Hiltunen, Mr Lassi Ilmari. Dept. of Chemistry, Helsinki U. of Techn., 02150 Espoo, Finland. (1943) MSc, inorganic chemistry (Helsinki U. of Techn., 1969). Lab. administrator. (tel. 0 4342591, fax. 0 465077, e-mail EKE-LH at FINHUT.BITNET).

Hocksell, Mr Veli Eerik. Metals Lab.,Ovako Steel Oy Ab, 55100 Imatra, Finland. (1951) MSc, physical metallurgy (Helsinki U. of Techn.,1976).Lab. eng. (tel. 54 6021, telex 5711 ovai sf, fax 54 63575). *Electron microscopy, microanalysis, metallography.*

Hölsä, Dr Jorma Pertti Kalervo. Dept. of Chemistry, Helsinki U. of Techn., 02150 Espoo, Finland. (1952) DTechn., spectroscopy of rare earth elements (Helsinki U. of Techn., 1983). Chief asst. (tel. 0 4342608, telex 125161, fax. 0 465077). *Luminescence, structure, thermal stability, rare earth elements, quantum mechanics.*

Hytönen, Dr Kai Kalevi Gabriel. Petrological Dept., Geological Survey of Finland, 02150 Espoo, Finland. (1925) PhD, geology and mineralogy (U. Turku, 1959). State geologist. (tel. 0 46931, ext. 263, telex 123185, fax 0 462205). *Rock-forming minerals in Finland, chemistry, structure, paragenesis.*

Järvinen, Dr Matti Johannes. Dept. of Information Technology, Lappeenranta U. of Techn., PO Box 20, 53851 Lappeenranta, Finland. (1934) PhD, physics (U. Helsinki, 1969). Lab. manager. (tel. 53 27570, ext. 2807). *X-ray diffraction, materials science.*

Johanson, Mr Bo Stefan. Geological Survey of Finland, 02150 Espoo, Finland. (1954) MSc, geology and mineralogy (U. Helsinki, 1984). Geologist. (tel. 0 46931). *Anorthosites/Gabbros, analytical mineralogy, microprobe.*

Kähkönen, Dr Heikki Antero. Dept. of Physical Sciences, U. of Turku, 20500 Turku, Finland. (1934) PhD, physics (U. Turku, 1968). Lect. (tel. 21 645671). *X-ray diffraction.*

Kallio, Mr Pekka Yrjö Juhani. Petrological Dept., Geological Survey of Finland, PO Box 237, 70101 Kuopio, Finland. (1937) MSc, geology and mineralogy (U. Helsinki, 1964). Geologist. (tel. 71 205111, fax. 71 228670). *Mineralogy, X-ray diffraction.*

Kansikas, Dr Jarno Juhani. Dept. of Chemistry, U. of Helsinki, 00100 Helsinki, Finland. (1947) PhD., inorganic chemistry (U. Helsinki, 1985). Instructor. (tel. 0 650211, e-mail KANSIKAS at FINFUN.BITNET). *Metal complex crystal structures.*

Karvinen, Mrs Saila Marjatta. Dept. of Chemistry, Helsinki U. of Techn., 02150 Espoo, Finland. (1959) MSc, inorganic chemistry (1984). Asst. (tel. 0 4512758, telex 125161 htkk sf). *Crystal structures.*

Ketolainen, Prof. Pertti Pekka Juhani. Dept. of Physics, U. of Joensuu, 80130 Joensuu, Finland. (1937) PhD, physics (U. Turku, 1969). Assoc. prof. (tel. 73 28311, ext. 375). *Crystal lattice defects.*

Kettunen, Prof. Pentti Olavi. Inst. of Materials Science, Tampere U. of Techn., 33100 Tampere, Finland. (1932) DTechn., physical metallurgy (Helsinki U. of Techn., 1965). Prof. (tel. 31 162280, telex 22313 ttktr sf). *Plasticity, strain hardening, fatigue, electron microscopy, X-ray diffraction, small-angle scattering, microanalyzing.*

Kivekäs, Dr Raikko Terjo Ilari. Dept. of Chemistry, U. of Helsinki, 00100 Helsinki, Finland. (1944) PhD, inorganic chemistry (U. Helsinki, 1977). Instructor. (tel. 0 650211). *Inorganic and organoselenium crystal structures.*

Kivilahti, Prof. Jorma Kalevi. Dept. of Mining and Metallurgy, Helsinki U. of Techn., 02150 Espoo, Finland. (1943) DSc, physical metallurgy (Helsinki U. of Techn., 1976). Assoc. prof. (tel. 0 4566115, telex 125161). *X-ray metallography, physical metallurgy, titanium and alloys, metal-hydrogen systems.*

Klinga, Dr Martti Evert. Dept. of Chemistry, U. of Helsinki, 00100 Helsinki, Finland. (1942) PhD, inorganic chemistry (U. Helsinki, 1987). Instructor. (tel. 0 650211, e-mail KLINGA at FINFUN.BITNET). *Inorganic and organometallic crystal structures.*

Knuuttila, Mrs Hilkka Ritva-Liisa. Dept. of Chemistry, U. of Jyväskylä, 40100 Jyväskylä, Finland. (1946) PhD, inorganic chemistry (U. Jyväskylä, 1983). Asst. (tel. 41 291211). *Inorganic and organometallic crystal structures.*

Knuuttila, Mr Pekka Juhani. Dept. of Chemistry, U. of Jyväskylä, 40100 Jyväskylä, Finland. (1944) PhD, inorganic chemistry (U. Jyväskylä, 1982). Asst. (tel. 41 291211). *Inorganic and organometallic crystal structures.*

Komu, Mr Markku Eino Sakari. Wihuri Physical Lab., U. of Turku, 20500 Turku, Finland. (1946) MA, physics (U. Turku, 1973). Asst. (tel. 21 645944). *Nuclear magnetic resonance, ionic crystals.*

Korvenranta, Dr Jorma Artturi. Dept. of Chemistry, U. of Helsinki, 00100 Helsinki, Finland. (1938) PhD, inorganic chemistry (U. Helsinki, 1974). Lect. (tel. 0 650211).

Kurki-Suonio, Prof. Kaarle Veikko Johannes. Dept. of Physics, U. of Helsinki, 00170 Helsinki, Finland. (1933) PhD, physics (U. Helsinki, 1959). Prof., dept. head. (tel. 0 650211, ext. 298, telex 122229 nuphu sf, fax 0 174072, e-mail KURKISUONIO at FINUHCB.BITNET). *Charge density, spin and momentum density, deformation models in diffraction data analysis.*

Kyröläinen, Mr Antero Johannes. Outokumpu Oy, 95400 Tornio, Finland. (1951) Dipl.eng., physical metallurgy (U. Oulu, 1976). Res. eng. (tel. 80 4521, ext. 580). *Electron microscopy, stainless steels.*

Lähdeniemi, Dr Matti Juhani Iisakki. Technical U. of Tampere, Technical Inst. of Pori, PO Box 23, 28601 Pori, Finland. (1949) PhD, electronic structure of metals (U. Turku, 1982). Sr. Lect. (tel. 39 27011). *Electronic structure, metals and alloys, molecular and semiconductor surfaces, XPS, XFS, ARUPS, robotic sensors.*

Lahti, Dr Seppo Ilmari. Petrological Dept., Geological Survey of Finland, 02150 Espoo, Finland. (1947) PhD, geology and mineralogy (U. Helsinki, 1981).

Geologist. (tel. 0 4693248, fax. 0 462205). *Mineral crystal structures, single crystal X-ray methods, powder diffraction methods, structure determination, minerals.*

Laiho, Dr Reino Toivo Salomo. Wihuri Physical Lab., U. of Turku, 20500 Turku, Finland. (1941) PhD, physics (U. Turku, 1973). Res. scient. (tel. 21 645943). *Optical properties of solids.*

Laine, Dr Ensio Sulo Uolevi. Dept. of Physical Sciences, U. of Turku, 20500 Turku, Finland. (1940) PhD, physics (U. Turku, 1973). Lect. *X-ray crystallography and applications.*

Lehtinen, Dr Martti Kalevi. Dept. of Geology, U. of Helsinki, 00170 Helsinki, Finland. (1941) PhD, geology and mineralogy (U. Helsinki, 1976). Curator of minerals. (tel. 0 1914424). *Crystallography, mineralogy.*

Leiro, Dr Jarkko Albert. Dept. of Physical Sciences, U. of Turku, 20520 Turku, Finland. (1949) PhD, physics (U. Turku, 1982). Asst. (tel. 21 645698). *Electronic structure, X-ray spectroscopy, photoelectron spectroscopy.*

Leskelä, Dr Markku Antero. Dept. of Chemistry, U. of Turku, 20500 Turku, Finland. (1950) DTechn., inorganic chemistry (Helsinki U. of Techn., 1980). Prof. (tel. 21 645704). *Solid state chemistry.*

Levoska, Mr Pentti Juhani. Dept. of Techn. Physics, U. of Oulu, Linnanmaa, 90570 Oulu, Finland. (1942) Techn.Lic., techn. physics (U. Oulu, 1974). Sr. asst. (tel. 81 352521). *X-ray diffraction, diffuse scattering.*

Lindqvist, Mr Kristian Vilhelm. Geological Survey of Finland, 02150 Espoo, Finland. (1949) MSc, geology and mineralogy (U. Helsinki, 1981). Geologist. (tel. 0 46931, fax. 0 462205). *X-ray diffraction.*

Lindroos, Prof. Veikko Kalervo. Dept. of Mining and Metallurgy, Helsinki U. of Techn., 02150 Espoo, Finland. (1938) DSc, physical metallurgy (Helsinki U. of Techn., 1968). Prof. (tel. 0 4512610, telex 125161). *Dislocation theory, electron microscopy, X-ray metallography, phase transformation, silicon technology, semiconductor metallurgy, non- waste technology.*

Lindström, Dr Rauno. Dept. of Math. and Physics, U. of Joensuu, 80100 Joensuu, Finland. (1937) PhD, physics (U. Turku, 1973). Lect. (tel. 73 28311, ext. 287). *Crystal defects.*

Lumme, Prof. Paavo Olavi. Dept. of Inorganic Chemistry, U. of Helsinki, 00100 Helsinki, Finland. (1923) PhD, physical chemistry (U. Helsinki, 1956). Prof., dept. head. (tel. 0 662134). *Bioinorganics, complexes, magneto-chemistry, organometallics, spectroscopy, structural chemistry, thermal chemistry.*

Mäki, Mr Jouko Kalervo. Lab. of Electron Microscopy, U. of Turku, 20520 Turku, Finland. (1943) MSc, physics (U. Turku, 1970). Lab. eng. (tel. 21 513355, ext. 318, telex 62293 tyl sf). *Electron microscopy, long period order, binary alloys, structure, macromolecules.*

Manninen, Dr Seppo Olavi. Dept. of Physics, U. of Helsinki, 00170 Helsinki, Finland. (1944) PhD, physics (U. Helsinki, 1972). Lect. (tel. 0 650211, ext. 233, telex 122229 nuphu sf, fax 0 174072, e-mail SEMANNINEN at FINUHCB.BITNET). *Inelastic X-ray scattering, resonance phenomena, synchrotron radiation.*

Mansikka, Prof. Kauko Antti. Dept. of Physical Sciences, U. of Turku, 20500 Turku, Finland. (1932) PhD, physics (U. Turku, 1961). Prof. (tel. 21 645111, ext. 5680). *Quantum theory of solids, radiation - matter interaction.*

Martikainen, Mr Hannu Olavi. Dept. of Mining and Metallurgy, Helsinki U. of Techn., 02150 Espoo, Finland. (1952) MSc, physical metallurgy (Helsinki U. of Techn., 1976). Asst. (tel. 0 4554122). *Electron microscopy, dislocation theory, grain boundaries.*

Meisalo, Prof. Veijo Pauli Juhani. Dept. of Teacher Education, U. of Helsinki, 00120 Helsinki, Finland. (1938) PhD, physics (U. Helsinki, 1967). Assoc. prof. (tel. 0 12391). *X-ray diffraction, high pressure studies, phase transformations, physics education.*

Merisalo, Dr Matti Juhani. Dept. of Physics, U. of Helsinki, 00170 Helsinki, Finland. (1937) PhD, physics (U. Helsinki, 1967). Lab. manager, docent. (tel. 0 650211, telex 1002125 attn Merisala, fax. 0 174072). *Charge density, materials science.*

Minni, Dr Erkki Esa Kalervo. Turku Regional Inst. of Occupational Health, 20500 Turku, Finland. (1946) PhD, physics (U. Turku, 1982). Docent. (tel. 21 337970). *Materials science, surface science.*

Muhonen, Dr Heikki Juhani. Dept. of Chemistry, U. of Helsinki, 00100 Helsinki, Finland. (1946) PhD., inorganic chemistry (U. Helsinki, 1987). Instructor. (tel. 0 650211, ext. 515). *Crystal structures, magnetism, transition metal complexes, small organic compounds, crystal structure.*

Mutikainen, Dr Ilpo Pellervo. Div. of Inorganic Chemistry, U. of Helsinki, 00100 Helsinki, Finland. (1947) PhD, inorganic chemistry (U. Helsinki, 1988). Instructor. (tel. 0 650211, ext. 568, e-mail MUTIKAINEN at FINFUN.BITNET). *Crystal structures, metal complexes.*

Mutka, Dr Hannu Mika Ilmari. Reactor laboratory, Techn. Research Centre of Finland, 02150 Espoo, Finland. (1952) DSc, physics (U. de Paris-Sud (Orsay), 1982). Sr. Scient. (tel. 0 4566354). *Neutron diffraction, electron microscopy, electrical and magnetic properties of solids, defects in solids.*

Näsäkkälä, Dr Matti Eerik. Dept. of Chemistry, U. of Helsinki, 00100 Helsinki, Finland. (1944) PhD, inorganic chemistry (U. Helsinki, 1977). Instructor, docent. (tel. 0 650211, ext. 536). *Crystal structures, transition metal complexes.*

Näsänen, Prof. Reino Olavi. Dept. of Chemistry, U. of Helsinki, 00100 Helsinki, Finland. (1908) PhD, chemistry (U. Helsinki, 1939). Prof. emeritus. (tel. 0 650211). *Coordination chemistry.*

Nenonen, Mr Pertti Olavi. Metallurgy Lab., Techn. Res. Centre of Finland, 02150 Espoo, Finland. (1943) Techn.Lic., physical metallurgy (Helsinki U. of Techn., 1974). Sr. Res. Scien. (tel. 0 4561, fax 0 463118, telex 122972 vtthasf). *Electron microscopy.*

Nieminen, Dr Kari Veikko Juhani. Dept. of Chemistry, U. of Helsinki, 00100 Helsinki, Finland. (1946) PhD, inorg. chemistry (U. Helsinki, 1983). Special scient. (tel. 0 650211, ext. 567). *Crystal structure, complexes formed by metal ions and some inorganic compounds.*

Niinistö, Dr Lauri. Dept. of Chemistry, Helsinki U. of Techn., 02150 Espoo, Finland. (1941) DSc, inorganic chemistry (Helsinki U. of Techn., 1973). Prof., lab. head. (tel. 0 4342600, telex 125161, fax 0 465077, e-mail EKE-LNE at FINHUT.BITNET). *Structural inorganic chemistry, solid state chemistry, thin films.*

Oksman, Mr Pentti. Wihuri Physical Lab., Dept. of Chemistry, U. of Turku, 20500 Turku, Finland. (1948) MA, physics (U. Turku, 1973). Lab. eng. (tel. 21 513355, ext. 391). *NMR, ionic crystals, mass spectrometry, organic compounds.*

Orama, Dr Olli Antero. Dept. of Chemistry, U. of Helsinki, 00100 Helsinki, Finland. (1944) PhD, inorganic chemistry (U. Helsinki, 1976). Docent. (tel. 0 650211, ext. 567, e-mail OORAMA at FINFUN.BITNET). *Crystal structure, transition metal complexes, organometallic compounds.*

Paakkari, Prof. Timo Lauri Päiviö. Dept. of Physics, U. of Helsinki, 00170 Helsinki, Finland. (1937) PhD, physics (U. Helsinki, 1968). Assoc. prof. (tel. 0 650211, ext. 245, telex 122229 nuphu sf, fax 0 174072, e-mail PAAKKARI at FINUHCB.BITNET). *Charge density, momentum and spin density, structure, disordered materials.*

Paalassalo, Mr Pentti Olavi. Wihuri Physical Lab., Dept. of Physical Sci., U. of Turku, 20500 Turku, Finland. (1933) PhLic., physics (U. Turku, 1965). Lab. eng. (tel. 21 645653). *Electron microscopy, metals and alloys.*

Pajunen, Dr Aarne Veikko. Dept. of Chemistry, U. of Helsinki, 00100 Helsinki, Finland. (1939) PhD, inorganic chemistry (U. Helsinki, 1967). Assoc. prof. (tel. 0 650211, fax. 656591, e-mail PAJUNEN at FINFUN.BITNET). *Crystal structures, coordination compounds.*

Pakkanen, Dr Tapani Antti. Dept. of Chemistry, U. of Joensuu, 80120 Joensuu, Finland. (1949) PhD, chemistry (SUNY, Stony Brook, USA, 1977). Assoc. prof. (tel. 73 1513345, fax. 73 1513390, e-mail TAP at FINNJO.BITNET). *Cluster chemistry, surface chemistry, catalysis.*

Pakkanen, Dr Tuula Tellervo. Dept. of Chemistry, U. of Joensuu, 80120 Joensuu, Finland. (1949) PhD, chemistry (SUNY, Stony Brook, USA, 1978). Lect. (tel. 73 1513340). *Cluster chemistry, catalysis, organometallic compounds.*

Papunen, Prof. Heikki Tapani. Inst. of Geology and Mineralogy, U. of Turku, 20500 Turku, Finland. (1936) PhD, geology and mineralogy (U. Turku, 1971). Prof. (tel. 21 645480). *Mineralogy.*

Perkkiö, Mrs Seija Anneli. Dept. of Physics, U. of Helsinki, 00170 Helsinki, Finland. (1957) MSc, physics (U. Helsinki, 1983). Asst. (tel. 0 650211, ext. 250 or 233, telex 122229 nuphu sf, fax. 0 174072, e-mail PERKKIO at FINUHCB.BITNET). *Compton Scattering.*

Pessa, Prof. Viljo Markus. Dept. of Physics, Techn. U. of Tampere, 33100 Tampere, Finland. (1941) PhD, physics (U. Turku, 1971). Assoc. prof. (tel. 31 162111). *Electron and X-ray spectroscopy.*

Pitkänen, Dr Ilkka Pellervo. Dept. of Chemistry, U. of Jyväskylä, 40100 Jyväskylä, Finland. (1937) PhD, inorganic chemistry (U. Jyväskylä, 1980). Lect. (tel. 41 291211, fax. 41 292797, e-mail PITKANEN at FINJYU.BITNET). *Metal complexes, structural chemistry.*

Pitkänen, Dr Tuula Esteri. Dept. of Physics, U. of Helsinki, 00170 Helsinki, Finland. (1953) PhD., physics (U. Helsinki, 1988). Asst. (tel. 0 650211, ext. 233, telex 122229 nuphu sf, e-mail PITKANEN at FINUHCB.BITNET). *Compton scattering.*

Punkkinen, Dr Matti. Wihuri Physical Lab., U. of Turku, 20500 Turku, Finland. (1939) PhD, physics (U. Turku, 1967). Prof. (tel. 21 645947, fax. 21 331167). *Nuclear magnetic resonance in solids.*

Pyykkö, Prof. Veli Pekka. Dept. of Chemistry, U. of Helsinki, Et. hesperiank 4, 00100 Helsinki, Finland. (1941) PhD, physics (U. Turku, 1967). Prof. (tel. 0 4027455, telex 121199 seism sf). *Relativistic quantum chemistry, heavy element structures, magnetic resonance spectroscopy.*

Ranta, Mr Lasse Kosti. Dept. of Physical Sciences, U. of Turku, 20500 Turku, Finland. (1949) MSc, physics (U. Turku, 1975). Asst. (tel. 21 645698). *Crystallography, minerals.*

Rautioaho, Dr Risto Heikki. Inst. of Physical Metallurgy, U. of Oulu, Linnanmaa, 90570 Oulu, Finland. (1945) DTechn., materials science (U. Oulu, 1981). Assoc. prof. (tel. 81 353611). *Structure, ceramics.*

Rissanen, Mr Kari Tapani. Dept. of Chemistry, U. of Jyväskylä, 40100 Jyväskylä, Finland. (1959) PhLic, inorganic chemistry (U. Jyväskylä, 1987). Jr. Researcher. (tel. 41 291211, fax. 41 292797, e-mail RISNSANEN at FINJYU.BITNET). *Bioinorganic chemistry, structural chemistry.*

Seitsonen, Dr Sulo Iivari. Dept. of Physical Sciences, U. of Turku, 20500 Turku, Finland. (1935) PhD, physics (U. Turku, 1967). Lect. *X-ray diffraction.*

Serimaa, Mrs Ritva Elina. Dept. of Physics, U. of Helsinki, 00170 Helsinki, Finland. (1957) MSc, physics (U. Helsinki, 1982). Asst. (tel. 0 650211, ext. 250 and 233, telex 122229 nuphu sf, e-mail RSERIMAA at FINFUN.BITNET). *Uncompletely crystallized materials, cellulose, computing.*

Siivola, Prof. Jaakko Uolevi. Dept. of Geology and Mineralogy, U. of Helsinki, 00170 Helsinki, Finland. (1938) PhD, geology and mineralogy (U. Helsinki, 1971). Prof. (tel. 0 1914457). *Crystallography, mineralogy.*

Sivonen, Mr Seppo Juhani. Inst. of Electron Optics, U. of Oulu, Linnanmaa, 90570 Oulu, Finland. (1942) MSc, techn. physics (U. Oulu, 1967). Dir. (tel. 81 352633). *X-ray microanalysis.*

Smolander, Dr Kari Juhani. Dept. of Physics, U. of Helsinki, 00170 Helsinki, Finland. (1947) PhD, physics (U. Helsinki, 1980). Asst. (tel. 0 650211, ext. 256, telex 122229 nuphu sf). *Ferroelectricity, Monte Carlo simulations, phase transitions at high pressures.*

Smolander, Dr Kimmo Juhani Nils-Eric. Dept. of Chemistry, U. of Joensuu, 80101 Joensuu, Finland. (1944) PhD, inorganic chemistry (U. Helsinki, 1983). Lect. (tel. 73 1513364, telex 46183 joy sf). *Metal complexes.*

Stubb, Dr Arne Henrik. Semiconductor Lab., Techn. Res. Centre of Finland, 02150 Espoo, Finland. (1946) PhD, physics (U. Helsinki, 1979). Res. officer. (tel. 0 4561, fax. 0 4550115). *Electronic materials synthesis and characterization.*

Sundius, Dr Tom Robert. Dept. of Physics, U. of Helsinki, 00170 Helsinki, Finland. (1942) PhD, physics (U. Helsinki, 1981). Aman., docent. (tel. 0 650211, ext. 218, telex 122229 nuphu sf, e-mail SUNDIUS at FINUHCB.BITNET). *Molecular vibrations and force fields, computer programming.*

Sundström, Dr Lorna Jean. Dept. of Physics, U. of Helsinki, 00170 Helsinki, Finland. (1940) PhD, solid state physics (U. Helsinki, 1968). Docent. *Solid state theory, energy bands, collective excitations, luminescence.*

Suoninen, Prof. Eero Juhani. Dept. of Physical Sciences, U. of Turku, 20500 Turku, Finland. (1929) PhD, physics (U. Helsinki, 1957). Prof. (tel. 21 645694, fax. 21 331126, e-mail FYS-ES at FINTUVM.BITNET). *Electronic structure of crystals, materials research, instrumentation, electron spectroscopy, surface science.*

Suortti, Prof. Pekka. Dept. of Physics, U. of Helsinki, 00170 Helsinki, Finland. (1938) PhD, physics (U. Helsinki, 1967). Assoc. prof. (tel. 0 650211, ext. 249, telex 122229 nuphu sf). *Electron distribution in simple solids, synchrotron radiation.*

Tarna, Mr Toivo Mikael. Dept. of Physical Sciences, U. of Turku, 20500 Turku, Finland. (1940) PhLic., physics (U. Turku, 1976). Asst. *Electron density.*

Teuho, Mr Juhani Erkki Tapani. Wihuri Physical Lab., Dept. of Physical Sci., U. of Turku, 20500 Turku, Finland. (1948) PhLic., physics (U. Turku, 1973). Asst. (tel. 21 645950). *Electron microscopy, metals and alloys.*

Tiitta, Mr Antero Tapani. Reactor Lab., Techn. Res. Centre of Finland, 02150 Espoo, Finland. (1946) Techn.Lic., eng. dept. of techn. physics (Helsinki U. of Techn., 1970). Sr. res. scient. (tel. 0 4566350, telex 122972 vttin sf). *Neutron time-of-flight, neutron diffraction, instrument design.*

Tilli, Mr Markku Väinö Kalevi. Okmetic Ltd., PO Box 44, 02631 Espoo, Finland. (1950) MSc, physical metallurgy (Helsinki U. of Techn., 1974). Manager. (tel. 0 524744, telex 124802, fax 0 529594). *X-ray topography, semiconductor metallurgy, electron microscopy.*

Toivonen, Mr Jukka Tapio. Dept. of Chemistry, Helsinki U. of Techn., 02150 Espoo, Finland. (1951) MSc, inorganic chemistry (Helsinki U. of Techn., 1976). Asst. (tel. 0 4512758). *Inorganic crystal structures.*

Törnroos, Assoc. Prof. Ragnar Fredrik. Dept. of Geology, U. of Helsinki, 00170 Helsinki, Finland. (1943) PhD, geology and mineralogy (U. Helsinki, 1983). (tel. 0 1914416). *Analytical mineralogy, descriptive mineralogy.*

Tuomi, Prof. Turkka Olavi. Dept. of Techn. Physics, Helsinki U. of Techn., 02150 Espoo, Finland. (1939) DSc, techn. physics (Helsinki U. of Techn., 1968). Dir. (tel. 0 4512145, telex 125161 htkk sf, fax 0 465077, e-mail FYS-TR at FINHUT.BITNET). *Synchrotron X-ray topography, semiconductors, optical properties of solids.*

Turpeinen, Dr Urho Taneli. Dept. of Chemistry, U. of Helsinki, 00100 Helsinki, Finland. (1944) PhD, inorganic chemistry (U. Helsinki, 1977). Instructor. (tel. 0 650211, e-mail TURPEINEN at FINFUN.BITNET). *Crystal structures, metal complexes.*

Turunen, Dr Markus Johannes. Valmet Process Automation, PO Box 237, 33101 Tampere, Finland. (1947) DSc, physical metallurgy (Helsinki U. of Techn., 1975). Docent. (tel. 31 688111, fax. 31 650317, e-mail INSTRUME at OPMVAX.KPO.FI). *Solid state sensors, semiconductors, lattice defects.*

Uggla, Dr Rolf Åke Magnus. Dept. of Chemistry, U. of Helsinki, 00100 Helsinki, Finland. (1924) DTechn., physical chemistry (Helsinki U. of Techn., 1960). Docent. (tel. 0 650211, ext. 538, fax. 0 656591, e-mail UGGLA at FINFUN.BITNET). *Copper amine complexes, copper hexaoxoiodates, bin(diphenylphosphino)acetylene derivates.*

Unonius, Dr Lars-Olof. Dept. of Physics, U. of Helsinki, 00170 Helsinki, Finland. (1947) PhD., physics (U. Helsinki, 1987). Asst. (tel. 0 650211, ext. 225, telex 122229 nuphu sf). *X-ray scattering and attenuation.*

Vähäkangas, Mr Jouko Kaarlo. Inst. of Techn. Physics, U. of Oulu, Linnanmaa, 90570 Oulu, Finland. (1948) Techn.Lic., techn. physics (U. Oulu, 1980). Asst. (tel. 81 353611). *X-ray diffraction, diffuse scattering, surface structure.*

Vahvaselkä, Dr Aino Margit. Dept. of Physics, U. of Helsinki, 00170 Helsinki, Finland. (1942) PhD, physics (U. Helsinki, 1978). Aman. (tel. 0 650211, ext. 240, telex 122229 nuphu sf). *Deformation models, X-ray and neutron diffraction analysis.*

Vahvaselkä, Dr Kaarlo Sakari. Dept. of Physics, U. of Helsinki, 00170 Helsinki, Finland. (1942) PhD, physics (U. Helsinki, 1977). Asst. (tel. 0 650211, ext. 418, telex 122229 nuphu sf). *X-ray diffraction, solid and liquid metals.*

Valkonen, Prof. Jussi Uolevi. Dept. of Chemistry, U. of Jyväskylä, 40100 Jyväskylä, Finland. (1947) DTechn., inorganic chemistry (Helsinki U. of Techn., 1979). Prof. (tel. 41 291211, fax. 41 292797, e-mail J_VALKON at FINJYU.BITNET). *Inorganic chemistry, structural chemistry.*

Visapää, Mr Asko Edvard. Chemical Lab., Techn. Res. Centre of Finland, 02150 Espoo, Finland. (1921) MA, analytical chemistry (U. Helsinki). Res. officer. (tel. 0 4561 or 0 4565270). *Analytical chemistry, instrumental analysis, spectrometry, cellulose structure, X-ray diffractometry.*

Vorma, Prof., Dr Atso Ilmari. Dept. of Petrology, Geological Survey of Finland, 02150 Espoo, Finland. (1933) PhD, geology and mineralogy (U. Helsinki, 1963). Dept. chief, (tel. 0 46931, ext. 266, telex 123185 geolo sf). *Mineralogy, crystal structure determination.*

Ylinen, Dr Eero Elias. Wihuri Physical Lab., U. of Turku, 20500 Turku, Finland. (1944) PhD, physics (U. Turku, 1978). Res. physicist. (tel. 21 645944). *Nuclear magnetic resonance, ionic crystals.*

FRANCE

Sub-Editor: **Y. Epelboin**

Notes

1. International telephone country code - 33.

2. Degrees conferred by the French universities are the *doctorat-ès-sciences* (DSc) (four to six years study), the *agrégation* (Agr), the *diplome de docteur-ingénieur* (D.Ing), which is more technical, the *doctorat de 3e cycle* (D.3e cycle) (approximately between MSc and PhD). Since 1984 these diploma have been replaced by the *doctorat d'Université (DU)*. which is the equivalent to the PhD (three years study). It also exists lower level degrees: the *diplome d'études approfondies* (D.E.A.), the *diplome d'études supérieures* (D.E.S.), the *maîtrise-ès-sciences* (MSc) (approximately equivalent to BSc) and the *licence-ès-sciences* (LSc). The Grandes Ecoles in France confer a *diplome d'ingénieur* (Ing).

3. The functions at the Universities are *Professeur* (Professor), or *maître de conférences* (associate Professor). At the Centre National de la Recherche Scientifique the functions are *directeur-*, and *chargé- de recherche*. In industrial laboratories the function *ingénieur* (ing.) is generally used; further, *chercheur* (research scientist) is used.

4. The following abbreviations for institutions are used in the French list:

CEA - Commissariat à l'Energie Atomique
CEN - Centre d'Etudes Nucléaires
CENG - Centre d'Etudes Nucléaires Grenoble
CNET - Centre National d'Etudes des Télécommunications

CNRS - Centre National de la Recherche Scientifique
ILL - Institut Laue-Langevin
INRIA - Institut National de la Recherche Agronomique
LURE - Lab. pour l'utilisation du rayonnement électromagnétique

5. Additional abbreviations used in this list:

BP - Boîte postale
Miss - Mademoiselle

Mrs - Madame
Mr - Monsieur

Albouy Dr Pierre. Lab.Physique des Solides, Bat. 510, U. Paris Sud, 91405 Orsay Cedex, France. (1956) DSc, physics (U. Paris Sud). Chargé de rech. CNRS (tel. 1 + 69 41 60 99, fax 1 + 69 41 60 86). *Diffuse scattering, phase transistions, X-Ray diffraction*

Aleonard, Dr Suzanne. Lab. de Cristallographie, CNRS Grenoble, 25 avenue des Martyrs, BP 166X, Centre de Tri, 38042 Grenoble Cedex, France. (1926) DSc, physics (U. de Grenoble, 1965). Directeur de rech. CNRS (tel. 76 88 10 44). *Crystal chemistry, fluorine compounds.*

Allais, Prof. Gérard. U. de Caen, UER des Sciences, Esplanade de la Paix, 14032 Caen, France. (1935) DSc, physics (U. de Paris, 1967). Prof. (tel. 31 45 26 51). *Defect structures, microscopy.*

Andonov, Dr Paulette. Lab. de Magnétisme, CNRS, 1 place A.Briand, 92195 Meudon principal Cedex, France. (1932) DSc, physics (U. Paris Sud). Chargée de rech. CNRS. (tel. 1 + 45 34 75 50, ext. 20 76, fax 1 + 45 34 46 96). *Amorphous materials, liquids, structures.*

André, Dr Daniel. Lab. de Physicochimie Structurale, U. Paris 12, UFR des Sciences, av. du Gal de Gaulle, 94010 Créteil, France. (1946) DSc, physics (U. Paris Sud), 1975). Maître de Conf. (tel. 1 + 48 98 91 44, ext. 24 98). *Disorder, molecular crystals, X-ray diffraction*

Arnoux, Dr Bernadette Inst. de Chimie des Substances Naturelles, CNRS, 91190 Gif sur Yvette, France. (1952) DSc, chemistry (U. Paris 7, 1985). Chargé de rech. CNRS (tel. 1 + 69 82 30 18). *X-Ray diffraction, proteins.*

Authier, Prof. André Lab. de Minéralogie- Cristallographie, tour 16, U. Pierre et Marie Curie (Paris 6), 4 place Jussieu, 75252 Paris Cedex 05, France. (1932) DSc, physics (U. de Paris, 1961). Prof., (tel. 1 + 43 54 84 76, fax 1 + 43 54 40 97, email authier at frlmcp61.bitnet). *Crystal growth, defect structures, teaching, X-ray diffraction.*

Averbuch-Pouchot, Dr Marie-Thérèse. Lab. de Cristallographie, CNRS Grenoble, 25 avenue des Martyrs, BP 166X, Centre de Tri, 38042 Grenoble Cedex, France. (1940) DSc, crystal chemistry (U. de Grenoble, 1974). Chargé de rech. CNRS. (tel. 76 88 78 01). *Phosphates, structures.*

Ayroles, Dr René Centre de Recherches Rhone-Poulenc, Lab. de Microscopie Electronique, 14 rue des Gardinoux, 93308 Aubervilliers Cedex, France. (1936) Dsc, physics (U. de Toulouse, 1968). Resp. lab., (tel. 1 + 48 39 62 89, fax 1 + 48 39 61 00). *Electron microscopy, electron diffraction, materials science.*

Bagieu, Dr Muriel. Lab. de Cristallographie, CNRS, 166X, 38042 Grenoble Cedex, France. (1943) DSc, physics (U. de Grenoble, 1980). Maître de Conf. (tel. 76 88 78 01, fax 76 87 21 97). *Structural chemistry, synthesis, phosphates.*

Baldy, André. Lab. Chimie Inorganique Moléculaire, UA 126, Faculté des Sciences, Av. de l'escadrille Normandie-Niemen, 13397 Marseille Cedex 13, France. (1939) DU, physics (U. Aix-Marseille, 1971). Ingénieur de Rech. CNRS. (tel. 91 28 81 93). *Instrumentation, graphics.*

Bally, Dr Renée. Lab. de Minéralogie- Cristallographie, tour 16, U. Pierre et Marie Curie (Paris 6), 4 place Jussieu, 75230 Paris Cedex 05, France. (1928) DSc, physics (U. de Paris, 1966). Chargée de rech. CNRS. (tel. 1 + 43 36 25 25, ext. 52 41, fax 1 + 43 54 40 97, *Protein structures.*

Baruchel, Dr José ILL, Ave. des Martyrs, BP 156X, F-38042 Grenoble Cedex, France. (1947) DSc, physics (U. de Grenoble, 1980). Attaché de rech. CNRS.

(tel. 76 48 70 91). *Magnetism, Defect structures, neutron diffraction, phase transitions, scattering theory, X-ray diffraction.*

Bavoux, Dr Claude. Lab. de Minéralogie-Cristallographie, U. Claude Bernard 43 bld du 11 Novembre 1918, 69622 Villeurbanne Cedex, France. (1943) DSc, crystallography (U. C. Bernard, 1986). Maître de conf. (tel. 78 89 81 24, ext. 32 96) *Organic compounds, molecular crystals, polymorphism.*

Bentley, Dr Graham Arthur. Institut Pasteur, Unité d'Immunolgie Structurale, 25 rue du Dr Roux, 75724 Paris, France. (1946) PhD, chemistry (U. of Auckland, New Zealand, 1971). Chercheur. (tel. 1 + 45 68 86 07). *Molecular biology.*

Bertaut, Dr Erwin Félix. Lab. de Cristallographie, CNRS, 25 avenue des Martyrs, BP 166X, Centre de Tri, 38042 Grenoble Cedex, France. (1913) DSc, solid state physics (U. de Grenoble, 1949). Membre de l'Institut, (tel. 76 88 10 00, ext. 14 84, home tel. 76 96 36 75, telex 32 02 54 CNRSALP). *Cristallography, magnetism, phase transitions.*

Berthet-Colominas, Dr Carmen. Grenoble Outstation, European Molecular Biology Lab., Ave. des Martyrs, BP 156X, F-38042 Grenoble Cedex, France. (1940) Dr, solid state physics (U. de Grenoble, 1967). Responsable, rotating anode X-ray generators. (tel. 76 48 72 76, telex: 32 06 21F, fax 76 48 39 06, email berthet at frembl51.earn). *X-ray crystallography, adenovirus.*

Bessiere Dr Michel. LURE, Bat. 209D, Centre Universitaire Paris-Sud, 91405 Orsay Cedex, France. (1950) Ing. chemistry (ENSCP, 1974). Ing. CNRS. (tel. 1 + 64 46 81 25, fax 64 46 41 48F, email bessiere at frlure51.earn). *Diffractometry, phase transitions, diffuse scattering, disorder, instrumentation, materials science, metals, phase transitions, powder diffraction, quasicrystals, synchrotron radiation.*

Bideau, Prof. Jean-Pierre. Lab. de Cristallographie et Physique Cristalline, U. de Bordeaux 1, Faculté des Sciences, 351 cours de la Libération, 33405 Talence, France. (1937) DSc, physics (U. de Bordeaux, 1971). Professor. (tel. 56 80 61 59). *Biological structures.*

Bienfait, Prof. Michel. Dépt. of Physics, Faculté des Sciences de Luminy, 13288 Marseille, France. (1939) DSc, physics (U. de Nancy, 1965). Prof. (tel. 91 26 91 81, fax 91 41 89 16, email bienfait at frmop11.earn). *Surface physics, interfaces.*

Bodot, Prof. Hubert. Université de Provence, Centre Saint Jérome, Case 542, 13397 Marseille Cedex 13, France. (1933) DSc, physics (U. de Montpellier, 1957). Prof. (tel. 91 63 65 10, fax 91 02 05 50, email bodot at frmrs11.earn). *Disorder, organic compounds, structural chemistry.*

Bonnet, Prof. Jean-Jacques. Lab. de Chimie de Coordination, CNRS, 205 route de Narbonne, 31077 Toulouse Cedex, France. (1939) DSc, physical chemistry (U. de Toulouse, 1972). Prof., (tel. 61 33 31 71, fax 61 17 56 00). *Coordination compounds.*

Bonnet, Dr Michel. CENG, DRF/SPh-MDN, 85X, 38041 Grenoble Cedex, France. (1944) Dsc Physics, (U. d'Orsay, 1976). Staff member, (tel. 76 88 39 56). *Neutron diffraction, magnetism.*

Bonpunt, Dr Louis. U. de Bordeaux 1, Lab. de Cristallographie et de Physique Cristalline, 351 cours de la Libération, 33405 Talence, France. (1944) Dsc Physics, (U. de Bordeaux 1, 1981). Chargö de Recherches, (tel. 56 84 61 55, email hospital at frbdx11.earn). *Point defects, scattering.*

Boulesteix, Prof. Claude. Lab. Microscopie Electronique Appliquée, Université Aix-Marseille 3, av. Escadrille Normandie-Niemen, Case 542, 13397 Marseille Cedex 13, France. (1937) DSc, physics (U. de Paris, 1966). Prof. (tel. 91 28 83 90, fax 91 02 05 50). *Electron microscopy, oxides, phase transitions, superconductors. Disorder, organic compounds, structural chemistry.*

Bouraoui, Prof. Ahmed. Dépt. de Physique, Ecole Normale Supérieure, 7021 Bizerte, Tunisie. (1931) DSc, physics 5U. de Paris, 1965). *Radiocristallography, materials science.*

Brown, Dr Penelope Jane. Institut Laue Langevin, Avenue Des Martyrs BP 156X, Grenoble 38042, France. (1932) PhD,Physics (Cambridge,UK,1958) Senior Scientist. (tel. 33 + 76487040, telex. 320 621, Fax. 76483906, email Brown at Frill). *Neutron diffraction, spin density, diffractometry.*

Brückel, Dr Thomas. ILL, 156X, Centre de tri, 38042 Grenoble Cedex, France. (1957) Dr (U. Tübingen, 1988). physicist. (tel. 76 48 70 66, fax 76 48 39 06, email bruckel at frill51.earn). *Diffuse scattering, disorder, magnetism, neutron diffraction.*

Brunie, Dr Simone. Lab. de Biochimie, Ecole Polytechnique, 91128 Palaiseau Cedex, France. (1936) DSc, physics (U. de Lyon, 1970). Dir. de rech. CNRS. (tel. 1 + 69 41 82 00, ext. 28 40 email brunie at frpoly52.earn). *Macromolecules.*

Capelle, Dr Bernard. Lab. de Minéralogie- Cristallographie, tour 16, U. Pierre et Marie Curie (Paris 6), 4 place Jussieu, 75230 Paris Cedex 05, France. (1949) DSc, physics (U. de Paris, 1982). Chargé de rech. (tel. 1 + 43 36 25 25, ext. 52 17, fax 1 + 43 54 40 97, email capelle at frlmcp61.earn). *Phase transitions, defect structure, quartz.*

Caranoni, Dr Anny. Lab. de Physique Cristalline, CNRS, avenue Escadrille Normandie-Niemen, Case 151, 13397 Marseille Cedex 13, France. (1941) DSc, physics (U. d'Aix-Marseille 3, 1979). Maître de conf. (tel. 91 28 83 10, fax 91 28 80 30, email rahma at frmrs11.earn). *Powder diffraction, phase determination, crystallographic data.*

Chaillout, Dr Catherine. Lab. de Cristallographie, CNRS, 166X, 38042 Grenoble Cedex, France. (1961) DU, physics (U. Grenoble, 1986). Chargé de rech. CNRS (tel. 76 88 78 02, fax 76 87 21 97). *X-ray diffraction, electron diffraction, electron microscopy, powder diffraction.*

Champier, Prof. Georges. Lab. de Physique du Solide, Institut National Polytechnique de Lorraine, Ecole Nationale Supérieure des Mines de Nancy, Parc de Saurupt, 54042 Nancy, France. (1927) DSc, physics (U. de Nancy, 1958). Prof. (tel. 83 57 41 56). *Defect structures, metals, X-ray diffraction.*

Chattopadhyay, Dr Tapan Kumar. CENG, DRF, Groupe Magnétisme et Diffraction Neutronique, BP 85X, 38041 Grenoble Cedex, France. (1942) PhD, physics (Indian Inst. of Technology, Kharagpur, India 1972), physicist (tel. 76 88 31 44, fax 76 48 39 06, email chatt at frill51.earn) *Neutron diffraction, magnetism, phase transitions.*

Chevrier, Dr Lab. de Cristallographie Biologique, IBMC, 15 rue René Descartes, 67084 Strasbourg Cedex, France. (1941) Dsc, chemistry (U. de Strasbourg, 1971). Chercheur CNRS (tel. 88 41 70 94, email cris at fribcp51.earn). *Structural biology, molecular crystals, macromolecules, computing.*

Cockcroft, Dr Jeremy Karl. ILL, BP 156X, 38042 Grenoble Cedex, France. (1958) PhD, chemistry (U. Oxford, 1985). scientist (tel. 76 48 70 89, fax 76 48 39 06, email cockcroft at frill51.earn) *Neutron diffraction, powder diffraction, phase transitions, magnetism, molecular crystals, computing.*

Cohen-Addad, Dr Claudine. ILL, BP 156X, 38042 Grenoble Cedex, France. (1939) DSc, physics (U. de Grenoble, 1969). Dir. de rech. (tel. 76 48 71 11, fax 76 48 39 06, email addad at frill51.earn). *Organic compounds, proteins, structural chemistry, structural biology.*

Collin, Dr Gaston. Lab. de Physique des Solides, Bat. 510, Université Paris Sud, 91405 Orsay Cedex, France. DSc, physics (U. Paris Sud, 1971). Chercheur CNRS. (tel. 1 + 69 41 60 56, fax 1 + 69 41 60 86). *Crystal growth, diffuse scattering, inorganic crystals, materials science.*

Courseille, Prof. Christian. Lab. de Cristallographie, Université Bordeaux I, 351 Cours de la Libération, 33405 Talence Cedex, France. (1945) DSc, physics (U. de Bordeaux I, 1977). Professor (56 84 61 63). *X-ray diffraction, structural biology, organic compounds, proteins, macromolecules.*

Curie, Prof. Daniel. Lab. de Luminescence, tour 13, U. Pierre et Marie Curie (Paris 6), 4 place Jussieu, 75252 Paris Cedex 05, France. (1927) DSc, physics (U. de Paris, 1951). Prof. émer. (tel. 1 + 43 36 25 25, ext. 44 57). *Luminescence, inorganic crystals, semiconductors.*

Curien, Prof. Hubert. Lab. de Minéralogie- Cristallographie, tour 16, U. Pierre et Marie Curie (Paris 6), 4 place Jussieu, 75252 Paris Cedex 05, France. (1924) DSc, physics (U. de Paris, 1952). Prof. (tel. 1 + 43 36 25 25, ext. 5083, fax 1 + 43 54 40 97). *Structures, defect structures.*

Dahan, Dr Françoise. Lab. de Chimie de Coordination, CNRS, 205 route de Narbonne, 31077 Toulouse Cedex, France. (1941) DSc, physics (U. de Paris 6, 1980). Ing. (tel. 61 33 31 29, fax 61 55 30 03). *Coordination compounds, X-ray diffraction, computing.*

d'Anterroches, Dr Cécile. CNET, BP98, 38243 Meylan Cedex, France. (1952) DSc, physics, (U. Grenoble, 1982), Ing., (tel. 76 76 42 93, fax 76 90 34 43, telex: 98 07 27F). *Electron microscopy, materials science, crystal growth.*

Darces, Dr Jean-François. Lab. de Cristallographie et Chimie Minérale, Faculté des Sciences, 25030 Besancon Cedex, France. (1944) DSc, physics (U. de Franche

Comté, 1984). Ma&icitre de conf., (tel. 81 53 81 22). *X-ray diffraction, minerals, defect structures.*

Dartyge, Dr Elisabeth. LURE, Bat. 209D, U. Paris Sud, 91405 Orsay, France. (1941) DSc, physics (U. de Paris sud, 1979). Maître de conf. (tel. 1 + 64 46 80 20, fax 1 + 64 46 41 48, email fontaine at frlure51.earn). *EXAFS.*

Davidson, Mr Patrick. Lab. de Physique des Solides, Bat. 510, U. Paris Sud, 91405 Orsay Cedex, France. (1960) D.3e cycle, material Sc. (U. de Paris sud, 1984). Chercheur CNRS, (tel. 1 + 69 41 53 93). *X-ray diffraction, diffuse scattering, liquid crystals, polymers.*

Delapalme, Dr Alain. Lab. Léon Brillouin, CEA/CEN Saclay BP 2, 91191 Gif sur Yvette Cedex, France. (1930) DSc, solid state physics (U. de Grenoble, 1967). Chercheur. (tel. 1 + 69 08 82 61). *Spin density, molecular crystals.*

Delettré, Dr Jean. Lab. de Minéralogie- Cristallographie, tour 16, U. Pierre et Marie Curie (Paris 6), 4 place Jussieu, 75230 Paris Cedex 05, France. (1943) DSc, physics (U. de Paris 6, 1978). Maître de conf. (tel. 1 + 43 36 25 25, ext. 52 41, fax 1 + 43 54 40 97, email delettre at frlmcp61.earn) *Proteins, organic structures.*

Delord, Prof. Pierre. Lab. de Cristallographie, U. des Sciences et Techniques du Languedoc, place Eugène Bataillon, 34000 Montpellier, France. (1937) DSc, crystallography (U. de Montpellier, 1970). Prof. (tel. 67 63 91 44, ext. 501, fax 67 54 30 79, email delord at frmop11.earn). *Small-angle scattering, liquid crystals, colloids.*

Dénoyer, Dr. Françoise. Lab. de Physique des Solides, Bat. 510, Université Paris Sud, 91405 Orsay Cedex, France. (1945) DSc, physics (U. de Paris-Sud, 1977). Dir. de Rech. (tel. 1 + 69 41 60 86). *Diffuse scattering, neutron diffraction, X-ray diffraction, phase transitions, quasicrystals.*

Doucet, Dr Jean. Lab. de Physique des Solides, Bat. 510, Université paris-Sud, 91405 Orsay Cedex, France. (1947) DSc, physics (U. de Paris Sud 1978). Charg é de rech. (tel. 1 + 69 41 60 59, email doucet at frlure51.earn). *Proteins.*

Dubourg, Mr Antoine. Lab. de Physique Industrielle Pharmaceutique, Faculté de pharmacie, Av. Charles Flahaud, 34060 Montpellier, France. (1945) D.3e cycle, physics,(U. P.Sabatier, 1970). Maître de conf. (tel. 67 63 53 58, fax 67 61 16 22, email dubourg at frmop22.earn). *X-ray diffraction, computing, hydrogen-bonding.*

Ducruix, Dr Arnaud. Inst. de Chimie des Substances Naturelles, CNRS, Cristallochimie, 91198 Gif sur Yvette, France. (1947) DSc, physics (U. de Paris sud, 1976). Dir. de rech. (tel. 1 + 69 82 30 62, fax 1 + 69 07 72 47, email ducruix at frcgm51.earn). *X-ray diffraction, macromolecules, crystal growth.*

Durand, Dr. Dominique, Lab. de Physique des Solides, Bat. 510, .91405 Orsay Cedex, France. (1957) DSc, physics (U. de Paris Sud, 1987). Chargé de rech. CNRS (tel. 1 + 69 41 60 61, email durand at frsol11.earn). *Phase transitions, neutron diffraction, X-ray diffraction.*

Durif, Dr André Lab. de Cristallographie, CNRS Grenoble, 25 av. des Martyrs, BP 166X, Centre de Tri, 38042 Grenoble cedex, France. (1929) DSc, crystal chemistry (U. de Grenoble, 1958). Dir. scient. (tel. 76 88 10 45, email crystrx at frcpn11.earn). *Phosphates, minerals.*

Epelboin, Dr Yves. Lab. de Minéralogie- Cristallographie, tour 16, U. Pierre et Marie Curie (Paris 6), 4 place Jussieu, 75252 Paris Cedex 05, France. (1944) DSc, physics (U. P.M.Curie , 1974). Maître de conf. (tel. 1 + 43 36 25 25, ext. 52 16, fax 43 54 40 97, email epelboin at frlmcp61.earn). *X-ray diffraction, computing, defect structures.*

Espinat, Dr Didier. Lab. de Rayons X, Institut Français du Pétrole, 1 et 4 av. de Bois-Préau, 92506 Rueil-Malmaison Cedex, France. (1954) D.Ing., (1982). dir. of the lab., (tel. 1 + 47 52 62 05). *Minerals, X-ray diffraction, polymers, powder diffraction, small-angle scattering.*

Fauvet, Mr Gérard. Lab de Rayons X, ENRAF-NONIUS FRANCE, 28 ter av. de Versailles, 93320 Gagny, France. (1948) D.3e cycle (1977). Ing. (tel. 1 + 45 09 04 04, fax 1 + 45 09 22 87). *Apparatus.*

Fontaine, Dr Alain. LURE, Université Paris Sud, Bat. 209, 91405 Orsay, France. (1944) DSc, physics (U. de Paris Sud, 1975). Maître de conf. (tel. 1 + 64 46 80 18/81 44, fax 1 + 64 46 41 48, email fontaine at frlure51.earn). *Synchrotron, EXAFS, metals, X-ray diffraction, supraconductivity.*

Fontaine, Prof. Hubert. Lab. de Dynamique des Cristaux Moléculaires, Université de Lille I, UFR de Physique, BP 36, 59655 Villeneuve d'Ascq, France. (1936) DSc, solid state physics (U. de Paris, 1973). Prof. (tel. 20 43 47 74, fax 20 43 49 95). *Molecular crystals, disorder, phase transitions, X-ray diffraction.*

Fouret, Prof. René U. des Sciences et Techniques de Lille, UER de Physique Fondamentale, BP 36, 59655 Villeneuve d'Ascq, France. (1925) DSc, solid state physics (U. de Paris, 1963). Prof. (tel. 20 91 92 22, ext. 2173, fax 20 43 49 95). *X-ray diffraction, molecular crystals, lattice dynamics, disorder.*

Fontecilla-Camps, Dr Juan C. Lab. de cristallographie et de cristallisation des Macromolécules biologiques, Faculté de Medecine Secteur Nord, bld Pierre Dramard, 13236 Marseille. (1949) PhD, ristallography (U. of Alabama, 1979). Chargé de rech. CNRS. (tel. 91 65 79 47, fax 91 65 75 95). *Protein crystallography.*

Friedel, Prof. Jacques. Lab. de Physique des Solides, U. Paris Sud, (Paris 11), Bat. 510, 91405 Orsay, France. (1921) PhD, DSc, solid state physics (Bristol U., UK, 1951; U. de Paris, 1954). Prof. (tel. 1 + 69 28 59 86, ext. 33 66). *Metals, alloys, defect structures.*

Gabis, Prof. Victor Michel. Lab. d'Etude des Matériaux Minéraux, Ecole Sup&eb.rieure de l'Energie et des Matériaux, 45067 Orléans Cedex 2, France.

(1932) DSc, geochemistry, mineralogy (U. de Paris 6, 1964). Prof. (tel. 38 41 70 51). *Ceramics, minerals.*

Gasperin, Dr Madeleine. Lab. de Minéralogie- Cristallographie, tour 16, U. Pierre et Marie Curie (Paris 6), 4 place Jussieu, 75252 Paris Cedex 05, France. (1924) DSc, crystallography (U. de Paris, 1960). Dir. de rech. CNRS (tel. 1 + 43 36 25 25, ext. 50 82, fax 1 + 43 54 40 97, email gasperin at frlmcp61.earn). *Crystal growth, X-ray diffraction, structural chemistry, inorganic crystals, crystallographic data.*

Gatineau, Dr Lucien Charles. Centre de Recherche sur les Solides à Organisation Cristalline Imparfaite (CRSOCI), 1bis rue de la Férollerie, 45071 Orléans cedex 2, France. (1927) DSc, physics (U. de Paris, 1964). Dir. de rech. (tel. 38 63 39 37). *Disorder, structural chemistry, X-ray diffraction.*

Gauthier, Prof. Jean-Pierre. Lab. de Minéralogie- Cristallographie, U. de Lyon, 43 bd du 11 novembre 1918, 69622 Villeurbanne Cedex, France. (1943) DSc, crystallography (U. de Lyon, 1978). Professor. (tel. 78 89 81 24, ext. 32 49). *Polytypism, surfaces, interfaces, electron diffraction, gems.*

Geoffre, Mr Serge. Lab. de Cristallographie, Université Bordeaux I,rie, 351 Cours de la Libération, 33405 Talence, France. (1941) D. 3e cycle, crystallography (U. Bordeaux I, 1967), Ing. (tel. 56 84 61 49). *X-ray diffraction, crystal growth, macromolecules, proteins, viruses.*

George, Dr Amand. Lab. de Physique du Solide, Ecole des Mines de Nancy, Parc de Saurupt, 54042 Nancy Cedex, France. (1946) DSc, physics, (Institut National Polytechnique de Lorraine, 1977). Dir. de rech. (tel. 83 57 41 58) *Material science, defect structures, semiconductors, X-ray diffraction.*

Giegé, Dr Richard. Institut de Biologie Moléculaire et Cellulaire du CNRS, 15 rue René Descartes, 67084 Strasbourg, France. (1942) DSc, chemistry. Dir. de rech. (tel. 88 41 70 58, fax 88 60 75 50, email giege at fribcp51.earn). *Crystal growth, macromolecules.*

Ginderow, Dr Daria. Lab. de Minéralogie- Cristallographie, tour 16, U. P.M.Curie (Paris 6), 4 place Jussieu, 75252 Paris Cedex 05, France. (1933) DSc, (U. d'Orléans, 1969). Chargée de rech. CNRS. (tel. 1 + 43 36 25 25, ext. 52 41 fax 1 + 43 54 40 97, email ginderow at frlmcp61.earn). *Minerals, organometallics.*

Gleizes, Prof. Alain Nicolas. Lab. Dépots chimiques en couches minces, UA 445, ENSCT, Institut National Polytechnique de Toulouse, 118 route de Narbonne, 31077 Toulouse Cedex, France. (1943) DSc, chemistry (U. de Toulouse, 1974). Professor. (tel. 61 17 56 56). *Coordination compounds, material science, teaching, X-ray diffraction.*

Graf, Prof. René Lab. des Rayons X, Faculté des Sciences de Rouen, BP 118, 76134 Mont-Saint-Aignan Cedex, France. (1920) DSc, physics (U. de Paris, 1955). Prof. (tel. 35 14 66 33, fax 35 14 63 49). *Material science, X-ray diffraction.*

Grandjean, Prof. Daniel Lab. de Cristallochimie, U. de Rennes, Campus de Beaulieu, 35042 Rennes Cedex, France. (1939) DSc, physics (U. de Strasbourg, 1966) Professor (tel. 99 28 62 46, fax 99 38 34 87). *Coordination compounds, metals.*

Guinier, Prof. André Jean. Lab. de Physique des Solides, U. Paris Sud, Bat. 510, 91405 Orsay, France. (1911) DSc, physics (U. de Paris, 1939). Prof.émérite. (tel. 1 + 69 41 00 92, ext. 32 73). *X-ray diffuse scattering.*

Guitel, Mr Jean-Claude. Lab. de Cristallographie, CNRS, BP 166X, Centre de Tri, 38042 Grenoble Cedex, France. (1939) Master Physics, (U. de Grenoble, 1961). Maître de conf. (tel. 76 88 11 41, fax 76 88 11 61, email crystal at frcpn11.earn). *Computing, inorganic structures, phosphates, biology.*

Haget, Dr Yvette. Lab. de Cristallographie, U. de Bordeaux 1, 315 cours de la Libération, 33405 Talence, France. (1934) DSc, physics (U. de Bordeaux, 1968). Dir. de rech. (tel. 56 84 61 57). *Disorder, molecular crystals, molecular alloys, phase transitions.*

Hansen, Dr Niels K. Lab. Minéralogie-Cristallographie, Université de Nancy I, BP 239, 54506 Vandoeuvre-les-Nancy Cedex, France. (1950) LSc., chemistry (U. Aarhus, Denmark, 1978). Mître de conf. (tel. 83 91 22 65, fax 83 32 95 90, email niels at frciil71.earn). *Electron density, momentum density, X-ray diffraction.*

Hardy, Mrs Anne-Marie. Lab. de Cristallochimie minérale, Faculté des Sciences de Poitiers, 40 av. du recteur Pineau, 86022 Poitiers Cedex, France. (1938) D.ét.sup., inorganic chemistry (U. de Rennes, 1961). Ing. (tel. 49 46 20 74). *X-ray diffraction, inorganic crystals.*

Hardy, Prof. Antoine. Lab. de Cristallochimie Minérale, Faculté des Sciences de Poitiers, 40 av. du recteur Pineau, 86022 Poitiers, France. (1929) DSc, inorganic chemistry (U. de Bordeaux, 1962). Prof. (tel. 49 46 20 74). *Structural chemistry, materials science.*

Haser, Dr Richard Michel. LCCMB-CNRS, Faculté de Médecine-Nord, Bld Pierre Dramard, 13326 Marseille Cedex 15, France. (1942) DSc, physics (U. d'Aix-Marseille, 1972). Dir. de rech. (tel. 91 65 79 47, fax 91 65 75 95). *Proteins, crystal growth. X-Ray diffraction.*

Hewat, Dr A.W. ILL, 156X, 38042 Grenoble Cedex, France. (1942) PhD, physics (U. Melbourne, 1970). Staff scientist (tel. 76 48 72 13, fax 76 48 39 06, email hewat at frill51.earn). *Neutron diffraction, phase transitions, superconductors.*

Hewat, Dr Elizabeth. D. LETI,, CENG, 85X, 38041 Grenoble Cedex, France. (1947) PhD, physics (U. Oxford, 1975). Ing. (tel. 76 88 40 20, fax 76 88 51 57). *Electron diffraction, electron microscopy, semiconductors, proteins.*

Hodeau, Dr. Jean-Louis. Lab. de Cristallographie, CNRS, 166X, 38042 Grenoble Cedex, France. (1952) DSc, physics, (U. Grenoble, 1984). Chargé de rech. (tel.

76 88 11 40, fax 76 87 21 97). *X-ray diffraction, electron diffraction, phase transitions,*

Horn, Prof. Paul. Lab. de Biophysique Moléculaire, BP 239, 54506 Vandoeuvre-les-Nancy, France. (1925) DSc, physics (U. de Strasbourg, 1954). Prof. (tel. 83 91 21 05). *Structural biology.*

Hospital, Dr Michel. Lab. de Cristallographie, U. Bordeaux 1, 351 cours de la Lib ération, 33405 Talence, France. (1935) DSc, physics (U. Bordeaux 1, 1968). Dir. de rech. CNRS. Dir. du lab. (tel. 56 84 61 61, fax 56 80 08 37, email hospital at frbdx11.earn). *X-ray diffraction, molecular crystals, structural biology.*

Housty, Dr Jacques. Lab. de Cristallographie, U. de Bordeaux 1, 351 cours de la Libération, 33405 Talence, France. (1928) DSc, crystallography (U. de Bordeaux, 1966). Maître de conf. (tel. 56 84 61 61, fax 56 80 08 37). *Structural chemistry, crystal growth.*

Janin, Prof. Joel. Inst. de biochimie, U. Paris Sud, bat. 430, 91405 Orsay Cedex, France. (1943) DSc, physics (U. de Paris Sud, 1969). Prof. (tel. 1 + 69 41 79 73, fax 1 + 64 46 41 48, email janin at frlure51.earn). *Proteins, computing, enzymology.*

Janot, Prof. Christian. ILL, 165X, 38042 Grenoble Cedex, France. (1936) DSc, physics (U. Nancy I, 1963). Senior scientist. (tel. 76 48 73 27, fax 76 48 39 06, email janot at frill51.earn). *Defect structures, disorder, material science, metals, neutron diffraction, quasicrystals, small-angle scattering.*

Joel, Dr Nahum. 46 rue Fabert, 75007 Paris, France. (1924) PhD, physics (Cambridge U., UK, 1959). (tel. 1 + 45 51 97 85). *Optical, elastic properties of crystals, teaching.*

Joubert, Prof. Jean-Claude. Lab. de Matériaux et de Génie Physique, INPG, ENSPG, BP 46, Saint Martin d'Heres, France. (1939) DSc, physics (U. de Grenoble, 1965). Prof. (tel. 76 82 63 09, fax 76 82 63 01, email joubert at frinpg.earn). *Supraconductors, magnetism, ceramics, materials science, crystal growth, inorganic crystals.*

Jourdan, Dr Claude, René Centre de Recherche des Mécanismes de la Croissance Cristalline, campus de Luminy, 13288 Marseille Cedex 2, France. (1934) DSc, solid state physics (U. de Marseille, 1965). Dir. de rech. (tel. 91 41 01 52). *Metals, crystal growth, defect structures.*

Kern, Prof. Raymond. Centre de Recherche des Mécanismes de la Croissance Cristalline, CNRS, campus Luminy, case 913, 13288 Marseille Cedex 9, France. (1928) DSc, physics (U. de Strasbourg, 1953). Prof. (tel. 91 41 01 52, fax 91 41 89 16). *Crystal growth, surface structure, polymers.*

Kléman, Dr Maurice. Lab. de Physique des Solides, U. Paris Sud, (Paris 11), Bat. 510, 91405 Orsay, France. (1934) DSc, solid state physics (U. de Paris, 1967). Dir. de rech., CNRS. (tel. 1 + 69 41 82 50, ext. 33 31, fax 1 + 69 41 60 86, email kleman at frsol11.earn). *Defect structures, liquid crystals, magnetism, glass, quasicrystals.*

Labbé, Dr Philippe. Lab. CRISMAT, ISMRa, Bld du Maréchal Juin, 14032 Caen Cedex, France. (1939) DSc, physics (U. de Caen, 1978). Maître de conf. (tel. 31 45 26 12, fax 31 47 31 11). *X-ray diffraction, structural chemistry, oxides, teaching.*

Lajzerowicz, Prof. Janine. Université Joseph Fourier, Lab. de Spectrométrie Physique, BP 87, 38042 Grenoble Cedex, France. (1932) DSc, physics (U. de Grenoble, 1964). Prof. (tel. 76 51 47 61, fax 76 51 48 48). *Phase transitions.*

Lambert, Prof. Marianne. Lab. Léon Brillouin, CEN Saclay, 91191 Gif sur Yvette Cedex, France. (1932) DSc, physics (U. de Paris, 1958). Prof. (tel. 1 + 69 08 32 54, fax 1 + 69 08 82 61). *Phase transitions, order-disorder, lattice dynamics.*

Langlois d'Estaintot, Mme Bétrice. Lab. de Cristallographie et de Physique cristalline, Université de Bordeaux I, 351 Cours de la libération, 33405 Talence Cedex, France. (1960) D.3e cycle, physics (U. de Paris 6, 1983) Chargé de rech.(tel. 56 84 60 00, ext. 80 55, email hospital at frbdx11.earn). *Macromolecules.*

Laugier, Mr Jean. CENG, DRF-G/ Service de Physique, Groupe Structures, 85X, 38041 Grenoble Cedex, France. (1932) LSc, (1954). Ing. (tel. 76 88 41 46, fax 76 88 51 53). *X-ray diffraction.*

Le Bars, Dr Michéle, Lab. de Minéralogie-cristallographie, Université P.M.Curie, 4 place Jussieu, 75252 Paris Cedex 05, France. (1940) DSC, physics (U. Grenoble I, 1985). Maître de conf. (tel. 1 + 43 36 25 25, ext. 45 85, fax 1 + 43 54 40 97, email lebars at frlmcp61.earn). *Molecular crystals, proteins.*

Leclaire, Dr André Lab. de Cristallographie et Sciences des Materiauxs, CRISMAT-ISMRa, bld du Maréchal Juin, 14032 Caen, France. (1944) Dsc. (U. de Caen, 1976). Chargé de rech. (tel. 31 45 26 22, fax 31 47 31 11). *X-ray diffraction, inorganic crystals.*

Ledesert, Dr Mariannick. Lab. CRISMAT, ISMRa, bld du Maréchal Juin, 14032 Caen, France. (1934) DSc, physics (U. de Caen 1965) Chargé de rech. (tel. 31 45 26 13). *Inorganic crystals, X-ray diffraction.*

Lefaucheux, Dr Françoise. Lab. Minéralogie-Cristallographie, Université P.M.Curie, 4 place Jussieu, 75252 Paris Cedex05, France. (1938) DSc, physics (U. P.M. Curie, 1974) Maître de conf. (tel. 1 + 43 36 25 25 ext. 52 16, fax 1 + 43 54 40 97, email labo at frlmcp61.earn). *Crystal growth, defect structures, X-ray topography.*

Legros, Prof. Jean-Pierre. Lab. de chimie de coordination, CNRS, 205 route de Narbonne, 31400 Toulouse, France. (1943) DSc,(1976). Professor (tel. 61 33

31 00, fax 61 55 30 03, email ausoleil at frtou71.earn). *X-ray diffraction, structural chemistry, molecular crystals, materials science, coordination compounds.*

Lehmann, Dr Mogens. ILL, avenue des Martyrs, 38042 Grenoble Cedex, France. (1942) LSc, (U. de Aarhus, 1972) staff scientist. (tel. 76 48 73 82, fax 76 48 39 06, email lehmann at frill51.earn). *Neutrons, molecular crystals, instrumentation.*

Leligny, Mr Henri. Lab. CRISMAT, ISMRa, bld du Maréchal Juin, 14032 Caen, France. (1941) D.3e cycle, crystallography (U. de Paris 6, 1971). Maître de conf. (tel. 31 45 26 13) *X-ray diffraction.*

Lepicard, Dr Geneviève, Lab. de Minéralogie- Cristallographie, U. de Nancy 1, BP 239, 54056 Vandoeuvre Cedex, France. (1939) DSc, physics (U. de Paris, 1978) Maître de conf. (tel. 83 91 20 00, ext. 22 68). *Structures.*

Levelut, Dr Anne-Marie. Lab. de Physique des Solides, U. Paris Sud, (Paris 11), Bat. 510, 91405 Orsay, France. (1936) DSc, physics (U. de Paris, 1968). Dir. de rech. CNRS. (tel. 1 + 69 41 53 94, fax 1 + 69 41 60 86). *Liquid crystal, diffuse scattering, X-ray diffraction.*

Lewit-Bentley Dr Anita. LURE, bat. 209D, Université Paris Sud, 91405 Orsay Cedex, France. (1948) RNDr, physical chemistry (Charles U., Prague, Czechoslovakia, 1972). Sci. (tel. 1 + 64 46 41 48, fax1 + 64 46 41 48). *Structural biology.*

Lifchitz, Mr Alain. Lab. de Minéralogie- Cristallographie, tour 16, U. Pierre et Marie Curie (Paris 6), 4 place Jussieu, 75230 Paris Cedex 05, France. (1946) D.3e cycle, physics (U. de Paris 6, 1974). Chargé de rech. CNRS (tel. 1 + 43 36 25 25, ext. 45 84, fax 1 + 43 54 40 97, email lifchitz at frlmcp61.earn). *Stuctural biology, computing.*

Louer, Dr Daniel. Lab. de Cristallochimie, Université Rennes I, av. du Gén&e&.ral Leclerc, 35042 Rennes Cedex, France. (1942) DSc, physics (U. de Rennes, 1969). Dir. de rech. (tel. 99 28 62 48, fax 99 38 34 87, email louer at frcicb81.earn). *Powder diffraction, crystal chemistry.*

Loupias, Prof. Geneviève. Lab. de Minéralogie- Cristallographie, U. P.M.Curie, 4 place Jussieu, 75230 Paris Cedex 05, France. (1940) DSc, physics (U. de Paris 6, 1977). Professor. (tel. 1 + 43 36 25 25, ext. 45 84, fax 1 + 43 54 40 97, email labo at frlmcp61.earn). *Momentum density, electron density, inelastic scattering.*

Malgrange, Prof. Cécile. Lab. de Minéralogie- Cristallographie, U. Pierre et Marie Curie, 4 place Jussieu, 75252 Paris Cedex 05, France. (1939) DSc, physics (U. de Paris, 1967). Prof. (tel. 1 + 43 36 25 25, ext. 52 15, fax 1 + 43 54 40 97, email labo at frlmcp61.earn). *X-ray topography, scattering theory.*

Marezio, Dr Massimo. Lab. de Cristallographie, CNRS, BP 166X, 38042 Grenoble Cedex, France. Dir. de rech. Dir. du lab. (tel. 76 88 10 40, email crystrx at frcpn11.earn). *Structural chemistry, superconductors, materials science.*

Marsau, Prof. Pierre. Lab. de Cristallographie et Physique Cristalline, U. de Bordeaux 1, 351 cours de la Libération, 33405 Talence, France. (1937) DSc, physics (U. de Bordeaux 1, 1972). Prof. (tel. 56 84 61 52). *Organic compounds, computing.*

Mason, Dr Sax Anton. ILL, BP 156X, 38042 Grenoble Cedex, France. (1946) PhD, chemistry (Melbourne U., Australia, 1971). Staff scient. (tel. 76 48 70 67, fax 76 48 39 06, email mason at frill51.earn). *Macromolecules, neutron diffraction, instrumentation.*

McIntyre, Dr Garry, James. ILL, 156X, 38042 Grenoble Cedex, France. (1951) PhD, physics (U. de Melbourne, 1978). Staff scientist. (tel. 76 48 70 90, fax 76 48 39 06, email mcintyre at frill51.earn). *Diffractometry, electron density, thermal vibrations, magnetism, scattering theory, proteins.*

Mérigoux, Prof. Henri. Lab. de Cristallographie, U. de Franche Comté 25030 Besançon, France. (1935) DSc, physics (U. de Paris, 1967). Prof. (tel. 84 53 81 22). *Cristal growth, diffractometry, minerals, symmetry.*

Metz, Dr Bernard. Inst. de Chimie, 67008 Strasbourg, France. (1941) DSc, chemical crystallography (U. de Strasbourg, 1972). Maître de conf. (tel. 88 41 68 00). *Structural chemistry, coordination compounds, organic compounds.*

Michel, Prof. Pierre. Lab. de Minéralogie- Cristallographie, U. Claude Bernard (Lyon 1), 43 bd du 11 novembre 1918, 69622 Villeurbanne Cedex, France. (1925) DSc, physics (U. de Strasbourg, 1958). Prof., Dir. du lab. (tel. 78 89 81 24, ext. 32 21). *Materials science, minerals, teaching.*

Mimault, Prof. Jean. Lab. de Métallurgie Physique, Faculté des Sciences de Poitiers, 40 avenue du recteur Pineau, 86022 Poitiers Cedex, France. (1945) DSc, physics (U. de Poitiers, 1978). Prof., (tel. 49 46 26 30 ext. 360). *Phase determination, Exafs.*

Minari, Prof. Fernand. Lab. de Physique Cristalline, U. Aix-Marseille 3, avenue Escadrille Normandie-Niemen, 13397 Marseille Cedex 13, France. (1934) DSc, physics (U. d'Aix-Marseille, 1971). Prof. (tel. 91 28 83 11, fax 91 28 80 30, email rahma at frmrs11.earn). *Crystal defects, metals, semiconductors.*

Moras, Dr Dino. Lab. de Cristallographie Biologique, IBMC,CNRS, 15 rue Descartes, 67084 Strasbourg Cedex, France. (1944) DSc, chemistry (U. Louis Pasteur, 1971) Dir. de rech. (tel. 88 41 70 24, fax 88 60 75 50, email moras at fribcp51.earn). *Macromolecules, proteins, structural biology.*

Moreau, Prof. Jean-Michel. Lab. de Structure de la Matière, U. de Savoie, 9 rue de l'Arc en Ciel, BP 240, 74942 Annecy le Vieux Cedex, France. (1944) Dsc physics (U. de Grenoble, 1976) Prof. (tel. 50 23 29 93). *Metals, inorganic crystals, X-ray diffraction, magnetism.*

Moret, Dr Roger. Lab. de Physique des Solides, Université Paris Sud, 91405 Orsay Cedex, France. (1947) DSc, physics, (U. P.M.Curie, 1977). Chargé de rech.

(tel. 1 + 69 41 60 55, fax 1 + 69 41 60 86, email moret at frsol11.earn). *X-ray diffraction, diffuse scattering, phase transitions, high pressures.*

Mornon, Dr Jean-Paul. Lab. de Minéralogie- Cristallographie, U. P.M. Curie, 4 place Jussieu, 75252 Paris Cedex 05, France. (1942) DSc, crystallography (U. de Paris, 1969). Dir. de rech. (tel. 1 + 43 36 25 25, ext. 52 36, fax 1 + 43 54 40 97, email mornon at frlmcp61.earn). *Proteins, structural biology, computing.*

Mosset, Dr Alain. Lab. de Chimie de Coordination, CNRS, 205 route de Narbonne, 31030 Toulouse, France. (1946) DSc., physics (U. de Toulouse, 1979). Maître de conf. (tel. 61 33 31 28, fax 61 55 30 03). *Amorphous materials.*

Naudon, Dr André Lab. de Métallurgie Physique, Faculté des Sciences de Poitiers, 40 av. du recteur Pineau, 86022 Poitiers Cedex, France. (1939) DSc, solid state physics (U. de Poitiers, 1971). Dir. de rech. CNRS. (tel. 49 46 26 30, ext. 682). *Metals, glasses, X-ray diffraction.*

Pannetier, Dr Jean. ILL, avenue des Martyrs, BP 156X, 38042 Grenoble Cedex, France. (1947) DSc, physics (U. de Rennes, 1974) Staff scient. (tel. 76 48 70 91, fax 76 48 39 06, email pannetier at frill51.earn). *Neutron diffraction, X-ray diffraction, structural chemistry, materials science.*

Pascard, Dr Claudine. Inst. de Chimie des Substances Naturelles, CNRS, Cristallochimie, 91198 Gif sur Yvette Cedex, France. (1929) DSc, physics (U. de Paris, 1960). Dir. de rech., CNRS. (tel. 1 + 69 82 30 46, fax 1 + 69 07 72 47). *Proteins, structural chemistry.*

Pebay-Peyroula, Dr Eva. ILL, BP 156X, 38042 Grenoble Cedex, France. (1956) DU, physics (U. de Grenoble, 1986). Staff scientist (tel. 76 48 73 11, fax 76 48 39 06, email pebay at frill51.earn). *Structural biology, neutron diffraction, proteins.*

Perz, Dr Serge. Centre de Recherches Agro-Alimentaires, INRA, BP 527, 44026 Nantes Cedex 03, France. (1947) DSc, physics (U. de Grenoble, 1978). Dir. de rech. (tel. 40 67 50 43, fax 40 67 50 05, email perez at frinra72.earn). *Computing, electron diffraction, macromolecules, molecular crystals, polymers, X-ray diffraction.*

Perrin, Prof. Monique. Lab. de Minéralogie- Cristallographie, U. Claude Bernard (Lyon 1), 43 bd du 11 novembre 1918, 69622 Villeurbanne Cedex, France. (1936) DSc, crystallography (U. de Lyon, 1974). Maître de conf. (tel. 78 89 81 24). *Organic compounds, phase transitions, structural chemistry.*

Petiau, Prof. Jacqueline. Lab. de Minéralogie- Cristallographie, Université P.M. Curie, 4 place Jussieu, 75252 Paris Cedex 05, France. (1937) DSc, physics (U. de Paris, 1966). Prof. (tel. 1 + 43 36 25 25, ext. 45 84, fax 1 + 43 54 40 97, email labo at frlmcp61.earn). *Materials science, EXAFS.*

Petipas, Prof. Claude. UA 808, Faculté des Sciences, Université de Rouen, 76134 Mont Saint Aignan Cedex, France. (1936) DSc, physics (U. de Paris Sud, 1969). Prof. (tel. 35 98 28 50, fax 35 14 63 49). *Small angle scattering, colloids.*

Petroff, Prof. Jean-François. Lab. de Minéralogie- Cristallographie, Université P.M. Curie, 4 place Jussieu, 75252 Paris Cedex 05, France. (1935) DSc, physics (U. de Paris, 1971) Prof. Dir. du lab. (tel. 1 + 43 36 25 25, ext. 52 19, fax 1 + 43 54 40 97, email petroff at frlmcp61.earn). *Materials science, X-ray diffraction, surface structure.*

Pierrot, Dr Marcel. Service de Cristallochimie, Faculté des Sciences de Saint-Jerome, rue Henri Poincaré, 13397 Marseille Cedex 13, France. (1938) DSc, physics (U. d'Aix-Marseille, 1968). Dir. de rech. CNRS. (tel. 91 98 32 08, fax 91 28 80 30, email bigpdp at frmrs11.earn). *Structural chemistry, proteins, molecular crystals, inorganic crystals, diffractometry.*

Poljak, Prof. Roberto J. Lab. d'Immunologie Structurale, Inst. Pasteur, 25 rue du Dr Roux, 75724 Paris Cedex 15, France. (1932) DSc, (U. de la Plata, 1956). Prof. (tel. 1 + 43 68 86 10, fax 1 + 43 06 98 35, email poljak at pasteur). *Macromolecules, proteins, structural biology.*

Pouget, Dr. Lab. de Physique des Solides, U. Paris Sud, Bat. 510, 91405 Orsay Cedex, France. (1947) DSc, physics (U. Paris Sud, 1974), Dir. de rech., (tel. 1 + 69 41 60 46, fax 1 + 69 41 60 86). *Diffuse scattering, organic compounds, phase transitions, symmetry.*

Precigoux, Dr Gilles. Lab. de Cristallographie, U. de Bordeaux 1, 351 Cours de la Libération, 33405 Talence, France. (1946) Dsc physics, (U. de Bordeaux 1, 1978). Dir. de rech. (tel. 56 84 61 63, email hospital at frbdx11.earn). *X-ray diffraction, peptides.*

Primot, Mr Jacques. CNET, BAG-PMM, 196 av. Ravera, 92220 Bagneux, France. (1931) Ing. (tel. 1 + 45 29 52 06). *Materials science, optics.*

Protas, Prof. Jean. Lab. de Minéralogie- Cristallographie, U. de Nancy 1, BP 239, 54506 Vandoeuvre Cedex, France. (1932) DSc, physics (U. de Paris, 1959). Prof. (tel. 83 91 22 64). *X-ray diffraction, diffractometry, computing, chemical bonding.*

Quéré, Prof. Yves. Ecole Polytechnique, 91128 Palaiseau Cedex, France. (1931) DSc, Physical metallurgy (U. d'Orsay, 1954). Professor (tel. 1 + 69 41 82 00, ext. 2171). *Defect structures, radiation damage, channeling.*

Rantsordas, Dr Shasikante. Lab. de Minéralogie-Cristallographie, Université Claude Bernard, 43 bld du 11 Novembre 1918, 69622 Villeurbanne Cedex, France. (1941) DSc, physics, (U. de Lyon I, 1987) Maître de conf. (tel. 78 89 81 24, ext. 4093). *Molecular crystals, organic compounds, crystal growth.*

Raoux, Dr Denis. LURE, Bat. 209d, Université Paris Sud, 91405 Orsay Cedex, France. (1942) DSc, physics (U. de Paris Sud, 1973). Dir. de rech. CNRS. (tel.

1 + 64 46 80 02, fax 1 + 64 46 41 48, email raoux at frlure51.earn). *Diffuse scattering, disorder, materials science, instrumentation.*

Ravy, Dr Sylvain. Lab. de Physique des Solides, Bat. 510, Université Paris Sud, 91405 Orsay Cedex, France. (1961) DU, Solid state physics (U. de Paris Sud, 1988). Chargé de rech., (tel. 1 + 69 41 60 47). *X-ray diffraction, organic compounds, phase transitions.*

Renaud, Dr Anne. Lab. de Spectrométrie Physique, Université Joseph Fourier, BP 87, 38402 Saint Martin d'Heres Cedex, France. (1960) DU, physics (U. de Grenoble, 1986). Chargé de rech. (tel. 76 51 47 59, fax 76 51 48 48, email renault at frcicg71.earn). *Molecular crystals, X-ray diffraction, phase transition.*

Rérat, Dr Claude. Lab. de Cristallographie, CNRS, 1 place A. Briand, 92190 Bellevue, France. (1924) DSc, physics (U. de Paris, 1959). Dir. de rech. (tel. 1 + 45 34 75 50, ext. 22 28). *Proteins, molecular crystals.*

Risler, Dr Jean-Loup. Centre de Génétique Moléculaire, CNRS, 91190 Gif sur Yvette, France. (1943) DSc, biophysics (U. de Paris, 1973). Dir. de rech. (tel. 1 + 69 82 31 34, email risler at frcgm51.earn). *Proteins.*

Robert-Picard, Dr Marie-Claire. Lab. de Minéralogie-Cristallographie, Université P.M.Curie, 4 place Jussieu, 75252 Paris Cedex 05, France. (1937) DSc, physics (U. de Paris, 1969). Dir. de rech. (tel. 1 + 43 36 25 25, ext. 5216, fax 1 + 43 54 40 97, email labo at frlmcp61.earn). *Crystal growth, defect structures, X-ray topography.*

Rolland, Mr Guy. CEA-CENG, DMG/SEM/LECM, av. des Martyrs, BP 85X, 38041 Grenoble Cedex, France. (1947) D. 3e cycle, physics (U. de Grenoble, 1973). Ing. (tel. 76 88 34 11). *X-ray diffraction, defect structure, texture, instrumentation.*

Roth, Dr Michel. ILL, avenue des Martyrs, BP 166X, 38042 Grenoble Cedex, France. (1939) DSc, metal physics (U. de Grenoble, 1969) Staff scient. (tel. 76 48 71 80). *Macromolecules, neutron diffraction, X-ray diffraction, instrumentation, small-angle scattering, structural biology, proteins.*

Roult, Mr Georges. Dépt. de Rech. Fondamentales SPH/S, Centre d'Etudes Nucléaires de Grenoble (CEA-CENG), av. des Martyrs, BP 85X, 38041 Grenoble Cedex, France. (1930) D.ét.sup., crystal growth (U. de Paris, 1954). Head, TOF neutron diffraction group (tel. 76 88 44 00, ext. 39 30, fax 76 88 51 53). *Neutron diffraction, high pressures, ceramics.*

Rousseaux, Dr Françoise. Lab. de Microstructure et de Microelectronique, 196 av. Henri Ravera, 92220 Bagneux, France. (1940) DSc, physics (U. d'Orléns, 1975). Dir. de rech. CNRS. (tel. 1 + 45 29 53 74, fax 1 + 45 29 53 78). *X-ray diffraction, defect structures, intercalation compounds, phase transitions, microlithography.*

Sadoc, Prof. Jean-çois. Lab. de Physique des Solides, Bat. 510, Université Paris-Sud, 91405 Orsay Cedex, France. (1941) DSc, physics (U. de Paris-Sud, 1976) professor. (tel. 1 + 69 41 60 48, fax 1 + 69 41 60 86, email nicolis at frsol11.earn). *Defect structures, disorder, quasicrystals.*

Soubeyroux, Dr Jean-Louis. ILL, 156X, 38042 Grenoble Cedex, France. (1949) DSc, chemistry (U. de Bordeaux, 1981) researcher (tel. 76 48 70 89, fax 76 48 39 06, email soubeyroux at frill51.earn). *Neutron diffraction, powder diffraction, magnetism, diffuse scattering, materials science, phase transitions.*

Sauvage- Simkin, Dr Michèle. Lab. de Minéralogie- Cristallographie, U. Pierre et Marie Curie, 4 place Jussieu, 75252 Paris Cedex 05, France. (1941) DSc, physics (U. de Paris 6, 1968). Dir. de rech. (tel. 1 + 43 36 25 25, ext. 52 15, fax 1 + 43 54 40 97, email labo at frlmcp61.earn). *X-ray topography, defect structures, surfaces, synchrotron.*

Schlenker, Prof. Michel. Lab. Louis Néel, CNRS, BP 166, 38042 Grenoble Cedex, France. (1940) DSc, physics (U. de Grenoble, 1970). Prof. (tel. 76 88 10 92, fax 76 88 11 91). *Neutron diffraction, topographic methods, electron diffraction, magnetism, phase transitions.*

Sfez, Dr Gérard. Lab. de Physique, Faculté de Pharmacie, 4 avenue de l'Observatoire, 75006 Paris, France. (1940) DSc, physics, (U. de Paris) Ingénieur. (tel. 1 + 43 29 12 08 ext. 338). *X-ray diffraction, proteins, computing.*

Tardieu, Dr Annette. CGM-CNRS, 91198 Gif sur Yvette Cedex, France. (1943) DSc, physics (U. de Paris, 1972). Dir. de rech., CNRS. (tel. 1 + 69 82 32 11, email tardieu at frcgm51.earn). *Small-angle scattering, disorder, structural biology. probabilities, statistics, protein crystallography.*

Théobald, Prof. François. Lab. de Chimie du Solide Cristallin, Bat. 420, Université Paris-Sud, 91405 Orsay Cedex, France. (1942) DSc, physics (U. de Besançon, 1975). professor. (tel. 1 30 21 79 52). *Structural chemistry, catalysis, oxides.*

Thozet, Dr Alain Maurice. Lab. de Minéralogie- Cristallographie, U. Claude Bernard (Lyon 1), 43 bd du 11 novembre 1918, 69622 Villeurbanne Cedex, France. (1941) DSc, physics (U. de Lyon, 1981). Maître de conf. (tel. 78 89 81 24, ext. 12). *Organic compounds, computing, powder diffraction.*

Timmins, Dr Peter. ILL, 156X, 38042 Grenoble Cedex, France. (1946) PhD, crystallography, (U. of London, 1972). staff scientist. (tel. 76 48 72 63, fax 76 48 39 06, email timmins at frill51.earn). *Neutron diffraction, X-ray diffraction, small-angle scattering, macromolecules.*

Tougard, Dr Pierre. Dépt. d'immunologie, Inst. Pasteur, 25 rue du Dr Roux, 75015 Paris, France. (1942) DSc, crystal chemistry (U. de Paris 6, 1974). Chargé de rech. CNRS. (tel. 1 + 45 68 80 00). *X-ray diffraction, structural biology, proteins.*

Toupet, Mr Loic. Physique Cristalline (UA 804), Faculté des Sciences de Rennes, Campus de Beaulieu, 35042 Rennes Cedex, France. (1949) D.3e cycle (U. de Rennes, 1976). Ingénieur. (tel. 99 28 60 65). *Instrumentation, diffractometry, phase transitions, X-ray diffraction.*

Vadon, Dr Albert. Lab. de Métallurgie Physique et Chimique, Faculté des Sciences, Ile du Saulcy, 57045, Metz Cedex 1, France. (1933) DSc, physics (U. de Metz, 1981). Maî de conf. (tel. 87 32 53 05, fax 87 30 39 89, email vadon at frciil71.earn). *Texture, X-ray diffraction, computing, instrumentation.*

Vettier, Dr Christian. ILL, av. des Martyrs, 156X, 38042 Grenoble Cedex, France. (1946) DSc., physics (U. de Grenoble, 1975). Staff scientist. (tel. 76 48 72 56, fax 76 48 39 06, email vettier at frill51.earn). *Neutron diffraction, high pressures, magnetism, X-ray diffraction, phase transitions.*

Vidal, Dr Geneviève. Lab. de Dynamique des Phases Condensées, U. des Sciences et Techniques du Languedoc, place Eugène Bataillon, 34060 Montpellier Cedex, France. (1944) DSc, physics (U. de Montpellier, 1975). Maître de conf. (tel. 67 54 48 30, fax 67 54 30 79, email jpvidal at frmop11.earn). *Materials science, X-ray diffraction.*

Vidal, Dr Jean-Pierre. Lab. de Dynamique des Phases Condensées, U. des Sciences et Techniques du Languedoc, place Eugène Bataillon, 34060 Montpellier Cedex, France. (1940) DSc, physics (U. de Montpellier, 1975). Maître de conf. (tel. 67 54 48 30, fax 67 54 30 79, email jpvidal at frmop11.earn). *Materials science, X-ray diffraction.*

Vogt, Dr Thomas. ILL, 156X, 38042 Grenoble Cedex, France. (1958) DSc, chemistry (U. T&uu.bingen, 1987). Staff scientist (tel. 76 48 70 37, fax 76 48 39 06, email vogt at frill51.earn). *Powder diffraction, diffractometry, inorganic crystals, neutron diffraction, structural chemistry.*

Weigel, Prof. Dominique. Chimie Physique du Solide, Ecole Centrale, Grande Voie des Vignes, 92290 Chatenay-Malabry, France. (1929) DSc, physics (U. de Paris, 1960). Prof. (tel. 1 + 46 83 63 23, fax 1 + 46 60 36 10, email cpsduff at frecp11.earn). *Phase transitions, symmetry, incommensurate structures.*

Weiss, Prof. Raymond. Inst. Le Bel, U. Louis Pasteur, 4 rue Blaise Pascal, 67070 Strasbourg, France. (1929) DSc, inorganic chemistry (U. de Nancy, 1959). Prof. (tel. 88 41 60 64, fax 88 60 75 50). *Coordination compounds, structural chemistry.*

Weulersse, Prof. Philippe. U. de Technologie, BP 649, 60206 Compiègne, France. (1937) DSc, physics (U. de Paris Sud, 1970). Prof. (tel. 44 20 99 60). *X-ray diffraction, materials science.*

Whuler, Dr Annick. Lab. de Minéralogie- Cristallographie, U. P.M.Curie, 75252 Paris Cedex 05, France. (1946) DSc, physics (U. P.M.Curie, 1978) Maître de conf. (tel. 1 + 43 36 25 25, ext. 5082, fax 1 + 43 54 40 97, email whuler at frlmcp61.earn). *Structural chemistry, polymers.*

Willaime, Prof. Christian. Centre Armoricain d'étude structurale des socles, U. de Rennes, avenue du Général Leclerc, 35042 Rennes Cedex, France. (1940) DSc, physics (U. de Paris 6, 1972). Prof. (tel. 99 28 63 78, fax 99 28 67 00). *Minerals, defect structures, electron microscopy.*

Wintenberger, Dr Micheline. Lab. Chimie Minérale, Faculté de Pharmacie, 4 avenue de l'Observatoire, 75270 Paris Cedex 06, France. (1929) DSc, physics (U. de Paris, 1962). Dir. de rech. (tel. 1 + 43 29 12 08, ext. 174). *Magnetism, neutron diffraction.*

Wyart, Prof. Jean. Lab. de Minéralogie- Cristallographie, U. Pierre et Marie Curie, 4 place Jussieu, 75252 Paris Cedex 05, France. (1902) DSc, physics (U. de Paris, 1933). Prof. Honoraire. (tel. 1 + 43 54 82 36). *Crystallography, mineralogy.*

Zaccai, Dr Giuseppe. ILL, Ave. des Martyrs, 156X, 38042 Grenoble Cedex, France. (1947) PhD, physics (U. Edinburgh, UK, 1972). Dir. de rech., CNRS. (tel. 76 48 70 46, fax 76 48 39 06, email zaccai at frill51.earn). *Macromolecules, neutron diffraction, proteins, small-angle scattering, structural biology.*

Zarka, Dr Albert. Lab. de Minéralogie- Cristallographie, U. Pierre et Marie Curie, 4 place Jussieu, 75252 Paris Cedex 05, France. (1942) DSc, crystallography (U. de Paris, 1973). Chargé de rech. CNRS. (tel. 1 + 43 36 25 25, ext. 52 16, fax 1 + 43 54 40 97, email labo at frlmcp61.earn). *Defect structures, X-ray diffraction.*

Zelwer, Dr Charles. Centre de Génétique Moléculaire, CNRS, 91190 Gif sur Yvette, France. (1940) DSc, chemistry (U. de Paris, 1967). Dir. de rech. (tel. 1 + 69 82 30 30, ext. 3219, fax 1 + 69 82 33 23, email zelwer at frcgm51.earn). *Structural biology, phase determination.*

GERMAN DEMOCRATIC REPUBLIC
DDR

Sub-Editor: **P. Rudolph**

Notes

1.International telephone country code - 37.

2.Telephone of Secretariat of Vereinigung für Kristallographie (VFK) in der GGW der DDR: +2826490.

3.The degrees conferred by the universities are the *Doctor habilitatus* (Dr.habil.) (comparable to DSc), *Doctor scientiae naturalis* (Dr.sc.nat.) (comparable to DSc), *Doctor rerum naturalium* (Dr.rer.nat.) (equivalent to PhD), *Diplom-Chemiker* (Dipl.Chem.), *Diplom-Metallurge* (Dipl.Met.), *Diplom-Mineraloge* (Dipl.Min.), *Diplom-Physiker* (Dipl.Phys.), *Diplom-Kristallograph* (Dipl. Krist.), *Diplom-Mathematiker* (Dipl.Math.), *Diplom-Geologe* (Dipl.Geol.), (these seven equivalent to MSc). At the *Technische Universität (TU)* and *Technische Hochschulen (TH)* the degrees *Doctor Ingenieur* (Dr.Ing.), *Diplom Ingenieur* (Dipl.Ing.), and *Physik Ingenieur* (Phys.Ing.) can be obtained.

Abelmann, Mr Rolf-Ulrich. Akademie der Wissenschaften der DDR, Inst.für Festkörperphysik und Elektronenmikroskopie, DDR-4010 Halle, Am Weinberg 2, German Dem. Rep. (1945). Dipl.-Phys. (tel. +601312). *Crystallography.*

Adamski, Mr Hannelore. Vereinigung für Kristallographie (VFK) in der GGW, DDR-1040 Berlin, Invalidenstrasse 43, German Dem. Rep. (1948). Dipl.Krist. *Ceramics, phase analysis.*

Albrecht, Prof. Günter. Akademie der Wissenschaften der DDR, DDR-1080 Berlin, Otto- Nuschke-Str. 22-23, German Dem. Rep. (1930). Dr. habil. (tel. +20700). *Low temperature physics, solid state physics.*

Alex, Dr Volker. Akademie der Wissenschaften der DDR, Zentrum für Wissenschaftlichen Gerätebau, DDR-1199 Berlin, Rudower Chaussee 6, German Dem. Rep. (1939). Dr. rer. nat. (tel. +6744444). *X-ray diffraction.*

Alter, Dr Uwe. Vereinigung für Kristallographie (VFK) in der GGW, DDR-1040 Berlin, Invalidenstrasse 43, German Dem. Rep. (1946). Dr.rer.nat. *Real structures, crystal physics.*

Anders, Mr Rudolf. Vereinigung für Kristallographie (VFK) in der GGW, DDR-1040 Berlin, Invalidenstrasse 43, German Dem. Rep. (1932). Dipl.Phys. *Crystal growth, epitaxy.*

Andree, Mr Anneliese. Vereinigung für Kristallographie (VFK) in der GGW, DDR-1040 Berlin, Invalidenstrasse 43, German Dem. Rep. (1951). Dipl.Krist. *Mineralogy.*

Andreeff, Prof. Alexander. Bergakademie Freiberg, Sektion Physik, DDR-9200 Freiberg, Gustav-Zeuner-Strasse 5, German Dem. Rep. (1932). Dr.habil. *Radiation damage, implantation.*

Andrehs, Dr Gerhard. Akademie der Wissenschaften der DDR, Zentralinst. für Physik der Erde DDR-1500 Potsdam, Telegraphenberg, German Dem. Rep. (1933). Dr.rer.nat. *Technical petrography.*

Arnold, Prof. Heinrich. Technische Hochschule Ilmenau, Sektion Physik, DDR-63 Ilmenau, German Dem. Rep. (1933). Dr.habil. (tel. +74576) *Semiconductors.*

Arnold, Mr Rolf. Technische Hochschule Carl Schorlemmer Merseburg, DDR-4200 Merseburg, Geusärstrasse, German Dem. Rep. (1947). Dipl. Min. (tel. +462042) *Crystallography.*

Backhaus, Dr Karl-Otto. Akademie der Wissenschaften der DDR, Zentralinstitut für physikalische Chemie, DDR-1199 Berlin, Rudower Chaussee 5, German Dem. Rep. (1936). Dr.rer.nat. (tel. +6742355) *X-ray structure analysis.*

Balarin, Prof. Manfred. Akademie der Wissenschaften der DDR, Zentralinstitut für Kernforschung Rossendorf, DDR-8050 Dresden, PF-19, German Dem. Rep. (1934). Dr.sc. (tel. +34251+389) *Solid state physics.*

Barthel, Prof. Johannes. Akademie der Wissenschaften der DDR, Zentralinstitut für Festkörperphysik und Werkstofforschung, DDR-8032 Dresden, Helmholtzstrasse 20, German Dem. Rep. (1931). Dr.habil. *Single crystal growth.*

Baumbach, Mr Manfred. Vereinigung für Kristallographie (VFK) in der GGW, DDR-1040 Berlin, Invalidenstrasse 43, German Dem. Rep. (1947). Dipl.Krist. *Real structures.*

Bautsch, Prof. Hans-Joachim. Humboldt-Univ. zu Berlin, Museum für Naturkunde, DDR-1040 Berlin, Invalidenstrasse 43, German Dem. Rep. (1929). Dr.habil. (tel. +2897560) *Crystallography, petrography.*

Becherer, Prof. Gerhard. Wilhelm-Pieck-Univ. Rostock, Sektion Physik, DDR-2520 Rostock, Universitätsplatz 3, German Dem. Rep. (1915). Dr.sc. *X-ray and crystal physics.*

Becker, Dr Claus. Akademie der Wissenschaften der DDR, Institut für Halbleiterphysik, DDR-1211 Falkenhagen, German Dem. Rep. (1933). Dr.rer.nat. *Crystallography.*

Beckmann, Prof. Günter. Ingenieurhochschule Zittau, DDR-88 Zittau, Str. der Jungen Pioniere 2, German Dem. Rep. (1927). Dr.habil. *Crystal physics, X-ray structure analysis.*

Beier, Dr Wilfried. Martin-Luther Univ. Halle-Wittenberg, DDR-4020 Halle, Friedemann-Bach-Platz 6, German Dem. Rep. (1936). Dr.rer.nat. (tel. +21153). *Electron beam microanalysis.*

Berger, Dr Hans. Humboldt-Univ. zu Berlin, Sektion Physik, DDR-1040 Berlin, Invalidenstrasse 43, German Dem. Rep. (1940). Dr.sc.nat. *X-ray diffraction, real structure.*

Bergner, Mr Joachim. Vereinigung für Kristallographie (VFK) in der GGW, DDR-1040 Berlin, Invalidenstrasse 43, German Dem. Rep. (1930). Physik-Ing. *Crystallography.*

Bergunde, Mr Thomas. Humboldt-Univ. zu Berlin, Sektion Physik, DDR-1040 Berlin, Invalidenstr. 110, German Dem. Rep. (1959). Dipl. Krist. *Epitaxy.*

Berthold, Mr Lutz. Akademie der Wissenschaften der DDR, Inst. für Festkörperphysik und Elektronenmikroskopie, DDR-4010 Halle, Am Weinberg 2, German. Dem. Rep. (1958). Dipl. Phys. (tel. +601312). *Crystallography, solid state reactions.*

Bertram, Dr Marion. Technische Univ. Dresden, Sektion Physik, DDR-8027 Dresden, Mommsenstrasse 13, German Dem. Rep. (1949). Dipl.Krist. · *Real structures.*

Betzl, Dr Manfred. Akademie der Wissenschaften der DDR, Zentralinstitut für Kernforschung, DDR-8051 Dresden, Postfach 19, German Dem. Rep. (1933). Dr.rer.nat. *X-ray physics, structure analysis.*

Binas, Dr Horst. Akademie der Wissenschaften der DDR, Zentralinstitut für Physikalische Chemie, DDR-1199 Berlin, Rudower Chaussee 5, German Dem. Rep. (1934). Dr.rer.nat. *Crystal growth, interfaces, analysis.*

Blankenburg, Dr Hans-Joachim. Bergakademie Freiberg, Sektion Geowissenschaften, DDR-9200 Freiberg, Brennhausgasse 14, German Dem. Rep. (1938). Dr.rer.nat. (tel. +512001). *Mineralogy, phase analysis.*

Blau, Prof. Winfried. Wilhelm-Pieck-Univ. Rostock, Sektion Physik, DDR-2500 Rostock, Universitätsplatz 3, German Dem. Rep. (1942). Dr. rer. nat. (tel. +369465). *Crystallography.*

Boeck, Mr Torsten. Humboldt-Univ. zu Berlin, Sektion Physik, DDR-1040 Berlin, Invalidenstr. 110, German Dem. Rep. (1957). Dipl. Krist. (tel. +2803361). *Crystal growth, epitaxy.*

Bohm, Dr Joachim. Akademie der Wissenschaften der DDR, Zentralinstitut für Optik und Spektroskopie, DDR-1199 Berlin, Rudower Chaussee 5, German Dem. Rep. (1935). Dr.habil. (tel. +6743611). *Crystal growth, real structures.*

Bornmann, Dr Horst. Vereinigung für Kristallographie (VFK) in der GGW, DDR-1040 Berlin, Invalidenstrasse 43, German Dem. Rep. (1936). Dr.rer.nat. *Crystal physics.*

Bornmann, Dr Peter. Vereinigung für Kristallographie (VFK) in der GGW, DDR-1040 Berlin, Invalidenstrasse 43, German Dem. Rep. (1945). Dr.rer.nat. *X-ray structure analysis.*

Brand, Prof. Paul. Bergakademie Freiberg, Sektion Chemie, DDR-9200 Freiberg, Leipziger Strasse, German Dem. Rep. (1931). Dr.sc. (tel. +512050). *Crystal chemistry, structures.*

Brauny, Mr Siegfried. Technische Univ. Dresden, Sektion Physik, DDR-8027 Dresden, Salvador-Allende-Platz 3, German Dem. Rep. (1934). Dipl.Phys. (tel. +4834221) *X-ray physics.*

Broosch, Mr Erika. Vereinigung für Kristallographie (VFK) in der GGW, DDR-1040 Berlin, Invalidenstrasse 43, German Dem. Rep. (1953). Dipl.Krist. *Crystallography.*

Brückner, Dr Winfried. Akademie der Wissenschaften der DDR, Zentralinstitut für Festkörperphysik und Werkstofforschung, DDR-8032 Dresden, Helmholtzstrasse 20, German Dem. Rep. (1939). Dr.rer.nat. *Crystal physics, solid state physics.*

Brühl, Dr Hans-Gerd. Karl-Marx-Univ. Leipzig, DDR-7010 Leipzig, Linnestrasse 5, German Dem. Rep. (1944). Dr.rer.nat. (tel. +68580) *Crystallography.*

Brümmer, Prof. Otto. Martin-Luther-Univ. Halle-Wittenberg, DDR-4020 Halle, Friedemann-Bach-Platz 6, German Dem. Rep. (1920). Dr.habil. (tel. +37761) *X-ray physics, crystal physics.*

Bublitz, Mr Günter. Akademie der Wissenschaften der DDR, Inst.für Meereskunde, DDR-253 Warnemünde, Seestrasse 15, German Dem. Rep. (1934). Dipl.Min. *Sedimentation.*

Butter, Prof. Ehrenfried. Karl-Marx-Univ. Leipzig, Sektion Chemie, DDR-701 Leipzig, Liebigstr. 18, German Dem. Rep. (1931). Dr.sc. *Crystal growth, epitaxy.*

Christoph, Mr Arthur. Vereinigung für Kristallographie (VFK) in der GGW, DDR-1040 Berlin, Invalidenstrasse 43, German Dem. Rep. (1921). Dipl.Chem. *Crystal growth.*

Däweritz, Dr Lutz. Akademie der Wissenschaften der DDR, Zentralinstitut für Elektronenphysik, DDR-1086 Berlin, Hausvogteiplatz 5-7 German Dem. Rep. (1943). Dr.rer.nat. (tel. +2077359). *Interfaces, epitaxy.*

Damaschun, Dr Ferdinand. Humboldt-Univ. zu Berlin, Museum für Naturkunde, DDR-1040 Berlin, Invalidenstrasse 43, German Dem. Rep. (1950). Dr.rer.nat. (tel. +2897670). *Crystallography, mineralogy.*

Demus, Prof Dietrich. Martin-Luther-Univ. Halle-Wittenberg, DDR-4020 Halle, Mühlpforte 1, German Dem. Rep. (1935). Dr.habil. *Crystallography.*

Deus, Dr Peter. Bergakademie Freiberg, Sektion Physik, DDR-9200 Freiberg/Sa., Cottastr. 4, German Dem. Rep. (1938). Dr.sc.nat. (tel. +512866). *Crystal growth, X-ray analysis.*

Dietrich, Prof Burkhard. Akademie der Wissenschaften der DDR, Institut für Halbleiterphysik Frankfurt/Oder, DDR-1200 Frankfurt/Oder, Walter-Korsing-Str. 2, German Dem. Rep. (1939). Dr.sc.nat. (tel. +09303730). *Crystal physics, real structures.*

Dlubek, Prof Günter. Pädagogische Hochschule Halle, Kröllwitzer Str. 44, DDR-4020 Halle, German Dem. Rep. (1947). Dr.sc.nat. (tel. +37761). *Solid state physics, positron annihilation.*

Dörrfeld, Mr Hans-Georg. Akademie der Wissenschaften der DDR, Zentrum für wissenschaftlichen Gerätebau, DDR-1199 Berlin, Rudower Chaussee 6, German Dem. Rep. (1935). Dipl.Ing. (tel. +6702871). *Crystal growth.*

Dörschel, Dr Jürgen. Akademie der Wissenschaften der DDR, Inst. für Physik der Werkstoffbearbeitung, DDR-1166 Berlin, Seestrasse 82, German Dem. Rep. (1942). Dr.rer.nat. (tel. +6489671). *Crystal growth, real structures.*

Dräger, Dr Günter. Martin-Luther-Univ. Halle-Wittenberg, Sektion Physik, DDR-4020 Halle, Friedemann-Bach-Platz 6, German Dem. Rep. (1937). Dr.sc.nat. (tel. +37761). *Solid state physics, X-ray analysis.*

Dressler, Dr Ludwig. Friedrich-Schiller-Univ. Jena, Sektion Physik, DDR-69 Jena, Max - Wien - Platz 1, German Dem. Rep. (1944). Dr.rer.nat. (tel. +8226250). *X-ray physics, crystal growth.*

Driesel, Dr Wolfgang. Akademie der Wissenschaften der DDR, Inst.für Festkörperphysik und Elektronenmikroskopie, DDR-4000 Halle, Weinberg 2, German Dem. Rep. (1946). Dr.rer.nat. (tel. +4861+203). *Real structures.*

Dünkel, Mr Lothar. Akademie der Wissenschaften der DDR, Zentralinstitut für physikalische Chemie, DDR-1199 Berlin, Rudower Chaussee 5, German Dem. Rep. (1942). Dipl.-Chem. *X-ray structure analysis.*

Eichhorn, Dr Gerd. Technische Hochschule Ilmenau, Sektion Physik, DDR-63 Ilmenau, German Dem. Rep. (1939). Dr.rer.nat. (tel. +74621). *Crystal growth, epitaxy.*

Eichler, Dr Klaus. Technische Univ. Dresden, Sektion Physik, DDR-8027 Dresden, Salvador-Allende-Platz 3, German Dem. Rep. (1941). Dr.rer.nat. (tel. +4832536). *Crystal growth.*

Eichler, Dr Wolfgang. Martin-Luther-Univ. Halle-Wittenberg, DDR-4020 Halle, Bachstrasse 6, German Dem. Rep. (1938). Dr.rer.nat. (tel. +37761). *Real structures, crystal growth.*

Eisenschmidt, Mr Christian. Martin-Luther-Univ. Halle-Wittenberg, Sektion Physik, DDR-4020 Halle, Friedemann-Bach-Platz 6, German Dem. Rep. (1953). Dipl. Phys. (tel. +37761). *X-ray-physics.*

Elbinger, Dr German. Akademie der Wissenschaften der DDR, Physikalisch-Technisches Institut Jena, DDR-6900 Jena, Helmholtzweg 4, German Dem. Rep. (1925). (tel. +27372). *Crystallography.*

Emons, Prof. Hans-Heinz. Akademie der Wissenschaften der DDR, DDR-1080 Berlin, Otto- Nuschke-Str. 22-23, German Dem. Rep. (1930). Dr.sc.nat. (tel. + 20700). *Inorganic and organic techniques, chemistry.*

Engel, Dr Aribert. Humboldt-Univ. zu Berlin, Sektion Physik, DDR-1040 Berlin, Invalidenstrasse 110, German Dem. Rep. (1934). Dr.rer.nat. (tel. +2803461). *Crystal growth, microscopy, surfaces.*

Engels, Prof. Siegfried. Technische Hochschule Carl Schorlemmer Merseburg, DDR-4200 Merseburg, Geusärstrasse, German Dem. Rep. (1932). Dr.habil. *Solid state chemistry.*

Fabian, Prof Eginhard. Ernst-Moritz-Arndt-Univ. Greifswald, DDR-2200 Greifswald, Demstr. 20, German Dem. Rep. (1935). Dr.sc.phil. (tel. +63336). *Crystallography, history.*

Falkenberg, Dr Wolfgang. Technische Hochschule Carl Schorlemmer Merseburg, DDR-4200 Merseburg, Geusärstrasse, German Dem. Rep. (1945). Dr.rer.nat. (tel. +462104). *Chemical crystallography, X-ray structure analysis.*

Fanter, Dr Detlef. Akademie der Wissenschaften der DDR, Inst.für Polymerenchemie, DDR-153 Teltow, Kantstrasse 55, German Dem. Rep. (1945). Dipl.Krist. *High polymer structures.*

Faust, Mr Wolfgang. Technische Univ. Karl-Marx-Stadt, Sektion Physik, DDR-9010 Karl-Marx-Stadt, Reichenhainerstrasse, German Dem. Rep. (1949). Dipl.Krist. *Crystallography.*

Feher, Mr Andreas. Vereinigung für Kristallographie (VFK) in der GGW, DDR-1040 Berlin, Invalidenstrasse 43, German Dem. Rep. (1953). Dipl.Krist. *Crystallography.*

Feher, Mr Elvira. Vereinigung für Kristallographie (VFK) in der GGW, DDR-1040 Berlin, Invalidenstrasse 43, German Dem. Rep. (1954). *Crystallography.*

Fehling, Mr Wolfgang. Vereinigung für Kristallographie (VFK) in der GGW, DDR-1040 Berlin, Invalidenstrasse 43, German Dem. Rep. (1941). Dipl.Min. *Crystal growth, crystal chemistry, materials science.*

Felbinger, Dr Adolf. Vereinigung für Kristallographie (VFK) in der GGW, DDR-1040 Berlin, Invalidenstrasse 43, German Dem. Rep. (1937). Dr.rer.nat. *Crystallization, crystallography.*

Feltz, Prof. Adalbert. Friedrich-Schiller-Univ. Jena, Sektion Chemie, DDR-69 Jena, August-Bebel-Strasse 2, German Dem. Rep. (1934). Dr.habil. (tel. +22578+209). *Solid state chemistry, semiconductor chemistry.*

Fichtner-Schmittler, Dr Helga. Akademie der Wissenschaften der DDR, Zentralinstitut für physikalische Chemie, DDR-1199 Berlin, Rudower Chaussee 5, German Dem. Rep. (1932). Dr.rer.nat. (tel. +6702841). *Crystal structures, powder methods, order-disorder phenomena.*

Fiedler, Dr Gustav. Vereinigung für Kristallographie (VFK) in der GGW, DDR-1040 Berlin, Invalidenstrasse 43, German Dem. Rep. (1932). Dr.rer.nat. *Mineralogy phase-analysis.*

Filscher, Mr Gerold. Friedrich-Schiller-Univ. Jena, Sektion Physik, DDR-69 Jena, Max - Wien - Platz 1, German Dem. Rep. (1933). *Crystallography.*

Finster, Dr Joachim. Wilhelm-Pieck-Univ. Rostock, Sektion Physik, DDR-2500 Rostock, Universitätsplatz 3, German Dem. Rep. (1936). Dr.sc.nat. (tel. +369465). *Amorphous states, surfaces.*

Fischer, Dr Karl. Akademie der Wissenschaften der DDR, Zentralinstitut für Anorganische Chemie, DDR-1199 Berlin, Rudower Chaussee 5, German Dem. Rep. (1934). Dipl.Min. (tel. +6702841). *Solid state physics, defect structures.*

Fitzl, Dr Günther. Karl-Marx-Univ. Leipzig, Sektion Chemie, DDR-701 Leipzig, Liebigstrasse 18, German Dem. Rep. (1950). Dipl.Krist. *X-ray structure analysis.*

Flade, Dr Tilo. Vereinigung für Kristallographie (VFK) in der GGW, DDR-1040 Berlin, Invalidenstr. 43, German Dem. Rep. (1942). Dr.ing. *Crystal growth.*

Flögel, Dr Peter. Akademie der Wissenschaften der DDR, Zentralinstitut für Elektronenphysik, DDR-1086 Berlin, Hausvogteiplatz 5-7, German Dem. Rep. (1926). Dr.rer.nat. (tel. +20770). *Crystal growth.*

Förster, Dr Eckhart. Friedrich-Schiller-Univ. Jena, Sektion Physik, DDR-69 Jena, Max - Wien - Platz 1, German Dem. Rep. (1944). Dr.rer.nat. (tel. +24036). *Real structures.*

Försterling, Dr Gerd. Technische Univ. Dresden, Sektion Physik, DDR-8027 Dresden, Mommsenstrasse 16, German Dem. Rep. (1940). Dr.rer.nat. (tel. +4633765). *Crystallography.*

Freudenberg, Mr Axel. Akademie der Wissenschaften der DDR, Zentralinstitut für organische Chemie, DDR-1040 Berlin, Invalidenstrasse 44, German Dem. Rep. (1952). Dipl.-Krist. (tel. +2262441). *Crystallography.*

Freydank, Mr Gisela-Christine. Vereinigung für Kristallographie (VFK) in der GGW, DDR-1040 Berlin, Invalidenstrasse 43, German Dem. Rep. (1944). Dipl.Krist. *Real structures, semiconductors.*

Frühauf, Dr Joachim. Technische Univ. Karl-Marx-Stadt, Sektion Physik, DDR-9010 Karl-Marx-Stadt, PSF 964, German Dem. Rep. (1943). Dr.Ing. *Crystal growth.*

Geidel, Mr Volkmar. Vereinigung für Kristallographie (VFK) in der GGW, DDR-1040 Berlin, Invalidenstr. 43, German Dem. Rep. (1944). Dipl.ing. *Crystal growth.*

Geil, Dr Werner. Vereinigung für Kristallographie (VFK) in der GGW, DDR-1040 Berlin, Invalidenstr. 43, German Dem. Rep. (1929). Dr.rer.nat. *Crystal growth.*

Geist, Dr Volker. Karl-Marx-Univ. Leipzig, Sektion Chemie, DDR-7010 Leipzig, Talstr. 35, German Dem. Rep. (1942). Dr.sc.nat. (tel. +310502). *Structure analysis, ion implantation.*

Gernat, Mr Christine. Akademie der Wissenschaften der DDR, Forschungszentrum für Molekularbiologie, DDR-1115 Berlin-Buch, Lindenberger Weg 70, German Dem. Rep. (1952). (tel. +3460). *Crystallography.*

Gesemann, Prof. Renate. Ingenieurhochschule Mittweida, Sektion Elektronischer Gerätebau, DDR-925 Mittweida, German Dem. Rep. (1936). Dr.rer.nat. (tel. +58370). *Structure in solid state.*

Geserick, Mr Sabine. Technische Univ. Karl-Marx-Stadt, Sektion Physik, DDR-9010 Karl-Marx-Stadt, PSF 964, German Dem. Rep. (1952). *Electron microscopy.*

Gille, Dr Peter. Humboldt-Univ. zu Berlin, Sektion Physik, DDR-1040 Berlin, Invalidenstr. 110, German Dem. Rep. (1954). Dr.rer.nat. (tel. +2083294). *Crystal growth.*

Göbel, Mr Ralf. Akademie der Wissenschaften der DDR, Zentralinstitut für Festkörperphysik und Werkstofforschung, DDR-8027 Dresden, Helmholzstrasse 20, German Dem. Rep. (1935). Dipl.Phys. (tel +23220). *Crystallography.*

Göcke, Dr Wolfhart. Wilhelm-Pieck-Univ. Rostock, Universitätsplatz 3, German Dem. Rep. (1940). Dr.rer.nat. (tel. +369353). *Crystallography.*

Görnert, Prof Peter. Akademie der Wissenschaften der DDR, Physikalisch-Technisches Institut, DDR-6900 Jena, Helmholtzweg 4, German Dem. Rep. (1943). Dr.sc.nat. (tel. +27372). *Crystal growth, crystallization.*

Götz, Dr Konrad. Friedrich-Schiller-Univ. Jena, Sektion Physik, DDR-69 Jena, Max - Wien - Platz 1, German Dem. Rep. (1938). Dr.rer.nat. (tel. +825220). *Interference theory.*

Götz, Dr Wolfgang. Friedrich-Schiller-Univ. Jena, Sektion Chemie, DDR-69 Jena, Sellierstrasse 6, German Dem. Rep. (1928). Dr.habil. (tel. +27122+346). *Crystallography, crystal chemistry.*

Gottschalch, Dr Volker. Karl-Marx-Univ. Leipzig, Sektion Chemie, DDR-701 Leipzig, Liebigstr. 18, German Dem. Rep. (1945). Dr.rer.nat. (tel. +7165+476). *Crystal growth, real structures.*

Gramlich, Prof Kurt. Ingenieurhochschule Köthen, DDR-4370 Köthen, Bernburger Str. 52-57, German Dem. Rep. (1941). Dr.sc.nat. (tel. +67220). *Mass crystallizazion, industrial crystallization.*

Grau, Mr Lutz. Akademie der Wissenschaften der DDR, Physikalisch-Technisches Institut Jena DDR-69 Jena, Helmholtzweg 4, German Dem. Rep. (1947). Dipl.Krist. (tel. +27372). *Crystallography.*

Gülzow, Mr Hansjürgen. Vereinigung für Kristallographie (VFK) in der GGW, DDR-1040 Berlin, Invalidenstrasse 43, German Dem. Rep. Dipl.Min. *Crystal growth.*

Günther, Prof. Fritz. Bergakademie Freiberg, Sektion Metallurgie und Werkstofftechnik, DDR-9200 Freiberg, Gustav-Zeuner-Strasse 5, German Dem. Rep. (1912). Dr.habil. (tel. +512601). *Real structure, crystal growth*

Gütt, Mr Rainer. Vereinigung für Kristallographie (VFK) in der GGW, DDR-1040 Berlin, Invalidenstrasse 43, German Dem. Rep. (1943). Dipl.Krist. *Crystallography.*

Hadan, Dr Marianne. Akademie der Wissenschaften der DDR, Institut für Chemische Technologie, DDR-1199 Berlin, Rudower Chaussee 5, German Dem. Rep. (1939). Dr.rer.nat. (tel. +6743431). *Silicates.*

Hähnert, Dr Irmela. Humboldt-Univ. zu Berlin, Sektion Physik, DDR-1040 Berlin, Invalidenstr. 110, German Dem. Rep. (1941). Dr.rer.nat. (tel. +2803361). *Crystal growth.*

Hähnert, Dr Manfred. Akademie der Wissenschaften der DDR, Zentralinstitut für Anorganische Chemie, DDR-1199 Berlin, Rudower Chaussee 5, German Dem. Rep. (1934). Dr.rer.nat. (tel. +67407+415). *Crystal growth, crystal chemistry.*

Hahne, Dr Bodo. Technische Hochschule Carl Schorlemmer Merseburg, DDR-4200 Merseburg, Geusärstrasse, German Dem. Rep. (1941). (tel. +462075). *Phase analysis.*

Hartmann, Dr Horst. Akademie der Wissenschaften der DDR, Zentralinstitut für Elektronenphysik, DDR-108 Berlin, Hausvogteiplatz 5-7, German Dem. Rep. (1934). Dr.rer.nat. (tel. +2000561+325). *Crystal growth, real structures.*

Hartung, Prof Helmut. Martin-Luther-Univ. Halle-Wittenberg, DDR-4020 Halle, Mühlpforte 1, German Dem. Rep. (1935). Dr.sc.nat. (tel. +25081). *X-ray structure analysis.*

Hegenbarth, Prof. Ernst. Technische U. Dresden, Sektion Physik, DDR-8027 Dresden, Mommsenstrasse 16, German Dem. Rep. (1933). Dr.habil. (tel. +4832459). *Solid state physics.*

Heide, Dr Klaus. Friedrich Schiller Univ. Jena, Otto Schott Institut, DDR-69 Jena, Sellierstrasse 6, German Dem. Rep. (1938). Dr.habil. (tel. +8225462). *Crystallography.*

Heim, Dr Joachim. Technische Univ. Karl-Marx-Stadt, Sektion Physik, DDR-9010 Karl-Marx-Stadt, PSF 964, German Dem. Rep. (1933). Dr.rer.nat. (tel. +5613037). *Crystallography, crystal growth.*

Hellmold, Prof Peter. Technische Hochschule Carl Schorlemmer Merseburg, DDR-4200 Merseburg, Geusärstrasse, German Dem. Rep. (1937). Dr.sc.nat. (tel. +462019). *Crystallography.*

Hennig, Prof. Klaus. Akademie der Wissenschaften der DDR, Zentralinstitut für Kernforschung Rossendorf, DDR-8050 Dresden, PF 19, German Dem. Rep. (1936). Dr.sc. (tel. +5912478). *Neutron diffraction.*

Herberger, Dr Jürgen. Technische Univ. Karl-Marx-Stadt, Sektion Physik, DDR-9010 Karl-Marx-Stadt, Strasse der Nation 62, German Dem. Rep. (1941). Dr.rer.nat. (tel. +88550). *Solid state physics, crystal growth, X-ray structure analysis.*

Hermoneit, Mr Bernd. Akademie der Wissenschaften der DDR, Zentralinstitut für Optik und Spektroskopie, DDR-1199 Berlin, Rudower Chaussee 5, German Dem. Rep. (1944). Dip.Krist. (tel. +6702871). *Crystal growth, crystallography.*

Herms, Dr Gerhard. Wilhelm-Pieck Universität, Sektion Physik, DDR-2520 Rostock, Universitätsplatz 3, German Dem. Rep. (1932). Dr.sc.nat. (tel. +23212).

Herrmann, Dr Frank-Peter. Akademie der Wissenschaften der DDR, Zentrum für Wissenschaftlichen Gerätebau, DDR-1199 Berlin, Rudower Chaussee 6, German Dem. Rep. (1940). Dr.sc.nat. (tel. +6744366). *Crystallography.*

Herrmann, Prof. Rudolf. Humboldt-Univ. zu Berlin, Sektion Physik, DDR-1040 Berlin, Invalidenstr. 110, German Dem. Rep. (1936). Dr.sc.nat. (tel. +2803330). *Solid state physics, superconductivity.*

Heydenreich, Prof Johannes. Akademie der Wissenschaften der DDR, Institut für Festkörperphysik und Elektronenmikroskopie, DDR-4000 Halle, Weinberg 2, German Dem. Rep. (1930). Dr.habil. (tel. +601312). *Boundary surface physics, solid state physics.*

Heymann, Prof Gunter. Humboldt-Universität-zu Berlin, Sektion Elektronik, DDR-1140 Berlin, Invalidenstr. 110, German Dem. Rep. (1940). Dr.sc.nat. (tel. +28030). *Crystallography, solid state physics.*

Hinz, Dr Dietrich. Akademie der Wissenschaften der DDR, Zentralinstitut für Festkörperphysik und Werkstofforschung, DDR-8027 Dresden, Helmholzstrasse 20, German Dem. Rep. (1945). Dr.rer.nat. (tel. +36092). *Crystallography.*

Höbler, Dr Hans-Joachim. Karl-Marx-Univ. Leipzig, Sektion Chemie, Wissenschaftsbereich Kristallographie, DDR-7030 Leipzig, Scharnhorststrasse 20, German Dem. Rep. (1951). Dr.rer.nat. (tel. +310502). *Crystal growth, mineralogy.*

Höche, Dr Hans-Reiner. Martin-Luther-Univ. Halle-Wittenberg, DDR-4020 Halle, Friedemann-Bach-Platz 6, German Dem. Rep. (1942). Dr.rer.nat. (tel. +832432). *Crystallography, X-ray diffraction.*

Höche, Dr Hellmut. Akademie der Wissenschaften der DDR, Inst.für Festkörperphysik und Elektronenmikroskopie, DDR-4010 Halle, Am Weinberg 2, German Dem. Rep. (1946). Dr.rer.nat. (tel. +601312). *Crystallography, crystal growth.*

Hofmeister, Dr Herbert. Akademie der Wissenschaften der DDR, Institut für Festkörperphysik und Elektronenmikroskopie, DDR-4010 Halle, Am Weinberg 2, German Dem. Rep. (1947). Dr.rer.nat. (tel. +691312). *Crystal growth, thin films, electron microscopy.*

Höhne, Prof. Ernst. Akademie der Wissenschaften der DDR, Zentralinstitut für Molekularbiologie, DDR-1115 Berlin, Lindenberger Weg 70, German Dem. Rep. (1927). Dr.habil. (tel. +3462270). *Structural chemistry, X-ray structure analysis.*

Hötzsch, Dr Günter. Bergakademie Freiberg, Sektion Metallurgie und Werkstofftechnik, DDR-9200 Freiberg, Gustav-Zeuner-Strasse 5, German Dem. Rep. (1932). Dr.habil. *Real structures, solid state physics.*

Hoff, von, Dr Siegfried. Vereinigung für Kristallographie (VFK) in der GGW, DDR-1040 Berlin, Invalidenstrasse 43, German Dem. Rep. (1940). Dr.rer.nat. *X-ray diffraction, thin film techniques.*

Holldorf, Dr Horst. Bergakademie Freiberg, Sektion Chemie, DDR-9200 Freiberg, Leipziger Strasse, German Dem. Rep. (1939). Dr.rer.nat. *Mass crystallization.*

Hoppe, Prof. Günter. Humboldt-Univ. zu Berlin, Museum für Naturkunde, DDR-1040 Berlin, Invalidenstrasse 43, German Dem. Rep. (1919). Dr.sc.nat. (tel. +28970). *Mineralogy, petrography.*

Hoyer, Dr Walter. Technische Hochschule Karl-Marx-Stadt, Sektion Physik, DDR-9010 Karl-Marx-Stadt, PSF 964, German Dem. Rep. (1944). Dr.rer.nat. (tel. +850525). *Crystallography.*

Hübner, Dr Manfred. Hochschule für Bauwesen, DDR-703 Leipzig, R.-Lehmannstrasse 32, German Dem. Rep. (1935). Dr.rer.nat. (tel. +398225). *Chemistry, boundary surfaces.*

Hultzsch, Mr Rainer. Vereinigung für Kristallographie (VFK) in der GGW, DDR-1040 Berlin, Invalidenstrasse 43, German Dem. Rep. (1940). Dipl. Phys. *Laser physics, crystal testing.*

Ickert, Dr Lars. Humboldt-Univ. zu Berlin, Sektion Physik, DDR-1040 Berlin, Invalidenstrasse 110, German Dem. Rep. (1934). Dr.habil. (tel. +2803461). *Crystal growth.*

Jacobs, Prof Klaus. Humboldt-Univ. zu Berlin, Sektion Physik, DDR-1040 Berlin, Invalidenstr. 110, German Dem. Rep. (1945). Dr.sc.nat. (tel. +2803417). *Epitaxy.*

Jährling, Mr Thomas. Akademie der Wissenschaften der DDR, Zentralinstitut für Elektronenphysik, DDR-108 Berlin, Hausvogteiplatz 5-7, German Dem. Rep. (1952). Dipl. Phys. (tel. +2000561). *Crystallography.*

Jegerlehner, Mr Kurt. Vereinigung für Kristallographie (VFK) in der GGW, DDR-1040 Berlin, Invalidenstrasse 43, German Dem. Rep. (1935). Dipl.Min. *Solid state physics.*

Jenichen, Dr Bernd. Akademie der Wissenschaften der DDR, Zentralinstitut für Elektronenphysik, DDR-1080 Berlin, Hausvogteiplatz 5-7, German Dem. Rep. (1953). Dipl.Phys. (tel. +2000561). *Crystallography, X-ray analysis.*

Johansen, Dr Heinrich. Akademie der Wissenschaften der DDR, Inst.für Festkörperphysik und Elektronenmikroskopie, DDR-4010 Halle, Am Weinberg 2, German Dem. Rep. (1939). Dr.rer.nat. (tel. +601312). *Electron microscopy.*

Jost, Dr Karlheinz. Akademie der Wissenschaften der DDR, Zentralinstitut für Anorganische Chemie, DDR-1199 Berlin, Rudower Chaussee 5, German Dem. Rep. (1927). Dr.habil. (tel. +6702841). *Inorganic structures, solid state chemistry.*

Jurisch, Dr Manfred. Akademie der Wissenschaften der DDR, Zentralinstitut für Festkörperphysik und Werkstofforschung, DDR-8032 Dresden, Helmholtzstrasse 20, German Dem. Rep. (1940). Dr.-Ing. (tel. +2322216). *Crystal growth, solidification.*

Kämmel, Dr Thomas. Martin-Luther-Univ. Halle-Wittenberg, DDR-4020 Halle, Domstr. 5, German Dem. Rep. (1931). Dr.habil. (tel. +37761). *Geological and geochemistry analysis.*

Kalinna, Mr Hartmut. Vereinigung für Kristallographie (VFK) in der GGW, DDR-1040 Berlin, Invalidenstrasse 43, German Dem. Rep. (1934). Dip.Min. *Crystal structures, semiconductors.*

Kalweit, Mr Harald. Akademie der Wissenschaften der DDR, Zentralinstitut für Anorganische Chemie, DDR-1199 Berlin, Rudower Chaussee 5, German Dem. Rep. (1945). Dipl. Min. (tel. +6702841). *Mineralogy, petrography.*

Kamprath, Mr Fred-Bodo. Akademie der Wissenschaften der DDR, Zentralinstitut für Anorganische Chemie, DDR-1199 Berlin, Rudower Chaussee 5, German Dem. Rep. (1948). Dipl.Krist. *Crystallography.*

Kanis, Dr Michael. Humboldt-Univ. zu Berlin, Sektion Physik, DDR-1040 Berlin, Invalidenstr. 110, German Dem. Rep. (1949). Dr.rer.nat. (tel. +2803287). *Crystal growth, crystallography.*

Katzschmann, Mr Kurt. Vereinigung für Kristallographie (VFK) in der GGW, DDR-1040 Berlin, Invalidenstrasse 43, German Dem. Rep. (1928). Dipl.Phys. *Crystal optics, thin film techniques.*

Kaufmann, Dr Thorsten. Technische Hochschule Ilmenau, Sektion Physik und technisch- elektronische Baülemente, DDR-63 Ilmenau, Weimarerstrasse, German Dem. Rep. (1941). Dr.rer.nat. (tel. +640+595). *Solid state chemistry.*

Keller, Dr Kurt Wolfgang. Akademie der Wissenschaften der DDR, Inst.für Festkörperphysik und Elektronenmikroskopie DDR-4010 Halle, Am Weinberg 2, German Dem. Rep. (1936). Dr.rer.nat. (tel. +601312). *Crystal growth, boundary surface physics.*

Kersten, Mr Friedrich. Ingenieurhochschule Mittweida, DDR-925 Mittweida, Platz der DSF, German Dem. Rep. (1934). Ing. (tel. +58219). *Crystal growth, real structures.*

Kiedrowski, von, Mr Hartmut. Akademie der Wissenschaften der DDR, Zentralinstitut für Elektronenphysik, DDR-108 Berlin, Hausvogteiplatz 5-7, German Dem. Rep. Dipl. Krist. (tel. +2000561). *Crystallography.*

Kies, Dr Jörg. Akademie der Wissenschaften der DDR, Zentralinstitut für Molekularbiologie, DDR-1115 Berlin, Lindenberger Weg 70, German Dem. Rep. (1938). Dr.rer.nat. (tel. +5697851). *Structure.*

Kiessling, Mr Frank-Michael. Humboldt-Univ. zu Berlin, Sektion Physik, DDR-1040 Berlin, Invalidenstr. 110, German Dem. Rep. (1960). Dipl. Krist. (tel. +2803334). *Crystal growth, crystallography.*

Kleinert, Dr Peter. Akademie der Wissenschaften der DDR, Physikalisch-Technisches Institut, DDR-6900 Jena, Helmholtzweg 4, German Dem. Rep. (1925). Dr.sc.nat. (tel. +27372). *Crystallography, crystal growth.*

Kleinstück, Prof. Karlheinz. Technische U. Dresden, Sektion Physik, DDR-8027 Dresden, Mommsenstrasse 16, German Dem. Rep. (1929). Dr.habil. *Crystallography.*

Kleint, Dr Christian. Karl-Marx-Univ. Leipzig, DDR-7010 Leipzig, Linnestrasse 5, German Dem. Rep. (1926). Dr.habil. (tel. +6858209). *Crystallography.*

Klimakow, Dr. Alexander. Humboldt-Univ. zu Berlin, Sektion Physik, DDR-1040 Berlin, Invalidenstr. 110, German Dem. Rep. (1949). Dr.rer.nat. (tel. +2803249). *Crystal growth.*

Klimanek, Dr Peter. Bergakademie Freiberg, Sektion Met. und Werkst. Technik, DDR-9200 Freiberg, Gustav Zeunerstrasse 5, German Dem. Rep. (1935). Dr.Ing. (tel. +513153). *Crystallography.*

Klimm, Dr Detlef. Karl-Marx-Universität Leipzig, Sektion Chemie, WB Kristallographie, DDR-7010 Leipzig, Talstr. 35, German Dem. Rep. (1957). Dr.rer.nat. (tel. +310502). *Real structure, computing.*

Klöss, Dr. Gerd. Karl-Marx-Universität Leipzig, Sektion Chemie, WB Kristallographie, DDR-7010 Leipzig, Talstr. 35, German Dem. Rep. (1956). Dr.rer.nat. (tel. +310502). *Mass density, microscopy.*

Kneschke, Dr Götz. Akademie der Wissenschaften der DDR, Forschungsinstitut für Aufbereitung, DDR-92 Freiberg, Strasse des Friedens 46, German Dem. Rep. (1937). Dr.rer.nat. (tel. +45230). *Real structures.*

Köhler, Dr Rolf. Akademie der Wissenschaften der DDR, Zentralinstitut für Elektronenphysik, DDR-1080 Berlin, Hausvogteiplatz 5-7, German Dem. Rep. (1942). Dr.rer.nat. (tel. +2000561). *Real structures, X-ray analysis.*

Köpernik, Mr Horst. Vereinigung für Kristallographie (VFK) in der GGW, DDR-1040 Berlin, Invalidenstrasse 43, German Dem. Rep. (1943). Dipl.Chem. *Zeolites.*

Kötitz, Dr Günther. Vereinigung für Kristallographie (VFK) in der GGW, DDR-1040 Berlin, Invalidenstrasse 43, German Dem. Rep. (1935). Dr.rer.nat. *Crystal growth, solid state physics.*

Kosche, Mr Ingeborg. Akademie der Wissenschaften der DDR, Zentralinstitut für Anorganische Chemie, DDR-1199 Berlin, Rudower Chaussee 5, German Dem. Rep. (1942). Dipl.Min. (tel. +226209). *Electron microscopy.*

Krause, Mr Christa. Akademie der Wissenschaften der DDR, Zentrum für Rechentechnik, DDR-1199 Berlin, Rudower Chaussee 5, German Dem. Rep. (1935). Dipl. Math. (tel. +6702841). *Crystallography.*

Krausse, Dr Joachim. Friedrich Schiller Univ. Jena, Sektion Chemie, DDR-69 Jena, Lessingstrasse 10, German Dem. Rep. (1933). Dr.rer.nat. *X-ray structure analysis.*

Kressner, Dr F.Harry. Akademie der Wissenschaften der DDR, Inst. für Physik der Werkstoffbearbeitung, DDR-1166 Berlin, Seestrasse 82, German Dem. Rep. (1936). Dr.rer.nat. (tel. +6589671). *Crystallography.*

Kretschmer, Dr Rolf-Günther. Akademie der Wissenschaften der DDR, Zentralinstitut für Molekularbiologie, DDR-1115 Berlin, Lindenberger Weg 70, German Dem. Rep. (1934). Dr.rer.nat. (tel. +4862270). *Computing methods in crystallography.*

Krohn, Prof Martin. Akademie der Wissenschaften der DDR, Inst. für Festkörperphysik und Elektronenmikroskopie, DDR-4010 Halle, Am Weinberg 2, German Dem. Rep. (1936). Dr.sc.nat. (tel. +601312). *Thin films.*

Krüger, Dr Albrecht. Akademie der Wissenschaften der DDR, Zentrum für Wissenschaftlichen Gerätebau, DDR-1199 Berlin, Rudower Chaussee 6, German Dem. Rep. (1947). Dr.rer.nat. (tel. +6745498). *Crystal growth.*

Kuban, Dr Ralf-Jürgen. Akademie der Wissenschaften der DDR, Zentralinstitut für physikalische Chemie, DDR-1199 Berlin, Rudower Chaussee 5, German Dem. Rep. (1953). Dr.rer.nat. (tel. +6742257). *Crystallography.*

Kühn, Prof Günther. Karl-Marx-Univ. Leipzig, Sektion Chemie, WB Kristallographie, DDR-7030 Leipzig, Scharnhorststrasse 20, German Dem. Rep. (1939). Dr.sc.nat. (tel. +310502). *Crystal growth, solid state chemistry.*

Kürsten, Dr Hans-Dieter. Akademie der Wissenschaften der DDR, Zentralinstitut für Optik und Spektroskopie, DDR-1199 Berlin, Rudower Chaussee 6, German Dem. Rep. (1945). Dr.rer.nat. (tel. +6702841). *Crystal growth, real structures.*

Kutschabsky, Dr Leo. Akademie der Wissenschaften der DDR, Zentralinstitut für Molekularbiologie, DDR-1115 Berlin, Lindenberger Weg 70, German Dem. Rep. (1933). Dr.rer.nat. (tel. +6743314). *X-ray structure analysis, crystal chemistry, crystallography.*

Lammert, Mr Barbara. Vereinigung für Kristallographie (VFK) in der GGW, DDR-1040 Berlin, Invalidenstrasse 43, German Dem. Rep. (1949). *Astronomy.*

Lebek, Dr Alexander. Akademie der Wissenschaften der DDR, Zentrum für Wissenschaftlichen Gerätebau, DDR-1199 Berlin, Rudower Chaussee 6, German Dem. Rep. (1926). Dr.rer.nat. (tel. +6702841). *Crystallography.*

Lehmann, Dr Gottfried. Humboldt Univ. zu Berlin, Sektion Physik, DDR-1040 Berlin, Invalidenstr. 110, German Dem. Rep. (1924). Dr.rer.nat. (tel. +2803250). *Crystallography.*

Leipner, Dr Hartmut. Martin-Luther-Univ. Halle-Wittenberg, Sektion Physik, DDR-4010 Halle, Friedemann-Bach-Platz 6, German Dem. Rep. (1958). Dipl.Phys. (tel. + 832432). *Electron microscopy, real structures.*

Leonhardt, Dr Albrecht. Akademie der Wissenschaften der DDR, Zentralinstitut für Werkstofforschung, DDR-8020 Dresden, Winterbergstrasse 20, German Dem. Rep. (1949). Dr.rer.nat. (tel. +34770+274). *Crystal growth.*

Leonhardt, Prof Gunter. Technische Universität Karl-Marx-Stadt, DDR-9010 Karl-Marx- Stadt, German Dem. Rep. (1939). Dr.rer.nat. (tel. +6742683). *Crystallography.*

Lilie, Mr Martin. Bergakademie Freiberg, Sektion Geowissenschaften, DDR-9200 Freiberg/Sa., German Dem. Rep. (1949). Dipl.Krist. *Mineralogy.*

Linke, Dr Dietmar. Akademie der Wissenschaften der DDR, Zentralinstitut für Anorganische Chemie, DDR-1199 Berlin, Rudower Chaussee 5, German Dem. Rep. (1940). Dr.rer.nat. (tel. +2362444). *Crystal and glass chemistry.*

Löffler, Prof Hans. Pädagogische Hochschule Halle, DDR-4020 Halle, Kröllwitzer Str. 44, German Dem. Rep. (1923). Dr.habil. (tel. +28211+33). *Metal physics, X-ray diffraction.*

Lösche, Prof. Artur. Karl-Marx-Univ. Leipzig, DDR-7010 Leipzig, Linnestrasse 5, German Dem. Rep. (1921). (tel. +68580). *Crystallography.*

Lutz, Mr Dieter. Vereinigung für Kristallographie (VFK) in der GGW, DDR-1040 Berlin, Invalidenstrasse 43, German Dem. Rep. (1951). Dipl.Krist. *Crystallography.*

Lux, Mr Bernd. Akademie der Wissenschaften der DDR, Zentrum für Wissenschaftlichen Gerätebau, DDR-1199 Berlin, Rudower Chaussee 6, German Dem. Rep. (1949). Dipl.Krist. (tel. +6702871). *Crystal growth.*

Marx, Prof. Günter. Technische Universität Karl Marx Stadt, DDR-9010 Karl-Marx-Stadt, Strasse der Nation 62, PS 964, German Dem. Rep. (1938). Dr.sc. (tel. +668371). *Crystallography.*

Masche, Mr Wolfgang. Vereinigung für Kristallographie (VFK) in der GGW, DDR-1040 Berlin, Invalidenstrasse 43, German Dem. Rep. (1947). Dipl.Min. *Crystallography.*

Meisel, Prof. Armin. Karl-Marx-Univ. Leipzig, Sektion Chemie, DDR-7010 Leipzig, Linnéstr. 2, German Dem. Rep. (1926). Dr.habil. (tel. +6858483). *X-ray spectroscopy.*

Meyer, Prof. Klaus. Akademie der Wissenschaften der DDR, Inst. für Chemische Technologie, DDR-1199 Berlin, Rudower Chaussee 5, German Dem. Rep. (1936). Dr.habil. (tel. +6743431). *Crystallography, chemistry, boundary surfaces.*

Michel, Prof Bernd. Akademie der Wissenschaften der DDR, Institut für Mechanik, DDR-9010 Karl-Marx-Stadt, PSF 408, German Dem. Rep. (1949). Dr.sc.nat. (tel. +42727). *Crystallography, crystallization.*

Möhling, Dr Werner. Akademie der Wissenschaften der DDR, Zentralinstitut für Elektronenphysik, DDR-1080 Berlin, Hausvogteiplatz 5-7, German Dem. Rep. (1934). Dr.rer.nat. (tel. +2000561). *Crystallography, real structures.*

Mohr, Dr Ulrich. Vereinigung für Kristallographie (VFK) in der GGW, DDR-1040 Berlin, Invalidenstrasse 43, German Dem. Rep. (1935). Dr.rer.nat. *Crystal growth, real structures.*

Mothes, Mr Heinrich. Vereinigung für Kristallographie (VFK) in der GGW, DDR-1040 Berlin, Invalidenstrasse 43, German Dem. Rep. (1935). Dipl.Phys. *Crystal growth, electronics.*

Mucha, Mr Christine. Vereinigung für Kristallographie (VFK) in der GGW, DDR-1040 Berlin, Invalidenstrasse 43, German Dem. Rep. (1948). Dipl.Krist. *Crystal chemistry, structure analysis.*

Mühlberg, Dr Manfred. Humboldt-Univ. zu Berlin, Sektion Physik, DDR-1040 Berlin, Invalidenstrasse 110, German Dem. Rep. (1949). Dr.rer.nat. (tel. +2803249). *Crystallography, crystal growth.*

Müller, Dr Bernd. Friedrich-Schiller-Univ. Jena, Sektion Chemie, DDR-69 Jena, Lessingstr. 10, German Dem. Rep. (1943). Dr.rer.nat. (tel. +8226053). *X-ray structure analysis.*

Müller, Dr Brigitte. Akademie der Wissenschaften der DDR, Zentralinstitut für Festkörperphysik und Werkstofforschung, DDR-8032 Dresden, Helmholtzstrasse 20, German Dem. Rep. (1933). (tel. +232200). *Crystallography.*

Müller, Dr Eberhard. Friedrich-Schiller-Univ. Jena, Sektion Chemie, DDR-69 Jena, Lessingstrasse 10, German Dem. Rep. (1942). Dr.rer.nat. *Boundary surface physics and chemistry.*

Muschner, Dr Wolfgang. Vereinigung für Kristallographie (VFK) in der GGW, DDR-1040 Berlin, Invalidenstrasse. 43, German Dem. Rep. (1934). *Crystal growth.*

Neels, Prof. Hermann. Karl-Marx-Univ. Leipzig, Sektion Chemie, Fachrichtung Kristallographie, DDR-7030 Leipzig, Scharnhorststrasse 20, German Dem. Rep. (1913). Dr.rer.nat. (tel. +32519). *Crystal growth, crystallization.*

Neumann, Prof Hans-Georg. Wilhelm-Pieck-Universität Rostock, Sektion Physik, DDR-2500 Rostock, Universitätsplatz 3, German Dem. Rep. (1939). Dr.sc.nat. (tel. +369465). *Amorphous states, real structures.*

Neumann, Dr Wolfgang. Akademie der Wissenschaften der DDR, Inst.für Festkörperphysik und Elektronenmikroskopie, DDR-4010 Halle, Am Weinberg 2, German Dem. Rep. (1944). Dipl.Min. (tel. +601312). *Electron diffraction.*

Nieber, Dr Johannes. Martin-Luther Univ. Halle-Wittenberg, DDR-4020 Halle, Friedemann-Bach-Platz 6, German Dem. Rep. (1948). Dr.rer.nat. (tel. +832432). *Crystallography.*

Niebsch, Dr Hans-Hermann. Humboldt-Univ. zu Berlin, Sektion Physik, DDR-1040 Berlin, Invalidenstrasse 110, German Dem. Rep. (1933). Dr.rer.nat. (tel. +2803361). *Crystal physics, crystal growth.*

Nitsche, Mr Walter. Vereinigung für Kristallographie (VFK) in der GGW, DDR-1040 Berlin, Invalidenstrasse 43, German Dem. Rep. (1948). Dipl.Krist. *Crystallography.*

Noack, Dr Joachim. Akademie der Wissenschaften der DDR, Zentralinstitut für Elektronenphysik, DDR-108 Berlin, Hausvogteiplatz 5-7, German Dem. Rep. (1929). Dr.rer.nat. (tel. +2000561). *Crystal growth.*

Oettel, Dr Heinrich. Bergakademie Freiberg, Sektion Metallurgie, DDR-9200 Freiberg, Zeunerstrasse 5, German Dem. Rep. (1940). Dr.Ing. (tel. +512617). *Crystallography.*

Oppermann, Mr Dieter. Karl-Marx-Univ. Leipzig, Sektion Chemie, Fachrichtung Kristallographie, DDR-70 Leipzig, Scharnhorststrasse 20, German Dem. Rep. (1938). Dipl. Krist. (tel. +310502). *Crystallography.*

Oppermann, Prof Heinrich. Technische Universität Dresden, Sektion Physik, DDR-8027 Dresden, Salvador-Allende-Platz 3, German Dem. Rep. (1934). Dr.sc.nat. (tel. +4834221). *Crystal growth, crystallization.*

Osterland, Mr Martina. Vereinigung für Kristallographie (VFK) in der GGW, DDR-1040 Berlin, Invalidenstrasse 43, German Dem. Rep. (1953). Dipl.Krist. *Petrography, precious stones.*

Parthier, Mr Lutz. Humboldt-Univ. zu Berlin, Sektion Physik, DDR-1040 Berlin, Invalidenstr. 110, German Dem. Rep. (1954). Dipl. Krist. (tel. +2803361). *Crystallography, crystal growth.*

Pätzke, Mr Nora. Vereinigung für Kristallographie (VFK) in der GGW, DDR-1040 Berlin, Invalidenstrasse 43, German Dem. Rep. (1950). Dipl.Krist. *Thermal analysis.*

Paufler, Prof. Peter. Karl-Marx-Univ. Leipzig, Sektion Chemie, DDR-7030 Leipzig, Scharnhorststrasse 20, German Dem. Rup. (1940). Dr.sc.nat., physics (TU Dresden, 1971) Prof. crystallography (1978), Prof. (tel. +310502). *Inorganic crystal structures, crystal physics, defects in crystals, mechanical properties.*

Pechstein, Mr Gisela. Vereinigung für Kristallographie (VFK) in der GGW, DDR-1040 Berlin, Invalidenstrasse 43, German Dem. Rep. (1949). Dipl.Krist. *Crystallography.*

Peibst, Dr Herbert. Akademie der Wissenschaften der DDR, Zentralinstitut für Elektronenphysik, DDR-108 Berlin, Hausvogteiplatz 5-7, German Dem. Rep. (1921). Dr.sc. *Crystallography.*

Penndorf, Dr Jürgen. Karl-Marx-Univ. Leipzig, Sektion Chemie, Fachrichtung Kristallographie, DDR-7030 Leipzig, Scharnhorststrasse 20, German Dem. Rep. (1950). Dipl.Krist. *Crystallography.*

Perthel, Dr Rolf. Akademie der Wissenschaften der DDR, Physikalisch-Technisches Institut, DDR-6900 Jena, Helmholtzweg 4, German Dem. Rep. (1929). Dr.sc.nat. (tel. +27372). *Solid state magnetism.*

Pietsch, Dr Ullrich. Karl-Marx-Universität Leipzig, Sektion Physik, DDR-7010 Leipzig, Linnéstr. 5, German Dem. Rep. (1952). Dr. rer. nat. (tel. +6858447). *Crystallography, X-ray diffraction.*

Pietzsch, Dr Claus. Bergakademie Freiberg, Sektion Physik, DDR-9200 Freiberg, Silbermannstr.1, German Dem. Rep. (1938). Dr.rer.nat. (tel. +512860). *Mössbaür spectroscopy.*

Preiss, Dr Henry, Akademie der Wissenschaften der DDR, Zentralinstitut für physikalische Chemie, DDR-1199 Berlin, Rudower Chaussee 5, German Dem. Rep. (1935). Dr.rer.nat. (tel. +6743154). *Inorganic chemistry, X-ray structure analysis, mass spectroscopy.*

Preuss, Dr Heinz. Ingenieurhochschule Zittau, DDR-8800 Zittau, Theodor-Körner- Allee 16, German Dem. Rep. (1934). Dr.rer.nat. (tel. +93031). *Crystallography, real structures.*

Pritzkow, Dr Wolfgang. Akademie der Wissenschaften der DDR, Zentralinstitut für Physikalische Chemie, DDR-1199 Berlin, Rudower Chaussee 5, German Dem. Rep. (1952). Dr.rer.nat. (tel. +6743310). *Crystallography.*

Puff, Mr Manfred. Ingenieurhochschule Zwickau, Abt.Mathematik, DDR-95 Zwickau, Dr.-Friedrich-Ring 2a, German Dem. Rep. (1929). Dipl. Phys. (tel. +4861+203). *Real structures, X-ray structure analysis.*

Raidt, Dr Helmut. Akademie der Wissenschaften der DDR, Zentralinstitut für Elektronenphysik, DDR-1086 Berlin, Hausvogteiplatz 5-7, German Dem. Rep. (1938). (tel. +2077359). *Crystallography.*

Rechenberg, Dr Ingrid. Akademie der Wissenschaften der DDR, Zentralinstitut für Optik und Spektroskopie, DDR-1199 Berlin, Rudower Chaussee 6, German Dem. Rep. (1938). Dr.rer.nat. *Crystal growth, crystallization, epitaxy.*

Reiche, Dr Manfred. Akademie der Wissenschaften der DDR, Inst.für Festkörperphysik und Elektronenmikroskopie, DDR-4010 Halle, Am Weinberg 2, German Dem. Rep. (1951). Dipl. Krist. (tel. +601312). *Crystallography.*

Reinecke, Mr Kriemhild. Akademie der Wissenschaften der DDR, Zentralinstitut für physikalische Chemie, DDR-1199 Berlin, Rudower Chaussee 5, German Dem. Rep. (1940). Dipl.Krist. (tel. +6743309). *X-ray structure analysis.*

Reinhold, Mr Ingrid. Vereinigung für Kristallographie (VFK) in der GGW, DDR-1040 Berlin, Invalidenstrasse 43, German Dem. Rep. (1935). Dipl.Min. *Crystal growth.*

Rentsch, Dr Harald. Karl-Marx-Univ. Leipzig, Sektion Chemie, Fachrichtung Kristallographie, DDR-7030 Leipzig, Scharnhorststrasse 20, German Dem. Rep. (1950). Dr.rer.nat. *Crystallography.*

Reuter, Mr Dietrich. Vereinigung für Kristallographie (VFK) in der GGW, DDR-1040 Berlin, Invalidenstrasse 43, German Dem. Rep. (1944). Dipl.Chem. *Crystallography.*

Richter, Dr Frank. Akademie der Wissenschaften der DDR, Institut für Halbleiterphysik Frankfurt/Oder, DDR-1211 Falkenhagen, German Dem. Rep. (1947). *Real structures, epitaxy.*

Richter, Mr Hans. Akademie der Wissenschaften der DDR, Zentralinstitut für Physik der Erde, DDR-1500 Potsdam, Telegraphenberg, German Dem. Rep. (1939). Dipl.-Met. *Crystallography.*

Richter, Dr Klaus. Akademie der Wissenschaften der DDR, Zentralinstitut für Physikalische Chemie, DDR-1199 Berlin, Rudower Chaussee 5, German Dem. Rep. (1935). Dr.rer.nat. (tel. +6702841). *X-ray analysis, catalysis.*

Richter, Dr Rainer. Karl-Marx-Univ. Leipzig, Sektion Chemie, DDR-701 Leipzig, Liebigstrasse 18, German Dem. Rep. (1946). Dr.rer.nat. (tel. +7165497). *Crystallography.*

Richter, Dr Waltraut. Akademie der landwirtschaft der DDR, Inst.für Düngungsforschung, DDR-7022 Leipzig, Gustaw-Kühn-Strasse 8, German Dem. Rep. (1944). Dr.rer.nat. *Crystallography.*

Ringel, Dr Lilli. Vereinigung für Kristallographie (VFK) in der GGW, DDR-1040 Berlin, Invalidenstrasse 43, German Dem. Rep. (1937). Dr.Ing. *Crystallography.*

Ritschel, Dr Manfred. Akademie der Wissenschaften der DDR, Zentralinstitut für Festkörperphysik und Werkstofforschung, DDR-8032 Dresden, Helmholtzstrasse 20, German Dem. Rep. (1947). Dr.rer.nat. (tel. +3477+272). *Crystal growth.*

Rosin, Dr Horst. Vereinigung für Kristallographie (VFK) in der GGW, DDR-1040 Berlin, Invalidenstr. 43, German Dem. Rep. (1943). Dr.rer.nat. *Real structures.*

Rossner, Dr Johannes. Vereinigung für Kristallographie (VFK) in der GGW, DDR-1040 Berlin, Invalidenstr. 43, German Dem. Rep. (1945). *Crystallography.*

Rost, Mr Jutta. Martin-Luther-Universität Halle-Wittenberg, Sektion Physik, DDR-4020 Halle, J.-Curie-Platz 26, German Dem. Rep. (1947). Dipl. Krist. *Interfaces, boundary physics, thin film techniques.*

Roestel, Dr Reinhard. Akademie der Wissenschaften der DDR, Inst. für Kosmosforschung, DDR-1199 Berlin, Rudower Chaussee 5, German Dem. Rep. (1951). Dr.rer.nat. *Crystal growth.*

Rudolph, Prof Peter. Humboldt-Univ. zu Berlin, Sektion Physik, DDR-1040 Berlin, Invalidenstrasse 110, German Dem. Rep. (1945). Dr.sc.nat. (tel. +2803334). *Crystallography, crystal growth.*

Ruscher, Prof. Christian. Karl-Marx-Universität Leipzig, Sektion Physik, DDR-7010 Leipzig, Linnéstr. 5, German Dem. Rep. (1928). Dr.habil. (tel. +6858276). *Polymerization, X-ray diffraction, thin films.*

Sarodnik, Dr Reinhard. Vereinigung für Kristallographie (VFK) in der GGW, DDR-1040 Berlin, Invalidenstrasse 43, German Dem. Rep. (1942). Dr.rer.nat. *Glasses.*

Sayfarth, Prof Hans-Heinz. Pädagogische Hochschule Erfurt-Mühlhausen, Sektion Chemie, DDR-5700 Mühlhausen, Schillerweg 59, German Dem. Rep. (1939). *Crystallography, chemical aspects.*

Schaal, Mr Joachim. Technische Hochschule Carl Schorlemmer Merseburg, DDR-4200 Merseburg, Geusärstrasse, German Dem. Rep. (1952). Dipl.Krist. (tel. +462155). *Polarization microscopy.*

Schäfer, Dr Peter. Humboldt-Univ. zu Berlin, Sektion Physik, DDR-1040 Berlin, Invalidenstrasse 110, German Dem. Rep. (1953). Dr.rer.nat. (tel. +28030). *Crystallography.*

Scharfenberg, Dr Rudolf. Akademie der Wissenschaften der DDR, Zentralinstitut für Festkörperphysik und Werkstofforschung, DDR-8032 Dresden, Helmholtzstrasse 20, German Dem. Rep. (1935). Dipl.Ing. (tel. +2322207). *Crystal growth.*

Schenk, Prof Manfred. Humboldt-Universität zu Berlin, Sektion Physik, DDR-1040 Berlin, Invalidenstr. 110, German Dem. Rep. (1934). Dr.habil. (tel. +2803361). *Crystallography.*

Schilling, Mr Hansjoachim. Akademie der Wissenschaften der DDR, Zentrum für Wissenschaftlichen Gerätebau, DDR-1199 Berlin, Rudower Chaussee 6, German Dem. Rep. (1938). Dipl.Ing. *Crystal growth.*

Schippel, Dr Erhard. Ingenieurhochschule Wismar, Sektion Technologie der Elektrotechnik und Elektronik, DDR-2500 Wismar, Phillip-Müller-Str., German Dem. Rep. (1935). Dr.rer.nat. *Crystallography, structure, germination.*

Schläfer, Dr Dietrich. Akademie der Wissenschaften der DDR, Zentralinstitut für Festkörperphysik und Werkstofforschung, DDR-8032 Dresden, Helmholtzstrasse 20, German Dem. Rep. (1939). Dr.Ing. (tel. +4659241). *Texture, metallography.*

Schläfer, Dr Ursula. Akademie der Wissenschaften der DDR, Zentralinstitut für Festkörperphysik und Werkstofforschung, DDR-8032 Dresden, Helmholtzstrasse 20, German Dem. Rep. (1938). Dr.Ing. (tel. +4659253). *Texture, metallography.*

Schmidt, Dr Peter. Akademie der Wissenschaften der DDR, Zentralinstitut für Optik und Spektroskopie, DDR-1199 Berlin, Rudower Chaussee 5, German Dem. Rep. (1952). Dr.rer.nat. *Crystallography, epitaxy.*

Schmidt, Prof. Günter. Martin-Luther-Univ. Halle-Wittenberg, DDR-4020 Halle, Friedemann-Bach-Platz 6, German Dem. Rep. (1921). Dr.sc. (tel. +37761). *Crystal and solid state physics.*

Schmidt, Prof. Werner. Pädagogische Hochschule, Sektion Physik, DDR-8060 Dresden, Wigardstrasse 17, German Dem. Rep. (1934). Dr.habil. *Real structures, crystal physics.*

Schmitz, Dr Werner. Karl-Marx-Univ. Leipzig, Sektion Chemie, Fachrichtung Kristallographie, DDR-7030 Leipzig, Scharnhorststrasse 20, German Dem. Rep. (1943). Dr.sc.nat. (tel. +310502). *Crystallography, crystal growth.*

Schneider, Dr Günter. Vereinigung für Kristallographie (VFK) in der GGW, DDR-1040 Berlin, Invalidenstr. 43, German Dem. Rep. (1934). Dr.habil. *Epitaxy.*

Schneider, Prof. Herbert. Bergakademie Freiberg, Section Physik, DDR-9200 Freiberg, Silberman Strasse 1, German Dem. Rep. (1926). Dr.habil. (tel. +512860). *Solid state physics, interfacial physics.*

Schott, Prof. Günter. Wilhelm-Pieck Univ. Rostock, Sektion Chemie, DDR-25 Rostock, Buchbinderstrasse 9, German Dem. Rep. (1921). Dr.habil. *Organic silicon chemistry.*

Schrader, Prof. Richard. Vereinigung für Kristallographie (VFK) in der GGW, DDR-1040 Berlin, Invalidenstrasse 43, German Dem. Rep. (1915). Dr.habil. *Inorganic chemistry, crystal chemistry.*

Schrauber, Dr Hannelore. Akademie der Wissenschaften der DDR, Zentralinstitut für Molekularbiologie, DDR-1115 Berlin, Robert-Roessle-Str. 10, German Dem. Rep. (1934). Dr.rer.nat. (tel. +3464433). *Crystallography, computer programming.*

Schreiter, Dr Peter. Karl-Marx-Univ. Leipzig, Sektion Chemie, Fachrichtung Kristallographie, DDR-7030 Leipzig, Scharnhorststrasse 20, German Dem. Rep. (1938). Dr.sc.nat. (tel. 310502). *Mineralogy, petrography.*

Schröder, Dr Winfried. Akademie der Wissenschaften der DDR, Zentrum für Wissenschaftlichen Gerätebau, DDR-1199 Berlin, Rudower Chaussee 6, German Dem. Rep. (1937). Dr.Ing. (tel. +6702871). *Crystal growth, real structures.*

Schubert, Dr Gernot. Krankenhaus Friedrichshain, Urologische Klinik, DDR-1017 Berlin, Leninallee 171, German Dem. Rep. (1945). Dr.rer.nat. (tel. +4367027). *X-ray structure analysis, order-disorder phenomena.*

Schubert, Mr Heike-Kristina. Vereinigung für Kristallographie (VFK) in der GGW, DDR-1040 Berlin, Invalidenstrasse 43, German Dem. Rep. (1945). Dipl.Krist. *Microscopy, structure analysis.*

Schulz, Dr Manfred. Vereinigung für Kristallographie (VFK) in der GGW, DDR-1040 Berlin, Invalidenstrasse 43, German Dem. Rep. (1934). Dr.rer.nat. *Crystal physics, semiconductor physics.*

Schulze, Prof Dietrich. Akademie der Wissenschaften der DDR, Zentralinstitut für Festkörperphysik und Werkstofforschung, DDR-8032 Dresden, Helmholtzstrasse 20, German Dem. Rep. (1922). Dr.habil. (tel. +4659230). *Solid state physics, electron microscopy.*

Schulze, Dr Günter. Technische Hochschule Carl Schorlemmer Merseburg, DDR-4200 Merseburg, Geusär Strasse, German Dem. Rep. (1932). Dr.rer.nat. (tel. +462790). *Crystallography.*

Schumann, Dr Bernd. Karl-Marx-Univ. Leipzig, Sektion Chemie, Fachrichtung Kristallographie, DDR-7030 Leipzig, Scharnhorststrasse 20, German Dem. Rep. (1947). Dr.rer.nat. (tel. +310502). *Real structures.*

Schunk, Dr Wolfgang. Martin-Luther-Univ. Halle, Sektion Chemie, DDR-4020 Halle, Domstrasse 4, German Dem. Rep. (1950). Dipl.Chem. (tel. +832441). *Crystallography.*

Sedlacek, Dr Paul. Akademie der Wissenschaften der DDR, Zentralinstitut für physikalische Chemie, DDR-1199 Berlin, Rudower Chaussee 5, German Dem. Rep. (1920). Dr.rer.nat. (tel. +6702841). *X-ray structure analysis.*

Seemann, Dr Hans. Akademie der Wissenschaften der DDR, Zentralinstitut für physikalische Chemie, DDR-1199 Berlin, Rudower Chaussee 5, German Dem. Rep. (1907). Dr.rer.nat. *Crystallography.*

Seidowski, Dr Eckart. Vereinigung für Kristallographie (VFK) in der GGW, DDR-1040 Berlin, Invalidenstr. 43, German Dem. Rep. (1946). Dr.sc. *Semiconductor physics and techniques.*

Seifert, Dr Wolfgang. Akademie der Wissenschaften der DDR, Zentralinstitut für Physik der Erde, DDR-1500 Potsdam, Telegrafenberg, German Dem. Rep. (1947). Dr.rer.nat. *Mineralogy.*

Sieler, Dr Joachim. Karl-Marx-Univ. Leipzig, Sektion Chemie, DDR-701 Leipzig, Liebigstrasse 18, German Dem. Rep. (1939). Dr. sc.nat. (tel. +7165497). *Crystallography.*

Soa, Dr Ernst-Adolf. Akademie der Wissenschaften der DDR, Zentralinstitut für Optik und Spektroskopie, DDR-69 Jena, Humboldtstrasse 11, German Dem. Rep. (1930). Dr.habil. *Thin film techniques.*

Sommer, Dr Joachim. TU Dresden, Sektion Physik, DDR-8027 Dresden, Mommsenstr. 13, German Dem. Rep. (1934). Dr.rer.nat. (tel. +4833887). *X-ray metallography.*

Sommermann, Mr Günter. Akademie der Wissenschaften der DDR, Zentralinstitut für physikalische Chemie, DDR-1199 Berlin, Rudower Chaussee 5, German Dem. Rep. (1931). Dipl.Phys. (tel. +6702841). *Spectroscopy, electron microscopy.*

Sorge, Prof Georg. Martin-Luther-Univ. Halle-Wittenberg, DDR-4020 Halle, Friedemann-Bach-Platz 6, German Dem. Rep. (1936). Dr.rer.nat. *Solid state physics.*

Spindler, Dr Herbert. Vereinigung für Kristallographie (VFK) in der GGW, DDR-1040 Berlin, Invalidenstrasse 43, German Dem. Rep. (1934). Dr.rer.nat. *X-ray structure analysis.*

Sprenger, Dr Heinz. Akademie der Wissenschaften der DDR, Zentralinstitut für Anorganische Chemie, DDR-1199 Berlin, Rudower Chaussee 5, German Dem. Rep. (1946). Dipl.Krist. *Crystallography.*

Starke, Dr Rainer. Bergakademie Freiberg, Sektion Geowissenschaften, DDR-9200 Freiberg, Brennhausgasse 14, German Dem. Rep. (1933). Dr.habil. (tel. +512660). *Mineralogy, crystal chemistry.*

Stecker, Prof. Kurt. Martin-Luther-Univ. Halle-Wittenberg, DDR-4020 Halle, Friedemann-Bach-Platz 6, German Dem. Rep. (1929). Dr.sc. (tel. +21153). *Semiconductor physics.*

Stegmann, Dr Eleonore. Technische Hochschule Carl Schorlemmer Merseburg, DDR-4200 Merseburg, Geusärstrasse, German Dem. Rep. (1929). Dr.rer.nat. (tel. +462071). *Chemistry, microscopy.*

Steil, Dr Helmut. Wilhelm-Pieck Univ. Rostock, Sektion Physik, DDR-2520 Rostock, Universitätsplatz 3, German Dem. Rep. (1949). Dr.rer.nat. *Crystallography.*

Steinbruch, Mr Uta. Vereinigung für Kristallographie (VFK) in der GGW, DDR-1040 Berlin, Invalidenstrasse 43, German Dem. Rep. (1949). Dipl.Chem. *Crystal growth.*

Steinicke, Prof. Ursula. Akademie der Wissenschaften der DDR, Zentralinstitut für physikalische Chemie, DDR-1199 Berlin, Rudower Chaussee 5, German Dem. Rep. (1935). Dr.rer.nat. (tel. +6742257). *Crystal growth, X-ray diffraction.*

Stelzner, Mr Sabine. Vereinigung für Kristallographie (VFK) in der GGW, DDR-1040 Berlin, Invalidenstrasse 43, German Dem. Rep. (1949). Dipl.Krist. *Solid state physics.*

Stephan, Dr Dieter. Akademie der Wissenschaften der DDR, Zentralinst. für Festkörperphysik und Werkstofforschung, DDR-8032 Dresden, Helmholtzstr. 20, German Dem. Rep. (1939). Dr.rer.nat. (tel. +4833488). *X-ray diffraction, real structures.*

Stephanik, Dr Heinz. Martin-Luther-Univ. Halle, Sektion Chemie, DDR-4020 Halle, Domstrasse 4, German Dem. Rep. (1925). Dr.rer.nat. *Crystallography.*

Storbeck, Prof. Fritz. TU Dresden, Sektion Physik, DDR-8027 Dresden, Mommsenstr. 13, German Dem. Rep. (1937). Dr.sc.nat. (tel. +4633249). *Crystallography.*

Stroppe, Prof Hilbert. Technische Hochschule Otto v. Guericke Magdeburg, DDR-3010 Magdeburg, PSF 124, German Dem. Rep. (1932). Dr.sc.nat. (tel. +592467). *Crystal physics, X-ray analysis.*

Stündel, Mr Peter. Vereinigung für Kristallographie (VFK) in der GGW, DDR-1040 Berlin, Invalidenstr. 43, German Dem. Rep. (1947). Dipl. Ing. *Crystal growth.*

Suessmann, Dr Hans. Martin-Luther-Univ. Halle-Wittenberg, Sektion Physik, DDR-4020 Halle, Friedemann-Bach-Platz 6, German Dem. Rep. (1930). Dr.sc.nat. (tel. +37761). *Crystal growth.*

Swillens, Mr Eckhard. Vereinigung für Kristallographie (VFK) in der GGW, DDR-1040 Berlin, Invalidenstrasse 43, German Dem. Rep. (1935). Dipl.Min. *Physical chemistry, crystallography.*

Syhre, Dr Hans. Akademie der Wissenschaften der DDR, Zentralinstitut für Werkstofforschung, DDR-8020 Dresden, Wittenbergstrasse 28, German Dem. Rep. *Crystallography.*

Szulzewsky, Dr Klaus. Akademie der Wissenschaften der DDR, Zentralinstitut für physikalische Chemie, DDR-1199 Berlin, Rudower Chaussee 5, German Dem. Rep. (1940). Dr.rer.nat. (tel. +6702841). *X-ray structure analysis.*

Tänzer, Dr Dietmar. Akademie der Wissenschaften der DDR, Zentralinstitut für Elektronenphysik, DDR-108 Berlin, Hausvogteiplatz 5-7, German Dem. Rep. (1940). Dipl.Phys. *Crystallography.*

Tempel, Dr Alfred. Technische UniversitäT Dresden, Sektion Physik, DDR-8027 Dresden, Salvador-Allende-Platz 3, German Dem. Rep. (1941). Dr.sc.nat. *Solid state physics, real physics, electron microscopy.*

Tempelhoff, Dr Klaus. Akademie der Wissenschaften der DDR, Zentralinstitut für Elektronenphysik, DDR-108 Berlin, Hausvogteiplatz 5-7, German Dem. Rep. (1938). Dr.rer.nat. *Crystal growth, ideal structures.*

Teresiak, Mr Angelika. Akademie der Wissenschaften der DDR, Zentralinstitut für Festkörperphysik und Werkstofforschung, DDR-8032 Dresden, Helmholtzstr. 20, German Dem. Rep. (1949). Dipl.Krist. *Crystallography.*

Thieme, Dr Wolfgang. Akademie der Wissenschaften der DDR, Zentralinstitut für Optik und Spektroskopie, DDR-1199 Berlin, Rudower Chaussee 6, German Dem. Rep. (1946). Dr.Ing. *Crystallography.*

Trempler, Mr Jörg. Technische Hochschule Merseburg, Sektion Werkstoffwissenschaften, DDR-4200 Merseburg, German Dem. Rep. (1946). (tel. +462543). *Electron microscopy.*

Trettin, Dr Reinhard. Akademie der Wissenschaften der DDR, Zentralinstitut für Anorganische Chemie, DDR-1199 Berlin, Rudower Chaussee 5, German Dem. Rep. (1951). Dipl.Krist. (tel. +6743366). *Real structures, surface analysis.*

Uecker, Mr Reinhard. Akademie der Wissenschaften der DDR, Zentralinstitut für organische Chemie, DDR-1199 Berlin, Rudower Chaussee 5, German Dem. Rep. (1951). Dipl.Krist. *Crystallography.*

Ulbricht, Prof. Heinz. Wilhelm-Pieck Univ. Rostock, Sektion Physik, DDR-2520 Rostock, Universitätsplatz 3, German Dem. Rep. (1931). Dr.sc.nat. (tel. +369327). *Crystallography.*

Ullrich, Prof Hans-Jürgen. Technische Univ. Dresden, Sektion Physik, DDR-8027 Dresden, Mommsenstr. 13, German Dem. Rep. (1938). Dr.sc. (tel. +4632366). *Real structures, X-ray structure analysis.*

Unangst, Prof Dietrich. Friedrich-Schiller-Univ. Jena, Sektion Physik, DDR-69 Jena, Max - Wien - Platz 1, German Dem. Rep. (1931). Dr.habil. (tel. +24036). *Crystallography, real structures.*

Unger, Prof. Konrad. Karl-Marx-Univ. Leipzig, DDR-7010 Leipzig, Linnestrasse 5, German Dem. Rep. (1934). Dr.sc. (tel. +6858219). *Crystallography.*

Velfe, Mr Hans Dieter. Akademie der Wissenschaften der DDR, Inst.für Festkörperphysik und Elektronenmikroskopie, DDR-4010 Halle, Am Weinberg 2, German Dem. Rep. (1942). Dipl.Phys. (tel. +601312). *Electron microscopy.*

Voigt, Dr Dieter. Akademie der Wissenschaften der DDR, Zentralinstitut für Festkörperphysik und Werkstofforschung, DDR-8032 Dresden, Helmholtzstr. 20, German Dem. Rep. (1938). Dr.rer.nat. *Crystallography.*

Voigt, Dr Rita. Akademie der Wissenschaften der DDR, Zentralinstitut für physikalische Chemie, DDR-1199 Berlin, Rudower Chaussee 5, German Dem. Rep. (1941). Dr.rer.nat. (tel. +6702841). *Real structures.*

Vollstädt, Prof Heiner. Akademie der Wissenschaften der DDR, Zentralinstitut für Physik der Erde, DDR-1500 Potsdam, Telegrafenberg, German Dem. Rep. (1939). Dr.rer.nat. (tel. +4551+384). *Mineralogy, geology.*

Wadewitz, Dr Heinz. Akademie der Wissenschaften der DDR, Zentralinstitut für Festkörperphysik und Werkstofforschung, DDR-8032 Dresden, Helmholtzstr. 20, German Dem. Rep. (1921). Dr.rer.nat. *X-ray diffraction, metal physics.*

Wäsch, Dr Elke. Humboldt-Universität zu Berlin, Museum für Naturkunde, DDR-1040 Berlin, Invalidenstr. 43, German Dem. Rep. (1942). Dr.rer.nat. (tel. +28970). *Mineralogy, crystallography.*

Wagner, Dr Gerald. Karl-Marx-Univ. Leipzig, Sektion Chemie, WB Kristallographie, DDR-7030 Leipzig, Scharnhorststrasse 20, German Dem. Rep. (1950). Dr.rer.nat. (tel. +71650). *Crystal growth, real structures.*

Wagner, Mr Gunther. Vereinigung für Kristallographie (VFK) in der GGW, DDR-1040 Berlin, Invalidenstrasse 43, German Dem. Rep. (1940). Dipl.Min. *Real structures, X-ray structure analysis.*

Wagner, Dr Günter. Akademie der Wissenschaften der DDR, Zentralinstitut für Optik und Spektroskopie, DDR-1199 Berlin, Rudower Chaussee 5, German Dem. Rep. (1953). Dr.rer.nat. (tel. +6743496). *Epitaxy.*

Wahner, Mr Bettina. Akademie der Wissenschaften der DDR, Zentralinstitut für Anorganische Chemie, Bereich Glas Keramik, DDR-1040 Berlin, Invalidenstrasse 44, German Dem. Rep. (1950). Dipl.Krist. *X-ray structure analysis, X-ray diffraction.*

Walther, Mr Christa. TH Carl Schorlemmer Merseburg, DDR-4200 Merseburg, Geusärstr., German Dem. Rep. (1934). Dipl.Phys. (tel. +462774). *X-ray structure analysis.*

Wappler, Dr Gert. Humboldt-Univ. zu Berlin, Museum für Naturkunde, DDR-1040 Berlin, Invalidenstrasse 43, German Dem. Rep. (1935). Dr.rer.nat. (tel. +28970). *Mineralogy, crystal structures.*

Wawra, Mr Herbert. Vereinigung für Kristallographie (VFK) in der GGW, DDR-1040 Berlin, Invalidenstrasse 43, German Dem. Rep. (1942). Dipl.Krist. *topiconductor and interface physics.*

Weis, Dr Josef. Vereinigung für Kristallographie (VFK) in der GGW, DDR-1040 Berlin, Invalidenstrasse 43, German Dem. Rep. (1933). Dr.rer.nat. *Crystal growth, X-ray diffraction.*

Weise, Dr Günter. Akademie der Wissenschaften der DDR, Zentralinstitut für Festkörperphysik und Werkstofforschung, DDR-8032 Dresden, Helmholtzstrasse 20, German Dem. Rep. (1930). Dr.habil. (tel. +3477395). *Crystallography.*

Weiss, Mr Hans-Georg. Akademie der Wissenschaften der DDR, Zentrum für Rechentechnik, DDR-1199 Berlin, Rudower Chaussee 5, German Dem. Rep. (1935). Dipl.Phys. (tel. +6702841). *Computing, computer programming.*

Weiss, Dr Helmut. Vereinigung für Kristallographie (VFK) in der GGW, DDR-1040 Berlin, Invalidenstrasse 43, German Dem. Rep. (1934). Dr.rer.nat. *Crystal growth, crystallography.*

Wendland, Mr Bettina. Vereinigung für Kristallographie (VFK) in der GGW, DDR-1040 Berlin, Invalidenstrasse 43, German Dem. Rep. (1950). Dipl.Krist. *Metals and material science.*

Werner, Dr Inge. Akademie der Wissenschaften der DDR, Zentralinstitut für Arbeitsmedizin der DDR, DDR-1134 Berlin, Nöldnerstrasse 40-42, German Dem. Rep. (1934). Dr.rer.nat. (tel. +5509901). *X-ray diffraction, phase analysis.*

Wieser, Prof Egbert. Akademie der Wissenschaften der DDR, Zentralinstitut für Kernforschung, DDR-8051 Dresden, Postfach 19, German Dem. Rep. (1938). Dr.sc.nat. (tel. +52541). *Solid state physics.*

Wilde, Dr Wolfgang. Akademie der Wissenschaften der DDR, Zentralinstitut für Anorganische Chemie, DDR-1199 Berlin, Rudower Chaussee 5, German Dem. Rep. (1935). Dr.rer.nat. *Crystal chemistry, X-ray structure analysis, synthetic chemistry.*

Wildner, Mr Günter. Vereinigung für Kristallographie (VFK) in der GGW, DDR-1040 Berlin, Invalidenstrasse 43, German Dem. Rep. (1932). Dipl.Min. *Petrography, polarization microscopy.*

Windsch, Prof. Wolfgang. Karl-Marx-Univ. Leipzig, DDR-7010 Leipzig, Linnestrasse 5, German Dem. Rep. (1931). (tel. +68580). *Solid state physics, ferroelectrics.*

Winkler, Dr Michael. Humboldt-Universität zu Berlin, Sektion Physik, DDR-1040 Berlin, Invalidenstr. 110, German Dem. Rep. (1955). Dr.rer.nat. (tel. +2803365). *Crystallography, epitaxy.*

Winzer, Dr Achim. TH Carl Schorlemmer Merseburg, DDR-4200 Merseburg, Geusärstr., German Dem. Rep. (1937). Dr.rer.nat. (tel. +462035). *Crystallography.*

Wolf, Mr Eberhard. Akademie der Wissenschaften der DDR, Zentrum für Wissenschaftlichen Gerätebau, DDR-1199 Berlin, Rudower Chaussee 6, German Dem. Rep. (1932). Dipl.Min. (tel. +6702871). *Crystal growth, X-ray diffraction.*

Wolf, Prof Dieter. Bergakademie Freiberg, Sektion Geowissenschaften, DDR-9200 Freiberg/Sa., PSF 47, German Dem. Rep. (1939). Dr.sc.nat. *Crystallography.*

Woltersdorf, Dr Jörg. Akademie der Wissenschaften der DDR, Institut für Festkörperphysik und Elektronenmikroskopie, DDR-4020 Halle, Am Weinberg, German Dem. Rep. (1944). Dr.rer.nat. (tel. +6013112). *Electron diffraction, interfaces, solid state physics.*

Worzala, Prof Horst. Akademie der Wissenschaften der DDR, Zentralinstitut für Anorganische Chemie, DDR-1199 Berlin, Rudower Chaussee 5, German Dem. Rep. (1938). Dr.rer.nat. (tel. +67402771). *X-ray structure analysis, structure chemistry.*

Wurl, Dr Bernd. Akademie der Wissenschaften der DDR, Zentralinstitut für Kybernetik und Informationsprozesse, DDR-1199 Berlin, Rudower Chaussee 5, German Dem. Rep. (1947). Dr.rer.nat. (tel. +674342). *Crystallography, crystal growth.*

Zahn, Dr Andreas. Karl-Marx-Universität Leipzig, Sektion Chemie, DDR-7010 Leipzig, Talstr. 35, German Dem. Rep. (1956). Dr.rer.nat. (tel. +71650). *Microscopy.*

Zahn, Dr Gernot. Karl-Marx-Universität Leipzig, Sektion Chemie, DDR-7010 Leipzig, Talstr. 35, German Dem. Rep. (1957). Dr.rer.nat. (tel. +71650). *X-ray structure analysis, computing.*

Zeigan, Dr Dieter. Akademie der Wissenschaften der DDR, Zentralinstitut für physikalische Chemie, DDR-1199 Berlin, Rudower Chaussee 5, German Dem. Rep. (1935). Dr.rer.nat. *X-ray structure analysis.*

Zickert, Mr Kurt. Wilhelm-Pieck Univ. Rostock, Sektion Physik, DDR-2520 Rostock, Universitätsplatz 3, German Dem. Rep. (1936). Dipl.Phys. (tel. +23212). *Amorphous solid state, X-ray physics.*

Ziemer, Dr Burkhard. Humboldt-Universität zu Berlin, Sektion Chemie, DDR-1040 Berlin, Hessische Str. 1-2, German Dem. Rep. (1945). Dr.rer.nat. (tel. +2838284). *Crystallography.*

Zschach, Dr Siegfried. Vereinigung für Kristallographie (VFK) in der GGW, DDR-1040 Berlin, Invalidenstrasse 43, German Dem. Rep. (1938). Dr.rer.nat. *Instrument manufacture.*

GERMANY, FED. REP.
BRD

Sub-Editors: **H. Burzlaff**

Notes

1. International telephone country code - 49. Exceptions are Austria 060 and Luxemburg 050 instead of 49. These numbers, e.g. 00949 for Denmark or 060 for Austria replace the first digit 0 of the area code.

2. Degrees conferred by the German Universities are the *Doctor honoris causae* (Dr h.c.) (an honorary degree), *Doctor philosophiae* (Dr phil.), *Doctor philosophiae naturalis* (Dr phil.nat.), *Doctor philosophiae rerum naturalium* (Dr phil.rer.nat.), *Doctor rerum naturalium* (Dr rer.nat.), and *Doctor scientiæ naturalis* (Dr sc.nat.) (all approximately equivalent to PhD), and the *Diplom-Chemiker* (Dipl.-Chem.), *Diplom-Mineraloge* (Dipl.-Min.) and *Diplom-Physiker* (Dipl.-Phys.) (these three are approximately equivalent to MSc). At the German Technische Hochschulen and Technical Universities the degrees *Doctor-Ingenieur* (Dr -Ing.), *Doctor rerum technicarum* (Dr rer.techn.), *Doctor scientiæ technicae* (Dr sc.techn.), and *Diplom-Ingenieur* (Dipl.-Ing.) can be obtained. To get the position of a *Dozent* (equivalent to reader) at a German University, Technische Hochschule, or Technical University the conditions of the *Habilitation* have to be fulfilled corresponding to the degree of *Dr habil.*

3. At a German University, Technische Hochschule or Techn. University, persons can be appointed in the following positions *Ord. Professor* (equivalent to full professor), *Wiss. Rat und Professor* (equivalent to associate professor), *ausserplanmässiger Professor* (equivalent to assistant professor), *Dozent* (equivalent to reader). Other positions are *Akad. Direktor, Akad. Oberrat, Akad. Rat, Kustos, Konservator, Oberassistent, Assistent (Asst.), Wiss. Angestellter (Ang.),* and *Wiss. Mitarbeiter (Mit.)* (equivalent to assistant professor, research associate, or research assistant).

4. Further abbreviations used in the German list are

Abt. - Abteilung (department)	Inst. - Institut
Ang. - Angestellter (employee)	i.R. - im Ruhestand (retired)
Ass. - Assistent (assistant)	Mit. - Mitarbeiter (co-worker)
akad. - akademisch (academic)	MPI - Max-Planck-Institut
angew. - angewandt (applied)	Ord. Prof. - ordentlicher Prof.
Apl. Prof. - ausserplanmässiger Prof.	Str. - Strasse (street)
em. - emeritiert (emeritus)	RWTH - Rheinisch-Westfälische Technische Hochschule at Aachen
FU - Freie Universität at Berlin	TH - Technische Hochschule
Geb. - Gebäude (building)	TU - Technische Universität
Ges. - Gesellschaft (society)	U. - Universität
GH - Gesamthochschule (University)	wiss. - wissenschaftlich (scientific)

5. The postal code consists of D- and a four-digit number, e.g. D-7500 and precedes the name of the town.

6. The telephone numbers included in the code list are normally those of a central switchboard. If the number ends in a '1', it is in general possible to bypass the central switchboard by dialing the extension number instead of the '1'. The first group of digits beginning with '0' is the area code.

7. The alphabetical order of the names in the list is according to German usage. So *von Schnering* is listed under S and vowels with an Umlaut (Ä, Ö, Ü) are placed as if they were written as AE, OE, and UE respectively.

8. The name of the 'present position' is given in English or German, in most cases according to the data presented by the crystallographer.

Abriel, Dr Walter. Inst. f. Anorg. Chemie, Universität, P.O.Box 397, D-8400 Regensburg, Germany, Fed. Rep. (1949) Dr rer.nat., inorganic chemistry (U. Regensburg, 1980). Prof. (tel. 0941 + 943, ext. 4543, telex 65658 unire d). *Inorganic solid state chemistry.*

Abs-Wurmbach, Dr Irmgard. Inst. f. Mineralogie, Universität, Lahnberge, D-3550 Marburg, Germany, Fed. Rep. (1938) Priv.-Doz. Dr rer.nat., mineralogy (U. Bonn, 1973). Wiss. Ang. (tel. 06421 + 28, ext. 5611). *Experimental petrology, experimental mineralogy, physical properties of minerals.*

Ackermann, Dr Lothar Peter Matthias. Dept. Kristalltechnologie, Hartmetallwerkzeugfabrik Andreas Maier GmbH+Co KG, Stegwiesen 2, D-7959 Schwendi(Hörenhausen), Germany, Fed. Rep. (1953) Dr rer.nat., crystal chemistry, spectroscopy (TU. Berlin, 1983). Chief of dept. (tel. 07347 + 61-0, ext. 260, fax 07347-7307, telex 712590 ham). *Crystal growth, crystal chemistry.*

Albers, Miss Ursula. An St. Albertus Magnus 31, D-4300 Essen, Germany, Fed. Rep. (1959)

Alexander, Prof. Dr Helmut. Abt. f. Metallphysik im II. Physik. Inst., Universität, Zülpicher Str. 77, D-5000 Köln 41, Germany, Fed. Rep. (1928) Prof., metal physics Wiss. Abteilungsvorsteher. (tel. 0221 + 4701, ext. 4200). *Electron microscopy, crystal defects, plasticity, electron paramagnetic resonance.*

Allmann, Prof. Dr Rudolf. Fachbereich Geowissenschaften, Universität, Lahnberge, D-3550 Marburg, Germany, Fed. Rep. (1931) Dr habil., mineralogy and crystallography (U. Marburg, 1968). Prof. (tel. 06421 + 281, ext. 3002). *Double layer structures, heteropoly acids, organic structures, electron density.*

D' Amour-Sturm, Dr Hedwig. Fachbereich Physik, Universität-GH, Warburgstr. 100, D-4790 Paderborn, Germany, Fed. Rep. (1948) Dr, crystallography (U. Marburg, 1974). Asst. (tel. 05251 + 602674). *Crystal structures, phase transitions, high pressure techniques.*

Amstutz, Prof. Dr Dr h.c. Christian Gerhard. Mineralogisch-Petrogr. Inst., Universität, Postfach 104040, D-6900 Heidelberg, Germany, Fed. Rep. (1922) Dr sc.nat. Dr h.c., mineralogy, petrology, geology (ETH Zürich, 1952). Prof. and Dir. (tel. 06221 + 561, ext. 2802). *Mineralogy, petrology, ore deposits, igneous and sedimentary petrology, history and philosophy of science, pseudomorphism.*

Angermund, Mr Klaus Peter. MPI f. Kohlenforschung, Lembkestr. 5, D-4330 Mülheim/Ruhr, Germany, Fed. Rep. (1958) Dipl.-Chem. (U. Düsseldorf, 1983). PhD student. (tel. 0208 + 306 ext. 491). *Organometallic compounds, electron deformation density distribution.*

Anselment, Dr Bernhard. Dept. RCK/K (Catalysts, Production & Development), BASF AG, Building A 520 D-6700 Ludwigshafen, Germany, Fed. Rep. (1955) Dipl.-Chem. (U. Karlsruhe, 1982). Developing chemist. (tel. 0621 + 60 ext. 21072). *Inorganic chemistry, solid state chemistry, phase transitions, crystallography.*

Arndt, Prof. Dr Jörg Friedrich. Mineralogisch-Petrogr. Inst., Universität, Wilhelmstr. 56, D-7400 Tübingen, Germany, Fed. Rep. (1938) Habilitation, mineralogy and crystallography (U. Tübingen, 1974). Prof. (tel. 07071 + 291, ext. 6802). *Mineralogy, crystallography, crystal chemistry, high pressure mineralogy, glasses - structure and properties, technical mineralogy, shock metamorphism, lunar and terrestrial materials.*

Arnold, Prof. Dr Heinrich Günther Alfred. Inst. f. Kristallographie, RWTH, Templergraben 55, D-5100 Aachen, Germany, Fed. Rep. (1930) Dr rer.nat., crystallography (U. Würzburg, 1964). Dozent. (tel. 0241 + 80, ext. 6901, telex 08/32704 THAC D). *Symmetry, phase transitions, lattice dynamics.*

Attig, Dr Rainer. Chemische Landesuntersuchungsanstalt, Hoffstr. 3, D-7500 Karlsruhe, Germany, Fed. Rep. (1944) Dr rer.nat., chemistry (TU Braunschweig, 1973). Reg. Chemierat. (tel. 0721 + 1351, ext. 3631). *Inorganic and organic crystal structures, neutron diffraction.*

Axmann, Dr Anton. Bereich C1, Hahn-Meitner-Inst. f. Kernforschung, Glienicker Str. 100, D-1000 Berlin 39, Germany, Fed. Rep. (1937) Dr rer.nat., physics

(RWTH Aachen, 1968). Asst. Group Leader. (tel. 030 + 80091, ext. 2756). *Instrumentation, neutron scattering.*

Baars, Dr -Ing. Jan Walter. Abt. Infrarotphysik, Fraunhofer Inst. f. Angew. Festkörperphysik, Eckerstr. 4, D-7800 Freiburg, Germany, Fed. Rep. (1931) Dr -Ing. solid state physics (TU Berlin, 1967). Branch Head. (tel. 0761 + 2714, ext. 247). *Crystal growth, epitaxy, inorganic crystal structures.*

Babel, Prof. Dr Dietrich. Fachbereich Chemie, Universität, Hans-Meerwein-Str., D-3550 Marburg, Germany, Fed. Rep. (1930) Prof., inorganic chemistry (U. Marburg, 1971). Prof. (tel. 06421 + 28, ext. 5625). *Solid state chemistry, inorganic crystal structures.*

Bade, Mr Dirk. Physik Dept., Techn. Universität, James-Franck-Str., D-8046 Garching, Germany, Fed. Rep. (1950) Dipl.-Phys., physics (TU München, 1974). Wiss. Ang. (tel. 089 + 3209, ext. 2509). *Protein structure analysis, Mössbauer - Rayleigh scattering interference method.*

Bärnighausen, Prof. Dr Hartmut. Inst. f. Anorganische Chemie, Universität, Engesserstr., Geb. Nr. 30.45, Postfach 6380, D-7500 Karlsruhe, Germany, Fed. Rep. (1933) Dr rer.nat., chemistry (U. Freiburg, 1959). Prof. (tel. 0721 + 608, ext. 3484). *Inorganic crystal structures, crystal chemistry, symmetry.*

Baier, Mr Hubert. Inst. f. Mineralogie, Universität, Corrensstr. 24, D-4400 Münster, Germany, Fed. Rep. (1959) Dipl.-Min., mineralogy (U. Münster). Doktorand. (tel. 0251 + 833489, telex 892529 unims d). *Crystal growth, electron microscopy, materials science, surface structure, teaching, X-ray diffraction.*

Bambauer, Prof. Dr Hans Ulrich. Inst. f. Mineralogie, Universität, Corrensstr. 24, D-4400 Münster, Germany, Fed. Rep. (1929) Dr rer.nat., mineralogy (U. Mainz, 1957). Full Prof. (tel. 0251 + 83, ext. 3450). *General and applied mineralogy, electron microscopy, materials science, microscopy, minerals, phase determination, phase transitions, powder diffraction, texture.*

Banner, Dr David William. Eur. Mol. Biol. Lab., Meyerhofstr. 1, D-6900 Heidelberg, Germany, Fed. Rep. (1946) D. phil., molecular biophysics (U. Oxford, UK, 1972). Sci. (tel. 06221 + 387, ext. 255). *Protein structure, nucleic acid structure, protein-nucleic acid interactions.*

Barbier, Mr Bruno. Mineralogisches Institut, Universität, Poppelsdorfer Schloss, D-5300 Bonn, Germany, Fed. Rep. (1952) DUT, physical technique of measurement (U. Rouen, France, 1974). Phys. Ass. (tel. 0228 + 73, ext. 3557). *Diffractometry, minerals, powder diffraction, symmetry, X-ray diffraction.*

Baresel, Dr -Ing. Detlef Wilhelm Berthold. Forschungszentrum, FCW, Robert Bosch GmbH, Postfach 50, D-7000 Stuttgart 1, Germany, Fed. Rep. (1926) Dr -Ing., inorganic chemistry (TU Berlin, 1962). Wiss. Referent. (tel. 0711 + 8111, ext. 6552). *Inorganic chemistry, physical chemistry, catalysis, surface chemistry, adsorption, chemisorption, solid state chemistry, surface structure, semiconductors, ionic conductors.*

Bartl, Prof. Dr Hans. Inst. f. Kristallographie, Universität, Senckenberg-Anlage 30, D-6000 Frankfurt/Main, Germany, Fed. Rep. (1933) Dr habil., crystallography and mineralogy (U. Frankfurt/Main, 1971). Prof. (tel. 069 + 798, ext. 2105, Telex 413932 Unif). *X-ray and neutron diffraction.*

Bats, Dr Jan, Willem. Inst. f. Organische Chemie, Universität, Niederurseler Hang, D-6000 Frankfurt/Main, Germany, Fed. Rep. (1949) Dr, X-ray and neutron diffraction (Twente U., Netherlands, 1976). Res. Asst. (tel. 069 + 5800, ext. 9128). *Organic crystal structures, structure determination, electron density studies.*

Bauer, Prof. Dr Ernst Georg. Physikalisches Inst., TU, Leibnizstr. 4, D-3392 Clausthal-Zellerfeld, Germany, Fed. Rep. (1928) Dr phil., physics (U. München, 1955). Ord. Prof. (tel. 05323 + 721, ext. 2249). *Surface physics, thin film growth and structure.*

Baum, Mrs Elke. Physikalisches Inst., TU, Michaelstr. 8, D-7500 Karlsruhe, Germany, Fed. Rep. (1960) Dipl.-Chem., chemistry (U. Marburg, 1985). (tel. 0721 + 814251). *X-ray structure determination, Mössbauer spectroscopy, magnetic properties, mineralogy.*

Baumann, Mr Jürgen Rudolf. Fakultt f. Physik, Lst. Prof. Bucher, Universität, Postfach 5560, D-7750 Konstanz, Germany, Fed. Rep. (1957) Dipl.-Min., crystallography and mineralogy (TU Clausthal-Zellerfeld, 1986). (tel. 07531 + 882036). *Computing, crystal growth, crystallographic data, diffractometry, inorganic crystals, materials science, metals, phase determination, powder diffraction, small angle scattering, texture, X-ray diffraction.*

Baur, Prof. Dr Werner H. Inst. f. Kristallographie u. Mineralogie, Universität Frankfurt, Senckenberganlage 30, D-6000 Frankfurt/M., Germany, Fed. Rep. (1931) Dr rer.nat., crystallography (U. Göttingen, 1956). Prof. (tel. 069+798, ext. 2100, email XBR1f11U at DDATHD21.EARN). *Powder diffraction, bond length prediction, computer simulation, solid state chemistry, zeolite studies, mineral and inorganic crystal structures, hydrogen bonding.*

Bayh, Prof. Dr Werner. Mineralogisches Inst., Universität, Wilhelmstr. 56, D-7400 Tübingen, Germany, Fed. Rep. (1928) Dr rer.nat., physics (U. Tübingen, 1962). Prof. (tel. 07071 + 29, ext. 2648). *Crystal growth, X-ray crystallography, electron microscopy.*

Beck, Prof. Dr Horst Philipp. Inst. f. Anorganische Chemie, Universität, Egerlandstr. 1, D-8520 Erlangen, Germany, Fed. Rep. (1941) Dr habil., inorganic chemistry (U. Karlsruhe, 1980). Prof. (tel. 09131 + 85, ext. 7353). *Crystal chemistry, high pressure.*

Becker, Prof. Dr Gerd. Inst. f. Anorganische Chemie, Universität, Pfaffenwaldring 55, D-7000 Stuttgart 80, Germany, Fed. Rep. (1940) Dr habil., inorganic chemistry (U. Karlsruhe, 1976). Ord. Prof. (tel. 0711 + 685, ext. 4172). *X-ray structure determination, compounds with elements of the main-groups (esp. P, As, Sb, Bi, Si).*

Behm, Dr Helmut Johannes Julius. TTC Microelectronic GmbH, D-3060 Stadthagen, Germany, Fed. Rep. (1947) Dr rer.nat., chemistry (U. Freiburg, 1976). (tel. 05721 + 75076, fax +6170). *Borates, crystal chemistry, Patterson methods, materials science.*

Behm, Prof. Dr Rolf Jürgen. Inst. f. Kristallographie u. Mineralogie, Universität, Theresienstr. 41, D-8000 München 2, Germany, Fed. Rep. (1949) Dr rer.nat., physical chemistry (U. München, 1980). Prof. (tel. 089 + 2394, ext. 4355, fax 089-5203-286, email KK401AC at DM0LRZ01EARN). *Scanning tunneling microscopy, surface structure, surface reaction. main-groups (esp. P, As, Sb, Bi).*

Behrens, Dr Heinrich. Fachinformationszentrum Energie Physik Mathematik GmbH, Kernforschungszentrum, D-7514 Eggenstein-Leopoldshafen 2, Germany, Fed. Rep. (1937) Dr rer.nat., physics (U. Karlsruhe, 1966). (tel. 07247 + 821, ext. 4554).

Behrens, Dr Peter. Fakultt f. Chemie, Universität, Universitätsstr. 10, D-7750 Konstanz, Germany, Fed. Rep. (1957) Dr rer.nat., chemistry (U. Hamburg, 1989). Wiss. Mit. (tel. 07531 + 88, ext. 2011, email CHFELS1 at DKNKURZ1). *X-ray and neutron diffraction, X-ray absorption spectroscopy, zeolites, graphite intercalation compounds.*

Behrens, Dr Ulrich Hermann. Inst. f. Anorg. u. Angew. Chemie, Universität, Martin-Luther-King-Platz 6, D-2000 Hamburg 13, Germany, Fed. Rep. (1946) Dr habil., inorganic chemistry (U. Hamburg 1975). Priv.-Doz. (tel. 040 + 4123, ext. 2894). *Organometallic and coordination chemistry, crystal structures.*

Behruzi, Dr Massoud. Inst. f. Kristallographie, RWTH, Templergraben 55, D-5100 Aachen, Germany, Fed. Rep. (1939) Dr, crystallography, mineralogy (RWTH Aachen, 1972). Wiss. Ang. (tel. 0241 + 80, ext. 6906, email BEHRUZI at DACTH51). *Crystal chemistry, X-ray diffraction, crystal growth.*

Bennett, Dr William, Samuel, Jr. Abteilung Wittmann, MPI f. Molekulare Genetik, Ihnestr. 63-73, D-1000 Berlin, Germany, Fed. Rep. (1949) PhD., molecular biology (Yale U., USA, 1978). Res. scient. (tel. 030 + 8307-1, ext. 344). *Protein structure and function, crystallographic computing.*

Bente, Prof. Dr Klaus Alexander. Mineralogisch-Kristallographisches Inst., Universität, Goldschmidtstr. 1, D-3400 Göttingen, Germany, Fed. Rep. (1946) Dr habil., mineralogy (U. Göttingen). Prof. *Crystal chemistry, synthesis, inorganic materials, physico-chemical crystallography.*

Benz, Prof. Dr Klaus-Werner. Kristallographisches Inst., Universität, Hebelstr. 25, D-7800 Freiburg, Germany, Fed. Rep. (1938) Dr Ing.habil. Prof.(tel. 07711 + 203 ext. 4295, fax +4369, telex 772740-65 uf d). *Crystal growth, defect structures, electron microscopy, inorganic crystals, materials science, microscopy, semiconductors, teaching, X-ray diffraction.*

Berg, Mrs Dr Lieselotte. Gmelin-Inst., Max-Planck-Ges., Varrentrappstr. 40/42, D-6000 Frankfurt/Main, Germany, Fed. Rep. (1933) Dr rer.nat., crystallography (TU Braunschweig, 1964). Wiss. Mit. (tel. 069 + 7917, ext. 239). *Crystallography, inorganic chemistry, metalorganic chemistry.*

Bergerhoff, Prof. Dr Guenter. Inst. f. Anorg. Chemie, Universität, Gerhard-Domagk-Str. 1, D-5300 Bonn 1, Germany, Fed. Rep. (1926) Dr rer.nat., inorganic chemistry (U. Bonn, 1954). Prof. (tel. 0228 + 73, ext. 2657, telex 886657, email UNC412 at DBNRHRZ1). *Crystallographic data, structural chemistry.*

Berking, Dr Bernhard. Fachbereich Seefahrt, Fachhochschule Hamburg, Rainvilleterrasse 4, D-2000 Hamburg 52, Germany, Fed. Rep. (1939) Dr habil., crystallography (U. Hamburg, 1975). Dozent. (tel. 040 + 3802848). *Mineralogy, organic biochemistry.*

Bernotat-Wulf, Mrs. Dr Hannelore. Inst. f. Kristallographie, Universität, Kaiserstr. 12, D-7500 Karlsruhe, Germany, Fed. Rep. (1940) Dr, crystallography (ETH Zürich, 1970). Wiss. Angest. (tel. 0721 + 6083927). *Structural chemistry, inorganic crystals, minerals.*

Bertelmann, Dr Dieter Wilhelm. E689, SIEMENS AG, Östliche Rheinbrückenstr. 50, D-7517 Karlsruhe, Germany, Fed. Rep. (1954) Dr rer.nat., mineralogy and crystallography (U. Karlsruhe, 1986). Product specialist X-ray diffraction. (tel. 0721 + 595, ext. 2425, fax +4506, telex 78255-69). *X-ray diffraction.*

Berthold, Prof. Dr Hans Joachim. Inst. f. Anorg. Chemie, Universität, Callinstr. 9, D-3000 Hannover, Germany, Fed. Rep. (1923) Dr phil., inorganic chemistry (U. Köln, 1952). Prof. C4, Institutsdirektor. (tel. 0511 + 762, ext. 2254). *Inorganic crystal structures, phase transitions.*

Berthold, Dr Thomas. ZFE F1 AMF 21, SIEMENS AG, Otto-Hahn-Ring 6, D-8000 München 83, Germany, Fed. Rep. (1955) Dr (tel. 089 + 636, ext. 2684, fax +48131). *X-ray diffraction, crystal growth, diffuse scattering, materials science.*

Betz, Mr Helmut. Arbeitsgruppe f. Chem. Kristallographie, MPI f. Biochemie, Am Klopferspitz, D-8033 Martinsried, Germany, Fed. Rep. (1956) Dipl.-Chem., chemistry (TU. München, 1983). (tel. 089 + 85782659). *X-ray structure analysis, organic compounds.*

Betzel, Dr Christian. European Molecular Biology Laboratory, c/o DESY, Notkestr. 85, D-2000 Hamburg 52, Germany, Fed. Rep. (1956) Dipl.-Phys., physics (U. Göttingen, 1982). Scientific staff member (tel. 040 + 890801, ext. 36, telex 215124, fax (040)890801-49). *Proteins, hydrogen bonding, macromolecules.*

Biedl, Dr Albrecht. Fachbereich Informatik, TU, Strasse des 17. Juni 135, D-1000 Berlin 12, Germany, Fed. Rep. (1938) Dr phil., mineralogy (U. Wien, 1963). Akad. Rat. (tel. 030 + 314, ext. 4893/4891). *Computer programming, symmetry.*

Bielen, Prof. Dr Helmut Josef. Neuhausstr. 15, D-6000 Frankfurt/Main 1, Germany, Fed. Rep. (1929) Prof., solid state physics (U.P.R., USA, 1963). Industrial Consultant. (tel. 069 + 593210). *Inorganic crystal structures, magnetic crystal structures.*

Bissert, Dr Elisabeth Gertrud. Mineralogisches Inst., Universität, Olshausenstr. 40, D-2300 Kiel, Germany, Fed. Rep. (1933) Dr, crystallography (U. Kiel, 1969). Akad. Direktor. (tel. 0431 + 880, ext. 2893). *Inorganic crystals, crystallographic data, X-ray diffraction.*

Blaschke, Prof. Dr Rochus Bruno Albert. Inst. f. Medizinische Physik, Universität, Hüfferstr. 68, D-4400 Münster, Germany, Fed. Rep. (1930) Dr rer.nat., mineralogy (U. Giessen, 1963). Prof. (tel. 0251 + 81799). *Electron microscopy, biomineralogy, applied mineralogy.*

Bleif, Dr Hans-Jürgen. Kernchemie und Reaktor, Hahn-Meitner-Inst. f. Kernforschung, Glienicker Str. 100, D-1000 Berlin 39, Germany, Fed. Rep. (1945) Dr rer.nat., solid state physics (U. Tübingen, 1978). Physicist. (tel. 030 + 8009, ext. 2758). *Phase transitions, diffuse scattering.*

Block, Prof. Dr Jochen Hermann. Fritz-Haber-Inst., Max-Planck-Ges., Faradayweg 4-6, D-1000 Berlin 33, Germany, Fed. Rep. (1929) Dr sc.h.c., physical chemistry. Direktor. (tel. 030 + 8305, ext. 411). *Surface chemistry, field ionization, mass spectrometry, fast chemical reactions.*

Blüthgen, Dipl-Min Waldemar. Fa. Ernst Leitz Wetzlar GmbH, Ernst-Leitz-Str. 30, Pf 2020, D-6330 Wetzlar, Germany, Fed. Rep. (1940) Diplom, crystallography, metal surfaces, epitaxy (U. Bonn, 1968). Patent Eng. (tel. 06441 + 292466, telex 483849 Leiz D). *Crystal optics, ceramics, optical glasses, microscopy, glass ceramics.*

Bode, Dr Wolfram. MPI f. Biochemie, D-8033 Martinsried, Germany, Fed. Rep. (1942) PhD, phys. chemistry (U. München, 1971). (tel. 089 + 8578, ext. 2676, fax +3777). *Macromolecules; proteins.*

Bögge, Dr Hartmut. Fak. f. Chemie, Universität, Universitätsstr., D-4800 Bielefeld, Germany, Fed. Rep. (1953) Dipl.-Chem., chemistry (U. Dortmund, 1977). (tel. 0521 + 106, ext. 2906). *Transition metal complexes.*

Boehm, Prof. Dr Hanns-Peter. Inst. f. Anorganische Chemie, Universität, Meiserstr. 1, D-8000 München 2, Germany, Fed. Rep. (1928) Dr rer.nat., inorganic chemistry (TH Darmstadt, 1953). Ord. Prof. (tel. 089 + 5902355). *Carbon and graphite chemistry, surface chemistry, catalysis.*

Böhm, Dr Horst. Inst. f. Geowissenschaften, Universität, Saarstr. 21, D-6500 Mainz, Germany, Fed. Rep. (1937) Dr, crystallography (ETH Zürich, 1969). Prof. (tel. 06131 + 39, ext. 2848). *Structure - physical properties relationships, phase transformations, modulated structures.*

Böhme, Prof. Dr Reinhild. Inst. f. Mineralogie, Ruhr-Universität, Universitätsstr. 150, D-4630 Bochum, Germany, Fed. Rep. (1942) Dr rer.nat.habil., crystallography (U. Erlangen-Nürnberg, 1982). Prof. (tel. 0234 + 700, ext. 3740). *Structure determination, superstructure effects, crystallographic computing.*

Boese, Dr Roland. Inst. f. Anorg, Chemie, Universität-GH, Universitätsstr. 3-5, D-4300 Essen, Germany, Fed. Rep. (1945) Dr, chemistry (U. Marburg, 1976). Akad. Oberrat. (tel. 0201 + 183, ext. 2416, fax +2151, email CAN010 at DE0HRZ1A). *Chemical bonding, electron density, phase transitions, molecular crystals, organic compounds, thermal vibrations, instrumentation, low temperature techniques.*

Böttcher, Dr Peter Ernst-August. Inst. f. Anorg. Chemie, Universität, Universitätsstr. 1, D-4000 Düsseldorf, Germany, Fed. Rep. (1939) Dr rer.nat., inorganic chemistry (RWTH Aachen, 1971). Prof. (tel. 0211 + 311, ext. 3147). *Solid state chemistry.*

Bohatý, Prof. Dr Ladislav. Inst. f. Kristallographie u. Mineralogie, Universität, Theresienstr. 41, D-8000 München, Germany, Fed. Rep. (1948) Dr rer.nat., mineralogy (U. Köln, 1975). Prof. (tel. 089 + 23944331). *Crystal physics, crystal growth.*

Bondza, Dr Harald Werner. Inst. f. Angew. Physik, Lehrstuhl f. Kristallogr., Universität, Bismarckstr. 10, D-8520 Erlangen, Germany, Fed. Rep. (1959) Dr rer.nat., physics (U. Erlangen-Nürnberg, 1989). *Dynamical theory, X-ray scattering.*

Bonse, Prof. Dr Dr h.c. Ulrich Karl Eberhard. Lehrstuhl f. Experimentelle Physik I, Universität, Otto-Hahn-Str. 4, Postfach 500500, D-4600 Dortmund 50, Germany, Fed. Rep. (1928) Dr rer.nat.habil., physics (U. Münster, 1963). Ord. Prof. (tel. 0231 + 755, ext. 3504, fax 0231-751532, telex 822445 unido d, email UPH038 at DDOHRZ11). *Solid state physics, X-ray and neutron diffraction esp. interferometry, microtomography, synchrotron radiation.*

Borchardt-Ott, Dr Walter. Inst. f. Mineralogie, Universität, Corrensstr. 24, D-4400 Münster, Germany, Fed. Rep. (1933) Dr rer.nat., mineralogy (U. Münster, 1964). Akad. Direktor. (tel. 0251 + 83, ext. 3453). *Crystal growth, teaching.*

Born, Dr Eberhard. Angew. Mineralogie u. Geochemie, Techn. Universität, Lichtenbergstr. 4, D-8046 Garching, Germany, Fed. Rep. (1939) Dr habil., crystallography (TU München, 1974). Prof. (tel. 089 + 3209, ext. 3226). *Crystal defects, X-ray topography, X-ray diffraction methods, texture, metals.*

Born, Dr Liborius. Zentralbereich ZF-D, Bayer AG, D-5090 Leverkusen-Bayerwerk, Germany, Fed. Rep. (1931) Dr, crystallography, physics (U. Marburg, 1961). Industrial Crystallographer. (tel. 0214 + 305852). *Organic crystal structures.*

Borrmann, Prof. Dr -Ing. Gerhard. Fritz-Haber-Inst., MPI, Faradayweg 4-6, D-1000 Berlin 33, Germany, Fed. Rep. (1908) O. Prof, physics (TH Danzig, 1936). Em.

Boysen, Dr Hans. Inst. f. Kristallographie und Mineralogie, Universität, Theresienstr. 41, D-8000 München 2, Germany, Fed. Rep. (1944) Dipl.-phys., physics (U. München, 1970). Wiss. Ang. (tel. 089 + 3209, ext. 4040, fax 089-5203286, telex 529815) *Neutron diffraction, disorder, powder diffraction, phase transitions, diffuse scattering, phase transitions.*

Bradaczek, Prof. Dr Hans Arthur. Inst. f. Kristallographie, Freie U. Berlin, Takustr. 6, D-1000 Berlin 33, Germany, Fed. Rep. (1930) Dr rer.nat., physics (FU Berlin, 1966). Dir. (tel. 030 + 838, ext. 3461). *Diffraction theory, small angle scattering, single crystal diffractometry, phase transitions, crystal surfaces.*

Brämer, Dr Wulf. Heraeus GmbH, Heraeusstr. 12-14, D-6450 Hanau/Main, Germany, Fed. Rep. (1946) Dipl.-Chem., inorganic chemistry (U. Münster, 1975). Leiter der metallurgischen Entwicklung. (tel. 06181 + 360, ext. 795). *Solid state physics, metals.*

Brandmüller, Prof. Dr Josef Karl August. Sektion Physik, LM Universität, Schellingstr. 4, D-8000 München, Germany, Fed. Rep. (1921) Prof., physics (U. München, 1955). O. Prof. (em.) (tel. 089 + 2180, ext. 3211). *Optical solid state spectroscopy, group and representation theory.*

Brandt, Ing Gernot. Abt. Kristallchemie, Fraunhofer Inst. f. Angew. Festkörperphysik, Eckerstr. 4, D-7800 Freiburg, Germany, Fed. Rep. (1936) Dr, physics (TU Berlin, 1969). (tel. 0761 + 2714, ext. 281). *Crystal growth, II-VI and III-V compounds, chalcopyrites.*

Brauer, Prof. Dr Georg Karl. Chemisches Laboratorium, Universität, Albertstr. 21, D-7800 Freiburg, Germany, Fed. Rep. (1908) Prof., Inorganic chemistry (U. Freiburg, 1932). Prof. Em. (tel. 0761 + 203, ext. 2894). *Inorganic solid state chemistry, crystal structure, metal oxides, fluorides, nitrides, carbides, intermetallic compounds.*

Braun, Dr Eckart. Inst. f. Mineralogie, FU, Takustr. 6, D-1000 Berlin 33, Germany, Fed. Rep.

Brauner, Mr Christian. Mineralogisches Institut, Universität, Meyerhofstr. 11, D-2300 Kiel, Germany, Fed. Rep. (1964) cand. inf., informatics, Hiwi. (tel. 0431 + 880, ext. 2569). *Computing.*

Breit, Dipl-Min Udo. Inst. f. Mineralogie, Universität, Corrensstr. 24, D-4400 Münster, Germany, Fed. Rep. (1939) Dipl.-Min., mineralogy (U. Münster, 1974). Postgrad. (tel. 0251 + 83, ext. 3405). *Instrumentation, silicate crystal structures.*

Breitinger, Prof. Dr Dietrich Karl. Inst. f. Anorg. Chemie, Universität, Egerlandstr. 1, D-8520 Erlangen, Germany, Fed. Rep. (1935) Dr rer.nat., inorganic and analytical chemistry (RWTH Aachen, 1964). Section leader, lect. (tel. 09131 + 85, ext. 7352). *Spectroscopy (vibrational - electronic - high pressure), inorganic solids, single crystal spectroscopy.*

Brill, Mr Wolfgang. Inst. f. Kristallographie, Universität, Im Stadtwald, D-6600 Saarbrücken, Germany, Fed. Rep. (1956) Dipl.phys., physics (U. Saarbrücken, 1982). Wiss. Mit. (tel. 0681 + 3022470). *Solid state physics, computer technology, X-ray diffraction.*

Brodalla, Dipl.-Chem Dieter. Inst. f. Anorg. Chemie und Strukturchemie, Universität, Universitätsstr. 1, D-4000 Düsseldorf, Germany, Fed. Rep. (1947) Dipl.-Chem., chemistry (U. Düsseldorf, 1979). Res. asst. (tel. 0211 + 311, ext. 3068). *Solid state chemistry, crystal growth, inorganic crystal structures, clathrate structures, low temperature crystal structure determination.*

Brodersen, Prof. Dr Klaus. Inst. f. Anorg. Chemie, Universität, Egerlandstr. 1, D-8520 Erlangen, Germany, Fed. Rep. (1926) Dr phil., inorganic and analytical chemistry (U. Greifswald, 1951). Full Prof. (tel. 09131 + 85, ext. 7350). *Structural chemistry, mercury compounds.*

Brokmeier, Dr Heinz-Günter. Inst. f. Physik, GKSS-Forschungszentrum, Max-Planck-Str., D-2054 Geesthacht, Germany, Fed. Rep. (1952) Dr rer.nat., crystallography (TU Clausthal-Zellerfeld, 1983). (tel. 04152 + 12257). *X-ray and neutron diffraction, textures, deformation and recrystallisation processes.*

Bronger, Prof. Dr Welf. Inst. f. Anorganische Chemie, RWTH, Prof. Pirlet-Str. 1, D-5100 Aachen, Germany, Fed. Rep. (1932) Dr rer.nat., chemistry (U. Münster, 1961). Ord. Prof. (tel. 0241 + 801, ext. 4643). *Inorganic solid state chemistry.*

Bruder, Dr Martin. AEG, Theresienstr. 2, D-7100 Heilbronn, Germany, Fed. Rep. (1955) Dr rer.nat. *Crystal growth, materials science, semiconductors.*

Buck, Prof. Dr Peter. Inst. f. Mineralogie, Universität, Lahnberge, D-3550 Marburg, Germany, Fed. Rep. (1939) Dr rer.nat., crystallography (U. Freiburg, 1967). Hochschullehrer. (tel. 06421 + 28, ext. 5500). *Crystal growth, defects in crystals, crystal physics.*

Buehner, Dr Manfred. Arbeitsgruppe Röntgenstrukturanalyse, Universität, Am Hubland, D-8700 Würzburg, Germany, Fed. Rep. (1940) Dr rer.nat., biochemistry (U. Konstanz, 1969). Head, Res. Group. (tel. 0931 + 888-386) *Structure analysis, proteins, enzymology.*

Buhl, Dr Josef-Christian. Inst. f. Mineralogie, Universität, Corrensstr. 24, D-4400 Münster, Germany, Fed. Rep. (1953) Dipl.-Min., mineralogy (FU Berlin, 1983). (tel. 0251 + 833459). *Crystal growth, high pressures, reaction pathways.*

Bunge, Prof. Dr Dr h.c. Hans-Joachim. Inst. f. Metallkunde und Metallphysik, Techn. Universität, Grosser Bruch 23, D-3392 Clausthal-Zellerfeld, Germany, Fed. Rep. (1929) Dr rer.nat.habil., physics (Humboldt U. Berlin, 1964). Prof.

(tel. 05323 + 72, ext. 2244). *Material science, X-ray and neutron diffraction, textures.*

Burschka, Dr Christian. Inst. f. Anorg. Chemie, Universität, Am Hubland, D-8700 Würzburg, Germany, Fed. Rep. (1946) Dr rer.nat., chemistry (RWTH Aachen, 1975). Akad. Rat (tel. 0931 + 888, ext. 286). *Solid state and inorganic chemistry, computer programming, direct methods, information retrieval.*

Burzlaff, Prof. Dr Hans. Inst. f. Angew. Physik, Lehrstuhl f. Kristallographie, Universität, Bismarckstr. 10, D-8520 Erlangen, Germany, Fed. Rep. (1932) Dr habil., crystallography (U. Marburg, 1968). Full Prof. (tel. 09131 + 85, ext. 2700). *Symmetry, structure determination, electron density, crystal physics.*

Buschmann, Dr Juergen Friedrich. Inst. f. Kristallographie, FU Berlin, Takustr. 6, D-1000 Berlin 33, Germany, Fed. Rep. (1939) Dr rer.nat., crystallography (FU Berlin, 1980). Scient. (tel. 030 + 838 ext. 3408, fax +6702). *Low temperature single crystal diffractometry, low temperature crystallization, small molecules, electron deformation density (X-N, X-X).*

Cammenga, Prof. Dr Heiko Karl. Inst. f. Theoretische Chemie, TU, Hans Sommer-Str. 10, D-3300 Braunschweig, Germany, Fed. Rep. (1938) Dr, physical chemistry (TU Braunschweig, 1967). Prof. (tel. 0531 + 391, ext. 5333). *Phase transitions, solid state reactions, heterogen. kinetics.*

Chen, Dr Jiang-Tsun. Inst. f. Kristallographie, FU Berlin, Takustr. 6, D-1000 Berlin 33, Germany, Fed. Rep. (1936) Dr, physical chemistry of polymers (1979). Scientist. (tel. 030 + 8382354). *Physical chemistry, charge-transfer complexes.*

Claus, Mr Karl Heinz. Röntgenlabor, MPI f. Kohlenforschung, Lembkestr. 5, D-4330 Mülheim/Ruhr, Germany, Fed. Rep. (1940) (tel. 0208 + 306, ext. 493). *Crystal data collection, powder diffractometry, instrumentation.*

Coll, Dr Miquel. Max-Plank Instut Fur Biochemie, Martinsried Bei Muchen, D-8033, Germany, Fed.Rep. (1955) PhD, Biology (Barcelona U., 1986) Postdoctoral asst. *DNA structure, protein structure.*

Cordier, Dr Gerhard. Inst. f. Anorg. Chemie, TH, Hochschulstr. 4, D-6100 Darmstadt, Germany, Fed. Rep. (1949) Dr, chemistry (TH Darmstadt, 1977). (tel. 06151 + 162492). *Structures, inorganic materials.*

Czank, Dr Michael. Mineralogisches Inst., Universität, Olshausenstr. 40-60, D-2300 Kiel, Germany, Fed. Rep. (1941) Dr rer.nat., crystallography (E.T.H. Zürich, 1973). (tel. 0431 + 880, ext. 2903). *Electron microscopy, realbau of minerals, silicate crystal chemistry.*

Dachs, Prof. Dr Hans. Hahn-Meitner-Inst. f. Kernforschung, Glienicker Str. 100, D-1000 Berlin 39, Germany, Fed. Rep. (1927) Dr rer.nat., crystallography (U. München, 1956). Prof. (tel. 030 + 8009, ext. 2741). *Neutron diffraction, magnetic structures, phase transitions.*

Dahlkamp, Dr Franz-Joses. Oelbergstr. 10, D-5307 Wachtberg-Liessem, Germany, Fed. Rep. (1931) Dr phil., Petrogr.-Geol. (U. Graz, Austria, 1958). Independent Consultant. (tel. 0228 + 341904). *Appl. mineralogy, exploration.*

Daniels, Mr Peter. Inst. f. Mineralogie, Universität, Universitätsstr. 150, D-4630 Bochum, Germany, Fed. Rep. (1959) Dipl.-Min., mineralogy (U. Bochum, 1987). Wiss.Mitarb. (tel. 0234 + 700, ext. 4379) *Crystal chemistry, crystal growth, silicates.*

Debaerdemaeker, Dr Tony. Sektion f. Röntgen- u. Elektronenbeugung, Universität, Oberer Eselsberg, D-7900 Ulm, Germany, Fed. Rep. (1945) Dr phil., physics (U. York, UK, 1971). Res. Asst. (tel. 0731 + 1761). *Direct methods, computer programming, crystal structures.*

Dederer, Dr Bernhard. Abt. f. Strukturforschung I, MPI f. Biochemie, Am Klopferspitz, D-8033 Martinsried, Germany, Fed. Rep. (1949) Dipl.-Chem., chemistry (TU München, 1975). Res. Assoc. (tel. 089 + 8585, ext. 661). *Structure analysis.*

Degenhardt, Dr Detlev. DESY - HASYLAB, Notkestr. 85, D-2000 Hamburg 52, Germany, Fed. Rep. (1960) Ph.D., surface physics (U.Kiel, 1988). Res. worker (tel. 040 + 8998, ext. 2603, fax +3282, email F41DEG at DHHDESY3). *Surface structure, phase transitions, X-ray diffraction.*

Deiseroth, Dr Hans-Jörg. Fachbereich 8, Anorg. Chemie, Universität-GH, Postfach 101240, D-5900 Siegen, Germany, Fed. Rep. (1945) Dr, inorganic chemistry (U. Giessen, 1972). Prof. (tel. 0271 + 740-4219). *Inorganic chemistry, crystal structures, computer programming, amorphous compounds, EXAFS.*

Depmeier, Prof Dr Wulf Helmut Heinz. Inst. f. Mineralogie u. Kristallographie, TU Berlin, BH1, Ernst-Reuter-Platz 1, D-1000 Berlin 12, Germany, Fed. Rep. (1944) Dr rer.nat.habil., crystallography, solid state chemistry (U. Hamburg, 1973). Prof. (tel. 030 + 314, ext. 23523, email 1171 at DB0TUZ01). *Inorganic crystal structures, phase transitions, modulated structures, superstructures.*

Dhupia, Mr Gursev Singh. Inst. f. Gesteinshüttenkunde, RWTH, Mauerstr. 5, D-5100 Aachen, Germany, Fed. Rep. (1946) Dipl.-Min., mineralogy (RWTH Aachen, 1979). Res. asst. (tel. 0241 + 804984). *Electron microscopy, refractories, ceramics.*

Diehl, Dr J. Inst. f. Werkstoffwiss., MPI f. Metallforschung, Seestr. 92, D-7000 Stuttgart 1, Germany, Fed. Rep.

Diehl, Dr Roland. Project Management, Fraunhofer Inst. f. Appl. Solid State Physics, Eckerstr. 4, D-7800 Freiburg, Germany, Fed. Rep. (1944) Dr rer.nat., crystallography (U. Freiburg, 1972). Project Manager. (tel. 0761 + 2714, ext. 286, fax +296, telex 0772510). *Crystal growth, III-V semiconductor technology, materials science.*

Dietrich, Mr. Andreas. Technische Univ. Berlin, Sekr.C2, Inst. fuer Analyt. Anorg. Chem., Strasse des 17 Juni 135, 1000 Berlin 12, Germany, Fed. Rep. (1962)

MSc, chemistry (1988) Res. Assoc. (tel. 30 + 37425729, email 1137 at DB0TUZ0) *Organometallic synthesis, small molecule interest group.*

Dietrich, Prof. Dr Hans Karl Ernst. Limastr. 21 A, D-1000 Berlin 37, Germany, Fed. Rep. (1923) Dr rer.nat., chemistry (U. Heidelberg, 1956). (tel. 030 + 8028715). *Structure, metallorganic compounds, electron density distribution, chemical bonding.*

Dittmar, Dr Günter. GID, Lyoner Str. 44-48, D-6000 Frankfurt/Main, Germany, Fed. Rep. (1936) Dr rer.nat., chemistry (TH Darmstadt, 1976). Wiss. Mit. *Inorganic crystal structures, crystal chemistry systematics, computer programming, symmetry.*

Dörr, Dr Friedrich Johannes. Instrumentelle Analytik + Mineralogie, Schott-Glaswerke, Hattenbergstr. 10, D-6500 Mainz, Germany, Fed. Rep. (1924) Dr rer.nat., mineralogy (U. Mainz, 1955). Leitender Wissenschaftler. (tel. 06131 + 663497, telex 4187920 sm). *X-ray diffractometry and fluorescence analysis, electron-probe microanalysis, structure and properties, silicates, glass systems.*

Dräger, Prof. Dr Martin. Inst. f. Anorg. und Analyt. Chemie, Universität, Johann-Joachim-Becher-Weg 24, D-6500 Mainz, Germany, Fed. Rep. (1940) Dr rer.nat., chemistry (U. Mainz, 1970). Prof. (tel. 06131 + 39, ext. 5757, fax +3382, telex 4187476 uni d). *Inorganic and organic crystal structures, conformational analysis, ring systems.*

Duerring, Mr Markus. Strukturforschung II, MPI f. Biochemie, Am Klopferspitz, D-8033 Martinsried, Germany, Fed. Rep. (1960) Dipl. Natw. ETH, biochemistry (ETH Zürich, Switzerland, 1986). Research student. (tel. 089 + 8578, ext. 2680, fax +3777, email DUERRING at DM0MPB51). *Proteins, structural biology, X-ray diffraction, crystal growth, crystallographic data.*

Durchschlag, Dr Helmut. Inst. f. Biophysik und Phys. Biochemie, Universität, Universitätsstr. 31, D-8400 Regensburg, Germany, Fed. Rep. (1944) Dr phil., physical chemistry (U. Graz, 1971). Akad.Rat. (tel. 0941 + 943, ext. 3041). *Small angle scattering, biopolymers.*

Eberhard, Prof. Dr Emil. Mineralogisches Inst., Universität, Welfengarten 1, D-3000 Hannover, Germany, Fed. Rep. (1928) Dr, mineralogy (U. Fribourg, Switzerland, 1954). Prof. (tel. 0511 + 762, ext. 2443). *Crystal chemistry.*

Eckerlin, Dr Peter. Philips GmbH Forschungslaboratorium, Weisshaus-Str., D-5100 Aachen, Germany, Fed. Rep. (1926) Dr, chemistry (TH Darmstadt, 1955). Res. Chemist. (tel. 0241 + 62071). *Ternary oxides, intermetallic compounds.*

Eckhardt, Prof. Dr Franz-Jörg. Abt. f. Mineralogie und Petrologie, Bundesanstalt f. Geowiss. und Rohstoffe, Postfach 510153, D-3000 Hannover 51, Germany, Fed. Rep. (1929) Dr phil.nat., mineralogy (U. Frankfurt/Main, 1957). Dir. and Prof. (tel. 0511 + 6468, ext. 559). *Crystallographic methods, mineralogy, petrography, layer silicates, geochemistry, crystal chemistry.*

Eckstein, PhD. Dipl. Ing. Juraj. Kristallogr. Inst., Universität, Hebelstr. 25, D-7800 Freiburg, Germany, Fed. Rep. (1927) PhD, chemistry (U. Prague, ČSSR, 1958). Res. asst. (tel. 0761 + 2034283). *Crystal growth, thermodynamics, analytics.*

Egert, Prof. Dr Ernst. Inst. f. Organische Chemie, Universität, Niederurseler Hang, D-6000 Frankfurt/M., Germany, Fed. Rep. (1949) Dr -Ing., chemistry (TH Darmstadt, 1979). Prof. (tel. 069 + 5800, ext. 9230). *Patterson search, structure-activity relationships, force-field calculations.*

Ehses, Dr Karl-Heinz. Inst. f. Kristallographie, Dep. f. Physik, Universität, Im Stadtwald, D-6600 Saarbrücken, Germany, Fed. Rep. (1943) Dr, physics (U. d. Saarlandes, 1973). Wiss. Mit. (tel. 0681 + 302, ext. 3460). *Solid state physics, X-ray diffraction.*

Eichhorn Dr Klaus-D. DESY-F41, HASYLAB, Notkestr. 85, D-2000 Hamburg 52, Germany, Fed. Rep. (1949) PhD, crystallography (U. Saarbrücken, 1982). Wiss. Ang. (tel. 040 + 8998, ext. 3901, fax +3282, telex 215124 DESY D). *Synchrotron radiation, X-ray diffraction, computing, teaching.*

Eisenmann, Mrs Dr Brigitte. Abt. II f. Anorganische Chemie, TH, Hochschulstr. 4, D-6100 Darmstadt, Germany, Fed. Rep. (1942) Dr, inorganic chemistry (U. München, 1971). Akad. Oberrätin. (tel. 06151 + 16, ext. 2492). *Inorganic crystal structures.*

Elf, Dr Frank. Mineralogisches Inst., Universität, in KFA Jülich, Postfach 1913, D-5170 Jülich, Germany, Fed. Rep. (1944) Dr rer.nat., crystallography (U. Bonn, 1989). (tel. 02461 + 61, ext. 6024, email IFF120 at DJUKFA11). *Neutron diffraction, magnetic and crystallographic structure analysis and refinement, profile analysis, synchrotron radiadion, high pressure, high temperature.*

Ellner, Dr Martin Oliver. Inst. f. Werkstoffwissenschaft, MPI f. Metallforschung, Seestr. 75, D-7000 Stuttgart, Germany, Fed. Rep. (1938) Dr rer.nat., chemistry (U. Stuttgart, 1971). Sen. res. sci. (tel. 0711 + 2095, ext. 244, fax 0711-225722, telex 723742 MPIM D). *Crystal chemistry, structure, intermetallic compounds, metastable crystalline and amorphous phases, instrumentation.*

Engel, Dr Walter. Fraunhoferinst. f. Treib- u. Explosivstoffe, Hummelberg, D-7507 Berghausen, Germany, Fed. Rep. (1935) PhD, inorg. chemistry (U. Giessen, 1962). (tel. 0721 + 46101) *Phase transitions.*

Engelhardt, Dr Günter. Fakultt f. Chemie, Universität, Universitätsstr. 10, D-7750 Konstanz, Germany, Fed. Rep. (1936) Dr rer.nat.habil., chemistry (Humboldt-U. Berlin, 1969). Wiss. Angest. (tel. 07531 + 88, ext. 2014, email CHFELS1 at DKNKURZ1) *Structural chemistry, solid-state NMR.*

Englisch, Mr Uwe-Franz. Abt. f. Chemie, MPI f. Experim. Medizin, Hermann-Rein-Str. 3, D-3400 Göttingen, Germany, Fed. Rep. (1954) (tel. 0551 + 303, ext. 356). *Protein crystallography, biochemistry.*

Ensling, Dr Jürgen. Inst. f. Anorg. und Analyt. Chemie, Universität, Jakob-Welder-Weg 11, D-6500 Mainz, Germany, Fed. Rep. (1940) Dr, Mössbauer-spectroscopy (TH Darmstadt, 1970). Res. Asst. (tel. 06131 + 39, ext. 2703). *Mössbauer spectroscopy, solid state chemistry and physics.*

Ermer, Prof. Dr Otto. Inst. f. Organische Chemie, Universität, Greinstr. 4, D-5000 Koe.ln 41, Germany, Fed. Rep. (1940) Dr sc.techn., chemistry (ETH. Zue.rich, Switzerland, 1970). Prof. (tel. 0221 + 470, ext. 4104). *Organic crystal chemistry, carboxylic acids, strained molecules.*

Esselborn, Dr Reiner Ferdinand. Zentrallabor f. Anorganische Chemie, E. Merck, Frankfurter Str. 250, D-6100 Darmstadt, Germany, Fed. Rep. (1927) Dr rer.nat., chemistry (U. Freiburg, 1958). Head of Res. & Dev. (tel. 06151 + 722296). *Solid state chemistry, thin films, crystal growth methods, properties, high purity materials, solid state technology, fibre optic technology.*

Eulenberger, Dr Günther Richard. Inst. f. Chemie, Universität Hohenheim, Garbenstr. 30, D-7000 Stuttgart, Germany, Fed. Rep. (1936) Dr phil., chemistry (U. Wien, Austria, 1963). Priv.-Doz. (tel. 0711 + 459, ext. 2166). *Solid state chemistry, inorganic crystal structures, computer programming.*

Euler, Dr Robert. Kanalstr. 13, D-6452 Hainburg 1, Germany, Fed. Rep. (1925) Dr phil., mineralogy (U. Marburg, 1958). Lehrbeauftragter. (tel. 06182 + 69231). *Technical mineralogy, metallurgical slags, refractories.*

Eysel, Prof. Dr Walter. Mineralogisch-Petrogr. Inst., Universität, Im Neuenheimer Feld 236, D-6900 Heidelberg, Germany, Fed. Rep. (1935) Dr, crystallography, mineralogy (RWTH Aachen, 1968). Lect. (tel. 06221 + 56, ext. 2807, fax +3111, telex 461745 UNIKL D, email EY0 at DHDURZ1). *X-ray, powder diffraction, thermal analysis, inorganic crystals, minerals, crystal chemistry, crystallographic data, phase transitions, teaching.*

Faber, Dr Peter. Anwendungstechnik (Elektrochemie), Rheinisch-Westf. Elektrizitätswerk AG, Kölner Str. 12, D-8757 Karlstein, Germany, Fed. Rep. (1924) Dr, electrochemistry (U. Strasbourg, France, 1976). Chief, Res. and Dev. Group. (tel. 06188 + 2197). *Solid state electrochemistry, structure, surface reactions.*

Feld, Dr Rainer Hans Helmut. Miat Ges. f. Informationssysteme, Nerobergstr. 1, D-6500 Mainz, Germany, Fed. Rep. (1950) Dr rer.nat., crystallography (U. Marburg, 1980). Managing Dir. (tel. 06131 + 681038). *Sensor and information technology, solid state physics, image processing.*

Felsche, Prof. Dr Jürgen. Fakultät f. Chemie, Universität, Universitätsstr. 10, Postfach 5560, D-7750 Konstanz, Germany, Fed. Rep. (1939) Prof., crystallography, crystal chemistry (1973). Full Prof. (tel. 07531 + 88, ext. 2025 or 2012, fax +3688, telex 0733359 univ d, email CHFELS1 at DKNKURZ1). *Structural chemistry, materials science, hydrogen bonding, phase transitions, reaction mechanisms.*

Fenske, Prof. Dr Dieter. Inst. f. Anorg. Chemie, Universität, Engesserstr. Geb. Nr. 30.45, D-7500 Karlsruhe, Germany, Fed. Rep. (1942) Dr rer.nat., chemistry (1972) Prof. (tel. 0721 + 6082086). *Complex chemistry (trans. metals), X-ray structure, phosphine ligands.*

Fischer, Dr Carl-Otto. Hahn-Meitner-Inst. f. Kernforschung, Glienicker Str. 100, D-1000 Berlin 39, Germany, Fed. Rep. (1938) Dr rer.nat., physics (FU Berlin, 1969). Physicist. (tel. 030 + 8009, ext. 2746). *Phase transitions, molecular compounds, liquid structure and dynamics.*

Fischer, Prof. Dr Karl. Inst. f. Kristallographie, Dept. f. Physik, Universität, Im Stadtwald, D-6600 Saarbrücken, Germany, Fed. Rep. (1925) Dr, chemistry (U. Erlangen, 1954). Full Prof. (tel. 0681 + 302, ext. 3410, fax +3900, telex 518-6817533). *Structure determination, X-ray and neutron diffraction, anomalous dispersion, instrumentation, synchrotron radiation.*

Fischer, Dr Reinhard X. Mineralogisches Inst. der Univ., Am Hubland, D-8700 Wurzburg, Germany, Fed. Rep. (1954) Dr rer.nat, mineralogy (Mainz, 1983) Akademischer Rat. (tel. 931+888410, telex 68671uniwbgd, fax 931+707012, email KRIS005 at DWUUNI21). *Neutron and X-ray structure analysis, zeolites, crystallographic computing.*

Fischer, Mrs Ute Eva-Maria. Inst. f. Kristallographie, Universität, Kaiserstr. 12, D-7500 Karlsruhe, Germany, Fed. Rep. (1957) Dipl.chem., chemistry (U. Karlsruhe, 1983).

Fischer, Prof. Dr Werner. Inst. f. Mineralogie, Universität, Hans-Meerwein-Str., D-3550 Marburg, Germany, Fed. Rep. (1931) Dr rer.nat.habil, crystallography and mineralogy (U. Marburg, 1970). Prof. (tel. 06421 + 28, ext. 5704). *Mathematical crystallography, crystal chemistry, crystal physics, crystal structure determination, crystallographic computing.*

Flörke, Prof. Dr Otto Wilhelm. Inst. f. Mineralogie, Ruhr-Universität, Universitätsstr. 150, Postfach 102148, D-4630 Bochum, Germany, Fed. Rep. (1926) Dr phil., crystallography (U. Marburg, 1951). Prof. (tel. 0234 + 700, ext. 3512). *Crystal chemistry, crystal physics, crystal growth, mineralogy, ceramics.*

Follner, Prof. Dr Heinz. Inst. f. Mineralogie u. Mineralische Rohstoffe, TU, Adolf-Roemer-Str. 2A, D-3392 Clausthal-Zellerfeld, Germany, Fed. Rep. (1938) Dr rer.nat., crystallography and mineralogy (TU Clausthal, 1971). Prof. (tel. 05323 + 722394). *Crystal growth, structure-physical properties relation.*

Foster, Dr Mark David. Polymer Physics, Max-Planck Inst. fuer Polymerforschung, Postfach 3148, D-6500 Mainz 1, Germany, Fed. Rep. (1959) PhD, Chem. Eng. (U. Minnesota, USA, 1987) Scientist. (tel. 06131 + 379-201) *SAXS, X-ray reflection, polymer surfaces and interfaces.*

Frahm, Dr Ronald Reinhard. HASYLAB at DESY, Notkestr. 85, D-2000 Hamburg 52, Germany, Fed. Rep. (1954) Dr rer.nat., experimental physics (U. Kiel, 1983). Staff member. (tel. 040 + 8998, ext. 3705, fax +3282, email F41FRA at DHHDESY3). *EXAFS, anomalous scattering, amorphous metals, disorder.*

Frank, Dr Walter. FR 13.1 Anorg. Chemie, Universität, Im Stadtwald, D-6600 Saarbrücken, Germany, Fed. Rep. (1957) Dr rer.nat., inorg. chemistry (TU Braunschweig, 1985). Wiss. Angest. (tel. 0681 + 302, ext. 2975). *Structural inorganic and organometallic chemistry.*

Freiburg, Dr Johann Christoph. Zentralabt. Chemische Analysen, Kernforschungsanlage Jülich, D-5170 Jülich, Germany, Fed. Rep. (1931) Dr, physical chemistry (U. Bonn, 1962). Scient. employee. (tel. 02461 + 61, ext. 3291). *Powder mixture identification, stoichiometry.*

Freund, Prof. Dr Friedemann. Code SSX MS 239-4, NASA Ames Research Center, CA 94035 Moffet Field, USA. (1933) Dr, mineralogy (U. Marburg, 1959). (tel. 415 + 694, ext. 5183, fax 415-585-7637).

Frey, Prof. Dr Friedrich. Inst. f. Kristallographie und Mineralogie, Universität, Theresienstr. 41, D-8000 München 2, Germany, Fed. Rep. (1942) Dr habil., crystallography (U. München, 1980). Prof. (tel. 089 + 2394, ext. 4332, fax 089-5203286, telex 529815). *Neutron diffraction, phase transitions, diffuse scattering, disorder, teaching.*

Frey, Mr Wolfgang Ulrich. Inst. f. Organische Chemie, Universität, Pfaffenwaldring 55, D-7000 Stuttgart 1, Germany, Fed. Rep. (1957) Dipl.-Chem., organic chemistry (1985). (tel. 0711 + 685, ext. 4335).

Freyhardt, Prof. Dr Herbert C. Inst. f. Metallphysik, Universität, Hospitalstr. 315, D-3400 Göttingen, Germany, Fed. Rep. (1941) Dr rer.nat.habil., metal physics (1976). Prof. (tel. 0551 + 39, ext. 4492, fax +9612, telex 96703). *Crystal growth, diffractometry, materials science, metals, powder diffraction, texture, X-ray diffraction.*

Fröhlich, Mr Armin. Inst. f. Mineralogie, Universität, Welfengarten 1, D-3000 Hannover 1, Germany, Fed. Rep. (1958) Dipl.-Min., mineralogy (U. Hannover, 1987). (tel. 0511 + 7622222). *Phase transitions, HREM investigations, crystal defects, non linear phenomena of crystals.*

Fröhlich, Dr Roland. Anorg.-Chem. Institut, Universität, Wilhelm-Klemm-Str.8, D-4400 Münster, Germany, Fed. Rep. (1952) Dr rer.nat., inorganic chemistry (U. Köln, 1982). (tel. 0251 + 833153). *Diffractometry, structural chemistry.*

Fuess, Prof. Dr Hartmut. Technische Hochschule, Darmstadt 6100, Petersenstr 20, Germany, Fed. Rep. (1941) Dr -Ing., structural chemistry (TH Darmstadt, 1968). Prof. (tel. 06151 + 16 22 98, fax + 16 54 89, telex 419 579). *Materials science, neutron diffraction, powder diffraction.*

Gahm, Dr Josef. Mikroskopisches Labor, Fa. Carl Zeiss, D-7082 Oberkochen, Germany, Fed. Rep. (1925) Dr, mineralogy, geology, physics (U. Tübingen, 1953). Wiss. Mit. (tel. 07364 + 20, ext. 3288). *Instrumentation, optics, computer programming, image analysis, microscopy (interference and polarisation).*

Gassmann, Dr Johann. MPI f. Biochemie, Am Klopferspitz, D-8033 Martinsried, Germany, Fed. Rep. (1934) Dr, crystallography (TU München, 1966). Res.asst. (tel. 089 + 8585, ext. 723). *Direct methods, crystallographic computing, electron microscopy, structure refinement, large molecules.*

Gebert, Dr Walter Richard. Inst. f. Mineralogie, Ruhr-Universität, Universitätsstr. 150, D-4630 Bochum, Germany, Fed. Rep. (1938) Dr, mineralogy (U. Bochum, 1970). Wiss. Ang. (tel. 0234 + 700, ext. 4380). *Inorganic crystal structures, crystal growth.*

Gebhardt, Prof. Dr Manfred Adolf Hermann. Mineralogisches Inst., Universität, Poppelsdorfer Schloss, D-5300 Bonn, Germany, Fed. Rep. (1934) Prof., mineralogy and crystallography (U. Bonn, 1972). (tel. 0228 + 73, ext. 3268). *Applied mineralogy, biomineralization, thin films.*

Gehlen, Prof. Kurt von. Inst. f. Geochemie, Universität, Senckenberganlage 28, D-6000 Frankfurt/Main, Germany, Fed. Rep. (1927) Dr rer.nat., mineralogy (U. Freiburg, 1952). Prof. (tel. 069 + 798, ext. 2102). *Ore deposits, mineralogy, geochemistry, isotope geochemistry, petrology.*

Geiger, Dr Charles Arthur. Bayerisches Forschungsinst. f. Experimentelle Geochemie u. Geophysik, Univesität Bayreuth, Postfach 101251, D-8580 Bayreuth, Germany, Fed. Rep. (1956) PhD, mineral physics - geochemistry (U. Chicago, USA, 1986). res. assoc. *crystal growth, high pressures, minerals, material science.*

Gentner, Mr Thomas. Abt. f. Kristallographie, Mineralog.-Petrograph. Inst. Universität, Im Neuenheimer Feld 236, D-6900 Heidelberg, Germany, Fed. Rep. (1960) Dipl.-Min., mineralogy (U. Heidelberg, 1989). Student. *Crystallography.*

Gering, Dr Erich. Europ. Inst. f. Transurane, EURATOM, Postfach 2340, D-7500 Karlsruhe, Germany, Fed. Rep. (1957) Dr rer.nat., (U. Karlsruhe, 1985). (tel. 07247 + 84368). *High pressure X-ray diffraction, elastic an inelastic neutron scattering, phonons.*

Gerold, Prof. Dr Volkmar. Inst. f. Werkstoffwissenschaften, MPI f. Metallforschung, Seestr. 92, D-7000 Stuttgart, Germany, Fed. Rep. (1922) Dr, physics (U. Stuttgart, 1953). Prof. (tel. 0711 + 2095, ext. 219). *Metal physics.*

Gertel, Ms Heike. Abt. Werkstofforschung, GKSS-Forschungszentrum Geesthacht, Max-Planck-Strasse, D-2054 Geesthacht, Germany, Fed. Rep. (1963) Dipl.-Min., mineralogy (U. Hamburg, 1988). Doktorandin. (tel. 04152 + 871327). *Neutron diffraction, texture, metals.*

Geyer, Mr Andreas. Mineralogisch-Petrogr. Inst., Universität, Im Neuenheimer Feld 236, D-6900 Heidelberg, Germany, Fed. Rep. (1958) Student. *Thermal analysis (TGA, DTA, DSC), crystal growth.*

Gies, Dr Hermann. Mineralogisches Inst., Universität, Olshausenstr. 60, D-2300 Kiel, Germany, Fed. Rep. (1952) Dr, chemistry (U. Kiel, 1982). (tel. 0431 + 8802912, email NMP34 at DKUNI0). *Crystal chemistry, crystal structure, zeolites synthesis.*

Glinnemann, Dr Jürgen Wilhelm Erich. Inst. f. Kristallographie, RWTH, Templergraben 55, D-5100 Aachen, Germany, Fed. Rep. (1951) Dr rer.nat, mineralogy (RWTH Aachen, 1987). (tel. 0241 + 80, ext. 6988). *Inorganic crystals, structural chemistry, high pressure, X-ray diffraction, symmetry, teaching.*

Goddard, Dr Richard John. MPI f. Kohlenforschung, Kaiser-Wilhelm-Platz 1, D-4330 Mülheim/Ruhr, Germany, Fed. Rep. (1952) PhD., chemistry (Bristol U., UK, 1977). Wiss. Mit. (tel. 0208 + 306, ext. 485, fax +407, telex 856615 kofo d, email GODDARD MPI-MUELHEIM.MPG.DBP.DE). *Structural chemistry, theoretical chemistry, electron density determination, organometallic compounds, molecular modelling.*

Göbel, Dr Herbert Ernst. ZFE FKE 42, SIEMENS AG, Otto-Hahn-Ring 6, D-8000 München 83, Germany, Fed. Rep. (1940) Dr rer.nat., gamma spectroscopy (TU München, 1969). Laboratory leader. (tel. 089 + 636, ext. 3274, telex 52109-11). *X-ray powder diffraction, materials science, high-temperature diffraction, thin films, position-sensitive detectors.*

Göttlicher, Prof. Dr Siegfried. Fachgebiet Strukturforschung, Inst. f. Phys. Chemie, Petersenstr. 20, D-6100 Darmstadt, Germany, Fed. Rep. (1929) Dr rer.nat., chemistry (TU Darmstadt, 1961). Prof. (tel. 06151 + 16, ext. 2893). *Electron density determination in crystals; structure determination.*

Gomm, Dr Martin. Inst. f. Angew. Physik, Lehrst. f. Kristallographie, Universität, Bismarckstr. 10, D-8500 Erlangen, Germany, Fed. Rep. (1943) Dr rer.nat., crystallography (U. Erlangen-Nürnberg, 1977). Akad. Oberrat (tel. 09131 + 85, ext. 2712). *Diffractometry, electron density, computing, mycology, honey bees.*

Gonschorek, Dr Walter. Inst. f. Physikalische Chemie der TH Darmstadt, Außtelle in KFA/ZFR, Postfach 1913, D-5170 Jülich, Germany, Fed. Rep. (1937) Dr, crystallography (RWTH Aachen, 1971). Priv.-Doz. (tel. 02461 + 61, ext. 3339). *Chemical bonding, lattice dynamics, physical properties, statistics.*

Graetsch, Dr Heribert. Inst. f. Mineralogie, Universität, Universitätsstr. 155, D-4630 Bochum, Germany, Fed. Rep. (1953) Dr rer.nat., mineralogy and crystallography (U. Bochum, 1986). Wiss. Mit. (tel. 0234 + 7003516). *Crystal chemistry, crystal physics.*

Graf, Dr Hans Anton. Dept. Neutron Scattering, Hahn-Meitner-Inst., Glienicker Str. 100, D-1000 Berlin 39, Germany, Fed. Rep. (1945) Dr rer.nat., chemistry (U. München, 1975). Res. assoc. (tel. 030 + 8009, ext. 2778, fax +2999, telex 01-85763, email CH6 at DB0HMI41). *Neutron scattering, magnetic structures, Jahn-Teller-compounds, order/disorder.*

Granzin, Dr Joachim. Inst. f. Kristallographie, FU Berlin, Takustr. 6, D-1000 Berlin 33, Germany, Fed. Rep. (1957) Dr rer.nat., mineralogy (U. Hamburg, 1987). (tel. 030 + 838, ext. 6909, fax 030-832-6561). *Chemical bonding, X-ray crystallography, inorganic crystals, microscopy, structural chemistry, biology.*

Greis, Dr Ortwin. Zentralbereich Elektronenmikroskopie, TU Hamburg-Harburg, Eissendorfer Str. 42, D-2100 Hamburg 90, Germany, Fed. Rep. (1941) Dr rer.nat.habil., inorganic chemistry, crystallography (U. Heidelberg, 1980). Leiter d. Zentralb. Elektronenmikroskopie. (tel. 040 + 7718, ext. 2434). *Inorganic solid state chemistry, X-ray microanalysis, materials science, mineralogy, superstructures, crystallography, X-ray powder methods, electron microscopy (TEM - AEM - REM).*

Grosse, Prof. Dr Peter. I. Physikalisches Inst., RWTH, Templergraben 55, D-5100 Aachen, Germany, Fed. Rep. (1932) Dr rer.nat.habil., physics (U. Köln, 1969). Ord. Prof. (tel. 0241 + 80, ext. 7155, telex 08/32704). *Solid state spectroscopy IR, FIR, preparation, semiconductors, semimetals.*

Grotepass-Deuter, Mrs Margit. Inst. f. Kristallogr., Dept. f. Physik, Universität, Im Stadtwald, D-6600 Saarbrücken, Germany, Fed. Rep. (1958) Dipl., mineralogy (RWTH Aachen, 1984). Grad. stud. (tel. 0681 + 3021). *Structure refinement, synchrotron radiation.*

Grube, Mr Jörg-Michael. Fak. f. Chemie, Universität, Postfach 5560, D-7750 Konstanz, Germany, Fed. Rep. (1963) (tel. 07531 + 883678).

Gruehn, Prof. Dr Reginald. Inst. f. Anorgan. u. Analyt. Chemie, Universität, Heinrich-Buff-Ring 58, D-6300 Giessen, Germany, Fed. Rep. (1929) Dr rer.nat.habil., inorg. and analyt. chemistry (U. Münster, 1969). Prof. (tel. 0641 + 7025670). *Transmission electron microscopy, inorganic crystal structures, chemical transport, materials science.*

Gütlich, Prof. Dr Philipp. Inst. f. Anorg. und Analyt. Chemie, Universität, Jakob-Welder-Weg 11, D-6500 Mainz, Germany, Fed. Rep. (1934) Prof., inorganic and analytical chemistry (U. Mainz, 1975). Ord. Prof. (tel. 06131 + 39, ext. 2373). *Transition metal chemistry, electronic structure, solid state chemistry, theoretical inorganic chemistry (ligand field and MO theories), magnetochemistry, spectroscopy.*

Gupta, Dr Amaresh. Bundesinstitut f. chem.-techn. Unters. beim BWB (BICT), Grosses Cent, D-5357 Swisttal 1, Germany, Fed. Rep. (1941) Dr rer.nat., crystallography (U. Bonn, 1975). Lab. Head. (tel. 02222 + 6008, ext. 384, telex 8869315 bict d, fax 02222-1852). *Crystal structures, high explosives, thermal properties.*

Guse, Prof. Dr Werner Joachim. Mineralog.-Petrogr. Inst., Universität, Grindelallee 48, D-2000 Hamburg 13, Germany, Fed. Rep. (1940) Prof., physical crystallography (U. Hamburg, 1987). Univ. Prof. (tel. 040 + 41232076). *Crystal growth, defect structures, diffuse scattering, disorder, electron diffraction, electron microscopy.*

Gussone, Dr Rainer Carl Leonard. Inst. f. Mineralogie und Lagerstättenlehre, RWTH, Wüllnerstr. 2, D-5100 Aachen, Germany, Fed. Rep. (1931) Dr -Ing., economic geology (RWTH Aachen, 1964). Akad. Oberrat. (tel. 0241 + 80, ext. 5759). *X-ray spectrometry, X-ray diffraction, mineral and rock diagnosis, ore microscopy.*

Guth, Dr Helmut Karl Richard. SFB 127 Universität Marburg, Inst. f. Angew. Kernphysik, KFZ., Postfach 3640, D-7500 Karlsruhe, Germany, Fed. Rep. (1952) Dr rer.nat. physics (U. Karlsruhe, 1979). Wiss. Mit. (tel. 07247 + 82, ext. 3438). *Neutron and X-ray diffraction, crystal structures, phase transitions.*

Haase, Prof. Dr Wolfgang. Inst. f. Physikalische Chemie, TH, Petersenstr. 20, D-6100 Darmstadt, Germany, Fed. Rep. (1936) Dr rer.nat., chemistry (U. Jena, 1964). Prof. (tel. 06151 + 16, ext. 3398, fax +5489, telex 419579 thd). *Liquid crystals, structure, higher ordered smectic phases, polymers, magnetig exchange effect, weak interactions, metalloproteins, solid state properties.*

Haase-Wessel, Dr Werner. Am Weidenfeld 2C, D-3352 Einbeck, Germany, Fed. Rep. (1941) Dr rer.nat., physical chemistry (U. Göttingen, 1973). Wiss. Mit. *Inorganic and organic crystal structures.*

Hädicke, Dr Erich Emil Hermann. Anorg. Chemie ZAA/S-M325, BASF AG, D-6700 Ludwigshafen, Germany, Fed. Rep. (1940) Dr rer.nat., chemistry (TU München, 1969). Lab. head. (tel. 0621 + 60, ext. 4805, telex ZA-TTX 6215934=basf). *Material science, X-ray diffraction, electron diffraction, magnetic materials, textures.*

Hafner, Prof. Dr Stefan S. Institut f. Mineralogie, Universität, Hans-Meerwein-Str., D-3550 Marburg, Germany, Fed. Rep. (1932) Dr sc.nat., petrography (Fed. Inst. of Techn. Switzerland, 1958). Prof. (tel. 06421 + 28, ext. 5617, telex 482372 umr d). *Inorganic crystal structures, spectroscopy, physical properties, minerals.*

Hahn, Prof. Dr Theodor. Inst. f. Kristallographie, RWTH, Templergraben 55, D-5100 Aachen, Germany, Fed. Rep. (1928) Dr rer.nat., mineralogy (U. Frankfurt/Main, 1952). Prof. (tel. 0241 + 80, ext. 6900, fax +4413, telex 0832704thac d, email KRISTA at DACTH01). *Inorganic crystal chemistry, crystal physics, symmetry.*

Hangleiter, Dr Thomas B. FB 6 - Physik, Universität, Warburger Str. 100 A, D-4790 Paderborn, Germany, Fed. Rep. (1948) Dr rer.nat., physics (U. Paderborn, 1979). Assoc. Prof. (tel. 05251 + 60, ext. 2761, fax +2519, telex 936776). *Crystal growth, inorganic crystals, semiconductors.*

Hanke, Dr Kurt. Mineralogisch-Kristallogr. Inst., Universität, Goldschmidtstr. 1, D-3400 Göttingen, Germany, Fed. Rep. (1935) Dr, mineralogy (U. Göttingen, 1965). Akad. Oberrat. (tel. 0551 + 39, ext. 3937). *Inorganic crystal structures.*

Harbrecht, Dr Bernd. Anorg. Chemie I, Universität, Otto-Hahn-Strasse, Postf. 500500, D-4600 Dortmund 50, Germany, Fed. Rep. (1950) Dr rer.nat., chemistry (RWTH Aachen, 1981). (tel. 0231 + 7553794, telex 822465). *Solid state chemistry.*

Harms, Dr Klaus. Inst. f. Anorg. Chemie, Universität, Tammannstr. 4, D-3400 Göttingen, Germany, Fed. Rep. (1953) Dr rer.nat., chemistry (U. Göttingen, 1984). Wiss. Angest. (tel. 0551 + 39, ext. 3073). *Structural organic chemistry, pericyclic reactions.*

Harr, Dr Albrecht Wolfgang Michael. Battelle-Institut, Wiesbadenerstr., D-6000 Frankfurt/Main, Germany, Fed. Rep. (1947) Dr rer.nat., physics (U. Göttingen, 1977). Princip. res. sci. (tel. 069 + 79082883). *Growth, compound and organic semiconductors.*

Hartl, Prof. Dr Hans. Inst. f. Anorg. u. Analyt. Chemie, Freie Universität, Fabeck-str. 34-36, D-1000 Berlin 33, Germany, Fed. Rep. (1940) Dr rer.nat., inorganic chemistry (TH München, 1969). Prof. (tel. 030 + 838, ext. 4003). *Inorganic chemistry (halogen-chemistry), structure determination, crystal chemistry.*

Hauck, Dr Jürgen. Inst. f. Festkörperforschung, Kernforschungsanlage Jülich, D-5170 Jülich, Germany, Fed. Rep. (1941) Dr, inorganic chemistry (U. Frankfurt/Main, 1968). Scient. (tel. 02461 + 61, ext. 4237). *Inorganic crystals, thin films, superconducting materials.*

Hausen, Dr Hans-Dieter. Inst. f. Anorg. Chemie, Universität, Pfaffenwaldring 55, D-7000 Stuttgart 80, Germany, Fed. Rep. (1937) Dr rer.nat., chemistry (U. Stuttgart, 1966). Akad. Oberrat. (tel. 0711 + 685, ext. 4220). *Inorganic and organometallic crystal structures.*

Haussühl, Prof. Dr Siegfried Georg. Inst. f. Kristallographie, Universität, Zülpicher Str. 49, D-5000 Köln 1, Germany, Fed. Rep. (1927) Dr rer.nat., natural sciences (U. Tübingen, 1956). Direktor. (tel. 0221 + 470, ext. 3194). *Crystal growth, thermal properties, mechanical properties, electrical properties, nonlinear optics, nonlinear acoustics, phase transitions, electron diffraction.*

Hecht, Dr Hans-Jürgen. Abt. Röntgenstrukturanalyse, Physiol.-Chem. Inst., Universität, Am Hubland, Zentralbau Chemie, D-8700 Würzburg, Germany, Fed. Rep. (1947) Dr rer.nat., chemistry (FU Berlin, 1976). Res. assoc. (tel. 0931 + 888, ext. 386). *X-ray crystallography, proteins, biologically interesting small molecules.*

Heide, Dr Helmut. Battelle Inst. e.V., Am Römerhof, D-6000 Frankfurt, Germany, Fed. Rep. (1934) Dr rer.nat., mineralogy (U. Bonn, 1969). Deputy chief. (tel.

069 + 7908, ext. 2764). *Materials research, biochemical engineering, ceramics, phase relations, sintering kinetics.*

Heim, Dr Harald Josef Robert. Mandelring 67, D-6730 Neustadt/Weinstr. 13, Germany, Fed. Rep. (1951) Dr rer.nat., chemistry (U. Karlsruhe, 1979). (tel. 06321 + 88103). *Inorganic crystal structures, symmetry, crystallography.*

Heime, Prof. Dr Klaus. Inst. f. Halbleitertechnik, RWTH, Templergraben 55, D-5100 Aachen, Germany, Fed. Rep. (1935) Dr rer.nat., physics (U. Darmstadt, 1967). Prof. and Dir. (tel. 0241 + 80, ext. 7745, fax +7751). *Crystal growth, crystallographic data, defect structures, electron microscopy, instrumentation, materials science, microscopy, semiconductors, surface structure, teaching, X-ray diffraction.*

Heinemann, Dr Udo. Inst. f. Kristallographie, FU Berlin, Takustr. 6, D-1000 Berlin 33, Germany, Fed. Rep. (1953) Dr rer.nat., chemistry (U. Göttingen, 1982). Hochschulass. (tel. 030 + 838, ext. 3456, fax 030-832-6561, email HEINEMANN at KRISTALL.CHEMIE.FU-BERLIN.DBP.DE). *Molecular biology, X-ray crystallography, proteins, nucleic acids.*

Hellner, Prof.em. Dr Erwin E. Inst. f. Mineralogie, Fachbereich Geowissenschaften, Universität, Lahnberge, D-3550 Marburg, Germany, Fed. Rep. (1920) Dr rer.nat., mineralogy (U. Göttingen, 1945). Prof.em. (tel. 06421 + 28, ext. 2045). *Theoretical crystallography, inorganic structure type symbology, high pressure research, charge density.*

Helmreich, Dr Dieter. HELIOTRONIC Forschungs- und Entwicklungsgesellschaft f. Solarzellengrundstoffe mbH. Joh.-Hess-Str. 24, P.O.Box 1129, D-8263 Burghausen, Germany, Fed. Rep. (1939) Dr rer.nat., physics (TU München, 1968). Director R&D crystals. (tel. 08677 + 83, ext. 2736, fax 08677-62171). *Crystal growth, materials science, semiconductors, rapid solidification.*

Henke, Dr Henning. Inst. f. Anorganische Chemie, Universität, Engesserstr., Geb. Nr. 30.45, Postfach 6380, D-7500 Karlsruhe, Germany, Fed. Rep. (1943) Dr rer.nat., chemistry (U. Karlsruhe, 1971). Wiss. Ang. (tel. 0721 + 608, ext. 2977). *Inorganic crystal structures, hydrates, hydrogen bonding.*

Henkel, Prof. Dr Gerald. Fachgebiet Anorg.-Chem./Festkörperchemie, Universität, Lotharstr. 1, D-4100 Duisburg 1, Germany, Fed. Rep. (1948) Dr rer.nat., chemistry (U. Bielefeld, 1976). Prof. (tel. 0203 + 379, ext. 3186, fax +3333, telex 855793 uni du d). *Structural chemistry, inorganic and organic compounds, biological structures.*

Hentschel, Dr Manfred Paul. Gruppe Parakristallforschung, BAM, Unter den Eichen 44/46, D-1000 Berlin 45, Germany, Fed. Rep. (1943) Dr rer.nat., physics (FU Berlin, 1981). Wiss. Mit. (tel. 030 + 8104, ext. 0912). *Paracrystals, glassy structures, composites, nondestructive testing, fibers, lipids, protein structure, molecular crystals.*

Herdtweck, Dr Eberhardt. Anorg.-chem. Institut, TU München, Lichtenbergstr. 4, D-8046 Garching, Germany, Fed. Rep. (1948) Dr rer.nat., chemistry (U. Tübingen, 1978). Akad. Rat. (tel. 089 + 3209, ext. 3143, fax +2727, telex 17898174) *Fluorides, oxifluorides, oxides and their hydrates, X-ray structures, organometallic compounds.*

Herres, Mr Nikolaus. IR-MA, Fraunhofer Inst. f. Angew. Festkörperphysik, Eckerstr. 4, D-7800 Freiburg, Germany, Fed. Rep. (1954) Dipl.-Min., mineralogy (RWTH Aachen, 1981). (tel. 0761 + 2714, ext. 374, fax +296, telex 772510 fhges d). *Diffractometry, instrumentation, microscopy, semiconductors, texture, X-ray topography.*

Herzberg, Dr Armin. Mineralog.-Petrogr. Inst., TU, Adolf-Roemer-Str. 2A, D-3392 Clausthal-Zellerfeld, Germany, Fed. Rep. (1948) Dr rer.nat., mineralogy (TU Clausthal, 1977). Wiss. Ass. (tel. 05323 + 72, ext. 2326). *Petrography, granites, autometasomatism, mineralogy, carbonates, chromites, micas, minerals analysis (RFA - AAS - ICP - EMPA).*

Hess, Prof. Dr Heinz. Mautenreutestr. 10, D-7425 Hohenstein, Germany, Fed. Rep. (1924) Dr rer.nat., chemistry (U. Stuttgart, 1958). Prof.(retired). (tel. 07387 + 1303). *Inorganic crystal structures, mineral structures.*

Hesse, Dr Karl-Friedrich. Mineralogisches Inst., Universität, Olshausenstr. 40-60, D-2300 Kiel, Germany, Fed. Rep. (1939) Dr, mineralogy (U. Kiel, 1973). Wiss. Mit. (tel. 0431 + 880, ext. 2901). *Crystal chemistry, silicates, X-ray crystallography, single crystal X-ray diffractometer.*

Hildebrandt, Prof. Dr Gerhard. Form. Fritz-Haber-Inst., Max-Planck-Ges., Faradayweg 4-6, D-1000 Berlin 33, Germany, Fed. Rep. (1922) Prof., physics (TU Berlin, 1975). (tel. 030 + 3655132). *X-ray diffraction, topography with X-rays, solid state physics.*

Hilgenfeld, Dr Rolf. Central Research/Biotechnology, HOECHST AG, Postfach 800320, D-6230 Frankfurt 80, Germany, Fed. Rep. (1954) Dr rer.nat., chemistry (FU Berlin, 1987). Group Leader. (tel. 069 + 305, ext. 6360, fax 069-307941). *Protein crystallography, molecular modelling, protein engineering.*

Hiller, Dr Wolfgang Paul. Inst. f. Anorg. Chemie, Universität, Auf der Morgenstelle 18, D-7400 Tübingen, Germany, Fed. Rep. (1953) Dr rer.nat., chemistry (U. Tübingen, 1981). Akad. Rat. (tel. 07071 + 29, ext. 6230, fax +5400, telex 17-7071-913, email bitnet CAHI001 at DTUZDV5A). *Crystal structure analysis, Cu-, Ag-, Au-complexes.*

Hinrichs, Dr Winfried. Inst. f. Kristallographie, FU Berlin, Takustr. 6, D-1000 Berlin 33, Germany, Fed. Rep. (1950) Dr rer.nat., chemistry (U. Hamburg, 1983). (tel. 030 + 838 ext 6326, fax 030-832-6561). *Chemical bonding, structural chemistry and biology, X-ray crystallography, semiconductors.*

Hinrichsen, Prof. Dr Georg. Inst. f. Nichtmetallische Werkstoffe, TU Berlin, Englische Str. 20, D-1000 Berlin 12, Germany, Fed. Rep. (1941) Dr rer.nat., physics (U. Mainz, 1970). Full prof. (tel. 030 + 31424464). *X-ray wide angle scattering, X-ray small angle scattering.*

Hinsch, Dr Thorsten Reinhard. Inst. f. Kristallographie, RWTH, Templergraben 55, D-5100 Aachen, Germany, Fed. Rep. (1959) Dr rer..nat., mineralogy (U. Hamburg, 1986). (tel. 0241 + 806903). *X-ray powder diffraction.*

Hinze, Prof. Dr Eckhard. Inst. f. Geowissenschaften und Lithosphärenforschung, Universität, Senckenbergstr. 3, D-6300 Giessen, Germany, Fed. Rep. (1934) Dr rer.nat.habil., mineralogy (U. Bonn, 1978). Prof. (tel. 0641 + 702, ext. 8282). *High pressure X-ray diffraction, crystal chemistry, physical properties, minerals, mantle minerals, phase equilibria, high pressure mineral synthesis.*

Höfer, Dr Hans Hermann. CORNING KERAMIK GmbH & CoKG, Dotzheimerstr. 168, D-6200 Wiesbaden, Germany, Fed. Rep. (1948) Dr, crystallography (RWTH Aachen, 1979). Quality assurance manager. (tel. 06121 + 804-140). *Thermodynamics and kinetics, solid state reactions; solid ion conductors.*

Höfler, Dr Sabine Ida Gerda. KFA Jülich, Postfach 1913, D-5170 Jülich, Germany, Fed. Rep. (1957) Dr rer.nat., crystallography (U. Bonn, 1989). (tel. 02461 + 616024). *Textures, structure, minerals, neutron diffraction.*

Höhling, Prof. Dr Hans Jürgen. Inst. f. Medizinische Physik, Universität, Hüfferstr. 68, D-4400 Münster, Germany, Fed. Rep. (1930) Dr rer.nat.habil., mineralogy (U. Münster, 1964). Prof. (tel. 0251 + 83, ext. 5191). *Crystallography, hard tissues (normal & diseased).*

Hönle, Dr Wolfgang. MPI f. Festkörperforschung, Heisenbergstr. 1, D-7000 Stuttgart, Germany, Fed. Rep. (1947) Dr, inorganic chemistry (U. Münster, 1975). Sci. coworker, assistant editor. (tel. 0711 + 6860, ext. 416 or 472, fax 0711-6874371, telex 7-255555 mpif-d, email HOENL at DS0MPI11). *Solid state chemistry, crystal structures, syntheses.*

Hoffmann, Prof. Dr Wolfgang. Inst. f. Mineralogie, Universität, Corrensstr. 24, D-4400 Münster, Germany, Fed. Rep. (1935) Dr rer.nat., mineralogy, crystallography (U. Hamburg, 1961). Direktor. (tel. 0251 + 83, ext. 3461). *Crystal structure analysis, physical properties, phase transitions.*

Hoffmeister, Dr Wolfgang. Material and Reliability Analysis, IBM Lab., Schoenaicher Str. 220, D-7030 Böblingen, Germany, Fed. Rep. (1930) Dr, nuclear chemistry (U. Köln, 1961). Chief chemist. (tel. 07031 + 660, ext. 8307). *Trace analysis, radiotracer application.*

Hofmeister, Dr Wolfgang. Inst. f. Geowissenschaften, Universität, Saarstr. 21, D-6500 Mainz, Germany, Fed. Rep. (1952) Dr rer.nat., mineralogy (U. Mainz, 1981). Hochschulassistent. (tel. 06131 + 39, ext. 4365). *Structure determination, crystal chemistry, minerals, mineralogy.*

Hohlwein, Dr Dietmar. Inst. f. Kristallographie, Universität, Charlottenstr. 33, D-7400 Tübingen, Germany, Fed. Rep. (1944) Dr habil., physics (U. Tübingen, 1981). Dozent. (tel. 07071 + 29, ext. 6058). *Neutron diffraction and instrumentation, disorder in crystals, diffuse scattering.*

Holinski, Dr Rüdiger. Res. and Dev., DOW CORNING GmbH, Pelkovenstr. 152, D-8000 München 50, Germany, Fed. Rep. (1939) Dr rer.nat., chemistry (TU Clausthal, 1968). Dept. Manager. (tel. 089 + 14860, ext. 265, telex 5215654). *Tribology, specialty lubrication, solid lubricants, metallurgy, solid state chemistry, analytical surface investigations.*

Holmes, Prof. Dr Kenneth Charles. Abt. f. Biophysik, MPI f. medizinische Forschung, Jahnstr. 29, D-6900 Heidelberg, Germany, Fed. Rep. (1934) Ph.d., biophysics (U. London, 1959). Dir. of res. (tel. 06221 + 486, ext. 270). *Macromolecular structures, crystallography, muscle physiology.*

Holzapfel, Prof. Dr Wilfried B. Fachbereich Physik, Universität, Warburger Str. 100, D-4790 Paderborn, Germany, Fed. Rep. (1938) Dr rer.nat., physics (U. Karlsruhe, 1966). Univ. Prof. (tel. 05251 + 602673, telex936776 unipb d). *X-ray diffraction on solids under pressure.*

Honigmann, Dr Bertold. Hauptlaboratorium, BASF-AG, Carl-Bosch-Str., D-6700 Ludwigshafen, Germany, Fed. Rep. (1921) Dr, physical chemistry (TU Berlin, 1948). Head, res. group. (tel. 0621 + 60, ext. 99481). *Crystal growth, pigments - crystal properties, dispersions - physical and chemical properties.*

Hoppe, Prof. Dr Dr hc Rudolf. Inst. f. Anorg. und Analytische Chemie, Justus-Liebig-Universität, Heinrich-Buff-Ring 58, D-6300 Giessen, Germany, Fed. Rep. (1922) Ord. Prof., inorganic chemistry (U. Giessen, 1965). Dir. (tel. 0641 + 702 ext. 5660). *Crystal chemistry, Madelung factors, metal oxides, synthesis, solid state chemistry, fluorine.*

Horst, Dr Wolfgang. Tucholskystr. 22, D-6000 Frankfurt/Main, Germany, Fed. Rep. (1948) Dr phil.nat., physics (U. Frankfurt/Main, 1981). (tel. 069 + 627260). *Modulated structures, rock forming minerals, satellite reflections.*

Horstmann, Prof. Dr Manfred. Universität Osnabrück, Neuer Graben, D-4500 Osnabrück, Germany, Fed. Rep. (1928) Dr rer.nat., experimental physics (U. Hamburg, 1960). Prof. (tel. 0541 + 608, ext. 4100). *Solid state physics, surface physics.*

Hosemann, Prof. Dr Dr hc Rolf. Gruppe Parakristallforschung, BAM, Unter den Eichen 44-46, D-1000 Berlin 45, Germany, Fed. Rep. (1912) Prof. Dr phil.nat.habil., physics (U. Freiburg, 1935). Res. Group Leader. (tel. 030 + 8104, ext. 0913). *Equilibrium states, thermodynamics, glass, melts, polymers, catalysts.*

Hoser, Dr Andrzej. Inst. f. Kristallographie, Universität, Charlottenstr. 33, D-7400 Tübingen, Germany, Fed. Rep. (1952) Dr, chemistry (U. Poznań, Poland,

1982). Wiss. Mit. (tel. 07071 + 296058). *X-ray and neutron structure analysis, molecular crystals, phase transitions.*

Hovestreydt, Dr Eric Robert. E 689 F2, SIEMENS AG, Östliche Rheinbrückenstr. 50, D-7500 Karlsruhe, Germany, Fed. Rep. (1957) Dr, crystallography (U. Genf, Switzerland, 1986). Application scientist. (tel. 0721 + 595, ext. 6573, fax +4506, telex 78255-69, email BJ02 at DKAUNI48). *Computing, crystallographic data, diffractometry, electron microscopy, high pressures, inorganic crystals, instrumentation, group theory, materials science, metals, phase transitions, powder diffraction, quasicrystals, scattering theory, structural chemistry, symmetry, X-ray diffraction.*

Hu, Ms Xiaorui. Inst. f. Mineralogie u. Kristallographie, TU Berlin, Ernst-Reuter-Platz 1, D-1000 Berlin 12, Germany, Fed. Rep. (1962) Dipl.-Min., mineralogy and crystallography (TU Berlin, 1988). (tel. 030.+ 31423225, email 1374 at DB0TUZ01). *Computing, crystal growth, crystallographic data, defect structures, inorganic crystals, group theory, neutron diffraction, phase transition, powder diffraction, quasicrystals, symmetry, X-ray diffraction.*

Hubbert, Mrs Dr Elisabeth. Anorg. Chemie I, Ruhr-Universität, Universitätsstr. 150, D-4630 Bochum, Germany, Fed. Rep. (1934) Dr, inorganic chemistry (U. Münster, 1966). Wiss. Ang. (tel. 0234 + 700, ext. 4153). *Crystal growth, inorganic crystal structures.*

Huber, Prof. Robert. Strukturforschung II, MPI f. Biochemie, Am Klopferspitz, D-8033 Martinsried, Germany, Fed. Rep. (1937) Prof., chemistry (TU München, 1976). Direktor. (tel. 089 + 8578, ext. 2677, fax +3777). *Biochemistry, protein crystallography.*

Hubert, Mr Georg. Analytical Systems, E689F, SIEMENS AG, Östliche Rheinbrückenstr. 50, D-7500 Karlsruhe, Germany, Fed. Rep. (1948) Diplom, physics (TH Karlsruhe, 1975). Product Manager X-ray Analysis. (tel. 0721 + 5954265, telex 78255-69). *X-ray analysis, instrumentation, software.*

Huch, Dr Volker. Inst. f. Anorg. Chemie, Universität, Im Stadtwald, D-6600 Saarbrücken, Germany, Fed. Rep. (1955) Dr, inorg. chemistry (TU Braunschweig, 1984). Akad. Rat. (tel. 0681 + 3022265). *Inorganic and organometallic crystal structures.*

Hübner, Mr Thomas. Arbeitsgruppe f. chem. Kristallogr., MPI f. Biochemie, Am Klopferspitz, D-8033 Martinsried, Germany, Fed. Rep. (1956) Diplom, chemistry (U. München, 1982). (tel. 089 + 8578, ext. 2659). *X-ray structure analysis, organic compounds, heterocyclic systems, organometallic molecules.*

Hümmer, Prof. Dr Kurt. Inst. f. Angew. Physik, Lehrst. f. Kristallographie, Universität, Bismarckstr. 10, D-8520 Erlangen, Germany, Fed. Rep. (1939) Dr rer.nat.habil., physics (U. Erlangen-Nürnberg, 1978). Univ.-Prof. (tel. 09131 + 85, ext. 2711, fax +2957, telex 629830 unierd, email MPKR00 at DERRZE). *Diffractometry, electron density, X-ray diffraction, multiple scattering, phase determination.*

Hummel, Dr Hans-Ulrich. Inst. f. Anorg. Chemie, Universität, Egerlandstr. 1, D-8520 Erlangen, Germany, Fed. Rep. (1954) Dr rer.nat., chemistry (1980). Akad. Rat. (tel. 09131 + 857393). *Preparation, structural investigation, inorganic solids, alkaline and earth-alkaline compounds, pseudohalogenides, pseudochalcogenides.*

Hund, Dr Franz Josef. AC-Forschung, Anorg. Wiss. Labor, Bayer AG, Rheinuferstr. 7-9, D-4150 Krefeld, Germany, Fed. Rep. (1916) Dr, inorganic chemistry (U. Heidelberg, 1945). Industrial chemist. (tel. 0215 + 52346). *Inorganic pigments, crystal structure, synthesis.*

Hundt, Dr Rudolf. Inst. f. Anorganische Chemie, Universität, Gerhard-Domagk-Str. 1, D-5300 Bonn 1, Germany, Fed. Rep. (1941) Dr rer.nat., inorg. chemistry (U. Bonn, 1973). OStR i. HDnst.. (tel. 0228 + 73, ext. 5337, email UNC404 at DBNRHRZ1). *Computing, inorganic crystals, crystallographic data.*

Ihringer, Dr Jörg-Peter, Inst. f. Kristallographie, Universität, Charlottenstr. 33, D-7400 Tübingen, Germany, Fed. Rep. (1943) Dr rer.nat.habil., physics (U. Tübingen). (tel. 07071 + 29, ext. 6394). *Phase transitions, X-ray powder diffraction, instrumentation, quasi crystals.*

Irngartinger, Prof. Dr Hermann. Organisch-Chem. Inst., Universität, Im Neuenheimer Feld 270, D-6900 Heidelberg, Germany, Fed. Rep. (1938) Dr habil., chemistry (U. Heidelberg, 1972). Prof. (tel. 06221 + 56, ext. 2422, fax +3111, email EARN A61 at DHDURZ2). *Organic crystal structures, organic solid state chemistry, electron density distribution.*

Isenberg, Dr Wilhelm. Analytical Systems, E689, SIEMENS AG, Östliche Rheinbrückenstr. 50, D-7500 Karlsruhe, Germany, Fed. Rep. (1953) Diplom, inorg. chemistry (U. Göttingen, 1979). Product manager X-ray diffraction. (tel. 0721 + 5952425, telex 78255-69). *Diffractometry theory, instumentation, software.*

Jacob, Dr Herbert. Res. Div., Wacker-Chemitronic GmbH, Postfach 1140, D-8263 Burghausen, Germany, Fed. Rep. (1928) Dr rer.nat., chemical thermodynamics (U. Münster, 1958). Forschungsbereichsleiter. (tel. 08677 + 83868). *Industrial crystal growth, epitaxy.*

Jacobi, Dr Hans. Inst. f. Mineralogie u. mineral. Rohstoffe, TU, Adolph-Roemer-Str. 2A, D-3392 Clausthal-Zellerfeld, Germany, Fed. Rep. (1934) Dr phil., mineralogy (U. Marburg, 1963). Akad. Oberrat. (tel. 05323 + 72, ext. 2392). *Modulated crystal structures, optics.*

Jacobs, Prof. Dr Herbert Ernst Hermann. Anorganische Chemie, Universität, Postfach 500500, Otto-Hahn-Str., D-4600 Dortmund 50, Germany, Fed. Rep.

(1936) Dr rer.nat., chemistry (U. Kiel, 1966). Univ.-Prof. (tel. 0231 + 7553802, telex 822465). *Solid state chemistry.*

Jäger, Dr Hans. Abt. Festkörperphysik, Battelle-Inst. e.V., Am Römerhof 35, D-6000 Frankfurt/Main, Germany, Fed. Rep. (1934) Dr rer.nat., solid state physics, metal physics (TH Stuttgart, 1963). (tel. 069 + 7908, ext. 2766). *Semiconductor physics, device development, crystal growth, epitaxy, X-ray detection, gamma-ray detection, infrared physics.*

Jäger, Dr Susanne Christine. Gmelin Inst., Joachim-Becher-Str. 2, D-6000 Frankfurt/Main, Germany, Fed. Rep. (1943) Dr rer.nat., chemistry (RWTH Aachen, 1982). Editorial assoc. (tel. 069 + 564122). *Solid state chemistry.*

Jagodzinski, Prof. Dr Dr hc Heinz Ernst. Inst. f. Kristallographie und Mineralogie, Universität, Theresienstr. 41, D-8000 München 2, Germany, Fed. Rep. (1916) Dr rer.nat., physics (U. Göttingen, 1941). Em. Ord. Prof. (tel. 089 + 2394, ext. 4357, telex 529815 univm d). *Diffraction (X-ray - neutron - electron), disorder, phase transitions, quasicrystals, surface structure.*

Jahn, Dr Irmin-Rudolf. Inst. f. Kristallographie, Universität, Charlottenstr. 33, D-7400 Tübingen, Germany, Fed. Rep. (1939) Dr rer.nat., physics (U. Tübingen, 1971). Wiss. Angest. (tel. 07071 + 29, ext. 6389). *Crystal physics, phase transitions, crystal optics.*

Jakobs, Dr Rüdiger-Hasko. VDO Adolf Schindling AG, Sodener Str. 9, D-6231 Schwalbach/Taunus, Germany, Fed. Rep. (1941) Dr, physics (TU Berlin, 1979). (tel. 06196 + 8012710). *General crystallography.*

Jansen, Prof. Dr Martin. Inst. f. Anorg.Chemie, Universität, Gerhard-Domagk-Str. 1, D-5300 Bonn 1, Germany, Fed. Rep. (1944) Dr rer.nat., inorganic and analytical chemistry (U. Giessen, 1973). Prof. (tel. 0228 + 73, ext. 3144 or 2708, telex 886675 unibo d). *Chemical bonding, crystal growth, crystallographic data, diffractometry, high pressures, inorganic crystals, materials science, phase transitions.*

Jarchow, Prof. Dr Otto. Mineralogisch-Petrogr. Inst., Universität, Grindelallee 48, D-2000 Hamburg, Germany, Fed. Rep. (1931) Dr rer.nat., mineralogy and crystallography (U. Saarbrücken, 1961). Dozent. (tel. 040 + 4123, ext. 2056). *Organic and inorganic crystal structures, crystal chemistry, crystal disorder, crystal growth, symmetry.*

Jauch, Dr Wolfgang. Arbeitsgruppe CI, Hahn-Meitner-Inst., Glienicker Str. 100, D-1000 Berlin 39, Germany, Fed. Rep. (1947) Dr rer.nat., solid state physics (TU Berlin, 1979). Wiss. Angest. (tel. 030 + 80091 ext. 2767, fax +2999, email CWJ at DB0HMI41, telex 185763). *Neutron and X-ray diffraction, structure and physical properties.*

Jeitschko, Prof. Dr Wolfgang. Anorg.-Chem. Inst., Universität, Wilhelm-Klemm-Str. 8, D-4400 Münster, Germany, Fed. Rep. (1936) Dr phil., chemistry (U. Wien, 1964). Dir. (tel. 0251 + 833121). *Structure and properties, inorganic and intermetallic crystals.*

Jessen, Mr Sven Michael. Mineralogisches Inst., Universität, Olshausenstr. 40, D-2300 Kiel, Germany, Fed. Rep. (1962) Dipl.-Chem., chemistry (U. Kiel). Ass. (tel. 0431 + 880, ext. 2932). *Hydrogen bonding, molecular crystals, neutron diffraction.*

Joest, Mr Stephan. Jägerstr. 14, D-5620 Velbert-Neviges, Germany, Fed. Rep. (1965) cand.rer.nat. *Crystal growth, materials science, metals, polymers.*

Joswig, Dr Werner. Inst. f. Kristallographie, Universität, Senckenberg-Anlage 30, D-6000 Frankfurt/Main, Germany, Fed. Rep. (1940) Dr, crystallography (U. Frankfurt/Main, 1972). Wiss. Mit. (tel. 069 + 798, ext. 3502, telex 413932). *Inorganic crystal structures.*

Jung, Dr Detlef. Abt. Forschung SIGRI GmbH, Werner-v.-Siemensstr., D-8901 Meitingen, Germany, Fed. Rep. (1937). (tel. 08271 + 3384).

Jung, Dr Volkhard. Inst. f. Angewandte Kernphysik, KFZ Karlsruhe GmbH, Postfach 3640, D-7500 Karlsruhe 1, Germany, Fed. Rep. (1934) Dr rer.nat., nuclear physics (U. Heidelberg, 1964). Physicist. (tel. 07247 + 82, ext. 3406). *Phase tranformations, austenitic steels, lattice parameter variation, stressed metals.*

Jung, Dr Walter. Inst. f. Anorg. Chemie, Universität, Greinstr. 6, D-5000 Köln 41, Germany, Fed. Rep. (1940) Dr rer.nat.habil., inorg. chemistry (U. Köln, 1979). Priv.-Doz. (tel. 0221 + 4703353). *Inorganic solid state chemistry, ternary borides.*

Kaat, te, Prof Dr Erich Heinz. Inst. f. Physik, Universität, Otto-Hahn-Str. D-4600 Dortmund, Germany, Fed. Rep. (1937) Dr, physics (1969). Prof. (tel. 0231 + 755, ext. 3506). *Electron microscopy, radiation damage, ion implantation.*

Kabs, Mr Michael. Abt. PCH-TEV, W.C.Heraeus GmbH, Heraeusstr. 12-14, D-6450 Hanau 1, Germany, Fed. Rep. (1952) Dr rer.nat., chemistry (TU Berlin, 1982). (tel. 06181 + 35, ext. 5122). *Slags, ceramics, material research.*

Kabsch, Dr Wolfgang. Abt. f. Biophysik, MPI f. medizinische Forschung, Jahnstr. 29, D-6900 Heidelberg, Germany, Fed. Rep. (1941) Dr rer.nat., physics (U. Heidelberg, 1972). Res. scient. (tel. 06221 + 486, ext. 276, telex 461505). *Protein structure, protein folding.*

Kaerlein, Mr Carsten-Peter. Arbeitsgr. f. chem. Kristallographie, MPI f. Biochemie, Am Klopferspitz, D-8033 Martinsried, Germany, Fed. Rep. (1942) Diplom, physics (U. München, 1981). (tel. 089 + 8578, ext. 2661). *X-ray structure analysis, organic compounds, biologically interesting molecules.*

Kambe, Dr Kyozaburo. Fritz-Haber-Inst., Max-Planck-Ges., Faradayweg 4-6, D-1000 Berlin 33, Germany, Fed. Rep. (1926) Dr Sci., physics (U. Tokyo, FU Berlin, 1961). Scient. (tel. 030 + 8305, ext. 527, fax +520, telex 185676 fhimp d) *Electron diffraction, electron microscopy, X-ray topography, diffraction theory, channeling, crystal defects, surface crystallography, electron density.*

Karl, Prof Dr Norbert. Physikalisches Inst., Teil 3, Universität, Pfaffenwaldring 57, D-7000 Stuttgart 80, Germany, Fed. Rep. (1939) Dr habil., crystal physics (U. Stuttgart, 1975). Apl. Prof. (tel. 0711 + 685, ext. 5195, fax +3500, telex 072-55445). *Crystal growth, organic crystals, high purity, electrical properties, organic laser crystals, epitaxial organic thin films, UHV-characterization techniques.*

Kassner, Mr Dethard. Inst. f. Kristallographie, Universität, Senckenberganlage 30, D-6000 Frankfurt/Main 1, Germany, Fed. Rep. (1949) Diplom, physics (U. Frankfurt, 1979). Wiss. Mit. (tel. 069 + 798-2105, telex 413932). *Crystallographic computing.*

Katscher, Dr Hartmut. Gmelin-Inst., Max-Planck-Ges., Varrentrappstr. 40/42, D-6000 Frankfurt/Main 90, Germany, Fed. Rep. (1933) Dr, chemistry (U. Würzburg, 1964). Editor. (tel. 069 + 79171). *Inorganic crystal structures.*

Kaub, Mr Jürgen. FB Chemie, Universität, Erwin-Schrödinger-Str., D-6750 Kaiserslautern, Germany, Fed. Rep. (1958) Diplom, chemistry (1984). Wiss. Mit. (tel. 0631 + 2052962). *Inorganic crystal structures.*

Kaus, Dr Gerhard. IBM, Dept. 0135-7032-47, Tübinger Allee, D-7032 Sindelfingen, Germany, Fed. Rep. (1941) Nat.Sc.D., mineralogy, crystallography (U. Mainz, 1969). Staff assoc. (tel. 07031 + 611, ext. 2269). *Surface analysis methods (SIMS - AES - ESCA), ceramics, semiconductors.*

Keesmann, Prof. Dr Karl-Ingo Ortwin. Fachbereich 22 - Geowissenschaften, Inst. f. Geowissenschaften, Saarstr. 21, D-6500 Mainz, Germany, Fed. Rep. (1937) Dr rer.nat., mineralogy (1968). Prof. (tel. 06131 + 39, ext. 2721/5934). *Archeometry, computing, materials science, metals, microscopy, minerals, teaching.*

Keller, Dr Egbert. Kristallogr. Inst., Universität, Hebelstr. 25, D-7800 Freiburg i.Br., Germany, Fed. Rep. (1950) Dr rer.nat., chemistry (U. Freiburg, 1978). Wiss. Angest. (tel. 0761 + 2034279). *X-ray structure analysis, computer graphics in crystallography.*

Keller, Priv.-Doz. Dr Hans-Lothar. Inst. f. Anorganische Chemie, Universität, Otto-Hahn-Platz 6/7, D-2300 Kiel, Germany, Fed. Rep. (1943) Dr rer.nat.habil., inorganic chemistry (U. Kiel, 1984). (tel. 0431 + 880, ext. 2096, fax +2072). *Inorganic crystal structures.*

Keller, Prof. Dr Heimo Jürgen. Anorganisch-Chemisches Inst., Universität, Im Neuenheimer Feld 270, D-6900 Heidelberg, Germany, Fed. Rep. (1935) Dr habil., inorganic chemistry (TU München, 1966). Full prof. (tel. 06221 + 56, ext. 2438). *Linear chain transition metal compounds, one-dimensional metals, crystal structure, coordination compounds.*

Keller, Prof Dr Paul. Inst. f. Mineralogie u. Kristallchemie, Universität, Pfaffenwaldring 55, D-7000 Stuttgart 80, Germany, Fed. Rep. (1940) Prof. (U. Stuttgart, 1973). (tel. 0711 + 6854112, telex 7255445). *Crystal chemistry, crystallography, mineralogy.*

Keller, Dr Wolfgang Ludwig. WDHLIZ, SIEMENS AG, Frankfurter Ring 152, D-8000 München 46, Germany, Fed. Rep. (1927) Dr rer.nat., experimental mineralogy (U. Tübingen, 1954). Lab. leader. (tel. 089 + 3500, ext. 778). *Crystal growth, float zone growth, silicon.*

Kemmler-Sack, Mrs Prof. Dr Sibylle. Inst. f. Anorganische Chemie, Universität, Auf der Morgenstelle 18, D-7400 Tübingen, Germany, Fed. Rep. (1934) Prof., inorganic chemistry (U. Tübingen, 1973). Prof. (tel. 07071 + 29, ext. 2439). *Noble metal compounds, complex oxides, transition metals, rare earths, optical properties.*

Kempa, Mr Paul-Bernd. Fak. f. Chemie, Universität, Postfach 5560, D-7750 Konstanz, Germany, Fed. Rep. (1957) Dipl.-Min., mineralogy (U. Clausthal-Zellerfeld, 1988). (tel. 07531 + 882012, email CHFELS1 at DKNKURZ1). *X-ray and neutron diffraction, zeolites, crystal structure determination.*

Keppler, Dr Ulrich Hermann. Schönbuch-Labor, Allemannenweg 9, D-7036 Schönaich, Germany, Fed. Rep. (1933) Dr rer.nat., crystallography (U. Hamburg, 1962). Manager. (tel. 07031 + 50086). *Biocrystallography, superconductivity, mineral identification.*

Kiel, Dr Gertrude Lina. Inst. f. Anorg. u. Analyt. Chemie, Universität, Johann-Joachim-Becher-Weg 24, D-6500 Mainz, Germany, Fed. Rep. (1937) Dr rer.nat., inorg. chemistry (U. Göttingen, 1967). Akad. Oberrat. (tel. 06131 + 392284). *X-ray structure determination.*

Kirfel, Dr Armin Harald. FR Kristallographie, Universität des Saarlandes, HASYLAB/DESY, Notkestr. 85, D-2000 Hamburg 52, Germany, Fed. Rep. (1943) Dr rer.nat.habil., crystallography (U. Bonn, 1982). (tel. 040 + 8998, ext. 2601). *Electron density, chemical bonding, diffractometry, X-ray diffraction, minerals.*

Klapper, Prof. Dr Helmut. Mineralog. Inst., Universität, . Poppelsdorfer Schloos, D-5300 Bonn, Germany, Fed. Rep. D-5100 Aachen, Germany, Fed. Rep. (1937) Dr Habil, crystallography & crystal physics. (RWTH Aachen, 1975). Prof. (tel. 0228 + 73, ext. 2769, fax +2770, telex UNIBO d 886657). *Crystal growth, crystal physics, X-ray topography, twinning, phase transitions.*

Klaska, Dr Karl-Heinz. Mineralog.-Petrogr. Inst., Universität, Grindelallee 48, D-2000 Hamburg 13, Germany, Fed. Rep. (1943) Dr rer.nat., mineralogy (U. Hamburg, 1974). Wiss. Mit. (tel. 040 + 41232063). *Organic and inorganic crystal structures, crystal chemistry.*

Klebe, Dr Gerhard, ZHV/D, BASF AG, D-6700 Ludwigshafen, Germany, Fed. Rep. (1954) Dr rer.nat., chemistry (U. Frankfurt, 1982). (tel. 0621 + 6041966). *Organic crystal structures, molecular modelling, structure prediction, structure and bonding.*

Klee, Prof. Dr Wilfrid Edgar. Inst. f. Kristallographie, Universität, Kaiserstr. 12, D-7500 Karlsruhe, Germany, Fed. Rep. (1935) Dr rer.nat., physical chemistry (U. Freiburg, 1960). Prof. (tel. 0721 + 608, ext. 2136). *Vibrational spectroscopy, apatites, graph theory.*

Klement, Dr Ulrich. FB Chemie/Pharmazie, Universität, Universitätsstr. 31, D-8400 Regensburg, Germany, Fed. Rep. (1930) Dr, crystallography (U. München, 1959). Akad. Dir. (tel. 0941 + 9434542) *Structure, inorganic and organic crystals.*

Klepp, Dr Kurt Otto. Inst. f. Anorg. Chemie, RWTH, Prof.-Pirlet-Str. 1, D-5100 Aachen, Germay, Fed. Rep. (1944) Ph.D., chemistry (U. Wien, Austria, 1971). Wiss. Angest. (tel. 0241 + 804643). *Structural chemistry, inorganic materials, solid state chemistry, mixed-valence compounds.*

Klessen, Mr Gerhard. Chemistry Division, KKW Gundremmingen Betr. Ges., Postfach 300, D-8871 Gundremmingen, Germany, Fed. Rep. (1949) Diplom, chemistry (RWTH Aachen, 1978). 2nd chemist. (tel. 08224 + 782192, telex 531143). *Intermetallic phases, corrosion products, oxide layers.*

Kniep, Prof Dr Rüdiger. Inst. f. Anorg. Chemie u. Strukturchemie, Universität, Universitätsstr. 1, D-4000 Düsseldorf, Germany, Fed. Rep. (1945) Dr rer.nat.habil., inorganic chemistry (U. Düsseldorf, 1978). Prof. (tel. 0211 + 311, ext. 3147). *Solid state chemistry, inorganic crystal structures.*

Knoch, Dr Falk A. Inst. f. Anorg. Chemie, Universität, Egerlandstr. 1, D-8520 Erlangen, Germany, Fed. Rep. (1953) Dr rer.nat., inorg. chemistry (U. Bonn, 1984). *Computers in chemistry, direct methods, crystal structures, phosphorous structural chemistry.*

Knöchel, Dr Claus-Dieter. V IC ELO, E.Merck, Frankfurter Str. 250, D-6100 Darmstadt, Germany, Fed. Rep. (1953) Dr rer.nat., physical chemistry (TH Darmstadt, 1979). (tel. 06151 + 72, ext. 3686, fax +3424, telex 419328-0). *Crystal growth, properties and applications.*

Knof, Dr Wolfgang Erich. Department of Physics, Kyushu University 33, Fukuoka 812, Japan. (1961) Dr rer.nat., crystallography (U. Saarbrücken, 1989). (fax 0081-92-631-4233). *Structure determination (theory).*

Knorr, Dr Klaus. Inst. f. Kristallographie, Universität, Charlottenstr. 33, D-7400 Tübingen, Germany, Fed. Rep. (1937) Dr rer.nat., physics (TU München, 1967). Akad. Oberrat. (tel. 07071 + 29, ext. 6058, fax +5400, telex 7-262867 utna d). *High pressures, instrumentation, neutron diffraction, phase transitions, powder diffraction, X-ray diffraction.*

Koch, Priv.-Doz. Dr Elke. Fachbereich Geowissenschaften, Inst. f. Mineralogie, Universität, Hans-Meerwein-Str., D-3550 Marburg/Lahn, Germany, Fed. Rep. (1943) Dr rer.nat.habil., crystallography (U. Marburg, 1987). Wiss. Ang. (tel. 06421 + 28, ext. 5610). *Mathematical crystallography, crystal chemistry, crystal physics, structure determination.*

Koch-Wallraf, Mrs Prof. Dr Maria. Am Prinzenrain 6, D-5300 Bonn, Germany, Fed. Rep. (1920) Dr rer.nat., physical chemistry (U. Bonn, 1950). Prof. *Inorganic crystal structures, computer programming, crystal growth.*

Kockel, Dr Andreas. Inst. f. Mineralogie, Ruhr-Universität, Universitätsstr. 150, D-4630 Bochum, Germany, Fed. Rep. (1932) Dr, mineralogy (U. Marburg, 1960). Akad. Oberrat. (tel. 0234 + 700, ext. 3514). *Thermal expansion, magnetic oxides formation.*

Koellner, Mrs Gertraud. Inst. f. Kristallographie, FU Berlin, Takustr. 6, D-1000 Berlin 33, Germany, Fed. Rep. (1960) Dipl.-Chem., chemistry (U. Marburg, 1988). (tel. 030 + 838, ext. 6325, fax 030-832-6561). *X-rax crystallography, proteins, structural chemistry, biology.*

König, Dr Burkhard. Bundesanstalt f. Geowiss. und Rohstoffe, Stilleweg 2, D-3000 Hannover 51, Germany, Fed. Rep. (1938) Dr rer.nat., crystallography, mineralogy (TU Clausthal, 1970). (tel. 0511 + 6468, ext. 652). *Inorganic crystal structures.*

Kokkinidis, Dr Michael. EMBL X-Ray Division, Meyerhof-Str. 1, D-6900 Heidelberg, Germany, Fed. Rep. (1952) Dr rer.nat., crystallography (TU München, 1981). Postdoc. (tel. 06221 + 387308). *Structure-activity relationships, biomolecules.*

Koller, Mr Hubert. Fakultät f. Chemie, Universität, Universitätsstr. 10, D-7750 Konstanz, Germany, Fed. Rep. (1962) (tel. 07531 + 882013).

Kopf, Dr Jürgen. Inst. f. Anorganische Chemie, Universität, Martin-Luther-King-Platz 6, D-2000 Hamburg 13, Germany, Fed. Rep. (1942) Dr rer.nat.habil., crystallography, X-ray structures analysis (U. Hamburg, 1973). Priv.-Doz. (tel. 040 + 4123, ext. 2897, email FC50060 at DHHUNI4). *X-ray diffraction, diffractometry, computing, structural chemistry, organic compounds.*

Kosten, Mr Klaus. Inst. f. Kristallographie, RWTH, Templergraben 55, D-5100 Aachen, Germany, Fed. Rep. (1949) Dipl., mineralogy (RWTH Aachen, 1978). *Phase transitions, solid state reactions, powder diffraction, synchrotron radiation.*

Krämer, Prof. Dr Volker. Kristallographisches Inst., Universität, Hebelstr. 25, D-7800 Freiburg, Germany, Fed. Rep. (1940) Dr rer.nat., (TH München, 1968). Prof. (tel. 0761 + 203, ext. 4277). *Crystal growth, thermal analysis, materials science, structure determination, inorganic crystal structures.*

Kramer, Dr Irmtraud. Batelle-Institut, Am Römerhof 35. D-6000 Frankfurt/Main, Germany, Fed. Rep. (1956) Dr rer.nat., physics (RWTH Aachen, 1982). Res. scient. (tel. 069 + 79082378). *Crystal growth (organic conductors).*

Krane, Mr Hans-Georg. Inst. f. Kristallographie, Dept. f. Physik, Universität, Im Stadtwald, D-6600 Saarbrücken, Germany, Fed. Rep. (1958) Dipl.-Min., crystallography (U. Saarbrücken, 1985). Grad. stud. (tel. 0681 + 3021). *Mineralogy, synchrotron radiation.*

Krause, Dr Christian. Kernforschungszentrum, Inst. f. nukleare Entsorgungstechnik, Postfach 3640, D-7500 Karlsruhe, Germany, Fed. Rep. (1955) Dr rer.nat. mineralogy (U. Münster, 1986). Wiss. Mit. (tel. 07247 + 824352). *Electron microscopy, materials science, metals, phase transitions.*

Krebs, Prof. Dr Bernt. Anorg.-Chem. Inst., Universität, Wilhelm-Klemm-Str. 8, D-4400 Münster, Germany, Fed. Rep. (1938) Dr rer.nat., chemistry (U. Göttingen, 1965). Prof. Inorganic Chemistry (tel. 0251 + 83, ext. 3131, fax +2090, telex 892529 unims d). *Structural chemistry, inorganic crystals, coordination compounds, chemical bonding, hydrogen bonding, macromolecules, electron density.*

Kreutz, Dr Ernst Wolfgang. Inst. f. Angew. Physik, Schlossgartenstr. 7, D-6100 Darmstadt, Germany, Fed. Rep. (1940) Dr rer.nat., physics (1969). Akad. Oberrat. (tel. 06151 + 161, ext. 2082). *Surface structure, laser lattices.*

Krogmann, Prof. Dr Klaus. Inst. f. Anorganische Chemie, Universität, Engesserstr., Geb. Nr. 30.45, Postfach 6380, D-7500 Karlsruhe, Germany, Fed. Rep. (1925) Dr, inorganic chemistry (U. Stuttgart, 1956). Prof. (tel. 0721 + 608, ext. 2980). *Inorganic crystal structures.*

Kroll, Prof. Dr Herbert. Inst. f. Mineralogie, Universität, Corrensstr. 24, D-4400 Münster, Germany, Fed. Rep. (1940) Dr rer.nat., mineralogy (U. Münster, 1971). Prof. (tel. 0251 + 83, ext. 3455). *Crystallography in earth sciences.*

Krüger, Prof. Dr Carl. Röntgenlabor, MPI f. Kohlenforschung, Lembkestr. 5, D-4330 Mülheim/Ruhr, Germany, Fed. Rep. (1933) Dr rer.nat., inorganic chemistry (RWTH Aachen, 1961). Prof., res. dir. (tel. 0208 + 306, ext. 487, fax +407, telex 856615 kofod, email KRUEGER at MPI-MUELHEIM.MPG.DBP.DE). *Chemical bonding, computing, crystallographic data, electron density, molecular crystals, reaction pathways, structural chemistry, molecular modelling.*

Krug, Prof. Dr Detlef. Laboratorium f. Anorganische Chemie, Universität, Auf der Morgenstelle 18, D-7400 Tübingen, Germany, Fed. Rep. (1936) Dr, inorganic chemistry (U. Tübingen, 1972). Prof. (tel. 07071 + 29, ext. 6227). *Inorganic solid state reactions, thermal analysis, X-ray investigations, chemical analysis.*

Kühlbrandt, Dr Werner. EMBL, Meyerhofstr. 1, D-6900 Heidelberg, Germany, Fed. Rep. (1951) PhD, molecular biologyy (U. Cambridge/U.K., 1981). Group leader. (tel. 06221 + 378-0, ext. 245, fax +306, telex 461613). *Electron microscopy, electron diffraction, structural biology, proteins.*

Kuehn, Prof. Dr Robert. Richard-Wagner-Str. 31, D-6901 Wilhelmsfeld, Germany, Fed. Rep. (1911) Dr phil., mineralogy (U. Kiel, 1938). Prof. (tel. 06220 + 8924). *Geosciences of salts and salt deposits.*

Küppers, Prof. Dr Horst. Mineralogisches Inst., Universität, Olshausenstr. 40, D-2300 Kiel, Germany, Fed. Rep. (1933) Dr rer.nat., physics (U. Freiburg, 1966). Prof. (tel. 0431 + 880, ext. 2897, email NMP26 at DKIUNI0). *Crystal physics, hydrogen bonding.*

Kuhs, Dr Werner Friedrich. Institut f. Kristallographie, Universität, Kaiserstr. 12, D-7500 Karlsruhe, Germany, Fed. Rep. (1952) Dr rer.nat., crystallography (U. Freiburg, 1978). Wiss. Ass. (tel. 0721 + 608, ext. 3317, fax +4290, telex 17-721166). *Hydrogen bonding, high pressure, instrumentation, neutron diffraction phase transitions, thermal vibrations.*

Kupčik, Prof. Dr Vladimir. Inst. f. Mineralogie und Kristallographie, Universität, V.M. Goldschmidtstr. 1, D-3400 Göttingen, Germany, Fed. Rep. (1934) Ph.D., mineralogy and crystallography (U. Bratislava, CSSR, 1965). Prof. (tel. 0551 + 39, ext. 3891). *Crystal structure analysis, crystal chemistry, sulfide compounds, synchrotron radiation, modulated structures.*

Kutoglu, Dr Ali. Fachbereich Geowissenschaften, Universität, Lahnberge, D-3550 Marburg, Germany, Fed. Rep. (1935) Dr rer.nat., geology (U. Kiel, 1963). Res. scient. (tel. 06421 + 28, ext. 2246). *Inorganic crystal structures, crystal structures, organometallic compounds.*

Labischinski, Dr Harald. Robert-Koch-Institut, Bundesgesundheitsamt, Nordufer 20, D-1000 Berlin, Germany, Fed. Rep. (1948) Dr rer.nat.habil., biocrystallography (FU Berlin, 1983). (tel. 030 + 4503406). *Biopolymer structure and function, small-angle X-ray scattering.*

Lacmann, Prof. Dr Ing. Rolf. Inst. f. Physikalische u. Theoretische Chemie, TU, Hans Sommer-Str. 10, D-3300/Braunschweig, Germany, Fed. Rep. (1927) Prof., physical chemistry (TU Braunschweig, 1974). Ord. Prof. (tel. 0531 + 391, ext. 5326, fax +4577, telex 0952526). *Crystal growth, nucleation, growth kinetic, growth form, thermodynamics of mixtures, materials science, phase transitions, surface structure, teaching.*

Ladenstein, Dr Rudolf. Abt. Strukturforschung, MPI f. Biochemie, Am Klopferspitz, D-8033 Martinsried, Germany, Fed. Rep. (1943) Dr, biochemistry Wiss. Mitarb. (tel. 089 + 8578, ext. 2704). *Crystal growth, macromolecules, structural biology, viruses, X-ray diffraction.*

Lamm, Mr Viktor Andreas. Abt. f. Strukturforschung, MPI f. Biochemie, Am Klopferspitz, D-8033 Martinsried, Germany, Fed. Rep. (1950) Dipl.-Phys., (TU München, 1977). (tel. 089 + 8585, ext. 660). *Structure analysis.*

Langbein, Prof Dr Werner Dieter. Battelle-Institut, Wiesbadenerstr., D-6000 Frankfurt/Main, Germany, Fed. Rep. (1932) Hon.Prof., physics (U. Frankfurt, 1972). Sen.Res.Sci. (tel. 069 + 79082539). *Convective heat and mass transport theory, stability of multicomponent systems.*

Lange, Mr Joachim Reinhard. Inst. f. Angew. Physik, Lst. f. Kristallographie, Universität, Bismarckstr. 10, D-8520 Erlangen, Germany, Fed. Rep. (1961) Dipl.-Phys., physics (U. Erlangen-Nürnberg, 1987). Wiss.Mit. (tel. 09131 + 852119). *Computing, diffractometry.*

Langer, Prof Dr Klaus. Inst. f. Mineralogie u. Kristallographie, TU, Ernst-Reuter-Platz 1, D-1000 Berlin 12, Germany, Fed. Rep. (1936) Dr rer.nat., inorg. chemistry (U. Kiel, 1965). Prof. (tel. 030 + 31425325). *Minerals, high pressures, spectroscopy, defect structures, disorder, hydrogen bonding.*

Lauck, Mr Rudolf. Kristall- und Materiallabor, Fak. f. Physik, Universität, Kaiserstr. 12, D-7500 Karlsruhe, Germany, Fed. Rep. (1947) Diplom, physics (U. Saarbrücken, 1974). Wiss. Mit. (tel. 0721 + 608, ext. 3551). *Crystal growth, characterization.*

Lehmann, Mr Christian Wolfgang. Inst. f. Kristallographie, FU Berlin, Takustr. 6, D-1000 Berlin 33, Germany, Fed. Rep. (1964) Dipl.-Chem., chemistry (FU Berlin, 1988). (tel. 030 + 838, ext. 3460, fax +6705, CHRISTIAN at KRISTALL. CHEMIE. FU-BERLIN. DBP. DE.). *Synchrotron radiation, deformation density.*

Lehmann, Prof. Dr Gerhard Rudolf. Inst. f. Physikalische Chemie, Universität, Schlossplatz 4, D-4400 Münster, Germany, Fed. Rep. (1935) Dr rer.nat., physical chemistry (U. Münster, 1963). Prof. (tel. 0251 + 83, ext. 3427). *Solid state spectroscopy, crystal growth, color centers, defects in solids, ligand field theory.*

Lehmpfuhl, Dr Gunter. Fritz-Haber-Inst., Max-Planck-Ges., Faradayweg 4-6, D-1000 Berlin 33, Germany, Fed. Rep. (1928) Dr, physics (FU Berlin, 1961). Wiss. Oberasst. (tel. 030 + 8305, ext. 261). *Electron diffraction, structure potential measurement, electron microscopy, direct observation of atomic steps.*

Lenz, Mrs Andrea. Inst. f. Kristallographie, FU Berlin, Takustr. 6, D-1000 Berlin 33, Germany, Fed. Rep. (1957) assessor, chemistry/mathematics (U.Heidelberg, 1984). (tel. 030 + 838, ext. 3456, fax 030-832-6561). *Proteins, chemical bonding, X-ray crystallography, computing, structural chemistry and biology.*

Lerf, Dr Anton Eduard. Zentralinstitut f. Tieftemperatur-Forschung, Bayerische Akademie der Wissenschaften, D-8046 Garching, Germany, Fed. Rep. (1948) Dr, chemistry (U. München, 1976). Wiss. Mit. (tel. 089 + 3209, ext. 5218). *Crystal chemistry, superconductors, organic solid state chemistry, intercalation reactions.*

Leute, Prof. Dr Volkmar. Inst. f, Physikalische Chemie, Universität, Schlossplatz 4, D-4400 Münster, Germany, Fed. Rep. (1938) Dr rer.nat., physical chemistry (U. München, 1969). Prof. (tel. 0251 + 83, ext. 3431). *Chalcogenide semiconductors, superstructures, phase diagrams, defects.*

Lewis, Dr James jr. Scientific and Technical Team, Army Material Command-STITEUR, IG-Farben-Hochhaus room 740, Grüneburgplatz, D-6000 Frankfurt/Main 1, Germany, Fed. Rep. (1940) PhD, crystallography (TU Clausthal, 1978). Material scient. (tel. 069 + 151, ext. 8263). *Crystal chemistry, electron density distribution, structure and physical properties, crystal physics, crystal defects.*

Liebau, Prof. Dr Friedrich. Mineralogisches Inst., Universität, Olshausenstr. 40, D-2300 Kiel, Germany, Fed. Rep. (1926) Dr rer.nat., chemistry (Humboldt U. Berlin, 1956). Prof.C4 (tel. 0431 + 880, ext. 2888). *Inorganic crystal chemistry (silicates, phosphates), minerals, materials science, phase transitions.*

Liebertz, Prof. Dr Josef. Inst. f. Kristallographie, Universität, Zülpicher Str. 49, D-5000 Köln 1, Germany, Fed. Rep. (1929) Dr rer.nat., mineralogy (U. Bonn, 1961). Prof. (tel. 0221 + 470, ext. 4420). *Crystal growth, crystal chemistry.*

Limper, Mr Wolfram. Inst. f. Kristallographie, Universität, Charlottenstr. 33, D-7400 Tübingen, Germany, Fed. Rep. (1961) Dipl.-Phys., physics (U. Kiel, 1988). Wiss.Ang. (tel. 07071 + 296389). *Instrumentation, phase transitions, powder diffraction, synchrotron radiation, X-ray diffraction.*

Lindemann, Prof. Dr Willi. Lehrstuhl f. Kristallstrukturlehre, Universität, Am Hubland, D-8700 Würzburg, Germany, Fed. Rep. (1921) Dr phil.nat., crystallography (U. Erlangen, 1951). Prof. (tel. 0931 + 881, ext. 430). *Mathematical crystallography, structure analysis, computer programming.*

Lipka, Mrs Dr Annegret. Analytisch-Chem. Labor, Sachtleben Bergbau GmbH, Meggener Str., D-5940 Lennestadt, Germany, Fed. Rep. (1948) Dr rer.nat., chemistry (U. Münster, 1975). Wiss. Ass. (tel. 02721 + 835275). *Crystal chemistry, antimony(III) compounds, stereochemistry.*

Löchner, Dr Ulrich. Inst. f. Kristallographie u. Mineralogie der Universität Frankfurt am DESY, HASYLAB, Notkestr. 85, D-2000 Hamburg 52, Germany, Fed. Rep. (1948). Dr chemistry (U. Karlsruhe, 1980). Res. assoc. (tel. 040 + 8998, ext. 2653, fax +3282). *Rare earths, catalysts, multiphase mixtures, textures, solid state, high temperature chemistry, synchrotron radiation.*

Löns, Dr Jürgen. Fachbereich Chemie, Universität, Corrensstr. 24, D-4400 Münster, Germany, Fed. Rep. (1939) Dr, crystallography (U. Hamburg, 1969). Akad. Oberrat. (tel. 0251 + 83, ext. 3456). *Crystallography, structure determination, crystal chemistry.*

Lorenz, Prof. Dr Wolfgang J. Inst. f. Physikalische Chemie und Elektrochemie, Universität, Kaiserstr. 12, D-7500 Karlsruhe, Germany, Fed. Rep. (1933) Prof., physical chemistry (TU, Clausthal-Cellerfeld, 1961). Prof. (tel. 0721 + 6083303). *Physics and chemistry of surfaces and interfaces, electrocrystallization.*

Luger, Prof. Dr Peter. Inst. f. Kristallographie, FU, Takustr. 6, D-1000 Berlin 33, Germany, Fed. Rep. (1943) Dr habil., crystallography (FU Berlin, 1974). Prof. (tel. 030 + 838, ext. 3411). *X-ray and neutron single crystal analysis, conforma-*

tional analysis, carbohydrates, drug design, computer graphics, low temperature diffraction.

Lutz, Prof. Dr Heinz Dieter. Anorganische Chemie I, Universität Siegen, Adolf-Reichwein-Strasse, D-5900 Siegen, Germany, Fed. Rep. (1934) Dr habil., inorganic chemistry (U.Köln, 1967). Prof. (tel. 0271 + 7404217). *Inorganic crystals, structural chemistry, X-ray diffraction, neutron diffraction, hydrogen bonding, phase transitions.*

Maas, Prof. Dr Gerhard. Department of Chemistry, Universität Kaiserslautern, Erwin-Schrödinger-Strasse, D-6750 Kaiserslautern, Germany, Fed. Rep. (1949) Dr, organic chemistry (U.Saarbrücken, 1974). Prof. (tel. 0631 + 2052957). *Organic compounds.*

Mages, Dr Gert Rudolf. FL ALE 3, Siemens AG, Guenther-Scharowsky-Str. 2, D-8520 Erlangen, Germany, Fed. Rep. (1939) Dr rer.nat., mineralogy, petrology (U. Erlangen-Nürnberg, 1970). Scient. (tel. 09131 + 22858). *X-ray analysis, small angle X-ray scattering, inorganic crystal structures.*

Maier, Dr Horst. AEG, FB Infrarot- u. Nachtsichtkomponenten, Theresienstr. 2, D-7100 Heilbronn, Germany, Fed. Rep. *Crystal growth, materials science, semiconductors.*

Mandelkow, Dr Eckard. Dept. of Biophysics, MPI f. Medical Research, Jahnstr. 29, D-6900 Heidelberg, Germany, Fed. Rep. (1943) Dr physics (U. Heidelberg, 1973). Res. scient. (tel. 06221 + 486277). *Fibre diffraction, small angle scattering, synchrotron radiation, electron microscopy, image reconstruction.*

Marie, Dr Alain Louis. Abt. f. Strukturforschung I, MPI f. Biochemie, Am Klopferspitz, D-8033 Martinsried, Germany, Fed. Rep. (1946) Ph.D., biochemistry (U. Sherbrooke, Quebec, Canada, 1973). Wiss. Ang. (tel. 089 + 8585, ext. 682). *Enzymology, protein sequencing, protein chemistry, crystallography.*

Mariolacos, Dr Konstantin. Mineralogisch-Kristallogr. Inst., Universität, V.M. Goldschmidtstr. 1, D-3400 Göttingen, Germany, Fed. Rep. (1936) Dr, crystal structures (U. Wien, 1972). Techn. Ang. (tel. 0551 + 39, ext. 3931). *Phase diagrams, crystal synthesis, crystal structures, sulpho-salts, sulphohalogenides.*

Marler, Mr Bernd. Mineralog. Inst., Universität, Olshausenstr. 40, D-2300 Kiel, Germany, Fed. Rep. (1956) Dipl.-Min., mineralogy (U. Kiel, 1985). Ass. (tel. 0431 + 880, ext. 2226). *Structural chemistry, zeolites, powder diffraction.*

Martin, Mrs Brigitte. Inst. f. Mineralogie, Universität, Universitätsstr. 150, D-4630 Bochum, Germany, Fed. Rep. (1957) Dipl.-Min., crystallography (U. Bochum, 1987). (tel. 0234 + 700, ext. 3881). *Microcrystalline SiO2, quartz, HTEM, FTIR.*

Massa, Prof. Dr Werner. Fachbereich Chemie, Universität, Hans-Meerwein-Str., D-3550 Marburg, Germany, Fed. Rep. (1944) Priv. Doz., inorganic chemistry (U. Marburg, 1982). Akad.Oberrat. (tel. 06421 + 28, ext. 5525). *Structure determination, inorganic solids, fluorometallates, hydrogen bonds, magnetism.*

Mateika, Dr Dieter. Philips GmbH Forschungslaboratorium, Postfach 540840, Vogt-Kölln-Str. 30, D-2000 Hamburg 54, Germany, Fed. Rep. (1935) Dr rer.nat., mineralogy (U. Bonn, 1969). Scient. (tel. 040 + 5493, ext. 553, telex 21331656). *Crystal growth.*

Mattes, Prof. Dr Rainer. Inst. f. Anorganische Chemie, Universität, Wilhelm-Klemm-Str. 8, D-4400 Münster, Germany, Fed. Rep. (1937) Dr habil., inorganic chemistry (U. Münster, 1970). Prof. (tel. 0251 + 83, ext. 3117). *Inorganic crystal structures, structure, oxo and thio complexes.*

Matz, Prof. Dr Günther. Ingenieurabteilung AP, Verfahrenstechnik, Bayer AG, Friedrich Ebert-Str. 217-319, D-5600 Wuppertal-Elberfeld, Germany, Fed. Rep. (1920) Dr, process eng. (U. Frankfurt/Main, 1950). Dept. head. (tel. 0202 + 36, ext. 7618). *Industrial crystallization, crystallization from solution, kinetics, crystallization.*

Matzat, Dr Eckhart. Kristallogr. Inst., Universität, Goldschmidtstr. 1, D-3400 Göttingen, Germany, Fed. Rep. (1938) Dr rer.nat., mineralogy (U. Göttingen, 1968). (tel. 0551 + 39, ext. 3893). *Structure determination, modulated structures.*

Mayer, Dr Hugo Werner Waldemar. IAK 1, Kernforschungszentrum Karlsruhe, Postfach 3640, D-7500 Karlsruhe, Germany, Fed. Rep. (1943) Dr rer.nat., chemistry (U. Stuttgart, 1979). Wiss. Ang. (tel. 07247 + 82, ext. 3438). *Crystal structures, phase transitions, elastic neutron scattering.*

Meier, Prof. Dr Hans. Staatl. Forschungsinstitut f. Geochemie, Concordiastr. 28, D-8600 Bamberg, Germany, Fed. Rep. (1927) Dr rer.nat., physical chemistry (U. Mainz, 1954). Inst. head. (tel. 0951 + 27280). *Organic semiconductors.*

Melzer, Mr Rolf. Inst. f. Mineralogie u. Kristallographie, BH1, TU Berlin, Ernst-Reuter-Platz 1, D-1000 Berlin, Germany, Fed. Rep. (1960) Dipl.-Min., crystallography (U. Hannover, 1987). Wiss. Mit. (tel. 030 + 314, ext. 26177, email 1171 at DB0TU401). *Defect structure, inorganic crystals, phase transitions.*

Mertin, Dr Wilhelm. Inst. f. Anorg. u. Analyt. Chemie, Universität, Heinrich-Buff-Ring 58, D-6300 Giessen, Germany, Fed. Rep. (1938) Dr rer.nat., inorg. chemistry (U. Münster, 1967). Akad. Oberrat. (tel. 0641 + 7025676). *Transmission electron microscopy, structural chemistry, materials science, analytical electron microscopy.*

Messerschmidt. Dr Albrecht. Strukturforschung II, MPI f. Biochemie, Am Klopferspitz 18 A, D-8033 Martinsried, Germany, Fed. Rep. (1945) Dr rer.nat., crystallography (Humboldt-U. Berlin, 1971). Res. scientist. (tel. 089 + 8578, ext. 2662, email MESSERSC at DM0MPB51). *Proteins, X-ray diffraction, diffractometry, macromolecules, structural biology.*

Messner, Dr Dieter. Physik-Labor, Kalle Niederlassung der Hoechst AG, Rheingaustr. 190, D-6200 Wiesbaden, Germany, Fed. Rep. (1928) Dr rer.nat.,

physics (TH Stuttgart, 1956). Physicist. (tel. 06121 + 68, ext. 6921). *Physics, high and low molecular weight materials, X-ray structures.*

Metcalf, Dr Peter. EMBL, Postfach 10.2209, Meyerhofstr. 1, D-6900 Heidelberg, Germany, Fed. Rep. (1952) PhD, electron microscopy (U. Auckland, New Zealand, 1981). Group leader. (tel. 06221 + 387355, email METCALF at EMBL). *Parasite surface antigens, proteins, viruses.*

Metter, Mr Joachim. Inst. f. Kristallographie, Universität, Senckenberganlage 30, D-6000 Frankfurt/Main 1, Germany, Fed. Rep. (1955) Dipl., inorg. chemistry (U. Würzburg, 1983). (tel. 069 + 7982105). *Rare earths, X-ray and neutron diffraction, group V elements, computer aided chemistry.*

Mewis, Prof. Dr Albrecht. Inst. f. Anorganische Chemie, Universität, Greinstr. 6, D-5000 Köln 41, Germany, Fed. Rep. (1942) Prof., inorg. chemistry (U. Köln, 1986). Prof. (tel. 0221 + 4703267). *Structural chemistry, crystallographic data, X-ray diffraction.*

Meyer, Dr Gerd Heinrich. Inst. f. Anorganische Chemie, Universität, Callinstr. 9, D-3000 Hannover, Germany, Fed. Rep. (1949) Dr rer.nat., inorganic chemistry (U. Giessen, 1976). Full Prof. (tel. 0511 + 762, ext. 3696, fax +3456). *Inorganic solid state chemistry, synthesis, structures.*

Meyer, Prof. Dr Hans-Jürgen. Mineralog. Inst., Universität, Poppelsdorfer Schloss, D-5300 Bonn, Germany, Fed. Rep. (1927) Dr habil., mineralogy and crystallography (U. Bonn, 1962). Wiss. Rat und Prof. (tel. 0228 + 73, ext. 2771). *Kinetics, crystal growth, physicochemical crystallography.*

Meyer-Ehmsen, Prof. Dr Gerhard. Fachbereich 4, Universität, Postfach 4469, D-4800 Osnabrück, Germany, Fed. Rep. (1932) Dr rer.nat., physics (U. Hamburg, 1961). Prof. (tel. 0541 + 608, ext. 2435). *Electron diffraction, surface science.*

Milius, Dr Wolfgang. MPI f. Festkörperforschung, Universität, Heisenbergstr. 1, D-7000 Stuttgart 80, Germany, Fed. Rep. (1958). Dr rer.nat., chemistry (U. Erlangen-Nürnberg). (tel. 0711 + 6860668).

Möller, Dr Manfred. Anorg. Chem. Inst., Universität, Corrensstr. 36, D-4400 Münster, Germany, Fed. Rep. (1953) Dr rer.nat., chemistry (U. Dortmund, 1983). Wiss. Ass. (tel. 0251 + 833116). *Structure and properties, inorganic and intermetallic crystals.*

Moh, Prof. Dr Günter Harald. Mineralogisch-Petrogr. Inst., Universität, Im Neuenheimer Feld 236, Postfach 104040, D-6900 Heidelberg, Germany, Fed. Rep. (1929) Dr habil., mineralogy and petrology (U. Heidelberg, 1967). Prof. (tel. 06221 + 562810). *Experimental mineralogy, sulfide petrology, crystal chemistry.*

Mootz, Prof. Dr Dietrich. Inst. f. Anorg. Chemie und Strukturchemie, Universität, Universitätsstr. 1, D-4000 Düsseldorf, Germany, Fed. Rep. (1933) Dr rer.nat., chemistry (TU Berlin, 1959). Prof. (tel. 0211 + 311, ext. 3135). *Inorganic and organic crystal structures, solid state chemistry, low temperatures, hydrogen bonding.*

Morgenroth, Mr. Wolfgang H. Inst. f. Mineralogie, Universität, Poppelsdorfer Schloss, D-5300 Bonn 1, Germany, Fed. Rep. (1962) cand.min., mineralogy and crystallography (tel. 0228 + 73, ext. 2761). *Diffractometry, phase transitions, powder diffraction, symmetry, X-ray diffraction.*

Moritz, Prof. Dr Wolfgang Otto. Inst. f. Kristallographie und Mineralogie, Universität, Theresienstr. 41, D-8000 München 2, Germany, Fed. Rep. (1943) Dr rer.nat.habil., Kristallographie (U. München, 1976). Prof. (tel. 089 + 2394, ext. 4336). *Surface structure, diffuse scattering, electron and X-ray diffraction.*

Müller, Prof. Dr Gerd. Inst. f. Mineralogie, TH, Schnittspahnstr. 9, D-6100 Darmstadt, Germany, Fed. Rep. (1942) Dr rer.nat., mineralogy (U. Karlsruhe, 1969). Prof. (tel. 06151 + 165280). *Crystal chemistry, material science, mineralogy.*

Müller, Prof. Dr Gerhard. Anorg.-chem. Inst., TU München, Lichtenbergstr. 4, D-8046 Garching, Germany, Fed. Rep. (1953) Dr rer.nat.habil., inorganic and analytical chemistry (TU. München, 1989). Priv.-Doz. (tel. 089 + 3209, ext. 3133, fax +2727). *Structural chemistry, organometallic chemistry, unusual bonding modes, weak interactions in chemistry, small molecule crystallography.*

Müller, Prof. Horst. Chemisches Laboratorium, Universität, Albertstr. 21, D-7800 Freiburg, Germany, Fed. Rep. (1929) Dr, inorganic chemistry (U. Freiburg, 1958). Prof. (tel. 0761 + 203, ext. 2915). *Solid state chemistry, radio and nuclear chemistry, hydrides.*

Müller, Dr Paul Hubert. Inst. f. Anorg. Chemie, RWTH, Prof.-Pirlet-Str. 1, D-5100 Aachen, Germany, Fed. Rep. (1951) Dr, inorg. chemistry (RWTH Aachen, 1980). (tel. 0241 + 804669). *Inorganic chemistry, neutron diffraction, magnetochemistry.*

Müller, Prof. Dr Ulrich. Fachbereich Chemie, Universität, Hans-Meerwein-Str., D-3550 Marburg, Germany, Fed. Rep. (1940) Dr, chemistry (U. Stuttgart, 1966). Prof. (tel. 06421 + 285686). *Structural chemistry, inorganic compounds, group theory, diffuse scattering and disorder.*

Müller, Prof. Dr Wolfgang Friedrich. Inst. f. Mineralogie, Techn. Universität, Schnittspahnstr. 9, D-6100 Darmstadt, Germany, Fed. Rep. (1939) Dr rer.nat., mineralogy (U. Tübingen, 1965). Prof. (tel. 06151 + 16, ext. 3380). *Crystal chemistry, petrography, meteorites, crystal defects, electron microscopy.*

Müller-Buschbaum, Prof. Dr Hanskarl. Inst. f. Anorganische Chemie, Universität, Olshausenstr. 40-60, D-2300 Kiel, Germany, Fed. Rep. (1931) Ord. Prof., inorganic chemistry (U. Kiel, 1969). Direktor. (tel. 0431 + 880, ext. 2410). *Solid state reactions, plasma and laser chemistry, high temperatures, solid state chemistry, inorganic crystal structures.*

Müller-Fahrnow, Mrs. Anke. Inst. f. Kristallographie, FU Berlin, Takustr. 6, D-1000 Berlin, Germany, Fed. Rep. (1960) Dipl., biochemistry (FU Berlin, 1985). (tel. 030 + 838, ext. 6599). *Proteins, macromolecules.*

Müller-Vogt, Dr German. Kristall- und Materiallabor, Fak. f. Physik, Universität, Kaiserstr. 12, D-7500 Karlsruhe, Germany, Fed. Rep. (1943) Dr rer.nat., physics (U. Karlsruhe, 1971). Leiter des Kristall- und Materiallabors. (tel. 0721 + 608, ext. 3470). *Crystal growth and characterization.*

Müllner, Dr Manfred. Inst. f. Kernphysik, Universität, August-Euler-Str. 6, D-6000 Frankfurt, Germany, Fed. Rep. (1928) Dr, physics (1966). (tel. 0611 + 798, ext. 4238). *Solid state physics, lattice dynamics, structure determination.*

Münninghoff, Dr Günter. E689, SIEMENS AG, Östliche Rheinbrückenstr. 50, D-7500 Karlsruhe 21, Germany, Fed. Rep. (1953) Dr crystallography (1978). Product manager single crystal diffractometry. (tel. 0721 + 595, ext. 6574, fax +4506, telex 78255-69). *Computing, coordination compounds, diffractometry, inorganic crystals, instrumentation, macromolecules, materials science, minerals, molecular crystals, neutron diffraction, organic compounds, polymers, proteins, small-angle scattering, structural chemistry, X-ray diffraction.*

Mundt, Dr Otto. Inst. f. Anorg. Chemie, Universität, Pfaffenwaldring 55, D-7000 Stuttgart, Germany, Fed. Rep. (1950) Dr rer.nat., chemistry (U. Karlsruhe, 1979). Akad.Rat. (tel. 0711 + 6854221). *Structural chemistry, organic compounds of main group elements.*

Murad, Dr Enver. Lehrstuhl f. Bodenkunde, TU München, D-8050 Freising-Weihenstephan, Germany, Fed. Rep. (1941) Dr phil.nat., mineralogy (U. Frankfurt, 1970). Wiss. Mit. (tel. 08161 + 71, ext. 735). *Crystal chemistry, Mössbauer spectroscopy.*

Mutter, Mrs Graciela. Electron Microscopy Group, MPI f. Metallforschung, Inst. f. Werkstoffwissenschaften, Seestr. 92, D-7000 Stuttgart 1, Germany, Fed. Rep. (1955) Dipl., mineralogy (U. Heidelberg, 1983). *Ceramics, cordierite, grain boundaries.*

Nägele, Dr Walter. Inst. f. Kristallographie, Universität, Charlottenstr. 33, D-7400 Tübingen, Germany, Fed. Rep. (1949) Dr rer.nat., physics (U. Tübingen, 1979). Wiss. Mit. (tel. 07247 + 82, ext. 3158). *Crystal structures, magnetic structures, neutron scattering, spin glasses.*

Neder, Dr Reinhard Bernhard. Inst. f. Kristallographie, Universität, Theresienstr. 41, D-8000 München 2, Germany, Fed. Rep. (1959) Dr rer.nat., crystallography (U. München, 1989). *Diffuse scattering, defect structures, electron diffraction, neutron diffraction.*

Neff, Prof. Dr Hans Josef. Abt. E63, Siemens AG, Rheinbrückenstr. 50, D-7500 Karlsruhe, Germany, Fed. Rep. (1920) Dr rer.nat., physics (U. Karlsruhe, 1951). Manager of application labs. (tel. 0721 + 595, ext. 2656). *Electron microscopy, X-ray diffraction, X-ray fluorescence analysis, gas and HP liquid chromatography.*

Neifeind, Dipl-Ing Axel. IBM, Dept. 0348, 7032-16, Tübinger Allee, D-7032 Sindelfingen, Germany, Fed. Rep. (1941) Dipl.-Ing., metallurgy (U. Stuttgart, 1969). Staff eng. (tel. 07031 + 611, ext. 3114). *Crystal structures, semi conductors, crystal defects, surface analysis, orientation methods, X-ray topography.*

Nelkowski, Prof. Dr Horst. Inst. f. Festkörperphysik, TU, Strasse des 17. Juni 135, D-1000 Berlin 12, Germany, Fed. Rep. (1921) Dr, experimental physics (TU Berlin, 1970). Dir. (tel. 030 + 314, ext. 2247). *Solid state physics, crystal growth (esp. II-VI- and I-III-V I2-compounds), surface physics, semiconductor physics, photoconductivity, luminescence (electroluminescence).*

Nesper, Dr Reinhard Friedrich. MPI f. Festkörperforschung, Heisenbergstr. 1, D-7000 Stuttgart, Germany, Fed. Rep. (1949) Dr rer.nat., chemistry (tel. 0711 + 6860320). *Solid state chemistry, crystallography, structure and properties, electronic structure of solids.*

Neubüser, Prof. Joachim Franz Friedrich Gerhard. Lehrstuhl D f. Mathematik, RWTH, Templergraben 64, D-5100 Aachen, Germany, Fed. Rep. (1932) Dr rer.nat., mathematics (U. Kiel, 1957). Prof. (tel. 0241 + 80, ext. 4543). *Crystallographic groups, computational group theory.*

Newesely, Prof. Dr Heinrich. Inst. f. Mineralogie und Kristallographie, TU, Ernst-Reuter-Platz 1, D-1000 Berlin, Germany, Fed. Rep. (1933) Dozent, crystal chemistry and micromorphology (TU Berlin, 1964). Prof. (tel. 030 + 314, ext. 2746). *Crystal chemistry, micromorphology (electron microscopy), biocrystallography, fine particle dusts research.*

Nitsche, Prof. Dr Rudolf. Kristallographisches Inst., Universität, Hebelstr. 25, D-7800 Freiburg, Germany, Fed. Rep. (1922) Ph.D., physical chemistry (U. Heidelberg, 1951). Em.Prof. (tel. 0761 + 203, ext. 4280). *Crystal growth, crystal chemistry, phase relations, crystal physics.*

Nitschmann, Dr Günter Max Alfred. Am Entenspiel 1, D-6330 Wetzlar, Germany, Fed. Rep. (1914) Dr, crystallography, mineralogy (U. Breslau, 1940). (tel. 06441 + 23748). *Crystal growth.*

Noltemeyer, Dr Mathias Rolf. Inst. f. Anorg. Chemie, Universität, Tammannstr. 4, D-3400 Göttingen, Germany, Fed. Rep. (1947) Dr, chemistry (U. Göttingen, 1977). Wiss. Angest. (tel. 0551 + 391). *X-ray apparatus.*

Nover, Dr Georg. Inst. f. Mineralogie, Universität, Poppelsdorfer Schloss, D-5300 Bonn 1, Germany, Fed. Rep. (1948) Dr rer.nat., mineralogy (U. Bonn, 1979). Wiss. Ang. (tel. 0228 + 73, ext. 2732). *Crystal structure analysis, inorganic and metallic phases.*

Nowack, Mrs Ellen Carla. Inst. f. Kristallographie, RWTH, Templergraben 55, D-5100 Aachen, Germany, Fed. Rep. (1956) Dipl.-Min., crystallography (1983). Wiss. Ang. (tel. 0241 + 806909, email KRISTA at DACTH01). *Electron density, chemical bonding, computing.*

Oefner, Mr Christian. X-Ray Division, EMBL, Meyerhofstr. 1, D-6900 Heidelberg, Germany, Fed. Rep. (1956) Dipl., physics (U. Heidelberg, 1982). Grad. student. (tel. 06221 + 387308). *Structure - activity relationships, biomolecules.*

Okrusch, Prof. Dr Martin. Mineralogisches Inst., Universität, Am Hubland, D-8700 Würzburg, Germany, Fed. Rep. (1934) Habilitation, mineralogy (U. Würzburg, 1968). Prof. *Rock forming minerals.*

Otten, Mr Peter. Gilbertstr. 27, D-2000 Hamburg, Germany, Fed. Rep. (1938) Dipl.-Min., crystallography (U. Bonn, 1967). Scientific consultant research vessel. (tel. 040 + 3193302). *X-ray diffractometry, phase transitions, instrument development.*

Otto, Dr Hans Hermann. Kristalllabor der NWF II-Physik, Universität, Universitätsstr. 31, Postfach 397, D-8400 Regensburg 2, Germany, Fed. Rep. (1938) Dr rer.nat., crystallography, mineralogy (TU Berlin, 1970). Priv.Doz., senior scientist of the crystal lab. (tel. 0941 + 943, ext. 2053). *Crystal structure analysis, physical properties of crystals, ferroics, high-Tc superconductors, quasicrystals.*

Overkott, Dr Paul Engelbert. Keramisches Laboratorium, Dr C. Otto und Comp. GmbH, Bochum, Nachtigallenstr. 30, D-5820 Gevelsberg, Germany, Fed. Rep. (1926) Dr rer.nat., mineralogy (U. Köln, 1959). Head, ceramic lab. (tel. 0234 + 4191, ext. 244). *Industrial ceramics, refractories, X-ray diffraction, REM-EDAX.*

Pachali, Dr Klaus Erich. Analytisches Labor, Deutsche Akzo Coatings GmbH, Magirusstr. 26, D-7000 Stuttgart 30, Germany, Fed. Rep. (1943) Dr rer.nat., inorganic chemistry (U. Stuttgart, 1970). Wiss. Mit. (tel. 0711 + 895, ext. 379). *Structures, ternary chalkogenides, polyphosphides, chelate complexes, inorganic crystal structures.*

Pai, Dr Emil Friedrich. Dept. of Biophysics, MPI f. Med. Res., Jahnstr. 29, D-6900 Heidelberg, Germany Fed. Rep. (1950) Dr rer.nat., chemistry (U. Heidelberg, 1978). Group leader (tel. 06221 + 486, ext. 275, fax +351, telex. 461505, email KABSCH at DHDEMBL5). *Macromolecules, structural biology, proteins, X-ray diffraction.*

Pal, Dr Gour Pada, Inst. f. Kristallographie, FU, Takustr. 6, D-1000 Berlin 33, Germany, Fed. Rep. (1953) PhD, protein chemistry (Calcutta U., India, 1980). Wiss. Mit. (tel. 030 + 838-6325). *Proteins, crystal growth, X-ray crystallography.*

Pannhorst, Dr Wolfgang. Res. and Dev., SCHOTT Glaswerke, Hattenbergstr. 10, D-6500 Mainz, Germany, Fed. Rep. (1942) Dr rer.nat.habil., crystallography (U. Karlsruhe, 1980). General manager. (tel. 06131 + 663676, telex. 4187920). *Glasses, glass ceramics, optical glasses, devitrification, nucleation, crystal growth, fibres, integrated optics, radiation defects, photochromism.*

Paulitsch, Prof. Dr Peter. Inst. f. Mineralogie, TH, Schnittspahnstr. 9, D-6100 Darmstadt, Germany, Fed. Rep. (1922) Dr rer.nat., mineralogy (U. Graz, Austria, 1944). Full prof. (tel. 06151 + 16, ext. 2180). *Experimental petrofabrics.*

Paulus, Dr Erich Friedrich. Angewandte Physik, Hoechst AG, Postfach 800320, D-6230 Frankfurt/Main 80, Germany, Fed. Rep. (1937) Dr, chemistry (U. München, 1965). Scient. (tel. 0611 + 3051, ext. 6360). *Crystal structures, organic and inorganic compounds, powder diffraction, fibres, high polymer structures.*

Pense, Prof. Dr Karl Eduard Jürgen. Institut f. Geowissenschaften, Universität, Postfach 3980, D-6500 Mainz, Germany, Fed. Rep. (1931) Prof., mineralogy, crystallography (U. Mainz, 1965). Prof. (tel. 06131 + 392256 or 381756, telex 4187408pensd). *Electron microscopy and diffraction, precious stones, crystal physics.*

Pentinghaus, Dr Horst. Dept. of Chemistry, Inst. f. Nuclear Waste Technology, KFZ, Postfach 3640, D-7500 Karlsruhe 1, Germany, Fed. Rep. (1932) Dr rer.nat., mineralogy (U. Münster, 1970). Head of the department. (tel. 07247 + 824476, telex 7826484). *Experimental crystallography, crystal chemistry.*

Petcov, Mr Alexe. FR Kristallographie, Universität d. Saarlandes, HASYLAB/DESY, Notkestr. 85, D-2000 Hamburg 52, Germany, Fed. Rep. (1952) Dipl.-Phys., solid state physics (U. Bucharest, Romania, 1976). Wiss.Mit. (tel. 040 + 8998, ext. 2601, fax +3282, telex 215124 desy d). *X-ray diffraction, scattering theory, instrumentation, diffractometry.*

Peterat, Dr Michael. HELIOTRONIC GmbH, Johannes-Hess-Str. 24, D-8263 Burghausen, Germany, Fed. Rep. (1946) Dipl., mineralogy (TU München, 1974). (tel. 08677 + 834026). *Unconventional crystallization techniques from Si-melt, solar-cells, order-disorder phenomena.*

Peters, Dr Dieter. Geschftsbereich A, HOECHST AG, Industriestrasse, D-5030 Hue.rth-Knapsack, Germany, Fed. Rep. (1956) Dr rer.nat., chemistry (tel. 02233 + 67203). *Powder diffractometry, nitrides, phosphates, high pressures.*

Peters, Dr Karl. MPI f. Festkörperforschung, Heisenbergstr. 1, D-7000 Stuttgart 80, Germany, Fed. Rep. (1940) Dr rer.nat., inorganic chemistry (U. Münster, 1971). Wiss. Ang. (tel. 0711 + 68601). *Crystal structure determination, organic and inorganic compounds.*

Petzoldt, Dr Jürgen Hugo Hans. Res. & Dev. Dept., Schott Glaswerke, Hattenberg Str. 10, D-6500 Mainz, Germany, Fed. Rep. (1935) Dr rer.nat., inorganic chemistry (U. Bonn, 1963). Executive Vice President Res. and Dev. (tel. 06131 + 66, ext. 3508, telex 4187920 sm d). *Glass ceramics, glass structure and properties.*

Pfefferkorn, Prof. Dr Gerhard Erich. Inst. f. Medizinische Physik, Universität, Hüfferstr. 68, D-4400 Münster, Germany, Fed. Rep. (1913) Dr rer.nat., physics (U. Berlin, 1938). Prof. (tel. 0251 + 831, ext. 5101). *Biocrystallography, electron microscopy, X-ray micro analysis.*

Pflugrath, Dr James William. MPI f. Biochemie, Am Klopferspitz, D-8033 Martinsried, Germany, Fed. Rep. (1957) PhD, biochemistry (Rice U., USA, 1984). Res. assoc. (tel. 089 + 8578-2681). *Biological macromolecules, structure and function, computational crystallography.*

Philipsborn, von, Prof. Dr Henning. Abt. f. Kristallographie, Universität, D-8400 Regensburg, Germany, Fed. Rep. (1934) Dr phil., crystallography (U. Zürich, 1964). Prof. (tel. 0941 + 943, ext. 2481, fax +2305). *Crystal chemistry, physics, electronic materials, radioactivity, symmetry.*

Pickardt, Prof Dr Joachim. Inst. f. Anorg. u. Analyt. Chemie, TU, Str. d. 17.Juni 135, D-1000 Berlin 12, Germany, Fed. Rep. (1939) Dr Ing., chemistry (TU Berlin, 1971). Apl. Prof. (tel. 030 + 3142469). *inorganic chemistry, structural chemistry, solid state chemistry.*

Pieper, Dr Gerhard. Inst. f. Kristallographie, Universität, Senckenberg-Anlage 30, D-6000 Frankfurt/Main, Germany, Fed. Rep. (1934) Dr phil.nat., mineralogy and crystallography (U. Frankfurt/Main, 1967). Akad. Oberrat. (tel. 069 + 798, ext. 104). *Inorganic crystal structures.*

Plesken, Prof. Wilhelm. Lehrstuhl B für Mathematik, RWTH Aachen, Templergraben 64, Aachen 5100, Germany, Fed. Rep. (1950) Dr rer.nat., mathematics (RWTH Aachen, 1974) Prof. (tel. 0241+80 4535, telex TH thac d 832704, email Plesken at dacth 01) *Algebra and group theory, crystallographic groups.*

Plies, Dr Volker. Inst. f. Anorg. u. Analyt. Chemie, Universität, Heinrich-Buff-Ring 58, D-6300 Giessen, Germany, Fed. Rep. (1945) Dr rer.nat., inorganic chemistry (U. Giessen, 1975). Akad. Rat. (tel. 0641 + 7025706). *Inorganic crystal structures, TEM.*

Ploog, Dr Klaus. MPI f. Festkörperforschung, Heisenbergstr. 1, D-7000 Stuttgart 80, Germany, Fed. Rep. (1941) Dr, inorganic chemistry (U. München, 1970). Project leader. (tel. 0711 + 6860383). *Double crystal X-ray diffraction, semiconductor superlattices, RHEED - grazing incidence, semiconductor (III-V) surfaces.*

Plust, Dr Heinz-Günther. Deutsche Automobilgesellschaft mbH, Geschäftsführung, Postfach 85, D-7300 Esslingen-Mettingen, Germany, Fed. Rep. (1927) Dr, solid state chemistry (TU Berlin, 1953). Geschäftsführer. (tel. 0711 + 3026877). *Crystal growth, inorganic crystal structures, instrumentation.*

Pöllmann, Dr Herbert Josef. Mineralog. Inst., Universität, Schlossgarten 5a, D-8520 Erlangen, Germany, Fed. Rep. (1956) Dr rer.nat., mineralogy (U. Erlangen-Nürnberg). (tel. 09131 + 853986). *Crystal chemistry, mineralogy.*

Pohl, Prof. Dr Dieter. Mineralog.-Petrogr. Inst., Universität, Grindelallee 48, D-2000 Hamburg 13, Germany, Fed. Rep. (1940) Dr rer.nat., physics (U. Hamburg, 1968). (tel. 040 + 41232482). *Crystal physics, inorganic crystals, minerals, X-ray diffraction.*

Pohl, Prof. Dr Siegfried. Fachbereich Chemie, Universität, Carl-von-Ossietzky Str., D-2900 Oldenburg, Germany, Fed. Rep. (1943) Dr, inorganic chemistry (U. Bielefeld, 1974). Prof. *Crystal chemistry, main group iodides, iron-sulfur-clusters.*

Polborn, Dr Kurt Volkmar. Inst. f. Anorg. Chemie, Universität, Meiserstr. 1, D-8000 München, Germany, Fed. Rep. (1946) Dr rer.nat. (U. München, 1975). Akad. Rat. (tel. 089 + 5902250). *Cluster compounds, superstructures.*

Poll, Dr Wolfgang. Inst. f. Anorg. Chemie und Strukturchemie, Universität, Universitätsstr. 1, D-4000 Düsseldorf, Germany, Fed. Rep. (1954) Dr rer.nat., chemistry (U. Düsseldorf, 1979). Akad. Rat. (tel. 0211 + 3113146). *Solid state chemistry, crystal structures, computer programming, instrumentation.*

Pollmann, Dipl-Min Siegfried. Chemisch-Phys. Labor, Steinmüller GmbH, Postfach 1949-1960, D-5270 Gummersbach, Germany, Fed. Rep. (1931) Dr rer.nat., mineralogy (U. Münster, 1964). Dept. head. (tel. 02261 + 85, ext. 2761). *Technical mineralogy, crystal chemistry.*

Postma, Dr Johannes Petrus Maria. Dept. Biocomputing, EMBL, Meyerhofstr. 1, D-6900 Heidelberg, Germany, Fed. Rep. (1951) Dr, theoretical chemistry (U. Groningen, 1979). Staff member. (tel. 06221 + 387, ext. 255, email POSTMA at DHDEMBL (EARN)). *Molecular dynamics, computer graphics, crystallographic refinement, drug design.*

Prandl, Prof. Dr Wolfram. Inst. f. Kristallographie, Universität Tübingen, Charlottenstr. 33, D-7400 Tübingen, Germany, Fed. Rep. (1935) Dr rer.nat.habil., crystallography, mineralogy (U. München, 1973). Prof. (tel. 07071 + 29, ext. 6058, fax +5400, telex 7-262867 utna d). *Elastic and inelastic neutron scattering, magnetism, molecular crystals, disorder, group theoretical methods.*

Press, Prof. Dr Werner. Inst. f. Experimentalphysik, Universität, Leibnizstr., D-2300 Kiel, Germany, Fed. Rep. (1942) Prof., solid state physicslogy Prof., (tel. 0431 + 8803852). *Disorder, molecular crystals, neutron diffraction, phase transitions, surface structure, thermal vibrations.*

Preut, Dr Hans. Anorganische Chemie, Universität, Otto-Hahn-Str., D-4600 Dortmund 50, Germany, Fed. Rep. (1940) Dr, inorganic chemistry (U. Dortmund, 1972). Wiss. Ang. (tel. 0231 + 755, ext. 3813, fax 0231-751532, email UCH002 at DDOHRZ11). *Inorganic and organic crystal structures, instrumentation, lattice energy calculations, molecular packing analysis.*

Puff, Prof. Dr Heinrich. Anorg.-Chemisches Inst., Universität, Gerhard-Domagk-Str. 1, D-5300 Bonn 1, Germany, Fed. Rep. (1921) Dr rer.nat., inorganic

chemistry (U. Kiel, 1956). Full prof. (tel. 0228 + 73, ext. 2661). *Inorganic crystal structures, computer programming.*

Punge-Witteler, Mrs Dr Barbara. Fachbereich Chemietechnik, Universität, Emil-Figge-Str., Postfach 500500, D-4600 Dortmund 50, Germany, Fed. Rep. (1956) Dr rer.nat., mineralogy, crystallography (U. Bochum, 1986). Wiss. Angest. (tel. 0231 + 7552352). *Surface cristallisation, metallic glasses, oxidation, corrosion, metallic materials.*

Queisser, Prof. Dr Hans Joachim. MPI f. Festkörperforschung, Heisenbergstr. 1, D-7000 Stuttgart 80, Germany, Fed. Rep. (1931) Dr rer.nat., semiconductor physics (U. Göttingen, 1958), Prof. and Direktor. (tel. 0711 + 6860, ext. 601, fax 0711-6874371, telex 7255555). *Crystal growth, defect structure, semiconductors.*

Rabenau, Prof. Dr Albrecht. MPI f. Festkörperforschung, Heisenbergstr. 1, D-7000 Stuttgart 80, Germany, Fed. Rep. (1922) Dr habil., solid state chemistry (RWTH Aachen, 1963), Prof. and Direktor. (tel. 0711 + 6860, ext. 720, fax 0711-6874371). *Solid state chemistry, materials research.*

Rager, Dr Helmut. FB Geowissenschaften, Inst. f. Mineralogie, Petrologie und Kristallographie, Lahnberge, D-3550 Marburg, Germany, Fed. Rep. (1941) Dr rer.nat., physical chemistry (U. Münster, 1973). Priv.-Doz. (tel. 06421 + 282232). *NMR, EPR, ligand field spectroscopy.*

Range, Prof. Dr Klaus-Jürgen. Inst. f. Chemie, Universität, Universitätstr. 31, D-8400 Regensburg, Germany, Fed. Rep. (1938) Dr rer.nat., inorganic chemistry (U. Heidelberg, 1966). Ord. Prof. (tel. 0941 + 943-1, ext. 4551, telex 065658 unire d). *Inorganic structural chemistry, general solid state chemistry, high pressure synthesis, crystal growth.*

Rath, Prof. Dr Robert. Mineralogisch-Petrogr. Inst., Universität, Grindelallee 48, D-2000 Hamburg, Germany, Fed. Rep. (1924) Prof., mineralogy (U. Hamburg, 1971). Prof. (tel. 040 + 4123, ext. 2053). *Crystal optics, applied mineralogy.*

Recker, Prof. Dr Kurt. Mineralog. Inst., Universität, Poppelsdorfer Schloss, D-5300 Bonn, Germany, Fed. Rep. (1924) Prof., crystallography and mineralogy (U. Bonn, 1970). Univ. Prof. (tel. 0228 + 73, ext. 2769). *Crystal growth, inorganic crystals, materials science, microscopy, minerals, phase transitions, teaching, texture.*

Rehfeldt-Oskierski, Dr Angeline. E689, SIEMENS AG, Östliche Rheinbrückenstr. 50, D-7500 Karlsruhe, Germany, Fed. Rep. (1959) Dr rer.nat., mineralogy (U. Münster, 1986). (tel. 0721 + 595, ext. 4503, fax +4506, telex 78255-69). *Diffractometry, diffuse scattering, electron diffraction, electron microscopy, microscopy, phase determination, powder diffraction, scattering theory, small-angle scattering, X-ray diffraction.*

Reimers, Dr Walter. Fachgebiet Qualitätskontrolle, Universität, Postfach 500 500, D-4600 Dortmund, Germany, Fed. Rep. (1954) Dr rer.nat., crystallography (U. Marburg, 1980). Hochschulass. (tel. 0231 + 755, ext. 4782, fax 0231-751532, telex 822445). *Structure determination, magnetic structure determination.*

Reinen, Prof. Dr Dirk. Fachbereich Chemie, Universität, Hans-Meerwein-Str., D-3550 Marburg, Germany, Fed. Rep. (1930) Dr, inorganic chemistry (U. Bonn, 1960). Ord. Prof. (tel. 06421 + 285668, telex 482372). *Structure and bonding, transition metal compounds, spectroscopy, vibronic coupling effects, Jahn-Teller effect.*

Reithmayer, Mr Klaus Thomas. Inst. f. Kristallographie u. Mineralogie, Universität, Theresienstr. 41, D-8000 München, Germany, Fed. Rep. (1962) Dipl.-Min., mineralogy (U. München, 1988). Wiss.Mit. (tel. 089 + 23944356). *High pressure, inorganic crystals.*

Reuber-Kürbs, Mrs Dr Ing Ellen. Fritz-Haber-Inst., Max-Planck-Ges. Faradayweg 4-6, D-1000 Berlin 33, Germany, Fed. Rep. Dr -Ing., physics (TU Berlin, 1953). Res. scient. (tel. 030 + 8305, ext. 346). *Electron microscopic structure analysis at atomic resolution, image reconstruction, biological structures, large molecules, instrumentation.*

Reuter, Dr Hans. Anorg. Chemisches Inst., Universizät, Gerhard-Domagk-Str. 1, D-5300 Bonn 1, Germany, Fed. Rep. (1952) Dr, chemistry (U. Bonn, 1987). (tel. 0228 + 735330). *Inorganic crystals, organic compounds, powder diffraction, X-ray diffraction.*

Richter, Dr Ursula. Morgengraben 1, D-5000 Köln 80, Germany, Fed. Rep. (1953) Dr rer.nat., mineralogy, crystallography, (U. Köln, 1986). (tel. 0221 + 664241). *Crystal growth, optical and electrical properties, crystal structure analysis.*

Rickert, Prof. Dr Hans. Physikalische Chemie, Universität, Postfach 500500, D-4600 Dortmund 50, Germany, Fed. Rep. (1928) Prof., physical chemistry (U. Dortmund, 1969). Prof. (tel. 0231 + 755, ext. 3900). *Physical chemistry of solids, chemical thermodynamics, transport reactions, electrochemistry, corrosion.*

Riedel, Prof. Dr Erwin. Inst. f. Anorg. u. Analyt. Chemie, TU, Strasse des 17. Juni 135, D-1000 Berlin 12, Germany, Fed. Rep. (1930) Dr -Ing., inorganic chemistry (TU Berlin, 1970). Prof. (tel. 030 + 314, ext. 3498). *Solid state chemistry.*

Robl, Dr Christian. Inst. f. Anorg. Chemie, Universität, Meiserstr. 1, D-8000 München 2, Germany, Fed. Rep. (1955) Dr rer.nat.habil., inorganic chemistry (U.München, 1988). Akad.Rat. (tel. 089 + 5902, ext. 371). *Neutron diffraction, X-ray diffraction, hydrogen bonding, coordination polymers, zeolites, intercalation reactions, solid state ionic conductors.*

Röller, Mr Klaus. Inst. f. Mineralogie, Universität, Universitätsstr. 150, D-4630 Bochum, Germany, Fed. Rep. (1960) Dipl.-Min., crystallography (U.Bochum, 1987). (tel. 0234 + 700, ext. 3881). *X-ray diffraction, quartz, FTIR.*

Rösch, Dr Heinrich. Abt. Geoch. u. Mineralogie, Bundesanst. f. Geowiss. u. Rohstoffe, Stilleweg 2, D-3000 Hannover 51, Germany, Fed. Rep. (1935) Dr phil., mineralogy and crystallography (U. Kiel, 1961). Sr. scient., sr. res. officer. (tel. 0511 + 6468, ext. 562). *Clay mineralogy, crystallography, silica modifications, quantitative X-ray diffraction analysis.*

Rohmer, Mr Christian. IBM, Dept. 0348, 7032-16, Tübinger Allee, D-7032 Sindelfingen, Germany, Fed. Rep. (1938) Technician. (tel. 07031 + 611, ext. 3963). *X-ray topography, silicon materials, X-ray analysis, single crystals, polycrystalline materials.*

Rossmanith, Mrs. Dr Elisabeth. Mineralogisch-Petrographisches Inst., Universität, Grindelallee 48, D-2000 Hamburg 13, Germany, Fed. Rep. (1943) Dr phil.habil., physics, crystallography (U. Hamburg, 1981). Wiss. Ass. (tel. 040 + 4123, ext. 2485, telex 214732unihhd). *Thermal vibrations, crystal physics.*

Rothammel, Mr Walter. Inst. f. Angew. Physik, Lst. f. Kristallographie, Universität, Bismarckstr. 10, D-8520 Erlangen, Germany, Fed. Rep. (1960) Dipl.-Phys., physics, crystallography (U. Erlangen-Nürnberg, 1987). Wiss.Mit. (tel. 09131 + 852119). *Computing, diffractometry, thermal vibrations.*

Rothbauer, Dr Richard. Weidenstr. 11, D-6234 Hattersheim 3, Germany, Fed. Rep. (1938) Dr phil.nat., crystallography (U. Frankfurt, 1971). *Methods, structure analysis, single crystal diffractometry, elastic neutron diffraction.*

Rott, Dr Volkwin. Schunk und Ebe GmbH, Rodheimer Str. 59-61, D-6300 Giessen, Germany, Fed. Rep. (1939) Dr, crystallography (TU Clausthal, 1970). Leiter der Patentabteilung. (tel. 0641 + 78081). *Carbon.*

Ruban, Prof. Dr Gerhard. FB 21, WE 5, Inst. f. Kristallographie, FU, Takustr. 6, D-1000 Berlin 33, Germany, Fed. Rep. (1926) Prof., crystallography (FU Berlin, 1971). Prof. (tel. 030 + 838, ext. 3462). *X-ray structure analysis, crystal chemistry.*

Rudert, Mr Rainer. Inst. f. Kristallographie, FU, Takustr. 6, D-1000 Berlin 33, Germany, Fed. Rep. (1956) Diplom, physics (FU Berlin, 1983). Physicist. (tel. 030 + 8382354). *X-ray structure analysis, electron density.*

Ruiz Perez, Mrs Catalina. Arbeitsgruppe f. chem. Kristallographie, MPI f. Biochemie, Am Klopferspitz, D-8033 Martinsried, Germany, Fed. Rep. (1957) Diplom, physics (U. Valencia, Spain, 1983). (tel. 089 + 85782661). *X-ray structure analysis, organic compounds.*

Ruppersberg, Prof. Dr Henner. Fachbereich Angewandte Physik, Universität, D-6600 Saarbrücken, Germany, Fed. Rep. (1933) Dr rer.nat., Prof. (tel. 0681 + 302, ext. 3448). *Amorphous phases, chemical short-range order, X-ray stress analysis.*

Saalfeld, Prof. Dr Horst. Mineralogisch-Petrogr. Inst., Universität, Grindelallee 48, D-2000 Hamburg, Germany, Fed. Rep. (1920) Prof., mineralogy (U. Saarbrücken, 1960). Direktor. (tel. 040 + 4123, ext. 2050). *Crystal chemistry, clay mineralogy.*

Sabrowsky, Prof. Dr Horst. Anorganische Chemie I, Ruhr-Universität, D-4630 Bochum, Germany, Fed. Rep. (1934) Prof. (tel. 0234 + 700, ext. 4151). *Solid state chemistry, structures, magneto chemistry, high-pressure thermogravimetry.*

Sänger, Dr Annette. Inst. f. Kristallographie, Universität, Kaiserstr. 12, D-7500 Karlsruhe, Germany, Fed. Rep. (1954) Dr rer.nat., mineralogy (U. Karlsruhe, 1985). (tel. 0721 + 6082976). *Hydrogen-bonding, disorder, defect structures, crystal growth, microscopy, minerals, phase transitions, powder diffraction, thermal vibrations, X-ray diffraction.*

Saenger, Prof. Dr Wolfram H. E. Inst. f. Kristallographie, FU, Takustr. 6, D-1000 Berlin 33, Germany, Fed. Rep. (1939) Prof., chemistry (Darmstadt, 1965). Prof. (tel. 030 + 838, ext. 3412, fax +6702, telex 184019). *Protein structures, nucleic acids, oligosaccharides, hydrogen bonding, inclusion complexes.*

Sahl, Prof. Dr Kurt. Inst. f. Mineralogie, Universität, Universitätsstr. 150, D-4630 Bochum, Germany, Fed. Rep. (1933) Dr rer. nat., crystallography (U. Göttingen, 1963). Prof. (tel. 0234 + 700, ext. 4381). *Inorganic crystal chemistry.*

Sahm, Prof. Dr Ing. Peter Rudolf. Giesserei Inst., RWTH, Intzestr. 5, D-5100 Aachen, Germany, Fed. Rep. (1934) Dr Ing., foundry technology (TU. Berlin, 1961). Director of the Foundry Institute. (tel. 0241 + 80, ext. 5880, fax +6726, telex 0832704 thac d). *Computing, crystal growth, materials science, metals teaching.*

Saur, Prof. Dr Ing Eugen. Inst. f. Angew. Physik, Universität, Leihgesterner Weg 106, D-6300 Giessen, Germany, Fed. Rep. (1910) Dr -Ing., physics (TU Stuttgart, 1936). Full prof. (tel. 0641 + 702, ext. 2791). *Superconductivity.*

Schäfer, Prof. Dr Herbert Leo. Abt. II f. Anorganische Chemie, TH, Hochschulstr. 4, D-6100 Darmstadt, Germany, Fed. Rep. (1933) Dr, inorganic chemistry (TH Darmstadt, 1971). Prof. (tel. 06151 + 162292). *Inorganic crystal structures.*

Schäfer, Dr Wolfgang. Mineralog. Inst., Universität Bonn, KFA, Postfach 1913, D-5170 Jülich, Germany, Fed. Rep. (1942) Dr, physics (U. Bonn, 1971). (tel. 02461 + 616024). *Structure analysis and refinement, neutron diffraction, magnetic ordering.*

Schanda, Dr Friedrich. Abt. f. Strukturforschung I, MPI f. Biochemie, Am Klopferspitz, D-8033 Martinsried, Germany, Fed. Rep. (1947) Diplom, physics, (TU München, 1975). Res. assoc. (tel. 089 + 8585, ext. 660). *X-ray and neutron structure analysis, bonding electron density studies.*

Schildberg, Dr Hans Peter. ZAA/F - M 320, BASF AG, D-6700 Ludwigshafen, Germany, Fed. Rep. (1957) Dr rer.nat., physics (U. Kiel, 1988). Physicist. (tel. 0621 + 60, ext. 56556).

Schimanski, Dr Uwe Lothar. Materials Lab. I, Dept. 4627, 65-05, IBM GmbH, Hechtsheimerstr. 2, D-6500 Mainz 1, Germany, Fed. Rep. (1953) Dr rer.nat., chemistry (U. Bielefeld, 1983). Specialist f. surface analysis. (tel. 06131 +

842038). *SEM, STEM, surface analysis, powder diffractometry, crystal structure determination.*

Schirmer, Dr Ulrich. VGB, Techn. Vereinigung der Gross-kraft-werks-betrei-ber, Klinkestr. 27-31, D-4300 Essen, Germany, Fed. Rep. (1949) Dr rer.nat., technical mineralogy (U. Münster, 1979). (tel. 0201 + 198, ext. 281). *Instrumentation, polymorphic transformations, ceramics, refractories, waste combustion.*

Schliephake, Dr Rolf-Werner. Abt. Mineralogie und Petrographie, Bergbau-Forschung GmbH, Frillendorfer Str. 351, D-4300 Essen 13, Germany, Fed. Rep. (1925) Dr rer.nat., chemistry (TH München, 1958). Leiter des Röntgenlabors. (tel. 0201 + 105, ext. 9381). *X-ray powder diffractometry, phase analysis, instrumentation.*

Schloemer, Prof. Dr Hermann J. Technische Mineralogie, Universität, Im Stadtwald, D-6600 Saarbrücken, Germany, Fed. Rep. (1923) Prof., mineralogy (U. Saarbrücken, 1968). Prof. (tel. 0681 + 302, ext. 2912). *Crystal growth, high pressure and temperature research, cement chemistry, mineral mobility, coal gasification.*

Schmetzer, Dr Karl. Mineralogisch-Petrogr. Inst., Universität, Im Neuenheimer Feld 236, D-6900 Heidelberg, Germany, Fed. Rep. (1951) Dr rer.nat., mineralogy (U. Heidelberg, 1978). (tel. 06221 + 56, ext. 2805). *Inorganic crystal structures, crystal chemistry, spectroscopy, colour of inorganic materials.*

Schmidt, Mr Bertram Felix Paul. Kristall-und Materiallabor, Fak. f. Physik, Universität (TH), Kaiserstr. 12, D-7500 Karlsruhe, Germany, Fed. Rep. (1953) Diplom, physics (U. Karlsruhe, 1979). Wiss. Mit. (tel. 0721 + 608, ext. 3471). *Crystal growth and characterization, X-ray topography.*

Schneider, Dr Hartmut. Forschungsinst. d. Feuerfest-Industr., An der Elisabethkirche 27, D-5300 Bonn 1, Germany, Fed. Rep. (1941) Dr rer.nat.habil., mineralogy Head mineralogical group. (tel. 0228 + 2110, ext. 51, fax +50, telex 886533). *Materials science (ceramics).*

Schneider, Dr Jochen Richard. Hahn-Meitner-Inst. f. Kernforschung, Glienicker Str. 100, D-1000 Berlin 39, Germany, Fed. Rep. (1941) Dr rer.nat., physics (U. Hamburg, 1973). Physicist. (tel. 030 + 8009, ext. 2768). *Electron charge, momentum density, metal hydrides, ionic conductors, semiconductors, imperfect single crystal diffraction, structural phase transitions.*

Schneider, Dr Julius. Inst. f. Kristallographie und Mineralogie, Universität, Theresienstr. 41, D-8000 München 2, Germany, Fed. Rep. (1942) Dr rer.nat., physics (T.U. München, 1975). Wiss. Angest. (tel. 089 + 23944354). *Neutron and X-ray diffraction, materials science, metals, diffuse scattering, disorder.*

Schneider, Dr Walter. Inst. f. Material- u. Festkörperforschung, KFZ, Postfach 3640, D-7500 Karlsruhe, Germany, Fed. Rep. (1936) Dr rer.nat., mineralogy (U. Göttingen, 1960). Gruppenleiter. (tel. 07247 + 824903). *Electron microscopy, electron diffraction, metals, radiation defects, precipitation in steels.*

Schnering, von, Prof. Dr Dr hc Hans Georg. MPI f. Festkörperforschung, Heisenbergstr. 1, D-7000 Stuttgart 80, Germany, Fed. Rep. (1931) Dr rer.nat.habil., inorganic chemistry (U. Münster, 1960). Direktor. (tel. 0711 + 68, ext. 60560, fax +74371, telex 7255555 mpif d). *Solid state chemistry, general structural chemistry, cluster compounds, structure and bonding in solids.*

Schnick, Dr Wolfgang. Anorg.-Chemisches Institut, Universität, Gerhard-Domagk-Str. 1, D-5300 Bonn 1, Germany, Fed. Rep. (1957) Dr rer.nat., inorganic solid state chemistry Hochschulass. (tel. 0228 + 73, ext. 3153, email UNC414 at DBNRHRZ1). *Chemical bonding, crystal growth, inorganic crystals, powder diffraction, structural chemistry, X-ray diffraction.*

Schöllhorn, Prof. Dr Robert. Inst. f. Anorganische u. Analytische Chemie, TU Berlin, Strasse des 17.Juni 135, D-1000 Berlin 12, Germany, Fed. Rep. (1935) Ph.D., inorganic chemistry (U. Heidelberg, 1963). Prof. (tel. 030 + 314 ext. 22740, fax +23222, telex 22777). *Chemical bonding, defect structures, diffractometry, disorder, hydrogen bonding, inorganic crystals, magnetism, materials science, phase determination, phase transitions, powder diffraction, structural chemistry.*

Scholz, Dr Heinz Werner. Philips GmbH Forschungslaboratorium, Weisshausstr., D-5100 Aachen, Germany, Fed. Rep. (1930) Dr chemistry (U. Münster, 1959). Res. chemist. (tel. 0241 + 62071). *Crystal growth.*

Schomburg, Priv.-Doz. Dr Dietmar. Ges. f. Biotechn. Forschung, Molekulare Strukturforschung, Mascheroder Weg 1, D-3300 Braunschweig, Germany, Fed. Rep. (1950) Dr rer.nat.habil., chemistry (TU Braunschweig, 1985). Head, molecular structure res. group. (tel. 0531 + 6181, ext. 350, fax +515). *Structural chemistry, hypervalent compounds, protein design, protein structure.*

Schramm, Dr Volker. FR 17.3 Kristallographie, Universität, D-6600 Saarbrücken 11, Germany, Fed. Rep. (1939) Dr, mineralogy (U. Saarbrücken, 1972). (tel. 0681 + 302, ext. 3470). *Crystal structures, organometallic complexes, computer programming, teaching.*

Schröcke, Prof. Dr Helmut. Inst. f. Kristallographie und Mineralogie, Universität, Theresienstr. 41, D-8000 München 2, Germany, Fed. Rep. (1922) Prof., physico-chemical mineralogy (U. München, 1951). Abteilungsvorstand. (tel. 089 + 2394, ext. 4331). *Physicochemical mineralogy, mineral deposits, mineral systematics.*

Schröder, Dr Friedrich Anton. Gmelin-Inst., Max-Planck-Ges., Varrentrappstr. 40/42, D-6000 Frankfurt/Main 90, Germany, Fed. Rep. (tel. 069 + 79171, fax +7917338). *Silicon compounds.*

Schröder, Mr Jens. GKSS Research Center, Max-Planck-Strasse, D-2054 Geesthacht, Germany, Fed. Rep. (1960) Dipl-Min., mineralogy (tel. 04152 +

87-0, ext. 1249, fax +1403, telex 218732). *Neutron diffraction, materials science, powder diffraction.*

Schröpfer, Dr Lothar Maximilian. Inst. f. Kristallographie, Universität, Senckenberg-Anlage 30, D-6000 Frankfurt/Main, Germany, Fed. Rep. (1941) Dr phil.nat., mineralogy (U. Frankfurt/Main, 1971). Akad. Oberrat. (tel. 069 + 7982103). *Crystal chemistry, inorganic crystal structures, electron microscopy.*

Schubert, Mr Helmut. Powder Metallurgy Lab., MPI f. Metallforschung, Heisenbergstr. 5, D-7000 Stuttgart 80, Germany, Fed. Rep. (1951) Dipl.-Min., materials science (RWTH Aachen, 1981). (tel. 0711 + 2095628). *Zirconia, powder preparation and characterisation, martensitic transformations, microstructural design.*

Schubert, Prof. Dr Konrad. Inst. f. Werkstoffwissenschaft, MPI f. Metallforschung, Seestr. 75, D-7000 Stuttgart, Germany, Fed. Rep. (1915) Prof., crystallography (U. Stuttgart, 1960). (tel. 0711 + 2095210). *Inorganic and metallic phases and structures, bonding, two-electron density matrix.*

Schülke, Prof Dr Winfried. Inst. f. Physik, Universität, Otto-Hahn-Str., D-4600 Dortmund 50, Germany, Fed. Rep. (1935) Dr rer.nat.habil. (U. Halle, 1971). Prof. (tel. 0231 + 7553507, telex 822465). *X-ray diffraction, inelastic X-ray scattering, gamma-Compton-scattering, dynamical X-ray diffraction theory, synchrotron radiation applications.*

Schüller, Prof. Dr Karl-Heinz. Fachbereich Werkstofftechnik, Ohm-FH, Kesslerplatz 12, D-8500 Nürnberg, Germany, Fed. Rep. (1928) Prof., technical mineralogy (U. Erlangen-Nürnberg, 1972). Prof. (tel. 0911 + 5880216). *Technical mineralogy, ceramic technology, ceramic microstructures, properties of ceramic raw materials.*

Schürmann, Dr Kay Uwe. Geosciences, University of Marburg, Lahnberge, D-3550 Marburg, Germany, Fed. Rep. (1939) Dr rer.nat., mineralogy (U. Marburg, 1966). Res. scient. (tel. 06421 + 282228). *Silicate structures, double-sheet structure, Mg-Fe-distribution (inter- and intracrystalline).*

Schultze-Rhonhof, Dr Ernst. KHD Humboldt-Wedag AG, Dr chenfelsweg 19, D-5300 Bonn 3, Germany, Fed. Rep. (1934) Dr rer.nat., chemistry (U. Bonn, 1964). Chemiker. (tel. 0228 + 465641). *Inorganic crystal structures, computer programming.*

Schulz, Dr Georg Eberhardt Bruno. Abt. f. Biophysik, MPI f. medizinische Forschung, Jahnstr. 29, D-6900 Heidelberg, Germany, Fed. Rep. (1939) Prof., biophysics (U. Heidelberg, 1973). Wiss. Ang. (tel. 06221 + 486274). *Protein structure, biochemistry.*

Schulz, Prof. Dr Heinz Hermann. Inst. f. Kristallographie u. Mineralogie, Universität, Theresienstr. 41, D-8000 München, Germany, Fed. Rep. (1935) Dr rer.nat., physics (U. Saarbrücken, 1964). Prof. (tel. 089 + 23944311, telex univm 529815). *Surface crystallography, high-pressure crystallography, fast ionic conductors, thermal motion.*

Schur, Dr Karl. Inst. f. Medizinische Physik, Universität, Hüfferstr. 68, D-4400 Münster, Germany, Fed. Rep. (1929) Dr rer.nat., mineralogy (U. Giessen, 1971). Dr (tel. 0251 + 83, ext. 5100). *Applied electron microscopy, crystal growth.*

Schuster, Prof. Dr Hans-Uwe. Inst. f. Anorganische Chemie, Universität, Greinstr. 6, D-5000 Köln 41, Germany, Fed. Rep. (1930) Prof., inorganic chemistry (U. Köln, 1971). Ord. Prof. (tel. 0221 + 470, ext. 3262). *Inorganic chemistry, solid state chemistry, structure determination, magnetism, conductivity, optical measurements, DTA.*

Schwabe, Prof. Dr Dietrich Gerhard. I. Physikalisches Inst., Universität, H.-Buff-Ring 16, D-6300 Giessen, Germany, Fed. Rep. (1942) Dr habil., experimental physics (tel. 0641 + 702, ext. 2715, fax +2099, telex 482391 phygi). *Crystal growth, defect structure, materials science.*

Schwahn, Dr Dietmar. IFF, KFA Jülich, Postfach 1913, D-5170 Jülich, Germany, Fed. Rep. (1943) (tel. 02461 + 61, ext. 6661, telex 833556). *Small-angle scattering, polymers.*

Schwarz, Dr Wolfgang. Inst. f. Anorg. Chemie, Universität, Pfaffenwaldring 55, D-7000 Stuttgart 80, Germany, Fed. Rep. (1939) Dr rer.nat., chemistry (U. Stuttgart, 1973). Akad. Oberrat. (tel. 0711 + 6854238). *Inorganic and organometallic crystal structures.*

Schwarzmann, Mrs Dr Sigrid. Mineralogische Sammlung, Staatl. Naturw. Sammlung, Theresienstr. 41, D-8000 München 2, Germany, Fed. Rep. (1928) Dr rer.nat., mineralogy, crystallography (U. Göttingen, 1956). Oberkonservatorin. (tel. 089 + 2394, ext. 4308). *Inorganic crystal structures, crystal optics, crystal growth.*

Schweda, Dr Eberhard. Inst. f. Anorganische Chemie, Universität, Auf der Morgenstelle 18. D-7400 Tübingen 1, Germany, Fed. Rep. (1928) Dr rer.nat., inorganic chemistry (U. Tübingen, 1981). Wiss. Angest. (tel. 07071 + 296217). *X-ray, neutron, electron diffraction, inorganic chemistry, high resolution electron microscopy.*

Schweinsberg, Dr Heinz Friedrich. Zentrale Technik, Thyssen Stahl AG, Postfach 110067, D-4100 Duisburg 11, Germany, Fed. Rep. (1936) Dr rer.nat., crystallography (U. Kiel, 1971). Res. manager. (tel. 0203 + 5224733). *Crystal chemistry, refractories.*

Sebastian, Dr Mailadil Thomas. Inst. Kristallographie, RWTH, Templergraben, D-5100 Aachen, Germany, Fed. Rep. (1952) PhD, physics (1983). A.v.Humboldt fellow. (tel. 0241 + 806917). *Crystal growth, defect structures, disorder, diffuse scattering, phase transitions, X-ray diffraction.*

Seidel, Dr Peter. Inst. f. Mineralogie, Universität, Corrensstr. 24, D-4400 Münster, Germany, Fed. Rep. (1938) Dr, crystallography (U. Münster, 1976). Wiss. Ang. (tel. 0251 + 833458). *Phase transitions, physical properties, instrumentation.*

Seifert, Prof. Dr Hans-Joachim. Inst. f. Anorg. Chemie, Gesamthochschule, Heinrich-Plett-Str. 40, D-3500 Kassel, Germany, Fed. Rep. (1930) Prof., inorganic chemistry (U. Giessen, 1969). Prof. (tel. 0561 + 8044760). *Inorganic crystal structures, thermochemistry.*

Seifert, Prof. Dr Karl-Friedrich. Mineralog. Inst., Universität, Poppelsdorfer Schloss, D-5300 Bonn, Germany, Fed. Rep. (1927) Dr rer.nat., mineralogy (U. Münster, 1957). Akad. Oberrat. (tel. 0228 + 732768). *High pressure, crystallography, mineralogy, physical properties, biomineralization.*

Serafin, Dr Michael. Inst. f. Anorg. u. Analyt. Chemie, Universität, Heinrich-Buff-Ring 58, D-6300 Giessen, Germany, Fed. Rep. (1949) Dr rer.nat., inorg. chemistry (U. Giessen, 1980). Akad. Rat. (tel. 0641 + 7025666, email U151420 at DGIHRZ01). *Solid state chemistry, inorg. crystal structures.*

Sheldrick, Prof. Dr George Michael. Inst. f. Anorg. Chemie, Universität, Tammannstr. 4, D-3400 Göttingen, Germany, Fed. Rep. (1942) PhD, chemistry (U. Cambridge, UK, 1966). Prof. (tel. 0551 + 39, ext. 3021, fax +9612, email GSHELDR at DGohGWDG1). *Structural chemistry, crystallographic computing.*

Sheldrick, Prof Dr William Stephen. FB Chemie, Universität, Erwin-Schrödinger-Str., D-6750 Kaiserslautern, Germany, Fed. Rep. (1945) PhD, inorganic chemistry. (U. Cambridge, UK, 1969). Prof. (tel. 0631 + 2052986). *Coordination compounds, inorganic crystals, reaction pathways, structural chemistry.*

Siebels, Mr Hansjörg. Abt. f. Strukturforschung I, MPI f. Biochemie, Am Klopferspitz, D-8033 Martinsried, Germany, Fed. Rep. (1950) Dipl.-Phys., physics (TU München, 1975). Res. assoc. (tel. 089 + 8585, ext. 660). *X-ray and neutron structure analysis.*

Sieber, Mr Norbert Hermann Wilhelm. Mineralogisches Inst., Universität, Am Hubland, D-8700 Würzburg, Germany, Fed. Rep. (1960) Dipl., mineralogy (U. Mainz, 1985). (tel. 0931 + 888431). *Crystal chemistry, crystal structure analysis, X-ray crystallography.*

Sieger, Mr Peter. Fak. f. Chemie, Universität, Universitätsstr. 10, D-7750 Konstanz, Germany, Fed. Rep. (1963) (tel. 07531 + 882013).

Sievers, Dr Rolf. Inst. f. Anorganische Chemie, Universität, Gerhard-Domagk-Str. 1, D-5300 Bonn, Germany, Fed. Rep. (1938) Dr, inorganic chemistry (U. Kiel, 1968). Akad. Oberrat. (tel. 0228 + 73, ext. 5336, UNC411 at DBNRHRZ1). *Computing, crystallographic data, diffractometry, inorganic crystals, instrumentation, molecular crystals, powder diffraction, teaching, X-ray diffraction.*

Simon, Prof. Dr Arndt. MPI f. Festkörperforschung, Heisenbergstr. 1, D-7000 Stuttgart, Germany, Fed. Rep. (1940) Dr, chemistry (U. Münster, 1966). Direktor. (tel. 0711 + 6860640, ext. 2640). *Inorganic crystal structures, preparation, crystal growth, structure and physical properties, instrumentation.*

Sirtl, Prof. Dr Erhard. R & D Heliotronic GmbH, Postfach 1129, D-8263 Burghausen, Germany, Fed. Rep. (1928) Apl. Prof., inorganic and physical chemistry (LMU München, 1968). R and D manager. (tel. 08677 + 83, ext. 2580). *Crystal growth, crystal defects.*

Sondermann, Dr Ulrich. Inst. f. Mineralogie, Universität, Lahnberge, D-3550 Marburg, Germany, Fed. Rep. (1936) Dr rer.nat., physics (U. Marburg, 1970). (tel. 06421 + 28, ext. 2226). *Magnetism, crystal physics.*

Sowa, Dr Heidrun. Inst. f. Kristallographie u. Mineralogie, Universität, Theresienstr. 41, D-8000 München, Germany, Fed. Rep. (1954) Dr rer.nat., mineralogy (U. Marburg, 1983). Akad. Rätin. (tel. 089 + 23944356, telex 529815). *High-pressure crystallography.*

Spaeth, Prof. Dr Johann-Martin. FB 6 - Physik, Universität, Warburger Str. 100 A, D-4790 Paderborn, Germany, Fed. Rep. (1937) Dr rer.nat., physics (U. Stuttgart, 1965). Full Prof. (tel. 05251 + 60, ext. 2742, fax +2519, telex 936776). *Defect structures, inorganic crystals, materials science, semiconductors.*

Speier, Mr Peter. ZFZ/WO Optoelectronic components, SEL-Research center, Lorenzstr. 10, D-7000 Stuttgart 40, Germany, Fed. Rep. (1958) Dipl.-Phys., physics (U. Stuttgart, 1983). Dept.Head. (tel. 0711 + 821, ext. 5850, fax +6355, telex 72526). *Crystal growth, electron microscopy, materials science, diffractometry, defect structures, semiconductors.*

Springer, Prof. Dr Tasso. Inst. f. Festkörperforschung, KfA Jülich GmbH, Postfach 1913, D-5170 Jülich, Germany, Fed. Rep. (1930) Dr physics (HT München, 1955). Director. (tel. 02461 + 61, ext. 4699, fax +5327, telex 833556-0 kf d). *Polymers, small angle scattering (SANS).*

Stadermann, Dr Gerd. Fachbereich Physik, WE1, FU Berlin, Arnimallee 14, D-1000 Berlin 33, Germany, Fed. Rep. (1947) Dr rer. nat., crystallography (Humboldt-U. Berlin, 1976). (tel. 030 + 8386088). *High temperature superconductors, crystal growth, surfacees, thin films, epitaxy, electron microscopy, optoelectronics.*

Stadler, Mr Maximilian Wolfgang. Inst. f. Anorg. Chemie, Universität, Universitätsstr. 31, D-8400 Regensburg, Germany, Fed. Rep. (1956).

Steeb, Prof. Dr Siegfried. Inst. f. Werkstoffwissenschaften, MPI f. Metallforschung, Seestr. 92, D-7000 Stuttgart, Germany, Fed. Rep. (1931) Prof. metal-physics (U. Stuttgart, 1969). Head, sci. dept. (tel. 0711 + 20951). *X-ray diffraction, electron and neutron diffraction, melts and amorphous solids.*

Steffen, Dipl. Chem Michael Georg. Inst. f. Anorg. Chemie u. Strukturchemie, Universität, Universitätsstr. 1, D-4000 Düsseldorf, Germany, Fed. Rep. (1950) Dipl.-Chem., chemistry (U. Düsseldorf, 1977). Res. asst. (tel. 0211 + 3113857). *Low temperature crystal growth, inorganic crystal structures.*

Steffer-Tun, Mr Wolfgang. Inst. f. Mineralogie, Universität, Corrensstr. 24, D-4400 Münster, Germany, Fed. Rep. (1960) Dipl.-Min., mineralogy (U. Münster, 1987). (tel. 0251 + 833489, telex 892529 unims d)). *Diffractometry, electron microscopy, powder diffraction, minerals.*

Stegemann, Dr Ing Jürgen. Nicolet Instrument GmbH, Senefelderstr. 162, D-6050 Offenbach/Main, Ger. Fed. Rep. (1947) Dr -Ing., X-ray structure analysis (TU Darmstadt, 1980). Application engineer. (tel. 0611 + 837001). *X-ray structure analysis, organic compounds, semi-empirical and force-field calculations, intra- and inter-molecular structures.*

Stegemeyer, Prof. Dr Horst. Inst. f. Physikalische Chemie, Universität, Warburger Str. 100, D-4790 Paderborn, Ger. Fed. Rep. (1931) Dr rer.nat., physical chemistry (U. Hannover, 1961). Prof. (tel. 05251 + 60, ext. 2156, fax +2519, telex 936776 unipb d). *Crystal growth, liquid crystals, materials science, microscopy, phase determination, phase transitions, small-angle scattering.*

Steiner, Prof. Dr Michael. Inst. f. Physik, Universität, Staudinger Weg 7, D-6500 Mainz, Germany, Fed. Rep. (1943) Dr rer.nat.habil., experimental physics (TU Berlin, 1978). Prof. (tel. 06131 + 39, ext. 3637 or 2282, fax +2964, telex 4187155phmsd). *Magnetism, phase transitions, neutron diffraction.*

Steiner, Mr Thomas Johann. Inst. f. Kristallographie, TU Berlin, Takustr. 6, D-1000 Berlin 33, Germany, Fed. Rep. (1961) Dipl.-Ing. (tel. 030 + 838, ext. 6599). *Hydrogen bonding, macromolecules, neutron diffraction.*

Steinmann, Dr Gerhard. Materiallaboratorien, IBM Deutschland GmbH, Tübinger Allee 49, Postfach 266, D-7032 Sindelfingen, Germany, Fed. Rep. (1954) Dr rer.nat., mineralogy, crystallography (U.Karlsruhe, 1984). Lab. assoc./proj. leader. (tel. 07032+91 ext. 3092). *Crystal growth, defect structure, diffractometry, group theory, X-ray/electron diffraction, materials science.*

Steuhl, Dr Hans Hermann. Inst. f. Mineralogie, Universität, Corrensstr. 24, D-4400 Münster, Germany, Fed. Rep. (1922) Dr rer.nat., physics (T.U. München, 1958). Prof. (tel. 02461 + 615774). *Neutron scattering, molecular solids, liquid crystals.*

Stezowski, Prof. Dr John J. Inst. f. Organische Chemie, Universität, Pfaffenwaldring 55, D-7000 Stuttgart 80, Germany, Fed. Rep. (1942) PhD, chemistry (Michigan State U., USA, 1969). Prof. (tel. 0711 + 685, ext. 4260, fax +3500). *Crystal structure analysis, biomolecules, proteins, protein design, organic molecules, temperature dependent studies.*

Steurer, Dr Walter. Inst. f. Kristallographie und Mineralogie, Universität, Theresienstr. 41, D-8000 München 2, Germany, Fed. Rep. (1950) Dr rer.nat.habil., crystallography and mineralogy (U. München, 1987). Lecturer. (tel. 089 + 23944314, fax 089-5203286, email UK4030 at DM0LRZ016 EARN). *Phase transitions, quasicrystals, structural chemistry, symmetry.*

Stöckelmann, Dr Diedrich. Inst. f. Mineralogie, Universität, Corrensstr. 24, D-4400 Münster, Germany, Fed. Rep. (1938) Dr ., mathematics, mineralogy (U. Münster, 1983). Wiss. Ass. (tel. 0251 + 833493, telex 892529 unims d). *Crystallographic computing.*

Strähle, Prof. Dr Joachim. Inst. f. Anorganische Chemie, Universität, Auf der Morgenstelle 18, D-7400 Tübingen, Germany, Fed. Rep. (1937) Ord. Prof., inorg. chemistry (U. Tübingen, 1976) Ord. Prof., chair of inorganic chemistry. (tel. 07071 + 29, ext. 6102, fax +5246, telex 177071913, email CAHI001 at DTUZDV5A). *Inorganic chemistry, synthesis, inorganic compounds, crystal structure determination.*

Strell, Mrs Dr Irmtraud. MPI f. Biochemie, Am Klopferspitz, D-8033 Martinsried, Germany, Fed. Rep. (1922) Dr rer.nat., physical chemistry (TU München, 1949). (tel. 089 + 8585697). *Organic molecules, crystal growth, biochemical derivatives.*

Strocka, Dr Bernhard. Philips GmbH Forschungslaboratorium, Vogt-Kölln-Str. 30, D-2000 Hamburg 54, Germany, Fed. Rep. (1943) Dr -Ing., solid state chemistry (TU Berlin, 1971). Wiss. Mit. (tel. 040 + 5493, ext. 550+568). *Precision measurement, lattice constants, X-ray topography, lattice defects, single-crystal characterization.*

Strübel, Prof. Dr Günter. Mineralogisch-Petrologisches Inst., Justus Liebig-Universität, Senckenbergstr. 3, D-6300 Giessen, Germany, Fed. Rep. (1932) Prof. Dr habil., natural sciences (Justus Liebig U. Giessen, 1962). Prof. (tel. 0641 + 7028372). *Technical mineralogy, applied mineralogy, hydrothermal mineralization, mineral raw materials.*

Strumpel, Dr Marianna Katona. Inst. f. Kristallographie, FU, Takustr. 6, D-1000 Berlin 33, Germany, Fed. Rep. (1939) Dr rer.nat., crystallography (FU Berlin, 1983) Wiss. Mit. (tel. 030 + 838-3457). *Small organic molecules, conformational calculations, programming, vector search techniques.*

Stubbs, Dr Milton. Strukturforschung II, MPI f. Biochemie, Am Klopferspitz, D-8023 Martinsried, Germany, Fed. Rep. (1962) Dr phil., biophysics (U.Oxford, UK, 1986). Res. assoc. (tel. 089 + 8578, ext. 2670, fax +3777, email STUBBS at DM0MPB51). *Proteins, structural biology, viruses, X-ray diffraction, computing.*

Stuhrmann, Prof. Dr Heinrich B. Macromolecular Structure Research, GKSS Research Centre, Max-Planck-Strasse, D-2054 Geesthacht, Germany, Fed. Rep. (1938) Prof., physical chemistry (U. Mainz, 1972). Head of Res. Group. (tel.

041520 + 87, ext. 1290, fax +1403, telex 1332). *Molecular biophysics, physical chemistry, macromolecules, neutron scattering, X-ray synchrotron radiation.*

Suck, Dr Dietrich. Eur. Mol. Biol. Lab., Postfach 102209, Meyerhofstr. 1, D-6900 Heidelberg, Germany, Fed. Rep. (1944) Dr structure analysis (Braunschweig U., 1971). Group leader. (tel. 06221 + 387307). *Macromolecules, proteins, structural biology, X-ray diffraction.*

Süsse, Prof. Dr Peter. Mineralogisch-Kristallogr. Inst., Universität, V.M.Goldschmidtstr. 1, D-3400 Göttingen, Germany, Fed. Rep. (1939) Dr phil., mineralogy (U. Göttingen, 1963). Prof. (tel. 0551 + 39-1, ext. 3936). *Mineralogy, mineral data files.*

Suhre, Miss Ursula. Anorg. Chem. Inst., Universität, Greinstr. 6, D-5000 Köln 41, Germany, Fed. Rep. (1956) Diplom, chemistry (1983). (tel. 0221 + 4702913). *Structure, coordination compounds, CO and PF3-complexes.*

Sussieck-Fornefeld, Dr Cornelia. Studienkreis, Grenzhöfer Str. 11, D-6830 Schwetzingen, Germany, Fed. Rep. (1951) Dr, hexaoxogermanates (U. Heidelberg, 1988). (tel. 06202 + 12260). *Synthesis, germanium and silica oxides, thermodynamics in mineralogy.*

Tapfer, Dr Leander. MPI f. Festkörperforschung, Heisenbergstr. 1, D-7000 Stuttgart 80, Germany, Fed. Rep. (1956) Dr rer.nat., physics (U. Stuttgart, 1984). MTS. (tel. 0711 + 68601, telex 7255555 mpif d). *Double crystal X-ray diffractometry, X-ray topography, structure and properties, semiconductors, molecular beam epitaxy.*

Taxer, Dr Karlheinz Jürgen. Hochschulrechenzentrum, Universität, Gräfstr. 38, D-6000 Frankfurt/Main, Germany, Fed. Rep. (1939) Dr (TU Berlin, 1970). (tel. 069 + 7988106). *Crystal structure analysis, crystal chemistry, order-disorder substructures, derivative structures.*

Tebbe, Prof. Dr Karl-Friedrich. Inst. f. Anorg. Chemie, Universität, Greinstr. 6, D-5000 Köln 41, Germany, Fed. Rep. (1941) Dr rer.nat., inorganic chemistry (U. Münster, 1970). Prof. (tel. 0221 + 407, ext. 3285, email AC002 at DK0RRZK0)). *Inorganic, crystal and solid state chemistry, structure determination, computer programming, halogen compounds.*

Tennyson, Mrs Prof. Dr Christel. Inst. f. Mineralogie u. Kristallographie, TU, Hardenbergstr. 42, D-1000 Berlin 12, Germany, Fed. Rep. (1925) Dr rer.nat., mineralogy (TU Berlin, 1953). Prof. (tel. 030 + 314, ext. 2192). *Crystal chemistry, inorganic crystal structures, mineral classification and structure.*

Teske, Dr Christoph Ludwig. Inst. f. Anorg. Chemie, Olshausenstr. 60, D-2300 Kiel, Germany, Fed. Rep. (1942) Dr rer.nat., chemistry (U. Giessen, 1970). Akad. Oberrat. (tel. 0431 + 880, ext. 2408). *Inorganic chemistry.*

Thewalt, Prof. Dr Ulf. Sektion Röntgenbeugung, Universität, Oberer Eselsberg, D-7900 Ulm, Germany, Fed. Rep. (1939) Dr rer.nat., chemistry (U. Heidelberg, 1965). Sektionsleiter. (tel. 0731 + 176, ext. 2554). *Coordination chemistry, organometallic chemistry, symmetry.*

Thiele, Prof. Dr Gerhard. Inst. f. Anorg. Chemie, Universität, Albertstr. 21, D-7800 Freiburg, Germany, Fed. Rep. (1935) Dr rer.nat., inorganic chemistry (RWTH Aachen, 1964). Prof. (tel. 0761 + 2032894). *Structural inorganic chemistry.*

Thurn, Dr Herbert. Inst. f. Anorganische Chemie, Universität, Pfaffenwaldring 55, D-7000 Stuttgart 80, Germany, Fed. Rep. (1937) Dr rer.nat., chemistry (U. Stuttgart, 1967). Oberasst. (tel. 0711 + 685, ext. 4237). *Inorganic crystal structures, solid state chemistry, symmetry.*

Tigges, Mr Hartmut R. RIGAKU EUROPE GmbH, Monschauer Str. 7, D-4000 Düsseldorf 11, Germany, Fed. Rep. (1945) Dipl.-Chem., chemistry (U. Münster, 1973). Sales manager. (tel. 0211 + 50, ext. 2186, fax +4496, telex 172114145). *X-ray diffraction, X-ray fluorescence.*

Tillmann, Dr Bruno. Mineralog.-Kristallogr. Inst., Universität, Goldschmidtstr. 1, D-3400 Göttingen, Germany, Fed. Rep. (1936) Dr rer.nat., physics (TU Hannover, 1967). Wiss. Ang. (tel. 0551 + 393892). *Inorganic and organic crystal structures, crystallographic computing.*

Tillmanns, Prof. Dr Ekkehart. Mineralog. Inst., Universität, Am Hubland, D-8700 Würzburg, Germany, Fed. Rep. (1941) Dr rer.nat., mineralogy (U. Bochum, 1968). Prof. (tel. 0931 + 888431, fax 0931-707012, email KRIS003 at DWUUNI21). *Crystal structure analysis, X-ray crystallography, crystal chemistry.*

Töpel-Schadt, Dr Jutta. Mineralog.-Petrogr. Inst., Universität, Zülpicherstr. 49, D-5000 Köln 1, Germany, Fed. Rep. (1952) Dr phil.nat., mineralogy (U. Frankfurt, 1980). Wiss. Mit. (tel. 0221 + 4702238). *Electron microscopy, real structures in crystals, phase transitions.*

Tolksdorf, Prof. Dr Wolfgang. Philips GmbH Forschungslaboratorium, Vogt-Kölln-Str. 30, D-2000 Hamburg 54, Germany, Fed. Rep. (1932) Dr phil.nat.habil., inorganic chemistry, applied mineralogy (U. Frankfurt, 1961). Group leader. (tel. 040 + 5493, ext. 548, fax + 210, telex 21331656 pfd). *Solid state chemistry, materials characterization, crystal growth.*

Trauth, Mr Jürgen. Research Lab., SIEMENS AG, Otto-Hahn-Ring 6, D-8000 München, Germany, Fed. Rep. (1959) Dipl.-Min., mineralogy (1987). (tel. 089 + 6362696). *Crystal growth.*

Treimer, Dr Wolfgang. Hahn-Meitner-Institut, Glienickerstr. 100, D-1000 Berlin 39, Germany, Fed. Rep. (1950) Dr phil., physics (U. Vienna, Austria, 1975). Asst. (tel. 030 + 8009221). *Neutron interferometry, dynamical diffraction, fundamental physics, magnetic domains and Bloch walls.*

Treutmann, Dr Werner. Inst. f. Mineralogie, Fachbereich Geowissenschaften, Universität, Hans-Meerwein-Str., D-3550 Marburg, Germany, Fed. Rep. (1939)

Dr rer.nat., physics (U. Marburg, 1968). Akad. Oberrat. (tel. 06421 + 285701). *Crystal growth and characterization, magnetic properties.*

Trömel, Prof. Dr Martin Gerhard. Inst. f. Anorganische Chemie, Universität, Niederurseler Hang, D-6000 Frankfurt/Main 50, Germany, Fed. Rep. (1934) Dr, inorganic chemistry (U. Frankfurt/Main, 1963). Prof. (tel. 069 + 5800, ext. 9159, fax +9494). *Inorganic crystal structures, crystal chemistry.*

Trost, Dr Friedrich Karl. Res.& Develop., Joh.Schaefer Kalkwerke, Brandenburgerstr. 38, D-6252 Diez/Lahn, Germany, Fed. Rep. (1931) Dr rer.nat. (U. Tübingen, 1964). (tel. 06432 + 2327). *Physics, chemistry, crystallography, calcium (carbonates - sulphates - silicates).*

Tsernoglou, Prof. Demetrius. Eur. Mol. Biol. Lab., Meyerhofstr. 1, D-6900 Heidelberg, Germany, Fed. Rep. (1935) PhD, biophysics (Yale U., 1966) Sr. scient. (tel. 06221 + 387270, telex embld 461613) *Proteins.*

Tschulena, Dr Guido. Electronic Systems Department, Battelle-Inst. e.V., Am Römerhof 35, D-6000 Frankfurt/Main, Germany, Fed. Rep. (1945) Dr phil., solid state physics (U. Vienna, Austria, 1970). Senior scientist. (tel. 069 + 7908, ext. 2221, fax + 80, telex 411966). *Electronic materials, sensors, thin film technology, screen printing.*

Untersteller, Mr Eugen. Inst. f. Mineralogie, Fachbereich Geowissenschaften, Universität, Hans-Meerwein-Str. D-3550 Marburg, Germany, Fed. Rep. (1953) Dipl.Min., mineralogy Wiss. Mit. (tel. 064219 + 283466). *Crystallography, magnetism, neutron diffraction.*

Urban, Dr Heinz. Nichtmetallische Werkstoffe, TU, Zehntner Str. 2A, D-3392 Clausthal-Zellerfeld, Germany, Fed. Rep. (1928) Dr habil., applied mineralogy (TU Clausthal, 1957). Prof. (tel. 05323 + 1091). *Applied mineralogy, structural ceramics.*

Urban, Prof. Dr Knut Wolf. Inst. f. Festkörperforschung, KfA Jülich, P.O.Box 1913, D-5170 Jülich, Germany, Fed. Rep. (1941) Dr, physics. Direktor, Prof. (tel. 02461 + 61, ext. 3153, fax +6444, telex 833556-0 kfd). *Defect structures, disorder, electron diffraction, metals, electron microscopy, phase transitions, quasicrystals, scattering theory, semiconductors.*

Urland, Prof. Dr Werner. Inst. f. Anorganische Chemie, Universität, Callinstr. 9, D-3000 Hannover, Germany, Fed. Rep. (1944) Dr rer.nat., inorganic chemistry (U. Giessen, 1971). Univ.-Prof. (tel. 0511 + 762, ext. 2567). *Inorganic chemistry, solid state chemistry, complex chemistry, theoretical chemistry.*

Vahrenkamp, Prof. Dr Heinrich. Inst. f. Anorganische Chemie, Universität, Albertstr. 21, D-7800 Freiburg, Germany, Fed. Rep. (1940) Dr rer.nat., inorganic chemistry (U. München, 1967). Full prof. (tel. 0761 + 203, ext. 2913). *Inorganic crystal structures.*

Valeton, Mrs Prof. Dr Ida Walburga Jakobine. Geologisch- Paläontologisches Inst., Universität, Bundesstr. 55, D-2000 Hamburg, Germany, Fed. Rep. (1922) Prof., (U. Hamburg, 1944). Head, sedimentology dept. (tel. 040 + 41235042). *Crystallography, clay minerals, glauconites, bauxite minerals (gibbsite - boehmite - diaspore).*

Veith, Prof. Dr Michael. Inst. f. Anorg. Chemie, Universität, Im Stadtwald, D-6600 Saarbrücken, Germany, Fed. Rep. (1944) Dr rer.nat., chemistry (U. München, 1971) Prof. (tel. 0681 + 3023415). *Structural inorganic and metalorganic chemistry, crystal structure analysis, low temperature crystal structures.*

Vielhaber, Dr Edmund Antonius. Inst. f. Anorg. und Analyt. Chemie, Universität, Heinrich-Buff-Ring 58, D-6300 Giessen, Germany, Fed. Rep. (1931) Dr rer.nat., inorganic chemistry (U. Münster, 1964). Akad. Oberrat. (tel. 0641 + 702, ext. 5693). *Crystal growth, inorganic crystal structures.*

Viswanathan, Prof. Dr Krishnamoorthy. Mineralog. Inst., TU, Gauss-Str. 29, D-3300 Braunschweig, Germany, Fed. Rep. (1936), Dr sci.nat., mineralogy (ETH Zürich, Switzerland, 1967). Prof. (tel. 0531 + 391-1, ext. 3628). *Inorganic crystal structures, minerals, crystal physics.*

Vorbach, Dr Angelika Irene. Siemensstr. 1A, D-7550 Rastatt, Germany, Fed. Rep. (1953) Dr rer.nat., mineralogy (U. Karlsruhe, 1983). (tel. 07222 + 25161). *Gem stones, experimental mineralogy, petrography, geochemistry.*

Vorderwisch, Dr Peter. Hahn-Meitner-Inst. f. Kernforschung, Glienicker Str. 100, D-1000 Berlin 39, Germany, Fed. Rep. (1940) Dr rer.nat., physics (FU Berlin, 1974). Physicist. (tel. 030 + 80092747). *Solid state physics, neutron scattering, lattice dynamics, molecular dynamics, crystal-field theory, spin dynamics.*

Wacker, Dr Friedel Klaus. Kristallogr. Inst., Universität, Hebelstr. 25, D-7800 Freiburg i. Br., Germany, Fed. Rep. (1954) Dr, crystallography (U. Marburg, 1983). (tel. 0761 + 2034287). *Physical properties of crystals.*

Wagner, Dr Ernst-Heinz. Fritz-Haber-Inst., Max-Planck-Ges., Faradayweg 4-6, D-1000 Berlin 33, Germany, Fed. Rep. (1925) Dr rer.nat., physics (TH Stuttgart, 1954). Wiss. Mit. (tel. 030 + 8305271). *Crystal physics.*

Walcher, Dr Herbert. Inst. f. Angew. Festkörperphysik, Fraunhofer Ges., Eckerstr. 4, D-7800 Freiburg i. Br., Germany, Fed. Rep. (1949) Dr, crystallography (U. Freiburg, 1982). Staff crystallographer. (tel. 0761 + 2714288, telex 07-72510). *Crystal growth, II-VI compounds, low-temperature methods, microgravity influence on crystal growth.*

Walker, Dr Nigel P. C. ZHV/D - A30, BASF AG, D-6700 Ludwigshafen, Germany, Fed. Rep. (1955) PhD, protein crystallography (U. Bristol, UK, 1983). Chemiker (tel. 0621 + 6042638). *Protein crystallography, molecular modelling, small molecule crystallography, computer programming.*

Wallis, Dr Julian Mark. Anorg.-Chem. Inst., TU, Lichtenbergstr. 4, D-8046 Garching, Germany, Fed. Rep. (1959) D. phil., organomet. chemistry (U. Oxford,

UK, 1984). Res. Assoc. (tel. 089 + 3130, telex 17898174). *Metal vapour synthesis, organometallic compounds, structures.*

Wallrafen, Dr Franz. Mineralog. Inst., Universität, Poppelsdorfer Schloss, D-5300 Bonn, Germany, Fed. Rep. (1938) Dr rer.nat., mineralogy and crystallography (U. Bonn, 1972). Wiss. Ang. (tel. 0228 + 732961). *Crystal growth.*

Wang, Dr Naiding. Mineralogisch-Petrogr. Inst., Universität, Im Neuenheimer Feld 236, D-6900 Heidelberg, Germany, Fed. Rep. (1935) Dr, crystallography (U. Heidelberg, 1968). Res. asst. (tel. 06221 + 562804). *Crystallography, sulfides, synthesis, phase relations.*

Wartchow, Dr Rudolf. Inst. f. Anorg. Chemie, Universität, Callinstr. 9, D-3000 Hannover, Germany, Fed. Rep. (1940) Dr rer.nat., inorganic chemistry (TU Hannover, 1975). Akad. Rat. (tel. 0511 + 7622216). *Inorganic crystal structures, computer programming, symmetry.*

Weber, Dr Hans-Jürgen. Inst. f. Physik, Universität, Otto-Hahn-Str., D-4600 Dortmund, Germany, Fed. Rep. (1943) Dr rer.nat.habil., crystallography (U. Köln, 1983). Priv.-Doz. (tel. 0231 + 7553526, telex 822445). *Crystal physics, crystal growth, crystal optics.*

Weber, Prof. Dr Kurt. Inst. f. Mineralogie u. Kristallographie, TU, Hardenbergstr. 42, D-1000 Berlin 12, Germany, Fed. Rep. (1926) Dr rer.nat.habil., crystallography (TU Berlin, 1967). Prof. (tel. 030 + 314, ext. 3919). *Crystal growth, physical properties, inorganic crystal structures.*

Weckert, Dr Edgar. Inst. f. Angew. Physik, Lehrst. f. Kristallographie, Universität, Bismarckstr. 10, D-8520 Erlangen, Germany, Fed. Rep. (1960) Dr rer.nat, crystallography (U. Erlangen-Nürnberg, 1988). Wiss. Mit. (tel. 09131 + 852705, email EDGARWE at DERRZE0). *Multiple diffraction, dynamic theory, materials science.*

Wegener, Dr Joachim Rolf. Chemie-Labor, Philips GmbH, Valvo RHW, Stresemannallee 101, D-2000 Hamburg 54, Germany, Fed. Rep. (1942) Dr rer.nat., inorganic chemistry (U. Göttingen, 1970). Chief Chemist. (tel. 040 + 5613297). *Silicon, SiO2, Si3N4, thin metals.*

Weiner, Dr Karl Ludwig. Inst. f. Kristallographie und Mineralogie, Universität, Theresienstr. 41, D-8000 München 2, Germany, Fed. Rep. (1922) Dr rer.nat., crystallography, mineralogy (U. Bonn, 1954). Akad. Direktor. (tel. 089 + 23944355). *Instrumentation, X-ray diffraction, thin films, high and low temperature, powder diffraction, applied crystallography, applied mineralogy, archaeometry.*

Weiss, Prof. Dr Alarich. Inst. f. Physikalische Chemie, TH, Petersenstr. 20, D-6100 Darmstadt, Germany, Fed. Rep. (1925) Dr rer.nat., physics (TH Darmstadt, 1955). Prof. (tel. 06151 + 162607). *Solid state physical chemistry.*

Weiss, Prof. Dr Erwin Ludwig. Inst. f. Anorg. und Angew. Chemie, Universität, Martin-Luther-King-Platz 6, D-2000 Hamburg, Germany, Fed. Rep. (1926) Prof., inorganic chemistry (TU München, 1965). Prof. (tel. 040 + 41233102). *Organometallic and coordination chemistry, crystal structures.*

Weitzel, Dr Hans. Inst. f. Physikal. Chemie, Strukturforschung, TH, Petersenstr. 20, D-6100 Darmstadt, Germany, Fed. Rep. (1941) Dr rer.nat., physics (U. Tübingen, 1969). Akad. Oberrat. (tel. 06151 + 162298). *Neutron diffraction, crystal structures, magnetic structures, phase transitions.*

Wendl, Dr Wolfgang. Kristall- und Materiallabor, Fak. f. Physik, Universität, Kaiserstr. 12, D-7500 Karlsruhe, Germany, Fed. Rep. (1943) Dr rer.nat., chemistry (U. Heidelberg, 1970). Wiss. Mit. (tel. 0721 + 6083558). *Crystal growth, chemical characterization of crystals.*

Wenig, Prof Dr Werner. Lab. f. Angew. Physik, Universität, Lotharplatz, D-4100 Duisburg, Germany, Fed. Rep. (1943) Dr rer.nat., physics (U. Ulm, 1974). Prof. (tel. 0203 + 3792386). *X-ray analysis, polymers, amorphous substructures.*

Weston, Mr Simon Allan. Dep. of Biological Structures, EMBL, Meyerhofstr 1, D-6900 Heidelberg, Germany, Fed. Rep. (1965) B.Sc., physical biochemistry (CNAA U.K., 1987). Predoctoral fellow of EMBL. (tel. 06221 + 387266). *Structural biology, macromolecules, X-ray diffraction.*

Wiebcke, Dr Michael Helmut. Fak. f. Chemie, Universität, Postfach 5560 D-7750 Konstanz, Germany, Fed. Rep. (1952) Dr rer.nat., chemistry (U. Düsseldorf, 1986). (tel. 07531 + 883678, email CHFELS1 at DKNKURZ1). *Crystal structure determination, X-ray and neutron diffraction, zeolites, salt hydrates.*

Wieder, Dr Thomas. Experimentalphysik III, FB Physik, GH-Universität, Wilhelmshher Allee 73, D-3500 Kassel, Germany, Fed. Rep. (1961) Dr rer.nat., physics (1985). Wiss. Mit. (tel. 0561 + 804, ext. 6437, fax +2330). *Thin film structure, powder diffraction, residual stress measurement.*

Wierenga, Dr Rikkert Klaas. Dept. f. Structure Programs, EMBL, Meyerhoffstr. 1, D-6900 Heidelberg, Germany, Fed. Rep. (1949) Dr, chemistry (U. Groningen, Netherlands, 1978). *Protein crystallography, drug design, enzyme design.*

Wildenburg, Mr. Jörg Werner. Lab. f. Ultrahartkeramik/Hochdruck, Mineralog. Inst., Universität, Poppelsdorfer Schloss, D-5300 Bonn, Germany, Fed. Rep. (1959) Dipl. Min., mineralogy (U. Bonn, 1985). Doctor fellow, Ass. (tel. 0228 + 732765). *Computing, diffractometry, electron microscopy, high pressures, microscopy, phase transitions, X-ray diffraction.*

Wilhelm, Dr Eberhard. Balthasar-Schönfelderstr. 2, D-8550 Forchheim, Germany, Fed. Rep. (1948) Dr rer.nat., chemistry (U. Saarbrücken, 1975). *Organic crystal structures, stereochemistry, NMR spectroscopy.*

Wilke, Prof. Dr Wolfgang. Abt. f. Exper. Physik, Universität, Oberer Eselsberg, D-7900 Ulm, Germany, Fed. Rep. (1934) Apl. Prof., physics (U. Ulm, 1974). Prof. (tel. 0731 + 1762510). *Crystal physics, polymer structure, defects in crystals.*

Wilken, Dr Gerdt. Exploration and Producing, Mobil Oil, Burggrafstr.1, D-3100 Celle, Germany, Fed. Rep. (1940) Dr rer.nat., physics (U. Saarbrücken, 1970). (tel. 5141 + 15228). *Crystal growth, X-ray analysis, equilibrium studies, precipitation from aqueous solutions.*

Will, Prof. Dr Georg. Mineralog. Inst., Universität, Poppelsdorfer Schloss, D-5300 Bonn, Germany, Fed. Rep. Dr rer.nat., crystallography (TU München, 1958). Prof. (tel. 0228 + 73, ext. 2760 or 2761, fax +5579, telex 886657 unibo d). *Chemical bonding, computing, diffractometry, high pressures, magnetism, materials science, minerals, neutron diffraction, phase transitions, powder diffraction, texture, X-ray diffraction.*

Winnacker, Prof. Dr Albrecht. ZFE F1 AMF 33, Corporate Research, SIEMENS AG, Paul-Gossen-Str. 100, D-8520 Erlangen, Germany, Fed. Rep. (1942) Dr rer.nat., physics (1970). Research scientist. (tel. 09131 + 732802). *Defect structures, inorganic crystals, semiconductors.*

Winter, Prof Dr Werner. Res. Center, Grünenthal GmbH, Zieglerstr. 6, D-5100 Aachen, Germany, Fed. Rep. (1943) Dr habil., chemistry (U. Tübingen, 1978). (tel. 0241 + 51022326). *Medicinal chemistry, molecular modelling.*

Witte, Prof. em. Dr Helmut Hermann Wolfgang. Inst. f. Physikalische Chemie, TH, Petersenstr. 20, D-6100 Darmstadt, Germany, Fed. Rep. (1909) Dr phil., physics (U. Göttingen, 1933). Em. ord. Prof. (tel. 06151 + 165107). *Inorganic crystal structures, physical chemistry, metals and alloys.*

Wögerbauer, Dr Rupert. RME, Lab. f. Röntgenographie, Mineralogie u. Edelsteinforschung, Am Schlehenbusch 15, D-8711 Mainstockheim, Germany, Fed. Rep. (1941) Dr rer.nat., crystallography (U. Würzburg, 1974). (tel. 09321 + 5628). *X-ray diffraction, crystal growth, crystal physics.*

Wölfel, Prof. Dr Erich Richard. STOE & Cie GmbH, Hilpertstr. 10, D-6100 Darmstadt, Germany, Fed. Rep. (1922) Dr rer.nat., chemistry (TH Darmstadt, 1952). Prof. (tel. 06151 + 891225 or 82174, telex 419336). *Instrumentation, valency electron distribution.*

Wolf, Dr Dieter. Inst. f. Kristallographie und Mineralogie, Universität, Theresienstr. 41, D-8000 München 2, Germany, Fed. Rep. (1939) Dr rer.nat.habil., physics (U. München, 1972). Priv.-Doz. (tel. 089 + 23944333, telex 529815). *LEED, surface crystallography, surface disorder.*

Wolmershäuser, Dr Gotthelf. Fachbereich Chemie, Universität Kaiserslautern, Erwin-Schrödinger-Strasse, D-6750 Kaiserslautern, Germany, Fed. Rep. (1945) Dr rer.nat., chemistry (U. Kaiserslautern, 1976). Akad. Oberrat. (tel. 0631 + 2052468, telex 04-5627). *Coordination compounds, inorganic crystals, materials science, metals, semiconductors.*

Wondratschek, Prof. Dr Hans. Inst. f. Kristallographie, Universität, Kaiserstr. 12, D-7500 Karlsruhe, Germany, Fed. Rep. (1925) Dr rer.nat., mineralogy (U. Bonn, 1953). Prof. (tel. 0721 + 6083320). *Symmetry, group theory, structural chemistry.*

Wroblewski, Dr Thomas. HASYLAB, DESY, Notkestr. 85, D-2000 Hamburg, Germany, Fed. Rep. (1955) Dr rer.nat., physics Wiss. Ang. (tel. 040 + 8998, ext. 3004, fax +3282, telex 215124 desy d). *Synchrotron radiation, instrumentation, powder diffraction.*

Wunderlich, Dr Hartmut. Inst. f. Anorg. Chemie u. Strukturchemie, Universität, Universitätsstr. 1, D-4000 Düsseldorf, Germany, Fed. Rep. (1938) Dr rer.nat., physics (TU Braunschweig, 1969). Akad. Oberrat. (tel. 0211 + 3113144). *Inorganic and organic crystal structures, computer programming, instrumentation, teaching.*

Zedler, Dr Achim. Inst. f. Kristallographie, Universität, Senckenberg-Anlage 30, D-6000 Frankfurt/M,, Germany, Fed. Rep. (1934) Dr rer.nat., physics (Akad.d.Wiss.d.DDR Berlin, 1973). (tel. 069 + 7982100). *Order-disorder crystal structures, OD-methods, structure analysis, OD-theory.*

Zeilinger, Prof. Dr Anton. Physik Department E21, TU München, D-8046 Garching, Germany, Fed. Rep. (1945) Dr rer.nat.habil., physics (U. Vienna, Austria, 1971). (tel. 089 + 3209 ext. 2476, fax 2112). *Neutron diffraction.*

Zemke, Mr Klaus Jürgen. Inst. f. Kristallographie, FU, Takustr. 6, D-1000 Berlin 33, Germany, Fed. Rep. (1962) M.Sc., chemistry (U.Cape Town/South Africa, 1985). (tel. 030 + 8386599). *Protein, macromolecules, computing.*

Zeyfang, Dr Rolf Robert. Forschungsinstitut, AEG-Telefunken, Goldsteinstr. 235, D-6000 Frankfurt/Main, Germany, Fed. Rep. (1938) Dr rer.nat., physics (U. Stuttgart, 1967). Group leader. (tel. 0611 + 6679221). *Inorganic crystal structures, crystal growth, ceramics, ferroelectrics, epitaxy.*

Ziegler, Prof. Dr Manfred Ludwig. Anorg.-Chem. Inst., Universität, Im Neuenheimer Feld 270, D-6900 Heidelberg, Germany, Fed. Rep. (1936). (tel. 06221 + 562468). *Metalorganic chemistry, X-ray structure analysis, molecular compounds.*

Zigan, Prof. Dr Franz Martinus. Inst. f. Kristallographie, Universität, Senckenberg-Anlage 30, D-6000 Frankfurt/Main, Germany, Fed. Rep. (1928) Dr phil., physics (U. Marburg, 1958). Prof. (tel. 069 + 7982293). *Crystal physics, neutron diffraction, inorganic crystal structures.*

Zimmermann, Dr Helmuth W. Mineralogisches Inst., Universität, Poppelsdorfer Schloss, D-5300 Bonn 1, Germany, Fed. Rep. (1945) Dr rer.nat.habil., crystallography (U. Erlangen-Nürnberg, 1982). Prof. (tel. 0228 + 732770). *Mathematical crystallography, X-ray crystallography, direct methods, crystallographic computing, crystal structures, teaching.*

Zobel, Dr Dieter. Inst. f. Kristallographie, FU Berlin, Takustr. 6, D-1000 Berlin 33, Germany, Fed. Rep. (1941) Dr rer.nat., crystallography (1968). Wiss.Ass. (tel. 030 + 8383453). *Crystal structures, physical properties, low temperatures, organic molecules.*

Zorn, Dr Gerhard. Res. Laboratories, SIEMENS AG, Otto-Hahn-Ring 6, D-8000 München 83, Germany, Fed. Rep. (1956) Dr, crystallography (tel. 089 + 6363274). *X-ray diffraction, diffraction software.*

Zulehner, Dr Werner. EK, Wacker-Chemitronic GmbH, Postfach 1140, D-8236 Burghausen, Germany, Fed. Rep. (1941) Dr rer.nat., physics (U. Saarbrucken, 1973). Dev. manager. (tel. 8677+832547). *Solid state phyics, semiconductor physics, crystal growth, crystallography.*

GHANA

Sub-Editor: **F. L. Phillips**

Notes

1. International telephone country code - 233.

Baeta, Dr Robert Domingo. Dept. of Physics, U. of Ghana, P.O. Box 63, Legon, Accra, Ghana. (1939) PhD, physics (U. Bristol, UK, 1969). Sr. lect. (tel. Accra + 75381, ext. 8427). *Electron microscopy, defects, X-ray topography.*

Phillips, Dr Frederick Lloyd. Dept. of Chemistry, U. of Ghana, Legon, Accra, Ghana. (1949) PhD, chemical crystallography (U. London, UK, 1975). Lect. (tel. Accra + 75381, ext. 8329). *Coordination complexes.*

GREECE

Sub-Editor: **P. J. Rentzeperis**

Notes

1. International telephone country code - 0030.

2. Greek Universities confer the following degrees: a) The basic degree of the corresponding field, at the end of the undergraduate studies (4 - 6 years), equivalent to BA or BSc. b) The *Doctor's* degree, equivalent to DSc or PhD. c) At some Universities a degree equivalent to MA or MSc, at the end of a postgraduate course in a special field, e.g. Electronics.

3. The regular teaching and research positions at Greek Universities are equivalent to *Professor, Associate Professor, Assistant Professor, Lecturer, Chief Assistant, Scientific Assistant* and *Teaching Assistant*. A specialist, holding a *Doctor's* degree, may be appointed temporarily to the position of a *Special Scientist* for teaching purposes.

Alexandropoulos, Prof. Nikos. Physics Dept., U. of Ioannina, Ioannina, Greece. (1934) PhD, Physics (Athens U., 1964). Prof. (tel. 0651 + 91-396, telex 0651 + 322160). *X-ray physics, electron distribution, dynamical diffraction, instrumentation.*

Alexandropoulos, Mrs Tina. Physics Dept., U. of Ioannina, Ioannina, Greece. (1934) BSc, Physics (Athens U., 1961). Instructor. (tel. 0651 + 91-950, telex 0651 + 322160). *X-ray physics.*

Alexopoulos, Prof. (Emeritus) Kessar. Spefsipou 7, Athens 139, Greece. (1909) DSc, physics (E.T.H. Zurich, Switzerland, 1935). Retired from Athens U. (tel. 01 + 738-442, telex 01 + 223815). *Crystal physics, structure defects.*

Antonopoulos, Prof. John. Physics Dept., U. of Thessaloniki, Thessaloniki, Greece. (1939) DSc, electron microscopy (Thessaloniki U., 1972). Prof. (tel. 031 + 99-2806, telex 031 + 412184, fax 031 + 206138) *Crystal structures, defects, elastic properties, fatigue.*

Basilakis, Mr Michael. Physics Lab., U. of Patras, Patras, Greece. (1947) BSc, physics (Patras U., 1973). Asst. (tel. 061 + 429-713, telex 061 + 312239). *Liquid crystals.*

Boyiatzis, Mr Ioannis. Physics Lab., U. of Patras, Patras, Greece. (1951) BSc, physics (Patras U., 1972). Asst. (tel. 061 + 429-713, telex 061 + 312239). *Diffuse X-ray scattering, inorganic crystal structures.*

Bozopoulos, Dr Anastasios Panayiotis. Appl. Physics Lab., U. of Thessaloniki, Thessaloniki, Greece. (1944) DSc, crystallography (Thessaloniki U., 1985). Lecturer. (tel. 031 + 99-2803, telex 031 + 412184, fax 031 + 206138, email caez04%grtheun1). *Inorganic and organic crystal structures.*

Calamiotou, Dr Maria. Physics Dept., U. of Athens, Athens, Greece. (1949) DSc, physics (Athens U., 1978). Lect. (tel. 01 + 363-3413, telex 01 + 223815). *Crystal physics, dynamical diffraction theory, archaeometry.*

Christidis, Prof. Panayiotis Chrysostomos. Appl. Physics Lab., U. of Thessaloniki, Thessaloniki, Greece. (1942) DSc, crystallography (Thessaloniki U., 1975). Assoc. prof. (tel. 031 + 99-2803, telex 031 + 412184, fax 031 + 206138, email caez02%grtheun1). *Inorganic and organic crystal structures, electron density distribution.*

Economou, Prof. Nicolaos Alkiviadis. Physics Lab., U. of Thessaloniki, Thessaloniki, Greece. (1926) PhD, physics (Wayne State U., USA, 1959). Prof. (tel. 031 + 99-1439, telex 031 + 412184, fax 214276). *Structure and defects, semiconductors.*

Euthymiou, Prof. Paraskevi. Physics Lab., U. of Athens, Athens, Greece. (1923) DSc, physics (Athens U., 1952). Asst. prof. (tel. 01 + 3611-927, telex 01 + 223815). *Semiconductor physics, transport phenomena in crystals.*

Evangelidou, Miss Christina. Electrical Eng. Dept., U. of Thrace, Xanthi, Greece. (1950) BSc, Physics (Thessaloniki U., 1973). Asst. (tel 0541 + 26-475, telex 0531 + 462205). *Crystal structures, magnetic structures, rare earths.*

Fanariotis, Mr Iakovos. Appl. Physics Lab., U. of Thessaloniki, Thessaloniki, Greece. (1951) BSc, chemistry (Thessaloniki U., 1974). Asst. (tel. 031 + 99-2803, telex 031 + 412184, fax 031 + 206138). *Inorganic crystal structures, crystal growth.*

Flevaris. Prof. Nikolaos. Physics Lab., U. of Thessaloniki, Thessaloniki, Greece. (1953) PhD, mater. sc. and eng. (Northwestern U.,Evanston USA, 1983). Asst.Prof. (tel. 031 + 99-1427, telex 031 + 412184, fax 206138). *Materials science.*

Filippakis, Dr Sophokles. Physics Lab., N. R. C. Demokritos, Athens, Greece. (1933) PhD, crystallography (Weizmann Inst. of Sci., Israel, 1966). Sr. Scient. (tel. 01 + 6513-111, telex 01 + 216199). *Crystal structures, crystal physics, structure of metals.*

Ftikos, Dr Christos. Chemistry Lab., Technical U. of Athens, Athens, Greece. (1949) DSc, inorg. chemistry (Techn. U. of Athens, Athens, 1972). Asst. prof. (tel. 01 + 369-1385, telex 01 + 221682). *X-ray diffraction, electron microscopy, cement chemistry and technology.*

Galinos, Prof. Andreas. Physics Lab., U. of Patras, Patras, Greece. (1925) DSc, inorganic chemistry (Athens U., 1955). Prof. *Inorganic crystal structures.*

Gountsidou, Mrs Vasiliki. Solid State Physics Dept., U. of Thessaloniki, Thessaloniki, Greece. (1955) MSc, Electronics, (Thessaloniki U., 1978). Res. Asst. (tel. 031 + 99-1434, telex 031 + 412184, fax 031 + 206138). *Measurements, semiconductors.*

Grigoriades, Prof. Panayotis. Physics Lab., U. of Thessaloniki, Thessaloniki, Greece. (1943) DSc, physics (Thessaloniki U., 1980). Asst. prof. (tel. 031 + 99-1400, telex 031 + 412184, fax 031 + 206138). *Crystal lattice defects.*

Hamodrakas, Dr Stavros. Physics Lab., N. R. C. Demokritos, Athens, Greece. (1947) PhD, biophysics (Leeds U., 1974). Res. asst. (tel. 01 + 6513-111, telex 01 + 216199). *Molecular biology, pharmacology, proteins.*

Hountas, Dr Athanasios. Physics Lab., N. R. C. Demokritos, Athens, Greece. (1945) DSc, X-ray physics (Thessaloniki U., 1977). Res. asst. (tel. 01 + 6513-111, telex 01 + 216199). *Crystal physics, dynamical diffraction theory.*

Kagarakis, Prof. Constantine. Electrical Eng. Dept., Techn. U. of Athens, Athens, Greece. (1933) DSc, chemical eng. (Techn. U. of Athens, 1955). Prof. (tel. 01 + 362-3770, telex 01 + 221682). *Semiconductors.*

Kambas, Prof. Kostas. Solid State Physics Dept., U. of Thessaloniki, Thessaloniki, Greece. (1945) PhD, solid state physics (Thessaloniki U., 1981). Asst. prof. (tel. 031 + 99-2850, telex 031 + 412184, fax 031 + 206138). *Lattice dynamics, Raman spectroscopy, superionic conductors, optical and electrical properties, thin films.*

Kanellis, Prof. George. Physics Lab., U. of Thessaloniki, Thessaloniki, Greece. (1942) DSc, physics (Thessaloniki U., 1977). Assoc. prof. (tel. 031 + 99-1421, telex 031 + 412184, fax 031 + 206138). *Lattice dynamics, Raman spectroscopy, superionic conductors, phase transitions.*

Katagas, Dr Christos. Geology Lab., U. of Patras, Patras, Greece. (1944) PhD, geology (U. of Manchester, U.K., 1975). Lect. (tel. 061 + 991-972, telex 061 + 312239). *Structure, minerals.*

Katsanos, Mr Demetrios Evangelos. Solid State Physics Dept., U. of Ioannina, Ioannina, Greece. (1959) BSc, Physics. Asst. (tel. 0651 + 91-234). *Incoherence X-ray spectroscopy.*

Kavounis, Dr Konstantinos. Appl. Physics Lab., U. of Thessaloniki, Thessaloniki, Greece. (1945) DSc, crystallography (Thessaloniki U., 1985). Lecturer. (tel. 031 + 99-2688, telex 031 + 412184, fax 031 + 206138, email caez03%grtheun1). *Organic crystal structures.*

Keramidas, Mr Konstantinos Georgios. Appl. Physics Lab., U. of Thessaloniki, Thessaloniki, Greece. (1960) BSc, physics (Thessaloniki U., 1981). Asst. (tel. 031 + 99-2803, telex 031 + 412184, fax 031 + 206138, email caez06%grtheun1). *Inorganic and organic crystal structures.*

Kokkou, Prof. Socrates Constantinos. Appl. Physics Lab., U. of Thessaloniki, Thessaloniki, Greece. (1940) DSc, crystallography (Thessaloniki U., 1975). Assoc. prof. (tel. 031 + 99-2688.tex.031 + 412184 fax 031 + 206138, email caez11%grtheun1). *Organic crystal structures, direct methods.*

Konguetsof, Dr Helen. Electrical Eng. Dept., U. of Thrace, Xanthi, Greece. (1938) PhD, Physics (Sussex U., England, 1.1.). Lect. (tel. 0541 + 26475, telex 0531 + 462205). *Crystal structures, magnetic structures, rare earths.*

Kosmopoulos, Dr John. Physics Lab., U. of Patras, Patras, Greece. (1945) DSc, physics (U. of Patras,1975). Chief Asst. (tel. 061 + 429-713, telex 061 + 312239). *Liquid crystals.*

Kotsanidis, Mr Panayotis. Electrical Eng. Dept., U. of Thrace, Xanthi, Greece. (1946) BSc, Physics (Thessaloniki U., 1970). Asst. (tel. 0541 + 26475, telex 0531 + 462205). *Crystal structures, magnetic structures, rare earths.*

Kotsis, Mr Konstantinos. Physics Dept., U. of Ioannina, Ioannina, Greece. (1959) BSc, Phycics (Thessaloniki U., 1980), Instructor. (tel. 0651 + 91-950, telex 0651 + 322160). *Dynamical diffraction theory.*

Koumelis, Dr Christos. Physics Lab., U. of Athens, Athens, Greece. (1931) DSc, physics (Athens U., 1963). Lect. (tel. 01 + 3633-412, telex 01 + 223815). *Inorganic crystal structures, solid state physics.*

Kountouris, Mr Costas. Physics Lab., U. of Patras, Patras, Greece. (1950) BSc, physics (Patras U., 1972). Asst. (tel. 061 + 429-713, telex 061 + 312239). *Diffuse X-ray scattering, inorganic crystal structures.*

Kyriakos, Prof. Demetrius. Physics Lab., U. of Thessaloniki, Thessaloniki, Greece. (1941) DSc, solid state physics (Thessaloniki U., 1978). Assoc. prof. (tel. 031 + 99-2807, telex 031 + 412184, fax 031 + 206138). *Transport phenomena in solids, crystal growth.*

Leventouri, Dr Dora. Physics Lab., U. of Athens, Athens, Greece. (1943) DSc, physics (Athens U., 1972). Teaching asst. (tel. 01 + 363-3413, telex 01 + 223815). *Solid state physics, X-ray spectroscopy.*

Loizos, Mr Zafiris. Chem. Eng. Dept., Techn. U. of Athens, Athens, Greece. (1950) BSc, chem. eng. (Techn. U. of Athens, 1973). Asst. (tel. 01 + 369-1244, telex 01 + 221682). *Inorganic crystal structures.*

Londos, Dr Charalampos. Physics Lab., U. of Athens, Athens, Greece. (1947) DSc, physics (Athens U., 1979). Asst. (tel. 01 + 363-3413, telex 01 + 223815). *Solid state physics, X-ray inelastic scattering.*

Manolikas, Prof. Konstantinos. Solid State Physics Dept., U. of Thessaloniki, Thessaloniki, Greece. (1944) PhD, Electron microscopy (Thessaloniki U., 1976). Prof. (tel. 031 + 99-1421, telex 031 + 412184, fax 031 + 206138). *Transmission electron microscopy.*

Mavridis, Prof. Aristides. Chemistry Dept., U. of Athens, Athens, Greece. (1944) PhD, crystallography (Michigan State U., USA, 1975). Asst. prof. (tel. 01 + 360-6529, telex 01 + 223815). *Structural chemistry, theoretical chemistry, small molecules.*

Miliotis, Prof. Demitrios Menelaos. Appl. Physics Lab., U. of Ioannina, Ioannina, Greece. (1931) DSc, physics (Athens U., 1965). Prof. (tel. 0651 + 22-956). *Inelastic X-ray scattering.*

Mourikis, Dr Stamatios. Physics Lab., U. of Athens, Athens, Greece. (1928) DSc, physics (Athens U., 1963). Chief asst. (tel. 01 + 363-3412, telex 01 + 223815). *Metal physics.*

Moustakali Mavridis, Dr Irene. Physics Lab., N. R. C. Demokritos, Athens, Greece. (1946) PhD, physical chemistry (U. Michigan, USA, 1975). Res. asst. (tel. 01 + 6513-111, telex 01 + 216199). *Organic and biological structures, solid state reactions.*

Panagos, Prof. Athanasios. Geology Lab., U. of Patras, Patras, Greece. (1926) PhD, crystallography (E.T.H. Zurich, Switzerland, 1960). Prof. (tel. 061 + 429-714, telex 061 + 312239). *Inorganic structures.*

Papadakis, Prof. Alexander. Mineralogy Lab., U. of Thessaloniki, Thessaloniki, Greece. (1928) DSc, mineralogy and petrology (Thessaloniki U., 1965). Asst. prof.Retired from Thessaloniki U. (tel. 031 + 99-2818, telex 031 + 412184, fax 031 + 206138). *Crystal optics.*

Papadimitraki Chlichlia, Prof. Helena. Physics Lab., U. of Thessaloniki, Thessaloniki, Greece. (1930) DSc, solid state physics (Thessaloniki U., 1960). Prof. (tel. 031 + 99-2405, telex 031 + 412184, fax 031 + 206138). *Electrical and thermal conductivity, hall effect, structure of materials.*

Papadopoulos, Mr Demetrius. Physics Lab., U. of Thessaloniki, Thessaloniki, Greece. (1945) BSc, physics (Thessaloniki U., 1969). Asst. (tel. 031 + 99-2851, telex 031 + 412184, fax 031 + 206138). *Electron microscopy, crystalline materials.*

Papathanassopoulos, Dr Constantinos. Physics Lab., N. R. C. Demokritos, Athens, Greece. (1934) DSc, physics (T. Hochschule, Braunschweig, 1964). Res. asst. (tel. 01 + 6513-111, telex 01 + 216199). *Radiation damage, crystal growth.*

Parissakis, Prof. George. Chemistry Lab., Technical U. of Athens, Athens, Greece. (1929) PhD, chemistry (E.T.H. Zurich, Switzerland, 1955). Prof. *X-ray diffraction, electron microscopy,cement chemistry and technology.*

Perdikatsis, Dr Basilios. Mineralogy Lab., Inst. of Geol. and Min. Res., Athens, Greece. (1942) PhD, crystallography (U. Erlangen, BRD, 1972). Head, X-ray dept. (tel. 01 + 779-8412, telex 01 + 223815). *Powder diffraction, microanalysis.*

Polychroniades, Prof. Efstathios. Physics Lab., U. of Thessaloniki, Thessaloniki, Greece. (1946) DSc, solid state physics (Thessaloniki U., 1972). Asst. prof. (tel. 031 + 99-1427, telex 031 + 412184, fax 031 + 206138). *Crystal lattice defects.*

Priftis, Prof. George. Physics Lab., U. of Patras, Patras, Greece. (1937) DSc, physics (Athens U., 1969). Asst. prof. *Diffuse X-ray scattering, inorganic crystal structures.*

Profi, Mrs Stella. Physics Lab., N. R. C. Demokritos, Athens, Greece. (1949) BSc, natural sciences (Athens U., 1972). Res. asst. (tel. 01 + 6513-111, telex 01 + 216199). *Inorganic and mineral structure, mineral chemistry.*

Rentzeperis, Prof. Panayiotis Ioannis. Appl. Physics Lab., U. of Thessaloniki, Thessaloniki, Greece. (1928) DSc, crystallography, mineralogy and petrography (Thessaloniki U., 1956). Prof. and head. (tel. 031 + 99-1444, telex 031 + 412184 fax 031 + 206138, email caez16%grtheun1). *Inorganic and organic crystal structures, crystal physics.*

Rigopoulos, Prof. Rigas. Physics Lab., U. of Patras, Patras, Greece. (1929) PhD, physics (Birmingham U., UK, 1963). Prof. (tel. 061 + 429-713, telex 061 + 312239). *Liquid crystals.*

Rocophyllou Agathonikou, Dr Elsa - Helena. Physics Lab., N. R. C. Demokritos, Athens, Greece. (1936) DSc, physical chemistry (U. Paris, France, 1964). Res. officer. (tel. 01 + 6513-111, telex 01 + 216199). *Radiation damage, metal physics, structure, metals.*

Roilos, Prof. Minas. Physics Lab., U. of Patras, Patras, Greece. (1930) DSc, solid state physics (Athens U., 1970). Prof. (tel. 061 + 99-1764, telex 061 + 312239). *Inorganic crystal structures, crystal growth, amorphous semiconductors.*

Sahalos, Prof. John. U. of Thessaloniki, Thessaloniki, Greece. (1943) PhD, Appl. electromagnetics (Thessaloniki U., 1974). Prof. (tel. 031 + 99-1400, telex 031 + 412184, fax 031 + 206138). *Magnetic materials, ferrites, microwave applications of ferrites.*

Sakellaridis, Prof. Paul. Chemistry Lab., Techn. U. of Athens, Athens, Greece. (1920) DSc, chemistry (U. Paris, France, 1953). Prof. *Inorganic crystal structures.*

Sakkopoulos, Dr Sotirios. Physics Lab., U. of Patras, Patras, Greece. (1945) DSc, solid state physics (Patras U., 1974). Chief asst. (tel. 061 + 429-713, telex 061 + 312239). *Electric properties, metals, semiconductors.*

Sandalaki, Dr Zefi. Physics Lab., U. of Athens, Athens, Greece. DSc, physics (Athens U., 1953). Chief asst. (tel. 01 + 363-3412, telex 01 + 223815). *Crystal physics.*

Semitelou, Mrs Julia. Electrical Eng. Dept., U. of Thrace, Xanthi, Greece. (1956) BSc, Physics (Thessaloniki U., 1980). Asst. (tel. 0541 + 26475, telex 0531 + 462205). *Crystal structures, magnetic structures, rare earths.*

Sianou, Miss Anna. Physics Lab., U. of Thessaloniki, Thessaloniki, Greece. (1945) MA, history of science (Cornell U., USA, 1970). Asst. (tel. 031 + 99-1456, telex 031 + 412184, fax 031 + 206138). *Electrical and thermal conductivity in crystals, structure materials.*

Siapkas, Prof. Demetrios John. Physics Lab., U. of Thessaloniki, Thessaloniki, Greece. (1934) DSc, physics (Thessaloniki U., 1970). Asst. prof. (tel. 031 + 99-1427, telex 031 + 412184, fax 031 + 206138). *F.I.R. and Raman spectroscopy, phase transition, ferroelectricity.*

Soldatos, Prof. Constantinos. Mineralogy Lab., U. of Thessaloniki, Thessaloniki, Greece. (1924) DSc, mineralogy and petrology (Thessaloniki U., 1955), DPhil, (E.T.H. Zurich, Switzerland, 1960). Prof. (tel. 031 + 99-1445, telex 031 + 412184, fax 031 + 206138). *Feldspars, crystal optics.*

Spyrellis, Dr Nicolaos. Chemistry Lab., Technical U. of Athens, Athens, Greece. (1943) DSc, appl. chemistry (U. P. et M. Curie, Paris, France, 1974). Lect. (tel. 01 + 369-1244, telex 01 + 221682). *Crystal growth, electrocrystallization.*

Spyridelis, Prof. John. Physics Lab., U. of Thessaloniki, Thessaloniki, Greece. (1929) DSc, solid state physics (Thessaloniki U., 1964). Prof. (tel. 031 + 99-1437, telex 031 + 412184, fax 031 + 206138). *Crystal physics, electron microscopy.*

Stergiou, Dr Anagnostis Charalambos. Appl. Physics Lab., U. of Thessaloniki, Thessaloniki, Greece. (1941) DSc, crystallography (Thessaloniki U., 1984). Assist.prof. (tel. 031 + 99-2643, telex 031 + 412184, fax 031 + 206138, email caez05%grtheun1). *Inorganic and organic crystal structures.*

Stergioudis, Dr Georgios Asterios. Appl. Physics Lab., U. of Thessaloniki, Thessaloniki, Greece. (1946) DSc, crystallography (Thessaloniki U., 1984). Assist.prof. (tel. 031 + 99-2803, telex 031 + 412184, fax 031 + 206138, email caez12%grtheun1). *Inorganic crystal structures.*

Stoimenos, Prof. John Nikolaos. Physics Lab., U. of Thessaloniki, Thessaloniki, Greece. (1934) DSc, physics (Thessaloniki U., 1969). Prof. (tel. 031 + 99-1427, telex 031 + 412184, fax 031 + 206138). *Crystal lattice defects.*

Terzis, Dr Aristides. Physics Dept., N. R. C. Democritos, Athens, Greece. (1941) PhD, Inorganic chemistry, (Princeton U., USA, 1970). Res. assoc. (tel. 01 + 651-3111, ext. 123, telex 21-6199, fax 01 + 216199). *Transition metal complexes, unidirectional compounds, low temperatures.*

Theodoridou, Dr Irini. Physics Dept., U. of Ioannina, Ioannina, Greece. (1948) PhD, physics (Ioannina U., 1983). Lect. (tel. 0651 + 91-951). *X-ray physics, electron distribution.*

Theodoropoulos, Prof. Dimitrios. Physics Lab., Techn. U. of Athens, Athens, Greece. (1926) DSc, chemistry (Athens U., 1951). Prof. (tel. 01 + 363-3412, telex 01 + 221682). *Organic crystal structures.*

Theodossiou, Prof. Alexandros. Physics Lab., U. of Patras, Patras, Greece. (1918) DSc, solid state physics (Athens U., 1950). Prof. *Crystal physics.*

Tsatis, Mr Demetrius. Physics Lab., U. of Patras, Patras, Greece. (1939) MSc, nuclear physics (Carleton U., Canada, 1968) Asst. (tel. 061 + 272-945, telex 061 + 312239). *Electric and thermal properties of crystals.*

Tsimberis, Mr Nikolaos. Physics Lab., U. of Patras, Patras, Greece. (1946) BSc, physics (Patras U., 1973). Asst. (tel. 061 + 429-713, telex 061 + 312239). *Liquid crystals.*

Tsoli Kataga, Dr (Mrs) Panayota. Geology Lab., U. of Patras, Patras, Greece. (1946) DSc, geology (Patras U., 1979). Lect. (tel. 061 + 991-972, telex 061 + 312239). *Structure, clay minerals.*

Tsoukalas, Prof. John. Physics Lab., U. of Thessaloniki, Thessaloniki, Greece. (1941) DSc, solid state physics (Thessaloniki U., 1975). Prof. (tel. 031 + 99-1456, telex 031 + 412184, fax 031 + 206138). *Hall effect, metals and alloys.*

Valassiades, Prof. Odysseus. Physics Lab., U. of Thessaloniki, Thessaloniki, Greece. (1944) DSc, physics (Thessaloniki U., 1976). Asst. prof. (tel. 031 +

99-1400, telex 031 + 412184, fax 031 + 206138). *Defects, semiconductors, transport phenomena.*

Venetopoulos, Dr Cleanthis. Appl. Physics Lab., U. of Thessaloniki, Thessaloniki, Greece. (1929) DSc, crystallography (Thessaloniki U., 1975). Res. assoc. (tel. 031 + 99-2803, telex 031 + 412184, fax 031 + 206138) *Inorganic crystal structures, computer programming.*

Vgenopoulos, Prof. Andreas. Mineralogy Dept., Techn. U. of Athens, Athens, Greece. (1940) PhD, mineralogy (Techn. U. Athens). Assoc. prof. (tel. 01 + 369-1307, telex 01 + 221682). *Powder diffraction.*

Voliotis, Prof. Stavros. Chemistry Lab., U. of Patras, Patras, Greece. (1934) DSc, chemistry (U. Paris VII, France, 1975). Prof. (tel. 061 + 99-1973, telex 061 + 312239). *Structural crystallography, crystal chemistry.*

Voutsas, Dr George Panayiotis. Appl. Physics Lab., U. of Thessaloniki, Thessaloniki, Greece. (1945) DSc, crystallography (Thessaloniki U., 1983). Assist.prof. (tel. 031 + 99-2803, telex 031 + 412184, fax 031 + 206138, email caez06%grtheun1). *Inorganic and organometallic crystal structures.*

Vradis, Mr Alexandros. Physics Lab., U. of Patras, Patras, Greece. (1950) BSc, physics (Patras U., 1974). Asst. (tel. 061 + 429-713, telex 061 + 312239). *Diffuse X-ray scattering, inorganic crystal structures.*

Yakinthos, Prof. John. Physics Lab., U. of Thrace, Xanthi, Greece. (1937) DSc, solid state physics (Grenoble U., France, 1971). Prof. (tel. 0541 + 26-944, telex 0531 + 462205). *Crystal structure, magnetic structure, Mossbauer, crystal fields,rare earths.*

Zardas, Mr George. Physics Lab., U. of Athens, Athens, Greece. (1944) BSc, physics (Athens U., 1970). Teaching asst. (tel. 01 + 363-3413, telex 01 + 223815). *Solid state physics, elastic properties of crystals.*

Zenginoglou, Mr Charalambos. Physics Lab., U. of Patras, Patras, Greece. (1946) BSc, physics (Athens U., 1970). Asst. (tel. 061 + 429-713, telex 061 + 312239). *Liquid crystals.*

HONG KONG

Sub-Editor: **T. F. Lai**

Notes

1. International telephone country code - 852.

2. Degrees conferred by the Hong Kong universities are the *Master of Science* (MSc), *Master of Philosophy* (MPhil) and *Doctor of Philosophy* (PhD).

3. The university appointments are similar to the UK equivalents.

Cheng, Mr Graham Cheng-hsun. Taching Petroleum Co. Ltd., 2B Boundary Crest, 177-177A Boundary St., Kowloon, Hong Kong. (1936) MSc, geology (U. of Manchester, UK, 1962). Director. (tel. 3 + 386143, fax 852 5 8902962). *Inorganic crystal structures, crystal physics, phase transitions, high temperature properties.*

Cheung, Dr Kung Kai. Chemistry Dept., U. of Hong Kong, Hong Kong. (1935) PhD, crystallography (U. of Glasgow, UK, 1965). Lect. (tel. 5 + 8592167, fax 852 5 479907). *Single-crystal and powder diffractometry.*

Hon, Dr Ping-Kay. Chemistry Dept., Chinese U. of Hong Kong, Shatin N.T., Hong Kong. (1936) PhD, chemistry (U. of Illinois, USA, 1964). Lect. (tel. 0 + 6952329, fax 852 0 6954234). *Organometallic compounds, crystal structures, analytical chemistry, instrumentation.*

Lai, Dr Ting Fong. Chemistry Dept., U. of Hong Kong, Hong Kong. (1930) DPhil, crystallography (U. of Oxford, UK, 1960). Sr. lect. (tel. 5 + 8592159, fax 852 5 479907). *Organic and organometallic structures.*

Mak, Prof. Thomas Chung-wai. Chemistry Dept., The Chinese U. of Hong Kong, Shatin, New Territories, Hong Kong. (1936) PhD, chemistry (U. of British Columbia, Canada, 1963). Prof. and chairman (tel. 0 + 6952343, fax 852 0 6954234). *Crystal structure analysis, inclusion compounds, hydrates, hydrogen bonded molecular adducts, metal complexes.*

Wong, Dr Yau-Shing. Forensic Div., Government Lab. of Hong Kong, 5th. Floor, 190 Argyle Street, Kowloon, Hong Kong. (1946) PhD, chemistry (U. of Waterloo, Canada, 1975). Sr. Chemist. (tel. 3 + 7149161, ext. 388, fax 852 3 7610531). *Amino acid - heavy metal complexes, Platinum group - thiocyanate complexes.*

HUNGARY

Sub-Editor: T. Ungár

Notes

1. International telephone country code - 36.

2. The university degrees *Dr. phil., Dr. techn.* and *Dr. rer. nat.* are approximately equivalent to PhD. The degrees *okl. fizikus, okl. mérnök, okl. tanár* and *okl. vegyész* refer to a university degree in physics, engineering, teaching or chemistry, respectively, and they are approximately equivalent to MSc. The degrees conferred by the Hungarian Academy of Sciences are *a MTA rendes tagja* (akadémikus) (member of the Hungarian Academy of Sciences), *a MTA levelező tagja* (lev. tag) (corresponding member of the Hungarian Academy of Sciences), *a ... tudomány doktora* (Dokt.) and *a ... tudomány kandidátusa* (Kand.). The latter two degrees are approximately equivalent to DSc and PhD, respectively. The abbreviations Dokt. and Kand. are not in general use in Hungary.

3. Special abbreviations used for institutions which confer degrees: BME - Budapesti Müszaki Egyetem (Technical University of Budapest), ELTE - Eötvös Loránd Tudományegyetem (L. Eötvös University, Budapest), MTA - Magyar Tudományos Akadémia (Hungarian Academy of Sciences).

Arató, Dr Péter. Dept. Metal Physics, Res. Inst. Techn. Physics, Hung. Acad. Sci., H-1325 Budapest P.O.B. 76, Hungary. (1941) Kand., physics (MTA, 1974). Res. scient. (tel. 1 + 692-100, facs 1+698-037). *X-ray diffraction, real structures.*

Argay, Mr Gyula. Dept. X-ray Diffraction, Central Res. Inst. Chemistry, Hung. Acad. Sci., H-1525 Budapest P.O.B. 17, Hungary. (1939) Okl. fizikus, physics (ELTE, 1962). Res. scient. (tel. 1 + 353-735, ext. 327). *Organic crystal structures, computer programming.*

Banizs, Dr Károly. Div. Physical Metallurgy, Res. Eng. Prime Contr. Centre, Hung. Aluminium Corp., ALUTERV-FKI, H-1509 Budapest P.O.B. 5, Hungary. (1940) Kand., physics (MTA, 1979). Res. scient. (tel. 1 + 669-311) *Electron microscopy, diffraction, Al-alloys, and metallurgy.*

Barna, Dr Péter. Dept. Thin Film Physics, Res. Inst. Techn. Physics, Hung. Acad. Sci., H-1325 Budapest, P.O.B. 76, Hungary. (1928) Kand., physics (MTA, 1966). Head. (tel. 1 + 692-100, facs 1+698-037). *Solid state physics, thin film physics, thin film growth, structure research.*

Bodor, Prof. Géza. Section Polymer Physics, Polymer Res. Inst., H-1950 Budapest, Hungary. (1930) Dokt., chemistry (MTA, 1969). Head. (tel. 1 + 634-200, ext. 146). *Polymer structures.*

Bognár, Dr László. Dept. Mineralogy, ELTE, Múzeum krt 4/A, H-1088 Budapest, Hungary. (1936) Dr.rer.nat. mineralogy (ELTE, 1969). Asst. lect. (tel. 1 + 189-833 ext. 322). *Structure determination, polycrystalline mineral-systems.*

Bottyán, Dr László. Central Research Inst.Physics, H-1525 Budapest P.O.B. 49, Hungary. (1951) Dr.rer.nat., physics (ELTE, 1978). Res. scient. (tel. 1 + 699-499,facs 1+853-894). *Powder diffraction methods, inorganic structures, computer science.*

Bőcskei, Mr Zsolt. Physical Chemistry Dept., Chinoin Pharmaceutical and Chemical Works Ltd., H-1325 Budapest P.O.B. 110, Hungary. (1962) Okl. vegyész,chemistry (ELTE, 1985).Res.scient. (tel. 1+692-500,ext.22-34) *Organic structures, pharmacological applications.*

Csanády-Bokody, Dr (Ms) Ágnes. Dept. Material Sci., Electronoptical Lab., Res. Eng.and Prime Contr. Centre, Hung. Aluminium Corp., ALUTERV-FKI, H-1389 Budapest P.O.B. 128, Hungary. (1935) Kand., chemistry (MTA,1983). Head of Dept. (tel. 1+850-153). *Electron beam methods, oxidation processes, metallurgy, aluminium and alloys, metallic and nonmetallic materials, physicochemical investigations.*

Csákvári, Dr (Ms) Éva. Structural Chem. Res. Group, Hung. Acad. Sci., Eötvös Univ., H-1431 Budapest P.O.B. 117, Hungary. (1950) Dr.rer. nat., chemistry (ELTE,1975). Res. scient. (tel.1+173-722). *Molecular structures, electron diffraction.*

Csordás, Dr László. Lab. of Surface and Interface Physics, ELTE, Múzeum krt. 6-8, H-1088 Budapest, Hungary. (1929) Kand., physics (MTA, 1968). Assoc. prof. (tel. 1 + 189-833). *Organic and inorganic structures.*

Csordás-Tóth, Dr (Ms) Anna. Dept. Material Sci.,Hungalu Eng. and Development Centre, Aluminium Corp., ALUTERV-FKI, H-1389 Budapest P.O.B.128, Hungary. (1947) Dr. rer. nat., physics (JATE, 1977). Res. scient. (tel. 1 + 669-311, ext. 353). *Real structures, non-metallic materials, scanning electron microscopy, microanalytical investigations, bauxites.*

Cziráki, Dr (Ms) Ágnes. Dept. Solid State Physics, Eötvös U., Múzeum krt 6-8, H-1088 Budapest, Hungary. (1944) Dr, physics (ELTE, 1972). Sr. asst. (tel. 1 + 189-833). *Electron microscopy, amorphous and crystalline materials.*

Czugler, Dr Mátyás. Dept. X-ray Diffraction, Central Res. Inst. Chemistry, Hung. Acad. Sci., H-1525 Budapest P.O.B. 17, Hungary. (1948) Dr. techn., chemistry (BME, 1977), Res. scient. (Assoc. prof.,Univ.Stockholm,1987). (tel. 1 + 353-735, ext. 327). *Organic crystal structures, inclusion compounds, proteins.*

Dankházi, Mr Zoltán. Lab. of Surface and Interface Physics, Eötvös U., Múzeum krt 6-8, H-1088 Budapest, Hungary. (1960) Okl. fizikus, physics (ELTE, 1978). Res. asst. (tel. 1 + 189-833). *X-ray fluorescent spectroscopy, X-ray diffraction.*

Dódony, Dr István. Dept. Mineralogy, ELTE, Múzeum krt. 4/a, H-1088 Budapest, Hungary. (1950) Dr. rer. nat., mineralogy (ELTE,1978). Asst. lect. (tel.1+189-833 ext.368). *Mineralogy, electron diffraction and microscopy.*

Farkas-Jahnke, Dr (Ms) Mária. X-ray Dept., Res. Inst. Techn. Physics, Hung. Acad. Sci., H-1325 Budapest P.O.B. 76, Hungary. (1932) Kand., physics (MTA, 1974). Head.(tel. 1 + 692-100, fax.1+692-037). *Real structure of crystals, phase transformations, polytypes.*

Farkas, Dr László. Dept. Material Sci., Res. Eng. and Prime Contr. Centre, Hung. Aluminium Corp., ALUTERV-FKI, H-1389 Budapest P.O.B. 128, Hungary. (1942) Kand., geology (MTA, 1981). Sr. res. scient. (tel. 1+669-028, ext.286). *Powder diffraction methods, mineral structures, material science.*

Fodor, Ms Krisztina. X-ray Lab., Alumina Plant & Aluminium Smelter, H-8400 Ajka, P.O.B.45, Hungary.

Frigyik, Mr Gábor. Dept. Mech. Techn., Miskolc U., H-3515 Miskolc, Hungary. (1948) Okl. mérnök, metallurgy (Miskolc U., 1967). Res. scient. (tel. 46 + 65111). *Fracture, X-ray diffraction.*

Fuchs, Prof. Dr Sc Techn Erik. Technical U. for Heavy Industry, Miskolc-Egyetemv áros H-3515 Hungary. (1930) Dokt., physics (MTA, 1973). Head. (tel. 46 + 65111). *Metallurgical investigation technology, properties, metals.*

Fülöp, Mr Vilmos. Dept. X-ray Diffraction, Central Res. Inst. Chemistry, Hung. Acad. Sci., H-1525 Budapest P.O.B. 17, Hungary. (1961) Okl. vegyész, chemistry (ELTE, 1985). Res. scient. (tel. 1+ 353-735). *Organic and organometallic crystal structures.*

Gaál, Dr István. Dept. Light Sources, Res. Inst. Techn. Physics, Hung. Acad. Sci., H-1325 Budapest P.O.B. 76, Hungary. (1936) Kand., physics (MTA, 1976). Res. scient. (tel. 1 + 692-100, fax. 1+ 698-037). *Physical metallurgy, X-ray scattering, lattice defects.*

Gadó, Dr Pál. MEOSZ, H-1553 Budapest P.O.B. 10, Hungary. (1933) Kand., physics (MTA, 1971). Secretary General.(tel. 1 + 298-087). *Reai structure, minerals, inorganic materials, powder diffractometry, electron microscopy (structural), phase analysis, automation.*

Geleji-Neubauer, Ms Irén. H-1037 Budapest, Seregély-köz 7. Hungary. (1938) Okl. fizikus, physics (ELTE, 1962). okl. mérnök, metallurgy (Miskolc U. 1967) Res. scient. (tel. 1 + 686-326). *Structure, inorganic crystals and metals, powder methods, inorganic phase transformations, crystal physics.*

Gergely, Dr Márton. Metallurgical Dept., Res. Inst. Ferrous Metallurgy, H-1509 Budapest P.O.B. 14, Hungary. (1943) Dr. rer. nat., metallurgy (Techn. U., Miskolc, 1972). Head of lab. (tel. 1 + 610-641). *Diffraction methods, phase transformations.*

Griger, Dr (Ms) Ágnes. Metal Sci. and Techn. Division, HUNGALU Engineering and Development Centre, ALUTERV-FKI, H-1389 Budapest, P.O.B. 128, Hungary. (1946) Dr. rer. nat., physics (ELTE, 1976). Head of Lab. (tel. 1 + 812-979). *Crystallographic computing, metal physics, composite materials, quantitative phase analysis.*

Grosz, Mr Tamás. X-ray Dept., Res. Inst. Techn. Physics, Hung. Acad. Sci., H-1325 Budapest P.O.B. 76, Hungary. (1949) Okl. fizikus, physics (ELTE, 1973). Res. scient. (tel. 1 + 692-100, fax. 1 + 698-037). *Crystal orientation, phase transformations, solid state reactions.*

Hajdu, Dr Ferenc. Dept. X-ray Diffraction, Central Res. Inst. Chemistry, Hung. Acad. Sci., H-1525 Budapest P.O.B. 17, Hungary. (1926) Kand., chemistry (MTA, 1976). Sr. res. scient. (tel. 1 + 353-735). *Diffraction methods, liquids, solutions, amorphous solids.*

Hajdu, Mr János. Lab. of Surface and Interface Physics, ELTE, Múzeum krt 4/A, H-1088 Budapest, Hungary. (1932) Okl. fizikus, physics (ELTE, 1958). Asst. lect. (tel. 1 + 336-316). *Metals and alloys.*

Hange, Mr Ferenc. Röntgen Lab. Tungsram Co. LTD. H-1340 Budapest, Váci út 77, Hungary. (1954) Okl. tanár, chemistry (ELTE, 1979). Head of Lab. (tel.1 + 692-800, ext. 16-63). *Powder diffractometry, spectrometry.*

Hargittai, Prof István. Structural Chemistry Res. Group Hung. Acad. Sci., Eötv&oes U., H-1431 Budapest P.O.B. 117, Hungary. (1941) lev. tag, chemistry (MTA, 1987). Head.(tel. 1 + 173-899). *Structural chemistry (exp. and theoretical), electron diffraction, symmetry.*

Hargittai, Dr (Ms) Magdolna. Structural Chemistry Res. Group Hung.Acad. Sci., Eötvös U., H-1431 Budapest P.O.B. 117, Hungary. (1945) Kand., chemistry (MTA, 1978). Sr. res. scient. (tel. 1 + 173-722). *Inorganic molecular structures, electron diffraction, symmetry.*

Harsányi, Dr László. Structural Chemistry Res. Group Hung. Acad. Sci., Eötvös U., H-1431 Budapest P.O.B. 117, Hungary. (1952) Dr. rer. nat., chemistry (ELTE, 1981). Res. scient. (tel. 1 + 173-722). *Molecular structures, quantum chemistry, electron diffraction.*

Hartmann, Dr Ervin. Res. Lab. Crystal Physics, Hung. Acad. Sci., Budaörsi út 45, H-1112 Budapest, Hungary. (1935) Kand., physics (Moscow Inst. of Steels and Alloys, USSR, 1968). Sr. res. scient., assoc. prof. (tel. 1 + 850-777, ext. 315). *Crystal growth, physical properties of crystals.*

Horváth, Dr (Ms) Ilona. Röntgen Lab. Tungsram Co. LTD. H-1340 Budapest, Váci út 77, Hungary. (1959) Dr.techn., chemistry (BME, 1987). Res.scient. (tel.1 + 692-800, ext. 16-63). *Powder diffractometry, spectrometry.*

Imre-Baán, Dr (Ms) Irén. Dept. Material Sci., Hungalu eng.and Development Centre, ALUTERV-FKI, H-1389 Budapest P.O.B. 128, Hungary. (1938) Dr techn., chemistry (BME, 1985). Res. scient. (tel. 1 + 669-311, ext. 353). *Microanylitical investigations, fused materials, aluminium alloys.*

Ipacs, Mr László Sci. and Techn. Div., HUNGALU Eng. and Dev. Centre, ALUTERV-FKI, H-1389 Budapest P.O.B. 128, Hungary. (1956) Okl. Fizikus, physics (ELTE, l984). Res. scient. (tel. 1+669-311, ext. 392). *Composite materials, metal physics, X-ray diffraction.*

Kádár, Dr György. Inst. Microelectronics, Central Res. Inst. Physics, Hung. Acad. Sci., H-1525 Budapest P.O.B. 49, Hungary. (1942) Kand., physics (MTA, 1978). Res. scient. (tel. 1 + 699-499, fax. 1+853-894) *Magnetic structures.*

Kálmán, Prof. Alajos. Dept. X-ray Diffraction, Central Res. Inst. Chemistry, Hung. Acad. Sci., H-1525 Budapest P.O.B. 17, Hungary. (1935) Dokt., chemistry (MTA, 1975). Head. (tel. 1 + 353-735, ext. 160). *Organic structures, clay minerals, X-ray diffraction.*

Kálmán, Dr Erika. Dept. X-ray Diffraction, Central Res. Inst. Chemistry, Hung. Acad. Sci., H-1525 Budapest P.O.B. 17, Hungary. (1942) Kand., chemistry (Techn. U. of Dresden, DDR, 1970) Sr. res. scient, (tel. 1 + 334-359, ext. 91). *Liquid structure, electron diffraction.*

Kardos, Ms Jutta. X-ray Dept. Res. Inst. Techn. Physics, Hung. Acad. Sci., H-1325 Budapest P.O.B. 76, Hungary. (1939) MSc, crystallography (Humboldt U., DDR, 1963). Head of res. group. (tel. 1 + 692-100). *Semiconductor single crystals, topography, orientation, diamond structures.*

Kertész, Dr László. Lab. of Surface and Interface Physics, ELTE, Múzeum krt 6-8, H-1088 Budapest, Hungary. (1925) Kand., physics (MTA, 1966). Assoc. prof. (tel. 1 + 336-316). *Metals and alloys.*

Kiss, Mr Sándor. Dept. X-ray Diffraction, Hungarian Hydrocarbon Institute, H-2443 Szászhalombatta, P.O.B. 32, Hungary. (1942) Electr. eng. (BME, 1970). Res. scient. (tel. 1+ 800-122, ext. 187). *Cement, clay minerals, X-ray powder diffraction.*

Klug, Dr (Ms) Annamária. National Technical Information Centre and Library, H-1428 Budapest, P.O.B.12, Hungary. (1947) PhD, chemistry (Stockholm U., Sweden, 1977). Res. scient. (tel. 1 + 139-851, fax. 1 + 137-822). *Real structures, inorganic materials, minerals, powder diffraction methods, X-ray spectrometry, information science, information policy.*

Koritsánszky, Mr Tibor. Dept. X-ray Diffraction, Central Res. Inst. Chemistry, Hung. Acad. Sci., H-1525 Budapest P.O.B. 17, Hungary. (1954) Okl. vegyész, chemistry (ELTE, 1979). Res. scient. (tel. 1 + 353-735). *Organic crystal structures, charge density.*

Kormány, Dr Teréz. Dept. Electronic Devices, BME, Goldmann Gy. tér 3, H-1521 Budapest, Hungary. (1936) Kand., technical sciences (MTA, 1969). Res. scient. (tel. 1 + 812-199). *Thin films, semiconductors, crystal defects.*

Köszegi, Mr László Inst. Solid State Physics, Central Res. Inst. Physics, Hung. Acad. Sci., H-1525 Budapest P.O.B. 49, Hungary. (1945) Okl. fizikus, physics (ELTE,1968). Res. scient. (tel. 1 + 699-499, ext. 14-69, fax. 1+ 853-894). *Neutron diffraction, amorphous structures.*

Krén, Dr Emil. Central Res. Inst. Physics, Hung. Acad. Sci., H-1525 Budapest P.O.B. 49, Hungary. (1935) Kand., physics (MTA, 1974). Director. (tel. 1 + 699-499, fax. 1 + 853-894). *Magnetic structures, phase transitions.*

Lovas, Dr György Antal. Dept. Mineralogy, ELTE, Múzeum krt 4/A, H-1088 Budapest, Hungary. (1947) Dr. rer. nat., chemistry (ELTE, 1985). Sr. asst. (tel. 1 + 189-833, ext.345). *Crystal and molecular structure determination, powder diffraction, minerals.*

Malisckó, Dr László. Res. Lab. Crystal Physics, Hung. Acad. Sci., Budaörsi út 45, H-1112 Budapest, Hungary. (1934) Kand., physics (MTA, 1968). Sr. res. scient, asst. prof. (tel. 1 + 850-777, ext. 363). *Crystal growth, real structure of crystals, electron microscopy.*

Mándy, Mr Tamás. Dept. Catalytic Processes, Hungarian Hydrocarbon Institute, H-1443 Szászhalombatta, P.O.B. 32, Hungary. (1929) Chem. eng. (BME, 1951). Head of dept. (tel. 26 + 52-817). *X-ray diffraction catalysis, zeolites.*

Medgyaszay, Dr Márton. Res. Lab. Crystallography, EGIS Pharmaceuticals, H-1475 Budapest, P.O.B.100, Hungary. (1936) Dr, pharm. chemist (BOTE, 1959). Head. (tel. 1 + 835-340). *Organic structures, powder diffractometry, morphology, technological properties, pharmaceutical crystallized products.*

Menczel, Dr György. Dept. Solid State Physics, ELTE, Múzeum krt. 6-8. H-1088 Budapest, Hungary. (1921) Kand., physics (MTA, 1973). Assoc. prof. (tel. 1 + 189-833). *Organic structures.*

Nemecz, Prof. Ernö. Dept. Mineralogy, Techn. U. Chemical Engineering, Veszprém, H-8201 Veszprém P.O.B. 158, Hungary. (1920) Dr. phil., chemistry (Pázmány Péter U. Sci.,1944). Head of dept. and rector of U. (tel. 80 + 22-022, ext. 370). *X-ray crystallography, clay and zeolite minerals.*

Pál, Ms Edith. X-ray Dept., Res. Inst. Techn. Physics, Hung. Acad. Sci., H-1325 Budapest P.O.B. 76, Hungary. (1948) Okl. fizikus, physics (ELTE, 1971). Res. scient. (tel. 1 + 692-100, fax. 1 + 698-037). *X-ray topography.*

Pálinkás, Dr Gábor. Dept. X-ray Diffraction, Central Res. Inst. Chemistry, Hung. Acad. Sci., H-1525 Budapest P.O.B. 17, Hungary. (1941) Dr. sci .chem. (MTA, 1986). Res. scient. (tel. 1 + 353-735). *Liquid structure, electron diffraction, X-ray diffraction.*

Papp, Mr Gábor. Dept. Mineralogy, ELTE, Múzeum krt. 4/A, H-1088 Budapest, Hungary. (1960) Okl. geológus, geology (ELTE, 1984). Res. asst. (tel. 1 + 189-833, ext. 345). *Mineralogy, electron diffraction and microscopy, X-ray diffraction.*

Párkányi, Dr László. Dept. X-ray Diffraction, Central Res. Inst. Chemistry, Hung. Acad. Sci., H-1525 Budapest P.O.B. 17, Hungary. (1940) Kand., chemistry (MTA, 1982). Head, res. group. (tel. 1 + 353-735, ext. 224). *Organic and organometallic crystal structures.*

Petrik, Ms Anna. Dept. X-ray Diffraction, Hungarian Hydrocarbon Institute, H-2443 Szászhalombatta, P.O.B. 32, Hungary. (1942) Chem. eng. (BME, 1971). Res. scient. (tel. 1 + 800-122, ext. 187). *Cement, clay minerals, X-ray diffraction.*

Péter, Ms Ágnes. Res. Lab. Crystal Physics, Hung. Acad. Sci., Budaörsi ut 45, H-1112 Budapest, Hungary. (1950) Okl. fizikus, physics (ELTE, 1973). Res. scient. (tel. 1 + 850-777, ext. 334). *Crystal defects, oxides, X-ray diffraction, crystal growth.*

Petrás, Mr László. X-ray Dept. Res. Inst. Techn. Phys., Hung. Acad. Sci., H-1325 Budapest P.O.B. 76, Hungary. (1957) Okl. fizikus, physics, (ELTE, 1981). Res. scient. (tel. 1 + 692-100, fax. 1 + 698-037). *Powder diffraction, phase transformations, X-ray topography.*

Pósfai, Mr Mihály. Dept. Mineralogy, ELTE, Múzeum krt. 4/A, H-1088 Budapest, Hungary. (1963) Okl. geológus, geology (ELTE, 1987). Res. asst. (tel. 1+189-833, ext. 368) *Sulphide minerals, electron diffraction and microscopy.*

Radnai, Dr Tamás. Dept. X-ray Diffraction, Central Res. Inst. Chemistry, Hung. Acad. Sci., H-1525 Budapest P.O.B. 17, Hungary. (1948) Dr. rer. nat., physics (ELTE, 1977). Res. scient. (tel. 1 + 353-735). *Liquid structure, small angle X-ray scattering, X-ray diffraction.*

Redler Mr László Dept. Silicate Chemistry, Central Res. and Design Inst. Silicate Industry, H-1300 Budapest P.O.B. 112, Hungary. (1954) Okl. fizikus, physics (Kossuth U. Debrecen, l979). Res. scient. (tel. 1+804-273). *Cement clinkers, X-ray diffraction.*

Rischák, Mr Géza. Mineralogical and Petrographical Dept., Hungarian State Geological Survey, H-1442 Budapest P.O.B. 106, Hungary. (1931) Okl. vegyész, chemistry (József Attila U., Szeged, 1955). Sr. res. scient. (tel. 1 + 837-940, ext. 106). *Powder diffraction methods, fluorescent spectroscopy, bauxite-, ore- and soil minerals.*

Rozsondai, Dr Béla. Stuctural Chemistry Res. Group Hung. Acad. Sci., Eötvös U., H-1431 Budapest P.O.B. 117, Hungary. (1934) Kand., chemistry (MTA, 1977). Sr. res. scient. (tel. 1 + 173-722). *Molecular structures, gas electron diffraction, computational methods.*

Sajó, Mr István. Dept. Mater. Sci., Res. Eng. and Prime Contr. Centre, Hung. Aluminium Corp., ALUTERV-FKI, H-1389 Budapest P.O.B. 128, Hungary. (1949) Okl. vegyész, chemistry (ELTE, 1973) Res. scient. (tel. 1+669-028, ext. 286). *Powder diffraction methods, mineral structures, materials science, phase analysis.*

Sasvári, Dr (Ms) Judit. Dept. Material Sci., Res. Eng. Prime Contr. Centre, Hung. Aluminium Corp., ALUTERV-FKI, H-1389 Budapest P.O.B. 128, Hungary. (1945) Dr. rer. nat., physics (ELTE, 1972). Head of lab. (tel. 1 + 669-028, ext. 286). *Powder diffraction methods, phase analysis, X-ray topography, computer science.*

Sasvári, Dr Kálmán. Dept. X-ray Diffraction, Central Res. Inst. Chemistry, Hung. Acad. Sci., H-1525 Budapest P.O.B. 17, Hungary. (1912) Dr phyl., physics (Pázmány Péter U., 1938). Retired (tel. 1 + 661-881) *Organic crystal structures, computer programming.*

Schultz, Dr György. Structural Chemistry Res. Group Hung. Acad Sci., Eötvös U., H-1431 Budapest P.O.B. 117, Hungary. (1938) Kand., chemistry (MTA, 1982). Sr. res. scient. (tel. 1 + 173-722). *Molecular structures, electron diffraction.*

Schuszter, Dr Ferenc. Lab. of Surface and Interface Physics, Eötvös U., Múzeum krt 6-8, H-1088 Budapest, Hungary. (1937) Dr. rer. nat., physics (ELTE, 1971).

Sr. asst. (tel. 1 + 189-833). *Electron microscopy, amorphous and crystalline materials.*

Simon, Dr Kálmán. Physical Chemistry Dept., Chinoin Pharmaceutical and Chemical, H-1325 Budapest P.O.B. 110, Hungary. (1946) Kand., chemistry (MTA, 1981). Head, res. group. (tel. 1 + 692-500, ext. 22-34). *Crystal structures, biological applications, direct methods.*

Soós, Mr Miklós. Dept. Mineralogy, ELTE, Múzeum krt. 4/A, H-1088 Budapest, Hungary. (1959) Okl. geológus, geology (ELTE, 1984). Res. asst. (tel. 1+189-833, ext. 368) *Mineralogy, electron diffraction and microscopy.*

Sváb, Dr (Ms) Erzsébet. Central Res. Int. Physics, Hung. Acad. Sci., H-1525 Budapest P.O.B. 49, Hungary. (1947) Kand., physics (MTA, 1986). Res. scient. (tel. 1+699-499, fax 1+853-894) *Powder diffraction methods, neutron diffraction, amorphous structures.*

Szemethy, Miss Andrea. Dept. X-ray Diffraction, Hungarian Geological Inst., H-1442 Budapest P.O.B. 106, Hungary. (1947) Okl. geol., geology (ELTE, 1972). Res. scient. (tel. 1 + 837-940, ext. 106). *X-ray phase analysis, zeolite structures, carbonates.*

Sztrókay, Prof. Kálmán. Dept. Mineralogy, ELTE, Múzeum krt 4/A, H-1088 Budapest, Hungary. (1907) Dokt., geology (MTA, 1958). Prof. (tel. 1 + 183-864). *Crystallography, structure and physical properties, mineralogy, phase composition.*

Tardy, Dr Pál. Dept. Phys. Metallurgy. Res. Dev. Enterprise for Steel Industry. H-1509 Budapest P.O.B. 14, Hungary. (1940) Kand., technical sci. (MTA, 1974). Head of Dept. (tel. 1 + 868-414). *Electron microscopy, ferrous metallurgy.*

Tarján, Prof. Imre. Dept. Biophysics, Semmelweiss Medical U., Budapest H-1088 Budapest Puskin utca 9, Hungary. (1912) Akadémikus, physics (MTA, 1976). Head. (tel. 1 + 339-599). *Crystal physics, crystal growth, biological macromolecules.*

Tasnády, Dr (Ms) Eleonóra. Hungarian Hydrocarbon Institute, H-2443 Százhalombatta, P.O.B. 32, Hungary. (1938) Dr. rer. nat., chemistry, (Chemical U. Veszprém, 1977) Sr. res. scient. (tel. 1+800-122, ext. 369) *Cement, clay minerals, X-ray diffraction.*

Tóth, Mr Lajos. Dept. Thin Films Physics, Res. Inst. Techn. Physics, Hung. Acad. Sci., H-1325 Budapest P.O.B. 76, Hungary. (1952) Okl. fizikus, physics (ELTE, 1975). Res. scient. (tel. 1 + 692-100, fax 1+698-037). *Thin films, structural research.*

Tremmel, Mr János. Structural Chemistry Res. Group Hung. Acad. Sci., Eötvös U., H-1431 Budapest P.O.B. 117, Hungary. (1927) Okl. mérnök, electrical eng. (BME, 1959). Res. scient. (tel. 1 + 173-722). *Electro optical techniques, high temperature, gas electron diffraction, instrumentation.*

Turmezey, Dr Tibor. Metal Sci. Techn. Div., HUNGALU, Eng. Developm. Centr., ALUTERV-FKI, H-1389 Budapest P.O.B. 128, Hungary. (1948) Dr. rer. nat.,

physics (ELTE, 1974). Head of Div. (tel. 1 + 810-181). *Material science, electron diffraction, metal physics.*

Ungár, Dr Tamás. Inst. for General Physics, ELTE, Múzeum krt. 6-8., H-1088 Budapest, Hungary. (1943) Kand., physics (MTA, 1980). Assoc. prof. (tel. 1 + 189-833 ext. 617). *Metals and alloys, high resolution diffractometry, line profile analysis, small angle scattering.*

Vajda, Dr Erzsébet. Structural Chemistry Res. Group Hung. Acad. Sci.,Eötv&oes U. H-1431 Budapest P.O.B. 117, Hungary. (1938) Dr. rer. nat., chemistry (ELTE, 1971). Res. scient. (tel. 1 + 173-722). *Molecular structures, electron diffraction.*

Varga, Dr László. Dept. Light Sources, Res. Inst. Techn. Physics, Hung. Acad. Sci., H-1325 Budapest P.O.B. 76, Hungary. (1935) Kand., techn. sci. (MTA, 1976). Res. scient. (tel. 1 + 692-100, fax. 1 + 698-037). *Metals and alloy structures.*

Varga, Mr László. Dept. Mech. Techn. Mat. Sci., BME, Goldmann Gy. tér 3, H-1111 Budapest, Hungary. (1944) Okl. mérnök, mechanical eng. (U. Heavy Industry, Miskolc, 1967). Res. scient. (tel. 1 + 664-011, ext. 28-83). *Crystal defects, stress analysis, metallurgy.*

Verö, Dr Balázs. Dept, Phys. Metallurgy Res. Dev. Enterprise for Steel Industry. H-1509 Budapest P.O.B. 14, Hungary. (1944) Kand., techn. sci. (MTA, 1980). Head of lab. (tel. 1 + 868-4149). *Ferrous metallurgy.*

Viczián, Dr István. Dept. X-ray Diffraction, Hungarian Geological Inst., H-1442 Budapest P.O.B. 106, Hungary. (1940) Kand., geology (MTA, 1976). Res. scient. (tel. 1 + 637-600). *X-ray phase analysis, clay mineralogy and petrology.*

Vízi, Dr Béla. Dept. Gen. Inorg. Chemistry, Veszprém U. of Chemical Engineering, H-8201 Veszprém P.O.B. 158, Hungary. (1936) Kand., chemistry (MTA, 1981). Sr. lect. (tel. 80+22-022). *Normal coordinate analysis, vibrational amplitudes.*

Weiszburg, Dr Tamás. Dept. Mineralogy, ELTE, Múzeum krt. 4/A, H-1088 Budapest, Hungary. (1956) Dr. rer. nat., mineralogy (ELTE, l987) Res. scient. (tel. 1+189-833, ext. 344). *Crystal chemistry, real structure of minerals, sulphides, silicates, electron diffraction and microscopy, X-ray diffraction.*

Zábráczki, Mr Jósef. Hungalu, Almásfüzitö Alumina Plant., H-2931 Almásfüzitö P.O.B. 2. Hungary. (1948) Dr. rer. nat., physics (ELTE, l980). Res. scient. (tel 34+14-364). *Powder diffraction methods, computer science, X-ray spectrometry.*

Zsoldos, Mrs Éva. Inst. Microelectronics, Central Res. Inst. Physics, Hung. Acad. Sci., H-1525 Budapest P.O.B. 49, Hungary. (1934) Okl. fizikus, physics (ELTE, 1957). Head, lab. (tel. 1 + 699-499, fax 1+853-894). *X-ray topography, magnetic crystals.*

Zsoldos, Dr Lehel. Dept. of Structure Res., Res. Inst. Techn. Physics, Hung. Acad. Sci., H-1325 Budapest P.O.B. 76, Hungary. (1931) Kand., physics (MTA, 1965). Head. (tel. 1 + 690-315). *Defects, phase transformations, instrumentation.*

ICELAND

Sub-Editor: **H. Kristmannsdóttir**

Notes

1. International telephone country code - 354.

Eiríksson, Dr Vésteinn Runi. Hamrahlíd Col., Reykjavík, Iceland. (1944) PhD, physics, crystallography (U. Edinburgh, UK, 1974). Lecturer. (tel. 91 + 44710). *Phase transitions and structure, ferroelectrics.*

Kristmannsdóttir, Cand.Real. Hrefna. National Energy Authority, Geothermal div., Grensásvegur 9, Reykjavík, Iceland. (1944) Cand.Real., mineralogy, petrology (U. Oslo, Norway, 1970). Head Dept. (tel. 91 + 83600, fax +354 1 688896, email hk(a)geysir.UUCP). *Clay minerals, feldspars, zeolites.*

Sigvaldason, Dr Gudmundur. The Nordic Volcanologic Institute, Reykjavík, Iceland. (1932) Dr.Rer.Nat., geochemistry, mineralogy (U. Göttingen, BRD, 1959). Director. (tel. 91 + 694491, e-mail KOSMOS: gesigvaldaso). *Crystal optics, silicate crystal structures, clay minerals.*

Steinthórsson, Dr Sigurdur. Dept. of Geology, U. of Iceland, Reykjavík, Iceland. (1940) PhD, geochemistry, petrology (Princeton U., USA, 1974). Assoc. prof. (tel. 91 + 694476). *Crystal optics, silicate crystal structures, ice crystal growth, ice chemistry.*

Tómasson, Cand.Real. Jens. National Energy Authority, Geothermal Div., Grensásvegur 9, Reykjavík, Iceland. (1925) Cand.Real., mineralogy, petrology (U. Oslo, Norway, 1962). Section leader. (tel. 91 + 83600, fax +354 1 688896, email jt(a)geysir.UUCP). *Low metamorphic minerals.*

INDIA

Sub-Editor: **J. K. Dattagupta**

Notes

1. International telephone country code - 91.

2. The Indian Universities and Institutes confer the degrees of DPhil, PhD, and DSc. The first two degrees are equivalent to the PhD degree of the UK Universities and the last one is a higher degree.

3. Abbreviations in this section include:

ATIRA - Ahmedabad Textile Industries Research Association
BARC - Bhabha Atomic Research Centre
C.G.C. - Crystal Growth and Characterization
C.S.S. - Central Scientific Services

I.A.C.S. - Indian Association for the Cultivation of Science
I.I.S. - Indian Institute of Science, Bangalore
I.I.T. - Indian Institute of Technology
SINP - Saha Inst. of Nuclear Physics

Agarwal, Dr Bhagwatiprasad. Physics Dept., U. Sch. of Sci., Ahmedabad 380009, India. (1943) PhD, physics (Sardar Patel U., 1972). Reader. (tel. 40929). *Growth of single crystals, defect properties.*

Agarwal, Dr Ramesh Chandra. Cntr. for Appl. Res. in Electronics, Indian Inst. of Techn., New Delhi 110029, India. (1947) PhD, electrical engineering (Rice U., USA, 1974). Res. Principal scient. officer. (tel. 665674) *Least-squares refinement, large structures, fast fourier transform applications, fast convolution techniques, digital signal processing applications.*

Aggarwal, Dr Prem Sarup. Central Glass & Ceramic Res. Inst., Calcutta 700032, India. (1934) PhD, chemistry (Poona U., 1958). Scient. (tel. 463496) *X-ray and electron diffraction, ceramics and raw materials, thermal analysis.*

Agrawal, Mr Jawahar Lal. Physics Dept., Ranchi U., Ranchi 834008, India. (1950) PhD, physics (Ranchi U., 1976). Res. scholar. *X-ray crystallography, organometallic compounds.*

Amarendra Kumar, Mr V. Dept. of Organic Chemistry, Indian Inst. of Sci., Bangalore 560012, India. (1962) MSc, chemistry (Sri Sathya Sai Inst. of Higher Learning, 1986). Res. scholar. *Crystallography, photochromic organic compounds.*

Amirthalingam, Dr V. Chemistry Div., BARC, Trombay, Bombay. 400085, India. (1929) PhD, physics (Wales U., UK, 1962). Scient. officer. *Biocrystallography, phase transition.*

Anantha Murthy, Dr Rayasa V. Crystal Growth and characterization Sec., Nat. Physical Lab., Dr. K.S. Krishnan Road, New Delhi. 110012, India. (1948) PhD, X-ray crystallography (I.I.T., Madras, 1976). Scient. (tel. 58 1733). *Crystal structure analysis, organic and metal complexes, crystal growth, characterization.*

Anantharaman, Prof. Tanjore Ramachandra. Inst. of Techn., Banaras Hindu U., Varanasi 221005, India. (1927) DPhil, metallurgy (Oxford U., 1954). Prof. & dir. *Structural changes, structural imperfections, metallic structures.*

Aravindakshan, Cheethambadi. Physical Res. Wing, Projects & Dev. India Ltd., Sindri, Dhanbad 828122, India. (1934) PhD, X-ray crystallography (Madras U., 1974). Addl. superintendent. *Metal physics, metals and alloys, corrosion, crystal structure, fertilizer materials.*

Arora, Dr Narinder Kumar. Materials Div., Nat. Physical Lab., Dr. K. S. Krishnan Road, New Delhi 110012, India. (1952) PhD, semiconductors (Delhi U., 1983). Scient. C.(tel. 5726058). *Growth of single crystals, silicon, solar cells.*

Asath Bahadur, Mr S. School of Physics, Madurai Kamaraj U., Madurai 625021, India. (1963) MPhil. (Madurai Kamaraj U., 1987). Res. scholar. *Crystal structure analysis.*

Awasthi, Dr Santosh Kumar. Radiochemistry Div., BARC, Trombay, Bombay 400085, India. (1944) PhD, chemistry (Bombay U., 1974). Scient. officer. (tel. 523321, ext. 456) *Inorganic and organometallic compounds, oxides, X-ray spectrometry.*

Balasingh, Dr C. Materials Sci. Div., Nat. Aeronautical Lab., Bangalore 560017, India. (1941) PhD, (I.I.S., 1981). Scient. E1. *Powder diffractometry, stress measurement, instrumentation.*

Balasubramanian, Dr R. Dept. of Crystallography and Biophysics, U. of Madras, A. C. C. Campus, Madras 600025, India. (1943) PhD, molecular biophysics (Madras U., 1973). Prof. (tel. 432248, ext. 16). *Theoretical studies on crystal structures.*

Bandopadhyay, Dr Mrs Tapati. Physics Dept., Calcutta U., 92 A.P.C. Rd., Calcutta 700009, India. (1950) PhD, physics (Calcutta U., 1985). Res. assoc. *Crystal structure, biologically important compounds.*

Banerjee, Dr Asok. Dept. of Biophysics, Bose Inst., P1/12 C.I.T. Scheme VIIM, Calcutta, India. (1942) PhD, physics (Calcutta U., 1974). Reader. (tel. 374734). *Molecular biophysics, macromolecular structure, function dynamics and molecular modelling.*

Banerjee, Dr Krishna. Physics Dept., Ranchi U., Ranchi, Bihar 834008, India. (1950) PhD, X-ray crystallography (Ranchi U., 1982). Res. scholar. *Crystal structure determination, X-ray diffraction techniques.*

Banerjee, Mr Rahul. Molecular Biophysics Unit, Indian Inst. of Sci., Bangalore 560012, India. (1965) MSc, physics (I.I.T., Madras, 1988). Res. student. (tel. 344411, ext. 2389). *Biomolecular crystallography.*

Banerjee, Dr Srikumar. Physical Metallurgy Div., Bhabha Atomic Res. Cntr., Trombay, Bombay 400085, India. (1946) PhD, metallurgy (I.I.T., Kharagpur, 1975). Head, Structural Metallurgy Sec. (tel. 5519949). *Phase transformation, electron microscopy, radiation damage.*

Bansigir, Prof. K. Goswami. Sch. of Studies in Physics, Jiwaji U., Gwalior, M. P., India. (1923) PhD, solid state physics (Osmania U., 1959). Prof. *Dislocations, X-ray topography, dissolution studies, colour centres, lattice vibrations, photo-elastic effect.*

Baruah, Mr Gagan Ch. Physics Dept., Biswanath C., Chariali 784176, India. (1945) MSc. Lect. *Natural fibres.*

Basak, Prof. Bejoysanker. Physics Dept., Presidency C., College St. Calcutta 700073, India. (1921) PhD, physics (Calcutta U., 1950). Principal(retd), Presidency C. *Crystal structure, organic compounds, bond lengths.*

Betal, Mr Badal Kumar. Physics Dept., Presidency C., Calcutta 700073, India. (1943) MSc, physics (Calcutta U., 1963). Lect. (tel. 341121). *Sterol structures.*

Bhagvantam, Prof. Suri. SriNiket Tarnaka, Hyderabad 500017, India. DSc(Hon.), physics (Andhra U., 1941; Osmania U., 1969). *Crystal physics.*

Bhagwat, Dr Vasant. Dept. of Chemistry, Vikram U., Ujjain, Madhya Pradesh 456001, India. (1947) PhD, Chemistry (Vikram U., 1970). Lect. *X-ray crystallographic studies, alkali and akaline earth complexes of macrocyclic compounds.*

Bhakay-Tamhane, Mrs Sandhya Nitin. Neutron Physics Div., Bhabha Atomic Res. Cntr., Trombay, Bombay 400085, India. (1954) PhD, physics (Bombay U., 1985). Scient. officer. (tel. 5519783, ext. 4392). *Structural phase transitions, X-ray and neutron diffraction, fractals.*

Bhaktapriya, Dr S. R. Y. Physics Dept., Ranchi U., Ranchi 834008, India. (1935) PhD, X-ray crystallography (Ranchi U., 1981) Prof. *Organo-metallic compounds, structures.*

Bharati Rao, Ms T. Dept. of Organic Chemistry, Indian Inst. of Sci., Bangalore 560012, India. (1965) MSc, chemistry (I.I.T., Madras, 1987). Res. scholar. *Structure-reactivity correlations.*

Bhatia, Dr Subhash Chander. Chemistry Dept., Kurukshetra U., Kurukshetra 132119, India. (1950) PhD, chemistry (Kurukshetra U., 1977). Lect. *Organometallic compounds.*

Bhatt, Dr Vaikunthray Promodray. Physics Dept., Faculty of Sci., M.S.U., Baroda 390002, India. (1929) PhD, solid state physics (M.S.U., Baroda, 1963). Reader. *Growth, dissolution and mechanical properties, metal and alloy single crystals.*

Bhattacharjee, Mrs Lilabati. Mineral Physics Div., Geological Survey of India, 29 Chowringhee Road, Calcutta 700016, India. MSc, physics (Calcutta U., 1951). Sr. mineralogist. (tel. 238321, ext. 8). *Structural crystallography, optical transform methods, computer programming, phase transformations, crystal growth, topography, instrumentation.*

Bhattacharya, Archana. Dept. of Physics, Jadavpur U., Calcutta 7000329, India. (1951) MSc, physics (Jadavpur U., 1974). Res. scholar. *X-ray crystallography.*

Bhattacharya, Dr Ramendranarayan. Dept. of Magnetism, Indian Assoc. for Cultivation of Sci., Calcutta 700032, India. (1934) PhD, physics (Calcutta U. 1975). Lect. *Crystal physics.*

Bhattacharyya, Dr Subodh Chandra. Biophysics Div., SINP, 37 Belgachia Rd., Calcutta 700037, India. (1933) PhD, physics (Calcutta U., 1975). Reader. *Structure and function, drug molecules.*

Bhattacherjee, Mr Santi Brata. Physics Dept., Serampore C., College Street, Serampore, W. Bengal, India. (1925) MSc, physics (Dacca U., Bangladesh, 1948). Prof. *Crystal structure, phase transformation, X-ray spectroscopy.*

Bhattacherjee, Dr Satyananda. Physics Dept., Indian Inst. of Techn., Kharagpur 721302, India. (1935) PhD, X-ray structure of matter (I.I.T., Kharagpur, 1970). Prof. *Polymer and blends, X-ray diffraction and structure property relation in solids.*

Bhawalkar, Dr Ramkrishna Haribhau. X-ray Group, Nat. Physical Lab., Hillside Road, New Delhi 110012, India. (1933) PhD, luminescence (Sagar U., 1963), Scient. (tel. 584179). *Powder diffractometry, X-ray fluorescence analysis, crystallography.*

Bist, Dr B. M. S. Physics Dept., Benaras Hindu U., Varanasi 221005, India. (1950) PhD, physics (Banaras Hindu U., 1974). Postdoctoral fellow. *Thin films, electron microscopy.*

Biswas, Goutam. Dept. of Biophysics, Bose Inst., P-1/12, C.I.T. Scheme Kankurgachi, Calcutta, India. (1960) MSc, physics (Calcutta U., 1982). Res. Scholar. *Organic crystal structure, biological macromolecules.*

Biswas, Dr Subhash Chandra. Dept. of Solid State Physics, Indian Assoc. for the Cultivation of Sci., Calcutta 700032, India. (1939) PhD, physics (Calcutta U., 1987). Lect. (B.K.C. College, Calcutta 700035). *Organic structures, disorder, energy calculations.*

Biswas, Dr Sundar Gopal. Physics Dept. Visva Bharati U., Santiniketan 731235, India. (1931) PhD, crystallography (Calcutta U., 1961). Lect. *Biologically important organic compounds.*

Bora, Dr M. N. Physics Dept., Gauhati U., Guwahati 781014, India. (1942) PhD (1974). Reader. *Clay minerals, natural fibres.*

Bose, Mr Shyamal Kumar. X-ray Crystallography Group, Nat. Metallurgical Lab., Jamshedpur 831007, India. (1936) MSc, physics (Allahabad U., 1957). Scient. 'C'. *Metals and alloys, structure and transformation.*

Chacko, Dr K. K. Dept. of Crystallography and Biophysics, U. of Madras, Guindy Campus, Madras 600025, India. (1942) PhD, physics (Madras U., 1971). Reader. (tel. 432248). *Biologically interesting compounds, medically interesting compounds, anomalous dispersion.*

Chadha, Asok Kumar. Fuel Chemistry Div., Bhabha Atomic Res. Centr., Trombay, Bombay 400085, India. MSc, chemistry (Delhi U., 1961). *Actinide complex, mixed oxides.*

Chadha, Dr Gopal Krishan. Dept. of Physics and Astrophysics, Delhi U., Delhi 110007, India. (1942) PhD, physics (Delhi U., 1968). Reader. *Crystal growth, X-ray diffration, defects in crystals, polytypism.*

Chakrabarti (Chatterjee), Dr (Mrs) Chandana. Crystallography & Molecular Biology Div., Saha Inst. of Nuclear Physics, Sector-1 Block-AF, Bidhannagar, Calcutta 700064, India. (1951) PhD, physics (Calcutta U. 1984). Pool officer. *Biologically important compounds, proteins.*

Chakrabarty, Dipak Kumar. Dept. of Physics, X-ray Lab., Jadabpur U., Calcutta 700032, India. (1953) MSc, physics (Calcutta U., 1976). Lect. *Crystal structure, lattice dynamics.*

Chakrabarty, Subhasis. Dept. of Biophysics, Bose Inst. P-1/12, C.I.T. Scheme, Kankurgach, Calcutta 700054, India. (1959) MSc, physics (Calcutta U., 1982). Res. scholar. *Molecular biophysics.*

Chakrabarty, Mr Sugoto. Physics Div., Bhabha Atomic Res. Cntr., Bombay 400085, India. (1957) MSc, physics (Delhi U., 1979). Scient. officer. (tel. 551-3848). *Protein crystallography.*

Chakraborty, Dipak. Physics Dept., Jadavpur U., Calcutta 700032, India. MSc, physics (Calcutta U., 1976). Lect. *Rigid body analysis, structure solution, computer simulation.*

Chakraborty, Dr Suchit Chandra. Physics Dept., U. of Burdwan, Burdwan, W. Bengal, India. (1928) PhD, physics (Allahabad U., 1958). Prof. *Crystal structure, thermal diffuse scattering, lattice vibrations, phase transitions.*

Chakravarti, Ms Lily. C.S.S. Dept., Indian Assoc. for the Cultivation of Sci., Calcutta 700032, India. (1960) MSc, physics (Calcutta U., 1983). Res. scholar. *Dynamical theory, X-ray diffraction.*

Chakravarty, Dr A. R. Inorganic and Physical Chemistry Dept., Indian Inst. of Sci., Bangalore 560012, India. (1953) PhD, chemistry (Calcutta U., 1982). Asst. Prof. *X-ray crystallography of transitition metal compounds.*

Chandrasekaran, Prof. Katuputhur Sarma. Physics Dept., Madurai U., Madurai 625021, india. (1925) PhD, X-ray crystallography (Madras U., 1956). Prof. and head. *Crystalline solid solutions, lattice defects, electron density, texture & crystal perfection.*

Chandrasekhar, Prof. Sivaramakrishna. Raman Res. Inst., Bangalore 560006, India. (1930) ScD physics (Cambridge U., UK). Prof. (tel. 30124). *X-ray diffraction, crystal optics, liquid crystals.*

Chandrasekharaiah, Dr M. N. Dept. of Metallurgical Eng., Inst. of Techn., Benaras Hindu U., Varanasi 221005, India. (1945) PhD physics (Benaras Hindu U., 1972). Res. assoc. *Crystal growth, geometrical crystallography, defects (planar).*

Chatterjee, Dr Amitava. Dept. of General Physics & X-ray, I.A.C.S., Calcutta 700032, India. (1948) PhD, physics (Calcutta U., 1978). Pool officer, CSI. *Macromolecular structure, conformational analysis.*

Chatterjee, Dr Sanat Kumar. Physics Dept., Regional Eng. C., Durgapur 731209, India. (1945) PhD, physics (Calcutta U., 1976). Asst. Prof. *Lattice imperfections, phase transformations, amorphous metals, mechanical properties.*

Chattopadhyay, Debasish. Crystallography & Molecular Biology Div., Saha Inst. of Nuclear Physics, 1/AF Bidhannagar, Calcutta 700064, India. (1957) MSc, biochemistry (Calcutta U., 1979). Res. fellow. *X-ray crystallography, biochemically interesting compounds, coordination chemistry, protein crystallography.*

Chaudhuri, Dr Ahindra Kumar. Solid State Lab., Central Glass & Ceramic Res. Inst., Calcutta 700032, India. (1941) PhD, physics (I.I.T., Kharagpur, 1974). Scient. *Nucleation, crystallisation, glass, order-disorder, thin films.*

Chaudhuri, Mr Siddhartha. RSIC,Bose Inst.,93 A.P.C. Rd.,Calcutta 700009, India. (1945) MSc, physics (Calcutta U., 1969). *Biologically important compounds, computing methods.*

Chawla, Dr Krishan Lal. Radiochemistry Div., Bhabha Atomic Res. Cntr., Bombay 400085, India. (1939) PhD, inorganic chemistry (Rajasthan U., 1968). Scient. officer. (tel. 523321, ext. 456). *Crystal chemistry, actinides.*

Chelliah, Dr Mahadevan. Dept. of Physics, S.T. Hindu College, Nagercoil, Tamilnadu 629002, India. (1958) PhD, physics (I.I.T. Madras, 1984). Asst. Prof. (tel. 3127). *X-ray crystallography, crystal, growth and characterization.*

Chetal, Prof. Amritlal R. Dept. of Applied Sci., Indian Sch. of Mines, Dhanbad 826004, India. (1938) PhD, physics (Poona U., 1966). Prof. *Crystal structure.*

Chidambaram, Dr Rajagopala. Physics Group, Bhabha Atomic Res. Cntr., Trombay, Bombay 400085, India. (1936) PhD, physics (I.I.S., Bangalore, 1962). Dir., physics group. (tel. 5518700). *Neutron diffraction, high pressure physics.*

Chinnakali, K. Dept. of Physics, Anna U., Madras 600025, India. (1962). Res. scholar. *X-ray crystallography.*

Chopra, Prof. Kasturilal. Physics Dept., Indian Inst. of Techn., Kharagpur 721302, India. (1933) PhD, solid state physics (British Columbia U., Canada, 1957). Director. (tel. 0322-386, telex 021-2760 ITKG IN). *Thin films, amorphous semiconductors, electron diffraction, electron microscopy, vacuum technology.*

Dadel, Mrs Snehlata. Physics Dept., Ranchi U., Ranchi 834008, India. (1939) MSc, physics (Ranchi U., 1962). Lect. *Crystallography, organo-metallic compounds.*

Das, Dr Birendra Nath. Physics Dept., Vivekananda C., D. H. Road, Calcutta 700063, India. (1941) PhD, physics (Calcutta U., 1981). Lect. (tel. 773434). *Biologically interesting organic compounds, structure and function.*

Das, Dr Indu Mohan. U.S.I.C., Gauhati U., Guwahati 781014, India. (1935) PhD, physics (Gauhati U., 1967). Prof. and Head. (tel. 88531). *X-ray fluorescence spectroscopy.*

Das, Pratap Kumar. Dept. of Solid State Physics, Indian Assoc. for Cultivation of Sci., Calcutta 700032, India. (1947) MSc, physics (Calcutta U., 1969). Lect. (Surendranath C., Calcutta 700009). *Organic structure, disorder, phase transition.*

Das, Dr Sabita. Physics Dept., Victoria Inst. (college branch), Calcutta 700009, India. (1934) PhD, physics (Calcutta U., 1969). Lect. *Theoretical crystallography, molecular biophysics.*

Das Gupta, Mr Prabal. C.S.S. Dept., Indian Assoc. for the Cultivation of Sci., Jadvpur, Calcutta 700032, India. (1952) MSc, chemistry (Jadavpur U., 1977). Techn. Superintendent. *Structure - properties relationships.*

Datta, Dr Amal Kumar. Physics Dept., Indian Inst. of Techn., Kharagpur 721302, India. (1953) PhD, X-ray diffraction (I.I.T., Kharagpur, 1986) Sr. Res. asst. *X-ray microscopy, small angle scattering, clays, fibres, polymers and blends.*

Dattagupta, Dr Jiban Kanti. Crystallography & Molecular Biology Div., Saha Inst. of Nuclear Physics, 1/AF Bidhannagar, Calcutta 700064, India. (1945) PhD, physics (Calcutta U., 1972). Assoc. Prof. (tel. 370659). *Structure & conformation, medicinal compounds, protein crystallography.*

Dayal, Dr Radha Raman. Glass Technology Div., Defence Sci. Cntr.,Metcalfe House, Delhi 110054, India. (1941) PhD, phase equilibria (Aberdeen U., UK, 1965). Scient. E. (tel. 2521325,ext. 335). *Phase equilibria, crystal growth, crystal chemistry, glass science and technology.*

De, Dr Amitabha. Physics Dept., Calcutta U., 92, A.P.C. Road, Calcutta 700009. India. (1958) PhD, Physics (Calcutta U.,1988). Res. scholar. *Biologically important compounds.*

De, Dr Madhusudan. Dept. of Materials Science, Indian Assoc. for the Cultivation of Sci., Calcutta 700032, India. (1941) PhD, physics (Calcutta U., 1970). *Metal physics, lattice imperfections, thermal expansion.*

Deopura, Dr B. L. Textile Techn. Dept., Indian Inst. of Techn., New Delhi 110016, India. (1946) PhD, physics (I.I.T., Kanpur, 1972). Asst. Prof. *Small angle X-ray scattering, polymers, fibres, X-ray diffraction.*

Dhanaraj, Dr V. Molecular Biophysics Unit, Indian Inst. of Sci., Bangalore 560012, India. (1957) PhD, X-ray crystallography (I.I.S., 1986). Res. Assoc. (tel. 344411, ext. 2389). *Biological crystallography, molecular biology.*

Dhaneshwar, Dr Narayandatta Nagesh. Physical Chemistry Div., Nat. Chemical Lab., Pashan Road, Poona 411008, India. (1936) PhD, X-ray crystallography (Poona U., 1973). Sr. scient. officer. (tel. 56451). *Low temperature crystallography, heterocyclic compounds, small organic molecules, natural products.*

Dhawan, Mrs Urmil. Specialized Techniques (X-rays) Div., Nat. Physical Lab., Hillside Road, New Delhi 110012, India. (1939) MSc, physics (Panjab U., 1964). Scient. (tel. 584179). *Inorganic structures, defects, diffraction theory, experimental methods & apparatus.*

Dutta, Dr Bishnu Pada. Physics Dept., Science C., Patna 800005, India. (1942) PhD, X-rays & structure of matter (Ranchi U., 1975). Reader. *X-ray crystallography.*

Dutta, Dr Sachindra Nath. Dept. of Education of Sci. & Math., Nat. Inst. of Education, Sri Arbindo Marg, New Delhi 110016, India. (1928) PhD, X-ray crystallography (Manchester U., UK, 1959). Reader. (tel. 79546, ext. 225). *Crystal structure, solid state physics, liquid physics.*

Dweltz, Dr Neville Edwin. Physics Div., ATIRA, Ahmedabad 380015, India. (1933) PhD, crystallography (Madras U., 1962). Deputy dir. *Fibre structure, macromolecular materials research, small angle scattering.*

Dwivedi, Dr Ganpat Lal. Mineral Physics Div., Geological Survey of India, B-8 Govind Marg, Adarsh Nagar, Jaipur, India. (1942) PhD, X-ray crystallography (I.I.T., Kanpur, 1971). Sr. Mineralogist. *Organic structures, inorganic and mineral crystal structures.*

Elango, Mr N. Dept. of Crystallography & Biophysics, Guindy Campus, Madras U, Madras 600025, India. (1957) MSc (Madras U.,1980). Res. scholar. *Theoretical crystallography, crystal structure analysis.*

Eswara Prasad, Dr Gummuluri. Physical Metallurgy Div., Bhabha Atomic Res. Cntr., Trombay, Bombay 400085, India. (1940) PhD, physical chemistry (Bombay U., 1978). Scient. officer. (tel. 5514910, ext. 2242). *Structure and properties, oxide ceramics, rapid solidification metals and alloys, failure analysis.*

Fernandes, Mr Jacob Richard. Liquid Crystals Lab., Raman Res. Inst., Bangalore 560006, India. (1951) MSc, physics (I.I.T., Kharagpur, 1973). Res. fellow. (tel. 30124). *Liquid crystals.*

Ganesan, V. Physics Dept., Indian Inst. of Techn., Madras, India. (1959) MSc, physics (Madras U.). *Lattice dynamics.*

Garg, Dr Ajau Kumar. Dept. of physics, N.A.S. College, Shanti Sadan, Sharda Road, Delhi Gate, Meerut, UP 25002, India. (1952) PhD,solid state physics (IIT, Delhi, 1975) Lect. (tel. 24528) *Crystal growth, crystallographic data, inorganic crystals, microscopy, phase transitions, powder diffraction, X-ray diffraction.*

Ghosh, Ms Manuja. Crystallography & Molecular Biology Div.,Saha Inst. of Nuclear Physics, 1/AF Bidhannagar,Calcutta 700064, India. (1959) MSc, physics (Calcutta U.,1981).Res. fellow. *Biologically important compounds.*

Ghosh, Dr Mrs Minakshi. Dept. of Physics, Presidency C., Calcutta 700073, India. (1948) PhD, physics (Calcutta U., 1981). Lect. *Organic and inorganic structures, phase transitions, disorder.*

Ghosh, Dr Sujit Kumar. Res. and Dev. Div., Projects & Dev., India Ltd., Sindri, Dhanbad 828122, India. (1932) PhD, X-ray crystallography (Indiana U., USA, 1960). Deputy General Manager. *Crystal structures, organic molecules, crystal chemistry, catalysts, line profile analysis, computer programming, mineral characterization and beneficiation, fertilizer materials.*

Ghosh, Ms Sutapa. Crystallography & Molecular Biology Div., Saha Inst of Nuclear Physics, Sector-1 Block-AF, Bidhannagar, Calcutta 700064, India. (1959) MSc, physics (Calcutta U., 1981). Res. scholar. *Crystal structure, biomolecules.*

Ghouse, Dr Khaja Mohd. X-ray Div., Regional Res. Lab., Hyderabad 500009, India. (1931) PhD. physics (Osmania U., 1967). Scient. (tel. 71351). *X-ray crystallography, organic and inorganic compounds.*

Giri, Mr Anit K. C.S.S Dept., Indian Assos. for the Cultivation of Sci., Jadavpur, Calcutta 700032, India. (1960) MSc, science (Jadavpur U.,1988).Res. scholar. *Structure-properties relationship, lattice defects, material characterization, amorphous materials.*

Giri, Mr Siba Narayan. Dept. of Physics, Bidhan Chandra Krishi Viswavidyalaya Mohanpur, Nadia, India. (1936) MSc, physics (Allahbad U., 1960). Reader. *Crystal structure, organic compounds.*

Girirajan, Dr K. S. Dept. of Biophysics and Crytallography, U. of Madras, Guindy Campus, Madras 600025, India. (1946) PhD, physics (I.I.T., Madras, 1974). Lect. *Thermal vibrations, crystallography.*

Godavarthi, Bhagavannarayana. C.G.C. Sect., Nat. Physical Lab., New Delhi 110012, India. (1955) MSc, physics (Andhra U., Waltair, 1979). Scient. *High resolution X-ray diffraction.*

Gomes, Albert Cardinal. Dept. of Biophysics, Bose Inst., P-1/12, C.I.T. Scheme 7 M, Kankurgachi, Calcutta 700054, India. (1947) MSc, molecular Biology (Calcutta U., 1972). Res. scholar. *Organic crystal structure, biological macromolecules.*

Goswami, Prof. Kaidar Nath. Physics Dept., Jammu U., Jammu 180001, India. (1934) PhD, solid state physics (Sardar Patel U., 1963). Prof. and Head. (tel. 48021 and 46340). *X-ray crystallography, structure determination and medicinal compounds, phase transition.*

Goswami, Dr (Mrs) S. N. N. Materials Characterization Div., Nat. Physical Lab. New Delhi, 110012, India. (1949) PhD, physics (Jiwaji U., Gwalior, 1975). Res assoc. *Single crystal characterization, perfection, X-ray diffraction.*

Guha, Dr (Mrs) Rommel. Mineral Physics Div., Geological Survey of India, 29 J.L.Nehru Road, Calcutta 700016, India. (1941) PhD, physics (Calcutta U., 1969).Sr.Mineralogist. (tel. 29-7645,ext.7). *Structure, minerals, biomolecules, powder diffractometry.*

Guha, Dr Sankarananda. Gayatri Chemicals, 37 Garia Park, Calcutta 700084, India. (1943) PhD, physics (Calcutta U., 1968). Proprietor. (tel. 72-2035). *Structure, biomolecules, ferroelectrices.*

Gupta, Dr Kinkar Prosad. Metallurgical Eng. Dept., Indian Inst. of Techn., Kanpur 208016, India. (1933) PhD, metallurgical engineering (Illinois U., USA, 1962) Prof. *Complex structures, phase transformations, theory & properties, metals and alloys, magnetic materials.*

Gupta, Mr Manoj Kumar. Physics Dept., Lucknow U., Lucknow 226007, India. (1955) MSc, physics (Kanpur U., 1973). Res. scholar. (tel. 25140). *Structure and conformation, polypeptides and biomolecules.*

Gupta, Prof. Manoranjan Prasad. Physics Dept., Ranchi U., Ranchi 834008, India. (1926) PhD, crystallography (London U., UK, 1953). Prof. and Head. (tel. 22336). *Crystal structure, crystal physics.*

Gupta, Dr Satish Chander. Neutron Physics Div., Bhabha Atomic Res. Cntr., Trombay, Bombay 400085, India. (1951) PhD, physics (Bombay U. 1980). Scient. officer. (tel. 5513848). *High pressure crystallography, shock wave phenomena in materials.*

Gupta, Miss Sunita. Physics Dept. Lucknow U., Lucknow 226007, India. (1952) MSc, physics (Lucknow U., 1972). Res. scholar. (tel. 25140). *Lattice dynamics, intermolecular forces.*

Gupta, Mr Vijai Prakash. Physics Dept., Lucknow U., Lucknow 226007, India. (1950) MSc, physics (Banaras Hindu U., 1972). Sr. res. fellow. (tel. 25140). *Conformation, polypeptide and biomolecular conformational transitions.*

Gupta, Prof. Vishwambhar Dayal. Physics Dept., Lucknow U., Lucknow 226007, India. (1934) DPhil, X-ray structure of cellulose (Allahabad U., 1958). Prof. (tel. 25140). *Biomolecules, small angle X-ray scattering, phase transitions, lattice dynamics.*

Gururow, Dr Tayur N. Physical Chem Div., Nat. Chem. Lab., Pune 411008, India. (1951) PhD, X-ray crystallography (I.I.S., 1976). Scient. *Crystal structures, organic and inorganic compounds, electron density, computer programming.*

Gyanchandani, Ms Jyoti S. Neutron Physics Div., Bhabha Atomic Res. Centr., Bombay 400085, India. (1963) MSc, physics (Nagpur U.,1985). Scient. officer. (tel 5513848).

Halder, Dr Sujit Kumar. Crystal Growth and Characterization Sect., Nat. Physical Lab., New Delhi 110012, India. (1948) PhD, Physics (Calcutta U., 1976). Scient. *X-ray diffraction, microstructure, alloys, thin films, crystal growth, characterization.*

Hariharan, Meena. Dept. of Crystallography and Biophysics, U. of Madras, Madras 60000, India. (1957) MSc, (U. Madras). Res. scholar. *Biocrystallography.*

Hemkar, Dr Mangla Prasad. Physics Dept., Allahabad U., Allahabad, UP, India. (1932) DPhil, lattice dynamics (Allahabad U., 1962). Lect. *X-ray crystallography, fibre structures, diffuse X-ray scattering, lattice dynamics.*

Hiremath, Mr Chaitanya, N. Molecular Biophysics Unit, Indian Inst. of Sci., Bangalore 560012, India. (1962) MSc, physics (Karnataka U., 1985). Res. student. (tel. 344411, ext. 2389). *Virus crystallography, molecular biophysics.*

Iyenger, Dr Leela. Physics Dept., C. of Sci., Osmania U., Hyderabad 500007, India. (1943) PhD, X-ray crystallography (Osmania U., 1970). Lect. *Thermal expansion, crystal structures.*

Jagannadham, Dr Adibhatla Vankata. Physics Dept., Government C., Ajmer 305001, India. (1924) PhD, crystalline field parameters by EPR (I.I.T., Kanpur, 1971). Postgrad. head. *X-ray crystallography, electron paramagnetic resonance, lasers.*

Jain, Dr Arun Kumar. Physics Div., Jute Techn. Res. Lab.,Indian Council of Agricultural Res., 12 Regent Park, Calcutta 700040, India. (1954) PhD, Fibre Science (I.I.T.,Delhi, 1988). Scientist S-2. (tel. 723192). *Fibre structure.*

Jain, Dr Prem Chand. Chemistry Dept., Kurukshetra Univ.,Kurukshetra 132119, India. (1931) PhD, chemistry (Saugar U., 1960). Prof. *Structure, biologically important compounds.*

Jakkal, Vasant Shankar. Water Chemistry Div, Bhabha Atomic Res Cntr., Trombay Bombay 400085, India. (1936) MSc, physics (Bombay U., 1961) Scient. officer. *Organic crystal structure.*

Jayadevan, Dr Naduviledath Chennuvittil. Radiochemistry Div., Bhabha Atomic Res. Cntr., Bombay 400085, India. (1936) PhD, inorgnic chemistry (McMaster U., Canada, 1968). Scient. officer. *Transition metal complexes, actinide complex, mixed oxides, mixed carbides.*

Jayanty, Dr Ashok Physics Dept., Ranchi U., Ranchi 834008, India. (1951) PhD, X-ray crystallography (Ranchi U., 1980) Lect. *Crystal srtucture determination.*

Jayashree, Ms A.N. Molecular Biophysics unit, Indian Inst. of Sci., Bangalore 560012, India. (1963) MSc, Physics (Karnatak U., 1984). Res. Student. *Biological crystallography.*

Joshi, Prof. Ramesh Vinayak. Applied Physics Dept., M.S.U., Baroda 390001, India. (1925) PhD, physics (Leeds U., UK, 1958). Prof. *Crystal defects.*

Joshi, Dr Shri Krishna. Physics Dept., Roorkee U., Roorkee 247667, India. (1935) DPhil, crystal physics (Allahabad U., 1962). Prof. & head. (tel. 743). *Solid state theory, electrons & phonons in disordered systems, magnetism.*

Kar(Roy), Dr (Mrs) Tanusree Dept. of Materials Science, Indian Assoc. for the Cultivation of Sci., Calcutta 700032, India. (1954) PhD, physics (Calcutta U., 1984). Res. Assoc. *Structures, organic and biological molecules, molecular conformation, crystalphysics, crystal growth, characterization, topography.*

Kalyanaraman, Dr A. R. Physics Dept., SITRA, Avinashi Road, Coimbatore 641014, India. (1937) PhD, X-ray crystallography (Madras U., 1967). Res. assoc. (tel. 28367, ext. 36). *Crystallography, fibre diffraction, computer programming, texture.*

Kannan, Dr Kazhiur Kothandapani. Neutron Physics Div., Bhabha Atomic Res. Cntr., Bombay 400085, India. (1939) PhD, X-ray crystallography (I.I.S., 1966).

Scient. (tel. 5513848, Telex 011-71017 BARC IN). *Macromolecular crystallography.*

Kashyap, Dr Ram Prasad. Chemistry Dept., Guru Nanak Dev U., Amritsar 143005, India. (1947) PhD, crystallography (Kurukshetra U., 1974). Lect. *X-ray crystallography.*

Kasturi, Prof Tirumali R. Organic Chemistry Dept., Indian Inst. of Sc., Bangalore 560012, India. (1931) PhD, organic chemistry (Bombay U., 1957). Prof.; Chairman, div. of chemical & biological sci. *Crystal structure, organic molecules.*

Kini, Mr Ullal Devappa. Liquid Crystals Lab., Raman Res. Inst., Bangalore 560006, India. (tel. 340124). *Liquid crystals (hydrodynamics).*

Kodandapani, Mr R. Molecular Biophysics Unit, Indian Inst of Sci., Bangalore 560012, India. (1961) MSc, physics (Karnataka U., 1983). Res. student. (tel. 344411, ext. 2389) *Protein crystallography, biophysics.*

Kohli, Dr Vijay Kumar. Crystal Growth & Characterization Sect., Nat. Physical Lab., New Delhi 110012, India. (1948) PhD, physics (Delhi U., 1984). Scient. *Crystal growth, topography, instrumentation, X-ray diffraction techniques.*

Krishna, Prof. Padmanabhan. Rajghat Education Centre, Krishnamurti Foundation India, Rajghat Fort, Varanasi - 221001, India. (1938) PhD, physics (Banaras Hindu U., 1962). Rector. (tel. 62717). *Crystal growth and perfection, phase transformation, polytypism.*

Krishna Rao, Prof. K. V. Physics Dept., C. of Sci., Osmania U., Hyderabad 500007, India. (1920) PhD, X-ray crystallography (London U., UK, 1958). Prof. *X-ray crystallography, photoelastic effect, solid state physics.*

Krishnaiah, Mr Musali. Physics Dept., Sri Venkateswara U., Tirupati, A. P. 517502, India. (1950) PhD, physics (Sri Venkateswara U., 1985). Lect. (tel. 8574 2781, ext. 274). *X-ray structure analysis.*

Kumar, Dr Rajendra. Regional Res. Lab. (CSIR), Bhopal 462026, India. (1929) PhD, metallurgy (Sheffield U., 1956). Dir. (tel. 65690). *Structure, rapidly solidified metallic phases, liquid metals, solidification, aluminium conductors, high temperature materials, crystal structure.*

Kumar, Mr Vinay Neutron Physics Div., Bhabha Atomic Res. Cntr., Bombay 400085, India. (1960) MSc, chemistry (Guru Nanak Dev U., 1981). Scient. officer. *Macromolecular crystallography.*

Kundra, Dr Krishan Dev. Materials Characterization (X-rays) Div., Nat. Physical Lab., Hillside Road, New Delhi 110012, India. (1934) PhD, X-ray crystallography (Delhi U., 1977). Scient. (tel. 584179). *Characterization of materials, X-ray diffraction, fluorescence, experimental methods & apparatus.*

Kundu, Dr Mohanlal. Physical Res. Wing, Projects & Dev. India Ltd., Sindri 828122, Dhanbad, India. (1942) PhD, Physics (Burdwan U.,1979).Deputy Superintendent. *Organic crystal structures, powder diffraction, catalysts, structural characterization, defect structures, minerals, fertilizer compounds, computer programming.*

Kuppuswamy, Dr Nagarajan. Res. & Dev., Searle (India) Ltd., Thane-Bhelapur Road, Thane 400601, India. (1930) PhD, (Madras U., 1954). Dir. *Organic crystal structure.*

Lahiri, Dr Barendra Nath. Physics Dept., Burdwan U., Burdwan, India. (1938) PhD, X-ray crystallography (Calcutta U, 1973). Lect. (tel. 341121). *X-ray crystallography, silicate minerals, textile fibres.*

Lakshminarayanan, Muthuvijayan. Inorganic and Physical Chemistry Dept., Indian Inst. of Sci., Bangalore 560012, India. (1960) MSc, chemistry (Madurai Kamaraj U.,1985).Res.scholar. *Crystallography, bio-coordination compounds.*

Lal, Dr Krishan. Nat. Physical Lab., Hillside Road, New Delhi 110012, India. (1941) PhD, (Delhi U., 1969) Deputy dir. *Crystal growth, lattice imperfections, high resolution X-ray diffraction techniques, topography, multi-crystal X-ray diffractometry, diffuse X-ray scattering.*

Lele, Prof. Shrikant Dept. of Metallurgical Eng., Banaras Hindu U., Varanasi 221005, India. (1943) PhD, physical metallurgy (Banaras Hindu U., 1967). Prof., physical metallurgy. (tel. 64491, ext. 250). *Phase transitions.*

Lokanatha, Mr S. Physics Dept., Indian Inst. of Techn., Kharagpur 721302, India. (1958) MSc (Bangalore U., 1981). Lect. (Physics Dept., SJRC C., Bangalore) *Structure - properties relationship, minerals.*

Mahanta, Dr Bhubaneswar. Physics Dept., Ranchi U., Ranchi 834008, India. (1928) PhD, X-ray crystallography (Ranchi U., 1979) Res. scholar. *X-ray crystallography, organo-metallic compounds.*

Madhusudan, Mr. Molecular Biophysics Unit, Indian Inst. of Sci., Bangalore 560012, India. (1962) MSc, physics (Delhi U.,1985).Res. student. (tel. 344411, ext. 2389). *Protein crystallography, biophysics.*

Madhusudana, Dr N. V. Liquid Crystals Lab., Raman Res. Inst., Bangalore 560006, India. (1944) PhD, liquid crystals (Mysore U., 1971). Scient. (tel. 30124). *Liquid crystals.*

Mahata, Dr Akhil Physics Dept., Ranchi U., Ranchi 834008, India. (1939) PhD, X-ray crystallography (Ranchi U., 1978) Res. scholar. *X-ray crystallography, organo-metallic compounds.*

Maiti, Dr Gobinda Chandra. Physical Res. Wing, Projects & Dev. India Ltd., Sindri 828122, Dhanbad, India. (1944) Phd, chemistry(Calcutta U.,1977).Asst. Superintendent. *Catalysis, crystal chemistry, heterogeneous catalysts, solid solution, structural changes.*

Majumdar, Dr Sunil Kumar. Crystallography & Molecular Biology Div., Saha Inst. of Nuclear Physics, 92 A. P. C. Road, Calcutta 700009, India. (1933) PhD, X-

ray crystallography (Madras U., 1964). Prof. (tel. 354281). *X-ray crystallography, proteins, nucleic acid, biological molecules.*

Mandal, Dr Pradip Kumar. North Eastern Hill U., Bijni Complex, Bhagyakul, India. (1958) PhD, physics (North Bengal U.,1986). Lect. *Liquid crystal, str. determination.*

Mande, Prof. Chintamani. Physics Dept., Nagpur U., Nagpur 440010, India. (1925) DSc, X-ray spectroscopy (Paris U., France, 1958). Prof. and head. (tel. 31946). *X-ray spectroscopy, intermetallic compounds, chemical bonding, spinel structures.*

Mande, Mr Sekhar Chintamani. Molecular Biophysics Unit, Indian Inst. of Sci., Bangalore 560012, India. (1962) MSc, physics (Nagpur U., 1984). Res. student. (tel. 344411, ext. 2389). *Protein crystallography, molecular biophysics.*

Mani, Mr A. Electrochemical Res. Inst., Karaikudi 623006, India. (1950) MSc, materials sci. (Anna U., 1981). Scient. *Materials science, structure and properties, thin-flim deposits, phase transitions, organic structures, X-ray crystallographic methods and apparatus.*

Manickkavachgam, Dr Ramanathan. Christian C., Madras, India. (1955) PhD, (Madurai Kamaraj U., 1988). Lect. *Crystal structure, phase transition.*

Manohar, Prof Hattikud²r. Dept. of Inorganic & Physical Chemistry, Indian Inst. of Sci., Bangalore 560012, India. (1929) PhD, physics (I.I.S., Bangalore, 1963). Prof. (tel. 364411, ext. 383). *Crystallography, bio-coordinated compounds, cyclophosphazenes, organometallics; solid state reactions.*

Mathews, Irimpan I. Inorganic and Physical Chemistry Dept., Indian Inst. of Sci., Bangalore 560012, India. (1962) MSc, chemistry (Calicut U.,1984).Res.scholar. *X-ray crystallography, bio-coordination compounds.*

Mathur, Dr Balbir Kumar. Physics Dept., Indian Inst. of Techn., Kharagpur 721302, India. (1948) PhD, physics (I.I.T., Kharagpur, 1978). Sr. scient. asst. *Lattice defects, optical transforms, computer programming.*

Misra, Prof Nirmal Kumar. Dept. of Physics & Meteorology, Indian Inst. of Techn., Kharagpur 721302, India. (1939) PhD, physics (I.I.T., Kharagpur, 1967). Prof. *Crystal growth, defect crystal structure, thin films, metallic glass, disorder materials.*

Misra, Prof. Somnath. Principal, Regional Eng. C., Rourkela 769008, India. (1936) ScD, metallurgy (MIT, USA, 1963). Principal. (tel. 5050). *Intermetallic compounds, order-disorder transformations.*

Misra, Dr Tripurari. Dept. of Physics, Regional Eng. C., Rourkela 769008, India. (1933) PhD, physics (Sambalpur U., 1973). Prof. and Head. *Molecular biophysics.*

Mitra, Prof. Girija Bhushan. C.S.S. Dept., Indian Assoc. for the Cultivation of Sci., Calcutta 700032, India. (1923) DSc, X-ray studies of lattice defects (Calcutta U., 1967). Em. scient. *Lattice defects, lattice vibrations, dynamical diffraction theory, material characterization, structure, instrumentation, small angle X-ray scattering, structure - properties relationships.*

Mohanlal, Dr Sembu Krishnaiyer. School of Physics, Madurai Kamaraj U., Madurai 625021, India. (1940) PhD, X-ray crystallography (Madurai U., 1971). Prof. (tel. 26514). *X-ray diffraction, solid solutions, lattice defects, electron density distribution, instrumentation.*

Mukherjee, Dr Alok Kumar. Dept. of Physics, Jadavpur U., Calcutta 700032, India. (1950) PhD, physics (Visva-Bharati U., 1978). Lect. *Organic and organometallic structures, disorder.*

Mukherjee, Dr Amal Bikash. Dept. of Geology & Geophysics, Indian Inst. of Tech., Kharagpur 721302, India. (1935) PhD, geology (I.I.T., Kharagpur, 1961). Prof. *Crystal growth, structure, minerals.*

Mukherjee, Dr Biswanath. X-ray Crystallography Dept., Central Glass & Ceramic Res. Inst., Calcutta 700032, India. (1936) PhD, physics (Calcutta U., 1965). Asst. dir. *Material science, crystal stability, structure.*

Mukherjee(mondal), Dr (Mrs) Monika. Dept. of Magnetism, Indian Assoc. for the Cultivation of Sci., Calcutta 700032, India. (1947) PhD, physics (Calcutta U., 1977). Res Asst. *Organic and organometallic structure, disorder, phase transition.*

Mukherjee, Dr Partha Sarathi. Materials Div., Regional Res. Lab., Trivandrum 695019, India. (1952) PhD, physics (I.I.T., Kharagpur, 1982). Scient. *Characterization, semicrystalline materials, polymers, minerals, high temp. superconductor.*

Mukhopadhyay, Dr (Mrs) Anuradha. Dept. of Solid State Physics, Indian Assoc. for the Cultivation of Sci., Calcutta 700032, India. (1955) PhD, physics (Calcutta U., 1987). Res. assoc. *Coordination complexes, organic compounds, disorder, phase transition, crystallographic statistics.*

Mukhopadhyay, Ashis. X-rays and Crystallograohic Unit, Physics Dept., Jadavpur U., Calcutta 700032, India. MSc, physics (I.I.T.,Kharagpur,1987).Res.scholar. *Biologically important compounds, drugs, lattice dynamics, computer modelling, computer simulation.*

Mukhopadhyay, Mr Bishnu Prasad. Crystallography & Molecular Biology Div., Saha Inst. of Nuclear Physics, 1/AF Bidhannagar, Calcutta 700064, India. (1954) MSc, chemistry (Burdwan U., 1977). Res. fellow. *Biomolecular structure and conformation, adrenergics.*

Mukhopadhyay, Dr Pradip. Physical Metallurgy Div. Bhabha Atomic Res. Cntr., Bombay 400085, India. (1943) PhD, physics (Bombay U., 1980). Scient. officer. (tel. 5514910 ,ext. 2435, 2242). *Phase transformation, defect structure properties correlation, electron microscopy and diffraction, electron spectroscopy.*

Munirathinam, Nethaji. Dept. of Crystallography & Biophysics, Madras U., Madras 60025, India. (1958) MSc (Madras U., 1981). Res. scholar. *Crystal structure, biomolecules.*

Munshi, Mr Sanjeev Kumar. Molecular Biophysics Unit, Indian Inst. of Sci., Bangalore 560012, India. (1962) MSc, biochemistry (Kashmir U., 1984). Res. student. (tel. 344411, ext. 2389). *Molecular biophysics, virus crystallography.*

Muralidharan, Mr K. V. Chemistry Div., Bhabha Atomic Res. Cntr., Bombay 400085, India. (1937) MSc, physics (Bombay U., 1971). Scient. officer. (tel. 523321, ext. 288). *Biologically important structures.*

Murthy, Dr M.R. N. Molecular Biophysics Unit, Indian Inst. of Sci., Bangalore 560012, India. (1950) PhD, X-ray crystallography (I.I.S., 1977). Asst. prof. (tel. 344411, ext. 2389, telex 0845-8349 IISC IN). *Biological crystallography, molecular evolution.*

Nag, Dr Dilip Kumar. Mineral Physics Div., Geological Survey of India, 29 Chowringhee Road, Calcutta 700016, India. (1945) PhD, X-ray crystallography (Calcutta U., 1974). Mineralogist. (tel. 238321, ext. 8) *Organic and inorganic structures, mineral crystal structures, instrumentation, crystal physics.*

Nag, Miss Jhumjhumi. Dept. of General Physics & X-rays, Indian Assoc. for the Cultivation of Sci., Calcutta 700032, India. (1955) MSc physics (Calcutta U., 1976). Res. scholar. *Structural characterization, amorphous materials, thin films crystallography, X-ray diffraction.*

Nagabhushana Rao, Mr Chemboli. Metallurgy Div., Bhabha Atomic Res. Cntr., Bombay 400085, India. (1941) MTech, metallurgy (I.I.T., Kanpur, 1973). Scient. officer. *X-ray metallography, texture studies.*

Nagpal, Dr Kailash Chander. X-ray Sect., Nat. Physical Lab., Dr. K. S. Krishnan Road, New Delhi 110012, India. (1932) PhD, X-ray crystallography (Delhi U., 1976). Scient. (tel. 584179, 5714624). *X-ray crystallography, diffractometry, X-ray fluorescence analysis, experimental methods & apparatus.*

Naik, Dr (Mrs) Uma Murlidhar. Metallurgy Div., Bhabha Atomic Res. Cntr., Bombay 400085, India. (1940) Scient. off. *X-ray metallography, electron microscopy.*

Nandi, Asok Kumar. X-ray Sect., Central Glass & Ceramic Res. Inst., Calcutta 700032, India. (1944) MSc, (Calcutta U., 1965). Scient. *Crystal structure, organic compounds, amorphous materials.*

Nandi, Dr Ranjan Kumar. 2E Neelamber, 28B Shakespeare Sarani, R & D Cntr. Steel Authority of India Ltd., Calcutta 700017, India. (1950) PhD, physics (Calcutta U., 1978). Principal Res. Eng. *Crystal imperfections, thin films EXAFS, supported metal catalysts, textures, electrical steels, deep drawing steels.*

Narasimhamurthy, Narasappa. Inorganic and Physical Chemistry Dept., Indian Inst. of Sci., Bangalore 560012, India. (1960) MSc, chemistry (Mysore U.,1983).Res. scholar. *Crystallography of organometallic compounds.*

Narasimhan, Dr P. Physics Dept., A. M. Jain C., Madras 600114, India. (1944) PhD, physics (Madras U., 1984). Prof. *Crystal structure, organic compounds, dipeptides, small molecules.*

Narayanan, Prof. Palamadi Sundaram. Physics Dept., Indian Inst. of Sci., Bangalore, Karanataka 560012, India. (1926) PhD, crystal physics (Madras U., 1951). Prof., chairman, div. of physics and math. sciences. (tel. 34411, ext., 469). *Ferroelectricity, structural phase transitions, crystal growth.*

Natarajan, Dr S. Physics Dept., Anna U., Madras 600025, India. (1941) PhD, physics (I.I.T., Madras 1969). Prof. and Head. *Crystal structure, high pressure X-ray diffraction.*

Natarajan, Dr Subramanian. Sch. of Physics, Madurai Kamaraj U., Madurai 625021, India. (1949) PhD, physics (Madurai Kamaraj U., 1979). Reader. (tel. 85492). *Crystal structure, amino acid complexes, crystal growth.*

Natesan, Mr Elango. Dept. of Crystallography and Biophysics, Madras U., Madras 60025, India. (1957) MSc (Madras U., 1980). Res. scholar. *Theoretical X-ray crystallograpy, structures.*

Nath, Mr Kashi. Physics Dept., Lucknow U., Lucknow 226007, India. (1951) MSc, physics (Lucknow U., 1973). Res. fellow. (tel. 25140). *Conformation and conformational transitions, polypeptides and biomolecules.*

Nigam, Gur Dayal. Physics Dept., Indian Inst. of Techn., Kharagpur 721302, India. (1939) PhD (I.I.T., Madras, 1969). Asst. prof. *Direct methods, crystallographic statistics, crystallographic symmetry.*

Nigli, Selina. Dept. of Physics & Astrophysics, U. of Delhi, Delhi 110007, India. (1949) MPhil (Delhi U., 1982). Res. scholar. *Crystal growth, X-ray diffraction, defects, physical properties.*

Noor, Sahina Begum. Inorganic and Physical Chemistry Dept., Indian Inst. of Sci., Bangalore 560012, India. (1960) MSc (Bangalore U., 1982). Res. scholar. *X-ray crystallography, metal nucleotide complexes.*

Pan, Dr Nitya Ranjan. Physics Dept., Calcutta U., 92 A. P. C. Road, Calcutta 700009, India. (1935) PhD, physics (Calcutta U., 1975). Lect. (tel. 359186). *Electret orientation & related properties.*

Pandey, Dr Dhananjai. Sch. of Material Sci. & Techn., Banaras Hindu U., Varanasi 221005, India. (1952) PhD, physics (Banaras Hindu U., 1976). Reader. *Structural imperfections, solid state transformations, ferroelectrics.*

Pandya, Prof. Janardhan Rameshchandra. Dept. of Physics, M.S.U., Baroda 390002, India. (1934) PhD, physics (M.S.U., 1961). Prof. *Crystal growth, dissolution (etch phenomenon), hardness, electrical conductivity.*

Pant, Dr Arun Kumar. Physics Dept. Gorakhpur U., Gorakhpur 273009, India. (1940) PhD, X-ray crystallography (Calcutta U., 1964). Reader. *Organic and inorganic structures, lattice dynamics, electron density distribution.*

Pant, Dr Lalit Mohan. Physical Chemistry Div., Nat. Chemical Lab., Pashan Road, Poona 411008, India. (1928) PhD, crystallography (London U., UK, 1958). Scient. *Structure analysis, disorder in crystals.*

Papavinasam, Dr E. Dept. of Physics, Thiagarajar C., Madurai 625009, India. (1943) PhD, X-ray crystallography (Madurai Kamaraj U., 1986). Prof. *Crystal growth, structure analysis.*

Parasnis, Prof. Arawind Shripad. Physics Dept., Indian Inst. of Techn., Kanpur 208016, India. (1928) PhD, crystal physics (Bristol U., UK, 1960). Prof. (tel. 214151). *Intercrystalline interfaces, surface free energy, dislocations, electron microscopy.*

Parthasarathy, Dr S. Dept. of Crystallography & Biophysics, Guindy Campus, Madras U., Madras 600025, India. (1940) PhD, crystallography (Madras U., 1967). Prof. (tel. 415367, ext. 16). *Theoretical crystallography, crystal structure analysis.*

Patel, Prof. Ambalal Ranchhodbhai. Dept. of Physics, Sardar Patel U., Vallabh Vidyanager, Gujrat, India. (1917) PhD, physics (London U., UK, 1958). Retired prof. *Mineral crystals.*

Patel, Dr Prabhudas Revandas. Silical Lab., Sandhporepardi, Valsad 396001, India. (1938) PhD, liquid crystals (M.S.U., Baroda, 1967). Partner. (tel. 2281, 2154). *Structural changes, organic molecules, liquid crystals.*

Patel, Dr Ranjan Prafulbhai. Chemistry Div., Bhabha Atomic Res. Cntr., Bombay 400085, India. (1953) PhD,inorganic chemistry (Bombay U.,1988). Scient. officer. *Synthesis, spectroscopy, structure, transition metal-organic compounds.*

Patel, Dr Tankadhar. Physics Dept., Regional Eng. C., Rourkela 769008, Orissa,India. (1943) PhD, physics (I.I.T., Bombay, 1982). Asst. Prof. *Crystal structure, organic and inorganic substances.*

Pathak, Prof. Pushkarrai Dalpatram. Physics Dept., U. Sch. of Sci., Ahmedabad 380009, India. (1916) MSc, physics (Bombay U., 1946). Prof. (tel. 40929). *X-ray diffraction, defects, thermal properties.*

Pathinettam, Dr Padiyan. D.V.H.N.S.N. College, Virudhunagar 626002, India. (1956) PhD (Madurai Kamaraj U., 1988). Lect. *Crystalline solid solutions, X-ray diffraction, bonding, semiconductors, anharmonicity.*

Pattabhi, Dr (Mrs) Vasantha. Dept. of Crystallography & Biophysics, U. of Madras, A.C.C. Campus, Madras 600025, India. (1944) PhD, X-ray crystallography (Madras U., 1972). Lect. (tel. 415367, ext. 16). *Crystallography, biomolecules, conformational analysis.*

Podder, Dr (Mrs) Aloka. Crystallography & Molecular Biology Div., Saha Inst. of Nuclear Physics, 1/AF Bidhannagar, Calcutta 700064, India. (1937) PhD, physics (Calcutta U., 1976) Reader. *Biomolecular structure and conformation.*

Poojary, Dr M. Damodara Inorganic and Physical Chemistry Dept., Indian Inst. of Sci., Bangalore 560012, India. (1955) PhD, physics (I.I.S., 1984). Scient. Asst. *X-ray crystallography, metal-nucleotide complexes.*

Pradhan, Dr Dukhabandhu. Dept. of Physics, Deogarh C., Sambalpur, India. (1948) PhD, (I.I.T., Kharagpur, 1987). Reader. *Crystal structure, organic compounds, intensity statistics.*

Prasad, Dr Narayan, Physics Dept., Ranchi U., Ranchi 834008, India. (1944) PhD, X-ray crystallography (Ranchi U., 1978) Lect. *X-ray crystallography, organometallic compounds, hydrogen bonds.*

Prasad, Dr Ravindra. Geological Survey of India, B-192 Niralanagar, Lucknow, India. (1943) PhD, defects (Banaras Hindu U., 1971). Sr. Mineralogist. *Defects, polytypic crystals, crystal growth, structure, phase transformations, stacking faults.*

Prasad, Dr Satya Murti. Physics Dept., Ranchi U., Ranchi 834008, India. (1943) PhD, X-ray crystallography (Ranchi U., 1970). Prof. *X-ray crystallography, organic and organo-metallic compounds,proteins.*

Prasad, Dr Y. R. Ananth. Materials Div., Nat. Aeronautical Lab., PO Box No. 1779, Bangalore 560017, India. (1942) PhD, semiconductors (I.I.T., Delhi, 1970). Fellow NAL (tel. 573351, ext. 248). *Single crystal growth, ferro-electric crystals, diffusion technology, solar energy ion-implantation and microscopy.*

Puranik, Dr (Mrs) Vedavati Gururaj. Physical Chemistry Div., Nat. Chemical Lab., Poona 411008, India. (1953) PhD, physics (Bangalore U., 1983). Res. Assoc. *Crystal structure, organic and inorganic molecules, organometallic molecules, biologically important molecules.*

Raghavacharyulu, Dr Iyyunni Venkata Veera. Nuclear Physics Div., Bhabha Atomic Res. Cntr., Bombay 400085, India. (1934) DSc, group theory & representation of space groups (Andhra U., 1958). Scientific Officer (E). (tel. 523321, ext. 352). *Mathematical crystallography, solid state physics.*

Raghunatha, Chary. Dept. of Physics, Indian Inst. of Sci., Bangalore 560012, India. (1955) MPhil, physics (Hyderabad U., 1979). Res. student. *Crystal growth, crystal structure.*

Raghurama, Mr G. Physics Dept., Indian Inst. of Sci., Bangalore 560012, India. (1958) MSc, physics (I.I.T., Madras, 1980). Res. scholar. *Crystal structure, crystal binding.*

Rajagopal, Mr Hariharasubramonia Iyer. Neutron Physics Div., Cirus Bhabha Atomic Res. Cntr., Bombay 400085, India. (1941) BSc, physics (Kerala U., 1962). Scient. officer. *Neutron diffraction, computer programming, software development.*

Rajan, Mr R. D. Dept. of Physics, Anna U., Madras 600025, India. (1945) MPhil, physics (Anna U., 1984). Lect. *Crystal structure, organic compounds.*

Rajan, Dr S. S. Crystallography & Biophysics Dept., Madras U., Madras 60025, India. (1949) PhD, physics (Madras U., 1977). Reader. *Crystal structure analysis, organic compounds, proteins.*

Rajaram, Dr Ramasamy Karunandam. Sch. of Physics, Madurai Kamaraj U., Madurai 625021, India. (1951) PhD (Madurai Kamaraj U., 1978). Reader. *X-ray crystallography, hydrogen bonding, phase transitions.*

Rajasekharan, Dr T. Defence Metallurgical Res. Lab., Hyderabad 500258, India. PhD, physics (I.I.T., Madras,1978).Scient. *Structure, struc.-phys. props. correlations, intermetallic phases.*

Raju, Dr K. S. Dept. of Crystallography & Biophysics, Guindy Campus, Madras U., Madras 600025, India. (1937) PhD (Sardar Patel U., 1969). Reader. *Crystal growth, defects and characterisation.*

Raju, Dr I. V. K. Bhagavan. Dept. of Physics, Kakatiya U., Warangal 506009, India. (1946) PhD, physics (Osmania U., 1973). Reader. *Crystal growth, lattice defects,plastic flow, materials.*

Ram, Dr Purushottam. Physics Dept., Ranchi U., Ranchi 834008, India. (1948) PhD, physics (Ranchi U., 1979). Lect. *X-ray crystallography, organo-metallic compounds.*

Ram Kishore, Dr. Materials Div., Nat. Physical Lab., Krishnan Road, New Delhi 110012, India. (1948) PhD, physics (Agra U., 1983). Scient. B. (tel. 5726058). *Single crystal growth, multicrystalline silicon ingot technology.*

Rama Rao, Dr B. X-ray Crystallography Sect., Regional Res. Lab., Hyderabad 500009, India. (1930) Dr.rer.nat., mineralogy & crystallography (Göttingen, BRD, 1961). Asst. dir. *Organic and inorganic crystal structures, X-ray powder analysis.*

Ramachandran, Prof. Gopalasamudram N. Mathematical Philosophy Group, Indian Inst. of Sci., Bangalore 560012, India. (1922) DSc, crystallography (Madras U., 1949). INSA Albert Einstein Prof. *Theoretical crystallography.*

Ramakrishnan, Dr Chandrasekhara. Molecular Biophysics Unit, Indian Inst. of Sci., Bangalore 560012, India. (1939) PhD, biophysics (Madras U., 1966). Asst. prof. (tel. 34411, ext. 460). *X-ray crystallography, bio-molecules, computer programming.*

Ramakumar, Dr Suryanarayana Rao. Physics Dept., Indian Inst. of Sc., Bangalore 560012, India. (1941) PhD (I.I.S., 1977). Asst. prof. *Crystallography, biomolecules, drug molecules, computational crystallography, databases.*

Ramanadham, Dr Muthyala. Neutron Physics Div., Bhabha Atomic Res. Cntr., Trombay, Bombay 400085, India. (1945) PhD, physics (Bombay U., 1975). Scient. officer (SF). (tel. 5513848). *X-ray crystallography, biological macromolecules, neutron diffraction studies on hydrogen bonded systems, crystallographic computations, macromolecular refinement and graphics.*

Ramaswamy, Dr Krishnamachari. Physics Dept., Annamalai U., Annamalai Nagar 608002, India. (1935) PhD, molecular spectroscopy (Annamalai U., 1961). Head. *Bio-molecular structures, Raman spectroscopy, infrared spectroscopy, nuclear magnetic resonance, X-ray techniques.*

Ramaswamy, Mr S. Molecular Biophysics Unit, Indian Inst. of Sci., Bangalore 560012, India. (1964) MSc, physics (Bharatidasan U., 1987). Res. student. (tel.344411, ext. 2389). *Biological crystallography.*

Ranganath, Dr G. S. Liquid Crystals Lab., Raman Res. Inst., Bangalore 560006, India. (1944) PhD, crystal optics (Bangalore U., 1974). Scient. (tel. 30124). *Crystal optics, liquid crystals.*

Ranganathan, Prof. Srinivasa. Dept. of Metallurgy, Indian Inst. of Sci., Bangalore 560012, India. (1941) PhD, metallurgy (Cambridge U., UK, 1965). Prof. *Quasicrystals.*

Rao, Dr Keshavamurthy Narayana Swamy. Physics Dept. Indian Inst. of Techn., Kanpur 208016, India. (1934) PhD, crystal physics (Kanpur U., 1974). Sr. res. asst. (tel. 40066, ext. 30). *Crystal growth, characterization, crystal defects, intercrystalline boundaries.*

Rao, Prof. P. Rama. Defense Metallurgical Res. Lab., Hyderabad, India. (1942) PhD, metallurgical eng. (Banaras Hindu U., 1968). Dir. *Structure, intermetallic phases.*

Ratna, Miss B. R. Liquid Crystals Lab., Raman Res. Inst., Bangalore 560006, India. (1949) MTech, physical eng. (I.I.S., 1972). Res. fellow. (tel. 30124). *Liquid crystals.*

Ravichandran, Mrs V. Nuclear Physics Dept., U. of Madras, Madras 600025, India. (1959) MSc, physics (Madras U.,1981). Res. scholar. *Molecular biophysics.*

Ray, Dr (Mrs) Gouri. Radon House (P) Ltd., 7 Sirdar Sankar Road, Calcutta 700026, India. (1935) PhD, physics (Calcutta U., 1966). Managing dir. (tel. 461773). *Crystal physics, instrumentation.*

Ray, Dr Pankaj Narayan. Crystallography & Molecular Biology Div., Saha Inst. of Nuclear Physics, 1/AF Bidhannagar, Calcutta 700064, India. (1934) PhD, physics (Calcutta U., 1974). Reader. *Amino acids, biocompounds.*

Ray, Dr Pradip Kumar. Physics Div., Jute Techn. Res. Lab., Indian Council of Agricultural Res., 12, Regent Park, Calcutta 700040, India. (1938) PhD, Fibre structure (Calcutta U., 1968). Senior Scient. (tel. 723192). *Cellulose fibre structure.*

Ray, Prof. Siddhartha. Dept. of Solid State Physics, Indian Assoc. for the Cultivation of Sci., Calcutta 700032, India. (1932) DSc, physics (Calcutta U., 1974). Prof. *Inorganic and organic structures, phase transitions, disorder.*

Roychowdhury, Dr Priyobroto. Physics Dept., Calcutta U., 92 A.P.C. Road, Calcutta 700009, India. (1940) PhD, physics (Calcutta U., 1974). Reader. *X-ray crystallography, biomolecules, proteins.*

Roychowdhury, Dr (Mrs) S. X-ray lab., Dept. of Physics, Presidency C., Calcutta, India. PhD. Lect. *Small molecule crystallography.*

Saha, Mr Ajay Prakash. Physics Dept., Ranchi U., Ranchi 834008, India. (1951) MSc, physics (Ranchi U., 1972). Lect. *X-ray crystallography, organic compounds.*

Saha, Mr Bishwa Nath. Physics Dept., Ranchi U., Ranchi 834008, India. (1936) MSc, physics (Bihar U., 1960). Res, Scholar. *X-ray crystallography, organic and organometallic compounds.*

Sahani, Dr Jiwan Lal. Surgery Dept., Inst. of Medical Sciences, Banaras Hindu U., Varanasi 221005, India. (1950) MS, surgery (Banaras Hindu U., 1976). Sr. resident. *Crystallographic studies, gallstones.*

Sahu, Dr Bholanath. Physics Dept., Ranchi U., Ranchi 834008, India. (1943) PhD, X-ray crystallography (Ranchi U., 1970). Lect. *X-ray crystallography, organic structures.*

Sahu, Dr Mahendra. Physics Dept., Ranchi U., Ranchi 834008, India. (1928) PhD, X-ray crystallography (Ranchi U., 1970). Lect. *X-ray crystallography, organic compounds.*

Sahu, Dr N. C. Dept. of Physics, Post Graduate Regional Eng. C., Rourkela 796008, India. (1937) PhD, physics (Sambalpur U., 1972). Prof. and head. *Crystal structure, fibres, molecular biophysics, molecular structure, small angle X-ray scattering.*

Sahu, Dr Ram Gopal. Dept. of Physics, G.A.L.C., Daltonganj 822102, India. (1934) PhD, physics (Ranchi U., 1968). Reader. *Crystal growth, crystal structure, organic crystals.*

Sahu, Dr Ramdhani. Physics Dept., Ranchi U., Ranchi 834008, India. (1927) PhD, X-ray crystallography (Ranchi U.1982). Reader. *Crystal structure determination.*

Sahaymary, Mrs J. James Dept. of Biophysics & Crystallography, A.C. C., Madras U., Madras 60025, India. (1958) MSc, physics (Madras U., 1981). Res. scholar. *Molecular biophysics.*

Salunke, Dr Dinakar M. National Inst. of Immunology, Shahid Jeet Singh Marg, New Delhi 110067, India. (1955) PhD, molecular biophysics (I.I.S., 1983). Principal Scient. officer. *Protein crystallography, electron microscopy and image processing, structural biology.*

Samanta, Chitra. Dept. of Physics, Jadavpur U., Calcutta 700032, India. (1949) MSc, Applied mathematics (Jadavput U., 1971). Res. scholar. *X-ray crystallography, molecular lattice dynamics, computer simulation.*

Samantaray, Dr Biswas Kumar. Dept. of Physics, Indian Inst. of Techn., Kharagpur 721302, India. (1947) PhD, physics (I.I.T., Kharagpur, 1977). Res asst. *Applied crystallography, thin films, mineralogy.*

Sanjeeviraja, Dr C. Alagappa U., Karaikudi 623003, India. (1954) PhD (Madurai Kamaraj U., 1985). Lect. *Crystalline solid solutions, X-ray diffraction, instrumentation.*

Sankar, Mr B. N. Dept. of Physics, C. of Eng., Anna U., Madras 60025, India. (1951) MPhil physics (Anna U., 1983).Lect. *Crystal structure, organic compounds.*

Sarkar, Chitrita. Dept. of Physics, Jadavpur U., Calcutta 700032, India. (1951) MSc, physics (Jadavpur U., 1975). Res. scholar. *X-ray crystallography.*

Sarkar, Dr Satyabrata. Dept. of physics Jadavpur U., Calcutta 700032, India. (1944) PhD, physics (Jadavpur U., 1983). Techn. Superintendent. *X-ray crystallography, diffuse scattering, lattice dynamics, structure solution, computer simulation.*

Sankaran, Ms Hema. Neutron Physics Div., Bhabha Atomic Res. Centr., Bombay 400085, India. (1962) MSc, physics (I.I.T., Bombay,1983).Scient.officer. (tel. 5513848). *Crystal structure, solid state physics under high pressure.*

Saravanan, Mr R. School of Physics, Madurai Kamaraj U., Madurai 625021, India. (1964) MPhil (Madurai Kamaraj U., 1988). Res. scholar. *X-ray diffraction, defect characterization.*

Sasisekharan, Prof. Viswanathan. Molecular Biophysics Unit, Indian Inst. of Sci., Bangalore 560012, India. (1933) PhD, crystallography & biophysics (Madras U., 1959). Prof. and Divisional Chairman, Div. of Biological Sciences. (tel. 344411, ext. 458). *Crystal structure, conformation theory, quantum chemistry.*

Sastry, Dr Bommakanti Sri Rama. Industrial Ceramics Div., Regional Res. Lab., Hyderabad 500009, India. (1923) PhD, ceramic techn. (Pennsylvania State U., USA, 1958). Asst. dir. (tel. 71351). *High temperature technology, industrial ceramics, material science, phase equilibria.*

Sastry, Dr G. V. S. Dept. of Metallurgical Eng., Banaras Hindu U., Varanasi, 221005, India. (1952) PhD, metallurgical eng. (Banaras Hindu U., 1982) Reader. *Crystal structure, alloy metastable phase, electron difraction*

Sastry, Dr Medury Dattatreya. Radiochemistry Div., Bhabha Atomic Res. Cntr., Bombay 400085, India. (1942) PhD, physics (I.I.T., Kanpur, 1967). Scient. Officer. (tel. 523321, ext. 456). *Crystallographic studies, magnetic resonance, Mössbauer spectroscopy, X-ray diffraction.*

Satyanarayana Murthy, Dr Keta. Physics Dept., Nizam C., Hyderabad 500001, India. (1940) PhD, X-ray crystallography (Osmania U., 1971). Lect. (tel. 34231). *X-ray crystallography, solid state physics, biophysics.*

Savithramma, Miss K. L. Liquid Crystals Lab., Raman Res. Inst., Bangalore 560006, India. (1952) MSc, theoretical physics (Mysore U., 1974). Res. fellow. (tel. 30124). *Liquid crystals.*

Seal, Dr Alpana. Magnetism Dept., Indian Assoc. for the Cultivation of Sci., Calcutta 700032, India. (1955) PhD, physics (Calcutta U., 1985). Res. scholar. *Co-orination complexes, organic compounds, disorder, phase transition. liquid crystals, neutron protein crystallography.*

Seal, Prof. Arun Kumar. Bengal Eng. C., Howrah 711103, India. (1928) PhD, physical metallurgy (Sheffild U., 1956). Principal. *Crystal structure, metals.*

Seetharaman, Mr Venkataramakrishanan. Metallurgy Div., Bhabha Atomic Res. Cntr., Bombay 400085, India. (1950) BTech, metallurgy (I.I.T., Madras, 1971). Scient. officer. (tel. 523321, ext. 242). *Phase transformations, electron microscopy, microstrain, radiation damage.*

Sekar, Mr K. Dept. of Crystallography & Biophysics, Guindy Campus, Madras U., Madras 600025, India. (1961) MSc, biophysics (Madras U., 1984). Res. scholar. (tel. 415367, ext. 16). *Theoretical crystallography, crystal structure analysis.*

Selladurai, S. Dept. of Physics, Anna U., Madras 600025, India. (1961). Res. fellow. *X-ray crystallography.*

Sen, Dr Deb Kumar. Mineral Physics Div., Geological Survey of India, 29 Jawaharlal Nehru Road, Calcutta 700016, India. (1942) PhD, X-ray crystallography (Calcutta U., 1972). Mineralogist (Sr.). (tel. 297645, ext. 7). *X-ray crystallography, organic compounds, minerals, electron microprobe analysis.*

Sen, Miss Mina. Physics Dept., Presidency C., Calcutta 700073, India. (1947) MSc, physics (Calcutta U., 1970). Lecturer in Vidyasagar C. (tel. 341121). *X-ray crystallography, organic and bio- molecules.*

Sen, Dr Ranjit Kumar. Dept. of General Physics & X-rays, Indian Assoc. for the Cultivation of Sci., Calcutta 700032, India. (1919) DSc, physics (Allahabad U., 1956). Em. scient. *Organic and organometallic structures, lattice vibrations, inelastic X-ray - phonon scattering.*

Sen, Dr (Mrs) Suchitra. Central Glass & Ceramic Res. Inst., Calcutta 700032, India. (1950) PhD, physics (Calcutta U., 1977). Scient. *Lattice imperfections, thin films, ceramic materials, X-ray diffraction, electron microscopy.*

Sen Gupta, Dr Amitava. Physical Res. Wing, Projects & Div. India Ltd., Sindri 828122, Dhanbad, India. (1944) PhD, physics (Jadavpur U., 1981).Deputy superintendent. *Powder diffraction, metals and alloys, corrosion products, catalysts, mineral compounds.*

Sen Gupta, Prof. Siba Prasad. Dept. of Materials Science, Indian Assoc. for the Cultivation of Sci., Calcutta 700032, India. (1941) DPhil, physics (Calcutta U., 1968). Prof. and Head. (tel. 469371, telex 0215501 IACS IN). *Organic and biological structures, lattice defects, thin films, crystal growth, characterization, topography, electron microscopy, amorphous materials.*

Sequeira, Dr Anisbert Stanislaus. Neutron Physics Div., Bhabha Atomic Res. Cntr., Bombay 400085, India. (1938) PhD, crystallography (Bombay U., 1970). Scient. SG. (tel. 5519783, ext. 4244). *Neutron diffraction, biological crystallography, automation.*

Sharma, Mr Braj Bhushan. Solid State Physics Lab., Ministry of Defence, Lucknow Road, Delhi 110007, India. (1941) MSc, physics (Agra U., 1961). Sr. scient. officer. (tel. 226919). *Semiconductors, X-ray topography.*

Sharma, Dr Surinder Dutt. Crystal Growth Characterization Sect., Nat. Physical Lab., New Delhi 110012, India. (1947) PhD, physics (Allahabad U., 1975). Scient. *Crystal growth, X-ray topography.*

Shashidhar, Dr R. Liquid Crystals Lab., Raman Res. Inst., Bangalore 560006, India. (1946) PhD, liquid crystals (Mysore U., 1972). Scient. (tel. 30124). *Liquid crystals.*

Shivaprakash, Dr N. C. Instrumentation & Service Unit, Indian Inst. of Sci., Bangalore 560012, India. (1955) PhD, physics (Mysore U., 1982). Scient. *Liquid crystals, structure.*

Shrivastava, Prof. Hari Narayan. Chemistry Dept., Indian Inst. of Techn., Bombay 400076, India. (1927) PhD (Glasgow U., UK, 1960). Prof. (tel. 581421, ext. 416). *X-ray crystallography, natural products, acid salts, electron spin resonance, infrared spectroscopy.*

Sikka, Dr Satinder Kumar. Neutron Physics Div., Bhabha Atomic Res. Cntr., Bombay 400085, India. (1942) PhD, physics (Bombay U., 1970). Scient. SG. (tel. 5513848). *High pressure crystallography, phase problem, neutron diffraction.*

Singh, Dr Anil Kumar. Materials Sci. Div., National Aeronautical Laboratory, Bangalore 560017, India. (1939) PhD, Physics (I.I.T., Madras, 1966). Scient. *Instrumentation, high pressure crystallography.*

Singh, Dr Bachchan. Laser Div., Defence Sci. Lab., Matcalfe House, Delhi 110006, India. (1930) PhD, physics (Pennsylvania State U., USA, 1962). Principal Scient. (tel. 221521, ext. 28). *Crystal growth & evaluation.*

Singh, Dr Bhanu Pratap. High Pressure Techn. Div., Nat. Physical Lab., New Delhi 110012, India. (1947) PhD, physics (Delhi U., 1980). Scient. *Defect characterization, diffuse X-ray scattering, high pressure phase transformation.*

Singh, Dr Govind. Physics Dept., Banaras Hindu U., Varanasi 221005, India. (1940) PhD, physics (Banaras Hindu U., 1967). Prof. *X-ray crystallography, crystal growth & imperfections, electron microscopy.*

Singh, Mr K. D. P. Atomic Minerals Div., Dept. of Atomic Energy, X-ray Diffraction Lab., AMD complex, Begumpet, Hyderabad 500016, India. (1944) MSc, (Banaras Hindu U., 1967). Scient. officer. *Inorganic crystal structure.*

Singh, Dr S.N. Materials Div., Nat. Physical Lab., Dr. K.S. Krishnan Road, New Delhi 110012, India. (1947) PhD, physics (Agra U., 1975). Scient. C. (tel. 585039). *Single crystal growth, polycrystalline silicon.*

Singh, Mr Surendra Prakash. Physics Dept., Ranchi U., Ranchi 834008, India. (1954) MSc, physics (Ranchi U., 1976). Res. scholar. *X-ray crystallography, organo-metallic compounds.*

Singh, Prof T.P. Biophysics Dept., All India Institute of Medical Sciences, New Delhi 110029, India. (1949) PhD (I.I.S., 1975). Prof and Head. (tel. 661123, ext. 201). *Crystallography, peptides, proteins.*

Singru, Prof. Ramesh Madhao. Physics Dept., Indian Inst. of Techn., Kanpur 208016, India. (1935) PhD, physics (Purdue U., USA, 1963). Prof. (tel. 40066). *Electron momentum density.*

Sinha, Dr Umesh Chandra. Physics Dept., Indian Inst. of Techn., Bombay 400076, India. (1936) DPhil, X-ray crystallography (Allahabad U., 1961). Asst. prof. *X-ray diffraction, crystallography, instrumentation.*

Sirdeshmukh, Dr Dinker. Physics Dept., Kakatiya U., Warangal 506009, India. (1935) PhD, X-ray crystallography (Osmania U., 1964). Prof. (tel. 7701). *X-ray crystallography, crystal growth, defects, thermal & mechanical properties.*

Sivakumar, Dr K. Dept. of Physics, Anna U., Madras 600025, India. (1960) PhD, physics (Anna U., 1988). Visiting lect. *Crystal structure determination.*

Soman, Ms Jayashree. Molecular Biophysics Unit, Indian Inst. of Sci., Baangalore 560012, India. (1963) MSc, physics(Karnataka U., 1984).Res. student (tel. 344411, ext. 2389). *Biological crystallography.*

Sridhar Prasad, Mr G. Molecular Biophysics Unit, Indian Inst. of Sci., Bangalore 560012, India. (1963) MSc, chemistry (Bangalore U., 1985). Res. student. (tel. 344411, ext. 2389). *Biomolecular crystallography, molecular biophysics.*

Srinivasa, Dr Vishwanathapuram Kalasa. Physical Res. Wing, Projects & Dev. India Ltd., P.O Sindri, Dhanbad 828122, India. (1934) PhD, physics (Indian Sch. of Mines, Dhanbad, 1978). Deputy superintendent. (tel. 2613, telex 0629-215-216 FPDIL) *High & Low temperature X-ray diffraction, powder diffraction techniques, X-ray cameras and instrumentation, fertilizer materials, nitrogen fixation, solid solutions, double salts, adducts, stress analysis, phase transformation.*

Srinivasan, Prof. R. Dept. of Crystallography & Biophysics, U. of Madras, A.C.C. Campus, Madras 600025, India. (1933) PhD, physics (Madras U., 1958). Sr. prof. and head. (tel. 432248, ext. 16). *X-ray diffraction, statistical application, protein crystallography, nuclear magnetic resonance (wide-line).*

Srinivasan, Dr Sampat. Radiochemistry Div., Bhabha Atomic Res. Cntr., Bombay 400085, India. (1943) PhD, inorganic chemistry (I.I.T., Madras, 1971). Scient. officer. (tel. 523321, ext. 456). *Crystal chemistry, actinides, oxides and carbides.*

Srivastava, Dr Ramesh Chandra. Physics Dept., Indian Inst. of Techn., Kanpur 208016, India. (1936) DPhil, physics (Allahabad U., 1960). Asst. prof. *Organic & inorganic structures, metal oxides, diffuse X-ray scattering.*

Srivastava, Dr Surendra Nath. Physics Dept., Allahabad U., Allahabad 211002, India. (1932) DPhil, X-ray crystallography (Allahabad U., 1963). Lect. *Liquid state.*

Subhadra, Dr K. G. Physics Dept., Kakatiya U., Warangal 506009, India. (1947) PhD, physics (Osmania U., 1976). Reader. *Crystal growth, inorganic crystal structure, chemical crystallography.*

Subramanian, Prof. Easwara. Dept. of Physics, Crystallography & Biophysics, U. of Madras, Madras, Tamilnadu 600025, India. (1938) PhD, physics (Madras U., 1965). Prof. (tel. 415367). *X-ray diffraction, protein crystallography, structural studies on peptides.*

Subramanian, Dr K. Physics Dept., C. of Eng., Anna U., Madras 60025, India. (1947) PhD, physics (Madras U., 1983). Lect. *Crystal structure determination.*

Subramanya, Mr H. S. Molecular Biophysics Unit, Indian Inst. of Sci., Bangalore 560012, India. (1963) MSc, physics (Mysore U.,1986). Res. student. (tel. 344411, ext. 2389). *Biological crystallography.*

Suguna, Dr K. Molecular Biophysics Unit, Indian Inst. of Sci., Bangalore 560012, India. (1956) PhD, physics (I.I.S., 1982). Pool officer (tel. 344411, ext. 2389, telex 0845-8349 IISCIN). *Protein crystallography.*

Sundaramoorthy, Mr M. Molecular Biophysics Unit, Indian Inst. of Sci., Bangalore 560012, India. (1960) MSc, material science (Anna U., 1983). Res. student. (tel. 344411, ext. 2534). *Molecular biophysics, fibre diffraction.*

Suresh, Mr C. G. Molecular Biophysics Unit, Indian Inst. of Sci., Bangalore 560012, India. (1955) post MSc, Dip. biophysics (I.I.S., 1980). Res. scholar. *Molecular biophysics.*

Suresh, Mr K. A. Liquid Crystals Lab., Raman Res. Inst., Bangalore 560006, India. (1948) MSc, solid state physics (Mysore U., 1969). Res. fellow. (tel. 30124). *Liquid crystals.*

Suri, D. K. X-ray Sect., Material Characterization Div., Nat. Physical Lab., New Delhi 110012, India. (1953) MSc, physics (Meerut U., 1972). Scient. *Crystal structure, phase transformation, inorganic materials.*

Suryanarayana, Dr Challapalli. Dept. of Metallurgical Eng., Banaras Hindu U., Varanasi 221005, India. (1945) PhD, physical metallurgy (Banaras Hindu U., 1970). Reader. (tel. 54290, ext. 250). *Crystallography, electron microscopy, defects.*

Suryanarayana, Dr Shambhuni V. Physics Dept., C. of Sci., Osmania U., Hyderabad 500007, India. (1943) PhD, X-ray crystallography (Osmania U., 1971). Lect. (tel. 71251, ext. 242). *X-ray crystallography, thermal expansion, Debye temperatures, alloy phases.*

Suta, Elizabeth Dept. of Physics, Indian Inst. of Sci., Bangalore 560012, India. (1959) MSc, physics (Kerala U., 1983) Scient. asst. *Crystal growth, characterization.*

Talapatra, Dr S. K. Dept. of Physics, Jadavpur U., Calcutta 700032, India. (1930) PhD, physics (Calcutta U., 1968). Reader. *X-ray crystallography, molecular lattice dynamics, computer simulations.*

Talukdar, Dr Amarendra Nath. Dept. of Physics, Gauhati U., Gauhati 781014, India. (1934) PhD, physics (Gauhati U., 1974). Reader. *Crystal structure analysis.*

Tavale, Dr Sudam Shankar. Physical Chemistry Div., Nat. Chemical Lab., Pushan Road, Poona 411008, India. (1934) PhD, X-ray crystallography (Poona U., 1964). Scient. 'C'. *X-ray crystallography, organic & bio-compounds.*

Tewari, Dr Raghavendra. Computer Cntr., Aligarh Muslim U., Aligarh 202001, India. (1945) PhD, X-ray crystallography (I.I.T., Kanpur, 1973). Lect. (tel. 4809). *X-ray crystallography.*

Trigunayat, Prof. Govind Chandra. Dept. of Physics, U. of Delhi, Delhi 110007, India. (1936) PhD, physics (Delhi U., 1960). Prof. *Crystal growth, defects, polytypism.*

Vasanth, Dr K. L. Chemistry Dept., PSG C. of Techn., Coimbatore 641004, India. (1940) PhD, chemistry (M.S.U., Baroda, 1968). Prof. and head. *Liquid crystals, metal & alloy electrodeposit structures.*

Veerapandian, Dr B. Physics Dept., Bharathidasan U., Trichy 620023, India. (1951) PhD, protein crystallography (I.I.S., 1985). Lect. (tel. 28416). *Protein crystallography, structures, pesticides, drugs, teaching methods.*

Venkatesan, Prof. Kailasam. Dept.of Organic Chemistry, Indian Inst. of Sci., Bangalore 560012, India. (1932) PhD, X-ray crystallography (Madras U., 1959). Prof. (tel. 344411, ext. 578). *Structure analysis, organic compounds, structure-reactivity correlations.*

Venkatasubramanian, Dr K. Diffraction and Computer Facilities, Central Salt & Marine Chemicals Res. Inst., Bhavnagar 364002, India. (1938) PhD, (Calcutta U., 1984). Scient. *Crystal structure, organic and organometallic coordination compounds, theoretical methods, molecular biophysics.*

Venkobarao, Mr H. N. Instrumentation Div., Central Electrochemical Res. Inst., Karaikudi 623006, India. (1925) MSc, physics (Mysore U., 1949). Scient. in charge. *X-ray crystallography, instrumentation, solid state physics.*

Venudhar, Dr Y. C. Physics Dept., Osmania U., Hyderabad 500007, India. (1950) PhD, X-ray crystallography (Osmania U., 1981). Reader. *X-ray crystallography, solid state physics, materials science.*

Venugopalan, Mr P. Organic Chemistry Dept., Indian Inst. of Sci., Bangalore 560012, India. (1963) MSc, Ed. chemistry (Mysore U., 1986). Res. scholar. *Structure-reactivity correlations.*

Verma, Dr Ajit Ram Nat. Physical Lab., Hillside Road, New Delhi 110012, India. (1921) DSc, physics (London U., 1969). *Em. scient. Crystal growth, lattice imperfection.*

Vijayakar, Mr Suresh Jaywant. Metallurgy Div., Bahbha Atomic Res. Cntr., Bombay 400085, India. (1931) MSc, chemistry (Bombay U., 1968). Scient. officer. (tel. 523321, ext. 242). *Phase transformations, electron microscopy.*

Vijayan, Dr (Mrs) Kalyani. Materials Sci. Div., Nat. Aeronautical Lab., Bangalore 560017, India. (1942) PhD, X-ray crystallography (I.I.S., 1969). Scient. E1. (tel. 570098). *Crystallographic methods, organic & bio-structures, liquid crystals, polymer diffraction.*

Vijayan, Prof. Mamannamana. Molecular Biophysics Unit, Indian Inst. of Sci., Bangalore 560012, India. (1941) PhD, X-ray crystallography (I.I.S., 1967). Prof. (tel. 344411, ext. 2460, telex 0845-8349 IISC IN). *Biological crystallography, molecular biophysics.*

Viswamitra, Prof. Mysore Ananthamurthy. Physics Dept., Indian Inst. of Sci., Bangalore 560012, India. (1932) PhD, X-ray crystallography (I.I.S., 1960). Prof. *Biological crystallography, high temperature crystallography.*

Vora, Dr Rasiklal Amulakhbhai. Appl. Chemistry Dept., Faculty of Techn. & Eng., M.S.U., Baroda 390001, India. (1938) PhD, liquid crystals (M.S.U., Baroda, 1975). Prof. and Head. *Liquid crystals.*

Vyas, Mr K. Inorganic & Physical Chem. Dept., Indian Inst. of Sci., Bangalore 560012, India. (1958) MSc, Ed. chemistry (Mysore U., 1980). Res. scholar. *Crystallography, small molecules, topochemistry, solid state reactions.*

Wadhawan, Dr Vinod Kumar. Neutron Physics Div., Bhabha Atomic Res. Cntr., Bombay 400085, India. (1944) PhD, physics (Bombay U., 1976). Scient. officer. (telex 01171017 BARC IN). *Crystallography, crystal physics, phase transitions, ferroic crystals, composites.*

Yadav, Dr Asheshwar. Dept. of Physics, Patna U., Patna 800005, India. (1945) PhD, physics (Patna U., 1983). Reader. *Crystal structure determination, X-ray diffraction methods.*

Yadav, Dr Vijay Singh. Neutron Physics Div., Bhabha Atomic Res. Cntr., Bombay 400085, India. (1942) PhD, crystallography (Bombay U., 1974). Scient. officer. (tel. 523321, ext. 265). *Organic & bio-molecular crystallography, X-ray & neutron diffraction.*

Yadav, Dr Tapaswi. Physics Dept., G.J. College, Rambagh, Bihta (Patna) 801103, India. (1939) PhD, X-ray crystallography (Ranchi U., 1984). Reader. *Crystal structure determination, X-ray diffraction techniques.*

Yadava, Dr Bishwanath Physics Dept., Ranchi U., Ranchi 834008, India. (1935) PhD, X-ray crystallography (Ranchi U., 1981) Res. scholar. *Crystallography.*

INDONESIA

Sub-Editor: **W. Loeksmanto**

Notes

1. International telephone country code - 62.

Loeksmanto, Dr Waloejo. Dept. of Physics, Bandung Inst. of Techn., Jalan Ganesya 10, P.O.Box 273, Bandung, Indonesia. (1933) D.3e cycle, structural chemistry (U. Sci. et Techn., Languedoc, France, 1979). Assoc. prof. *Structure, electrical conductivity, one dimensional conductors, solid electrolytes - LISICON.*

IRAN

Sub-Editor: E. Arzi

Notes

1. International telephone country code - 98.

2. At the Iranian universities persons with an academic training can be appointed to the following academic positions: *Professor* (Prof.), *Associate Professor* (Assoc. prof.), *Assistant Professor* (Asst. prof.) and *Instructor*.

3. Degrees conferred by Iranian Universities in science and technology are BSc, MSc and PhD. Degrees conferred by other countries such as the UK, Germany, etc. are quoted in their original form. The descriptions of such other degrees can be found in the *Notes* of the appropriate country.

Abed Esfahani, Mr Abbas. Chemistry, Esfahan Uni., University St., Esfahan 81744, Iran. (1959) Master of Science, analytical chemistry (Esfahan U., 1989) (tel. 031 71071-09, ext. 533) *X-ray diffraction, powder diffraction.*

Alalee, Dr. Mohammad Sadegh. Dept. of Physics, U. of Tehran, Amirabad Shomali Avenue, Tehran, Iran. (1945) PhD, Physics (Brunel U.). Asst. Prof. (tel. 021+635776, telex 213944 IBBIR). *Crystal growth, electron microscopy, materials science, semiconductors, surface science, X-ray diffraction.*

Alavi, Prof. Mehdi. Chemistry Dept., Fac. of Sci., U. of Esfahan, University Street, Esfahan 81744, Iran. (1937) Dr rer.nat., mineralogy and crystallography (Techn. U. München, BRD, 1969). Assoc. prof. (tel. 031 + 710 71-9). *Powder diffraction, diffractometry, minerals, symmetry, teaching, computing.*

Alemi, Prof. Abdol Ali. Chemistry, U. of Tabriz, Azerbaijan, Tabriz, Iran. (1949) Dr, Inorganic Chemistry. Prof. (tel. 041 30081-9 ext. 479). *Inorganic crystals, powder diffraction, solid state reaction.*

Aliani, Dr. Farhad. Geology, U. Bou Ali Sina, Shariati, Hamadan, Iran. (1942) Dr. Mineralogy (Johannes Gutenburg Mainze, W. Germany, 1984) (tel. 0261 + 21900 - 38) *Microscopy, minerals, powder diffraction, symmetry.*

Amighian, Prof. Jamshid. Physics Dept., Fac. of Sci., U. of Isfahan, University Street, Isfahan, Iran. (1945) PhD, solid state physics (U. Durham, UK, 1975). Assoc. prof. (tel. 031 + 71071-59 ext. 356). *Crystal growth, magnetism, materials science, powder diffraction, X-ray diffraction.*

Arzi, Dr Ezatollah. Physics Dept., U. of Tehran, P.O. Box 11365-7693, Tehran, Iran. (1944) PhD, solid state physics (Queen Mary C., U. of London, UK, 1975). Editor-in-Chief, Journal of Science. (tel. 021 + 635776, ext. 19) *Diode lasers, fibre optics, teaching.*

Azarnia, Dr Nezhat. Dept. of Basic Sci., Tehran Polytechnic, Hafez Avenue, Tehran, Iran. (1945) PhD, crystallography (U. Pittsburgh, USA, 1971). Asst. prof., chemistry. *Carbohydrates, biological molecules, vitamins, heavy atom molecules.*

Banaii, Dr Nasser. Atomic Energy Organization of Iran, Nuclear Research Center, P.O. Box 11365-8486, Tehran, Iran. (1941) PhD, Solid State Physics (Ghent State U., Belgium, 1976) Asst. prof., spectroscopy div. head. (tel. 021 + 638025-28, ext. 284-279, telex 212165). *Color centers, crystal growth, X-ray diffraction, topography, rocking curve.*

Barghi, Mr Mohammad Ali Barghi. Materials Div., Materials and Energy Res. Centre, Tehran, P.O. Box 14155-4777, Iran. (1951) M.S.C. Crystallography, mineralogy (Meunster U., West Germany) Res. (tel 021 + 685581-7, telex 222720) *Crystal growth, structural chemistry, minerals, crystallographic data.*

Beglari, Dr Ing. Parviz. Geology Dept., Fac. of Sci., U. of Azerbaijan, Tabriz, Iran. (1934) Dr.Ing., petrography (T.H. Aachen, BRD, 1975). Asst. prof. *Petrography.*

Etemadi-Abdolabadi, Dr Bijan. Geology Dept., College of Sci., Shiraz U., Shiraz, Iran. (1951) PhD, crystal structure (Birkbeck C., U. London, UK, 1981). Asst. prof. *Crystal growth, inorganic crystals, minerals, molecular crystals, powder diffraction, symmetry, X-ray diffraction.*

Farrahi, Dr Gholam-Hossein. Fac. of Eng., Bu-Ali sina U., Hamadan, Iran. (1956) Dr.Eng. Mechanical metallurgy (ENSAM, France, 1981) Dean of Eng. (tel. 0261 + 20940 or 27096) *X-ray diffraction, diffractometry, electron microscopy, materials science, phase determination, powder diffraction, surface structure, metals.*

Feiz, Dr Sayid Mohammad Hasan. Physics Dept., Fac. of Sci., U. of Isfahan, Hazarjerib Street, Isfahan, Iran. (1944) PhD, Electrical and Optical (St. Andrews U., UK, 1981). Assoc. prof., Dean, Fac. of Sci. (tel. 031 + 71071-79 ext. 356). *Semiconductors.*

Foroughi, Dr Ali-Assghar. Geology Dept., Fac. of Sci., U. of Azerbaijan, Tabriz, Iran. (1935) Dr rer.nat, geology (U. Mainz, BRD, 1974). Asst. prof. *SiO_2-Al_2O_3-P_2O_5 system.*

Gourang, Dr Mansour. Geology Dept., Fac. of Sci., U. of Isfahan, Hezardjarib, Isfahan, Iran. (1938) PhD, mineralogy and petrography (U. Vienna, Austria, 1972). Asst. prof. (tel. 031 + 71071-9 ext. 531,532). *X-ray crystallography, differential thermal analysis.*

Janghorban, Dr Kamal. Materials Sci. & Eng. Dept., Shiraz U., Shiraz, Iran. (1948) PhD, Materials Science (1981) Asst. prof. (tel. 071 + 36104 ext. 2679). *Microscopy, crystallographic data, phase determination, electron diffraction, microscopy, powder diffraction, instrumentation.*

Keshmiri, Seyyed-Hosein. Physics Dept., Sch. of Sci., Mashhad U., Masshad 91384, Iran. (1946) PhD, Solid State Science, (Penn. State U., USA) Asst. prof. (tel. 51 + 32021 ext. 63). *Semiconductors, crystal growth, defect structures, materials science, thin films, characterization.*

Khakzar, Dr Ahmad. Geology Dept., Shahid Beheshti U., Evin St., Tehran, 19834 Iran. (1944) Dr rec nat, Appl. Mineralogy, Crystallography (1977) Ass. prof. (tel. 021 + 2141655). *Minerals, materials science, powder diffraction, teaching.*

Khandar-Shahabad, Dr Ali-Akbar. Chemistry Dept., Fac. of Sci., Tabriz U., Daneshgah Street, Tabriz, 51666 Iran. (1951) PhD, X-ray crystallography (U. East Anglia, UK, 1980). Lecturer (tel. 041 + 30081, ext. 237). *X-ray diffraction, crystal structure analysis, phase problem.*

Kianvash, Dr Abbas. Dept. Mechanical Eng., Tabriz U., 29 Bahman Bulvar, Tabriz, Azerbaijan, Iran. (1953) PhD, Phys. Metallurgy & Materials Sci. (U. Birmingham, UK, 1981). Senior Lecturer (tel. 041 + 30081-9, ext. 646). *Magnestism, materials science, metals, microscopy.*

Kohsari, Dr Amir Hossein. Mining Eng. Dept., Isfahan U. of Techn., Isfahan, Iran. (1959) PhD Economic Geology (Indian Inst. Techn., Bombay, 1986). Assoc. prof. (tel. 03291 + 4589). *Electron microscopy, diffraction, microscopy, minerals, phase determination, powder diffraction, texture, thermal vibration, x-ray diffraction.*

Manutchehr-Danai, Dr Mohsen. Geology Dept., Fac. of Sci., U. of Meshed, Asrar, Meshed, Iran. (1939) Dr, mineralogy (U. Mainz, BRD, 1968). Prof. (tel. 051 + 35468). *Inorganic crystal structures.*

Modjtahedi, Dr Mansour. Geology Dept., Fac. of Sci., U. Azerbaijan, Tabriz, Iran. (1946) PhD, geology (U. Vienna, Austria, 1973). Asst. prof. *Mineralogy, crystal optics, microscopy.*

Mohajeri Moghaddam, Dr Hamid Reza. Physics Dept. Ferdowski U., Mashhad, 91384 Iran. (1954) Solid State Physics (U. Hull, UK, 1983). Lecturer (tel. 051 + 32021-24 ext. 85). *Crystal growth, II-VI compounds, photon effect, diffusion in IR detectors, semiconductors, x-ray diffraction, surface effects, thin films, epitaxy.*

Nakissa, Dr Manutschehr. Res. & Dev. Div. (Mineralogy), NICICO, Orchide 6/6, Townsite, Rafsanjan, 144 Iran. (1943) Dr rer nat Mineralogy & Crystallography (Techn. Hochsch. Darmstadt, U. Heidelberg FRG, 1975). Head of Mineralogy (Tel. 03431 + 4010, 4011-279). *X-ray diffraction, computing, instrumentation, crystal growth, microscopy, electron microscopy, diffraction, minerals, metals, materials science, structural chemistry, viruses, quasicrystals.*

Nani, Dr Rahim. MERC. Research Centre, Karaj, Iran. (1924) PhD, Solid state physics (U. Bakou, USSR,1978) Deputy Dir. (tel. Karaj 664361). *Crystal growth, optical, electrical and magnetic crystal properties, low and high temperature techniques.*

Nazarboland, Mr Abbas Ali. Materials Sci. & Eng. Dept., Shiraz U., Shiraz, Iran. (1948) MS, Materials Sci. (1986). Instructor (tel. 071 + 36104 ext. 2630). *Crystal growth, powder diffraction, liquid crystals.*

Noorbehesht, Dr Iradj. Geology Dept., Fac. of Sci., U. of Isfahan, Hezardjerib, Isfahan 81744, Iran. (1943) Dr rer nat, mineralogy, crystallography (Darmstadt, BRD, 1975). Ass. prof. (tel. 031 + 71071-9 ext. 530). *X-ray diffraction, inorganic crystals.*

Pourghazi, Prof Azam. Physics Dept., Fac. of Sci., U. of Isfahan, Hezarjarib, Isfahan 81744, Iran. (1946) PhD, Theoretical Solid State Physics (Durham U., UK, 1978). Professor (Tel. 031 + 71071-79 ext. 369). *Group theory, materials science, scattering theory, semiconductors, symmetry, teaching.*

Rostami, Dr Farzaneh. Geology Dept., Fac. of Sci., U. of Azerbaijan, Khomeini Street, Tabriz, Iran. (1935) PhD, geology (U. Vienna, Austria, 1972). Asst. prof. (tel. ext. 22397). *Mineralogy, crystallography.*

Saadinam, Dr Abolfazl. Chemistry Dept., Fac. of Sci., U. of Birjand, Khorasan, Iran. (1933) D.3e, Inorganic Chemistry (U. of Bordeaux, France, 1974). Asst. prof. (tel. 0561 + 2429). *Sb-V-O phases, AgTiS and MSbO systems, antitumour complexes.*

Safa, Dr Mehdi. Dept. of Physics, Isfahan U. of Techn., Isfahan, Iran. (1946) PhD, physics (U. Durham, UK, 1978). Lecturer (tel. 03291 + 3811). *Solid state physics, magnestism, crystal growth, X-ray topography.*

Salagegheh, Mehdi. Res. & Dev. Div. (Mineralogy), NICICO, Yas 1/4 Townsites, Rafsanjan, Kerman 144, Iran. (1950) BS Geology (U. Tehran, 1975). Mineralogist (Tel. 03431 +4010/4011 -279). *Crystal growth, diffractometry, electron diffraction, microscopy, minerals, molecular crystals, powder diffraction, teaching.*

Taeb Dr Abbas. Chemical Eng. Dept., Iran U. of Sci. & Techn., Normak, Tehran 16844, Iran. (1949) PhD Structure analysis (Techn. U. of Graz, Austria, 1986). Overassistant (tel. 021 + 798082), Telex I.U.S.T. 212965 IR. *Materials science, minerals, powder diffraction, structural chemistry.*

Tajabor, Dr Nasser. Physics Dept., Fac. of Sci., U. of Ferdowski, Mashhad, Iran. (1947) PhD, Solid State Physics (U. Essex, UK, 1976). Asst. prof. (tel. 051 + 32021-24, ext. 64). *Crystal growth, defect structures, magnetism, powder diffraction, semiconductors.*

Tourchi, Dr Mohammad. Dept. Geology, Shaheed-ba-Honar U., P.O. Box 736, Kerman, Iran. (1943) PhD Crystal Chemistry (U. of Vienna, Austria, 1976). Asst. prof. *Mineralogy, petrology.*

Valizadeh, Dr. Mohammad-vali. Dept. of Geology, Fac. of Sci., U. Tehran, Enghelab Avenue, Tehran, Iran. (1940) Doctorat, petrology (Academie de Clermont, France) Assoc. Prof. (tel. 021+6113266) *Geochronology, igneous and metamorphic petrology.*

Zadeh Kashani, Mr Hassan. Non-destructive testing Laboratory, National Iranian Steel Corporation of Esfahan, 45km of Shar cord Rd. Esfahan, Iran. (1940) MSc Metallurgy (U. Sheffield, UK, 1981). Chief of N.D.T. (Mechanical) Lab. (tel 031 + 75051-57 ext. 3195-3175), Telex 312832 NI 8P - IR *X-ray diffraction, powder diffraction, coordination compounds, crystal growth, disorder, diffuse scattering, liquid crystals, metals, quasicrystals, reaction pathways.*

Zanjanchi, Dr Mohammad Ali. Chemistry Dept., U. of Gilan, P.O. Box 451 Rasht, Iran. (1954) PhD X-ray Diffraction (1980). Sr. Lect. (tel. 0231 + 30035-7 ext.23). *Crystallographic computing, diffractometry, electron diffraction, powder diffraction, structural chemistry, x-ray diffraction.*

IRAQ

Sub-Editor: **N.N. Rammo**

Notes

1. International telephone country code - 964.

Alabdulla, Dr Ihsan Abdul Ghani. Chemistry Dept., C. of Sci., U. of Mosul, Mosul, Iraq. (1944) PhD, inorganic structural chemistry (Leeds U., UK, 1974). Lect. *Structure determination, transition metal complexes, X-ray diffraction methods.*

Al-Karaghouli, Dr Abdulrazzak Hamoudi. Chemistry Dept., Nuclear Res. Inst., Tuwaitha, Baghdad, Iraq. (1941) PhD, inorganic chemistry (Southampton U., UK, 1970). Res. *Structural chemistry, X-ray diffraction, neutron diffraction, metal ion complexes, biologically interesting compounds.*

Ibrahim, Dr T. K. Atomic Energy Commission, P.O. Box 765, Tuwaitha-Baghdad, Iraq.

Mahmoud, Dr Mouyed Mohamed. Physics Dept., C. of Sci., Al-Mustansiryha U., Baghdad, Iraq. (1947) PhD, physics (Nottingham U., UK, 1976). Lect. *Crystallography, crystal structure analysis.*

Obaid, Dr Yassin Najim. Physics Dept., C. of Sci., Al-Mustansiryha U., Baghdad, Iraq. (1937) D.3e cycle, physics (U. de Bordeaux, France, 1976). Asst. lect. *Crystallography.*

Rammo, Dr Nabil Naim. Physics Dept., C. of Sci., U. of Baghdad, Jadryia - Baghdad, Iraq. (1946) PhD, polymer crystallography (Manchester U., UK, 1977). Lect. (tel. 01 + 20001, ext. 210). *Polymer crystallography, molecular biology.*

IRELAND

Sub-Editor: **B. J. Hathaway**

Notes

1. International telephone country code - 353.

Byrne, Mr Peter G. Analytical Chemistry Dept., Inst. for Industrial Res. and Standards, Ballymum Road, Dublin 9, Ireland. (1946) MSc, analytical chemistry (London U., UK, 1972). Sr. scient. officer. (tel. 370101, ext. 621, telex 25449). *X-ray analysis.*

Cardin, Dr Christine Janet. Dept. of Chemistry, Trinity C., Dublin 2, Ireland. (1947) DPhil., organometallic synthesis and structures (Sussex U., 1973). Lect. (tel. 1-772941, ext. 1357) *Organometallic and coordination chemistry, bioinorganic chemistry, single crystal X-ray diffraction studies.*

Cunningham, Dr Patrick Desmond. Dept. of Chemistry, Inorganic Div., U.C. Galway, Galway, Ireland. (1942) PhD, inorganic/physical chemistry (Nat. U., U. C. Dublin, 1966). Lect. (tel. 091 + 24411, ext. 483). *Tin chemistry (structural and synthetic), Mössbauer spectroscopy.*

Hathaway, Prof. Brian John. Dept. of Chemistry, U.C. Cork, Cork, Ireland. (1929) PhD, inorganic chemistry (Nottingham U., UK, 1954). Prof. (tel. 276871, ext. 2270, fax 021 277194). *Electronic properties, stereochemistry, copper complexes, X-ray crystallography.*

Higgins, Dr. Timothy. Dept. of Physical Sci., Regional Techn. C. Galway, Galway, Ireland. (1955) PhD, organometallic, physical chemistry (Nat. U., U.C. Galway, Ireland,1981). Lect. (tel. 091+53161, ext. 272). *Crystallographic computing, molecular modelling.*

Kelly, Dr Thomas C. Ceramics Research Centre, EOLAS, Dublin 9, Ireland. (1954) PhD, inorganic, physical chemistry (Nat. U., U.C. Galway, 1980) Sr. scient. officer. (tel. 01-30101, ext. 2642, fax 01-379620). *Inorganic materials, ceramic materials, X-ray identification, crystallite size and strain measurements.*

Kiely, Dr Patrick Vincent. The Agricultural Inst., Johnstown Castle, Wexford, Ireland. (1934) PhD, soil science (Wisconsin U., USA, 1964). Res. officer. (tel. 053 + 22888) *Clay minerals, crystal optics.*

McArdle, Dr Patrick. Crystallography Lab., Dept. of Chemistry, Inorganic Div., U.C. Galway, Galway, Ireland. (1945) PhD, inorganic/physical chemistry (Nat. U., U. C. Dublin, 1968). Lect. (tel. 091 + 24411, ext. 487, fax 353-91-25700, email CHEMCARDLE at CS8700.UCG.IE). *Organometallic iron chemistry, metal-metal bonded rhenium compunds, structure, synthesis.*

Strogen, Mr Peter. Dept. of Geology, U. C. Dublin, Nat. U. of Ireland, Belfield, Stillorgan Road, Dublin 4, Ireland. (1937) BSc, geology (London U., UK, 1958). Lect. (tel. 693244, ext. 244). *X-ray analysis, natural mineral composition.*

ISRAEL

Sub-Editor: **M. Harel**

Notes

1. International telephone country code - 972.

Agmon, Mrs Ilana. Chemistry Dept., Technion, Israel Inst. of Techn., Haifa 32000, Israel. (1949) MSc, crystallography (Technion, 1980). Res. assoc. (tel. 04 + 293716). *Crystal structure, phase transitions.*

Arad Mr Talmon. Structural Chemistry Dept., Weizmann Inst., Rehovot 76100, Israel. (1938). (tel. 08 + 482631, email csarad at weizmann). *Electron microscopy.*

Atzmony, Prof. Uzi. Physics Dept., Nuclear Res. Center-Negev, P.O.B. 9001, Beer-Sheva 84190, Israel. (1937) PhD, physics (Hebrew U., 1967). Assoc. prof. and group leader. (tel. 057 + 968030). *Magnetic and electric crystal fields, magnetism, gas-solid interactions, surface properties.*

Azmon, Prof. Emanuel. Dept. of Geology and Mineralogy, Ben-Gurion U., P.O.B. 653, Beer-Sheva 84105, Israel. (1929) PhD, geology (U. Southern California, USA, 1960). Assoc. prof. (tel. 057 + 461378). *Sedimentology, clay mineralogy, clay ceramics.*

Ben Yair, Dr Moshe Pinhas. The Standards Inst. of Israel, 42 University st., Tel Aviv 69977, Israel. (1922) PhD, geochemistry (Hebrew U., 1957). (tel. 03 + 5454154). *Crystal growth, clay minerals.*

Benghiat, Dr Victor. Biological Services, Weizmann Inst., Rehovot 76100, Israel. (1932) PhD, crystallography (Weizmann Inst., 1971). Head of Electron Microscopy Unit. (tel. 08 + 483223). *Electron microscopy.*

Berkovitch-Yellin, Prof. Ziva. Structural Chemistry Dept., Weizmann Inst., Rehovot 76100, Israel. (1946) PhD, structural chemistry (Weizmann Inst., 1976). Sr. sci. (tel. 08 + 482631, email csberkov at weizmann). *Drug design, crystal growth, ribosomes.*

Bernstein, Prof. Joel. Chemistry Dept., Ben-Gurion U., P.O.B. 653, Beer-Sheva 84105, Israel. (1941) PhD, physical chemistry (Yale U., USA, 1967). Prof. (tel. 057 + 461187, telex 5253 UNASI-IL, email geba100 at bgunos). *Chemical crystallography, organic compounds, solid state chemistry, polymorphism, crystal forces, molecular conformation.*

Bino, Prof. Avi. Dept. of Inorganic and Analytical Chemistry, Hebrew U., Jerusalem 91904, Israel. (1949) PhD, inorganic chemistry (Hebrew U., 1978). Assoc. prof. (tel. 02 + 585722). *Inorganic compounds.*

Cohen, Dr Shmuel. Dept. of Inorganic Chemistry, Hebrew U., Jerusalem 91904, Israel. (1944) PhD, chemistry (Hebrew U., 1976). Crystallographer. (tel. 02 + 585482). *Solid state chemistry.*

Dariel, Prof. Moshe Pierre. Materials Eng. Dept., Ben-Gurion U., Beer-Sheva 84501, Israel. (1935) PhD, crystallography (Weizmann Inst., 1968). Prof. (tel. 057 + 967500). *Metals, intermetallic compounds, hydrogen absorbers, solid state amorphization.*

Deutsch, Prof. Moshe. Physics Dept., Bar-Ilan U., Ramat-Gan 52100, Israel. (1946) PhD, physics (Bar-Ilan U., 1979). Assoc. prof. (tel. 03 + 5318476). *Mathematical methods, X-ray physics, X-ray optics, liquid crystals, surface structure, monolayers.*

Eisenberg, Prof. Henryk. Polymer Dept., Weizmann Inst., Rehovot 76100, Israel. (1921) PhD, chemistry (Hebrew U., 1951). Prof. (tel. 08 + 483252, telex 381300 WIX IL, fax 8-466966, email bpeisenb at weizmann). *Nucleic acids, enzymes, halophilic proteins, nucleoproteins, chromatin, small angle scattering, neutron scattering, light scattering.*

Eisenstein, Dr Miriam. Structural Chemistry Dept., Weizmann Inst., Rehovot 76100, Israel. (1951) PhD, structural chemistry (Weizmann Inst., 1981). Assoc. (tel. 08 + 482672, email cseisen at weizmann). *DNA, interaction energy, electron density.*

Erez, Prof. Gidon. Physics Dept., Nuclear Res. Center-Negev, P.O.B. 9001, Beer-Sheva 84190, Israel. DSc, physics (Technion, 1960). Prof. (tel. 057 + 57111). *Metals.*

Felsteiner, Prof. Joshua. Physics Dept., Technion, Israel Inst. of Techn., Haifa 32000, Israel. (1938) PhD, physics (Toronto U., Canada, 1967). Assoc. prof. (tel. 04 + 293869). *Momentum density, magnetism.*

Fraenkel, Prof. Benjamin. Racah Inst. of Physics, Hebrew U., Jerusalem, Israel. (1923) PhD, physics (Hebrew U., 1955). Chairman, Plasma Section. (tel. 02 + 584335). *Highly ionized atomic spectroscopy, X-ray (2D) and V.U.V. spectroscopy, accelerator diagnostics, plasma diagnostics.*

Frolow, Dr Felix. Chemical Services, Weizmann Inst., Rehovot 76100, Israel. (1947) PhD, crystallography (Weizmann Inst., 1980). (tel. 08 + 483421, email cvfrolow at weizmann). *X-ray diffraction, neutron diffraction, physical crystallography, macromolecules.*

Goldberg, Prof. Israel. Sch. of Chemistry, Tel-Aviv U., Ramat Aviv 69978, Israel. (1945) PhD, physical chemistry (Tel-Aviv U., 1974). Assoc. prof. (tel. 03 + 5450258, email d66 at taunos). *Organic compounds, inclusion compounds, clathrates, structure - properties relationships.*

Goshen, Dr Shmuel Yehuda. Physics Dept., Nuclear Res. Center-Negev, P.O.B. 9001, Beer-Sheva 84190, Israel. (1933) PhD, nuclear physics (Weizmann Inst., 1966). *Symmetry, phase transitions, simulations.*

Grünbaum, Prof. Enrique. Sch. of Eng., Tel-Aviv U., Tel Aviv 69978, Israel. (1926) MSc, physics (London U., UK, 1957). Prof. (tel. 03 + 5450138). *Growth and structure and molecular beam epitaxy of metal and semiconductor films, electron diffraction, electron microscopy.*

Gurewitz, Dr Eitan. Physics Dept., Nuclear Res. Center-Negev, P.O.B. 9001, Beer-Sheva 84190, Israel. (1938) PhD, physics (Weizmann Inst., 1976). (tel. 057 + 967357). *Magnetism, symmetry, neutron diffraction, phase transitions.*

Harel, Dr Michal. Structural Chemistry Dept., Weizmann Inst., Rehovot 76100, Israel. (1940) PhD, crystallography (Weizmann Inst., 1975). (tel. 08 + 482647, email csharel2 at weizmann). *Proteins.*

Heller-Kallai, Prof. Lisa. Geology Dept., Hebrew U., Jerusalem, Israel. (1926) PhD, crystallography (London U., UK, 1951). Prof. (tel. 02 + 584883). *Clay mineralogy.*

Herbstein, Prof. Frank Herzl. Chemistry Dept., Technion, Israel Inst. of Techn., Haifa 32000, Israel. (1926) DSc, physical and chemical crystallography (Cape Town U., S. Africa, 1967). Prof. (tel. 04 + 293715). *Crystal structures, phase transitions, solid state reactions.*

Hirshfeld, Prof. Fred. Structural Chemistry Dept., Weizmann Inst., Rehovot 76100, Israel. (1927) PhD, crystallography (Hebrew U., 1956). Prof. (tel. 08 + 483320, email cshirshf at weizmann). *Organic compounds, electron density, computing.*

Joshua-Tor, Ms Leemor. Structural Chemistry Dept., Weizmann Inst., Rehovot 76100, Israel. (1961) BSc, chemistry (Tel Aviv U., 1982). PhD student. (tel. 08 + 482541, email csleemor at weizmann). *DNA, proteins.*

Kaftory, Prof. Menahem. Chemistry Dept., Technion, Israel Inst. of Techn., Haifa 32000, Israel. (1943) DSc, chemistry (Technion, 1973). Assoc. prof. (tel. 04 + 293761, email chr13mk at technion). *Chemical crystallography, solid state chemistry, reaction pathways.*

Kalb(Gilboa), Dr Aaron Joseph. Biophysics Dept., Weizmann Inst., Rehovot 76100, Israel. (1937) PhD, chemistry (U. California, USA, 1963). Sr. sci. (tel. 08 + 483621). *Proteins.*

Kalman, Prof. Zwi Heinrich. Racah Inst. of Physics, Hebrew U., Jerusalem 91904, Israel. (1924) PhD, physics (Hebrew U.,1965). Assoc. prof. (tel. 02 + 584646). *X-ray diffraction, materials science, solid state physics.*

Kapon, Dr Moshe. Chemistry Dept., Technion, Israel Inst. of Techn., Haifa 32000, Israel. (1941) DSc, chemistry (Technion, 1974). Dept. crystallographer. (tel. 04 + 293716, email chr03kp at technion). *Organic polyiodides, thermal decomposition, inclusion complexes.*

Katz, Dr Gerald. 21 Ra'anan St., Haifa 34384, Israel. PhD, solid state science (Pennsylvania State U., USA, 1965). *Crystal growth, topotaxy, epitaxy, optical spectroscopy, rare earth doped glasses.*

Kazinets, Dr Maria. SEM Lab., Ben-Gurion U., P.O.B. 1025, Beer-Sheva 84110, Israel. (1935) PhD, solid state physics (Inst. of Physics, Azerb. USSR, 1969). Eng. (tel. 057 + 461244). *Crystal growth, phase transitions, thin films, electron microscopy, diffraction.*

Kimmel, Prof. Giora. Materials Eng. Dept., Technion, Israel Inst. of Techn., Haifa 32000, Israel. (1939) DSc, materials science (Technion, 1973). Assoc. prof. (tel. 04 + 294580, telex TECLIIL 46650, email mtrgk01 at vmsa.technion.ac.il). *Alloy phases, powders, applied crystallography, thin films, coating, metal-metal composites.*

Kisch, Prof. Hanan Josef. Dept. of Geology and Mineralogy, Ben-Gurion U., P.O.B. 653, Beer-Sheva 84105, Israel. (1935) PhD, geology and mineralogy (Amsterdam U., Netherlands, 1962). Prof. (tel. 057 + 461290). *Very low grade metamorphism, burial diagenesis, clay minerals, coalification, fluid inclusions, amphiboles, X-ray diffraction, metamorphic microfabrics, regional metamorphism, orogenic belts.*

Leiserowitz, Prof. Leslie. Structural Chemistry Dept., Weizmann Inst., Rehovot 76100, Israel. (1934) PhD, crystallography (Hebrew U., 1965). Prof. (tel. 08 + 482631). *Packing modes, electron density, solid state, surface crystallography.*

Livnah, Mr Oded. Structural Chemistry Dept., Weizmann Inst., Rehovot 76100, Israel. (1958) BSc, chemistry (Ben-Gurion U., 1986). PhD student. (tel. 08 + 482541, email csoded at weizmann). *Proteins.*

Low, Prof. William Zev. Div. of Microwave Physics, Hebrew U., Jerusalem, Israel. (1922) PhD, physics (Columbia U., USA, 1950). Prof. (tel. 02 + 585848). *Solid state physics.*

Mayer, Prof. Itzchak. Chemistry Dept., Hebrew U., Jerusalem, Israel. (1927) PhD, chemistry (Hebrew U., 1960). Assoc. prof. (tel. 02 + 585214). *Inorganic compounds, apatites, powder diffraction.*

Melamud, Dr Mordechai. Physics Dept., Nuclear Res. Center-Negev, P.O.B. 9001, Beer-Sheva 84190, Israel. (1943) PhD, physics (Weizmann Inst., 1976). Researcher. (tel. 057 + 967468). *Neutron diffraction.*

Minkoff, Prof. Isaac. Materials Eng. Dept., Technion, Israel Inst. of Techn., Haifa 32000, Israel. DSc, metallurgy (M.I.T., USA, 1957). Prof. (tel. 04 + 294582, telex 46406). *Crystal growth.*

Nathan, Dr Yaacov. Geochemistry Dept., Geological Survey of Israel, 30 Malchei Israel St., Jerusalem, Israel. (1931) PhD, geology (Hebrew U., 1969). Sr. sci. (tel. 02 + 208225, telex 26362 ENERG IL). *Inorganic compounds, clay minerals, phosphates.*

Pinto, Mr Haim. Physics Dept., Nuclear Res. Center-Negev, P.O.B. 9001, Beer-Sheva 84190, Israel. (1937) BSc, physics (Ben-Gurion U., 1976). Researcher. (tel. 057 + 967468). *Neutron diffraction.*

Rabin, Mr Baruch. Dept. 1600, Israel Aircraft Industries, Lod 70100, Israel. (1929) MSc, physics of metals (Ural Polytechnical Inst., USSR, 1965). Manager, manufacturing. (tel. 03 + 973940). *Metals, phase transitions.*

Rabinovich, Prof. Dov. Structural Chemistry Dept., Weizmann Inst., Rehovot 76100, Israel. (1929) PhD, crystallography (Hebrew U., 1964). Assoc. prof. (tel. 08 + 482672, email csrabin1 at weizmann). *Methods for structure solution, flash X-ray diffraction, DNA.*

Reisner, Dr George. Chemistry Dept., Technion, Israel Inst. of Techn., Haifa 32000, Israel. (1943) PhD, chemistry (Technion, 1966). Sr. res. fellow. (tel. 04 + 293716).

Sariel, Mr Joseph. Metallurgy Dept., Nuclear Res. Center-Negev, P.O.B. 9001, Beer-Sheva 84190, Israel. (1947) MSc, materials science (Ben-Gurion U., 1981). Metallurgist. (tel. 057 + 968786). *Alloy phases, powder diffraction, crystal structure.*

Schieber, Prof. Michael. Sch. of Applied Sci., Hebrew U., Jerusalem, Israel. (1928) PhD, material sci. (Hebrew U., 1962). Prof. (tel. 02 + 584364, fax 972 2 666 804). *Crystal growth, electronic materials, radiation detectors, superconductors.*

Shaanan, Dr Boaz. Structural Chemistry Dept., Weizmann Inst., Rehovot 76100, Israel. (1946) PhD, physical chemistry (Tel-Aviv U., 1979). Sr. sci. (tel. 08 + 482647). *Biological macromolecules, lattice dynamics.*

Shaked, Prof. Hagai. Physics Dept., Ben-Gurion U., P.O.B. 653, Beer-Sheva 84105, Israel. (1931) PhD, eng. sci. (U. California, Berkeley, USA, 1963). Prof. (tel. 057 + 461567). *Neutron diffraction, magnetism, phase transitions, symmetry, group theory.*

Shakked, Prof. Zippora. Structural Chemistry Dept., Weizmann Inst., Rehovot 76100, Israel. (1943) PhD, crystallography (Weizmann Inst., 1975). Assoc. prof. (tel. 08 + 482672, email cshaked2 at weizmann). *Organic compounds, DNA, protein-DNA interactions.*

Shamir, Dr Noah. Physics Dept., Nuclear Res. Center-Negev, P.O.B. 9001, Beer-Sheva 84190, Israel. (1947) PhD, physics (Weizmann. Inst., 1977). Res. assoc. (tel. 057 + 967795). *Surface structure and reactions, gas surface interactions.*

Shmueli, Prof. Uri. Sch. of Chemistry, Tel Aviv U., Tel Aviv 69978, Israel. (1928) PhD, crystallography (Weizmann Inst., 1966). Assoc. prof. (tel. 03 + 5450258, telex 342171 VERSY IL, email d22 at taunos). *Molecular crystals, intensity statistics, direct methods.*

Shoham, Dr Gil. Dept. of Inorganic Chemistry, Hebrew U., Jerusalem 91904, Israel. (1952) PhD, crystallography (Harvard U., USA, 1984). Lect. (tel. 02 + 585611,585610, email gil at hujivms). *Biological compounds, macromolecules.*

Shoham, Dr Menachem. Structural Chemistry Dept., Weizmann Inst., Rehovot 76100, Israel. (1944) PhD, chemistry (Weizmann Inst., 1979). Sr. sci. (tel. 08 + 482647). *Proteins, nucleic acids.*

Shpigler, Mr Bilu. Metalurgical Lab., Israel Inst. of Metals, Technion, Haifa 32000, Israel. (1932) MSc, metallurgy (Technion, 1963). Lab. head. (tel. 04 + 235104). *Metals microstructure, instrumentation, electron optics, fracture mechanics, color metallography, failure analysis, composite materials, hot isostatic pressing.*

Stein, Mrs Zafra. Sch. of Chemistry, Tel Aviv U., Tel Aviv 69978, Israel. (1953) MSc, chemical eng. (Technion, 1980). Crystallographer. (tel. 03 + 5450258). *Structure determination, refinement, computing.*

Steinberger, Prof. Itzhak T. Racah Inst. of Physics, Hebrew U., Jerusalem 91904, Israel. (1928) PhD, physics (Hebrew U., 1957). Prof. (tel. 02 + 584648). *Inorganic compounds, crystal physics, simple fluid electronic and optical properties.*

Sussman, Prof. Joel Leonard. Structural Chemistry Dept., Weizmann Inst., Rehovot 76100, Israel. (1943) PhD, biophysics (M.I.T., USA, 1972). Assoc. prof. (tel. 08 + 482638, telex 381300 WIXIL, email csjoel at weizmann). *Proteins, nucleic acids, computer graphics, model building, structure determination, refinement.*

Traub, Prof. Wolfie. Structural Chemistry Dept., Weizmann Inst., Rehovot 76100, Israel. (1927) PhD, physics (London U., UK, 1955). Prof. (tel. 08 + 482668). *Biological macromolecules, mineralized tissues.*

Wolf, Ms Sharon. Structural Chemistry Dept., Weizmann Inst., Rehovot 76100, Israel. (1958) MSc, structural chemistry (Weizmann Inst., 1986). PhD student. (tel. 08 + 482631, email csgrayer at weizmann). *Langmuir monolayers, two-dimensional X-ray diffraction, order-disorder.*

Yahalom, Prof. Joseph. Materials Eng. Dept., Technion, Israel Inst. of Techn., Haifa 32000, Israel. (1933) PhD, metallurgy (Cambridge U., UK, 1963). Prof. corrosion lab. head. (tel. 04 + 294578, telex 46406). *Oxides, corrosion, electrodeposition.*

Yonath, Prof. Ada. Structural Chemistry Dept., Weizmann Inst., Rehovot 76100, Israel. PhD, crystallography (Weizmann Inst., 1969). Prof. (tel. 08 + 482541, telex 08 466966, email csyonath at weizmann). *Biological macromolecules and assemblies, ribosomes.*

Zeldes, Mr Nathan. Intel Electronics, P.O.B. 3173, Jerusalem 91031, Israel. MSc, physics (Hebrew U., 1977). Quality assurance group leader. (tel. 02 + 897111, ext. 7496). *Electron microscopy, silicon, microelectronics fabrication.*

Zelingher, Mr Naphtaly. Mechanical Eng. Dept., Ben-Gurion U., P.O.B. 653, Beer-Sheva 84105, Israel. (1927) MSc, physics (C.I.Parhon U., Romania, 1958). Head, mech. eng. lab. (tel. 057 + 461358). *Whisker growth, crystal layers from superfluid solutions, computing.*

Zevin, Prof. Lev. Inst. for Applied Res., Ben-Gurion U., P.O.B. 1025, Beer-Sheva 84110, Israel. (1930) PhD, applied crystallography (Moscow Inst. for Crystallography, USSR, 1963). Assoc. prof. (tel. 057 + 461244). *Powder diffraction, thin films, modulated structures, intermetallic compounds, applied crystallography.*

ITALY

Sub-Editor: **M. Mammi**

Notes

1. International telephone country code - 39.

2. The first degree awarded by Italian Universities for scientific subjects is *Dottore* (Dr). It is obtained after a 4 year course, with the exceptions of the degrees in chemistry (5 years), engineering (5 years), and medicine (6 years). The previous training is given by a 5 + 3 + 5 year course at the elementary, secondary and higher schools, respectively. At some Universities various diplomas for special training may be awarded to Dr after courses of instruction. The doctorate *(Dottore di Ricerca)* is obtained after regular courses and research lasting three-four years beyond the Dr, and is awarded by the Ministry of Education.

3. Positions and functions of teachers and researchers at an Italian University are at present *approximately* equivalent to: *Professore ordinario* (Prof.) - Full Professor; *Professore associato* (Prof. assoc.) - Associate Professor; *Ricercatore* - Researcher; *Professore Incaricato* (Prof. inc.) - Lecturer; *Assistente* (Asst.) - Research Assistant; *Tecnico Laureato* (Tecn. laur.) - Technical Research Assistant; *Borsista* - Postdoctoral Research Fellow. New appointments cannot be made to the position of *Libero Docente* (Lib. doc.) - (Reader) since 1970.

4. At the Consiglio Nazionale delle Ricerche (CNR - National Research Council) and other public institutions, persons can be appointed: *Dirigente di ricerca* (Dirig. ric.) - Research leader; *Primo ricercatore* (Primo ric.) - First researcher; *Ricercatore* Researcher, and *Borsista* - Postdoctoral Research Fellow.

5. Other functions in the above and other institutions and companies are *Direttore* , director (of Institute, Div., Centre, etc.); *Capo-reparto* - Head (of Dept., Group, Lab., etc.); *Ricercatore* - Research Scientist.

6. Abbreviations used in this section include: Dip. for Dipartimento, Tecn. for Tecnologia, and Ist. for Istituto.

Abbona, Prof. Francesco. Dip. di Scienze della Terra, U. della Calabria, Castiglione Cosentino Staz., 87030 Cosenza, Italy. (1937) Dr, chemistry (U. di Torino, 1961). Prof. (tel. 0984 + 838118). *Crystal growth.*

Accorsi, Prof. Carla Alberta. Dip. di Chimica, U. di Ferrara, Via Luigi Borsari 46, 44100 Ferrara, Italy. (1941) Dr, chemistry (U. di Ferrara, 1966). Prof. assoc. (tel. 0532 + 36374). *Crystal growth, molecular crystals.*

Ajó, Dr David. Ist. di Chimica e Tecn. dei Radioelementi, C.N.R., Corso Stati Uniti, 35100 Padova, Italy. (1946) Dr, chemistry (U. di Roma, 1970). Ricercatore. (tel. 049 + 845373, telex 430302 CNRPDI, Fax 049 + 845449, email DAVIDE at PDCNR2.INFN.IT). *Coordination compounds, magnetism, materials science, molecular crystals, organic compounds, structural biology, X-ray diffraction, conformation.*

Albano, Prof. Vincenzo Giulio. Ist. Chimico 'Ciamician', U. di Bologna, Via F. Selmi 2, 40126 Bologna, Italy. (1937) Dr, chemistry (U. di Bari, 1960). Prof., inorganic chemistry. (tel. 051 + 235292). *Coordination compounds.*

Alberti, Prof. Alberto. Ist. Geologico Mineralogico, U. di Sassari, Corso G. M. Angioj 10, 07100 Sassari, Italy. (1938) Dr, physics (U. di Modena, 1962). Prof., mineralogy. (tel. 079 + 231250). *Materials science, minerals, X-ray diffraction.*

Albinati, Prof. Alberto. Ist. di Chimica Farmaceutica, U. di Milano, Viale Abruzzi 42, 20131 Milano, Italy. (1945) Dr, chemistry (U. di Milano, 1970) Prof. assoc. (tel. 02 + 209197, telex 320484 UNIMII, email ALBINATI at IMICLVM.BITNET). *Coordination compounds, neutron diffraction, powder diffraction, structural chemistry, X-ray diffraction.*

Alietti, Prof. Andrea. Ist. di Mineralogia e Petrografia, U. di Modena, Via S. Eufemia 19, 41100 Modena, Italy. (1923) Dr, chemistry (U. di Pavia, 1947). Prof., mineralogy. (tel. 059 + 218062). *Clay minerals.*

Allegra, Prof. Giuseppe. Ist. di Chimica, Politecnico di Milano, Piazza L. Da Vinci 32, 20133 Milano, Italy. (1933) Dr, chemical eng. (Politecnico di Milano, 1958). Prof., chemistry. (tel. 02 + 23993023, telex 333467 POLIMI I, fax 02 + 23992206, email RCHIGE06 at IMIPOLI.BITNET). *Defect structures, polymers, powder diffraction.*

Amicarelli, Prof. Vincenzo. Ist. di Chimica Applicata, Facoltà di Ingegneria, U. di Bari, Via Re David 200, 70125 Bari, Italy. (1930) Dr, Industrial chemistry (U. di Napoli, 1956). Prof., appl. chemistry. (tel. 081 + 228014). *Materials science.*

Andreetti, Prof. Giovanni Dario. Ist. di Strutturistica Chimica, U. di Parma, Viale delle Scienze, 43100 Parma, Italy. (1939) Dr, chemistry (U. di Parma, 1963). Prof., organic chemical crystallography. (tel. 0521 + 580548 or 580552, fax 0521 + 580542, email STRUCHI at UNIVPR.BITNET). *Computing, crystal growth, diffuse scattering, disorder, electron diffraction, instrumentation, liquid crystals, magnetism, microscopy, neutron diffraction, phase transitions, powder diffraction, small-angle scattering, surface structure, texture.*

Antolini, Prof. Luciano. Dip. di Chimica, U. di Modena, Via Campi 183, 41100 Modena, Italy. (1942) Dr, chemistry (U. di Modena, 1967). Prof. assoc. (tel. 059 + 378422). *Coordination compounds, proteins.*

Antonini, Prof. Marcello. Dip. di Fisica, U. di Modena, Via Campi 213, 41100 Modena, Italy. (1939) Dr, physics (U. di Rome, 1962). Prof. assoc. (tel. 059 + 361142 ext.83). *Defect structures, metals, microstructure determination, glasses.*

Antonione, Prof. Carlo. Ist. di Chimica Generale ed Inorganica, Fac. di Farmacia, U. di Torino, Via P. Giuria 9, 10125 Torino, Italy. (1931) Dr, chemistry (U. di Torino, 1958). Prof. assoc. (tel. 011 + 657000). *Solid state chemistry, structural chemistry., metals.*

Aquilano, Prof. Dino. Dip. di Scienze della Terra, U. di Torino, Via S. Massimo 24, 10123 Torino, Italy. (1940) Dr, physics (U. di Torino, 1963). Prof. assoc. (tel. 011 + 832193). *Crystal growth, defect structures.*

Armigliato, Dr Aldo. Ist. LAMEL, C.N.R., Via de' Castagnoli 1, 40126 Bologna, Italy. (1940) Dr, physics (U. di Padova, 1965). Ricercatore. (tel. 051 + 287932, Telex 511350 CNR BO, Fax 051 + 229702). *Defect structures, disorder, electron diffraction, electron microscopy, materials science, semiconductors.*

Artioli, Dr Gilberto. Ist. di Mineralogia e Petrologia, U. di Modena, Via S. Eufemia 19, 41100 Modena, Italy. (1957) PhD, Geological sciences (U. Chicago, USA, 1987). Ricercatore (tel. 059 + 218062, email ARTIOLI at IMOVX2.INFNET). *Diffractometry, hydrogen bonding, instrumentation, minerals, neutron diffraction, powder diffraction.*

Bachechi, Dr Fiorella. Ist. di Strutturistica Chimica 'Giordano Giacomello', C.N.R., C.P. 10 - 00016 Monterotondo Stazione, Roma, Italy. (1939) PhD, crystallography (Toronto U., Canada, 1970). Promo ricercatore. (tel. 06 + 9005641, ext 616). *Coordination compounds, organic compounds.*

Balzarotti, Prof. Adalberto. Dip. di Fisica, U. di Roma II 'Tor Vergata', Via O. Raimondo, 00173 Roma, Italy. (1939) Dr, physics (U. di Pavia, 1961). Prof., physics. (tel. 06 + 79791 ext. 2305, telex 626382 FIUNTV I, fax 06 + 6133074, email BALZA at VAXTOV.INFNET). *Optical properties of solids, synchrotron radiation, X-ray spectroscopy.*

Bandoli, Prof.Giuliano. Dip. di Scienze Farmaceutiche, U. di Padova, Via Marzolo 5, 35131 Padova, Italy. (1941) Dr, chemistry, pharmacy (U. di Padova, 1966, U. di Ferrara, 1975). Prof. assoc. (tel. 049 + 831634, Fax 049 + 831660). *Coordination compounds, organic compounds.*

Barbieri, Prof. Renato. Dipartimento di Chimica Inorganica, U. di Palermo, 28 Via Archirafi, 90123 Palermo, Italy. (1930) Dr, chemistry (U. di Padova, 1956). Prof. (tel. 091 + 6161474, email BARBIERI at IPACUC.BITNET). *Proteins, structural biology, structural chemistry.*

Bardi, Prof. Renato. Dip. di Chimica Organica, U. di Padova, Via Marzolo 1, 35100 Padova, Italy. (1924) Dr, chemistry (U. di Padova, 1957). Prof. assoc. (tel. 049 + 831111 or 831311). *Organic compounds, structural chemistry.*

Bartolucci, Dr Cecilia. Ist. di Strutturistica Chimica 'Giordano Giacomello, C.N.R., C.P. 10, 00016 Monterotondo Stazione, Roma, Italy. (1987) Dr, chemistry (U. di Roma, 1987). Ricercatore. (tel. 06 + 9005142, Fax 06 + 9005849). *Organic compounds.*

Basso, Prof. Riccardo. Dipartimento di Scienze della Terra, Sez. di Mineralogia, U. degli Studi, Palazzo delle Scienze, Corso Europa 26, 16132 Genova, Italy. (1947) Dr, mathematics (U. di Genova, 1971). Prof., mineralogia applicata. (tel. 010 + 3538311, fax 010 + 352169 DISTER GE, email MINERAL at IGECUNIV.BITNET). *Crystallographic data, minerals, X-ray diffraction.*

Battaglia, Prof. Luigi Pietro. Ist. di Chimica Generale ed Inorganica, U. di Parma, Viale delle Scienze, 43100 Parma, Italy. (1943) Dr, chemistry (U. di Parma,

1970). Prof. assoc. (tel. 0521 + 580416). *Coordination compounds, molecular crystals, organic compounds, structural chemistry.*

Battezzati, Prof. Livio. Dipartimento di Chimica Inorganica, Chimica Fisica e Chimica de, materiali, U. di Torino, Via P. Giuria 9, 10125 Torino, Italy. (1950) Dr, chemistry (U. di Torino, 1974). Prof. assoc. (tel. 011 + 657000). *Electron microscopy, materials science, metals, quasicrystals, X-ray diffraction.*

Bedarida, Prof. Federico. Dipartimento di Scienze della Terra, Sez. Mineralogia, U. di Genova, Corso Europa (Palazzo delle Scienze), 16100 Genova, Italy. (1924) Dr, physics and mathematics (U. di Genova, 1958). Prof. (tel. 010 + 35383 ext. 01, Fax 010 + 352169). *Crystal growth, coherent light crystallographic applications.*

Bellon, Prof. Pier Luigi. Ist. di Chimica Strutturistica Inorganica, U. di Milano, 21, Via Venezian, 20133 Milano, Italy. (1931) Dr, industrial chemistry (U. di Padova, 1955). Prof., general chemistry. (tel. 02 + 2361410). *Electron microscopy, instrumentation, small-angle scattering.*

Benedetti, Prof. Ettore. Dip. di Chimica, U. di Napoli, Via Mezzocannone 4, 80134 Napoli, Italy. (1940) Dr, chemistry (U. di Napoli, 1965). Prof. (tel. 081 + 206800). *Macromolecules, molecular crystals, structural biology, structural chemistry, biomolecules.*

Benetollo, Dr Franco. Ist. di Chimica e Tecn. dei Radioelementi, C.N.R., Corso Stati Uniti, 35100 Padova, Italy. (1947) Dr, pharmacy (U. di Padova, 1981) Ricercatore. (tel. 049 + 845111 or 845368, Telex 430302 CNRI, fax 049 + 845449, email FRANCO at PDCNR2.INFN.IT). *Coordination compounds, organic compounds, structural chemistry, X-ray diffraction.*

Benna, Dr Piera. Dip. di Scienze della Terra, U. di Torino, Via S. Massimo 24, 10123 Torino, Italy. (1954) Dr, geological sciences (U. di Torino, 1978) Ricercatore. (tel. 011 + 832193). *Structural chemistry, minerals.*

Berti, Dr Giovanni. Dip. di Scienze della Terra, U. di Pisa, Via S. Maria 53, 56100 Pisa, Italy. (1949) Dr, physics (U. di Pisa) Ricercatore. (tel. 050 + 501457). *Structural chemistry, minerals.*

Bertolasi, Prof. Valerio. Dip. di Chimica, U. di Ferrara, Via L. Borsari 46, 44100 Ferrara, Italy. (1949) Dr, chemistry (U. di Ferrara, 1973). Prof. assoc. (tel. 0532 + 33522, email VG3A at ICINECA). *Chemical bonding, coordination compounds, hydrogen bonding, molecular crystals, organic compounds, reaction pathways, structural biology, structural chemistry, X-ray diffraction.*

Biagini Cingi, Prof. Marina. Ist. di Chimica Generale ed Inorganica, U. di Parma, Viale delle Scienze 78, 43100 Parma, Italy. (1925) Dr, chemistry (U. di Parma, 1949). Prof., general chemistry. (tel. 0521 + 580419, Telex 530327 UNIV PR I, fax 0521 + 580542, email CHIMIC3 at IPRUNIV.BITNET). *Coordination compounds, electron density, inorganic crystals, structural chemistry.*

Bianchi, Dr Riccardo. Centro di Studio per la Relazione tra Struttura e Reattività Chimica, C.N.R., Via Golgi 19, 20133 Milano, Italy. (1942) Dr, mathematics (U. di Milano, 1972). Ricercatore. (tel. 02 + 231226, email VALRAY at IMISIAM.BITNET) *Electron density, molecular crystals, X-ray diffraction.*

Bianchi Orlandini, Dr Annabella. Ist. Stereochimica, C.N.R., Via Jacopo Nardi 39, 50132 Firenze, Italy. (1944) Dr, chemistry (U. di Firenze, 1968). Ricercatore. (tel. 055 + 243990). *Structural chemistry, metal complexes.*

Bigi Dr Adriana. Dip. di Chimica 'G. Ciamician', U. di Bologna, Via Selmi 2, 40126 Bologna, Italy. (1951), Dr, chemistry (U. di Bologna, 1975). Ricercatore. (tel. 051 + 259552, fax 051 + 259546, email VS3BODI3 at ICINECA.BITNET). *Inorganic crystals, powder diffraction, proteins, small-angle scattering, X-ray diffrraction.*

Bigi Dr Simona. Istituto di Mineralogia e Petrologia, U. di Modena, Largo S. Eufemia 19, 41100 Modena, Italy. (1961), Dr, geological Scinces (U. di Modena, 1985). Dottorando. (tel. 059 + 218062). *Electron diffraction, electron microscopy, minerals.*

Bigoli, Prof. Francesco. Ist. di Chimica Generale ed Inorganica, U. di Parma, Viale delle Scienze 78, 43100 Parma, Italy. (1936) Dr, chemistry (U. di Parma, 1964). Prof. assoc. (tel. 0521 + 580428). *Coordination compounds, molecular crystals, organic compounds, structural chemistry.*

Bisi Castellani, Prof. Carla. Dip. di Chimica Generale, U. di Pavia, Via Taramelli 12, 27100 Pavia, Italy. (1931), Dr, chemistry (1954). Prof. assoc. (tel 0382 + 392343). *Coordination compounds.*

Blasi, Prof. Achille. Dip. di Scienze della Terra, U. degli Studi, Via Botticelli 23, 20133 Milano, Italy. (1940) Dr, geological sciences (U. di Milano, 1968). Prof. (tel. 02 + 2663994, Fax 02 + 2364393). *Minerals, X-ray diffraction.*

Bocchi, Dr Claudio. MASPEC Ist, materiali Speciali per Elettronica , magnetismo, Via Chivari 18A, 43100 Parma, Italy. (1948) Dr, physics (U. di Parma, 1986). Ricercatore. (tel. 0521 + 96841, Fax 0521 + 96315). *Defect structures, diffractometry, diffuse scattering, materials science, semiconductors, thermal vibrations. X-ray diffraction.*

Bocelli, Dr Gabriele. Centro di Studio per la Strutturistica Diffrattometrica, C.N.R., Viale delle Scienze, 43100 Parma, Italy. (1942) Dr, chemistry (U. di Parma, 1970). Ricercatore. (tel. 0521 + 580448). *Coordination compounds, inorganic crystals, materials science, organic compounds, semiconductors, X-ray diffraction.*

Bolis, Dr Vera Maria. Ist. di Chimica Generale ed Inorganica, Fac. di Farmacia, U. di Torino, via P. Giuria 9, 10125 Torino, Italy. (1950) Dr, chemistry (U. di Torino, 1974) Ricercatore. (tel. 011 + 657000). *Surface structure, adsorption, catalysis.*

Bolognesi, Prof. Martino. Sezione di Cristallografia,Dip. di Genetica e Microbiologia, U. di Pavia, Via Taramelli 16, 27100 Pavia, Italy. (1951) Scuola perfezionamento, chemistry-biochemistry (U. di Pavia, 1974-78). Prof. assoc. (tel. 0382 + 392528, Fax 0382 + 422286, email HOBBIT at IPV85). *Computing, crystal growth, proteins.*

Bombieri, Prof. Gabriella. Ist. di Chimica Farmaceutica, U. di Milano, viale Abruzzi 42, I-20131 Italy. (1936) Dr, chemistry (U. di Padova, 1962). Prof. (tel. 02 + 222041 ext 602, telex 320484UNIMII). *Inorganic crystals, pharmaceutical compounds, actinide and lanthanide derivatives stereochemistry.*

Bonamartini-Corradi, Prof. Anna. Ist. di Chimica Generale ed Inorganica, U. di Parma, Viale delle Scienze, 43100 Parma, Italy. (1942) Dr, chemistry (U. di Parma, 1967). Prof. assoc. (tel. 0521 + 580416). *Structural chemistry, inorganic crystals, organic crystals.*

Bonamico, Dr Mario. Ist. Teoria e Struttura Elettronica e Comportamento Spettrochimico dei Composti di Coordinazione, C.N.R., Area della Ricerca di Roma, 00016 Monterotondo Stazione, Roma, Italy. (1930) Dr, chemistry (U. di Roma, 1956). Direttore di Ricerca. (tel. 06 + 9005849, telex 624809 CNRMLI). *Coordination compounds, inorganic crystals, materials science, molecular crystals, structural chemistry.*

Bonazzi, Dr Paola. Dip. di Scienze della Terra, U. di Firenze, Via La Pira 4, 50121 Firenze, Italy. (1960) Dr (U. di Firenze, 1984). Dottorando. (tel. 055 + 287140). *Crystal structures, crystal chemistry, minerals, borates.*

Bovio, Prof. Bruna. Dip. di Chimica Generale, U. di Pavia, Viale Taramelli 12, 27100 Pavia, Italy. (1928) Dr, chemistry (U. di Pavia, 1960). Prof. assoc. (tel. 0382 + 29270). *Coordination compounds, organic compounds.*

Braga, Prof. Dario. Dip. di Chimica 'G. Ciamician', via Selmi 2, 40126 Bologna, Italy. (1953) Dr, chemistry (U. di Bologna, 1977). Prof. Assoc. (tel. 051 + 259555, fax 051 + 259456, email VN3BODG1 at ICINECA). *Coordination compounds, thermal vibrations.*

Braibanti, Prof. Antonio. Ist. di Chimica Farmaceutica, Sezione Chimica Fisica, U. di Parma, Via La Spezia 73, 43100 Parma, Italy. (1927) Dr, chemistry (U. di Parma, 1951). Prof., physical chemistry. (tel. 0521 + 592207). *Structures, metal complexes, drugs, equilibria in solutions, thermodynamics.*

Bresciani Pahor, Prof. Nevenka. Dip. di Scienze Chimiche, U. di Trieste, Piazzale Europa 1, 34127 Trieste, Italy. (1949) Dr, chemistry (U. di Trieste, 1973). Prof. assoc. (tel. 040 + 5603111, Fax 040 + 5603176). *Coordination compounds, organic compounds, structural chemistry.*

Brigatti, Dr Maria Franca. Ist. di Chimica, U. della Basilicata, Via N. Sauro 85, 85100 Potenza, Italy. (1946) Dr, geological sci. (U. di Modena, 1970). Prof. Assoc. (tel. 0971 + 334111, telex 812492 UNSTBAI). *Crystallographic data, minerals, powder diffraction, X-ray diffraction, crystal chemistry.*

Brückner, Prof. Sergio. Dip. di Chimica, Politecnico di Milano, Piazza L. da Vinci 32, 20133 Milano, Italy. (1943) Dr, chemistry (U. di Trieste, 1967) Prof. assoc. (tel. 02 + 23993022, fax 02 + 23992206, email RCHIGE01 at CIMIPOLI). *Liquid crystals, macromolecules, powder diffraction.*

Bruno, Prof. Emiliano. Dip. di Scienze della Terra, U. di Torino, Via S. Massimo 24, 10123 Torino, Italy. (1938) Dr, geological sci. (U. di Genova, 1962). Prof. (tel. 011 + 832193). *Crystal chemistry, experimental mineralogy.*

Bruno, Prof. Giuseppe. Dip. di Chimica Inorganica e Struttura Molecolare, U. di Messina, C.da Papardo, 98100 Messina, Italy. (1952) Dr, chemistry (U. di Messina, 1976). Prof. assoc. (tel. 090 + 393537, email BRUNO at IMEUNIV.BITNET). *Coordination compounds, X-ray diffraction.*

Bruzzone, Prof. Giacomo. Ist. di Chimica Fisica, U. di Genova, Corso Europa (Palazzo delle Scienze), 16132 Genova, Italy. (1931) Dr, chemistry (U. di Genova, 1959). Prof., physical chemistry. (tel. 010 + 3538221). *Chemical bonding, hydrogen bonding, high pressure, inorganic crystals, materials science.*

Burla, Dr Maria Cristina. Dip. di Scienze della Terra, Sezione di Cristallografia, U. di Perugia, Piazza Università, 06100 Perugia, Italy. (1953) Dr, mathematics (U. di Perugia, 1976). Ricercatore (tel. 075 + 4693215, telex 662078 UNIPG I, email SIR2 at IPGUNIV). *Computing, direct methods.*

Busetti, Prof. Vilma. Dip. di Chimica Organica, U. di Padova, Via Marzolo 1, 35100 Padova, Italy. (1934) Dr, chemistry (U. di Padova, 1960). Prof. assoc. (tel. 049 + 831273). *Organic compounds, inorganic compounds.*

Buttinelli, Prof. Dante. Dip. di Ingegneria Chimica, de, materiali, dell, materie prime e Metallurgia, U. di Roma 'La Sapienza', Via Eudossiana 18, 00184 Roma, Italy. (1935) Dr, chemistry (U. di Roma, 1958). Prof. stab. (tel. 06 + 464286). *Electron diffraction, electron microscopy, metals, phase determination, phase transitions, texture.*

Caglioti, Prof. Dr Giuseppe. Laboratori, materiali, CESNEF - Ist. Ingegneria Nucleare del Politecnico, Via Ponzio 34/3, 20133 Milano, Italy. (1931) Dr, physics (U. di Roma, 1953). Prof., solid state physics. (tel. 02 + 23996310, telex 333467 POLIMI I, Fax 23992206). *Materials science, metals, symmetry, thermal vibrations.*

Calestani, Dr Gianluca. Ist. di Strutturistica Chimica, U. di Parma, Viale delle Scienze, 43100 Parma, Italy. (1952) PhD (U. Saarlandes, BRD, 1980). Ricercatore. (tel. 0521 + 580548 or 580447, Fax 0521 + 580542). *Computing, crystal growth, inorganic crystals, magnetism, materials science, powder diffraction, semiconductors, X-ray diffraction.*

Callegari, Dr Athos. Dip. di Scienze della Terra, U. di Pavia, Via A. Bassi 4, 27100 Pavia, Italy. (1958) Dr, geological sci. (U. di Pavia, 1984). Museum curator.

(tel. 382 + 392269, email CSCSPV at IPVIAN.BITNET). *Crystal chemistry, minerals.*

Calleri, Prof. Mariano Bernardino. Dip. di Scienze della Terra, U. di Torino, Via San Massimo 24, 10123 Torino, Italy. (1934) Dr, chemistry (U. di Torino, 1958). Prof. (tel. 011 + 832193, email U137F at ITOXIP). *Materials science, molecular crystals, phase determination.*

Calligaris, Prof. Mario. Dip. di Scienze Chimiche, U. di Trieste, Piazzale Europa 1, 34127 Trieste, Italy. (1939) Dr, chemistry (U. di Trieste, 1964). Prof. (tel. 040 + 5603113, Fax 040 + 5603176). *Coordination compounds, structural chemistry, organometallic compounds.*

Camalli, Dr Mercedes. Ist. di Strutturistica Chimica 'G. Giacomello', C.N.R., C.P. 10, Monterotondo Stazione, 00016 Roma, Italy. (1947) Dr, chemistry (Buenos Aires U., Argentina, 1974, U. di Napoli, 1978). Ricercatore. (tel. 06 + 9005142, Fax 06 + 9005849, email STRUTA at IRMIAS). *Computing, coordination compounds, diffractometry, instrumentation, phase determination.*

Cameroni, Prof. Riccardo. Dip. di Scienze Farmaceutiche, U. di Modena, Via S. Eufemia 19, 41100 Modena, Italy. (1926) Dr, chemistry and pharmacy (U. di Modena, 1950 and 1954). Prof., appl. farmaceutical chemistry. (tel. 059 + 219093). *Powder diffraction, structural chemistry, X-ray diffraction.*

Campanelli, Dr Anna Rita. Dip. di Chimica, U. di Roma 'La Sapienza', Piazzale A. Moro 5, 00185 Roma, Italy. (1953) Dr, chemistry (U. di Roma, 1978) Ricercatore. (tel. 06 + 49913322, Fax 06 + 4958251, email CAMPANELLI at VAXRMA.INFNET). *Structural biology, X-ray diffraction, inclusion compounds, micellar aggregates.*

Candeloro De Sanctis, Prof. Sofia. Dip. di Chimica, U. di Roma 'La Sapienza', Piazzale A. Moro, 00185 Roma, Italy. (1942) DPhil, crystallography (U. di Oxford, UK, 1970). Prof. assoc. (tel. 06 + 49913322, Fax 06 + 4958251, email CANDELORO at VAXRMA.INFNET). *Micellar aggregates, inclusion compounds, molecular crystals, structural biology.*

Cannas, Prof. Mario. Dip. di Scienze Chimiche, U. di Cagliari, Via Ospedale 72, 09124 Cagliari, Italy. (1926) Dr, chemistry (U. di Cagliari, 1951). Prof., structural chemistry. (tel. 070 + 668047). *Coordination compounds, structural chemistry, teaching.*

Cannillo, Dr Elio. Centro di Studio per la Cristallografia Strutturale, C.N.R., Via A. Bassi 4, 27100 Pavia, Italy. (1938) Dr, chemistry (U. di Pavia, 1962). Ricercatore. (tel. 0382 + 392267, Fax 0382 + 392252, email CSCSPV at IPVIAN.BITNET). *Computing, minerals.*

Capasso, Prof. Sante. Dip. Chimico, U. di Napoli, Via Mezzocannone 4, 80134 Napoli, Italy. (1944) Dr, chemistry (U. di Napoli, 1969). Prof. assoc. (tel. 081 + 205730). *Organic compounds, proteins, X-ray diffraction.*

Capotorto, Prof. Concetta. Dip. di Ingegneria Chimica, de, materiali, dell, materie prime e Metallurgia, U. di Roma 'La Sapienza', Via Eudossiana 18, 00184 Roma, Italy. (1936) Dr, chemistry (U. di Roma, 1967). Prof. assoc. (tel. 06 + 464286). *Crystallographic data, electron diffraction, electron microscopy, materials science, metals, microscopy, phase determination, phase transition, quasicrystals, texture, X-ray diffraction.*

Carbonin, Dr Susanna. Ist. di Mineralogia e Petrologia, U. di Padova, Corso Garibaldi 37, 35100 Padova, Italy. (1948) Dr, natural sci. (U. di Padova, 1972). Ricercatore. (tel. 049 + 663122, Fax 049 + 831711) *Minerals.*

Cartení-Farina, Prof. Maria. Ist. di Biochimica dell, macromoleculecole, U. di Napoli, Via Costantinopoli 16, 80138 Napoli, Italy. (1945) Dr, chemistry (U. di Napoli, 1969). Prof. assoc. (tel. 081 + 217052-213668, Fax 081 + 217052). *Chemistry, sulfonium compounds, polyamine biosynthesis, structure and function, enzymes.*

Carugo, Dr Oliviero. Dip. di Chimica Generale, U. di Pavia, Viale Taramelli 12, 27100 Pavia, Italy. (1963), Dr, chemisty (U. Pavia, 1986). Tecnico laureato. (tel. 0382 + 392343). *Structural chemistry, coordination compounds. NMR.*

Caruso, Dr Francesco. Ist. di Strutturistica Chimica 'G. Giacomello', C.N.R., C.P. 10, Monterotondo Stazione, 00016 Roma, Italy. (1947), Dr, chemisty (U. di Buenos Aires, 1974, U. of Napoli, 1977). Ricercatore. (tel. 06 + 9005142). *Structural chemistry, coordination compounds. NMR.*

Casalone, Dr Gianluigi. Centro di Studio per le Relazioni tra Struttura e Reattività Chimica C.N.R., Via Golgi 19, 20133 Milano, Italy. (1939) Dr, industrial chemistry (U. di Milano, 1965). Ricercatore. (tel. 02 + 2663805, fax 02 + 2362946, email MI0354 at IMISIAM). *Electron diffraction, surface structure.*

Cascarano, Dr Giovanni Luca. Dip. Geomineralogico, U. di Bari, Trav. 200 via Re David 4, Campus Universitario, 70121 Bari, Italy. (1953) Dr, computer sci. (U. di Bari, 1979). (tel. 080 + 242597) *Computing, direct methods.*

Casellato, Dr Umberto. Ist. di Chimica e Tecn. dei Radioelementi, C.N.R., Corso Stati Uniti, 35100 Padova, Italy. (1942) Dr, chemistry (U. di Padova, 1970). Primo ric. (tel. 049 + 845359, telex 430302, Fax 049 + 845449, email BERTO at ICTR02.INFN.IT) *Coordination compounds, organic compounds. X-ray diffraction.*

Catti, Prof. Michele. Dip. di Chimica Fisica ed Elettrochimica, U. di Milano, Via Golgi 19, 20133 Milano, Italy. (1945) Dr, chemistry (U. di Torino, 1969). Prof., crystal chemistry. (tel. 02 + 2664329). *Disorder, inorganic crystals, materials science, minerals, phase transitions.*

Caucia, Dr Franca. Dip. di Scienze della Terra, U. di Pavia, Via A. Bassi 4, 27100 Pavia, Italy. (1957) Dr, geological sci. (U. di Pavia, 1980). Ricercatore. (tel. 0382 + 392269, email CSCSPV at IPVIAN.EARN). *Minerals, crystal chemistry.*

Cavalca, Prof. Luigi. Via G. C. Ferrarini 2, 43100 Parma, Italy. (1911) Dr, chemistry (U. di Parma, 1945) Prof., structural chemistry. (tel. 0521 + 241827). *Inorganic compounds, organic compounds.*

Cellai, Dr Luciano. Ist. di Strutturistica Chimica 'G. Giacomello', C.N.R., C.P. 10 - 00016 Monterotondo Stazione, Roma, Italy. (1944) Dr, chemistry (U. di Roma, 1969). Ricercatore. (tel. 06 + 9005142, fax 06 + 9005849, email STRUTA at IRMIAS). *Organic compounds.*

Celotti, Dr Giancarlo. Ist. di Chimica e Tecn. de, materiali e Componenti per l'Elettronica (LAMEL), C.N.R., Via de' Castagnoli 1, 40126 Bologna, Italy. (1943) Dr, physics (U. di Bologna, 1966). Ricercatore. (tel. 051 + 519593, ext. 218). *Defect structures, semiconductors, inorganic compounds, organic compounds, precise lattice parameter determination, perturbed structures, X-ray diffraction.*

Cerrini, Dr Silvio. Ist. di Strutturistica Chimica 'G. Giacomello', C.N.R., C.P. 10 - 00016 Monterotondo Stazione, Roma, Italy. (1937) Dr, physics (U. di Roma, 1963). Primo ric. (tel. 06 + 9005142, fax 06 + 9005849, email STRUTA at IRMIAS). *Computing, diffractometry, molecular crystals, organic compounds, structure-activity relationships.*

Cesari, Prof. lib. doc. Marco. ENIRICERCHE, Via F. Maritano 26, 20097 San Donato Milanese, Milano, Italy. (1930) Dr, chemistry (U. di Parma, 1954). Capo-reparto, lib. doc. at U. of Parma. (tel. 02 + 5205897). *Organometallic compounds, inorganic compounds, structure and morphology, polymers.*

Chiari, Prof. Giacomo. Dip. di Scienze della Terra, U. di Torino, Via San Massimo 24, 10123 Torino, Italy. (1943) Dr, chemistry (U. di Torino, 1967). Prof. assoc. (tel. 011 + 832193, email U137 at ITOCSIP). *Computing, diffractometry, hydrogen bonding, minerals, organic compounds, teaching, X-ray diffraction.*

Chiesi-Villa, Prof. Angiola. Ist. di Strutturistica Chimica, U. di Parma, Via M. D'Azeglio 85, 43100 Parma, Italy. (1940) Dr, chemistry (U. di Parma, 1964). Prof. assoc. (tel. 0521 + 580449, email STRUCHI at IPRUNIV). *Chemical bonding, coordination compounds, structure-activity relationships.*

Ciajolo, Prof. Maria Rosaria. Dipartimento di Chimica, U. di Napoli, Via Mezzocannone 4, 80134 Milano, Italy. (1946) Dr, chemistry (U. di Napoli, 1970). Prof. assoc. (tel. 081 + 206800). *Liquid crystals.*

Ciani, Prof. Gianfranco. Ist. di Chimica Generale ed Inorganica, U. degli Studi di Milano, Via G. Venezian 21, 20133 Milano, Italy. (1944) Dr, chemistry (U. di Milano, 1968). Prof. assoc. (tel. 02 + 235120). *Chemical bonding, coordination compounds, materials science, structural chemistry, X-ray diffraction.*

Cimino, Prof. Alessandro. Genera, inorganic Chemistry, U. of Rome, Città U., 00100 Roma, Italy. (1926) Dr, chemistry (U. di Roma, 1948). Prof. (tel. 06 + 4991, ext. 3582). *Materials science, surface chemistry.*

Cini, Prof. Renzo. Dip. di Chimica, U. di Siena, Pian dei Mantellini 44, 53100 Siena, Italy. (1949) Dr, chemistry (U. di Pisa, 1974). Prof. assoc. (tel. 0577 + 47054, fax 0577 + 280405, email SICHIM at ICNUCEVM). *Coordination compounds, structural biology, structural chemistry.*

Cipriani, Prof. Curzio. Museo di Mineralogia, U. di Firenze, Via La Pira 4, 50121 Firenze, Italy. (1927) Dr, chemistry (U. di Firenze, 1950). Prof. (tel. 055 + 216936). *Crystallographic data, diffractometry, inorganic crystals, minerals, phase determination, phase transitions. X-ray diffraction.*

Cirafici, Prof. Salvino S. Ist. di Chimica Fisica, U. di Genova, Corso Europa (Palazzo delle Scienze), 16132 Genova, Italy. (1944) Dr, chemistry (U. di Genova, 1971). Prof. assoc. (tel. 010 + 3538225, fax 010 + 515076). *Materials science, metals, powder diffraction.*

Clemente, Prof. Dore Augusto. Dipartimento di Scienza de, materiali, U. di Lecce, Via Monteroni, 73100 Lecce, Italy. (1940) Dr, chemistry (U. di Padova, 1966). Prof. (tel. 0832 + 621529). *Computing, electron density, structural chemistry.*

Coda, Prof. Alessandro. Dip. di Genetica e Microbiologia, Sez. di Cristallografia, U. di Pavia, Via Taramelli 16, 27100 Pavia, Italy. (1936) Dr, chemistry (U. di Pavia, 1959). Prof., crystallography. (tel. 0382 + 422394). *Organic compounds, bio- or pharmacologically interesting structures.*

Coiro, Dr Vincenza Maria. Ist. di Strutturistica Chimica 'G. Giacomello', C.N.R., C.P. 10 - 00016 Monterotondo Stazione, Roma, Italy. (1937) Dr, (U. di Napoli, 1961). Primo ric. (tel. 06 + 9005142, fax 06 + 9005849, email STRUTA at IRMIAS). *Organic compounds, structural chemistry, micellar aggregates.*

Cojazzi, Dr Gianna. Centro di Studio per la Fisica dell, macromoleculecole, C.N.R., Via Selmi 2, 40126 Bologna, Italy. (1935) Dr, chemistry (U. di Padova, 1961). Ricercatore. (tel. 051 + 259552, fax 051 + 259456). *Macromolecules, polymers, powder diffraction, X-ray diffraction.*

Cola, Prof. Mario Luigi. Dip. di Chimica Generale, U. di Pavia, Viale Taramelli 12, 27100 Pavia, Italy. (1922) Dr, chemistry (U. di Pavia, 1950). Prof. assoc. (tel. 0382 + 392331). *Inorganic crystals, powder diffraction.*

Colapietro, Prof. Marcello. Dipartimento di Chimica, U. di Roma 'La Sapienza', Piazz.le A. Moro 5, 00185 Roma, Italy. (1937) Dr, physics (U. di Roma, 1962). Prof. Assoc. (tel. 06 + 49913715, fax 06 + 4958251). *Computing, diffractometry, instrumentation, phase determination, scattering theory, structural chemistry, X-ray diffraction.*

Colombo, Dr Arturo. Ist. di Chimica dell, macromoleculecole, C.N.R., Via Bassini 15, 20133 Milano, Italy. (1935) Dr, chemistry (U. di Pavia, 1960). Sr. scient. (tel. 02 + 2666071, email KRIS at IMISIAM). *Computing, structural chemistry.*

Corchía, Dr Massimo. Divisione Scienza de, materiali, E.N.E.A. - C.R.E. Casaccia, S. P. Anguillarese 301, Casella Postale N. 2400, 00100 Roma, Italy. (1944) Dr, physics (U. di Pavia, 1967). Ricercatore. (tel. 06 + 69484355, telex ENEACAI 613296). *Metals, solid state physics, inorganic crystals.*

Corradini, Prof. Paolo. Dip. di Chimica, U. di Napoli, Via mezzocannone 4, 80134 Napoli, Italy. (1930) Dr, chemistry (U. di Roma, 1951). Prof. (tel. 081 + 205730, ext. 121). *Structural chemistry, macromolecules, structure - property relationships, statistical thermodynamics, polymers.*

Cremona, Ing. Luigi. Assing S.p.A., Via A. de Pretis 70, 00184 Roma, Italy. (1945) Dr, eng. (U. di Rome, 1969). Techn. manager. (tel. 06 + 4750341, fax 06 + 460576). *Crystal growth, electron diffraction, electron microscopy, powder diffraction, semiconductors.*

Cubiotti, Prof. Gaetano. Ist. di Fisica Teorica, U. di Messina, P.O. Box 50, S. Agata di Messina, 98166 Messina, Italy. (1939) Dr, physics (U. di Messina, 1962). Prof., physics. (tel. 090 + 714492, email CUBIOTTI at IMEUNIV). *Electron density, metals, scattering theory, surface structure.*

Dall'Aglio, Dr Gian Antonio. Dipartimento di Scienze della Terra, sezione di Mineralogia, Corso Europa 26, 16136 Genova, Italy. (1959) Dr, geological sci. (U. di Genova, 1986). Ricercatore. (tel. 010 + 3538303, fax 010 + 352169). *Crystal growth, holographic interferometry.*

Dal Negro, Prof. Alberto. Dipartimento di Mineralogia e Petrologia, U. di Padova, Corso Garibaldi 37, 35100 Padova, Italy. (1941) Dr, chemistry (U. di Pavia, 1964). Prof., crystallography. (tel. 049 + 663122). *Minerals.*

Dapporto, Prof. Paolo. Dip. di Energetica, U. di Firenze, Via Santa Marta 3, 50139 Firenze, Italy. (1942) Dr, chemistry (U. di Firenze, 1966). Prof. (tel. 055 + 4796209, telex 580681 UNFING I). *Structural chemistry.*

Davoli, Dr Paolo. Ist. di Mineralogia e Petrologia, U. di Modena, Via S. Eufemia 19, 41100 Modena, Italy. (1958) Dr, Physics (U. di Modena, 1982). (tel. 059 + 218062). *Computing, inorganic crystals.*

De Angelis, Prof. Giuseppe. Dip. di Scienze della Terra, U. di Roma 'la Sapienza', Piazzale A. Moro 5, 00185 Roma, Italy. (1944) Dr, geological sci. (U. di Roma, 1970). Asst. (tel. 06 + 49914621). *Computing, diffractometry, X-ray diffraction, Mössbauer spectrometry, Mössbauer methods, crystallographic computer applications.*

Della Giusta, Prof. Antonio. Dipartimento di Mineralogia e Petrologia, U. di Padova, Corso Garibaldi 37, 35100 Padova, Italy. (1941) Dr, geology (U. di Genova, 1964). Prof. (tel. 049 + 663122). *Computing, crystallographic data, diffractometry, inorganic crystals, minerals, teaching, X-ray diffraction.*

Del Monte, Prof. Marco Emiliano. Dipartimento di Scienze Geologiche, Via Zamboni 67, 40126 Bologna, Italy. (1939) Dr, geological sci. (U. di Bologna, 1964). Prof., mineralogy and geology (tel. 051 + 228810, telex 511350, fax 051 + 229702). *Crystal growth, diffractometry, microscopy, minerals, powder diffraction, X-ray diffraction.*

Del Piero, Dr Gastone. ENIRICERCHE, via F. Maritano 26, 20097 San Donato Milanese, Milano, Italy. (1943) Dr, chemistry (U. di Padova, 1968). Ricercatore. (tel. 02 + 52024904) *Computing, diffractometry, electron diffraction, electron microscopy, inorganic crystals, liquid crystals, materials science, minerals, molecular crystals, phase determination, polymers, powder diffraction, small-angle scattering, structural chemistry, surface structure, organometallic compounds.*

Del Pra, Prof. Antonio. Ist. Chimico Farmaceutico e Tossicologico, U. di Milano, Viale Abruzzi 42, 20131 Milano, Italy. (1932) Dr, chemistry (U. di Padova, 1962). Prof. (tel. 02 + 222041, ext 25) *Coordination compounds, diffuse scattering, structural biology.*

Demartin, Prof. Francesco. Ist. di Chimica Strutturistica Inorganica, U. di Milano, Via G. Venezian 21, 20133 Milano, Italy. (1953), Dr, chemistry (U. di Milano, 1978). Prof. Assoc. (tel. 02 + 235120, email STINCH at IMISIAM). *Coordination compounds, structural chemistry, X-ray diffraction.*

Demontis, Dr Pierfranco. Dipartimento di Chimica, U. di Sassari, Via Vienna 2, 07100 Sassari, Italy. (1954) Dr, Chemistry (U. di Sassari, 1979). Ricercatore. (tel. 079 + 218415) *Disorder, high pressure, minerals, thermal vibrations.*

De Munno, Prof. Giovanni. Dipartimento di Chimica, U. della Calbria, Arcavacata di Rende, 87030 Cosenza, Italy. (1951) Dr, Chemistry (U. di Roma, 1975). Prof. Assoc. (tel. 098 + 4839321, email 2201DEM at ICSUNIV) *Coordination compounds, liquid crystals, magnetism.*

Depero, Dr Laura Eleonora. Dipartimento di Ingegneria Meccanica, U. di Brescia, Via D. Valotti 9, 25060 Brescia, Italy. (1960) Dr, Physics (U. di Milano, 1984). Borsista. (tel. 030 + 398461, ext 272, telex 304116 UNIVI) *Inorganic crystals, materials science, accurate intensity measurements.*

De Pol Blasi, Prof. Carla. Dip. di Scienze della Terra, U. degli Studi, Via Botticelli 23, 20133 Milano, Italy. (1936) Dr, geological sci. (U. di Milano, 1960). Prof. assoc. (tel. 02 + 2663994, fax 02 + 2364393). *Minerals, X-ray diffraction.*

De Santis, Prof. Pasquale. Dip. di Chimica, U. di Roma, Piazzale A. Moro 5, 00100 Roma, Italy. (1935) Dr, chemistry (U. di Bari, 1959). Prof., physical chemistry. (tel. 06 + 4952993). *Computing, hydrogen bonding, macromolecules, molecular crystals, polymers, proteins, small-angle scattering, structural biology, structural chemistry.*

Destro, Prof. Riccardo. Dip. di Chimica Fisica ed Elettrochimica, U. degli Studi di Milano, Via Golgi 19, 20133 Milano, Italy. (1940) Dr, industrial chemistry (U. di Milano, 1964). Prof. assoc. (tel. 02 + 2663805, telex 331370 MICHIM I, email MI0354 at IMISIAM). *Chemical bonding, electron density, molecular crystals, electrostatic properties.*

Di Blasio, Prof. Benedetto. Dip. di Chimica, U. di Napoli, Via Mezzocannone 4, 80134 Napoli, Italy. (1945) Dr, physics (U. di Napoli, 1971). Prof. assoc. (tel. 081 + 206800). *Structural biology, X-ray diffraction.*

Di Lorenzo, Dr Guido. Dip. di Chimica, U. di Napoli, Via Mezzocannone 4, 80134 Napoli, Italy. (1948) Dr, chemistry (U. di Napoli, 1975). (tel. 081 + 206800). *Proteins, X-ray diffraction.*

Di Vaira, Prof. Massimo. Dip. di Chimica, U. di Firenze, Via Maragliano 77, 50144 Firenze, Italy. (1940) Dr, chemistry (U. di Firenze, 1963). Prof. (tel. 055 + 354841, telex 570123 CHIMFI, fax 055 + 244102, email DIVAIRA at IFICHIM.BITNET). *Chemical bonding, coordination compounds, structural chemistry.*

Domeneghetti, Dr Maria Chiara. Centro di Studio per la Cristallografia Strutturale, C.N.R., Via Bassi 4, 27100 Pavia, Italy. (1954) Dr, natural sci. (U. di Pavia, 1978). Ricercatore. (tel. 0382 + 392265, fax 0382 + 392252, email CSCS at IPVIAN.EARN). *Minerals, phase transitions, X-ray diffraction, crystal chemistry.*

Domenicano, Prof. Aldo. Dipartimento di Chimica, Ingegneria Chimica , materiali, U. dell'Aquila, 67100 L'Aquila, Italy. (1938) Dr, chemistry (U. di Roma, 1965). Prof. (tel. 0862 + 25387). *Electron diffraction, organic compounds, structural chemistry, X-ray diffraction.*

Domenici, Dr Marcello. DN, electroni, materials S.p.A., Viale Gherzi 31, 28100 Novara, Italy. (1932) Dr, physics (U. di Pisa, 1958). R & D director. (tel. 0321 + 442, ext. 313, telex 200486, fax 0321 + 440054). *Crystal growth, materials science, powder diffraction, small-angle scattering, texture, X-ray diffraction.*

Domiano, Prof. Paolo. Ist. di Strutturistica, U. di Parma, Viale delle Scienze, 43100 Parma, Italy. (1936) Dr, chemistry (U. di Parma, 1960). Prof. assoc. (tel. 0521 + 580446, telex 530327 UNIVPR I, fax 0521 + 580542). *Computing, crystallographic data, molecular crystals, organic compounds, structural chemistry, teaching, X-ray diffraction.*

Donati, Dr Donato. Ist. di Chimica Organica, U. di Siena, Pian dei Mantellini 44, 53100 Siena, Italy. (1946) Dr, chemistry (U. di Firenze, 1970). Prof. assoc. (tel. 0577 + 47054). *Organic compounds.*

Dovesi, Prof. Roberto. Dipartimento di Chimica Inorganica, Chimica Fisica e de, materiali, U. di Torino, Via P. Giuria 5, 10125 Torino, Italy. (1947) Dr, chemistry (U. di Torino, 1971). Prof. assoc. (tel. 011 + 657274, email U107 at ITOCSIP). *Solid state theoretical chemistry.*

Duca, Dr Dario. Dipartimento di Chimica Inorganica, U. di Palermo, Via Archirafi 26, 90123 Palermo, Italy. (1961) Dr, chemistry (U. di Palermo, 1985). (tel. 091 + 6161502). *Catalysis, morphology, disorder, structural chemistry, small-angle scattering.*

Emiliani, Prof. Francesco. Ist. di Mineralogia, U. di Parma, Viale delle Scienze 78, 43100 Parma, Italy. (1923) Dr, chemistry (U. di Bologna, 1949). Prof., mineralogy. (tel. 0521 + 580321, fax 0521 + 580342). *Diffractometry, minerals, powder diffraction, teaching.*

Fagherazzi, Prof. Giuliano. Dip. di Chimica Fisica, U. Cà Foscari di Venezia, D.D. 2137, 30123 Venezia, Italy. (1938) Dr, physics (U. di Padova, 1961). Prof., solid state physical chemistry. (tel. 041 + 5227554, telex 410638 UNIVVE I, fax 041 + 5210112). *Small-angle scattering, inorganic polycrystalline and amorphous materials, X-ray diffraction.*

Fagnani, Prof. Gustavo. Dip. di Scienze della Terra, U. degli Studi, Via Botticelli 23, 20133 Milano, Italy. (1917) Dr, natural sciences (U. di Milano, 1942). Prof. assoc. (tel. 02 + 2663994). *Minerals.*

Fanfani, Prof. Luca. Dip. di Scienze della Terra, U. di Cagliari, Via Trentino 51, 09100 Cagliari, Italy. (1941) Dr, chemistry (U. di Firenze, 1964). Prof., mineralogy. (tel. 070 + 2006, ext 219). *Minerals.*

Fares, Dr Vincenzo. Ist. Teoria e Struttura Elettronica e Comportamento Spettrochimico dei Composti di Coordinazione, C.N.R., Area della Ricerca di Roma, 00016 Monterotondo Stazione, Roma, Italy. (1942) Dr, chemistry (U. di Roma, 1966). Ricercatore. (tel. 06 + 90020285, telex 624809 CNRMLI). *Coordination compounds, inorganic crystals, materials science, structural chemistry.*

Fedeli, Prof. Walter. Dip. di Chimica, Ingegneria Chimica , materiali, U. dell'Aquila, Via Assergi 4, 67100 L'Aquila, Italy. (1933) Dr, chemistry (U. di Roma, 1961). Prof., general chemistry. (tel. 0862 + 25387). *Biologically important substances.*

Ferracini, Prof. Elena. Dip. di Chimica 'G. Ciamician', U. di Bologna, Via Selmi 2, 40126 Bologna, Italy. (1938) Dr, industrial chemistry (U. di Bologna, 1962). Prof. assoc. (tel. 051 + 259552, fax 051 + 259456). *Disorder, macromolecules, polymers, small-angle scattering, teaching, X-ray diffraction.*

Ferrari, Dr Claudio. MASPEC C.N.R., Via Chiavari 18/A, 43100 Parma, Italy. (1958) Dr, physics (U. di Parma). Ricercatore. (tel. 0521 + 96841, telex 531639, fax 0521 + 96315). *Defect structures, diffractometry, diffuse scattering, materials science, semiconductors, X-ray diffraction.*

Ferrari-Belicchi, Prof. Marisa. Ist. di Chimica Generale, U. di Parma, Viale delle Scienze 78 , 43100 Parma, Italy. (1942) Dr, chemistry (U. di Parma, 1966). Prof. assoc. (tel. 0521 + 580420, fax 0521 + 580542, email CHIMIC3 at IPRUNIV). *Coordination compounds, organic compounds, structural chemistry, X-ray diffraction.*

Ferraris, Prof. Giovanni. Dip. di Scienze della Terra, U. di Torino, Via S. Massimo 24, 10123 Torino, Italy. (1937) Dr, physics (U. di Torino, 1960). Prof., crystallography. (tel. 011 + 832193, email U137 at ITOCSIP). *Chemical bonding, hydrogen bonding, minerals, neutron diffraction, inorganic crystal chemistry.*

Ferrero Rognoni, Prof. Adele. Ist. Chimico 'G. Ciamician', U. di Bologna, via F. Selmi 2 40126 Bologna, Italy. (1925), Dr, chemistry (U. di Bologna, 1950). Prof. assoc. (tel. 03951 + 259450). *Defect structures, diffuse scattering, disorder, macromolecules, polymers, small-angle scattering, X-ray diffraction.*

Ferretti, Dr Valeria. Dip. di Chimica, U. di Ferrara, Via L. Borsari 46, 44100 Ferrara, Italy. (1958) Dr, chemistry (U. di Ferrara, 1981). Ricercatore. (tel. 0532 + 33522, ext. 15, email VG3A at ICINECA). *Chemical bonding, coordination compounds, hydrogen bonding, molecular crystals, organic compounds, reaction pathways.*

Fichera, Dr Anna Maria. Centro di Studio per la Fisica dell, macromoleculecole, C.N.R., Via Selmi 2, 40126 Bologna, Italy. (1939) Dr, chemistry (U. di Bologna, 1965). Ricercatore. (tel. 051 + 259552, fax 051 + 259456, email VS3BODI3 at ICINECA). *Macromolecules, polymers, powder diffraction, X-ray diffraction.*

Filippini, Dr Giuseppe. Centro di Studio per le Relazioni tra Struttura e Reattività Chimica, C.N.R., Via Golgi 19, 20133 Milano, Italy. (1940) Dr, industrial chemistry (U. di Milano, 1968). Ricercatore. (tel. 02 + 231226, telex 331370 MICHIM I, fax 02 + 2663030, email LATTDINA at IMISIAM). *Computing, molecular crystals, thermal vibrations.*

Foresti, Prof. Elisabetta. Dip. di Chimica 'G. Ciamician', U. di Bologna, Via Selmi 2, 40126 Bologna, Italy. (1943) Dr, chemistry (U. di Bologna, 1968). Prof. assoc. (tel. 051 + 259450). *Crystallographic data, diffractometry, inorganic crystals, organic compounds, powder diffraction, structural chemistry. X-ray diffraction.*

Fornasini, Prof. Maria Luisa. Ist. di Chimica Fisica, U. di Genova, Corso Europa 26 (Palazzo delle Scienze), 16132 Genova, Italy. (1941) Dr, chemistry (U. di Genova, 1965). Prof. assoc. (tel. 010 + 3538225, email CFMET at IGECUNIV). *Metals, structural chemistry.*

Forni, Prof. Flavio. Dip. di Scienze Farmaceutiche, U. di Modena, Via S. Eufemia 19, 41100 Modena, Italy. (1951) Dr, pharmacy and biological sci. (U. di Modena, 1973, 1978). Prof. assoc. (tel. 059 + 219093). *Crystallographic data, structural chemistry, X-ray diffraction.*

Forsellini, Dr Eleonora. Ist. di Chimica e Tecn. dei Radioelementi, C.N.R., Corso Stati Uniti, 35100 Padova, Italy. (1934) Dr, chemistry (U. di Padova, 1965). Primo ric. (tel. 049 + 845369, telex 430302 CNR I, fax 049 + 845449, email NORA at ICTR02.BITNET). *Coordination compounds, structural chemistry, X-ray diffraction.*

Forti, Prof. Paolo, Ist. di Geologia, U. di Bologna, Via Zamboni 67, 40127 Bologna, Italy. (1945) Dr, chemistry (U. di Bologna, 1969). Prof. assoc. (tel. 051 + 242251). *Minerals, powder diffraction, X-ray diffraction.*

Franceschi, Prof. Enrico. Ist. di Chimica Fisica, U. di Genova, Corso Europa (Palazzo delle Scienze), 16132 Genova, Italy. (1942) Dr, chemistry (U. di Genova, 1965). Prof. assoc. (tel. 010 + 3538224, fax 010 + 515076). *Structure and physical properties, intermetallic compounds.*

Franchini, Prof. Marinella. Dip. di Scienze della Terra, U. di Torino, Via San Massimo 24, 10123 Torino, Italy. (1936) Dr, natural sci. (U. di Torino, 1959). Prof. assoc. (tel. 011 + 832193). *Crystal growth, minerals.*

Franzini, Prof. Marco. Dip. di Scienze della Terra, U. di Pisa, Via S. Maria 53, 56100 Pisa, Italy. (1938) Dr, geological sci. (U. di Pisa, 1960). Prof., mineralogy. (tel. 050 + 501457). *Optical crystallography, low energy radiation interaction with condensed matter.*

Franzosi, Dr Paolo. MASPEC Ist, materiali Speciali per Elettronica e Magnetismo, C.N.R., Via Chiavari 18/A, 43100 Parma, Italy. (1948) Dr, physics (U. di Parma, 1972). Ricercatore. (tel. 0521 + 96841, 0521 + 96315). *Defect structures, diffractometry, electron microscopy, materials science, semiconductors, X-ray diffraction.*

Frigeri, Dr Cesare. Ist. MASPEC, C.N.R., Via Chiavari 18/A, 43100 Parma, Italy. (1950) Dr, Physics (U. di Bologna). Ricercatore. (tel. 0521 + 96841, fax 0521 + 96315). *Defect structures, electron diffraction, electron microscopy, materials science, powder diffraction, semiconductors, texture.*

Fumi, Prof. Fausto Gherardo. Dip. di Fisica, U. di Genova, Viale Dodecaneso 33, 16146 Genova, Italy. (1924) Dr, physics (U. di Genova, 1948). Prof., solid state physics. (tel. 010 + 59931, ext. 222, telex 211154 INFNGE I). *Crystal bonding, materials science, symmetry, thermal vibrations.*

Galli, Prof. Ermanno. Ist. di Mineralogia e Petrologia, U. di Modena, Via S. Eufemia 19, 41100 Modena, Italy. (1937) Dr, geological sci. (U. di Modena, 1963). Prof., mineralogy (tel. 059 + 218062, email MINERALOGIA at IMOVX2.INFNET). *Zeolites, silicates, inorganic crystal structures.*

Ganazzoli, Dr Fabio. Dip. di Chimica, Politecnico di Milano, Piazza L. da Vinci 32, 20133 Milano, Italy. (1955) Dr, chemistry (U. di Milano, 1978). Ricercatore. (tel. 02 + 23993024, email RCHIGE06 at IMIPOLI). *Coordination compounds, organic compounds, polymers.*

Garbassi, Dr Fabio. Ist. G. Donegani S.p.A., Via Fauser 4, 28100 Novara, Italy. (1942) Dr, chemistry (U. di Trieste, 1965). Senior scientist. (tel. 0321 + 447284). *Electron microscopy, materials science, polymers, surface chemistry, X-ray diffraction.*

Gasparri-Fava, Prof. Giovanna. Ist. di Chimica Generale, U. di Parma, Viale delle Scienze 78, 43100 Parma, Italy. (1930) Dr, chemistry (U. di Parma, 1955). Prof., general chemistry. (tel. 0521 + 580420, fax 0521 + 580542, email CHIMIC3 at IPRUNIV). *Coordination compounds, organic compounds, structural chemistry, X-ray diffraction.*

Gastaldi, Dr Leonardo. Ist. Teoria e Struttura Elettronica e Comportamento Spettrochimico dei Composti di Coordinazione, C.N.R., Area della Ricerca di Roma, 00016 Monterotondo Stazione, Roma, Italy. (1944) Dr, chemistry (U. di Roma, 1969). Ricercatore. (tel. 06 + 90020343). *Crystal growth, defect structures, X-ray diffraction.*

Gatti, Dr Giuseppina. Dip. di Genetica e Microbiologia, Sez. di Cristallografia, U. di Pavia, via Taramelli 16, 27100 Pavia, Italy. (1954) Dr, chemistry (U. di Pavia, 1978). Tecn. laur. (tel. 0382 + 392529, fax 0382 + 32234, email HOBBIT at IPV85.INFNET). *Crystal growth, proteins.*

Gauzzi, Prof. Franco. Dip. di Ingegneria Meccanica, U. di Roma Tor Vergata, Via O. Raimondo, 00173 Roma, Italy. (1931) Dr, chemistry (U. di Roma, 1955). Prof. (tel. 06 + 79792610, telex 611462 UNIVRM I, fax 06 + 6131757). *Martensitic transformations, metallic and non metallic materials, composite materials, quantitative measurements, defects.*

Gavezzotti, Dr Angelo. Dip. di Chimica Fisica ed Elettrochimica, U. di Milano, Via Golgi 19, 20133 Milano, Italy. (1944) Dr, chemistry (U. di Milano, 1968). Prof. (tel. 02 + 231226, telex 331370 MICHIM I, email MI0354 at IMISIAM). *Molecular crystals, structural chemistry.*

Gavuzzo, Dr Enrico. Ist. di Strutturistica Chimica 'G. Giacomello', C.N.R., C.P. 10 - 00016 Monterotondo Stazione, Roma, Italy. (1943) Dr, chemistry (U. di Roma, 1972). Ricercatore. (tel. 06 + 9005142, fax 06 + 9005849, STRUTA at IRMIAS). *Structural chemistry.*

Gazzano, Dr Massimo. Dipartimento di Chimica 'G. Ciamician', U. di Bologna, Via Selmi 2, 40126 Bologna, Italy. (1958) Dr, chemistry (U. di Bologna, 1982) (tel. 051 + 259450, fax 051 + 259456, email VS3BODI3 at ICINECA). *Inorganic crystals, powder diffraction, X-ray diffraction.*

Gazzoni, Dr Giuseppe. Dip. di Scienze della Terra, U. di Torino, Via San Massimo 22, 10123 Torino, Italy. (1937) Dr, physics (U. di Torino, 1963). Primo ric. (tel. 011 + 832193). *Inorganic crystals, phase transitions, structural chemistry.*

Gervasio, Prof. Giuliana. Dip. di Chimica Inorganica, Chimica Fisica e Chimica de, materiali, U. di Torino, Via P. Giuria 7, 10125 Torino, Italy. (1943) Dr, chemistry (U. di Torino, 1967). Prof. assoc. (tel. 011 + 655832-3, email U213 at ITOCSIP). *Coordination compounds, molecular crystals, neutron diffraction, X-ray diffraction, organometallic compounds.*

Ghezzi, Prof. Carlo. Dip. di Fisica, U. di Parma, Viale delle Scienze, 43100 Parma, Italy. (1940) Dr, physics (U. di Milano, 1963). Prof. (tel. 0521 + 580270, 0521 + 580223). *Defect structures, diffuse scattering, semiconductors, X-ray diffraction.*

Ghilardi, Dr Carlo Alfredo. Ist. Stereochimica, C.N.R., Via I. Nardi 39, 50132 Firenze, Italy. (1943) Dr, chemistry (U. di Firenze, 1968). Ricercatore. (tel. 055 + 243990). *Coordination compounds.*

Giacomazzo, Prof. Carmelo. Dip. Geomineralogico, U. di Bari, Via Amendola, 70121 Bari, Italy. (1940) Dr, physics (U. di Bari, 1965). Prof., mineralogy. (tel. 080 + 242590 or 242597). *Phase determination.*

Giglio, Prof. Edoardo. Dip. di Chimica, U. di Roma, Piazzale A. Moro 5, 00185 Roma, Italy. (1931) Dr, chemistry (U. di Bari, 1955). Prof., physical chemistry. (tel. 06 + 4952993). *Macromolecules, phase determination, small-angle scattering, micelles, inclusion compounds.*

Gilli, Prof. Gastone. Dip. di Chimica, U. di Ferrara, Via L. Borsari 46, 44100 Ferrara, Italy. (1937) Dr, chemistry (U. di Ferrara, 1962). Prof., physical chemistry. (tel. 0532 + 33522, ext. 20, M38A at ICINECA). *Chemical bonding, coordination compounds, hydrogen bonding, molecular crystals, organic compounds, reaction pathways, structural biology, structural chemistry, teaching, X-ray diffraction.*

Giordani, Prof. Marino. Ist. di Chimica, Facoltà di Ingegneria, U. di Genova, Fiera del Mare, Padiglione D, Piazzale J.F. Kennedy, 16129 Genova, Italy. (1944) Dr, chemistry (U. di Genova, 1968). Prof. assoc. (tel. 010 + 566427). *Crystal growth, materials science, phase transitions.*

Giordano, Prof. Federico. Dip. di Chimica, U. di Napoli, Via Mezzocannone 4, 80134 Napoli, Italy. (1939) Dr, chemistry (U. di Napoli, 1964). Prof. assoc. (tel. 081 + 206800). *Chemical bonding, organic compounds, proteins.*

Giordano Orsini, Prof. Paolo. Dip. di Ingegneria ,U. di Trento, Mesiano, 38050 Trento, Italy. (1926) Dr, industrial chemistry (U. di Napoli, 1949). Prof., chemistry. (fax 0461 + 881999) *Crystal growth, crystallographic data, diffractometry, electron microscopy, materials science, microscopy, phase transition, powder diffraction, surface structure, texture, X-ray diffraction.*

Giunchi, Dr Giovanni. Keramont Italia S.p.A., Via Fauser 4, 28100 Novara, Italy. (1946), Dr, physics (U. di Bologna 1970). Manager R & D. (tel 0321 + 24701, fax 0321 + 447378). *Diffractometry, disorder, inorganic crystals, materials science, polymers, powder diffraction, structural chemistry..*

Giuseppetti, Prof. Giuseppe. Dip. di Scienze della Terra, U. di Pavia, Via A. Bassi 4, 27100 Pavia, Italy. (1923) Dr, chemistry and pharmacy (U. di Camerino, 1947 and 1949). Prof., mineralogy. (tel. 0382 + 392259). *Inorganic crystals, minerals. organic compounds, powder diffraction, X-ray diffraction.*

Gramaccioli, Prof. Carlo Maria. Dip. di Scienze della Terra, U. di Milano, Via Botticelli 23, 20133 Milano, Italy. (1935) Dr, industrial chemistry (U. di Milano, 1959). Prof., physical chemistry (tel. 02 + 293994, telex 320484 UNIMI I, fax 02 + 2364393, email XLAMIN at IMISIAM). *Minerals, molecular crystals, thermal vibrations.*

Graziani, Prof. Rodolfo. Ist. di Chimica Generale, U., Via Loredan 4, 35100 Padova, Italy. (1937) Dr, chemistry (U. di Padova, 1964). Prof. assoc. (tel. 049

+ 831359). *Chemical bonding, coordination compounds, crystallographic data, diffractometry, teaching, X-ray diffraction.*

Grepioni, Dr Fabrizia. Dip. di Chimica 'G. Ciamician', U. di Bologna, Via Selmi 2, 40132 Bologna, Italy. (1960) Dr, chemistry (U. di Bologna, 1985). Borsista. (tel. 0521 + 259555, fax 051 + 259456, email ICINECA at VN3BOD41). *Coordination compounds, thermal vibrations.*

Guastini, Prof. Carlo. Ist. di Strutturistica Chimica, U. di Parma, Viale delle Scienze, 43100 Parma, Italy. (1931) Dr, chemistry (U. di Parma, 1962). Prof. assoc. (tel. 0521 + 580548 or 580449). *Coordination compounds, structural chemistry, X-ray diffraction.*

Guerreschi, Dr Luigi Giuseppe. Dip. di Chimica, Città Universitaria, 00185 Roma, Italy. (1921) Dr, chemistry (U. di Roma, 1949). Ricercatore, Prof. inc. stab. (tel. 06 + 4991, ext. 297).

Guzman, Dr Luis Alberto. Ist. per la Ricerca Scientifica e Tecn. (I.R.S.T.), 38050 Povo, Trento, Italy. (1947) Dr, Physics (U. Geneva, Switzerland, 1972). Sr. scient. (tel. 0461 + 810105, fax 0461 + 810851). *Materials science, metals, phase determination, surface chemistry. X-ray diffraction.*

Iandelli, Prof. Aldo. Ist. di Chimica Fisica, U. di Genova, Corso Europa 26 (Palazzo delle Scienze), 16132 Genova, Italy. (1912) Dr, chemistry (U. di Firenze, 1933). Prof., physical chemistry, inst. dir. (tel. 010 + 3538220). *Metals, structural chemistry.*

Ianelli, Dr Sandra. Ist. di Chimica Generale, U. di Parma, Viale delle Scienze, 43100 Parma, Italy. (1951) Dr, Chemistry (U. di Parma, 1976). Ricercatore. (tel. 0521 + 580425, fax 0521 + 580542, email CHIMIC3 at IPRUNIV). *Coordination compounds, organic compounds, structural chemistry, X-ray diffraction.*

Iannelli, Dr Pio. Dip. di Fisica, U. di Palermo, Strada Provinciale Baronissi-Penta, 84081 Baronissi (SA), Italy. (1958) Dr, chemistry (U. di Napoli, 1981). Ricercatore. (tel. 089 + 822245, telex 722582 USADIF, fax 089 + 958291). *Computing, diffractometry, liquid crystals, macromolecules, polymers, structural chemistry, X-ray diffraction.*

Immirzi, Prof. Attilio. Dip. di Fisica, U. di Salerno, Via Baronissi, 84100 Salerno, Italy. (1938) Dr, industrial chemistry (U. di Napoli, 1961). Prof. (tel. 089 + 822, ext 231, telex 722582 USADIF I, fax 089 + 958291) *Polymers, fibers, Rietveld method.*

Imperatori, Dr Patrizia. Ist. Teoria e Struttura Elettronica e Comportamento Spettrochimico dei Composti di Coordinazione, C.N.R., Area della Ricerca di Roma, 00016 Monterotondo Stazione, Roma, Italy. (1957) Dr, chemistry (U. di Roma, 1981). Ricercatore. (tel. 06 + 90020346, telex 624809 CNR MLI, fax 06 + 9005849). *Coordination compounds, inorganic crystals. materials science, structural chemistry.*

Ioppolo, Prof. Salvatore. Ist. di Scienze della Terra, Facoltà di Scienze, Salita Papardo - Casella postale 54, 98166 Sant'Agata (Messina), Italy. (1939) Dr, chemistry (U. di Messina, 1965). Prof. assoc. (tel. 090 + 392333). *Diffractometry, electron microscopy, powder diffraction, X-ray diffraction.*

Isetti, Prof. Giovanni. Dip. di Scienze della Terra, U. di Genova, Corso Europa, Palazzo delle Scienze, 16132 Genova, Italy. (1925) Dr, chemistry (U. di Genova, 1949). Prof., mineralogy. (tel. 010 + 3538300 or 311). *Crystal physics, defect structures.*

Ivaldi, Dr Gabriella. Dip. di Scienze della Terra, U. di Torino, Via S. Massimo 24, 10123 Torino, Italy. (1950) Dr, natural sci. (U. di Torino, 1974). Prof. Assoc. (tel. 011 + 832193). *Minerals, neutron diffraction, X-ray diffraction.*

Kovács, Prof. Alessandro L. Dip. di Medicina Sperimentale, U. di Roma, Città Universitaria, 00100 Roma, Italy. (1936) Dr, physics (U. di Bologna, 1961). Prof., physics. (tel. 06 + 493971). *Computing, symmetry, molecular dynamics, symmetry breaking.*

Krajewski, Dr Adriano. Dip. di Ceramiche Classiche, Ist. di Ricerche Tecnologiche per la Ceramica, C.N.R., Via Granarolo 64, 48018 Faenza (Ravenna), Italy. (1947) Dr, chemistry (U. di Bologna, 1971). (tel. 0546 + 46147, ext. 87, telex 583462 IRTEC I, fax 0546 + 46381). *Structure - activity correlation in solid state, numerical analysis, crystal growth, sintering, bioceramics, refractories, electronic ceramics, ancient ceramics.*

Lagomarsino, Dr Stefano. Ist. di Elettronica dello Stato Solido, C.N.R., Via Cineto Romano 42, 00156 Roma, Italy. (1948) Dr, physics (U. di Roma). Ricercatore. (tel. 06 + 4124345). *X-ray diffraction, materials characterization.*

Lamba, Dr Doriano. Ist. di Strutturistica Chimica 'G. Giacomello', C.N.R., C. P. 10, 00016 Monterotondo Stazione, Roma, Italy. (1955) Dr, chemistry (U. di Roma, 1980). Ricercatore. (tel 06 + 9005142, telex 624809 CNR MLI, fax 06 + 9005849, email STRUTA at IRMIAS)). *Organic compounds, structural chemistry, structure-activity relationships, carbohydrates.*

Lanfranchi, Dr Maurizio. Ist. di Chimica Generale ed Inorganica, U. di Parma, Viale delle Scienze 78, 43100 Parma, Italy. (1942) Dr, chemistry (U. di Parma, 1985). Tecn. laur. (tel. 0521 + 580424, email CHIMIC3 at IPRUNIV). *Coordination compounds, molecular crystals, organic compounds, structural chemistry.*

Leccabue, Dr Fabrizio. MASPEC, C.N.R., Via Chiavari 18/A, 43100 Parma, Italy. (1947) Dr, chemistry (U. di Parma, 1972). Ricercatore. (tel. 0521 + 96841, fax 0521 + 96315). *Crystal growth, inorganic crystals, magnetism, materials science, semiconductors.*

Leoni, Prof. Leonardo. Dip. di Scienze della Terra, U. di Pisa, Via S. Maria 53, 56100 Pisa, Italy. (1944) Dr, geological sci. (U. di Pisa, 1967). Prof. (tel. 050 + 501457). *Microscopy. minerals, powder diffraction.*

Leporati, Prof. Enrico. Ist. di Chimica Generale ed Inorganica, U. di Parma, Viale delle Scienze, 43100 Parma, Italy. (1934) Dr, chemistry (U. di Bologna, 1964). Prof. assoc. (tel. 0521 + 580426, ext. 94064, fax 0521 + 580542). *Structural chemistry, inorganic crystals, organic compounds, organometallic compounds.*

Licci, Dr Francesca. MASPEC - C.N.R., Via Chiavari 18/A, 43100 Parma, Italy. (1945) Dr, chemistry (U. di Bologna, 1969). Ricercatore. (tel. 0521 + 96841, telex 531639 MASPEC, fax 0521 + 96315) *High TC superconductors.*

Licheri, Prof. Giovanni. Dip. di Scienze Chimiche, U. di Cagliari, Via Ospedale 72, 09100 Cagliari, Italy. (1939) Dr, physics (U. di Cagliari, 1962). Prof. (tel. 070 + 668047). *Disorder, materials science, powder diffraction, X-ray diffraction.*

Liquori, Prof. Alfonso Maria. ICTB, Palazzo Loredan, Campo Santo Stefano 2945, 30124 Venezia, Italy. (1926) Dr, chemistry (U. di Roma, 1948). Prof., president. (tel. 041 + 5203121, fax 041 + 5209457). *Molecular conformation, macromolecules, conformational transition, tertiary structure, proteins, crystal packing.*

Locchi, Prof. Stelio Giovanni. Dip. di Chimica Generale, U. di Pavia, Viale Taramelli 12, 27100 Pavia, Italy. (1929) Dr, chemistry (U. di Pavia, 1952). Prof. assoc. (tel. 0382 + 29270). *Organic compounds.*

Loreto, Prof. Lucio. Dip. di Scienze della Terra, U. di Roma 'la Sapienza', Piazzale delle Scienze, Città Universitaria, 00185 Roma, Italy. (1937) Dr, geological sci. (U. di Roma, 1965). Prof. assoc. (tel. 06 + 49914686, ext. 490329). *Computing, crystal growth, minerals, symmetry, teaching.*

Lucchetti, Prof. Gabriella. Dip. di Scienze della Terra, Corso Europa 26, 16132 Genova, Italy. (1947) Dr, natural sci. (U. di Genova, 19760). Prof. Assoc. (tel. 010 + 3538304, telex 352169 DISTER GE). *Crystallographic data, electron microscopy, minerals, powder diffraction.*

Magini, Dr Mauro. ENEA - Progett, materiali Amorfi, settore chimico, Via Anguillarese, 00060 CRE-Casaccia, Roma, Italy. (1941) Dr, chemistry (U. di Roma, 1965). Res. (tel. 06 + 30483391, fax 06 + 30483327). *Solutions, liquids, amorphous structures.*

Malpezzi, Dr Luciana. Dip. di Chimica, Politecnico di Milano, Piazza L. da Vinci 32, 20133 Milano, Italy. (1946) Dr, Chemistry (U. di Bologna, 1970). Ricercatore. (tel. 02 + 23993000, ext. 3022). *Chemical bonding, hydrogen bonding, macromolecules, molecular crystals, organic compounds, polymers, structural chemistry, teaching, X-ray diffraction.*

Malta, Dr Viscardo. Centro di Studio per la Fisica dell, macromoleculecole, C.N.R., Via Selmi 2, 40126 Bologna, Italy. (1938) Dr, chemistry (U. di Bologna, 1968). Ricercatore. (tel. 051 + 259552, fax 051 + 259456, email VS3BODI3 at ICINECA). *Macromolecules, polymers, powder diffraction, X-ray diffraction.*

Mammi, Prof. Mario. Dip. di Chimica Organica, U. di Padova, Via Marzolo 1, 35131 Padova, Italy. (1932) Dr, chemistry (U. di Padova, 1956). Prof., structural chemistry, dir., Biopolymer Res. Centre of C.N.R. (tel. 049 + 663736 and 831-234 or 232 or 111, telex 430176 UNPADU I, fax 049 + 831222, email CHIM09 at UNIPAD.INFN.IT or CHIM09 at IPDUNIVX.BITNET). *Macromolecules, proteins, structural biology, structural chemistry, teaching.*

Manassero, Prof. Mario. Ist. di Chimica Strutturistica Inorganica, U. degli Studi di Milano, Via G. Venezian 21, 20133 Milano, Italy. (1944) Dr, chemistry (U. di Milano, 1968). Prof. assoc. (tel. 02 + 235120). *Chemical bonding, coordination compounds, structural chemistry.*

Mancini, Dr Annamaria. Dip. di Scienza de, materiali, U. di Lecce, Via Arnesano, 73100 Lecce, Italy. (1944) Dr, chemistry (U. di Bari, 1970). Prof. assoc. (tel. 0832 + 627247). *Crystal growth, structural and physical characterization.*

Mangani, Dr Stefano. Dip. di Chimica, U. di Siena, Pian dei Mantellini 44, 53100 Siena, Italy. (1951) Dr, chemistry (U. di Firenze, 1977). Ricercatore. (tel. 0577 + 47054, email CRYSTL at IFICHIM). *Coordination compounds, proteins, EXAFS.*

Mangia, Prof. Alessandro. Ist. di Chimica Generale ed Inorganica, U. di Parma, Via D'Azeglio 85, 43100 Parma, Italy. (1941) Dr, chemistry (U. di Parma, 1965). Prof. assoc. (tel. 0521 + 34400). *Coordination compounds.*

Manotti-Lanfredi, Prof. Anna Maria. Ist. di Chimica Generale ed Inorganica, U. di Parma, Viale delle Scienze, 43100 Parma, Italy. (1933) Dr, chemistry (U. di Parma, 1957). Prof. (tel. 0521 + 580417). *Coordination compounds, inorganic crystals.*

Mantovani, Prof. Giorgio. Dip. di Chimica, U. di Ferrara, Via Luigi Borsari 46, 44100 Ferrara, Italy. (1925) Dr, chemistry (U. di Ferrara, 1948). Prof., industrial chemistry. (tel. 0532 + 36374). *Crystal growth.*

Marchetti, Dr Fabio. Dip. di Chimica e Chimica Industriale, U. di Pisa, Via Risorgimento 35, 56126 Pisa, Italy. (1950) Dr, chemistry (U. di Pisa, 1974). Ricercatore. (tel. 050 + 587220, email MINER at ICNUCEVM). *Chemical bonding, coordination compounds, molecular crystals, structural chemistry.*

Marega, Dr Carla. Dip. di Chimica Inorganica, Organometallica ed Analitica, U. di Padova, Via Loredan 4, 35131 Padova, Italy. (1961) Dr, physics (U. di Padova, 1985) Borsista. (tel. 049 + 831210, email CHIMIC05 at UNIPAD.INFN.IT). *Macromolecular physics, macromolecules, polyolefins polyammides, fluorinated polymers.*

Marigo, Prof. Antonio. Dip. di Chimica Inorganica, Organometallica ed Analitica, U. di Padova, Via Loredan 4, 35100 Padova, Italy. (1950) Dr, chemistry (U. di Padova, 1974) Prof. assoc. (tel. 049 + 831210, email CHIMICA05 at UNIPAD.INFN.IT). *Computing, disorder, inorganic crystals, macromolecules, phase transitions, powder diffraction.*

Marongiu, Prof. Giaime. Dip. di Scienze Chimiche, U. di Cagliari, Via Ospedale 72, 09100 Cagliari, Italy. (1939) Dr, chemistry (U. di Cagliari, 1963). Prof., inorganic chemistry. (tel. 070 + 668047). *Materials sciences, powder diffraction.*

Martelli, Dr Stefano. TIB-IMA-MAT-MIC, ENEA - Casaccia, Via Anguillarese Km 1+300 C.P. 2400, 00100 Roma, Italy. (1955) Dr, physics and chemistry (U. di Bologna and Freie Universitat, Berlin). (tel. 06 + 3048, ext. 4728, telex 613296 ENEACA I, fax 06 + 38484729). *Interfacial reaction, residual stress.*

Martinelli, Prof. Giuliano. Dip. di Fisica, U. di Ferrara, Via Paradiso 12, 44100 Ferrara, Italy. (1938) Dr, physics (U. di Ferrara, 1969). Prof. assoc. (tel. 0532 + 760023, fax 0532 + 762057). *Crystal growth, electron microscopy, semiconductors.*

Martorana, Dr Antonino. Dip. di Chimica Inorganica, U. di Palermo, Via Archirafi 26-28, 90123 Palermo, Italy. (1952) Dr, physics (U. di Padova, 1979). Prof. Assoc. (tel. 091 + 6161502, email CRIC1 at IPACUC.BITNET). *Computing, disorder, macromolecules, phase determination, powder diffraction.*

Masciocchi, Dr Norberto. Ist. di Chimica Strutturistica Inorganica, U. di Milano, Via Venezian 21, 20133 Milano, Italy. (1959) Dr, chemistry. (tel. 02 + 235120, email STINCH at IMISIAM.EARNET). *Chemical bonding, computing, coordination compounds, powder diffraction, structural chemistry, X-ray diffraction.*

Masi, Mr Dante. Ist. poer lo Studio della Stereochimica ed Energetica, dei Composti di Coordinazione, C.N.R., Via J. Nardi 39, 50132 Firenze, Italy. (1948) Diploma. Technical assistant. (tel. 055 + 243990, email ISSEC at IFIIDG). *Coordination compounds.*

Massarotti, Prof. Vincenzo. Dip. di Chimica Fisica, U. di Pavia, Viale Taramelli 16, 27100 Pavia, Italy. (1944) Dr, chemistry (U. di Pavia, 1968). Prof. assoc. (tel. 0382 + 27082). *Crystal growth, defect structures, diffractometry, disorder, materials science, phase transitions, powder diffraction.*

Mattia, Prof. Carlo. Dip. di Chimica, U. di Napoli, Via Mezzocannone 4, 80134 Napoli, Italy. (1946) Dr, chemistry (U. di Napoli, 1970). Prof. assoc. (tel. 081 + 206800). *Proteins, structural biology.*

Mattias, Prof. Pierpaolo. Ist. di Geologia appl. e Giacimenti Minerari, Fac. di Ingegneria, Via Eudossiana 18, 00184 Roma, Italy. (1936) Dr, geological sci. (U. di Roma, 1960). Prof. assoc. (tel. 06 + 461810 or 465913). *Minerals.*

Mazza, Dr Fernando. Ist. di Strutturistica Chimica 'G. Giacomello', C.N.R., C.P. 10 - 00016 Monterotondo Stazione, Roma, Italy. (1938) Dr, chemistry (U. di Roma, 1964). Prof. assoc. (tel. 06 + 9005641). *Structural chemistry.*

Mazzarella, Prof. Lelio. Dip. di Chimica, U. di Napoli, Via Mezzocannone 4, 80134 Napoli, Italy. (1938) Dr, chemistry (U. di Napoli, 1961). Prof. (tel. 081 + 206800). *Organic compounds, proteins, X-ray diffraction.*

Mazzi, Prof. Fiorenzo. Dip. di Scienze della Terra, U. di Pavia, Via A. Bassi 4, 27100 Pavia, Italy. (1924) Dr, chemistry (U. di Firenze, 1947). Prof., mineralogy. (tel. 0382 + 392254, fax 0382 + 392252, email CSCSPV at IPVIAN.EARN). *Crystallographic data, minerals, symmetry.*

Mealli, Prof. Carlo. Ist. per lo Studio della Stereochimica ed Energetica dei Composti di Coordinazione, ISSECC - C.N.R., Via J. Nardi 39, 50132 Firenze, Italy. (1946) Dr, chemistry (U. di Firenze, 1969). Ricercatore. (tel. 055 + 243990, email ISSECC at IFIIDG). *Chemical bonding, coordination compounds, structure-activity relationships.*

Mellini, Prof. Marcello. Dip. di Scienze della Terra, Piazza Università, 06100 Perugia, Italy. (1949) Dr, chemistry (U. di Pisa, 1974). Prof., mineralogy. (tel. 075 + 4693214, telex 662078 UNIPG I, fax 075 + 4692067, email CRYSTAL at IPGUNIV). *Electron microscopy, high pressure, minerals, X-ray diffraction.*

Menchetti, Prof. Silvio. Dip. di Scienze della Terra, U. di Firenze, Via La Pira 4, 50121 Firenze, Italy. (1937) Dr, geological sci. (U. di Firenze, 1961). Prof. of mineralogy (tel. 055 + 287140, fax 055 + 2757430)). *Crystallographic data, diffractometry, inorganic crystals, minerals, X-ray diffraction.*

Menzinger, Prof. Filippo. Dip. di Fisica, U. di Roma II, via O. Raimondo, 00173 Roma, Italy. (1937) Dr, physics (U. di Roma, 1961). Prof. (tel. 06 + 24990441, telex 06 + 626382 FIUNTV, fax 06 + 6133074). *Neutron diffraction, spin density, thermal vibrations.*

Meriani, Dr Sergio. Ist. di Chimica Applicata, U. di Trieste, Via Valerio 2, 34127 Trieste, Italy. (1941) Dr, inorganic chemistry (U. di Trieste, 1966). Prof. (tel. 040 + 5603705, fax 040 + 5603176, email TK7T01 at CHIMMAP.BITNET). *Materials scince, phase determination, phase transitions.*

Merlini, Dr Alfonso Enrico. Physics Div., C.C.R. Euratom (Joint Res. Center of the European Community), 21020 Ispra, Varese, Italy. (1926) PhD, metallurgical engineering (Illinois U., USA, 1954). Head, physics div. (tel. 0332 + 789809, telex 380042,380058 EUR I). *Dynamical diffraction theory, EXAFS, Synchrotron light applications.*

Merlino, Prof. Stefano. Dip. di Scienze della Terra, U. di Pisa, Via S. Maria 53, 56100 Pisa, Italy. (1938) Dr, chemistry (U. di Pisa, 1962). Prof., crystallography. (tel. 050 + 501457, email MINER at ICNUCEVM). *Disorder, electron diffraction, electron microscopy, minerals.*

Merlo, Prof. Franco. Ist. di Chimica Fisica, U. di Genova, Corso Europa 26 (Palazzo delle Scienze), 16132 Genova, Italy. (1940) Dr, chemistry (U. di Genova, 1963). Prof. assoc. (tel. 010 + 3538225). *Materials science, metals, powder diffraction.*

Millini, Dr Roberto. ENIRICERCHE, Via F. Maritano 26, 20097 San Donato Milanese, Milano, Italy. (1959) Dr, chemistry (U. di Pavia, 1983). Ricercatore. (tel. 02 + 5207543) *Coordination compounds, diffractometry, disorder, electron microscopy, inorganic crystals, organic compounds, powder diffraction, small-angle scattering, surface structure, texture, X-ray diffraction.*

Molin, Prof. Gianmario. Dip. di Mineralogia e Petrologia, U. di Padova, Corso Garibaldi 37, 35100 Padova, Italy. (1948) Dr, geological sci. (U. di Padova, 1975). Prof. assoc. (tel. 049 + 663122, fax 049 + 831711) *Crystallographic data, diffractometry, minerals, teaching, X-ray diffraction.*

Monaco, Dr Hugo Luis. Dip. di Genetica e Microbiologia, Sez. di Cristallografia, U. di Pavia, Via Taramelli 16, 27100 Pavia, Italy. (1947) Ph.D., chemistry (Harvard U., USA, 1978). Prof. assoc. (tel. 0382 + 392528, email MONACO at IPV85.INFNET). *Macromolecules, proteins, structural biology.*

Monari, Dr Magda. Dip. di Chimica 'G. Ciamician', U. di Bologna, Via Selmi 2, 40126 Bologna, Italy. (1955) Dr, industrial chemistry (U. di Bologna, 1982). (tel. 051 + 259450, fax 051 + 259456). *Coordination compounds, crystallographic data, inorganic crystals, structural chemistry.*

Mongiorgi, Prof. Romano. Ist. di Mineralogia, U. di Bologna, Piazza di Porta S. Donato 1, 40127 Bologna, Italy. (1940) Dr, geological sci. (U. di Bologna, 1969). Prof. assoc. (tel. 051 + 243556). *Crystal growth, electron microscopy, inorganic crystals, biominerals.*

Montenero, Prof. Angelo. Ist. di Strutturistica Chimica, U. di Parma, Viale delle Scienze, 43100 Parma, Italy. (1944) Dr, chemistry (U. di Parma, 1968). Prof. assoc. (tel. 0521 + 580553, fax 0521 + 580542). *Crystal growth, electron microscopy, materials science, phase determination, powder diffraction, X-ray diffraction.*

Morandi, Prof. Noris. Dip. di Scienze Mineralogiche, U. di Bologna, Piazza San Donato 1, 40127 Bologna, Italy. (1938), Dr, geology (U. di Bologna, 1961). Prof. assoc. (tel. 051 + 243556). *Minerals, diffractometry, phase determination, powder diffraction, teaching.*

Moret, Dr Massimo. Ist. di Chimica Strutturistica Inorganica, U. di Milano, Via Venezian 21, 20123 Milano, Italy. (1961) Dr, chemistry. (tel. 02 + 235120, email STINCH at IMISIAM.EARNET). *Chemical bonding, computing, coordination compounds, reaction pathways, structural chemistry, X-ray diffraction.*

Motta, Dr Nunzio. Dip. di Fisica, II U. di Roma, Via E. Carnevale, 00173 Roma, Italy. (1957), Dr, physics (U. di Roma, 1981). Ricercatore. (tel. 06 + 24990, ext. 438, fax 06 + 6133074, email MOTTA at VAXTOV.INFNET). *Semiconductors, metals, superconductors, thermodynamic, EXAFS, XPS.*

Mottana, Prof. Annibale. Dip. di Scienze della Terra, U. di Roma 'La Sapienza', Piazzale A. Moro 5, 00185 Roma, Italy. (1940) Dr, mineralogy. Prof., mineralogy. (tel. 06 + 490844, fax 06 + 4952824 or 4746067). *Crystal growth, diffractometry, high pressure, materials science, minerals, phase transitions, powder diffraction, X-ray diffraction.*

Moze, Dr Oscar. MASPEC, C.N.R., Via Chiavari 18/A, 43100 Parma, Italy. (1954) Phd, physics (Monash University). Ricercatore. (tel. 0521 + 96841, fax 0521 + 93615). *Diffuse scattering, disorder, magnetism, materials science, metals, neutron diffraction.*

Mugnoli, Prof. Angelo. Ist. di Chimica Fisica, U. di Genova, Corso Europa 26, 16132 Genova, Italy. (1933) Dr, industrial chemistry (U. di Milano, 1958). Prof., structural chemistry (tel. 010 + 3538235, fax 010 + 515076, email CHIFISI at IGECUNIV). *Computing, molecular crystals, organic compounds, reaction pathways, structural chemistry, teaching.*

Mura, Dr Pasquale. Ist. di Strutturistica Chimica 'G. Giacomello', C.N.R., C.P. 10 - 00016 Monterotondo Stazione, Roma, Italy. (1944) Dr, chemistry (U. di Roma, 1975). Ricercatore. (tel. 06 + 9005142, fax 06 + 9005849). *Coordination compounds.*

Musatti, Prof. Amos. Ist. di Strutturistica Chimica, U. di Parma, Viale delle Scienze, 43100 Parma, Italy. (1935) Dr, chemistry (U. di Parma, 1962). Prof. assoc. (tel. 0521 + 580548 or 580450, fax 0521 + 580542, email STRUCHI at UNIVPR.EARNET). *Computing, coordination compounds, organic compounds.*

Napolitano, Prof. Roberto. Ist. di Chimica, U. della Basilicata, Via N. Sauro 85, 85100 Potenza, Italy. (1949) Dr, chemistry (U. di Napoli, 1974). Prof. assoc. (tel. 0971 + 334246). *Computing, crystallographic data, macromolecules, powder diffraction, proteins, X-ray diffraction.*

Nardelli, Prof. Mario. Ist. di Chimica Generale ed Inorganica, U. di Parma, Viale delle Scienze 76, 43100 Parma, Italy. (1922) Dr, chemistry (U. di Parma, 1946). Prof., general chemistry. (tel. 0521 + 580433, telex 530327 UNIV PR I, fax 0521 + 580542, email NARDELLI at IPRUNIV). *Chemical bonding, computing, coordination compounds, crystallographic data, hydrogen bonding, molecular crystals, organic compounds, structural chemistry.*

Nardin, Prof. Giorgio. Dip. di Scienze Chimiche, U. di Trieste, Piazzale Europa 1, 34127 Trieste, Italy. (1940) Dr, chemistry (U. di Trieste, 1964). Prof. assoc. (tel. 040 + 5064111, fax 040 + 5603176). *Coordination compounds, minerals, ceramic superconductors.*

Navarra, Dr Gabriele. Dip. di Scienze Chimiche, U. di Cagliari, Via Ospedale 72, 09124 Cagliari, Italy. (1954) Dr, chemistry (U. di Cagliari, 1980). Ricercatore. (tel. 070 + 668047). *Computing, materials science, powder diffraction, X-ray diffraction.*

Nicolo, Dr Francesco. Dip. di Chimica Organica, U. di Padova, Via Marzolo 1, 35131 Padova, Italy. (1959) Dr, chemistry (U. di Messina, 1983). Borsista. (tel. 049 + 831293, fax 049 + 831222, email CHIM09 at UNIPAD.INFN.IT). *Coordination compounds, computing, macromolecules.*

Nunzi, Prof. Antonio. Dip. di Scienze della Terra, Sez. di Cristallografia, U. di Perugia, Piazza Università, 06100 Perugia, Italy. (1943) Dr, chemistry (U. di Perugia, 1966). Prof. assoc. (tel. 075 + 4693215, telex 075 + 662078 UNIPG I, fax 075 + 4692067, email PG2 at ICNUCEVM). *Computing, direct methods.*

Oberti, Dr Roberta. Centro di Studio per la Cristallografia Strutturale, C.N.R., Via Bassi 4, 27100 Pavia, Italy. (1952) Dr, chemistry (U. di Pavia, 1976). Ricercatore. (tel. 0382 + 392267, fax 0382 + 392252, email CSCSPV at IPVIAN.EARN). *Crystal-chemistry, minerals. organic compounds, X-ray diffraction.*

Olcese, Prof. Giorgio L. Ist. di Chimica Fisica, U. di Genova, Corso Europa (Palazzo delle Scienze), 16132 Genova, Italy. (1933) Dr, chemistry (U. di Genova, 1957). Prof., physical chemistry. (tel. 010 + 3538221). *Crystal growth, crystallographic data, high pressure, inorganic crystals, magnetism, materials science, metals.*

Orioli, Prof. Pierluigi. Dip. di Chimica, U. di Firenze, Via Capponi 7, 50132 Firenze, Italy. (1933) Dr, chemistry (U. di Firenze, 1956). Prof., structural chemistry. (tel. 055 + 2476949). *Coordination compounds, structural biology, X-ray diffraction.*

Palenzona, Prof. Andrea. Ist. di Chimica Fisica, U. di Genova, Corso Europa (Palazzo delle Scienze), 16132 Genova, Italy. (1935) Dr, chemistry (U. di Genova, 1961). Prof., inorganic chemistry. (tel. 010 + 3538221 or 3538223). *Crystallographic data, materials science, metals, minerals, phase determination, powder diffraction, X-ray diffraction.*

Pani, Dr Marcella. Ist. di Chimica Fisica, U. di Genova, Corso Europa 26, 16132 Genova, Italy. (1959) Dr, industrial chemistry (U. di Genova, 1984). Funzionario tecnico. (tel. 010 + 3538225). *Crystallographic data, inorganic crystals, structural biology, structural chemistry, X-ray diffraction.*

Paorici, Prof. Carlo. Dip. di Fisica, U. di Parma, Viale delle Scienze, 43100 Parma, Italy. (1936) Dr, chemistry (U. di Roma, 1962). Prof. assoc. (tel. 0521 + 580271, telex 531639 MASPEC I, fax 0521 + 96315). *Crystal growth, defect structures, semiconductors, teaching.*

Paris, Dr Eleonora. Dip. di Scienze della Terra, U. di Camerino, Via Betti 1, 62032 Camerino, Italy. (1958) Dr, geological sci. (U. di Pisa, 1981). Ricercatore. *High pressure, minerals, X-ray diffraction, synchrotron radiation, EXAFS.*

Pasero, Dr Marco. Dip. di Scienze della Terra, U. di Pisa, Via S. Maria 53, 56126 Pisa, Italy. (1958) PhD, mineralogy (U. di Pisa, 1988). (tel. 050 + 501457). *Diffractometry, disorder, electron diffraction, electron microscopy, minerals.*

Passaglia, Prof. Elio. Ist. di Mineralogia, U. di Ferrara, Corso Ercole I D'Este 32, 44100 Ferrara, Italy. (1941) Dr, geological sci. (U. di Modena, 1965). Prof. of mineralogy (tel. 0532 + 32987). *Minerals, powder diffraction, X-ray diffraction.*

Pasti, Mr Fabio. Ital Structures, Area Industriale, 38066 Riva del Garda, Trento, Italy. (1937) Diploma, electrotechnique (Higher Sch., 1957). Maker of scientific instruments. (tel. 0464 + 553426, fax 0464 + 555270). *Diffractometry, instrumentation, powder diffraction, texture, X-ray diffraction.*

Pavel, Dr Nicolae Viorel. Dip. di Chimica, U. di Roma 'La Sapienza', Piazzale A. Moro 5, 00185 Roma, Italy. (1949) Dr, chemistry (U. fo Roma, 1973). Prof. Assoc. (tel. 06 + 49913652, fax 06 + 4958251, email PAVEL at VAXRMA.INFNET or PAVEL at IRMLNF.BITNET). *Macromolecules, polymers, small-angle scattering, micelles, inclusion compounds.*

Pedone, Prof. Carlo. Ist. Chimico, U. di Napoli, Via Mezzocannone 4, 80134 Napoli, Italy. (1938) Dr, chemistry (U. di Napoli, 1961). Prof. (tel. 081 + 205730). *Macromolecules, structural biology, structural chemistry, bio-molecules.*

Pelizzi, Prof. Corrado. Ist. di Chimica Biologica, U. di Sassari, Via Muroni 23/A, Sassari, Italy. (1942) Dr, chemistry (U. di Parma, 1966). Prof., general chemistry. (tel. 079 + 238582). *Coordination compounds, structural chemistry.*

Pelizzi, Prof. Giancarlo. Ist. di Chimica Generale ed Inorganica, U. di Parma, Viaale delle Scienze, 43100 Parma, Italy. (1939) Dr, chemistry (U. di Parma, 1964). Prof., general and inorganic chemistry. (tel. 0521 + 580421). *Coordination compounds, organic compounds, structural chemistry, X-ray diffraction.*

Pellinghelli, Prof. Maria Angela. Ist. di Chimica Generale ed Inorganica, U. di Parma, Viale delle Scienze 78, 43100 Parma, Italy. (1943) Dr, chemistry (U. di Parma, 1966). Prof. assoc. (tel. 0521 + 580428, email CHIMIC3 at IPRUNIV). *Coordination compounds, molecular crystals, organic compounds, structural chemistry.*

Penco, Prof. Anna Maria. Dip. di Scienze della Terra, U. di Genova, Corso Europa 26, 16132 Genova, Italy. (1927) Dr, natural sci. (U. di Genova, 1953). Prof., mineralogy. (tel. 010 + 3538311, telex 352169 DISTER GE). *Optical crystallography, structural chemistry.*

Perego, Dr Giovanni. ENIRICERCHE, Via F. Maritano 26, 20097 San Donato Milanese, Milano, Italy. (1938) Doctorat d'Université, sciences (U. de Strasbourg, France, 1971). Manager. (tel. 02 + 5207543). *Structural chemistry, organometallic compounds, inorganic crystals, polymers.*

Petraccone, Prof. Vittorio. Dip. di Chimica, U. di Napoli, Via Mezzocannone 4, 80134 Napoli, Italy. (1943) Dr, chemistry (U. di Napoli, 1966). Prof. assoc. (tel. 081 + 205730 or 206450). *Computing, crystallographic data, defect structures, diffuse scattering, disorder, liquid crystals, macromolecules, materials science, phase transitions, polymers, small-angle scattering, texture, X-ray diffraction.*

Peyronel, Prof. Giorgio. Dip. di Chimica, Sez. Inorganica, U. di Modena, Via G. Campi 183, 41100 Modena, Italy. (1913) Dr, chemistry (U. di Milano, 1936). Prof., general and inorganic chemistry. (tel. 059 + 378415). *Chemical bonding,*

coordination compounds, crystallographic data, diffractometry, disorder, hydrogen bonding, inorganic crystals, instrumentation, magnetism, metals, organic compounds, powder diffraction, structural chemistry, teaching, X-ray diffraction, coordination compounds.

Piazzesi, Prof. AnnaMaria. Dip. di Chimica Organica, U. di Padova, Via Marzolo 1, 35100 Padova, Italy. (1927) Dr, chemistry (U. di Padova, 1963). Prof. assoc. (tel. 049 + 831311). *Organic compounds, structural chemistry.*

Pifferi, Dr Augusto. Ist. di Strutturistica Chimica 'G. Giacomello', C.N.R., 00016 Monterotondo Stazione, Roma, Italy. (1955) Dr, physics (U. di Roma, 1979). Ricercatore. (tel. 06 + 9005142, fax 06 + 9005849). *Synchrotron radiation, methodology, powder diffraction.*

Pignedoli, Prof. Anna. Dip. di Chimica Generale, sez. Inorganica, U. di Modena, Via G. Campi 183, 41100 Modena, Italy. (1924) Dr, chemistry (U. di Modena, 1947). Prof. assoc. (tel. 059 + 362243). *Structural chemistry, inorganic crystals, coordination compounds.*

Pilati, Dr Tullio. Centro di Studio per le relazioni tra Struttura e reattività Chimica, C.N.R., via Golgi 19, 20133 Milano, Italy. (1946) Dr, chemistry (U. di Milano, 1971). Ricercatore. (tel. 02 + 2663805, fax 02 + 2362946, email MI0354 at IMISIAM). *Computing, diffractometry, electron density, minerals, molecular crystals, organic compounds, X-ray diffraction.*

Pirozzi, Prof. Beniamino. Dip. di Chimica, U. di Napoli, Via Mezzocannone 4, 80134 Napoli, Italy. (1947) Dr, chemistry (U. di Napoli, 1970). Prof. assoc. (tel. 081 + 206450). *Computing, crystallographic data, liquid crystals, polymers, powder diffraction, symmetry, X-ray diffraction.*

Pisani, Prof. Cesare. Dip. di Ch. Inorganica, Ch. Fisica e Ch. de, materiali, U. di Torino, Via P. Giuria 5, 10125 Torino, Italy. (1938) Dr, physics (U. di Milano, 1963). Prof. (tel. 011 + 657274, email U107 at ITOCSIP). *Solid state theoretical chemistry.*

Pochetti, Dr Giorgio. Ist. di Strutturistica Chimica 'G. Giacomello', C.N.R., C.P. 10 - 00016 Monterotondo Stazione, Roma, Italy. (1957) Dr, chemistry (U. di Roma, 1981). Ricercatore. (tel. 06 + 9005142, ext. 633, fax 06 + 9005849, email STRUTA at IRMIAS). *Computing, organic compounds, peptides.*

Polidori, Dr Giampiero. Dip. di Scienze della Terra, Sez. Cristallografia, U. di Perugia, Piazza Università, 06100 Perugia, Italy. (1950) Dr, mathematics (U. di Perugia, 1973). Ricercatore. (tel. 075 + 4693213, telex 662078 UNIPG I, fax 075 + 4692067, email SIRI at IPGUNIV). *Computing, powder diffraction, direct methods.*

Polo, Dr Adriano. Ist. di Chimica e Tecn. dei radioelementi, C.N.R., Corso Stati Uniti 4, 35100 Padova, Italy. (1956), Dr, chemistry (U. di Padova, 1982). Ricercatore. (tel. 049 + 845111 or 845368, telex 430302 CNR I, fax 049 + 845449, email ADRIANO at PDCNR2.INFNET). *Coordination compounds, structural chemistry, X-ray diffraction.*

Pompa, Dr Francesco. Divisione Scienza de, materiali, ENEA - C.R.E. Casaccia, S. P. Anguillarese - C. P. 2400, 00100 Roma, Italy. (1931) Dr, chemistry (U. di Bari, 1958). Ricercatore (tel. 06 + 3048, ext. 3117, telex ENEACA I 613296, fax 06 + 30484729). *Instrumentation, computing, inorganic crystals, metals.*

Poppi, Prof. Luciano. Ist. di Mineralogia e Petrografia, U. di Bologna, P.zza S. Donato 1, 40127 Bologna, Italy. (1940) Dr, geology (U. di Modena, 1968). Prof. assoc. (tel. 051 + 231961, ext. 57). *Minerals, powder diffraction, crystal chemistry.*

Porta, Prof. Piero. Dip. di Chimica, U. di Roma 'La Sapienza', Piazzale A. Moro 5, 00185 Roma, Italy. (1933) Dr, chemistry (U. di Bari, 1958). Prof. (tel. 06 + 49913378). *Defect structures, inorganic crystals, powder diffraction, surface structure.*

Portalone, Dr Gustavo. Dip. di Chimica, U. di Roma 'La Sapienza', Città Universitaria, P.le A. Moro 5, 00185 Roma, Italy. (1952) Dr, pharmaceutical chemistry and technology (U. di Roma, 1977). Ricercatore. (tel. 06 + 4991, ext. 3715, fax 06 + 4958251, email PORTALONE at VAXRMA). *Chemical bonding, electron diffraction, organic compounds, structural chemistry, thermal vibrations.*

Porzio, Dr William. Ist. di Chimica dell, macromoleculecole, C.N.R., Via E. Bassini 15, 20133 Milano, Italy. (1951) Dr, industrial chemistry (U. di Milano, 1976). Ricercatore. (tel. 02 + 235310, email NMR at IMISIAM). *Materials science, polymers, powder diffraction, structural chemistry, X-ray diffraction.*

Puliti, Dr Raffaella. Ist. per la Chimica di Molecole di Interesse Biologico, C.N.R., Via Toiano 2, 80072 Arco Felice, Napoli, Italy. (1936) Dr, chemistry (U. di Bari, 1961). Primo ric. (tel. 081 + 8601444 or 205730). *Organic compounds, structural biology, X-ray diffraction.*

Quagliata, Prof. Claudio. Dip. di Chimica, U. di Roma, Piazzale A. Moro 5, 00185 Roma, Italy. (1945) Dr, chemistry (U. di Roma, 1970). Prof. assoc. (tel. 06 + 4952993). *Potential energy calculations, inclusion compounds, conformational analysis, micelles.*

Quartieri, Dr Simona. Ist. di Mineralogia e Petrologia, U. di Modena, Via S. Eufemia 19, 41100 Modena, Italy. (1955) Dr, chemistry (U. di Modena, 1979). (tel. 059 + 218062, ext. 16, email VEZZALINI at IMOVX2.INFNET). *Diffractometry, materials science, minerals, X-ray diffraction.*

Randaccio, Prof. Lucio. Dip. di Scienze Chimiche, U. di Trieste, Piazzale Europa 1, 34127 Trieste, Italy. (1940) Dr, chemistry (U. di Napoli, 1963). Prof. (tel. 040 + 5603111, telex 460865 UNIVTS I, fax 040 + 5603176). *Coordination compounds, minerals. structural chemistry, ceramic superconductors.*

Ravaglioli, Dr Antonio. Dip. di Ceramiche Classiche ed Avanzate, Ist. di Ricerche Tecnologiche per la Ceramica, C.N.R., Via Granarolo 64, 48018 Faenza (Ravenna), Italy. (1938) Dr, chemistry (U. di Bologna, 1967). Ricercatore. (tel. 0546 + 46147). *Crystal growth, sintering, bioceramics, glazes, refractories, ceramic raw materials, electronic ceramics, ancient ceramics.*

Riganti, Prof. Vincenzo. Dip. di Chimica Generale, U. di Pavia, Viale Taramelli 12, 27100 Pavia, Italy. (1932) Dr, chemistry (U. di Pavia, 1955). Prof. (tel. 0382 + 392345). *Diffractometry, materials science.*

Rigault de la Longrais, Prof. Germain. Dip. di Scienze della Terra, U. di Torino, Via San Massimo 24, 10123 Torino, Italy. (1930) Dr, chemistry (U. di Torino, 1953). Prof., mineralogy. (tel. 011 + 832193). *Group theory, microscopy, symmetry, teaching.*

Rinaldi, Prof. Romano. Dip. di Scienze della Terra, U. di Cagliari, Via Trentino 51, 09127 Cagliari, Italy. (1944) Dr, geological sci. (U. di Modena, 1969). Prof., mineralogy. (tel. 070 + 2006212 or 4, email MINERALOGIA at IMOVX2.INFNET). *Electron diffraction, electron microscopy, instrumentation, minerals, teaching, X-ray diffraction.*

Rinaudo, Dr Caterina. Dip. di Scienze della Terra, U. di Torino, Via S. Massimo 24, 10123 Torino, Italy. (1951) Dr, natural sci. (U. di Torino, 1974). Ricercatore. (tel. 011 + 832193). *Crystal growth, minerals.*

Ripamonti, Prof. Alberto. Dip. di Chimica 'G. Ciamician', U. di Bologna, Via Selmi 2, 40126 Bologna, Italy. (1930) Dr, chemistry (U. di Roma, 1953). Prof., general chemistry. (tel. 051 + 259551, fax 051 + 259456, email VS3BOOI3 at ICINECA). *Inorganic crystals, powder diffraction, proteins, small-angle scattering, X-ray diffraction.*

Ripamonti, Dr Carlo. Dip. di Fisica, U. di Genova, Via Dodecaneso 33, 16146 Genova, Italy. (1950) Dr, physics (U. di Genova, 1975). Ricercatore. *Crystal physics.*

Riva di Sanseverino, Prof. Lodovico. Dip. di Scienze Mineralogiche, U. di Bologna, Piazza di Porta S. Donato 1, 40126 Bologna, Italy. (1939) Dr, chemistry (U. di Firenze, 1962). Prof., mineralogy. (tel. 051 + 243513 or 56, ext. 50, fax 051 + 247244, email T54BOM12 at ICINECA). *Organic compounds, structural chemistry, teaching.*

Riva, Dr Fernando. Istituto Polimeri, C.N.R., Via Mezzocannone 4, 80134 Napoli, Italy. (1933) Dr, chemistry (U. di Pavia, 1958). Ricercatore. (tel. 081 + 205730). *Macromolecules, materials science, polymers, powder diffraction, small-angle scattering, X-ray diffraction.*

Rizzoli, Dr Corrado. Ist. di Strutturistica Chimica, U. di Parma, Viale delle Scienze, 43100 Parma, Italy. (1957) Dr, chemistry (U. di Parma, 1981). Tecnico laureato. (tel. 0521 + 580447). *Computing, crystal modelling, structural chemistry.*

Rosa, Dr Rodolfo. Ist. LAMEL, C.N.R., Via de' Castagnoli 1, 40126 Bologna, Italy. (1944) Dr, physics and philosophy (U. di Bologna, 1968, 1977). Primo ric. (tel. 051 + 287912, fax 051 + 229702). *Computing, electron microscopy.*

Rossi, Dr Marco. Dip. di Energetica, U. di Roma 'La Sapienza', Via Scarpa 14, 00161 Roma, Italy. (1961) Dr, engineering (U. di Roma). Consultant. (tel. 06 + 4941108, fax 06 + 4270183). *Defect structures, electron microscopy semiconductors.*

Roveri, Prof. Norberto. Dip. di Chimica 'G. Ciamician', U. di Bologna, Via Selmi 2, 40126 Bologna, Italy. (1947) Dr, chemistry (U. di Bologna, 1972). Prof. assoc. (tel. 051 + 259552, fax 051 + 259456, email VS3BODI3 at ICINECA). *Inorganic crystals, powder diffraction, proteins, small-angle scattering, X-ray diffraction.*

Rubbo, Prof. Marco. U. della Calabria, 87030 Castiglione Cosentino sc., Italy. (1946) Dr, chemistry (U. di Torino, 1971). Prof. assoc. (tel. 011 + 832193). *Crystal growth, minerals, molecular crystals.*

Sabat, Dr Michal. Ist. per lo Studio della Stereochimica ed Energetica dei Composti di Coordinazione, C.N.R., Via Guerrazzi 27, 50132 Firenze, Italy. (1947) PhD, Chemistry (U. Wroclaw, Poland, 1976). Ricercatore. (tel. 055 + 243990). *Structural chemistry, molecular orbital methods, coordination compounds, organometallic and bioinorganic compounds.*

Sabatino, Dr Piera. Dip. di Chimica 'G. Ciamician', U. di Bologna, Via Selmi 2, 40126 Bologna, Italy. (1950) Dr, pharmacy (U. di Bologna, 1974). Ricercatore. (tel. 051 + 259555, fax 051 + 259456, email VN3BODG1 at ICINECA1). *Coordination compounds, organic compounds.*

Sabbioni, Dr Cristina. Ist. FISBAT, C.N.R., Via Castagnoli 1, 40126 Bologna, Italy. (1954) Dr, physics (U. di Bologna, 1978). Ricercatore. (tel. 051 + 287093, fax 051 + 229702). *Diffractometry, electron microscopy, microscopy.*

Sabelli, Dr Cesare. Centro di Studio per la Mineralogia e la Geochimica dei Sedimenti, C.N.R., Via La Pira 4, 50121 Firenze, Italy. (1934) Dr, geological sci. (U. di Firenze, 1959). Ricercatore. (tel. 055 + 287140). *Crystallographic data, minerals.*

Sacerdoti, Prof. Michele. Ist. di Mineralogia, U. di Ferrara, Corso Ercole I d'Este 32, 44100 Ferrara, Italy. (1935) Dr, geological sci. (U. di Padova, 1959). Prof. assoc. (tel. 0532 + 32987, email M83FEM12 at ICINECA). *Electron microscopy, hydrogen bonding, minerals. X-ray diffraction.*

Salviati, Dr Giancarlo. Ist. MASPEC, C.N.R., Via Chiavari 18/A, 43100 Parma, Italy. (1950) Dr, physics (U. di Parma). Ricercatore. (tel. 0521 + 96841, fax 0521 + 96315). *Defect structures, electron diffraction, electron microscopy, materials science, semiconductors.*

Sansoni, Prof. Mirella. Ist. di Chimica Strutturistica Inorganica, U. degli Studi di Milano, Via G. Venezian 21, 20148 Milano, Italy. (1939) Dr, physics (U. di Milano, 1964). Prof. assoc. (tel. 02 + 235120). *Chemical bonding, coordination compounds, structural chemistry.*

Santini, Dr Antonello. Dip. di Chimica, U. di Napoli, Via Mezzocannone 4, 80134 Napoli, Italy. (1958) (tel. 081 + 206800). *Computing, crystallographic data, diffractometry, X-ray diffraction, biocrystallography, structure of peptides.*

Sartori, Prof. Franco. Dip. di Scienze della Terra, U. di Pisa, Via S. Maria 53, 56100 Pisa, Italy. (1938) Dr, geological sci. (U. di Pavia, 1962). Prof. assoc. (tel. 050 + 501663). *Diffractometry, minerals.*

Scandale, Prof. Eugenio. Dip. Geomineralogico, U. di Bari, Campus, Via Salvemini, 70124 Bari, Italy. (1943) Dr, physics (U. di Bari, 1971). Prof. assoc. (tel. 080 + 242585). *Crystal growth, defect structures, minerals.*

Scapin, Dr Giovanna. Dip. di Chimica Organica, U. di Padova, Via Marzolo 1, 35100 Padova, Italy. (1961) Dr, chemistry (U. di Padova, 1985). Dottorando. (tel. 049 + 831350, fax 049 + 831222, email CHIM09 at UNIPAD.INFN.IT). *Macromolecules, proteins.*

Scaramuzza, Dr Lucio. Ist. Teoria e Struttura Elettronica e Comportamento Spettrochimico dei Composti di Coordinazione, C.N.R., Area della Ricerca di Roma, 00016 Monterotondo Stazione, Roma, Italy. (1935) Dr, chemistry (U. di Roma, 1965). Ricercatore. (tel. 06 + 4991, ext. 322). *Coordination compounds, instrumentation.*

Scatturin, Prof. Vladimiro. Ist. di Chimica Strutturistica Inorganica, U. degli Studi di Milano, Via G. Venezian 21, 20133 Milano, Italy. (1922) Dr, chemistry (U. di Padova, 1946). Prof., general chemistry. (tel. 02 + 235288). *Quasicrystals, structural chemistry. teaching.*

Schiavinato, Prof. Giuseppe. Dip. di Scienze della Terra, U. degli Studi, Via Botticelli 23, 20133 Milano, Italy. (1915) Dr, geological sci. (U. di Padova, 1939). Prof., mineralogy. (tel. 02 + 2663994). *Minerals, X-ray diffraction.*

Scordari, Prof. Fernando. Dip. Geomineralogico, U. di Bari, Campus Universitario, 70100 Bari, Italy. (1944) Dr, geological sci. (U. di Bari, 1968). Prof. (tel. 080 + 242587). *Disorder, inorganic crystals, minerals, X-ray diffraction.*

Secco, Dr Luciano. Dip. di Mineralogia e Petrologia, U. di Padova, Corso Garibaldi 37, 35100 Padova, Italy. (1955) Dr, geological sci. (U. di Padova, 1980). Ricercatore. (tel 049 + 663122). *Crystallographic data, diffractometry, X-ray diffraction, crystal chemistry.*

Sersale, Prof. Riccardo. Dip. di Ingegneria de, materiali e della Produzione, U. di Napoli, P.le Tecchio, 80125 Napoli, Italy. (1921) Dr, chemistry (U. di Napoli, 1943). Prof., appl. chemistry. (tel. 081 + 7682395, telex INGENA I 722392, fax 081 + 7682362). *Crystallographic data, defect structures, disorder, electron diffraction, electron microscopy, inorganic crystals, materials science, microscopy, minerals, phase determination, powder diffraction, semiconductors, surface structure, texture, X-ray diffraction.*

Servidori, Dr Marco. Ist. LAMEL, CNR, Via de' Castagnoli 1, 40126 Bologna, Italy. (1943) Dr, industrial chemistry (U. di Bologna, 1967). Ricercatore. (tel. 051 + 287932, telex 511350 CNRBO, fax 051 + 229702). *X-ray topography, electron microscopy, electronics materials, multiple crystals diffraction, X-ray diffraction.*

Sgarabotto, Prof. Paolo. Ist. di Strutturistica Chimica, U. degli Studi di Parma, Viale delle Scienze, 43100 Parma, Italy. (1940) Dr, chemistry (U. di Parma, 1967). Prof. assoc. (tel. 0521 + 580448). *Crystallographic data, diffractometry, powder diffraction, structural chemistry, X-ray diffraction.*

Sgarlata, Prof. Francesco. Ist. di Mineralogia e Petrografia, U. di Roma, Piazzale A. Moro, Città Universitaria, 00185 Roma, Italy. (1926) Dr, physics (U. di Roma, 1949). Prof., crystallography. *Solid state physics, X-ray diffraction, gamma resonance spectrometry, phases in natural heterogeneous systems.*

Sgualdino, Dr Giulio. Dip. di Chimica, U. di Ferrara, Via L. Borsari 46, 44100 Ferrara, Italy. (1947) Dr, chemistry (U. di Ferrara, 1974). Ricercatore. (tel. 0532 + 36374). *Crystal growth.*

Sica, Dr Filomena. Dip. di Chimica, U. di Napoli, Via Mezzocannone 4, 80134 Napoli, Italy. (1959) PhD, chemistry (U. di Napoli, 1988). (tel. 081 + 206800). *Organic compounds, proteins, X-ray diffraction.*

Sironi, Prof. Angelo. Ist. Chimica Strutturistica Inorganica, U. degli Studi di Milano, Via G. Venezian 21, 20133 Milano, Italy. (1948) Dr, chemistry (U. di Milano, 1972). Prof. assoc. (tel. 02 + 235120, email STINCH at IMISIAM.EARNET). *Chemical bonding, computing, coordination compounds, powder diffraction, structural chemistry, X-ray diffraction.*

Spadon, Prof. Paola. Dip. di Chimica Organica, U. di Padova, Via Marzolo 1, 35131 Padova, Italy. (1947) Dr, chemistry (U. di Padova, 1971). Prof. assoc. (tel. 049 + 831327 or 831236, fax 049 + 831222, email CHIM09 at UNIPAD.INFN.IT). *Macromolecules, proteins, structural biology.*

Spagna, Dr Riccardo. Ist. di Strutturistica Chimica 'G. Giacomello', C.N.R., C.P. 10 - 00016 Monterotondo Stazione, Roma, Italy. (1941) Dr, chemistry (U. di Roma, 1967). Primo ric. (tel. 06 + 90020, ext. 614, fax 06 + 9005849, email STRUTA at IRMIAS). *Computing, diffractometry, instrumentation, phase determination, X-ray diffraction.*

Stasi, Dr Francesca. Dip. Geomineralogico, U. di Bari, Via Amendola, 70121 Bari, Italy. (1939), Dr, physics (U. di Bari, 1971). Tecn. laur. (tel. 080 + 242615). *Defect structures, electron microscopy, inorganic crystals, minerals, organic compounds.*

Suffritti, Prof. Giuseppe Baldovino. Dip. di Chimica, U. di Sassari, Via Vienna 2, 07100 Sassari, Italy. (1947) Dr, physics (U. di Milano, 1972). Prof. assoc. (tel 079 + 218415). *Disorder, materials science, minerals, thermal vibrations.*

Tadini, Prof. Carla. Dip. di Scienze della Terra, U. di Pavia, Via A. Bassi 4, 27100 Pavia, Italy. (1924) Dr, chemistry (U. di Pavia, 1956). Prof. assoc. (tel. 0382 + 392254). *Inorganic crystals, minerals. organic compounds, X-ray diffraction.*

Tazzoli, Prof. Vittorio. Dip. di Scienze della Terra, U. di Pavia, Via A. Bassi 4, 27100 Pavia, Italy. (1938) Dr, chemistry (U. di Pavia, 1964). Prof. assoc. (tel. 0382 + 392265, fax 0382 + 392252, email CSCSPV at IPVIAN.EARN). *Minerals, phase transitions, X-ray diffraction, crystal chemistry.*

Tieghi, Prof. Giuseppe. Dip. di Chimica Industriale ed Ingegneria Chimica, Politecnico di Milano, Piazza L. da Vinci 32, 20133 Milano, Italy. (1943) Dr, chemical engineering. (U. di Milano, 1968). Prof. assoc. (tel. 02 + 23993218). *Macromolecules, materials science, polymers, powder diffraction, symmetry.*

Tiripicchio, Prof. Antonio. Ist. di Chimica Generale ed Inorganica, U. di Parma, Viale delle Scienze, 43100 Parma, Italy. (1936) Dr, chemistry (U. di Parma, 1959). Prof., general chemistry. (tel. 0521 + 580418). *Coordination compounds, inorganic crystals, structural chemistry, organometallic compounds.*

Tiripicchio-Camellini, Prof. Marisa. Ist. di Chimica Generale ed Inorganica, U. di Parma, Viale delle Scienze, 43100 Parma, Italy. (1938) Dr, natural sci. (U. di Parma, 1962). Prof. assoc. (tel. 0521 + 580417). *Coordination compounds, inorganic crystals, structural chemistry, organometallic compounds.*

Tomasicchio, Dr Michele. Dip. di Chimica Organica, U. di Padova, Via Marzolo 1, 35131 Padova, Italy. (1959) Dr, chemistry (U. di Padova, 1985). Dottorando. (tel. 049 + 831293, email CHIM09 at UNIPAD.INFN.IT). *Macromolecules, proteins.*

Tomassini, Dr Marco. Dip. di Scienze della Terra, Sez. di Cristallografia, U. di Perugia, Piazza U., 06100 Perugia, Italy. (1949) Dr, chemistry (U. di Perugia, 1974). Prof. inc. (tel. 075 + 23822). *Computing, potential energy calculations in crystals.*

Tosi, Prof. Giorgio. Dip. di Scienze de, materiali e della Terra, Fac. di Ingegneria, Via Brecce Bianche, 60100 Ancona, Italy. (1942) Dr, Industrial chemistry (U. di Bologna, 1966). Prof. assoc. (tel. 071 + 5893723, telex 561838, fax 071 + 5893714). *Chemical and physical relationship, clusters.*

Trosti-Ferroni, Dr Renza. Dip. di Scienze della Terra, U. di Firenze, Via La Pira 4, 50121 Firenze, Italy. (1947) Dr, natural sci. (U. di Firenze, 1969). Prof. assoc. (tel. 055 + 287140). *Inorganic crystals, structural chemistry, minerals.*

Tuzi, Dr Angela. Dip. di Chimica, U. di Napoli, Via Mezzocannone 4, 80134 Napoli, Italy. (1954) Dr, chemistry (U. di Napoli, 1978). Prof. assoc. (tel. 081 + 205730). *Liquid crystals, polymers, X-ray diffraction.*

Ugliengo, Dr Piero. Dip. di Chimica Inorganica, Chimica Fisica e de, materiali, U. di Torino, Via P. Giuria 7, 10125 Torino, Italy. (1957) Dr, chemistry (U. di Torino, 1981). Ricercatore. (tel. 011 + 6505102, ext. 15, email U102 at ITOCSIP). *Chemical bonding, computiing, structural chemistry.*

Ugozzoli, Dr Franco. Ist. di Strutturistica Chimica, Viale delle Scienze, 43100 Parma, Italy. (1947) Dr, physics (U. di Parma, 1983). Tecn. laur. (tel. 0521 + 580548 or 580450, fax 0521 + 580542, email CHIMORG at IPRUNIV). *Structural chemistry, organic compounds, X-ray spectroscopy, Mossbauer spectroscopy, amorphous systems, computing.*

Ungaretti, Prof. Luciano. Dip. di Scienze della Terra, U. di Pavia, Via A. Bassi 4, 27100 Pavia, Italy. (1942) Dr, chemistry (U. di Pavia, 1965). Prof., mineralogy. (tel. 0382 + 392266, fax 0382 + 392252, email CSCSPV at IPVIAN.EARN). *Chemical bonding, crystallographic data, diffractometry, high pressure, instrumentation, materials science, minerals, phase transitions, structural chemistry, thermal vibrations, X-ray diffraction.*

Vaccari, Dr Giuseppe. Dip. di Chimica, U. di Ferrara, Via Luigi Borsari 46, 44100 Ferrara, Italy. (1948) Dr, chemistry (U. di Ferrara, 1972). Prof. inc. (tel. 0532 + 36374). *Crystal growth.*

Vaciago, Prof. Alessandro. Dip. di Chimica, U. di Roma, Città Universitaria, 00100 Roma, Italy. (1931) Dr, physics (U. di Milano, 1953). Prof., structural chemistry. (tel. 06 + 49913715, telex 613255 INFNRO I). *Instrumentation, structural chemistry.*

Valdrè, Prof. Ugo. Dip. di Fisica, U. di Bologna, Via Irnerio 46, 40126 Bologna, Italy. (1926) Dr, physics (U. di Bologna, 1954). Professor of Physics. (tel. 051 + 241134 or 244190, ext. 217 or 108, telex 211664 INFN BO, fax 051 + 247244). *Electron microscopy, instrumentation, semiconductors, superconductors, low temperatures, organic compounds, nanoanalysis.*

Valigi, Prof. Mario. Dip. di Chimica, U. di Roma, Piazzale A. Moro 5, 00100 Roma, Italy. (1936) Dr, chemistry (U. di Roma, 1961). Prof. (tel. 06 + 4991, ext. 23377). *Magnetism, materials science, phase transitions, powder diffraction, surface structure, X-ray diffraction.*

Valle, Dr Giovanni. Centro di Studio sui Biopolimeri, C.N.R., Dip. di Chimica Organica, Via Marzolo 1, 351310 Padova, Italy. (1930) Dr, industrial chemistry (U. di Padova, 1960). Primo ric. (tel. 049 + 831229, fax 049 + 831222). *Organic compounds, structural biology.*

Venturello, Porf. Giovanni. Ist. di Chimica Generale, Fac. di Farmacia, U. di Torino, Via P. Giuria 9, 10125 Torino, Italy. (1912) Dr, chemistry (U. di Torino, 1935). Prof. emerito. (tel 011 + 657000). *Metals, phase transitions, quasicrystals, surface structure.*

Verdini, Prof. Brunella. Dip. di Ingegneria Chimica, de, materiali, dell, materie Prime e Metallurgia, 00184 Roma, Italy. (1931) Dr, physics (U. di Roma, 1960). Prof. assoc. (tel. 06 + 4687350). *Crystallographic data, diffractometry, electron microscopy, metals, microscopy, phase determination, phase transitions, teaching, texture, X-ray diffraction.*

Vezzalini, Dr Maria Giovanna. Ist. di Mineralogia e Petrologia, U. di Modena, Via S. Eufemia 19, 41100 Modena, Italy. (1951) Dr, geological sci. (U. di Modena, 1974). Ricercatore. (tel. 059 + 218062, ext. 16, email VEZZALINI at IMOVX2.INFNET). *Diffractometry, minerals, X-ray diffraction.*

Vitali, Dr Francesca. Ist. di Chimica Generale, U. di Parma, Viale delle Scienze, 43100 Parma, Italy. (1957) Dr, Pharmaceutical chemistry (U. di Parma, 1983). Ricercatore. (tel. 0521 + 580421, fax 0521 + 580542, email CHIMIC3 at IPRUNIV). *Coordination compounds, structural chemistry, X-ray diffraction.*

Vitali, Prof. Gianfranco. Dip. di Energetica, U. di Roma 'La Sapienza', Via A. Scarpa 14, 00161 Roma, Italy. (1934) Dr, physics. (tel. 06 + 425787, fax 06 + 4270183). *Electron microscopy, semiconductors.*

Viterbo, Prof. Davide Lazzaro Marco. Dip. di Chimica, U. della Calabria, 87030 Arcavacata di Rende, Cosenza, Italy. (1939) Dr, chemistry (U. di Torino, 1962). Prof., Physical chemistry. (tel. 0984 + 839321, email 2201VIT at ICSUNIC). *Computing, direct methods, structural chemistry.*

Zagari, Prof. Adriana. Dip. di Chimica, U. di Napoli, Via Mezzocannone 4, 80134 Napoli, Italy. (1946) Dr, chemistry (U. di Napoli, 1969). Prof. assoc. (tel. 081 + 205730). *Molecular crystals, organic compounds, proteins.*

Zanazzi, Prof. Pier Francesco. Dip. di Scienze della Terra, Sez. di Cristallografia, U. di Perugia, Piazza U., 06100 Perugia, Italy. (1939) Dr, chemistry (U. di Firenze, 1962). Prof., crystallography. (tel. 075 + 4693212, telex 662078 UNIPG I, fax 075 + 4692067, email CRYSTAL at IPGUNIV). *High pressure, inorganic crystals, minerals, crystal-chemistry.*

Zangrando, Dr Ennio. Dip. di Scienze Chimiche, U. di Trieste, P.le Europa 1, 34127 Trieste, Italy. (1950) Dr, chemistry (U. di Trieste, 1974). Ricercatore. (tel 040 + 5603111, fax 040 + 5603176). *Structures, inorganic chemistry.*

Zannetti, Prof. Roberto. Dip. di Chimica Inorganica, Metallorganica ed Analitica, U. di Padova, Via Loredan 4, 35100 Padova, Italy. (1929) Dr, chemistry (U. di Padova, 1953). Prof., industrial chemistry. (tel. 049 + 831257). *Disorder, macromolecules, inorganic crystals, phase transitions, powder diffraction.*

Zanotti, Prof. Giuseppe. Dip. di Chimica Organica, U. di Padova, Via Marzolo 1, 35131 Padova, Italy. (1950) Dr, chemistry (U. di Padova, 1974). Prof. assoc. (tel. 049 + 831229, fax 049 + 831222, email CHIM09 at UNIPAD.INFN.IT or ZANOTTI at PDCHOR.INFN.IT). *Macromolecules, proteins, structural biology.*

Zanotti, Dr Lucio. Ist. MASPEC, C.N.R., Via Chiavari 18/A, 43100 Parma, Italy. (1944) Dr, chemistry (U. di Bologna, 1969). Ricercatore. (tel. 0521 + 96841, fax 0521 + 96315). *Crystal growth, defect structures, diffractometry, electron microscopy, inorganic crystals, magnetism, materials science, semiconductors.*

Zappia, Prof. Vincenzo. Ist. di Biochimica dell, macromoleculecole, U. di Napoli, Via Costantinopoli 16, 80138 Napoli, Italy. (1939) Dr, medicine (U. di Napoli, 1963). Prof., biological chemistry. (tel. 081 + 217052-213668, fax 081 + 217052). *Structural chemistry, sulfonium compounds, polyamine biosynthesis, structure and function, enzymes.*

Zefiro, Dr Livilo. Dip. di Scienze della Terra, U. degli Studi, Corso Europa 26, Palazzo delle Scienze, 16132 Genova, Italy. (1948) Dr. physics (U. di Genova, 1972). Ricercatore. (tel. 010 + 3538303, telex 352169 DISTER GE, email MINERAL at IGECUNIV). *Crystal growth, coherent optics in crystallography, holographic interferometry, hydrodynamics of solution.*

Zerbi, Prof. Giuseppe. Dip. di Chimica Industriale, Politecnico di Milano, Piazza L. da Vinci 32, 20133 Milano, Italy. (1933) Dr, chemistry (U. di Pavia, 1956). Prof., appl. chemistry (tel. 02 + 23993235, fax 02 + 23992206). *Defect structures, disorder, liquid crystals, macromolecules, materials science, polymers, semiconductors, thermal vibrations.*

Zocchi, Prof. Marcello. Dip. di Ingegneria Meccanica, U. di Brescia, Via D. Valotti 9, 25060 Brescia, Italy. (1929) Dr, chemistry (U. di Roma, 1956). Prof., chemistry. (tel. 030 + 398461, ext. 272, telex 304116 UNIVI). *Inorganic stereochemistry, materials science, accurate intensity measurements.*

Zosi, Prof. Gianfranco Luigi. Ist. di Fisica Generale, U. di Torino, Corso M. D'Azeglio 46, 10125 Torino, Italy. (1940) Dr, physics (U. di Torino, 1962). Prof. assoc. (tel. 011 + 65271, ext. 426, fax 011 + 6699579, email 2051 at VAXTO). *X-ray diffraction.*

IVORY COAST

Sub-Editor: N. Ebby

Notes

1. International telephone country code - 225.

Bonnaud, Dr Bernard Henri. Dépt. de Physique, Faculté des Sciences et Techniques, U. d'Abidjan, 22 BP 582 Abidjan 22, Côte d'Ivoire. (1946) D.3e cycle, structural sci. (U. de Marseille-Provence, France, 1985). Asst. (tel. 225 + 43 90 00, ext. 3125, telex RECTUCI 26138). *Organic structures, liquid crystals.*

Bonny, Dr Roger. Dépt. de Physique, Faculté des Sciences et Techniques, U. d'Abidjan, 22 BP 582 Abidjan 22, Côte d'Ivoire. (1955) D.3e cycle, électronique(USTL, Montpellier, France, 1984). Asst. (tel. 225 + 43 90 00, ext. 3129, telex RECTUCI 26138). *Organic structures, X-ray diffraction.*

Briard, Mrs Pierrette. Dépt. de Physique, Faculté des Sciences et Techniques, U. d'Abidjan, 22 BP 582 Abidjan 22, Côte d'Ivoire. (1940) D.3e cycle, metallurgy (U. de Toulouse, France, 1973). Maître-asst. (tel. 225 + 43 90 00, ext. 3125, telex RECTUCI 26138). *Organic structures, X-ray diffraction.*

Cossu, Dr Michéle Josette. Dépt. de Physique, Faculté des Sciences et Techniques, U. d'Abidjan, 22 BP 582 Abidjan 22, Côte d'Ivoire. (1943) DSc, physics (U. de Provence, France, 1983). Maître de conf. (tel. 225 + 43 90 00, ext. 3129, telex RECTUCI 26138). *Organic structures, strained molecules.*

Ebby, Dr N' Dédé. Dépt. de Physique, Faculté des Sciences et Techniques, U. d'Abidjan, 22 BP 582 Abidjan 22, Côte d'Ivoire. (1941) DSc, physics (U. d'Abidjan, 1980). Prof. (tel. 225 + 93 90 00, ext. 3124, telex RECTUCI 26138). *Organic structures, X-ray diffraction.*

Mansilla - Koblavi, Mrs Frédérica Gbédégbé Marie Sonia. Dépt. de Physique, Faculté des Sciences et Techniques, U. d'Abidjan, 22 BP 582 Abidjan 22, Côte d'Ivoire. (1958) Doctorat Spec., molecular chemistry (U. de Toulouse Paul Sabtier, France). Asst. (tel. 225 + 43 90 00, ext. 3129, telex RECTUCI 26138). *Organometallic structures, X-ray diffraction.*

Tenon, Mr Abodou Jules. Dépt. de Physique, Faculte des Sciences et Techniques, U. d'Abidjan, 22 BP 582 Abidjan 22, Côte d'Ivoire. (1955) DEA, structural science (U. d'Abidjan, 1986). Asst. (tel. 225 + 43 90 00, ext. 3125, telex RECTUCI 26138). *Organic structures, X-ray diffraction.*

Toure, Dr Siaka. Dépt. de Physique, Faculte des Sciences et Techniques, U. d'Abidjan, 22 BP 582 Abidjan 22, Côte d'Ivoire. (1947) DSc, physics (U. d'Abidjan, 1979). Prof. (tel. 225 + 43 90 00, ext. 3122, telex RECTUCI 26138). *Organic compounds, structure determination, X-ray diffraction, natural products.*

JAPAN

Sub-Editor: M. Tokonami

Notes

1. International telephone country code - 81.

2. At the Universities, persons with an academic training can be appointed in various functions ranging from *professor, assistant professor, lecturer* to *research associate.* At some Universities the use of the English translations *associate professor* and *research assistant* instead of assistant professor and research associate, respectively, is preferred although in Japanese these differences do not occur.

Abe, Prof. Hideo. 2-18-14 Sasuke Kamakura, Kanagawa 248, Japan. (1924) Dr.honoris causa, eng. (Tokyo U., 1960). Em. prof. (tel. 0467 + 24-8210). *Texture, recrystallization in metals.*

Abe, Prof. Ryuji. Electrical Eng., Kanazawa Inst. of Techn., 7-1 Oogigaoka Nonoichi, Kanazawa, Ishikawa 921, Japan. (1922) DSc, physics (Nagoya U., 1959). Prof. (tel. 0762 + 48-1100 ext 463). *Phase transitions.*

Abe, Mr Shuich. (1927) BSc, physics (Tohoku U., 1951). *X-ray crystallography, solid state physics.*

Abe, Dr. Takao. SEH R and D Center, Shin-Etsu Handotai Co., 2-13-1 Isobe Annaka, Gunma 379-01, Japan. (1936) DEng.(Hokkaido U., 1964). Dupty Dir. (tel. 0273 + 85-2511, fax +81 273 85 4995). *Semiconductor Si crystals.*

Abe, Mr. Yoshio. Ceramics Res. and Deveropement, Murata MFG. Co.,Ltd., 2-26-10 Nagaokakyo-shi, Kyoto 617, Japan. (1957) MSc., mineralogy.(1984). (tel. 075 + 955-9379, fax +81 75 954 7720). *Chemical bonding, defect structures, diffuse scattering, disorder, electron microscopy, phase transitions, powder diffraction, surface structure, X-ray diffraction.*

Achiwa, Prof. Norio. Dep. of Physics, Kyushu U., 6-10-1 Hakozaki Higashi-ku Fukuoka, Fukuoka 812, Japan. (1940) DSc.(Kyoto U., 1969). Assoc. prof. (tel. 092 + 641-1101 ext 4176). *High pressure, magnetism, neutron diffraction, inorganic crystals, phase transition, powder diffraction, structural chemistry, symmetry, X-ray diffraction.*

Adachi, Prof. Kengo. Fac. of Eng., Nagoya U., Furo-cho Chikusa-ku, Nagoya 464, Japan. (1926) DSc, physics (Tohoku U., 1958). Prof. of physics. (tel. 052 + 781-5111 ext 3567). *Magnetism, order-disorder phenomena, alloys.*

Aihara, Prof. Ariyuki. 11-404 Sakura-Josui 4-1 Setagaya-ku, Tokyo 156, Japan. (1920) DSc, chemistry (Tokyo U., 1954). Em. prof., U. of Electro-Communication (tel. 03 + 303-2702). *Disorder, hydrogen bonding, phase transition.*

Aikawa, Dr Nobuyuki. Dept. of Geosciences, Fac. of Sci., Osaka City U., 459 Sugimoto-cho Sumiyoshi-ku, Osaka 558, Japan. (1944) DSc, mineralogy (Tokyo U., 1975). Assoc. prof. (tel. 06 + 605-2587). *Crystal structure, minerals, diffuse X-ray reflections.*

Aizaki, Mr Naoaki. Fundamental Res. Labs., Nippon Electric Co. Ltd., 4-1-1 Miyazaki Miyamae-ku Kawasaki, Kanagawa 213, Japan. (1945) MEng, applied physics (Tokyo U., 1971). Res. member. (tel. 044 + 856-2186, fax +81 44 856 2214). *Semiconductors, crystal growth, electron diffraction, surface structure, X-ray diffraction.*

Akagi, Dr. Yoshiro. Materials Res. and Analysis Center, Sharp Co. Ltd. Eng. Center, 2613-1 Ichinomoto Tenri, Nara 632, Japan. (1949) DSc., physical chem.(Osaka U., 1983). (tel. 07436 + 5-1321 ext 3090, fax +81 7436 5 1131). *Molecular crystals, surface structure, materials science.*

Akao, Dr Masaru. Inst. for Med. Dent. Eng., Tokyo Med. Dent. U., 2-3-10 Kanda Surugadai Chiyoda-ku, Tokyo 101, Japan. (1945) DEng, chemical eng. (Tokyo Inst. Techn., 1975). (tel. 03 + 291-3721, fax +81 3 291 3727). *Minerals, bioceramics.*

Akimitsu, Prof. Jun. Dept. of Physics, Aoyama Gakuin U., 6-2-3 Chitosedai Setagaya-ku, Tokyo 157, Japan. (1940) DSc, physics (Tokyo U., 1970). Prof. (tel. 03 + 307-2888 ext 253, fax +81 3 326 3235). *Magnetism, X-ray diffraction, neutron diffraction.*

Akimoto, Dr. Koichi. Fundamental Res. Lab., NEC Co., 1-1 Miyazaki 4-chome Miyamae-ku Kawasaki, Kanagawa 213, Japan. (1958) DEng, applied physics.(1985). Superviser. (tel. 044 + 856-2184, fax +81 44 856 2214). *X-ray diffraction, surface structure, interface structure, semiconductors.*

Akimoto, Dr. Toshio. Res. Management Dept., Chugai Pharmaceutical Co., Ltd., 135 Komakado 1-chome Gotenba-shi, Shizuoka 412, Japan. (1941) DPharm, pharmaceutical Sci. (Tokyo U., 1970). staff manager. (tel. 0550 + 87-3411, fax +81 550 87 1960).

Akizuki, Dr Mizuhiko. Inst. of Mineralogy, Petrology and Economic Geology, Tohoku U., Aramaki Aza-Aoba, Sendai 980, Japan. (1937) DSc, mineralogy (Tohoku U., 1968). Asst. prof. (tel. 0222 + 22-1800 ext 4227). *Texture, minerals, twining.*

Amemiya, Dr Yoshiyuki. Photon Factory, Nat. Lab. for High Energy Physics, Oho-machi, Tsukuba, Ibaraki 305, Japan. (1952) DEng, applied physics (Tokyo U., 1979) Res. assoc. (tel. 0298 + 64-1171 ext 5025). *Synchrotron radiation, instrumentation, small angle scattering, time-resolved X-ray measurement, position sensitive X-ray detector.*

Ando, Dr Masami. Photon Factory, Nat. Lab. for High Energy Physics, Oho-machi, Tsukuba, Ibaraki 305, Japan. (1942) DEng., applied physics (Tokyo U., 1974). Prof. (tel. 0298 + 64-1171 ext 5040). *Synchrotron radiation, instrumentation.*

Ando, Dr Yoshinori. Dept. of Physics, Meijo U., 1-501 Shiogamaguchi, Tenpaku-ku, Nagoya 468, Japan. (1942) DEng, applied physics (Nagoya U., 1970). Asst. prof. (tel. 052 + 832-1151 ext 5280). *Crystal growth, X-ray and electron diffraction, electron microscopy, ceramics.*

Annaka, Dr. Shoichi. Tokyo U. of Mercantile Marine, 48-6 Nukui 3-chome Nerima-ku, Tokyo 176, Japan. (1924) DSc, physics (Tokyo U. of Education, 1958). Prof. Em. (tel. 03 + 970-6124). *Materials science.*

Aoki, Dr Katsuyuki. Inst. of Physical and Chemical Res., 1 Hirosawa 2-chome Wako, Saitama 351, Japan. (1945) DPharm, chemistry (Tokyo U., 1978). Res. (tel. 0484 + 62-1111, telex 02962818 RIKEN J, fax +81 484 62 1111). *Inorganic structural biochemistry, organometallic structural chemistry.*

Aoki, Dr Yoshikazu. Dept. of Geology, Fac. of Sci., Kyushu U., 10-1 Hakozaki 6-chome, Higashi-ku, Fukuoka 812, Japan. (1939) DSc, mineralogy (Kyushu U., 1977). Assoc. prof. (tel. 092 + 641-1101 ext 4312, fax +81 92 631 4233). *Mineralogy, crystal growth.*

Asada, Prof. Eiichi. Dept. of Materials Sci., Toyohashi U. of Techn., 1-1 Hibarigaoka Tempaku-cho Toyohashi 440 Japan. (1924) DEng, applied chemistry (Tokyo U., 1979). Prof. (tel. 0532 + 47-0111 ext 442, telex 4322201JPNTUT, fax +81 532 45 0480). *Powder diffraction, structural chemistry, phase determination.*

Asai, Dr Takeshi. Inst. of Scient. and Industrial Res., Osaka U., Mihogaoka Ibaraki, Osaka 567, Japan. (1942) DSc, chemistry (Osaka U., 1971). Res. assoc. (tel. 06 + 877-5111 ext 3551, telex 5286213, fax +81 6 877 4977). *Inorganic crystals, materials science, structural chemistry.*

Ashida, Dr Sakichi. Production and Dev. Div., Nihon Dempa Kogyo Co. Ltd., 1275-2 Kamihirose Sayama, Saitama 350-13, Japan. (1935) DSc, mineralogy (Tohoku U., 1965). Deputy Div. Dir. (tel. 0429 + 52-7211 ext 1403, fax +81 429 54 3968). *Crystal growth, piezoelectric materials, solid state physics.*

Ashida, Prof. Tamaichi. Dept. of Applied Chemistry, Fac. of Eng., Nagoya U., Furo-cho Chikusa-ku, Nagoya 464, Japan. (1933) DSc, chemistry (Osaka U., 1964). Prof. (tel. 052 + 781-5111 ext 3339, telex 4477355 ENUNAG J, fax +81 52 781 4895). *Proteins, computing, organic compounds.*

Ashizawa, Mr Kazuhide. Physical Chemistry, Central Res. Lab., Eisai Company, Ltd., 1-3 Tokodai 5-chome Tsukuba, Ibaraki 300-26, Japan. (1956) (Kitasato U.). Res. (tel. 029747 + 2211 ext 417, fax +81 29747 8489). *Physical properties, crystal characterization, phase transitions, polymorphism, solid state chemistry, pharmaceutical sicences, organic crystal structures.*

Azuma, Dr Nagao. Fac. of General Education, Ehime U., 3 Bunkyo-cho Matsuyama, Ehime 790, Japan. (1943) DSc, science (Kyoto U., 1982). Asst. prof. (tel. 0899 + 24-7111 ext 3892). *Coordination compounds, magnetism, reaction.*

Bunno, Dr Michiaki. Dept. of Geology, Geological Survey of Japan, 1-1-3 Higashi Tsukuba, Ibaraki 305, Japan. (1942) DSc, mineralogy (Tokyo U., 1980). (tel. 0298 + 54-3752). *Descriptive mineralogy.*

Chen Mr Jie. Dept. of Applied chemistry, Osaka U., Yamadakami Suita, Osaka 565, Japan. (1958) MEng, chemistry (Osaka U., 1988). Graduate student (tel. 06 + 877-5111 ext 4322). *X-ray crystallography, organometallic structural chemistry.*

Chikaura, Dr Yoshinori. Fac. of Eng., Dept. of Phys., Kyushu Inst. of Techn., 1 Sensui-cho Tobata-ku Kitakyushu, Fukuoka 804, Japan. (1946) DEng., applied crystallography.(Tokyo Inst. of Techn., 1973). Prof. (tel. 093 + 871-1931 ext 488, fax +81 92 582 4201). *X-ray topography, defect structures, scattering radiography, crystal growth, magnetic domains.*

Chikawa, Dr Jun-ichi. Photon Factory, Nat. Lab. for High Energy Physics, Oho-machi, Tsukuba, Ibaraki 305, Japan. (1930) DSc, physics (Kyoto U., 1961). Prof. (tel. 0298 + 64-1171, telex 3652534, fax +81 298 64 2801). *X-ray topography, magnetic domains, crystal growth, crystal imperfection, phase transition.*

Daimon, Dr Hiroshi. Dept. of Physics, Fac. of Sci., Tokyo U. 7-3-1 Hongo Bunkyo-ku, Tokyo 113, Japan. (1953) DSc (Tokyo U., 1983). Asst. prof. (tel. 03 + 812-2111 ext 4209, telex UTPHYSIC J-23472, fax +81 3 814 9717). *Surface physics, surface structure, photoemission, RHEED, synchrotron radiation.*

Dohi, Prof. Shoso. Dept. of Applied Physics, National Defense Academy, 10-20 Hashirimizu 1-chome Yokosuka, Kanagawa 239, Japan. (1927) DSc, physics (Hiroshima U., 1961). Prof. (tel. 0468 + 42-3810 ext 540). *Physical properties, less-common metals.*

Doi, Dr Kenji. Dept. of Appl. Physics, Fac. of Eng., Ibaraki U., 4-12-1 Narusawa-cho Hitachi, Ibaraki 316, Japan. (1929) DSc, mineralogy, solid state physics, (Tokyo U., 1959). Sr. physicist. (tel. 0294 + 35-6101). *X-ray and neutron diffraction, neutron diffraction topography, electron microscopy.*

Doi, Dr Mitsunobu. Dept. of Physical Chemistry, Osaka U. of Phermaceutical Sci., 2-10-65 Kawai Matsubara, Osaka 580, Japan. (1957) Ph.D., molecular biology.(Osaka U., 1988). Asst. (tel. 0723 + 32-1015 ext 215, fax +81 723 32 9929). *Enkephalin, peptides, nucleic acid, X-ray structure analysis.*

Fujii, Dr Satoshi. Fac. of Pharmaceutical Sci., Osaka U., 1-6 Yamada-oka Suita, Osaka 565, Japan. (1946) DPharm, pharmacy (Osaka U., 1978). Assoc. prof. (tel. 06 + 877-5111 ext 6212, fax +81 6 877 4489). *Biophysical sciences, organic crystal structure.*

Fujii, Dr Tetsuo. Advanced Materials Res. Lab., Tosoh Corp., 2743-1 Hayakawa Ayase-shi, Kanagawa 252, Japan. (1957) DSc, materials science (Tokyo Inst. of Techn., 1986). Res. (tel. 0467+ 77-2211 ext 4447, fax +81 467 78 5385). *Non-crystalline materials.*

Fujii, Prof. Yasuhiko. Inst. of Material Sci., Fac. of Sci., 1-1-1 Tennodai Tsukuba, Ibaraki 305, Japan. (1943) DSc, physics (Osaka U., 1973). Assoc. prof. (tel. 0298 + 53-5308, fax +81 298 53 5205). *X-ray diffraction, neutron scattering, phase transitions, high pressure physics, nonequilibrium physics.*

Fujime, Dr Satoru. Biophysics Div., Mitsubishi-Kasei Inst. of Life Sciences, 11 Minamiooya Machida, Tokyo 194, Japan. (1936) DSc, physics (Tohoku U., 1967). Div. chief. (tel. 0427 + 24-6288 ext 288, fax +81 427 29 1252). *Dynamical properties, biological macromolecules.*

Fujimoto, Prof. Fuminori. Inst. of Scient. and Industrial Res., Osaka U., 8-1 Mihogaoka Ibaraki, Osaka 567, Japan. (1928) DSc, physics (Tokyo U., 1959). Prof. (tel. 06 + 877-5111 ext 3495, telex 5286213 ISIROU-J, fax +81 6 877 4977). *Electron diffraction, electron channeling, channeling radiation, high energy electron microscopy, ion beam physics, nuclear and high energy physics - crystal applications.*

Fujimoto, Prof. Hirofumi. 1-9-51 Kashiwagi Sendai, Miyagi 981, Japan. (1923) DSc, physics (Tohoku U., 1961). Prof. Em., Miyagi U. of Education. (tel. 0222 + 34-4009). *X-ray diffraction.*

Fujino, Dr Kiyoshi. Dept. of Earth Sci., Fac. of Sci., Ehime U. Bunkyo-cho 2-5 Matsuyama, Ehime 790, Japan. (1945) DSc, mineralogy (Tokyo U., 1974). Assoc. prof. (tel. 0899 + 24-7111 ext 3597, fax +81 899 23 2545). *X-ray crystallography, electron microscopy, rock forming minerals.*

Fujino, Dr Nobukatsu. Fundamental Res. Section, Central Res. Labs., Sumitomo Metal Industries Ltd., 3 Nishinagasu Hondori 1-chome Amagasaki, Hyogo 660, Japan. (1935) DSc, chemistry (Osaka U., 1967). Sr. res. eng. (tel. 06 + 401-6201 ext 311). *X-ray spectroscopy, electron spectroscopy, X-ray diffraction, metallurgical applications.*

Fujita, Prof. Francisco Eiichi. Dept. of Material Sci., Fac. of Eng. Sci., Osaka U., Machikaneyama 1-1 Toyonaka, Osaka 560, Japan. (1925) DSc, physics (Hokkaido U., 1959). Prof. of metal physics. (tel. 06 + 844-1151 ext 4660). *Metal physics, lattice defect, Mössbauer spectroscopy, electron microscopy.*

Fujita, Prof. Hiroshi. Res. Center, Ultra- High Voltage Electron Microscopy, Osaka U., Yamada-kami Suita, Osaka 565, Japan. (1926) DEng, metal physics (Osaka U., 1958). Prof. (tel. 06 + 877-5111 ext 4131, 4133). *Metal physics, lattice imperfection behavior, electron microscopy applications, materials science.*

Fujiwara, Prof. Hiroshi. Fac. of Sci., Hiroshima U., 1-89 Higashisendamachi 1-chome, Hiroshima 730, Japan. (1928) DSc, physics (Tokyo U., 1959). Prof. (tel. 0822 + 41-1221 ext 438). *Materials under high pressure, high pressure X-ray diffraction, thin film crystallography, magnetic properties, metals and alloys.*

Fujiwara, Prof. Kunio. Inst. of Phys., C. of Arts and Sci., U. of Tokyo Komaba 3-8-1 Meguro-ku, Tokyo 153, Japan. (1932) DSc, Phys. (Tohoku U., 1962). Prof. (tel. 03 + 467-1171 ext 556, fax +81 3 485 2904). *Positrons in crystals.*

Fujiwara, Prof. Takaji. Fac. of Sci., Shimane U., 1060 Nishi-Kawatsu Matsue, Shimane 690, Japan. (1936) DSc, chemistry (Osaka U., 1971). Asst. prof. (tel. 0852 + 21-7100 ext 585, fax +81 852 31 0812). *Structural biology, proteins, organic compounds, reaction pathways, disorder, hydrogen bonding.*

Fujiyoshi, Dr Yoshinori. The First Res. Dept., Protein Eng. Res. Inst., 6-2-3 Furuedai Suita, Osaka 565, Japan. (1948) DSc, chemistry (Kyoto U., 1982). Res. Dir. (tel. 06 + 872-8209 Direct, telex 06-872-8200, fax +81 6 872 8210). *Electron microscopy, proteins, structural biology.*

Fukamachi, Dr Tomoe. Dept. of Electronic Eng., Saitama Inst. of Techn., 1690 Fusaiji Okabe, Saitama 369-02, Japan. (1943) DSc, physics (Tokyo U., 1976). Prof. (tel. 0485 + 85-2521). *Energy-dispersive diffractometry, anomalous scattering, Compton scattering.*

Fukano, Mr Tatsuo. Center of Optronics Product, Hoya Corp., 3-1 Musashino 3-chome Akishima, Tokyo 196, Japan. (1960) MEng, materials science (Tokyo Inst. of Techn., 1985). Eng. (tel. 0425 + 46-2736, fax +81 425 46 2708). *Optics, thin film, solid state physics.*

Fukano, Prof. Yasushige. C. of General Education, Nagoya U., Furo-cho Chikusa-ku, Nagoya 464, Japan. (1927) DSc, physics (Nagoya U., 1961). Prof. (tel. 052 + 781-5111 ext 4846). *Physics, fine particles, metals, alloys, semiconductors.*

Fukuda, Prof. Tsuguo. Res. Inst. for Iron, Steel and Other Metals, Tohoku U., 1-1 Katahira 2-chome Sendai, Miyagi 980, Japan. (1939) DSc, crystallography (Tokyo U., 1971). Prof. (tel. 0222 + 27-6200 ext 2546). *Crystal growth, ferroelectrics, semiconductors, optics, crystallography, mineralogy.*

Fukuhara, Dr Akira. Advanced Res. Lab., Hitachi Ltd., 280 Higashi-Koigakubo 1-chome Kokubunji, Tokyo 185, Japan. (1933) DSc, physics (Tokyo U., 1961). (tel. 0423 + 23-1111, fax +81 423 26 0880). *Electron and X-ray diffraction.*

Fukuyama, Dr Keiichi. Dept. of Biology, Fac. of Sci., Osaka U., 1-1 Machikaneyama-cho Toyonaka, Osaka 560, Japan. (1949) DSc, chemistry (Osaka U., 1979). Asst. prof. (tel. 06 + 844-1151 ext 4296). *Proteins, viruses.*

Fukuyama, Dr Tsutomu. Atmospheric Environment Div., Nat. Inst. for Environmental Studies, 16-2 Onogawa Tsukuba, Ibaraki 305, Japan. (1942) DSc, chemistry (Tokyo U., 1971) Sect. chief. (tel. 0298 + 51-6111 ext 393, fax +81 298 51 4732). *Aerosol chemistry, nucleation.*

Furusaki, Dr Akio. Dept. of Chemistry, Fac. of Sci., Hokkaido U., Kita 10-jo Nishi 8-chome Kita-ku, Sapporo 060, Japan. (1939) DSc, chemistry (Osaka U., 1968). Asst prof. (tel. 011 + 716-2111 ext 2703, telex 932510HOKUSC J, fax +81 11 717 9394). *Organic compounds, phase determination, structural chemisty.*

Goto, Mr Akira. Applied Physics. Hokkaido U. North13 West8 Sapporo, Hokkaido 060, Japan. (1963) MEng. (Hokkaido U., 1988). Res. Assoc. (tel. 011 + 716-2111 ext 6637, telex 932302 HOKUEN-J, fax +81 11 717 4745). *Hydrogen bonding, electron density, inorganic crystals, molecular crystals, X-ray diffraction.*

Goto, Prof. Masaru. Fac. of Eng., Oita U., 700 Dannoharu, Oita 870-11, Japan. (1932) DSc, physics (Hiroshima U., 1965). Prof. of physics. (tel. 0975 + 69-3311 ext 426, 425). *Mechanical properties (elastic and non-elastic), metals, defects, dislocations, internal friction.*

Goto, Dr Yoshiaki. Dept. of Chemistry, Fac. of Techn., Gunma U., 5-1 Tenjin-cho 1-chome Kiryu, Gunma 376, Japan. (1940) MSc, chemistry (Gunma U., 1967). Res. assoc. (tel. 0277 + 22-3181, fax +81 277 43 6556).

Gyobu, Mr Atsuo. Dept. of Mechanical Eng., Niihama Nat. C. of Techn., 7-1 Yakumo-cho Niihama, Ehime 792, Japan. (1945) BEng, Res. assoc. (tel. 0897 + 37-1240, fax +81 897 37 1245). *Materials science.*

Haga, Dr Nobuhiko. Mineralogical Inst., Fac. of Sci., Tokyo U., 3-1 Hongo 7-chome Bunkyo-ku, Tokyo 113, Japan. (1945) DSc, mineralogy (Tokyo U., 1973). Res. assoc. (tel. 03 + 812-2111 ext 4547). *Minerals, X-ray diffraction.*

Haisa, Prof. Masao. Dept. of Chemistry, Fac. of Sci., Okayama U., Tsushima Okayama 700, Japan. (1923) DSc, chemistry (Osaka U., 1961). Prof. (tel. 0862 + 52-1111 ext 441). *Organic crystal structures, large molecules, crystal physics, polymer chemistry.*

Hakoshima, Dr Toshio. Pharm. Sci. Osaka U. 1-6 Yamadaoka Suita, Osaka 565, JAPAN. (1954) DPharm.Chem.(Osaka U., 1982). Res.Assoc. (tel. 06 + 877-5111 ext 6213, fax +81 6 877 4489). *Structural biology, macromolecules, proteins, reaction pathways.*

Hamamura, Prof. Kenji. Dept. of Materials Sci., Inst. of Voc. Training, 1960 Aihara Sagamihara 229, Japan. (1941) DEng, metallurgy (Tokyo Inst. of Techn., 1975). Prof. (tel. 0427 + 61-2111 ext 346, fax +81 427 61 8293). *Materials science, metals, crystal growth, X-ray diffraction.*

Hamauzu, Prof. Yoshihiro. Dept. of Material Techn., Numazu C. of Techn., 3600 Ooka Numazu, Shizuoka 410, Japan. (1944) DSc, physics (Hokkaido U., 1974) Prof. (tel. 0559 + 21-2700 ext 455). *Surface science.*

Harada, Prof. Jimpei. Dept. of Applied Physics, Fac. of Eng., Nagoya U., Chikusa-ku, Nagoya 464-01, Japan. (1931) DSc, physics (Tokyo Inst. of Techn., 1964). Prof. (tel. 052 + 781-5111 ext 4464, fax +81 52 782 2129). *Diffuse scattering, X-ray diffraction, neutron diffraction, phase transition, disorder, thermal vibration, surface structure.*

Harada, Dr Shigeharu. Applied Chemistry Fac. of Eng. Osaka U., Yamadaoka Suita, Osaka 565, Japan. (1954) DSc, (Osaka U., 1982). Res. Assoc. (tel. 06 + 877-5111 ext 4322). *Protein crystallography.*

Harada, Dr Zyunpei. 4426 Ikuta Tama-ku, Kawasaki 214, Japan. (1898) DSc, mineralogy (Tokyo U., 1939). Em. prof. (tel. 044 + 911-8177). *Mineralogy, physical crystallography.*

Hariya, Prof. Yu. Dept. of Geology and Mineralogy, Hokkaido U., Kita 10-jo Nishi 8-chome Kita-ku, Sapporo 060, Japan. (1929) DSc, mineralogy (Hokkaido U., 1961). Prof. (tel. 011 + 716-2111 ext 2728, fax +81 11 717 9394). *Geochemistry of manganese, phase equilibrium, minerals at high pressures.*

Hashimoto, Prof. Hatsujiro. Dept. of Mechanical Eng., Okayama U. of Sci., 1-1 Ridaicho, Okayama 700, Japan. (1921) DSc, physics (Kyoto U., 1953). Prof. (tel. 0862 + 52-3161 ext 4553, fax (office)+81 862 55 3611, (home)+81 775 23 0118). *Crystal growth, electron diffraction, electron microscopy, materials science, phase transformation, scattering theory.*

Hashimoto, Mr Hideki. Res. Toray Res. Cntr. Inc., Sonoyama, Otsu, Shiga 520, Japan. (1958) MSc, X-ray topography (Nagoya U., 1983). (tel. 0775 + 37-0700 ext 4133). *EXAFS.*

Hashizume, Prof. Hiroo. Res. Lab. of Eng. Materials, Tokyo Inst. of Techn., Nagatsuta, Midori-ku, Yokohama 227, Japan. (1940) DSc(eng.), applied physics (Tokyo U., 1970). Assoc. prof. (tel. 03 + 260-4271 ext 439, fax +81 45 921 1015, email HHASHIZU at TITNCA.NC.TITECH.JUNET%UTOKYO-RELAY.CSNET at RELAY.CS.NET). *X-ray diffraction, materials science, inorganic crystals, surface structure, powder diffraction, instrumentation, X-ray optics characterization.*

Hasiguti, Prof. Ryukiti. Fac. of Eng., The Science U. of Tokyo, Kagurazaka Shinjuku-ku, Tokyo 162, Japan. (1914) DEng, metallurgy (Tokyo U., 1953). Prof. of materials science. (tel. 03 + 260-4271 ext 439). *Crystal lattice defects, radiation damage, crystal growth.*

Hata, Dr Yasuo. Inst. for Protein Res., Osaka U., 3-2 Yamadaoka Suita, Osaka 565, Japan. (1951) DSc, polymer science (Osaka U., 1979). Res. sciencent. (tel. 06 + 877-5111 ext 3837, fax +81 6 876 2533). *Structural biology, proteins, X-ray diffraction.*

Hatta, Dr Tamao. Marginal Land Res. Div. Tropical Agriculture Res. Center. 1-2 Owashi Tukuba, Ibaraki 305, Japan. (1956) DSc, geoscience (Tukuba U., 1986). Res. (tel. 02975 + 6-6360, fax +81 2975 6 6316). *Weathering.*

Hayakawa, Dr Kazunobu. Advanced Res. Lab., Hitachi Ltd., 280 Higashi-Koigakubo 1-chome, Kokubunji Tokyo 185, Japan. (1936) DSc, physics (Tokyo U., 1970). Sr. res. (tel. 0423 + 23-1111 ext 2811). *Surface physics, spin analysis.*

Hayakawa, Dr Motozo. Mechanical Eng. Tottori U. Koyama-minami Tottori, Tottori 680, Japan. (1943) PhD, materials science (1973) Assoc. prof. (tel. 0857 + 28-0321 ext 4371, fax +81 857 28 1092). *Materials science, metal, ceramics, X-ray diffraction, electron microscopy.*

Hayashi, Dr Kooya. Dept. of Chemistry, Fac. of Sci., Okayama U. of Sci., 1 Ridai-cho 1-chome, Okayama 700, Japan. (1947) DEng, solid state chemistry (Tokyo Inst. of Techn., 1975). Assoc. prof. (tel. 0862 + 52-3161 ext 315, fax +81 862 55 3611). *Inorganic synthetic chemistry, structure analysis.*

Hayashi, Prof. Mituhiko. Physics Dept., Toyama Medical and Pharm. U. 2630 Sugitani Toyama, Toyama 930-01, Japan. (1930) DSc, physics (Tokyo Inst. of Techn., 1971). Prof. (tel. 0764 + 34-2281 ext 2720, fax +81 764 34 1463). *Mössbauer spectroscopy, small particles, magnetism, minerals, inorganic crystals.*

Hidaka, Mr Tsuneo. New Ceramics Div., Asahi Optical Co. Ltd., 36-9 Maeno-cho 2-chome, Itabashi-ku, Tokyo 174, Japan. (1940) MSc, ceramics (MIT, USA, 1974). General manager. (tel. 03 + 960-5161, fax +81 3 960 5226). *Optics, materials science, bio-materials.*

Higashi, Prof. Akira. Div. of Natural Sci., International Christian U., 3-10-2 Osawa Mitaka-shi, Tokyo 181, Japan. (1922) DSc, physics (Hokkaido U., 1951). Prof.(also Em. prof. of Hokkaido U.) (tel. 0422 + 33-3337, dial in. fax +81 422 33 9887). *Crystal growth, lattice defects, electronmicroscope, ice crystals.*

Higuchi Dr Yoshiki. Basic Res. Lab. Himeji Inst. of Techn., 2167 Shosha Himeji, Hyogo 671-22, Japan. (1956) DSc, chemistry (Osaka U., 1984). Lect. (tel. 0792 + 66-1661 ext 402, fax +81 792 668868). *Protein crystallography.*

Higuchi, Prof. Taiichi. Dept. of Chemistry, Fac. of Sci., Osaka City U., 3-3-138 Sugimoto-cho, Sumiyoshi-ku, Osaka 558, Japan. (1929) DSc, chemistry (Osaka City U., 1967). Assoc. prof. (tel. 06 + 605-2557 ext 3244, fax +81 6 605 2522). *Large molecules, proteins, enzymes, organic and organometallic structures, solid state chemistry, bio-mimetic chemistry.*

Hirabayashi, Prof. Makoto. Inst. for Materials Res., Tohoku U., 1-1 Katahira 2-chome, Sendai 980, Japan. (1925) DSc, metal physics (Tohoku U., 1959). Dir. (tel. 022 + 227-6200 ext 2901, telex 852238 KINKENJ, fax +81 22 264 7984). *Electron microscopy, materials science.*

Hiragi, Dr Yuzuru. Physical Chemistry of Enzyme. Inst.for Chemical Res., Kyoto U.. Gokasho Uji, Kyoto 611, Japan. (1939) DSc, (1973). Asst.Prof. (tel. 0774 + 32-3111 ext 2161, telex 5453638 UCLKU-J, fax +81 774 33 1247). *Proteins, scattering theory, small-angle scattering, structural biology, viruses.*

Hirahara, Dr Eiji. 2-6-17 Yagiyamahoncho Sendai, Miyagi 982, Japan. (1912) DSc, physics (Hiroshima U., 1949). Em. prof. of Tohoku U., (tel. 022 + 229-1544). *Solid state physics, magnetism, semiconductors, phase transitions.*

Hirano, Prof. Shin-ichi. Dept. of Applied Chemistry., Fac. of Eng., Nagoya U., Furocho Chikusaku, Nagoya 464, Japan. (1942) D. Eng, applied chemistry (Nagoya U., 1970). Prof. (tel. 052 + 781-5111 ext 3343, fax +81 52 782 5170). *Hydrothermal crystal growth, crystallization, hydrolysis, organometallic compounds, ceramic processing.*

Hirayama, Dr Noriaki. Tokyo Res. Lab., Kyowa Hakko Kogyo Co. Ltd., 3-6-6 Asahimachi Machida, Tokyo 194, Japan. (1948) DSc, chemistry (Tokyo Inst. of Techn., 1981). Sr. res. fellow. (tel. 0427 + 25-2555 ext 256). *Drug - macromolecule interaction, protein crystallograpy, polymorphism.*

Hirota, Dr Fumio. X-ray analysis. Dental Reseach Institute. Nippon Dental U., 1-9-20 Fujimi Chiyoda-ku, Tokyo 102, Japan. (1943) DDs, dentistry (Nippon Dental U., 1986). Lect. (tel. 03 + 261-8311 ext 268). *Structural biology, texture, X-ray diffraction.*

Hirotsu, Dr Ken. Dept. of Chemistry, Fac. of Sci., Osaka City U., Sugimoto, Sumiyoshi-ku, Osaka 558, Japan. (1942) DSc, chemistry (Osaka City U., 1974). Lect. (tel. 06 + 692-1231 ext 3244). *Organic crystal structures, large molecules.*

Hirotsu, Dr Yoshihiko. Nagaoka U. of Techn., Kamitomioka-cho, Nagaoka 940-21, Japan. (1945) DEng (Tokyo Inst. of Techn., 1974). Asst. prof. (tel. 0258 + 46-6000). *Structure, metals and alloys, electron diffraction, electron microscopy.*

Homma, Mr Shigeru. Nat. Inst. for Inorganic Materials Res., 1-1 Namiki Tsukuba, Ibaraki 305, Japan. (1936) BSc, mineralogy (Hokkaido U., 1962). Head res. (tel. 0298 + 51-3351 ext 280). *X-ray diffraction topography, crystal growth.*

Homma, Mr Teiichi. Inst. of Industrial Science, U. of Tokyo, 22-1 Roppongi 7 Minato-ku, Tokyo 106, Japan. (1931) DEng, metals (Tokyo U., 1961). Prof. (tel. 03 + 402-6231 ext 2160, telex 0242-3216 IISTYOJ, fax +81 3 402 6375). *Defect structure, surface structure, metals.*

Horioka, Dr Keiji. ULSI Res. Center, Toshiba. 1 Komukai-Toshiba-cho Saiwaiku Kawasaki, Kanagawa 210, Japan. (1955) DSc, chemistry(Osaka U.,1983). (tel. 044 + 549-2189). *Semiconductors, materials science.*

Horiuchi, Dr Hiroyuki. Mineralogical Inst., Fac. of Sci., Tokyo U., Hongo Bunkyo-ku, Tokyo 113. Japan. (1940) DSc, mineralogy (Tokyo U., 1969). Assoc. prof. (tel. 03 + 812-2111 ext 4542, fax +81 3 816 5714). *Structure of minerals, X-ray diffraction, instrumentation.*

Horiuchi, Dr Shigeo. Nat. Inst. for Inorganic Materials Res., Sakura-mura Niihari-gun, Ibaraki 305, Japan. (1939) DEng, metallurgy (Tokyo U., 1967). Supervising res. (tel. 0298 + 51-3351 ext 283). *Structure analysis, high resolution electron microscopy.*

Hosaka, Dr Masahiro. Gemmology, Gemmology and Jewelry Arts of Yamanashi, Tokojicho 1955-1, Kofu Yamanashi 400, Japan. (1944) PhD, crystal growth, inorganic chemistry (Yamanashi U., 1967). Asst. prof. (tel. 0552 + 32-6672). *Hydrothermal growth, inorganic chemistry.*

Hoshino, Prof. Sadao. Inst. of Applied Physics, Tsukuba U., 1-1-1 Tennodai Tsukuba, Ibaraki 305, Japan. (1926) DSc, physics (Osaka U., 1958). Prof. (tel. 0298 + 53-5308, fax +81 298 53 5205). *X-ray and neutron diffraction, phase transitions.*

Hosoya, Prof. Masahiko. Dept. of Physics, U. of the Ryukyus, Senbaru-1 Nisihara, Okinawa 903-01, Japan. (1943) MSc, Physics (Hokkaido U., 1967). Assoc. prof. (tel. 09889 + 5-2221 ext 2635, fax +81 9889 5 2247). *Computing, group theory, phase transitions, symmetry.*

Hosoya, Prof. Sukeaki. 154-8 Meguridamachi Kodaira, Tokyo 187, Japan. (1924) PhD, physics (Wales U., UK, 1958) DSc, physics (Tokyo U., 1961). Em. prof. (Tokyo U.). (tel. 0423 + 23-7463). *Teaching crystallography, anomalous dispersion, Compton scattering, molecular biology.*

Ibata, Dr Koichi. Techn. Div., Graphtec Corp., 3-19-6 Nishishinagawa Shinagawa-ku, Tokyo 141, Japan. (1947) DSc, chemistry (Tokyo Inst. of Techn., 1975). Chief res. (tel. 03 + 491-0655, fax +81 3 492 1072). *Chemistry, applied physics, microelectronics, astronomy, biology.*

Ichida, Dr Hikaru. Dept. of Chemistry, Fac. of Sci., U. of Tokyo, 7-3-1 Hongo Bunkyo-ku, Tokyo 113, Japan. (1953) DSc, chemistry (Tokyo U., 1982). Res. assoc. (tel. 03 + 812-2111 ext 4361, telex UTYOSCI J33659, fax +81 3 814 2627). *Inorganic crystals, chemical bonding, structural chemistry, metal oxides, co-ordination compounds.*

Ichikawa, Mr Kimio. Fujinomiya Res. Lab. Fuji Photo Firm Co.Ltd., Oonakazato 200 Fujinomiya, Shizuoka 418, Japan. (1957) MSc, polymer science(Hokkaido U., 1982). (tel. 0544 + 26-7631, fax +81 544 26 7641). *Hydrogen bonding, liquid crystals, macromolecules, phase transitions, Mechanical and thermal properties of polymers.*

Ichikawa, Dr Mizuhiko. Dept. of Physics, Fac. of Sci. Hokkaido U., Kita 10-jo Nishi 8-chome, Kita-ku, Sapporo 060, Japan. (1940) DSc, chemistry (Osaka U., 1979). Res. assoc. (tel. 011 + 716-2111 ext 5427, fax +81 11 717 9304, telex 932510HOKUSC J). *Hydrogen bonding, inorganic crystals, phase transitions, X-ray diffraction.*

Ichimiya, Prof. Ayahiko. Dept. of Applied Physics, Nagoya U., Furo-cho Chikusa-ku, Nagoya 464-01, Japan. (1940) DSc, physics (Nagoya U., 1966). Assoc. Prof. (tel. 052 + 781-5111 ext 4459, fax +81 52 782 2129). *Crystal growth, electron diffraction, semiconductors, surface structure.*

Ichimura, Mr Takeo. Glass Manufacturing Div., Nippon Kogaku K.K., 1773 Asamizodai Sagamihara, Kanagawa 228, Japan. (1929) BSc, applied chemistry (Tokyo U., 1954). (tel. 0427 + 45-3311).

Ichinokawa, Prof. Takeo. Dept. of Applied Physics, Waseda U., 3-4-1 Ohkubo Shinjuku-ku, Tokyo 160, Japan. (1928) DSc, physics (Waseda U., 1958). Prof. (tel. 03 + 209-3211 ext 3559). *Electron microscopy, surface science.*

Iida, Dr Atsuo. Photon Factory, Nat. Lab. for High Energy Physics, 1-1 Oho Tsukuba, Ibaraki 305, Japan. (1948) DEng, crystal characterization (Tokyo U., 1977). Assoc. prof. (tel. 0298 + 64-1171). *X-ray spectrometry, microscopy, X-ray diffraction, instrumentation.*

Iijima, Dr Kinya. Dept. of Chemistry, Shizuoka U., 836 Oya, Shizuoka 422, Japan. (1941) DSc, chemistry (Hokkaido U., 1976). Res. assoc. (tel. 0542 + 37-1111 ext 552). *Electron diffraction, structural chemistry, chemical bonding.*

Iijima, Dr Sumio. Fundamental Res. Labs. NEC Corp., 4-1-1 Miyazaki Miyamae-ku Kawasaki, Kanagawa 213, Japan. (1939) PhD, physics (Tohoku U., 1968). Sr. Res. Manager (tel. 044 + 856-2049, fax +81 44 856 2213). *Defect structures, diffuse scattering, disorder, electron diffraction, electron micrscopy, inorganic crystals, surface structure, semiconductors.*

Iijima, Prof. Takao. Fac. of Sci., Gakushuin U., 1-5-1 Mejiro, Toshima-ku, Tokyo 171 Japan. (1934) DSc, chemistry (Tokyo U., 1962). Prof. (tel. 03 + 986-0221 ext 424).

Iimura, Mr Yasuhiro. The Inst. of Phys. and Chem. Res., 2-1 Hirosawa Wako, Saitama 351-01, Japan. (1942) BEng, elect.eng. (Shibaura Inst. of Techn., 1966). Techn. scient. (tel. 0484 + 62-1111 ext 3381, telex 02962818 RIKEN J, fax +81 484 62 1449). *Structural chemistry, X-ray diffraction.*

Iishi, Dr Kazuaki. Dept. of Mineralogical Sci. and Geology, Fac. of Sci., Yamaguchi U., 1677-1 Yoshida, Yamaguchi 753, Japan. (1942) DSc, mineralogy (Hiroshima U., 1973). Asst. prof. (tel. 0839 + 22-6111 ext 385, fax +81 839 32 2041). *Crystal growth, electron microscopy, minerals, phase transitions, structural chemistry.*

Iitaka, Prof. Yoichi. Sch. of Medicine, Biophysical lab., Teikyo U., 359 Ohtsuka Hachioji, Tokyo 192-03, Japan. (1927) DSc, crystallography (Tokyo U., 1959). Prof. (tel. 0426 + 76-8211). *Organic crystal structures, instrumentation, large molecules, biological substances, computer programming.*

Iizuka, Dr Masakatsu. Fac. of Education, Bunkyo U., 3337 Minamiogishima Koshigaya, Saitama 343, Japan. (1933) DSc, structural crystallography (Tokyo

U. of Education, 1972). Lect. (tel. 0489 + 74-8811). *Crystal structure, crystal chemistry.*

Iizumi, Dr Masashi. Office of Planning, Japan Atomic Energy Res. Inst., 2-2-2 Uchisaiwai-cho Chiyoda-ku, Tokyo 100, Japan. (1935) DSc, Physics (Tokyo U., 1974). Deputy dir. (fax +81 3 592 2119). *Neutron scattering, phase transitions.*

Ikawa, Dr Hiroyuki. Dept. of Inorganic Materials, Tokyo Inst. of Techn., 2-12-1 Ookayama Megro-ku, Tokyo 152, Japan. DEng. Assoc. Prof. (tel. 03 + 726-1111 ext 2517, fax +81 3 729 0393). *Inorganic crystals, materials science, structural chemistry.*

:hp2, **Imai**, Mr Katsuhiro. Chemistry, Tukuba U., Tukuba, Ibaraki 305, Japan. (1965) BSc, chemistry(1988). (tel. 0298 + 53-4520). *Inorganic crystals, powder diffraction, X-ray diffraction, structural chemistry.*

Imanishi, Dr Yasuhiro. 5-19-17 Fukasawa Setagaya-ku, Tokyo 158, Japan. (1947) DSc, mineralogy (Tokyo U., 1977). (tel. 03 + 702-2976). *Phase transformation, minerals, crystal growth, semiconductor materials, physical properties.*

Imura, Prof. Toru. Dept. of Metallurgy, Fac. of Eng., Nagoya U., Furo-cho Chikusa-ku, Nagoya 464, Japan. (1924) DSc, physics (Osaka U., 1957). Prof. (tel. 052 + 781-5111 ext 3350). *Metal physics, electron and X-ray diffraction, electron microscopy, crystal growth, crystal characterization, deformation and strength, amorphous metal structure and stability.*

Ino, Prof. Shozo. Dept. of Physics, Fac. of Sci., Tokyo U., 7-3-1 Hongo Bunkyo-ku, Tokyo 113, Japan. (1936) DSc, science (Tohoku U., 1970). Prof. (tel. 03 + 812-2111 ext 4208, telex UTPHYSIC J-23472, fax +81 3 814 9717) *Surface structure, electron diffraction, electron microscopy, semiconductor, metals.*

Ino, Dr Tadashi. Dept. of Physics, Fac. of Sci., Osaka City U., Sugimoto-cho Sumiyoshi-ku, Osaka 558, Japan. (1923) DSc, physics (Nagoya U., 1961). Prof. (tel. 06 + 692-1231 ext 3232). *Gas molecular structures, amorphous substances, microcrystals.*

Inokuchi, Prof. Hiroo. Inst. for Molecular Sci., Okazaki Nat. Res. Inst., 38 Saigo Naka Myodaiji-cho Okazaki, Aichi 444, Japan. (1927) DSc (Tokyo U., 1956). Dir. (tel. 0564 + 54-1111 ext 2209). *Molecular science.*

Inoue, Mr Hironao. Designing, MAC Sci.Co.,Ltd. 2-25-16 Nakanokami-cho Hachioji, Tokyo 192, Japan. (1946) BSc, (Shizuoka U., 1970). System Eng. (tel. 0426 + 24-2201, fax +81 426 24 2723). *Instrumentation.*

Inoue, Dr Morio. Kyoto Res. Lab. Matsushita Electronics Corp., 19 Nishikujo-Kasuga-cho, Minami-ku, Kyoto 601, Japan. (1937) DSc, chemistry (Kyoto U., 1972). General Manager. (tel. 075 + 681-3181 ext 227, fax +81 75 681 0705). *Semiconductors, materials science, defect structures, crystal growth.*

Inoue, Dr Zenzaburo. Third Res. Group, Nat. Inst. for Inorganic Materials Res., 1-1 Namiki Tsukuba, Ibaraki 305, Japan. (1940) DSc, mineralogy (Tokyo U., 1974). Sr. res. officer. (tel. 0298 + 51-3351 ext 373, telex 0298 51 3351, fax +81 298 52 7449). *Polytypism, phase transition in crystals, high temperature X-ray crystallography, ceramic crystal structures, crystal growth.*

Ishibashi, Prof. Yoshihiro. Synthetic Crystal Res. Lab., Fac. of Eng., Nagoya U., Furo-cho Chikusa-ku, Nagoya 464, Japan. (1935) DSc, physics (Tokyo U., 1963). Prof. (tel. 052 + 781-5111 ext 3597, fax +81 52 782 9209). *Ferroelectricity, phase transition, crystal growth.*

Ishida, Prof. Kohtaro. Phys. of Physics, Fac. of Sci. and Techn., Sci. U. of Tokyo, Noda-shi, Chiba 278, Japan. (1940) DSc, physics (Kyoto U., 1970). Prof. (tel. 0471 + 24-1501 ext 3209, fax +81 471 23 9361). *X-ray and electron diffraction, crystal growth, superconductors.*

Ishida, Dr Toshimasa. Physical Chemistry, Osaka U. of Pharmaceutical Sci. 2-10-65 Kawai Matsubara, Osaka 580, Japan. (1946) Dr, pharmacy (Osaka U., 1977). Assoc. prof. (tel. 0723 + 32-1015 ext 298, fax +81 723 32 9929). *Structure - function relationships, bioactive substances, drugs.*

Ishihara, Mr Nobukazu. Dept. of Physics, C. of Humanities and Sci., Nihon U., 25-40 Sakurajosui 3-chome Setagaya-ku, Tokyo 156, Japan. (1933) BSc, chemistry (1956). Asst. prof. (tel. 03 + 302-8131 ext 283). *Crystal growth, polymer crystal, electron microscopy.*

Ishikawa, Dr Tetsuya. Applied Physics Tokyo U. 7-3-1 Hongo Bunkyo-ku, Tokyo 113, Japan. (1954) DEng.(Tokyo U., 1982). Asst. Prof. (tel. 03 + 812-2111). *X-ray optics, synchrotron radiation, dynamical X-ray diffraction, X-ray topography.*

Ishizawa, Dr Nobuo. Res. Lab. of Eng. Materials, Tokyo Inst. of Techn., 4259 Nagatsuta Midori-ku Yokohama, Kanagawa 227, Japan. (1949) DEng.(Tokyo Inst. of Techn., 1979). Assoc. Prof. (tel. 045 + 922-1111 ext 2309, fax +81 45 921 1015). *Structural chemistry.*

Ishizuka, Dr Kazuo. Lab. of Crystal and Powder Chemistry, Inst. for Chemical Res., Kyoto U., Gokasho Uji 611, Japan. (1947) DSc, chemistry (Kyoto U., 1978). Sr res. (tel. 0774 + 32-3111 ext 2072, fax +81 774 33 1247). *Electron diffraction, electron microscopy, electron energy loss spectroscopy.*

Isobe, Dr Mitsumasa. Nat. Inst. for Inorganic Materials Res., 1-1 Namiki Tsukuba, Ibaraki 305, Japan. (1944) DEng, chemistry (Tokyo Inst. of Techn., 1973). Res. (tel. 0298 + 51-3351 ext 254). *Crystal chemistry.*

Itai, Dr Akiko. Fac. of Pharmaceutical Sci., Tokyo U., 7-3-1 Hongo Bunkyo-ku, Tokyo 113, Japan. (1941) Dr, chemistry (Tokyo U., 1969). Res. assoc. (tel. 03 + 812-2111 ext 4841). *X-ray crystallography, computer drug design.*

Ito, Prof. Masatoki. Chemistry Dept., Fac. of Sci. and Eng., Keio U. 3-14-1 Hiyoshi, Kohoku-ku, Yokohama 223, Japan. (1942) DSc, chemistry (Tokyo U., 1970) Assoc. prof. (tel. 044 + 63-1141 ext 3911). *Surface physics, catalysis.*

Ito, Prof. Tetsuzo. Dept. of Chemical Process Eng., Kanagawa Inst. of Techn., 1030 Simoogino Atsugi, Kanagawa 243-02, Japan. (1936) DSc, chemistry (Tokyo U., 1965). Prof. (tel. 0462 + 41-1211 ext 3192, fax +81 462 42 3737). *Molecular crystals, electron density.*

Ito, Mr Yoshiaki. Inst. of Scient. and Industrial Res. 8-1 Mihogaoka Ibaraki, Osaka 567, Japan. (1954) PhD, (Osaka U.). (tel. 06 + 877-5111 ext 3546). *Ionic conductor, materials science, X-ray diffraction and spectrometry.*

Ito, Dr Yuji. Inst. for Solid State Physics, Tokyo U., Roppongi Minato-ku, Tokyo 106, Japan. (1936) PhD, physics (MIT, USA, 1967). Assoc. prof. (tel. 03 + 402-6231 ext 688). *Neutron scattering, solid state physics, biophysics.*

Ito, Dr Yukio. Central Res. Lab. Hitachi Ltd. 1-280 Higashikoigakubo Kokubunji, Tokyo 185, Japan. (1947) DSc, Chemistry(1981). Res. (tel. 0423 + 23-1111 ext 3931, telex 2832522 CHUKEN J). *Superconductors, piezoelectric materials, solid electrolyte.*

Itoh, Prof. Kazuyuki. Materials Sci., Fac. of Sci. Hiroshima U. Higashisenda-machi Naka-ku, Hiroshima 730, Japan. (1941) DSc, Physics(Hokkaido U., 1972). Assoc.Prof. (tel. 082 + 241-1221 ext 3802, fax +81 82 242 7454). *Disorder, phase transitions, thermal vibrations, X-ray diffraction.*

Iwanaga, Dr Hiroshi. Fac. of Liberal Arts, Nagasaki U., 1-14 Bunkyo-machi, Nagasaki 852, Japan. (1938) DSc (Tokyo Inst. of Techn., 1978). Prof. (tel. 0958 + 47-1111 ext 3258, fax +81 958 43 1379). *Twin morphology polarity, dislocation, stacking faults, electron diffraction, X-ray diffraction, etching.*

Iwasaki, Prof. Fujiko. Dept. of Applied Physics and Chemistry, Electro-Communications U., 1-5-1 Chofugaoka Chofu, Tokyo 182, Japan. (1937) DSc, chemistry (Tokyo U., 1966). Prof. (tel. 0424 + 83-2161 ext 3822, fax +81 424 84 4518). *Structural chemistry, molecular crystals, chemical bonding, reaction pathways, electron density.*

Iwasaki, Dr Hiroshi. Res. Div., Nippon Telegraph and Telelphone Corp., 9-11 Midori-cho 3-chome Musashino, Tokyo 180, Japan. (1933) DSc, chemistry (Osaka U., 1963). (tel. 0422 + 59-2616) *Ferroelectric crystals, crystal growth.*

Iwasaki, Prof. Hiroshi. Photon Factory, Nat. Lab. for High Energy Physics, 1-1 Oho Tsukuba, Ibaraki 305, Japan. (1933) DSc, physics (Tohoku U., 1966). Prof. (tel. 0298 + 64-1171 ext 5002, telex 3652 534 KEKOHO, fax +81 298 64 2801). *Phase transitions, high pressures, materials science.*

Iwasaki, Dr Hitoshi. Crystal Physics Lab., Inst. of Physical and Chemical Res., 1 Hirosawa 2-chome Wako, Saitama 351-01, Japan. (1935) DSc, chemistry (Tokyo U., 1965). Chief scient. (tel. 0484 + 62-1111 ext 3341, fax +81 484 62 1554). *Inorganic crystals, molecular crystals, symmetry.*

Iwata, Mrs Miyuki. Inst. for Solid State Physics, Tokyo U., 22-1 Roppongi 7-chome Minato-ku, Tokyo 106, Japan. (1940) MSc, chemistry (Tokyo U., 1964). (tel. 03 + 402-6231 ext 663). *Accurate electron density determination, diffraction in general, molecular science.*

Iwata, Dr Yutaka. Res. Reactor Inst., Kyoto U., Kumatori-cho Sennan-gun, Osaka 590-04, Japan. (1937) DSc, physics (Kyushu U., 1977). Asst. Prof. (tel. 0724 + 52-0901 ext 2322). *Neutron diffraction, phase transitions.*

Izui, Dr Kazuhiko. Chemistry Div., Japan Atomic Energy Res. Inst., 2-4 Shirakata-Shirane Tokaimura Nakagun, Ibaraki 319-11, Japan. (1929) DSc, physics (Hiroshima U., 1965). Principal sciencent. (tel. 02928 + 2-5523). *Crystal structure, lattice defects, radiation damage, electron microscopy.*

Izumi, Prof. Takatoshi. Dept. of Eng. Physics, Fac. of Eng., Chubu U., 1200 Matsumoto-cho, Kasugai Aichi 487, Japan. (1938) DEng, BSc, physics (Osaka Gakugei U., 1966). Prof. (tel. 0568 + 51-1111 ext 446). *X-ray crystallography, ferroelectrics.*

Izumi, Dr Yoshinobu. Dept. of Polymer Sci., Fuculty of Sci., Hokkaido U. Kita10 Nishi8 Kita-ku Sapporo, Hokkaido 060, Japan. (1944) PhD, polymer(Hokkaido U., 1972). Instr. (tel. 011 + 716-2111 ext 3810, fax +81 11 717 9394). *Macromolecules, phase transitions, diffuse scattering, small-angle scattering.*

Kageyama, Dr Hiroyuki. Inorganic Material Dept., Government Industrial Res. Inst., Osaka 1-8-31 Midorigaoka Ikeda, Osaka 563, Japan. (1955) DEng.(Osaka U., 1984). Res. (tel. 0727 + 51-8351 ext 1153). *Defect structures, inorganic crystals, molecular crystals, reaction pathways, structural chemistry.*

Kagotani, Dr Toshio. Materials Sci. Tohoku U., Fac. of Eng. Aramaki aza Aoba Sendai, Miyagi 980, Japan. (1941) DEng, technology (Tohoku U.). Instr. (tel. 022 + 222-1800 ext 4453, fax +81 22 268 2949). *Crystal growth, defect structure, inorganic crystals, magnetism, materials science, semiconductors, chemical bonding, structural chemistry, X-ray diffraction, quasicrystals.*

Kai, Prof. Yasushi. Dept. of Applied Chemistry, Osaka U., 2-1 Yamadaoka Suita, Osaka 565, Japan. (1943) DEng, chemistry (Osaka U., 1973). Assoc. prof. (tel. 06 + 877-5111 ext 4322, fax +81 6 877 0178). *Proteins, organic compounds, organometallic compounds, molecular crystals.*

Kainuma, Prof. Yoshiro. General Education, Nagoya U. of Foreign Studies, 57 Iwasaki Takenoyama Nisshin-cho Aichi-gun, Aichi 470-01, Japan. (1922) DSc, physics (Nagoya U., 1955). Prof. (tel. 0568 + 91-3738). *Symmetry, electron diffraction, X-ray diffraction.*

Kaito, Dr Chihiro. Dept. of Electronics and Information Sci., Kyoto Inst. of Techn. Matsugasaki, Sakyo-ku, Kyoto 606, Japan. (1943) BSc, Physics (Ritsumeikan

U., 1965). Asst. prof. (tel. 075 + 791-3211 ext 445). *Crystal growth, electron microscopy, materials science, minerals, defect structure, X-ray diffraction.*

Kakudo, Prof. Masao. 18-18 Ohara Ashiya, Hyogo 659, Japan. (1918) DSc, chemistry (Osaka U., 1953). Em. prof. (tel. 0797 + 22-7137). *Proteins, biological substances, structure.*

Kamijo, Dr Nagao. Gov. Ind. Res. Inst., Osaka, 1-8-31 Midorigaoka Ikeda, Osaka 563, Japan. (1936) DSc, chemistry, solid state physics (Kwansei Gakuin U., 1978). Sr. res. (tel. 0727 + 51-8351). *X-ray crystallography, solid state physics, phase transition.*

Kamiya, Dr Kazuhide. Chemical Res. Labs., Central Res. Div., Takeda Chemical Industries Ltd., Jusohonmachi Yodogawa-ku, Osaka 532, Japan. (1939) DSc, crystallography (Tokyo U., 1980). Res. (tel. 06 + 301-1231 ext 2401). *X-ray crystallography.*

Kamiya, Dr Nobuo. X-ray Crystallography Lab., RIKEN - Inst. of Physical and Chemical Res., Hirosawa 2-1 Wako, Saitama 351-01, Japan. (1953) DSc, chemistry (Nagoya U., 1984). (tel. 0484 + 62-1111 ext 3342, fax +81 484 62 1449). *Protein X-ray crystallography, structural biology, synchrotron radiation.*

Kamiya, Prof. Yoshihiro. Toyota Technological Inst., Hisakata Tenpaku Nagoya, Aichi 468, Japan. (1932) DSc. (Nagoya U., 1960). Prof. (tel. 052 + 802-1111 ext 315, fax +81 52 802 6069). *Electron diffraction, electron microscopy. un2*

Kanamaru, prof. Fumikazu. Inst. of Sci. and Industry, Osaka U., 8-1 Mihogaoka Ibaraki, Osaka 567, Japan. (1932) DSc., structural chemistry (Osaka U., 1963). Prof. (tel. 06 + 877-5111 ext 3545, telex 5286213 ISIROU-J, fax +81 6 877 4977). *Inorganic crystals, materials science, structural chemistry.*

Kanehisa, Miss Nobuko. Fac. of Eng., Osaka U. Yamadaoka Suita, Osaka 565, Japan. (1943) BSc, chemistry (Osaka City U., 1966). Techn. official (tel. 06 + 877-5111 ext 4322, fax +81 6 876 6484). *Structural chemistry.*

Kasai, Prof. Nobutami. Dept. of Applied Chemistry, Osaka U., 2-1 Yamadoka Suita, Osaka 565, Japan. (1929) DEng, chemistry (Osaka U., 1962). Prof. (tel. 06 + 877-5111 ext 4321). *Organic and organometallic compounds, macromolecules, polymers, proteins, instrumentation, teaching.*

Kashino, Dr Setsuo. Dept. of Chemistry, Fac. of Sci., Okayama U., Tsushima Okayama 700, Japan. (1937) DSc, chemistry (Osaka U., 1973). Asst. prof. (tel. 0862 + 52-1111 ext 391). *Organic crystal structures, solid phase organic reaction.*

Kashiwase, Dr Yasuji. Dept. of Physics, C. of General Education, Nagoya U., Furo-cho Chikusa-ku, Nagoya 464, Japan. (1932) DSc, physics (Nagoya U., 1965). Prof. (tel. 052 + 781-5111 ext 4845). *X-ray diffuse scattering, X-ray crystallography, dynamical diffraction, solid state physics.*

Kasuga, Prof. Masanobu. Dept. of Electronics, Fac. of Eng., Yamanashi U., 3-11 Takeda 4-chome, Kofu 400, Japan. (1941) DEng, electronics (Nagoya U., 1971). Prof. (tel. 0552 + 52-1111 ext 5245, fax +81 552 51 6828). *MCZ.*

Katada, Prof. Kinya. Dept. of Physics, U. of Osaka Prefecture, Mozu Umemachi 4-chome Sakai, Osaka 591, Japan. (1925) DSc, chemistry (Osaka U., 1961). Prof. (tel. 0722 + 52-1161). *Structure, phase stability, metals and alloys.*

Katagawa, Dr Takeshi. ULVAC Corp., 2500 Hagisono Chigasaki, Kanagawa 253, Japan. (1941) DEng, applied physics (Nagoya U., 1977). Res. asst. *X-ray crystallography.*

Katayama, Dr Chuji. Res. Lab., MAC Sci. Co. Ltd., 2-25-16 Nakanokami-cho Hachioji, Tokyo 192, Japan. (1944) DSc, chemistry (Nagoya U., 1985). Dir. (tel. 0426 + 24-2201, fax +81 426 24 2723). *Instrumentation, computing, crystallographic data.*

Katayama, Mr Kenichi. R and D Showa Denko Silicon 1505 Shimokagemori Chichibu, Saitama 369-18, Japan. (1958) MSc, phys.(Tohoku U., 1984). (tel. 0494 + 23-6115, fax +81 494 22 5700). *Crystal growth, defect structures, materials science, semiconductors.*

Katayama, Prof. Mikio. Dept. of Informatics, Teikyo U. of Techn. Uruido Ichihara, Chiba 290-01, Japan. (1926) DSc, chemistry(Tokyo U.,1957). Prof. (tel. 0436 + 74-5511 ext 2300, fax +81 436 74 7551). *Structural chemistry, laser spectroscopy.*

Katayanagi, Mr Katsuo. 1st Dept., Protein Eng. Res. Inst., 6-2-1 Suita, Osaka 565, Japan. (1961) MEng, materials science (Tokyo Inst. of Techn., 1986). Res. Scient. (tel. 06 + 872-8201, fax +81 6 872 8210). *Protein crystal structure analysis.*

Kato, Dr Akira. Dept. of Geology, Nat. Sci. Museum, 23-1 Hyakunin-cho 3-chome Shinjuku, Tokyo 160, Japan. (1931) DSc, geology (Tokyo U., 1959). Res. officer. (tel. 03 + 364-2311 ext 331). *Descriptive mineralogy, mineral classification.*

Kato, Mr Ichiro. Consumer Goods Testing Lab., Tokyo Shibaura Electric Co., 14-8 Omori-nishi 1-chome Ota-ku, Tokyo 143, Japan. (1931) MSc, mineralogy (Tokyo U. 1955). Manager. (tel. 03 + 762-6844). *Applied physics, crystallography, perfect crystals.*

Kato, Dr Katsuo. Nat. Inst. for Inorganic Materials Res., 1-1 Namiki Tsukuba, Ibaraki 305, Japan. (1938) Dr.rer.nat. habil., mineralogy (U. Hamburg, FRG, 1968, 1972). Group leader. (tel. 0298 + 51-3351 ext 375, fax +81 298 52 7449). *Inorganic crystals.*

Kato, Prof. Masanori. Dept. of Inorganic Materials, Fac. of Eng., Tokyo Inst. of Techn., 12-1 Oh-okayama 2-chome Meguro-ku, Tokyo 152, Japan. (1928) DEng, chemistry (Tokyo Inst. of Techn., 1966). Prof. (tel. 03 + 726-1111 ext 2518). *Inorganic materials.*

Kato, Prof. Norio. Dept. of Physics, Fac. of Sci. and Techn., Meijo U., Tenpaku-cho Tenpaku-ku Nagoya, Aichi 468, Japan. (1923) DSc, physics (Nagoya U., 1954).

Prof. (tel. 052 + 832-1151 ext 5278). *Diffraction theory, crystal perfection, crystal growth.*

Kato, Prof. Toshio. Inst. of Earth Sci., Yamaguchi U., 1677-1 Yoshida, Yamaguchi 753, Japan. (1931) DSc, mineralogy (Tokyo U., 1958). Prof. (tel. 0839 + 22-6111 ext 518). *X-ray diffraction, minerals.*

Kato, Mr Yoshihiro. Dept. of Physics, Osaka Kyoiku U., 43 Minami-Kawahori-cho Tennoji-ku, Osaka 543, Japan. (1934) BEd, physics (Osaka Kyoiku U., 1958). Asst. prof. (tel. 06 + 771-8131 ext 295). *Structural disorder, phase transitions, organic crystals.*

Katoh, Mr Ichiro. Ultra Precision Techn. Group, Manufacturing Eng. Lab., Toshiba Corp., 8 Shinsugita-cho Isogo-ku Yokohama, Kanagawa 235, Japan. Expert (tel. 045 + 756-2752, fax +81 45 775 0719). *Crystal growth, defect structures, semiconductors.*

Katsube, Prof. Yukiteru. Inst. for Protein Res., Osaka U. Yamada-oka 3-2 Suita, Osaka 565, Japan. (1930) DSc, chemistry (Osaka U., 1963). Prof. (tel. 06 + 877-5111 ext 3836, fax +81 6 876 2533). *Proteins, organic compounds, instrumentation.*

Katsuya, Mr Yoshio. Inst. for Protein Res. Osaka U. 3-2 Yamadaoka Suita, Osaka 565, Japan. (1962) MSc.(Osaka U., 1986). Grad. Student (tel. 06 + 877-5111 ext 3838, fax +81 6 876 2533). *Proteins, macromolecules.*

Kawado, Dr Seiji. Res. Center, Sony Corp., 174 Fujitsuka-cho, Hodogaya-ku, Yokohama 240, Japan. (1940) DEng, applied physics (Tokyo U., 1982). Chief res. scient. (tel. 045 + 334-6910, fax +81 45 352 3169). *X-ray topography, crystal imperfections, semiconductors.*

Kawahara, Prof. Akira. Dept. of Earth Sci., Fac. of Sci., Okayama U., Tsushima Okayama 700, Japan. (1932) DSc, mineralogy (Tokyo U., 1962). Prof. (tel. 0862 + 52-1111 ext 439). *Crystal structure, minerals.*

Kawai, Mr Takatoshi. Tukuba Res. Labs. Eisai Co., Ltd. 1-3 Tokodai 5-chome Tukuba, Ibaraki 300-26, Japan. (1960)MAg (tel. 0297 + 47-2211 ext 401, fax +81 297 47 2037). *Organic chemistry, computational chemistry.*

Kawai, Mr Toshiaki. Res. Inst. of Electronics, Shizuoka U., 3-5-1 Jyohoku Hamamatsu, Shizuoka 432, Japan. (1942) MEng, crystal growth (Shizuoka U., 1974). (tel.0534 + 71-1171). *Crystal growth, semiconductor physics, infrared physics, X-ray diffraction.*

Kawaminami, Prof. Masaru. Dept. of Physics, C. of Liberal Arts, Kagoshima U., 21-30 Korimoto 1-chome, Kagoshima 890, Japan. (1941) DSc, physics (Kyushu U., 1970). Prof. (tel. 0992 + 54-7141 ext 5793, fax +81 992 58 4866). *Phase transitions, precision lattice constants.*

Kawamori, Miss Asako. School of Sci., Kwansei Gakuin U., Uegahara Nishinomiya, Hyogo 662, Japan. (1935) DSc, chemical physics (Osaka U., 1962). Prof. (tel. 0798 + 53-6111 ext 5279, fax +81 798 51 0914). *Magnetic resonance, solid state physics, biophysics.*

Kawamura, Dr Tsutomu. Res. and Development Group, Nippon Electric Co. Ltd., 34 Miyukigaoka Tsukuba, Ibaraki 305, Japan. (1931) DSc, mineralogy (Tokyo U., 1966). Res. fellow. (tel. 029757 + 1111, fax +81 44 856 2213). *Crystal growth, crystal defects, inorganic crystal structures.*

Kawano, Prof. Shigeaki. Kyushu Jogakuin Junior C., 3-12-16 Kurokami, Kumamoto 860, Japan. (1931) BSc, physics (Kyushu U., 1954). Prof. (tel. 096 + 343-3246). *X-ray crystallography, crystallographic database system, universal program system, solid state physics.*

Kawata, Dr Hiroshi. Photon Factory, Nat. Lab. For High Energy Physics, Oho-machi, Tsukuba, Ibaraki 305, Japan. (1955) DSc, physics (Tokyo Inst. of Techn., 1982). Asst. prof. (tel. 0298 + 64-1171 ext 5030). *X-ray crystallography, ferroelectrics.*

Kifune, Dr Kouichi. Dept. of Natural Sci., Osaka Women's U. Daisen-cho 2-1 Sakai, Osaka 590, Japan. (1958) DSc. (Hiroshima U., 1988). Asst. (tel. 0722 + 22-4811 ext 335, fax +81 722 38 5539). *Materials science, phase transitions, inorganic crystals, electron microscopy, X-ray diffraction.*

Kihara, Dr Kuniaki. Dept. of Earth Sci., Kanazawa U., 1-1 Marunouchi, Kanazawa 920, Japan. (1943) DSc, mineralogy (Tokyo U. of Education, 1972). Assoc. prof. (tel. 0762 + 62-4281 ext 576). *X-ray crystallography.*

Kikuchi, Dr Makoto. Dept. of Metallurgical Eng., Tokyo Inst. of Techn., 12-1 Oh-okayama 2-chome Meguro-ku, Tokyo 152, Japan. (1935) DEng, metallurgy (Tokyo Inst. of Techn., 1966). Prof. (tel. 03 + 726-1111 ext 3138, telex 2466360 TITECHJ fax +81 3 729 0393) *Electron microscopy, materials science, metals.*

Kikuta, Prof. Seishi. Applied Physics Dept., Fac. of Eng., Tokyo U., 3-1 Hongo 7-chome Bunkyo-ku, Tokyo 113, Japan. (1938) DSc, physics (Tokyo U., 1970). Prof. (tel. 03 + 812-2111 ext 6825, telex 2722111 FEUT J, fax +81 3 816 7805). *Diffractometry, instrumention, materials science, neutron diffraction, semiconductors, surface structure, X-ray diffraction.*

Kimura, Prof. Masao. Dept. of Chemistry, Fac. of Sci., Hokkaido U., Kita 10-jo Nishi 8-chome Kita-ku, Sapporo 060, Japan. (1921) DSc, chemistry (Nagoya U., 1949). Prof. (tel. 011 + 716-2111 ext 3501). *Gas electron diffraction.*

Kimura, Mr Masao. R and D Labs.-I Nippon Steel Corp. 1618 Ida Nakahara-ku Kawasaki, Kanagawa 211, Japan. (1962) MSc,chemistry(Kyoto U., 1987). (tel. 044 + 777-4111 ext 471, fax +81 44 751 2142). *X-ray diffraction, structural chemistry.*

Kiriyama, Prof. Hideko. 1-12 Imajuku Ogaki, Gifu 503, Japan. (1923) DSc, chemistry (Osaka U., 1960). (tel. 0584 + 78-5702). *Coordination compounds, inorganic crystals.*

Kiriyama, Prof. Ryoiti. 1-12 Imajuku Ogaki, Gifu 503, Japan. (1913) DSc, chemistry (Osaka U., 1949). Prof.Em. (tel. 0584 + 78-5702). *Inorganic crystals, materials science.*

Kishi, Dr Kiyoshi. Dept. of Applied Physics, Sci. U. of Tokyo, 1-3 Kagurazaka, Tokyo 162, Japan. (1933) DSc, physics (Sci. U. of Tokyo, 1974). Asst. prof. (tel. 03 + 260-4271). *Crystal growth, surface physics, electronic instrumentation.*

Kishino, Prof. Seigo. Electronics Himeji Inst. of Techn., 2167 Shosha Himeji, Hyogo 671-22, Japan. (1938) DSc, X-ray diffraction(Tokyo U., 1972). Prof. (tel. 0792 + 66-1661 ext 302, fax +81 792 66 8868). *Defect structures, materials science, semiconductor, surface structure.*

Kitagawa, Dr Yasuyuki. Res. Center for Protein Eng., Inst. for Protein Res., Osaka U., 3-2 Yamadaoka Suita, Osaka 565, Japan. (1959) DSc,macromolecules(Osaka U., 1987). Instr. (tel. 06 + 877-5111 ext 3912, fax +81 6 876 2533). *Protein crystallography.*

Kitahama, Dr Katsuki. Inst. of Scient. and Industrial Res., Osaka U., Yamadakami Suita, Osaka 565, Japan. (1941) DSc,inorganic chemistry (Osaka U., 1972). Res. assoc. (tel. 06 + 877-5111 ext 3551, fax +81 6 877 4977). *Crystal growth, electron diffraction, hydrogen bonding, inorganic crystals, materials science, neutron diffraction, phase transition, structural chemistry, X-ray diffraction.*

Kitano, Dr Yasuyuki. Dept. Material Sci., Hiroshima U. Higashi-senda-machi Naka-ku Hiroshima, Hiroshima 730, Japan. (1940) DEng,(Osaka U., 1980). Assoc. Prof. (tel. 082 + 241-1221 ext 3654, fax +81 82 242 7454). *Diffuse scattering, electron diffraction, electron microscopy, metals, X-ray diffraction.*

Kitano, Mr Yukishige. Pioneering Res. and Dev. Labs., Toray Industries Inc., 2-1 Sonoyama 3-chome Otsu, Shiga 510, Japan. (1941) BSc, chemistry (Osaka U., 1965). (tel. 0775 + 37-0600). *Organic crystal structures, large molecules.*

Kobayashi, Dr Akiko. Dept. of Chemistry, Fac. of Sci., Tokyo U., 3-1 Hongo 7-chome Bunkyo-ku, Tokyo 113, Japan. (1943) DSc, chemistry (Tokyo U., 1972). Res. assoc. (tel. 03 + 812-2111, telex UTYOSCI J33659, fax +81 3 814 2627). *Inorganic chemistry, solid state chemistry.*

Kobayashi, Dr Hayao. Dept. of Chemistry, Toho U., 542 Miyama-cho Funabashi, Chiba 274, Japan. (1942) DSc, chemistry (Tokyo U., 1970). Asst. prof. (tel. 0474 + 72-1141). *Solid state chemistry.*

Kobayashi, Prof. Jinzo. Dept. of Applied Physics, Waseda U., 3-4-1 Ohkubo, Shinjuku-ku, Tokyo 160, Japan. (1925) DEng, applied physics (Waseda U., 1960). Prof. (tel. 03 + 203-4141 ext 73-3564,3566, fax +81 474 23 2375). *Phase transitions, optical activity, ferro electricity.*

Kobayashi, Mr Masaaki. Electron Devices Div. Fujitsu Labs., 10-1 Morinosato-Wakamiya Atsugi, Kanagawa 243-01, Japan. (1935) BSc,applied phys.(Kyoto U., 1958). General Manager (tel. 0462 + 48-3111 ext 2800, fax +81 462 48 3896). *Semiconductors, crystal growth.*

Kobayashi, Prof. Nobuyuki. Applied Physics Toyama U., Gofuku 3190, Toyama 930, Japan. (1942) DEng, crystal growth (Nagoya U., 1972). Prof. (tel. 0764 + 41-1271 ext 837). *Crystal growth, fluid mechanics, heat transfer, mass transfer, numerical analysis, computer simulation, materials science, solid state physics, thermal stress.*

Kobayashi, Prof. Tadashi. Dept. of Pharm., Hokuriku U., 3 Ho Kanagawa-machi, Kanazawa 920-11, Japan. (1944) DSc, physics (Kyushu U., 1972). Prof. (tel. 0762 + 29-1161 ext 227). *X-ray diffraction, phase transitions, electron spin resonance(ESR).*

Kobayashi, Dr Takaaki. Tokyo Inst. of Techn., 4259 Nagatsuta Yokohama, Kanagawa 227, Japan. (1944) DSc, chemistry (Tohoku U., 1973). Assoc. prof. (tel. 045 + 922-1111 ext 2451, fax +81 45 921 1089). *Crystal growth, electron microscopy, semiconductors.*

Kobayashi, Dr Takashi. Inst. for Chemical Res., Kyoto U., Gokasho Uji 611, Japan. (1938) DSc, chemistry (Kyoto U., 1970). Prof. (tel. 0774 + 32-3111 ext 2071, telex 5453-638 UCLKU-J). *Crystal growth, defect structures, disorder, electron diffraction, electron microscopy, electron energy loss spectrpscopy, structural chemistry, reaction pathways, organic crystal.*

Koda, Dr Shigetaka. Analytical Res. Labs., Fujisawa Pharmaceutical Co. Ltd., 1-6 Kashima 2-chome Yodogawa-ku, Osaka 532, Japan. (1946) Ph.D.,physical chemistry (Hiroshima U., 1987). Sr. scient. (tel. 06 + 390-1169 ext 532, telex FUJIPA J 64724, fax +81 6 308 5006). *Coordination compounds, powder diffraction, structural chemistry, electron microscopy.*

Kohra, Prof. Kazutake. 1-28-2 Higashinakano Nakano-ku, Tokyo 164, Japan. (1921) DSc, physics (Kyushu U., 1954). (tel. 03 + 371-7574, fax +81 3 227 1482). *Dynamical diffraction, X-ray optics.*

Koide, Prof. Tsutomu. Dept. of Chemistry, Osaka Kyoiku U., 43 Minami-Kawahori-cho Tennoji-ku, Osaka 543, Japan. (1930) DSc, chemistry (Osaka U., 1955). Asst. prof. (tel. 06 + 771-8131). *Phase transition, organic compounds, phase transition effects of crystal particle size.*

Koizumi, Dr Hideo. Res. Div., Musashino Electrical Communication Lab., Nippon Telegraph and Telephone Corp., 9-11 Midori-cho 3-chome Musashino, Tokyo 180, Japan. (1928) DSc, physics (Osaka U., 1962). Sr. scient. (tel. 0422 + 59-3120). *Crystal structure analysis, crystal physics, structure, laser materials, ferroelectric materials.*

Koizumi, Prof. Mitsue. Inst. of Sci. and Eng., Ryukoku U., 67 Fukakusatsukamoto-cho Fushimi, Kyoto 612, Japan. (1923) DSc, High pressure synthesis (Tokyo U., 1958). Prof. (tel. 075 + 642-1111). *High pressure, ceramics, zeolites.*

Kojima, Dr Seiji. Inst. of Applied Physics, Tsukuba U. 1-1-1 Tennodai Tsukuba, Ibaraki 305, Japan. (1951) DSc,phys.(Tokyo U., 1979). Asst.Prof. (tel. 0298 + 53-5307, telex 3652580 UNTUKU J, fax +81 298 53 5205). *Phase transitions, microscopy.*

Komatsu, Prof. Hiroshi. Inst. for Materials Res., Tohoku U.,1-1 Katahira 2-chome, Sendai 980, Japan. (1935) DSc, mineralogy (Tokyo U. of Education, 1964). Prof. (tel. 0222 + 27-6200 ext 2908, telex 852238(KINKEN J), fax +81 22 264 7984). *Crystal growth, surface microtopography, interferometry, optical microscopy.*

Komura, Prof. Yukitomo. Fac. of Integrated Arts & Sci., Hiroshima U., 1-89 Higashisendamachi 1-chome, Hiroshima 730, Japan. (1933) DSc, physics (Tokyo U., 1967). Prof. (tel. 082 + 241-1221 ext 3565, fax +81 82 244 5170). *X-ray diffraction, neutron diffraction, phase transitions, magnetic materials, alloys, superconductors, liquid crystals, biomembranes.*

Komura, Prof. Yukitomo. Dept. of Materials Sci., Fac. of Sci., Hiroshima Inst. of Techn., 2-1-1 Miyake Saeki-ku, Hiroshima 731-51, Japan. (1924) DSc, physics (Osaka U., 1961). Prof. (tel. 0829 + 21-3121 ext 541, fax +81 829 22 1480). *Defect structures, disorder, electron microscopy, materials science, metals, X-ray diffraction.*

Konaka, Prof. Shigehiro. Dept. of Chemistry, Fac. of Sci., Hokkaido U., Kita 10-jo Nishi 8-chome, Kita-ku, Sapporo 060, Japan. (1939) DSc, chemistry (Hokkaido U., 1969). Prof. (tel. 011 + 716-2111 ext 3501, fax +81 11 717 9394). *Electron diffraction, molecular structure, electron density, small-angle scattering, structural chemistry.*

Konno, Dr Michiko. Dept. of Chemisty, Ochanomizu U., 2-1-1 Otsuka Bunkyo-ku, Tokyo 112, Japan. (1946) DSc, chemistry (Tokyo U., 1974). Assoc. prof. (tel. 03 + 943-3151 ext 745, fax +81 3 942 2815). *Phase transitions, proteins, structural chemistry, X-ray diffraction.*

Kotani, Prof. Masao. 2nd Div. Japan Academy, Ueno Park, Daito-ku, Tokyo 110, Japan. (1906) DSc, (U. Tokyo, 1942). Member, Japan Acad. (tel. 03 + 822-2101). *Protein structure.*

Koto, Prof. Kichiro. Dept. of Materials Sci., Fac. of Integrated Arts and Sci., Tokushima U., Minami-Josanjima,Tokushima 770, Japan. (1936) DSc, mineralogy (Tokyo U., 1969). Prof. (tel. 0886 + 23-2311 ext 2330, telex 5862256TSKL J, fax +81 886 55 2108). *Crystal structure, EXAFS, inorganic crystals, materials science, minerals, structural chemistry, superionic conductors, superstructure, thermal vibrations.*

Koyama, Prof. Hirozo. Fac. of Sci. & Techn., Kinki U., 3-4-1 Kowakae Higashi-Osaka, Osaka 577, Japan. (1924) DSc, chemistry (Osaka U., 1962). Prof. (tel. 06 + 721-2332 ext 4006). *Structure analysis methods, organic structures.*

Koyama, Dr Yasumasa. Dept. of Metallurgy, Fac. of Eng., Tokyo Inst. of Techn., 2-12-1 Ookayama Meguro-ku, Tokyo 152, Japan. (1952) DEng, metallurgy (Tokyo Inst. of Techn., 1981). Res. assoc. (tel. 03 + 726-1111 ext 3145). *Diffuse scattering, phase transitions, thermal vibrations.*

Koyano, Mr Kazuo. Central Res. Inst., Teijin Limited, 3-2 Asahigaoka 4-chome Hino, Tokyo 191, Japan. (1932) MSc, chemistry. Res. assoc. (tel. 0425 + 81-4321). *Structure, biopolymers.*

Kozaki, Prof. Shigeru. The Inst. of Vocational Training, 1960 Aihara Sagamihara, Kanagawa 229, Japan. (1934) DEng, applied physics (Tokyo U., 1974). Prof. (tel. 0427 + 61-2111, fax +81 427 61 8293). *X-ray diffraction.*

Kubo, Prof. Teruichiro. Dept. of Chemistry, Musashi Inst. of Techn., 28-1 Tamazutsumi 1-chome Setagaya-ku, Tokyo 158, Japan. (1907) DSc, chemistry (Tokyo Inst. of Techn., 1940). Prof.; Em. prof. (Tokyo Inst. of Techn.). (tel. 03 + 703-3111 ext 490). *Mechanochemistry, powder technology, solid state chemistry.*

Kuchitsu, Prof. Kozo. Chemistry Dept., Nagaoka U. of Techn., 1603-1 Kamitomioka Nagaoka, Niigata 940-21, Japan. (1927) DSc, chemistry (Tokyo U., 1958). Prof., prof. em. of Tokyo. (tel. 0258 + 46-6000 ext 3203, telex 3232123 TUNJ, fax +81 258 46 6507). *Gas electron diffraction, vibration - rotation spectroscopy, chemical processes by electronic and atomic impact.*

Kudoh, Dr Yasuhiro. Mineralogical Inst., Fac. of Sci., Tokyo U., 3-1 Hongo 7-chome Bunkyo-ku, Tokyo 113, Japan. (1947) DSc, mineralogy (Tokyo U., 1975). Res. assoc. (tel. 03 + 812-2111 ext 4548). *Inorganic crystal structures, high pressure X-ray diffraction.*

Kumagawa, Prof. Masashi. Res. Inst. of Electronics, Shizuoka U., 3-5-1 Johoku Hamamatsu, Shizouka 432, Japan. (1938) DEng, materials science (Tohoku U., 1967). Prof. (tel. 0534 + 71-1171 ext 434, telex 4225-280SHZDDK-J, fax +81 534 74 0630). *Crystal growth, semiconductors.*

Kumao, Prof. Akihiro. Electronics and Information Sci., Kyoto Inst. of Techn., Matsugasaki, Sakyo-ku, Kyoto 606, Japan. (1941) DSc, physics (Hiroshima U., 1983). Prof. (tel. 075 + 791-3211 ext 446, fax +81 75 711 9483) *Electron microscopy, crystal growth.*

Kurahashi, Dr Masayasu. Nat. Chemical Lab. for Industry, 1-1 Higashi Tsukuba, Ibaraki 305, Japan. (1943) DSc, chemistry (Osaka City U., 1975). Sr. res. officer. (tel. 0298 + 54-4626). *Inorganic and organic crystal structures, LEED,*

Kuribayashi, Mr Shunsuke. Government Industrial Res. Inst., 1-8-31 Midorigaoka Ikeda, Osaka 563, Japan. (1923) BSc, chemistry (Osaka U., 1949). Res. (tel. 0727 + 51-8351 ext 279). *Structure, polymers and related compounds.*

Kuroda, Prof. Haruo. Dept. of Chemistry, Fac. of Sci., Tokyo U., 3-1 Hongo 7-chome Bunkyo-ku, Tokyo 113, Japan. (1931) DSc, chemistry (Tokyo U., 1958). Prof. (tel. 03 + 812-2111 ext 2447). *Physical properties, crystal structure, molecular complexes, radical salts, X-ray photoelectron spectroscopy.*

Kuroya, Dr Hisao. Dept. of Chemistry, Okayama U. of Sci., 1-1 Ridai-cho Okayama, Okayama 700, Japan. (1916) DSc, chemistry (Osaka U., 1949). Prof. (tel. 0862 + 52-3161 ext 4263, fax +81 862 55 7700). *Coordination compounds, structural chemistry.*

Kushi, Dr Yoshihiko. The Inst. of Chemistry, C. of General Education, Osaka U. 1-1 Machikaneyama Toyonaka, Osaka 560, Japan. (1937) DSc,chemistry(1968). Prof. (tel. 06 + 844-1151 ext 5270). *Coordination compounds, inorganic crystal, structural chemistry.*

Kusunoki, Dr Masami. Res. Center for Protein Eng., Osaka U., Yamadaoka 3-2 Suita, Osaka 565, Japan. (1953) DSc, macromolecular science (Osaka U., 1980). Res. asst. (tel. 06 + 877-5111 ext 3912, fax +81 6 876 2533). *Proteins, structural biology.*

Kyotani, Dr Mutsumasa. Dept. of Polymer Physics, Res. Inst. for Polymers and Textiles, 1-1-4 Higashi Tsukuba, Ibaraki 305, Japan. (1938) DEng., polymer eng. (tel. 0298 + 54-6297). *Crystal growth, crystallographic data, electron microscopy, liquid crystals, polymers, electron diffraction, X-ray diffraction.*

Little, Mr Thomas W. Res. & Dev. Dept., Seiko Epson Corp., 3-5, Owa 3-Chrome, Suwa-shi, Ngano-ken 392, Japan. (1962) MS, materials science & eng. (U. of Illinois at Urbana-Champaign, 1987). Res. eng. (tel. 0266 + 52-3131, fax 0266 + 52-4927). *Liquid crystal display.*

Maeda, Dr Hironobu. Chemistry, Okayama U. Tsushima-naka Okayama, Okayama 700, Japan. (1943) DSc.(Hiroshima U., 1978). Res. Assoc. (tel. 0862 + 52-1111 ext 391, fax +81 862 52 6601). *EXAFS, structural chemistry, phase transition.*

Mannami, Dr Michihiko. Dept. of Eng. Sci., Kyoto U. Sakyo-ku, kyoto 606, Japan. (1935) DSc,physics.(Kyoto U.) Prof. (tel. 075 + 753-5196, fax +81 75 771 7286). *Electron diffraction, surface structure.*

Maruha, Prof. Juro. C. of Liberal Arts, Kanazawa U., 1-1 Marunouchi, Kanazawa 920, Japan. (1921) DSc, chemistry (Tohoku U., 1960). Prof. (tel. 0762 + 62-4281 ext. 648). *Physical chemistry.*

Marukawa, Prof. Kenzaburo. Dept. of Applied Physics, Fac. of Eng., Hokkaido U., Kita 13-jo Nishi 8-chome Kita-ku, Sapporo 060, Japan. (1937) DSc, physics (Kyoto U., 1967). Prof. (tel. 011 + 716-2111 ext 6643, telex 932302, fax +81 11 717 4745). *Lattice defects, metal crystals.*

Marumo, Prof. Fumiyuki. Res. Lab. of Eng. Materials, Tokyo Inst. of Techn., Nagatsuta-machi 4259 Midori-ku, Yokohama 227, Japan. (1931) DSc, mineralogy (Tokyo U., 1960). Prof. (tel. 045 + 922-1111 ext 2312, telex 3823553 TITNAG J, fax +81 45 921 1015) *Structural chemistry of inorganic materials, Electron density in inorganic crystals.*

Maruse, Prof. Susumu. Dept. of Electronics, Nagoya U., Furo-cho Chikusa-ku, Nagoya 464, Japan. (1926) DEng, electrical eng. (Nagoya U., 1962). Prof. (tel. 052 + 781-5111 ext 4436). *Electron optics.*

Maruyama, Prof. Saiyu. Div. of Physics, Dept. of Natural Sci., Osaka Women's U., Daisen-cho Sakai, Osaka 590, Japan. (1923) DSc, physics (Kyoto U., 1961). Prof. (tel. 0722 + 22-4811 ext 335). *Long period structure, thin films, electron microscopy.*

Masaki, Dr Norio. Fac. of Pharm. Sci., Kyoto U., Yoshida Shimoadachi-cho Sakyo-ku, Kyoto 606, Japan. (1931) DSc, physics (Osaka U., 1961). Asst. prof. (tel. 075 + 753-4533 ext 4533, telex 05422693LIBKYUJ, fax +81 75 761 2698, email A50206 at JPNKUDPC.BITNET). *Structural biology, X-ray diffaction, electron microscopy.*

Masakuni, Dr Mayumi. First Dept. of Biochemistry, Nat. Defense Medical C., 2-3 Namiki, Tokorozawa, Saitama 359, Japan. (1934) MD, biochemistry (Nihon U., 1983). Asst. (tel. 0429 + 95-1211 ext 2293). *Protein structure and function, enzymes.*

Masuda, Dr Hideki. Coord.Chemistry Labs. Inst. for Molecular Sci. Myodaiji Okazaki, Aichi 444, Japan. (1950) DPharm, pharm. science (Kyoto U., 1982). Res. Assoc. (tel. 0564 + 54-1111 ext 3451, telex 4537475 KOKKEN J, fax +81 564 54 2254, email MASUDA at JPNONRI) *Bioinorganic chemistry, coordination chemistry, semiconductors.*

Masuko, Dr Akiyoshi. Central Techn. Res. Lab. Nippon Oil Co. Ltd., 8 chidori-cho Nakaku Yokohama, Kanagawa 231, Japan. (1935) DEng.(Waseda U., 1989). Sr. Res. Physicist (tel. 045 + 212-7261, telex J27237 NIPOIL, fax +81 45 212 7270). *Electron microscopy, powder diffraction, small-angle scattering, surface structure, X-ray diffraction.*

Matsubara, Dr Ikuo. Res. Div. Toray Res. Center Inc., 1-1 1-chome Sonoyama Otsu, Shiga 520, Japan. (1931) Ph.D., chemistry (Kyushu U., 1962). Managing Director (tel. 0775 + 37-0700 ext 4370). *Macromolecules, structural chemistry.*

Matsubara, Prof. Takeo. Dept. of Applied Physics, Okayama U. of Sci. 1-1 Ridai-cho, Okayama 700, Japan. (1921) DSc,physics(Osaka U., 1951). Prof. (tel. 0862 + 751-3161 ext 4129). *Lattice dynamics, phase transition, random systems, surface physics.*

Matsuda, Prof. Hidehiko. Dept. of Applied Physics, Kyushu Inst. of Techn., 1 Sensuicho Tobata-ku Kitakyushu, Fukuoka 804, Japan. (1931) DEng, metallurgy (Kyushu U., 1972). Prof. (tel. 093 + 871-1931 ext 467). *Physics, metals and alloys.*

Matsui, Dr Masanori. Chemical Lab., Kanazawa Medical U., Uchinada, Kahoku-gun, Ishikawa 920-02, Japan. (1949) DSc (Kwansei Gakuin U., 1982). Lect. (tel. 0762 + 86-2211 ext 7108, fax +81 762 86 0224). *Modeling crystal structures, high temperature, high pressure.*

Matsui, Mr Toshiro. Electron Device Eng. Lab., Toshiba Corp., 8 Shin-Sugita-cho Isogo-ku Yokohama, Kanagawa 235, Japan. (1943) BSc, mineralogy (Tokyo U., 1967). Res. (tel. 045 + 756-2542, fax +81 45 773 5978). *Liquid crystals.*

Matsui, Dr Yoshio. Nat. Inst. for Inorganic Materials Res., 1-1 Namiki Tsukuba, Ibaraki 305, Japan. DSc.(Tokyo U., 1984). Sr. Res. (tel. 0298 + 51-3351 ext 283, fax +81 298 52 7449). *Electron microscopy, materials science, superconductors, ceramics, radiation damage.*

Matsui, Prof. Yoshiro. Inst. for the Study of Earth's Interior, Okayama U. 827 Yamada Misasa Tohaku-gun, Tottori 682-02, Japan. (1931) DSc,chemistry(1961). Prof. (tel. 0858 + 43-1215 ext 504,740, fax +81 858 43 2184). *Modelling of crystal structures, interatomic forces, computational physics of silica and silicate crystals, ultrahigh-pressure mineralogy.*

Matsumoto, Mr Osamu. Fac. of Pharm. Sci., Kyoto U. Sakyo-ku, Kyoto 606, Japan. (1960) Master (tel. 075 + 753-4533, fax +81 75 761 2698, email A50359 at JPNKUDPC). *Proteins, structural biology.*

Matsumoto, Prof. Takeo. Dept. of Earth Sci., Fac. of Sci., Kanazawa U., 1-1 Marunouchi Kanazawa, Ishikawa 920, Japan. (1932) DSc, mineralogy (U.of Tokyo, 1962). Prof. (tel. 0762 + 62-4281 ext 560, fax +81 762 64 1059). *Inorganic crystal structures, mathematical crystallography, symmetry.*

Matsuo, Dr Munetsugu. Fundamental Res. Labs., Nippon Steel Corp., 1618 Ida Nakahara-ku Kawasaki, Kanagawa 211, Japan. (1936) DEng, metallurgy (Tokyo U., 1967). Sr. res. (tel. 044 + 777-4111). *Metal physics, texture.*

Matsushima, Dr Norio. School of Allied Health Proffessions, Sapporo Medical College, Minami 3 Nishi 8 Chuo-ku, Sapporo 060, Japan. (1948) DSc., polymer phisics (Hokkaido U., 1976). Assoc. prof. (tel. 011 + 611-2111 ext 2834). *Small-angle scattering, biomineralllzation, protein structure, macromolecules.*

Matsushita, Dr Tadashi. Photon Factory, Nat. Lab. for High Energy Physics, Oho, Tsukuba, Ibaraki 305, Japan. (1945) DEng., applied physics (Tokyo U., 1972). Prof. (tel. 0298 + 64-1171 ext 5039). *X-ray optics, instrumentation, synchrotron radiation.*

Matsuura, Dr Yoshiki. Inst. for Protein Res., Osaka U.,. Yamada-oka Suita, Osaka 565, Japan. (1943) DSc, chemistry (Osaka U. 1976). Assoc. prof. (tel. 06 + 877-5111 ext 3837, fax +81 6 876 2533). *Protein crystal structure analysis and its methods.*

Matsuzaki, Dr Takao. Res. Center, Mitsubishi Kasei Corp, 1000 Kamoshida Midori-ku, Yokohama 227, Japan. (1945) DPharm. chemistry (Tokyo U., 1988). Res. assoc. (tel. 045 + 963-3156,3157 ext 3312, fax +81 45 961 6561). *Proteins, structural biology.*

Mihama, Prof. Kazuhiro. Dept. of Applied Physics, Fac. of Eng., Nagoya U., Furo-cho Chikusa-ku, Nagoya 464, Japan. (1927) DSc. (Universite de Paris, France, 1960). Prof. (tel. 052 + 781-5111 ext 4457). *High resolution electron microscopy, crystal growth, thin films.*

Miida, Mr Rokuro. Dept. of Physics, Tohoku U., Aramaki Aza-Aoba, Sendai 980, Japan. (1938) BSc, physics (Sci. U. of Tokyo, 1961). Res. assoc. (tel. 0222 + 22-1800 ext 5345). *Structure, alloys.*

Miki, Dr Kunio. Dept. of Applied Chemistry, Fac. of Eng., Osaka U., Yamadakami Suita, Osaka 565, Japan. (1952) DEng, chemistry (Osaka U., 1981). Res. assoc. (tel. 06 + 877-5111 ext 4322, telex 5286227 FEOUJ J, fax +81 6 876 6484). *Protein crystallography, protein structure and function, organometallic crystal structures.*

Min, Dr Eungi. Central Res. Lab., Sumitomo Cement Co., Ltd., 585 Toyotomicho Hunabashi, Chiba 274, Japan. (1949) DSc, crystallography (Tokyo U., 1980). Res. eng. (tel. 0474 + 57-0745). *Materials science, electron microscopy.*

Minagawa, Dr Teruaki. Dept. of Physics, Osaka Kyoiku U., 43 Minami-Kawahori-cho Tennoji-ku, Osaka 543, Japan. (1942) DSc, physics (Osaka City U., 1971). Asst. prof. (tel. 06 + 771-8131 ext 456). *Disorder, phase transitions, thermal vibrations, materials science.*

Minami, Dr Nobuyuki. Physics, Kinki U. 3-4-1 Kowakae Higashi-osaka, Osaka 577, Japan. (1940) DSc.(Osaka City U., 1985). Asst.Prof. (tel. 06 + 721-2332 ext 4057). *Electron diffraction, disorder, quasicrystals, scattering theory, X-ray diffraction.*

Minami, Mr. Takashi. Japan. (1964) Student *Inorganic crystals, magnetism.*

Minato, Prof. Hideo. 5-37-17 Kugayama Suginami-ku, Tokyo 167, Japan. (1921) DSc, mineralogy (Tokyo U., 1952). Prof. em.(Tokyo U.) (tel. 03 + 334-6231). *Mineralogy, mineral chemistry, clay mineralogy, descriptive mineralogy, analytical chemistry, minerals, clay minerals, zeolites.*

Minomura, Prof. Shigeru. Fac. of Sci., Okayama U. of Sci., 1-1 Ridai-cho, Okayama 700, Japan. (1923) DSc, chemistry (Kyoto U.). Prof. of chemistry. (tel. 0862 + 52-3161). *High pressure physics and chemistry.*

Mitsuda, Dr. Hiromichi. Materials Characterization Dept., Matsushita Technoresearch,Inc. 1006 Kadoma, Osaka 571, Japan. (1935)

DSc,physics(Osaka U., 1967). Manager (tel. 06 + 908-1291 ext 311, fax +81 6 906 1748). *X-ray structure analysis, X-ray spectroscopy.*

Mitsuda, Dr. Takeshi. Lab. of Ceramic Techn., Nagoya Inst. of Techn., 10 Asahigaoka Tajimi, Gifu 507, Japan. (1931) DSc, mineralogy (Hokkaido U., 1961). Asst. prof. (tel. 0572 + 27-6811). *Chemistry, cements and allied materials.*

Mitsui, Prof. Toshio. Dept. of Biophysical Eng., Fac. of Eng. Sci., Osaka U., 1 Machikaneyama-cho 1-chome Toyonaka, Osaka 560, Japan. (1926) DSc, physics (Hokkaido U., 1960). Prof. (tel. 06 + 844-1151 ext 4765, telex 5286110 OUFES J, fax +81 6 843 9354). *Structural biology, quasicrystals, small-angle scattering, muscles, biomembranes.*

Mitsui, Prof. Yukio. Fac. of Eng., Nagaoka U. of Techn., Kamitomioka Nagaoka, Niigata 940-21, Japan. (1938) DPharm, chemistry (Tokyo U., 1966). Prof. (tel. 0258 + 46-6000 ext 3121, fax +81 258 46 6504). *Proteins, structural biology, phase determination, X-ray diffraction computing, instrumentation, teaching.*

Mitsuishi, Prof. Tomokuni. Dep. of Management and Information Sci., Jobu U., 270 Shinmachi Tano-gun, Gunma 370-13, Japan. (1917) DSc, physics (Tokyo U., 1967). Prof. (tel. 0270 + 32-1011). *Semiconductors.*

Miura, Dr Yasuhiro. Dept. of Metallurgy, Fac. of Eng., Kyushu U., 10-1 Hakozaki 6-chome Higashi-ku, Fukuoka 812, Japan. (1941) PhD, physical metallurgy (U. California, Berkeley, USA,1970). Assoc. prof. (tel. 092 + 641-1101 ext 5709, fax +81 92 632 0434). *Materials science, microstructure, plastic deformation, strength of materials.*

Miura, Dr Yasunori. Dept. of Mineralogical Sci. and Geology, Fac. of Sci., Yamaguchi U., 1677-1 Yoshida, Yamaguchi 753, Japan. (1946) DSc, mineralogy (Tohoku U., 1976). Assoc. prof. (tel. 0839 + 22-6111 ext 382, fax +81 839 32 2041). *Ion and electron microprobe analyses, inorganic crystal structure, X-ray crystallography, electron microscopy, crystal optics, computing, minerals, instrumentation.*

Miyaji, Dr. Hideki. Dept. of Physics, Kyoto U. Kitashirakawa Oiwake-cho, Kyoto 606, Japan. (1941) DSc,physics(Kyoto U., 1975). Asst.Prof. (tel. 075 + 753-3754). *Polymers, surface structure, high pressure, crystal growth, X-ray diffraction, phase transitions, small-angle scattering.*

Miyake, Prof. Shizuo. 2-29-1 Minamiogikubo Suginami-ku, Tokyo 167, Japan. (1911) DSc, physics (Tokyo U., 1942).Em. prof. (tel. 03 + 334-7180). *Crystal physics, dynamical diffraction (HEED - LEED - X-rays).*

Miyake, Dr Yasuhiro. Dept. of Polymer Sci., Fac. of Sci., Hokkaido U., Kita 10-jo Nishi 8-chome Kita-ku, Sapporo 060, Japan. (1925) DSc, physics (Hokkaido U., 1963). Prof. (tel. 011 + 716-2111). *Polymers, biophysics.*

Miyamae, Dr Hiroshi. Fac. of Sci., Josai U., Keyakidai 1-1, Sakado, Saitama 350-02, Japan. (1950) DSc, chemistry (Tokyo U., 1978). Lect. (tel. 0492 + 86-2233 ext 525). *X-ray crystallography, transition metal complexes, intercalation.*

Miyamoto, Dr. Masamichi. Pure and Applied Sci., C. of Arts and Sci., Tokyo U., 3-8-1 Komaba, Meguro-ku, Tokyo 153, Japan. (1949) Res. assoc. (tel. 03 + 467-1171 ext 402, fax +81 3 485 2904). *High pressure, inorganic crystals, minerals, phase transitions, computing, crystallographic data, X-ray diffraction.*

Miyano, Mr. Toshio. Maizuru C. of Techn. Shiraya 234 Maizuru, Kyoto 625, Japan. (1953) Master Lect. (tel. 0773 + 62-5600 ext 229, fax +81 773 62 5558). *Electron microscopy, inorganic crystals, defect structures.*

Miyata, Dr. Takeshi. Yamanashi Inst. of Gemmology & Jewelry Arts, 1955-1 Tokojicho, Kofu Yamanashi 400, Japan. (1949) PhD, mineralogy (Tohoku U. 1980). Assoc. prof. (tel. 0552 + 32-6671 ext 36). *Crystal growth, morphology, gemmology.*

Miyazawa, Dr. Shintaro. Device Physics and Techn. Lab., N.T.T. LSI Labs, 3-1 Morinosato Wakamiya Atsugi-shi, Kanagawa 243-01, Japan. (1942) DEng, electronics (Tohoku U., 1978). Executive res. eng. (tel. 0462 + 40-2720, telex +72 0387 2110 ECLATGJ, fax +81 462 40 2872). *Crystal growth, materials science.*

Miyoshi, Dr Tadahiko. Materials Res. Dept., Hitachi Res. Lab., Hitachi Ltd., 4026 Kuji-cho Hitachi-shi, Ibaraki 319-12, Japan. (1943) DSc, chemistry (Tokyo U., 1971). Res. (tel. 0294 + 52-5111 ext 284). *Metal oxide semiconductors, ceramics.*

Mizota, Dr Tadato. Dept. of Mining and Mineral Eng., Fac. of Eng., Yamaguchi U., Tokiwadai Ube, Yamaguchi 755, Japan. (1941) DSc. (Tohoku U., 1975). Assoc. prof. (tel. 0836 + 31-5100 ext 236, fax +81 836 33 4404). *Mineralogy, crystallography, mineral processing.*

Mizuki, Dr Junichiro. Semiconductor Res. Lab. NEC Corp., 1-1 Miyazaki 4-chome Miyamae-ku Kawasaki, Kanagawa 213, Japan. (1950) DSc,solid state physics(1980). Res. Manager (tel. 044 + 856-2184(dial in), fax +81 44 856 2214). *Semiconductor, magnetism, surface structure, spin density, neutron diffraction, X-ray diffraction, phase transition.*

Mizuno, Dr Hiroshi. Dept. of Molecular Biology, Nat. Inst. of Agrobiological Resources, Kan-nondai 2-1-2, Yatabe Tsukuba Sci. City, Ibaraki 305, Japan. (1943) DPharm, chemistry (Osaka U., 1972). Sr. res. (tel. 02975 + 6-7014, fax +81 2975 6 7408). *Macromolecular crystallography, molecular biology.*

Mizuno, Prof. Joji. Dept. of Electronics, Tohoku Inst. of Techn., 19 Koeji, Nagamachi Sendai 982, Japan. (1918) DSc, crystallography (Tohoku U., 1962). Prof. (tel. 0222 + 29-1151 ext 253). *Inorganic crystal structures.*

Morikawa, Mr Hideki. Res. Lab. of Eng. Materials, Tokyo Inst. of Techn., Nagatsuta 4259, Midori-ku, Yokohama 227, Japan. (1942) DEng, ceramics

(Tokyo Inst. of Techn. 1973). Assoc. prof. (tel. 045 + 922-1111 ext 2627, telex 3823553 TITNAG J, fax +81 45 921 1015). *Amorphous structure.*

Morikawa, Dr. Hiroshi. Dept. of Materials Sci. and Eng., Nagoya Inst. of Techn., Gokiso-cho Showa-ku, Nagoya 466, Japan. (1942) DSc, chemistry (Kyoto U., 1972). Assoc. prof. (tel. 052 + 732-2111 ext 2561, fax +81 52 735 0487). *Field ion microscopy, electron microscopy, surface structure, crystal growth, thin film.*

Morimoto, Dr. Jun. Dept. of Applied Physics, Nat. Defense Academy, 1-10-20 Hashirimizu Yokosuka, Kanagawa 239, Japan. (1950) DEng, information processing (Tokyo Inst. of Techn. 1983). Assoc. prof. (tel. 0468 + 41-3810 ext 2459, fax +81 468 43 6236). *Crystal growth, semiconductor physics.*

Morimoto, Prof. Nobuo. C. of General Education, Osaka Sangyo U., 3-1-1 Nakagaito Daito-shi, Osaka 574, Japan. (1925) DSc, mineralogy (Tokyo U., 1954). Prof. (tel. 0720 + 75-3001 ext 4241). *Electron microscopy, minerals, phase transition, materials science.*

Morimoto, Dr. Yukio. Basic Research Lab., Himeji Inst. of Tech., 2167 Shosha Himeji, Hyogo 671-22, Japan. (1958) Dr. Sci., chemistry (Osaka U., 1987). Research assoc. (tel. 0792 + 66-1661 ext 402, fax +81 792 66 8868). *Proteins, small-angle scattering, structural biology, X-ray diffraction.*

Morinaga, Dr Masahiko. Production Systems Eng. Toyohashi U. of Techn., Tempaku-cho Toyohashi, Aichi 440, Japan. (1946) PhD, materials science (Northwestern U., 1978). Assoc. Prof. (tel. 0532 + 47-0111 ext 623, telex 4322-201 JPNTUT, fax +81 532 47 2688). *X-ray diffraction, diffuse scattering, defect structures, materials science, phase transitions, chemical bonding.*

Morino, Prof. Yonezo. Sagami Chemical Res. Center, 4-1 Nishi-Ohnuma 4-chome Sagamihara, Kanagawa 229, Japan. (1908) DSc, chemistry (Tokyo U., 1937). President, Em. prof. (Tokyo U.). (tel. 0427 + 42-4791). *Molecular structure.*

Moritani, Mr Yoshimitsu. Material Sci. & Analysis Labs. Mitsui Petrochemical Industries Ltd., Waki-town Kuga-gun, Yamaguchi 740, JAPAN. (1961) MSc.(Okayama U., 1986). Res. Chemist (tel. 08275 + 3-2311 ext 3923, fax +81 8275 3 6203). *Chemical bonding, structural chemistry, liquid crystals, materials science, organic compounds, polymers, powder diffraction, small-angle scattering, symmetry, texture, X-ray diffraction.*

Motegi, Prof. Hiroshi. Maritime Safety Academy, 1-1 Wakaba-cho Kure, Hiroshima 737, Japan. (1939) DSc, physics (Tokyo U., 1973). Prof. (tel. 0823 + 21-4961). *Ferroelectricity.*

Mukai, Prof. Tadasuke. Physics Lab., Fukuoka U. of Education, Akama Munakata-machi, Fukuoka 811-14, Japan. (1917) DSc, physics (Hiroshima U., 1959). Prof. (tel. 09403 + 2-2381 ext 360). *Crystal growth, plastic deformation, metals.*

Murakami, Dr Takashi. Dept. of Environmental Safety Res., Japan Atomic Energy Res. Inst., Shirakata Tokai, Ibaraki 319-11, Japan. (1951) DSc, Mineralogy (Tokyo U., 1980) Res. scient. (tel. 0292 + 82-5872). *High level waste management.*

Muraoka, Dr Hisashi. Purex, 735 Nippa-cho Kohoku-ku Yokohama, Kanagawa 222, Japan. (1924) DSc, mineralogy (Tokyo U., 1956). Corporate fellow. (tel. 045 + 541-9493). *Semiconductors, crystal growth.*

Murata, Prof. Yoshitada. Inst. for Solid State Physics, U. of Tokyo, 7-22-1 Roppongi Minato-ku, Tokyo 106, Japan. (1935) DSc, chemistry (Tokyo U., 1964). Prof. (tel. 03 + 478-6811 ext 5301, telex ISSP UT J 32469, fax +81 3 401 5169). *Surface science, low energy electron diffraction.*

Nagai, Prof. Ryutaro. Dept. of Physics, Tokyo Gakugei U., 1-1 Nukuikita-machi 4-chome Koganei, Tokyo 184, Japan. (1916) DSc, physics (Hiroshima U., 1961). Prof. (tel. 0423 + 21-1741 ext 343). *Crystal physics, metal physics.*

Nagakura, Prof. Saburo. Okazaki Nat. Res. Inst., 38 Myodaiji-cho Saigo Naka Okazaki, Aichi 444, Japan. (1920) DSc, chemistry (Tokyo U., 1953). President. (tel. 0564 + 52-9771). *Structures, optical properties, molecular crystals.*

Nagakura, Prof. Sigemaro. Dept. of Mechanical Eng., Nagaoka U. of Techn. Kamitomioka Nagaoka, Niigata 940-21, Japan. (1926) DSc, physics (Kyoto U., 1959). Prof. (tel. 0258 + 46-6000 ext 7121, fax +81 258 46 6972). *Structure, metals and alloys, crystal growth, electron diffraction, electron microscopy, X-ray diffraction, topography.*

Nagano, Dr Kozo. Fac. of Pharm. Sci., Tokyo U., 7-3-1 Hongo Bunkyo-ku, Tokyo 113, Japan. (1933) DPharm. (Tokyo U., 1962). Assoc. prof. (tel. 03 + 812-2111 ext 4841). *Tertiary structure prediction, biological macromolecules.*

Nagashima, Dr. Seiichi. Dept. of Physics, C. of Eng., Nihon U. 1 Nakagawara Tokusada, Tamura-cho Koriyama, Fukushima 963, Japan. (1949) DEng, electrical eng. (Nihon U., 1977). Lect. (tel. 0249 + 44-1300 ext 312, fax +81 249 43 3146). *Epitaxial growth, thin films, LEED.*

Nagata, Dr Fumio. 4th Dept., Central Res. Lab., Hitachi Ltd., 280 Higashi-Koigakubo 1-chome, Kokubunji Tokyo 185, Japan. (1940) DEng, applied physics (Nagoya U., 1971). (tel. 0423 + 23-1111). *Electron microscopy.*

Naiki, Prof. Toshio. Dept. of Electronics and Information Sci., Kyoto Inst. of Techn., Matsugasaki Sakyo-ku, Kyoto 606, Japan. (1925) DSc, physics (Kyoto U., 1962). Prof. (tel. 075 + 791-3211 ext 441). *Crystal growth, electron microscopy, thin films.*

Nakagawa, Dr Atsushi. Photon Factory Nat. Lab. for High Energy Physics, 1-1 Oho Tsukuba, Ibaraki 305, Japan. (1961) MSc,chemistry(Osaka U., 1985). Res. Assoc. (tel. 0298 + 64-1171 ext 5027, fax +81 298 64 2801, email NAKAGAWA at JPNKEKVX.BITNET). *Protein crystallography.*

Nakahigashi, Dr Kiyotaka. Fac. of General Education, Osaka Prefecture U., Mozu Umemaci 4-chome Sakai, Osaka 591, Japan. (1941) BSc, physics (Osaka City U., 1964). Lect. (tel. 0722 + 52-1161). *Structure, metals and alloys.*

Nakai, Dr Hisayoshi. Dept. of Chemistry, Hyogo C. of Medicine, 1 Mukogawa 1-chome Nishinomiya, Hyogo 663, Japan. (1942) DSc, chemistry (Osaka City U., 1971). Asst. prof. (tel. 0798 + 45-6442). *Coordination chemistry, inorganic biochemistry.*

Nakai, Dr. Izumi. Chemistry Tsukuba U. Tsukuba, Ibaraki 305, Japan. (1953) DSc,chemistry(Tsukuba U., 1980). Asst. Prof. (tel. 0298 + 53-4520 ext 1, fax +81 298 53 6503). *Inorganic crystals, minerals, structural chemistry, powder diffraction, X-ray diffraction.*

Nakai, Dr. Yasuo. Dep. of Physics, Nagoya U. Furocho Chikusa-ku Nagoya, Aichi 464, Japan. (1941) DSc (Nagoya U., 1970) Assoc. prof. (tel. 052 + 781-5111 ext 2465, telex 447-7323, fax +81 52 782 9794). *Electron diffraction., electron microscopy, instrumentation, scattering theory, surface structure.*

Nakajima, Dr. Tetuo. Photon Factory, Nat. Lab. For High Energy Physics, 1-1 Uehara Oho Tsukuba, Ibaraki 305, Japan. (1935) DSc, metal physics (Kyoto U., 1964) Assoc. prof. (tel. 0298 + 64-1171 ext 5027). *Low temperature physics, phase transition, neutron diffraction, X-ray diffraction, magnetic substances, synchrotron radiation.*

Nakajima, Dr Yoshiharu. Materials Res. and Analysis center Sharp Corp., 2613-1 Ichinomoto Tenri-shi, Nara 632, Japan. (1946) DSc, inorganic and physical chemistry (Osaka U., 1975). General Manager. (tel. 07436 + 5-1321 ext 3090, telex 5522-364SHAPEL J, fax +81 7436 5 1131). *Crystal growth, surface structure, structural chemistry, electron microscopy.*

Nakamura, Dr Kazuo. Fac. of Pharm. Sci. Tokyo U., 3-1 Hongo 7-chome Bunkyo-ku, Tokyo 113, Japan. (1945) DPharm, pharmaceutical science (Tokyo U., 1974). Res. assoc. (tel. 03 + 812-2111 ext 4842, email A32281%tansei.cc.u-tokyo. junet at relay.cs.net). *Protein crystallography, computer programming, microcomputers.*

Nakamura, Dr Naotake. Dept. of Chemistry, Ritsumeikan U., 56-1 Tojiin-Kitamachi Kita-ku, Kyoto 603, Japan. (1943) DEng, chemistry (Ritsumeikan U.). Assoc.prof. (tel. 075 + 463-1131 ext 3638, telex 5423171 RITSUU J). *Diffuse scattering, liquid crystals, macromolecules, molecular crystals, organic compounds, phase transitions, polymers, small-angle scattering*

Nakamura, Dr Osamu. Inorganic Material Dept., Gov. Indust. Res. Inst., Osaka, 8-31 Midorigaoka 1-chome Ikeda, Osaka 563, Japan. (1946) DSc. chemistry (Osaka Univ., 1977). Manager, Powder Material Section (tel. 0727 + 51-8351 ext 1154, fax +81 727 53 4660). *Defect structure, inorganic crystals, materials science, surface structure.*

Nakamura, Prof. Terutaro. Tokai U. School of Eng., 1117 Kitakaname Hiratsuka, Kanagawa 259-12, Japan. (1923) DSc, physics (Tokyo U., 1961). Prof. (tel. 0463 + 58-1211). *Dielectric physics, ferroelectric phase transition, light scattering in solids, amorphous state.*

Nakanishi, Dr Hachiro. Polymer Chemistry Div., Res. Inst. for Polymers and Textiles, 1-1-4 Higashi Tsukuba, Ibaraki 305, Japan. (1942) PhD,chemistry. Molecular Eng. Lab. Chief (tel. 0298 + 54-6310, telex 3652570 AIST J). *Macromolecules, organic compounds, crystallographic data, materials science, structural chemistry.*

Nakanishi, Prof. Norihiko. Dept. of Chemistry, Fac. of Sci., Konan U., 9-1 Okamoto 8-chome, Kobe 658, Japan. (1928) DSc, chemistry. Prof. (tel. 078 + 431-4341 ext 653). *Solid state chemistry, physical metallurgy, powder metallurgy.*

Nakano, Prof. Shigeru. Dept. of Physics, Fac. of Sci., Chiba U., 33 Yayoicho 1-chome, Chiba 260, Japan. (1929) DSc, physics (Tokyo U., 1965). Prof. (tel. 0472 + 51-1111 ext 2610, fax +81 472 56 5793). *X-ray diffraction, superconductivity.*

Nakashima, Prof. Shin-ichi. Dep. of Appl. physics, Osaka U., 2-1 Yamadaoka Suita, Osaka 565, Japan. (1935) DSc, physics (1966). Assoc. prof. (tel. 06 + 877-5111 ext 4668, fax +81 6 877 2900) *Raman spectroscopy, semiconductors, disorder, phase transition.*

Nakata, Mr. Kazuaki. Dept. of Physics, Osaka Kyoiku U., 43 Minami-Kawahori-cho Tennoji-ku, Osaka 543, Japan. (1946) MEd, physics (Osaka Kyoiku U., 1971). Res. assoc. (tel. 06 + 771-8131 ext 216). *Disorder, phase transition, organic crystals.*

Nakatsu, Prof. Kazumi. Dept. of Chemistry, Fac. of Sci., Kwansei Gakuin U., Uegahara Nishinomiya, Hyogo 662, Japan. (1928) DSc, chemistry (Osaka U., 1961). Prof. (tel. 0798 + 53-6111 ext 5278, fax +81 798 51 0914). *Materials science, molecular crystals, proteins, structural chemistry, chemical bonding.*

Nakayama, Mr Kan. Quantum Metrology Dept., Nat. Res. Lab., 1-4 1-chome Umezono Tsukuba, Ibaraki 305, Japan. (1943) DEng,(Tokyo U., 1971). Chief, Nanometrology Section (tel. 0298 + 54-4152 (dial in), fax +81 298 54 4135). *X-ray diffraction, instrumentation.*

Nakazawa, Dr Hiromoto. Nat. Inst. for Inorganic Materials Res., 1 Namiki Tsukuba-shi, Ibaraki 305, Japan. (1940) DSc, inorg. chemistry (Osaka U.). Supervising res. sci. (tel. 0298 + 51-3351 ext 289). *X-ray diffraction, materials science, minerals, instrumentation, microscopy.*

Nakazumi, Dr Yoshihide. Nakazumi Crystal Lab., 3-1 Sugahara-cho Ikeda, Osaka 563, Japan. (1919) DEng, chemistry (Osaka U., 1961). President of Nakazumi Crystals (tel. 0727 + 51-8832, fax +81 727 52 2003). *Crystal growth, fine ceramics, geology.*

Namba, Dr Yoshiyuki. Dept. of Chemistry, Osaka Kyoiku U., 43 Minami-Kawahori-cho Tennoji-ku, Osaka 543, Japan. (1924) DSc, physics (Osaka U., 1962). Prof. (tel. 06 + 771-8131). *Coordination compounds.*

Namikawa, Prof. Kazumichi. Physics, Tokyo Gakugei U. Nukui-kitamachi 4-1-1 Koganei, Tokyo 184, Japan. (1944) DSc,physics(1981). Assoc. Prof. (tel. 0423 + 25-2111 ext 638). *Nonlinear optics, X-ray scattering, magnetic scattering, surface physics, magnetism.*

Narita, Dr. Hajime. Techn. Dev. Dept., I&E Div., Nihon Philips Corp., Philips Building 3-13-37 Konan Minato-ku, Tokyo 100, Japan. Acting chief (tel. 03 + 448-5579 Direct, telex NIPHILT J 26388, fax +81 3 448 5585).

Narita, Mr. Masaki. Konan U., 8-9-1 Okamoto Higashi-Nada Kobe, Hyogo 658, Japan. (1964) MSc, inorganic chemistry post-graduate (tel. 078 + 431-4341 ext 652). *Inorganic crystals, phase transition.*

Nawata, Dr Yoshiharu. Res. Labs., Chugai Pharmaceutical Co. Ltd., 41-8 Takada 3-chome, Toshima-ku, Tokyo 171, Japan. (1934) DPharm, pharamcy (Tokyo U., 1984). Sr. scient. (tel. 03 + 987-7111 ext 219, fax +81 3 980 3578). *X-ray crystallography, molecular structure analysis, natural organic products.*

Naya, Prof. Shigeo. Dept. of Physics, Fac. of Sci., Kwansei Gakuin U., 1-155 Uegahara 1-bancho Nishinomiya, Hyogo 662, Japan. (1927) DSc, physics (Osaka U., 1959). Prof. (tel. 0798 + 51-3301). *Phase transition, crystal physics.*

Niimura, Dr. Nobuo. Lab. of Nuclear Sci., Fac. of Sci., Tohoku U., Mikamine Sendai, Miyagi 982, Japan. (1942) DSc, physics (Tokyo U., 1970). Assoc. Prof. (tel. 022 + 245-2151 ext17, fax +81 22 243 0965). *Neutron diffraction, small-angle scattering, structural biology, instrumentation.*

Nishi, Prof. Fumito. Saitama Inst. of Techn., 1690 Fusaiji Okabe-cho Osato-gun, Saitama 369-02, Japan. (1949) DSc, mineralogy (Tokyo U., 1978). Prof. (tel. 0485 + 85-0886). *Inorganic crystal structures.*

Nishida, Dr Isao. Physics Lab., Dept. of General Education, Nagoya U., Furo-cho Chikusa-ku, Nagoya 464, Japan. (1933) DSc, physics (Tokyo Inst. of Techn., 1971). Prof. (tel. 052 + 781-5111 ext 4846, fax +81 52 782 8261). *Metal fine particles, electron microscopy, crystal morphology.*

Nishida, Prof Takashi. C. of Arts and Sci., Chiba U., 1-33 Yayoi-cho Chiba-city, Chiba 260, Japan. (1938) DSc, mineralogy (Tokyo U., 1970). Prof. (tel. 0472 + 51-1111 ext 2282). *Inorganic crystal structures, polytypism, twinning, crystal growth.*

Nishikawa, Dr Masana. Eng. & Dev. Div., Mitsubishi Atomic Power Ind. Inc., 297 Kitabukuro 1-chome Omiya, Saitama 330, Japan. (1942) DSc, geophysics (Tokyo U., 1971). Chief. (tel. 0486 + 41-5111 ext 292, 370). *Nuclear fusion experiment facility design, high energy particle - solid interactions.*

Nishinaga, Prof. Tatau. Dept. of Electronics, Tokyo U., 7-3-1 Hongo Bunkyo-ku, Tokyo 113, Japan. (1939) DEng, electronics (Nagoya U., 1967). Prof. (tel. 03 + 812-2111 ext 6673, +81 3 818 5706). *Crystal growth, semiconductor.*

Nishino, Dr. Yoichi. Materials Sci. and Eng. Dept., Nagoya Inst. of Techn., Gokiso-cho Showa-ku, Nagoya 466, Japan. (1955) DEng, materials science (Nagoya U., 1983). Asst. prof. (tel. 052 + 732-2111 ext 2525, fax +81 52 735 0487). *Materials science, lattice defects, semiconductors, X-ray diffraction, metals.*

Nishiyama, Dr Tsutomu. Natural Sci. Lab., Toyo U., 28-20 Hakusan 5-chome Bunkyo-ku, Tokyo 112, Japan. (1939) DSc, mineralogy (Tokyo U. of Education, 1971). Asst. prof. (tel. 03 + 945-7392). *Clay minerals.*

Nishiyama, Prof. Zenji. 3-21-7 Shimoda-cho Kohoku-ku, Yokohama 223, Japan. (1901) DSc, physics (Tohoku U., 1932). Em. prof. (Osaka U.). (tel. 044 + 61-7774). *Metals, phase transition.*

Nittono, Dr. Osamu. Dept. of Metallurgy, Tokyo Inst. of Techn., 12-1 Oh-okayama 2-chome Meguro-ku, Tokyo 152, Japan. (1941) DEng, metallurgy (Tokyo Inst. of Techn., 1970). Prof. (tel. 03 + 726-1111 ext 3145, telex 2466360, fax +81 3 729 0393). *X-ray and electron diffraction, crystal growth and characterization, materials science.*

Noda, Prof. Tokiti. 15-3 Shikannonmichi Nishi, Tashiro-cho Chikusa-ku, Nagoya 464, Japan. (1903) DEng, applied chemistry (Tokyo U., 1940). Em. prof. (Nagoya U. and Mie U.). (tel. 052 + 711-2959). *Crystal growth, inorganic materials, carbon and graphite.*

Noda, Prof. Yasutoshi. Dept. of Materials Sci., Fac. of Eng., Tohoku U., Aoba Aramaki, Sendai 980, Japan. (1942) DEng, materials science (Tohoku U., 1970). Assoc. prof. (tel. 0222 + 22-1800 ext 4464, telex 852246 THUCOM, fax +81 22 268 2949). *Chemical bonding, crystal growth, inorganic crystals, materials science, semiconductors, X-ray diffraction.*

Noda, Dr. Yukio. Fac. of Eng. Sci., Osaka U., 1 Machikaneyama-cho 1-chome Toyonaka, Osaka 560, Japan. (1948) PhD,physics(Osaka U.,1977). Res. Assoc. (tel. 06 + 844-1151 ext 4686, telex 5286110 OUFES J, fax +81 6 845 4632). *Diffuse scattering, materials science, phase transition.*

Nukui, Dr Akihiko. Group 9, Nat. Inst. for Inorganic Materials Res., 1-1 Namiki Tsukuba, Ibaraki 305, Japan. (1944) DEng, materials science (Tokyo Inst. of Techn., 1973). Sr. res. (tel. 0298 + 51-3351 ext 249, fax +81 298 52 7449). *Glass structure, phase transitions, X-ray diffraction.*

Oda, Prof. Tsutomu. Dept. of Chemistry, Osaka Kyoiku U., 43 Minami-Kawahori-cho Tennoji-ku, Osaka 543, Japan. (1916) DSc, physical chemistry (Osaka U., 1945). Prof. (tel. 06 + 771-8131). *Crystal chemistry.*

Ogata, Dr Kiyoshi. 2nd Dept., Production Eng. Res. Lab., Hitachi Ltd., 292 Yoshida-cho Totsuka-ku Yokohama, Kanagawa 244, Japan. (1956)

DSc,mineralogy(Tokyo U., 1986). Res. (tel. 045 + 881-1241 ext 3024, telex 3822470, fax +81 45 864 5721).

Ogawa, Dr Katsumi. Materials Processing Dept., Nat. Res. Inst. for Pollution and Resources, Onogawa Tsukuba, Ibaraki 305, Japan. (1950) DEng, mineralogy (Waseda U., 1978). (tel. 0298 + 54-3057, fax +81 298 54 3049).

Ogawa, Prof. Kazuhide. 1-7-31 Fukaekita-machi Higashinada-ku Kobe, Hyogo 658, Japan. (1922) DSc, physics (Osaka U., 1962). Em. prof. *Inorganic and organic crystal structures.*

Ogawa, Dr Keiichiro. Dept. of Chemistry, C. of Arts & Sci. U. of Tokyo, Komaba Meguro-ku, Tokyo 153, Japan. (1952) DSc, chemistry(Tokyo U., 1983). Instr. (tel. 03 + 467-1171 ext 593, telex: 2426728 TODAIK J, fax +81 3 485 2904, email ogawa%komaba.c.u-tokyo.junet at relay.cs.net). *Molecular crystals, organic compounds, physical organic chemistry.*

Ogawa, Prof. Shiro. 2-14-1 Midorigaoka Sendai, Miyagi 982, Japan. (1912) DSc, physics (Tohoku U., 1946). Prof. Em. of Tohoku U. (tel. 022 + 248-9106). *Electron diffraction, electron microscopy, metals and alloys, structure of films and surfaces.*

Ogawa, Prof. Tomoya. Dept. of Physics, Fac. of Sci., Gakushuin U., 5-1 Mejiro 1-chome Toshima-ku, Tokyo 171, Japan. (1930) DSci, crystal growth (Tohoku U., 1986). Prof. (tel. 03 + 986-0221 ext 459, fax +81 3 590 2602). *Crystal growth, texture.*

Ogura, Prof. Iwao. C. of Eng., Nihon U., Tamura-machi Koriyama, Fukushima 963, Japan. (1922) DSc, physics (Hiroshima U., 1961). Prof. (tel. 0249 + 44-1300 ext 314). *Thin films, surface science, materials science and engineering, electron microscopy.*

Ohachi, Prof. Tadashi. Dept. of Electrical Eng., Doshisha U., Karasuma-Imadegawa, Kyoto 602, Japan. (1941) DEng, electrical eng. (Doshisha U.,1975). Prof. (tel. 075 + 251-3784, fax +81 75 251 3737). *Crystal growth, phase transition, semiconductors, surface structure, teaching.*

Ohama, Prof. Nobuhiko. Div. of Childhood Education, Dept. of Literature, Seinan Gakuin U. Nishijin 6-chome 2-92 Sawara-ku Fukuoka, Fukuoka 814, Japan. (1940) DSc.(Kyushu U., 1971). Prof. (tel. 092 + 841-1311 ext 3415, fax +81 92 843 9337). *X-ray diffraction, diffractometry, phase transitions, crystal physics, instrumentation.*

Ohashi, Prof. Yuji. Dep. of Chemistry, Ochanomizu U., 1-1 Otsuka 2-chome Bunkyo-ku, Tokyo 112, Japan. (1941) DSc, chemistry (Tokyo U., 1974). Prof. (tel. 03 + 943-3151 ext 551, fax +81 3 942 2815). *Reaction pathways, molecular crystals.*

Ohba, Dr. Shigeru. Dept. of Chemistry, Fac. of Sci. & Techn., Keio U. Hiyoshi 3, Kohoku-ku, Yokohama 223, Japan. (1953) DSc, chemistry (Tokyo U., 1981). Asst Prof. (tel. 044 + 63-1141 ext 3912, fax +81 44 63 3421). *Electron density distribution, transition metal complexes.*

Ohba, Dr. Takuya. Inst. of Materials Sci., Tsukuba U. Tennoudai Tsukuba, Ibaraki 305, Japan. (1953) DSc (Hiroshima U., 1984). Res. Asst. (tel. 0298 + 53-5470). *Inorganic crystals, metals, X-ray diffraction.*

Ohgaki, Mr Masataka. Res. Lab. of Eng. Materials, Tokyo Inst. of Techn., 4259 Nagatsuta-cho, Midori-ku, Yokohama Kanagawa 227, Japan. (1959) MSc, materials science (Tokyo Inst. of Techn., 1985). Postgraduate student (tel. 045 + 922-1111 ext 2311). *Crystal structure analysis, electron density determination.*

Ohmasa, Dr Masaaki. Inst. of Materials Sci., Tsukuba U., Sakura-mura Niihari-gun, Ibaraki 305, Japan. (1935) DSc, mineralogy (Tokyo U., 1964). Asst. prof. (tel. 0298 + 53-5012). *Crystal structures, phase transition.*

Ohno, Dr Tamotsu. Physics Dept., Aichi Medical U., Yazako Nagakute-cho, Aichi 480-11, Japan. (1937) DEng, physics (Nagoya U., 1975). Prof. (tel. 05616 + 2-3311 ext 2055). *Electron diffraction, electron microscopy, macromolecules.*

Ohsato, Dr. Hitoshi. Materials Sci. and Eng., Nagoya Inst. of Techn., Gokiso-cho Showa-ku Nagoya, Aichi 466, Japan. (1944) DSc.(Tokyo U., 1984). (tel. 052 + 732-2111 ext 2520, fax +81 52 732 2925). *Inorganic crystal structure, phase relation, crystal growth, optical microscopy.*

Ohshima, Dr Ken-ichi. Inst. of Applied Physics, Tsukuba U., 1-1-1 Ten-nodai Tsukuba, Ibaraki 305, Japan. (1946) DSc, physics (Tokyo U., 1975). Assoc. prof. (tel. 0298 + 53-5300,5049, fax +81 298 53 5205). *Diffuse scattering, X-ray diffraction*

Ohsumi, Dr Kazumasa. Photon Factory, Nat. Lab. for High Energy Physics, Oho Tsukuba, Ibaraki 305, Japan. (1943) DSc, mineralogy (Tokyo U., 1971). Assoc. prof. (tel. 0298 + 64-1171). *Symmetry, inorganic crystal structures, phase relation.*

Ohta, Dr Takao. Toshiba Res. and Dev. Center, Tokyo Shibaura Electric Co. Ltd., 1 Komukai Toshiba-cho Saiwai-ku, Kawasaki Kanagawa 210, Japan. (1941) DSc, mineralogy (Tokyo U., 1970). Res. (tel. 044 + 511-2111 ext 2291). *Inorganic crystal structure, crystal growth.*

Ohta, Dr. Tsutomu. Mineralogical Inst., Fac. of Sci., Tokyo U., 3-1 Hongo 7-chome Bunkyo-ku, Tokyo 113, Japan. (1950) PhD. mineralogy (Tokyo U., 1978). (tel. 03 + 812-2111 ext 3728). *Crystal chemistry.*

Ohtani, Prof. Eiji. Inst. of Mineralogy - Petrology and Economic Geology, Tohoku U., Aoba Sendai, Miyagi 980, Japan. (1950) Dr. Sci., earth science (Nagoya U., 1979). Assoc. prof. (tel. 022 + 222-1800 ext 3442, fax +81 22 262 6609). *Crystallography, experimental petrology, geophysics, seismology, igneous petrology, planetology, high pressure, phase transitions.*

Ohtsuka, Mr Yasukuni. Dept. of Physics, Tohoku Inst. of Techn., 35-1 Kasumi-cho Yagiyama Sendai, Miyagi 982, Japan. (1935) MSc, physics (Tohoku U., 1961). Prof. (tel. 022 + 229-1151 ext 529). *Crystal growth.*

Okabe, Dr. Toshio. Dept. of Physics, Fac. of Literature and Sci., Toyama U., 3190 Gofuku, Toyama 930, Japan. (1942) DSc, physics (Kyoto U., 1974). Asst. prof. (tel. 0764 + 41-1271 ext 2314, fax +81 764 41 2972). *Amorphous semiconductor, crystallization, electron microscopy.*

Okada Prof. Masakazu. Dept. of Applied Physics, Fac. of Applied Biology Sci., Hiroshima U. Midorimachi Fukuyama, Hiroshima 720 Japan. (1928) DSc, physics (Tokyo Sci.U.,1956) Prof. (tel. 0849 + 24-6211 ext 320). *Crystal physics, electron crystallography.*

Okada, Mr Kenji. System Products Marketing Div. Ricoh Co. Ltd., 2-38-5 Nishi-Shinbashi,Minato-ku, Tokyo 105 Japan. (1942) BSc, applied chemistry (Sci. U. of Tokyo, 1965). Computer graphics planning manager. (tel. 03 + 578-3086, telex 246-6201, fax +81 3 578 3086). *Computing, phase determination, organic compounds.*

Okada, Dr. Kiyoshi. Dept. of Inorganic Materials, Fac. of Eng., Tokyo Inst. of Techn., 12-1 Oh-okayama 2-chome, Meguro-ku, Tokyo 152, Japan. (1948) DEng, inorganic chemistry (Tokyo Inst. of Techn., 1976). Assoc. prof. (tel. 03 + 726-1111 ext 2524, fax +81 3 729 0393). *Crystal chemistry, clay mineralogy, amorphous substances.*

Okada, Prof. Masakazu. Applied Physics Dept., Fac. of Applied Biology Sci., Hiroshima U., Midorimachi, Fukuyama-city 720, Japan. (1928) DSc, physics (Tokyo Sci. U., 1956). Prof. (tel. 0849 + 24-6211 ext 320). *Crystal physics, electron crystallography.*

Okada, Mr Yasumasa. Fundamental Sci. Div., Electrotechnical Lab., Umezono Tukuba-shi, Ibaraki 305, Japan. (1940) BSc, solid state physics (Sci. U. of Tokyo, 1965). Sr. res. (tel. 0298 + 54-5136, telex 3652570 AIST J, fax +81 298 54 5156). *Defect structures, microscopy, semiconductors, X-ray diffraction.*

Okamura, Dr Fujio Peter. 15th Res. Group, Nat. Inst. for Inorganic Materials Res., Namiki 1-1 Tsukuba, Ibaraki 305, Japan. (1939) DSc, mineralogy (Tokyo U., 1969). Sr. res. officer (tel. 0298 + 51-3351 ext 289). *Crystal chemistry, phase prediction, CAD of materials.*

Okawa, Dr. Tokio. Dept. of Electronics, Fac. of Eng. The Inst. of Vocational Training, 1960 Aihara Sagamihara, Kanagawa 229, Japan. (1934) DEng. (Osaka U., 1984). Prof. (tel. 0427 + 61-2111 ext 306, fax +81 427 61 8293). *Electron microscopy, X-ray microscopy, X-ray diffraction.*

Okazaki, Prof. Atsushi. Dept. of Physics, Kyushu U., 10-1 Hakozaki 6-chome Higashi-ku, Fukuoka 812, Japan. (1931) DSc, physics (Kyushu U., 1961). Prof. (tel. 092 + 641-1101 ext 4177, fax +81 92 631 4233). *X-ray diffraction, phase transitions, instrumentation.*

Okuno, Dr. Masayuki. Dept. of Earth Sci., Fac. of Sci. Kanazawa U., 1-1 Marunouchi Kanazawa, Ishikawa 920, Japan. (1955) DSc, mineralogy(Tokyo Inst. of Techn., 1983). Res. Assoc. (tel. 0762 + 62-4281 ext 576, fax +81 762 64 1059). *Mineralogy, melt and glass structures, plasticity.*

Okunuki, Mr Masahiko. Semiconductor Equipment Dev. Div. Canon Ltd., Morinosato Wakamiya 5 Atsugi, Kanagawa 243-01, Japan. (1946) MSc, physics (Gakushuin U., 1971). (tel. 0462 + 47-2111). *X-ray diffraction, X-ray lithography, focused ion beam, micro lithography.*

Okuyama, Prof. Kenji. Polymer Eng. Tokyo U. of Agriculture and Techn., Nakamachi 2-24-16 Koganei, Tokyo 184, Japan. (1946) DSc, polymer chemistry (Osaka U., 1977). Prof. (tel. 0423 + 81-4221 ext 300, telex 2832663 TUATT J, fax +81 423 84 3804) *Biological substances, structure.*

Onodera, Dr. Akira. Physics, Hokkaido U. N10W8 Sapporo, Hokkaido 060, Japan. (1950) DSc.(Hokkaido U., 1979). (tel. 011 + 716-2111 ext 3218, telex 932510 HOKUSCJ). *Crystallographic data, phase transitions, X-ray diffraction.*

Onuma, Dr Shigeki. Dept. of Chemistry, Shizuoka U., 836 Oya, Shizuoka 422, Japan. (1937) DSc, chemistry (Tohoku U., 1966). Asst. prof. (tel. 0542 + 37-1111 ext 5601).

Ookawa, Prof. Akiya. Dept. of Physics, Fac. of Sci., Gakushuin U., 1-5 Mejiro Toshima-ku, Tokyo 171, Japan. (1918) DSc, physics (Tokyo U., 1958). Prof. (tel. 03 + 986-0221 ext 485). *Crystal growth, statistical thermodynamics.*

Osaka, Dr. Toshiaki. Materials Sci. and Eng. Waseda U. 3-4-1 Ohkubo Shinjuku-ku, Tokyo 169, Japan. (1941) DEng, thin films(Waseda U., 1972). Prof. (tel. 03 + 203-4141 ext 74-2100, fax +81 3 203 1353). *Growth, structure and electronic states of thin films, electron microscopy, electron diffraction, metal, alloy and semiconductors.*

Osakabe, Mr Nobuyuki. Advanced Res. Lab.Hitachi 1-280 Higashikoigakubo Kokubunji-shi, Tokyo 185, Japan. (1955) MSc, physics (Tokyo Inst. of Techn., 1980). Res. (tel. 0423 + 23-1111, fax +81 423 26 0880, email OSAKABE at HARL186.HARL.HITACH.JUNET). *Surface structure, magnetism, electron microscopy.*

Osaki, Dr Kenji. 5-9-6 Nampei-dai Takatsuki, Osaka 569, Japan. (1920) DSc, physics (Osaka U., 1958). Em. prof. (Kyoto U.). (tel. 0726 + 95-3116). *Crystal structures, molecular interactions, crystallographic information system.*

Osano, Ms. Yasuko. Res. Center Mitsubishi Kasei Corp. 1000 Kamoshida-cho Midoriku Yokohama, Kanagawa 227, Japan. (1960) MSc, chemistry(Tokyo Inst. of Techn., 1984). (tel. 045 + 963-3156, fax +81 45 961 6561). *Structural chemistry, structural biology.*

Otsuka, Prof. Kazuhiro. Inst. of Materials Science, U. of Tsukuba, 1 Tennodai Tsukuba, Ibaraki 305, Japan. (1937) DEng, metallurgy (Tokyo U., 1972). Prof. (tel. 0298 + 53-5294(direct), telex 3652580 UNTUKUJ, fax +81 298 53 5208). *Materials science, physical metallurgy, martensitic transformations, shape memory alloys.*

Otsuka, Prof. Ryohei. Dept. of Mineral Resources Eng., Waseda U., 170 Nishiokubo 4-chome Shinjuku-ku, Tokyo 169, Japan. (1922) DEng, mineralogy (Waseda U., 1957). Prof. (tel. 03 + 203-4141 ext 73-3211, telex 232-5115 WARIKO, Fax: +81 3 200 2567). *Minerals.*

Oyanagi, Dr Hiroyuki. Fundamental Sci. Div., Electrotechn. Lab., 1-1-4 Umezono Tsukuba, Ibaraki 305, Japan. (1952) PhD, physical chemistry (Tokyo U. 1976). Chief res. (tel. 0298 + 54-5112). *Synchrotron radiation, X-ray spectroscopy, EXAFS, organic conductors, thin films.*

Ozawa, Dr Tohru. Mineralogical Inst., Fac. of Sci., Tokyo U., 7-3-1 Hongo Bunkyo-ku, Tokyo 113, Japan. (1940) DSc, mineralogy (Tokyo U., 1968). Lect. (tel. 03 + 812-2111 ext 4546, fax +81 3 816 5714). *Crystal chemistry, minerals, electron microscopy, inorganic materials.*

Sadanaga, Prof. Ryoichi. 4-1-4 Suimeidai Kawanishi-shi, Hyogo 666-01, Japan. (1920) DSc, mineralogy (Tokyo U., 1953). Em. prof. (Tokyo U.); Member, Japan Academy. (tel. 0727 + 92-5100, fax +81 727 92 5200). *Mathematical crystallography, inorganic and organic crystal structures.*

Saito, Prof. Norio. Chemical Lab., Meiji-gakuin U., 2-37 Shiroganedai 1-chome Minato-ku, Tokyo 108, Japan. (1935) DSc, biochemistry (Tokyo U. of Education, 1964). Prof. (tel. 03 + 443-8231 ext 254). *Plant pigments, biosynthesis.*

Saito, Prof. Yoshihiko. Dept. of Chemistry, Fac. of Sci. & Techn., Keio U., 14-1 Hiyoshi 3-chome, Kohoku-ku, Yokohama 223, Japan. (1920) DSc, Chemistry (Osaka U., 1952). Prof. (tel. 044 + 63-1141 ext 3910). *teaching, history of science.*

Saito, Dr Yoshio. Dept. of Electronics and Info. Sci., Kyoto Inst. of Techn., Matsugasaki, Sakyo-ku, Kyoto 606, Japan. (1944) BEE, electronics (Osaka Inst. of Techn., 1968). Asst. (tel. 075 + 791-3211 ext 448). *Crystal growth, computing, electron microscopy, coordination compounds, molecular crystals, semiconductors.*

Saito, Mr Yoshiyuki. Res. Lab., Kawasaki Steel Corp., 8-4 Ritsurincho 1-chome, Takamatsu 760, Japan. (1948) MSc, physics (1974). (tel. 0878 + 62-2853).

Saka, Dr Takashi. New Materials Res. Lab., Daido Steel Co. Ltd., 2-30 Daido-cho, Nagoya 457, Japan. (1946) DEng, applied physics (Nagoya U., 1974). Chief res. eng. (tel. 052 + 611-2511 ext 2773). *Materials science.*

Sakabe, Dr. Noriyoshi. Photon Factory, Nat. Lab. for High Energy Physics, Oho Tsukuba, Ibaraki 305, Japan. (1934) DSc, chemistry (Nagoya U., 1966). Prof. (tel. 0298 + 64-1171). *Protein crystallography, molecular biophysics.*

Sakai, Mr. Katsura. New Materials Res. Lab., , Nisshin Steel Co. Ltd., 7-1 Koya-shinmachi Ichikawa, Chiba 272, Japan. (1962)applied physics Student of master course (tel. 0473 + 28-1211 ext 670). *Semiconductors, surface structure, X-ray diffraction.*

Sakaki, Prof. Yoneichiro. 105 Yayoigaoka Tempaku-ku, Nagoya 468, Japan. (1913) DEng, electron microscopy (Nagoya U., 1951). Prof. em. - Nagoya U. (tel. 052 + 831-1787). *Electron optics.*

Sakamoto, Dr Yosio. SAKAMOTO's Niggli-&-Born Inst. for Studying Crystal Structure Types, in Hirosima 15-5 Higasikasumi-machi Minami-ku, Hirosima 734, Japan. (1919) DSc, chemistry (Hirosima U. of Lit. & Sci., 1957). Res. crystallographer, former Prof. Hiroshima U. (tel. 082 + 282-6513). *Codification, crystal structure types, lattice & point complex method crystal lattice potential.*

Sakata, Dr Makoto. Dept. of Applied Physics, Fac. of Eng., Nagoya U., Furo-cho, Chikusa-ku, Nagoya Aichi 464, Japan. (1944) PhD, chemistry (Nagoya U. of Education, 1974). Assoc. prof. (tel. 052 + 781-5111 ext 4453, fax +81 52 782 2129). *Solid state physics, X-ray crystallography, neutron diffraction.*

Sakuma, Dr Takashi. Dept. of Physics, Fac. of Sci., Ibaraki U., Mito 310, Japan. (1951) DSc, physics, (Tokyo U., 1978). Assoc.prof. (tel. 0292 + 26-1621 ext 707, fax +81 292 31 5740). *X-ray and neutron diffraction.*

Sakurai, Dr Junji. Dept. of Materials Sci., Fac. of Sci., Hiroshima U., 1-89 Higashisendamachi 1-chome, Hiroshima 730, Japan. (1936) DSc, chemistry (Kyoto U., 1964). Asst. prof. (tel. 0822 + 41-1221 ext 654). *Magnetism.*

Sakurai, Prof. Tosio. Fac. of Education, Shinshu U. Nishinagano, Nagano 380, Japan. (1926) DSc, physics (Tokyo U., 1962). Prof. (tel. 0262 + 32-8106 ext 359). *Structural chemistry.*

Sasada, Prof. Yoshio. 885-36 Higashihongo Midori-ku, Yokohama 226, Japan. (1926) DSc, chemistry (Osaka U., 1958). Prof. (tel. 045 + 473-0674). *Structural chemistry, organic compounds, solid state reaction, biological molecules.*

Sasaki, Prof. Akio. Dept. of Electrical Eng., Kyoto U., Kyoto 606, Japan. (1932) PhD, (U. California, Berkley, USA, 1966) Prof. (tel. 075 + 753-5296, 5270 ext 5296, telex 5422455 DEEKYU J, fax +81 75 751 1576). *Solid state electronics.*

Sasaki, Dr Kyoyu. C. of Medical Techn., 1-1-20 Daiko-minami, Higashi-ku, Nagoya 461, Japan. (1940) DSc, chemistry (Nagoya U., 1973). Assoc. Prof. (tel. 052 + 723-1111 ext 241, fax +81 52 723 0290). *Protein crystallography, computing techniques.*

Sasaki, Dr Satoshi. Photon Factory, Nat. Lab. For High Energy Physics, Oho, Tsukuba, Ibaraki 305, Japan. (1951) DSc, mineralogy (Tokyo U., 1979). Res. assoc. (tel. 0298 + 64-1171 ext 5023, telex 3652-534, fax +81 298 64 2801, email BITNET: SASAKIS at JPNKEKVM). *X-ray diffraction, synchrotron radiation research, crystallography, mineralogy.*

Sasaki, Prof. Yukiyoshi. Dept. of Chemistry, Fac. of Sci., Tokyo U., 7-3-1 Hongo Bunkyo-ku, Tokyo 113, Japan. (1928) DSc, chemistry (Tokyo U., 1960). Prof. (tel 03 + 812-2111 ext 4359). *Polyanion, inorganic structural chemistry.*

Sato, Mr Hiroki. Musashi Works, Hitachi Ltd., 1450 Josui-honmachi Kodaira, Tokyo 187, Japan. (1944) MSc, mineralogy (Tohoku U., 1970). Eng. (tel. 0423 + 43-8623). *Surface or interface science.*

Sato, Dr Mamoru. Protein Crystallography Inst. for Protein Res. Yamada-oka 3-2 Suita, Osaka 565, Japan. (1956) DEng.(Osaka U., 1983). Instr. (tel. 06 + 877-5111 ext 3837, fax +81 6 876 2533). *Small-angle scattering, proteins, X-ray diffraction, macromolecules.*

Sato, Prof. Mitsuo. Dept. of Chemistry, Fac. of Techn., Gunma U., Tenkincho 1-5-1, Kiryu, Gunma 376, Japan. (1932) DSc, mineralogy (Tokyo Kyoiku U., 1962). Prof. (tel. 0277 + 22-3181 ext 429, fax +81 277 43 6556). *Structural chemistry, inorganic crystals, minerals.*

Sato, Prof. Shin'ichi. 3-3-3-17 Hassamu Nishi-ku, Sapporo 063, Japan. (1925) DSc, physics (Osaka U., 1962). Em. prof. (tel. 011 + 661-0498). *Diffraction crystallography, martensitic phase transformation.*

Sato, Dr Shoichi. Inst. for Solid State Physics, Tokyo U., 7-22-1 Roppongi Minato-ku, Tokyo 106, Japan. (1930) DSc, chemisrty (Tokyo U., 1980). Res. assoc. (tel. 03 + 478-6811 ext 5976, telex ISSPUT J32469, fax +81 3 401 5169). *Coordination compounds, electron density, inorganic crystals, phase transition, structural chemistry.*

Sato, Mr Tomohiro. Shionogi Res. Labs., Shionogi and Co. Ltd., Fukushima-ku, Osaka 553, Japan. (1939) MPharm. (Tokyo U., 1965). Res. (tel 06 + 458-5861). *X-ray crystallography.*

Satow, Prof. Yoshinori. Fac. of Pharm. Sci., U. of Tokyo, 7-3-1 Hongo Bunkyo-ku, Tokyo 113, Japan. (1949) DPharm, pharmaceutical sci. (Tokyo U., 1977). Prof. (tel. 03 + 812-2111 ext 4840). *Diffractometry, instrumentation, macromolecules, proteins, synchrotron radiation, X-ray diffraction.*

Sawada, Prof. Akikatsu. Synthetic Crystal Res. Lab., Nagoya U., Furo-cho Chikusa-ku, Nagoya 464-01, Japan. (1941) DEng, physics (Nagoya U., 1968). Assoc. prof. (tel. 052 + 781-5111 ext 3596, fax +81 52 782 9209). *Ferroelectrics, phase transitions, lattice vibration.*

Sawada, Dr Haruo. Photon Factory Nat. Lab. for High Energy Physics 1-1 Oho Tsukuba, Ibaraki 305, Japan. (1956) DSc.(Tokyo U., 1986). Visiting Res. (tel. 0298 + 64-1171 ext 072, telex (o)3652 534 (International)). *Crystallographic data, electron density, inorganic crystals, minerals, phase transitions, structural chemistry, X-ray diffraction, chemical bonding, coordination compounds.*

Sawada, Dr Toshiyuki. Electron Device Eng. Lab., Toshiba Corp., 8 Shinsugita-cho Isogo-ku Yokohama, Kanagawa 235, Japan. (1949) MSc, mineralogy (Tokyo U., 1975). (tel. 045 + 756-2522). *Crystal structures.*

Sawada, Mr Yasuaki. Dept. of Info. Sci., Kanazawa Inst. of Techn., 7-1 Ogigaoka Nonoichi-machi, Ishikawa 921, Japan. (1941) MSc, mathematics (Kanazawa U., 1967). Asst. prof. (tel. 0762 + 48-1100). *Symmetry.*

Sawaguchi, Prof. Etsuro. Dept. of Physics, Fac. of Sci., Hokkaido U., Kita 10-jo Nishi 8-chome Kita-ku, Sapporo 060, Japan. (1925) DSc, physics (Tokyo U., 1960). Prof. (tel. 011 + 716-2111 ext 2680). *Phase transitions, materials science.*

Seki, Prof. Syuzo. 3-7 Tsukiwaka-cho Ashiya, Hyogo 659, Japan. (1915) DSc, chemistry (Osaka U., 1945).Em. Prof. of Osaka U., Member of Japan Academy. (tel. 0797 + 22-2082). *Physical chemistry, thermodynamical studies, phase transitions in solids.*

Sekido, Dr Kiyotane. Chemistry, The Nat. Defense Academy, Hashirimizu 1-10 Yokosuka, Kanagawa 239, Japan. (1937) DSc. (Tsukuba U., 1986). Lect. (tel. 0468 + 41-3810 ext2413, fax +81 468 43 6236). *Coordination compounds, organic compounds, X-ray diffraction, chemical bonding, hydrogen bonding, strutural chemistry.*

Sekizaki, Prof. Masao. C. of Liberal Arts, Kanazawa U., 1-1 Marunouchi, Kanazawa 920, Japan. (1941) DSc, chemistry (Nagoya U., 1972). Prof. (tel. 0762 + 62-4281 ext 648). *Structures, transition metal complexes.*

Shibata, Prof. Noboru. Fac. of Liberal Arts, Nagasaki U., 1-14 Bunkyo-machi, Nagasaki 852, Japan. (1924) BSc, physics (Osaka U., 1946). Prof. (tel. 0958 + 47-1111, fax +81 958 43 1379). *Crystal growth, 2-6 and 3-5 compounds, etching behavior.*

Shibata, Prof. Shuzo. Dept. of Chemistry, Shizuoka U., 836 Oya, Shizuoka 422, Japan. (1924) DSc, chemistry (Nagoya U., 1958). Prof. (tel. 0542 + 85-1171). *Gas electron diffraction, inorganic crystal structure.*

Shibuya, Prof. Iwao. Div. of Slow Neutron Physics, Res. Reactor Inst., Kyoto U., 1052 Noda Kumatori-cho, Osaka 590-04, Japan. (1930) DSc, solid state physics (Kyushu U., 1961). Prof. (tel. 07245 + 2-0901 ext 2322, 2282). *Neutron diffraction studies, ferroelectrics.*

Shichiri, Dr Takaki. Dept. of Physics, Fac. of Sci., Osaka City U., Sugimoto-cho Sumiyoshi-ku, Osaka 558, Japan. (1933) DTechn., applied physics (Nagoya U., 1972). Asst. prof. (tel. 06 + 692-1231 ext 3232). *Crystal growth.*

Shigenari, Prof. Takeshi. Dept. of Applied Physics and Chemistry, U. of Electro-Communications, 5-1 Chofugaoka 1-chome Chofu, Tokyo 182, Japan. (1939) DEng, physics (Tokyo U., 1970). Prof. (tel. 0424 + 83-2161 ext 3921). *Phase transition, thermal vibrations.*

Shimanouchi, Dr Hirotaka. Dept. of Life Sci., Tokyo Inst. of Techn., 4259 Nagatsuta Midori-ku Yokohama, Kanagawa 227, Japan. (1939) DSc, chemistry (Tohoku U., 1967). Asst. prof. (tel. 045 + 922-1111 ext 2388, fax +81 45 922 2432). *Macromolecules, proteins, polymers, structural biology, liquid crystals.*

Shimaoka, Prof. Kohji. Dept. of Math. and Physics, Ritsumeikan U., 28-1 Tojiin-Kitamachi Kita-ku, Kyoto 603, Japan. (1928) DSc, physics (Tokyo Inst. of Techn., 1958). Prof. (tel. 075 + 463-1131 ext 326). *Crystal structures, neutron diffraction, phase transition.*

Shimazu, Dr Masaji. Group 10, Nat. Inst. for Inorganic Materials Res., 1-1 Namiki Tsukuba, Ibaraki 305, Japan. (1930) DSc, mineralogy (Tokyo U. of Education, 1965). Nonlinear optics. (tel. 0298 + 51-3351, fax +81 298 52 7449). *Crystal growth, crystallographic data, diffractometry, inorganic crystals, materials science, microscopy, powder diffraction, X-ray diffraction, structural chemistry.*

Shimizu, Prof. Ken'ichi. Inst. of Scient. and Industrial Res., Osaka U., 8-1 Mihoga-oka Ibaraki, Osaka 567, Japan. (1928) DSc, physics (Nagoya U., 1962). Prof. (tel. 06 + 877-5111 ext 3555, telex 5286213 ISIROU J, fax +81 6 877 4977). *Physical metallurgy, crystallography, metals.*

Shimoi, Dr Mamoru. Dept. of Chemistry, Fac. of Sci. Tohoku U. Aramaki Aoba Sendai, Miyagi 980, Japan. (1946) DSc, chemistry (1976). Assoc. Prof. (tel. 022 + 222-1800 ext 3350, fax +81 22 262 6609). *Coordination compounds.*

Shimura, Mr Fumio. Material Res. Lab., Central Res. Labs., Nippon Electric Co. Ltd., 1753 Shimonumabe Nakahara-ku Kawasaki, Kanagawa 211, Japan. (1948) MEng, crystallography (Nagoya Inst. of Techn., 1974). Res. member. (tel. 044 + 433-1111 ext 2839). *Growth theory of crystals, crystal growth, crystals grown for electronics.*

Shinnaka, Prof. Yasuhiro. Dept. of Physics, Fac. of Sci., Yamaguchi U., 1677-1 Yoshida, Yamaguchi 753, Japan. (1926) DSc, physics (Kyoto U., 1960). Prof. (tel. 0839 + 22-6111 ext 371). *Phase transition, X-ray diffraction.*

Shintani, Dr Ryuichi. Dept. of Physics, Fac. of Sci., Kwansei Gakuin U., 1-155 Uegahara 1-bancho Nishinomiya, Hyogo 662, Japan. (1928) DSc, chemistry (Osaka U., 1960). Prof. (tel. 0798 + 51-0912). *Organic semiconductors.*

Shiojiri, Prof. Makoto. Dept. of Electronics and Info. Sci., Kyoto Inst. of Techn., Matsugasaki Sakyo-ku, Kyoto 606, Japan. (1936) DSc, chemistry (Kyoto U., 1967). Prof. (tel. 075 + 791-3211 ext 444). *Crystal growth, electron microscopy.*

Shiro, Dr Motoo. Dept. of Physical Chemistry, Shionogi Res. Lab., 12-4 Sagisu 5-chome Fukushima-ku, Osaka 553, Japan. (1931) DSc, chemistry (Osaka City U., 1979). Sr. res. chemist. (tel. 06 + 458-5861 ext 271savelex SHIOGI OSAKA, fax +81 6 458 0987) *Crystal structure analysis, organic compounds, physicochemistry, biochemistry.*

Shoda, Prof. Tokugoro. 2-2-7 Kobinata Bunkyo-ku, Tokyo 112, Japan. (1913) DSc, mineralogy (Tokyo U., 1957). Prof. (tel. 03 + 947-4286). *Optical crystallography.*

Soejima, Dr. Yuji. Dept. of Physics Kyushu U. Hakozaki 6-10-1 Fukuoka, Fukuoka 814, Japan. (1957) DSc.(Kyushu U., 1986). Res. Assoc. (tel. 092 + 641-1101 ext 4184, fax +81 92 631 4233). *Diffractometry, phase transition, powder diffraction, X-ray diffraction.*

Somiya, Prof. Shigeyuki. Teikyo U., 2-11-1 Kaga Itabashi-ku, Tokyo 173, Japan. (1928) Prof.; dir. (Hydrothermal Syn. Lab.). (fax +81 3 415 6619). *Hydrothermal synthesis, phase equilibria.*

Sonoike, Prof. Sanemi. Dept. of Physics, Chuo U., Kasuga Bunkyo-ku, Tokyo 112, Japan. (1922) DEng, applied physics (Tokyo U., 1958). Prof. (tel. 03 + 813-4175 ext 363). *Defects in crystalline solids, physical properties of silver halide.*

Sudo, Prof. Toshio. Geol. & Mineral. Inst., Fac. of Sci., Tokyo U. of Ed. (retired), 14-27 Amishimadai Kouhoku-ku, Yokohama 223, Japan. (1911) DSc, mineralogy (Tokyo U., 1944). Em. prof. retired from Tokyo U. of Education (tel. 045 + 544-9762). *Minerals, structural chemistry.*

Sueno, Prof. Shigeho. Inst. of Geoscience, Tsukuba U., Tennoudai Tsukuba, Ibaraki 305, Japan. (1937) DSc, mineralogy (Tokyo U., 1966). Prof. (tel. 0298 + 53-4427, telex 3652580UNTUKU J, fax +81 298 53 4012). *Inorganic crystal chemistry.*

Suga, Prof. Hiroshi. Dept. of Chemistry, Osaka U., Toyonaka, Osaka 560, Japan. (1930) DSc. Prof. (tel. 06 + 844-1151 ext 4200, telex 05286207 OSAKU J, fax +81 6 855 8139). *Disorder, neutron diffraction, phase transition.*

Sugaike, Dr Suezo. Sci.-Techn Res. Co. Ltd., 5-1 Yuurakucho 1-chome Chiyoda-ku, Tokyo 100, Japan. (1921) DSc, mineralogy (Tokyo U., 1957). Techn. consultant. (tel. 03 + 501-1942) *Inorganic crystals, semiconductors.*

Sugawara, Dr Yoko. X-ray Crystallography Lab., RIKEN - Inst. of Physical and Chem. Res., Wako Saitama 351-01, Japan. (1952) DPharm, physical chemistry (Tokyo U., 1980). Scient. (tel. 0484 + 62-1111 ext 3342, telex 02962818 RIKEN J, fax +81 484 62 1449) *Biophysical chemistry.*

Sugihara, Dr Akio. Dept. of Biochemistry, Osaka Municipal Techn. Res. Inst., 1-6-50 Morinomiya Jyoto-ku, Osaka 536, Japan. (1943) DSc, chemistry (1970). Res. worker. (tel. 06 + 969-1031 ext 509). *Protein structure and function.*

Sugio, Dr Shigetoshi. Res. Div., The Green Cross Corp., Shodai-ohtani 2-1180-1 Hirakata, Osaka 573, Japan. (1958) DPharm, structural biology (Osaka U., 1985). (tel. 0720 + 56-9204). *Structural biology, macromolecules, proteins, X-ray diffraction, protein crystallography.*

Suito, Dr Eiji. Electron microscopy, 30 Kamiikeda-cho Kitashirakawa Sakyo-ku, Kyoto 606, Japan. (1912) DSc, chemistry (Kyoto U., 1942). (tel. 075 + 781-2737). *Electron microscopy, microcrystals (colloid or powder).*

Sunagawa, Prof. Ichiro. 3-54-2 Kashiwa-cho Tachikawa, Tokyo 190, Japan. (1924) DSc, mineralogy (Hokkaido U., 1957). Prof. Em. (Tohoku U.) (tel. 0425 + 36-2564, fax +81 425 35 3637). *Crystal growth mechanism, crystal morphology, characterization, natural minerals.*

Suzuki, Prof. Hideji. Dept. of Electronic Eng., Tokyo Eng. U., 1404-1 Katakura Hachioji, Tokyo 192, Japan. (1924) DSc, physics (Tohoku U., 1955). Prof. (tel. 0426 + 37-2111 ext 2103, telex 2862-558 TEUN J, fax +81 426 37 2118). *X-ray topography, lattice defects, mechanical properties, quantum crystals, phase transitions.*

Suzuki, Prof. Ikuo. Dept. of Electrical and Computer Eng., Nagoya Inst. of Techn., Gokiso-cho Showa-ku, Nagoya 466, Japan. (1941) DEng, physics (Nagoya U., 1968). Asst. prof. (tel. 052 + 732-2111 ext 2570, fax +81 52 733 6589). *Phase transitions, computing.*

Suzuki, Prof. Kazuo. Dept. of Applied Materials Sci., Muroran Inst. of Techn., 27-1 Mizumoto-cho Muroran, Hokkaido 050, Japan. (1925) DSc, physics(Tokyo U., 1962). Prof. (tel. 0143 + 44-4181 ext 2533, fax +81 143 47 3300). *Solid state physics, ionic crystals, mixed crystals, defects, X-ray diffuse scattering.*

Suzuki, Dr Michio. 1-11-9 Kunimi, Sendai 981, Japan. (1924) DSc, physics (Tohoku U., 1961). Prof. em. (tel. 022 + 271-5070). *X-ray scattering, X-ray lasers.*

Suzuki, Dr Shigeo. Tsukuba Res. Center Sanyo Electric Co. Ltd. 2-1 Koyadai Tsukuba, Ibaraki 305, Japan. (1939) DSc, physics(Tokyo Inst. of Techn., 1969). (tel. 02975 + 5-1151 ext 521, fax +81 2975 5 0164). *Semiconductors.*

Suzuki, Prof. Tadasu. Physics Sophia U. Kioi-cho 7 Chiyoda-ku, Tokyo 102, Japan. (1925) DSc, physics(1950). Chairman, Grad. Div. (tel. 03 + 238-3428, fax +81 3 238 3885). *Electron, momentum and spin density, electron and X-ray diffraction, scattering theory.*

Suzuki, Dr Toshimasa. Dept. of System Eng., Fac. of Eng., Nippon Inst. of Techn., 4-1 Gakuendai Miyashiro, Minami-Saitama, Saitama 345, Japan. (1948) DEng, applied physics (Tokyo Inst. of Techn., 1979). Assoc. prof. (tel. 0480 + 34-4111 ext 576, fax +81 480 34 2941). *Crystal growth, semiconductors, superconductors.*

Suzuki, Prof. Yoshio. Inst. of Geoscience, Tsukuba U., Tennodai Tsukuba, Ibaraki 305, Japan. (1927) DSc, petrology (Hokkaido U., 1958). Prof. (tel. 0298 + 53-4243). *Petrology, crystal growth, materials science.*

Tabata, Dr Hideyo. Res. Planning Office, Gov. Ind. Res. Inst. of Nagoya, 1-1 Hirate-machi, Kita-ku, Nagoya Aichi 462, Japan. (1941) DSc, chemistry (Osaka U., 1976). Dir. (tel. 052 + 911-2111 ext 780, telex NAGINBTH J 59500(Ref.NAG-128), fax +81 52 916 2802). *Crystal growth, defect structures.*

Tabira, Mr Yasunori. Res. Lab. of Eng. Materials, Tokyo Inst. of Techn., 4259 Nagatsuda Midori-ku Yokohama, Kanagawa 227, Japan. (1963) MSc, materials science (Tokyo Inst. of Techn., 1988). Grad. Student (tel. 045 + 922-1111 ext 2311, fax +81 45 921 1015). *Minerals, X-ray diffraction, electron microscopy, EXAFS.*

Tada, Dr Toshiji. Analytical Res. Labs. Fujisawa Pharm. Co. Ltd., 1-6 2-chome Kashima Yodogawa-ku Osaka, Osaka 532, Japan. (1949) PhD, chemistry (Hiroshima U., 1982). Chief (tel. 06 + 390-1173 ext 2866, telex FUJIPA J 64724, fax +81 6 308 5006). *Coordination compounds, organic compounds, powder diffraction, proteins, structural chemistry.*

Tadaki, Dr Tsugio. Inst. of Scient. and Industrial Res., Osaka U., 8-1 Mihogaoka Ibaraki, Osaka 567, Japan. (1943) DEng, metallurgy(1978). Assoc. Prof. (tel. 06 + 877-5111 ext 3557, fax +81 6 877 4977). *Electron diffraction, electron microscopy, materials science, metals, phase transitions.*

Tadokoro, Prof. Hiroyuki. 3-9-9 Ishibashi Ikeda, Osaka 563, Japan. (1920) DSc, chemistry (Osaka U., 1959). Em. prof. (tel. 0727 + 62-7554). *Structure, crystalline polymers, X-ray diffraction, infrared spectroscopy, Raman spectroscopy, energy calculations.*

Taga, Dr Tooru. Fac. of Pharm. Sci., Kyoto U. Sakyo-ku Shimoadachi-cho, Kyoto 606, Japan. (1938) DSc.(Osaka U., 1969). Lect. (tel. 075 + 753-4533, fax +81 75 761 2698 email A50950 at JPNKUDPC). *Computer simulation, natural products.*

Tagai, Dr Tokuhei. Mineralogical Inst., Fac. of Sci., U. of Tokyo, 7-3-1 Hongo Bunkyo-ku, Tokyo 113, Japan. (1943) DSc, mineralogy (Tokyo U., 1972). Asst. prof. (tel. 03 + 812-2111 ext 4544, fax +81 3 816 5714). *Real structure, extra-terrestrial minerals.*

Takagi, Dr Mieko. 1-10-6 Tsurumaki Setagaya-ku, Tokyo 154, Japan. (1919) DSc, physics (Osaka U., 1956). (tel. 03 + 429-9636). *X-ray topography, ferroelectrics.*

Takagi, Prof. Satio. 1-10-6 Tsurumaki Setagaya-ku, Tokyo 154, Japan. (1916) DSc, physics (Tokyo U., 1958). Prof. (tel. 03 + 429-9636). *Diffraction theory, X-ray topography, electron diffraction, electron microscopy.*

Takagi, Prof. Yutaka. 2-8 Ryokuenkita Kagamihara, Gifu 509-01, Japan. (1914) DSc, physics (Tokyo U., 1944). Em. prof. of Nagoya U. (tel.0583 + 84-8149). *Phase transition, ferroelectricity, ferroelasticity, mechanical properties of solids.*

Takahashi, Mr Shoichi. Res. Lab., Toshiba Ceramics Co. Ltd., 30 Soya Hatano, Kanagawa 257, Japan. (1938) BSc, mineralogy (Tokyo U., 1962). Chief res. (tel. 0463 + 81-8407). *Mineralogy, crystallography, ceramics, crystal growth.*

Takahashi, Dr Toshio. Inst. for Solid State Physics, Tokyo U., 7-22-1 Roppongi Minato-ku, Tokyo 106, Japan. (1950) DEng, applied physics (Tokyo U., 1979). Assoc. prof. (tel. 03 + 478-6811 ext 5621, fax +81 3 401 5169) *X-ray diffraction, surface structure, neutron diffraction.*

Takahashi, Prof. Yasuhiro. Dept. of Macromolecular Sci., Fac. of Sci., Osaka U. Machikaneyama Toyonaka, Osaka 560, Japan. (1941) DSc, macromolecular science (Osaka U., 1973). Assoc. prof. (tel. 06 + 844-1151 ext 4252). *Macromolecules, disorder, phase transformation, molecular motion, X-ray crystallography.*

Takai, Dr Mitsuo. Dept. of Applied Chemistry, Fac. of Eng., Kita 13-jo Nishi 8-chome Kita-ku Sapporo, Hokkaido 060, Japan. (1941) DEng, physical chemistry (Hokkaido U., 1969). Assoc.prof. (tel. 011 + 716-2111 ext 6567, telex 932302 HOKUEN-J, fax +81 11 717 4745). *Fine structure, biosynthesis, cellulose.*

Takaki, Dr Yoshito. Dept. of Physics, Osaka Kyoiku U., 43 Minami-Kawahori-cho Tennoji-ku, Osaka 543, Japan. (1931) DSc, physics (Osaka U., 1963). Prof. (tel. 06 + 771-8131 ext 228). *Computing, powder diffraction, disorder.*

Takama, Dr Toshihiko. Dept. of Applied Physics, Fac. of Eng., Hokkaido U., Kita Chuo-ku Sapporo, Hokkaido 060, Japan. (1941) DEng.(Hokkaido U., 1984). Assoc. Prof. (tel. 011 + 716-2111 ext 6644, telex 932-302 HOKUEN-J, fax +81 11 717 4745). *X-ray diffraction, electron density, defect structures, diffractometry.*

Takamura, Prof. Jin-ichi. Dept. of Metal Sci. and Techn., Fac. of Eng., Kyoto U., Yoshida Honmachi Sakyo-ku, Kyoto 606, Japan. (1921) DEng, metallurgy (Kyoto U., 1953). Prof. (tel. 075 + 751-2111 ext 5461). *Lattice defects, strength of metals, metal physics.*

Takano, Dr Tsunehiro. Fac. of Pharm. Sci., Setsunan U., 45-1 Nagao-toge cho, Hirakata, Osaka 573-01, Japan. (1936) DSc, physical chemistry (Osaka U., 1965). Prof. (tel. 0720 + 68-7000 ext 349). *Proteins, nucleic acids.*

Takano, Prof. Yasumasa. Dept. of Mechanical Eng., Okayama U. of Sci., 1 Ridai-cho 1-chome, Okayama 700, Japan. (1921) DSc, physics (Hiroshima U., 1961). Chief prof. of Mechanical Fundamental Study (tel. 0862 + 52-3161 ext 4557). *Radiation detection, crystal physics, X-ray chemical analysis.*

Takano, Dr Yukio. 5-14-3-202 Yoyogi Shibuya-ku, Tokyo 151, Japan. (1924) DSc, mineralogy (Tokyo U., 1959). (tel. 03 + 460-5820). *Crystal morphology, crystal growth, inorganic crystal structures.*

Takasu, Dr Shin-ichiro. Res. Lab. Toshiba Ceramics Co. Ltd., 30 Soya Hatano, Kanagawa 257, Japan. (1928) DSc, mineralogy (Tokyo U., 1958). Sr. res. member. (tel. 0463 + 81-8407). *Crystal growth, morphology, X-ray crystallography, optical crystallography, silicon, semiconducting materials, inorganic materials.*

Takayanagi, Prof. Kunio. Materials Sci. & Eng. Dept., Tokyo Inst. of Techn., 4259 Nagatsuta Midori-ku, Yokohama, 227, Japan. (1947) DSc, surface physics, electron microscopy Prof. (tel. 045 + 922-1111 ext 2619, fax +81 45 922 5173). *Surface structure, surface phase transition, crystal growth, thin films, electron diffraction, electron microscopy, X-ray crystallography.*

Takazawa, Mr Hiroyuki. Dept. of Chemistry, Keio U. Hiyoshi 3 Kohoku-ku Yokohama, Kanagawa 223, Japan. (1961) MSc, chemistry(Keio U., 1987). (tel. 044 + 63-1141 ext 3912). *Electron density, heavy metal complexes.*

Takeda, Dr Hirofumi. Kansai R&D Center, Dainippon Ink and Chemicals Inc., 3-1 Takasago Takaishi-shi, Osaka 592, Japan. (1950) PhD, chemistry (Osaka U., 1978). (tel. 0722 + 68-3745, fax +81 722 68 3789). *Macromolecules, organic compounds, polymers, structural chemistry, surface structure.*

Takeda, Prof. Hiroshi. Mineralogical Inst., Fac. of Sci., U. of Tokyo, 3-1 Hongo 7-chome Bunkyo-ku, Tokyo 113, Japan. (1934) DSc, mineralogy (Tokyo U., 1962). Prof. (tel. 03 + 812-2111 ext 4543, fax +81 3 816 5714). *Minerals, polymorphism, polytypism, twinning, crystal chemistry, meteoritic and lunar minerals, crystallographic data.*

Takeda, Prof. Takayoshi. Fac. of Integrated Arts and Sci. Hiroshima U., 1-1-89 Higashi-senda-machi Naka-ku, Hiroshima 730, Japan. (1944) DSc, physics (Tohoku U., 1971). Assoc. Prof. (tel. 082 + 241-1221 ext 2204, telex 652712 HIFT J, fax +81 82 244 5170). *Small-angle scattering, neutron diffraction, magnetism, phase transition, structural biology, liquid crystals.*

Takenaka, Dr Akio. Fac. of Sci., Tokyo Inst. of Techn., Nagatsuta, Midori-ku, Yokohama 227, Japan. (1942) DSc, chemistry (Kwansei Gakuin U., 1971). Res. Assoc. (tel. 045 + 922-1111, fax +81 45 922 2432). *Structural chemistry, structural biology, molecular crystals, hydrogen bonding, computing, macromolecules.*

Takenaka, Mr Yasuyuki. Res. Lab. of Eng. Materials, Tokyo Inst. of Techn., Midori-ku Nagatsuda-cho 4259 Yokohama, Kanagawa 227, Japan. (1964) BEng, techn. (Tokyo Inst. of Techn., 1987). (tel. 045 + 922-1111 ext 2627, fax +81 45 921 1015, email ytakenak at titnca.nc.titech.junet). *Chemical bonding, computing, diffractometry, electron density, scattering theory, thermal vibrations, X-ray diffraction.*

Takeoka, Mr Yoshikatsu. Toshiba Res. and Dev. Center, Tokyo Shibaura Electric Co. Ltd., 1 Komukai Toshiba-cho Saiwai-ku, Kawasaki Kanagawa 210, Japan. (1945). (tel. 044 + 511-2111 ext 2329).

Takeuchi, Prof. Yoshio. Dept. of Earth Sci., Nihon U., 25-40 Sakurajosui 3-chome Setagaya-ku, Tokyo 156, Japan. (1924) DSc, mineralogy (Tokyo U., 1953). Prof. (tel. 03 + 329-1151 ext 288, fax +81 3 303 9899). *Structural chemistry of minerals and inorganic crystals.*

Tamada, Dr Osamu. Inst. of Earth Sci., C. of General Arts, Kyoto U., Yoshida Nihonmatsu-cho Sakyo-ku, Kyoto 606, Japan. (1944) DSc, mineralogy (Kyoto U., 1980). Assoc. prof. (tel. 075 + 753-6869 ext 6865). *Chemical bonding, electron density, minerals, inorganic crystals, X-ray diffraction.*

Tamura, Dr Chihiro. Analytical & Metabolic Res. Lab., 2-58 Hiromachi 1-chome, Shinagawa-ku, Tokyo 140, Japan. (1930) DSc, chemistry (Tokyo Metropolitan U., 1963). Res. officer. (tel. 03 + 492-3131 ext 599). *Organic crystal structures, biological molecular structures.*

Tanaka, Dr Isao. Macromolecular Div., Fac. of Sci, Hokkaido U., 10-Jyo Kita Kita-ku, Sapporo 060, Japan. (1948) MSc, chemistry (Osaka U., 1973). Assoc. Prof. (tel. 011 + 716-2111 ext 3221). *Proteins.*

Tanaka, Prof. Jiro. Dept. of Chemistry, Fac. of Sci., Nagoya U., Furo-cho Chikusa-ku, Nagoya 464, Japan. (1929) DSc, chemistry (Tokyo U., 1957). Prof. (tel. 052 + 781-5111 ext 2481). *Intermolecular interaction, electronic spectra, molecular crystals, absolute configuration, chiral molecules.*

Tanaka, Dr Kiyoaki. Res. Lab. of Eng. Materials, Tokyo Inst. of Techn., 4259 Nagatsuta, Midori-ku, Yokohama 227, Japan. (1946) DSc, chemistry (Tokyo U., 1975). Res. assoc. (tel. 045 + 922-1111 ext 2312). *Crystal structure analysis, accurate electron density determination.*

Tanaka, Dr Michiyoshi. Dept. of Physics, Fac. of Sci., Tohoku U., Aramaki Aza-Aoba, Sendai 980, Japan. (1938) DSc, physics (Tokyo Inst. of Techn., 1965). Assoc. prof. (tel. 022 + 222-1800 ext 3296, fax +81 22 225 1891). *Electron diffraction, electron microscopy, condensed matter.*

Tanaka, Dr Nobuo. Dept. of Life Sci., Tokyo Inst. of Techn., Nagatsuda 4259 Midori-ku Yokohama, kanagawa 227, Japan. (1941) DSc, chemistry (Osaka U., 1971). Prof. (tel. 045 + 922-1111 ext 2386, telex 3823553 TITNAG J, fax +81 45 922 2432). *Protein crystallography.*

Tanaka, Dr Nobuo. Dept. of Applied Physics, Sch. of Eng., Nagoya U., Chikusa-ku Nagoya, Aichi 464-01, Japan. (1949) DEng, electron microscopy (1978). Res. Assoc. (tel. 052 + 781-5111 ext 4457, telex 4477355 ENUNAG-J, fax +81 52 782 2129, email A41263A at NUCC.NAGOYA-U.JUNET at U-TOKYO-RELAY.CSNET) *High resolution electron microscopy, thin film physics, crystal growth, surface physics, diffuse scattering, quasi-crystals.*

Tanemura, Dr Sakae. Dept. of Ceramics Sci., Gov. Ind. Res. Inst., Nagoya, 1-1 Hirate-machi Kita-ku, Nagoya 462, Japan. (1943) DEng, applied physics (Nagoya U., 1971). Div. head. (tel. 052 + 911-2111 ext 570, fax +81 52 916 2802). *Materials science, thin film, electron and optical spectroscopy, X-ray diffraction, surface structure, semiconductors, superconductors.*

Tani, Dr Katsuhiko. R and D Center Ricoh Co. Ltd., 16-1 Shin'ei-cho Kohoku-ku, Yokohama 223, Japan. (1944) DSc, mineralogy (Tokyo U., 1976). (tel. 045 + 593-3411, fax +81 45 593 3482). *Symmetry, X-ray diffraction.*

Tanigaki, Mr Takeshige. Semiconductor Group Sony Corp., 4-14-1 Asahi-cho Atsugi, Kanagawa 243, Japan. (1944) MEng, metallurgy (Tokyo Inst. of Techn., 1970). Manager of Characterization Center (tel. 0462 + 30-6206, fax +81 462 30 6233). *Semiconductors, characterization, FIB, surface analysis, SIMS, crystal, X-ray diffraction, electron diffraction, focused ion beam.*

Taniguchi, Mr Tomohiko. Dept. of Physics, Osaka Kyoiku U., 43 Minami-Kawahori-cho Tennoji-ku, Osaka 543, Japan. (1936) BEd, physics (Osaka Kyoiku U., 1962). Assoc. prof. (tel. 06 + 771-8131 ext 295). *Phase transformation, molecular crystals.*

Tanisaki, Prof. Sigetosi. Dept. of Liberal Arts Yahata U., Edamitsu 5-9-1 Yahata Higashi-ku Kitakyushu, Fukuoka 805, Japan. (1922) DSc, physics (Kyoto U., 1960). Prof. (tel. 093 + 671-8923 ext 370, fax +81 93 671 8995). *Ferroelectricity, X-ray diffraction.*

Tanishiro, Mr Yasumasa. Physics Dept., Tokyo Inst. of Techn., Oh-okayama Meguro-ku, Tokyo 152, Japan. (1955) MSc, physics (Tokyo Inst. of Techn., 1980). Res. assoc. (tel. 03 + 726-1111 ext 2079, telex 03-246-6360, fax +81 3 729 0042). *Surface physics, crystal growth, electron microscopy, electron spectroscopy.*

Taoka, Prof. Tadami. Dept. Fundamental Eng., Union Optical Co. Ltd. 2-20-9 Shimura Itabashi-ku, Tokyo 174, Japan. (1916) DSc, physics (Tokyo U., 1957). Prof. (tel. 03 + 968-8521, email 433-1-1004, ICHINOMIYA, TAMA-SHI, TOKYO 206). *Metal physics.*

Tate, Dr Isao. 1380-67 Yoshida Ueda, Nagano 386-01, Japan. (1923) DEng, industrial chemistry (Nagoya U., 1971). Em. prof. (tel. 0268 + 36-3890). *Applied mineral chemistry, crystal growth.*

Tatsuzaki, Prof. Itaru. Fac. of Eng., Hokkaido-Tokai U., 1-1-1 5-jo Minamizawa Minami-ku, Sapporo 061-21, Japan. (1925) DSc, physics (Hokkaido U., 1960). Prof. *Phase transition, ferroelectrics, dielectrics.*

Terauchi, Dr Hikaru. Fac. of Sci., Kwansei Gakuin U., 1-155 Uegahara 1-bancho Nishinomiya, Hyogo 662, Japan. (1942) DSc, physics (Kwansei Gakuin U., 1972). Prof. (tel. 0798 + 53-6111 ext 5280, fax +81 798 51 0914). *Phase transitions, artificial superlattices.*

Togawa, Dr Sen-ichi. Tokyo U. of Mercantile Marine, 1-6 Etchujima 2-chome Koto-ku, Tokyo 135, Japan. (1926) DSc, physics (Tokyo U., 1965). Prof. (tel. 03 + 641-1171). *Solid state physics.*

Tokonami, Prof. Masayasu. Mineralogical Inst., Fac. of Sci., U. of Tokyo, 7-3-1 Hongo Bunkyo-ku, Tokyo 113, Japan. (1933) DSc, mineralogy (Tokyo U., 1966). Prof. (tel. 03 + 812-2111 ext 4541, telex UTYOSCIJ33659, fax +81 3 816 5714). *Inorganic crystal structures, crystal physics.*

Tokushita, Mr Motoyuki. Explosive Plant, Asahi Chemical Industry, 6-1 Mizushiri-cho Nobeoka, Miyazaki 882, Japan. (1936) BSc, physics (Hokkaido U., 1960). Asst. manager. (tel. 09823 + 33-6141). *Analytical chemistry, metallurgy.*

Tomeoka, Mr Kazushige. Mineralogical Inst., Fac. of Sci., Tokyo U., 3-1 Hongo 7-chome Bunkyo-ku, Tokyo 113, Japan. (1952) PhD, mineralogy (Tokyo U., 1975). Res. Assoc. (tel. 03 + 812-2111 ext 4548, fax +81 3 816 5714). *Crystal chemistry, cosmic mineralogy, electron microscopy.*

Tomimitsu, Mr Hiroshi. Solid State Physics - 1, Japan Atomic Energy Res. Inst., 2-4 Shirakata-Shirane Tokaimura, Nakagun Ibaraki 319-11, Japan. (1942) MSc, physics (Tokyo U., 1969). Res. staff. (tel. 02928 + 2-5466). *X-ray crystallography, neutron diffraction, electron diffraction.*

Tomita, Dr Katsutoshi. Dept. of Geology & Mineralogy, Fac. of Sci., Kyoto U., Oiwake-cho Kitashirakawa Sakyo-ku, Kyoto 606, Japan. (1934) DSc, mineralogy (Kyoto U., 1965). Lect. (tel. 075 + 753-4162, telex 5422302SCIKYU, fax +81 75 753 4189). *Crystal structure, minerals, micro-texture, rock forming minerals.*

Tomita, Prof. Ken-ichi. Fac. of Pharm. Sci., Osaka U., 1-6 Yamadaoka Suita, Osaka 565, Japan. (1928) DSc, physical chemistry (Osaka U., 1959). Prof. (tel. 06 + 877-5111 ext 6211, fax +81 6 877 4489, email BITNET LBA2014 at JPNOSAKA). *Organic crystal structures, large molecules, biological substances.*

Tonomura, Dr Akira. Advanced Res. Lab., Hitachi Ltd., 280 Higashi-koigakubo 1-chrome, Kokubunji Tokyo 185, Japan. (1942) DEng, applied physics. (Nagoya U., 1973). Chief res. (tel. 0423 + 23-1111 ext 2820, fax +81 423 26 0880) *Electron holography, electron microscopy, electron diffraction.*

Toraya, Dr Hideo. Ceramic Eng. Res. Lab., Nagoya Inst. of Techn. 10-6-29, Asahigaoka Tajimi, Gifu 507, Japan. (1949) DSc, materials science (Tokyo Inst. of Techn., 1980). Assoc. prof. (tel. 0572 + 27-6811, fax +81 572 27 6812). *X-ray crystallography, crystal structure analysis, powder diffraction.*

Toriumi, Dr Koshiro. Inst. for Molecular Sci., Okazaki Nat. Res. Inst., Myodaiji Okazaki Aichi 444, Japan. (1949) DSc, chemistry (Tokyo U., 1978). Res. assoc.. (tel. 0564 + 54-1111 ext 3432, telex 4537475 KOKKEN J, fax +81 564 54 2254). *X-ray diffraction, coordination compounds, materials science.*

Toyoda, Prof. Koichi. Res. Inst. of Electronics, Shizuoka U., 5-1 Johoku 3-chome, Hamamatsu 432, Japan. (1933) DSc, physics (Kyoto U., 1986). Assoc. prof. (tel. 0534 + 71-1171 ext 431, telex 4225280SHZDDK J, fax +81 534 74 0630). *Ferroelectric crystals, information storage and retrieval, solid state data.*

Tsuda, Mr Noritoshi. Dept. of Math. and Physics, Ritsumeikan U., 28-1 Tojiin-Kitamachi Kita-ku, Kyoto 603, Japan. (1940) MSc, physics (Ritsumeikan U., 1966). Res. assoc. (tel. 075 + 463-1131 ext 326). *Inorganic crystal structures, phase transition.*

Tsuji, Mr Kazuhiko. Semiconductor Res. Center Matsushita Electric Industrial Co. Ltd., 3-15 Yagumo-Nakamachi Moriguchi, Osaka 570, Japan. (1947) MEng, physics (Nagoya U.,). Sr. res. eng. (tel. 06 + 909-1121 ext 2716, telex 06 529 5737, fax +81 6 906 0177). *Semiconductors, crystal growth, defect structures, metals, surface structure.*

Tsuji, Mr Koji. Wabco Co. Ltd. 14-2 Fukusaki Kogyo Danchi Saiji, Fukusaki-cho Kanzaki-gun, Hyogo 679-22, Japan. (1944) MSc, chemistry (Kwansei Gakuin U., 1969). Dir. (tel. 0790 + 22-6000, fax +81 790 22 0120).

Tsukihara, Dr Tomitake. Fac. of Eng., Tottori U., 1-1 Koyama, Tottori 680, Japan. (1944) DSc, biochemistry (Osaka U., 1974). Lect. (tel. 0857 + 28-0321 ext 4220, fax +81 857 28 1092). *Protein crystallography.*

Tsukimura, Dr Katsuhiro. Mineralogy Div. Geological Survey of Japan, 1-1-3 Higashi Tsukuba, Ibaraki 305, Japan. (1953) DSc, mineralogy (Tokyo U.). Sr. Res. (tel. 0298 + 54-3632). *High pressures, group theory, minerals, powder diffraction, structural chemistry, X-ray diffraction.*

Tsunekawa, Dr Shin. Inst. for Materials Res. Tohoku U., Katahira, Sendai, Miyagi 980, Japan. (1943) DSc, chemistry (Nagoya U., 1979). Res. asst. (tel. 022 + 227-6200 ext 2335, 2852, fax +81 22 264 7984). *Crystal growth, inorganic crystals, semiconductors, structural chemistry.*

Uechi, Mr Tetsuo. General Education Dept., Kyushu U., 2-1 Ropponmatsu 4-chome Chuo-ku, Fukuoka 810, Japan. (1936) Res. assoc. (tel. 092 + 771-4161). *Crystal structure analysis.*

Ueda, Prof. Ikuhiko. C. of General Education, Kyushu U., 2-1 Ropponmatsu 4-chome Chuo-ku, Fukuoka 810, Japan. (1921) DSc, physics (Kyushu U., 1963). Prof. (tel. 092 + 771-4161 ext 277). *Structural analysis of complex molecules, heterocyclic and chelate compounds.*

Uefuji, Mr Tateki. Electronics Dept., Fac. of Sci. & Eng., Saga U., 1 Honjomachi, Saga 840, Japan. (1945) MSc, physics (1970). Res. assoc. (tel. 09522 + 4-5191). *Electron diffraction, lattice defects, metals.*

Ueki, Dr Tatzuo. Dept. of Biophysical Eng., Fac. of Eng. Sci., Osaka U., 1 Machikaneyama-cho 1-chome Toyonaka, Osaka 560, Japan. (1940) DSc, physical chemistry (Osaka U., 1968). Assoc. prof. (tel. 06 + 844-1151 ext 4766, fax +81 6 843 9354). *Structural biology, solution X-ray scattering, macromolecules, proteins, membranes.*

Ueno, Mr Tsunehisa. Semiconductor Div., Tokyo Shibaura Electric Co. Ltd., 1 Komukai Toshiba-cho Saiwai-ku, Kawasaki Kanagawa 210, Japan. (1948) MEng, mineralogy (Tohoku U., 1973). (tel. 044 + 511-3111 ext 691, 473). *Crystallography, applied physics, mineralogy, spectroscopy, electronics, photographic engineering.*

Uesu, Prof. Yoshiaki. Dept. of Physics, Waseda U., Okubo 3-chome Shinjuku-ku, Tokyo 169, Japan. (1942) DSc, physics (Waseda U., 1971). Prof. (tel. 03 + 203-4141 ext 3661, 3662, telex 03 232 5115, fax +81 3 200 2567). *Materials science, phase transitions, X-ray diffraction.*

Ukaji, Prof. Takeshi. Dept. of Process Chemical Eng., Ikutoku Techn. U., 1030 Shimo-ogino Atsugi, Kanagawa 243-02, Japan. (1922) DSc, chemistry (Tokyo U. of Education, 1958). Prof. (tel. 0462 + 41-1211 ext 254). *Molecular structure determination, gas electron diffraction.*

Umegaki, Prof. Yoshiharu. 38 Uenota Oyamazaki-cho Otokunj-gun, Kyoto 618, Japan. (1909) DSc, mineralogy (Kyoto U., 1952). Em. prof. (Hiroshima U.). (tel. 075 + 961-0156). *Crystallographic data, electron diffraction, neutron diffraction, X-ray diffraction.*

Umeno, Prof. Masataka. Dept. of Precision Eng., Fac. of Eng., Osaka U., 133-1 Yamadakami Suita, Osaka 565, Japan. (1939) DEng, applied physics (Osaka U., 1967). Prof. (tel. 06 + 877-5111 ext 4616, fax +81 6 878 3819). *Crystal growth, defect structures, electron microscopy, materials science, metals, semiconductors, X-ray diffraction.*

Uno, Prof. Ryosei. Dept. of Physics, C. of Humanities and Sci., Nihon U., 25-40 Sakurajosui 3-chome Setagaya-ku, Tokyo 156, Japan. (1924) DSc, physics (Kyushu U., 1947). Prof. (tel. 03 + 329-1151 ext 283, fax +81 3 303 9899). *Semiconductors, powder diffraction, electron density.*

Unoki, Dr Hiromi. Superconductive Material Section Electrotechnical Lab. 1-1-4 Umezono Tsukuba, Ibaraki 305, Japan. (1932) DSc, solid state physics(1971). Section Chief (tel. 0298 + 54-5202, fax +81 298 55 1729). *Dielectrics, crystal growth, superconductivity.*

Uragami, Prof. Takuyuki. General Education, Okayama U. of Sci., 1-1 Ridai-cho, Okayama 700, Japan. (1938) DSc, physics (Tokyo U., 1971). Prof. (tel. 0862 + 52-3161 ext 3118, fax +81 862 55 3847). *X-ray diffraction.*

Uyeda, Dr Natsu. 3-30-11 Senriyama-Nishi Suita, Osaka 565, Japan. (1924) DSc, chemistry (Kyoto U., 1958). Prof. Em., Kyoto U. (tel. 06 + 385-3165). *Epitaxial growth, thin crystalline films, organic semiconductors, high resolution electron microscopy in atomic order.*

Uyeda, Prof. Ryozi. Dept. of Physics, Meijo U., Tempaku, Nagoya 468, Japan. (1911) DSc, physics (Tokyo U., 1944). Prof. (tel. 052 + 832-1151 ext 570). *Electron diffraction, electron microscopy.*

Wada, Mr Takeo. Res. Labs., Chemical Products Div., Takeda Chemical Industries Ltd., 17-85 Juso-honmachi 2-chome, Yodogawa-ku, Osaka 532, Japan. (1934) DSc, inorganic and physical chemistry (Osaka U., 1963). Chief res. (tel. 0727 + 99-3108, telex TAKEDA J63484, fax +81 6 300 6565). *Clay minerals.*

Wakabayashi, Dr Katsuzo. Dept. of Biophysical Eng., Fac. of Eng. Sci., Osaka U., 1 Machikaneyama-cho 1-chome Toyonaka, Osaka 560, Japan. (1943) DSc, biophysics (Hokkaido U., 1971). Res. assoc. (tel. 06 + 844-1151 ext 4768, telex 5286110, fax +81 6 843 9354). *Small angle X-ray scattering, X-ray diffraction, structural biology.*

Wakino, Dr Kikuo. Murata MFG. Co. Ltd. 2-26-10 Tenjin Nagaokakyo, Kyoto 617, Japan. (1925) DEng, electronics (Osaka U., 1980). Sr. Executive (tel. 075 + 951-9111 ext 3005). *Inorganic crystals, crystallographic data, phase transitions, powder diffraction.*

Wakoh, Prof. Shinya. U. for Library and Info. Sci., Tsukuba, Ibaraki 305, Japan. (1938) DSc (Tokyo U., 1966). Prof. (tel. 0298 + 52-3969 ext 312, fax +81 298 52 4326). *Electron density, magnetism, metals, momentum density, spin density.*

Watanabe, Dr Akiteru. Fourth Group, Nat. Inst. for Inorganic Materials Res., 1-1 Namiki Tsukuba, Ibaraki 305, Japan. (1945) DEng, crystal chemistry (Tokyo Inst. Techn., 1983). Sr. res. (tel. 0298 + 51-3351 ext 209, fax +81 298 52 7449). *Inorganic crystals, phase transitions, materials science, structural chemistry.*

Watanabe, Prof. Denjiro. Dept. of Physics, Fac. of Sci., Tohoku U., Aramaki Aza-Aoba, Sendai 980, Japan. (1926) DSc, physics (Tohoku U., 1960). Prof. (tel. 0222 + 22-1800 ext 3295, fax +81 22 225 1891). *Metal and alloy structures, electron diffraction, electron microscopy.*

Watanabe, Dr Mamoru. 7th Group, Nat. Inst. for Inorganic Materials Res., 1-1 Namiki Tsukuba, Ibaraki 305, Japan. (1945) PhD, inorganic chemistry (Osaka U., 1974). Sr. Res. Officer (tel. 0298 + 51-3351 ext279, fax +81 298 52 7449). *Inorganic crystals, materials science, structural chemistry.*

Watanabe, Dr Masaru. Electron Optics Div. JEOL Ltd., 1418 Nakagami Akishima, Tokyo 196, Japan. (1922) DEng, physics (Tokyo Inst. of Techn., 1962). Dir., General Manager Electron Optics Div. (tel. 0425 + 43-1111 ext 417). *Instrumentation, electron optics, electron diffraction.*

Watanabe, Prof. Takashi. Geoscience, Joetsu U. of Education, Yamayashiki Joetsu, Niigata 943, Japan. (1940) DSc, mineralogy (Tokyo U. of Educ., 1970). Prof. (tel. 0255 + 22-2411 ext 442). *Clay mineralogy.*

Watanabe, Mr Yasunari. X-ray Crystallography Lab., Inst. of Physical and Chemical Res., 1 Hirosawa 2-chome Wako, Saitama 351-01, Japan. (1936) BSc, physics (Sci. U. of Tokyo, 1963). Scient. (tel. 0484 + 62-1111 ext 3343, fax +81 484 62 1449). *Quasicrystals.*

Watanabe, Prof. Yasuyoshi. Dept. of Physics, Fac. of Sci., Chiba U., 33 Yayoicho 1-chome, Chiba 280, Japan. (1920) DSc, physics (Tokyo U. of Educ., 1956). Prof. (tel. 0472 + 51-1111 ext 2632). *Electron diffraction, crystal growth, LEED.*

Watari, Dr Fumio. Fac. of Dentistry, Dept. of Dental Techn. I, Tokyo Medical and Dental U., 1-5-45 Yushima Bunkyo-ku, Tokyo 113, Japan. (1949) PhD., eng. (Tokyo U.). Assoc. prof. (tel. 03 + 813-6111 ext 5147). *Electron diffraction, electron microscopy, inorganic crystals, materials science, reaction pathways, X-ray diffraction.*

Yagi, Prof. Katsumichi. Physics Dept., Tokyo Inst. of Techn., Oh-okayama Meguro-ku, Tokyo 152, Japan. (1939) DSc, physics (Tokyo Inst. of Techn.,

1967). Prof. (tel. 03 + 726-1111 ext 2078-9, telex 03 2466360 TIT TECH J, fax +81 3 729 0042). *Surface physics, thin film growth, phase transition.*

Yakushi, Mr Kyuya. Dept. of Chemistry, Fac. of Sci., Tokyo U., 3-1 Hongo 7-chome Bunkyo-ku, Tokyo 113, Japan. (1945) MSc, chemistry (Tokyo U., 1970). Res. assoc. (tel. 03 + 812-2111). *Structure and physical properties, molecular crystals.*

Yamada, Dr Yukio. Physics C. of General Education, Tohoku U. Kawauchi Sendai, Miyagi 980, Japan. (1945) DSc.(Tohoku U., 1986). Asst. Prof. (tel. 022 + 222-1800 ext 5287, fax +81 22 262 2429). *Crystal growth, electron diffraction, electron microscopy, materials science, phase transition, surface structure.*

Yamagata, Dr Yuriko. Fac. of Pharm. Sci., Osaka U. 1-6 Yamadaoka Suita, Osaka 565, Japan. (1952) DParm, chemistry(Osaka U., 1980). Res. Assoc. (tel. 06 + 877-5111 ext 6214, fax +81 6 877 4489). *Organic compounds, structural biology.*

Yamaguchi, Mr Hiroshi. Inst. for Protein Res., Osaka U. 3-2 Yamadaoka Suita, Osaka 565, Japan. (1963) MSc (Osaka U., 1988). Grad. Student. (tel. 06 + 877-5111 ext 3837, fax +81 6 876 2533). *Proteins, macromolecules.*

Yamaguchi, Dr Toshio. Dept. of Chemistry, Fac. of Sci., Fukuoka U. Nanakuma Jonan-ku, Fukuoka 814-01, Japan. (1949) DSc.(1978). Assoc. Prof. (tel. 092 + 871-6631 ext 6224, fax +81 92 865 6030, email BITNET A71206G at JPNCCKU). *X-ray diffraction, neutron diffraction, amorphous materials.*

Yamaguti, Prof. Tasaburo. B-208 353-1 Eda Midori-ku, Yokohama Kanagawa 227, Japan. (1904) DSc, physics (Tokyo U., 1938). Em. prof. (Chiba U.). (tel. 045 + 911-8596). *Crystal growth, diffusion, surfaces, films, deposition.*

Yamamoto, Dr Akiji. Nat. Inst. for Inorganic Materials Res., Namiki Tsukuba, Ibaraki 305, Japan. (1945) DSc, mineralogy and physics (Kyoto U., 1981). Sr. res. (tel. 0298 + 51-3351 ext 371, fax +81 298 52 7449). *Incommensurate structure, quasicrystals.*

Yamamoto, Dr Atsushi. Techn. Res. Div. Kawasaki Steel Div. 1 Kawasaki-cho Chiba, Chiba 260, Japan. (1951) DEng.(Nagoya U., 1980). Sr. Res. (tel. 0472 + 62-2455, fax +81 472 62 2061). *Electron microscopy.*

Yamamoto, Dr Naoki. Dept. of Sci. and Math., Nagaoka U. of Techn. 1603-1 Kamitomioka Nagaoka, Niigata 940-21, Japan. (1950) DSc, physics (Tokyo Inst. of Techn., 1979). Asst. Prof. (tel. 0258 + 46-6000 ext 3106, fax +81 258 46 6504). *Electron microscopy, electron diffraction, ferroelectricity, phase transition, luminescence.*

Yamamoto, Mr Shinichi. Chemical Techn. Asahi Chemical Industry Co. Ltd. 1-3-2 Yako Kawasaki-ku Kawasaki, Kanagawa 210, Japan. (1958) BEng., industrial chemistry (1982). (tel. 044 + 271-2352, fax +81 44 271 2355). *Catalysts.*

Yamamoto, Prof. Takashi. Electrical Eng. Nat. Defense Academy, Hashirimizu 1-10-20 Yokosuka, Kanagawa 239, Japan. (1948) DEng, electronics(Kyoto U., 1980). Assoc. Prof. (tel. 0468 + 41-3810 ext2586). *Electronics, electronics ceramics, thin films.*

Yamanaka, Prof. Takamitsu. C. of General Education Osaka U., 1-1 Machikaneyama Toyonaka, Osaka 560, Japan. (1942) DSc, mineralogy (Tokyo U., 1971). Assoc. prof. (tel. 06 + 844-1151 ext 5049, 5300, telex 528607 OSAKU J, fax +81 6 855 7621). *Structure chemistry, physical property.*

Yamane, Dr Takashi. Dept. of Applied Chemistry, Fac. of Eng., Nagoya U., Furo-cho Chikusa-ku, Nagoya 464, Japan. (1946) DSc, chemistry (Osaka U., 1975). Res. assoc. (tel. 052 + 781-5111 ext 3342, fax +81 52 781 4895). *Organic compounds, proteins, structural chemistry, X-ray diffraction*

Yamashita, Mr Shuji. Physics Div. Tokyo Metropolitan Isotope Res. Centre, 11-1 Fukazawa 2-chome Setagaya-ku, Tokyo 158, Japan. *Crystal physics.*

Yamazaki, Dr Yohtaro. Grad. School at Nagatsuda Tokyo Inst. of Techn., 4259 Nagatsuda Midori-ku Yokohama, Kanagawa 227, Japan. (1945) DEng. (Tokyo Inst. of Techn., 1974). Assoc. Prof. (tel. 045 + 922-1111 ext 2436). *Crystal growth, crystallographic data, computer graphics, magnetism, teaching.*

Yase, Dr Kiyoshi. Fac. of Appl. Biological Sci., Hiroshima U., Shitami Saijo-cho Higashi-Hiroshima, Hiroshima 724, Japan. (1954) DSc, crystal chemistry (Kyoto U., 1985). Res. Assoc. (tel. 0824 + 22-7111 ext 4064, fax +81 824 22 7067). *Crystal growth, electron diffraction, electron microscopy, molecular crystals, organic compounds, macromolecules, polymers, surface structure, X-ray diffraction.*

Yasuda, Prof. Yukio. Dept. of Eng. Nagoya U., Chikusa-ku Nagoya, Aichi 464-01, Japan. (1940) DEng, thin film growth (Nagoya U., 1973). Prof. (tel. 052 + 781-5111 ext 6481, telex 4477355 ENUNAG J, fax + 81 52 782 6948). *Defect structures, electron microscopy, semiconductor, surface structure.*

Yasuoka, Prof. Noritake. Basic Res. Lab. Himeji Inst. of Techn., 2167 Shosha Himeji, Hyogo 671-22, Japan. (1936) DSc, macromolecules (Osaka U., 1968). Prof. (tel. 0792 + 66-1661 ext 401, fax +81 792 66 8868). *Proteins, organic compounds, crystallographic data, computing.*

Ye, Ms Jinhua. Mineralogical Inst. U. of Tokyo, 7-3-1 Hongo Bunkyo-ku, Tokyo 113, Japan. (1963) MSc, crystallography (Tokyo U., 1987). PhD cand. (tel. 03 + 812-2111 ext 4550, fax +81 3 816 5714). *Crystal growth, materials science, phase transitions, X-ray diffraction.*

Yoda, Dr Osamu. Takasaki Res. Est. Japan Atomic Energy Res. Inst., 1233 Watanuki-machi Takasaki, Gunma 370-12, Japan. (1944) DEng, chemistry (Hokkaido U., 1979). Res. scient. (tel. 0273 + 46-1211 ext 382). *X-ray and neutron diffraction, polymers.*

Yokomori, Dr Yoshinobu. Dept. of Chemistry, Nat. Defense Academy, 1-10-20 Hashirimizu, Yokosuka, Kanagawa 239, Japan. (1950) DEng, chemistry (Tokyo

U., 1978). Lect. (tel. 0468 + 41-3810 ext 2127). *Organic compounds, oligopeptides.*

Yoshiasa, Dr Akira. Inst. of Geology and Mineralogy, Fac. of Sci. Hiroshima U. Higashi-sendamachi, Hiroshima 730, Japan. (1957) DSc, inorganic and physical chemistry (Osaka U., 1988). Res. Assoc. (tel. 082 + 241-1221 ext 3519, fax +81 82 242 7454). *Crystal structure, inorganic compounds, minerals, EXAFS.*

Yoshida, Dr Kentaro. Fac. of Eng., Kobe U., 1 Rokkodai-machi Nada-ku, Kobe 657, Japan. (1935) DSc, physics (Kyoto U., 1973). Assoc. prof. (tel. 078 + 881-1212 ext 5261, fax +81 78 881 1346). *Electron microscopy, electron diffraction, materials science, metals and alloys, phase transition, surface and inter-face structure.*

Yoshimatsu, Prof. Mitsuru. Applied Physics, Fac. of Sci., Fukuoka U., 8-19-1 Nanakuma, Johnan-ku, Fukuoka 814-01, Japan. (1926) DSc, physics (Kyushu U., 1962). Prof. (tel. 0552 + 52-1111 ext 571). *Crystal growth, X-ray diffraction.*

Yoshimura, Dr Junichi. Inst. of Inorganic Synthesis, Yamanashi U., 3-11 Takeda 4-chome Kofu, Yamanashi 400, Japan. (1943) DEng, applied physics (Tokyo U., 1975). Assoc. prof. (tel. 0552 + 52-1111 ext 5427, fax +81 552 51 6828). *X-ray diffraction, materials science, defect structures, crystal growth.*

Yoshimura, Mr Yukio. Dept. of Math. and Physics, Ritsumeikan U., 28-1 Tojiin-kitamachi Kita-ku, Kyoto 603, Japan. (1941) BSc, physics (Ritsumeikan U., 1964). Res. assoc. (tel. 075 + 463-1131 ext 326). *Crystal physics, inorganic crystal structures, phase transition.*

Yoshioka, Prof. Hide. Sch. of Medicine, Fujita-Gakuen Health U., Kutsukake-cho Toyoake, Aichi 470-11, Japan. (1922) DSc, physics (Nagoya U., 1958). Prof. *Low temperature physics.*

Yoshizawa, Mr Masami. Computer Center, Saitama Inst. of Techn., Fusaiji 1960 Okabe, Saitama 369-02, Japan. (1956) (fax +81 485 85 2523). *X-ray diffraction, instrumentation.*

KOREA

Sub-Editor: **W. Shin**

Notes

1. International telephone country code - 82.

Ahn, Prof. Dr Choong Tai. Chemistry Dept., Hankuk U. of Foreign Studies, Yongin-gun, Kyunggi-do 170, Korea. (1931) PhD, physical chem (Seoul Nat. U., 1979). Prof. (tel. Seoul 02 + 359-8632). *X-ray diffraction, organic compounds.*

Cho, Prof. Sung-Il. Chemistry Dept., City C. of Seoul, Jeonnong-dong, Dongdaemun-gu, Seoul 131, Korea. (1947) PhD, physical chemistry (Seoul Nat. U., 1981). Asst. prof. (tel. 02 + 245-8111). *X-ray diffraction, biological compounds (small and medium size).*

Chung, Prof. Dr Su Jin. Inorganic Materials Eng. Dept., Seoul Nat. U., Sinrim-dong, Gwanag-gu, Seoul 151-742, Korea. (1938) Dr.rer.nat., Crystallography (Tech. Hochschule Aachen, BRD, 1972). Prof. (tel. 02 + 880-5635). *Inorganic crystal chemistry, X-ray diffraction, electron diffraction, symmetry.*

Jeong, Dr Jong Hwa. Inorganic Chem. Lab., Chemistry Div., Korea Advanced Inst. of Sci. and Tech., Seoul, Korea. (1955) PhD, Inorganic Chem.(U. Hawaii,1987). Snr. Inv. (tel. 02 + 967-8801, ext 3293, fax 963-4013). *Inorganic crystal structure analysis.*

Kim, Dr Hoon Sup. P.O. Box 35, Daejun, Chungnam, Korea. (1935) PhD, crystallography (Pittsburgh U., USA, 1970). *X-ray diffraction, biological compounds (small and medium size).*

Kim, Prof. Ho Sung. Inorganic Materials Eng. Dept., Chonnam Nat. U., Kwangju, Chonnam 500-757, Korea. (1959)PhD, Crystallography (Seoul Nat. U.,1989) Asst. Prof. *X-ray diffraction, electron diffraction, symmetry, computer application.*

Kim, Dr Key Soo. Electro-Optics Lab., Korea Atomic Energy Res. Inst., 170-2 Gongneung-dong, Dobong-gu, Seoul, Korea. (1929) DSc, solid state physics (Seoul Nat. U., 1972). Head. (tel. 96 5083). *X-ray crystallography, electron microscopy, crystal growth, radiation damage, nuclear materials.*

Kim, Prof. Kimoon. Chemistry Dept., Pohang Inst. of Sci. and Tech., Pohang, P.O. Box 125, Korea. (1954) PhD, Inorganic Chem. (Standford U.,1986) Asst. prof. (tel. 0562 + 79-2113, fax 0562+79-2099). *Inorganic crystal structure.*

Kim, Prof. Moon Il. Dept. Metallugical engineering, Yonsei U., Suhdaemun-ku, Seoul, 120-749, Korea. (1929) DSc, Crystallography (Seoul Nat. U.,1970) Prof. (tel. 02 + 392-0131, ext. 2353). *Surface hardening, crystal growth.*

Kim, Prof. Moon-Jib. Physics Dept., Soon Chun Hyang U., Onyang, Chungnam 331, Korea. (1954) PhD, physics (Chungnam Nat. U., 1988). Prof. (tel. 0418 + 2-4751). *Solid state physics.*

Kim, Dr Sangsoo. Protein Eng. Lab., Lucky Central Res. Inst., Sci. Town, PO box 10 Dae Deog Dan Ji, Dae Jeon Chung-Nam 302-343, Korea. (1958) PhD, Physical Chemistry (Iowa State U.,1986) Snr. Chemist. (tel. 042 + 861-9981, fax (8242)294-2918, telex LUCKYR K45580) *Protein Crystallography.*

Kim, Prof. Dr Soo Jin. Geological Sciences Dept., Seoul Nat. U., Sinrim-dong, Gwanag-gu, Seoul 151-742, Korea. (1939) PhD, mineralogy (Seoul Nat. U., 1971), Dr.rer.nat. (U. Heidelberg, BRD, 1979). Prof. (tel. 02 + 880-5480). *X-ray diffraction, crystal chemistry, minerals, crystal optics.*

Kim, Prof. Dr Yang. Chemistry Dept., Pusan Nat. U., Dongrae-gu, Pusan 607, Korea. (1940) PhD, chemistry (Hawaii U., USA, 1979). Prof. (tel. 051 + 56-0171). *Crystal structure analysis, zeolites.*

Kim, Prof. Dr Yang Bae. Manufacturing Pharmacy Dept., Seoul Nat. U., Sinrim-dong, Gwanag-gu, Seoul 151-742, Korea. (1940) PhD, pharmacy (Seoul Nat.

U., 1974). Prof. (tel. 02 + 886-0101 Ext. 2884). *Crystal structure analysis, biological compounds.*

Kim, Dr Yoonho. Fine ceramics lab., Korea Advanced Inst. of Sci. and Tech., Chungryangri, P.O. Box 131, Seoul, Korea. (1942) Dr. Eng. (E.N.S.C.I. France,1982). Lab. manager. (tel. 02 + 967-8801 ext 3593, fax 82-2-963-4013). *Ceramics, material research.*

Koo, Prof. Dr Chung Hoe. Chemistry Dept., Seoul Nat. U., Sinrim-dong, Gwanag-gu, Seoul 151-742, Korea. (1922) DSc, crystallography (Seoul Nat. U., 1960). Emeritus Prof. *Crystal structure analysis, organic compounds, X-ray diffraction.*

Namgung, Prof. Dr Hae. Chemistry Dept., Kukmin U., Sungbuk-gu, Seoul 132, Korea. (1942) Dr.rer.nat. (U. Bonn, BRD, 1978) Prof. (tel. 02 + 914-3141) *Crystal structure, crystal chemistry, crystal physics.*

Park, Dr Byung Kyu. Refractory lab., Energy Dept., Res. Inst. of Industrial Sci. and Tech., Pohang, Kyeongnam, Korea. (1956) PhD,crystallography (Seoul Nat. U.,1987) Snr. res. (tel. 0562 + 71-6370). *Crystal growth, inorganic crystal chemistry, X-ray diffraction, symmetry.*

Park, Prof. Young Ja. Chemistry Dept., Sook-Myung Women's U., Yongsan-ku, Seoul 140-742, Korea. (1942) PhD, crystallography (Pittsburgh U., USA, 1970). Prof. (tel. 02 + 713-4528). *X-ray diffraction, molecular mechanics, biological compounds.*

Shin, Prof. Hyun So. Chemistry Dept., Dong-Guk U., Pil-dong, Chung-gu, Seoul 110, Korea. (1937) PhD, crystallography (Seoul Nat. U., 1979). Prof. (tel. 02 + 261-8131). *X-ray diffraction, organic compounds.*

Shin, Prof. Dr Whanchul. Chemistry Dept., Seoul Nat. U., Sinrim-dong, Gwanag-gu, Seoul 151-742, Korea. (1950) PhD, crystallography (Pittsburgh U., USA, 1978). Assoc. prof. (tel. 02 + 886-0101, ext 3318, fax 82+2+889-1568) *X-ray crystallography, biologically active small molecules, proteins, crystallographic computing.*

Suh, Prof. Dr Ill-Hwan. Physics Dept., Chungham Nat. U., Daejun, Chungnam 302-764, Korea. (1936) PhD, crystallography (Korea U., 1976). Prof. (tel. 042 + 822-0101). *Crystal structure determination, crystal physics.*

Suh, Prof. Jung Sun. Chemistry Dept., Myong-Ji U., Yongin-kun, Korea. (1942) PhD, physical chemistry (Seoul Nat. U., 1979). Prof. (tel. 0335 + 32-4001). *X-ray diffraction, pharmacological compounds.*

Suh, Prof. Se Won. Chemistry Dept., Seoul Nat. U., Sinrim-dong, Gwanag-gu, Seoul 151-742, Korea. (1951) PhD, chemistry (UCLA, USA, 1980). Assoc. prof. (tel. 02 + 886-0101, ext 3249, fax 82+2+889-1568). *Protein structure and function.*

Yeon, Prof. Younghee. Chem. Dept., Korea Air Force Academy, Chungwon, 363-849- Korea. (1949) PhD. Crystallography(U.Pitsburgh, 1988) Assoc. Prof. (tel. 0431 + 53-9165 ext. 3112). *X-ray diffraction, superconductivity materials.*

Yoon, Prof. Choon Sup. Physics Dept., Korea Inst. of Tech., Dae Duck Sci. Town, Daejon, Korea. (1950)PhD, Physics (U. Strathclyde, Glasgow, Scotland,1987). Assoc. prof. (tel. 042 + 861-1234 ext 430, fax 042-861-5636). *Solid state physics, X-ray topography, non-linear optic materials.*

LIBYA

Sub-Editor: **M. S. Ellid**

Notes

1. International telephone country code - 218.

Ajaal, Mr Tawfik Taher. Material Sci. Dept., Tajura Nuclear Res. Center, P.O. Box 30878, Tajura - Tripoli, Libya. (1958) BSc, material science (Alfateh U., 1983). Eng. (telex 13615) *Neutron diffractometry, solid state research, corrosion.*

Almajdub, Mr Musbah Meftah. Solid State Physics Lab., Tajura Res. Center, T.N.R.C, P.O. Box 397, Tripoli, Libya. (1954) BSc, physics (Tennessee Techn. U., USA, 1980) Res. (telex 20792 TAJ RC LY) *Neutron diffractometry, crystal structure, crystal dynamics, inelastic neutron scattering, phonon resonance, alloys.*

El-Azizi, Mr Ibrahim M. Physics and Material Sci., Tajura Nuclear Res. Center, P.O. Box 30878, Tripoli, Libya. (1961) BSc, solid state physics (Tbilisi U., USSR, 1989) Res. eng. solid state physics. (tel. 0021821 607023, telex 20792 TAJ RC LY) *Superconductivity, martensitic transformations.*

Ellid, Dr Mohamed S. Solid State Physics Lab, Tajura Res. Centre, P.O. Box 30878, Tajura, Tripoli, Libya. (1947) PhD, Mössbauer effect (U. West Virginia, USA, 1958). Res. (telex 20792 TAJ RC LY). *Mössbauer effect, catalysis, crystal structure investigation, neutron diffraction.*

El-Mashri, Dr S.M. Solid State Physics and Material Science Dept., Tajura Res. Centre, P.O. Box 397, Tripoli, Libya. (1950) PhD, Surface physics (U. Warwick, UK, 1984). Res. (telex 20792 TAJ RC LY). *X-ray diffraction, electron microscopy, surface physics.*

Elzawi, Mr Rajab Abdulla. Dept. of Physics, Solid State Lab., Tajura Res. Center, T.N.R.C, P.O. Box 30878, Tajura-Tripoli, Libya. (1954) BSc, physics (U. Oklahoma, USA, 1980) Res. asst. (tel. 607052, telex 20792 TAJ RC LY) *Neutron diffraction, magnetic structure analysis, nuclear structure, Mössbauer spectroscopy.*

Khalf, Mr Fuad M. Solid State Physics Lab, Tajura Res. centre, P.O. Box 30878, Tajura, Tripoli, Libya. (1958) BSc, Electrical engineering (U. Washington, USA, 1980) Res. Asst. (telex 20792 TAJ RC LY). *Crystallogrphic Computing, computer programming, determination of crystal structure.*

Mrayed, Mr Yaseen S. Solid State Physics Lab., Tajura Res. Centre, P.O. Box 30878, Tajura, Tripoli, Libya. (1958) BSc, Physics (U Minnesota-Duluth,USA,1985) Res. Asst. (telex 20792 TAJ RC LY). *neutrons spectrometry, martensite transformation, alloys, invar like behaviour.*

Sherfad. Mr Mohamed E. . Physics and Materials Sci., Tajura Nuclear Res. Center, P.O. Box 30878, Tripoli, Libya. (1958) BSc, nuclear eng. (Al-Fateh U., 1983) Res. eng. solid state physics. (tel. 0021821 607023, ext. 491, telex 20792 TAJ RC LY) *Superconductivity, martensitic transformations.*

Shihub, Mr Salahedin I. Solid State Physics Lab., Tajura Nuclear Res. Center, P.O. Box 397, Tripoli, Libya. (1954) BSc, physics (U. Washington, USA, 1980) Res. asst. (telex 20792 TAJ RC LY) *Neutron spectrometry, crystal dynamics.*

MALAYSIA

Sub-Editor: **Abdul Hamid Othman**

Notes

1. International telephone country code - 60

2. The science degrees conferred by Malaysian universities are the PhD, MSc, and BSc and correspond to British degrees.

3. The following abbreviations have been used in this section:

U.M. - *Universiti Malaya* (University of Malaya).
U.S.M. - *Universiti Sains Malaysia* (University of Science of Malaysia).
U.K.M. - *Universiti Kebangsaan Malaysia* (National University of Malaysia).
U.P.M. - *Universiti Pertanian Malaysia* (University of Agriculture of Malaysia).
U.T.M. - *Universiti Teknologi Malaysia* (University of Technology of Malaysia).

M.I.T. - Mara Institute of Technology.
M.A.R.D.I. - Malaysian Agricultural Research and Development Institute.
S.I.R.I.M. - Standards and Industrial Research Institute of Malaysia.
P.O.R.I.M. - Palm Oil Research Institute of Malaysia.

Abdul Aziz, Mr Abdul Halim bin. Sch. of Physics, U.S.M., 11800 Penang, Malaysia. (1958) MSc, X-ray crystallography (U. London, UK, 1983). Lect. (tel. 04 + 883822, ext. 654). *X-ray crystal structure determination.*

Abu Bakar, Mr Ismail. Central Res. Labs. Div.,M.A.R.D.I., P.O.Box 12301, Kuala Lumpur 01-02, Malaysia. (1952) MSc, mineralogy (Ghent U, Belgium, 1981). Res. officer. (tel. 03 + 356601, ext. 478). *Malaysian soils, mineralogy, soil survey, micromorphology.*

Almashoor, Dr Syed Sheikh. Dept. of Geology, U.K.M., 43600 UKM Bangi, Selangor, Malaysia. (1944) PhD, igneous petrology (Penn. State U., U.S.A., 1983). Lect. (tel. 03 + 8250001, ext. 2390). *Mineral, X-ray diffraction.*

Ang, Dr Ha Ming. Chemical Eng. Dept., U.M., Lembah Pantai, 59100 Kuala Lumpur,Malaysia. (1946) PhD, bulk crystallization (U. London, UK, 1973). Assoc. prof. (tel. 03 + 553466, ext. 293). *Crystal characteristics, bulk crystallization.*

Baba, Mrs Jasmin. Mechanical Eng. Fac., U.T.M., 81300 Sekudai, Johor, Malaysia. (1954) MSc, material science (Cranfield Inst. of Techn., UK, 1981). Lect. (tel. 03 + 929033, ext. 533). *Electron microscopy, X-ray diffraction techniques.*

Cheang, Dr Kok Keong. School of Materials & Mineral Resources Eng., U.S.M., Perak Branch Campus, Jln. Bandaraya, 30000 Ipoh, Perak, Malaysia. (1949) PhD, geology, (U. Georgia, USA, 1982). Sr. lect. (tel. 05 + 503131, ext. 2330). *X-ray crystallography, electron microscopy, ore petrology, economic geology, exploration and isotope geochemistry, optical minerology*

Chen, Dr Wei. Chemistry Dept., U.M., Lembah Pantai, 59100 Kuala Lumpur, Malaysia. (1948) PhD, X-ray crystallography (U. New South Wales, Australia,

1976). Assoc. prof. (tel. 03 + 7555466). *Single crystal diffractometry, structure, small molecules, structural chemistry, symmetry.*

Faqir, Dr Gul. Sch. of Eng., M.I.T., 40450 Shah Alam, Selangor, Malaysia. (1951) PhD, metallurgical eng. (U. Liverpool, UK, 1982). Lect. (tel. 03 + 362311). *Mechanical metallurgy, materials.*

Fun, Dr Hoong Kun. Sch. of Physics, U.S.M., 11800 USM Penang, Malaysia. (1946) PhD, experimental solid state physics (Purdue U., USA, 1974). Assoc. prof. (tel. 04 + 883822, ext. 652, telex USMLIB MA40254, fax +604 871526). *Magnetic resonance, structure analysis, laser physics.*

Gan, Mr Ah Sai. Mineralogy and Petrology Div., Geological Survey of Malaysia, 31400 Scrivenor Road, Ipoh, Perak, Malaysia. (1945) BSc(Hons), applied geology (U.M., 1970). Sr. geologist. (tel. 05 + 557644, ext. 116). *Mineralogy, ore microscopy, powder diffraction.*

Hassan, Dr Wan Fuad. Geology Dept., U.K.M., 43600 UKM Bangi, Selangor, Malaysia. (1948) PhD, mineralogy and geochemistry (U. Leeds, UK, 1982). Lect. (tel. 03 + 8250001, ext. 2657). *Microscopy, minerals, inorganic crystals, teaching, X-ray diffraction.*

Hutchison, Prof. Dr Charles Strachan. Geology Dept., U.M., Lembah Pantai, 59100 Kuala Lumpur, Malaysia. (1933) PhD, petrology and mineralogy (U.M., 1966). Prof. (tel. 03 + 555466, ext. 203). *Mineral chemistry, petrology.*

Lee, Dr Chnoong Kheng, Chemistry Dept., U.P.M., 43400 UPM Serdang, Selangor, Malaysia. (1948) PhD, inorganic chemistry (U. Aberdeen, UK, 1972). Lect. (tel. 03 + 9486101, ext. 3613). *Minerals, phase determination, powder diffraction, structural chemistry, disorder.*

Mohamad, Dr Hamzah. Geology Dept., U.K.M., 43600 UKM Bangi, Selangor, Malaysia. (1951) PhD, geochemistry, petrology (U. Strathclyde, Glasgow, UK, 1980). Lect. (tel. 03 + 8250001, ext. 2664). *Minerals, powder diffraction, X-ray diffraction.*

Ng, Dr Wee Lam. Chemistry Dept., U.M., Lembah Pantai, 59100 Kuala Lumpur, Malaysia. (1943) PhD, solid state chemistry (U. Western Ontario, Canada, 1971). Assoc. prof. (tel. 03 + 555466, ext. 249). *Electrical conduction, thermal decomposition, crystallization, structural defects, crystal structures.*

Oh, Mr Chuan Ho, Flingoh. Chemistry and Techn. Div., P.O.R.I.M., 6 Persiaran Institusi, Bandar Baru Bangi, 40300 Kajang, Selangor, Malaysia. (1947) MSc, crystallization (U.M., 1980). Sr. res. officer. (tel. 03 + 8259775, ext. 1086, 1084, telex MA31609, fax 603 8259446). *Crystallization, oil and fat, powder diffraction.*

Othman, Dr Abdul Hamid bin. Chemistry Dept., U.K.M., 43600 UKM Bangi, Selangor, Malaysia. (1948) PhD, X-ray crystallography (U. Reading, UK, 1977). Assoc. prof., dy. dean (tel. 03 + 8250001, ext. 2439, 3330, telex UNIKEB MA 31496, fax: 603 825 6484). *Coordination compounds, molecular crystals, computing,X-ray diffraction.*

Othman, Dr Radzali. Sch. of materials and mineral resourses Eng., U.S.M., Perak Branch Campus , Jln. Bandaraya, 30000 Ipoh, Perak, Malaysia. (1954) PhD, ceramics (U. Sheffield, UK, 1982). Dean (tel. 05 + 503131, ext. 2317, telex USMKCP MA 45080). *Electron microscopy, material science, X-ray diffraction.*

Ong, Mr Yeoh Han. Minerals Clearance Project, Geological Survey Dept., Scrivenor Road, Ipoh, Perak, Malaysia. (1948) BSc, geology (U.M., 1972). Geologist. (tel. 05 + 557644, ext.7). *Mineralogy, gemology, X-ray methods.*

Rao, Mr Nutakki Nageswara. Materials Eng. Div., Sch. of Eng. Sci. and Industrial Techn., U.S.M., 11800 Penang, Malaysia. (1949) MTechn., metallurgy (I.I.T., Madras, India, 1973). Lect. (tel. 04 + 883822, ext. 613). *Powder metallurgy, fracture analysis, welding, corrosion.*

Salleh, Dr Mansor bin Haji. S.I.R.I.M., P.O.Box 35, 40000 Shah Alam, Selangor, Malaysia. (1944) PhD, corrosion (U. Manchester, UK, 1978). Controller. (tel. 03 + 592635). *Metallurgy, materials, corrosion.*

Salleh, Dr Mohammad Nawi. Sch. of Eng., M.I.T., Shah Alam, Selangor, Malaysia. (1946) PhD, metallurgical eng. (Colorado Sch. of Mines, USA, 1979). Sr. lect. (tel. 03 + 362464). *Mechanical metallurgy, metal forming.*

Silong, Dr Sidik bin. Chemistry Dept., U.P.M., 43400 UPM Serdang, Selangor, Malaysia. (1953) PhD, inorganic chemistry (U. Reading, UK, 1982). Lect. (tel. 03 + 356101, ext. 546). *Coordination compounds, reaction pathways, structural chemistry, X-ray diffraction.*

Singh, Dato' Prof. Dr Chatar. Sch. of Physics, U.S.M., 1180 Penang, Malaysia. (1929) PhD, X-ray crystallography and physics (Cambridge U., UK, 1961). Prof. (tel. 04 + 883822, ext. 659). *X-ray diffraction, organic compounds, hydrogen bonding, teaching.*

Teh, Dr Guan Hoe. Geology Dept., U.M., Lembah Pantai, 59100 Kuala Lumpur 22-11, Malaysia. (1946) Dr.rer.nat., mineralogy, geomicrobiology, experimental petrology (U. Heidelberg, BRD, 1979). Lect. (tel. 03 + 560022). *Economic geology, mineralogy, experimental petrology, analytical methods.*

Teh, Dr Ser Kok. Mechanical Eng. Dept., U.M., Lembah Pantai, 59100 Kuala Lumpur, Malaysia. (1947) PhD, materials science (U. London, UK, 1976). Lect. (tel. 03 + 553466, ext. 265). *Defects structure, electron diffraction, electron microscopy, materials science, ceramics, alumina, metals, creep.*

Teoh, Mr Lay Hock. Geological Survey of Malaysia, 20th floor, Tabung Haji Building, Jalan Tun Razak, 50736 Kuala Lumpur, Malaysia. (1950) BSc(Hons), geology and geochemistry (U. Victoria, New Zealand, 1972). Principal geologist. (tel. 05 + 557644). *Minerals, X-ray diffraction.*

Tuan Sarif, Mr Tuan Besar. Mineral Resources Eng. Div., Sch. of Eng. Sci. and Industrial Techn., U.S.M., 11800 Penang, Malaysia. (1957) MSc, geochemistry, (Iowa State U., U.S.A., 1983). Lect. (tel. 04 + 883822, ext. 614). *Mineralogy, crystallography, geochemistry, mining, geotechnical engineering.*

Yahaya, Dr Muhamad. Physics Dept., U.K.M., 43600 UKM Bangi, Selangor, Malaysia. (1947) PhD, electronic properties (Monash U., Australia, 1979). Assoc. prof. (tel. 03 + 8250001, ext. 2418). *Electron density, group theory, semi-conductors, thermal vibrations.*

Yong, Mr Swee Kee. Technical Support Services, Geological Survey Dept., Scrivenor Road, 31400 Ipoh, Perak, Malaysia. (1942) BSc, geology (U. Adelaide, Australia, 1964). Geologist. (tel. 05 + 557644, ext. 7). *Mineralogy, mineragraphy, petrography, economic geology.*

MEXICO

Sub-Editor: A.E. Cordera-Borboa

Notes

1.International telephone country code - 52.

2. Degrees conferred by the Mexican Universities are the *doctor en ciencias* (dr.c.)(equivalent to phd. *maestro en ciencias* (m.c.)(equivalent to msc), and *ingeniero* (ing.) or *licenciado en ciencias* (lic.), both equivalent to bsc.

3. At a Mexican University, persons can be appointed in the pollowing positions not belonging to the regular formation of personnel: *profesor* (titular, asociado, ayudante), *(investigador)* (Titular, Asociado, Ayudante), and *Tecnico Academico* (Titular, Asociado, Ayudante).

Aguilera Herrera, Prof. Nicolás. Depto. de Biologia, Facultad de Ciencias, U. Nac. Aut. de México, Circuito Exterior, Ciudad Universitaria, D.F. 04510 México. (1920) MSc, soil science (U. Wisconsin, USA, 1953). Prof. (tel.550-5913). *Clay, soil, mineralogy, edaphology.*

Bosch Giral, Dr Pedro. Depto. De Catálisis, UAM, Iztapalapa, A.P. 55-532, D.F. 13 México. (1948) DSc, crystal chemistry and catalysis (U. Claude Bernard,France,1946). Investigador cientifico. (tel. 5815206) *Catalysis, X-ray diffraction.*

Cabrera Bravo, Prof. Enrique. Depto. Materia Condensada, Inst. de Física, UNAM A.P. 20-364, Ciudad Universitaria, D.F. 04510 México. (1946) Dr.C. theoretical physics (Fac. de Ciencias,UNAM, 1980) (tel. 5488192). *Electron microscopy, crystallography, diffusion.*

Carrillo-Hoyo, Dr. Eduardo. Depto. de Estado Sólido, Inst. de Física, UNAM A.P. 20-364, México D.F. 01000, México. (1948) Dr.C. (Fac.de Ciencias, UNAM 1986). Investigador. (telex UNAMME 17 74523, tel. 5505936, fax +91 5 5483111). *Metal physics.*

Castellanos Guzman, Dr A. Guillermo. Centro de Investigación en Ciencias Básicas, U. de Colima, Ap. Post. 2-1694, Colima 28000, México. (1939) PhD, physics (London U., UK, 1981). Scient. (tel. 331 + 2-58-18, telex. UCOLMEX 62248) *Crystal growth, dialectric and optical properties, polar crystals, X-ray structure determination.*

Castellanos Román, Mrs Maria Asunción. Div. Estudios de Postgrado, Fac. de Quimica, U. Nac. Aut. de México, Ciudad Universitaria, D.F. 04510 México. (1943) MSc, (Aberdeen U., UK, 1979). Jefe lab. Rayos X (tel. 5-48-82-10). *Crystal chemistry, oxide complexes.*

Chapela Castañares, Dr Víctor Manuel. Inst. de Investigaciones de Materiales, U. Nac. Aut. de México, Ap. Post. 70-360, Ciudad Universitaria, D.F. 04510 México. (1944) PhD, (Imperial C. London U., UK, 1979). Jefe depto. bajas temperaturas. (tel. 550-5215, ext. 4735). *Intercalation of aminoacids in dichalcogenides.*

Cano Corona, Dr Octavio. Inst. de Fisica, U. Nac. Aut. de México, Ap. Post. 70-364, Ciudad Universitaria, D.F. 04510 México. (1921) PhD, physics (Pennsylvania State U., USA, 1954). Acad. techn. (tel. 550-5215, ext. 5940). *X-ray powder diffraction, crystal studies.*

Cordero-Borboa, Dr Adolfo. Inst. de Fisica, UNAM. A.P. 20-364, D.F. 01000, M éxico. (1953)Dr.C. (Fac. de Ciencias, UNAM, 1983) Investigador. (tel. 5488192, telex 1774523 - UNAMME, fax +905-5483111). *Crystallography of minerals, man-made materials and proteins.*

Cota Araiza, Mr Leonel Susano. Depto. Estado Solido, Inst. de Fisica, U. Nac. Aut. de México, Ap. Post. 20-364, Ciudad Universitaria, D.F. 04510 México. (1944) MPh, surface physics (Warwick U., UK, 1974). Res. assoc. (tel. 5-48-81-92, ext. 332). *Low energy electron diffraction, Auger electron spectroscopy, photoemission.*

De Pablo Galan, Dr Linerto. Inst. de Geologia, U. Nac. Aut. de México, Ap. Post. 70-296, Ciudad Universitaria, D.F. 04510 México. (1934) PhD, (Ohio State U., USA). Jefe del Depto. de Geoquimica. (tel. 550-5215, ext. 4268). *Mineralogy, crystallography.*

Domínguez Esquivel, Dr José Manuel. Instituto Mexicano del Petróleo, Ap. Post. 14-805, Ave. 100 Metros 152, D.F. 14 México. (1948) DSc, physics (U. Claude Bernard, France, 1977). Investigador cientifico. (tel. 5-67-66-00, ext. 2377) *Catalysis, surface science.*

Echavarri Hernandez, Dr Ariel. Direccion de Mineria, Geologia Y Energeticos, Gobierno del Estado de Sonora, Paseo de la Arboleda 30, Hermosillo Sonora 83000, México. (1939) Dr, petrology (U. de Paris, France, 1967). Dir. (tel. 621 + 31-968). *Mineralogy, petrology.*

Fernández González, Dr Alonso. Rectory, U. Aut. Metropolitana, Unidad Iztapalapa, Ap. Post. 55-532, D.F. 13 México. (1927) PhD, solid state physics (Manchester U., UK, 1958). Prof. (tel. 5-81-52-06). *Crystal growth,electrical breakdown,conductivity in crystals,semiconductors.*

Gómezdaza Almendaro, Mr Mariano. Inst. de Investigaciones de Materiales, U. Nac. Aut. de México, Ap. Post. 70-360, Ciudad Universitaria, D.F. 04510 México. (1944) BSc, chemical eng. (Fac. Quimica, UNAM, 1969). Investigador. (tel. 550-5215, ext. 4747). *Semiconductors, sulfides.*

Gomez Ramirez, Dr Ricardo. Depto. Estado Solido, Inst. de Fisica, U. Nac. Aut. de México, Ap. Post. 20-364, Ciudad Universitaria, D.F. 04510 México. (1944) PhD, materials science (Stanford U., USA, 1971). Investigador titular. (tel. 5-48-81-92). *Plastic deformation, creep, nucleation in solid-solid deformation.*

Gomez Rodriguez, Dr. Alfredo. Depto. Materia Condensada, UNAM A.P. 20-364, D.F. 01000 México. (1952) PhD,physics (U. Warwick, U.K.,1989). Investigador (tel. 5505215, ext. 3981). *Crystal diffraction, quasi-crystals.*

Herrera-Becerra, Dr. Raul. Depto, de Materia Condensada, Inst. de Física, UNAM. A.P. 20-364, D.F. 01000 México. (1952) Dr.C. (CECESE, Mexico,1989) Investigador (tel. 5488192). *Electron Diffraction.*

Huanosta Tera, Mr Alfonso. Inst. de Investigaciones de Materiales, U. Nac. Aut. de México, Ap. Post. 70-360, Ciudad Universitaria, D.F. 04510 México. (1944) MSc, (Fac. de Ciencias, UNAM, 1978) Investigador. (tel. 550-5215, ext. 4746). *Electron microscopy, alloys of Cu-Al.*

José Yacaman, Dr Miguel. Inst. de Fisica, U. Nac. Aut. de México, Ap. Post. 20-364, Ciudad Universitaria, D.F. 04510 México. (1946) DrSc, (Fac. de Ciencias, UNAM, 1972). Head Director. (tel. 550-5215, ext. 5940). *Crystallography, microcrystals.*

Lara Magaña, Mrs María Eugenia. Ciencia e Ingenieria de Materiales al Servico de la Industria, Ojito no.34 Coyoacán, D.F. 04000 México. (1950) BSc, pharmacy (U. Texas, USA, 1975). Investigadora. (tel. 554-5945) *Crystal chemistry, organic materials obtained from plants.*

Lee Moreno, Dr José Luis. Consejo de Recursos Minerales., Ninos Heroes 139, D.F. 7 México. (1939) PhD, geological eng. (U. Arizona, USA, 1972). Manager of special studies. (tel. 5-78-59-42). *geochemistry, minerals exploration, fluid occlusions, computer applications.*

Muñoz Picone, Dr Eduardo. Depto. Materia Condensada, Inst. de Fisica, UNAM A.P. 20-364, Ciudad Universitaria, D.F. 04510 México. (1937) PhD, physics (UNAM, 1970) Investigador, prof. (tel. 5488192). *Crystal growth.*

Murrieta Sánchez, Dr. Héctor, Depto. Estado Sólido, UNAM. A.P. 20-364, D.F. 01000, México. (1945) PhD,physics (UNAM, 1978) Investigador (tel. 5488192). *Optical properties in solids.*

Piña de Noyola, Prof. Maria Cristina. Depto. Estado Solido, Inst. de Fisica, U. Nac. Aut. de México, Ap. Post. 20-364, Ciudad Universitaria, D.F. 04510 México. (1946) MSc, Physics (U. Nac. Aut. de México, 1976). Res. worker. (tel.5-48-81-92). *Thermodynamics of solids, diffusion, metallurgy.*

Quintana Owen, Mrs Patricia. Division Estudios de Postgrado. Fac. de Quimica, U. Nac. Aut. de México, Ciudad Universitaria, D.F. 04510 México. (1951) MSc, chemistry (UNAM, 1977). Investigadora. (tel. 5-48-52-10). *Phase diagrams, oxide system Li-Zr-Si.*

Ramos Bernal, Dr Sergio. Centro de Estudios Nucleares, U. Nac. Aut. Mexico, Circuito exterior, Ciudad Universitaria, D.F. 04510 México. (1945) PhD, (U. Manchester, UK, 1947). Secretario academico. (tel. 5-48-45-69). *Radiation damage, magnetic materials.*

Rendón Diaz Mirón, Dr Luis Emilio. Inst. de Investigaciones de Materiales, U. Nac. Aut. de México, Ap. Post. 70-360, Ciudad Universitaria, D.F. 04510 México. (1946) PhD, materials sci. and eng. (U. Texas, USA, 1977). Jefe proyecto materiales para electronica.(tel. 550-5215, ext. 4747). *Crystal chemistry, transition metal sulfides and selenides.*

Rios Jara, Dr David. Inst. de Investigaciones de Materiales, U. Nac. Aut. de México, Ap. Post. 70-360, Ciudad Universitaria, D.F. 04510 México. (1950) Dr, (Inst. Nac. des Ciences Appl. de Lyon, France). Res. (tel. 550-5215, ext. 4746). *Phase transformation, electron microscopy, alloys.*

Rivera Moras, Mr Vicente. Inst. de Investigaciones de Materiales, U. Nac. Aut. de México, Ap. Post. 70-360, Ciudad Universitaria, D.F. 04510 México. (1948) MSc, (Fac. de Ciencias, UNAM, 1979). Investigador. (tel. 550-5215, ext. 4746). *Electron microscopy (lorentz), magnetic materials.*

Riveros Rotgé, Dr Héctor Gerardo. Depto. Materia Condensada, Inst. de Fisica, UNAM A.P. 20-364, Ciudad Universitaria, D.F. 04510 México. (1940) Dr, solid state (U. Nac. Aut. de México, 1973). Investigador titular. (tel. 5488192). *Crystal growth.*

Romero, Dr Miguel. Romero S. Hnos. S. A., Calle 7 Norte 356, Tehuacan, Puebla, México. (1925) Dr, organic chemistry (U. Nac. Aut. Mexico, 1964). General director. (tel. 238ý-15-80). *Chemistry, mineralogy, crystallography, geology, nutrition.*

Rouffignac, Dr Eric de. Instituto Mexicano del Petróleo, Ap. Post. 14-805, Ave. 100 Metros 152, D.F. 14 México. (1945) PhD (U. Texas, USA, 1978). Jefe depto. de adsorcion, IBP. (tel. 5-67-81-67) *Thermodynamic properties of crystals, adsorption, chemisorption.*

Ruiz Mejia, Dr Carlos. Depto. Estado Solido, Inst. de Fisica, UNAM A.P. 20-364, Ciudad Universitaria, D.F. 04510 México. (1939) DrC, crystallography (U. Nac. Aut. de México, 1964). Investigador, Prof. (tel. 5488192). *Formation energies, optics, defects in crystals.*

Reyes Chumacero, Mr Antonio. Div. Estudios de Postgrado, Fac. de Quimica, U. Nac. Aut. de México, Ciudad Universitaria, D.F. 04510 México. (1940) Ing, (Fac. de Quimica, UNAM, 1964). Jefe Depto. fisico-quimica. (tel. 5-48-02-49). *Thermodynamics, condensed phases.*

Romero Romo, Dr Mario. Depto. Ciencia de Materiales, U. Autonoma. Metropolitana, Av. San Pablo 180, D.F. 02200 México. (1947) PhD, (U. Liverpool). Res. (tel. 382-5000, ext. 235). *Phase transformation, properties, oxides, alloys.*

Salas, Dr Guillermo Armando. Depto. de Geologia, U. de Sonora, Hermosillo, Sonora, México. (1942) PhD, geology, (Stanford U. USA, 1971). Chairman. (tel. 621-4390, ext. 149). *Geology, mineralogy, geochemistry, ore deposts.*

Solorio Munguía, Prof. José Gregorio. Inst. de Geologia, U. Nac. Aut. de México, Ap. Post. 70-296, Ciudad Universitaria, D.F. 04510 México. (1922) QuimMet, metallurgy (U. Nac. Aut. de México, 1958). Invest. Prof. (tel. 550-5215, ext. 4270). *Mineralogy, minerals separation, crystallography, geochemistry.*

Soriano Garcia, Dr Manuel. Inst. de Quimica, U. Nac. Aut. de México, Circuito Exterior, Delagación Coyoacán, D.F. 04510, México. (1947) PhD, biophysics (SUNY at Buffalo, USA, 1976) Res. prof. (tel. 5-50-5215, ext. 2456). *Crystal structure determination, biologically interesting compounds, protein crystallography, crystal chemistry.*

Téllez Ortiz, Mrs Minerva Estela. Div. Estudios de Postgrado, Fac. de Quimica, U. Nac. Aut. de México, Ciudad Universitaria, D.F. 04510 México. (1943) MSc, (Fac. de Quimica, UNAM, 1977). Tecnico academico. (tel 5-48-82-10) *X-ray diffraction, X-ray spectroscopy.*

Torres Villaseñor, Dr Gabriel. Inst. de Investigaciones de Materiales, U. Nac. Aut. de México, Ap. Post. 70-360, Ciudad Universitaria, D.F. 04510 México. (1944) PhD, (Case Western U., USA, 1972). Project chief. (tel. 550-5215, ext. 4746). *Alloys of Cu, domains in Cu gamma phases.*

Toscano, Mr Ruben Alfredo, Inst. de Quimica, U. Nac. Aut. de México, Circuito Exterior, Ciudad Universitaria, D.F. 04510 México. (1958) BSc (U. Nac. Aut. de México) Acad. Techn. (tel. 548-5448). *X-ray structure determination, infrared spectroscopy.*

Valenzuela Monjarás, Dr Raúl Alejandro. Inst. de Investigaciones de Materiales, U. Nac. Aut. de México, Ap. Post. 70-360, Ciudad Universitaria, D.F. 04510 México. (1946) DSc, (Fac. des Sciencies, Paris, 1974). Jefe depto. ceramica y metalurgia. (tel. 550-5215, ext. 4746). *Magnetic properties, ceramic materials.*

Vera Calderón, Mrs Gloria. Div. Estudios de Postgrado, Fac. de Quimica, U. Nac. Aut. de México, Ciudad Universitaria, D.F. 04510 México. (1938) Qumica Industrial (U. de Guanajuato, 1962). Tecnico academico. (tel. 5-8210). *X-ray diffraction, organics.*

Villafuerte Castrejón, Mrs María Elena. Inst. de Investigaciones de Materiales, U. Nac. Aut. de México, Ap. Post. 70-360, Ciudad Universitaria, D.F. 04510 México. (1948) MSc, inorganic chemistry (Fac. de Quimica, UNAM, 1979). Investigadora. (tel. 550-5215, ext. 4747). *Crystal chemistry, ceramic materials.*

NETHERLANDS

Sub-Editors: R. Olthof-Hazekamp & I. Preistnall

Notes

1. International telephone country code - 31.

2. Degrees conferred by the Netherlands universities are *doctor* (Dr) (approximately equivalent to PhD at British universities), *doctorandus* (Drs) and *ingenieur* (Ir) (these latter two between MSc and PhD).

3. At universities, persons with an academic training can be appointed in various positions: *hoogleraar* (professor), *universitair hoofddocent, universitair docent, research scientist* and *research assistent* (approximately equivalent to assistent lecturer).

Admiraal, Dr Gerrit. Lab. voor Kristallografie, U. of Amsterdam, Nieuwe Achtergracht 166, 1018 WV Amsterdam, The Netherlands. (1953), Dr, chemistry (U. Groningen, 1981). Res. scient. (tel. 020 + 5257039). *Peptide, DNA structures, Patterson methods, computer programming.*

Aerts, Dr Jozef. Corporate Res. Dept., AKZO, Velperweg 76, P. O. Box 9300, 6800 SB Arnhem, The Netherlands. (1957) PhD, chemistry (U. Leuven, Belgium, 1983). Res. scient. (tel. 085 + 662669). *Structures and physical properties of polymers.*

Altona, Prof. Cornelis. Dept. of Organic Chemistry, Gorlaeus Lab., P. O. Box 9502, 2300 RA Leiden, The Netherlands. (1931) Dr, organic chemistry (U. Leiden, 1964). Prof. (tel. 071 + 274329 or 274505). *Conformational analysis, biomolecules (nucleotides).*

Baak, Ing. Leonardus Cornelis. Res. and Development Lab., ENRAF-NONIUS, Röntgenweg 1, P. O. Box 483, 2600 AL Delft, The Netherlands. (1944) Ing., software product specialist. (tel. 015 + 698500, ext 678, telex 38083, fax 015 + 619574). *Computing, diffractometry, X-ray diffraction.*

Bastin, Dr Ir Guillaume F. Dept. of Physical Chemistry, Eindhoven U. of Techn., P. O. Box 513, 5600 MB Eindhoven, The Netherlands. (1944) Dr, chemistry (Eindhoven U. of Techn., 1972). Universitair hoofddocent. (tel. 040 + 473049, telex 51163, fax 040 + 445619). *Powder diffraction, texture, metals.*

Bennema, Prof. Dr Pieter. Science Dept., U. of Nijmegen, Toernooiveld, 6525 ED Nijmegen, The Netherlands. (1932) Dr, crystal growth (Delft U. of Techn., 1965). Prof. (tel. 080 + 613070, telex 48226 WINA NL, fax 080 + 553450). *Crystal growth, morphology.*

Berg, van den, Ir Adrianus Johannes. Afd. Technische Natuurkunde, Delft U. of Techn., Lorentzweg 1, 2628 CJ Delft, The Netherlands. (1940) Ir, physical chemistry (Delft U. of Techn., 1966). Docent. (tel. 015 + 782481). *Crystal physics, powder diffraction.*

Berger, Dr Rolf Anders. Lab. voor Anorganische Chemie, U. of Groningen, Nijenborgh 16, 9747 AG Groningen, The Netherlands. (1946) Dr, chemistry (Uppsala U., Sweden, 1978). Res. scient. (tel. 050 + 634414). *Transition metal alloys, chalcogenides.*

Bergsma, Dr Jitze. Physics Dept., Netherlands Energy Res. Foundation, ECN, P. O. Box 1, 1755 ZG Petten NH, The Netherlands. (1932) Dr, physics (U. Leiden, 1970). Head physics dept. (tel. 02246 + 4949). *Neutron diffraction, neutron inelastic scattering, crystal dynamics.*

Beurskens, Dr Gezina. Lab. voor Kristallografie, U. of Nijmegen, Toernooiveld, 6525 ED Nijmegen, The Netherlands. (1936) Dr, chemistry (U. Utrecht, 1961). (tel. 080 + 612842 or 612875, telex 48228 WINA NL, fax 080 + 553450, email U625004 at HNYKUN11.EARN). *Crystal structure determination, direct methods.*

Beurskens, Prof. Dr Paul T. Lab. voor Kristallografie, U. of Nijmegen, Toernooiveld, 6525 ED Nijmegen, The Netherlands. (1934) Dr, chemistry (U. Utrecht, 1965). Prof. (tel. 080 + 612188 or 612875, telex 48228 WINA NL, fax 080 + 553450, email U625004 at HNYKUN11.EARN). *Crystal structure determination, direct methods, Patterson methods, computer programming, automation.*

Beyer, Dr Ir Jenö. Dept. of Mechanical Engineering, U. of Twente, P. O. Box 217, 7500 AE Enschede, The Netherlands. (1942) Dr, mat. sci. (U. Twente, 1982). Universitair docent. (tel. 053 + 894232). *Kinetics, crystallography, martensitic transformations.*

Boer, de, Dr Jan Louwert. Lab. voor Anorganische Chemie, U. of Groningen, Nijenborgh 16, 9747 AG Groningen, The Netherlands. (1936) Dr, chemistry (U. Groningen, 1970). Universitair hoofddocent. (tel. 050 + 634424, fax 050 + 634200, email JANDEBOER at HGRRUG52.EARN). *Modulated crystal structures, diffuse scattering, instrumentation.*

Bolhuis, van, Mr Fré. Nijenborgh 16, 9747 AG Groningen, The Netherlands. (1934). Crystallographer. (tel. 050 + 634368 or 634370). *X-ray diffraction, instrumentation.*

Boom, Dr Geert. Vakgroep Techn. Fysica, U. of Groningen, Nijenborgh 18, 9747 AG Groningen, The Netherlands. (1933) Dr, crystallography (U. Groningen,

1966). Universitair hoofddocent. (tel. 050 + 634896). *Scanning electron microscopy, transmission electron microscopy, microprobe elemental analysis.*

Bosman, Drs Wilhelmus P. J. H. Lab. voor Kristallografie, U. of Nijmegen, Toernooiveld, 6525 ED Nijmegen, The Netherlands. (1937) Drs, chemistry (U. Nijmegen, 1969). Res. scient. (tel. 080 + 612591, telex 48228 WINA NL, fax 080 + 553450, email U625008 at HNYKUN11.EARN). *Computer programming, inorganic crystal structures, direct methods.*

Bouwmeester, Dr Henny J.M. Dept. Chemical Technology, U. of Twente, P. O. Box 217, 7500 AE Enschede, The Netherlands. (1954) Dr, inorganic chemistry (U. Groningen, 1988). Universitair docent. (tel. 053 + 892992, fax 053 + 356024). *Inorganic structures, oxides, ionic conductors.*

Braam, Dr Adrianus Wilhelmus Maria. Dept. of Physical Chemistry, CRO-DSM, P. O. Box 16, 6160 MD Geleen, The Netherlands. (1952) Dr, chemistry and physics (U. Groningen, 1981). Head of X-ray dept. (tel. 04494 + 65307 or 66782). *X-ray crystallography, inter- and intra-molecular interactions, morphology, polymers, small angle X-ray scattering.*

Braun, Dr Poul Bernard. Bremdreef 6, 5571 AD Bergeyk, The Netherlands. (1917) Dr, crystallography (U. Amsterdam, 1956). Sr. sci. (tel. 04975 + 1814). *Organic and inorganic crystal structures, Patterson methods.*

Bronsema, Dr Klaas Derk. Corporate Res. Dept. CRL, Arla, AKZO, Velperweg 76, P. O. Box 9300, 6800 SB Arnhem, The Netherlands. (1957) Dr, chemistry (U. Groningen, 1985). Res. scient. (tel. 085 + 663924, telex 45204, fax 085 + 662669). *X-ray diffraction, modulated structures (commensurate and incommensurate), neutron diffraction.*

Bruggen, van, Dr Christiaan Frans. Lab. voor Anorganische Chemie, Materialen Studie Centrum, U. of Groningen, Nijenborgh 16, 9747 AG Groningen, The Netherlands. (1934) Dr, chemistry (U. Groningen, 1969). Universitair hoofddocent. (tel. 050 + 634435). *Transition element solid compounds, structure - physical properties relation.*

Bruins Slot, Dr Hilbert Jan. CAOS/CAMM Center, U. of Nijmegen, Toernooiveld, 6525 ED Nijmegen, The Netherlands. (1956) Dr, chemistry (U. Nijmegen, 1986). Res. scient. (tel. 080 + 613387, telex 48228 WINAT NL, fax 080 + 553450, email CAOS at HNYKUN11.EARN) *Computing, structural chemistry, phase determination, molecular modelling.*

Buschow, Prof. Dr Kurt H. J. Philips Res. Lab., Ned. Philips Bedrijven B.V., P. O. Box 80000, 5600 JA Eindhoven, The Netherlands. (1934) Dr, physical chemistry (Free U. Amsterdam, 1963). Sr. res. (tel. 040 + 743552, telex 35000 PHTC NL). *Intermetallic compounds, crystal structure, magnetic properties, intermetallics, ternary hydrides, amorphous alloys.*

Bijen, Dr Jan M.J.M. INTRON B.V., Inst. for Material and Environmental Res. B.V., Het Rondeel 18, 6219 PG Maastricht, The Netherlands. and: Dept. Material Science, Delft U. of Techn., P. O. Box 5048, 2600 GA Delft, The Netherlands. (1948) Dr, chemistry (U. Utrecht, 1974). Director INTRON / Prof. civil engeneering Delft. (tel. 043 + 254577, ext. 10, telex 56886, fax 043 + 253923). *Gas electron diffraction, microwave spectroscopy, crystallographic structures, inorganic bonding materials.*

Daams, Mr Johannes L.C. Metals Dept., Philips Res. Lab., Ned. Philips Bedrijven B.V., P. O. Box 80000, 5600 JA Eindhoven, The Netherlands. (1942) HBO. (tel. 040 + 742664, fax 040 + 743352). *Crystal structure analysis, intermetallic compounds, powder diffraction, computer aided modelling.*

Deblieck, Dr Rudy André Cornelis. FA-OM, DSM Research, P. O. Box 18, 6160 MD Geleen, The Netherlands. (1956) PhD, physical chemistry (V.U. Brussel, Belgium, 1986). Res. scient. (tel. 04490 + 61661, telex 36777 DSM NL, fax 04490 + 67244, email JOOSTEN at HUTRUU51.EARN). *Polymers, electron microscopy, structure and morphology.*

Delhez, Dr Ir Robert. Fac. der Scheik. Techn. en der Materiaalk., Delft U. of Techn., Rotterdamseweg 137, 2628 AL Delft, The Netherlands. (1940) Dr, chemistry (Delft U. of Techn., 1978). Universitair docent. (tel. 015 + 786730). *Powder diffraction, instrumentation, line profile analysis, materials science.*

Doesburg, Dr Hendrikus M. Lab. Automation, Duphar B.V., C. J. van Houtenlaan 36, P. O. Box 2, 1380 AA Weesp, The Netherlands. (1953) Dr, chemistry (U. Nijmegen, 1984). Automation expert. (tel. 02940 + 77489, telex 14232, fax 02940 + 80253). *Computer programming, crystal structure determination, development of information systems.*

Drenth, Prof. Dr Jan. Lab. voor Chemische Fysica, U. of Groningen, Nijenborgh 16, 9747 AG Groningen, The Netherlands. (1925) Dr, chemistry (U. Groningen, 1957). Prof. (tel. 050 + 634382, telex 53935, fax 050 + 634200). *Structure and action, biological macromolecular systems.*

Driessen, Drs René A. J. Lab. voor Kristallografie, U. of Amsterdam, Nieuwe Achtergracht 166, 1018 WV Amsterdam, The Netherlands. (1959) Drs, chemistry (U. Leiden, 1982). Res. scient. (tel. 020 + 5257040). *Direct methods.*

Duisenberg, Drs Albert Jozef Maria. Lab. voor Kristal- en Structuurchemie, U. of Utrecht, Padualaan 8, P. O. Box 80050, 3508 TB Utrecht, The Netherlands. (1935) Dr, chemistry (U. Utrecht, 1964). Sr. sci. (tel. 030 + 533127, fax 030 + 521877, email DUISENBERG at HUTRUU54.EARN). *Instrumentation, computer programming, organic crystal structures.*

Dijk, van, Dr Cornelis. Physics Dept., Netherlands Energy Res. Foundation, ECN, P. O. Box 1, 1755 ZG Petten NH, The Netherlands. (1934) Dr, physics (U. Leiden, 1970). Group leader. (tel. 02246 + 4576, fax 02246 + 4480). *Static and dynamic structure determination, neutron diffraction, small angle neutron scattering.*

Dijkstra, Dr Bauke Wiepke. Lab. voor Chemische Fysica, U. of Groningen, Nijenborgh 16, 9747 AG Groningen, The Netherlands. (1948) Dr, chemistry (U. Groningen, 1980). Universitair hoofddocent. (tel. 050 + 634381 or 634378, fax 050 + 634200, email BAUKE at HGRRUG52.EARN). *Macromolecules, proteins.*

Edmonds, Dr James William. Philips Export, B.V., Lelweg 1, 7602 EA Almelo, The Netherlands. (1943) PhD,chemistry (Rice U., 1968) International sales manager. (tel. 31 5490 39429) *X-ray powder diffraction, computer automated diffractometry, CD-Rom databases, X-ray flourescence spectroscopy, optical emission spectroscopy.*

Enckevort, van, Dr Wilhelmus Johannus Petrus. Res. Div., Drukker Internationaal, Beversestraat 20, 5431 SH Cuyk, The Netherlands. (1952) Dr, chemistry (U. Nijmegen, 1976). Group leader. (tel. 08850 + 95700, fax 08850 + 16104). *Research on diamonds, crystal growth, defect structures, opto-electrical properties.*

Feil, Prof. Dr Dirk. Chemical Physics Lab., U. of Twente, P. O. Box 217, 7500 AE Enschede, The Netherlands. (1933) Dr, chemistry (U. Utrecht, 1961). Prof. (tel. 053 + 892949, fax 053 + 356024, email FEIL at HENUT5.EARN). *Electron density, molecular interaction, quantum chemistry.*

Felius, Dr Robert Onno. Inst. voor Aardwetenschappen, U. of Utrecht, Budapestlaan 4, P. O. Box 80021, 3508 TA Utrecht, The Netherlands. (1938) Dr, geology (U. Leiden, 1976). Universitair docent. (tel. 030 + 535097, telex 40704 VMLRU NL). *Minerals, materials science, electron microscopy, diffraction.*

Fleischmann, Dr Klaus Dietrich. Scientific Instruments Div., ENRAF-NONIUS, Röntgenweg 1, P. O. Box 483, 2600 AL Delft, The Netherlands. (1937) Dr, physical chemistry (Munich U. of Techn., FRD, 1970). Vice president. (tel. 015 + 698500, ext. 525, telex 38083). *Instrumentation, organic structures.*

Frikkee, Dr Evert. Physics Dept., Netherlands Energy Res. Foundation, ECN, P. O. Box 1, 1755 ZG Petten NH, The Netherlands. (1934) Dr, physics (U. Leiden, 1973). Res. scient. (tel. 02246 + 4527, fax 02246 + 4480). *Magnetism, neutron diffraction, phase transitions.*

Geerestein, van, Dr Vincent Johan. CMC Dept., Organon International B.V., P. O. Box 20, 5340 BH Oss, The Netherlands. (1959) Dr, chemistry (U. Utrecht, 1988). Res. scient. (tel. 04120 + 63092). *Bio-molecular structures, structure - activity relation.*

Gelder, de, Drs René. Chemistry Dept., U. of Leiden, Einsteinweg 5, 2333 CC Leiden, The Netherlands. (1965) Drs, chemistry (U. Leiden, 1988). Res. asst. (tel. 071 + 274414, email ASRGELDER at HLERUL52.EARN). *Direct methods, small structures.*

Gellings, Prof. Dr Paul Johann. Dept. of Inorganic Chemistry and Materials Science, U. of Twente, P. O. Box 217, 7500 AE Enschede, The Netherlands. (1927) Dr, chemistry (U. Amsterdam, 1963). Prof. (tel. 053 + 892861, telex 44200, fax 053 + 893360). *Structure and bonding, coordination compounds, transition metals, catalists, oxidation products.*

Goedkoop, Prof. Dr Jacob A. de Rougemont - Nes 1, 1862 AB Bergen (NH), The Netherlands. (1921) Dr, chemistry (U. Amsterdam, 1952). Prof. Em., U. Leiden (tel. 02208 + 13450). *Neutron diffraction.*

Gorter, Drs Ing. Sybout. Vakgroep ASKA, Gorlaeus Lab., U. of Leiden, P. O. Box 9502, 2300 RA Leiden, The Netherlands. (1940) Drs, chemistry (U. Leiden, 1974). Universitair docent. (tel. 071 + 274415, fax 071 + 274537, email ASKASG at HLERUL2.EARN). *Programming, measurement of intensities, inorganic and organic structures, development of apparatus.*

Goubitz, Drs Kees. Lab. voor Kristallografie, U. of Amsterdam, Nieuwe Achtergracht 166, 1018 WV Amsterdam, The Netherlands. (1953) Drs, chemistry (U. Amsterdam, 1981). Res. scient. (tel. 020 + 5257038). *Crystal structure determination.*

Graaff, de, Dr Rudolf Adriaan Gerard. Vakgroep ASKA, Gorlaeus Lab., U. of Leiden, P. O. Box 9502, 2300 RA Leiden, The Netherlands. (1941) Dr, X-ray crystallography (U. Leiden, 1974). Universitair hoofddocent. (tel. 071 +

274211, email ASRKINNIG at HLERUL52.EARN). *Direct methods, accurate structure factor determination, computer programming, organic crystal structures.*

Grampel, van de, Prof. Dr Johan Christoph. Lab. voor Polymeer Chemie, U. of Groningen, Nijenborgh 16, 9747 AG Groningen, The Netherlands. (1934) Dr, chemistry (U. Groningen, 1967). Associate Prof. (tel. 050 + 634442, telex 53935). *Inorganic polymers, structure, bonding, main group elements.*

Halen, van, Drs Cornelis Jozef Gerardus. FA-OM, DSM Research, P. O. Box 18, 6160 MD Geleen, The Netherlands. (1961) Drs, chemistry (U. Nijmegen, 1986). Res. scient. (tel. 04490 + 61625, telex 36777, fax 04490 + 67244). *Polymer science, orientational studies, WAXS, SAXS.*

Harkema, Dr Sybolt. Chemical Physics Lab., U. of Twente, P. O. Box 217, 7500 AE Enschede, The Netherlands. (1940) Dr, chemistry (U. Twente, 1971). Universitair hoofddocent. (tel. 053 + 893080 or 892950, email HARKEMA at HENUT5.EARN). *Electron density, structure, crown-ethers.*

Hartman, Prof. Dr Piet. Inst. voor Aardwetenschappen, U. of Utrecht, Budapestlaan 4, P. O. Box 80021, 3508 TA Utrecht, The Netherlands. (1922) Dr, crystallography (U. Groningen, 1953). Prof. Em. (tel. 030 + 535092). *Crystal growth, inorganic and mineral structures.*

Havinga, Dr Edsko Enno. Philips Res. Lab., Ned. Philips Bedrijven B.V., P. O. Box 80000, 5600 JA Eindhoven, The Netherlands. (1932) Dr, physical chemistry (U. Groningen, 1957). Sr. sci. (tel. 040 + 742547). *Structure - physical properties relation, organic molecules, organic crystals.*

Haije, Dr Willem Gerrit. Solid-State Physics, Netherlands Energy Res. Foundation, ECN, P. O. Box 1, 1755 ZG Petten, The Netherlands. (1959) Dr, theoretical inorganic chemistry (U. Leiden, 1988). Res. scient. (tel. 02246 + 4548). *Electron density, inorganic crystals, group theory, magnetism, neutron diffraction, phase transitions, symmetry, thermal vibrations.*

Heinerman, Dr Jacobus Johannes Leonardus. Chemical Div. AKZO, Nieuwendammerkade 1-3, P. O. Box 15, 1000 AA Amsterdam, The Netherlands. (1951) Dr, chemistry (U. Utrecht, 1977). Res. manager. (tel. 020 + 212812, fax 341103). *Crystal structure determination, direct methods, catalysis, catalytic cracking, hydrotreating.*

Helmholdt, Dr Robert Barteld. Physics Dept., Netherlands Energy Res. Foundation, ECN, P. O. Box 1, 1755 ZG Petten NH, The Netherlands. (1943) Dr, chemistry (U. Groningen, 1975). Res. scient. (tel. 02246 + 4529, fax 02246 + 4480, email ESU0105 at HPEENR51.EARN). *Neutron diffraction, powder and single crystal diffraction, texture and stress analysis, computing.*

Heijnen, Drs Wilhelmus Marinus Maria. Inst. voor Aardwetenschappen, U. of Utrecht, Budapestlaan 4, P. O. Box 80021, 3508 TA Utrecht, The Netherlands. (1956) Drs, geology (U. Leiden, 1980). Res. scient. (tel. 030 + 535062). *Crystal growth, morphology, calcium oxalates and carbonates, X-ray crystallography.*

Hol, Prof. Dr Wim G. J. Lab. voor Chemische Fysica,, U. of Groningen, Nijenborgh 16, 9747 AG Groningen, The Netherlands. (1945) Dr, protein crystallography (U. Groningen, 1971). Prof. (tel. 050 + 634378, telex 53935 CHRUG NL, fax 050 + 634200). *Biostructural crystallography, molecular modelling, protein engineering, drug design.*

Hornstra, Drs Jan. Heesakkerstraat 22, 5616 SL Eindhoven, The Netherlands. (1927) Drs, chemistry (U. Groningen, 1952). Sr. sci. (tel. 040 + 552448). *Crystal structure determination, diffractometry, lattice defects.*

Huiszoon, Dr Cornelis. Chemical Physics Lab., U. of Twente, P. O. Box 217, 7500 AE Enschede, The Netherlands. (1933) Dr, microwave spectroscopy (U. Nijmegen, 1966). Universitair docent. (tel. 053 + 893081). *Intermolecular forces, quantum mechanics.*

Hummel, van, Ing. Gerrit Jan. Chemical Physics Lab., U. of Twente, P. O. Box 217, 7500 AE Enschede, The Netherlands. (1945). Sr. res. asst. (tel. 053 + 893082, fax 053 + 356024, email HUMMEL at HENUT5.EARN). *Electron density, organic structures, computer programming, instrumentation.*

Janner, Prof. Dr Aloysio. Inst. voor Theoretische Fysica, U. of Nijmegen, Toernooiveld 1, 6525 ED Nijmegen, The Netherlands. (1928) PhD, philosophy (U. Zürich, Switzerland, 1962). Prof. (tel. 080 + 613408 or 612981, fax 080 + 553450, email U633002 at HNYKUN11.EARN or ALO at WN2.SCI.KUN.NL.UUCP). *Solid state physics, group theory, incommensurate structures.*

Jansen, Drs Jacob, Lab. voor Kristallografie, U. of Amsterdam, Nieuwe Achtergracht 166, 1018 WV Amsterdam, The Netherlands. (1959) Drs, theoretical physics (U. Amsterdam). Res. asst. (tel. 020 + 5257030). *Direct methods.*

Jellinek, Prof. Dr Franz. Lab. voor Anorganische Chemie, U. of Groningen, Nijenborgh 16, 9747 AG Groningen, The Netherlands. (1925) Dr, chemistry (U. Utrecht, 1957). Prof. Em., U. Groningen. (tel. 050 + 344332). *Structure - physical properties - composition relation, inorganic structures, modulated structures.*

Kalk, Mr Kornelis Harm. Lab. voor Chemische Fysica, U. of Groningen, Nijenborgh 16, 9747 AG Groningen, The Netherlands. (1944). Crystallographer (tel. 050 + 634239, fax 050 + 634200, email BAUKE at HGRRUG52.EARN). *Biologically important molecules, instrumentation.*

Kamphuis, Dr Irenus Gerhardus. ENR Computing Services for Techn., Netherlands Energy Res. Foundation, ECN, P. O. Box 1, 1755 ZG Petten (NH), The Netherlands. (1952) Dr, protein crystallography (U. Groningen, 1983). Res. scient. (tel. 02246 + 4099, fax 02246 + 1864). *Protein crystallography.*

Kanters, Dr Jan. Lab. voor Kristal- en Structuurchemie, U. of Utrecht, Padualaan 8, P. O. Box 80050, 3508 TB Utrecht, The Netherlands. (1928) Dr, chemistry (U. Utrecht, 1958). Sr. sci. (tel. 030 + 533410, fax 030 + 521877, email KANTERS at HUTRUU54.EARN). *Hydrogen-bond pattern, conformation, saccharides, organic acids.*

Keulen, Dr Evert. Applic. Lab. X-ray Diffraction, Philips Analytical, I & E Division, Lelyweg 1, 7602 EA Almelo, The Netherlands. (1932) Dr, chemistry (U. Groningen, 1969). Res. scient. (tel. 05490 + 39445, telex 36591 XLCALAA, fax 05490 + 39598). *Instrumentation, precision structure analysis, powder diffraction.*

Keijser, de, Dr Ir Thomas Henri. Fac. der Scheik. Techn. en der Materiaalk., Delft U. of Techn., Rotterdamseweg 137, 2628 AL Delft, The Netherlands. (1937) Dr, chemistry (Delft U. of Techn., 1977). Universitair hoofddocent. (tel. 015 + 784105). *Crystallography, diffraction, phase transformations, thin layers.*

Kiers, Dr Conradus Theodorus. Res. and Develepment Lab., ENRAF-NONIUS, Röntgenweg 1, P. O. Box 483, 2600 AL Delft, The Netherlands. (1947) Dr, chemistry (U. Groningen, 1976). Software product specialist. (tel. 015 + 698500, ext. 440, telex 38083, fax 015 + 619574). *Computing, diffractometry, phase determination.*

Kinneging, Dr Albertus Jacobus. Philips Res. Lab., Ned. Philips Bedrijven B.V., P. O. Box 80000, 5600 JA Eindhoven, the Netherlands. (1959) Dr, chemistry (U. Leiden, 1986). Res. scient. (tel. 040 + 742293, telex 35000 PHTC NL, fax 040 + 743783). *Diffraction physics.*

Klop, Drs Enno Anton. Lab. voor Kristal- en Structuurchemie, U. of Utrecht, Padualaan 8, P. O. Box 80050, 3508 TB Utrecht, The Netherlands. (1960) Drs, chemistry (U. Utrecht, 1984). Res. asst. (tel. 030 + 532865, fax 030 + 521877, email KLOP at HUTRUU54.EARN). *Anomalous scattering, X-ray crystal structure determination.*

Koch, Dr Beatrix. Lab. voor Kristallografie, Werkgroep Röntgenfluorescentie Spectrometrie en Poederdiffractie, U. of Amsterdam, Nieuwe Achtergracht 166, 1018 WV Amsterdam, The Netherlands. Dr, crystallography (U. Amsterdam, 1975). Sr. staff member. (tel. 020 + 5256574). *Mineralogical crystallography, powder diffraction, X-ray spectroscopy, carotenoid structures.*

Kolster, Prof. Dr Ir Benjamin Harry. Dept. of Mechanical Engg., U. of Twente, P. O. Box 217, 7500 AE Enschede, The Netherlands. and: Stichting Geavanceerde Metaalkunde, P. O. Box 8039, 7550 KA Hemgelo, The Netherlands. (1938) Dr, material science (Delft. U. of Techn., 1985). Prof. (tel. 053 + 892472 or 892599, fax 053 + 338135). *Corrosion, structure - mechanical properties relation, powder metallurgy, non-waste technology.*

Koningsveld, van, Dr Hendrikus. Afd. der Technische Natuurkunde, Delft U. of Techn., Lorentzweg 1, 2628 CJ Delft, The Netherlands. (1942) Dr, physical chemistry (U. Utrecht, 1970). Universitair hoofddocent (tel. 015 + 782605). *Crystal structures.*

Koopmans, Prof. Dr Kasper. Fac. of Mining and Petroleum Eng., Delft U. of Techn., Mijnbouwstraat 120, 2628 RX Delft, The Netherlands. (1927) Dr, techn. sciences (Eindhoven U. of Techn., 1971). Prof. (tel. 015 + 785001). *Powder diffraction, ores, minerals, rocks.*

Kooijman, Drs Huub. Lab. voor Kristal- en Structuurchemie, U. of Utrecht, Padualaan 8, P. O. Box 80050, 3508 TB Utrecht, The Netherlands. (1964) Drs, chemistry (U. Utrecht, 1987). Res. asst. (tel. 030 + 533407, fax 030 + 521877, email HUUB at HUTRUU54.EARN). *Structural chemistry.*

Krabbendam, Drs Hendrik. Lab. voor Kristal- en Structuurchemie, U. of Utrecht, Padualaan 8, P. O. Box 80050, 3508 TB Utrecht, The Netherlands. (1934) Drs, chemistry (U. Utrecht, 1959). Universitair hoofddocent (tel. 030 + 533414, fax 030 + 521877, email SCHREURS at HUTRUU54.EARN). *Phase determination, organic compounds, macromolecules. protein crystal structures.*

Krever, Dr Maarten. Lab. voor Kristallografie, U. of Amsterdam, Nieuwe Achtergracht 166, 1018 WV Amsterdam, The Netherlands. (1955) Dr, chemistry (U. Leiden, 1989). Res. scient. (tel. 020 + 5257039). *Direct methods, crystal structure determination.*

Kronenburg, Drs Martinus Johannes. Lab. voor Kristallografie, U. of Amsterdam, Nieuwe Achtergracht 166, 1018 WV Amsterdam, The Netherlands. (1964) Drs, physics (U. Utrecht, 1988). Res. asst. Asst. (tel. 020 + 5257032). *Direct methods.*

Kroon, Prof. Dr Jan. Lab. voor Kristal- en Structuurchemie, U. of Utrecht, Padualaan 8, P. O. Box 80050, 3508 TB Utrecht, The Netherlands. (1937) Dr, chemistry (U. Utrecht, 1964). Prof. (tel. 030 + 532383 or 533209, fax 030 + 521877, email KROON at HUTRUU54.EARN). *X-ray diffraction, struturedetermination methods, molecular conformation, hydrogen bonding, computer simulations, biological structure-activity relation.*

Kroon-Batenburg, Dr Louise Maria Johanna. Lab. voor Kristal- en Structuurchemie, U. of Utrecht, Padualaan 8, P. O. Box 80050, 3508 TB Utrecht, The Netherlands. (1956) Dr, chemistry (U. Utrecht, 1985). Res. scient. (tel. 030 + 532533, fax 030 + 521877, email BATE at HUTRUU54.EARN). *Crystal and molecular structure, carbohydrates, hydrogen bonds, molecular mechanics, molecular dynamics.*

Kummer, Drs Ernst Albertus. Fysisch Geografisch en Bodemkundig Lab., U. of Amsterdam, Dapperstraat 115, 1093 BS Amsterdam, The Netherlands. (1932) Drs, physical geography (U. Amsterdam, 1962). Clay mineralogist. (tel. 020 + 5257410 or 5257439). *X-ray diffraction, clay mineralogy, mineralogy, weathering and soilforming processes.*

Loopstra, Prof. Dr Bert Onno. Lab. voor Kristallografie, U. of Amsterdam, Nieuwe Achtergracht 166, 1018 WV Amsterdam, The Netherlands. (1928) Dr, crystallography (U. Amsterdam, 1958). Prof. Em. (tel. 020 + 5257030). *Neutron diffraction, crystal structure determination.*

Lugt, van der, Dr Willem. Solid State Physics Dept.,U. of Groningen, Melkweg 1, 9718 EP Groningen, The Netherlands. (1929) Dr, physics (U. Leiden, 1961). Prof. (tel. 050 + 115427). *Liquid metals, structural research. nuclear magnetic resonance.*

Maaskant, Prof. Dr Willem Johannes Albert. Lab. voor Anorganische Chemie, Gorlaeus Lab., U. of Leiden, P. O. Box 9502, 2300 RA Leiden, The Netherlands. (1932) Dr, theoretical chemistry (U. Leiden, 1963). Prof. (tel. 071 + 274214 or 274450). *Theoretical physics, solid state physics, solid state chemistry, physical and theoretical chemistry, organic and inorganic chemistry, crystallography, applied mathematics.*

Macgillavry, Prof. Dr Carolina H. Mensinge 63, 1083 HE Amsterdam, The Netherlands. (1904) Dr, natural sciences (U. Amsterdam, 1937). Em., U. Amsterdam. (tel. 020 + 443999). *General crystallographic interest.*

Mahy, Dr Jan Willem Gaston. Corporate Res. Dept. CRL, Surface Analysis, AKZO, Velperweg 76, P. O. Box 9300, 6800 SB Arnhem, The Netherlands. (1959) Dr Sc, physics (U. Antwerpen, Belgium, 1987). Res. scient. (tel. 085 + 663924, telex 45204, fax 085 + 662669). *Surface spectroscopy (SIMS, XPS), electron microscopy, electronic materials, inorganic-organic interfaces.*

Malssen van, Drs Kees Frederik. Lab. voor Kristallografie, U. of Amsterdam, Nieuwe Achtergracht 166, 1018 WV Amsterdam, The Netherlands. (1964) Drs, chemistry (U. Amsterdam, 1988). Res. asst. (tel. 020 + 5257030). *Crystal growth, powder diffraction.*

Meetsma, Drs Auke. Afd. Participatie onderzoek, U. of Groningen, Nijenborgh 16, 9747 AG Groningen, The Netherlands. (1946) Drs, physical chemistry (U. Groningen, 1980). Res. asst. (tel. 050 + 634368, fax 050 + 634200, email AMEET at HGRRUG5.EARN). *Crystal structure determination.*

Meurs, van, Dr Ir Frank. Scientific Instruments Div., ENRAF-NONIUS, Röntgenweg 1, P. O. Box 483, 2600 AL Delft, The Netherlands. (1946) Dr, chemistry (Delft U. of Techn., 1978). Head of div. (tel. 015 + 698507, telex 38083, fax 015 + 619574). *Computing, diffractometry, instrumentation, X-ray diffraction, structural chemistry.*

Mittemeijer, Prof. Dr Ir Eric Jan. Lab. of Metallurgy, Delft U. of Techn., Rotterdamseweg 137, 2628 AL Delft, The Netherlands. (1950) Dr, physical chemistry (Delft U. of Techn., 1978). Prof. (tel. 015 + 782207, fax 015 + 786730). *X-ray diffraction, line profile analysis, electron diffraction, electron microscopy, diffusion, thin films, phase transformations, surface coatings, stress analysis.*

Mijlhoff, Dr Frans Cornelis. Vakgroep ASKA, Gorlaeus Lab., U. of Leiden, P. O. Box 9502, 2300 RA Leiden, The Netherlands. (1932) Dr, chemistry (U. Amsterdam, 1964). Universitair hoofddocent. (tel. 071 + 274213). *Gas phase molecular structure, electron microscopy, EXAFS.*

Noordik, Dr Jan Hendrik. CAOS/CAMM Center, U. of Nijmegen, Toernooiveld, 6525 ED Nijmegen, The Netherlands. (1944) Dr, chemistry (U. Nijmegen, 1971). Res. scient. (tel. 080 + 613386, telex 48228, email CAOS at HNYKUN52.EARN). *Crystal structure determination, computer programming, automation in chemistry.*

Northolt, Dr Ir Maurits Gerhard. Corporate Res. Dept., AKZO, Velperweg 76, P. O. Box 9300, 6800 SB Arnhem, The Netherlands. (1939) Dr, crystallography (U. Amsterdam, 1968). Res. scient. (tel. 085 + 664056). *Polymer crystal structures, diffraction, structure - mechanical properties relation.*

Nijveldt, Dr Ir Dick. Dept. of Measurement Development, Delft Hydraulics, Rotterdamseweg 185, 2600 MH Delft, The Netherlands. (1951) Dr, physics (U. Groningen, 1985). Experimental physicist. (tel. 015 + 569353, fax 015 + 619674). *Accurate electron density, structure determination.*

Oen, Prof. Dr Ing Soen. Inst. voor Aardwetenschappen, Free U. of Amsterdam, de Boelelaan 1085, 1081 HV Amsterdam, The Netherlands. (1928) Dr, geology (U. Amsterdam, 1958). Prof. (tel. 020 + 5482915). *Minerals, phase determinations, phase transitions, crystal growth, texture.*

Olthof, Dr Gerrit Jan. Afd. Epidemologie en Informatica, Ministerie van WVC, P.O. Box 5406, 2280 HK Rijswijk, The Netherlands. (1950) Dr, chemistry (U. Amsterdam, 1981). Informatica adv. (tel. 070 + 407190). *Direct methods.*

Olthof-Hazekamp, Drs Roeli. Lab. voor Kristal- en Structuurchemie, U. of Utrecht, Padualaan 8, P. O. Box 80050, 3508 TB Utrecht, The Netherlands. (1937) Drs, chemistry (U. Groningen, 1963). Sr. sci. (tel. 030 + 532865 or 532869, fax 030 + 521877, email ROH at HUTRUU54.EARN). *Computer programming.*

Peerdeman, Prof. Dr Antonius Franciscus. Prof. L. Fuchslaan 41, 3571 HE Utrecht, The Netherlands. (1921) Dr, X-ray crystallography (U. Utrecht, 1955). Prof. Em., U. Utrecht. (tel. 030 + 712460). *X-ray crystallography, apparatus, direct methods, anomalous scattering, molecular conformations, intermolecular interactions, thermodynamics.*

Perdok, Prof. Dr Wiepko Gerhardus. Dental School, Lab. of Material Technica, Ant. Deusinglaan 1, 9713 AV Groningen, The Netherlands. (1914) Dr, chemistry, crystallography (U. Groningen, 1942). Prof. Em., U. Groningen. (tel. 050 + 633141). *Crystal optics, crystal growth, X-ray diffraction applications.*

Peschar, Dr René. Lab. voor Kristallografie, U. of Amsterdam, Nieuwe Achtergracht 166, 1018 WV Amsterdam, The Netherlands. (1956) Dr, chemistry (U. Amsterdam, 1987). Res. scient. (tel. 020 + 5257040). *Direct methods, computing methods.*

Peterse, Ir Wilhelmus J. A. M. Afd. der Technische Natuurkunde, Delft U. of Techn., Lorentzweg 1, 2628 CJ Delft, The Netherlands. (1934) Ir, physics (Delft U. of Techn., 1963). Universitair hoofddocent. (tel. 015 + 782405 or 784267, email INMSWJP at HDETUD1.EARN). *X-ray diffraction, inorganic crystals, phase transitions, instrumentation, computing.*

Plas, van der, Prof. Dr Leendert. Dept. of Soil Science and Geology, Wageningen U. of Agriculture, Duivendaal 10, P. O. Box 37, 6700 AA Wageningen, The Netherlands. (1928) Dr, petrography and mineralogy (U. Leiden, 1959). Prof. (tel. 08370 + 84415). *Clay mineralogy, feldspar properties, clay geochemistry, zeolites, salt minerals.*

Pontenagel, Dr Wilbert M. G. F. Dept. Material Techn., DSM Research B.V., P. O. Box 18, 6160 MD Geleen, The Netherlands. (1954) Dr, crystallography (U. Utrecht, 1983). Res. manager. (tel. 04490 + 61768, fax 04490 + 67244). *X-ray crystal structure determination, polymer science.*

Popma, Prof. Dr Theo Johan August. Dept. of Applied Physics and Electrical Engeneering, U. of Twente, P. O. Box 217, 7500 AE Enschede, The Netherlands. (1941) Dr, solid state chemistry (U. Groningen, 1970). Prof. (tel. 053 + 892747). *Materials science.*

Preistnall, Mr. Ian. Publisher D Reidel Publishing Company, Spuiboulevard 50, PO Box 17, 3300 AA Dordrecht, The Netherlands.

Prick, Dr Petrus Antonius Johannes. Hiddemaheerd 16, 9737 JN Groningen, The Netherlands. (1952) Dr, chemistry (U. Nijmegen, 1979). *Direct methods, refinement, protein structures.*

Reiss, Drs Céleste A. Lab. voor Kristallografie, U. of Amsterdam, Nieuwe Achtergracht 166, 1018 WV Amsterdam, The Netherlands. (1952) Drs, chemistry (U. Amsterdam, 1979). Res. scient. (tel. 020 + 5257038). *Direct methods, crystal structure determination.*

Ridder, de, Drs Dirk. Lab. voor Kristallografie, U. of Amsterdam, Nieuwe Achtergracht 166, 1018 WV Amsterdam, The Netherlands. (1959) Licentiaat (U. Antwerpen, Belgium, 1986). Res. scient. (tel. 020 + 5257036). *Crystal structure determination, structure-activity relation, odor compounds.*

Rieck, Prof. Dr Gerard Daniel. Mecklenburglaan 5, 5583 AG Waalre, The Netherlands. (1911) Dr, chemistry (U. Utrecht, 1945). Prof. Em., Eindhoven U. of Techn. (tel. 04904 + 13768). *Reactions, recrystallization, grain growth, textures, metals, oxide solids.*

Rietveld, Dr Hugo Marie. Dienst Technische en Wetenschappelijke Informatie, Netherlands Energy Res. Foundation, ECN, P. O. Box 1, 1755 ZG Petten, The Netherlands. (1932) Dr, physics (U. Western Aust., Australia, 1964). Head. (tel. 02246 + 4365). *Data base technology, information retrieval.*

Romers, Prof. Dr Cornelis. Nachtegaallaan 17, 2172 JP Sassenheim, The Netherlands. (1919) Dr, chemistry (U. Amsterdam, 1948). Prof. Em., U. Leiden.

Rutten-Keulemans, Drs Elisabeth Wilhelmina Maria. Vakgroep ASKA, Gorlaeus Lab., U. of Leiden, P. O. Box 9502, 2300 RA Leiden, The Netherlands. (1932) Drs, chemistry (U. Leiden, 1959). Sr. sci. (tel. 071 + 123923 or 274211). *Computer programming.*

Schagen, Dr Jan-Dirk. Scientific Instruments Div., ENRAF-NONIUS, Röntgenweg 1, P. O. Box 483, 2600 AL Delft, The Netherlands. (1951) Dr, crystallografphy (U. Amsterdam, 1986). Applic. scient. (tel. 015 + 698678, telex 38083, fax 015 + 619574). *Computer programming.*

Schapink, Dr Frederik Willem. Lab. of Metallurgy, Delft U. of Techn., Rotterdamseweg 137, 2628 AL Delft, The Netherlands. (1931) Dr, physics (Delft U. of Techn., 1969). Universitair hoofddocent (tel. 015 + 782272). *Electron diffraction, electron microscopy, X-ray diffraction, physical metallurgy.*

Schenk, Prof. Dr Hendrik. Lab. voor Kristallografie, U. of Amsterdam, Nieuwe Achtergracht 166, 1018 WV Amsterdam, The Netherlands. (1939) Dr, chemistry (U. Amsterdam, 1969). Prof. (tel. 020 + 5257035, fax 020 + 5255802, email U00082 at HASARA5.EARN). *Direct methods, crystal structure determination.*

Schierbeek, Dr Abraham Johan. Scientific Instruments Div., ENRAF-NONIUS, Röntgenweg 1, P. O. Box 483, 2600 AL Delft, The Netherlands. (1955) Dr,protein crystallography (U. Groningen, 1988). Appl. scient. (tel. 015 + 698500, ext. 678, telex 38083, fax 015 + 619574). *Protein crystallography, area detectors, structural biology.*

Schoone, Prof. Dr Jean C. Groenlinglaan 80, 3722 VB Bilthoven, The Netherlands. (1919) Dr, chemistry (U. Utrecht, 1950). Prof. Em., U. Utrecht. (tel. 030 + 293799). *Computer programming.*

Schooneveld, van, Ing. Marinus. Scientific Instruments Div., ENRAF-NONIUS, Röntgenweg 1, P. O. Box 483, 2600 AL Delft, The Netherlands. (1943) Ing., electrical engineering. Product specialist. (tel. 015 + 698503, telex 38083, fax 015 + 619574). *Instrumentation.*

Schouten, Mr Arie. Lab. voor Kristal- en Structuurchemie, U. of Utrecht, Padualaan 8, P. O. Box 80050, 3508 TB Utrecht, The Netherlands. (1956). Techn. asst. (tel. 030 + 533122, fax 030 + 521877, email SCHOUTEN at HUTRUU54.EARN). *X-ray diffraction, powder diffraction, structural chemistry,* *instrumentation, computing, crystallographic data, symmetry, organic compounds, crystal growth, phase transitions.*

Schreuder, Dr Herman Antony. Lab. voor Chemische Fysica, U. of Groningen, Nijenborgh 16, 9747 AG Groningen, The Netherlands. (1958) Dr, biochemistry (U. Groningen, 1983). Res. scient. (tel. 050 + 634378, fax 050 + 634200). *Protein crystallography.*

Schreurs, Drs Antonius Mathias Maria. Lab. voor Kristal- en Structuurchemie, U. of Utrecht, Padualaan 8, P. O. Box 80050, 3508 TB Utrecht, The Netherlands. (1955) Drs, chemistry (U. Utrecht, 1981). Res. scient. (tel. 030 + 533902 or 532869, fax 030 + 521877, email SCHREURS at HUTRUU54.EARN). *Computer programming, computer graphics, crystal structure statistics.*

Schutte, Drs Willy. Lab. voor Anorganische Chemie, U. of Groningen, Nijenborgh 16, 9747 AG Groningen, The Netherlands. (1960) Drs, physical and inorganic chemistry (U. Utrecht, 1985). Res. asst. (tel. 050 + 634457, fax 050 + 634200, email WANGO at HGRRUG5.EARN). *X-ray diffraction, modulated structures, diffuse scattering, inorganic crystals, materials science, computing, minerals, phase transitions, symmetry.*

Seal, Dr Michael. D. Drukker & ZN. N.V., P.O. Box 15120, Amsterdam 1001 MG, The Netherlands. (1930) PhD, physics (Cambridge U., UK, 1957). Res. dir. (tel. 020 + 267321, telex 14143, fax 020 + 248650). *Crystal growth, electron microscopy, high pressures, instrumentation, materials science, microscopy, surface structure.*

Sluis, van der, Drs Paul. Lab. voor Kristal- en Structuurchemie, U. of Utrecht, Padualaan 8, P. O. Box 80050, 3508 TB Utrecht, The Netherlands. (1962) Drs, chemistry (U. Utrecht, 1985). Res. asst. (tel. 030 + 533407 or 532869, fax 030 + 521877, email SLUP at HUTRUU54.EARN). *Crystal growth, organic compounds.*

Smaalen, van, Dr Sander. Lab. voor Anorganische Chemie, U. of Groningen, Nijenborgh 16, 9747 AG Groningen, The Netherlands. (1958) Dr, chemistry (U. Groningen, 1985). Res. fellow. (tel. 050 + 634457, fax 050 + 636200, email SMASH at HGRRUG50.EARN). *Modulated structures, quasi crystals, physical properties of solids.*

Smeets, Drs Wilberthus J. J. Lab. voor Kristal- en Structuurchemie, U. of Utrecht, Padualaan 8, P. O. Box 80050, 3508 TB Utrecht, The Netherlands. (1958) Drs, chemistry (U. Utrecht, 1985). Res. scient. (tel. 030 + 532533 or 532869, fax 030 + 521877, email SMEW at HUTRUU54.EARN). *X-ray structure determinations.*

Smit, Dr Paul H. I & E, Ned. Philips Bedrijven B.V., Bldg TQ III, P. O. Box 80000, 5600 MD Eindhoven, The Netherlands. (1949) Dr, chemistry (U. Utrecht, 1978). Marketing manager. (tel. 040 + 757289). *Signal processing.*

Smits, Mr Johannes Martinus Maria. Lab. voor Kristallografie, U. of Nijmegen, Toernooiveld, 6525 ED Nijmegen, The Netherlands. (1948). Techn. asst. (tel. 080 + 612030, telex 48228 WINA NL, fax 080 + 553450, email U625021 at HNYKUN11.EARN). *Inorganic and organic crystal structures, computer programming.*

Smoorenburg, Mr Henricus C. A. M. Philips Res. Lab., Ned. Philips Bedrijven B.V., P. O. Box 80000, 5600 JA Eindhoven, The Netherlands. (1951). Res. scient. (tel. 040 + 742623). *X-ray diffraction, structure, texture.*

Smout, Ing. Scientific Instruments Div., ENRAF-NONIUS, Röntgenweg 1, P. O. Box 483, 2600 AL Delft, The Netherlands. Product specialist. (tel. 015 + 698500, ext. 698, telex 38083, fax 015 + 619574). *Instrumentation, X-ray diffraction, powder diffraction, texture.*

Sonneveld, Mr Eduard Jan. X-ray Dept., Technisch Physische Dienst TNO-TH, Stieltjesweg 1, P. O. Box 155, 2600 AD Delft, The Netherlands. (1945). Res. asst. (tel. 015 + 787005). *Powder diffraction.*

Spek, Dr Anthony Louis. Lab. voor Kristal- en Structuurchemie, U. of Utrecht, Padualaan 8, P. O. Box 80050, 3508 TB Utrecht, The Netherlands. (1944) Dr, chemistry (U. Utrecht, 1975). Universitair docent. (tel. 030 + 532538, fax 030 + 521877, email SPEA at HUTRUU54.EARN). *Direct methods, automation, computer programming, computer graphics, organic and organometallic structures.*

Stam, Dr Casper Hendrik. Lab. voor Kristallografie, U. of Amsterdam, Nieuwe Achtergracht 166, 1018 WV Amsterdam, The Netherlands. (1925) Dr, chemistry (U. Amsterdam, 1963). Universitair hoofddocent. (tel. 020 + 5257033). *Crystal structure determination, crystal optics.*

Stouten, Drs Pieter F. W. Lab. voor Kristal- en Structuurchemie, U. of Utrecht, Padualaan 8, P. O. Box 80050, 3508 TB Utrecht, The Netherlands. (1959) Drs, chemistry (U. Utrecht, 1985). Res. asst. (tel. 030 + 532866, fax 030 + 521877, email PIER at HUTRUU54.EARN). *Hydrogen bonding, conformational analysis, molecular dynamics, computer simulations.*

Straver, Drs Leonardus Hendrikus. Scientific Instruments Div., ENRAF- NONIUS, Röntgenweg 1, P. O. Box 483, 2600 AL Delft, The Netherlands. (1954) Drs, chemistry (U. Utrecht, 1980). Application scient. (tel. 015 + 698678, telex 38083, fax 015 + 619574). *Computer programming, instrumentation, organometallics, conformational analysis.*

Struikmans, Drs Rink. Fac. der Techn. Natuurkunde, Delft U. of Techn., Lorentzweg 1, 2628 CJ Delft, The Netherlands. (1941) Drs, physics (Free U. of Amsterdam, 1970). Universitair hoofddocent (tel. 015 + 784098, telex 38151 BUTUD, fax 015 + 78251). *Phase transitions, optical crystallography, crystal physics, instrumentation.*

Thijsse, Dr Barend Jan. Fac. der Scheik. Techn. en der materiaalk., Delft U. of Techn., Rotterdamseweg 137, 2628 AL Delft, The Netherlands. (1950) Dr, physics (U. Leiden, 1978). Universitair docent. (tel. 015 + 782221). *X-ray diffraction, neutron diffraction, non-crystalline solids, metallic glasses, structural relaxation.*

Timmers, Drs Jacob. Lab. voor Technische Natuurkunde, Delft U. of Techn., Lorentzweg 1, 2628 CJ Delft, The Netherlands. (1961) Drs, physics (U. Leiden, 1985). Res. scient. (tel. 015 + 781914, fax 015 + 783251). *X-ray diffractometry apparatus, powder diffraction.*

Tuinstra, Prof. Dr Ir Fokke. Afd. der Technische Natuurkunde, Delft U. of Techn., Lorentzweg 1, 2628 CJ Delft, The Netherlands. (1934) Dr, physics (Delft U. of Techn., 1967). Prof. (tel. 015 + 786112 or 784276). *Structure - properties relation, crystal physics.*

Veen, van der, Dr Adriaan Hendrik. Inst. voor Aardwetenschappen, U. of Utrecht, Budapestlaan 4, P. O. Box 80021, 3508 TA Utrecht, The Netherlands. (1925) Dr, mineralogy, petrology (Delft U. of Techn., 1963). Added res. (tel. 030 - 535090, telex 40704). *Quantitative X-ray diffraction (compositions), high temperature X-ray diffraction, differential thermal analysis, thermal gravimetric analysis, quantitative microscopy (stereology), petrology, mineralogy, ore-microscopy, geochemistry, economic geology.*

Veld, In 't, Ir Gerard Adriaan. R & D PHI, ENRAF-NONIUS, P. O. Box 483, 2600 AL Delft, The Netherlands. (1956) Ir, electronics (Delft U. of Techn., 1982). Head product specific group. (tel. 015 + 698432, fax 619574). *Hardware development, computers, software.*

Versteeve, Dr Abraham Jan. Inorg. Materials Dept., Billiton Res. B.V., Westervoortsedijk 67, P. O. Box 40, 6800 AA Arnhem, The Netherlands. (1942) Dr, petrology (U. Utrecht, 1974). Dept. head. (tel. 085 + 654370, telex 75026 BIRES NL, fax 085 + 640041). *Ceramics, refractories, industrial minerals.*

Verwer, Drs Paul. Lab. voor Kristal- en Structuurchemie, U. of Utrecht, Padualaan 8, P. O. Box 80050, 3508 TB Utrecht, The Netherlands. (1965) Drs, chemistry (U. Utrecht, 1988). Res. asst. (tel. 030 + 532865, fax 030 + 521877, email VERWER at HUTRUU54.EARN). *Structure determination methods.*

Visser, Drs Jan Willem. Henry Dunantlaan 81, 2614 GL Delft, The Netherlands. (1925) Drs, chemistry (U. Amsterdam, 1955). Retired. (tel. 015 + 123593). *Powder diffraction, automation.*

Visser, Dr Rudolph Joseph Jacobus. Lab. voor Techn. Natuurkunde, Delft U. of Techn., Lorentzweg 1, 2628 CJ Delft, The Netherlands. (1953) Dr, chemistry (U. Groningen, 1984). Software engineer. (tel. 015 + 781914, fax 015 + 783251). *Powder diffraction, instrumentation, software development.*

Vlak, Dr Wim. Physics Dept., Netherlands Energy Res. Foundation, ECN, P. O. Box 1, 1755 ZG Petten, The Netherlands. (1957) Dr, physics (U. Utrecht). Res. scient. (tel. 02246 + 4532, fax 02246 + 4480). *Small-angle scattering, neutron diffraction, materials science, macromolecules, computing.*

Vonk, Dr Christ Gijsbertus. Beatrixlaan 12, 6165 CX Geleen, The Netherlands. (1925) Dr, physical chemistry (U. Groningen, 1957). (tel. 04490 + 42153). *Polymers, small angle X-ray scattering.*

Voort, van der, Drs Elisabeth. Inst. voor Aardwetenschappen, U. of Utrecht, Budapestlaan 4, P. O. Box 80021, 3508 TA Utrecht, The Netherlands. (1962) Drs, chemistry (U. Utrecht, 1986). Res. asst. (tel. 030 + 535062, email WGDKRI3 at HUTRUU0.EARN). *Crystal growth, surface structure.*

Vos-Looyenga, Dr Aafje. Roland Holstlaan 908, 2624 JK Delft, The Netherlands. (1928) Dr, structural chemistry (U. Groningen, 1952). (tel. 015 + 566590). *Accurate structure determinations, oligopeptides.*

Vries, de, Drs Johan Louis. Humperdincklaan 51, 5654 PB Eindhoven, The Netherlands. (1920) Drs, chemistry (U. Amsterdam, 1950). (tel. 040 + 522102). *Instrumentation, powder diffraction in industry.*

Vucht, van, Dr Johannes Hendrikus Nicolaas. Isodorusweg 23, 5624 KD Eindhoven, The Netherlands. (1924) Dr, technical sciences (Eindhoven U. of Techn., 1963). Sr. res. (tel. 040 + 446196). *intermetallic compounds, inorganic compounds, structure - physical properties relation.*

Waal, van de, Benjamin Willem. Dept. of Physics, CT1324, U. of Twente, P. O. Box 217, 7500 AE Enschede, The Netherlands. (1936) Ir, physics (Delft U. of Techn., 1966). Sr. sci. (tel. 053 + 892954, fax 053 + 356024). *Molecular packing, intermolecular forces, atomic and molecular clusters.*

Wagner, Dr Anton Johan. Lab. voor Chemische Fysica, U. of Groningen, Nijenborgh 16, 9747 AG Groningen, The Netherlands. (1933) Dr, chemistry (U. Groningen, 1966). Universitair hoofddocent. (tel. 050 + 634376). *X-ray diffraction, molecular structure, quantum chemistry.*

Wal, van der, Dr Robert Jan. SARA, Ondersteuning, U. of Amsterdam, Kruislaan 415, 1098 SJ Amsterdam, The Netherlands. (1955) Dr, chemistry (U. Groningen, 1982). Sr. system progr. *X-ray crystallography, accurate electron density determination, electron density distribution interpretation, supercomputers.*

Wiebenga, Prof. Dr Eelco Herman. 5 Allée des Cimes (la Pinède), 83420 La Croix-Valmer, France. (1913) Dr, chemistry (U. Utrecht, 1940). Prof. Em., U. Groningen. (tel. 94 + 796853). *Quantum chemistry.*

Wiegers, Dr Gerrit Adriaan. Lab. voor Anorganische Chemie, U. of Groningen, Nijenborgh 16, 9747 AG Groningen, The Netherlands. (1930) Dr, chemistry (U. Groningen, 1963). Universitair hoofddocent. (tel. 050 + 634433, tel 050 + 634200). *Inorganic crystal structures, solid state chemistry. phase transitions, powder diffraction.*

Wit, de, Drs Martin. Lab. voor Kristallografie, U. of Amsterdam, Nieuwe Achtergracht 166, 1018 WV Amsterdam, The Netherlands. (1963) Drs, chemistry (U. Amsterdam, 1987). Res. asst. (tel. 020 + 5257036). *Powder diffraction, computing, phase transitions.*

With, de, Prof. Dr Gijsbertus. Philips Res. Lab., Ned. Philips Bedrijven B.V., P. O. Box 80000, 5600 JA Eindoven, The Netherlands. (1950) Dr, chemical physics (U. Twente, 1978). Prof. (tel. 040 + 742132, fax 040 + 744282). *Oxides, nitrides, carbides, mechanical and chemical properties.*

Woensdregt, Drs Cornelis Franciscus. Inst. voor Aardwetenschappen, U. of Utrecht, Budapestlaan 4, P. O. Box 80021, 3508 TA Utrecht, The Netherlands. (1937) Drs, geology (U. Leiden, 1963). Universitair hoofddocent. (tel. 030 + 535070, email WGDKRI1 at HUTRUU0.EARN). *Crystal growth, crystal morphology, electron microscopy, X-ray diffraction, computer programming.*

Wolff, de, Dr Pieter Maarten. Meermanstraat 126, 2614 AM Delft, The Netherlands. (1919) Dr, physics (Delft U. of Techn., 1951). Prof. Em., Delft U. of Techn. (tel. 015 + 120396). *Symmetry, phase transitions.*

Zeedijk, Ir Hendrik Bastiaan. Metaalinst. TNO, Laan van Westenenk 501, 7334 DT Apeldoorn, The Netherlands. (1936) Ir, physical chemistry (Delft U. of Techn., 1960). Sr. scent. (tel. 055 + 493493). *Electron microscopy, metallography, micro-analysis.*

Zoutberg, Drs Martinus C. Lab. voor Kristallografie, U. of Amsterdam, Nieuwe Achtergracht 166, 1018 WV Amsterdam, The Netherlands. (1960) Drs, chemistry (U. Amsterdam, 1986). Res. asst. (tel. 020 + 5257041). *Direct methods.*

Zwaan, Dr Pieter Cornelis. National Museum of Geology and Mineralogy, Hooglandse Kerkgracht 17, 2312 HS Leiden, The Netherlands. (1928) Dr, mineralogy (U. Leiden, 1955). Universitair hoofddocent. keeper of minerals. (tel. 071 + 143844). *X-ray diffraction, instrumentation, gem minerals.*

NEW ZEALAND

Sub-Editor: **C.E.F. Rickard**

Notes

1. International telephone country code - 64

2. Degrees conferred by New Zealand universities are generally similar to British Degrees.

3. Department of Scientific and Industrial Research is abbreviated to D.S.I.R.

Aldridge, Dr Laurence Philip. Chemistry Div., DSIR, Private Bag, Petone, New Zealand. (1945) PhD, chemistry (Otago U., 1971). Res. Chemist. (tel. 4 + 666-919, ext. 470). *Zeolites, cement.*

Anderson, Dr Bryan Frederick. Chemistry and Biochemistry Dept., Massey U., Palmerston North, New Zealand. (1937) PhD, chemistry (Auckland U., 1967). Snr. Research officer. (tel. 63 + 69-099, ext. 7551, fax (63) 62140, e-mail B.ANDERSON at NZ.AC.MASSEY). *Protein structures.*

Baker, Dr Edward Neill. Chemistry and Biochemistry Dept., Massey U., Palmerston North, New Zealand. (1942) PhD, chemistry (Auckland U., 1968). Sr. lect. (tel. 63 + 69-099, ext. 7773, fax (63) 62140, e-mail T.BAKER at NZ.AC.MASSEY). *Protein structures.*

Bates, Prof. Richard Heaton Tunstall. Dept. of Electrical Eng., U. of Canterbury, Private Bag, Christchurch, New Zealand. (1929) PhD, engineering (London U., UK, 1972). Prof. (tel. 3 + 482-009, ext. 7278, fax (0064 3) 430326). *Image processing, computationally-orientated diffraction theory, computer modelling, applied Fourier theory.*

Beckingsale, Mr Peter Gerard. Chemistry Dept., U. of Auckland, Private Bag, Auckland, New Zealand. (1952) MSc, chemistry (Auckland U., 1976). Technician. (tel. 9 + 792-300, ext. 9274). *Computing, organic and organometallic compounds.*

Brown, Dr Kevin Laurie. Geothermal Research Centre, D.S.I.R., Private Bag, Taupo, New Zealand. (1946) PhD, chemistry (Auckland U., 1972). Officer in Charge. (tel. 074 + 48-211, fax (074) 48199). *Inorganic and organic crystal structures.*

Childs, Dr Cyril Walter. Soil Bureau, D.S.I.R., Private Bag, Lower Hutt, New Zealand. (1941) PhD, chemistry (Otago U., 1967). Sci. (tel. 4 + 673-119, ext. 8872). *Soil mineralogy, Mossbauer spectroscopy.*

Churchman, Dr Gordon John. Soil Bureau, D.S.I.R., Private Bag, Lower Hutt, New Zealand. (1944) PhD, chemistry (Otago U., 1970). Res. chemist. (tel. 4 + 673-119, ext. 876). *Clay mineralogy, surface chemistry.*

Claridge, Dr Graeme Geoffrey. Soil Bureau, D.S.I.R., Private Bag, Lower Hutt, New Zealand. (1931) PhD, chemistry (Auckland U., 1955). Res. chemist. (tel. 4 + 673-119). *Clay mineral formation in soils.*

Clark, Dr George Raymond. Chemistry Dept., U. of Auckland, Private Bag, Auckland, New Zealand. (1942) PhD, chemistry (Auckland U., 1968). Assoc. prof. (tel. 9 + 737999, ext. 8294, fax (0064 9) 33429, email CHX119R at AUCC1.AUKUNI.AC.NZ). *Organometallic structures, bioinorganic structures.*

Coombs, Prof. Douglas Saxon. Geology Dept., U. of Otago, P.O. Box 56, Dunedin, New Zealand. (1924) Hon. DSc, (Geneva U., Switzerland, 1974). Prof. (tel. 24 + 797-520, fax (024) 741607). *Mineralogy, metamorphic and volcanic rocks, zeolites.*

Cooper, Dr Alan Frederick. Geology Dept., U. of Otago, P.O. Box 56, Dunedin, New Zealand. (1945) PhD, petrology (Otago U., 1970). Lect. (tel. 24 + 791-100, ext. 7515, fax (024) 741607). *Mineralogy, metamorphic petrology, lamprophyre and carbonatite mineralogy and petrology.*

Couldwell, Dr Margaret Claire. Computer Centre, Massey U., Palmerston North, New Zealand. (1950) PhD, chemistry (Canterbury U., 1974). Computer Scient. (tel. 63 + 69099, ext. 8564). *Structural chemistry, organometallic compounds, molecular complexes.*

Cutfield, Dr John Franklin. Biochemistry Dept., U. of Otago, P.O. Box 56, Dunedin, New Zealand. (1945) PhD, chemistry (Auckland U., 1970). Sr. lect. (tel. 24 + 771-640, ext. 618). *Protein structure-function studies.*

Evans, Mr David Lindsay. 3a Snowdon Road, Christchurch, New Zealand. (1949) MSc, chemistry (Canterbury U., 1981). Medical scientist. (tel. 3 + 516-134). *Charge transfer complexes, structure and physical properties, computer management techniques, data analysis.*

Freeman, Dr Alan George. Chemistry Dept., Victoria U. of Wellington, Private Bag, Wellington, New Zealand. (1935) PhD, chemistry (Aberdeen U., UK, 1962). Sr. lect. (tel. 4 + 721-000, ext. 772). *Solid state reactions, layer structures, defect structures.*

Gainsford, Dr Graeme John. Chemistry Div., D.S.I.R., Private Bag, Petone, New Zealand. (1945) PhD, chemistry (Canterbury U., 1969). Scientist. (tel. 4 + 666-919, ext. 682, fax (04) 694500, email SRGCMGR at GRV.GOVT.NZ). *Computing techniques, crystal structure, powder diffraction.*

Hall, Dr David. U. Grants Committee, P.O. Box 12-348, Wellington, New Zealand. (1928) DSc, chemistry (Auckland U., 1969). Chairman. (tel. 4 + 728-600). *Molecular packing, conformation.*

Jones, Dr Tony Cristofer. Chemistry Dept., U. of Auckland, Auckland, New Zealand. (1955) PhD, chemistry (York U., UK, 1980). Techn. officer. (tel. 9 + 737-999, ext. 8274). *Computing, pulsed NMR, metal hydrides, hydrogen energy.*

Kawachi, Dr Yosuke. Geology Dept., U. of Otago, P.O. Box 56, Dunedin, New Zealand. (1932) PhD, geology (Otago U., 1970). Sr. res. officer. (tel. 24 + 79-7507, fax (024) 741607). *Petrology, metamorphism, mineralogy.*

Kirkman, Dr John Henry. Soil Sci. Dept., Massey U., Palmerston North, New Zealand. (1938) PhD, mineralogy (Aberdeen U., UK, 1965). Reader. (tel. 63 + 69-099). *Clay mineralogy.*

Lyons, Dr Karen. Food Techn. Dept., Massey U., Private Bag, Palmerston North, New Zealand. (1955) PhD, chemistry (Auckland U., 1981). Res chemist (tel. 63 + 69-099, ext. 2448). *Organometallic structures.*

Lyons, Mr Paul John. Computer Sci. Dept., Massey U., Palmerston North, New Zealand. (1953) MSc, chemistry (Auckland U., 1977). Lect. (tel. 63 + 69-099). *Molecular packing analysis, artificial intelligence applied to chemistry, protein structure, programming languages.*

March, Dr Frank Conroy. Computer Services Centre, Victoria U., Private Bag, Wellington, New Zealand. (1944) PhD, crystallography (Canterbury U., 1970). Director. (tel. 4 + 721-000, fax (4) 712070, email director at RS1.VUW.AC.NZ). *Organometallic structure.*

Maslen, Dr Hugh Stafford. Chemistry Dept., U. of Auckland, Private Bag, Auckland, New Zealand. (1924) PhD, chemistry (Auckland U., 1974). Sr. lect. (tel. 9 + 737999, ext. 8291). *Structural chemistry, molecular conformation.*

McKee, Dr Vickie. Chemistry Dept., U. of Canterbury, Private Bag, Christchurch, New Zealand. (1955) PhD, Inorganic Chemistry (Queens U., Belfast, 1979). Sr. Lect. (tel 3 + 667-001, ext. 7443, fax (0064 3) 483308, email V.MCKEE%CANTERBURY.AC.NZ%RELAY.CS.NET). *Macrocyclic complexes, bioinorganic structure analysis.*

Nicholson, Dr Brian Kenneth. Chemistry Dept., U. of Waikato, Hamilton, New Zealand. (1947) PhD, chemistry (U. of Otago, 1973). Sr. Lect. (fax (071) 60135, email B.NICHOLSON at NZ.AC.WAIKATO). *Inorganic and organometallic structures.*

Norris, Dr Gillian Erna. Chemistry and Biochemistry Dept., Massey U., Palmerston North, New Zealand. (1948) PhD, chemistry(Massey U., 1982). Research officer. (tel. 63 + 69-099, ext. 7551, fax (63) 62140, email G.NORRIS at NZ.AC.MASSEY). *Protein crystallography.*

Oliver, Dr Peter John. N.Z. Geological Survey, P.O. Box 30368, Lower Hutt, New Zealand. (1948) PhD, geochemistry (Canterbury U., 1977). Geologist. (tel. 4 + 699-059). *Petrology and geochemistry, volcanic rocks, hydrothermal geochemistry and crystal chemistry, paleomagnetism.*

Page, Dr Campbell Thomas. Textile Chemistry Dept., Wool Research Organisation of N.Z., Private Bag, Christchurch, New Zealand. (1951) PhD, chemistry (Canterbury U., 1979). Res. scientist. (tel. 3 + 252-421). *Structure, transition metal complexes containing sulphur donor ligands.*

Parry, Prof. David Anthony Dougall. Physics and Biophysics Dept., Massey U., Palmerston North, New Zealand. (1948) DSc, biophysics (London U., UK, 1982). Prof. (tel. 63 + 69-099, ext. 7551, fax (063) 62140, email D.PARRY at MASSEY.AC.NZ.CSNET). *Structural and functional studies, proteins, muscle, collagen, intermediate filaments.*

Penfold, Prof. Bruce Russell. Chemistry Dept., U. of Canterbury, Private Bag, Christchurch, New Zealand. (1927) PhD, chemistry (Cambridge U., UK, 1952). Prof. (tel. 3 + 482-009). *Organometallic and inorganic structures, computer aided teaching, information retrieval.*

Percival, Dr Henry Joseph. NZ Soil Bureau, DSIR, Eastern Hutt Rd., Taita, Private Bag, Lower Hutt, New Zealand. (1943) PhD, chemistry (Victoria U., 1970). Scientist. (tel. 4 + 673-119, ext. 857, fax (04) 673114). *Soils - physical chemistry, clay mineralogy.*

Rickard, Dr Clifton Edward Frank. Chemistry Dept., U. of Auckland, Private Bag, Auckland, New Zealand. (1941) PhD, chemistry (Auckland U., 1967). Sr. lect. (tel. 9 + 737999, ext. 8289, fax (0064 9) 33429, email CHX117T at AUCC1.AUKUNI.AC.NZ). *Inorganic and organic structures.*

Robinson, Dr Ward Thomas. Chemistry Dept., U. of Canterbury, Private Bag, Christchurch, New Zealand. (1937) PhD, chemistry (Canterbury U., 1964). Reader. (tel. 3 + 482-009, ext. 294, fax (0064 3) 483308, email W.ROBINSON%CANTERBURY.AC.NZ%RELAY.CS.NET). *Structure analysis, symmetry, model building, computing.*

Rodgers, Dr Kerry Anthony. Geology Dept., U. of Auckland, Private Bag, Auckland, New Zealand. (1942 PhD, geology (Auckland U., 1972). Assoc. prof. (tel. 9 + 737-999 ext 7414). *Crystal chemistry, minerals.*

Rumball, Dr Sylvia Vine. Chemistry Dept., Massey U., Palmerston North, New Zealand. (1939) PhD, chemistry (Auckland U., 1966). Sr. lect. (tel. 63 + 69-089, ext. 7958). *Protein three-dimensional structure.*

Shelley, Dr David. Geology Dept., U. of Canterbury, Private Bag , Christchurch, New Zealand. (1940) PhD, geology (Bristol U., UK, 1964). Sr. lect. (tel. 3 + 667-001, ext. 7723). *Crystal growth, silicate mineralogy.*

Smale, Mr David. N.Z. Geological Survey, U. of Canterbury, Private Bag, Christchurch, New Zealand. (1939) MSc, geology (Auckland U., 1962). Sedimentary petrologist. (tel. 3 + 482-009, ext. 7769). *Detrital mineralogy, sedimentary petrology.*

Smalley, Dr Ian James. Soil Bureau, D.S.I.R., Private Bag, Lower Hutt, New Zealand. (1936) PhD, materials science (City U., London, UK, 1966). Sci. (tel. 4 + 673-119, ext. 883). *Clay minerals, quartz particles in sediments.*

Stewart, Dr Robert Bruce. Soil Sci. Dept., Massey U., Palmerston North, New Zealand. (1951) PhD, soil science (Massey U, 1983). Lect. (tel. 63 + 69-089, ext. 2454). *Mineralogy, weathering processes, isotope distribution.*

Waters, Dr Joyce Mary. Dept. of Chem. and Biochem., Massey U., Palmerston North, New Zealand. (1931) PhD, chemistry (New Zealand U., 1960). Snr. Res. Fellow. (tel. 063 + 69-099, ext. 8691, fax (64 63) 62140). *Inorganic and organic structures, large molecules.*

Waters, Dr Thomas Neil Morris. Massey U., Palmerston North, New Zealand. (1931) DSc, (Auckland U., 1969). Vice-Chancellor. (tel. 63 + 69-099). *Inorganic and organic structures, large molecules.*

Watters, Dr William Asher. N.Z. Geological Survey, P.O. Box 30368, Lower Hutt, New Zealand. (1926) PhD, petrology (Cambridge U., UK, 1956). Chief petrologist. (tel 4 + 699-059). *Mineralogy of rock-forming minerals.*

Weaver, Dr Stephen Donald. Geology Dept., U. of Canterbury, Christchurch, New Zealand. (1947) PhD, geology (London U., UK, 1973). Lect. (tel. 3 + 482-009, ext. 569). *Silicate minerals.*

Whimp, Dr Peter Olaf. Physics and Eng. Lab., D.S.I.R., Private Bag, Lower Hutt, New Zealand. (1942) PhD, chemistry (Victoria U. of Wellington, 1967). Sci. (tel. 4 + 666-919, ext. 547). *Coordination chemistry, organometallics, computing.*

NIGERIA

Sub-Editor: O.O. Adewoye

Notes

1. International telephone country code - 234.

Adetunji, Dr Jacob. Physics Dept., Ahmadu Bello U., Main Campus, Zaria, Kaduna State, Nigeria. (1944) PhD, physics (Essex U., UK, 1976). Lect. *Electron microscopy, radiation damage in crystalline solids.*

Adewoye, Dr Olusegun Oyeleke. Metallurgical and Materials Eng. Dept., U. of Ife, Faculty of Techn., Ile-Ife, Oyo State, Nigeria. (1947) PhD, materials science (Cambridge U., UK, 1976). Sr. lect. (tel. Ife 2290-2299, telex IFEVASITY IFE). *Structure determination, crystallography, ceramics, deformation modes, coal diffraction.*

Aladekomo, Prof. Johnson Bandele. Physics Dept., U. of Ife, Ife Campus, Ile-Ife, Oyo State, Nigeria. (1938) PhD, physics (Manchester U., UK, 1965). Prof. *Exciton diffusion, molecular crystals, luminescence effects, crystal impurities.*

Dubey, Dr Ram Janam. Chemistry Dept., U. of Maiduguri, P.M.B. 1069, Maiduguri, Nigeria. (1941) DSc, crystallography (Laval U., Canada, 1973). Asst. prof. *Synthesis, structure, bonding, organometallics, protein structure, impurities.*

Koshy, Dr Jacob. Physics Dept., U. of Ibadan, 1 Parry Road, Ibadan, Nigeria. (1942) PhD, solid state physics (Sardar Patel U., 1970). Sr. lect. (tel. 462550, ext. 1042). *Crystal imperfections, crystal growth, electron microscopy, epitaxial growth, thin films.*

Onyeagocha, Dr Anthony Chukwuma. Geology Dept., U. of Nigeria, Nsukka Campus, Nsukka, Anambra State, Nigeria. (1942) PhD, geology (U. Washington, USA, 1973). Lect. *Petrology, crystal chemistry, phase equilibria, geochemistry, mineralogy.*

Sanni, Dr Bamidele. Chemistry Dept., U. of Benin, Benin City, Nigeria. (1943) PhD, chemistry (Ibadan U., 1974). Lect. (tel. Benin City 343). *Large molecules, organic structures.*

Sharma, Dr (Mrs) Aysel. Physics Dept., U. of Benin, B28 Ugbowo Campus, Benin City, Bendel State, Nigeria. (1941) PhD, Physics (St. Andrews U., UK, 1973). Lect. *Crystal structure analysis, organic compounds, X-ray and neutron diffraction.*

Sharma, Dr Vinod Chander. Physics Dept., U. of Benin, B28 Ugbowo Campus, Benin City, Bendel State, Nigeria. (1940) PhD, Physics (St. Andrews U., UK, 1973). Sr. lect. *Kinematic and dynamic X-ray diffraction, topography.*

NORWAY

Sub-Editors: A. Hordvik & K. Maartmann-Moe

Notes

1. International telephone country code - 47.

2. Degrees conferred by the Norwegian universities are the *Doctor philosophiae* (Dr philos.) and *Doctor technicae* (Dr techn.)(both approximately equivalent to the English DSc), *Doctor scientiarium* (Dr scient.) and *Doctor ingenieur* (Dr ing.)(both equivalent to PhD), *Candidatus realium* (Cand. real.) and *Magister scientiarium* (Mag. scient.)(both range between PhD and MSc), *Candidatus scientiarium* (Cand.scient.) and *Sivilingeniör* (siv.ing.)(approximately equivalent to MSc).

3. The position *förstelektor* and *försteamanuensis* both correspond to senior lecturer, whereas *amanuensis* corresponds to lecturer.

Almenningen, Mr Arne. Dept. of Chemistry, U. of Oslo, N-0315 Oslo 3, Norway. (1921) Cand. real., physics (U. Oslo, 1952). Försteamanuensis. (tel. 02 + 455408, fax 02 + 455441). *Electron diffraction, instrumentation.*

Alver, Mr Eyvind. Dept. of Chemistry, U. of Bergen, N-5007 Bergen, Norway. (1922) Cand. real., chemistry (U. Oslo, 1954). Sr. lect. (tel. 05 + 213544, fax 05 + 329058). *Structural chemistry, philosophy of scientific method.*

Andersen, Dr Per. Dept. of Chemistry, U. of Oslo, N-0315 Oslo 3, Norway. (1919) Dr philos., chemistry (U. Oslo, 1968). Försteamanuensis. (tel. 02 + 455458, fax 02 + 455441). *Electron diffraction, free radicals.*

Andresen, Dr Arne Fridtjof. Dept. of Neutron Physics, Inst. for Energy Techn., P.O. Box 40, N-2007 Kjeller, Norway. (1926) Dr philos., physics (U. Oslo, 1972). Asst. div. head. (tel. 06 + 812560 ext. 294, fax 06 + 815553). *Neutron diffraction, inorganic crystal structures, magnetic properties.*

Bastiansen, Prof. Otto Christian Astrup. Dept. of Chemistry, U. of Oslo, N-0315 Oslo 3, Norway. (1918) Dr philos., physical chemistry (U. Oslo, 1949). Prof. emer. of physical chemistry. (tel. 02 + 455401, fax 02 + 455441). *Electron diffraction, molecular structure.*

Bremer, Dr Johannes. Dept. of Physics and Mathematics, U. of Trondheim-NTH, N-7034 Trondheim, Norway. (1949) Dr ing., (U. Trondheim-NTH, 1979). Försteamanuensis. (tel. 07 + 593582). *Spectroscopy, scattering effects.*

Bye, Dr Erik. Inst. of Occupational Health, P.O. Box 8149, N-0033 Oslo 1, Norway. (1945) Dr philos., chemistry (U. Oslo, 1976). Scient. (tel. 02 + 466850 ext. 784, fax 02 + 603276). *Biologically active molecules, powder diffraction, mineral dust, inorganic dust, chemometrics.*

Dahl, Dr Tor. Dept. of Chemistry, Inst. of Mathematical and Physical Sci., U. of Tromsö, P.O. Box 953, N-9001 Tromsö, Norway. (1938) Dr philos., chemistry (U. Tromsö, 1976). Sr. lect. (tel. 083 + 44075, fax 083 + 55418, email kjtd at es.uit.uninett). *Crystal structures, organic charge transfer compounds.*

Fjaer, Dr Erling. IKU SINTEF Group, N-7034 Trondheim, Norway. (1951) Dr ing., physics (U. Trondheim-NTH, 1983). Sr. scient. (tel. 07 + 920611, fax 07 + 920924, email fjaer at iku.uninett). *X-ray diffraction, modulated crystal structures.*

Fjellvåg, Mr Helmer. Dept. of Chemistry, U. of Oslo, N-0315 Oslo 3, Norway. (1954) Cand. real. (U. Oslo, 1978). Sr. lect. (tel. 02 + 455562, fax 02 + 455441). *X-ray and neutron diffraction, oxides, transition metal compounds, magnetic structures.*

Foss, Prof. Olav. Dept. of Chemistry, U. of Bergen, N-5007 Bergen, Norway. (1918) Dr techn., chemistry (Norges Tekniske Högskole, 1947). Prof. emer. (tel. 05 + 213563, fax 05 + 329058). *Inorganic crystal structures.*

Furuseth, Mrs Sigrid. Dept. of Chemistry, U. of Oslo, N-0315 Oslo 3, Norway. (1939) Cand. real., chemistry (U. Oslo, 1964). Försteamanuensis. (tel. 02 + 455561, fax 02 + 455441). *Alloy structures.*

Gjönnes, Prof. Jon Kjell. Dept. of Physics, U. of Oslo, N-0316 Oslo 3, Norway. (1931) Dr philos., physics (U. Oslo, 1967). Prof. (tel. 02 + 456490). *Electron diffraction, microscopy, inorganic materials.*

Görbitz, Mr Carl Henrik. Dept. of Chemistry, U. of Oslo, N-.0315 Oslo 3, Norway. (1961) Cand. scient., chemistry (U. Oslo, 1985). Res. fellow. (tel. 02 + 455460, fax 02 + 455441, email m__goerbitz__c at use.uio.uninett). *Structure, peptides.*

Grjotheim, Prof. Kai. Dept. of Chemistry, U. of Oslo, N-0315 Oslo 3, Norway. (1919) Dr techn., chemistry (Norges Tekniske Högskole, 1956). Prof. (tel. 02 + 455039, fax 02 + 455441). *Inorganic crystal structures.*

Groth, Mr Per Arne. Dept. of Chemistry, U. of Oslo, N-0315 Oslo 3, Norway. (1934) Cand. real., chemistry (U. Oslo, 1960). Res. fellow. (tel. 02 + 455692, fax 02 + 455441). *Organic crystal structures, programming.*

Grönvold, Prof. Fredrik. Dept. of Chemistry, U. of Oslo, N-0315 Oslo 3, Norway. (1924) MSc, metallurgy (U. Michigan, USA, 1951). Prof. (tel. 02 + 455599, fax 02 + 455441). *Inorganic structures and transitions.*

Gundersen, Prof. Grete. Dept. of Chemistry, U. of Oslo, N-0315 Oslo 3, Norway. (1940) Cand. real., chemistry (U. Oslo, 1967). Prof. (tel. 02 + 455406, fax 02 + 455441, email m__gunder__g at use.uio.uninett). *Electron diffraction, molecular structure.*

Haaland, Prof. Arne. Dept. of Chemistry, U. of Oslo, N-0315 Oslo 3, Norway. (1936) Dr philos., chemistry (U. Oslo, 1969). Prof. (tel. 02 + 455407, fax 02 + 455441). *Molecular structure, organometallic compounds.*

Hadler, Mrs Eva. Dept. of Pharmacy, U. of Oslo, N-0316 Oslo 3, Norway. (1921) Cand. real., chemistry (U. Oslo, 1950). Försteamanuensis. (tel. 02 + 456580). *Inorganic and organic crystal structures, large molecules and molecular biology.*

Hagen, Dr Kolbjörn. Dept. of Chemistry, U. of Trondheim, N-7055 Dragvoll, Norway. (1943) Dr philos., physical chemistry (U. Trondheim, 1979). Sr. lect. (tel. 07 + 596223, fax 07 + 593337, email k__hagen at avh.unit.uninett). *Structure determination, conformation, gas-phase molecules, electron diffraction.*

Hansen, Mr Lars Kristian. Dept. of Chemistry, Inst. of Mathematical and Physical Sci., U. of Tromsö, P.O. Box 953, N-9001 Tromsö, Norway. (1944) Cand. real., (U. Bergen, 1971). Res. fellow. (tel. 083 + 44079, fax 083 + 55418, email kjlkh at es.uit.uninett). *Organic crystal structures, protein crystallography.*

Hauback, Mr Björn Christian. Dept. of Physics and Mathematics, U. of Trondheim-NTH, N-7034 Trondheim, Norway. (1957) Siv. ing., physics (U. Trondheim-NTH, 1981). Res. asst. (tel. 07 + 593584). *X-ray diffraction, spectroscopy, electron density distribution.*

Hauge, Dr Sverre. Dept. of Chemistry, U. of Bergen, N-5007 Bergen, Norway. (1932) Dr philos., chemistry (U. Bergen, 1978). Sr. lect. (tel. 05 + 213566, fax 05 + 329058). *Inorganic complexes.*

Hordvik, Prof. Asbjörn. Dept. of Chemistry, Inst. of Mathematical and Physical Sci., U. of Tromsö, P.O. Box 953, N-9001 Tromsö, Norway. (1928) Dr philos., (U. Bergen, 1968). Prof. (tel. 083 + 44072, fax 083 + 55418, email kjah.at es.uit.uninett). *Inorganic and organic structures, protein crystallography.*

Hough, Dr Edward. Dept. of Chemistry, Inst. of Mathematical and Physical Sci., U. of Tromsö, P.O. Box 953, N-9001 Tromsö, Norway. (1941) PhD, X-ray crystallography (London U., UK, 1975). Sr. lect. (tel. 083 + 44073, fax 083 + 55418, email kjprotein at es.uit.uninett). *Protein crystallography.*

Husebye, Prof. Steinar. Dept. of Chemistry, U. of Bergen, N-5007 Bergen, Norway. (1933) PhD, chemistry (Tulane U., USA, 1963), Dr philos., chemistry (Bergen, 1970) (tel. 05 + 213551, fax 05 + 329058). *Complexes with central Se and Te, inorganic and organic structures.*

Höier, Prof. Ragnvald. Dept. of Physics and Mathematics, U. of Trondheim-NTH, N-7034 Trondheim, Norway. (1938) Dr philos., physics (U. Oslo, 1973). Prof. (tel. 07 + 593588, fax 07 + 592886, email hoier at norunit). *Electron diffraction, electron microscopy.*

Jynge, Mr Knut. Dept. of Chemistry, Inst. of Mathematical and Physical Sci., U. of Tromsö, P.O. Box 953, N-9001 Tromsö, Norway. (1933) Cand. real., chemistry (U. Tromsö, 1976). Res. asst. (tel. 083 + 44071, fax 083 + 55418, email kjprotein at es.uit.uninett). *Physical chemistry, biologically important structures, protein crystallography.*

Kjekshus, Prof. Arne. Dept. of Chemistry, U. of Oslo, N-0315 Oslo 3, Norway. (1932) Dr philos., (U. Oslo, 1971). Prof. (tel. 02 + 455560, fax 02 + 455441). *Inorganic crystal structures, metal structures, magnetic structures.*

Klewe, Mr Bernt. Dept. of Chemistry, U. of Oslo, N-0315 Oslo 3, Norway. (1933) Cand. real., chemistry (U. Oslo, 1961). Sr. lect. (tel. 02 + 455460, fax 02 + 455441, email m__klewe__b at use.uio.uninett). *Organic crystal structures.*

Maartmann-Moe, Mr Knut. Dept. of Chemistry, U. of Bergen, N-5007 Bergen, Norway. (1928) Cand. real., chemistry (U. Bergen, 1961). Försteamanuensis (tel. 05 + 213446, fax 05 + 329058). *X-ray diffractometry, computing, crystal structures.*

Marthinsen, Dr Knut. Div. of Applied Physics, SINTEF, N-7034 Trondheim, Norway. (1956) Dr ing., physics (U. Trondheim-NTH, 1986). Res. scient. (tel. 07 + 593473, fax 07 + 592886). *Electron diffraction, electron microscopy, dynamical diffraction effects.*

Maröy, Dr Kjartan. Dept. of Chemistry, U. of Bergen, N-5007 Bergen, Norway. (1930) Dr philos., inorganic chemistry (U. Bergen, 1976). Sr. lect. (tel. 05 + 213565, fax 05 + 329058). *Tellurium complexes, polythionates.*

Mo, Prof. Frode. Dept. of Physics and Mathematics, U. of Trondheim-NTH, N-7034 Trondheim, Norway. (1937) Dr techn., crystallography (U. Trondheim-NTH, 1980). Prof. (tel. 07 + 593585, fax 07 + 592886, email nthmo at norunit.earn). *Multiple scattering, X-ray phase determination, accurate structures, organic chalcogenides, biopolymer components.*

Mostad, Mr Arvid. Dept. of Chemistry, U. of Oslo, N-0315 Oslo 3, Norway. (1929) Cand. real., chemistry (U. Oslo, 1959). Sr. lect. (tel. 02 + 455415, fax 02 + 455441, email m__mostad__a at use.uio.uninett). *Molecular structure - biological activity relationships.*

Nicholson, Dr David Graham. Dept. of Chemistry, U. of Trondheim, N-7055 Dragvoll, Norway. (1944) PhD, inorganic chemistry (London U., UK, 1969). Sr. lect. (tel. 07 + 596204, fax 07 + 593337, email d__nicholson at avh.unit.uninett). *Inorganic crystal structures.*

Nordenson, Mr Svein. Dept. of Chemistry, U. of Oslo, N-0315 Oslo 3, Norway. (1949) Cand. real., chemistry (U. Oslo, 1976). Res. fellow. (tel. 02 + 455452, fax 02 + 455441). *Organic crystal structures.*

Norman, Prof. Nico. Dept. of Physics, U. of Oslo, N-0316 Oslo 3, Norway. (1919) Dr philos., (U. Oslo, 1956). Prof. (tel. 02 + 456434, fax 02 + 456422). *Crystal structures, imperfections.*

Olsen, Dr Arne. Dept. of Physics, U. of Oslo, N-0316 Oslo 3, Norway. (1944) Dr philos., physics (U. Oslo, 1978). Försteamanuensis. (tel. 02 + 456432, fax 02 + 456422). *Electron diffraction, microscopy, inorganic materials.*

Pedersen, Dr Berit Fjærtoft. Dept. of Pharmacy, U. of Oslo, N-0316 Oslo 3, Norway. (1933) Dr philos., physical chemistry (U. Oslo, 1969). Sr. lect. (tel. 02 + 455694, fax 02 + 455441, email m__peders__bf at use.uio.uninett). *Organic and inorganic crystal structures, conformations.*

Pedersen, Prof. Björn. Dept. of Chemistry, U. of Oslo, N-0315 Oslo 3, Norway. (1933) Dr philos., chemical physics (U. Oslo, 1964). Prof. (tel. 02 + 455690, fax 02 + 455441). *Crystal structures, atomic motion, nuclear magnetic resonance.*

Raade, Mr Gunnar. Mineralogical-Geological Museum, U. of Oslo, Sarsgate 1, 0562 Oslo 5, Norway. (1944) Cand. real., mineralogy (U. Oslo, 1973). Curator of minerals. (tel. 02 + 686960 ext. 147). *Minerals.*

Riste, Prof. Tormod. Dept. of Physics, Inst. for Energy Techn., P.O. Box 40, N-2007 Kjeller, Norway. (1925) Dr philos., physics (U. Oslo, 1961). Prof. (tel. 06 + 812560, fax 06 + 816356). *Neutron diffraction, solid state physics, statistical physics.*

Rosenqvist, Prof. Ivan Thoralf. Dept. of Geology, U. of Oslo, N-0316 Oslo 3, Norway. (1916) Dr philos., mineralogy (U. Oslo, 1945). Prof. (tel. 02 + 456652). *Clay mineralogy, rock forming minerals.*

Römming, Prof. Christian. Dept. of Chemistry, U. of Oslo, N-0315 Oslo 3, Norway. (1928) Dr philos., chemistry (U. Oslo, 1968). Prof. (tel. 02 + 455403, fax 02 + 455441, email m__roemming__c.use.uio.uninett). *Molecular structure - properties relationships.*

Röst, Mr Erling. Dept. of Chemistry, U. of Oslo, N-0315 Oslo 3, Norway. (1924) Cand. real., chemistry (U. Oslo, 1954). Sr. lect. (tel. 02 + 455613, fax 02 + 455441). *Alloy structures, phase equilibria.*

Samdal, Mr Svein. Oslo College of Engineering, Cort Adelersgate 30, N-0254 Oslo 2, Norway. (1945) Cand. real., chemistry (U. Oslo, 1973). Försteamanuensis. (tel. 02 + 553000). *Electron diffraction, conformational analysis.*

Samuelsen, Prof. Emil J. Dept. of Physics and Mathematics, U. of Trondheim-NTH, N-7034 Trondheim, Norway. (1937) Dr philos., physics (U. Oslo, 1971). Prof. (tel. 07 + 593412, fax 07 + 592886). *Partly disordered solids, x-ray and neutron scattering, Raman spectroscopy.*

Semmingsen, Dr Dag. Högskolesenteret i Rogaland, P.O. Box 2557 Ullandhaug, N-4004 Stavanger, Norway. (1940) Dr philos., chemistry (U. Oslo, 1976). Sr. lect. (tel. 04 + 874248, fax 04 + 874300). *Structure determinations by X-rays and neutrons, organic and inorganic compounds, phase transitions and cooperative phenomena.*

Skjerpe, Mr Per Martin. Dept. of Physics, U. of Oslo, N-0316 Oslo 3, Norway. (1953) Cand. real., physichs (U. Oslo, 1980). Res. asst. (tel. 02 + 455049). *Transmission electron microscopy in metallurgy.*

Sletten, Dr Einar. Dept. of Chemistry, U. of Bergen, N-5007 Bergen, Norway. (1939) Dr philos., chemistry (U. Bergen, 1979). Sr. lect. (tel. 05 + 213352, fax 05 + 329058). *Transition metal complexes, nucleic acid constituents.*

Sletten, Prof. Jorunn. Dept. of Chemistry, U. of Bergen, N-5007 Bergen, Norway. (1941) Dr philos., chemistry (U. Bergen, 1976). Prof. (tel. 05 + 213562, fax 05 + 329058). *Transition metal complexes.*

Strand, Dr Tor Gogstad. Dept. of Chemistry, U. of Oslo, N-0315 Oslo 3, Norway. (1934) Dr philos., (U. Oslo, 1968). Sr. lect. (tel. 02 + 455411, fax 02 + 455441). *Electron diffraction.*

Stölevik, Prof. Reidar. Dept. of Chemistry, U. of Trondheim, N-7055 Dragvoll, Norway. (1938) Dr philos., physical chemistry (U. Oslo, 1975). Prof. (tel. 07 + 596224). *Structure and conformation, electron diffraction, molecular mechanics calculations.*

Svinning, Dr Torgeir. Div. of Materials and Processes, SINTEF, N-7034 Trondheim, Norway. (1948) Dr ing., crystallography (U. Trondheim-NTH, 1978). Division director. (tel. 07 + 592037, fax 07 + 597043). *Structure and mechanical properties, metal working.*

Sæthre, Mr Leif Jarle. Dept. of Chemistry, Inst. of Mathematical and Physical Sci., U. of Tromsö, P.O. Box 953, N-9001 Tromsö, Norway. (1945) Cand. real., (U. Bergen, 1971). Sr. lect. (tel. 083 + 44078, fax 083 + 55418, email kjls at es.uit.uninett). *Molecular structures, ESCA, Auger spectroscopy.*

Thorkildsen, Dr Gunnar. Högskolesenteret i Rogaland, P.O. Box 2557 Ullandhaug, N-4004 Stavanger, Norway. (1953) Dr ing., crystallography (U. Trondheim-NTH, 1983). Försteamanuensis. (tel. 04 + 874257, fax 04 + 874300). *X-ray diffraction and spectroscopy, charge density in solids, solid state physics.*

Tibballs, Dr John Earl. Senter for Industriforskning (SI), P. O. Box 124 Blindern, N-0314 Oslo 3, Norway. (1947) PhD, physics (Melbourne U., Australia, 1974). Sen. res. scient. (tel. 02 + 452978, fax 02 + 452040). *Phase diagrams, electron microscopy, welds, diffuse scattering, intermetallic phases, neutron diffraction.*

Trætteberg, Prof. Marit. Dept. of Chemistry, U. of Trondheim, N-7055 Dragvoll, Norway. (1930) Dr philos., physical chemistry (U. Trondheim, 1970). Prof. (tel. 07 + 596225, email m__tratteberg at avh.unit.uninett). *Physical organic chemistry, molecular structure, unsaturated compunds, conformational analysis, gas electron diffraction.*

PAKISTAN

Sub-Editor: **S.S.H. Rizvi**

Notes

1. International telephone country code - 92.

2. Abbreviations used for the Pakistan entries include:
   PCSIR - Pakistan Council of Scientific and Industrial Research
   NPSL - National Physical and Standards Laboratory
   GSP - Geological Survey of Pakistan

Ahmad, Dr Zulfiqar. Centre of Excellence in Mineralogy, U. of Baluchistan, Sariab Road, Quetta, Pakistan. (1945) PhD, geology (London U., UK, 1982). Prof. and Dir. *Minerals, structural chemistry, microscopy.*

Akhtar, Mr Mohammad. GSP, 22-Ali Block, New Garden Town, Lahore-16, Pakistan. (1944) MSc, geology (Sind U., 1965). Dpty. dir. (tel. 042 + 855922). *Minerals.*

Akhter, Mr Javed. GSP, 22-Ali Block, New Garden Town, Lahore-16, Pakistan. (1955) MSc, geophysics (Punjab U., 1978). Asst. geophysicist. (tel. 042 + 855816). *Geophysics.*

Ali, Dr Syed Wajahat. Solar Energy Research Center,PCSIR Latifabad unit 2, Autobahn, Hyderabad, Pakistan. (1937) PhD, physical chemistry (Karachi U., 1981). Dir. (tel. 0221+83040). *Structural chemistry.*

Anwar, Mr Muhammad. GSP, 16-G Model Town, Lahore, Pakistan. (1953) MSc, geology (Punjab U., 1976). Asst. dir. (tel. 042 + 855922) *Minerals, geology.*

Baqri, Dr Syed Rafiqul-Hassan. Earth Sci. Div., Pakistan Museum of Natural History, Al-Markaz, F-7/2, Islamabad, Pakistan. (1945) PhD, X-ray diffraction (Southhampton U., UK, 1977). Dir. (tel. 82439). *X-ray diffraction.*

Bhatti, Mr Muhammad Akram. GSP, Ministry of Petroleum and Natural resources, 83-D Model Town, Lahore, Pakistan. (1947) MSc, geology (Punjab U., 1969). Dpty. dir. (tel.042+855232). *Minerals, geology.*

Butt, Dr Khurshid Alam. Atomic Energy Mineral Centre, Ferozepur Road, Lahore, Pakistan. (1947) PhD, petrology (New Brunswick U., Canada, 1976). Head. (tel. 042 + 870237, ext. 07). *Minerals, X-ray diffraction.*

Butt, Mr Muhammad Hafeez. GSP, Quetta, Pakistan. (1945) MSc. geology (Punjab U., 1968). Geophysicist. (tel. 042 + 855232). *Microscopy, geophysics.*

Butt, Dr Noor Mohammad. PINSTECH, Nilore, Islamabad, Pakistan. (1936) PhD, solid state physics (Birmingham U., UK, 1965). Assoc. dir. (tel.051+840151). *Materials, neutron diffraction.*

Chaudhary, Dr Abdul Majid. Nuclear Materials Div., Pakistan Inst. of Nuclear Sci. and Techn., Nilore, Islamabad, Pakistan. (1945) PhD, amorphous solids (New England U., Australia, 1978). Principal scient. officer. (tel.840103,ext.240). *X-ray diffraction, diffuse scattering.*

Choudhry, Dr Muhammad Iqbal. X-ray Diffraction facilities, H.E.J. Res. Inst. of chemistry, U. of Karachi-32, Karachi, Pakistan. (1959) PhD, organic chemistry (Karachi U.,1987). Incharge. (tel.463414). *Structural biology, X-ray diffraction.*

Chuadry, Dr Fazal Muhammad. Investment Promotion Bureau, Ministry of Industries, Kandawala building, M.A.Jinaah Rd, Karachi, Pakistan. (1932) PhD, high temperature X-ray diffraction (Strathclyde U., UK, 1966). Dpty. dir. general. (tel.021+714295). *Structural chemistry, X-ray diffraction, metals.*

Chaudhary, Mr G. Sarwar Alam. GSP, (Punjab Div.), 14-Canal Park, Gulberg, Lahore-11, Pakistan. (1944) MSc, mineralogy (Punjab U., 1967). Dpty. dir. (tel. 042+ 881192). *Minerals.*

Chaudhry, Mr Mohammad Anwar. GSP, 14-Canal Park, Gulberg, Lahore-11, Pakistan. (1944) MSc, geology (Peshawar U., 1972). Asst. dir. (tel. 042 + 881192). *Minerals.*

Chauhan, Mr Ehsanul Haq. Chem. Div., GSP, Sariab Road, Quetta, Pakistan. (1931) MS, geology (Idaho U., USA, 1968). Chief chemist. (tel. 081 + 72617). *Structural chemistry, microscopy, minerals.*

Elahi, Mr Manzoor. Metallurgy Group, Defence Sci. and Techn. Organization, Chaklala, Rawalpindi, Pakistan. (1937) M.Tech., metallurgical quality control (Brunel U., UK, 1980). Res. officer. (tel. 64746, ext. 52). *Metals, structural chemistry.*

Fatmi, Prof. Ali Nasir. GSP, Ministry of Petroleum and Natural Resources, Sariab Road, Quetta, Pakistan. (1930) PhD, stratigraphy and paleontology (Wales U., 1968). Dpty. dir. general. (tel. 081+73564). *Minerals.*

Gilani, Mr Jamshed Ali. GSP, 157-Sher Shah Block, New Garden Town, Lahore, Pakistan. (1956) MSc, geochemistry (Karachi U., 1977). Asst. dir. (tel. 081 + 78519). *Minerals.*

Habib, Mr Syed Abbas. GSP, 22-Ali Block, New Garden Town, Lahore-16, Pakistan. (1936) MSc, sedimentology (Victoria U., New Zealand, 1971). Dpty. dir. (tel. 042 + 855923). *Minerals.*

Hasan, Dr Faizul. Metallurgical Eng. Dept., U. of Eng. and Techn., G.T. Road, Lahore, Pakistan. (1948) PhD, metallurgy (Manchester U., 1984). Assoc. prof. (tel. 042 + 339207) *X-ray diffraction, electron diffraction, structural chemistry.*

Haq, Dr Anwarul. Material Science/Metallurgy Div., Dr.A.Q.Khan Res.Labs., Kahuta, P.O. Box 502, Rawalpindi, Pakistan. (1947) PhD, thin films (Karlsruhe U., W.Germany, 1982). Principal scient. officer. *Phase transitions, metals.*

Hussain, Dr Khadim. Centre For Solid State Physics, Punjab U. New Campus, Lahore- 20, Pakistan. (1947) PhD, physics (Victoria U., Manchester, UK, 1981). Asst. prof. (tel 042+854113). *Structural chemistry, X-ray diffraction.*

Ikram, Dr Nazma. Centre For Solid State Physics, Punjab U., Lahore, Pakistan. (1949) PhD, theoretical solid state physics (Cambridge U., U.K, 1976). Asst. prof. (tel. 854113). *Structural chemistry, electron diffraction, X-ray diffraction.*

Iqbal, Mr Mir Waseluddin Ahmad. GSP, 22-Ali Block, New Garden Town, Lahore-16, Pakistan. (1926) MSc, geology (U. California, USA, 1964). Dir. (tel. 852547). *Minerals.*

Jafry Mr Syed Qamar Abbas. GSP, 16-G, Model Town, Lahore, Pakistan. (1953) MSc,geology (Punjab U., 1977). Asst. dir. (tel. 042 + 852826). *Minerals, geology, computing.*

Khalid, Mr Mohammad. NPSL Centre, PCSIR Labs., Off University Road, Karachi, Pakistan. (1940) MSc, physics (Karachi U., 1963). Sr. scient. officer. (tel. 460101, ext. 76). *Structural chemistry, X-ray spectroscopy.*

Khan, Dr Abdul Quadeer. Material Science/Metallurgy Div., Dr.A.Q.Khan Res.Labs., Kahuta, P.O.Box 502, Rawalpindi, Pakistan. (1936) PhD, physical metallurgy (Leuven U.,Beigium, 1972). Project dir. *Phase transitions, structural chemistry, material science.*

Khan, Dr Ainul Hassan. R and D Div. PCSIR, Shahrahe-Kamal Attaturk, Karachi, Pakistan. (1932) PhD, geochemistry (Manchester U., U.K, 1965). Dir. (tel. 213454). *X-ray diffraction.*

Khan, Mr Mohammad Afaq. GSP, Ministry of Petroleum and Natural Resources, 16-G Model Town, Lahore, Pakistan. (1951) MSc, geology (Punjab U., 1976). Asst. dir. *Minerals, geophysics.*

Khawaja, Mr Mahmood-ul-Hassan. GSP, 14-Canal Park, Gulberg, Lahore-11, Pakistan. (1944) MSc, geology (Punjab U., 1966). Dpty. dir. (tel. 042 + 881192). *Minerals.*

Khwaja, Dr Farid Akhtar. Physics Dept., Quaid-i-Azam U., Islamabad, Pakistan. (1949) PhD, solid state physics (Moscow State U., USSR, 1976). Asst. prof. (tel. 29472). *Disorder, X-ray and neutron diffraction, materials science.*

Mahmood, Mr Khursheed. Dept. of Mech. Eng., NED U. of Eng. and Techn., University Road, Karachi, Pakistan. (1952) MSc, metallurgical eng. (Cranfield Inst. of Tech., UK, 1978). Asst. prof. (tel. 461866). *Surface structure, materials science.*

Maqsood, Dr Asghari. Dept. of Physics, Quaid Azam U., Islamabad, Pakistan. (1947) PhD, materials science (Goteborg U.,Sweden,1982). Assoc. Prof. (tel. 829472). *X-ray diffraction, metals.*

Mian, Dr Mohammad Ashraf. Inst. of Applied Geology, Azad Jammu and Kashmir U., Muzaffarabad, Azad Kashmir, Pakistan. (1938) PhD, Geochemistry (Punjab U., 1976). Teacher, appl. geology. (tel. 058 + 2706). *Minerals, X-ray diffraction.*

Mian, Mr Muhammad Asghar. GSP, 22-Ali Block, New Garden Town, Lahore-16, Pakistan. (1951) MSc, geology (Punjab U., 1976). Geophysicist. (tel. 042 + 855816). *Geophysics.*

Mir, Mr Jan Mohammad. Dept. of Geology, U. of Karachi, University Road, Karachi, Pakistan. (1937) MSc, sedimentology, mineralogy (Manitoba U., Canada, 1972). Asst. prof. (tel. 460211, ext. 95). *Crystallographic data, minerals.*

Munir, Mr Mohammad. Centre of Excellence in Mineralogy, U. of Baluchistan, Sariab Road, Quetta, Pakistan. (1950) MSc, geology (Baluchistan U., 1974). Lect. *Structural chemistry, minerals.*

Naqvi, Dr Syed Ali Anwar. Dept. of Urology, Dow Medical C., Karachi, Pakistan. (1947) MS, urology (Karachi U., 1984). Sr. registrar. (tel. 219551, ext. 247). *X-ray diffraction.*

Nasreen, Miss Shagufta. Mineral Res. Div., PCSIR Labs., Jamrud Road, Peshawar, Pakistan. (1955) MSc, physical chemistry (Peshawar U., 1979). Res. officer. (tel. 8817). *X-ray diffraction.*

Pathan, Dr Muhammad Taqee. Geology Dept. U. of Sind, Jamshoro, Pakistan. (1938) PhD, geology, mineralogy (Moscow State U., USSR, 1972). Prof. (tel. 71291, ext. 16). *Structural chemistry, minerals.*

Pervaiz, Mr Rashed. GSP, 22-Ali Block, New Garden Town, Lahore-16, Pakistan. (1952) MSc, geophysics (Quad-i-Azam U. Islamabad, 1975). Geophysicist. (tel. 042 + 855816) *Geophysics.*

Qaiser, Mr Mohammad Ali. Mineral Res. Div., PCSIR Labs., Jamrud Road, Peshawar, Pakistan. (1940) MSc, physics (Bihar U., India, 1962). Sr. scient. officer. (tel.0521+41191, ext. 19). *Structural chemistry, minerals.*

Qurashi, Dr Mazhar Mahmood. Pakistan Academy of Sciences, Editor and Secretary-general, Constitution Avenue G-5, Islamabad, Pakistan. (1925) DSc, physics, X-ray crystallography (Manchester U., UK, 1962). Editor and Sec.Gen. (tel.824843). *Structural chemistry, minerals, metals, x-ray diffraction.*

Qureshi, Mr Khalid Mahmood. Mineralogy Div., Atomic Energy Mineral Centre, Ferozpur Road, Lahore, Pakistan. (1953) MSc, physics (Punjab U., 1980). Sci. officer. (tel. 99 + 870276, ext. 04). *X-ray diffraction.*

Qureshi, Mr Mohammed Kaleem Akhtar. GSP, Ministry of Petroleum and Natural resources, 83-D Model Town, Lahore 54700, Pakistan. (1945) MSc, geology (Punjab U., 1968). Dpty. dir. (tel. 042 + 852826). *Geology, minerals.*

Rahman, Mr Mohammad Abdul. Geology Div., Pakistan Atomic Energy Commission, D.G.Khan, Pakistan. (1941) MSc, mineralogy (Aberdeen U., UK, 1979). Principal geologist. *Geology, minerals.*

Rana, Mr Riaz Ahmad. GSP, 14-Canal Park, Gulberg, Lahore-11, Pakistan. (1950) MSc, geology (Punjab U., 1977). Asst. dir. (tel. 042 + 881192). *Minerals.*

Rizvi, Prof Adibul Hasan. Dept. of Urology, Dow Medical C., Karachi, Pakistan. (1938) FRCS, surgery/urology (Royal C. of Surgeons, UK, 1967). Prof. (tel. 219551, ext. 247). *X-ray diffraction.*

Rizvi, Dr Syed Sadrul Hassan. NPSL Centre, PCSIR Labs., Off University Road, Karachi, Pakistan. (1933) PhD, physics (Manshester U., UK, 1962). Dir. (tel. 466308). *Structural chemistry, X-ray diffraction, instrumentation.*

Russell, Mr Nazirullah. GSP, 22-Ali Block, New Garden Town, Lahore-16, Pakistan. (1948) MSc, geophysics (Punjab U., 1970). Geophysicist. (tel. 042 + 855816). *Geophysics.*

Saeed, Mr Syed Mohammad. NPSL Centre, PCSIR Labs., Off University Road, Karachi, Pakistan. (1943) MSc, physics (Karachi U., 1971). Sr. scient. officer. (tel. 460101, ext. 76). *Instrumentation, X-ray diffraction, computing.*

Saghir, Mr Ahmad. NPSL Centre, PCSIR Labs., Off University Road, Karachi, Pakistan. (1947) MSc, physics (Karachi U., 1968). Sr. scient. officer. (tel. 460101, ext. 76). *Structural chemistry, powder diffraction, computing.*

Shah, Prof Muzaffar Ali. Centre for Solid State Physics, Punjab U. New Campus, Lahore 20, Pakistan. PhD, physics (U. C., London U., UK, 1965). Prof. (tel. 854113). *Phase transition, semiconductors.*

Shahi, Mr Ghulam Nabi. Geology Dept., U. of Baluchistan, Sariab Road, Quetta, Pakistan. (1953) MSc, geology (Baluchistan U., 1976). Asst. prof. (tel. 081 + 73484). *Minerals.*

Shaikh, Mr Mohammad Iqbal. GSP, 16-G, Model Town, Lahore- 16, Pakistan. (1951) MSc, geology (Punjab U., 1971). Asst. dir. (tel. 042 + 855922). *Minerals.*

Shaikh, Mr Qameruddin. NPSL Centre, PCSIR Labs., Off University Road, Karachi, Pakistan. (1947) MSc, physics (Karachi U., 1974). Exp. officer. (tel. 460101, ext. 76). *X-ray diffraction, instrumentation, computing.*

Shaikh, Mr Mohammad Sualehin. NPSL Centre PCSIR Labs., Off University Road, Karachi, Pakistan. (1939) MSc, physics (Manchester U., UK, 1966). Sr. scient. officer. (tel. 460101, ext. 80). *Structural chemistry.*

Shuja, Mr Tauqir Ahmad. GSP, Natural resources Div., 53 Plaza blue area, Islamabad, Pakistan. (1943) MSc, geology (Punjab U., 1967). Dpty. dir. (tel. 051 + 825779). *Geology.*

Siddiqui, Mr Jawed Ahmad. Centre of Excellence in Mineralogy, U. of Baluchistan, Sariab Road, Quetta, Pakistan. (1950) MSc, geology (Karachi U., 1977). Lect. *Geology.*

Siddiqui, Dr Rafiq Ahmad. Training Div., PCSIR, Shahrah-e-Kamal Attaturk, Karachi, Pakistan. (1930) PhD, physical chemistry (British Columbia U., Canada, 1961). Dir. (tel. 871349). *Crystal growth.*

Tauqir, Dr Anjum. Material Science/Metallurgy Div., Dr.A.Q.Khan Res.Labs.,Kahuta, P.O. Box 502, Rawalpindi, Pakistan. (1953) PhD, metallurgy (Connecticut U.,USA,1986). Sr. eng. *Materials science, metals.*

Yousufzai, Mr Inayatullah Khan. X-ray and Microscopy Div., Pakistan Inst. of Cotton Res. and Techn., Moulvi Tameezuddin Khan Road, Karachi, Pakistan. (1938) MSc, fibre science (Strathcylde U., UK, 1966). Sr. res. officer. (tel. 552007). *Microscopy, X-ray diffraction.*

Yusaf, Dr Mohammad. Glass and Ceramics Div., PCSIR Labs., Roomi Road, Lahore, Pakistan. (1940) PhD, analytical chemistry (Charles U., Czechoslovakia, 1973). Principal scient. officer. (tel. 870324). *Minerals, X-ray diffraction.*

PERU

Sub-Editor: R. Salazar Orrego

Notes

1. International telephone country code - 51.

Asmat, Dr Humberto. Physics Dept., U. Nac. de Ingenieria, Ave. Tupac Amaru s/n Rimac, Lima 25, Peru. (1944) D. 3e cycle, solid state physics (U. Scient. et Medicale de Grenoble, France, 1977). Prof. (tel. 811070, ext. 136). *Solid state physics, magnetism.*

Avalos, Dr Jaime. Physics Dept., U. Nac. de Ingenieria, Ave. Tupac Amaru s/n Rimac, Lima 25, Peru. (1947) D. 3e cycle, solid state physics (U. Scient. et Medicale de Grenoble, France, 1977). Prof. (tel. 811070, ext. 137). *Magnetic resonance.*

Cavero Ghersi, Dr César Augusto. Lab. de Quimica Inorganica, U. Nac. de Ingenieria, Ave. Tupac Amaru s/n Rimac, Lima 25, Peru. (1942) Dr, structural chemistry (U. Scient. et Medicale de Grenoble, France, 1975). Assoc. prof. (tel. 811070, ext. 216). *Structural crystallography, crystal chemistry.*

Cisneros Ramos, Prof. Luis. Facultad de Ingenieria Quimica y Manufacturera, U. Nac. de Ingenieria, Brema Lima 5, Lima, Peru. (1945) Dr, physics (U. Scient. et Medicale de Grenoble, France, 1975). *Solid state physics, mineralogy, ceramics.*

Espinoza, Prof. Odon. Chemistry Dept., U. Nac. Mayor de San Marcos, Ave. Venezuela s/n Lima, Lima, Peru. (1920) chemistry - chemical engineering (U. Nac. de San Marcos, 1958). Prof. (tel. 525635). *Mineralogy.*

Horn, Prof. Manfred Josef. Dept. de Fisica, U. Nac. de Ingenieria, Casilla 1301, Lima, Peru. (1938) PhD, physics (British Columbia U., Canada, 1971). Assoc. prof. (tel. 811070, ext. 281). *Solid state physics, phase transitions, mineralogy, solar energy conversion (selective surfaces), X-ray diffraction.*

Linares, Dr Jorge. Dept. of Sci., Pontifica U. Catolica, Ave. Bolivar s/n Pueblo Libre, Lima, Peru. (1949) D. 3e cycle, solid state physics (U. Scient. et Medicale de Grenoble, France, 1978). Prof. (tel. 622540, ext. 239). *Magnetism, crystallography, Mössbauer spectroscopy, X-rays.*

Salazar Orrego, Prof. Ramon. Dept. de Fisica, U. Nac. de Ingenieria, Casilla 1301, Lima, Peru. (1932) MSc, physics (U. Nac. de Ingenieria, 1977). Principal prof. (tel. 811070, ext. 137). *Solid state physics, electron paramagnetic resonance.*

Valera, Dr Anibal Abel. Physics Dept., U. Nac. de Ingenieria, Ave. Tupac Amaru s/n Rimac, Lima, Peru. (1950) Dr.Rer.Nat., solid state physics (U. Stuttgart, BRD, 1979). Prof. (tel. 811070, ext. 136). *Solar energy, photovoltaic cells.*

Vega, Prof. Juan. Physics Dept., U. Nac. de Ingenieria, Ave. Tupac Amaru s/n Rimac, Lima, Peru. (1944) MSc, solid state physics (U. Nac. de Ingenieria, 1978). Prof. (tel. 811070, ext. 137). *Magnetic resonance.*

Velasquez, Prof. Jaime. Physics Dept., U. Nac. de Ingenieria, Ave. Tupac Amaru s/n Rimac, Lima, Peru. (1943) Lic(BS), solid state physics (U. Nac. de Ingenieria, 1974). Prof. (tel. 811070, ext. 136). *Solar energy, photovoltaic, cells, selective absorbtion coatings.*

PHILIPPINES

Sub-Editor: **B. S. Austria**

Notes

1. International telephone country code - 63.

Austria, Dr Benjamin Suarez. Nat. Inst. of Geological Sci., U. of The Philippines, Diliman, Quezon City 1101, Philippines. (1946) PhD, geology (Harvard U., USA, 1975). Dir. (tel. 98-17-26, fax. 6328179807). *Ore mineralogy.*

Cejalvo, Prof. Flor. Dept. of Mathematics, U. of The Philippines, Diliman, Quezon City 1101, Philippines. (1933) MS, mathematics (Stanford U., USA, 1964). Prof. (tel. 95-14-71). *Crystallographic groups.*

Felix, Dr Rene P.. Dept. of Mathematics, U. of The Philippines, Diliman, Quezon City 1101, Philippines. (1950) PhD, mathematics (U. Philippines, 1980). Prof. (tel. 95-14-71). *Crystallographic groups.*

Fernandez, Prof. Aurora Reyes. Dept. of Mathematics, U. of The Philippines, Diliman, Quezon City 1101, Philippines. (1938) MS, mathematics (U. Detroit, USA, 1963). Prof. (tel. 95-14-71). *Crystallographic groups.*

Llaguno, Dr Elma Caballes. Institute of Chemistry, U. of The Philippines, Diliman, Quezon City 1101, Philippines. (1947) PhD, physical chemistry (Illinois U., USA, 1973). Prof. (tel. 97-60-61, ext. 410). *Organic crystal structures, enviromental chemistry.*

Mallari-Kaballo, Mrs Paz P. Dept. of Mathematics, U. of The Philippines, Diliman, Quezon City 1101, Philippines. (1950) MS, statistics (U. Philippines, 1976). Asst. prof. (tel. 95-14-71). *Crystallographic groups.*

Nochefranca, Dr Luz R. Dept. of Mathematics, U. of The Philippines, Diliman, Quezon City 1101, Philippines. (1953) PhD, mathematics (U. Philippines, 1988) Prof. (tel. 95-14-71). *Crystallographic groups.*

Patalinghug, Dr Wyona Cruz. Chemistry Dept., De la Salle University, 2401 Taft Ave., Manila, Philippines. (1950) PhD, chemistry (U. Hawaii, 1982). Prof. (tel. 50-46-11). *Small molecules.*

Quibilan, Mr Edelmiro I., Solar Energy Sect., Philippine Nat. Oil Co., Energy Res. and Dev. Center, Don M. Marcos Ave., Diliman, Quezon City, Philippines. (1946) MS, physics (Pennsylvania State U., USA, 1973). Analyst. *Solid state physics, nuclear magnetic resonance, photovoltaics.*

Rapanut, Dr Teofina Axibal. University of the Philippines, U.P. College - Baguio, Baguio City, Philippines. (1947) PhD, mathematics (U. Philippines,1988) Asst. Prof. *Crystallographic groups.*

Soriano-Calix, Mrs Virginia B. Physics Res. Div., Philippine Atomic Energy Commission, Don M. Marcos Ave., Diliman, Quezon City, Philippines. (1944) MS, physics (U. Kansas, USA, 1970). Res. assoc. III. *Solid state physics, electron paramagnetic resonance.*

Trance, Dr Aurora Serrana. Dept. of Mathematics, De La Salle University, Taft Ave., Manila, Philippines. (1944) PhD, mathematics (Ateneo de Manila U., 1980). Prof. (tel. 95-14-71). *Crystallographic groups.*

Valencia, Dr Iluminado G. Araneta University Foundation, Victoria Park, Malabon, Metro Manila, Philippines. (1926) PhD, mineralogy (U. Wisconsin, USA, 1962). President. (tel. 361-90-56). *Clay mineralogy, fission product adsorption, clay analysis and identification.*

Victorio-Gervasio, Mrs Visitacion. Nat. Inst. of Geological Sci., U. of The Philippines, Diliman, Quezon City 1101, Philippines. (1929) BS, chemistry (U. Philippines, 1951). Instructor. *Mineralogy, petrography, petrology.*

POLAND

Sub-Editor: **A. Pietraszko**

Notes

1. International telephone country code - 48.

2. The Polish equivalent of MSc is *magister* (Mgr), which is also used as a title. *Adiunkt* is equivalent to reader and *docent* to assistant Professor. The degree of *doktor habilitowany* (Dr hab.) is next higher than PhD.

Adamiak, Dr Dorota Anna. Polish Acad. of Sci., Inst. of Bioorganic Chemistry, Noskowskiego 12/14, Poznań 61-704, Poland. (1948) doctor, crystallography (U. A.Mickiewicza, 1975). Adiunct.(tel. 061+526877, telex 0413600PANPL). *Crystallization, X-ray investigation, biologicaly important compounds, nucleic acids and protein components.*

Anulewicz, Dr Romana. Dept. of Chemistry, U. of Warsaw, Pasteura 1, Warsaw 02-093, Poland. (1937) doctor, chemistry (Warsaw, 1979). (tel. 022+222892, telex 815439UWPL). *Organic crystal chemistry, intermolecular interactions in solids, substituent effect on molecular properties.*

Barcik, Dr Jan. Inst. of Physics and Chemistry of Metals, Silesian U., Bankowa 12, Katowice 40-007, Poland. (1933) PhD, solid state physics (Silesian U., 1968). Adiunkt. (tel. 40007). *X-ray metallography, electron microscopy (TEM & SEM), metal physics, phase transformation.*

Barszcz, doc. Edward. Physics Metallurgy, Inst. of Ferrous Metallurgy, K.Miarki 12, Gliwice 44-100, Poland. (1936) PhD, physics (Silesian U., 1969). Adiunkt. (tel. 914051, ext.327). *Phase transformations, metals.*

Bartczak, Dr hab. Tadeusz Jan. Inst. of General Chemistry, Techn. U. of Łódź, Żwirki 36, Łódź 90-924, Poland. (1935) Dr hab., chemistry (Techn. U. of Łódź, 1986). Docent. (tel. 0-42+365522, ext.598, telex 866136PL). *X-ray crystal structure analysis, metalloporphyrins, heterocycles containing phosphorus.*

Bąk, Dr Jadwiga. Central Lab., X-ray and Electron Microscopy, Inst. of Physics, Pol. Acad. of Sci., Al.Lotników 32/46, Warszawa 02-668, Poland. (1942) PhD, physics (Pol.Acad. of Sci., 1974). Scient. worker. (tel. 436034). *Real crystal structure, X-ray studies.*

Bednarski, Dr Stanisław. Solid State Physics, Inst. of Nuclear Res., Świerk 05-400, Poland. (1930) MSc, physics (Wrocław U., 1955). Adiunkt. (tel. 798, ext.684). *Crystal growth.*

Bedyńska, Dr hab. Teresa. Inst. of Physics, Polish Acad. of Sci., Al.Lotników, 32/46, Warszawa 02-668, Poland. (1932) Dr hab., physics (Inst. of Physics PAS, 1978). Adiunct. (tel. 22+437001, ext.301, telex 812468). *Dynamical theory, X-ray diffraction.*

Blinowski, Dr Konrad. Solid State Physics, Inst. of Nuclear Res., Świerk Res. Est., Świerk 05-400, Poland. (1928) PhD, physics (Świerk Inst. of Nuclear Res., 1965). Adiunkt. (tel. 798, ext.805). *Neutron diffraction, magnetic structures.*

Bogucka-Ledóchowska, Dr Maria. Dept. of Pharm. Techn. and Biochem., Techn. U., Majakowskiego 11/12, Gdańsk 80-952, Poland. (1927) PhD, organic chemistry (Techn. U. Gdańsk, 1965). Res. worker. (tel. 539652). *X-ray structure analysis, biologically active compounds, biological activity - chemical structure relation.*

Bojarski, Prof. Dr Zbigniew. Inst. of Pysics and Chemistry of Metals, Silesian U., Bankowa 12, Katowice 40-007, Poland. (1921) Dr (Silesian Techn. U., 1956). Professor. (tel. 596929, telex 0315584USKPL). *Materials science, crystallography, structure of metals.*

Bołd, Dr Tadeusz. Physical Metallurgy, Inst. of Ferrous Metallurgy, K.Miarki 12, Gliwice 44-100, Poland. (1934) PhD, physical metallurgy (Inst. of Ferrous Metallugry 1972). Head of X-ray Lab. (tel. 914051, ext.327). *Phase transformations, metals and alloys.*

Borowiak, assoc. Prof. Teresa. Fac. of Chemistry, A.Mickiewicz U., Dept. of Crystallography, Grunwaldzka 6, Poznań 60-780, Poland. (1939) assoc. Prof., chemistry (A.Mickiewicz U., 1975). Assoc. Prof. (tel. 48-61+699181, ext.374).

Structure organic compounds, organic crystal chemistry, biologically active compounds.

Borzecka-Prokop, Dr Barbara. Fac. of Chemistry, Jagiellonian U., Karasia 3, Kraków 30-060, Poland. (1953) PhD, chemistry (Jagiellonian U., 1985). Chemistry. (tel. 04-81+2336377, ext.268). *X-ray powder diffraction, phase transitions.*

Bronowska, Dr Wiesława. Inst. of Physics, Techn. U. of Wrocław, Wybrzeże Wyspiańskiego 27, Wrocław 50-370, Poland. (1944) dr, physics (Inst. for Low Temp. and Structure Res., 1978). Adiunkt. (tel. 203278). *Ferroelectrics phase transitions, crystal structures and physical properties, powder diffractometry.*

Brzozowski, Dr Andrzej Marek. Dept. of Crystallography, Inst. of Chemistry, U. of Łódź, Nowotki 18, Łódź 91-416, Poland. (1953) PhD, chemistry (Łódź U., 1976). Adiunkt. (tel. 332365). *Protein structure, crystallography, biological active compounds.*

Bukowska-Strzyżewska, Dr hab. Maria. Dept. of Structure Res. and Crystal Chem., Inst. of General Chemistry, Techn. U. of Łódź, Żwirki 36, Łódź 90-924, Poland. (1929) Dr hab., chemistry (Techn. U. Łódź, 1976). Profesor. (tel. 65522). *X-ray crystallography, organic and complex compounds.*

Chełkowski, Prof. August Jan. Physics Dept., Silesian U., Uniwersytecka 4, Katowice 40-007, Poland. (1927) Dr hab., solid state physics (A.Mickiewicz U., 1959). Prof. (tel. 598764). *Metal physics, molecular physics.*

Ciechanowicz-Rutkowska, Dr Maria. X-ray Div., Regional Lab., Physicochem. Analysis and Structural Res. JU, Karasia 3, Kraków 30-060, Poland. (1941) PhD, chemistry (Imperial C. London, UK, 1971). dr. (tel. 048-12+336377, ext.267). *Crystal structures, organic compounds.*

Ciszak, Dr Ewa. Dept. of Crystallography, A.Mickiewicz U., Grunwaldzka 6, Poznań 60-780, Poland. (1960) MSc, chemistry (A.Mickiewicz U., 1984). Postgraduate student. (tel. 61+699181, ext.489, telex 0413260UAMPL). *Organic crystal structures, X-ray analysis.*

Ciunik, Dr Zbigniew. Inst. of Chemistry, U. of Wrocław, Joliot-Curie 12, Wrocław 50-383, Poland. (1949) dr, asst. (Wrocław U., 1979). *X-ray crystallography, structural chemistry, amino acids, peptides.*

Czachor, Dr hab. Andrzej. Dept. of Solid State Physics, Inst. of Nuclear Res., Świerk Res. Est., Otwock 05-400, Poland. (1934) Dr hab., physics (Świerk Nuclear Res. Inst., 1975). Assi.Prof. (tel. 798649). *Crystal lattice dynamics theory.*

Czerwonka, Mr Janusz. Fac. of Chemistry, Jagiellonian U., Karasia 3, Kraków 30-060, Poland. (1954) MSc (Jagiellonian U., 1978). Assistent. (tel. 0-48-12+336377, ext.268, telex 322297PLUJ). *Phase transitions.*

Damm, Prof. Józef Zbigniew. Dept.of Crystal Defects, Inst. for Low Temp. and Structure Res. pl.Katedralny 1, Wrocław 50-950, Poland. (1924) Dr hab., chemistry (Inst. of Physical Chemistry PAS, 1969). Head. (tel. 221071, ext.29). *Structural defects in ionic crystals.*

Dauter, Dr Zbigniew. Dept. of Pharm. Techn. and Biochem., Techn. U., Majakowskiego 11/12, Gdańsk 80-952, Poland. (1948) PhD, organic chemistry (Techn. U. Gdańsk, 1975). Adiunkt. (tel. 471618). *Stereochemistry, biological activity-molecular structure relations.*

Dobrowolska, Dr Wanda. Dept. of Structure Res., Inst. of General Chemistry, Techn. U. of Łódź, Żwirki 36, Łódź 90-924, Poland. (1945) PhD, chemistry (Inst. of General Chemistry, 1976). Adiunkt. (tel. 65522). *Organic crystal structures.*

Dobrzyński, Dr hab. Ludwik. Solid State Physics, Inst. of Nuclear Res., Świerk Res.Est., Otwock 05-400, Poland. (1941) Dr hab., physics (Świerk Inst. of Nuclear Res., 1975). Adiunkt. (tel. 798, ext.805). *Neutron diffraction, magnetic structures.*

Durski, Dr Stanisław. Dept. of General Chem. & Inorg., Techn. U. of Warszawa, Noakowskiego 3, Warszawa 00-662, Poland. (1931) PhD, chemistry (Techn. U. of Warszawa, 1970). Adiunkt. *Inorganic and organic crystal structures, X-ray crystallography.*

Dynowska, Mrs Elżbieta. Central Lab., X-ray and Electron Microscopy, Inst. of Physics PAS, al.Lotników 32/46, Warszawa 02-668, Poland. (1944) MSc, physics of solid state (Warszawa U., 1969). Asst. (tel. 436034). *X-ray diffraction methods, crystal studies.*

Figas, Mgr Elżbieta Teresa. Dept. of Chemistry, Adam Mickiewicz U., Grunwaldzka 6, Poznań 60-780, Poland. (1949) MSc (Adam Mickiewicz Univ., 1977). Specjalista. (tel. 699181, ext.443). *Molecular crystals.*

Figielski, Dr hab. Tadeusz. Inst. of Physics PAS, al.Lotników 32/46, Warszawa 00-681, Poland. Dr hab., physics (Inst. of Physics). Asst. Prof. *Crystal physics.*

Gałązka, Dr hab. Robert. Inst.of Physics PAS, al.Lotników 32/46, Warszawa 00-681, Poland. Dr hab., physics (Inst. of Physics). Asst. Prof. (tel. 436034). *Crystal growth.*

Gałdecka, Dr Ewa Renata. Dept. of Crystallography, Inst. for Low Temp. and Structure Res., PAS, pl.Katedralny 1, Wrocław 50-950, Poland. (1947) dr, crystallographer (Inst. for Low Temp. and Structure Res., 1981). Adiunkt. (tel. 221071, ext.44). *Mathematical and statistical methods, crystallographic data interpretation, experimental optimization, computer programing.*

Gałdecki, Prof. Zdzisław. Dept. of Structure Res. and Crystal Chem., Inst. of General and Inorganic Chemistry, T.U. Łódź, Żwirki 36, Łódź 90-924, Poland. (1924) PhD, chemistry (Techn. U. of Łódź, 1960). Head. (tel. 65522, ext.524,). *Crystal structure analysis, crystal chemistry, biologically active compounds, crystallographic computer programming.*

Gawron, Dr Marian. Dept. of Crystallography, Fac. of Chemistry, A.Mickiewicz U., Grunwaldzka 6, Poznań 60-780, Poland. (1950) dr, chemistry (A.Mickiewicz U., Poznań, 1982). Adiunkt. (tel. 48-61+699181, ext.242). *Organic crystal chemistry, structure, organic and organometallic compounds, biologically active compounds.*

Gdaniec, Dr Maria. Dept. of Crystallography, A.Mickiewicz U., Grunwaldzka 6, Poznań 60-780, Poland. (1951) dr, chemistry (A.Mickiewicz U., 1978). Adiunkt. (tel. 61+699181, ext.489, telex 0413260UAMPL). *Crystal chemistry of natural products.*

Giebułtowicz, Dr Tomasz. Nuclear Methods, Solid State Physics Dept., Inst. of Exp. Physics, Warszawa U., Hoża 69, Warszawa 00-681, Poland. (1945) PhD, physics (Warszawa U., 1975). Adiunkt. (tel. 283031, ext.166). *Neutron studies of solids.*

Gilski, Mr Mirosław. Dept. of Crystallography, A.Mickiewicz U., Grunwaldzka 6, Poznań 60-780, Poland. (1963) MSc, physics (A.Mickiewicz U., 1987). Physicist. (tel. 61+699181, ext.489, telex 0413260UAMPL). *Computer programming.*

Gluziński, Dr Przemysław. Inst. of Organic Chemistry, Polish Acad. of Sci., Kasprzaka 44, Warszawa 01-224, Poland. (1924) PhD, chemistry (Techn. U. Warszawa, 1968). Adiunkt. (tel. 323221, ext.142, telex 817097ICHFPL). *X-ray structural investigation, organic compounds, crystallographic computer programming.*

Głowiak, Dr hab. Tadeusz. Inst. of Chemistry, U. of Wrocław, Joliot-Curie 14, Wrocław 50-383, Poland. (1935) PhD, X-ray crystallography (Wrocław U., 1969). Dept. head. *X-ray crystallography, crystal chemistry, coordination compounds, structural chemistry.*

Główka, Dr Marek. Inst. of General Chemistry, Tech. Univ. of Łódź, Żwirki 36, Łódź 90-924, Poland. (1948) Dr hab., crystal chemistry (Tech. Univ. of Łódź, 1983). Assistant Professor. (tel. 365522, ext.595). *Crystallography, therapeutic agents, modeling, receptor binding sites, receptor-drug interactions.*

Godwod, MSc Krzysztof Jan. Central Lab., X-ray and Electron Microscopy, Inst. of Physics PAS, Al.Lotników 32/46, Warszawa 02-668, Poland. (1938) MSc, exp. physics (Warszawa U., 1962). Special. in X-ray spectroscopy (tel. 436034). *X-ray diffraction dynamical effects, dynamical theory, X-ray spectrometry, X-ray diffractometry.*

Goliński, Dr Bohdan. Dept. of Structure Res., Inst. of General Chemistry, Tech. U., Żwirki 36, Łódź 90-539, Poland. (1931) PhD, chemistry (Techn. U. of Łódź, 1965). Adiunkt. (tel. 65-522). *Crystal structure, computer programming.*

Górkiewicz, Mgr Zbigniew. Dept. of Structure Res., Inst. of General Chemistry, Techn. U., Żwirki 36, Łódź 90-924, Poland. (1936) MSc, textile technology (Techn. U. of Łódź, 1968). Asst. (tel. 65522).

Górski, Dr Ludwik. Plasma Physics and Techn., Inst. of Nuclear Res, Świerk Res. Est., Otwock-Świerk 05-400, Poland. (1936) MSc, physical chemistry (Warszawa U., 1960). Scient. worker. (tel. 799612). *Phase transitions, thermal and pressure treatment, nonstoichiometric and unstable phases, plasma spray and similar processes, organic structures with biological activity, X-ray and neutron diffraction, hydrogen bonds.*

Grabowski, Asst. Prof. Mieczysław Jerzy. Dept. of Crystallography, Inst. of Chemistry, U. of Łódź, Nowotki 18, Łódź 91-416, Poland. (1928) doctor hab., crystallography (Łódź U., 1969). Dept. head. (tel. 332365). *Crystal structure and properties, X-ray analysis.*

Grochowski, Dr Jacek Mathias. X-ray Div., Regional Lab. of Physicochemical Res., Jagiellonian U., Karasia 3, Kraków 30-060, Poland. (1943) dr, chemistry (Jagiellonian U., 1975). Head, X-ray Div. (tel. 012 + 33 63 77, ext. 267, telex 0322-297). *Anomalous dispersion, small molecule structure analysis, materials research.*

Grochulski, Dr Paweł Inst. of Physics, Techn. U. of Łódź, Wólczańska 219, Łódź 93-005, Poland. (1955) dr, physics (Techn. U. of Łódź, 1988). Adiunct. (tel. 42+365522, ext.894). *X-ray structure analysis.*

Gronkowski, Dr Jerzy. Dept. of Structure Res., Inst. of Exp. Physics, U. of Warsaw, Hoża 69, Warszawa 00-681, Poland. (1949) PhD, physics (U. of Warsaw, 1979). Adiunkt. (tel. 22+283031, ext.137, telex 815548UWPHYPL). *Dynamical X-ray diffraction, real cystal structure, X-ray topography.*

Grylicki, Dr Mirosław. Inst. of Inorganic Chemistry and Technology, Gdańsk Techn. U., Majakowskiego 11/12, Gdańsk 80-952, Poland. (1926) PhD, chem.tech. (Acad. of Mining and Metallurgy, Kraków, 1963). Assoc.Prof. (tel. 58+472065, telex 512302PL). *Phase analysis, X-ray diffraction (powder methods), mineralogical analysis, solid state chemistry, materials science, materials engineering, ceramic materials technology.*

Grzymek, Prof. Jerzy. Inst. of Building and Refractory Materials, Acad. of Mining and Metallurgy, Al.Mickiewicza 30, Kraków 30-059, Poland. (1908) Prof.dr, physicochemistry of Silicate. Prof. (tel. 012+339595, telex 0322203PL). *Physicochemistry, silicate crystallization, polymorphic phase transformation, technological processes, crystallization phenomena during phase transformation.*

Habla, Dr Halina. Inst. of Physics and Chemistry of Metals, Silesian U., Bankowa 12, Katowice 40-007, Poland. (1939) dr, solid state physics (Silesian U., 1970). Adiunkt. (tel. 588211, ext.471). *Structure, solid state, X-ray microanalysis.*

Hodorowicz, Prof. Stanisław. Fac. of Chemistry, Jagiellonian U., Karasia 3, Kraków 30-060, Poland. (1941) Dr hab., solid state chem. (Jagiellonian U., 1979). Prof. (tel. 04-812+336377, ext.267, telex 322297UJ). *Crystal chemistry,* *isopolymolybdates, crystal structure, phase transitions, superconductors, biologically active compounds.*

Holas, Dr hab. Andrzej. Dept. of Solid State Physics, Inst. of Nuclear Res., Świerk Res. Est., Otwock 05-400, Poland. (1940) Dr hab., physics (Inst. of Nuclear Res., 1980). Adiunkt. (tel. 798649). *Crystal lattice dynamics theory.*

Horn, Dr Jerzy. Inst. of Inorganic Chemistry, Techn. U. of Wrocław, Wybrzeże Wyspiańskiego 27, Wrocław 50-370, Poland. (1935) PhD, chemistry (Inst. for Low Temp. and Structure Res., 1972). Adiunkt. *Inorganic crystal structures.*

Horyń, Dr hab. Roman. Inst. for Low Temp. and Structure Res., Pol. Acad. of Sci., Pl.Katedralny 1, Wrocław 50-950, Poland. (1936) Dr hab., inorganic chemistry (Techn. U., 1967). (tel. 221071, ext.280). *Phase equilibria, crystal structure, crystal growth, intermetallics, inorganic materials.*

Isakow, Dr Zofia. Inst. of Physics and Chemistry of Metals, Silesian U., Bankowa 12, Katowice 40-007, Poland. (1947) dr, physics (Silesian U., 1981). Doctor. (tel. 845+596929). *Phase analysis, structure research, applied crystallography.*

Iwanicka, Mrs Iwona Ewa. Inst. of General Chemistry, Techn. U. of Łódź, Zwirki 36, Łódź 90-924, Poland. (1957) Mrs, physics (Techn. U. of Łódź, 1980). Assistant. (tel. 365522, ext.595). *Crystallography, biologically active compounds.*

Jackowski, Dr Józef Wojciech. Zakład Metaloznawstwa, Instytut Metalurgii AGH, Mickiewicza 30, Kraków 30-059, Poland. (1942) PhD, metal science (Acad. of Mining and Metallurgy, 1971). Adiunkt. (tel. 38100, ext.2628). *Deformation mechanism, recrystalization, metals and alloys, texture formation.*

Janecka, inż Maria. Inst. of Organic Mat.Sc., Chrobrego 27, Radom 26-600, Poland. (1952) inż, chemistry (TU Radom, 1979). Chemistry. (tel. 48+40031, ext.162). *Metallorganic compounds chemistry.*

Janecki, Mgr inż Hieronim Piotr. Inst. of Organic Sc., Chrobrego 27, Radom 26-600, Poland. (1951) Mgr inż, chemistry (TU Radom, 1978). Asst. (tel. 48+40031, ext.215). *Metal surface chemistry, morphology, structure, metals and elements.*

Janik, Prof. Jerzy. Inst. of Nuclear Physics, Radzikowskiego 152, Kraków 31-342, Poland. (1927) Dr hab., physics (U. Jagiellonian, 1950). Dept. head. *Molecular crystal dynamics.*

Janko, Dr Andrzej. Dept. of Catalysis on Metals, Inst. of Physical Chemistry PAS, Kasprzaka 44/52, Warszawa 01-224, Poland. (1925) PhD, physics (PAS, 1965). Adiunkt. (tel. 323221, ext.284). *Metals, structures, phase processes, metal catalysis, instrumentation.*

Jaskólski, Dr Mariusz. Dept. of Crystallography, A.Mickiewicz U., Grunwaldzka 6, Poznań 60-780, Poland. (1952) Dr hab., chemical crystallography (A.Mickiewicz U., 1985). Docent. (tel. 61+699181, ext.489, telex 0413260UAMLP). *Stuctural chemistry, nucleic acids, constituents, protein crystallography, hydrogen bond.*

Joachimiak, Dr Andrzej. Polish Acad. of Sci., Inst. of Bioorganic Chemistry, Noskowskiego 12/14, Poznań 61-704, Poland. (1951) PhD, chemistry (A.Mickiewicz U. Poznań, 1979). Adiunct. (tel. 061+528503, telex 0413600PANPL). *Crystallization, X-ray analysis, biologicaly important proteins, nucleic acids, structure - function relation.*

Kajzar, Dr Franciszek. Dept. of Condensed Phase Physics, Inst. of Physics and Nuclear Techn., Ac. of Mining and Metall., Al.Mickiewicza 30, Kraków 30-059, Poland. (1942) PhD, theoretical physics (Jagiellonian U. Kraków, 1970). Asst. Prof. (tel. 39100, ext.2955). *Magnetic interaction theory.*

Kalicińska-Karut, Mgr Jarosława. Dept. of Crystallography, Inst. for Low Temp. and Structure Res., pl.Katedralny 1, Wrocław 50-950, Poland. (1941) Mgr, chemistry (Techn. U. of Wrocław, 1967). Chemistry. (tel. 221071, ext.54). *Ferroelectric crystal structures, phase transitions.*

Kałuski, Prof. Dr hab. Zygmunt. Chemistry Dept.U.Poznań, Grunwaldzka 6, Pozna ń 60-780 Poland. (1920) D.Sc., chemistry (A.Mickiewicz U. Poznań, 1967). Profesor. (tel. 699181, ext.443). *Crystal chemistry, organic and organometallic compounds, bis-quinolizidine alkaloids.*

Karniewicz, Dr Jan. Inst. of Physics, Techn. U. of Łódź, Wólczańska 219, Łódź 93-005, Poland. (1928) dr, physics (Techn. U. of Łódź, 1966). Physicist. (tel. 363139). *Crystal growth.*

Karolak-Wojciechowska, Dr Janina. Chemistry Dept., Inst. of General Chemistry, Techn. U., Żwirki 36, Łódź 90-924, Poland. (1942) Dr hab., chemistry (Techn. U. of Łódź, 1986). Docent. (tel. 365522). *Crystallography, small molecules, drug action, molecular mechanics.*

Karp, Prof. Dr Jan. Instytut Metalografii, Akademia Górniczo-Hutnicza, Al.Mickiewicza 30, Kraków 30-059, Poland. (1922) Prof., physical metallurgy (Akademia Górniczo-Hutnicza, 1961). Prof. (tel. 33823). *Texture, metals, relative orientations in metals.*

Kartusiak, Dr Andrzej Szczepan. Dept. of Chemistry, Adam Mickiewicz U., Grunwaldzka 6, Poznań 60-780, Poland. (1955) PhD (Adam Mickiewicz Univ., 1983). Adiunkt. (tel. 699181, ext.443). *High-pressure X-ray crystallography, statistics applications in crystallography, bis-quinolizidine alkaloids.*

Kaszkur, Dr Zbigniew. Dept. of Catalysis on Metals, Inst. of Physical Chemistry, PAS, Kasprzaka 44/52, Warszawa 01-224, Poland. (1954) PhD, physics (Instutute of Physical Chemistry PAS, 1987). Adiunkt. (tel. 323221, ext.284). *Amorphous state, radial distribution function, structure, solid carbons.*

Keller, Dr hab. Włodzimierz. Inst. of Material Science, Warsaw Techn. U., Narbutta 85, Warszawa, Poland. (1929) Dr hab., crystallography (Warsaw Techn. U.,

1972). Docent. (tel. 499929). *Thermal motion in crystals, dynamical diffraction theory, precise parameter measurement.*

Kociński, Prof. Dr Jerzy. Inst. of Physics, Politechnika Warszawska, Koszykowa 75, Warszawa 00-628, Poland. *Phase transitions.*

Kołakowski, Dr Bogdan Józef. Magnetic Div., Inst. of Physics, Pol. Acad. of Sci., al.Lotników 32/46, Warszawa 02-668, Poland. (1933) PhD, crystal physics (Inst. of Physics, Pol.Acad.Sci.1968). Adiunkt. *Symmetry theory, phase transitions, organic crystal structure.*

Kołakowski, Dr Andrzej. Magnetic Div., Inst. of Physics, Pol. Acad. of Sci., al.Lotników 32/46, Warszawa 02-668, Poland. (1941) MSc, crystal physics (Łódź U., 1975). Physicist. *Organic crystal structure, computer programming.*

Konitz, Mr Antoni. Dept. of Pharm. Techn. and Biochem., Techn. U., Majakowskiego 11/12, Gdańsk 80-952, Poland. (1948) MSc, organic chemistry (Techn. U. of Gdańsk, 1972). Res. worker. (tel. 471618). *Organic chemistry, biochemistry, crystallography.*

Kosturkiewicz, Prof. Dr Zofia. Dept. of Crystallography, A.Mickiewicz U., Grunwaldzka 6, Poznań 60-780, Poland. (1928) Dr hab., chem. crystallography (A.Mickiewicz U., 1969). Professor. (tel. 699181, ext.488). *Crystal chemistry, organic compounds, crystallography.*

Kowalski, Dr Grzegorz. Dept.of Structure Res., Inst. of Exp. Physics, U. of Warsaw, Hoża 69, Warszawa 00-681, Poland. (1951) PhD, physics (U. of Warsaw, 1984). (tel. 22+294229, telex PHYPL815548UW). *Dynamical X-ray diffraction, real crystal structure, X-ray topography.*

Kozioł, Dr Anna E. Inst. of Chemistry, Maria Curie-Skłodowska U., pl.Marii Curie-Skłodowskiej 3, Lublin 20-031, Poland. (1951) PhD, chemistry (A.Mickiewicz U., 1980). Adiunkt. (tel. 081+375662). *Stereochemistry, natural products, structure, small organic molecules, metal complexes.*

Kozłowska, Mrs Krystyna. Dept. of Structure Res., Inst. of General Chemistry, Techn. U., Żwirki 36, Łódź 90-539, Poland. (1948) Mgr, matematics (Łódź U., 1972). Sr.asst. (tel. 65522). *Computer programming, structure analysis.*

Krajewski, Dr Janusz. Inst. of Organic Chemistry, Polish Acad. of Sci., Kasprzaka 44/52, Warszawa 01-224, Poland. (1929) PhD, phys.chem. (U. of Warsaw, 1964). Adiunkt. (tel. 323221, ext.142, telex 817097ICHFPL). *X-ray analysis.*

Krawiec, Mr Mariusz. Dept. of Chemistry, U. of Warsaw, Pasteura 1, Warsaw 02-093, Poland. (1963) chemistry (1987). (tel. 022+222892, telex 815439UWPL).

Krukowski, Dr Marek. Dept. of Physico-chemistry of Solid State, Inst. of Physical Chemistry, Pol.Acad.Sci., Kasprzaka 44/52, Warszawa 01-224, Poland. (1946) MSc, light organic technology (Warszawa Techn. U., 1969). Scient. co-worker. (tel. 323221, ext.262). *Structural and physico-chemical properties in solid state, metals and alloys, high pressures, high temperature, low temperature.*

Krygowski, prof. Tadeusz Marek. Dept. of Chemistry, U. of Warsaw, Pasteura 1, Warsaw 02-093, Poland. (1937) DSci, chemistry (Warsaw, 1973). Professor. (tel. 022+222892, telex 815439UWPL). *Intermolecular interactions, substituent effects, pi-electron systems statistics in chemistry, organic crystal chemistry.*

Kubiak, Dr hab. Ryszard. Dept. of Crystallography, Inst for Low Temp. and Structure Res., pl.Katedralny 1, Wrocław 50-950, Poland. (1944) PhD, chemistry (Inst. for Low Temp. and Structure Res. PAS, 1974). Docent. (tel. 221071). *Low temperature X-ray analysis, metal compounds.*

Kubiak, Dr Maria. Dept. of Chemistry, U. of Wrocław, Joliot-Curie 14, Wrocław 50-383, Poland. (1945) dr, chemistry (tel. 229283). *Applied crystallography, palladium and platinum chemistry, S.N-heterocyclic ligands.*

Kubicki, MSc Maciej. Dept. of Crystallography, A.Mickiewicz U., Grunwaldzka 6, Poznań 60-780, Poland. (1963) MSc, physics (A.Mickiewicz U. Poznań, 1986). (tel. 48-61+699181, ext.242). *Structure, organic compounds, organic crystal chemistry, biological active compounds.*

Kucab, Dr Marian. Dept. of Condensed Phase Physics, Inst. of Physics and Nuclear Techn., Acad. of Mining and Metall., al.Mickiewicza 30, Kraków 30-059, Poland. (1946) PhD, theoretical physics (Jagiellonian U. Kraków, 1973). Asst. Prof. (tel. 39100, ext.2955). *Group theory applications, magnetic structures.*

Kucharczyk, Dr Damian. Dept. of Crystallography, Inst. for Low Temp. and Structure Res. PAS, pl.Katedralny 1, Wrocław 50-950, Poland. (1951) dr, physics (Inst. for Low Temp. and Structure Res. PAS, 1978). Adiunkt. (tel. 221071, ext.44). *Phase transitions, ferroelectrics materials, modulated structure.*

Kusz, Mgr Joachim. Silesian U., Inst. of Physics, Uniwersytecka 4, Katowice 40-007, Poland. (1955) MSc, physics (1979). Mgr. (tel. 588211, ext.114). *Crystallographic computing, modulated structures.*

Kwiatkowski, Mgr inż Witold. Dept. of Crystallography, Techn. U., Inst. of General Chemistry, Żwirki 36, Łódź 90-924, Poland. (1961) Mgr, physics (Techn. U. of Łódź, 1985). Assistent. (tel. 365522, ext.524). *Crystallography, drug action, molecular mechanics.*

Lappa, Dr Ryszard. Dept. of Chemistry, U. Teachers C., 3 Maja 54, Siedlce 08-110, Poland. (1920) PhD, solid state physics (Inst. of Physics PAS, 1969). Adiunkt. (tel. 431645). *Crystal growth physics, nucleation phenomena, experimental & theoretical studies, molecular beam technology.*

Leciejewicz, Prof. Dr Janusz. Inst. of Nuclear Chemistry and Techn., Dorodna 16, Warszawa 03-195, Poland. (1928) Dr hab., physics (Inst. of Nuclear Research, 1975). Inst. dir. (tel. 110656). *X-ray and neutron crystallography, magnetic materials, f-electron complex compounds.*

Lefeld-Sosnowska, Prof. Dr hab. Maria. Dept. of Structure Res., Inst. of Exp. Physics, U. of Warsaw, Hoża 69, Warszawa 00-681, Poland. (1934) Dr hab., physics (U. of Warsaw, 1979). Prof. (tel. 283031, ext.135). *Dynamical X-ray diffraction, real crystal structure, X-ray topography.*

Lewiński, Dr Krzysztof. Fac. of Chemistry, Jagiellonian U., Karasia 3, Kraków 30-060, Poland. (1954) PhD, chemistry (Jagiellonian U., 1983). Adiunkt. (tel. 048-12+336377, ext.270, telex 322297PLUJ). *Crystal structure analysis, metal ion binding, organometallic compounds, proteins.*

Ligenza, Asst. Prof. Sylwester. Solid State Physics Lab., Inst. of Atomic Energy, Świerk 05-400, Poland. (1935) PhD, physics (A.Mickiewicz U. Poznań, 1958). Asst. Prof. (tel. 798642). *Neutron diffraction, magnetic and crystal structures, neutron spectroscopy, Mossbauer effect.*

Lipkowska, Dr Zofia. Inst. of Organic Chemistry, Polish Acad. of Sci., Kasprzaka 44/52, Warsaw 01-224, Poland. (1946) PhD, crystallogr (Inst. of Organic Chemistry, PAS, 1975). Adiunkt. (tel. 323221, ext.142, telex 817097ICHF PL). *Organic crystal structures, structure-activiy relationships, molecular modelling.*

Lipkowski, Asst. Prof. Janusz. Inst. of Physical Chemistry, Polish Acad. of Sci., Kasprzaka 44/52, Warszawa 01-224, Poland. (1943) Dr hab., physical chemistry (Polish Acad. of Sci., 1983). Docent. (tel. 022+322159, telex 817097ICHFPL). *Clathrate(s), inclusion phenomena, crown compounds, cryptates, cyclodextrin(s), structure and chemical behaviour.*

Lis, Dr Tadeusz. Dépt. of Chemistry, U. of Wrocław, F.Joliot-Curie 14, Wrocław 50-383, Poland. (1947) dr.hab., chemistry (IU, 1980). Docent. (tel. 229283). *Applied crystallography, carbohydrate chemistry, rhenium and manganese chemistry.*

Łagiewka, Dr Eugeniusz. Inst. of Physics and Chemistry of Metals, Silesian U., Bankowa 12, Katowice 40-007, Poland. (1939) Dr hab., physics (PAS, INTiBS, Wrocław, 1983). Docent. (tel. 596929). *Metal physics, structure research, applied crystallography.*

Lasocha, Dr Wiesław. Dept. of Chemistry, Jagiellonian U., Karasia 3, Kraków 30-060, Poland. (1957) PhD, solid state chem. (Jagiellonian U., 1986). Assistant. (tel. 048-12+336377, ext.268, telex 322297PLUJ). *Crystal chemistry, isopolymolybdates, superconductors.*

Łukaszewicz, Prof. Kazimierz. Dept. of Crystallography, Inst. for Low Temp. and Structure Res., pl.Katedralny 1, Wrocław 50-950, Poland. (1927) Dr hab., physics (Inst.of Physical Chemistry PAS, 1968). Prof. (tel. 48-071+225739, telex 0712777INTPL). *Crystal structure analysis, phase transitions, precision lattice parameters.*

Maciosowski, Dr Andrzej. Physical Metallurgy, Inst. of Ferrous Metallurgy, K.Miarki 12, Gliwice 44-101, Poland. (1944) PhD, physical metallurgy (Acad. Metallurgy Kraków, 1973). Adiunkt (tel. 914051, ext.288). *Crystallography, phase transformations in solid state, structure - properties relation, steel, texture.*

Majchrzak, Dr Stanisław. Dept. of Catalysis on Metals, Inst. of Chemistry PAS, Kasprzaka 44/52, Warszawa 01-224, Poland. (1925) PhD, chemistry (Inst. of Physical Chem. PAS, 1969). Adiunkt. (tel. 323221, ext.284). *High pressure X-ray diffraction, hydrides, clathrates, phase transformations, instrumentation.*

Majewska, Miss Katarzyna. Inst. of Physics, Silesian U., Uniwersytecka 4, Katowice 40-007, Poland. (1961) MSc, physics (Silesian U., 1985). Assistant. (tel. 588211, ext.501). *Solid state physics.*

Malinowski, Dr Mariusz. Dept. of Crystallography, Inst. for Low Temp. and Structure Res., pl. Katedralny 1, Wrocław 50-950, Poland. (1950) dr, physics (Inst. for Low Temperature and Structure Res., 1980). Adiunkt. (tel. 221071, ext.44). *Phase transitions, ferroelectric materials.*

Maliszewski, Dr Edward. Solid State Physics, Inst. of Nuclear Res., Świerk Res. Est., Otwock 05-400, Poland. (1930) PhD, physics (Świerk Inst. of Nuclear Res., 1967). Adiunkt. (tel. 798, ext.303). *Crystal lattice dynamics, neutron scattering.*

Maluszyńska, Dr Hanna. Chemistry Dept., A.Mickiewicz U. Poznań, Grunwaldzka 6, Poznań 60-780, Poland. (1947) MSc, physics (A.Mickiewicz U., 1971). Asst. (tel. 699181, ext.443). *X-ray crystallography, organic compounds.*

Matyja, Dr Elżbieta. Central Lab., X-ray and Electron Microscopy, Inst. of Physics, PAS, al.Lotników 32/46, Warszawa 02- 668, Poland. (1938) PhD, techn.sci. (Polytechnic of Warszawa, 1972). Scient. worker. (tel. 437001, ext.143). *Crystal defects, electron microscopy.*

Matyja, Dr Przemysław. Inst. of Physics and Chemistry of Metals, Silesian U., Bankowa 12, Katowice 40-007, Poland. (1939) PhD, physics of metals (Silesian U. of Katowice, 1976). Adiunkt. (tel. 587231, ext.442). *X-ray metallography.*

Modrzejewski, Dr hab. Antoni. E-VI, Inst. of Atomic Energy, Świerk 05-400, Poland. (1931) Dr hab., technology (MSc in Physics, 1975). Doc.Dr hab. (tel. 799644, telex 813244). *Crystal growth.*

Morawiec, Prof. Henryk. Inst. of Physics and Chemistry of Metals, Silesian U., Bankowa 12, Katowice 40-007, Poland. (1933) Dr Sc.PhD (1967). Dean of Fac. (tel. 596929). *Crystallography, phase transformation, martensitic transformation, X-ray methods.*

Murasik, Dr Andrzej. Solid State Physics, Inst. of Nuclear Res., Świerk Res.Est., Otwock 05-400, Poland. (1931) PhD, physics (Świerk Inst. of Nuclear Res., 1970). Adiunkt. (tel. 798, ext.324). *Neutron diffraction, magnetic structures.*

Nizioł, Dr Stanisław. Dept. of Condensed Phase Physics, Inst. of Physics and Techniques, Acad. of Mining and Metallurgy, A.Mickiewicza 30, Kraków 30-059, Poland. (1941) PhD, exp. physics (Acad. of Mining and Metallurgy,

Kraków, 1972). Asst. Prof. (tel. 39100, ext.2960). *Magnetic structure determination, neutron diffraction method.*

Olech, Mr Andrzej. Fac. of Chemistry, Jagiellonian U., Karasia 3, Kraków 30-060, Poland. (1956) MSc, chemistry (Jagiellonian U., 1980). Teaching Assistant. (tel. 048-12+336377, ext.270, telex 322297PLUJ). *Crystal growth kinetics.*

Olejnik, Dr Stanisław. Dept. of Crystallography, Inst. for Low Temp. and Structure Res., pl.Katedralny 1, Wrocław 50-950, Poland. (1947) PhD, chemistry (Inst. for Low Temp. and Structure Res. PAS, 1975). Adiunkt. (tel. 221071). *Solid state, ferroelectrics, crystal structure analysis.*

Oleksyn, Dr Barbara. Fac. of Chemistry, Jagiellonian U., Karasia 3, Kraków 30-060, Poland. (1940) PhD, chemistry (Jagiellonian U., 1972). Adiunkt. (tel. 048-12+336377, ext.267, telex 322297PLUJ). *Crystal structure analysis, biologically active compounds.*

Oleś, Prof. Andrzej. Inst. of Physics and Nuclear Techn., Acad. of Mining and Metallurgy, al.Mickiewicza 30, Kraków 30-059, Poland. (1923) D.Ph., solid State Phys. (Jagiellonian U.Kraków, 1968). Chief of Dept. (tel. 012+344482, telex 322203PL). *Neutron diffraction, magnetic structures.*

Paciorek, Dr Włodzimierz. Crystallography, Inst.for Low Temp. & Structure Res. PAS, pl.Katedralny 1, Wrocław 50-950, Poland. (1953) Dr, physics (1987). Crystallographer. (tel. 48-71+221071, ext.44, telex 712777INTPL). *Crystallographic computing, modulated structures, crystallographic computing.*

Pająk, Mgr Lucjan. Inst. of Physics and Chemistry of Metals, Silesian U., Bankowa 12, Katowice 40-007, Poland. (1947) Mgr, chemistry (Jagiellonian U., 1970). Asst. (tel. 588211, ext.442). *Small angle X-ray scattering.*

Pawlak, Dr Stanisław. Physical Metallurgy, Inst. of Ferrous Metallurgy, K.Miarki 12, Gliwice 44-101, Poland. (1944) PhD, physics of metals and physical metallurgy (Acad. of Mining and Metallurgy, Kraków, 1973). Adiunkt. (tel. 914051, ext.752). *Crystallography, martensitic transformation, metals, texture, anisotropic properties, structure - properties relations, high strength steels.*

Pielaszek, Dr Jerzy. Dept. Catalysis on Metals, Inst. of Physical Chemistry, Kasprzaka 44/52, Warszawa 01-224, Poland. (1941) PhD, physics (Polish Acad. of Sci., 1972). Adiunkt. (tel. 323221, ext.284). *Structure, metals, XRD, catalysts, amorphous solids.*

Pietraszko, Mrs Donata. Instytut Materiałoznawstwa, Politechnika Wrocławska, Smoluchowskiego 25, Wrocław 50-370, Poland. (1941) Mgr, physics (Wrocław U.1966). Sr. asst. *Inorganic crystal structures.*

Pietraszko, Dr Adam. Dept. of Crystallography, Inst. for Low Temp. and Structure Res. PAS, pl.Katedralny 1, Wrocław 50-950, Poland. (1943) PhD, physics (Inst. for Low Temp. and Structure Res. PAS, 1974). Adiunkt. (tel. 221071, ext.50). *X-ray crystal structure analysis, phase transitions, ferroelectrics, subconductors.*

Ratajczak-Sitarz, Mgr Małgorzata Agnieszka. Dept. of Chemistry, Adam Mickiewicz U., Grunwaldzka 6, Poznań 60-780, Poland. (1958) MSc, chemistry (Poznań Techn. Univ., 1982). Assist. (tel. 699181, ext.443). *Molecular crystals.*

Ratuszek, Dr Wiktoria Maria. Inst. of Physical Metallurgy, Acad. of Mining and Metallurgy, al.MIckiewicza 30, Kraków 30-059, Poland. (1936) PhD, metal science (Acad. of Mining and Metallurgy, 1971). Adiunkt. (tel. 38100, ext.2628). *Stacking fault energy, deformation, recrystallization, metals and alloys, texture formation.*

Ratuszna, Dr Alicja. Inst. of Physics, Silesian U., Uniwersytecka 4, Katowice 40-007, Poland. (1947) doctor, physics (Silesian U., 1968). Adiunkt. (tel. 588211, ext.501). *Condensed matter physics.*

Rychlewska, Dr Urszula. Dept. of Crystallography, A.Mickiewicz U., Grunwaldzka 6, Poznań 60-780, Poland. (1948) dr, chemistry (A.Mickiewicz U., Poznań, 1976). Adiunkt. (tel. 61+699181, ext.489, telex 0413260UAMPL). *X-ray structure analysis, natural products, conformation, organic molecules, crystallography, crystal chemistry.*

Sawka-Dobrowolska, Dr Wanda. Dept. of Chemistry, U. of Wrocław, F. Joliot - Curie 14, Wrocław 50-383, Poland. (1945) dr, chemistry. Adiunkt. (tel. 229283). *Applied crystallography, aminophosphonic & aminophosphinic acid chemistry.*

Serda, Mr Paweł Reg. Lab. of Physicochemical Analysis & Structural Res., Jagiellonian U., Karasia 3, Kraków 30-060, Poland. (1955) MSc, physics (Jagiellonian U., 1979). Research fellow. (tel. 048-12+336377, ext.267, telex 322297PLUJ). *Crystal structure analysis, crystallographic computing.*

Sikora, Dr Wiesława. Dept. of Condensed Phase Physics, Inst. of Physics and Nuclear Techn., Acad. of Mining and Metallurgy, A.Mickiewicza 30, Kraków 30-059, Poland. (1945) PhD, theoretical physics (Jagiellonian U. Kraków, 1974). Adiunkt. (tel. 39100, ext.2955). *Magnetic group theory, symmetry, magnetic structures.*

Skowerenda, Dr Jolanta. Dept. of Structure Res., Inst. of General Chemistry, Techn. U., Żwirki 36, Łódź 90-539, Poland. (1945) PhD, chemistry (Techn. U. of Łódź, 1976). Adiunkt. (tel. 65522). *Organic crystal structures.*

Skrzat, Dr Zofia. Dept. Crystallography and Mineralogy, Inst. Chemistry, N.Copernicus U., Gagarina 7, Toruń 87-100, Poland. (1919) drNat.Sc. (Nicolaus Copernicus U., 1960). Retired academic teacher. *Structure, organic compounds.*

Sosnowska, Dr hab. Izabela. Nuclear Methods, Solid State Physics Dept., Inst. of Exp. Physics, Warszawa U., Hoża 69, Warszawa 00-681, Poland. (1939) Dr hab., physics (Warszawa U., 1973). Asst. Prof. and head. (tel. 287252). *Structure and dynamics, crystal lattices, neutron scattering.*

Sosnowski, Dr hab. Jerzy. Solid State Dept., E-8, Inst. of Atomic Energy, Otwock-Świerk 05-400, Poland. (1936) D.Sc., physics (1975). (tel. 798303, telex 813244). *Lattice dynamics, neutron diffraction.*

Stadnicka, Dr Katarzyna. Fac. of Chemistry, Jagiellonian U., Karasia 3, Kraków 30-060, Poland. (1943) PhD, chemistry (Jagiellonian U., 1973). Adiunkt. (tel. 048-12+336377, ext.270, telex 322297PLUJ). *Crystal structure analysis, inorganics, organics, biologically active compounds, structure - properties relation.*

Staliński, Prof. Bohdan. Inst.for Low Temp. and Structure Res., PAS, pl.Katedralny 1, Wrocław 50-950, Poland. (1924) Dr hab., chemistry (Techn. U. Wrocław, 1956). Head. (tel. 221071). *Crystal structures and physical properties, magnetic compounds.*

Stanisz, Mgr Grzegorz. Inst. of Physics, Jagiellonian U., Reymonta 4, Kraków 30-059, Poland. MSc, physics (tel. 336377, ext.576). *Profile analysis, phase transitions.*

Stępień, Dr Andrzej. Dept. of Crystallography, Inst. of Chemistry, U. of Łódź, Nowotki 18, Łódź 91-416, Poland. (1945) PhD, crystallography (Łódź U., 1973). Adiunkt. (tel. 332365). *Organic crystal structures, X-ray analysis.*

Stępień-Damm, Dr Julia. Dept. of Crystallography, Inst. for Low Temperature and Structure Res., P.A.Sc., pl.Katedralny 1, Wrocław 50-950, Poland. (1945) dr, physics (MSc 1969, 1980). Adiunkt. (tel. 48-71+221071, ext.50). *Crystal structure, lattice defects, superconductivity.*

Stróż, Dr Danuta. Inst. of Physics and Chemistry of Metals, Silesian U., Bankowa 12, Katowice 40-007, Poland. (1951) dr, phys.metall. (Inst. of Metallurgy, Polish Acad.of Sci., 1984). *Electron microscopy, phase transformations, shape memory effect.*

Surowiec, Dr Marian Ryszard. Dept. of Technic, Inst. of Physics & Chemistry of Metals, Bankowa 12, Katowice 40-007, Poland. (1948) dr, physics (Silesian U., 1977). Doctor. (tel. 832+596929, ext.855, telex 0315584USKPL). *X-ray topography, real structure, semiconductors.*

Suwińska, Dr Kinga. Inst. of Physical Chemistry, Polish Acad. Sci., Kasprzaka 44/52, Warszawa 01-224, Poland. (1954) PhD, chemistry (Inst. of Physical Chemistry, PAS, 1984). Adiunkt. (tel. 022+323221, ext.254, telex 817097 ICHFPL). *Organic crystal structures, inclusion compounds, clathrates, molecular mechanics.*

Szarras, Dr Stanisław. Solid State Physics, Inst. of Nuclear Res., Świerk Res. Est., Otwock 05-400, Poland. (1924) PhD, chemistry (Techn. U. Warszawa, 1966). Adiunkt. (tel. 798, ext.648). *Crystal perfection.*

Szmid, Dr Zofia. Solid State Physics, Inst. of Nuclear Res., Świerk Res. Est., Otwock 05-400, Poland. (1921) PhD, physics (Świerk Inst. of Nuclear Res., 1967). Adiunkt. (tel. 798, ext.648). *Crystal perfection.*

Szummer, Dr Andrzej. Inst. of Materials Eng., Techn. U. of Warszawa, Nowowiejska 24, Warszawa 00-665, Poland. *Phase transformation, metals.*

Szurgot, Dr Marian. Inst. of Physics, Techn. U. of Łódź, Wólczańska 219, Łódź 93-005, Poland. (1946) dr, physics (Techn. U. of Łódź, 1987). Adiunkt. (tel. 362512, ext.14). *Crystal growth, etching, defects, morphology, surface micromorphology.*

Ślebarski, Dr Andrzej. Silesian U., Inst. of Physics, Uniwersytecka 4, Katowice 40-007, Poland. (1950) MSc, physics (Silesian U., 1973). Asst. (tel. 588211). *X-ray physics, solid state physics.*

Ślepowroński, Mr Marek. E-VI, Inst. of Atomic Energy, Świerk 05-400, Poland. (1951) MSc, physics (MSc in Physics, 1975). Assistant. (tel. 799644, telex 813244). *Crystal growth.*

Śliwiński, Mr Jan. Fac. of Chemistry, Jagiellonian U., Karasia 3, Kraków 30-060, Poland. (1952) MSc, chemistry (Jagiellonian U., 1976). Research fellow. (tel. 048-12+336377, ext.268, telex 322297PLUJ). *Crystal structure analysis, biologically active compounds.*

Tomaszewski, Dr Paweł. Dept. of Crystallography, Inst. for Low Temp. and Structure Res., pl.Katedralny 1, Wrocław 50-950, Poland. (1952) PhD, physics (Wrocław U., 1985). Adiunkt. (tel. 221071, ext.54, telex 712777INTPL). *Phase transitions, X-ray crystallography, ferroelectrics.*

Tosik, Dr Anita. Dept. of Structure Res., Inst. of General Chemistry, Techn. U. of Łódź, Żwirki 36, Łódź 90-539, Poland. (1947) MSc, chemistry (Łódź U., 1970). Sr.asst. (tel. 65522). *Inorganic crystal structures.*

Turowska-Tyrk, MSc Ilona. Dept. of Chemistry, U. of Warsaw, Pasteura 1, Warsaw 02-093, Poland. (1959) master, chemistry (Warsaw 1983). (tel. 022+222892, telex 815439UWPL). *Substituent effect on molecular properties, intermolecular interactions in solids, pi-electron system stability, organic crystal chemistry.*

Tykarska, Dr Ewa Maria. Dept. of Chemistry, A.Mickiewicz U., Grunwaldzka 6, Poznań 60-780, Poland. (1957) PhD, chemistry (A.Mickiewicz U., 1984). Dr - adiunkt. (tel. 69918, ext.489). *Crystallography, molecular biology, organic crystal structures.*

Urbańczyk, Prof. Dr hab. Grzegorz. Inst. of Fiber Physics and Textile Finishing, Techn. U. of Łódź, Żwirki 36, Łódź 90-924, Poland. (1928) Prof.Dr hab., techn.sci. (Techn. Universite of Łódź, 1963). Inst. Director. (tel. 362762, telex 8800136). *Fiber physics, methods, fiber fine structure investigation.*

Uszyński, Dr Ignacy. Crystallography, Inst. for Low Temp. & Structure Res., P.Ac. Sc., pl.Katedralny 1, Wrocław 50-950, Poland. (1950) dr, physics (MSc 1973, 1977). Crystallographer. (tel. 48-71+221071, ext.50, telex 712777INTPL). *Crystallographic computing, modulated structures, crystallographic symmetry.*

Wajsman, Dr Elżbieta. Dept. of Crystallography, Inst. of Chemistry U. of Łódź, Nowotki 18, Łódź 91-416, Poland. (1938) PhD, crystallography (Łódź U., 1975). Adiunkt. (tel. 332365). *Crystal structures, X-ray analysis.*

Warchoł, Dr Stanisław. E-VI, Inst. of Atomic Energy, Świerk 05-400, Poland. (1948) dr, technology (MSc in Physics, 1987). Adjunct. (tel. 799644, telex 813244). *Crystal growth.*

Warczewski, doc. Jerzy. Dept. of Condensed Phase Physics, Inst. of Physics and Nuclear Techn., Acad. of Mining and Metallurgy, al.Mickiewicza 30, Kraków 30-059, Poland. (1939) PhD, exp. physics (Acad. of Mining and Metallurgy, Kraków, 1969). Asst. Prof. (tel. 39100, ext.2955). *Crystal structure determination, phase transitions in crystals, displacive modulation in crystals, powder diffraction data error analysis.*

Waśkowska, Dr Alicja. Dept. of Crystallography, Inst. for Low Temperature & Structure Res., Polish Acad. Sci., pl.Katedralny 1, Wrocław 50-950, Poland. (1942) dr, physics (MSc, 1975). Adiunct. (tel. 221071, ext.54). *Crystal structure - property relationship, phase transitions, ferroelectricity, hydrogen bonding.*

Węglowski, Dr Stanisław. Inst. of Inorganic Chemistry, Techn. U. of Wrocław, Wybrzeże Wyspiańskiego 27, Wrocław 50-370, Poland. (1929) PhD, chemistry (Techn. U. of Wrocław, 1963). Adiunkt. *Inorganic crystal chemistry.*

Wawrzak, Dr Zdzisław. Inst. of Physics, Techn. U. of Łódź, Wólczańska 219, Łódź 93-005, Poland. (1955) dr, physics (Techn. U. of Łódź, 1988). Adiunct. (tel. 42+365522, ext.524). *X-ray structure analysis.*

Wieczorek, Dr Michał Dept. of Structure Res., Inst. of General Chemistry, Techn. U. of Łódź, Żwirki 36, Łódź 90-539, Poland. (1937) PhD, chemistry (Inst. of General Chemistry, 1970). Adiunkt. (tel. 65522). *Organic crystal structures.*

Wieteska, Dr Krzysztof. Dept. of Solid State Physics, Inst. of Nuclear Res., Świerk Res. Est., Otwock 05-400, Poland. (1946) dr, physics (Świerk Inst. of Nuclear Res., 1980). Adiunkt. *Crystal defects, X-ray topography.*

Wiewióra, Prof. Dr hab. Andrzej. Inst. of Geological Sci., Polish Acad. Sci., Al.Żwirki i Wigury 93, Warszawa 02-089, Poland. (1933) Prof.natural sci. (Warszawa U., 1981). Prof. (tel. 223051, ext.142). *Polytypism, phyllosilicates, X-ray methods, polytypes.*

Wokulski, Dr Zygmunt. Inst. of Physics and Chemistry of Metals, U. of Silesia, Bankowa 12, Katowice 40-007, Poland. (1940) PhD, solid state physics (U. of Silesia, 1978). Lecturer. (tel. 48+582441, ext.521, telex 0315584USKPL). *Crystal growth, Czochralski method, Bridgman method, CVD and PVD techniques, morphology, whiskers, dislocation structure, carbides, nitrides.*

Wolska, Dr Irena. Fac. of Chemistry, Dept. of Crystallography, A.Mickiewicz U., Grunwaldzka 6, Poznań 60-780, Poland. (1950) dr, chemistry (A.Mickiewicz U., Poznań, 1985). (tel. 48-61+699181, ext.374). *Structure, organic compounds, organic crystal chemistry, biologically active compounds.*

Wołcyrz, Dr Marek. Dept. of Crystallography, Inst. for Low Temperature & Structure Res., PAS, pl.Katedralny 1, Wrocław 50-950, Poland. (1952) dr, cryst. (Inst. for Low Temp. & Structure Res., PAS, 1982). Adiunkt. (tel. 221071, ext.54). *Crystal structure analysis, phase transitions, powder diffractometry.*

Woźniak, Mr Krzysztof. Dept. of Chemistry, Warsaw U., Pasteura 1, Warsaw 02-093, Poland. (1961) master, chemistry (Warsaw, 1986). Assistant. (tel. 022+222892, telex 815439UWPL). *Statistics applications, organic crystal chemistry, substituent effects, molecular properties.*

Zielińska-Rohozińska, Dr hab. Elżbieta. Dept. of Structure Res., Inst. of Exp. Physics, U. of Warsaw, Hoża 69, Warszawa 00-681, Poland. (1938) Dr hab., physics (U. of Warsaw, 1984). Adiunkt. (tel. 22+294229, telex 815548 UWHYPL). *Dynamical X-ray diffraction, real crystal structure, X-ray topography.*

Ziołowski, Dr Zbigniew. Dept. of Structure Res., Inst. of Iron Metallurgy, Miarki 12/14, Gliwice 44-100, Poland. (1919) PhD, techn.sci. (Techn. U. of Gliwice, 1960). Asst. Prof. *Metals and metallurgical materials - structure.*

Ziółowska, Dr Blanka. Inst. of General Chem. & Inorg. Techn., Techn. U. of Warszawa, Noakowskiego 3, Warszawa 00-664, Poland. (1928) PhD, chemistry (Techn. U. of Warszawa). Adiunkt. *Organic crystal structures.*

Żmija, Prof. Józef. Techn. Acad. of Military, Lazurowa, Warszawa, Poland. Dept. head. (tel. 366661, ext.3331). *Crystal growth.*

PORTUGAL

Sub-Editor: J. Lima-de-Faria

Notes

1. International telephone country code - 351.

2. Degrees conferred by the Portuguese Universities are *Doutor* (Dr), approximately equivalent to PhD at British Universities, and *Licenciado* (Lic. or Eng. for engineering), approximately equivalent to MSc at British universities.

3. University positions include (with approximate British equivalents added in parentheses), *Professor catedrático* (professor), *Professor associado* (reader), *Professor auxiliar* (lecturer), *Assistente* (demonstrator). A research career parallel to the university career includes *investigador coordenador* (equivalent to *professor caterdrático*), *investigador principal* (equivalent to *professor associado*), and *investigador auxiliar* (equivalent to *professor auxiliar*). *Naturalista* corresponds to curator.

Aires-Barros, Prof. Luis. Secção de Mineralogia, Instituto Superior Técnico, Av. Rovisco Pais, 1096 Lisboa Codex, Portugal. (1931) Dr, mineralogy (Instituto Superior Técnico, Lisboa, 1964). Prof. catedrático. (tel. 1 + 800806, fax +351-1-899242). *Mineralogical applications, X-ray crystallography, clay minerals, crystal optics.*

Almeida, de, Prof. Maria José Marques. Depto. de Física, Fac. de Ciencias e Tecnologia, 3000 Coimbra, Portugal. (1946) PhD, physics (Cambridge U., UK, 1975). Prof. associado. (tel. 39 + 23675, ext. 254, fax +351-39.29158). *Alloy structures, electron and spin densities.*

Alte da Veiga, Prof. Luis. Depto. de Física, Fac. de Ciencias e Tecnologia, 3000 Coimbra, Portugal. (1932) PhD, crystallography (Cambridge U., UK, 1964). Prof. catedrático. (tel. 39 + 23675, ext. 251, fax +351-39.29158). *Alloy structures, electron densities.*

Andrade, Dr Lourdes Rodrigues. Depto. de Física, Fac. de Ciencias e Tecnologia, 3000 Coimbra, Portugal. (1954) Dr., physics (Coimbra U., 1986). Prof. auxiliar. (tel. 39 + 23675, ext. 244, fax +351-39-29158). *Inorganic crystal structures, electron densities.*

Basto, Mrs Maria João. Laboratório de Mineralogia e Petrologia, Instituto Superior Técnico, Av. Rovisco Pais, 1096 Lisboa Codex, Portugal. (1945) Chemical Engineer (Instituto Superior Técnico, Lisboa, 1968). Investigador auxiliar. (tel. 1 + 800111, telex 63423 istutl p, fax 351-1-899242). *X-ray diffraction, crystal chemistry, sulfide minerals.*

Borges, Prof. Frederico. Depto. de Mineralogia e Geologia, Fac. de Ciencias, 4000, Porto, Portugal. (1942) Dr, geology (London U., UK, 1978). Prof. catedrático.(tel. 2 + 310290). *Recrystallization, intracrystalline deformation, electron microscopy.*

Bravo, Prof. Manuel. Depto. de Geologia, Fac. de Ciencias e Tecnologia, Univ. Nova de Lisboa, Quinta da Torre, 2825 Monte de Caparica, Portugal. (1933) PhD, experimental petrology (Edinburgh U., UK, 1973). Prof. associado. (tel. 1 + 2954464). *X-ray crystallography, teaching crystallography.*

Carrondo, Prof. Maria Arménia. Centro de Química Estrutural, Complexo Interdisciplinar, Instituto Superior Tecnico, Av. Rovisco Pais, 1096 Lisboa codex, Portugal. (1948) PhD, chemical crystallography (London U., UK, 1978). Prof. associado. (tel. 351-1+572616, fax +351-1-899242, telex 63423ISTUTLP) *Coordination compounds, macromolecules, proteins.*

Correia dos Santos, Dr António. Centro de Química-Física e Radioquímica, Faculdade de Ciências de Lisboa, R. Escola Politécnica, 1200 Lisboa, Portugal. (1948) Dr, Chemistry (Lisboa U.,1987). Prof. auxiliar. (tel. 1 + 608932, fax 351-1-758.31.87). *Solid state chemistry, crystal structures.*

Costa, Prof. M. Margarida Ramalho. Depto. de Física, Fac. de Ciencias e Tecnologia, 3000 Coimbra, Portugal. (1945) PhD, physics (Cambridge U., UK, 1974). Prof. catedrático. (tel. 39 + 29252, ext. 256, fax +351-39.29158). *Alloy structures, electron and spin densities.*

Domingos, Dr Angela. Depto. de Química, Lab. Nacional de Engenharia e Tecnologia Industrial, Estrada Nacional no. 10, 2685 Sacavém, Portugal. (1939) PhD, chemical crystallography (Cambridge U., UK, 1973) Investigador principal. (tel. 1 + 2510021, fax +351-1-899242). *X-ray structural characterization, inorganic and organometallic compounds.*

Duarte, M. Teresa. Centro de Química Estrutural, Complexo Interdisciplinar, Instituto Superior Técnico, Av. Rovisco Pais, 1096 Lisboa codex, Portugal. (1958) BSc,Chem.Eng.(IST,1983) Asst.IST. (tel. 351-1+572096, ext. 282, fax +351-1-899242, telex 63423ISTUTLP, email D1992 at ETA.IST.RCCN.PT) *Coordination compounds, structural chemistry.*

Figueiredo, Dr Maria Ondina. Centro de Cristalografia e Mineralogia, Inst. de Investigação Científica Tropical, Alameda D. Afonso Henriques 41-4Esq., 1000 Lisboa, Portugal. (1938) Dr, geology (Techn. U. of Lisbon, 1980). Investigador coordenador. (tel. 1 + 534596). *Inorganic structure systematics, mathematical crystallography, phase transformations.*

Fortes, Prof. Manuel Amaral. Depto. de Engenharia de Materiais, Instituto Superior Técnico, Av. Rovisco Pais, 1000 Lisboa, Portugal. (1938) PhD, physics (Cambridge U., UK, 1969). Prof. catedrático. (tel. 1 + 802045). *Structure and properties, interfaces, topological aspects, structures, random structures.*

Gama Carvalho, Dr Frederico. Lab. Nacional de Engenharia e Tecnologia Industriais, Estrada Nacional N-10, 2685 Sacavem, Portugal. (1936) Dr, physics (U. Karlsruhe, BRD, 1967). Investigador. (tel. 1 + 2510021, fax +351-1-899242). *Crystal structure, neutron diffraction, solid state physics, statistical physics.*

Lima-de-Faria, Dr José. Centro de Cristalografia e Mineralogia, Inst. de Investigação Científica Tropical, Alameda D. Afonso Henriques 41-4Esq., 1000 Lisboa, Portugal. (1925) PhD, crystallography (Cambridge U., UK, 1962). Investigador coordenador. (tel. 1 + 534596). *Inorganic structure systematics, condensed models, phase transformations, minerals.*

Lopes-Vieira, Prof. António. Centro de Química Estrutural, Complexo Interdisciplinar, Inst. Superior Técnico, Av. Rovisco Pais, 1000 Lisboa, Portugal. (1929) PhD, mineralogy (Oxford U., UK, ·1967). Prof. associado. (tel 1 + 534596, fax +351-1-899242, email JNET%"IST_1322 at PTIFM"LOPESVIEIRA Bitnet). *Inorganic crystal structures, minerals.*

Margarido, Eng. Fernanda. M.R, Cruz. Depto. Mater. Eng., Instituto Superior Técnico, Techn. U. of Lisbon, 1000 Lisboa, Portugal. (1957) Metal Eng. (Techn. U. Lisbon.,1979). Asst. (tel. 1 + 801210). *Crystal chemistry, alloys, X-ray diffraction.*

Matias, Dr Pedro. Cetro Qiímica Estrutural, Complexo Interdisciplinar, Instituto Superior Técnico, Av.Rovisco Pais, 1096 Lisboa codex, Portugal. (1959) PhD, Crystallography (U. of Pittsburgh, 1986). Res. Fellow. (tel. 315 + 1 572056, ext. 282, telex 63423ISTUTLP, fax 351-1-899742, email D1892 at ETA.IST.RCCN.PT) *Macromolecules, protein.*

Matos Beja, Dr Ana Maria. Depto. de Fisica, Fac. de Ciencias e Tecnologia, 3000 Coimbra, Portugal. (1949) Dr, physics (Coimbra U., 1988). Investigador auxiliar. (tel. 39 + 23675, fax +351-39.29158). *Alloy structures, electron densities.*

Montenegro de Andrade, Prof. Miguel. Depto. de Mineralogia e Geologia, Fac. de Ciencias, 4000 Porto, Portugal. (1918) Dr, petrology (Coimbra U., 1955). Prof. catedrático. (tel. 2 + 21208). *Crystal optics.*

Quadrado, Prof. Ricardo. Depto. de Mineralogia e Geologia, Fac. de Ciencias, Rua da Escola Politécnica, 1200 Lisboa, Portugal. (1920) Dr, crystallography (Madrid U., Spain, 1967). Prof. catedrático. (tel. 1 + 605850). *Crystal physics, teaching crystallography.*

Salgado, Dr José. Lab. Nacional de Engenharia e Tecnologia Industriais, Estrada Nacional N-10, 2685 Sacavem, Portugal. (1940) Dr, physics (U. Karlsruhe, BRD, 1974). Investigador principal. (tel. 1 + 2550021, fax +315-1-2550117). *Neutron diffraction, solid state physics, structures, dynamics of solids.*

Salvado Canelhas, Mrs Maria da Graça. Depto. de Geologia, Fac. de Ciencias, Univ. de Lisboa, Bloco C2 - No.5 Piso, 1700 Lisboa, Portugal. (1940) Lic., geology (Lisboa U., 1964). Naturalista. (tel. 1 + 7583141). *Mineral identification.*

ROMANIA

Sub-Editor: J. Ionescu

Notes

1. International telephone country code - 40.

2. Degrees conferred by the Romanian universities are the *doctor docent în ştiinţe* (Dr.doc.st.), *doctor în ştiinţe* (Dr.st.), and *licentiat în ştiinţe* (Lic.st.), equivalent approximately to DSc, MSc and BSc respectively.

3. Other abbreviations used are: Acad. - academician, member of the Academia R.S.R.; Acad. R.S.R. - Academia Republucii Socialistă Romănia; and Conf. - conferenţiar universitar (assistant professor).

Anton, Mr Liviu. Ministry of Mines & Geology, Geological & Geophysical Inst. of Romania, 1 Caransebes Street, Bucarest 8, Romania. (1942) Scient. res. (tel. 657530). *Crystallography, mineralogy, geochemistry, experimental works.*

Apostolescu, Eugenia Rodica. Crystallography and Mineralogy Lab., Inst. Politehnic 'Gh. Gheorghiu-Dej', Str. Polizu 1, Bucureşti, Romania. (1930) Ing., geology (Mining and Geology Inst., Bucureşti, 1953). Asst. (tel. 139440). *Crystallography, minerals.*

Balan, Mr Mihai. Ministry of Mines & Geology, Geological & Geophysical Inst. of Romania, 1 Caransebes Street, Bucarest 8, 7000 Romania. (1940) Dr, geological sciences (U. Bucarest, 1975). Scient. res. (tel. 657530). *Crystallography, mineralogy, geochemistry, experimental works.*

Bally, Conf. Dorel. Comitetul de Stat pentru Energia Nucleară- Bucureşti, Inst. of Physics, Bucureşti, Romania. (1923) Dr.st., physics (Moscow State U., USSR, 1953). Dr.doc.st. (U. de Bucureşti, 1967). Div. chief, Neutron Physics Lab., Căsuta Poştală 35, Bucureşti. (tel. 23 68 60). *Neutron diffraction, metal physics.*

Baltă, Conf. Petru. Glass Chemistry and Techn. Lab., Institutul Politehnic 'Gh. Gheorghiu-Dej', Calea Grivitei 132, Bucureşti, Romania. (1930) Dr.Ing., silicate chemistry (Polytechnic Inst. Bucureşti, 1956). Conf. glass techn. (tel. 139440). *Glass, silicates.*

Cioflica, Prof. Graţian. Dept. of Geology & Geography, Blvd. N. Bălcescu 1, Universitatea Bucureşti, Bucureşti, Romania. (1927) Dr.st., mineralogy and petrography (U. de Bucureşti, 1958). Prof. of ore deposits. *Crystal optics.*

Constantinescu, Mr Radu. Ministry of Mines and Geology, Geological & Geophysical Inst. of Romania, 1 Caransebes Street, Bucarest 8, 7000 Romania. (1945) Scient. res. (tel. 657530). *Crystallography, mineralogy, geochemistry, experimental works.*

Cruceanu, Mr Eugen. Comitetul de Stat pentru Energia Nucleară, Inst. of Physics, Bucureşti, Romania. (1931) Dr.st., semiconductor crystals and crystallography (Inst. of Metals 'M. Kalinin', Moscow, USSR, 1960). Res. scient. (tel. 16 66 50). *Crystal growth, crystal structures.*

Dinescu, Prof. Radu. Ceramics and Refractory Materials, Fac. of Chemical Industry, Institutul Politehnic 'Gh. Gheorghiu-Dej', Calea Grivitei 132, Bucureşti, Romania. (1917) Dr.Ing., ceramics (Polytechnic Inst., Bucureşti, 1943). Prof. (tel. 139440). *Minerals, ceramics and refractory materials.*

Draghici, Mr Iosif. Mineralogy Dept., U. Bucharest, 1 N. Balcescu, Bucureşti, Romania. (1930) BSc, Geology and Geography (U. Bucharest). Reader. *Hydrothermal deposits.*

Dumitrescu, Aurelia. Crystallography & Mineralogy Lab., Inst. Politehnic 'Gh. Gheorghiu-Dej', Str. Polizu 1, Bucureşti, Romania. (1928) Mining engineer (Mining Inst., Bucureşti, 1953). Asst. (tel. 139440). *Thermal analysis.*

Giuşcă, Prof. Dan. Dept. of Geology & Geography, Blvd. N. Bălcescu 1, Universitatea Bucureşti, Bucureşti, Romania. (1904) Dr.doc.st., crystal chemistry (U. Cluj, 1927). Acad.; prof., petrology. *Crystal chemistry.*

Ianovici, Prof. Virgil. Ministerul Minelor, Petrolului şi Geologei, Institutul Geologic, Str. Mendeleev 36, Bucureşti 1, Romania. (1900) Dr.doc.st., mineralogy and petrography (U. Iaşi, 1929). Acad.; prof., crystallography, mineralogy. (tel. 333187). *Crystallography, mineralogy.*

Imreh, Mr Iosif. Mineralogy Lab., Universtatea Babeş-Bolyai, Str. Rogălniceanu, 1 Cluj, Romania. (1924) Dr.st., mineralogy and crystallography (U. Iaşi, 1957). Reader. (tel. 3001). *Morphology - structure relationship.*

Ionescu, Mrs Jeana. Mineralogical Lab., Ministerul Minelor, Petrolului şi Geologei, Institutul Geologic, Str. Caransebes 1, Bucureşti, Romania. (1924) Dr.doc.st., crystallography and mineralogy (Acad.R.S.R., 1960). Chief. (tel. 657530). *Inorganic crystal structures, crystal optics, clay minerals.*

Jude, Dr (Mrs) Lidia. Dept. of Geology and Geography, Universitatea Bucureşti, Blvd. N. Balcescu 1, Bucureşti 78344, Romania. (1930) Dr.doc.st., mineralogy and geology (U. Bucharest, 1962). (tel. 90 + 468673). *Crystal optics, crystal physics, crystal growth.*

Kissling, Mr Alexandru. Inst. of Oil and Nat. Gas, Ploieşti, Romania. (1921) Dr.st., mineralogy (U. de Bucureşti, 1964). Conf. *Radiocrystallography, crystal chemistry.*

Lazar, Mr Constantin. Inst. of Geology and Geophysics, Str. Caransebes 1, Bucure şti 78344, Romania. (1935) Trainer for Dr, mineralogy (U. Bucharest). Sr. res. scient. (tel. 90 + 657530). *Mineralogy, ore minerals, crystallography, geochemistry.*

Mănăilă, Mrs Rodica. X-ray Lab., Comitetul de Stat pentru Energia Nucleară, Inst. of Physics, Bucureşti, Romania. (1935) Lic.st., physics (U. de Bucureşti, 1957). Res. scient. (tel. 166550). *Semiconductor crystal structure, amorphous semiconductor structures.*

Mastacan, Prof. Gheorghe. Dept. of Geology and Geography, Universitatea Bucure şti, Blvd N. Bălcescu 1, Bucureşti, Romania. (1907) Dr.st., mineralogy and petrography (U. Iaşi, 1948). Reader. (tel. 127796 and 230754). *Crystal physics, structures, clay minerals, crystal optics.*

Mirzu-Ghergariu, Mrs Lucretia. Mineralogy Lab., Universitatea Babeş-Bolyai, Str. Kogălniceanu, 1 Cluj, Romania. (1932) Lic.st., chemistry, crystallography and mineralogy (U. Cluj, 1954). Asst. (tel. 3001). *Crystal optics, clay minerals.*

Petreus, Mr Ion. Mineralogy Lab., U. 'Al. I. Cuza', Calea 23 August 20A, Iaşi, Romania. (1939) Lic.st., geology (U. de Bucureşti). Asst. *Crystal morphology, clay minerals.*

Popescu, Conf. Ion. Dept. of Geology and Geography, Universitatea Bucureşti, Blvd. N. Bălcescu 1, Bucureşti, Romania. (1906) Dr.st., chemistry (U. de Bucure şti, 1943). Reader. (tel. 151798). *Inorganic crystal structures, clay minerals, crystal optics.*

Radulescu, Prof. Dan. Sedimentary Petrography Lab., Dept. of Geology and Geography, Universitatea Bucureşti, Blvd. N. Bălcescu 1, Bucureşti, Romania. (1928) Dr.doc.st., mineralogy (U. de Bucureşti, 1957). Prof. *Crystal optics.*

Rosca, Mr Liviu. Crystallography and Mineralogy Lab., Inst. Politehnic 'Gh. Gheorghiu-Dej', Str. Polizu 1, Bucureşti, Romania. (1914) Lic.st., (U. Cernăuti, 1941). Reader. (tel. 139440). *Crystallography, mineralogy.*

Segal, Mr Eugen. Physical Chemistry Lab., Universitatea Bucureşti, Blvd. Republicii 13, Bucureşti, Romania. (1933) Dr.st., chemistry (U. de Bucureşti, 1964). Asst. prof. (tel. 157980).

Stiopol, Prof. Victoria. Crystallography and Geochemistry Lab., Universitatea Bucureşti, Bucureşti, Romania. (1928) Dr.st., mineralogy (U. de Bucureşti, 1960). Prof. (tel. 156713). *Crystallography, mineralogy.*

Stoicovici, Prof. Eugen. Mineralogy and Geochemistry Lab., Universitatea Babeş-Bolyai, Str. Kogălniceanu, 1 Cluj, Romania. (1906) Dr.doc.st., mineralogy and crystallography (U. Cluj, 1954). Prof. (tel. 3001). *Inorganic crystal structures, crystal optics, crystal physics, metals structure, clay minerals.*

Udubasa, Dr Gheorghe. Institutul de Geologie şi Geofizica, Str. Caransebes nr. 1, Bucureşti 7000, Romania. (1938) Dr, mineralogy (U. Heidelberg, DDR, 1972). Scient. res. (tel. 657530). *Mineralogy, geochemistry, structure, ore minerals.*

Vanghelie, Mr Iulian. Inst. de Geologie şi Geofizica, Str. Caransebes nr. 1, Bucure şti 7000, Romania. (1946) Lic.st., chemistry (U. de Bucureşti, 1969). Scient. res. (tel. 90 + 657530). *X-ray crystallography, inorganic and organic crystal structures, X-ray diffraction.*

Vlad, Dr Serban-Nicolae. Inst. de Geologie şi Geofizica, Str. Caransebes nr. 1, Bucureşti 78344, Romania. (1941) Dr, mineralogy (U. de Bucureşti, 1971). Sr. res. scient. (tel. 90 + 657530). *Skarn silicates, ore minerals, mineralogy, crystallography.*

SAUDI ARABIA

Sub-Editor: **M.S. Hussain**

Notes

1. International telephone country code - 966.

Al-Shanti, Prof. Ahmed Mahmoud. Economic Geology Dept., Fac. of Earth Sci., King Abdulaziz U., P.O. Box 1744, Jeddah 21441, Saudi Arabia. (1932) PhD, mining geology (Imp. C., U. London, UK, 1973). Chairman. (tel. Jeddah - 6653735, Telex 401141 KAUNI). *Crystal optics (incident light on opaque minerals).*

Avci, Dr Recep. Physics Dept., King Fahd U. of Petroleum and Minerals, UPM Box 2018, Dhahran 31261, Saudi Arabia. (1950) PhD, solid state physics (U. Illinois, Urbana, USA, 1978). Asst. prof. (tel. 3 + 860-2292, telex 601060 UPMSI, fax 860-3306). *Photoelectron spectroscopy, LEED, Auger spectroscopy, EELS, Photoemmision (inverse) spectroscopy, interfaces, materials research.*

Bakr, Dr Abdel Razak. Geology Dept., Fac. of Sci., King Abdulaziz U., P.O. Box 1540, Jeddah, Saudi Arabia. PhD, petrology (U. Leeds, UK, 1973). Head. *Crystal optics in transmitted light.*

El-Mahdi, Dr Omar. Inst. of Applied Geology, King Abdulaziz U., P.O. Box 1744, Jeddah, Saudi Arabia. (1938) PhD, economic geology (U. Utah, USA, 1966). Assoc. Prof. (tel. Jeddah - 24263). *Crystal optics (incident light on opaque materials).*

Haque, Prof. Mazhar-ul. Chemistry Dept., King Fahd U. of Petroleum and Minerals, KFUPM Box 165, Dhahran 31261, Saudi Arabia. (1936) PhD, chemistry (Imperial C., U. London, UK, 1964). Prof. (tel. 03 + 860-2378, fax 860-3306). *Inorganic and organic crystal structures, natural products, phosphetans.*

Horne, Mr William. King Fahd U. of Petroleum and Minerals, KFUPM Airport Box 82, Dhahran, Saudi Arabia. (1945) BSc(Hons), chemistry (U. Leeds, UK, 1966). Lect. (tel. 3 + 860-3827, fax 860-3306). *Computer programming, crystal structures, phosphetans, phosphorinanes.*

Hussain, Dr M. Sakhawat. Chemistry Dept., King Fahd U. of Petroleum and Minerals, KFUPM Box 1830, Dhahran 31261, Saudi Arabia. (1939) PhD, chemistry, chemical crystallography (U. California, Davis, USA, 1968). Prof. (tel. 3 + 860-3821, fax 860-3306). *Organometallic compounds, crystal structures, complexes with short hydrogen bonds, biologically significant compounds, superconducting ceramics.*

Hussain, Dr Zahid. Physics Dept., King Fahd U. of Petroleum and Minerals, KFUPM Box 580, Dhahran 31261, Saudi Arabia. (1949) PhD, chemical physics (U. Hawaii, USA, 1979). Assoc. prof. (tel. 3 + 860-2292, fax 860-3306).

Electron spectroscopy, XPS, UPS, LEED, Auger spectroscopy, EELS, synchrotron radiation instrumentation, EXAFS, surface crystallography, photoelectron diffraction.

Keith, Dr Vepan. Physics Dept., King Fahd U. of Petroleum and Minerals, KFUPM box 1622, Dhahran 31261, Saudi Arabia. (1946) PhD, Solid State Physics (U. of Waterloo, Canada, 1977). Assoc. Prof. (fax 3 + 860-3306). *Specific heat, superconductivity, electron tunnelling, magnetic suceptibility and resistivity of semi-magnetic semiconductors.*

Kenaan, Dr Feisal. Directorate General of Mineral Resources, P.O. Box 345, Jeddah, Saudi Arabia. PhD, petrology (Colorado School of Mines, USA, 1976). *Crystal optics in transmitted light.*

Khattak, Dr Guldad Khattak. Physics Dept., King Fahd U. of Petroleum and Minerals, KFUPM Box 1854, Dhahran 31261, Saudi Arabia. (1948) PhD, solid state physics (Purdue U., USA, 1978). Assoc. prof. (tel. 3 + 860-2260, fax 860-3306). *Specific heat, magnetic susceptibility, resistivity, low temperature investigation, semimagnetic semiconductors, oxides, inorganic complexes, superconductors.*

Khawaja, Dr Ehsan Ellahi. Physics Dept., King Fahd U. of Petroleum and Minerals, KFUPM Box 1987, Dhahran 31261, Saudi Arabia. (1945) PhD, solid state physics (U. Adelaide, Australia). Asst. prof. (tel. 3 + 860-2267, fax 860-3306). *Semiconductor physics, thin film structure, X-ray diffraction, electron diffraction.*

Koyama, Dr Kazutoshi. Inst. of Applied Geology, King Abdulaziz U., P.O. Box 1744, Jeddah, Saudi Arabia. (1946) PhD, X-ray crystallography (U. Tokyo, Japan, 1976). Assoc. UNESCO Expert. (tel. Jeddah - 24263). *Crystal structures.*

Naseif, Dr Abdulah. King Abdulaziz U., P.O. Box 1540, Jeddah, Saudi Arabia. PhD, petrology (U. Leeds, UK, 1972). Deputy President. *Crystal optics in transmitted light, minerals.*

Saif, Dr Saiful-Islam. Dept. of Earth Sciences, King Fahd U. of Petroleum and Minerals, KFUPM Box 1853, Dhahran 31261, Saudi Arabia. (1944) PhD, Massive sulfides (U. of New Brunswick, Canada, 1977). Assoc. Prof. (tel 3+860-2621, telex 801060 kfupm sj, fax 860-3306). *Economic minerals, gem minerals.*

Tahoun, Prof. Salah. Inst. of Applied Geology, King Abdulaziz U., P.O. Box 1744, Jeddah, Saudi Arabia. (1937) PhD, clay mineralogy (Michigan State U., USA, 1965). Prof. of soil mineralogy. (tel. Jeddah - 24263). *Structure and identification, clay minerals.*

SINGAPORE

Sub-Editor: **L. L. Koh**

Notes

1. International telephone country code - 65

2. The science degrees conferred by the National University of Singapore are the PhD, MSc and BSc and correspond to British degrees.

Chowdari, Prof. B.V.R. Physics Dept., Nat. U. of Singapore, Kent Ridge, Singapore 0511, Singapore. (1943) PhD, physics (I.I.T., Kanpur, India, 1968). Assoc. Prof. (tel. 7722956). *Glasses, solid state batteries, thin films, materials science.*

Chung, Dr Mui-Fatt. Physics Dept., Nat. U. of Singapore, Lower Kent Ridge Rd., Singapore 0511, Singapore. (1936) PhD, surface physics (U. New South Wales, Australia, 1966). Sr. lect. (tel. 7722618, telex UNISPO-RS33943, fax 7774279). *Surface properties, surface treatment, electron diffraction, auger and other electron sprectroscopy.*

Fong, Dr Hock Sun. Mechanical Eng., Nat. U. of Singapore, Kent Ridge, Singapore 0511, Singapore. (1941) PhD, physical metallurgy, materials science (U. Birmingham, UK, 1969). Assoc. prof. (tel. 7756666, ext. 2211). *Crystallography, phase transformations.*

Hosea, Dr Thomas Jeffrey Cockburn. Physics Dept., Nat. U. of Singapore, Kent Ridge, Singapore 0511, Singapore. (1952) PhD, solid state physics (U. Edinburgh, Scotland, 1978). Lect. (tel. 7756666, ext. 2629). *Raman and brillouin spectroscopy, phase transitions.*

Koh, Dr Lip Lin. Dept. of Chemistry, Nat. U. of Singapore, Kent Ridge, Singapore 0511, Singapore. (1935) PhD, physical chemistry (Boston U., USA, 1964). Assoc. Prof. (tel. 7722847). *Crystallography.*

Kuok, Dr Meng Hau. Physics Dept., Nat. U. of Singapore, Kent Ridge, Singapore 0511, Singapore. (1951) PhD, solid state physics (U. Canterbury, New Zealand,1978). Sr. Lect. (tel. 7722609, fax 7774279). *Raman spectroscopy, infrared spectroscopy, fluoresence spectroscopy.*

Kwik, Dr. Whei Lu. Chemistry, Nat. U. of Singapore, 10 Kent Ridge Crescent, Singapore 0511, Singapore. (1942) PhD, Inorganic Chemistry (SUNY, Stony Brook,1973) Assoc. Prof. (tel. 7722820). *Coordination chemistry, organometallic chemistry, analytical chemistry.*

Mok, Dr Kum-fun. Dept. of Chemistry, Nat. U. of Singapore, Kent Ridge, Singapore 0511, Singapore. PhD, Inorganic Chemistry (Victora U., New Zealand,1965) Assoc. Prof. (tel. 7722669). *Coordination compounds, structural chemistry.*

Ng, Prof. Ser Choon. Physics Dept., Nat. U. of Singapore, Kent Ridge, Singapore 0511, Singapore. (1937) PhD, solid state physics (McMaster U., Canada, 1967). Assoc. Prof. (tel. 7722610). *X-ray diffraction, neutron diffraction, laser light scattering.*

Rajaratnam, Prof. Arthur. Physics Dept., Nat. U. of Singapore, Kent Ridge, Singapore 0511, Singapore. (1927) PhD, spectroscopy (U. London, UK, 1958). Prof. (tel. 7756666, ext. 2628). *Spectroscopy, structure of matter.*

Tan, Dr Hock Siew. Physics Dept., Nat. U. of Singapore, Kent Ridge, Singapore 0511, Singapore. (1950) PhD, solid state physics (U. Rochester, USA, 1980). Lect. (tel. 7756666, ext. 2614). *Structure of solids.*

Teh, Dr Hung Chuan. Information Systems and Computer Sci. Dept., Nat. U. of Singapore, Kent Ridge, Singapore 0511, Singapore. (1941) PhD, solid state physics (McMaster U., Canada, 1972). Sr. lect. (tel. 7722912, fax 7794580). *Neutron diffraction and scattering, microcomputer systems and applications, computer graphics.*

SOUTH AFRICA

Sub-Editor: **S.M. Dobson**

Notes

1. International telephone country code - 27.

Alberts, Prof. Hermanus Lambertus. Physics Dept., Rand Afrikaans U., P.O. Box 524, Johannesburg 2000, South Africa. (1941) PhD, physics (Rand Afrikaans U., 1970). Prof. (tel. 011 + 489-2330, fax +11 726 7723). *Magnetic and elastic properties.*

Allen, Mrs Christine Corinna. Chemistry Dept., U. of the Witwatersrand, P.O. Wits, Johannesburg 2050, South Africa. (1960) BSc, chemistry (U. Witwatersrand, 1985). Res. Tech. (tel. 011 + 716-4097, fax +11 403 1926). *Disorder.*

Archer, Dr Steven James. Chemistry Dept., U. of Cape Town, Private Bag, Rondebosch 7700, Cape Town, South Africa. (1958) PhD, chemistry (U. Cape Town, 1984). Temp. Lect. (tel. 021 + 650-2530, fax +21 650 3726). *Inorganic and organometallic structures.*

Auf der Heyde, Dr Thomas Paul Edwin. Chemistry Dept., U. of the Western Cape, Private Bag X17, Bellville 7530, South Africa. (1958) PhD, physical-inorganic chemistry (U. Cape Town, 1988). Sen. Lect. (tel. 021 + 959-2263, fax +21 959 2779). *Inorganic crystals and structural chemistry.*

Ball, Prof. Anthony. Dept. of Materials Eng., U. of Cape Town, Private Bag, Rondebosch 7700, Cape Town, South Africa. (1939) DEng, physical metallurgy (U. Birmingham, UK, 1987). Head. (tel. 021 + 650-3173, fax +21 650 3726). *Materials deformation, tribology and microstructure.*

Basson, Dr Stephen Smuts. Chemistry Dept. U. of the Orange Free State, P.O. Box 339, Bloemfontein 9300, South Africa. (1942) DSc, chemistry (U. Orange Free State, 1969). Prof. (tel. 051 + 401-2348, fax +51 8279). *Complex cyanides, rhodium and iridium chemistry.*

Beukes, Prof Dr Gerhardus Johannes. Geology Dept. U. of the Orange Free State, P.O. Box 339, Bloemfontein 9300, South Africa. (1943) DSc, geology (U. Orange Free State, 1973). Assoc. prof. (tel. 051 + 401-2393, fax +51 8279). *X-ray diffractometry, X-ray fluorescence spectrometry, applied mineralogy.*

Billing, Mr David Gordon. Chemistry Dept. U. of the Witwatersrand, P.O. Wits, Johannesburg 2050, South Africa. (1966) BSc(Hons), chemistry (U. Witwatersrand, 1989). Res. Tech. (tel. 011 + 716-2169, fax +11 403 1926). *Low temperature crystallography, charge density studies.*

Boeyens, Prof. Jan Christoffel Antonie. Chemistry Dept. U. of the Witwatersrand, P.O. Wits, Johannesburg 2050, South Africa. (1934) DSc, physical and theoretical chemistry (U. Pretoria, 1964). Prof. (tel. 011 + 716-2076, fax +11 403 1926). *Disorder, structural theory, molecular mechanics, phase transitions.*

Bond, Dr Dianne Ruth. Chemistry Dept. U. of Cape Town, Private Bag, Rondebosch 7700, Cape Town, South Africa. (1958) PhD, chemistry (U. Cape Town, 1986). Temp. Lect. (tel. 021 + 650-2562, fax +21 650 3726). *Liquid clathrates, organic clathrates.*

Boonstra, Prof. Eelco Gerrit. U. of the Orange Free State, P.O. Box 339, Bloemfontein 9300, South Africa. (1935) PhD, physics (U. Natal, 1966). Vice Rector. (tel. 051 + 401-2661, fax +51 8279). *Crystal structures, automation, computing, surface structures.*

Bourne, Miss Susan Ann. Chemistry Dept. U. of Cape Town, Private Bag, Rondebosch 7700, South Africa. (1965) BSc(Hons), chemistry (U. Cape Town, 1987). Teaching Assist. (tel. 021 + 650-2562, fax +21 650 3726). *Organic clathrates.*

Brown, Prof Michael Ewart. Chemistry Dept. Rhodes U, P.O. Box 94, Grahamstown 6140, South Africa. (1938) PhD, physical chemistry (Rhodes U., 1966). Prof. (tel. 0461 + 22023, ext. 258, fax +461 25049). *Solid phase reactions, kinetics and mechanisms.*

Caira, Prof Mino Rodolfo. Chemistry Dept., U. of Cape Town, Private Bag, Rondebosch 7700, South Africa. (1949) PhD, chemistry (U. Cape Town, 1975). Assoc. Prof. (tel. 021 + 650-3071, fax +21 650 3726). *Clathrates, drug structure-reactivity relationships, drug polymorphism.*

Caveney, Dr Robert John. De Beers Industrial Diamond Div., P.O. Box 916, Johannesburg 2000, South Africa. (1941) PhD, physics (U. Witwatersrand,

1970). Deputy res. dir.. (tel. 011 + 835-3232, ext. 2101, fax +11 835 2337). *Crystal growth, defects, diamond-structure, composite materials.*

Clark, Dr James Brian. Research, Development and Implementation, CSIR, P.O. Box 395, Pretoria 0001, South Africa. (1949) DSc, physics (U. Pretoria, 1973). Group Executive. (tel. 012 + 841-2429, fax +12 86 9167). *Inorganic crystals, phase transitions.*

Comins, Dr Neville Raymond. Speciality Metals, DMST, CSIR, P.O. Box 395, Pretoria 0001, South Africa. (1945) PhD, physics (Cambridge U., UK, 1971). Programme Manager. (tel. 012 + 841-3420, fax +12 841 4395). *Electron microscopy, physical metallurgy.*

Copperthwaite, Dr Richard George. Chemistry Dept, U. of the Witwatersrand, P.O. Wits, Johannesburg 2050, South Africa. (1945) PhD, chemistry (U.of London, UK, 1971). Sr. lect. (tel. 011 + 716-2262, fax +11 403 1926). *Solid state and surface chemistry, heterogeneous catalysis.*

Coville, Prof. Neil John. Chemistry Dept, U. of the Witwatersrand, P.O. Wits, Johannesburg 2050, South Africa. (1945) PhD, chemistry (McGill U., Canada, 1973). Prof. (tel. 011 + 716-2371, fax +11 403 1926). *Organometallic chemistry.*

Crawford, Mr John Lawrence. Physics Dept., U. of the Witwatersrand, P.O. Wits, Johannesburg 2050, South Africa. (1937) BSc Hons., physics (U. Witwatersrand, 1959). Sr. Lect. (tel. 011 + 716-2287, fax +11 403 1926). *Electron microscopy, defect structures.*

Davies, Dr Geoffrey John. De Beers Diamond Res. Lab., P.O. Box 916, Johannesburg 2000, South Africa. (1948) PhD, physics (U. Reading, UK, 1972). Assist. Res. Manager. (tel. 011 + 835-3232, fax +11 835 2337). *High pressures, diamond physics.*

Davies, Dr Gladstone. Div. Building Tech., CSIR, P.O.Box 395, Pretoria 0001, South Africa. (1952) PhD, geology (U. Witwatersrand, 1983). Ch. res. (tel. 012 + 841-2507, fax +12 841 4680). *Materials science, powder diffraction.*

Denner, Dr Louis. Res. and Dev. Dept., H.T.P. (Pty) Ltd, P.O. Box 19823, Pretoria West 0117, South Africa. (1959) PhD, conformational analysis (U. Witwatersrand, 1987). Sen. Scientist. (tel. 012 + 296-3775, fax +12 296 3767). *Powder diffraction, electron deformation density.*

De Villiers, Dr. Johan Pieter Roos. Mineralogy Division, MINTEK, Private Bag X3015, Randburg 2125, South Africa. (1942) PhD, mineralogy (U. Illinois, USA, 1969). Director. (tel. 011 + 793-3511, ext. 459, fax +11 793 2413). *Minerals, inorganic phase determination.*

Dillen, Dr Jan Louis Maria. Chemistry Dept., University of Pretoria, Pretoria 0002, South Africa. (1955) PhD, physical chemistry (U. Antwerp, Belgium,1981). Sr. Lect. (tel. 012 + 420-3089, fax +12 342 2453). *Conformational analysis, molecular mechanics, inter- and intramolecular forces.*

Dobson, Dr Susan Mary. Chemistry Dept, U. of the Witwatersrand, P.O. Wits, Johannesburg 2050, South Africa. (1959) PhD, inorganic chemistry (U. Witwatersrand, 1986). Res. Assoc. (tel. 011 + 716-2176, fax +11 403 1926). *Coordination compounds, macrocyclic ligands, conformational analysis.*

Eales, Prof. Hugh Victor. Geology Dept., Rhodes U., P.O. Box 94, Grahamstown 6140, South Africa. (1929) PhD, geology (Rhodes U., 1961). Prof., dept. head. (tel. 0461 + 22023, ext. 310, fax +461 25049). *Mineralogy, spinel group minerals, petrology, geochemistry, basic rocks, Bushveld Complex, XRF spectrometry, microprobe techniques, reflected light microscopy.*

Engel, Prof. Dennis Walter. Physics Dept., U. of Durban-Westville, Private Bag X54001, Durban 4000, South Africa. (1939) Dr. rer. nat., physics (Tech. U. München, BRD, 1971). Prof. (tel. 031 + 820-2226, fax +31 820 2383). *Anomalous scattering, powder diffraction.*

English, Dr Robert Bertram. Chemistry Dept., Rhodes U., P.O.Box 94, Grahamstown 6140, South Africa. (1948) PhD, organometallic chemistry (U. Cape Town, 1977). Sr. lect. (tel. 0461 + 22023, fax +461 25049). *Metal cluster*

chemistry, platinum group metal nitrosyl chemistry, alkoxides, carboxylates (Zr,Ti,Al,Pb).

Field, Prof. John Stainer. Chemistry Dept., U. of Natal, P.O. Box 375, Pietermaritzburg 3200, South Africa. (1946) PhD, X-ray crystallography (Cambridge U., UK, 1973). Assoc. Prof. (tel. 0331 + 63320). *Structural organometallic chemistry.*

Fourie, Dr Jacobus Theodor. Div. Mat. Sci. and Tech., CSIR, P.O. Box 395, Pretoria 0001, South Africa. (1930) DSc, physics (U. Pretoria, 1956). Chief specialist sci. (tel. 012 + 841-3386, fax + 12 841 4395). *Transmission electron microscopy, plastic deformation, metals.*

Glasser, Prof. Leslie. Chemistry Dept., U. of the Witwatersrand, P.O. Wits, Johannesburg 2050, South Africa. (1935) PhD D.I.C., chemical eng. (Imperial Coll., U. of London, 1960). Prof. of phys. chem. (tel. 011 + 716-2219, fax +11 403 1926). *Electrical properties, materials, hydrogen bonding, chemometrics, computer modelling.*

Heckroodt, Prof. Renier Oelof. Dept. of Civil Eng., U. of Cape Town, Private Bag, Rondebosch 7700, Cape Town, South Africa. (1935) DSc, geology (U. Pretoria, 1968). Prof. (tel. 021 + 650-3176, fax +21 650 3726). *Clay mineralogy.*

Heyns, Prof. Anton Michal. Chemistry Dept., U. of Pretoria, Hatfield, Pretoria 0002, South Africa. (1939) PhD, chemistry (U. South Africa, 1968). Prof. (tel. 012 + 420-2516, fax +12 342 2453). *IR spectroscopy, Raman spectroscopy, X-ray powder diffraction, ionic solids.*

Horsfield, Mr Edgar Charles. Physics Dept., U. of Durban-Westville, Private Bag X54001, Durban 4000, South Africa. (1942) MSc, crystallography (U. Natal, 1969). Lect. (tel. 031 + 820-2662). *Organometallic complexes.*

Irving, Dr Anne. Chemistry Dept., U. of Cape Town, Private Bag, Rondebosch 7700, Cape Town, South Africa. (1940) PhD, X-ray crystallography (U. Leeds, UK, 1969). Sr. Lect. (tel. 021 + 650-2564, fax +21 650 3726). *Organic and inorganic structures.*

Johnson, Ms Louise. Chemistry Dept. University of Cape Town, Private Bag, Rondebosch 7700, South Africa. (1965) BSc(Hons), chemistry (U. Cape Town, 1987). Teaching Assist. (tel. 021 + 650-2562, fax +21 650 3726). *Organic clathrates.*

Jones, Miss Elizabeth Louise. Chemistry Dept. University of Cape Town, Private Bag, Rondebosch 7700, South Africa. (1965) BSc(Hons), biochemistry (U. Cape Town, 1986). Teaching Assist. (tel. 012 + 650-2564, fax +21 650 3726). *Biological clathrates.*

Kruger, Dr Gert Jacobus. Chemistry dept., Rand Afrikaans U., P.O.Box 524, Johannesburg 2000, South Africa. (1943) DSc, chemistry (Potchefstroom U., 1970). Prof. (tel. 011 + 489-2368, fax +11 726 7723). *Crystallographic computing, direct methods, powder diffraction.*

Laing, Dr Mary Elizabeth. 61 Baines Road, Durban 4001, South Africa. (1935) PhD, chemistry (U. California, Los Angeles, USA, 1964). Part-time lect. (tel. 031 + 25-1951). *Organic and inorganic structures.*

Laing, Prof. Michael John. Chemistry Dept., U. of Natal, King George V Ave, Durban 4001, South Africa. (1937) PhD, inorganic chemistry (U. California, Los Angeles, USA, 1965). Assoc. prof. (tel. 031 + 816-3103, fax +31 816 2214). *Coordination compounds, strained and aromatic organic compounds, polymorphism in organic crystals.*

Leipoldt, Prof. Johann Gotlieb. Chemistry Dept., U. of the Orange Free State, P.O. Box 339, Bloemfontein 9300, South Africa. (1940) DSc, inorganic chemistry (U. Orange Free State, 1969). Prof. (tel. 051 + 401-2497, fax +51 8279). *Crystal structure, transition metal complexes.*

Le Roux, Dr Johannes Hendrik. Res. and Dev. Dept, SASTECH, P.O. Box 1, Sasolburg 9570, South Africa. (1934) PhD, physical chemistry (U. South Africa, 1968). Instrumental Techniques Manager (tel. 016 + 708-2904). *Waxes, explosives, catalysts, minerals, industrial problem solving.*

Le Roux, Dr Stephanus David. Atomic Energy Corporation, P.O. Box 582, Pretoria 0001, South Africa. (1947) PhD, solid state physics (Purdue U., USA, 1975). Division head. (tel. 012 + 324-2811, ext. 533, fax +12 579 1515). *Materials science, diffraction, texture analysis.*

Levendis, Dr Demetrius Christos. Chemistry Dept, U. of the Witwatersrand, P.O. Wits, Johannesburg 2050, South Africa. (1957) PhD, crystallography (U. Witwatersrand, 1984). Lect. (tel. 011 + 716-2348, fax +11 403 1926). *Disorder, phase transitions, electron density, conducting molecular crystals.*

Liles, Mr David Charles. Div. of Proc. and Chem. Manufacturing Tech., CSIR, P.O.Box 395, Pretoria 0001, South Africa. (1950) BSc(Hons), chemistry (Loughborough U., 1973). Chief res. (tel. 012 + 841-2628, fax +12 841 2689). *Structure, transition metal complexes.*

Lombard, Dr Anthonie van Altena. Chemistry dept., Rand Afrikaans U., P.O.Box 524, Johannesburg 2000, South Africa. (1958) PhD, chemistry (Rand Afrikaans U., 1985). Res. officer. (tel. 011 + 489-2838, fax +11 726 7723). *Gold chemistry.*

Maske, Prof. Siegfried. Geology Dept., U. of the Witwatersrand, P.O. Wits, Johannesburg 2050, South Africa. (1928) DSc, geology (Stellenbosch U., 1964). Prof. of mining geology. (tel. 011 + 716-2799, fax +11 403 1926). *Ore genesis, mineralogy, geochemistry, crystal structure, sulphide minerals.*

Nabarro, Prof. Frank Reginald Nunes. U. of the Witwatersrand, P.O. Wits, Johannesburg 2050, South Africa. (1916) DSc, FRS, metallurgy(dislocation

theory) (U. Birmingham, UK, 1953). Prof. em. (tel. 011 + 716-2175, fax +11 403 1926). *Crystal defects.*

Nassimbeni, Prof. Luigi Renzo. Chemistry Dept., U. of Cape Town, Private Bag, Rondebosch 7700, Cape Town, South Africa. (1939) PhD, physical chemistry (U. Cape Town, 1969). Prof. (tel. 021 + 650-2569, fax +21 650 3726). *Organic and inorganic crystal structures, clathrates.*

Niven, Dr Margaret Lillian. Chemistry Dept, U. of Cape Town, Private Bag, Rondebosch 7700, Cape Town, South Africa. (1954) PhD, chemistry (U. Cape Town, 1980). Sr. res. Officer. (tel. 021 + 650-2570, fax +21 650 3726). *Small molecules, disorder.*

O'Neill, Ms Françoise Marcelle. Chemistry Dept, U. of the Witwatersrand, P.O. Wits, Johannesburg 2050, South Africa. (1965) BSc(Hons) chemistry (U. Witwatersrand 1987). Jr. lect. (tel. 011 + 716-3826, fax +11 403 1926). *Bond orders, dimetal centres.*

Paige-Green, Mr Philip. DRTT, CSIR, P.O. Box 395, Pretoria 0001, South Africa. (1952) MSc, geology (U. Natal, 1975). Eng. geologist. (tel. 012 + 841-2924, fax +12 841 3232). *X-ray diffraction analysis, minerals, clays, road materials.*

Pipkin, Dr Noel John. Physics Dept., De Beers Diamond Res. Lab., P.O. Box 916, Johannesburg 2000, South Africa. (1942) PhD, metallurgy (U. Newcastle-upon-Tyne, UK, 1967). Ass. Res. Manager. (tel. 011 + 835-3232, ext. 240, fax +11 835 2337). *Graphite, diamonds, cubic boron nitride, characterisation.*

Pretorius, Dr Jan Andries. Analytical Group, Res. Dept., AECI Explosives and Chemicals (ltd), P.O.North Rand 1645, South Africa. (1949) PhD, chemistry (U. South Africa, 1978). Chief res. officer. (tel. 011 + 605-9111). *Powder diffraction, X-ray fluorescence, automation, crystallographic computing.*

Retief, Dr Johannes Jacobus. Res. Div., Sasol Technology, P.O. Box 1, Sasolburg 9570, South Africa. (1941) PhD, physics (U. Orange Free State, 1978). Principal res. sci. (tel. 016 + 708-2940). *Structure, catalysts, waxes, carbons, clay mineral identification.*

Reynhardt, Prof. Eduard Christiaan. Physics Dept., U. of South Africa, P.O. Box 392, Pretoria 0001, South Africa. (1944) PhD, physics U. of South Africa, 1971). Prof. (tel. 012 + 429-8062, fax +12 429 3221). *Molecular reorientation in solids, phase transitions in solids.*

Richter, Dr Paul Wilhelm. Div. Mat. Sci. Tech., CSIR, P.O. Box 395, Pretoria 0001, South Africa. (1946) PhD, physical chemistry (U. South Africa, 1971). Chief specialist Res. (tel. 012 + 841-2434, fax +12 841 4395). *Materials research, high pressure, thermal analysis, ceramics, crystal growth.*

Rodgers, Prof. Allen Lawrence. Chemistry Dept., U. of Cape Town, Private Bag., Rondebosch 7700, Cape Town, South Africa. (1946) PhD, chemistry (U. Cape Town, 1974). Assoc. prof. (tel. 021 + 650-2572, fax +21 650 3726). *X-ray powder diffraction, scanning electron microscopy, calculi.*

Rutherford, Prof. John Stewart. Chemistry Dept., Univ. of Transkei, Private Bag X1, UNITRA, Umtata, Transkei, South Africa. (1938) PhD,physical chemistry (McMaster U., Canada, 1967) Prof. and head. (tel. 0471 + 26811, telex 734TT) *Small molecules, bonding in solids, discrete mathematics applications.*

Schoch, Dr Aylva Ernest. Geology Dept., U. of the Orange Free State, Bloemfontein 9301, South Africa. (1933) DSc, igneous petrology (Stellenbosch U., 1972). Res. Prof. (tel. 051 + 401-2593, fax +51 8279). *Order-disorder relations and non-linear variations applicable to rock-forming minerals.*

Schöning, Prof. Friedrich Richard Ludwig. Physics Dept., U. of the Witwatersrand, P.O. Wits, Johannesburg 2050, South Africa. (1923) PhD, physics (U. Witwatersrand, 1959). Reader in crystallography. (tel. 011 + 716-2132, fax +11 403 1926). *Diffraction physics, crystal defects, non-crystalline materials.*

Schutte, Prof. Casper Jan Hendrik. Chemistry Dept., U. of South Africa, P.O. Box 392, Pretoria 0001, South Africa. (1934) Dr, physical chemistry (U. Amsterdam, Netherlands, 1960). Prof. (tel. 012 + 429-8008, fax +12 429 3221). *IR spectroscopy, Raman spectroscopy, molecular vibrations, solids.*

Sewell, Dr Bryan Trevor. Biochemistry Dept., U. of Cape Town, Rondebosch 7700, Cape Town, South Africa. (1953) PhD, protein crystallography (London, UK, 1981). Assoc. prof. (tel. 021 + 650-2405, fax +21 685 5931). *Computer graphics, macromolecular structure, chromatin, E.M. tomography.*

Shaw, Dr Martin Philip. Dept. of Materials Eng., U. of Cape Town, Private Bag, Rondebosch 7700, South Africa. (1948) PhD, materials science (U. of Exeter, UK, 1977). Sr. lect. (tel. 021 + 650-3177, fax +21 650 3726). *Structure-property relationships, electron diffraction, electron microscopy.*

Sommerville, Mrs Polly Baker Melville. Chemistry Dept., U. of Natal, King George V Ave., Durban 4001, South Africa. (1924) MSc, chemistry (U. Natal, 1970). Res. asst. (tel. 031 + 816-3090). *Organic structures.*

Spalding, Dr Dennis Raymond. Physics Dept., U. of Natal, King George V Ave., Durban 4001, South Africa. (1942) PhD, physics (Cambridge U., UK, 1969). Sr. lect. (tel. 031 + 816-2775, fax +31 816 2214). *High temperature superconductors.*

Strydom, Dr Ockert Andries Wilhelm. Building B-E1, A.E.C, Private Bag 285, Pelindaba Pretoria 0001, South Africa. (1933) PhD, physics (Rand Afrikaans U., 1969). Div. head. (tel. 012 + 324-2811, ext. 524, fax +12 S79 1515). *Materials science.*

Subramony, Mr Loganathan. Sch. of Health Sci., M L Sultan Technikon, P.O. Box 1334, Durban 4000, South Africa. (1946) MSc, physics (U. Durban-Westville, 1984). Lect. (tel. 031 + 31-6681, ext. 2203). *Organometallic compounds.*

Taylor, Mr Michael William. MINTEK, Private bag X3015, Randburg 2125, South Africa. (1959) BSc(Hons), chemistry (U. Cape Town, 1984). Res. Officer. (tel. 011 + 793-3511). *Inclusion compounds.*

Thackeray, Dr Michael Makepeace. Div. of Mat. Sci. and Tech., CSIR, P.O. Box 395, Pretoria 0001, South Africa. (1949) PhD, chemistry (U. Cape Town, 1977). Programme Manager, Battery Technology. (tel. 012 + 841-3304, fax +12 841 4395). *Solid electrolytes, solid solution electrodes, non-stoichiometric compounds battery systems.*

Van Dyk, Dr Martha Sophia. Chemistry Dept., U. of the Orange Free State, P.O.Box 339, Bloemfontein 9300, South Africa. (1959) PhD, organic chemistry (Rand Afrikaans U., 1986). Post doc. Res. (tel. 051 + 401-2495, fax +51 8279). *Organic structures, NMR analysis, flavonoid structures.*

Van Rooyen, Prof. Petrus Hendrik. Chemistry Dept., U. of Pretoria, Hatfield, Pretoria 0002, South Africa. (1949) PhD, chemistry (Rand Afrikaans U., 1979). Assoc. prof. (tel. 012 + 420-2519, fax +12 342 2453). *Organic and organometallic structures, structure-activity relationships.*

Van Schalkwyk, Prof. Theunis Gabriel Dirkse. Chemistry Dept., U. of Cape Town, Private Bag, Rondebosch 7700, South Africa. (1920) MSc, physics (Stellenbosch U., 1943). Prof. (tel. 021 + 650-2568, fax +21 650-3726). *Organic and inorganic structures.*

SPAIN

Sub-Editor: X. Solans

Notes

1. Degrees confered by Spanish Universities are *Doctor*(DSc), *Graduado*(Grad), and *Licenciado*(Msc).

2. The occupational titles in University are in decreasing order: *Catedratico, Prof. Titular, Prof. Asociado, Prof. Ayudante, Laboral,* and *Becario.*

3. The occupational titles in CSIC are in decreasing order: *Prof. Investigacion, Investigador, Colaborador, Tecnico Superior,* and *Becario.*

Aguilo, Dr Magdalena. Dep. de Quimica Univ. de Barcelona., Pza. Imperial Tarraco 1, 43005-Tarragona, Spain. (1953) DSc, physic (Univ. Barcelona, 1983). Prof. Titular. (tel. 34 + 77 225254, ext 2205) *Crystal growth, morphology-structure relationships.*

Alamo-Serrano, Dr Jaime. Dep. de Quimica Inorganica, Univ. de Valencia, Avda. Dr Moliner 50, 46100-Burjassot, Spain. (1946) DSc, Chemistry (Univ. Valencia, 1971). Prof. Titular. (tel. 34 + 6 3630011, ext 369) *Powder diffraction, properties-structure relationships, thermal expansion, ion transport.*

Alcobe, Mr Xavier, Servicio Cientifico-Tecnico, Univ. de Barcelona Marti y Franques s/n, 08028-Barcelona, Spain. (1962) Grad, Geology (Univ. Barcelona, 1985). Laboral 1. (tel. 34 + 3 3307311, ext 1958) *Powder diffraction, phase transitions and molecular alloys.*

Alonso, Dr Jose Antonio. Inst. Quimica Inorganica Elhuyar, CSIC, Serrano 113, 28006-Madrid, Spain. (1958) DSc, Chemistry (Univ. Complutense Madrid, 1984). Becario. (tel. 34 + 1 4111772) *Structural inorganic chemistry.*

Alvarez, Dr Aurelio. Dep. Geologia, Univ. Autonoma de Barcelona 08193-Bellaterra, Spain. (1935) DSc, Geology (Univ. Barcelona, 1974). Prof. Titular. (tel. 34 + 3 5811611) *Minerals.*

Alvarez, Dr M. Angeles. Dep. Geologia, Univ. Sevilla Apdo. 553. 41071-Sevilla, Spain. (1951) DSc, Chemistry (Univ. Autonome de Madrid, 1976) Prof. Titular. (tel. 34 + 54 625060) *Minerals.*

Amigo, Prof. Jose-Maria. Dep. Geologia, Univ. Valencia Avda. Dr Moliner, 50, 46100-Valencia, Spain. (1940) DSc, Geology (Univ. Barcelona, 1966) Catedratico. (tel. 34 + 6 3630011) *Minerals, crystal structure determination.*

Apreda, Dr M. Carmen. Inst. Quimica Fisica Rocasolano, CSIC, Serrano 119, 28006-Madrid, Spain. (1943) DSc, Physics (Univ. Nacional de la Plata, Argentina, 1976) Contratado. (tel. 34 + 1 2619400) *Crystal structure determination.*

Aragon de la Cruz, Prof. Francisco. Inst. Quimica Inorganica, CSIC, Serrano 113, 28006-Madrid, Spain. (1933) DSc, Chemistry (Univ. Complutense Madrid, 1960) Prof. Investigacion (tel. 34 + 1 4111772) *Inorganic crystals.*

Arana, Prof. Rafael. Dep. Quimica Agricola, Geologia y Edafologia, Univ. de Murcia Sto. Cristo 1, 30001-Murcia, Spain. (1942) DSc, Geology (Univ. Granada, 1972) Catedratico (tel. 34 + 68 236207) *Optical crystallography, minerals, X-ray diffraction, computing, mathematical crystallography.*

Arrieta, Dr Juan Manuel. Dep. Quimica, Univ. del Pais Vasco Apdo. 644 48080-Bilbao, Spain. (1952) DSc, Chemistry (Univ. Pais Vasco) Prof. Titular (tel. 34 + 4 4641000 Ext. 348) *Crystal structure determination.*

Arriortua, Dr Maribel. Dep. Cristalografia y Mineralogia, Univ. Pais Vasco Apdo. 644, 48080-Bilbao, Spain. (1950) DSc, Chemistry (Univ. Pais Vasco, 1981) Prof. Titular (tel. 34 + 4 4641000, ext. 316,318) *Crystal structure determination.*

Balcazar, Dr Jose Luis. Univ. Alcala de Henares ALcala de Henares (Madrid), Spain. (1929) DSc, Chemistry (Univ. Vallalodid) Prof. Titular (tel. 34 + 1 8890400) *Crystal structure determination, minerals.*

Bastida, Dr Joaquin. Dep. Geologia, Univ. Valencia Avda. Dr Moliner, 50, 46100-Valencia, Spain. (1955) DSc, Geology (Univ. Autonoma de Barcelona) Prof. Titular. (tel. 34 + 6 3478101) *Teaching, X-ray diffraction, minerals.*

Bayon, Dr J. Carlos. Dep. Quimica, Univ. Autonoma de Barcelona 08193-Bellaterra, Spain. (1954) DSc, Chemistry (Univ. Autonoma de Barcelona, 1981) Prof. Titular (tel. 34 + 3 5811889) *Structural chemistry.*

Bellver, Dr Consuelo. Dep. Optica, Univ. de Sevilla Appdo. 1065, 41080-Sevilla, Spain. (1957) DSc, Physic (Univ. Seville, 1985) Prof. Colaborador. (tel. 34 + 54 616615) *Crystal structure determination.*

Bermudez-Polonio, Dr Joaquin. CSIC, Serrano 113, 28006-Madrid, Spain. (1930) DSc, Chemistry (Univ. Complutense de Madrid, 1964) Investigador Cientifico. (tel. 34 + 1 4111772) *X-ray spectroscopy, relationship between physical properties and structure, X-ray diffraction, inorganic crystals.*

Bernalte, Prof. Antoni. Univ. Nac. de Educacion a distancia Appdo 50487, 28080-Madrid, Spain. (1927) DSc, Physics (U. of California, Berkeley, USA, 1968) Catedratico. (tel. 34 + 1 2439431) *Defect structures, mathematical crystallography.*

Blanco, Dr Marta. Dep. Geologia, Univ. de Oviedo Arias de Velasco s/n, 33005-Oviedo. Spain. (1950) DSc, Geology. (Univ. Oviedo, 1986) Prof. Ayudante. (tel. 34 + 85233200, Ext. 150) *Minerals, polymorphism.*

Brianso, Prof. Jose Luis. Dep. Geologia, Univ. Autonoma de Barcelona 08193-Bellaterra, Spain. (1944) DSc, Geology (Univ. de Barcelona, 1972) Catedratico. (tel. 34 + 3 5811611) *Crystal structure.*

Brime, Dr Covadonga. Dep. Geologia, Univ. de Oviedo Arias de Velasco s/n, 33005-Oviedo. Spain. (1950) DSc, Geology. (Univ. Oviedo, 1978) Prof. Titular. (tel. 34 + 85233200, Ext. 150) *Minerals.*

Caballero, Prof. M. Antonio. Dep. Estructura y Propiedades de los Materiales, Univ. de Cadiz Appdo. 40, 11510 Pto. Real. Spain. (1942) DSc, Geology. (Univ. Complutense de Madrid, 1972) Catedratico. (tel. 34 + 56850210) *X-ray diffraction.*

Calvet, Mr. Teresa. Dep. Cristalografia, Mineralogia y Depositos Minerales, Univ. de Barcelona, Marti y Franques s/n, 08028-Barcelona, Spain. (1960) Grad, Geology (Univ. Barcelona, 1985). Becario. (tel. 34 + 3 3307311, ext 1973) *X-ray diffraction, syncrystallization.*

Cardellach, Dr Esteve. Dep. Geologia, Univ. Autonoma de Barcelona 08193-Bellaterra, Spain. (1949) DSc, Geology (Univ. de Barcelona) Prof. Titular. (tel. 34 + 3 5811611) *Minerals.*

Carriedo, Dr Gabino-Alejandro. Dep. Quimica Organometalica, Univ. Oviedo 33071-Oviedo, Spain. (1952) DSc, Chemistry (Univ. de Vallalodid, 1981) Prof. Titular. (tel. 34 + 3 5811611) *Organometalic structural chemistry.*

Casabo, Prof. Jaume. Dep. Quimica, Univ. Autonoma de Barcelona 08193-Bellaterra, Spain. (1941) DSc, Chemistry (Univ. de Barcelona, 1972) Catedratico. (tel. 34 + 3 5811369) *Structural chemistry.*

Cascales, Dr Concepcion. Inst. Quimica Inorganica Elhuyar, CSIC, Serrano 113, 28006-Madrid, Spain. (1959) DSc, Chemistry (Univ. Complutense de Madrid, 1986). Becario. (tel. 34 + 1 4111772). *Structural chemistry.*

Castiñeiras, Dr Alfonso. Dep. Quimica Inorganica Univ. de Santiago de Compostela, Campus Universitario, 15706-Santiago, Spain. (1942) DSc, Chemistry (Univ. Santiago, 1974). Prof. Titular. (tel. 34 + 81 594636). *Structural chemistry, crystal structure determination.*

Castro, Dr Alicia. Inst. Quimica Inorganica Elhuyar, CSIC, Serrano 113, 28006-Madrid, Spain. (1958) DSc, Chemistry (Univ. Complutense de Madrid, 1984). Becario. (tel. 34 + 1 4111772). *Structural chemistry.*

Claramunt, Prof. Rosa Maria. Univ. Nacional de Educacion a distancia Ciudad Universitaria, 28040-Madrid, Spain. (1948) DSc, Chemistry (Univ. Barcelona, 1973) Catedratico. (tel. 34 + 1 4493600) *Structural chemistry.*

Colacio, Dr Enrique. Dep. Quimica Inorganica, Univ. de Granada Fuentenueva s/n, 18071-Granada, Spain. (1957) DSc, Chemistry (Univ. Granada) Prof. Titular. (tel. 34 + 58 202212, ext. 322) *Structural chemistry.*

Conde, Prof. Alejandro. Dep. Fisica de la Materia Condensada, Univ. de Sevilla Appdo. 1065, 41080-Sevilla, Spain. (1947) DSc, Physics (Univ. Sevilla, 1972) Catedratico. (tel. 34 + 54 616615) *Crystal structure determination, amorphous materials.*

Conde, Dr Clara Francisca. Dep. Fisica de la Materia Condensada, Univ. de Sevilla Appdo. 1065, 41080-Sevilla, Spain. (1952) DSc, Physics (Univ. Sevilla, 1981) Prof. Titular. (tel. 34 + 54 616615) *Crystal structure determination, amorphous materials.*

Coy-Yll, Prof. Ramon. Dep. Cristalografia y Mineralogia, Univ. Complutense de Madrid Ciudad Universitaria, 28040-Madrid, Spain. (1940) DSc, Geology (Univ. Barcelona, 1964) Catedratico. (tel. 34 + 1 2437195) *Minerals, dynamical crystal properties.*

Cuevas-Diarte, Dr Miquel Angel. Dep. Cristalografia, Mineralogia y Depositos Minerales, Univ. de Barcelona, Marti y Franques s/n, 08028-Barcelona, Spain. (1948) Grad, Geology (Univ. Barcelona, 1979). Prof. Titular. (tel. 34 + 3 3307311, ext 1973) *X-ray diffraction, syncrystallization, thermal analysis.*

Cumbrera, Dr Francisco Luis. Dep. Fisica de la Materia Condensada, Univ. de Sevilla Appdo. 1065, 41080-Sevilla, Spain. (1954) DSc, Physics (Univ. Sevilla, 1982) Catedratico. (tel. 34 + 54 616615) *X-ray diffraction, amorphous materials.*

Diaz, Dr Francesc. Dep. de Quimica, Univ. de Barcelona. Pza. Imperial Tarraco 1, 43005-Tarragona, Spain. (1953) DSc, physic (Univ. Barcelona, 1982). Prof. Titular. (tel. 34 + 77 225254, ext 2220) *Crystal growth.*

Dianez-Millan, Dr M. Jesus. Dep. Fisica de la Materia Condensada, Univ. de Sevilla Appdo. 1065, 41080-Sevilla, Spain. (1950) DSc, Physics (Univ. Sevilla, 1985) Prof. Associado. (tel. 34 + 54 616615) *X-ray diffraction, amorphous materials.*

Domenech, Dr M. Victoria. Dep. Cristalografia, Mineralogia y Depositos Minerales, Univ. de Barcelona, Marti y Franques s/n, 08028-Barcelona, Spain. (1940) Grad, Chemistry (Univ. Oviedo, 1981). Colaborador. (tel. 34 + 3 3307311, ext 1979) *Crystal structure determination.*

Dominguez, Dr Esther. Dep. Quimica, Univ. del Pais Vasco Apdo. 644 48080-Bilbao, Spain. (1947) DSc, Chemistry (Univ. Pais Vasco, 1975) Prof. Titular. (tel. 34 + 4 4641000 Ext. 510) *Crystal structure determination.*

Espinet, Prof. Pablo. Dep. Quimica Inorganica, Univ. de Vallaodid, Prado de la Magdalena s/n, 47005-Vallalodid, Spain. (1949) DSc, Chemistry (Univ. Zaragoza, 1975). Catedratico (tel. 34 + 48 251713). *Structural chemistry.*

Esteban-Calderon, Dr M. Carmen. Esc. Univ. Ingenieria Tecnica Industrial, Univ. de Vallalodid. Joaquin Velasco Martin s/n, 47014-Vallalodid, Spain. (1953) DSc, Chemistry (Univ. Madrid, 1981). Prof. Asociado (tel. 34 + 48 337244). *Crystal structure determination.*

Estop, Dr Eugenia. Dep. Geologia, Univ. Autonoma de Barcelona 08193-Bellaterra, Spain. (1950) DSc, Geology (Univ. de Barcelona, 1980) Prof. Titular. (tel. 34 + 3 5811611) *Molecular alloys, X-ray diffraction.*

Estrada, DrM. Dolores. Dep. Fisica de la Materia Condensada, Univ. de Sevilla Appdo. 1065, 41080-Sevilla, Spain. (1953) DSc, Physics (Univ. Sevilla, 1984) Prof. Titular. (tel. 34 + 54 616615) *Crystal structure determination.*

Faus-Paya, Prof. Juan. Dep. Quimica Inorganica, Univ. de Valencia, Avda. Dr Moliner, 50, 46100-Valencia, Spain. (1946) DSc, Chemistry (Univ. Valencia, 1971) Catedratico. (tel. 34 + 6 3630011, ext 365) *Structural chemistry.*

Fayos, Prof. Jose. Inst. Rocasolano, CSIC, Serrano, 119, 28006-Madrid, Spain. (1940) DSc, Physics (Univ. Complutense de Madrid) Prof. Investigacion. (tel. 34 + 1 2619400) *Crystal structure determination.*

Fenoll, Prof. Purificacion. Dep. Mineralogia y Petrologia, Univ. de Granada Fuentenueva s/n, 18002-Granada, Spain. (1957) DSc, Chemistry (Univ. Granada, 1966) Catedratico. (tel. 34 + 58 202212) *X-ray diffraction, minerals.*

Fernandez, Dr Carlos Jose. Dep. Geologia, Universidad de Oviedo Arias de Velasco s/n, 33005-Oviedo. Spain. (1951) DSc, Geology. (Univ. Oviedo, 1982) Prof. Titular. (tel. 34 + 85233200) *Minerals.*

Fita, Dr Ignacio. Dep. Ingenieria Quimica, Univ. Politecnica de Catalunya Diagonal 647, 08028-Barcelona. Spain. (1953) DSc, Biology. (Univ. Autonoma de Barcelona, 1981) Prof. Titular. (tel. 34 + 3 2495800, ext. 236) *Structural biology.*

Florencio, Dr Feliciana. Inst. Rocasolano, CSIC, Serrano, 119, 28006-Madrid, Spain. (1924) DSc, Chemistry (Univ. Complutense de Madrid) Investigador. (tel. 34 + 1 2619400) *Crystal structure determination.*

Foces-Foces, Dr Concepcion. Inst. Rocasolano, CSIC, Serrano, 119, 28006-Madrid, Spain. (1946) DSc, Physics (Univ. Complutense de Madrid, 1974) Investigador. (tel. 34 + 1 2619400) *Crystal structure determination, computing.*

Fonseca, Dr Isabel. Inst. Rocasolano, CSIC, Serrano, 119, 28006-Madrid, Spain. (1933) DSc, Chemistry (Univ. Complutense de Madrid) Colaborador. (tel. 34 + 1 2619400) *Crystal structure determination.*

Font-Altaba, Prof. Manuel. Dep. Cristalografia, Mineralogia y Depositos Minerales, Univ. de Barcelona, Marti y Franques s/n, 08028-Barcelona, Spain. (1923) DSc, Chemistry (Univ. Barcelona, 1954). Prof. Emeritus. (tel. 34 + 3 3307311, ext 1959) *Crystal structure determination, crystal growth, minerals.*

Font-Bardia, Mr. Merce. Dep. Cristalografia, Mineralogia y Depositos Minerales, Univ. de Barcelona, Marti y Franques s/n, 08028-Barcelona, Spain. (1956)

Grad., Pharmacy (Univ. Barcelona, 1984). Becario. (tel. 34 + 3 3307311, ext 1959) *Crystal structure determination.*

Forteza, Mr Matilde. Dep. Geologia, Univ. de Sevilla Appdo. 553, 41071-Sevilla, Spain. (1954) Grad. Pharmacy (Univ. Sevilla, 1978) Prof. Ayudante. (tel. 34 + 54 625060) *Minerals.*

Fuente-Cullel, Dr Carlos. Dep. Cristalografia, Mineralogia y Depositos Minerales, Univ. de Barcelona, Marti y Franques s/n, 08028-Barcelona, Spain. (1941) DSc, Geology (Univ. Barcelona). Prof. Titular. (tel. 34 + 3 3307311, ext 1971) *Minerals.*

Fuertes, Dr Amparo. Inst. de Ciencia de Materiales de Barcelona, CSIC, Marti y Franques s/n, 08028-Barcelona, Spain. (1959) DSc, Chemistry (Univ. Valencia, 1986). Colaborador. (tel. 34 + 3 3302716) *Structural chemistry, crystal structure determination.*

Gaete, Dr Walter. Dep. Quimica, Univ. Autonoma de Barcelona 08193-Bellaterra, Spain. (1934) DSc, Chemistry (Univ. Sta. Maria, Chile, 1960) Prof. Titular. (tel. 34 + 3 5811010) *Structural chemistry.*

Galan, Prof. Emilio. Dep. Geologia, Univ. Sevilla Apdo. 553. 41071-Sevilla, Spain. (1942) DSc, Geology (Univ. de Madrid, 1972) Catedratico. (tel. 34 + 54 625060) *Minerals.*

Gali, Dr Salvador. Dep. Cristalografia, Mineralogia y Depositos Minerales, Univ. de Barcelona, Marti y Franques s/n, 08028-Barcelona, Spain. (1949) DSc, Geology (Univ. Barcelona, 1976). Prof. Titular. (tel. 34 + 3 3307311, ext 1970) *Crystal structure determination.*

Galindo, Dr Agustin. Dep. Quimica Inorganica, Univ. Sevilla 41012-Sevilla, Spain. (1961) DSc, Chemistry (Univ. de Sevilla, 1986) Becario. (tel. 34 + 54 629061) *Structural chemistry.*

Garcia-Blanco, Prof. Severino. Inst. Rocasolano, CSIC, Serrano, 119, 28006-Madrid, Spain. (1922) DSc, Chemistry (Univ. Complutense de Madrid) Prof. Emeritus. (tel. 34 + 1 2619400) *Crystal structure determination.*

Garcia-Casado, Dr Pedro. Facultad de Farmacia, Univ. de Navarra, 31080-Pamplona, Spain. (1942) DSc, Chemistry (Univ. Navarra, 1986) Prof. Titular. (tel. 34 + 48 252150) *Crystal structure determination.*

Garcia-Rodriguez, Prof. Antonio. Dep. Quimica Inorganica, Univ. de Granada Fuentenueva s/n, 18071-Granada, Spain. (1945) DSc, Chemistry (Univ. Granada, 1972) Catedratico. (tel. 34 + 58 202212, ext. 322) *Structural chemistry.*

Garcia-Ruiz, Dr Joaquin. Dep. Fisica de la Materia Condensada, Univ. de Zaragoza, Pza. S. Francisco, 50009-Zaragoza, Spain. (1951) DSc, Physics (Univ. Zaragoza, 1981) Prof. Titular. (tel. 34 + 76 353557) *X-ray absorption spectroscopy.*

Gervilla, Mr Fernando. Dep. Mineralogia y Petrologia, Univ. de Granada, Fuentenueva s/n, 18002-Granada, Spain. (1961) Grad. Geology (Univ. Granada, 1985) Becario. (tel. 34 + 58 202212) *X-ray diffraction, minerals.*

Gimeno, Prof. Jose. Dep. Quimica Organometalica, Univ. Oviedo 33071-Oviedo, Spain. (1947) DSc, Chemistry. (Univ. de Zaragoza, 1972) Catedratico. (tel. 34 + 85 239516) *Structural chemistry.*

Gomez-Sal, Dr M. Pilar. Dep. Quimica Inorganica, Univ. Alcala de Henares Campus Universitario, Madrid, Spain. (1951) DSc, Chemistry. (Univ. Complutense de Madrid, 1978) Prof. Titular. (tel. 34 + 1 8890400) *Crystal structure determination.*

Gonzalez-Calbet, Dr Jose M. Dep. Quimica Inorganica, Univ. Complutense de Madrid Campus Universitario, 28040-Madrid, Spain. (1952) DSc, Chemistry. (Univ. Complutense de Madrid, 1979) Prof. Titular. (tel. 34 + 1 4491850) *Solid state chemistry.*

Gregorkiewitz, Dr Miguel. Inst. de Ciencia de Materiales, CSIC, Serrano 115 bis, 28006-Madrid, Spain. (1946) DSc, Natural Sciences (Tech. Hochschule, Darmstadt, FRD, 1980) Investigador. (tel. 34 + 1 2624526) *Crystal structure determination, X-ray diffraction.*

Gutierrez-Puebla, Dr Enrique. Inst. de Quimica Inorganica, CSIC, Serrano 113, 28006-Madrid, Spain. (1952) DSc, Chemistry. (Univ. Complutense de Madrid, 1978) Colaborador. (tel. 34 + 1 4111772) *Crystal structure determination.*

Gutierrez-Zorrilla, Dr Juan Manuel. Dep. Quimica Inorganica, Univ. del Pais Vasco Apddo. 644, 48080-Bilbao, Spain. (1957) DSc, Chemistry. (Univ. Pais Vasco, 1984) Prof. Titular. (tel. 34 + 44 4641000, ext 259) *Crystal structure determination, structural chemistry.*

Hernandez-Cano, Prof. Felix. Inst. Rocasolano, CSIC, Serrano, 119, 28006-Madrid, Spain. (1941) DSc, Physics (Univ. Complutense de Madrid) Prof. Investigacion. (tel. 34 + 1 2619400) *Crystal structure determination, computing, teaching.*

Herreras, Mr M. Luisa. Inst. de Ciencia de Materiales de Barcelona, CSIC, Marti y Franques s/n, 08028-Barcelona, Spain. (1961) Grad. Chemistry (Univ. Vallalodid, 1985). Contratado. (tel. 34 + 3 3302716) *Crystal growth.*

Iglesias, Prof. Juan Eugenio. Inst. de Ciencia de Materiales, CSIC, Serrano 115 bis, 28006-Madrid, Spain. (1942) DSc, Chemical Eng. (U. of Texas, USA, 1971) Investigador. (tel. 34 + 1 2624526) *Structural chemistry, X-ray diffraction.*

Julve-Oliva, Dr Miguel. Dep. Quimica Inorganica, Univ. de Valencia, Avda. Dr Moliner, 50, 46100-Valencia, Spain. (1953) DSc, Chemistry. (Univ. Valencia, 1981) Prof. Titular. (tel. 34 + 6 3630011, ext 369) *Structural chemistry.*

Lahoz, Dr Fernando J. Dpto. Quimica Inorganica, Univ. de Zaragoza, 50009-Zaragoza, Spain. (1958) DSc, Chemistry (Univ. Zaragoza, 1983) Prof.

Ayudante. (tel. 34 + 76 454559) *Crystal structure determination, structural chemistry.*

Liso-Rubio, Prof. M. Jesus. Universidad de Extremadura Carretera de Portugal, 06071-Badajoz, Spain. (1944) DSc, Pharmacy (Univ. Complutense de Madrid, 1969) Catedratico. (tel. 34 + 24 238800) *Minerals.*

Lloveras, Dr Joaquim. Dep. Chemical Engineering, Univ. Politecnica de Catalunya Diagonal 647, 08028-Barcelona, Spain. (1946) DSc, Engineering (Univ. Politecnica de Catalunya, 1974) Prof. Titular. (tel. 34 + 3 2495800, ext. 236) *Fiber diffraction, small-angle scattering, structural biology.*

Lopez-Acevedo, Dr M. Victoria. Dep. Cristalografia y Mineralogia, Univ. Complutense de Madrid 28040-Madrid, Spain. (1953) DSc, Geology (Univ. Complutense de Madrid, 1983) Prof. Titular. (tel. 34 + 1 2433468) *Minerals.*

Lopez-Castro, Prof. Amparo. Inst. de Ciencia de Materiales de Sevilla, CSIC, Reina Mercedes s/n, 41080-Sevilla, Spain. (1928) DSc, Physics (Univ. Complutense de Madrid, 1954) Prof. Investigacion. (tel. 34 + 54 616615) *Crystal structure determination, X-ray diffraction.*

Lopez de Lerma, Dr Julian. Inst. Rocasolano, CSIC, Serrano, 119, 28006-Madrid, Spain. (1928) DSc, Chemistry. (Univ. Complutense de Madrid) Investigador. (tel. 34 + 1 2619400) *Crystal structure determination.*

Lopez-Galindo, Dr Alberto. Dep. Mineralogia y Petrologia, Univ. de Granada, Fuentenueva s/n, 18002-Granada, Spain. (1960) DSc, Geology (Univ. Granada, 1986) Becario. (tel. 34 + 58 202212) *X-ray diffraction, minerals.*

Lopez-Gonzalez, Prof. Juan de D. Dep. Quimica Inorganica Univ. Nacional Educacion a distancia, Campus Universitario, 28040-Madrid, Spain. (1924) DSc, Chemistry. (Univ. Complutense de Madrid, 1949). Catedratico. (tel. 34 + 1 2439431). *Structural chemistry.*

Lopez-Soler, Dr Angel. Inst. Jaime Almera, CSIC, Marti y Franques s/n, 08028-Barcelona, Spain. (1940) DSc, Geology (Univ. Barcelona, 1968). Colaborador. (tel. 34 + 3 3302716). *Minerals, optical properties - solids.*

Madariaga, Dr Gotzon. Dep. Fisica, Univ. del Pais Vasco Apddo. 644, 48080-Bilbao, Spain. (1959) DSc, Physics (Univ. Pais Vasco, 1985) Prof. Titular. (tel. 34 + 44 4641000, ext 275) *Phase transitions.*

Marcano-Fermin, Dr Cenis M. Univ. de Oriente, Av. Universidad, Cerro Colorado, Cumana, Venezuela. (1949) DSc, Chemistry (Univ. Complutense de Madrid, 1986) Prof. Agregado. (tel. 58 + 93 662244) *Structural chemistry.*

Marcos, Dr Celia. Dep. Geologia, Univ. de Oviedo, Arias de Velasco s/n, 33005-Oviedo, Spain. (1954) DSc, Geology (Univ. Oviedo, 1985) Prof. Ayudante. (tel. 34 + 85 233200, ext 152) *Minerals.*

Marin-Elena, Dr Jose Manuel. Dep. Quimica Inorganica, Univ. de Sevilla, 41012-Sevilla, Spain. (1955) DSc, Chemistry (Univ. Sevilla, 1981) Prof. Titular. (tel. 34 + 54 629061) *Structural chemistry.*

Marquez-Delgado, Prof. Rafael. Dep. Fisica de la Materia Condensada, Univ. de Sevilla Reina Mercedes s/n, 41080-Sevilla, Spain. (1929) DSc, Physics (Univ. Complutense de Madrid, 1957) Catedratico. (tel. 34 + 54 616615) *Crystal structure determination, X-ray diffraction, electron diffraction.*

Marti-Artoy, Mr Xavier. Dep. Geologia, Univ. Autonoma de Barcelona 08193-Bellaterra, Spain. (1963) Grad, Geology (Univ. Autonoma de Barcelona, 1986) Becario. (tel. 34 + 3 5811611) *Structural chemistry.*

Martin-Ramos, Dr Jose Daniel. Dep. Mineralogia y Petrologia, Univ. de Granada, Fuentenueva s/n, 18002-Granada, Spain. (1949) DSc, Geology (Univ. Granada, 1977) Prof. Titular. (tel. 34 + 58 202212, ext 340) *Crystal structure determination, powder methods, texture.*

Martin-Vivaldi, Dr Juan Luis. Dep. Cristalografia y Mineralogia, Univ. Complutense de Madrid 28040-Madrid, Spain. (1950) DSc, Geology (Univ. Complutense de Madrid, 1983) Prof. Titular. (tel. 34 + 1 2433468) *Minerals.*

Martinez, Mr Benjamin. Inst. de Ciencia de Materiales de Barcelona, CSIC, Marti y Franques s/n, 08028-Barcelona, Spain. (1960) Grad. Physics (Univ. Barcelona, 1984). Becario. (tel. 34 + 3 3302716) *Crystal structure determination.*

Martinez, Dr Francisco. Dpto. Quimica Inorganica, Univ. de Zaragoza, 50009-Zaragoza, Spain. (1945) DSc, Chemistry (Univ. Zaragoza, 1975) Prof. Titular. (tel. 34 + 76 452347) *Crystal structure determination, structural chemistry.*

Martinez-Carrera, Prof. Sagrario. Inst. Rocasolano, CSIC, Serrano, 119, 28006-Madrid, Spain. (1925) DSc, Chemistry. (Univ. Complutense de Madrid) Prof. Investigacion. (tel. 34 + 1 2619400) *Crystal structure determination.*

Martinez-Ripoll, Prof. Martin. Inst. Rocasolano, CSIC, Serrano, 119, 28006-Madrid, Spain. (1946) DSc, Chemistry. (Univ. Complutense de Madrid) Prof. Investigacion. (tel. 34 + 1 2619400) *Crystal structure determination, computing.*

Millan-Muñoz, Dr Maria. Dep. Fisica de la Materia Condensada, Univ. de Sevilla, Reina Mercedes s/n. 41080-Sevilla, Spain. (1950) DSc, Physics (Univ. Sevilla, 1980). Prof. Titular. (tel. 34 + 54 616615). *Structural chemistry, amorphous crystals.*

Miravitlles, Prof. Carlos. Inst. de Ciencia de Materiales de Barcelona, CSIC, Marti y Franques s/n, 08028-Barcelona, Spain. (1942) DSc, Pharmacy (Univ. Barcelona, 1971). Prof. Investigacion. (tel. 34 + 3 3302716) *Crystal structure determination, materials science.*

Molins, Dr Elies. Inst. de Ciencia de Materiales de Barcelona, CSIC, Marti y Franques s/n, 08028-Barcelona, Spain. (1957) DSc, Physics (Univ. Barcelona, 1985). Colaborador. (tel. 34 + 3 3302716) *Crystal structure determination, materials science.*

Monge, Dr M. Angeles. Inst. de Quimica Inorganica, CSIC, Serrano 113, 28006-Madrid, Spain. (1951) DSc, Chemistry. (Univ. Complutense de Madrid, 1978) Colaborador. (tel. 34 + 1 4111772) *Crystal structure determination.*

Moreiras, Dr Damaso. Dep. Geologia, Univ. de Oviedo, Arias de Velasco s/n, 33005-Oviedo, Spain. (1947) DSc, Geology (Univ. Oviedo, 1980) Prof. Titular. (tel. 34 + 85 233200, ext 126) *Minerals, crystal structure determination, X-ray diffraction.*

Moreno-Carretero, Dr Miguel. Dep. Quimica Inorganica, Univ. de Granada Fuentenueva s/n, 18071-Granada, Spain. (1958) DSc, Chemistry. (Univ. Granada, 1983) Prof. Titular. (tel. 34 + 58 202212, ext. 322) *Structural chemistry.*

Moreno-Echevarria, Dr M. Esperanza. Inst. de Ciencia de Materiales de Sevilla, CSIC, Reina Mercedes s/n, 41080-Sevilla, Spain. (1925) DSc, Physics (Univ. Sevilla, 1960) Colaborador. (tel. 34 + 54 616615) *Crystal structure determination, X-ray diffraction.*

Moron, Dr M. Carmen. Inst. Ciencia de Materiales de Aragon, CSIC, 50009-Zaragoza, Spain. (1960) DSc, Physics (Univ. Zaragoza, 1988) Becario. (tel. 34 + 76 353557) *Structural chemistry.*

Mzayek, Mr Elias. Inst. de Quimica Inorganica, CSIC, Serrano 113, 28006-Madrid, Spain. (1957) Grad, Chemistry. (Univ. Alepo, Syria, 1981) Becario. (tel. 34 + 1 4111772) *Structural chemistry.*

Navarro, Dr Carmen. Dep. Quimica Inorganica, Univ. Autonoma de Madrid. 28049-Madrid, Spain. (1947) DSc, Chemistry (Univ. Autonoma de Madrid, 1976) Prof. Titular. (tel. 34 + 1 3974356) *Structural chemistry.*

Nieto-Garcia, Dr Fernando. Dep. Mineralogia y Petrologia, Univ. de Granada, Fuentenueva s/n, 18002-Granada, Spain. (1955) DSc, Geology (Univ. Granada, 1982) Prof. Titular. (tel. 34 + 58 202212, ext 342) *Minerals.*

Nogues-Carulla, Dr Joaquim. Dep. Cristalografia, Mineralogia y Depositos Minerales, Univ. de Barcelona, Marti y Franques s/n, 08028-Barcelona, Spain. (1946) DSc, Geology (Univ. Barcelona, 1976). Prof. Titular. (tel. 34 + 3 3307311, ext 1972) *Minerals.*

Ortega-Huertas, Dr Miguel. Dep. Mineralogia y Petrologia, Univ. de Granada, Fuentenueva s/n, 18002-Granada, Spain. (1949) DSc, Geology (Univ. Granada, 1978) Prof. Titular. (tel. 34 + 58 202212) *Minerals, X-ray diffraction.*

Otero-Diaz, Dr L. Carlos. Dep. Quimica Inorganica, Univ. Complutense de Madrid Campus Universitario, 28040-Madrid, Spain. (1951) DSc, Chemistry. (Univ. Complutense de Madrid, 1979) Prof. Titular. (tel. 34 + 1 4491850) *Solid state chemistry.*

Palacio, Dr Fernando. Inst. Ciencia de Materiales de Aragon, CSIC, 50009-Zaragoza, Spain. (1944) DSc, Chemistry (Univ. Zaragoza, 1974) Investigador. (tel. 34 + 76 353557) *Structural chemistry, magnetism.*

Palomo-Delgado, Dr M. Inmaculada. Dep. Mineralogia y Petrologia, Univ. de Granada, Fuentenueva s/n, 18002-Granada, Spain. (1957) DSc, Geology (Univ. Granada, 1987) Prof. Titular. (tel. 34 + 58 202212, ext 343) *Minerals, X-ray diffraction.*

Perales, Dr Aurea. Inst. Rocasolano, CSIC, Serrano, 119, 28006-Madrid, Spain. (1929) DSc, Chemistry. (Univ. Valencia, 1952) Investigador. (tel. 34 + 1 2619400) *Crystal structure determination, computing.*

Perez-Garcia, Mr Virginia. Dep. Quimica Inorganica, Univ. Complutense de Madrid Campus Universitario, 28040-Madrid, Spain. (1960) Grad, Chemistry. (Univ. Complutense de Madrid, 1984) Becario. (tel. 34 + 1 4491850) *Crystal structure determination.*

Perez-Garrido, Dr Simeon. Dep. Optica, Univ. de Sevilla, Reina Mercedes s/n, 41080-Sevilla, Spain. (1943) DSc, Physics (Univ. Sevilla, 1971) Prof. Titular. (tel. 34 + 54 616615) *Crystal structure determination, X-ray diffraction.*

Perez-Mato, Dr Juan Manuel. Dep. Fisica, Univ. del Pais Vasco Apddo. 644, 48080-Bilbao, Spain. (1952) DSc, Physics (Univ. Pais Vasco, 1980) Prof. Titular. (tel. 34 + 44 4641000, ext 524) *X-ray diffraction.*

Piniella, Dr Juan Francisco. Dep. Geologia, Univ. Autonoma de Barcelona 08193-Bellaterra, Spain. (1952) DSc, Chemistry. (Univ. Autonoma de Barcelona, 1985) Prof. Titular. (tel. 34 + 3 5811611) *Structural chemistry.*

Plana-Llevat, Dr Feliciano. Inst. Jaime Almera, CSIC, Marti y Franques s/n, 08028-Barcelona, Spain. (1946) DSc, Geology (Univ. Barcelona, 1974). Colaborador. (tel. 34 + 3 3302716). *X-ray diffraction, powder diffraction, minerals.*

Polvorinos, Dr Angel Jesus. Dep. Geologia, Univ. Sevilla Apdo. 553. 41071-Sevilla, Spain. (1952) DSc, Geology (Univ. de Sevilla, 1981) Prof. Titular. (tel. 34 + 54 625060) *Minerals.*

Prieto, Dr Manuel. Dep. Cristalografia y Mineralogia, Univ. Complutense de Madrid 28040-Madrid, Spain. (1950) DSc, Geology (Univ. Complutense de Madrid) Prof. Titular. (tel. 34 + 1 2433468) *Minerals.*

Puigjaner, Prof. Luis. Dep. Quimica Macromolecular Univ. Politecnica de Catalunya, Diagonal 647, 08028-Barcelona, Spain. (1935) DSc, Engineering (Univ. Politecnica de Madrid, 1968). Catedratico. (tel. 34 + 3 2496400). *Structural biology, computing, X-ray diffraction.*

Rasines, Prof. Isidoro. Inst. de Quimica Inorganica, CSIC, Serrano 113, 28006-Madrid, Spain. (1927) DSc, Chemistry. (Univ. Complutense de Madrid, 1970) Prof. Investigacion. (tel. 34 + 1 4111772) *Crystal structure determination, powder diffraction.*

Rausell-Colom, Prof. Jose Antonio. Inst. de Ciencia de Materiales, CSIC, Serrano 115 bis, 28006-Madrid, Spain. (1932) DSc, Chemistry. (Univ. Complutense de Madrid) Prof. Investigacion. (tel. 34 + 1 4111772) *Structural chemistry, minerals, small-angle scattering.*

Ribas, Prof. Joan. Dep. Quimica Inorganica, Univ. de Barcelona, Diagonal 647. 08028-Barcelona, Spain. (1943) DSc, Chemistry (Univ. Barcelona, 1974) Catedratico. (tel. 34 + 3 3308813) *Structural chemistry.*

Riera, Prof. Victor. Dep. Quimica Organometalica, Univ. Oviedo 33071-Oviedo, Spain. (1936) DSc, Chemistry. (Univ. de Oviedo) Catedratico. (tel. 34 + 85 232526) *Structural chemistry.*

Rius, Dr Jordi. Inst. de Ciencia de Materiales de Barcelona, CSIC, Marti y Franques s/n, 08028-Barcelona, Spain. (1954) DSc, Natural Sciences (Univ. Marburg, DFR, 1980) Colaborador. (tel. 34 + 3 3302716) *Crystal structure determination, minerals, X-ray diffraction.*

Rodriguez-Carvajal, Dr Juan. Inst. de Ciencia de Materiales de Barcelona, CSIC, Marti y Franques s/n, 08028-Barcelona, Spain. (1953) DSc, Physics (Univ. Barcelona, 1984) Colaborador. (tel. 34 + 3 3302716) *X-ray diffraction, neutron diffraction, powder diffraction.*

Rodriguez-Clemente, Dr Rafael. Inst. de Ciencia de Materiales de Barcelona, CSIC, Marti y Franques s/n, 08028-Barcelona, Spain. (1948) DSc, Geology (Univ. Barcelona) Investigador. (tel. 34 + 3 3302716) *Crystal growth.*

Rodriguez-Gallego, Prof. Manuel. Dep. Mineralogia y Petrologia, Univ. de Granada, Fuentenueva s/n, 18002-Granada, Spain. (1935) DSc, Geology (Univ. Granada, 1960) Catedratico. (tel. 34 + 58 202212, ext 339) *X-ray diffraction, crystal structure determination, small-angle scattering, defect structures.*

Rodriguez-Gordillo, Dr Jose. Dep. Mineralogia y Petrologia, Univ. de Granada, Fuentenueva s/n, 18002-Granada, Spain. (1946) DSc, Geology (Univ. Granada, 1976) Prof. Titular. (tel. 34 + 58 202212) *X-ray diffraction, minerals.*

Rodriguez-Roldan, Dr Ana. Dep. Quimica Inorganica, Univ. Complutense de Madrid Campus Universitario, 28040-Madrid, Spain. (1951) DSc, Chemistry. (Univ. Complutense de Madrid, 1983) Becario. (tel. 34 + 1 4491850) *Crystal structure determination.*

Roman, Dr Pascual. Dep. Quimica Inorganica, Univ. del Pais Vasco Apddo. 644, 48080-Bilbao, Spain. (1947) DSc, Chemistry (Univ. Pais Vasco, 1976) Prof. Titular. (tel. 34 + 44 4641000,ext 304) *Structural chemistry, crystal structure determination..*

Rojo, Dr Teofilo. Dep. Quimica Inorganica, Univ. del Pais Vasco Apddo. 644, 48080-Bilbao, Spain. (1951) DSc, Chemistry (Univ. Pais Vasco, 1981) Prof. Titular. (tel. 34 + 44 4641000) *Structural chemistry.*

Romero, Dr Antonio. Inst. de Quimica Inorganica, CSIC, Serrano 113, 28006-Madrid, Spain. (1958) DSc, Chemistry. (Univ. Complutense de Madrid) Becario. (tel. 34 + 1 4111772) *Crystal structure determination.*

Romero-Molina, Dr M. Angustias. Dep. Quimica Inorganica, Univ. de Granada Fuentenueva s/n, 18071-Granada, Spain. (1950) DSc, Chemistry. (Univ. Granada) Prof. Titular. (tel. 34 + 58 202212, ext. 322) *Structural chemistry.*

Ros, Dr Josep. Dep. Quimica, Univ. Autonoma de Barcelona 08193-Bellaterra, Spain. (1952) DSc, Chemistry. (Univ. Autonoma de Barcelona, 1981) Prof. Titular. (tel. 34 + 3 5811889) *Structural chemistry.*

Ruiz, Mr Xavier. Dep. Quimica, Univ. de Barcelona, Pza. Imperial Tarraco, 1, 43005-Tarragona, Spain. (1953) Grad, Physics (Univ. Barcelona, 1985) Prof. Ayudante. (tel. 34 + 77 225254, ext 2200) *Crystal growth, computing.*

Ruiz-Perez, Dr Catalina. Inst. Universitario de Quimica Organica, CSIC, Cta. de la Esperanza, 2, 28206-La Laguna, Spain. (1957) DSc, Physycs (Univ. Valencia, 1987) Prof. Titular. (tel. 34 + 22 250723) *Structural chemistry, crystal structure determination.*

Ruiz-Valero, Dr Caridad. Inst. de Quimica Inorganica, CSIC, Serrano 113, 28006-Madrid, Spain. (1957) DSc, Chemistry. (Univ. Complutense de Madrid, 1982) Becario. (tel. 34 + 1 4111772) *Crystal structure determination.*

Salas-Aparicio, Dr Juan Manuel. Dep. Quimica Inorganica, Univ. de Granada Fuentenueva s/n, 18071-Granada, Spain. (1952) DSc, Chemistry. (Univ. Granada, 1979) Prof. Titular. (tel. 34 + 58 202212, ext. 322) *Structural chemistry.*

Sanchez-Aparicio, Dr Purificacion. Dep. Quimica Inorganica, Univ. de Granada Fuentenueva s/n, 18071-Granada, Spain. (1958) DSc, Chemistry. (Univ. Granada, 1984) Prof. Asociado. (tel. 34 + 58 202212, ext. 322) *Structural chemistry.*

Sanchez-Navas, Mr Antonio. Dep. Mineralogia y Petrologia, Univ. de Granada, Fuentenueva s/n, 18002-Granada, Spain. (1961) Grad. Geology (Univ. Granada, 1985) Becario. (tel. 34 + 58 202212) *X-ray diffraction, structural biology.*

Sanz-Aparicio, Dr Juliana. Inst. Rocasolano, CSIC, Serrano, 119, 28006-Madrid, Spain. (1959) DSc, Chemistry. (Univ. Complutense de Madrid) Colaborador. (tel. 34 + 1 2619400) *Crystal structure determination.*

Sanz-Aparicio, Prof. Francisco. Dep. Termologia, Univ. de Valencia, Avda. Dr Moliner, 50, 46100-Valencia, Spain. (1941) DSc, Physics (Univ. Complutense de Madrid, 1969) Catedratico. (tel. 34 + 6 3630011, ext 280) *Crystal structure determination. amorphous diffraction.*

Sebastian, Dr Eduardo Manuel. Dep. Mineralogia y Petrologia, Univ. de Granada, Fuentenueva s/n, 18002-Granada, Spain. (1949) DSc, Geology (Univ. Granada, 1979) Prof. Titular. (tel. 34 + 58 202212, ext 340) *X-ray diffraction, crystal structure determination.*

Sola, Dr Joan. Dep. Quimica, Univ. Autonoma de Barcelona 08193-Bellaterra, Spain. (1952) DSc, Chemistry. (Univ. Autonoma de Barcelona, 1982) Prof. Titular. (tel. 34 + 3 5811372) *Structural chemistry.*

Solans, Prof. Joaquim. Dep. Cristalografia, Mineralogia y Depositos Minerales, Univ. de Barcelona, Marti y Franques s/n, 08028-Barcelona, Spain. (1940) DSc, Geology (Univ. Barcelona, 1966). Catedratico. (tel. 34 + 3 3307311, ext 1975) *Structural chemistry, physical properties.*

Solans, Prof. Xavier. Dep. Cristalografia, Mineralogia y Depositos Minerales, Univ. de Barcelona, Marti y Franques s/n, 08028-Barcelona, Spain. (1949) DSc, Physics (Univ. Barcelona, 1977). Catedratico. (tel. 34 + 3 3307311, ext 1959, email d3cmxsh0 at eb0ub011). *Crystal structure determination, crystal growth, phase transitions, structural chemistry, relationship between structure and properties.*

Suades, Dr Joan. Dep. Quimica, Univ. Autonoma de Barcelona, 08193-Bellaterra, Spain. (1954) DSc, Chemistry. (Univ. Autonoma de Barcelona, 1983) Prof. Titular. (tel. 34 + 3 5811010) *Structural chemistry.*

Subirana, Prof. Juan A. Dep. Quimica Macromolecular, Univ. Politecnica de Catalunya Diagonal 647, 08028-Barcelona, Spain. (1936) DSc, Chemistry. (Univ. Complutense de Madrid, 1960) Catedratico. (tel. 34 + 3 5811010) *Fiber diffraction, structural biology.*

Tauler, Dr Esperanza. Dep. Cristalografia, Mineralogia y Depositos Minerales, Univ. de Barcelona, Marti y Franques s/n, 08028-Barcelona, Spain. (1953) DSc, Geology (Univ. Barcelona, 1983). Prof. Titular. (tel. 34 + 3 3307311, ext 1973) *X-ray diffraction, molecular alloys.*

Teixidor, Dr Francesc. Inst. de Ciencia de Materiales de Barcelona, CSIC, Marti y Franques s/n, 08028-Barcelona, Spain. (1952) DSc, Chemistry (Univ. Barcelona, 1979) Investigador. (tel. 34 + 3 3302716) *Structural chemistry.*

Tomas, Dr Milagros. Dep. Quimica Inorganica, Univ. de Zaragoza, 50009-Zaragoza, Spain. (1954) DSc, Chemistry (Univ. Zaragoza, 1979) Prof. Titular. (tel. 34 + 76 452347) *Structural chemistry.*

Torres-Ruiz, Dr Jose. Dep. Mineralogia y Petrologia, Univ. de Granada, Fuentenueva s/n, 18002-Granada, Spain. (1952) DSc, Geology (Univ. Granada, 1980) Prof. Titular. (tel. 34 + 58 202212, ext 340) *Minerals.*

Traveria-Cros, Dr Adolfo. Inst. Jaime Almera, CSIC, Marti y Franques s/n, 08028-Barcelona, Spain. (1928) DSc, Geology (Univ. Barcelona, 1964) Investigador. (tel. 34 + 3 3302716) *X-ray diffraction, powder diffraction.*

Turrillas, Dr Javier. Facultad de Farmacia, Univ. de Navarra, 31080-Pamplona, Spain. (1955) DSc, Chemistry. (Univ. Navarra, 1986) Becario. (tel. 34 + 1 5895111) *Crystal structure determination, powder diffraction.*

Valin, Dr Mariluz. Dep. Geologia, Univ. de Oviedo, Arias de Velasco s/n, 33005-Oviedo, Spain. (1956) DSc, Geology (Univ. Oviedo, 1986) Prof. Titular. (tel. 34 + 85 233200, ext 151) *Minerals, crystal structure determination.*

Vallet-Regi, Dr Maria. Dep. Quimica Inorganica, Univ. Complutense de Madrid Campus Universitario, 28040-Madrid, Spain. (1946) DSc, Chemistry. (Univ. Complutense de Madrid, 1975) Prof. Titular. (tel. 34 + 1 4491850) *Solid state chemistry.*

Vega, Dr Rosario. Dep. Optica, Univ. de Sevilla, Reina Mercedes s/n, 41080-Sevilla, Spain. (1921) DSc, Physics (Univ. Complutense de Madrid, 1952) Prof. Titular. (tel. 34 + 54 616615) *Crystal structure determination, X-ray diffraction.*

Vegas, Dr Angel. Inst. de Quimica Inorganica, CSIC, Serrano 113, 28006 Madrid, Spain. (1947) DSc, Chemistry. (Univ. Complutense de Madrid, 1975) Colaborador. (tel. 34 + 1 4111772) *Crystal structure determination, electron density.*

Veintemillas, Mr Sabino. Inst. de Ciencia de Materiales de Barcelona, CSIC, Marti y Franques s/n, 08028-Barcelona, Spain. (1957) DSc, Physics (Univ. Complutense de Madrid) Becario. (tel. 34 + 3 3302716) *Crystal growth.*

Velilla, Dr Nicolas. Dep. Mineralogia y Petrologia, Univ. de Granada, Fuentenueva s/n, 18002-Granada, Spain. (1952) DSc, Geology (Univ. Granada, 1983) Prof. Titular. (tel. 34 + 58 202212) *Minerals.*

Vendrell, Dr Marius. Dep. Cristalografia, Mineralogia y Depositos Minerales, Univ. de Barcelona, Marti y Franques s/n, 08028-Barcelona, Spain. (1949) DSc, Geology (Univ. Barcelona, 1978). Prof. Titular. (tel. 34 + 3 3307311, ext 1972) *Minerals.*

Vicente, Prof. Jose. Dep. Quimica Inorganica, Univ. de Murcia, 30001-Murcia, Spain. (1943) DSc, Chemistry. (Univ. Zaragoza, 1973) Catedratico. (tel. 34 + 68 239959) *Structural chemistry.*

Vila, Dr Eladio. Inst. de Quimica Inorganica, Serrano 113, 28006-Madrid, Spain. (1953) DSc, Chemistry. (Univ. Complutense de Madrid, 1980) Titulado Superior. (tel. 34 + 1 4111772) *X-ray spectroscopy.*

Zuñiga, Dr Fco. javier. Depto. Fisica, Univ. del Pais Vasco, Appdo. 644, Bilbao, Spain. (1953) DSc, Physics (Univ. Complutense de Madrid, 1980) Prof. Titular. (tel. 34 + 4 4641000, ext 524) *Phase transitions, crystal structure determination.*

SRI LANKA

Sub-Editor: H. W. Dias

Notes

1. International telephone country code - 94.

Dias, Dr Hanwellage Wijayapala. Chemistry Dept., U. of Peradeniya, Peradeniya, Sri Lanka. (1936) PhD, inorganic chemistry (Leeds U., UK, 1964). Prof. (tel. 08 + 88018). *X-ray crystallography, mineral chemistry.*

Gunawardane, Dr Richard Pemasiri. Chemistry Dept., U. of Peradeniya, Peradeniya, Sri Lanka. (1945) PhD, inorganic chemistry (Aberdeen U., UK, 1974). Assoc. prof. (tel. 08 + 88018). *Silicate chemistry.*

SUDAN

Sub-Editor: S. el D. Hamad

Ali, Dr E. M. Dept. of Physics, Faculty of Education, U. of Khartoum, P.O. Box 406, Omdurman, Sudan. (1936) PhD, physics (U. of Cambridge, UK, 1970). Assoc. prof. (tel. 72271, ext. 298-299). *Crystallography.*

Hamad, Dr Sa'ad El Din. Dept. of Geology, U. of Khartoum, P.O. Box 321, Khartoum, Sudan. (1936) PhD, mineralogy (U. of Cambridge, UK, 1970). Assoc. prof. (tel. 72271, ext. 298-299). *Experimental mineralogy, crystallography, low temperature hydration-dehydration, minerals.*

SWEDEN

Sub-Editor: Y. Andersson

Notes

1. International telephone country code - 46. Omit the zero in the area code after the country code.

2. Degrees conferred by Swedish Universities are *filosofie doktor* (fil.dr.)(approximately equivalent to PhD), *filosofie licentiat* (fil.lic.)(approximately equivalent to MSc) *filosofie kandidat* (fil.kand.) and *högskoleexamen* (högsk.ex.)(approximately equivalent to BSc) at Faculties of Science and *medicine doktor* (med.dr.) and *medicine kandidat* (med.kand.) at Faculties of Medicine and at the Institutes of Technology *teknologie doktor* (tekn.dr.) and *civilingenjör* (civ.ing.) or *bergsingenjör* (bergsing.)(at the School of Mines). The older degree *filosofie magister* (fil.mag.) is no longer given but is approximately equivalent to BSc.

3. *Docent* is either a title given by a faculty to a person with a scientific competence well above the doctor's level or a position for a person performing independent academic research. *Research associate* and *research assistant* are positions for research at lower levels and are often combined with teaching elementary courses.

Åberg, Dr Märtha M. Dept. of Inorganic Chemistry, Royal Inst. of Techn., S-10044 Stockholm, Sweden. (1942) Tekn.dr., chemistry (Royal Inst. of Techn., 1971). Lect. (tel. 08-7908150). *Coordination compounds, inorganic crystals, molecular crystals, structural chemistry, teaching, X-ray diffraction.*

Adelsköld, Dr Volrath. Dept. of Structural Chemistry, Arrhenius Lab., U. of Stockholm, S-10691 Stockholm, Sweden. (1911) Fil.lic., chemistry (Stockholm U., 1939). Sr. sci. (tel. 08-162393). *Inorganic crystal structures.*

Adlerborn, Dr Jan. ABB Cerama AB, S-91500 Robertsfors, Sweden. (1938) Fil.lic., inorganic chemistry (Göteborg U., 1971). Manager R & D. (tel. 0934-10510, telex 54014, fax +46 934 10139).

Ahlzén, Mr Per-Johan. Dept. of Inorganic Chemistry, U. of Uppsala, Box 531, S-75121 Uppsala, Sweden. (1962) Fil. kand., chemistry (Uppsala U., 1986). Res. assoc. (tel. 018-183736, fax +46 18 108542). *Materials science, X-ray diffraction, neutron diffraction, magnetism.*

Albertsson, Dr Jörgen. Div. of Inorganic Chemistry 2, Chemical Center, U. of Lund, Box 124, S-22100 Lund, Sweden. (1939) Fil.dr., chemistry (Lund U., 1972). Docent. (tel. 046-108223, fax +46 46 146030). *Solid state chemistry, instrumentation.*

Aleby, Dr Stig E. Dept. of Inorganic Chemistry, U. of Göteborg, S-41296 Göteborg, Sweden. (1932) Fil.dr., structural chemistry (Göteborg U., 1969). Sen. lect. (tel. 031-722853). *Diffractometry, molecular crystals, phase transitions, structural chemistry, teaching.*

Alfredsson, Miss Viveka. Div. of Inorganic Chemistry 2, Chemical Center, U. of Lund, Box 124, S-22100 Lund, Sweden. (1963) Civ. ing., chemical eng. (Lund U., 1988). Res. asst. (tel. 046-108233, fax +46 46 146030). *Electron microscopy.*

Al Karadaghi, Dr Salam. Dept. of Structural Chemistry, Arrhenius Lab., U. of Stockholm, S-10691 Stockholm, Sweden. (1953) Fil.dr., biophysics (Moscow U.,USSR, 1982). Res. assoc. (tel. 08-162380). *Electron microscopy, image processing, biological macromolecules.*

Andersson, Dr Inger A. Dept. of Molecular Biology, Swedish U. of Agricultural Sciences, Uppsala Biomedical Center, Box 590, S-75124 Uppsala, Sweden. (1949) Dr.rer.nat., biochemistry (U. des Saarlandes, Germany, Fed. Rep., 1980). Res. assoc. (tel. 018-174523). *Biological macromolecules, enzyme catalysis.*

Andersson, Dr Staffan. Dept. of Inorganic Chemistry, Chalmers U. of Techn., S-41296 Göteborg, Sweden. (1948) Fil.dr., inorganic chemistry (Chalmers U. of Techn., 1988). Res. assoc.(tel. 031-722850). *Coordination compounds, chemical bonding, structural chemistry.*

Andersson, Prof. Sten. Div. of Inorganic Chemistry 2, Chemical Center, U. of Lund, Box 124, S-22100 Lund, Sweden. (1931) Fil.dr., chemistry (Stockholm U., 1967). Prof. (tel. 046-108227, fax +46 46 146030). *Inorganic chemistry.*

Andersson, Dr Yvonne. Dept. of Inorganic Chemistry, U. of Uppsala, Box 531, S-75121 Uppsala, Sweden. (1947) Fil.dr., inorganic chemistry (Uppsala U., 1983). Docent. (tel. 018-183726, telex 76088 TSLISV S, fax +46 18 108542, email bitnet: yba at semax51). *Materials science, crystal growth, structural chemistry, phase determination, magnetism.*

Andersson-Söderberg, Mrs Margaretha. Dept. of Inorganic Chemistry, U. of Uppsala, Box 531, S-75121 Uppsala, Sweden. (1963) Högsk.ex., materials chemistry (Uppsala U., 1986). Res. asst. (tel. 018-183723, telex 76088 TSLISV S, fax +46 18 108542). *Materials science, semiconductors, metals, structural chemistry, phase determination.*

Aronsson, Prof. Bertil. Dept. of Inorganic Chemistry, U. of Uppsala, Box 531, S-75121 Uppsala,Sweden. (1929) Fil.dr., chemistry (Uppsala U., 1960). Adjunct.prof. (tel. 018-183719). *Metal physics and structure.*

Åsbrink, Dr Gudrun. Dept. of Structural Chemistry, Arrhenius Lab., U. of Stockholm, S-10691 Stockholm, Sweden. (1930) Fil.lic., chemistry (Stockholm U., 1971). Res. asst. (tel. 08-162387). *Small angle scattering.*

Åsbrink, Dr Stig. Dept. of Inorganic Chemistry, Arrhenius Lab., U. of Stockholm, S-10691 Stockholm, Sweden. (1929) Fil.dr., chemistry (Stockholm U., 1973). Docent. (tel. 08-162387 fax +46 8 152187). *Phase transitions, high pressures, inorganic crystals, structural chemistry, X-ray diffraction.*

Åström, Prof. Hans U. Dept. of Solid State Physics, Royal Inst. of Techn., S-10044 Stockholm, Sweden. (1926) Tekn.dr., metal physics (Royal Inst. of Techn., 1958). Prof. (tel. 08-7907300). *Low temperature physics, magnetism, metals, phase transitions, defect structures.*

Aurivillius, Prof. Bengt. Div. of Inorganic Chemistry 2, Chemical Center, U. of Lund, Box 124, S-22100 Lund, Sweden. (1918) Fil.dr., inorganic chemistry (Stockholm U., 1951). Prof. em. (tel. 046-108230, fax +46 46 146030). *Inorganic crystal structures, heavy metals, block structures.*

Bäckerud, Dr Lennart S. Dept. of Structural Chemistry, Arrhenius Lab., U. of Stockholm, S-10691 Stockholm, Sweden. (1932) Fil.dr., inorganic chemistry (Uppsala U., 1968). Adjunct prof. (tel. 08-162383). *Nucleation, crystal growth.*

Blum, Dr Zoltan. Div. of Inorganic Chemistry 2, Chemical Center, U. of Lund, Box 124, S-22100 Lund, Sweden. (1952) Fil.dr., organic chemistry (Lund U. 1981). Res. assoc. (tel. 046-108232, fax +46 46 146030). *Biochemistry, inclusion compounds, inorganic chemistry.*

Boström, Dr N. Dan. Dept. of Inorganic Chemistry, U. of Umeå, S-90187 Umeå, Sweden. (1954) Fil.dr., inorganic chemistry (Umeå U., 1988). Res. assoc. (tel. 090-165264, fax +46 90 136310). *Crystal growth, minerals, inorganic crystals.*

Bovin, Dr Jan-Olov. National Center for HREM, Inorganic Chemistry 2, Chemical Center, U. of Lund, Box 124, S-22100 Lund, Sweden. (1943) Fil.dr., inorganic chemistry (Lund U., 1975). Docent. (tel. 046-104769, fax +46 46 146030). *Structural chemistry, electron microscopy.*

Brändén, Prof. Carl-Ivar. Dept. of Molecular Biology, Uppsala Biomedical Center, Box 590, S-75124 Uppsala, Sweden. (1934) Fil.dr., chemistry (Uppsala U., 1964). Prof. (tel. 018-174478, fax +46 18 151759, email bitnet: branden at semax51). *Biological macromolecules, enzyme catalysis, computer programming.*

Bryntse, Mrs Ingrid. Dept. of Inorganic Chemistry, Arrhenius Lab., U. of Stockholm, S-10691 Stockholm, Sweden. (1956) ämneslär.ex., chemistry (Linköping U., 1982). Res. asst. (tel. 08-162368, fax +46 8 152187). *High temp. superconductors, inorganic crystals, electron microscopy.*

Calais, Dr Jean-Louis. Quantum Chemistry Group, U. of Uppsala, Box 518, S-75120 Uppsala, Sweden. (1932) Fil.dr., quantum chemistry (Uppsala U., 1965). Res. position at the Swedish Natural Sciences Res. Council. (tel. 018-183264). *Solid state theory, electronic structure, ionic crystals, transition metal compounds.*

Cassel, Dr Anders Ö. Surface Chemistry Div., Berol Kemi AB, Box 851, S-44401 Stenungsund, Sweden. (1946) Fil.dr., inorg. chemistry (Lund U., 1979). R&D group manager. (tel. 0303-85554). *Metal halide phosphines, solid state structures, sulfide mineral flotation.*

Collini, Prof. Bengt H. E. Inst. of Geology, U. of Uppsala, Box 555, S-75122 Uppsala, Sweden. (1917) Fil.lic., mineralogy and petrology (Uppsala U., 1943), fil.dr.h.c. (Uppsala U., 1983). Prof. em. (tel. 018-182557). *Sedimentary petrology, meteorite impact geology.*

Cowan, Dr Sandra Wendy. Dept. of Molecular Biology, Biomedical Center, U. of Uppsala, Box 590, S-75124 Uppsala, Sweden. (1962) Ph.D., biochemistry (Melbourne U., Australia, 1988). Res. asst. (tel. 018-174566, fax +46 18 171559, email bitnet: sandra at semax51). *Protein crystallography.*

Csöregh, Dr Ingeborg. Dept. of Structural Chemistry, Arrhenius Lab., U. of Stockholm, S-10691 Stockholm, Sweden. (1942) Fil.dr., structural chemistry (Stockholm U., 1983). Docent. (tel. 08-162381, telex 8105199, fax +46 8 152187). *Organic and organometallic crystal structures, X-ray diffraction.*

Dagerhamn, Dr Tore. Dept. of Inorganic Chemistry, Arrhenius Lab., U. of Stockholm, S-10691 Stockholm, Sweden. (1930) Fil.lic., chemistry (Stockholm U., 1965). Lect. (tel. 08-162352). *Inorganic crystal structures, metals structure.*

Delaplane, Dr Robert G. Inst. of Chemistry, U. of Uppsala, Box 531, S-75121 Uppsala, Sweden. (1942) PhD, phys. chemistry (Northwestern U., USA, 1969). Docent. (tel. 018-183773, fax +46 18 108542). *Neutron diffraction, amorphous materials.*

Edström, Mrs Kristina. Inst. of Chemistry, U. of Uppsala, Box 531, S-75121 Uppsala, Sweden. (1958) Högsk.ex., chemistry (U. of Uppsala). Res.asst. (tel. 018-183775, fax +46 18 108542). *Crystal structure - physical properties relationships.*

Engström, Dr Ingvar O. J. Inst. of Chemistry, U. of Uppsala, Box 531, S-75121 Uppsala, Sweden. (1934) Fil.dr., chemistry (Uppsala U., 1970). Docent. (tel. 018-183740, fax +46 18 108542). *Structural chemistry, microelectronics.*

Ericsson, Mr Thomas. Dept. of Inorganic Chemistry, U. of Göteborg, S-41296 Göteborg, Sweden. (1964) Högsk.ex., chemistry (Göteborg U., 1988). Res. asst. (tel. 031-722866). *Inorganic crystals.*

Eriksson, Dr Anders. Inst. of Chemistry, U. of Uppsala, Box 531, S-75121 Uppsala, Sweden. (1945) Fil.dr., chemistry (Uppsala U., 1981).Docent. (tel.

018-183766, telex 76088 TSLISV S, fax +46 18 108542, email bitnet: ae at semax51). *Vibrational spectroscopy.*

Eriksson, Dr Birgitta. Dept. of Structural Chemistry, Arrhenius Lab., U. of Stockholm, S-10691 Stockholm, Sweden. (1945) Fil.dr., structural chemistry (Stockholm U., 1982). Res. assoc. (tel. 08-163730, fax + 46 8 152187). *Coordination compounds, structural chemistry, teaching.*

Eriksson, Dr A. Elisabeth. Dept. of Molecular Biology, Biomedical Center, U. of Uppsala, Box 590, S-75124 Uppsala, Sweden. (1959) Fil.dr., molecular biology (Uppsala U., 1988). Res. assoc. (tel. 018-174566, email bitnet: lizzie at semax51). *Biological macromolecules, enzyme catalysis.*

Eriksson, Mr Lars. Dept. of Structural Chemistry, Arrhenius Lab., U. of Stockholm, S-10691 Stockholm, Sweden. (1960) Res. asst. (tel. 08-162393, fax +46 8 152187). *Computing, powder diffraction, group theory.*

Eriksson, Mr Sten. Dept. of Inorganic Chemistry, U. of Göteborg, S-41296 Göteborg, Sweden. (1958) Högsk.ex., chemistry (Göteborg U., 1983). Res. asst. (tel. 031-722860). *Inorganic crystals, solid state chemistry, powder diffraction.*

Ersson, Dr Nils Olov. Inst. of Chemistry, U. of Uppsala, Box 531, S-75121 Uppsala, Sweden. (1942) Fil.lic., chemistry (Uppsala U., 1985). Res. asst. (tel. 018-183728). *Computing, diffractometry, instrumentation.*

Ertan, Mrs Anne. Dept. of Structural Chemistry, Arrhenius Lab., U. of Stockholm, S-10691 Stockholm, Sweden. (1957) Högsk.ex., chemistry (Stockholm U., 1984). Res. asst. (tel. 08-162381, telex 8105199, fax +46 8 152187). *Organic and biological structures, X-ray diffraction, structural chemistry, structural biology.*

Fischer-Hjalmars, Prof. Inga M. Dept. of Physics, U. of Stockholm, Vanadisvägen 9, S-11346 Stockholm, Sweden. (1918) Fil.dr., molecular quantum mechanics (Stockholm U., 1952). Prof. em., theoretical physics. (tel. 08-164608, fax +46 8 347817). *Chemical bonding, coordination compounds.*

Flodmark, Dr Stig. Fysikum, U. of Stockholm, Vanadisvägen 9, S-11346 Stockholm, Sweden. (1926) Fil.dr., solid state theory (Stockholm U., 1959). Ass. prof. (tel. 08-164620, telex 15433 FYSTUS, email sesuf51). *Symmetry, group, solid state, quantum, earth rotation, Chandler wobbles, geodynamics, geomagnetism, supermagnetism, superconductivity, theoretical mechanics, hydroelectrodynamics.*

Forslund, Dr S. Bertil. Dept. of Inorganic Chemistry, Arrhenius Lab., U. of Stockholm, S-10691 Stockholm, Sweden. (1943) Fil.dr., inorg. chemistry (Stockholm U., 1984). Res. assoc. (tel. 08-162353, fax +46 8 152187). *Inorganic materials, crystal growth, high pressures.*

Frostäng, Mr F. Sten E. Dept. of Structural Chemistry, Arrhenius Lab., U. of Stockholm, S-10691 Stockholm, Sweden. (1951) Högsk.ex., chemistry (Stockholm U., 1984). Res. asst. (tel. 08-162365, fax +46 8 152187). *Materials science, powder diffraction, ionic solid state conductors.*

Glaser, Dr Julius. Dept. of Inorganic Chemistry, Royal Inst. of Techn., S-10044 Stockholm, Sweden. (1948) Fil.dr., inorganic chemistry (Royal Inst. of Techn., 1981). Sen. lect. (tel. 08-7908151). *Coordination chemistry, equilibria, structure and dynamics for complex ions in solution, multinuclear NMR, X-ray diffraction.*

Glehn, von, Dr Marianne. Swedish Board for Space Activities, Box 4006, S-17104 Solna, Sweden. (1941) Fil.dr., chemistry (Stockholm U., 1971). Docent. (tel. 08-7336483, telex 17128 SPACELO S, fax +46 8 7335014). *Proteins, liquid crystals, structural biology.*

Grenthe, Prof. Ingmar. Dept. of Inorganic Chemistry, Royal Inst. of Techn., S-10044 Stockholm, Sweden. (1933) Fil.dr., inorg. and phys. chemistry (Lund U., 1964). Prof. (tel. 08-7906000). *Structure and bonding, coordination compounds, reactivity in solids.*

Grins, Dr Jekabs. Dept. of Inorganic Chemistry, Arrhenius Lab., U. of Stockholm, S-10691 Stockholm, Sweden. (1952) Fil.dr., chemistry (Stockholm U., 1980). Res. assoc. (tel. 08-162365, fax +46 8 152187). *Solid state chemistry, materials science.*

Gullman, Dr Jan O. The Central Board of Antiquities, Box 5405, S-11484 Stockholm, Sweden. (1943) Fil.lic., inorganic chemistry (Uppsala U., 1987). Avd. dir. (tel. 08-7839340). *Corrosion, inorganic crystal structures.*

Gustafsson, Dr Torbjörn. Inst. of Chemistry, U. of Uppsala, Box 531, S-75121 Uppsala, Sweden. (1949) Fil.dr., chemistry (Uppsala U., 1987). Res. eng. (tel. 018-183767, fax +46 18 108542). *Inorganic crystal structures, liquid structures, hydrogen bonding, instrumentation.*

Hansen, Dr Staffan S. National Center for HREM, Inorganic Chemistry 2, Chemical Center, U. of Lund, Box 124, S-22100 Lund, Sweden. (1954) Tekn.dr., chemistry (Lund U., 1985). Res. assoc. (tel. 046-108233, telex 33533 LUNIVER S, fax +46 46 146030). *Inorganic crystals, electron microscopy.*

Hansson, Dr Arne E. Inst. of Chemistry, U. of Uppsala, Box 531, S-75121 Uppsala, Sweden. (1926) Fil.lic., chemistry (Uppsala U., 1959). Director of studies in chemistry. (tel. 018-183713). *Inorganic and organic crystal structures.*

Håkansson, Mr Mikael. Dept. of Inorganic Chemistry, Chalmers U. of Techn., S-41296 Göteborg, Sweden. (1957) Civ.ing., chemical eng. (Chalmers U. of Techn., 1984). Res. asst. (tel. 031-722871). *Coordination compounds, organometallic chemistry.*

Hårsta, Dr Anders. Inst. of Chemistry, U. of Uppsala, Box 531, S-75121 Uppsala, Sweden. (1952) Fil.dr., chemistry (Uppsala U., 1985). Res. assoc. (tel. 018-183729, fax +46 18 108542). *Materials science.*

Hassler, Dr Eivind. Inst. of Chemistry, U. of Uppsala, Box 531, S-75121 Uppsala, Sweden. (1939) Fil.lic., chemistry (Uppsala U., 1970). Res. assoc. (tel. 018-183729, fax +46 18 108542). *Inorganic crystal structures.*

Hebert, Dr Hans. Dept. of Medical Biophysics, Karolinska Inst., S-10401 Stockholm, Sweden. (1951) Tekn.dr., medical biophysics (Karolinska Inst., 1979). Res. assoc. (tel. 08-340560, ext. 1561). *Biological macromolecules, electron microscopy, image processing.*

Hegedüs, Dr Zsolt. Dept. of Inorganic Chemistry, Arrhenius Lab., S-10691 Stockholm, Sweden. (1948) Fil.dr. (Babes-Bolyai U., Cluj-Napoca, Romania, 1971). Res. asst. (tel. 08-162417, fax +46 8 152187). *Inorganic crystals, phosphors, superconducting substances.*

Helgesson, Mr Göran. Dept. of Inorganic Chemistry, Chalmers U. of Techn., S-41296 Göteborg, Sweden. (1953) Civ.ing., chemical eng. (Chalmers U. of Techn., 1986). Res. asst. (tel. 031-722871, email bitnet: col47 at seguc21). *Coordination compounds, X-ray diffraction.*

Herbertsson, Dr B. Harald. Eng. Materials, Luleå U.of Techn., S-95187 Luleå, Sweden. (1940) Fil.dr., inorganic chemistry (Royal Inst. of Techn., 1980). Res. assoc. (tel. 0920-91000, ext. 233, telex: 80447 LUH S). *High performance ceramics, high temperature X-ray diffraction, materials science.*

Hermansson, Dr Kersti. Inst. of Chemistry, U. of Uppsala, Box 531, S-75121 Uppsala, Sweden. (1951) Fil.dr., inorganic chemistry (Uppsala U., 1984). Docent. (tel. 018-183767, telex 76088 TSLISV S, fax +46 8 108542, email bitnet: kersti at semax51). *Molecular dynamics, intermolecular interactions, electron distribution.*

Hermansson, Dr Leif Å. G. DOXA CERTEX AB, Box 33, S-51045 Sparsör, Sweden. (1947) Tekn.dr., chemistry (Chalmers U. of Techn., 1977). Managing dir. (docent). (tel. 033-60200, fax +46 33 69580). *Inorganic crystal structures, high performance ceramics, bioceramics.*

Hermodsson, Dr Yngve. Inst. of Chemistry, U. of Uppsala, Box 531, S-75121 Uppsala, Sweden. (1929) Fil.dr., chemistry (Uppsala U., 1969). Docent. (tel. 018-183701, fax +46 18 108542). *Inorganic crystal structures.*

Hesse, Dr Rolf S. Inst. of Chemistry, U. of Uppsala, Box 531, S-75121 Uppsala, Sweden. (1923) Fil.dr., chemistry (Uppsala U., 1963). Docent. (tel. 018-135829). *Coordination compounds, crystal structures.*

Hjertén, Dr Inger. Dept. of Structural Chemistry, Arrhenius Lab., U. of Stockholm, S-10691 Stockholm, Sweden. (1937) Fil.lic., chemistry (Stockholm U., 1972). Res. asst. (tel. 08-162381). *Inorganic crystal structures.*

Hong, Dr Sam-Hyo. R. & D. Dept., Ericsson Components AB, Isafjordsgatan 16, Kista, S-16481 Stockholm, Sweden. (1940) Fil.dr., inorg. chemistry (Stockholm U., 1982). Res. assoc. (tel. 08-7574690, telex 8125008 COMEKA S, fax +46 8 7575040). *Inorganic crystal structures, phase transitions.*

Hovmöller, Dr Sven. Dept. of Structural Chemistry, Arrhenius Lab., U. of Stockholm, S-10691 Stockholm, Sweden. (1947) Fil.dr., chemistry (Stockholm U., 1980). Docent. (tel. 08-162380). *Electron microscopy and image processing, membrane proteins, inorganic crystals, direct methods.*

Humble, Dr Sten G. Dept. of Solid-State Physics, Royal Inst. of Techn., S-10044 Stockholm, Sweden. (1925) Tekn.dr., solid-state physics (Royal Inst. of Techn., 1971). Lect. (tel. 08-7907303). *Metal physics, metals structure.*

Ingri, Prof. Nils. Dept. of Inorganic Chemistry, U. of Umeå, S-90187 Umeå, Sweden. (1929) Fil.dr., chemistry (Stockholm U., 1963). Prof. (tel. 090-165260). *Inorganic crystal structures, computer programming.*

Ivarsson, Dr Gun J. M. Dept. of Inorganic Chemistry, U. of Umeå, S-90187 Umeå, Sweden. (1943) Fil.dr., chemistry (Umeå U., 1983). Sen. lect. (tel. 090-165467, fax +46 90 136310). *Inorganic and organic crystal structures, teaching.*

Jagner, Dr Susan. Dept. of Inorganic Chemistry, Chalmers U. of Techn. and U. of Göteborg, S-41296 Göteborg, Sweden. (1940) Fil.dr., chemistry (Göteborg U., 1970). Docent. (tel. 031-722852, fax +46 31 167194, email bitnet: col40 at seguc21). *Coordination compounds, chemical bonding, structural chemistry, disorder.*

Jahnberg, Dr Lena. Dept. of Inorganic Chemistry, Arrhenius Lab., U. of Stockholm, S-10691 Stockholm, Sweden. (1937) Fil.dr., chemistry (Stockholm U., 1972). Res. assoc. (tel. 08-162368, fax +46 8 152187). *Inorganic crystals, structural chemistry, electron microscopy, X-ray diffraction.*

Jansson, Dr Kjell. Dept. of Inorganic Chemistry, Arrhenius Lab., U. of Stockholm, S-10691 Stockholm, Sweden. (1959) Fil.dr., chemistry (Stockholm U., 1988). Res. asst. (tel. 08-162372). *Structure, amorphous metals, electron microscopy.*

Jennische, Dr Per. Nat. Swedish Lab. for Agricultural Chemistry, Box 7004, S-75007 Uppsala, Sweden. (1943) Fil.dr., inorg. chemistry (Uppsala U., 1976). Director. (tel. 018-673000, fax +46 18 302753). *Coordination chemistry, analytical chemistry.*

Johansson, Dr Georg. Dept. of Inorganic Chemistry, Royal Inst. of Techn., S-10044 Stockholm, Sweden. (1925) Tekn.dr., inorg. chemistry (Royal Inst. of Techn., 1963). Docent. (tel. 08-7908295). *Complexes, structure in solution, crystal structures.*

Johansson, Mr Karl-Erik. Dept. of Structural Chemistry, Arrhenius Lab., U. of Stockholm, S-10691 Stockholm, Sweden. Res. eng. (tel. 08-162389). *Instrumentation.*

Jones, Prof. T. Alwyn. Dept. of Molecular Biology, Biomedical Center, U. of Uppsala, Box 590, S-75124 Uppsala, Sweden. (1947) PhD, biophysics (London U., UK, 1973). Prof. (tel. 018-174566, fax +46 18 151759, email bitnet: alwyn at semax51). *Computing, macromolecules, proteins, structural biology, viruses, X-ray diffraction.*

Jönsson, Dr Per-Gunnar. Inst. of Chemistry, FOA, Cementväg. 20, S-90182 Umeå, Sweden. (1937) Fil.dr., chemistry (Uppsala U., 1973). Ass. dir. of res. (tel. 090-189230). *Structural chemistry, computing.*

Josefsson, Miss Magdalena. Dept. of Inorganic Chemistry, Chalmers U. of Techn. and U. of Göteborg, S-41296 Göteborg, Sweden. (1965) Civ.ing., chemical eng. (Chalmers U. of Techn., 1988). Res. asst.(tel. 031-722869). *Structural chemistry, coordination compounds.*

Käll, Mr Per-Olov. Dept. of Inorganic Chemistry, Arrhenius Lab., U. of Stockholm, S-10691 Stockholm, Sweden. (1947) Fil.kand., chemistry (Stockholm U., 1981). Res. asst. (tel. 08-162365, fax +46 8 152187). *Materials science, phase determination, powder diffraction.*

Karlsson, Dr Bengt E. Natural Science Research Council, Box 6711, S-11385 Stockholm, Sweden. (1948) Fil.dr., chemistry (Stockholm U., 1978). Docent. (tel. 08-151580 ext:174, fax +46 8 301386). *Organic crystal structures, X-ray analysis, biological specimen structure, electron microscopy and image reconstruction.*

Karppinen, Dr Markku. Inst. of Chemistry, U. of Uppsala, Box 531, S-75121 Uppsala, Sweden. (1948) Fil.dr., inorganic chemistry (Uppsala U., 1988). Res. assoc. (tel. 018-183770). *Solid state chemistry, physical properties,electron density.*

Kierkegaard, Prof. Peder. Dept. of Structural Chemistry, Arrhenius Lab., U. of Stockholm, S-10691 Stockholm, Sweden. (1928) Fil.dr., chemistry (Stockholm U., 1962). Prof. (tel. 08-162385, fax +46 8 152187). *Inorganic and organic structures, protein and vitreous structures, instrumentation.*

Kihlborg, Prof. Lars. Dept. of Inorganic Chemistry, Arrhenius Lab., U. of Stockholm, S-10691 Stockholm, Sweden. (1930) Fil.dr., chemistry (Uppsala U., 1964). Prof. (tel. 08-162370, telex 8105199 UNIVER S, fax +46 8 152187, email larsk at inorg.su.se). *Inorganic structures, solid state chemistry, structural defects, high temperature superconductors, electron microscopy.*

Knight, Mr Stefan D. Dept. of Molecular Biology, Swedish U. of Agricultural Sciences, Uppsala Biomedical Center, Box 590, S-75124 Uppsala, Sweden. (1957) Högsk.ex., chemistry (Uppsala U., 1984). Res. asst. (tel. 018-174524). *Biological macromolecules, protein engineering.*

Kritikos, Mr Mikael. Dept. of Structural Chemistry, Arrhenius Lab., U. of Stockholm, S-10691 Stockholm, Sweden. (1961) Högsk.ex., chemistry (Stockholm U., 1985). (tel. 08-162382, fax +46 8 152187). *Coordination compounds, neutron diffraction, X-ray diffraction, structural chemistry.*

Langer, Dr Vratislav. Dept. of Inorganic Chemistry, Chalmers U. of Techn. and U. of Göteborg, Kemigården 3, S-41296 Göteborg, Sweden. (1949) Fil.dr., physics (Charles U., Praha, Czeckoslovakia, 1978). Docent. (tel. 031-722860, fax +46 31 167194, email bitnet: colp6 at seguc21). *Inorganic crystals, powder diffraction, proteins, diffractometry, computing.*

Larsson, Miss Ann-Kristin. Div. of Inorganic Chemistry 2, Chemical Center, U. of Lund, Box 124, S-22100 Lund, Sweden. (1963) Civ.ing., chemical eng. (Lund U., 1988). Res. asst. (tel. 046-108112, fax +46 46 146030). *X-ray diffraction, crystal growth, inorganic crystals, electron density.*

Larsson, Dr Sven. Dept. of Physical Chemistry, Chalmers U. of Techn. and U. of Göteborg, S-41296 Göteborg, Sweden. (1941) Fil.dr., quantum chemistry (Uppsala U., 1972). Docent. (tel. 031-723058, fax +46 31 167194). *Metalloproteins, solid state chemistry, superconductivity.*

Lenner, Dr Magnus. Dept. of Physical Chemistry, Chalmers U. of Techn., S-41296 Göteborg, Sweden. (1944) Fil.dr., chemistry (Göteborg U., 1980). Res. assoc. (tel. 031-723047, fax +46 31 167194, email bitnet: ckaml at seguc21). *Actinide complexes - structure.*

Lidin, Mr Sven. Div. of Inorganic Chemistry 2, Chemical Center, U. of Lund, Box 124, S-22100 Lund, Sweden. (1961) Civ.ing., chemical eng. (Lund U., 1986). (tel. 046-108232, fax +46 46 146030, email bitnet: 002sl at seldc51). *Computing, inorganic crystals, symmetry.*

Liem, Dr D. Hay. Dept. of Inorganic Chemistry, Royal Inst. of Techn., S-10044 Stockholm, Sweden. (1932) Tekn.dr., chemistry (Royal Inst. of Techn., 1971). Docent, ass. prof. (tel. 08-7878330). *Metallo-organic compounds, complexes.*

Liljas, Prof. Anders. Dept. of Molecular Biophysics, Chemical Center, U. of Lund, Box 124, S-22100 Lund, Sweden. (1939) Fil.dr., chemistry (Uppsala U., 1971). Prof. (tel. 046-104681, telex 33533 LUNIVER S, fax +46 46 146030, email bitnet: mbfanders at seldc51). *Macromolecules, proteins, X-ray diffraction.*

Liljas, Dr Lars. Dept. of Molecular Biology, Biomedical Center, U. of Uppsala, Box 590, S-75124 Uppsala, Sweden. (1947) Fil.dr., biochemistry (Uppsala U., 1977). Docent. (tel. 018-174000, fax +46 18 151759, email bitnet: lars at semax51). *Biological macromolecules.*

Liminga, Prof. Rune. Inst. of Chemistry, U. of Uppsala, Box 530, S-75121 Uppsala, Sweden. (1932) Fil.dr., chemistry (Uppsala U., 1968). Prof. (tel. 018-183770). *Crystal structure - physical properties relationships, materials.*

Lindahl, Mr Martin. Dept. of Molecular Biophysics, Chemical Center, U. of Lund, Box 124, S-22100 Lund, Sweden. (1963) H&oegsk.ex., chemistry (Lund U., 1986). Res.asst. (tel. 046-104692, telex 33533 LUNIVER S, fax +46 46 146030, email bitnet: mbfmartin at seldc51). *Macromolecules, proteins, X-ray diffraction.*

Lindahl, Dr Tommie. CEA AB, Box 174, S-64523 Strängnäs, Sweden. (1937) Fil.lic., chemistry (Stockholm U., 1969). (tel. 0152-12930) *X-ray diffraction film.*

Lindner, Dr Peter W. Dept. of Quantum Chemistry, U. of Uppsala, Box 518, S-75120 Uppsala, Sweden. (1937) Fil.dr., quantum chemistry (Uppsala U., 1970). Docent. (tel. 018-183501). *X-ray scattering theory, interaction between X-rays and molecules.*

Lindqvist, Prof. Ingvar. The Royal Swedish Academy of Sciences, Box 50005, S-10405 Stockholm, Sweden. (1921) Fil.dr., inorg. chemistry (Uppsala U., 1951). Prof.em. (tel. 08-150430 and 018-301221). *Structural biochemistry.*

Lindqvist, Prof. Oliver. Dept. of Inorganic Chemistry, Chalmers U. of Techn. and U. of Göteborg, S-41296 Göteborg, Sweden. (1943) Fil.dr., chemistry (Göteborg U., 1973). Prof. (tel. 031-722862, fax +46 31 167194). *Inorganic crystal structures, glass and liquid structures.*

Lindqvist, Dr Ylva Ch. Dept. of Molecular Biology, Swedish U. of Agricultural Sciences, Uppsala Biomedical Center, Box 590, S-75124 Uppsala, Sweden. (1947) Fil.dr, chemistry (Swedish U. of Agricultural Sciences, 1981). Res. assoc. (tel. 018-174535, telex 76132 BIOMED, fax +46 18 151759). *Protein crystallography, photosynthesis.*

Ljungström, Dr Evert B. Dept. of Inorganic Chemistry, Chalmers U. of Techn. and U. of Göteborg, S-41296 Göteborg, Sweden. (1949) Fil.dr., chemistry (Göteborg U., 1979). Docent. (tel. 031-722880, fax +46 31 167194, email bitnet: col86 at seguc21). *Instrumentation, inorganic and organometallic crystal structures.*

Löfgren, Dr Percy. Dept. of Inorganic Chemistry, Arrhenius Lab., U. of Stockholm, S-10691 Stockholm, Sweden. (1927) Fil.dr., chemistry (Stockholm U., 1974). Res. assoc. (tel. 08-162353, fax +46 8 152187). *Inorganic crystal structures.*

Lundberg, Dr Bruno K. S. Dept. of Inorganic Chemistry, U. of Umeå, S-90187 Umeå, Sweden. (1939) Fil.dr., chemistry (Umeå U., 1972). Docent. (tel. 090-165155). *Coordination compounds, teaching.*

Lundberg, Dr Monica. Dept. of Inorganic Chemistry, Arrhenius Lab., U. of Stockholm, S-10691 Stockholm, Sweden. (1938) Fil.dr., chemistry (Stockholm U., 1971). Docent. (tel. 08-162368, fax +46 8 152187, email bitnet: mol at seqz51). *Defect structures, electron diffraction, electron microscopy, inorganic crystals, materials science, powder diffraction, X-ray diffraction.*

Lundgren, Mr Lennart. National Inst. of Occupational Health, Aerosol Division, S-17184 Solna, Sweden. (1950) Fil.kand., chemistry (Uppsala U., 1973). Sen. res. (tel. 08-7319100, ext. 9742). *Powder diffraction, aerosol chemistry.*

Lundström, Dr Torsten. Inst. of Chemistry, U. of Uppsala, Box 531, S-75121 Uppsala, Sweden. (1929) Fil.dr., inorg. chemistry (Uppsala U., 1969). Docent. (tel. 018-183722, fax +46 18 108542). *Instrumentation, crystal growth, structure and properties.*

Lyxell, Mr Dan-Göran. Dept. of Inorganic Chemistry, U. of Umeå, S-90187 Umeå, Sweden. (1945) Fil.mag., physics (Umeå U., 1969). Lect. (tel. 090-165445, fax +46 90 136310). *Large angle X-ray scattering, liquids, inorganic crystal structures.*

Magnéli, Prof. Arne. Dept. of Inorganic Chemistry, Arrhenius Lab., U. of Stockholm, S-10691 Stockholm, Sweden. (1914) Fil.dr., chemistry (Uppsala U., 1950). Prof. em. (tel. 08-162417, also 018-118650). *Inorganic crystals, defect structures, materials science.*

Malm, Dr Jan-Olle. Div. of Inorganic Chemistry 2, Chemical Center, U. of Lund Box 124, S-12200 Lund, Sweden. (1962) Tekn.lic., inorganic chemistry (Lund U., 1988). Res. asst. (tel. 046-108231, fax +46 46 146030, email bitnet: 002jom at seldc51). *Electron microscopy.*

Marinder, Dr Bengt-Olov. Dept. of Inorganic Chemistry, Arrhenius Lab., U. of Stockholm, S-10691 Stockholm, Sweden. (1927) Fil.dr., inorg. chemistry (Stockholm U., 1986). Docent. (tel. 08-162417, fax +46 8 152187). *Electron microscopy, inorganic crystals, structural chemistry, X-ray diffraction.*

Nenner, Miss Ann-Marie. Dept. of Inorganic Chemistry, U. of Umeå, S-90187 Umeå, Sweden. (1953) Fil.kand., chemistry (Umeå U., 1975). Res. asst. (tel. 090-166327, fax +46 90 136310). *Inorganic crystal structures.*

Nilsson, Dr Karin I. Materials Lab., Alfa Laval Thermal AB, Box 74, S-22100 Lund, Sweden. (1954) Fil.dr., inorganic chemistry (Lund U., 1986). Chemist. (tel. 046-106785). *Materials science, metals.*

Nilsson, Dr Rolf O. Div. Kemiteknik, Boliden Kemi AB, Box 902, S-25109 Helsingborg, Sweden. (1927) Tekn.dr., inorg. chemistry (Chalmers U. of Techn., 1958). Chief chemist (docent). (tel. 042-171200). *Liquid structures.*

Noläng, Dr Bengt. Dept of Inorganic Chemistry, U. of Uppsala, Box 531, S-75121 Uppsala, Sweden. (1949) Fil.dr., inorganic chemistry (Uppsala U., 1984). Res.assoc. (tel. 018-183726, telex 76088 TSLISV S, fax +46 18 108542, email bitnet: bin at semax51). *Chemical bonding, computing, electron density, materials science.*

Nord, Dr Anders G. Dept. of Structural Chemistry, Arrhenius Lab., S-10691 Stockholm, Sweden. (1942) Fil.dr., inorganic chemistry (Stockholm U., 1974). Scientist (docent).(tel. 08-7839000, ext. 9339). *Inorganic and mineral structures, solid solutions, powder diffraction.*

Noreland, Mr Jakob. Dept. of Inorganic Chemistry, U. of Uppsala, Box 531, S-75121 Uppsala, Sweden. (1964) Högsk.ex., chemistry (Uppsala U., 1986). Res. asst. (tel. 018-183770, fax +46 18 108542). *Structural chemistry.*

Norén, Dr Nils Bertil. Div. of Inorganic Chemistry 1, Chemical Center, U. of Lund, Box 124, S-22100 Lund, Sweden. (1931) Fil.dr., chemistry (1970). Docent. (tel. 046-107000, ext. 8109). *Metallo-organic crystal structures, coordination compounds.*

Noréus, Dr Dag. Dept. of Structural Chemistry, Arrhenius Lab., U. of Stockholm, S-10691 Stockholm, Sweden. (1951) Tekn.dr., reactor physics (Royal Inst. of Techn., 1982). Docent. (tel. 08-162391, fax +46 8 152187, email bitnet: dag at seqz51). *Neutron scattering, X-ray diffraction, metal hydrides.*

Norin, Dr Rolf. Dept. of Chemistry, Linköping U., S-58183 Linköping, Sweden. (1930) Fil.dr., inorganic chemistry (Göteborg U., 1970). Sen.lect. (tel. 013-281380). *Inorganic crystal structures, chemical education.*

Nygren, Prof. Mats. Dept. of Inorganic Chemistry, Arrhenius Lab., U. of Stockholm, S-10691 Stockholm, Sweden. (1938) Fil.dr., chemistry (Stockholm U., 1972). Prof. (tel. 08-162366, fax +46 8 152187). *Solid state chemistry, preparation, sol-gel.*

Ojamäe, Mr Lars. Inst. of Chemistry, U. of Uppsala, Box 531, S-75121 Uppsala, Sweden. (1964) Fil.kand., chemistry (Uppsala U., 1988). Res.asst. (tel. 018-183772, telex 76088 TSLISV S, fax +46 18 108542, email bitnet: lars at semax51). *Computational chemistry.*

Olovsson, Dr Gunnar. Inst. of Chemistry, U. of Uppsala, Box 531, S-75121 Uppsala, Sweden. (1953) Fil.lic., chemistry (Uppsala U., 1987). Res. asst. (tel. 018-183775, telex 76088 TSLISV S, fax +46 18 108542, email bitnet: kemgo at semax51). *Accurate structure analysis, X-ray and neutron diffraction, hydrogen bonding, cation radical salts ("organic metals").*

Olovsson, Prof. Ivar. Inst. of Chemistry, U. of Uppsala, Box 531, S-75121 Uppsala, Sweden. (1928) Fil.dr., chemistry (Uppsala U., 1960). Prof. (tel. 018-183721, telex 76088 TSLISV S, fax +46 18 108542). *Accurate structure analysis, X-ray diffraction, neutron diffraction, hydrogen bonding, electron density, reaction pathways.*

Olson, Mrs Solveig. Dept. of Inorganic Chemistry, Chalmers U. of Techn. and U. of Göteborg, S-41296 Göteborg, Sweden. (1944) Res. asst. (tel. 031-722881). *Coordination compounds, structural chemistry.*

Olsson, Mr Per-Olof. Dept. of Inorganic Chemistry, Arrhenius Lab., U. of Stockholm, S-10691 Stockholm, Sweden. (1957) Högsk.ex., chemistry (Stockholm U., 1983). Res. asst. (tel. 08-162372). *Defect structures, electron microscopy, inorganic crystals, structural chemistry, materials science.*

Oskarsson, Dr Åke. Dept. of Inorganic Chemistry, Royal Inst. of Techn., S-10044 Stockholm, Sweden. (1942) Fil.dr., chemistry (Lund U., 1974). Sen.lect. (tel. 08-7908157). *Coordination compounds.*

Österberg, Prof. Ragnar. Dept. of Chemistry, Swedish U. of Agricultural Sciences, Box 7015, S-75007 Uppsala, Sweden. (1932) Med.dr., biochemistry (Göteborg U., 1966). Prof. (tel. 018-301485). *Macromolecular interactions, small angle X-ray and neutron scattering.*

Pascher, Dr Irmin. Dept. of Structural Chemistry, Fac. of Medicine, U. of Göteborg, Box 33031, S-40033 Göteborg, Sweden. (1935) Phil.Dr., chemistry (Graz U., Austria, 1963). Sen.lect. (tel. 031-853451, email bitnet: medpg at secthf51). *Structural biology, X-ray diffraction, membrane lipids, structure and function.*

Permér, Miss Lotta. Dept. of Inorganic Chemistry, Arrhenius Lab., U. of Stockholm, S-10691 Stockholm, Sweden. (1964) Högsk.ex., chemistry (Stockholm U., 1986). Res.asst. (tel. 08-162368, fax +46 8 152187). *Defect structures, disorder, electron diffraction, electron microscopy, inorganic crystals, materials science, powder diffraction.*

Persdotter, Dr Ingeborg M. Lerums Gymnasieskola, Box 301, S-44301 Lerum, Sweden. (1956) Fil.dr., chemistry (Göteborg U., 1988). Lect. (tel. 0302-50535). *Inorganic crystal structures.*

Persson, Miss Jeanette. Dept. of Inorganic Chemistry, Arrhenius Lab., U. of Stockholm, S-10691 Stockholm, Sweden. (1962) Högsk.ex., chemistry (Stockholm U., 1988). Res.asst. (tel. 08-162368, fax +46 8 152187). *Ceramics, oxidation, microscopy.*

Pilotti, Dr Anne-Marie. Dept. of Structural Chemistry, Arrhenius Lab., U. of Stockholm, S-10691 Stockholm, Sweden. (1942) Fil.dr., chemistry (Stockholm U., 1971).Docent. (tel. 08-162284). *Organic crystal structures.*

Ribbing, Dr Carl-Gustaf. Dept. of Solid State Physics, Inst. of Techn., Box 534, S-75121 Uppsala, Sweden. (1942) Fil.dr., solid state physics (Uppsala U., 1973). Lect. (docent). (tel. 018-183133, fax +46 18 155095). *Optical coatings, scattering, optical selectivity.*

Rundqvist, Prof. Stig O. Inst. of Chemistry, U. of Uppsala, Box 531, S-75121 Uppsala, Sweden. (1929) Fil.dr., chemistry (Uppsala U., 1963). Prof. (tel. 018-183718, telex 76088 TSLISV S, fax +46 18 108542). *Materials science, crystal growth, structural chemistry.*

Sahle, Dr Wubeshet. Aerosol, National Inst. of Occupational Health, S-17184 Solna, Sweden. (1949) Fil.dr., inorganic chemistry (Stockholm U., 1983). Res. assoc. (tel. 08-162368). *Electron microscopy, electron diffraction, minerals.*

Sandström, Dr Magnus K. E. Dept. of Inorganic Chemistry, Royal Inst. of Techn., S-10044 Stockholm, Sweden. (1945) Tekn.dr., inorg. chemistry (Royal Inst. of Techn., 1978). Sen.Lect. (tel. 08-7908156, telex: 10389 KTHB S). *Chemical bonding, coordination compounds, neutron diffraction, X-ray diffraction on solutions.*

Schneider, Dr Gunter. Dept. of Molecular Biology, Swedish U. of Agricultural Sciences, Uppsala Biomedical Center, Box 590, S-75124 Uppsala, Sweden.

(1953) Dr.rer.nat., chemistry (Saarbrücken U., Germany, Fed. Rep., 1983). Res. assoc. (tel. 018-174524). *Biological macromolecules, protein crystallography.*

Sjöberg, Prof. Bo. Dept. of Medical Biochemistry, U. of Göteborg, Box 33031, S-40033 Göteborg, Sweden. (1941) Fil.dr., chemistry (Göteborg U., 1974). Prof. (tel. 031-853458). *Biomolecular structure, hydration and dynamics in solution; small-angle X-ray and neutron scattering.*

Sjöberg, Mr Jörgen. Dept. of Inorganic Chemistry, Chalmers U. of Techn. and U. of Göteborg, S-41296 Göteborg, Sweden. (1959) Tekn.lic., inorganic chemistry (Chalmers U. of Techn., 1988). Res.asst. (tel. 031-722882). *Powder diffraction, ceramic materials, single crystals, neutron diffraction, thermal analysis.*

Sjölin, Dr H. Lennart G. Dept. of Inorganic Chemistry, Chalmers U. of Techn. and U. of Göteborg, S-41296 Göteborg, Sweden. (1949) Fil.dr., chemistry (Göteborg U., 1979). Docent. (tel. 031-721000, fax +46 31 167194). *Protein crystallography, neutron diffraction.*

Sjövall, Mr Rune. Div. of Inorganic Chemistry 2, Chemical Center, U. of Lund, Box 124, S-22100 Lund, Sweden. (1960) Civ.ing., chemical eng. (Lund U., 1985). Res. asst. (tel. 046-108224, fax +46 46 146030). *Inorganic chemistry, X-ray diffraction.*

Ståhl, Dr Kenny. Div. of Inorganic Chemistry 2, Chemical Center, U. of Lund, Box 124, S-22100 Lund, Sweden. (1953) Tekn.dr., inorg. chemistry (Lund U., 1983). Docent. (tel. 046-108117, fax +46 46 146030). *Inorganic structures, X-ray diffraction, neutron diffraction, X-ray synchroton radiation.*

Stålhandske, Dr Claes. Div. of Inorganic Chemistry 2, Chemical Center, U. of Lund, Box 124, S-22100 Lund, Sweden. (1941) Tekn.dr., inorg. chemistry (Lund U., 1980). Docent. (tel. 046-108234, fax +46 46 146030). *Inorganic crystal structures.*

Stefanidis, Mr Theodoros. Dept. of Structural Chemistry, Arrhenius Lab., U. of Stockholm, S-10691 Stockholm, Sweden. (1955) Fil.kand., chemistry (Stockholm U., 1978). Res. asst. (tel. 08-162382). *Phosphates - crystal studies.*

Stenberg, Dr Lars. Div. of Inorganic Chemistry 2, Chemical Center, U. of Lund, Box 124, S-22100 Lund, Sweden. (1949) Tekn.dr., inorganic chemistry (Lund U., 1983). Res. assoc. (tel. 046-108233, fax +46 46 146030). *Electron microscopy, structural chemistry.*

Stensland, Dr Birgitta. Dept. of Structural Chemistry, Arrhenius Lab., U. of Stockholm, S-10691 Stockholm, Sweden. (1938) Fil.lic., chemistry (Stockholm U., 1970). Res. assoc. (tel. 08-162381, email bitnet: bsd at seq51). *Biologically active molecules.*

Stomberg, Dr Rolf. Dept. of Inorganic Chemistry, Chalmers U. of Techn. and U. of Göteborg, S-41296 Göteborg, Sweden. (1933) Tekn.dr., inorg. chemistry (Chalmers U. of Techn., 1965). U. lect. (docent). (tel. 031-722874). *Teaching, inorganic crystal structures.*

Strandberg, Prof. Bror E. Dept. of Molecular Biology, Biomedical Center, U. of Uppsala, Box 590, S-75124 Uppsala, Sweden. (1930) Fil.dr., chemistry (Uppsala U., 1967). Prof. (tel. 018-113453 or 174475, fax +46 18 151759). *Structure and function, proteins, viruses, protein-nucleic acid interaction.*

Strandberg, Dr Rolf A.G. Inorganic Chemistry Dept., U. of Umeå, S-90187 Umeå, Sweden. (1938) Fil.dr., chemistry (Umeå U., 1974). Sen.lect. (tel. 090-165467, fax +46 90 136310). *Inorganic crystal structures.*

Strid, Prof. Karl-Gustav. Biomaterials group, Dept. of Handicap Res., U. of Göteborg, Brunnsgatan 2, S-41312 Göteborg, Sweden. (1940) Tekn.dr., physics (Chalmers U. of Techn., 1976). Prof. of biomaterials res.(tel. 031-410919). *Biomaterials, image science, instrumentation, materials science.*

Stridh, Mr Kjell. Dept. of Inorganic Chemistry, Chalmers U. of Techn. and U. of Göteborg, S-41296 Göteborg, Sweden. (1958) Tekn.lic., inorganic chemistry (Chalmers and Göteborg U., 1988). Res.asst. (tel. 031-722866). *Flotation, mineral surfaces, sulphide minerals, orientation.*

Su, Mr Xiao-dong. Dept. of Molecular Genetics, Karolinska Inst., Box 60400, S-10401 Stockholm, Sweden. (1963) B.Sc., physics, biophysics (1985). Res. asst. (tel. 08-340560 ext. 1251). *Electron microscopy, X-ray diffraction, structural biology, viruses, quasicrystals, molecular crystals, hydrogen bonding, crystal growth.*

Sundberg, Dr Margareta. Dept. of Inorganic Chemistry, Arrhenius Lab., U. of Stockholm, S-10691 Stockholm, Sweden. (1944) Fil.dr., inorg. chemistry (Stockholm U., 1981). Docent. (tel. 08-162368, fax +46 8 152187). *Inorganic crystals, defect structures, electron microscopy.*

Svensson, Dr L. Anders. Dept. of Inorganic Chemistry, Chalmers U. of Techn. and U. of Göteborg, S-41296 Göteborg, Sweden. (1959) Fil.dr., chemistry (Göteborg U., 1989). Res. assoc. (tel. 031-722857, email bitnet: colp4 at seguc21). *Protein crystallography.*

Svensson, Dr Christer. Div. of Inorganic Chemistry 2, Chemical Center, U. of Lund, Box 124, S-22100 Lund, Sweden. (1945) Tekn.dr., inorg. chemistry (Lund U., 1978). Docent. (tel. 046-108117, fax +46 46 146030). *Crystal structures, properties, diffraction methods, computer programming.*

Svensson, Mr Göran. Div. of Inorganic Chemistry 2, Chemical Center, U. of Lund, Box 124, S-22100 Lund, Sweden. (1955) Civ.ing., chemistry (The Lund Inst. of Techn., 1983). Res. asst. (tel. 046-108112, fax +46 46 146030). *Crystal structures, crystal growth, coordination compounds.*

Svensson, Mr Gunnar. Dept. of Inorganic Chemistry, Arrhenius Lab., U. of Stockholm, S-10691 Stockholm, Sweden. (1960) Högsk.ex., chemistry (Stockholm U., 1984). Res.asst. (tel. 08-162365, fax +46 8 152187). *Electron microscopy, inorganic crystals, powder diffraction.*

Svensson, Dr Ing-Britt A. Dept. of Chemical Eng., Chalmers U. of Techn., S-41296 Göteborg, Sweden. (1942) Fil.lic., chemistry (Göteborg U., 1971). Director of studies. (tel. 031-810100, ext. 1303). *Inorganic crystal structures.*

Tegman, Dr Ragnar. Swedish Inst. of Production Eng. Res., Regnbågsallén, S-95187 Luleå, Sweden. (1943) Fil.dr., chemistry (Umeå U., 1974). Docent. (tel. 0920-91000, ext. 770, fax +46 920 97288). *Inorganic crystal structures, nitrides, non-oxide ceramics, materials research, hot isostatic pressing (HIP).*

Tellgren, Dr I. G. Roland. Inst. of Chemistry, U. of Uppsala, Box 531, S-75121 Uppsala, Sweden. (1930) Fil.dr., chemistry (Uppsala U., 1975). Docent. (tel. 018-183776, fax +46 18 108542, email bitnet: rte at semax51). *Neutron diffraction, metal hydrides, magnetic structures, powder diffraction.*

Thomas, Prof. John O. (Josh). Inst. of Chemistry, U. of Uppsala, Box 531, S-75121 Uppsala, Sweden. (1944) Ph.D., crystallography (London U., UK, 1969). Prof. (tel. 018-183763, email bitnet: josh at semax51). *Materials science, disorder, solid ionics, polymers, molecular dynamics, powder diffraction.*

Thomasson, Mr Ronnie. Div. of Inorganic Chemistry 2, Chemical Center, U. of Lund, Box 124, S-22100 Lund, Sweden. (1959) Fil.kand., inorganic chemistry (Lund U., 1984). Res. asst. (tel. 046-108229, fax +46 46 146030). *Zeolites, silicate chemistry, adsorption, inorganic crystal structures.*

Törnroos, Mr Karl Wilhelm. Dept. of Structural Chemistry, Arrhenius Lab., U. of Stockholm, S-10691 Stockholm, Sweden. (1956) Högsk.ex., structural chemistry (Stockholm U., 1983). Res. asst. (tel. 08-162379, fax +46 8 152187, email bitnet: kwtrose at seqz51). *Coordination compounds, organic compounds, proteins.*

Unge, Dr K. Torsten. Dept. of Molecular Biology, Biomedical Center, U. of Uppsala, Box 590, S-75124 Uppsala, Sweden. (1945) Fil.dr., biochemistry (Uppsala U., 1979). Docent. (tel. 018-174000, fax +46 18 151759). *Biological macromolecules.*

Vannerberg, Prof. Nils-Gösta. Eka Nobel AB, S-44501 Surte, also, Dept. of Inorganic Chemistry, Chalmers U. of Techn. and U. of Göteborg, S-41296 Göteborg, Sweden. (1930) Fil.dr., inorganic chemistry (Chalmers U. of Techn., 1959). Prof., Dir. of R & D. (tel. 0303-98000, ext. 209, telex: 2435, fax +46 31 981774). *Inorganic crystal structures.*

Wahlberg, Dr Anders. Inst. of Chemistry, U. of Uppsala, Box 531, S-75121 Uppsala, Sweden. (1945) Fil.dr., chemistry (Uppsala U., 1985). Lect. (tel. 018-183669, telex 76088 TSLISV S, fax +46 18 108542). *Crystal structure types, amphiphilic interactions, 1:1 salts.*

Wahlberg, Dr Olof. Dept. of Structural Chemistry., Arrhenius Lab., U. of Stockholm, S-10691 Stockholm, Sweden. (1936) Fil.dr., bio-inorganic solution chemistry (Stockholm U., 1971). Docent.(tel. 08-162390, fax +46 8 152187). *Coordination compounds, bio-inorganic chemistry, geochemistry.*

Wahlström, Dr Ebba B. Dept. of Inorganic Chemistry, Arrhenius Lab., U. of Stockholm, S-10691 Stockholm, Sweden. (1938) Fil.dr., chemistry (Stockholm U., 1970). Res. assoc. (tel. 08-162417, fax +46 8 152187). *Materials science, inorganic crystals.*

Wallenberg, Dr L. Reine. Div. of Inorganic Chemistry 2, Chemical Center, U. of Lund, Box 124, S-22100 Lund, Sweden. (1957) Tekn.dr., inorganic chemistry (Lund U., 1987). Res. assoc. (tel. 046-108233, fax +46 46 146030, email bitnet: 002rw at seldc51). *Electron microscopy, surface structure.*

Waller, Prof. Ivar. Dept. of Theoretical Physics, U. of Uppsala, Thunbergsvägen 3b, S-75238 Uppsala, Sweden. (1898) Fil.dr., theoretical physics (Uppsala U., 1925). Prof. em. (tel. 018-115159). *X-ray scattering, neutron scattering, crystal physics, liquid physics.*

Waltersson, Dr Kjell. CEA AB, Box 174, S-64523 Strängnäs, Sweden. (1947) Fil.dr., chemistry (Stockholm U., 1976). Docent. (tel. 0152-12930). *Inorganic crystal structures, solid state chemistry.*

Wang, Dr Da Neng. Dept. of Structural chemistry, Arrhenius Lab., U. of Stockholm, S-10691 Stockholm, Sweden. (1961) Fil.dr., chemistry (1988). Res.asst. (tel. 08-162380, telex 8105199 UNIVER S, fax +46 8 152187, email bitnet: wang at embl). *Electron diffraction, electron microscopy, biological macromolecules, structural biology.*

Werner, Prof. Per-Erik. Dept. of Structural Chemistry, Arrhenius Lab., U. of Stockholm, S-10691 Stockholm, Sweden. (1931) Fil.dr., chemistry (Stockholm U., 1971). Prof. (tel. 08-162393, fax +46 8 152187, email bitnet: werner at seqz51). *Powder diffraction, computer programming, inorganic and organic structures.*

Westdahl, Miss Marianne. Dept. of Structural Chemistry, Arrhenius Lab., U. of Stockholm, S-10691 Stockholm, Sweden. (1951) Res. asst. (tel. 08-162393, fax +46 8 152187). *Powder diffraction, crystal structures, computer programming.*

Westman, Mrs Anna-Karin. ABB Cerama AB, S-91500 Robertsfors, Sweden. (1957) Tekn.lic., inorganic chemistry (Chalmers U. of Techn., 1987). Manager materials R & D. (tel. 0934-10510, telex: 54014, fax +46 934 10139). *High performance ceramics.*

Westman, Dr Sven. Dept. of Structural Chemistry, Arrhenius Lab., U. of Stockholm, S-10691 Stockholm, Sweden. (1933) Fil.dr., inorg. chemistry (Stockholm U., 1972). Docent. (tel. 08-162390). *Chemical education, inorganic structures.*

Wickman, Prof. Frans Erik. Dept. of Geology, U. of Stockholm, S-10691 Stockholm, Sweden. (1915) Fil.dr., mineralogy (U. of Stockholm). Prof. em., mineralogy, petrology and geochemistry. (tel. 046-112045). *High pressures, inorganic crystals, minerals.*

Ymén, Dr B. Ingvar. Technical Res. & Dev. Dept., SUPRA AB, Box 516, S-26124 Landskrona, Sweden. (1954) Fil.dr., inorg. chemistry (Lund U., 1983). Res. assoc. (tel. 0418-76100). *Inorganic crystals, powder diffraction, microscopy.*

Zou, Ms Xiao-dong, Dept. of Structural Chemistry, Arrhenius Lab., U. of Stockholm, S-10405 Stockholm, Sweden. (1964) Solid state physics (Beijing U. of Iron and Steel Techn.,1986). (tel. 08-162380). *Electron microscopy, inorganic crystals, materials science, phase determination, quasicrystals.*

SWITZERLAND

Sub-Editor: **M. Dobler**

Notes

1. International telephone country code - 0041.

2. The degrees *Dr.sc.nat., Dr.sc.techn., Dr.sc.math., Dr.phil.* and *Dr.phil.nat.* are approximately equivalent to a PhD. *Dipl.Ing.Chem., Dipl.Chem., Dipl.Phys.* and *Dipl.Nat.* are first degrees requiring at least 4 years of course work including some research, approximately equivalent to MSc. *Dipl.Ing.HTL* (french *Ing.ETS*) are degrees from Technical Schools, approximately equivalent to Polytechnical Schools in Britain.

3. Appointments at universities other than professors are *Oberassistent* (french *Maître assistent*) (oberasst.), about equivalent to lecturer. *Chargé de cours* or *Chargé de recherche* is equivalent to assistant. The title *PD* (Privatdozent) is about equivalent to reader.

4. Abbreviations:
   Inst. - Institut (department)
   Lab. - Laboratorium (as Institut)

   ETH (french EPF) - Eidgenössische Technische Hochschule
   HTL (french ETS) - Höhere Technische Lehranstalt

Altermatt, Dr Urs Daniel. Messtechnik und Automation, CIBA-GEIGY AG, R-1055.4.46, CH-4002 Basel, Switzerland. (1955) Dr.sc.nat., crystallography (ETH Zürich, 1983). (tel. 061 + 6975722). *Computing, diffractometry, drivers.*

Arend, Prof. Hanns. Inst. für Quantenelektronik, ETH Hönggerberg, CH-8093 Zürich, Switzerland. (1922) Dr, physical chemistry (U. Prague, Czechoslovakia, 1952). Acad. guest. (tel. 01 + 3772329, fax: + 3720630). *Crystal growth, electrooptic materials, dielectrics, ferroelectrics.*

Armbruster, PD Thomas Michael Ludwig. Lab. für chemische und mineralogische Kristallographie, U. Bern, Freiestr. 3, CH-3012 Bern, Switzerland. (1950) PD, mineralogical crystallography (U. Bern, 1986). Lect. (tel. 031 + 654281, fax + 654499, email u422 at cbebda3t). *Crystal optics, crystal chemistry, minerals.*

Banner, Dr David William. ZFE, Bau 65/102, F. Hoffmann-LaRoche, Grenzacherstr., CH-4002 Basel, Switzerland. (1946) D.Phil., protein crystallography (U. Oxford, UK, 1972). res. scient. (tel 061 + 6887587, email banner at dhdembl5). *Proteins, drug design, protein design.*

Bärlocher, Dr Christian. Inst. für Kristallographie und Petrographie, ETH Zentrum, CH-8092 Zürich, Switzerland. (1944) PhD, physical chemistry (London U., UK, 1973). wiss. Adjunkt (tel. 01 + 2563749, fax + 2527008, email u5567 at czheth5a). *Rietveld refinement, powder diffraction, powder structure determination, zeolites, computing.*

Bayer, Prof. Gerhard. Inst. für Kristallographie und Petrographie, ETH Zentrum, CH-8092 Zürich, Switzerland. (1923) Dr.sc.nat., crystallography (ETH Zürich, 1961). Prof. (tel. 01 + 2563735). *Crystal chemistry, oxides, silicates, crystallization, glasses, ceramics, industrial minerals.*

Bernardinelli, Dr Gérald. Lab. de Cristallographie aux Rayons X, U. de Genève, 24 quai Ernest-Ansermet, CH-1211 Genève, Switzerland. (1945) PhD, crystallography (U. de Genève, 1977). Maître d'enseignement et de recherche. (tel. 022 + 219355, ext. 2372). *Crystallography, organic compounds.*

Bialek, Mr Roland. Inst. für Kristallographie und Petrographie, ETH Zentrum, CH-8092 Zürich, Switzerland. (1962) Dipl.chem.ETH, chemistry (ETH Zürich, 1985). Asst. (tel. 01 + 2563775). *Zeolites.*

Blanc, Mr Eric Charles-Henri. Inst. de Cristallographie, U. de Lausanne, B.S.P. Dorigny, CH-1015 Lausanne, Switzerland. (1963) Ing. phys., physics (EPF Lausanne, 1987). Asst. (tel. 021 + 6922360, email eblanc at clsuni51). *Computing, quasicrystals.*

Bohac, Dr Petr. Ind. Res. unit, ETH Hönggerberg, CH-8093 Zürich, Switzerland. (1942) Dr.sc.techn., inorganic chemistry (ETH Zürich, 1979). Res. scient. (tel. 01 + 3772174). *Ceramics, crystal growth.*

Brinkmann, Prof. Detlef. Physik Inst., U. Zürich, Schönberggasse 9, CH-8001 Zürich, Switzerland. (1931) Dr, exp. physics (U. Zürich, 1961). Prof. (tel. 01 + 2572930). *NMR, crystallography, minerals, solid state physics.*

Bührer, Dr Willi. Lab. für Neutronenstreuung, ETH Zürich, CH-5303 Würenlingen, Switzerland. (1938) Dr.sc.nat., solid state physics (ETH Zürich, 1970). Res. scient. (tel. 056 + 992086, fax: + 982327). *Neutron diffraction, phase transitions.*

Bukowiecki, Dr Stanislaw. Pharmaceutical R + D, Bldg. 340, Sandoz Ltd., CH-4002 Basel, Switzerland. (1948) Dr.sc.nat., crystallography (ETH Zürich, 1975). Head R + D. (tel. 061 + 245195). *Polymorphism, crystal engineering.*

Bürgi, Prof. Hans-Beat. Lab. für chemische und mineralogische Kristallographie, U. Bern, Freiestr. 3, CH-3012 Bern, Switzerland. (1942) Dr, chemistry (ETH Zürich, 1969). Prof. (tel. 031 + 654282, fax + 654499, email u422 at cbebda3t). *Structure analysis, organic and inorganic compounds, reaction paths, structural correlations, molecular mechanics.*

Burkhard, Dr Andreas. Physics Dept. Klybeck, CIBA-GEIGY AG, CH-4002 Basel, Switzerland. (1947) Dr Phil., mineralogy (U. Basel, 1977). Res. scient. (tel. 061 + 6964014). *Powder diffraction, microscopy, alpine minerals.*

Bürki, Dr Hans. Viktoriastr. 69, CH-3013 Bern, Switzerland. (1921) Dr.phil.nat., chemistry (U. Bern, 1950). (tel. 031 + 420593). *Structure determination, crystal structures, biological structures.*

Busch, Prof. Georg Adolf. Lab. für Festkörperphysik, ETH Hönggerberg, CH-8093 Zürich, Switzerland. (1908) Dr, solid state physics (ETH Zürich, 1938). Prof. em. (tel. 01 + 3772240). *Condensed matter physics.*

Chapuis, Prof. Gervais Constant. Inst. de Cristallographie, U. de Lausanne, B.S.P. Dorigny, CH-1015 Lausanne, Switzerland. (1944) Dr, crystallography (ETH Zürich, 1971). Prof. (tel. 021 + 6922350). *Modulated structures, phase transitions, symmetry, structure modeling.*

Chollet, Dr Lucien-Francois. Materials Science and Micromechanics, Centre Suisse d'electronique et de microtechnique, CSEM, rue Breguet 2, CH-2000 Neuchatel, Switzerland. (1931) Dr, physics (U. de Neuchatel, 1960). Scient. (tel. 038 + 245566). *Structure, metals, alloys.*

Daly, Dr John Joseph. ZFE, F. Hoffmann-LaRoche AG, CH-4002 Basel, Switzerland. (1931) PhD, chemistry (Leeds U., UK, 1959). Group leader. (tel. 061 + 6886046). *Structural chemistry, organic compounds, proteins.*

Delaloye, Prof. Michel. U. de Genève, 13, rue des Maraichers, CH-1211 Genève, Switzerland. (1936) Dr, mineralogy (U. de Genève, 1966). Prof. (tel. 022 + 219355). *Geochemistry, geochronometry.*

Dobler, Prof. Max. Lab. für organische Chemie, ETH Zentrum, CH-8092 Zürich, Switzerland. (1937) Dr.sc.techn., chemical crystallography (ETH Zürich, 1963). Prof. (tel. 01 + 2564509, email dobler at czheth5a). *Biological molecules, molecular modeling.*

Donatz, Mrs Martina. Inst. für Molekularbiologie und Biophysik, ETH Hönggerberg, CH-8093 Zürich, Switzerland. (1962) Dipl.nat.ETH, macromolecular crystallography (ETH Zürich, 1986). PhD student. (tel. 01 + 3772461). *Proteins.*

Doudin, Mr Bernard. Inst. de Cristallographie, U. de Lausanne, B.S.P. Dorigny, CH-1015 Lausanne, Switzerland. (1961) Dipl. phys., physics (U. de Lausanne, 1985). Asst. (tel. 021 + 6922360, email bdoudin at clsuni51). *Structure analysis, phase transitions, modulated phases.*

Dubler, Prof. Erich. Anorganisch-chemisches Inst., U. Zürich, Winterthurerstr. 190, CH-8057 Zürich, Switzerland. (1939) Dr.phil., chemistry (U. Zürich, 1970). Prof. (tel. 01 + 2574621). *Bioinorganic chemistry, crystal structures, thermal analysis.*

Dunitz, Prof. Jack David. Lab. für organische Chemie, ETH Zentrum, CH-8092 Zürich, Switzerland. (1923) PhD, chemistry (Glasgow U., UK, 1947). Prof. (tel. 01 + 2562892, fax + 2514633, email jdd at czheth5a). *Crystal and molecular structure.*

Duroc-Danner, Mr Jean-Marie. Gemgrading, rue Albert-Gos 4, CH-1206 Genève, Switzerland. (1947) Dipl. IUED (IUED Genève, 1977). Expert. (tel. 022 + 466061, fax: + 473676). *Syntheses, diamonds, beryls.*

Egli, Dr Martin. Lab. für organische Chemie, ETH Zentrum, CH-8092 Zürich, Switzerland. (1961) Dr.sc.nat., crystallography (ETH Zürich, 1988). Asst. (tel. 01 + 2564510, email egli at czheth5a). *Crystal structures.*

Emmenegger, Prof. Franzpeter. Inorganic Chemistry Inst., U. de Fribourg, Pérolles, CH-1700 Fribourg, Switzerland. (1935) Dr, chemistry (ETH Zürich, 1963). Prof. (tel. 037 + 826422). *Crystal growth, chemical transport.*

Engel, Dr Nora. Dept. de Minéralogie, Muséum d'Histoire naturelle, 1, rte. de Malagnou, CH-1211 Genève, Switzerland. (1953) PhD, crystallography (U. de Genève, 1986). Chargée de recherche. (tel. 022 + 359130). *Minerals, computing, classification.*

Epprecht, Prof. Willfried Th. Ottenbergstr. 45, CH-8049 Zürich, Switzerland. (1918) Dr.sc.nat. (ETH Zürich). Prof. em. (tel. 01 + 3421386). *Physical metallurgy, nuclear reactor materials, creep, fatigue.*

Estermann, Mr Michael. Inst. für Kristallographie und Petrographie, ETH Zentrum, CH-8092 Zürich, Switzerland. (1961) Dipl.phil II, physics (U. Zürich, 1987). PhD student. (tel. 01 + 2563727). *Crystallography, maximum entropy.*

Fischer, Dr Peter. Lab. für Neutronenstreuung, ETH Zürich, CH-5303 Würenlingen, Switzerland. (1937) Dr, neutron diffraction (ETH Zürich, 1966). Physicist. (tel. 056 + 992094, fax: + 827417). *Neutron diffraction, solid state physics, structure research, magnetism.*

Flack, Dr Howard David. Lab. de Cristallographie aux Rayons X, U. de Genève, 24 Quai Ernest-Ansermet, CH-1211 Genève, Switzerland. (1943) PhD, crystallography (London U., UK, 1968). Chargé de recherche. (tel. 022 + 219355, ext. 2249, fax: + 812192, email flack at cgeuge52). *Crystallography.*

Forster, Dr Martin. Res. dep., Cerberus AG, alte Landstr. 411, CH-8708 Männedorf, Switzerland. (1945) Dr.sc.nat., natural sciences (ETH Zürich, 1978). Project leader. (tel. 01 + 9226476). *Single crystals, ceramics, metal oxides.*

Galetti, Prof. Giulio. Inst. de Minéralogie, U. de Fribourg, Pérolles, CH-1700 Fribourg, Switzerland. (1937) Dr, geochemistry (U. di Padova, Italy, 1971). Prof. (tel. 037 + 826268). *X-ray diffraction, inorganic crystals.*

Gelato-Volders, Dr (Mrs) Louise Marie. Lab. de Cristallographie aux Rayons X, U. de Genève, 24 quai Ernest-Ansermet, CH-1211 Genève, Switzerland. (1935) Dr, mathematics (U. di Bologna, Italy, 1965). Asst. (tel. 022 + 219355, ext. 2236, email gelato at cgeuge11 or cgeuge52). *Computing.*

Gerdil, Prof. Raymond. U. de Genève, 30 quai Ernest Ansermet, CH-1211 Genève 4, Switzerland. (1929) Dr.sc.techn., organic physical chemistry (ETH Zürich, 1957). Prof. (tel. 022 + 219355). *Clathrates, stereoselectivity, host-guest interaction, reactivity, organic compounds.*

Giovanoli, Prof. Rudolf. Inorganic Chemistry Inst., U. Bern, Freiestr. 3, CH-3000 Bern 9, Switzerland. (1936) Dr, chemistry (U. Bern, 1965). Prof. (tel. 031 + 654317). *Finely divided solids, metal oxides and oxidehydroxides, interconversion reactions, topotaxy, structure-texture relationships.*

Girgis, Dr Kamal. Inst. für Kristallographie und Petrographie, ETH Zentrum, CH-8092 Zürich, Switzerland. (1936) Dr.sc.nat., crystallography (ETH Zürich, 1969). Head materials res. group. (tel. 01 + 2563770). *Crystal chemistry, crystal physics, structure, intermetallic compounds, crystal growth, superconductivity, magnetism.*

Glumoff, Mr Tuomo. Lab. für Biochemie I, ETH Zentrum, CH-8092 Zürich, Switzerland. (1961) MSc, biochemistry (U. Turku, Finland, 1987). PhD student. (tel. 01 + 2563141, fax + 2526323, email u4610 at czheth5a). *Proteins, structure-function relationship, protein engineering.*

Gotthardt, Dr Rolf. Inst. Génie Atomique, EPF Lausanne, CH-1015 Lausanne, Switzerland. (1941) Dr.rer.nat., physics (U. Stuttgart, FRG, 1977). Physicist. (tel. 021 + 6933392, fax + 6934444, email gotthardt at clsepf51). *Dislocation mobility, radiation damage, martensitic transformation, electron microscopy.*

Gramlich, Dr Rahel. Inst. für Kristallographie und Petrographie, ETH Zentrum, CH-8092 Zürich, Switzerland. (1953) Dr.sc.nat., crystallography (ETH Zürich, 1981). *Zeolites.*

Gramlich, PD Volker. Inst. für Kristallographie und Petrographie, ETH Zentrum, CH-8092 Zürich, Switzerland. (1941) Dr, crystallography. Oberass. (tel. 01 + 2563756). *Superstructures, quasicrystals, maximum entropy.*

Gränicher, Prof. Walter Hans Heini. Lab. für Festkörperphysik, ETH Hönggerberg, CH-8093 Zürich, Switzerland. (1924) Dr.sc.nat., solid state physics (ETH Zürich, 1959). Prof. (tel. 01 + 3772330). *Phase transitions, ferroelectrics, hydrogen bonding.*

Grimmer, Dr Hans. Lab. für Materialwissenschaften, Paul Scherrer Institut, CH-5303 Würenlingen, Switzerland. (1941) PhD, mathematical physics (Edinburgh U., UK, 1969). Wiss. Adjunkt. (tel. 056 + 992421, fax: + 982327, email grimmer at cageir5a). *Structure and properties, interfaces.*

Grubenmann, Dr Arnold. Forschungszentrum Marly, CIBA-GEIGY AG, CH-1701 Fribourg, Switzerland. (1933) Dr.phil II, chemistry (U. Zürich, 1968). Res. chemist. (tel. 037 + 214972). *Crystal growth, organic compounds, nonlinear optics.*

Guenter, Prof. John Ralph. Anorganisch-chemisches Inst., U. Zürich, Winterthurerstr. 190, CH-8057 Zürich, Switzerland. (1943) Dr, chemistry (U. Zürich, 1970). Asst. Prof. (tel. 01 + 2574646). *Inorganic crystals, crystal chemistry, topotactic reactions, electron microscopy.*

Hepp, Dr Alfred. Trutztobel, CH-7074 Malix, Switzerland. (1939) Dr.phil., crystallography (U. Zürich, 1981). *Powder diffraction, computing, silicate structures.*

Hintermann, Dr Hans-Erich. Materials Science and Micromechanics, Centre Suisse d'electronique et de microtechnique, CSEM, rue A.L. Breguet 2, CH-2000 Neuchatel, Switzerland. (1929) Dr.sc.nat., physical chemistry (ETH Zürich, 1957). Director. (tel. 038 + 245566). *Composite materials, metallurgy, surface structure, tribology, corrosion, catalysis, high temperature materials.*

Hulliger, Dr Jürg. Inst. für Quantenelektronik, ETH Hönggerberg, CH-8093 Zürich, Switzerland. (1953) Dr.phil., (U. Zürich). Oberasst. (tel. 01 + 3772329, fax: + 590630). *Crystal growth, materials science, dielectrics, ferroelectrics, optoelectronics.*

Hummel, Mr Wolfgang. Lab. für chemische und mineralogische Kristallographie, U. Bern, Freiestr. 3, CH-3012 Bern, Switzerland. (1958) Dipl.min., mineralogy (U. Tübingen, FRG, 1985). PhD student. (tel. 031 + 654272, email u422 at cbebda3t). *Inorganic crystals, minerals, glasses, structural correlations, molecular graphics.*

Irwin, Mr John Joseph. Lab. für organische Chemie, ETH Zentrum, CH-8092 Zürich, Switzerland. (1963) MSc, chemistry (U. of Toronto, Canada, 1987). PhD student. (tel. 01 + 2564507, email j2i at czheth5a) *Molecular mechanics.*

Kaldis, Prof. Emanuel. Lab. für Festkörperphysik, ETH Hönggerberg, CH-8093 Zürich, Switzerland. (1931) Dr.rer.nat., physical chemistry (U. München, 1962). Head materials res. group. (tel. 01 + 3772251). *Solid state chemistry and physics, crystal growth, high temperature chemistry, ultra pure materials, superconductors.*

Kallen, Dr Jörg. Abt. Strukturbiologie, Biozentrum der Universität Basel, Klingelbergstr. 70, CH-4056 Basel, Switzerland. (1957) Dr, biophysics (U. Basel, 1985). Postdoctoral fellow. (tel. 061 + 253880, ext. 274, email kallen at urz.unibas.ch). *Proteins, linear dichroism.*

Karlsson, Dr Rolf. Abt. Mikrobiologie, Biozentrum der Universität Basel, Klingelbergstr. 70, CH-4056 Basel, Switzerland. (1942) PhD, chemistry (U. Stockholm, 1977). Asst. *Proteins.*

Kellenberger, Prof. Eduard. Abt. Mikrobiologie, Biozentrum der Universität Basel, Klingelbergstr. 70, CH-4056 Basel, Switzerland. (1920) Dr, biophysics (U. de Genève, 1953). Prof. (tel. 061 + 253880, ext. 284). *Supramolecular structures, electron microscopy.*

Keller, Dr Eva Barbara. Inst. für Kristallographie und Petrographie, ETH Zentrum, CH-8092 Zürich, Switzerland. (1956) Dr.sc.nat., crystallography (ETH Zürich, 1987). Asst. (tel. 01 + 2563768). *Zeolites, zeolite-like materials.*

Kostorz, Prof. Gernot. Inst. für angewandte Physik, ETH Hönggerberg, CH-8093 Zürich, Switzerland. (1941) Dr.rer.nat., metal physics (U. Göttingen, FRG, 1968). Prof. (tel. 01 + 3773399). *Alloys, defect structures, plasticity, short-range order, phase transitions, surface structure, diffuse, small-angle, neutron, X-ray diffraction.*

Kunz, Mr Martin. Mineralogisches Inst., U. Bern, Baltzerstr. 1, CH-3012 Bern, Switzerland. (1963) Lic.phil.nat., (U. Bern, 1988). (tel. 031 + 658798). *Feldspars, order-disorder.*

Lévy, Prof. Francis. Inst. de physique appliquée, EPF-Lausanne, Ecublens, CH-1015 Lausanne, Switzerland. (1940) Dr.sc.nat., physics (ETH Zürich, 1969). Adj. scient. (tel. 021 + 471111). *Synthesis, crystal growth, thin films, surface structure, interfaces, semiconductors, physical characterization, electronic properties, spectroscopy.*

Ludi, Prof. Andreas. Inorganic Chemistry Inst., U. Bern, Freiestr. 3, CH-3012 Bern, Switzerland. (1936) Dr.phil., inorganic chemistry (U. Bern, 1962). Prof. (tel. 031 + 654244, fax + 654499, email u30g at cbebda3t). *Structure, reactivity, coordination compounds, electron transfer, ruthenium chemistry.*

McCusker, Dr Lynne Bridget. Inst. für Kristallographie und Petrographie, ETH Zentrum, CH-8092 Zürich, Switzerland. (1951) PhD, chemistry (U. of Hawaii, 1980). Res. assoc. (tel. 01 + 2563762, fax + 2527008, email u5569 at czheth5a). *Zeolite crystallography, powder diffraction.*

Meier, Prof. Walter M. Inst. für Kristallographie und Petrographie, ETH Zentrum, CH-8092 Zürich, Switzerland. (1926) D.Sc., physical chemistry (London U., UK, 1983). Prof. (tel. 01 + 2563730). *Inorganic crystals, zeolites, microporous materials, powder diffraction, model constructions, teaching aids.*

Mez, Dr Hans-Christian. Messtechnik und Automation, CIBA-GEIGY AG, Schwarzwaldallee 215, CH-4002 Basel, Switzerland. (1935) Dr.sc.techn., chemical crystallography (ETH Zürich, 1961). Group leader. (tel. 061 + 6974950). *Social contacts.*

Moeckli, Dr Pedro. Lab. de ceramique, Dept. des matériaux, EPF Lausanne, av. des Bains 37-39, CH-1007 Lausanne, Switzerland. (1940) Dr.sc.nat., crystallography (ETH Zürich, 1983). Ing. (tel. 021 + 474954). *Powder diffraction, texture.*

Moor, Dr Robert. Swiss federal propellant plant, CH-3752 Wimmis, Switzerland. (1948) Dr.sc.nat., crystallography (ETH Zürich, 1983). Consulting eng. (tel. 033 + 552396). *Theoretical crystallography.*

Muller, Prof. Jean. Dép. de physique de la matière condensée, U. de Genève, 24 quai Ernest Ansermet, CH-1211 Genève 4, Switzerland. (1929) Dr, physics (ETH Zürich, 1958). Prof. (tel. 022 + 219355, fax: + 812192). *Metals physics, superconductivity.*

Müller, Dr Rudolf O. CIBA-GEIGY AG, Physik Klybeck, CH-4000 Basel, Switzerland. (1929) Dr.phil.nat., mineralogy and petrology (U. Bern, 1958). Scient. expert. (tel. 061 + 6964235). *Polymorphism, identification.*

Nissen, Prof. Hans-Ude. Lab. für Festkörperphysik, ETH Hönggerberg, CH-8093 Zürich, Switzerland. (1932) Dr.rer.nat., geology (U. Münster, FRG, 1960). (tel.

01 + 3772262). *Electron microscopy, minerals, inorganic crystals, minerals, petrology.*

Oberhänsli, Dr Willi E. Central Res. Units, F. Hoffmann-LaRoche AG, Grenzacherstr. 124, CH-4002 Basel, Switzerland. (1930) PhD, chemistry (U. of Wisconsin, USA, 1964). (tel. 061 + 6883564). *Direct methods, organic compounds.*

Parthé, Prof. Erwin. Lab. de Cristallographie aux Rayons X, U. de Genève, 24 quai Ernest-Ansermet, CH-1211 Genève, Switzerland. (1928) Dr.phil., chemistry (U. Wien, Austria, 1954). Prof. (tel. 022 + 219355, ext. 2208). *Inorganic crystals, metals, crystal chemistry.*

Patscheider, Mr Jörg. Anorganisch-chemisches Inst., U. Zürich, Winterthurerstr. 190, CH-8057 Zürich, Switzerland. (1956) Dipl.chem., chemistry (U. Zürich, 1984). PhD student. (tel. 01 + 2574648). *Plasma chemistry.*

Petcher, Dr Trevor James. Preclinical Res. Dept., Sandoz AG, CH-4002 Basel, Switzerland. (1943) PhD, chemical crystallography (Sheffield U., UK, 1967). Mol. pharmacologist, part time crystallographer. (tel. 061 + 241111, ext. 4851). *Structure-activity relationships, organic compounds, drug design, conformational analysis, molecular pharmacology, receptor binding studies.*

Petter, Prof. Walter. Inst. für Kristallographie und Petrographie, ETH Zentrum, CH-8092 Zürich, Switzerland. (1926) Dr.sc.nat., physics (ETH Zürich, 1969). Prof. (tel. 01 + 2563752, fax + 2527008, email kristall at czheth5a). *Crystal chemistry, inorganic crystals.*

Priestle, Dr John P. Abt. Strukturbiologie, Biozentrum der Universität Basel, Klingelbergstr. 70, CH-4056 Basel, Switzerland. (1954) PhD, biochemistry (U. of Texas, Austin, USA, 1982). Postdoctoral fellow. (tel. 061 + 253880, ext. 256, email priestle at urz.unibas.chunet). *Biological macromolecules, computing, macromolecular structure refinement.*

Raselli, Mr Andrea-Raeto. Lab. für chemische und mineralogische Kristallographie, U. Bern, Freiestr. 3, CH-3012 Bern, Switzerland. (1961) Dipl.chem., chemistry (U. Bern, 1986). PhD student. (tel. 031 + 654272, email w434 at cbebda3t). *Crystal structure analysis.*

Reller, Dr Armin. Anorganisch-chemisches Inst., U. Zürich, Winterthurerstr. 190, CH-8057 Zürich, Switzerland. (1952) Dr, chemistry (U. Zürich, 1981). Res. scient. (tel. 01 + 2574617, fax: + 817251). *Structural mechanism, kinetics, reversible heterogeneous solid state reactions, mixed metal oxides, metal carbonates.*

Rieger, Dr Hans Wolfhart. TKT-Metoxit AG, CH-8240 Thayngen, Switzerland. (1939) Dr.phil., chemistry (U. Wien, Austria, 1965). Managing director. (tel. 053 + 391050, fax: + 393965). *Ceramics.*

Riesen, Dr Andreas. Inst. für anorganische Chemie, U. Basel, Spitalstr. 51, CH-4056 Basel, Switzerland. (1955) PhD, chemistry (U. Basel, 1984). Asst. (tel. 061 + 571557). *Structure, metal complexes.*

Rossi, Mr Franco Antonio. Lab. für Biochemie I, ETH Zentrum, CH-8092 Zürich, Switzerland. (1955) Dipl.bot., plant physiology (U. Zürich, 1982). Asst. (tel. 01 + 2563138). *Macromolecular crystallography, protein structure and function.*

Rüegg, Dr Andreas. Dep. UR-D, Ugimag Recoma AG, Industriestr. 297, CH-5242 Lupfig, Switzerland. (1946) Dr.sc.nat., crystallography (ETH Zürich, 1977). Dev. head. (tel. 056 + 949066, fax: + 949081). *Inorganic crystals, phase transitions, ceramics.*

Sakellariou, Dr Evangelos. Gebr. Sulzer AG, CH-8401 Winterthur, Switzerland. (1953) Dr.sc.nat., crystallography (ETH Zürich, 1986). (tel. 052 + 811122). *Hydrogen bonding, structures, anion complexes, crystallisation.*

Scheel, Prof. Hans J. Inst. of Micro- and Optoelectronics DP, Swiss Fed. Inst. of Technology, CH-1015 Lausanne, Switzerland. (1937) Res. assoc. (tel. 21 + 693, ext. 4452, fax +4444, telex 454478 epfv ch). *Crystal growth technology, epitaxy, charecterization, semiconductors, high-temperature semiconductors.*

Schenk, Dr Kurt Johann. Inst. de Cristallographie, U. de Lausanne, B.S.P. Dorigny, CH-1015 Lausanne, Switzerland. (1951) Dr, crystallography (U. de Lausanne, 1984). Res. asst. (tel. 021 + 6922354, fax: + 462307, email kschenk at clsuni51). *Phase transitions, disordered and modulated structures, structure and physical properties, computing.*

Schmalle, Dr Helmut Willi. Anorganisch-chemisches Inst., U. Zürich, Winterthurerstr. 190, CH-8057 Zürich, Switzerland. (1941) Dr, crystallography (U. Hamburg, FRG, 1977). Res. asst. (tel. 01 + 2574650,4624, email schmalle at czhrzu1a). *Crystal structure determination, organic compounds, inorganic crystals, allergenic compounds.*

Schmelczer, Dr Robert. Physiklabor, Huber + Suhner AG, CH-8330 Pfäffikon, Switzerland. (1946) Dr, crystallography (U. de Lausanne, 1984). Res. asst. (tel. 01 + 9522449). *Metal physics, small-angle scattering, structure and physical properties.*

Schmid, Prof. Hans. Dép. de chimie minérale, analytique et appliquée, U. de Genève, 30, quai Ernest-Ansermet, CH-1211 Genève, Switzerland. (1931) Prof., inorganic chemistry (U. de Genève, 1977). Applied chemistry lab. head. (tel. 022 + 219355, ext. 2405). *Magnetically ordered ferroelectrics, ferroics.*

Schobinger-Papamantellos, Dr (Mrs) Penelope. Inst. für Kristallographie und Petrographie, ETH Zentrum, CH-8092 Zürich, Switzerland. (1937) Dr.phil.chem., physical chemistry and crystallography (U. Wien, Austria, 1962). Res. asst. (tel. 01 + 2563773). *Magnetic structures, neutron diffraction, X-ray diffraction.*

Schönholzer, Mr Peter. ZFE, F. Hoffmann-LaRoche AG, Grenzacherstr. 124, CH-4002 Basel, Switzerland. (1937) Dipl.Phys., physics (U. Bern, 1968). (tel. 061 + 6882902). *Organic compounds, single crystals, powder diffraction.*

Schwarzenbach, Prof. Dieter. Inst. de Cristallographie, U. de Lausanne, B.S.P. Dorigny, CH-1015 Lausanne, Switzerland. (1936) Dr, crystallography (ETH Zürich, 1965). Prof. (tel. 021 + 6922349). *Crystal structures, electron density, refinement procedures.*

Schweizer, Dr Wolfhard Bernd. Lab. für organische Chemie, ETH Zentrum, CH-8092 Zürich, Switzerland. (1947) Dr.sc.nat., chemical crystallography (ETH Zürich, 1977). Res. Assoc. (tel. 01 + 2564507, fax: + 2514633, email schweizer at czheth5a). *Reaction pathways.*

Seiler, Mr Paul. Lab. für organische Chemie, ETH Zentrum, CH-8092 Zürich, Switzerland. (1945) Res. Assoc. (tel. 01 + 2564508, email seiler2 at czheth5a). *Crystallography.*

Simmen, Mr André. Inst. für Kristallographie und Petrographie, ETH Zentrum, CH-8092 Zürich, Switzerland. (1960) Dipl.chem.ETH (ETH Zürich, 1985). Asst. (tel. 01 + 2563721). *Crystal chemistry, structural chemistry, zeolites, model construction.*

Smit, Dr Jan Derk Geert. Lab. für Biochemie, ETH Zentrum, CH-8092 Zürich, Switzerland. (1945) Dr, chemistry (U. Groningen, Netherlands, 1973). wiss. Adjunkt. (tel. 01 + 2563141, fax: + 2526323, email u4610 at czheth5a). *Proteins, structure and function, molecular evolution.*

Speziali, Mr Nivaldo Lucio. Inst. de Cristallographie, U. de Lausanne, B.S.P. Dorigny, CH-1015 Lausanne, Switzerland. (1953) MSc, solid state physics (U. Federal de Minas Gerais, Brazil, 1981). PhD student. (tel. 021 + 6922352, email nspeziali at clsuni51). *Structure determination, phase transitions, modulated phases.*

Staehlin, Dr Walter. Gfeller Telecommunications, Bruennenstr. 66, CH-3018 Bern, Switzerland. (1942) Dr, chemistry (U. Zürich, 1970). Head. (tel. 031 + 505111, fax: + 554841). *Materials science.*

Stoeckli-Evans, Prof. (Mrs) Helen Margaret. Inst. de Chimie, U. de Neuchatel, av. de Bellevaux 51, CH-2000 Neuchatel, Switzerland. (1944) PhD, spectroscopy (Salford U., UK, 1969). Prof. assoc. (tel. 038 + 252815, ext. 46). *Crystal structure analysis, inorganic crystals, organic compounds, molecular modeling.*

Strickler, Dr Peter. Abt. Chemie, Kantonsschule Zürcher Oberland, CH-8620 Wetzikon, Switzerland. (1936) Dr, chemical crystallography (ETH Zürich, 1966). Insr. (tel. 01 + 9321933). *Inorganic crystals, organic compounds, structural chemistry.*

Veprek, PD Stanislav. Anorganisch-chemisches Inst., U. Zürich, Winterthurerstr. 190, CH-8057 Zürich, Switzerland. (1939) Dr.phil., physics and chemistry (U. Zürich, 1972). PD. (tel. 01 + 2574651). *Crystal growth, thin films, plasma chemistry, mass spectrometry, photoelectron spectrometry, surface chemistry, surface processes.*

Very, Dr Jean-Michel. Dep. of morphology, University Medical Center, CH-1211 Genève, Switzerland. (1942) PhD, geology and mineralogy (U. de Genève, 1971). Maitre d'enseignement et de recherche. (tel. 022 + 473300, email very at cgecmu51). *biological systems with crystals, bone metabolism, arthropatic disease crystals.*

Villars, Dr. P. Villars' Intermetallic Phases Data Bank, Postfach, CH-5628 Aristau, Switzerland. (1949) Dr.sc.nat., chemistry (ETH Zürich, 1982). Editor CRYSTMET. (tel. 057 + 444386). *Solid state chemistry, intermetallic phases.*

Vincent, Dr Beverly Robert. Lab. für organische Chemie, U. Zürich, Winterthurerstr. 190, CH-8057 Zürich, Switzerland. (1961) PhD, chemistry (Dalhousie U., Canada, 1988). Dep. crystallographer. (tel. 01 + 2574228, email k468920 at czhrzu1a). *Hydrogen bonding, disorder, phase transitions, medium rings.*

Waldmann, Dr Hans. Rheinstr. 4, CH-4127 Birsfelden, Switzerland. (1906) Dr.sc.nat., chemistry and physics (ETH Zürich, 1935). *Chemical microscopy, optical identification methods, chemical crystallography, instrumental optics.*

Walkinshaw, Dr Malcolm Douglas. Preclinical Res., Pharmaceutical Div., Sandoz AG, CH-4002 Basel, Switzerland. (1950) PhD, physical chemistry (Edinburg U., UK, 1975). Res. scient. (tel. 061 + 2411111). *Proteins, peptides, protein-drug interactions.*

Weber, Prof. Hans-Peter. Inst. de Cristallographie, U. de Lausanne, B.S.P. Dorigny, CH-1015 Lausanne, Switzerland. (1941) PhD, crystal physics (U. of Chicago, USA, 1971). Prof. (tel. 021 + 6922354, fax + 6922307, email hweber at clsuni51). *Diffractometry, electron density, minerals, bioorganic structures, phase transitions.*

Weber, Dr Hans Peter. Preclinical Res., Bldg. 503/560, Sandoz AG, CH-4002 Basel, Switzerland. (1936) Dr.sc.nat., X-ray analysis (ETH Zürich, 1964). Head. (tel. 061 + 244343, fax: + 248001). *X-ray analysis, biological molecules, molecular modeling, drug design, molecular mechanics, quantum chemistry.*

Werk, Dr Margit L. Inst. de Cristallographie, U. de Lausanne, B.S.P. Dorigny, CH-1015 Lausanne, Switzerland. (1956) Dr.rer.nat., crystallography (U. Hamburg, FRG, 1984). Prem. asst. (tel. 021 + 6922352, email mwerk at clsuni51). *Inorganic crystals, phase transitions.*

Winkler, Dr Fritz Karl. ZFE, F. Hoffmann-LaRoche, Grenzacherstr., CH-4002 Basel, Switzerland. (1944) Dr.sc.techn., structural organic chemistry (ETH Zürich, 1973). Scient. *Macromolecular crystallography, drug design.*

Wüest, Mr Hermann. Inst. für Quantenelektronik, ETH Hönggerberg, CH-8093 Zürich, Switzerland. (1952) Technician. (tel. 01 + 3772336). *Crystal growth.*

Yvon, Prof. Klaus. Lab. de Cristallographie aux Rayons X, U. de Genève, 24 quai Ernest-Ansermet, CH-1211 Genève, Switzerland. (1943) PhD, structural chemistry (U. Wien, Austria, 1967). Prof. (tel. 022 + 219355, ext. 2231). *Structural chemistry, condensed matter physics.*

Zbinden, Mr Peter. Lab. für organische Chemie, ETH Zentrum, CH-8092 Zürich, Switzerland. (1958) Dipl.chem.ETH, chemistry (ETH Zürich, 1986). PhD student. (tel. 01 + 2564510, email peter at czeth5a) *Molecular modeling, enzyme-substrate/inhibitor complexes, proteins.*

Zehnder, PD (Mrs) Margareta. Inst. für anorganische Chemie, U. Basel, Spitalstr. 51, CH-4056 Basel, Switzerland. (1942) PhD, (1973). Akad. Adjunkt. (tel. 061 + 571557, ext. 273). *Structure, metal complexes.*

Zhao, Mr Jing Tai. Lab. de Cristallographie aux Rayons X, U. de Genève, 24 quai Ernest-Ansermet, CH-1211 Genève, Switzerland. (1962) MSc, solid state physics (Academia Sinica, 1984). Asst. (tel. 022 + 219355, ext. 2372, email zhaoj at cgeuge52). *Intermetallic compounds.*

Zimmermann, Dr Ulrich. Paul Scherrer Institut, ETH Zürich, CH-5303 Würenlingen, Switzerland. (1944) PhD, physics (ETH Zürich, 1978). (tel. 056 + 992466, fax: + 982327). *Powder diffraction, small-angle scattering, positron annihilation, metal physics, radiation damage.*

Zschokke-Gränacher, Prof. (Mrs) Iris. Inst. für Physik, U. Basel, Klingelbergstr. 82, CH-4056 Basel, Switzerland. (1933) Dr, physics (U. Basel, 1960). Prof. (tel. 061 + 442040). *Molecular solids with nonlinear optical properties.*

SYRIAN ARAB REP.

Sub-Editor: S.E. Ali

Notes

1. International telephone country code - 90.

Abou Ghaloun, Dr Omar Farouk. Chemistry Dept., Faculty of Sci., Aleppo U., Aleppo, Syrian Arab Republic. (1950) PhD, physical chemistry (Nantes U., France, 1983). Lect. *Ionic conductors, semiconductors, crystal structure.*

Ali, Prof. Shams El Din. Physics Dept., Faculty of Sci., University Homs, Syrian Arab Republic. (1939) PhD, X-ray crystallography (Hull U., UK, 1969). Prof. *Solid state physics, crystallography.*

TAIWAN

Sub-Editor: L. K. Liu

Notes

1. International telephone country code - 886.

Chang, Dr Chin-Pu. Inst. of Materials Sci. and Eng., Nat. Sun Yat-Sen U., Kaoshiung, Taiwan 80424, ROC. (1956) PhD, Metallurgy and Materials (U. Birmingham, UK, 1987). Assoc. Prof. (tel. 07 + 5316171, ext. 435, fax 07 + 5614914). *Structure and defect analysis, phase transformation.*

Chang, Dr Hou-Cheng. Chemical Systems Res. Div., CSIST, 2 Chung Shan Rd., Chia An Village, Lung-tan, Taiwan 32500, ROC. (1946) PhD, structure chemistry (Weizmann Inst. of Sci., Israel, 1980). Assoc. scient., (tel. 02 + 9253733, fax 03 + 4250085). *Structure analysis, chemical analysis.*

Chang, Prof. Shih-Lin. Physics Dept., Nat. Tsing-Hua U., 101 Kuang-Fu Rd., Sec. 2, Hsinchu, Taiwan 30043, ROC. (1946) PhD, physics (Polytech. Inst. New York, USA, 1975) Chairman and director. (tel. 035 + 719037, fax 035 + 710776, email c47b0006 at twnmoe10) *X-ray multi-beam dynamical diffraction, x-ray phase problem, x-ray optics.*

Chang, Mr Tien-Show. Mining and Metallurgical Eng. Dept., Taipei Inst. of Techn., 3 Shin-Sheng S. Rd., Sec.1, Taipei, Taiwan 10626, ROC. (1941) MSc, earth science (Pennsylvania State U., USA, 1974). Prof. (tel. 02 + 7712171). *Mineralogy, petrology.*

Chen, Prof. Pei-Yuan. Earth Sci. Dept., Nat. Taiwan Normal U., 88 Roosevelt Rd., sec. 5, Taipei, Taiwan 11718, ROC. (1919) PhD, geological sci. (U. Texas, Austin, USA, 1968). Prof. emeritus NTU. (tel. 02 + 9347120-office, 7329513-home). *Clay mineralogy, x-ray mineralogy.*

Chen, Prof. Ruey-Hong. Physics Dept., Nat. Taiwan Normal U., 88 Roosevelt Rd., sec. 5, Taipei, Taiwan 11718, ROC. (1947) PhD, crystallography (U. Pittsburgh, USA, 1977). Prof. (tel. 02 + 9317511, ext. 252). *Solid state physics, diffraction physics, valence charge density.*

Chen, Mrs Yueh-Hua. Metering and Testing Lab., Taiwan Power Co., 198 Roosevelt Rd., Sec. 4, Taipei, Taiwan 107, ROC. (1948) BSc, chemistry (Nat. Cheng Kung U., 1970). Chemist (tel. 02 + 3218121 ext. 37). *X-ray powder diffraction, fluorescence analysis.*

Chiu, Prof. Kuan-Cheng Physics Dept., Chung Yuan Christian U., Chung-Li, Taiwan 32023, ROC. (1954) PhD, physics (U. Utah, USA, 1986). Assoc. Prof. (tel. 03 + 4566228). *Fluid dynamics, vapor growth, semi-conductor compounds.*

Chuang, Mr Kung-Chou. Exploration and Development Res. Center, Chinese Petroleum Corporation, 1 Ta Yuan, Wen Shan, Miaoli, Taiwan 36010, ROC. (1951) MSc, Marine geology (Chinese Culture U., 1978). Geologist. (tel. 037 + 356150, ext. 206). *Clay mineralogy, reservoir study.*

Chung, Prof. Being-Tau. Physics Dept., Chung Yuan Christian U., Chung-Li, Taiwan 32023, ROC. (1944) PhD, applied science (Southern Methodist U., USA, 1974). Prof. (tel. 03 + 4563168, ext. 259). *Electronic properties of materials, crystal structure.*

Houng, Prof. Kun-Huang. Agricultural Chemistry Dept., Nat. Taiwan U., Taipei, Taiwan 10764, ROC. (1932) PhD, soil science (U. Hawaii, USA, 1964). Prof. (tel. 02 + 7808467). *Clay mineralogy, soil science.*

Hseu, Prof. Shu-En. Materials Sci. Dept., Nat. Taiwan U., Taipei, Taiwan 10764, ROC. (1929) PhD, materials science (Stanford U., USA, 1972). Prof. (tel. 02 + 9329693). *X-ray crystallography.*

Hseu, Prof. Tzong-Hsiung. Inst. of Life Sci., Nat. Tsing Hua U., 101 Kuang Fu Rd., Sec. 2, Hsinchu, Taiwan 30043, ROC. (1941) PhD, Physical chemistry (U. Washington, USA, 1972). Prof. (tel. 035 + 715131, ext. 261, fax 035 + 715934). *Structure analysis, molecular biology.*

Huang, Mrs Chi-Yung. Power Res. Inst., Taiwan Power Co., 198 Roosevelt Rd., Sec. 4, Taipei, Taiwan 107, ROC. (1948) BEng., metallurgy (Nat. Cheng Kung U., 1971). Metallurgist (tel. 02 + 6815424 ext. 242). *X-ray powder diffraction, fluorescence analysis, metallography.*

Huang, Mr Tung-Woo. Materials R & D Center, Chung-Shan Inst. of Sci. and Techn., P.O.Box 1-26, Lung-Tang, Taiwan 32526, ROC. (1948) MSc, mineralogy (Chinese Culture U., 1974). Res. scient.. (tel. 02 + 3145384, ext. 4831). *Inorganic compounds, superconductor, structure analysis.*

Jan, Prof. Gwo-Jen. Physics Dept., Nat. Taiwan U., Taipei, Taiwan 10764, ROC. (1946) MSc, Physics. (Nat. Taiwan U., 1970). Lect. (tel. 02 + 3630231). *Solid state physics, x-ray instrumentation.*

Jan, Mr Fong-Yee. Chemical Systems Res. Div., CSIST, 2 Chung Shang Rd., Chia An Village, Lung-Tan, Taiwan, 32500, ROC. (1960) MSc, organic chemistry (Natinoal Tsing Hua U., 1985). Asst. scient., (tel. 02 + 5221475, fax 034 + 250085). *Natural product analysis, structure analysis.*

Juang, Prof. Tzo-Chuan. Res. Inst. of Soil Sci., Nat. Chung Hsing U. 250 Kuo-Kuang Rd., Taichug, Taiwan, 40227, ROC. (1931) PhD, soil science (U. Hawaii, USA, 1970). Prof. (tel. 04 + 2840375). *Clay mineralogy, x-ray crystallography.*

Lee, Prof. Wang Chihming. Geology Dept., Nat. Taiwan U., Taipei, Taiwan 10764, ROC. (1932) PhD, mineralogy (U. Bochum, BRD, 1968). Prof. (tel. 02 + 3630231, ext. 2387). *Mineralogy, crystal chemistry, petrology.*

Lii, Dr Kwang-Hwa. Inst. of Chemistry, Acad. Sinica, 128 Yen-Chiu-Yuan Rd., Sec. 2, Nankang, Taipei, Taiwan 11529, ROC. (1954) PhD, inorganic chemistry (Iowa State University, USA, 1985). Assoc. res. fellow (tel. 02 + 7821889, fax 02 + 7831237) *Inorganic solid state chemistry, early transition metal phosphates, oxides, crystal structure analysis, magnetochemistry.*

Lin, Dr Hsi-Che. Materials Criterion Lab., Inst. of Industrial Techn., 1021 Kuang-Fu Rd., Hsinchu, Taiwan 300, ROC. (1936) PhD, mineralogy (Ohio State U., USA, 1967). Res. scient.. (tel. 035 + 966100). *Phase equilibria, x-ray crystallography, fine ceramics, mineral synthesis.*

Lin, Prof. Szu-Bin. Geology Dept., Nat. Taiwan U., 245 Choushan Rd., Taipei, Taiwan 10764, ROC. (1938) PhD, crystallography and mineralogy (McMaster U., Canada, 1971). Prof. (tel. 02 + 3630231, ext. 2343). *X-ray fluorescence analysis, crystal structure analysis, mineralogical sciences.*

Lin, Prof. Tsang-Lang. Nuclear Eng. Dept., Nat. Tsing-Hua U., Hsin-Chu, Taiwan, 30043, ROC. (1954) PhD, nuclear eng., appl. radiation physics. (MIT, USA, 1986). Assoc. prof. (tel. 035 + 715131 ext. 652, fax 035 + 716770). *Small-angle neutron, x-ray scattering; colloid science.*

Liu, Prof. Ling-Kang. Inst. of Chemistry, Acad. Sinica, 128 Yen-Chiu-Yuan Rd., Sec. 2, Taipei, Taiwan 11529, ROC. (1950) PhD, physical chemistry (U. Texas, Austin, USA, 1978). Res. fellow. (tel. 02 + 7821889, fax 02 + 7831237, email wc7a0001 at twnmoe10). *Structure analysis, organic and organometallic compounds, phase problem.*

Lu, Prof. Tian-Huey. Physics Dept., Nat. Tsing Hua U., 101 Kuang-Fu Rd., Sec. 2, Hsinchu, Taipei, Taiwan 30043, ROC. (1939) MSc, Physics (Nat. Tsing Hua U., 1965). Prof. (tel. 035 + 715131, ext. 430, fax 035 + 710776, email c47b0005 at twnmoe10). *Structure analysis, protein crystallography, coordination compounds, computer programming.*

Ma, Dr Che-Bao. Inst. of Nuclear Energy Res., A.E.C., P.O. Box 3, Lung-tan, Tao-yuan, Taiwan 32526, ROC. (1941) PhD, geological sciences (Harvard U., USA, 1973). Sr. scient. (tel. 02 + 3145384, ext. 2310, fax 03 + 4706519). *Phase equilibria and structure analysis.*

Peng, Prof. Shie-ming. Chemistry Dept., Nat. Taiwan U., Taipei, Taiwan 10764, ROC. (1949) PhD, inorganic chemistry (U. Chicago, USA, 1975). Prof. (tel. 02 + 3630231 ext. 2111, fax 02 + 3935359). *Inorganic chemistry, crystal structure analysis.*

Shen, Prof. Pooyan. Inst. of Materials Sci. and Eng., Nat. Sun Yat-Sen U., Kaosiung, Taiwan 80424, ROC. (1952) PhD, geophysics (Cornell U., USA, 1982). Prof. (tel. 07 + 5316171, ext. 456). *Mineral physics, intermetallic compounds, electron diffraction, electron microscopy.*

Shiu, Prof. Kom-Bei. Chemistry Dept., Nat. Cheng Kung U., Tainan, Taiwan 70101, ROC. (1951) PhD, inorganic chemistry (U. Michigan, USA, 1984). Assoc. prof. (tel. 06 + 2361111 ext. 839). *Structure analysis, inorganic and organometallic chemistry.*

Tang, Dr Chia-Pin. Chemistry Dept., Chung-Shan Inst. of Sci. and Techn., P.O.Box 1-4, Lung-Tan, Taiwan 32526, ROC. (1946) PhD, structural chemistry (Weizmann Inst. of Sci., Israel, 1979). Assoc. sci. (tel. 02 + 3931621). *Structure analysis, organic and inorganic compounds, solid state chemistry.*

Tseng, Prof. Poh-Kun. Physics Dept., Nat. Taiwan U., Taipei, Taiwan 10764, ROC. (1930) PhD, nuclear eng. (U. Michigan, USA, 1968). Prof. (tel. 02 + 3630231 ext. 2321). *Solid state physics, x-ray instrumentaion.*

Ueng, Prof. Chuen-Her. Chemistry Dept., Nat. Taiwan Normal Univ., 88 Roosevelt Rd., Sec. 5, Taipei, Taiwan, 11718, ROC. (1948) PhD, inorganic chemistry (Nat. Taiwan U., 1987) Assoc. Prof. (tel. 02 + 2536935). *Structure analysis and electron density maps, organic, inorganic, and coordination compounds.*

Wang, Mr. Jinnlung. Chemical Systems Res. Div., CSIST, 2 Chung Shang Rd., Chia An Village, Lung-Tan, Taiwan, 32500, ROC. (1952) MSc, structure chemistry (Weizmann Inst. of Sci., Israel, 1983). Asst. scient., (tel. 03 + 4713755, fax 034 + 250085). *Structure analysis.*

Wang, Prof. Ju-Chun. Chemistry Dept., Soochow U., Taipei, Taiwan, 11102, ROC. (1958) PhD, inorg. chemistry (Louisiana State U., USA, 1987). Assoc. Prof. (tel.02 + 8819471, ext. 329, 339, email scut001 at twnmoe10, bit005 at twnscu10). *Solids state physics and chemistry, quantum chemistry.*

Wang, Prof. Sue-Lein. Chemistry Dept., Nat. Tsing Hua U., Hsinchu, Taiwan, 30043, ROC. (1953) PhD, physical chemistry (Iowa State U., USA, 1985). Assoc. Professor (tel. 035 + 716640, fax 035 + 711082). *Crystal structure analysis, powder XRD, Rietveld analysis, EXAFS techniques.*

Wang, Prof. Yu. Chemistry Dept., Nat. Taiwan U., Taipei, Taiwan 10764, ROC. (1943) PhD, chemistry (U. Illinois, USA, 1973). Prof. (tel. 02 + 3630231, ext. 2325, fax 02 + 3935359, email ac7b0001 at twnmoe10). *Computational crystallography, metal alloys, transition metal complexes, computerized data file base.*

Wu, Prof. Nan-Chung. Material Sci. Dept., Nat. Cheng Kung U., Taiwan 70101, ROC. (1936) MSc, physics (Tohoku U., Japan, 1972). Assoc. prof. (tel. 06 + 2361111 ext. 556). *X-ray diffraction, electronic ceramics.*

Wu, Prof. Shyi-Kaan. Graduate Inst. of Materials Eng., Nat. Twiwan U., 1 Roosevelt Rd., sec. 4, Taipei, Taiwan 10764, ROC. (1950) PhD, physical metallurgy (U. Illinois, Urbana-champaign, USA, 1986). Professor. (tel. 02 + 3630231 ext. 2701, fax 02 + 3514562). *Martensitic transformation, structure analysis, by transmission electron microscopy .*

Yang, Prof. Houng-Yi. Earth Sci. Dept., Nat. Cheng Kung U., Ta-Hsue Rd., Tainan, Taiwan 70101, ROC. (1938) PhD, mineralogy (Ohio State U., USA, 1970). Prof. (tel. 06 + 2361111, ext. 408). *Mineralogy, petrology.*

Yang, Prof. Tse Chun. Soil Sci. Dept., Nat. Chung Hsing U., 250 Kuo-Kuang Rd., Taichung, Taiwan 40227, ROC. (1932) PhD, soil sci. (Michigan State U., USA, 1970). Prof. (tel. 04 + 2840373). *Soil physics, physical chemistry, mineralogy.*

Yang, Prof. Yui-Whei. Chemistry Dept., Chung-yuan Christian U., Chung-Li, Taiwan 32023, ROC. (1948) PhD, chemistry (SUNY-Buffalo, USA, 1976). Prof. (tel. 03 + 4563171, ext. 258). *X-ray crystallography, electron density.*

Yu, Prof. Shu-Cheng. Earth Sci. Dept., Nat. Cheng Kung U., Ta-Hsue Rd., Tainan, Taiwan 70101, ROC. (1942) PhD, mineralogy (Pennsylvania State U., USA, 1976). Prof. (tel. 06 + 2361111, ext. 409). *Structure, microstructure, minerals, ceramic sciences.*

THAILAND

Sub-Editor: **S. Pramatus**

Notes

1. International telephone country code - 66.

2. At a Thai University, persons can be appointed to the following academic positions: *lecturer, assistant professor, associate professor* and *professor*. The functions are approximately equivalent to the US system, but with *lecturer* equivalent to *instructor.*

Anantachai, Mrs Suda Yasarawana. Physics Dept., Chiangmai U., Huey Keow Road, Chiangmai 50002, Thailand. (1940) MSc, crystallography (London U., UK, 1973). Lect. (tel. 053 + 22-1934, ext. 51). *Damage in semiconductors.*

Anugul, Mrs Surang. Chemistry Dept., Chulalongkorn U., Phya Thai Road, Bangkok 10330, Thailand. (1935) MS, inorganic chemistry (Oregon State U., USA, 1961). Assoc. prof. (tel. 02 + 252-7019). *Inorganic structures, alloys.*

Busaracome, Mr Suwin. Geology Dept., Khon Kaen U., Khon Kaen 40002, Thailand. (1948) MSc, geochemistry (Victoria U. of Wellington, New Zealand, 1978). Lect. (tel. 043 + 23-6199, ext. 1320). *Crystal growth, mineralogy, petrology, gemmology, X-ray crystallography.*

Chaichit, Dr Narongsak. Physics Dept., Silpakorn U., Nakorn Pathom 73000, Thailand. (1947) PhD, inorganic chemistry - X-ray crystallography (Monash U., Australia, 1982). Lect. (tel. 034 + 25-5093, ext. 15, fax +66 34 255099). *Natural-products, organic and organometallic structures.*

Chaikum, Dr Nitirampai Latavalya. Chemistry Dept., Mahidol U., Rama 6 Road, Bangkok 10400, Thailand. (1947) PhD, X-ray crystallography (The Flinders U. of South Australia, Australia, 1976). Asst. prof. (tel. 02 + 246-1360, ext. 156). *Crystal structure analysis.*

Chaikum, Dr Nopadol. Chemistry Dept., Mahidol U., Rama 6 Road, Bangkok 10400, Thailand. (1949) PhD, geochemistry (Otago U., New Zealand, 1976). Assoc. prof. (tel. 02 + 246-1360, ext. 124). *Clays and clay minerals.*

Choosang, Mrs Pilai. Chemistry Dept., Chulalongkorn U., Phya Thai Road, Bangkok 10330, Thailand. (1937) BSc, chemistry (Chulalongkorn U., 1959). Asst. prof. (tel. 02 + 252-7019). *Alloy structures.*

Hoonnivathana, Mr Ekachai. Physics Dept., Kasetsart U., Phaholythin Road, Bangkok 10900, Thailand. (1956) MSc, physics (Chulalongkorn U.,1984). Lect. (tel. 02 + 579-5529). *Inorganic and organometallic crystal structures, physical metallurgy.*

Jinawath, Dr Supatra. Materials Sci. Dept., Chulalongkorn U., Phya Thai Road, Bangkok 10330, Thailand. (1945) PhD, mineral science (Leeds U., UK, 1974). Assoc. prof. (tel. 02 + 251-1954, fax 02 + 662 2155523). *Mineral science, ceramics, cement chemistry.*

Jirajesda, Mr Jate. Geological Survey Div., Dept. of Mineral Resources, Rama 6 Road, Bangkok 10400, Thailand. (1955) BSc, chemistry (Khon Kaen U., 1980). Scient. (tel. 02 + 282-1164). *Powder diffraction, fluorescence.*

Kamolchote, Mr Poonsak. Chemistry Dept., Silpakorn U., Nakorn Pathom 73000, Thailand. (1953) MSc, physical chemistry (Chiangmai U., 1977). Lect. (tel. 034 + 24-2072). *Polymer structures.*

Keankeo, Miss Watcharaporn. Geology Dept., Khon Kaen U., Khon Kaen 40002, Thailand. (1958) BSc, geology (Chiangmai U., 1980). Lect. (tel. 043 + 23-6199, ext. 1320). *Mineral crystallography.*

Keow-kam-nerd, Dr Kanchana. Chemistry Dept., Chiangmai U., Huey Keow Road, Chiangmai 50002, Thailand. (1937) PhD, chemical technology - inorganics (U. de Besancon, France, 1970). Assoc. prof. (tel. 053 + 22-1934, ext. 21). *Ceramics, silicate technology.*

Khantaprab, Dr Chaiyudh. Geology Dept., Chulalongkorn U., Phya Thai Road, Bangkok 10330, Thailand. (1942) PhD, sedimentology (geology) (Imperial C., London U., UK, 1972). Asst. prof. (tel. 02 + 252-5931). *Sedimentology, sediment crystallography, environmental geology.*

Kritayakirana, Mrs Rungsri. Physics Dept., Chulalongkorn U., Phya Thai Road, Bangkok 10330, Thailand. (1940) MS, physics (Northeastern U., USA, 1969). Asst. prof. (tel. 02 + 252-9987, fax 02 + 662 2155523). *Powder diffraction.*

Mitrprachachon, Dr Pachanee. Chemistry Dept., Khon Kaen U., Khon Kaen 40002, Thailand. (1950) PhD, inorganic chemistry - X-ray crystallography (Bristol U., UK, 1980). Lect. (tel. 043 + 23-7606, ext. 1269). *Crystal structure, crystallographic computing.*

Nimgirawath, Mrs Kloy. Physics Dept., Silpakorn U., Nakorn Pathom 73000, Thailand. (1944) MSc, X-ray crystallography (U. New South Wales, Australia, 1975). Asst. prof.(tel. 034 + 24-2072). *Organic and inorganic structures.*

Padmasuta, Mrs Soontari. Geological Survey Division, Dept. of Mineral Resources, Rama 6 Road, Bangkok 10400, Thailand. (1939) BSc, physics (Chulalongkorn U., 1963). Scient. (tel. 02 + 282-1164). *Mineral identification, powder diffraction, fluorescence.*

Pakawatchai, Dr Chaveng. Chemistry Dept., Prince of Songkla U., Haad Yai, Songkla 90112, Thailand. (1951) PhD, crystallography (U. Western Australia, Australia, 1984). Lect. (tel. 074 + 32-5800, ext. 2260, telex 62168). *Organic and inorganic structures.*

Phaovibul, Dr Orapin. Chemistry Dept., Mahidol U., Rama 6 Road, Bangkok 10400, Thailand. (1941) Dr.rer.nat., physics (The Free U., Berlin, BRD, 1971). Assoc. prof. (tel. 02 + 246-1360, ext. 147). *Physical properties, liquid crystals, polymers, structure - properties relations.*

Phavanantha, Dr Phathana. Physics Dept., Chulalongkorn U., Phya Thai Road, Bangkok 10330, Thailand. (1942) PhD, crystallography (Imperial C., London U., UK, 1970). Assoc. prof. (tel. 02 + 252-9987, fax 02 + 662 2155523). *Natural-products, organic and inorganic structures, applied crystallography.*

Pisutha-Arnond, Dr Visut. Geology Dept., Chulalongkorn U., Phya Thai Road,Bangkok 10330, Thailand. (1951) PhD, inorganic geochemistry (Penn State U., USA, 1982). Asst. prof. (tel. 02 + 252-5931). *Ore mineralogy, ore deposit research.*

Pongsapich, Dr Wasant. Geology Dept., Chulalongkorn U., Phya Thai Road, Bangkok 10330, Thailand. (1942) PhD, geology (U. Washington, USA, 1974). Asst. prof. (tel. 02 + 252-7989). *Chemical analyses, mineralogical determination.*

Pontchour, Miss Cha-on. Chemistry Dept., Chulalongkorn U., Phya Thai Road, Bangkok 10330, Thailand. (1937) BSc, chemistry (Chulalongkorn U., 1959). Asst. prof. (tel. 02 + 252-7019). *Inorganic structures, alloys.*

Pramatus, Miss Supanich. Physics Dept., Chulalongkorn U., Phya Thai Road, Bangkok 10330, Thailand. (1933) MSc, crystallography (U.C., London U., UK, 1968). Assoc. prof. (tel. 02 + 252-9987, fax 02 + 662 2554441). *Inorganic structures, powder diffraction.*

Puttajakr, Mrs Taswal. Physics Dept., King Mongkut's Inst. of Techn. Thonburi, Suksawad Road, Bangkok 10140, Thailand. (1955) MSc, high energy physics (Chulalongkorn U., 1982). Asst. prof. (tel. 02 + 427-0039, ext. 6203). *X-ray diffraction, gemmology, crystallization of materials.*

Ratanasthien, Dr Benjavun. Geological Sci. Dept., Chiangmai U., Huey Keow Road, Chiangmai 50002, Thailand. (1946) PhD, geochemistry (Aston U., Birmingham, UK, 1975). Assoc. prof. (tel. 053 + 22-1699, ext. 129). *Clays, clay minerals, X-ray and electron diffraction, crystallography.*

Rukvichal, Mr Surapol. Applied Physics Dept., King Mongkut's Inst. of Techn., Bangkok 10520, Thailand. (1950) MSc, physics (Chulalongkorn U., 1976). Asst. prof. (tel. 02 + 326-7320-9, ext. 395, 396). *Inorganic structures, alloys.*

Sangariyavanich, Mrs Archara. Physics Dept., Office of Atomic Energy for Peace, Vibhavados Road, Bangkok 10900, Thailand. (1948) MSc, solid state physics (Chulalongkorn U., 1973). Nuclear physicist (tel. 02 + 579-0138-9, ext. 332). *Material science.*

Satittada, Miss Gannaga. Physics Dept., King Mongkut's Inst. of Techn. Thonburi, Suksawad Road, Bangkok 10140, Thailand. (1953) MSc, physics (Chulalongkorn U., 1977). Asst. prof. (tel. 02 + 427-0039, ext. 714). *X-ray crystallography, powder diffraction.*

Silskulsuk, Mr Buncha. Physics Dept., Srinakharinwirot U., Sukumwit soi 23, Bangkok 10110, Thailand. (1955) MSc, physics (Chulalongkorn U., 1984). Instr. (tel. 02 + 258-3989). *Alloys.*

Siripaisarnpipat, Dr Sutatip. Chemistry Dept., Kasetsart U., Phaholyothin Road, Bangkok 10900, Thailand. (1947) PhD, inorganic chemistry (U. of Missouri-Columbia, USA, 1981). Asst. Prof. (tel. 02 + 579-0658). *Crystal structure, organic compounds, coordination compounds.*

Siripitayanon, Dr Jintana. Chemistry Dept., Fac. of Sci., Chiangmai U., Chiangmai 50002, Thailand. (1954) PhD, crystallography (Australian Nat. U., Australia, 1986). Lect. (tel. 053 + 22-1699, ext. 3331). *X-ray diffraction, polymers, superconductors, diffuse scattering.*

Siriratwatanakul, Mr Narin. Physics Dept., Chiangmai U., Huey Keow Road, Chiangmai 50002, Thailand. (1952) MSc, physics (Chiangmai U., 1981). Asst. prof. (tel. 053 + 22-1934, ext. 51). *Damage in semiconductors.*

Suddhiprakarn, Dr Anohsloe. Dept. of Soils, Kasetsart U., Bankok 10903, Thailand. (1948) PhD, minerology (U. of Western Australia at Perth, Australia, 1978). Asst. prof. (tel. 02 + 579-2028, fax +66 2 5799538). *Minerology.*

Sukapaddhanadhi, Mr Narong. Eng. Techn. Service, Thai Oil Refinery Co. Ltd., Km 124 1/2 Sukhumvit Road, Au Udom, Sriracha, Chonburi 20210, Thailand. (1941) MSc, X-ray crystallography (London U., UK, 1970). Sr. Metallurgist. (tel. 038 + 31-1070, ext. 1511). *Metallography, electron diffraction, dislocations, polycrystalline texture.*

Thanomkul, Dr Srinuan Chaiwasie. Physics Dept. Chulalongkorn U., Phya Thai Road, Bangkok 10330, Thailand. (1936) Dr ing., X-ray crystallography (U. Trondheim, Norway, 1974). Assoc. prof. (tel. 02 + 252-9987, fax 02 + 662 2554441). *Organic and inorganic structures.*

Thinapong, Dr Pongchan Chananont. Chemistry Dept., Mahidol U., Rama 6 Road, Bangkok 10400, Thailand. (1948) PhD, chemistry (Birmingham U., UK, 1981). Lect. (tel. 02 + 246-1360, ext. 154). *Structure and activity, biologically active compounds.*

Tontrakoon, Mr Jeerapong. Physics Dept., Chiangmai U., Huey Keow Road, Chiangmai 50002, Thailand. (1950) MSc, physics (Chiangmai U., 1978). Asst. prof. (tel. 053 + 22-1934, ext. 51). *Damage in semiconductors.*

Tooptakong, Dr Uncharee Methong. Chemistry Dept., Silpakorn U., Nakorn Pathom 73000, Thailand. (1954) PhD, X-ray crystallography (Australian Nat. U., 1985). Lect. (tel. 034 + 25-5797, fax +66 34 255099). *Organic and inorganic structures.*

Trechairusma, Mr Kamchai, Physics Dept., Silpakorn U., Nakorn Pathom 73000, Thailand. (1955) MSc, physics (Chulalongkorn U., 1985). Lect. (tel. 034 + 25-5093). *Inorganic structures.*

Tunkasiri, Dr Tawee. Physics Dept., Chiangmai U., Huey Keow Road, Chiangmai 50002, Thailand. (1943) PhD, physics (U. Surrey, UK, 1975). Assoc. prof. (tel. 053 + 22-1934, ext. 51). *Damage in semiconductors.*

Uttamasil, Dr Lek. Res. Inst. of Metal and Material Sci., Chulalongkorn U., Phya Thai Road, Bangkok 10330, Thailand. (1943) PhD, ceramic eng. (Ohio State U., USA, 1971). Asst. prof., dir. (tel. 02 + 251-1954, fax 02 + 662 2155523). *Clay minerals, high temperature materials.*

Wongshaiboon, Dr Sajee. Physics Dept., Chulalongkorn U., Phya Thai Road, Bangkok 10330, Thailand. (1946) PhD, chemistry (Uppsala U., Sweden, 1981). Asst. prof. (tel. 02 + 252-9987, fax 02 + 662 2554441). *General crystallography, crystal structure, physical properties.*

TUNISIA

Sub-Editor: **M. Ghedira**

Notes

1. International telephone country code - 216.

2. The degrees granted and functional positions at Tunisian universities are patterned upon the French academic designations.

Amara, Dr Mongi. Faculté de Pharmacie, Dépt de Chimie, 5000 Monastir, Tunisia. Dr d'Etat (U. Bordeaux, France) Maître de conf. *Structures, phase transitions.*

Ariguib, Dr Najia. Directrice I.N.R.S.T. Borj Cédria, B.P. 95, Hammam-Lif, Tunisia. (1937) DSc, chemistry (U. Paris, France). Prof. *Structures, phase transitions.*

Belhadj, Dr Ali. Dépt. de Physique, Ecole Normale Supérieure, 7029 Bizerte, Tunisia. (1938) DSc, physics (U. Paris, France, 1974). Prof. *Clays, structure.*

Bel Hassen, Miss Dalila. Dépt. de Chimie, Ecole Normale Supérieure, 7029 Bizerte, Tunisia. (1955) Dipl., chemistry (Ecole Normale Supérieure de Tunis). Asst. *Structures, phase transitions.*

Belkheria, Mr Salah. Dépt. de Chimie, Faculté des Sci. et Techniques, 5000 Monastir, Tunisia. DEA (U. Tunis). Asst. *Structures.*

Ben Amor, Dr. Dépt. de Chimie, Ecole Nat. d'Ingénieurs, 6029 Gabès, Tunisia. Dr 3e cycle. Maître-asst.

Ben Brahim, Dr Jemaïel. Dépt. de Physique, Ecole Normale Supérieure, 7029 Bizerte, Tunisia. (1954) Dr 3e cycle, physique (Faculté des Sci. de Tunis). Maître-asst. *Clays, structure.*

Ben Ghozlane, Dr Hédi. Dépt. de Physique, Ecole Nat. d'Ingénieurs de Sfax, Route de la Soukra, B.P.W. 3038 Sfax, Tunisia. (1951) DSc, physique (Faculté des Sci. de Tunis). Maître de conf. *Complexes, structure.*

Ben Romdhane, Dr. Dépt. de Chimie, Ecole Nat. d'Ingénieurs, 6029 Gabès, Tunisia. Dr d'Etat (France). Maître-asst.

Ben Salah, Dr Abdelhamid. Dépt. de Chimie, Ecole Nat. d'Ingénieurs de Sfax, Route de la Soukra, B.P.W. 3038 Sfax, Tunisia. (1950) DSc, physique. Maître de conf. (tel. 216 - 42088, ext. 04). *Structure determination, phase transitions.*

Billiet, Prof. Yves. Dépt. de Physique, Ecole Nat. d'Ingénieurs, Route de la Soukra, B.P.W. 3038 Sfax, Tunisia. (1936) DSc, sciences physiques (U. Paris Sud, France, 1969). Prof. (tel. 216 - 4 41430). *Symmetry, group theory, phase transition, teaching.*

Bizid, Dr Abdelmalek dit Youssef. Dépt. de Physique, Faculté des Sci. de Tunis, Tunis, Tunisia. (1940) DSc, physics (U. Paris VI, France). Prof. *Structures.*

Bouraoui, Prof. Ahmed. Dépt. de Physique, Ecole Normale Supérieure, 7029 Bizerte, Tunisia. (1931) DSc, physiques (U. Paris, France, 1965). *Radiocrystallography, solid state physics.*

Boutiba, Dr Samia. Dept. Physique, Ecole Normale Supérieure, 5 avenue Tah Hassis, Tunis, Tunisia. Dr 3e cycle (U. Tunis) (tel. (01) 491272). *Structures, composés pharmaceutiques.*

Cheikhrouhou, Dr Abdelwaheb. Dépt. de Physique, Ecole Nat. D'Ingénieurs, Route de la Soukra, B.P.W. 3038 Sfax, Tunisia. (1948) Dr d'Etat es Sci., physique (Faculté de Sci. de Tunis). Maître de conf. *Complexes, structure.*

Dabbabi, Dr Mongi. Dépt. de Chimie, Faculté des Sci. et Techniques, 5000 Monastir, Tunisia. Dr d'Etat (U. Paris, France). Prof. *Structure determination.*

Damak, Dr Mabrouk. Dépt. de Physique, Ecole Nat. d'Ingénieurs de Sfax, Route de la Soukra, B.P.W. 3038 Sfax, Tunisia. (1943) Dr Ing (U. Dijon, France). Maître asst. *Structures.*

Daoud, Prof. Abdelaziz. Dépt. de Chimie, Ecole Nat. d'Ingénieurs, Route de la Soukra, B.P.W. 3038 Sfax, Tunisia. (1939) Dr d'Etat, crystallography (U. Dijon, France, 1970). Prof. (tel. 216 - 42088, ext. 04). *Structures, phase transitions.*

Driss, Dr Ahmed. Dépt. de Chimie, Fac. des Sci. de Tunis, Tunis, Tunisia. (1950) Dr 3e cycle, chimie (Fac. des Sci. de Tunis). Maître-asst. *Structure determination.*

Fakhar, Miss Noura. Dépt. de Chimie, Fac. des Sci. de Tunis, Tunis, Tunisia. (1955) MSc, chemistry (Fac. des Sci. de Tunis). Asst. *Structure determination.*

Ghedira, Dr Mounir. Dépt. de Physique, Faculté des Sci. et Techniques, 5000 Monastir, Tunisia. (1951) Dr d'Etat es Sci., physique (U. Sci. et Medicale, Grenoble, France) Maître de conf. (tel. 216 + 361766). *Structure determination, phase transitions, charge localization.*

Harzallah, Miss Besma. Dépt Physique, Faculté des Sciences, Monastir 5000, Monastir, Tunisia. Asst. *Structures.*

Jouini, Dr Amor. Dépt. de Chimie, Faculté des Sci. et Techniques, 5000 Monastir, Tunisia. Dr 3e cycle (U. Dijon, France). Maître-asst. *Structure determination.*

Jouini, Dr Noureddine. Ecole Normale Supérieure Tunis, 5 Avenue Tah Hassin, Tunisia. Dr d'Etat, Maître de conf. *Structure.*

Jouini, Dr Tahar. Dépt. de Chimie, Faculté des Sci. de Tunis, Tunis, Tunisia. (1939) DSc, chemistry (France). Prof. *Structure determination.*

Kallal, Dr Ahmed. Dépt. de Physique, Faculté des Sci. de Tunis, Tunis, Tunisia. (1937) DSc, physics (U. Grenoble, France). Prof. *Structures.*

Kamoun, Dr Slaheddine. Dépt de Physique, Faculté des Sciences, 5000 Monastir, Tunisia. Dr 3r cycle(U. Paris VI France) *Structures, phosphates organiques.*

Maaref, Dr. Saida. Dépt. de Chimie, Ecole Nat. d'Ingénieurs, 6029 Gabès, Tunisia. Dr 3e Cycle. Maître-asst. *Radio-crystallography.*

Mhiri, Dr Tahar. Dépt. de Chimie, Ecole Nat. d'Ingénieurs de Sfax, Route de la Soukra, B.P.W. 3038 Sfax, Tunisia. Dr 3e cycle (U. Paris, France). Maître asst. *Structures.*

Mlik, Dr Youssef. Dépt. de Physique, Ecole Nat. d'Ingénieurs de Sfax, Route de la Soukra, B.P.W. 3038 Sfax, Tunisia. (1950) Dr d'Etat es Sci., physique (Faculé des Sci. de Tunis). Maître de conf. *Complexes, structure.*

Oumezzine, Dr Belgacem. Dépt. de Chimie, Faculé des Sci. de Tunis, 1060 Tunis, Tunisia. (1943) Dr 3e cycle, chimie (Ecole Normale Supérieure). Asst. *Structures, phase transitions.*

Oumezzine, Dr Mohamed. Dépt de Physique, Faculté des Sciences, Monastir, Tunisia. Dr 3e cycle (Paris VI, France). *Structures, composés amorphes.*

Omrani, Dr Hédi. Dépt. de Chimie, Faculté des Sci. et Techniques, 5000 Monastir, Tunisia. Dr 3e cycle (U. Besacon, France) Maître-asst. *Structure determination.*

Rzaigui, Dr Mohamed. Dépt. de Chimie, Ecole Normale Supérieure, 7029 Bizerte, Tunisia. (1948) Dr d'Etat, chemistry (France). Maître-asst. *Structures, phase transitions.*

Soua, Dr Moncef. Dept chimie Institut Textiles, 5000 Ksarhellal DEA, Faculté des Sci., Tunis, Tunisia. *Structures.*

Trabelsi, Dr Malika. Dépt. de Chimie, Ecole Normale Supérieure, 7029 Bizerte, Tunisia. (1947) DSc, chemistry (France). Prof. *Structures, phase transitions.*

Zemni, Dr Sadok. Dept. Physique, Faculté des Sci., Monastir 5000, Tunisia. Dr 3e cycle (U. Grenoble, France). *Structures, macles.*

TURKEY

Sub-Editor: **N. Armağan**

Notes

1. International telephone country code-90.

2. O.D.T.Ü. (Orta Doğu Teknik Üniversitesi) is the Turkish equivalent of M.E.T.U. (Middle East Technical University).

Adigüzel, Dr Osman. Fizik Böl., Firat Ü. Fen-Ed. Fak., Elazığ, Turkey. (1952) PhD, physics (Diyarbakir U., 1980). Asst. prof. (tel.811-12407), ext.259). *Phase transformations, metal and shape memory alloys, crystal structure, powder diffraction.*

Akalan, Prof. İlhan. Toprak Bilimi Böl., Ankara Ü. Ziraat Fak., Ankara, Turkey. (1955) Prof, soil science (Ankara U., 1964). Lect. (tel. 4-13472100, ext.182). *Clay mineralogy.*

Akgül, Mr Muharrem. Jeoloji Müh. Böl., Firat Ü. Müh-Mim. Fak., Elazığ, Turkey. (1962). MSc, petrography and petrology (Karadeniz U.). Asst. (tel. 811-11904, ext.244). *Mineral deposits, ore microscopy, crystallography, X-ray diffraction.*

Akgün, Dr İlhan. Fizik Böl., Gazi Ü. Fen-Ed. Fak., Ankara, Turkey. (1955) PhD, physics (Firat U., 1981). Asst. Prof. (tel. 4-1135538). *Phase transformations, metals.*

Akkurt, Yrd. Doç. Dr Mehmet. Fizik Böl., Erciyes Ü. Fen-Ed. Fak., Talas Yolu, 38039 Kayseri, Turkey. (1958) PhD, solid state physics (Erciyes U.). Lect. (tel.351-74901). *X-ray diffraction, crystal structure determination.*

Aksoy, Yrd. Doç. Dr İlhan. Fen Bilimleri Eğitimi Böl., İnönü Ü. Eğitim Fak. Kampus, 44000 Malatya, Turkey. (1958) PhD, physics (İnöü U.). Lect. (tel. 821-21871). *Clay minerals, X-ray diffraction.*

Akyüz, Dr Tanil. Spektro-Kimya, M.T.A., Ankara, Turkey. (1941) MSc, X-rays (Ankara U., Faculty of Science). Engineer (tel.4-12873430, ext.669). *XRF, XRD, IR, clay minerals.*

Alpaslan, Mr Musa. Jeoloji Müh. Böl., Cumhuriyet Ü., 58140 Sivas, Turkey. (1962) MSc, geology (Cumhuriyet U., 1987). Asst. (tel. 477-61527, ext.118, fax477-61513). *Mineralogy, petrography, X-ray diffraction.*

Alpaut, Prof. Dr Okyay. Kimya Böl., Hacettepe Ü. Müh. Fak., Ankara, Turkey. (1928) Prof, physical metallurgy (Hacettepe U.,1971). Dean. (tel.4-2235164). *Physical chemistry, metals.*

Arikan, Doç. Dr Rafet. Makina Böl., Gazi Ü. Müh-Mim. Fak., Maltepe, Ankara, Turkey. (1948) PhD, material science (ODTU). Lect. (tel. 4-12317400, ext. 72). *Phase transformations, fracture.*

Armağan, Nizamettin. Fizik Böl., Selçuk Ü. Fen-Ed. Fak., 42049 Kampus, Konya, Turkey. (1942) PhD, physics (Wales U., UK, 1970). Prof. (tel. 33-182663, ext. 412). *Crystal structure, thermal vibrations, powder diffraction, crystallographic programming, clay minerals.*

Artunç, Dr Ekrem. Fizik Boe.l., Firat Ü. Fen-Ed. Fak., 23169 Elazığ, Turkey. (1955) PhD, physics (Firat U., 1988). Lect. (tel. 811-12407, ext.66). *Crystallographical analysis, phase transformations.*

Aslaner, Prof. Dr Mustafa. Jeoloji Müh., Karadeniz Teknik Ü. 61080 Trabzon, Turkey. (19340 Prof, mineralogy, petrography (K.T.U.). Lect. (tel. 031-16920). *Mineralogy, petrography, optic mineralogy, geochemistry.*

Atasoy, Doç. Dr Ö. Aydin. Metalurji Müh. Böl., İstanbul Teknik Ü. Maslak, İstanbul, Turkey. (1950) PhD, materials science (Manchester, UK). Lect. (tel. 1-1763387, telex 28186 ITU TR.). *Solidification, crystal growth, transmission electron microscopy.*

Aydinol, Doç. Dr Mahmut. Fizik Böl., Dicle Ü. Fen-Ed. Fak., Diyarbakir, Turkey. (1948) PhD, atomic and molecular physics (Stirling U., Scotland, UK, 1980). Lect.(tel. 831-13581, ext. 227). *Multipurpose electron-atom (molecule) crossed beam system, X-ray spectroscopic methods, ESCA, AES, XRFS.*

Aydinuraz, Prof. Dr Arsin. Fizik Böl., Ankara Ü. Fen Fak., Beşevler, Ankara, Turkey. (1941) Prof., physics (Ankara U.,1987). Lect. (tel. 4-1236550, ext 145). *X-ray diffraction analysis.*

Ayhan, Doç. Dr Ahmet. Jeoloji Müh. Böl., Selçuk Ü. Müh-Mim. Fak., 42040 Konya, Turkey. (1950) PhD, mineral deposits-geochemistry (Ruprecht-Karlu U., Heidelberg, BRD).Lect. (tel. 33-115952). *Mineral deposits, mineralogy-petrography, geochemistry.*

Aypar, Prof. Dr Abidin. Fizik Böl., Selçuk Ü., Fen-Ed. Fak., 42049 Kampus, Konya, Turkey. (1943) PhD, physics (Durham U., UK.). Head of Physics Dept. (tel. 33-182663, ext. 411). *Electron spin resonance, crystal growth, thermoluminescence dosimetry, piezoelectric ceramic, transducers, optical properties.*

Aytaş, Doç. Dr S. Işik. Fizik Böl., Marmara Araş. Enst., Gebze, İzmit, Turkey. (1942) PhD, physics (Surrey U., UK.,1971). Res. scient. *Lattice defects.*

Bayhan, Doç. Dr Hasan. Jeoloji Müh. Böl., Hacettepe Ü. Müh. Fak., 06532 Beytepe, Ankara, Turkey. (1950) PhD, geology (Hacettepe U., 1980). Lect. (tel. 4-2873060, ext.1113, telex 42237 htk tr). *Igneous petrology, mineralogy.*

Bayvas, Dr Fehime. Fizik Böl., TAEK, ANAEM, 06100 Beşevler, Ankara, Turkey. (1935) PhD, physics (İstanbul U., 1981). Res. scient. (tel. 4-2126230, ext. 95). *Crystallography, X-ray diffraction, neutron diffraction.*

Bingöl, Doç. Dr A. Feyzi. Jeoloji Müh. Böl., Firat Ü., 23300 Elazığ, Turkey. (1952) PhD, mineralogy-petrography (Strassbourg-Louis Pasteur U., France).Lect. (tel. 811-11904, ext.257). *Magmatic rocks, mineralogy.*

Birsoy, Doç. Dr Rezan. Jeoloji Müh. Böl., Dokuz Eylül Ü., 35102 Bornova, İzmir, Turkey. (1947) Doç, mineralogy (Dokuz Eylül U.). Lect. (tel. 51-180110, ext.2426). *Crystal defects, physical chemistry, mineral phases.*

Boztuğ, Yrd. Doç. Dr Durmuş. Jeoloji Müh. Böl., Cumhuriyet Ü., 58140 Sivas, Turkey. (1957) PhD, geology (Hacettepe U., 1980). Lect. (tel. 477-61527, ext.112,fax477-61513). *Mineralogy,petrography, clay minerals, X-ray diffraction.*

Büyükgüngör, Dr Orhan. Fizik Böl., Ondokuz Mayis Ü. Fen-Ed. Fak., Samsun, Turkey. (1954) PhD, physics (hacettepe U.,1983). Asst. Prof. (tel. 361-19680, ext. 656). *Crystallography.*

Ceylan, Doç. Dr Mehmet. Fizik Böl., Firat Ü. Fen-Ed. Fak., 23169 Elazığ, Turkey. (1950) PhD, physics (Diyarbakir U.,1980).Lect. (tel. 811-12407, ext. 60). *Crystal structure, austenite-martensite phase transformation, dislocations, electron microscopy.*

Ceylan, Dr Kazim. Fizik Böl., Gazi Ü. Fen-Ed. Fak., Beşevler, Ankara, Turkey. (1952) PhD, physics (A.D.M.M.A., 1981). Lect. *Phase transformations, metals.*

Çakmak, Mr Seyfettin. Fizik Böl., Firat Ü. Fen-Ed. Fak., 23169 Elazığ,Turkey. (1961) MSc, physics (Firat U., 1987). Asst. (tel. 811-12407, ext. 66). *Phase transformation, shape memory effect.*

Çalişkan, Dr Nezihe. Fizik Böl., İnönü Ü. Fen-Ed. Fak., Malatya, Turkey. (1956) PhD, physics (İnönü U., 1983). Asst. Prof. (tel. 821-21871, ext. 274). *Clay minerals, X-ray diffraction.*

Çolakoğlu, Doç. Dr Kemal. Fizik Böl., Gazi Ü. Fen-Ed. Fak., Ankara, Turkey. (1947) PhD, physics (Diyarbakir U., 1978). Lect. (tel. 4-1135538). *Crystallography.*

Danaci, Dr Süheyla. Fizik Müh. Böl., Hacettepe Ü. Müh. Fak., 06532 Beytepe,Ankara, Turkey. (1957) PhD, physics (Hacettepe U., 1988). Asst. (tel. 4-223039, telex 422237 htk tr). *Crystal structure analysis, biological small molecules, powder diffraction.*

Dikici, Doç. Dr Mustafa. Fen Bilimleri Eğtimi Böl., Ondokuz Mayis Ü. Eğitim Fak., 55200 Atakum, Samsun, Turkey. (1946) PhD, physics (Firat U., 1982). Lect. (tel. 361-12586). *Phase transformations, elastic fields.*

Dinçer, Dr Muharrem. Fizik Böl., Ondokuz Mayis Ü. Fen-Ed. Fak., Samsun, Turkey. (1955) PhD, physics (Firat U., 1981). Asst. Prof. (tel. 361-19680). *Martensitic phase transformations, alloys, crystal structure, powder diffraction.*

Doğan, Dr Ali. Fizik Boe.l., Firat Ü. Fen-Ed. Fak., 23169 Elazığ, Turkey. (1952) PhD, physics (Firat U., 1982). Lect. (tel. 811-12407, ext. 68). *Phase transformations, kinetics, nucleation, growth, defects, thermodynamics.*

Dora, Prof. Dr Özcan. Jeoloji Müh. Böl., Dokuz Eylül Ü., 35102 İzmir, Turkey. (1935) Prof. feldspar mineralogy (Ege U.). Lect. (tel. 51-180110. ext. 2429). *X-ray diffraction analysis, metamorphic petrology, feldspar mineralogy.*

Durlu, Prof. Dr Tahsin Nuri. Fizik Boe.l., Ankara Ü. Fen Fak., 06100 Tandoğan, Ankara, Turkey. (1945) PhD, physical metallurgy (Oxford U., UK., 1974). Prof. (tel. 4- 2126720, exr. 1245). *Electron densities, crystal structure.*

Dündar, Yrd. Doç. Dr Sacit. Metalurji Müh., İstanbul Teknik Ü. Sakarya Müh. Fak., Adapazari Turkey. (1951) PhD, material science (Leeds U., UK.). Lect. (tel. 261-16044). *X-ray diffraction, texture, electron microscopy, metals and alloys.*

Elmali, Mr Ayhan. Fizik Müh. Böl., Ankara Ü. Fen Fak., 06100 Tandoğan, Ankara, Turkey. (1965) MSc, physics (Ankara U., 1986).Asst. (tel. 4-2127620, ext. 1215). *Crystal structure determination.*

Elerman, Doç. Dr Yalçin. Fizik Müh. Böl., Ankara Ü. Fen Fak., 06100 Tandoğan, Ankara, Turkey. (1951) PhD, physics (Ankara U., 1978). Assoc. Prof. (tel. 4-2126720, ext. 1245). *Crystal structure, electron densities.*

Erdönmez, Dr Ahmet. Fizik Böl., Ondokuz Mayis Ü. Fen-Ed. Fak., Samsun, Turkey. (1950) PhD, physics (Hacettepe U., 1980). Asst. Prof. (tel. 361-19747). *Crystal structure, powder diffraction, crystallographic computer programs.*

Ergin, Yrd. Doç. Dr Ömer. Fen Bilimleri Eğitimi Böl., Uludağ Ü. Necati Bey Eğitim Fak., Kasaplar Mah., Soma Cad., 10100 Balikesir, Turkey. (1949) PhD, crystal structure analysis (Atatürk U.). Asst. Prof. (tel. 661-12762). *Structure determination.*

Gedikoğlu, Prof. Dr Atasever. Jeoloji Müh. Böl., Akdeniz Ü. İsparta Müh. Fak., Kampus, Cünür, 32260 İsparta, Turkey. (1942) Prof., petrography (Akdeniz U.).Lect. (tel. 327-15628, ext. 14, fax 327-13514). *Mineral deposits, petrography, optical mineralogy, X-ray fluorescence, X-ray diffraction.*

Gezci, Doç. Dr Sami. Fizik Böl., İ.T.Ü. Fen-Ed. Fak., 80626 Maslak, İstanbul, Turkey. (1940) Doç., solid state physics (İ.T.U., 1981).Lect. (tel. 91-1763205, telex 28186 itu tr). *Crystal growth, II-VI compounds, X-ray diffraction, lattice defects, thermoluminescence, thin films.*

Girgin, Doç. Dr İsmail. Maden Müh., Hacettepe Ü., 06532 Beytepe, Ankara,Turkey. (1950) Doç., mineral processing-chemical mining (Hacettepe U.). Lect. (tel. 4-2873060, ext. 1148). *Mineral processing, chemical mining, X-ray diffraction.*

Güçer, Prof. Dr Şeref. Kimya Boe.l., İnönü Ü. Fen-Ed. Fak., Malatya, Turkey. (1946) Prof., analytical chemistry (İnönü U.). lect. (tel. 821-21871, ext. 2353, fax 66140). *Atomic spectroscopy, X-ray diffraction, micro methods, trace analysis.*

Güler, Yrd. Doç. Dr Hülya. Kimya Böl., Cumhuriyet Ü. Fen-Ed. Fak., 58140 Sivas, Turkey. (1950) PhD, DTA (Hacettepe U.). Lect. (tel. 477-61724). *Clay analysis, physical metallurgy.*

Günel, Doç. Dr Gülten. Fizik Böl., Çukurova Ü. Fen-Ed. Fak., 01330 Balcali, Adana, Turkey. (1943) PhD, solid state physics (Ankara U.). Lect. (tel. 71-133394, ext. 2470). *X-ray crystallography, crystal structure analysis.*

Güven, Doç. Dr Olgun. Kimya Böl., Hacettepe Ü., 06532 Beytepe, Ankara, Turkey. (1947) PhD, physicalchemistry (Hacettepe U.). Lect. (tel. 4-2873060, ext. 1339, telex 42237). *Polymer crystals, crystallisation kinetics, morphology.*

Hökelek, Dr Tuncer. Fizik Müh. Böl., Hacettepe Ü. Müh. Fak., 06532 Beytepe, Ankara, Turkey. (1957) PhD, physics (Hacettepe U., 1986). Asst. (tel. 4-2230391, telex 42237 htk tr). *Crystal structure, organometallic compounds, powder diffraction, thin films.*

İnce, Prof. Dr Faruk. Toprak Böl., Dicle Ü. Ziraat Fak., 63 Şanliurfa, Turkey. (1944) Prof., soil sciences (Y.ÖK.). Prof. (tel. 12829-21). *Earth micromorphology, clay mineralogy, earth genesis and classification.*

İşci, Doç. Dr Coşkun. Elektrik ve Elektronik Böl., Dokuz Eylül Ü. Müh-Mim. Fak., Bornova, İzmir, Turkey. (1950) PhD, solid state physics (Hull U., UK., 1977). Lect. (tel. 51-182880). *Crystal growth, crystal structure, material science, phase transitions.*

Karapinar, Mr Ridvan. Fizik Böl., Yüzüncü Yil Ü. Fen-Ed. Fak., Van, Turkey. (1966) (tel. 061-18374). Asst. *Phase transformations, liquid crystals, optical properties, nematic liquid crystals.*

Kendi, Doç. Dr Engin. Fizik Müh. Böl., Hacettepe Ü. Müh. Fak., 06532 Beytepe, Ankara, Turkey. (1945) PhD, physics (Hacettepe U., 1974). Assoc. Prof. (tel. 4-2230391). *Crystal structure analysis, biological small molecules, powder diffraction.*

Kimyongür, Yrd. Doç. Dr Nurettin. Jeoloji Böl., Firat Ü. Müh-Mim. Fak., Elaziğ, Turkey. (1955) PhD, industrial minerals (Hull U., UK.). Lect. (tel. 811-11904). *Clay mineralogy and other industrial minerals.*

Kökçe, Dr Ali. Fizik Böl., Firat Ü. Fen-Ed. Fak., 23169 Elaziğ, Turkey. (1953) PhD, physics (Firat U., 1982). Asst. (tel. 811-12407, ext. 68). *Phase transformations, nucleation and growth, crystal defects, X-ray diffraction topography.*

Kumru, Dr Mustafa. Fizik Böl., Firat Ü. Fen-Ed. Fak., 23169 Elaziğ, Turkey. (19555) PhD, physics (Firat U., 1985). Asst. (tel. 811-12407, ext. 32). *Molecular crystallography, IR- spectroscopy.*

Kun, Yrd. Doç. Dr Nejat. Jeoloji Müh. Böl., Dokuz Eylül Ü., 35102 Bornova, İzmir, Turkey. (1951) PhD, metamorphic petrography (Dokuz Eylül U.). Lect. *Optical mineralogy, metamorphic petrography.*

Kuzucu, Mr Veysel. Fizik Böl., Firat Ü. Fen-Ed. Fak., 23169 Elaziğ, Turkey. (1960) MSc, physics (Firat U.,1987). Asst. (tel. 811-12407, ext. 66). *phonon softening, phase transformations.*

Küçükçelebi, Mr Hayrettin. Fizik Böl., Selçuk Ü. Fen-Ed. Fak., 42049 Kampus, Konya, Turkey. (1964) MSc, physics (Selçuk U., 1988). Asst. (tel. 331-182663, ext. 412). *Clay minerals, X-ray crystallography.*

Külcü, Doç. Dr Nevzat. Kimya Böl., Erciyes Ü. Fen-Ed. Fak., 39039 Kayseri, Turkey. (1950) PhD, X-ray emission spectrometry (J.W.Goethe U., Frankfurt, BRD). Lect. (tel. 351-74901,ext. 421). *X-ray emission spectrometry, X-ray diffractometry, crystal structure determination, differential thermal analysis, phase transitions.*

Munsuz, Prof. Nuri. Toprak Bilimleri Böl., Ankara Ü. Ziraat Fak., Ankara, Turkey. (1933) Prof., soil sciences (Ankara U., 1972). Lect. (tel. 4-1472100, ext. 162). *Clay mineralogy.*

Özkaplan, Dr Habib. Fizik Böl., Ondokuz Mayis Ü. Fen-Ed. Fak., Samsun, Turkey. (1945) PhD, physics (Louisville U., USA.,1978). Asst. Prof. (tel. 361-19680, ext. 657). *Semiconductors, crystal structure, thin films.*

Özpozan, Mrs Nilgün. Kimya Böl., Erciyes Ü. Fen-Ed. Fak., Talas Yolu, 38039 Kayseri, Turkey. (1959) MSc, inorganic chemistry (Erciyes U.). Asst. (tel. 351-74901). *X-ray powder diffraction, solid state reactions.*

Öztunali, Prof. Dr Önder. Jeoloji Müh. Böl., İstanbul Ü. Müh. Fak., 34459 Vezneciler, İstanbul, Turkey. (1935) Prof., mineralogy-mineral deposits (İstanbul U.). Lect. (tel. 1-5118480, ext. 317). *Mineral deposits, mineral chemistry, X-ray analysis methods.*

Soylu, Prof. Dr Hüseyin. Fizik Böl., Erciyes Ü. Fen-Ed. Fak., 38039 Talas Yolu, Kayseri, Turkey. (1933) PhD, physics (Hacettepe U.). Lect. (tel. 351-74901, ext. 417). *Crystal structure determination, solid state physics.*

Tarimci, Doç. Dr Çelik. Fizik Müh. Böl., Ankara Ü. Fen Fak., 06100 Tandoğan, Ankara, Turkey. (1945 PhD, crystallography (Pittsburg U., USA., 1975). Assoc. Prof. (tel. 4-2126720, ext. 1205). *Small molecules, structures, NQR spectroscopy.*

Taşer, Mr Mehmet. Fizik Böl., Selçuk Ü. Fen-Ed. Fak., 42049 Kampus, Konya, Turkey. (1961) MSc, physics (Selçuk U., 1988). Asst. (tel. 33- 182663, ext. 412). *Clay minerals, X-ray crystallography, crystallographic programming.*

Tokay, Yrd. Doç. Dr Nesrin. Kimya Böl., Hacettepe Ü. Müh. Fak., 06532 Ankara, Turkey. (1951) PhD, chemistry (Hacettepe U.). Lect. (tel. 4-2973060, ext. 1335, telex 42237 htk tr). *X-ray diffraction, physical metallurgy, crystallinity in polymers.*

Tokel, Doç. Dr Selçuk. Jeoloji Müh. Böl., Karadeniz Teknik Ü. Müh-Mim. Fak., Trabzon, Turkey. (1941) PhD, ore deposits-geochemistry (Karadeniz Teknik U.). Lect. (tel. 031-16920, ext. 2749). *Ore deposits, petrology, geochemistry, X-ray fluorescence spectroscopy.*

Tunç, Yrd. Doç. Dr Cemil. Fizik Böl., Yildiz Ü. Fen-Ed. Fak., 87270 Şişli, İstanbul, Turkey. (1940) PhD, physics (Karadeniz Teknik U.). Lect. (tel. 1-1468100, ext. 86). *Crystal growth-crystal defects, crystallography-metallurgy.ESR (EPR).*

Unan, Doç. Dr Coşkun. Jeoloji Müh. Böl., Orta Doğu Teknik Ü., İnönü Cad., 06532 Ankara, Turkey. (1936) PhD, mineralogy-petrography (M.E.T.U.). Lect. (tel. 4-2237100, ext. 2678, telex 42761, fax 2233054). *Mineralogy, ore microscopy, geochemistry, instrumentation.*

Üçişik, Prof. A. Hikmet. Protez Malzeme Anabilim Dali, Boğaziçi Ü., P.K. 2, 80815 Bebek, İstanbul, Turkey. (1945) PhD, material science (İstanbul T. U.). Lect. (tel. 1-5213959). *Materials, biomaterials, ceramics, mechanical properties.*

Ülkü, Prof. Dr Dinçer. Fizik Böl., Hacettepe Ü. Müh. Fak., 06532 Beytepe, Ankara, Turkey. (1940) Dr.rer.nat, crystallography (Munchen U., BRD, 1965). Prof. (tel. 4-2230391), telex 42237 htk tr). *Crystal structure analysis.*

Ünal, Doç. Dr Narin. Teknik Programlar, Akdeniz Ü. Antalya Meslek Y. O., Çalli, 07050 Antalya, Turkey. (1947) PhD, ceramic metarials (O.D.T.Ü.). Lect. (tel. 31-151330). *Phase tansformations, X-ray diffractometry.*

Yağbasan, Dr Rahmi. Fizik Böl., İnönü Ü. Fen-Ed. Fak., Malatya, Turkey. (1949) PhD, physics (Hacettepe U.,1980). Asst. Prof. (tel. 821-21871, ext. 276). *Crystal structure, powder diffraction, clay minerals.*

Yağci, Yrd. Doç. Dr Osman. Sağlik Programi Böl., Akdeniz Ü. Meslek Y.O., Çevre Yolu, Antalya, Turkey. (1942) PhD, X-ray spectroscopy (Leicester U., UK.). Lect. (tel. 31-151330). *Emission spectrometry.*

Yalçin, Yrd. Doç. Dr Hüseyin. Jeoloji Müh. Böl., Cumhuriyet Ü, 58140 Sivas, Turkey. (1957) PhD, geology (Hacettepe U., 1988). Lect. (tel. 477-61527, ext. 118, fax 477-61513). *Mineralogy, petrography, clay minerals, X-ray diffraction.*

Yaman, Doç. Dr Y. Macit. Metalurji Araştirma Ens., Anadolu Ü., 36030 Eskişehir, Turkey. (1946) PhD, manufacturing processes (Anadolu U.). Lect. (tel. 221-18550, ext. 90, telex 35147 tr, fax (221) 53616). *Material science, amorphous metals, solidification.*

Zor, Prof. Dr Muhsin. Fizik Böl., Anadolu Ü. Fen-Ed. Fak., 26470 Eskişehir, Turkey. (1947) PhD, solid state physics (W. London U.,Brunel, 1977). Lect. (tel. 221-50124). *Compound semiconductors, film structures, powder diffraction.*

UNION OF SOVIET SOCIALIST REPUBLICS

Sub-Editor: **E. N. Belova**

Notes

1. International telephone country code - 7.

2. The degree *doctor* (Dr) and *candidate* (Cand.) are approximately equivalent to DSc and PhD, respectively.

3. Degrees can be conferred by the Universities, by Institutes belonging to the Academy of Sciences of the USSR (abbreviated as Acad. Sci. USSR) or belonging to Academies of Sciences of the individual Soviet Socialist Republics, and by other Scientific Institutes. The following abbreviations are used to indicate where the degrees were obtained: GORNY - Leningrad Mining Institute; INCRYS - Institute of Crystallography, Academy of Sciences of the USSR; INEOS - Institute of Elemento-Organic Compounds, Academy of Sciences of the USSR; IONCH - Institute of General and Inorganic Chemistry, Academy of Sciences of the USSR; IMGRE - Institute of Mineralogy, Geochemistry, Crystal Chemistry of Rare Elements, Academy of Sciences of the USSR; IGEM - Institute of Geology, Mineralogy and Petrography of Ore Minerals, Academy of Sciences of the USSR; IGPM - Institut of Geochemistry and Physics of Minerals, Academy of Sciences, Ukrainian SSR; KARPOV - Karpov Physical Chemistry Institute; LGU - Leningrad State University; MGU - Moscow State University.

4. It should be remarked that more than one transliteration of names, from the non-Latin characters used in the USSR, is often possible. For example, names which in this Edition (as well as in the previous editions) have 'y' at the end may alternatively terminate in 'ii' as in the English translation of the journal *Kristallografiya*.

Abdullaev, Dr Abdulkhamid Aliyevich. Inst. of Crystallography, Acad. Sci. USSR, Leninsky pr. 59, Moscow 117333, USSR. (1936) Cand, physics and mathematics (INCRYS, 1971). Sr. scient. *Crystal growth, optical properties.*

Afanasyev, Dr Igor' Ivanovich. Leningrad Mining Inst., 21st Liniya 2, Leningrad 199026, USSR. (1935) Cand, geology and mineralogy (GORNY, 1966). Sr. scient. *Crystal growth, crystallography, crystal physics.*

Akchurin, Dr Marat Shikhapovich. Inst. of Crystallography, Acad. Sci. USSR, Leninsky pr. 59, Moscow 117333, USSR. (1947) Cand, physics and mathematics (INCRYS, 1983).Sr. scient. *Real structure, mechanical properties.*

Akhmetov, Dr Spartak Fatykhovich. Res. Inst. for Synthesis of Mineral Raw Materials, Institutskaya St. 1, Alexandrov 601600, Vladimirskaya Oblast', USSR. (1938) Cand, geology and mineralogy (Kazakh Polytechnic Inst., 1965). Sr. scient. *Crystal growth, crystal optics, X-ray structure analysis.*

Akhmetova, Dr Galina Leonidovna. Res. Inst. for Synthesis of Mineral Raw Materials, Institutskaya St. 1, Alexandrov 601600, Vladimirskaya Oblast', USSR. (1937) Cand, technics (Inst. of Metallurgy, Acad. Sci. Kazakh SSR, 1966). Sr. scient. *Synthesis of single crystals, optical methods, X-ray diffraction methods, physical chemistry, melts.*

Akselrud, Dr Lev Grigoryevich. Dept. of Chemistry, Lvov State U., University St. 1, Lvov 290602, USSR. (1948) Cand, chemistry (Lvov U., 1980). Sr. scient. *X-ray structure analysis, crystal chemistry, intermetallic compounds, computing.*

Aksenov, Dr Viktor Lazarevich. Joint Inst. for Nuclear Reasearch. Moskovskaya obl., Dubna 141980, USSR. (1947) Dr, physics and mathematics (Steklov Inst. of Mathematics, Moscow, 1985). Asst. dir. *Structure investigations, single crystals, polycrystals, neutron diffraction.*

Aleshina, Dr Lyudmila Aleksandrovna, Petrozavodsk State U., Prospekt Lenina 33, Petrozavodsk 185640, USSR. (1942) Cand, physics and mathematics (MGU, 1983), Sr. lect. *Structure, amorphous metals, polycrystal X-ray analysis.*

Al'shits, Dr Vladimir Iosifovich. Inst. of Crystallography, Acad. Sci. USSR, Leninsky pr. 59, Moscow 117333, USSR. (1941) Dr, physics and mathematics (INCRYS, 1977). Lab. head. *Dislocation theory.*

Alaverdova, Dr Olga Georgiyevna. Kharkov Polytechnic Inst., Frunze St. 21, Kharkov 310002, USSR. (1938) Cand, technics (Kharkov Polytechnic Inst., 1975). Docent. *Structure, real crystals, thin films.*

Aldoshin, Dr Sergey Mikhailovich. Branch Inst. of Chemical Physics, Acad. Sci. USSR, Chernogolovka 142432, Noginsky Rayon, Moskovskaya Oblast', USSR. (1953) Dr, chemistry (Inst. of Chemical Physics, 1986). Sr. scient. *Structural photochemistry, superconductors, structure.*

Aleshko-Ozhevsky, Dr Oleg Pavlovich. Inst. of Crystallography, Acad. Sci. USSR, Leninsky pr. 59, Moscow 117333, USSR. (1934) Cand, physics and mathematics (INCRYS, 1969). Sr. scient. *Real structure.*

Alexandrov, Prof. Alexander Danilovich. Mathematical Inst., Acad. Sci. USSR, Siberian Dept., Novosibirsk 630090, USSR. (1912) Full member, Acad. Sci. USSR; Dr, physics and mathematics (LGU, 1936). *Geometrical crystallography.*

Alexandrov, Prof. Kirill Sergeyevich. Inst. of Physics, Acad. Sci. USSR Siberian Dept., Akademgorodok, Krasnoyarsk 660036, USSR. (1931) Full member, Acad. Sci. USSR; Dr, physics and mathematics (INCRYS, 1967). Dir. *Crystal physics, phase transitions in crystals, ferroelectricity.*

Alexandrov, Dr Vladimir Borisovich. Inst. of Crystallography, Acad. Sci. USSR, Leninsky pr. 59, Moscow 11733, USSR. (1931), Cand, geology and mineralogy (MGU, 1964). Sr. scient. *X-ray structure analysis, crystal chemistry, minerals.*

Alexandrova, Dr Inga Petrovna. Inst. of Physics, Acad. Sci. USSR, Siberian Dept., Akademgorodok, Krasnoyarsk 660036, USSR. (1934) Dr, physics and mathematics (INCRYS,1988) Sr. scient. *Ferroelectricity, phase transitions.*

Aliev, Dr Fazil Isa ogly. Inst. of Physics, Acad. Sci. Azerbaidzhan SSR, Narimanov Pr. 33, Baku 370143, USSR. (1937) Cand, physics and mathematics (Inst. of Physics, Acad. Sci. Azerbaidzhan SSR, 1969). Sr. scient. *Amorphous films, crystallization, structure.*

Aliev, Dr Zainutdin Gasanovich. Branch Inst. of Chemical Physics, Acad. Sci. USSR, Chernogolovka 142432, Noginsky Rayon, Moskovskaya Oblast', USSR. (1939) Cand, physics and mathematics (Inst. of Chemical Physics, Acad. Sci. USSR, 1974). Sr. scient. *Structure of organic and inorganic compaunds.*

Alikhanov, Dr Ruben Abramovich. Inst. of High Pressure Physics, Acad. Sci. USSR, Akademgorodok, Podol'sky Rayon, Moskovskaya Oblast' 142092, USSR. Cand, physics and mathematics (Inst. of Physical Problems, Acad. Sci. USSR, 1959). Sr. scient. *High pressure, low temperature, structure, magnetic crystals, molecular crystals.*

Alyavdin, Dr Vladimir Fedorovich. Leningrad Mining Inst., 21st Liniya 2, Leningrad 199026, USSR. (1913) Cand, geology and mineralogy (Inst. of Geology, Acad. Sci. USSR, 1947). *Crystal morphology, goniometry.*

Amiraslanov, Dr Imameddin Radzhabali ogly. Inst. of Physics, Acad. Sci Azerbaidzan SSR, Prospekt Narimanova 33, Baku 370143, USSR. (1948) Cand, physics and mathematics (Inst. Appl. Phys., Kishinev,1979). Sr. scient. *Crystal chemistry, inorganic compounds, inclusion and intercalate compounds.*

Amirov, Dr Savalan Teimur ogly. Inst. of Inorganic and Physical Chemistry, Acad. Sci. Azerbaidzhan SSR, Narimanov Prospekt 29, Baku 370143, USSR. (1939) Cand, chemistry (INCRYS, 1968). Sr. scient. *X-ray structure analysis, crystal chemistry, silicates.*

Andreeva, Dr Nataliya Sergeyevna. Inst. of Molecular Biology, Acad. Sci. USSR, Vavilov St. 32, Moscow 117312, USSR. Dr, physics and mathematics (INCRYS, 1970). Sr. scient. *Protein crystallography, fibrillous structures.*

Andrianov, Dr Valery Ivanovich. Inst. of Crystallography, Acad. Sci. USSR, Leninsky pr. 59, Moscow 117333, USSR. (1938) Cand, physics and mathematics (INCRYS, 1969). Sr. scient. *Computing methods, structure analysis.*

Andrianova, Dr Mariya Egorovna. Inst. of Crystallography, Acad. Sci. USSR. Leninskii Prospekt 59, Moscow 117333, USSR. Cand, physics and mathematics (INCRYS, 1987). Sr. scient. *Protein crystallography, automation, diffractometry.*

Andrushevskii, Dr Nikolai Matveevich. Computing Dept., Moscow State U., Leninskie Gory, Moscow 117234, USSR. (1943) Cand, physicvs and mathemaatics (Inst. Appl. Physics, Kishinev, 1982).Sr. scient. *Mathematical problems, structure analysis.*

Anikin, Dr Igor' Nikolayevich. Moscow State U., Dept. of Geology, Leninskiye Gory, Moscow 117234, USSR. (1929) Cand, geology and mineralogy (MGU, 1955). Sr. scient. *Crystal growth, crystal formation, instrumentation, physical chemistry, melts.*

Anisimova, Vera Nikolaevna. Inst. of Crystallography, Acad. Sci. USSR. Leninskii Prospekt 59, Moscow 117333, USSR. (1947) Cand, physics and mathematics (INCRYS, 1986) Sr. scient. *Dielectric physics, phase transitions.*

Anistratov, Dr Anatoly Tikhonovich. Inst. of Physics, Acad. Sci. USSR, Siberian Dept., Akademgorodok, Krasnoyarsk 660036, USSR. (1935) Cand, physics and mathematics (Krasnoyarsk State Pedagogic Inst., 1967). Sr. scient. *Crystal optics, phase transitions.*

Antipin, Dr Mikhail Yuvenaliyevich. Inst. of Elemento-Organic Compounds, Acad. Sci. USSR, Vavilov St. 28, Moscow 117813, USSR. (1951) Cand, chemistry

(INEOS, 1980). Sr. scient. *Organic and elemento-organic compounds, crystal structure, low temperature X-ray structure analysis.*

Antonov, Dr Petr Iosifovich. Physico-Techn. Inst., Acad. Sci. USSR, Politekhnicheskaya St. 26, Leningrad 194021, USSR. (1935) Dr, physics and mathematics (Physico-Techn. Inst., Acad. Sci. USSR, 1988). Sr. scient. *Crystal growth, growth from melts by Stepanov method, crystal structure and properties.*

Antsishkina, Dr Alla Sergeevna. Inst. of General and Inorganic Chemistry, Acad. Sci. USSR, Leninsky pr. 31, Moscow 117071, USSR. (1926) cand, chemistry (IONCH, 1959). Sr. scient. *Crystal chemistry, stereochemistry, coordination compounds.*

Apinitis, Dr Smuidris Karlovich. Riga Polytechnic Inst., Ul. Aizenes 14, Riga 226048, USSR. (1933) Cand, chemistry (Riga Polytechic Inst., 1970). Docent. *X-ray structure analysis.*

Arakcheeva, Dr Alla Vladimirovna. Inst. of Metallurgy, Acad. Sci. USSR, Leninskii Prospekt 49, Moscow 117334, USSR. (1954) Cand, chemistry (INCRYS, 1988). Scient. *Crystal chemistry, inorganic compounds, ferrites, high-temperature supercon superconductivity,polytypysm, isomorphism.*

Argunova, Dr Tatyana Sergeevna. Physical-Technical Inst., Acad. Sci. USSR, Zapovednaya 51, Leningrad 194037, USSR. (1957) Cand, physics and mathematics (Physical-Technical Inst., 1986),Scient. *Structural defects in single-crystal layers.*

Arinkin, Dr Aleksandr Viktorovich. Polytechnical Inst., Frunze 21, Kharkov 310002, USSR. (1948) Cand, physics and mathematics (Kharkov Polytechnical Inst., 1978). Sr. scient. *X-ray analysis, imperfect crystals, crystal lattice defects.*

Arkhipenko, Dr Diana Konstantinovna. Inst. of Geology and Geophysics, Acad. Sci. USSR, Siberian Dept., Prospekt Nauki 3, Novosibirsk 630090, USSR. (1928) Dr, physics and mathematics (INCRYS, 1983). Lab. head. *Mineral structures, X-ray diffraction methods, infrared spectroscopy.*

Arzumanyan, Dr Gennadiy Ashotovich. Inst. of Crystallography, Acad. Sci. USSR, Leninsky pr. 59, Moscow 117333, USSR. (1953) Cand, chemistry (Moscow Chem. Techn. Inst., 1980). Scient. *Physical chemistry, high temperature oxide crystallization processes, mass spectrometry.*

Asadchikov, Dr Viktor Evgen'yevich. Inst. of Crystallography, Acad. Sci. USSR, Leninsky pr. 59, Moscow 117333, USSR. (1948) Cand, physics and mathematics (INCRYS, 1982). Sr. scient. *X-ray small angle scattering, apparatus, experimental methods.*

Asadov, Prof. Yusif Gazanfar ogly. Inst. of Physics, Acad. of Sci. Azerbaidzhan SSR, Narimanov Prospekt 33, Baku 370143, USSR. (1934) Cand, physics and mathematics (Azerbaidzhan State U., 1964). Lab. head. *Crystal growth, phase transformations, structure, semiconductors.*

Ashirov, Dr Aman. Physical Technical Inst., Acad. Sci. Turkmen SSR, Golgol St. 15, Ashkhabad 74400, USSR. (1935) Cand, physics and mathematics (INCRYS, 1963). Lab. head. *X-ray structure analysis.*

Askhabov, Dr Askhab Magomedovich. Inst. of Geology, Acad. Sci. USSR Komi Dept., Kommunisticheskaya 28, Syktyvkar 167000, USSR. (1948) Cand, geology and mineralogy (GORNY, 1977). Lab. head. *Crystallography.*

Aslanov, Prof. Leonid Aleksandrovich. Moscow State U., Dept. of Chemistry, Leninskiye Gory, Moscow 117234, USSR. (1938) Dr, chemistry (MGU, 1973). Prof. *X-ray structure analysis, crystal chemistry, instrumentation.*

Atovmyan, Prof. Lev Oganovich. Branch Inst. of Chemical Physics, Acad. Sci. USSR, Chernogolovka 142432, Noginsky Rayon, Moskovskaya Oblast', USSR. (1928) Dr, chemistry (IONCH, 1971). Lab. head. *Crystal chemistry, coordination compounds.*

Avdiyenko, Dr Klavdiya Ilyinishna. Inst. of Semiconductor Physics, Acad. Sci. USSR, Siberian Dept., Prospekt Nauki 13, Novosibirsk 630090, USSR. Cand, physics and mathematics (Inst. of Semiconductor Physics, Acad. Sci. USSR, Siberian Dept., 1970). Sr. scient. *Crystal growth, structure - physical properties relationship.*

Avilov, Dr Anatoly Sergeevich. Inst. of Crystallography, Acad. Sci. USSR, Leninsky pr. 59, Moscow 117333, USSR. (1943) Cand, physics and mathematics (INCRYS, 1973). Sr. scient. *Electron diffraction, structure analysis, crystal chemistry.*

Babareko, Dr Alesya Adamovna. Inst. of Metallurgy, Acad. Sci. USSR, Leninsky Prospekt 49, Moscow 117334, USSR. (1928) Cand, techn. (Inst. of Metallurgy, 1978). Sr. scient. *Plastic deformation, crystal growth.*

Bagdasarov, Dr, Khachik Saakovich. Inst. of Crystallography, Acad. Sci. USSR, Leninsky pr. 59, Moscow 117333, USSR. (1929) Dr, physics and mathematics (INCRYS, 1972). Lab. head. *Crystal growth, high temperature.*

Bakakin, Dr Vladimir Vasilyevich. Inst. of Inorganic Chemistry, Acad. Sci. USSR, Siberian Dept., Pr. Lavrent'eva 3, Novosibirsk 630090, USSR. (1933) Dr, geology and mineralogy (MGU, 1963). Sr. scient. *Crystal chemistry, inorganic compounds, minerals, teaching crystallography.*

Bakovets, Dr Vladimir Viktorovich. Inst. Inorganic Chemistry, Siberian Dept.of the USSR Acad.Sci., Prospekt Akademika Lavrent'eva 3, Novosibirsk 630090, USSR. (1942) Cand, chemistry (INCh, 1972). Sr. scient. *Crystal growth, growth forms.*

Balagurov, Mr Anatoly Mikhailovich. Joint Inst. for Nuclear Res., Dubna 141980, Moskovskaya Oblast', USSR. (1945). Scient. *Neutron physics.*

Balakirev, Dr Vladimir Georgiyevich. Res. Inst. for Synthesis of Mineral Raw Materials, Institutskaya St. 1, Alexandrov 601600, Vladimirskaya Oblast', USSR.

Cand, geology and mineralogy (IGEM, 1977). Sr. scient. *Solid state physics, defects in crystals.*

Balitsky, Dr Vladimir Sergeyevich. Res. Inst. for Synthesis of Mineral Raw Materials, Institutskaya St. 1, Alexandrov 601600, Vladimirskaya Oblast', USSR. (1932) Dr, geology and mineralogy (IGEM, 1971). Dept. head. *Crystal growth in hydrothermal systems.*

Balyunis, Dr Lyubov' Evgenievna. Physics Inst., Rostov State U., Prospekt Stachki 194, Rostov-on-Don 344104, USSR. (1955) Cand, physics and mathematics (Rostov U., 1984). Sr. scient. *Crystal structure, extremal conditions.*

Barabanov, Prof. Vladimir Fedorovich. Leningrad State U., University Emb. 7/9, Leningrad 199164, USSR. (1918) Dr, geology and mineralogy (LGU, 1961). Head of Chair. *Crystallography, morphology, typomorphism, solid state physics.*

Baranov, Dr Anatoliy Ivanovich. Inst. of Crystallography, Acad. Sci. USSR, Leninsky pr. 59, Moscow 117333, USSR. (1947) Cand, physics and mathematics (INCRYS, 1973). Sr. scient. *Structural phase transitions, ferroelectricity, superionic conductivity.*

Baranskii, Prof Konstantin Konstantinovich. Physics Dept., Moscow State U., Leninskie Gory, Moscow 117234, USSR. (1921) Dr, physics and mathematics (MGU, 1981). Prof. *Physical acoustics, ferroelectricity.*

Barsukova, Dr Marina Leonidovna. Inst. of Crystallography, Acad. Sci. USSR, Leninsky Prospekt 59, Moscow 117333, USSR. (1943) Cand, chemistry (INCRYS, 1980). Scient. *Crystal growth.*

Bartoshinsky, Prof. Zinovy Vladislavovich. Lvov State U., Dept. of Mineralogy, Shcherbakova St. 4, Lvov 290005, USSR. (1929) Dr, geology and mineralogy (Inst. of Minerals Raw Materials 1983). Prof. *Crystal morphology, minerals, diamond crystallography and mineralogy.*

Bataliyeva, Dr Nataliya Glebovna. Inst. of Mineralogy, Geochemistry and Crystal Chemistry of Rare Elements, Sadovnicheskaya Emb. 71, Moscow 113127, USSR. (1931) Cand, geology and mineralogy (MGU, 1971). Sr. scient. *Crystal chemistry, minerals, synthetic compounds, rare earth compounds.*

Batsanov, Dr Andrey Stepanovich. Inst. of Elemento-Organic Compounds, Acad. Sci. USSR, Vavilova 28, Moscow 117813, USSR. (1955) Cand, chemistry (INEOS, 1983). Scient. *X-ray structure analysis, structural chemistry, organometallic compounds, coordination compounds, bioactive compounds.*

Bekesha, Dr Sergei Nikolaevich. Geology Dept, Lvov State U., Shcherbakova 4, Lvov 290005, USSR. (1953) Cand, geology and mineralogy (Lvov U., 1967) Sr. scient. *Crystal morphology, physics, minerals.*

Bekrenev, Dr Anatoly Nikolayevich. Kuibyshev Polytechnic Inst., Pervomaiskaya St. 18, Kuibyshev 443002, USSR. (1944) Dr, physics and mathematics (Kharkov U., 1971). Prof. *Crystal structure, deformed crystals.*

Belan, Dr Bogdana Dmitrievna. Chemistry Dept., Lvov State U., Universitetskaya 1, Lvov 290602.USSR. (1953) Cand, chemistry (Lvov U., 1988). Scient. *Crystal chemistry, intermetallic compounds.*

Bel'sky, Dr Vitaly Konstantinovich. Karpov Physical Chemistry Inst., Obukha St. 10, Moscow 107120, USSR. (1943) Cand, chemistry (MGU, 1969) Sr. scient. *Crystal chemistry, organic compounds, symmetry, X-ray structure analysis.*

Belikova, Dr Galina Sergeyevna. Inst. of Crystallography, Acad. Sci. USSR, Leninsky pr. 59, Moscow 117333, USSR. (1928) Cand, chemistry (INCRYS, 1968). Sr. scient. *Organic crystals, growth.*

Belokoneva, Dr Elena Leonidovna. Moscow State U., Dept. of Geology, Leninskiye Gory, Moscow 117234, USSR. Cand, geology and mineralogy (MGU, 1975). Sr. scient. *Crystal chemistry, inorganic compounds, X-ray structure analysis.*

Belova, Dr Elizaveta Nikolayevna. Inst. of Crystallography, Acad. Sci. USSR, Leninsky pr. 59, Moscow 117333, USSR. Cand, physics and mathematics (INCRYS, 1949). Sr. scient. *Structure analysis.*

Belugina, Dr Nataliya Vasilyevna. Inst. of Crystallography, Acad. Sci. USSR, Leninsky pr. 59, Moscow 117333, USSR. (1941) Cand, physics and mathematics (INCRYS, 1978). Scient. *Real structure, crystal growth - real structure relations.*

Belyaeva, Mrs Klara Fedorovna. Inst. of Applied Physics, Acad. Sci. Moldavian SSR, Akademicheskaya 5, Kishinev 277028, USSR. (1933). *X-ray structure analysis methods, coordination compounds.*

Belyakov, Dr Sergei Vasil'evich. Polytechnic Inst., Aizenes st.14, Riga 226048, USSR. (1961) Cand, physics and mathematics (Inst. Appl. Physics, Kishinev, 1967). Scient. *Computational methods, X-ray analysis, crystal chemistry, organic compounds.*

Belyustin, Dr Aleksey Vsevolodovich. Gorky State U., Gagarin Pr. 23, Gorky 603022, USSR. (1913) Cand, physics and mathematics (Gorky U., 1945). Docent. *Crystal growth from solutions.*

Bendeliani, Dr Nikolay Alexandrovich. Inst. of High Pressure Physics, Acad. Sci. USSR, Troitsk, Podol'sky Rayon, Moskovskaya Oblast' 142092, USSR. Dr, chemistry (MGU, 1982). Lab. Head. *High pressure, crystal structure.*

Beresnev, Dr Leonid Alekseyevich. Inst. of Crystallography, Acad. Sci. USSR, Leninsky pr. 59, Moscow 117333, USSR. (1947) Cand, physics and mathematics (Inst. of Solid State Physics, Acad. Sci. USSR, 1979). Sr. scient. *Structure and properties, ferroelectric liquid crystal systems.*

Berezhkova, Dr Galina Vasilyevna. Inst. of Crystallography, Acad. Sci. USSR, Leninsky pr. 59, Moscow 117333, USSR. (1933) Cand, physics and mathematics (INCRYS, 1964). Sr. scient. *Defects in crystals, mechanical properties.*

Berezkin, Dr Vladimir Viktorovich. Inst. of Crystallography, Acad. Sci. USSR,. Leninskii Prospekt 59, Moscow 117333, USSR. (1938) Cand, eng.

(Agrophysical Inst., Leningrad, 1974). Sr. scient. *Nuclear filters, purification, liquid and gaseous crystallization media.*

Berezyuk, Dr Dar'ya Aleksandrovna. Dr Chemistry Dept., Lvov State U., Universitetskaya 1, Lvov 290602, USSR. (1953) Cand, chemistry (Lvov U., 1986).Scient. *Crystal chemistry, intermetallic compounds.*

Bershov, Dr Leonid Viktorovich. Inst. of Geology, Mineralogy and Petrography (IGEM), Acad. Sci. USSR, Staromonetny 35, Moscow 109017, USSR. (1935) Dr, geology and mineralogy (IGEM, 1973). Lab. head. *Minerals (physics), crystal field theory, electron paramagnetic resonance.*

Bersuker, Prof. Isaak Borukhovich. Inst. of Chemistry, Acad. Sci. Moldavian SSR, Akademicheskaya 3, Kishinev 277028, USSR. (1928) Corresp. member Acad. Sci. USSR, Dr, physics and mathematics (LGU, 1964). Dept. head. *Crystal chemistry, ferroelectricity, structural phase transitions.*

Berzinya, Dr Inese Rudol'fovna. Inst. of inorganic chemistry, Acad. Sci. Latvian SSR, Miera st.34, Rizhskii raion, Salaspils 1, Latvian SSR 229021, USSR. (1948) Cand, chemistry (Inst. Inorg. Chem., Riga, 1988). Scient. *Crystal chemistry, complex compounds.*

Betsofen, Dr Sergey Yakovlevich. Inst. of Metallurgy, Acad. Sci. USSR, Leninsky pr. 49, Moscow 117334, USSR. (1946) Cand, technics (Inst. of Metallurgy, Acad. Sci. USSR, 1978) Sr. scient. *Applied crystallography, structure, amorphous materials, diffraction apparatus.*

Beznosikov, Dr Boris Valeriyanivich. Inst. of Physics, Acad. Sci. USSR, Siberian Dept., Akademgorodak, Krasnoyarsk 660036, USSR. (1930) Cand, physics and mathematics (Krasnoyarsk Inst. of Physics, Acad. Sci. USSR, 1978). Sr. scient. *Crystal chemistry, crystal growth.*

Bichurin, Dr Rinnat Chingizkhanovich. Inst. of Crystallography, Acad. Sci. USSR, Leninsky pr. 59, Moscow 117333, USSR. (1952) Cand, chemistry (Inst. of Chemistry, Acad. Sci. Tadzhik SSR). Scient. *Crystal growth, properties, ferroelectrics.*

Biyushkin, Dr Victor Nikolayevich. Inst. of Applied Physics, Moldavian SSR, Akademicheskaya 5, Kishinev 277028, USSR. (1935) Cand, physics and mathematics (INCRYS, 1969). Sr. scient. *X-ray structure analysis methods, organic compounds.*

Bleidelis, Dr Yanis Yazepovich. Inst. of Organic Synthesis, Acad. Sci. Latvian SSR, Aizkraukles St. 23, Riga 226006, USSR. (1926) Cand, chemistry (IONCH, 1957). Sr. scient. *Crystal chemistry, organic and elemento-organic compounds.*

Blinov, Dr Lev Mikhailovich. Inst. of Crystallography, Acad. Sci. USSR, Leninsky pr. 59, Moscow 117333, USSR. (1939) Dr, physics and mathematics (INCRYS, 1977). Lab. head. *Liquid crystals, structure and properties.*

Blinov, Dr Victor Aleksandrovich. Inst. of Mineralogy- Geochemistry & Crystal Chemistry of Rare Elements, Sadovnicheskaya nab.71, Moscow 113035, USSR. (1946) Cand, geology and mineralogy (MGU, 1975). Sr. scient. *Crystal chemistry, minerals, X-ray diffractometry.*

Blistanov, Prof Aleksandr Aleksandrovich. Moscow Inst. of Steel and Alloys. Leninskii Prospekt 4, Moscow 117049, USSR. (1937) Doctor, physics and mathematics (Inst. of Steel and Alloys, Moscow, 1972). Head of chair. *Crystal physics, defect theory.*

Blokhin, Prof. Mikhail Arnol'dovich. Rostov State U., Dept. of Physics, Stachki pr. 192, Rostov-on-Don 344061, USSR. (1908) Dr, physics and mathematics (Kiev U., 1955). *Electronic energy-producing structure of crystals.*

Bodak, Prof. Oksana Ivanovna. Lvov State U., Dept. of Chemistry, University St. 1, Lvov 290602, USSR. (1942) Dr, chemistry (Lvov U., 1981). Prof. *X-ray structure analysis, crystal chemistry, intermetallic compounds.*

Bogdanov, Dr Gennadii Evgen'evich. Inst. of Geology, Ural Div. Acad. Sci. USSR, Oplesnina 2, Syktyvkar 167610, USSR. (1958) Cand, geology and mineralogy (MGU, 1987). Lab. Head. *Crystallography, crystal growth.*

Boiko, Prof. Boris Timofeyevich. Kharkov Polytechnik Inst., Frunze St. 21, Kharkov 310002, USSR. (1930) Dr, physics and mathematics (Kharkov U., 1971). Prof. *Crystal structure, metals and alloys, defects in crystal structure.*

Boikova, Dr Alexandra Ivanovna. Inst. of Silicate Chemistry, Acad. Sci. USSR, Makarov Emb. 2, Leningrad 199164, USSR. (1926) Cand, chemistry (Inst. of Silicate Chemistry, Acad. Sci. USSR, 1955). Sr. scient. *Crystal chemistry, natural and synthetic minerals.*

Bokii, Prof. Georgy Borisovich. Inst. of Geology, Mineralogy and Petrography (IGEM), Acad. of Sci. USSR, Staromonetny 35, Moscow 109017, USSR. (1909) Corresp. member, Acad. Sci. USSR, Dr, chemistry (IONCH, 1942). Lab. head. *Crystal chemistry, inorganic compounds, minerals.*

Bondar', Dr Anatolii Mikhail;ovich. Inst. of Metallurgy, Acad. Sci. USSR. Leninskii Prospekt 49, Moscow 117334, USSR. (1940) Cand., physics and mathematics (INCRYS, 1973).Scient. *Spectroscopy (NMR,NPR, EPR, etc.) and electrophysical properties properties of materials, structural chemistry, phase transformations, hydrogen dynamics.*

Bondar', Dr Iraida Adamovna. Inst. of Silicate Chemistry, Acad. Sci. USSR, Makarov Emb. 2, Leningrad 199164, USSR. Dr, chemistry (Inst. of Silicate Chemistry, Acad. Sci. USSR, 1967). Lab. head. *Crystal chemistry, crystal growth.*

Bondars, Dr Bruno Yanovich. Inst. of Inorganic Chemistry, Acad. Sci. Latvian SSR, Ul. Miera 34,Rizhskii rayon, Saalspils 1, Latv. SSR, 229021, USSR. (1951) Cand, chemistry (Inst. of Inorganic Chemistry, Acad. Sci. Latvian SSR, 1981). Sr. scient. *Powder X-ray diffraction.*

Borisanova, Dr Lidiya Mikhailovna. Moscow State U., Dept. of Chemistry, Leninskiye Gory, Moscow 117234, USSR. (1940) Cand, chemistry (MGU, 1971). Sr. scient. *Crystal chemistry.*

Borisov, Dr Stanislav Vasilyevich. Inst. of Inorganic Chemistry, Acad. Sci. USSR, Siberian Dept., Pr. Lavrent'eva 3, Novosibirsk 630090, USSR. (1930) Dr, physics and mathematics (INCRYS, 1974). Lab. Head. *Structure determination, methods, inorganic compounds, superstructures.*

Borisov, Dr Vsevolod Vasilyevich. Inst. of Crystallography, Acad. Sci. USSR, Leninsky Prospekt 59, Moscow 117333, USSR. (1937) Cand, physics and mathematics (INCRYS, 1975). Sr. scient. *Protein crystallography.*

Boyarskaya, Prof. Yuliya Stanislavovna. Inst. of Applied Physics, Acad. Sci. Moldavian SSR, Akademicheskaya 5, Kishinev 277028, USSR. (1928) Dr, physics and mathematics (INCRYS, 1974). Sr. scient. *Mechanical properties, crystal lattice defects.*

Brainin, Dr Boris Matveyevich. Petrozavodsk State U., Lenin pr. 33, Petrozavodsk 185018, USSR. (1937) Cand, physics and mathematics (Petrozavodsk State U., 1967). Docent. *Structure, real crystals.*

Brovkin, Dr Anatoly Afanasyevich. Inst. of Mineral Raw Materials, Staromonetny 29, Moscow 109017, USSR. (1937) Cand, geology and mineralogy (Yakutsk Branch of Siberian Dept., Acad. Sci. USSR, 1966). Sr. scient. *Crystal chemistry, borates, X-ray diffraction, quantitative phase analysis.*

Bublik, Prof Valdimir Timofeevich. Moscow Inst. of Steel and Alloys, Leninskii Prospekt 4, Moscow 117049, USSR. (1934) Doctor, physics and mathematics (Moscow Inst. of Steel and Alloys,1980). Professor. *Physics, defect crystals.*

Bud'ko, Dr Ivetta Alexandrovna. Inst. 'Mekhanobr', 21st Liniya 8a, Leningrad 199026, USSR. (1932) Cand, geology and mineralogy (LGU, 1968). Sr. scient. *X-ray analysis, crystal chemistry.*

Bukin, Dr Alexander Sergeyevich. Inst. of Geology, Acad. Sci. USSR, Pyzhevsky per. 7, Moscow 109017, USSR. (1947) Cand, physics and mathematics (MGU, 1975). Scient. *Polytypism, layer silicates, order-disorder, isomorphically substituted silicates.*

Bukvetsky, Dr Boris Vladimirovich. Inst. of Chemistry, Far East Scient. Center, Acad. Sci. USSR, Pr. of the 100th Aniv. of Vladivostok 159, Vladivostok 690022, USSR. (1944) Cand, physics and mathematics (INCRYS, 1977). Sr. scient. *Crystal chemistry, inorganic compounds, hydrogen bonds, phase transitions.*

Bulakh, Dr Andrey Glebovich. Leningrad State U., Dept. of Geology, University Emb. 7/9, Leningrad 199164, USSR. (1933) Dr, geology and mineralogy (LGU, 1978). Chair. Head. *Crystal morphology, goniometry.*

Bulgarovskaya, Dr Irina Vsevolodovna. Karpov Physical-Chemical Inst., Obukha 10, Moscow 103064, USSR. (1941) Cand, physics and mathematics (KARPOV, 1981). Scient. *X-ray analysis, crystal chemistry, photonics, organic molecular crystals.*

Bunina, Dr Ol'ga Alekseevna. Inst. of Physics, Rostov State U., Prospekt Stachki 194, Rostov-on-Don 344104, USSR. (1955) Cand., physics and mathematics (Rostov-U., 1988). Sr. scient. *Structure, oxides compounds, textures.*

Burdina, Dr Valentina Ivanovna. Inst. of Crystallography, Acad. Sci. USSR, Leninsky pr. 59, Moscow 117333, USSR. (1927) Cand, physics and mathematics (Inst. of Mathematics, Acad. Sci. USSR, 1953). Scient. *Computing methods in crystallography.*

Burnasheva, Dr Veniana Venediktovna. Inst. of New Chemical Problems, Acad. Sci. USSR, Chernogolovka 142432, Moskovskaya Oblast', USSR. (1940) Cand, chemistry (Lvov State U., 1970). *Crystal structure, intermetallic compounds, hydride phases.*

Burova, Dr Elena Mikhailovna. Computing Dept., Moscow State U., Leninskie Gory, Moscow 117234, USSR. (1952) Cand, physics and mathematics (MGU, 1981). Scient. *Computing methods in powder crystallography.*

Burshtein, Dr Izya Fridelevich. Inst. of Applied Physics, Acad. Sci. Moldavian SSR, Akademicheskaya 5, Kishinev 277028, USSR. (1942) Cand, physics and mathematics (Inst. of Applied Physics, Moldavian Acad. Sci., 1977). Sr. scient. *Computing methods, crystal chemistry, coordination compounds, bioinorganic compounds.*

Bushuev, Dr Vladimir Alekseevich. Physics Dept., Moscow State U., Leninskie Gory, Moscow 117234, USSR. (1947) Cand, physics and mathematics (MGU, 1975). Sr. scient. *X-ray diffraction in perfect and distorted crystals, inelastic and diffuse X-ray scattering.*

Butaev, Dr Boris Savel'evich. Dept. of Chemistry, Moscow State University, Leninskie Gory, Moscow 117234, USSR. (1948) Cand, chemistry (MGU, 1979). Sr. scient. *Gas electron diffraction, inorganic structures.*

Butman, Dr Lev Abramovich. Inst. of General and Inorganic Chemistry, Acad. Sci. USSR, Leninsky pr. 31, Moscow 117071, USSR. (1930) Cand, physics and mathematics (INCRYS, 1971). Sr. scient. *Crystal chemistry, stereochemistry, coordination compounds, X-ray diffractometry, high temperature superconductivity.*

Bykov, Aleksei Borisovich. Inst. of Crystallography, Acad. Sci. USSR, Leninskii Prospekt 59, Moscow 117333, USSR. (1952) Cand, chemistry (INCRYS, 1982). Sr. scient. *Crystal growth from melts and solutions.*

Chaban, Dr Nadezhda Fedorovna. Lvov State U., Dept. of Chemistry, University St. 1, Lvov 290602, USSR. (1942) Cand, chemistry (Lvov U., 1973). Scient. *X-ray structure analysis, crystal chemistry, intermetallic compounds.*

Chashchinov, Dr Yury Mikhailovich. Leningrad Mining Inst., 21st Liniya 2, Leningrad 199026, USSR. (1939) Cand, geology and mineralogy (GORNY, 1972). Sr. scient. *Crystal growth, defects.*

Cheremskoy, Dr Petr Grigoryevich. Kharkov Polytechnic Inst., Frunze St. 21, Kharkov 310002, USSR. (1942) Cand, technics (Kharkov Polytechnic Inst., 1973). Sr. scient. *Metals and alloys, structure, crystal structure defects, small-angle scattering*

Cherepanova, Dr Tamara Alekseyevna. Latvian State U., Boulevard Rainisa 19, Riga 226050, USSR. (1944) Dr, physics and mathematics(Inst. of Physics, Akad. Sci. Latvian SSR, 1985). Lab. head. *Crystal growth, crystallography theory.*

Cherkezyan, Dr Suren Asaturovich. Inst. of Crystallography, Acad. Sci. USSR, Leninskii Prospekt 59, Moscow 117333, USSR. (1949) Cand, physics and mathematics (INCRYS, 1988). Scient. *Magnetic properties of crystals.*

Chernega, Dr Aleksandr Nikolaevich. Inst. of Inorganic Chemistry, Ukrainian SSR Acad. Sci., Murmanskaya 5, Kiev 152660, USSR. (1955) Cand, chemistry (INEOS, 1985). Head of group. *X-ray analysis, crystal chemistry, organic and elemento-organic compounds.*

Cherner, Dr Yakov Efremovich. Inst. of Physics, Rostov State U., Prospekt Stachki 194, Rostov-on-Don 344104, USSR. (1949) Cand, physics and mathematics (Rostov U., 1981). Sr. scient. *Powder X-ray diffraction, structural phase transitions.*

Chernov, Dr Aleksandr Nikolaevich. Dzerzhinsk Branch, Moscow Inst. of Qualification Improvement, Gagarin st.3,Gor'kovskaya obl., Dzerzhinsk 606000, USSR. (1944) Cand, physics and mathematics (INCRYS, 1970). Docent. *Crystal chemistry, organic and metal-organic compounds, X-ray analysis.*

Chernov, Prof. Alexander Alexandrodich. Inst. of Crystallography, Acad. Sci. USSR, Leninsky pr. 59, Moscow 117333, USSR. (1931) Corresp. member, Acad. Sci. USSR, Dr, physics and mathematics (INCRYS, 1970). Lab. head. *Crystal growth, surface phenomena, solid state theory*

Chernyshev, Dr Vladimir Vasil'evich. Dept. of Chemistry, Moscow State U., Leninskie Gory, Moscow 117234, USSR. (1955) Cand, physics and mathematics (Inst.Applied.Physics,Kishinev, 1988).Scient. *Precision X-ray diffractometry, single crystals, computing methods.*

Chernysheva, Dr Marina Alexandrovna. Inst. of Crystallography, Acad. Sci. USSR, Leninsky pr. 59, Moscow 117333, USSR. (1911) Cand, physics and mathematics (INCRYS, 1955). Sr. scient. *Real structure, mechanical properties.*

Chernysheva, Mrs Valentina Fedorovna. Leningrad State U., Dept. of Geology, University Emb. 7/9, Leningrad 199164, USSR. (1933) Asst. *Crystallography, crystal optics, crystal chemistry, goniometry.*

Chertanova, Dr Lyubov' Fedorovna. Inst. of Organic and Physical Chemistry, Kazan Branch of the USSR Acad.Sci., acad. Arbuzov st.8, Kazan 420083, USSR. (1958) Cand, chemistry (Inst. of Chemical Engineering, Kazan, 1988). Scient. *Organic crystal chemistry, molecular and crystal structure, organic compounds.*

Chetkina, Dr Larisa Arkadyevna. Karpov Physical Chemistry Inst., Obukha St. 10, Moscow 103064, USSR. (1932) Cand, physics and mathematics (INCRYS, 1966). Sr. scient. *X-ray structure analysis, crystal physics, crystal chemistry, organic compounds.*

Chiragov, Dr Mamed Isa ogly. Azerbaidzhan State U., Patrisa Lumumba St. 23, Baku 370073, USSR. (1937) Cand, geology and mineralogy (MGU, 1969). Docent. *X-ray structure analysis, crystal chemistry, silicates.*

Chirgadze, Dr Yury Nikolayevich. Inst. of Protein, Acad. Sci. USSR, Pushchino, Serpukhovsky Rayon, Moskovskaya Oblast' 142292, USSR. (1935) Dr, biology (Inst. of Chemical Physics, 1983). Lab. head. *X-ray structure analysis, globular proteins.*

Chudinov, Prof. Sergei Mikhailovich. Physics Dept., Moscow State U., Leninskie Gory, Moscow 117234, USSR. (1935) Dr, physics and mathematics (MGU, 1979).Prof. *Low-dimensional crystals (intercalated layer compounds, superlattices, Wigner crystals, quasi-unidimensional crystals).*

Chudinova, Dr Svetlana Alekseyevna. Petrozavodsk State U., Lenin St. 33, Petrozavodsk 185018, USSR. (1939) Cand, physics and mathematics, (Ural U.). Docent. *Metal oxides, structure, order-disorder transformations, solid solutions.*

Chukhovsky, Dr Felix Nikolaevich. Inst. of Crystallograrhy, Acad. Sci. USSR, Leninsky pr. 59, Moscow 117333, USSR. (1940) Dr, physics and mathematics (INCRYS, 1986). Sr. scient. *X-ray crystal optics.*

Chumakova, Dr Svetlana Petrovna. Inst. of Crystallography, Acad. Sci. USSR, Leninskii Prospekt 59, Moscow 117333, USSR. Cand, physics and mathematics (INCRYS, 1982) Scient. *Liquid crystals.*

Chuprunov, Dr Evgeniy Vladimirovich. Gorky State U., Sverdlova 37, Gorky 603000, USSR. (1951) Cand, physics and mathematics (Inst. of Applied Physics, Acad. Sci. Moldavian SSR, 1979). Docent. *X-ray structure analysis, symmetry theory.*

Chvalun, Dr, Sergei Nikolaevich. Karpov Physical-Chemical Institute, Obukha 10, Moscow 103064, USSR. (1955) Cand, physics and mathematics (KARPOV, 1981).Sr. scient. *X-ray analysis, polymers.*

D'yachenko, Dr Oleg Anatolyevich. Branch Inst. of Chemical Physics, Acad. Sci. USSR, Chernogolovka 142432, Noginsky Rayon, Moskovskaya Oblast', USSR. (1939) Dr, chemistry (Bransh Inst. of Chem. Physics, 1986). Sr. scient. *Crystal chemistry, organic compounds.*

D'yakon, Dr Ivan Andreyevich. Inst. of Applied Physics, Acad. Sci. Moldavian SSR, Akademicheskaya 5, Kishinev 277028, USSR. (1934) Cand, physics and mathematics (INCRYS, 1970). Sr. scient. *X-ray structure analysis, instrumentation, inorganic compounds, structure.*

Darinskaya, Dr Elena Vladimirovna. Inst. of Crystallography, Acad. Sci. USSR, Leninskii Prospekt 59, Moscow 117333, USSR. Cand, physics and mathematics,(INCRYS, 1984). Scient. *Real structure, mechanical properties.*

Datt, Dr Igor Daudovich. Mendeleev Inst.of Chemical Technology, Miusskaya sq. 9, Moscow 125820, USSR. (1941) Cand, physics and mathematics (KARPOV, 1973) Docent. *Structure analysis, neutron diffraction.*

Davydchenko, Dr Anatoliy Georgiyevich. Res. Inst. for Synthesis of Mineral Raw Materials, Institutskaya St. 1, Alexandrov 601600, Vladimirskaya Oblast', USSR. (1934) Cand, geology and mineralogy (IGEM, 1966). Dept. head. *Crystal growth, natural and artificial crystal formation, physical chemistry.*

Dedukh, Dr Leonid Mikhailovich. Inst. of Solid State Physics, Acad. Sci. USSR, Moskovskaya obl.,Chernogolovka 142432, USSR. (1934) Doctor, phsics and mathematics (Inst. of Solid State Physics, Chernogolovka, 1986). Sr. scient. *Magnetic properties, defects, amorphous state.*

Degtyareva, Dr Valentina Feognievna. Inst. of Solid State Physics, Acad. Sci. USSR, Moskovskaya obl., Chernogolovka 142432, USSR. Cand, physics and mathematics (Inst. of Steel and Alloys, Moscow 1975).Sr. scient. *Phase transformations in solids, high pressure, amorphous state.*

Dem'yanets, Dr Lyudmila Nikolayevna. Inst. of Crystallography, Acad. Sci. USSR, Leninsky pr. 59, Moscow 117333, USSR. (1939) Cand, chemistry (INCRYS, 1966). Sr. scient. *Crystal chemistry, crystal growth.*

Dembo, Dr Alexander Teodorovich. Inst. of Crystallography, Acad. Sci. USSR, Leninsky Pr. 59, Moscow 117333, USSR. (1939) Cand, biology (Inst. of Mol. Biology and Genetics, Acad. Sci. Ukrainian SSR, 1977). Sr. scient. *Structure, biopolymers, viruses, small-angle scattering, X-ray scattering, neutron scattering.*

Denisenko, Dr Georgy Alexandrovich. Inst. of Crystallography, Acad. Sci. USSR, Leninsky pr. 59, Moscow 117333, USSR. (1945) Cand, physics and mathematics (Kazan' U., 1975). Scient. secretary. *Crystal spectroscopy.*

Deyanov, Dr Ramil' Zinyatullovich. Computing Dept., Moscow State U., Leninskie Gory, Moscow 117234, USSR. (1953) Cand, physics and mathematics (Moscow Physical-Technical Inst., 1988). Scient. *Diffraction data processing.*

Dimitrova, Dr Ol'ga Vladimirovna. Moscow State U., Dept. of Geology, Leninskiye Gory, Moscow 119899, USSR. (1948) Cand, geology and mineralogy (MGU, 1977). Scient. *Crystal chemistry, rare earth compounds.*

Dmitrieva, Dr Tatyana Vladimirovna. Inst. of Crystallography, Acad. Sci. USSR, Leninsky pr. 59, Moscow 117333, USSR. (1933) Cand, physics and mathematics (INCRYS, 1975). Scient. *Magnetic properties, Mössbauer effect.*

Dmitriyeva, Dr Margarita Timofeyevna. Inst. of Geology, Mineralogy and Petrography, Acad. Sci. USSR, Staromonetny 35, Moscow 109017, USSR. (1932) Cand, geology and mineralogy (IGEM, 1977). Sr. scient. *X-ray structure analysis, minerals.*

Dodokin, Dr Anatoly Petrovich. Inst. of Crystallography, Acad. Sci. USSR, Leninsky pr. 59, Moscow 117333, USSR. (1943) Cand, physics and mathematics (Physico-Technical Inst., Acad. Sci. USSR, 1972). Sr. scient. *Magnetic properties, Mössbauer effect.*

Dolivo-Dobrovol'skaya, Dr Galina Ilyinishna. Leningrad Mining Inst., 21st Liniya 2, Leningrad 199026, USSR. (1935) Cand, geology and mineralogy (GORNY, 1964). *Crystal morphology, defects.*

Dolivo-Dobrovol'skaya, Mrs Elena Maximovna. Leningrad State U., Dept. of Geology, University Emb. 7/9, Leningrad 199164, USSR. (1927). Scient. *Crystal chemistry, X-ray structure analysis, symmetry.*

Domnitskaya, Dr Yaroslava Fedorovna. Chemistry Dept., Lvov State U., Universitetskaya 1, Lvov 290602, .USSR. (1952) Cand, chemistry (Lvov U., 1982) Asst. *Crystal chemistry, metal phosphides.*

Dorfman, Dr Moisey Davydovich. Fersman Mineralogical Museum, Leninsky pr. 14/16, Moscow 117071, USSR. (1908) Dr, geology and mineralogy (IGEM, 1962). Sr. scient. *Crystal chemistry, minerals.*

Dorokhova, Dr Galina Igorevna. Moscow State U., Dept. of Geology, Leninskiye Gory, Moscow 119899, USSR. (1952) Cand, geology and mineralogy (MGU, 1983). Asst. *X-ray structure analysis, crystal chemistry, inorganic compounds, morphology, minerals.*

Doroshinsky, Dr Alexander Leibovich. Inst. of New Chemical Problems, Acad. Sci. USSR, Chernogolovka 142432, Moskovskaya Oblast', USSR. (1933) Cand, chemistry (KARPOV, 1973). Scient. *Crystal chemistry, coordination compounds.*

Drits, Dr Victor Anatolyevich. Inst. of Geology, Acad. Sci. USSR, Pyzhevsky per. 7, Moscow 109017, USSR. (1932) Dr, geology and mineralogy (IGEM, 1905). Lab. head. *Layer minerals, structure, diffraction methods.*

Drozdov, Dr Yury Nikolayevich. Gorky State U., Gagarin Prospekt 23, Gorky 603022, USSR. (1947) Cand, physics and mathematics (Gorky U., 1974). Sr. scient. *X-ray structure analysis.*

Dubov, Dr Petr Lvovich. Leningrad State U., University Emb. 7/9, Leningrad 199164, USSR. (1943) Cand, geology and mineralogy (LGU, 1971). Sr. lect. *Symmetry theory and history, geometrical crystallography.*

Dubravina, Dr Aida Nikolaevna. Moscow Inst. of Steel and Alloys. Leninskii Prospekt 4, Moscow 117049, USSR. (1935) Cand, physics and mathematics

(Moscow Inst. of Steel and Alloys,1966). Docent. *Phase and structural transitions in solids.*

Dudarev, Dr Vasily Yakovlevich. Karpov Physical Chemistry Inst., Obukha St. 10, Moscow 107120, USSR. (MGU, 1963). *Structure analysis (X-ray - neutrons - electrons).*

Duderov, Dr Nikolay Grigoryevich. Inst. of Crystallography, Acad. Sci. USSR, Leninsky pr. 59, Moscow 117333, USSR. (1945) Cand, chemistry (INCRYS, 1975). Scient. *Crystal growth, crystal chemistry.*

Dudkevich, Prof. Vladimir Petrovich. Rostov State U., Dept. of Physics, Prospekt Stachki 192, Rostov-on-Don 344061, USSR. (1935) Dr, physics and mathematics (Rostov U., 1980). Prof. *Defect structures.*

Dukova, Dr Elena Dmitriyevna. Inst. of Crystallography, Acad. Sci. USSR, Leninsky pr. 59, Moscow 117333, USSR. (1925) Cand, physics and mathematics (INCRYS, 1956). *Crystal morphology, crystal growth.*

Dvorkin, Dr Alexander Arkadyevich. Inst. of Applied Physics, Acad. Sci. Moldavian SSR, Akademicheskaya 5, Kishinev 277028, USSR. (1947) Cand, physics and mathematics (Inst. of Applied Physics, Acad. Sci. Moldavian SSR, 1975). Scient. *X-ray structure analysis, crystal chemistry.*

Dyuzheva, Dr Tat'yana Ivanovna. Inst. of High Pressure Physics, Acad. Sci. USSR, Moskovskaya obl., Troitsk 142092, USSR. (1954) Cand, geology and mineralogy (MGU, 1979). Sr. scient. *Crystal structure, high pressure.*

Dzhafarov, Dr Kara Mustafa ogly. Inst. of Physics, Acad. Sci. Azerbaidzan SSR, Prospekt Narimanova 33, Baku 370143, USSR. (1954) Cand, physics and mathematics (Inst. of Physics, Acad. Sci. Azerbaidzan SSR, 1986). Sr. scientist. *Structure, semiconductors, crystal chemistry, inorganic compounds.*

Dzyabchenko, Dr Aleksandr Valentinovich. Karpov Physical-Chemical Inst., Obukha 10, Moscow 103064, USSR. (1948) Cand, chemistry (KARPOV, 1981). Scient. *Structure, molecular crystals, symmetry, crystal data banks.*

Efendiev, Dr El'dar Gusein ogly. Inst. of Physics, Acad. Sci. Azerbaidzan SSR, Prospekt Narimanova 33, Baku 370143, USSR. (1947) Cand, physics and mathematics (Inst. of Physics, Baku, 1976). Sr. scient. *Thin films, phase transitions.*

Efimov, Dr Aleksandr Vasil'evich. Inst. of Proteins, Acad. Sci. USSR. Moskovskaya obl., Pushchino 142292, USSR. (1954) Cand, chemistry (Inst. of Proteins, Pushchino, 1963). Sr. scient. *Crystallography, proteins, nucleic acids.*

Efremov, Dr Valery Alexandrovich. Inst. of Chemical Reagents and Pure Substances, Acad. Sci. USSR, Bogorodsky Val 3, Moscow 107258, USSR. (1950) Cand, chemistry (MGU, 1976). Sr. scient. *Crystal chemistry, inorganic compounds.*

Efremova, Dr Elena Pavlovna. Inst. of Crystallography, Acad. Sci. USSR, Leninskii Prospekt 59, Moscow 117333, USSR. (1949) Cand, chemistry (INCRYS, 1982) Scient. *Crystal growth, morphology, physical chemistry, aqueous growth systems.*

Egorov-Tismenko, Dr Yury Klavdiyevich. Moscow State U., Dept. of Geology, Leninskiye Gory, Moscow 117234, USSR. (1938) Cand, geology and mineralogy (MGU, 1973). Docent. *Crystal chemistry, inorganic compounds, X-ray structure analysis.*

Em, Dr Vyacheslav Terent'evich. Inst. of Nuclear Physics, Uzbek SSR Acad. Sci. Ulugbek, Tashkent 702132, USSR. (1945) Cand, physics and mathematics (Joint Scient. Council of the Uzbek SSR Acad. Sci., Tashkent, 1975) Scient. *Crystallography, interstitiaal phases.*

Entin, Dr Il'ya Ruvimovih. Inst. of Solid State Physics, Acad. Sci. USSR, Moskovskaya obl., Chernogolovka 142432, USSR. (1946) Doctor, physics and mathematics (Inst.of Solid State Physics,Chernogolovka,1987) Sr. scientist. *X-ray optics and acoustics.*

Eremkin, Dr Vladimir Vasil'evich. Inst. of Physics, Rostov State U., Prospekt Stachki 194, Rostov-on-Don, USSR. (1955) Cand, physics and mathematics (Rostov U., 1984). Scient. *X-ray analysis, inorganic compounds, crystal growth.*

Esipova, Dr Nataliya Georgiyevna. Inst. of Molecular Biology, Acad. Sci. USSR, Vavilov St. 32, Moscow 117312, USSR. Cand, physics and mathematics (Inst. of Biophysics, Acad. Sci. USSR). Sr. scient. *Fibrillous structures.*

Farber, Dr Boris Yakovlevich. Inst. of Solid State Physics, USSR Acad,Sci., Moskovskaya obl., Chernogolovka 142432, USSR. (1954) Cand, physics and mathematics (Inst. of Steel and Alloys,Moscow, 1983). Sr. scient. *Real structure, dynamical properties, crystal lattice.*

Fedorenko, Prof. Anatoly Ivanovich. Kharkov Polytechnic Inst., Frunze St. 21, Kharkov 310002, USSR. (1937) Dr, technics (Kharkov Polytechnic Inst., 1978). Docent. *Thin film structure, defects, methods for defect study.*

Fedorov, Dr Aleksandr Aleksandrovich. Inst. of Molecular Biology, Acad. Sci. USSR, Vavilov st.32, Moscow 117984, USSR. (1946) (Inst. Molec. Biolog., 1980). Sr. scient. *Protein crystallography, X-ray structure analysis theory.*

Fedorov, Dr Boris Alexandrovich. Inst. of Protein, Acad. Sci. USSR, Pushchino 142292, Serpukhovsky Rayon, Moskovskaya Oblast', USSR. (1939) Cand, physics and mathematics (Inst. of High Molecular Compounds, Acad. Sci. USSR, 1966). Sr. scient. *X-ray structure analysis, proteins, diffuse X-ray scattering, macromolecules in solution.*

Fedorov, Dr Pavel Pavlovich. Inst. of Crystallography, Acad. Sci. USSR, Leninsky pr. 59, Moscow 117333, USSR. (1950) Cand, chemistry (Moscow Inst. of Fine Chem. Techn., 1977). Sr. scient. *Crystal growth, phase diagrams.*

Fedotov, Dr Alexander Fedorovich. Inst. of Geology - Mineralogy and Petrography, Acad. Sci. USSR, Staromonetny 35, Moscow 109017, USSR. (1924) Cand,

geology and mineralogy (IGEM, 1974). Sr. scient. *Structural mineralogy, electron diffraction.*

Fedyna, Dr Mikhail Fedorovich. Chemistry Dept, Lvov State U., Universitetskaya 1, Lvov 290602, USSR. (1960) Cand, chemistry (Lvov U., 1988). Scient. *Crystal chemistry, intermetallic compounds.*

Feigin, Dr Lev Abramovich. Inst. of Crystallography, Acad. Sci. USSR, Leninsky pr. 59, Moscow 117333, USSR. (1927) Dr, physics and mathematics (INCRYS, 1976). Lab. Head. *Biopolymer structures, X-ray diffraction.*

Fesenko, Prof. Evgeny Grigoryevich. Rostov State U., Dept. of Physics, Prospekt Stachki 192, Rostov-on-Don 344061, USSR. (1918) Dr, physics and mathematics (Rostov U., 1973). Dir., Physics Inst. *Crystal chemistry, complex oxides, structure, physical properties.*

Fesenko, Dr Oleg Evgenyevich. Rostov State U., Prospekt Stachki 192, Rostov-on-Don 344090, USSR. (1950) Cand, physics and mathematics (Rostov U., 1978). Lab. head. *Ferroelectric crystal physics.*

Fetisov, Dr Gennadii Vladimirovich. Chemistry Dept., Moscow State U., Leninskie Gory, Moscow 117234, USSR. (1947) Cand, engineering (Moscow Physical Eng. Inst., 1973). Sr. scient. *Precision X-ray diffractometry, single crystals, instrumentation.*

Filatov, Dr Stanislav Konstantinovich. Leningrad State U., Dept. of Geology, University Emb. 7/9, Leningrad 199164, USSR. (1940) Dr, geology and mineralogy (LGU, 1987). Docent. *Crystallography, crystal chemistry, X-ray structure analysis.*

Filip'yev, Dr Victor Semenovich. Rostov State U., Dept. of Physics, Prospekt Stachki 192, Rostov-on-Don 344061, USSR. (1937) Cand, physics and mathematics (Rostov U., 1966). Docent. *Crystal chemistry, complex oxides, structure and physical properties, crystals.*

Filipenko, Dr Olga Savelyevna. Branch Inst. of Chemical Physics, Acad. Sci. USSR, Chernogolovka 142432, Noginsky Rayon, Moskovskaya Oblast', USSR. (1940) Cand, chemistry (MGU, 1971). Sr. scient. *Crystal chemistry, organic compounds.*

Finkel'shtein, Dr Aleksey Vital'yevich. Inst. of Protein, Acad. Sci. USSR, Pushchino 142292, Serpukhovsky Rayon, Moskovskaya Oblast', USSR. (1947) Cand, physics and mathematics (Moscow Physico-Techn. Inst., 1976). Sr. scient. *Structure, proteins, nucleic acids.*

Flerov, Dr Igor' Nikolayevich. Inst. of Physics, Acad. Sci. USSR, Siberian Dept., Akademgorodak, Krasnoyarsk 660036, USSR. (1942) Cand, physics and mathematics (Krasnoyarsk Inst. of Physics, Acad. Sci. USSR, 1978). Sr. scient. *Thermal properties, phase transitions (structural).*

Fofanov, Dr Anatolii Dmitrievich. Petrozavodsk State U., Prospekt Lenina 33, Petrazavodsk 185640, USSR. (1946) Cand, chemistry (Chernovtsy U., 1974) Docent. *Structure, amorphous materials, X-ray analysis, polycrystals.*

Fotchenkov, Dr Anatoly Andreyevich. Res. Inst. for Synthesis of Mineral Raw Materials, Institutskaya St. 1, Alexandrov 601600, Vladimirskaya Oblast', USSR. (1925) Cand, physics and mathematics (INCRYS, 1960). Dept. head. *Physical properties, piezoelectric properties,dialectric properties, acoustic and elastic properties, optical properties.*

Frank-Kamenetskaya, Dr Olga Victorovna. Leningrad State U., Dept. of Geology, University Emb. 7/9, Leningrad 199164, USSR. (1945) Cand, geology and mineralogy (LGU, 1973). Scient. *Structure analysis, crystal chemistry.*

Frank-Kamenetsky, Prof. Victor Al'bertovich. Leningrad State U., University Emb. 7/9, Leningrad 199164, USSR. (1915) Dr, geology and mineralogy (GORNY, 1962) Prof. *Crystallography, crystal chemistry, X-ray structure analysis, layer silicates.*

Franke, Dr Valeriya Dmitriyevna. Leningrad State U., University Emb. 7/9, Leningrad 199164, USSR. (1945) Cand, geology and mineralogy (LGU, 1982). Sr. scient. *Crystal growth, morphology.*

Fridkin, Prof. Vladimir Mikhailovich. Inst. of Crystallography, Acad. Sci. USSR, Leninsky pr. 59, Moscow 117333, USSR. (1929) Dr, physics and mathematics (INCRYS, 1963). Sr. scient. *Physical properties, phase transitions.*

Fundamensky, Mr Vladimir Semenovich. NPO Burevestnik, Stakhanovtsev St. 1, Leningrad 195112, USSR. (1946) Lab. head. *Methods, X-ray structural analysis, crystal chemistry, organic and complex compounds.*

Furmanova (Bokii), Dr Nina Georgievna. Inst. of Crystallography, Acad. Sci. USSR, Leninsky pr. 59, Moscow 117333, USSR. (1939) Cand, chemistry (INEOS, 1968). Sr. scient. *Crystal chemistry, organic and organometallic compounds.*

Fursenko, Dr Boris Alexandrovich. Inst. of Geology and Geophysics, Acad. Sci. USSR, Siberian Dept., University Prospekt 3, Novosibirsk 630090, USSR. (1946) Cand, geology and mineralogy (Inst. of Geology and Geophysics, Acad. Sci. USSR, Siberian Dept., 1973). Sr. scient. *Properties, X-ray structure analysis, high pressure.*

Fykin, Dr Leonid Efimovich. Branch Karpov Physical Chemistry Inst., Obninsk, Kaluzhskaya oblast',249020, USSR. (1936) Cand, physics and mathematics (Moscow Inst. of Eng. and Physics, 1972). Sr. scient. *Neutron diffraction, instrumentation, methods, structure analysis.*

Gabuda, Prof. Svyatoslav Petrovich. Inst. of Inorganic Chemistry, Acad. Sci. USSR, Siberian Dept., Lavrent'yev Prospekt 3, Novosibirsk 630090, USSR. (1936) Cand, physics and mathematics (Acad. Sci. USSR, Siberian Dept., 1970). Lab. head. *Non-diffraction structure methods, NMR, crystal chemistry, hydrogen.*

Gagarina, Dr Elena Stanislavovna. Inst. of Physics, Rostov State U., Pr. Stachki 194, Rostov-on-Don 344104 , USSR. (1949) Cand, physics and mathematics (Rostov U., 1988). Scient. *Structure, inorganic crystals, phase transitions, twinning.*

Galitsky, Dr Nikolay Mikhailovich. Inst. of Bio-organic Chemistry, Acad. Sci. Belorussian SSR, Leninsky pr. 68, Minsk 220600, USSR. (1950) Cand, chemistry (Inst. of Bio-organic Chemistry Belorussian Acad. Sci., 1978). Group head. *X-ray structure analysis, peptide-protein compounds.*

Galiulin, Dr Ravil Vagizovich. Inst. of Crystallography, Acad. Sci. USSR, Leninsky pr. 59, Moscow 117333, USSR. (1940) Dr, physics and mathematics (INCRYS, 1978). Sr. scient. *Crystallography fundamentals, mathematical crystallography.*

Galstyan, Dr Viktor Gaikovich. Inst. of Crystallography, Acad. Sci. USSR, Leninsky pr. 59, Moscow 117333, USSR. (1940) Dr, physics and mathematics (MGU, 1971). Sr. scient. *Surfaces, real structure.*

Gamarnik, Dr Moisei Yanvelevich. Inst. of Geochemistry and Physics of Minerals, Ukrainian SSR Acad. Sci., Prospekt Palladina 34, Kiev 252680, USSR. (1936) Cand, physics and mathematics (Kharkov U., 1984). Scient. *X-ray crystallography, ultradisperse systems.*

Garashina, Dr Lyudmila Solomonovna. Inst. of Crystallography, Acad. Sci. USSR, Leninsky pr. 59, Moscow 117333, USSR. (1939) Cand, chemistry (IONCH, 1969). Scient. *Isomorphism.*

Gavrilova, Dr Irina Vladimirovna. Inst. of Crystallography, Acad. Sci. USSR, Leninsky pr. 59, Moscow 117333, USSR. (1924) Cand, physics and mathematics (INCRYS, 1973). Scient. *Crystal growth.*

Gavrilova, Dr Nadezhda Dmitrievna. Physics Dept., Moscow State U., Leninskie Gory, Moscow 117234, USSR. (1937) Cand, physics and mathematics (MGU, 1965) Scient. *Ferro- and pyrroelectrics.*

Gavrilyachenko, Dr Victor Georgievich. Physics Dept., Rostov State U., Pr. Stachki 192, Rostov-on-Don, USSR. (1935) Cand,physics and mathematics (Rostov U., 1971). Docent. *Phase transitiions, domain structure, ferroelectrics, antiferroelectrics, ferroelastic crystals.*

Geguzina, Dr Galina Alexandrovna. Rostov State U., Dept. of Physics, Prospekt Stachki 192, Rostov-on-Don 344061, USSR. (1945) Cand, physics and mathematics (Rostov U., 1975). Sr. scient. *Crystal chemistry, complex oxides, structure and physical properties.*

Gel'man, Dr Yury Alexandrovich. Inst. of Crystallography, Acad. Sci. USSR, Leninsky pr. 59, Moscow 117333, USSR. (1938) Cand, physics and mathematics (INCRYS, 1978). Sr. scient. *Crystal growth from gas phase, sublimation, surface studies.*

Genin, Dr Yakov Vladimirovich. Inst. of Elemento-Organic Compounds, Acad. Sci. USSR, Vavilov St. 28, Moscow 117813, USSR. (1941) Cand, physics and mathematics (INCRYS, 1976). Scient. *Polymer X-ray diffraction, polymer physics.*

Genkina, Dr Elena Aleksandrovna. Inst. of Crystallography, Acad. Sci. USSR, Leninskii Prospekt 59, Moscow 117333, USSR. (1954) Cand, geology and mineralogy (MGU, 1987). Scient. *Structure analysis, inorganic compounds.*

Gerasimov, Dr Viktor Ivanovich. Mordovian State U., Bol'shevistskaya 68a, Saransk 430000, USSR. (1943) Cand, physics and mathematics.(Inst. Appl. Physics, Kishinev, 1979).Docent. *X-ray analysis.*

Geras'kin, Dr Valerii Vasil'evich. Moscow Inst. of Steel and Alloys. Leninskii Prospekt 4, Moscow 117049, USSR. (1940) Cand, physics and mathematics (Moscow Inst. of Steel and Alloys,1970). Docent. *Crystal physics and optics.*

Gevork'yan, Dr Svetlana Vasil'evna. Inst. of Geochemistry and Physics of Minerals, Ukrainian SSR Acad. Sci., Prospekt Palladina 34, Kiev 252680, USSR. (1934) Cand, geology and mineralogy (IGFM, 1977) Sr. scient. *Infrared spectroscopy of crystals.*

Gilinskaya, Dr Emma Abramovna. Inst. of Scient. and Techn. Information, Acad. Sci. USSR, Baltiyskaya St. 14, Moscow 125219, USSR. (1922) Cand, chemistry (IONCH, 1953). *Crystal chemistry, coordination compounds.*

Givargizov, Dr Evgeny Inviyevich. Inst. of Crystallography, Acad. Sci. USSR, Leninsky pr. 59, Moscow 117333, USSR. (1934) Dr, physics and mathematics (INCRYS, 1976). Lab. Head. *Crystal growth, whisker-crystal growth from vapor phase.*

Gladkikh, Dr Liliya Ivanovna. Kharkov Polytechnic Inst., Frunze St. 21, Kharkov 310002, USSR. (1934) Cand, technics (Kharkov Polytechnic Inst., 1966). Docent. *Metal and alloy structures, defects, methods.*

Gladky, Dr Vsevolod Vladimirovich. Inst. of Crystallography, Acad. Sci. USSR, Leninsky pr. 59, Moscow 117333, USSR. (1934) Dr, physics and mathematics (INCRYS, 1985). Sr. scient. *Ferroelectrics (physics), phase transitions (ferroelectric).*

Gladyshevsky, Prof. Evgeny Ivanovich. Lvov State U., University St. 1, Lvov 290602, USSR. (1924) Dr, chemistry (MGU, 1967). Head of Chair. *X-ray structure analysis, crystal chemistry, intermetallic compounds.*

Glazov, Dr Aleksey Ivanovich. Leningrad Mining Inst., 21st Liniya 2, Leningrad 199026, USSR. (1942) Cand, geology and mineralogy (GORNY, 1976). Sr. scient. *Crystal morphology, X-ray diffraction.*

Glikin, Mr Arkady Eduardovich. Leningrad State U., Dept. of Geology, University Emb. 7/9, Leningrad 199034, USSR. (1943) Cand, geology and mineralogy (LGU, 1978). Sr. scient. *Crystal growth, crystal morphology.*

Godovikov, Prof. Alexander Alexandrovich. Fersman Mineralogical Museum, Leninsky pr. 18, Moscow 117071, USSR. (1927) Dr, geology and mineralogy (MGU, 1970). Dir. *Theoretical and experimental mineralogy, mineral synthesis.*

Goilo, Dr Eduard Al'bertovich. Leningrad State U., Dept. of Geology, University Emb. 7/9, Leningrad 199034, USSR. (1941) Cand, geology and mineralogy (LGU, 1970). Sr. scient. *X-ray and electron diffraction, layer silicates, crystal chemistry.*

Golovachev, Dr Vladimir Pavlovich. Gorky State U., Gagarin Prospekt 23, Gorky 603022, USSR. (1930) Cand, physics and mathematics (INCRYS, 1972). Docent. *X-ray structure analysis.*

Golovastikov, Dr Nikolay Ivanovich. Inst. of Crystallography, Acad. Sci. USSR, Leninsky pr. 59, Moscow 117333, USSR. (1915) Cand, physics and mathematics (INCRYS, 1953). Sr. scient. *Structure analysis.*

Golovin, Dr Andrei Leonidovich. Inst. of Crystallography, Acad. Sci. USSR, Leninsky Prospekt 59, Moscow 117333, USSR. (1956) Cand, physics and mathematics (INCRYS, 1985) Scient. *Real structure, subsurface crystal layers.*

Golovina, Dr Nina Ivanovna. Branch Inst. of Chemical Physics, Acad. Sci. USSR, Chernogolovka, Noginsky Rayon, Moskovskaya Oblast' 142432, USSR. (1934) Dr, chemistry (Branch Inst. of Chemical Physics, 1988). Sr. scient. *Metallo-organic compounds, structure.*

Golovko, Dr Yurii Ilarionovich. Inst. of physics, Rostov State U., Pr. Stachki 194, Rostov-on-Don 344104, USSR. (1947) Cand, physics and mathematics (Rostov U., 1980). Sr. scient. *Structure, thin films, complicated oxides.*

Golubev, Dr Alexander Mikhailovich. Inst. of Crystallography, Acad. Sci. USSR, Leninsky pr. 59, Moscow 117333, USSR. (1948) Cand, chemistry (MGU, 1975). Sr. scient. *Precision structures, inorganic compounds.*

Golubinskii, Dr Aleksei Vladimirovich. Chemistry Dept., Moscow State U., Leninskie Gory, Moscow 117234, USSR. (1941). Cand, chemistry (MGU, 1978). Scient. *Gas electron diffraction.*

Golubkov, Dr Alexander Vasilyevich. Physico-Techn. Inst., Acad. Sci. USSR, Zapovednaya 51, Leningrad 194047, USSR. Cand, chemistry (Inst. of Silicate Chemistry, Acad. Sci. USSR, 1969). Sr. sci. *Rare earth element compounds.*

Golyshev, Dr Vladimir Mikhailovich. Mordovian State U., Bol'shevistskaya 68a, Saransk 430000, USSR. Cand, physics and mathematics (INCRYS, 1971).Docent. *X-ray analysis.*

Goncharov, Dr Aleksandr Vasil'evich. Vladimir State Pedagogical Inst., Prospekt Stroitelei 11, Vladimir 600024, USSR. (1949) Cand, physics and mathematics (INCRYS, 1975). Docent. *X-ray analysis, organic compounds.*

Goncharov, Dr Georgy Nikolayevich. Leningrad State U., Dept. of Geology, University Emb. 7/9, Leningrad 199034, USSR. (1941) Cand, geology and mineralogy (LGU, 1968). Docent. *Physics and chemistry, minerals, Mössbauer spectroscopy.*

Gorbunova, Dr Yuliya Efimovna. Inst. of General and Inorganic Chemistry, Acad. Sci. USSR, Leninsky pr. 31, Moscow 117071, USSR. (1932) Cand, chemistry (IONCH, 1971). Scient. *Crystal chemistry, inorganic compounds.*

Gordiyenko, Dr Vladimir Vasilyevich. Leningrad State U., University Emb. 7/9, Leningrad 199034, USSR. (1934) Dr, geology and mineralogy (LGU, 1988). Sr. scient. *Mineralogical crystallography, geochemistry, X-ray studies, minerals.*

Gorelik, Prof. Semen Samuilovich. Moscow Inst. of Steel and Alloys. Leninskii Prospekt 4, Moscow 117049, USSR. (1911) Doctor, physics and mathematics (Moscow Inst. of Steel and Alloys,1962). Professor. *Structure defects, textures.*

Gorina, Dr Iza Ivanovna. Inst. of Crystallography, Acad. Sci. USSR, Leninsky pr. 59, Moscow 117333, USSR. (1936) Cand, chemistry (Inst. of Oil-Chem. Synthesis, Acad. Sci. USSR, 1966). Sr. scient. *Liquid crystals, physical chemistry.*

Gorogotskaya, Dr Lydumila Ivanova. Inst. of Geochemistry and Physics of Minerals, Acad. of Sci. Ukrainian SSR, Palladin Prospekt 34, Kiev 252068, USSR. (1935) Cand, geology and mineralogy (IGEM, 1967). Sr. scient. *Crystal chemistry, crystal structure, minerals.*

Gorshkov, Prof. Anatoly Ivanovich. Inst. of Geology, Mineralogy and Petrography (IGEM), Staromonetny 35, Moscow 109017, USSR. (1929) Dr, geology and mineralogy (IGEM, 1971). Lab. head. *Mineral structure, morphology, phase transformation.*

Gorskaya, Dr Marina Gennad'evna. Geology Dept., Leningrad State U., Universitetskaya Nab.7/9, Leningrad 199034, USSR. (1950) Cand, geology and mineralogy (LGU, 1985).Scient. *Crystal chemistry, tourmalins, volcanic minerals.*

Grebenshchikov, Prof. Roman Georgiyevich. Inst. of Silicate Chemistry, Acad. Sci. USSR, Makarov Emb. 2, Leningrad 199164, USSR. (1929) Dr, chemistry (Inst. of Silicate Chemistry, Acad. Sci. USSR, 1967). Lab. head. *Crystal chemistry, X-ray crystallography, inorganic compounds, silicates.*

Grechushnikov, Prof. Boris Nikolayevich. Inst. of Crystallography, Acad. Sci. USSR, Leninsky pr. 59, Moscow 117333, USSR. (1925) Dr, physics and mathematics (INCRYS, 1986). Lab. head. *Crystal optics, optical spectroscopy, radio spectroscopy.*

Grigorov, Dr Sergey Nikolayevich. Kharkov Polytechnic Inst., Frunze St. 21, Kharkov 310002, USSR. (1944) Cand, technics (Kharkov Polytechnic Inst., 1971). Docent. *Crystal structure defects, thin film growth, electron microscopy.*

Grigoryev, Prof. Dimitry Pavlovich. Leningrad Mining Inst., 21st Liniya 2, Leningrad 199026, USSR. (1909) Dr, geology and mineralogy (IGEM, 1943). Prof. *Crystal growth.*

Grigoryeva, Dr Tamara Nikolayevna. Inst. of Geology and Geophysics, Acad. Sci. USSR, Siberian Dept., Prospekt Nauki 3, Novosibirsk 630090, USSR. (1933)

Cand, physics and mathematics (Irkutsk U., 1971). Scient. *Mineral structures, X-ray structure analysis.*

Grin', Dr Yury Nikolayevich. Dept. of Chemistry, Lvov State U., University St. 1, Lvov 742388, USSR. (1955) Cand, chemistry (Lvov U., 1980). Asst. *X-ray structure analysis, crystal chemistry, intermetallic compounds.*

Grinberg, Dr Svetlana Arnol'dovna, Inst. of Crystallography, Acad. Sci. USSR, Leninsky pr. 59, Moscow 117333, USSR. (1944) Cand, physics and mathematics (INCRYS, 1979). Scient. *Crystal growth.*

Grunskii, Dr Oleg Sergeevich. Geology Dept., Leningrad State U., Universitetskaya Nab. 7/9, Leningrad 199034, USSR. (1960) Cand, geology and mineralogy (LGU, 1988). Scient. *Crystal growth and crystal forms.*

Gurin, Dr Vladimir Nikolayevich. Physico-Techn. Inst., Acad. Sci. USSR, Politekhnicheskaya St. 26, Leningrad 194021, USSR. (1936) Cand, chemistry (Physico-Techn. Inst., Acad. Sci. USSR, 1968). Sr. scient. *Crystal growth.*

Gurskaya, Dr Galina Victorovna. Inst. of Molecular Biology, Acad. Sci. USSR, Vavilov St. 32, Moscow 117312, USSR. Cand, physics and mathematics (INCRYS, 1964). Sr. scient. *Biological materials, structure.*

Guseinov, Dr Gakhraman Gusein ogly. Inst. of Physics, Acad. Sci. Azerbaidzhan SSR, Narimanov Prospekt 33, Baku 370143, USSR. (1937) Cand, chemistry (Azerbaidzhan Inst. of Inorg. and Physical Chemistry, 1968). Sr. scient. *Crystal structure, phase transformations, semiconductor compounds.*

Guseinova, Dr Maya Kara kyzy. Inst. for Petrochemical Processes, Acad. Sci. Azerbaidzhan SSR, Telnov St. 30, Baku 370025, USSR. *X-ray structure analysis, crystal chemistry, complex and organic compounds.*

Gushchina, Mrs. Alla Evgen'evna. Inst. of Molecular Biology, Acad. Sci. USSR, Vavilov st. 32, 117984 Moscow, USSR. (1948) Scient. *Protein crystallography.*

Harutunyan, Dr Emil' Haikovich. Inst. of Crystallography, Acad. Sci. USSR, Leninsky pr. 59, Moscow 117333, USSR. (1935) Dr, chemistry (INCRYS, 1984). Sr. scient. *Protein structure, X-ray analysis.*

Ikornikova, Prof. Nina Yuryevna. Inst. of Crystallography, Acad. Sci. USSR, Leninsky pr. 59, Moscow 117333, USSR. (1913) Dr, geology and mineralogy (INCRYS, 1970). *Hydrothermal crystal synthesis, solutions - physical and chemical studies.*

Il'inets, Dr Aleksei Mikhailovich. Central Res. Inst., Organization- Mechanization and Techn. Assistance of Building Eng. Dmitrovskoe shosse 9, Moscow 127434, USSR. (1948) Cand, physics and mathematics (INCRYS, 1985). Lab. Head. *Crystal chemistry, silicates.*

Ilyinsky, Dr Alexander Lvovich. Moscow State U., Dept. of Chemistry, Leninskiye Gory, Moscow 117234, USSR. (1937) Cand, chemistry (MGU, 1975). Sr. scient. *X-ray structure analysis, instrumentation, crystal chemistry.*

Ilyushin, Dr Alexander Sergeyvich. Moscow State U., Dept. of Physics, Leninskiye Gory, Moscow 119899, USSR. (1943) Cand, physics and mathematics (MGU, 1971). Head Chair. *Phase transitions, low-temperature X-ray diffraction.*

Imamov, Dr Rafik Mamedovich. Inst. of Crystallography, Acad. Sci. USSR, Leninsky pr. 59, Moscow 117333, USSR. (1938) Dr, physics and mathematics (INCRYS, 1978). Lab. head. *Electron diffraction, structure analysis, crystal chemistry, real structure.*

Indenbom, Prof. Vladimir Lvovich. Inst. of Crystallography, Acad. Sci. USSR, Leninsky pr. 59, Moscow 117333, USSR. (1924) Dr, physics and mathematics (Leningrad Physical Techn. Inst., 1964). Lab. head. *Real crystals (physics), phase transition theory, X-ray optics, internal strain theory, dislocations.*

Ionov, Dr Pavel Victorovich. Inst. of Crystallography, Acad. Sci. USSR, Leninsky pr. 59, Moscow 117333, USSR. (1941) Cand, physics and mathematics (INCRYS, 1975). Scient. *Photo-stimulated solid processes.*

Ionov, Dr Vladislav Mikhailovich. Moscow State U., Dept. of Chemistry, Leninskiye Gory, Moscow 117234, USSR. (1934) Cand, chemistry (MGU, 1973). Sr. scient. *X-ray structure analysis, crystal chemistry.*

Ishchenko, Dr Anatolii Aleksandrovich. Chemistry Dept., Moscow State U., Leninskie Gory, Moscow 117234, USSR. (1948) Cand, chemistry (MGU, 1974). Sr. scient. *Electron diffraction, large-amplitude molecular vibrations.*

Iskhakova, Dr Lyudmila Dmitriyevna. Inst. of Chemical Reagents and Pure Substances, Bogorodsky Val 3, Moscow 107258, USSR. (1942) Cand, chemistry (Moscow Inst. of Fine Chem. Eng., 1970). Sr. scient. *Crystal chemistry, inorganic compounds.*

Ismailov, Dr Dzhabir Ibragim ogly. Inst. of Physics, Acad. Sci. Azerbaidzan SSR, Prospekt Narimanova 33, Baku 370143, USSR. (1948) Cand, physics and mathematics (Inst. of Physics, Acad. Sci. AzSSR, 1986). Scient. *Thin films, ageing processes, phase transformations.*

Ivanitskii, Dr Vladimir Pavlovich. Inst. of Geochemistry and Physics of Minerals, Ukrainian SSR Acad. Sci., Prospekt Palladina 34, Kiev 252680, USSR. (1937) Cand, geology and mineralogy (IGFM, 1977). Sr. scient. *Structure defects.*

Ivanov, Dr Arkadii Aleksandrovich. Chemistry Dept., Moscow State U., Leninskie Gory, Moscow 117234, USSR. (1937) Cand, chemistry (MGU, 1976). Scient. *Electron diffraction, instrumentaation.*

Ivanov, Dr Nikolay Rafailovich. Inst. of Crystallography, Acad. Sci. USSR, Leninsky pr. 59, Moscow 117333, USSR. (1938) Cand, physics and mathematics (INCRYS, 1967). Sr. scient. *Optical properties, phase transitions, ferroelectricity, ferroelasticity.*

Ivanov, Dr Sergei Aleksandrovich. Karpov Physical-Chemical Inst., Obukha 10, Moscow 103064, USSR. (1951) Cand, physics and mathematics (KARPOV,

1981). Sr. scient. *X-ray analysis, polycrystals, crystal chemistry, complicated oxides, phase transitions, ferroelectrics.*

Ivanova, Dr Irina Vadimovna. Inst. of Physics, Acad. Sci. Azerbaidzhan SSR, Narimanov Prospekt 33, Baku 370143, USSR. (1936) Cand, physics and mathematics (Inst. of Physics, Acad. Sci. Azerbaidzhan SSR, 1981). Sr. scient. *Semiconductor films, thin film formation, structure and physical properties.*

Kabalkina, Prof. Sara Samsonovna. Inst. of High Pressure Physics, Acad. Sci. USSR, Troitsk, Podol'sky Rayon, Moskovskaya Oblast' 142092, USSR. Dr, physics and mathematics (INCRYS, 1975). Sr. scient. *High pressure effects, crystal structure.*

Kabalov, Dr Yurii Konstantinovich. Geology Dept., Moscow State U., Leninskie Gory, Moscow 117234, USSR. (1936) Cand, geology and mineralogy (MGU, 1972). Sr. scient. *Powder diffractometry, crystal chemistry, X-ray analysis, inorganic structures, structural mineralogy.*

Kachalov, Dr Oleg Viktorovich. Inst. of Crystallography, Acad. Sci. USSR, Leninsky Pr. 59, Moscow 117333, USSR. (1942) Cand, physics and mathematics (INCRYS, 1973). Sr. scient. *Lattice dynamics, vibrational spectroscopy.*

Kachinsky, Dr Vitol'd Nikolayevch. Inst. of Crystallography, Acad. Sci. USSR, Leninsky pr. 59, Moscow 117333, USSR. (1931) Cand, physics and mathematics (INCRYS, 1964). Sr. scient. *Electronic structure, phase transitions, high pressures.*

Kaganer, Dr Vladimir Mikhailovich. Inst. of Crystallography, Acad. Sci. USSR, Leninsky pr. 59, Moscow 117333, USSR. (1956) Cand, physics and mathematics (INCRYS, 1984). Sr. scient. *Dynamic diffraction, X-ray and electron diffraction, neutron diffraction.*

Kaidalova, Dr Taisiya Alexandrovna. Inst. of Chemistry, Far East Scient. Center, Acad. Sci. USSR, Pr. of the 100th Aniv. of Vladivostok 159, Vladivostok 690022, USSR. (1940) Cand, chemistry (IONCH, 1974). Sr. scient. *Crystal chemistry, inorganic compounds, physical properties.*

Kalanov, Dr Makhmud. Inst. of Nuclear Physics, Uzbek SSR Acad. Sci., Ulugbek, Taskent 702132, USSR. (1942) Cand., physics and mathematics (Inst. of Nuclear Physics, Leningraad, 1974). Sr. scient. *Crystallography, carbon, layered materials.*

Kalikhman, Dr Vyacheslav Mikhailovich. Irkutsk State U., Karl Marx st.1, Irkutsk 664003, USSR. (1943) Cand, physics and mathematics (Irkutsk U., 1972). Docent. *Crystal physics, physical properties, mica.*

Kalinchenko, Dr Anatolii Mikhailovich. Inst. of Geochemistry and Physics of Minerals of the Ukrainian SSR Acad. Sci., Prospekt Palladina 34, Kiev 252680, USSR. (1941) Cand, geology and mineralogy (IGFM, 1975) Sr. scient. *Defects - structural and nonstructural, minerals.*

Kalinin, Dr Vadim Rodionovich. Gorky State U., Gagarin Prospect 23, Gorky 603600, USSR. (1950) Cand, physics and mahtematics (INCRYS, 1979). Head Lab. *X-ray structure analysis, symmetry.*

Kalinin, Dr Vladimir Ivanovich. Inst. of Crystallography, Acad. Sci. USSR, Leninsky pr. 59, Moscow 117333, USSR. (1948) Cand, physics and mathematics (MGU, 1977). Scient. *Electrical conductivity, physical properties - crystal growth relationship.*

Kalinkina, Dr Irina Nikolayevna. Inst. of Crystallography, Acad. Sci. USSR, Leninsky pr. 59, Moscow 117333, USSR. (1930) Cand, physics and mathematics (Inst. of Physical Problems, Acad. Sci. USSR, 1963). Sr. scient. *Spectroscopy.*

Kalychak, Dr Yaroslav Mikhailovich. Lvov State U., Universitetskaya St. 1, Lvov 290602, USSR. (1947) Cand, chemistry (Lvov State U., 1977). Docent. *Crystal chemistry, intermetallic compounds.*

Kamentsev, Dr Igor' Evgenyevich. Leningrad State U., Dept. of Geology, University Emb. 7/9, Leningrad 199034, USSR. (1933) Dr, geology and mineralogy (LGU, 1988). Docent. *Crystal chemistry, X-ray structure analysis.*

Kaminsky, Dr Alexander Alexandrovich. Inst. of Crystallography, Acad. Sci. USSR, Leninsky pr. 59, Moscow 117333, USSR. (1934) Dr, physics and mathematics (INCRYS, 1974). Sr. scient. *Crystal physics, laser crystals.*

Kaminsky, Dr Vladimir Fedorovich. Branch Inst. of Chemical Physics, Acad. Sci. USSR, Chernogolovka 142432, Noginsky Rayon, Moskovskaya Oblast', USSR. (1941) Cand, physics and mathematics (Inst. of Applied Physics, Acad. Sci. Moldavian SSR, 1974). Scient. *Organic compounds, direct methods, structure determination.*

Kanevskii, Dr Vladimir Mikhailovich. Inst. of Crystallography, Acad. Sci. USSR, Leninskii Prospekt 59, Moscow 117333, USSR. (1948) Cand, physsics and mathematics (INCRYS, 1987). Scient. *Real crystal physics.*

Kantor, Dr Matvey Matveyevich. Inst. of Metallurgy, Acad. Sci. USSR, Leninsky pr. 49, Moscow 117234, USSR. (1936) Cand, technics (Inst. of Metallurgy, Acad. Sci. USSR, 1970). Sr. scient. *Electron microscopy, electron diffraction.*

Kaplunnik, Dr Lidiya Nikolayevna. Dept. of Geology, Moscow State U., Leninskiye Gory, Moscow 117234, USSR. (1947) Cand, geology and mineralogy (MGU, 1978). Asst. *Crystal chemistry, inorganic compounds, X-ray structure analysis.*

Karapetyan, Arutyun Arshaluisovich. Inst. of Fine Organic Chemistry, Acad. Sci. Armenian SSR, Prosp. Azatutyan 26, Erevan 375014, USSR. (1954) Cand, physics and mathematics (Inst. Appl. Physics, Kishinev, 1984). Sr. scient. *X-ray studies, biologically active compounds.*

Kardashev, Dr Boris Konstantinovich. Physico-Techn. Inst., Acad. Sci. USSR, Politekhnicheskaya St. 26, Leningrad 194021, USSR. (1941) Dr, physics and mathematics (Physico-Techn. Inst., Acad. Sci. USSR, 1988). Sr. scient. *Crystal structure defects, elasticity and plasticity, crystal physics.*

Karpinsky, Dr Oleg Georgiyevich. Inst. of Metallurgy, Acad. Sci. USSR, Leninsky pr. 49, Moscow 117334, USSR. (1923) Cand, physics and mathematics (Moscow Inst. of Eng. and Physics, 1958). Sr. scient. *Inorganic compounds, structure.*

Karyakina, Dr Tatyana Alexandrovna. Leningrad Mining Inst., 21st Liniya 2, Leningrad 199026, USSR. (1939) Cand, geology and mineralogy (GORNY, 1969). Sr. scient. *Mineralogical crystallography, crystal growth, defects, quartz.*

Kashaev, Dr Anvar Akhyarovich. State Pedagogical Inst., Nizhnyaya Naberezhnaya 6, Irkutsk 664011, USSR. (1932) Cand, physics and mathematics (Irkutsk State U., 1968). Lab. Head. *Crystal chemistry, X-ray structure analysis, inorganic compounds.*

Kas'yanenko, Dr Evgenii Vasil'evich. Leningrad Mining Inst., 21st Liniya 2, Leningrad 199026, USSR. (1950) Cand, physics and mathematics (LGU, 1977). Docent. *Physics, minerals.*

Kataeva, Ol'ga Nikolaevna. Inst. of Organic and Physical Chemistry, Kazan Branch Acad. Sci. USSR, acad. Arbuzov st.8, Kazan 420083, USSR. (1957) Cand, chemistry (Inst. of Chemical Engineering, Kazan, 1984). Scient. *Organic crystal chemistry.*

Kats, Dr Moisey Sukherovich. Inst. of Applied Physics, Acad. Sci. Moldavian SSR, Grosula St. 5, Kishinev 277028, USSR. (1941) Cand, physics and mathematics (Physico-Techn. Inst. Acad. Sci. USSR, 1976). Sr. scient. *Mechanical properties, defects.*

Katsnel'son, Prof. Al'bert Anatolyevich. Moscow State U., Dept. of Physics, Leninskiye Gory, Moscow 117234, USSR. (1930) Dr, physics and mathematics (MGU, 1968). Prof. *Solid state physics, X-ray crystallography, X-ray scattering theory.*

Kayushina, Dr Renata Lvovna. Inst. of Crystallography, Acad. Sci. USSR, Leninsky pr. 59, Moscow 117333, USSR. Cand, physics and mathematics (INCRYS, 1965). Sr. scient. *Biopolymer structural studies, X-ray diffraction method.*

Kazeeva, Dr Lyudmila Pavlovna. Inst. of Inorganic Chemistry, Siberian Dept. Acad. Sci. USSR, Prospekt Akademika Lavrent'eva 3, Novosibirsk 630090, USSR. (1942) Cand, chemistry (Council of Chemical Sciences, Siberian Dept. Acad. Sci. USSR, 1972) Sr. scient. *Crystal growth, morphology and growth conditions.*

Kemme, Dr Andrey Andreyevich. Inst. of Organic Synthesis, Acad. Sci. Latvian SSR, Aizkraukles St. 21, Riga 226006, USSR. (1941) Cand, chemistry (Inst. of Organic Synthesis, Acad. Sci. Latvian SSR, 1977). Scient. *Organic crystal chemistry.*

Kessenikh, Dr Galina Georgiyevna. Inst. of Crystallography, Acad. Sci. USSR, Leninsky pr. 59, Moscow 117333, USSR. (1938) Cand, physics and mathematics (INCRYS, 1972). Sr. scient. *Phase transitions, elastic properties.*

Khadzhi, Dr Valentin Evstafyevich. Res. Inst. for Synthesis of Mineral Raw Materials, Institutskaya St. 1, Alexandrov 601600, Vladimirskaya Oblast', USSR. (1932) Cand, geology and mineralogy (INCRYS, 1968). Dept. head. *Crystal growth, hydrothermal solutions.*

Khaikin, Dr Leonid Solomonovich. Moscow State U., Dept. of Chemistry, Leninskiye Gory, Moscow 117234, USSR. (1937) Cand, chemistry (MGU, 1969). Sr. scient. *Elemento-organic compounds, structure, electron diffraction by gases.*

Kharchenko, Dr Lyudmila Yulianovna. Inst. of Inorganic Chemistry, Acad. Sci. USSR, Siberian Dept., Prospekt Nauki 3, Novosibirsk 630090, USSR. (1937) Cand, chemistry (Inst. of Inorganic Chemistry, Acad. Sci. USSR, Siberian Dept., 1968). Sr. scient. *Crystal growth, crystal chemistry, inorganic compounds.*

Kharchenko, Dr Olga Ivanovna. Dept. of Chemistry, Lvov State U., University St. 1, Lvov 290602, USSR. (1946) Cand, chemistry (Lvov U., 1978). Sr. scient. *X-ray structure analysis, crystal chemistry, intermetallic compounds.*

Kharitonov, Dr Yury Alexandrovich. Inst. of Crystallography, Acad. Sci. USSR, Leninsky pr. 59, Moscow 117333, USSR. (1940) Cand, geology and mineralogy (MGU, 1971). Scient. *Crystal structure,crystal growth, structure - properties relationships.*

Khasanov, Dr Salavat Salim'yanovich. Inst. of Solid State Physics, Acad. Sci. USSR, Moskovskaya obl., Chernogolovka 142432, USSR. (1956) Cand, physics and mathematics (Inst. of Solid State Physics., Chernogolovka, 1938). Sr. scient. *Phase transitions, incommensurate structures.*

Khatanova, Dr Nina Abdulovna. Moscow State U., Dept. of Physics, Leninskiye Gory, Moscow 117234, USSR. Cand, physics and mathematics (MGU, 1968). Asst. *X-ray crystallography, phase transformations.*

Kheiker, Prof. Daniel' Moiseyevich. Inst. of Crystallography, Acad. Sci. USSR, Leninsky pr. 59, Moscow 117333, USSR. (1930) Dr, physics and mathematics (INCRYS, 1972). Lab. head.. *X-ray structure analysis, instrumentation.*

Kheirov, Dr Mamed Bekovch. Azerbaidzhan Res. Inst., Oil Petrolium Industry, Aganeimatulla St. 39, Baku 370033, USSR. (1925) Dr, geology and mineralogy (Azerbaidzhan Inst. of Oil Chemical Industry, 1975). Sr. scient. *Crystal chemistry, clay minerals, structure, transformation.*

Khidirov, Dr I. Inst. of Nuclear Physics, Uzbek SSR Acad. Sci., Ulugbeck, Tashkent 702132, USSR. (1947) Cand, physics and mathematics (Ural U., Sverdlovsk, 1983). Sr. scient. *Crystallography of interstitial phases.*

Khimich, Dr Tamara Andranikovna. Inst. of Crystallography, Acad. Sci. USSR, Leninskii Prospekt 59, Moscow 117333, USSR. (1941) Cand, physics and mathematics (U. of Voronezh, 1972). Scient. *Magnetic properties.*

Khisina, Dr Nataliya Rafailovna. Inst. of Geochemistry and Analytical Chemistry, Acad. Sci. USSR, Kosygin pr. 19, Moscow 117334, USSR. (1945) Cand, geology and mineralogy (Inst. of Geochemistry and Analytical Chemistry, Acad. Sci. USSR, 1978). Scient. *Mineral solid solution decay, structure, cation ordering.*

Khitrova, Dr Valentina Ivanovna. Inst. of Crystallography, Acad. Sci. USSR, Leninsky pr. 59, Moscow 117333, USSR. (1928) Cand, physics and mathematics (INCRYS, 1963). Sr. scient. *Electron diffraction, structure analysis, crystal chemistry.*

Khodashova, Dr Tatyana Semenovna. Inst. of General and Inorganic Chemistry, Acad. Sci. USSR, Leninsky pr. 31, Moscow 117071, USSR. (1928) Cand, chemistry (IONCH, 1963). Sr. scient. *Crystal chemistry, stereochemistry, coordination compounds.*

Khokhlov, Prof Aleksandr Fedorovich. Gorky State U., Prospekt Gagrina 23, Gorky 603600, USSR. (1945) Dr, physics and mathematics (1982). Rector. *Physics, crystalline and noncrystalline materials, real structure, crystal physics.*

Kholov, Dr Alimakhmad. Physical-Techn. Inst., Tadzik SSR Acad. Sci. Akademgorodok, Dushanbe 734063 , USSR. (1950) Cand., chemistry (INCRYS, 1980). Lab. Head. *Crystal growth.*

Khotsyanova, Dr Tatyana Lvovna. Inst. of Elemento-Organic Cmpounds, Acad. Sci. USSR, Vavilov St. 28, Moscow 117312, USSR. (1924) Cand, chemistry (INCRYS, 1952). Scient. *Nuclear quadrupole resonance, structure determination, X-ray structure analysis.*

Khundzhua, Dr Andrei Georgievich. Physics Dept., Moscow State U., Leninskie Gory, Moscow 117234, USSR. (1949) Cand, physics and mathematics (MGU, 1980) Sr. lect. *Low-temperature X-ray diffraction. structural phase transitions in inhomogeneous solid solutions.*

Khurshudyan, Dr Era Khristoforovna. Inst. of Geological Sciences, Acad. Sci. Armenian SSR, Barekamutyan St. 24a, Erevan 375200, USSR. (1934) Cand, geology and mineralogy (Erevan State U., 1972). Lab. head. *Crystal chemistry, X-ray structure analysis.*

Kinzhibalo, Dr Vladimir Vasil'evich. Chemistry Dept., Lvov State U., Universitetskaya 1, Lvov 290502, USSR. (1947) Cand, chemistry (Lvov U., 1982). Docent. *Crystal chemistry, intermetallic compounds.*

Klosse, Dr Georgy Alexandrovich. Inst. of Applied Physics, Acad. Sci. Moldavian SSR, Akademicheskaya 5, Kishinev 277028, USSR. (1932) Cand, physics and mathematics (Inst. of Applied Physics, Acad. Sci. Moldavian SSR, 1975). Sr. scient. *X-ray structure analysis, crystal chemistry, inorganic compounds, phase transitions.*

Kirichenko, Dr Valentina Vasilyevna. Inst. of Crystallography, Acad. Sci. USSR, Leninsky pr. 59. Moscow 117333, USSR. (1932) Cand, physics and mathematics (INCRYS, 1975). Scient. *Crystal lattice defects.*

Kirikov, Dr Vladimir Arkadyevich. Inst. of Crystallography, Acad. Sci. USSR, Leninsky pr. 59, Moscow 117333, USSR. (1941) Cand, physics and mathematics (INCRYS, 1976). Scient. *Ferroelectrics (physics), phase transitions (ferroelectric).*

Kirkinsky, Dr Vitaly Alekseyevich. Inst. of Geology and Geophysics, Acad. Sci. USSR, Siberian Dept., Novosibirsk 630090, USSR. (1937) Dr, geology and mineralogy (Inst. of Geochemistry and Analytical Chemistry, Acad. Sci. USSR, 1984). Lab. head. *Isomorphism, polymorphism, high-pressure crystal chemistry.*

Kirpichnikova, Dr Lyubov' Fedorovna. Inst. of Crystallography, Acad. Sci. USSR, Leninsky pr. 59. Moscow 117333, USSR. (1944) Cand, physics and mathematics (INCRYS, 1972). Scient. *Structural phase transitions, ferroelectrics, ferroelastics.*

Kir'yanova, Dr Elena Viktorovna. Geology Dept., Leningrad State U., Universitetskaya Nab. 7/9, Leningrad 199034, USSR. (1956) Cand, geology and mineralogy (LGU, 1986). Scient. *Crystal growth, crystal forms.*

Kiselev, Dr Nikolay Andreyevich. Inst. of Crystallography, Acad. Sci. USSR, Leninsky pr. 59, Moscow 117333, USSR. (1928) Coresp. member, Acad. Sci. USSR; Dr, biology (Inst. of Biochemistry, Acad. Sci. USSR, 1964). Lab. head. *Protein structures, nucleic acids, viruses.*

Kislovsky, Dr Lev Dmitriyevich. Inst. of Crystallography, Acad. Sci. USSR, Leninsky pr. 59, Moscow 117333, USSR. (1924) Cand, physics and mathematics (State Optical Inst., 1960). Sr. scient. *Infrared spectroscopy, quantum crystal chemistry, inorganic compounds.*

Klassen-Neklyudova, Prof. Marina Victorovna. Inst. of Crystallography, Acad. Sci. USSR, Leninsky pr. 59, Moscow 117333, USSR. (1904) Dr, physics and mathematics (Leningrad Physico-Techn. Inst., 1936). Consulting scient. *Mechanical properties.*

Klechkovskaya, Dr Vera Vsevolodovna. Inst. of Crystallography, Acad. Sci. USSR, Leninsky pr. 59, Moscow 117333, USSR. (1938) Cand, physics and mathematics (INCRYS, 1974). Sr. scient. *Structure analysis, electron diffraction, crystal chemistry.*

Kleshchinsky, Dr Leonid Innokentyevich. Irkutsk Inst. of Railway Engineers, Kurchatova St. 10, Irkutsk 664028, USSR. (1934) Cand, physics and mathematics (Leningrad Pedagogical Inst., 1968). Head of Chair. *Electron density, momentum density, real crystals (X-ray diffraction), powder diffractometry.*

Klevtsov, Dr Petr Vasilyevich. Inst. of Inorganic Chemistry, Acad. Sci. USSR, Siberian Dept., Pr. Lavrent'ev 3, Novosibirsk 630090, USSR. (1930) Cand, physics and mathematics (INCRYS, 1955). Sr. scient. *Crystal growth, crystal chemistry, inorganic compounds, polymorphism.*

Klevtsova, Dr Rimma Fedorovna. Inst. of Inorganic Chemistry, Acad. Sci. USSR, Siberian Deprtment, Prospekt Lavrent'ev 3, Novosibirsk 630090, USSR. (1928) Cand, physics and mathematics (INCRYS, 1954). Sr. scient. *Crystal chemistry, inorganic compounds, X-ray structure analysis, polymorphism.*

Klimova, Dr Anna Yuryevna. Inst. of Crystallography, Acad. Sci. USSR, Leninsky pr. 59, Moscow 117333, USSR. Cand, chemistry (INCRYS, 1976). Scient. *Optical properties, optical activity of crystals.*

Kliya, Dr Maya Ottovna. Inst. of Crystallography, Acad. Sci. USSR, Leninsky pr. 59, Moscow 117333, USSR. (1927) Cand, geology and mineralogy (INCRYS, 1952). Sr. scient. *Crystal growth, crystal morphology.*

Klyavin, Dr Oleg Vladimirovich. Physico-Techn. Inst., Acad. Sci. USSR, Politekhnicheskaya St. 26, Leningrad 194021, USSR. (1931) Dr, physics and mathematics (Leningrad Polytechnic Inst., 1972). Sr. scient. *Defects, plasticity physics, strength.*

Knab, Dr Galina Grigoryevna. Inst. of Crystallography, Acad. Sci. USSR, Leninsky pr. 59, Moscow 117333, USSR. (1934) Cand, physics and mathematics (INCRYS, 1974). Scient. *Defects, optical properties, mechanical properties.*

Kobzareva, Dr Svetlana Alekseyevna. Inst. of Crystallography, Acad. Sci. USSR, Leninsky pr. 59, Moscow 117333 USSR. Cand, chemistry (INCRYS, 1966). Scient. *Crystal nucleation, epitaxy, electron microscopy.*

Kocharov, Dr Alexander Georgiyevich. Inst. of Crystallography, Acad. Sci. USSR, Leninsky pr. 59, Moscow 117333, USSR. (1945) Cand, physics and mathematics (INCRYS, 1972). Lab. Head. *Neutron diffraction, crystal structures, magnetic structures.*

Kolesova, Dr Rimma Vladimirovna. Rostov State U., Dept. of Physics, Prospekt Stachki 192, Rostov-on-Don 344061, USSR. Cand, physics and mathematics (Rostov U., 1967). Docent. *Crystal structure.*

Kolin, Dr Nikolai Georgievich . Obninsk Branch, Karpov Physical-Chemical Inst., Kaluzhskaya obl.,Obninsk 249020, USSR. (1948) Cand, physics and mathematics (Inst. of Steel and Alloys, 1986) Sr. scient. *Impurities, defects, semiconductor properties, dielectric properties.*

Kolobyanina, Dr Tatyana Nikolayevna. Inst. of High Pressure Physics, Acad. Sci. USSR, Troitsk, Podol'sky Rayon, Moskovskaya Oblast' 142092, USSR. Cand, physics and mathematics (MGU, 1974). Lab. Head. *High pressure, crystal structure.*

Kolodiyeva, Dr Svetlana Vasilyevna. Res. Inst. for Synthesis of Mineral Raw Materials, Institutskaya St. 1, Alexandrov 601600, Vladimirskaya Oblast', USSR. (1936) Cand, physics and mathematics (MGU, 1979). Sr. scient. *Defects - physical properties interrelation, real crystals.*

Kolontsova, Dr Ekaterina Vasilyevna. Moscow State U., Dept. of Physics, Leninskiye Gory, Moscow 117234, USSR. Dr, physics and mathematics (MGU, 1970). Sr. scient. *Diffuse X-ray scattering, radiation effects, phase transformations, single crystals.*

Kolpakov, Dr Andrey Vasilyevich. Moscow State U., Dept. of Physics, Leninskiye Gory, Moscow 117234, USSR. (1941) Cand, physics and mathematics (MGU, 1968). Asst. *X-ray diffraction theory, group theory.*

Kon, Dr Aviv Yuliseyevich. Kishinev Polytechnic Inst., Lenin pr. 168, Kishinev 277004, USSR. (1929) Cand, physics and mathematics (Gorky U., 1967). Docent. *X-ray structure analysis methods, crystal chemistry, coordination compounds.*

Kondrashev, Dr Yury Dmitriyevich. State Inst. of Applied Chemistry, Vatny Ostrov 2, Leningrad, USSR. (1916) Cand, chemistry (State Inst. of Applied Chemistry, 1949). Sr. scient. *Organic crystal structure, structure determination methods.*

Kondratenko, Dr Lyudmila Konstantinovna. Inst of Metallurgy, Acad. Sci. USSR, Leninskii Prospekt 49,Moscow 117334, USSR. (1946) Cand, eng. (Inst. of Metallurgy, 1976). Scient. *Crystallography, martensitic transformations, crystal chemistry, inorganic compounds.*

Kondratyeva, Dr Victoria Victorovna. Leningrad State U., Dept. of Geology, University Emb. 7/9, Leningrad 199164, USSR. (1932) Cand, geology and mineralogy (LGU, 1966). Scient. *Crystallography, crystal chemistry, X-ray structure analysis, borates, borosilicates.*

Konstantinova, Dr Alisa Fedorovna. Inst. of Crystallography, Acad. Sci. USSR, Leninsky pr. 59, Moscow 117333, USSR. (1936) Dr, physics and mathematics (INCRYS, 1988). Scient. *Crystal optics, optical activity.*

Koptsik, Prof. Vladimir Alexandrovich. Moscow State U., Dept. of Physics, Leninskiye Gory, Moscow 117234, USSR. (1924) Dr, physics and mathematics (MGU, 1963). Prof. *Symmetry, crystal physics.*

Koreshkov, Dr Boris Dmitriyevich. Kolomna Pedagogical Inst., Zelenaya 30, Kolomna 140410, Moskovskaya Oblast', USSR. (1940) Cand, physics and mathematics (Moscow State Pedagogical Inst., 1968). Rector. *Organic molecular crystals, thermodynamic properties.*

Kornev, Dr Aleksey Nikolayevich. Inst. of Biological Physics, Acad. Sci. USSR, Pushchino 142292, Moskovskaya Oblast', USSR. (1944) Cand, physics and mathematics (INCRYS, 1973). Sr. scient. *Structure analysis, biologically active compounds.*

Korsukova, Dr Mariya Mikhailovna. Physico-Techn. Inst., Acad. Sci. USSR, Politekhnicheskaya St. 26, Leningrad 194021, USSR. (1945) Cand, chemistry (Lvov State U.'1976). Sr. scient. *Crystal growth.*

Koryagin, Mr. Vyacheslav Filippovich. Inst. of Crystallography, Acad. Sci. USSR, Leninsky pr. 59, Moscow 117333, USSR. (1921). Scient. *Spectroscopy, magnetic resonance.*

Kosevich, Prof. Vadim Markovich. Kharkov Polytechnic Inst., Frunze St. 21, Kharkov 310002, USSR. (1931) Dr, physics and mathematics (Physico-Techn. Inst. of Low Temperatures, Acad. Sci. Ukrainian SSR, 1969). Head of Chair. *Real structure, imperfections, crystal structure, thin films.*

Kosmachev, Dr Sergey Mikhailovich. Kharkov Polytechnic Inst., Frunze St. 21, Kharkov 310002, USSR. (1946) Cand, technics (Kharkov Polytechnic Inst., 1974). Docent. *Crystal structure defects, thin film growth, electron microscopy.*

Kosova, Dr Tatyana Borisovna. Moscow State U., Dept. of Geology, Leninskiye Gory, Moscow 117234, USSR. (1940) Cand, geology and mineralogy (MGU, 1973). Scient. *Non-diffraction methods, crystal growth.*

Kosterin, Dr Evgeny Andreyevich. Dept. of Physics, Ivanovo State Medical Inst., Engels St. 8, Ivanovo 153462, USSR. (1935) Cand, physics and mathematics (Ivanovo State Medical Inst.). Head of chair. *Liquid crystals, structure, properties.*

Kotel'nikova, Dr Elena Nikolayevna. Leningrad State U., Dept. of Geology, University Emb. 7/9, Leningrad 199164, USSR. (1945) Cand, geology and mineralogy (LGU, 1982). Scient. *Crystal chemistry, laminated minerals.*

Kotov, Prof. Nikolay Vladimirovich. Leningrad State U., Dept. of Geology, University Emb. 7/9, Leningrad 199034, USSR. (1935) Dr, geology and mineralogy (LGU, 1974). Sr. scient. *Crystal chemistry, high pressure synthesis, mineral structures, transformations, high pressure.*

Kotur, Dr Bogdan Yaroslavovich. Dept. of Chemistry, Lvov State U., University St. 1, Lvov 290602, USSR. (1952) Cand, chemistry (Lvov U., 1978). Docent. *X-ray structure analysis, crystal chemistry, intermetallic compounds.*

Koval'chuk, Dr Mikhail Valentinovich. Inst. of Crystallography, Acad. Sci. USSR, Leninsky pr. 59, Moscow 117333, USSR. (1946) Dr, physics and mathematics (Inst. of Solid State Physics, 1988). Lab. Head. *Dynamic X-ray scattering, real structure.*

Kovda, Prof. Leonid Mikhailovich. Dept. of Chemistry, Moscow State U., Leninskiye Gory, Moscow 117234, USSR. (1932) Cand, chemistry (MGU, 1971). Prof. *Crystal chemistry, inorganic compounds.*

Kovyev, Dr Ernest Konstantinovich. Inst. of Crystallography, Acad. Sci. USSR, Leninsky pr. 59, Moscow 117333, USSR. (1941) Cand, physics and mathematics (MGU, 1974). Sr. scient. *X-ray diffraction, real structure.*

Koz'ma, Dr Alexander Alekseyevich. Kharkov Polytechnic Inst., Frunze St. 21, Kharkov 310002, USSR. (1939) Cand, technics (Kharkov Polytechnic Inst., 1968). Sr. scient. *Metals and alloys, structure, defects, methods for defect study.*

Koz'min, Dr Petr Alekseyevich. Inst. of General and Inorganic Chemistry, Acad. Sci. USSR, Leninsky pr. 31, Moscow 117071, USSR. (1929) Cand, chemistry (IONCH, 1965). Sr. scient. *Crystal chemistry, inorganic compounds.*

Kozlenkov, Dr Alexander Ivanovich. Inst. of Metallurgy, Acad. Sci. USSR, Leninsky pr. 49, Moscow 117334, USSR. (1932) Cand, physics and mathematics (Rostov U., 1965). Sr. scient. *Electronic states, lattice symmetry - electronic states relation, spectroscopy of crystals.*

Kozlova, Dr Olga Gerasimovna. Moscow State U., Dept. of Geology, Leninskiye Gory, Moscow 117234, USSR. (1927) Cand, geology and mineralogy (MGU, 1957). Docent. *Crystal growth, crystal morphology.*

Krasnikov, Dr Vladimir Vladimirovich. Inst. of Inorganic Chemistry, Acad. Sci. Latvian SSR. Miera st.34, Rizhskii raion, Salaspils 1, Latvian SSR 229021, USSR. (1953) Cand, chemistry (Inst. Inorg. Chem., Riga, 1983). Sr. scient. *Structure, polycrystals.*

Krivandina, Dr Elena Alekseyevna. Inst. of Crystallography, Acad. Sci. USSR, Leninsky pr. 59, Moscow 117333, USSR. (1938) Cand, physics and mathematics (INCRYS, 1980). Scient. *Crystal growth from melt, fluoride - single crystal growth.*

Krivenko, Dr Vladimir Georgievich. Inst. of Crystallography, Acad. Sci. USSR, Leninskii Prospekt 59, Moscow 117333, USSR. (1939) Cand, physics and mathematics (Inst. of Biochemistry, Moscow, 1968). Scient. *High temperature superconductivity.*

Krivokoneva, Dr Galina Kirillovna. Inst. of Mineral Raw Materials (VIMS), Staromonetny 29, Moscow 109017, USSR. (1938) Cand, geology and mineralogy (VIMS, 1971). *X-ray structure analysis, polycrystals, crystal chemistry, minerals.*

Krochuk, Dr Vasilii Maksimovich. Inst. of Geochemistry and Physics of Minerals, Ukrainian SSR Acad. Sci., Prospekt Palladina 34, Kiev 252680, USSR. (1954) Cand, geology and mineralogy (IGFM, 1982). Sr. scient. *Crystallography, minerals, twinning.*

Krol', Dr Inna Mikhailovna. NPO IREA, Bogorodskii val 3, Moscow 107258, USSR. (1946) Cand, chemistry (Inst. of Chemical Reagents, Moscow, 1982) Scient. *Crystal chemistry, organic and coordination compounds.*

Kruglik, Dr Anatoliy Ivanovich. Inst. of Physics, Acad. Sci. USSR, Siberian Dept., Akademgorodok, Krasnoyarsk 660036, USSR. (1947) Cand, physics and mathematics (Inst. of Physics, Acad. Sci. USSR, Siberian Dept., 1981). Sr. scient. *Structure, ferroelectrics, ferroelastics, X-ray and neutron diffraction.*

Krutova, Dr Glafira Ivanova. Dept. of Geology, Moscow State U., Leninskiye Gory, Moscow 117234, USSR. (1931) Cand, geology and mineralogy (MGU, 1971). Scient. *Crystal chemistry, inorganic compounds, X-ray diffraction.*

Krymov, Dr Vladimir Mikhailovich. Physical Techn. Inst., Acad. Sci. USSR, Politekhnicheskaya St. 26, Leningrad 194021, USSR. (1948) Cand, physics and mathematics (Physical Techn. Inst., Acad. Sci. USSR, 1979). Sr. scient. *Semiconductors, dielectrics, crystal physics, crystallization from melt.*

Kryshtop, Dr Viktor Mikhailovich. Physics Dept., Rostov State U., Zorge 5, Rostov-on-Don, 344104, USSR. (1945) Cand, physics and mathematics (Rostov U., 1980). Docent. *Structure, physical properties, complicated oxides.*

Kudryavtseva, Dr Galina Petrova. Dept. of Geology, Moscow State U., Leninskiye Gory, Moscow 117234, USSR. (1947) Cand, geology and mineralogy (MGU, 1973). Sr. scient. *Crystal chemistry, real crystals.*

Kudryavtseva, Dr Rimma Vasil'evna. Gorky State U., Prospekt Gagarina 23,Gorky, 603600, USSR. (1934) Cand, physics and mathematics (Gorky U., 1972) Docent. *Crystal growth, real structure.*

Kukharenko, Prof. Alexander Alexandrovich. Leningrad State U., Dept. of Geology, University Emb. 7/9, Leningrad 199164, USSR. (1914) Dr, geology and mineralogy (LGU, 1954). Prof. *Crystal chemistry.*

Kukina, Dr Galina Alexandrovna. Inst. of General and Inorganic Chemistry, Acad. Sci. USSR, Leninsky pr. 31, Moscow 117071, USSR. (1927) Cand, chemistry (IONCH, 1963). Scient. *Crystal chemistry, stereochemistry, coordination compounds, X-ray structure analysis techniques.*

Kukuy, Dr Anatoly Lvovich. Leningrad Mining Inst., 21st Liniya 2, Leningrad 199026, USSR. (1939) Cand, geology and mineralogy (GORNY, 1970). Scient. *Morphology, crystal growth.*

Kuntsevich, Dr Tamara Serafimovna. Gorky State U., Gagarin Prospekt 23, Gorky 603022, USSR. (1932) Cand, physics and mathematics (Gorky U., 1970). Sr. scient. *X-ray structure analysis, symmetry groups.*

Kupriyanov, Dr Mikhail Fedotovich. Rostov State U., Dept. of Physics, Prospekt Stachki 192, Rostov-on-Don 344061, USSR. (1937) Cand, physics and mathematics (Rostov U., 1968). Docent. *Phase transitions, complex oxides, crystal chemistry, ferroelectrics.*

Kuranova, Dr Inna Petrovna. Inst. of Crystallography, Acad. Sci. USSR, Leninsky pr. 59, Moscow 117333, USSR. (1933) Cand, chemistry (MGU, 1963). Sr. scient. *Protein structures.*

Kurazhkovskaya, Dr Victoriya Semenovna. Moscow State U., Dept. of Geology, Leninskiye Gory, Moscow 117234, USSR. (1944) Cand, geology and mineralogy (MGU, 1971). Scient. *Crystal chemistry, intermetals, inorganic compounds.*

Kurbanov, Dr Khakim Mamadaliyevich. Physico-Techn. Inst., Acad. Sci. Tadzhik SSR, Akademgorodok, Dushanbe, USSR. (1935) Cand, physics and mathematics (INCRYS, 1964). Lab. head. *Crystal structure, crystal growth.*

Kurdyumov, Prof. Georgy Vyacheslavovich. Inst. of Metal Physics, 2nd Bauman St. 9/23, Moscow 107005, USSR. (1902) Full member, Acad. Sci. USSR, Dr, physics and mathematics (1937). Dir. *Metal physics, real structure.*

Kurkutova, Prof. Evdokiya Nikitichna. Vladimir State Pedagogical Inst., Prospekt Stroiteley 11, Vladimir 600024, USSR. (1930) Dr, physics and mathematics (INCRYS, 1978). Prof. *X-ray structure analysis.*

Kuz'ma, Prof. Yury Bogdanovich. Lvov State U., Dept. of Chemistry, University St. 1, Lvov 290602, USSR. (1934) Dr, chemistry (Lvov U., 1974). Head of Chair. *X-ray structure analysis, crystal chemistry, intermetallic compounds.*

Kuz'min, Prof. Eduard Alekseyevich. Gorky State U., Gagarin Prospekt 23, Gorky 603022, USSR. (1939) Dr, physics and mathematics (Rostov U., 1974). *X-ray structure analysis.*

Kuz'min, Dr Ivan Ivanovich. Obninsk Branch, Karpov Physical-Chemical Inst., Kaluzhskaya obl.,Obninsk 249020, USSR. (1932) Cand, physics and mathematics (INCRYS, 1970). Lab. Head. *Composition, structure, lattice dynamics, electric properties, crystals, radiation-induced defects, impurities, crystal properties.*

Kuz'min, Prof. Runar Nikolayevich. Moscow State U., Dept. of Physics, Leninskiye Gory, Moscow 117234, USSR. (1932) Dr, physics and mathematics (MGU, 1970). Prof. *Crystallography, solid state.*

Kuz'mina, Dr Irina Pavlovna. Inst. of Crystallography, Acad. Sci. USSR, Leninsky pr. 59, Moscow 117333, USSR. (1932) Cand, chemistry (INCRYS, 1968). Sr. scient. *Crystal growth, crystal chemistry.*

Kuz'mina, Dr Lyudmila Georgievna. Inst. of General and Inorganic Chemistry, Acad. Sci. USSR, Leninskii Prospekt 31, Moscow, USSR. (1945) Cand, chemistry (INEOS, 1978). Scientist. *Crystallography, elemento-organic compounds, organic compounds, coordination compounds.*

Kuz'mina, Dr Mariya Anatol'evna. Geology Dept., Leningrad State U., Universitetskaya Nab. 7/9, Leningrad 199034, USSR. (1962) Cand, geology and mineralogy (LGU, 1987). Scient. *Crystal growth.*

Kuznetsov, Dr Alexander Victorovich. Petrozavodsk State U., Lenin pr. 33, Petrozavodsk 185018, USSR. (1932) Cand, physics and mathematics (MGU, 1963). Docent. *Dynamical theory, X-ray scattering.*

Kuznetsov, Prof. Fedor Andreyevich. Inst. of Inorganic Chemistry, Acad. Sci. USSR, Siberian Dept., Prospekt Nauki 3, Novosibirsk 630090, USSR. (1932) Full member, Acad. Sci. USSR, Dr, chemistry (Scient. Council for Chemical Sci., Acad. Sci. USSR, Siberian Dept., 1972). Dir. *Crystal growth, thin film growth.*

Kuznetsov, Dr Gennadii Vasil'evich. Inst. of Geochemistry and Physics of Minerals, Ukrainian SSR Acad. Sci. Prospekt Palladina 34, Kiev 252680, USSR. (1937)

Cand, geology and mineralogy (IGFM, 1977). Sr. scient. *Crystal physics and chemistry.*

Kuznetsov, Prof. Vasily Grigoryevich. Inst. of General and Inorganic Chemistry, Acad. Sci. USSR, Leninsky pr. 31, Moscow 117071, USSR. (1906) Dr, chemistry (IONCH, 1969). *Crystal chemistry, coordination compounds, semiconductors.*

Kuznetsov, Dr Victor Andreyevich. Inst. of Crystallography, Acad. Sci. USSR, Leninsky pr. 59, Moscow 117333, USSR. (1938) Cand, geology and mineralogy (INCRYS, 1967). Sr. scient. *Crystal growth, crystal chemistry.*

Kuznetsov, Dr Yurii Georgievich. Inst. of Crystallography, Acad. Sci. USSR, Leninskii Prospekt 59, Moscow 117333, USSR. (1944) Cand, physics and mathematics (INCRYS, 1988). Scient. *Crystal growth, aqueous solutions, X-ray topography, defects.*

Kuzyukevich, Dr Anatolii Anatol'evich. Inst. of Inorganic Chemistry, Acad, Sci. Latvian SSR, Miera st.34, Rizhskii raion, Salaspils 1, Latvian SSR 229021, USSR. Sr. scient. *Electron diffraction structure analysis, microdiffraction.*

Kvasnitsa, Dr Victor Nikolayevich. Inst. of Geochemistry and Physics of Minerals, Acad. Sci. Ukrainian SSR, Palladina Pr. 34, Kiev 252680, USSR. (1942) Cand, geology and mineralogy (Inst. of Geochemistry and Physics of Minerals, Acad. Sci. Ukrainian SSR, 1974). Sr. scient. *Mineralogical crystallography.*

Kyutt, Dr Reginal'd Nikolayevich. Physical Techn. Inst., Acad. Sci. USSR, Zapovednaya St. 51, Leningrad 194037, USSR. (1944) Cand, physica and mathematics (Physical Techn. Inst., Acad. Sci. USSR, 1979). Scient. *Defects, diffractometry - single crystal.*

Lapergauz, Dr Il'ya Samuilovich. Inst. of Nuclear Physics, Uzbek SSR Acad. Sci., Ulugbeck, Tashkent 702132, USSR. (1950) Cand., physics and mathematics (Inst. of Nuclear Physics, Tashkent, 1986) Scient. *Crystallography, interstitial phases.*

Lazarenkov, Prof. Vadim Grigor'yevich. Leningrad Mining Inst., 21st Liniya 2, Leningrad 199026, USSR. (1933) Dr, geology and mineralogy (GORNY, 1980). Prof. *Mineralogical crystallography.*

Lazarev, Dr Eduard Mikhailovich. Inst. of Metallurgy, Acad. Sci. USSR, Leninsky pr. 49, Moscow 117334, USSR. (1937) Cand, technics (Inst. of Metallurgy, Acad. Sci. USSR, 1967). Sr. scient. *Inorganic compounds, structure.*

Lazarev, Dr Valerii Georgievich. Inst. of Crystallography, Acad. Sci. USSR, Leninskii Prospekt 59, Moscow 117333, USSR. (1950) Cand, physics and mathematics (INCRYS, 1985) Sr. scient. *Photoelectric and optical phenomena, pyro- and ferroelectrics, transfer phenomena.*

Lebedev, Prof. Vasily Ilyich. Leningrad State U., Dept. of Geology, University Emb. 7/9, Leningrad 199034, USSR. (1911) Dr, geology and mineralogy (LGU, 1955). Lab. head. *Structural crystallography, crystal chemistry, mineralogy, geochemistry.*

Lebedeva, Dr Marina Vladimirovna. Kharkov Polytechnic Inst., Frunze St. 21, Kharkov 310002, USSR. (1941) Cand, technics (Kharkov Polytechnic Inst., 1973). Sr. lect. *Metals and alloys, structure, defects, methods for defect study.*

Leonyuk, Dr Lidiya Ivanovna. Moscow State U., Dept. of Geology, Leninskiye Gory, Moscow 117234, USSR. (1950) Cand, geology and mineralogy (MGU, 1978). Scient. *Crystal growth, morphology.*

Leonyuk, Dr Nikolay Ivanovich. Moscow State U., Dept. of Geology, Leninskiye Gory, Moscow 117234, USSR. (1941) Dr, geology and mineralogy (MGU, 1985). Docent. *Crystal growth, crystal chemistry, borates.*

Levanyuk, Prof. Arkady Petrovich. Inst. of Crystallography, Acad. Sci. USSR, Leninsky pr. 59, Moscow 117333, USSR. (1933) Dr, physics and mathematics (INCRYS, 1977). Lab. Head. *Polymorphic transformations, ferroelectrics.*

Lim, Dr Valery Irovich. Inst. of Protein, Acad. Sci. USSR, Pushchino, Serpukhovsky Rayon, Moskovskaya Oblast' 142292, USSR. (1943) Dr, biology (MGU, 1986). Assos. dir. *Biopolymer crystallography.*

Lindeman, Dr Sergei Vital'evich. Inst. of Elementoorganic Compounds, Acad. Sci. USSR, Vavilov st. 28, Moscow 117334, USSR. (1959) Cand, chemistry (INEOS, 1988). Scient. *Biologically active compounds, structure modeling, polymer chains.*

Lindin', Dr Lauma Felixovna. Riga Polytechnical Inst., Kronvald Boulevard 4, Riga 226828, USSR. (1935) Cand, technics (Riga Polytechn. Inst., 1972). Sr. scient. *Crystal chemistry, crystal phase formation, inorganic oxide systems.*

Liopo, Dr Valery Alexandrovich. University, Ozheshhko St. 12, Grodno 230023, USSR. (1939) Cand, physics and mathematics (Irkutsk State U., 1968). Prorector. *Crystallographic group theory, teaching, structure and properties of crystals.*

Lisoivan, Dr Vladimir Ivanovich. Inst. of Inorganic Chemistry, Acad. Sci. USSR, Siberian Dept., Prospekt Lavrent'ev 3, Novosibirsk 630090, USSR. (1936) Cand, physics and mathematics (Inst. of Semiconductor Physics, Acad. Sci. USSR, Siberian Dept., 1971). Sr. scient. *Crystal lattice defects, X-ray structure analysis.*

Litovchenko, Dr Anatolii Stepanovich. Inst. of Geochemistry and Physics of Minerals, Ukrainian SSR Acad. Sci., Prospekt Palladina 34, Kiev 252680, USSR. (1941) Cand, geology and mineralogy (Physical-Techn. Inst., Donetsk, 1974). Sr. scient. *Real structure, minerals.*

Litvin, Dr Alexander Lukich. Inst. of Geochemistry and Physics of Minerals, Acad. Sci. Ukrainian SSR, Palladina pr. 34, Kiev 252680, USSR. (1927) Dr, geology and mineralogy (Kiev U., 1978). Lab. Head. *Structural typomorphism, rock-forming minerals.*

Litvin, Dr Boris Nikolayevich. Moscow State U., Dept. of Geology, Leninskiye Gory, Moscow 117234, USSR. (1934) Dr, chemistry (IONCH, 1978). Sr. scient. *Crystal growth, crystal chemistry.*

Litvinov, Dr Igor' Anatol'evich. Inst. of Organic and Physical Chemistry, Kazan Branch of the Acad. Sci. USSR,acad. Arbuzov st. 8, Kazan' 420083, USSR. (1953) Cand, chemistry (Inst. of Chemical Engineering, Kazan, 1984). Sr. scient. *Organic crystal chemistry, molecular and crystal structure, organic compounds.*

Litvinskaya, Mrs Galina Petrovna. Moscow State U., Dept. of Geology, Leninskiye Gory, Moscow 117234, USSR. (1920) Sr. lect. *Geometrical crystallography, crystal chemistry, inorganic compounds, structure.*

Lityagina, Dr Lyudmila Mitrofanovna. Inst. of High Pressure Physics, Acad. Sci. USSR, Troitsk, Podol'sky Rayon, Moskovskaya Oblast' 142091, USSR. Cand, geology and mineralogy (MGU, 1976). Scient. *High pressure effects, crystal structure.*

Lobkovsky, Dr Emil' Borisovich. Moscow State U., Dept. of Chemistry, Leninskiye Gory,Moscow 117234, USSR. (1941) Cand, chemistry (MGU, 1974). Sr. scient. *Crystal chemistry, transition metal hydrides, tetrahydroborates.*

Lomonov, Dr Vladimir Alekseevich. Inst. of Crystallography, Acad. Sci. USSR, Leninskii Prospekt 59, Moscow 117333, USSR. (1948) Cand, chemistry (Chemical-Technological Inst., Moscow, 1979). Sr. scient. *Crystal growth.*

Lomov, Dr Andrei Aleksandrovich. Inst. of Crystallography, Acad. Sci. USSR, Leninskii Prospekt 59, Moscow 117333, USSR. (1954) Cand, physics and mathematics (INCRYS, 1987). Scient. *Real structure, surface layers.*

Loshmanov, Dr Arkady Andreyevich. Inst. of Crystallography, Acad. Sci. USSR, Leninsky pr. 59, Moscow 117333, USSR. (1929) Cand, physics and mathematics (Inst. of Metallurgy of Ferrous Metals, 1967). Sr. scient. *Neutron diffraction, non-crystalline and crystalline state.*

Lube, Dr Emil' Lvovich. Inst. of Crystallography, Acad. Sci. USSR, Leninskii Prospekt 59, Moscow 117333, USSR. (1936) Cand, physics and mathematics (INCRYS, 1971). Sr. scient. *Growth from melt, crystal growth, modelling.*

Lunin, Dr Vladimir Yur'evich. Res. Computational Center, Acad. Sci. USSR, Moskovskaya obl., Pushchino 142292, USSR. (1951) Cand, physics sand mathematics (MGU, 1977). Sr. scient. *Theoretical problems, structure determination, biological molecules, computational problems, crystallography.*

Lvov, Dr Yury Mikhailovich. Inst. of Crystallography, Acad. Sci. USSR, Leninsky pr. 59, Moscow 117334, USSR. (1952) Cand, physics and mathematics (MGU, 1978). Sr. scient. *Small angle X-ray scattering.*

Lyakhovitskaya, Dr Vera Aronovna. Inst. of Crystallography, Acad. Sci. USSR, Leninsky pr. 59, Moscow 117333, USSR. (1929) Cand, chemistry (INCRYS, 1967). Sr. scient. *Crystal growth from melt and vapour phase.*

Lyubalin, Dr Mark Dmitriyevich. Leningrad Mining Inst., 21st Liniya 2, Leningrad 199026, USSR. (1937) Cand, technics (GORNY, 1970). Sr. scient. *Morphology, crystal growth.*

Lyubimov, Dr Vasily Nikolayevich. Karpov Physical Chemistry Inst., Obukha St. 10, Moscow 107120, USSR. (1936) Dr, physics and mathematics (INCRYS, 1988). Sr. scient. *Theoretical crystal physics.*

Lyubitov, Dr Yury Naumovich. Inst. of Crystallography, Acad. Sci. USSR, Leninsky pr. 59, Moscow 117333, USSR. (1932) Cand, physics and mathematics (INCRYS, 1978). *Crystal growth.*

Lyubushkina, Dr, Lyudmila Mikhailovna. Irkutsk Pedagogical Inst., Nizhnyay nab. 6, Irkutsk 664654, USSR. (1951) Cand, physics and mathematics (Solid State Inst., Acad. Sci. Belorussian SSR, 1985). Sr. scient. *Crystal chemistry,inorganic compounds, alloys.*

Lyubutin, Dr Igor' Savelyevich. Inst. of Crystallography, Acad. Sci. USSR, Leninsky pr. 59, Moscow 117333, USSR. (1938) Dr, physics and mathematics (INCRYS, 1975). Sr. scient. *Magnetism, Mössbauer spectroscopy.*

Lyutin, Dr Vladimir Ivanovich. Res. Inst. for Synthesis of Mineral Raw Materials, Institutskaya St. 1, Alexandrov 601600, Vladimirskaya Oblast', USSR. (1948) Cand, physics and mathematics (INCRYS, 1974). Sr. scient. *Crystal growth, X-ray structure analysis, crystal chemistry.*

Lyutzau, Prof. Vsevolod Grigoryevich. Machinery Inst., Acad. Sci. USSR, Griboyedov St. 4, Moscow 101000, USSR. (1922) Dr, technics (Inst. of Steel and Alloys, 1972). Lab. head. *Real crystals (substructure), lattice defects.*

Magomedova, Dr Nina Samuilovna. Karpov Physic₂l-CHemical Inst., Obukha 10, Moscow 10364, USSR. (1945) Cand,chemistry (KARPOV, 1984). Scient. *X-ray analysis, crystal chemistry, organic and inorganic compounds.*

Maiyer, Prof. Alexander Artemyevich. Mendeleev Inst. of Chemical Technology, Miusskaya Square 9, Moscow 125820, USSR. (1927) Dr, chemistry (Moscow Chemical Techn. Inst., 1970). *Crystallography, crystal chemistry, selenites, chromites, molybdates, tungstates.*

Makarenko, Dr Igor' Nikolayevich. Inst. of Crystallography, Acad. Sci. USSR, Leninsky pr. 59, Moscow 117333, USSR. (1938) Cand, physics and mathematics (INCRYS, 1971). Sr. scient. *High pressure phase transitions.*

Makarov, Prof. Evgeny Sergeyevich. Inst. of Geochemistry and Analytical Chemistry, Acad. Sci. USSR, Kosygin pr. 19, Moscow 117334, USSR. (1911) Dr, chemistry (IONCH, 1954). Consulting prof. *X-ray diffraction, crystal chemistry, inorganic compounds.*

Makhmudova, Dr Nailiya Kamilovna. Inst. of Chemistry, Uzbek SSR Acad. Sci., Prospekt Gor'kogo 77, Tashkent 700170, USSR. (1949) Cand.chemestry (IONCH, 1986). Scient. *Crystal chemistry, inorganic and coordination compounds.*

Maksimova, Dr Nadezhda Vasil'evna. Inst. of Mineralogy, Geochemistry, and Crystal Chemistry of Rare Elements, Sadovnicheskaya nab.71, Moscow 113035, USSR. (1932) Cand., Geology and mineralogy (MGU, 1983). Scient. *Crystal chemistry, inorganic compounds.*

Malakhova, Dr Lyudmila Fedorovna. Inst. of Crystallography, Acad. Sci. USSR, Leninsky pr. 59, Moscow 117333, USSR. (1941) Cand, physics and mathematics (INCRYS, 1976). Sr. scient. *X-ray diffractometry, apparatus and methods.*

Malinenko, Dr Inna Avramovna. Petrozavodsk State U., Lenin Prospekt 33, Petrozavodsk 185018, USSR. (1941) Cand, physics and mathematics (Moscow Inst. of Steel and Alloys, 1971). Docent. *Point defects, lattice dynamics in crystals.*

Malinovsky, Dr Stanislav Tadeushevich. Inst. of Applied Physics, Acad. Sci. Moldavian SSR, Akademicheskaya 5, Kishinev 277028, USSR. (1949) Cand, physics and mathematics (INCRYS, 1978). Scient. *Structural chemistry, coordination compounds, organic and bio-organic compounds.*

Malinovsky, Dr Yury Alexandrovich. Inst. of Crystallography, Acad. Sci. USSR, Leninsky pr. 59, Moscow 117333, USSR. (1947) Cand, geology and mineralogy (MGU, 1976). Sr. scient. *Crystal structure.*

Malinovsky, Prof. Tadeush Iosifovich. Inst. of Applied Physics, Acad. of Sciences Moldavian SSR, Akademicheskaya 5, Kishinev 277028, USSR. (1921) Full member, Acad. Sci. Moldavian SSR, Dr, physics and mathematics (INCRYS, 1967). Lab. head. *Crystallography, crystal chemistry, X-ray crystal structure analysis.*

Mal'tsev, Dr Yurii Fedorovich. Physics Dept., Rostov State U., Zorge 5, Rostov-on-Don, 344104, USSR. (1945) Cand, physics and mathemtics (Rostov U., 1974). Docent. *Dynamical theory, X-ray diffraction, multicrystal diffractometry.*

Malyushitskaya, Dr Zinaida Vladimirovna. Inst. of High Pressure Physics, Acad. Sci. USSR, Troitsk, Podol'sky Rayon, Moskovskaya Oblast' 142092, USSR. Cand, chemistry (MGU, 1974). Scient. *High pressure effect on crystal structure.*

Man, Dr Lucia Ivanovna. Inst. of Crystallography, Acad. Sci. USSR, Leninsky pr. 59, Moscow 117333, USSR. Cand, physics and mathematics (INCRYS, 1970). Scient. *Electron diffraction, semiconductors, crystal chemistry, structure analysis.*

Marfunin, Dr Arnol'd Sergeyevich. Inst. of Geology- Mineralogy and Petrography, Acad. Sci. USSR, Staromonetny 35, Moscow 109017, USSR. (1926) Corresp. member, Acad. Sci. USSR, Dr, geology and mineralogy (IGEM, 1962). Lab. head. *Minerals (physics), crystal field theory, electron paramagnetic resonance, electron-hole centers.*

Martyshev, Dr Yury Nikolayevich. Inst. of Crystallography, Acad. Sci. USSR, Leninsky pr. 59, Moscow 117333, USSR. (1931) Cand, physics and mathematics (INCRYS, 1971). Scient. *Optical properties.*

Marusin, Dr Evgenii Petrovich. Chemistry Dept., Lvov State U., Universitetskaya 1, Lvov 290602, USSR. (1949) Cand, chemistry (Lvov U., 1982). Sr. scient. *Crystal chemistry, intermetallic compounds.*

Maslennikov, Dr Aleksei Vladimirovich. Inst. of Geology and Geochronology of Precambrian of the Acad. Sci. USSR, Nab. Makarova 2, Leningrad 199034, USSR. (1945) Cand, geology and mineralogy (LGU, 1979) Scient. *Crystal chemistry, minerals.*

Maslov, Dr Andrei Viktorovich. Inst. of Crystallography, Acad. Sci. USSR, Leninskii Prospekt 59, Moscow 117333, USSR. (1952) Cand, physics and mathematics (INCRYS, 1988). Scient. *Real structure, subsurface layers.*

Massalimov, Dr Ismail Aleksandrovich. Inst. of Chemistry, Uzbek Acad. Sci., Prospekt Gor'kogo 77, Tashkent 700170, USSR. (1952) Cand, physics and mathematics (MGU, 1983).Sr. scient. *Precision X-ray investigations, electron density.*

Mastryukov, Dr Vladimir Saidovich. Moscow State U., Dept. of Chemistry, Leninskiye Gory, Moscow 117234, USSR. (1935) Cand, chemistry (MGU, 1966). Sr. scient. *Structure, elemento-organic compounds, electron diffraction by gases.*

Matkovsky, Dr Orest Ilyarovich. Lvov State U., Shcherbakov St. 4, Lvov 290005, USSR. (1929) Dr, geology and mineralogy (IGPM, 1979). Head of Chair. *Mineralogical crystallography.*

Matveeva, Dr Ol'ga Petrovna. Leningrad Mining Inst., 21st Liniya 2, Leningrad 199026, USSR. (1951) Cand, physics and mathematics (LGU, 1985). Sr. scient. *Mineral physics.*

Matveeva, Mrs Rimma Georgiyevna. Inst. of Crystallography, Acad. Sci. USSR, Leninsky pr. 59, Moscow 117333, USSR. Scient. *X-ray structure analysis, computing.*

Matyash, Prof Ivan Vasil'evich. Inst. of Geochemistry and Physics of Minerals, Ukrainian SSR Acad. Sci., Prospekt Palladina 34, Kiev 252680, USSR. (1930) Corresponding Member, Ukrainian SSR Acad. Sci., Dr, physics and mathematics (U. of Kiev, 1972) Lab. Head. *Real structure, minerals.*

Mavlonov, Dr Sharaf. Physical Techn. Inst., Acad. Sci. Tadzhik SSR, Akademgorodok, Dushanbe 734630, USSR. (1935) Cand, physics and mathematics (Azerbaidzhan U., 1962). Lab. head. *Semiconductors, crystal structure.*

Maximov, Dr Boris Alekseyevich. Inst. of Crystallography, Acad. Sci. USSR, Leninsky pr. 59, Moscow 117333, USSR. (1941) Cand, physics an mathematics (INCRYS, 1969). Sr. scient. *X-ray structure analysis.*

Mazus, Dr Mark Davidovich. Inst. of Applied Physics, Acad. of Sciences Moldavian SSR, Akademicheskaya 5, Kishinev 277028, USSR. (1937) Cand, physics and mathematics (Inst. of Applied Physics, Acad. Sci. Moldavian SSR, 1974). Sr. scient. *X-ray structure analysis, coordination and organic compounds.*

Mchedlishvili, Dr Boris Victorovich. Inst. of Crystallography, Acad. Sci. USSR, Leninskii Prospekt 59, Moscow 117333, USSR. (1944) Cand, physics and mathematics (Polytechnical Inst., Lenin- grad, 1980). Lab. Head. *Nuclear filters, purification, liquid and gaseous media.*

Mel'nikov, Dr Oleg Konstantinovich. Inst. of Crystallography, Acad. Sci. USSR, Leninsky pr. 59, Moscow 117333, USSR. (1940) Cand, geology and mineralogy (INCRYS, 1968). Sr. scient. *Crystal growth, crystal chemistry.*

Mel'nikov, Dr Vitaly Alexandrovich. Inst. of Crystallography, Acad. Sci. USSR, Leninsky pr. 59, Moscow 117333, USSR. (1937). Cand, physics and mathematics (INCRYS, 1982). Computing center head. *Automation, crystallographic investigations.*

Mel'nikov, Dr Vladimir Stepanovich. Inst. of Geochemistry and Physics of Minerals, Ukrainian SSR Acad. Sci., Palladina 34, Kiev 252680, USSR. (1931) Cand, geology and mineralogy (IGFM, 1973). Sr. scient. *Structure analysis.*

Mel'nikova, Dr Alina Mikhailovna. Inst. of Crystallography, Acad. Sci. USSR, Leninskii Prospekt 59, Moscow 117333, USSR. (1940) Cand, physics and mathematics (INCRYS, 1982). Scient. *Crystal growth.*

Melekh, Dr Bernard Abu-Talibovich. Physical Techn. Inst., Acad. Sci. USSR, Zapovednaya St. 51, Leningrad 194037, USSR. (1937) Cand, chemistry (Moscow Inst. of Steel and Alloys, 1967). Sr. scient. *Thermoelectric materials, oxide compounds.*

Meleshina, Dr Valentina Alexandrovna. Inst. of Crystallography, Acad. Sci. USSR, Leninsky pr. 59, Moscow 117333, USSR. Cand, physics and mathematics (INCRYS, 1967). Scient. *Real structure, optical microscopy, electron microscopy.*

Melik-Adamyan, Dr Vil'yam Rafailovich. Inst. of Crystallography, Acad. Sci. USSR, Leninsky pr. 59, Moscow 117333, USSR. (1937) Cand, physics and mathematics (INCRYS, 1969). Sr. scient. *Protein structures.*

Merinov, Dr Boris Vladimirevich. Inst. of Crystallography, Acad. Sci. USSR, Leninsky pr. 59, Moscow 117333, USSR. (1954) Cand, physics and mathematics (INCRYS, 1981). Sr. scient. *X-ray structure analysis, solid electrolytes.*

Metsik, Prof. Mikhail Stepanovich. Irkutsk State U., Dept. of Physics, Karl Marx St. 1, Irkutsk 664003, USSR. (1918) Dr, physics and mathematics (Physical Chemistry Inst., 1965). Prof. *Layer silicates, structure, thin water films.*

Mikhailov, Dr Al'bert Mikhailovich. Inst. of Crystallography, Acad. Sci. USSR, Leninsky pr. 59, Moscow 117333, USSR. (1939) Cand, physics and mathematics (INCRYS, 1971). Sr. scient. *Protein structures, virus structures, X-ray structure analysis, electron microscopy.*

Mikhailov, Dr Igor' Fedorovich. Kharkov Polytechnic Inst., Frunze St. 21, Kharkov 310002, USSR. (1949) Cand, technics (Kharkov Polytechnic Inst., 1975). Sr. scient. *Metals, structure, semiconductors, defects, methods for defect study.*

Mikhailov, Dr Vladimir Ivanovich. Inst. of Crystallography, Acad. Sci. USSR, Leninskii Prospekt 59, Moscow 117333, USSR. (1944) Cand, physics and mathematics (INCRYS, 1978).Sr. scient. *Crystal growth, electron microscopy.*

Mikhailov, Dr Yuriy Nikolayevich. Inst. of General and Inorganic Chemistry, Acad. Sci. USSR, Leninsky pr. 31, Moscow 117071, USSR. (1932) Cand, chemistry (IONCH, 1969). Sr. scient. *Inorganic structures, coordination compounds (chemistry).*

Mikhalenko, Dr Svetlana Ivanovna. Chemistry Dept., Lvov State U., Universitetskaya 1, Lvov 290602, USSR. (1946) Cand, chemistry (Lvov U., 1976) Sr. scient. *Crystal chemistry, intermetallic compounds, borides.*

Mikheeva, Dr Irina Victorovna. Inst. 'Mekhanobr', 21st Liniya 8a, Leningrad 199026, USSR. (1921)Cand, geology and mineralogy (GORNY, 1952). Sr. scient. *X-ray studies, crystal chemistry.*

Millionova, Dr Margarita Ivanovna. Inst. of Molecular Biology, Acad. Sci. USSR, Vavilov St. 32, Moscow 117312, USSR. Cand, physics and mathematics (Inst. of Biophysics, Acad. Sci. USSR, 1964). Scient. *Fibrillous structures.*

Minacheva, Dr Lidiya Khabibovna. Inst. of General and Inorganic Chemistry, Acad. Sci. USSR, Leninsky pr. 31, Moscow 117071, USSR. (1938) Cand, chemistry (IONCH, 1971). Scient. *Crystal chemistry, stereochemistry, coordination compounds*

Mineeva, Dr Rimma Mikhailovna. Inst. of Geology- Mineralogy and Petrography, Acad. Sci. USSR, Staromonetny 35, Moscow 109017, USSR. (1938) Cand, physics and mathematics (Kazan U., 1967). Sr. scient. *Minerals (physics), crystal field theory, electron paramagnetic resonance.*

Mints, Prof. Rafail Isaakovich. Ural Polytechnic Inst., Sverdlovsk 620002, USSR. (1931) Dr, technics (Ural Polytechnic Inst., 1965). Dept. head. *Structure phase transformations, metastable states, ordered condensed systems.*

Mishnev, Dr Anatolii Fedorovich. Inst. of Organic Synthesis, Acad. Sci. Latvian SSR, Aizkraukles st.23, Riga 226006, USSR. (1952) Cand, physics and mathematics (Inst. Appl. Physics, Kishinev, 1982). Sr. scient. *Phase problem in structure analysis, X-ray analysis, biologically active substances.*

Misyul', Dr Sergei Vladimirovich. Inst. of Physics, Siberian Dept. Acad. Sci. USSR, Akademgorodok, Krasnoyarsk 660036, USSR. (1949) Cand., physics and mathematics (Inst.of Physics, Siberian Dept.,Acad. Sci. USSR, 1986). Scient. *Phase transformations, structure, inorganic dielectrics.*

Miuskov, Dr Vasily Fedorovich. Inst. of Crystallography, Acad. Sci. USSR, Leninsky pr. 59, Moscow 117333, USSR. (1909) Cand, physics and mathematics (Inst. of Crystallography, 1941). Sr. scient. *X-ray topography, X-ray moire, crystal growth, defects in crystals.*

Mogilevskii, Dr Leonid Yur'evich. Inst. of Crystallography, Acad. Sci. USSR, Leninskii Prospekt 59, Moscow 117333, USSR. (1948) Cand, physics and mathematics (INCRYS, 1986) Scient. *Small-angle scattering, overatomic structure, biopolymers and amorphous metals.*

Mokeeva, Dr Valentina Ivanovna. Inst. of Geochemistry and Analytical Chemistry, Acad. Sci. USSR, Kosygin pr. 19, Moscow 117334, USSR. (1923) Cand, physics and mathematics (INCRYS, 1952). Scient. *Inorganic compounds, structure, isomorphism.*

Mokhov, Dr Andrei Vladimirovich. Inst. of Geology of Ore Deposits- Petrography-Mineralogy and Geochemistry, Acad. Sci. USSR, Staromonetnii Per.35, Moscow 109017, USSR. (1953) Cand, geology and mineralogy (IGEM, 1987) Scient. *Crystal chemistry, minerals.*

Mokraya, Dr Ivanna Romanovna. Chemistry Dept., Lvov State U. Universitetskaya 1, Lvov 290602, USSR. (1950) Cand, chemistry (Lvov U., 1979). Sr. scient. *Crystal chemistry, intermetallic compounds.*

Molchanov, Dr Vladimir Nikolayevich. Inst. of Crystallography, Acad. Sci. USSR, Leninsky pr. 59, Moscow 117333, USSR. (1952) Cand, chemistry (INCRYS, 1982). Sr. scient. *X-ray structure analysis, high pressure.*

Moroz, Dr Ella Mikhailovna. Inst. of Catalysis, Siberian Dept. Acad. Sci. USSR, Prospekt Lavrent'ev 3, Novosibirsk 630090, USSR. (1939) Dr, chemistry (Inst. of Catalysis, 1989). Scient. *Inorganic compounds, structure, phase transformation.*

Moshkin, Dr Sergei Vladimirovich. Geology Dept., Leningrad State U., Universitetskaya Nab. 7/9, Leningrad 199034, USSR. (1952) Cand, geology and mineralogy (LGU, 1982). Sr. scient. *Crystal growth.*

Moskvin, Dr Valentin Vasilyevich. Inst. of Crystallography, Acad. Sci. USSR, Leninsky pr. 59, Moscow 117333, USSR. (1941) Cand, chemistry (INCRYS, 1979). Scient. *Nucleation, thin films, epitaxy, electron microscopy.*

Movchan, Dr Nikolai Prokof'evich. Inst. of Geochemistry and Physics of Minerals, Ukrainian SSR Acad. Sci., Pr. Palladina 34, Kiev 252680, USSR. (1938) Cand, physics and mathematics (IGFM, 1971). Lab. Head. *Crystallography, layer silicates.*

Mukhamedzhanov, Dr Enver Khamzyaevich. Inst. of Crystallography, Acad. Sci. USSR, Leninskii Prospekt 59, Moscow 117333, USSR. (1955) Cand., physics and mathematics (INCRYS, 1986). Sr. Scient. *Real structure, thin films.*

Mul:htarova, Dr Nina Nikolayevna. Inst. of Nuclear Physics, Acad. Sci. Uzbek SSR, Ulugbek, Tashkent 702132, USSR. Cand, physics and mathematics (INCRYS, 1981). Scient. *Structural crystallography, crystal chemistry, inorganic compounds.*

Musaev, Dr Aidyn Alipanakh ogly. Inst. of Physics, Acad. Sci. Azerbaidzan SSR, Prospekt Narimanova 33, Baku 370143, USSR. (1954) Cand, chemistry (MGU, 1983). Scient. *Crystal chemistry, structure, inorganic compounds.*

Musayev, Dr Faig Nasib ogly. Inst. of Inorganic and Physical Chemistry, Acad. Sci. Azerbaidzhan SSR, Narimanov Pr. 29, Baku 370143, USSR. (1952) Cand, chemistry (Tbilisi State U., 1979). Sr. scient. *Structure, coordination compounds.*

Mustafayev, Dr Nariman Mustafa ogly. Inst. of Inorganic and Physical Chemistry, Acad. Sci. Azerbaidzhan SSR, Narimanov Pr. 29, Baku 370143, USSR. (1929) Cand, geology and mineralogy (INCRYS, 1966). Sr. scient. *Hydrothermal synthesis, crystal chemistry, silicates.*

Myasnikova, Dr Rimma Mikhailovna. Inst. of Elemento-Organic Compounds, Acad. Sci. USSR, Vavilov St. 28, Moscow 117312, USSR. (1932) Dr, physics and mathematics (INCRYS, 1981). Sr. scient. *Crystal structure and properties, binary organic systems.*

Mys'kiv, Dr Mar'yan Grigoryevich. Lvov State U., Dept. of Chemistry, University St. 1, Lvov 290602, USSR. (1947) Cand, chemistry (Lvov U., 1973). Asst. *X-ray structure analysis, crystal chemistry, intermetallic compounds.*

Myshlyayev, Prof. Mikhail Mikhailovich. Inst. of Solid State Physics, Acad. Sci. USSR, Chernogolovka, Noginsky Rayon, Moskovskaya Oblast' 142432, USSR. (1934) Dr, physics and mathematics (INCRYS, 1982). Sr. scient. *Electron microscopy (contrasting defects), defects, plastic deformation mechanisms.*

Naboka, Dr Marat Nikolayevich. Kharkov Polytechnic Inst., Frunze St. 21, Kharkov 310002, USSR. (1933) Cand, technics (Kharkov, Polytechnic Inst., 1969). Sr. scient. *Metals and alloys, structures, defects, measurement methods.*

Nadezhina, Dr Tamara Nikolayevna. Moscow State U., Dept. of Geology, Leninskiye Gory, Moscow 117234, USSR. (1946) Cand, geology and mineralogy (MGU, 1975). Scient. *X-ray structure analysis, minerals and synthetic compounds, geocrystal chemistry.*

Nagaitsev, Dr Yury Valeryevich. Leningrad State U., Dept. of Geology, University Emb. 7/9, Leningrad 199164, USSR. (1937) Cand, geology and mineralogy (LGU, 1965). Sr. scient. *Crystal chemistry, metamorphic minerals.*

Nardov, Dr Andrei Vladimirovich. Geology Dept., Leningrad State U., Universitetskaya Nab.7/9, Leningrad 119034, USSR. (1955) Cand, geology and mineralogy (LGU, 1986). Sr. scient. *Crystal growth.*

Naumova, Dr Inessa Ivanovna. Physics Dept., Moscow State U., Leninskie Gory, Moscow 117234, USSR. (1939) Cand, geology and mineralogy. Sr. eng. *Ferroelectricity, photorefraction, phase transitions, scattering, of light, crystal growth.*

Nefedova, Dr Elena Vasilyevna. Leningrad Mining Inst., 21st Liniya 2, Leningrad 199026, USSR. (1934) Cand, geology and mineralogy (GORNY, 1971). Sr. scient. *Crystal growth, morphology.*

Neronova, Dr Nina Nikolayevna. All-Union Correspondence Inst. of Eng. and Construction, Lesnaya St. 5, Kostroma 156021, USSR. Cand, physics and mathematics (INCRYS, 1965). Docent. *Symmetry, teaching.*

Nesterov, Dr Vladimir Nikolaevich. Voroshilovgrad Pedagogical Inst., Oboronnaya 2, Voroshilovgrad 348011, USSR. (1959) Cand, chemistry (IOCH, 1988). Sr. scient. *Organic crystal chemistry.*

Nesterova, Dr Yaroslava Mikhailovna. Moscow State U., Dept. of Chemistry, Leninskiye Gory, Moscow 177234, USSR. (1937) Cand, chemistry (MGU, 1973). Scient. *Crystal chemistry, coordination compounds, X-ray structure analysis.*

Nevskaya, Dr Natalia Aleksandrovna. Inst. of Proteins, Acad. Sci. USSR, Moskovskaya obl., Pushchino 142293, USSR. (1944) Cand, physics and mathematics (Inst. of Proteins, Pushchino, 1978). Scient. *Structure, proteins.*

Nevskii, Dr Nukolai Nikolaevich. NPO IREA, Bogorodskii val 3, Moscow 107258, USSR. (1949) Cand, physics and mathematics (INCRYS, 1960).Sr. scient. *Mathematical methods, structure determination.*

Nikanorov, Dr Stanislav Prokhorovich. Physico-Techn. Inst., Acad. Sci. USSR, Politekhnicheskaya St. 26, Leningrad 194021, USSR. (1928) Dr, physics and mathematics (Phsico-Techn. Inst., Acad. Sci. USSR, 1978). Lab. Head. *Crystal growth, structure imperfections, elasticity and plasticity, crystal physics.*

Nikiforov, Prof. Igor' Yakovlevich. Rostov Inst. of Agricultural Mashinery , Pl. Gagarina 1, Rostov-on-Don 344010, USSR. (1930) Doctor, physics and mathematics (Rostov U., 1985). Head of chair. *Perfection, X-ray spectrometry.*

Nikishova, Dr Lidiya Vasilyevna. Inst. of Geology, Acad. Sci. USSR, Yakutsk Branch of Siberian Dept., Lenin Pr. 39, Yakutsk 677007, USSR. (1938) Cand, physics and mathematics (Irkutsk U., 1976). Scient. *Isomorphism, polymorphism, polytypism, electron diffraction, microdiffraction.*

Nikitenko, Prof. Valerian Ivanovich. Inst. of Solid State Physics, Acad. Sci. USSR, Chernogolovka 142432, Noginsky Rayon, Moskovskaya Oblast', USSR. (1937) Dr, physics and mathematics (Inst. of Solid State Physics, 1972). Lab. head. *Defects, crystal structure, physical properties - defects relationship.*

Nikol'skaya, Dr Larisa Viktorovna. Leningrad Mining Inst., 21st Liniya 2, Leningrad 199026, USSR. (1949) Cand, geology and mineralogy (All-Union Research Inst.for Synthesis of Mineral Raw Materials, 1976). Sr. scient. *Mineral physics.*

Nikol'skaya, Dr Natal'ya Kimovna. Geology Dept., Leningrad State U., Universitetskaya Nab. 7/9, Leningrad 119034, USSR. (1952) Cand, geology and mineralogy (LGU, 1987). Scient. *Crystal chemistry, minerals, X-ray structure analysis.*

Nikonov, Dr Stanislav Valdimirovich. Inst. of Proteins, Acad. Sci. USSR, Moskovskaya obl., Pushchino 142292, USSR. (1938) Cand, physics and mathematics (INCRYS, 1985). Sr. scient. *Structure, proteins.*

Nizamutdinov, Dr Nazym Minsafovich. Kazan State U., Lenina 18, Kazan 420008, USSR. (1939) Cand, physics and mathematics (Kazan U., 1977). Docent. *Crystal growth, symmetry.*

Novozhilov, Dr Alexander Ivanovich. Res. Inst. for Synthesis of Mineral Raw Materials, Institutskaya St. 1, Alexandrov 601600, Vladimirskaya Oblast', USSR. (1939) Cand, physics and mathematics (INCRYS, 1974). Sr. scient. *Optical spectroscopy, electron paramagnetic resonance.*

Nuriyev, Dr Idayat Ragim ogly. Inst. of Physics, Acad. Sci. Azerbaidzhan SSR, Narimanov Pr. 33, Baku 370143, USSR. (1941) Cand, physics and mathematics (Inst. of Physics, Acad. Sci. Azerbaidzhan SSR, 1971). Lab. head *Thin film growth, structure and physical properties, epitaxial films, semiconductors.*

Obodovskaya, Dr Alla Efimovna. NPO IREA, Bogorodskii val 3, Moscow 107258, USSR. (1941) Cand, chemistry. (Inst. of Chemical Reagents, Moscow, 1967). Scient. *Crystal chemistry, organic and coordination compounds.*

Okhrimenko, Dr Tatyana Mikhailovna. Inst. of Crystallography, Acad. Sci. USSR, Leninskii Prospekt 59, Moscow 117333, USSR. (1944) Cand, chemistry (INCRYS, 1987) Scient. *Crystal growth and morphology, geometric crystallization.*

Oleinikov, Dr Vladimir Aleksandrovich. Inst. of Crystallography, Acad. Sci. USSR, Leninskii Prospekt 59, Moscow 117333, USSR. (1949) Cand, physics and mathematics (Physical-Eng. Inst., Moscow, 1980) Sr. scient. *Solid state physics, surface physics.*

Oliinik, Dr Vladimir Vladimirovich. Chemistry Depart., Lvov State U., Universitetskaya 1, Lvov 290602, USSR. (1959) Cand, chemistry (Lvov U., 1985). Sr. scient. *Crystal chemistry, X-ray analysis, transition metals.*

Onishchina, Dr Ninel' Mitrofanovna. Leningrad Mining Inst., 21st Liniya 2, Leningrad 199026, USSR. (1935) Cand, geology and mineralogy (GORNY, 1975). Sr. teacher. *Crystal morphology, crystal growth.*

Opekunov, Dr Victor Nikolayevich. Inst. of Crystallography, Acad. Sci. USSR, Leninsky pr. 59, Moscow 117333, USSR. (1950) Cand, physics and mathematics (MGU, 1982). Scient. *Real structure, defects, mechanical properties.*

Orekhova, Dr Valentina Petrovna. Inst. of Crystallography, Acad. Sci. USSR, Leninsky pr. 59, Moscow 117333, USSR. (1940) Cand, physics and mathematics (INCRYS, 1975). Scient. *Optical spectroscopy, doped crystals.*

Organova, Dr Nataliya Ivanovna. Inst. of Geology-, Mineralogy and Petrography, Acad. Sci. USSR, Staromonetny 35, Moscow 109017, USSR. (1929) Dr,

geology and mineralogy (IGEM, 1987). Sr. scient. *Minerals, structure, atomic ordering, hydrogen bonding.*

Orishin, Dr Stepan Vasil'evich. Chemistry Dept, Lvov State U., Universitetskaya 1, Lvov 290602, USSR. Cand, chemistry (Lvov U., 1984). Lab. Head. *Crystal chemistry, phosphates.*

Osipov, Dr Mikhail Alekseevich. Inst. of Crystallography, Acad. Sci. USSR, Leninskii Prospekt 59, Moscow 117333, USSR. (1955) Cand, physics and mathematics (MGU, 1982) Sr. scient. *Liquid crystals, biologically ordered structures, phase transition theory.*

Osip'yan, Prof. Yury Andreyevich. Inst. of Solid State Physics, Acad. Sci. USSR, Chernogolovka, Noginsky Rayon, Moskovskaya Oblast' 142432, USSR. (1931) Full member, Acad. Sci. USSR; Dr, physics and mathematics; Director. *Crystal structure, semiconductors, dielectrics.*

Ostanevich, Dr Yury Mechislavovich. Joint Inst. for Nuclear Res., Dubna 141980, Moskovskaya Oblast', USSR. (1936) Dr, physics and mathematics (Joint Inst. for Nuclear Res., 1972). Dept. head. *Neutron physics.*

Ostapenko, Prof Georgii Tikhonovich. Inst. of Geochemistry and Physics of Minerals, Ukrainian SSR Acad. Sci. Pr. Palladina 34, Kiev 252680, USSR. (1935) Doctor, geology and mineralogy (IGFM, 1979). Lab. Head. *Crystal growth, isomorphism, crystallization pressures.*

Otroshchenko, Dr Lyudmila Petrovna. Inst. of Crystallography, Acad. Sci. USSR, Leninsky pr. 59, Moscow 117333, USSR. (1940) Cand, geology and mineralogy (MGU, 1981). Scient. *X-ray structure analysis.*

Ostrovskaya, Dr Anna Borisovna. Inst. of Geochemistry and Physics of Minerals, Ukrainian SSR Acad. Sci., Pr. Palladina 34, Kiev 252680, USSR. (1935) Cand, physics and mathematics (Kiev U., 1964) Sr. scient. *Mineralogy, crystal chemistry, clays.*

Ostrovskii, Dr Boris Isaakovich. Inst. of Crystallography, Acad. Sci. USSR, Leninskii Prospekt 59, Moscow 117333, USSR. (1949) Cand, physics and mathematics (MGU, 1981) Sr. scient. *Liquid crystals, phase transitions.*

Ovchinnikov, Dr Yurii Eduardovich. Novosibirsk Pedagogical Inst., Vilyuiskaya 28., Novosibirsk 630126, USSR. (1951) Cand, physics and mathematics (Moscow Pedagogical Inst., 1986). Sr. lecturer. *Crystallography, organic compounds, chain molecules including polymers.*

Ozerin, Dr Alexander Nikiforovich. Karpov Physical Chemistry Inst., Obukha St. 10, Moscow 107120, USSR. (1952) Cand, physics and mathematics (Moscow Physico-Techn Inst., 1977). Sr. scient. *X-ray structure analysis, polymers.*

Ozerov, Prof. Ruslan Pavlovich. Mendeleev Inst. of Chemical Technology, Miusskaya sq. 9, Moscow 125820, USSR. (1926) Dr, physics and mathematics (INCRYS, 1969). Head of Chair. *X-ray diffraction, neutron diffraction, solid state physics.*

Ozola, Dr Astrida Davovna. Inst. of Inorganic Chemistry, Acad. Sci. Latvian SSR, Ul Miera 34, Rizhsky Rayon, Saalspils 1, LAtv. SSR, 229021, USSR. Cand, chemistry (Inst. of Inorganic Chemistry, Acad. Sci. Latvian SSR, 1977). Sr. scient. *Crystal chemistry.*

Ozolin'sh, Dr Gerkhard Vladimirovich. Inst. of Inorganic Chemistry, Acad. Sci. Latvian SSR, Ul Miera 34, Rizhsky Rayon, Saalspils 1, LAtv. SSR, 229021, USSR. (1934) Cand, physics and mathematics (Latvian State U., 1969). Lab. head. *Instrumentation and techniques for X-ray crystallography, lattice parameter precision determination.*

Ozols, Dr Yan Karlovich. Inst. of Inorganic Chemistry, Acad. Sci. Latvian SSR, Ul Miera 34, Rizhsky Rayon, Saalspils 1, LAtv. SSR, 229021, USSR. (1915) Cand, chemistry (Latvian State U., 1949). Sr. scient. *Crystal chemistry, borates, chelate compounds.*

Pakhomov, Dr Vladimir Ivanovich. Inst. of General and Inorganic Chemistry, Acad. Sci. USSR, Leninsky pr.31, Moscow 117071, USSR. (1932) Dr, chemistry (IONCH, 1973). Sr. scient. *Crystal chemistry, inorganic compounds.*

Palatnik, Prof. Lev Samoilovich. Kharkov Polytechnik Inst., Frunze St. 21, Kharkov 310002, USSR. (1909) Dr, physics and mathematics (Kharkov U., 1952). Head of Chair. *Metals, structure, sub-structure, semiconductors, dielectrics, films (single- & poly-crystal), amorphous films.*

Pal'chik, Dr Nadezhda Arsent'evna. Inst. of Geology and Geophysics, Siberian Dept. Acad. Sci. USSR, Universitetskii Prospekt 3, Novosibirsk 630090, USSR. (1946) Cand, geology and mineralogy (IGEM, 1981). Sr. scient. *Structure, minerals, diffraction methods, non-diffraction structure methods.*

Palistrant, Prof. Alexander Filippovich. Kishinev U., Sadovaya 60, Kishinev 277003, USSR. (1933) Dr, physics and mathematics (INCRYS, 1984). Prof. *Symmetry theory (generalization and applications).*

Palkina, Dr Kapitolina Kapitonovna. Inst. of General and Inorganic Chemistry, Acad. Sci. USSR, Leninsky pr. 31, Moscow 117071, USSR. (1932) Cand, chemistry (IONCH, 1963). Sci. *Crystal chemistry, coordination compounds, semiconductors.*

Panchekha, Dr Petr Alekseyevich. Kharkov Polytechnic Inst., Frunze St. 21, Kharkov 310002, USSR. (1938) Cand, technics (Kharkov Polytechnic Inst., 1968). Sr. scient. *Metals and alloy structures, defects, defect investigation methods.*

Parvov, Mr Vladimir Fedorovich. Inst. of Crystallography, Acad. Sci. USSR, Leninskii Prospekt 59, Moscow 117333, USSR. (1920) Eng. *Apparatus, growth, water-soluble crystals, microfilming.*

Pashaev, Dr El'khan Mekhrali-Ogly. Inst. of Crystallography, Acad. Sci. USSR, Leninskii Prospekt 59, Moscow 117333, USSR. (1949) Cand, physics and mathematics (INCRYS, 1985). Scient. *Real structure, subsurface layers, semiconductor materials.*

Patmalnieks, Dr Aloisii Alekseevich. Inst. of Solid State Physics, Latvian State U., Kengaraga st.8, Riga 226063, USSR. (1926) Cand, physics and mathematics (Inst. of Physics, Riga, 1974). Lab. Head. *Growth, structure, transition metal oxides.*

Pavlishin, Dr Vladimir Ivanovich. Inst. of Geochemistry and Physics of Minerals, Acad. Sci. Ukrainian SSR, Palladina Prospekt 34, Kiev 252680, USSR. (1940) Cand, geology and mineralogy (Lvov U., 1966). Sr. scient. *Crystallography, minerals, silicates.*

Pavlovsky, Dr Alexander Grigoryevich. Inst. of Crystallography, Acad. Sci. USSR, Leninsky pr. 59, Moscow 117333, USSR. (1947) Cand, chemistry (Inst. of Molecular Biology, Acad. Sci. USSR, 1979). Sr. scient. *Protein crystallography.*

Pavlyuk, Dr Anatolii Alekseevich . Inst. of Inorganic Chemistry, Siberian Dept. Acad. Sci. USSR, Prospekt Akademika Lavrent'eva 3, Novosibirsk 630090, USSR. (1939) Cand, chemistry (Council of Chemical Sciences, Siberian Dept., USSR Acad. Sci., 1975). Sr. scient. *Crystal growth from melt and flux.*

Pech, Dr Lucia Yanovna. Inst. of Inorganic Chemistry, Acad. Sci. Latvian SSR, Ul Miera 34, Rizhsky Rayon, Saalspils 1, LAtv. SSR, 229021, USSR. Cand, chemistry (Inst. of Inorganic Chemistry Acad. Sci. Latvian SSR, 1977). Sr. scient. *Crystal chemistry.*

Pecharskii, Dr Vitalii Konstantinovich. Chemistry Dept., Lvov State U., Universitetskaya 1, Lvov 290602, USSR. (1954) Cand, chemistry (Lvov U., 1979). Docent. *Crystal chemistry, intermetallic compounds, computation.*

Perekalina, Dr Tatyana Mikhailovna. Inst. of Crystallography, Acad. Sci. USSR, Leninsky pr. 59, Moscow 117333, USSR. (1922) Dr, physics and mathematics (INCRYS, 1973). Sr. scient. *Magnetic properties, magnetic phase transitions.*

Perekalina, Dr Zoya Borisovna. Inst. of Crystallography, Acad. Sci. USSR, Leninsky pr. 59, Moscow 117333, USSR. (1929) Cand, physics and mathematics (INCRYS, 1969). Sr. scient. *Optical properties, optical activity.*

Perelomova, Dr Nataliya Vladislavovna. Moscow Inst. of Steel and alloys, Leninskii Prospekt 4, Moscow 117049, USSR. (1936) Cand, physics and mathematics (Moscow Inst. of Steel and Alloys, 1972). Docent. *Crystal physics and acoustics, data banks.*

Pershin, Dr Vitaly Konstantinovich. Ural Polytechnic Inst., Sverdlovsk 620002, USSR. Cand, physics and mathematics (Ural Polytechnic Inst., 1976). Asst. *Structural phase transformations, metastable states, organic systems.*

Perstnev, Dr Petr Petrovich. Inst. of Crystallography, Acad. Sci. USSR, Leninskii Prospekt 59, Moscow 117333, USSR. (1951) Cand, physics and mathematics. (INCRYS, 1983). Sr. scient. *Real structure, mechanical properties.*

Pertsin, Dr Alexander Iosifovich. Inst. of Elemento-Organic Compounds, Acad. Sci. USSR, Vavilov St. 28, Moscow 117813, USSR. (1948) Dr, physics and mathematics (Inst. of Chemical Physics, Acad. Sci. USSR, 1987). Lab. Head. *Phase transitions, organic crystal thermodynamics.*

Peskin, Dr Vladimir Fedorovich. Inst. of Crystallography, Acad. Sci. USSR, Leninsky pr. 59, Moscow 117333, USSR. (1937) Cand, geology and mineralogy (MGU, 1981). Scient. *Crystal growth.*

Petropavlov, Dr Nikolay Nikolayevich. Inst. of Biological Physics, Acad. Sci. USSR, Pushchino 142292, Moskovskaya Oblast', USSR. (1938) Cand, physics and mathematics (Moscow State Pedagogical Inst., 1971). Sr. scient. *Crystal structure and properties, biologically important substances, phase transitions.*

Petrov, Dr Thomas Georgiyevich. Leningrad State U., Dept. of Geology, University Emb. 7/9, Leningrad 199034, USSR. (1931) Cand, geology and mineralogy (GORNY, 1962). Sr. scient. *Crystallography, crystal growth, defects in crystals.*

Petrova, Dr Irina Vladimirovna. Dept. of Geology, Moscow State U., Leninskiye Gory, Moscow 117234, USSR. (1949) Cand, geology and mineralogy (MGU, 1980). Scient. *Crystal chemistry, inorganic compounds, X-ray diffraction.*

Petrova, Dr Valentina Vasil'evna. Petrozavodsk State U., Prospekt Lenina 33, Petrozavodsk 185640, USSR. (1938) Cand, chemistry (Khar'kov, 1968).Docent. *Defects in crystalline and amorphous materials, X-ray analysis of polycrystals.*

Petrovskii, Dr Vitalii Aleksandrovich. Inst. of Geology, Ural Dept., Acad. Sci. USSR, Oplesnina 2, Syktyvkar 167610, USSR. (1946) Cand, geology and mineralogy (Moscow Inst. of Geological Survey,1981). Sr. scient. *Crystal morphology, genesis, environmental effects.*

Petrunina, Dr Alla Anastas'yevna. Inst. of Geochemistry and Physics of Minerals, Acad. of Sci. Ukrainian SSR, Palladina pr. 34, Kiev 252068, USSR. (1930) Cand, physics and mathematics (INCRYS, 1972). *Structural typomorphism, rock-forming minerals.*

Petukhov, Dr Boris Vladimirovich. Inst. of Crystallography, Acad. Sci. USSR, Leninsky pr. 59, Moscow 117333, USSR. (1941) Dr, physics and mathematics (INCRYS, 1988). Sr. scient. *Crystal lattice defects, strength and plasticity.*

Pidzhyan, Prof. Grigory Oganesovich. Inst. of Geology, Acad. of Sciences Armenian SSR, Barekamutyan St. 24a, Erevan 375200, USSR. (1919) Dr, geology and mineralogy (GORNY, 1969). Dept. head. *Crystal chemistry, sulphides and sulphosalts.*

Pikin, Dr Sergey Alekseyevich. Inst. of Crystallography, Acad. Sci. USSR, Leninsky pr. 59, Moscow 117333, USSR. (1941) Dr, physics and mathematics (Moscow Inst. of Eng. and Physics, 1978). Sr. scient. *Theoretical physics, liquid crystal theory, phase transition theory.*

Pinsker, Dr Garry Zinov'yevich. Inst. of General and Inorganic Chemistry, Acad. Sci. USSR, Leninsky pr. 31, Moscow 117071, USSR. (1929) Dr, physics and mathematics (Inst. of Physics, Acad. Sci. Latvian SSR, 1984). Scient. *Short range order, symmetry, diffraction, structure, amorphous bodies.*

Pisarevsky, Dr Yury Vladimirovich. Inst. of Crystallography, Acad. Sci. USSR, Leninsky pr. 59, Moscow 117333, USSR. (1940) Cand, physics and mathematics (INCRYS, 1974). Sr. scient. *Acoustic and elastic properties, piezoelectric properties.*

Plakhov, Dr Gennady Fedorovich. Moscow State U., Dept. of Geology, Leninskiye Gory, Moscow 117234, USSR. (1938) Cand, geology and mineralogy (MGU, 1976). Scient. *Crystal chemistry, inorganic compounds.*

Plastinina, Dr Marina Arkad'evna. Intitute of Geochemistry and Physics of Minerals, Ukrainian SSR Acad. Sci., Pr.Palladina 34, Kiev 252680, USSR. (1941) Cand, geology and mineralogy (IGFM, 1971). Sr. scient. *Structure, minerals.*

Platonov, Prof Aleksei Nikolaevich. Inst. of Geochemistry and Physics of Minerals, Ukrainian SSR Acad. Sci., Pr. Palladina 34, Kiev 252680, USSR. (1937) Doctor, physics and mathematics (IGEM, 1974). Lab. Head. *Optical spectroscopy, crystals.*

Pletnev, Dr Vladimir Zakharovich. Inst. of Bioorganic Chemistry, Acad. Sci. USSR, Vavilov St. 32, Moscow 117312, USSR. (1944) Dr, chemistry (INCRYS, 1988). Sr. scient. *Peptide-protein related structures.*

Plyasova, Dr Lyudmila Mikhailovna. Inst. of Catalysis, Acad. Sci. USSR, Siberian Dept., Prospekt Lavrent'ev 5, Novosibirsk 630090, USSR. Cand, physics and mathematics (Gorky U., 1967). Sr. scient. *Crystal chemistry, inorganic compounds.*

Plyusnina, Dr Inga Ivanovna. Moscow State U., Dept. of Geology, Leninskiye Gory, Moscow 117234, USSR. (1931) Cand, geology and mineralogy (MGU, 1959). Sr. scient. *Infrared spectroscopy, minerals, inorganic compounds.*

Pobedimskaya, Dr Elena Alexandrovna. Moscow State U., Dept. of Geology, Leninskiye Gory, Moscow 117234, USSR. (1925) Cand, geology and mineralogy (MGU, 1962). Docent. *X-ray structure analysis, minerals, synthetic compounds, geocrystal chemistry, elements.*

Podberezskaya, Dr Nina Vasilyevna. Inst. of Inorganic Chemistry, Acad. Sci. USSR, Siberian Dept., Prospekt LAvrent'ev 3, Novosibirsk 630090, USSR. (1938) Cand, physics and mathematics (Gorky U., 1971). Sr. scient. *Structure determination methods, crystal chemistry, inorganic compounds, structure.*

Polchovskaya, Dr Tatyana Mikhailovna. Inst. of Crystallography, Acad. Sci. USSR, Leninsky pr. 59, Moscow 117333, USSR. (1944) Cand, physics and mathematics (INCRYS, 1975). Scient. *Crystal growth from melt.*

Poltev, Prof. Valerii Ivanovich. Inst. of Biophysics, Acad. Sci. USSR, Moskovskaya obl., Pushchino 142292, USSR. (1939) Doctor, physics and mathematics (MGU, 1985). Sr. scient. *Structure of fragments of nucleic acids and other bioorganic compounds.*

Polyakova, Dr Irina Nikolaevna. NPO IREA, Bogorodskii val 3, Moscow 107258, USSR. (1951) Cand, chemistry (MGU, 1979). Scient. *Crystal chemistry, organic and complex compounds.*

Polyanskaya, Dr Tamara Mikhailovna. Inst. of Inorganic Chemistry, Acad. Sci. USSR, Siberian Dept., Prospekt Lavrent'ev 3, Novosibirsk 630090, USSR. (1942) Cand, physics and mathematics (Gorky U., 1971). Scient. *Structure determination methods, crystal chemistry, inorganic compounds.*

Polyansky, Dr Evgeny Vasilyevich. Res. Inst. for Synthesis of Mineral Raw Materials, Institutskaya St. 1, Alexandrov 601600, Vladimirskaya Oblast', USSR. (1944) Cand, geology and mineralogy (MGU, 1973). Sr. scient. *Crystal growth, crystal optics.*

Polynova, Dr Tamara Nikitichna. Moscow State U., Dept. of Chemistry, Leninskiye Gory, Moscow 117234, USSR. (1930) Cand, chemistry (MGU, 1963). Docent. *Crystal chemistry, coordination compounds, X-ray structure analysis.*

Ponomarev, Dr Vasily Ivanovich. Branch Inst. of Chemical Physics, Acad. Sci. USSR, Chernogolovka 142432, Noginsky Rayon Moskovskaya Oblast', USSR. (1940) Cand, chemistry (INCRYS, 1971). Sr. scient. *Crystallography, crystal physics.*

Ponyatovsky, Prof. Evgeny Genrikhovich. Inst. of Solid State Physics, Acad. Sci. USSR, Chernogolovka, Noginsky Rayon, Moskovskaya Oblast' 142432, USSR. (1930) Dr, physics and mathematics. Lab. head. *Phase transformations at high pressures, high pressure synthesis, metal hydrides.*

Popik, Dr Mikhail Vasil'evich. Chemistry Dept., Moscow State U., Leninskie Gory, Moscow 117234, USSR. (1946) Cand., chemistry (MGU, 1979) Scient. *Gas electron diffraction.*

Popolitov, Dr Vladislav Ivanovich. Inst. of Crystallography, Acad. Sci. USSR, Leninsky pr. 59, Moscow 117333, USSR. (1938) Dr, chemistry (INCRYS, 1987). Scient. *Crystal growth.*

Popov, Dr Aleksandr Nikolaevich. Inst. of Crystallography, Acad. Sci. USSR, Leninskii Prospekt 59, Moscow 117333, USSR. (1953) Cand, Physics and mathematics (INCRYS, 1984). Sr. Scient. *X-ray diffractometry.*

Popova, Dr Anastasiya Arsentyevna. Inst. of Crystallography, Acad. Sci. USSR, Leninsky pr. 59, Moscow 117333, USSR. (1916) Cand, chemistry (INCRYS, 1963). *Crystal growth from melt.*

Porai-Koshits, Prof. Mikhail Alexandrovich. Inst. of General and Inorganic Chemistry, Acad. Sci. USSR, Leninsky pr. 31, Moscow 117071, USSR. (1918) Corresp. member, Acad. Sci. USSR, Dr, physics and mathematics (INCRYS,

1960). Lab. head. *Crystal chemistry, stereochemistry, coordination compounds, methods, structure analysis.*

Porai-Koshits, Prof. Evgeny Alexandrovich. Inst. of Silicate Chemistry, Acad. Sci. USSR, Makarov Emb. 2, Leningrad 199164, USSR. (1907) Dr, physics and mathematics (LGU, 1953). *Disordered systems, glass structure, glass-like state.*

Portnov, Dr Vadim Nikolayevich. Gorky State U., Gagarin Prospekt 23, Gorky 603022, USSR. (1934) Cand, physics and mathematics (Gorky U., 1966). Docent. *Crystal growth.*

Potekhin, Dr Konstantin Al'bertovich. Vladimir State Pedagogical Inst., Prospekt Stroitelei 11, Vladimir 600024, USSR. (1954) Cand, physics and mathematics (INCRYS, 1980). Docent. ' *Patterson methods, molecule packing in crystals.*

Prevarsky, Dr Anatoly Petrovich. Lvov State U., Dept. of Chemistry, University St. 1, Lvov 290602, USSR. (1924) Cand, chemistry (Lvov U., 1973). Sr. scient. *X-ray structure analysis, crystal chemistry, intermetallic compounds.*

Prikhod'ko, Dr Leonid Vasilyevich. Inst. of Crystallography, Acad. Sci. USSR, Leninsky pr. 59, Moscow 117333, USSR. (1938) Cand, physics and mathematics (INCRYS, (1971). Scient. *High temperature crystallization.*

Provotorov, Dr Mikhail Viktorovich. Moscow Chemico-Techn. Inst., Miusskaya sq. 9, Moscow 125047, USSR. (1946) Cand, chemistry (Moscow Chemico-Techn. Inst., 1976). Docent. *Synthesis, new crystalline phases, crystal growth from melts.*

Ptitsyn, Prof. Oleg Borisovich. Inst. of Protein, Acad. Sci. USSR, Pushchino, Serpukhovskoy Rayon, Moskovskaya Oblast' 142292, USSR. (1929) Dr, physics and mathematics (Inst. of High Molecular Compounds, Acad. Sci. USSR, 1962). Assoc. dir. *X-ray structure analysis, proteins, diffuse X-ray scattering, biological macromolecules.*

Pudovkina, Dr Zoya Vasilyevna. Inst. of Mineralogy- Geochemistry and Crystal Chemistry of Rare Elements, Sadovnicheskaya Emb. 71, Moscow 113127, USSR. (1933) Cand, geology and mineralogy (MGU, 1968). Sr. scient. *Crystal chemistry, X-ray structure analysis, inorganic compounds, minerals.*

Pugachev, Prof. Anatoly Tarasovich. Kharkov Polytechnic Inst., Frunze St. 21, Kharkov 310002, USSR. (1940) Dr, technics (Leningrad Phys. Techn. Inst., 1982). Prof. *Metals and alloy structures, defects, methods for defect study.*

Punin, Dr Yury Olegovich. Leningrad State U., University Emb. 7/9, Leningrad 199034, USSR. (1941) Cand, geology and mineralogy (LGU, 1970). Scient. *Crystal growth, defect formation.*

Pushcharovsky, Dr Dmitry Yuryevich. Moscow State U., Dept. of Geology, Leninskiye Gory, Moscow 117234, USSR. (1944) Dr, geology and mineralogy (IGEM, 1984). Docent. *Crystal chemistry, silicates and analogs, physical properties, crystal structure.*

Pyatenko, Dr Yury Andreyevich. Inst. of Mineralogy- Geochemistry and Crystal Chemistry of Rare Elements, Acad. Sci. USSR, Sadovnicheskaya Emb. 71, Moscow 113127, USSR. (1928) Dr, geology and mineralogy (IGEM, 1969). Lab. head. *Crystal chemistry, minerals.*

Radautsan, Prof. Sergey Ivanovich. Inst. of Applied Physics, Acad. Sci. Moldavian SSR, Akademicheskaya 5, Kishinev 277028, USSR. (1926) Full member, Acad. Sci Moldavian SSR, Dr, technics (Leningrad Polytechnic Inst., 1966). Lab. head. *Crystal growth, semiconductor materials, glasses, amorphous semiconductors.*

Rakin, Dr Vladimir Ivanovich. Inst. of Geology, Ural Dept., Acad. Sci. USSR, Oplesnina 2, Syktyvkar 167610, USSR. (1956) Cand, geology and mineralogy (MGU, 1985). Scient. *Crystal growth.*

Rakova, Dr Elena Vasilyevna. Inst. of Crystallography, Acad. Sci. USSR, Leninsky pr. 59, Moscow 117333, USSR. (1941) Cand, chemistry (INCRYS, 1978). Scient. *Thin film growth and structure.*

Ramans, Dr Guntis Mirvaldovich. Inst. of Solid State Physics, Latvian State U., Kengaraga st.8, Riga 226063, USSR. (1950) Cand, physics and mathematics (Inst. of Physics, Riga, 1986). Sr. scient. *Crystal and amorphous structure of transition metal oxides.*

Rashkovich, Prof. Leonid Nikolaevich. Physics Dept., Moscow State U., Leninskie Gory, Moscow 117234, USSR. (1931)Dr, eng. (MGU, 1981) Sr. cient. *Crystal growth.*

Rastsvetaeva, Dr Ramiza Kerarovna. Inst. of Crystallography, Acad. Sci. USSR, Leninsky pr. 59, Moscow 117333, USSR. (1936) Cand, geology and mineralogy (MGU, 1971). Scient. *X-ray structure analysis, inorganic crystals.*

Rau, Dr Tamara Fedorovna. Vladimir Pedagogical Inst., Prospekt Stroiteley 11, Vladimir 600024, USSR. (1940) Cand, physics and mathematics (Gorky U., 1972). Docent. *X-ray crystal structure analysis.*

Rau, Prof. Valery Georgiyevich. Vladimir Pedagogical Inst., Prospekt Stroiteley 11, Vladimir 600024, USSR. (1940) Dr, physics and mathematics (INCRYS, 1985). Head Chair. *X-ray crystal structure analysis.*

Rebrov, Dr Aleksandr Nikolaevich. Karpov Physical-Chemical Inst., Obukha 10, Moscow 103064, USSR. (1950) Cand, physics and mathematics (KARPOV, 1988). Scient. *X-ray analysis, organic and inorganic compounds, programming.*

Regel', Dr Vadim Robertovich. Inst. of Crystallography, Acad. Sci. USSR, Leninsky pr. 59, Moscow 117333, USSR. (1917) Cand, physics and mathematics (INCRYS, 1965). Sr. scient. *Mechanical properties, surface physics, strength.*

Revkevich, Dr Galina Panteleimonovna. Physics Dept., Moscow State U., Leninskie Gory, Moscow 117234, USSR. (1933) Cand, physics and mathematics (MGU, 1967).Assistant. *Defect crystals.*

Rez, Prof. Iosif Solomonovich. All-Union Gosstandart Research Center of Materials and Substances. Krasnoproletarskaya 35, Moscow 103030, USSR. (1915) Doctor, chemistry (IONCH, 1970). Consultant. *Crystal chemistry, crystal growth.*

Rogacheva, Dr Evelina Danilovna. Gorky Agricultural Inst., Gagarin Prospekt 97, Gorky 603078, USSR. (1928) Cand, physics and mathematics (Gorky U., 1966). Head of Chair. *Crystal formation from solutions.*

Roginskaya, Dr Yuliana Eremeyevna. Physical Chemistry Inst., Obukha St. 10, Moscow 107120, USSR. (1937) Cand, chemistry (KARPOV, 1965). Sr. scient. *Crystal chemistry, complex metal oxides.*

Romaka (Komarovskaya), Dr Lyubov' Petrovna. Chemistry Dept., Lvov State U., Universitetskaya 1, Lvov 290602, USSR. (1958) Cand, chemistry (Lvov U.,1984). Sr. scient. *Crystal structures, ternary stannides, rare earth and transition metal.*

Romanov, Dr Gennady Vasilyevich. Moscow State U., Dept. of Chemistry, Leninskiye Gory, Moscow 117234, USSR. (1937) Cand, chemistry (MGU, 1967). Sr. scient. *Inorganic compounds, structure, electron diffraction by gases.*

Rozenberg, Dr Yurii Aleksandrovich. Inst. of Railway Eng., Chernyshevsky st.15, Irkutsk 664074, USSR. (1949) Cand, physics and mathematics (Irkutsk U., 1977). Docent. *Dynamical theory, inelastic X-ray scattering.*

Rozhansky, Dr Vladimir Nikolayevich. Inst. of Crystallography, Acad. Sci. USSR, Leninsky pr. 59, Moscow 117333, USSR. (1923) Dr, physics and mathematics (Inst. of Metal Physics, 1969). Lab. head. *Real structure.*

Rozhdestvenskaya, Dr Ira Vasilyevna. NPO Burevestnik, Malookhtensky Prospekt 78, Leningrad 195112, USSR. (1938) Cand, geology and mineralogy (LGU,1975). Sr. scient. *X-ray crystal structure analysis, crystal chemistry, silicates.*

Rubina, Dr Elena Borisovna. Inst. of Metallurgy, Acad. Sci. USSR, Leninskii Prospekt 49, Moscow 117334, USSR. Cand, physics and mathematics (Moscow Aviation Inst., 1984). Scient. *Structure, amorphous metal alloys, deformation mechanisms, hexagonal single- and polycrystal alloys.*

Rubinina, Dr Nataliya Mikhailovna. Physics Dept., Moscow State U., Leninskie Gory, Moscow 117234, USSR. (1934) Cand, chemistry (MGU, 1977). Sr. eng. *Crystal growth , physical properties as a function of composition and growth conditions.*

Rudenko, Mr Sergei Sergeevich. Leningrad Mining Inst., 21st Liniya 2, Leningrad 199026, USSR. (1950). Scient. *Crystallorgraphy, crystal chemistry, pegmatites.*

Rumanova, Dr Iskra Mikhailovna. Inst. of History of Science and Technics, Acad. Sci. USSR, Staropansky 1/5, Moscow 103012, USSR. Dr, physics and mathematics (INCRYS, 1971). Sr. scient. *Inorganic crystal structures, methodology, structure - properties relationships.*

Rusanovskii, Dr Mikhail Evstaf'evich. Polytechnical Inst., Lenina 168, Kishinev 277012, USSR. (1941) Cand, physics and mathematics (Inst. Appl.Phys., Kishinev, 1985).Sr. Lecturer. *Structure, coordination compounds, semiconductor materials.*

Russo, Dr Galina Vladimirovna. Geology Dept., Leningrad State U., Universitetskaya Nab. 7/9, Leningrad 119034, USSR. (1951) Cand, geology and mineralogy (LGU, 1986). Sr. scient. *Crystal growth, crystal defects, crystal morphology.*

Ruvimov, Dr Segei Sergeevich. Physical-Technical Inst., Acad. Sci. USSR, Zapovednaya 51, Leningrad 194035, USSR. (1953) Cand., physics and mathematics (Physical-Technical Inst., 1957). Scient. *Real structure.*

Ryabchenkov, Dr Vladimir Vasil'evich. Inst. of Crystallography, Acad. Sci. USSR, Leninskii Prospekt 59, Moscow 117333, USSR. (1955) Cand, physics and mathematics (INCRYS, 1988) Scient. *Solid-state lasers, laser spectroscopy.*

Ryabchikov, Dr Sergei Aleksandrovich. Physics Dept., Moscow State U., Leninskie Gory, Moscow 117234, USSR. (1947) Cand, physics and mathematics (MGU, 1985). *Theory, structural phase transitions, crystal lattice dynamics.*

Ryaboshapka, Dr Karl Petrovich. Inst. of Metallophysics, Acad. Sci. Ukrainian SSR, Vernadsky Prospekt 36, Kiev 252142, USSR. (1931) Cand, physics and mathematics (Joint Scient. Council Acad. Sci. Ukranian SSR, 1964). Sr. scient. *X-ray scattering, dislocation distorted crystals.*

Ryadnov, Dr Sergei Nikolaevich. Inst. of Crystallography, Acad. Sci. USSR, Leninskii Prospekt 59, Moscow 117333, USSR. (1957) Cand, physics and mathematics (INCRYS,1988). Scient. *Growth from melt, impurities.*

Rybakov, Dr Victor Borisovich. Chemistry Dept., Moscow State U., Leninskie Gory, Moscow 117234, USSR. (1950) Cand, chemistry (MGU, 1985). Scient. *Inorganic and organic crystal chemistry, instrumentation.*

Ryvkin, Dr Viktor Adol'phovich. Inst. of Crystallography, Acad. Sci. USSR, Leninskii Prospekt 59, Moscow 117333, USSR. (1950) Cand., physics and mathematics (INCRYS, 1987). Scient. *Phase transition physics.*

Ryzhenkov, Dr Alexander Pavlovich. Kolomna Pedagogical Inst., Zelenaya St. 30, Kolomna 140410, Moskovskaya Oblast', USSR. (1935) Cand, physics and mathematics (V.I. Lenin Moscow State Pedagogical Inst., 1969). Docent. *Organic crystals, structure, physical properties.*

Sabirov, Dr Vakhobzhon Khusanovich. Tashkent Pedagogical Inst., Dimitrov st., kvartal 48, Tashkentskaaya obl., Angren 702500, USSR. (1950) Cand, chemistry (Inst. of Chemistry, Tashkent, 1986). Sr.Lect. *Structure, coordination metal compounds, bioactive ligands.*

Sadikov, Dr Georgiy Georgievich. Inst. of General and Inorganic Chemistry, Acad. Sci. USSR, Leninsky pr. 31, Moscow 117071, USSR. (1932) Cand, chemistry (IONCH, 1969). Sr. scient. *X-ray crystal structure analysis, fluorides.*

Sadova, Dr Nina Ivanovna. Moscow State U., Dept. of Chemistry, Leninskiye Gory, Moscow 117234, USSR. (1937) Cand, chemistry (MGU, 1968). Sr. scient. *Elemento-organic compounds, structure, electron diffraction by gases.*

Sadybakasov, Dr Bolot Kemelovich. Inst. of Inorganic Chemistry, Kirghiz SSR Acad. Sci., Leninskii Prospekt 267, Frunze 720071, USSR. (1950) Cand., physics and mathematics (MGU, 1978). Lab Head. *Structure, reactivity, crystal chemistry, sugars.*

Safro, Dr Mark Grigor'evich. Inst. of Molecular Biology, Acad. Sci. USSR. Vavilov st.32, Moscow 117984, USSR. (1946) Cand, physics and mathematics (INCRYS, 1980). Sr. scient. *X-ray crystallography, biomolecules.*

Saf'yanov, Dr Yurii Nikolaevich. Gorky State U., Prospekt Gagarina 23, Gorky 603600, USSR. (1949) Cand, physics and mathematics (Gorky U., 1976) Scient. *X-ray analysis, metalloorganic compounds.*

Sakharov, Dr Boris Alexandrovich. Inst. of Geology, Acad. Sci. USSR, Pyzhevsky per. 7, Moscow 109017, USSR. (1944) Cand, geology and mineralogy (Inst. of Geology Acad. Sci. USSR, 1974). Scient. *Diffraction applications, defects, minerals.*

Sal'dau, Dr El'ga Petrovna. Leningrad Mining Inst., 21st Liniya 2, Leningrad 199026, USSR. (1930) Cand, geology and mineralogy (GORNY, 1958). Docent. *Crystal chemistry, X-ray structure analysis.*

Samarskaya, Dr Valentina Dmitriyevna. Kolomna Pedagogical Inst., Zelenaya 30, Kolomna 140410, Moskovskaya Oblast', USSR. (1936) Cand, physics and mathematics (Moscow Pedagogical Inst., 1971). Docent. *Crystallization processes, physical properties and phase state, organic molecular crystals.*

Samoilovich, Dr Lidiya Alexandrovna. Res. Inst. for Synthesis of Mineral Raw Materials, Institutskaya St. 1, Alexandrov 601600, Vladimirskaya Oblast', USSR. (1934) Cand, geology and mineralogy (MGU, 1968). Sr. scient. *Crystal growth, real structure, crystal chemistry.*

Samoilovich, Prof. Mikhail Isaakovich. Res. Inst. for Synthesis of Mineral Raw Materials, Institutskaya St. 1, Alexandrov 601600, Vladimirskaya Oblast', USSR. (1937) Dr, physics and mathematics (Kazan State U., 1973). Lab. head. *Crystal growth, real structure, crystal chemistry.*

Samotin, Dr Nikoloai Dmitrievich. Inst. of Geology of Ore Deposits- Petrography-Mineralogy and Geochemistry, Acad. Sci. USSR, Staromonetny 35, Moscow 109017, USSR. (1935) Cand, geology and mineralogy (IGEM, 1974). Sr. scient. *Growth, morphology, structure and properties.*

Samus', Dr Ivan Dmitriyevich. Kishinev Polytechnic Inst., Lenin Pr. 168, Kishinev 277004, USSR. (1926) Cand, physics and mathematics (INCRYS, 1967). Head of Chair. *X-ray structure analysis methods, crystal chemistry, coordination compounds.*

Samusina, Mrs. Svetlana Nikolaevna. Leningrad Mining Inst., 21st Liniya 2, Leningrad 199026, USSR. (1931) Sr. lecturer. *Crystal morphology, crystal growth.*

Sanadze, Prof. Vladimir Vladimirovich. Polytechnic Inst., Lenin St. 77, Tbilisi 380015, USSR. (1920) Dr, physics and mathematics (INCRYS, 1962). Head of Chair. *Phase transformations, structure analysis, metals and alloys.*

Sannikov, Dr Daniil Grigoryevich. Inst. of Crystallography, Acad. Sci. USSR, Leninsky pr. 59, Moscow 117333, USSR. (1931) Dr, physics and mathematics (INCRYS, 1986). Sr. scient. *Ferroelectrics (physics), phase transitions (ferroelectric).*

Sarkisov, Dr Stepan Ervandovich. Inst. of Crystallography, Acad. Sci. USSR, Leninsky pr. 59, Moscow 117333, USSR. (1948) Cand, physics and mathematics (INCRYS, 1979). Scient. *Crystal spectroscopy, crystal growth from melt.*

Sarin, Dr Victor Anatol'yevich. Branch Karpov Physical Chemistry Inst., Obninsk, Kaluzhskaya oblast', 249020, USSR. (1947) Cand, physics and mathematics (Inst. of Appl. Physics, Moldavian SSR, 1978). Lab. head *Neutron diffraction, structures.*

Sedmalis, Prof. Uldis Yanovich. Riga Polytechnic Inst., Kronvalda Boulevard 4, Riga 226828, USSR. (1933) Dr, technics (Byelorussian Polytechnic Inst., 1970). Prof. *Geometrical crystallography, crystallo-optical and X-ray phase analyses.*

Semenchev', Dr Aleksandr Fedorovich. Physics Dept., Rostov State U., Zorge 5, Rostov-on-Don 344104, USSR. (1947) Cand, physics and mathematics (Rostov U., 1982) Docent. *Structural phase transitions, mechanical twinning in crystals.*

Semenova, Dr Tat'yana Fedorovna, Leningrad State U., Dept. of Geology, University Emb. 7/9, Leningrad 199034, USSR. (1951) Cand, geology and mineralogy (LGU, 1978). Sr. scient. *Crystal chemistry, structure analysis, laminated silicates.*

Serdyuk, Dr Igor' Nikolayevich. Inst. of Protein, Acad. Sci. USSR, Pushchino, Serpukhovsky Rayon, Moskovskaya Oblast' 142292, USSR. (1939) Cand, physics and mathematics (Inst. of High Molecular Compounds, Acad. Sci. USSR, 1968). Sr. scient. *Diffuse X-ray scattering, biological macromolecules.*

Serebryanaya, Dr Nadezhda Ruvimovna. Inst. of High Pressure Physics, Acad. Sci. USSR, Akademgorodok, Podol'sky Rayon, Moskovskaya Oblast' 142092, USSR. Cand, chemistry (INCRYS, 1970). Scient. *High pressure effects, crystal structure.*

Sereda, Dr Sergei Vladimirovich. Inst. of Organic Chemistry, Ukrainian SSR Acad. Sci., Murmanskaya 5, Kiev 252660, USSR. (1959) Cand, chemistry (INEOS, 1987). Scient. *X-ray analysis and structural chemistry of organic compounds.*

Sergeev, Dr Yurii Vladimirovich. Inst. of Proteins, Acad. Sci. USSR, Moskovskaya obl., Pushchino 142292, USSR. (1949) Cand, physics and mathematics (Inst. of Proteins, Pushchino, 1982) Scient. *Structure, proteins.*

Sergienko, Dr Vladimir Semenovich. Inst. of General and Inorganic Chemistry, Acad. Sci. USSR, Leninsky pr. 31, Moscow 117071, USSR. (1941) Cand, chemistry (IONCH, 1973). Sr. sci. *Crystal chemistry, stereochemistry, coordination compounds.*

Sevast'yanov, Dr Boris Konstantinovich. Inst. of Crystallography, Acad. Sci. USSR, Leninsky pr. 59, Moscow 117333, USSR. (1930) Cand, physics and mathematics (Moscow Physical Techn. Inst., 1962). Assoc. dir. *Optical spectroscopy, magnetic properties, doped ionic crystals.*

Shafizade, Prof. Rafik Bekhbud ogly. Inst. of Physics, Acad. Sci. Azerbaidzhan SSR, Narimanov Prospekt 33, Baku 370143, USSR. (1934) Dr, physics and mathematics (Inst. of Physics, Acad. Sci. Azerbaidzhan SSR, 1985). Lab. head. *Semiconductor thin film structure, thin films (phase formation and transformation).*

Shafranovsky, Prof. Ilarion Ilarionovich. Leningrad Mining Inst., 21st Liniya 2, Leningrad 199026, USSR. (1910) Dr, geology and mineralogy (LGU, 1942). Prof. *Morphology, geometrical crystallography.*

Shaldin, Dr Yury Vitalyevich. Inst. of Crystallography, Acad. Sci. USSR, Leninsky pr. 59, Moscow 117333, USSR. (1935) Cand, physics and mathematics (Inst. of Steel and Alloys, 1967). Sr. scient. *Non-linear crystal properties.*

Shamburov, Dr Vladimir Alekseyevich. Inst. of Crystallography, Acad. Sci. USSR, Leninsky pr. 59, Moscow 117333, USSR. (1920) Cand, physics and mathematics (Machinery Inst., Acad. Sci. USSR, 1950). Sr. scient. *Ferroelectrics (physics), crystal optics.*

Shamray, Dr Vladimir Fedorovich. Inst. of Metallurgy, Acad. Sci. USSR, Leninsky pr. 49, Moscow 117334, USSR. (1937) Cand, technics (Inst. of Metallurgy, Acad. Sci. USSR, 1970). Lab. Head. *X-ray structure analysis, superconducting compounds.*

Sharipov, Dr Khasan Turabochiv. Inst. of Chemistry, Uzbek SSR Acad. Sci., Prospekt Gor'kogo 77, Tashkent 700170, USSR. (1947), Cand., chenmistry (Lomonosov Inst. of chemical Technology, 1974). Head lab. *Crystal chemistry, inorganic and coordination compounds.*

Shashkin, Dr Dmitry Petrovich. Inst. of Chemical Physics, Acad. Sci. USSR, Vorobyevskoye Chaussee 2-b, Moscow 117334, USSR. (1936) Cand, geology and mineralogy (MGU, 1970). Sr. scient. *Crystal structure analysis.*

Shchedrin, Dr Boris Mikhailovich. Moscow State U., Computing Center, Leninskiye Gory, Moscow 117234, USSR. (1934) Cand, physics and mathematics (INCRYS, 1966). Scient. *Computing methods, structure analysis.*

Shchepitil'nikov. Dr Boris Vladimirovich. Inst. of Crystallography, Acad. Sci. USSR, Leninskii Prospekt 59, Moscow 117333, USSR. (1945) Cand, physics and mathematics (INCRYS, 1986). Scient. *Crystal acoustics, phase transitions.*

Shcherbakova, Dr Mira Yakovlevna. Inst. of Geology and Geophysics, Acad. Sci. USSR, Siberian Dept., Novosibirsk 630090, USSR. (1926) Cand, physics and mathematics (Tomsk Polytechnic Inst., 1961). Lab. head. *Electron paramagnetic resonance, defects and impurities, minerals, synthetic materials.*

Shebanov, Dr Leonid Anatol'evich. Inst. of Solid State Physics, Latvian State U., Kengaraga st.8, Riga 226063, USSR. (1949) Cand, physics and mathematics (Inst. of Physics, Riga, 1980), Sr. scient. *Phase transitions, ferroelectrics and related materials.*

Shekhtman, Dr Veniamin Sholomovich. Inst. of Solid State Physics, Acad. Sci. USSR, Chernogolovka, Noginsky Rayon, Moskovskaya Oblast' 142432, USSR. (1929) Cand, technics (Moscow Inst. of Steel and Alloys, 1962). Lab. head. *X-ray structure analysis, phase and structural changes.*

Shepelev, Dr Yury Fedorovich. Inst. of Silicate Chemistry, Acad. Sci. USSR, Makarov Emb. 2, Leningrad 199164, USSR. (1939) Cand, physics and mathematics (LGU, 1971). Sr. scient. *Crystal chemistry, inorganic compounds.*

Shibaeva, Dr Rimma Pavlovna. Branch Inst. of Chemical Physics, Acad. Sci. USSR, Chernogolovka 142432, Noginsky Rayon, Moskovskaya Oblast', USSR. Dr, physics and mathematics (INCRYS, 1977). Sr. scient. *Organic and metallo-organic compounds, structure analysis, electrical and magnetic properties, direct methods.*

Shishova, Dr Tatyana Gennadiyevna. Gorky Agricultural Inst., Gagarin Prospekt 97, Gorky 603078, USSR. (1949) Cand, physics and mathematics (INCRYS, 1977). Asst. *X-ray structure analysis, organic compounds, crystal growth.*

Shivrin, Dr Oleg Mikolayevich. Petrozavodsk State U., Lenin Prospekt 33, Petrozavodsk 185018, USSR. (1923) Cand, physics and mathematics (MGU, 1961). Docent. *Order-disorder transformations, radiation defects, crystal lattice dynamics.*

Shklover, Dr Valery Efimovich. Inst. of Elemento-Organic Compounds, Acad. Sci. USSR, Vavilov St. 28, Moscow 117312, USSR. (1946) Dr, chemistry (INEOS, 1986). Sr. sci. *Structural chemistry, metallo-organic compounds, organic and bio-organic compounds, X-ray structure analysis.*

Shkol'nikova, Dr Larisa Mikhailovna. Inst. of Chemical Reagents and Pure Substances, Bogorodsky Val 3, Moscow 107258, USSR. (1930) Cand, chemistry (KARPOV, 1960). Sr. scient. *X-ray structure analysis, complex and organic compounds.*

Shlenskii, Dr Aleksei Leonidovich. Inst. of Crystallography, Acad. Sci. USSR, Leninskii Prospekt 59, Moscow 117333, USSR. (1955) Cand, physics and mathematics (INCRYS, 1986) Scient. *Optical and photoelectric properties, pyro- and piezoelectrics.*

Shnulin, Dr Anatoly Nikolayevich. Inst. of Inorganic and Physical Chemistry, Acad. Sci. Azerbaidzhan SSR, Narimanov Pr. 29, Baku 370143, USSR. (1936)

Cand, physics and mathematics (Azerbaidzhan State U., 1960). Sr. scient. *Structure analysis, complexes, biological compounds.*

Shmyt'ko, Dr Ivan Mikhailovich. Inst. of Solid State Physics, Acad. Sci. USSR, Chernogolovka 142432, Noginsky Rayon, Moskovskaya Oblast', USSR. (1946) Cand, physics and mathematics (Moscow Physico-Techn. Inst., 1976). Sr. scient. *Crystallography, phase transitions, X-ray diffraction optics, real crystals.*

Shternberg, Dr Aleksey Alexandrovich. Inst. of Crystallography, Acad. Sci. USSR, Leninsky pr. 59, Moscow 117333, USSR. (1911) Dr, physics and mathematics (INCRYS, 1969). Consultant. *Crystal growth.*

Shul'pina, Dr Iren Leonidovna. Physico-Techn. Inst.,Acad. Sci. USSR, Zapovednaya St. 51, Leningrad 194037, USSR. (1936) Dr, physics and mathematics (Inst. of Semiconductors, Acad. Sci. USSR, 1983). Sr. scient. *Crystal lattice defects, X-ray methods, dynamic scattering.*

Shulakov, Dr Evgeniy Vladimirovich. Inst. of Solid State Physics, Acad. Sci. USSR, Chernogolovka 142432, Noginsky Rayon, Moskovskaya Oblast', USSR. (1949) Cand, physics and mathematics (Inst. of Solid State Physics, Acad. Sci USSR, 1978). Sr. scient. *X-ray diffraction optics, perfect crystals, real crystals; diffraction pattern modelling.*

Shumyatskaya, Dr Ninel' Grigoryevna. Inst. of Mineralogy- Geochemistry and Crystal Chemistry of Rare Elements, Sadovnicheskaya Emb. 71, Moscow 113127, USSR. (1930) Cand, geology and mineralogy (MGU, 1974). Scient. *Crystal chemistry, minerals, synthetic compounds.*

Shustov, Dr Alexander Vsevolodovich. Leningrad Mining Inst., 21st Liniya 2, Leningrad 199026, USSR. (1937) Cand, geology and mineralogy (GORNY, 1967). Sr. scient. *Crystal morphology, optical crystallography.*

Shuvalov, Dr Aleksandr L'vovich. Inst. of Crystallography, Acad. Sci. USSR, Leninskii Prospekt 59, Moscow 117333, USSR. (1955) Cand, physics and mathematics (INCRYS, 1985). Scient. *Theoretical crystallography.*

Shuvalov, Prof. Lev Alexandrovich. Inst. of Crystallography, Acad. Sci. USSR, Leninsky pr. 59, Moscow 117333, USSR. (1923) Dr, physics and mathematics (INCRYS, 1972). Lab. head. *Physical properties, structural phase transitions, ferroelectricity and ferroelasticity.*

Shvelashvili, Prof. Arsen Eristovich. Inst. of Physical and Inorganic Chemistry, Acad. Sci. Georgian SSR, Dzhikiya St. 5, Tbilisi 380086, USSR. (1935) Dr, chemistry (Tbilisi State U., 1974). Lab. head. *Stereochemistry, coordination compounds.*

Sichevich, Dr Olga Mikhailovna. Chemistry Dept, Lvov State U., Universitetskaya 1, Lvov 290602, USSR. (1956) Cand, chemistry (Lvov U., 1986). Sr. scient. *Crystal chemistry, intermetallic compounds.*

Sidorenko, Dr Galina Alexandrovna. Inst. of Mineral Raw Materials (VIMS), Staromonetny 29, Moscow 109017, USSR. (1926) Dr, geology and mineralogy (VIMS, 1976). Lab. head. *Crystal chemistry, minerals, X-ray crystallography, polycrystals.*

Sigayev, Dr Vladimir Nikolayevich. Inst. of Crystallography, Acad. Sci. USSR, Leninsky pr. 59, Moscow 117333, USSR. (1945) Cand, physics and mathematics (INCRYS, 1975). Scient. *Neutron diffraction, liquids and glasses.*

Silin', Dr Elga Yanovna. Inst. of Inorganic Chemistry, Acad. Sci. Latvian SSR, Miyera St. 34, Salaspils 229021, Riga District, USSR. (1941) Cand, chemistry (Inst. of Inorganic Chemistry, Acad. Sci. Latvian SSR, 1978). Sr. scient. *Crystal chemistry.*

Sil'vestrova, Dr Iraida Mikhailovna. Inst. of Crystallography, Acad. Sci. USSR, Leninsky pr. 59, Moscow 117333, USSR. (1924) Cand, physics and mathematics (INCRYS, 1963). Sr. scient. *Physical properties, piezoelectricity, elasticity, acoustic properties.*

Simonov, Dr Mikhail Alexandrovich. Moscow State U., Dept. of Geology, Leninskiye Gory, Moscow 117234, USSR. (1940) Cand, physics and mathematics (INCRYS, 1969). Docent. *Crystal chemistry, inorganic compounds, X-ray structure analysis.*

Simonov, Dr Valentin Ivanovich. Inst. of Crystallography, Acad. Sci. USSR, Leninsky pr. 59, Moscow 117333, USSR. (1930) Dr, physics and mathematics (INCRYS, 1972). Assoc. dir. *Structure analysis methods, computing.*

Simonov, Dr Yury Alexandrovich. Inst. of Applied Physics, Acad. Sci. Moldavian SSR, Akademicheskaya 5, Kishinev 277028, USSR. (1937) Cand, physics and mathematics (Gorky U., 1967). Sr. scient. *X-ray structure analysis, methods, inorganic compounds.*

Sirota, Dr Mikhail Isaakovich. Inst. of Crystallography, Acad. Sci. USSR, Leninsky pr. 59, Moscow 117333, USSR. (1945) Cand, physics and mathematics (INCRYS, 1975). Sr. scient. *Computer programming, structure analysis.*

Sizova, Dr Nataliya Leonidovna. Inst. of Crystallography, Acad. Sci. USSR, Leninsky pr. 59, Moscow 117333, USSR. (1937) Cand, physics and mathematics (INCRYS, 1974). Sr. scient. *Crystal lattice defects, mechanical properties of crystals.*

Skakov, Prof. Yurii Aleksandrovich. Moscow Inst. of Steel and Alloys, Leninskii Prospekt 4, Moscow 117049, USSR. (1925) Doctor, physics and mathematics (Moscow Inst, of Steel and Alloys, 1967). Head of chair. *Crystal chemistry, structure, metal alloys.*

Skolozdra, Dr Roman Vladimirovich. Lvov State U., Dept. of Chemistry, University St. 1, Lvov 290602, USSR. (1941) Cand, chemistry (Lvov U., 1967). Docent. *X-ray structure analysis, crystal chemistry, intermetallic compounds.*

Smetannikova, Dr Olga Gennadiyevna. Leningrad State U., Dept. of Geology, University Emb. 7/9, Leningrad 199034, USSR. (1947) Cand, geology and mineralogy (LGU, 1974). Scient. *Crystal chemistry, X-ray structure analysis.*

Slovokhotov, Dr Yurii Leonidovich. Inst. of Elementoorganic Compounds, Acad. Sci. USSR. Vavilov st. 28, Moscow 117334, USSR. (1955) Cand, chemistry (INEOS, 1986). Scient. *Theoretical chemistry, structure, metalloorganic compounds, cluster chemistry.*

Smirnov, Dr Aleksei Evgenievich. Inst. of Crystallography, Acad. Sci. USSR, Leninskii Prospekt 59, Moscow 117333, USSR. (1946) Cand, chemistry (INCRYS, 1982) Scient. *Electromagnetic effects, real structure, high temperature superconductivity.*

Smirnov, Prof. Yury Mstislavovich. Kalinin State U., Zhelyabova St. 33, Kalinin 170013, USSR. (1932) Dr, technics (GORNY, 1986). Head. chair. *Crystal growth, morphology.*

Smirnova, Dr Nina Lvovna. Moscow State U., Dept. of Geology, Leninskiye Gory, Moscow 117234, USSR. (1926) Cand, chemistry (INCRYS, 1961). Sr. scient. *Crystal chemistry.*

Smolin, Dr Yury Ivanovich. Inst. of Silicate Chemistry, Acad. Sci. USSR, Makarov Emb. 2, Leningrad 199164, USSR. (1930) Dr, physics and mathematics (INCRYS, 1974). Sr. scient. *Crystal chemistry, inorganic compounds.*

Smotrakov, Dr Valery Georgiyevich. Rostov State U., Prospekt Stachki 192, Rostov-on-Don 344090, USSR. (1944) Cand, chemistry (Rostov U., 1971). Sr. scient. *Crystal growth.*

Sobolev, Dr Boris Pavlovich. Inst. of Crystallography, Acad. Sci. USSR, Leninsky pr. 59, Moscow 117333, USSR. (1936) Dr, chemistry (INCRYS, 1978). Lab. Head. *Crystal growth, crystal chemistry, inorganic compounds.*

Sobolev, Dr Chingis Sergeyevich. Leningrad Mining Inst., 21st Liniya 2, Leningrad 199026, USSR. (1931) Cand, geology and mineralogy (GORNY, 1965). Scient. *Crystallography, crystal morphology.*

Soboleva, Dr Lidiya Victorovna. Inst. of Crystallography, Acad. Sci. USSR, Leninsky pr. 59, Moscow 117333, USSR. (1927) Cand, chemistry (IONCH, 1954). Scient. *Inorganic crystal growth.*

Soboleva, Dr Svetlana Vsevolodovna. Inst. of Geology- Mineralogy and Petrography, Acad. Sci. USSR, Staromonetny 35, Moscow 109017, USSR. (1937) Dr, geology and mineralogy (IGEM, 1988). Sr. scient. *Structural mineralogy, polytypism, electron diffraction.*

Sokol, Dr Anatoly Afanasyevich. Kharkov Polytechnic Inst., Frunze St. 21, Kharkov 310002, USSR. (1937) Cand, technics (Kharkov Polytechnic Inst., 1970). Docent. *Structure, defects, thin film growth, electron microscopy.*

Sokol, Dr Valentina Ivanovna. Inst. of General and Inorganic Chemistry, Acad. Sci. USSR, Leninsky pr. 31, Moscow 117071, USSR. (1927) Cand, chemistry (IONCH, 1965). Scient. *Crystal chemistry, stereochemistry, coordination compounds.*

Sokolov, Dr Yury Alexandrovich. Inst. of Crystallography, Acad. Sci. USSR, Leninsky pr. 59, Moscow 117333, USSR. (1940) Cand, physics and mathematics (INCRYS, 1979). Scient. *Crystal spectroscopy, nuclear magnetic resonance, vibrational spectroscopy.*

Sokolova, Dr Elena Vadimovna. Geology Dept., Moscow State U., Leninskie Gory, Moscow 117234, USSR. (1953) Cand, geology and mineralogy (MGU, 1980). Scient. *Crystal chemistry, X-ray analysis, inorganic compounds, structural mineralogy.*

Sokolova, Dr Nataliya Gavrilovna. Leningrad Mining Inst., 21st Liniya 2, Leningrad 199026, USSR. (1939) Cand, geology and mineralogy (GORNY, 1969). Sr. scient. *X-ray crystallography, morphology.*

Soldatov, Dr Evgeniy Alexandrovich. Gorky State U., Sverdlova 37, Gorky 603000, USSR. (1949) Cand, physics and mathematics (INCRYS, 1979). Docent. *Mathematical methods, structure analysis, symmetry theory.*

Solotchina, Dr Emiliya Pavlovna. Inst. of Geology and Geophysics, Siberian Dept. Acad. Sci. USSR, Universitetskii Prospekt 3, Novosibirsk 630090, USSR. (1946) Cand, geology and mineralogy (Inst. of Geology and Geophysics, Siberian Dept. Acad. Sci. USSR, 1982). Sr. scient. *Real structure, layer minerals,clay, by diffraction and nondiffraction methods.*

Solo'vyev, Dr Sergey Petrovich. Branch Moscow Phys.-Eng. Inst., Lenin Pr. 71, Obninsk Kaluzhskaya oblast',249020, USSR. (1932) Dr, physics and mathematics (Physico-Energetic Inst., 1976). Lab. head. *X-ray and neutron structure analysis, crystal lattice dynamics.*

Solovyeva, Dr Lidiya Pavlovna. Inst. of Catalysis, Siberian Dept.Acad. Sci. USSR, Lavrent'yeva pr. 5, Novosibirsk 630090, USSR. (1935) Cand, geology and mineralogy (MGU, 1965). Sr. scient. *X-ray diffraction, methods, programming, structure, zeolites, layered compounds.*

Sonin, Prof. Anatoliy Stepanovich. Res. Inst. of Semi-Products and Dye-Stuffs, Bol'shaya Sadovaya 1, Moscow 103787, USSR. (1931) Dr, physics and mathematics (Dnepropetrovsk State U., 1972). Sr. scient. *Symmetry, crystallo-physics, liquid crystals.*

Sorokin, Dr Aleksandr Alekseevich. Inst. of Crystallography, Acad. Sci. USSR, Leninskii Prospekt 59, Moscow 117333, USSR. (1954) Cand, physics and mathematics (INCRYS, 1988) Scient. *Structure, partially ordered systems, superconductivity.*

Sorokin, Dr Lev Mikhailovich. Physico-Techn. Inst.,Acad. Sci. USSR, Zapovednaya St. 51, Leningrad 194037, USSR. (1937) Cand, physics and mathematics (Inst. of Semiconductors, Acad. Sci. USSR, 1968). Sr. scient. *Defect structure, defect dynamics, methods for defect study, dynamic electron and X-ray scattering.*

Sorokina, Dr Natalya Ivanovna. Inst. of Crystallography, Acad. Sci. USSR, Leninskii Prospekt 59, Moscow 117333, USSR. (1952) Cand, chemitry (INCRYS, 1986) Scient. *Electron density by X-ray and neutron diffraction.*

Sosfenov, Dr Nikita Ilyich. Inst. of Crystallography, Acad. Sci. USSR, Leninsky pr. 59, Moscow 117333, USSR. (1932) Cand, physics and mathematics (INCRYS, 1972). Sr. scient. *Protein crystallography, instrumentation for X-ray structure analysis.*

Spiridonov, Prof. Victor Pavlovich. Moscow State U., Dept. of Chemistry, Leninskiye Gory, Moscow 117234, USSR. (1931) Dr, chemistry (MGU, 1969). Lab. head. *Inorganic compounds, structure, electron diffraction by gases.*

Spitsyna, Dr Valentina Danilovna. Inst. of Crystallography, Acad. Sci. USSR, Leninsky pr. 59, Moscow 117333, USSR. (1941) Cand, chemistry (INCRYS, 1975). Scient. *Crystallization processes in multicomponent systems.*

Starikova, Dr Zoya Alexandrovna. Inst. of Chemical Reagents and Pure Substances, Bogorodsky Val 3, Moscow 107258, USSR. (1934) Cand, chemistry (IONCH, 1968). Sr. scient. *Crystal chemistry, complexes, organic compounds.*

Starostina, Dr Lyudmila Sergeyevna. Inst. of Crystallography, Acad. Sci. USSR, Leninsky pr. 59, Moscow 117333, USSR. (1933) Cand, physics and mathematics (INCRYS, 1964). Sr. scient. *Spectroscopy, crystal growth.*

Stepanova, Dr Alla Nikolayevna. Inst. of Crystallography, Acad. Sci. USSR, Leninsky pr. 59, Moscow 117333, USSR. (1934) Cand, physics and mathematics (INCRYS, 1974). Scient. *Crystal growth, thin film growth, whisker growth.*

Stepanova, Dr Nataliya Stepanovna. Gorky State U., Prospekt Gagarina 23, Gorky 603600, USSR. (1934) Cand, physics and mathematics (Gorky U., 1970). Sr. scient. *Crystal growth from aqueous solutions.*

Stepantsov, Dr Evgenii Arkad'evich. Inst. of Crystallography, Acad. Sci. USSR, Leninskii Prospekt 59, Moscow 117333, USSR. (1951) Cand., physics and mathematics (INCRYS, 1960). Sr. scient. *Growth, properties, crystalline composites.*

Stishov, Dr Sergey Mikhailovich. Inst. of Crystallography, Acad. Sci. USSR, Leninsky pr. 59, Moscow 117333, USSR. Dr, physics and mathematics (INCRYS, 1974). Lab. head. *High pressure crystallography.*

Struchkov, Prof. Yury Timofeyevich. Inst. of Elemento-Organic Compounds, Acad. Sci. USSR, Vavilov St. 28, Moscow 117312, USSR. (1926) Dr, chemistry (INEOS, 1978). Lab. head. *Structural chemistry, organometallic compounds, organic and bio-organic compounds, X-ray analysis.*

Strukov, Prof. Boris Anatolyevich. Moscow State U., Dept. of Physics, Leninskiye Gory, Moscow 117234, USSR. (1935) Dr, physics and mathematics (MGU, 1975). Prof. *Ferroelectricity, structural phase transitions, Raman scattering.*

Sultanov, Dr Rafik Mukhadastovich. Inst. of Physics, Acad. Sci. Azerbaidzan SSR, Prospekt Narimanova 33, Baku 370143, USSR. (1939) Cand, physics and mathematics (Inst. of Physics, Acad. Sci. AzSSR, 1987) Scient. *Thin films, ageing, phase transformations.*

Sumin, Dr Vyacheslav Vasil'evich. Obninsk Branch, Karpov Physical-Chemical Inst., Kaluzhskaya obl., Obninsk 249020, USSR. (1946) Cand, eng. (Inst. of Metallurgy, Acad. Sci. USSR, 1974) Scient. *Structure and dynamics of defect crystals.*

Suvorov, Dr Ernest Vitalyevich. Inst. of Solid State Physics, Acad. Sci. USSR, Chernogolovka, Noginsky Rayon, Moskovskaya oblast 142432, USSR. (1937) Dr, physics and mathematics. Assoc. dir. *Dynamic X-ray scattering, real crystals.*

Svergun, Dr Dmitry Ivanovich. Inst. of Crystallography, Acad. Sci. USSR, Leninsky pr. 59, Moscow 117333, USSR. (1954) Cand, physics and mathematics (INCRYS, 1982). Scient. *X-ray and neutron small angle scattering, diffraction theory, methods.*

Sviridov, Prof. Dmitry Timofeyevich. Inst. of Crystallography, Acad. Sci. USSR, Leninsky pr. 59, Moscow 117333, USSR. (1931) Dr, physics and mathematics (INCRYS, 1973). Sr. scient. *Crystal optics, spectroscopy crystals, crystal structures.*

Tafeenko, Dr Viktor Aleksandrovich. NPO of Organic Products and Dies, Bolshaya Sadovaya 1, korp.3, Moscow 103787, USSR. (1953) Cand, chemistry (MGU, 1981) Scient. *Organic crystal chemistry.*

Takhodzhaev, Dr Bakhodirhodzha. Inst. of Chemistry, Uzbek SSR Acad. Sci., Prospekt Gor'kogo 77, Tashkent 700170, USSR. (1948) Cand, chemistry (MGU, 1977). Sr. scient. *Organic crystal chemistry, computation, structural chemistry.*

Tamasyan, Dr Rafael Arshamovich. Inst. of Crystallography, Acad. Sci. USSR, Leninskii Prospekt 59, Moscow 117333, USSR. (1958) Cand, physics and mathematics (INCRYS,1987). Scient. *Twinning crystals and incommensurate phases by X-ray analysis.*

Tarashchan, Prof Arkadii Nikolaevich. Inst. of Geochemistry and Physics of Minerals, Ukrainian SSR Acad. Sci. Pr. Palladina 34, Kiev 252680, USSR. (IGEM, 1974). Head lab. *Real structure, crystal properties.*

Tarasov, Dr Yurii Igorevich. Chemistry Dept., Moscow State U., Leninskie Gory, Moscow 117234, USSR. (1959) Cand, chemistry (MGU, 1986). Scient. *Gas electron diffraction, high-temperature superconductivity.*

Tarkhova, Dr Tatyana Nikolayevna. Gorky State U., Gagarin Prospekt 23, Gorky 603022. USSR. Cand, physics and mathematics (INCRYS, 1949). Docent. *X-ray structure analysis, crystal chemistry.*

Tarnopol'sky, Dr Boris Lvovich. Branch Inst. of Chemical Physics, Acad. Sci. USSR, Chernogolovka 142432, Noginsky Rayon, Moskovskaya Oblast', USSR. (1924) Cand, physics and mathematics (INCRYS, 1965). Sr. scient. *Computing methods, structure analysis.*

Tatarchenko, Dr Vitaly Antonovich. Inst. of Solid State Physics, Acad. Sci. USSR, Chernogolovka, Noginsky Rayon, Moskovskaya Oblast' 142432, USSR. (1938) Cand, physics and mathematics. Dept. head. *Crystallization from melt.*

Tatarinova, Dr Lyudmila Ivanovna. Inst. of Crystallography, Acad. Sci. USSR, Leninsky pr. 59, Moscow 117333, USSR. (1903) Cand, physics and mathematics (INCRYS, 1953). *Synthetic polypeptide structures, amorphous substances, electron diffraction, X-ray diffraction.*

Tatarsky, Prof. Vitaly Borisovich. Leningrad State U., Dept. of Geology, University Emb. 7/9, Leningrad 199164, USSR. (1907) Dr, geology and mineralogy (LGU, 1953). Emeritus Prof. *Crystal optics, microscopic phase-analysis, goniometry.*

Telegina, Dr Inna Vasilyevna. Moscow State U., Dept. of Physics, Leninskiye Gory, Moscow 117234, USSR. Cand, physics and mathematics (MGU, 1968). Scient. *Diffuse X-ray scattering, small angle X-ray scattering, radiation effects in crystals.*

Terent'ev, Dr Evgenii Mikhailovich. Inst. of Crystallography, Acad. Sci. USSR, Leninskii Prospekt 59, Moscow 117333, USSR. (1959) Cand, physics and mathematics (INCRYS, 1985). Scient. *Liquid crystals, static theory, crystal strength and plasticity.*

Teslenko, Dr Valery Fedorovich. Kolomna Pedagogical Inst., Zelenaya 30, Kolomna 140410, Moskovskaya Oblast', USSR. (1937) Cand, physics and mathematics (Moscow Pedagogical Inst., 1967). Docent. *Crystallization processes, physical properties and phase state, organic molecular crystals.*

Tikhomirova, Dr Nataliya Alexandrovna. Inst. of Crystallography, Acad. Sci. USSR, Leninsky pr. 59, Moscow 117333, USSR. (1932) Cand, physics and mathematics (INCRYS, 1966). Sr. scient. *Phase transitions in solids, liquid crystals, high pressure properties, ferroelectrics.*

Tikhonova, Dr Anna Andreyevna. Inst. of Crystallography, Acad. Sci. USSR, Leninsky pr. 59, Moscow 117333, USSR. (1934) Cand, physics and mathematics (INCRYS, 1973). Scient. *Thin film growth and structure.*

Timchenko, Dr Tamara Iosifovna. Moscow State U., Dept. of Geology, Leninskiye Gory, Moscow 117234, USSR. (1930) Cand, physics and mathematics (IGEM, 1962). Sr. scient. *Crystal growth, synthesis.*

Timofeeva, Dr Valentina Alexandrovna. Inst. of Crystallography, Acad. Sci. USSR, Leninsky pr. 59, Moscow 117333, USSR. (1923) Cand, chemistry (Inst. of Chemistry, Acad. Sci. Kazakh SSR, 1949). Sr. scient. *Crystal growth from fluxed melts.*

Timofeeva, Dr Tat'yana Vladimirovna. Inst. of Elemento-Organic Compounds, Acad. Sci. USSR, Vavilov St. 28, Moscow 117813, USSR. (1947) Cand, chemistry (INEOS, 1982). Scient. *Molecular packing energy calculations, conformational analysis.*

Tishchenko, Dr Galina Nikolayevna. Inst. of Crystallography, Acad. Sci. USSR, Leninsky pr. 59, Moscow 117333, USSR. Dr, chemistry (INCRYS, 1984). Sr. scient. *Structure analysis, proteins and cyclic peptides.*

Tkachev, Dr Valery Vladimirovich. Branch Inst. of Chemical Physics, Acad. Sci. USSR, Chernogolovka 142432, Moskovskaya Oblast', USSR. (1943) Cand, physics and mathematics (Inst. of Chemical Physics, Acad. Sci. USSR). Scient. *Crystal chemistry, ion-conducting compounds, coordination compounds, crystal structure, phospho-organic compounds.*

Tobelko, Mr Konstantin Ivanovich. Inst. of Geochemistry and Analytical Chemistry, Acad. Sci. USSR, Kosygin pr. 19, Moscow 117334, USSR. (1923). Scient. *Isomorphism, minerals, inorganic compounds, X-ray analysis.*

Tolstikhina, Dr Alla Leonidovna. Inst. of Crystallography, Acad. Sci. USSR, Leninskii Prospekt 59, Moscow 117333, USSR. Cand, physics and mathematics (INCRYS, 1988). Scient. *Electron diffraction, electron microscopy, thin films.*

Tomashpol'sky, Dr Yury Yakovlevich. Karpov Physical Chemistry Inst., Obukha St. 10, Moscow 103064, USSR. (1937) Cand, chemistry (KARPOV, 1965). Lab. head. *Complex oxides, structure, surfaces, thin layers.*

Topor, Dr Nikolay Dmitriyevich. Moscow State U., Dept. of Geology, Leninskiye Gory, Moscow 117234, USSR. (1915) Cand, geology and mineralogy (MGU, 1946). Sr. scient. *Geocrystal chemistry, thermal properties, minerals.*

Tovbis, Dr Alexander Borisovich. Inst. of Crystallography, Acad. Sci. USSR, Leninsky pr. 59, Moscow 117333, USSR. (1940) Cand, physics and mathematics (INCRYS, 1971). Sr. scient. *Computing methods, structure analysis.*

Treivus, Dr Evgeny Borisovich. Leningrad State U., Dept. of Geology, University Emb. 7/9, Leningrad 199164, USSR. (1934) Cand, geology and mineralogy (LGU, 1965). Sr. scient. *Crystal growth, crystal morphology.*

Tret'yakov, Dr Vyacheslav Nukolaevich. Leningrad Mining Inst., 21st Liniya 2, Leningrad 199026, USSR. (1933) Cand, geology and mineralogy (GORNY, 1978). Lab. Head. *Morphology and crystal growth.*

Treushnikov, Dr Evgeniy Nikolayevich. Dept. of Chemistry, Moscow State U., Leninskiye Gory, Moscow 117234, USSR. Cand, physics and mathematics (INCRYS, 1970). Sr. scient. *Electron density distribution (diffraction method).*

Triodina, Dr Nina Sergeyevna. Inst. of Crystallography, Acad. Sci. USSR, Leninsky pr. 59, Moscow 117333, USSR. (1941) Cand, chemistry (INCRYS, 1979). Scient. *Crystal growth.*

Troyanov, Dr Sergei Igorevich. Chemistry Dept., Moscow State U., Leninskie Gory, Moscow 117234, USSR. (1936) Cand, physics and mathematics (MGU, 1969) Sr. Scient. *Ferroelectricity, high-temperature superconductivity, crystal growth.*

Trubkin, Dr Nikolai Viktorovich. Inst. of Geology of Ore Deposits- Petrography-Mineralogy and Geochemistry, Acad. Sci. USSR, Staromonetnii 35, Moscow 109017, USSR. (1949) Cand, geology and geochemistry (IGEM, 1985). Scient. *Structure, minerals.*

Trunov, Dr Vadim Konstantinovich. Inst. of Chemical Reagents and Pure Substances, Bogorodsky Val 3, Moscow 107258, USSR. (1936) Dr, chemistry (MGU, 1972). Dept. head. *Crystallography, oxide compounds, double salts.*

Tseitlin, Dr Mikhail Nevakhovich. Physico-Technical Inst., Acad. Sci. Tadzhik SSR, Akademgorodok, Dushanbe 773630, USSR. (1945) Cand, chemistry (INCRYS, 1974). Sr. scient. *Crystal growth, crystal chemistry.*

Tsikhotsky, Dr Evgeny Stanislavovich. Rostov State U., Dept. of Physics, Prospekt Stachki 192, Rostov-on-Don 344061, USSR. (1946) Cand, physics and mathematics (Rostov U., 1975). Sr. scient. *Perfection studies in crystals.*

Tsinober, Dr Leonid Iosifovich. Res. Inst. for Synthesis of Mineral Raw Materials, Institutskaya St. 1, Alexandrov 601600, Vladimirskaya Oblast', USSR. (1924) Cand, geology and mineralogy (INCRYS, 1962). Lab. head. *Real structure, structural mineralogy, X-ray crystallography, electron microscopy.*

Tsintsadze, Prof. Givi Vasilyevich. Polytechnic Inst., Lenin St. 77, Tbilisi 360015, USSR. (1933) Dr, chemistry (Tbilisi U., 1971). Head of Chair. *Crystal chemistry, coordination compounds.*

Tsirel'son, Dr Valadimir Grigoryevich. Moscow Chemical Techn. Inst., Miusskaya Square 9, Moscow 125820, USSR. (1948) Cand, physics and mathematics (Inst. of Applied Physics, Acad. Sci. Moldavian SSR). Scient. *Precision diffraction measurement, electron density distribution.*

Tsuprun, Dr Vladimir Lvovich. Inst. of Crystallography, Acad. Sci. USSR, Leninsky pr. 59, Moscow 117333, USSR. (1948) Cand, physics and mathematics (Inst. of Proteins, Acad. Sci. USSR, 1979). Scient. *Protein crystallography.*

Tsvankin, Dr Daniel' Yakovlevich. Inst. of Elemento-Organic Compounds, Acad. Sci. USSR, Vavilov St. 28, Moscow 117312, USSR. (1929) Dr, physics and mathematics (Inst. of Macromolecular Compounds, Acad. Sci. USSR, 1971). Sr. scient. *Polymer physics, small angle X-ray scattering.*

Tsyganov, Dr Evgeny Matveyevich. Res. Inst. for Synthesis of Mineral Raw Materials, Institutskaya St. 1, Alexandrov 601600, Vladimirskaya Oblast', USSR. (1919) Cand, geology and mineralogy (Lvov State U., 1951). Lab. head. *Crystal synthesis.*

Tumanyan, Dr Vladimir Gayevich. Inst. of Molecular Biology, Acad. Sci. USSR, Vavilov St. 32, Moscow 117312, USSR. (1938) Dr, physics and mathematics (MGU, 1986). Sr. scient. *Fibrillous structures, conformational calculations.*

Tyapunina, Dr Nataliya Alexandrovna. Dept. of Physics, Moscow State U., Leninskiye Gory, Moscow 117234, USSR. (1922) Dr, physics and mathematics (MGU, 1972). Sr. scient. *Defects, physical properties.*

Tyvanchuk, Dr Anna Teodorovna. Lvov State U., Dept. of Chemistry, University St. 1, Lvov 290602, USSR. (1947) Cand, chemistry (Lvov State U., 1980). Scient. *Crystal chemistry, intermetallic compounds.*

Udalova, Dr Valentina Vasilyevna. Inst. of Crystallography, Acad. Sci. USSR, Leninsky pr. 59, Moscow 117333, USSR. (1932) Cand, physics and mathematics (INCRYS, 1974). Sr. scient. *Electron diffraction, structure analysis.*

Ukraintsev, Dr Vladimir Alekseevich. Inst. of Crystallography, Acad. Sci. USSR, Leninskii Prospekt 59, Moscow 117333, USSR. (1955) Cand, physics and mathemtics (INCRYS, 1985). Scient. *Surface physics, hetorophase processes.*

Umansky, Prof. Mark Moiseyevich. Moscow State U., Dept. of Physics, Leninskiye Gory, Moscow 117234, USSR. (1906) Dr, physics and mathematics (INCRYS, 1957). *X-ray structure analysis, instrumentation.*

Urusov, Dr Vadim Sergeyevich. Moscow State U., Dept. of Geology, Leninskiye Gory, Moscow 119899, USSR. (1936) Dr, chemistry (Inst. of Geochemistry and Analytical Chemistry, Acad. Sci. USSR, 1975). Head of chair. *Crystal chemistry theory, energetic crystal chemistry, isomorphism, polymorphism.*

Urusovskaya, Dr Aida Alexandrovna. Inst. of Crystallography, Acad. Sci. USSR, Leninsky pr. 59, Moscow 117333, USSR. (1929) Dr, physics and mathematics (INCRYS, 1981). Sr. scient. *Defects, mechanical properties.*

Urzhumtsev, Dr Aleksandr Georgievich. Research Computational Center, Acad. Sci. USSR, Moskovskaya obl., Pushchino 142292, USSR. (1956) Cand, physics and mathematics (INCRYS, 1985). Scient. *Computation, crystallography.*

Usov, Dr Oleg Alekseyevich. Physical Techn. Inst.,Acad. Sci. USSR, Zapovednaya St. 51, Leningrad 194037, USSR. (1936) Cand, physics and mathematics (Inst. of Semiconductors, Acad. Sci. USSR, 1967). Sr. scient. *Structural and dynamical properties, crystallographic computing methods.*

Uyukin, Dr Evgeny Mikhailovich. Inst. of Crystallography, Acad. Sci. USSR, Leninsky pr. 59, Moscow 117333, USSR. (1946) Cand, physics and mathematics (INCRYS, 1980). Sr. scient. *Optical properties, electrical properties.*

Vainshtein, Prof. Boris Konstantinovich. Inst. of Crystallography, Acad. Sci. USSR, Leninsky pr. 59, Moscow 117333, USSR. (1921) Full member, Acad. Sci. USSR, Dr, physics and mathematics (INCRYS, 1955). Dir. *X-ray 'crystallography, electron microscopy, biological macromolecules, diffraction theory, crystal structure analysis theory.*

Val'kovskaya, Dr Margarita Ivanovna. Inst. of Applied Physics, Acad. Sci. SSR, Akademicheskaya 5, Kishinev 277028, USSR. (1938) Cand, physics and mathematics (Kishinev U., 1966). Sr. scient. *Physical crystallography, mechanical properties.*

Val'ter, Prof Anton Antonovich. Inst. of Geochemistry and Physics of Minerals, Ukrainian SSR Acad. Sci., Pr. Palladina 34, Kiev 252680, USSR. (1933) Doctor, geology and mineralogy (Inst. of Mineral Raw Materials, Moscow, 1980) Lab. Head. *Crystallography, minerals.*

Varnek, Dr Aleksandr Aleksandrovich. Mendeleev Inst. of Chemical Technology, Miusskaya pl.9, Moscow 125820, USSR. (1955) Cand, chemistry (IONCH, 1985) Docent. *Electron density distribution in molecules and crystals.*

Vasil'ev, Dr Aleksandr Dmitrievich. Inst. of Physics, Siberian Dept. Acad. Sci. USSR, Akademgorodok, Krasnoyarsk 660036, USSR. (1947) Cand, physics and mathematics (INCRYS, 1986). Scient. *Phase transformations, structure, inorganic and low-molecular organic compounds.*

Vasil'ev, Dr Yan Vladimirovich. Inst. of Inorganic Chemistry, Siberian Dept., Acad. Sci. USSR, Prospekt Lavrent'eva 3, Novosibirsk 630090, USSR. (1936) Cand, chemistry (LGU, 1965). Sr. scient. *Crystal growth, morphology, growing surfaces, morphology and physical properties as a function of growth conditions.*

Vasil'yev, Dr Alexander Borisovich. Inst. of Crystallography, Acad. Sci. USSR, Leninsky pr. 59, Moscow 117333, USSR. (1951) Cand, physics and mathematics (INCRYS, 1979). Sr. scient. *Optics of solids.*

Vasilyev, Dr Evgeny Konstantinovich. Inst. of Earth Crust, Acad. Sci. USSR, Siberian Dept., Lermontov St. 128, Irkutsk 664033, USSR. (1922) Cand, physics and mathematics (Irkutsk State U., 1966). Sr. scient. *Crystal chemistry, X-ray structure analysis, inorganic compounds, isomorphism, instrumentation.*

Veispals, Dr Aris Arvidovich. Inst. of Solid State Physics, Latvian State U., Kengaraga st.8, Riga 226063, USSR. (1943)Cand, physics and mathematics (Inst. of Physics, Riga, 1980). Sr. scient. *Growth, structure, oxide crystals.*

Velikodnyi, Dr Yury Andreyevich. Inst. of Chemical Reagents and Pure Substances, Bogorodsky Val 3, Moscow 107258, USSR. (1941) Cand, chemistry (MGU, 1975). Sr. scient. *Crystal chemistry, transition metals, double salts.*

Venevtsev, Prof. Yury Nikolayevich. Karpov Physical Chemistry Inst., Obukha St. 10, Moscow 103064, USSR. (1926) Dr, physics and mathematics (Inst. of Physics, Acad. Sci. USSR, 1970). Lab. head. *Crystallography, crystal chemistry, ferroelectrics.*

Veremeichik, Dr Tamara Fedorovna. Inst. of Crystallography, Acad. Sci. USSR, Leninsky pr. 59, Moscow 117333, USSR. (1945) Cand, physics and mathematics (INCRYS, 1977). Scient. *Crystalline field theory, spectroscopy, impurities.*

Verkhovskaya, Dr Kira Alexandrovna. Inst. of Crystallography, Acad. Sci. USSR, Leninsky pr. 59, Moscow 117333, USSR. (1940) Cand, physics and mathematics (INCRYS, 1968). Scient. *Ferroelectrics, optical properties.*

Vilkov, Prof Lev Vasilyevich. Moscow State U., Dept. of Chemistry, Leninskiye Gory, Moscow 117234, USSR. (1931) Dr, chemistry (MGU, 1969). Prof. *Elemento-organic compounds, structure, electron diffraction by gases.*

Vinokurov, Prof. Vladimir Mikhailovich. Kazan' State U., Lenina 18, Kazan' 420008, USSR. (1921) Dr, geology and mineralogy (IGEM, 1966). Head of Chair. *Radiospectroscopy, physical properties.*

Vistin', Dr Leonard Kazimirovich. Inst. of Crystallography, Acad. Sci. USSR, Leninsky pr. 59, Moscow 117333, USSR. (1933) Dr, physics and mathematics (INCRYS, 1988). Sr. scient. *Crystal physics, liquid crystals.*

Vlasov, Dr Vasily Platonovich. Inst. of Crystallography, Acad. Sci. USSR, Leninsky pr. 59, Moscow 117333, USSR. (1941) Cand, physics and mathematics (INCRYS, 1979). Scient. *Surface physicochemical properties, nucleation, epitaxy, electron and ionic spectroscopy.*

Voitsekhovsky, Dr Vladimir Nikolayevich. Leningrad Mining Inst., 21st Liniya 2, Leningrad 199026, USSR. (1931) Cand, geology and mineralogy (GORNY, 1966). Sr. scient. *Crystal growth, crystal morphology.*

Vol'kenshtein, Prof. Mikhail Vladimirovich. Inst. of Molecular Biology, Acad. Sci. USSR, Vavilov St. 32, Moscow 117312, USSR. (1912) Corresp. member, Acad. Sci. USSR; Dr, physics and mathematics (Tomsk U., 1942). Lab. head. *Polymers, macromolecular compounds.*

Volk, Dr Tat'yans Rafailovna. Inst. of Crystallography, Acad. Sci. USSR, Leninsky pr. 59, Moscow 117333, USSR. (1942) Cand, physics and mathematics (INCRYS, 1972). Scient. *Phase transitions, ferroelectrics, radiation effects, crystal properties.*

Volkova. Dr Olga Leonidovna. Leningrad Mining Inst., 21st Liniya 2, Leningrad 199026, USSR. (1960) Cand, physics and mathematics (Leningrad Polytechnical Inst., 1988). Eng. *Crystal physics, structure, real crystals.*

Volodin, Dr Alexander Petrovich. Inst. of Crystallography, Acad. Sci. USSR, Leninsky pr. 59, Moscow 117333, USSR. (1950) Cand, physics and mathematics (Inst. of Physical Problems, Acad. Sci. USSR, 1978). Scient. *EPR spectroscopy, activated single crystals, low temperatures.*

Volodina, Dr Galina Fedorovna. Inst. of Applied Physics, Acad. Sci. Moldavian SSR, Akademicheskaya 5, Kishinev 277028, USSR. (1935) Cand, physics and mathematics (INCRYS, 1964). Sr. scient. *X-ray structure analysis methods, coordination and inorganic compounds.*

Voloshin, Dr Aleksei Eduardovich. Inst. of Crystallography, Acad. Sci. USSR, Leninskii Prospekt 59, Moscow 117333, USSR. (1960) Cand, physics and mathematics (Inst. of Fine Chemical Technology, Moscow, 1986). Scient. *Real structure, structural defects, diffraction methods.*

Voronkova, Dr Valentina Ivanovna, Physics Dept., Moscow State U., Leninskie Gory, Moscow 117234, USSR. (1936) Cand, physics and mathematics (MGU, 1969). Sr. scient. *Ferroelectricity, high-temperature superconductivity, crystal growth.*

Voronova, Dr Alexandra Alekseyevna. Inst. of Crystallography, Acad. Sci. USSR, Leninsky pr. 59, Moscow 117333, USSR. Cand, physics and mathematics (INCRYS, 1971). *X-ray structure analysis, proteins.*

Voskresenskaya, Dr Inna Evgenyevna. Inst. of Crystallography, Acad. Sci. USSR, Leninsky pr. 59, Moscow 117333, USSR. Cand, geology and mineralogy (MGU, 1968). Sr. scient. *Crystal growth.*

Voznyak, Dr Dmitry Konstantinovich. Inst. of Geochemistry and Physics of Minerals, Acad. Sci.Ukrainian SSR, Palladina Prospekt 34, Kiev 252068, USSR. (1938) Cand, geology and mineralogy (Inst. of Geology, Acad. Sci. Ukrainian SSR, 1971). Sr. scient. *Mineralogical crystallography.*

Vozzhennikov, Dr Valery Mikhailovich. Karpov Physical Chemistry Inst., Obukha St. 10, Moscow 107120, USSR. (1936) Cand, chemistry (KARPOV, 1970). Sci. *Crystal chemistry, organic compounds.*

Vrublevskaya, Dr Zoya Vasilyevna. Inst. of Geology, Mineralogy and Petrography,Acad. Sci. USSR, Staromonetny 35, Moscow 109017, USSR. (1940) Cand, geology and mineralogy (IGEM, 1974). Scient. *Structural mineralogy, polytypism, electron diffraction.*

Yakhontova, Prof. Liya Konstantinovna. Moscow State U., Dept. of Geology, Leninskiye Gory, Moscow 117234, USSR. (1925) Dr, geology and mineralogy (MGU, 1973). Docent. *Structural mineralogy, crystal chemistry, minerals.*

Yakovenko Dr Sergey Sergeyevich. Inst. of Crystallography, Acad. Sci. USSR, Leninsky pr. 59, Moscow 117333, USSR. (1945) Cand, physics and mathematics (INCRYS, 1979). Scient. *Optical properties, inhomogeneous media.*

Yakovlev, Dr Viktor Alekseevich. Inst. of Crystallography, Acad. Sci. USSR, Leninskii Prospekt 59, Moscow 117333, USSR. (1955) Cand, physics and mathematics (INCRYS, 1986). Scient. *Crystal growth, thin films, surface physics.*

Yakubovich, Dr Ol'ga Vselodovna. Moscow State U., Dept. of Geology, Leninskiye Gory, Moscow 119899, USSR. (1950) Cand, geology and mineralogy (MGU, 1978). Scient. *X-ray structure analysis, crystal chemistry, inorganic compounds.*

Yakushkin, Dr Evgenii Dmitrievich. Inst. of Crystallography, Acad. Sci. USSR, Leninskii Prospekt 59, Moscow 117333, USSR. (1952) Cand, physics and mathematics (INCRYS, 1988). Scient. *Crystal acoustics, physics, phase transitions.*

Yamnova, Dr Nataliya Arkadyevna. Moscow State U., Dept. of Geology, Leninskiye Gory, Moscow 117234, USSR. (1950) Cand, geology and mineralogy (MGU, 1976). Scient. *Crystal chemistry, inorganic compounds.*

Yanovskii, Dr Aleksandr Il'ich. Inst. of Elementoorganic Compounds, Acad. Sci. USSR, Vavilov st. 28. Moscow 117334, USSR. (1957) Cand, chemistry (INEOS, 1963). Sr. scient. *Structural chemistry, elemento- and metalloorganic compounds, computation methods, programming, X-ray analysis.*

Yanovskii, Dr Vladimir Karlovich. Physics Dept., Moscow State U., Leninskie Gory, Moscow 117234, USSR. (1931) Cand, physics and mathematics (MGU, 1963) Sr. scient. *Ferroelectricity, high-temperature superconductivity, crystal growth.*

Yanson, Dr Tamara Ivanovna. Lvov State U., Dept. of Chemistry, University St. 1, Lvov 290602, USSR. (1947) Cand, chemistry (Lvov U., 1975). Scient. *X-ray structure analysis, crystal chemistry, intermetallic compounds.*

Yanulov, Dr Kirill Paskalyevich. Inst. of Geology, Acad. Sci. USSR, Komi Dept., Kommunisticheskaya St. 28, Siktivkar 167007, USSR. (1920) Cand, geology and mineralogy (LGU, 1950). *Isomorphism, epitaxial growth.*

Yanulova, Dr Lyudmila Alekseevna. Inst. of Geology, Ural Dept.of the Acad. Sci. USSR, Oplesnina 2, Syktyvkar 167610, USSR. (1942) Cand, geology and mineralogy (Inst. of Geology and Geochemistry, Sverdlovsk). Scient. *Crystal chemistry, sulphides.*

Yanusova, Dr Lyudmila Germanovna. Inst. of Crystallography , Acad. Sci. USSR, Leninskii Prospekt 59, Moscow 117333, USSR. (1949) Cand, physics and mathematics (Inst. of Steel and Alloys, Moscow, 1980). Scient. *Langmuir-Blodgett films, small-angle X-ray scattering.*

Yarmolyuk, Dr Yaroslav Petrovich. Lvov State U., Dept. of Chemistry, University St. 1, Lvov 290602, USSR. (1942) Cand, chemistry (Lvov U., 1972). Docent. *X-ray structure analysis, crystal chemistry, intermetallic compounds.*

Yasinskaya, Dr Angelina Andreyevna. Lvov State U., Dept. of Geology, Shcherbakov St. 4, Lvov 290005, USSR. (1922) Cand, geology and mineralogy (Lvov State U., 1951). Docent. *Mineralogical crystallography.*

Yufit, Dr Dmitrii Sergeevich. Inst. of Elementoorganic compounds, Acad. Sci. USSR, Vavilov st. 28, Moscow 117334, USSR. (1955) Cand, chemistry (MGU, 1988). Scient. *Crystal chemistry, organic compounds.*

Yurin, Dr Vladimir Alexandrovich. Inst. of Crystallography, Acad. Sci. USSR, Leninsky pr. 59, Moscow 117333, USSR. (1927) Cand, physics and mathematics (INCRYS, 1964). Sr. scient. *Ferroelectrics, phase transitions.*

Yushin, Dr Yury Yakovlevich. Inst. of Crystallography, Acad. Sci. USSR, Leninsky pr. 59, Moscow 117333, USSR. (1937) Cand, physics and mathematics (Inst. of High Energy Physics, 1968). Scient. *Electromagnetic radiation interaction with crystals.*

Yushkin, Prof. Nikolay Pavlovich. Inst. of Geology, Acad. Sci. USSR, Komi Department, Kommunisticheskaya 28, Syktyvkar 167000, USSR. (1936) Dr, geology and mineralogy (GORNY, 1968). Dept. head. *Earth's crust, minerological crystallography, crystallogeny.*

Zadorozhnaya, Dr Lyudmila Alexandrovna. Inst. of Crystallography, Acad. Sci. USSR, Leninsky pr. 59, Moscow 117333, USSR. (1944) Cand, geology and mineralogy (MGU, 1977). Scient. *Crystal growth.*

Zagal'skaya, Dr Yudif' Gertsevna. Moscow State U., Dept. of Geology, Leninskiye Gory, Moscow 117234, USSR. (1921) Cand, geology and mineralogy (MGU, 1966). Docent. *Geometrical crystallography, crystal chemistry, inorganic compounds.*

Zaitsev, Dr Sergey Mikhailovich. Rostov State U., Prospekt Stachki 192, Rostov-on-Don 344090, USSR. (1951) Cand, physics and mathematics (Rostov U., 1979). Scient. *X-ray structure analysis.*

Zaitseva, Dr Mariya Panteleimonovna. Inst. of Physics, Acad. Sci. USSR, Siberian Dept., Akademgorodok, Krasnoyarsk 660036, USSR. (1930) Cand, physics and mathematics (Inst. of Physics, Acad. Sci. USSR, Siberian Dept., 1968). Sr. scient. *Crystal physics, phase transitions, ferroelectrics, nonlinear electromechanical properties.*

Zakharchenko, Dr Irina Nikolayevna. Rostov State U., Inst. of Physics, Stachki pr. 194, Rostov-on-Don 344090, USSR. (1946) Cand, physics and mathematics (Rostov State U., 1978). Sr. scient. *X-ray structure analysis, defects, phase transitions, two-dimensional crystals.*

Zakharov, Dr Lev Nikolaevich. Inst. of Metalloogrganic Chemistry, Acad. Sci. USSR, Tropinina 49, Gorky 603600, USSR. (1950) Cand, physics and mathematics (Inst. Applied Physics, Kishinev,1981). Sr. scient. *X-ray single crystal analysis, growth, structure, thin films.*

Zakharov, Dr Nikolay Dmitriyevich. Inst. of Crystallography, Acad. Sci. USSR, Leninsky pr. 59, Moscow 117333, USSR. (1944) Cand, physics and mathematics (INCRYS, 1976). Scient. *Real structure, crystals, electron microscopy.*

Zakharova, Prof. Mariya Ivanovna. Moscow State U., Dept. of Physics, Leninskiye Gory, Moscow 117234, USSR. (1904) Dr, physics and mathematics (MGU, 1949). Prof. *X-ray crystallography, phase transformations.*

Zalessky, Dr Andrey Vladimirovich. Inst. of Crystallography, Acad. Sci. USSR, Leninsky pr. 59, Moscow 117333, USSR. (1930) Dr, physics and mathematics (INCRYS, 1985). Sr. scient. *Magnetic properties,, nuclear magnetic resonance.*

Zalutsky, Dr Ivan Ilyich. Lvov State U., Dept. of Chemistry, University St. 1, Lvov 290602, USSR. (1935) Cand, chemistry (Lvov U., 1968). Docent. *X-ray structure analysis, crystal chemistry, intermetallic compounds.*

Zamorzayev, Dr Alexander Mikhailovich. Kishinev U., Sadovaya 60, Kishinev 277003, USSR. (1927) Dr, physics and mathematics (INCRYS, 1971). Prof. *Symmetry theory (generalization and applications).*

Zarechnyuk, Dr Oleg Safonovich. Lvov State U., Dept. of Chemistry, University St. 1, Lvov 290602, USSR. (1923) Cand, chemistry (Lvov U., 1968). Docent. *X-ray structure analysis, crystal chemistry, intermetallic compounds.*

Zasorin, Dr Evgeny Zotikovich. Moscow State U., Dept. of Chemistry, Leninskiye Gory, Moscow 117234, USSR. (1934) Cand, chemistry (MGU, 1966). Sr. scient. *Structure, inorganic compounds, electron diffraction by gases.*

Zavalii, Dr Petro Yuliyanovich. Chemistry Dept., Lvov State U., Universitetskaya 1, Lvov 290602, USSR. (1957) Cand, chemistry (Lvov U., 1983). Asst. *Crystal chemistry, intermetallic compounds, computation methods, structure analysis.*

Zavodnik, Dr Valerii Efimovich. Karpov Physical-Chemical Inst., Obukha 10, Moscow 103064, USSR. (1941) Sr. scient. *X-ray analysis, crystal chemistry.*

Zav'yalova, Anna Arkadyevna. Inst. of Crystallography, Acad. Sci. USSR, Leninsky pr. 59, Moscow 117333, USSR. (1937) Cand, geology and mineralogy (INCRYS, 1970). Scient. *Electron diffraction, thin film structures, crystal chemistry.*

Zayakina, Dr Nadezhda Viktorovna. Inst. of Geology, Acad. Sci. USSR, Siberian Dept., Lenin Pr. 39, Yakutsk 677982, USSR. (1943) Cand, geology and mineralogy (MGU, 1976). Sr. scient. *Crystal structure determination, crystal chemistry, silicates, typomorphism, minerals.*

Zhdanov, Prof. German Stepanovich. Moscow State U., Dept. of Physics, Leninskiye Gory, Moscow 117234, USSR. (1906) Dr, physics and mathematics MGU, 1941). Prof. *Crystal structure, physical properties.*

Zheludev, Prof. Ivan Stepanovich. Inst. of Crystallography, Acad. Sci. USSR, Leninsky pr. 59, Moscow 117333, USSR. (1921) Dr, physics and mathematics (MGU, 1961). Sr. scient. *Ferroelectrics (physics), phase transitions (ferroelectric), symmetry in physics.*

Zheludeva, Dr Svetlana Ivanovna. Inst. of Crystallography, Acad. Sci. USSR, Leninsky pr. 59, Moscow 117333, USSR. (1948) Cand, physics and mathematics (MGU, 1976). Scient. *Thin film growth, thin film properties.*

Zhidkov, Dr Nikolay Petrovich. Moscow State U., Computing Center, Leninskie Gory, Moscow 117234, USSR. (1918) Cand, physics and mathematics (MGU, 1949). Docent. *Computing methods in structure analysis.*

Zhmurova, Dr Zinaida Ivanovna. Inst. of Crystallography, Acad. Sci. USSR, Leninsky pr. 59, Moscow 117333, USSR. (1930) Cand, physics and mathematics (INCRYS, 1970). Scient. *Crystal growth from melt.*

Zhukhlistov, Dr Anatoliy Pavlovich. Inst. of Geology, Mineralogy and Petrography,Acad. Sci. USSR, Staromonetny 35, Moscow 109017, USSR. (1938) Cand, geology and mineralogy (IGEM, 1977). Sr. scient. *Structural crystallography, mineralogy, polytipy, electron diffraction, electron microscopy.*

Zhuze, Prof. Vladimir Panteleimonovich. Physico-Techn. Inst.,Acad. Sci. USSR, Zapovednaya St. 51, Leningrad 194037, USSR. (1904) Dr, physics and mathematics (Inst. of Semiconductors, Acad. Sci. USSR, 1955). Sr. scient. *Phase transitions, non-stoichiometric compounds.*

Zinenko, Dr Victor Ivanovich. Inst. of Physics, Acad. Sci. USSR, Siberian Dept., Akademgorodok, Krasnoyarsk 660036, USSR. (1942) Dr, physics and mathematics (Inst. of Physics, Acad. Sci. USSR, Siberian Dept.). Scient. *Phase transitions in crystals.*

Zolotoi, Dr Aleksandr Borisovich. Branch of the Inst. of Chemical Physics, Acad. Sci. USSR. Moskovskaya obl., Chernogolovka 142432, USSR. (1953) Cand, physics and mathematics (Inst. Chem. Physics, Chernogolovka, 1981).Sr. scient. *Structure, organic molecules and crystals.*

Zorky, Prof. Petr Markovich. Moscow State U., Dept. of Chemistry, Leninskiye Gory, Moscow 117234, USSR. (1933) Dr, chemistry (MGU, 1973). Lab. head. *Symmetry, crystal chemistry.*

Zubenko, Dr Vasily Vasilyevich. Moscow State U., Dept. of Physics, Leninskiye Gory, Moscow 117234, USSR. (1930) Cand, physics and mathematics (MGU, 1968). Sr. scient. *X-ray crystallography, instrumentation and methods.*

Zubov, Dr Yuriy Alexandrovich. Karpov Physical Chemistry Inst., Obukha St. 10, Moscow 107120, USSR. (1932) Dr, chemistry (KARPOV, 1976). Sr. scient. *X-ray structure analysis, polymers.*

Zvezdinskaya, Dr Larisa Vsevolodovna. Inst. of Geology, Mineralogy and Petrography,Acad. Sci. USSR, Staromonetny 35, Moscow 109017, USSR.

(1948) Cand, geology and mineralogy (MGU, 1978). Scient. *Crystal chemistry, classification, inorganic compounds, minerals.*

Zviedre, Dr Irena Ilyinichna. Inst. of Inorganic Chemistry, Acad. Sci. Latvian SSR, Meistaru St. 10, Riga 226934, USSR. (1938) Cand, chemistry (Dept. of Biological and Chemical Sciences, Acad. Sci. Latvian SSR). Sr. scient. *Crystal chemistry, borates.*

Zvinchuk, Dr Rostislav Alekseyevich. Leningrad State U., Dept. of Chemistry, 14th Liniya 29, Leningrad 199178, USSR. (1929) Cand, chemistry (LGU, 1964). Docent. *Diffuse phase transitions.*

Zvirgzede, Dr Yulia Vil'gel'movna. Polytechnical Inst. Lenina 1, 226003 Riga, USSR. (1940) Cand., physics and mathematics (Inst. of Physics, Riga, 1977). Docent. *Ionic thermal vibrations, phase transition, ferroelectrics.*

Zvirgzede, Dr Yurii Al'fredovich. Inst. of Solid State Physics, Latvian State U. Kengaraga st.8, Riga 226063, USSR. (1942) Cand, physics and mathematics (Latvian State U., 1975). Lab. Head. *Ionic thermal vibrations, phase transitions, ferroelectrics.*

Zvyagin, Dr Boris Borisovich. Inst. of Geology, Mineralogy and Petrography, Acad. Sci. USSR, Staromonetny 35, Moscow 109017, USSR. (1921) Dr, physics and mathematics (INCRYS, 1963). Lab. head. *Structural crystallography, structural mineralogy, polytypism, electron diffraction.*

UNITED KINGDOM

Sub-Editor: **J.A.K. Howard**

Notes

1. International telephone country code - 44

2. Unless otherwise stated, the exchange name for the telephone number is the place name in the address. Most exchanges now have direct dialling (STD) codes (separated from the rest of the number by '+'). The STD code may not apply to relatively short distance calls (e.g. London to Welwyn Garden); consult the local booklet of dialling codes.

3. In the biographic data, the position held (e.g. Head, Prof.) relates to the Div. or Dept. in the address, unless further details are given.

4. In general the bachelor's degree (BA, BSc) is the first degree awarded by universities in the U.K.; the holder of an Oxford or Cambridge BA may proceed to the MA after the passage of time and payment of money. Qualifications of first degree standard awarded by other bodies include Dip. Tech. (Diploma in Technology), HND (Higher National Diploma) and graduateship, licentiateship or membership of various professional institutions. The Higher National Certificate (HNC) is similar to HND but with a narrower range of subjects. Higher degrees include master's degrees (MSc, MPhil) and various diplomas awarded after courses of instruction or research lasting one or two years. The doctorate (PhD, DPhil) is obtained after research lasting (normally) three years. The senior doctorate (DSc, ScD) is awarded on the basis of published contributions to knowledge.

5. Colleges of the University of London include:

Bedford	King's (KQC)
Birkbeck	Queen Mary (QMC)
Chelsea	University (UC)
Imperial (Imp.)	

6. Abbreviations used for counties or regions are:

Beds. - Bedfordshire	Middx. - Middlesex
Berks. - Berkshire	Northants. - Northamptonshire
Bucks. - Buckinghamshire	Oxon - Oxfordshire
Hants. - Hampshire	Staffs.- Staffordshire
Herts. - Hertfordshire	Wilts. - Wiltshire
Lancs. - Lancashire	Worcs. - Worcestershire
Leics. - Leicestershire	Yorks. - Yorkshire

7. The following are grades in the Scientific Civil Service and in some universities:

CSO - Chief Scientific Officer	SSO - Senior Scientific Officer
DCSO - Deputy Chief Scientific Officer	HSO - Higher Scientific Officer
SPSO - Senior Principal Scientific Officer	EO - Experimental Officer
PSO - Principal Scientific Officer	SEO - Senior Experimental Officer

8. Other abbreviations used include:

AERE - Atomic Energy Research Establishment	Off. - Officer
AWRE - Atomic Weapons Research Establishment	plc - public liability company
BR - British Rail	RAE - Royal Aircraft Establishment
CEGB - Central Electricity Generating Board	RMCS - Royal Military College of Science
CERL - Central Electricity Research Laboratories	RSRE - Royal Signals and Radar Establishment
CNAA - Council for National Academic Awards	Roy. Soc. - Royal Society
GEC - General Electric Company	Sect. - Section
ICI - Imperial Chemical Industries	SERC - Science and Engineering Research Council
Lect. - Lecturer	UKAEA - United Kingdon Atomic Energy Authority
MRC - Medical Research Council	UMIST - University of Manchester Inst. of Science and Technology

Abell, Dr John Stuart. Dept. of Metallurgy and Materials, U. of Birmingham, PO Box 363, Birmingham B15 2TT, England. (1944) PhD, solid state physics (U. Surrey, 1969). Res. fellow. (tel. 021 + 472-1301, ext. 3446). *Single crystal growth, rare earths, intermetallic compounds, opto-electronic device materials, metals, structure and physical properties, Nb-H dilute alloys.*

Acharya, Dr Ravindra. Lab. of Molecular Biophysics, U. of Oxford, S. Parks Rd., Oxford OX1 3QU, England. (1955) PhD, X-ray crystallography (U. Bangalore, India, 1982). Res. sci. (tel. 0865 + 275385; fax 0865 + 510454; email RAVI at UK.AC.OX.BIOP). *Macromolecular crystallography.*

Adam, Dr Jerzy. Timber Top, Brightwell-cum-Sotwell, Wallingford, Oxon OX10 0RG, England. (1918) PhD, physics (U. St. Andrews, 1949). Retired. (tel. 0491 + 36045). *X-ray crystallography, structure, nuclear reactor materials, computing.*

Adams, Dr Margaret Joan. Lab. of Molecular Biophysics, Zoology Dept., Oxford U., Rex Richards Building, South Parks Rd., Oxford OX1 3QU, England. (1939) DPhil, protein crystallography (U. Oxford, 1968). Fellow, tutor in chemistry, Somerville C. (tel. 0865 + 56733, ext. 426). *X-ray methods, protein structures, adenine nucleotide dependent enzymes, carboxylation enzymes.*

Ainsworth, Mr Leonard Ralph. Springfields Nuclear Power Dev. Lab., UKAEA (N. Div.), Preston PR4 0RR, England. (1932) CChem, FRSC, Chemistry (Royal Society of Chemistry, 1960). SSO (tel. 0772 + 728262, ext. 31284). *X-ray powder diffraction, computer methods, vitreous state.*

Alcock, Dr Nathaniel Warren. Dept. of Chemistry, U. of Warwick, Coventry CV4 7AL, England. (1939) PhD, chemistry (U. Cambridge, 1963). Reader. (tel. 0203 + 523228; fax 0203 + 461606; email MSRBB at UK.AC.WARWICK.CU). *Chemical bonding, coordination compounds, main group inorganic compounds, inorganic crystals.*

Allen, Dr Andrew John. Materials Phys. and Metall. Div., Building 521, UKAEA Harwell, Didcot, Oxon OX11 0RA, England. (1955) PhD, neutron scattering (U. Birmingham, 1981). Professional grade 2. (tel. 0235 + 24141, ext. 5171; telex 83135 ATOMHAG; fax 0235 + 432726). *Neutron scattering, small angle scattering, ASAXS, double crystal diffraction, ultra high resolution diffraction, SAS.*

Allen, Dr Frank Harmsworth. Cambridge Crystallographic Centre, U. Chemical Lab., Lensfield Rd., Cambridge CB2 1EW, England. (1944) PhD, Physical chemistry (U. London, 1968). Principal Scientist (tel. 0223 + 336425; telex 81240 CAMSPL G; fax 0223 + 336362; email FHA1 at UK.AC.CAM.PHX or FHA1 at UK.AC.CAM.CHEMCRYS). *Organic compunds, crystallographic databases, computer programming, molecular systematics.*

Angel, Dr Ross John. Dept. of Geological Sciences, U. C. London, Gower St, WCIE GBT, London, England. (1959) PhD, Minerals (Cambridge, 1985). Resfilellow (Tel. 01-387 7050 ext. 2427). *Mineralogy, mineral physics, mineral chemistry, high pressure X-ray diffraction.*

Angelova-Tiurkedjieva, Dr Maia Nikolova. Dept. of Chemistry, U. of Kent, Canterbury, Kent, CT2 7NH, England. (1958) DPhil, Corepresentations of Magn. Groups (U. Sofia, 1988). Visiting res. fellow. (tel. 0277 + 764000, ext. 3995; telex 965449; fax 0227 + 459025; email MNAT at UK.AC.UKC). *Theoretical crystallography, crystallographic space groups, magnetic space groups, representations, corepresentations, structures, magnetic materials, phase transitions.*

Anwar, Mr Jamshed. Pharmacy, Kings College London, Chelsea campus, Manresa Rd, London SW3 6LX, England. (1957) MSc, Pharmacy (London U., 1982) Lecturer. (tel. 01-351 2488 ext. 2504, email UDKJ060 at KCL.CC.ELM) *Phase transformation, polymorhism, molecular solids, computer simulation.*

Arndt, Dr Ulrich Wolfgang. Structural Studies Div., MRC Lab. of Molecular Biology, Hills Road, Cambridge CB2 2QH, England. (1924) PhD, physics (U. Cambridge, 1949). Scient. staff, MRC. (tel. 0223 + 248011; telex 81532; fax 0224 + 213556). *Instrumentation, diffractometry, proteins.*

Arnott, Prof. Struther. U. of St Andrews, College Gate, St. Andrews KY16 9AJ, Scotland. (1934) PhD, Chemistry (Glascow U., 1960) Vice Chancellor. (tel. 0334-76161 ext. 237, fax 76213). *Fibrous structures, nucleic acids, polysaccharides.*

Artymiuk, Dr Peter Joseph. Dept. of Biochemistry, Sheffield U., Sheffield, S10 2TN, England. (1952) DPhil, Molecular biophysics (U. Oxford, 1979). (tel. 0742 + 768555, ext. 4190; fax 0742 + 727949; email BI1PA at UK.AC.SHEF.PRIMEA). *Macromolecules, X-ray diffraction, structural biology, computing, crystal growth, diffractometry, diffuse scattering, hydrogen bonding.*

Ashwell, Dr Geoffrey Joseph. Dept. of Chemistry, Sheffield City Poly., Pond Street, Sheffield S1 1WB, England. (1947) PhD, organic semiconductors (U. Nottingham, 1972). Lect. (tel. 0742 + 20911, ext. 380). *One-dimensional metals, organic semiconductors, molecular rectifiers.*

Attfield, Dr John Paul. Chemical Crystallography Lab., U. of Oxford, 9 Parks Rd., Oxford, OX1 3PD, England. (1962) DPhil, Chemistry (U. Oxford, 1988). Junior res. fellow. (tel 0865 + 270823). *Inorganic crystals, magnetism, neutron diffraction, powder diffrction, structural chemistry, X-ray diffraction.*

Aupers, Dr John Henry. Chemistry Div., R.I.H.E. Whitelands college, West Hill, Putney, London SW15 3SN, England. (1942) PhD, Crystallography (London, 1983) Sr. Lect. (tel. 01-788 8268, ext. 5296) *Coordination compounds, organic compounds.*

Bacon, Prof. George Edward. Windrush Way, Guiting Power, Cheltenham, GL54 5US, England. (1917) ScD, physics and crystallography (U. Cambridge, 1964). Em. prof., physics, U. Sheffield. (tel. 04515 + 631). *Neutron diffraction.*

Bagley, Mr Arthur George. Director, Hiltonbrooks Ltd., Yew Tree Cottage, Knutsford Rd., Cranage, Holmes Chapel, Cheshire CW4 8EP, England. (1947). Dir. (tel. 0477 + 32687). *Analytical X-ray equipment, X-ray diffraction.*

Bailey, Prof. David Kenneth. Dept. of Geology, U. of Reading, Whiteknights, Reading RG6 2AB, England. (1931) PhD, geology (U. London, 1959). Prof. (tel. 0734 + 875123, ext. 7871). *Petrology, mineralogy, geochemistry.*

Bailey, Dr Neil Anthony. Dept. of Chemistry, U. of Sheffield, Brook Hill, Sheffield S3 7HF, England. (1940) PhD, chemistry (Imp. C., U. of London, 1964). Lect. (tel. 0742 + 78555, ext. 4464). *X-ray crystal structure analysis, binucleating and compartmental acyclic and macrocyclic ligands, metal complexes, organometallic molybdenum, tungsten structural chemistry.*

Bain, Dr Derek Charles. Dept. of Mineral Soils, Macaulay Inst. for Soil Res., Craigiebuckler, Aberdeen AB9 2QJ, Scotland. (1944) PhD, geology (U. Aberdeen, 1974). PSO. (tel. 0224 + 38611). *Soil mineralogy and weathering, X-ray fluorescence spectroscopy.*

Bainbridge, Mr John Evelyn. Materials Physics and Metallurgy Div., AERE Harwell, Didcot, Oxon OX11 0RA, England. (1933) AIM, metallurgy (Inst. of Metallurgists, 1965). SSO. (tel. 0235 + 24141, ext. 4134). *Scanning electron microscopy with analysis, transmission electron microscopy.*

Baker, Dr Patrick Julian. Biochemistry Dep., U. of Sheffield, Western Bank, Sheffield S10 2TN, England. (1961) PhD, Protein structure (Sheffield U., 1988). Res. assoc. (tel. 0742 + 768555, ext. 4242). *Protein structure, protein function, dehydrogenases, molecular modelling.*

Baker, Mr R.W. U. of London Computer Centre, 20 Guildford St., London WC1 1DZ, England.

Baker, Mr Thomas Wilfred. Tirrold Scientific and Technical Services Ltd., Berry Croft, Spring Lane, Aston Upthorpe, Didcot, Oxfordshire OX11 9EH, England. (1923) BSc, physics (U. Manchester, 1949). Dir. & consultant, X-ray diffraction. (tel. 0235 + 850264). *X-ray diffraction applications, materials, powder diffraction, line broadening, preferred orientation, high and low temperatures, high precision, automatic control, computer data processing, X-ray diffraction instruments.*

Balchin, Dr Anthony Arthur. Dept. of Physical Sci., Brighton Polytechnic, Lewes Rd., Moulsecoomb, Brighton BN2 4GJ, England. (1932) PhD, crystallography (U. London, 1968). Sr. lect. (tel. 0273 + 60090, ext. 2498). *Layer compounds, semiconductors.*

Balyuzi, Dr Hushang H.M. Physics Dept., King's C. (KQC), Strand, London WC2R 2LS, England. (1942) PhD, X-ray diffraction (U. London, 1970). Lect. (tel. 01 + 836-5454, ext. 2145). *Amorphous systems, structure, X-ray diffraction.*

Barnes, Dr John Conquest. Dept. of Chemistry, U. of Dundee, Dundee DD1 4HN, Scotland. (1935) DSC, chemistry (U. Dundee, 1985). Sr. lect. (tel. 0382 + 23181, ext. 4705). *Coordination compounds, molecular crystals, structural chemistry.*

Barnes, Dr Paul. Dept. of Crystallography, Birkbeck C., Malet St., London WC1E 7HX, England. (1942) PhD, physics (U. Cambridge, 1968). Sr. lect. (tel. 01 + 631-6417; fax 01 + 436-8918). *Powder diffraction, X-ray diffraction and phase transitions, microscopy, computer simulation.*

Barron, Dr Hugh Wilson Taylor. Dept. of Natural Philosophy, U. of Aberdeen, Aberdeen AB9 2UE, Scotland. (1943) PhD, physics (U. Aberdeen, 1971). Lect. (tel. 0224 + 40241, ext. 284). *Phonon - X-ray scattering, high field transport, semiconductors.*

Barrow, Dr Michael John. Dept. of Appl. Chemical Sci., Napier Polytechnic of Edinburgh, 10 Colinton Rd, Edinburgh EH10 5DT, Scotland. (1946) PhD, chemistry (U. Manchester, 1972). Sr. Lect. (tel. 031 + 444-2266, ext. 2621 or 2206, fax 031-445-7209). *Structural chemistry, molecular crystals, computing, X-ray diffraction.*

Basak, Dr Ajit Kumar. Lab. of Molecular Biophysics, Dept. of Zoology, U. of Oxford, South Parks Rd., Oxford, OX1 3QV, England. (1955) PhD, X-ray crystallography (Calcutta, 1987). Res. assoc. (tel. 0865 + 275392). *Structure and function of biological compounds, crystallographic data, macromolecules, organic compounds, proteins, structural biology, structural chemistry. X-ray diffraction.*

Bassett, Prof. David Clifford. Physics Dept., U. of Reading, Whiteknights, PO Box 220, Reading, Berks RG6 2AF, England. (1937) ScD, polymer physics (U. Cambridge, 1981). Prof., physics. (tel. 0734 + 875123, ext. 369, telex 847813). *Polymer crystallization, crystal growth, high pressures.*

Bassett, Mr G. Alan. Dept. of Physics, U. of Warwick, Coventry CV4 7AL, England. (1926) MSc, physics (U. London, 1950). Sr. tutor., physics. (tel. 0203 + 24011, ext. 2375). *Crystal surfaces, electron microscopy, decoration, mechanical properties, crack propagation.*

Batchelder, Dr David Neville. Dept. of Physics, Queen Mary C., Mile End Rd., London E1 4NS, England. (1938) PhD, physics (U. Illinois, USA, 1965). Lect. (tel. 01 + 980-4811, ext. 4004). *Structure and properties, polymer single crystals, conjugated polymers.*

Bates, Dr David Ronald. Analytical Div., British Gas Corp., London Res. Station, Michael Road, London SW6 2AD, England. (1955) PhD, physical chemistry (U. Nottingham, 1982). Res. Scient. (tel. 01 + 736-3344, ext. 4059). *X-ray diffraction, industrial applications; automatic powder diffractometry; crystallite size studies, catalysis.*

Bates, Dr Peter Arthur. Div. of Physics and Astronomy, Lancashire Polytechnic, Corporation St., Preston, Lancs. PR1 2TQ, England. (1945) PhD, physics (U. of Wales). Sr. lect. in physics. (tel. 0772 + 22141, ext. 2182). *X-ray diffraction, instrumentation, magnetic fluids, fine particle magnetic systems.*

Battle, Dr Peter David. Sch. of Chemistry, U. of Leeds, Leeds, LS2 9JT, England. (1954) DPhil, crystallography (U. Oxford, 1980). Lect. (tel. 0532 + 336412). *Solid state chemistry, mixed metal oxides, magnetism.*

Baxter, Mr Colin. Materials Res. & Dev. Lab., Rolls-Royce Ltd., PO Box 31, Derby DE2 8BJ, England. (1949) HND, metallurgy, (Derby and District C. of Techn., 1971). Principal Eng. (tel. 0332 + 42424, ext. 520). *Aerospace industry applications, qualitative analysis, preferred orientation measurement, single-crystal orientation measurement, residual stress measurement.*

Beagley, Dr Brian. Dept. of Chemistry, UMIST, PO Box 88, Manchester M60 1QD, England. (1936) DSc, structural chemistry (U. Birmingham, 1981). Reader. (tel. 061 + 236-3311, ext. 2567/2535; telex 666094; fax 061 + 228 1249; email MCDSSBB at UK.AC.UMRCC.CMS). *Structural chemistry, electron diffraction, X-ray crystallography, EXAFS, chemical bonding, computing, crystallographic data.*

Beamson, Dr G. 34 Westbourne Grove, Goole, North Humberside DN14 6NB, England.

Beddell, Dr Christopher Raymond. Physical Chemistry Dept., Wellcome Res. Labs., Langley Court, Beckenham, Kent BR3 3BS, England. (1944) DPhil, molecular biophysics (U. Oxford, 1971). Res. (tel. 01 + 658-2211, ext. 406). *Macromolecule structure and function, macromolecule - ligand interactions, pharmacophore definition.*

Beddoes, Mr Roy L. Dept. of Chemistry, U. of Manchester, Brunswick St., Manchester M13 9PL, England. (1938) MA, physics (U. Cambridge, 1965). Sr. exp. off. (tel. 061 + 275-4688) *Structure, computation, apparatus.*

Beevers, Dr Cecil Arnold. Dept. of Chemistry, U. of Edinburgh, 128 Blackford Avenue, Edinburgh EH9 3HH, Scotland. (1908) DSc, X-ray crystallography (U. Liverpool, 1943). Em. Res. Fellow (tel. 031 + 667-3843). *Crystallographic data.*

Begley, Dr Michael John. Dept. of Chemistry, U. of Nottingham, University Park, Nottingham NG7 2RD, England. (1944) PhD, crystallography (U. Surrey, 1970). Lect. (tel. 0602 + 484848, ext. 2392). *Organic compounds, solid state reactions, structural chemistry, X-ray diffraction.*

Bell, Mr Anthony Martin Thomas. X-ray Analysis Section, UKAEA, Springfields, Preston PR4 0RR, Lancs, England. (1963) BSc, chemistry (U. Sheffield, 1984). Materials Scientist. (tel. 0772 + 728262, ext. 31460/31450; telex 67545; fax 0772 + 736248). *Powder diffraction, powder diffraction, materials science.*

Bellamy, Mr Brian Arthur. Materials Dev. Div., AERE Harwell, Didcot, Oxon OX11 0RA, England. (1936) MInstP (Inst. of Physics, 1936). SSO. (tel. 0235 + 24141, ext. 4524). *X-ray diffraction applications, materials.*

Bellard, Dr Sharon Ann. Dept. of Organic and Inorganic Chemistry, U. of Cambridge, Lensfield Rd., Cambridge CB2 1EW, England. (1950) PhD, chemistry (U. Cambridge, 1979). Res. asst. (tel. 0223 + 66499, ext. 314). *Informa-*

tion storage and retrieval, data base management, organic and organometallic crystal structures, low temperature crystallography.

Bennett, Dr Pauline Mary. MRC Cell Biophysics unit, 26-29 Drury Lane, London, WC2B 5RL, England. (1944) PhD, Biophysics (U. London, 1977). Sci. Staff (tel. 01 + 836-8851). *Electron microscopy, electron diffraction, X-ray diffraction, structural biology, proteins, macromolecules.*

Bevis, Prof. M.J. Dept of Materials Tech., Brunel U., Uxbridge, Middx. UB8 3PH, England.

Bishop, Dr Arthur Clive. Dept. of Mineralogy, British Museum (Natural History), Cromwell Rd., London SW7 5BD, England. (1930) PhD, petrology (U. London, 1954). Deputy dir., Keeper, mineralogy. (tel. 01 + 589-6323, ext. 226). *Mineralogy, petrology.*

Blake, Dr Alexander John. Dept. of Chemistry, U. of Edinburgh, West Mains Rd., Edinburgh EH9 3JJ, Scotland. (1954) PhD, structural chemistry (U. Aberdeen, 1980). Res. Fellow. (tel. 031 + 667-1081, ext. 3414; telex 727442 UNIVED G; fax 031 + 662-4054; email A.J.BLAKE at UK.AC.EDINBURGH). *Low-melting compounds, powder diffraction, X-ray diffraction, disorder, low-temperature crystallography.*

Blake, Dr Antony Brian. Dept. of Chemistry, U. of Hull, Hull HU6 7RX, England. (1933) PhD, chemistry (U. London, 1960). Lect. (tel. 0482 + 46311, ext. 7433). *Transition-metal chemistry, magnetic properties, polynuclear complexes, exchange interactions.*

Bland, Dr J.A. Coach House, Matson Drive, Remenham, Henley-on-Thames, Oxford, England.

Bloomer, Dr Anne Christine, MRC Molecular Biology Lab., Hills Rd., Cambridge, CB2 2QH, England. (1945) DPhil, Molecular Biophysics (U. Oxford, 1972). Sci. Staff (tel. 0223 + 248011; telex 81532; fax 0223 + 213556; email ACB1 at UK.AC.CAM.MRC-LMB). *Structural biology, macromolecules, proteins, viruses, instrumentation.*

Bloor, Dr David. Dept. of Physics, Queen Mary C., Mile End Rd., London, E1 4NS, England. (1937) PhD, physics (U. London, 1961). Reader. (tel. 01 + 980-4811, ext. 334). *Structure, properties, preparation, diacetylene monomer, polymer single crystals.*

Blow, Prof David Mervyn. Blackett Lab., Imperial C. of Sci. and Techn., London SW7 2BZ, England. (1931) PhD, physics (U. Cambridge, 1957). Prof., biophysics. (tel. 01 + 589-5111, ext. 6721; fax 01 + 589-9463). *Crystal growth, proteins, macromolecules, X-ray diffraction, enzymes, protein engineering.*

Blundell, Dr David James. Petrochemicals and Plastics Div., ICI plc, PO Box 90, Wilton, Middlesborough, Cleveland, TS6 8JE, England. (1940) PhD, physics (U. Bristol, 1967). Sr. res. physicist. (tel. 0642 + 455522, ext. 2007, telex 587461). *Polymer morphology, small-angle X-ray scattering, wide-angle X-ray scattering, thermal analysis.*

Blundell, Prof. Thomas Leon. Lab. of Mol. Biology, Dept. of Crystallography, Birkbeck C., Malet St., London WC1E 7HX, England. (1942) DPhil, crystallography (U. Oxford, 1967). Prof. (tel. 01 + 631-6284). *Protein crystallography, molecular biology, protein modelling and design.*

Boles, Dr Michael Owen. Dept. of Env. Sci., Polytechnic South West, Drake Circus, Plymouth, Devon PL4 8AA, England. (1941) PhD, physics (U. Exeter, 1962). Sr. lect. (tel. 0752 + 600600) *X-ray diffraction, powder diffraction, materials science.*

Borkakoti, Dr Nivedita Neera. Physical Methods, Roche Products Ltd., Broadwater Rd., Welwyn Garden City, AL7 3AY, England. (1949) PhD, crystallography (U. London, 1978). Res. sci. (tel. 0707 + 328128, ext. 2436). *Protein crystallography, computing, molecular modelling.*

Bowen, Dr Alun Wynne Materials and structures Dept, RAE, Farnborough, Hants GU14 6TD, England. (1942) PhD, Metallurgy, solid state physics (U. Manchester, 1968). Section leader. (tel. 0252 + 24461, ext. 2208; telex 858134; fax 0252 + 24461, ext. 2173). *Electron diffraction, electron microscopy, materials science, metals, microscopy, phase transitions, powder diffraction, texture, X-ray diffraction, residual stress.*

Bowen, Dr David Keith. Dept. of Eng., U. of Warwick, Coventry CV4 7AL, England. (1940) DPhil, metallurgy (U. Oxford, 1967). Reader. (tel. 0203 + 523133, ext. ddi, telex 311904 uniwk, fax 0203 + 418922, email DBK at UK.AC.WARWICK.EAGLF). *X-ray topography, diffractometry and interferometry, synchrotron radiation, materials science, X-ray metrology, electronic materials.*

Bowen-Jones, Dr J. Lower Bawdon Farm, Charley, Loughborough, Leics LE12 9XL, England.

Bown, Dr Michael George. Dept. of Earth Sci., U. of Cambridge, Downing St., Cambridge CB2 3EQ, England. (1928) PhD, crystallography (U. Cambridge, 1955). Lect. (tel. 0223 + 333400). *Minerals.*

Boyle, Dr Lewis Laurence. University Chemical Lab., Canterbury, Kent CT2 7NH, England. (1942) DPhil, chemistry with supp. mineralogy (U. Oxford, 1966). Sr. lect. (tel. 0227 + 66822, ext. 584, telex 965449). *Group theory, phase transitions, mineral structures, space group character tables.*

Boys, Dr Cecil William Gordon. Dept. of Biophysics, U. of Leeds, Leeds, LS2 9JJ, England. (1954) DPhil, protein crystallography (U. Oxford, 1987). Res. fellow (tel. 0532 + 333029). *Prothrombin fragment 1, molecular recognition, DNA binding, proteins, crystallisation, structure determination.*

Brady, Dr Robert Leo. Dept. of Chemistry, U. of York, Heslington, York, YO1 5DD, England. (1961) DPhil, Protein Crystallography (York, 1988). Res. fellow (tel. 0904 + 432566; fax 0904 + 410519; email BRADY at UK.AC.YORK.YORVIC). *Crystal growth, proteins, structural biology, viruses.*

Briant, Dr Clive Edward. 131 Drayton Rd., Sutton Courtenay, Oxon OX14 4HA, England. (1954) PhD, chemistry (U. Bradford, 1983). *Structural chemistry, organic and organometallic compounds, polycyclics, nucleosides, metal clusters.*

Brice, Dr John Chadwick. 20 Kitsmead, Copthorne Bank, Crawley, West Sussex, RH10 3PW, England. (1934) PhD, materials science (U. Cambridge, 1969). Consultant. (tel. 0342 + 712825). *Crystal growth, and properties of electronic materials.*

Bright, Dr Alan Aubrey Samuel. Chemical Synthesis Group, FBC Ltd., Chesterford Pk. Res. Stn., Nr. Saffron Walden, Essex CB10 1XL, England. (1948) PhD, chemistry (U. Dundee, 1974). Team Leader (Spectroscopy/Physical Chemistry). (tel. 0799 + 30123). *Molecular structure, biologically important chemicals, pesticides, agrochemicals.*

Britton, Ms Karen Linda. Biochemistry Dept., Sheffield U., Western Bank, Sheffield, S10 2TN, England. (1965) BSc, Biophysics (U. Leeds, 1987). Postgrad. (tel. 0742 + 768555, ext. 4242). *Protein structure, dehydrogenases, molecular modelling, protein function.*

Brown, Miss Betty Rosina. Crystallography Group, Materials Characterisation Div., Materials Sci. Lab., GEC Res. Labs., Hirst Res. Centre, East Lane, Wembley Middx HA0 7PP, England. (1926) BSc, physics and mathematics (U. Reading, 1945). Sci. res. staff. (tel. 01 + 904-1262, ext. 281). *Crystal orientation, multiple diffraction, epitaxial layers.*

Brown, Dr Cedric John. Dept. of Metallurgy and Materials, City of London Polytechnic, Central House, Whitechapel High St., London E1 7PF, England. (1915) DSc, crystallography (U. Birmingham, 1955). Res. fellow. (tel. 01 + 283-1030, ext. 479). *Organic structures, diffuse scattering.*

Brown, Mr D. Stoe & Cie GmbH, 21 Dorset Ave, Southall, Middx. UB2 4HF, England.

Brown, Dr David Summers. Dept. of Chemistry, Loughborough U. of Techn., Loughborough, Leics. LE11 3TU, England. (1937) PhD, crystallography (U. Nottingham, 1962). Sr. lect. (tel. 0509 + 222558). *Structural chemistry, small angle X-ray scattering.*

Bruce, Dr P.G. Dept. of Chemistry, Heriot-Watt U., Riccarton, Edinburgh, Scotland.

Bryant, Mr P.K. 43 Fraser St, Bilston, W Midlands WV14 7PD, England.

Buckley, Dr Christopher Paul. Polymer Eng. Div., Dept. of Mechanical Eng., UMIST, PO Box 88, Manchester M60 1QD, England. (1946) DPhil, eng. sci. (U. Oxford, 1968). Lect. (tel. 061 + 236-3311, ext. 2716). *Polymers, structure, properties, processing.*

Bullen, Dr G.J. Tanglewood, Church Lane, Lexden, Colchester CO3 4DX, England.

Bullen, Mr Henry Eric. Res. and Dev. Lab. (U.K.), Gillette, 454 Basingstoke Rd., Reading RG2 0QE, England. (1930) MSc, crystallography (U. London, 1957). Res. physicist. (tel. 0734 + 875222, ext. 326). *Structures, steels, polymers.*

Bunn, Dr Charles William. 6 Pentley Park, Welwyn Garden City, Herts. AL8 7RU, England. (1905) DSc, chemistry (U. Oxford, 1953). Retired res. fellow, Royal Inst. (tel. 07073 + 23581). *Crystal growth, X-ray diffraction, structures, organic crystals, macromolecules, physical properties - structure relationship.*

Bunning, Dr John David. Dept. of Physical Sciences, Trent Polytechnic, Clifton Lane, Nottingham, NQ11 8NS, England. (1949) PhD, structre of liquid crystals (U. Leeds, 1980). Sr. Lect. (tel. 0602 + 418248, ext. 3112). *Liquid crystals.*

Burge, Prof. Ronald Edgar. Dept. of Physics, King's C. London, Strand, London WC2R 2LS, England. (1932) DSc, physics and biophysics (U. London, 1975). Head. (tel. 01 + 836-5454, ext. 2514). *Structure and properties, natural and synthetic polymers, polymer associations, scattering theory, image analysis, electron microscopy instrumentation, soft X-ray microscopy.*

Bush, Dr Michael Anthony. Wilkinson Sword Ltd., Totteridge Rd., High Wycombe, Bucks. HP13 6EJ, England. (1943) PhD, chemistry (U. Bristol, 1967). Marketing manager. (tel. 0494 + 33300). *Structural inorganic chemistry.*

Bushnell-Wye, Dr Graham. Dept. of Crystallography, Birkbeck C., Malet St., London WC1E 7HX, England. (1950) PhD, crystallography (U. London, 1983). Res. Officer. (tel. 01 + 801-9356, ext. 2136; fax 01 + 436-8918; email UBCG022 at UK.AC.BBK.CU). *Computing, crystal growth, defect structures, diffuse scattering, disorder, instrumentation, materials science, scattering theory, teaching, X-ray diffraction.*

Butler, Dr Barry Conrad Milne. Dept. of Earth Sci., U. of Oxford, Parks Rd., Oxford OX1 3PR, England. (1932) PhD, petrology (U. Cambridge, 1960). Lect. (tel. 0865 + 54511). *Mineralogy, industrial materials.*

Butler, Dr Stephen Andrew. Plates Dept., British Steel Technical, Swinden House, Moorgate, Rotherham, South Yorks, S60 3DD, England. (1946) PhD, crystallographic mineralogy (U. Cambridge, 1972). Sr. Investigator. (tel. 0709 + 820166, ext. 3283; telex 547279). *Diffractometry, electron diffraction, electron microscopy, metals, minerals, powder diffraction, texture, X-ray diffraction.*

Cahn, Prof. R.W. 6 Storeys Way, Cambridge CB3 0DT, England.

Cain, Mr Peter Maurice. General Chemistry Div., Metropolitan Police Forensic Science Lab., 109 Lambeth Rd., London SE1 7LP, England. (1944) BSc, chemistry (U. Reading, 1966). HSO. (tel. 01 + 230-6243). *Computing, diffractometry, instrumentation, powder diffraction.*

Cameron, Dr Allan Forbes. Dept. of Chemistry, U. of Glasgow, Glasgow G12 8QQ, Scotland. (1943) PhD, crystallography (U. Glasgow, 1968). Sr. lect. (tel. 041 + 339-8855, ext. 7133). *Structure, organic compounds, ylides, biologically active molecules, structure and properties (chemical - physical - biological).*

Campbell, Dr John Wilson. SBT Divn., SERC Daresbury Lab., Keckwick Lane, Daresbury, Warrington, Cheshire WA4 4AD, England. (1944) PhD, chemistry (U. Edinburgh, 1969). Computational Scientist. (tel. 0925 + 603000, ext. 3528; email JWC at UK.AC.DL.DLVD). *Computing, crystallographic data, macromolecules, proteins, structural biology, viruses, X-ray diffraction, Laue diffraction.*

Cardyn, Dr. Christine Janet. Chemistry Dept., Trinity College, Dublin 2, Ireland. (1947) PhD, Chemistry (Sussex, 1973) Lect. (tel. 01-772941, ext. 2026, email CCARDIN at VAXI.TCD.IE). *Organometallics, clusters.*

Carlisle, Prof. Charles Harold. 12 Marney Rd., London SW11 5EP, England. (1911) DPhil, crystallography (U. Oxford, 1943). Retired Head, Dept. of Crystallography, Birkbeck C., London. *Organic structures, biologically important molecules.*

Carter, Mr Trevor John. Analytical Services Dept., BICC Res. and Eng. Ltd., 38 Wood Lane, London W12 7DX, England. (1938) BSc, physics (U. Leicester, 1959). Technical off. (tel. 01 + 743-1212, ext. 385). *Structure, metals, minerals, polymers.*

Cartwright, Dr Michael. R.M.C.S., Shrivenham, Swindon, Wilts., England. (1940) PhD, chemistry (U. London, 1974). *Inorganic structural chemistry, defect structures, radiation damage.*

Cebula, Dr D. 18 Blakes Way, Bushmead Fields, Eaton Socon, Cambs, England.

Cernik, Dr Robert Joseph. Process Analysis Dept., Ferranti Electronics Ltd., Fields New Rd., Chadderton, Oldham OL9 8NP, England. (1954) PhD, physics (U. Wales, 1984). Res. physicist. (tel. 061 + 682-6844, ext. 226). *Crystal structure analysis, lattice defects, semiconductors.*

Champion, Dr John Anthony. Div. of Materials Applications, Nat. Physical Lab., Teddington, Middx. TW11 0LW, England. (1930) PhD, physics (U. London, 1961). SPSO. (tel. 01 + 977-3222, ext. 4284, telex 262344). *Crystal physics, surface effects, crystal growth.*

Champness, Dr John Norman. Physical Sciences Dept., Wellcome Foundation Ltd., Langley Court, Beckenham, Kent BR3 3BS, England. (1943) PhD, biophysics (U. London, 1968). Sr. res. scientist. (tel. 01 + 658-2211, ext. 5272; fax 01 + 658-2278). *Macromolecules (structure, function and interaction with ligands), Biological activity - structure relation, computing, proteins, viruses.*

Champness, Dr Pamela Eileen. Dept. of Geology, U. of Manchester, Oxford Rd., Manchester M13 9PL, England. (1942) PhD, mineralogy (U. Cambridge, 1969). Sr. lect. (tel. 061 + 275-3808; telex 666517 UNIMAN G; fax 061 + 275-5584). *Electron diffraction & microscopy, phase transformations, minerals.*

Cheetham, Prof. Anthony Kevin. Chemical Crystallography Lab., U. of Oxford, 9 Parks Rd., Oxford OX1 3PD, England. (1946) DPhil, chemistry (U. Oxford, 1971). Lect. (tel. 0865 + 57387, ext. 66). *Solid-state chemistry, X-ray and neutron diffraction, electron microscopy.*

Cherns, Dr David. H.H. Wills Physics Lab., Bristol U., Tyndall Ave., Bristol BS8 1TL, England. (1948) PhD, metal physics (U. Cambridge, 1974). Lect. (tel. 0272 + 303030, ext. 3646; fax 0272 + 732657). *Electron microscopy, electron diffraction, defect structures, interfaces, metals, semiconductors, materials science.*

Chisholm, Dr James Edwin. Dept. of Mineralogy, British Museum (Natural History), Cromwell Rd., London SW7 5BD, England. (1945) PhD, mineralogy (U. Manchester, 1973). SSO. (tel. 01 + 938-8816) *Minerals, X-ray diffraction, powder diffraction, electron diffraction, electron microscopy, defect structures.*

Christian, Prof. John Wyrill. Dept. of Metallurgy and Sci. of Materials, U. of Oxford, Parks Rd., Oxford OX1 3PH, England. (1926) DPhil, metallurgy (U. Oxford, 1949). Prof. of physical metallurgy. (tel. 0865 + 59981, ext. 227). *Phase transformations, metals and alloys, lattice defects.*

Clay, Mrs Kathleen. Materials Res., Rolls-Royce PLC, PO Box 31, Derby, DE2 8BJ, England. (1960) MA, Metallurgy and Materials Sci. (Cambridge, 1986) Res. Metallurgist (tel. 0332 + 240210). *X-ray diffraction, texture, metals, materials science, metal phase determination, single crystal orientation, residual stress.*

Clegg, Dr William. Dept. of Inorganic Chemistry, The U., Newcastle upon Tyne NE1 7RU, England. (1949) ScD, Chemistry (U. of Cambridge, 1989). Reader. (tel. 091 + 222-6649; telex 53654; fax 091+222-6929; email W.CLEGG at UK.AC.NEWCASTLE). *Coordination compounds, computing, diffractometry, structural chemistry, X-ray diffraction, organic compounds.*

Clewer, Mr Peter John. Jules Thorn Lighting Lab., Chemical Res. Dept., Thorn EMI Lighting Ltd., Gt. Cambridge Rd., Enfield, Middx. EN1 1UL, England. (1930) MSc, physics (U. London, 1959). Res. physicist (tel. 01 + 363-5353, ext. 2418). *Phase identification, inorganic systems, X-ray diffraction techniques.*

Cochran, Prof. William. Dept. of Physics, U. of Edinburgh, Mayfield Rd., Edinburgh EH9 3JZ, Scotland. (1922) PhD, chemistry (U. Edinburgh, 1946). Prof. (tel. 031 + 667-1081, ext. 2771). *Solid state physics.*

Cohen, Dr L. 9 Limewood Close, London W13 8HL, England.

Cook, Dr David Stanley. Dept. of Metallurgy and Materials Eng., City of London Polytechnic, Central House, Whitechapel High St., London E1 7PF, England. (1938) PhD, crystallography (U. London, 1968). Sr. lect. (tel. 01 + 283-1030, ext. 470). *X-ray crystal structure analysis, computing, teaching.*

Cooper, Dr Malcolm John. Dept. of Physics, U. of Warwick, Coventry CV4 7AL, England. (1944) PhD, physics (U. Cambridge, 1967). Reader. (tel. 0203 + 523523, ext. 3379; telex 312331; fax 0203 + 692016). *Electron density, momentum density, Compton scattering.*

Cooper, Dr Martyn John. Materials Physics and Metallurgy Div., Building 521.2, AERE Harwell, Didcot, Oxon OX11 0RA, England. (1935) PhD, physics (U. Cambridge, 1962). PSO. (tel. 0235 + 24141, ext. 5184, telex 83135). *Composite properties, high-accuracy diffraction techniques, reliability, techniques.*

Costello, Mr Bernard Anthony de Lacy. Chem. Eng. and Chem. Tech., Imperial C., Prince Consort Rd., London, SW7 2BY, England. (1954) MSc, crystallography (U. London, 1986). Res. Chemist. (tel. 01 + 589-5111, ext. 4348). *Macromolecules, materials science, polymers, structural chemistry.*

Cousins, Mr Christopher Stanley George. Dept. of Physics, U. of Exeter, Stocker Rd., Exeter EX4 4QL, England. (1934) MA, physics (U. Oxford, 1958). Sr. lect. (tel. 0392 + 264116, email COUSINS.CSG at UK.AC.EXTER). *Structure, stress, electric field effects, electron density, synchrotron radiation applications.*

Cowlam, Dr Neil. Dept. of Physics, The Hicks Building, U. of Sheffield, Sheffield S3 7RH, England. (1941) PhD, magnetism (U. Sheffield, 1968). Lect. (tel. 0742 + 78555, ext. 4295, telex Unilib Sheff. 54348). *Magnetic crystallography, neutron diffraction, X-ray diffraction, metals and alloys, metallic glasses.*

Cox, Dr Philip John. Sch. of Pharmacy, Robert Gordon's Inst. of Techn., Schoolhill, Aberdeen AB9 1FR, Scotland. (1947) PhD, crystallography (U. Glasgow, 1972). Lect., physical pharmaceutical chemistry. (tel. 0224 + 633611, ext. 495). *Natural products, sesquiterpenoids, drug molecules, steroids, peptides.*

Craig, Mr G.R. 44 Litherland Park, Litherland, Merseyside L21 9HR, England.

Crennell, Mrs K.M. Atlas Centre, Rutherford Appleton Lab., Chilton, Didcot, Oxon OX11 0QX, England.

Crennell, Ms. Susan Jane. Chemical Crystallography Lab., U. of Oxford, 9 Parks Rd., Oxford, OX1 3PD, England. (1966) BSc, chemistry (U. Bristol, 1988). Postgrad. (tel 0865 + 270823). *Inorganic crystals, solid state lasers, neutron diffraction, X-ray diffraction, structural chemistry.*

Cressey, Dr Barbara Anne. 19 Crendon Court, Caversham, Reading RG4 8BE, England. (1953) PhD, electron microscopy (U. Manchester, 1977). (tel. 0734 + 483335). *Hydrous silicate structures.*

Cressey, Dr Gordon. Dept. of Mineralogy, British Museum (Natural History), Cromwell Rd., London SW7 5BD, England. (1952) PhD, mineralogy (U. Manchester, 1979). Sr. res. fellow. (tel. 01 + 589-6323, ext. 385). *Silicate garnets, epidotes.*

Crocker, Prof. Alan Godfrey. Dept. of Physics, U. of Surrey, Guildford, Surrey GU2 5XH, England. (1935) DSc, physics (U. London, 1971). Reader. (tel. 0483 + 71281, ext. 553). *Theory, crystal defects, metals, crystallography, deformation twinning, phase transformations, computer simulation, porosity.*

Cruickshank, Prof. Durward William John. Dept. of Chemistry, UMIST, PO Box 88, Manchester M60 1QD, England. (1924) ScD, crystallography (U. Cambridge, 1961). Em. prof. (tel. 061 + 236-3311, ext. 2647; fax 061 + 228-1249) *Synchrotron methods, macromolecules, structural chemistry.*

Cullen, Mr F.L. Building 393, MDD, AERE Harwell, Didcot, Oxon OX11 0RA, England.

Cummings, Dr Stewart, Dept. of Physics, U. of Manchester, Brunswick St., Manchester, M13 9PL, England. (1953) PhD, Liquids (U. Bristol, 1979). Env. Off. (tel. 061 + 275-4059). *X-ray optics/interferometry, energy dispersive X-ray diffraction.*

Dacombe, Mr Michael H. International Union of Crystallography, 5 Abbey Square, Chester CH1 2HU, England. (1950) BSc, chemistry & earth sciences (U. Leeds, 1972). Tech. Editor. (tel. 0244 + 342878; telex 669755 OFFICE G, Attn. UNICRYSTAL; fax 0244 + 314888). *Information science.*

Darlington, Dr Charles Nicholas Wright. Dept. of Physics, U. of Birmingham, Birmingham B15 2TT, England. (1945) PhD, physics (U. Cambridge, 1971). Res. fellow. (tel. 021 + 472-1301, ext. 2545 or 3470). *Phase transitions, scattering, Mössbauer gamma rays in X-ray crystallography, critical phenomena.*

Davies, Dr John Edward. Cambridge Crystallographic Data Centre, Lensfield Rd., Cambridge CB2 1EW, England. (1947) PhD, chemistry (Monash U., Australia, 1973). Res. asst. (tel. 0223 + 336442; telex 82140 CAMSPL G; fax 0223 + 336362; email JED2 at UK.AC.CAM.CHEMCRYS).

de Meester, Dr Patrice. Imperial College, London SW7, England. (1943) PhD, crystallography (London, 1973) Academic visitor. (tel. 01 + 589-5111) *Chemical crystallography, protein crystallography.*

Dekker, Mr Henri. Enraf-Nonius Ltd., High View House, 165-7 Station Rd., Edgware, Middx. HA8 7JU, England. (1945). (Tel. 01 + 952-1643 or 952-7255 extn 24, fax 01-951 1463). *Instrumentation.*

Delf, Dr Brian William. Dept. of Physics, U. C. Cardiff, PO Box 78, Cardiff CF1 1XL, Wales. (1935) PhD, physics (U. Wales, 1962). Sr. lect. (tel. 0222 + 44211, ext. 2379). *X-ray powder diffractometry, thin films.*

Dent Glasser, Dr Lesley Scott. Dept. of Chemistry, U. of Aberdeen, Old Aberdeen AB9 2UE, Scotland. (1932) DSc, crystallography (U. Aberdeen, 1972). Reader. (tel. 0224 + 40241, ext. 5658, telex 73458 UNIABN G). *Inorganic structures, silicates and aluminates, teaching crystallography.*

Derewenda, Mrs Urszula. Chemistry, U. of York, Heslington, York, Nth Yorkshire Y01 5DD, England. (1957) MSc, Chemistry (Łódź, Poland, 1983). (tel. 904 + 430 000 ext. 2570). *Proteins, macromolecules, structural biology, X-ray diffraction.*

Derewenda, Dr. Zygmunt. Chemistry, U. of York, Heslington, York, Nth Yorkshire Y01 5DD, England. (1953) PhD, Chemistry (Łódź, Poland, 1981). Res. fellow. (tel. 904 + 403 000, ext. 2570). *Proteins, structural biology, macromolecules, viruses, phase determination, crystallographic data, diffractometry, X-ray diffraction.*

Derwent, Mr Frank William. 70 London Rd., Liphook, Hampshire, GU30 7TA, England. (1912) MSc, physics (U. London, 1933). Retired. (tel. 0428 + 722784). *Electron diffraction, electron microscopy, surface structure.*

Diamond, Dr Robert. MRC Lab. of Molecular Biology, Hills Rd., Cambridge CB2 2QH, England. (1929) PhD, physics (U. Cambridge, 1956). Sr. scient. staff. (tel. 0223 + 248011, ext. 210; telex 81532; fax 0223 + 213556; email RD10 at UK.AC.CAM.MRC-LMB). *Proteins, mathematical methods, thermal vibrations.*

Dineen, Mr C. GEC Res. Labs., Hirst Res. Centre, East Lane, Wembley HA9 7PP, England.

Dingley, Dr David Joseph. Dept. of Physics, U. of Bristol, Tyndall Ave., Bristol BS8 1TL, England. (1939) PhD, metallurgy (U. London, 1965). Reader. (tel. 0272 + 303607; fax 0272 + 732657). *Micro diffraction, electron microscopy, thin films, magnetism.*

Dobson, Dr Peter James. Dept. of Eng. Sci., U. of Oxford, Parks Rd., Oxford, OX1 3PJ, England. (1942) PhD, physics (U. Southampton, 1969). Lect. (tel. 0865 + 273000; telex 83295 NUCLOX G; fax 0865 + 273010). *Surface structure, electron diffraction, semiconductors.*

Dodson, Mrs Eleanor Joy. Dept. of Chemistry, U. of York, Heslington, York YO1 5DD, England. (1936) BA(hons), mathematics (New Zealand, 1958). Res. fellow. (tel. 0904 + 432520, fax 0904 410519, email XTAL at UK.AC.YORK.YORVIC and CHEM1 at UK.AC.YORK.VAXB) *Computing, crystallographic data, electron density, phase determination, symmetry.*

Dodson, Prof George Guy. Dept. of Chemistry, U. of York, Heslington, York YO1 5DD, England. (1937) PhD, biochemistry (U. New Zealand, 1962). Prof. (tel. 0904 + 432520. fax 0904 410519, email CHEM1 at UK.AC.YORK.VAXB). *Hydrogen bonding, macromolecules, phase determination, proteins, viruses, X-ray diffraction.*

Doidge-Harrison, Mrs Solange Maria Silva Veloso. Sch. of Appl. Sci., Robert Gordon's Inst. of Techn., St. Andrew St., Aberdeen, AB1 1HG, Scotland. (1948) BA, Chemistry (Open U.) Res. Asst./Lab. Techn. (tel. 0224 + 633611, ext. 527). *Coordination compounds, chemical bonding, structural chemistry.*

Donaldson, Prof John Dallas. Dept. of Chemistry, The City U., Northampton Square, London EC1V 0HB, England. (1935) DSc, chemistry (U. London, 1970). Prof. (tel. 01 + 253-4399). *Structures, main group elements (compounds in lower oxidation states).*

Dougill, Dr Maryon W. 5 Fraser Ave., Horsforth, Leeds LS18 5EA, England. PhD, crystallography (U. Leeds, 1953). Retired. (tel. 0532 + 582472). *Structures, inorganic compounds.*

Dover, Dr Stanley David. Dept. of Biophysics, King's C., 26-29 Drury Lane, London WC2B 5RL, England. (1943) PhD, biophysics (U. London, 1968). Lect. (tel. 01 + 836-8851). *Electron microscopy, three-dimensional reconstruction, image processing, electron microscope tomography, holography.*

Downie, Mr George. Dept. of Geology and Petroleum Geology, U. of Aberdeen, Marischal C., Broad St., Aberdeen AB9 1AS, Scotland. (1931) BSc, geology and mineralogy (U. Aberdeen, 1954). Lect. (tel. 0224 + 273064). *Electron diffraction, teaching, X-ray diffraction.*

Doyle, Mr Michael Joseph. CCDC Cambridge, Lensfield Rd., Cambridge, CB2 1EW, England. (1964) BA, chemistry (U. Cambridge, 1986). Res. worker. (tel. 0223 + 336425; email MD39 at UK.AC.CAMBRIDGE) *Computing, crystallographic data, inorganic crystals, proteins.*

Drew, Dr Michael George Brindley. Dept. of Chemistry, U. of Reading, Whiteknights, Reading, Berks. RG6 2AD, England. (1941) PhD, chemistry (U. London, 1966). Lect. (tel. 0734 + 85123, ext. 7952). *X-ray crystallography.*

Duckett, Mr G.R. 21 Glen Ave, Springboig, Glasgow G32 0DL, Scotland.

Duke, Mr J.R.C. Heatherley, Gypsy Lane, Great Amwell, Ware, Herts, England.

Dunham, Prof A.C. Dept of Geology, U. of Hull, Cottingham Rd, Hull HU6 7RX, England.

Dyson, Mr David John. Electron Metallography Div., British Steel Corp., Moorgate, Rotherham, S. Yorks. S60 3AR, England. (1939) BSc, physics (U. Durham, 1960). Div. head. (tel. 0709 + 60166, ext. 3290, telex 547279). *X-ray and electron diffraction methods, transmission, scanning and high voltage electron microscopy, iron and steel technology, quantitive phase analysis, dust analysis, texture studies, automatic data collection and reduction.*

Edmondson, Mr M. 22 Passmonds Crescent, Rochdale, Lancs. OL11 5AW, England.

Edwards, Dr Anthony John. Dept. of Chemistry, U. of Birmingham, P.O. Box 363, Birmingham B15 2TT, England. (1936) DSc, chemistry (U. Birmingham, 1973). Sr. lect. (tel. 021 + 414-4381). *Inorganic fluorine chemistry, structure determination, structural chemistry, X-ray diffraction.*

Edwards, Dr Ian Arthur Samuel. Northern Carbon Res. Labs., Dept. of Physical Chemistry, U. of Newcastle upon Tyne, Newcastle upon Tyne NE1 7RU, England. (1941) PhD, crystallography (U. Newcastle upon Tyne, 1968). Sr. experimental asst. (tel. 091 + 2328511). *Structure determination, X-ray methods, surface studies, metals, carbon deposits, LEED, AES; teaching crystallography; carbon science; structure and properties, coals, cokes, graphites and carbons.*

Eeles, Mr Wilfred Trefor. Electricity Council Res. Centre, Capenhurst, Chester CH1 6ES, England. (1930) MSc, physics (U. Wales, 1953). Head, materials sci. sect. (tel. 051 + 339-4181, ext. 309). *Powder diffraction identification, defect structure diffraction, graphite and its intercalates.*

Elder, Dr D.P. 37 Bartongate Drive, Barnton, Edinburgh EH4 8BE, Scotland.

Elias, Mr E.E. Biophysics Dept., U. of Leeds, Leeds LS2 9JT, England.

Eliopoulos, Dr Elias Edward. Dept. of Biophysics, U. of Leeds, Woodhouse Lane, Leeds LS2 9JT, England. (1958) PhD, physics and biophysics (U. Leeds, 1986). Post-doc. fellow. (tel. 0532 + 333042; email ELIAS at UK.AC.LEEDS.BIOVAX). *X-ray diffraction, macromolecular crystallography, crystallographic computing, mechanics, graphics, electron density, structural biology.*

Elliott, Dr Gerald Frank. Biophysics Group, Open U. Res. Unit, Foxcombe Hall, Boars Hill, Oxford, England. (1931) PhD, biophysics (U. London, 1960). Prof. of physics. (tel. 0865 + 730031, ext. 285). *Low angle X-ray and neutron diffraction, biological polyelectrolytes - muscle and cornea.*

Elliott, Dr James Cornelis. Child Dental Health, The London Hospital Medical C., Turner St., London E1 2AD, England. (1937) PhD, crystallography (U. London, 1964). Reader in biophysics. (tel. 01 + 377-7000, ext. 3088; fax 01 + 377-7677). *Instrumentation, powder diffraction, minerals.*

Elliott, Dr Robert Brian. Dept. of Geology, U. of Nottingham, University Park, Nottingham NG7 2RD, England. (1921) PhD, petrology (U. Nottingham, 1952). Reader(retired). (tel. 0602 + 56101, ext. 3159). *Minerals, volcanic rocks and metabasites, X-ray diffractometry.*

Embrey, Mr Peter Godwin. Dept. of Mineralogy, British Museum (Natural History), Cromwell Rd., London SW7 5BD, England. (1929) MA, chemistry, mineralogy (U. Oxford, 1954). Curator of minerals (PSO). (tel. 01 + 589-6323, ext. 567). *Descriptive mineralogy, optical and morphological crystallography, history of science.*

Evans, Dr Anthony Meredith. Dept. of Geology, University, Leicester LE18 7RH, England. (1929) PhD, petrology and mining geology (Queen's U., Ontario, 1962). Sr. lect. (tel. 0533 + 554455, ext. 148). *Ore microscopy, industrial mineralogy, mining geology.*

Evans, Dr E.M.H. 3 Waverley Close, Llandough, Penarth, S Glamorgan, Wales.

Evans, Dr John Hedley. Materials Development Div., Harwell Lab., Didcot, Oxon OX11 0RA, England. (1941) DSc, physics (U. Wales). Professional grade 1. (tel. 0235 + 24141, ext. 4166; fax 0235 + 432620). *Metals, radiation damage, inert gases, positron annihilation, electron misroscopy, ion implantation.*

Evans, Dr P.A. Dept. of Ceramics, U. of Leeds, Clarendon Rd., Leeds, LS2 9JT, England.

Evans, Mr R.R. Rhos Dawel, Trerhyngyll, Nr Cowbridge, S Glamorgan, Wales.

Fairclough, Dr D.P. Nicolet Instruments Ltd, Budbrooke Rd, Warwick CV34 5XH, England.

Faerman, Dr Carlos Hugo. Dept. of Theoretical Chemistry, U. Chemical Lab., Lensfield Rd., Cambridge CB2 1EW, England. (1954) PhD (U. Toronto, Canada, 1987). Postdoctoral res. asst. (tel. 0223 + 336504, email CF106 at UK.AC.CAM.PHX). *Intermolecular forces, molecular crystals.*

Falshaw, Dr C.P. Chemistry Dept, U. of Sheffield, Sheffield S3 7HF, England.

Farrar, Prof. Roy Alfred. Dept. of Mechanical Eng., U. of Southampton, University Rd., Southampton SO9 5NH, England. (1939) PhD, metallurgy (U. London, 1967). Prof. (tel. 0703 + 559122, ext. 2891). *Non-stoichiometry in II-VI compounds, phase transformations, metallic systems, intermetallic phases in weld metals.*

Farrugia, Dr L.J. Chemistry Dept, U. of Glasgow, Glasgow G12 8QQ, Scotland.

Faruqi, Dr Abdul Raffey. MRC Molecular Biology Lab., Hills Rd, Cambridge CB2 2QH, England. (1940) PhD, physics (U. London, 1965). Senior sci. (tel. 0223 + 248011, ext. 219). *X-ray detectors, time resolved measurements, structure and function, muscle, fibre diffraction.*

Fawcett, Dr John. Chemistry Dept., Leicester U., University Road, Leicester LE1 7RH, England. (1947) PhD, Prep. and struct. chemistry (U. Leicester, 1980). EO. (tel. 0533 + 522110; fax 0533 + 522200). *Fluorides, noble gases.*

Fejer, Miss Eleonora Eva. Dept. of Mineralogy, British Museum (Natural History), Cromwell Rd., London SW7 5BD, England. (1927) SSO, head, X-ray sect. (tel. 01 + 589-6323, ext. 447/274). *Crystallography, mineralogy.*

Fenn, Dr Ruth Helen. Applied Physics and Physical Electronics, Portsmouth Polytechnic, King Henry I St., Portsmouth PO1 2DZ, England. (1938) PhD, crystallography (U. London, 1964). Principal lect. (tel. 0705 + 827681, ext. 2157). *X-ray and neutron diffraction, materials science.*

Ferguson, Dr Ian Forster. X-ray Analysis Section, UKAEA Northern Res. Labs., Springfields, Preston PR4 0RR, England. (1931) PhD, chemistry (U. London, 1961). Sect. leader (tel. 0772 + 728262, ext. 31219/31450; telex 67545; fax 0772 + 736248) *Computing, crystal growth, crystallographic data, defect structures, inorganic crystals, instrumentation, materials science, texture.*

Fewster, Dr Paul Frederick. Solid State Electronics Div., Philips Res. Labs., Cross Oak Lane, Redhill, Surrey RH1 5HA, England. (1950) PhD, crystallography (U. London, 1977). Head of X-ray diffraction. (tel. 0293 + 785544, ext. 312; telex 877261; fax 0293 + 776495). *Scattering theory, semiconductors, diffractometry, computing, X-ray diffraction, diffuse scattering, defects.*

Fielding, Mr W.D. 5 Openfield Croft, Water Orton, Warwicks B46 1RE, England.

Finney, Prof. John Leslie. Neutron Sci. Div., Rutherford Appleton Lab., Chilton, Didcot, OX11 0QX, England. (1943) PhD, crystallography (U. London, 1968). Head, Neutron Sci. Div. (tel. 0235 + 446287; telex, 83159 RUTHLB G; fax,

0235 + 445720; email JLF%RLDE). *Defect structures, disorder, hydrogen bonding, high pressures, macromolecules, material science, neutron diffraction.*

Fisher, Mr Graham Richard. Aplied Physics Div., GEC Avionics Ltd., Elstree Way, Borehamwood, Herts WD6 1RX, England. (1951) BSc, physics (U. Salford, 1973). Chief Eng. (tel. 01 + 953-2030, ext. 6739). *X-ray topography, X-ray physics, material science.*

Fitzgerald, Dr Alexander Grant. Dept. of Physics, U. of Dundee, Dundee DD1 4HN, Scotland. (1939) PhD, physics (U. Cambridge, 1964). Lect. (tel. 0382 + 23181, ext. 297). *Electron spectroscopy.*

Fleet, Dr Stephen George. Downing College, U. of Cambridge, Cambridge CB2 1 DQ, England. (1936) PhD, crystallography (U. of Cambridge, 1962). Registrary. (tel. 0223 + 332294). Member: Finance Committee of IUCr. *Crystal structure, electron microscopy.*

Fletcher, Dr Steven Reginald. Technical Dept., ICI plc, Mond Div., P.O. Box 8, The Heath, Runcorn, Cheshire WA7 4QD, England. (1946) PhD, chemical crystallography (U. London, 1972). Sr. res. scient. (tel. 0928 + 513445). *Powder diffraction, line profile analysis, powder structure analysis.*

Flewitt, Prof. Peter Edwin John. Scientific and Technical Branch (south), CEGB South Eastern Region, Canal Rd., Gravesend, Kent DA12 2RS, England. DSc, metallurgy (U. London, 1980). Sect. Manager Materials Inspection and Structural Analysis. (tel. 0474 + 351122, ext. 505). *Defect structures, electron diffraction, electron microscopy, metals, materials science, X-ray diffraction.*

Fones, Mr M.D. Building 393, MDD, AERE Harwell, Didcot, Oxon OX11 0RA, England.

Ford, Dr Geoffrey Charles. Molecular Biology and Biotechnology, U. of Sheffield, Western Bank, Sheffield, S. Yorks S10 2TN, England. (1942) DPhil, mathematics and crystallography (U. Oxford, 1969). Reader and Wellcome Trust sen. lect. (tel. 0742 + 768555, ext. 4241; telex 547216 ULSHEF G; fax 0742 + 727949; email BI1GF at UK.AC.SHEF.IBM). *Computing, crystal growth, diffractometry, instrumentation, macromolecules, phase determination, proteins, reaction pathways, structural biology, teaching, X-ray diffraction.*

Forsyth, Prof. John Bruce. Neutron Div., SERC Rutherford Lab., Chilton, Didcot, Oxon OX11 0QX, England. (1932) PhD, physics (U. Cambridge, 1959). SPSO. (tel. 0235 + 446116). *Neutron diffraction, magnetization densities, magnetic photon scattering, chemical bonding, computing, diffractometry, magnetism, spin density.*

Forsyth, Mr V.T. Physics Dept, Keele U., Keele, Staffs, England.

Fox, Mr Bruce Edward. Analytical R and D., Raychem Ltd, Faraday Rd, Dorcan, Swindon SN3 5HH, England. (1945) BTech, chemistry (U. Loughbrough, 1968) Professional Specialist. (tel. 0793 + 482413). *X-ray diffraction, materials science, polymers, powder diffraction.*

Francis, Mr John Godfrey. Dept. of Mineralogy, British Museum (Natural History), Cromwell Rd., London SW7 5BD, England. (1941) BSc, geology (U. London, 1966). SSO. (tel. 01 + 938-9274). *Mineralogy, X-ray powder diffraction, mineral identification and structures.*

Frank, Prof. Sir Frederick Charles. H.H. Wills Physics Lab., U. of Bristol, Royal Fort, Bristol BS8 1TL, England. (1911) DPhil, chemistry (U. Oxford, 1937). Em. prof. (tel. 0272 + 681708). *Crystal growth, crystal defects.*

Franks, Dr Albert. Mechanical and Optical Metrology Div., Nat. Physical Lab., Teddington, Middx. TW11 0LW, England. DSc, physics (U. London, 1976). DCSO. (tel. 01 + 977-3222, ext. 3515). *X-ray optics, X-ray microscopy, instrumentation.*

Franks, Dr Joseph. Ion Tech Ltd., 2 Park St., Teddington, Middx. TW11 0LT, England. (1924) PhD, physics (U. London, 1952). Managing dir. (tel. 01 + 977-9306). *Ion equipment, X-ray optics.*

Freeman, Mr Walter Gerard. Crystallography Grp., GEC, East Lane, Wembley, Middx, HA9 7PP, England. (1956) MSc, crystallography (U. London, 1981). Res. scient. (tel. 01 + 908-9000, ext. 2135). *Powder diffraction, crystallographic data, diffractometry, materials science, X-ray diffraction, semiconductors, superconductors.*

Freer, Dr Andrew Aloysius. Dept. of Chemistry, U. of Glasgow, University Ave., Glasgow, G69 6ET, Scotland. (1946) PhD, direct methods (U. Glasgow, 1980). Res. asst. (tel. 041 + 339-8855, ext. 5495). *Phase determination, direct methods, structure, biologically active compounds, hexa-host clathrates.*

Freundlich, Dr A. Biophysics Sect., Physics Dept, Imperial C., London SW7 2BZ, England.

Fuller, Prof. W. Physics Dept, Keele U, Staffs STY5 5BG, England.

Fulton, Dr William Stephen. H.H. Wills Physics Lab., U. of Bristol, Tyndall Ave, Bristol BS8 1TL, England. (1953) PhD, Biophysics (U. Bristol, 1980). Res. assoc. (Tel. 0272 + 24161, ext. 105) *Biopolymers, polymers, crystallography, molecular modelling.*

Gale, Dr Brian. Div. of Mechanical and Optical Metrology, Nat. Physical Lab., Teddington, Middx. TW11 0LW, England. (1928) PhD, physics (U. Bristol, 1952). PSO. (tel. 01 + 977-3222, ext. 4295). *Electromagnetic scattering theory, optical instrument design theory, crystal surface theory, interface structure theory.*

Gallagher, Dr Kevin Joseph. Dept. of Chemistry, University C. of Swansea, Singleton Park, Swansea SA2 8PP, Wales. (1928) PhD, crystallography (Queen's U. Belfast, 1953). Sr. lect. (tel. 0792 + 205678, ext. 5272; telex 48358 ULSWAN; fax 0792 + 295618). *Solid state reactions, hydrogen bonding, diffusion, new inorganic materials (oxides and hydroxides).*

Galloy, Dr Jean. University Chemical Lab., Lensfield Rd., Cambridge CB2 1EW, England. (1949) PhD, chemistry (U. de Louvain, Belgium, 1975). Res. associate. (tel. 0223 + 336416; email JJG5 at UK.AC.CAM.CHEMCRYS). *Computer programming, chemical graphics.*

Gard, Dr John Alan. Dept. of Chemistry, U. of Aberdeen, Old Aberdeen AB9 2UE, Scotland. (1919) DSc, chemistry (U. Aberdeen, 1973). Res. lect. (tel. 0224 + 40241, ext. 5661). *Electron microscopy, Electron & X-ray diffraction, minerals, inorganic phases, zeolites, calcium silicates.*

Gare, Mr Terence. Alternative Processes Dept., British Steel Technical., Teesside Labs., PO Box 11, Grangetown, Middlesbrough, Cleveland TS6 6UB, England. (1949) MSc, crystallography (London, 1974). Res. officer. (tel. 0642 + 467144, ext. 2411; telex 58347; fax 0642 + 467144, ext. 2527). *Materials science, metals, reaction pathways, surface structure.*

Garner, Prof. Christopher David. Dept. of Chemistry, U. of Manchester, Brunswick St., Manchester M13 9PL, England. (1941) PhD, chemistry (U. Nottingham, 1966). Prof. inorg. chem. (tel. 061 + 275-4653). *Bioinorganic chemistry, coordination chemistry, EXAFS, crystallographic data, inorganic crystals, minerals, proteins, structural chemistry and biology, surfaces.*

Geddes, Dr Alexander John. Astbury Dept. of Biophysics, U. of Leeds, Leeds LS2 9JT, England. (1941) PhD, biophysics (U. Leeds, 1966). Sr. lect. (tel. 0532 + 333041). *Macromolecules, proteins.*

Giles, Mr Raymond Richard. Powder characterisation, Alcan Chemicals LTD., Chalfont Park, Gerrards Cross, Bucks 5L9 0QB, England. (1944) HNC, app. Physics (1966) Sr. physicist. (tel. 0753+887373, ext. 341, telex 847343, fax 0753+889667). *X-ray diffraction, electron & light microscopy, inorganic crystals.*

Gilmartin, Mr M.G.M. Flat 2D, Lennox Court, 14 Sutherland Ave, Mosshead, Bearsden, Glasgow, Scotland.

Gilmore, Dr Christopher John. Dept. of Chemistry, U. of Glasgow, Glasgow G12 8QQ, Scotland. (1946) PhD, chemical crystallography (U. Bristol, 1971). Reader. (tel. 041 + 339-8855, ext. 5497; email GACA41 at UK.AC.GLASGOW.VME). *Phase determination, computing, powder diffraction, electron microscopy.*

Glasser, Prof. Frederik Paul. Dept. of Chemistry, U. of Aberdeen, Meston Building, Meston Walk, Old Aberdeen, AB9 2UE, Scotland. (1929) DSc, chemistry (U. Aberdeen, 1969). Prof. (tel. 0224 + 272906). *Inorganic structures, property-composition-structure relationships, crystal chemistry.*

Glasson, Dr Douglas Royston. Dept. of Environmental Sci., Polytechnic South-West, Drake's Circus, Plymouth PL4 8AA, England. (1926) PhD, chemistry (U. London, 1949). Head of solid-state res. group. (tel. 0752 + 221312, ext. 5387). *Industrial, materials, building, ceramics, refractories, carbons, minerals, crystallinity, microstructure, reactivity.*

Glazer, Dr Anthony Michael. Clarendon Lab., U. of Oxford, Parks Rd., Oxford OX1 3PU, England. (1943) PhD, crystallography (U. London, 1968). Lect. physics. (tel. 0865 + 272290; telex 83154; fax 0865 + 272400; email WONDRE at UK.AC.OXFORD.VAX). *Diffuse scattering, disorder, materials science, phase transitions, symmetry, X-ray diffraction, optical properties, instrumentation, absolute structures.*

Glazier, Mr Edward James. Chemistry I Div., Metropolitan Police Forensic Sci. Lab., 109 Lambeth Rd., London, SE1 7LP, England. (1948) MSc, Chemical Analysis (Thames Poly., 1976). SSO (tel. 01 + 230-6243). *Computing, powder diffraction.*

Glen, Mr John Wallington. School of Phys. and Space Res., U. of Birmingham, PO Box 363,Birmingham B15 2TT, England. (1927) DSc, physics (U. Birmingham, 1981). Reader, ice physics. (tel. 021 + 414-4675; telex 338938 SPAPHYG; fax 021 + 414-6709) *Ice physics.*

Glidewell, Dr Christopher. Dept. of Chemistry, U. of St. Andrews, North Haugh, St. Andrews, Fife KY16 9ST, Scotland. (1944) PhD, chemistry (U. Cambridge, 1970). Reader. (tel. 0334 + 76161, ext. 8393; fax 0334 + 78292; email: CHSCG at UK.AC.ST-AND.SAVA). *Hydrogen bonding, inorganic crystals, molecular crystals.*

Glover, Dr Ian David. Dept. of Physics, U. of Keele, Keele, Staffs., ST5 5BG, England. (1957) PhD, Protein crystallography (U. London, 1984). Lect. (tel. 0782 + 621111, ext. 3942; email MA6 at UK.AC.DARESBURY.DLVD). *Protein crystallography, anomalous scattering, diffuse scattering, biological structures, protein dynamics.*

Goaman, Dr Llawenydd Constance Gwynne. Dept. of Physics, Portsmouth Polytechnic, King Henry I St., Portsmouth PO1 2DZ, England. PhD, physics (U. Wales, 1962). Sr. lect. (tel. 0705 + 827681, ext. 102). *Biologically interesting organic molecules.*

Godden, Ms Manuela Joanna. Internat. Union of Crystallography, 5 Abbey Sq., Chester, CH1 2HU, England. (1966) BSc, Industrial and Nat. Resource Chemistry (Brunel, 1988). Editorial asst. (tel. 0244 + 342878; telex 669755 OFFICE G attn Unicrystal; fax 0244 + 314888)

Goodfellow, Dr Julia Mary. Dept. of Crystallography, Birkbeck C., U. of London, Malet St., London WC1E 7HX, England. (1951) PhD, biophysics (Open U., 1975). Lect. (tel. 01 + 631-6368; fax 01 + 436-8918; email UBCG08A at UK.AC.BBK.CU). *Macromolecular interaction modelling.*

Goodman, Prof. C.H.L. 5 Hollies End, Mill Hill Village, London NW7 2RY, England.

Gould, Dr Robert Ozburn. Dept. of Chemistry, U. of Edinburgh, West Mains Rd., Edinburgh EH9 3JJ, Scotland. (1938) PhD, chemistry (U. St. Andrews, 1963). Sr. lect. (tel. 031 + 667-1081, ext. 3649; telex 727442 UNIVED G; fax 031 + 662-4053). *Molecusar crystals, structural chemistry, coordination compounds.*

Gould, Dr Sheila Elizabeth Buchan. Beevers Miniature Models Unit, Chemistry Dept, U. of Edinburgh, Edinburgh EH9 3JJ, Scotland. (1940) PhD, chemistry (U. Edinburgh, 1965). Director. (tel. 031 + 667-1081, ext. 3405; telex 727442 UNIVED G; fax 031 + 662-4054). *Biological structures, model making.*

Gover, Dr Sheila. Lab. of Molecular Biophysics, U. of Oxford, Rex Richards Bldng., South Parks Rd., Oxford, OX1 3QU, England. (1948) PhD, crystallography (U. Cambridge, 1976). Res. asst. (tel 0865 + 275392). *X-ray diffraction, computing, proteins, structural biology, electron density, macromolecules.*

Grant, Dr Douglas Frank. Warlawbank Cottage, Reston, Berwickshire TD14 5LW, Scotland. (1923) PhD, physics (U. Durham, 1950). Hon. Sr. lect. in physics, St Andrews U. (tel. 03904 + 385). *Crystallographic computing.*

Green, Mr R.S. GTP Eng. Ltd., Station Industrial Estate, Sheppard St, Swindon, Wilts, England.

Grimes, Dr N.W. 246 May Lane, Kings Heath, Birmingham 14, England.

Gutteridge, Mr Walter Alfred. British Cement Assocn., Wexham Springs, Slough SL3 6PL, England. (1931) MSc, crystallography (Birkbeck C., U. of London, 1967). Principal Scient. (tel. 02816 + 2727, ext. 490). *Powder diffraction, minerals, X-ray diffracton, diffractometry, materials science, microscopy.*

Habash, Dr Jarjis. Dept. of Chemistry, U. of Manchester, Manchester, M13 9QL, England. (1943) PhD, X-ray crystallography (U. Sheffield, 1981). Res. associate. *Proteins, X-ray diffraction.*

Hails, Dr J.E. 16 Redewater Rd, Fenham, Newcastle-upon-Tyne NE4 9UD, England.

Halfpenny, Dr Joan Christine. Dept. of Physical Sci., Trent Polytechnic, Clifton La., Nottingham, NG11 8NS, England. (1954) PhD, crystallography (U. Lancaster, 1978). Sr. lect. (tel. 0602 + 418248, extn. 3312). *X-ray diffraction, structural chemistry.*

Hall, Dr Ivan Harold. Dept. of Pure and Appl. Physics, UMIST, PO Box 88, Sackville St., Manchester M60 1QD, England. (1928) PhD, physics (U. London, 1965). Sr. lect. (tel. 061 + 236-3311, ext. 2969). *Crystalline polymer structures, interaction between structure and bulk properties.*

Hall, Mr Norman Michael. Res. and Dev. Lab. (U.K.), Gillette, 454 Basingstoke Rd., Reading RG2 0QE, England. (1939) MA, science of metals (U. Oxford, 1960). Principal scient. (tel. 0734 + 875222, ext. 355). *Precipitation phenomena, steels, thin film structures, electron diffraction.*

Halliwell, Mrs Mary Ann Griffiths. Group RT2312, British Telecom Res. Labs, Martlesham Heath, Ipswich IP10 0SS, England. (1942) MSc, solid state physics (U. of London, 1967). Head of group. (tel. 0473 + 646805; fax 0473 + 646885) *Multiple crystal diffractometry, heteroepitaxial growth, perfection, semiconductors, X-ray topography.*

Hammond, Dr Christopher. Dept. of Materials, U. of Leeds, Leeds LS2 9JT, England. (1942) PhD, metallurgy (U. Leeds, 1968). Sr. lect. (tel. 0532 + 332382). *Materials, electron diffraction, X-ray diffraction, phase transformations.*

Hammonds, Mr T.G. Green Lane, Churt, Farnham, Surrey GU10 2LT, England.

Hamor, Dr Thomas Andrew. Dept. of Chemistry, U. of Birmingham, Birmingham B15 2TT, England. (1930) DSc, chemistry (U. Birmingham, 1974). Lect. (tel. 021 + 414 4434, fax. 5954). *Coordination compounds, organic compounds, structural biology, structural chemistry, X-ray diffraction.*

Hannon, Mrs Rosemary Anne. Dept. of Biochemistry- Molecular Biology and Biotech., U. of Sheffield, Western Bank, Sheffield, S10 2TN, England. (1949) BSc, Biochemistry (U. Belfast, 1971). Res. asst. (tel. 0742 + 768555, ext. 4241; fax 0742 + 727949; email BI1RAH at UK.AC.SHEF.IBM). *Computing, proteins, crystal growth.*

Harding, Mr J.W. BR 131B Railway Techn. Ctr., London Rd, Derby DE2 8UP, England.

Harding, Dr Marjorie Mary. Dept. of Chemistry, U. of Liverpool, Donnan Labs., Grove St., Liverpool, L69 3BX, England. (1934) DPhil, crystallography (U. Oxford, 1961). Sr. lect. (tel. 051 + 794-3535; telex 627095 UNILPL G; fax 051 + 708-6502; email: PXLI at UK.AC.DL.DLVD). *Synchrotron radiation, structural chemistry, proteins, molecular crystals, zeolites, crystal growth.*

Harding, Dr Roger Robertson. Geological Museum, Exhibition Rd., London SW7 2DE, England. (1938) DPhil, geology (U. Oxford, 1962). PSO. (tel. 01 + 589-3444). *Gemmology, mineralogy.*

Hardy, Dr Andrew David. Chemistry Dept., Sultan Qaboos U., PO Box 32486, Al-Khod, Sultanate of Oman. (1946) DPhil, crystallography (U. Sussex, 1971). Asst. Prof. (tel. OMAN 515473). *X-ray powder diffraction (geological samples), X-ray single-crystal diffraction (organic molecules).*

Hargreaves, Dr A. 26 Ley Hey Rd, Marple, Cheshire, England.

Harper, Mr W.H. 4 Katherine Drive, Woodview, Toton, Nottingham, England.

Harries, Mr J.E. 9 New Hey Rd, Cheadle, Cheshire SK8 2AQ, England.

Harris, Miss Deborah. Dept. of Chemistry, U. of Glasgow, Glasgow, G12 8QQ, Scotland. (1964) BA, Chemistry (U. Oxford, 1986). Res. asst. (tel. 041 + 339-8855). *Proteins.*

Harris, Dr Gillian Wendy. Biotechnology and Enzymology Dept., AFRC Inst. of Food Res., Shinfield, Reading, RG2 9AT, England. (1960) PhD (U. Witwatersrand, 1986). SSO. (tel 0734 + 883103; telex 265871; fax 0734 +

884763) *Computing, diffuse scattering, macromolecules, proteins, structural biology, thermal vibrations, X-ray diffraction.*

Harrison, Prof. Pauline May. Dept. of Molecular Biology and Biotechnology, U. of Sheffield, Western Bank, Sheffield S10 2TN, England. (1926) DPhil, crystallography (U. Oxford, 1947). Prof. (tel. 0742 + 768555, ext. 4843; telex 547216 UGSHEF G; fax 0742 + 727949; email BI1PMH at UK.AC.SHEF.IBM). *Protein structure, metalloproteins, iron metabolism, macromolecular assembly, protein crystallisation.*

Hart, Mr Derrik Gordon. Plessey Res. (UK) Ltd., Allen Clark Res. Centre, Caswell, Towchester, Northants, NN12 8EQ, England. (1942) LRIC, analytical chemistry (RIC, 1966) SPSO. (tel 0327 + 50581, ext. 43464; telex 31572; fax 0327 + 53410). *X-ray diffraction, electronic materials.*

Hart, Prof. Michael. Dept. of Physics, Manchester U., Manchester M13 9PL, England. (1938) DSc, physics (U. Bristol, 1970). Prof. (tel. 061 + 275-4105) Joint appointment as Res. Program Coordinator, SERC Daresbury Lab., Warrington WA4 4AD, England. (tel. 0925 + 65000; telex 629609). *Bragg reflection X-ray and neutron optics, X-ray interferometry, polarimetry, synchrotron radiation.*

Harvey, Dr T.A. 57 Guildford Park Ave, Guildford, Surrey GU2 5NN, England.

Hatt, Mr B.A. 23 Tancred Rd, High Wycombe, Bucks, England.

Hatton, Dr Peter David. Dept. of Physics, U. of Edinburgh, Kings Bldngs., Mayfield Rd., Edinburgh, EH9 3JZ, Scotland. (1957) PhD, Chemistry (Leicester U., 1983). Lect. (tel. 031 + 667-1081, ext. 2821). *High-pressure X-ray and neutron diffraction, phase transitions, powder diffraction, surface structure.*

Häusermann, Mr Daniel. Industrial Materials Group, Dept. of Crystallography, Birkbeck C., Malet St., London, WC1E 7HX, England. (1955) PhD, X-ray optics/diffraction (U. London, 1987). Res. off. (tel. 01 + 580-6622, ext. 2453; fax 01 + 436-8918). *X-ray diffraction, diffractometry, instrumentation, metals, material science, high pressures, phase transitions, powder diffraction.*

Hawley, Dr D.M. Bankhead of Keir, Keir, Thornhill, Dumfriesshire DG3 5EB, Scotland.

Haworth, Dr Colin William. School of Materials, U. of Sheffield, Mappin St., Sheffield, S1 3JH, England. (1932) DPhil, metallurgy (U. Oxford, 1958). Hon. sr. lect., res. associate. (tel. 0742 + 768555, ext. 5509; fax 0742 + 768496). *Microscopy, X-ray diffraction, metals.*

Hay, Dr James Neilson. Chemistry Dept., U. of Birmingham, Birmingham B15 2TT, England. (1935) DSc, chemistry (U. Birmingham, 1955). Sr. lect. (tel. 021 + 472-1301, ext. 2719). *Polymer crystal structure, melting and crystallization rate studies, amorphous materials.*

Heavens, Prof. Oliver Samuel. Dept. of Physics, U. of York, Heslington, York YO1 3JX, England. (1922) DSc, physics (U. London, 1964). Prof. (tel. 0904 + 432240; fax 0904 + 432335) *LEED, epitaxy, microcroscopy.*

Helliwell, Prof. John Richard. Dept. of Chemistry, U. of Manchester, Manchester, M13 9PL, England. (1953) DPhil, physics (U. York, 1974). Prof. (tel. 061 + 275-4686; telex 666517 UNIMAN; fax 061 + 275-5082; email MBDSCJH at UK.AC.UMRCC.CMS). *X-ray diffraction, macromolecules, structural chemistry.*

Henderson, Dr Christopher Michael Bradford. Dept. of Geology, U. of Manchester, Manchester M13 9PL, England. (1938) PhD, geochemistry (U. London, 1964). Reader (tel. 061 + 275-3812; fax 061 + 275-5584) *High pressures, minerals, phase transitions, powder diffraction, small angle scattering, X-ray diffraction.*

Henderson, Dr Richard. MRC Lab. of Molecular Biology, Hills Rd., Cambridge CB2 2QH, England. (1945) PhD, Molecular Biology (Cambridge, 1970) Scient. (tel. 0223+248011 extn 215, telex. 81532, fax. 0223+213556, email RHIS at UK.AC.CAM.MRC-LMB). *Proteins, membrane protein structure.*

Henderson, Dr Robert Keith. Chemistry Dept., U. of Glasgow, University Ave., Glasgow, G12 8QQ, Scotland. (1957) DPhil, Physics (U. York, 1987). Res. asst. (tel. 041 + 339-8855 ext. 6579). *Computing, Phase determination, powder diffraction, X-ray diffraction.*

Highcock. Dr Rona Margaret. Chemical Analysis Dept., Glaxo Group Res. Ltd., Greenford Rd., Greenford, Middx., WB6 0HE, England. (1958) PhD, Chemistry (Bristol U., 1982). Sr. res. analyst. (tel. 01 + 422-3434, extn. 3992). *Organic compounds, structural chemistry.*

Hill, Mr Christopher Peter. Chemistry Dept., U. of York, Heslington, York YO1 5DD, England. (1958) BA, Chemistry (U. York, 1980). Res. asst. (tel. 0904 + 59861, ext. 337). *Protein structure, enzyme catalysis.*

Hilleard, Dr Ronald James. Dept. of Physical Sci., The Polytechnic, Wulfruna St., Wolverhampton WV1 1LY, England. (1936) PhD, physics (U. Aston, 1973). Sr. lect. (tel. 0902 + 27371, ext. 126). *Crystallography, magnetic properties, ferrites.*

Hillman, Dr Harold. Unity Lab., U. of Surrey, Guildford, Surrey GU2 5XH, England. (1930) PhD, physiology (U. London, 1958), biochemistry (U. London, 1963). Lab. dir., Reader, physiology. (tel. 0483 + 571281, ext. 573; telex 859331). *Membrane - ionic reactions, tissue - ion affinity.*

Hine, Dr Raymond. Dept. of Physics, U. C. Cardiff, PO Box 78, Cardiff CF1 1XL, Wales. (1934) PhD, crystallography (U. Wales, 1958). Sr. lect. (tel. 0222 + 44211, ext. 2125). *Phase changes, computer programming.*

Hirsch, Prof. Sir Peter. Metallurgy Dept., Oxford U., Parks Road, Oxford, England.

Hitchcock, Dr P.B. Sch. of Molecular Sci., Sussex U., Falmer, Brighton BN1 9QJ, England.

Hockly, Dr M. Group R3 1 3 British Telecom Res. Lab., Martlesham Heath, Ipswich, Suffolk IP5 7RE, England.

Hodgkin, Prof. Dorothy Mary Crowfoot. Chemical Crystallography Lab., U. of Oxford, 9 Parks Rd., Oxford OX1 3PD, England. (1910) PhD, X-ray crystallography (U. Cambridge, 1936). Em. prof. (tel. 0865 + 53387, ext. 293). *Large molecules, biologically interesting molecules.*

Hodgkinson, Mr R.A. QMC, London U., Mile End Rd, London E1 4NS, England.

Hogg, Mr C. I. & A.P., Building 347-2, AERE Harwell, Didcot, Oxon OX11 0RA, England.

Hogg, Dr Joshua Herbert Christopher. Dept. of Physics, U. of Hull, Cottingham Rd., Hull HU6 7RX, England. (1940) PhD, crystallography (U. Hull, 1966). Lect. (tel. 0482 + 46311, ext. 7389 or 7819). *Inorganic and organic crystal structures, liquid crystals, X-ray topography.*

Holben, Mr John. Enraf Nonius Ltd., Highview Hse., 165/7 Station Rd., Edgware, HA8 7JU, England. (tel. 01+952 1643/7255 extn 22, fax. 01+951 1463).

Holland, Miss S. Lab. of Molecular Biophysics, Richards Building, Oxford OX1 3QU, England.

Holt, Dr Ronald Stanley. Dept. of Physics, U. of Warwick, Coventry CV4 7AL, England. (1953) PhD, physics (U. Warwick, 1978). Sr. postdoctoral fellow. (tel. 0203 + 24011, ext. 2377). *Compton X-ray & gamma-ray scattering, Compton scatter imaging & densitometry applications.*

Hornung, Dr George. Dept. of Earth Sci., U. of Leeds, Leeds LS2 9JT, England. (1934) PhD, mineralogy (U. Leeds, 1961). Sr. lect. (tel. 0532 + 431751, ext. 6472). *Optical mineralogy, X-ray diffraction, X-ray spectrometry, electron microprobe, mineral deposits, volcanic rocks.*

Howard, Dr Judith Ann Kathleen. Dept. of Inorganic Chemistry, U. of Bristol, Cantocks Close, Bristol BS8 1TS, England. (1945) DSc, chemistry (U. Bristol, 1986). Sr. res. fellow. (tel. 0272 + 303689; fax 0272 + 251259; email HOWARDJ at UK.AC.BRISTOL.CSA). *Chemical crystallography, neutron diffraction, deformation density studies, cryogenic applications, transition metal polyhydride complexes.*

Howie, Dr Robert Alan. Dept. of Chemistry, U. of Aberdeen, Meston Walk, Aberdeen AB9 2UE, Scotland. (1940) PhD, Inorganic chemistry (U. Aberdeen, 1972). Sr. res. officer. (tel. 0224 + 272907; telex 73458 UNIABN G; fax 0224 + 272921; email: CHE24 at UK.AC.ABERDEEN). *Diffractometry, powder diffraction, structural chemistry, inorganic crystals, organic compounds, computing.*

Howie, Prof. Robert Andrew. Dept. of Geology, King's C., Strand, London WC2R 2LS, England. (1923) ScD, mineralogy (U. Cambridge, 1974). Prof. of mineralogy. (tel. 01 + 836-5454, ext. 2521). *Mineralogy, composition-cell parameter relationships, rock-forming minerals.*

Howlin, Dr Brendan James. Dept. of Chemistry, U. of Surrey, Stag Hill, Guildford, GU2 5XH, England. (1959) PhD, Chemistry (U. Essex, 1984). Lect. (tel. 0483 + 571281, ext. 2592; telex 859331; email CHM096 at UK.AC.SURREY.SYSI). *Computing, coordination compounds, macromolecules, proteins, structural biology, teaching, thermal vibrations, X-ray diffraction.*

Hriljac, Dr Joseph A. Chemical Crystallography Lab., U. of Oxford, 9 Parks Rd., Oxford, OX1 3PD, England. (1960) PhD, inorganic chemistry (Northwestern U., 1986). Postdoc. (tel 0865 + 270823). *Inorganic crystals, materials science, neutron diffraction, powder diffraction, structural chemistry, X-ray diffraction.*

Hubbard, Dr Roderick Eliot. Chemistry Dept., U. of York, Heslington, York YO1 5DD, England. (1956) DPhil,Chemistry(York,1981) Lect. (tel. 0904+432569, fax. 0904+410519, email REH2 at UK.AC.YORK.VAXA) *Computing, proteins, hydrogen bonding, molecular graphics.*

Hudd, Mr R.C. 2 Looks Court, Porthcawl, Mid Glamorgan CF36 3JJ, Wales.

Hughes, Dr Antony Elwyn. Materials Physics and Metallurgy Div., AERE Harwell, Didcot, Oxon OX11 0RA, England. (1941) DPhil, physics (U. Oxford, 1966). Division Head. (tel. 0235 + 24141, ext. 5273). *Defect structure, radiation damage, superionic conductors, oxides, diffusion, positron annihilation.*

Hughes, Dr David Lewis. AFRC-IPSR Nitrogen Fixation Lab. U. of Sussex, Brighton BN1 9RQ, England. (1941) PhD, chemical crystallography (U. British Columbia, Canada, 1971). Crystallographer. (tel. 0273 + 678198). *Structural chemistry, X-ray diffraction, transition-metal complexes.*

Hughes, Mr Thomas Ernest. British Nuclear Fuels plc, Springfields Works, Salwick, Preston PR4 0XJ, England. (1929) HNC, chemistry (Lancs. and Cheshire Insts., 1953). HSO. (tel. 0772 + 728262, ext. 31158). *X-ray powder diffraction - all aspects, powder cameras, automation, electron microprobe analysis.*

Hukin, Dr David Ainsworth. Clarendon Lab., U. of Oxford, Parks Rd., Oxford OX1 3PU, England. (1936) DPhil, crystal growth (U. Oxford, 1966). Sr. res. off. (tel. 0865 + 59291, ext. 271). *Crystal growth, rare earth metals, alloys, intermetallic compounds, high temperature crystal growth techniques, high temperature equipment design and development.*

Hukins, Dr David William Laurence. Dept. of Medical Biophysics, U. of Manchester, Oxford Rd., Manchester M13 9PT, England. (1947) PhD, biophysics (King's C., U. of London, 1972). Sr. lect. (tel. 061 + 275-5139; fax 061 + 275-5145). *Structural biology.*

Hull, Dr Stephen Edward. Fraser Williams (Scientific Systems) Ltd., London House, London Road South, Poynton, Cheshire SK12 1YP, England. (1948) PhD, chemistry (U. Sheffield, 1972). Computer Systems Consultant. (tel. 0625 + 871126, ext. 8). *Chemical and crystallographic data bases.*

Hulme, Mr Ralph. Dept. of Chemistry, U. of St. Andrews, North Haugh, St. Andrews, Fife KY16 9ST, Scotland. (1924) MA, chemistry (U. Oxford, 1948) Sr. lect.(semi-retired). (tel. 0334 + 76161, ext. 8210). *Organometallic structures, antimony and iodine chemistry.*

Humphreys, Prof. Colin John. Dept. of Materials Sci. and Eng., U. of Liverpool, PO Box 147, Liverpool L20 4JS, England. (1941) PhD, physics (U. Cambridge, 1968). Prof. of Materials Eng., Dept. Head. (tel. 051 + 794-2000, ext. 4664/4665; telex 62095 UNILPL G; fax 051 + 794-4655) *Electron diffraction, electron microscopy, electron density, material science, semiconductors.*

Hurley, Mr Patrick Walter. Analytical Dept., Pye Unicam, York St., Cambridge CB1 2PX, England. (1937) HNC, chemistry (Birmingham C. of Advanced Techn., 1959). Sales Manager. (tel. 0223 + 358866, ext. 335). *Analytical chemistry, X-ray diffraction, X-ray spectrometry, optical emission spectrometry.*

Hursthouse, Dr Michael Barry. Dept. of Chemistry, Queen Mary C., Mile End Rd., London E1 4NS, England. (1941) PhD, chemistry (U. London, 1965). Reader, structural chemistry. (tel. 01 + 980-4811, ext. 3717; telex 893750). *X-ray structure analysis, inorganic complexes, complexes with bulky ligands, natural and synthetic organic compounds, inorganic molecular mechanics.*

Husain, Dr Jasmine. Dept. of Crystallography, Birkbeck C., Malet St., London WC1E 7HX, England. (1951) PhD, crystallography (U. London, 1981) Res. off. (tel. 01 + 631-6448). *Protein crystallography, molecular complexes, molecular dynamics, molecular complexes, drug/hormone receptor interactions.*

Hutchings, Dr M.T. Materials Physics Div., Building 521.2, AERE Harwell, Didcot, Oxon OX11 0RA, England.

Hutton, Dr Alan Thomas. Dept. of Chemistry, Queens U., Belfast, BT9 5AJ, N. Ireland. (1953) PhD, chemistry (U. Capetown, South Africa, 1980) *Chemical crystallography, organometallic and coordination chemistry.*

Huxley, Dr Hugh Esmor. Dept. of Structural Studies, MRC Lab. of Molecular Biology, Hills Rd., Cambridge CB2 2QH, England. (1924) ScD, biology (U. Cambridge, 1964). Scient. staff, MRC. (tel. 0223 + 48011). *Molecular biology, physiology.*

Iball, Prof. John. Dept. of Chemistry, U. of Dundee, Dundee DD1 4HN, Scotland. (1907) DSc, X-ray crystallography (U. Wales. 1939). Hon. res. fellow. (tel. 0382 + 23181, ext. 284). *Crystallography, organic compounds.*

Ingram, Mrs Lorna. 'Kenmore', 40 Manor Place, Cults, Aberdeen AB1 9QN, Scotland. (1943) MSc, chemistry/crystallography (U. Aberdeen, 1968). Res. fellow (U. Aberdeen). (tel. 0224 + 861183). *Structure determination, transition-metal compounds.*

Isaac, Dr D.H. Dept of Metallurgy, U. C. of Swansea, Swansea SA2 8PP, Wales.

Isaacs, Prof. Neil William. Dept. of Chemistry, U. of Glasgow, Glasgow, G12 8QQ, Scotland. (1945) PhD, chemistry (QLD, 1970). Prof. of protein crystallography. (tel. 041 + 339-8855, ext. 5945). *Proteins, structural biology, computing.*

Isherwood, Dr Brian James. Materials Sci. Lab., GEC Hirst Res. Centre, East Lane, Wembley, Middx. HA9 7PP, England. (1941) PhD, physics (Brunel U., 1970). Lab. Manager. (tel. 01 + 904-1262, ext. 291). *X-ray diffraction, characterization, growth, perfection, single crystals, polycrystals, epitaxial layers.*

Isherwood, Mrs S.A. 19 Ilmington Rd, Kenton Harrow, Middx, England.

Jack, Prof. Kenneth Henderson. 147 Broadway, Cullercoats, North Sheilds, Tyne and Wear, NE30 3TA, England. (1918) ScD, materials science (U. Cambridge, 1978). Consultant, Cookson Group plc. (tel. 091 + 262-2211, ext. 264; 091 + 257-3664 (home); telex 537357 ZIRCON G; fax 091 + 263-3847). *Nitrogen ceramics, oxynitride glasses, nitrogen steels, interstitial alloys.*

Jakubovics, Dr John Paul. Dept. of Metallurgy and Sci. of Materials, U. of Oxford, Parks Rd, Oxford OX1 3PH, England. (1938) PhD, physics (U. Cambridge, 1965) U. lect., metallurgy. (tel. 0865 + 59981). *Electron microscopy, magnetic materials, micromagnetism, field ion microscopy, atom probe microanalysis.*

Jaswon, Prof. Maurice Arthur. Dept. of Mathematics, The City U., St. John St., London EC1V 0HB, England. (1922) PhD, applied mathematics (U. Birmingham, 1949). Prof. *Dislocations, space groups, elastomers.*

Jayaweera, Dr Shanath Amarasiri Arumabadu. Chemistry Dept., Teeside Polytechnic, Borough Road, Middlesbrough, Cleveland TS1 3BA, England. (1938) PhD, chemistry (U. London, 1969). Head of inorganic chemistry sect. (tel. 0642 + 218121, extn. 4188). *Inorganic structures, thermal decomposotions, electron microscopy, inorganic crystals, materials science, minerals, powder diffraction, teaching, X-ray diffraction.*

Jeffreys, Dr John Alexander David. Dept. of Chemistry, U. of Strathclyde, 295 Cathedral St.,Glasgow G1 1XL, Scotland. (1927) DPhil, chemistry (U. Oxford, 1952). Lect. (tel. 041 + 552-4400, ext. 2259). *Computing, organic compounds, X-ray diffraction.*

Jeyaratnam, Mr Mailoo. Occupational Medicine and Hygiene Lab., Health and Safety Exec., 403-405 Edgeware Rd., London, NW2 6LN, England. (1929) MSc, Forensic sci (U. Strathclyde, 1967). SSO. (tel. 01 + 450-8911, ext. 359). *X-ray diffractometry, X-ray spectroscopy, microscopy, crystallographic data, inorganic crystals, instrumentation, minerals, phase determination, powder diffraction.*

Johnson, Dr David Julian. Dept. of Textile Industries, U. of Leeds, Leeds LS2 9JT, England. (1936) PhD, textile physics (U. Leeds, 1965). Reader, textile physics. (tel. 0532 + 431751, ext. 6026). *Fibre structures, high modulus organic and inorganic fibres, profile analysis, diffraction patterns, high resolution electron microscopy.*

Johnson, Dr Louise Napier. Lab. of Molecular Biophysics, Dept. of Zoology, U. of Oxford, South Parks Rd., Oxford OX1 3QU, England. PhD, biophysics (U. London, 1965). Lect. (tel. 0865 + 275371). *Protein crystallography.*

Johnson, Dr Michael William. Neutron Div., SERC Rutherford Lab., Chilton, Didcot, Oxon. OX11 0QX, England. (1944) PhD, crystallography (U. London, 1971). Res. scient. (tel. 0235 + 21900, ext. 5418). *Molecular solids, phase transitions, liquid structure-dynamics.*

Johnson, Dr N.P. Chemistry Dept, Portsmouth Polytechnic, White Swan Rd, Portsmouth, Hants, England.

Johnson, Mr Owen. Cambridge Crystallographic Data Centre, Lensfield Rd., Cambridge, CB2 1EW, England. (1958) PhD, Chemical crystallography (U. Bradford, 1985). SO (tel. 0223 + 336442; email OJ100 at UK.AC.CAM.CHEMCRYS). *Crystallographic data, instrumentation, X-ray diffraction, organic compounds.*

Johnson, Dr Peter Anthony Victor. Fuel and Materials Div., UKAEA Springfields Nuclear Power Dev. Labs., Salwick, Nr Preston, Lancs. PR4 0RR, England. (1953) PhD, physics (U. Reading, 1981). Res. chemist. (tel. 0772 + 72862, ext. 31012). *Oxide glasses, structure, neutron diffraction.*

Jones, (Milne) Dr Angela Alice. Dept. of Soil Sci., U. of Reading, London Rd., Reading, Berks. RG1 5AQ, England. (1927) PhD, geology and mineralogy (U. Aberdeen, 1952). Lect. (tel. 0734 + 875123, ext. 7896; telex 847813; fax 0734 + 314404). *Clay minerals, poorly crystalline hydrated silica, alumina and iron oxide, biogenic crystals.*

Jones, Mr D. Nicolet Instr. Ltd, Budbrooke Rd, Warwick CB34 5HX, England.

Jones, Prof. Derry Wynn. Dept. of Chemistry and Chemical Techn., U. of Bradford, Bradford BD7 1DP, England. (1928) DSc, chemistry (U. Bradford, 1980). Hon. Prof., appl. structural chemistry. (tel. 0274 + 733466, ext. 317 or 480; telex 51309 UNIBFD G; fax 0274 + 305340). *X-ray and neutron diffraction, NMR, fuel constituents, carcinogenic polycyclics, nucleosides, carbides.*

Jones, Dr G.R. E707 RSRE, St Andrews Rd, Great Malvern, Worcs WR14 3PS, England.

Jones, Dr William. Chemistry Dept., U. of Cambridge, Lensfield Rd, Cambridge CB2 1EW, England. (1949) PhD, Chemistry (U. Wales, 1974). Asst. Res. Dir. (tel 0223 + 336468; telex 81240; fax 0223 + 336362; email WJ10 at UK.AC.CAM.PHX). *Electron microscopy, materials science, molecular crystals, reaction pathways, structural chemistry.*

Kamminga, Dr H. 178 Sturton St, Cambridge CB1 2QF, England.

Kay, Dr Herbert Frederick. Dept. of Physics, U. of Bristol, Tyndall Ave., Bristol BS8 1TL, England. (1923) PhD, physics (U. Manchester, 1947). Sr. lect. (tel. 0272 + 24161, ext. 103). *Electrical and magnetic materials, liquids, physics of sailing, electronics.*

Kelly, Dr Anthony. U. of Surrey, Guildford, Surrey GU2 5XH, England. (1929) ScD, physics (U. Cambridge, 1967). Vice chancellor. (tel. 0483 + 571281, ext. 696). *Ceramics, metals.*

Kelly, Mr Eric. Electricity Council Res. Centre, Capenhurst, Chester CH1 6ES, England. (1939) BA, earth science (The Open U., 1976). Technologist. (tel. 051 + 339-4181, ext. 266). *Industrial crystallographic applications, corrosion.*

Kempster, Dr Charles John Edgar. Dept. of Pure and Appl. Physics, UMIST, PO Box 88, Manchester M60 1QD, England. (1932) PhD, physics (U. Cambridge, 1958). Lect. (tel. 061 + 228-7040, ext. 3913; te.ex 666094; fax 061 + 228-7040). *X-ray diffraction and small angle scattering from polymers, texture.*

Kennard, Dr Olga. Cambridge Crystallographic Data Centre, University Chemical Lab., Lensfield Rd., Cambridge CB2 1EW, England. (1924) ScD, crystallography (U. Cambridge, 1971). Dir., Crystallographic Data Centre; external staff, MRC. (tel. 0223 + 336408; fax 0223 + 336362). *Crystallographic data, macromolecules, structural biology, structural chemistry.*

Kennedy, Miss D.A. Chem. Res., Glaxo Group Res. Ltd., Ware, Herts., SG12 0DJ, England.

Kennedy, Mr John Matthew. Dept. of Chemistry, U. of Kent, Giles Lane, Canterbury, CT2 7NH, England. (1964) BSc, Chemistry (U. Kent, 1986). Postgrad. (tel. 0227 + 764000, ext. 3995; telex 965449; fax 0227 + 459025; email: CHG404 at UK.AC.UKC.SATURN). *Representation theory, group theory, phase transitions.*

Kerr, Mr Ian Segrave. Dept. of Chemistry, Imperial C., London SW7 2AY, England. (1929) PhD, electron diffraction (U. London, 1955). Lect. (tel. 01 + 589-5111, ext. 4508). *Structure, alumino-silicates, organometallic complexes and clathrates.*

Killean, Dr Reginald Cameron Gordon. Sch. of Physical Sci., U. of St. Andrews, North Haugh, St. Andrews, Fife KY16 9SS, Scotland. (1934) PhD, physics (U. St. Andrews, 1962). Sr. lect. (tel. 0334 + 76161, ext. 8402). *Solid state physics.*

King, Dr James Newington. International Union of Crystallography, 5 Abbey Square, Chester CH1 2HU, England. (1937) PhD, physics (Imp. C., London, 1963). Executive secretary. (tel. 0244 + 342878; telex 669755 OFFICE G, attn. UNICRYSTAL; fax 0244 + 314888).

Kipling, Miss Susan Jane. Catalyst Group Res. Dept., ICI Agricultural Div., Billingham, Cleveland, England. MA, mineralogy (U. Cambridge, 1970). Tech. off. (tel. 0642 + 553601, ext. 5419). *Catalysts, line profile analysis.*

Klug, Dr Aaron. Div. of Structural Studies, MRC Lab. of Molecular Biology, Hills Rd., Cambridge CB2 2QH, England. (1926) PhD, physics (U. Cambridge, 1952). Scient. staff, MRC. (tel. 0223 + 48011). *Biological macromolecular structures, viruses, nucleic acids and chromatin; image reconstruction techniques, electron microscopy.*

Knight, Mr Kevin Stephen. Spectroscopy Branch, BP Res. Centre, Chertsey Rd., Sunbury-on-Thames, Middlesex, TW16 7LN, England. (1956) MSc, crystallography (U. London, 1982). (tel. 093276 + 3276). *X-ray topography, mineralogy, structure solution.*

Knight, Mr Robert. Dept. of Physics, U. of Hull, Cottingham Rd., Kingston on Hull, N. Humberside HU6 7RX, England. (1956) BSc, chemistry (U. Hull, 1978). Res. technician. (tel. 0482 + 497389). *Organic and inorganic crystal structures, liquid crystals, polmer liquid crystals, topography.*

Knott, Mr P.R. 68 Dorset Ave, Great Baddow, Chelmsford, Essex CM2 9UA, England.

Körber, Dr. Fritjof Carl Friedrich. Dept. Chemistry, U. of York, York YO1 5DD, England. (1952) PhD,Biophysics(Leeds U.,1984) Res. Asst. (tel. 0904+432590). *X-ray diffraction, proteins, crystal growth, crystallographic data, instrumentation, macromolecules, structural biology.*

Krohn, Dr A. 36 Park View Gardens, London NW4, England.

Kuroda, Dr Reiko. Dept. of Biophysics, King's C., 26-29 Drury Lane, London WC2B 5RL, England. (1947) PhD, chemistry (U. Tokyo, 1975). Hon. lect., res. fellow. (tel. 01 + 836-8851). *Crystal and molecular structures, chiral compounds, oligonucleotides, biologically active compounds, chiroptical spectroscopy, chiral discrimination, drug-nuclei acid interactions.*

Ladd, Dr Marcus Frederick Charles. Sub-Dept. of Chemical Physics (Chemistry), U. of Surrey, Stag Hill, Guildford, Surrey GU2 5XH, England. (1926) DSc, solid state chemistry (U. London, 1979). Reader (head of sub-dept.). (tel. 0487 + 71281, ext. 427, telex 859331). *Solid state chemistry, crystallographic computing, structure and bonding, ionic crystals, crystal chemistry.*

Laker, Mr Thomas James. Analytical Div., British Gas Corp., London Res. Station, Michael Rd., London SW6 2AD, England. (1945) HNC, chemistry (Borough Poly., 1967). Sr. scient. (tel. 01 + 736-3344, ext.4059). *Industrial X-ray diffraction applications, automatic powder diffractometry, crystallite size studies, LIMS.*

Lang, Prof. Andrew Richard. H.H. Wills Physics Lab., U. of Bristol, Tyndall Ave., Bristol BS8 1TL, England. (1924) PhD, crystallography (U. Cambridge, 1953). Emeritus Prof. (tel. 0272 + 303030, ext. 3626; telex 445938, fax 0272 732657). *Crystal growth, materials science, dynamical diffraction theory.*

Langford, Dr John Ian. Dept. of Physics and Space Res., U. of Birmingham, PO Box 363, Birmingham B15 2TT, England. (1935) PhD, crystallography (U. Wales, 1965). Res. fellow. (tel. 021 + 414-4662, telex 338938 SPAPHY G, fax 414 6709, email XT-1 at UK.AC.BHAM.PH.I). *Powder diffractometry, crystal imperfections, profile fitting, total pattern size-strain analysis, industrial applications.*

Lappin, Miss Alison. Analytical Support, BP Res., Chertsey Rd., Sunbury-on-Thames, Middlesex TW16 7LN, England. (1966) BSc, chemistry (Paisley C. of Techn., 1987). Scient. Asst. (tel. 0932 + 76-3728). *X-ray powder diffraction.*

Last, Mr Paul Edward. Analytical Chemistry Dept., Central Res., Pfizer Ltd, Ramsgate Road, Sandwich, Kent CT13 9NJ, England. (1954) GRSC (Royal Society of Chemistry, 1979). Sr. res. scient. (tel. 0304 + 616621). *Pharmaceutical crystallographic applications, polymorphism, phase changes.*

Lawson, Mr David Mark. Dept. of Biochemistry, U. of Sheffield, Western bank, Sheffield, S. Yorkshire S10 2TN, England. (1963) BSc, biochemistry (Sheffield, 1985). Res. asst. (tel. 0742 + 768555, ext. 4190; telex 547216 UGSHEF G, fax 742 727949). *Computing, crystal growth, macromolecules, proteins, structural biology, X-ray diffraction.*

Lawson, Mr Robert Ian. Petrographical Dept., British Geological Survey, Murchison House, West Mains Rd., Edinburgh EH9 3LA, Scotland. (1925) BSc, geology (U. Edinburgh, 1952). Principal geologist. (tel. 031 + 667-1000). *X-ray and electron probe, rocks, minerals.*

Leadbetter, Prof. Alan James. SERC Daresbury Lab., Daresbury, Warrington, Cheshire WA4 4AD, England. (1934) DSc, chemistry (U. Bristol, 1971). Dir. (tel. 0925 + 603119, fax 0925-603100). *Diffractometry, disorder, molecular crystals, liquid crystals, phase transitions and X-ray diffraction.*

Leake, Dr John Anthony. Dept. of Metallurgy and Materials Sci., U. of Cambridge, Pembroke St., Cambridge CB2 3QZ, England. (1939) PhD, crystallography (U. Cambridge, 1965). Lect. (tel. 0223 + 334331, telex 81240 CAMSPL G, fax 0223 334748). *Alloys, intermetallic compounds, ordering, rapid-quenching, amorphous materials, ceramics - technical, teaching.*

Lee, Dr John David. Dept. of Chemistry, Loughborough U. of Techn., Loughborough, Leics. LE11 3TU, England. (1931) PhD, chemistry, crystallography (U. Nottingham, 1959). Sr. lect. (tel. 0509 + 263171, ext. 362). *Crystal structure determination, computer programming.*

Leslie, Dr Andrew Greig William. M.R.C. Lab. of Molecular Biology, Hills Rd., Cambridge, CB2 2QH, England. (1949) PhD, crystal structure (U. Manchester, 1975). Scientific Staff. (tel. 0223 + 248011, ext. 212, telex 81532, fax 213556). *Protein crystallography*

Lewis, Dr Eric Leslie Vallance. Dept. of Physics, U. of Leeds, Leeds LS2 9JT, England. (1943) PhD, physics. (U. Birmingham, 1967). Res. fellow. (tel. 0532 + 333861). *Wide and small angle scattering, structure and phase transitions, organic crystals and polymers, computer programming.*

Lewis, Prof. Michael Harold. Dept. of Physics, U. of Warwick, Coventry CV4 7AL, England. (1938) DPhil, materials science (U. Oxford, 1964). Prof, physics. (tel.

0203 + 523392, fax 0203 692016). *Ceramics, electron microscopy, electron diffraction, solid state NMR spectroscopy.*

Li, Dr Jade. Structural Studies Division, Medical Res. Council, Molecular Biology Lab., Hills Rd., Cambridge CB2 2 QH, England. Ph.D., Biophysics (Harvard U., 1978) Scientific staff. (tel. 0223 + 245133 ext. 223, telex. 81532, fax 0223 + 245133, email: JL at UK.AC.CAM.MRC-LMB) *Macromolecules, structural biology, proteins, X-ray diffraction, electron microscopy.*

Liddell, Dr Katharine. Dept. of Mechanical Materials, U. of Newcastle, Newcastle Upon Tyne NE1 7RU, England. (1949) PhD, crystal chemistry (U. of Newcastle, 1984). Res. asst. (tel. 091 + 2328511, ext. 7206; telex 53654 UNINEW G, fax 091 2611182). *Nitrogen ceramics, sialons, phase relationships, X-ray techniques, powders.*

Lindley, Dr Peter Frank. Dept. of Crystallography, Birkbeck C., Malet St., London WC1E 7HX, England. (1942) PhD, chemistry (U. Bristol, 1966). Reader. (tel. 01 + 631-6422, fax 01 436 8918). *Proteins, X-ray diffraction, structural biology, structural chemistry, teaching.*

Lipson, Prof. Henry, FRS. Dept. of Physics, UMIST, PO Box 88, Manchester M60 1QD, England. (1910) DSc, physics (U. Liverpool, 1930). Em. prof. (tel. 061 + 236-3311, ext. 2743). *Crystal structure determination, X-ray optics.*

Lorimer, Dr Gordon Winston. Materials Sci. Centre, U. of Manchester, Grosvenor St., Manchester M20 0AJ, England. (1941) PhD, metallurgy (U. Cambridge, 1968). Reader. (tel. 061 + 236-3311, ext. 2390; telex 666094, fax 061-228-7040). *Electron microscopy, electron diffraction, convergent beam diffraction, phase transformations.*

Lovell, Mrs S.E. 131B Railway Technical Centre, London Rd, Derby DE2 8UP, England.

Low, Dr John Nicolson. Depts. of Physics and Chemistry, U. of Dundee, Nethergate, Dundee DD1 4HN, Scotland. (1940) PhD, crystallography (U. Dundee, 1982). SSO. (tel. 0382 + 23181, ext. 4562). *Plant virus structures, nucleoside structures, nucleotide structures.*

Lowde, Dr Raymond Douglas. Materials Physics Div., AERE Harwell, Didcot, Oxon. OX11 0RA, England. (1923) DSc, solid state physics (U. London, 1968). SPSO. (tel. 0235 + 24141, ext. 5101). *Solid state physics, metal physics, magnetism, phase transitions, neutron scattering.*

Lowe, Dr Philip Richard. Dept. of Pharmaceutical Sci., U. of Aston in Birmingham, Goster Green, Birmingham B4 7ET, England. (1948) PhD, crystallography (U. Aston, Birmingham, 1984). Dept. Superintendent. (tel. 021 + 359-3611, ext. 4190; telex 336997 UNIAST G, fax 021-333-3172). *X-ray diffraction, structural chemistry, organic compounds, hydrogen bonding.*

Lowe, Ms Susan Elizabeth. International Union of Crystallography, 5 Abbey Square, Chester CH1 2HU, England. (1950) BSc, biophysics (U. Leeds, 1974). Asst. Tech. Editor. (tel. 0244 + 342878, telex 669755 OFFICE G attn. UNICRYSTAL, fax 0244 314888). *Editing.*

Lydon, Dr John Ennis. Astbury Dept. of Biophysics, U. of Leeds, Leeds LS2 9JT, England. (1940) PhD, chemistry (U. Leeds, 1966). Lect. (tel. 0532 + 431751, ext. 6443). *Structural investigations, liquid crystal systems, mesogenic compounds.*

Lyons, Dr Michael Hamilton. R3.1.2, British Telecom Res. Labs., Martlesham Heath, Ipswich, Suffolk IP5 7RE, England. (1953) PhD, chemistry (U. London, 1982). Executive Eng. (tel. 0473 + 642619). *Crystal growth, semiconductors, dynamical theory.*

Mackay, Prof. Alan Lindsay. Dept. of Crystallography, Birkbeck C., Malet St., London WC1E 7HX, England. (1926) PhD, physics-crystallography (U. London, 1951), DSc . Prof. (tel. 01 + 580-6622, fax 01 631 6370, email: UBCG04M at UK.AC.BBK.CU). *Quasi-crystals, electron microscopy, structural chemistry, symmetry.*

Mackie, Dr Fiona L. Wrexham Techn. Centre, BICC Cables plc., Wrexham Ind. Estate, Wrexham, Clwyd LL13 9XP, Wales. (1959) PhD, chemistry (U. of Dundee, 1981). Crystallographer. (tel. 0978 + 662345, ext. 2515). *X-ray diffraction, polymers.*

Main, Dr Peter. Dept. of Physics, U. of York, Heslington, York YO1 5DD, England. (1939) PhD, physics (U. Manchester, 1963). Sr. lect. (tel. 0904 + 432265, fax. 0904+432335, email PMI%UK.AC.YORK.VAXB). *Direct methods, crystal structure determination.*

Mallinson, Dr Paul Raymond. Dept. of Chemistry, U. of Glasgow, Glasgow G12 8QQ, Scotland. (1943) PhD, chemistry (U. Essex, 1970). Computer programmer. (tel. 041 + 339-8855, ext. 4409; telex. 777070 UNIGLA, fax 041 330-4808, email P.R.MALLINSON at UK.AC.GLASGOW.VME). *Chemical bonding, computing, electron density, hydrogen bonding, molecular crystals, organic compounds, structural chemistry, X-ray diffraction.*

Malone, Dr John Francis. Dept. of Chemistry, Queen's U., Stranmillis Rd., Belfast BT9 5AG, N. Ireland. (1944) PhD, chemical crystallography (U. Leeds, 1969). Lect. (tel. 0232 + 661111, ext. 4423; telex QUBADM 74487, fax 0232-247895). *Structural chemistry, molecular crystals, X-ray diffraction, molecular modelling.*

Manning, Mr David Charles. Enraf-Nonius Ltd., Highview House, 165/7 Station Road, Edgware, Middlesex HA8 7JU, England. Technical Manager. (tel. 01+952 1643/7255 extn 26, fax. 01+951 1463).

Manojlović-Muir, Dr Ljubica. Dept. of Chemistry, U. of Glasgow, Glasgow G12 8QQ, Scotland. (1931) PhD, crystallography (U. Belgrade, Yugoslavia, 1962). Reader. (tel. 041 + 339-8855, ext. 4506). *X-ray diffraction, neutron diffraction,*

coordination chemistry, structural chemistry, chemical bonding, reaction pathways, hydrogen bonding.

Martin, Dr John Wilson. Dept. of Metallurgy and Sci. of Materials, U. of Oxford, Parks Rd., Oxford OX1 3PH, England. (1926) ScD, metallurgy (U. Cambridge, 1978). Lect. (tel. 0865 + 273711). *Electron microscopy, electron diffraction.*

Marvin, Dr Donald Arthur. Dept. of Biochemistry, U. of Cambridge, Tennis Court Rd., Cambridge CB2 1QW, England. (1934) PhD physics (U. of London, 1960). Sr. Res. asst. (tel. 0223 + 333662 or 353090). *Macromolecules, structural biology, viruses, X-ray diffraction.*

Mason, Dr Kenneth George. Dept. of Chemistry, Loughborough U. of Techn., Loughborough, Leics. LE11 3TU, England. (1928) PhD, organic chemistry (U. Nottingham, 1955). Lect. (tel. 0509 + 263171, ext. 603). *Organic structures.*

Mason, Prof. Sir Ronald, FRS. Sch. of Molecular Sci., U. of Sussex, Brighton, Sussex BN1 9QJ, England. (1930) PhD, crystallography (U. London, 1953). Prof. of chemistry. (tel. 0273 + 606755). *Structural chemistry, surface chemistry.*

Mason, Prof. Stephen Finney. Dept. of Chemistry, King's C., Strand, London WC2R 2LS, England. (1923) DSc, chemistry (U. Oxford, 1945). Prof. (tel. 01 + 836-5454, ext. 2259). *Crystal and molecular structure, chiral compounds, chiroptical spectroscopy, chiral discrimination.*

Matthew, Prof. James Andrew Davidson. Dept. of Physics, U. of York, Heslington, York YO1 5DD, England. (1938) DSc, physics (U. Aberdeen, 1985). Head of Physics. (tel. 0904 + 432000). *Electron scattering theory, surface structure, spin polarised electron diffraction, surface magnetism.*

Mazey, Dr David John. B393 Materials Dev. Div., AERE Harwell, Didcot, Oxon OX11 0RA, England. (1929) DPhil, physical metallurgy (U. Salford, 1975). PSO. (tel. 0235 + 24141, ext 4592). *Electron microscopy, TEM, STEM, SEM, microanalysis, microdiffraction, defects, precipitates, radiation damage, metals.*

McAllister, Mr Patrick Brian. Wrexham Tech. Centre, BICC Cables, Wrexham, Clwyd, LL13 9XP, Wales. (1934) chartered eng. (M. Inst. Metal., 1979). *Structure, materials, metals, polymers.*

McDonald, Dr Walter Stanley. Dept. of Inorganic and Structural Chemistry, U. of Leeds. Leeds LS2 9JT, England. (1933) PhD, crystallography (U. Glasgow, 1965). Sr. lect. (tel. 0532 + 31751, ext. 6062). *Structural chemistry.*

McHardy, Dr William James. Dept. of Mineral Soils, Macaulay Inst. for Soil Res., Craigiebuckler, Aberdeen AB9 2QJ, Scotland. (1936) PhD, chemistry (U. Aberdeen, 1962). PSO. (tel. 0224 + 38611, ext. 255). *Electron microscopy (SEM & TEM), electron probe microanalysis, soil minerals.*

McKie, Dr Christine Hilary. Dept. of Earth Sci., U. of Cambridge, Downing St., Cambridge CB2 3EQ, England. (1931) PhD, crystallography (U. Cambridge, 1958). Lect. (tel. 0223 + 335463, ext. 279). *Mineral structures, solid-state transformations.*

McKie, Dr Duncan. Dept. of Earth Sci., U. of Cambridge, Downing St., Cambridge CB2 3EQ, England. (1930) PhD, mineralogy (U. Cambridge, 1962). Lect. (tel. 0223 + 335463, ext. 294). *Mineral structures, polymorphism, solid state reactions.*

McMahon, Mr Brian. International Union of Crystallography, 5 Abbey Square, Chester CH1 2HU, England. (1956) MA, physics (Oxford, 1978). Editorial asst. (tel. 0244 + 342878, fax 0244 314888). *Publishing*

McPartlin, Dr Mary. Dept. of Chemistry, The Polytechnic of North London, Holloway, London N7 8DB, England. PhD, crystallography (U. New South Wales, Australia, 1966). Sr. lect. (tel. 01 + 607-2789, ext. 2142). *Chemical crystallography.*

Meek, Dr Keith Michael Andrew. Oxford Res. Unit, Open University, Foxcombe Hall, Berkeley Rd., Boars Hill, Oxford OX1 5HR, England. (1951) PhD, biophysics (U. of Manchester, 1976). Lect. (tel. 0865 + 730031, extn. 287). *X-ray diffraction, structural biology, proteins, macromolecules, electron microscopy.*

Megaw, Dr Helen Dick. 22 Dunamallaght Rd., Ballycastle, Co. Antrim BT54 6PB, N. Ireland. (1907) ScD, crystallography (U. Cambridge, 1967). Retired. (tel. 002657 + 62729) *Crystal structures.*

Mendelssohn, Dr Monica Jutta. Crystallography Unit, Geology Dept., U. C. London, Gower St., London WC1E 6BT, England. (1943) PhD, physics (U. London, 1971). Res. fellow. (tel. 01 + 387-7050, ext. 445). *Divergent beam measurements, diamonds, computing.*

Mercer, Dr William Duncan. Dept. of Biochemistry, Medical Biology Centre, Queen's U. of Belfast, Belfast BT9 7BL, N. Ireland. (1947) PhD, biochemistry (U. Bristol, 1972). Lect. (tel. 0232 + 29241, ext. 2795). *Enzymes, structure determination, structure display, protein structure prediction and analysis.*

Merriman, Mr Richard James. Mineral Sci. Res. Group, British Geological Survey, Keyworth, Nottingham NG12 5GG, England. (1943) BSc, geology (U. London, 1970). PSO. (tel. 06077 + 6111, ext. 3130). *X-ray diffraction, electron diffraction (TEM).*

Metcalfe, Dr Edward. Materials Div., Central Electricity Res. Labs., Kelvin Ave., Leatherhead KT22 7SE, England. (1947) PhD, physics, metallurgy (U. Cambridge, 1973). Res. officer. (tel. 037 23 + 74488, ext. 167). *High temperature oxidation, surface analytical techniques, order-disorder phenomena.*

Michell, Mr Ernest William John. Cookson Group plc, 7 Wadsworth Road, Perivale, Greenford, Middx. UB6 7JQ, England. (1935) MSc, crystallography (U. London, 1965). Head of crystallography. (tel. 01 + 997-5635, ext. 15). *X-ray powder diffraction, crystalline phase composition analysis.*

Middleton, Dr Andrew Philip. Res. Labs., British Museum, Gt. Russell St., London WC1B 3DG, England. (1949) DPhil, mineralogy (U. Oxford, 1974). Scient.

(tel. 01 + 636-1555, ext. 282). *X-ray diffraction, scanning microscopy, optical microscopy, archeological materials.*

Milburn, Prof. George Henry William. Dean of Sci. Faculty, Napier Polytechnic., Colinton Rd., Edinburgh EH10 5DT, Scotland. (1934) PhD, crystallography (U. Leeds, 1963). Dean. (tel. 031 + 444-2266, ext. 2463). *Diacetylenes, optoelectronic materials, liquid crystal polymers, molecular crystals.*

Milledge, Dr H. Judith. Crystallography Unit, Dept. of Geology, U. C. London, Gower St., London WC1E 6BT, England. (1927) DSc, crystallography (U. London, 1963). Reader, crystallography. (tel. 01 + 387-7050, ext. 431). *Automation, structure analysis, oscillation photographs, unstable crystals, solid state reactions, high pressure diffraction, high temperature diffraction, diamonds.*

Miller, Prof. Andrew. Dept. of Biochemistry, U. of Edinburgh, Hugh Robson Building, George Square, Edinburgh EH8 9XD, Scotland. (1936) PhD, X-ray crystallography (U. Edinburgh, 1962). Prof. (tel. 031 + 667-1011, ext. 2336; telex 727442 UNIVED G). *Structural biology, small-angle scattering, synchrotron radiation.*

Mills, Mr Owen S. Dept. of Chemistry, U. of Manchester, Manchester M13 9PL, England. BSc, chemistry (U. Liverpool, 1945). Reader. (tel. 061 + 273-7121, ext. 5285). *Chemical crystallography, data retrieval, computer graphics.*

Mitchell, Dr Gary Findlater. Software Development, Cambridge Crystallographic Data Centre, Lensfield Rd., Cambridge CB2 1EW, England. (1961) PhD, crystallography (U. of Edinburgh, 1988). Programmer. (tel. 0223 + 336425, email GFM10 at UK.AC.CAM.CHEMCRYS). *Chemical bonding, computing, crystallographic data, electron density.*

Mitchell, Dr Geoffrey Robert. J.J. Thomson Physical Lab., U. of Reading, Whiteknights, Reading RG6 2AF, England. (1949) PhD, material science, (CNAA). Lect. (tel. 0734 + 318573, telex 847813, fax 0734 750203, email JANET, SPSMITLL at UK.AC.RDG.AM). *Liquid crystals, X-ray diffraction, neutron diffraction, conformational analysis, non-crystalline materials.*

Moody, Dr Peter Charles Edmund. Chemistry Dept., U. of York, Heslington, York YO1 5DD, England. (1956) PhD biophysics (U. of London, 1984). Res. asst. (tel. 0904 + 432589, telex 57933 YORKUL, fax 410519, email MOODY at UK.AC.YORKYORVIC). *Structural biology, proteins, macromolecules, X-ray diffraction.*

Moore, Miss Alice Elizabeth. Tinkers Coppice, Cliff Bridge, Shanklin, Isle of Wight PO37 6QL, England. (1921) MSc, crystallography (Birkbeck C., U. of London, 1954). Retired from Cement and Concrete Res. Assoc. (tel. 0983 + 864270). *Cement mineralolgy, anhydrous and hydrated cement.*

Moore, Dr Anthony Moreton. Dept. of Physics, Royal Holloway and Bedford New C., U. of London, Egham, Surrey TW20 0EX, England. (1943) PhD, physics (U. Bristol, 1973). Sr. lect. (tel. 0784 (Egham) + 439941, telex 935504, fax 0784 437520). *Crystal growth, minerals, semiconductors, X-ray diffraction, X-ray topography.*

Moore, Dr John Carlton. Dept. of Sci. and Techn., Slough C. of Higher Education, Wellington St., Slough, Berks SL1 1YG, England. (1934) PhD, crystallography (U. London, 1985). Sr. lect., appl. physics. (tel. 0753 + 34585, ext. 34). *Solid-state phase changes.*

Moore, Mr Peter Leonard. Techn. Centre, British Steel Corp. Tubes Div., Corby Works, Corby, Northants. NN17 1UA, England. (1941) HNC, applied physics (Northern Polytechnic, London, 1964). Superintendent. (tel. 053 66 + 2121, ext. 4583). *Powder diffraction techniques, quantitative and qualitative analysis, occupational hygiene, electron microscopy.*

Morffew, Dr Andrew James. Molecular Graphics Project, IBM UK Scient. Centre, Athelstan House, St Clement St., Winchester, Hants SO23 9UT, England. (1950) PhD, protein crystallography (U. London, 1981). Sr. Assoc. scient. (tel. 0962 + 68191, ext. 234). *Molecular graphics, refinement, restrained least squares, protein conformation analysis, peptides (small) structures.*

Morgan, Dr Colin Harris. Dept. of Computer Studies, U. of Hull, Cottingham Rd., Hull HU6 7RX, England. (1937) PhD, crystallography (U. St. Andrews, 1964). Lect. (tel. 0482 + 46311, ext. 7295). *Crystallographic computing.*

Morris, Dr Donald Frank Charles. Nuclear Science, Chemistry Dept., Brunel U., Uxbridge, Middx. UB8 3PH, England. (1928) DPhil, chemistry (U. Oxford, 1953). Reader. (tel. 0895 + 37188, ext. 537). *Structure and thermodynamics, inorganic crystals, radiochemistry.*

Moseley, Dr Patrick. Materials Dev. Div., AERE Harwell, Didcot, Oxon OX11 0RA, England. (1943) PhD, chemical crystallography (U. Durham, 1968). Res. staff. (tel. 0235 + 24141, ext. 4262). *Structure, solid state inorganic chemistry.*

Moss, Dr David Stanley. Dept. of Crystallography, Birkbeck C., Malet St., London WC1E 7HX, England. (1941) PhD, chemical crystallography (U. London, 1967). Sr. lect. (tel. 01 + 580-6622, ext. 368). *Protein structure and dynamics, computing, statistics.*

Motherwell, Dr William David Samuel. Automation Office, Cambridge U. Library, West Rd., Cambridge CB3 9DR, England. (1941) PhD, chemistry (U. St. Andrews, 1967). Automation Officer. (tel. 0223 + 61441, ext. 234). *Computer programming, information retrieval, crystal structures, molecular packing, packing energy calculation.*

Muir, Dr Kenneth Walter. Dept. of Chemistry, U. of Glasgow, Glasgow G12 8QQ, Scotland. (1941) PhD, crystallography (U. Glasgow, 1967). Reader. (tel. 041 + 339-8855, ext. 5345; telex 777070 UNIGLA, fax 330 4808). *Transition metals, crystallographic computing, inorganic and organometallic structures.*

Muirhead, Dr Hilary. Dept. of Biochemistry, U. of Bristol, Bristol BS8 1TD, England. (1937) PhD, protein crystallography (U. Cambridge, 1964). Reader. (tel. 0223+ 24161, ext. 1127). *Protein crystallography, enzyme structure and function.*

Murray-Rust, Dr Judith. Dept. of Crystallography, Birkbeck C., Malet St., London WC1E 7HX, England. (1946) PhD, chemistry (U. Stirling, 1971). Res. Asst. (tel. 01 + 580 6622, extn. 2364; email UBCG09J at UK.AC.BBK.CR). *Protein structures.*

Murray-Rust, Dr Peter. Dept. of Chemistry, Glaxo Group Res., Greenford Rd., Middx. UB6 0HE, England. (1941) DPhil, chemistry (U. Oxford, 1967). Head, molecular graphics sect. (tel. 01 + 422-3434). *Molecular geometry analysis, organic molecules, crystallographic data file use, structural correlations, reaction pathways, drug design.*

Nave, Dr Colin. SERC Daresbury Lab., Warrington, Cheshire WA4 4AD, England. (1949) PhD, crystallography (U. London, 1974). Head of Protein Crystallography Project Team. (tel. 0925 + 603265, telex 692609, fax 603100, email CN at UK.AC.DL.DLVD). *Structural biology, viruses, synchrotron radiation, protein crystallography, fibre diffraction.*

Nawaz, Dr Rab. Geology Dept., Ulster Museum, Botanic Gardens, Belfast BT9 5AB, N. Ireland. (1940) PhD, mineralogy and petrology (Queen's U., Belfast, 1974). Curator. (tel. 0232 + 381251, ext. 272). *Computing, crystallographic data, zeolites, diffractometry, inorganic crystals, microscopy, powder diffraction, symmetry, gemstones, minerals.*

Neidle, Dr Stephen. CRC Biomolecular Structure Unit, Inst. of Cancer Res., U. of London, Cotswold Rd., Sutton, Surrey SM2 5ng, England. (1946) PhD, chemical crystallography (Imp. C., London, 1970). Dir. (tel. 01 + 643-8901, ext. 4251; fax 642 9598, email: STEVE at UK.AC.LON.ICR.VAXF). *Nucleic acid structures, anti-cancer drug molecular design, drug-DNA interactions.*

Nelmes, Dr Richard J. Dept. of Physics, U. of Edinburgh, Mayfield Rd., Edinburgh EH9 3JZ, Scotland. (1943) DSc, physics (U. Cambridge, 1982). Sr. lect. (tel. 031 + 667-1081, ext. 2743; telex 727442 UNIVED G, fax 031 662 4712). *Diffuse scattering, disorder, hydrogen bonding, high pressures, inorganic crystals, phase transitions, thermal vibrations, X-ray and neutron diffraction.*

Nicholas, Mr David Michael. Div. of Materials Sci., Thames Polytechnic, Wellington St., London SE18 6PF, England. (1936) MSc, crystallography (U. London, 1965). Sr. lect. (tel. 01 + 316-8427). *Powder diffraction, materials science, phase transitions, teaching.*

Nicholls, Dr Reginald A. Analytical Dept., Pye-Unicam, Philips, York St., Cambridge CB1 2PX, England. (1952) PhD, clay mineral geochemistry (U. Bristol, 1979). Sr. appl. XRD. (tel. 0223 + 358866, ext. 322; telex 817331). *Clay mineral diffraction, computation in diffraction.*

Nieduszynski, Dr Ian Alexander. Dept. of Biological Sci. (Biochemistry), U. of Lancaster, Bailrigg, Lancaster LA1 4YQ, England. (1944) PhD, biophysics (U. Leeds, 1969). Lect. (tel. 0524 + 65201, ext. 4662). *X-ray fibre diffraction, physico-chemical studies, polysaccharides, glycosaminoglycans.*

North, Prof. Anthony Charles Thomas. Astbury Dept. of Biophysics, U. of Leeds, Leeds LS2 9JT, England. (1931) PhD, biophysics (U. London, 1955). Prof. (tel. 0532 + 333023, telex 556473 UNILDS G, fax 0532 336017, email ACTN at UK.AC.LEEDS.BIOVAX). *Proteins, structural biology, computing, X-ray diffraction.*

Norval, Dr Stephen Vynne. Res. and Techn. Dept., ICI Petrochemicals and Plastics Div., PO Box 90, Wilton, Middlesborough, Cleveland TS6 8JE, England. (1949) PhD, chemistry (Glasgow U., 1976). Res. scient. (tel. 0642 + 455522, ext. 2005; telex 587461). *X-ray diffraction, electron microscopy, heterogeneous catalysts.*

Nowell, Dr Ian William. Dept. of Chemistry, R.G.I.S.T., St Andrews St., Aberdeen, AB1 1HG, Scotland. (1944) PhD, chemistry (U. Leicester, 1969). *Coordination complexes, macrocyclic complexes, organometallic complexes.*

Nyburg, Prof. Stanley C. Dept. of Chemistry, King's College, Strand, London WC2R 2LS, England. (1924) DSc, crystallography (London). Hon. Sr. Res. Fellow. (tel. 01 + 836 5454, ext. 2257). *Molecular crystals, intermolecular forces.*

Nye, Prof. John Frederick. H.H. Wills Physics Lab., U. of Bristol, Tyndall Ave., Bristol BS8 1TL, England. (1923) PhD, physics (U of Cambridge, 1948). Emeritus. Prof. (tel. 0272 + 303030, ext. 3680). *Physical properties, tensor properties, ice, optics of crystals.*

O'Connor, Dr Denis Arthur. Dept. of Physics, U. of Birmingham, PO Box 363, Birmingham B15 2TT, England. (1927) DSc, physics (U. Birmingham, 1968). Reader, crystal physics. (tel. 021 + 472-1301, ext. 2545). *Crystal phase transitions, dynamical diffraction theory, Mössbauer effect.*

Orpen, Dr Anthony Guy. Dept. of Inorganic Chemistry, Bristol U., Cantock's Close, Bristol BS8 1TS, England. (1955) PhD, inorganic chemistry (U. Cambridge, 1979). Lect. (tel. 0272 + 303699, fax 0272 251295, email CSD at UK.AC.BRIS.CSA). *Crystallographic databases, neutron diffraction, X-ray diffraction, structural chemistry, neutron diffraction, reaction pathways.*

Owston, Dr Philip George. Afton Cottage, Berks Hill, Chorleywood, Herts. WD3 5AJ, England. (1921) DSc, crystallography, structural chemistry (U. London, 1976). Retired. (tel. 09278 + 3708). *Coordination compounds, structural chemistry, materials science, chemical bonding.*

Page, Dr James Ernest. 127 Northumberland Rd., Harrow, Middx. HA2 7RB, England. (1915) DSc, chemistry (U. London, 1958). Retired (tel. 01 + 866-8871). *Organic compounds.*

Page, Prof. Trevor Francis. Dept. Mechanical Materials, U. Newcastle Upon Tyne, NE1 7RU, England. (1946) PhD, metallurgy-materials science (U. Cambridge, 1971). Prof. (tel. 0223 + 65151, ext. 388 or 375). *Ceramics, silicon carbide polytypes, silicon nitride, alumina, electron microscopy (TEM, HREM, SEM, EDX), hardness, tribology, microstructure, ion implantation, coatings.*

Palmer, Dr Rex Alfred. Dept. of Crystallography, Birkbeck C., Malet St., London WC1E 7HX, England. (1936) PhD, crystallography (U. London, 1962). Sr. lect. (tel. 01 + 580-6622, ext. 330). *Biological molecules, structure determination, methods, teaching crystallography, drug design, quantum calculations, pharmacological activity.*

Pamplin, Dr Brian Randall. Sch. of Physics, U. of Bath, Claverton Down, Bath BA2 7AY, England. (1933) PhD, physics (U. Cambridge, 1960). Sr. lect., (tel. 0225 + 61244, ext. 445). *Semiconducting compounds, crystal growth, characterization.*

Papiz, Dr Miroslav Zenko. SERC Daresbury Lab., Warrington WA4 4AD, England. (1955) PhD, biophysics (CNAA - Napier C., Edinburgh, 1982). SSO. (tel. 0925 + 603388, telex 629609, email KUM%UK.AC.DARESBURY.DLVD). *Synchrotron radiation, area detectors, light-harvesting protein.*

Parker, Dr Andrew. Postgrad. Res. Inst. for Sedimentology, U. of Reading, Whiteknights, Reading RG6 2AB, England. (1941) PhD, geochemistry (U. Reading, 1969). Assoc. Director. (tel. 0734 + 318944). *Clay mineralogy, quantitative powder diffraction.*

Parker, Dr Michael William. European Molecular Biology Lab., Meyerhofstrasse 1, D-6900, Heidelberg, Postfach 10 22 09, West Germany. (1959) DPhil protein crystallography (Oxford, 1986). Staff scient. (tel. 6221 387269, telex 461613 embld, fax 6221 387306, email PARKER at EMBL). *Proteins, structural biology.*

Parkinson, Dr Gordon Michael. Analytical Service and Res. Div., BP Res., Sunbury Res. Centre, Chertsey Rd., Sunbury on Thames, Middlesex TW16 7LN, England. (1951) PhD chemistry (U. London, 1972). Res. Chem. (tel. 0932 + 762419, telex 296041, fax 0932-762999). *Solid state chemistry, catalysis, electron microscopy, electron diffraction, gas-solid reactions, in-situ chemistry, clays.*

Parpia, Dr Dawood Yusuf. Dept. of Physics, U. of Durham, South Road, Durham DH1 3LE, England. (1945) PhD, physics (Cambridge U., 1974). Sr. res. asst. (tel. 0385 + 64971, ext. 244). *X-ray topography, crystal growth, crystal defects, ferroelectric domains.*

Parsons, Prof. Ian. Dept. of Geology and Geophysics, U. of Edinburgh, West Mains Rd., Edinburgh EH9 3JW, Scotland. (1939) PhD, petrology (U. Durham, 1963). Prof. (tel. 031 + 6670181, ext. 3572; telex 727442 UNIVEDG). *Mineralogy, exsolution, microstructures, petrology.*

Pashley, Prof. Donald William. Dept. of Metallurgy and Materials Sci., Imperial C., Prince Consort Rd., London SW7 2AZ, England. (1927) PhD, physics (Imp. C., London, 1950). Prof. of materials, Dept. head. (tel. 01 + 789-5111, ext. 5901). *Electron microscopy, electron diffraction, thin film growth, epitaxy, alloys, crystal imperfections.*

Paster, Mr Simeon. Physics, King's College, Strand, London WC2R 2LS, England. (1965) BSc, physics (Liverpool, 1986). Res. Student. (tel. 01 + 836 5454, ext. 2718; email SP2 at UK.AC.KCL.PH.IPG). *Magnetic structure, 1st transition phosphates, single-crystal neutron diffraction.*

Paton, Mr John Dennis. Dept. of Chemistry, U. of Dundee, Dundee DD1 4HN, Scotland. (1946) MSc, crystallography (U. Dundee, 1979). SO. (tel. 0382 + 23181, ext. 4705). *Powder diffraction, X-ray structure analysis.*

Pawley, Prof. G. Stuart. Dept. of Physics, U. of Edinburgh, Mayfield Rd., Edinburgh EH9 3JZ, Scotland. (1937) PhD, physics (U. Cambridge, 1962). Prof. (tel. 031 + 667-1081, ext. 2699; fax 662 4712, email GSP at UK.AC.EDINBURGH). *Molecular dynamics, molecular lattice dynamics, massively parallel computation, plastic crystals, polymer folding.*

Pearce, Mr I.R. CERL, Kelvin Ave, Leatherhead, Surrey KT22 7SE, England. (1956) BSc, chemistry (CNAA, 1979). Second eng. (tel. 0372 + 374488, ext. 2404; telex 917338). *Electron microscopy, TEM, STEM, SEM, AUGER, electron diffraction, XRD, materials science, metallurgy, EELS, image processing.*

Perry, Dr Alan Leonard. Lab. of Molecular Biophysics, Oxford U., South Parks Rd., Oxford OX1 3QU, England. (1957) PhD, Crystallography (U. of London, 1987). Postdoc. (tel. 0865 + 275369). *Proteins, macromolecules, structural biology.*

Perutz, Dr Max Ferdinand. MRC Lab. of Molecular Biology, Hills Rd., Cambridge CB2 2QH, England. (1914) PhD, crystallography (U. Cambridge, 1940). (tel. 0223 + 248011, telex 81532, fax 223-213556). *Protein crystallography.*

Phillips, Prof. Sir David Chilton. Lab. of Molecular Biophysics, Dept. of Zoology, U. of Oxford, The Rex Richards Building, South Parks Rd., Oxford OX1 3QU, England. (1924) PhD, physics, crystallography (U. Wales, 1951). Prof. (tel. 0865 + 275365, fax 0865 510454). *Proteins, structural biology, instrumentation.*

Phillips, Dr Simon Edward Victor. Dept. of Biophysics, U. of Leeds, Leeds LS2 9JT, England. (1950) PhD, chemistry (U. London, 1974). Lect. (tel. 0532 + 431751, ext. 7581). *Protein crystallography.*

Pierce-Butler, Dr Melanie Anne. Crystallography Sect., Royal Armaments Res. and Dev. Est., Powdermill Lane, Waltham Abbey, Essex EN9 1BP, England. (1949) PhD, chemistry (U. Warwick, 1975). Sect. head. (tel. 0992 (Lea Valley) + 713030, ext. 305). *Crystallography, explosives, structural chemistry, crystallisation.*

Pigram, Dr William John. Physics Dept, U. of Keele, Keele, Staffs ST5 5BG, England. (1943) PhD, physics (U. of London, 1968). Lect. (tel. 0782 + 621111, ext. 3912). *X-ray and neutron diffraction, fibres, DNA, RNA, drug binding, industrial polymers, conformational transitions, anomalous scattering, molecular modelling.*

Pirie, Dr John Douglas. Dept. of Physics, U. of Aberdeen, Old Aberdeen AB5 0AB, Scotland. (1939) PhD, physics (U. Aberdeen, 1965). Lect. (tel. 0224 + 272500, telex 73458 UNIABN G). *Thermal X-ray scattering, diffuse scattering.*

Plant, Dr John Stewart. Computing Centre, Keele U., Keele, Staffs ST5 5BG, England. (1945) PhD, neutron diffraction (U. Sheffield, 1970). Res. systems analyst. (tel. 0782 + 621111, ext. 3307, email CCA08 at KL.SEQ1). *Computing, electron density, hydrogen bonding, magnetism, neutron diffraction.*

Pointer, Dr David John. Dept. of Chemistry, Teesside Polytechnic, Borough Rd., Middlesborough, Cleveland TS1 3BA, England. (1937) PhD, chemistry (U. Wales, 1964). Principal lect. (tel. 0642 + 218121, ext. 4179). *Organic molecules, biological activity, neuromuscular blocking agents.*

Pollard, Dr David Ronald. Dept. of Education and Sci., Elizabeth House, York Rd., London SE1 7PH, England. (1942) PhD, crystallography (U. Glasgow, 1968). Principal. (tel. 01 + 934-0740). *Computing methods, organic structures.*

Pond, Dr Robert Charles. Dept. of Metallurgy and Materials Sci., U. of Liverpool, PO Box 147, Liverpool L69 3BX, England. (1946) PhD, materials science (U. Bristol, 1973). Sr. lect. (tel. 051 + 709-6022, ext. 2028, telex 627095). *Electron microscopy, structure and properties, interfaces, semiconductors.*

Potter, Dr Reginald. Dept. of Physics, U. C. Cardiff, PO Box 78, Cardiff CF1 1XL, Wales. (1931) PhD, physics, crystallography (U. Wales, 1962). Sr. lect. (tel. 0222 + 44211, ext. 2342). *Instrumentation, accurate intensities.*

Povey, Dr David Christopher. Dept. of Chemistry, U. of Surrey, Stag Hill, Guildford, Surrey GU2 5XH, England. (1946) PhD, X-ray crystallography (U. Surrey, 1974). Lect. (tel. 0483 + 571281, ext. 2591; telex 859331, fax 0483 300803). *Coordination chemistry, instrumentation, proteins, teaching, structural chemistry.*

Powell, Prof. Herbert Marcus. 46 Davenant Rd., Oxford OX2 8BY, England. (1906) MA, chemistry (U. Oxford, 1927). Em. Prof. (tel. 0865 + 54951). *Chemistry, crystallography.*

Price, Dr Geoffrey David. Dept. of Geology, U. C. London, Gower St., London WC1E 6BT, England. (1956) PhD, mineralogy (U. Cambridge, 1980). Res. fellow. (tel. 01 + 387-7050, ext. 433). *Silicates, phase transformations, deformation, high pressures.*

Prout, Dr Charles Keith. Chemical Crystallography Lab., U. of Oxford, 9 Parks Rd., Oxford OX1 3PD, England. (1934) DPhil, chemistry (U. Oxford, 1959). Lect., fellow of Oriel C. (tel. 0865 + 270820). *Structural chemistry.*

Putnis, Dr Andrew. Dept. of Earth Sci., U. of Cambridge, Downing St., Cambridge CB2 3EW, England. (1947) PhD, mineralogy and petrology (U. Cambridge, 1976). Lect. (tel. 0223 + 333400, ext. 3431; fax 333450). *Electron microscopy, phase transformations, order-disorder, NMR spectroscopy.*

Puttick, Prof. Keith Ernest. Dept. of Physics, U. of Surrey, Guildford, Surrey GU2 5XH, England. (1926) PhD, physics (U. Bristol, 1957). Prof. (tel. 0483 + 571281, ext. 741, telex 859331). *Mechanical properties (scaling laws, hardness, fracture), ceramics, particulate crystals, polymers, composites, metals.*

Puxley, Dr David Charles. Analytical Div., British Gas Corp., London Res. Station, Michael Rd., London SW6 2AD, England. (1946) PhD, inorganic chemistry (U. London, 1972). Principal scient. (tel. 01 + 736-3344, ext. 4051). *Industrial X-ray diffraction, automatic powder diffractometry, crystallite size studies, catalysis, position-sensitive detectors.*

Rabinowich, Prof. D. University Chemical Lab, Lensfield Rd, Cambridge CB2 1EW, England.

Rae, Dr Alan William James Melville. Technical Ceramics Div., Anzon Ltd., Cookson House, Willington Quay, Wallsend NE28 6UQ, England. (1951) PhD, metallurgy (U. Newcastle, 1976). General Manager. (tel. 0632 + 622211, telex 537537). *Technical ceramics, inorganic refractory compounds.*

Rae, Dr Alastair Ian Maxwell. Dept. of Physics, U. of Birmingham, PO Box 363, Birmingham B15 2TT, England. (1938) PhD, physics (U. W. Australia, 1963). Lect. (tel. 021 + 472-1301, ext. 3460). *Crystallography, intermolecular forces, phase changes.*

Raftery, Dr James. Dept. of Chemistry, U. of Manchester, Manchester, M13 9PL, England. (1952) PhD, crystallography (U. London, 1981). *Parallel computers, proteins, data bases.*

Raithby, Dr Paul Robert. Dept. of Inorganic Chemistry, U. of Cambridge, Lensfield Rd., Cambridge CB2 1EW, England. (1951) PhD, chemistry (U. London, 1976). Res. off. (tel. 0223 + 66499, ext. 220). *X-ray crystallography, cluster compounds, osmium chemistry.*

Ralph, Prof. Brian. Dept. of Metallurgy and Materials Sci., U. C. Cardiff, Newport Road, Cardiff CF2 1TA, Wales. (1939) ScD, materials science (U. Cambridge, 1980). Dept. head. (tel. 0222 + 44211, ext. 7003; telex 498635). *Grain boundaries, interfaces, electron diffraction, electron microscopy.*

Ramdas, Dr S. Analytical Div., BP Res. Centre, Chertsey Rd, Sunbury-on Thames, Middx. TW16 7LN, England.

Rawas, Dr Ahmad. Dept. of Biochemistry, U. of Bristol, Univ. Walk, Bristol BS8 1TD, England. (1952) PhD, crystallography (U. of Hull, 1985). Res. asst. (tel.

0272 + 303030, ext. 3777). *Computing, crystallographic data, proteins, polymers, liquid crystals, powder diffraction, small-angle scattering, X-ray diffraction, teaching.*

Redhouse, Dr Alan David. Dept. of Chemistry and Appl. Chemistry, U. of Salford, Salford M5 4WT, England. (1940) PhD, inorganic chemistry (U. Bristol, 1964). Lect. (tel. 061 + 736-5843, ext. 643). *Crystal structure analysis.*

Reid, Dr John Sinclair. Dept. of Physics, U. of Aberdeen, Fraser Noble Building, Meston Walk, Old Aberdeen AB9 2UE, Scotland. (1942) PhD, physics (U. Aberdeen, 1970). Lect. (tel. 0224 + 272507, telex 73458, fax 487048). *Diffuse X-ray scattering, lattice dynamics, teaching.*

Rendle, Dr David Forbes. Chemistry I Div., Metropolitan Police Forensic Sci. Lab., 109 Lambeth Rd., London SE1 7LP, England. (1946) PhD, chemistry (U. Guelph, Canada, 1972). SSO. (tel. 01 + 230-6246, telex 892733). *Forensic applications, powder diffraction methods, instrumentation.*

Reynolds, Dr C.D. Biophysics Lab, Dept of Physics, Liverpool Poly., Liverpool L3 3AF, England.

Rhodes, Prof. Rene George. Dept. of Eng., U. of Warwick, Coventry CV4 7AL, England. (1916) PhD, physics (U. London, 1950). Prof. (tel. 0203 + 24011, ext. 2128). *Semiconductors, superconductors, magnetic levitation.*

Rice, Dr David William. Dept. of Molecular Biology and Biotechn., U. of Sheffield, Western Bank, Sheffield S10 2TN, England. (1952) DPhil, X-ray crystallography of proteins (U. Oxford, 1979). Sen. Res. Fellow. (tel. 0742 + 768555, ext. 4242, fax 727949, email BI1DWR%UK.AC.SHEF.PA. at AC.UK). *Protein crystallography, crystallization, macromolecules, structural homology.*

Richards, Dr Brian Peter. Materials Science Div., GEC Hirst Res. Centre, East Lane, Wembley, Middx. HA9 7PP, England. (1939) PhD, crystallography (U. London, 1974). Manager Materials Characterization Div. (tel. 01 + 904-1262, ext. 280). *Characterisation, crystalline materials, amorphous materials, carbons and graphites, defects in semiconductor materials.*

Richards, Dr John Philip Gerald. Dept. of Physics, U. C. Cardiff, PO Box 78, Cardiff CF1 1XL, Wales. (1932) PhD, physics (U. Wales, 1960). Sr. lect. (tel. 0222 + 44211, ext. 2360). *Molecular motions, phase transitions, thin films, surfaces.*

Richardson, Dr Robert Melville. Sch. of Chemistry, Cantock's Close, Bristol, Avon BS8 1TS, England. (1952) PhD,liquid crystals (U. Bristol,1977). Lect. (tel. 0272 303688, fax 250612). *Liquid crystals, thin films, colloids, porous materials, small-angle scattering, powder diffraction, neutron diffraction, disorder, surface structure, macromolecules.*

Rickards, Mr A.L. T and N Materials Res. Ltd., Dell Rd., Rochdale, Lancs., OL12 6BY, England.

Roberts, Dr Kevin John. Dept. of Pure and Appl. Chemistry, U. of Strathclyde, 295 Cathedral St., Glasgow G1 1XL, Scotland. (1950) PhD, material science, crystallography (Portsmouth Poly, 1979). Lect. (tel. 041 + 552-4400, ext. 2265). *Crystal growth (theoretical & experimental studies), molecular crystals, surface crystallography, X-ray diffraction, novel uses of synchrotron radiation.*

Robertson, Dr John Harry. Sch. of Chemistry, U. of Leeds, Leeds LS2 9JT, England. (1923) PhD, crystallography (U. Edinburgh, 1949). Retired. (tel. 0532 + 431751, ext. 6406). *Inorganic, biological and organic structures, physical properties of the crystalline state, philosophy of scientific method.*

Robertson, Prof. John Monteath, FRS. 11A Eriskay Rd., Inverness IV2 3LX, Scotland. DSc, chemistry (U. Glasgow, 1933). Em. prof. (tel. 0463 + 225561). Gregorie Aminoff Gold Medal, Royal Swedish Academy, June 1982. *Chemistry.*

Roebuck, Dr Peter Hamish Athey. Production Dept., Kanthal Ltd., Inveralmond Industrial Estate, Perth PH1 3EE, Scotland. (1952) PhD, materials science - ceramics (U. Newcastle, 1978). Technical & Production Manager. (tel. 0738 + 20931, ext. 124; telex 76460, fax 0738 20936). *Silicon carbide, high temperature materials (electrical).*

Rogers, Prof. Donald. 11 Salvington Crescent, Bexhill on Sea, E Sussex, TN39 3NP, England. (1921) PhD, physics (U. London, 1944) Em. Prof. (tel. 0424 + 222157). *Absolute configuration assignment, automated setting of single crystals.*

Rogers, Dr K.D. App. Physics and Electro-Optics Group, Royal Military C. of Sci., Shrivenham, Swindon, Wiltshire SN6 8LA, England. (1959) PhD, physics, crystallography (U. of Wales, 1985). Lect. (tel. 0793 + 782551, ext. 2399). *Thin films, phase transitions, Rietveld methods.*

Rollett, Dr John Sydney. Computing Lab., U of Oxford, 8-11 Keble Rd., Oxford OX1 3QD, England. (1927) PhD, chemistry (U. Leeds, 1952). U. lect. in numerical maths. (tel. 0865 + 54141, ext. 317). *Crystallographic computing, numerical analysis (optimization, linear algebra).*

Ross, Dr Donald Keith. Dept. of Physics, U. of Birmingham, Birmingham B15 2TT, England. (1939) PhD, physics (U. Birmingham, 1972). Lect. (tel. 021 + 472-1307, ext. 3467 or 2078). *Hydrogen in metals, neutron diffraction, quasielastic scattering, clay-water systems, intermetallic systems.*

Ross, Dr. Nancy L. Geological Sci. Dept., U. C. London, Gower St., London WC1E 6BT, England. PhD, geology (Arizona State U.,1985). Lect. (tel. 44-(0) 1-387-7050, ext. 2423). *High pressure crystallography, minerals, thermodynamic properties.*

Rout, Ms Joanne Elizabeth. Res. and Techn. Div., ICI Chemicals and polymers, The Heath, Runcorn, Cheshire WA7 4QD, England. (1961) BSc, materials science (U. of Sussex, 1982). Sen. res. scient. (tel. 0928 + 513742, telex 629655

ICIMOH G, fax 576376, email 1253 at DL.DLVB). *Crystal growth, proteins, inorganic crystals, materials science, structural chemistry, phase transitions.*

Rowley, Dr Colin Raymond. Dept. of Geology, Portsmouth Polytechnic, Burnaby Rd., Portsmouth PO1 3QL, England. (1938) PhD, geology (U. Durham, 1965). Principal lect. (tel. 0705 + 827681, ext. 254). *Crystal structures and chemistry, clay minerals, rock-forming carbonates.*

Rule, Dr Stephen A. SERC Daresbury Lab., Warrington, Cheshire WA4 4AD, England. (1955) BSc, physics (U. Keele, 1981). Post-doc. Res. asst. (tel. 0925 + 65000, ext. 388). *Synchrotron radiation, protein crystallography.*

Russell, Dr David Robin. Dept. of Chemistry, U. of Leicester, Leicester LE1 7RH, England. (1939) PhD, inorganic chemistry (U. Glasgow, 1963). Sr. lect. (tel. 0533 + 554455). *Chemical crystallography.*

Sampson, Mr Christopher. Materials Physics Div., AERE Harwell, Didcot, Oxon OX11 0RA, England. (1935) Nat. Cert., applied physics (1960). HSO. (tel. 0235 + 24141, ext. 5280). *Lattice parameter determination, powder cameras, powder diffractometry, low temperature distortions, single crystals and powders.*

Sanderson, Dr Mark Rutherford. CRC Biomolecular Structure Group, Inst. of Cancer Res., Block F, 15 Cotswold Rd., Sutton, Surrey SM2 5NG, England. (1956) PhD, biophysics, (U. of London, 1981). Staff Scient. (tel. 01 + 6438901, ext. 4536; fax 01 642 9598). *Nucleic acids, anti-tumour drug structure, DNA binding proteins.*

Sándor, Dr Endre Elek. 1 Fairlawn Drive, Woodford Green, Essex IG8 9AW, England.

Sawyer, Dr Lindsay. Dept. of Biochemistry, U. of Edinburgh, Hugh Robson Building, George Square, Edinburgh EH8 9XD, Scotland. (1944) PhD, protein crystallography (U. Edinburgh, 1971). Sen. Lect. (tel. 031 + 667-1011, ext. 2363; telex 727442 UNIVED G, email LINDSAY at UK.AC.EDINBURGH.BIOVAX). *Molecules of biological interest, proteins, structural biology.*

Schwalbe, Dr Carl Hellmuth Walter. Dept. of Pharmaceutical Sci., U. of Aston, Gosta Green, Birmingham B4 7ET, England. (1942) PhD, chemistry (U. Harvard, 1970). Sr. lect. (tel. 021 + 359-3611, ext. 4201, telex 336997, fax 333 3172, email UK.AC.ASTON.MAIL SCHWALBECH). *Crystal structure determination, drug molecules, molecular graphics, molecular orbital calculations.*

Schwarzenberger, Mrs D.R. Eng. Dept., Warwick U., Coventry CV4 7AL, England.

Scouloudi, Dr Helen. Lab. of Molecular Biophysics, Dept. of Zoology, U. of Oxford, The Rex Richards Building, South Parks Rd., Oxford OX1 3QU, England. PhD, physics (U. London, 1951). Res. officer. (tel. 0865 + 275368). *Protein structure, biological macromolecules, X-ray diffraction.*

Scrimgeour, Dr Sheelagh Nicoll. Dept. of Chemistry, U. of Dundee, Nethergate, Dundee DD1 4HN, Scotland. (1947) PhD, chemistry and crystallography (U. Dundee, 1973). Hon. Res. fellow. (tel. 0382 + 23181, ext. 4682). *Small molecules, biologically interesting molecules, solid state NMR.*

Seddon, Dr John Michael. Dept. of Chemistry, The University, Southampton SO9 5NH, England. (1953) PhD, Biophysics (U. London, 1980). Lect. (tel. 0703 + 595000, ext. 2193; telex 47661, fax 0703 593939, email CHR009 at IBM.SOTON). *Liquid crystals, X-ray diffraction, neutron diffraction, small angle scattering.*

Shackleton, Miss Judith Mary. Analytical Dept., Philips Scientific, York St., Cambridge, CB1 2PX, England. (1958) MSc, physical methods of analysis (U. Aston in Birmingham, 1984). Applications Specialist. (tel. 0223 + 58866, ext. 323; telex 817331 PHISCI, fax 0223 312764). *Powder diffraction methods, small angle scattering, metals, minerals, X-ray diffraction.*

Shah, Dr Jitendra Shantilal. H.H. Wills Physics Lab., U. of Bristol, Royal Fort, Bristol BS8 1TL, England. (1939) PhD, physics (U. Bath, 1968). (tel. 0272 + 303030, ext. 3635; fax 0272 732657). *Fibre diffraction, small-angle scattering, powder diffraction.*

Sharp, Prof. David William Arthur. Chemistry Dept., U. of Glasgow, Glasgow G12 8QQ, Scotland. (1931) PhD, chemistry (U. Cambridge, 1957). Prof. (tel. 041 + 330-5290, telex 777070 UNIGLA, fax 041-330-4920). *Inorganic chemistry.*

Sharpe, Miss Andrea Jane. International Union of Crystallography, 5 Abbey Square, Chester CH1 2HU, England. (1958) BSc, biological chemistry (U. of Essex, 1980). Editorial Asst. (tel. 0244 + 342878, telex 669755 OFFICE G, fax 0244 314888).

Shaw, Mr Andrew. Dept. of Biochemistry, U. of Sheffield, Western Bank, Sheffield S10 2TN, England. (1966) BSc, biochemistry (U. of York, 1988). Res. Student. (tel. 768555, ext. 4241). *Flavodoxins, crystallisation, nitrogen-fixation.*

Shaw (née Gözen), Dr Leylâ Süheylâ. Queen Mary C., Mile End Road, London E1 4NS, England. (1951) PhD, chemical crystallography (CNAA, 1982). Hon. res. fellow. (tel. 01 + 980-4811, ext. 3711). *Structure - property relationship, NMR, NQR, small molecules, medium-sized molecules, phosphorus-nitrogen compounds, organic nitrogen heterocycles.*

Sheldrick, Dr Bernard. Astbury Dept. of Biophysics, U. of Leeds, Leeds LS2 9JT, England. (1929) PhD, chemical crystallography (U. Leeds, 1964). Sr. lect. (tel. 0532 + 31751, ext. 6104). *Biological molecular structures, computer methods.*

Sherwood, Prof. John Neil. Dept. of Pure and Appl. Chemistry, U. of Strathclyde, 295 Cathedral St., Glasgow G1 1XL, Scotland. (1933) DSc, chemistry (U. Durham, 1976). Prof. (tel. 041 + 552-4400, ext. 2797; telex 77472 UNSLIB

G, fax 041 552 0775). *Lattice defects in solids, crystal growth (defect influence), chemical properties, physical properties.*

Silver, Dr Jack. Dept. of Chemistry, U. of Essex, Wivenhoe Park, Colchester, Essex, England. (1948) PhD, inorganic chemistry (U. London, 1973). Sen. Lect. (tel. 0206 + 872097, telex 98440 UNILIB G, fax 0206 873598). *Iron sandwich compounds, metalloporphyrins, metallophthalocyanines.*

Sim, Prof. George Andrew. Dept. of Chemistry, U. of Glasgow, Glasgow G12 8QQ, Scotland. (1929) PhD, chemistry (U. Glasgow, 1955). Prof. (tel. 041 + 339-8855, ext. 419). *Conformation, organic molecules, molecular mechanics.*

Sinn, Prof. Ekkehard. Chemistry Dept., U. of Hull, Hull HU6 7RX, England. (1945) PhD (U. New South Wales, Australia,1968). (tel. 0482 + 466353, fax 0482 + 466410, email ES10 at seq.hull.ac.uk) *Electronic - molecular structure relation, transition metal complexes; magnetic exchange interactions; bio-inorganic chemistry.*

Skapski, Dr Andrzej Czeslaw. Chemical Crystallography Lab., Dept. of Chemistry, Imperial C., London SW7 2AY, England. (1938) PhD, chemistry (U. London, 1963). Lect. (tel. 01 + 589-5111, ext. 4609). *Inorganic crystal structures, metal binding sites, nucleic acid components.*

Skarnulis, Dr Anthony Jerome. Dept. of Chemical Crystallography, U. of Oxford, 9 Parks Rd., Oxford OX1 3PD, England. (1948) PhD, chemistry (Arizona State U., 1975). Res. assoc. (tel. 0865 + 53424, ext. 295). *Crystallography, electron microscopy, electron diffraction, X-ray diffraction, computer graphics.*

Skellett, Mr C.A. Appl. Physics Div., GEC Avionics Ltd, Borehamwood, Herts WD2 1RX, England.

Small, Dr Ronald W.H. Dept. of Chemistry, U. of Lancaster, Bailrigg, Lancaster LA1 4YA, England. (1921) DSc, chemistry (U. Birmingham, 1982). Reader. (tel. 0524 + 65201, ext. 4033). *Polymorhism, phase studies, molecular conformation.*

Smart, Dr Lesley Elizabeth. Dept. of Chemistry, Open U., Walton Hall, Milton Keynes, MK7 6AA, England. (1947) PhD, chemistry (U. Southampton, 1971). Lect. (tel. 0908 + 653191). *Organometallic structures, halides, fluorides and oxide-fluorides of transition and non-transition elements.*

Smith, Dr Arnold John. Dept. of Chemistry, U. of Sheffield, Brook Hill, Sheffield S3 7HF, England. (1931) PhD, inorganic chemistry (U. London, 1957). Lect. (tel. 0742 + 768555, ext. 4476, fax 0742-739826, telex 54348 ULSHEF G, Email CH1AJS at UK.AC.SHEFFIELD.PRIMEA). *Inorganic crystal structures, lanthanide and actinide compounds, oxide phases, high coordination numbers.*

Smith, Dr Bryan Edward. Dept. of Mechanical Eng., Brunel U., Uxbridge, Middx. UB8 3PH, England. (1936) PhD, physical metallurgy (Brunel U., 1974). Lect. (tel. 0895 + 37188, ext. 417). *Lattice defects, line broadening X-ray techniques, solar collectors, surfaces, coatings, high temperature materials, microstructure - mechanical properties relationships.*

Smith, Mr Gallienus William. Dept. of Chemistry, U. of Surrey, Guildford, Surrey, England. (1924) MSc, crystallography (U. London, 1953). Sen. scient. (tel. 0483 + 571281, ext. 2591; telex 859331). *Organic/inorganic crystal structure determination, teaching.*

Smith, Mr John Michael Andrew. Dept. of Biochemistry, U. of Sheffield, Sheffield S10 2TN, England. (1956) PhD, crystallography (U. Sheffield, 1981). Res. contract staff. (tel. 0742 + 768555, ext..4241; telex 547216 UGSHEF G, fax 739826, email BI1JMS%UK.AC.SHEF.IBM at AC.UK). *Macromolecular crystallography, small molecule crystallography, diffractometry, instrumentation, crystallographic computing.*

Smith, Dr Robert Carr. Director, Kingston Polytechnic, Penrhyn Rd., Kingston upon Thames, Surrey KT1 2EE, England. (1935) PhD, physics (U. London, 1961). Dir. (tel. 01 + 549-1366, ext. 200). *Nonlinear optics, ternary semiconducting compounds.*

Smith, Dr Sidney Herbert. Clarendon Lab., U. of Oxford, Parks Rd., Oxford OX1 3PU, England. (1938) PhD, physics (U. London, 1970). Res. support II. (tel. 0865 + 59291, ext. 275). *Crystal growth - flux.*

Soar, Mr Martin. Dept. of Spectroscopy, BP Res., Chertsey Rd., Sunbury on Thames TW16 7LN, England. (1959) GRSC(1) chemistry (Sheffield Polytech., 1980). Chemist. (tel. 0932 + 763724). *XRPD, catalysis, in-situ XRD.*

Sprackling, Dr Michael Thomas. Dept. of Physics, King's C. London, Strand, London WC2R 2LS, England. (1934) PhD, physics (U. Bristol, 1959). Sr. lect. (tel. 01 + 836-5454, ext. 2119). *Dislocations, ionic crystals, photography.*

Spratt, Mr S.B.D. Johnson Matthey Res. Centre, Blounts Court, Sonning Common, Reading, Berks RG4 9NH, England.

Spreadborough, Dr John. John Spreadborough and Co. Ltd., 30 Clarence Rd., Windsor, Berks SL4 5AQ, England. (1933) DPhil, physical metallurgy (U. Oxford, 1958). Managing dir. (tel. 07535 + 61552). *Physical metallurgy, X-ray diffraction, crystallography, crystal structure.*

Spriggs, Dr Paul Humphrey. Dept. of Metallurgy, U. of Manchester, Grosvenor St., Manchester M1 7HS, England. (1931) PhD, metallurgy (U. Manchester, 1967). Fellow. (tel. 061 + 236-3311, ext. 2234). *Binary and ternary compounds, transition metals, transition metal alloy theory.*

Squire, Dr John Michael. Biophysics Sect., Blackett Lab., Imperial C., London SW7 2BZ, England. (1945) PhD, biophysics (U. London, 1969). Reader. (tel. 01 + 589-5111, ext. 6741; fax 01 589 9463). *Electron microscopy, macromolecules, neutron diffraction, polymers, proteins, small angle scattering, structural biology, X-ray diffraction.*

Stadler, Dr Hans Peter. Crystallography Lab., U. of Newcastle upon Tyne, Newcastle upon Tyne NE1 7RU, England. (1921) PhD, crystallography (U. Leeds, 1948). Sr. lect. (tel. 0632 + 328511, ext. 3203). *Crystal structure determination, transform methods, radiation protection.*

Stammers, Dr David Kingsley. Biochemistry Dept., Wellcome Res. Labs., Langley Court, Beckenham, Kent BR3 3RS, England. (1949) PhD, crystallography (U. Bristol, 1974). Sr. scient. (tel. 01 + 658-2211, ext. 263). *Protein structure and function.*

Stanford, Mr Michael John. 92 Peregrine Rd., Sunbury-on-Thames, Middx. TW16 6JP, England. (1949) MSc, crystallography (U. Cape Town, S. Africa, 1979). Continuing student. (tel. 01 + 768-3473). *Symmetry, phase problem, crystallographic computing, molecular biology.*

Stansfield, Dr Robert Frank David. Merrell Dow Res. Inst., Strasbourg Res. Centre, 16 Rue D' Ankara, 67084 Strasbourg Cedex, France. (1954) PhD, chemistry (U. of Bristol, 1979). Manager Comp. Services. (tel. 88 61 48 89, ext. 460; telex 890 252, fax 88 60 37 98). *Crystal structures, small molecules, proteins, X-ray and neutron diffraction, position sensitive detectors.*

Steeds, Prof. John Wickham, FRS. Dept. of Physics, U. of Bristol, Tyndall Av., Bristol BS8 1TL, England. (1940) PhD, electron microscopy (U. Cambridge, 1967). Prof. (tel. 0272 + 303597, fax 0272-732657). *Electron diffraction, phase transformation, incommensurate structures.*

Steigmann, Dr Gottfried Albert. Dept. of Physics, U. of Hull, Cottingham Rd., Hull HU6 7RX, England. (1938) PhD, crystallography (U. Hull, 1963). Lect. (tel. 0482 + 46311, ext. 7545 or 7389). *Crystal structures, inorganic compounds, computation.*

Steward, Prof. Edward George. Dept. of Physics, The City U., Northampton Sq., London EC1V 0HB, England. (1923) DSc, crystallography (U. London, 1975). Prof. (tel. 01 + 253-4399, ext. 4403). *Structure - property relationships, molecular medicine.*

Stillman, Dr Timothy James. Dept. of Biochemistry, U. of Sheffield, Western Bank, Sheffield S10 2TN, England. (1961) PhD, crystallography (U. of Lancaster, 1988). Post-Doc. (tel. 0742 + 768555, ext. 4242). *Proteins, dehydrogenases, molecular modelling, molecular graphics.*

Stothart, Dr Philip Hamilton. Food Res. Inst., Shinfield, Reading, Berks. RG2 9AT, England. (1946) PhD, physics (U. Reading, 1978). Res. physicist. (tel. 0734 + 883103). *Small-angle X-ray scattering, small-angle neutron scattering, protein structure.*

Strickland, Mr Peter R. International Union of Crystallography, 5 Abbey Square, Chester CH1 2HU, England. (1956) BSc, chemistry (U. Sheffield, 1977). Asst. Tech. Editor. (tel. 0244 + 342878, telex 669755 OFFICE G attn. UNICRYSTAL, fax 0244 314888). *Editing.*

Sullivan, Dr Richard Arthur. Dept. of Physics, U. of Bath, Claverton Down, Bath BA2 7AY, England. (1936) PhD, crystallography (U. Manchester, 1960). Lect. (tel. 0225 + 61244). *Geophysics, X-ray diffraction.*

Sundaresan, Dr Thiagarajan. Dept. of Sci. Math. and Computing, Leigh C., Marshall St., Leigh WN7 4HX, England. (1939) PhD, crystallography (U. Nottingham, 1972). Lect. (tel. 0942 + 608811, ext. 222). *Fibre diffraction, synthetic fibres.*

Sutherland, Dr Hector Howieson. Dept. of Physics, U. of Hull, Cottingham Rd., Hull HU6 7RX, England. (1935) PhD, crystallography and mineralogy (U. St. Andrews, 1962). Sr. lect. (tel. 0482 + 46311, ext. 7389 or 7820). *Inorganic and organic crystal structures, liquid crystals, topography.*

Sutor, Dr Dorothy June. Dept. of Crystallography, Birkbeck C., Malet St., London WC1E 7HX, England. (1929) PhD, chemistry (Auckland U.C., New Zealand, 1954) PhD, crystallography (U. Cambridge, 1958). Res. fellow. (tel. 01 + 631-6420). *Pathological crystals, powder diffraction.*

Sutton, Dr A.L. Appl. Physics Dept, The City U., St John St, London EC1V 0HB, England.

Sutton, Dr Brian John. Biomolecular Scinces Division, King's College London, 26 Drury Lane, London WC2B 5RL, England. (1954) DPhil, molecular biophysics (U. Oxford, 1980). Lect. (tel. 01-836 + 8851, ext. 223; fax 497 9078). *Molecular immunology, antibody structure, metalloenzymes, enzymology, antibiotic resistance mechanisms, XRD.*

Sutton, Mr J.D. Oxford Instruments Ltd, Osney Mead, Oxford OX2 0DX, England.

Swallow, Dr Arnold Graham. Slough C. of Further Education, Wellington St., Slough SL1 1YG, England. (1934) PhD, chemistry (U. Leeds, 1961).

Swindells, Dr David Campbell Neil. Production Dept., IOP Publishing Ltd., Techno House, Redcliffe Way, Bristol BS1 6NX, England. (1952) PhD, chemistry (U. New Brunswick, Canada, 1981). Sen. editorial asst. (tel. 0272 + 297481, ext. 217; telex 449149, fax 0272 294318, email IOPPL at UK.AC.RL.GB).

Tait, Dr John Mervyn. Dept. of Mineral Soils, Macaulay Inst. for Soil Res., Craigiebuckler, Aberdeen AB9 2QJ, Scotland. (1947) PhD, chemistry (U. Aberdeen, 1973). SSO. (tel. 0224 + 38611, ext. 255). *Transmission electron microscopy, electron diffraction, soil minerals.*

Tanner, Dr Brian Keith. Dept. of Physics, U. of Durham, South Rd., Durham DH1 3LE, England. (1947) DPhil, metallurgy (U. Oxford, 1972). Reader. (tel. 091 + 374-2137, telex 537351 DURLIB G, fax 091 374-3749). *X-ray diffractometry, instrumentation, magnetic materials, defect structures.*

Tarling, Dr Stephen Edward. Industrial Materials Group, Dept. of Crystallography, Birkbeck C., Malet St, London WC1E 7HX, England. (1955) PhD, crystallography (London ,1984). Lect.(tel. 01 + 631-6513, fax 01 436 8918). *Powder diffraction, materials science, phase transitions, teaching.*

Tasker, Mr Michael Peter. XM2 Div., RARDE Fort Halstead, Sevenoaks, Kent TN14 7BP, England. (1939) MSc, crystallographic and spectroscopic techniques (CNAA, 1976). HSO. (tel. 0959 + 32222, ext. 3381). *Materials science.*

Tasker, Dr Peter Anthony. Sch. of Chemistry, The Polytechnic of North London, Holloway, London N7 8DB, England. (1944) DPhil, chemistry (U. York, 1968). Sr. lect. (tel. 01 + 607-2789, ext. 2153). *Coordination chemistry, metal ions, biological inorganic chemistry, catalysis, complexation, metals extraction.*

Tate, Dr Cecil. Physics Dept., U. of York, Heslington, York YO1 5DD, England. (1934) PhD, mathematical physics (U. London, 1960). Res. fellow. (tel. 0904 + 430000, ext. 2208). *Direct methods, X-ray crystallography.*

Taylor, Prof. Charles Alfred. 9 Hill Deverill, Warminster, Wilts BA12 7EF, England. (1922) DSc, physics (U. Manchester, 1959). Em. prof. (tel. 0985 + 40574). *Teaching crystallography, X-ray diffraction.*

Taylor, Dr Derek. Fairey Tecramics Limited, Filleybrooks, Stone, Staffs ST15 0PU, England. (1939) PhD, mineralogy (U. Manchester, 1966). Managing. Dir. (tel. 0785 + 813241, ext. 273; telex 36277, fax 0785-816903). *Thermal expansion, framework structures, ceramic science.*

Taylor, Prof. Harry Francis West. Dept. of Chemistry, U. of Aberdeen, Old Aberdeen AB9 2UE, Scotland. (1923) DSc, chemistry (U. London, 1957). Prof. (tel. 0224 + 40241, ext. 481). *Silicates, cement chemistry, mineralogical chemistry.*

Taylor, Dr R. Sartoria, Bruisyard Rd, Rendham, Saxmundham, Suffolk IP17 2AH, England.

Tempest, Dr Paul Anthony. Res. Div. CEGB, Berkeley Nuclear Labs., Berkeley, Gloucestershire GL11 5PH, England. (1947) PhD, materials science (U. London, 1974). Res. off. (tel. 0453 + 810451, ext. 146). *X-ray absorption, preferred orientation, stainless steel oxidation products, hyperstoichiometric uranium dioxide, spinels, sructure.*

Thatcher, Mr J.S. GTP Eng. Ltd., Station Ind. Est., Sheppard St, Swindon, Wilts, England.

Theocharis, Dr C.R. Chemistry Dept, Brunal U., Uxbridge, Middx. UB8 3PH, England.

Thomas, Dr Charles Richard. Dept. of Eng., U. of Warwick, Coventry, West Midlands CV4 7AL, England. (1947) D.Phil, eng. science (U. of Oxford, 1975). Lect. (tel. 0203 + 523139). *X-ray diffraction, materials science, semiconductors.*

Thomas, Mr David Huw. Dept. of Biochemistry, U. of Sheffield, Western Bank, Sheffield S10 2TN, England. (1964) BSc, biochemistry (U. of Sheffield, 1985). Res. asst. (tel. 0742 + 768555, ext. 4242). *Protein structure, bacterial toxins.*

Thomas, Mrs Elizabeth Ann. Analytical Div., BP Res., Chertsey Rd, Sunbury on Thames, Middlesex TW16 7LN, England. (1964) BSc chemistry (U. of Nottingham, 1985). Res. asst. (tel. 0932 + 763728). *Powder diffraction, zeolites, polymers.*

Thomas, Prof John Meurig, FRS. The Director, Royal Institution, 21 Albermale St., London, W1X 4BS, England. Cambridg(1932) DSc, chemistry (U. C. Swansea, 1954). *Solid-state chemistry, catalysis (heterogeneous), imperfections in solids, surface characterization techniques, solid state NMR.*

Thomas, Dr Pamela Anne. Clarendon Lab., Oxford U., Parks Rd., Oxford OX1 3PU, England. (1962) DPhil, physics, crystallography (U. of Oxford, 1987). Post-Doc. (tel. 0865 + 272200 ext. 72334). *Physical properties, non-linear optical materials, structure-property relationships, XRD.*

Thompson, Dr Derek Parr. Wolfson Lab., Dept.of Metallurgy and Eng. Materials, U. of Newcastle upon Tyne, Newcastle upon Tyne NE1 7RU, England. (1945) PhD, mineralogy, crystallography (U. Cambridge, 1972). Lect. (tel. 091 + 2328511, ext. 7202; telex 53654 UNINEW, fax 091 232 9259). *Powder diffraction, materials science, phase determination, crystal chemistry, minerals, nitrogen ceramics.*

Thornton-Pett, Dr M.A. 4 Westfield Court, Westfield Rd, Leeds, England.

Tickle, Dr Ian James. Dept. of Crystallography, Birkbeck C., Malet St., London WC1E 7HX, England. (1947) DPhil, crystallography (U. Oxford, 1972). Res. Technologist. (tel. 01 + 631-6485, fax 01 436 8918, email TICKLE at UK.AC.BBK.CR). *Computing, diffractometry, macromolecules, XRD.*

Tofield, Dr Bruce C. Materials Dev. Div., AERE Harwell, Didcot, Oxon OX11 0RA, England. (1943) DPhil, chemistry (U. Oxford, 1965). Group Leader - Materials and surface chemistry group. (tel. 0235 + 24141, ext. 4453; telex 83135). *Solid state chemistry, gas detectors, lithium batteries, oxidation, surface analysis.*

Tollin, Dr Patrick. Carnegie Lab. of Physics, U. of Dundee, Dundee DD1 4HN, Scotland. (1938) PhD, crystallography (U. Cambridge, 1963). Reader. (tel. 0382 + 23181, ext. 4561). *Structural biology, phase determination, macromolecules.*

Tomkeieff, Mr Michael Vamime. 3 Osgathorpe Drive, Pitsmoor, Sheffield, S. Yorks S4 7AP, England. (1934) MA, mineralogy and crystallography (U. Cambridge, 1962). (tel. 0742 + 388560). *Crystallography, metallurgy, mineralogy, petrology, statistics, electron beam instruments.*

Townsend, Dr Stephen Phillip. Dept. of Computing Sci., U. of Aberdeen, Dunbar St., Old Aberdeen AB9 2TY, Scotland. (1948) DPhil, numerical analysis (U. Oxford, 1977). Lect. (tel. 0224 + 40241, ext. 6417). *Numerical analysis, theoretical crystallography.*

Toy, Dr Mark. Dept. of Physics, Liverpool Polytechnic, Byron St., Liverpool L3 3AD, England. (1932) PhD, physics (UMIST, 1961). Ex. lect. (tel. 051 + 207 3581, ext. 2058). *Biological crystallography.*

Truter, Prof. Mary Rosaleen. Dept. of Chemistry, U. C. London, 20 Gordon St., London WC1H 0AJ, England. (1925) DSc, chemistry (U. London, 1965). Visiting prof. (tel. 01 + 387-7050, ext. 4657; fax 01 380 7463). *Molecular structure, coordination chemistry.*

Vickers, Miss Mary Elizabeth. X-ray Diffraction Group, BP International, Chertsey Rd., Sunbury-on-Thames, Middx. TW16 7LN, England. (1953) BSc, chemical physics (U. Bristol, 1975). Physicist. (tel. 09327 + 62079, telex 296041, fax 0932762999). *Polymers, orientation studies, X-ray diffraction, small-angle scattering, materials science.*

Vrielink, Ms Alice. Biophysics Section, Blackett Lab., Imperial C. of Sci. and Tech., London SW7 2BZ, England. (1959) MSc, chemistry (U. Calgary, Canada, 1986). PhD student. (tel. 02 + 589-5111, ext. 6733). *Protein crystallography, biological structures.*

Walker, Dr Peter Jonathan. Clarendon Lab., U. of Oxford, Parks Rd., Oxford OX1 3PU, England. (1947) PhD, chemistry (U. Liverpool, 1971). Res. chemist. (tel. 0865 + 59291, ext. 330). *Purification, preparation, halide single crystals, Bridgman-Stockbarger method, Czochralski method, oxygen-17 enriched materials.*

Wallis, Dr John Douglas. The Chemical Lab., U. of Kent, Canterbury, Kent CT2 7NH, England. (1954) D.Phil medicinal chemistry (U. of Oxford, 1979). Lect. (tel. 0227 + 764000, ext. 3547; fax 459025, email JDW at UK.AC.UKC). *Organic compounds, materials science, structural chemistry.*

Wallwork, Dr Stephen Collier. 'Inglewood', 15 Elm Ave., Beeston, Nottingham, NG6 1BV, England. (1925) DPhil, chemical crystallography (U. Oxford, 1950). Retired. *Organic complexes, radical ion salts and complexes, anhydrous metal nitrates, organonitrogen-metal complexes.*

Ward, Mr Roger Charles Chavannes. Marconi Infrared Devices Lab., GEC Hirst Res. Centre, East Lane, Wembley, Middx. HA9 7PP, England. (1948) DPhil, physics (U. Oxford, 1981). Res. assoc. (tel. 01 + 904-1262, ext. 499). *X-ray diffraction techniques, thin film growth and analysis, IR materials, crystal growth.*

Waring, Dr J.R.S. 3 Fop St, Uley, Nr Dursley, Glos, England.

Warner, Ms Joanne Kathleen. Chemical Crystallography Lab., U. of Oxford, 9 Parks Rd., Oxford OX1 3PD, England. (1961) B.App.Sci. applied chemistry (NSW Inst. Tech., 1987). DPhil Student. (tel. 0865 + 270823, email KRYST7 at OX.VAX). *Neutron diffraction, electron microscopy, X-ray diffraction.*

Watkin, Dr David John. Chemical Crystallography Lab., U. of Oxford, 9 Parks Rd., Oxford OX1 3PD, England. (1942) PhD, crystallography (U. Birmingham, 1967). Res. asst. (tel. 0865 + 270826). *Computing, instrumentation, teaching.*

Watson, Dr David Gilfillan. University Chemical Lab., U. of Cambridge, Lensfield Rd., Cambridge CB2 1EW, England. (1934) PhD, chemical crystallography (U. Glasgow, 1960). Asst. dir. of res. (tel. 0223 + 66499, ext. 317). *Organic crystal structures, crystallographic data storage and retrieval.*

Watson, Dr Herman Charles. Dept. of Biochemistry, U. of Bristol, University Walk, Bristol BS8 1TD, England. (1933) DSc, physics (U. Manchester, 1972). Reader. (tel. 0272 + 303030, ext. 3734). *X-ray diffraction, proteins, structural biology, diffractometry, reaction pathways, instrumentation, computing, graphics.*

Watts, Dr Bernard Enrico. Clarendon Lab., Oxford U., Parks Rd, Oxford OX1 3PU, England. (1954) M.A. Chemistry(Oxford, 1976) Res. Asst. (tel. 0865+272312). *Crystal growth, Inorganic crystals, materials science, phase determination, structural chemistry, optical properties.*

Weakley, Dr Timothy John Ruffer. Dept. of Chemistry, U. of Dundee, Dundee DD1 4HN, Scotland. (1933) DPhil, chemistry (U. Oxford, 1959). Lect. (tel. 0382 + 23181, ext. 284). *Polyoxoanions, V and Mo sub-groups, inorganic crystal chemistry.*

Webster, Dr Michael. Dept. of Chemistry, U. of Southampton, Southampton SO9 5NH, England. (1938) PhD, chemistry (U. London, 1962). Lect. (tel. 0703 + 559122). *Chemical crystallography.*

Weiss, Prof. Richard J. Physics, Kings College London, Strand WC2R 2LS, London, England. (1923) PhD, physics (New York U., USA, 1950). Prof. physics. *Electron distribution, polymers, thermodynamics, X-ray physics, compton scattering, optical fibre sensors.*

Welch, Dr Alan Jeffrey. Chemistry Dept, Edinburgh U, Edinburgh EH9 3JJ, Scotland. (1949) PhD, chemistry (U. of London, 1974). Sen. Lect. (tel. 031 + 667 1081, ext. 3406; telex 727442 UNIVED G, fax 662 4054, email EDCK01 at UK.AC.ED.EMAS). *Chemical bonding, electron density, molecular crystals, structural chemistry.*

Welch, Dr Dorothy Ann. Dept. of Computer Sci., Edinburgh U, Edinburgh EH9 3JZ, Scotland. (1960) PhD, inorganic chemistry (U. of Cambridge, 1985). Computing Officer. (tel. 031 + 6671081, ext. 2710; telex 727442 UNIVED G, fax 031 662 4712, email DAW at UK.AC.EDINBURGH). *Graphics, integrated project support environments, HCI.*

West, Dr Anthony Roy. Dept. of Chemistry, U. of Aberdeen, Old Aberdeen AB9 2UE, Scotland. (1947) DSc, solid state chemistry (U. Aberdeen, 1984). Reader. (tel. 0224 + 272918). *Crystallographic data, defect structures, disorder, diffraction and microscopy, magnetism, materials science, phase transitions, semiconductors, structural and solid state chemistry.*

West, Dr N.G. Health and Safety Executive, 403 Edgware Rd, London NW2 6LN, England.

Wheatley, Dr Peter Jaffrey. Dept. of Physical Chemistry, U. of Cambridge, Lensfield Rd., Cambridge CB2 1EP, England. (1921) DPhil, physical chemistry (U. Oxford, 1950). Lect., fellow of Queens' C. (tel. 0223 + 66499, ext. 428). *Chemical crystallography.*

Whelan, Dr Michael John. Dept. of Metallurgy and Sci. of Materials, U. of Oxford, Parks Rd., Oxford OX1 3PH, England. (1931) PhD, physics (U. Cambridge, 1958). Reader. (tel. 0865 + 273654, telex 83295 NUCLOX G, fax 0865-270708). *Electron microscopy, electron diffraction.*

Whiston, Dr Clive David. Sch. of Appl. Sci., The Polytechnic, Wulfruna St., Wolverhampton WV1 1LY, England. (1937) PhD, crystallography (U. Sheffield, 1963). Sr. lect. in inorganic chemistry. (tel. 0902 + 27371, ext. 129). *Crystal structures, anti-cancer drugs.*

Whitaker, Dr Alan. Dept. of Physics, Brunel U., Kingston Lane, Uxbridge, Middx. UB8 3PH, England. (1932) PhD, crystallography (Birkbeck C., London, 1965). Lect. (tel. 0895 + 74000, ext. 2406). *Organic crystal structures, pigments.*

White, Dr David Nathaniel James. Dept. of Chemistry, U. of Glasgow, Glasgow G12 8QQ, Scotland. (1946) DPhil, chemical crystallography (U. Sussex, 1970). Reader. (tel. 041 + 339-8855, ext. 7168). *Molecular conformation and mechanics, computer graphics, polypeptides, proteins, drug design.*

White, Dr Janice Larraine. Dept. of Biochemistry, U. of Sheffield, Western Bank, Sheffield, S. Yorks S10 2TN, England. (1948) PhD, biology, protein crystallography (Purdue U.,USA, 1976). Post-doc. res. asst. (tel. 0742 + 768555, ext. 4241, telex 547216 UGSHEF G, fax 727949, email BI1JW%UK.AC.SHEF.IBM. at AC.UK). *Computing, crystal growth, crystallographic data, electron density, hydrogen bonding, instrumentation, macromolecules, phase determination, structure prediction.*

Whittaker, Dr Eric James William. Dept. of Earth Sci., U. of Oxford, Parks Rd., Oxford OX1 3PR, England. (1921) PhD, crystallography (U. London, 1956). Retired. (tel. 0865 + 54511). *Silicate structures, disordered structures, four-dimensional crystallography.*

Whitworth, Dr Robert William. Dept. of Physics and Space Res., U. of Birmingham, Birmingham B15 2TT, England. (1932) PhD, physics (U. Cambridge, 1958). Sr. lect. (tel. 021 + 414-3344, fax 021-414-6709). *Dislocations, lattice defects, topography, ionic crystals, ice.*

Wierzchowski, Dr Wojciech Krzysztof. Dept. of Physics, RHBNC London, Egham Hill, Egham, Surrey TW20 0EX, England. (1948) PhD solid state physics (U. of Warsaw, 1982). Post Doc. (tel. 0784 + 434455, ext. 3499; telex 935504, fax 0784-437520). *X-ray diffraction, X-ray topography, crystal growth, defect structures, semiconductors, minerals, diamonds.*

Wilford, Dr John Bernard. Dept. of Computer Sci., Teesside Polytechnic, Borough Rd., Middlesborough TS1 3BA, England. (1940) PhD, chemistry (U. Bristol, 1966). Sr. lect. (tel. 0642 + 218121). *Crystal structures, industrially interesting compounds, computer programming (scientific and educational).*

Wilkinson, Dr Anthony Joseph. Dept. Chemistry, York U. Heslington, York YO1 5DD, England. (1960) PhD. Chemistry (London U.,1984) Lect. (tel. 0904+432589, fax YORK 410519, email WILKINSON at UK.AC.YORK.YORVIC). *Proteins, Structural biology.*

Wilkinson, Dr Clive. Dept. of Physics, King's C., Strand, London WC2R 2LS, England. (1941) PhD, physics (U. Cambridge, 1965). Lect. (tel. 01 + 836-5454, ext. 2586; fax 836-1799). *Neutron diffraction, position sensitive detectors, magnetism.*

Wilkinson, Dr D. Exsell Tech Comms Ltd., Dock Office, Trafford Rd., Salford, M5 2XB, England.

Willis, Prof. Bertram Terence Martin. Chemical Crystallography Lab., U. of Oxford, 9 Parks Rd., Oxford OX1 3PD, England. (1927) DSc, physics (U. London, 1968). Sr. res. fellow. (tel. 0865 + 270833, fax 0865 270708). *Diffraction physics, neutron scattering.*

Wilson, Prof. Arthur James Cochran. University Chemical Lab., Lensfield Rd., Cambridge CB2 1EW, England. (1914) PhD, physics (MIT, 1938; U. of Cambridge, 1942). Em. Prof. (tel. 0223 + 333655, email AJCW at CAM.PHX) *Crystallographic statistics, International Tables for Crystallography, data, information.*

Wilson, Prof. Herbert Rees. Physics Dept., U. of Stirling, Stirling FK9 4LA, Scotland. (1929) PhD, physics (U. Wales, 1952). Prof. and Head of Dept. (tel. 0786 + 73171, ext. 2008; telex 777557 STUNIVG, fax 0786 63000). *Structural biology, viruses, X-ray diffraction.*

Wilson, Dr Keith Sanderson. Dept. of Physics, U. of York, Heslington, York YO1 5DD, England. (1949) DPhil, Chemistry (U. Oxford, 1971). Lect. (tel. 0904 + 59861, ext. 5507). *Protein crystallography, macromolecular structure, biophysics.*

Wilson, Dr Michael Jeffrey. Dept. of Mineral Soils, Macaulay Inst. for Soil Res., Craigiebuckler, Aberdeen AB9 2QJ, Scotland. (1937) DSc, geology and soil science (U. Wales, 1984). Head of Dept. (tel. 0224 + 38611, ext. 241). *Mineralogy, rocks and soils, clay mineralogy.*

Windle, Dr Alan Hardwick. Dept. of Metallurgy and Materials Sci., U. of Cambridge, Pembroke St., Cambridge CB2 3QZ, England. (1942) PhD, metallurgy (U. Cambridge, 1966). Lect. (tel. 0223 + 65151, ext. 333). *Polymer physics, structured liquids, metal-polymer adhesion, composites.*

Windsor, Dr Colin George. Materials Physics and Metallurgy Div., B418, AERE Harwell, Oxon OX11 0RA, England. (1938) DPhil, physics (U. of Oxford. 1960). Group leader, neutron physics. (tel. 0235 + 24141, ext. 4025; telex 83135). *Neutron scattering.*

Winter, Dr Marcus John. X-ray Analytical Systems, Siemens Hse., Varey Rd., Congleton, Cheshire CW12 1PH, England. (1956) PhD, chemical physics (U. of Southampton, 1980). Sales eng. (tel. 0260 + 283441, ext. 3441; telex 8951091). *X-ray diffractometry techniques.*

Wonacott, Dr Alan John. Biophysics Sect., Blackett Lab., Imperial C., Prince Consort Rd., London SW7 2BZ, England. (1941) PhD, biophysics (U. London, 1966). Res. fellow. (tel. 01 + 589-5111, ext. 6725) *Protein crystallography.*

Wondre, Mr Friedrich Rudolf. Clarendon Lab., Oxford U., Parks Rd., Oxford OX1 3PU, England. (1943) G.Inst.P applied physics (Oxford Poly., 1969). Res. support. (tel. 0865 + 272312). *X-ray crystallography.*

Wood, Mr Dermott. X-ray Lab. Analytical Branch, British Petroleum Co. Ltd., Chertsey Rd., Sunbury-on-Thames, Middx. TW16 7LN, England. (1940) BSc, general (Nat. U. of Ireland, 1961). Technologist. (tel. 093 27 + 85533, ext. 8033). *Catalysts, poorly crystalline materials, powder diffraction, line profile analysis.*

Wood, Dr Ian George. Soils and Plant Nutrition Dept., Rothamsted Exp. Station, Harpenden, Herts AL5 2JQ, England. (1952) PhD, crystallography (U. London, 1977). HSO. (tel. 05827 + 63133, ext. 311). *Mineralogy, phase transitions, disorder, powder diffraction.*

Wood, Dr Raymond Maurice. Dept. of Appl. Physics, Sheffield City Polytechnic, Pond St., Sheffield S1 1WB, England. (1927) PhD, metallurgy (U. Sheffield, 1973). Sr. lect. (tel. 0742 + 20911, ext. 228). *Diffusionless transformations, structures, liquids, liquid crystals.*

Woods, Dr Geoffrey Steward. CSO Valuations Ltd, 17 Charterhouse St., London EC1N 6RA, England. (1939) PhD, physics (U. Witwatersrand, South Africa, 1971). Res. physicist. (tel. 01 + 404-4444, ext. 3181). *Diamonds, electron microscopy, infrared spectroscopy, optical spectroscopy, defects in solids, diffraction methods, radiation damage.*

Woolfson, Prof. Michael Mark. Dept. of Physics, U. of York, Heslington, York YO1 5DD, England. (1927) DSc, physics (U. Manchester, 1961). Prof. (tel. 0904 + 432230, fax 0904 432335). *Direct methods, small biological molecules.*

Wooster, Mr Antony Martin. Christie and Wooster Res. Ltd., 91 North St., Burwell, Cambridge CB5 0BB, England. (1935) Dir. (tel. 0638 + 741315). *Protein structures, automatic diffractometers, X-ray detectors.*

Wright, Dr Helen. Dept. of Computing Sci., U. of York, Heslington, York YO1 5DD, England. (1957) DPhil, crystallography (U. York, 1983). Appl. programmer. (tel. 0904 + 433809, email HW1 at YORK). *Direct methods, random phasing methods.*

Wright, Dr John Albert. Dept. of Civil Eng. and Construction, U. of Aston in Birmingham, Gosta Green, Birmingham B4 7ET, England. (1936) PhD, metallurgy (U. Sheffield, 1961). Lect. (tel. 021 + 359-3611, ext. 5195). *Environmental cracking, hydrogen embrittlement, safe-life predictions of structures, damage-tolerant design.*

Wright, Dr John Dalton. University Chemical Lab., U. of Kent, Canterbury, Kent CT2 7NH, England. (1941) DPhil, chemistry (U. Oxford, 1965). Lect. (tel. 0227 + 66822, ext. 519). *Crystal structure, electrical properties and spectra, molecular complexes, molecular crystals.*

Yates, Dr Paul Christopher. Dept. of Pharmaceutical Sci., Aston U., Aston Triangle, Birmingham B4 7ET, England. (1961) PhD, chemical physics (U. of Reading, 1987). Lect. (tel. 021 + 3593611, ext. 5223; telex 336997 UNIAST G, email YATESPC at UK.AC.ASTON.CLUST). *Computing, metals, X-ray diffraction, proteins, structural chemistry.*

Yewdall, Dr Stephen John. Dept. of Biochemistry, U. of Sheffield, Western Bank, Sheffield S10 2TN, England. (1961) PhD, biophysics (U. of Leeds, 1988). Post-Doc. (tel. 0742 + 768555, ext. 4190). *Computing, crystal growth, diffractometry, phase determination, proteins, structural biology, X-ray diffraction.*

Young, Mr Brian Raymond. Burnbank, New Road, Shiplake, Henley-on-Thames, Oxfordshire RG9 3LG, England. (1927) Retired. (tel. 0734 + 402142). *Mineralogy, clay mineralogy.*

Zussman, Prof. Jack. Dept. of Geology, U. of Manchester, Manchester M13 9PL, England. (1924) PhD, crystallography (U. Cambridge, 1952). Prof. (tel. 061 + 275-3804,). *Electron diffraction, electron microscopy, microscopy, minerals, powder diffraction, X-ray diffraction.*

UNITED STATES OF AMERICA

Sub-Editor: **Robert C. Taylor**

Notes

1. International telephone country code - 101. In the ten digit telephone numbers given in the entries, the first three are the regional area code; local calls within a regional area require only the last seven digits. Extension (ext.) numbers are used within an institution or company.

2. In the references to universities at which degrees were conferred, the following special abbreviations are used:

CIT - California Institute of Technology
MIT - Massachusetts Institute of Technology
PSU - Pennsylvania State University
PUN - Polytechnic University of New York (formerly: Polytechnic Institute of Brooklyn)

SUNY - State University of New York
UCB - University of California at Berkeley
UCLA - University of California at Los Angeles
USC - University of Southern California

3. The following acronyms are used in addresses for specific institutions:
NCI - National Cancer Institute
NIH - National Institutes of Health
HHMI - Howard Hughes Medical Institute

4. In the addresses, the following two-letter abbreviations are used for states and territories:

AL Alabama	IA Iowa	MT Montana	RI Rhode Island
AK Alaska	ID Idaho	NC North Carolina	SC South Carolina
AR Arkansas	IL Illinois	ND North Dakota	SD South Dakota
AZ Arizona	IN Indiana	NE Nebraska	TN Tennessee
CA California	KS Kansas	NH New Hampshire	TX Texas
CO Colorado	KY Kentucky	NJ New Jersey	UT Utah
CT Connecticut	LA Louisiana	NM New Mexico	VA Virginia
CZ Canal Zone	MA Massachusetts	NV Nevada	VI Virgin Islands
DC District of Columbia	MD Maryland	NY New York	VT Vermont
DE Delaware	ME Maine	OH Ohio	WA Washington
FL Florida	MI Michigan	OK Oklahoma	WI Wisconsin
GA Georgia	MN Minnesota	OR Oregon	WV West Virginia
GU Guam	MO Missouri	PA Pennsylvania	WY Wyoming
HI Hawaii	MS Mississippi	PR Puerto Rico	

5. The three degrees usually awarded by U.S.A. colleges and universities to graduates in scientific subjects are BS (four year program), MS (additional one or two years of courses which may include research), and PhD (three to five years beyond the BS, including research and dissertation; in some universities the MS degree may be an intermediate requirement). Occasionally the BA degree is awarded to batchelors' graduates in science. The DSc or ScD degree often, although not always, is an honorary degree not necessarily indicating professional training in science.

Aarif, Dr. Atta M. Chemistry Dept., U. of Utah, Salt Lake City, UT 84112, USA. (1953) PhD crystallography (U. of London, Queen Mary C., 1983). Staff crystallographer. (tel. 801 + 581-5320). *Structure determinations, small molecules, software, programming, modification.*

Abad-Zapatero, Dr. Celerino. Protein Crystallography Lab., D-47E, AP-9A, Abbott Labs., Abbott Park, IL 60064, USA. (1947) PhD biological sciences (U. of Texas at Austin, 1978). Res. scient., crystallogr. (tel. 312 + 937-0294, fax 312 + 937-0294). *Macromolecular crystallography, direct methods, protein structure refinement, crystallographic symmetry, interactive display, proteins, viruses, structure - function relation, drug design.*

Abdel-Meguid, Dr. Sherin S. Monsanto, 700 Chesterfield Village Pkwy., Chesterfield Village, MO 63198, USA. (1946) PhD chemistry (U. of Nebraska, 1977). Sr. group leadr. (tel. 314 + 537-6395, fax 314 + 537-6806). *Proteins, nucleic acids, viruses.*

Abel, Mr. James E. 265 W. Shore Trail, Sparta, NJ 07871, USA. (1915) MS chemistry (Stevens Inst. of Techn., 1954). Retired (tel. 201 + 729-5503).

Abola, Dr. Enrique E. Chem. Dept., Bldg. 555, Brookhaven Nat. Lab., Upton, NY 11973, USA. (1947) PhD crystallography (U. of Pittsburgh, 1973). Assoc. chemist. (tel. 516 + 282-4383, fax 516 + 282-5815, email Bitnet ABOLA at BNLCHM). *Information retrieval, scientific databases, crystallographic computing.*

Abola, Dr. Jaime Esteva. Crystallography Dept., U. of Pittsburgh, Fifth Ave., Pittsburgh, PA 15260, USA. (1947) PhD crystallography (U. of Pittsburgh, 1973). (tel. 412 + 624-9300). *Crystallography, proteins.*

Abraham, Prof. Donald James. Medicinal Chem. Dept., U. of Pittsburgh, 737 Salk Hall, Pittsburgh, PA 15261, USA. (1936) PhD org. chemistry (Purdue U., 1963). Prof. and chairman. (tel. 412 + 624-3261). *Medicinal chemistry, sickle cell anemia, Alzheimer's disease, drug design, drug - protein interactions, neurochemistry, X-ray crystallography.*

Abrahams, Dr. Sidney Cyril. Inst. f. Krystallographie der Univ. Tubingen, Charlottenstr. 33, D-7400 Tübingen, GERMAN FED. REP., Addr. fall 1990: Physics Dept., Southern Oregon State C., Ashland, OR 97520, USA. (1924) DSc crystallography (U. of Glasgow, U.K., 1957). Retired from AT&T Labs.

(tel. 49-07071-29-6058, telex 7262714 AITD, fax 49-7071-31153). *Condensed matter physics, atomic field displacements, crystallographic accuracy.*

Achari, Dr. Aniruddha. Genex Corp., 16020 Industrial Dr., Gaithersburg, MD 20877, USA.

Acosta Ramos, Mr. Oscar. Calle 23 00-6 Villa Guadalupe, Caguas, PR 00625, USA.

Adams, Prof. Richard Darwin. Chemistry Dept., U. of South.Carolina, Columbia, SC 29208, USA. (1947) PhD inorg. chemistry (MIT, 1973). Prof. (tel. 803 + 777-5104). *Inorganic chemistry, clusters, catalysts.*

Adams, Dr. Walter Wade. Polymer Branch, AFWAL/MLBP, AF Wright Aeronautical Labs., Wright-Patterson AFB, OH 45433-6533, USA. (1946) PhD Polymer Sci. & Eng. (U. of Massachusetts, 1984). Material res. eng. (tel. 513 + 255-9148, fax 513 + 255-5375, email ADAMS at WPAFB-SEVAX.ARPA or ADAMS at ADAWC.WPAFB.AF.MIL). *Polymer morphology, structure determination, fiber diffraction.*

Adhya, Dr. Sankar. Mol. Biology Lab., NCI, Nat. Inst. of Health, Bethesda, MD 20892, USA. (1937) PhD biochemistry (U. of Wisconsin, 1966). Head, Dev. Genetics Sect. (tel. 301 + 496-2495, fax 301 + 496-0260). *Structure, protein, DNA.*

Adler, Mr. George. 21 Harvard Rd., Shoreham, NY 11786, USA. (1920) MA phys. chemistry (Brooklyn C., 1952). Retired (tel. 516 + 744-2549). *Organic solid state chemistry, polymers.*

Adman, Dr. Elinor Thomson. Biological Structure Dept., Sch. of Medicine, U. of Washington - SM20 Seattle, WA 98195, USA. (1941) PhD phys. chemistry (Brandeis U., 1967). Res. assoc. prof. (tel. 206 + 543-6589, fax 206 + 543-1524, email ADMAN at XRAY0.BCHEM.WASHINGTON.EDU). *Macromolecular structures, electron transfer proteins.*

Afshar, Mrs. Carol E. Inst. for Cancer Res., Fox Chase Cancer Ctr., 7701 Burholme Ave., Philadelphia, PA 19111, USA. (1956) BS biochemistry (Spring Garden C., 1980). Res. asst. (tel. 215 + 728-2548).

Agard, Prof. David Andrew. Biochemistry Dept., U. of Calif. at San Francisco, Parnassus Ave., San Francisco, CA 94143, USA. (1953) PhD biological chemistry (CIT, 1980). Assoc. prof. (tel. 415 + 476-2521). *X-ray crystallography, 3-D image reconstruction, chromosome structure, protein structure.*

Aggarwal, Prof. Aneel Kumar. Biochem. & Mol. Biophys. Dept., C. Phys. & Surg., Columbia U., 630 W. 168th. St., New York, NY 10032, USA. (1957) PhD biophysics (U. of London, UK, 1984). Asst. prof. *DNA recognition, transcription control.*

Agron, Mr. Paul A. Chemistry Div., Oak Ridge Nat. Lab., P.O. Box X, Oak Ridge, TN 37831, USA. (alt. addr.: 102 Wilderness Ln., Oak Ridge, TN 37830, USA). (1915) MS phys. chemistry (PUN, 1946). Consultant. (tel. 615 + 574-5039, fax 615 + 626-2912). *Inorganic crystal structure determination, X-ray, neutrons, surface science, VVPES, ESCA, XPS.*

Akella, Dr. Radha. Whistler Ctr. Carbohydrate Res., Purdue U., Smith Hall, West Lafayette, IN 47907, USA. (1960) PhD crystallography (Indian Inst. Techn., Madras, 1988). Postdoct. res. assoc. (tel. 317 + 494-4924, email JPJ at MACE.CC.PURDUE.EDU). *Macromolecular crystallography, fiber diffraction, small angle X-ray scattering.*

Akers, Dr. Charles Kenton. 73 Oakgrove Dr., Williamsville, NY 14221, USA. (1942) PhD biophysics (SUNY at Buffalo, 1972). Sr. chemist. (tel. 716 + 634-3697). *Biomedical science, surface science, X-ray scattering (small angle).*

Alber, Dr. Tom. Biochemistry Dept., U. of Utah Sch. of Med., 50 N. Medical Dr., Salt Lake City, UT 84132, USA. (1954) PhD biology (MIT, 1981). Asst. prof. (tel. 801 + 581-2797 or 581-2117). *Protein crystallography, protein stability, protein folding, enzyme activity.*

Albert, Mr. Charles W. Glidden Co., 3901 Hawkins Point Rd., Baltimore, MD 21226, USA.

Alexander, Prof. Leroy. 68401 Hill St., Sturgis, MI 49091, USA. (1910) PhD phys. chemistry (U. of Minnesota, 1943). Retired. (tel. 616 + 651-2850). *Organic and polymer structures, X-ray diffraction.*

Alexander, Mr. Richard Scott. Chemistry Dept., U. of Pennsylvania, 34th. & Spruce St., Philadelphia, PA 19401, USA. (1965) BS chemistry (Philadelphia C. Textiles & Sci., 1987). Grad. student. (tel. 215 + 898-2227, fax 215 + 898-2037, email ALEX at XTAL.CHEM.UPENN.EDU). *Protein crystallography, biological chemistry.*

Alkire, Dr. Randy W. Nat.l Synchrotron Light Source, Bldg. 725, Beamline X8, Upton, NY 11973, USA. (1953) PhD chemistry (U. of Missouri at Rolla, 1982). Scient. specialist II. (tel. 516 + 282-5608). *Crystallography, instrumentation.*

Allen, Mr. Joseph H. 3629 Swallow Lane, Irving, TX 75062, USA. (tel. 214 + 255-8175).

Allersma, Mr. Ties. Glass Res. Center, PPG Industries, Box 11472, Pittsburgh, PA 15238, USA. (1936) Ir physics (Techn. U. of Delft, Netherlands, 1965). Sr. res. assoc. (tel. 412 + 665-8500). *Crystallization, glass-ceramics, optics as related to glass.*

Amin, Dr. Ahmed A. 22 Village Way, North Attleboro, MA 02760, USA. (1945) PhD solid state science (PSU, 1979). (tel. 617 + 699-1094). *Ferroelectrics, structure - properties relationship, grain boundary phenomena, device materials applications, instrumentation, automatic data acquisition.*

Amma, Prof. Elmer Louis. Chemistry Dept., U. of South Carolina, Columbia, SC 29208, USA. (1929) PhD phys. chemistry (Case Inst. of Tech., 1952). Prof. (tel. 803 + 777-2542). *Inorganic structural chemistry, solid state metal NMR, protein crystallography.*

Ammon, Prof. Herman L. Dept. of Chem. & Biochem., U. of Maryland, College Park, MD 20742, USA. (1936) PhD chemistry (U. of Washington, 1963). Prof. (tel. 301 + 454-2634, email AMMON at XRAY.UMD.EDU or HERMA at UMD2.UMD.EDU). *Protein crystallography, small molecules.*

Amzel, Dr. Leon Mario. Biophysics Dept., Johns Hopkins Sch. of Med., 725 N. Wolfe St., Baltimore, MD 21205, USA. (1942) Doc.Univ. phys. chemistry (U. of Buenos Aires, Argentina, 1968) Prof. (tel. 301 + 955-3955, email MARIO at JHUIGF). *X-ray diffraction, proteins, peptides, conformational studies, immunoglobulins, binding proteins.*

Anderson, Ms. Christine Alexis Francis Chemistry Dept., Hofstra U., Hempstead, NY 11550, USA. (1955) MS mineralogy (PSU, 1980). Manager, Chem. Labs. (tel. 516 + 560-5541). *Uranium mineralogy, crystal growth, hydrothermal alterations.*

Anderson, Dr. Daniel Horacio. Mol. Biology Inst., U. of Calif. at Los Angeles, Los Angeles, CA 90024-1570, USA. (1956) PhD biochemistry (U. of Calif. at San Diego, 1986). Postdoct. fel. (tel. 213 + 825-8901). *Protein crystallography, AIDS proteins, protein crystal growth, data collection.*

Anderson, Dr. Gary Don. Chemistry Dept., Marshall U., Huntington, WV 25701, USA. (1943) PhD org. chemistry (Florida State U., 1972). Assoc. prof. (tel. 304 + 696-6594). *Organic synthesis, natural products chemistry, computers in chemistry, X-ray crystallography.*

Anderson, Dr. John E. Cold Spring Harbor Lab., Bungtown Rd., P.O. Box 100, Cold Spring Harbor, NY 11724, USA. (1952) PhD biophysics (Harvard U., 1985). Sr. staff investig. (tel. 516 + 367-8822, email Bitnet ANDERSON at CSHLAB). *Nucleic acid binding proteins, signal transduction.*

Anderson, Prof. Oren Paul. Chemistry Dept., Colorado State U., Fort Collins, CO 80523, USA. (1942) PhD chemistry (Northwestern U., 1968). Prof. (tel. 303 + 491-6339, email OPA at CSUGREEN). *Chemistry, coordination, bioinorganic, polydentate chelates, mixed-valence compounds.*

Andrews, Dr. Lawrence Charles. Chemistry Dept., U. of Washington, BG-10, Seattle, WA 98125, USA. (1941) PhD chemistry (U. of Washington, 1975). Mgr., spectral serv. (tel. 206 + 364-9564).

Angilello, Mr. Joseph. Res. Div., IBM T. J. Watson Res. Center, Box 218, Yorktown Hgts., NY 10598, USA. (1929). Sr. res. eng. (tel. 914 + 945-1509, telex 137456, fax 914 + 945-2141, email ANGILEL at IBM.COM). *X-ray diffraction, topography, instrumentation.*

Ansell, Dr. Gerald Brian. 3 Windsor Ct., Annandale, NJ 08801, USA. (1936) PhD crystallography (U. of Essex, UK, 1966). *Diffraction, single crystal, powder, catalysts, inorganics, organometallics, organic small molecules.*

Antal, Dr. John Joseph. Materials Reliability Div., Army Matls. Techn. Lab. SLCMT-MRM, Arsenal St., Watertown, MA 02172-0001, USA. (1926) PhD physics (Saint Louis U., 1952). Sup. res. physicist. (tel. 617 + 923-5454). *Neutron scattering, material characterization by neutrons, instrumentation.*

Anthony, Dr. John W. P.O. Box 40726, Tucson, AZ 85717, USA. (1920) PhD geology (Harvard U., 1965). Prof. (tel. 602 + 621-2973). *Mineralogy, crystal structures, epitaxy, oxidation zone minerals.*

Antonio, Dr. Mark Ricci. Analyt., Env. & Appl. Sci. Div., BP Res. Internatl., 4440 Warrensville Ctr. Rd., Cleveland, OH 44128-2837, USA. (1954) PhD inorg. chemistry (Michigan State U., 1983). Project leader. (tel. 216 + 581-6754, fax 216-581-5621). *Synchrotron radiation diffraction, X-ray absorption spectroscopy, (XANES/EXAFS), Mossbauer spectroscopy.*

Appleman, Dr. Daniel E. Mineral Sci. Dept., Smithsonian Inst., NHB 119, Washington, DC 20560, USA. (1931) PhD geology, crystallography (Johns Hopkins U., 1956). Crystallographer. (tel. 202 + 357-2632). *Crystal chemistry, silicates, crystal structures, complex polytypic minerals.*

Aragón, Prof. Ricardo. Chem. Eng., Matls. Sci. Prog., U. of Delaware, Colburn Lab., Newark, DE 19716, USA. (1947) PhD geochemistry (Purdue U., 1979). Assoc. prof. (tel. 302 + 451-1132, fax 302-451-1048). *Transition metal oxides, electrical, magnetic properties, critical phenomena, incommensurate phases.*

Arai, Dr. Gerda Johanna. Eng. Dept., Zenith Electronics Corp., 2407 North Ave., Melrose Park, IL 60160, USA. PhD crystallography (U. of Leiden, Netherlands, 1960). Sect. mgr., analyt. chem. group (tel. 312 + 450-8380). *Inorganic structures, quantitative analysis, X-ray fluorescence.*

Araki, Mr. Takaharu. 97 Fitz-Henry Blvd., Columbus, OH 43214, USA. (Alt.: Chemical Abstracts Service, 43210 Columbus, OH 43210, USA). (1929) DSc mineralogy and crystallography (Kyoto U., Japan, 1961). Assoc. editor (tel. 614 + 447-3600 (ofc)). *Mineral and inorganic structures, computer programming.*

Archer, Dr. Ronald D. Chemistry Dept., U. of Massachusetts, Amherst, MA 01003, USA. PhD chemistry (U. of Illinois at Urbana, 1959). Prof., dept. head. (tel. 413 + 545-2291). *Structures, coordination compounds.*

Arem, Dr. Joel Edward. Joel E. Arem Inc., P.O. Box 5056, Laytonsville, MD 20879, USA. (1943) PhD mineralogy (Harvard U., 1970). Pres. (tel. 301 + 977-0335, telex 510-600-7531, fax 301 + 869-5898). *Geology, mineralogy, crystal growth, synthetic gemstones, computer sciences.*

Arents, Dr. Gina. Biology Dept., Johns Hopkins U., Charles and 34th Sts., Baltimore, MD 21218, USA. (1944) PhD biophysics (Johns Hopkins U., 1987). Postdoct. fel. (tel. 301 + 338-8590). *Protein crystallography, hemoglobins, DNA binding proteins.*

Arevalo, Mr. Jairo H. Scripps Clinic, 10666 N. Torrey Pines Rd., La Jolla, CA 92037, USA.

Arif, Dr. Atta Mahmood. Chemistry Dept., U. of Utah, HEB Bldg., Salt Lake City, UT 84112, USA. (1953) PhD chem. crystallography (London U., Queen Mary C., GB, 1983). Res. staff crystallographer. (tel. 801 + 581-5320). *Structures, inorganic, coordination, organometallic compounds, computer program applications.*

Armendarez, Prof. Peter X. Physics Dept., Brescia C., Owensboro, KY 42301, USA.

Armstrong, Prof. Ronald William. Mechanical Eng. Dept., U. of Maryland, College Park, MD 20742, USA. (1934) PhD metallurgical eng. (Carnegie Mellon U., 1958). Prof. (tel. 301 + 454-8881). *Dislocations, X-ray diffraction topography, single crystal strength properties, polycrystal plasticity, metals, ceramic structures, electronic semiconductors, energetic molecular crystals.*

Arnold, Dr. Edward Van Dyke. CABM at Rutgers U., Waksman Inst. Microbiology, P.O. Box 759, Piscataway, NJ 08855-0759, USA. (1957) PhD org. chemistry (Cornell U., 1982). Asst. prof. chem. (tel. 201 + 932-5182, email ARNOLD at BIOVAX). *Molecules in living systems, structure - function, viruses, viral proteins, drug design, viral polymerase, viruses, common cold, human immunodeficiency.*

Arnone, Prof. Arthur. Biochemistry Dept., U. of Iowa, Iowa City, IA 52242, USA. (1942) PhD phys. chemistry (MIT, 1970). Prof. (tel. 319 + 335-7882, email Bitnet CMDAAAVA at UIAMVS). *Macromolecular crystallography.*

Arora, Dr. Satish Kumar. Drug Dynamics Inst., C. of Pharmacy, U. of Texas at Austin, Austin, TX 78712, USA. (1942) PhD chemistry (U. of Poona, India, 1970). Sr. scient. (tel. 512 + 471-9267). *Biological structures.*

Arrington, Mr. Wendell. 175 E. Kenilworth, Newton Square, PA 19073, USA. (tel. 215 + 341-5680).

Artz, Ms. P. 461 Spring St., Pottstown, PA 19464, USA. (tel. 215 + 970-5309).

Aruffo, Dr. Alejandro A. Mol. Biology Dept., Massachusetts General Hospital, Fruit St., Boston, MA 02114, USA. (1959) PhD biophysics (Harvard U., 1988). Res. fellow. (tel. 617 + 726-5972). *Protein structure and function.*

Ashfaquzzaman, Mr. Syed. 82 Orion Walk, Holbrook, NY 11741, USA. (tel. 516 + 589-2885).

Atassi, Prof. M. Zouhair. Biochemistry Dept., Baylor C. of Med., One Baylor Plaza, Houston, TX 77030, USA. (1934) PhD chemistry (U. of Birmingham, England, 1960). Robt. A. Welch Prof. of Chem. & Biochem. (tel. 713 + 798-6050). *Proteins, binding sites, synthetic peptides, enzymes, antigens, antibodies, acetylcholine receptors, neurotoxins, lysozyme, trypsin, hemoglobin, myoglobin, urokinase, ragweed allergen.*

Athappily, Dr. Francis Kuriakose. Biochem. & Molec. Biophys. Dept., C. Phys. & Sur., Columbia U., 630 W. 168th. St., New York, NY 10032, USA. (1956) PhD chemistry (Calicut U., 1984). Postdoct. res. scient. (tel. 212 + 305-2219). *X-ray crystallography, biomolecules, molecular biology, protein engineering, protein structure, protein folding.*

Athey, Dr. Brian D. Phoenix Biotech, LTD., 309 N. Ashley, Ann Arbor, MI 48103, USA. (1957) PhD cellular/molecular biology (U. of Michigan, 1989). Vice-pres., R. & D. (tel. 313 + 663-9366, fax 313 + 669-9356). *Structures, protein, virus.*

Atkinson, Dr. David. Biophysics Dept., Boston U. Sch. of Medicine, 80 E. Concord St., Boston, MA 02118, USA. (1944) PhD biophysics (Council Nat. Academic Awards - UK, 1975). Assoc. prof., medicine & biochem. (tel. 617 + 638-4015). *Scattering, X-ray, neutron, biological macromolecular diffraction, biophysics, lipids, proteins, lipoproteins, membranes.*

Atoji, Dr. Masao. Basic Industry Res. Lab., Northwestern U., 1801 Maple Ave., Evanston, IL 60201-3135, USA. DSc physical chem. & crystallog. (Osaka U., Japan, 1956). Prin. res. scient. (tel. 312 + 985-1248 or 491-5030, fax 312 + 491-4486). *Semiconductors, electronics materials, diffraction, X-ray, neutron, magnetic structures, metals and alloys, crystal growth.*

Attard, Dr. Alfred E. 5434 Phelps Luck Dr., Columbia, MD 21045, USA. (1926) PhD physics (Illinois Inst. of Techn., 1962). *Solid state physics, chemistry, optics, electro-optics.*

Atwood, Prof. Jerry Lee. Chemistry Dept., U. of Alabama, Tuscaloosa, AL 35487, USA. (1942) PhD inorg. chemistry (U. of Illinois, 1968). U. res. prof. (tel. 205 + 348-8447). *Organometallic chemistry, molecular recognition, separations.*

Augustin, Mr. Rolf M. Polaroid Corporation, 575 Technology Square, Cambridge, MA 02139, USA.

Austerman, Mr. Stanley Boone. Austerman Associates, 17853 Santiago Blvd., #107-118, Villa Park, CA 92667, USA. (1922) BA physics (Purdue U., 1949). Consultant (tel. 714 + 639-2742). *X-ray crystallography, topography, crystal growth, characterization (thermal - electrical - physical).*

Averbach, Prof. Benjamin Lewis. Materials Sci. & Eng. Dept., Massachusetts Inst. of Techn., 77 Massachusetts Ave., Cambridge, MA 02139, USA. (1919) ScD metallurgy (MIT, 1947). Prof. (tel. 617 + 253-3320, telex 92-1473 MITCAM). fax 617 + 253-8000). *Amorphous structures, scattering, small angle X-ray, neutron.*

Ayers, Mr. John E. 6002 Jonsson Eng. Ctr., Rensselaer Polytechnic Inst., Troy, NY 12180, USA.

Azaroff, Prof. Leonid V. Inst. of Materials Sci., U. of Connecticut, Storrs, CT 06268, USA. (1926) PhD crystallography (MIT, 1954). Dir., prof. of physics (tel. 203 + 486-4623,4). *Structure, liquid crystals, LC polymers, diffraction studies, solids.*

Babich, Prof. Michael Wayne. Chemistry Dept., Florida Inst. of Techn., 150 W. University Blvd., Melbourne, FL 32901-6988, USA. (1945) PhD chemistry (U. of Nevada, 1974). Prof. and Head. (tel. 407 + 768-8000, Ext. 8046). *Chemistry, structural, solid state, coordination, thermal analysis, solid phase reaction kinetics.*

Babu, Dr. Yarlagadda Sudhakar. Ctr. for Macromolecular Cryst., Rm. 262, BHS Box 79 THT, U. of Alabama in Birmingham, Birmingham, AL 35294, USA. (1952) PhD X-ray crystallography (Indian Inst. of Sci., 1980) (tel. 205 + 934-7974, fax 205 + 934-0480). *X-ray crystallography, drug design, enzyme structure - function, calcium binding proteins.*

Badger, Dr. John. Rosenstiel Basic Med. Sci. Res. Ctr., Brandeis U., P.O. Box 9110, Waltham, MA 02254-9110, USA. (1962) PhD protein crystallography (York U., 1986). Postdoct. fel. (tel. 617 + 736-2495, email Bitnet BADGER at BRANDEIS). Virus structure, molecular dynamics, phase determination, protein refinement, water, drug design.

Baenziger, Prof. Norman C. Chemistry Dept., U. of Iowa, Iowa City, IA 52242, USA. (1922) PhD phys. chemistry (Iowa State U., 1948). Prof. (tel. 319 + 335-1374). *Crystal structures, X-ray diffraction.*

Bailey, Prof. Marcia F. Chemistry Dept., Central Michigan U., Mt. Pleasant, MI 48859, USA. (1939) PhD phys. chemistry (U. of Wisconsin at Madison, 1965). Assoc. prof., temp. (tel. 517 + 774-3956). *Crystal structures, organometallic, organic, structural chemistry.*

Bailey, Prof. Sturges Williams. Geology and Geophysics Dept., U. of Wisconsin, Weeks Hall, 1215 W Dayton St., Madison, WI 53706, USA. (1919) PhD crystallography (U. of Cambridge, UK, 1955). Prof. (tel. 608 + 262-1806). *Layer silicate structures, feldspars.*

Baird, Dr. Herbert Wallace. P.O. Box 849, Walkertown, NC 27051, USA. Walkertown, NC 27051, USA. (1936) PhD phys. chemistry (U. of Wisconsin at Madison, 1963). Consultant (tel. 919 + 595-4825). *Crystal and molecular structures.*

Baker, Dr. Kenneth Neil. Chemistry Dept., U. of Dayton, Matls. Lab. Off., Bldg. 654, 300 College Park, Dayton, OH 45469, USA. (1957) PhD solid state

chemistry (U. of Mississippi, 1986). Postdoct. res. assoc. (tel. 513 + 255-9164 or 255-9133). *Crystallography, X-ray diffraction, oligomers, polymers, non-linear optical materials.*

Baker, Dr. R. J. Chemistry Dept., Cleveland State U., Cleveland, OH 44115, USA. (1952) PhD phys. chemistry (U. of New Orleans, 1982). Adjunct asst. prof. (tel. 216 + 687-3974). *X-ray diffraction, chemotherapeutic agents, charge-transfer complexes.*

Baldwin, Mr. Eric T. Crystallog. Lab., BRI-Basic Res. Prg., NCI, Bldg. 539, Frederick Cancer Res. Facil., Frederick, MD 21701, USA. (1960) BS biochemistry (Purdue U., 1982). Sr. res. techn. (tel. 301 + 698-5031, fax 301 + 698-5991). *Macromolecular crystallography, DNA-protein interactions.*

Baldwin, Mr. Kenneth John. Mineral Physics Inst., ESS Dept., SUNY at Stony Brook, Stony Brook, NY 11794-2100, USA. (1948) MS geochemistry (SUNY at Stony Brook, 1973). Res. assoc. (tel. 516 + 632-8196, email Bitnet KBALDWIN at SBCCMAIL). *Crystallography, instrumentation, automation, X-ray diffraction.*

Bale, Prof. Harold D. Physics Dept., U. of North Dakota, Grand Forks, ND 58202, USA. (1927) PhD physics (U. of Missouri, 1959). Prof. (tel. 701 + 777-3529). *Small angle X-ray scattering, liquids, non-crystalline solids.*

Bales, Mr. Howard E. 247 Edison Blvd., Xenia, OH 45385, USA. (1912) BSc education (Wilmington C., 1934). Retired (tel. 513 + 372-3688). *Education, chemistry, instrumental analysis, research funding.*

Ball, Dr. Richard George. Biophysical Chem. Dept., Merck Sharp & Dohme Res. Labs., P.O. Box 2000, Rahway, NJ 07065, USA. (1950) PhD chemical crystallography (U. of Western Ontario, 1978). Res. fellow. (tel. 201 + 594-5341, fax 201 + 594-6645, email XRAY%ROB at UUNET). *Biological small molecules, drug design, crystallographic computing.*

Banaszak, Prof. Leonard J. Biochemistry Dept., U. of Minnesota, 435 Delaware St. SE, Minneapolis, MN 55455, USA. (1933) PhD biochemistry (Loyola U., 1961). Prof. (tel. 612 + 625-6907). *Protein crystallography, lipid-protein systems.*

Banko, Mr. Brad. 1223 Waterford Dr., Apt. C, Columbus, OH 43220, USA. (1961) BS physics (Brown U., 1983). Engr. (tel. 614 + 875-7912, email banko at ohstpy.mps.ohio-state.edu). *Sputtering, thin films.*

Banks, Prof. Ephraim. Chemistry Dept., Polytechnic U., 333 Jay St., Brooklyn, NY 11201, USA. (1918) PhD inorg. chemistry (PUN, 1949). Prof. (tel. 718 + 260-3285). *Crystal growth, chemistry, spectra, properties (magnetic - electrical - luminescent), oxide superconductors.*

Barber, Dr. Ann M. Lab. of Mathematical Biol., NCI, BG 469, Frederick, MD 21701, USA. (1951) MD medicine (Northwestern U., 1981). Sr. staff fel. (tel. 301 + 698-5576, fax 301 + 698-5598, email BARBER at NCIFCRF.GOV). *GTP-binding proteins, reverse transcriptase.*

Barber, Prof. Patrick George. Natural Sci. Dept., Longwood C., Farmville, VA 23901, USA. (1942) PhD Physical chemistry (Cornell U., 1969). Prof. of Chemistry. (tel. 804 + 392-9351). *Liquid crystals, crystal growth.*

Barcza, Dr. Sandor. CADD 403, Sandoz Res. Inst., Route 10, East Hanover, NJ 07936, USA. (tel. 201 + 503-8108). *Molecular structure determination, computer assisted drug design, structure - activity relations, molecular modeling, information storage, retrieval.*

Bardhan, Mr. Pronob. Sci. Prod. Portfolio, R&D Labs., Corning Glass Works, Corning, NY 14830, USA.

Barkigia, Dr. Kathleen M. 63 Silver St., Patchogue, NY 11772, USA. (1951) PhD Chemistry (Georgetown U., 1978). Assoc. sci. (tel. 516 + 282-4382). *X-ray diffraction, photosynthetic pigments, modeling.*

Barnes, Dr. Charles Leslie. Chemistry Dept., U. of Missouri, Columbia, MO 65211, USA. (1949) PhD biochemistry (U. of Tennessee, 1980). Dept. crystallographer, res. investig. (tel. 314 + 882-2962). *Crystallography, natural products, peptide conformation, structure - function relationships.*

Barnett, Dr. Bobby L. Miami Valley Res. Labs., Procter & Gamble Co., P.O. Box 398707, Cincinnati, OH 45239-8707, USA. (1939) PhD phys. chemistry (U. of Texas at Austin, 1970). Group leader. (tel. 513 + 245-2321, fax 513 + 235-1612). *Molecules, computer aided design, protein engineering.*

Barney, Ms. Elsa Pauline. 312 Burton St., Bath, NY 14810, USA. (1922) MS geology (MIT, 1947). Tax map tech. (tel. 607 + 776-4161). *Mineral structures.*

Barnhart, Dr. David Merle. Physical Sci. Dept., Eastern Montana U., Billings, MT 59101, USA. (1933) PhD phys. chemistry (Oregon State U., 1964). Prof. (tel. 406 + 657-2341). *Organometallic complexes.*

Barrett, Prof. Charles Sanborn. Eng. Dept., U. of Denver, Denver, CO 80208, USA. (Winter months: 115 E. Los Arcos, Green Valley, AZ 85614, USA) (1902) PhD physics (U. of Chicago, 1928). Sr. res. scient., adjunct prof. phys. (tel. 303 + 871-3515 (ofc)). *X-ray, diffraction, instrumentation, imaging, metals, polymerics, composites, stress analysis, phase identification, editing.*

Barrick, Dr. James Clinton. 564 S. Selby Blvd., Worthington, OH 43085, USA. (1940) PhD inorg. chemistry (McMaster U., Canada, 1972). Information scient. (tel. 614 + 471-3558). *X-ray crystallography, single crystal, powder, materials research, alloys, organic compounds, cobalt complexes, computer program applications.*

Bartell, Prof. Lawrence Sims. Chemistry Dept., U. of Michigan, Ann Arbor, MI 48109, USA. (1923) PhD chemistry (U. of Michigan, 1951). Prof. (tel. 313 + 764-7375, fax 313 + 747-4865, email L.S.BARTELL at UB.CC.UMICH.EDU). *Electron diffraction, holography, distributions, molecular* *structure, vibrations, force fields, isotope effects, quantum chemistry, liquid structure, molecular clusters.*

Bartlett, Prof. Neil. Chemistry Dept., U. of Calif. at Berkeley, Berkeley, CA 94720, USA. (1932) DSc inorganic chemistry (U. Newcastle Upon Tyne, UK, 1958). Prof. (tel. 415 + 642-7259, fax 415 + 486-7000). *Solid state chemistry, fluorine chemistry, noble-gas chemistry, graphite intercalations, novel graphite relatives.*

Barton, Dr. C. J. Dorr-Oliver Inc., 77 Havemeyer Lane, P.O. Box 9312, Stamford, CT 06904, USA. (1936) PhD metallurgy (Rensselaer Polytechnic Inst., 1966). President. (tel. 203 + 358-3200). *Structure - properties relationships.*

Barton, Dr. Randolph Jr. Fibers Dept., Experimental Station, Box 80302, E.I. du Pont de Nemours Co., Wilmington, DE 19880-0302, USA. (1941) PhD phys. chemistry (Johns Hopkins U., 1968). Res. assoc. (tel. 302 + 695-2578, fax 302 + 695-1717). *Fiber structure, morphology.*

Basu, Dr. Sankar Prasad. Lab. for the Structure of Matter, Naval Res. Lab., Overlook Ave., Washington, DC 20375, USA. (1941) PhD biophysics (U. of Oklahoma, 1977). Res. scient. (tel. 202 + 767-0656). *Macromolecular crystallography.*

Bates, Prof. Robert Brown. Chemistry Dept., U. of Arizona, Tucson, AZ 85721, USA. (1933) PhD chemistry (U. of Wisconsin, 1957). Prof. (tel. 602 + 621-6317). *Organic chemistry, natural products, carbanions.*

Batterman, Prof. Boris William. Sch. of Appl. & Eng. Physics, CHESS-Synchrotron Radiation Lab., Cornell U., Ithaca, NY 14853, USA. (1930) PhD physics (MIT, 1956). Dir. (tel. 607 + 256-5161, telex MCI16713054, fax 607 + 255-8062, email Bitnet BATTERMAN at CRNLCHES). *Synchrotron radiation, diffraction, X-ray, neutron, dynamical X-ray, solid state physics, anharmonic vibrations.*

Bau, Prof. Robert. Chemistry Dept., U. of Southern California, Exposition Blvd., Los Angeles, CA 90007, USA. (1944) PhD inorg. chemistry (UCLA, 1968). Prof. (tel. 213 + 743-8800, telex 674-803, fax 213 + 747-4176, email Bitnet BAU at RAMOTH). *Neutron diffraction, structure determination, transition metal hydride complexes, metal-nucleotide complexes.*

Baughman, Mr. Richard Joseph. Div. 1144, Sandia Nat. Lab., Albuquerque, NM 87185, USA. (1927) BS chemistry-biology (Mount Union C., 1950). Techn. staff member. (tel. 505 + 844-6337). *Crystal growth, materials preparation, electron microscopy.*

Baughman, Prof. Russell George. Div. Sci., Northeast Missouri State U., Kirksville, MO 63501, USA. (1946) PhD phys. chemistry (Iowa State U., 1977). Assoc. prof. (tel. 816 + 785-4627). *Pesticides, charge transfer complexes, crystallography education.*

Baures, Mr. Paul William. Physical & Struct. Chem., Smith Kline & French Labs., 709 Swedeland Rd., King of Prussia, PA 19406, USA. (1963) MS organic chemistry (U. of Minnesota at Minneapolis, 1988). Scient. (tel. 215 + 270-4263). *Co-crystals, co-crystallization, hydrogen bonding, triphenylphosphine oxide.*

Bear, Prof. Richard Scott. 1515 E. Franklin St., Unit 43, Chapel Hill, NC 27514, USA. (1908) PhD chemistry (UCB, 1933). Prof. emeritus. (tel. 919 + 929-8337). *Natural fibers, membranes, structures, optics, X-ray diffraction.*

Beard, Mr. Donald W. Siemens Analyt. X-ray Instruments, 6300 Enterprise Ln., Madison, WI 53719-1173, USA. (1928) BS chemistry (Allegheny C., 1950). Prod. mgr., XRD & XRF. (tel. 800 + 552-9729, fax 608 + 276-3015). *X-ray diffraction, spectroscopy, instrumentation.*

Beasley, Prof. Wayne Machon. Materials Sci. Div., Mechanical Eng. Dept., U. of New Hampshire, Durham, NH 03824, USA. (1922) SM ceramics (MIT, 1965). Prof. emeritus. (tel. 603 + 332-6375). *Inorganic crystal structures, quantitative stereology.*

Becker, Prof. Joseph Whitney. Developmental & Mol. Biol. Dept., Rockefeller U., 1230 York Ave., New York, NY 10021-6399, USA. (1943) PhD chemistry (Stanford U., 1970). Assoc. prof. (tel. 212 + 570-8183, telex 7105814146, fax 212 + 570-7974, email Bitnet EAU at ROCKVAX). *Protein crystallography, image reconstruction.*

Bedarkar, Dr. Sudhir. Biophysics Dept. Rm. 612, Johns Hopkins Sch. of Med., 725 N. Wolfe St., Baltimore, MD 21205, USA. (1951) PhD protein crystallography (U. of London, UK, 1982). Postdoctoral Fellow (tel. 301 + 955-8715). *Structure & function, proteins, nucleic acids, protein crystallography, biochemistry.*

Bednowitz, Dr. Allan Lloyd. Res. Div., IBM T. J. Watson Res. Center, P.O. Box 218, Yorktown Hgts., NY 10598, USA. (1939) PhD chemical physics (PUN, 1966). Res. staff member. (tel. 914 + 945-1529, telex 137456, fax 914 + 945-2141, email BENWITZ at IBM.COM). *Computer graphics, programming, automation, direct determination.*

Bedzyk, Dr. Michael James. CHESS, Wilson Lab., Cornell U., Ithaca, NY 14853, USA. (1951) PhD physics (SUNY at Albany, 1982). Sr. res. assoc. (tel. 607 + 255-0920, telex MCI6713054, fax 607 + 255-8062, email Bitnet BEDZYK at CRNLCHES). *Dynamical X-ray diffraction, synchrotron radiation, structure determination, surface, interface, solid state physics.*

Beem, Dr. Karl Michael. 3088 Boxwood Dr., NE., Atlanta, GA 30345, USA. (1943) PhD chemistry (U. of Michigan, 1969). Software scient. (tel. 404 + 496-1603). *Crystallography, inorganic chemistry, NMR.*

Beese, Ms. Lorena S. Mol. Biophysics & Biochem. Dept., Yale U., 260 Whitney Ave., New Haven, CT 06511, USA. (1962) Assoc. res. scient. (tel. 203 + 432-5611, email Bitnet BEESE at YALEVMS). *Macromolecular crystallography, protein-nucleic acid interactions, direct methods.*

Beiter, Mr. Thomas Albert. Chemistry Dept., 320 Hughes Labs., Miami U., Oxford, OH 45056, USA. (1947) MS physics (Miami U., 1984). Grad. student. (tel. 513 + 529-2834). *Crystallography, structure determination, nuclear structure, logic and mathematical foundations.*

Bell, Dr. Jeffrey A. Inst. of Mol. Biology, U. of Oregon, Eugene, OR 97403, USA. (1954) PhD biochemistry (Cornell U., 1985). Res. assoc. (tel. 503 + 686-5176, email JEFF at UOXRAY.UOREGON.EDU). *Enzymes, protein structure - function, thermostability.*

Belt, Dr. Roger F. Airtron, Litton Industries Inc., 200 E. Hanover Ave., Morris Plains, NJ 07950, USA. (1929) PhD phys. chemistry (State U. of Iowa, 1956). Res. dir. (tel. 201 + 539-5500, ext. 501). *Crystal growth, structure, perfection of crystals, X-ray techniques, magnetic materials, superconductors.*

Bené, Prof. Robert William. Elec. & Computer Eng. Dept., U. of Texas at Austin, ENS 143, Austin, TX 78712, USA. (1939) PhD electrical eng. (Stanford U., 1968). Prof. (tel. 512 + 471-1225). *Interface structure, thin film kinetics, nucleation, metastable structure.*

Bennett, Prof. Dennis W. Chemistry Dept., U. of Wisconsin at Milwaukee, 3210 Cramer St., Milwaukee, WI 53201, USA. (1946) PhD inorg. chemistry (U. of Utah, 1978). Assoc. prof. (tel. 414 + 229-5276). *Synthesis, structure, bonding, transition metal and metal-organic compounds, crystallography, chemical, computational.*

Bennett, Dr. John Michael. Central Scientific Labs., Tarrytown Techn. Center, Union Carbide Corp., Tarrytown, NY 10591, USA. (1939) PhD chemistry (U. of Aberdeen, 1966). Sr. res. assoc. (tel. 914 + 789-3604). *Molecular sieve materials, diffraction, X-ray, electron, neutron, powder.*

Beno, Dr. Mark A. Chemistry Div., Bldg. 200 C101, Argonne Nat. Lab., 9700 Cass Ave., Argonne, IL 60439, USA. (1951) PhD phys. chemistry (Ohio State U., 1979). Chemist. (tel. 312 + 972-3507, fax 312 + 972-4470, email Bitnet BENO at ANLCHM). *X-ray and neutron diffraction, instrumentation, low temperature diffractometry.*

Benson, Mr. James Edward. Materials Characterization Lab., General Electric Co., 21800 Tungsten Rd., Cleveland, OH 44117, USA. (1933) MS phys. chemistry (Iowa State U., 1963). Assoc chemist. (tel. 216 + 266-8370). *X-ray diffraction, refractory metals, automation.*

Beres, Mr. John J. 6014 Echodell, NW, North Canton, OH 44720, USA. (tel. 216 + 798-4015).

Bergmann, Mrs. Margot E. Physics Dept., Polytechnic Inst. of New York, 640 Riverside Dr., New York, NY 10031, USA. (1913) MS phys. chemistry (Rutgers U., 1940). Res. consultant. (tel. 718 + 643-4984). *Diffraction, X-ray, electron, electron microscopy, X-ray camera development, polymers, biological materials.*

Berliner, Prof. Lawrence J. Chemistry Dept., Ohio State U., 120 W. 18th. Ave., Columbus, OH 43210, USA. (1941) PhD phys. chemistry (Stanford U., 1967). Prof. (tel. 614 + 292-0134, telex 332 911, fax 614 + 292-1685, email Bitnet BERLINER at OHSTCH or TS0288 at OHSTVMA). *Protein structure - function, imaging, NMR.*

Berliner, Dr. Ronald Richard. Res. Reactor, U. of Missouri, Res. Park, Columbia, MO 65211, USA. (1943) PhD physics (U. of Illinois, 1973). Sr. res. scient. (tel. 314 + 882-5235, telex MURR COMA 434199, fax 314 + 882-3443). *Phase transformations, crystal defects, instrumentation.*

Berman, Dr. Helen M. Inst. for Cancer Res., Fox Chase Cancer Ctr., 7701 Burholme Ave., Philadelphia, PA 19111, USA. (1943) PhD crystallography (U. of Pittsburgh, 1967). Sr. member. (tel. 215 + 728-2548). *Crystallography, biological molecules, nucleic acids, interactions, conformation.*

Bernal, Prof. Ivan. Chemistry Dept., U. of Houston, Cullen Blvd., Houston, TX 77204-5641, USA. (1931) PhD chemical physics (Columbia U., 1963). Prof. (tel. 713 + 749-2618). *Absolute configurations, organometallics, coordination compounds, resolutions, spontaneous, chiral, conglomerate crystallizations, energetics.*

Bernheim, Dr. Marguerite Mary. Computer Sci. Dept., Pennsylvania State U., 333 Whitmore Lab., University Park, PA 16802, USA. (1946) PhD inorganic chem., crystallog. (PSU, 1976). Prof. (tel. 814 + 863-0897). *Molecular graphics, computer modeling, bridged metallic compounds, organometallic systems, structure determination methods.*

Bernstein, Ms. Frances C. Protein Data Bank, Chemistry Dept., Brookhaven Nat. Lab., Upton, NY 11973, USA. (1942) MS mathematics (New York U., 1965). Computer analyst. (tel. 516 + 282-4382, telex 6852516 BNL DOE, fax 516 + 282-5815, email Bitnet BERNSTEIN at BNLCHM). *Macromolecular data bases, macromolecular structures, computer programming.*

Bernstein, Dr. Herbert Jacob. Courant Inst. of Math. Sci., New York U., 251 Mercer St., New York, NY 10012, USA. (1944) PhD mathematics (New York U., 1968). Sr. res. sci. (tel. 212 + 998-3038, email YAYA at NYU.EDU). *Scientific computing, theoretical crystallography, data acquisition.*

Berry, Dr. Chester Ridlon. 37 Heritage Dr., S. Orleans, MA 02662, USA. (1919) PhD physics (Cornell U., 1946). Consultant (tel. 617 + 255-6206). *Mechanisms, microcrystalline nucleation and growth, crystal imperfection methods, small particle optical behavior, photographic sensitivity theories.*

Bertrand, Prof. Joseph Aaron. Sch. of Chemistry, Georgia Inst. of Techn., Atlanta, GA 30332, USA. (1933) PhD inorg. chemistry (Tulane U., 1961). Prof. (tel. 404 + 894-4050). *Transition metal complexes, X-ray diffraction, heterogeneous catalysis.*

Bertraud, Mr. J. A. Sch. of Chemistry, Georgia Inst. of Techn., Atlanta, GA 30332, USA. (tel. 404 + 894-2186).

Bethge, Dr. Paul Herman. Physiology & Biophysics Dept., Washington U. Med. Sch., 660 S. Euclid Ave., St. Louis, MO 63110, USA. (1945) PhD chemistry (Harvard U., 1973). Res. instructor. (tel. 314 + 362-3354). *Crystallography, biological macromolecules.*

Betts, Dr. Foster. Res. Lab., Hosp. for Special Surg., Cornell U. Med. C., 535 E. 70th St., New York, NY 10021, USA. (1932) PhD electrical eng. (Stanford U., 1972). Assoc. scient. *Amorphous materials.*

Betts, Ms. Laurie. Biochemistry Dept., U. of North Carolina, 405 FLOB 231 H, Chapel Hill, NC 27514, USA. Grad. student. (tel. 919 + 966-3263).

Bhandary, Dr. Krishna K. Oral Biology Dept., SUNY at Buffalo, 109 Foster Hall, Buffalo, NY 1421463, USA. (1946) PhD chemistry (Indian Inst of Sci., India, 1974). Asst. prof. (tel. 716 + 831-2114, fax 716 + 831-3942). *Cyclic peptides, cardiotonic agents, ionophores, salivary proteins.*

Bhattacharjee, Mr. Sarama. 36 Constantine Pl., Apt. 53, Summitt, NJ 07901, USA.

Bienenstock, Prof. Arthur Irwin. Stanford Synchrotron Radiat. Lab., Stanford U., P.O. Box 4349, Bin 69, Stanford, CA 94309-0210, USA. (1935) PhD applied physics (Harvard U., 1962). Prof., dir. (tel. 415 + 926-3153, fax 415 + 926-4100, email A at SSRL750). *Non-crystalline solids, poorly crystallized solids, physical properties, atomic arrangements, synchrotron radiation.*

Bigelow, Prof. Wilbur Charles. Matls. & Metallurg. Eng. Dept., Dow Building, North Campus, U. of Michigan, Ann Arbor, MI 48109, USA. (1923) PhD phys. chemistry (U. of Michigan, 1951). Prof. (tel. 313 + 764-3321). *Electron diffraction, microscopy.*

Bilderback, Dr. Donald Heywood. Sch. of Appl. & Eng. Physics, Cornell U., Ithaca, NY 14853, USA. (1947) PhD solid state physics (Purdue U., 1975). Staff Scient. for CHESS (tel. 607 + 255-7163, email Bitnet BILDERBACK at CRNLCHES). *Diffraction physics, multiwire X-ray detectors, X-ray mirrors, monochromators, Laue diffraction, synchrotron radiation.*

Billman, Mr. John F. U. of Washington, Benson Hall BF-10, Seattle, WA 98195, USA.

Bingman, Mr. Craig A. Biochemistry Dept., U. of Wisconsin, 420 Henry Mall, Madison, WI 53706, USA.

Binnie, Dr. William Polson. The Carborundum Co., BP America, P.O. Box 832, Niagara Falls, NY 14302, USA. (1924) PhD chemistry (U. of Glasgow, UK, 1948). Scient. (tel. 716 + 278-2565). *Materials science.*

Birdsall, Prof. William John. Chemistry Dept., Albright C., 13th & Exeter Sts., Reading, PA 19612-5234, USA. (1944) PhD inorg. chemistry (PSU, 1971). Prof. (tel. 215 + 921-2381). *Metal ion complexes, purines, indoles, pyrroles.*

Birgeneau, Prof. Robert J. Physics Dept., Rm. 13-2114, Mass. Inst. of Techn., Cambridge, MA 02139, USA. (tel. 617 + 253-4937).

Birks, Mr. L. S. 306 Lands End Rd., Chapin, SC 29036, USA. (1919) MS physics (U. of Maryland, 1951). Retired, part-time consultant *Spectrometer crystals, crystals, defects, diffraction theory.*

Birktoft, Dr. Jens Julius. Biochem. & Mol. Biophys. Dept., Washington U. Sch. of Med., 660 S. Euclid Ave., Box 8094, St. Louis, MO 63110, USA. (1942) Cand.scient. biochemistry (U. of Copenhagen, 1967). (tel. 314 + 362-3340, fax 314 + 362-7183, email BIRKTOFT at WUMS). *Protein crystallography, enzyme mechanism, evolution.*

Bish, Dr. David Lee. Earth and Space Sci. Div., Los Alamos Nat. Lab., MS D469, Los Alamos, NM 87545, USA. (1952) PhD mineralogy, (PSU, 1977). Staff mineralogist. (tel. 505 + 667-1165). *Mineralogy, clay mineralogy, X-ray powder diffraction.*

Biswas, Dr. Amit. Central Res. & Dev. Dept., E.I. duPont de Nemours Co., P.O. Box 80356, Wilmington, DE 19880-0356, USA. (1960) PhD macromolecular science (Case Western Reserve U., 1988). Visiting scient. (tel. 302 + 695-1848). *Polymers, structure, conformation, liquid crystals, computer modeling.*

Black, Dr. Shaun D. Med. Chem. & Pharmacognosy Dept., Ohio State U., 500 W. 12th. Ave., Columbus, OH 43210-1291, USA. (1954) PhD biological chemistry (U. of Michigan, 1982). Asst. prof. (tel. 614 + 292-3925, fax 614 + 292-2435, email BLACK at OHSTPHRM). *Crystallization, topology, dynamics, membrane proteins, cytochromes, P-450.*

Blake, Prof. Jerry Wayne. Veterinary Sci. Dept., U. of Kentucky, 103 Animal Path., Med. Ctr. Dr., Lexington, KY 40546, USA. (1936) PhD mineralogy (Ohio State U., 1968). Prof. (tel. 606 + 257-4826).

Blanton, Mr. Thomas Nelson. Bldg. 49, Res. Labs., Eastman Kodak Co., Rochester, NY 14652, USA. (1959) MS analyt. chemistry (Emory U., 1981). Res. chemist, suprv. X-ray sect. (tel. 716 + 722-3323). *X-ray diffraction, solid state, thin films.*

Blessing, Dr. Robert Harry. Mol. Biophysics Dept., Med. Fndn. of Buffalo, 73 High St., Buffalo, NY 14203, USA. (1941) PhD chemistry (Ohio U., 1971). Sr. res. scient. (tel. 716 + 856-9600, fax 716 + 852-4846, email Bitnet WPRHB%MFB at UBVMS). *Small biological molecules, hydrogen bonding, accurate structure analysis, electron density distributions.*

Blevins, Dr. Richard A. Merck & Co., P.O. Box 2000, Rahway, NJ 07065, USA. PhD chemistry (Michigan State U., 1984). Sr. res. chemist. (tel. 201 + 594-7329). *Biological macromolecules, computer graphics.*

Bliss, Dr. Mary. Neutron Devices Dept., General Electric Co., P.O. Box 2908, Largo, FL 34649, USA. (1959) PhD solid state science (PSU, 1989). Adv. matls. eng. (tel. 813 + 541-8001).

Block, Dr. Stanley. Ceramics Div., Nat. Inst. Stnds. & Techn., Gaithersburg, MD 20899, USA. (1926) PhD chemistry (Johns Hopkins U., 1955). Res. chemist. (tel. 301 + 975-5733, fax 301 + 975-2128). *Inorganic compounds, ceramics, high pressures, structures, identification, powder, single crystal.*

Blount, Dr. Alice M. Newark Museum, P.O. Box 540, Newark, NJ 07101, USA. (1942) PhD geology (U. of Wisconsin, 1970). Curator, earth sciences. (tel. 210 + 596-6672). *Mineralogy, layer silicates.*

Blount, Dr. John Franklin. Hoffman-La Roche Inc., Nutley, NJ 07110, USA. (1937) PhD chemistry (U. of Wisconsin, 1965). Dir., phys. chem. (tel. 201 + 235-3580). *Crystal structures, organic, organometallic, automated diffractometer software, computer programming, crystal structure analysis.*

Bly, Dr. Donald D. Expt. Sta., E328/123, E.I. du Pont de Nemours Co., Wilmington, DE 19898, USA. (tel. 302 + 695-2791).

Boehme, Mr. Richard Frederick. Res. Div., IBM T. J. Watson Res. Ctr., P. O. Box 218 Yorktown Heights, NY 10598, USA. (1950) MA chemistry-biology (Boston U.). Sr. assoc. eng. (tel. 914 + 945-1820, telex 137456, fax 914 + 945-2141, email BOEHM at IBM.COM). *Electron density determination, modeling, low temperature crystallography, EXAFS, atomic structure.*

Boggs, Dr. Rita Rose. American Res. & Testing Inc., 14934 S. Figueroa St., Gardena, CA 90248, USA. (1938) PhD phys. chemistry (U. of Pennsylvania, 1973). Pres. (tel. 213 + 538-9709). *Crystal structure, small molecules, direct methods.*

Bohlen, Mr. David S. U. of Minnesota, 421 Washington Ave., SE, Minneapolis, MN 55455, USA.

Bolin, Prof. Jeffrey T. Biological Sci. Dept., Lilly Hall, Purdue U., West Lafayette, IN 47907, USA. (1952) PhD chemistry (U. of Calif. at San Diego, 1982). Asst. prof. (tel. 317 + 494-4922). *Biochemistry, protein structure, enzymology, crystallography.*

Bonaventura, Dr. Joseph. Marine Biomed. Ctr., Duke U. Marine Lab., Pivers Island, Beaufort, NC 28516, USA. (1942) PhD biochemistry (U. of Texas at Austin, 1968). Assoc. prof. (tel. 919 + 728-2111, fax 919 + 728-2514). *Hemoproteins, protein design, protein engineering, structure - function.*

Bonham, Prof. Russell Aubrey. Chemistry Dept., Indiana U., Bloomington, IN 47401, USA. (1931) PhD phys. chemistry (Iowa State U., 1958). Prof. (tel. 812 + 855-4843, fax 812 + 855-6611, email BONHAM at IUBACS). *Spectroscopy, electron impact, secondary electron, electron beam time of flight, negative ion resonance, electron diffraction, charge and momentum determination, X-ray incoherent scattering factor determination.*

Boo, Prof. William O. J. Chemistry Dept., U. of Mississippi, University, MS 38677, USA. (1933) PhD chemistry (U. of Chicago, 1966). Prof. (tel. 601 + 232-7301). *Solid state chemistry.*

Borhani, Dr. David. Biochemistry Dept., Harvard U., 7 Divinity Ave., Cambridge, MA 02138, USA. (1960) PhD org. chemistry (MIT, 1986). (tel. 617 + 495-5043). *Protein crystallography.*

Borie, Dr. Bernard Simon. 13 Brookside Dr., Oak Ridge, TN 37830, USA. (1924) PhD physics (MIT, 1956). Prof. (tel. 615 + 974-5336). *Diffraction crystallography.*

Boris, Mrs. Linda Jo. Life Sci. Res. Lab., Eastman Kodak Co., Bldg. 82, Flr. 1, Rochester, NY 14650-2136, USA. (1967) BS chemical eng. (U. of Virginia, 1988). Scient. (tel. 716 + 477-2864). *Proteins, crystallization, crystallography, rational drug design.*

Boskey, Dr. Adele Ludin. Res. Div., Hosp. for Special Surg., Cornell U. Med. C., 535 East 70th St., New York, NY 10021, USA. (1943) PhD chemistry (Boston U., 1970). Dir., Lab. for Ultrastruct. Biochem. (tel. 212 + 606-1453). *Calcification mechanisms, biologic calcification.*

Boss, Dr. James William. 134 North 54th St., Philadelphia, PA 19139, USA. (1932) PhD phys. chemistry (Ohio State U., 1966). (tel. 215 + 472-1137). *X-ray diffraction, electron spin resonance, disordered structures, glasses.*

Bota, Prof. Kofi B. Chemistry Dept., Atlanta U., 223 J.P. Brawley Dr., Atlanta, GA 30314, USA. (tel. 404 + 653-8595).

Bourne, Dr. Philip Eric. Health Sci. Computing Fac., Columbia U., 630 W. 168th. St., New York, NY 10032, USA. (1953) PhD chemistry (Flinders U. of South Australia, 1980). Sr. assoc. (tel. 212 + 305-3657, fax 212 + 305-7379, email SYSTEM at CUMBG). *Computational methods.*

Boviatsis, Dr. Ioanis. Chemistry Dept., SUNY at Buffalo, Buffalo, NY 14214, USA. (tel. 716 + 831-3263).

Bowman, Mr. Allen L. 10 Encino, Los Alamos, NM 87544, USA. (tel. 505 + 662-9560).

Box, Mr. Harold C. Roswell Park, Mem. Inst., 666 Elm St., Buffalo, NY 14263, USA. (tel. 716 + 845-3135).

Boyd, Dr. Donald B. Res. Labs., Lilly Corporate Ctr., Eli Lilly and Co., Indianapolis, IN 46285, USA. (1941) PhD phys. chemistry (Harvard U., 1968). Res. scient. (tel. 317 + 276-4232, fax 317 + 276-5431). *Computational chemistry, molecular modeling.*

Boyington, Mr. Jeffrey C. Biophysics Dept., Johns Hopkins Sch. of Med., 725 N. Wolfe St., Baltimore, MD 21205, USA. (1965) MS biophysics (U. of

Connecticut, 1987). Grad. student. (tel. 301 + 955-8715). *Protein crystallography.*

Boyko, Prof. Edward Raymond. Chemistry Dept., Providence C., Providence, RI 02918, USA. (1930) PhD phys. chemistry (Rutgers U., 1956). Prof. (tel. 401 + 865-2108). *Coordination compounds, organic crystal structures.*

Boyle, Mr. Paul. Chemistry Dept., U. of Minnesota, 207 Pleasant St., SE, Minneapolis, MN 55455, USA.

Braden, Dr. Bradford Carl. Biophysics Dept., Johns Hopkins U., 34th and Charles St., Baltimore, MD 21218, USA. (1951) PhD biophysics (Indiana U., 1978). Res. assoc. (tel. 301 + 338-7250). *Protein structure, peptide ionophores.*

Bradshaw, Mr. Joseph Earl. Chemistry Dept., Rice U., P.O. Box 1892, Houston, TX 77251, USA.

Brady, Mr. George W. Div. Labs. and Res., NY State Dept. Health, New Scotland Ave., Albany, NY 12201, USA. (tel. 518 + 474-0671).

Bragg, Prof. Robert Henry. Matls. Sci. & Mineral Eng. Dept., U. of Calif. at Berkeley, Hearst Mem. Mining 370, Berkeley, CA 94720, USA. (1919) PhD physics (Illinois Inst. of Techn., 1960). Prof. emeritus. (tel. 415 + 642-7393). *Structure, electrical properties, carbon materials, graphite intercalation compounds, diffraction physics, small-angle X-ray and neutron scattering.*

Brammer, Dr. Lee. Chemistry Dept., Brookhaven Nat. Lab., Upton, NY 11973, USA. (1963) PhD inorg. chemistry (U. of Bristol, U.K., 1987). NATO postdoct. fel. (tel. 516 + 282-4375, telex 6852516 BNL DOE, fax 516 + 282-5815, email Bitnet ARAMMER at BNLCHM). *Organometallic structures, diffraction, X-ray, neutron, charge densities, crystallographic data bases.*

Brandhuber, Dr. Barbara. Synergen, 1885 33rd. St., Boulder, CO 80301, USA. (1959) PhD chemistry (U. of Colorado at Boulder, 1988). Res. fel. (tel. 303 + 938-6200, ext. 233, 252). *Protein crystallography.*

Brathovde, Prof. James Robert. P.O. Box 2299, Camp Verde, AZ 86322, USA. (1926) PhD phys. chem., X-ray crystallog. (U. of Washington, 1956). Prof. emeritus, N. Ariz. U. (tel. 602 + 567-9493). *Geothermal.*

Bray, Prof. Diana D. Chemistry Dept., Fordham U., Bronx, NY 10458, USA. (1945) PhD inorg. chemistry (Fordham U., 1976). Assoc. prof. (tel. 212 + 579-2604).

Brech, Mr. Frederick. P. O. Box 145, Dover, MA 02030, USA. (tel. 508 + 785-0091).

Brennan, Dr. Richard Gerald. Biochemistry Dept., Oregon Health Sci. U., 3181 SW Sam Jackson Park Rd., Portland, OR 97201, USA. (1955) PhD biochemistry (U. of Wisconsin at Madison, 1984). Asst. prof. (tel. 503 + 279-7781). *Nucleic acids, protein complexes, modifications of, transcription.*

Brennan, Dr. Sean Michael. Stanford Synchrotron Rad. Lab., Stanford U., 2575 Sandhill Rd., Menlo Park, CA 94025, USA. (1955) PhD materials sci. & engin. (Stanford U., 1982). Sr. res. assoc. (tel. 415 + 926-3173, fax 415 + 926-4100, email BREN at SSRL750). *Grazing incidence X-ray scattering, surface scattering, GIXS, DEXTER, SEXAFS.*

Brennan, Dr. Thomas Francis. 512 General Booth Blvd., Virginia Beach, VA 23451, USA. (1943) PhD inorganic chem., crystallog. (SUNY at Stony Brook, 1970) Asst. dir. *Neuropeptide structure - function, catechol estrogens, DNA-drug interactions, hormone-receptor binding models.*

Brenner, Mr. Stephen A. Naval Res. Lab., Code 6030, Washington, DC 20375, USA. (1937) MA phys. chemistry (Boston U., 1962). Res. chemist. (tel. 202 + 767-2735). *Mathematics, chemistry, physics, computer science.*

Brese, Mr. Nathaniel E. Chemistry Dept., Arizona State U., Tempe, AZ 85287, USA. (1965) BA chemistry (Northwestern U., 1987). Grad. student. (tel. 602 + 965-6570, email AGNEB at ASUACVAX). *Structure, single crystal X-ray, powder neutron, transmission electron microscopy, crystal chemistry.*

Brickenkamp, Dr. Carroll Shelton. 12405 Beall Spring Rd., Potomac, MD 20854, USA. (1945) PhD crystallography (U. of Pittsburgh, 1970). *Moisture measurement, materials, compliance sampling, prepackaged commodities, borophosphates, structure, chemistry.*

Bright, Dr. William M. P.O Box 25210 Washington, DC 20007, USA. PhD chemistry (Georgetown U., 1974).

Briguglio, Dr. James. 177 W. 18th. St., Bayonne, NJ 07002, USA. (1954) PhD inorg. chemistry (U. of Pennsylvania, 1985). Sr. scient. (tel. 201 + 437-9073). *Analytical chemistry, crystallography.*

Britton, Prof. Doyle. Chemistry Dept., U. of Minnesota, Minneapolis, MN 55455, USA. (1930) PhD chemistry (CIT, 1955). Prof. (tel. 612 + 625-9535). *Intermolecular interactions.*

Broach, Dr. Robert William. Phys. Chem. & Surface Sci. Dept., UOP Res. Centr., 50 E. Algonquin Rd., Des Plaines, IL 60016, USA. (1949) PhD chemistry (U. of Wisconsin at Madison, 1977). Assoc. dir. (tel. 312 + 391-3313). *X-ray and neutron diffraction, EXAFS, catalysts and adsorbents (heterogeneous - petrochemical - exhaust gas conversion).*

Brock, Prof. Carolyn Pratt. Chemistry Dept., U. of Kentucky, Lexington, KY 40506-0055, USA. (1946) PhD chemistry (Northwestern U., 1972). Prof. (tel. 606 + 257-1959, email CPBROCK at UKCC). *Molecular packing, thermal motion.*

Brooks, Dr. Charles L. Chemistry Dept., Carnegie-Mellon U., 4400 5th. Ave., Pittsburgh, PA 15213, USA. (1956) PhD phys. chemistry (Purdue U., 1982). Prof. (tel. 412 + 268-3176, email BROOKS at CMCHEM). *Chemical physics, computational statistical mechanics, macromolecular structure, dynamics.*

Brooks, Prof. Frederick Phillips Jr. Computer Sci. Dept., U. of North Carolina, Campus Box 3175, Chapel Hill, NC 27599, USA. (1931) PhD applied mathematics (Harvard U., 1956). Kenan Prof. of Computer Sci. (tel. 919 + 962-1931, fax 919 + 962-1799, email BROOKS at CS.UNC.EDU). *Molecular graphics.*

Brown, Prof. Bruce Elliot. Geosciences Dept., U. of Wisconsin at Milwaukee, Milwaukee, WI 53201, USA. (1930) PhD geology (U. of Wisconsin, 1960). Assoc. prof. (tel. 414 + 963-4972). *Mineralogy, geochemistry, limnology.*

Brown, Prof. Bruce Willard. Chemistry Dept., Portland State U., Portland, OR 97207-0751, USA. (1927) PhD chemistry (U. of Washington, 1961). Prof. (tel. 503 + 464-3811, email Bitnet BABB at PSUORVM). *Coordination compounds, computer programming.*

Brown, Dr. George Marshall. Bldg. 4500-N, MS 6197, Oak Ridge Nat. Lab., P.O. Box 2008, Oak Ridge, TN 37831-6197, USA. (1921) PhD phys. chemistry (Princeton U., 1949). Res. staff member. (tel. 615 + 574-4989). *X-ray and neutron diffraction, crystal structure analysis.*

Brown, Prof. Glenn H. Liquid Crystal Inst., Kent State U., Kent, OH 44242, USA. (1915) PhD chemistry (Iowa State U., 1951). Dir., Regents prof. of chemistry. (tel. 216 + 672-2654). *Liquid crystals, structure - properties relationship.*

Brown, Prof. Gordon Edgar Jr. Geology Dept., Stanford U., Stanford, CA 94305, USA. (1943) PhD mineralogy (Virginia Polytechnic Inst., 1970). Prof. of mineral. & crystallog., Co-Dir. Ctr. Matls. Res. (tel. 415 + 497-3518). *X-ray crystallography, absorption spectroscopy, crystal chemistry, mineralogy, silicate glass/melt systems, structure, properties, sorption reactions, oxide surfaces, synchrotron radiation, applications to earth materials.*

Brown, Dr. Katherine Anne. Chemistry Dept., B-017, U. of Calif. at San Diego, La Jolla, CA 92093, USA. (1963) PhD biophysics, crystallography (Imperial C., U. of London, UK, 1988). Res. chemist. (tel. 619 + 534-2153, telex 188929, fax 619 + 534-0058, email KBROWN at UCSD.EDU). *Protein crystallography, protein engineering, computer graphics.*

Brown, Dr. Leo Dale. Exxon Res. & Devel. Labs., P.O. Box 2226, Baton Rouge, LA 70821, USA. (1948) PhD inorg. chemistry (UCB, 1974). Res. assoc. (tel. 504 + 359-4519). *Structural inorganic chemistry, silicate minerals, crystal chemistry, X-ray powder diffraction, high and low temperature.*

Brown, Mr. Raymond Sidney. HHMI, Harvard U., 7 Divinity Ave., Cambridge, MA 02138, USA. (1946) M.Phil. struct. molec. biology (London U., UK, 1975). Res. assoc., instr. microbiol. (tel. 617 + 495-3332, fax 617 + 495-9613, email Bitnet RAYBROWN at HUXTAL). *Interactions, RNA - protein, DNA - protein.*

Bruccoleri, Dr. Robert E. Squibb Inst. for Med. Red., D-4116, P.O. Box 4000, Princeton, NJ 08543-4000, USA. (1956) PhD biochemistry (Harvard U., 1984). Res. group leader. (tel. 609 + 683-6165). *Macromolecular modeling, protein folding, antibodies.*

Bruck, Dr. Michael Allen. Chemistry Dept., U. of Arizona, Tucson, AZ 85721, USA. (1953) PhD inorg. chemistry (USC, 1983). Asst. staff scient. (tel. 216 + 621-4168, email Bitnet MBRUCK at ARIZRVAX).

Brumberger, Prof. Harry. Chemistry Dept., U. of Syracuse, Syracuse, NY 13244-1200, USA. (1926) PhD chemistry (PUN, 1955). Prof. (tel. 315 + 443-5923). *Small angle scattering, catalysts.*

Bryan, Prof. Robert Finlay. Chemistry Dept., U. of Virginia, McCormick Rd, Charlottesville, VA 22901, USA. (1933) PhD chemistry (U. of Glasgow, UK, 1957). Prof. (tel. 804 + 924-3619). *Liquid crystals and precursors, structure determinations, crystallographic computing.*

Bryant, Dr. Stephen Howard. Wadsworth Center for Labs. & Res., NY State Dept. of Health, Empire State Plaza, Albany, NY 12201, USA. (1954) PhD biophysics (Johns Hopkins U., 1982). Res. scient. (tel. 516 + 474-1518). *Proteins, crystallography, data bases, structural analysis.*

Bryden, Prof. John Heilner. Chemistry Dept., Calif. State U. at Fullerton, 800 N. State College Blvd., Fullerton, CA 92634, USA. (1920) PhD phys. chemistry (UCLA, 1951). Prof. (tel. 714 + 773-3833). *Crystal structures, organic, inorganic, computer programming, crystallography.*

Buchanan, Prof. David R. Textile Eng., Chem. and Sci. Dept., North Carolina State U., P.O. Box 8301, Raleigh, NC 27695-8301, USA. (1934) PhD phys. chemistry (Ohio State U., 1962). Prof. (tel. 919 + 737-3058). *Polymers, fibers, structure - properties relationship, small angle X-ray scattering.*

Buchanan, Prof. Robert Martin. Chemistry Dept., U. of Louisville, Brook St., Louisville, KY 40929, USA. (1951) PhD inorg. chemistry (U. of Colorado, 1980). Assoc. prof. (tel. 502 + 588-6580). *Spectroscopy, EPR, NMR, bioinorganic chemistry, polyimidazole cluster complexes, Mn, Fe, mixed valence complexes, electrochemistry, magnetic properties, alkane oxidation via binuclear Fe complexes.*

Buchner, Mr. Robert. Immunetech Pharmaceuticals, 11045 Roselle St., Suite 9A, San Diego, CA 92129, USA. (1958) MS biology (San Francisco State U., 1983). Mgr., res. & resource anal. (tel. 619 + 457-2553, ext. 348). *Protein analysis, peptides.*

Budai, Dr. J. D. Solid State Div., Oak Ridge Nat. Lab., P.O. Box 2008, Oak Ridge, TN 37831, USA. (1952) PhD physics (Cornell U., 1982). Res. scient. (tel. 615 + 576-6721). *Crystal defects, interfaces, quasicrystals, liquid crystal phase transitions.*

Bugg, Prof. Charles E. Biochemistry Dept., U. of Alabama, University Station, Birmingham, AL 35233, USA. (1941) PhD chemistry (Rice U., 1965). Prof. (tel. 205 + 934-5329). *Biological crystallography.*

Bunick, Dr. Gerard J. Biology Div., MS 8077, Oak Ridge Nat. Lab., P.O. Box 2009, Oak Ridge, N 37830-8077, USA. (1947) PhD chemistry (U. of Pennsylvania, 1975). Res. staff mem. (tel. 615 + 576-2685, fax 615 + 574-1274, email Bitnet BUNICKGJ at ORNLSTC or BUNICKGJ at STD10.CTD.ORNL.GOV). *X-ray crystallography, small angle scattering, X-ray, neutron, nucleic acid - protein interactions, biological macromolecules.*

Buranda, Mr. Tione. Chemistry Dept., Wayne State U., Detroit, MI 48202, USA. (1960) MS inorganic chemistry (U. of Toledo, 1987). Grad. student. (tel. 313 + 577-3087). *Crystallography, small molecule, inorganic photochemistry.*

Burbank, Mr. Robinson D. 45 Woodland Ave., Summit, NJ 07901, USA. (1921) PhD inorganic chem. (MIT, 1950). Retired from Bell Labs. (tel. 201 + 582-3236). *X-ray crystallography, inorganic compounds, interhalogen compounds, noble gas compounds, phase transformations, thin films.*

Burling, Mr. Temple. Biophysics Dept., U. of Rochester Med. Ctr., Rochester, NY 14642, USA. (1963) MS physics (Iowa State U., 1987). Grad. fel. (tel. 716 + 275-2571, email FTBU at UORDBV).

Burnett, Prof. Roger MacDonald. The Wistar Inst., 3601 Spruce St., Philadelphia, PA 19104, USA. (1941) PhD protein crystallography (Purdue U., 1970). Prof. (tel. 215 + 898-2201, fax 215 + 898-3868). *Macromolecular structure, recognition, interaction, virus structure and assembly.*

Burnham, Prof. Charles Wilson. Earth and Planetary Sci. Dept., Harvard U., 20 Oxford St., Cambridge, MA 02138, USA. (1933) PhD mineralogy and cystallography (MIT, 1961). Prof. mineralogy. (tel. 617 + 495-2484). *Minerals, crystal structure analysis, crystal chemistry, physics, high temperature and pressure, physical properties - structure relations.*

Burns, Dr. John Howard. Chemistry Div., Oak Ridge Nat. Lab., P.O. Box 2008, Oak Ridge, TN 37831-6119, USA. (1930) PhD phys. chemistry (Rice Inst., 1955). Sr. res. staff memb. (tel. 615 + 574-5018). *Structure, coordination complexes, neutron diffraction, molecular mechanics.*

Burton, Dr. Benjamin Paul. Inst. Materials Sci. & Eng., Nat. Inst. Stnds. & Techn. Gaithersburg, MD 20899, USA. (1949) PhD earth sci. (SUNY at Stony Brook, 1982). Physical scient. (tel. 301 + 975-6043, fax 301 + 975-2128, email Bitnet BURTON at NBS). *Phase transitions, order-disorder, mineralogy.*

Bush, Dr. Bruce L. Merck & Co. Inc., R80-M101, P.O. Box 2000, Rahway, NJ 07065, USA. (1965) PhD physics (Stanford U., 1974). Res. fel. (tel. 201 + 594-6758). *Protein biophysics, molecular modeling, numerical methods.*

Bush, Mr. Stewart Fowler. P.O. Box 127, Newell, NC 28126, USA.

Busing, Dr. William Richard. Chemistry Div., Oak Ridge Nat. Lab., P.O. Box 2008, Oak Ridge, TN 37831-6197, USA. (1923) PhD phys. chemistry (Princton U., 1949). Sr. res. staff memb. (tel. 615 + 574-4976). *Neutron diffraction, crystallographic computing, molecular modeling, crystal modeling, fiber crystallography.*

Butcher, Prof. Raymond John. Chemistry Dept., Howard U., College St., Washington, DC 20059, USA. (1945) PhD chemistry (U. of Canterbury, New Zealand, 1974). Assoc. prof. (tel. 202 + 636-6886, email Bitnet ALACH09 at HUMAIN). *Structure, magnetism, polynuclear complexes, bio-inorganic chemistry, molybdenum, enzymes containing copper (model complexes), iron (III) complexes, magnetic properties.*

Butler, Mr. Brent D. Materials Sci. Dept., Northwestern U., Techn. Inst., Evanston, IL 60201, USA. (1962) MS (U. of Missouri, 1986). Res. asst. (tel. 312 + 491-7677).

Butler, Dr. William M. Chemistry Dept., Rm. 1094, U. of Michigan, Ann Arbor, MI 48109-1055, USA. (1943) PhD inorg. chemistry (U. of Arizona, 1972). Dept. crystallographer. (tel. 313 + 763-2009, fax 313 + 747-4865). *X-ray crystallography, small molecules, computer aided instruction, computer graphics.*

Butler, Mr. William O. Analytical Services, Ricerca Inc., P.O. Box 1000, Painesville, OH 44077, USA. (1939) BS physics (Miami U., 1961). Res. assoc. (tel. 216 + 357-3388). *Materials, surface analysis techniques, microscopy, electron (scanning and transmission), optical, X-ray diffraction, spectroscopy, ESCA, Auger electron spectroscopy.*

Buttrey, Prof. Douglas James. Chemical Engin. Dept., U. of Delaware, Colburn Lab., Newark, DE 19716, USA. (1954) PhD chemistry (Purdue U., 1984). Asst. prof. (tel. 302 + 451-2034, fax 302 + 451-1048). *Structure, magnetism, electronic properties, complex transition metal oxides, synthesis, crystal growth.*

Byram, Mrs. Susan Katherine. Siemens Analytical X-ray Instr., 6300 Enterprise Ln., Madison, WI 53719, USA. (1945) MSc crystallography (U. of Toronto, Canada, 1970). Product mgr., crystallog. systs. (tel. 800 + 552-XRAY, fax 608 + 276-3015). *Crystallographic programming, real time instrument control.*

Byrn, Prof. Stephen Robert. Med. Chem. & Pharmacognosy Dept., Sch. of Pharm. and Pharmacol. Sci., Purdue U., West Lafayette, IN 47906, USA. (1944) PhD chemistry (U. of Illinois, 1971). Prof. (tel. 317 + 494-1460). *Mechanisms, solid state reactions, pharmaceutical chemistry, drug-nucleic acid interactions.*

Cady, Dr. Howard Hamilton. M-1, Mail Stop C920, Los Alamos Nat. Lab., P.O. Box 1663, Los Alamos, NM 87545, USA. (1931) PhD chemistry (UCB, 1957). Staff member. (tel. 505 + 667-4992). *Optical and X-ray crystallography, explosives, solid state phase studies, physical chemistry.*

Caffrey, Prof. Martin. Chemistry Dept., Ohio State U., 120 W. 18th. Ave., Columbus, OH 43210, USA. (1950) PhD biochemistry (Cornell U., 1982). Asst. prof. (tel. 614 + 292-8437, fax 614 + 292-2928, email TS7074 at OHSTVMA). *Lipids, membranes.*

Cahn, Dr. John Werner. Inst. Matls. Sci. & Eng., Nat. Inst. Stnds. & Techn., Bldg. 223, Rm. A153, Gaithersburg, MD 20899, USA. (1928) PhD chemistry (UCB, 1953). Sr. fel. (tel. 301 + 975-6154, fax 301 + 975-2128, email CAHN at NBSENH). *Quasicrystals, phase transitions, intermetallics.*

Calabrese, Dr. Joseph C. Central Res. Div., Expt. Sta. E356/247, E.I. du Pont de Nemours Co., P.O. Box 80228, Wilmington, DE 19898-0288, USA. (1943) PhD chemistry (U. of Wisconsin, 1971). Res. sci. (tel. 302 + 695-3952, fax 302 + 695-9183, email CALABRES%ESVAX at DUPONT). *Inorganic, organometallic, organic, macromolecular compounds, structure - function, computing, data bases, instrumentation.*

Calandra, Mr. Peter M. Innovative Tech. Inc., 205 Willow St., South Hamilton, MA 01982, USA. (1939) BS chemistry (Cansius C., 1962). President. (tel. 508 + 468-3543, telex 510 101 2674). fax 508 + 468-1101,

Camerman, Prof. Arthur. Dept. of Medicine (Neurology), RG-27, U. of Washington, Seattle, WA 98195, USA. (1939) PhD chemistry (U. of British Columbia, Canada, 1964). Prof. (tel. 206 + 543-2340). *Drug action mechanisms, drug design, structure - activity relationships, biological molecules.*

Cameron, Mr. Robert P. 5013 Iberville St., #D, New Orleans, LA 70119, USA.

Campana, Dr. Charles F. Siemens Analytical X-ray Instr. Inc., 6300 Enterprise Ln., Madison, WI 53719, USA. (1947) PhD inorg. chemistry (U. of Wisconsin at Madison, 1976). Sr. appl. scient. (tel. 608 + 276-3000). *X-ray crystallography, inorganic chemistry, organometallic chemistry.*

Compobasso, Mr. Nino. 5514 S. University Ave., Rm. 1702, Chicago, IL 60637, USA. (tel. 312 + 702-0286).

Cantrell, Prof. Joseph Sires. Chemistry Dept., Miami U., Oxford, OH 45056, USA. (1932) PhD phys. chemistry (Kansas State U., 1961). Prof. (tel. 513 + 529-2834, fax 513 + 529-3841). *X-ray crystallography, organic, biological, metal hydrides (powder), electron diffraction, thermal stability studies (DSC).*

Capano, Mr. Michael A. Room 13-4077, MIT, 77 Massachusetts Ave., Cambridge, MA 02139, USA. (tel. 617 + 253-6999).

Cargill, Dr. George Slade III. Res. Div., IBM T. J. Watson Res. Ctr., P. O. Box 218, Yorktown Heights, NY 10598, USA. (1943) PhD applied physics (Harvard U., 1969). Sr. mgr., Struct. of Materials. (tel. 914 + 945-1958, telex 137456, fax 914 + 945-2141, email CARGILL at IBM.COM). *Amorphous solids, defects, semiconductors, EXAFS, X-ray scattering.*

Carlson, Dr. Ernest Howard. Geology Dept., Kent State U., Kent, OH 44242, USA. (1933) PhD geology (McGill U., Canada, 1966). Assoc. prof. (tel. 216 + 672-3778). *X-ray crystallography, mineralogy, exploration geochemistry.*

Carlson, Mrs. Vanice Aparecida Perin. Los Alamos Nat. Lab., P-LANSCE, H805, Los Alamos, NM 87545, USA. (1952) MS nuclear physics (U. de Sao Paulo, Brazil, 1978). Staff res. asst. (tel. 505 + 667-6069, email CARLSON at WNRVAX.LANL.GOV). *Heusler alloys, magnetic structures, glass metals, amorphous metals.*

Carlson, Dr. William D. Cardiac Unit, Massachusetts Genl. Hosp., Fruit St., Boston, MA 02114, USA. (1946) MD/PhD molecular biophysics (Yale U., 1976). Asst. prof (Harvard U.) (tel. 617 + 726-7790). *Renin inhibitors, molecular graphics, drug design, molecular dynamics.*

Carnahan, Dr. Gary Ellis. Biochemistry Dept., Washington U. Sch. of Med., 660 S. Euclid Ave., St. Louis, MO 63110, USA. (1950) MD,PhD biochemistry (Vanderbilt U., 1982). (tel. 314 + 362-3340).

Caron, Dr. Aimery Pierre. Office of Continuing Education, U. of the Virgin Islands, St. Thomas, VI 00801, USA. (1930) PhD chemistry (USC, 1962). Dir. continuing educ. (tel. 809 + 776-9200, ext. 1230, telex 3470102 UVI, fax 809 + 776-2399). *Molecular structures, X-ray diffraction.*

Carpenter, Dr. Donald Allmand. Martin Marietta Energy Systs., Box 2009, Bldg. 9203, MS 8084, Oak Ridge, TN 37831-8084, USA. (1941) PhD inorg. chemistry (Georgia Inst. of Techn., 1968). Dev. chemist. (tel. 615 + 574-0931). *Materials science, ceramics, metallurgy.*

Carpenter, Prof. Gene B. Chemistry Dept., Brown U., Providence, RI 02912, USA. (1922) PhD phys. chemistry (Harvard U., 1947). Prof. Emeritus. (tel. 401 + 863-3389, fax 401 + 863-2594, email CH408000 at BROWNVM). *Crystal and molecular structures, X-ray diffraction techniques.*

Carperos, Dr. William E. La Jolla Cancer Res. Fndn., 10901 N. Torrey Pines Rd., La Jolla, CA 92037, USA. (1958) PhD chemistry (Emory U., 1985). Res. assoc. (tel. 619 + 455-6480, ext. 590, fax 619 + 455-0181). *Macromolecular crystallography.*

Carr, Dr. Martin J. Division 1822, Sandia Nat. Labs., Albuquerque, NM 87185, USA. (1949) PhD materials eng. (Rensselaer Polytechnic Inst., 1976). Tech. staff membr. (tel. 505 + 846-1405). *Electron diffraction, materials science.*

Carrano, Mr. Carl J. Chemistry Dept., U. of Vermont, Burlington, VT 05405, USA. (tel. 802 + 656-0197).

Carrell, Dr. Horace L. The Inst. for Cancer Res., Fox Chase Cancer Center, 7701 Burholme Ave., Philadelphia, PA 19111, USA. (1940) PhD chemistry (USC, 1966). Sr. res. assoc. (tel. 215 + 728-2220). *Biological activity - molecular structure relationships, X-radiation effects, organic molecules.*

Carrithers, Mr. Charles H. Enraf-Nonius Service Corp., 390 Central Ave., Bohemia, NY 11716, USA. (tel. 516 + 589-2885).

Carroll, Mr. Patrick J. 15 Anders Dr., Cherry Hill, NJ 08003, USA. (tel. 215 + 898-3505).

Carson, DMr. William Michael. Ctr. for Macromolec. Cryst., U. of Alabama at Birmingham, UAB Sta., Box THT-79, Birmingham, AL 35294, USA. (1951) PhD chemistry (U. of Texas, 1980). Assoc. scient. (tel. 205 + 934-1983, fax 205 + 934-0480). *Computational crystallography, molecular graphics.*

Carter, Prof. Charles Williams. Biochemistry Dept., U. of North Carolina, Fac. Lab. & Office Bldg. 231H, Chapel Hill, NC 27514, USA. (1945) PhD biology (U. of Calif. at San Diego, 1972). Prof. (tel. 919 + 966-3263, fax 919 + 966-6923, email CARTER at MED.UNC.EDU). *Structure, macromolecular, function, evolution, genetic molecular apparatus, electron transport systems.*

Carter, Dr. Daniel Clark. Microgravity Sci. Div., Biophys. Br., NASA, Marshall Space Flt. Ctr., Bldg. 4481, Code ES76, Huntsville, AL 35812, USA. (1854) PhD crystallography (U. of Pittsburgh, 1984). Senior scient. (tel. 205 + 544-5492). *Protein, crystal growth, crystallography, serum albumin.*

Cartz, Dr. Louis. C. of Eng., Marquette U., 1515 W. Wisconsin Ave., Milwaukee, WI 53233, USA. (1926) PhD crystallography (U. London, UK, 1954). Prof. *X-ray crystallography, radiation damage, minerals, ceramics.*

Case, Mr. J. A. M. 1812 N. 74th. St., Milwaukee, WI 53213, USA. (tel. 414 + 444-5735).

Caslavsky, Dr. Jaroslav L. US Army Matls. & Techn. Lab., Watertown, MA 02172-0001, USA. (1928) PhD, RNDr solid state chem. (PSU 1969, Charles U., Prague, 1957). Prin. scient. (tel. 617 + 923-5352). *Crystal growth, defects, solid state reactions at high temperatures.*

Caspar, Prof. Donald L. D. Rosenstiel Basic Med. Sci. Ctr., Brandeis U., 415 South St., Waltham, MA 02254, USA. (1927) PhD biophysics (Yale U., 1955). Prof. (tel. 617 + 736-2465). *Structural biology, viruses, membranes, X-ray diffraction, electron microscopy, macromolecular assemblies.*

Casson, Mr. Lawrence P. Molec. Biology Dept., Princeton U., Princeton, NJ 08544, USA. (1959) BS biology (MIT, 1983). Grad. student. (tel. 609 + 987-2835).

Caughlan, Prof. Charles N. Chemistry Dept., Montana State U., Bozeman, MT 59717, USA. (1915) PhD chemistry (U. of Washington, 1941). Prof. emeritus. (tel. 406 + 994-5385). *X-ray diffraction structures, accurate parameters, thio-urea complexes, structure - chemical reactivity relations, organophosphorus compounds.*

Cava, Dr. Robert J. AT&T Bell Labs., Solid State Chem. Res., 600 Mountain Ave., Murray Hill, NJ 07974, USA. (1951) PhD ceramics (MIT, 1978). Disting. memb. technical staff. (tel. 201 + 582-2180). *High temperature superconductivity.*

Cavin, Mr. Odis Burl. Metals and Ceramics, Bldg. 4515, Oak Ridge Nat. Lab., P.O. Box 2008, Oak Ridge, TN 37831-6064, USA. (1929) MS metallurgy (U. of Tennessee, 1959). Res. assoc. (tel. 615 + 574-5121). *X-ray diffraction studies, powders, single crystals.*

Celikel, Dr. Reha. Biology Dept., U. of Calif. at San Diego, Mayer Hall, BB-017, La Jolla, CA 92093, USA. (1949) PhD molecular biophysics (Leeds U., England, 1979). Postdoct. res. biologist. (tel. 619 + 534-4241). *Structure/function relationships, biological molecules.*

Chakoumakos, Dr. Bryan Charles. Solid State Div., Bldg. 2000, Oak Ridge Nat. Lab., P.O. Box 2008, MS 6056, Oak Ridge, TN 37831-6056, USA. (1955) PhD geological sci. (Virginia Inst. Tech., 1984). Post-doctoral fellow (tel. 615 + 574-5495, email Bitnet KOU at ORNLSTC). *X-ray diffraction, powder, single crystal, crystal chemistry, physics, mineralogy, materials science.*

Chamberland, Prof. Bertrand L. Chemistry Dept., U. of Connecticut, 215 Glenbrook Rd., Storrs, CT 06268, USA. (1934) PhD chemistry (U. of Pennsylvania, 1960). Prof. (tel. 203 + 486-5381, fax 203 + 486-5381). *Structures, oxides, nitrides, fluorides, inorganic compounds.*

Champion, Mr. William C. Chemistry Dept., Colorado C., Colorado Springs, CO 80903, USA. (tel. 303 + 473-2233).

Chandra, Prof. Dhanesh. Chem. & Metallurg. Dept., Mackay Sch. of Mines, U. of Nevada at Reno, Reno, NV 89557-0047, USA. (1944) PhD metallurgy, materials sci. (U. of Denver, 1976). Assoc. prof. (tel. 702 + 784-1300, fax 702 + 784-1300). *Phase transformations, diagrams, metal hydrides, solid state energy storage materials, high temperature X-ray diffractometry.*

Chandrasekaran, Prof. Rengaswami. Whistler Ctr. Carbohydrate Res., Smith Hall, Purdue U., West Lafayette, IN 47907, USA. (1939) PhD X-ray crystallography (U. of Madras, India, 1966). Assoc. prof. (tel. 317 + 494-4923, telex 276147-AGAD-PU-LAF, fax 317 + 494-7953, email EOK at MACE.CC.PURDUE.EDU). *X-ray fiber diffraction, theory, polymers, structure - property - function relationships, nucleic acids, structure, polysaccharides, nucleic acid - drug interactions, computer graphics.*

Chandrasekhar, Dr. K. Crystallography Dept., U. of Pittsburgh, 304 Thaw Hall, Pittsburgh, PA 15260, USA. (1945) PhD physics (U. of Madras, India, 1979). Res. assoc. (tel. 412 + 683-3000, ext. 517). *Macromolecules, direct methods, structural crystallography, reaction paths.*

Chandross, Dr. Ronald Jay. Physics Dept., Physics Bldg., McCormick Rd., Charlottesville, VA 22901, USA. (1935) PhD phys. chemistry (MIT, 1961). Sr. scient. (tel. 804 + 924-6803). *Proteins, small angle scattering, instrumentation.*

Chaney, Dr. Michael Owen. Physical Chemistry Dept., Eli Lilly Res. Lab., 307 E. McCarty St., Indianapolis, IN 46206, USA. (1943) PhD biochemistry (Indiana

U., 1969). Res. scient. (tel. 317 + 276-4135). *X-ray crystallographic studies, biologically important molecules, structure - activity relationships, molecular modeling.*

Chang, Dr. Chong-Hwan. Biological & Med. Res. Div., Argonne Nat. Lab., 9700 S. Cass Ave., Argonne, IL 60439, USA. (1950) PhD crystallography (U. of Pittsburgh, 1982). (tel. 312 + 972-3887). *Macromolecular structure determination, immunoglobin, photo-reaction centers, interactive computer graphics.*

Chang, Prof. Shih-Chi. Physics Dept., Duquesne U., Pittsburgh, PA 15282, USA. (1933) PhD physics (Kansas State U., 1963). Prof. (tel. 412 + 434-6353). *Structure analysis, X-ray diffraction.*

Chapman, Dr. Michael Stewart. Lilly Hall of Biol. Sci., Purdue U., West Lafayette, IN 47907, USA. (1961) PhD biochemistry (UCLA, 1987). Postdoct. res. fel. (tel. 317 + 494-6766, fax 317 + 494-0876, email 640 at MAU.CC.PURDUE.EDU). *Crystallography, protein A virus.*

Chasen, Prof. Edith. Physics Dept., St. John's U., Grand Central & Utopia Parkways, Jamaica, NY 11439, USA. (1947) MA geology, crystallography (Boston U., 1970). Instructor (tel. 718 + 990-6161, ext. 6289). *Mineralogy, crystallography, computer use in education.*

Chastain, Dr. Roger Vernon Jr. Res. Information Management, US Army Med. Res. & Dev. Cmd., Fort Detrick, Frederick, MD 21701-5012, USA. (1938) PhD chemistry (U. of Washington, 1965). Lieut. Col. (tel. 301 + 349-5159). email CHASTAIN at DETRICK-EMHI.ARMY.MIL, *X-ray crystallography, micro computers, data communications.*

Chaudhuri, Dr. Jharna. Mech. Eng. Dept. Box 35, Wichita State U., Wichita, KS 67208, USA. PhD (1982). Asst. Prof. *Applied crystallography.*

Cheatam, Ms. Linda J. 28 Gilbert St., Watertown, MA 02172, USA. (tel. 617 + 671-8321).

Cheer, Prof. Clair James. Chemistry Dept., U. of Rhode Island, Kingston, RI 02881, USA. (1937) PhD org. chemistry (Wayne State U., 1964). Prof. (tel. 401 + 792-2103). *Organic crystallography, structurally novel substances, small biologically interesting .*

Chen, Dr. Cheng-San. Syst. Res. & Eng., High Perform. Group, Digital Equipment Corp., MROH/L126, 200 Forest St., Marlboro, MA 01752, USA. (1943) PhD biophysics (SUNY at Buffalo, 1977). Prin. eng. (tel. 508 + 462-2290). *Molecules, small and macro, crystal structures, biological activity, magnetic recording, molecular modeling, simulation, high performance computers for modeling, simulation, crystallographic computations.*

Chen, Prof. Haydn H. Matls. Sci. & Eng. Dept., U. of Illinois, 1304 W. Green St., Urbana, IL 61801, USA. (1948) PhD material sci. & eng. (Northwestern U., 1977). Prof. (tel. 217 + 333-7636). *Applied crystallography, double crystal diffractometry, structures, kinetics of ordering transformation, neutron scattering, synchrotron radiation, topography.*

Chen, Mr. Liqing. Crystallography Dept., U. of Pittsburgh, Pittsburgh, PA 15260, USA. (1956) MS crystallography (Fujian Inst. Struct. Matter, PRC, 1984). Grad. student. (tel. 412 + 624-9300). *Crystallography, macromolecular, applied, molecular biology, biochemistry, biophysics.*

Chen, Dr. Longyin. Cell Biol. & Physiology Dept., Washington U. Med. Sch., 660 S. Euclid Ave., St. Louis, MO 63110, USA. (1947) PhD biochemistry (Indiana U., 1988). Res. assoc. (tel. 314 + 362-1079). *Protein crystallography, molecular pharmacology.*

Chen, Miss Ying. Appl. & Eng. Physics Dept., Cornell U., Clark Hall, Ithaca, NY 14853, USA. (1963) BS physics (Beijing U., PRC, 1985). Grad. student. (tel. 607 + 255-2174, email Bitnet CPVY at CORNELLA). *Protein crystallography.*

Chiang, Dr. Michael Yen-Nan. Chemistry Dept., Columbia U., P.O. Box 968, New York, NY 10027, USA. (1954) PhD chemistry (USC, 1984). Manager, X-ray lab. (tel. 212 + 280-8402). *Diffraction, X-ray, neutron, small molecules.*

Chidester, Ms. Connie. Phys. & Analyt. Res., The Upjohn Co., 301 Henrietta St., Kalamazoo, MI 49001, USA. (1937) MA mathematics (Western Michigan U., 1968). Res. sci. (tel. 616 + 385-7624). *Organic, biological crystal structures, computer programming.*

Childs, Dr. Jerry D. Halliburton Services, Res. and Dev. Ctr., Duncan, OK 73536, USA. (1943) PhD phys. chemistry (U. of Oklahoma, 1972). Chemist (tel. 405 + 251-3907). *Cement, spectroscopy, thermodynamics.*

Chipman, Dr. David Randolph. 54 Captains Ln., South Dartmouth, MA 02748, USA. (1928) ScD metallurgy (MIT, 1955). Res. physicist *Charge and momentum density, crystal imperfections, amorphous metal alloys.*

Chirino, Mr. Arthur. Chem. & Biochem. Dept., U. of Calif. at Los Angeles, 405 Hilgard Ave., Los Angeles, CA 90024, USA. (tel. 213 + 825-8901).

Chirlian, Dr. L. E. 2 Morton Ct., Lawrenceville, NJ 08648, USA. (tel. 609 + 896-3259).

Chiu, Ms. Celia C. 10137 Pasture Gate Ln., Columbia, MD 21044, USA.

Chiu, Prof. Wah. Biochemistry Dept., Baylor C. of Med. One Baylor Plaza, Houston, TX 77030, USA. (1947) PhD biophysics (UCB, 1975). Prof. (tel. 713 + 798-6985, fax 713 + 796-8133, email WAH%BCM.TMC.EDU at TMS.EDU). *Structural biophysics, electron crystallography, macromolecules.*

Chlebowski, Dr. Jan F. Biochem. & Biophysics Dept., Med. C. Box 614, MCV Station, Richmond, VA 23298, USA. (1943) PhD chemistry (Case Western Reserve U., 1969). Assoc. prof. (tel. 804 + 786-9762). *Proteins, structure and function, crystallography, alkaline phosphatase, metalloenzymes.*

Chodosh, Dr. Daniel F. Distributed Chemical Graphics, 1326 Carol Rd., Meadowbrook, PA 19046, USA. (1952) PhD chemistry (SUNY at Buffalo, 1977). (tel. 215 + 885-3706, fax 215 + 885-5278).

Choi, Dr. Chang Sun. Reactor Div., Nat. Inst. Stnds. & Techn., Bldg. 235, Gaithersburg, MD 20899, USA. (1926) PhD physics (Kyung Pook U., Korea, 1968). Res. physicist (tel. 301 + 975-6225). *Diffraction, X-ray, neutron, applied crystallography, crystal structures, texture.*

Choi, Dr. Hok-Kin. Chemistry Dept., U. of Southern Calif., Los Angeles, CA 90089, USA. (tel. 213 + 743-7527).

Christianson, Dr. David William. Chemistry Dept., U. of Pennsylvania, 231 S. 34th. St., Philadelphia, PA 19104-6323, USA. (1961) PhD chemistry (Harvard U., 1987). Asst. prof. (tel. 215 + 898-5714). *Protein crystallography, complexes, enzyme - substrate, enzyme - inhibitor, catalysis.*

Christofferson, Dr. Glen D. Chevron Res. Co., 576 Standard Ave:, Richmond, CA 94802, USA. (1931) PhD phys. chemistry (UCLA, 1958). Sr. res. assoc. (tel. 415 + 620-2837). *Crystal chemistry, powder diffraction.*

Christoph, Dr. Gary Gordon. Group C-B, Los Alamos Nat. Lab., P.O. Box 1663, Mail Stop B294, Los Alamos, NM 87545, USA. (1945) PhD chemical physics (U. of Chicago, 1971). Sr. scient. (tel. 505 + 667-3709). *Diffraction, neutron, X-ray, catalysis, computational techniques, structural theory.*

Chu, Prof. Shirley Shan-C. Electrical Eng. Dept., U. of South Florida, 4202 E. Fowler Ave., Tampa, FL 33620, USA. (1929) PhD phys. chem., crystallog. (U. of Pittsburgh, 1961). Prof. (tel. 813 + 974-3939, fax 813 + 974-3651). *X-ray crystal structure determination, organic, inorganic compounds, semiconductor materials, devices, compound semiconductor films, epitaxial growth, characterization, photovoltaic solar energy conversion.*

Chuknyisky, Dr. Peter Peterson. Lab. of Cellular & Molec. Biol., NIH, Gerontology Res. Ctr., 4940 Eastern Ave., Baltimore, MD 21224, USA. (1942) PhD biophysics (Bulgarian Acad. of Sci., 1976). Sr. res. scient. (tel. 301 + 550-1802). *Structures, NMR, EPR, proteins, nucleic acids, RNA polymerase, transcription, regulation, interactions.*

Chung, Prof. Y. W. Matls. Sci. & Eng. Dept., Northwestern U., 2145 Sheridan Rd., Evanston, IL 60208, USA. (1950) PhD physics (UCB, 1977). Prof. (tel. 312 + 491-3112, fax 312 + 491-4133). *Surface physics, chemistry, tribology, fatigue.*

Chung, Mr. Yong Je. Crystallography Dept., U. of Pittsburgh, Pittsburgh, PA 15260, USA. (tel. 412 + 624-4641).

Churchill, Prof. Melvyn Rowen. Chemistry Dept., SUNY at Buffalo, Buffalo, NY 14214, USA. (1940) PhD inorg. chemistry (Imperial C., London, UK, 1964). Prof. (tel. 716 + 831-3906). *Organo-transition metal structural chemistry.*

Clancy, Ms. Laura Lee. Phys. & Analyt. Chem., 7255-209-1, Upjohn International Inc., 301 Henrietta St., Kalamazoo, MI 49001, USA. (1950) MS crystallography (U. of Pittsburgh, 1986). Chemist. (tel. 616 + 384-9794). *Protein crystallography, immunobiology, bioenergetics.*

Clarage, Mr. James Braun II. Physics Dept., Brandeis U., 415 South St., Waltham, MA 02254, USA. (1963) BA physics, mathematics (Illinois Wesleyan U., 1985). Grad. student. (tel. 617 + 736-2475). *X-ray crystallography, protein dynamics, diffuse (variational) scattering.*

Clardy, Prof. Jon Christel. Chemistry Dept., Baker Lab., Cornell U., Ithaca, NY 14853-1301, USA. (1943) PhD chemistry (Harvard U., 1969). Prof. and Chairman. (tel. 607 + 255-7583, fax 607 + 255-4137). *Natural products, organic chemistry.*

Clark, Ms. Connie M. 66K/301, Gen. Prods. Div., IBM Corp., 9000 S. Rita Rd., Tucson, AZ 85744, USA. (1945) MS geochem., computer sci. (PSU, 1974). Staff programmer. (tel. 602 + 629-4664). *Computer programming, geochemistry, mineralogy.*

Clark, Prof. Edward Shannon. Polymer Eng., U. of Tennessee, Knoxville, TN 37996, USA. (1930) PhD chemistry (UCB, 1956). Prof. (tel. 615 + 974-5340). *Crystalline polymers, structure - properties relationship.*

Clark, Mrs. Joan Robinson. 56 Citation Dr., Los Altos, CA 94022-7136, USA. (1920) PhD crystallography (Johns Hopkins U., 1958). Retired (tel. 415 + 960-0628). *Crystallography, mineralogy.*

Clarke, Mr. Frank H. RR3, Box 510, Califon, NJ 07830, USA. (1927) PhD org. chemistry (Harvard U., 1954). Disting. Res. Fel., Ciba-Geigy Corp. (tel. 201 + 277-7525). *X-ray crystallography, small molecule, molecular graphics, drug design.*

Clarke, Prof. Roy. Physics Dept., U. of Michigan, Ann Arbor, MI 48109-1120, USA. (1947) PhD physics (U. of London, 1973). Prof., Dir. appl. phys. (tel. 313 + 764-4466, email Bitnet CLARKE at UMIPHYS). *Molecular beam epitaxy, metallic superlattices, semiconductor superlattices, X-ray synchrotron radiation.*

Claus, Mr. Albert C. 15928 W. Woodbine Cir., Mundelein, IL 60060, USA.

Clawson, Mr. David K. 104 N. Blair St., Crawfordsville, IN 47933, USA. (tel. 317 + 276-4216).

Clayton, Dr. William Rex. Miller Brewing Co., 3939 West Highland Blvd., Milwaukee, WI 53201, USA. (1938) PhD phys. chemistry (Texas A.& M. U., 1971). Res. chemist (tel. 414 + 931-3789). *Inorganic structures*

Clearfield, Prof. Abraham. Chemistry Dept., Texas A.& M. U., College Station, TX 77843, USA. (1927) PhD inorg. chem., crystallog. (Rutgers U., 1952). Prof., Assoc. Dean for Res. (tel. 409 + 845-2936, fax 409 + 845-4719). *Diffraction, powder, X-ray, neutron, inorganic ion exchangers, solid electrolytes, catalysis.*

Cline, Dr. James P. 1882 Columbia Rd., NW, Washington, DC 20009, USA.

Clinger, Dr. Kent. Chemistry Dept., Bethany College, Richardson Hall 217, Bethany, WV 26032, USA. (1955) PhD chemistry (U. of Texas at Austin, 1985). Asst. prof. (tel. 304 + 829-7754). *Protein crystallography, structure.*

Cochran, Prof. Todd G. Sch. of Pharmacy, U. of Montana, Missoula, MT 59812, USA. (1943) PhD pharmaceutical chemistry (U. of Washington, 1970). Assoc. prof. (tel. 406 + 243-4941). *Biological small molecules.*

Cody, Dr. Vivian. Mol. Biophysics Dept., Med. Fndn. of Buffalo, 73 High St., Buffalo, NY 14203-1196, USA. (1943) PhD chemistry (U. of Cincinnati, 1969). Sr. res. scient. (tel. 716 + 856-9600, ext. 322, fax 716 + 852-4846, email Bitnet WPVC%MFB at UBVMS). *Crystallography, protein, small molecules, molecular endocrinology, thyroid hormones, antifolates, drug design, protein - drug complexes, computer graphics.*

Cohen, Prof. Carolyn. Rosenstiel Basic Med. Sci. Res. Ctr., Brandeis U., 415 South St., Waltham, MA 02154, USA. (1929) PhD biophysics (MIT, 1954). Prof. (tel. 617 + 736-2466, fax 617 + 736-2405, email Bitnet CCOHEN at BRANDEIS). *Structural biology, protein assemblies in the cell, fibrous proteins.*

Cohen, Dr. Gerson H. Nat. Insts. of Health, Bldg. 2, Rm. 312, 9000 Wisconsin Ave., Bethesda, MD 20892, USA. (1939) PhD chemistry (Cornell U., 1965). Res. chemist (tel. 301 + 496-4295). *Proteins, crystallography, structure refinement, computer graphics, structure display, crystallographic computing, minicomputers, automated diffractometry, densitometry.*

Cohen, Ms. Janet Paula. 21 Sycamore St., Bronxville, NY 10708, USA. (1951) JD law (Pace U., 1982). Lawyer (tel. 212 + 715-0633). *Patent law.*

Cohen, Prof. Jerome Bernard. Materials Sci. and Eng., Northwestern U., 2145 Sheridan Road, Evanston, IL 60208, USA. (1932) ScD metallurgy (MIT, 1957). Prof. (tel. 312 + 491-5220). *X-Ray Diffraction, neutron diffraction, deformation, phase transformations, residual stresses, local order, clustering, catalysis, oxides, alloys.*

Cole, Dr. Henderson. Technical Computing Systs., IBM, 472 Wheelers Farms Rd., Milford, CT 06460, USA. (1924) PhD physics (MIT, 1952). Scient. specialist. (tel. 203 + 797-0527). *X-ray diffraction, laboratory automation, large scale scientific computing.*

Cole, Dr. L. Brent. HHMI, U. of Texas Health Ctr., 5323 Harry Hines Blvd., Dallas, TX 75235, USA. (1960) PhD org. chemistry (Oklahoma State U., 1986). Assoc. (tel. 214 + 689-5059). *Crystallography, X-ray, protein.*

Colella, Prof. Roberto. Physics Dept., Purdue U., West Lafayette, IN 47907, USA. (1935) PhD physics (U. of Milan, Italy, 1958). Prof. (tel. 317 + 494-3029, telex 493-0593 PHERLUI, fax 317 + 494-0706, email COLELLA at NEWTON.PHYSICS.PURDUE.EDU). *Diffraction physics, perfect crystals, phonons, diffuse scattering, charge density waves, phase problem.*

Colgate, Prof. Samuel Oran. Chemistry Dept., U. of Florida, 218 Leigh Hall, Gainesville, FL 32611, USA. (1933) PhD chemistry (MIT, 1959). Prof. (tel. 904 + 392-5876). *Chemical vapor deposition, MOCVD, volatile metal compounds, synthetic diamond, diamond-like carbon.*

Collins, Dr. Douglas MacPherson. Lab. for the Struct. of Matter, Naval Res. Lab., Code 6030, Bldg. 35, Washington, DC 20375-5000, USA. (1939) PhD chemistry (Rutgers U., 1966). Sr. scient. (tel. 202 + 767-2735, fax 202 + 767-6874, email COLLINS1 at CCF3.NRL.NAVY.MIL). *Structural chemistry, large biological molecules, electron density representations, chemical bonding, crystal structure determination theory.*

Collins, Dr. Richard C. 1997 McFarlane Ave., Lake City, FL 32055, USA. (1946) PhD analytical chemistry (U. of Texas at Austin, 1977). (tel. 904 + 397-8194). *Solid state chemistry, X-ray methods of analysis.*

Comey, Mr. Paul Van A. 340 Eastern Promenade, Apt. 151, Portland, ME 04101, USA. (1898) BChE (NYU, 1954). Retired, consultant (tel. 207 + 774-2874).

Conant, Dr. John W. P.O. Box 159, Tesuque, NM 87574, USA. (1924) PhD chemistry (U. of Iowa, 1955).

Connolly, Dr. Michael Lee. Chemistry Dept., New York U., Washington Square, New York, NY 10003, USA. (1951) PhD biophysics (UCB, 1981). Res. scient. (tel. 212 + 998-8458). *Proteins, theoretical molecular biophysics, differential geometry, molecular graphics.*

Cook, Dr. William Joseph. Pathology Dept., U. of Alabama at Birmingham, University Station, Birmingham, AL 35294, USA. (1949) MD, PhD medicine (U. of Alabama at Birmingham, 1974, 1976). Prof. (tel. 205 + 934-4880, fax 205 + 934-0480). *Biological crystallography.*

Cook, Dr. William R. Jr. 684 Quilliams Rd., Cleveland Heights, OH 44121, USA. (1927) PhD geology (Case Western Reserve U., 1971). Corporate sec., Cleveland Crystals Inc. (tel. 216 + 381-9003, fax 216 + 486-6103). *Solid state chemistry, piezoelectricity, ferroelectricity, mineralogy.*

Cooley, Mr. James W. Res. Div., IBM Thomas J. Watson Res. Ctr., P.O. Box 218, Yorktown Heights, NY 10598, USA. (tel. 914 + 945-2550, telex 137456, fax 914 + 945-2141, email COOLEY at IBM.COM).

Cooper, Mrs. Ann S. Solid State Res. Dept., AT&T Bell Labs., Rm. 1C-211, 600 Mountain Ave., Murray Hill, NJ 07974, USA. (1929) BS chemistry (St. Lawrence U., 1950). Member techn. staff. (tel. 201 + 582-6921). *Powder X-ray diffraction, crystal growing.*

Cooper, Dr. Brian J. Geology Dept., Sam Houston State U., Huntsville, TX 77341, USA. (1953) PhD geology (Virginia Polytechnic Inst., 1988). Asst. prof., geol. prog. coordin. (tel. 409 + 294-1566). *Optical crystallography, crystal chemistry, mineralogy.*

Cooper, Dr. John Neale. Chemistry Dept., Bucknell U., Lewisburg, PA 17837, USA. (1938) PhD chemistry (UCB, 1964). Prof. (tel. 717 + 524-3673, email COOPER at BUCKNELL). *Physical inorganic chemistry, transition metal complexes.*

Copeland, Prof. Richard Franklin. Div. of Sci. and Mathematics, Bethune-Cookman C., Daytona Beach, FL 32015, USA. (1938) PhD chemistry (Texas A&M U., 1965). Prof. (tel. 904 + 255-1401, ext. 468). *Organometallic complexes, scientific and technical information, crystallographic computing.*

Copley, Dr. John R. D. Reactor Radiation Div., Nat. Inst. Stnds. & Techn., Gaithersburg, MD 20899, USA. (1944) PhD physics (McMaster U., Canada, 1970). physicist (tel. 301 + 975-5133, fax 301 + 921-9847, email Bitnet COPLEY at NBSENH). *Structure, simple liquids, neutron guide instrumentation.*

Coppens, Prof. Philip. Chemistry Dept., Acheson Hall, SUNY at Buffalo, Buffalo, NY 14214, USA. (1930) PhD phys. chemistry (U. of Amsterdam, Netherlands, 1960). Prof. (tel. 716 + 831-3911, fax 716 + 831-2960). *Charge density distribution, chemical bonding, modulated structures, low temperature crystallography, accurate measurements, crystallographic computing, low dimensional conductors, applications of synchrotron radiation.*

Cordes, Prof. A. Wallace. Dept. of Chem. & Biochem., U. of Arkansas, Fayetteville, AR 72701, USA. (1934) PhD inorg. chemistry (U. of Illinois, 1959). Prof. (tel. 501 + 442-6608). *Inorganic ring and cage molecules, sulfur chemistry, sulfur-nitrogen structures.*

Corfield, Prof. Peter William Reginald. Chemistry Dept., The Kings C., Briarcliff Manor, NY 10510, USA. (1937) PhD chemistry (Durham U., UK, 1963). Prof. and chairman. (tel. 914 + 941-7200, EXT. 203). *Crystallography, protein, small molecule, computer software.*

Corliss, Dr. Lester Myron. Fearington Post 252, Pittsboro, NC 27312, USA. (1919) PhD chemical physics (Harvard U., 1949). Sr. chemist. *Neutron scattering, magnetism, phase transformations, critical phenomena, diffuse motion.*

Correll, Mr. Carl Clayton. Biophysics Res. Div., 232 IST Bldg., U. of Michigan, North Campus, Ann Arbor, MI 48109, USA. (1962) BS chemistry (U. of Michigan, 1984). Grad. student. (tel. 313 + 763-2199). *Flavo-metallo proteins, membrane proteins.*

Coulter, Dr. Charles L. Res. Facilities Improvement Prog., Div. of Res. Resources, Nat. Insts. of Health, Bethesda, MD 20892, USA. (1933) PhD chemistry (UCLA, 1960). Dir. (tel. 301 + 496-8482, fax 301 + 496-0019, email Bitnet CCO at NIHCO). *Structural biochemistry, instrumentation.*

Cowan, Dr. Paul L. Quantum Metrology Div., Nat. Inst. Stnds. & Techn., A141 Physics Bldg., Gaithersburg, MD 20899, USA. (1950) PhD physics (PSU, 1977). Physicist. (tel. 301 + 975-4846, telex 197674 NBS UT, fax 301 + 869-7761, email COWAN at NBSENH). *X-ray physics, diffraction, surface science.*

Cowley, Prof. John M. Physics Dept., Arizona State U., Tempe, AZ 85287-1504, USA. (1923) DSc physics (U. of Adelaide, Australia, 1957). Prof. (tel. 602 + 965-6459). *Electron diffraction, microscopy, crystals, defects, disorder, surface structure.*

Cowley, Miss Margaret. Molec. Biol. Dept., Expt. Sta. E228/316, E.I. du Pont de Nemours Co., Wilmington, DE 19898, USA. (tel. 302 + 695-9060).

Cox, Dr. David Ernest. Physics Dept., Brookhaven Nat. Lab., Upton, NY 11973, USA. (1934) PhD inorg. chemistry (Royal C. of Sci., London, UK, 1959) Physicist (tel. 516 + 282-3818, telex 96-7703, fax 516 + 282-2739, email Bitnet COX at BNLUX0). *Diffraction, X-ray, neutron, synchrotron, powder, Rietveld analysis, magnetic structures, solid state chemistry, zeolites, complex oxides, high T(c) oxide superconductors.*

Cox, Ms. Jane. E.I. duPont de Nemours Co., E228-340 Experimental Sta., Wilmington, DE 19898, USA.

Cox, Ms. Mary Beth. 7600 Hollow Hollow, #1307, Austin, TX 78731, USA.

Cramer, Prof. Roger Earl. Chemistry Dept., U. of Hawaii, 2545 The Mall, Honolulu, HI 96822-2275, USA. (1943) PhD inorg. chemistry (U. of Illinois, 1969). Prof. and chairman. (tel. 808 + 948-7480, fax 808 + 949-3025, email Bitnet ROGERS at UHCCUX). *Organoactinide chemistry, ionophores, vitamin B1 coordination chemistry.*

Craston, Mr. Dennis F. 10834 N. 44th St., Phoenix, AZ 85028, USA. (1925) MSc analytical chem. (Fordham U.). Sr. physicist (tel. 602 + 996-4115). *Forensic toxicology, spectroscopy, X-ray diffraction, drug identification.*

Craven, Prof. Bryan Maxwell. Crystallography Dept., U.of Pittsburgh, Pittsburgh, PA 15260, USA. (1932) PhD chemistry (U. of New Zealand, 1957). Prof. (tel. 412 + 624-9300, fax 412 + 624-1882, email CRAVEN at PITTVMS). *Biological structure, diffraction, X-ray, neutron, charge density distribution.*

Creswick, Dr. Michael William. Texas Operations B-1217, Dow Chemical USA, Freeport, TX 77541, USA. (1953) PhD chemistry (U. of Houston, 1981). Proj. leader. (tel. 409 + 238-3765).

Crist, Prof. Buckley, Jr. Materials Sci. and Eng. Dept., Northwestern U., Tech. Inst., Evanston, IL 60208, USA. (1941) PhD chemistry (Duke U., 1966). Prof. (tel. 312 + 491-3279). *Polymers, morphology, mechanical, optical properties, X-ray diffraction, small-angle X-ray, neutron scattering.*

Critchlow, Dr. Susan C. Chemistry Dept., BG-10, U. of Washington, Seattle, WA 98195, USA. (1954) PhD inorg. chemistry (Iowa State U., 1983). Res. assoc. (tel. 206 + 543-0210). *Chemistry, inorganic, solid state, structure determinations.*

Croft, Dr. William J. Materials Sci. Branch, Army Materials Techn. Lab., Arsenal St., Watertown, MA 02172, USA. (1926) PhD crystallography (Columbia U., 1954). Chemist. (tel. 617 + 923-5358). *Diffraction, X-ray, electron, ceramic materials.*

Cromer, Dr. Don Tiffany. INC-4, MS C346, Los Alamos Nat. Lab., P.O. Box 1663, Los Alamos, NM 87545, USA. (1923) PhD chemistry (U. of Wisconsin, 1953). Staff member. (tel. 505 + 667-2424). *Intermetallic structures, charge density, computer programming, scattering factors, anomalous dispersion.*

Cueto, Ms. Maria A. Chemistry Dept., Rutgers U., Graduate Sch. New Brunswick, Piscataway, NJ 08854, USA. (tel. 201 + 932-5630).

Cuff, Ms. Marianne Elaine. Biochem. and. Mol. Biophys. Dept., Columbia U., 630 W. 168th St., New York, NY 10032, USA. (1963) M.Phil. Biochem., mol. biophys. (Columbia U., 1988). Grad. student. (tel. 212 + 305-1846, ext. F-1846, email CUFF at CUMBG). *Protein structure - function, invertebrate oxygen carriers, macromolecular assembly.*

Cullen, Prof. David Lawrence. Chemistry Dept., Connecticut C., New London, CT 06320, USA. (1940) PhD chemistry (U. of Washington, 1969). Assoc. prof. (tel. 203 + 447-1911). *Porphyrin structures, bile pigments, biologically interesting molecules, coordination chemistry.*

Curtin, Prof. David Yarrow. Chemistry Dept., U. of Illinois, 1209 W. California, Urbana, IL 61801, USA. (1920) PhD chemistry (U. of Illinois, 1945). Prof. emeritus. (tel. 217 + 333-0797). *Chemistry, organic compounds.*

Curtis, Prof. M. David. Chemistry Dept., U. of Michigan, Ann Arbor, MI 48109-1055, USA. (1937) PhD chemistry (Northwestern U., 1964). Prof. and chairman. (tel. 313 + 764-7314, fax 313 + 747-4865, email Bitnet USERK26G at UMICHUM). *Catalysis, homogeneous, heterogeneous, metal clusters, organometallic chemistry, early transition metals.*

Czerwinski, Prof. Edmund William. Div. of Biochemistry, U. of Texas Med. Branch, Galveston, TX 77550, USA. (1940) PhD biochemistry (Indiana U., 1971). Asst. prof. (tel. 409 + 761-3287). *Protein crystallography, biologically important small molecules.*

Dahl, Prof. Lawrence F. Chemistry Dept., U. of Wisconsin, Madison, WI 53706, USA. (1929) PhD phys. chemistry (Iowa State U., 1956). Prof. (tel. 608 + 262-5859). *Synthesis, structure, bonding, transition metal compounds, organometallics, metal cluster systems.*

Dalley, Prof. Nelson Kent. Chemistry Dept., Brigham Young U., Provo, UT 84602, USA. (1935) PhD chemistry (U. of Texas at Austin, 1968). Prof. (tel. 801 + 378-3434). *Crystal structures, cyclic polyethers (derivatives and cation complexes), nucleosides, biologically interesting small molecules.*

Daniels, Dr. Lee M. Chemistry Dept., Iowa State U., Ames, IA 50011, USA. (1956) PhD inorg. chemistry (Texas A&M U., 1984). X-ray crystallographer. (tel. 515 + 294-7956, email Bitnet DANIELS at ALISUVAX).

Daniels, Mr. R. E. R.E. Daniels Inc., 36 Rolling Hill Dr., Morristown, NJ 07960, USA.

Dann, Mr. Jeffrey Neil. Chem. and Metallurgical Div., GTE Products Corp., Hawes St., Towanda, PA 18848, USA. (1946) MS physics (PUN, 1970). Sr. proj. eng. (tel. 717 + 265-2121, ext. 2425, telex 834610, fax 717 + 265-1430). *X-ray powder diffraction, computer applications.*

Dantonio, Mr. Peter. Code 6030, U. S. Naval Res. Lab., 4555 Overlook Ave., SW, Washington, DC 20375, USA. (tel. 202 + 767-2735).

Darby, Dr. Willie L. Chemistry Dept., Hampton U., P.O. Box 6594, Hampton, VA 23668, USA. PhD inorg. chemistry (Auburn U., 1982). Asst. prof. (tel. 804 + 727-5249).

Darling, Dr. Stephen D. Chemistry Dept., U. of Akron, Akron, OH 44325, USA. (tel. 216 + 375-7366).

Darovskikh, Mr. Alexander. X3 Nat. Synchrotron Light Source, Brookhaven Nat. Lab., Upton, NY 11973, USA. (1950) MS physics (Moscow Inst. Phys. & Engin., 1974). Postdoct. res. assoc. (tel. 516 + 282-3770). *Crystallography, phase transitions, instrumentation, X-ray, neutron.*

Das, Dr. Badri Narayan. 12 Thurston Dr., Upper Marlboro, MD 20772, USA. (1927) PhD metallurgy (Illinois Inst. of Techn., 1964). (tel. 202 + 767-2714). *Materials processing, crystal growth, structure, properties, physical, magnetic, superconducting.*

David, Mr. Peter Rensis. Chemistry Dept., Harvard U., 12 Oxford St., Cambridge, MA 02138, USA. (1959) BA chemistry, mathematics (Oberlin C., 1981). Grad. student. (tel. 617 + 495-4097, email DAVID at HARVSC3). *Protein crystallography, detector, generator design, computational methods.*

Davies, Dr. David R. Lab. of Mol. Biology, Nat. Inst. of Health, Bldg. 2, Rm. 316, Bethesda, MD 20892, USA. (1927) PhD chemical crystallography (Oxford U., UK, 1952). Chief, Mol. struct. sect. (tel. 301 + 496-4925, fax 301 + 496-0201, email DRD at NIHKLMB). *Enzymes, antibodies, nucleic acids.*

Davies, Dr. Jay Franklin, II. Chemistry Dept., U. of Calif. at San Diego, La Jolla, CA 92093, USA. (1955) PhD chemistry (U. of Calif. at San Diego, 1989). Res. Asst. (tel. 619 + 534-2011, email Bitnet JFD at UCSD). *Macromolecular crystallography.*

Davies, Dr. Julian Anthony. Chemistry Dept., U. of Toledo, W. Bancroft St., Toledo, OH 43606, USA. (1955) PhD chemistry (U. of London, England, 1979). Prof. chem., and med. & biol. chem. (tel. 419 + 536-2254, email UOFT01 at FAC1074). *Chemistry, organometallic, coordination.*

Davis, Dr. Briant LeRoy. Inst. of Atmospheric Sci., South Dakota Sch. of Mines & Techn., Rapid City, SD 57701, USA. (1936) PhD geology (UCLA, 1964). Asst. to V.P. for Grad. Studies. (tel. 605 + 394-2291). *Cloud physics, nucleation processes, air pollution, chemistry, physics.*

Davis, Prof. Phillip Howard. Chemistry Dept., U. of Tennessee at Martin, Martin, TN 38238, USA. (1946) PhD chemistry (U. of Illinois, 1972). Prof. (tel. 901 + 587-7456). *Magnetic resonance, X-ray crystallography, molecular and electronic structure, inorganic materials.*

Davis, Prof. Raymond Edward. Chemistry Dept., U. of Texas at Austin, Austin, TX 78712, USA. (1938) PhD chemistry (Yale U., 1965). Prof. (tel. 512 + 471-3097). *Molecular structure determination, organometallic compounds, organic structures.*

Davison, Prof. Daniel B. Biochem. & Biophys. Sci. Dept., U. of Houston, 4800 Calhoun, Houston, TX 77204-5500, USA. (1955) PhD biological sci. (SUNY at Stony Brook, 1985). Asst. prof. (tel. 713 +749-2801, email BCHSO at UHNIX2.UH.EDU). *Protein structure prediction, distance geometry.*

Day, Miss Catherine L. Ames Lab., Iowa State U., 40 Spedding Hall, Ames, IA 50011, USA. (1964) BS chemistry (Iowa State U., 1986). Grad. student. (tel. 515 + 294-8477).

Day, Dr. Cynthia Ann Secauer. Crystalytics Co., P.O. Box 82286, Lincoln, NE 68501, USA. (1952) PhD inorg. chemistry (U. of Nebraska, 1978). President. (tel. 402 + 421-2797).

Day, Mr. Michael W. 7730 Sedan Ave., Canoga Park, CA 91304, USA.

Day, Prof. Roberta Ogilvie. Chemistry Dept., U. of Massachusetts, Amherst, MA 01003, USA. (1941) PhD phys. chemistry (MIT, 1971). Assoc. prof. (tel. 413 + 545-2375).

Day, Prof. Victor Warren. Chemistry Dept., U. of Nebraska, Lincoln, NE 68588, USA. (1943) PhD phys. chemistry (Cornell U., 1969). Prof. (tel. 402 + 472-3540). *Structure, bonding, organic, inorganic, organometallic compounds, structure - function relationships, biological macromolecules.*

De Camp, Dr. Wilson H. II. Food & Drug Admin., HFD-520, Div. Anti-Infective Drug. Prod., 5600 Fishers Ln., Rockville, MD 20857, USA. (1936) PhD phys. chemistry (U. of Maryland, 1970). Chemist. (tel. 301 + 443-6714). *Crystal structures, absolute configuration, conformational analysis, organic compounds, natural products, crystallographic computing, powder diffraction, drug polymorphism.*

De Fontaine, Prof. Robert Didier. Materials Sci. Dept., Hearst Mining Bldg., U. of Calif. at Berkeley, Berkeley, CA 94720, USA. (1931) PhD materials sci. (Northwestern U., 1967). Prof. (tel. 415 + 642-8177). *Phase transformations, alloys, thermodynamics.*

De Haven, Dr. Patrick William. Z/40E, IBM Corp., Route 52, Hopewell Junction, NY 12533, USA. (1949) PhD phys. chemistry (Iowa State U., 1976). Advisory eng. (tel. 914 + 894-6859). *Diffraction, high temperature, small area, thin film, solid state kinetics.*

De Jarnette, Miss F. Elaine. Mol. Biophysics Dept., Med. Fndn. of Buffalo, 73 High St., Buffalo, NY 14203, USA. (1931) BS education (SUNY at Buffalo, 1970). Asst. res. sci. (tel. 716 + 856-9600). *Crystallography.*

De la Camp, Prof. Ulrich Otto. Chemistry Dept., California State U., Dominguez Hills, Carson, CA 90747, USA. (1929) PhD chemistry (U. of Calif. at Davis, 1966). Prof. (tel. 213 + 516-3417). *Structure, conformation, small organic molecules, anomalous scattering of X-rays.*

De Lucia, Dr. Mary Lou. 61D-300/2 Kimberly-Clark Corp., 1400 Holcomb Bridge Rd., Roswell, GA 30076, USA. (1947) PhD phys. chemistry (SUNY at Buffalo, 1978). Res. scient. (tel. 413 + 587-8420). *Paper, structure, performance, biological structure - function.*

De Maggio, Mr. Gregory B. 1834 Spring Grove, Bloomfield Hills, MI 48013, USA. (tel. 313 + 853-7830).

De Rosier, Prof. David John. Biology Dept., Brandeis U., 215 South St., Waltham, MA 02254, USA. (1939) PhD biophysics (U. of Chicago, 1965). Prof. (tel. 617 + 736-2426, fax 617 + 736-2405, email Bitnet DEROSIER at BRANDEIS). *Actin structure, bacterial flagella, image analysis, electron microscopy.*

De Titta, Dr. George Thomas. Mol. Biophysics Dept., Med. Fndn. of Buffalo, 73 High St., Buffalo, NY 14203, USA. (1947) PhD biochemistry-crystallography (U. of Pittsburgh, 1973). Assoc. res. scient. (tel. 716 + 856-9600). *Biocrystallography, prostaglandins, biotin vitamins, lipid related molecules.*

De Vos, Dr. Abraham Martien. Mol. Biochem. Dept., Genentech Inc., 460 Pt. San Bruno Blvd., South San Francisco, CA 94080, USA. (1955) Dr chemistry (U. of Utrecht, The Netherlands, 1985). (tel. 415 + 266-2523). *Protein crystallography, macromolecular structure - function.*

De Vries, Dr. Adriaan. Res. Office, Kent State U., Kent, OH 44242, USA. (1931) PhD phys. chem., X-ray crystallog. (State U. of Utrecht, Netherlands, 1963). Asst. dean. (tel. 216 + 672-2251). *Organic crystal structures, liquid crystals, liquids (complex molecules).*

Deadwyler, Mr. Daniel A. Chemistry Dept., U. of North Carolina, Charlotte, NC 28223, USA. (tel. 704 + 547-4437).

Dean, Mr. Johnny Clyde. MOBAY Synthetics Corp., 8701 Park Place Blvd., Houston, TX 77017, USA. (1928) BS chemistry (U. of Southwestern Louisiana, 1952). Staff chemist. (tel. 713 + 946-9352). *X-ray diffraction, polycrystalline materials.*

DeBoer, Dr. Barry Goodwin. GTE Products Corp., 60 Boston St., Salem, MA 01970, USA. (1942) PhD chemistry (UCB, 1968). Eng. specialist (tel. 508 + 741-9568). *Applications of structural information, powder X-ray diffraction, solid state: simulations, chemistry, physics, phosphors, luminescent materials.*

Deganello, Prof. Sergio. SCL-Box 42, U. of Chicago, 5735 S. Ellis, Chicago, IL 60637, USA. (Alt.: Instituto Mineralogia, U. di Palermo, Via Archirafi 36, Palermo, 90123, Italy). PhD mineralogy, crystallography (U. of Chicago, 1971). Prof. (tel. 312 + 702-7062 (Italy:91-6161516)).

Deisenhofer, Prof. Johan. Biochem. Dept. and HHMI. U. Texas SW Med. Center at Dallas, 5323 Harry Hines Blvd., Y4.206, Dallas, TX 75235-9050, USA. (1943) PhD physics (Techn. U. of Munich, 1974). Prof., Investigator, HHMI. (tel. 214 + 689-5089, fax 214 + 689-5066, email Bitnet DEISENHO at UTSW). *X-ray crystallography, biological macromolecules, proteins, structure - function.*

Delaney, Prof. Matthew S. Office of the Academic Dean, Mount St. Mary's C., 12001 Chalon Rd., CA 90049, USA. (1927) PhD mathematics (Ohio State U., 1971). Academic dean. (tel. 213 + 476-2237). *Mathematical crystallography.*

Delatore, Ms. Diana L. Structural Biol. Lab., Brandeis U., Rosenstiel Res. Ctr., Waltham, MA 02254, USA.

Delgado, Mr. Jose Miguel. MIT, Room 13-4069, 77 Massachusetts Ave., Cambridge, MA 02139, USA. (tel. 617 + 253-6894).

DeLucas, Dr. Lawrence James. Ctr. for Macromolec. Cryst. U. of Alabama at Birmingham, Box THT-79, Room 270, Birmingham, AL 35294, USA. (1950) PhD biochem., crystallography (U. of Alabama at Birmingham, 1982). Assoc. dir. (tel. 205 + 934-3802, fax 205 + 934-0480). *Protein purification, crystallization, structure.*

Depalma, Mr. Vincent M. Zip AM1, IBM Corp., East Fishkill, Route 52, Hopewell Junction, NY 12533, USA.

Depp, Mr. Mark. Siemens Analytical X-ray Co., 5225-1 Verona Rd., Madison, WI 53711, USA. (1954) BS chemistry (U. of Wisconsin at Milwaukee, 1975). Eng. mgr. (tel. 608 + 271-3333, ext. 2534, telex 910/286/2736, fax 608 + 273-5046). *Instrumentation, computing methods, computer graphics.*

Deroski, Ms. Betty Rolfs. Suffolk County Community C., 553 College Rd., Selden, NY 11784, USA.

Desiraju, Dr. Gautam R. Central R. & D., Exptl. Sta. E356/131, E.I. du Pont de Nemours Co., Wilmington, DE 19880, USA. (1952) PhD chemistry (U. of Illinois at Urbana, 1976). Visiting scient. (tel. 302 + 695-9495). *Solid state chemistry, organic, Cambridge database, intermolecular interactions.*

Desper, Dr. C. Richard. Army Materials Mech. Res. Ctr., SLCMT-EMX, Watertown, MA 02172, USA. (1937) PhD chemistry (U. of Massachusetts at Amherst, 1966). Res. chemist. (tel. 617 + 923-5391). *Synthetic polymers.*

Dewan, Prof. John C. Chemistry Dept., New York U., 4 Washington Pl., New York, NY 10003, USA. (1949) PhD inorg. chemistry (U. of Western Australia, 1975). Res. prof. of chemistry. (tel. 212 + 998-8400, telex 235128NYU, email DEWAN at NYUACF (BN)). *Crystallography, macromolecular, small-molecule, low temperature, bio-inorganic chemistry.*

Dexter, Dr. David D. Computing Center, Alma C., Alma, MI 48801, USA. (1940) PhD chemistry (Georgetown U., 1968). (tel. 517 + 463-7303).

Dhere, Dr. Ashok G. Techn. Sect., Fibers Dept., E. I. duPont de Nemours Co., 4501 N. Access Rd., Chattanooga, TN 37415, USA. (1948) PhD metallurg. eng., matls. sci. (U. of Kentucky, 1982). Res. eng. (tel. 615 + 875-7768). *Structural studies, polymers, carbon, nylon fibers.*

Dickerson, Prof. Richard Earl. Mol. Biology Inst., U. of California at Los Angeles, Los Angeles, CA 90024, USA. (1931) PhD phys. chemistry (U. of Minnesota, 1957). Prof. and Director. (tel. 213 + 825-5864). *DNA structure, DNA-drug binding, protein-DNA recognition, molecular evolution.*

Dickinson, Dr. Charles. Res. Dept., White Oak Lab., Naval Surface Weapons Ctr., Silver Spring, MD 20903-5000, USA. (1937) PhD chemistry (U. of Maryland, 1972). Dir., Indep. res. off. (tel. 202 + 394-1259, email CDICKIN at NSWC.OAS). *Computing systems, small molecules, structure - property relationships.*

Dickman, Dr. Michael H. 4515 Willard Ave., Apt. 1414 S., Chevy Chase, MD 20815, USA. (1953) PhD chemistry (U. of Calif. at Irvine, 1984). (tel. 301 + 652-5066). *Small molecules, inorganic complexes.*

Dixon, Ms. Melinda M. Inst. of Mol. Biology, U. of Oregon, Eugene, OR 97401, USA. (tel. 503 + 686-5176).

Djebli, Mr. Abdellah. Chemistry Dept., Case Western Reserve U., Box 110, Cleveland, OH 44106, USA. (tel. 216 + 368-2598).

Dodd, Dr. Charles Gardner. 581-B N. Trail, Stratford, CT 06497, USA. (1915) PhD phys. chemistry (U. of Michigan, 1948). (tel. 203 + 375-5015). *Surface chemistry, physics, microanalysis, soft X-ray spectrometry, chemical bonding, ion implantation.*

Dodge, Prof. Richard Patrick. Chemistry Dept., U. of the Pacific, 3601 Pacific Ave., Stockton, CA 95211, USA. (1932) PhD phys. chemistry (UCB, 1958). Prof. (tel. 209 + 946-2268). *Molecular structure, computer applications, quantum chemistry.*

Doedens, Prof. Robert John. Chemistry Dept., U. of California at Irvine, Irvine, CA 92717, USA. (1937) PhD chemistry (U. of Wisconsin, 1965). Prof. (tel. 714 + 856-6605, email Bitnet RDOEDENS at UCI). *Structural chemistry, inorganic, organometallic, coordination.*

Dollase, Prof. Wayne A. Earth & Space Sci. Dept., U. of California at Los Angeles, Los Angeles, CA 90024, USA. (1938) PhD crystallography (MIT, 1966). Prof. (tel. 213 + 825-3823). *Chemistry, crystal, solid state, mineralogy.*

Doniach, Mr. S. Applied Physics Dept., Stanford U., Stanford, CA 94305, USA. (tel. 415 + 497-4786).

Dorset, Dr. Douglas Lewis. Electron Diffraction Dept., Med. Fndn. of Buffalo, 73 High Street, Buffalo, NY 14203, USA. (1942) PhD biophysics (U. of Maryland, 1971). Prin. res. scient., dept. head (tel. 716 + 856-9600, ext. 475). *Electron diffraction, crystal structure analysis, biomembrane structure, phase transitions, polymer crystal growth.*

Doscher, Dr. Marilynn S. Biochemistry Dept., Wayne State U., 540 E. Canfield, Detroit, MI 48201, USA. (tel. 313 + 577-1295).

Downing, Dr. Kenneth H. Donner Lab., 1-326, Lawrence Berkeley Lab., Berkeley, CA 94720, USA. (1945) PhD applied physics (Cornell U., 1974). Staff scient. (tel. 415 + 486-5941, email KDOWNING at LBL). *Structural biology, electron microscopy, crystallography.*

Downs, Dr. James Winston. Geology and Mineralogy Dept., The Ohio State U., 291 Watts Hall, 104 W. 19th St., Columbus, OH 43210, USA. (1952) PhD geology (Virginia Poly. Inst. & State U., 1983). Asst. prof. (tel. 614 + 292-6290, email Bitnet TS1790 at OHSTVMA). *Electron density distributions, minerals, crystal chemistry, mineralogy.*

Dowty, Mr. Eric. 196 Beechwood Ave., Bogota, NJ 07603, USA. (tel. 201 + 487-3254).

Doyle, Prof. John Robert. Chemistry Dept., U. of Iowa, Iowa City, IA 52242, USA. (1924) PhD chemistry (Tulane U., 1955). Prof. (tel. 319 + 353-3585). *Inorganic, organometallic compounds, chemistry, structures.*

Doyne, Prof. Thomas H. Chemistry Dept., Villanova U., Villanova, PA 19085, USA. (1927) PhD biochemistry (PSU, 1957). Prof. of biochemistry, chairman (tel. 215 + 645-4874, email TDOYNE at VILLVM). *Peptides, biologically interesting small molecules.*

Dragsdorf, Prof. Russell Dean. Physics Dept., Cardwell Hall, Kansas State U., Manhattan, KS 66506, USA. (1922) PhD physics (MIT, 1948). Prof. emeritus. (tel. 913 + 539-9277). *Diffraction, X-ray, electron, defect structure.*

Drendel, Dr. William Bruce. 3225 Debbie Dr., Hendersonville, NC 28739, USA. (1953) PhD biochemistry (U. of Wisconsin at Madison, 1986). (tel. 704 + 696-9546). *Macromolecular crystallography, protein structure, sequence homology.*

Drickman, Dr. Myra Vivian. 1114 Princeton St., Apt. 5, Santa Monica, CA 90403, USA. (1942) MD medicine (New York U., 1976). *Medical imaging systems.*

Druyan, Prof. Mary Ellen. Biochemistry Dept., Loyola U. Sch. of Dentistry, 2160 South First Ave., Maywood, IL 60153, USA. (1938) PhD biochemistry (U. of Chicago, 1972). Asst. prof. (tel. 312 + 531-3578). *Molecular model building, low angle scattering, biologically relevant systems.*

Duax, Dr. William Leo. Med. Fndn. of Buffalo, 73 High St., Buffalo, NY 14203, USA. (1939) PhD phys. chemistry (U. of Iowa, 1967). Res. dir. (tel. 716 + 856-9600, ext. 61, fax 716 + 852-4846, email Bitnet ACAMJV%MFB at VBVMS). *Conformation, hormones, drugs, antibiotics, biological function - structure relationships, direct methods, data storage, data retrieval and analysis, computer graphics.*

Duchamp, Dr. David James. Pharmaceutical Res. and Dev., The Upjohn Co., 301 Henrietta St., Kalamazoo, MI 49001, USA. (1939) PhD chemistry (CIT, 1965). Dir., phys., analyt. chem. res. (tel. 616 + 385-7766). *Organic, biological crystal structures, crystallographic computing, molecular mechanics, potential energy calculations.*

Duesler, Dr. Eileen N. Chemistry Dept., U. of New Mexico, Albuquerque, NM 87131, USA. PhD chemistry (UCB, 1973). Res. scient. (tel. 505 + 277-0505, email Bitnet DUESLER at UNMB). *Structural chemistry.*

Duke, Dr Norma Edith. Dept. of molecular Biophyics and Biochemistry, Yale U., 260 Whitney Ave. (JWG 402), New Haven, CT 06511, USA. (1959) PhD, chemistry (U. Calgary,Canada,1988). Postdoctoral Fellow. *Small molecule crystallography, drug design, proteins, nucleotides.*

Dumke, Prof. Warren Lloyd. RR 1, Box 148, Chesapeake, OH 45619, USA. (1928) PhD phys. chemistry (U. of Nebraska, 1965). *Diffraction, X-ray, neutron, quantum mechanical calculations.*

Dunn, Dr. Deborah Anne. Chemical Design, Inc., 200 Rte. 17, South Suite 120, Mahwah, NJ 07430, USA. (1950) PhD inorg. chemistry (Florida State U., 1978). Exec. vice-pres. (tel. 201 + 529-3323, fax 201 + 529-2443). *Molecular modeling, chemistry.*

Dunn, Mr. Karl L. 1044 Joe Quick Rd., Hazel Green, AL 35750, USA. (tel. 205 + 882-4127).

Dunsieth, Mr. Dana G. 12804 Antioch Rd., R.R.2, Leesburg, OH 45135, USA. Leesburg, OH 45135, USA. (1951).

Dutremez, Mr. Sylvain G. Chemistry Dept., U. of Toledo, 2801 W. Bancroft St., Toledo, OH 43606, USA. (1962) Eng. chemistry (U. of Lille, France, 1985). Grad. student. (tel. 419 + 537-2109, ext. 5983). *X-ray structure, small molecules, solid state NMR, organometallic chemistry, ceramics, semiconductors.*

Dwiggins, Dr. Claudius William. 1211 S. Keeler St., Bartlesville, OK 74003, USA. (1933) PhD phys. chemistry (U. of Arkansas, 1958). Res. chemist. (tel. 918 + 336-8546). *Small angle X-ray scattering, scattering theory, non-crystalline materials, ultracentrifuge, X-ray fluorescence.*

Dwight, Mr. Austin Elbert. 8720 Lemont Rd., Downers Grove, IL 60516, USA. (1919) MS metallurgical eng. (U. of Michigan, 1950). *Intermetallic structures, alloy crystal structure, chemistry, alloys, intermetallic compounds.*

Dyar, Dr. M. Darby. Geological Sci. Dept., U. of Oregon, Eugene, OR 97403, USA. (1958) PhD geochemistry (MIT, 1985). Asst. prof. (tel. 503 + 686-4552, fax 503 + 686-3127, email MATSCI at OREGON). *Mineral physics, crystallography, mineralogy, Moessbauer spectroscopy, crystal chemistry.*

Dytrych, Mr. William J. Geophysical Sci. Dept., U. of Chicago, 5734 S. Ellis Ave., Chicago, IL 60637-1434, USA. MS geology (Virginia Polytech. Inst. & State U., 1983). Grad. student. (tel. 312 + 684-2964). *Molecular sieve materials, characterization, theoretical framework structures.*

Ealick, Dr. Steven Edward. Ctr. for Macromolecular Crystallog., U. of Alabama in Birmingham, 1918 University Blvd., Birmingham, AL 35294, USA. (1951) PhD phys. chemistry (U. of Oklahoma, 1976). Assoc. Dir. Ctr. Macromol. Cryst. (tel. 205 + 934-7277, fax 205 + 934-0480, email EALICK at UABCMC). *Protein crystallography, drug design, synchrotron radiation, X-ray instrumentation.*

Eanes, Dr. Edward David. Bone Res. Br., Bldg. 30, Rm. 106, Nat. Inst. of Dental Res., Nat. Insts. of Health, Bethesda, MD 20892, USA. (1934) PhD phys. chem., crystallog. (Johns Hopkins U., 1961). Chief, min. chem. & struct. sect. (tel. 301 + 975-6832). *Crystal chemistry, calcium phosphates, biological mineralization.*

Earnest, Dr. Thomas. Biochem. & Biophysics Dept., Sch. of Medicine, U. of Calif. at San Francisco, San Francisco, CA 94143, USA. (1950) PhD physics (Boston U., 1987). Postdoct. fel. (tel. 415 + 665-6932, email TNE at MSE.OCSF.EDU). *Macromolecular crystallography, cryo-crystallography, electron diffraction.*

Ebinger, Mr. Michael H. Agronomy Dept., Purdue U., West Lafayette, IN 47907, USA.

Eck, Mr. Michael J. HHMI, U. of Texas - SW, 5323 Harry Hines Blvd., Dallas, TX 75235, USA. (tel. 214 + 689-5039).

Eddy, Prof. Lowell Perry. Chemistry Dept., Western Washington U., Bellingham, WA 98225, USA. (1920) PhD chemistry (Purdue U., 1952). Assoc. prof. (tel. 206 + 676-3070). *Inorganic coordination chemistry, compounds, structure.*

Eddy, Mr. Michael. Chemistry Dept., U. of Calif. at Santa Barbara, Santa Barbara, CA 93106, USA.

Edmondson, Dr. Stephen P. Molec. Biology Dept., Vanderbilt U., P.O. Box 1820, Sta. B, Nashville, TN 37235, USA. (tel. 615 + 322-2012).

Edmundson, Prof. Allen Broderick. Biology Dept., U. of Utah, 410 Chipeta Way, Salt Lake City, UT 84112, USA. (1932) PhD biochemistry (Rockefeller U., 1961). Prof. (tel. 801 + 581-3997, ext. 3997/3996, fax 801 + 581-4668). *Protein crystallography.*

Edwards, Dr. Brian Francis Peregrin. Biochemistry Dept., Wayne State U., 540 E. Canfield, Detroit, MI 48201, USA. (1947) PhD chemistry (Harvard U., 1975). Assoc. prof. (tel. 313 + 577-5107, fax 313 + 577-2765). *Protein structure, blood coagulation proteins, proteins from thermophiles.*

Eggleston, Dr. Drake Stephen. Physical & Struct. Chem. Dept., Smith Kline & French Labs., 600 Allendale Rd., King of Prussia, PA 19406, USA. (1954) PhD chemistry (U. of North Carolina at Chapel Hill, 1983). Sr. investigator. (tel. 215 + 270-6690, email EGGLESTON at SMITHKLINE.COM). *Peptides, structure, X-ray crystallography, NMR, inorganic chemistry.*

Eick, Prof. Harry A. Chemistry Dept., Michigan State U., East Lansing, MI 48824-1322, USA. (1929) PhD inorg. chemistry (U. of Iowa, 1956). Prof. (tel. 517 + 353-4511). *X-ray diffraction, single crystal, Rietveld powder, solid state inorganic chemistry, lanthanide chemistry, lower oxidation states, preparatory procedures, chalcogens, halogens and oxygen.*

Eigenbrot, Dr. Charles Weaver. 5918 Dover St., Oakland, CA 94609, USA. (1954) PhD chemistry (UCB, 1981). *Diffraction, single crystal, X-ray, neutrons, small, macromolecular.*

Eilerman, Dr. Donna Paige. 2020 Van Roo Ave., Merrick, NY 11566, USA. (Alt.: Nassau Community C., Garden City, NY). (1951) PhD chemistry-crystallography (Adelphi U., 1979). Asst. prof. (tel. 516 + 868-8826). *Orientationally disordered plastic crystals, gel-sol transformations, gelatin mixtures.*

Einck, Dr. James J. Quality Control Dept., Steris Labs. Inc., Box 23160, Phoenix, AZ 85363, USA. (1945) PhD org. chemistry (Arizona State U., 1976). Chem. control mgr. (tel. 602 + 278-1400. ext. 178). *Small molecules, organics, natural products, software developments.*

Einspahr, Dr. Howard Martin. Physical and Analytical Chem., The Upjohn Co., 301 Henrietta St., Kalamazoo, MI 49001, USA. (1943) PhD chemistry (U. of Pennsylvania, 1970). Sen. res. scient. (tel. 616 + 385-5492). *Macromolecular crystallography.*

Eisenberg, Prof. David. Chemistry & Biochem. Dept., U. of Calif. at Los Angeles, 405 Hilgard Ave., Los Angeles, CA 90024, USA. (1939) D.Phil. chemistry (Oxford U., 1964). Prof. (tel. 213 + 825-3754). *Proteins, structure - function, protein folding, water and aqueous solutions.*

Eisenberg, Prof. Richard. Chemistry Dept., U. of Rochester, River Campus, Rochester, NY 14627, USA. (1943) PhD chemistry (Columbia U., 1967). Prof. (tel. 716 + 275-5573). *Inorganic chemistry, homogeneous catalysis, structure - reactivity relationships, organometallic systems, catalytically active complexes.*

Eisenman, Prof. George. Physiology Dept., Sch. of Med., U. of Calif. at Los Angeles, Los Angeles, CA 90024-1751, USA. (1929) MD biophysics (Harvard U., 1953). Prof. (tel. 213 + 825-7138, fax 213 + 206-5661, email IMN5FJB at UCLAMVS). *Ion binding, channels, bioenergetics.*

El-Kabbani, Mr. Ossama A. L. Biological Res., Bldg. 202 Rm. A-109, Argonne Nat. Lab., Argonne, IL 60439, USA.

Elder, Prof. Richard C. Chemistry Dept., U. of Cincinnati, Mail Location 172, Cincinnati, OH 45221, USA. (1939) PhD chemistry (MIT, 1964). Prof. (tel. 513 + 556-9224). *Inorganic chemistry, X-ray absorption spectroscopy (EXAFS, XANES, high energy EXAFS), X-ray diffraction, WAXS, DAS, single crystal, gold-based arthritis pharmaceuticals, biological chemistry, transition metal complexes, structure, bonding.*

Ely, Dr. Kathryn R. La Jolla Cancer Res. Fndn., 10901 N. Torrey Pines Rd., La Jolla, CA 92037, USA. (1944) PhD biochemistry (U. of Utah, 1981). Staff scient. (tel. 619 + 455-6480, ext. 591). *Protein - nucleic acid interactions, protein crystallography, computer graphics.*

Emerson, Prof. Kenneth. Chemistry Dept., Montana State U., Bozeman, MT 59715, USA. (1931) PhD phys. chemistry (U. of Minnesota, 1961). Prof. (tel. 406 + 994-5393). *Complex halide salts, transition metals, synthesis, structures, low dimensional compounds.*

Emerson, Mr. Merle T. Chemistry Dept., U. of Alabama at Huntsville, Huntsville, AL 35899, USA. (tel. 205 + 895-6153).

Emge, Dr. Thomas James. Corp. Res. Div., Procter & Gamble Co., Miami Valley Labs., Box 398707, Cincinnati, OH 45239, USA. (1955) PhD phys. chemistry (The Johns Hopkins U., 1981). Memb., X-ray staff. *X-ray and neutron diffraction, analytical studies, surfactants, clays, celluloses, liquid crystals, single crystal studies, small molecules.*

Enemark, Prof. John Henry. Chemistry Dept., U. of Arizona, Tucson, AZ 85721, USA. (1940) PhD chemistry (Harvard U., 1966). Prof. (tel. 602 + 621-2245). *Inorganic and bio-inorganic chemistry, molybdenum compounds, heteronuclear NMR.*

Eng-Wilmot, Dr. David Lawrence. Chemistry Dept., Rollins C., P.O. Box 2743, Winter Park, FL 32789, USA. (1947) PhD inorg. chemistry (U. of South Florida, 1978). Assoc. prof. (tel. 305 + 647-2045). *Coordination compounds, iron transport, siderophores, ionophores, molecular structure, small molecules, bioinorganics, conformational analysis, stereochemistry.*

Enwall, Dr. Eric Lee. Chemistry Dept., U. of Oklahoma, 620 Parrington Oval, Norman, OK 73019, USA. (1940) PhD chemistry (Montana State U., 1969). Dir., analytical services center (tel. 405 + 325-2843, email ENWALL at AARDVARK.UCS.UOKNOR.EDU). *Structural inorganic chemistry, chemical instrumentation.*

Eppelsheimer, Dr. Daniel Snell Jr. Energy Lab., Room 8-115 MIT, 77 Massachusetts Ave., Cambridge, MA 02139, USA. (1941) Dr.rer.nat. mineralogie (U. Heidelberg, Germany, 1981). Postdoct. assoc. (tel. 617 + 253-5069). *Crystal physics.*

Epperson, Dr. John Ernest. Materials Sci. Div., MST-212, Argonne Nat. Lab., 9700 S. Cass Ave., Argonne, IL 60439, USA. (1933) PhD metallurgical eng. (U. of Tennessee, 1968). Metallurgist. (tel. 312 + 972-4971). *Diffuse scattering studies, short range order, small angle scattering, alloy decomposition.*

Erickson, Dr. John. Protein Crystallography Lab., Abbott Labs., D-47E, Abbott Park, IL 60064, USA. PhD virology (1978). (tel. 312 + 937-0268). *Macromolecular structure and function, drug design, antiviral research.*

Eriks, Prof. Klaas. Chemistry Dept., Boston U., 685 Commonwealth Ave., Boston, MA 02215, USA. (1922) PhD chemistry (U. of Amsterdam, Netherlands, 1952). Prof. (tel. 617 + 353-2497). *Amino acid, polypeptide complexes, phosphates, inorganic, organic, organometallic complexes with phosphine ligands.*

Ernst, Dr. Stephen Richard. Chemistry Dept., U. of Texas, Austin, TX 78712, USA. (1939) PhD phys. chemistry (U. of Utah, 1972). Res. scient. (tel. 512 + 471-1105, email ERNST at UTADNX). *Structure - activity relations, crystallographic computing.*

Estes, Dr. Eva Dixon. Environmental Chemistry Dept., Res. Triangle Inst., Bldg. 6, P.O. Box 12194, Ressearch Triangle Park, NC 27709, USA. (1949) PhD inorg. chemistry (U. of North Carolina at Chapel Hill, 1975). Res. chemist. (tel. 919 + 541-6283). *Ion chromatography, environmental source assessment, air monitoring.*

Etter, Prof. Margaret E. Cairns. Chemistry Dept., 78 Kolthoff Hall, U. of Minnesota, 207 Pleasant St., SE, Minneapolis, MN 55455, USA. (1943) PhD org. chemistry (U. of Minnesota, 1974). Assoc. prof. (tel. 612 + 624-5217, fax 612 + 624-6369). *Solid-state organic chemistry, crystallography, solid state NMR, hydrogen bond interactions, organic non-linear optical materials.*

Evans, Prof. David Robert. Biochemistry Dept., Wayne State U. Sch. of Med., 540 E. Canfield St., Detroit, MI 48201, USA. (1941) PhD biochemistry (Wayne State U., 1968). Prof. (tel. 313 + 577-1016, fax 313 + 577-2765). *Protein structure, enzymology, site directed mutagenesis, proteins, multifunctional, multidomain.*

Evans, Dr. Doris Louise. RD #1, Box 276, Watkins Glen, NY 14891, USA. (1923) PhD crystallography (London U., UK, 1969). Retired. (tel. 607 + 292-3205). *Structure - properties relationships, phase transformations, silicates, glasses.*

Evans, Ms. Eloise Humez. 1621 Gruenther Ave., Rockville, MD 20851, USA. (1921) BSc mathematics (MIT, 1942). (tel. 301 + 921-2921). *Standard X-ray diffraction powder patterns, powder patterns (experimental and calculated).*

Evans, Dr. Howard Tasker Jr. U.S. Geological Survey, Nat. Center Stop 959, Reston, VA 22092, USA. (1919) PhD inorg. chemistry (MIT, 1948). Res. chemist. (tel. 703 + 648-6762). *Inorganic and mineral crystal structures.*

Ezell, Dr. Edward F. BOC Group Technical Ctr., 100 Mountain Ave., Murray Hill, NJ 07974, USA. (1945) PhD physics (Stevens Inst. of Techn., 1979). Section leader. (tel. 201 + 771-6373). *Crystallography, diffraction, powder, scattering, X-ray, small angle, spectroscopy, surfaces, gases.*

Faber, Dr. John, Jr. Materials Sci. Div., Argonne Nat. Lab., 9700 S. Cass Ave., Argonne, IL 60439, USA. (1941) PhD materials science (Marquette U., 1973). Staff scient. (tel. 312 + 972-4969, email Bitnet FABER at ANLPNS). *Structure - properties relationships, high temperature materials, nonstoichiometric effects, low temperature magnetic properties, phase transitions.*

Fackler, Prof. John Paul, Jr. C. of Sci., Texas A&M U., College Station, TX 77843, USA. (1934) PhD chemistry (MIT, 1960). Dean, C. of Sci., Distin. prof. of chem. (tel. 713 + 845-7364). *Transition metal organometallics, sulfur compounds, coordination compounds, metal-containing polymers.*

Fahey, Prof. James A. Chem. Dept., Bronx Community C., 181 St., W. University Ave., Bronx, NY 10453, USA. (1941) PhD chemistry (U. of Tennessee at Knoxville, 1971). Prof. (tel. 212 + 220-6218, ext. 6903). *Radial-distribution studies, nuclear waste glasses, powder structural analysis, lanthanide, actinide compounds.*

Fair, Dr. Carolyn Kay. Enraf-Nonius Co., Res. & Dev., 390 Central Ave., Bohemia, NY 11716, USA. (1945) PhD inorg. chemistry (U. of Arkansas, 1973). Software developer. (tel. 516 + 589-2885, fax 516 + 589-2068). *Neutron diffraction, software development.*

Fairchild, Ms. Beatrice M. Columbia U., St. Luke's Hospital, 114 St. Amsterdam Ave., New York, NY 10025, USA. (tel. 212 + 870-6156).

Fajer, Dr. Jack. Applied Sci. Dept., Brookhaven Nat. Lab., 75 Rutherford Dr., Upton, NY 11973, USA. (1936) PhD phys. chemistry (Brandeis U., 1962). Sr. scient. (tel. 516 + 282-4521, email Bitnet FAJERJ at BNLUXO). *Diffraction, X-ray, neutron, EXAFS, structure - function correlations, porphyrins, hydroporphyrins.*

Faller, Prof. John William (Jack). Chemistry Dept., Yale U., P.O. Box 6666, New Haven, CT 06511-8118, USA. (1942) PhD chemistry (MIT, 1967). Prof. (tel. 203 + 432-3954, fax 203 + 432-6144, email Bitnet FALLER at YALEVMS). *Chemistry, inorganic, organometallic, coordination, metal-oxo complexes.*

Falvello, Dr. Lawrence R. Chemistry Dept., Texas A&M U., College Station, TX 77843-3255, USA. (1954) PhD chemistry (Cambridge U., UK, 1979). Res. scient. (tel. 409 + 845-3726, fax 409 + 845-4719, email FALVELLO at TAMLMSB). *Inorganic chemistry.*

Fanchon, Dr. Eric. Biochemistry Dept., BB 523, Columbia U., 630 W. 168th. St., New York, NY 10032, USA. (tel. 212 + 305-2219).

Fang, Prof. Jen Ho. Geology Dept., U. of Alabama, Tuscaloosa, AL 35487, USA. (1929) PhD geochemistry (PSU, 1961). Prof. and Chairman. (tel. 205 + 348-5097). *Powder diffraction.*

Farber, Dr. Gregory K. Biochemistry Dept., U. of Wisconsin, 420 Henry Mall, Madison, WI 53706-1569, USA. (1962) PhD phys. chemistry (MIT, 1988). Postdoct. fel. *Protein crystallography, Laue method, synchrotron radiation.*

Farmer, Dr. Barry L. Materials Sci. Dept., U. of Virginia, Thornton Hall, Charlottesville, VA 22901, USA. (1947) PhD macromol. sci. (Case Western Reserve U., 1974). Assoc. prof. (tel. 804 + 924-0605, email BLF2V at VIRGINIA). *Polymers, structure, properties, molecular modeling, simulations.*

Fasman, Prof. Gerald David. Biochemistry Dept., Brandeis U., Waltham, MA 02254, USA. (1925) PhD org. chemistry (CIT, 1952). Prof. (tel. 617 + 736-2370, fax 617 + 736-2349). *Structure, protein, nucleic acid.*

Faulk, Mr. John Warren. Rte #2, Box 1430, Sulphur, LA 70663, USA. (1940) BS physics (McNeese State, 1964). Administrative mgr. (tel. 318 + 583-7554). *X-ray diffraction.*

Fawcett, Dr. Timothy G. Analytical Sci., Dow Chemical Co., 1897 Bldg., Midland, MI 48640, USA. (1953) PhD chemistry (Rutgers U., 1979). Group leader. (tel. 617 + 636-7786).

Fay, Prof. Robert Clinton. Chemistry Dept., Cornell U., Ithaca, NY 14853, USA. (1936) PhD inorg. chemistry (U. of Illinois, 1962). Prof. (tel. 607 + 255-3636). *Stereochemistry, configurational rearrangements, metal chelate compounds.*

Feigelson, Prof. Robert Saul. Ctr. for Matls. Res., Stanford U., McCullough Bldg., Stanford, CA 94305-4045, USA. (1935) PhD material sci. (Stanford U., 1974). Prof., res. (tel. 415 + 723-4007, telex 348 402 STANFORD, fax 415 + 723-0100, email FEIGELSON at SIERRA). *Crystal growth.*

Feldman, Dr. Robert Edward. 425 Lincoln Blvd., Apt. 2U, Hauppauge, NY 11788-2913, USA. (Alt.: Physics Dept., NY Inst. of Techn., Carleton Ave., Central Islip, NY, 11722, USA) (1939) PhD physics (PUN, 1969). Asst. prof. (tel. 516 + 348-3053). *X-ray physics, crystal structures.*

Feldmann, Mr. Richard Joseph. Div. of Computer Res. & Techn., Nat. Inst. of Health, Bldg. 12A, Room 2008, Bethesda, MD 20014, USA. (1939) MS electrical eng. (PUN, 1962). Computer specialist. (tel. 301 + 496-1100, fax 301 + 496-4005, email Bitnet RJF at NIHCUDEC). *Protein crystallography, molecular structure data retrieval, macromolecular surface representation.*

Femec, Dr. Douglas Anthony. Dept. Chemistry, Arizona State U., Tempe, AZ 85287-1604, USA. (1958) PhD inorg. chemistry (Cornell U., 1987). Postdoct. res. assoc. (tel. 602 + 965-7164, email Bitnet ATDAF at ACVAX or FEMEC at XRAY). *Structure - activity relationships.*

Fenderson, Dr. Faith. Biological Struct. Dept., SM-20, U. of Washington, Seattle, WA 98196, USA. (1956) PhD biochemistry (U. of Washington, 1988). Sr. fellow. (tel. 206 + 543-4496). *Macromolecular structure.*

Fenna, Dr. Roger Edward. Biochemistry Dept., U. of Miami Med. Sch., Box 016129, Miami, FL 33101, USA. (1947) D.Phil. protein crystallog. (Corpus Christi C., Oxford, UK, 1973). Asst. prof. (tel. 305 + 547-6564). *Structure - function, biological macromolecules, photosynthesis.*

Fernando, Prof. Quintus. Chemistry Dept., U. of Arizona, Tucson, AZ 85721, USA. (1926) PhD analytical chemistry (U. of Louisville, 1953). Prof. (tel. 602 + 621-2105). *Metal complexes, trace element analysis.*

Ferrara, Dr. Joseph David. Mol. Structure Corp., 3200A Research Forest Dr., The Woodlands, TX 77381, USA. (1961) PhD chemistry (Case Western Reserve U., 1988). (tel. 713 + 363-1033, telex 55-9314, fax 713 + 292-2472).

Fidelis, Dr. Krzysztof Andrze. Chemistry and Biochem. Dept., U. of Oklahoma, 620 Parrington Oval, Norman, OK 73019-0370, USA. (1957) PhD biophysics (U. of Oklahoma, 1989). Res. asst. (tel. 405 + 325-7612, email AA3024 at UOKMVSA). *Molecular dynamics, mechanics, drug design, experimental electron density determination.*

Filman, Dr. David Jeffrey. Mol. Biology Dept. MB13, Res. Inst. Scripps Clinic, 10666 N. Torrey Pines Rd., La Jolla, CA 92037, USA. (1949) PhD biochemistry (U. of Calif. at San Diego, 1981). Asst. member. (tel. 619 + 554-8271). *Crystallography, protein, macromolecular, virus.*

Finer-Moore, Dr. Janet Sue. Biochemistry Dept., U. of Calif. Sch. of Med., Box S964, San Francisco, CA 94143, USA. (1951) PhD chemistry (Iowa State U., 1978). Asst. res. biochemist. (tel. 415 + 476-3937). *Protein crystallography, membrane protein, structure - function.*

Finger, Dr. Larry W. Geophysical Lab., 2801 Upton St., NW, Washington, DC 20008-3898, USA. (1940) PhD geology (U. of Minnesota, 1967). Crystallographer. (tel. 202 + 966-0334, fax 202 + 895-4225). *Mineral structures, high temperature and pressure, crystallographic computing.*

Fink, Prof. Anthony L. Chemistry Dept., U. of Calif. at Santa Cruz, Santa Cruz, CA 95064, USA. (1943) PhD chemistry (Queen's U., Canada, 1968). Prof. (tel. 408 + 429-2744, fax 408 + 5429-0146, email Bitnet ENZYME at UCSCC). *Enzyme mechanisms, protein folding.*

Fink, Dr. William LaVilla. 410 Emerson Ave., Aspinwall, Pittsburgh, PA 15215, USA. (1896) PhD chemistry, metallurgy (U. of Michigan, 1926). Retired. (tel. 412 + 781-4216). *Chemistry, metallurgy, X-ray diffraction, crystallography.*

Finzel, Dr. Barry C. Phys. & Analyt. Chem., The Upjohn Co., 301 Henrietta St., Kalamazoo, MI 49001, USA. (1956) PhD chemistry (U. of Calif. at San Diego, 1983). Res. scient. (tel. 616 + 384-9744). *Heme enzymes, protein/protein recognition, protein structure.*

Fischer, Mr. Gerhard Richard. R.D.& E. DIV., SP-FR-1-7, Corning Glass Works, Corning, NY 14831, USA. (1926) BSc physics (U. of Jena, DDR, 1951). Physicist. (tel. 607 + 974-3342). *High temperature X-ray diffraction, structure - properties relationships, automated powder diffraction.*

Fisher, Dr. Richard G. Advanced Methods Group, Kraft Techn. Ctr., 801 Waukegan Rd., Glenview, IL 60025, USA. (1952) PhD molecular biology (UCLA, 1980). Sr. res. scient. II (tel. 312 + 998-7378, fax 312 + 558-3864). *Computer graphics, molecular modeling, structural predictions, macromolecular and peptide techniques, computer methods.*

Fisher, Dr. Robert M. Center for Advanced Materials, Lawrence Berkeley Lab., 1 Cyclotron Rd., Berkeley, CA 94720, USA. (1927) PhD metallurgy (U. of Cambridge, UK, 1962). Acting prof. (tel. 415 + 486-4760). *High voltage electron diffraction.*

Fitzgerald, Dr. Paula Marie Dean. Merck Sharp & Dohme Res. Labs., R80M203, P.O. Box 2000, Rahway, NJ 07065, USA. (1949) PhD biophysics (Johns Hopkins U., 1977). Res. fellow. (tel. 201 + 594-5510). *Biological macromolecular structures, molecular replacement.*

Fletterick, Dr. Robert J. Biochemistry Dept. U. of Calif. at San Francisco, 513 Parnassus St., S-964, San Francisco, CA 94143-0448, USA. (1943) PhD phys. chemistry (Cornell U., 1970). Prof. (tel. 415 + 476-5080,or 476-5051, fax 415 + 476-0943, email FLETT at CGL.ECSF.EDU). *Proteins, structure - function.*

Flippen-Anderson, Ms. Judith Lee. Lab. for the Struct. of Matter, Naval Res. Lab., Code 6030, Washington, DC 20375, USA. (1941) MS chemistry (Arizona State U., 1966). X-ray crystallographer. (tel. 202 + 767-3463, fax 202 + 767-6874, email FLIPPEN at LSMNIC.NAVY.NRL.MIL).

Florio, Dr. John Victor. 31 Marion Dr., North Haven, CT 06473, USA. (1925) PhD phys. chemistry (Iowa State U., 1952). *Surface structure, chemistry, liquid phase epitaxy.*

Folting-Streib, Mrs. Kirsten. Chemistry Dept., Mol. Structure Center, Indiana U., Bloomington, IN 47405, USA. (1932) Lic.pharm. pharmacy (Royal Danish S. of Pharmacy, Denmark, 1964). Staff crystallographer (tel. 812 + 335-6604). *Organic crystal structures.*

Foord, Mr. Eugene E. U.S. Geological Survey, MS 905, Denver Federal Ctr., Lakewood, CO 80225, USA. (tel. 303 + 236-4755).

Foreman, Prof. Dennis W. Jr. Oral Biology Dept., Ohio State U., Wendell Postle Hall, 305 W. 12th. Ave., Columbus, OH 43210, USA. (1929) PhD mineralogy (Ohio State U., 1966). Prof. (tel. 614 + 292-6316).

Forest, Mrs. Katrina. Molec. Biology Dept., Princeton U., Princeton, NJ 08544, USA. (1966) BS applied biology (MIT, 1987). Grad. student. (tel. 609 + 987-2826). *Biological macromolecules.*

Foris, Ms. Catherine M. Central Res. Bldg. 356, E.I. du Pont de Nemours Co., Exptl. Sta., P.O. Box 80356, Wilmington, DE 19880-0356, USA. (tel. 302 + 695-3687, telex 835420, fax 302 + 695-1664, email FORIS at ESVAX%DUPONT.COM@RELAY.CS.NET). *Inorganic crystal chemistry, powder diffraction techniques, Guinier cameras.*

Fornoff, Mr. Mario M. P.O. Box 4262, Wilmington, DE 19807, USA. (1932) ChE chemical eng. (U. of Cincinnati, 1957). *Powder diffraction, XRF, electron optics, microanalysis.*

Foster, Dr. Mark David. 5021 Tudor Cir., Carmel, IN 46032-9321, USA. (Addr. until 11/89: Arbeitsgruppe Fischer, Max-Planck-Institut f. Polymerforschung, Postfach 3148, D-6500 Mainz 1, German Fed. Rep.). (1959) PhD chem. eng. (U. of Minnesota, 1986). Res. scient. (tel. 06131 + 379-201, FRG, email FOSTER at MAX.MPIP-MAINZ.MPG.DBP.DE). *X-ray reflection, scattering, surface, small angle, neutron reflection, thin polymer films, polymer surfaces, porous solids.*

Foundling, Dr. Stephen Ian. Central Res. & Dev., E.I. duPont de Nemours Co., Bldg. 328, Rm. 350A, Wilmington, DE 19880-0328, USA. (1958) PhD crystallography (U. of London, UK, 1987). Visiting scient. (tel. 302 + 695-4128, fax 302 + 695-9183). *Polyproteins, virus, structure - function, proteases, maturation, inhibitor complexes, specialist fields, inhibitor design, modeling, macromolecular.*

Fox, Prof. Robert O. Jr. Mol. Biophys. & Biochem. Dept., Yale U., 260 Whitney Ave., New Haven, CT 06511, USA. (1954) PhD mol. biophysics and biochem. (Yale U., 1981). Asst. prof. (tel. 203 + 432-5645, fax 203 + 432-3282, email FOX at YALEVMS). *Protein crystallography, protein folding, protein engineering.*

Foxman, Prof. Bruce Mayer. Chemistry Dept., Brandeis U., South St., Waltham, MA 02254-9110, USA. (1942) PhD inorg. chemistry (MIT, 1968). Prof. (tel. 617 + 736-2532, email Bitnet FOXMAN1 at BRANDEIS). *Solid state reactions, structural chemistry, sterically crowded molecules.*

Franzen, Prof. Hugo Friedrich. Chemistry Dept., Iowa State U., Ames, IA 50011, USA. (1934) PhD phys. chemistry (U. of Kansas, 1962). Prof. (tel. 515 + 294-5773). *Phase transitions, vacancy ordering, heterogeneous equilibria, transition metal chalcogenide structures.*

Fratini, Prof. Albert V. Chemistry Dept., U. of Dayton, Dayton, OH 45469, USA. (1939) PhD chemistry (Yale U., 1965). Prof. (tel. 513 + 229-2849, email FRATINI at DAYTON). *Structure and morphology, ordered polymers, organic compounds, nonlinear optical materials.*

Frazer, Dr. Benjamin Chalmers. Div. of Matls. Sci., Off. of Basic Energy Sci., US Dept. of Energy, ER 132, GTN, Washington, DC 20545, USA. (1922) PhD physics (PSU, 1952). Branch chief, solid st. & matls. chem. (tel. 301 + 353-3426). *X-ray and neutron scattering, phase transitions, magnetism, ferroelectricity, synchrotron radiation research, nuclear research reactors.*

Fredrich, Mr. Michael F. P.O. Box 14466, Fremont, CA 94539, USA.

Freed, Prof. Robert Lowell. Geology Dept., Trinity U., 715 Stadium Drive, San Antonio, TX 78284, USA. (1938) PhD mineralogy (U. of Michigan, 1966). Prof., chairman. (tel. 512 + 736-7609). *Inorganic structures, clay minerals.*

Freer, Dr. Stephan T. Crystallography Dept., Agouron Pharmaceuticals Inc., 11025 N. Torrey Pines Rd., La Jolla, CA 92037, USA. (1933) PhD biochemistry (U. of Washington at Seattle, 1964). Res. scient. (tel. 619 + 535-5548). *Macromolecules.*

Frelinger, Dr. Andrew Lawrence. Immunology Dept., Imm 15, Scripps Clinic & Res. Fndn., 10666 N. Torreu Pines Rd., La Jolla, CA 92037, USA. (1953) PhD biology (Case Western Reserve U., 1984). Sr. res. assoc. (tel. 619 + 455-9100, ext. 2766). *Adhesion, platelets, integrins, receptors, signal transduction.*

French, Dr. Alfred Dexter. Southern Regional Res. Center, USDA, P.O. Box 19687, New Orleans, LA 70179, USA. (1943) PhD phys. chemistry (Arizona State U., 1971). Res. chemist. (tel. 504 + 286-4250, fax 504 + 286-4419). *Polysaccharide crystal structures, computer modeling.*

French, Dr. Robert D. 40 Reed St., Lexington, MA 02173, USA. (tel. 617 + 923-3471).

Frenz, Dr. Bertram Anton. B. A. Frenz and Associates, Inc., 1140 E. Harvey Rd., College Station, TX 77840, USA. (1945) PhD chemistry (Northwestern U., 1971). Pres. (tel. 409 + 764-3999, fax 409 + 764-8945). *Computer programming.*

Frevel, Dr. Ludo Karl. 1205 W. Park Dr., Midland, MI 48640, USA. (1910) PhD phys. chemistry (Johns Hopkins U., 1934). Fellow by courtesy J.H.U. (tel. 517 + 832-8983). *X-ray crystallography, catalysis, applied mathematics.*

Friedlander, Dr. Peter H. 18 Hunting Ridge Rd., Greenwich, CT 06831, USA. (tel. 203 + 629-3989).

Fritchie, Prof. Charles Julius Jr. Chemistry Dept., Tulane U., 6823 St. Charles Ave., New Orleans, LA 70118, USA. (1936) PhD chemistry (CIT, 1962). Prof. (tel. 504 + 865-5573). *Charge-transfer complexes, structure, spectra, organometallic crystallography.*

Fronczek, Dr. Frank R. Chemistry Dept., Louisiana State U., Baton Rouge, LA 70803, USA. (1948) PhD chemistry (CIT, 1975). Res. assoc., dept. crystallogr. (tel. 504 + 388-8270). *Crystal structures, macrocycles, transition metal complexes, natural products.*

Frueh, Prof. Alfred Joseph. 23 Bundy Lane, Storrs, CT 06268, USA. (1919) PhD mineralogy (MIT, 1949). Prof. emeritus. (tel. 203 + 429-1045). *Mineral and inorganic structures, order-disorder imperfections.*

Fry, Dr. David C. Physical Chem. Dept., Hoffmann-La Roche Inc., 340 Kingsland St., Nutley, NJ 07110, USA. (tel. 201 + 235-3709). *NMR.*

Fullam, Mr. Ernest F. Ernest F. Fullam Inc., 2217 Stoneridge Rd., Schenectady, NY 12309, USA. (1910) AB chemistry (Cornell U., 1937). Retired. (tel. 518 + 374-2049). *Chemical microscopy, micro-probe analysis, diffraction, X-ray, electron.*

Fullenwider, Dr. Malcolm Allen. R.K. Labs., 2970 MacArthur Rd., P.O. Box 2, Whitehall, PA 18052, USA. (1940) PhD electrochemistry (U. of Pennsylvania, 1969). Consultant. (tel. 215 + 435-1452).

Fuoss, Dr. Paul Henry. AT&T Bell Labs., 4C-316, Crawford Corner Rd., Holmdel, NJ 07733, USA. (1953) PhD materials sci. (Stanford U., 1980). Member tech. staff. (tel. 201 + 949-3581). *Surface crystallography.*

Furey, Dr. William F. Biocrystallography Lab., V.A. Med. Center, P.O. Box 12055, Pittsburgh, PA 15240, USA. (1952) PhD crystallography, phys. chem. (Rutgers U., 1977). Adjunct assoc. prof. (tel. 412 + 683-3000, ext. 517). *Macromolecular crystallography, molecular dynamics, conformational energy analysis, direct methods, crystallographic computing, computer graphics.*

Furnas, Dr. Thomas Coleman Jr. Molecular Data Corp., 2869 Scarborough Rd., Cleveland Heights, OH 44118, USA. (1922) PhD physics (MIT, 1952). Pres. (tel. 216 + 321-4173). *X-ray instrumentation applications, analytical, industrial, medical, X-ray diffraction, single crystal, powder, small angle scattering, spectrographic systems, dispersive systems, monochromators, energy, wave length.*

Fuzek, Dr. John Frank. 4603 Mitchell Rd., Kingsport, TN 37664, USA. (1921) PhD phys. chemistry (U. of Tennessee, 1947). (tel. 615 + 229-4183). *Fiber structure, polymer structure.*

Gaier, Dr. James Richard. Electro-Physics Section, NASA Lewis Res. Center, 21000 Brookpark Rd., Cleveland, OH 44135, USA. (1952) PhD chemistry (Michigan State U., 1983). Physicist. (tel. 216 + 433-2311). *Intercalated graphite, protein-protein interactions, high-temperature superconductors, conducting polymers.*

Gaines, Dr. James Matthew. Physics & Matls. Dept., Philips Labs., 345 Scarborough Rd., Briarcliff Manor, NY 10510, USA. (1958) PhD electrical eng. (U. of Colorado, 1986). Sr. memb. res. staff. (tel. 914 + 945-6541). *Structure - properties relationships.*

Gall, Prof. William Einar. The Rockefeller U., 1230 York Ave., New York, NY 10021, USA. (1942) PhD biochemistry (The Rockefeller U., 1969). Assoc. prof. (tel. 212 + 570-8975). *Protein structure, electron microscopy, three-dimensional reconstruction techniques.*

Gallacher, Mr. Anthony. 2002 Wyndhurst Rd., Toledo, OH 43607, USA.

Gallagher, Mr. Travis. Chemistry Dept., U. of Texas, Austin, TX 78712, USA. (1957) BS electrical eng., computer sci. (MIT, 1979). Grad. student. (tel. 512 + 471-3625). *Structure determination, macromolecular, protein folding, enzymatic catalysis.*

Gallucci, Dr. Judith Chlastawa. Chemistry Dept., The Ohio State U., 120 W. 18th Ave., Columbus, OH 43210, USA. (1953) PhD inorg. chemistry (U. of Massachusetts at Amherst, 1979). Dept. crystallographer. (tel. 614 + 292-4039). *Chemical crystallography, small molecule structures.*

Gandour, Prof. Richard David. Chemistry Dept., Louisiana State U., Baton Rouge, LA 70803-1804, USA. (1945) PhD chemistry (Rice U., 1972). Prof. (tel. 504 + 388-1529, email CHGAND at LSUVM). *Molecular recognition, bioorganic chemistry.*

Gantzel, Dr. Peter Kellogg. 8308 Paseo Del Ocaso, La Jolla, CA 92037, USA. (1934) PhD phys. chemistry (UCLA, 1962). Retired. (tel. 619 + 459-6440). *X-ray diffraction, powder, single crystal, atomic geometry, microstructure via electron microscopy, diffraction.*

Garafalo, Mr. Alfred R. Massachusetts C. of Pharmacy, 179 Longwood Ave., Boston, MA 02115, USA. (tel. 617 + 732-2949).

Garavito, Dr. R. Michael. Biochem. and Mol. Biol. Dept., U. of Chicago, 920 East 57th. St., Chicago, IL 60637, USA. (1952) PhD biochem./biophys. (Purdue U., 1978). Asst. prof. (tel. 312 + 702-9481, email GARAVITO at BIOVAX.UCHICAGO.EDU). *Proteins, X-ray crystallography, structure - function, crystallization, macromolecules.*

Garbauskas, Dr. Mary Frances. Corporate Res. and Development, General Electric, Bldg. K-1, P.O. Box 8, Schenectady, NY 12345, USA. (1953) PhD chemistry (UCLA, 1979). Staff scient. (tel. 518 + 387-5797, fax 518 + 387-7597, email GARBAUSKAS at GE-CRD.ARPA). *X-rays, powder diffraction, spectrometry, structure determination, inorganic, organic, crystallinity, polymer systems.*

Gardner, Dr. Kenncorwin Hancock. Central Res. and Dev. Dept., E. I. du Pont de Nemours Co., Box 80356, Wilmington, DE 19880-0356, USA. (1947) PhD macromolecular sci. (Case Western Reserve U., 1974). Res. scient. (tel. 302 + 695-2408). *Fibrous polymers, diffraction, structure - properties - morphology relationship, polymers.*

Garvey, Prof. Roy George. Chemistry Dept., North Dakota State U., Fargo, ND 58105, USA. (1941) PhD chemistry (U. of Utah, 1966). Assoc prof. (tel. 701 + 237-8697). *Inorganic, coordination chemistry.*

Geckle, Mr. Raymond J. AMP Inc., 425 Prince St., Harrisburg, PA 17105, USA.

Geib, Dr. Steven J. Biochemistry Dept., Vanderbilt U., Nashville, TN 37211, USA. (1959) PhD chemistry (U. of Delaware, 1988). Res. assoc. (tel. 615 + 322-6352).

Geiger, Dr. David K. Chemistry Dept., SUNY at Geneseo, Geneseo, NY 14454, USA. (1956) PhD inorg. chemistry (U. of Notre Dame, 1983). Asst. prof. (tel. 716 + 245-5319).

Geil, Prof. Phillip Herbert. Polymer Div., Matls. Sci. & Eng. Dept., U. of Illinois, 1304 W. Green St., Urbana, IL 61801, USA. (1930) PhD physics (U. of Wisconsin, 1956). Prof. (tel. 217 + 333-0149). *Structure - properties - morphology relationships, synthetic macromolecules, biological macromolecules.*

Geiser, Dr. Urs W. Chemistry Div., Argonne Nat. Lab., 9700 S. Cass Ave., Argonne, IL 60439, USA. (1956) PhD chemistry (Washington State U., 1985). Asst. chemist. (tel. 312 + 972-3509, fax 312 + 972-4470, email Bitnet GEISER at ANLCHM). *Solids, electrical, magnetic properties, organic conductors, structure - properties relationships.*

Geisinger, Ms. Karen L. Spring Ponds Apts., 47 Indian Pipe, Painted Post, NY 14870, USA. (tel. 607 + 974-6290).

Geiss, Dr. Roy Howard. Dept. K34/802, IBM Almaden Res. Ctr., 650 Harry Rd., San Jose, CA 95120, USA. (1937) PhD applied physics (Cornell U., 1967). Res. staff member. (tel. 408 + 927-2445, telex 31 69 69, fax 408 + 927-2100). *Electron microscopy (scanning and transmission), diffraction, inorganic materials, thin films, metals and alloys, corrosion, magnetic properties, defects.*

Geller, Prof. Seymour. Elec. & Computer Eng. Dept., U. of Colorado, Campus Box 425, Boulder, CO 80309-0425, USA. (1921) PhD phys. chemistry (Cornell U., 1949). Prof. (tel. 303 + 492-7157). *Structure - properties relationship, magnetic materials, structures, solid electrolytes, superconducting materials, pressure induced phases, phase transformations.*

George, Dr. Clifford F. Naval Res. Labs., 4555 Overlook Ave., SW, Washington, DC 20375, USA.

Gerdes, Dr. Reiner Josef. Continental Telecom Labs., 270 Scientific Dr., Techn. Park - Atlanta, Norcross, GA 30092, USA. (1935) Dr.rer.nat. phys. chemistry (Technical U., Hannover, BRD, 1963). Dir. (tel. 404 + 448-2206). *Structure - properties relationship, surface structure, catalysis.*

Gerhard, Mr. F. Bruce. 38 Spring Ln., Canton, MA 02021, USA. (tel. 617 + 828-3678).

Gerkin, Prof. Roger Estlick. Chemistry Dept., Ohio State U., 120 W. 18th. Ave., Columbus, OH 43210, USA. (1931) PhD phys. chemistry (UCB, 1960). Prof. (tel. 614 + 292-6053). *Single crystals, structure, electron paramagnetic resonance.*

Getzoff, Dr. Elizabeth Dickinson. Mol. Biology Dept., MB5, Res. Inst. of Scripps Clinic, 10666 North Torrey Pines, La Jolla, CA 92037, USA. (1954) PhD biochemistry (Duke U., 1982). Asst. member. (tel. 619 + 554-2878, telex 697168 SCRF SCRF, fax 619 + 554-8841, email GETZOFF at SCRIPPS.EDU). *Protein structure determination, protein engineering, protein analysis, macromolecular interactions, conformational change, antibody-antigen recognition, computer graphics.*

Gheith, Prof. Mohamed A. Geology Dept., Boston U., 675 Commonwealth Ave., Boston, MA 02215, USA. (1925) PhD geochemistry and mineralogy, (U. of Minnesota, 1952). Dir., spec. ext. progs., Middle East (tel. 617 + 353-2616). *Geological education, economic geology, geochemistry, ore deposits, sulfide mineralogy.*

Ghose, Prof. Subrata. Geological Sci. Dept., AJ-20, U. of Washington, Seattle, WA 98195, USA. (1932) PhD mineralogy, crystallography (U. of Chicago, 1959). Prof. (tel. 206 + 543-7378). *Structural and magnetic phase transitions, lattice and molecular dynamics, crystal chemistry, rock-forming silicates.*

Ghosh, Dr. Debashis. Mol. Biophysics, Med. Fndn. of Buffalo, 73 High St., Buffalo, NY 14203, USA. (1952) PhD crystallography (U. of Pittsburgh, 1981). Res. scient. (tel. 716 + 856-9600, fax 716 + 852-4846). *Macromolecular crystallography, diffraction physics.*

Ghosh, Mr. Partho. Biochem. & Biophys. Dept., U. of Calif. at San Francisco, San Francisco, CA 94142-0448, USA. (1962) BS biophysics, biochemistry (Yale U., 1985). Grad. student. (tel. 415 + 476-3937). *Membrane proteins.*

Giese, Mr. Rossman F. Jr. 298 Schenck St., North Tonawanda, NY 14120, USA. (tel. 716 + 831-3051).

Giessen, Prof. Bill Cormann. Chemistry Dept., Northeastern U., 360 Huntington Ave., Boston, MA 02115, USA. (1932) Dr.sc.nat. metallurgy (U. of Goettingen, BRD, 1958). Prof., Assoc. Dir. Barnett Inst. Chem. Anal. & Matl. Sci. (tel. 617 + 437-2827). *X-ray diffraction, electron microscopy, ceramics, intermetallic phases, amorphous metals, superconductors, metastable alloys, alloy chemistry, surface structures.*

Gilardi, Dr. Richard Dean. Lab. for the Struct. of Matter, Naval Res. Lab., Code 6030, Washington, DC 20375, USA. (1940) PhD phys. chemistry (U. of Maryland, 1966). Res. chemist. (tel. 202 + 767-2624, fax 202 + 767-6874, email GILARDI at NRL3.ARPA). *Diffraction analysis methods, conformation energy calculations.*

Gilfrich, Mr. John Valentine. 8710 Lowell St., Bethesda, MD 20817-3218, USA. (1927) BA chemistry (American International C., 1949). Consulting res. chemist. (tel. 301 + 365-5070). *X-rays, spectroscopy (fluorescence), diffraction, physics.*

Gilje, Prof. John W. Chemistry Dept., U. of Hawaii, 2545 The Mall, Honolulu, HI 96822-2275, USA. (1939) PhD inorg. chemistry (U. of Michigan, 1965). Prof.

(tel. 808 + 948-7389, fax 808 + 949-8025, email Bitnet GILJE at UHCCUX). *Chemistry, inorganic, organometallic, f-elements.*

Gillette, Dr. Paul Calvin. Res. Center, Matls. Sci. Div., Hercules, Inc., Hercules & Lancaster Rds., Wilmington, DE 19894, USA. (1956) PhD macromolecular sci. (Case Western Reserve U., 1983). Sr. res. chemist. (tel. 302 + 995-3815, telex 83-5479, fax 302 + 995-4135). *Polymers, crystallography, infrared spectroscopy, diffusion, film morphology, molecular modeling.*

Gillies, Dr. Donald Chalmers. Microelectronics Ctr., McDonnell Douglas Corp., 8905 Airport Rd., Berkeley, MO 63134, USA. (1939) PhD crystallography (U. of London, UK, 1969). Lead eng. (tel. 314 + 234-8072). *Crystal growth, electronic materials, space processing.*

Gilliland, Dr. Gary Lynn. Ctr. Adv. Res. Biotechn., Nat. Inst. Stnds. & Techn., 9600 Gudelsky Dr., Rockville, MD 20850, USA. (1948) PhD (1979). Res. chemist. (tel. 301 + 975-2981). *Protein crystallography.*

Ginell, Prof. Robert. Chemistry Dept., Brooklyn C. of CCNY, Avenue H and Bedford Ave., Brooklyn, NY 11210, USA. (1912) PhD chemistry (PUN, 1943). Prof. emeritus. (tel. 718 + 780-5753). *Liquids, solids, equations of state, association theory, nucleation.*

Ginell, Dr. Stephan Lawrence. Chem. Dept., Rutgers U., P.O. Box 939, Piscataway, NJ 08855, USA. (1949) PhD biophysics (SUNY at Buffalo, Roswell Park Div., 1980). Res. assoc. (tel. 201 + 932-2618). *Structure - function, proteins, nucleic acids, drugs, hydration, biological systems, radiation damage.*

Giordano, Mr. Joseph. 43 Deerfield Rd., Parlin, NJ 08859, USA. (tel. 201 + 596-3555).

Giorgi, Dr. Angelo Louis. MST-5, MS G734, Los Alamos Nat. Lab., Los Alamos, NM 87545, USA. (1917) PhD phys. chemistry (U. of New Mexico, 1957). Staff member. (tel. 505 + 667-5815). *High temperature chemistry, superconductivity.*

Giranda, Mr. Vincent. 1239 Fuller St., Philadelphia, PA 19111, USA. (tel. 215 + 728-3617).

Glaeser, Prof. Robert Martin. Biophysics Dept., U. of Calif. at Berkeley, Berkeley, CA 94720, USA. (1937) PhD biophysics (UCB, 1964). Prof. (tel. 415 + 643-8874, email RMGLAESER at LBL). *Electron diffraction, microscopy, membrane structure.*

Glass, Dr. Howard L. Electro-Optical Ctr., 031-BD14, Rockwell International, P.O. Box 3105, Anaheim, CA 92803, USA. (1942) PhD physics (Rutgers U., 1969). Techn. staff member. (tel. 714 + 632-3691, telex 678437). *Crystal growth, magnetism, epitaxy, X-ray diffraction.*

Gleason, Dr. William Bourke. Central Res. Lab., 201-1W-31, Minn. Mining & Mfg. Co., St. Paul, MN 55144, USA. (1945) PhD org. chemistry (U. of Minnesota, 1974). Sr. res. chemist. (tel. 612 + 733-0720). *Organic, organometallic crystal structures.*

Glick, Prof. Milton Don. Office of the Provost, Iowa State U., Ames, IA 50011, USA. (1937) PhD chemistry (U. of Wisconsin at Madison, 1965). Provost. (tel. 515 + 294-0071). *Structural chemistry, X-ray crystallography, computing.*

Glinka, Dr. Charles Joseph. Reactor Radiation Div., Nat. Inst. Stnds. & Techn., Gaithersburg, MD 20899, USA. (1947) PhD solid state physics (U. of Maryland, 1975). Physicist. (tel. 301 + 975-6242, fax 301 + 921-9847).

Glucksman, Dr. Marc J. Fishberg Ctr. for Neurobiology, Mount Sinai Sch. of Med., 1 Gustave Levy Pl., New York, NY 10029, USA. (1956) PhD biochem., molec. biophys. (Columbia U., 1988). Res. assoc. (tel. 212 + 241-9230, email GLUX at CUMBG). *Diffraction, X-ray, fiber, DNA - protein interactions, macromolecular assemblies, neurobiology, knowledge based structure prediction schemes.*

Glusker, Dr. Jenny Pickworth. Mol. Structure Dept., The Inst. for Cancer Res., 7701 Burholme Ave., Philadelphia, PA 19111, USA. (1931) DPh chemistry (Oxford U., UK, 1957). Sr. member (tel. 215 + 728-2220, fax 215 + 728-2839 or 215 + 728-3574). *Enzyme mechanisms, chemical carcinogenesis, mutagenesis, metal chelation.*

Go, Miss Kuan Tee. Biophysics Dept., Roswell Park Mem. Inst., 666 Elm St., Buffalo, NY 14263, USA. MS (1962). Res. scient. (tel. 716 + 845-8295).

Godden, Mr. Jeff. Biochemistry Dept. SJ-70, U. of Washington, Seattle, WA 98195, USA. (tel. 206 + 543-6047).

Goddette, Dr. Dean. Molecular Genetics Dept., Henkel Res. Corp., 2330 Circadian Way, Santa Rosa, CA 95407, USA. (1958) PhD biomedical sci. (Washington U. of St. Louis, 1985). Sr. res. scient. (tel. 707 + 575-7155, ext. 281, fax 707 + 575-7833). *Proteases, lipases.*

Godycki, Mr. L. Edward. 1060 Granada Ave., San Marino, CA 91108, USA. (tel. 714 + 732-2065).

Goehner, Mr. Raymond Philip. Siemens Analyt. X-ray Instr., Inc., 6300 Enterprise Ln., Madison, WI 53719, USA. (1945) MSc physics (Rensselaer Polytechnic Inst., 1971). Vice-pres., marketing. (tel. 608 + 276-3000). *X-ray diffraction, fluorescence.*

Goland, Dr. Allen N. Applied Sci. Dept., Brookhaven Nat. Lab., Bldg. 815, Upton, NY 11973, USA. (1930) PhD physics (Northwestern U., 1956). Sr. physicist, Deputy chrm. (tel. 516 + 282-3819, telex 6852516 BNL DOE, fax 516 + 282-3000). *Atomic defects, diffuse scattering, radiation - matter interaction.*

Gold, Dr. Karen Walter. Res. Triangle Inst., P.O. Box 12194, Research Triangle Park, NC 27709, USA. (1943) PhD chemistry (Harvard U., 1971). Environ. chemist. (tel. 919 + 541-5840). *Powder X-ray diffraction, asbestos, air pollutants, porphyrins.*

Goldberg, Prof. Stephen Z. Chemistry Dept., Adelphi U., Garden City, NY 11530, USA. (1947) PhD chemistry (UCB, 1973). Prof. chem. (tel. 516 + 294-8700, ext. 7519). *Structural inorganic chemistry.*

Goldenberg, Ms. Maria. Chemistry Dept., Hunter C. of CUNY, 695 Park Ave., New York, NY 10021, USA.

Goldish, Dr. Elihu. 4821 Ostrom Ave., Lakewood, CA 90713, USA. (1928) PhD chemistry (CIT, 1956). *X-ray diffraction, spectrometry, powder diffractometry.*

Goldman, Prof. Adrian. Waksman Inst., P.O. Box 759, Hoes Ln., Piscataway, NJ 08855, USA. (1958) PhD X-ray crystallography (Yale U., 1985). Asst. prof. (tel. 201 + 932-5204, fax 201 + 932-5735, email GOLDMAN at BIOVAX.RUTGERS.EDU). *Protein X-ray crystallography, structure - function relationships, enzyme mechanisms.*

Goldsmith, Dr. Elizabeth Jane. Biochemistry Dept., U. of Texas S.W. Med. Center, 5323 Harry Hines Blvd., Dallas, TX 75235-9050, USA. (1945) PhD chemistry (UCLA, 1971). Asst. prof. (tel. 214 + 689-5009, fax 214 + 689-5066, email Bitnet GOLD02 at UTSW). *Protein crystallography, allosteric proteins, glycogen phosphorylase, insulin receptor, protease inhibitors, carboxypeptidase.*

Goldsmith, Prof. Julian Royce. Geophysical Sci. Dept., U. of Chicago, 5734 Ellis Ave., Chicago, IL 60637, USA. (1918) PhD geochemistry (U. of Chicago, 1947). Charles E. Merriam Distin. Serv. Prof. Emeritus. (tel. 312 + 962-8155). *Geochemistry, phase equilibria, crystal chemistry, silicates and carbonates.*

Goldstein, Dr. Barry Michael. Biophysics Dept., U. of Rochester Med. Ctr., 601 Elmwood Ave., Rochester, NY 14642, USA. (1952) MD, PhD medicine (U. of Rochester, 1982). Asst. prof. (tel. 716 + 275-5095, email GOLDSTEIN_B at UORDBV). *Structure - function relationships, pharmacologically active agents, X-ray diffraction, NMR (solution structure), computational chemistry.*

Goldstone, Dr. Joyce. Neutron Scattering Ctr., Los Alamos Nat. Lab., MS H805, Los Alamos, NM 87545, USA. (1949) PhD materials sci. (SUNY at Stony Brook, 1978). Staff member. (tel. 505 + 667-3629, telex 660495, fax 585 + 665-2676, email GOLDSTONE%WNRVAX at LANL.GOV). *Neutron scattering, phase transformations, hydrides, actinides.*

Golikeri, Mr. Ganesh D. Res. Center, Lever Bros. Co., 45 River Rd., Edgewater, NJ 07020, USA. (tel. 201 + 943-7100).

Gong, Dr. Ping-Po +. Detector and Eng. Dept., 4115, EG&G/EM, Inc., P.O. Box 1912, Las Vegas, NV 89125, USA. (1950) PhD physics (PUN, 1983). Scient. specialist (tel. 702 + 295-3353). *X-ray, measurements, optics, fluorescence, diffraction, systems.*

Goodenough, Prof. John Bannister. Matls. Sci. and Eng. Ctr., U. of Texas at Austin, ETC 5.150, Austin, TX 78712-1084, USA. (1922) PhD physics (U. of Chicago, 1952). Virginia H. Cockrell Centennial Prof. of Eng. (tel. 512 + 471-1646, fax 512 + 471-8727). *Transition metal oxides, fast ionic conductors, catalytic electrodes, magnetism, superconductivity, phase transitions.*

Goonesekere, Mr. Nalin. Chemistry Dept., Princeton U., Washington Rd., Princeton, NJ 08540, USA. (tel. 609 + 987-2827).

Gordon, Ms. Janice T. 1250 Upper Gulph Rd., Radnor, PA 19087, USA. (tel. 215 + 688-8422).

Gougoutas, Dr. Jack Zanos. Squibb Inst. Med. Res., P.O. Box 4000, Princeton, NJ 08540, USA. (1939) PhD chemistry (Harvard U., 1963). Sr. res. fellow. (tel. 609 + 921-4562). *Solid state reactions, topotaxy, organic compounds, structure - properties relationships.*

Graeber, Dr. Edward J. Dept. 5214, Sandia Nat. Lab., Albuquerque, NM 87185, USA. (1934) PhD geology (U. of New Mexico, 1970). Staff member. (tel. 505 + 844-1621). *Structures, mineral, inorganic.*

Graves, Dr. Bradford J. Physical Chemistry Dept., Hoffman-La Roche Inc., Bldg. 76/15, Nutley, NJ 07110, USA. (1954) PhD chemistry (U. of North Carolina, 1980). Sr. scient. (tel. 201 + 235-5815). *Structure - function relationships, biological macromolecules.*

Gray, Prof. Terry Mitchell. Chemistry Dept., Calvin C., Grand Rapids, MI 49506, USA. (1958) PhD molecular biology (U. of Oregon, 1985). Asst. prof. (tel. 616 + 957-7187). *Proteins, protein folding, protein stability, protein engineering.*

Greenberg, Dr. Berton Laurence. Matls. Characterization Res., Philips Labs., 345 Scarborough Rd., Briarcliff Manor, NY 10510, USA. (1940) PhD materials sci. (Stevens Inst. of Techn., 1979). Sr. mem. res. staff. (tel. 914 + 945-6074, telex 646326 PHILAB BRRF, email BLG at PHILABS.PHILIPS.COM). *X-ray powder diffraction, epitaxial structures, single crystal structure analysis.*

Greenblatt, Prof. Martha. Chemistry Dept., Rutgers U., P.O. Box 939, Piscataway, NJ 08854, USA. (1941) PhD inorg. chemistry (PUN, 1967). Prof. (tel. 201 + 932-3277, fax 201 + 932-5312). *Solid state chemistry, crystal growth, structure - properties relationships, bronzes, solid electrolytes, insertion compounds, high T(c) superconductors.*

Greene, Dr. Andrea Claire. AiResearch Dept., Allied-Signal Aerospace, 2525 W. 190th. St., MS T42, Torrance, CA 90509, USA. (1957) PhD analytical chemistry (Louisiana State U., 1988). Matls. analyst. (tel. 213 + 512-4111). *Superconductors, powder diffraction, ceramics, composites.*

Greenhouse, Mr. Harold M. Dept. 480, Bendix Communications Div., East Joppa Rd., Baltimore, MD 21204, USA. (1924) MS phys. chemistry (Ohio State U., 1951). Sr. staff eng. (tel. 301 + 583-4102). *Materials science, microelectronics.*

Greer, Dr. Jonathan. Dept. 47E, Abbott Labs., Abbott Park, IL 60064, USA. (1943) PhD molecular biology (U. of Cambridge, UK, 1970). Sr. project leader.

(tel. 312 + 937-6933). *Structure - function, proteins, biological macromolecules, protein - ligand interactions, comparative molecular modeling.*

Gregg, Mr. R. Q. 3207 Henrietta, Bartlesville, OK 74003, USA.

Gregory, Mr. Don. Chemistry Dept., U. of Calif. at Santa Barbara, Santa Barbara, CA 93106, USA. (1952) BS biochemistry (U. of Calif. at Santa Barbara, 1979). Grad. student. (tel. 805 + 961-2761, email Bitnet GREGORY at SBIDP). *Proteins, molecular mechanics, dynamics.*

Gremillion, Prof. Alcuin Florian. Chemistry Dept., U. of Arkansas at Little Rock, 2801 S. University Ave., Little Rock, AR 72204, USA. (1925) PhD inorg. chemistry (Tulane U., 1958). Assoc. prof. (tel. 501 + 569-8827). *Inorganic structures.*

Gress, Dr. Mary Edith. Chemical Sci. Div., ER-141 GTN, Office of Basic Energy Sci., Dept. of Energy, Washington, DC 20545, USA. (1946) PhD phys. chemistry (Iowa State U., 1973). Prog. mgr., photochem. & rad. sci. (tel. 301 + 353-5820). *Crystal structures, inorganic, organic.*

Grev, Prof. Dennis Merle. Chemistry Dept., Columbia C., 10th and Rogers, Columbia, MO 65216, USA. (1935) MS chemistry (U. of Missouri at Columbia, 1963). Prof. (tel. 314 + 875-7633). *Crystal structures, environmental analytical chemistry, methodology.*

Gribskov, Dr. Michael Ray. Cryst. Lab., Bionetics Res. Inc., BRP, NCI, Frederick Cancer Fac. P.O. Box B Frederick, MD 21701-1013, USA. (1958) PhD molecular biology (U. of Wisconsin, 1985). Scient. assoc. (tel. 301 + 698-5031, fax 301 + 698-5991, email GRIBSKOV at NCIFCRF.GOV). *Proteins, structure, nucleic acid interactions, sequence relationships, molecular graphics.*

Griffin, Dr. Jane Flanigen. Mol. Biophysics Dept., Med. Fndn. of Buffalo, 73 High St., Buffalo, NY 14203, USA. (1933) PhD chemistry (SUNY at Buffalo, 1974). Res. scient. (tel. 716 + 856-9600). *Structure - activity relationships, cardiac glycosides, opiates, steroid structure, crystallographic information dissemination.*

Griffith, Dr. Elizabeth Ann Hall. Chemistry Dept., U. of South Carolina, Columbia, SC 29205, USA. (1935) PhD phys. chemistry (U. of South Carolina, 1970). Asst. res. prof. (tel. 803 + 777-6108). *X-ray crystallography, physical methods, structural chemistry, biologically significant compounds, non-linear optical materials.*

Grill, Mr. Charles M. Separations Techn., P.O. Box 63, Wakefield, RI 02879, USA.

Grossie, Dr. David Alan. Chemistry Dept., Wright State U., Dayton, OH 45435, USA. (1954) PhD chemistry (Texas Christian U., 1982). Asst. prof. (tel. 513 + 873-2210, email Bitnet DGROSSIE at WSU).

Groy, Dr. Thomas. Chemistry Dept., Arizona State U., Tempe, AZ 85287-1604, USA. (1954) PhD phys. chemistry (Arizona State U., 1982). Res. specialist. (tel. 602 + 965-1511). *Solid state reaction kinetics, multilayer films.*

Grzeskowiak, Dr. Kazmierz. Molec. Biology Inst., U. of Calif. at Los Angeles, Los Angeles, CA 90024, USA. (1942) PhD chemistry (A. Mickiewicz U., Poland, 1974). Molecular biologist. (tel. 213 + 206-8270). *Nucleic acids, structure, crystallography, spectroscopy, synthesis, DNA-drug complexes.*

Gschneidner, Prof. Karl A. Ames Lab., Dept. of Matls. Sci. and Eng., 255 Spedding, Iowa State U., Ames, IA 50011, USA. (1930) PhD phys. chemistry (Iowa State U., 1957). Distinguished prof. Sci. & Humanities (tel. 515 + 294-7931, telex 269266, fax 515 + 294-3226, email Bitnet RIC at ALISUVAX). *Metallic systems, metallurgy, physics, alloys, intermetallic compounds, low temperature heat capacity, magnetic susceptibility, electrical resistivity, high purity metals, preparation, crystal growth, rare earth metals.*

Gude, Mr. Arthur James III. 845 Dudley St., Lakewood, CO 80215, USA. (1917) MSc geology (mineralogy) (Colorado S. of Mines, 1949). Retired, US Geol. Surv. (tel. 303 + 237-7560). *Geology, mineralogy, authigenic zeolites, low temperature, low pressure minerals, zeolites, silicate minerals.*

Guerra, Mr. Ralph. Dow Chemical Co., Bldg. 1225, Freeport, TX 77541, USA. (1952) BS chemistry (U. of Texas at Austin, 1974). Res. leader. (tel. 409 + 238-1228). *Catalysts, refractories, metals, microstructure, crystallite size, amorphous materials.*

Guggenheim, Prof. Stephen. Geological Sci. Dept., U. of Illinois at Chicago Circle, Box 4348, Chicago, IL 60680, USA. (1948) PhD geology (U. of Wisconsin at Madison, 1976). Prof. (tel. 312 + 996-3263). *X-ray and electron diffraction methods, geologic problems, crystal structure, crystal chemistry, stability.*

Guttmann, Dr. Geoffrey D. Center for X-Ray Optics, MS 80-101, Lawrence Berkeley Lab., One Cyclotron Rd., Berkeley, CA 94720, USA. (1954) PhD medical physics (UCB, 1989). Res. asst. (tel. 415 + 486-6839, email GDGUTTMAN at LBL). *X-ray lithography, microscopy, optics, imaging methods.*

Guven, Prof. Necip. Geosciences Dept. Texas Tech U., P.O. Box 4109, Lubbock, TX 79409, USA. (1936) Dr.rer.nat. mineralogy (U. of Goettingen, BRD, 1962). Prof., geology. (tel. 806 + 742-3110). *Diffraction, X-ray, electron, clay minerals, micas.*

Guy, Dr. Joseph Thomas Jr. E.I. duPont de Nemours Co., P.O. Box 267, Brevard, NC 28712, USA. (1958) PhD inorg. chemistry (U. of Wisconsin at Milwaukee, 1985) Res. chemist. (tel. 704 + 885-5488).

Hackert, Prof. Marvin LeRoy. Chemistry Dept., U. of Texas, Austin, TX 78712, USA. (1944) PhD chemistry (Iowa State U., 1970). Prof. (tel. 512 + 471-1105). *Protein crystallography, protein structure - function, PRP, Prv enzymes, instrumentation.*

Haeffner, Mr. Dean R. Materials Sci. Dept., Northwestern U., 2145 Sheridan Ave., Evanston, IL 60201, USA.

Haendler, Prof. Helmut M. Chemistry Dept., Parsons Hall, U. of New Hampshire, Durham, NH 03824, USA. (1913) PhD inorg. chemistry (U. of Washington, 1940). Prof. emeritus. (tel. 603 + 659-3942). *Inorganic chemical structure applications.*

Hage, Mr. Frank W. Biochemistry Dept., U. of N. Carolina at Chapel Hill, CB #7260 FLOB, Chapel Hill, NC 27514, USA. Grad. student. (tel. 919 + 966-3263).

Hager, Dr. Gordon. NCI, NIH, Bldg. 37, Rm. 3C-19, Bethesda, MD 20892, USA. (1942) PhD genetics (U. of Washington, 1971). Chief, Hormone Act. & Oncogenesis Sect. (tel. 301 + 496-9867, fax 301 + 496-0734). *Structure, protein, DNA, genome, protein/DNA interactions, genetic engineering, gene therapy, expression, transcription, genetics, virology, microbiology, neurobiology, chromatin, molecular biology.*

Hagler, Dr. Arnold T. Biophysics Dept., Agouron Inst., 505 Coast Blvd., South, La Jolla, CA 92117, USA. (1942) PhD biophysics (Cornell U., 1970). Chairman. (tel. 619 + 456-1623). *Peptide hormones, proteins, structure, design, molecular dynamics, peptide design, crystal structure and dynamics.*

Haldar, Dr. Pradeep. Res. & Devel., Johnson Matthey Co., E. 15128 Euclid Ave., Spokane, WA 99216, USA. (1962) PhD materials sci. (Northeastern U.). Dev. eng. (tel. 509 + 922-8725, fax 509 + 922-8628). *Metallurgy, solid-state chemistry, superconductivity, alloy phase diagrams, rapid solidification, electronic materials.*

Hale, Dr. Danforth Rawson. Box 23, Aurora, OH 44202, USA. (1901) PhD phys. chemistry (Cornell U., 1928). Consultant. (tel. 216 + 562-6275). *Quartz, crystal growth, electronic properties, optical properties.*

Haller, Dr. Kenneth James. Oneida Res. Services, Inc., 1 Halsey Rd., Whiteboro, NY 13492, USA. (1951) PhD chemistry (U. of Arizona, 1978). Dir., X-ray crystallography. (tel. 315 + 736-3050). *Transition metal chemistry.*

Halpern, Dr. B. David. RTD, Polysciences/NL, 400 Valley Rd., Warrington, PA 18976, USA. (1921) PhD org. chemistry (Notre Dame U., 1949). President. (tel. 215 + 343-6484). *Photographic emulsions for crystallography.*

Haltiwanger, Mr. Ralph Curtis. Chemistry and Biochem. Dept., Campus Box 215, U. of Colorado, Boulder, CO 80309-0215, USA. (1947) MS chemistry (U. of Virginia, 1971). Res. chemist. (tel. 303 + 492-7239). *X-ray crystallography, organic, inorganic compounds, computer programming.*

Hamill, Dr. Gregory Prince. Materials Characterization Dept., GTE Labs., Inc., 40 Sylvan Rd., Waltham, MA 02154, USA. (1949) PhD applied phys. - matl. sci. (CIT, 1978). Techn. staff member. (tel. 617 + 466-2748, fax 617 + 890-9320). *X-rays, powder diffraction, topography, phase transitions, automation, computer programming, defect analysis, structure - properties relationships.*

Hamilton, Prof. Jean Allan. Biochemistry Dept., Indiana Sch. of Medicine, 635 Barnhill Dr., Indianapolis, IN 46223, USA. (1938) PhD X-ray crystallography (Glasgow U., Scotland, 1962). Prof. (tel. 317 + 274-7544). *X-ray structure determination, biochemically interesting molecules, beta-cyclodextrin complexes, ion-transporting antibiotic complexes, proteins.*

Hamilton, Prof. Robert David. 1042 Xenophon St., Golden, CO 80401, USA. (1942) PhD geology (Colorado School of Mines, 1978). (tel. 303 + 273-3817). *Inorganic structures, structure - properties relationships.*

Hamlin, Dr. Ronald Craig. 7776 Camino Glorita San Diego, CA 92122, USA. (1946) PhD physics, biophysics (U. of Calif. at San Diego, 1975). (tel. 714 + 452-2565). *Position sensitive X-ray detectors, macromolecular structures.*

Han, Dr. Fusen. Physical & Analyt. Chem. Res., The Upjohn Co., Kalamazoo, MI 49001, USA. (1946) PhD crystallography (Chinese Acad. of Sci., 1983). Res. scient. (tel. 616 + 385-5271, fax 616 + 385-7522). *Crystallography, small molecules, direct methods.*

Hannick, Dr. Linda I. Lab. for the Struct. of Matter, Code 6030, Naval Res. Lab., Washington, DC 20375, USA. (1947) PhD phys. chem., crystallog. (U. of New Orleans, 1986). Postdoct. fel. (tel. 202 + 767-0657, telex 7108 220147, fax 202 + 767-9046, email hannick at lsm.nrl.navy.mil). *Crystallography, protein, small molecule, electron density determination, area detectors.*

Hansen, Mr. Harly. Biochem. & Biophysics Dept., Texas A&M U., College Station, TX 77843, USA.

Hanson, Dr. Jonathan C. Chemistry Dept., Brookhaven Nat. Lab., Upton, NY 11973, USA. (1941) PhD chemistry (U. of Michigan, 1969). Sr. computer analyst (tel. 516 + 282-4378). *Molecular graphics, computer control.*

Hardcastle, Prof. Kenneth Irvin. Chemistry Dept., Calif. State U. at Northridge, 18111 Nordhoff, Northridge, CA 91330, USA. (1931) PhD inorganic and physical chem. (USC, 1961). Prof. (tel. 818 + 885-3381). *Structures, metal hydrides, organometallics, metal cluster compounds, small biological molecules.*

Hardgrove, Prof. George Lind Jr. Chemistry Dept., St. Olaf C., Northfield, MN 55057, USA. (1933) PhD chemistry (UCB, 1959). Prof. (tel. 507 + 663-3404). *Organic structures, diffraction, magnetic resonance methods.*

Hardiman, Ms. June. Fox Chase Cancer Ctr., 7701 Burholme Ave., Philadelphia, PA 19111, USA. (tel. 215 + 728-3660).

Hardman, Mr. Karl D. Protein Eng. Div., Genex Corp., 16020 Industrial Rd., Gaithersburg, MD 20877, USA. (tel. 301 + 258-0552).

Harker, Dr. David. Mol. Biophysics Dept., Med. Fndn. of Buffalo, 73 High St., Buffalo, NY 14203, USA. (1906) PhD chemistry (CIT, 1936). Res. sci. emeritus. (tel. 716 + 886-2666). *Crystal, molecular structure, color symmetry, structural chemistry.*

Harlow, Dr. Richard Leslie. Central Res. and Dev., E228/316D E. I. du Pont de Nemours Co., Wilmington, DE 19711, USA. (1942) PhD chemistry (Syracuse U., 1971). Group supr. (tel. 302 + 695-2097). *Single crystal structure determination, X-ray powder diffraction, powder diffraction software.*

Haromy, Dr. Tuli Patrick. 1716 Kendall Ave., Madison, WI 53705, USA. (1958) PhD biochemistry (U. of Wisconsin at Madison, 1982). (tel. 608 + 238-7809).

Harper, Prof. Richard A. Physics Dept., Rensselaer Polytechnic Inst., Eighth St., Troy, NY 12181, USA. (1936) PhD physics (New York U., 1970). Assoc. prof., Dir. crystallog. biophys. (tel. 518 + 270-6434). *Programming applications, linear and quadratic techniques, direct determination, crystal structure refinement, powder diffraction data analysis.*

Harris, Prof. David R. Computer Sci. Dept., California State U. at Chico, Chico, CA 95926, USA. (1932) PhD phys. chemistry (U. of Colorado, 1963). Prof. (tel. 916 + 895-5884). *Organic, biological structures, computer applications, artificial intelligence.*

Harris, Dr. Lester. Cancer Resarch Lab., Abbott Northwestern Hosp., 800 E. 28th. St., Minneapolis, MN 55407, USA. (1939) PhD molec. biol., med. microbiol. (Ohio State U., 1975). Tech. dir., Cancer Res. Lab. (tel. 612 + 863-4000, ext. 4617,4439, fax 612 + 863-4507). *Structure prediction, sequence analysis, supercomputer information, molecular dynamics simulation, protein/DNA complexes, model building, gene regulation, mouse mammary tumor virus.*

Harris, Mr. Mark. Computer Sci. Dept., U. of North Carolina, Chapel Hill, NC 27514, USA. (1960) PhD biophysics (U. of Leeds, England, 1985). Dir., GRIP Mol. Graphics Proj. (tel. 919 + 962-1753, fax 919 + 962-1799, email harris at cs.unc.edu). *Protein structures, molecular modeling software.*

Harrison, Dr. Robert Wilson. 8258 Blackhaw Cir., Frederick, MD 21701, USA. (1957) PhD biophysics (Yale U., 1985). *Protein crystallography, computational methods, enzyme structure, molecular calculations.*

Harrison, Prof. Stephen Coplan. Biochemistry Dept. and HHMI, Harvard U., 7 Divinity Ave., Cambridge, MA 02138, USA. (1943) PhD biophysics (Harvard U., 1968). Prof., biochemistry. (tel. 617 + 495-4090). *Macromolecular structure and assembly, viruses, protein - nucleic acid interactions, methods development, non-crystallographic symmetry, very large unit cells.*

Harrison, Dr. William Thomas Alexander. Chemistry Dept., U. of Calif. at Santa Barbara, Santa Barbara, CA 93102, USA. (1960) D.Phil. chemistry (Oxford U., England, 1986). Postdoct. res. asst. (tel. 805 + 961-8126, email WTAH at SBITP). *Solid state chemistry, neutron diffraction, powder methods, computing.*

Hart, Mr. Donald W. Shell Development Co., P.O. Box 1380, Houston, TX 77001, USA.

Hart, Dr. Haskell Vincent. Mol. Spectroscopy Dept., Shell Development Co., P.O. Box 1380, Houston, TX 77251, USA. (1943) PhD chemistry (Harvard U., 1973). Res. mgr. (tel. 713 + 496-9229). *Mineralogy, diffraction, powder, electron, crystallographic databases.*

Hartsuck, Dr. Jean Ann. Lab. of Protein Studies, Oklahoma Med. Res. Fndn., 825 N.E. 13th St., Okla. City, OK 73104, USA. (1939) PhD chemistry (Harvard U., 1964). Assoc. member. (tel. 405 + 271-7293). *Protein crystallography, enzyme structure - function.*

Hassell, Ms. Anne M. Macromol. Sci., Smith Kline & French R & D Labs., L110, P.O. Box 1539, King of Prussia, PA 19406-0939, USA. (1946) MS microbiol., biochem. (U. of Georgia, 1972). Res. scient. (tel. 215 + 270-4265). *Protein crystallography.*

Hastings, Dr. Jerome Biller. Nat. Synchrotron Light Source, Brookhaven Nat. Lab., Bldg. 725D, Upton, NY 11973, USA. (1948) PhD applied physics (Cornell U., 1975). Assoc. physicist. (tel. 516 + 282-3930, email Bitnet HASTINGS at BNLCL1). *X-ray physics, synchrotron radiation diffraction applications, phase transitions.*

Hastings, Dr. Julius Mitchell. 65 Ketcham Ave., Patchogue, NY 11772, USA. (1920) PhD phys. chemistry (Cornell U., 1945). *Neutron scattering, phase transitions.*

Hatada, Mr. Marcos H. Hoffman - La Roche, Bldg. 76/15, Nutley, NJ 07110, USA. (1955) PhD chemistry (Michigan State U., 1982). Sr. scient. (tel. 201 + 235-5385). *Protein crystallography.*

Hau, Dr. Herbert H. Restorative Dentistry, Tufts School of Dental Medicine, 1 Kneeland St., Boston, MA 02111, USA. (1941) PhD, DMD chemistry, dentistry (Boston U. 1970, Harvard U., 1977). Asst. Clinical Prof. (tel. 617 + 423-5655). *Crystal structure, inorganic molecules.*

Haupt, Mr. Gary Robert. Ames Lab., Iowa State U., 16 Wilhelm Hall, Ames, IA 50011, USA. (1953) MS metallurgy (Iowa State U., 1982). Assoc. metallurgist. (tel. 515 + 294-2564). *Physical metallurgy.*

Hauptman, Prof. Herbert Aaron. Med. Fndn. of Buffalo, 73 High St., Buffalo, NY 14203, USA. (1917) PhD mathematics (U. of Maryland, 1955). President. (tel. 716 + 856-9600, ext. 307, fax 716 + 852-4846). *Direct methods, organic crystal structures.*

Hayden, Dr. Thomas Day. Construction Products Div., W. R. Grace and Co., 62 Whittemore Ave., Cambridge, MA 02140, USA. (1944) PhD chemistry (Boston U., 1976). Res. assoc. (tel. 617 + 876-1400, ext. 3133). *Structural chemistry, microstructure, performance, silicates, aluminates, portland cement, experimental design.*

Hazen, Dr. Robert Miller. Geophysical Lab., Carnegie Inst. of Washington, 2801 Upton St., NW, Washington, DC 20008, USA. (1948) PhD mineralogy-

crystallography (Harvard U., 1975). Exp. mineralogist (tel. 202 + 966-0334, fax 202 + 895-4225). *Crystal structure, variation with pressure - temperature - composition, physical properties.*

He, Dr. Xiao-Min. ES76 Marshall Space Flight Ctr., NASA, Huntsville, AL 35812, USA. (1944) PhD crystallography (U. of Pittsburgh, 1984). Visit. scient. (tel. 205 + 544-5531). *Molecular thermal motion, charge density, molecular structure determination, crystallographic computing.*

Heath, Mr. James R. Chemistry Dept., Rice U., P.O. Box 1892, Houston, TX 77251, USA.

Hedberg, Prof. Kenneth Wayne. Chemistry Dept., Oregon State U., Corvallis, OR 97331, USA. (1920) PhD phys. chemistry (CIT, 1948). Prof. (tel. 503 + 754-2081). *Electron diffraction (gases), molecular dynamics, force fields.*

Hedman, Dr. Gun-Britt Margareta. Stanford Synchrotron Radiat. Lab., Stanford U., SLAC Bin 69, P.O. Box 4349, Stanford, CA 94309, USA. (1949) PhD chemistry (Umeå U., 1978). Sr. res. assoc. (tel. 415 + 926-3052, telex STANFRD STNU 348402, fax 415 + 926-4100, email Bitnet HEDMAN at SSRL750). *Bio-inorganic chemistry, synchrotron radiation, X-ray absorption spectroscopy, protein crystallography using anomalous dispersion.*

Heeg, Dr. Mary Jane. Chemistry Dept., Wayne State U., Detroit, MI 48202, USA.

Hegde, Dr. Rashmi. Mol. Biophys. & Biochem. Dept., HHMI, Yale U., 260 Whitney Ave., JWG 402, New Haven, CT 06511, USA. (1962) PhD med. chemistry, crystallography (U. of Pittsburgh, 1989). Postdoct. fel. (tel. 203 + 432-5096, fax 203 + 432-5175). *Protein crystallography, drug design.*

Heinz, Prof. Dion Larson. Geophysical Sci. Dept., U. of Chicago, 5734 S. Ellis Ave., Chicago, IL 60637, USA. (1958) PhD geology (UCB, 1986). Asst. prof. (tel. 312 + 702-3046, fax 312 + 702-1225, email heinz at geovax.uchicago.edu). *Ultra-high pressure diamond anvil studies, mineral physics.*

Hejna, Ms. Carolyn I. Corporate R&D, Bldg. K-1 2C28, General Electric Co., Schenectady, NY 12301, USA.

Held, Dr. Glenn A. Res. Div., IBM T.J. Watson Res. Ctr., P.O. Box 218, Yorktown Heights, NY 10598, USA. (1957) PhD physics (UCB). Res. staff memb., physical Sci. (tel. 914 + 945-2609, telex 137456, fax 914 + 945-2141, email HELD at IBM.COM). *Experimental condensed matter physics, surface phase transitions, surface crystallography.*

Hemily, Dr. Philip Wright. Off. of International Affairs, Nat. Academy of Sci., 2101 Constitution Ave., Washington, DC 20418, USA. (1922) Doctorat phys. chemistry (U. de Paris, France, 1953). Consultant. (tel. 202 + 334-2807, fax 202 + 334-3094).

Hempel, Dr. Judith C. 539 Westminster Ave., Swarthmore, PA 19081, USA. (tel. 215 + 270-6528).

Hendricks, Prof. Robert Wayne. Materials Eng. Dept., VA Polytech. Inst. & State U. 210 Holden Hall, Blacksburg, VA 24061, USA. (1937) PhD materials sci. (Cornell U., 1964). Prof. (tel. 703 + 231-6917). *Position-sensitive detectors, X-ray analysis, small-angle scattering, physics, diffuse scattering, electronic materials, residual stress, phase transformations, polymers.*

Hendrickson, Prof. Wayne A. Biochem. & Mol. Biophysics Dept., Columbia U., New York, NY 10032, USA. (1941) PhD biophysics (Johns Hopkins U., 1968). Prof. (tel. 212 + 305-3456, fax 212 + 305-7379, email Bitnet HENDW at CUMBG). *Diffraction methods, biological macromolecules.*

Hendrixson, Mr. Thomas L. Ames Lab., 43 Spedding Hall, Iowa State U., Ames, IA 50011, USA. (1960) BS chemistry (Wichita State U., 1982). Grad. student. (tel. 515 + 294-8477). *Patterson methods, metal complexes.*

Henley, Mr. Christopher Lee. Physics Dept., Clark Hall, Cornell U., Ithaca, NY 14853, USA. (tel. 607 + 255-9688).

Henslee, Dr. Walter Warren. Analytical Sci., A915, Dow Chemical Co., Freeport, TX 77541, USA. (1946) PhD chemistry (U. of Texas at Austin, 1974). Group leader. (tel. 409 + 238-4531). *X-ray powder diffraction, ceramics, catalysts, metals, mixed metal oxides and hydroxides.*

Heo, Dr. Nam Ho. Chemistry Dept., Boston C., 1400 Commonwealth Ave., Chestnut Hill, MA 02167, USA. (1956) PhD phys. chemistry (1987). Postdoct. fel. (tel. 617 + 552-8481, email NAMHO at BCCHEM). *Crystallography, macromolecular, zeolites.*

Hepp, Mr. Mark. Chemistry Dept., Georgetown U., 37th. & O Sts. NW, Washington, DC 20057, USA. (tel. 202 + 687-5704).

Herbette, Prof. Leo Gerard. Radiology Dept., U. of Connecticut Health Center, 263 Farmington Ave., Farmington, CT 06032, USA. (1953) PhD biophysics (U. of Pennsylvania, 1980). Assoc. prof. (tel. 203 + 679-2545, fax 203 + 679-2518). *Diffraction, X-ray, neutron, drug design methods.*

Herman, Prof. Herbert. Materials Sci. Dept., SUNY at Stony Brook, Stony Brook, NY 11794-2275, USA. (1934) PhD materials sci. (Northwestern U., 1961). Prof. (tel. 516 + 632-8480, fax 516 + 632-6252). *Powder diffractometry, stress analysis.*

Hermans, Prof. Jan. Biochemistry Dept., U. of North Carolina, Chapel Hill, NC 27599-7260, USA. (1933) PhD chemistry (U. of Leiden, 1958). Prof. (tel. 919 + 966-4644, email hermans at med.unc.edu). *Proteins, modeling, conformation change.*

Herriott, Prof. Jon Roger. Biochemistry Dept., U. of Washington, Seattle, WA 98195, USA. (1937) PhD biophysics (The Johns Hopkins U., 1967). Assoc. prof. (tel. 206 + 543-9484). *Protein structure, X-rays, NMR.*

Herron, Dr. James Nelson. Pharmaceutics Dept., U. of Utah, 421 Wakara Way, Suite 316, Salt Lake City, UT 84108, USA. (1954) PhD microbiology (U. of Illinois, 1981). Asst. prof. (tel. 801 + 581-7303, 3997, fax 801 + 581-7880, email bi.herron at science.utah.edu). *Biophysics, immunology, macromolecular crystallography.*

Herzberg, Dr. Osnat. CARB, 9600 Gudelsky Dr., Rockville, MD 20850, USA. (1949) PhD chemistry (Weizmann Inst. of Sci., Israel, 1982). Assoc. staff member. (tel. 301 + 975-2980, fax 301 + 258-1115). *Protein crystallography, protein structure - function.*

Heuer, Prof. Arthur H. Matls. Sci. & Eng. Dept., Case Western Reserve U., 10900 Euclid Ave., Cleveland, OH 44106, USA. (1936) DSc physical ceramics (U. of Leeds, England, 1977). Kyocera Prof. of Ceramics, (tel. 216 + 368-3868).

Hibbard, Dr. Lyndon Stanley. Neurology & Neurolog. Surg. Dept., Washington U. Med. Ctr., P.O. Box 8111, St. Louis, MO 63110, USA. (1947) PhD phys. chemistry (Michigan State U., 1977). Res. assoc. prof., Dir. Neuro-Imaging Lab. (tel. 314 + 362-3390, email lyn at loni.wustl). *Digital image analysis, three dimensional brain image construction.*

Higgins, Prof. John Britt. Central Res. Lab., Mobil Res. and Devel. Corp., P.O. Box 1025, Princeton, NJ 08540, USA. (1947) PhD mineralogy (geo. sci.) (Virginia Polytech. Inst., 1978). Res. chemist. (tel. 609 + 737-4215, fax 609 + 737-5217). *Zeolites, crystallography, silicate crystal chemistry.*

Hinch, Mr. Ralph J. Jr. Walter C. McCrone Assoc., Inc., 850 Pasquinelli Dr., Chicago, IL 60559, USA. (1926) BS chemistry (Elmhurst C., 1950). Res. scient. (tel. 312 + 887-7100, fax 312 + 887-7417). *Crystallography, optical, X-ray, analysis, X-ray diffraction, optical microscopy.*

Hinckley, Prof. Conrad Cutler. Chem. and Biochem. Dept., Southern Illinois U., Carbondale, IL 62901, USA. (1934) PhD chemistry (U. of Texas at Austin, 1964). Prof. (tel. 618 + 453-5721). *Inorganic chemistry, coordination complexes, transition metals, iron congeners, oxygen chemistry, osmium compounds, medicinal applications, coal chemistry.*

Hingerty, Dr. Brian Edward. Health & Safety Res. Div., Oak Ridge Nat. Lab., P.O. Box 2009, MS 8077, Oak Ridge, TN 37831, USA. (1948) PhD physics, biophysics (Princeton U., 1974). Res. staff. (tel. 615 + 574-0844). *Biologically important structures, nucleic acids, polysaccharides, proteins, supercomputers.*

Hirshfield, Mr. Jordan M. Biophysics Dept., Merck and Co., Inc., P.O. Box 2000, Rahway, NJ 07065, USA. (tel. 201 + 574-5510).

Hirth, Prof. John Price. Mech. and Matls. Eng. Dept., Washington State U., Pullman, WA 99164-2920, USA. (1930) PhD metallurgical eng.g (Carnegie-Mellon U., 1957). Prof. (tel. 509 + 335-4971). *Metal physics, dislocation theory, surfaces.*

Hite, Prof. Gilbert James. Medicinal Chemistry, School of Pharmacy, U-92, U. of Connecticut, Storrs, CT 06268, USA. (1931) PhD medicinal chemistry (U. of Wisconsin, 1959). Prof. (tel. 203 + 486-3350). *Organic crystal structures, structure - reaction mechanism - stereochemistry relationships, drug action, optical rotatory dispersion.*

Ho, Prof. Chien. Biological Sci. Dept., Carnegie Mellon U., 4400 Fifth Ave., Pittsburgh, PA 15213, USA. (1934) PhD chemistry (Yale U., 1961). Prof. (tel. 412 + 268-3395, fax 412 + 268-7083). *Structure - function relationships, proteins, enzymes, lipids, NMR in biology, biochemistry, biophysics, molecular biology.*

Ho, Dr. Douglas M. Chemistry Dept., U. of Cincinnati, Mail Location 172, Cincinnati, OH 45221, USA. (1951) PhD inorg. chemistry (USC, 1981). Staff crystallographer. (tel. 513 + 556-9250).

Hoard, Prof. James Lynn. 42 Cornell St., Ithaca, NY 14850, USA. (1905) PhD chemistry (CIT, 1932). Prof. emeritus. (tel. 607 + 273-7243). *Structural chemistry.*

Hoard, Dr. Laurence Graham. 68 Southern Lane, Warwick, NY 10990, USA. (1940) PhD chemistry (U. of Michigan, 1977). (tel. 914 + 986-1275).

Hobbs, Prof. Linn Walker. Matls. Sci. & Eng. Dept., Massachusetts Inst. of Techn., Rm. 13-4062, Cambridge, MA 02139, USA. (1944) D.Phil. materials sci. (Oxford U., UK, 1972). Prof.of ceramics & matl. sci. (tel. 617 + 253-6835, telex 921473 MIT C fax 617 + 253-8000). *Non-stoichiometric compounds, extended defect structures, high resolution transmission electron microscopy, structures, aperiodic, quasi-periodic.*

Hodgson, Prof. Derek John. Chemistry Dept., U. of Wyoming, Box 3838, Laramie, WY 82071-3838, USA. (1942) PhD chemistry (Northwestern U., 1969). Prof. & head. (tel. 307 + 766-2434, fax 307 + 766-2271). *Magnetically condensed systems, dimer and cluster complexes, metal peptide complexes.*

Hodgson, Prof. Keith Owen. Chemistry Dept., Stanford U., Stanford, CA 94305, USA. (1947) PhD inorg. chemistry (UCB, 1972). Prof. (tel. 415 + 723-1328, fax 415 + 926-4100, email Bitnet HODGSON at SSRL750). *Bio-inorganic chemistry, anomalous dispersion, synchrotron radiation, X-ray absorption spectroscopy.*

Hodsdon, Dr. John Marshall. Longridge Farm, Meredith, NH 03253, USA. (1938) PhD biochemistry (UCB, 1970). (tel. 603 + 279-6126). *Proteins, structure, small molecule - protein interactions, refinement methods, accuracy, molecular graphics.*

Hoff, Mr. Henry A. Matls. Sci. & Techn., Code 6325, Naval Res. Lab., 4555 Overlook Ave., SW, Washington, DC 20375-5000, USA. (1946) AB, BA geology, math. (U. of Illinois, 1969, C. of Grt. Falls, MT, 1973). Physical scient.

(tel. 202 + 767-2743). *Superconductivity, crystallography, optical, X-ray, mineralogy.*

Hoggins, Dr. James Thomas. Diamond Techn. Center, Norton Christensen, Inc., 2532 South 3270 West, Salt Lake City, UT 84119, USA. (1942) PhD materials sci. (U. of Texas at Austin, 1975). Sr. scient. (tel. 801 + 972-3140). *Diamonds, X-ray crystallography, ceramics.*

Hogle, Dr. James Martin. Mol. Biology Dept. MB13, Res. Inst. Scripps Clinic, 10666 N. Torrey Pines Rd., La Jolla, CA 92037, USA. (1951) PhD biochemistry (U. of Wisconsin at Madison, 1978). Member. (tel. 619 + 554-9705, fax 619 + 554-6105). *Virus structure and function, macromolecular X-ray crystallography.*

Holbrook, Dr. Stephen Roy. Lab. of Chemical Biodynamics, Lawrence Berkeley Lab., Bldg. 3, U. of Calif. at Berkeley, Berkeley, CA 94720, USA. (1948) PhD phys. chemistry (U. of Oklahoma, 1974). Staff scient. (tel. 415 + 486-4304). *Nucleic acid, structure - dynamics, crystallographic refinement, molecular modeling.*

Holden, Dr. Hazel Marguerite. Chemistry Dept., Inst. for Enzyme Res., 1710 University Ave., Madison, WI 53705, USA. (1955) PhD biochemistry (Washington U., 1982). Asst. prof. (tel. 608 + 262-4988). *Transport proteins, electron, lipid, protein crystallography.*

Holden, Dr. James Richard. 10414 Glenmore Dr., Adelphi, MD 20783-1203, USA. (1928) PhD phys. chemistry (State U. of Iowa, 1955). Res. chemist, retired. (tel. 301 + 439-2807). *Organic crystal structures, crystal density prediction.*

Holland, Dr. Hans J. Technical Staff Div., Corning Glass Works, Sullivan Park FR-18, Corning, NY 14830, USA. (1929) PhD chemistry (U. of Utah, 1963). Res. assoc. (tel. 607 + 974-3675). *Glass ceramic structures, service crystallography, automation (XRD).*

Hollander, Dr. Frederick J. Chemistry Dept., U. of California at Berkeley, Berkeley, CA 94720, USA. (1946) PhD phys. chemistry (UCB, 1972). Assoc. specialist. (tel. 415 + 642-8444). *Small molecules, service crystallography.*

Holmes, Mr. Francis Edward. Drawer D, Accokeek, MD 20607, USA. (tel. 301 + 283-3376).

Holomany, Mr. Mark A. JCPDS, Internatl. Ctr. Diffraction Data, 1601 Park Ln., Swarthmore, PA 19081, USA.

Holt, Dr. Elizabeth Manners. Chemistry Dept., Oklahoma State U., Stillwater, OK 74078, USA. (1939) PhD chemistry (Brown U., 1966). Prof. (tel. 405 + 744-5949). email CHEMEMH at OSUCC, *Fluorescent Cu(I) systems, calcium-allergen interactions, metal cluster systems.*

Holtzberg, Dr. Frederic. Res. Div., IBM T. J. Watson Res. Ctr., P.O. Box 218, Yorktown Hgts., NY 10598, USA. (1922) PhD phys. chemistry (PUN, 1952). Res. staff memb., condensed matter. (tel. 914 + 945-1045, telex 137456, fax 914 + 945-2141, email at IBM.COM)). *Solid state physics, magnetism, phase transitions, rare earth compounds, materials research, high temperature superconductors, heavy fermions.*

Holtzman, Ms. Susan. Molecular Architects Corp., 231 S. Bemiston, Suite 800, Clayton, MO 63105, USA. (tel. 314 + 862-7767).

Hom, Dr. Tommy. Eng. Dept., Philips Electronic Instruments, Inc., 85 McKee Dr., Mahwah, NJ 07430, USA. (1949) PhD physics (PUN, 1979). Scient. *Diffraction physics, X-ray optics, automated instrumentation, computer programming.*

Honzatko, Prof. Richard E. Biochem. and Biophysics Dept., Gilman Hall, Iowa State U., Ames, IA 50011, USA. (1954) PhD phys. chemistry (Harvard U., 1982). Asst. prof. (tel. 515 + 294-7103). *Macromolecular structure determination, catalysis mechanisms, allostery, anomalous scattering.*

Hoogsteen, Dr. Karst. Biophys. and Pharmacol. Dept., Merck Inst. for Therapeutic Res., P.O. Box 2000, Rahway, NJ 07065, USA. (1923) PhD crystallography (U. of Groningen, Netherlands, 1957). Sr. investigator. (tel. 201 + 232-4108). *Crystal structures, organic, inorganic.*

Hope, Prof. Håkon. Dept. of Chemistry, U. of California, Davis, Davis, CA 95616, USA. (1930) Cand. real. chemistry (U. of Oslo, Norway, 1958). Prof. (tel. 916 + 752-0957, telex 910-531-0785, fax 916 + 752-6363, email HHOPE at UCDAVIS). *X-ray diffraction methods, rapid structure determination, instrumentation, biological macromolecules, cryocrystallography.*

Hopkins, Prof. Robert C. Chemistry Dept., U. of Houston - Clear Lake, 2700 Bay Area Blvd., Houston, TX 77058, USA. (1937) PhD phys. chemistry (Harvard U., 1965). Prof. (tel. 713 + 488-9394). email Bitnet HOPKINS at UHCL2, *Molecular biophysics, nucleic acid structure, protein - nucleic acid interactions.*

Horrigan, Ms. Jane Akerlund. RADC - ESM, Hanscom AFB, Bedford, MA 01731, USA. (1937) MS chemistry (Northeastern U., 1965). Res. chemist. (tel. 617 + 377-2215). *Defect structures, X-ray topography, materials characterization.*

Horrocks, Prof. William DeWitt Jr. Chemistry Dept., Pennsylvania State U., 152 Davey Lab., University Park, PA 16802, USA. (1934) PhD inorg. chemistry (MIT, 1960). Prof. (tel. 814 + 865-1191, fax 814 + 865-3314, email WDH2 at PSUVM). *Lanthanide ion chemistry, calcium binding proteins, metalloenzymes, metal ion active sites, molecular mechanics simulation of X-ray structures.*

Horsey, Dr. Richard Stephen. Res. and Dev., Keystone Carbon Co., 1935 State St., St. Marys, PA 15857, USA. (1950) PhD ceramic sci. (PSU, 1981). Mgr. R&D, thermistor div. (tel. 814 + 781-1591). *Spinels, thick films, thermistors, titanates, ionic conductors.*

Horton, Mr. John R. Biochem. & Biophys. Dept., Columbia U., Rm. 523 Black Bldg., 630 W. 168th. St., New York, NY 10032, USA. (1962) M.Phil. Biochem., molec. biophys (Columbia U.). Grad. student. (tel. 212 + 305-1846, fax 212 +

305-7379, email HORTON at CUMBG). *Biological macromolecules, X-ray diffraction, 2-D NMR.*

Hosmane, Prof. Narayan S. Chemistry Dept., Southern Methodist U., Airline & Daniel Sts., Dallas, TX 75275, USA. (1948) PhD inorg. chemistry (U. of Edinburgh, UK, 1974). Prof. (tel. 214 + 692-2950, 4141, fax 214 + 692-4138, email HADR1001 at SMUVM1). *Crystallography, electron diffraction, main group chemistry, carboranes, metallacarboranes, silicon hydrides, boron hydrides, sulfur-fluorine chemistry.*

Hossain, Dr. M. Bilayet. Chemistry Dept., Oklahoma U., 620 Parrington Oval, Norman, OK 73019, USA. (1937) PhD X-ray crystallography (London U., UK, 1965). Res. scient. (tel. 405 + 325-5831). *Structures, biomolecules, siderophores, peptides, marine natural products, anti-cancer compounds.*

Houska, Prof. Charles Robert. Materials Eng., Holden Hall, Virginia Polytechnic Inst., Blacksburg, VA 24061, USA. (1927) ScD metallurgy (MIT, 1957). Prof. (tel. 703 + 951-5652). *X-ray diffraction, atomic diffusion, physical metallurgy.*

Howard, Mr. Andrew Jay. Protein Eng. Dept., Genex Corp., 16020 Industrial Dr., Gaithersburg, MD 20877, USA. (1954) PhD physics (U. of Calif. at San Diego, 1981). Prin. res. scient. (tel. 301 + 258-0552, ext. 336, telex 908775, fax 301 + 926-1221). *Protein crystallographic software, protein engineering, structure - function relationships, diffraction methods.*

Howard, Dr. Scott A. Ceramics Dept., U. of Missouri at Rolla, B47 McNutt, Rolla, MO 65401, USA. (1958) PhD ceramic sci. (New York State C. of Ceramics, 1984). Asst. prof. (tel. 314 + 341-4403). *Powder diffraction, X-ray, neutron, crystallographic computing.*

Howatson, Prof. John. Box 3342, Laramie, WY 82071, USA. (1920) PhD chemistry (U. of Wisconsin, 1950). (tel. 307 + 766-4370). *Inorganic and mineral structures.*

Howe, Ms. Donna-Beth. 10104 Gardiner Ave., Silver Spring, MD 20902, USA. (tel. 301 + 681-5998).

Howe, Prof. James M. Met. Eng. and Matl. Sci. Dept., Carnegie Mellon Inst., 5000 Forbes Ave., Pittsburgh, PA 15213, USA. (1955) PhD materials sci. & eng. (UCB, 1985). Alcoa Asst. Prof. (tel. 412 + 268-2683, fax 412 + 268-6421, email JH3P at CMC). *Solid state phase transformations, high resolution electron microscopy, atomic structure, kinetics, interfaces.*

Howell, Dr. Peter Adam. I.S.D., Bldg. 230-1S-07, 3M Center, Minn. Mining & Mfg. Co., St. Paul, MN 55144, USA. (1928) PhD phys. chemistry (U. of Minnesota, 1955). Sr. res. specialist. (tel. 612 + 733-9007). *Alumino silicates, zeolites, crystal growth, polymers, structures, glass.*

Howes, Ms. Catherine. Immunetech Pharmaceuticals, 11045 Roselle St., San Diego, CA 92121, USA. (1961) BS chemistry (Michigan State U., 1985). Res. resources coord. (tel. 619 + 457-2553, ext. 334, telex 75-4566, fax 619 + 452-7524). *Proteins, sequence analysis, structural predictions, immunology, peptides.*

Hsu, Miss Barbara T. Chemistry Dept., U. of Calif. at Los Angeles, Los Angeles, CA 90024, USA. (1957) BS chemistry (CIT, 1979). Grad. student. (tel. 213 + 825-8901).

Hsu, Mr. Hsyh-Min. Superconductivity Dept., Microelectronic & Computer Techn. Corp., 12100 Technology Blvd., Austin, TX 78727, USA. (1961) MS Matl. sci. & eng. (U. of Texas at Austin, 1988). Tech. staff membr. (tel. 512 + 250-2754, fax 512 + 250-2895). *Superconductors, thin film technology, electronic materials.*

Hsu, Prof. I-Nan. Chemistry Dept., Calif. State U. at Northridge, 18111 Nordhoff St., Northridge, CA 91330, USA. (1939) PhD phys. chemistry (U. of Oklahoma, 1971). Assoc. prof. (tel. 818 + 885-3366). *Protein structure - function, molecular and metal complexes.*

Hsu, Dr. Leh-Yeh Ruth. Chemistry Dept., Ohio State U., 120 W. 18th. Ave., Columbus, OH 43210, USA. (1948) PhD phys. chemistry (U. of Louisville, 1980). Sr. res. assoc. (tel. 614 + 292-7827, fax 614 + 292-1685, email HSUS at OHSTCH). *Structure determination, computer molecular modeling, intermolecular forces.*

Hu, Dr. Hengliang. Electron Diffr. Dept., Med. Found. of Buffalo, 73 High St., Buffalo, NY 14203, USA. (1942) BS structural chemistry (Fudan U., PRC, 1966). Res. assoc. (tel. 716 + 856-9600, ext. 354, email edhweng%mfb at ubyms). *Electron diffraction, thermal analysis, crystal structure, supermolecular structure, polymers, polymer materials, fibers.*

Huang, Mr. De-Bin. Crystallography Dept., U. of Pittsburgh, Pittsburgh, PA 15260, USA. (1951) MS phys. chemistry (Fuzhou U., PRC, 1981). Grad. student. (tel. 412 + 624-9305). *X-ray crystal structures, clusters, proteins, liquid crystals.*

Huang, Mr. Kuei-Shang. 1194 W. 30th. St., Apt. 1, Los Angeles, CA 90007, USA. (tel. 213 + 743-7527).

Huang, Dr. Ting Chun. Almaden Res. Center, K34/802, IBM Res. Div., 650 Harry Rd., San Jose, CA 95120-6099, USA. (1942) PhD physics (PUN, 1972). Res. staff member. (tel. 408 + 927-2375, telex 316969, fax 408 + 927-2100). *X-rays, powder diffraction, thin film analysis, fluorescence, electron microprobe, multiple diffraction, double crystal diffraction, applied crystallography, computer techniques, dynamical theory, laboratory automation.*

Hubbard, Dr. Camden Richards. Ctr. for Matls. Sci., A257, MATL, Nat. Inst. Stnds. & Techn., Gaithersburg, MD 20899, USA. (1944) PhD phys. chemistry (Iowa State U., 1971). Res. chemist. (tel. 301 + 975-6121, fax 615 + 574-4913,

email Bitnet HUH at ORNLSTC). *Powder diffraction, materials characterization, data evaluation.*

Hubbard, Dr. Stevan R. Biochem. & Molec. Biophysics Dept., Columbia U. C. P&S, 630 W. 168th. St., New York, NY 10032, USA. (1957) PhD applied physics (Stanford U., 1988). Postdoct. res. scient. (tel. 212 + 305-2219, email HUBBARD at CUMBG). *Calcium binding proteins, anomalous dispersion, synchrotron radiation.*

Hubbell, Mr. John Howard. Ionizing Radiation Div., Nat. Inst. Stnds. & Techn. Gaithersburg, MD 20899, USA. (1925) MSc eng. physics (U. of Michigan, 1950). Consultant, X-ray interaction data. (tel. 301 + 975-5550, telex 197674 *X-rays, attenuation coefficients, interactions with atoms, Compton scattering, Rayleigh coherent scattering, atomic form factors, incoherent scattering functions.*

Huddle, Prof. Benjamin Paul. Chemistry Dept., Roanoke C., Salem, VA 24153, USA. (1941) PhD phys. chemistry (U. of North Carolina, 1968). Prof. and Chair. (tel. 703 + 375-2440). *Small molecule structures, inorganic complexes, computer methods.*

Hudgens, Mr. Claude R. Mound Facility, E. G. & G., Miamisburg, OH 45343, USA. (1921) PhD chemistry (U. of Illinois, 1950). Res. specialist. (tel. 513 + 865-3303). *Crystallography, general X-ray applications.*

Huffman, Dr. John Curtis. Mol. Structure Center, Chemistry Dept., Indiana U., Bloomington, IN 47405, USA. (1941) PhD chemistry (Indiana U., 1974). Sr. scient. (tel. 812 + 855-6742, email HUFFMAN at IUBACS). *Instrumentation, computer graphics, low temperature crystallography.*

Hughes, Prof. John M. Geology Dept., Miami U., Oxford, OH 45056, USA. (1952) PhD mineralogy (Dartmouth U., 1981). Assoc. prof. (tel. 513 + 529-3218). *Mineralogy, crystallography, minerals, apatite, vanadium.*

Hughes, Dr. Robert Edward. Associated Universities Inc., 1717 Massachusetts Ave., Washington, DC 20036, USA. (1924) PhD phys. chemistry (Cornell U., 1953). (tel. 202 + 462-1676). *Structures, inorganic, biochemical, macromolecular.*

Hughes, Prof. William Eugene. Physics and Astronomy Dept., U. of Southern Mississippi, Southern Sta. Box 5046, Hattiesburg, MS 39406, USA. (1932) PhD physics (U. of Alabama, 1963). Chairman. (tel. 601 + 266-4934). *Magnetic resonance, astronomy, crystallography.*

Humblet, Dr. Christine. Parke-Davis Res. Labs., 2800 Plymouth Rd., Ann Arbor, MI 48105, USA. (1953) PhD chemistry (FNDP NAMUR, Belgium, 1975). Sr. mgr., comp. assisted drug design. (tel. 313 + 996-7034). *Molecular modeling, drug design, 3-D structures, X-ray, NMR.*

Hunt, Dr. Lois T. Nat. Biomed. Res. Fndn., Georgetown U. Med. Ctr., 3900 Reservoir Rd. NW, Washington, DC 20007, USA. (1933) PhD zoology (U. of Maryland at College Park, 1968). Sr. res. scient. (tel. 202 + 687-2121, fax 202 + 687-1662, email Bitnet PIRMAIL at GUNBRF). *Protein sequences, data bases, identification, comparison, predictions, computer analyses, theoretical molecular biology.*

Hunt, Mr. Richard E. 7609 Range Rd., Alexandria, VA 22306, USA. (1931) BS chemistry (Wagner C., 1956). Manufacturers repr. (tel. 703 + 768-0836). *Materials analysis.*

Hurst, Prof. Vernon James. Geology Dept., GGS Building, U. of Georgia, Athens, GA 30601, USA. (1923) PhD geology (Johns Hopkins U., 1954). Res. prof. (tel. 404 + 542-2400). *X-ray crystallography, petrology (experimental), crystal growth.*

Hutchings, Mr. Alan E. Barbeau-Hutchings, Inc., 10 S. Franklin Turnpike, Ramsey, NJ 07446, USA. (1937) AB geology (U. of Arizona, 1958). Vice-pres. (tel. 201 + 327-6611, fax 201 + 327-8862). *Analytical instruments, microelectronics.*

Hutchins, Mr. Charles. 36071 N. Grand Oaks Ct., Apt. G3, Gurnee, IL 60031, USA.

Hybl, Prof. Albert. Biophysics Dept., U. of Maryland Sch. of Medicine, 660 W. Redwood St., Baltimore, MD 21201, USA. (1932) PhD chemistry and mathematics (CIT, 1961). Assoc. prof. (tel. 301 + 528-7940). *Biological structures, membrane structure - function, computer applications.*

Hyde, Dr. C. Craig. Lab. of Mol. Biology, Bldg. 2, Rm. 316, Nat. Insts. of Health, Bethesda, MD 20829, USA. (1956) PhD biochemistry (U. of Iowa, 1985). Sr. staff fel. (tel. 301 + 496-4295, fax 301 + 496-0201, email CCH at NIHKLMB). *X-ray crystallography, biological macromolecules, protein crystallography.*

Ibers, Prof. James A. Chemistry Dept., Northwestern U., Evanston, IL 60208, USA. (1930) PhD chemistry (CIT, 1954). Prof. (tel. 312 + 491-5449, fax 312 + 491-4133, email ibers at nuacc.acns.nwu.edu). *Structures, coordination compounds, organometallic compounds, ternary chlcogenides, metalloporphyrins.*

Ice, Dr. Gene Emery. ORNL Beamline X14, NSLS, Bldg. 725A, Brookhaven Nat. Lab., Upton, NY 11973, USA. (1950) PhD physics (U. of Oregon, 1977). Staff scient. (tel. 516 + 282-5632). *Sscattering, X-ray, anomalous, synchrotron radiation, inner shell cross sections.*

Inniss, Mr. Daryl. 3283 Sepulveda Blvd., #4, Los Angeles, CA 90034, USA.

Inouye, Dr. Hideyo. Neuroscience Dept., Children's Hosp., Harvard Med. Sch., 300 Longwood Ave., Boston, MA 02115, USA. (1951) PhD biophysical sci. (1979). Instr. neuropathology. (tel. 617 + 735-6102). *Biomolecules.*

Iyere, Mr. Peter Abeta. Chemistry Dept., Brandeis U., P.O. Box 9110, Waltham, MA 02254-9110, USA. (1956) MS chemistry (U. of Abadan, Nigeria, 1981).

Grad. student. (tel. 617 + 736-2000, ext. 2535). *X-ray crystallography, solid state reactions.*

Jackson, Dr. Kenneth Arthur. Physics of Matls., Rm. 1A-157, AT&T Bell Labs., 600 Mountain View, Murray Hill, NJ 07974, USA. (1930) PhD applied physics (Harvard U., 1956). (tel. 201 + 582-4188, fax 201 + 582-2783). *Crystal growth, phase transformations, growth morphologies, crystalline defects, novel crystalline materials.*

Jacobs, Prof. Gerald Daniel. Chemistry Dept., Northern Michigan U., Marquette, MI 49855, USA. (1935) PhD phys. chemistry (Michigan State U., 1961). Prof., head. (tel. 906 + 227-2912). *Crystallography, thermodynamics, kinetics.*

Jacobson, Mr. Bruce L. HHMI, Baylor C. Med., Houston, TX 77030, USA. (1962) BS chemistry & mol. biology (U. of Wisconsin at Madison, 1984). Grad. student. (tel. 713 + 799-6564). *Protein structure - function, protein - ligand interactions.*

Jacobson, Prof. Robert Andrew. Chemistry Dept., Ames Lab., 42 Spedding, Iowa State U., Ames, IA 50011, USA. (1932) PhD chemistry (U. of Minnesota, 1959). Prof., sr. chemist. (tel. 515 + 294-1144, email XRA at ALISUVAX). *Crystallographic computing, automation, metal complexes, Patterson methods.*

Jain, Mr. Sanjeev. Biochemistry Dept., U. of Wisconsin, 420 Henrry Mall, Madison, WI 53706, USA.

Jain, Dr. Shri C. Biophys. Dept., Sch. of Medicine and Dentistry, U. of Rochester, Rochester, NY 14642, USA. (1940) PhD physical chem., crystallog. (Poona U., India, 1968). Scient. (tel. 716 + 275-1458). *Crystal structure determination, biological and organic molecules, drug-nucleic acid interactions, nucleic acids, structure, function, computer programming.*

James, Prof. William J. Graduate Ctr. Materials Res., U. of Missouri, Rolla, MO 65401, USA. (1922) PhD chemistry (Iowa State U., 1953). Prof., sr. investigator. (tel. 314 + 341-4324). *Neutron diffraction, magnetic structures, rare earth alloys.*

Jameson, Prof. Geoffrey Brind. Chemistry Dept., Georgetown U., 37 and O Sts., NW, Washington, DC 20057, USA. (1952) PhD chemistry (Canterbury U., New Zealand, 1977). Assoc. prof. (tel. 202 + 687-4027, email JAMESON at GUVAX). *Bio-inorganic chemistry, twinning, topotaxy.*

Janakiraman, Dr. M. N. Biochem. and Biophys. Dept., Iowa State U., 360 Gilman Hall, Ames, IA 50011, USA. (1955) PhD phys. chemistry (Iowa State U., 1988). Postdoct. fel. (tel. 515 + 294-0567). *Crystallography, structure - activity relationships, macromolecules, structure, organic synthesis.*

Janakiraman, Dr. Vijayalakshmi. Biochem. & Biophys. Dept., Texas A&M U., College Station, TX 77843-2128, USA. (1950) PhD X-ray crystallography (Madras U., India, 1987). Postdoct. res. assoc. (tel. 409 + 846-1744). *Crystallography, protein, small molecule, protein folding, theoretical studies, small molecules, size, shape analysis by radius of gyration, proteins, statistical data analysis, programming.*

Jandacek, Dr. Ronald James. Procter and Gamble Co., Miami Valley Labs., P.O. Box 39175, Cincinnati, OH 45239-8707, USA. (1942) PhD chemistry (U. of Texas, 1968). Chemist. (tel. 513 + 245-2767). *Lipid chemistry.*

Jap, Dr. Bing Kiat. Cell & Mol. Biol. Div., Lawrence Berkeley Lab., Donner Lab. Rm. 212, U. of Calif., Berkeley, CA 94720, USA. (1945) PhD biophysics (UCB, 1975). Staff scient. III. (tel. 415 + 486-7104, fax 415 + 486-5454, email jap%rhoda.hepnet at lbl.arpa). *Electron crystallography, image processing, electron multiple scattering theory, spectroscopy, infrared, circular dichroic, membrane proteins, purification, crystallization.*

Jasinski, Prof. Jerry P. 12 Orchard Ln., Springfield, VT 05156, USA. (alt.: Chem. Dept., Keene State C., Keene, NH 03431, USA). (1940) PhD chemistry (U. of Wyoming, 1974). Assoc. prof. chem. (tel. 802 + 885-9297). *Chemistry, inorganic, physical, polymers, molecular modeling, X-ray crystallography.*

Jeffrey, Dr. George Alan. Crystallography Dept., U. of Pittsburgh, 304 Thaw Hall, Pittsburgh, PA 15260, USA. (1915) DSc chemistry (U. of Birmingham, UK, 1953). Prof. emeritus. (tel. 412 + 624-9300, telex 812466 (dom) 199126(internatl.), fax 412 + 624-1882). *Structure, small molecules, carbohydrates, hydrates, hydrogen bonding.*

Jendrek, Dr. Eugene F. E.G.& G. Mound Applied Techn., P.O. Box 3000, Miamisburg, OH 45343-0987, USA. (1949) PhD phys. chemistry (U. of Maryland, 1979). Sr. res. chemist. (tel. 513 + 865-4205). *Crystallographic computing, inorganic crystal chemistry.*

Jeng, Dr. Tzyy-Wen. Consumer Diagnostics, Abbott Labs., Abbott Park, IL 60064, USA. (1947) PhD comparative biochemistry (UCB, 1978). Biochemist. (tel. 312 + 937-8880). *Macromolecular structure, electron scattering, crystallography.*

Jenkins, Dr. Ron. JCPDS, Internatl. Ctr. Diffraction Data, 1601 Park Ln., Swarthmore, PA 19081, USA. (1932) PhD chemical physics (PUN, 1981). Principal scient. (tel. 215 + 328-9403). *X-ray diffraction, powder diffractometry.*

Jennings, Dr. Laurence Duane. SLCMT-OMM, Army Matls. Tech. Lab., Watertown, MA 02172-0001, USA. (1929) PhD physics (MIT, 1955). Retired. (tel. 617 + 923-5375). *Diffraction techniques, characterization.*

Jensen, Prof. Lyle Howard. Biological Structure Dept., U. of Washington, Seattle, WA 98195, USA. (1915) PhD chemistry (U. of Washington, 1943). Prof. emeritus. (tel. 206 + 543-1983). *Structure and function, biological molecules, accurate molecular parameters.*

Jensen, Prof. William Phelps. Chemistry Dept., South Dakota State U., Brookings, SD 57006, USA. (1937) PhD inorg. chemistry (U. of Iowa, 1964). Prof. (tel. 605 + 688-5151). *Coordination compounds, small organic molecules.*

Jesser, Prof. William Augustus. Materials Sci. Dept., Thornton Hall, U. of Virginia, Charlottesville, VA 22901, USA. (1939) PhD physics (U. of Virginia, 1966). Prof. (tel. 804 + 924-6349, fax 804 + 924-7553). *Thin films, electron microscopy, epitaxy, electronic materials.*

Ji, Mr. Xinhua. Chemistry Dept., U. of Oklahoma, 620 Parrington Oval, Rm. 208, Norman, OK 73019, USA. (1948) BS (Inner Mongolia Teachers C., PRC 1982). Grad. student. (tel. 405 + 325-7612). *Single crystal X-ray diffraction, structure determination, natural products, anticancer agents.*

Jiang, Mr. Fan. Chemical Biodynamics Lab., U. of Calif. at Berkeley, 1 Cyclotron Rd., Bldg. 3, Berkeley, CA 94720, USA. (1962) BA physics (Peking U., PRC, 1984). Grad. student. (tel. 415 + 486-5449, email Bitnet F-JIANG at LBL). *X-ray crystallography, proteins, structure analysis, prediction, folding, modeling, design, statistical mechanics.*

Jircitano, Dr. Alan John. Chemistry Dept., Pennsylvania State U. - Behrend C., Station Rd., Erie, PA 16563, USA. (1955) PhD inorg. chemistry (U. of Kansas, 1982). Asst. prof. (tel. 814 + 898-6400, email A0J at PSUVM). *Bioinorganic chemistry, transition metal complexes, macrocyclic ligands.*

Joesten, Prof. Raymond. Geology & Geophys. Dept., U. of Connecticut, 345 Mansfield Rd., Storrs, CT 06268-2045, USA. (1944) PhD geological sci. (CIT, 1974). Prof. (tel. 203 + 486-4434). *Electron microscopy, transmission, crystal chemistry, silicates, oxides.*

Johnson, Prof. Alan Arthur. Room 311, Ernst Hall, U. of Louisville, Louisville, KY 40292, USA. (1930) PhD,Physics (London U.,UK, 1960) Prof. (tel. 502+588 ext. 6338) *Metals, microscopes, phase transitions, teaching.*

Johnson, Dr. Carroll K. 344 East Dr., Oak Ridge, TN 37830, USA. (1929) PhD biophysics (MIT, 1959). (tel. 615 + 574-4975). *Crystallographic computing, computer graphics, thermal motion, modulated structures, neutron diffraction, artificial intelligence.*

Johnson, Dr. Frank Bacchus. Chemical Pathology Dept., Armed Forces Inst. of Pathology, Washington, DC 20306, USA. (1919) MD medicine (Howard U., 1944). Chairman. (tel. 202 + 829-2211). *Inorganic crystal structures, microstructure.*

Johnson, Prof. Gerald Glenn Jr. Computer Sci. Dept., 164 MRL, Pennsylvania State U., University Park, PA 16801, USA. (1939) PhD solid state sci. (PSU, 1965). Assoc. prof. (tel. 814 + 865-1637, telex 842510, fax 814 + 865-2326, email Bitnet JOHNSON at PSUCES). *Powder diffraction, search and match techniques, indexing, laboratory automation, computer graphics, pattern recognition, VAX/VMS software development.*

Johnson, Prof. John Emil. Biological Sci. Dept., Lilly Hall of Life Sci., Purdue U., W. Lafayette, IN 47907, USA. (1945) PhD phys. chemistry (Iowa State U., 1972). Prof. (tel. 317 + 494-5911). *Virus structure, macromolecular assembly, protein crystallography, biophysical chemistry.*

Johnson, Prof. Michael E. Med. Chem. & Pharmacognosy Dept., U. of Illinois at Chicago, P.O. Box 6998, Chicago, IL 60680, USA. (1945) PhD physics, biophysics (Northwestern U., 1973). Prof., Assoc. Dean. (tel. 312 + 996-0796, email Bitnet U40190 at UICVM). *Molecular modeling, graphics, computer aided molecular design, NMR spectroscopy, protein structure.*

Johnson, Dr. Paul Lorentz. Internatl. Energy Dev. Progs., Nat. Energy Software Ctr., Argonne Nat. Lab., 9700 S. Cass Av., Argonne, IL 60439, USA. (1941) PhD phys. chemistry (Washington State U., 1968). Computer scient. (tel. 312 + 972-4043, email Bitnet B24559 at ANLNESC). *XRD, neutron diffraction, scientific computer applications, computer program exchange.*

Johnson, Dr. Quintin C. Chemistry & Matls. Sci. Dept., Lawrence Livermore Lab., L-370, Livermore, CA 94550, USA. (1935) PhD chemistry (UCB, 1961). (tel. 415 + 422-6390). *Powder pattern analysis, materials characterization, micro-organized materials.*

Johnson, Mrs. Ruth Jeannette Beach. Rochelle Crystal Corp., 2004 Randolph Ave. at the C. of St. Catherine, St. Paul, MN 55105, USA. (1950) BS chemistry, math., English (U. of Montevallo, Alabama, 1972 Corporate scient. (tel. 612 + 698-1161 or 698-6125). *Crystallography, crystal growth, solid state organic chemistry.*

Jones, Prof. Daniel Silas. Chemistry Dept., U. of North Carolina at Charlotte, UNCC Station, Charlotte, NC 28223, USA. (1943) PhD phys. chemistry (Harvard U., 1971). Assoc. prof. (tel. 704 + 547-4438). *Transition metal complexes.*

Jones, Mr. Glover A. Central Res. & Dev. Dept., Expt. Sta., 228/320-d(D), E.I. duPont de Nemours Co., Wilmington, DE 19898, USA. (1938) MS phys. chemistry (Tuskegee U., 1966). Physical chemist. (tel. 302 + 695-3935). *Materials science, catalysts, structure - property relationships, instrumentation, high/low temperature diffraction studies.*

Jones, Dr. Noel Duane. Physical Chemistry Res. Div., Lilly Res.saves., Lilly Corporate Center, Indianapolis, IN 46285, USA. (1937) PhD chemistry (CIT, 1964). Sr. res. scient. (tel. 317 + 276-4668, fax 317 + 276-5431). *Automation, protein crystallization, crystal structures, macromolecules, small molecules, synchrotron radiation.*

Jordan, Dr. Steven R. Agouron Pharmaceuticals, 11025 N. Torrey Pines Rd., P.O. Box 12209, La Jolla, CA 92037, USA. (1952) PhD chemistry (U. of Arizona,

1983). Crystallographer. (tel. 619 + 535-5500). *Protein - DNA interactions, macromolecular interactions, drug design.*

Jordan, Prof. Truman H. Chemistry Dept., Cornell C., Mount Vernon, IA 52314, USA. (1937) PhD chemistry (Harvard U., 1964). Prof. (tel. 319 + 895-8811). *Tin (II) chemistry (phosphates), titanium phosphates.*

Jorgensen, Dr. James D. Matls. Sci. Div., Bldg. 223, Argonne Nat. Lab., 9700 S. Cass Ave., Argonne, IL 60439, USA. (1948) PhD solid state physics (Brigham Young U., 1975). Physicist, Group Leader, neutron & X-ray scattering. (tel. 312 + 972-3308, fax 312 + 972-3308, email JORGENSEN at ANLPNS). *Superconducting materials, fast ion conductors, high pressure phase transitions, neutron diffraction.*

Julian, Prof. Maureen O'Donnell. Geological Sci. Dept., Virginia Polytech. Inst. & State U., Blacksburg, VA 24061, USA. PhD phys. chemistry (Cornell U., 1966). Adjunct prof. (tel. 703 + 231-9191 (h.703/961-6521, email IRELAND at VTVM1). *Silicon nitrides, ab initio calculations including heat capacity, diamond elastic constants.*

Jurnak, Prof. Frances Anne. Biochemistry Dept., U. of California at Riverside, Riverside, CA 92521, USA. (1946) PhD chemistry (UCB, 1973). Assoc. prof. (tel. 714 + 787-4245). *Macromolecular crystallography, protein elongation factors, complexes, antibiotics, proteins.*

Jaidong, Dr. Ko. Mineral Phys. Inst., ESS Dept., SUNY at Stony Brook, Stony Brook, NY 11794-2100, USA. (1955) PhD geochemistry (SUNY at Stony Brook, 1988). Postdoct. res. assoc. (tel. 516 + 632-8058). *Crystallography, high pressure, temperature.*

Kaduk, Dr. James Albert. Amoco Res. Ctr., Amoco Corp., P.O. Box 400, Naperville, IL 60566, USA. (1952) PhD inorg. chemistry (Northwestern U., 1977). Sr. res. chemist. (tel. 312 + 420-4547). *Catalysts, minerals, zeolites.*

Kaler, Prof. Eric W. Chem. Eng. Dept. BF-10, U. of Washington, Seattle, WA 98195, USA. Prof. (tel. 206 + 545-2324).

Kalnik, Dr. Matthew Walter. Mol. Biol. Dept., MB-2, Res. Inst. of Scripps Clinic, 10666 N. Torrey Pines Rd., La Jolla, CA 92037, USA. (1962) PhD biochem., mol. biophys. (Columbia U., 1989). Postdoct. fel. (tel. 619 + 554-2801, email kalnik at scripps.edu). *NMR, two-dimensional, proteins, DNA.*

Kamer, Mr. Greg. Lilly Hall, Purdue U., West Lafayette, IN 47907, USA.

Kampf, Mr. Jeff W. Chemistry Dept., Wayne State U., Detroit, MI 48202, USA. (tel. 313 + 577-3076).

Kanatzidis, Dr. Mercouri G. Chemistry Dept., Michigan State U., East Lansing, MI 48824-1322, USA. (1957) PhD (U. of Iowa, 1984). Asst. prof. *Bioinorganic chemistry, Fe-S proteins, sulfur chemistry, low dimensional materials.*

Kantardjieff, Prof. Katharine Ann. Chemistry and Biochem. Dept., Calif. State U., 800 State College Blvd., Fullerton, CA 92634, USA. (1957) PhD phys. chemistry (UCLA, 1988). Asst. prof. (tel. 714 + 773-3621, email KATHYK at UCLAVE). *Macromolecules, proteins, toxins, protein translocation.*

Kapecki, Dr. Jon Alfred. Res. Labs., Eastman Kodak Co., Rochester, NY 14650, USA. (1942) PhD phys. org. chem., X-ray cryst. (U. of Illinois, 1969). Head, photo. res. lab. (tel. 716 + 477-5629). *Cycloaddition processes, molecular orbital theory, structure - reactivity relationships, sterically crowded molecules, reaction mechanisms, computer modeling, applications.*

Kaplan, Mr. David B. Pittsburgh NMR Inst., 3260 Fifth Ave., Pittsburgh, PA 15213, USA. (tel. 412 + 647-6674).

Karcher, Dr. Barbara Ann. Computer Sci. Dept., U. of Wisconsin at Oshkosh, Oshkosh, WI 54901, USA. (1953) PhD phys. chemistry (Iowa State U., 1981). Prof. (tel. 414 + 424-1260). *Crystallographic computing.*

Karipides, Prof. Anastas. Chemistry Dept., Miami U., Oxford, OH 45056, USA. (1937) PhD chemistry (U. of Illinois, 1964). Prof. (tel. 513 + 529-7274). *Chemistry, structural, coordination, fluoro-organic compounds, crystal energetics, non-bonded interactions.*

Karle, Dr. Isabella L. Code 6030, Naval Res. Lab., Washington, DC 20375, USA. (1921) PhD phys. chemistry (U. of Michigan, 1944). Head, X-ray crystallog. (tel. 202 + 767-2624). *Structure analysis methods, biologically interesting molecules, polypeptides, ionophores, photo rearrangement products.*

Karle, Dr. Jean Marianne. Pharmacology Dept., Walter Reed Army Inst. of Res., Washington, DC 20307-5100, USA. (1950) PhD chemistry (Duke U., 1976). Res. chemist. (tel. 301 + 427-5177, fax 301-427-6569). *Conformational analysis, antimalarials, organic molecules.*

Karle, Prof. Jerome. Lab. for the Structure of Matter, Code 6030, Naval Res. Lab., Washington, DC 20375, USA. (1918) PhD phys. chemistry (U. of Michigan, 1944). Chief scient. (tel. 202 + 767-2665, Telex. TWX7108220147, fax 202-767-6874, email KARLE2 at NRL). *Structure analysis methods, diffraction applications, X-ray, electron, neutron.*

Kasper, Dr. John S. 18 Cumberland Pl., Scotia, NY 12302, USA. (1915) PhD chemistry (Johns Hopkins U., 1941). Physical chemist. (tel. 518 + 399-3213). *Crystal structures, inorganic compounds, metals and alloys, high pressure phases, magnetic structures.*

Kastner, Dr. Margaret Ellen. Chemistry Dept., Bucknell U., Lewisburg, PA 17837, USA. (1950) PhD inorg. chemistry (U. of Notre Dame, 1979). Asst. prof. (tel. 717 + 524-3258, email KASTNER at BKNLVMS). *Small molecules, inorganic, transition metal complexes, lathanide complexes.*

Katz, Mr. Henry. Mol. Structure Dept., Inst. for Cancer Res., 7701 Burholme Ave., Philadelphia, PA 19111, USA. (1927) MS chemistry (U. of Pennsylvania,

1955). Res. specialist. (tel. 215 + 728-2220, email katzh at rm.fccc.edu). *Small molecule structures, instrumentation.*

Katz, Prof. J. Lawrence. Biomedical Eng. Dept., Rensselaer Polytechnic Inst., Troy, NY 12181, USA. (1927) PhD physics (PUN, 1957). Prof., biophys. & biomed. eng. (tel. 518 + 270-6547, fax 518-276-6003). *Biomechanical properties, structure, calcified and connective tissues, ultrasonic studies, bone and teeth, scanning microscopy, electron, acoustic, biomedical materials, strain, biological inorganic crystals, X-ray diffraction.*

Katz, Prof. Lewis. Chemistry Dept., U. of Connecticut, Storrs, CT 06269, USA. (1923) PhD chemistry (U. of Minnesota, 1951). Prof. emeritus. (tel. 203 + 429-4122). *Inorganic crystal structures, complex metal oxides, molecular crystals.*

Katz, Dr. Robert. Metabolic Diseases Res. Prog., NIH, Westwood Bldg., 5333 Westbard Ave., Bethesda, MD 20892, USA.

Kaufman, Prof. Hershall William. Oral Biology Dept., Sch. of Dental Medicine, SUNY at Stony Brook, Stony Brook, NY 11794-8702, USA. (1940) PhD oral biology (U. of Manitoba, Canada, 1967). Prof. (tel. 516 + 632-8925). *Calcium phosphate crystallography, bone, tooth crystal structure.*

Kay, Dr. Mortimer I. 70 Oak Shade Rd., Gaithersburg, MD 20878, USA. (1930) PhD phys. chemistry (U. of Connecticut, 1958). (tel. 301 + 353-4861). *Ferroelectrics, time resolved effects, materials, energy conversion, ocean thermal energy.*

Keder, Dr. Nancy Lynn. Chemistry Dept., U. of Calif. at Santa Barbara, Santa Barbara, CA 93106, USA. (1955) PhD chemistry (UCLA, 1984). Staff crystallog. (tel. 805 + 961-2399, email Bitnet NLKEDER at SBITP). *Crystallographic computing.*

Keefe, Ms. Lisa J. Molec. Biol. & Genetics Dept., Johns Hopkins Sch. of Med., 725 N. Wolfe St., Baltimore, MD 21205, USA. (tel. 301 + 955-2731, email KEEFE at JHUIGF). *Molecular structure and function, protein - DNA interactions.*

Keefe, Prof. William Edward. Biostatistics Dept., Med. C. of Virginia, P.O. Box 678, Med. C. Station, Richmond, VA 23220, USA. (1923) PhD biophysics (Med. C. of Virginia, 1967). Assoc. prof. (tel. 804 + 786-9824). *Structure, biological molecules, image analysis techniques.*

Keem, Dr. John Edward. Ovionic Synthetic Materials Co. Inc., 1788 Northwood, Troy, MI 48084, USA. (1948) PhD solid state physics (Purdue U., 1977). Dir., R & D. (tel. 313 + 362-1290, ext. 3516, fax 313-362-4043, email keem at ssrl750). *Rietveld powder profile analysis, X-ray, neutron, multilayered structures.*

Keller, Dr. Ludwig. CAMET Res. Inc., 14809 Calvert St., Van Nuys, CA 91411, USA. (1949) Dr.rer.nat. mineralogy (Tech. U. of München, FRG, 1978). Pres. (tel. 818 + 376-0341). *X-ray diffractometry, powder, double crystal.*

Kelly, Mrs. Carol J. Central Lab. Services, Ford Motor Co., 15000 Century Dr., Dearborn, MI 48210, USA. MS management (U. of Michigan, 1982). Manager. (tel. 313 + 322-1676, fax 313-322-1614). *X-ray diffraction.*

Kelly, Dr. Judith Ann. Molec. and Cell Biol. Dept., U. of Connecticut, 75 N. Eagleville Rd., Storrs, CT 06269, USA. (1944) PhD biophysics (U. of Connecticut, 1977). Assoc. prof. (tel. 203 + 486-4622 or 486-4353, Telex. 994484/UCONNCOOP, email Bitnet KELLY at UCONNVM). *Enzymes, structure, function, drug/protein interactions, interactive computer graphics.*

Kemp, Dr. Nantelle Smith Pantaleo. Amoco Production Co., P.O. Box 3385, Tulsa, OK 74102, USA. (1947) PhD phys. chemistry (Emory U., 1971). Staff res. scient. (tel. 918 + 660-3147). *Small molecules, clay chemistry, geochemistry, thermodynamics of electrolyte solutions.*

Keszler, Prof. Douglas A. Chemistry Dept., Oregon State U., Corvallis, OR 97331, USA. (1957) PhD inorg. chemistry (Northwestern U., 1984). Asst. prof. (tel. 503 + 754-2081). *Solid state chemistry.*

Khachaturyan, Mr. Armen G. Lawrence Berkeley Lab., U. of California HMB, Berkeley, CA 94720, USA.

Khalil, Miss Safia. Chemistry Dept., U. of Oklahoma, 620 Parrington Oval, Norman, OK 73019, USA. (1963) MS chemistry (U. of Karachi, India, 1985). Grad. student. (tel. 405 + 325-7612). *Protein crystallography, structure determination, biological compounds.*

Khan, Mr. A. Chemical Eng. Dept., U. of Florida, Gainesville, FL 32611, USA.

Khan, Dr. Masood. Chemistry Dept., U. of Oklahoma, Norman, OK 73019, USA. (1947) PhD chemistry (U. of Victoria, Canada, 1976). Asst. prof., staff crystallographer. (tel. 405 + 325-4542).

Kiefer, Dr. Charles R. Cell & Molec. Biol. Dept., Med. C. of Georgia, Augusta, GA 30912, USA. (1947) PhD microbiology (Med. C. of Georgia, 1981). Asst. res. scient. (tel. 404 + 721-3757). *Acquired immunodeficiency syndrome, antibodies, monoclonal, binding sites, antibody, hemoglobin, immunoassay, immunochemistry, immunology.*

Kim, Ms. Eunice E. Molec. Biophys. & Biochem. Dept., Yale U., P.O. Box 6666, New Haven, CT 06510, USA.

Kim, Ms. Heasook. Biophysics Dept., Roswell Park Memorial Inst., 666 Elm St., Buffalo, NY 14263, USA. (1956) MS nuclear phys., crystallog. (1980). Grad. student. (tel. 716 + 845-8298). *Crystallography, ESR, ENDOR, DNA, radiation biophysics.*

Kim, Prof. Jung Ja P. Biochemistry Dept., Med. C. of Wisconsin, 8701 Watertown Plank Rd., Milwaukee, WI 53226, USA. (1941) PhD chemistry (Cornell U., 1969). Assoc. prof. (tel. 414 + 257-8479, email JJKIM at MEDCOLWI). *Protein structure, protein-nucleic acid interaction.*

Kim, Mr. Kyung Hyun. Biochem. & Biophys. Dept., Iowa State U., Ames, IA 50010, USA. (1957) biophysics. Grad. student. (tel. 515 + 294-0567). *Macromolecular structure.*

Kim, Dr. Nancy Ellen Kime. Div. of Environ. Health Assessment, New York State Dept. of Health, 2 University Pl., Albany, NY 12203-3313, USA. (1942) PhD inorg. chemistry (Northwestern U., 1969). Dir. (tel. 518 + 458-6438, Telex. 518-458-6434). *Bioinorganic crystallography, structure - activity correlations, toxicology.*

Kim, Prof. Peter S. Whitehead Inst., Mass. Inst. of Techn., Nine Cambridge Ctr., Cambridge, MA 02142, USA. (1958) PhD biochemistry (Stanford U., 1985). Asst. prof., biology. (tel. 617 + 258-5184, fax 617-258-5061). *Protein folding, peptides, NMR, coiled coils.*

Kim, Mr. Sukyoung. Chem. and Matls. Eng. Dept., U. of Vermont, Votey Bldg., Rm. 213, Burlington, VT 05404, USA. (1954) MS materials sci. (Alfred U., 1985). Grad. student (tel. 802 + 656-8399). *X-ray diffraction.*

Kim, Prof. Sung-Hou. Chemistry Dept., U. of Calif. at Berkeley, Berkeley, CA 94720, USA. (1937) PhD chemistry (U. of Pittsburgh, 1966). Prof. (tel. 415 + 642-8270, fax 415-642-8368). *Structure - function relationships, biological molecules.*

Kimball, Dr. Martha R. 200 Cabrini Blvd., Apt. 3, New York, NY 10033, USA. (1945) PhD crystallography (U. of Cambridge, UK, 1974). (tel. 212 + 795-7813). *Crystallography, biological macromolecules.*

King, Dr. Hubert Ellis. Exxon Res. and Eng. Co., Route 22 East, Annandale, NJ 08801, USA. (1949) PhD earth and space sci. (SUNY at Stony Brook, 1979). Sr. physicist. (tel. 201 + 730-2888, email heking at erenj). *Pressure and temperature effects upon materials, lattice dynamics, mineralogy.*

King, Dr. Murray Vernon. Wadsworth Ctr. Labs. & Res., New York State Dept. of Health, Albany, NY 12201-0509, USA. (1922) PhD phys. chemistry (U. of Minnesota, 1949). Res. scient. III (tel. 518 + 486-4971). *Structural molecular biology, contractile systems, electron microscopy, diffraction, electron, X-ray, design, materials, devices.*

Kingma, Ms. Kathleen J. Geology & Mineralogy Dept., Ohio State U., 125 S. Oval Mall, Columbus, OH 43210, USA. (1962) MS geology (Ohio State U., 1989). Grad. res. asst. (tel. 614 + 263-0464). *Mineralogy, crystallography.*

Kingman, Dr. Priscilla Ward. Warhead Mechanics Br., USABRL, APG, 1115 High Country Rd., Towson, MD 21204, USA. (1934) PhD mechanics, metallurgy (Johns Hopkins U., 1982). Metallurgist. (tel. 301 + 825-0321, email patk at brl.mil). *Imperfections, twinning, diffraction physics.*

Kirchhoff, Ms. Pamela Moore. FP&PR, 1710 Bldg. Dow Chemical Co., Midland, MI 48640, USA. (tel. 517 + 636-5090).

Kirchner, Prof. Richard Martin. Chemistry Dept., Manhatten C., 4513 Manhatten College Pky., Riverdale, NY 10471, USA. (1941) PhD chemistry (U. of Washington, 1971). Prof. (tel. 212 + 920-0206). *Crystallography, inorganic chemistry, characterization, microcrystalline zeolite-like materials, coordination compounds.*

Kirn, Mr. J. F. 4324 Southampton Rd., Richmond, VA 23235, USA. (tel. 804 + 272-5784).

Kirschner, Dr. Daniel A. Neurology Res., Children's Hospital Corp., 300 Longwood Ave., Boston, MA 02115, USA. (1944) PhD Biophysics (Harvard U., 1972). Assoc. Prof. & Sr. Res. Assoc. (tel. 617 + 735-6102,3). *Membrane structure, neuropathology, macromolecular assemblies, electron microscopy, diffraction, X-ray, neutron.*

Kirz, Prof. Janos. Physics Dept., SUNY at Stony Brook, Stony Brook, NY 11794, USA. (1937) PhD physics (UCB, 1963). Prof. (tel. 516 + 632-8106, email Bitnet KIRZ at SUNYSBNP). *X-ray optics, microscopy, high energy physics.*

Kiss, Dr. Klara. Matls. & Struct. Analysis, AKZO Chemicals Inc., Livingstone Ave., Dobbs Ferry, NY 10522, USA. (1930) PhD analyt. chemistry (U. of Budapest for Sci. and Tech., 1982). Sr. res. scient. (tel. 914 + 693-1200, ext. 2096). *Materials characterization, ceramics, plastics, microbeam analysis, X-ray crystallography, physical testing, instrumental analysis, materials science, electron diffraction.*

Kissinger, Dr. Charles R. Molec. Biol. & Genetics Dept., Johns Hopkins U., Sch. of Med., 725 N. Wolfe St., Baltimore, MD 21205, USA. (1956) PhD biological structure (U. of Washington, 1989). Res. assoc.

Kissinger, Mr. Homer Everett. 1733 Horn Ave., (also: Battelle Pacific Northwest Labs., P.O. Box 999), Richland, WA 99352, USA. (1923) MS physics (Kansas State U., 1950). Retired, consultant. (tel. 509 + 946-8236). *X-ray diffraction, general.*

Kistenmacher, Dr. Thomas John. MS Eisenhower Res. Ctr. 2-115, Appl. Phys. Lab., Johns Hopkins U., Johns Hopkins Rd., Laurel, MD 20707, USA. (1943) PhD chemistry (U. of Illinois, 1970). Chemist, prin. prof. staff. (tel. 301 + 953-6215). *Synthetic organic metals, amorphous materials, phase transition phenomena, structure - property relationships, semiconductor superlattices.*

Klanderman, Prof. Kent Arlen. Chemistry Dept., SUNY at Cortland, Cortland, NY 13045, USA. (1936) PhD phys. chemistry (U. of Wisconsin, 1965). Assoc. prof. (tel. 607 + 753-2908). *Bonding and structure, transition metal complexes, organometallic compounds.*

Klein, Prof. Cheryl Lynn. Chemistry Dept., Xavier U., 7325 Palmetto St., New Orleans, LA 70125, USA. (1956) PhD phys. chemistry (U. of New Orleans,

1982). Assoc. prof. (tel. 504 + 486-7411, ext. 7377). *Crystallography, drug molecules, neuroleptics, analgesics, charge density, magnetism.*

Klinger, Ms. Alexandra. 814 Hinton Ave., Charlottesville, VA 22901, USA. (tel. 804 + 924-7039).

Klooster, Dr. W. Crystallography Dept., U. of Pittsburgh, Pittsburgh, PA 15260, USA. (tel. 412 + 624-9300).

Klug, Prof. Harold Philip. Bethany Homes, 201 S. University Dr., Fargo, ND 58103, USA. (1902) PhD phys. chemistry (Ohio State U., 1928). Retired. (tel. 701 + 239-3068). *X-ray crystallography, organic, inorganic.*

Knighton, Mr. Daniel R. Chemistry Dept. D-006, U. of Calif. at San Diego, La Jolla, CA 92093], USA. (1962) BS chemistry (U. of the Pacific, 1984). Res. asst. (tel. 619 + 534-4241, email Bitnet DKNIGHTO at UCSD). *Macromolecular crystallography, protein structure/function, protein kinases.*

Knobler, Dr. Carolyn B. Chemistry Dept., U. of Calif. at Los Angeles, 405 Hilgard Ave., Los Angeles, CA 90024, USA. (1934) PhD chemistry (PSU, 1959). Assoc. res. chemist. (tel. 213 + 474-4708, email Bitnet KNOBLER at UCLACH). *Crystal structures.*

Knorr, Dr. David B. Matls. Eng. Dept., Rensselear Polytechnic Inst., Troy, NY 12180, USA. (1952) ScD metallurgy (MIT, 1981). Asst. prof. (tel. 518 + 276-2890). *Metallurgy, ceramics, texture.*

Knox, Prof. James Russell. Mol. and Cell Biology Dept., U. of Connecticut, U-125, Storrs, CT 062698, USA. (1941) PhD phys. chemistry (Boston U., 1967). Prof. (tel. 203 + 486-3133 or 4622, fax 203-486-5381, email Bitnet KNOX at UCONNVM). *Protein crystallography, penicillin-binding enzymes, low angle scattering.*

Knox, Dr. Ralph David. Microelectronic Res. Ctr., Iowa State U., 1925 Scholl Rd., Ames, IA 50011, USA. (1961) PhD solid state physics (Iowa State U., 1988). Assoc. Scient. (tel. 515 + 294-7732, fax 515-294-9584).

Koch, Prof. Stephen A. Chemistry Dept., SUNY at Stony Brook, Stony Brook, NY 11794, USA. (1948) PhD inorg. chemistry (MIT, 1975). Assoc. prof. (tel. 516 + 632-7944, fax 516-632-7960, email Bitnet SKOCK at SBCCMAIL). *Chemistry, solid-state, inorganic, bioinorganic.*

Koeppe, Dr. Roger E. II. Chemistry Dept., U. of Arkansas, Fayetteville, AR 72701, USA. (1949) PhD chemistry and biochemistry (CIT, 1976). Prof. (tel. 501 + 575-4601, email R642 at UAFMUSA). *Protein structure and function.*

Koetzle, Dr. Thomas Frederick. Chemistry Dept., Bldg. 555, Brookhaven Nat. Lab., Upton, NY 11973, USA. (1943) PhD chemistry (Harvard U., 1970). Chemist. (tel. 516 + 282-4384, Telex. 6852516 BNL DOE, fax 516-282-3000, email Bitnet KOETZLE at BNLCHM). *Neutron diffraction, organometallic compounds, metal hydrides, synchrotron radiation studies.*

Kohn, Prof. Jack Arnold. 65 Wigwam Rd., Locust, NJ 07760, USA. (1925) PhD mineralogy (U. of Michigan, 1950). Adjunct prof., electronic materials. (tel. 201 + 872-2295). *Crystallography, electronic materials, twinning, polytypism, polymorphism.*

Koknat, Prof. Friedrich Wilhelm. Chemistry Dept., Youngstown State U., 410 Wick Ave., Youngstown, OH 44555, USA. (1938) Dr.rer.nat. chemistry (U. of Giessen, BRD, 1965). Prof. (tel. 216 + 742-3668). *Inorganic crystal structures.*

Kokotailo, Dr. George Thomas. 98 North American St., Woodbury, NJ 08096, USA. (1919) PhD physics (Temple U., 1955). Consultant. (tel. 609 + 845-6508). *Crystal structures, catalyst characterization, crystal growth, zeolite synthesis.*

Kolatkar, Mr. Anand Ratnakar. Biochemistry Dept., Rice U., Houston, TX 772521, USA. (1965) BA chemistry, biology (Augustana C., 1987). Grad. student. (tel. 713 + 527-8101, ext. 3346, email anand at crysvax.rice.edu). *X-ray crystallography, diffuse scattering, proteins, protein dynamics.*

Kolatkar, Mr. Prasanna R. Chemistry Dept., U. of Texas, Austin, TX 78712, USA. (1963) BA chemistry, biology (Augustana C., 1985). Res. asst. (tel. 512 + 471-3625, email cmdj354 at chpc.brc.utexas.edu).

Kolks, Dr. Gary. Chemistry Dept., Manhatten C., Riverdale, NY 10471, USA. (tel. 212 + 920-0322).

Kolpak, Dr. Francis John. Materials Sci. Div., Hercules Inc., Res. Center, Lancaster Pike & Hercules Rd., Wilmington, DE 19894, USA. (1950) PhD macromolecular sci. (Case Western Reserve U., 1977). Sr. res. chemist. (tel. 302 + 995-3623). *Structure/property relationships, polymers, macromolecular crystallography.*

Komiya, Dr. Hiromi. Mol. Biology Inst., U. of Calif. at Los Angeles, 405 Hilgard Ave., Los Angeles, CA 90024-1570, USA. (1956) PhD biochemistry (1987). Postdoct. fel. (tel. 213 + 825-8901, email komiya%uclasp.span at star.stanford.edu). *Crystallography, structure, biochemical function, proteins, biomacromolecules.*

Kong, Dr. Eric Siu Wai. Matls. Sci. & Techn. Inst., 936 Bluebonnet Dr., Sunnyvale, CA 94086, USA. (1953) PhD polymer chemistry (Rensselaer Polytechnic Inst., 1978). Res. dir. (tel. 408 + 739-8065). *Polymers, materials science, composite materials.*

Konnert, Dr. John H. Lab. for the Struct. of Matter, Naval Res. Lab., Code 6030, 4555 Overlook Ave., Washington, DC 20375, USA. (1941) PhD phys. chemistry (U. of Minnesota, 1967). Res. chemist. (tel. 202 + 767-2735). *Glassy materials, macromolecules.*

Konnert, Mrs. Judith A. U.S. Geological Survey, National Center, Stop 959, Sunrise Valley Dr., Reston, VA 22092, USA. (1941) BA chemistry (C. of Wooster, 1963). Chemist. (tel. 703 + 648-6763).

Kopelman, Prof. Raoul. Chemistry Dept., U. of Michigan, Ann Arbor, MI 48109-1055, USA. (1933) PhD chemistry (Columbia U., 1960). Prof. (tel. 313 + 764-1541, fax 313-747-4865, email raoul_kopelman at ub.cc.umich.edu). *Molecular crystals, excitons, phonons, internal rotation.*

Kopka, Mrs. Mary Lou. 259A Mol. Biol. Inst., U. of Calif. at Los Angeles, 405 Hilgard, Los Angeles, CA 91024, USA. (1938) BS chemistry. Res. assoc. (tel. 213 + 206-8278). *DNA, proteins, cancer drugs.*

Korp, Prof. James Douglas. Chemistry Dept., U. of Houston, University Park, Houston, TX 77204-5641, USA. (1950) PhD analytical chemistry (U. of Texas at Austin, 1975). Asst. prof., dept. crystallographer (tel. 713 + 749-2108). *Biologically active small molecules, charge transfer compounds, porphyrins, organometallics.*

Kosel, Mr. George Eugene. American Gas & Chemical Co. Ltd., 220 Pegasus Ave., Northvale, NJ 07656, USA. (1923) MS pharmacology (U. of Rochester, 1951). Chief chemist. (tel. 201 + 767-7300). *Electrophotography, inorganic crystallography.*

Kostiner, Prof. Edward Stephen. Chemistry Dept. U-60, U. of Connecticut, Storrs, CT 06269-3060, USA. (1940) PhD chemistry (PUN, 1960). Prof. and Dept. Head. (tel. 203 + 486-3214). *Solid state inorganic chemistry, crystal chemistry, growth.*

Koszelak, Mr. Stanley N. Lab. of Protein Studies, Oklahoma Med. Res. Fndn., 825 Northeast 13th St., Oklahoma City, OK 73104, USA. (1953) BS microbiology (U. of Oklahoma, 1976). Grad. student. (tel. 405 + 235-8331). *Macromolecular structure and function.*

Kounts, Dr. Dennis James. Centr. Res. & Dev. Dept., Supercond. Prod. Concept Ctr., E. I. du Pont de Nemours Co., Box 80304, Wilmington, DE 19880-0304, USA. (1956) PhD phys. chemistry (Ohio State U., 1984). Sr. chemist. (tel. 302 + 695-4256). *High temperature superconductivity, ceramics.*

Kraatz, Dr. Paul. Northrup Res. & Techn. Ctr., One Research Park, Palos Verdes Peninsula, CA 90274, USA. (1940) PhD geology, mineralogy (U. of Minnesota, 1972). Res. techn. staff member (tel. 213 + 377-4271). *Crystal growth, thin films, structure - properties - growth relationship, physical properties.*

Krasner, Prof. Saul. Science Dept., U.S. Coast Guard Academy, New London, CT 06320-4195, USA. (1929) PhD physics (PUN, 1970). Prof. (tel. 203 + 444-8638). *Solid state physics, education.*

Krause, Dr. Jeannette Alice. Chemistry Dept., The Ohio State U., 140 W. 18th. Ave., Columbus, OH 43210, USA. (1960) PhD inorg. chem., crystallog. (Ohio State U., 1989). Asst. Deptl. Crystallog. (tel. 614 + 292-4388).

Krause, Dr. Stephen. Chemical Eng. Dept., Arizona State U., Tempe, AZ 85287, USA. (tel. 602 + 965-2050).

Kraut, Prof. Joseph. Chemistry Dept., U. of Calif. at San Diego, La Jolla, CA 92093, USA. (1926) PhD chemistry (CIT, 1954). Prof. (tel. 619 + 534-3366, fax 619-534-0058, email jkraut at ucsd.edu). *Biological macromolecules (structure - function - evolution).*

Krawitz, Prof. Aaron David. Mech. and Aerospace Eng. Dept., U. of Missouri, 1006 Engineering Bldg., Columbia, MO 65211, USA. (1943) PhD materials sci. (Northwestern U., 1972). Prof. (tel. 314 + 882-7671). *Neutron diffraction, engineering materials, diffraction stress measurements, composites, alloys, cemented carbide composites.*

Kretsinger, Prof. Robert Harvey. Biology Dept., U. of Virginia, Charlottesville, VA 22901, USA. (1937) PhD biophysics (MIT, 1964). Prof. (tel. 804 + 924-7039). *Protein structure, evolution, data measurement by multiwire proportional counters.*

Krimm, Prof. Samuel. Physics Dept., 729 Dennison 1090, U. of Michigan, Ann Arbor, MI 48109-1090, USA. (1925) PhD phys. chemistry (Princeton U., 1950). Prof., Dir., Protein Struct. & Design Prog. (tel. 313 + 764-1146). *Vibrational spectroscopy, polymers, proteins, macromolecules.*

Krueger, Ms. Joanna Katherine. Chemistry Dept., Princeton U., Princeton, NJ 08544, USA. (1963) MA chemistry (Princeton U., 1987). Res. asst. (tel. 609 + 987-2826). *Protein biochemistry, structure, function, chemotaxis.*

Kuchenmeister, Dr. Mark Edward. Chemistry Dept., Texas A&M U., College Station, TX 77843, USA. (1958) PhD phys. chemistry (Michigan State U., 1989). Postdoct. fel.

Kudoh, Dr. Yasuhiro. Geophysical Lab., Carnegie Inst. of Washington, 2801 Upton St., NW, Washington, DC 20008, USA. (1947) PhD mineralogy (U. of Tokyo, 1975). Res. assoc. (tel. 202 + 966-0334, fax 202-895-4225). *Mineral structures, high temperatures and pressures.*

Kullnig, Dr. Rudolph Karl. Box 424 McClellan Rd., RD 1, Nassau, NY 12123, USA. (1918) PhD chemistry (U. of Ottawa, Canada, 1958). (tel. 518 + 766-3827). *Structure determination, X-ray diffraction, spectroscopy, theoretical chemistry.*

Kumosinski, Mr. Thomas F. Eastern Regional Res. Ctr., USDA, 600 E. Mermaid Ln., Philadelphia, PA 19118, USA. (tel. 215 + 233-6475).

Kuo, Prof. Lawrence C. Chemistry Dept., Boston U., 590 Commonwealth Ave., Boston, MA 02129, USA. PhD biophysics (U. of Chicago, 1981). Asst. prof. *Crystallography, enzymology, enzyme regulation, metalloenzymes.*

Kuriyama, Dr. Masao. Center for Matls. Sci., Natl. Inst. Stnds. & Techn., Rm. A161, Bldg. 223, Gaithersburg, MD 20899, USA. (1931) DSc physics (U. of Tokyo, Japan, 1958). Physicist. (tel. 301 + 975-5974). *X-rays, dynamical diffraction, topography, inelastic scattering, synchrotron radiation, real time imaging.*

Kuriyan, Dr. John. Box 3, Rockefeller U., 1230 York Ave., New York, NY 10021, USA. (1960) PhD chemistry (MIT, 1986). Asst. prof. (tel. 212 + 570-8342). *Protein crystallography, molecular dynamics, X-ray refinement.*

Kvick, Dr. Ake H. Chemistry Dept., Bldg. 555, Brookhaven Nat. Lab., Upton, NY 11973, USA. (1942) PhD chemistry (U. of Uppsala, Sweden, 1974). Chemist. (tel. 516 + 282-4381). *Diffraction, X-ray, neutron, synchrotron radiation, molecular sieves, dielectric structures, hydrogen bonding.*

Kwong, Mr. Peter. Biochem. & Mol. Biophys. Dept., Columbia U., BB523, 630 W. 168th. St., New York, NY 10032, USA. (tel. 212 + 305-2219).

LaBean, Mr. Thomas Henry. Biochem. & Biophys. Dept., U. of Pennsylvania, 37th. & Hamilton Walk, Philadelphia, PA 19104, USA. (1963) BS biochemistry (Michigan State U., 1985). Grad. fel. (tel. 215 + 898-8731, fax 215-898-4217, email THOM at AMY.MED.UPENN.EDU). *Protein structure, folding, space.*

La Prade, Dr. Marie Douglas. 192 Lakeside Blvd., Hamilton, NJ 08610, USA. (1942) PhD chemistry (MIT, 1969). (tel. 201 + 932-3762). *Organometallic structures, crystallographic computing.*

Ladell, Dr. Joshua. Matls. Characteriz. Res. Group, Philips Labs., P.O. Box 198 Briarcliff Manor, NY 10510, USA. (1923) PhD physics (PUN, 1954). Sr. res. scient. (tel. 914 + 945-6332). *Computer control, X-ray instrumentation, experimental phase determination, crystal diffractometry.*

Laderman, Dr. Stephen Stromberg. Hewlett-Packard Laboratories, 3500 Deer Creek Rd., Palo Alto, CA 94304, USA. (1955) PhD materials sci. (Stanford U., 1983). Member tech. staff. (tel. 415 + 857-1501). *Crystalline semiconductors, superconductors, amorphous materials, diffraction methods, X-ray topography, synchrotron radiation.*

Ladner, Dr. Robert Charles. Protein Eng. Corp., 765 Concord Ave., Cambridge, MA 02138, USA. (1944) PhD theoretical chemistry (CIT, 1972). CEO, Sci. Dir. (tel. 617 + 868-0868, fax 617-868-0898). *Molecular biology.*

Lai, Mr. Hsin-Hsi. Chemistry Dept., U. of Wisconsin, Madison, WI 53706, USA. (1960) MS chemistry (U. of N. Carolina at Charlotte, 1988). Grad. student. (tel. 608 + 262-1483).

Lake, Prof. James A. Mol. Biology Inst., U. of Calif. at Los Angeles, Los Angeles, CA 90024, USA. (1941) PhD physics, molecular biology (U. of Wisconsin, 1967). Prof. (tel. 213 + 825-2546). *Biological structures, evolution of structure.*

Lalancette, Prof. Roger A. Chemistry Dept., Rutgers U., Olson Labs., 73 Warren St., Newark, NJ 07102, USA. (1939) PhD analytical chemistry (Fordham U., 1967). Assoc. prof. (tel. 201 + 648-5646). *Macromolecules, small peptides, proteins, hormones, small molecules.*

LaLonde, Ms. Judith. 2322 Spruce St., Philadelphia, PA 19103, USA. (tel. 215 + 728-3617).

Lam, Dr. Chiu Tin. Res. & Dev. Dept., AKZO Chemical Inc., 13000 Bay Park Rd., Pasadena, CA 77507, USA. (1945) PhD inorg. chemistry (Columbia U., 1977). Sr. scient. (tel. 713 + 474-2864, ext. 418, Telex. 02-3223, fax 713-474-7539). *Structures, inorganic materials, biological polymers.*

Lambert, Mr. Guy. Mol. Biology Dept. MB-13, Res. Inst. Scripps Clinic, 10666 N. Torrey Pines Rd., La Jolla, CA 92037, USA. (tel. 619 + 455-9100).

Lando, Prof. Jerome B. Macromolecular Sci. Dept., Case Western Reserve U., University Circle, Cleveland, OH 44106, USA. (1932) PhD phys. chemistry (PUN, 1963). Prof. (tel. 216 + 368-6366). *Solid state reactions, polymers, crystal structures, pyroelectric, piezoelectric properties, conformation transitions, solution behavior.*

Landy, Dr. Richard Allen. Res. Dept., North American Refractories Co., The Halle Bldg., 1228 Euclid Ave., Cleveland, OH 44115, USA. (1931) PhD mineralogy, petrology (PSU, 1961). Genl. mgr., corporate techn. (tel. 216 + 621-5200). *Refractory materials.*

Langridge, Prof. Robert. Computer Graphics Lab., U. of Calif. at San Francisco, 926 Med. Sci., San Francisco, CA 94143, USA. (1933) PhD crystallography (U. of London). Prof. (tel. 415 + 666-2630). *Molecular structure, computer graphics, drug design.*

Langs, Dr. David Alan. Mol. Biophysics Dept., Med. Fndn. of Buffalo, 73 High St., Buffalo, NY 14203, USA. (1941) PhD inorg. chemistry (SUNY, Buffalo, 1968). Assoc. res. scient. (tel. 716 + 856-9600). *Drug/hormone receptor binding, calcium channel drugs, prostaglandins, direct methods, molecular replacement methods.*

Larson, Dr. Allen C. LANSCLE, MS-H805, Los Alamos Nat. Lab., Los Alamos, NM 87545, USA. (1928) PhD chemistry (Washington U., 1956). Staff member. (tel. 505 + 667-2942, email larson at wnrvax.lanl.gov). *Structural chemistry, computational crystallography, materials science, computer applications.*

Larson, Dr. Bennett Charles. Solid State Div., Oak Ridge Nat. Lab., P.O. Box 2008, Oak Ridge, TN 37831, USA. (1941) PhD physics (U. of Missouri, 1970). Group leader. (tel. 615 + 574-5506, fax 615-574-4143). *X-ray scattering, diffuse, time resolved, X-ray topography, radiation damage in metals, defects.*

Larson, Dr. Elizabeth Margaret. Lawrence Livermore Nat. Lab., P.O. Box 808, L-356, Livermore, CA 94550, USA. (1951) PhD inorg. chemistry (Arizona State U., 1985). Res. fel. (tel. 415 + 423-0175). *X-ray diffraction, single crystal, powder, EXAFS, Rietveld methods, neutron scattering, vibrational spectroscopy, molecular mechanics calculations, solid state materials.*

Larson, Dr. Steven Bland. Analytical Instrumentation, ICN - Nucleic Acid Res. Inst., 3300 Hyland Ave., Costa Mesa, CA 92626, USA. (1949) PhD analytical chemistry (Brigham Young U., 1980). Dept. head. (tel. 714 + 641-7232).

Crystallography, small molecules, macrocyclic compounds, nucleoside and nucleotide analogs.

Lashewycz-Rubycz, Dr. Romana Alexandra. Chemistry Dept., Hobart and William Smith C., Geneva, NY 14456, USA. (1952) PhD inorg. chemistry (SUNY at Buffalo, 1979). Assoc. prof. (tel. 315 + 789-5500, ext. 555). *Chemistry, organo-transition metal, bioinorganic, structural and kinetic studies, X-ray diffraction, spectroscopic techniques.*

Laswick, Prof. Patty Hall. Chemistry Dept., Clarion U. of Pennsylvania, Greenville Ave., Clarion, PA 16214, USA. (1937) PhD phys. chemistry (U. of Michigan, 1969). Prof. (tel. 814 + 226-2567). *Spectroscopy, infrared, Raman, vibrational analysis, X-ray crystallography, boron-nitrogen complexes.*

Lattman, Prof. Eaton Edward. Biophysics Dept., John Hopkins Sch. of Med., Baltimore, MD 21205, USA. (1940) PhD biophysics (Johns Hopkins U., 1969). Assoc. prof. (tel. 301 + 955-1210, email LATTMAN at JHUIGF). *X-ray diffraction analysis, large biological molecules, protein structure and function.*

Laudise, Dr. Robert Alfred. Materials Chem. Res. Lab., Rm. #1A-264, AT&T Bell Labs., 600 Mountain Ave., Murray Hill, NJ 07974, USA. (1930) PhD chemistry (MIT, 1956). Dir. res., phys. & inorg. chem. (tel. 201 + 582-6220). *Crystal growth, hydrothermal chemistry, materials science, quartz.*

Laughlin, Prof. David Eugene. Met. Eng. and Matls. Sci. Dept., Carnegie-Mellon U., 5000 Forbes Ave., Pittsburgh, PA 15213, USA. (1947) PhD materials sci. (MIT, 1973). Prof. (tel. 412 + 268-2706). *Phase transitions, magnetic materials, electron microscopy, symmetry change, ordering.*

Lauher, Dr. Joseph. Chemistry Dept., SUNY at Stony Brook, Stony Brook, NY 11794, USA. (tel. 516 + 632-7925).

Lawless, Prof. Kenneth Robert. Materials Sci. Dept., Thornton Hall, U. of Virginia, Charlottesville, VA 22901, USA. (1922) PhD phys. chemistry (U. of Virginia, 1951). Prof. (tel. 804 + 924-6335). *Surface structure, properties, electron microscopy, diffraction, electron, X-ray, oxidation, epitaxy, magnetic properties.*

Lawson, Dr. Andrew Cowper. Physical Metallurgy Group, Los Alamos Nat. Lab., Los Alamos, NM 87545, USA. (1946) PhD physics (U. of Calif. at San Diego, 1972). Staff mem. (tel. 505 + 667-8844). *Neutron diffraction, phase transitions, actinites.*

Lawton, Mr. Stephen Latham. Res. Dept., Mobil Res. and Dev. Corp., Paulsboro, NJ 08066, USA. (1939) MS chemistry (Iowa State U., 1966). Assoc. (tel. 609 + 423-1040). *Crystallography, aluminosilicate and inorganic crystal structures, computer science.*

Lawyer, Dr. Carl Henry. TEMM Software Div., Rewal Associates, 10320 Fontainebleau, Meguon, WI 53092, USA. (1946) MD chemistry, BA 1968, medicine (U. of Colorado, 1972). Mol. biologist, pharmacologist (tel. 414 + 242-1312, 4048, email CARL on MACNET). *Protein, DNA, RNA structure, computer analysis, 3-D modeling, homology, tertiary protein folding, pharmacology receptor modeling, NMR, crystallography, protein - nucleic acid interactions, drug development, software.*

Le, Mr. Liang-Hsien. 17 Marvin Ln., Piscataway, NJ 08854, USA. (tel. 201 + 932-5374).

Le Trong, Mrs. Isolde. Biological Struct. Dept., SM-20, U. of Washington, Seattle, WA 98195, USA. (1949) Diplombiol. biochem., genetics, virology (U. of Hohenheim, W. Res. technologist. (tel. 206 + 543-4496).

Lebioda, Prof. Lukasz. Chemistry Dept., U. of South Carolina, Columbia, SC 29208, USA. (1943) PhD chemistry (Jagiellonian U., Poland, 1972). Assoc. prof. (tel. 803 + 777-2140, email Bitnet D130011 at UNIVSCVM). *Protein crystallography, glycolytic enzymes.*

Ledbetter, Dr. Hassel. Fracture & Deformation Div. (430), Natl. Inst. Stnds. & Techn., Boulder, CO 80303, USA. (1937) PhD metallurgy (U. of Illinois at Urbana, 1969). Res. metallurgist. (tel. 303 + 497-3443, fax 303-497-5030). *Phase transformations, elastic properties, physical properties, stacking faults, superconductors.*

Lee, Dr. Byungkook. DCRT/PSL, Room 2007, Bldg. 12A, Natl. Insts. of Health, Bethesda, MD 20892, USA. (1941) PhD phys. chemistry (Cornell U., 1967). (tel. 301 + 496-1135). *Protein structures, solvent role, computer graphics.*

Lee, Prof. James Ching. Biochemistry Dept., St. Louis U., 1402 S. Grand Blvd., St. Louis, MO 63104, USA. (1941) PhD biochemistry (Case Western Reserve U., 1971). Prof. (tel. 314 + 577-8144). *Protein - nucleic acid interaction, macromolecular assembly.*

Lee, Mrs. Katharine Darby. Analytical & Eng. Sci., Dow Chemical Co., Bldg. B-1225, Freeport, TX 77541, USA. (1961) BS chemistry (Lamar U., 1983). Res. chemist. (tel. 409 + 238-3799). *X-ray diffraction, polymer scattering.*

Lee, Mr. Peter L. Chemistry Dept., SUNY at Buffalo, Buffalo, NY 14214, USA.

Lee, Mr. Xuye. Biochemistry Dept., U. of Wisconsin at Madison, 420 Henry Hall, Madison, WI 53706, USA. (1962) MS chemical crystallography (Chinese Acad. of Sci., 1985). Grad. student. (tel. 608 + 262-3019). *Crystallography, protein, nucleic acid, structure and function, biomolecules.*

Lefebvre, Mr. Kevin R. S55 W229550 Windcrest Dr., Waukesha, WI 53188, USA. (tel. 414 + 968-2226).

LeGeros, Dr. Racquel Z. Dental Materials Sci., New York U., 345 East 24th. St. New York, NY 10010, USA. (1935) PhD biochemistry (NYU, 1967). Prof., project dir. (tel. 212 + 998-9580). *Calcium phosphates and apatites, synthetic and biological.*

Lenhert, Prof. P. Galen. Physics Dept., Vanderbilt U., Box 1807, Station B, Nashville, TN 37235, USA. (1933) PhD biophysics (Johns Hopkins U., 1960). Prof. (tel. 615 + 343-6045). *Crystal structure, inorganic, biological, computer programming for data collection, polymer diffraction.*

Leon-Escamilla, Mr. Efigenio Alejandro. Chemistry Dept., Georgetown U., Washington, DC 20057, USA. inorg. chemistry (Universidad Autonoma de Puebla). Grad. student. (tel. 202 + 687-5925). *Solid state, inorganic chemistry, materials science.*

Leonowicz, Dr. Michael Edward. Res. Dept., Mobil Res. and Dev. Corp., Paulsboro, NJ 08066, USA. (1949) PhD phys. chemistry (Cornell U., 1976). Sr. Res. chemist. (tel. 201 + 423-1040). *Inorganic and zeolite structures, electron diffraction, microscopy, powder diffraction.*

Lessinger, Prof. Leslie. Chemistry Dept., Barnard C., 3009 Broadway, New York, NY 10027-6598, USA. (1943) PhD chemistry (Harvard U., 1972). Assoc. prof. (tel. 212 + 854-8461). *Direct methods, natural products, biologically active small molecules, hydration.*

Leung, Dr. Peter C. 201-2E-08, Sci. Res. Labs., Minn. Mining & Mfg. Co. Ctr., St. Paul, MN 55144, USA. (1949) PhD chemistry (SUNY at Buffalo, 1982). Sr. res. chemist. (tel. 612 + 736-0483). *Organic conductors and superconductors, non-linear optics, charge density.*

Levan, Mr. Keith R. Systems Software & Tech., Hughes Aircraft Co., 28622 Mt. Rose Rd., Rancho Palos Verdes, CA 90732, USA.

Levien, Dr. Louise. Exxon Production Res. Co., P.O. Box 2189, Houston, TX 77001, USA. (1952) PhD earth and space sci. (SUNY at Stony Brook, 1979). Res. specialist. (tel. 713 + 966-6041). *Mineralogy, crystallography, mineral elasticity.*

Levy, Dr. Henri A. 116 Meadow Rd., Oak Ridge, TN 37830, USA. (1913) PhD chemistry (CIT, 1938). Consultant, adjunct prof. U. of Tenn. (tel. 615 + 483-9567). *Neutron diffraction, small molecules, structural chemistry, electron microscope tomography.*

Lewis, Mr. Michael D. 18 Maple Ave., North Brunswick, NJ 08902, USA. (tel. 201 + 932-3811).

Leyerle, Mr. Richard W. Allied-Signal Res. Ctr., 50 E. Algonquin Rd., Des Plaines, IL 60017, USA. (tel. 312 + 391-3312).

Li, Dr. Chi-Tang. Res. Labs., Dow Corning Corp., MAIL C41D01, Midland, MI 48686-0994, USA. (1934) PhD phys. chemistry (Montana State U., 1964). Res. specialist. (tel. 517 + 496-6058). *Ceramic fiber, glass-ceramics, fuel cells, material science, ESCA.*

Li, Mr. Hui-Ying. Chemistry Dept., U. of South Carolina, Columbia, SC 29208, USA. (1951) BS chemistry (Shanxi U., PRC, 1976). Grad. student. (tel. 803 + 777-2682). *Small molecule crystallography, Cd(113) NMR.*

Li, Dr. Naiyin. Med. Fndn. of Buffalo, 73 High St., Buffalo, NY 14203, USA. (1955) PhD phys. chemistry (SUNY st Buffalo, 1989). Postdoct. fel. (tel. 716 + 856-9600). *Charge density studies, biomolecules.*

Li, Mr. Tien-Hsiung. Chemistry Dept., U. of South Florida, Tampa, FL 33620, USA. (1957) chemistry. Grad. student. (tel. 813 + 974-2365).

Li, Dr. Yong Ji. Dept. of Chemistry. U. of Puerto Rico, Rio Piedras, PR 00931, USA. (1933) PhD chemistry (SUNY at Buffalo, 1985). Assoc. Prof. (tel. 809-756-6686). *Charge density distribution, accurate measurement, crystallographic computing, biological structure, organometallics.*

Licklider, Mr. Robert A. Missle Simulation Br., Naval Weapons Center, China Lake, CA 93555, USA. (1956) MS chem. eng. (MIT, 1979). Supr., operations res. anal. (tel. 619 + 939-3800, email licklider at nwc.arpa). *Scattering, small angle, neutron.*

Liebman, Dr. Michael N. Bioinformation Group, Amoco Techn. Co., Warrenville Rd., P.O. Box 400, Naperville, IL 60566, USA. (1947) PhD phys. chemistry (Michigan State U., 1977). Sr. scient. (tel. 312 + 961-7050, fax 312-961-7668). *Macromolecular modeling, structure - function analysis, design and structure prediction, artificial intelligence.*

Lim, Dr. Louis W. Phys. Sci. Ctr., BB4K, MONSANTO, 700 Chesterfield Village Pkwy., St. Louis, MO 63198, USA. (1950) PhD mol. biology (Washington U., St. Louis, 1979). Res. specialist. (tel. 314 + 537-6225). *Macromolecular structure - function, protein engineering.*

Lin, Mr. Jar-Shyong. Solid State Div., Oak Ridge Nat. Lab., P.O. Box X, Oak Ridge, TN 37830, USA. (tel. 615 + 574-4534).

Lind, Dr. Maurice David. Sci. Center, Rockwell International, 1049 Camino Dos Rios, Thousand Oaks, CA 91360, USA. (1934) PhD chemistry (Cornell U., 1962). Techn. staff member. (tel. 805 + 373-4190). *X-ray crystallography, crystal growth.*

Lindenmeyer, Dr. Paul Henry. 165 Lee St., Seattle, WA 98109, USA. (1921) PhD phys. chemistry (Ohio State U., 1951). Consultant. (tel. 206 + 284-1283). *Materials research.*

Lingafelter, Prof. Edward Clay. Chemistry Dept., U. of Washington, BG-10, Seattle, WA 98195, USA. (1914) PhD chemistry (UCB, 1939). Prof. emeritus. (tel. 206 + 543-1686). *Structures, inorganic, coordination compounds.*

Lippard, Prof. Stephen James. Chemistry Dept., Mass. Inst. of Techn., 77 Massachusetts Ave., Cambridge, MA 02139, USA. (1940) PhD chemistry (MIT, 1965). Prof. (tel. 617 + 253-1892). *Chemistry, inorganic, organometallic, transition metal bimetallic centers, higher coordinate metal complexes, synthesis,* structure determination, chemistry and biology, heavy atom labeling, biopolymers, antitumor platinum drugs, coupling reactions.

Lippert, Dr. Ernest L. Monarch Analytical Labs. Inc., P.O. Box 2990, Toledo, OH 43606, USA. (1931) PhD inorg. and struct. chemistry (U. of Leeds, UK, 1965). Chief, analytical chemistry. (tel. 419 + 535-1780). *Scattering theory, scientific computing, X-ray diffraction, fluorescence, analytical instrumentation.*

Lippman, Dr. Robert. Chemistry Dept., Adelphi U., Garden City, NY 11530, USA. (1943) PhD chemistry (Adelphi U., 1977). (tel. 516 + 294-8700). *Low temperature X-ray diffraction instrumentation, crystal structure analysis.*

Lipscomb, Prof. William Nunn. Chemistry Dept., Harvard U., 12 Oxford St., Cambridge, MA 02138, USA. (1919) PhD chemistry (CIT, 1946). Abbott & James Lawrence prof. (tel. 617 + 495-4098, fax 617-495-1792). *Structures, enzymes, proteins, inorganic and organic compounds, low temperatures.*

Litvin, Prof. Daniel Bernard. Physics Dept., Pennsylvania State U., Berks Campus, P.O. Box 7009, Reading, PA 19610-6009, USA. (1940) PhD physics (Technion - Israel Inst. of Techn., 1971). Prof. (tel. 215 + 320-4856). *Solid state physics, mathematical crystallography, phase transitions.*

Liu, Mr. Hansong. Molec. & Cell Biology Dept., U. of Connecticut, U-125, Storrs, CT 06268, USA. (1963) MS biophysics (U. of Connecticut, 1987). Grad. student. (tel. 203 + 486-4622). *Microcrystallography.*

Liu, Mr. Hung-Yu. 2007 Brandeis Dr., Richardson, TX 75081, USA.

Liu, Dr. Xing. Earth Sci. Dept., SUNY at Stony Brook, Stony Brook, NY 11794, USA. (1959) PhD crystallography (SUNY at Stony Brook, 1988). Postdoct. assoc. (tel. 516 + 632-8196). *Phase transitions, crystal chemistry, X-ray diffraction, high temperature and pressure, inorganic synthesis.*

Lo, Mr. Chi-Fung. 769 Bevier Rd., Piscataway, NJ 08854, USA.

Loeb, Dr. Arthur L. Visual and Env. Studies Dept., Carpenter Center, Harvard U., Cambridge, MA 02138, USA. (1923) PhD chemical physics (Harvard U., 1949). Sr. lect., curator, trustee, Radcliffe C. (tel. 617 + 495-3216). *Mathematical crystallography, design science.*

Loehlin, Prof. James Herbert. Chemistry Dept., Wellesley C., Wellesley, MA 02181, USA. (1934) PhD phys. chemistry (MIT, 1960). Prof. (tel. 617 + 235-0320, ext. 3043). *Intermolecular hydrogen bonding, small molecule crystals.*

Loghry, Dr. Ray Allen. Chemical Res. Dept., Haliburton Services Co., Box 1431, Duncan, OK 73536, USA. (1949) PhD analyt. chem., crystallog. (U. of Texas, 1976). Sr. dev. chemist. (tel. 405 + 251-3950). *Powder diffraction studies, characterization, industrial scales, iron phosphides.*

Lok, Mr. Charles. P.O. Box 524, Centereach, NY 11720, USA. (tel. 516 + 751-7438).

Loll, Mr. Patrick J. Biophysics Dept., Johns Hopkins Sch. of Medicine, 725 N. Wolfe St., Baltimore, MD 21205, USA. (1958) BChE chemical eng. (Catholic U. of America, 1981). Grad. student. (tel. 301 + 995-8388). *Protein crystallography, site-directed mutagenesis.*

Long, Dr. Gabrielle Gibbs. Bldg. 223, Rm A256, Natl. Inst. Stnds. & Techn., Gaithersburg, MD 20899, USA. PhD physics (Polytechnic Inst. of NY, 1972). Physicist. (tel. 301 + 975-5975). *Dynamical diffraction, inelastic scattering, disordered materials, small angle scattering.*

Love, Dr. R. A. Biological Sci. Dept., U. of Pittsburgh, Pittsburgh, PA 15260, USA. (tel. 412 + 624-4638).

Love, Prof. Warner Edwards. Thomas C. Jenkins Biophys. Dept., Johns Hopkins U., 3400 N. Charles St., Baltimore, MD 21218, USA. (1922) PhD physiology (U. of Pennsylvania, 1951). Prof. (tel. 301 + 338-7250). *Protein crystal structures, hemoglobin, histone, hemocyanin.*

Lovell, Dr. Frederick Mauri. 128 Lakeview Ave., Leonia, NJ 07605, USA. (1930) PhD physics (U. of Wales, UK, 1960). *Organic structures, pharmaceutically interesting structures, computer programming.*

Low, Prof. Barbara Wharton. Biochem. and Mol. Biophys. Dept., Columbia U., 630 West 168th St., New York, NY 10032, USA. (1920) DPhil chemistry (Oxford U., UK, 1948). Prof. (tel. 212 + 305-3895, 3896). *Structure and function, snake venom postsynaptic neurotoxins, cytotoxins, protein conformation theory, acetylcholine receptor - neurotoxin complexes.*

Lowe-Ma, Dr. Charlotte Kathryn. Chemistry Div., Code 3854, Naval Weapons Ctr., China Lake, CA 93555, USA. (1951) PhD chemistry (CIT, 1979). Res. chemist. (tel. 619 + 939-1607). *Small-molecule crystallography, energetic materials, X-ray powder diffraction, solid state chemistry.*

Lublin, Mr. Paul. 16 Montgomery Dr., Framingham, MA 01701, USA. (1924) MS phys. chemistry (Purdue U., 1949). Retired. (tel. 508 + 877-7879). *X-ray diffraction analysis, powder, single crystal, X-ray spectroscopy, electron probe, automation, electron microscopy (scanning, transmission), spectroscopy, surface analysis.*

Ludwig, Prof. Martha L. Biophysics Res. Div., IST Bldg., U. of Michigan, 2200 Bonisteel Blvd., Ann Arbor, MI 48109, USA. (1931) PhD biochemistry (Cornell U. Med. Sch., 1956). Prof., res. biophysicist. (tel. 313 + 763-2199). *Protein crystallography.*

Luecke, Mr. Hartmut. Biochemistry Dept., Rice U., P.O. Box 1892, Houston, TX 77251, USA. (1962) BS chemistry, comp. sci., phys. (Heidelberg C., 1984). (tel. 713 + 799-6564, email hudel at rice.edu). *Protein crystallography, molecular dynamics.*

Luft, Mr. Joseph. Med. Fndn. of Buffalo, 73 High St., Buffalo, NY 14203, USA. (tel. 716 + 856-9600).

Luisi, Dr. Ben Francesco. Molec. Biophys. & Biochem. Dept., Yale U., 260 Whitney Ave., JSW 402, New Haven, CT 06511, USA. (1959) PhD molec. biology (Cambridge U., UK, 1986). Postdoct. fel. *Molecular biology.*

Luo, Mr. Ming. Biological Sci. Dept., Purdue U., West Lafayette, IN 47906, USA.

Luss, Mr. Henry Richard. Analytical Techn. Div., Eastman Kodak Co., Bldg. 49, Kodak Park, Rochester, NY 14652-3712, USA. (1939) BS chemistry (Rochester Inst. of Techn., 1969). Crystallographer. (tel. 716 + 722-0138). *Small molecule crystallography.*

Lustig, Mr. Stanley. Viskase Corp., 6733 W. 65th. St., Chicago, IL 60638, USA. (1933) BS chemistry (U. of Toledo, 1958). Techn. manager. (tel. 312 + 496-4672). *High polymer structure and properties.*

Lynch, Dr. V. M. Chemistry Dept., U. of Texas at Austin, Austin, TX 78712, USA. (1952) PhD chemistry (U. of Florida, 1982). Staff crystallographer. (tel. 512 + 471-4042).

Ma, Mr. Cheuk Ki. 3214 Cedar Trail, Middleton, WI 53562, USA.

MacCrone, Prof. Robert Kirsten. Materials Eng., Rensselear Polytechnic Inst., Troy, NY 12180, USA. DPhil physics (U. of Oxford, UK). Prof. (tel. 518 + 276-6047). *Magnetism, EPR, glasses, oxides.*

Mack, Mr. John. Mack Computer Services, 15 Pine Tree Rd., Huntington Station, NY 11746, USA. (1947) M.Phil. biochemistry (Columbia U., 1973). Director, software prod. management. (tel. 516 + 385-1294). *Structures, models, computer graphics, software, proteins, DNA.*

Mack, Dr. Joseph Philip Grant. Crystallography, FCRF-NCI, P.O. Box B, Frederick, MD 21701, USA. (1947) PhD biochemistry (U. of Sydney, Australia, 1974). Scient. assoc. (tel. 301 + 698-5033). *Enzyme kinetics, DNA binding proteins, genetic recombination.*

Mackie, Dr. Paul E. 5695 Salem Rd., Lithonia, GA 30058, USA. (1942) PhD physics (Georgia Inst. of Techn., 1972). Res. scient. (tel. 404 + 894-3455). *Crystal physics, powder, single-crystal diffractometry, computer systems programming.*

Mackinnon, Dr. Ian Donald. Geology Dept., U. of New Mexico, Albuquerque, NM 87131, USA. PhD geochemistry (James Cook U., Australia, 1979). Sr. res. scient. (tel. 505 + 277-7536). *Silicate mineralogy, cosmochemistry.*

MacRae, Dr. Alfred U. AT&T Bell Labs., Murray Hill, NJ 07974, USA. (1932) PhD physics (Syracuse U., 1960). Lab. dir. (tel. 201 + 949-6722). *Low energy electron diffraction, surfaces.*

Madden, Prof. John Joseph. Depts. of Psychiatry & Biochem., Emory U. Sch. of Medicine, Box Af, Atlanta, GA 30322, USA. (1943) PhD biochemistry (Emory U., 1968). Assoc. prof. (tel. 404 + 894-5951). *Nucleic acid structures, mutagenesis, DNA repair.*

Madejski, Ms. Julie. 4341 Autumn Trail, Clarence, NY 14031, USA. (tel. 716 + 856-9600).

Magnus, Dr. Karen A. Biochemistry Dept., Case Western Reserve Sch. of Med., 2119 Abington Rd., Cleveland, OH 44106, USA. (1952) PhD biophysics (Johns Hopkins U., 1980). (tel. 216 + 368-4666). email MAGNUS at CWRU. *Biological macromolecules, DNA-binding proteins, oxygen transport proteins.*

Magnuson, Dr. Vincent Richard. Chemistry Dept., U. of Minnesota at Duluth, Duluth, MN 55812, USA. (1942) PhD inorg. chemistry (U. of Illinois, 1968). Prof. (tel. 218 + 726-7591). *Chemistry, coordination, organometallic, physical-inorganic.*

Majeste, Prof. Richard J. Chemistry Dept., Southern U. of New Orleans, 6400 Press Dr., New Orleans, LA 70126, USA. (1943) PhD phys. chemistry (Louisiana State U. at New Orleans, 1969). Prof. (tel. 504 + 282-4401, ext. 250). *X-ray crystallography, opiates.*

Makinen, Prof. Marvin W. Biochem. & Molec. Biol. Dept., U. of Chicago, 920 E. 58th. St., Chicago, IL 60637, USA. (1939) MD, D.Phil. medicine, mol. biophys. (U. of Pennsylvania, 1968, Oxford U., England, 1976). Prof. & Chairman. (tel. 312 + 702-1080, fax 312-702-0439). *Protein structure, function, enzyme mechanisms, molecular dynamics, molecular modeling.*

Makowski, Prof. Lee. Physics Dept., Boston U., 590 Commonwealth Ave., Boston, MA 02215, USA. (1949) PhD molecular biophysics (MIT, 1976). Assoc. prof. *Macromolecular structure, membranes, viruses, fiber diffraction, diffraction theory.*

Malley, Ms. Mary F. Sci. Information Dept., E.R. Squibb and Sons, Box 4000, Princeton, NJ 08543-4000, USA. (1953) BA chemistry (Rutgers U., 1975). Res. investigator. (tel. 609 + 921-4986). *Molecular structure, conformation, X-ray diffraction.*

Mallory, Dr. Chester L. 1073 Lucot Way, Campbell, CA 95008, USA. (1952) PhD ceramic eng. (Alfred U., 1979). (tel. 408 + 379-7542). *Computer automation, analytical instrumentation, crystallographic computing.*

Mandel, Prof. Gretchen Sue. Res. Service #151, Med. C. of Wisconsin, VA Hosp., 5000 West National Ave., Milwaukee, WI 53295, USA. (1946) PhD X-ray crystallography (U. of Pennsylvania, 1972). Assoc. prof. (tel. 414 + 384-2000, ext. 2498). *Biological - medical X-ray structural analysis, crystal-induced diseases.*

Mandel, Prof. Neil Stanley. Res. Service #151, Med. C. of Wisconsin, VA Hosp., 5000 West National Ave., Milwaukee, WI 53295, USA. (1947) PhD X-ray crystallography (U. of Pennsylvania, 1971). Prof. (tel. 414 + 384-2000, ext. 2494,2498). *Biological - medical X-ray structural analysis, crystal-induced diseases.*

Manghnani, Mr. Murli H. Hawai Inst. Geophysics, U. of Hawaii, 2525 Corea Rd., Honolulu, HI 96822, USA.

Manor, Dr. Philip C. Gordon & Breach, 50 W. 23rd. St., New York, NY 10010, USA. (1944) PhD phys. chemistry (MIT, 1973). Sci. editor. (tel. 212 + 206-8900). *Crystallography, protein, nucleic acid.*

Maquire, Ms. Theresa. JCPDS, Internatl. Ctr. Diffraction Data, 1601 Park Ln., Swarthmore, PA 19081, USA.

Margulis, Prof. Thomas N. Chemistry Dept., U. of Massachusetts, Boston, MA 02125, USA. (1937) PhD chemistry (UCB, 1962). Prof. (tel. 617 + 929-7539). *Organic structure, small rings, drugs.*

Mariano, Dr. Anthony N. 48 Page Brook Rd., Carlisle, MA 01741, USA. (1930) PhD geology (Boston U., 1968). Consultant, mineral explor. (tel. 508 + 369-9242). *Mineralogy, petrology, economic geology, crystallography.*

Marians, Ms. Carol. Rm. 13-4053, Mass. Inst. of Techn., Cambridge, MA 02139, USA. (tel. 617 + 253-0560).

Markgraf, Mr. Steven A. Materials Res. Lab., Pennsylvania State U., University Park, PA 16802, USA. (1959) PhD solid state sci. (PSU, 1987). Res. assoc. (tel. 814 + 865-7102). *Phase transitions, structure - property relationships, ferroic crystals, crystal growth.*

Markley, Prof. John L. Biochemistry Dept., U. of Wisconsin, 420 Henry Mall, Madison, WI 53706, USA. (1941) PhD biophysics (Harvard U., 1969). Prof. (tel. 608 + 263-9349, fax 608-262-3453, email Bitnet MARKLEY at WISCMACC). *Protein structure, dynamics, folding.*

Marko, Mr. Eric. Spellman High Voltage, 7 Fairchild Ave., Plainview, NY 11803, USA.

Marks, Prof. Laurence Daniel. Materials Sci. Dept., Northwestern U., Evanston, IL 60208, USA. (1954) PhD physics (Cambridge U., UK, 1980). Assoc. prof. (tel. 312 + 491-4133, fax 312-491-4133, email LDM05055 at NUACC). *Electron microscopy, diffraction, small particles, surfaces, radiation damage.*

Marsh, Dr. Charles P. Materials Sci. Dept., Metallurgy Bldg. Rm. 201, U. of Illinois, Urbana, IL 61801, USA. (1957) PhD materials sci. (U. of Illinois at Urbana, 1989). Res. asst. (tel. 217 + 333-0789). *Coarsening, Ostwald ripening, precipitate morphology, phase kinetics, Ni-base superalloys.*

Marsh, Mr. Philip. AT&T Bell Labs., 600 Mountain Ave., Murray Hill, NJ 07974, USA.

Marsh, Dr. Richard Edward. Chemistry Dept., Calif. Inst. of Techn., 1201 East California Blvd., Pasadena, CA 91125, USA. (1922) PhD chemistry (UCLA, 1950). Sr. res. assoc. (tel. 213 + 356-6526). *Refinement techniques, biological, inorganic structures.*

Marshall, Prof. Garland R. Pharmacology Dept., Washington U. Med. Sch., St. Louis, MO 63110, USA. (1940) PhD biochemistry (Rockefeller U., 1966). Prof. (tel. 314 + 362-2286). *Peptide chemistry, viral proteases, peptaibols, inhibitors, drug design, computational chemistry.*

Martin, Mr. Bruce Alan. M/S B02-106, Grumman Aircraft Systems, Bethpage, NY 11714, USA. (1944) BS applied mathematics (PUN, 1964). Prin. software eng., adj. prof. BPI (tel. 516 + 577-1426). *Crystallographic computing, real-time online scattering experiments, computer control, mathematical crystallography, group theory.*

Martin, Mr. David C. 701 Grad. Res. Tower, U. of Massachusetts, Amherst, MA 01003, USA. (1961) MS materials eng. (U. of Michigan, 1985). Res. asst. (tel. 413 + 545-0044). email MILTY at UMAS). *Polymers, disorder, information theory.*

Martin, Mr. George William. 1671 Yale Dr., Mountain View, CA 94040, USA. (1932) MS material sci. (Stanford U., 1968). Res. scient. (tel. 415 + 969-4694). *Electron microprobe, X-ray diffraction, crystallography, spectroscopy, SEM, TEM, Auger.*

Martin, Dr. Kimberly Ann. MST-5, MS-6730, Los Alamos Nat. Lab.., Los Alamos, NM 87545, USA. (1956) PhD chemistry (U. of Wisconsin at Madison, 1986). Staff memb. (tel. 505 + 667-4341). *Chemistry, materials, organometallic.*

Martin, Dr. Philip D. Biochemistry Dept., Wayne State U., 540 E. Canfield, Detroit, MI 48201, USA. (1942) PhD biophysics (Wayne State U., 1978). Res. assoc. (tel. 313 + 577-1506). *X-ray crystallography, macromolecules.*

Martin, Dr. Yvonne C. D-47E AP9, Abbott Labs., Abbott Park, IL 60064, USA. (1936) PhD chemistry (Northwestern U., 1964). Sr. proj. leader, computer asst. mol. design. (tel. 312 + 937-4981, fax 312 + 937-1511).

Martinez, Mr. Sergio E. Biological Sci. Dept., Purdue U., West Lafayette, IN 47907, USA. (1964) BS physics (Cornell U., 1986). Res. asst. (tel. 317 + 494-9247). *Proteins, anomalous dispersion.*

Mason, Mr. John T. Ames Lab., 232 Metallurgy Bldg., Iowa State U., Ames, IA 50011, USA. (1933) MS phys. chemistry (Tufts U., 1956). Assoc. metallurgist. (tel. 515 + 294-6529). *Alloy structure, surface phenomena.*

Mason, Prof. Paul Robert. Div. of Physics & Appl. Optics, Rose-Hulman Inst. of Techn., 5500 Wabash Ave., Terre Haute, IN 47803, USA. (1934) MS physics (Indiana U., 1958). Assoc. prof: (tel. 812 + 877-1511, ext. 305). *Solid state physics, X-ray diffraction, polycrystalline materials.*

Massa, Prof. Louis. Chemistry Dept. City U. of New York, 695 Park Ave., New York, NY 10021, USA. (1940) PhD physics (Georgetown U., 1966). Prof., chem. and phys. (tel. 212 + 772-5330, fax 212 + 772-4941). *Chemistry, mathematics, physics.*

Massalski, Prof. Thaddeus B. Met. Eng. and Matl. Sci. Dept., Carnegie-Mellon U., 3303 Wean Hall, Pittsburgh, PA 15213-3890, USA. (1926) PhD physical met-allurgy (U. of Birmingham, UK, 1954). Prof. (tel. 412 + 268-2708). *Alloy phases, phase diagrams, metallic glasses, solid state transformations.*

Mastropaolo, Dr. Donald. Div. of Neurology, RG-27, U. of Washington, Seattle, WA 98195, USA. (1945) PhD chemistry (Rutgers U., 1974). Sr. res. assoc. (tel. 206 + 543-2340). email Bitnet 57023 at MAX). *X-ray crystal structure determination, biological molecules, structure - activity relationships.*

Matassa, Dr. Victor Giulio. Biomedical Res. Dept., ICI Pharmaceutical Group, New Murphy Rd., Wilmington, DE 19897, USA. (1951) PhD org. chemistry (Glasgow U., Scotland, 1978). Prin. chemist. (tel. 302 + 575-2579). *Medicinal chemistry, conformation, biomacromolecules, small molecules, enzyme mechanism, receptors.*

Mathew, Mr. M. Ada Health Fndn., Natl. Inst. Stnds. & Technol. Gaithersburg, MD 20899, USA. (tel. 301 + 975-6824).

Mathews, Prof. F. Scott. Cell Biology & Physiology Dept., Washington U. Med. School, 4566 Scott Ave., St Louis, MO 63110, USA. (1934) PhD chemistry (U. of Minnesota, 1969). Prof. (tel. 314 + 362-1080). email MATHEWS_S at WUMS). *Protein crystallography, molecular graphics, biologically interesting structures.*

Matias, Mr. Pedro M. 1416 Oxford St., Berkeley, CA 94704, USA.

Matthews, Prof. Brian W. Inst. of Mol. Biology, U. of Oregon, Eugene, OR 97403, USA. (1938) PhD physics (U. of Adelaide, Australia, 1964). Prof. of physics, res. assoc. (tel. 503 + 686-5151, fax 503 + 686-3127). *Macromolecular crystallography.*

Matthews, Dr. David Allan. Crystallography Dept., Agouron Pharmaceuticals, 11025 N. Torrey Pines Rd., La Jolla, CA 92024, USA. (1943) PhD chemistry (U. of Illinois at Urbana, 1971). Dir., protein crystalog. (tel. 619 + 535-5516). *Biological macromolecules, structure - function.*

Matyi, Prof. Richard James. Matls. Sci. & Eng. Dept., U. of Wisconsin at Madison, 1509 University Ave., Madison, WI 53706, USA. (1953) PhD materials sci. & eng. (Northwestern U., 1983). Asst. prof. (tel. 608 + 262-0711). *X-ray diffraction analysis, semiconductors, molecular beam expitaxy.*

Maulik, Dr. Prakas R. Biophysics Dept., Boston U. Sch. of Med., 80 East Concord St., Boston, MA 02118, USA. (1950) PhD X-ray crystallography (Calcutta U., India, 1982). Res. assoc. (tel. 617 + 638-4007). *Crystallography, protein, small molecule, protein crystallization, membrane, membrane binding, structure and function, lipids, biological membranes.*

Maverick, Dr. Andrew William. Chemistry Dept., Louisiana State U., Baton Rouge, LA 70803, USA. (1955) PhD inorg. chemistry (CIT, 1982). Asst. prof. (tel. 504 + 388-4415). email CHMAV at LSUVAX). *Transition metal chemistry, redox reactions, photochemistry, structure of metal complexes, metalloenzyme model systems.*

Maverick, Prof. Emily Fisch. Chemistry Dept., Los Angeles City C., 855 N. Vermont Ave., Los Angeles, CA 90029, USA. (1929) PhD analytical chemistry (UCLA, 1972). Prof. (tel. 213 + 669-4224). *Strained molecules, conformational energy, thermal motion, disorder.*

Mayo, Mr. William Edward. Mechanical & Matls. Sci. Dept., Rutgers U., Piscataway, NJ 08854, USA. (tel. 201 + 932-3666).

Mazany, Dr. Anthony Michael. Corporate Res., The BF Goodrich Co., 9921 Brecksville Rd., Brecksville, OH 44141-3289, USA. (1954) PhD inorg. chemistry (Case Western Reserve U., 1984). Res. & Dev. chemist. (tel. 216 + 447-5559). *Compounds, organometallic, gold, iron, molybdenum, inorganic, cluster, dimer, sulfur, thio, tin, phosphorus.*

McAlister, Dr. John Paul. Tripos Associates Inc., 1699 S. Hanley, Suite 303, St. Louis, MO 63144, USA. (1948) PhD biochemistry (U. of Wisconsin, Madison, 1978). Pres. (tel. 314 + 647-1099, fax 314 + 647-9241, email johnmc at tripos.com). *Software design, molecular modeling, graphics, crystallographic computing.*

McAtee, Prof. James L. Chemistry Dept., Baylor U., Waco, TX 76703, USA. (1924) PhD phys. chemistry (Rice U., 1951). Prof. emeritus. (tel. 817 + 755-3311). *Clay minerals, organic, inorganic complexes.*

McCallum, Prof. Malcolm Ernest. Earth Resources (Geology) Dept., Colorado State U., Fort Collins, CO 80523, USA. (1934) PhD geology (U. of Wyoming, 1964). Prof., res. geologist, US Geol. Survey. (tel. 303 + 491-6250). *Mineralogy, geochemistry, petrology, structure, mineral exploration, Rocky Mountain Precambrian crystalline rocks, kimberlite and included upper mantle-lower crustal nodules, diamonds from diatremes.*

McCammon, Prof. James Andrew. Chemistry Dept., U. of Houston, Houston, TX 77204-5641, USA. (1947) PhD chemical physics (Harvard U., 1976). M.D.Anderson Prof. of Chemistry. (tel. 713 + 749-7351, fax 713 + 749-1407, email Bitnet MCCAMMON at UHRCC2). *Simulations, computer, molecular dynamics, protein structure, dynamics, molecular design.*

McCarthy, Prof. Gregory Joseph. Chemistry and Geology Dept., North Dakota State U., Fargo, ND 58105, USA. (1943) PhD solid state sci. (PSU, 1969). Prof. (tel. 701 + 237-7193). *Solid state chemistry, geochemistry, X-ray powder diffraction.*

McCauley, Dr. James Weymann. US Army Materials Tech. Lab., Materials Science Br., Arsenal St., Watertown, MA 02172, USA. (1940) PhD solid state sci. (PSU, 1968). Chief, matls. sci. br. (tel. 617 + 923-5238, fax 617 + 923-5477). *Materials science, ceramics, crystallography, mineralogy.*

McClune, Mr. W. Frank. JCPDS, Internatl. Ctr. Diffraction Data, 1601 Park Ln., Swarthmore, PA 19081, USA.

McClure, Dr. Richard James. Medicinal Chemistry Dept., U. of Pittsburgh, 720 Salk Hall, Pittsburgh, PA 15261, USA. (1942) PhD chemistry (Georgia Inst. of Techn., 1969). Res. asst. prof. (tel. 412 + 648-8549). *Crystal structure, biological molecules, proteins - small molecule binding.*

McCollor, Mr. Donald P. 1703 1st. Ave. N., Grand Forks, ND 58201, USA. (tel. 701 + 772-2464).

McConnell, Prof. Duncan. 4312 S. 31st. St., Apt. 129, Temple, TX 76502, USA. (1909) PhD mineralogy (U. of Minnesota, 1937). Prof. emeritus. (tel. 817 + 771-1810). *Bone phosphate minerals, biominerals, calcification, crystal chemistry, apatite, francolite, dahllite, phosphorite, clay minerals.*

McCrone, Dr. Walter C. McCrone Res. Inst. Inc., 2820 S. Michigan Ave. Chicago, IL 60616, USA. (1916) PhD chemistry (Cornell U., 1942). Pres. (tel. 312 + 842-7100). *Optical crystallography, polymorphism.*

McDonald, Dr. Robert Charles. Whittaker-Yardney Power Sources, 520 Winter St., Waltham, MA 02154, USA. (1946) PhD chemistry (Boston U., 1975). Mgr., R & D. (tel. 617 + 890-3040, ext. 37312). *Biological structures, thionyl chloride chemistry, tin.*

McFarlane III, Dr. Samuel H. Materials Characterization Group, David Sarnoff Res. Center, Princeton, NJ 08543, USA. (1937) PhD physics (Brown U., 1967). Techn. staff member. (tel. 609 + 734-2206). *X-ray topography, crystal imperfections.*

McGuire, Dr. Nancy K. Chemistry Dept., Texas A&M U., College Station, TX 77843, USA. (1956) PhD chemistry, phys.(solid state) (Arizona State U., 1985). Res. assoc. (tel. 409 + 845-2936). email MCGUIRE at TAMCHEM). *Inorganic structures, materials science.*

McKay, Dr. David Bruce. Chemistry Dept., U. of Colorado, Campus Box 215, Boulder, CO 80309, USA. (1946) PhD biophysics (U. of Chicago, 1976). (tel. 303 + 492-6641). *Protein crystallography.*

McKenzie, Dr. Thomas Charles. Immunetech Pharmaceuticals, San Diego, CA 92121, USA. (1945) PhD org. chemistry (Columbia U., 1971). Sr. chemist. (tel. 619 + 457-2553). *Organic chemistry, synthesis, direct methods.*

McKeown, Dr. David Alexander. Semiconductor Electronics Div., Natl. Inst. Stnds. & Technol., Bldg. 225, Rm., A305, Gaithersburg, MD 20899, USA. (1957) PhD geology (Stanford U., 1985). Physicist. (tel. 301 + 975-3095). *Amorphous materials, Raman, EXAFS spectroscopies, X-ray scattering, diffraction.*

McKinstry, Prof. Herbert Alden. Material Sci. Dept., Pennsylvania State U., University Park, PA 16802, USA. (1925) PhD physics (PSU, 1960). Assoc. prof. (tel. 814 + 865-1614). *Ceramic powders, characterization, mechanical, thermal properties, computer graphics.*

McLaren, Prof. Eugene Herbert. Chemistry Dept., SUNY at Albany, 1400 Washington Ave., Albany, NY 12222, USA. (1924) PhD chemistry (Washington U., 1955). Prof. (tel. 518 + 457-8399). *Inorganic molecular structures, atmospheric chemistry, geochemistry.*

McLean, Dr. W. John. 2519 N. Walnut, Tucson, AZ 85712, USA. (1937) PhD earth and planetary sci. (U. of Pittsburgh, 1968). *Applied mineralogy, crystal structures, minerals.*

McMillan, Dr. Joyce A. Materials Res. Lab., U. of Illinois, Urbana, IL 61801, USA. (1938) PhD phys. chemistry (U. of Illinois, 1964). Dept. crystallographer. (tel. 217 + 333-1612). *Crystallography.*

McMullan, Dr. Richard K. Chemistry Dept. 555, Brookhaven Nat. Lab.., Upton, NY 11973, USA. (1929) PhD chemistry (Iowa State U., 1956). Assoc. chemist. (tel. 515 + 282-4380). *Structure determination, diffraction, X-ray, neutron, hydrogen bonded systems, clathrates, inclusion compounds.*

McMurdie, Mr. Howard Francis. JCPDS - Internatl. Ctr. Diffraction Data, Natl. Inst. Stnds. & Technol., Gaithersburg, MD 20899, USA. (1905) BS chemistry (Northwestern U., 1928). Consultant. (tel. 301 + 975-5792). *Powder diffraction, phase equilibria.*

McNulty, Mr. Thomas. Nicolet Instrument Corp., 5225-1 Verona Rd., P.O. Box 4508, Madison, WI 53711, USA. (tel. 608 + 271-3333).

McPhail, Prof. Andrew Tennent. Chemistry Dept., Duke U., Durham, NC 27706, USA. (1937) PhD chemistry (U. of Glasgow, UK, 1963). Prof. (tel. 919 + 684-2889). *Crystal structures, organic, biologically important molecules, transition metal complexes.*

McPherson, Prof. Alexander. Biochemistry Dept., U. of Calif. at Riverside, Watkins Dr., Riverside, CA 92521, USA. (1944) PhD biological science (Purdue U., 1970). Prof. & Chair, Biochem. (tel. 714 + 787-5391, fax 714 + 7873590). *X-ray diffraction, protein crystallization, proteins, nucleic acids, enzymes, macromolecules.*

McPherson, Mr. William G. 2201 Cedar, Duncan, OK 73533, USA.

McRee, Dr. Duncan E. Res. Inst. Scripps Clinic, MB5, 10666 N. Torrey Pines Rd., La Jolla, CA 92137, USA. (1957) PhD biochemistry (Duke U., 1984). Asst. member. (tel. 619 + 554-2806, telex 697168 SCRF SCRF, fax 619 + 554-8841). *Proteins, structure, crystallography.*

McTigue, Ms. Michele A. Chemistry Dept. B-017, U. of Calif. at San Diego, La Jolla, CA 92092, USA. (tel. 619 + 534-2011).

McWhan, Dr. Denis Bayman. Bldg. 510E, Brookhaven Nat. Lab.s, Upton, NY 11973, USA. (1935) PhD chemistry (UCB, 1961). Tech. staff member. (tel. 516 + 282-3927, fax 516 + 282-4462, email MCWHAN at BNL). *Condensed matter physics, synchrotron radiation.*

Meagher, Dr Edward Patrick. 516 Briar Road, Bellingham, WA 98225, USA. (1939) PhD,mineralogy (Penn State U., 1967). (tel. 206 + 647-1571). *Crystal chemistry, bonding theory.*

Medrud, Dr. Ronald Curtis. Analytical Res. & Services Div., Chevron Res. Co., 576 Standard Ave., Richmond, CA 94802, USA. (1934) PhD phys. chemistry (State U. of Iowa, 1963). Sr. res. chemist. (tel. 415 + 620-4090, fax 415 + 620-4647). *Powder diffraction, materials characterization, crystal chemistry, molecular sieves, molecular modeling.*

Meehan, Prof. Edward Joseph Jr. Chemistry Dept., U. of Alabama, Huntsville, AL 35899, USA. (1950) PhD biochemistry (U. of Alabama at Birmingham, 1978). Assoc. prof. (tel. 205 + 895-6188). *Protein crystallography, crystal growth.*

Meinke, Ms. Gretchen. Sch. of Chemistry and Biochem., Georgia Inst. Tech., Boggs Bldg., Atlanta, GA 30332-0400, USA. (1962) BS chemistry (Eckerd C., 1984). Grad. student. (tel. 404 + 894-8338, fax 404 + 894-7452, email meinke at max.gatech.edu). *Protein crystallography, molecular dynamics.*

Mel, de, Mr. Vidanalage S. J. Chemistry Dept., Wayne State U., Detroit, MI 48202, USA. (tel. 313 + 577-3076).

Mercola, Prof. Daniel Anthony. Pathology Dept., V-151, U. of Calif. at San Diego, La Jolla, CA 92093, USA. (1940) PhD pathology, molec. biophys. (U. of Calif., 1969). Asst. clin. prof. (tel. 619 + 453-7500). *Regulatory proteins, structure - function, c-fos, Troponin-C, Insulin, oncogenes, PDGF, sis, jun.*

Merritt, Dr. Ethan Allen. Biological Struct. Dept., SM-20, U. of Washington, Seattle, WA 98195, USA. (1952) PhD molecular biology (U. of Wisconsin, 1980). Res. assoc. (tel. 206 + 543-8865). email merritt at xray.bchem.washington.edu). *Anomalous scattering, artificial intelligence, macromolecular structure determination, synchrotron radiation, computer applications.*

Merritt, Prof. Lynne Lionel Jr. Chemistry Dept., Indiana U., Bloomington, IN 47405, USA. (1915) PhD analyt. chemistry (U. of Michigan, 1940). Prof. emeritus. (tel. 812 + 335-7368). *Instrumental analysis, structure, chelate compounds, X-ray crystal structure determinations.*

Mertes, Prof. Kristin Bowman. Chemistry Dept., Malott Hall, U. of Kansas, Lawrence, KS 66045, USA. (1946) PhD inorg. chemistry (Temple U., 1974). Prof. (tel. 913 + 864-3669). *Transition metal complexes, synthetic macrocyclic ligands, intermolecular interactions.*

Messick, Mr. Julian. JCPDS, Internatl. Ctr. for Diffraction Data, 1601 Park Lane, Swarthmore, PA 19081, USA. (1933) BS chemistry (Widener C., 1955). Corporate sec. & gen. manager. (tel. 215 + 328-9404, telex 847 170, fax 215 + 328-2503). *Diffraction analysis.*

Metzger, Prof. Robert Melville. Chemistry Dept., U. of Alabama, Tuscaloosa, AL 35487-0336, USA. (1940) PhD chemistry (CIT, 1969). Prof. (tel. 205 + 348-5952, fax 205 + 348-5051, email RMETZGER at UALVM). *Energies, Madelung, lattice, organic crystals, TCNQ salts, semi-empirical molecular orbital calculations, organic unimolecular rectifiers, EPR, triplet spin excitons, high temperature ceramic oxide superconductors.*

Meyer, Prof. Edgar Frederich. Biochem. and Biophysics Dept., Texas A&M U. College Station, TX 77843-2128, USA. (1935) PhD chemistry (U. of Texas at Austin, 1963). Prof. (tel. 409 + 845-1744, fax 409 + 845-9274, email MEYER at TAMBIGRF). *Interactive computer graphic modeling, protein crystallography, small molecule/large molecule interactions.*

Meyer, Prof. Frank Henry. 1103 15th Ave., S.E., Minneapolis, MN 55414, USA. (alt.: Internatl. Soc. Unified Sci. Inc., 1680 E. Atkins Ave., Salt Lake City, UT, USA). (1915) MS, MA physics and philosophy (PUN). Prof. emeritus, Pres., ISUS (tel. 612 + 331-6086). *Solid cohesion, forces in solids, interatomic closest approach theory, cohesive energy theoretical derivation, element crystals, binary compounds.*

Meyers, Dr. Bernard Lee. Analytical Div., Amoco Res. Center, P.O. Box 400, Naperville, IL 60566, USA. (1934) PhD analytical chemistry (U. of Illinois, 1960). Sr. res. assoc. (tel. 312 + 420-5226). *Catalysts, X-ray diffraction, physical characterizations.*

Meyers, Prof. Edward Arthur. Chemistry Dept., Texas A. & M. U., College Station, TX 77840, USA. (1930) PhD chemistry (U. of Minnesota, 1955). Prof. (tel. 713 + 845-2544). *Crystal structures, organic, inorganic.*

Michel, Dr. David John. U. S. Naval Res. Lab., Code 6320, 4555 Overlook Ave., S.W., Washington, DC 20375, USA. (1942) PhD metallurgy (PSU, 1968). Metallurgist. (tel. 202 + 767-2621). *Radiation effects, intermetallic compounds, high temperature properties.*

Mighell, Dr. Alan D. Reactor Radiation Div., Natl. Inst. Stnds. & Technol., Gaithersburg, MD 20899, USA. (1935) PhD chemistry (Princeton U.). Res. chemist. (tel. 301 + 975-6255). *Crystallographic data bases, crystal structure analysis.*

Mikkola, Prof. Donald Emil. Metallurgical Eng. Dept., Michigan Technological U., Houghton, MI 49931, USA. (1938) PhD materials science (Northwestern U., 1964). Prof. (tel. 906 + 487-2636). *Structure - properties relationships, X-ray diffraction.*

Milberg, Dr. Morton Edwin. 5448 E. Placita Apan, Tucson, AZ 85718, USA. (1926) PhD phys. chemistry (Cornell U., 1949). Prin. res. scient. (tel. 313 + 323-1724). *Noncrystalline solids, high temperature ceramics.*

Milburn, Mr. Michael Vance. MCL, Rm. 108, U. of Calif. at Berkeley, Berkeley, CA 93720, USA. (1964) MS biophys. chemistry (UCB, 1988). Res. asst. (tel. 415 + 486-4349). *Macromolecular crystallography.*

Mildner, Mr. David F. R. Res. Reactor Facility, U. of Missouri, Columbia, MO 65211, USA. (tel. 314 + 882-4211).

Mililllo, Prof. Frank. Mechanical Eng. Dept., Union C., Schenectady, NY 12308, USA. (1943) PhD physical metallurgy (PUN, 1974). Prof. (tel. 518 + 370-6264). *Phase transformations, interstitial atom ordering.*

Millane, Prof. Rick P. Whistler Ctr. Carbohydrate Res., Purdue U., Smith Hall, West Lafayette, IN 47907, USA. (1954) PhD electrical eng. (U. of Canterbury, New Zealand, 19 Asst. prof. (tel. 317 + 494-9272, telex 276147-AGAD-PU-LAF, fax 317 + 494-7953, email ojg at mace.cc.purdue.edu). *Fiber diffraction, theory, X-ray techniques, applications, structures, polysaccharides, nucleic acids, data collection and processing, phase retrieval, computing.*

Miller, Prof. Donald P. Physics and Astronomy Dept., Clemson U., Clemson, SC 29631, USA. (1927) PhD physics (PUN, 1962). Prof. (tel. 803 + 656-5314, telex 703648, fax 803 + 656-4040, email XRAY1 at CLEMSON). *Crystal and molecular structure, fibrous polymers, electron crystallography, image analysis.*

Miller, Dr. Kristine Elaine. CAMD - Monsanto Co., 700 Chesterfield Village Pkwy., Chesterfield, MO 63017, USA. (1960) PhD phys. chemistry (U. of Notre Dame, 1986). Sr. res. scient. (tel. 314 + 537-6491, fax 314 + 537-6480). *Chemistry, theoretical, computational, molecular modeling, proteins and peptides, structure - function.*

Miller, Dr. Lance L. Solid State Phys. Dept., Ames Lab., Iowa State U., Ames, IA 50011, USA. (1961) PhD chemistry (Iowa State U., 1988). Asst. physicist. (tel. 515 + 294-6816). *Crystallography, solid state physics, materials science.*

Miller, Dr. Maria. BRI Crystallography Lab., Nat. Cancer Inst., Frederick Cancer Res. Facil., Frederick, MD 21701, USA. PhD biophysics (U. of Warsaw, 1979). *Proteins, nucleic acids, structure, interactions.*

Miller, Mr. Mark L. Geology Dept., U. of New Mexico, Northrup Hall, Albuquerque, NM 87131, USA. (1956) MS geology (Virginia Polytechnic Inst., 1986). Grad. student. (tel. 505 + 277-9447, fax 505 + 277-0090, email CRYSTAL at UNMB). *Crystal growth, chemistry, radiation damage.*

Miller, Dr. Richard Wayne. Fiber Res. Dept., Monsanto Chemical Co., P.O. Box 12830, Pensacola, FL 32575, USA. (1947) PhD phys. chem. (Duke U., 1976). Res. fel. (tel. 904 + 968-8358). *Fiber diffraction, morphology, process statistical control.*

Miller, Prof. Robert Llewellyn. Michigan Molecular Inst., 1910 W. St. Andrews Drive, Midland, MI 48640, USA. (1929) PhD chemical physics (Brown U., 1954). Sr. res. scient. (tel. 517 + 832-5555). *Macromolecules, structure, properties.*

Millner, Dr. Ozra Elmo. Computational Chem., Becton Dickinson Res. Ctr., 21 Davis Dr., P.O. Box 12016, Research Triangle Park, NC 27709, USA. (1946) PhD medicinal chemistry (U. of Tennessee, 1973). Sr. computational chemist. (tel. 919 + 549-8641, ext. 105, fax 919 + 549-7572). *Protein structure, molecular mechanics, modeling, dynamics, quantum chemistry.*

Minkin, Mrs. Jean Albert. Geologic Div., Mail Stop 929, Eastern Mineral Resources Br., U.S. Geological Survey, Reston, VA 22092, USA. (1925) BA physics (Bryn Mawr C., 1947). Res. physicist. (tel. 703 + 648-6461). *Mineralogy, petrology, fine-grained mineral deposits, trace element geochemistry, nondestructive micro analysis techniques.*

Mirsky, Dr. Kira. 290 Skycrest Dr., Ashland, OR 97520, USA. (1935) PhD physics, mathematics (Academy of Sciences, USSR, 1967). Assoc. res. chemist. (tel. 503 + 482-5858). *Interatomic and intermolecular interactions, conformation analysis, organic chemical crystallography.*

Modrick, Ms. Michelle Ann. Corporate Res., Analyt. Sci. Lab., Exxon Res. and Eng. Co., US Rte. 22 East, Annandale, NJ 08801, USA. Assoc. res. chemist. (tel. 201 + 730-2108).

Moffat, Prof. John Keith. Section of Biochemistry, Mol. & Cell Biology, Cornell U., Ithaca, NY 14853, USA. (1943) PhD protein crystallography (Cambridge U., UK, 1970). Prof. (tel. 607 + 255-4677). email MOFFAT at CRNLCHES). *Macromolecular crystallography, synchrotron radiation, biophysics, polypeptide hormones, calcium binding proteins, metalloproteins.*

Moini, Dr. Ahmad. Chemistry Dept., Michigan State U., East Lansing, MI 48824, USA. (1963) PhD chemistry (Texas A&M U., 1986). Res. assoc. (tel. 517 + 353-9196). *Solid state chemistry, catalysis, inorganic chemistry, small angle scattering, crystallography.*

Mokren, Dr. James David. 1 Mitchell St., Jackson, OH 45640, USA. (1937) PhD inorg. chemistry (Ohio State U., 1974). (tel. 614 + 682-7083). *Inorganic crystal structures.*

Molea, Mr. Frank N. Div. of Eng. and Applied Physics, Harvard U., 9 Oxford St., Cambridge, MA 02138, USA. (1934) BBA elec. engin., bus. management (Northeastern U., 1960). Techn. assoc. (tel. 617 + 495-4469). *Amorphous structure determination, small angle X-ray scattering, amorphous metals, glasses.*

Momany, Mr. Cory. Chemistry Dept., U. of Texas, Austin, TX 78712, USA. (1961) BA biochemistry (Rice U., 1983). Res. asst. (tel. 512 + 471-3625). *Protein crystallography, structure - function relationships, biochemistry.*

Moncrief, Dr. J. William. Office, V.P. for Academic Affairs, Presbyterian C., Clinton, SC 29325, USA. (1941) PhD chemistry (Harvard U., 1966). Vice-

Pres. (tel. 803 + 833-2820). *Liquid crystals, crystal and molecular structure determination.*

Mondragón, Prof. Alfonso. Biochem., Mol. Biol. & Cell Biol. Dept., Northwestern U., 2153 Sheridan Rd., Evanston, IL 60208, USA. (1958) PhD biophysics (U. of Cambridge, UK, 1985). Asst. prof. (tel. 312 + 491-4254, fax 312 + 491-5211, email Bitnet MONDRAGON at NUALL). *X-ray crystallography, proteins, nucleic acids.*

Monroe, Prof. Eugene A. Ceramics Dept., SUNY at Alfred, Alfred, NY 14802, USA. (1934) PhD, mineralogy (1961), DDS, dentistry (1973). Assoc. prof. (tel. 607 + 871-2231). *Crystallography, mineralogy, dental materials, prosthesis.*

Montfort, Dr. William R. Biochem. & Biophys. Dept. 964-S, U. of Calif. at San Francisco, San Francisco, CA 94143, USA. (1958) PhD biochemistry (U. of Texas at Austin, 1985). Postdoct. fel. (tel. 415 + 665-6932). *Macromolecular crystallography.*

Montgomery, Mr. Thomas S. 648 Berkshire Dr., State College, PA 16803, USA.

Moore, Mr. Donald L. Instruments & Techn. Inc., 112 Water St., Naperville, IL 60540, USA. (1928) Met. eng. (U. of Cincinnati, 1952). Pres. (tel. 312 + 355-7748, telex 324597). *Scientific instrumentation.*

Moore, Mrs. Elizabeth J. Weichel. P.O. Box 63, Underhill Center, VT 05490, USA. (1925) AB geology (Radcliffe C., 1946). Retired from IBM. (tel. 802 + 899-4788). *Optical crystallography, crystal physics, microscopy.*

Moore, Prof. Paul Brian. Geophysical Sci. Dept., U. of Chicago, 5734 S. Ellis Ave., Chicago, IL 60637, USA. (1940) PhD geophysical sciences (U. of Chicago, 1965). Prof. (tel. 312 + 753-8111). *Systematology of atomic arrangements, minerals, inorganic crystals, mineral paragenesis, plane and space partitioning, convex polyhedra theory, crystal chemical homologies.*

Moreland, Dr. James Andrew. Wacker Siltronic Corp., P.O. Box 03180, Portland, OR 97203, USA. (1946) PhD inorganic crystallography (U. of Calif. at Irvine, 1974) Dir. (tel. 503 + 243-2020). *Crystal growth, crystal defects, electronic materials.*

Morgan, Dr. Joseph. 10153 Oakton Terrace Rd., Oakton, VA 22124, USA. (1909) PhD physics (MIT, 1937). Retired. (tel. 703 + 255-6136). *Structure, liquids.*

Moriarty, Dr. John Lawrence Jr. Materials Sci. Div. Eng. Directorate, Rock Island Arsenal, Rock Island, IL 61299, USA. (1932) PhD phys. chemistry (U. of Iowa, 1960). Sr. metallurgist. (tel. 319 + 323-4686). *Applied crystallography, composites, intermetallic compounds.*

Morimoto, Dr. Carl Noboru. 4003 Hamilton Park Dr., San Jose, CA 95130, USA. (1942) PhD phys. chem., crystallog. (U. of Washington, 1970). (tel. 408 + 943-7921). *Crystallographic computing, computer graphics, biochemical crystal structures.*

Moring, Dr. Jill. Biomolecular Structure Anal. Ctr., U. of Connecticut Health Ctr., Farmington, CT 06032, USA. (1943) PhD chemistry (solid state) (U. of Connecticut, 1986). Postdoct. fel. (tel. 203 + 679-4315). *Small-angle scattering, crystallography, protein, small molecule.*

Morosin, Dr. Bruno. Shock Wave & Explosive Physics-1131, Sandia Nat. Lab.s., Albuquerque, NM 87185, USA. (1934) PhD phys. chemistry (U. of Washington, 1959). Div. supervisor. (tel. 505 + 844-8169, fax 505 + 846-2009). *Crystal physics, structures.*

Morosoff, Dr. Nicholas C. Polymer Res. Lab., Res. Triangle Inst., P.O. Box 12194, Research Triangle Park, NC 27709, USA. (1937) PhD phys. chemistry (PUN, 1965). Sr. res. physical chemist. (tel. 919+ 541-6866, telex 802509 RTIRTPK). *Polymer morphology, polymerization, plasma, solid state, small angle X-ray scattering.*

Morris, Mrs. Marlene Cook. JCPDS, Internatl. Ctr. Diffraction Data, 1601 Park Ln., Swarthmore, PA 19081, USA. (1933) BS chemistry (Howard U., 1955). Dir., JCPDS assoc. at NBS (tel. 301 + 921-2921). *X-ray powder diffraction, inorganic crystal structures.*

Morris, Ms. Nancy L. Chemistry Dept., Georgetown U., 37th & O St. NW, Washington, DC 20057, USA. (1959) PhD chemistry (Georgetown U., 1989). (tel. 202 + 687-4353). email Bitnet N__MORRIS at GUVAX). *Diffraction from liquid crystals, NMR, paramagnetic molecules.*

Morrison, Mr. George. 810 John St., Wauuakee, WI 53597, USA. (tel. 608 + 849-4917).

Morrow, Prof. John Charles III. Chemistry Dept., 263 Venable Hall, CB#3290, U. of North Carolina, Chapel Hill, NC 27599-3290, USA. (1924) PhD phys. chemistry (MIT, 1949). Prof. (tel. 919 + 962-0165). *X-ray diffraction, molecular structure determination.*

Morrow, Dr. Scott Imlay. US Army Res., Devel. & Eng. Cmd., SMCAR AEE WW B 3022 Picatinny Arsenal, NJ 07806-5000, USA. (1920) PhD inorg. chemistry (Case-Western Reserve U., 1951). Res. chemist, microscopist. (tel. 201 + 724-4703). *Microscopy, polarizing, scanning electron, phase rule studies, polymorphism and isomorphism, explosives, propellants, environmental studies and analysis, pollution.*

Moss, Mr. Simon C. Physics Dept., U. of Houston, University Park, Houston, TX 77004, USA. (tel. 713 + 749-2840).

Moudrianakis, Prof. Evangelos N. Biology Dept., Johns Hopkins U., Charles & 34th Sts., Baltimore, MD 21218, USA. (1939) PhD mol. biology/biophysics (Johns Hopkins U., 1964). Prof. (tel. 301 + 338-7305). *Self-assembling systems, chromosomes, chromatin, histone - DNA complexes, photosynthetic membranes, enzymology, energy coupling.*

Moudy, Ms. Lavada Ann. 2830 Madonna Dr., Fullerton, CA 92635, USA. (1926) MS ceramics (U. of Washington, 1959). (tel. 714 + 762-6144). *Thin film structure and properties, crystal defects, crystal growth, X-ray diffraction and fluorescence, scanning electron microscopy.*

Mowbray, Dr. Sherry Lynn. HHMI, U. of Texas Health Ctr., 5323 Harry Hines Blvd., Dallas, TX 75235, USA. (1954) PhD chemistry (MIT, 1983). Asst. investigator. (tel. 214 + 689-5018). *Protein crystallography.*

Mozzi, Dr. Robert Lewis. Res. Div., Raytheon Co., 131 Spring St., Lexington, MA 02173, USA. (1931) PhD physics (MIT, 1967). Consulting scient. (tel. 617 + 860-3095). *Crystal defects, semiconductors, integrated circuit technology, microlithography.*

Mrose, Miss Mary E. 114 N. Wayne St., Apt. 2, Arlington, VA 22201, USA. (1910) MA geography (Boston U., 1944). Geologist-mineralogist. (tel. 703 + 860-6670). *Mineralogical investigations.*

Muchmore, Ms. Christine R. A. 216 Stuart Ave., Apt. 3, Kalamazoo, MI 49007, USA. (tel. 616 + 385-7780).

Muchmore, Mr. Steven W. 7255-209-1, Upjohn Pharmaceutical Co., 301 Henrietta St., Kalamazoo, MI 49001, USA. (1958) BA science (U. of Oklahoma, 1980). Chemist. (tel. 616 + 385-7501). *Crystallography, macromolecular, low temperature, structure refinement, accurate data collection.*

Mucker, Prof. Kenneth. Physics and Astronomy Dept., Bowling Green State U., Bowling Green, OH 43403, USA. (1939) PhD phys. chemistry (Ohio State U., 1966). Assoc. prof. (tel. 419 + 372-8707). *Applied crystallography.*

Mueller, Dr. Melvin H. IPNS, Bldg. 360, Argonne Nat. Lab.s, 9700 S. Cass Ave., Argonne, IL 60439, USA. (1918) PhD chemistry (U. of Illinois, 1949). Sr. scient. (tel. 312 + 972-3554 or 832-4056). *Neutron diffraction pulsed sources, metal hydrides, actinide compounds, instrumentation.*

Mueser, Mr. Timothy C. 3053 T St., Lincoln, NE 68503, USA. (tel. 402 + 472-1470).

Muir, Prof. James Alexander. Physics Dept., U. of Puerto Rico, Rio Piedras, PR 00931, USA. (1938) PhD physics (Northwestern U., 1966). Prof. (tel. 809 + 764-2258). email j__muir at uprenet). *Crystal structures, transition metal complexes, semiconductors.*

Muldawer, Prof. Leonard. Physics Dept., Temple U., Philadelphia, PA 19122, USA. (1920) PhD physics (MIT, 1948). Prof. (tel. 215 + 787-7668). *Phase transformations, alloys, diffuse X-ray scattering, line profile analysis, small angle neutron scattering, Fourier methods.*

Mullica, Dr. Donald Foster. Chemistry Dept., Baylor U., Waco, TX 76703, USA. (1928) PhD phys. chemistry (Baylor U., 1977). Res. crystallographer. (tel. 817 + 755-3311, ext. 40). *Crystallography, lanthanide and actinide chemistry, physical chemistry.*

Murali, Mr. Ramachandran. Molec. Biophys. & Biochem. Dept., Columbia U., 630 W. 168th St., New York, NY 10032, USA.

Murfitt, Mr. Robert R. 229 Old Westford Rd., Chelmsford, MA 01824, USA. (tel. 617 + 256-0906).

Murmann, Prof. R. Kent. Chemistry Dept., U. of Missouri, Columbia, MO 65211, USA. (1927) PhD chemistry (Northwestern U., 1953). Prof. (tel. 314 + 882-2826). *Complex inorganic ions.*

Murthy, Dr. Krishna H. M. Life Sci. Res. Labs., Eastman Kodak Co., Bldg. 82-B, 1999 Lake Ave., Rochester, NY 14650-2118, USA. (1952) PhD mol. biophysics (Indian Inst. of Sci., Bangalore, 1981). Sr. res. scient. (tel. 716 + 722-4960, fax 716 + 722-1675). *Macromolecular crystallography, structure - function relationships, anomalous scattering methods, enzyme action mechanism, refinement methods.*

Murthy, Dr. N. Sanjeeva. Corporate Res., Allied-Signal Inc., Columbia Rd., P.O. Box 1021R, Morristown, NJ 07960, USA. (1949) PhD materials science (U. of Connecticut, 1976). Res. assoc. (tel. 201 + 455-3764). *Polymer structure and morphology, structure - property relationships, disordered materials, small angle scattering.*

Myer, Prof. George Henry. Geology Dept., 303 Beury Hall, Temple U., 13th. and Norris Sts., Philadelphia, PA 19122, USA. (1937) PhD geology (Yale U., 1965). Assoc. prof. (tel. 215 + 787-7173). *Crystallography, mineralogy, petrology, ceramics.*

Mylvaganam, Dr. Sangari Eshwari. Mol. Biology Dept. MB5, Scripps Clinic & Res. Fndn., 10666 N. Torrey Pines Rd., La Jolla, CA 92037, USA. (1961) PhD crystallography (U. of London, England, 1987). Postdoct. fel. (tel. 619 + 554-2807). *Structural studies, antibodies.*

Namboodiri, Dr. Krishnan. Bio / Molec. Eng. Branch, Naval Res. Lab., Code 6190, Washington, DC 20375, USA. (1953) PhD org. chemistry (U. of Madras, India, 1983). Res. scient. (tel. 202 + 767-1154, 767-1679). *Protein crystallography, data bases, automated structure searching, protein engineering, molecular modeling.*

Nanni, Mr. Raymond. Crystallography Dept., U. of Pittsburgh, 305 Thaw Hall, Pittsburgh, PA 15260, USA. (tel. 412 + 624-9305).

Narayanan, Prof. V. Anantha. Physics Dept., Savannah State C., P.O. Box 20473, Savannah, GA 31404, USA. (1936) PhD physics (Indian Inst. of Sci., 1962). Prof. (tel. 912 + 356-2317). *Crystals, vibrational spectra, hydrogen bond vibrations, crystal field calculations, structure parameters, molecular spectra, crystals.*

Narendra, Dr. Narayana. Inst. for Cancer Res., Fox Chase Cancer Ctr., 7701 Burholme Ave., Philadelphia, PA 19111, USA. (1959) PhD crystallography

(Indian Inst. of Sci., Bangalore, 1986). Postdoct. assoc. (tel. 215 + 728-3617). *Biological molecular crystallography.*

Nassau, Dr. Kurt. Physics of Matls. Res., AT&T Bell Labs., Rm. 6D-205, 600 Mountain Ave., Murray Hill, NJ 07974, USA. (1927) PhD phys. chemistry (U. of Pittsburgh, 1959). Distinguished mem. techn. staff. (tel. 201 + 582-2589). *Crystal chemistry, crystal growth, glass.*

Nathan, Dr. Robert. Jet Propulsion Lab., Caltech., 4800 Oak Grove Dr., Pasadena, CA 91109, USA. (Alt.: 1125 Rexford Av., Pasadena, CA 91107, USA). (1927) PhD phys. chem., crystallography (CIT, 1955). Member techn. staff. (tel. 818 + 351-8606). *Image processing, computer architecture, gerontology.*

Navia, Dr. Manuel Alberto. Merck Inst. for Therapeutic Res., Merck and Co., (R80M203) P.O. Box 2000. Rahway, NJ 07065, USA. (1946) PhD biophysics (U. of Chicago, 1974). Sr. res. biophysicist. (tel. 201 + 594-7256, fax 201 + 574-4773). *X-ray crystallography, macromolecular, electron microscopy, enzyme-inhibitor complexes, enzyme mechanisms.*

Neidhart, Dr. David James. Computer Assisted Mol. Des. D-47E, AP 9A/LL. Abbott Labs., Abbott Park, IL 60064, USA. (1962) PhD biological chemistry (MIT, 1989). Postdoct. fel. *Protein crystallography, molecular pharmacology.*

Neilson, Dr. George Francis. Matls. Sci. & Eng. Dept., U. of Arizona, 4715 E. Fort Lowell Rd., Tucson, AZ 85712, USA. (1930) PhD phys. chemistry (Ohio State U., 1962). Res. prof. (tel. 602 + 322-2996). *X-ray scattering, small angle, phase separation, microstructure characterizatiion, nucleation, crystallization, space processing, glass.*

Nelson, Mr. A. Dwayne. 822 S. 2nd St., Stillwater, MN 55082, USA. (tel. 612 + 439-7490).

Nemiroff, Dr. Michael. Crystal Dynamics, 14024 Rue D'Antibes, Del Mar, CA 92014, USA. (1940) PhD inorg. chemistry (PUN, 1972). Pres. (tel. 619 + 481-9242).

Neuman, Mr. Melvin A. 4002 Sternberg, Schofield, WI 54476, USA.

Newman, Dr. Robert Alan. Analytical Sci., Bldg. 1897, Dow Chemical Co., Midland, MI 48667, USA. (1952) PhD inorg. chemistry (Iowa State U., 1981). Proj. leader. (tel. 517 + 636-4001).

Newnham, Prof. Robert Everest. Materials Res. Lab., Pennsylvania State U., University Park, PA 16802, USA. (1929) PhD physics (Cambridge U., UK, 1960). Prof., solid state sci. (tel. 814 + 865-1612). *Crystal physics, chemistry, electroceramics, composite materials.*

Newsam, Dr. John Michael. Exxon Res. and Eng. Co., Route 22 East, Annandale, NJ 08801, USA. PhD chemistry (Oxford U., UK, 1980). Staff chemist. (tel. 201 + 730-2901). *Catalysis, neutron scattering, synchrotron X-ray diffraction, superionic conductors, zeolites, oxide superconductors, computer modeling.*

Nguyen, Dr. Khe Thanh. Peptide Chem., ProCyte Corp., 2893 152nd. Ave., NE, Redmond, WA 98052, USA. (1957) PhD org. chemistry (U. of Washington, 1986). Sr. scient. (tel. 206 + 867-1820). *Drug design, NMR, computer applications, synthesis.*

Nguyen, Mr. Thao A. Student Rm. 13-4053, Mass. Inst. of Techn., Cambridge, MA 02139, USA.

Nichols, Mr. Monte Carl. Sandia Nat. Lab.., Div. 8313, Livermore, CA 94550, USA. (1938) MS phys. chemistry (U. of Arizona, 1962). Member tech. staff. (tel. 415 + 422-2906). *Mineralogy, X-ray crystallography, spectroscopy.*

Nicklow, Dr. Robert Merle. Solid State Div., MS 031, Oak Ridge Nat. Lab.., Bldg. 3025, P.O. Box 2008, Oak Ridge, TN 37831-6031, USA. (1936) PhD physics (Georgia Inst. of Techn., 1964). Sr. staff scient./Mgr. neutron fac. (tel. 615 + 574-5240, fax 615 + 574-4143, email RMN at ORNLSTC). *Lattice dynamics, magnetic excitations in solids, neutron spectrometry.*

Nicolosi, Dr. Joseph Anthony. Eng. Dept., Philips Electronic Instruments, 85 McKee Dr., Mahwah, NJ 07430, USA. (1950) PhD physics, crystallography (PUN, 1982). Development mgr. (tel. 201 + 529-3800, ext. 456). *X-ray diffraction, materials characterization, X-ray spectroscopy by EDS and WDS, EDAX micro-analytical instrumentation, research, development, applied research.*

Nielsen, Mr. Christopher Pine. Biology Dept., B-014, U. of Calif. at San Diego, La Jolla, CA 92093, USA. (1953) MA applied mathematics (U. of Calif. at San Diego, 1977). Programmer/analyst. (tel. 619 + 534-2565). email cn at chem.ucsd.edu). *Area detector software.*

Nilakantan, Dr. Ramaswamy. Med. Res. Div., Dept. 908, Lederle Labs., American Cyanamid Co., N. Middletown Rd., Pearl River, NY 10965, USA. (1956) PhD molecular biophysics (Indian Inst. of Sci., Bangalore, Res. biologist. (tel. 914 + 732-3773). *Chemical information, chemical graph theory, databases.*

Noether, Dr. Herman Dietrich. Textile Res. Inst., P.O. Box 625, Princeton, NJ 08542, USA. (Alt.: 20 Greenbriar Dr., Summit, NJ 07901, USA). (1912) PhD phys. chemistry (Harvard U., 1943). Res. assoc., consultant. (tel. 201 + 522-1653). *Polymers, fibers, plastics, structure - properties - morphology relationships, X-ray diffraction, scattering, small angle X-ray scattering, polycrystalline materials, polymeric materials.*

Nolta, Miss Kathleen Virginia. Biochem. & Mol. Biol. Dept., U. of Chicago, 920 E. 58th. St., Chicago, IL 60637, USA. (1966) BS chemistry, cellular biol. (U. of Michigan, 1988). Grad. student (tel. 312 + 363-7328). *Protein crystallography, crystallization, membrane proteins.*

Nomura, Dr. Glenn S. Centers for Disease Control, CEHIC, Toxicology, MS F-17, Atlanta, GA 30333, USA. (1947) PhD org. chemistry (Georgia Inst. of Techn.,

1985). Res. chemist. (tel. 404 + 488-4176, fax 404 + 488-4141). *Molecular modeling, organic synthesis, toxicology.*

Nordman, Prof. Christer Eric. Chemistry Dept., U. of Michigan, Ann Arbor, MI 48109-1055, USA. (1925) PhD phys. chemistry (U. of Minnesota, 1953). Prof. (tel. 313 + 764-7326, fax 313 + 747-4865, email Bitnet USERN767 at UMICHUB). *Organic, biological crystal structures, computer programming.*

Norton, Dr. Michael Louis. Chemistry Dept., U. of Georgia, Athens, GA 30602, USA. (1955) PhD solid state chemistry (Arizona State U., 1982). Asst. prof. (tel. 404 + 542-2626). *Superconductivity, synthetic metals, bronzes, superlattices, crystal growth, electrochemical deposition, layered compounds.*

Novotny, Dr. Jiri. Macromol. Modeling Dept., Squibb Inst. Med. Res., P.O. Box 4000, Princeton, NJ 08543, USA. (1943) PhD biochemistry (Charles U., Prague, 1970). Dir. (tel. 609 + 683-6209, fax 609 + 683-6280, email novotny at squibb.com.csnet). *Proteins, structure, computing.*

Nowotny, Prof. Dr. Hans. Inst. of Material Sci., U. of Connecticut, Storrs, CT 06268, USA. (1911) Dr.techn. physics (Techn. U. Vienna, 1934). Prof. Emeritus. (tel. 203 + 486-4619). *Inorganic crystal structures, alloy chemistry.*

Noyan, Dr. Ismail Cevdet. Res. Div., IBM T.J. Watson Res. Center., P.O. Box 218, Yorktown Heights, NY 10598, USA. (1956) PhD materials sci. and eng. (Northwestern U., 1984). Res. staff member (tel. 914 + 945-3941, telex 137456, fax 914 + 945-2141, email NOYAN at IBM.COM). *Stress/strain determination, X-ray diffraction, synchrotron radiation.*

Nunes, Prof. Anthony Charles. Physics Dept., U. of Rhode Island, Kingston, RI 02881, USA. (1942) PhD physics (MIT, 1969). Prof. (tel. 401 + 792-2048). *Diffraction, X-ray, neutron, polarized neutron, powder, small angle scattering, magnetic colloids, superconductivity, biophysics.*

Nunn, Dr. Christine M. Chemistry Dept., U. of Texas at Austin, Austin, TX 78705, USA. (1961) PhD inorg. chemistry (U. of Bristol, England, 1986). Res. assoc. (tel. 512 + 471-7710). *Small molecule X-ray crystallography, main-group organometallic compounds.*

O'Donnell, Dr. Terence J. 1307 W. Byron St., Chicago, IL 60613, USA. (1951) PhD phys. chemistry (U. of Illinois at Chicago, 1980). Consultant. (tel. 312 + 327-9390). *Molecular modeling, computer graphics.*

O'Keeffe, Prof. Michael. Chemistry Dept., Arizona State U., Tempe, AZ 85287, USA. (1934) PhD chemistry (U. of Bristol, UK, 1959). Prof. (tel. 602 + 965-3670). *Chemistry, crystal, solid state.*

O'Rourke, Mr. John A. MST-5, MS-G734, Los Alamos Nat. Lab.., P.O. Box 1663, Los Alamos, NM 87545, USA. (1924) MS metallurgical eng. (Colorado School of Mines, 1950). Staff member. (tel. 505 + 667-4033). *Crystallography, textures, mechanical properties.*

Oatley, Dr. Susan Alaine. VA Med. Ctr., V-151, 3350 La Jolla Village Dr., San Diego, CA 92161, USA. DPhil biophysics (Oxford U., UK, 1985). Res. assoc. (tel. 619 + 453-7500, ext. 3370, email S0oatley at ucsd.edu). *Protein crystallography, oncogenes, antibodies.*

Ogata, Dr. Craig. Biochemistry Dept., Columbia U., 630 W. 168th. St., New York, NY 10032, USA. (tel. 212 + 305-2219).

Ohashi, Dr. Yoshikazu. Exploration and Prod. Res. Ctr., ARCO Oil and Gas, 2300 W. Plano Parkway, Plano, TX 75075, USA. (1941) PhD geology (Harvard U., 1973). Sr. res. geologist. (tel. 214 + 754-6510). *Silicates, minerals, crystal structures.*

Ohlendorf, Dr. Douglas Henry. Experimental Station, E.I. duPont de Nemours Co., Bldg. 228, Rm. 316F, Wilmington, DE 19898, USA. (1950) PhD biochemistry (Washington U., 1978). Prin. investig. (tel. 302 + 695-7389). *Structure, macromolecules, protein - DNA interaction, oxygenases.*

Ohrt, Miss Jean Marie. RPMI - Crystallography, NYS Dept. of Health, 33 Tamarack St., Buffalo, NY 14220, USA. (1923) BA biology (U. of Buffalo, 1949). Cancer res. scient., retired. (tel. 716 + 825-5113). *Steroid structure, structure - function relationships, biologically active compounds, carcinogens, carcinostats.*

Okaya, Prof. Yoshi Haru. Chemistry Dept., SUNY at Stony Brook, Stony Brook, NY 11794-3400, USA. (1927) PhD chemistry (U. of Osaka, Japan, 1956). Prof. (tel. 516 + 632-7911). *Computer controlled diffractometry, synchrotron radiation, solid state physics, absolute configuration determination.*

Oki, Mr. Aderemi Rasaq. Chemistry Dept., U. of Wyoming, Laramie, WY 82071, USA. (1958) MS chemistry (U. of Ilorin, Nigeria, 1983). Grad. student. (tel. 307 + 766-5367). *Inorganic complexes, synthesis, characterization.*

Olafson, Dr. Barry. Biodesign Inc., 199 S. Los Robles, #270, Pasadena, CA 91101, USA. (1949) PhD chemical physics (CIT, 1978). Chief sci. off. (tel. 818 + 793-3600, fax 818 + 793-8098). *Computational chemistry.*

Oliver, Dr. Joel Day. Miami Valley Res. Lab., Procter & Gamble Co., P.O. Box 398707, Cincinnati, OH 45239-8707, USA. (1945) PhD phys. chemistry (U. of Texas at Austin, 1971). X-ray group leader. (tel. 513 + 245-1437, fax 513 + 741-9154). *Macromolecular crystallography, protein crystallization (automated).*

Ollis, Dr. David L. Biochemistry Dept., Northwestern U., 2153 Sheridan Rd., Evanston, IL 60208, USA. (1952) PhD chemistry (U. of Sydney, Australia, 1980). Asst. prof. (tel. 312 + 491-4253). *Proteins, structure - function.*

Olmstead, Dr. Marilyn Morgan. Chemistry Dept., U. of Calif. at Davis, Davis, CA 95616, USA. (1943) PhD chemistry (U. of Wisconsin at Madison, 1969). Staff res. assoc. (tel. 916 + 752-6668, email Bitnet MMOLMSTEAD at UCDAVIS). *Chemistry, inorganic, organometallic.*

Olsen, Prof. Kenneth Wayne. Chemistry Dept., Loyola U. of Chicago, 6525 N. Sheridan Rd., Chicago, IL 60626, USA. (1944) PhD biochemistry (Duke U., 1972). Assoc. prof. (tel. 312 + 508-3121). *Protein crystallography, stability, structure prediction, evolution.*

Olson, Dr. Arthur Jules. Mol. Biology Dept., Res. Inst. of Scripps Clinic, 10666 N. Torrey Pines Rd., La Jolla, CA 92087, USA. (1946) PhD phys. chemistry (UCB, 1975). Sr. staff scient. (tel. 619 + 554-9702, email olson at scripps.edu). *Computation, computer graphics, macromolecular modeling, macromolecular interactions, supramolecular assemblies.*

Olson, Dr. David Harold. Central Res. Lab., Mobil Res. & Dev. Corp., P.O. Box 1025, Princeton, NJ 08540, USA. (1937) PhD phys. chemistry (Iowa State U., 1963). Res. scient. (tel. 609 + 734-4253). *Zeolite crystal chemistry, heterogeneous catalyst characterization.*

Onan, Prof. Kay Denise. Chemistry Dept., Northeastern U., 360 Huntington Ave., Boston, MA 02115, USA. (1949) PhD phys. chem., X-ray crystallog. (Duke U., 1975). Assoc. prof. (tel. 617 + 437-2847). *Structure determination, biologically significant organic molecules, transition metal complexes.*

Ordway, Dr. Fred. ARTECH Corp., 14554 Lee Road, Chantilly, VA 22021, USA. (1922) PhD phys. chemistry (CIT, 1949). Executive vice-pres. (tel. 703 + 378-7263, fax 703 + 378-7267). *Amorphous structures, computer applications.*

Ortega, Dr. Richard B. Siemens Analytical X-ray Instruments Inc., 6300 Enterprise Ln., Madison, WI 53719-1173, USA. (1953) PhD chemistry (U. of New Mexico, 1980). Sr. appl. scient. (tel. 608 + 276-3000). *Data collection, macromolecular compounds, low temperature hardware, area detector applications.*

Osgood, Mr. Brian Clair. 2302 Pine Log Rd., Aiken, SC 29801, USA. (1954) BS ceramic eng. (Alfred U., 1976). Sr. chemist. (tel. 803 + 725-2173). *Powder diffraction, automation.*

Ostrofsky, Mr. Bernard. 23 River Rd., Naperville, IL 60566, USA. (1922) BSc phys. chemistry (City C. of New York, 1945). Consultant. (tel. 312 + 420-7314). *Instrumentation, crystal physics, X-ray spectroscopy, liquid structures.*

Ott, Miss J. E. Chemistry Dept., U. of Pennsylvania, 231 S. 34th. St., Philadelphia, PA 19104-6323, USA. (tel. 214 + 898-4785).

Ou, Dr. Chia-Chih. Construction Products Div., W. R. Grace and Co., 62 Whittemore Ave., Cambridge, MA 02140, USA. (1945) PhD phys. chemistry (Rutgers U., 1976). Res. Group Leader (tel. 617 + 876-1400). *Inorganic crystal structures, clay minerals.*

Pabo, Prof. Carl Ogren. Molec. Biology Dept. John Hopkins Med. School, 725 North Wolfe St., Baltimore, MD 21205, USA. (1952) PhD biochemistry (Harvard U., 1980). Assoc. prof. (tel. 301 + 955-3933, fax 301 + 955-4857, email Bitnet PABO at JHUIGF). *Structural molecular biology, protein design, folding, protein - DNA interactions.*

Pabst, Prof. Adolf. Geology and Geophysics Dept., U. of Calif. at Berkeley, Berkeley, CA 94720, USA. (1899) PhD geology, mineralogy (UCB, 1928). Prof. emeritus. (tel. 415 + 642-1878). *Mineralogy, crystallography, crystal chemistry, minerals.*

Pace, Ms. Helen C. 426 Dickinson St., Philadelphia, PA 19147, USA. (tel. 215 + 270-4265).

Padlan, Dr. Eduardo Agustin. Lab. Mol. Biol., NIDDK, Nat. Insts. of Health,. Bldg. 2, Rm. 323, Bethesda, MD 20892, USA. (1940) PhD biophysics (Johns Hopkins U., 1968). Visit. scient. (tel. 301 + 496-4295). *Biological macromolecules.*

Padmanabhan, Mr. K. Chemistry Dept., U. of Illinois, Urbana, IL 61801, USA.

Pagoaga, Dr. Katherine. Centre for Computing and Appl. Math., Natl. Bureau of standards, Mail Stop 713, 325 Broadway, Boulder, CO 80303-3328, USA. (1952) PhD geochemistry (U. of Maryland, 1983). Computer special. (tel. 303 + 497-5104, email Bitnet PAGOAGA at NBS, kpagoga at umd2.umd.edu). *Uraniam minerals. crystallographic programming, programming.*

Pähler, Dr. Arno. Biochem. & Molec. Biophysics Dept., Columbia U., 630 W. 168th St., New York, NY 10032, USA. (1949) PhD physics (Götingen, FRG, 1983). Postdoc. res. scient. (tel. 212 + 305-1846, email Bitnet PAHLER at CUMBG). *Protein crystallography.*

Palenik, Prof. Gus Joseph. Chemistry Dept., U. of Florida, Gainesville, FL 32611, USA. (1933) PhD chemistry (USC, 1960). Prof. (tel. 904 + 392-6734, email CRYSTAL at UFFSC). *Volatile metal complexes, aluminum complexes, complexes with unusual coordination numbers, synthesis, structure, characterization.*

Pallai, Dr. Peter V. Med. Chem. Dept., Boehinger Intelheim Pharmaceuticals,Inc. P.O. Box 368, Ridgefield, CT 06877, USA. (1945) PhD org. chemistry (Jozsef A. U., Hungary) (tel. 203 + 798-5133, fax 202 + 790-6815).

Palmer, Dr. Kenneth James. 1134 Mesters Dr., Pebble Beach, CA 93953, USA. (1910) PhD chemistry (CIT, 1938). Retired. (U.S.D.A.) (tel. 408 + 373-5154). *X-ray crystallography, diffraction, organic crystals, fibers, membranes.*

Pangborn, Prof. Robert Northrup. Eng. Sci. and Mechanics Dept., Pennsylvania State U., 227 Hammond Bldg., University Park, PA 16802, USA. (1951) PhD mechanics, materials science (Rutgers U., 1979). Assoc. prof. (tel. 814 + 863-0721). *Mechanical behavior, stress analysis, defect characterization.*

Pangborn, Mr. Walter. Med. Fndn. of Buffalo, 73 High St., Buffalo, NY 14203, USA. (tel. 716 + 856-9600).

Paretzkin, Mr. Boris. Natl. Inst. Stnds. & Technol. Materials A209, Gaithersburg, MD 20899, USA. (1922) AB chemistry (New York U., 1944). Res. assoc. (tel. 301 + 975-5789). *X-ray powder patterns, superconductors.*

Parge, Dr. Hans Erich. Mol. Biology Dept., MB5, Res. Inst. of Scripps Clinic, 10666 N. Torrey Pines Rd., La Jolla, CA 92037, USA. (1955) PhD chemistry, crystallog. (Trinity C., Dublin, 1982). Res. assoc. (tel. 619 + 554-2806, telex 697168 SCRF SCRF, fax 619 + 554-8841, email parge at scripps.edu). *Macromolecules, crystallography, recognition, DNA, Drug - protein interactions.*

Parise, Dr. John B. Central Res. & Dev., Bldg.356, E. I. du Pont de Nemours Co., P.O. Box 80356 Wilmington, DE 19880-0356, USA. (1953) PhD (James Cook U., Australia, 1981). (tel. 302 + 695-9565 or 695-3602, telex 835420, fax 302 + 695-1664, email parisejb%esvax%dupont.com at relay.cs.net). *Crystal chemistry, physics, oxides, diffractometry, X-ray, neutron.*

Park, Dr. Chang Hoon. Protein Crystallography Lab., D47E, Abbott Labs., AP9Am Abbott Park, IL 60064, USA. (1947) PhD chemistry (U. of Kansas, 1982). Protein crystallog. (tel. 312 + 937-0488, fax 312 + 937-1511). *Biological macromolecules.*

Parker, Dr. Robert Louis. Metallurgy Div., Natl. Inst. Stnds. & Technol., Gaithersburg, MD 20899, USA. (1929) PhD solid state physics (U. of Maryland, 1960). Physicist, consultant. (tel. 301 + 975-6166). *Crystal growth, solidification, morphological stability, ultrasonics.*

Parker, Dr. Sidney Glenn. Materials Sci. Lab., Texas Instruments Inc., P.O Box 655936, M.S. 147, Dallas, TX 75265, USA. (1925) PhD inorg. chemistry (U. of Texas, 1951). Techn. staff member. (tel. 214 + 995-4190). *Electronic materials, Si solar cells, crystal growth, II-VI compounds, infrared detector materials.*

Parks, Ms. Elizabeth Annette Heady. 1860 Glendmere Dr., Birmingham, AL 35216, USA. (1958) PhD chemistry (U. of Texas at Austin, 1984). (tel. 205 + 979-6541). *Biological macromolecules.*

Parris, Dr. Kevin Delos. Mol. Biol. Lab., Natl. Inst. of Health, Bldg. #2, Rm. 323., Bethesda, MD 20892, USA. (1960) PhD org. chemistry (U. of Pittsburgh, 1988). Postdoct. fel. (tel. 301 + 496-4295). *Protein crystallography, molecular recognition.*

Parrish, Dr. William. Physical Sci. Dept., IBM Almaden Res. Ctr. K31/802, 650 Harry Rd., San Jose, CA 95120-6099, USA. (1914) PhD mineralogy & crystallography (MIT, 1940). Memb., res. staff. (tel. 408 + 927-2350). *X-ray diffractometry, synchrotron radiation, materials characterization.*

Parsons, Dr. Donald Frederick. Div. Labs. & Res., N. Y. State Dept. of Health, Empire State Plaza, Albany, NY 12201, USA. (1928) MD, DSc biophysics (U. of London, UK, 1956). Res. physician III. (tel. 518 + 474-7047). *Electron diffraction, high voltage electron microscopy, small angle X-ray diffraction, cancer research.*

Parthasarathy, Dr. R. Cntr. for Crystallographic Res., Roswell Park Memorial Inst., 666 Elm Street, Buffalo, NY 14263, USA. (1936) PhD physics (U. of Madras, India, 1962). Cancer res. scient. VI. (tel. 716 + 845-5819). *Biological molecules, structure and function, stereochemistry, X-ray diffraction physics, conformational analysis, nuclear magnetic resonance.*

Partin, Mr. Daniel Edward. Chemistry Dept., Arizona State U., Tempe, AZ 85287, USA. (1965) BS chemistry (Oregon State U., 1987). Grad. student. *Powder diffraction, X-ray, neutron.*

Parvez, Dr. Masood. Chemistry Dept., Pennsylvania State U., 152 Davey Lab., University Park, PA 16802, USA. (1947) PhD org. chemistry (Queen's U., Belfast, 1977). Dir. (tel. 814 + 865-1554, email Bitnet PYS at PSUVM). *Crystal, molecular structures, drugs, natural products, synthetic intermediates, inorganic and organometallic complexes, polyphosphazene derivatives, electron density distribution, drugs, organometallics.*

Patel, Dr. Jamshed R. Natls. Interface Res. Dept., AT&T Bell Labs., 600 Mountain Ave., Murray Hill, NJ 07974, USA. (1925) ScD physical metallurgy (MIT, 1954). Techn. staff member. (tel. 201 + 582-6698, fax 201 + 582-4228). *Defects, X-rays, topography, standing waves, dynamical diffraction, surface structure, interfaces.*

Pattabiraman, Dr. Nagarajan. Lab. for the Struct. of Matter, Code 6030, Naval Res. Lab., 4555 Overlook Ave., Washington, DC 20375-5000, USA. (1951) PhD biophysics (Indian Inst. of Sci., Bangalore, 1979). Sr. scient. (tel. 202 + 767-0657, fax 202 + 767-6874, email pattabiraman at nrl.arpa). *Molecular modeling, mechanics, computer graphics, nucleic acids, peptides, proteins, protein folding, X-ray crystallography, drug design.*

Pattanayek, Dr. Rekha. Molec. Biology Dept., Vanderbilt U., Box 1820-B, Station B, Nashville, TN 37235, USA. (1953) PhD X-ray crystallography (Calcutta U., India, 1986). Res. assoc. (tel. 615 + 322-2012). *X-ray diffraction, fiber, protein crystallography, protein - DNA interactions, macromolecular assemblies.*

Pattridge, Ms. Katherine A. Biophysics Res. Div., U. of Michigan, 2200 Bonisteel Blvd., Ann Arbor, MI 48109-2099, USA. (1948) MS geological sci. (U. of Michigan, 1976). Res. assoc. (tel. 313 + 763-2199). *Crystallographic computing, graphics, refinement, macromolecular structures.*

Paturle, Mr. Antoine. Chemistry Dept., Acheson Hall, SUNY at Buffalo, Buffalo, NY 14214, USA. (1964) MS physics (U. of Grenoble, France, 1986). Grad. student. (tel. 716 + 831-3263). *Synchrotron radiation, non-linear optical effects.*

Paul, Miss Elizabeth A. Chemistry Dept., Brandeis U., P.O. Box 9110, Waltham, MA 02154-9110, USA. (1958) MA chemistry (Brandeis U., 1988). Grad. student. (tel. 616 + 736-2534, email PAUL at BBRANDEI). *X-ray*

Paul, Prof. Iain Campbell. Chemistry Dept., U. of Illinois, 505 S. Mathews, Urbana, IL 61801, USA. (1938) PhD chemistry (U. of Glasgow, UK, 1962). Prof. (tel.

217 + 333-3007). *Chemical crystallography, solid state org. chemistry, X-ray structure analysis, molecular geometry.*

Pauling, Prof. Linus. 440 Page Mill Rd., Palo Alto, CA 94306, USA. (1901) PhD chemistry, physics, math. (CIT, 1925). (tel. 415 + 327-4064). *Physics, chemistry, crystallography, biology, medicine.*

Pavalow, Prof. Melvin. Physics Dept., Hofstra U., Hempstead, NY 11550, USA. (1923) PhD physics (Adelphi U., 1972). Assoc. prof. (tel. 516 + 560-5583). *X-ray crystallography, solid state physics, molecular vibrational amplitudes.*

Pavkovic, Prof. Stephen Frank. Chemistry Dept., Loyola U. of Chicago, 6525 N. Sheridan Rd., Chicago, IL 60626, USA. (1932) PhD inorg. chemistry (Ohio State U., 1964). Prof. and Chairman. (tel. 312 + 508-3100). *Single crystal structure determination.*

Peacor, Prof. Donald R. Geological Sci. Dept., U. of Michigan, Ann Arbor, MI 48109-1063, USA. (1937) PhD mineralogy (MIT, 1962). Prof. (tel. 313 + 764-1452). *Mineralogy, TEM.*

Pecoraro, Prof. Vincent L. Chemistry Dept., U. of Michigan, Ann Arbor, MI 48109-1055, USA. (1956) PhD chemistry (UCB, 1981). Asst. prof. (tel. 313 + 763-1519). *Photosynthesis, enzymes, manganese, vanadium, electron transfer.*

Pedersen, Prof. Lee G. Chemistry Dept., U. of North Carolina CB #3290, Chapel Hill, NC 27514, USA. (1938) PhD phys. chemistry (U. of Arkansas, 1965). Prof. (tel. 919 + 962-1578, email PEDERSEN at NIEHS). *Structure and function, proteins, peptides, nucleic acids.*

Peiser, Dr. Herbert Steffen. 638 Blossom Dr., Rockville, MD 20850, USA. (1917) MA chemistry (Cambridge U., UK, 1943). Retired. (tel. 301 + 762-6860, Telex. 904059). *Symmetry, atomic weights, metrology, crystal growth.*

Penner-Hahn, Prof. James E. Chemistry Dept., U. of Michigan, Ann Arbor, MI 48109, USA. (1957) PhD chemistry (Stanford U., 1984). Asst. prof. (tel. 313 + 764-7324, fax 313 + 747-4865, email Bitnet USERGB9A at UMICHUB). *X-ray excited optical luminescence, XANES, metal site structure in inorganic systems.*

Pennington, Prof. William T. Chemistry Dept., Hunter Res. Lab., Clemson U., Clemson, SC 29634, USA. (1955) PhD inorg. chemistry (U. of Arkansas, 1983). Dir., Mol. struct. facility. (tel. 803 + 656-4200). *Solid state chemistry, physics, analytical crystallography, structure - properties relationship.*

Perkins, Mr. Herbert O. Vista Chemical, P.O. Box 500, Ponca City, OK 74602, USA.

Perozzo, Ms. Mary Ann. Lab. for the Struct. of Matter, Naval Res. Lab., Code 6030, 4555 Overlook Ave., Washington, DC 20375-5000, USA. (1961) BS chemistry (U. of Wisconsin at Parkside, 1983). Res. chemist. (tel. 202 + 767-2735, Telex. twx 7108 220147, fax 202 + 767-6874, email perozzo at lsm.nrl.navy.mil). *Protein crystallography, automated protein crystallization.*

Pessen, Dr. Helmut. Eastern Regional Res. Center, U.S. Dept. of Agriculture, 600 E. Mermaid Lane, Philadelphia, PA 19118, USA. (1921) PhD phys. chemistry (Temple U., 1961). Res. scient. (tel. 215 + 233-6490, 6491). *Small angle X-ray scattering, NMR, biopolymers, structure - function relationships, biophysics, phys. chemistry.*

Peters, Mr. Charles Richard. 16051 Middlebury Dr., Dearborn, MI 48120, USA. (1934) MS phys. chemistry (U. of Michigan, 1958). Res. scient. (tel. 313 + 323-1533). *X-ray crystallography, single crystal studies, defect structures, temperature dependent phase transitions in solids.*

Petersen, Dr. Donald Ralph. Analytical Res., MS-23, Dow Corning Corp., Midland, MI 48686-0995, USA. (1929) PhD phys. chemistry (CIT, 1955). Res. scient. (tel. 517 + 496-5015). *Inorganic structures, powder diffraction, X-ray instrumentation.*

Petersen, Prof. Jeffrey L. Chemistry Dept., West Virginia U., P.O. Box 6045, Morgantown, WV 26506, USA. (1947) PhD phys. chemistry (U. of Wisconsin at Madison, 1974). Prof. (tel. 304 + 293-3435). *Diffraction, neutron, X-ray, metal hydrides, bis(cyclopentadienyl) metal chemistry, transition metal complexes, homogeneous catalysis, CO activation.*

Peterson, Prof. John Robert. Pharmacognosy Dept., School of Pharmacy, U. of Mississippi, University, MS 38677, USA. (1953) PhD chemistry (U. of Minnesota, 1984). Asst. prof. (tel. 601 + 232-7486). *Drug design, computer assisted, drug synthesis, structure, anticancer agents, antivirals, antagonists/agonists drug binding.*

Peterson, Dr. Selmer W. 6222 Wehner Way, San Jose, CA 95135, USA. (1917) PhD phys. chemistry (U. of Maryland, 1942). Retired. *Neutron diffraction, inorganic complexes, one-dimensional conductors.*

Petsko, Prof. Gregory Anthony. Chemistry Dept., Room 2-202, Mass. Inst. of Technol., Cambridge, MA 02139, USA. (1948) PhD molecular biophysics (Oxford U., UK, 1973). Prof. (tel. 617 + 253-1837, fax 617 + 258-7500). *X-ray crystallography, enzymatic catalysis, Laue diffraction of proteins, site-directed mutagenesis.*

Pett, Dr. Virginia B. Chemistry Dept., College of Wooster, Wooster, OH 44691, USA. (1941) PhD inorg. chemistry (Wayne State U., 1979). Assoc. prof. (tel. 216 + 263-2114). *Organo-cobalt complexes as coenzyme B12 models, Crnucleotide adducts, peptide structure, modeling.*

Petz, Mr. John Ignatius. Alumina Dept., Reynolds Metals, 103 Hawthorne, Portland, TX 78374, USA. (1935) MS physics (U. of Arkansas, 1959). Sr. staff member. (tel. 512 + 643-6531, ext. 2394). *Liquid state.*

Pflaum, Mr. Wolfgang Richard. Crystal Devices Bus. Unit, Hughes Aircraft Co., P.O. Box H, 500 Superior Ave., Newport Beach, CA 92658-8903, USA. (1944)

BS phys. chemistry (Harvey Mudd C., 1966). Member techn. staff. (tel. 714 + 759-2283). *Inorganic structures.*

Pfluger, Prof. Clarence Eugene. Chemistry Dept., Syracuse U., Syracuse, NY 13244, USA. (1930) PhD chemistry (U. of Texas at Austin, 1958). Prof. (tel. 315 + 423-3920). *Coordination compound structures, low temperature structure determinations, solid state photochemistry.*

Pflugrath, Dr. James W. Cold Spring Harbor Lab., P.O. Box 100, Cold Spring Harbor, NY 11724, USA. (1957) PhD biochemistry (Rice U., 1984). Sr. staff investigator. (tel. 516 + 367-8821, email Bitnet PFLUGRATH at CSHLAB). *Biological macromolecules, computational crystallography.*

Phillips, Prof. George Neal Jr. Dept. Biochem. & Cell Biol., Rice U., P.O. Box 1892, Houston, TX 77251, USA. (1952) PhD biochemistry (Rice U., 1977). Assoc. prof. *Protein structure, macromolecular assemblies, diffuse X-ray scattering.*

Phillips, Prof. James Christopher. NSLS, Brookhaven National Lab., Upton, NY 11973, USA. (1952) PhD applied physics (Stanford U., 1979). Res. assoc. prof., SUNY X3 Tech. Dir. (tel. 516 + 282-3770, fax 516 + 282-4745). *Structural studies, synchrotron radiation.*

Phillips, Mr. Michael W. Geology Dept., U. of Toledo, Toledo, OH 43606, USA.

Phillips, Dr. Susan R. School of Chemistry & Biochem., Georgia Inst. Techn., Atlanta, GA 30332, USA. (1959) PhD phys. chemistry (Georgia Inst. of Technology, 1986). Postdoct. fel. (tel. 404 + 894-2186, email CMFLSSP at GITVM1). *Protein crystallography.*

Phillips, Mr. T. J. 1750 S. Flannery Rd., Baton Rouge, LA 70816, USA. Prof. emeritus (chem.)

Phillips, Dr. Theodore II. 1 Manito Dr., Cambridge, MD21613, USA. (1938) PhD inorganic-phys. chemistry (U. of Kentucky, 1968). *Structures, biologically interesting compounds, crystallographic computing, small molecules.*

Phillips, Dr. Walter C. Rosenstiel Center, Brandeis U., Waltham, MA 02254, USA. (1936) PhD physics (MIT, 1964). Sr. scient. (tel. 617 + 736-2452, fax 617 + 735-2405, email PHILLIPS at BRANDEIS). *X-ray detectors, small angle scattering, macromolecules.*

Phizackerley, Dr. Richard Paul. Stanford Synchrotron Radiat. Lab., Stanford U., P.O. Box 4349, Bin 69, Stanford, CA 94309, USA. (1945) PhD physics (U. of Cambridge, U.K., 1971). Sr. res. assoc. (tel. 415 + 926-3431, fax 415 + 926-4100, email PHIZ at SSRL750). *Structure and function, biological macromolecules, anomalous scattering, synchrotron radiation applications, protein crystallography, instrumentation.*

Pichert, Mr. Jerome. 3 Idle Day Dr., Centerport, NY 11721, USA. BSE (Bridgeport Eng. Inst., 1948). Consultant. (tel. 516 + 261-5648).

Picot, Mr. Daniel. U. of Chicago, CLSC 321, 920 E. 58th St., Chicago, IL 60637, USA. (tel. 312 + 702-0286).

Piermarini, Dr. Gasper John. Ceramics Div., Inst. Matls. Sci. & Eng., Natl. Inst. Stnds. & Technol., Gaithersburg, MD 20899, USA. (1933) PhD phys. chemistry (American U., 1971). Res. scient. (tel. 301 + 975-5734). *High pressure X-ray crystallography, diamond anvil cells, optical measurements at high pressure, ruby fluorescent measurement techniques.*

Pierpont, Prof. Cortlandt G. Dept. of Chem. & Biochem., U. of Colorado, Boulder, CO 80309, USA. (1942) PhD chemistry (Brown U., 1971). Prof. (tel. 303 + 492-8420). *Inorganic chemistry.*

Pignataro, Mrs. Edith H. 230 Jay St., Brooklyn, NY 11201, USA. (1925) MS physics (PUN, 1954). (tel. 718 + 858-7561). *X-ray crystallography.*

Pinkerton, Prof. Andrew Alan. Chemistry Dept., U. of Toledo, 2801 W. Bancroft St., Toledo, OH 43606, USA. (1943) PhD inorg. chemistry (U. of Alberta, Canada, 1971). Prof. (tel. 419 + 537-4568). *Lanthanide and actinide chemistry, NMR of paramagnetics, small molecule crystallography, electron density determination.*

Pique, Mr. Michael E. Mol. Biology, MB-5, Res. Inst. Scripps Clinic, 10666 N. Torrey Pines Rd., La Jolla, CA 92037, USA. (1951) MS computer sci. (U. of N. Carolina at Chapel Hill, 1980). (tel. 619 + 554=9775, email mp at scripps.edu). *Computer graphics, scientific computing.*

Pjura, Dr. Phillip Edward. Inst. of Mol. Biology, U. of Oregon, Eugene, OR 97403, USA. (1956) PhD chemical biology (CIT, 1987). Postdoct. res. assoc. (tel. 503 + 686-5192).

Pletcher, Dr. James F. Biocrystallography Lab., VA Med. Center, P.O. Box 12055, Pittsburgh, PA 15240, USA. (1935) PhD biochemistry (Columbia U., 1965). Res. chemist. (tel. 412 + 371-0875). *Biological systems - structure and function, thiamine, enzymatic and non-enzymatic catalysis.*

Pluth, Dr. Joseph John. Geophys. Sci. Dept. & Matls. Res. Lab., U. of Chicago, 5734 S. Ellis Ave., Chicago, IL 60637, USA. (1943) PhD chemistry (U. of Washington, 1971). Sr. res. assoc. (tel. 312 + 702-8109, email pluth at geovax.uchicago.edu). *Zeolites, mineral sieves, mineral and inorganic structures, heterogeneous catalysis, diffraction, X-ray, neutron, synchrotron radiation.*

Poe, Dr. Martin. Enzymology Dept., Merck & Co. Inc., 126 E. Lincoln Ave., Rahway, NJ 07076, USA. (1942) PhD biophysics (U. of Pennsylvania, 1968). Sr. investigator. (tel. 201 + 594-5931). *Protein NMR, lynphocyte proteases, renin, RAS oncogene product, dihydrofolate reductase.*

Poland, Mr. Virgil Laverne. 315 Glenvale, Youngstown, NY 14174, USA. (1928) BS chemistry (Illinois Wesleyan U., 1953). (tel. 716 + 745-3089). *X-ray diffraction, spectroscopy, crystallography, chemistry.*

Pollack, Dr. Sidney Solomon. Pittsburgh Energy Techn. Ctr., U.S. Dept. of Energy, Box 10940, Pittsburgh, PA 15236, USA. (1929) PhD soil mineralogy (U. of Wisconsin, 1956). Res. chemist. (tel. 412 + 892-6108). *X-ray diffraction, fluorescence, scattering, surface area measurement, minerals, coal, chars, catalysts, synthetic oil and gas products, small angle X-ray scattering.*

Pon, Mr. George W. Chemistry Dept., Washington State U., Pullman, WA 99164, USA. (1966) BS chemistry (1989). Grad. student. (tel. 509 + 332-8343). *Copper crystals, structure, magnetic properties.*

Poojary, Dr. Maradamoole Damodara. Rosenstiel Basic. Med. Sci. Res. Ctr., Brandeis U., 415 South St., Waltham, MA 02254, USA. (1953) PhD inorg. chem., crystallog. (Indian Inst. of Sci., Bangalore, 1984). Res. assoc. (tel. 617 + 736-2495). *X-ray crystallography, electron microscopy, protein structure, inorganic chemistry.*

Porter, Dr. Leigh Christopher. Chem. & Matls. Sci. Div., Argonne National Lab., 9700 S. Cass Ave., Argonne, IL 60439, USA. (1955) PhD inorg. chemistry (U. of Calif. at Irvine, 1984). (tel. 312 + 972-3461). *Chemistry, structural inorganic, solid state, organometallic, small molecule crystallography.*

Posner, Prof. Aaron Sidney. 2 Longview Dr., Scardsale, NY 10583, USA. (1920) PhD phys. chemistry (U. of Liège, Belgium, 1954). Dir. of res. (tel. 212 + 535-7908). *Ultrastructure, bone and teeth, tissue mineralization mechanism, calcium phosphate structures.*

Post, Prof. Benjamin. Physics Dept., Polytechnic U. of New York, 333 Jay St., Brooklyn, NY 11201, USA. (1911) PhD phys. chemistry (PUN, 1949). Prof. emeritus. (tel. 718 + 643-8804). *Dynamical diffraction theory, X-ray instrumentation, precise measurements, simultaneous diffraction effects.*

Post, Mr. Jeffrey E. Smithsonian Institution, NHB119, Washington, DC 20560, USA. (1954) PhD chemistry (Arizona State U., 1981). Mineralogist. (tel. 202-357-4009). *Mineralogy, crystallography, geochemistry.*

Potenza, Prof. Joseph Anthony. Chemistry Dept., Rutgers U., New Brunswick, NJ 08904, USA. (1941) PhD chemistry (Harvard U., 1967). Prof. (tel. 201 + 932-2115). *Chemistry, physical, inorganic.*

Potter, Mr. Stephen Anthony. Mol. Biophysics, Med. Fndn. of Buffalo, 73 High St., Buffalo, NY 14203, USA. (1946) BA physics (SUNY at Buffalo, 1969). Res. assoc. (tel. 716 + 856-9600). *Direct methods, macromolecules.*

Poulos, Mr. Thomas L. CARB, 9600 Gudelsky Dr., Rockville, MD 20850, USA.

Powell, Dr. Douglas R. Sch. of Geology & Geophys., U. of Oklahoma, 830 Van Vleet Oval, Norman, OK 73019, USA. (1953) PhD Physical chemistry (Iowa State U., 1980). Res. staff membr. (tel. 405 + 325-2362). *Crystallographic computing, instrumentation.*

Predecki, Mr. Paul K. 4830 E. Harvard Ln., Denver, CO 80222, USA. (tel. 303 + 871-2102).

Prendergast, Prof. Franklyn G. Biochem. & Mol. Biol. Dept., Mayo Fndn., Rochester, MN 55905, USA. (1945) MD/PhD medicine/biochemistry (U. of Minnesota, 1968, 1977). Prof. (tel. 507 + 284-4081, fax 507 + 284-9439, email prendergast at mayo). *Protein structure, dynamics.*

Pressprich, Mr. Mark R. Chemistry Dept., Washington State U., Pullman, WA 99164, USA. (tel. 509 + 335-9125).

Preston, Ms. Kimberly. Ceramic Eng. Dept., U. of Missouri, 222 McNutt Hall, Rolla, MO 65401, USA. (1960) BS ceramic engineering (U. of Missouri at Rolla, 1988). Grad. student. (tel. 314 + 341-2585).

Prevey, Mr. Paul S. Lambda Res. Inc., 1111 Harrison Ave., Cincinnati, OH 45214-1801, USA. (1948) BS physics (Case Inst. of Technology, 1970). President. (tel. 513 + 621-3933, fax 513 + 621-3935). *Powder diffraction, qualitative, quantitative, residual stress measurement, pole figure determination.*

Prewitt, Dr. Charles Thompson. Geophysical Lab., 2801 Upton St., NW, Washington, DC 20008, USA. (1933) PhD mineralogy and crystallography (MIT, 1962). Dir. (tel. 202 + 966-0334, fax 202 + 895-4225, email Bitnet GEOLAB at GWUVM). *Crystallography, mineralogy, crystal chemistry.*

Price, Ms. Rebecca Alexis. Chemistry Dept., West Virginia U., P.O. Box 6045, Morgantown, WV 26506, USA. (1951) MS (U. of Oregon, 1985). Res. asst. (tel. 304 + 296-5299). *Macromolecular structure, enzymes.*

Prince, Dr. Edward. Reactor Radiation Div., Natl. Inst. Stnds. & Technol., Rm. A106, Bldg. 235, Gaithersburg, MD 20899, USA. (1928) PhD physics (U. of Cambridge, UK, 1952). Res. physicist. (tel. 301 + 975-6230, Telex. 197674 NBS, fax 301 + 921-9847, email PRINCE at NBSENH). *Neutron diffraction, instrumentation, refinement techniques.*

Privé, Dr. Gilbert Gérard. Chem. Biodynanics Div., Lawrence Berkeley Lab., U. of Calif. at Berkeley, Berkeley, CA 94720, USA. (1960) PhD biochemistry (UCLA, 1988). Postdoct. fel. (tel. 415 + 486-4311). *Macromolecular crystallography, structure, DNA, protein, molecular recognition.*

Profeta, Dr. Salvatore Jr. Computational Div., Glaxo Inc., Five Moore Dr., Research Triangle Park, NC 27709, USA. (1951) PhD (U. of Georgia, 1978). Sr. consult./Head, theoret. chem. (tel. 919 + 248-7224). *Computational chemistry, X-ray structure determination, small molecules, macromolecules, computer graphics and drug design.*

Pulliam, Dr. Curtis R. Chemistry Dept., Utica C. of Syracuse U., Utica, NY 13502, USA. (1957) PhD inorg. chemistry (U. of Wisconsin at Madison, 1986). Asst. prof. (tel. 315 + 792-3140, 792-3028). *Chemistry, inorganic, organometallic, metal clusters, crystallographic computing, electrochemistry.*

Punzi, Dr. John Stephen. Pharmacology Dept., U. of Rochester, 601 Elmwood Ave., Rochester, NY 14642, USA. (1956) PhD biochemistry (SUNY at Buffalo, 1988). Postdoct. fel. (tel. 716 + 275-1681). *Crystal growth, protein crystallography.*

Purdy, Mr. Samuel M. National Steel Corp., 1745 Fritz Dr., Trenton, MI 48183, USA. (1926) BS metallurgical eng. (Lehigh U., 1948). Sr. Res. Assoc. (tel. 313 + 676-2682, fax 313 + 676-2030). *Metallography, structure of metals, materials science.*

Pyrros, Dr. Nikos P. Hercules Res. Center, Bldg. 8136, R259, Wilmington, DE 19894, USA. PhD physics (McMaster U., Canada, 1972). Sr. res. scientist. (tel. 302 + 995-3405). *X-ray diffraction, crystal structures, polymers.*

Quicksall, Dr. Carl O. 113 Woodbine Dr., Terre Haute, IN 47803, USA. (1941) PhD chemistry (Princeton U., 1971). (tel. 812 + 877-2820). *Inorganic chemistry, crystallography, X-ray diffraction.*

Quigley, Prof. Gary Joseph. Chemistry Dept., Hunter C., CUNY, 695 Park Ave., New York, NY 10021, USA. (1942) PhD chemistry (SUNY, C. of Env. Sci. and Forestry, 1969). Prof. chem. and biochem. (tel. 212 + 772-5377, email GJQHC at CUNYVM). *Nucleic acid structure and function, macromolecular crystallography, macromolecular modeling.*

Quintana, Mr. John P. Materials Sci. & Eng. Dept., The Techn. Inst., Northwestern U., Evanston, IL 60201, USA.

Quiocho, Prof. Florante A. HHMI, Baylor C. of Medicine, One Baylor Plaza, Houston, TX 77030, USA. (1937) PhD biochemistry (Yale U., 1966). Prof. and investigator. (tel. 713 + 799-6565, fax 713 + 797-6718). *Biochemistry, biophysics, proteins, small molecules, crystallography.*

Rabenberg, Dr. Llewellyn K. Mech. Engineering Dept., Ctr. Matls. Sci. & Eng., U. of Texas at Austin, Austin, TX 78712, USA. (1956) PhD Matls. sci. & eng. (UCB, 1983). Asst. Prof. (tel. 512 + 471-3178, fax 512 + 471-8727). *Transmission electron microscopy, electron diffraction, materials science.*

Rabinowitz, Dr. Israel Nathan. Virgilan Co., 2534 Foothill Rd., Santa Barbara, CA 93105, USA. (1935) PhD biochemistry (Rutgers U., 1965). Pres. (tel. 805 + 687-0047). *Crystal growth, biological calcification, industrial crystallization.*

Radhakrishnan, Dr. R. Biochem. & Biophys. Dept., Texas A&M U., College Station, TX 77843, USA. (1947) PhD physics (Madras U., India, 1983). Asst. res. scient. (tel. 409 + 845-1744, email RADHA at TAMBIGRF). *Crystallography, energy calculations, computer graphics, drug design, molecular modeling.*

Radonovich, Prof. Lewis J. Chemistry Dept., U. of North Dakota, Grand Forks, ND 58202, USA. (1944) PhD phys. chemistry (Wayne State U., 1970). Prof. (tel. 701 + 777-2541). *Structural chemistry, biologically important molecules, inorganic compounds.*

Rafalko, Ms. Patrice White. Analytical Dept., The BOC Group Inc., 100 Mountain Ave., Murray Hill, NJ 07974, USA. (1951) MA chemistry, crystallography (U. of Texas as Austin, 1977). Assist. scient. (tel. 201 + 771-6380). *Diffraction, single crystal, powder, small organic molecules.*

Raines, Prof. Ronald T. Biochemistry Dept., U. of Wisconsin, 420 Henry Mall, Madison, WI 53706-1569, USA. (1958) PhD chemistry (Harvard U., 1986). Asst. prof.

Rajeswaran, Dr. Manju. 6 Camborne Cir., Fairport, NY 14450, USA. (1956) PhD biophysics (SUNY at Buffalo, 1983). (tel. 716 + 377-4353). *Crystallography, biologically important molecules, proteins.*

Ramalingam, Dr. Veerappa Pillai. Biochem. & Biophysics Dept., U. of Calif. at San Francisco, San Francisco, CA 94143, USA. (1953) PhD biochem., X-ray crystallogr. (U. of Miami, 1986). Postdoct. res. assoc. (tel. 415 + 476-0953, email cgl ramu at msg.ucsf.edu). *Protein structure - function, crystallography, macromolecular, laue methods, rational drug design.*

Randall, Dr. Clive Alan. Materials Res. Lab., Pennsylvania State U., University Park, PA 16802, USA. (1960) PhD physics (U. of Essex, UK, 1987). Res. assoc., (tel. 814 + 863-1328). *TEM, phase transitions, ferroelectricity, ferroelastics, domain structures.*

Rao, Dr. Jejjala Krishna Mohana. Crystallography Lab., NCI-FCRF, BRI Basic Res. Prog., P.O. Box B, Frederick, MD 21701, USA. (1943) PhD crystallography (Indian Inst. of Science, Bangalore, 1971). Scient. assoc. (tel. 301 + 698-5031). *Proteins, viruses, secondary structures, programming, small molecules.*

Rao, Dr. S. Narasinga. Physics Dept., Central State U., 100 N. University Dr., Edmond, OK 73034-0177, USA. (1938) PhD biophysics, X-ray crystallog. (SUNY at Buffalo, 1973). Prof. (tel. 405 + 341-2980, ext. 5472, fax 405 + 341-4964). *Structure, function, conformation, proteins, biological small molecules, protein biochemistry.*

Rao, Dr. Sambhorao Thyagaraja. Biochemistry Dept., U. of Wisconsin, Madison, WI 53706, USA. (1937) PhD physics (U. of Madras, India, 1966). Sr. scient. (tel. 608 + 262-3019, email RAOST at WISCMACC). *X-ray crystallography, biologically significant molecules.*

Rao, Mr. Siram N. V. CABM at Rutgers U., P.O. Box 759, Piscataway, NJ 08855, USA. (tel. 201 + 932-3199).

Rao, Dr. Sudharkara. Rosenstiel Med. Sci. Res. Ctr., Brandeis U., 415 South St., Waltham, MA 02256, USA. (1958) PhD chemistry (Indian Inst. of Science, 1986). Postdoct. assoc. (tel. 617 + 736-2495, email RAO at BRANDEIS). *Protein crystallography, molecular dynamics, structural biology.*

Rao, Ms. Usha. Chemistry Dept., Boston C., Chestnut Hill, MA 02167, USA. (tel. 617 + 552-8481). *Macromolecular crystallography, molecular dynamics, structure prediction, toxins.*

Rappaport, Prof. Harry P. Biology Dept., Temple U., Philadelphia, PA 19122, USA. (1927) PhD physics (Yale U., 1956). Prof. (tel. 215 + 787-8875). *Protein structure, function, nucleic acids.*

Raptis, Dr. Raphael. Chemistry Dept., Texas A&M U., College Station, TX 77843-3257, USA. (1957) PhD chemistry (Texas A&M U., 1988). Res. assoc. (tel. 409 + 845-4837). *Chemistry, inorganic, organometallic.*

Rardin, Mr. R. Lynn. Rm. 18-127, Mass. Inst. of Technol., 77 Massachusetts Ave., Cambridge, MA 02139, USA. (1963) AB chemistry (Kenyon C., 1985). Grad. student. (tel. 617 + 253-1823).

Rasmussen, Dr. Bjarne. Chemistry Dept., Mass. Inst. of Technol., Rm. 2-202, 77 Massachusetts Ave., Cambridge, MA 02139, USA. (1952) Cand.Scient. Crystallography (U. of Copenhagen, 1983). Visiting Scient. (tel. 617 + 253-7146). *Protein crystallography, enzyme mechanism.*

Rastinejad, Mr. F. 4114 Spruce St., Apt. 3R, Philadelphia, PA 19104, USA. (tel. 215 + 387-3126).

Rath, Mr. Nigam Prasad. Chemistry Dept., U. of Notre Dame, Notre Dame, IN 46556, USA. (1958) PhD chemistry (Oklahoma State U., 1985). Asst. Fac. Fel. (tel. 219 + 239-6220). *X-ray crystallography, chemistry, organometallic, transition metal coordination.*

Rath, Ms. Virginia. Biochem. & Biophys. Dept., S-960, U. of Calif. at San Francisco, 3rd. & Parnassus Aves., San Francisco, CA 94143, USA. (1956) MS biological sciences (Stanford U., 1983). Grad. student. (tel. 415 + 476-5051). *Biological macromolecules, X-ray crystallography, allosteric mechanisms, protein structure.*

Rau, Mr. Robert C. 2542 Fleetwood Ave., Cincinnati, OH 45211, USA. (1935) MSc materials science (U. of Cincinnati, 1965). Consultant. (tel. 513 + 721-6383). *Inorganic crystal structures, powder diffraction analysis.*

Ravichandran, Mr. K. G. Chemistry Dept., Michigan State U., East Lansing, MI 48824, USA. (1960) MS chemistry (Indian Inst. Technol., Madras, 1983). Grad. student. (tel. 517 + 353-4505, email RAVI at MSUCEM). *Macromolecular crystallography, protein modeling, molecular dynamics.*

Ravikumar, Dr. Krishnan. Chemistry Dept. & Ctr. Biophysics, Rensselaer Polytechnic Inst., Troy, NY 12180, USA. (1958) PhD physics (U. of Madras, India, 1987). Postdoct. res. assoc. (tel. 518 + 276-2835, email csm.ravikuma at cicgj.rpi.edu). *Crystallography, small molecule, protein, natural products, drug - protein interactions.*

Ray, Prof. Alden Earl. Metals and Ceramics Div., U. of Dayton Res. Inst., 300 College Park Ave., Dayton, OH 45469, USA. (1931) PhD physical metallurgy (Iowa State U., 1959). Prof. & sr. metallurgist. (tel. 513 + 229-3529, fax 513 + 229-3433). *Metallurgy, rare earth-transition metal alloys, phase diagrams, phase stability, crystal structures, permanent magnets, processing, heat treatment.*

Raykhtsaum, Mr. Grigory. Technology Dept., Leach and Garner Co., P.O. Box 2018, Attleboro, MA 02703, USA.

Rayment, Prof. Ivan. Inst. for Enzyme Res., U. of Wisconsin, 1710 University Ave., Madison, WI 53705, USA. (1951) PhD chemistry (U. of Durham, England, 1975). Assoc. prof. (tel. 608 + 262-0437). *Crystallography, protein, macromolecular, structure and function, muscle proteins, motility.*

Raymond, Prof. Kenneth Norman. Chemistry Dept., U. of Calif. at Berkeley, Berkeley, CA 94720, USA. (1942) PhD inorg. chemistry (Northwestern U., 1968). Assoc. prof. (tel. 415 + 642-7219). *Coordination isomers, metal ion substitution, biological transport, structure and bonding, coordination complexes, lanthanide, actinide organometallic compounds, transuranium sequestering agents.*

Reddy, Dr. B. Swamintha. ST Systems Corp., 4400 Forbes Blvd., Lanham, MD 20706, USA. (1946) PhD physics (Indian Inst. of Sci., Bangalore, 1974). Dept. manager. (tel. 301 + 236-5319). *Astrophysics, biophysics.*

Reeber, Dr. Robert Richard. Matls. Science Div., Army Res. Office, P.O. Box 12211, Research Triangel Park, NC 27709, USA. (1937) PhD industrial mineralogy (Ohio State U., 1968). Materials eng. (tel. 919 + 549-0641, ext. 318). *Phase transformations, thermal expansion, crystal chemistry, diffraction methods.*

Reed, Dr. A. Thomas. Analytical Services, Anchor Hocking Glass Co., 1749 W. Fair Ave., Lancaster, OH 43130, USA. (1946) PhD inorg. chemistry (Miami U., Ohio, 1975). Chemist. (tel. 614 + 687-2804). *Inorganic chemistry, X-ray fluorescence spectrometry, powder diffraction.*

Reed, Dr. Larry L. Argonne National Lab., 9700 S. Cass Ave., Argonne, IL 60439, USA. (1949) PhD org. chemistry (U. of Arizona, 1971). Computer scient. (tel. 312 + 972-7585). *Chemistry, computer science.*

Reed, Dr. R. A. Biophysics Dept., Boston U. School of Med., 80 E. Concord St., Boston, MA 02118, USA. (1961) PhD biochemistry (Boston U., 1987). Instructor. (tel. 617 + 638-4008). *Protein structure determination via X-ray diffraction, electron microscope techniques.*

Reeder, Prof. Richard James. Earth & Space Sci. Dept., SUNY at Stony Brook, Stony Brook, NY 11794, USA. (1953) PhD geochemistry (UCB, 1980). Prof. (tel. 516 + 632-8208, email Bitnet RJREEDER at SBCCMAIL). *Crystal chemistry, mineralogy.*

Reeke, Prof. George Norman Jr. Developmental and Mol. Biol. Dept., The Rockefeller U., 1230 York Ave., New York, NY 10021, USA. (1943) PhD chemistry (Harvard U., 1969). Assoc. prof. (tel. 212 + 570-8183, fax 212 +

570-7974, email Bitnet CDRNI at CUNYVM). *Protein crystallography, crystallographic computing, pattern recognition, nervous system models.*

Rees, Prof. Douglas Charles. Chemistry and Biochem. Dept., U. of Calif. at Los Angeles, 405 Hilgard Ave., Los Angeles, CA 90024, USA. (1952) PhD biophysics (Harvard U., 1980). Assoc. prof. (tel. 213 + 206-1166). *Macromolecular crystallography.*

Reibenspies, Dr. Joseph. Chemistry Dept., Texas A&M U., College Station, TX 77843, USA. (1958) PhD chemistry (Colorado State U., 1987). Res. instrum. spec. (tel. 409 + 845-9125, fax 409 + 845-4719, email REIBENSPIES at TAMCHEM).

Reid, Dr. Austin Henry Jr. Chemicals & Pigments Dept., E.I. duPont de Nemours Co., Edge Moor Res. & Dev., Hay Rd., Edge Moor, DE 19809, USA. (1957) PhD inorg. chemistry (Auburn U., 1982). Dev. Supervisor. (tel. 302 + 761-2464). *Chemistry, inorganic, polymer, small molecule crystallography, X-ray spectroscopy.*

Reid, Ms. Susan Sarah. Chemistry Dept., Ohio State U., 140 W. 18th. Ave., Columbus, OH 43210, USA. (1965) BA chemistry, physics (St. Olaf C., 1987). Grad. student. (tel. 614 + 292-2251, ext. 3587).

Reidinger, Mr. Franz. CRL 206, Allied Corp., P.O. Box 1021R, Morristown, NJ 07960, USA. (tel. 201 + 455-3685).

Reiher, Dr. Walter E. III. BioDesign Inc., 199 S. Los Robles Ave., Suite 540, Pasadena, CA 91101, USA. (1955) PhD chemistry (Harvard U., 1985). Sr. scient. (tel. 818 + 793-3600, fax 818 + 793-8098). *Proteins, molecular modeling, computation.*

Reis, Dr. Arthur Henry Jr. Office of the Provost, Brandeis U., Irving 104, Waltham, MA 02254, USA. (1946) PhD inorg. chemistry (Harvard U., 1972). Assoc. dean resources & planning. (tel. 617 + 736-2105, email Bitnet REIS at BRANDEIS). *Low dimensional interactions, pulsed neutron diffraction.*

Rek, Ms. Zofia. SSRL Bin #69, Stanford U., 2575 Sand Hill Rd., Menlo Park, CA 94025, USA.

Remington, Prof. Stephen James. Physics Dept., U. of Oregon, Eugene, OR 97403, USA. (1950) PhD biophysics (U. of Oregon, 1977). Asst. prof. (tel. 503 + 686-5151, fax 503 + 686-3127, email jun at uoxray.uoregon.edu). *Macromolecules, crystallography, structure/function relationships, crystallographic methods, enzyme mechanisms.*

Reppart, Dr. William James. Analytical Chem. Dept., Shell Development Co., P.O. Box 481, Houston, TX 77001, USA. (1958) PhD phys. chemistry (The Ohio State U., 1980). Assoc. res. chemist. (tel. 713 + 663-2790). *X-ray powder diffraction, rare earth chemistry, clay and minerals.*

Restori, Dr. Renzo. Chemistry Dept., SUNY at Buffalo, Acheson Hall 218, Buffalo, NY 14214, USA. (tel. 716 + 831-3263).

Rey, Dr. Felix Augusto. Biochem. & Mol. Biol. Dept., Harvard U., 7 Divinity Ave., Cambridge, MA 02138, USA. (1957) PhD biochemistry (U. de Paris Sud, France, 1988). Postdoct. fel. (tel. 617 + 495-4091, fax 617 + 495-9613). *Crystallography, biological macromolecules.*

Reynolds, Dr. Ross Anthony. Physics Dept., Colby C., Eustis Pkwy., Waterville, MA 04901, USA. (1951) PhD physics (U. of Oregon, 1983). Asst. prof. (tel. 207 + 872-3599, fax 207 + 872-3555). *Protein crystallography.*

Rheingold, Dr. Arnold Lange. Chemistry Dept., U. of Delaware, Newark, DE 19716, USA. (1940) PhD chemistry (U. of Maryland, 1969). Prof. (tel. 302 + 451-8720, fax 302 + 451-6335, email ALQ01424 at UDACSVM). *Organometallic chemistry, compounds, structure.*

Rhodes, Prof. Gale. Chemistry Dept., 96 Falmouth St., U. of Southern Maine, Portland, ME 04103, USA. (1943) PhD chemistry (U. of N. Carolina at Chapel Hill, 1971). Prof. (tel. 207 + 780-4734, email RHODES at PORTLAND). *Refinement, malate dehydrogenase, nitrogenase, education, chemistry, biochemistry, biophysics.*

Rhyne, Ms. Kay A. 14521 Pebble Hill Ln., Gaithersburg, MD 20878, USA. (tel. 301 + 975-6222).

Ribbe, Prof. Paul H. Geological Sci. Dept., Virginia Polytechnic Inst. & State U., Blacksburg, VA 24061, USA. (1935) PhD physics, crystallog. (U. of Cambridge, UK, 1963). Prof., mineralogy (tel. 703 + 231-6880). *Silicate crystal chemistry.*

Ricci, Prof. John S. Jr. Chemistry Dept., U. of Southern Maine, Portland, ME 04103, USA. (1940) PhD chemistry (SUNY at Stony Brook, 1969). Prof. (tel. 207 + 780-4232). *Neutron diffraction, transition metal complexes, molecular structure.*

Rice, Dr. Catherine Ellen. AT&T Bell Labs., Rm. 1D-345, 600 Mountain Ave., Murray Hill, NJ 07974, USA. (1951) PhD inorg. chemistry (Purdue U., 1976). Techn. staff member. (tel. 201 + 582-4942). *Solid state chemistry.*

Rich, Prof. Alexander. Biology Dept., Mass. Inst. of Technol., 77 Massachusetts Ave., Cambridge, MA 02139, USA. (1924) MD medicine (Harvard Med. Sch., 1949). Sedgwick prof. (tel. 617 + 253-4715). *Molecular structure, proteins, nucleic acids, mechanism of protein synthesis, origin of life.*

Richards, Prof. Frederic Middlebrook. Mol. Biophysics & Biochem. Dept., Yale U., 260 Whitney Ave., New Haven, CT 06520, USA. (1925) PhD biophys. chemistry (Harvard U., 1952). Henry Ford II prof. (tel. 203 + 432-5620, fax 203 + 432-3282, email RICHARDS at YALEVMS). *Proteins, enzymes, structure and function.*

Richards, Dr. Gerald F. 850 Siemers St., Platteville, WI 53818, USA. (1935) PhD phys. chemistry (U. of Iowa, 1964).

Richardson, Dr. David Claude. Biochemistry Dept., Duke U., 210B Nanaline Duke Bldg., Durham, NC 27710, USA. (1940) PhD inorg. chemistry (MIT, 1967). Assoc. prof. (tel. 919 + 684-6010, fax 919 + 684-8885, email DXRAY at TUCC). *Protein crystallography, metalloenzymes, molecular graphics, protein design.*

Richardson, Dr. James Wyman Jr. IPNS Div., Argonne National Lab., 9700 S. Cass Ave., Argonne, IL 60439, USA. (1955) PhD crystallography (Iowa State U., 1984). Asst. chemist. (tel. 312 + 972-3554). *Powder diffraction, X-ray, neutron, zeolites and molecular sieves, amorphous diffraction.*

Richardson, Mrs. Jane Shelby. Biochemistry Dept., Duke U., 213 Nanaline Duke Bldg., Durham, NC 27710, USA. (1941) DSc (honorary) (Swarthmore C., 1986). Assoc. medical res. prof. (tel. 919 + 684-6010, fax 919 + 684-8885, email DXRAY at TUCC). *Comparison and classification of protein structures, protein design, crystallography, folding.*

Richardson, Dr. John Frederick. Chemistry Dept., U. of Louisville, Louisville, KY 40292, USA. (1954) PhD chemistry (U. of Western Ontario, 1981). (tel. 502 + 588-7069). *X-ray crystal structures of small molecules.*

Richman, Prof. Marc Herbert. Eng. Materials Sci. Dept., Brown U., Providence, RI 02912, USA. (1936) ScD metallurgy (MIT, 1963). Prof. (tel. 401 + 863-2317). *Diffraction, X-ray, electron, field ion microscopy, crystal structures, inorganic compounds, bulk materials, thin deposited films.*

Allen, Dr. Richon. Research Computing, Glaxo Inc., Five Moore Dr., Research Triangle Park, NC 27709, USA. (1951) PhD chemistry (U. of North Carolina, 1977). Mgr. (tel. 919 + 248-7223). *Molecular modeling, chemical graphics, structural chemistry, protein crystallography.*

Riegert, Mr. Richard Paul. QUAD Group, 331 Palm Ave., Santa Barbara, CA 93101, USA. (1927) MS ceramic eng., crystal chem. (Alfred U., 1957). Pres. & techn. dir. (tel. 805 + 965-1041). *Surface analysis, thin film morphology, crystal chemistry.*

Riess, Prof. John Karlem. 17 Audubon Blvd., New Orleans, LA 70118, USA. (1913) PhD physics (Brown U., 1943). Prof. emeritus. (tel. 504 + 861-9872). *Biophysics.*

Ringe, Dr. Dagmar. Chemistry Dept. 4-449, Mass. Inst. of Technol., Cambridge, MA 02139, USA. (1942) PhD chemistry (Boston U., 1968). Sr. Lect. (tel. 617 + 253-4526). *Proteases, inhibition, engineering, protein structure/function, protein crystallography.*

Rini, Dr. James. Molec. Biology Dept. MB-13, Scripps Clinic & Research Fndn., 10666 N. Torrey Pines Rd., La Jolla, CA 92037, USA. (1958) PhD medical biophysics (U. of Toronto, 1986). Postdoct. fel. (tel. 619 + 554-2456, email rini%scr.sdscnet at sdsc).

Rios Steiner, Mr. Jorge L. Chemistry Dept., U. of Puerto Rico, Rio Piedras, PR 00931, USA.

Robbins, Mr. Carl Richard. 6220 Winnebago Rd., Bethesda, MD 20816, USA. MA mineralogy-geochemistry (U. of Missouri, 1952). Res. chemist. (tel. 301 + 975-5786). *Inorganic structural chemistry, oxides, silicates, germanates, aluminates, environments, anhydrous, hydro-thermal, phase equilibria, structures, X-ray powder, single crystal.*

Roberts, Dr. Michael Mark. Lab. Mol. Virology & Carcinogenisis, NCI - Frederick Cancer Res. Fac. P.O. Box B, Bldg. 560, Frederick, MD 21701, USA. (1956) PhD Structural inorg. chemistry (U. of Warwick, England, 1981). Scient. assoc. (tel. 301 + 698-5832). *Protein crystallography, retrovirus structure.*

Roberts, Dr. Sue A. Chemistry Dept., U. of Arizona, Tucson, AZ 85721, USA. PhD (Washington State U., 1978). (tel. 602 + 621-6335).

Roberts, Dr. Victoria Anne. Mol. Biology Dept., Res. Inst. of Scripps Clinic, 10666 N. Torrey Pines Rd., La Jolla, CA 92037, USA. (1953) PhD synthetic org. chemistry (U. of Calif. at San Diego, 1 Res. assoc. (tel. 619 + 554-2806, fax 619 + 554-8841, email vickie%scr.sdscnet at sdsc). *Computational chemistry, molecular dynamics, electrostatics, conformational search.*

Robertson, Dean B. Ken. Chemistry Dept., U. of Missouri - Rolla, Rolla, MO 65401, USA. (1938) PhD chemistry (Texas A&M U., 1965). (tel. 314 + 341-4292).

Robertson, Prof. James David. Dept. of Anatomy, Box 3209, Duke Med. Center, Duke U., Durham, NC 27710, USA. (1922) MD, PhD biochemistry (Harvard Med. S. (1945), MIT, 1952). James B. Duke prof. of neurobiology. (tel. 919 + 684-5136). *Molecular structure, membranes, nervous system, structure, function.*

Robertus, Prof. Jon David. Chemistry Dept., U. of Texas at Austin, Welch 5.266, Austin, TX 78712, USA. (1945) PhD biochemistry (U. of Calif. at San Diego, 1972). Prof. (tel. 512 + 471-3175, Telex. 910-874-1305, fax 512 + 471-8696). *Structure and action, proteins, enzymes (antitumor & antiviral), site directed mutagenesis, protein engineering.*

Robinson, Dr. Ian Keith. AT&T Bell Labs., 1E445, 600 Mountain Ave., Murray Hill, NJ 07974, USA. (1955) PhD physics (Harvard U., 1981). Member tech. staff. (tel. 201 + 582-6056). *Surfaces, reconstruction, thin films, viruses, proteins, interfaces, phase transitions.*

Robinson, Mr. John C. 21699 Terrace Dr., Cupertino, CA 95014 USA. (tel. 408 + 253-8925).

Robinson, Prof. William Robert. Chemistry Dept., Purdue U., West Lafayette, IN 47907, USA. (1939) PhD inorg. chemistry (MIT, 1966). Prof. (tel. 317 +

494-5453). *Solid state chemistry, synthesis, structure, metal phosphates, oxides, silicates, sulfides, transition metal compounds.*

Roderick, Dr. Steven L. Inst. of Mol. Biology, U. of Oregon, Eugene, OR 97403-1229, USA. (tel. 503 + 686-5176). *Protein crystallography.*

Rodgers, Dr. David William. Biochem. & Molec. Biol. Dept., Harvard U., 7 Divinity Ave., Cambridge, MA 03138, USA. (1958) PhD biochemistry (Cornell U., 1987). Postdoct. fel. (tel. 617 + 495-5043).

Roe, Dr. Alfred Lawrence. Washington Res. Ctr., W.R. Grace & Co., 7379 Route 32, Columbia, MD 21045, USA. PhD chemistry (Cornell U., 1984). Res. chemist. (tel. 301 + 531-4272).

Roettgers, Mr. Wolbert. Henry L. Mattin Laboratories, The Mearl Corp., 217 North Highland Ave., Ossining, NY 10562, USA. (1926) MSc physics (Fairleigh Dickinson, U., 1966). Res. physicist. (tel. 914 + 941-7450, ext. 25). *Crystallography, clay minerals, crystal growth, epitaxy.*

Rogers, Prof. Robin Don. Chemistry Dept., Northern Illinois U., DeKalb, IL 60115, USA. (1957) PhD chemistry (U. of Alabama, 1982). Assoc. prof. (tel. 815 + 753-6885, email T40RDR1 at NIU). *Organometallic compounds, complexes (f-element - crown ether).*

Rognlie, Mr. David G. Blake Industries Inc., 660 Jerusalem Rd., Scotch Plains, NJ 07076, USA. (1934) BSEE electrical engineering (U. of North Dakota, 1956). Pres. (tel. 201 + 233-7240). *Diffraction instrumentation.*

Rohrbaugh, Dr. Wayne Joseph. Res. Dept., Mobil Res. & Dev. Corp., Paulsboro Res. Lab., Paulsboro, NJ 08066, USA. (1948) PhD phys. chemistry (Iowa State U., 1977). Project leader, matls. structure res. (tel. 609 + 423-1040, ext 2796, T *Zeolite structures, catalysis, X-ray methods, single crystal, powder, structure - properties relationship.*

Rohrer, Dr. Douglas C. The Upjohn Co., 301 Henrietta St., Kalamazoo, MI 49001, USA. (1942) PhD chemistry (Case-Western Reserve U., 1970). Sr. res. scient. (tel. 616 + 385-7729). *Drug/DNA interactions, molecular mechanics, dynamics, molecular structure from NMR data.*

Roldan, Dr. Luis-Gonzalez. 124 Becky Don Dr., Greer, SC 29651, USA. (1925) DSc chemistry-crystallography (U. of Sevilla, Spain, 1957). (tel. 803 + 877-0904). *Polymers, organic crystal structures, structure - properties - morphology relationships, crystal nucleation, polymorphism.*

Roof, Dr. Raymond Bradley. Chemistry Div., MS-G740, Los Alamos National Lab., P.O. Box 1663, Los Alamos, NM 87545, USA. (1929) PhD mineralogy-crystallography (U. of Michigan, 1955). Staff member. (tel. 505 + 667-2296). *Intermetallic compounds, line broadening, instrumentation, high pressure, computer programming, mineral synthesis.*

Rories, Dr. Charles C. P. Hormone Act. & Oncogenesis Sect., NCI, NIH, Bldg. 37, Rm. 3C-19, Bethesda, MD 20892, USA. (1955) PhD biochemistry (U. of N. Carolina at Chapel Hill, 1986). Postdoct. fel. (tel. 301 + 496-9867). *Protein - DNA interactions.*

Rosario, Mr. Nery Rivera. Calle Pio Baroja, #307 El Senorial, Rio Piedras, PR 00926, USA.

Rose, Prof. George D. Biological Chem. Dept., Hershey Med. Center, Pennsylvania State U., Hershey, PA 17033, USA. (1939) PhD biochem. and biophys. (Oregon State U., 1976). Prof. and chairman. (tel. 717 + 531-8271, fax 717 + 531-7072, email ROSE at PSUHMED). *Protein conformation, folding.*

Rose, Mr. Jon Patrick. Crystallography Dept., U. of Pittsburgh, 304 Thaw Hall, Pittsburgh, PA 15260, USA. (1953) PhD chemistry (Rutgers U., 1980). Mgr., X-ray diffr. fac. (tel. 412 + 624-3219, fax 412 + 624-1882, email ROSE at PITTVMS). *Protein crystallography, interactions, area detectors.*

Rosenberg, Prof. John M. Crystallography Dept., U. of Pittsburgh, Pittsburgh, PA 15260, USA. (1945) PhD physics (MIT, 1974). Prof. (tel. 412 + 624-4636, fax 412 + 624-1882, email ROSENBRG at PITTVMS). *Structures, protein, nucleic acid, DNA-protein interactions, macromolecular assemblies.*

Rosenstein, Prof. Robert Daniel. 6350 Genesee, Ave., Apt. 122A, San Diego, CA 92122, USA. (1922) Fil. Lic. chemistry (U. of Uppsala, Sweden, 1961). *Crystal structures.*

Ross, Ms. Dawn L. Rte. 4, Columbia, MO 65203, USA. (tel. 314 + 657-9235).

Ross, Dr. Frederick Keith. Res. Reactor Facility, U. of Missouri, Res. Park, Columbia, MO 65211, USA. (1942) PhD chemistry (U. of Illinois at Urbana, 1969). Sr. res. scient. (tel. 314 + 882-4211, email Bitnet CO365FR at UMVMA). *Structure and bonding, valence electron densities, diffraction, neutron, low temperature, phase transitions.*

Rossi, Dr. Miriam. Chemistry Dept., Vassar C., Box 484 Poughkeepsie, NY 12601, USA. (1952) PhD chemistry (Johns Hopkins U., 1978). Asst. prof. (tel. 914 + 437-5746, email Bitnet ROSSI at VASSAR). *Crystal structure, small molecules.*

Rossmann, Prof. Michael G. Biological Sci. Dept., Purdue U., Lilly Hall of Life Sci., West Lafayette, IN 47907, USA. (1930) PhD chemistry (U. of Glasgow, UK, 1956). Hanley prof. of biology. (tel. 317 + 494-4911). *Viruses, enzymes, protein folding, molecular evolution, crystallographic techniques, methods and theory, computing.*

Rotella, Dr. Frank J. IPNS Div., Bldg. 360., Argonne National Lab., 9700 S. Cass Ave., Argonne, IL 60439, USA. (1949) PhD chemistry (SUNY at Buffalo, 1979). Asst. chemist. (tel. 312 + 972-5785, fax 312 + 972-4163, email ROTELLA at ANLPNS). *Diffraction, single crystal, neutron powder, Rietveld analysis, structural studies, organometallic compounds, metal hydrides.*

Roth, Dr. Robert Sidney. Ceramics Div., Room B214, Materials Bldg., National Inst. Stnds. & Technol., Gaithersburg, MD 20899, USA. (1926) PhD geology (U. of Illinois, 1951). Res. chemist. (tel. 301 + 975-6116). *Powder and single crystal diffraction, non-metallic and inorganic materials.*

Roth, Prof. Walter Lester. 1552 Baker Ave., Schenectady, NY 12309, USA. (1917) PhD chemistry (UCB, 1941). (tel. 518 + 442-4498). *Crystal structure, diffraction, X-ray, neutron, EXAFS, superionic conductors, properties of materials.*

Roy, Prof. Rustum. Materials Res. Lab., Pennsylvania State U., University Park, PA 16802, USA. (1924) PhD ceramics (PSU, 1948). Dir., Evan Pugh Prof. of the Solid State. (tel. 814 + 865-3421). *Materials, preparation, characterization, crystal chemistry, non-metallic systems, synthesis, stability, phase equilibria, crystal growth, ultrahigh pressure solid reactions, non-crystalline solids (chemistry and physics).*

Royer, Dr. William Edward, Jr. Molec. Biophys. & Biochem. Dept., Columbia U., 630 168th St., New York, NY 10032, USA. (1954) PhD biophysics (Johns Hopkins U., 1984). Postdoct. res. scient. (tel. 212 + 305-1846). *Macromolecular crystallography, protein structure and evolution.*

Rozwarski, Ms. Denise Andrea. Chem. & Biochem Dept., U. of Texas at Austin, WEL 5.262, Austin, TX 78712, USA. (1964) BS chemistry, biochemistry (Purdue U., 1986). Grad. student. (tel. 512 + 471-3625). *Protein crystallography, proteins, structure, function.*

Ruben, Mrs. Helena W. 651 Vincente Ave., Berkeley, CA 94707, USA. AB physics (UCB, 1935). Retired. (tel. 415 + 526-1897). *Single crystals, diffraction, structure.*

Rubin, Dr. Ben Z. 124-H Highland St., Manchester, CT 06040, USA. (1917) PhD biophysics (U. of Michigan, 1976). Retired. (tel. 203 + 646-3067). *X-ray diffraction analysis, biological materials.*

Rubin, Dr. Byron. Protein Engineering Group, Eastman Kodak Co., 1999 Lake Ave., Rochester, NY 14650-2118, USA. (1943) PhD chemistry (Duke U., 1971). Res. scient./proj. leader. (tel. 716 + 722-4960, fax 716 + 722-2327). *Protein structures, glycoproteins, lipases.*

Rubin, Dr. John Ronald. Crystallography Lab., NCI - Frederick Cancer Res. Fac., P.O. Box B., Frederick, MD 21701, USA. (1947) PhD molecular biology (U. of Wisconsin at Madison, 1974). Group leader. (tel. 301 + 698-5035, fax 301 + 698-5991). *Structure, proteins, nucleic acids, protein - nucleic acid interactions, molecular pharmacology, drug interactions.*

Ruble, Dr. John Rollo. Crystallography Dept., U. of Pittsburgh, 304 Thaw Hall, Pittsburgh, PA 15260, USA. (1946) PhD crystallography (U. of Pittsburgh, 1975). Staff crystallographer (tel. 412 + 624-9309, fax 412 + 624-1882). *Small molecules, charge density.*

Ruderman, Dr. I. Warren. INRAD Inc., 181 Legrand Ave., Northvale, NJ 07647, USA. (1920) PhD chemical physics (Columbia U., 1949). Pres. and CEO. (tel. 201 + 767-1910, Telex. 13-5248, fax 201 + 767-9644). *Crystal growth, X-ray spectroscopy, electro-optics, acousto-optics.*

Rudman, Prof. Reuben M. Chemistry Dept., Adelphi U., Garden City, NY 11530, USA. (1937) PhD chemistry (PUN, 1966). Prof. (tel. 516 + 294-8700, ext. 7519). *Low temperature X-ray diffraction instrumentation, crystal structure analysis, phase transitions, molecular crystals.*

Rudnick, Dr. Suzanne Ellen. Chemistry Dept., Manhatten C., Manhatten College Parkway, Riverdale, NY 10471, USA. (1951) PhD chemistry (Boston U., 1979). Assoc. prof. (tel. 212 + 920-0211). *Protein structure and function.*

Rudnik, Mr. Paul J. 1233 W. Jarvis, Chicago, IL 60626, USA. (tel. 312 + 761-2063).

Rudolf, Dr. Philip Reinhold. Analyt. Sciences, Inorg. Grp., The Dow Chemical Co., 1897 Bldg., Midland, MI 48667, USA. (1955) PhD chemistry (Texas A. & M. U., 1983). Proj. leader. (tel. 517 + 636-0565). *Powder diffraction, X-ray, neutron, X-ray methods, software and graphics, powder structure solutions.*

Ruiz, Mr. Bienvenido. P.O. Box 8298, Humaco, PR 00661, USA.

Russell, Dr. Thomas Paul. K91/802, IBM Almaden Res. Ctr., 650 Harry Rd., San Jose, CA 95120, USA. (1952) PhD polymer science (U. of Massachusetts, 1979). Res. staff member. (tel. 408 + 927-1638). *Polymers, mixtures, scattering, X-ray, neutron, time resolved, synchrotron radiation, X-ray and neutron reflectivity.*

Ruszala, Mr. Ferdinand A. 1 Lebanon Ave., Colchester, CT 06415, USA.

Ruud, Prof. Clayton Olaf. Materials Res. Lab., 159 MRL, Pennsylvania State U., University Park, PA 16802, USA. (1934) PhD materials science (U. of Denver, 1970). Assoc. prof., Asst. Dir. Matls. Res.Lab. (tel. 814 + 863-2843). *X-ray powder diffraction, residual stresses, electron microprobe, materials characterization, X-ray spectroscopy, metallurgy, material science.*

Ryan, Prof. Clarence A. Inst. of Biol. Chem., Washington State U., Pullman, WA 99164-6340, USA. (1931) PhD chemistry (Montana State U., 1959). Prof. (tel. 509 + 335-3304). *Proteinase, inhibitors, structure, function, plant proteinase inhibitor genes, regulation.*

Ryan, Dr. Robert Reynolds. Group CNC-4, MS 346, Los Alamos National Lab., Los Alamos, NM 87545, USA. (1936) PhD chemistry (Oregon State U., 1965). Res. chemist (tel. 505 + 667-6045). *Crystallography, activation of small molecules, chemistry, coordination, actinide, continuous phase changes, vibrational spectroscopy, gas phase electron diffraction.*

Ryba, Prof. Earle Richard. Materials Science and Eng., Pennsylvania State U., 304 Steidle, University Park, PA 16802, USA. (1934) PhD physical metallurgy (Iowa State U., 1960). Assoc. prof. (tel. 814 + 865-3760). *Crystal structure,* *chemistry, intermetallic compounds, microstrains, synchrotron radiation, quasicrystalline materials.*

Rydel, Mr. Timothy John. Chemistry Dept., Michigan State U., East Lansing, MI 48824, USA. (1959) MS phys. chemistry (Michigan State U., 1983). Grad. student. (tel. 517 + 353-7298, fax 517 + 353-1793). *Protein crystallography, crystallization, structure of blood clotting cascade proteins.*

Rypniewski, Dr. Wojciech Robert. Inst. for Enzyme Res., U. of Wisconsin, 1710 University Ave., Madison, WI 53705, USA. (1958) PhD X-ray crystallography (U. of Cambridge, England, 1987). Postdoct. fel. (tel. 608 + 262-0529). *Protein crystallography.*

Sabat, Dr. Michael. Chemistry Dept., Northwestern U., 2145 Sheridan Rd., Evanston, IL 60208, USA. (1947) PhD chemistry (U. of Wroclaw, Poland, 1976). Lect., Mgr. cryst. & mol. graphics. (tel. 312 + 491-2950, fax 312 + 491-4133). *X-ray crystallography, small and biological macromolecules, molecular graphics, modeling, chemistry, structural, bioinorganic, organometallic.*

Sack, Dr. John Stuart. HHMI, One Baylor Plaza, Houston, TX 77030, USA. (1953) PhD biophysics (Johns Hopkins U., 1981). Res. Assoc. (tel. 713 + 799-6563). *Protein crystallography.*

Sadowski, Mr. Lucian M. US Army Armament Res., Developm. and Eng. Ctr., SMCAR CCL FA, Picatinny Arsenal, NJ 07806-5000, USA. (1954) MS engineering management (Florida Inst. Technol., 1982). Mech. eng./physicist (tel. 210 + 724-7932, email sadowski at ardec.arpa). *Crystallography, ballistics, optics, astronomy, aerodynamics, electromechanics, metallurgy.*

Sakore, Dr. Tukaram D. P. O. Box R.B. & B., Medical Center, U. of Rochester, Rochester, NY 14642, USA. PhD crystallography (U. of Poona, India, 1966). Sr. scient. (tel. 716 + 334-7770). *Structure, crystallographic studies, nucleic acids, drug-oligonucleotide complexes, biologically important molecules.*

Saldin, Dr Dilano Kerzaman. Physics Dept.,Uni. of Wisconsin-Milwaukee, P.O. Box 413, Wisconsin WI 53201, USA. (1949) DPhil,physics (Oxford,UK,1976) Asst. prof. (tel. 414 + 229 6423, fax 414+229 5589, email DKSALDIN at CSD4.MILW.WISC.EDU) *Electron diffraction, electron microscopy, surface structure, scattering theory.*

Salemme, Dr. Francis Raymond. Central Res. & Dev. Dept., E.I. du Pont de Nemours Co., Expt. Sta., P.O. Box 80228, Wilmington, DE 19880-0228, USA. (1945) PhD chemistry (U. of Calif. at San Diego, 1972). Res. leader. (tel. 302 + 695-1877, fax 302 + 695-9183). *Biomolecular structure, electron transport, molecular dynamics, protein crystallography.*

Salkind, Prof. Alvin J. Bioeng. Div., Surgery Dept., UMDNJ - Robert Wood Johnson Med. Sch., 675 Hoes Lane, Piscataway, NJ 08854, USA. (1927) DChE chemical engineering (PUN, 1958). Chief. (tel. 201 + 463-4799). *Electrochemically active surfaces, batteries, catalysts, implantable materials.*

Samson, Dr. Sten. Chemistry Dept., Calif. Inst. of Techn., 1201 East California Blvd., Pasadena, CA 91125, USA. (1916) Fil Dr. chemistry (U. of Stockholm, Sweden, 1968). Sr. res. assoc. emeritus. (tel. 818 + 356-6528, ext. 6528). *Intermetallic compounds (complex structures), organic conductors, quasicrystals, instrumentation, low temperature diffractometry.*

Samudzi, Mr. Cleopas Tendai. Biol. Sciences Dept., U. of Pittsburgh, 5th. & Ruskin Aves., Pittsburgh, PA 15260, USA. (1958) MS biocrystallography (U. of Pittsburgh, 1983). Grad. student. (tel. 412 + 624-4638, email CLEO at PITTVMS). *Macromolecular crystallography.*

Samuels, Prof. Robert J. School of Chem. Engineering, Georgia Inst. of Techn., Atlanta, GA 30332-0100, USA. (1931) PhD polymers (U. of Akron, 1961). Prof. (tel. 404 + 894-2885). *Polymer engineering, physics, chemistry, structure - properties - process relationship, diffraction, X-ray, neutron, small and wide angle.*

Sands, Prof. Donald E. Office of Academic Affairs, U. of Kentucky, 207 Administration Bldg., Lexington, KY 40506, USA. (1929) PhD phys. chemistry (Cornell U., 1955). Vice Chancellor, academic affairs (tel. 606 + 257-1961, email Bitnet SANDS at UKCC). *Crystals, tensor properties, motion in, thermodynamics.*

Saper, Dr. Mark A. Biochemistry Dept. and HHMI, Harvard U., 7 Divinity Ave., Cambridge, MA 02138, USA. (1954) PhD biochemistry (Rice U., 1983). Res. Assoc. (tel. 617 + 495-9613, fax 617 + 495-9613, email Bitnet SAPER at HUXTAL). *Protein crystallography, molecular immunology, computer graphics.*

Sappenfield, Dr. Eric L. Chemistry Dept., Baylor U., Waco, TX 76798, USA. (1949) PhD chemistry (Baylor U., 1987). (tel. 817 + 755-3311, ext. 4823).

Sarko, Prof. Anatole. Chemistry Dept., C. of Env. Sci. & Forestry, SUNY Syracuse, NY 13210, USA. (1930) PhD polymer chemistry (SUNY at Syracuse, 1966). Prof. and chairman. (tel. 315 + 470-6824, Telex. 7105410555, fax 315 + 470-6779, email Bitnet ASARKO at SUVM). *Polymers, crystallography, molecular mechanics, dynamics, quantum mechanics, polysaccharides, computer methods.*

Sarma, Prof. Raghupathy. Biochemistry Dept., SUNY at Stony Brook, Stony Brook, NY 11794, USA. PhD physics (U. of Madras, India, 1963). Assoc. prof. (tel. 516 + 632-8558, fax 516 + 632-8575, email RSARMA at SBCCMAIL). *Proteins, structure and function.*

Sass, Prof. Stephen Louis. Materials Sci. and Eng. Dept., Bard Hall, Cornell U., Ithaca, NY 14853, USA. (1940) PhD materials science (Northwestern U., 1966). Prof. (tel. 607 + 255-5239). *Grain boundary structure, diffraction, X-ray, electron, crystal defects, electron microscopy.*

Satyshur, Dr. Kenneth A. Biochemistry Dept., U. of Wisconsin, 420 Henry Mall, Madison, WI 53706, USA. (1948) PhD biochemistry (U. of Wisconsin at Madison, 1978). Asst. scient. (tel. 608 + 262-3019). *Macromolecular crystallography.*

Satz, Mr. Ronald Wayne. Transpower Corp., 1 Oak Drive, Parkerford, PA 19457, USA. (1951) MS engineering (Rensselaer Polytechnic Ins., 1974). President, systs. eng. (tel. 215 + 495-6362). *Theoretical physics, interatomic distances, liquids, solids, operations research, mechanical engineering, system dynamics, control.*

Sawzik, Ms. Patricia. 406 N. Neville Ave., Pittsburgh, PA 15213, USA.

Sax, Dr. Martin. Biocrystallography Lab., VA Hospital, University Drive C, Pittsburgh, PA 15204, USA. (1919) PhD chemistry, crystallography (U. of Pittsburgh). (tel. 412 + 683-3000, ext. 517 or204). *Antibody structures, stereochemistry, thiamine catalysis, biologically important structures, stereoelectronic features of biochemical reactions.*

Sayler, Prof. Alice Ann. Chemistry Dept., Bloomfield C., Franklin St., Bloomfield, NJ 07003, USA. (1946) PhD chemistry (Worcester Polytechnic Inst., 1974). Prof. chem. & Chrm. Natl. Sci. & Math. (tel. 201 + 736-5744). *X-ray studies, transition metal complexes.*

Sayre, Dr. David. Res. Div., IBM T. J. Watson Res. Cntr., P.O. Box 218, Yorktown Heights, NY 10598, USA. (1924) PhD chemical crystallography (Oxford U., UK, 1951). Retired. (tel. 914 + 945-1040). *Structure determination methods, X-ray microscopy, diffraction analysis methods.*

Scarbrough, Dr. Frank Edward. Office of Nutrition & Food Sci., Food and Drug Administration, 200 "C" St., Washington, DC 20204, USA. (1942) PhD phys. chemistry (Harvard U., 1971). Deputy dir. (tel. 202 + 245-1561, fax 202 + 426-1658). *Structure - flavor - odor correlations, structure - biological activity correlations.*

Scaringe, Dr. Raymond P. Chemistry Div., Res. Labs., Eastman-Kodak Co., 1669 Lake Ave., Rochester, NY 14650, USA. (1950) PhD phys.-inorg. chemistry (U. of North Carolina, 1976). Sr. res. scient. (tel. 716 + 477-7052). *Packing in molecular solids, small molecule - polymer interactions, structure determinations.*

Schaefer, Dr. William Palzer. Div. of Chemistry 127-72, Calif. Inst. of Techn., 1201 E. California St., Pasadena, CA 91125, USA. (1931) PhD chemistry (UCLA, 1960). Sr. res. assoc. (tel. 818 + 356-6567). *Inorganic chemistry, transition metal complexes.*

Schäfer, Prof. Lothar. Chemistry Dept., U. of Arkansas, Fayettesville, AR 72701, USA. (1939) Dr.rer.nat. inorg. chemistry (U. of Münich, FRG, 1965). Prof. (tel. 501 + 575-4601). *Structural studies, electron diffraction, vibrational analysis and theoretical procedures.*

Scheetz, Dr. Barry E. Materials Res. Lab., Pennsylvania State U., University Park, PA 16802, USA. (1946) PhD geochemistry, mineralogy (PSU, 1976). Assoc. prof., solid state sci. (tel. 814 + 865-3539, fax 814 + 865-2326). *Waste management, nuclear, hazardous, vibrational spectroscopy, crystal chemistry, cement chemistry.*

Scheidt, Prof. W. Robert. Chemistry Dept., U. of Notre Dame, Notre Dame, IN 46556, USA. (1942) PhD chemistry (U. of Michigan, 1968). Prof. (tel. 219 + 239-5939, email FNFTY9 at IRISHMVS). *Structures, inorganic compounds, macrocycles, porphyrins.*

Schellman, Dr. F. Charlotte. Inst. of Molec. Biology, U. of Oregon, Eugene, OR 97403, USA. (1922) PhD chemistry (Stanford U., 1950). Adjunct assoc. prof. (tel. 503 + 686-4633). *Protein conformation, folding.*

Schevitz, Mr. Richard. Biochem. & Molec. Biology Dept., U. of Chicago, 920 East 58th St., Chicago, IL 60637, USA.

Schiffer, Dr. Marianne. Biol. Environ, & Medical Res. Div., Argonne National Lab., 9700 South Cass Ave., Argonne, IL 60439, USA. (1935) PhD biochemistry (Columbia U., 1965). Sr. scient. (tel. 312 + 972-3883, Telex. 687-1701 DOE-ANL, fax 312 + 972-2206, email SCHIFFER at ANLPHY). *Protein structures, immunoglobulins, photosynthetic reaction centers.*

Schildekamp, Dr. Wilfried. 281 Wilson Lab., Cornell U., Ithaca, NY 14853, USA. (1949) Dr.rer.nat. crystallography (U. of Saarbrücken, FRG, 1975). Sr. res. assoc. (tel. 607 + 255-0916, Telex. MCI 6713054, fax 607 + 255-8062). *X-ray optics, synchrotron radiation instrumentation, macromolecular crystallography.*

Schioler, Dr. Liselotte Jensen. Directorate of Electronic & Matl. Sci., Air Force Off. Scient. Res., AFOSR/NE, Rolling AFB, DC 20332-6448, USA. (1950) ScD ceramic science (MIT, 1983). Prog. manager. (tel. 202 + 767-4933, fax 202 + 767-4977). *Structures, neutron diffraction.*

Schirber, Dr. James Emmanuel. Dept. 1150, Sandia Nat. Labs., Albuquerque, NM 87115, USA. (1931) PhD physics (Iowa State U., 1960). Manager, solid state res. dept. (tel. 505 + 844-8134). *Low temperature structures, pressure dependence, atomic positional parameters, thermal expansion, electronic structure - properties relationship.*

Schlemper, Prof. Elmer Otto. Chemistry Dept., U. of Missouri, Columbia, MO 65211, USA. (1939) PhD inorg. chemistry (U. of Minnesota, 1965). Prof. (tel. 314 + 445-1340). *Hydrogen bonding, neutron diffraction, inorganic compounds, structure.*

Schmid, Dr. Michael Francis. Biochemistry Dept., Baylor C. of Med., 1 Baylor Plaza, Houston, TX 77030, USA. (1947) PhD biochemistry (U. of Washington, 1974). Res. asst. prof. (tel. 713 + 798-6984). *Biological macromolecules, biological assemblies, X-ray crystallography, electron microscopy, diffraction, image processing.*

Schmidt, Dr. Melvin C. Techn. Dept., Amoco Techn. Co., P.O. Box 400, Naperville, IL 60566, USA. (1953) PhD physics (Purdue U., 1986). Res.

physicist. (tel. 312 + 420-3694). *X-ray diffraction, crystallography, semiconductors.*

Schmidt, Prof. Paul Woodward. Physics Dept., U. of Missouri - Columbia, Columbia, MO 65211, USA. (1926) PhD physics (U. of Wisconsin at Madison, 1953). Prof. (tel. 314 + 882-8241, fax 314-882-4195). *Small angle scattering, X-ray, neutron, biophysics, physics of fluids, critical phenomena, fractals, porous solids.*

Schmidt, Dr. William Charles Jr. 304 Cedarfield Ln., West Columbia, SC 29169, USA. (1948) PhD biochemistry, crystallography (U. of Virginia, 1975). (tel. 803 + 356-1210). *Protein structure, crystallography, refinement techniques, macromolecular crystallography.*

Schneider, Dr. Diane M. Inst. for Cancer Res., 7701 Burholme Ave., Philadelphia, PA 19111, USA. (tel. 215 + 728-3159).

Schneider, Dr. Dieter K. Biology Dept., Brookhaven Nat. Lab., Upton, NY 11973, USA. (1947) PhD biophysics (U. of Basel, Switzerland). Biophysicist. (tel. 516 + 282-3423). *Neutron scattering.*

Schoenborn, Dr. Benno Paul. Biology Dept., Brookhaven Nat. Lab., Upton, NY 11973, USA. (1936) PhD physics (U. of New South Wales, Australia, 1962). Sr. Sci., BNL, Prof., Columbia Med. Ctr. (tel. 516 + 282-3421, email SCHOENBORN at BNLCL1). *Biological structures, scattering, X-ray, neutron, structure - function relationships, pharmacological agents, proteins.*

Schomaker, Prof. Verner. Jan. 1-June 30: 1066 San Pasqual, #8, Pasadena, CA 91106, USA. Jul. 1-Dec. 31: 13224 42nd. Av., NE, Seattle, WA 98125, USA. (1914) PhD chemistry (CIT, 1938). Prof. emeritus, U. Wash., Vis. Res. Assoc. CIT. *Gaseous electron diffraction, X-ray structure determination methods.*

Schrader, Ms. Patricia Ann. Chemistry Dept., U. of Pennsylvania, 34th. & Spruce Sts., Philadelphia, PA 19104, USA. (1961) BS chemistry (Chestnut Hill C., 1983). Grad. student. (tel. 215 + 898-4886). *Solution NMR of proteins.*

Schroeder, Dr. LeRoy William. Mech. & Matls. Sci. Div., Food & Drug Administration, 12200 Wilkens Ave., Rockville, MD 20857, USA. (1943) PhD phys. chemistry (Northwestern U., 1969). Res.chemist. (tel. 301 + 443-7003). *Molecular structure, dynamics, membrane phenomena.*

Schulte, Ms. Gayle K. Chemistry Dept., Yale U., 225 Prospect St., New Haven, CT 06511, USA.

Schultz, Dr. Arthur Jay. Chemistry Dept., Argonne Nat. Lab., 9700 S. Cass Ave., Argonne, IL 60439, USA. (1947) PhD chemistry (Brown U., 1973). Chemist. (tel. 312 + 972-3465). *Diffraction, X-ray, single crystal, time-of-flight neutron, data analysis techniques, instrumentation, structure - property relationships, superconductivity.*

Schultz, Prof. Jerold Marvin. Chemical Eng. Dept., U. of Delaware, Newark, DE 19716, USA. (1935) PhD metallurgical eng. (Carnegie Inst. of Techn., 1965). Prof. (tel. 302 + 451-8145). *Structure - properties relationship, polymeric materials, phase transitions.*

Schuman, Mr. Clifford Alan. 149 Buckhill Rd., Canterbury, CT 06331, USA.

Schuster, Dr. Sanford Lee. Dept. of Physics, Mankato State U., Mankato, MN 56001, USA. (1938) PhD physics (U. of Nebraska at Lincoln, 1969). Prof. and chairman. (tel. 507 + 389-5743). *Solid state physics, lattice dynamics.*

Schutt, Prof. Clarence Ernest. Chemistry Dept., Princeton U., Washington Rd., Princeton, NJ 08544, USA. (1945) PhD applied mathematics (Harvard U., 1976). Assoc. prof. (tel. 609 + 452-4435, fax 609-987-6746). *Protein crystallography, actin, cytoplasmic proteins.*

Schwartz, Dr. Lyle Howard. Inst. Matls. Sci. & Eng., Natl. Inst. Stnds. & Techn., Bldg. 223, Rm. B309, Gaithersburg, MD 20899, USA. (1936) PhD materials science (Northwestern U., 1964). Dir., IMSE. (tel. 301 + 975-5658, telex 197674NISTUT, fax 301-926-8349). *Mossbauer spectroscopy, diffraction, neutron, X-ray, phase transformations in solids.*

Sclar, Prof. Charles Bertram. Geological Sci. Dept., Lehigh U., Williams Hall, Bldg. 31, Bethlehem, PA 18015, USA. (1925) PhD geology (Yale U., 1951). Prof. (tel. 215 + 861-3660). *Minerals, synthetic analogues, structure, stability, equilibrium, high pressure phases, synthesis, phase transformations.*

Scott, Prof. Alastair Ian. Chemistry Dept., Texas A&M U., C. Station, TX 77843, USA. (1928) D.Sc. org. chemistry (U. of Glasgow, Scotland, 1963). Davidson Prof. of Science, Dir. Ctr. for Biolog. NMR. (tel. 409 + 845-3243, fax 409-845-4719, email Bitnet SCOTTGROUP at TAMCHEM). *Enzyme mechanism, biosynthesis, NMR.*

Scott, Mr. Brian. Chemistry Dept., Washington State U., Pullman, WA 99164, USA. (1961) MS phys. chemistry (Washington State U., 1987). Grad. student (tel. 509 + 335-9125, email Bitnet WILLETT at WSUVM1). *Chemistry, solid state, magneto, 1-dimensional solids, mixed valence compounds, band calculations.*

Scott, Mr. Donald Lee. Technical Div., Martin Marietta Energy Systs. Inc., P.O. Box 628, Piketon, OH 45661, USA. (1931) electronic eng. (C.R.E.I., 1959). Sr. physicist. (tel. 614 + 897-2331, ext. 5777, fax 614-897-2985). *X-ray diffraction, spectroscopy, optical, electron microscopy, scientific photography.*

Seabaugh, Mr. Pyrtle W. 7795 Raintree Rd., Centerville, OH 45459, USA. (tel. 513 + 433-6925).

Seale, Dr. Steven Keith. Dept. of Chem. Research, Tennessee Valley Authority, T207-NFDC, Muscle Shoals, AL 35660, USA. (1944) PhD phys.-inorg. chemistry (U. of Alabama, 1974). Dept. head. (tel. 205 + 386-2214, email CIS 74326,1555). *Computational methods, instrumentation.*

Seaton, Prof. Barbara A. Physiology Dept., Boston U. Sch. of Medicine, 80 E. Concord St., Boston, MA 02118, USA. (1952) PhD chemistry (MIT, 1983). Asst. res. prof. (tel. 617 + 638-5061). *X-ray crystallography, proteins, structure - function, regulatory & control mechanisms.*

Sedzik, Dr. Jan. Neuroscience Dept., Childrens Hosp., Harvard Med. Sch., 300 Longwood Ave., Boston, MA 02115, USA. (1946) PhD biophysics, neurochem. (Poznan Medical Acad., Poland, 1978). Asst. prog. (tel. 617 + 735-7187 or 735-6102). *Structures, protein, membrane.*

Seeholzer, Dr. Steven H. Inst. for Cancer Res., 7701 Burholme Ave., Philadelphia, PA 19111, USA. (tel. 215 + 728-3159).

Seely, Prof. Oliver Jr. Chemistry Dept., California State C., Dominguez Hills, CA 90747, USA. PhD. Prof.

Seeman, Prof. Nadrian Charles. Chemistry Dept., New York U., New York, NY 10003, USA. (1945) PhD biochemistry-crystallography (U. of Pittsburgh, 1970). Prof. (tel. 212 + 998-8463, email SEEMAN at NYUACF1). *Nucleic acid branched junctions, macromolecular design.*

Seff, Prof. Karl. Chemistry Dept., U. of Hawaii, 2545 The Mall, Honolulu, HI 96822-2275, USA. (1938) PhD chemistry (MIT, 1964). Prof. (tel. 808 + 948-7665, fax 808-949-8025, email keff at helium.hawaii.chem.edu). *Intrazeolitic chemistry, structure, transition metal complexes, small organic molecules.*

Segmüller, Dr. Armin Paul. Res. Div., IBM, T. J. Watson Res. Cntr., P.O. Box 218, Yorktown Heights, NY 10598, USA. (1924) Dr.phil.nat. crystallography (U. Erlangen-Nürnberg, BRD, 1954). Res. staff member. (tel. 914 + 945-1287, telex 910-240-0632). *Crystal physics, diffraction physics, phonons, superlattices, laboratory automation, multilayer structures.*

Sehnke, Mr. Paul C. 203 Montefiore St., Apt. 216, Lafayette, IN 47905, USA.

Seitz, Dr. Frederick. Rockefeller U., 1230 York Ave., New York, NY 10021, USA. (1911) PhD physics (Princeton U., 1934). retired (tel. 212 + 570-8423).

Sekharudu, Dr. Chandra Y. Biochemistry Dept., U. of Wisconsin at Madison, Madison, WI 53706, USA. (tel. 608 + 262-3019).

Sen Gupta, Prof. Pradip Kumar. Geology Dept., Memphis State U., Memphis, TN 38152, USA. Memphis, TN 38152, USA. (1936) PhD geology, mineralogy (Washington U., 1964). Prof. (tel. 901 + 678-4361). *Structural mineralogy, organo-metallic complexes, inorganic structures.*

Senechal, Prof. Marjorie Lee. Mathematics Dept., Smith C., Northampton, MA 01063, USA. (1939) PhD mathematics (Illinois Inst. of Techn., 1965). Prof. (tel. 413 + 584-3862, email MSENECHAL at SMITH). *Mathematical crystallography, history of crystallography.*

Sennett, Ms. Judith Brodkin. Chemistry Dept., Brandeis U., 415 South St., Waltham, MA 02254-9110, USA.

Serra, Mr. Michael A. Biochem. & Biophysics, Iowa State U., Gilman Hall, Rm. 397, Ames, IA 50011, USA. (1962) BS biology and mathematics (Adrian C., 1984). Grad. student (tel. 515 + 294-0567). *Protein crystallography.*

Servos, Prof. Kurt. Geology Dept., Menlo C., Menlo Park, CA 94025, USA. (1928) MS geology (Yale U., 1954). Prof. (tel. 415 + 322-1245). *Crystal symmetry, morphological, geometrical crystallography, mineral crystal structures.*

Shafer, Ms. Donna. 77 Clinton St., Apt. M4, New York Mills, NY 13417, USA. (tel. 315 + 736-3050).

Shaffer, Prof. Lawrence B. Physics Dept., Anderson U., Anderson, IN 46012, USA. Anderson, IN 46012, USA. (1937) PhD physics (U. of Wisconsin, 1964). Prof. and chairman. (tel. 317 + 641-4375). *Small angle X-ray scattering, synchrotron radiation, alloy structure.*

Shaffner, Dr. Thomas Jackson. Central Res. Labs., Texas Instruments Inc., 13500 N. Central Exprwy., Dallas, TX 75265, USA. (1941) PhD physics (Vanderbilt U., 1969). Branch mgr. (tel. 214 + 995-6764, ext. 6764). *X-ray crystallography, microprobe analysis, microscopy, scanning electron, Auger electron, X-ray photoelectron, materials characterization, scientific data processing.*

Shafiee, Dr. Fathieh. Enzyme Inst., U. of Wisconsin at Madison, 1710 University Ave., Madison, WI 53705, USA. (1955) PhD inorg. chemistry (U. of Wisconsin at Madison, 1985). Staff res. asst. (tel. 608 + 262-0401). *X-ray crystallography, macromolecules, small molecules.*

Shannon, Dr. Robert Day. Central Res. & Dev., Exp. Sta., E. I. du Pont de Nemours Co., Wilmington, DE 19898, USA. (1935) PhD ceramic eng. (UCB, 1964). Res. chemist. (tel. 302 + 695-1024). *Crystallography, synthesis, crystal growth, structure - properties relationships, dielectric properties of oxides.*

Sharma, Prof. Brahama Datta. Chemistry Dept., Calif. State U. at Los Angeles, Los Angeles, CA 90032, USA. (1931) PhD chemistry (USC, 1961). Prof. (tel. 818 + 347-0551, ext. 264, fax 818-710-9844). *Structural chemstry, proteins, DNA, chemistry, inorganic, physical, analytical, molecular biology, number theory, solid state.*

Sharp, Prof. Paul R. Chemistry Dept., U. of Missouri, Columbia, MO 65211, USA. (1952) PhD inorg. chemistry (MIT, 1980). Assoc. prof. (tel. 314 + 882-7715). *Chemistry, synthetic, structural, inorganic, organometallic, catalysis, small molecule activation.*

Sharrah, Prof. Paul Chester. Physics Dept., U. of Arkansas, Fayetteville, AR 72701, USA. (1914) PhD physics (U. of Missouri, 1942). Prof. emeritus. (tel. 501 + 575-2506). *Diffraction, X-rays, neutrons, liquids.*

Shea, Dr. Madeline A. Biology Dept., Johns Hopkins U., Mudd Hall, Baltimore, MD 21218, USA. (1956) PhD biophysics (Johns Hopkins U., 1984). Assoc. res. scient. (tel. 301 + 338-7239). *Structure - function relationships, interactions,* *protein-protein, protein-DNA, thermodynamics, kinetics, regulatory mechanisms, calcium binding, EF hands.*

Shea, Mr. Michael E. 2199 NW Everett, Apt. 601, Portland, OR 97210, USA.

Shefter, Dr. Eli. Genentech Inc., 460 Pt. San Bruno Blvd., South San Francisco, CA 94080, USA. (1936) PhD pharmaceutics (U. of Wisconsin, 1963). Sr. scient. (tel. 415 + 266-2255). *Drug delivery, pharmocokinetics, structure - activity relationships, formulation design.*

Sheldon, Dr. Robert Isaly. Physical Metallurgy MST-5, MS G730, Los Alamos Nat. Lab., Los Alamos, NM 87545, USA. (1945) PhD phys. chemistry (U. of Kansas, 1976). Staff member. (tel. 505 + 665-0144). *Diffraction, small molecules, macromolecules, single crystals, powder, direct methods, actinides, deformation twinning.*

Shen, Dr. Ming-Shing. Applied Science Div., U. S. Dept. of Energy, P.O. Box 880, Morgantown, WV 26505, USA. (1940) PhD materials science (U. of Pittsburgh, 1974). Chemical eng. (tel. 304 + 291-4112). *Fossil fuel utilization, oil shale process technology, hot gas stream contaminant cleanup.*

Shen, Dr. Qun. CHESS, Wilson Lab., Cornell U., Ithaca, NY 14853, USA. (1959) PhD physics (Purdue U., 1987). Postdoct. res. assoc. (tel. 607 + 255-0923, fax 607-255-8062, email Bitnet SHEN at CRNLCHES). *Phase problem in crystallography, X-rays, diffraction, multi-beam, dynamical theory of, surface.*

Shenkin, Prof. Peter S. Chemistry Dept., Barnard C., Columbia U., New York, NY 10027, USA. (1947) PhD chemistry (Princeton U., 1979). Asst. prof. (tel. 212 + 854-1418, email shenkin at cubmol.bio.columbia.edu). *Modeling, structure prediction, minimization, optimization, physical biochemistry.*

Sheridan, Mr. Robert E. Chemistry Dept., U. of Wisconsin, Madison, WI 53705, USA.

Sheriff, Dr. Steven. Macromolecular Crystallography, Squibb Inst. Med. Res., P.O. Box 4000, Princeton, NJ 08543-4000, USA. (1951) PhD biochemistry (U. of Washington, 1979). (tel. 609 + 921-5934, fax 609-683-6280, email sheriff at squibb.com.csnet). *Macromolecular crystallography, structure - function, crystallographic computing.*

Sherry, Mrs. Elizabeth Ann Gebert. 1370 Blue Lilac Ln. Alpine, CA 92001, USA. (1926) BS chemistry (Ball State U., 1947). Retired. (tel. 619 + 445-3736). *Inorganic crystal structures.*

Shieh, Dr. Huey-Sheng. BB4K, MONSANTO, 700 Chesterfield Village Pkwy., Chesterfield, MO 63198, USA. (1946) PhD chemistry (U. of Pennsylvania, 1975). Sr. res. specialist. (tel. 314 + 537-6025). *Protein structure, design, energy calculations, molecular graphics, structure - function relationships.*

Shinn, Dr. Dennis Burton. Sylvania Lighting Center, GTE Sylvania, 100 Endicott St., Danvers, MA 01923, USA. (1939) PhD chemistry (Michigan State U., 1968). Dir. R&D, HID Labs. (tel. 508 + 750-2163). *Lamp materials, chemistry, phase equilibria, ceramics.*

Shiono, Prof. Ryonosuke. Crystallography Dept., U. of Pittsburgh, Pittsburgh, PA 15260, USA. (1923) DSc physics (Osaka U., Japan, 1960). Assoc. prof. (tel. 412 + 624-9302). *Organic crystal structures, computer programming.*

Shipley, Prof. G. Graham. Biophysics Dept., Boston U. Sch. of Medicine, 80 East Concord St., Boston, MA 02118, USA. (1937) PhD X-ray crystallography (U. of Nottingham, UK, 1963). Prof., biochemistry. (tel. 617 + 638-4009). *Lipids, lipoproteins, membranes, scattering, X-ray, neutron.*

Shoemaker, Prof. Clara Brink. 3453 NW Hayes Ave., Corvallis, OR 97330, USA. (1921) PhD chem. crystallog. (U. of Leiden, Netherlands, 1950). Prof. emeritus (tel. 503 + 752-7069). *Structures, crystal, inorganic, metal, alloy, quasicrystal related alloys.*

Shoemaker, Prof. David Powell. Chemistry Dept., Oregon State U., Corvallis, OR 97331, USA. (1920) PhD phys. chemistry (CIT, 1947). Prof. emeritus. (tel. 503 + 754-2081, telex 5105960682.OSU.COVS, fax 503-754-2400, email Bitnet CHEMOF1 at ORSTATE). *Tetrahedrally close packed (t.c.p.) metal phases, zeolites, alloy hydrides.*

Shoham, Mr. Menachem. Central R&D Dept., E.I. duPont de Nemours Co., Exptl. Sta. Bldg. 228 Rm. 320, Wilmington, DE 19898, USA.

Shoja, Dr. Massud. Chemistry Dept., Fordham U., Bronx, NY 10458, USA. (1948) PhD phys. chemistry (Fordham U., 1979). Lab. manager. (tel. 212 + 579-2581).

Shomaly, Mr. Walid. Chemistry Dept., Northeastern U., Boston, MA 02115, USA. (1957) MS polymer science (U. of Bradford, England, 1979). Grad. student. (tel. 617 + 437-2854).

Shon, Mr. Ki-Joon. Chemistry Dept., U. of Pennsylvania, 231 S. 34th. St., Philadelphia, PA 19104, USA. (tel. 215 + 898-4886).

Short, Dr. Michael Arthur. SRS Technologies, 24832 Weyburn Dr., Laguna Hills, CA 92653, USA. (1930) PhD X-ray diffraction (PSU, 1961). Mgr., X-ray technology. (tel. 714 + 250-4206, fax 714-250-7468). *X-rays, instrumentation, physics, diffraction, fluorescence, electron microprobe analysis.*

Short, Dr. Michael R. Central Res. & Devel., E320/264, E. I. duPont de Nemours Co., Exptl. Sta., Wilmington, DE 19880-0320, USA. (1944) PhD chemistry (U. of Texas at Austin, 1971). Consultant. (tel. 302 + 675-1942).

Shull, Prof. Clifford G. Physics Dept. Rm. 13-2154, Mass. Inst. of Technol., Cambridge, MA 02139, USA. (1915) PhD physics (New York U., 1941). Prof. emeritus. (tel. 617 + 253-4812). *Neutron diffraction, physics, solid state physics.*

Siegel, Dr. Lester Aaron. 44 Strawberry Hill Ave., Apt. 10E, Stamford, CT 06902, USA. (1925) PhD physics (MIT, 1948). (tel. 203 + 359-2030). *X-ray diffraction, spectroscopy, solid state physics.*

Siegel, Dr. Stanley. Chemical Techn. Div., D205, Argonne Nat. Lab., 9700 S. Cass Ave., Argonne, IL 60439, USA. (1915) PhD physics (U. of Chicago, 1941). Sr. physicist (tel. 312 + 972-4348). *Structure, crystal chemistry.*

Siegrist, Dr. Theo. AT&T Bell Labs., ID-348, 600 Mountain Ave., Murray Hill, NJ 07974, USA. (1955) PhD solid state physics (Eidgenössische Tech. Hochschüle, 1982). (tel. 201 + 582-5253, email allwiseltsi at uucp).

Sieker, Dr. Larry C. Biological Structure Div., U. of Washington, Seattle, WA 98195, USA. (1931) PhD biological structures (U. of Washington, 1981). Res. asst. prof. (tel. 206 + 543-6541). *Macromolecules, structure - function relationships, crystallization techniques, data acquisition, proteins, interactions.*

Sigalovsky, Dr. Julia I. Barnett Inst., Northeastern U., Boston, MA 02115, USA. (1955) PhD crystallography (Vernadsky Inst. of Geochem., USSR, 1986). Postdoct. fel. (tel. 617 + 437-5067). *Crystal structures, X-ray analysis, thermodynamics of disordering in crystals.*

Sigler, Prof. Paul Benjamin. Mol. Biophys. & Biochem. Dept. & HHMI, Yale U., 260 Whitney Ave., New Haven, CT 06511, USA. (1934) MD, PhD biochemistry (Columbia U., 1959, Cambridge U., UK, 1967). Prof./Investigator (HHMI). (tel. 203 + 432-5096, fax 203-432-5175). *Biological macromolecules, genetic regulation, proteins, membranes, interactions.*

Silcox, Prof. John. Sch. of Appl. & Eng. Physics, Cornell U., 235 Clark Hall, Ithaca, NY 14853, USA. (1935) PhD physics (Cambridge U., UK, 1961). David E. Burr Prof. of Eng. (tel. 607 + 255-3332, fax 607-255-7658). *Electron microscopy, spectroscopy, diffraction.*

Sills, Mr. Denis. 361 North Ridgewood Rd., South Orange, NJ 07079, USA.

Silverman, Ms. Sheryl. Inst. for Cancer Res., 7701 Burholme Ave., Philadelphia, PA 19111, USA. (tel. 215 + 728-2220).

Silverton, Ms. Enid. Natl. Insts. of Health, Bldg. 2, Room 312, Bethesda, MD 20205, USA. (tel. 301 + 496-4295).

Silverton, Dr. James V. Natl. Insts. of Health, Bldg. 10, Rm. 7N-316, Bethesda, MD 20014, USA. (1934) PhD chemical crystallography (U. of Glasgow, UK, 1958). Scient. (tel. 301 + 496-1515). *Direct methods, natural products, high molecular weight molecules (500-1500 Daltons), chemical crystallography (general), diffraction techniques.*

Simard, Dr. Roger Gerard. P.O. Box 242, Swarthmore, PA 19081, USA. (1909) PhD phys. chemistry (MIT, 1939). Consultant (tel. 215 + 543-7417). *Powder diffraction, solid state physics, molecular structure.*

Sime, Prof. Rodney J. Chemistry Dept., Calif. State U., Sacramento, CA 95819, USA. Sacramento, CA 95819, USA. (1931) PhD phys. chemistry (U. of Washington, 1959). (tel. 916 + 454-6659). *Crystal, molecular structure, transition element complexes, structure, activity, biologically important molecules.*

Sime, Dr. Ruth Lewin. Chemistry Dept., Sacramento City C., Sacramento, CA 95822, USA. (1939) PhD phys. chemistry (Harvard U., 1964). Instructor. (tel. 916 + 929-7356). *Biologically important molecular structures, history of science.*

Simmins, Dr. John J. Advanced Matls. Div., Titan Corp., 5130 Evans Ave., Valparaiso, IN 46383, USA. (1961) PhD ceramic science (Alfred U., 1989). Sr. ferrite scient. (tel. 219 + 462-4141, ext. 60, fax 219-462-0376).

Simmons, Dr. Charles J. Chemistry Dept., Brigham Young U., Hawaii Campus, Laie, HI 96762, USA. (1948) PhD phys.-inorg. chemistry (U. of Hawaii, 1980). Assoc. prof. (tel. 808 + 293-3803). *X-ray structural characterization, pseudo Jahn-Teller complexes, cobalt dioxygen complexes.*

Simmons, Prof. Ralph O. Physics Dept., U. of Illinois, 1110 W. Green St., Urbana, IL 61801, USA. (1928) PhD physics (U. of Illinois at Urbana, 1957). Prof. (tel. 217 + 333-4170, fax 217-333-4990, email SIMMONS at UIUCMRL). *Crystal defects, momentum distributions, noble gas solids, liquids.*

Simon, Dr. David Eugene. Chem. Res. & Devel., Halliburton Services, P.O. Box 1431-NRC, Duncan, OK 73536-0438, USA. (1943) PhD geology (Iowa State U., 1972). Res. geologist. (tel. 405 + 251-4208). *X-ray diffraction analysis, clay reactions with organics, quantitative analysis of mixtures.*

Simonsen, Prof. Stanley Harold. Chemistry Dept., WEL 3.148, U. of Texas at Austin, Austin, TX 78712, USA. (1918) PhD chemistry (U. of Illinois, 1949). Prof. (tel. 512 + 471-5755). *Transition metal coordination compounds, small ring heterocycles, fused rings, biochemically significant compounds.*

Simpson, Dr. Dale R. Geol. Sciences Dept., Lehigh U., Bethlehem, PA 18015, USA. (1930) PhD geology (CIT, 1960). Prof. (tel. 215 + 758-3664). *Mineralogy, crystallography, petrology.*

Simpson, Dr. Howard D. 3772 Hamilton St., Irvine, CA 92714, USA. (1937) PhD chemical eng. (U. of Texas, 1969). (tel. 714 + 528-7201). *Catalysis, chemical bonding.*

Sinclair, Dr. Alison. Neuroscience Dept., Children's Hosptal, 320 Longwood Ave., Boston, MA 02115, USA. (tel. 617 + 735-6443).

Singh, Dr. Phirtu. Chemistry Dept., North Carolina State U., Box 8204, Raleigh, NC 27695, USA. (1933) PhD chemistry (U. of Colorado, 1965). Dir., X-ray crystallography. (tel. 919 + 737-7362). *Biochemical molecules, proteins, nucleic acids, interactions, conformational analysis.*

Singh, Dr. Sarjant. P.O. Box 426, Del Mar, CA 92014, USA. PhD chemistry (U. of Calif. at San Diego, 1968). (tel. 619 + 755-6270). *Electron microscopy, diffraction, X-ray crystallography, biochemical systems.*

Sinha, Dr. Sunil K. Exxon Res. & Eng. Co., Route 22 East, Annandale, NJ 08801, USA. (1939) PhD physics (Cambridge U., England, 1964). Sr. res. assoc. (tel. 201 + 730-2875, telex 136140 EXXONRES CLN, fax 201-730-3042). *Diffraction, X-ray, neutron, structure, surface, interface, high T(c) materials, magnetic structures, inelastic neutron scattering, critical phenomena.*

Skarstad, Dr. Paul Michael. Promeon Div., Medtronic Inc., 6700 Shingle Creek Parkway, Minneapolis, MN 55430, USA. (1942) PhD phys. chemistry (Cornell U., 1971). Corporate res. fel. (tel. 612 + 574-6380). *Solid state chemistry, electrochemistry.*

Skelton, Dr. Earl Franklin. Cond. Matter & Radiation Sci. Div. U. S. Naval Res. Lab., Code 4683, Washington, DC 20375, USA. (1940) PhD physics (Rensselaer Polytechnic Inst., 1967). Supervisory res. physicist. (tel. 202 + 767-3014, fax 202-767-4868, email skelton%cmrsd.decnet. at nrl.arpa). *Phase transformations, high pressure, diffraction physics, synchrotron radiation, EXAFS.*

Skinner, Mr. Doyle P. Battelle Inst., 505 King Ave., Columbus, OH 43201, USA.

Skinner, Mr. Matthew M. Chemistry Dept., U. of Calif. at San Diego, 4434 Mayer Hall, La Jolla, CA 92093, USA. (1962) MS biochemistry (U. of Calif. at San Diego, 1988). Grad. student. (tel. 619 + 534-4241). *Crystallography, protein structure - function.*

Skrzypczak-Jankun, Dr. Ewa. Chemistry Dept., Michigan State U., East Lansing, MI 48824, USA. (1948) PhD chemistry (A. Mickiewicz U., Poland, 1976). Res. assoc. (tel. 517 + 353-7298). *X-ray crystallography, single crystal, proteins, structure, crystal growth, alkaloids, small molecules, organic, metallo-organic.*

Slagle, 3831 Riverside Dr., Tulsa, OK 74105, USA. (1962) MS phys. chemistry (Kansas State U., 1987). Res. chemist. (tel. 918 + 744-1712). *Single crystal X-ray diffraction, small biologically important molecules.*

Slaughter, Prof. Maynard. Chemistry & Geochemistry Dept., Colorado Sch. of Mines, Golden, CO 80401, USA. (1934) PhD X-ray crystallography (U. Pittsburgh, 1962). Prof. (tel. 303 + 273-3648, fax 303-273-3278). *Structures, zeolites, clay minerals.*

Sleight, Prof. Arthur William. Chemistry Dept., Oregon State U., Gilbert Hall, Corvallis, OR 97331, USA. (1939) PhD inorg. chemistry (U. of Connecticut, 1963). Milton Harris Prof. of Matls. Sci. (tel. 503 + 754-2081). *Solid state inorganic chemistry, inorganic structures, catalysts, powder structure analysis, superconductors.*

Sliva, Mr. Paul. Materials Research Lab., Pennsylvania State U., University Park, PA 16802, USA. (tel. 814 + 865-1114).

Sloan, Dr. Gilbert J. Polymer Prods. Dept., Exptl. Sta., E. I. du Pont de Nemours Co., P.O. Box 80323, Wilmington, DE 19880-0323, USA. (1928) PhD chemistry (U. of Michigan, 1954). Sr. res. fel. (tel. 302 + 695-2552). *Organic crystal growth, impurity microdistribution, zone refining.*

Sly, Prof. William Glenn. Chemistry Dept., Harvey Mudd C., Claremont, CA 91711, USA. (1922) PhD chemistry (CIT, 1955). Prof. (tel. 714 + 621-8000). *Organic and metal-organic structures, crystallographic computing.*

Smart, Dr. Bruce E. 22 Beethoven Dr., Wilmington, DE 19807, USA. (tel. 302 + 695-4179).

Smelik, Mr. Eugene. Earth & Planetary Sci. Dept., Johns Hopkins U., Charles & 34th Sts., Baltimore, MD 21218, USA. (1956) MS geology (U. of N. Carolina at Chapel Hill, 1987). Grad. student. (tel. 301 + 338-8342). *Mineralogy, crystallography, amphiboles, crystal chemistry.*

Smith, Mr. Albert Edward. 72 San Mateo Rd., Berkeley, CA 94707, USA. (1908) MS chemistry (UCB, 1935). Retired. (tel. 415 + 524-3697). *Compounds, organic, organometallic, structure/property relationships, metal oxides.*

Smith, Mr. Allan L. Chemistry Dept., Drexel U., 32 & Chestnut Sts., Philadelphia, PA 19104, USA. (tel. 215 + 895-2667).

Smith, Dr. Craig Daniel. Ctr. Macromol. Crystallog., Rm. BHS 265, U. of Alabama at Birmingham, THT 79 University Station, Birmingham, AL 35294, USA. (1954) PhD biophysical sci. (U. of Alabama at Birmingham, 1986). Mgr., X-ray cryst. fac., Cancer ctr. (tel. 205 + 934-6003). *Protein structure, area detectors, crystal growth, drug design.*

Smith, Prof. David John. Center for Solid State Sci., Arizona State U., Tempe, AZ 85287, USA. (1948) ScD physics (U. Melbourne, Australia, 1988). Prof. (tel. 602 + 965-4540, fax 602-965-2012). *High resolution electron microscopy, defect structures, surfaces.*

Smith, Prof. Deane Kingsley Jr. Dept. of Geosciences, Pennsylvania State U., 239 Deike Bldg., University Park, PA 16802, USA. (1930) PhD geology (U. of Minnesota, 1956). Prof. of mineralogy. (tel. 814 + 865-5782, telex 842510, fax 814-865-2326, email DKS1 at PSUVM). *Powder diffraction methods, crystal chemistry, minerals, mineral-like compounds.*

Smith, Dr. Dennis H. Molecular Design Ltd., 2132 Farallon Dr., San Leandro, CA 94577, USA. PhD chemistry (UCB, 1967). Vice-pres., prod. dev. (tel. 415 + 895-1313). *Computational chemistry, chemical databases, geometric searching.*

Smith, Dr. Douglas Lee. Life Sciences Res. Labs., Eastman Kodak Co., Building 82, Kodak Park, Rochester, NY 14650, USA. (1937) PhD phys. chemistry (U. of Wisconsin, 1962). Res. assoc. (tel. 716 + 722-3892). *Protein crystal structures.*

Smith, Dr. Francine. Biochemistry Dept., U. of North Carolina, CB #7260, FLOB, Chapel Hill, NC 27599-7260, USA. (1958) PhD biology (Johns Hopkins U., 1985). Res. instr. (tel. 919 + 966-3263, email fran at med.unc.edu). *X-ray crystallography, energetics, macromolecules, interactions, hemoglobin.*

Smith, Dr. George David. Mol. Biophysics Dept., Med. Fndn. of Buffalo, 73 High St., Buffalo, NY 14203, USA. (1941) PhD phys. chemistry (Ohio U., 1968). Sr. res. scient. (tel. 716 + 856-9600, ext. 430). *Polypeptides, polypeptide hormones, ion-transport antibiotics, protein structures, insulin action mode, structure - function relationships, molecular modeling, graphics.*

Smith, Dr. Gordon Stuart. Chemistry Dept., L-370, Lawrence Livermore Nat. Lab., P.O. Box 808, Livermore, CA 94550, USA. PhD phys. chemistry (Cornell U., 1957). Staff scient. (tel. 415 + 422-8008). *Powder diffraction, crystal structures, inorganic, intermetallic, instrumentation.*

Smith, Dr. Graham Monro. Research Labs., W42-3, Merck Sharp & Dohme, West Point, PA 19486, USA. (1947) PhD syn. org. chemistry (SUNY at Buffalo, 1974). Sr. investigator. (tel. 215 + 661-7620). *Structural chemistry, molecular modeling, medicinal chemistry.*

Smith, Dr. Harold Glenn. Solid State Div., Oak Ridge Nat. Lab., P.O. Box X, Oak Ridge, TN 37830, USA. (1927) PhD physics (Iowa State U., 1957). Sr. scient. (tel. 615 + 574-5243). *Crystallography, neutron scattering, X-ray diffraction, lattice dynamics.*

Smith, Prof. Janet Louise. Biological Sci. Dept., Purdue U., West Lafayette, IN 47907, USA. (1951) PhD biochemistry (U. of Wisconsin at Madison, 1978). Asst. prof. (tel. 317 + 494-9246, fax 317-494-0876, email SMITHJ at PURCCVM). *Protein crystallography.*

Smith, Prof. John Francis. Materials Science & Eng. Dept., Iowa State U., 122 Metallurgy Bldg., Ames, IA 50010, USA. (1923) PhD phys. chemistry (Iowa State U., 1953). Prof. emeritus, consultant. (tel. 515 + 294-5083). *Intermetallic phases, bonding, structure - properties relationships.*

Smith, Prof. Joseph Victor. Geophysical Sci. Dept., U. of Chicago - M9S, Chicago, IL 60637, USA. (1928) PhD crystallography (Cambridge U., UK, 1951). Louis Block Prof. of Physical Sci. (tel. 312 + 702-8110). *Mineral crystal structures, mineralogy, petrology, geochemistry.*

Smith, Prof. Phillip Ross. Cell Biology Dept., NYU Med. Ctr., 550 First Ave., New York, NY 10016, USA. (1945) PhD physics (U. of Cambridge, UK, 1971). Assoc. prof. (tel. 212 + 340-5356, fax 212-340-7190, email Bitnet SMITH at NYUMED). *Electron microscopy, image processing, structure, membrane protein.*

Smith, Mr. Robert David. Structural Biol. Lab. Rm. Y4.320, HHMI, 5323 Harry Hines Blvd., Dallas, TX 75235, USA. (1958) MS chemistry (U. of Virginia, 1986). Res. tech. (tel. 214 + 689-5000, ext. 5043, email Bitnet SMITH01 at UTSW). *Structural biology, protein crystallography, chemical physics, quantum mechanics, theoretical spectroscopy.*

Smith, Dr. Steven Sidney. Surgery Div., City of Hope Natl. Med. C. Shapiro Bldg., Rm. S101, 1500 E. Duarte Rd., Duarte, CA 91010, USA. (1946) PhD molecular biology (UCLA, 1974). Assoc. res. scient (tel. 818 + 301-8316). *DNA, methylation, structure, dynamics, protein conformation, catalysis.*

Smith, Dr. Ward Whitlock. Agouron Pharmaceuticals, Suite 120, 11025 N. Torrey Pines Rd., La Jolla, CA 92037, USA. (1949) PhD biological chemistry (U. of Michigan, 1977). *Biological macromolecules, structure - function, diffraction methods, instrumentation.*

Smoluchowski, Prof. Roman. Physics and Astronomy Dept., U. of Texas at Austin, Austin, TX 78712, USA. (1910) PhD (U. Groningen, Netherlands, 1935). Prof. (tel. 512 + 471-1305). *Crystal defects, amorphous surfaces, phase transitions.*

Snow, Dr. Mark Edward. Scient. Computation Group, U. of Michigan Computing Ctr., 535 W. William St., Ann Arbor, MI 48103, USA. (1959) PhD biophysics (Johns Hopkins U., 1986). Asst. res. scient. (tel. 313 + 763-6054, email smokey at um.cc.umich.edu). *Molecular structure and function, computational chemistry, computer programming.*

Snyder, Prof. Robert L. NYS C. of Ceramics, Alfred U., Alfred, NY 14802, USA. (1941) PhD chemistry (Fordham U., 1968). Prof., ceramic sci. (tel. 607 + 871-2438, fax 607-871-3469, email Bitnet SNYDER at CERAMICS). *Powder diffraction, crystal structure analysis, superconductors.*

Soltis, Mr. S. Michael. Chemistry Dept., U. of Calif. at Los Angeles, 405 Hilgard Ave., Los Angeles, CA 90024, USA.

Soltzberg, Prof. Leonard Jay. Chemistry Dept., Simmons C., 300 The Fenway, Boston, MA 02115, USA. (1944) PhD phys. chemistry (Brandeis U., 1969). Prof. (tel. 617 + 738-2181). *Phase transformations, crystal optics, computer graphics, on-line instrument control.*

Soman, Dr. Kizhake Variyath. Molec. Biology Dept., MB1, Res. Inst. of Scripps Clinic, 10666 N. Torrey Pines Rd., La Jolla, CA 92037, USA. (1953) PhD mol. biophysics (Indian Inst. of Science, Bangalore, 1984). Res. fel. (tel. 619 + 554-9957, email soman%scr.sdscnet at sdsc). *Proteins, theoretical/computational studies, dynamics, folding.*

Soni, Dr. Sulil-Datta. Oral Biology Dept., SUNY at Buffalo, 109 Foster Hall, Buffalo, NY 14203, USA. (1958) PhD biophysics (SUNY at Buffalo, 1989). Postdoct. fel. (tel. 716 + 831-2114). *NMR, synthesis, anti-viral peptides, nucleic acids.*

Spangler, Dr. Brenda D. Biol., Environ. & Med. Red. Div., Argonne Nat. Lab., 9700 S. Cass Ave., Argonne, IL 60439, USA. (1939) PhD biochemistry (Northern Illinois U., 1984). Postdoct. fel. (tel. 312 + 972-3238). *Proteins, crystallography, structure - function, microbial toxins.*

Spanton, Mr. Stephen. 17539 Cottonwood Ct., Grayslake, IL 60030, USA. (tel. 312 + 937-0869).

Sparks, Dr. Cullie James Jr. Metals and Ceramics Div., Oak Ridge Nat. Lab., P.O. Box 2008, Oak Ridge, TN 37831-6117, USA. (1929) PhD metallurgy (U. of Kentucky, 1957). Res. metallurgist. (tel. 615 + 574-6996). *X-ray diffraction physics, spectroscopy, diffuse X-ray scattering, imperfect materials, synchrotron radiation.*

Sparks, Dr. Robert Allen. 2085 Sandalwood Ct., Palo Alto, CA 94303, USA. (1928) PhD phys. chemistry (UCLA, 1958). Consultant. (tel. 415 + 321-1014). *Instrumentation, automation, methods of structure determination, computing methods, image science.*

Spence, Prof. John Charles Howorth. Physics Dept., Arizona State U., Tempe, AZ 85287, USA. (1946) PhD physics (U. of Melbourne, Australia, 1973). Prof. (tel. 602 + 965-6486). *Diffraction, electron, surface.*

Spielberg, Prof. Nathan. Physics Dept., Kent State U., Kent, OH 44242, USA. (1926) PhD physics (Ohio State U., 1952). Prof. (tel. 216 + 672-2881, email NSPIELBE at KENTVM). *X-ray physics, interferometry, ultra-soft X-rays, liquid crystals, instrumentation.*

Spooner, Prof. Stephen. Solid State Div., Oak Ridge Nat. Lab., Oak Ridge, TN 37831, USA. (1937) ScD metallurgy (MIT, 1965). (tel. 615 + 574-4535). *Phase transformations, scattering, X-ray, neutron, small angle, structure of solids.*

Sprang, Dr. Stephen Robert. HHMI, U. of Texas Southwestern, 5323 Harry Hines Blvd., Dallas, TX 75235, USA. (1949) PhD biochemistry (U. of Wisconsin at Madison, 1977). Asst. prof., asst. investig. (HHMI) (tel. 214 + 689-5008). *Macromolecular crystallography, recognition processes, allostery, structural basis, enzyme systems.*

Springer, Dr. James Patrick. Biophysical Chem. Dept., R80M203, Merck Sharp & Dohme Res. Lab., P.O. Box 2000, Rahway, NJ 07065, USA. (1950) PhD chemistry (Iowa State U., 1976). Assoc. Dir. (tel. 201 + 593-5496). *Macromolecular crystallography, molecular structure and function, ligand binding, NMR.*

Sproul, Prof. Gordon Duane. Chemistry Dept., U. of South Carolina, 800 Carteret St., Beaufort, SC 29902, USA. (1944) PhD inorg. chemistry (U. of Illinois, 1971). Prof. (tel. 803 + 524-7112, ext. 28). *Chemical education, coordination compounds, transition metal complexes.*

Spruiell, Prof. Joseph Earl. Materials Sci. and Eng. Dept., U. of Tennessee, 434 Dougherty Eng. Bldg., Knoxville, TN 37996-2200, USA. (1935) PhD metallurgical eng. (U. of Tennessee, 1963). Prof. and Head. (tel. 615 + 974-5336, fax 615-974-2669). *Polymers, morphology, structure, mechanical, physical properties, processing.*

Spurlino, Dr. John C. Structural Biol. Dept., HHMI, Baylor C. Med., One Baylor Plaza, Houston, TX 77030, USA. (1958) PhD biochemistry (Rice U., 1988). Res. assoc. (tel. 713 + 798-6564). *Protein crystallography, neurological receptor structure.*

Squattrito, Dr. Philip John. Chemistry Dept., Texas A&M U., College Station, TX 77843, USA. (1960) PhD inorg. chemistry (Northwestern U., 1987). Postdoct. assoc. (tel. 409 + 845-3812, email SQUATTRITO at TAMCHEM). *Chemistry, inorganic, solid state, transition metal chalcogenides, framework structures.*

Srikrishnan, Dr. Thamarapu. Biophysics Dept., Ctr. Cryst. Res., Roswell Park Memorial Inst., 666 Elm St., Buffalo, NY 14263, USA. (1943) PhD physics (U. of Madras, India, 1969). Cancer res. scient. (tel. 716 + 845-8302 or 8301). *X-ray diffraction techniques, structures, biological molecules, immuno modulators, carcinogens, compounds of chemotherapeutic interest, direct methods, applications.*

Srivastava, Dr. Alok Mani. Chemistry Dept., Polytechnic U., 333 Jay St., Brooklyn, NY 11201, USA. (1960) PhD inorg. chemistry (PUN, 1986). Postdoct. fel. (tel. 718 + 260-3742). *Crystal structures, inorganic materials, superconductors, solid state luminescence.*

Srivastava, Dr. Ramesh C. Crystallography Dept., U. of Pittsburgh, Pittsburgh, PA 15260, USA. (tel. 412 + 624-9300).

Stalick, Dr. Judith Kay. Reactor Radiation Div., Natl. Inst. Stnds. & Techn., A106 Reactor Bldg., Gaithersburg, MD 20899, USA. (1943) PhD inorg. chemistry (Northwestern U., 1969). Res. chemist. (tel. 301 + 975-6223, fax 301-921-9847). *Neutron diffraction, magnetic structures, Rietveld refinement techniques, phase identification methods.*

Stallings, Dr. William C. Monsanto Corp., BB4K, 700 Chesterfield Pkw., St. Louis, MO 63198, USA. (1947) PhD chemistry (U. of Pennsylvania, 1974). *Biological structures.*

Stanfield, Dr. Robyn L. Mol. Biology Dept., Res. Inst. of Scripps Clinic, 10666 N. Torrey Pines Rd., La Jolla, CA 92037, USA. (1958) PhD chemistry (U. of Texas at Austin, 1986). Postdoct. fel. (tel. 619 + 455-2456). *Protein crystallography.*

Stanko, Prof. Joseph Anthony. Chemistry Dept., SCA 211, U. of South Florida, 4202 Fowler Ave., Tampa, FL 33620, USA. (1941) PhD inorg. chemistry (U. of Illinois at Urbana, 1966). Assoc. Prof. (tel. 813 + 974-2373). *Diffraction, single crystal, powder X-ray, platinum group metals, coordination compounds, cancer chemotherapeutic drugs, one-dimensional electrical conductors, superconductors.*

Stanley, Prof. George Geoffrey. Chemistry Dept., Louisiana State U., Baton Rouge, LA 70803-1804, USA. (1953) PhD inorg. chemistry (Texas A&M U., 1979). Assoc. prof. (tel. 504 + 388-3471). *Chemistry, inorganic, organometallic, catalysis, bimetallics, cluster compounds.*

Stanton, Mr. Martin. Rosenstiel Center, Brandeis U., Waltham, MA 02254, USA. (1957) MS biochemistry (Brandeis U., 1984). Res. assoc. (tel. 617 + 736-2424, email Bitnet STANTON at BRANDEIS).

Staudenmann, Dr. Jean-Louis. Synchrotron Resource, HHMI, NSLS-X4, Bldg. 725, Brookhaven Natl. Lab., Upton, NY 11973-5000, USA. (1940) PhD solid state physics (U. of Geneva, 1976). Sr. assoc. scient. (tel. 516 + 282-5466, fax 516-282-5466). *Alloys, electron density maps, thermal properties, X-ray and neutron diffraction, instrumentation.*

Stauffacher, Dr. Cynthia Vianne. Biological Sciences Dept., Purdue U., Lilly Hall of Life Sciences, West Lafayette, IN 47907, USA. (1948) PhD phys. chemistry (UCLA, 1977). Asst. prof. (tel. 317 + 494-4937). *Macromolecular structure and assembly, X-ray crystallography, structural biology, viruses, enzymes, membrane proteins.*

Stearns, Mr. Frederick Stanley. 632 Lemke Dr., Placentia, CA 92670, USA. (1926) BS physics (West Coast U., 1971). (tel. 714 + 528-9775). *Crystal growth, inorganics.*

Stec, Dr. Boguslaw. Chemistry Dept., U. of South Carolina, Columbia, SC 29208, USA. (1956) PhD theoret. physics (Jagellonian U., Poland, 1986). Res. assoc. prof. (tel. 803 + 777-2140). *Protein crystallography, atomic physics - quantum mechanics.*

Steele, Dr. Ian McKay. Geophysical Sci. Dept., U. of Chicago, 5734 S. Ellis Ave., Chicago, IL 60637, USA. (1944) PhD geology-crystallography (U. of Illinois, 1971). Sr. res. assoc. (tel. 312 + 702-8109, email steele at geovax.uchicago.edu). *Mineralogy, mineral analysis, X-ray diffraction, crystal structures.*

Stein, Prof. Richard Stephen. Polymer Res. Inst., U. of Massachusetts, Grad. Res. Center, Tower A, Amherst, MA 01002, USA. (1925) PhD phys. chemistry (Princeton U., 1949). Prof. (tel. 413 + 545-2727). *Polymer texture, small angle X-ray scattering, crystalline polymers, blends, optical properties of polymers.*

Steinfink, Prof. Hugo. Materials Sci. and Eng., ETC 9.46, U. of Texas, Austin, TX 78712, USA. (1924) PhD phys. chemistry (PUN, 1954). Prof. (tel. 512 + 471-5233, fax 512-471-5233, email Bitnet CHAA013 at UTA3081). *Inorganic structures, physical properties.*

Steinrauf, Prof. Larry King. Biochemistry Dept., Indiana U. Sch. of Medicine, 635 Barnhill Dr., Indianapolis, IN 46223, USA. (1931) PhD biochemistry (U. of Washington, 1957). Prof. (tel. 317 + 274-7544). *Membrane transport, heavy metal poisoning, microprocessor computers, athletic medicine.*

Steitz, Prof. Thomas Arthur. Mol. Biophys. & Biochem. Dept., Yale U., P.O. Box 6666, New Haven, CT 06511, USA. (1940) PhD molecular biology and biochem. (Harvard U., 1966). Prof./ Investigator, HHMI. (tel. 203 + 432-5617, fax 203-432-3282). *Proteins, nucleic acids, structure - function, replication, recombination, translation, enzyme mechanisms.*

Stenkamp, Dr. Ronald Eugene. Biological Structure Dept., SM-20, U. of Washington, Seattle, WA 98195, USA. (1948) PhD chemistry (U. of Washington, 1975). Assoc. prof. (tel. 206 + 545-1721, fax 206-543-1524, email STENKAMP at UWALOCKE). *Structural studies, small molecules of biological interest, biological macromolecules.*

Stephens, Dr. Anthony E. Silicon Products Dept., Texas Instruments, Sherman, TX 75090, USA. (1940) PhD physics, solid state (U. of North Texas, 1975). Sr. memb. tech. staff. (tel. 214 + 868-5741). *Silicon, X-ray topography, oxygen precipitation.*

Stephenson, Dr. Gregory Brian. Res. Div., IBM Corp., T.J. Watson Res. Ctr., P.O. Box 218, Yorktown Heights, NY 10598, USA. (1957) PhD materials science (Stanford U., 1983). Mgr., phase transitions group. (tel. 914 + 945-3008, email gbs at yktvmz). *Phase transitions, synchrotron radiation, X-ray scattering, kinetics.*

Stern, Ms. Charlotte. U. of Illinois, 505 S. Mathews, Urbana, IL 61801, USA.

Stevens, Prof. Edwin David. Chemistry Dept., U. of New Orleans, Lakefront, New Orleans, LA 70148, USA. (1947) PhD chemistry (U. of Calif. at Davis, 1973). Prof. (tel. 504 + 286-6856, email Bitnet EDSCM at UNO). *Electron density distributions, accurate X-ray intensity measurements, synchrotron radiation, EXAFS.*

Stevens, Dr. Raymond Charles. Chemistry Dept., #309, Harvard U., 12 Oxford St., Cambridge, MA 02138, USA. (1963) PhD chemistry (USC, 1988). Postdoct. res. assoc. (tel. 617 + 495-40998). *Protein crystallography, diffraction, neutron, synchrotron radiation.*

Stewart, Prof. James McDonald. Chemistry Dept., U. of Maryland, C. Park, MD 20742, USA. (1931) PhD phys. chemistry (U. of Washington, 1958). Prof. (tel. 301 + 864-8275, email JSTEW at UMD2.UMD.EDU). *Crystallographic software development, crystal structure determination.*

Stewart, Dr. Martin Van Buren. Chemistry and Physics Dept., Middle Tennessee State U., Box 123, Murfreesboro, TN 37132, USA. (1944) PhD org. chemistry (U. of Georgia, 1979). Assoc. prof., chemistry (tel. 615 + 898-2949). *Organic synthesis, stereochemistry, physical organic chemistry, surface chemistry, spectroscopy, analytical methods, X-ray crystal structure analysis.*

Stewart, Prof. Robert Farrell. Chemistry Dept., Carnegie-Mellon U., 4400 Fifth Ave., Pittsburgh, PA 15213, USA. (1936) PhD chemistry (CIT, 1963). Prof. (tel. 412 + 268-3165). *Electrostatic properties, molecules, crystals, accurate diffraction data, Rayleigh scattering by X-rays, theory.*

Sthanam, Dr. V. L. Narayana. Cancer Center, U. of Alabama at Birmingham, 262 BHS, Birmingham, AL 35209, USA. (1952) PhD chem. crystallog. (Indian Inst. of Techn., Bombay, 1980). Sr. res. assoc. (tel. 205 + 934-7974). *Protein crystallography, rational drug design, structure solution methods.*

Stock, Dr. Ann. Chemistry Dept., Princeton U., Princeton, NJ 08544, USA. (tel. 609 + 4152-6112).

Stock, Prof. Stuart R. Sch. of Matls. Eng., Georgia Inst. of Techn., Atlanta, GA 30332-0245, USA. (1955) PhD metallurgical eng. (U. of Illinois at Urbana, 1983). Asst. prof. (tel. 404 + 894-6882, fax 404-894-3120). *X-ray diffraction, microtomography, stress/strain measurement, plastic deformation.*

Stodola, Mr. Robert. Computer Res. Facilities, Fox Chase Cancer Ctr., 7701 Burholme Ave., Philadelphia, PA 19111, USA. (tel. 215 + 728-3660).

Stollstorff, Mr. Gregory R. ENRAF-NONIUS Co., 390 Central Ave., Bohemia, NY 11716, USA. (tel. 516 + 589-2068, telex 960250).

Stouch, Dr. Terry Richard. Lab. for the Struct. of Matter, Code 6030, Naval Research Lab., Washington, DC 20375, USA. (1956) PhD chemistry (PSU, 1985). Res. chemist. (tel. 202 + 767-3463, telex TWX7108220147, fax 202-767-6874, email stouch at ccf3.nrl.navy.mil). *X-ray crystallography, structure - activity relationships, drug design, molecular simulations, modeling, dynamics, graphics, QSAR, computational chemistry, computer aided molecular design.*

Stout, Dr. Charles David. Molec. Biology Dept., Scripps Clinic Res. Inst., 10666 N. Torrey Pines Rd., La Jolla, CA 92037, USA. (1947) PhD biochemistry, crystallography (U. of Wisconsin, 1976). Assoc. member. (tel. 619 + 554-8738). *Protein crystallography, structure, metalloproteins.*

Stout, Dr. George H. Biological Struct. Dept. SM-20, U. of Washington, Seattle, WA 98195, USA. (1932) PhD chemistry (Harvard U., 1956). Res. prof. (tel. 206 + 543-1421). *Proteins, structure determination, phasing, computation.*

Stout, Mr. Thomas J. Chemistry Dept., Baker Lab., Box 116, Cornell U., Ithaca, NY 14853, USA. (1964) MS chemistry (Cornell U., 1989). Grad. student, Sea grant res. fel. (tel. 607 + 255-6145, email PXCY at CRNLVAXS). *Biologically active natural products.*

Streib, Dr. William E. Chemistry Dept., Indiana U., Bloomington, IN 47405, USA. (1931) PhD phys. chemistry (U. of Minnesota, 1962). Res. crystallographer. (tel. 812 + 335-6604). *Instrumentation, structures, organic, inorganic.*

Strong, Mr. Roland Kirk. Chemistry Dept., Rm. 2-212, Mass. Inst. of Techn., 77 Mass. Ave., Cambridge, MA 02139, USA. (1963) BS biophysics (U. of Michigan, 1984). Visit. scient. (tel. 617 + 253-0056). *Macromolecular structure, protein crystallography.*

Stroud, Prof. Robert Michael. Biochem. and Biophysics Dept., U. of Calif. at San Francisco, San Francisco, CA 94143-0448, USA. (1942) PhD crystallography (U. of London, UK, 1968). Prof. (tel. 415 + 476-4224 or 3937, fax 415-476-0961). *Structure - function, biological macromolecules.*

Strouse, Prof. Charles Earl. Chemistry and Biochem. Dept., U. of Calif. at Los Angeles, 405 Hilgard Ave., Los Angeles, CA 90024-1569, USA. (1944) PhD phys. chemistry (U. of Wisconsin at Madison, 1969). Prof. (tel. 213 + 825-1811). *Solid state equilibria, porphyrin chemistry, single crystal spectroscopy, transitions, inclusion compounds.*

Stubbs, Dr. Gerald James. Mol. Biology Dept., Vanderbilt U., Box 1820, Station B, Nashville, TN 37235, USA. (1947) PhD biophysics (Oxford U., 1972). Assoc. prof. (tel. 615 + 322-2018, email Bitnet STUBBSGJ at VVCTRVAX). *Macromolecular assemblies, structures, virus, protein, nucleic acid, fiber diffraction, protein crystallography.*

Stults, Dr. Bailey Ray. Battelle-Pacific N.W. Labs., P.O. Box 999, Richland, WA 99352, USA. (1948) PhD inorg. chemistry (U. of Nebraska, 1974). Mgr., chem. sci. dept. (tel. 509 + 375-2687). *Small molecule crystallography, EXAFS, structure - function relations, catalysts (homogeneous and heterogeneous), NMR.*

Sturcken, Dr. Edward Francis. P.O. Box 6846, North Augusta, SC 29841, USA. (1927) PhD nuclear physics (St. Louis U., 1953). (tel. 803 + 725-2790). *X-ray diffraction, materials science, electron microscopy, applied physics.*

Sturkey, Mr. Lorenzo. 1081 Mohr Lane, #317, Concord, CA 94518, USA. MS physics (U. of Kentucky, 1939). Retired. (tel. 415 + 680-4322). *Electron diffraction, magnesium metallurgy, intermetallic phases, dynamic theory of electron diffraction.*

Su, Dr. Shu-Chun. Analyt. Sci. Div., Res. Centr., MS 8136-258, Hercules Inc., Hercules Rd., Wilmington, DE 19894, USA. (1940) PhD mineralogy (Virginia Polytech. Inst. & State U., 1986). Res. microscopist. (tel. 302 + 995-3498, telex 4994538 HERCL, fax 302-995-4315). *Crystallography, X-ray, optical, mineralogy, crystal chemistry of silicates.*

Su, Mr. Ying-Zhong. Chemistry Dept., Ames Lab., 38 Spedding Hall, Iowa State U., Ames, IA 50010, USA. (1943) MS phys. chemistry (Zhongshan U., PRC, 1983). Grad. student. (tel. 515 + 294-8477). *Biological molecules, powder methods.*

Subramanian, Dr. Sethuraman. Biotechnology Res. & Dev., Miles Inc., P.O. Box 932, Elkhart, IN 46515, USA. (1940) PhD phys. chemistry (Indian Inst. of Techn., 1969). Staff scient. (tel. 219 + 262-7611, fax 219-264-8399). *Proteins, structure, chemistry, function, stability, folding, sequencing, engineering, enzymology.*

Sudarsanam, Mr. Padmanaban. Biophysics Dept., Roswell Park Memorial Inst., 666 Elm St., Buffalo, NY 14263, USA.

Suddath, Prof. Frederick L. Sch. of Chemistry, Georgia Inst. of Techn., Atlanta, GA 30332, USA. (1942) PhD chemistry (Georgia Inst. of Techn., 1970). Prof. (tel. 404 + 894-4028, fax 404-894-7452, email CMFLSBS at GITVMI). *Protein crystallography.*

Sugar, Prof. Istvan Peter. Biomathematical Sciences Dept., Mount Sinai Med. Ctr., New York, NY 10029, USA. (1947) PhD theoretical physics (Roland Eotvos, Budapest, 1973). Assoc. prof. (tel. 212 + 241-5851). *Phase transitions, lipid membranes, cell membranes, electroporation, electrofusion, protein structure prediction, NMR, two dimensional, theory.*

Suitch, Dr. Paul Raymond. Anglo-American Clays Corp., P.O. Box 471, Sandersville, GA 31082, USA. (1958) PhD ceramic eng. (Georgia Inst. of Techn., 1987). Prod. dev. eng. (tel. 912 + 552-6136, ext. 291). *Clay mineralogy, ceramic and oxide chemistry, polymers.*

Sullenger, Dr. Don Bruce. E.G.& G. Mound Appl. Techn., P.O. Box 3000, Miamisburg, OH 45343, USA. (1929) PhD phys. chemistry (Cornell U., 1969). Sr. Res. specialist. (tel. 513 + 865-3665). *Applied crystallography, crystal chemistry, inorganic, organic, materials science.*

Sundaralingam, Prof. Muttaiya. Biochemistry Dept., U. of Wisconsin at Madison, 420 Henry Mall, Madison, WI 53706, USA. (1931) PhD chemistry (U. of Pittsburgh, 1961). Steenbock Prof. of Biomolec. Struct., Chair, Biophys. prog. (tel. 608 + 262-1448, fax 608-262-3453). *Crystallography, nucleic acids, muscle proteins, conformational analyses, protein folding mechanisms, protein hydration.*

Sunshine, Dr. Steven A. AT&T Bell Labs., 600 Mountain Ave., Murray Hill, NJ 07974, USA. (1960) PhD chemistry (Northwestern U., 1986). Tech. staff member. (tel. 201 + 582-3446, fax 201-582-2783). *Chemistry, solid state, materials.*

Supper, Mr. Lee R. Charles Supper Co., 15 Tech Circle, Natick, MA 01760, USA. (1938) economics (Dickinson C., 1960). Pres. (tel. 508 + 655-4610). *Instrumentation, diffraction instrument manufacture, cameras and accessories.*

Sutton, Mr. Paul A. Analytical Services, Sandoz Res. Inst., Route 10, East Hanover, NJ 07936, USA. (1955) PhD medicinal chemistry (Purdue U., 1984). Sr. res. chemist. (tel. 201 + 503-7305, fax 201-503-8487). *Crystallography, small molecule, polymorphism, X-ray powder diffraction, compounds of pharmaceutical interest.*

Swain, Dr. Amy L. Cryst. Lab., Bionetics Res. Inc., BRP, NCI, Frederick Cancer Fac., P.O. Box B, Bldg. 539, Frederick, MD 21701, USA. (1961) PhD biochemistry (U. of South Carolina, 1988). Postdoct. fel. (tel. 301 + 698-5031, fax 301-698-5991). *Macromolecular crystallography.*

Swaminathan, Dr. P. Med. Chem. - Drug Design, Wyeth/Ayerst Res., 567 Ridge Rd., Monmouth Junction, NJ 08542, USA. (1948) PhD biophysics (U. of Madras, India, 1975). Res. scient. (tel. 201 + 274-4617). *Macromolecular crystallography, molecular modeling, computational chemistry.*

Swaminathan, Dr. S. Crystallography Dept., U. of Pittsburgh, Pittsburgh, PA 15260, USA. (tel. 412 + 683-3000).

Swaminathan, Dr. Vijayalakshmi. Off. of Computing Services, Drexel U., 32nd. & Chestnut Sts., Philadelphia, PA 19104, USA. (1948) PhD physics, crystallography (U. of Madras, India, 1975). Analyst, programmer. (tel. 215 + 895-6872). *Crystallography, biological molecules, nucleic acids, nucleoside antibiotics.*

Swank, Prof. Duane Douglas. Chemistry Dept., Pacific Lutheran U., Tacoma, WA 98447, USA. (1942) PhD phys. chemistry (Montana State U., 1969). Prof. (tel. 206 + 535-7556, fax 206-535-8320, email Bitnet SWANK at PLU). *Solid state spectroscopy, magnetic phenomena, inorganic solids, inorganic crystal structures.*

Swanson, Dr. Rosemarie. Biochem. & Biophys. Dept., Texas A&M U., College Station, TX 77843, USA. (1942) PhD chemistry (Stanford U., 1965). Res. scient. (tel. 409 + 846-4504, email ROSEMAR at TAMBIGRF). *Protein crystallography, crystal growth, entropy.*

Swanson, Dr. Stanley M. Biochemistry Dept., Texas A&M U., College Station, TX 77843, USA. (1938) PhD theoret. physics (Stanford U., 1968). Res. scient. (tel. 409 + 845-1744, email Bitnet STAN at TAMBIGRF). *Quantum theory, computer graphics and algorithms, electron density representation.*

Swartzendruber, Mr. John K. Eli Lilly Co., Indianapolis, IN 46285, USA.

Sweet, Dr. Robert M. Biology Dept., Brookhaven Nat. Lab., Upton, NY 11973, USA. (1943) PhD phys. chemistry (U. of Wisconsin at Madison, 1970). Scient. (tel. 516 + 282-3401, email Bitnet SWEET at BNLCL1). *Macromolecular crystallography, synchrotron radiation.*

Swenson, Dr. Dale Carl. Mol. Biophysics Dept., Med. Fndn. of Buffalo, 73 High St., Buffalo, NY 14203, USA. (1951) PhD phys. chemistry (U. of Iowa, 1979). Res. scient. (tel. 716 + 856-9600, ext. 327). *Packing forces, charge density studies.*

Swepston, Dr. Paul Nathan. Molecular Structure Corp., 3200 A Research Forest Dr., The Woodlands, TX 77381, USA. (1954) PhD chemistry (U. of Arkansas, 1981). Vice-pres. (tel. 713 + 363-1033, telex 55-9314(MOLECULAR), fax 713-292-2472). *Chemical crystallography, crystallographic computing.*

Swerdlow, Mr. Max. 5704 Lenox Rd., Bethesda, MD 20817, USA. (1915) BA physics (Brooklyn U., 1938). Program manager. (tel. 301 + 229-6084). *Microstructure, crystallography, electronic structure, optical properties of solids, superconductivity, magnetism, nonlinear optics.*

Swink, Dr. Laurence Nim. 1617 E. Julie Dr., Tempe, AZ 85283, USA. (1934) PhD chemistry (Brown U., 1969). Technical Dir. (tel. 602 + 839-9489). *Applied crystallography.*

Swinnea, Dr. John Steven. Materials Sci. & Eng. Dept., U. of Texas, ETC 9.114 Austin, TX 78712, USA. (1953) PhD chemical eng. (U. of Texas at Austin, 1981). Res. assoc. (tel. 512 + 471-3173, fax 512-471-8727, email Bitnet CHAA013 at UTA3081). *Chemical, physical properties of crystals, solid state chemistry.*

Switendick, Dr. Alfred Carl. Div. 1151, Sandia Nat. Lab., P.O. Box 5800, Albuquerque, NM 87185, USA. (1931) PhD solid state physics (MIT, 1963).

Staff member, solid state theory div. (tel. 505 + 846-2288, telex 1690, fax 505-846-2009, email acswite at sandia.gov). *Intermetallic compounds, interstitial compounds, hydrides, borides, electronic properties, structure.*

Syed, Dr. Ashfaquzzaman. Physical X-Ray Div., ENRAF-NONIUS, 390 Central Ave., Bohemia, NY 11716, USA. (1952) PhD physics (U. Gorakhpur, India, 1979). Product specialist. (tel. 516 + 589-2885, ext. 19, telex 960250, fax 516-589-2068). *Structures, organic, inorganic, instrumentation, system integration, accurate intensity measurements, electron density, crystallographic computing.*

Syed, Dr. Rashid. Mol. Biology Dept. MB-13, Res. Inst. Scripps Clinic, 10666 N. Torrey Pines Rd., La Jolla, CA 92037, USA. (1960) PhD chemistry (U. of Toledo, 1987). Res. assoc. (tel. 619 + 554-2456, fax 619-554-6105). *Protein crystallography.*

Szebenyi, Dr. Doletha M. E. Biochem., Mol. & Cell Biol. Dept., 209 Biotechn. Bldg., Cornell U., Ithaca, NY 14853, USA. (1947) PhD phys. chemistry (U. of Connecticut, 1972). Res. assoc. (tel. 607 + 255-2174). *Protein structure, calcium-binding proteins, Laue diffraction.*

Tai, Dr. Douglas Leung-Tak. Radiology Dept., U. of Tennessee at Memphis, 865 Jefferson Ave., Memphis, TN 38163, USA. (1940) PhD phys. chemistry (Cornell U., 1969). Assoc. prof. (tel. 901 + 575-7730). *Crystallography, radiation dosimetry, oncology physics.*

Tainer, Dr. John Arthur. Mol. Biology Dept., MB5, Res. Inst. of Scripps Clinic, 10666 N. Torrey Pines Rd., La Jolla, CA 92037, USA. (1951) PhD biochemistry (Duke U., 1982). Asst. member. (tel. 619 + 554-8119, telex 697168 SCRF SCRF, fax 619-554-8841, email tainer at scripps.edu). *Macromolecules, structure - function and design, proteins, interactions, molecular computer graphics, protein engineering, antigenic recognition.*

Takagi, Dr. Shozo. Paffenbarger Res. Cntr., American Dental Assn. Health Fndn., Natl. Inst. Stnds. & Techn., Gaithersburg, MD 20899, USA. (1943) PhD crystallography (U. of Pittsburgh, 1971). Res. assoc. (tel. 301 + 975-6825). *Structure analysis, carbohydrates, X-ray and neutron diffraction.*

Takei, Dr. William J. Res. Lab., Westinghouse Electric Corp., Beulah-Churchill Borough, Pittsburgh, PA 15235, USA. (1931) PhD chemistry (CIT, 1957). Chemistry fellow. (tel. 412 + 256-1992). *Defects, crystal growth, thin films.*

Takusagawa, Dr. Fusao. Chemistry Dept., U. of Kansas, Malott Hall, Lawrence, KS 66045, USA. (1946) PhD chemistry (Osaka City U., Japan, 1974). Dir., X-ray crystallography. (tel. 913 + 864-4727). *Proteins, nucleic acids, biologically interesting small molecules.*

Tang, Dr. Sunny C. Shell Dev. Co., P.O. Box 1380, Houston, TX 77251-1380, USA. PhD inorg. chemistry (MIT, 1975).

Tanner, Mr. Lee Elliot. Chem. & Matls. Sci. Dept., Lawrence Livermore Nat. Lab., P.O. Box 808/L-355, Livermore, CA 94550, USA. (1931) MME metallurgy (U. of Pennsylvania, 1958). Prin. investigator. (tel. 415 + 423-2653). *Physical metallurgy, phase transformations, electron microscopy.*

Tao, Dr. Yuan Kai. Texas Ctr. Superconductivity, Sci. & Res. One, U. of Houston, Houston, TX 77204-5506, USA. (1960) PhD matls. sci. & eng. (U. of Texas at Austin, 1989). Postdoct. fel. (tel. 713 + 749-3948).

Tatsch, Dr. Clinton Eugene. Res. Triangle Inst., Res. Triangle Park, NC 27709, USA. (1940) PhD phys. chemistry (U. of Oklahoma, 1972). (tel. 919 + 541-6930). *Biological macromolecules, protein model compounds, protein sequences, structure - function relationships.*

Tauber, Dr. Arthur. US Army Electronics, 927 Woodgate Ave., Elbron, NJ 07740, USA. (1928) PhD chemistry (PUN, 1972). Res. physical scient. (tel. 201 + 222-8156). *Solid state, magnetic materials, hydrides.*

Taylor, Ms. Denise A. HHMI, One Baylor Plaza, Houston, TX 77030, USA. (1962) BA biochemistry (Rice U., 1985). Lab. mgr. (tel. 713 + 799-6563). *Protein crystallography.*

Taylor, Dr. Ivan Fate Jr. User Services Group, P.O. Box 32883, Texas Christian U., Fort Worth, TX 76129, USA. (1944) PhD phys. chemistry (U. of South Carolina, 1971). Sr. user services consultant. (tel. 817 + 921-7695, ext. 6856, email Bitnet CC071CC at TCUAMUS). *Crystallographic software development and conversion, diffraction theory, direct methods.*

Taylor, Prof. Jean Ellen. Mathematics Dept., Rutgers U., New Brunswick, NJ 08903, USA. (1944) PhD mathematics (Princeton U., 1973). Prof. (tel. 201 + 932-2473, fax 201 + 932-5530, email taylor at fermat.rutgers.edu). *Grain boundaries, surfaces, surface energy minimization, quasicrystals.*

Taylor, Prof. Richard T. Chemistry Dept., Miami U., Oxford, OH 45056, USA. (1950) PhD chemistry (Ohio State U., 1977). Assoc. prof. (tel. 513 + 529-2826). *Chemistry, organic, polymer.*

Taylor, Prof. Robert Cooper. Chemistry Dept., U. of Michigan, Ann Arbor, MI 48109-1055, USA. (1917) PhD chemistry (Brown U., 1947). Prof. emeritus (tel. 313 + 747-2120, fax 313 + 747-4865, email Bitnet USERN708 at UMICHUM). *Spectroscopy, infrared, Raman, crystals, vibrational spectra, force constants - molecular structure relationships.*

Teeter, Dr. Martha Mary. Chemistry Dept., Boston C., Chestnut Hill, MA 02167, USA. (1944) PhD inorg. chemistry (PSU, 1973). Assoc. prof. biochem. (tel. 617 + 552-3615, email Bitnet TEETER at BCCHEM). *High resolution protein structure, water structure in crystalline proteins, structure prediction, homologous proteins.*

Teller, Dr. Raymond. Analytical Chem. Div., BP America Res. Internatl., 4440 Warrensville Center Rd., Cleveland, OH 44128, USA. (1946) PhD chemistry

(USC, 1978). Res. scient./group leader. (tel. 216 + 581-5953). *Solid state chemistry, catalysis, neutron diffraction.*

Templeton, Prof. David Henry. Chemistry Dept., U. of Calif. at Berkeley, Berkeley, CA 94720, USA. (1920) PhD chemistry (UCB, 1947). Prof. (tel. 415 + 486-5615, fax 415 + 642-8369, email lilo at lbl.gov). *Crystal structure, crystal chemistry, anomalous scattering of X-rays.*

Templeton, Dr. Lieselotte K. Chemistry Dept., U. of Calif. at Berkeley, Berkeley, CA 94720, USA. PhD chemistry (UCB, 1950). Res. scient. (tel. 415 + 486-5615, fax 415 + 642-8369, email lilo at lbl.gov). *Anomalous X-ray scattering, crystallographic computing.*

Ten Eyck, Dr. Lynn Forest. San Diegeo Supercomputer Ctr., D005, U. of Calif. at San Diego, La Jolla, CA 92093, USA. (1942) PhD biochemical sci. (Princeton U., 1970). email teneyckl at sds.sdsc.edu). *Macromolecular crystallography, molecular biology, enzyme catalysis mechanism.*

Tench, Dr. Alan Howard. 2044 Walnut St., Boulder, CO 80302, USA. (1926) PhD phys. chemistry (U. of Colorado, 1972). Self-employed. (tel. 303 + 444-1730). *Direct methods, nucleosides, pyrenes and adducts, humic compounds.*

Teng, Dr. Siu Tjeng. Finnigan Corp., 355 River Oaks Pkwy., San Jose, CA 95134, USA. (1944) Dr. crystallography (U. Frankfurt/M., FRG, 1983). Staff chemist. (tel. 408 + 433-4800, ext. 2934).

Terwilliger, Dr. Thomas C. Biochem. & Mol. Biol. Dept., U. of Chicago, 920 E. 58th. St., Chicago, IL 60637, USA. (1956) PhD mol. biology (UCLA, 1981). Asst. prof. (tel. 312 + 702-1346). *Protein folding, stability, X-ray crystallography.*

Thailambal, Dr. Vadakkanthara G. Whistler Ctr., Smith Hall, Purdue U., West Lafayette, IN 47907, USA. (1959) PhD physics/crystallography (U. of Madras, India, 1987). Postdoct. res. assoc. (tel. 317 + 743-4924, email e05 at mace.cc.purdue.edu). *Crystallography, macromolecules, small molecules, fiber diffraction.*

Thatcher, Dr. Walter E. Corporate Res. Lab. 201-BW-09, Minn. Mining & Mfg. Res. Cntr., St. Paul, MN 55144, USA. (1927) PhD chem., mineralogy (U. of Illinois, 1955). Sr. res. spec. (tel. 612 + 733-4437).

Thathachari, Mr. Y. T. Materials Sci. & Eng. Dept., Stanford U., Stanford, CA 94305, USA. (tel. 415 + 497-0806).

Thayer, Dr. Maria Julie McLeod. Chem. & Biochem. Dept., U. of Colorado, Boulder, CO 80026, USA. (1961) PhD protein crystallography (U. of Colorado at Boulder, 1989 Res. asst. (tel. 303 + 492-8331). *Protein crystallography, proteases, enzymes.*

Thielke, Mr. Harry G. 212 Westhaven Rd., Greenville, NC 27834, USA. (1929) MS chemistry (U. of Delaware, 1962). Retired. (tel. 919 + 756-8141). *X-ray powder diffraction.*

Thomas, Dr. Robert. Office of Educ. Progs., Brookhaven Nat. Lab., Bldg. 555A, Upton, NY 11973, USA. (1934) PhD phys. chemistry (Boston U., 1965). Scient. (tel. 516 + 282-4385, email Bitnet THOMAS at BNLCHM). *Instrumentation, parallel data collection, diffraction studies, X-ray, neutron.*

Thompson, Prof. Doris M. 8867 Highland Rd., Suite 124, Baton Rouge, LA 70808, USA. *X-ray crystal structure determination, metal complexes, sterically interesting molecules.*

Thota, Mr. Narasaiah V. Whistler Ctr. Carbohydrate Res., Smith Hall, Purdue U., West Lafayette, IN 47907, USA. (1958) PhD crystallography (Indian Inst. Techn., Kharagpur, 1987) Postdoct. res. assoc. (tel. 317 + 494-4914, email eoi at mace.cc.purdue.edu). *Fiber diffraction, structural phase transitions.*

Thudium, Mr. Richard N. Tire Res., Goodyear Tire & Rubber Co., 142 Goodyear Blvd., Akron, OH 44305, USA. (1932) MS chemical physics (Kent State U., 1969). Sr. res. physicist. (tel. 216 + 796-4218). *Polymer structures, carbon black structure.*

Tibbitts, Mr. Thomas Theodore. Rosenstiel Ctr. 441, Brandeis U., P.O. Box 9110, Waltham, MA 02254-9110, USA. (1960) BS biophysics (Oregon State U., 1983). Res. scient. (tel. 617 + 736-2495, email Bitnet TIBBITTS at BRANDEIS). *Fiber diffraction, computer graphics, membrane proteins, diffraction, refinement, gap junctions.*

Tillinger, Dr. Martin H. 422 Concord St., Cresskill, NJ 07626, USA. (1943) PhD physics (PUN, 1976). Consultant. (tel. 201 + 569-0311). *Computing techniques, measurement.*

Tilton, Dr. Robert F. Jr. Miles Res. Ctr., Miles Inc., 400 Morgan Lane, West Haven, CT 06516, USA. (1957) PhD pharmaceutical chemistry (U. of Calif. at San Francisco, Sr. staff scient. (tel. 203 + 937-2829). *Protein structure, dynamics, low temperature - high pressure crystallography.*

Tischler, Dr. Jonathan Zachary. Solid State Div., Oak Ridge Nat. Lab., P.O. Box 2008, Oak Ridge, TN 37831-6024, USA. (1953) PhD physics (Cornell U., 1982). Res. staff. (tel. 615 + 574-6505, email Bitnet ZZT at ORNLSTC). *Diffraction, Mössbauer spectra, dynamical theory, defects, thin films.*

Titus, Prof. Donald Dean. Chemistry Dept., Temple U., Broad & Montgomery Sts., Philadelphia, PA 19122, USA. (1944) PhD inorg. chemistry (CIT, 1971). Assoc. prof. (tel. 215 + 787-7127). *Organic selenium/tellurium compounds, charge transfer complexes.*

Toby, Dr. Brian H. Lab. for Res. Struct. Matter, U. of Pennsylvania, 3231 Walnut Ave., Philadelphia, PA 19104-6272, USA. (1958) PhD phys. chemistry (CIT, 1986). Res. assoc. (tel. 215 + 898-2919, email toby at petvax.lrsm.upenn.edu). *Disordered structures, diffraction, neutron, powder.*

Todaro, Mr. Louis J. Hoffmann-La Roche Inc., Nutley, NJ 07110, USA. (tel. 201 + 235-2923).

Tomchick, Ms. Diana R. Chemistry Dept., U. of Wisconsin at Madison, Madison, WI 53706, USA. (1960). (tel. 608 + 262-0414, fax 608 + 262-6707, email tomchick at bert.chem.wisc.edu). *Metal clusters, crystallography, small molecules, biological macromolecules.*

Tomlin, Mr. David W. Chemistry Dept., Miami U., Oxford, OH 45056, USA. (1963) BS chemistry (Miami U., 1985). Grad. student. (tel. 513 + 529-1620). *Small organic molecules.*

Tomlinson, Dr. Gail Elizabeth. 4103 Avondale, #6, Dallas, TX 75219, USA. (1952) MD, PhD biochemistry. *Biological molecules, pharmacologically active small molecules.*

Tong, Mr. Liang. Lab. Chemical Biodynamics, U. of Calif. at Berkeley, Berkeley, CA 94720, USA. (1963) BS biophys. chemistry (UCB, 1983). Res. asst. (tel. 415 + 486-4349, email Bitnet L.TONG at LBL). *Macromolecular crystallography.*

Topiol, Dr. Sidney. Medicinal Chem. Dept., Berlex Labs. Inc., 110 E. Hanover Ave., Cedar Knolls, NJ 07927, USA. (1949) PhD phys. chemistry (New York U., 1976). Sr. scient. (tel. 201 + 292-8080, telex 135354, fax 201 + 540-9046). *Chemistry, medicinal, theoretical, physical, bio.*

Torardi, Dr. Carmine Charlie. Central Res. & Dev., E.I. du Pont de Nemours Co., Expt. Sta., P.O. Box 80356, Wilmington, DE 19880-0356, USA. (1951) PhD solid state inorg. chemistry (Iowa State U., 1981). Staff scient. (tel. 302 + 695-2236). *Crystallography, solid state, oxides, transition metal.*

Torre, Prof. Louis Peter. Library of Sci. &d Medicine, Rutgers U., Busch Campus, New Brunswick, NJ 08904, USA. (1941) PhD phys. chemistry (U. of Washington, 1971). Asst. prof. (tel. 201 + 932-3526). *Information retrieval, data banks, computers.*

Towns, Prof. Robert Lee Roy. Chemistry Dept., Cleveland State U., 24th. and Euclid Sts. Cleveland, OH 44115, USA. (1940) PhD phys. chemistry (U. of Texas at Austin, 1969). Prof. (tel. 216 + 687-2468). *Molecular structures, organic heterocycles, biologically interesting molecules, trace metal analysis in biological specimens, X-ray fluorescence analysis.*

Treasurywala, Dr. Adi M. Med. Chem. Dept., Sterling Res. Group/Rensselaer, 81 Columbia Turnpike, Rensselaer, NY 12144, USA. (1947) PhD org. chemistry (U. of Brit. Columbia, Canada 1973). Sect. head, computational chem. (tel. 518 + 445-7042, email sterling at uhrcc2).

Trefonas, Prof. Louis Marco. Office of the Dean, Div. of Graduate Studies, U. of Central Florida, Orlando, FL 32816, USA. (1931) PhD phys. chemistry (U. of Minnesota, 1959). Grad. dean. (tel. 305 + 275-2197). *Biologically interesting molecular structures, coordination compounds, proteins.*

Treharne, Dr. Anne-Marie. Expt. Sta., Bldg. 228/316, E.I. du Pont de Nemours Co., P.O. Box 80228, Wilmington, DE 19880-0228, USA. (1960) PhD crystallography (U. of London, UK, 1987). Visit. res. scient. (tel. 302 + 695-9059). *Crystallography, proteins, macromolecules, theoretical simulations, molecular dynamics.*

Troup, Dr. Jan Marshall. Molecular Structure Corp., 3200A Research Forest Dr., The Woodlands, TX 77381, USA. (1946) PhD chemistry (Texas A&M U., 1974). Pres. (tel. 713 + 363-1033, telex 55-9314 (MOLECULAR), fax 713 + 292-2472). *Inorganic structures, X-rays, data collection techniques, powder structures, experimental electron density.*

Trueblood, Prof. Kenneth N. Chemistry Dept., U. of Calif. at Los Angeles, Los Angeles, CA 90024, USA. (1920) PhD chemistry (CIT, 1947). Prof. (tel. 213 + 825-1259). *Organic structures, molecular motion.*

Trus, Dr. Benes Louis. Nat. Insts. of Health, Bldg. 12A, Rm. 2053, Bethesda, MD 20892, USA. (1946) PhD chemistry (CIT, 1972). Res. chemist. (tel. 301 + 496-2250, telex 248232 NIHUR, email Bitnet TRU at NIHCUDEC). *Macromolecular structure, electron microscopy, image processing.*

Tsai, Prof. Chun-che. Chemistry Dept., Kent State U., Kent, OH 44242, USA. (1937) PhD phys. chemistry (Indiana U., 1968). Assoc. prof. (tel. 216 + 672-2989). *Biological crystallography, drugs, nucleic acid, interactions, structure - function, drug design.*

Tucker, Mr. Craig Anthony. 3548 Turtle Cove Ct., Marietta, GA 30067, USA. (1961) MS chemistry (N. Carolina State U. at Rayleigh, 1984). Chemist, Food & Drug Admin. (tel. 404 + 347-4005). *Macromolecules.*

Tulinsky, Prof. Alexander. Chemistry Dept., Michigan State U., East Lansing, MI 48824-1322, USA. (1928) PhD phys. chemistry (Princeton U., 1956). Prof. (tel. 517 + 353-4511). *Protein crystallography, structure - function, blood proteins, enzymes.*

Tuomi, Dr. Donald. 626 S. Kaspar, Arlington Heights, IL 60005, USA. (1920) PhD phys. chemistry (Ohio State U., 1952). (tel. 312 + 259-2374). *Structural chemistry, solid state physics, semiconductors, polymers.*

Turano, Dr. August M. Data Processing Dept., Med Chek Labs Inc., 11700 Althea Rd., Pittsburgh, PA 15235, USA. (1956) PhD crystallography (U. of Pittsburgh, 1982). (tel. 412 + 241-5739). *Biomolecules, charge density, protein crystallography, database methodology.*

Turley, Dr. June Williams. Regulatory & Legislative Issues, Dow Chemical Co., Building 1803, Midland, MI 48674, USA. (1929) PhD biological chemistry (PSU, 1957). Regulatory manager. (tel. 517 + 636-5443). *Structure - activity*

relationships (SAR), biologically active molecules, toxicology, risk assessment, data base management.

Turley, Mr. Stewart. Biological Struct. Dept., SM-20, U. of Washington, Seattle, WA 98195, USA. (1950) MS chemistry (Leeds U., UK, 1973). Res. technician. (tel. 206 + 543-4496, email turley at max.acs.washington.edu).

Turner, Dr. Ralph Waldo. Chemistry Dept., Florida A&M U., Tallahassee, FL 32307, USA. (1938) PhD phys. chemistry (U. of Pittsburgh, 1965). Prof. (tel. 904 + 599-3638). *Structural chemistry, metal-ion organic complexes, proteins, enzymes.*

Twigg, Mrs. Pamela Dekle. Space Sci. Lab., Marshall Space Flight Ctr., ES-76, Huntsville, AL 35812, USA. (1960) MS biomedical eng. (U. of Alabama at Birmingham, 1986). Visit. scient. (tel. 205 + 544-5494). *Protein crystallography, crystal growth.*

Tzeng, Dr. Shih-Ying. Analytical Chem. Div., Alcoa Tech. Ctr., 7th. St. Rd.- Rt. 780, Alcoa Center, PA 15069, USA. (1941) PhD geology (U. of Pittsburgh, 1976). Staff scient. (tel. 412 + 337-5803). *High temperature X-ray diffraction, small angle X-ray scattering.*

Uberbacher, Mr. Edward C. Oak Ridge Grad. Sch. of Biomed. Sci., U. of Tennessee, Biology Div. - Oak Ridge Nat. Lab., Oak Ridge, TN 37831-8077, USA. (1951) PhD chemistry (U. of Pennsylvania, 1979). Res. asst. prof. (tel. 615 + 576-2685). *Neutron scattering, macromolecules, chromatin, DNA.*

Ulmer, Dr. Kevin M. seQ Ltd., 30 Margin St., Cohasset, MA 02025, USA. (1951) PhD Biol. oceanography (MIT (Woods Hole Ocean. Inst.), 1978). President (tel. 617 + 383-2858). *Protein engineering, molecular modeling, molecular electronics.*

Ultsch, Mr. Mark. Genentech Inc., MS-27, 460 Pt. San Bruno Blvd., South San Francisco, CA 94080, USA. (1956) MA immunology, biol. sci. (Calif. State U. at Sacramento, Res. assoc. (tel. 415 + 266-1197).

Underwood, Dr. Dennis John. Merck Sharp & Dohme Res. Labs., P.O. Box 2000, Rahway, NJ 07065-0900, USA. (1952) PhD physical org. chemistry (Adelaide U., Australia, 1982). Sr. res. chemist. (tel. 201 + 594-7221). *Crystallography, chemistry, structural, theoretical, computational, physical organic, biochemistry.*

Unwalla, Dr. Rayomand. MIS, Glaxo Inc., Five Moore Dr., Research Triangle Park, NC 27713, USA. (1956) PhD org. chemistry (Louisiana State U., 1986). Postdoct. fel. (tel. 919 + 248-7385). *Crystallography, spectroscopy, molecular modeling, force field development, organic synthesis.*

Urdy, Prof. Charles Eugene. Chemistry Dept., Huston-Tillotson C., 1820 East 8th. St., Austin, TX 78723, USA. (1933) PhD chemistry (U. of Texas at Austin, 1962). Prof. (tel. 512 + 476-7421, ext. 273). *Crystal structure analysis.*

Valente, Dr. Edward Joseph. Chemistry Dept., Mississippi C., Box 4065, Clinton, MS 39058, USA. (1949) PhD chemistry (U. of Washington, 1977). Prof. (tel. 601 + 925-3424). *Structure, conformation, bio-medical compounds.*

Van der Helm, Prof. Dick. Chemistry Dept., U. of Oklahoma, 620 Parrington Oval, Norman, OK 73069, USA. (1933) DSc chemistry (U. of Amsterdam, Netherlands, 1960). George Lynn Cross Res. Prof. (tel. 405 + 325-5831). *Molecular structure, conformation, natural products, siderophores, peptides, peptide chelates, anticancer agents, membrane proteins.*

Van Duyne, Dr. Gregory D. Chemistry Dept., Baker Lab., Cornell U., Ithaca, NY 14853, USA. (1960) PhD chemistry (Cornell U., 1988). Dir., X-ray diffr. facil. (tel. 607 + 255-6145). *Structure/activity in biologically active compounds.*

Van Engen, Dr. Donna. Analytical Div., Exxon Res. and Eng. Co., Route 22 East, Annandale, NJ 08801, USA. (1954) PhD chemistry (Cornell U., 1981). Sr. chemist. (tel. 201 + 730-2107). *Natural products, organometallic complexes.*

Van Hove, Dr. Michel Andre. Lawrence Berkeley Lab., Berkeley, CA 94720, USA. (1947) PhD physics (Cambridge U., UK, 1974). Staff sr. scient. (tel. 415 + 486-6160, fax 415 + 486-4995, email Bitnet VANHOVE at LBL). *Surface crystallography, electron diffraction, low energy.*

Van Nordstrand, Mr. Robert Alexander. 520 Montecillo Rd., San Rafael, CA 94903, USA. (1917) MS chemistry (U. of Michigan, 1939). (tel. 415 + 472-7062). *Catalyst structures, alumina, clay minerals, X-ray absorption edges, zeolites.*

Van Opdenbosch, Dr. Nicole Marie. TRIPOS Associates, 1699 S. Hanley, Suite 303, St. Louis, MO 63144, USA. (1953) PhD chemistry (U. of Namur, Belgium, 1980). Mgr. cust. training. (tel. 314 + 647-1099, fax 314 + 647-9241). *Molecular modeling, drug design, computer chemistry, graphics.*

Van Roey, Dr. Patrick M. Mol. Biophysics Dept., Med. Fndn. of Buffalo, 73 High St., Buffalo, NY 14203, USA. (1952) PhD chemistry (U. of Calgary, Canada, 1979). Res. scient. (tel. 716 + 856-9600). *Protein structures, biological molecules.*

Vance, Dr. T. Blake Jr. Crop Sci. Dept., Hawaiian Sugar Planters' Assocn., 99-193 Aiea Hts. Dr., Aiea, HI 96701, USA. (1950) PhD inorg. chemistry (U. of Wyoming, 1983). (tel. 808 + 486-5401). *X-ray structural chemistry, coordination compounds.*

Vandenberg, Dr. Joanna Maria. Physics Res., AT&T Bell Labs., 600 Mountain Ave., Murray Hill, NJ 07974, USA. (1938) PhD inorg. chemistry (Leiden State U., Netherlands, 1964) Techn. staff member. (tel. 201 + 582-4186). *High resolution X-ray diffraction, thin film interfaces, epitaxial films, superlattices.*

Vander Sande, Prof. John Bruce. Rm. 13-5025, Matls. Sci. Dept., Mass. Inst. of Technol., 77 Massachusetts Ave., Cambridge, MA 02139, USA. (1944) PhD materials sci. (Northwestern U., 1971). Prof. (tel. 617 + 253-6933). *Materials* science, metallurgy, electron microscopy, diffraction, energy dispersive X-ray analysis, defect structures, phase transformations.

Vanderah, Ms. Terrell Ann. Res. Dept., Chemistry Div., Naval Weapons Ctr., Code 3854, China Lake, CA 93555, USA.

Vanderhoff, Dr. Peggy A. Health & Safety Res. Div., Oak Ridge Nat. Lab., P.O. Box 2009, Oak Ridge, TN 37831-8077, USA. (1959) PhD chemistry (Rutgers U., 1989). Postdoct. fel. (tel. 615 + 574-0844). Hydrogen bonding, keto-carboxylic acids, neutron diffraction.

VanDerveer, Dr. Donald G. Sch. of Chemistry, Georgia Inst. of Techn., Atlanta, GA 30332, USA. (1947) PhD chemistry (Brown U., 1974). Res. scient. (tel. 404 + 894-8517). Single crystal X-ray structure determination.

Vandlen, Dr. Richard L. Protein Chem. Dept., Genentech Inc., 460 Pt. San Bruno Blvd., South San Francisco, CA 94080, USA. (1947) PhD phys. chemistry (Michigan State U., 1972). Dir. (tel. 415 + 266-2648, fax 415 + 266-2739). Proteins, structure - function, hormone receptors, membrane proteins, bound enzymes, lipid metabolism, protein sequencing, structure determination via mass spectrometry.

Varughese, Dr. Kottayil Iype. Biology Dept., MC-B017, U. of Calif. at San Diego, La Jolla, CA 92093, USA. (1946) PhD X-ray crystallography (U. of Madras, India, 1974). Asst. res. biologist. (tel. 619 + 534-6423, email Bitnet KVARUGHESE at UCSD). Protein crystallography.

Vasilevsky, Dr. Igor V. Chemistry Dept., BG-10, U. of Washington, Seattle, WA 98195, USA. (1951) PhD chemistry (U. of Washington, 1988). Res. assoc. (tel. 206 + 543-6913). Structures, coordination compounds.

Veblen, Prof. David Rodli. Earth & Planetary Sci. Dept., Johns Hopkins U., Baltimore, MD 21218, USA. (1947) PhD geological sci. (Harvard U., 1976). Prof. (tel. 301 + 338-8487). Crystallography, mineralogy, electron transmission microscopy, diffraction, electron, X-ray.

Velsko, Dr. Stephan Paul. Y Division, Lawrence Livermore Nat. Lab., P.O. Box 5508, Livermore, CA 94550, USA. (1955) PhD chemistry (U. of Chicago, 1981). Res. chemist. (tel. 415 + 423-0191). Optical crystallography, non-linear optics.

Versic, Dr. Ronald J. Ronald T. Dodge Co., P.O. Box 630, Dayton, OH 45459, USA. (1942) PhD materials eng. (Ohio State U., 1969). President. (tel. 513 + 439-4497, telex 910-250-1073, fax 513 + 439-1704). Mineralogy, crystal chemistry, growth.

Vezie, Miss Deborah L. P.O. Box 223, Laveen, AZ 85339, USA. (tel. 602 + 965-3313).

Via, Mr. Grayson Hall. 740 Crescent Parkway, Westfield, NJ 07090, USA.

Vijay-Kumar, Prof. Senadhi. Comprehensive Cancer Clinic, U. of Alabama at Birmingham, 248 BHS, Birmingham, AL 35294, USA. (1952) PhD X-ray crystallog. (Indian Inst. of Techn., Bombay, 1980). Asst. prof. (tel. 205 + 934-1972, fax 205 + 934-0480). Proteins, nucleic acids, crystallization, structural work, structure refinement, protein - drug binding.

Vitali, Dr. Jacqueline. Biology Dept., U. of Virginia, Gilmer Hall, Charlottesville, VA 22901, USA. (1950) PhD biophysical sci. (SUNY at Buffalo, 1986). Res. assoc. (tel. 804 + 924-7039, email VITALI at BIOVAX or JV4V at VIRGINIA). Structures, small molecules, moderately complex macromolecules, crystallographic computing, crystallographic results interpretation.

Vlasse, Dr. Marcus. Solidification Sci., ES74, NASA, Marshall Space Flight Ctr., Huntsville, AL 35812, USA. (1933) PhD, DSc crystallog., matls. sci. (U. Pittsburgh 1965, U. of Bordeaux, France, 1980). Sr. scient. (tel. 205 + 544-7781, fax 205 + 544-8762). Crystal growth, organic, inorganic materials, crystallographic analysis, powder, single crystal, characterization, materials processing in space.

Voet, Prof. Donald Herman. Chemistry Dept., U. of Pennsylvania, 231 S. 34th. St., Philadelphia, PA 19104, USA. (1938) PhD chemistry (Harvard U., 1966). Prof. (tel. 215 + 898-6457, fax 215 + 898-2037, email voet%c.chem at penn.edu). Structures, biological macromolecules, proteins, nucleic acids.

Vold, Mr. Carl L. Material Sci. and Techn. Div., Code 6320, Naval Res. Lab., Washington, DC 20375, USA. (1932) MS metallurgy (Iowa State U., 1959). solid state physicis (tel. 202 + 767-2440). Lattice dynamics, anisotropic elasticity, elastic constants.

Volz, Mr. Karl W. Microbiol. Immunology Dept., U. of Illinois, MC790, Box 6998, Chicago, IL 60680, USA. (tel. 312 + 996-2314).

Von Dreele, Dr. Robert Bruce. 3841 Villa St., Los Alamos, NM 97544, USA. (1943) PhD chemistry (Cornell U., 1971). (tel. 505 + 667-3630, email vondreele%wnrvax at lanl.gov). Powder diffraction, neutron scattering, solid state chemistry.

Vranka, Mr. Robert G. Westec Services Inc., 1221 State St., Suite 200, Santa Barbara, CA 93101, USA.

Vyas, Dr. Meenakshi N. HHMI, One Baylor Plaza, Houston, TX 77030, USA. (1949) PhD X-ray crystallography (Indian Inst. of Techn., Bombay Assoc. X-ray crystallography.

Vyas, Dr. Nand Kishore. HHMI, One Baylor Plaza, Houston, TX 77030, USA. (1949) PhD chem., X-ray crystallog. (Indian Inst. of Techn., Bombay, 1977). Assoc. (tel. 713 + 798-6563). X-ray crystallography, structure determination, biological macromolecules.

Waerstad, Mr. Kjell Robert. Chemical Res., Tennessee Valley Authority, Muscle Shoals, AL 35660, USA. (1940) Cand. real. phys. chemistry (U. of Oslo,

Norway, 1965). Res. chemist. (tel. 205 + 383-2964). X-ray crystallography, materials characterization, fertilizer development.

Wagner, Prof. Christian Nikolaus Johan. Materials Sci. & Eng. Dept. U. of Calif. at Los Angeles Los Angeles, CA 90024-1595, USA. (1927) Dr.rer.nat. physical metallurgy (U. of the Saar, FRG, 1957). Prof. (tel. 213 + 825-6265). Amorphous and liquid alloy structures, plastic deformation, alloys, biomaterials, thin films.

Walker, Dr. Christopher Bland. 22 Baskin Rd., Lexington, MA 02173, USA. (1925) PhD physics (MIT, 1951). Retired. (tel. 617 + 862-6943). Metal physics, imperfections in crystals.

Walker, Mr. Donald L. 104 Mill Pond, Andover, MA 01845, USA. (tel. 617 + 470-2785).

Wallace, Dr. Bonnie Ann. Chemistry Dept. & Ctr. Biophys., Rensselaer Polytechnic Inst., Troy, NY 12180-3590, USA. (1951) PhD biophysics (Yale U., 1977). Prof. (tel. 518 + 276-2814, email cms at wallace.cicgj.rpi.edu). Membrane protein, structure - function, biophysics, macromolecular crystallography.

Wallace, Mr. John C. Biophysics Dept., Roswell Park Memorial Inst., 666 Elm St., Buffalo, NY 14203, USA. (tel. 716 + 845-2361).

Wallace, Mr. Peter Lindsay. Lawrence Livermore Nat. Lab., L-360, P.O. Box 808, Livermore, CA 94550, USA. (1937) BS metallurgical eng. (U. of Arizona, 1959). Metallurgist. (tel. 415 + 423-1679). Applied crystallography, X-ray spectrometry, actinide metallurgy.

Walter, Mr. Mark Harrison. Biochem., Mol., & Cell. Biology Dept., Northwestern U., 2153 Sheridan Rd., Hogan 2-100, Evanston, IL 60208, USA. (1953) BA history (U. of Northern Colorado, 1976). Grad. student (tel. 312 + 491-7140, email MWC02202 at NWACC). X-ray crystallography, biological macromolecules.

Walter, Mr. Norman Macmillan. P.O. Box 417, Round Lake, NY 12151-0417, USA. (1923) MS physics (PSU, 1954). Retired. (tel. 518 + 899-5726). Structure - properties relationships, residual stresses in materials, microprobe analysis.

Wan, Dr. Cheng. 1111 NE 130th. St., Seattle, WA 98125, USA. (1935) PhD chemistry (McMaster U., Canada, 1970). Res. scient. Inorganic crystal structures.

Wang, Prof. Andrew Hwei-Jiu. Physiology & Biophys. Dept., U. of Illinois, 407 S. Goodwin Ave., Urbana, IL 61801, USA. (1945) PhD chemistry (U. of Illinois, 1974). Prof., biophysics. (tel. 217 + 244-6637). X-ray crystallography, biological macromolecules, DNA structure, antitumor drug - DNA interactions.

Wang, Prof. Bi-Cheng. Crystallography Dept., U. of Pittsburgh, Pittsburgh, PA 15260, USA. (1938) PhD chemistry (U. of Arkansas, 1968). Prof. (tel. 412 + 624-9310, fax 412 + 624-1882). Macromolecular structure - function, diffraction methods.

Wang, Dr. Frederick E. Innovative Techn. Int. Inc., 10747-3 Tucker St., Beltsville, MD 20705, USA. (1932) PhD phys. chemistry, matl. sci. (Syracuse U., 1960). President. (tel. 301 + 937-3688, fax 301 + 595-5409). NITINOL technology (shape memory alloy), superconduction materials.

Wang, Mr. Gwo-Ching. Physics Dept., Rensselaer Polytech. Inst., Troy, NY 12180-3590, USA. (1946) PhD physics (Rensselaer Polytechnic Inst., 1978). Prof. (tel. 518 + 266-8387). Surfaces, structures, phase transitions, kinetics of ordering.

Wang, Dr. Po-Wen. 6850 Santa Teresa Rd., San Jose, CA 95119, USA. (1943) PhD physics (PUN, 1979). X-ray physics, magnetic recording heads, powder diffractometry.

Wang, Dr. Shou-Dao. Biomolec. Struct. Anal. Ctr., U. of Connecticut, 263 Farmington Ave., Farmington, CT 06032, USA. (1939) PhD crystallography (1985). Assoc. prof., Head, cryst. prog. (tel. 203 + 679-3979, telex 701-423-5 Drug/lipid crystal structures.

Ward, Dr. Donald Leslie. Chemistry Dept., Michigan State U., East Lansing, MI 48824-1322, USA. (1943) PhD chemistry (Montana State U., 1972). Dept. crystallographer (tel. 517 + 353-4511). Crystal structure determination, computer programming.

Ward, Dr. Keith Bolen. Lab. for the Structure of Matter, Code 6030, Naval Res. Lab., Washington, DC 20375-5000, USA. (1943) PhD biophysics (Johns Hopkins U., 1974). Res. biophysicist. (tel. 202 + 767-2735, telex TWX 7108 220147, fax 202 + 767-6874, email ward at lsm.nrl.navy.mil). Protein crystallography, molecular modeling, robotics.

Warren, Prof. Bertram Eugene. Physics Dept. Room 6-108, Mass. Inst. of Techn., 77 Massachusetts Ave., Cambridge, MA 02139, USA. (1902) ScD physics (MIT, 1929). Prof. emeritus (tel. 617 + 253-4351). X-ray diffraction, amorphous materials, imperfections, order-disorder.

Warren, Dr. Stephen George. Geophysics Dept., AK-50, U. of Washington, Seattle, WA 98195, USA. (1945) PhD phys. chemistry (Harvard U., 1973). Assoc. prof. (tel. 206 + 543-7230). Glaciology.

Waser, Prof. Jurg. 6120 Terryhill Dr., La Jolla, CA 92037, USA. (1916) PhD chemistry (CIT, 1944). Retired from CIT. (tel. 619 + 454-5622). Crystal structure.

Watenpaugh, Dr. Keith Donald. Physical and Analyt. Chem., The Upjohn Co., 7255-209-1, 301 Henrietta St., Kalamazoo, MI 49007, USA. (1939) PhD chemistry (Montana State U., 1967). Sr. scient. (tel. 616 + 385-5481, fax 616 + 385-7373). Macromolecular crystallography, computing methods.

Watkins, Prof. Steven F. Chemistry Dept., Louisiana State U., Baton Rouge, LA 70803, USA. (1940) PhD phys. chemistry (U. of Wisconsin at Madison, 1967).

Assoc. prof. (tel. 504 + 388-1525). *Chemistry, inorganic, organometallic, natural products, self consistent field calculations, computing techniques.*

Watson, Mr. Douglas J. A-9 Materials Res. Lab., Pennsylvania State U., University Park, PA 16803, USA. (tel. 814 + 865-2434).

Watson, Prof. William Harold. Chemistry Dept., Texas Christian U., TCU Station, Fort Worth, TX 76129, USA. (1931) PhD chemistry (Rice U., 1957). Prof. (tel. 817 + 921-7195, fax 817 + 921-7110, email RKAA at TCUAMUS). *Synthetic metals, electronic materials, molecular recognition.*

Watt, Mr. William. Phys. & Analyt. Chem. Res. Div., The Upjohn Co., 301 Henrietta St., Kalamazoo, MI 49007, USA. (1955) MS analytical chemistry (U. of Michigan, 1981). Res. chemist. (tel. 616 + 385-4585). *Protein crystallography, high resolution refinement, biologically interesting small molecules, computer graphics.*

Watterson, Dr. D. Martin. Pharmacology Dept., Vanderbilt U. Sch. of Med., 723 Light Hall, Nashville, TN 37232-0295, USA. (1946) PhD biochemistry (Emory U., 1975). Prof. (tel. 615 + 322-4403, fax 615 + 343-4726, email WATTERMAN at CTRVx1). *Calcium binding proteins, protein kinases.*

Watts, Mrs. Ethel Jean. Matls. Sci. Lab., Analysis Dept., The AEROSPACE Corp., 2350 E. El Segundo Blvd., El Segundo, CA 90245-4691, USA. (1928) MS physics (PSU, 1952). Techn. staff member. (tel. 213 + 336-0030). *Thin films, surfaces, interfaces, analytical electron microscopy, particulate and NVR contamination on spacecraft.*

Waychunas, Dr. Glenn A. Center for Materials Res., 105 McCullough Bldg., Stanford U., Stanford, CA 94305, USA. (1948) PhD geochemistry (UCLA, 1979). Lab. dir., X-ray diffr. (tel. 415 + 723-0187). *X-ray absorption spectroscopy, anomalous scattering, Mössbauer spectroscopy, order-disorder, surface diffraction.*

Weakley, Dr Timothy John Ruffer. Dept. of Chemistry, U. of Oregon, Eugene OR 97403-1210, USA. (1933) DPhil,chemistry (Oxford, UK, 1959) Res. Assoc. (tel. 503+686 4615) *Coordination compounds, inorganic crystals, molecular crystals, structural chemistry.*

Weaver, Dr. Larry H. Inst. of Mol. Biology, U. of Oregon, Rm. 205 Science II, Eugene, OR 97403, USA. (1942) PhD physics (U. of Oregon, 1975). Res. assoc. (tel. 503 + 686-5176). *Protein crystallography.*

Webb, Dr. Lawrence Edward. Applied Cellular Biology, Baxter Healthcare, Rte. 120 & Wilson Rd., Round Lake, IL 60073, USA. (1936) PhD phys. chemistry (U. of Chicago, 1965). Sr. res. assoc. (tel. 312 + 546-6311). *Protein crystallography, clinical chemistry, peptide structures.*

Webb, Prof. Thomas. Chemistry Dept., Auburn U., Auburn, AL 35849, USA. (tel. 205 + 826-4043).

Weber, Dr. Irene Teresa. Cryst. Lab., Bionetics Res. Inc., BRP, NCI, Frederick Cancer Fac., P.O. Box B, Bldg. 539, Frederick, MD 21701, USA. (1953) D. Phil. mol. biophysics (Oxford U., UK, 1978). Group leader. (tel. 301 + 698-5034, fax 301 + 698-5991). *Protein crystallography, protein - nucleic acid interaction, enzyme structure - function.*

Weber, Dr. Lawrence D. Pall Corp., 30 Sea Cliff Ave., Glen Cove, NY 11542, USA. (1950) PhD phys. chemistry (Michigan State U., 1978). Sr. staff scient. (tel. 516 + 671-4000, ext. 388). *Biopolymers, structure, physical chemistry, filter media - process fluid interaction.*

Weber, Dr. Patricia C. Central R&D Dept., E.I. duPont de Nemours Co., Exptl. Sta., P.O. Box 80228, Wilmington, DE 19880-0288, USA. (1952) PhD chemistry (U. of Arizona, 1979). Prin. investig. (tel. 302 + 695-3762, fax 302 + 695-9183). *Proteins, crystallization, structure analysis, engineering.*

Wechsler, Dr. Barry Andrew. Optical Physics Dept., MS RL65, Hughes Res. Labs., 3011 Malibu Canyon Rd., Malibu, CA 90265, USA. (1951) PhD geochemistry (SUNY at Stony Brook, 1981). Member techn. staff (tel. 213 + 317-5639). *Crystal chemistry, crystal growth, optical materials.*

Weeks, Dr. Charles M. Mol. Biophysics Dept., Med. Fndn. of Buffalo, 73 High St., Buffalo, NY 14203, USA. (1944) PhD biophysics (SUNY at Buffalo, 1970). Sr. res. scient. (tel. 716 + 856-9600, ext. 333, fax 716 + 852-4846, email Bitnet WPCMW%MFB at UBVMS). *Biologically interesting compounds, direct methods, crystallographic computing.*

Weertman, Dr. Julia Randall. Materials Sci. Dept., Northwestern U., Evanston, IL 60201, USA. (1926) DSc physics (Carnegie-Mellon U., 1951). Walter P. Murphy Prof. Matls. Sci. & Eng (tel. 312 + 491-7823). *Metals, alloys, mechanical behavior, dislocation effects, high temperature microstructural changes, neutron scattering.*

Wehrenberg, Prof. John Patteson. Geology Dept., U. of Montana, Sci. Complex, Missoula, MT 59801, USA. (1927) PhD geology (U. of Illinois, 1956). Prof. (tel. 406 + 243-2341). *Mineralogy, crystallography, infrared spectroscopy of solids.*

Wei, Dr. Chin Hsuan. Biology Div., Oak Ridge Nat. Lab., P.O. Box Y, Oak Ridge, TN 37830, USA. (1926) PhD chemistry (U. of Wisconsin, 1962). Biophysicist. (tel. 615 + 574-1253). *Biological structures.*

Weininger, Prof. Marc Sterling. Chemistry Dept., Florida A&M U., Tallahassee, FL 32301, USA. (1943) PhD chemistry (U. of South Carolina, 1972). Assoc. prof. (tel. 904 + 599-3638). *Biological nitrogen fixation, enzyme structure - function, metalloenzymes.*

Weintraub, Dr. Herschel J. R. Chemical Sci. Dept., Merrell Dow Res. Inst., 2110 E. Galbraith Rd., Cincinnati, OH 45215, USA. (1948) PhD macromolecular sci. (Case Western Reserve U., 1975). Group leader. (tel. 513 + 948-6089, fax 513 + 948-7982). *Molecular modeling, drug design, intermolecular interactions, protein homology.*

Weis, Dr. William I. Biochem. & Molec. Biophys. Dept., Columbia U., 630 W. 168th. St., New York, NY 10032, USA. (1959) PhD biochemistry (Harvard U., 1987). Postdoct. res. assoc. (tel. 212 + 305-2219, email Bitnet WEIS at CUMBG). *Crystallography, protein structure, cell surface, immunology, biochemistry, biophysics.*

Weiser, Mr. Calvin H. Corporate Res. Center, A02-26, Grumman Aerospace Corp., Bethpage, NY 11714, USA. (Alt.: 140-20 Debs Place, Bronx, NY 10475, USA). (1932) MS physics (PUN, 1960). *Spectroscopy, ultra-violet, infrared, rapid scanning, hot gas physics, surface physics.*

Wiess, Dr. George H. Div. Computer Res. & Techn., Nat. Insts. of Health, 9000 Rockville Pke, Bldg. 12A, Rm. 2007, Bethesda, MD 20892, USA. (1930) PhD applied mathematics (U. of Maryland, 1958). Chief, phys. sci. lab. (tel. 101 + 496-1135, fax 301 + 496-2443, email GHW at NIHCV). *Applied probability, random walks.*

Weissman, Dr. Larry. 8620 221 Ave., NE, Redmond, WA 98053, USA. (1947) PhD biochemistry (UCLA, 1979). (tel. 206 + 868-7310). *Macromolecular crystallography.*

Weissmann, Prof. Sigmund. Mechanics & Matls. Sci. Dept., C. of Eng., Rutgers U., Piscataway, NJ 08854, USA. (1917) PhD phys. chemistry (PUN, 1952). Prof. (tel. 201 + 932-3664). *Lattice defects, defects - physical properties relationship, X-ray topography, dynamical diffraction theory, physical metallurgy, structure.*

Welsh, Mr. William James. Chemistry Dept., U. of Missouri, St. Louis, MO 63121, USA.

Wenk, Prof. Hans-Rudolf. Geology & Geophys. Dept., U. of Calif. at Berkeley, Berkeley, CA 94720, USA. (1941) PhD mineralogy, crystallog. (U. of Zurich, Switzerland, 1965). Prof. (tel. 415 + 642-7431). *Crystal chemistry, minerals, electron microscopy, structural geology, texture analysis.*

West, Ms. Elizabeth M. Chemistry Dept., Brown U., Providence, RI 02912, USA. (1962) BS chemistry (Brown U., 1984). Grad. student. (tel. 401 + 863-3963). *Chemistry, organic, organometallic, inorganic, polymer, catalysis.*

West, Mr. Presbury B. 3283 Colfax St., Evanston, IL 60201, USA. (tel. 312 + 965-7500).

Westbrook, Dr. Edwin M. Div. of Biol. & Med. Res., Argonne Nat. Lab., 9700 Cass Ave., Argonne, IL 60439, USA. (1948) PhD, MD biophysics (U. of Chicago, 1981). Biophysicist. (tel. 312 + 972-3983). *Protein structure - function, steroid-protein, membrane-protein interaction, synchrotron radiation.*

Wester, Dr. Dennis Wayne. NeoRx Inc., 410 W. Harrison, Seattle, WA 98119, USA. (1949) PhD chemistry (U. of Florida, 1975). Sr. scient. *Radiochemistry, chemistry, inorganic, technetium.*

Weston, Dr. Norman Ernest. Micron Inc., 3815 Lancaster Pike, Wilmington, DE 19805, USA. (1930) PhD phys. chemistry (MIT, 1957). Vice pres. (tel. 302 + 998-1184). *Optical and electron microscopy, X-ray diffraction, electron probe microanalysis.*

Wheeler, Dr. George L. Dept. of Chem. & Chem. Eng., U. of New Haven, 300 Orange Ave., West Haven, CT 06516, USA. (1944) PhD phys. chemistry (U. of Maryland, 1973). J.F.Buckman Prof. of Chem. (tel. 203 + 932-7171). *Biophysics, environmental chemistry.*

White, Prof. Joe L. Agronomy Dept., Purdue U., Life Sci. Building, West Lafayette, IN 47907, USA. (1921) PhD soil chemistry (U. of Wisconsin, 1947). Prof. (tel. 317 + 494-8052). *Relationships, structure, spectra, composition, reactivity, micas, chlorites, aluminum hydroxide gels.*

White, Dr. John Greville. 50 Scott Ln., Princeton, NJ 08540, USA. (1922) PhD chemistry (Glasgow U., 194). Prof. emeritus. (Fordham U.) *Crystal structures, organic, inorganic.*

White, Dr. Peter Sutherland. Dept. of Chemistry, CB#3290, Venable & Kenan Labs., U. of North Carolina, Chapel Hill, NC 27599-3290, USA. (1947) PhD, inorganic chemistry (Dalhousie U., Canada,1973). X-ray crystallographic specialist. *Computing methods, microcomputers, direct methods, inorganic and organic structures.*

White, Dr. Steven P. Mol. Biophys. & Biochem. Dept., Yale U., 260 Whitney Ave., New Haven, CT 06511, USA. (1958) PhD biochemistry (U. of Illinois at Urbana, 1988). Postdoct. fel.

Whitlow, Dr. Marc David. Protein Eng. Dept., Genex Corp., 16020 Industrial Dr., Gaithersburg, MD 20877, USA. (1956) PhD biochemistry (Boston U., 1986). Res. scient. (tel. 301 + 258-0552, telex 908775, fax 301 + 926-1221). *Protein engineering, protein structure - function, macromolecular structure predictions.*

Whittle, Mr. Robert. 56 Oxford Rd., New Hartford, NY 13413, USA.

Wickersham, Dr. Charles Edward Jr. TOSOH SMD Inc., 3515 Grove City Rd., Grove City, OH 43123, USA. (1951) PhD metallurgical eng. (U. of Illinois, 1978). Engin. mgr. (tel. 614 + 875-7912, telex 246552 TSMDIUD, fax 614 + 875-0031). *Sputtering, thin films.*

Wickham, Dr. Donald Guy. Eng. Dept., Ampex Computer Products Div., 200 N. Nash St., El Segundo, CA 90245, USA. (1922) PhD inorg. chemistry (MIT, 1954). Manager, memory core dev. (tel. 213 + 416-1255). *Chemistry, inorganic, crystal, magnetic oxides.*

Wiff, Dr. Donald Ray. Corporate Res., GenCorp Inc., 2990 Gilchrist Rd., Akron, OH 44305, USA. (1936) PhD theoret. physics (Texas A. & M. U., 1967). Sect. leader, polymer phys. (tel. 216 + 794-6339). *Polymers, liquid crystal, electrically* *conducting, molecular composite, structure-properties relationship, theoretical predictions.*

Wignall, Dr. George Denis. Solid State Div., Oak Ridge Nat. Lab., Bldg. 3025, Oak Ridge, TN 37831-6031, USA. (1941) PhD physics (Sheffield U., England, 1966). Dir., Nat. Ctr. Small Angle Scatt. Res. (tel. 615 + 574-5237). *Polymer physics, scattering, neutron, X-ray, small angle.*

Wilchinsky, Dr. Zigmond Walter. P.O. Box 188, Linden, NJ 07036, USA. (1915) PhD physics (MIT, 1942). Retired. (tel. 201 + 862-3861). *Polymers, structural characteristics, physical behavior, molecular orientation, X-ray diffraction, crack propagation, rubbers, birefringence.*

Wildman, Dr. Gary C. Materials & Devices Res., Schering Plough, 3030 Jackson Ave., Memphis, TN 38151, USA. (1942) PhD phys. chemistry (Duke U., 1969). Pres. (tel. 901 + 320-5053). *Polymer physics, morphology, structure, material science.*

Wiley, Prof. Don Craig. Biochem. & Mol. Biol. Dept., HHMI, Harvard U., 7 Divinity Ave., Cambridge, MA 02138, USA. (1944) PhD biophysics (Harvard U., 1971). Prof. (tel. 617 + 495-1808). *Membrane surface antigens, X-ray diffraction, influenza virus membrane glycoproteins, human histocompatability antigens, trypanosome variable surface glycoproteins.*

Wilkes, Dr. Glenn Richard. Film Sensitizing Div., Eastman Kodak Co., 1669 Lake Ave., Kodak Park, Rochester, NY 14650, USA. (1937) PhD inorg.-phys. chemistry (U. of Wisconsin, 1965). Chemist, tech. assoc. (tel. 716 + 427-3568). *X-ray diffraction, inorganic compounds, catalysis, photography.*

Wilkinson, Dr. Michael Kennerly. Solid State Div., Oak Ridge Nat. Lab., P.O. Box 2008, MS 6024, Oak Ridge, TN 37831-6024, USA. (1921) PhD physics (MIT, 1950). (tel. 615 + 574-2592). *Neutron scattering, magnetic properties, low temperature physics.*

Willett, Prof. Roger DuWayne. Chemistry Dept., Washington State U., Pullman, WA 99164-4620, USA. (1936) PhD chemistry and physics (Iowa State U., 1962). Prof. (tel. 509 + 335-3925). *Phase transition dynamics, magnetic properties, transition metal salts, crystal structure studies, EPR, NMR, Moessbauer studies, incommensurate systems.*

Williams, Dr. Arthur Ray. Los Alamos Nat. Lab., MS H805, Los Alamos, NM 87545, USA. (1954) PhD physics (CIT, 1981). Staff member. (tel. 505 + 667-8841). *Powder diffraction, X-ray, neutron, Rietveld refinement.*

Williams, Prof. Donald Elmer. Chemistry Dept., U. of Louisville, Louisville, KY 40292, USA. (1930) PhD chemistry (Iowa State U., 1964). Prof. (tel. 502 + 588-6798). *Organic crystal structures, molecular packing, non-bonded interactions.*

Williams, Dr. Graheme John Bramald. ENRAF-NONIUS, 390 Central Ave., Bohemia, NY 11716, USA. (1942) PhD biochemistry (U. of Alberta, Canada, 1972). Vice Pres. (tel. 516 + 589-2885, fax 516 + 589-2068). *Crystallography methodology, instrumentation, structures.*

Williams, Dr. Ian Duncan. Matls. Res. Lab., Pennsylvania State U., University Park, PA 16802, USA. (1959) PhD inorg. chemistry (U. of Bristol, UK, 1985). Res. assoc. (tel. 814 + 865-3624, fax 814 + 865-2326, email IDW1 at PSUVM). *Solid state reactions, topotacticity, piezo electricity, non-linear optics, ferroelectric materials, non-centrosymmetry testing, polar materials.*

Williams, Dr. Jack Marvin. Chemistry Div., Bldg. 200-A113, Argonne Nat. Lab., 9700 Cass Ave., Argonne, IL 60439, USA. (1938) PhD phys.-inorg. chemistry (Washington State U., 1966). Sr. chemist, group leader. (tel. 312 + 972-3464). *Neutron diffraction, inorganic and organometallic structures, organic superconductors, pulsed neutron sources, instrumentation.*

Williams, Dr. John A. Circon ACMI, 300 Stillwater Ave., Stamford, CT 06902, USA. (1940) PhD ceramic sci. (Penn. State U., 1970). Dir., matls. techn. (tel. 203 + 328-8663). *Glass structure, small angle X-ray scattering.*

Williams, Prof. Rickey J. Chemistry Dept., Midwestern U., 3400 Taft, Wichita Falls, TX 76308, USA. (1942) PhD phys. chemistry (Texas Christian U., 1968). Prof. (tel. 817 + 692-6611, ext. 4187). *Inorganic crystal structures.*

Williams, Dr. Roger L. Baker Lab., Box 43, Cornell U., Ithaca, NY 14853, USA. (tel. 607 + 255-0556).

Williams, Dr. Roger M. MIC 303-310, Jet Propulsion Lab., 4800 Oak Grove Dr., Pasadena, CA 91109, USA. (1953) PhD chemistry (CIT, 1979). Group leader. (tel. 818 + 354-2727). *Solid electrolytes, high temperature electrochemistry, surface diffusion, order-disorder, anisotropic structures, properties.*

Williamson, Dr. Michael Mario. Biology Dept., U. of Calif. at San Diego, La Jolla, CA 92093-0317, USA. (1957) PhD inorg. chemistry (Emory U., 1987). Postgrad. res. biologist. (tel. 619 + 534-6423, fax 619 + 534-0053). *X-ray crystallography, crystal structure, proteins, macromolecules.*

Williard, Prof. Paul Gregory. Chemistry Dept., Brown U., Box H, Providence, RI 02912, USA. (1950) PhD chemistry (Columbia U., 1976). Assoc. prof. (tel. 401 + 863-3589, fax 401 + 863-2584, email CH412000 at BROWNVM). *Chemistry, organic, marine natural products, X-ray crystallography, small molecule.*

Wilson, Prof. Richard C. Chem. Engin. Dept., U. of Houston, Houston, TX 77204-4792, USA. (1959) PhD chem. eng. (MIT, 1988). Asst. prof., adjunct asst. prof. biochem. (tel. 713 + 749-2415, fax 713 + 747-6323). *Molecular recognition, immunoglobins, cytochromes.*

Wilsdorf, Prof. Heinz Gerhard Friedrich. Materials Sci. Dept., U. of Virginia, Thornton Hall, Charlottesville, VA 22901, USA. (1917) PhD metallurgy (Göttingen U., FRG, 1947). Wm. G. Reynolds Prof., Dir., Light Metals Ctr.

(tel. 804 + 924-6354 or 924-781). *Electron diffraction, diffraction contrast, electron microscopy.*

Wilson, Mr. Charles. Biochem. & Biophys. Dept., U. of Calif. at San Francisco, San Francisco, CA 94143-0448, USA. (1964) MA biophysics (Boston U., 1986). Grad. student. (tel. 415 + 476-5143, fax 415 + 476-0961, email wilson at msg.ucsf.edu). *Proteins, structure - function relationships, folding.*

Wilson, Mr. David Krause. Structural Biol. Lab., HHMI, One Baylor Plaza, Houston, TX 77030, USA. (tel. 713 + 799-6564, email wookie at fred.qci.bioc.bcm.tmc.edu).

Wilson, Dr. Frank Charles. 1410 Emory Rd., Green Acres, Wilmington, DE 19803, USA. (1927) PhD phys. chemistry (MIT, 1957). (tel. 302 + 478-6619). *Polymers, small angle X-ray scattering.*

Wilson, Dr. Ian Andrew. Mol. Biology Dept. MB7, Res. Inst. Scripps Clinic, 10666 N. Torrey Pines Rd., La Jolla, CA 92037, USA. (1949) D.Phil. mol. biophysics (Oxford U., UK, 1976). Assoc. memb. (tel. 619 + 554-9706, fax 619 + 554-6105). *X-ray crystallography, protein structure, antibody - antigen interactions, lymphokines.*

Wilson, Dr. Robert L. Rigaku/USA, Northwoods Business Park, 200 Rosewood Dr., Suite 190, Danvers, MA 01923, USA. (1946) PhD analyt. chemistry (Oregon State U., 1976). Applications lab. mgr. (tel. 508 + 777-2446, fax 508 + 777-3594). *X-ray fluorescence, diffraction.*

Wilson, Dr. Roxy B. Sch. of Chemical Sci., U. of Illinois, 601 S. Mathews Ave., Urbana, IL 61801, USA. (1951) PhD inorg. chemistry (U. of North Carolina, 1978). Asst. dir., Gen. Chem. (tel. 217 + 333-6274). *Crystallography education, organic solid state.*

Wilson, Mr. Scott R. U. of Illinois, 505 S. Mathews Ave., Urbana, IL 61801, USA.

Winborne, Ms. Evon L. 2028 Mount Royal Terrace, Apt. 105, Baltimore, MD 21217, USA. (tel. 301 + 975-2163).

Winchell, Prof. Horace. Geology & Geophysics Dept., Kline Geology Lab., 210 Whitney Ave., P.O. Box 6666 New Haven, CT 06511, USA. (alt.: 40 Deep Wood Dr., Hamden, CT 06517, USA). (1915) PhD mineralogy, petrography (Harvard U., 1941). Prof. emeritus. (tel. 203 + 776-2642 (home)). *Optical mineralogy, crystallography, systematics.*

Wing, Prof. Richard M. Chemistry Dept., U. of Calif. at Riverside, Riverside, CA 92502, USA. (1936) PhD phys. chemistry (SUNY at Buffalo, 1965). Prof. (tel. 714 + 787-3503, email DICKWING at UCRVMS). *Organometallics, DNA, reaction mechanisms, molecular transduction.*

Wingert, Dr. Lavinia Meinzer. Chemistry Dept., Carnegie Mellon U., 4400 Fifth Ave., Pittsburgh, PA 15213, USA. (1943) PhD crystallography (U. of Pittsburgh, 1986). Res. Chemist. (tel. 412 + 268-4978). *Electrostatic properties, molecules and crystals, liquid crystals, non-linear optical properties.*

Wink, Prof. Donald J. Chemistry Dept., New York U., New York, NY 10003, USA. (1958) PhD chemistry (Harvard U.). Asst. prof. (tel. 212 + 998-8475). *Organometallic synthesis, catalysis.*

Winslow, Mr. Douglas N. Sch. of Civil Eng., Purdue U., West Lafayette, IN 47907, USA. (tel. 317 + 494-5019).

Winter, Prof. William Thomas. Chemistry Dept., SUNY - Environ. Sci. and Forestry, 315 Baker Lab., Syracuse, NY 13210, USA. (1944) PhD phys. chemistry (SUNY C. of Env. Sci. and Forestry, 1974). Assoc. prof. (tel. 315 + 470-6870). *Diffraction, X-ray fiber, electron, polymer structure, solid state NMR, molecular modeling.*

Wintergrass, Mr. David J. 4830 Bernal Ave. #F, Pleasanton, CA 94566, USA. (1961) MS phys. chemistry (Iowa State U., 1987). Chemist, Syntex Corp. (tel. 415 + 855-5712). *Structure - property relationships, small molecules.*

Wismer, Prof. Robert Kingsley. Chemistry Dept., Millersville U., Millersville, PA 17551, USA. (1945) PhD phys. chemistry (Iowa State U., 1972). Prof. (tel. 717 + 872-3661). *Crystal structure, small molecules, structure solution methods.*

Witters, Prof. Robert Dale. Chemistry & Geochem. Dept., Colorado Sch. of Mines, Golden, CO 80401, USA. (1929) PhD phys. chemistry (Montana State U., 1964). Prof. (tel. 303 + 273-3632). *Molecular structure determination, X-ray diffraction.*

Wlodawer, Dr. Alexander. Basic Res. Program, NCI, Frederick Cancer Res. Fac., Frederick, MD 21701, USA. (1946) PhD molecular biology (UCLA, 1974). Lab. dir. (tel. 301 + 698-5036). *Protein crystallography, structure, enzyme, nucleic acid.*

Wold, Prof. Aaron. Chemistry Dept., Brown U., Room 610, Barus and Holley Bldg., Providence, RI 02912, USA. (1927) PhD solid state inorganic chem. (PUN, 1952). Prof. (tel. 401 + 863-2857). *Solid state chemistry, growth, structure - properties relationship, chalcopyrite oxide crystals, platinum metal chalcogenides.*

Wolff, Dr. Gunther Arthur. G.A. Consultants, NPO, 3776 Northampton Rd., Cleveland Heights, OH 44121, USA. (1918) ScD phys. chemistry (Techn. U. of Berlin, BRD, 1948). Consultant. (tel. 216 + 381-7284). *Crystals, growth, morphology (face, shape), point defects, physics, chemistry, surface and bulk kinetics, thermodynamics, solid state science, materials engineering, device applications, semiconduction, gallium phosphide/arsenide, electroluminescence.*

Wong, Dr. Ricky Ngok-Shun. Molec. Biology Dept., Bissendorf Biosciences Inc., P.O. Box 832406, Richardson, TX 75083, USA. (1954) PhD molecular biology (U. of Oklahoma, 1984). Staff mol. biologist. (tel. 214 + 907-6200, fax 214 +

907-9966). *Proteins, structure - function relationships, aspartic proteases, yeast expression system.*

Wong, Ms. Rosalind Y. U.S. Dept. of Agriculture, Res. Serv., Western Regional Res. Center, 800 Buchanan St., Albany, CA 94710, USA. (1940) BS chemistry (San Jose State U., 1965). Res. chemist. (tel. 415 + 559-5842, fax 415 + 559-5777). *Molecular structure, biological functions, structure - activity relationships, agriculturally interesting compounds.*

Wong-Ng, Dr. Winnie Kwai-Wah. Ceramics Div., Nat. Inst. Stnds. & Techn., A329 Matls., Gaithersburg, MD 20899, USA. (1947) PhD chemistry (Louisiana State U., 1974). Res. chemist. (tel. 301 + 975-5791). *Crystal chemistry, structures, powder diffraction, intermolecular forces, electron density, molecular orbital calculations, data bases, superconductors.*

Wood, Dr. Elizabeth Armstrong. 17 Alston Ct., Red Bank, NJ 07701, USA. (1912) PhD geology (Bryn Mawr C., 1939). Retired from Bell Labs. (tel. 201 + 747-2863). *Ferroelectrics, epitaxy, teaching.*

Wood, Prof. John Stanley. Chemistry Dept., U. of Massachusetts, Amherst, MA 01003, USA. (1936) PhD physical chem., crystallog. (U. of Manchester, UK, 1962) Prof. (tel. 413 + 545-2375). *Structure, charge density distributions, small inorganic systems, coordination compounds, EPR, magnetic, spectroscopic methods, molecular orbital calculations.*

Wood, Mr. Kenneth J. Med. Fndn. of Buffalo, 73 High St., Buffalo, NY 14203, USA. (tel. 716 + 856-9600).

Woode, Prof. Moses Kwamena. Sch. of Medicine, U. of Virginia, Box 446, Charlottesville, VA 22908, USA. (1947) PhD chemical crystallography (Imperial C., U. of London, 1978 Asst. dean, Assoc. prof. (tel. 804 + 924-2189, email kaw4f at virginia.edu). *Structure determination, biologically important molecules.*

Woon, Dr. Tai Chin. Chemistry Dept., U. of Wyoming, P.O. Box 3838, Laramie, WY 82071, USA. (1952) PhD inorg. chemistry (La Trobe U., Australia, 1981). Postdoct. fel. (tel. 307 + 766-5367). *Chemistry, bio-inorganic, coordination, enzyme models, cytochrome P450, C oxidase, proteins with type (III) copper, manganese complex in photosystem II.*

Worcester, Prof. David L. Biology Div., U. of Missouri, Columbia, MO 65211, USA. (1944) PhD physics (Harvard U., 1971). Assoc. prof. (tel. 314 + 882-6864, fax 314 + 882-5217). *Biological membranes, membrane components.*

Workman, Mr. Samuel Thomas. Siemens Analyt. X-ray Instruments Inc., 6300 Enterprise Ln., Madison, WI 53719, USA. (1926) BS physics (St. Josephs C., 1950). Pres. and CEO. (tel. 608 + 276-3000). *X-ray instrumentation.*

Worthington, Mr. C. R. Mellon Inst., Carnegie-Mellon U., 4400 Fifth Ave., Pittsburgh, PA 15213, USA.

Wright, Dr. Christine Gerda Schubert. Biochem. & Mol. Biophys. Dept., MCV, Virginia Commonwealth U., Box 614, MCV Station, Richmond, VA 23298, USA. (1940) PhD chemistry (U. of Calif. at San Diego, 1969). Res. asst. prof. (tel. 804 + 786-6139). *Macromolecular structure, carbohydrate binding proteins, evolution.*

Wright, Dr. Harlan Tonie. Biochemistry Dept., MCV, Virginia Commonwealth U., Box 614, MCV Station, Richmond, VA 23298, USA. (1941) PhD chemistry (U. of Calif. at San Diego, 1968). Assoc. prof. (tel. 804 + 786-6139). *Macromolecular structure.*

Wright, Dr. William Vaughn. Computer Sci. Dept., UNC, IBM Communications Div., 104 Campbell Lane, North Carolina, NC 27514, USA. (1931) PhD computer sci. (U. of N. Carolina at Chapel Hill, 1972). Adjunct prof. and Sr. eng. (tel. 919 + 942-1144). *Interactive computer graphics, molecular graphics, computational models, crystallographic computing, molecular structures.*

Wu, Dr. Ellen L. Mobil Res. & Dev. Corp., Paulsboro, NJ 08066, USA. PhD physical chem. (U. of Minnesota, 1962). Sr. consultant. (tel. 609 + 423-1040). *Catalysis, catalyst characterization, X-ray diffraction, infrared spectroscopy.*

Wu, Mrs. Jin. Biochemistry Dept., Med. C. of Wisconsin, Milwaukee, WI 53226, USA.

Wu, Mr. Jong-Chang. Chemistry Dept., U. of New Orleans, New Orleans, LA 70148, USA.

Wu, Mr. Jong-Ching. Physics Dept., U. of Oregon, Eugene, OR 97403, USA. (1959) MS physics (1988). Grad. student. (tel. 503 + 686-9952).

Wuensch, Prof. Bernhardt John. Materials Sci. and Eng. Dept., Mass. Inst. of Techn., 77 Massachusetts Ave., Cambridge, MA 02139, USA. (1933) PhD crystallography (MIT, 1963). TDK Prof. Matls. Sci. & Eng., Dir. Matls. Sci. Eng. Ctr. (tel. 617 + 253-6889). *Point defects, diffusion, inorganic crystal chemistry, fast ion conductors, sulfides, sulfosalts.*

Wyckoff, Prof. Harold Winfield. Molec. Biophys. & Biochem. Dept., Yale U., P.O. Box 6666, New Haven, CT 06520, USA. (1926) PhD biophysics (MIT, 1955). Prof. (tel. 203 + 432-5621, fax 203 + 432-3282). *Protein structure - function, instrumentation, X-ray diffraction, enzymes.*

Wyckoff, Prof. Ralph W. G. 4741 E. Cherry Hills Dr., Tucson, AZ 85718, USA. (1897) PhD chemistry (Cornell U., 1919). Prof. emeritus. (tel. 602 + 299-6029). *Crystal structures, biological macromolecular crystals.*

Xu, Dr. Zhang-Bao. Biol., Environ., & Med. Res., Argonne Nat. Lab., 9700 S. Cass Ave., Argonne, IL 60439, USA. (1964) PhD crystallography (U. of Pittsburgh, 1986). Postdoct. fel. (tel. 312 + 972-3887, email B36487 at ANLCVI). *X-ray diffraction, data collection, direct methods for data refinement, structure determination, proteins, small molecules.*

Xuong, Prof. Nguyen-huu. Physics Dept., U. of Calif. at San Diego, La Jolla, CA 92093, USA. (1933) PhD physics (UCB, 1962). Prof., Phys., Chem., Biol. (tel. 619 + 534-2501, fax 619 + 534-0053). *Protein structures, high speed data collection systems.*

Yakel, Dr. Harry Leonard. 129 Westlook Cir., Oak Ridge, TN 37830, USA. (1929) PhD chemistry (CIT, 1952). (tel. 615 + 483-7907). *Structures, transformations, alloys, inorganic compounds, synchrotron radiation.*

Yalkovsky, Dr. Rafael. Nat. Assoc. of Sci. Writers, P.O. Box 398, Grand Island, NY 14072, USA. (1917) PhD geology (U. of Chicago, 1956). Gemologist, free lance sci. writer. (tel. 716 + 773-2747). *Marine geology, mineralogy, international law of the sea, public policy.*

Yan, Mr. Gao. SUNY at Buffalo, 164 Acheson Hall, Buffalo, NY 14214, USA.

Yang, Miss Wei. BB #523, Columbia U., 630 W. 168th. St., New York, NY 10032, USA. (tel. 212 + 305-1846).

Yates, Dr. John Harry. Chemistry Dept., U. of Pennsylvania, Philadelphia, PA 19104, USA. (1948) PhD phys. chemistry (Ohio State U., 1976). Dir., Computer Facility. (tel. 215 + 898-4714). *Computer code development, molecular properties (experimental - theoretical.*

Yau, Mr. Jen-Kuan. P.O. Box 1899, Rolla, MO 65401, USA.

Yazici, Dr. Rahmi M. Matls. and Metallurg. Eng. Dept., Stevens Inst. of Techn., Castle Point, Hoboken, NJ 07030, USA. (1949) PhD materials sci. (Rutgers U., 1982). Asst. prof. (tel. 201 + 420-5261). *Strain analysis (micro and macro), ceramics, metals, thin films, coatings, general crystallography.*

Yeates, Dr. Todd O. Mol. Biology Dept., Res. Inst. of Scripps Clinic, 10666 N. Torrey Pines Rd., La Jolla, CA 92037, USA. (1961) PhD biochemistry (UCLA, 1988). Postdoct. fel. (tel. 619 + 753-0285, email yeatesU!riscc1 at sdcsvax.ucsd.edu). *Macromolecular crystallography, computation, theory, virus structure.*

Yelon, Dr. William B. U. of Missouri, Res. Reactor, Columbia, MO 65211, USA. (1944) PhD physics (Carnegie Mellon U., 1970). Group leader and prof. of physics. (tel. 314 + 882-5236, fax 314 + 882-3443). *Scattering, neutron, gamma ray, Mossbauer, diffraction, X-ray, powder, magnetic ordering, charge density extinction.*

Yoel, Mr. David William. Hetherington Inc., 4171 Market St., Ventura, CA 93003, USA. (1955) MS physics (Utah State U., 1983). Product mgr. (tel. 805 + 658-8111, fax 805 + 658-8985). *Crystals, vapor growth, growth equipment, materials processing in space.*

Yoon, Dr. Hyo Sub. Biomedical Eng. Sect., Veterans Admin. Med. Center, Middleville Rd., Northport, NY 11768, USA. (1935) PhD solid state sci. (PSU, 1971). Chief. (tel. 516 + 261-4400, ext. 7429). *Biomedical ultrasonics, crystal physics, biomaterials, biomechanics.*

Young, Dr. Frederick Walter Jr. Solid State Div., Oak Ridge Nat. Lab., P.O. Box 2008, Oak Ridge, TN 37831-6033, USA. (1924) PhD chemistry (U. of Virginia, 1950). Div. dir. (tel. 615 + 574-6151). *Highly perfect crystals, dynamical diffraction, X-ray topography.*

Young, Prof. R. A. Sch. of Physics, Georgia Inst. of Techn., Atlanta, GA 30332, USA. (1921) PhD physics (PUN, 1959). Prof. emeritus. (tel. 404 + 894-5208, telex 542507, fax 404 + 894-3120, email PH268RY at GITVM1). *Crystal physics, atomic-scale mechanisms, diffraction theory, applications.*

Young, Dr. Victor Gordon Jr. Chemistry Dept., Arizona State U., Tempe, AZ 85287-1604, USA. (1956) PhD inorg. chemistry (Arizona State U., 1985). Postdoct. res. assoc. (tel. 602 + 965-3641). *Powder diffraction, neutron, X-ray.*

Youngs, Dr. Wiley J. Chemistry Dept., Case Western Reserve U., Cleveland, OH 44106, USA. (tel. 216 + 368-5060).

Ysern, Dr. Xavier. Biophysics Dept., Johns Hopkins Sch. of Med., 725 Wolfe St., Baltimore, MD 21205, USA. (1947) PhD (Johns Hopkins U., 1977). (tel. 301 + 955-8715, email Bitnet XAVIER at JHUJGF). *Protein structure.*

Yuhasz, Dr. Stacieann. Biophysics Dept., Johns Hopkins Sch. of Med., 725 N. Wolfe St., Baltimore, MD 21205, USA. (1958) PhD biophysics (Johns Hopkins U., 1987). Postdoct. fel. (tel. 301 + 955-8715). *Protein - ligand recognition, interaction.*

Zabel, Dr. Volker. Biology Div., Oak Ridge Nat. Lab., P.O. Box 2009, MS 8077, Oak Ridge, TN 37831, USA. (1944) PhD crystallography (Freie Universitaet, Berlin, 1976). Sr. res. assoc. (tel. 615 + 574-0844). email VAZ at ORNLSTC). *Protein crystallography, neutron diffraction, direct methods, anomalous scattering, polypeptides, hydrogen bonding.*

Zacharias, Dr. David Edward. Mol. Structure Dept., Inst. for Cancer Res., 7701 Burholme Ave., Philadelphia, PA 19111, USA. (1926) PhD X-ray crystallography (U. of Pittsburgh, 1969). Sr. res. assoc. (tel. 215 + 728-2220, fax 215 + 728-3574, email zacharias at curie.rm.fccc.edu). *Organic molecular structure determination.*

Zahrt, Dr. John D. MS F669, Los Alamos Nat. Lab., Los Alamos, NM 87545, USA. (1941) PhD chemistry (Arizona State U., 1967). Staff member. (tel. 505 + 667-3410). *X-ray spectroscopy, neutron scattering, crystallography, powder diffraction.*

Zalkin, Dr. Allan. Molec. & Chem. Sci. Div., Lawrence Berkeley Lab., 70A-3307, U. of Calif. at Berkeley, Berkeley, CA 94720, USA. (1926) PhD chemistry (UCB, 1951). Sr. scient. (tel. 415 + 486-5762). *Structures, inorganic, organometallic, actinide, actinide chemistry, crystallographic computation, programming.*

Zaluzec, Mr. Eugene J. Chemistry Dept., Loyola U. of Chicago, 6525 N. Sheridan Rd., Chicago, IL 60626, USA. (1956) BS chemistry (Southern Illinois U., 1984). Grad. student. (tel. 312 + 868-3095).

Zaslow, Prof. Bertram. Chemistry Dept., Arizona State U., Tempe, AZ 85287-1604, USA. (1924) PhD chemistry (Iowa State U., 1956). Prof. (tel. 602 + 965-3685). *Polymer crystallography, symmetry.*

Zauhar, Dr. Randy Jay. Biocomputing Ctr., Biotechn. Inst., Penn. State U., 519 Wartik Lab., University Park, PA 16802, USA. (1959) PhD molecular, cell biology (PSU, 1986). Coordinator, comp. ctr. (tel. 814 + 865-1099). *Electrostatics, molecular dynamics.*

Zegenhagen, Dr. Jorg. Matls. & Interface Res. Div., AT&T Bell Labs., 600 Mountain Ave., Murray Hill, NY 07974, USA. (1950) PhD physics (U. of Hamburg, FRG, 1984). Techn. staff membr. (tel. 201 + 582-7005, fax 201 + 582-4228, email arpa!jorg%allwise at att.com). *Surfaces, interfaces.*

Zhang, Dr. Rongguang. Dept. of Biochem. & Mol. Biol., U. of Chicago, 920 E. 58th St., Chicago, IL 60637, USA. (1949) PhD crystallography (1985). Res. assoc. (tel. 312 + 702-1087). *Biomolecular crystallography.*

Zhang, Miss Xiaohua. Chemistry Dept., Wright-Rieman Lab., Rutgers U., Piscataway, NJ 08854, USA. (1956) MS phys. chem. (Rutgers U., 1989). Grad. student. (tel. 201 + 932-2115). *Crystallography, small molecules, macromolecules, electron charge transfer, synthesis, ruthenium complexes.*

Zhao, Dr. Haiching. Biochem. & Mol. Biol. Dept., Harvard U., 7 Divinity Ave., Cambridge, MA 02138, USA. (1956) PhD biophysics (U. of Connecticut, 1988). Postdoct. res. fel. (tel. 617 + 495-5043, fax 617 + 495-9613). *Protein crystallography, viruses, X-ray structure.*

Zhou, Mr. Rongsheng. P.O. Box 1242, Alfred, NY 14802, USA. (tel. 607 + 587-8415).

Ziller, Dr. Joseph W. Chemistry Dept., U. of Calif. at Irvine, Irvine, CA 92717, USA. (1959) PhD inorg. chem., crystallogr. (SUNY at Buffalo, 1986). Staff crystallographer. (tel. 714 + 856-4091). email Bitnet JZILLER at UCIUMSA). *Crystallography, chemistry, inorganic, organometallic.*

Ziolo, Dr. Ronald F. Webster Res. Center, Xerox Corp., 0114-39D, 800 Phillips Rd., Webster, NY 14580, USA. (1944) PhD inorg. chemistry (Temple U., 1970). Sr. res. scient. (tel. 716 + 422-3341, fax 716 + 422-2126). *Inorganic solid state compounds, properties (electrical, magnetic, optical), structure - properties relationships, chemistry.*

Zolensky, Dr. Michael Ewing. SN2/NASA, Johnson Space Center, Houston, TX 77058, USA. (1955) PhD geochemistry, mineralogy (PSU, 1983). Space scient. (tel. 713 + 483-5128). *Mineralogy, meteorites, powder diffraction.*

Zolliker, Dr. Peter. Brookhaven Nat. Lab., Bldg. 510B, Upton, NY 11973, USA. (tel. 516 + 282-4791).

Zoltai, Prof. Tibor Z. Geology and Geophys. Dept., U. of Minnesota, Minneapolis, MN 55455, USA. (1925) PhD mineralogy & crystallography (MIT, 1959). Prof. (tel. 612 + 624-7370). *Structures, minerals, crystal, surface.*

Zompa, Dr. Leverett Joseph. Office of the Provost, U. of Massachusetts at Boston, Harbor Campus, Boston, MA 02125, USA. (1938) PhD inorg. chemistry (Boston C., 1964). Acting vice-chancellor acad. affairs, acting provost. (tel. 617 + 929-7300). *Coordination compounds, complexes, transition metal, anion.*

Zuccola, Mr. Harmon J. Chemistry Dept., Georgia Inst. of Techn., Atlanta, GA 30332, USA.

Zuk, Dr. William. Naval Res. Lab., Code 6030, Washington, DC 20375-5000, USA. (tel. 202 + 767-2735).

Zupko, Ms. Sue A. ZAIAZ International Corp., 2610 Artie St., Huntsville, AL 35805, USA. (tel. 205 + 534-0050).

Zwell, Mr. Leo. JCPDS, Internat. Center for Diffraction Data, 117 S. Chester Rd., Swarthmore, PA 19081, USA. (1915) BS physics (Brooklyn C., 1934). Consultant. (tel. 215 + 328-0617). *Powder diffraction, X-ray metallography.*

Zwick, Prof. Martin. Systems Sci. Ph.D. Prog., Portland State U., P.O. Box 751, Portland, OR 97207, USA. (1939) PhD biophysics (MIT, 1968). Prof. (tel. 503 + 464-4960, fax 503 + 464-4882, email Bitnet HWMZ at PSUORVM). *Structure, macromolecular, phase problem, density modification, phase determination, refinement methods.*

URUGUAY

Sub-Editor: **R. A. Mariezcurrena**

Notes

1. International telephone country code - 598.
   Fax - 5982233738

2. Degrees conferred by the University of Uruguay include *Doctor en Química Farmacéutica* (Dr.Q.F.).

Fornaro, Prof Laura. Catedra de Fisica, Facultad de Química, Gral. Flores 2124-Casilla de Correa 1157, Montevido, Uruguay. (1954) Chem. Eng. (U. Mayor de la Republica,1982). Asst. Prof. (tel. 5982+206 736, ext. 40, fax 233738) *Organic crystal structure, crystal growth.*

Gomes, Mr Osvaldo. Crystallography Lab., Facultad de Química, Gral. Flores 2124-Casilla de Correo 1157, Montevideo, Uruguay. (1953) Chem. Eng. (U. Mayor de la Republica,1983). Res. asst. (tel. 2-206736, ext. 40, fax 233738) *Organic crystal structures, protein crystallization.*

Gonzalez, Mr. Oscar. Catedra de Fisica, Facultad de Química, Gral. Flores 2124-Casilla de Correo 1157, Montevideo, Uruguay. (1955) Res. Asst. (tel. 5982+206736, ext. 40, fax 233738) *Crystal structure determination.*

Mariezcurrena, Prof Raul Alfredo. Facultad de Química, Gral. Flores 2124-Casilla de Correo 1157, Montevideo, Uruguay. (1939) Dr Q.F., chemistry (U. Mayor de la República, 1975). Res. prof., lect. in physics. (tel. 2-206736, ext. 40, fax 233738). *Organic crystal structures, protein crystallography.*

Michaelis de Saenz, Dr Irene Marianne. Cristalofisica Aplicada, Facultad de Química, Gral. Flores 2124-Casill de Correo 1157, Montevideo, Uruguay. (1927) Dr. en Química, Chemistry(U. de la República, 1966) Res. *Polarizing microscopy, phase transformations in solids, non-metallic minerals, feldspars, twinning.*

Pessi Albisu, Malena Silvia. Cristalofisica Aplicada, Facultad de Química, Gral. Flores 2124-Casilla de Correo 1157, Montevideo, Uruguay. (1960) BSc, Geology(U. de la República, 1983) Asst. Res. *Macroscopic and microscopic examination, minerals.*

VENEZUELA

Sub-Editor: **M.V. Capparelli**

Notes

1. International telephone country code - 58.

2. The following abbreviations have been used :
   IVIC - Instituto Venezolano de Investigaciones Cientificas UCV - U. Central de Venezuela UDO - U. de Oriente ULA - U. de Los Andes USB - U. Simon Bolivar Depto. - Departamento Esc. - Escuela Fac. - Facultad

Arnstein, Prof. Gustavo. Esc. de Quimica, Fac. de Ciencias, UCV, Apartado 47102, Caracas 1041-A, Venezuela. (1942) PhD, Eng. materials (Maryland U., U.S.A., 1972). Prof. (tel. 02 + 662-6592, fax 02 + 662-7121). *X-ray topography, defects.*

Bello, Dr Alfredo. Depto. de Fisica, USB, Apartado 89000, Caracas 1081-A, Venezuela. (1957) PhD, physics (Case Western Reserve U., USA, 1983) Prof. agregado. (tel. 02 + 963-3022, ext. 7313, fax 02 + 962-1695). *Defects, EPR, computational methods, ionic crystals.*

Capparelli, Prof. Marlo Vicente. Esc. de Quimica, Fac. de Ciencias, UCV, Apartado 47102, Caracas 1041-A, Venezuela. (1937) PhD, crystallography (U. London, UK, 1984). Prof. (tel. 02 + 662-6592, fax 02 + 662-7121). *Chemical crystallography, crystal structures, computer programming.*

Delgado Quinones, Prof. Jose Miguel. Depto. de Quimica, Fac. de Ciencias, ULA, Venezuela. (1956) PhD, materials sci. (MIT, U.S.A., 1988) Prof. Agregado. (tel. 074-447379, fax 074-441723). *Solid state chemistry, single crystal diffraction, powder diffraction.*

Dunia, Prof. Emery. Esc. de Fisca y Matematica, Fac. de Ciencias, UCV, Apartado 20513, Caracas 1020-A, Venezuela. (1947) DSc, physics (U. Paris, France, 1980). Prof agregado. (tel. 02 + 662-9734, fax 02 + 662-7121). *X-ray topography, crystal defects.*

Gomez de Anderez, Prof. Dora Maria. Depto. de Quimica, Fac. de Ciencias, ULA, Merida, Venezuela. (1955) Lic., chemistry (U. de Los Andes,1980). Prof. Agregado. (tel. 074-447379, fax 074-441723). *Crystal structure determinations.*

Gonzalez R., Prof. Oscar V. Depto. de Fisica, Esc. de Ciencias, UDO, Cerro Colorado, Cumana, Venezuela. (1939) Dr, Solid state physics (U. de Strasbourg, 1977). Assoc. Prof. (tel. 093-653612 al 17, ext. 320). *Solid state physics, synthesis, characterization, graphite Intercalation compounds.*

Khan, Prof. Ali. Depto. de Quimica, Esc. de Ciencias, UDO, Nucleo de Sucre, Apartado 245, Cumana 6101-A, Venezuela. (1936) PhD, chemistry (Rensselear Poly. Inst., USA, 1968). Prof. titular. (tel. 093 + 65-3612, ext. 402). *Crystal growth, crystallography, electro-optics, solid state electrochemistry.*

Laredo, Prof. Estrella. Depto. de Fisica, USB, Apartado 89000, Caracas 1081-A, Venezuela. (1940) Dr (3rd cycle) crystallography (U. de Paris, France, 1965). Prof. (tel. 02 + 963-3022, ext 7313, fax 02 + 962-1695). *Defects in crystals, powder techniques, dielectric techniques, EPR, ionic crystals, ionic conductivity.*

Mateu, Dr. Leonardo. Centro de Biofisica y Bioquimica, IVIC, Apartado 21827, Caracas 1020-A, Venezuela. (1939) ScD, molecular biology (U. de Paris, France, 1974). Investigador titular. (tel. 02 + 74-9543, fax 02 + 571-3143). *Membrane structure, lipoprotein structure, small angle X-ray scattering, small angle neutron scattering.*

Mora Rodriguez, Prof. Asilloe Jasmina. Depto. de Quimica, Fac. de Ciencias, ULA, Merida, Venezuela. (1959) Lic, chemistry (Universidad de Los Andes, 1984) Asst. prof. (tel. 074-447379, fax 074-441723). *Crystal structure determination.*

Murgich, Dr Juan. Centro de Quimica, IVIC, Apartado 21827, Caracas 1020-A, Venezuela. DSc, physics (U. Cordoba, Argentina, 1974). Investigador titular. (tel. 02 + 74-5054, fax 02 + 571-3143). *Solid state physics, incommensurate phases, structural phase transitions, lattice dynamics.*

Padron, Dr. Raul. Centro de Biofisica y Bioquimica, IVIC, Apartado 21827, Caracas 1020-A, Venezuela. (1950) PhSc, biology - physiology, biophysics (IVIC, 1980). Investigador associado. (tel. 02 + 74-9543, fax 02 + 571-3143) *Molecular biology, biophysics, physiology.*

Rivera Ocando, Prof. Angela Valentina. Depto. de Quimica, Fac. de Ciencas, ULA, Merida, Venezuela. (1945) PhD, chemistry (Cambridge U., UK, 1979) Assoc. prof. (tel. 074-447379, fax 074-441723). *Crystal structure determination.*

Rodulfo de Gil, Prof. Eldrys Emilia. Depto. de Quimica, Fac. de Ciencias, ULA, Merida, Venezuela. (1936) PhD, chemistry (Wisconsin U., USA, 1968). Prof. (tel. 074-447379, fax 074-441723). *Crystal structure determination, organometallic chemistry.*

Schorin, Dr Hasso. Centro de Ingenieria, IVIC, Apartado 21827, Caracas 1020-A, Venezuela. Dr.rer.nat. (U. Tubingen, FRG, 1972). Investigador asociado. (tel. 02 + 74-8821, fax 02 + 571-3143). *Geochemistry, mineralogy, bauxites, powder X-ray diffraction.*

Suarez, MSc Nery. Depto. de Fisica, USB, Apartado 89000, Caracas 1081-A, Venezuela. (1948) MSc (U. Manchester, UK, 1977). Prof. asociado. (tel. 02 + 963-3022, ext. 7313, fax 02 + 962-1695). *Defects, polarization/depolarization techniques, ionic crystals, ionic conductivity.*

Urbina de Navarro, Dr Caribay. Centro de Microscopia Electronica, Fac. de Ciencias, UCV, Apartado 47114, Caracas 1041-A, Venezuela. (1952) Dr, chemistry (UCV, Venezuela, 1988) Prof. asistente. (tel. 02 + 662-4752, fax 02 + 662-7121). *Electron diffraction pattern, high resolution electron microscopy, heterogeneous catalysis.*

VIETNAM

Sub-Editor: **Lecong Dzurong**

Notes

1. International telephone country code - 84.

2. The Science degrees in the socialist countries: (a) the Cand. nauk, Dr.-Ing., Dr.rer.nat. (b) the Dr. nauk, Dr.sc.

Dao Cong Ngoan, Dr. Inorganic Chem. Dept., Nat. U. Hanoi, Truong DH Tonghop Hanoi, Vietnam. (1953) Dr.rer.nat., crystallography (U. Wroslav, Poland, 1988). Lect. *Structure, crystallography.*

Do Minh Nghiep, Dr. Dept. of Metal Physics and Materials, Nat. Polytechn. Inst. of Hanoi, Truong DH Bachkhoa Hanoi, Vietnam. (1945) Dr.rer.nat., metal physics (TU Dresden, GDR, 1980). Lect. *Plastic deformation, structure, magnetic materials, superconductors.*

Lecong Dzuong, Prof. Dept. Metal Physics and Materials, Nat. Polytechn. Inst. of Hanoi, Truong DH Bachkhoa, Hanoi, Vietnam. (1940) Dr.sc.nat, metal physics (PH Halle, GDR, 1987). Prof. (tel. 56116). *Physical metallurgy, X-ray and electron diffraction, structure, rapidly solidified materials.*

Le Cong Quy, Dr. Magnetic Materials and Superconductors Lab., Inst. of Physics, Vien Vatly, Vien KH Vietnam, Nghiado, Hanoi, Vietnam. (1949) Cand.nauk, crystallography (Inst. of Crystallography, Acad. Sci. USSR, 1986). *Crystallography.*

Le Nguyen Soc, Prof. Structure Analysis Lab., Res. Inst. of Industrial Chem., Vien Hoahoc Congnghiep Hanoi, Vietnam. (1937) MSc, crystallography (U. Hanoi, 1964). Prof. *Structure, metals.*

Le Van Huan, Dr. Magnetic Materials &d Superconductors Lab., Inst. of Physics, Vien Vatly, Vien KH Vietnam, Nghiado, Hanoi, Vietnam. (1950) Cand.nauk, solid state physics (Inst. Solid State Physics Moscow, USSR, 1986). *Structure, semiconductors.*

Nguy Tuyet Nhung, Dr. Geography Dept., Nat.U. Hanoi, Truong DH Tonghop Hanoi, Vietnam. (1946) Cand.nauk, crystallography (U. Leningrad, USSR, 1979). Lect. *Crystal structure.*

Nguyen An, Prof. Physics Dept., Nat. U. of Hanoi, Truong DH Tonghop Hanoi, Vietnam. (1938) Dr.sc.nat., metal physics (Humboldt-U. Berlin, GDR, 1984). Prof. (tel. 44529). *Metal physics, crystal structures.*

Nguyen Hong Quyen, Dr. Magnetic Materials and Superconductors Lab., Inst. of Physics, Vien Vatly KH Vietnam, Nghiado, Hanoi, Vietnam. (1950) Dr.rer.nat., metal physics (ZFW Dresden, GDR, 1980). *Structure, magnetic materials, superconductors.*

Nguyen Huu Chu, Dr. Technology Dept., Danang Polytechn. Inst., Truong DH Bachkhoa Danang, Vietnam. (1940) Dr.Ing., physical metallurgy (Bergakademie Freiberg, GDR, 1975). Lect. *Structure, alloys.*

Nguyen Khoa Phuc, Dr. Structure Analysis Lab., Res. Inst. of Technology, Vien Cong nghe, Tongcuc Kythuat, Hanoi, Vietnam. (1939) Dr.Ing, physical metallurgy (Bergakademie Freiberg, GDR, 1974). *Structure, alloys.*

Nguyen Van Chi, Dr. Dept. Metal Physics and Materials, Nat. Polytechn. Inst. of Hanoi, Truong DH Bachkhoa, Hanoi, Vietnam. (1942) Dr.rer.nat., metal physics (Bergakademie Freiberg, GDR, 1977). Lect. *Diffusion, structure, metals.*

Nguyen Van Vuong, Dr. Magnetic Materials and Superconductors Lab., Inst. Physics, Vien Vatly, Vien KH Vietnam, Nghiado, Hanoi, Vietnam. (1950) Dr.rer.nat., semiconductors (Humboldt-Uni, Berlin, GDR, 1984). *Structure, semiconductors.*

Phan Ving Phuc, Dr. Structure Analysis Lab., Inst. Physics, Vien Vatly, Vien KH Vietnam, Nghiado, Hanoi, Vietnam. (1938) Dr.rer.nat., chemistry (Acad.of Sciences Berlin, GDR, 1976). *X-ray fluorescence, structure.*

Quang Han Khang, Prof. Physics Dept., U. of Hochiminh-City, Truong DH Tonghop TP Ho Chi Minh, Vietnam. (1937) Cand.nauk, crystallography (U. of Leningrad, USSR, 1966). Prof. *Crystallography, structure, minerals.*

Ta Van That, Prof. Dept. of Metal Physics and Materials, Nat. Polytechn. Inst. of Hanoi, Truong DHBK Hanoi, Vietnam. (1938) physical metallurgy (Krakow Polytechn. Inst., Poland, 1972). Prof. *Structure, alloys, phase transformation.*

Tran Hai Huynh, Dr. Physics Dept., Nat.U. of Hanoi, Truong DH Tonghop Hanoi, Vietnam. (1937) Cand.nauk, metal physics (Nat U. of Moscow, USSR, 1975). Lect. *Metal physics, structure, solids.*

Tran Quoc Thang, Dr. Dept. of Metal Physics and Materials, Nat. Polytechn. Inst. of Hanoi, Truong DH Bachkhoa Hanoi, Vietnam. (1949) Dr, metal physics (INP Grenoble, France, 1987). Lect. *Electron microscopy, structure, metals and alloys.*

Trinh Han, Prof. Geography Dept., Nat.U. Hanoi, Truong DH Tonghop Hanoi, Vietnam. (1937) Cand.nauk, crystallography (Nat. U. of Moscow, USSR, 1969). Prof. *Structure, minerals.*

Trinh Le Thu, Dr. X-ray Lab., Res. Inst. of Geology and Mineralogy, Vien Diachat Khoangsan Hanoi, Vietnam. (1944) Cand.nauk, crystal chemistry (Nat. U. of Moscow, USSR, 1984). *Structure, minerals.*

Vo Vong, Dr. Magnetic Materials and Superconductors Lab., Inst. Physics, Vien Vatly, Vien KH Vietnam, Nghiado, Hanoi, Vietnam. (1941) Dr.rer.nat., metal physics (Centre Electron Micro. Res. Halle, GDR, 1974). *Structure.*

YUGOSLAVIA

Sub-Editor: **D. Matković-Čalogović**

Notes

1. International telephone country code - 0038.

2. The following degrees conferred in Yugoslavia: (a) by universities: *Doctor of Physical Sciences, Doctor of Chemical Sciences,* etc. (in the list all abbreviated as Dr) (approximately equivalent to PhD); (b) by universities and some faculties: postgraduate degrees *Master of Physical Sciences, Master of Chemical Sciences,* etc. (all abbreviated as Magistar) (approximately equivalent to MSc); (c) by Faculties of Sciences and various Technical Faculties: diplomas in physics, chemistry, physical chemistry, geology, etc. (abbreviated as *Dipl. fiz., Dipl. kem.,* etc. or *Dipl. inž. fiz., Dipl. inž. kem.,* etc. and *Dipl. inž.*) (approximately equivalent to BSc or BA).

3. At the universities persons with an academic training can be appointed in various positions ranging from *redovni profesor* (Prof.) (equivalent to Professor) via *izvanredni profesor* (Izv. prof.), *docent* to *asistent* (approximately equivalent to reader, lecturer and res. assoc. or demonstrator, respectively). The equivalent positions in the scientific institutes are: *scientific adviser, senior scientific associate, scientific associate* and *research associate* or *postgraduate student.*

Alujević - Stipanov, Dr Višnja. Farmaceutsko - biokemijski fakultet, Sveučilište u Zagrebu, ul. Ante Kovačića 1, 41000 Zagreb, Yugoslavia. (1948) Dr, chemistry (Zagreb U., 1981). Asistent. (tel. 041 - 445311, ext. 36). *X-ray diffraction analysis, polycrystalline systems.*

Arhar, Magistar Andrej. VTOZD Kemija in kem. tehnologija, FNT, Univerza E. Kardelja v Ljubljani, Murnikova 6, P.O.Box 537, 61001 Ljubljana, Yugoslavia. (1942) Magistar, chemistry (Ljubljana U., 1975). Sr. res. assoc. (tel. 061 - 214444). *Inorganic chemistry.*

Babič, Dr Danilo. Rudarsko-geološki fakultet, Univerzitet u Beogradu, Đušina 7, 11000 Beograd, Yugoslavia. (1948) Dr, mineralogy (Beograd U., 1984). Docent. (tel. 011 - 180111, ext. 710). *Crystal growth, crystal physics, mineral genesis, surface energy.*

Balić Žunić, Dr Tonči. Mineraloško-petrografski zavod, Prirodoslovno-matematički fakultet, Sveučilište u Zagrebu, Demetrova ul. 1, 41000 Zagreb, Yugoslavia. (1952) Dr, geology (Zagreb U., 1984). Docent. (tel. 041 - 422136). *Minerals, X-ray diffraction, teaching.*

Balzar, Mr Davor. Metalurški fakultet, Sisak, Institut za metalurgiju, Aleja narodnih heroja 1, 44000 Sisak, Yugoslavia. (1957) Magistar, physics (Zagreb U., 1987). Asistent. (tel. 044 - 32865). *Metals.*

Ban, Prof. Dr Zvonimir. Zavod za opću i anorgansku kemiju, Prirodoslovno-matematički fakultet, Sveučilište u Zagrebu, ul. Soc. revolucije 8, 41000 Zagreb, Yugoslavia. (1934) Dr, chemistry (Zagreb U., 1963). Prof. (tel. 041 - 442823). *Metals and alloys, structural chemistry, powder diffraction.*

Bermanec, Magistar Vladimir. Mineraloško-petrografski zavod, Prirodoslovno-matematički fakultet, Sveučilište u Zagrebu, Demetrova 1, 41000 Zagreb, Yugoslavia. (1955) Magistar, geology (Zagreb U., 1987). Asistent. (tel. 041 - 422136). *Microscopy, X-ray diffraction, minerals.*

Bezjak, Prof. Dr Aleksandar. Farmaceutsko - biokemijski fakultet, Sveučilište u Zagrebu, ul. Ante Kovačića 1, 41000 Zagreb, Yugoslavia. (1928) Dr, chemistry (Zagreb U., 1964). Prof. (tel. 041 - 446061). *X-ray diffraction analysis, polycrystalline systems.*

Blažič, Magistar Boris. FNT - VTOZD Kemija in kem. tehnologija, Univerza E. Kardelja v Ljubljani, Murnikova 6, P.O.Box 537, 61001 Ljubljana, Yugoslavia. (1961) Magistar, chemistry (Ljubljana U., 1988). Res. assistant. (tel. 061 - 214444). *Biologically interesting small molecules.*

Blažina, Dr Želimir. Institut 'Ruđer Bošković', Bijenička c. 54, P.O.Box 1016, 41001 Zagreb, Yugoslavia. (1946) Dr, chemistry (Zagreb U., 1979). Scient. assoc. (tel. 041 - 435111, ext. 482). *Intermetallic compounds.*

Bonefačić, Prof. Dr Antun. Institut za fiziku Sveučilišta, Bijenička c. 46, P.O.Box 304, 41001 Zagreb, Yugoslavia. (1925) Dr, physics (Zagreb U., 1963). Prof. (tel. 041 - 271211). *Metals.*

Brenčič, Prof. Dr Jurij. VTOZD Kemija in kem. tehnologija, FNT, Univerza E. Kardelja v Ljubljani, Murnikova 6, P.O.Box 537, 61001 Ljubljana, Yugoslavia. (1940) Dr, chemistry (Ljubljana U., 1969). Prof. (tel. 061 - 214444). *Coordination compounds (chromium - molybdenum - tungsten).*

Bruvo, Magistar Milenko. Zavod za opću i anorgansku kemiju, Prirodoslovno-matematički fakultet, Sveučilište u Zagrebu, ul. Soc. revolucije 8, 41000 Zagreb, Yugoslavia. (1934) Magistar, chemistry (Zagreb U., 1973). Asistent. (tel. 041 - 416023). *Inorganic crystals, coordination compounds.*

Bukovec, Prof. Dr Peter. VTOZD Kemija in kem. tehnologija, FNT, Univerza E. Kardelja v Ljubljani, Murnikova 6, P.O.Box 537, 61001 Ljubljana, Yugoslavia. (1946) Dr, chemistry (Ljubljana U., 1972). Prof. (tel. 061 - 214444). *Inorganic chemistry.*

Bukovec, Dr Nataša. VTOZD Kemija in kem. tehnologija, FNT, Univerza E. Kardelja v Ljubljani, Murnikova 6, P.O.Box 537, 61001 Ljubljana, Yugoslavia. (1946) Dr, chemistry (Ljubljana U., 1978). Docent. (tel. 061 - 214444). *Inorganic chemistry.*

Bulc, Mrs Nada. VTOZD Kemija in kem. tehnologija, FNT, Univerza E. Kardelja v Ljubljani, Murnikova 6, P.O.Box 537, 61001 Ljubljana, Yugoslavia. (1933)

Dipl. inž., chemistry (Ljubljana U., 1962). Sr. res. assoc. (tel. 061 - 214444). *Inorganic crystal structures.*

Colombo, Dr Lidija. Institut 'Ruđer Bošković', Bijenička c. 54, P.O. Box 1016, 41001 Zagreb, Yugoslavia. (1922) Dr, physics (Zagreb U., 1961). Scient. adviser. (tel. 041-435111, ext. 239). *Molecular crystals, organic compounds.*

Čeh, Dr Boris, VTOZD Kemija in kem. tehnologija, FNT, Univerza E. Kardelja v Ljubljani, Murnikova 6, P.O. Box 537, 61001 Ljubljana, Yugoslavia. (1951) Dr, chemistry (Ljubljana U., 1987). Asistent. (tel. 061 - 214444). *Coordination compounds, chemical bonding, inorganic crystal structures.*

Cvetković, Mr Ljubiša. Institut 'Mihajlo Pupin', OOUR 'Kristali', Volgina ul. 15, 11000 Beograd, Yugoslavia. (1931) Dipl. inž., geology (Beograd U., 1961). Adviser in OOUR 'Kristali'. (tel. 011 - 771373). *Piezoelectric crystals.*

Cvetković, Mrs Miroslava. Istraživačko - razvojni institut, Elektronska industrija, Batajnički drum 23, 11080 Zemun Polje, Yugoslavia. (1935) Dipl. inž. min., geology (Beograd U., 1961). Res. assoc. (tel. 011 - 696423). *X-ray diffraction analysis, polycrystalline materials, ceramics.*

Demšar, Magistar Alojz. VTOZD Kemija in kem. tehnologija, FNT, Univerza E. Kardelja v Ljubljani, Murnikova 6, P.O.Box 537, 61001 Ljubljana, Yugoslavia. (1955) Magistar, chemistry (Ljubljana U., 1979). Asistent. (tel. 061 - 214444). *Inorganic crystal structures.*

Despotović, Mr Zlatko. RO Chromos Centar za kemijska istraživanja i razvoj, Žitnjak bb, 41000 Zagreb, Yugoslavia. (1934) Dipl. inž., chemistry (Zagreb U., 1958). Sr. res. assoc. (tel. 041 - 229800, ext. 395). *Thermal properties, materials.*

Dimitrijević, Magistar Radovan. Rudarsko - geološki fakultet, Univerzitet u Beogradu, Đušina ul. 7, 11000 Beograd, Yugoslavia. (1947) Magistar, mineralogy (Beograd U., 1978). Asistent. (tel. 011 - 180111, ext. 701). *Inorganic crystal structures, computer programming.*

Divjaković, Prof. Dr Vladimir. Institut za fiziku Prirodno-matematičkog fakulteta, ul. Ilije Đuričića 4, 21000 Novi Sad, Yugoslavia. (1946) Dr, crystallography (Bern U., Switzerland, 1976). Prof. (tel. 021 - 55318). *Coordination compounds, minerals, polymers, X-ray diffraction, electron microscopy, small-angle scattering.*

Djinović, Miss Kristina. VTOZD Kemija in kem. tehnologija, FNT, Univerza E. Kardelja v Ljubljani, Murnikova 6, P.O.Box 537, 61001 Ljubljana, Yugoslavia. (1963) Dipl. inž, chemistry (Ljubljana U. 1986). Res. assistant. (tel. 061 - 214 444, ext. 44). *Hydrogen bonding, X-ray diffraction.*

Djordjević, Prof. Dr Slobodan. Metalurški fakultet, ul. Matije Gupca 1, 72000 Zenica, Yugoslavia. (1929) Dr, metallurgy (Sarajevo U., 1976). Izv. prof. (tel. 072 - 21831). *Metals, structure.*

Djurić, Dr Stevan. Rudarsko-geološki fakultet, Univerzitet u Beogradu, Đušina ul. 7, 11000 Beograd, Yugoslavia. (1931) Dr, geology (Beograd U., 1980). Docent. (tel. 011 - 180111, ext. 713) *Instrumentation, phase determination of clay minerals, materials science.*

Drašner, Magistar Antun. Institut 'Ruđer Bošković', Bijenička c.54, P.O. Box 1016, 41001 Zagreb, Yugoslavia. (1955) Magistar, chemistry (Zagreb U., 1984). Res. assoc. (tel. 041 - 435111). *Intermetallic compounds.*

Dubček, Mr Pavo. Fizički zavod, Prirodoslovno-matematički fakultet, Sveučilište u Zagrebu, Marulićev trg 19, 41000 Zagreb, Yugoslavia. (1960) Dipl. inž. fiz., physics (Zagreb U., 1983). Jr. res. assoc. (tel. 041 - 446242). *Metal physics, small-angle X-ray scattering.*

Dužević, Dr Davor. Zavod za tehnologiju, Sektor za razvoj proizvoda, 'Rade Končar' - Elektrotehnički institut, Baštijanova b.b., 41000 Zagreb, Yugoslavia. (1936) Dr, physics (Zagreb U., 1979). Senior scient. assoc. (tel. 041 - 312222, ext. 2015). *Materials science, metals.*

Fazlić, Magistar Refik. Elektrotehnički fakultet, Ozrenskog odreda 2, 75000 Tuzla, Yugoslavia. (1942) Magistar, physics (Beograd U., 1974). Docent. (tel. 075 - 35935). *Computing, coordination compounds, X-ray diffraction.*

Gabela, Prof. Dr Fikret. Medicinski fakultet, Univerzitet Sarajevo, ul. Moše Pijade 5, 71000 Sarajevo, Yugoslavia. (1936) Dr, physics (Sarajevo U., 1977). Izv. prof. (tel. 071 - 38244, ext. 53). *Inorganic crystal structures.*

Galešić, Dr Nikola. Institut 'Ruđer Bošković', Bijenička c. 54, P.O.Box 1016, 41001 Zagreb, Yugoslavia. (1937) Dr, chemistry (Zagreb U., 1971). Scient. assoc. (tel. 041 - 435111, ext. 335). *Coordination compounds, organic compounds.*

Girt, Prof. Dr Egvin. Prirodno - matematički fakultet, Vojvode Putnika 43, 71000 Sarajevo, Yugoslavia. (1936) Dr, physics (Zagreb U., 1977). Prof. (tel. 071 - 646755). *Metal physics.*

Gladić, Mr Jadranko. Institut za fiziku Sveučilišta, Bijenička c. 46, P.O.Box 304, 41000 Zagreb, Yugoslavia. (1959) Dipl. inž. fiz., physics (Zagreb U., 1983). Jr. res. assoc. (tel. 041 - 271211, ext. 307). *Superionic conductors, structure.*

Golič, Prof. Dr Ljubo. VTOZD Kemija in kem. tehnologija, FNT, Univerza E. Kardelja v Ljubljani, Murnikova 6, P.O.Box 537, 61001 Ljubljana, Yugoslavia. (1932) Dr, chemistry (Ljubljana U., 1965). Prof. (tel. 061 - 214444). *Inorganic crystal structures, hydrogen bond structures, oxalates and thiooxalates.*

Grdenić, Prof. Dr Drago. Zavod za opću i anorgansku kemiju, Prirodoslovno-matematički fakultet, Sveučilište u Zagrebu, ul. Soc. revolucije 8, 41000 Zagreb, Yugoslavia. (1919) Dr, chemistry (Zagreb U., 1951). Prof.retired. (tel. 041 - 442823). *Structural chemistry.*

Gržeta, Dr Biserka. Institut 'Ruđer Bošković', Bijenička c. 54, P.O. Box 1016, 41001 Zagreb, Yugoslavia. (1949) Dr, physics (Zagreb U., 1980). Scient. assoc. (tel. 041 - 435111, ext. 320). *X-ray diffraction, phase determination, phase transitions, materials science.*

Gspan, Dr Primož. Zavod SRS za varstvo pri delu, Bohoričeva 22 A, 61000 Ljubljana, Yugoslavia. (1934) Dr, physics (Zagreb U., 1980). Head, Dept. of Ecology and Toxicology. (tel. 061 - 320853). *Occupational safety, environment protection.*

Hečimović, Mrs Ljerka. RO 'Jugokeramika', OOUR Razvoj, Srebrnjak 169, 41000 Zagreb, Yugoslavia. (1949) Dipl. inž. min., mineralogy (Zagreb U., 1974). Res. assistant. (tel. 041 - 220344). *Minerals - clay and raw materials, microscopy - polarised-light microscopy (ceramics and porcelain).*

Herak, Dr Rajna. Institut za nuklearne nauke 'Boris Kidrič' - Vinča, Laboratorija za fiziku čvrstog stanja i radijacionu hemiju, P.O.Box 522, 11000 Beograd, Yugoslavia. (1930) Dr, technical sciences (Beograd U., 1969). Scient. adviser. (tel. 011 - 458222, ext. 417). *Crystal structures, inorganic and coordination compounds.*

Herceg, Dr Marija. Institut 'Ruđer Bošković', Bijenička c. 54, P.O.Box 1016, 41001 Zagreb, Yugoslavia. (1938) Dr, chemistry (Zagreb U., 1970). Scient. assoc. (tel. 041 - 435111, ext. 335). *Crystal structures, biologically interesting synthetic compounds.*

Hergold-Brundić, Dr Antonija. Zavod za opću i anorgansku kemiju, Prirodoslovno-matematički fakultet, Sveučilište u Zagrebu, ul. Soc. revolucije 8, 41000 Zagreb, Yugoslavia. (1942) Dr, chemistry (Zagreb U., 1980). Asistent. (tel. 041 - 416023, ext. 8005). *Structural chemistry.*

Horvatić, Magistar Davor. Institut 'Ruđer Bošković', Bijenička 54, P.O.Box 1016, 41001 Zagreb, Yugoslavia. (1957) Magistar, mathematics (Zagreb U., 1984). Asistent. (tel. 041 - 435111, ext. 264) *Computing.*

Hurem, Magistar Zihnija. Institut za opštu i fizičku hemiju, Studentski trg 12/V, PP 551, 11000 Beograd, Yugoslavia. (1960) Magistar, physical chemistry (Beograd U., 1988). Res. assistant. (tel. 011 - 628074). *Powder diffraction (zeolites).*

Janjić, Prof. Dr Svetislav. Rudarsko-topioničarski basen Bor, Institut za bakar-Bor, ul. AVNOJ-a 35, 19210 Bor, Yugoslavia. (1931) Dr, geology (Beograd U., 1979). Prof. (tel. 030 - 32299, ext. 792). *Minerals, powder diffraction, X-ray diffraction.*

Jelenić-Bezjak, Dr Ivanka. Savska c. 104, 41000 Zagreb, Yugoslavia. (1937) Dr, chemistry (Zagreb U., 1970). Retired. (tel. 041 - 517094). *Silicate structures.*

Jenček, Mr Ladislav August. Institut za materiale, Zavod za raziskavo materiala in konstrukcij, Dimičeva 12, 61000 Ljubljana, Yugoslavia. (1918) Dipl. inž., electronics (Techn. U. of Praha, Czechoslovakia, 1943). Res. adviser. (tel. 061 - 344061). *Natural and synthetic fibres.*

Jordanovska, Prof. Dr (Mrs) Vera. Institut za hemija, Prirodno-matematički fakultet, Univerzitet 'Kiril i Metodij', Arhimedova 5, 91000 Skopje, Yugoslavia. (1938) Dr, chemistry (Ljubljana U., 1982). Prof. (tel. 091 - 221033). *Coordination compounds.*

Jovanovski, Prof. Dr Gligor. Institut za hemija, Prirodno-matematički fakultet, Univerzitet 'Kiril i Metodij', Arhimedova 5, 91000 Skopje, Yugoslavia. (1945) Dr, chemistry (Zagreb U., 1981). Head, Inst. of Chem. (tel. 091 - 221033). *Structural chemistry.*

Jurković, Prof. Dr Ivan. Rudarsko-geološko-naftni fakultet, Sveučilište u Zagrebu, Pierottijeva ul 6, 41000 Zagreb, Yugoslavia. (1917) Dr, geology (Zagreb U., 1956). Prof. emeritus, Academician. (tel. 041 - 445658). *Crystal optics (ore microscopy), mineral genesis.*

Kaitner, Dr Branko. Zavod za opću i anorgansku kemiju, Prirodoslovno- matematički fakultet, Sveučilište u Zagrebu, ul. Soc. revolucije 8, 41000 Zagreb, Yugoslavia. (1942) Dr, chemistry (Zagreb U., 1979). Asistent. (tel. 041 - 416023, ext. 8005). *Structural chemistry, coordination compounds, organic compounds, inorganic crystals.*

Kamenar, Prof. Dr Boris. Zavod za opću i anorgansku kemiju, Prirodoslovno-matematički fakultet, Sveučilište u Zagrebu, ul. Soc. revolucije 8, 41000 Zagreb, Yugoslavia. (1929) Dr, chemistry (Zagreb U., 1960). Prof. (tel. 041 - 442823). *Inorganic and organic crystal structures.*

Kapor-Nahlovski, Dr Agneš. Institut za fiziku Prirodno-matematičkog fakulteta, ul. Ilije Đuričića 4, 21000 Novi Sad, Yugoslavia. (1950) Dr, physics (Novi Sad U., 1981). Prof. (tel. 021 - 55318). *Organic compounds, molecular crystals, structural chemistry.*

Karanović, Dr Ljiljana. Lab. za kristalografiju, Rudarsko-geološki fakultet, Univerzitet u Beogradu, Đušina ul. 7, 11000 Beograd, Yugoslavia. (1950) Dr, mineralogy (Beograd U., 1985). Res. assoc. (tel. 011 - 180111, ext. 701). *Organic and inorganic crystal structures, powder diffraction.*

Kaučič, Dr Venčeslav. Raziskovalna skupnost SR Slovenije, Jadranska 21, 61000 Ljubljana, Yugoslavia. (1950) Dr, chemistry (Ljubljana U., 1977). Assoc. Prof., Secretary general. (tel. 061 - 211931). *Inorganic crystals.*

Kerenović, Magistar Midhat. Pedagoška akademija, ul. Mirka Višnjića 1, 78000 Banja Luka, Yugoslavia. (1930) Magistar, physics (Zagreb U., 1973). Prof. of physics. (tel. 078 - 35625). *Metal physics.*

Kirin, Dr (Mrs) Ankica. Medicinski fakultet, Sveučilište u Zagrebu, Šalata 3b, 41000 Zagreb, Yugoslavia. (1929) Dr, physics (Zagreb U., 1973). Docent. (tel. 041 - 271188, ext. 379). *Solid state physics.*

Kojić-Prodić, Dr Biserka. Institut 'Ruđer Bošković', Bijenička c. 54, P.O.Box 1016, 41001 Zagreb, Yugoslavia. (1938) Dr, chemistry (Zagreb U., 1968). Scient. adviser. (tel. 041 - 435111, ext. 529). *Structural chemistry, inorganic crystals, organic compounds.*

Kosovinc, Prof. Dr Ivan. Katedra za metalografijo Fakultete za naravoslovje in tehnologijo, Univerza E. Kardelja v Ljubljani, Aškerčeva 20, 61000 Ljubljana, Yugoslavia. (1922) Dr, technology (Ljubljana U., 1968). Prof. (tel. 061 - 212121). *Physical metallurgy, metallography.*

Kranjc, Prof. Dr Katarina. Institut za fiziku Sveučilišta, Bijenička c. 46, P.O.Box 304, 41001 Zagreb, Yugoslavia. (1915) Dr, physics (Zagreb U., 1954). Prof. (tel. 041 - 271211). *Small angle X-ray scattering, crystal imperfections.*

Kraševec, Dr Viktor. Institut 'Jožef Stefan', Univerza E. Kardelja v Ljubljani, Jamova 39, 61000 Ljubljana, Yugoslavia. (1932) Dr, physics (Ljubljana U., 1974). Scient. assoc. (tel. 061 - 214399). *Microstructure, ceramic materials.*

Krstanović, Prof. Dr Ilija. Lab. za kristalografiju, Rudarsko-geološki fakultet, Univerzitet u Beogradu, Đušina ul. 7, 11000 Beograd, Yugoslavia. (1927) Dr, mineralogy (Beograd U., 1961). Prof. of crystallography. (tel. 011 - 180111, ext. 701). *Inorganic crystals, materials science.*

Kunstelj, Dr Dragan. Institut za fiziku Sveučilišta, Bijenička c. 46, P.O.Box 304, 41001 Zagreb, Yugoslavia. (1941) Dr, physics (Zagreb U., 1979). Docent. (tel. 041 - 271211). *Solid state physics, metal physics, electron microscopy.*

Lahodny-Šarc, Prof. Dr (Mrs) Olga. Zavod za kemiju RGN fakulteta, Sveučilište u Zagrebu, Pierottijeva 6, 41000 Zagreb, Yugoslavia. (1928) Dr, physical chemistry (Zagreb U., 1962). Prof. (tel. 041 - 442409). *Silicate chemistry, corrosion science.*

Lazar, Mr Dušan. Institut za fiziku Prirodno - matematičkog fakulteta, ul. Ilije Đuričića 4, 21000 Novi Sad, Yugoslavia. (1951) Dipl. fiz., physics (Novi Sad U., 1974). Res. asst. (tel. 021 - 55318). *Organic crystal structures.*

Lazarini, Prof. Dr Franc. VTOZD Kemija in kem. tehnologija, FNT, Univerza E. Kardelja v Ljubljani, Murnikova 6, P.O.Box 537, 61001 Ljubljana, Yugoslavia. (1940) Dr, chemistry (Ljubljana U., 1971). Prof. (tel. 061 - 214444). *Inorganic crystals, coordination compounds, structural chemistry, X-ray diffraction.*

Leban, Prof. Dr Ivan. VTOZD Kemija in kem. tehnologija, FNT, Univerza E. Kardelja v Ljubljani, Murnikova 6, P.O.Box 537, 61001 Ljubljana, Yugoslavia. (1947) PhD, physics (York U., UK, 1973). Izv. prof. (tel. 061 - 214444). *Crystal structure determination.*

Leovac, Prof. Dr Vukadin. Institut za hemiju, Prirodno-matematički fakultet, Ilije Đuričića 4, 21000 Novi Sad, Yugoslavia. (1943) Dr, chemistry (Novi Sad U., 1978). Izv. prof. (tel. 021 - 54065). *Coordination compounds, synthesis, structure.*

Luić, Magistar Marija. Institut 'Ruđer Bošković', Bijenička c. 54, P.O.Box 1016, 41001 Zagreb, Yugoslavia. (1953) Magistar, geology (Zagreb U., 1981). Res. assoc. (tel. 041 - 435111, ext. 319). *Structure analysis, direct methods.*

Lugomer, Dr Stjepan. Elektrotehnički fakultet, Meštrovićeva ul. 4, 78000 Banja Luka, Yugoslavia. (1944) Dr, physics (Zagreb U., 1974). Docent. (tel. 078 - 24097). *Metal thin films.*

Maksić, Prof. Dr Zvonimir. Institut 'Ruđer Bošković', Bijenička c. 54, P.O.Box 1016, 41001 Zagreb, Yugoslavia. (1938) Dr, chemistry (Zagreb U., 1967). Prof. of theoretical chemistry. (tel. 041 - 435111, ext. 502). *Molecular structure.*

Marinković, Prof. Dr Velibor. Institut 'Jožef Stefan', Univerza E. Kardelja v Ljubljani, Jamova 39, 61000 Ljubljana, Yugoslavia. (1929) Dr, chemistry (Ljubljana U., 1965). Scient. adviser. (tel. 061 - 214399, telex 31296 JOSTIN YU). *Materials science.*

Marinković, Magistar (Mrs) Živka. Institut za fiziku, Maksima Gorkog 118, 11080 Zemun-Beograd, Yugoslavia. (1932) Magistar, physical chemistry (Beograd U., 1967). Scient. assoc. (tel. 011 - 212219). *Thin films, intermetallic compound formation.*

Marković, Magistar Berislav. OOUR TENEZ, Institut "Ruđer Bošković" Bijenička 54, P.O.Box 1016, 41001 Zagreb, Yugoslavia. (1957) Magistar, chemistry (Zagreb U., 1985). Res. assistant. (tel. 041 - 435111, ext. 314). *Crystal structures, powder diffraction, phase determination.*

Marković, Magistar Desimir. Tehnički fakultet-Bor, JNA 12, 19210 Bor, Yugoslavia. (1950) Magistar, metallurgy (Beograd U., 1982). Asistent. (tel. 030 - 24555). *Metal physics, metals structure.*

Marković, Dr Vesna. Institut za opštu i fizičku hemiju, Studentski trg 12/V, PP 551, 11000 Beograd, Yugoslavia. (1946) Dr, physical chemistry (Beograd U., 1985). Res. assoc., Head, X-ray diffraction lab. (tel. 011 - 185243, telex 11176 YU IOFH). *Materials science, powder diffraction (X-ray).*

Mašić, Magistar Nikola. Institut 'Ruđer Bošković', Bijenička c. 54, P.O.Box 1016, 41001 Zagreb, Yugoslavia. (1948) Magistar, physics (Zagreb U., 1975). Res. assoc. (tel. 041 - 435111). *Polymers, structure and properties.*

Matijašić, Dr (Mrs) Ivanka. Zavod za organsku kemiju i biokemiju, Prirodoslovno-matematički fakultet, Sveučilište u Zagrebu, Strossmayerov trg 14, 41000 Zagreb, Yugoslavia. (1944) Dr, chemistry (Zagreb U., 1984). Scient. asst. (tel. 041 - 432580). *Organic compounds.*

Matković, Dr Boris. Institut 'Ruđer Bošković', Bijenička c. 54, P.O.Box 1016, 41001 Zagreb, Yugoslavia. (1927) Dr, chemistry (Zagreb U., 1961). Scient. adviser. (tel. 041 - 435111, ext. 335, telex 21383 YU). *Materials science, structural chemistry.*

Matković, Dr Prosper. Metalurški fakultet, Sisak, Institut za metalurgiju, Aleja narodnih heroja 1, 44000 Sisak, Yugoslavia. (1945) Dr, chemistry (Stuttgart U., BRD, 1977). Docent. (tel. 044 - 32044). *Inorganic crystal structures, physical metallurgy.*

Matković, Dr Tanja. Metalurški fakultet, Sisak, Institut za metalurgiju, Aleja narodnih heroja 1, 44000 Sisak, Yugoslavia. (1948) Dr, chemistry (Stuttgart U., BRD, 1977). Docent. (tel. 044 - 32044). *Structure, metals and alloys, inorganic compounds, physical metallurgy.*

Matković-Čalogović, Magistar (Mrs) Dubravka. Zavod za opću i anorgansku kemiju, Prirodoslovno-matematički fakultet, Sveučilište u Zagrebu, ul. Soc. revolucije 8, 41000 Zagreb, Yugoslavia. (1957) Magistar, chemistry (Zagreb U., 1985). Asistent. (tel. 041 - 416023, ext. 83). *Coordination compounds, inorganic crystals, organic compounds.*

Međimorec, Magistar Stanislav. Mineraloško - petrografski zavod, Prirodoslovno-matematički fakultet, Sveučilište u Zagrebu, Demetrova ul. 1, 41000 Zagreb, Yugoslavia. (1939) Magistar, mineralogy (Zagreb U., 1977). Asistent. (tel. 041 - 445315). *Physicochemical methods, mineral analysis.*

Mészáros, Mr Csaba, Prirodno-matematički fakultet, Institut za fiziku, ul. Ilije Đuričića 4, 21000 Novi Sad, Yugoslavia. (1962) Dipl. fiz., physics (Novi Sad U., 1987). Asistent. (tel 021 - 55318). *Inorganic crystals.*

Milat, Magistar Ognjen. Institut za fiziku Sveučilišta, Bijenička c. 46, P.O.Box 304, 41001 Zagreb, Yugoslavia. (1949) Magistar, physics (Zagreb U., 1978). Asistent. (tel. 041 - 271211, ext. 334). *Metal physics, electron microscopy.*

Milićev, Prof. Dr Svetozar. VTOZD Kemija in kem. tehnologija, FNT, Univerza E. Kardelja v Ljubljani, Murnikova 6, P.O.Box 537, 61001 Ljubljana, Yugoslavia. (1934) Dr, chemistry (Ljubljana U., 1972). Izv. prof. (tel. 061 - 214444). *Vibration spectroscopy.*

Milinski, Prof. Dr Nikola. Institut za fiziku Prirodno - matematičkog fakulteta, ul. Ilije Đuričića 4, 21000 Novi Sad, Yugoslavia. (1936) Dr, physics (Novi Sad U., 1976). Izv. prof. (tel. 021 - 55622). *Solid state physics, crystal physics.*

Mirčeva, Magistar Aneta. Hemiski institut, Prirodno-matematički fakultet, Univerzitet 'Kiril i Metodij', Arhimedova 5, 91000 Skopje, Yugoslavia. (1946) Magistar, chemistry (Skopje U., 1979). Asistent. (tel. 091 - 221033). *Crystal structures.*

Moguš-Milanković, Magistar Andrea. Institut 'Ruđer Bošković', Bijenička c. 54, P.O. Box 1016, 41001 Zagreb, Yugoslavia. (1953) Magistar, chemistry (Zagreb U., 1982). Res. assoc. (tel. 041-435111). *Electrical properties.*

Morvaj, Magistar (Mrs) Jasmina. Zavod za opću i anorgansku kemiju, Prirodoslovno-matematički fakultet, Sveučilište u Zagrebu, ul. Soc. revolucije 8, 41000 Zagreb, Yugoslavia. (1957) Magistar, chemistry (Zagreb U., 1985). Res. assoc. (tel. 041 - 416023, ext. 9). *Inorganic crystals, powder diffraction.*

Mrvoš-Sermek, Magistar (Mrs) Draginja. Zavod za opć i anorgansku kemiju, Prirodoslovno-matematički fakultet, Sveučilište u Zagrebu, ul. Soc. revolucije 8, 41000 Zagreb, Yugoslavia. (1961) Magistar, chemistry (Zagreb U.,1987). Res. assistant. (tel. 041 - 416023, ext. 8005). *Organic compounds, coordination compounds, structural chemistry.*

Nagl, Dr Antun. Zavod za opću i anorgansku kemiju, Prirodoslovno-matematički fakultet, Sveučilište u Zagrebu, ul. Soc. revolucije 8, 41000 Zagreb, Yugoslavia. (1942) Dr, chemistry (Bern U., Switzerland, 1979). Asistent. (tel. 041 - 416023, ext. 8005). *Structural chemistry.*

Nančovski, Mr Kosta. Građežen institut 'Makedonija'-Skopje, ul. Drezdenska 52, 91000 Skopje, Yugoslavia. (1934) Dipl. hem., chemistry (Beograd U., 1960). Asst. director. (tel. 091 - 253929). *Instrumentation.*

Nigović, Miss Biljana. Institut 'Ruđer Boškovi&ca', Bijenička c. 54, P.O.Box 1016, 41001 Zagreb, Yugoslavia. (1963) Dipl. inž, pharmacy (Zagreb U., 1987). Res. assistant. (tel. 041 - 435111, ext. 264). *Synthesis, crystal growth, organic crystal structures.*

Nikolić, Prof. Dr Pantelija. Elektrotehnički fakultet, Univerzitet u Beogradu, Bulevar Revolucije 73, 11000 Beograd, Yugoslavia. (1928) Dr, physics (Nottingham U., UK, 1969). Prof. (tel. 011 - 329212, ext. 384). *Semiconductor crystal physics, optical properties.*

Novosel Radović, Dr Vjera. MK 'Željezara Sisak', RO Institut za metalurgiju, OOUR Metalurški fakultet, Aleja narodnih heroja 1, 44000 Sisak, Yugoslavia. (1937) Dr, chemistry (Zagreb U., 1983). Chief organizer in Lab. for investigation of structure and properties of materials. (tel. 044 - 30444, ext. 395). *Powder diffraction.*

Očko, Magistar Miroslav. TVA KOV, Ilica 256, 41000 Zagreb, Yugoslavia. (1947) Magistar, physics (Zagreb U., 1976). Lect. (tel. 041 - 579666). *Amorphous materials.*

Osterc, Prof. Dr (Mrs) Valerija. Odsek za geologijo, Oddelek za montanistiko, Fakulteta za naravoslovje in tehnologijo, Univerza E. Kardelja v Ljubljani, Aškerčeva 20, 61000 Ljubljana, Yugoslavia. (1924) Dr, ceramics (Rheinisch - Westfälische Technische Hochschule Aachen, BRD, 1967). Prof. (tel. 061 - 212121). *Clay minerals, X-ray diffraction.*

Paljević, Dr Matija. Institut 'Ruđer Bošković', Bijenička c. 54, P.O.Box 1016, 41001 Zagreb, Yugoslavia. (1943) Dr, chemistry (Zagreb U., 1978). Res. assoc. (tel. 041 - 435111, ext. 274). *Gas-solid reactions.*

Pećina, Miss Planinka. Fizički zavod, Prirodoslovno-matematički fakultet, Sveučilište u Zagrebu, Marulićev trg 19, 41000 Zagreb, Yugoslavia. (1957) Dipl. inž. fiz., physics (Zagreb U., 1983). Jr. res. assoc. (tel. 041 - 446242). *Metal physics, electron microscopy.*

Penavić, Dr Maja. Zavod za opću i anorgansku kemiju, Prirodoslovno-matematički fakultet, Sveučilište u Zagrebu, ul. Soc. revolucije 8, 41000 Zagreb, Yugoslavia. (1941) Dr, chemistry (Zagreb U., 1977). Asistent. (tel. 041 - 416023, ext. 8005). *Structural chemistry.*

Petrović, Dr Dragoslav. Institut za fiziku Prirodno - matematičkog fakulteta, ul. Ilije Đuričića 4, 21000 Novi Sad, Yugoslavia. (1949) Dr, physics (Novi Sad U., 1980). Docent. (tel. 021 - 55318). *Solid state physics.*

Pocev, Prof. Dr Stefan. Tehnološki fakultet, Univerzitet 'Kiril i Metodij', Ruđer Bošković 16, 91000 Skopje, Yugoslavia. (1940) Dr, chemistry (Zagreb U., 1978). Izv. prof. (tel. 091 - 259725). *Powder diffraction.*

Poharc-Logar, Magistar Vesna. Rudarsko-geološki fakultet, Univerzitet u Beogradu, Đušina 7, 11000 Beograd, Yugoslavia. (1949) Magistar, mineralogy (Beograd U., 1979). Asistent. (tel. 011 - 183814). *Clay minerals.*

Polanc, Magistar Ivan. Visoka tehniška šola Maribor, Univerza v Mariboru, Smetanova 17, 62000 Maribor, Yugoslavia. (1949) Magistar, chemistry (Ljubljana U., 1980). Asistent. (tel. 062 - 25461). *Inorganic chemistry.*

Poleti, Dr Dejan. Tehnološko-metalurški fakultet, Karnegijeva 4, 11000 Beograd, Yugoslavia. (1952) Dr, chemistry (Beograd U., 1988). Docent. (tel. 011 - 328721, ext. 624). *Coordination compounds, transition metals.*

Popović, Prof. Dr Stanko. Institut 'Ruđer Bošković', Bijenička c. 54, P.O.Box 1016, 41001 Zagreb, Yugoslavia. (1938) Dr, physics (Zagreb U., 1968).Prof., Scient. adviser. (tel. 041 - 435111, ext. 320). *Powder diffraction, diffractometry, phase determination, phase transition.*

Prelesnik, Dr Bogdan. Institut za nuklearne nauke 'Boris Kidrič' - Vinča, Laboratorija za fiziku čvrstog stanja i radijacionu hemiju, P.O.Box 522, 11000 Beograd, Yugoslavia. (1938) Dr, crystallography (Bern U., Switzerland, 1975). Scient. assoc. (tel. 011 - 4440871, ext. 417). *Inorganic and organic crystal structures, neutron diffraction, ice nucleation.*

Prodan, Dr Albert. Institut 'Jožef Stefan', Univerza E. Kardelja v Ljubljani, Jamova 39, 61000 Ljubljana, Yugoslavia. (1944) Dr, physics (Zagreb U., 1974). Scient. assoc. (tel. 061 - 214399, ext. 444, telex 31296 JOSTIN YU). *Electron diffraction, electron microscopy, defect structures.*

Radaković, Magistar Aleksandra. Rudarsko-geološki fakultet, Univerzitet u Beogradu, Đušina ul. 7, 11000 Beograd, Yugoslavia. (1956) Magistar, geology (Beograd U., 1988). Res. assoc. (tel. 011 - 180111, ext. 701). *Inorganic crystals, materials science.*

Radivojević, Magistar Pavle. Institut za fiziku Univerziteta u Novom Sadu, Dr. Ilije Đuričića 4, 21000 Novi Sad, Yugoslavia. (1935) Magistar, chemistry (Novi Sad U., 1976). Asistent. (tel. 021 - 55318). *Inorganic crystals.*

Radmilović, Magistar Velimir. Tehnološko-metalurški fakultet, Katedra za fizičku metalurgiju, Karnegijeva ul. 4, 11000 Beograd, Yugoslavia. (1948) Magistar, metallurgy (Beograd U., 1980). Asistent. (tel. 011 - 328671). *Metal structures, electron diffraction, metal physics.*

Radukić, Prof. Dr Gordana. Rudarsko-geološki fakultet, Univerzitet u Beogradu, Đušina ul. 7, 11000 Beograd, Yugoslavia. (1931) Dr, geology (Beograd U., 1977). Prof. (tel. 011 - 180111, ext. 771). *Microscopy, materials science, minerals, phyllosilicates.*

Rebić, Mr Milenko. Institut za materiale, Zavod za raziskavo materiala in konstrukcij, Dimičeva 12, 61000 Ljubljana, Yugoslavia. (1938) Dipl. inž. fiz., physics (Ljubljana U., 1969). Res. fellow. (tel. 061 - 344061). *Metal physics.*

Ribár, Prof. Dr Béla. Institut za fiziku Prirodno - matematičkog fakulteta, ul. Ilije Đuričića 4, 21000 Novi Sad, Yugoslavia. (1930) Dr, crystallography (Bern U., Switzerland, 1969). Prof. (tel. 021 - 55318). *Inorganic crystals, organic compounds.*

Rogić, Prof. Dr Vinko. Građevinski fakultet, Univerzitet "Džemal Bijedić" Mostar, ul. A. Zuanića 14, 88000 Mostar, Yugoslavia. (1945) Dr, chemistry (Zagreb U., 1975). Prof. (tel. 088 - 416737). *Materials science, structural chemistry.*

Rogulić, Prof. Dr Mileva. Tehnološko-metalurški fakultet, Univerzitet u Beogradu, Karnegijeva ul. 4, 11000 Beograd, Yugoslavia. (1928) Dr, metallurgy (Cambridge U., UK, 1964). Izv. prof. (tel. 011 - 328721). *Physical metallurgy, metal and alloy structures.*

Runje, Dr Vesna. Farmaceutsko-biokemijski fakultet, Sveučilište u Zagrebu, ul. Ante Kovačića 1, 41000 Zagreb, Yugoslavia. (1947) Dr, chemistry (Zagreb U., 1981). Asistent. (tel. 041 - 445311, ext. 36). *X-ray diffraction analysis, polycrystalline systems.*

Ružić-Toroš, Dr Živa. Institut 'Ruđer Bošković', Bijenička c. 54, P.O.Box 1016, 41001 Zagreb, Yugoslavia. (1944) Dr, chemistry (Zagreb U., 1974). Sr. scient. assoc. (tel. 041 - 435111, ext. 529). *Structural chemistry, inorganic crystals, organic compounds.*

Sijarić, Dr (Mrs) Galiba. Odsjek za geografiju, katedra za geologiju, Prirodno-matematički fakultet, Vojvode Putnika 43, 71000 Sarajevo, Yugoslavia. (1939) Dr, geology (Sarajevo U., 1975). Docent. (tel. 071 - 659377). *Minerals, powder diffraction.*

Sikirica, Prof. Dr Milan. Zavod za opću i anorgansku kemiju, Prirodoslovno-matematički fakultet, Sveučilište u Zagrebu, ul. Soc. revolucije 8, 41000 Zagreb, Yugoslavia. (1934) Dr, chemistry (Zagreb U., 1963). Prof. (tel. 041 - 442823). *Inorganic crystal structures.*

Simić, Dr Vojislav. Institut za fiziku, Maksima Gorkog 118, 11080 Zemun-Beograd, Yugoslavia. (1926) Dr, chemistry (Zagreb U., 1960). Scient. adviser. (tel. 011 - 212219). *Thin solid films, reactions at room temperature.*

Slovenec, Dr Dragutin. Rudarsko - geološko - naftni fakultet, Sveučilište u Zagrebu, Pierottijeva ul. 6, 41000 Zagreb, Yugoslavia. (1941) Dr, geology (Zagreb U., 1980). Docent. (tel. 041 - 440422, ext. 425). *Mineralogy.*

Srdanov, Magistar Gordana. Institut za nuklearne nauke 'Boris Kidrič' - Vinča, Laboratorija za fiziku čvrstog stanja i radijacionu hemiju, P.O.Box 522, 11000 Beograd, Yugoslavia. (1950) Magistar, chemistry (Zagreb U., 1979). Res. assoc. (tel. 011 - 458222, ext. 417). *Crystal structures, inorganic and organometallic compounds.*

Stanković, Dr (Mrs) Slobodanka. Institut za fiziku Prirodno- matematičkog fakulteta, ul. Ilije Đuričića 4, 21000 Novi Sad, Yugoslavia. (1941) Dr, physics (Novi Sad U., 1980). Prof. (tel. 021 - 55318). *Organic crystal structures.*

Stefanović, Magistar Aleksandar. Zavod za opću i anorgansku kemiju, Prirodoslovno-matematički fakultet, Sveučilište u Zagrebu, ul. Soc. revolucije 8, 41000 Zagreb, Yugoslavia. (1957) Magistar, chemistry (Zagreb U., 1985). Res. assoc. (tel. 041 - 416023, ext. 8005). *Coordination compounds, structural chemistry.*

Stipančić, Prof. Dr Mladen. Elektrotehnički fakultet, Dr. Vase Butozana 3, 78000 Banja Luka, Yugoslavia. (1935) Dr, physics (Zagreb U., 1983). Prof., Electr. Engeneering Dept. Vice chairman. (tel. 078 - 24408) *Materials science.*

Stojadinović, Prof. Dr Slobodan. CIRM "Energoinvest", STUP, Tvornička 3, 71000 Sarajevo, Yugoslavia. (1947) Dr, metallurgy (Beograd U., 1980). Sr. res. (tel. 071 - 628929, ext. 980, telex 41204). *Metals, metal physics, metals structure.*

Stojanović, Mr Dobrica. RO 'MAGNOHROM'- Institut za vatrostalne materijale, P.O.Box 17, 36001 Kraljevo, Yugoslavia. (1931) Dipl. inž. geol., geology (Beograd U., 1957). Principal res. officer. (tel. 036 - 331322, ext. 513, telex 17630). *Materials science (raw and refractory materials), microscopy, phase determination, X-ray diffraction.*

Strukan, Dipl. inž. Neven. Zavod za opću i anorgansku kemiju, Prirodoslovno-matematički fakultet, Sveučilište u Zagrebu, ul. Soc. revolucije 8, 41000 Zagreb, Yugoslavia. (1961) Dipl. inž., chemistry (Zagreb U., 1986). Res. assistant. (tel. 041 - 416023, ext. 8005) *Coordination compounds, structural chemistry.*

Stubičar, Dr Mirko. Fizički zavod, Prirodoslovno-matematički fakultet, Sveučilište u Zagrebu, Marulićev trg 19, P.O.Box 162, 41001 Zagreb, Yugoslavia. (1940) Dr, physics (Zagreb U., 1986). Docent. (tel. 041 - 446242). *Materials science, X-ray diffraction, small-angle scattering.*

Šćavničar, Prof. Dr Stjepan. Mineraloško-petrografski zavod, Prirodoslovno-matematički fakultet, Sveučilište u Zagrebu, Demetrova ul. 1, 41000 Zagreb, Yugoslavia. (1923) Dr, chemistry (Zagreb U., 1956). Prof., head, dept. of mineralogy and petrography, Faculty of Science, Zagreb. (tel. 041 - 428610). *X-ray diffraction, inorganic crystals, minerals.*

Šegedin, Dr Primož. VTOZD Kemija in kem. tehnologija, FNT, Univerza E. Kardelja v Ljubljani, Murnikova 6, P.O.Box 537, 61001 Ljubljana, Yugoslavia. (1948) Dr, chemistry (Ljubljana U., 1987). Asistent. (tel. 061 - 214444). *Inorganic crystal structures, coordination compounds, chemical bonding.*

Šljukić, Prof. Dr Momčilo. Tehnički fakultet, Univerzitet 'Veljko Vlahović' Titograd, Kruševac b.b., 81000 Titograd, Yugoslavia. (1936) Dr, chemistry (Zagreb U., 1968). Prof. (tel. 081 - 52111). *Inorganic crystal structures.*

Šmit, Dr Ivan. Institut 'Ruđer Bošković', Bijenička c. 54, P.O.Box 1016, 41001 Zagreb, Yugoslavia. (1948) Dr, chemistry (Zagreb U., 1979). Res. assoc. (tel. 041 - 435111). *Polymers, structure and properties.*

Šoptrajanov, Prof. Dr Bojan. Hemiski institut, Prirodno-matematički fakultet, Univerzitet 'Kiril i Metodij', Arhimedova 5, 91000 Skopje, Yugoslavia. (1937) Dr, chemistry (Skopje U., 1973). Prof. (tel. 091 - 221033). *Structural chemistry.*

Šoptrajanova, Prof. Dr Gorica. Rudarsko-geološki fakultet, 92000 Štip, Yugoslavia. (1929) Dr, geology (Beograd U., 1967). Prof. (tel. 092 - 21379). *Mineral structures.*

Tibljaš, Mr Darko. Mineraloško-petrografski zavod, Prirodoslovno-matematički fakultet, Sveučilište u Zagrebu, Demetrova ul. 1, 41000 Zagreb, Yugoslavia. (1957) Magistar, geology (Zagreb U., 1987). Asistent. (tel. 041 - 422136). *X-ray diffraction.*

Tkalčec, Prof. Dr Emilija. Tehnološki fakultet, Sveučilište u Zagrebu, Marulićev trg 20, 41000 Zagreb, Yugoslavia. (1931) Dr, chemistry (Zagreb U., 1975). Izv. prof. (tel. 041 - 446439). *X-ray diffraction analysis, polycrystalline systems, silicates.*

Tonejc, Dr (Mrs) Anđelka. Fizički zavod, Prirodoslovno - matematički fakultet, Sveučilište u Zagrebu, Marulićev trg 19, 41000 Zagreb, Yugoslavia. (1942) Dr, physics (Zagreb U., 1980). Asistent, scient. assoc. (tel. 041 - 446211). *Metal physics, crystal imperfections, phase transitions, X-ray fiffraction, electron diffraction, electron microscopy.*

Tonejc, Dr Anton. Fizički zavod, Prirodoslovno - matematički fakultet, Sveučilište u Zagrebu, Marulićev trg 19, 41000 Zagreb, Yugoslavia. (1942) Dr, physics (Zagreb U., 1972). Docent, sr. scient. assoc. (tel. 041 - 446211). *Metal physics, metals structure.*

Topić, Dr Mladen. Institut 'Ruđer Bošković', Bijenička c. 54, P.O.Box 1016, 41001 Zagreb, Yugoslavia. (1934) Dr, chemistry (Zagreb U., 1965). Sr. scient. assoc. (tel. 041 - 435111, ext. 275). *Electrical properties.*

Trojko, Magistar Rudolf. Institut 'Ruđer Bošković', Bijenička c. 54, P.O.Box 1016, 41001 Zagreb, Yugoslavia. (1942) Magistar, chemistry (Zagreb U., 1974). Res. assoc. (tel. 041 - 435111, ext. 319). *Intermetallic compounds.*

Trubelja, Prof. Dr Fabijan. Prirodno-matematički fakultet, Univerzitet u Sarajevu, Vojvode Putnika 43, 71000 Sarajevo, Yugoslavia. (1927) Dr, geology (Zagreb U., 1958). Prof. (tel. 071 - 659377). *Mineralogy.*

Tudja, Dr Marijan. SOUR Chromos RZ Kemijski istraživački centar, Žitnjak bb, 41000 Zagreb, Yugoslavia. (1942) Dr, chemistry (Zagreb U., 1974). Res. assoc. (tel. 041 - 210200, ext. 426). *Electron microscopy.*

Ungar, Dr Goran. Institut 'Ruđer Bošković', Bijenička c. 54, P.O.Box 1016, 41001 Zagreb, Yugoslavia. (1948) Dr, chemistry (Bristol U., UK, 1979). Scient. assoc. (tel. 041 - 435111). *Polymer structure and reactivity.*

Vasić, Magistar Pavle. Prirodno - matematički fakultet Priština, ul. Maršala Tita, 38000 Priština, Yugoslavia. (1939) Magistar, physics (Beograd U., 1975). Lecturer in physics. (tel. 038 - 25855). *Inorganic crystal structures, crystal physics.*

Vene, Mrs Nada. Institut 'Jožef Stefan', Univerza E. Kardelja v Ljubljani, Jamova 39, 61000 Ljubljana, Yugoslavia. (1931) Dipl. mat., mathematics (Ljubljana U., 1954). Res. assoc. (tel. 061 - 214399). *Inorganic crystal structures.*

Vicković, Dr Ivan. Zavod za opću i anorgansku kemiju, Prirodoslovno-matematički fakultet, Sveučilište u Zagrebu, ul. Soc. revolucije 8, 41000 Zagreb, Yugoslavia. (1945) Dr, physics (Zagreb U., 1977). Sr. scient. assoc. (tel. 041 - 416023, ext. 5). *Computing, structural chemistry.*

Vladimirov, Dr Sote. Zavod za farmaceutsku hemiju, Farmaceutski fakultet, Dr. Subotića 8, P. Fah 146, 11000 Beograd, Yugoslavia. (1951) Dr, pharmaceutical chemistry (Beograd U., 1985). Docent. (tel. 011 - 685930). *Structural chemistry.*

Vukasović, Magistar Momčilo. Institut za geološko-rudarska istraživanja i ispitivanja nuklearnih i drugih mineralnih sirovina, Rovinjska ul. 12, 11000 Beograd, Yugoslavia. (1923) Magistar, geology (Beograd U., 1973). Res. adviser. (tel. 011 - 480506). *Minerals structure, clay minerals.*

Zajc, Magistar Andrej. Institut za materiale, Zavod za raziskavo materiala in konstrukcij, Dimičeva 12, 61000 Ljubljana, Yugoslavia. (1938) Magistar, physical chemistry (Ljubljana U., 1979). Dept. Dir. (tel. 061 - 344061, telex 31449). *Materials science.*

Zec, Magistar Slavica. Institut za nuklearne nauke 'Boris Kidrič' - Vinča, Institut za materijale 'IM', P.O.Box 522, 11000 Beograd, Yugoslavia. (1953) Magistar, technology (Beograd U., 1977). Independent investigator. (tel. 011 - 458222, ext. 594, telex 11563 YU). *Powder diffraction.*

ZAMBIA

Sub-Editor: **M. E. Kamwaya**

Notes

1. International country code - 260

Kamwaya, Dr Mombo E. Dept. of Physics, U. of Zambia, P.O. Box 32379, Lusaka, Zambia. (1941) PhD (Free Univ. Berlin, BRD, 1980). Lect. (tel. 218474, ext. 1706). *X-ray crystallography, solid state physics, computing.*

Nugteren, Mr Hendrik W. Sch. of Mines, U. of Zambia, P.O. Box 32379, Lusaka, Zambia. (1952) MSc, geology (Free U. Amsterdam, Neterlands, 1978). Lect. (tel. 213221). *X-ray diffraction, X-ray fluoresence, mineralogy, exploitation geology.*

Tembo, Mr Francis. Sch. of Mines, U. of Zambia, P.O. Box 32379, Lusaka, Zambia. (1958) MSc, geology (U. Zambia, 1986). Lect. (tel. 213221, ext. 1401). *X-ray diffraction, X-ray fluoresence, geochemistry.*

NAME INDEX

Anantachai, Mrs Suda Yasarawana. *Thailand.*
Anantha Murthy, Dr Rayasa V. *India.*
Anantharaman, Prof. Tanjore Ramachandra. *India.*
Anders, Mr Rudolf. *DDR.*
Andersen, Mr Erik Krogh. *Denmark.*
Andersen, Mrs Inger Grete Krogh. *Denmark.*
Andersen, Dr Niels Hessel. *Denmark.*
Andersen, Dr Per. *Norway.*
Andersen, Dr Peter. *Denmark.*
Andersen, Dr Stig Kjær. *Denmark.*
Anderson, Dr Bryan Frederick. *New Zealand.*
Anderson, Ms. Christine Alexis Francis *USA.*
Anderson, Dr. Daniel Horacio. *USA.*
Anderson, Dr. Gary Don. *USA.*
Anderson, Dr. John E. *USA.*
Anderson, Prof. Oren Paul. *USA.*
Andersson-Söderberg, Mrs Margaretha. *Sweden.*
Andersson, Dr Inger A. *Sweden.*
Andersson, Dr Staffan. *Sweden.*
Andersson, Prof. Sten. *Sweden.*
Andersson, Dr Yvonne. *Sweden.*
Ando, Dr Masami. *Japan.*
Ando, Dr Yoshinori. *Japan.*
Andonov, Dr Paulette. *France.*
Andrade, Dr Lourdes Rodrigues. *Portugal.*
André, Dr Daniel. *France.*
Andree, Mr Anneliese. *DDR.*
Andreeff, Prof. Alexander. *DDR.*
Andreetti, Prof. Giovanni Dario. *Italy.*
Andreeva, Dr Nataliya Sergeyevna. *USSR.*
Andrehs, Dr Gerhard. *DDR.*
Andresen, Dr Arne Fridtjof. *Norway.*
Andrews, Dr. Lawrence Charles. *USA.*
Andrianov, Dr Valery Ivanovich. *USSR.*
Andrianova, Dr Mariya Egorovna. *USSR.*
Andrushevskii, Dr Nikolai Matveevich. *USSR.*
Ang, Dr Ha Ming. *Lumpur, Malaysia.*
Angel, Dr Ross John. *UK.*
Angelova-Tiurkedjieva, Dr Maia Nikolova. *UK.*
Angermund, Mr Klaus Peter. *BRD.*
Angilello, Mr. Joseph. *USA.*
Anikin, Dr Igor' Nikolayevich. *USSR.*
Anisimova, Vera Nikolaevna. *USSR.*
Anistratov, Dr Anatoly Tikhonovich. *USSR.*
Annaka, Dr. Shoichi. *Japan.*
Ansell, Dr. Gerald Brian. *USA.*
Anselment, Dr Bernhard. *BRD.*
Anstis, Dr Geoffrey Richard. *Australia.*
Antal, Dr. John Joseph. *USA.*
Anthony, Dr. John W. *USA.*
Antipin, Dr Mikhail Yuvenaliyevich. *USSR.*
Antolini, Prof. Luciano. *Italy.*
Anton, Mr Liviu. *Romania.*
Antonini, Prof. Marcello. *Italy.*
Antonio, Dr. Mark Ricci. *USA.*
Antonione, Prof. Carlo. *Italy.*
Antonopoulos, Prof. John. *Greece.*
Antonov, Dr Petr Iosifovich. *USSR.*
Antsishkina, Dr Alla Sergeyevna. *USSR.*
Anugul, Mrs Surang. *Thailand.*
Anulewicz, Dr Romana. *Poland.*
Anwar, Mr Jamshed. *UK.*
Anwar, Mr Muhammad. *Pakistan.*
Anwar, Dr Yehia. *Egypt.*
Aoki, Dr Katsuyuki. *Japan.*
Aoki, Dr Yoshikazu. *Japan.*
Apinitis, Dr Smuidris Karlovich. *USSR.*
Apostolescu, Eugenia Rodica. *Romania.*
Apostolov, Prof. Andrei. *Bulgaria.*
Apostolov, Mr Anton. *Bulgaria.*
Appleman, Dr. Daniel E. *USA.*
Apreda, Dr M. Carmen. *Spain.*
Aquilano, Prof. Dino. *Italy.*
Arad Mr Talmon. *Israel.*
Arafa, Prof. Salah Arafa Mohamed. *Egypt.*
Aragon de la Cruz, Prof. Francisco. *Spain.*
Aragón, Prof. Ricardo. *USA.*
Arai, Dr. Gerda Johanna. *USA.*
Arakcheeva, Dr Alla Vladimirovna. *USSR.*
Araki, Mr. Takaharu. *USA.*

Arana, Prof. Rafael. *Spain.*
Arató, Dr Péter. *Hungary.*
Aravindakshan, Cheethambadi. *India.*
Archer, Dr. Ronald D. *USA.*
Archer, Dr Steven James. *South Africa.*
Arduz, Mr Marcelo. *Bolivia.*
Arellano, Mr J. *Bolivia.*
Arem, Dr. Joel Edward. *USA.*
Arend, Prof. Hanns. *Switzerland.*
Arents, Dr. Gina. *USA.*
Arevalo, Mr. Jairo H. *USA.*
Argay, Mr Gyula. *Hungary.*
Arguello, Dr Zoraide Primerano. *Brasil.*
Argunova, Dr Tatyana Sergeevna. *USSR.*
Arhar, Magistar Andrej. *Yugoslavia.*
Arif, Dr. Atta Mahmood. *USA.*
Ariguib, Dr Najia. *Tunisia.*
Arikan, Doç. Dr Rafet. *Turkey.*
Arinkin, Dr Aleksandr Viktorovich. *USSR.*
Arkhipenko, Dr Diana Konstantinovna. *USSR.*
Armağan, Nizamettin. *Turkey.*
Armbruster, PD Thomas M. L. *Switzerland.*
Armendarez, Prof. Peter X. *USA.*
Armigliato, Dr Aldo. *Italy.*
Armstrong, Prof. Ronald William. *USA.*
Arndt, Prof. Dr Jörg Friedrich. *BRD.*
Arndt, Dr Ulrich Wolfgang. *UK.*
Arnold, Dr. Edward Van Dyke. *USA.*
Arnold, Prof. Dr Heinrich Günther Alfred. *BRD.*
Arnold, Prof. Heinrich. *DDR.*
Arnold, Mr Rolf. *DDR.*
Arnone, Prof. Arthur. *USA.*
Arnott, Prof. Struther. *UK.*
Arnoux, Dr Bernadette *France.*
Arnstein, Prof. Gustavo. *Venezuela.*
Aronsson, Prof. Bertil. *Uppsala, Sweden.*
Arora, Dr Narinder Kumar. *India.*
Arora, Dr. Satish Kumar. *USA.*
Arrieta, Dr Juan Manuel. *Spain.*
Arrington, Mr. Wendell. *USA.*
Arriortua, Dr Maribel. *Spain.*
Arruda, Prof. Moacir Rabelo. *Brasil.*
Artioli, Dr Gilberto. *Italy.*
Artunç, Dr Ekrem. *Turkey.*
Artymiuk, Dr Peter Joseph. *UK.*
Artz, Ms. P. *USA.*
Aruffo, Dr. Alejandro A. *USA.*
Arzi, Dr Ezatollah. *Iran.*
Arzumanyan, Dr Gennadiy Ashotovich. *USSR.*
Asada, Prof. Eiichi. *Japan.*
Asadchikov, Dr Viktor Evgen'yevich. *USSR.*
Asadov, Prof. Yusif Gazanfar ogly. *USSR.*
Asai, Dr Takeshi. *Japan.*
Asath Bahadur, Mr S. *India.*
Ashfaquzzaman, Mr. Syed. *USA.*
Ashida, Dr Sakichi. *Japan.*
Ashida, Prof. Tamaichi. *Japan.*
Ashirov, Dr Aman. *USSR.*
Ashizawa, Mr Kazuhide. *Japan.*
Ashry, Dr Mamdouh. *Egypt.*
Ashwell, Dr Geoffrey Joseph. *UK.*
Askhabov, Dr Askhab Magomedovich. *USSR.*
Aslaner, Prof. Dr Mustafa. *Turkey.*
Aslanian, Dr Selma. *Bulgaria.*
Aslanov, Prof. Leonid Alexandrovich. *USSR.*
Asmat, Dr Humberto. *Peru.*
Atanassov, Dr Vassil. *Bulgaria.*
Atasoy, Doç. Dr Ö. Aydin. *Turkey.*
Atassi, Prof. M. Zouhair. *USA.*
Athappilly, Dr. Francis Kuriakose. *USA.*
Athey, Dr. Brian D. *USA.*
Atkinson, Dr. David. *USA.*
Atoji, Dr. Masao. *USA.*
Atovmyan, Prof. Lev Oganovich. *USSR.*
Attard, Dr. Alfred E. *USA.*
Attfield, Dr John Paul. *UK.*
Attig, Dr Rainer. *BRD.*
Atwood, Prof. Jerry Lee. *USA.*
Atzmony, Prof. Uzi. *Israel.*
Auf der Heyde, Dr Thomas Paul Edwin. *South Africa.*

Augustin, Mr. Rolf M. *USA.*
Aupers, Dr John Henry. *UK.*
Aurivillius, Prof. Bengt. *Sweden.*
Austerman, Mr. Stanley Boone. *USA.*
Austria, Dr Benjamin Suarez. *Philippines.*
Authier, Prof. André *France.*
Avalos, Dr Jaime. *Peru.*
Avci, Dr Recep. *Saudi Arabia.*
Avdiyenko, Dr Klavdiya Ilyinishna. *USSR.*
Averbach, Prof. Benjamin Lewis. *USA.*
Averbuch-Pouchot, Dr Marie-Thérèse. *France.*
Avey, Dr Hugh Philip. *Australia.*
Avila-Salinas, Mr Waldo. *Bolivia.*
Avilov, Dr Anatoly Sergeyevich. *USSR.*
Avramov, Dr Isak. *Bulgaria.*
Awasthi, Dr Santosh Kumar. *India.*
Axmann, Dr Anton. *BRD.*
Aydinol, Doç. Dr Mahmut. *Turkey.*
Aydinuraz, Prof. Dr Arsin. *Turkey.*
Ayers, Mr. John E. *USA.*
Ayhan, Doç. Dr Ahmet. *Turkey.*
Aypar, Prof. Dr Abidin. *Turkey.*
Ayroles, Dr René *France.*
Aytaş, Doç. Dr S. Işik. *Turkey.*
Azarnia, Dr Nezhat. *Iran.*
Azaroff, Prof. Leonid V. *USA.*
Azer, Dr Nazmi. *Egypt.*
Azmon, Prof. Emanuel. *Israel.*
Azuma, Dr Nagao. *Japan.*
Baak, Ing. Leonardus Cornelis. *Netherlands.*
Baars, Dr -Ing. Jan Walter. *BRD.*
Baba, Mrs Jasmin. *Malaysia.*
Babareko, Dr Alesya Adamovna. *USSR.*
Babel, Prof. Dr Dietrich. *BRD.*
Babič, Dr Danilo. *Yugoslavia.*
Babich, Prof. Michael Wayne. *USA.*
Babu, Dr. Yarlagadda Sudhakar. *USA.*
Bachechi, Dr Fiorella. *Italy.*
Backhaus, Dr Karl-Otto. *DDR.*
Bacon, Prof. George Edward. *UK.*
Bade, Mr Dirk. *BRD.*
Badger, Dr. John. *USA.*
Badr, Dr Yehia Abd-El Hamid. *Egypt.*
Badurek, Prof. Dr Dipl.-Ing. Gerald. *Austria.*
Bäckerud, Dr Lennart S. *Sweden.*
Baele, Miss Ingrid Albertina Frans Mariette. *Belgium.*
Baenziger, Prof. Norman C. *USA.*
Bärlocher, Dr Christian. *Switzerland.*
Bärnighausen, Prof. Dr Hartmut. *BRD.*
Baeta, Dr Robert Domingo. *Ghana.*
Bagchi, Prof. Subodh Nath. *Canada.*
Bagdasarov, Dr, Khachik Saakovich. *USSR.*
Baggio, Dr Ricardo. *Argentina.*
Baggio, Dr Sergio. *Argentina.*
Bagieu, Dr Muriel. *France.*
Bagley, Mr Arthur George. *UK.*
Bai, Mr Chun-li. *China.*
Baier, Mr Hubert. *BRD.*
Bailey, Mr David Eric. *Australia.*
Bailey, Prof. David Kenneth. *UK.*
Bailey, Prof. Marcia F. *USA.*
Bailey, Dr Neil Anthony. *UK.*
Bailey, Prof. Sturges Williams. *USA.*
Bain, Dr Derek Charles. *UK.*
Bainbridge, Mr John Evelyn. *UK.*
Baird, Dr. Herbert Wallace. *USA.*
Bąk, Dr Jadwiga. *Poland.*
Bakakin, Dr Vladimir Vasilyevich. *USSR.*
Baker, Dr Anthony Thomas. *Australia.*
Baker, Dr Edward Neill. *New Zealand.*
Baker, Dr. Kenneth Neil. *UK.*
Baker, Dr Patrick Julian. *UK.*
Baker, Dr. R. J. *USA.*
Baker, Mr R.W. *UK.*
Baker, Mr Thomas Wilfred. *UK.*
Bakovets, Dr Vladimir Viktorovich. *USSR.*
Bakr, Dr Abdel Razak. *Saudi Arabia.*
Bakshi, Dr Edward. *Australia.*
Balagurov, Mr Anatoly Mikhailovich. *USSR.*
Balakirev, Dr Vladimir Georgiyevich. *USSR.*

Balan, Mr Mihai. *Romania.*
Balarin, Prof. Manfred. *DDR.*
Balasingh, Dr C. *India.*
Balasubramanian, Dr R. *India.*
Balcazar, Dr Jose Luis. *Spain.*
Balchin, Dr Anthony Arthur. *UK.*
Baldrian, Dr Josef. *CSSR.*
Baldwin, Mr. Eric T. *USA.*
Baldwin, Mr. Kenneth John. *USA.*
Baldy, André. *France.*
Bale, Prof. Harold D. *USA.*
Bales, Mr. Howard E. *USA.*
Balić Žunić, Dr Tonči. *Yugoslavia.*
Balitsky, Dr Vladimir Sergeyevich. *USSR.*
Balkanov, Mr Ivan. *Bulgaria.*
Ball, Prof. Anthony. *South Africa.*
Ball, Dr. Richard George. *USA.*
Bally, Conf. Dorel. *Romania.*
Bally, Dr Renée. *France.*
Baltă, Conf. Petru. *Romania.*
Balyunis, Dr Lyubov' Evgenievna. *USSR.*
Balyuzi, Dr Hushang H.M. *UK.*
Balzar, Mr Davor. *Yugoslavia.*
Balzarotti, Prof. Adalberto. *Italy.*
Bambauer, Prof. Dr Hans Ulrich. *BRD.*
Ban, Prof. Dr Zvonimir. *Yugoslavia.*
Banaii, Dr Nasser. *Iran.*
Banaszak, Prof. Leonard J. *USA.*
Bandoli, Prof.Giuliano. *Italy.*
Bandopadhyay, Dr Mrs Tapati. *India.*
Banerjee, Dr Asok. *India.*
Banerjee, Dr Krishna. *India.*
Banerjee, Mr Rahul. *India.*
Banerjee, Dr Srikumar. *India.*
Bang, Dr Eva Henriette. *Denmark.*
Banizs, Dr Károly. *Hungary.*
Banko, Mr. Brad. *USA.*
Banks, Prof. Ephraim. *USA.*
Banner, Dr David William. *BRD.*
Banner, Dr David William. *Switzerland.*
Bansigir, Prof. K. Goswami. *India.*
Bao, Prof. Guang-hong. *China.*
Baptista, Mr Augusto. *Brasil.*
Baptista, Mrs Neysa Rocha. *Brasil.*
Baqri, Dr Syed Rafiqul-Hassan. *Pakistan.*
Barabanov, Prof. Vladimir Fedorovich. *USSR.*
Barakat, Prof. Nayel. *Egypt.*
Baran, Dr Zbigniew. *Brasil.*
Baranov, Dr Anatoliy Ivanovich. *USSR.*
Baranskii, Prof Konstantin Konstantinovich. *USSR.*
Baranyi, Dr Anthony David. *Canada.*
Barbagelata, Mr Franco. *Chile.*
Barber, Dr. Ann M. *USA.*
Barber, Prof. Patrick George. *USA.*
Barbier, Mr Bruno. *BRD.*
Barbieri, Prof. Renato. *Italy.*
Barcik, Dr Jan. *Poland.*
Barcza, Dr. Sandor. *USA.*
Bardhan, Mr. Pronob. *USA.*
Bardi, Prof. Renato. *Italy.*
Barelli, Dr Nilso. *Brasil.*
Baresel, Dr -Ing. Detlef Wilhelm Berthold. *BRD.*
Barghi, Mr Mohammad Ali Barghi. *Iran.*
Barkigia, Dr. Kathleen M. *USA.*
Barna, Dr Péter. *Hungary.*
Barnea, Dr Zwi. *Australia.*
Barnes, Dr. Charles Leslie. *USA.*
Barnes, Dr John Conquest. *UK.*
Barnes, Dr Paul. *UK.*
Barnett, Dr. Bobby L. *USA.*
Barney, Ms. Elsa Pauline. *USA.*
Barnhart, Dr. David Merle. *USA.*
Barrett, Prof. Charles Sanborn. *USA.*
Barrick, Dr. James Clinton. *USA.*
Barriga Villalba, Prof. Antonio María. *Colombia.*
Barrington-Leigh Dr John. *Canada.*
Barrios, Mr Nelson. *Chile.*
Barron, Dr Hugh Wilson Taylor. *UK.*
Barrow, Dr Michael John. *UK.*
Barsukova, Dr Marina Leonidovna. *USSR.*

Barszcz, doc. Edward. *Poland.*
Barta, Ing Čestmír. *CSSR.*
Bartczak, Dr hab. Tadeusz Jan. *Poland.*
Bartell, Prof. Lawrence Sims. *USA.*
Barthel, Prof. Johannes. *DDR.*
Bartl, Prof. Dr Hans. *BRD.*
Bartlett, Dr Michael William. *Canada.*
Bartlett, Prof. Neil. *USA.*
Bartolucci, Dr Cecilia. *Italy.*
Barton, Dr. C. J. *USA.*
Barton, Dr. Randolph Jr. *USA.*
Barton, Prof. Richard J. *Canada.*
Bartoshinsky, Prof. Zbignev Vladislavovich. *USSR.*
Baruah, Mr Gagan Ch. *India.*
Baruchel, Dr José *France.*
Basak, Dr Ajit Kumar. *UK.*
Basak, Prof. Bejoysanker. *India.*
Basha, Dr Ahmed Fouad. *Egypt.*
Basilakis, Mr Michael. *Greece.*
Bassett, Prof. David Clifford. *UK.*
Bassett, Mr G. Alan. *UK.*
Bassignana, Dr Isabella C. *Canada.*
Bassiouny, Dr Mohamed Khafagi. *Egypt.*
Basso, Prof. Riccardo. *Italy.*
Basson, Dr Stephen Smuts. *South Africa.*
Bastiansen, prof. Otto Christian Astrup. *Norway.*
Bastida, Dr Joaquin. *Spain.*
Bastin, Dr Ir Guillaume F. *Netherlands.*
Basto, Mrs Maria João. *Portugal.*
Basu, Dr. Sankar Prasad. *USA.*
Bataliyeva, Dr Nataliya Glebovna. *USSR.*
Batchelder, Dr David Neville. *UK.*
Batchelor, Dr Raymond John. *Canada.*
Bates, Dr David Ronald. *UK.*
Bates, Dr Peter Arthur. *UK.*
Bates, Prof. Richard Heaton Tunstall. *New Zealand.*
Bates, Prof. Robert Brown. *USA.*
Bats, Dr Jan, Willem. *BRD.*
Batsanov, Dr Andrey Stepanovich. *USSR.*
Battaglia, Prof. Luigi Pietro. *Italy.*
Batterman, Prof. Boris William. *USA.*
Battezzati, Prof. Livio. *Italy.*
Battle, Dr Peter David. *UK.*
Bau, Prof. Robert. *USA.*
Bauduin, Mrs Anne-Marie G. G. *Belgium.*
Bauer, Doc. Ing Jaroslav. *CSSR.*
Bauer, Prof. Dr Ernst Georg. *USA.*
Bauer, Prof. Dr Günther Ernst. *Austria.*
Baughman, Mr. Richard Joseph. *USA.*
Baughman, Prof. Russell George. *USA.*
Baum, Mrs Elke. *BRD.*
Baumann, Mr Jürgen Rudolf. *BRD.*
Baumbach, Mr Manfred. *DDR.*
Baumgartner, Prof. Dr Dipl.-Ing. Oswald. *Austria.*
Baur, Prof. Dr Werner H. *BRD.*
Baures, Mr. Paul William. *USA.*
Bautsch, Prof. Hans-Joachim. *DDR.*
Bavoux, Dr Claude. *France.*
Baxter, Mr Colin. *UK.*
Bayer, Prof. Gerhard. *Switzerland.*
Bayh, Prof. Dr Werner. *BRD.*
Bayhan Doç. Dr Hasan. *Turkey.*
Bayliss, Prof. Peter. *Canada.*
Bayon, Dr J. Carlos. *Spain.*
Bayvas, Dr Fehime. *Turkey.*
Beagley, Dr Brian. *UK.*
Beale, Dr John Phillip. *Australia.*
Beamson, Dr G. *UK.*
Bear, Prof. Richard Scott. *USA.*
Beard, Mr. Donald W. *USA.*
Beasley, Prof. Wayne Machon. *USA.*
Beauchamp, Prof. Dr André. *Canada.*
Becherer, Prof. Gerhard. *DDR.*
Becherer, Dr Karl. *Austria.*
Beck, Prof. Dr Horst Philipp. *BRD.*
Becker, Dr Claus. *DDR.*
Becker, Prof. Dr Gerd. *BRD.*
Becker, Prof. Joseph Whitney. *USA.*
Beckingsale, Mr Peter Gerard. *New Zealand.*
Beckmann, Prof. Günter. *DDR.*

Bedarida, Prof. Federico. *Italy.*
Bedarkar, Dr. Sudhir. *USA.*
Beddell, Dr Christopher Raymond. *UK.*
Beddoes, Mr Roy L. *UK.*
Bednarski, Dr Stanisław. *Poland.*
Bednowitz, Dr. Allan Lloyd. *USA.*
Bedyńska, Dr hab. Teresa. *Poland.*
Bedzyk, Dr. Michael James. *USA.*
Beem, Dr. Karl Michael. *USA.*
Beese, Ms. Lorena S. *USA.*
Beevers, Dr Cecil Arnold. *UK.*
Beglari, Dr Ing. Parviz. *Iran.*
Begley, Dr Michael John. *UK.*
Behm, Dr Helmut Johannes Julius. *BRD.*
Behm, Prof. Dr Rolf Jürgen. *BRD.*
Behrens, Dr Heinrich. *BRD.*
Behrens, Dr Peter. *BRD.*
Behrens, Dr Ulrich Hermann. *BRD.*
Behruzi, Dr Massoud. *BRD.*
Beier, Dr Wilfried. *DDR.*
Beiter, Mr. Thomas Albert. *USA.*
Bekesha, Dr Sergei Nikolaevich. *USSR.*
Bekrenev, Dr Anatoly Nikolayevich. *USSR.*
Bel Hassen, Miss Dalila. *Tunisia.*
Bel'sky, Dr Vitaly Konstantinovich. *USSR.*
Belan, Dr Bogdana Dmitrievna. *290602.USSR.*
Belanger-Gariepy, Mme Francine. *Canada.*
Belhadj, Dr Ali. *Tunisia.*
Belikova, Dr Galina Sergeyevna. *USSR.*
Belkheria, Mr Salah. *Tunisia.*
Bell, Mr Anthony Martin Thomas. *UK.*
Bell, Dr. Jeffrey A. *USA.*
Bellamy, Mr Brian Arthur. *UK.*
Bellard, Dr Sharon Ann. *UK.*
Bello, Dr Alfredo. *Venezuela.*
Bellon, Prof. Pier Luigi. *Italy.*
Bellver, Dr Consuelo. *Spain.*
Belokoneva, Dr Elena Leonidovna. *USSR.*
Belova, Dr Elizaveta Nikolayevna. *USSR.*
Belt, Dr. Roger F. *USA.*
Beltrán Abrego, Dr José Ramón. *Brasil.*
Belugina, Dr Nataliya Vasilyevna. *USSR.*
Belyaeva, Mrs Klara Fedorovna. *USSR.*
Belyakov, Dr Sergei Vasil'evich. *USSR.*
Belyustin, Dr Aleksey Vsevolodovich. *USSR.*
Ben Amor, Dr. *Tunisia.*
Ben Brahim, Dr Jemaïel. *Tunisia.*
Ben Ghozlane, Dr Hédi. *Tunisia.*
Ben Romdhane, Dr. *Tunisia.*
Ben Salah, Dr Abdelhamid. *Tunisia.*
Ben Yair, Dr Moshe Pinhas. *Israel.*
Bendeliani, Dr Nikolay Alexandrovich. *USSR.*
Bender, Dr Hugo J.M.R. *Belgium.*
Bené, Prof. Robert William. *USA.*
Benedetti, Prof. Ettore. *Italy.*
Benetollo, Dr Franco. *Italy.*
Benghiat, Dr Victor. *Israel.*
Bengochea, Dr Amado Leandro. *Argentina.*
Benna, Dr Piera. *Italy.*
Bennema, Prof. Dr Pieter. *Netherlands.*
Bennett, Prof. Dennis W. *USA.*
Bennett, Dr. John Michael. *USA.*
Bennett, Dr Pauline Mary. *UK.*
Bennett, Dr William, Samuel, Jr. *BRD.*
Beno, Dr. Mark A. *USA.*
Bensimon, Dr Corinne. *Canada.*
Benson, Mr. James Edward. *USA.*
Bente, Prof. Dr Klaus Alexander. *BRD.*
Bentley, Dr Graham Arthur. *France.*
Benyacar, de, Lic María Angélica R. *Argentina.*
Benz, Prof. Dr Klaus-Werner. *BRD.*
Beran, Prof. Dr Anton. *Austria.*
Beres, Mr. John J. *USA.*
Beresnev, Dr Leonid Alekseyevich. *USSR.*
Beretka, Mr Julius. *Australia.*
Berezhkova, Dr Galina Vasilyevna. *USSR.*
Berezkin, Dr Vladimir Viktorovich. *USSR.*
Berezyuk, Dr Dar'ya Aleksandrovna. *USSR.*
Berg, Mrs Dr Lieselotte. *BRD.*
Berg, van den, Ir Adrianus Johannes. *Netherlands.*

Berger, Dr Hans. *DDR.*
Berger, Dr Rolf Anders. *Netherlands.*
Bergerhoff, Prof. Dr Guenter. *BRD.*
Bergmann, Mrs. Margot E. *USA.*
Bergner, Mr Joachim. *DDR.*
Bergsma, Dr Jitze. *Netherlands.*
Bergunde, Mr Thomas. *DDR.*
Berking, Dr Bernhard. *BRD.*
Berkovitch-Yellin, Prof. Ziva. *Israel.*
Berliner, Prof. Lawrence J. *USA.*
Berliner, Dr. Ronald Richard. *USA.*
Berman, Dr. Helen M. *USA.*
Bermanec, Magistar Vladimir. *Yugoslavia.*
Bermudez-Polonio, Dr Joaquin. *Spain.*
Bernal de Ramírez, Prof. Inés. *Colombia.*
Bernal, Prof. Ivan. *USA.*
Bernalte, Prof. Antoni. *Spain.*
Bernardinelli, Dr Gérald. *Switzerland.*
Bernheim, Dr. Marguerite Mary. *USA.*
Bernotat-Wulf, Mrs. Dr Hannelore. *BRD.*
Bernstein, Ms. Frances C. *USA.*
Bernstein, Dr. Herbert Jacob. *USA.*
Bernstein, Prof. Joel. *Israel.*
Berry, Dr. Chester Ridlon. *USA.*
Bershov, Dr Leonid Viktorovich. *USSR.*
Bersuker, Prof. Isaak Borukhovich. *USSR.*
Bertaut, Dr Erwin Félix. *France.*
Bertelmann, Dr Dieter Wilhelm. *BRD.*
Berthet-Colominas, Dr Carmen. *France.*
Berthold, Prof. Dr Hans Joachim. *BRD.*
Berthold, Mr Lutz. *DDR.*
Berthold, Dr Thomas. *BRD.*
Berti, Dr Giovanni. *Italy.*
Bertolasi, Prof. Valerio. *Italy.*
Bertram, Dr Marion. *DDR.*
Bertrand, Prof. Joseph Aaron. *USA.*
Bertraud, Mr. J. A. *USA.*
Berzinya, Dr Inese Rudol'fovna. *USSR.*
Besoaín, Dr Eduardo. *Chile.*
Bessiere Dr Michel. *France.*
Besso, Ms Karyne. *Canada.*
Betal, Mr Badal Kumar. *India.*
Bethge, Dr. Paul Herman. *USA.*
Betsofen, Dr Sergey Yakovlevich. *USSR.*
Betts, Dr. Foster. *USA.*
Betts, Ms. Laurie. *USA.*
Betz, Prof. Dr Gerhard. *Austria.*
Betz, Mr Helmut. *BRD.*
Betzel, Dr Christian. *BRD.*
Betzl, Dr Manfred. *DDR.*
Beukes, Prof Dr Gerhardus Johannes. *South Africa.*
Beurskens, Mr Gezina. *Netherlands.*
Beurskens, Prof. Dr Paul T. *Netherlands.*
Bevan, Prof David John Martin. *Australia.*
Bevis, Prof. M.J. *UK.*
Beyer, Dr Ir Jenö. *Netherlands.*
Bezerra, Dr George Humberto. *Brasil.*
Bezjak, Prof. Dr Aleksandar. *Yugoslavia.*
Beznosikov, Dr Boris Valeriyanivich. *USSR.*
Bhagvantam, Prof. Suri. *India.*
Bhagwat, Dr Vasant. *India.*
Bhakay-Tamhane, Mrs Sandhya Nitin. *India.*
Bhaktapriya, Dr S. R. Y. *India.*
Bhandary, Dr. Krishna K. *USA.*
Bharati Rao, Ms T. *India.*
Bhatia, Dr Subhash Chander. *India.*
Bhatt, Dr Vaikunthray Promodray. *India.*
Bhattacharjee, Mrs Lilabati. *India.*
Bhattacharjee, Mr. Sarama. *USA.*
Bhattacharya, Archana. *India.*
Bhattacharya, Dr Ramendranarayan. *India.*
Bhattacharyya, Dr Subodh Chandra. *India.*
Bhattacherjee, Mr Santi Brata. *India.*
Bhattacherjee, Dr Satyananda. *India.*
Bhatti, Mr Muhammad Akram. *Pakistan.*
Bhawalkar, Dr Ramkrishna Haribhau. *India.*
Bi, Prof. Ru-chang. *China.*
Bi, Prof. Yu-run. *China.*
Biagini Cingi, Prof. Marina. *Italy.*
Bialek, Mr Roland. *Switzerland.*

Bianchet, Dr Mario Antonio. *Argentina.*
Bianchi Orlandini, Dr Annabella. *Italy.*
Bianchi, Dr Riccardo. *Italy.*
Bichurin, Dr Rinnat Chingizkhanovich. *USSR.*
Bideau, Prof. Jean-Pierre. *France.*
Biedl, Dr Albrecht. *BRD.*
Bielen, Prof. Dr Helmut Josef. *BRD.*
Bienenstock, Prof. Arthur Irwin. *USA.*
Bienfait, Prof. Michel. *France.*
Bigelow, Prof. Wilbur Charles. *USA.*
Bigi Dr Adriana. *Italy.*
Bigi Dr Simona. *Italy.*
Bigoli, Prof. Francesco. *Italy.*
Bijen, Dr Jan M.J.M. *Netherlands.*
Bilderback, Dr. Donald Heywood. *USA.*
Billiet, Prof. Yves. *Tunisia.*
Billing, Mr David Gordon. *South Africa.*
Billman, Mr. John F. *USA.*
Binas, Dr Horst. *DDR.*
Bingman, Mr. Craig A. *USA.*
Bingöl, Doç. Dr A. Feyzi. *Turkey.*
Binnie, Dr. William Polson. *USA.*
Bino, Prof. Avi. *Israel.*
Bird, Prof. Peter Hans. *Canada.*
Birdsall, Prof. William John. *USA.*
Birgeneau, Prof. Robert J. *USA.*
Birks, Mr. L. S. *USA.*
Birktoft, Dr. Jens Julius. *USA.*
Birnbaum, Dr George I. *Canada.*
Birsoy, Doç. Dr Rezan. *Turkey.*
Bish, Dr. David Lee. *USA.*
Bishop, Dr Arthur Clive. *UK.*
Bisi Castellani, Prof. Carla. *Italy.*
Bissert, Dr Elisabeth Gertrud. *BRD.*
Bist, Dr B. M. S. *India.*
Biswas, Dr. Amit. *USA.*
Biswas, Goutam. *India.*
Biswas, Dr Mohammad Alim. *Bangladesh.*
Biswas, Dr Subhash Chandra. *India.*
Biswas, Dr Sundar Gopal. *India.*
Biyushkin, Dr Victor Nikolayevich. *USSR.*
Bizid, Dr Abdelmalek dit Youssef. *Tunisia.*
Black, Dr. Shaun D. *USA.*
Blake, Dr Alexander John. *UK.*
Blake, Dr Antony Brian. *UK.*
Blake, Prof. Jerry Wayne. *USA.*
Blake, Mr Ronald George. *Australia.*
Blanc, Mr Eric Charles-Henri. *Switzerland.*
Blanco, Dr Marta. *Spain.*
Bland, Dr J.A. *UK.*
Blankenburg, Dr Hans-Joachim. *DDR.*
Blanton, Mr. Thomas Nelson. *USA.*
Blaschke, Prof. Dr Rochus Bruno Albert. *BRD.*
Blaschko, Dr Oskar. *Austria.*
Blasi, Prof. Achille. *Italy.*
Blaton, Dr Norbert Louis. *Belgium.*
Blau, Prof. Winfried. *DDR.*
Blažič, Magistar Boris. *Yugoslavia.*
Blažina, Dr Želimir. *Yugoslavia.*
Bleidelis, Dr Yanis Yazepovich. *USSR.*
Bleif, Dr Hans-Jürgen. *BRD.*
Blessing, Dr. Robert Harry. *USA.*
Blevins, Dr. Richard A. *USA.*
Blinov, Dr Lev Mikhailovich. *USSR.*
Blinov, Dr Victor Aleksandrovich. *USSR.*
Blinowski, Dr Konrad. *Poland.*
Bliss, Dr. Mary. *USA.*
Blistanov, Prof Aleksandr Aleksandrovich. *USSR.*
Bliznakov, Prof. Georgi. *Bulgaria.*
Block, Prof. Dr Jochen Hermann. *BRD.*
Block, Dr. Stanley. *USA.*
Blokhin, Prof. Mikhail Arnol'dovich. *USSR.*
Blomberg, Miss Merja Kristiina. *Finland.*
Bloomer, Dr Anne Christine. *UK.*
Bloor, Dr David. *UK.*
Blount, Dr. Alice M. *USA.*
Blount, Dr. John Franklin. *USA.*
Blow, Prof David Mervyn. *UK.*
Blüthgen, Dipl-Min Waldemar. *BRD.*
Bluhm, Dr Terry Lee. *Canada.*

Blum, Dr Zoltan. *Sweden.*
Blundell, Dr David James. *UK.*
Blundell, Prof. Thomas Leon. *UK.*
Bly, Dr. Donald D. *USA.*
Bocchi, Dr Claudio. *Italy.*
Bocelli, Dr Gabriele. *Italy.*
Bodak, Prof. Oksana Ivanovna. *USSR.*
Bode, Dr Wolfram. *BRD.*
Bodor, Prof. Géza. *Hungary.*
Bodot, Prof. Hubert. *France.*
Boeck, Mr Torsten. *DDR.*
Böcskei, Mr Zsolt. *Hungary.*
Bögge, Dr Hartmut. *BRD.*
Boehm, Prof. Dr Hanns-Peter. *BRD.*
Böhm, Dr Horst. *BRD.*
Boehm, Dr James M. *Australia.*
Böhme, Prof. Dr Reinhild. *BRD.*
Boehme, Mr. Richard Frederick. *USA.*
Boer, de, Dr Jan Louwert. *Netherlands.*
Boese, Dr Roland. *BRD.*
Böttcher, Dr Peter Ernst-August. *BRD.*
Boeyens, Prof. Jan Christoffel Antonie. *South Africa.*
Bogdanov, Dr Gennadii Evgen'evich. *USSR.*
Boggs, Dr. Rita Rose. *USA.*
Bognár, Dr László. *Hungary.*
Bogucka-Ledóchowska, Dr Maria. *Poland.*
Bohac, Dr Petr. *Switzerland.*
Bohatý, Prof. Dr Ladislav. *BRD.*
Bohlen, Mr. David S. *USA.*
Bohm, Dr Joachim. *DDR.*
Bohr, Dr Jakob. *Denmark.*
Boiko, Prof. Boris Timofeyevich. *USSR.*
Boikova, Dr Alexandra Ivanovna. *USSR.*
Bojarski, Prof. Dr Zbigniew. *Poland.*
Bokii, Prof. Georgy Borisovich. *USSR.*
Bołd, Dr Tadeusz. *Poland.*
Boles, Dr Michael Owen. *UK.*
Bolhuis, van, Mr Fré. *Netherlands.*
Bolin, Prof. Jeffrey T. *USA.*
Bolis, Dr Vera Maria. *Italy.*
Boller, Prof. Dr Herbert. *Austria.*
Bolognesi, Prof. Martino. *Italy.*
Bombieri, Prof. Gabriella. *Italy.*
Bonamartini-Corradi, Prof. Anna. *Italy.*
Bonamico, Dr Mario. *Italy.*
Bonaventura, Dr. Joseph. *USA.*
Bonazzi, Dr Paola. *Italy.*
Bond, Dr Dianne Ruth. *South Africa.*
Bondar', Dr Anatolii Mikhail;ovich. *USSR.*
Bondar', Dr Iraida Adamovna. *USSR.*
Bondars, Dr Bruno Yanovich. *USSR.*
Bondza, Dr Harald Werner. *BRD.*
Bonefačić, Prof. Dr Antun. *Yugoslavia.*
Bonetto, Miss Rita Dominga., *Argentina.*
Bonev, Dr Ivan. *Bulgaria.*
Bonham, Prof. Russell Aubrey. *USA.*
Bonnaud, Dr Bernard Henri. *Côte d'Ivoire.*
Bonnet, Prof. Jean-Jacques. *France.*
Bonnet, Dr Michel. *France.*
Bonny, Dr Roger. *Côte d'Ivoire.*
Bonpunt, Dr Louis. *France.*
Bonse, Prof. Dr Dr Dr h.c. Ulrich Karl Eberhard. *BRD.*
Boo, Dr. William O. J. *USA.*
Boom, Dr Geert. *Netherlands.*
Boonstra, Prof. Eelco Gerrit. *South Africa.*
Boorman, Dr Philip Michael. *Canada.*
Booth, Dr Andrew Donald. *Canada.*
Bora, Dr M. N. *India.*
Borchardt-Ott, Dr Walter. *BRD.*
Borges, Prof. Frederico. *Portugal.*
Borhani, Dr. David. *USA.*
Borie, Dr. Bernard Simon. *USA.*
Boris, Mrs. Linda Jo. *USA.*
Borisanova, Dr Lidiya Mikhailovna. *USSR.*
Borisov, Dr Stanislav Vasilyevich. *USSR.*
Borisov, Dr Vsevolod Vasilyevich. *USSR.*
Borkakoti, Dr Nivedita Neera. *UK.*
Born, Dr Eberhard. *BRD.*
Born, Dr Liborius. *BRD.*
Bornmann, Dr Horst. *DDR.*

Bornmann, Dr Peter. *DDR.*
Borowiak, assoc. Prof. Teresa. *Poland.*
Borrmann, Prof. Dr -Ing. Gerhard. *BRD.*
Borzecka-Prokop, Dr Barbara. *Poland.*
Bosch Giral, Dr Pedro. *México.*
Bose, Mr Shyamal Kumar. *India.*
Boskey, Dr. Adele Ludin. *USA.*
Bosman, Drs Wilhelmus P. J. H. *Netherlands.*
Boss, Dr. James William. *USA.*
Bostanov, Mr Vesselin. *Bulgaria.*
Boström, Dr N. Dan. *Sweden.*
Bota, Prof. Kofi B. *USA.*
Bottomley, Prof. Frank. *Canada.*
Bottyán, Dr László. *Hungary.*
Boulesteix, Prof. Claude. *France.*
Bouraoui, Prof. Ahmed. *Tunisia.*
Bouraoui, Prof. Ahmed. *Tunisie.*
Bourne, Dr. Philip Eric. *USA.*
Bourne, Miss Susan Ann. *South Africa.*
Bouška, Doc. Dr Vladimír. *CSSR.*
Boutiba, Dr Samia. *Tunisia.*
Bouwmeester, Dr Henny J.M. *Netherlands.*
Boviatsis, Dr. Ioanis. *USA.*
Bovin, Dr Jan-Olov. *Sweden.*
Bovio, Prof. Bruna. *Italy.*
Bowen-Jones, Dr J. *UK.*
Bowen, Dr Alun Wynne *UK.*
Bowen, Dr David Keith. *UK.*
Bowman, Mr. Allen L. *USA.*
Bown, Dr Michael George. *UK.*
Box, Mr. Harold C. *USA.*
Boyarskaya, Prof. Yuliya Stanislavovna. *USSR.*
Boyd, Dr. Donald B. *USA.*
Boyiatzis, Mr Ioannis. *Greece.*
Boyington, Mr. Jeffrey C. *USA.*
Boyko, Prof. Edward Raymond. *USA.*
Boyle, Dr Lewis Laurence. *UK.*
Boyle, Mr. Paul. *USA.*
Boys, Dr Cecil William Gordon. *UK.*
Boys, Dr Daphne. *Chile.*
Boysen, Dr Hans. *BRD.*
Bozopoulos, Dr Anastasios Panayiotis. *Greece.*
Boztuğ, Yrd. Doç. Dr Durmuş. *Turkey.*
Braam, Dr Adrianus Wilhelmus Maria. *Netherlands.*
Bracke, Mr Benedikt Rachel Frans. *Belgium.*
Bradaczek, Prof. Dr Hans Arthur. *BRD.*
Braden, Dr. Bradford Carl. *USA.*
Brádler, Ing Jaroslav. *CSSR.*
Bradshaw, Mr. Joseph Earl. *USA.*
Brady, Mr. George W. *USA.*
Brady, Dr Robert Leo. *UK.*
Brämer, Dr Wulf. *BRD.*
Brändén, Prof. Carl-Ivar. *Sweden.*
Braga, Prof. Dario. *Italy.*
Bragg, Prof. Robert Henry. *USA.*
Braibanti, Prof. Antonio. *Italy.*
Brainin, Dr Boris Matveyevich. *USSR.*
Brammer, Dr. Lee. *USA.*
Brand, Prof. Paul. *DDR.*
Brandhuber, Dr. Barbara. *USA.*
Brandmüller, Prof. Dr Josef Karl August. *BRD.*
Brandon, Dr James Kenneth. *Canada.*
Brandstätter, Dr Franz. *Austria.*
Brandt, Ing Gernot. *BRD.*
Brasseur, Prof. Dr Henri A. L. *Belgium.*
Brathovde, Prof. James Robert. *USA.*
Brauer, Prof. Dr Georg Karl. *BRD.*
Braun, Dr Eckart. *BRD.*
Braun, Dr Poul Bernard. *Netherlands.*
Brauner, Mr Christian. *BRD.*
Brauny, Mr Siegfried. *BRD.*
Bravo, Prof. Manuel. *Portugal.*
Bray, Prof. Diana D. *USA.*
Brayer, Prof. Gary David. *Canada.*
Brech, Mr. Frederick. *USA.*
Brehm, Dr Lotte. *Denmark.*
Breit, Dipl-Min Udo. *BRD.*
Breiter, Prof. Dr Manfred. *Austria.*
Breitinger, Prof. Dr Dietrich Karl. *BRD.*
Bremer, Dr Johannes. *Norway.*

Brenčič, Prof. Dr Jurij. *Yugoslavia.*
Brennan, Dr. Richard Gerald. *USA.*
Brennan, Dr. Sean Michael. *USA.*
Brennan, Dr. Thomas Francis. *USA.*
Brenner, Mr. Stephen A. *USA.*
Bresciani Pahor, Prof. Nevenka. *Italy.*
Brese, Mr. Nathaniel E. *USA.*
Březina, Ing Bohuslav. *CSSR.*
Brianso, Prof. Jose Luis. *Spain.*
Briant, Dr Clive Edward. *UK.*
Briard, Mrs Pierrette. *Côte d'Ivoire.*
Brice, Dr John Chadwick. *UK.*
Brickenkamp, Dr. Carroll Shelton. *USA.*
Brierley, Mr Cameron. *Canada.*
Brieva, Prof. Jorge Alfonso. *Colombia.*
Brigatti, Dr Maria Franca. *Italy.*
Bright, Dr Alan Aubrey Samuel. *UK.*
Bright, Dr. William M. *USA.*
Briguglio, Dr. James. *USA.*
Brill, Mr Wolfgang. *BRD.*
Brime, Dr Covadonga. *Spain.*
Brinkmann, Prof. Detlef. *Switzerland.*
Brisse, Dr François. *Canada.*
Brisson, Dr Josée. *Canada.*
Bristoti, Dr Anildo. *Brasil.*
Britten, Dr James Francis. *Canada.*
Britton, Prof. Doyle. *USA.*
Britton, Ms Karen Linda. *UK.*
Broach, Dr. Robert William. *USA.*
Brock, Prof. Carolyn Pratt. *USA.*
Brodalla, Dipl.-Chem Dieter. *BRD.*
Brodersen, Prof. Dr Klaus. *BRD.*
Broesby-Olsen, Mr Finn. *Denmark.*
Brokmeier, Dr Heinz-Günter. *BRD.*
Bronger, Prof. Dr Welf. *BRD.*
Bronowska, Dr Wiesława. *Poland.*
Bronsema, Dr Klaas Derk. *Netherlands.*
Brooks, Dr. Charles L. *USA.*
Brooks, Prof. Frederick Phillips Jr. *USA.*
Broosch, Mr Erika. *DDR.*
Broul, Ing Miroslav. *CSSR.*
Brovkin, Dr Anatoly Afanasyevich. *USSR.*
Brown, Miss Betty Rosina. *UK.*
Brown, Prof. Bruce Elliot. *USA.*
Brown, Prof. Bruce Willard. *USA.*
Brown, Dr Cedric John. *UK.*
Brown, Mr D. *UK.*
Brown, Dr David Summers. *UK.*
Brown, Dr. George Marshall. *USA.*
Brown, Prof. Glenn H. *USA.*
Brown, Prof. Gordon Edgar Jr. *USA.*
Brown, Dr Ian David. *Canada.*
Brown, Dr. Katherine Anne. *USA.*
Brown, Dr Kevin Laurie. *New Zealand.*
Brown, Dr. Leo Dale. *USA.*
Brown, Prof Michael Ewart. *South Africa.*
Brown, Dr Penelope Jane. *France.*
Brown, Mr. Raymond Sidney. *USA.*
Brown, Dr Roger Norman. *Australia.*
Browne, Mr Ian Bruce. *Australia.*
Bruccoleri, Dr. Robert E. *USA.*
Bruce, Dr P.G. *UK.*
Bruck, Dr. Michael Allen. *USA.*
Bruder, Dr Martin. *BRD.*
Brückel, Dr Thomas. *France.*
Brückel, Prof. Sergio. *Italy.*
Brückner, Dr Winfried. *DDR.*
Brühl, Dr Hans-Gerd. *DDR.*
Brümmer, Prof. Otto. *DDR.*
Bruggen, van, Dr Christiaan Frans. *Netherlands.*
Bruins Slot, Dr Hilbert Jan. *Netherlands.*
Brumberger, Prof. Harry. *USA.*
Brunie, Dr Simone. *France.*
Bruno, Prof. Emiliano. *Italy.*
Bruno, Prof. Giuseppe. *Italy.*
Bruvo, Magistar Milenko. *Yugoslavia.*
Bruzzone, Prof. Giacomo. *Italy.*
Bryan, Prof. Robert Finlay. *USA.*
Bryant, Mr P.K. *UK.*
Bryant, Dr. Stephen Howard. *USA.*

Bryden, Prof. John Heilner. *USA.*
Bryntse, Mrs Ingrid. *Sweden.*
Brzozowski, Dr Andrzej Marek. *Poland.*
Bubáková, Dr Růžena. *CSSR.*
Bublik, Prof Valdimir Timofeevich. *USSR.*
Bublitz, Mr Günter. *DDR.*
Buchanan, Prof. David R. *USA.*
Buchanan, Prof. Robert Martin. *USA.*
Buchner, Mr. Robert. *USA.*
Buchwald, Dr Vagn Fabritius. *Denmark.*
Buck, Prof. Dr Peter. *BRD.*
Buckley, Dr Christopher Paul. *UK.*
Bud'ko, Dr Ivetta Alexandrovna. *USSR.*
Budai, Dr. J. D. *USA.*
Budevski, Prof. Evgeni. *Bulgaria.*
Budurov, Prof. Stoyan. *Bulgaria.*
Buehner, Dr Manfred. *BRD.*
Bührer, Dr Willi. *Switzerland.*
Bürgi, Prof. Hans-Beat. *Switzerland.*
Bürki, Dr Hans. *Switzerland.*
Büyükgüngör, Dr Orhan. *Turkey.*
Bugg, Prof. Charles E. *USA.*
Buhl, Dr Josef-Christian. *BRD.*
Bukin, Dr Alexander Sergeyevich. *USSR.*
Bukovec, Dr Nataša. *Yugoslavia.*
Bukovec, Prof. Dr Peter. *Yugoslavia.*
Bukowiecki, Dr Stanislaw. *Switzerland.*
Bukowska-Strzyżewska, Dr hab. Maria. *Poland.*
Bukvetsky, Dr Boris Vladimirovich. *USSR.*
Bulakh, Dr Andrey Glebovich. *USSR.*
Bulc, Mrs Nada. *Yugoslavia.*
Bulgarovskaya, Dr Irina Vsevolodovna. *USSR.*
Bulhões, Mrs Iseli Angelica M. *Brasil.*
Bullen, Dr G.J. *UK.*
Bullen, Mr Henry Eric. *UK.*
Bunge, Prof. Dr Dr h.c. Hans-Joachim. *BRD.*
Bunick, Dr. Gerard J. *USA.*
Bunina, Dr Ol'ga Alekseevna. *USSR.*
Bunn, Dr Charles William. *UK.*
Bunning, Dr John David. *UK.*
Bunno, Dr Michiaki. *Japan.*
Buranda, Mr. Tione. *USA.*
Buras, Prof. Bronislaw. *Denmark.*
Burbank, Mr. Robinson D. *USA.*
Burdina, Dr Valentina Ivanovna. *USSR.*
Burge, Prof. Ronald Edgar. *UK.*
Burkhard, Dr Andreas. *Switzerland.*
Burla, Dr Maria Cristina. *Italy.*
Burling, Mr. Temple. *USA.*
Burnasheva, Dr Veniana Venediktovna. *USSR.*
Burnett, Prof. Roger MacDonald. *USA.*
Burnham, Prof. Charles Wilson. *USA.*
Burns, Dr. John Howard. *USA.*
Burova, Dr Elena Mikhailovna. *USSR.*
Burschka, Dr Christian. *BRD.*
Burshtein, Dr Izya Fridelevich. *USSR.*
Bursill, Dr Leslie Arthur. *Australia.*
Burton, Dr. Benjamin Paul. *USA.*
Burzlaff, Prof. Dr Hans. *BRD.*
Busaracome, Mr Suwin. *Thailand.*
Busch, Prof. Georg Adolf. *Switzerland.*
Buschmann, Dr Juergen Friedrich. *BRD.*
Buschow, Prof. Dr Kurt H. J. *Netherlands.*
Busetti, Prof. Vilma. *Italy.*
Bush, Dr. Bruce L. *USA.*
Bush, Dr Michael Anthony. *UK.*
Bush, Mr. Stewart Fowler. *USA.*
Bushnell-Wye, Dr Graham. *UK.*
Bushnell, Prof. Gordon William. *Canada.*
Bushuev, Dr Vladimir Alekseevich. *USSR.*
Busing, Dr. William Richard. *USA.*
Butaev, Dr Boris Savel'evich. *USSR.*
Butcher, Prof. Raymond John. *USA.*
Butler, Dr Barry Conrad Milne. *UK.*
Butler, Mr. Brent D. *USA.*
Butler, Dr Stephen Andrew. *UK.*
Butler, Dr. William M. *USA.*
Butler, Mr. William O. *USA.*
Butman, Dr Lev Abramovich. *USSR.*
Butt, Dr Khurshid Alam. *Pakistan.*

Butt, Mr Muhammad Hafeez. *Pakistan.*
Butt, Dr Noor Mohammad. *Pakistan.*
Butter, Prof. Ehrenfried. *DDR.*
Buttinelli, Prof. Dante. *Italy.*
Buttrey, Prof. Douglas James. *USA.*
Buyers, Dr William James Leslie, FRSC. *Canada.*
Bye, Dr Erik. *Norway.*
Bykov, Aleksei Borisovich. *USSR.*
Byram, Mrs. Susan Katherine. *USA.*
Byrn, Prof. Stephen Robert. *USA.*
Byrne, Mr Peter G. *Ireland.*
Caballero, Prof. M. Antonio. *Spain.*
Cabrera Bravo, Prof. Enrique. *México.*
Cady, Dr. Howard Hamilton. *USA.*
Caffrey, Prof. Martin. *USA.*
Caglioti, Prof. Dr Giuseppe. *Italy.*
Cahn, Dr. John Werner. *USA.*
Cahn, Prof. R.W. *UK.*
Cai, Prof. Jing-hua. *China.*
Cain, Mr Peter Maurice. *UK.*
Caira, Prof Mino Rodolfo. *South Africa.*
Çakmak, Mr Seyfettin. *Elaziǧ,Turkey.*
Calabrese, Dr. Joseph C. *USA.*
Calais, Dr Jean-Louis. *Sweden.*
Calamiotou, Dr Maria. *Greece.*
Calandra, Mr. Peter M. *USA.*
Calderón, Prof. Gómez Eduardo. *Colombia.*
Calestani, Dr Gianluca. *Italy.*
Çalişkan, Dr Nezihe. *Turkey.*
Callegari, Dr Athos. *Italy.*
Calleri, Prof. Mariano Bernardino. *Italy.*
Calligaris, Prof. Mario. *Italy.*
Calos, Mr Nicholas James. *Australia.*
Calvet, Mr. Teresa. *Spain.*
Calzadilla Amava, Mr Octąvio. *Cuba.*
Camalli, Dr Mercedes. *Italy.*
Camerman, Prof. Arthur. *USA.*
Camerman, Dr Norman. *Canada.*
Cameron, Dr Allan Forbes. *UK.*
Cameron, Mr. Robert P. *USA.*
Cameron, Prof. Theodore Stanley. *Canada.*
Cameroni, Prof. Riccardo. *Italy.*
Cammenga, Prof. Dr Heiko Karl. *BRD.*
Campana, Dr. Charles F. *USA.*
Campanelli, Dr Anna Rita. *Italy.*
Campbell, Dr John Wilson. *UK.*
Campbell, Dr Robert Laurence. *Canada.*
Campelo Farias, Prof. Carlinda. *Brasil.*
Campos, Dr Cícero. *Brasil.*
Candeloro De Sanctis, Prof. Sofia. *Italy.*
Canepa, Dr Horacio Ricardo. *Argentina.*
Cannas, Prof. Mario. *Italy.*
Cannillo, Dr Elio. *Italy.*
Cano Corona, Dr Octavio. *México.*
Cantrell, Prof. Joseph Sires. *USA.*
Cao, Prof. Ming-zhong. *China.*
Cao, Prof. Zheng-min. *China.*
Capano, Mr. Michael A. *USA.*
Capasso, Prof. Sante. *Italy.*
Capelle, Dr Bernard. *France.*
Čapková, Dr Pavla. *CSSR.*
Capotorto, Prof. Concetta. *Italy.*
Capparelli, Prof. Marlo Vicente. *Venezuela.*
Caranoni, Dr Anny. *France.*
Carbonin, Dr Susanna. *Italy.*
Cardellach, Dr Esteve. *Spain.*
Cardin, Dr Christine Janet. *Ireland.*
Cardyn, Dr. Christine Janet. *UK.*
Cargill, Dr. George Slade III. *USA.*
Carlisle, Prof. Charles Harold. *UK.*
Carlson, Dr. Ernest Howard. *USA.*
Carlson, Dr Sirkka Liisa. *Finland.*
Carlson, Mrs. Vanice Aparecida Perin. *USA.*
Carlson, Dr. William D. *USA.*
Carnahan, Dr. Gary Ellis. *USA.*
Caron, Dr. Aimery Pierre. *USA.*
Carpenter, Dr. Donald Allmand. *USA.*
Carpenter, Prof. Gene B. *USA.*
Carperos, Dr. William E. *USA.*
Carr, Dr. Martin J. *USA.*

Carrano, Mr. Carl J. *USA.*
Carrell, Dr. Horace L. *USA.*
Carriedo, Dr Gabino-Alejandro. *Spain.*
Carrillo-Hoyo, Dr. Eduardo. *México.*
Carrithers, Mr. Charles H. *USA.*
Carroll, Mr. Patrick J. *USA.*
Carrondo, Prof. Maria Arménia. *Portugal.*
Carson, DMr. William Michael. *USA.*
Carteni-Farina, Prof. Maria. *Italy.*
Carter, Prof. Charles Williams. *USA.*
Carter, Dr. Daniel Clark. *USA.*
Carter, Mr Trevor John. *UK.*
Cartwright, Dr Michael. *UK.*
Carty, Prof. Arthur John. *Canada.*
Cartz, Dr. Louis. *USA.*
Carugo, Dr Oliviero. *Italy.*
Caruso, Dr Francesco. *Italy.*
Carvalho da Silva, Prof. Jair. *Brasil.*
Casabo, Prof. Jaume. *Spain.*
Casalone, Dr Gianluigi. *Italy.*
Casanova, Lic Jorge Ramón. *Argentina.*
Cascales, Dr Concepcion. *Spain.*
Cascarano, Dr Giovanni Luca. *Italy.*
Case, Mr. J. A. M. *USA.*
Casellato, Dr Umberto. *Italy.*
Cashion, Assoc. Prof. John Dixon. *Australia.*
Caslavsky, Dr. Jaroslav L. *USA.*
Caspar, Prof. Donald L. D. *USA.*
Cassedane, Dr Jeannine. *Brasil.*
Cassel, Dr Anders Ö. *Sweden.*
Casson, Mr. Lawrence P. *USA.*
Castellano, Dr Eduardo Ernesto. *Brasil.*
Castellanos Guzman, Dr A. Guillermo. *México.*
Castellanos Román, Mrs Maria Asunción. *México.*
Castiñeiras, Dr Alfonso. *Spain.*
Castro, Dr Alicia. *Spain.*
Caticha Alfonso, Mr Ariel. *Brasil.*
Caticha Ellis, Prof. Stephenson. *Brasil.*
Catti, Prof. Michele. *Italy.*
Caucia, Dr Franca. *Italy.*
Caughlan, Prof. Charles N. *USA.*
Cava, Dr. Robert J. *USA.*
Cavalca, Prof. Luigi. *Italy.*
Caveney, Dr Robert John. *South Africa.*
Cavero Ghersi, Dr César Augusto. *Peru.*
Cavin, Mr. Odis Burl. *USA.*
Cebula, Dr D. *UK.*
Čech, Prof. Dr František. *CSSR.*
Čeh, Dr Boris, *Yugoslavia.*
Cejalvo, Prof. Flor. *Philippines.*
Celikel, Dr. Reha. *USA.*
Cellai, Dr Luciano. *Italy.*
Celotti, Dr Giancarlo. *Italy.*
Čermák, Dr Jan. *CSSR.*
Čeřňanský, Ing Marian. *CSSR.*
Cernik, Dr Robert Joseph. *UK.*
Černohorský, Doc. Dr Martin. *CSSR.*
Černý, Dr Petr. *Canada.*
Černý, Dr Radovan. *CSSR.*
Cerrini, Dr Silvio. *Italy.*
Červeň, Doc. Dr Ivan. *CSSR.*
Červinka, Dr Ladislav. *CSSR.*
Cesari, Prof. lib. doc. Marco. *Italy.*
Ceylan, Doç. Dr Mehmet. *Turkey.*
Ceylan, Dr Kazim. *Turkey.*
Chaban, Dr Nadezhda Fedorovna. *USSR.*
Chacko, Dr K. K. *India.*
Chadha, Asok Kumar. *India.*
Chadha, Dr Gopal Krishan. *India.*
Chaichit, Dr Narongsak. *Thailand.*
Chaikum, Dr Nitirampai Latavalya. *Thailand.*
Chaikum, Dr Nopadol. *Thailand.*
Chaillout, Dr Catherine. *France.*
Chakoumakos, Dr. Bryan Charles. *USA.*
Chakrabarti (Chatterjee), Dr (Mrs) C. *India.*
Chakrabarty, Dipak Kumar. *India.*
Chakrabarty, Subhasis. *India.*
Chakrabarty, Mr Sugoto. *India.*
Chakraborty, Dipak. *Lect.*
Chakraborty, Dr Suchit Chandra. *India.*

Chakravarti, Ms Lily. *India.*
Chakravarty, Dr A. R. *India.*
Chalupa, Ing Bohumil. *CSSR.*
Chamberland, Prof. Bertrand L. *USA.*
Champier, Prof. Georges. *France.*
Champion, Dr John Anthony. *UK.*
Champion, Mr. William C. *USA.*
Champness, Dr John Norman. *UK.*
Champness, Dr Pamela Eileen. *UK.*
Chandra, Prof. Dhanesh. *USA.*
Chandrasekaran, Prof. Katuputhur Sarma. *India.*
Chandrasekaran, Prof. Rengaswami. *USA.*
Chandrasekhar, Dr. K. *USA.*
Chandrasekhar, Prof. Sivaramakrishna. *India.*
Chandrasekharaiah, Dr M. N. *India.*
Chandross, Dr. Ronald Jay. *USA.*
Chaney, Dr. Michael Owen. *USA.*
Chang, Dr Chin-Pu. *Taiwan.*
Chang, Dr. Chong-Hwan. *USA.*
Chang, Dr Hou-Cheng. *Taiwan.*
Chang, Dr Shih- Lin. *Brasil.*
Chang, Prof. Shih-Chi. *USA.*
Chang, Prof. Shih-Lin. *Taiwan.*
Chang, Mr Tien-Show. *Taiwan.*
Chang, Dr Wei-Jui. *Australia.*
Chang, Mr Wen-rui. *China.*
Chao, Dr George Y. *Canada.*
Chapela Castañares, Dr Víctor Manuel. *México.*
Chapman, Dr. Michael Stewart. *USA.*
Chapuis, Prof. Gervais Constant. *Switzerland.*
Charland, Dr Jean-Pierre. *Canada.*
Chasen, Prof. Edith. *USA.*
Chashchinov, Dr Yury Mikhailovich. *USSR.*
Chastain, Dr. Roger Vernon Jr. *USA.*
Chatterjee, Dr Amitava. *India.*
Chatterjee, Dr Sanat Kumar. *India.*
Chattopadhyay, Debasish. *India.*
Chattopadhyay, Dr Tapan Kumar. *France.*
Chaudhary, Dr Abdul Majid. *Pakistan.*
Chaudhary, Mr G. Sarwar Alam. *Pakistan.*
Chaudhry, Mr Mohammad Anwar. *Pakistan.*
Chaudhuri, Dr Ahindra Kumar. *India.*
Chaudhuri, Dr. Jharna. *USA.*
Chaudhuri, Mr Siddhartha. *India.*
Chauhan, Mr Ehsanul Haq. *Pakistan.*
Chawdhury, Prof. Sadruddin Ahmed. *Bangladesh.*
Chawla, Dr Krishan Lal. *India.*
Cheang, Dr Kok Keong. *Malaysia.*
Cheary, Dr Robert Winston. *Australia.*
Cheatam, Ms. Linda J. *USA.*
Cheer, Prof. Clair James. *USA.*
Cheetham, Prof. Anthony Kevin. *UK.*
Cheikhrouhou, Dr Abdelwaheb. *Tunisia.*
Chełkowski, Prof. August Jan. *Poland.*
Chelliah, Dr Mahadevan. *India.*
Chen Mr Jie. *Japan.*
Chen, Mr Ben-ming. *China.*
Chen, Dr. Cheng-San. *USA.*
Chen, Prof. Dai-zhang. *China.*
Chen, Mr Guo-ying. *China.*
Chen, Prof. Haydn H. *USA.*
Chen, Dr Jiang-Tsun. *BRD.*
Chen, Mr Jing-yi. *China.*
Chen, Mr Jing-zhong. *China.*
Chen, Prof. Kuang-yuan. *China.*
Chen, Prof. Li-quan. *China.*
Chen, Mr. Liqing. *USA.*
Chen, Dr. Longyin. *USA.*
Chen, Prof. Pei-Yuan. *Taiwan.*
Chen, Prof. Ruey-Hong. *Taiwan.*
Chen, Mrs Shi-zhi. *China.*
Chen, Dr Wei. *Malaysia.*
Chen, Prof. Xian-qiu. *China.*
Chen, Miss Ying. *USA.*
Chen, Mr Yuan-zhu. *China.*
Chen, Mrs Yueh-Hua. *Taiwan.*
Chen, Mr Zhi-xue. *China.*
Chen, Dr Zhong-guo. *China.*
Cheng, Mr Graham Cheng-hsun. *Hong Kong.*
Cheng, Mr Min-chin. *China.*

Cheng, Dr Pei-Tak. *Canada.*
Cheremskoy, Dr Petr Grigoryevich. *USSR.*
Cherepanova, Dr Tamara Alekseyevna. *USSR.*
Cherkezyan, Dr Suren Asaturovich. *USSR.*
Chernega, Dr Aleksandr Nikolaevich. *USSR.*
Cherner, Dr Yakov Efremovich. *USSR.*
Chernov, Dr Aleksandr Nikolaevich. *USSR.*
Chernov, Prof. Alexander Alexandrodich. *USSR.*
Cherns, Dr David. *UK.*
Chernyshev, Dr Vladimir Vasil'evich. *USSR.*
Chernysheva, Dr Marina Alexandrovna. *USSR.*
Chernysheva, Mrs Valentina Fedorovna. *USSR.*
Chertanova, Dr Lyubov' Fedorovna. *USSR.*
Chetal, Prof. Amritlal R. *India.*
Chetkina, Dr Larisa Arkadyevna. *USSR.*
Cheung, Dr Kung Kai. *Hong Kong.*
Chevrier, Dr *France.*
Chiang, Prof. Liang-jun. *China.*
Chiang, Dr. Michael Yen-Nan. *USA.*
Chiari, Prof. Giacomo. *Italy.*
Chidambaram, Dr Rajagopala. *India.*
Chidester, Ms. Connie. *USA.*
Chieh, Prof. Chung (Peter). *Canada.*
Chiesi-Villa, Prof. Angiola. *Italy.*
Chikaura, Dr Yoshinori. *Japan.*
Chikawa, Dr Jun-ichi. *Japan.*
Childs, Dr Cyril Walter. *New Zealand.*
Childs, Dr. Jerry D. *USA.*
Chinnakali, K. *India.*
Chipman, Dr. David Randolph. *USA.*
Chiragov, Dr Mamed Isa ogly. *USSR.*
Chirgadze, Dr Yury Nikolayevich. *USSR.*
Chirino, Mr. Arthur. *USA.*
Chirlian, Dr. L. E. *USA.*
Chisholm, Dr James Edwin. *UK.*
Chiu, Ms. Celia C. *USA.*
Chiu, Prof. Kuan-Cheng *Taiwan.*
Chiu, Prof. Wah. *USA.*
Chlebowski, Dr. Jan F. *USA.*
Cho, Prof. Sung-Il. *Korea.*
Chodosh, Dr. Daniel F. *USA.*
Choi, Dr. Chang Sun. *USA.*
Choi, Mr. Hok-Kin. *USA.*
Chollet, Dr Lucien-Francois. *Switzerland.*
Choosang, Mrs Pilai. *Thailand.*
Chopra, Prof. Kasturilal. *India.*
Chornik, Dr Boris. *Chile.*
Choudhary, Dr Muhammad Iqbal. *Pakistan.*
Choudhury, Mrs Shamima. *Bangladesh.*
Chowdari, Prof. B.V.R. *Singapore.*
Chowdhury, Prof. Fazlul Halim. *India.*
Christensen, Dr Axel Nørlund. *Denmark.*
Christian, Prof. John Wyrill. *UK.*
Christianson, Dr. David William. *USA.*
Christidis, Prof. Panayiotis Chrysostomos. *Greece.*
Christofferson, Dr. Glen D. *USA.*
Christoph, Mr Arthur. *DDR.*
Christoph, Dr. Gary Gordon. *USA.*
Chu, Prof. Shirley Shan-C. *USA.*
Chuadry, Dr Fazal Muhammad. *Pakistan.*
Chuang, Mr Kung-Chou. *Taiwan.*
Chudinov, Prof. Sergei Mikhailovich. *USSR.*
Chudinova, Dr Svetlana Alekseyevna. *USSR.*
Chukhovsky, Dr Felix Nikolayevich. *USSR.*
Chuknyisky, Dr. Peter Peterson. *USA.*
Chumakova, Dr Svetlana Petrovna. *USSR.*
Chung, Prof. Being-Tau. *Taiwan.*
Chung, Dr Mui-Fatt. *Singapore.*
Chung, Prof. Dr Su Jin. *Korea.*
Chung, Prof. Y. W. *USA.*
Chung, Mr. Yong Je. *USA.*
Chuprunov, Dr Evgeniy Vladimirovich. *USSR.*
Churchill, Prof. Melvyn Rowen. *USA.*
Churchman, Dr Gordon John. *New Zealand.*
Chvalun, Dr. Sergei Nikolaevich. *USSR.*
Cia, Mr Jin-hua. *China.*
Ciajolo, Prof. Maria Rosaria. *Italy.*
Ciani, Prof. Gianfranco. *Italy.*
Čičel, Ing Blahoslav. *CSSR.*
Cid, Dr Hilda. *Chile.*

Ciechanowicz-Rutkowska, Dr Maria. *Poland.*
Cimino, Prof. Alessandro. *Italy.*
Cini, Prof. Renzo. *Italy.*
Cioflica, Prof. Graţian. *Romania.*
Cipriani, Prof. Curzio. *Italy.*
Cirafici, Prof. Salvino S. *Italy.*
Cisneros Ramos, Prof. Luis. *Peru.*
Ciszak, Dr Ewa. *Poland.*
Ciunik, Dr Zbigniew. *Poland.*
Clancy, Ms. Laura Lee. *USA.*
Clarage, Mr. James Braun II. *USA.*
Claramunt, Prof. Rosa Maria. *Spain.*
Clardy, Prof. Jon Christel. *USA.*
Claridge, Dr Graeme Geoffrey. *New Zealand.*
Clark, Ms. Connie M. *USA.*
Clark, Prof. Edward Shannon. *USA.*
Clark, Dr George Raymond. *New Zealand.*
Clark, Dr James Brian. *South Africa.*
Clark, Mrs. Joan Robinson. *USA.*
Clark, Dr Malcolm John Roy. *Canada.*
Clarke, Mr. Frank H. *USA.*
Clarke, Prof. Roy. *USA.*
Claus, Mr. Albert C. *USA.*
Claus, Mr Karl Heinz. *BRD.*
Clausen, Dr Kurt N. *Denmark.*
Clawson, Mr. David K. *USA.*
Clay, Mrs Kathleen. *UK.*
Clayton, Dr. William Rex. *USA.*
Clearfield, Prof. Abraham. *USA.*
Clegg, Dr William. *UK.*
Clemente, Prof. Dore Augusto. *Italy.*
Clewer, Mr Peter John. *UK.*
Cline, Dr. James P. *USA.*
Clinger, Dr. Kent. *USA.*
Cobbledick, Dr Roger Ernest. *Canada.*
Cochran, Prof. Todd G. *USA.*
Cochran, Prof. William. *UK.*
Cockayne, Dr David John Hugh. *Australia.*
Cockcroft, Dr Jeremy Karl. *France.*
Coda, Prof. Alessandro. *Italy.*
Codding, Dr Penelope Wixson. *Canada.*
Cody, Dr. Vivian. *USA.*
Cohen-Addad, Dr Claudine. *France.*
Cohen, Prof. Carolyn. *USA.*
Cohen, Dr. Gerson H. *USA.*
Cohen, Ms. Janet Paula. *USA.*
Cohen, Prof. Jerome Bernard. *USA.*
Cohen, Dr L. *USA.*
Cohen, Dr Shmuel. *Israel.*
Coiro, Dr Vincenza Maria. *Italy.*
Cojazzi, Dr Gianna. *Italy.*
Cola, Prof. Mario Luigi. *Italy.*
Colacio, Dr Enrique. *Spain.*
Çolakoğlu, Doç. Dr Kemal. *Turkey.*
Colapietro, Prof. Marcello. *Italy.*
Cole, Dr. Henderson. *USA.*
Cole, Dr. L. Brent. *USA.*
Colella, Prof. Roberto. *USA.*
Colgate, Prof. Samuel Oran. *USA.*
Coll, Dr Miquel. *Fed.BRD.*
Collin, Dr Gaston. *France.*
Collin, Dr Sonia Bertha Josepha. *Belgium.*
Collini, Prof. Bengt H. E. *Sweden.*
Collins, Dr. Douglas MacPherson. *USA.*
Collins, Dr. Richard C. *USA.*
Colman, Dr Peter Malcolm. *Australia.*
Colmanet, Dr Silvano Francesco. *Australia.*
Colombo, Dr Arturo. *Italy.*
Colombo, Dr Lidija. *Yugoslavia.*
Colyvas, Mr Kim. *Australia.*
Comey, Mr. Paul Van A. *USA.*
Comins, Dr Neville Raymond. *South Africa.*
Compobasso, Mr. Nino. *USA.*
Conant, Dr. John W. *USA.*
Conde, Prof. Alejandro. *Spain.*
Conde, Dr Clara Francisca. *Spain.*
Connolly, Dr. Michael Lee. *USA.*
Constantinescu, Mr Radu. *Romania.*
Cook, Prof Alan Cecil. *Australia.*
Cook, Dr David Stanley. *UK.*

Cook, Dr. William Joseph. *USA.*
Cook, Dr. William R. Jr. *USA.*
Cooley, Mr. James W. *USA.*
Coombs, Prof. Douglas Saxon. *New Zealand.*
Cooper, Dr Alan Frederick. *New Zealand.*
Cooper, Mrs. Ann S. *USA.*
Cooper, Dr. Brian J. *USA.*
Cooper, Dr. John Neale. *USA.*
Cooper, Dr Malcolm John. *UK.*
Cooper, Dr Martyn John. *UK.*
Copeland, Prof. Richard Franklin. *USA.*
Copley, Dr. John R. D. *USA.*
Coppens, Prof. Philip. *USA.*
Copperthwaite, Dr Richard George. *South Africa.*
Corbett, Dr Madeline. *Australia.*
Corchia, Dr Massimo. *Italy.*
Cordero-Borboa, Dr Adolfo. *México.*
Cordes, Prof. A. Wallace. *USA.*
Cordier, Dr Gerhard. *BRD.*
Corfield, Prof. Peter William Reginald. *USA.*
Corliss, Dr. Lester Myron. *USA.*
Corradini, Prof. Paolo. *Italy.*
Correia dos Santos, Dr António. *Portugal.*
Correia Neves, Prof. José Marques. *Brasil.*
Correll, Mr. Carl Clayton. *USA.*
Cortelezzi, Dr César Rafael. *Argentina.*
Cortés, Dr Abdón. *Colombia.*
Cossu, Dr Michéle Josette. *Côte d'Ivoire.*
Costa Gouveia, Prof. Albany H. *Brasil.*
Costa Viana, Prof. Carlos Sergio da. *Brasil.*
Costa, Prof. M. Margarida Ramalho. *Portugal.*
Costamagna, Dr Juan Alberto. *Chile.*
Costello, Mr Bernard Anthony de Lacy. *UK.*
Cota Araiza, Mr Leonel Susano. *México.*
Couldwell, Dr Margaret Claire. *New Zealand.*
Coulter, Dr. Charles L. *USA.*
Cour, la, Dr Troels Frederik Marstrand. *Denmark.*
Courseille, Prof. Christian. *France.*
Cousins, Mr Christopher Stanley George. *UK.*
Coville, Prof. Neil John. *South Africa.*
Cowan, Dr. Paul L. *USA.*
Cowan, Dr Sandra Wendy. *Sweden.*
Cowie, Prof. Martin. *Canada.*
Cowlam, Dr Neil. *UK.*
Cowley, Prof. John M. *USA.*
Cowley, Miss Margaret. *USA.*
Cox, Dr David Ernest. *USA.*
Cox, Ms. Jane. *USA.*
Cox, Ms. Mary Beth. *USA.*
Cox, Dr Philip John. *UK.*
Coy-Yll, Prof. Ramon. *Spain.*
Coyle, Mr Richard Alan. *Australia.*
Craievich, Dr Aldo Felix. *Brasil.*
Craig, Mr Donald Chadwick. *Australia.*
Craig, Mr G.R. *UK.*
Cramer, Prof. Roger Earl. *USA.*
Craston, Mr. Dennis F. *USA.*
Craven, Prof. Bryan Maxwell. *USA.*
Crawford, Mr John Lawrence. *South Africa.*
Creagh, Prof Dudley Cecil. *Australia.*
Creek, Mr Russell Charles. *Australia.*
Cremona, Ing. Luigi. *Italy.*
Crennell, Mrs K.M. *UK.*
Crennell, Ms. Susan Jane. *UK.*
Cressey, Dr Barbara Anne. *UK.*
Cressey, Dr Gordon. *UK.*
Creswick, Dr. Michael William. *USA.*
Crist, Prof. Buckley, Jr. *USA.*
Critchlow, Dr. Susan C. *USA.*
Crocker, Prof. Alan Godfrey. *UK.*
Croft, Dr. William J. *USA.*
Cromer, Dr. Don Tiffany. *USA.*
Cruceanu, Mr Eugen. *Romania.*
Cruickshank, Prof. Durward William John. *UK.*
Cruz Gandrilla, Mr Francisco. *Cuba.*
Cruz Inclan, Dr.Sc. Carlos. *Cuba.*
Csákvári, Dr (Ms) Éva. *Hungary.*
Csanády-Bokody, Dr (Ms) Ágnes. *Hungary.*
Csöregh, Dr Ingeborg. *Sweden.*
Csordás-Tóth, Dr (Ms) Anna. *Hungary.*

Csordás, Dr László. *Hungary.*
Cubiotti, Prof. Gaetano. *Italy.*
Cuchý, Ing Zdeněk. *CSSR.*
Cueto, Ms. Maria A. *USA.*
Cuevas-Diarte, Dr Miquel Angel. *Spain.*
Cuff, Dr Christopher. *Australia.*
Cuff, Ms. Marianne Elaine. *USA.*
Cui, Prof. Wen-yuan. *China.*
Cullen, Prof. David Lawrence. *USA.*
Cullen, Mr F.L. *UK.*
Cumbrera, Dr Francisco Luis. *Spain.*
Cummings, Dr Stewart. *UK.*
Cunningham, Dr Patrick Desmond. *Ireland.*
Curbelo Ramirez, Mr. Ciro. *Cuba.*
Curie, Prof. Daniel. *France.*
Curien, Prof. Hubert. *France.*
Curtin, Prof. David Yarrow. *USA.*
Curtis, Prof. M. David. *USA.*
Curzon, Prof. Albert Edward. *Canada.*
Cusatis, Dr Cesar. *Brasil.*
Cutfield, Dr John Franklin. *New Zealand.*
Cvetković, Mr Ljubiša. *Yugoslavia.*
Cvetković, Mrs Miroslava. *Yugoslavia.*
Cygler, Dr Mirek. *Canada.*
Czachor, Dr hab. Andrzej. *Poland.*
Czank, Dr Michael. *BRD.*
Czerwinski, Prof. Edmund William. *USA.*
Czerwonka, Mr Janusz. *Poland.*
Cziráki, Dr (Ms) Ágnes. *Hungary.*
Czugler, Dr Mátyás. *Hungary.*
D' Amour-Sturm, Dr Hedwig. *BRD.*
d'Anterroches, Dr Cécile. *France.*
D'yachenko, Dr Oleg Anatolyevich. *USSR.*
D'yakon, Dr Ivan Andreyevich. *USSR.*
Daams, Mr Johannes L.C. *Netherlands.*
Dabbabi, Dr Mongi. *Tunisia.*
Dachs, Prof. Dr Hans. *BRD.*
Dacombe, Mr Michael H. *UK.*
Dadel, Mrs Snehlata. *India.*
Däweritz, Dr Lutz. *DDR.*
Dagerhamn, Dr Tore. *Sweden.*
Dahan, Dr Françoise. *France.*
Dahl, Prof. Lawrence F. *USA.*
Dahl, Dr Tor. *Norway.*
Dahlkamp, Dr Franz-Joses. *BRD.*
Dai, Mr Jin-bi. *China.*
Daimon, Dr Hiroshi. *Japan.*
Dal Negro, Prof. Alberto. *Italy.*
Dall'Aglio, Dr Gian Antonio. *Italy.*
Dalley, Prof. Nelson Kent. *USA.*
Daly, Dr John Joseph. *Switzerland.*
Damak, Dr Mabrouk. *Tunisia.*
Damaschun, Dr Ferdinand. *DDR.*
Damm, Prof. Józef Zbigniew. *Poland.*
Danaci, Dr Süheyla. *Turkey.*
Dance, Prof Ian Gordon. *Australia.*
Daniels, Dr. Lee M. *USA.*
Daniels, Mr Peter. *BRD.*
Daniels, Mr. R. E. *USA.*
Danielsen, Dr Jacob. *Denmark.*
Dankházi, Mr Zoltán. *Hungary.*
Dann, Mr. Jeffrey Neil. *USA.*
Dantonio, Mr. Peter. *USA.*
Dao Cong Ngoan, Dr. *Vietnam.*
Daoo Morales, Dr.Sc. Anoel. *Cuba.*
Daoud, Prof. Abdelaziz. *Tunisia.*
Dapporto, Prof. Paolo. *Italy.*
Darby, Dr. Willie L. *USA.*
Darces, Dr Jean-François. *France.*
Dariel, Prof. Moshe Pierre. *Israel.*
Darinskaya, Dr Elena Vladimirovna. *USSR.*
Darling, Dr. Stephen D. *USA.*
Darlington, Dr Charles Nicholas Wright. *UK.*
Darovskikh, Mr. Alexander. *USA.*
Dartyge, Dr Elisabeth. *France.*
Das Gupta, Mr Prabal. *India.*
Das, Dr. Badri Narayan. *USA.*
Das, Dr Birendra Nath. *India.*
Das, Dr Indu Mohan. *India.*
Das, Pratap Kumar. *India.*

Das, Dr Sabita. *India.*
Datt, Dr Igor Daudovich. *USSR.*
Datta, Dr Amal Kumar. *India.*
Dattagupta, Dr Jiban Kanti. *India.*
Dauter, Dr Zbigniew. *Poland.*
David, Mr. Peter Rensis. *USA.*
Davidson, Mr Patrick. *France.*
Davies, Dr. David R. *USA.*
Davies, Dr Geoffrey John. *South Africa.*
Davies, Dr Gladstone. *South Africa.*
Davies, Dr. Jay Franklin, II. *USA.*
Davies, Dr John Edward. *UK.*
Davies, Dr. Julian Anthony. *USA.*
Davis, Dr. Briant LeRoy. *USA.*
Davis, Mr Paul Christopher. *Australia.*
Davis, Prof. Phillip Howard. *USA.*
Davis, Prof. Raymond Edward. *USA.*
Davis, Dr Ronald Lindsay. *Australia.*
Davison, Prof. Daniel B. *USA.*
Davoli, Dr Paolo. *Italy.*
Davydchenko, Dr Anatoliy Georgiyevich. *USSR.*
Day, Miss Catherine L. *USA.*
Day, Dr. Cynthia Ann Secauer. *USA.*
Day, Mr. Michael W. *USA.*
Day, Prof. Roberta Ogilvie. *USA.*
Day, Prof. Victor Warren. *USA.*
Dayal, Dr Radha Raman. *India.*
De Angelis, Prof. Giuseppe. *Italy.*
De Bondt, Mr Hendrik Leon Augusta Jozef. *Belgium.*
De Camp, Dr. Wilson H. II. *USA.*
De Fontaine, Prof. Robert Didier. *USA.*
De Gryse, Dr Roger Marc. *Belgium.*
De Haven, Dr. Patrick William. *USA.*
De Jarnette, Miss F. Elaine. *USA.*
De la Camp, Prof. Ulrich Otto. *USA.*
De Lucia, Dr. Mary Lou. *USA.*
De Maggio, Mr. Gregory B. *USA.*
de Meester, Dr Patrice. *UK.*
De Munno, Prof. Giovanni. *Italy.*
De Pablo Galan, Dr Linerto. *México.*
De Pol Blasi, Prof. Carla. *Italy.*
De Ranter, Prof. Dr Camiel Joseph. *Belgium.*
De Rosier, Prof. David John. *USA.*
De Santis, Prof. Pasquale. *Italy.*
De Smedt, Mr Jozef Maria A. L. *Belgium.*
De Titta, Dr. George Thomas. *USA.*
De Villiers, Dr. Johan Pieter Roos. *South Africa.*
De Vos, Dr. Abraham Martien. *USA.*
De Vries, Dr. Adriaan. *USA.*
De Winter. Mr Hans Louis Jos *Belgium.*
De, Dr Amitabha. *India.*
De, Dr Mabhusudan. *India.*
Deadwyler, Mr. Daniel A. *USA.*
Dean, Dr Christopher. *Australia.*
Dean, Mr. Johnny Clyde. *USA.*
Debaerdemaeker, Dr Tony. *BRD.*
Deblieck, Dr Rudy André Cornelis. *Netherlands.*
DeBoer, Dr. Barry Goodwin. *USA.*
Declercq, Prof. Jean Paul. *Belgium.*
Dederer, Dr Bernhard. *BRD.*
Dedukh, Dr Leonid Mikhailovich. *USSR.*
Deganello, Prof. Sergio. *USA.*
Degenhardt, Dr Detlev. *BRD.*
Degtyareva, Dr Valentina Feognievna. *USSR.*
Deguire, Mrs Suzanne. *Canada.*
Deisenhofer, Prof. Johan. *USA.*
Deiseroth, Dr Hans-Jörg. *BRD.*
Dekeyser, Prof. Dr Willy Clement. *Belgium.*
Dekker, Mr Henri. *UK.*
Del Monte, Prof. Marco Emiliano. *Italy.*
Del Piero, Dr Gastone. *Italy.*
Del Pra, Prof. Antonio. *Italy.*
Delaey, Prof. Luc J. M. A. E. *Belgium.*
Delaloye, Prof. Michel. *Switzerland.*
Delaney, Prof. Matthew S. *USA.*
Delaney, Dr William Timothy. *Australia.*
Delapalme, Dr Alain. *France.*
Delaplane, Dr Robert G. *Sweden.*
Delatore, Ms. Diana L. *USA.*
Delbaere, Dr Louis Theophil Joseph. *Canada.*

Delettré, Dr Jean. *France.*
Delf, Dr Brian William. *UK.*
Delgado Quinones, Prof. Jose Miguel. *Venezuela.*
Delgado, Mr. Jose Miguel. *USA.*
Delhez, Dr Ir Robert. *Netherlands.*
Deliens, Dr Michel. *Belgium.*
Delineshev, Dr Svetoslav. *Bulgaria.*
Della Giusta, Prof. Antonio. *Italy.*
Delord, Prof. Pierre. *France.*
Deltour, Prof. Robert. *Belgium.*
DeLucas, Dr. Lawrence James. *USA.*
Dem'yanets, Dr Lyudmila Nikolayevna. *USSR.*
Demartin, Prof. Francesco. *Italy.*
Dembo, Dr Alexander Teodorovich. *USSR.*
Demontis, Dr Pierfranco. *Italy.*
Demšar, Magistar Alojz. *Yugoslavia.*
Demus, Prof Dietrich. *DDR.*
Denisenko, Dr Georgy Alexandrovich. *USSR.*
Denner, Dr Louis. *South Africa.*
Dénoyer, Dr. Françoise. *France.*
Dent Glasser, Dr Lesley Scott. *UK.*
Deopura, Dr B. L. *India.*
Depalma, Mr. Vincent M. *USA.*
Depero, Dr Laura Eleonora. *Italy.*
Depmeier, Prof Dr Wulf Helmut Heinz. *BRD.*
Depp, Mr. Mark. *USA.*
Derewenda, Mrs Urszula. *UK.*
Derewenda, Dr. Zygmunt. *UK.*
Deroski, Ms. Betty Rolfs. *USA.*
Deruyttere, Prof. Dr André. *Belgium.*
Derwent, Mr Frank William. *UK.*
Desiraju, Dr. Gautam R. *USA.*
Desper, Dr. C. Richard. *USA.*
Despotović, Mr Zlatko. *Yugoslavia.*
Destro, Prof. Riccardo. *Italy.*
Deus, Dr Peter. *DDR.*
Deutsch, Prof. Moshe. *Israel.*
Dewan, Prof. John C. *USA.*
Dexter, Dr. David D. *USA.*
Deyanov, Dr Ramil' Zinyatullovich. *USSR.*
Dhanaraj, Dr V. *India.*
Dhaneshwar, Dr Narayandatta Nagesh. *India.*
Dhawan, Mrs Urmil. *India.*
Dhere, Dr. Ashok G. *USA.*
Dhupia, Mr Gursev Singh. *BRD.*
Di Blasio, Prof. Benedetto. *Italy.*
Di Lorenzo, Dr Guido. *Italy.*
Di Vaira, Prof. Massimo. *Italy.*
Diamond, Dr Robert. *UK.*
Dianez-Millan, Dr M. Jesus. *Spain.*
Dias Rodrigues, Mrs Ana Maria Gonçalves. *Brasil.*
Dias, Dr Hanwellage Wijayapala. *Sri Lanka.*
Díaz Peraza, Prof. José Milcíades. *Colombia.*
Diaz, Dr Francesc. *Spain.*
Dichmann, Dr Klaus. *Canada.*
Dickerson, Prof. Richard Earl. *USA.*
Dickinson, Dr. Charles. *USA.*
Dickman, Dr. Michael H. *USA.*
Dideberg, Dr Otto. *Belgium.*
Diehl, Dr J. *BRD.*
Diehl, Dr Roland. *BRD.*
Dietrich, Mr. Andreas. *BRD.*
Dietrich, Prof Burkhard. *DDR.*
Dietrich, Prof. Dr Hans Karl Ernst. *BRD.*
Dijk, van, Dr Cornelis. *Netherlands.*
Dijkstra, Dr Bauke Wiepke. *Netherlands.*
Dikici, Doç. Dr Mustafa. *Turkey.*
Dillen, Dr Jan Louis Maria. *South Africa.*
Dimitrijević, Magistar Radovan. *Yugoslavia.*
Dimitrova, Dr Ol'ga Vladimirovna. *USSR.*
Dimov, Mr Vergil. *Bulgaria.*
Dinçer, Dr Muharrem. *Turkey.*
Dineen, Mr C. *UK.*
Dinescu, Prof. Radu. *Romania.*
Dingley, Dr David Joseph. *UK.*
Diodati, Dr Francisco Piero. *Argentina.*
Dion, Mrs Chantal. *Canada.*
Dirl, Prof. Dr Rainer. *Austria.*
Dittmar, Dr Günter. *BRD.*
Divjaković, Prof. Dr Vladimir. *Yugoslavia.*

Dixon, Ms. Melinda M. *USA.*
Djarova, Dr Maria. *Bulgaria.*
Djebli, Mr. Abdellah. *USA.*
Djinovič, Miss Kristina. *Yugoslavia.*
Djordjević, Prof. Dr Slobodan. *Yugoslavia.*
Djurić, Dr Stevan. *Yugoslavia.*
Dlouhá, Ing Maja. *CSSR.*
Dlubek, Prof Günter. *DDR.*
Dmitrieva, Dr Tatyana Vladimirovna. *USSR.*
Dmitriyeva, Dr Margarita Timofeyevna. *USSR.*
Do Minh Nghiep, Dr. *Vietnam.*
Dobler, Prof. Max. *Switzerland.*
Dobrev, Dr Dobri. *Bulgaria.*
Dobrowolska, Dr Wanda. *Poland.*
Dobrzyński, Dr hab. Ludwik. *Poland.*
Dobson, Dr Peter James. *UK.*
Dobson, Dr Susan Mary. *South Africa.*
Dodd, Dr. Charles Gardner. *USA.*
Dodge, Prof. Richard Patrick. *USA.*
Dodokin, Dr Anatoly Petrovich. *USSR.*
Dódony, Dr István. *Hungary.*
Dodson, Mrs Eleanor Joy. *UK.*
Dodson, Prof George Guy. *UK.*
Doedens, Prof. Robert John. *USA.*
Dörr, Dr Friedrich Johannes. *BRD.*
Dörrfeld, Mr Hans-Georg. *DDR.*
Dörschel, Dr Jürgen. *DDR.*
Doesburg, Dr Hendrikus M. *Netherlands.*
Doğan, Dr Ali. *Turkey.*
Dohi, Prof. Shoso. *Japan.*
Doi, Dr Kenji. *Japan.*
Doi, Dr Mitsunobu. *Japan.*
Doidge-Harrison, Mrs Solange Maria S. V. *UK.*
Dolivo-Dobrovol'skaya, Mrs Elena M. *USSR.*
Dolivo-Dobrovol'skaya, Dr Galina Ilyinishna. *USSR.*
Dollase, Prof. Wayne A. *USA.*
Dolling, Dr Gerald. *Canada.*
Domenech, Dr M. Victoria. *Spain.*
Domeneghetti, Dr Maria Chiara. *Italy.*
Domenicano, Prof. Aldo. *Italy.*
Domenici, Dr Marcello. *Italy.*
Domiano, Prof. Paolo. *Italy.*
Domingos, Dr Angela. *Portugal.*
Domínguez Esquivel, Dr José Manuel. *México.*
Dominguez, Dr Esther. *Spain.*
Domnitskaya, Dr Yaroslava Fedorovna. *USSR.*
Donaldson, Prof John Dallas. *UK.*
Donati, Dr Donato. *Italy.*
Donatz, Mrs Martina. *Switzerland.*
Dong, Mr Ji-he. *China.*
Dong, Prof. Yi-cheng. *China.*
Doniach, Mr. S. *USA.*
Donnay, Prof. Joseph Désiré Hubert. *Canada.*
Donoso, Mr Eduardo. *Chile.*
Dora, Prof. Dr Özcan. *Turkey.*
Dorfman, Dr Moisey Davydovich. *USSR.*
Dorokhova, Dr Galina Igorevna. *USSR.*
Doroshinsky, Dr Alexander Leibovich. *USSR.*
Dorset, Dr. Douglas Lewis. *USA.*
Doscher, Dr. Marilynn S. *USA.*
Dou, Mr Shi-qi. *China.*
Doucet, Dr Jean. *France.*
Doudin, Mr Bernard. *Switzerland.*
Dougill, Dr Maryon W. *UK.*
Dover, Dr Stanley David. *UK.*
Dovesi, Prof. Roberto. *Italy.*
Downie, Mr George. *UK.*
Downing, Dr. Kenneth H. *USA.*
Downs, Dr. James Winston. *USA.*
Dowty, Mr. Eric. *USA.*
Doyle, Prof. John Robert. *USA.*
Doyle, Mr Michael Joseph. *UK.*
Doyne, Prof. Thomas H. *USA.*
Dräger, Dr Günter. *DDR.*
Dräger, Prof. Dr Martin. *BRD.*
Draganova, Dr Dragana. *Bulgaria.*
Draghici, Mr Iosif. *Romania.*
Dragsdorf, Prof. Russell Dean. *USA.*
Drake, Prof. John E. *Canada.*
Drašner, Magistar Antun. *Yugoslavia.*

Drenck, Prof. Kaj. *Denmark.*
Drendel, Dr. William Bruce. *USA.*
Drennan, Dr John. *Australia.*
Drenth, Prof. Dr Jan. *Netherlands.*
Dressler, Dr Ludwig. *DDR.*
Drew, Dr Michael George Brindley. *UK.*
Drickman, Dr. Myra Vivian. *USA.*
Driesel, Dr Wolfgang. *DDR.*
Driessen, Drs René A. J. *Netherlands.*
Driss, Dr Ahmed. *Tunisia.*
Dristas, Dr Jorge Anastasio. *Argentina.*
Drits, Dr Victor Anatolyevich. *USSR.*
Drozdov, Dr Yury Nikolayevich. *USSR.*
Druyan, Prof. Mary Ellen. *USA.*
Duarte, M. Teresa. *Portugal.*
Duax, Dr. William Leo. *USA.*
Dubček, Mr Pavo. *Yugoslavia.*
Dubey, Dr Ram Janam. *Nigeria.*
Dubler, Prof. Erich. *Switzerland.*
Dubourg, Mr Antoine. *France.*
Dubov, Dr Petr Lvovich. *USSR.*
Dubravina, Dr Aida Nikolaevna. *USSR.*
Duca, Dr Dario. *Italy.*
Duchamp, Dr. David James. *USA.*
Duckett, Mr G.R. *UK.*
Ducruix, Dr Arnaud. *France.*
Dudarev, Dr Vasily Yakovlevich. *USSR.*
Duderov, Dr Nikolay Grigoryevich. *USSR.*
Dudkevich, Prof. Vladimir Petrovich. *USSR.*
Dündar, Yrd. Doç. Dr Sacit. *Turkey.*
Dünkel, Mr Lothar. *DDR.*
Duerring, Mr Markus. *BRD.*
Duesler, Dr. Eileen N. *USA.*
Duffy, Dr Douglas Neil. *Australia.*
Duisenberg, Drs Albert Jozef Maria. *Netherlands.*
Duke, Mr J.R.C. *UK.*
Duke, Dr Norma Edith. *USA.*
Dukova, Dr Elena Dmitriyevna. *USSR.*
Dumitrescu, Aurelia. *Romania.*
Dumke, Prof. Warren Lloyd. *USA.*
Dunaj-Jurčo, Doc. Ing Michal. *CSSR.*
Dunham, Prof A.C. *UK.*
Dunia, Prof. Emery. *Venezuela.*
Dunitz, Prof. Jack David. *Switzerland.*
Dunn, Dr. Deborah Anne. *USA.*
Dunn, Mr. Karl L. *USA.*
Dunsieth, Mr. Dana G. *USA.*
Duoue Rodriguez, Mr. Julio. *Cuba.*
Dupont, Dr Leon. *Belgium.*
Durand, Dr. Dominique. *France.*
Durant, Prof. François Victor. *Belgium.*
Ďurčanská, Dr Edita. *CSSR.*
Durchschlag, Dr Helmut. *BRD.*
Durif, Dr André. *France.*
Durlu, Prof. Dr Tahsin Nuri. *Turkey.*
Duroc-Danner, Mr Jean-Marie. *Switzerland.*
Ďurovič, Ing Slavomil. *CSSR.*
Durski, Dr Stanisław. *Poland.*
Dutremez, Mr. Sylvain G. *USA.*
Dutta, Dr Bishnu Pada. *India.*
Dutta, Dr Sachindra Nath. *India.*
Dužević, Dr Davor. *Yugoslavia.*
Dvorkin, Dr Alexander Arkadyevich. *USSR.*
Dweltz, Dr Neville Edwin. *India.*
Dwiggins, Dr. Claudius William. *USA.*
Dwight, Mr. Austin Elbert. *USA.*
Dwivedi, Dr Ganpat Lal. *India.*
Dyar, Dr. M. Darby. *USA.*
Dynowska, Mrs Elżbieta. *Poland.*
Dyson, Mr David John. *UK.*
Dytrych, Mr. William J. *USA.*
Dyuzheva, Dr Tat'yana Ivanovna. *USSR.*
Dzhafarov, Dr Kara Mustafa ogly. *USSR.*
Dzyabchenko, Dr Aleksandr Valentinovich. *USSR.*
Eales, Prof. Hugh Victor. *South Africa.*
Ealick, Dr. Steven Edward. *USA.*
Eanes, Dr. Edward David. *USA.*
Earnest, Dr. Thomas. *USA.*
Ebby, Dr N' Dédé. *Côte d'Ivoire.*
Eberhard, Prof. Dr Emil. *BRD.*

Ebinger, Mr. Michael H. *USA.*
Echavarri Hernandez, Dr Ariel. *México.*
Eck, Mr. Michael J. *USA.*
Eckerlin, Dr Peter. *BRD.*
Eckhardt, Prof. Dr Franz-Jörg. *BRD.*
Eckstein, PhD. Dipl. Ing. Juraj. *BRD.*
Economou, Prof. Nicolaos Alkiviadis. *Greece.*
Eddy, Prof. Lowell Perry. *USA.*
Eddy, Mr. Michael. *USA.*
Eder, Prof. Dr Dipl.-Ing. Otto Josef. *Austria.*
Edmonds, Dr James William. *Netherlands.*
Edmondson, Mr M. *UK.*
Edmondson, Dr. Stephen P. *USA.*
Edmundson, Prof. Allen Broderick. *USA.*
Edström, Mrs Kristina. *Sweden.*
Edwards, Dr Anthony John. *UK.*
Edwards, Dr. Brian Francis Peregrin. *USA.*
Edwards, Dr Ian Arthur Samuel. *UK.*
Edwards, Dr William Donald. *Canada.*
Eeles, Mr Wilfred Trefor. *UK.*
Efendiev, Dr El'dar Gusein ogly. *USSR.*
Effenberger, Prof. Dr Herta. *Austria.*
Efimov, Dr Aleksandr Vasil'evich. *USSR.*
Efremov, Dr Valery Alexandrovich. *USSR.*
Efremova, Dr Elena Pavlovna. *USSR.*
Egert, Prof. Dr Ernst. *BRD.*
Eggleston, Dr. Drake Stephen. *USA.*
Eggleton, Dr Richard Anthony. *Australia.*
Egli, Dr Martin. *Switzerland.*
Egorov-Tismenko, Dr Yury Klavdiyevich. *USSR.*
Ehses, Dr Karl-Heinz. *BRD.*
Eichhorn Dr Klaus-D. *BRD.*
Eichhorn, Dr Gerd. *DDR.*
Eichler, Dr Klaus. *BRD.*
Eichler, Dr Wolfgang. *DDR.*
Eick, Prof. Harry A. *USA.*
Eigenbrot, Dr. Charles Weaver. *USA.*
Eilerman, Dr. Donna Paige. *USA.*
Einck, Dr. James J. *USA.*
Einspahr, Dr. Howard Martin. *USA.*
Einstein, Prof. Frederick William Boldt. *Canada.*
Eiríksson, Dr Vésteinn Runi. *Iceland.*
Eisenberg, Prof. David. *USA.*
Eisenberg, Prof. Henryk. *Israel.*
Eisenberg, Prof. Richard. *USA.*
Eisenman, Prof. George. *USA.*
Eisenmann, Mrs Dr Brigitte. *BRD.*
Eisenschmidt, Mr Christian. *DDR.*
Eisenstein, Dr Miriam. *Israel.*
El Demerdash, Dr Saad. *Egypt.*
El Gabi, Dr Sami. *Egypt.*
El Naggar, Dr Mohamed. *Egypt.*
El Ramly, Mr Mohamed Fawzi. *Egypt.*
El Sayed, Prof. (Mrs) Karimat. *Egypt.*
El Shaabini, Prof. (Mrs) Aida Moustafa. *Egypt.*
El Shanshury, Dr Ismail. *Egypt.*
El Sharkawi, Dr Mohamed Abdel Hamid. *Egypt.*
El Shazli, Dr El Shazli Mohamed. *Egypt.*
El-Azizi, Mr Ibrahim M. *Libya.*
El-Kabbani, Mr. Ossama A. L. *USA.*
El-Mahdi, Dr Omar. *Saudi Arabia.*
El-Mashri, Dr S.M. *Libya.*
Elahi, Mr Manzoor. *Pakistan.*
Elango, Mr N. *India.*
Elbadri, Dr H. *Egypt.*
Elbinger, Dr German. *DDR.*
Elcombe, Dr Margaret Marion. *Australia.*
Elder, Dr D.P. *UK.*
Elder, Prof. Richard C. *USA.*
Elerman, Doç. Dr Yalçin. *Turkey.*
Elf, Dr Frank. *BRD.*
Elias, Mr E.E. *UK.*
Eliopoulos, Dr Elias Edward. *UK.*
Ellid, Dr Mohamed S. *Libya.*
Elliott, Dr Gerald Frank. *UK.*
Elliott, Dr James Cornelis. *UK.*
Elliott, Dr Robert Brian. *UK.*
Ellmeyer, Mag. Wolfgang. *Austria.*
Ellner, Dr Martin Oliver. *BRD.*
Elmali, Mr Ayhan. *Turkey.*

Elwan, Dr Ahmed Abdel Salam. *Egypt.*
Ely, Dr. Kathryn R. *USA.*
Elzawi, Mr Rajab Abdulla. *Libya.*
Em, Dr Vyacheslav Terent'evich. *USSR.*
Embrey, Mr Peter Godwin. *UK.*
Emerson, Prof. Kenneth. *USA.*
Emerson, Mr. Merle T. *USA.*
Emge, Dr. Thomas James. *USA.*
Emiliani, Prof. Francesco. *Italy.*
Emmenegger, Prof. Franzpeter. *Switzerland.*
Emons, Prof. Hans-Heinz. *DDR.*
Enckevort, van, Dr Wilhelmus J. P. *Netherlands.*
Enemark, Prof. John Henry. *USA.*
Eng-Wilmot, Dr. David Lawrence. *USA.*
Engel, Dr Aribert. *DDR.*
Engel, Prof. Dennis Walter. *South Africa.*
Engel, Dr Nora. *Switzerland.*
Engel, Dr Walter. *BRD.*
Engelhardt, Dr Günter. *BRD.*
Engels, Prof. Siegfried. *DDR.*
Englisch, Mr Uwe-Franz. *BRD.*
English, Dr Robert Bertram. *South Africa.*
Engström, Dr Ingvar O. J. *Sweden.*
Ensling, Dr Jürgen. *BRD.*
Entin, Dr Il'ya Ruvimovih. *USSR.*
Enwall, Dr. Eric Lee. *USA.*
Epelboin, Dr Yves. *France.*
Eppelsheimer, Dr. Daniel Snell Jr. *USA.*
Epperson, Dr. John Ernest. *USA.*
Epprecht, Prof. Willfried Th. *Switzerland.*
Erazo Plaza, Mr Antonio David. *Colombia.*
Ercit, Dr Timothy Scott. *Canada.*
Erdönmez, Dr Ahmet. *Turkey.*
Eremkin, Dr Vladimir Vasil'evich. *USSR.*
Erez, Prof. Gidon. *Israel.*
Ergin, Yrd. Doç. Dr Ömer. *Turkey.*
Erickson, Dr. John. *USA.*
Ericsson, Mr Thomas. *Sweden.*
Eriks, Prof. Klaas. *USA.*
Eriksson, Dr A. Elisabeth. *Sweden.*
Eriksson, Dr Anders. *Sweden.*
Eriksson, Dr Birgitta. *Sweden.*
Eriksson, Mr Lars. *Sweden.*
Eriksson, Mr Sten. *Sweden.*
Ermer, Prof. Dr Otto. *BRD.*
Ernst, Dr Gert. *Austria.*
Ernst, Dr. Stephen Richard. *USA.*
Ersson, Dr Nils Olov. *Sweden.*
Ertan, Mrs Anne. *Sweden.*
Escobar, Dr Carmen. *Chile.*
Esipova, Dr Nataliya Georgiyevna. *USSR.*
Espinat, Dr Didier. *France.*
Espinet, Prof. Pablo. *Spain.*
Espinoza, Prof. Odon. *Peru.*
Esselborn, Dr Reiner Ferdinand. *BRD.*
Esteban-Calderon, Dr M. Carmen. *Spain.*
Estermann, Mr Michael. *Switzerland.*
Estes, Dr. Eva Dixon. *USA.*
Estop, Dr Eugenia. *Spain.*
Estrada, DrM. Dolores. *Spain.*
Eswara Prasad, Dr Gummuluri. *India.*
Etemadi-Abdolabadi, Dr Bijan. *Iran.*
Etheridge, Ms Joanne. *Australia.*
Etter, Prof. Margaret E. Cairns. *USA.*
Ettmayer, Prof. Dr Dipl.-Ing. Peter. *Austria.*
Eulenberger, Dr Günther Richard. *BRD.*
Euler, Dr Robert. *BRD.*
Euthymiou, Prof. Paraskevi. *Greece.*
Evangelidou, Miss Christina. *Greece.*
Evans, Dr Anthony Meredith. *UK.*
Evans, Mr David Lindsay. *New Zealand.*
Evans, Prof. David Robert. *USA.*
Evans, Dr. Doris Louise. *USA.*
Evans, Dr E.M.H. *UK.*
Evans, Ms. Eloise Humez. *USA.*
Evans, Dr. Howard Tasker Jr. *USA.*
Evans, Dr John Hedley. *UK.*
Evans, Dr P.A. *UK.*
Evans, Mr R.R. *UK.*
Evrard, Prof. Guy Henri. *Belgium.*

Eysel, Prof. Dr Walter. *BRD.*
Ezell, Dr. Edward F. *USA.*
Faber, Dr. John, Jr. *USA.*
Faber, Dr Peter. *BRD.*
Fabian, Prof Eginhard. *DDR.*
Fabius, Mrs Birgit. *Denmark.*
Fackler, Prof. John Paul, Jr. *USA.*
Faerman, Dr Carlos Hugo. *UK.*
Fagherazzi, Prof. Giuliano. *Italy.*
Fagnani, Prof. Gustavo. *Italy.*
Fahey, Prof. James A. *USA.*
Fair, Dr. Carolyn Kay. *USA.*
Fairchild, Ms. Beatrice M. *USA.*
Fairclough, Dr D.P. *UK.*
Fajardo, Dr.Sc. Fabio. *Cuba.*
Fajer, Dr. Jack. *USA.*
Fakhar, Miss Noura. *Tunisia.*
Falk, Dr Michael. *Canada.*
Falkenberg, Dr Wolfgang. *DDR.*
Faller, Prof. John William (Jack). *USA.*
Fallon, Dr Gary David. *Australia.*
Falshaw, Dr C.P. *UK.*
Falvello, Dr. Lawrence R. *USA.*
Fan, Mr Guang-yu. *China.*
Fan, Prof. Hai-fu. *China.*
Fan, Mr Yu-guo. *China.*
Fan, Mr Zhao-chang. *China.*
Fanariotis, Mr Iakovos. *Greece.*
Fanchon, Dr. Eric. *France.*
Fanfani, Prof. Luca. *Italy.*
Fang, Prof. Jen Ho. *USA.*
Fanter, Mr Detlef. *DDR.*
Faqir, Dr Gul. *Malaysia.*
Farber, Dr Boris Yakovlevich. *USSR.*
Farber, Dr. Gregory K. *USA.*
Fares, Dr Vincenzo. *Italy.*
Farkas-Jahnke, Dr (Ms) Mária. *Hungary.*
Farkas, Dr László. *Hungary.*
Farmer, Dr. Barry L. *USA.*
Farrahi, Dr Gholam-Hossein. *Iran.*
Farrar, Prof. Roy Alfred. *UK.*
Farrell, Miss Yvonne. *Australia.*
Farrugia, Dr L.J. *UK.*
Faruqi, Dr Abdul Raffey. *UK.*
Fasman, Prof. Gerald David. *USA.*
Fatmi, Prof. Ali Nasir. *Pakistan.*
Faulk, Mr. John Warren. *USA.*
Faus-Paya, Prof. Juan. *Spain.*
Faust, Mr Wolfgang. *DDR.*
Fauvet, Mr Gérard. *France.*
Fawcett, Dr John Keith. *Canada.*
Fawcett, Dr John. *UK.*
Fawcett, Dr. Timothy G. *USA.*
Fay, Prof. Robert Clinton. *USA.*
Fayed, Dr (Mrs) Leila. *Egypt.*
Fayos, Prof. Jose. *Spain.*
Fazlić, Magistar Refik. *Yugoslavia.*
Fedeli, Prof. Walter. *Italy.*
Fedorenko, Prof. Anatoly Ivanovich. *USSR.*
Fedorov, Dr Aleksandr Aleksandrovich. *USSR.*
Fedorov, Dr Boris Alexandrovich. *USSR.*
Fedorov, Dr Pavel Pavlovich. *USSR.*
Fedotov, Dr Alexander Fedorovich. *USSR.*
Fedyna, Dr Mikhail Fedorovich. *USSR.*
Feher, Mr Andreas. *DDR.*
Feher, Mr Elvira. *DDR.*
Fehling, Mr Wolfgang. *DDR.*
Feidenhans'l, Dr Robert. *Denmark.*
Feigelson, Prof. Robert Saul. *USA.*
Feigin, Dr Lev Abramovich. *USSR.*
Feil, Prof. Dr Dirk. *Netherlands.*
Feiz, Dr Sayid Mohammad Hasan. *Iran.*
Fejer, Miss Eleonora Eva. *UK.*
Felbinger, Dr Adolf. *DDR.*
Feld, Dr Rainer Hans Helmut. *BRD.*
Feldman, Dr. Robert Edward. *USA.*
Feldmann, Mr. Richard Joseph. *USA.*
Felius, Dr Robert Onno. *Netherlands.*
Felix, Dr Rene P.. *Philippines.*
Felsche, Prof. Dr Jürgen. *BRD.*

Felsteiner, Prof. Joshua. *Israel.*
Feltz, Prof. Adalbert. *DDR.*
Femec, Dr. Douglas Anthony. *USA.*
Fenderson, Dr. Faith. *USA.*
Feneau-Dupont, Mrs Janine. *Belgium.*
Fenn, Dr Ruth Helen. *UK.*
Fenna, Dr. Roger Edward. *USA.*
Fenoll, Prof. Purificacion. *Spain.*
Fenske, Prof. Dr Dieter. *BRD.*
Ferguson, Dr George. *Canada.*
Ferguson, Dr Ian Forster. *UK.*
Ferguson, Prof. Robert Bury. *Canada.*
Fernandes, Mr Jacob Richard. *India.*
Fernández González, Dr Alonso. *México.*
Fernandez, Prof. Aurora Reyes. *Philippines.*
Fernandez, Dr Carlos Jose. *Spain.*
Fernandez, Ing Juan Carlos. *Argentina.*
Fernando, Prof. Quintus. *USA.*
Ferracini, Prof. Elena. *Italy.*
Ferran, Dr Gustan. *Brasil.*
Ferrara, Dr. Joseph David. *USA.*
Ferrari-Belicchi, Prof. Marisa. *Italy.*
Ferrari, Dr Claudio. *Italy.*
Ferraris, Prof. Giovanni. *Italy.*
Ferreira de Souza, Prof. Milton. *Brasil.*
Ferrero Rognoni, Prof. Adele. *Italy.*
Ferretti, Dr Valeria. *Italy.*
Fesenko, Prof. Evgeny Grigoryevich. *USSR.*
Fesenko, Dr Oleg Evgenyevich. *USSR.*
Fetisov, Dr Gennadii Vladimirovich. *USSR.*
Fewster, Dr Paul Frederick. *UK.*
Fiala, Dr. Jaroslav. *CSSR.*
Fichera, Dr Anna Maria. *Italy.*
Fichtner-Schmittler, Dr Helga. *DDR.*
Fidelis, Dr. Krzysztof Andrze. *USA.*
Fiedler, Dr Gustav. *DDR.*
Field, Dr Donald William. *Australia.*
Field, Prof. John Stainer. *South Africa.*
Fielding, Mr W.D. *UK.*
Fields, Mr Barry Arthur. *Australia.*
Fiermans, Dr Lucien Victor August. *Belgium.*
Figas, Mgr Elżbieta Teresa. *Poland.*
Figgis, Prof Brian Norman. *Australia.*
Figielski, Dr hab. Tadeusz. *Poland.*
Figlas, Lic Norma Debora. *Argentina.*
Figueiredo Neto, Dr Antonio Martins. *Brasil.*
Figueiredo, Dr Maria Ondina. *Portugal.*
Filatov, Dr Stanislav Konstantinovich. *USSR.*
Filip'yev, Dr Victor Semenovich. *USSR.*
Filipenko, Dr Olga Savelyevna. *USSR.*
Filippakis, Dr Sophokles. *Greece.*
Filippini, Dr Giuseppe. *Italy.*
Filizova, Dr Lyudmila. *Bulgaria.*
Filman, Dr. David Jeffrey. *USA.*
Filscher, Mr Gerold. *DDR.*
Finer-Moore, Dr. Janet Sue. *USA.*
Finger, Dr. Larry W. *USA.*
Fingerland, Dr Antonín. *CSSR.*
Fink, Prof. Anthony L. *USA.*
Fink, Dr. William LaVilla. *USA.*
Finkel'shtein, Dr Aleksey Vital'yevich. *USSR.*
Finlayson, Dr Trevor Roy. *Australia.*
Finney, Prof. John Leslie. *UK.*
Finster, Dr Joachim. *DDR.*
Finzel, Dr. Barry C. *USA.*
Fischer-Hjalmars, Prof. Inga M. *Sweden.*
Fischer, Dr Carl-Otto. *BRD.*
Fischer, Mr. Gerhard Richard. *USA.*
Fischer, Prof. Dr Karl. *BRD.*
Fischer, Mr Karl. *DDR.*
Fischer, Dr Peter. *Switzerland.*
Fischer, Dr Reinhard X. *BRD.*
Fischer, Dr Richard. *Austria.*
Fischer, Mrs Ute Eva-Maria. *BRD.*
Fischer, Prof. Dr Werner. *BRD.*
Fisher, Mr Graham Richard. *UK.*
Fisher, Dr. Richard G. *USA.*
Fisher, Dr. Robert M. *USA.*
Fita, Dr Ignacio. *Spain.*
Fitzgerald, Dr Alexander Grant. *UK.*

Fitzgerald, Dr. Paula Marie Dean. *USA.*
Fitzl, Dr Günther. *DDR.*
Fjaer, Dr Erling. *Norway.*
Fjellvåg, Mr Helmer. *Norway.*
Flack, Dr Howard David. *Switzerland.*
Flade, Dr Tilo. *DDR.*
Fleet, Dr Michael Edward. *Canada.*
Fleet, Dr Stephen George. *UK.*
Fleischmann, Dr Klaus Dietrich. *Netherlands.*
Flerov, Dr Igor' Nikolayevich. *USSR.*
Fletcher, Dr Neville Horner. *Australia.*
Fletcher, Dr Steven Reginald. *UK.*
Fletterick, Dr. Robert J. *USA.*
Flevaris. Prof. Nikolaos. *Greece.*
Flewitt, Prof. Peter Edwin John. *UK.*
Flippen-Anderson, Ms. Judith Lee. *USA.*
Flodmark, Dr Stig. *Sweden.*
Flögel, Dr Peter. *DDR.*
Flörke, Prof. Dr Otto Wilhelm. *BRD.*
Florencio, Dr Feliciana. *Spain.*
Florio, Dr. John Victor. *USA.*
Foces-Foces, Dr Concepcion. *Spain.*
Fodor, Ms Krisztina. *Hungary.*
Förster, Dr Eckhart. *DDR.*
Försterling, Dr Gerd. *DDR.*
Fofanov, Dr Anatolii Dmitrievich. *USSR.*
Folgueras Dominguez, Dr Sérvulo. *Brasil.*
Follner, Prof. Dr Heinz. *BRD.*
Folting-Streib, Mrs. Kirsten. *USA.*
Fones, Mr M.D. *UK.*
Fong, Dr Hock Sun. *Singapore.*
Fonseca, Dr Isabel. *Spain.*
Font-Altaba, Prof. Manuel. *Spain.*
Font-Bardia, Mr. Merce. *Spain.*
Fontaine, Dr Alain. *France.*
Fontaine, Dr Frederic Desiré Albert. *Belgium.*
Fontaine, Prof. Hubert. *France.*
Fontecilla-Camps, Dr Juan C. *Marseille.*
Foord, Mr. Eugene E. *USA.*
Ford, Dr Geoffrey Charles. *UK.*
Foreman, Prof. Dennis W. Jr. *USA.*
Forest, Mrs. Katrina. *USA.*
Foresti, Prof. Elisabetta. *Italy.*
Foris, Ms. Catherine M. *USA.*
Formoso, Dr Milton Luiz. *Brasil.*
Fornaro, Prof Laura. *Uruguay.*
Fornasini, Prof. Maria Luisa. *Italy.*
Forni, Prof. Flavio. *Italy.*
Fornoff, Mr. Mario M. *USA.*
Forsellini, Dr Eleonora. *Italy.*
Forslund, Dr S. Bertil. *Sweden.*
Forster, Dr Martin. *Switzerland.*
Forsyth, Prof. John Bruce. *UK.*
Forsyth, Mr V.T. *UK.*
Fortes, Prof. Manuel Amaral. *Portugal.*
Forteza, Mr Matilde. *Spain.*
Forti, Prof. Paolo. *Italy.*
Fortier, Dr Suzanne. *Canada.*
Foroughi, Dr Ali-Assghar. *Iran.*
Forwood, Dr Christopher Thomas. *Australia.*
Foss, Prof. Olav. *Norway.*
Foster, Dr Mark David. *BRD.*
Foster, Dr. Mark David. *USA.*
Fotchenkov, Dr Anatoly Andreyevich. *USSR.*
Foundling, Dr. Stephen Ian. *USA.*
Fouret, Prof. René *France.*
Fourie, Dr Jacobus Theodor. *South Africa.*
Fox, Mr Bruce Edward. *UK.*
Fox, Prof. Robert O. Jr. *USA.*
Foxman, Prof. Bruce Mayer. *USA.*
Fraenkel, Prof. Benjamin. *Israel.*
Frahm, Dr Ronald Reinhard. *BRD.*
Franceschi, Prof. Enrico. *Italy.*
Francesconi, Dr Ricardo. *Brasil.*
Franchini, Prof. Marinella. *Italy.*
Francis, Mr John Godfrey. *UK.*
Francisco, Miss Regina Helena Porto. *Brasil.*
Frank-Kamenetskaya, Dr Olga Victorovna. *USSR.*
Frank-Kamenetsky, Prof. Victor A. *USSR.*
Frank, Prof. Sir Frederick Charles. *UK.*

Frank, Dr Walter. *BRD.*
Franke, Dr Valeriya Dmitriyevna. *USSR.*
Franklin, Dr Kenneth James. *Canada.*
Franklin, Prof. Ursula Martius. *Canada.*
Franks, Dr Albert. *UK.*
Franks, Dr Joseph. *UK.*
Fransolet, Dr André-Mathieu. *Belgium.*
Franzen, Prof. Hugo Friedrich. *USA.*
Franzini, Prof. Marco. *Italy.*
Franzosi, Dr Paolo. *Italy.*
Fratini, Prof. Albert V. *USA.*
Frazer, Dr. Benjamin Chalmers. *USA.*
Fredrich, Mr. Michael F. *USA.*
Freed, Prof. Robert Lowell. *USA.*
Freeman, Dr Alan George. *New Zealand.*
Freeman, Prof Hans Charles. *Australia.*
Freeman, Mr Walter Gerard. *UK.*
Freer, Dr Andrew Aloysius. *UK.*
Freer, Dr. Stephan T. *USA.*
Freiburg, Dr Johann Christoph. *BRD.*
Freire D'aguiar, Mr Manoel Marcos. *Brasil.*
Frelinger, Dr. Andrew Lawrence. *USA.*
French, Dr. Alfred Dexter. *USA.*
French, Mr. Robert D. *USA.*
Frenz, Dr. Bertram Anton. *USA.*
Freudenberg, Mr Axel. *DDR.*
Freund, Prof. Dr Friedemann. *BRD.*
Freundlich, Dr A. *UK.*
Frevel, Dr. Ludo Karl. *USA.*
Frey, Prof. Dr Friedrich. *BRD.*
Frey, Mr Wolfgang Ulrich. *BRD.*
Freydank, Mr Gisela-Christine. *DDR.*
Freyhardt, Prof. Dr Herbert C. *BRD.*
Fridkin, Prof. Vladimir Mikhailovich. *USSR.*
Friedel, Prof. Jacques. *France.*
Friedlander, Dr. Peter H. *USA.*
Frigeri, Dr Cesare. *Italy.*
Frigyik, Mr Gábor. *Hungary.*
Frikkee, Dr Evert. *Netherlands.*
Friman, Dr Rauno Kalevi. *Finland.*
Fritchie, Prof. Charles Julius Jr. *USA.*
Fröhlich, Mr Armin. *BRD.*
Fröhlich, Dr Roland. *BRD.*
Frolow, Dr Felix. *Israel.*
Fronczek, Dr. Frank R. *USA.*
Frostäng, Mr F. Sten E. *Sweden.*
Frueh, Dr Alfred Joseph. *USA.*
Frühauf, Dr Joachim. *DDR.*
Fry, Dr. David C. *USA.*
Ftikos, Dr Christos. *Greece.*
Fu, Prof. Heng. *China.*
Fu, Prof. Ping-qiu. *China.*
Fu, Mr Zheng-min. *China.*
Fu, Mr Zhu-ji. *China.*
Fuchs, Prof. Dr Sc Techn Erik. *Hungary.*
Fülöp, Mr Vilmos. *Hungary.*
Fuente-Cullel, Dr Carlos. *Spain.*
Fuentes Betancourt, Prof.Dr.Sc. Juan E. *Cuba.*
Fuentes Cobas, Dr.Sc. Luis E. *Cuba.*
Fuertes, Dr Amparo. *Spain.*
Fuess, Prof. Dr Hartmut. *BRD.*
Fuith, Dr Mag. Armin. *Austria.*
Fujii, Dr Satoshi. *Japan.*
Fujii, Dr Tetsuo. *Japan.*
Fujii, Prof. Yasuhiko. *Japan.*
Fujime, Dr Satoru. *Japan.*
Fujimore, Prof. Kenkichi. *Brasil.*
Fujimoto, Prof. Fuminori. *Japan.*
Fujimoto, Prof. Hirofumi. *Japan.*
Fujino, Dr Kiyoshi. *Japan.*
Fujino, Dr Nobukatsu. *Japan.*
Fujita, Prof. Francisco Eiichi. *Japan.*
Fujita, Prof. Hiroshi. *Japan.*
Fujiwara, Prof. Hiroshi. *Japan.*
Fujiwara, Prof. Kunio. *Japan.*
Fujiwara, Prof. Takaji. *Japan.*
Fujiyoshi, Dr Yoshinori. *Japan.*
Fukamachi, Dr Tomoe. *Japan.*
Fukano, Mr Tatsuo. *Japan.*
Fukano, Prof. Yasushige. *Japan.*

Fukuda, Prof. Tsuguo. *Japan.*
Fukuhara, Dr Akira. *Japan.*
Fukuyama, Dr Keiichi. *Japan.*
Fukuyama, Dr Tsutomu. *Japan.*
Fulfaro, Dr Roberto. *Brasil.*
Fullam, Mr. Ernest F. *USA.*
Fullenwider, Dr. Malcolm Allen. *USA.*
Fuller, Prof. W. *UK.*
Fulton, Dr William Stephen. *UK.*
Fumi, Prof. Fausto Gherardo. *Italy.*
Fun, Dr Hoong Kun. *Malaysia.*
Fundamensky, Mr Vladimir Semenovich. *USSR.*
Fuoss, Dr. Paul Henry. *USA.*
Furey, Dr. William F. *USA.*
Furmanova (Bokii), Dr Nina Georgievna. *USSR.*
Furnas, Dr. Thomas Coleman Jr. *USA.*
Fursenko, Dr Boris Alexandrovich. *USSR.*
Furusaki, Dr Akio. *Japan.*
Furuseth, Mrs Sigrid. *Norway.*
Fuzek, Dr. John Frank. *USA.*
Fykin, Dr Leonid Efimovich. *USSR.*
Gaál, Dr István. *Hungary.*
Gabe, Dr Eric James. *Canada.*
Gabela, Prof. Dr Fikret. *Yugoslavia.*
Gabelli, Lic Sandra Beatriz Argentina.
Gabis, Prof. Victor Michel. *France.*
Gable, Dr Robert William. *Australia.*
Gabuda, Prof. Svyatoslav Petrovich. *USSR.*
Gad, Dr Gamal Mohamed. *Egypt.*
Gadó, Dr Pál. *Hungary.*
Gaete, Dr Walter. *Spain.*
Gagarina, Dr Elena Stanislavovna. *USSR.*
Gahm, Dr Josef. *BRD.*
Gaier, Dr. James Richard. *USA.*
Gaines, Dr. James Matthew. *USA.*
Gainsford, Dr Graeme John. *New Zealand.*
Gait, Dr Robert Irwin. *Canada.*
Gajhede, Dr Michael. *Denmark.*
Galan, Prof. Emilio. *Spain.*
Gałązka, Dr hab. Robert. *Poland.*
Gałdecka, Dr Ewa Renata. *Poland.*
Gałdecki, Prof. Zdzisław. *Poland.*
Gale, Dr Brian. *UK.*
Galešić, Dr Nikola. *Yugoslavia.*
Galetti, Prof. Giulio. *Switzerland.*
Gali, Dr Salvador. *Spain.*
Galindo, Dr Agustin. *Spain.*
Galinos, Prof. Andreas. *Greece.*
Galitsky, Dr Nikolay Mikhailovich. *USSR.*
Galiulin, Dr Ravil Vagizovich. *USSR.*
Gall, Prof. William Einar. *USA.*
Gallacher, Mr. Anthony. *USA.*
Gallagher, Dr Kevin Joseph. *UK.*
Gallagher, Mr. Travis. *USA.*
Galli, Prof. Ermanno. *Italy.*
Galloy, Dr Jean. *UK.*
Gallucci, Dr. Judith Chlastawa. *USA.*
Galstyan, Dr Viktor Gaikovich. *USSR.*
Galvis, Mr Jaime. *Colombia.*
Gama Carvalho, Dr Frederico. *Portugal.*
Gamarnik, Dr Moisei Yanvelevich. *USSR.*
Gan, Mr Ah Sai. *Malaysia.*
Ganazzoli, Dr Fabio. *Italy.*
Gandour, Prof. Richard David. *USA.*
Ganesan, V. *India.*
Ganev, Ing Nikolaj. *CSSR.*
Gantzel, Dr. Peter Kellogg. *USA.*
Gao, Mr Yi-guei. *China.*
Garafalo, Mr. Alfred R. *USA.*
Garaj, Prof. Dr Ján. *CSSR.*
Garashina, Dr Lyudmila Solomonovna. *USSR.*
Garavito, Dr. R. Michael. *USA.*
Garbassi, Dr Fabio. *Italy.*
Garbauskas, Dr. Mary Frances. *USA.*
Garcia-Blanco, Prof. Severino. *Spain.*
Garcia-Casado, Dr Pedro. *Spain.*
Garcia-Rodriguez, Prof. Antonio. *Spain.*
Garcia-Ruiz, Dr Joaquin. *Spain.*
Gard, Dr John Alan. *UK.*
Gardner, Dr. Kenncorwin Hancock. *USA.*

Gare, Mr Terence. *UK.*
Garg, Dr Ajau Kumar. *India.*
Garín, Dr Jorge. *Chile.*
Garland, Mrs María Teresa. *Chile.*
Garner, Prof. Christopher David. *UK.*
Garvey, Prof. Roy George. *USA.*
Gaspard, Prof. Jean-Pierre *Belgium.*
Gasparri-Fava, Prof. Giovanna. *Italy.*
Gasperin, Dr Madeleine. *France.*
Gassmann, Dr Johann. *BRD.*
Gastaldi, Dr Leonardo. *Italy.*
Gatehouse, Dr Bryan Michael K. C. *Australia.*
Gates, Dr Jeffrey Douglas. *Australia.*
Gatineau, Dr Lucien Charles. *France.*
Gatti, Dr Giuseppina. *Italy.*
Gaunt, Dr Paul. *Canada.*
Gauthier, Prof. Jean-Pierre. *France.*
Gauzzi, Prof. Franco. *Italy.*
Gavezzotti, Prof. Angelo. *Italy.*
Gavrilova, Dr Irina Vladimirovna. *USSR.*
Gavrilova, Dr Nadezhda Dmitrievna. *USSR.*
Gavrilyachenko, Dr Victor Georgievich. *USSR.*
Gavuzzo, Dr Enrico. *Italy.*
Gawron, Dr Marian. *Poland.*
Gay, Dra Hebe Dina. *Argentina.*
Gazzano, Dr Massimo. *Italy.*
Gazzoni, Dr Giuseppe. *Italy.*
Gdaniec, Dr Maria. *Poland.*
Gebert, Dr Walter Richard. *BRD.*
Gebhardt, Prof. Dr Manfred Adolf Hermann. *BRD.*
Geckle, Mr. Raymond J. *USA.*
Geddes, Dr Alexander John. *UK.*
Gedikoğlu, Prof. Dr Atasever. *Turkey.*
Geerestein, van, Dr Vincent Johan. *Netherlands.*
Geguzina, Dr Galina Alexandrovna. *USSR.*
Gehlen, Prof. Kurt von. *BRD.*
Geib, Dr. Steven J. *USA.*
Geidel, Mr Volkmar. *DDR.*
Geiger, Dr Charles Arthur. *BRD.*
Geiger, Dr. David K. *USA.*
Geil, Prof. Phillip Herbert. *USA.*
Geil, Dr Werner. *DDR.*
Geise, Prof. Dr Herman Joseph V. H. *Belgium.*
Geiser, Dr. Urs W. *USA.*
Geisinger, Ms. Karen L. *USA.*
Geiss, Dr. Roy Howard. *USA.*
Geist, Dr Volker. *DDR.*
Gel'man, Dr Yury Alexandrovich. *USSR.*
Gelato-Volders, Dr (Mrs) Louise Marie. *Switzerland.*
Gelder, de, Drs René. *Netherlands.*
Geleji-Neubauer, Ms Irén. *Hungary.*
Geller, Prof. Seymour. *USA.*
Gellings, Prof. Dr Paul Johann. *Netherlands.*
Genin, Dr Yakov Vladimirovich. *USSR.*
Genkina, Dr Elena Aleksandrovna. *USSR.*
Gentner, Mr Thomas. *BRD.*
Geoffre, Mr Serge. *France.*
George, Dr Amand. *France.*
George, Dr. Clifford F. *USA.*
Geras'kin, Dr Valerii Vasil'evich. *USSR.*
Gerasimov, Dr Viktor Ivanovich. *USSR.*
Gerdes, Dr. Reiner Josef. *USA.*
Gerdil, Prof. Raymond. *Switzerland.*
Gergely, Dr Márton. *Hungary.*
Gerhard, Mr. F. Bruce. *USA.*
Gering, Dr Erich. *BRD.*
Gerkin, Prof. Roger Estlick. *USA.*
Germain, Prof. Gabriel. *Belgium.*
Gernat, Mr Christine. *DDR.*
Gerold, Prof. Dr Volkmar. *BRD.*
Gertel, Ms Heike. *BRD.*
Gervasio, Prof. Giuliana. *Italy.*
Gervilla, Mr Fernando. *Spain.*
Gerward, Dr Leif. *Denmark.*
Gesemann, Prof. Renate. *DDR.*
Geserick, Mr Sabine. *DDR.*
Getzoff, Dr. Elizabeth Dickinson. *USA.*
Gevork'yan, Dr Svetlana Vasil'evna. *USSR.*
Geyer, Mr Andreas. *BRD.*
Gezci, Doç. Dr Sami. *Turkey.*

Ghebrial, Dr Mounir Guirgis. *Egypt.*
Ghedira, Dr Mounir. *Tunisia.*
Gheith, Prof. Mohamed A. *USA.*
Ghezzi, Prof. Carlo. *Italy.*
Ghilardi, Dr Carlo Alfredo. *Italy.*
Ghose, Prof. Subrata. *USA.*
Ghosh, Dr. Debashis. *USA.*
Ghosh, Ms Manuja. *India.*
Ghosh, Dr Mrs Minakshi. *India.*
Ghosh, Mr. Partho. *USA.*
Ghosh, Dr Sujit Kumar. *India.*
Ghosh, Ms Sutapa. *India.*
Ghouse, Dr Khaja Mohd. *India.*
Giacovazzo, Prof. Carmelo. *Italy.*
Gibbons, Dr Cyril Stephen. *Canada.*
Gibon, Dr Véronique Julie Jacques Laure. *Belgium.*
Giebułtowicz, Dr Tomasz. *Poland.*
Giegé, Dr Richard. *France.*
Gies, Dr Hermann. *BRD.*
Giese, Mr. Rossman F. Jr. *USA.*
Giessen, Prof. Bill Cormann. *USA.*
Giglio, Prof. Edoardo. *Italy.*
Gilani, Mr Jamshed Ali. *Pakistan.*
Gilardi, Dr. Richard Dean. *USA.*
Giles, Mr Raymond Richard. *UK.*
Gilfrich, Mr. John Valentine. *USA.*
Gilinskaya, Dr Emma Abramovna. *USSR.*
Gilje, Prof. John W. *USA.*
Gille, Dr Peter. *DDR.*
Gillette, Dr. Paul Calvin. *USA.*
Gilli, Prof. Gastone. *Italy.*
Gillies, Dr. Donald Chalmers. *USA.*
Gilliland, Dr. Gary Lynn. *USA.*
Gilmartin, Mr M.G.M. *UK.*
Gilmore, Dr Christopher John. *UK.*
Gilski, Mr Mirosław. *Poland.*
Gimeno, Prof. Jose. *Spain.*
Ginderow, Dr Daria. *France.*
Ginell, Prof. Robert. *USA.*
Ginell, Dr. Stephan Lawrence. *USA.*
Giordani, Prof. Marino. *Italy.*
Giordano Orsini, Prof. Paolo. *Italy.*
Giordano, Prof. Federico. *Italy.*
Giordano, Mr. Joseph. *USA.*
Giorgi, Dr. Angelo Louis. *USA.*
Giovanoli, Prof. Rudolf. *Switzerland.*
Giranda, Mr. Vincent. *USA.*
Girgin, Doç. Dr İsmail. *Ankara,Turkey.*
Girgis, Dr Kamal. *Switzerland.*
Giri, Mr Anit K. *India.*
Giri, Mr Siba Narayan. *India.*
Girirajan, Dr K. S. *India.*
Girt, Prof. Dr Egvin. *Yugoslavia.*
Giunchi, Dr Giovanni. *Italy.*
Giuşcă, Prof. Dan. *Romania.*
Giuseppetti, Prof. Giuseppe. *Italy.*
Givargizov, Dr Evgeny Inviyevich. *USSR.*
Gjønnes, Prof. Jon Kjell. *Norway.*
Gladić, Mr Jadranko. *Yugoslavia.*
Gladkikh, Dr Liliya Ivanovna. *USSR.*
Gladky, Dr Vsevolod Vladimirovich. *USSR.*
Gladyshevsky, Prof. Evgeny Ivanovich. *USSR.*
Glaeser, Prof. Robert Martin. *USA.*
Glaser, Dr Julius. *Sweden.*
Glass, Dr. Howard L. *USA.*
Glasser, Prof. Frederik Paul. *UK.*
Glasser, Prof. Leslie. *South Africa.*
Glasson, Dr Douglas Royston. *UK.*
Glatter, Dr Otto. *Austria.*
Glazer, Dr Anthony Michael. *UK.*
Glazier, Mr Edward James. *UK.*
Glazov, Dr Aleksey Ivanovich. *USSR.*
Gleason, Dr. William Bourke. *USA.*
Glehn, von, Dr Marianne. *Sweden.*
Gleizes, Prof. Alain Nicolas. *France.*
Glen, Dr John Wallington. *UK.*
Glick, Prof. Milton Don. *USA.*
Glidewell, Dr Christopher. *UK.*
Glikin, Mr Arkady Eduardovich. *USSR.*
Glinka, Dr. Charles Joseph. *USA.*

Glinnemann, Dr Jürgen Wilhelm Erich. *BRD.*
Glover, Dr Ian David. *UK.*
Głowiak, Dr hab. Tadeusz. *Poland.*
Główka, Dr Marek. *Poland.*
Glucksman, Dr. Marc J. *USA.*
Glumoff, Mr Tuomo. *Switzerland.*
Glusker, Dr. Jenny Pickworth. *USA.*
Gluziński, Dr Przemysław. *Poland.*
Go, Miss Kuan Tee. *USA.*
Goaman, Dr Llawenydd Constance Gwynne. *UK.*
Godavarthi, Bhagavannarayana. *India.*
Goddard, Dr Richard John. *BRD.*
Godden, Mr. Jeff. *USA.*
Godden, Ms Manuela Joanna. *UK.*
Goddette, Dr. Dean. *USA.*
Godovikov, Prof. Alexander Alexandrovich. *USSR.*
Godwod, MSc Krzysztof Jan. *Poland.*
Godycki, Mr. L. Edward. *USA.*
Göbel, Dr Herbert Ernst. *BRD.*
Göbel, Mr Ralf. *DDR.*
Göcke, Dr Wolfhart. *DDR.*
Goedkoop, Prof. Dr Jacob A. *Netherlands.*
Goehner, Mr. Raymond Philip. *USA.*
Görbitz, Mr Carl Henrik. *Norway.*
Görnert, Prof Peter. *DDR.*
Göttlicher, Prof. Dr Siegfried. *BRD.*
Götz, Dr Konrad. *DDR.*
Götz, Dr Wolfgang. *DDR.*
Götzinger, Dr Michael Alois. *Austria.*
Goilo, Dr Eduard Al'bertovich. *USSR.*
Goland, Dr. Allen N. *USA.*
Gold, Dr. Karen Walter. *USA.*
Goldberg, Prof. Israel. *Israel.*
Goldberg, Prof. Stephen Z. *USA.*
Goldenberg, Ms. Maria. *USA.*
Goldish, Dr. Elihu. *USA.*
Goldman, Prof. Adrian. *USA.*
Goldsmith, Dr. Elizabeth Jane. *USA.*
Goldsmith, Prof. Julian Royce. *USA.*
Goldstein, Dr. Barry Michael. *USA.*
Goldstone, Dr. Joyce. *USA.*
Golič, Prof. Dr Ljubo. *Yugoslavia.*
Golikeri, Mr. Ganesh D. *USA.*
Goliński, Dr Bohdan. *Poland.*
Golovachev, Dr Vladimir Pavlovich. *USSR.*
Golovastikov, Dr Nikolay Ivanovich. *USSR.*
Golovin, Dr Andrei Leonidovich. *USSR.*
Golovina, Dr Nina Ivanovna. *USSR.*
Golovko, Dr Yurii Ilarionovich. *USSR.*
Golubev, Dr Alexander Mikhailovich. *USSR.*
Golubinskii, Dr Aleksei Vladimirovich. *USSR.*
Golubkov, Dr Alexander Vasilyevich. *USSR.*
Golyshev, Dr Vladimir Mikhailovich. *USSR.*
Gomes, Albert Cardinal. *India.*
Gomes, Mr Osvaldo. *Uruguay.*
Gomes, Mr Samuel Irati Novaes. *Brasil.*
Gomez de Anderez, Prof. Dora Maria. *Venezuela.*
Gomez Ramirez, Dr Ricardo. *México.*
Gomez Rodriguez, Dr. Alfredo. *México.*
Gomez-Sal, Dr M. Pilar. *Spain.*
Gómezdaza Almendaro, Mr Mariano. *México.*
Gomm, Dr Martin. *BRD.*
Goncharov, Dr Aleksandr Vasil'evich. *USSR.*
Goncharov, Dr Georgy Nikolayevich. *USSR.*
Gong, Dr. Ping-Po +. *USA.*
Gong, Prof. Xia-sheng. *China.*
Gonschorek, Dr Walter. *BRD.*
Gonzalez R., Prof. Oscar V. *Venezuela.*
Gonzalez-Calbet, Dr Jose M. *Spain.*
González, Mrs Irma. *Chile.*
Gonzalez, Mr. Oscar. *Uruguay.*
González, Mr Yanko. *Chile.*
Goodenough, Prof. John Bannister. *USA.*
Goodfellow, Dr Julia Mary. *UK.*
Goodman, Prof. C.H.L. *UK.*
Goodman, Dr Peter. *Australia.*
Goonesekere, Mr. Nalin. *USA.*
Gopal, Dr Ramanathan. *Canada.*
Gorbunova, Dr Yuliya Efimovna. *USSR.*
Gordiyenko, Dr Vladimir Vasilyevich. *USSR.*

Hamilton, Prof. Robert David. *USA.*
Hamlin, Dr. Ronald Craig. *USA.*
Hammond, Dr Christopher. *UK.*
Hammond, Mr Lloyd Charles. *Australia.*
Hammonds, Mr T.G. *UK.*
Hamodrakas, Dr Stavros. *Greece.*
Hamor, Dr Thomas Andrew. *UK.*
Han, Mr Fu-son. *China.*
Han, Dr. Fusen. *USA.*
Han, Mr Shao-xu. *China.*
Han, Mr Yu-zhen. *China.*
Handlovič, Ing Milan. *CSSR.*
Hange, Mr Ferenc. *Hungary.*
Hangleiter, Dr Thomas B. *BRD.*
Hanic, Doc. Dr František. *CSSR.*
Hanke, Dr Kurt. *BRD.*
Hannick, Dr. Linda I. *USA.*
Hannon, Mrs Rosemary Anne. *UK.*
Hansen, Mr. Harly. *USA.*
Hansen, Mr Lars Kristian. *Norway.*
Hansen, Dr Niels K. *France.*
Hansen, Dr Staffan S. *Sweden.*
Hanson, Dr Alfred Wallace. *Canada.*
Hanson, Dr. Jonathan C. *USA.*
Hansson, Dr Arne E. *Sweden.*
Haq, Dr Anwarul. *Pakistan.*
Haque, Prof. Mazhar-ul. *Saudi Arabia.*
Harada, Prof. Jimpei. *Japan.*
Harada, Dr Shigeharu. *Japan.*
Harada, Dr Zyunpei. *Japan.*
Harbrecht, Dr Bernd. *BRD.*
Hardcastle, Prof. Kenneth Irvin. *USA.*
Hardgrove, Prof. George Lind Jr. *USA.*
Hardiman, Ms. June. *USA.*
Harding, Mr J.W. *UK.*
Harding, Dr Marjorie Mary. *UK.*
Harding, Dr Roger Robertson. *UK.*
Hardman, Mr. Karl D. *USA.*
Hardy, Dr Andrew David. *UK.*
Hardy, Mrs Anne-Marie. *France.*
Hardy, Prof. Antoine. *France.*
Harel, Dr Michal. *Israel.*
Harga, Dr Ahmed Amin. *Egypt.*
Hargittai, Prof István. *Hungary.*
Hargittai, Dr (Ms) Magdolna. *Hungary.*
Hargreaves, Dr A. *UK.*
Hariharan, Meena. *India.*
Hariya, Prof. Yu. *Japan.*
Harkema, Dr Sybolt. *Netherlands.*
Harker, Dr. David. *USA.*
Harle, Mrs Säde Pirjo Anneli. *Finland.*
Harlow, Dr. Richard Leslie. *USA.*
Harms, Dr Klaus. *BRD.*
Haromy, Dr. Tuli Patrick. *USA.*
Harper, Prof. Richard A. *USA.*
Harper, Mr W.H. *UK.*
Harr, Dr Albrecht Wolfgang Michael. *BRD.*
Harries, Mr J.E. *UK.*
Harris, Prof. David R. *USA.*
Harris, Miss Deborah. *UK.*
Harris, Dr Gillian Wendy. *UK.*
Harris, Dr. Lester. *USA.*
Harris, Mr. Mark. *USA.*
Harrison, Prof. Pauline May. *UK.*
Harrison, Dr. Robert Wilson. *USA.*
Harrison, Prof. Stephen Coplan. *USA.*
Harrison, Dr. William Thomas Alexander. *USA.*
Harsányi, Dr László. *Hungary.*
Hart, Mr Derrik Gordon. *UK.*
Hart, Mr. Donald W. *USA.*
Hart, Dr. Haskell Vincent. *USA.*
Hart, Prof. Michael. *UK.*
Hartl, Prof. Dr Hans. *BRD.*
Hartley, Dr Richard H. *Australia.*
Hartman, Prof. Dr Piet. *Netherlands.*
Hartmann, Dr Ervin. *Hungary.*
Hartmann, Dr Horst. *DDR.*
Hartsuck, Dr. Jean Ann. *USA.*
Hartung, Prof Helmut. *DDR.*
Harutunyan, Dr Emil' Haikovich. *USSR.*

Harvey, Dr T.A. *UK.*
Harzallah, Miss Besma. *Tunisia.*
Hasan, Dr Faizul. *Pakistan.*
Hašek, Dr Jindřich. *CSSR.*
Haser, Dr Richard Michel. *France.*
Hashimoto, Prof. Hatsujiro. *Japan.*
Hashimoto, Mr Hideki. *Japan.*
Hashizume, Prof. Hiroo. *Japan.*
Hasiguti, Prof. Ryukiti. *Japan.*
Hassan, Prof. Ishmael. *Canada.*
Hassan, Prof. Mohamed Youssef. *Egypt.*
Hassan, Dr Wan Fuad. *Malaysia.*
Hassell, Ms. Anne M. *USA.*
Hassler, Dr Eivind. *Sweden.*
Hastings, Dr. Jerome Biller. *USA.*
Hastings, Dr. Julius Mitchell. *USA.*
Hata, Dr Yasuo. *Japan.*
Hatada, Mr. Marcos H. *USA.*
Hathaway, Prof. Brian John. *Ireland.*
Hatt, Mr B.A. *UK.*
Hatta, Dr Tamao. *Japan.*
Hatton, Dr Peter David. *UK.*
Hau, Dr. Herbert H. *USA.*
Hauback, Mr Björn Christian. *Norway.*
Hauck, Dr Jürgen. *BRD.*
Hauge, Dr Sverre. *Norway.*
Haupt, Mr. Gary Robert. *USA.*
Hauptman, Prof. Herbert Aaron. *USA.*
Hausen, Dr Hans-Dieter. *BRD.*
Hausner, Dipl.-Ing. Robert. *Austria.*
Haussühl, Prof. Dr Siegfried Georg. *BRD.*
Havinga, Dr Edsko Enno. *Netherlands.*
Havlík, Ing Tomáš. *CSSR.*
Hawkins, Ms Kathleen Darrelle. *Australia.*
Hawley, Dr D.M. *UK.*
Haworth, Dr Colin William. *UK.*
Hawthorne, Dr Frank Christopher. *Canada.*
Hay, Dr David Gilbert. *Australia.*
Hay, Dr James Neilson. *UK.*
Hayakawa, Dr Kazunobu. *Japan.*
Hayakawa, Dr Motozo. *Japan.*
Hayashi, Dr Kooya. *Japan.*
Hayashi, Prof. Mituhiko. *Japan.*
Hayden, Dr. Thomas Day. *USA.*
Hazell, Dr Alan Charles. *Denmark.*
Hazell, Mrs Rita Grønbæk. *Denmark.*
Hazen, Dr. Robert Miller. *USA.*
He, Prof. Chong-fan. *China.*
He, Prof. Cun-heng. *China.*
He, Mr Rei-ling. *China.*
He, Dr. Xiao-Min. *USA.*
Healy, Dr Peter Conrad. *Australia.*
Heath, Mr. James R. *USA.*
Heavens, Prof. Oliver Samuel. *UK.*
Hebert, Dr Hans. *Sweden.*
Hecht, Dr Hans-Jürgen. *BRD.*
Hečímović, Mrs Ljerka. *Yugoslavia.*
Heckroodt, Prof. Renier Oelof. *South Africa.*
Hedberg, Prof. Kenneth Wayne. *USA.*
Hedman, Dr. Gun-Britt Margareta. *USA.*
Heeg, Dr. Mary Jane. *USA.*
Hegde, Dr. Rashmi. *USA.*
Hegedüs, Dr Zsolt. *Sweden.*
Hegenbarth, Prof. Ernst. *DDR.*
Heide, Dr Helmut. *BRD.*
Heide, Dr Klaus. *DDR.*
Heijnen, Drs Wilhelmus Marinus Maria. *Netherlands.*
Heim, Dr Harald Josef Robert. *BRD.*
Heim, Dr Joachim. *DDR.*
Heime, Prof. Dr Klaus. *BRD.*
Heinemann, Dr Udo. *BRD.*
Heinerman, Dr Jacobus J. L. *Netherlands.*
Heinz, Prof. Dion Larson. *USA.*
Hejna, Ms. Carolyn I. *USA.*
Held, Dr. Glenn A. *USA.*
Heleskivi, Dr Jouni Martti. *Finland.*
Helgesson, Mr Göran. *Sweden.*
Heller-Kallai, Prof. Lisa. *Israel.*
Helliwell, Prof. John Richard. *UK.*
Hellmold, Prof Peter. *DDR.*

Hellner, Prof.em. Dr Erwin E. *BRD.*
Helmholdt, Dr Robert Barteld. *Netherlands.*
Helmi, Prof. Mohamed Ezzeldin. *Egypt.*
Helmreich, Dr Dieter. *BRD.*
Hemily, Dr. Philip Wright. *USA.*
Hemkar, Dr Mangla Prasad. *India.*
Hempel, Dr Andrew. *Canada.*
Hempel, Dr. Judith C. *USA.*
Henderson, Dr Christopher Michael Bradford. *UK.*
Henderson, Dr Richard. *UK.*
Henderson, Dr Robert Keith. *UK.*
Hendricks, Prof. Robert Wayne. *USA.*
Hendrickson, Prof. Wayne A. *USA.*
Hendrixson, Mr. Thomas L. *USA.*
Henke, Dr Henning. *BRD.*
Henkel, Prof. Dr Gerald. *BRD.*
Henley, Mr. Christopher Lee. *USA.*
Hennig, Prof. Klaus. *DDR.*
Henríquez, Mr Fernando. *Chile.*
Henslee, Dr. Walter Warren. *USA.*
Hentschel, Dr Manfred Paul. *BRD.*
Heo, Dr. Nam Ho. *USA.*
Hepp, Dr Alfred. *Switzerland.*
Hepp, Mr. Mark. *USA.*
Herak, Dr Rajna. *Yugoslavia.*
Herberger, Dr Jürgen. *DDR.*
Herbertsson, Dr B. Harald. *Sweden.*
Herbette, Prof. Leo Gerard. *USA.*
Herbstein, Prof. Frank Herzl. *Israel.*
Herceg, Dr Marija. *Yugoslavia.*
Herdade, Dr Silvio B. *Brasil.*
Herdtweck, Dr Eberhardt. *BRD.*
Heredia, Lic Eduardo Armando. *Argentina.*
Hergold-Brundić, Dr Antonija. *Yugoslavia.*
Heritsch, Prof. em. Dr Haymo. *Austria.*
Herman, Prof. Herbert. *USA.*
Hermans, Prof. Jan. *USA.*
Hermansson, Dr Kersti. *Sweden.*
Hermansson, Dr Leif Å. G. *Sweden.*
Hermida, Lic Jorge Daniel. *Argentina.*
Hermodsson, Dr Yngve. *Sweden.*
Hermoneit, Mr Bernd. *DDR.*
Herms, Dr Gerhard. *DDR.*
Hernandez-Cano, Prof. Felix. *Spain.*
Hernández, Prof. Luis C. *Colombia.*
Herren, Lic Gustavo Guillermo. *Argentina.*
Herrera Palma, Mrs Victoria. *Cuba.*
Herrera-Becerra, Dr. Raul. *México.*
Herreras, Mr M. Luisa. *Spain.*
Herres, Mr Nikolaus. *BRD.*
Herriott, Prof. Jon Roger. *USA.*
Herrmann, Dr Frank-Peter. *DDR.*
Herrmann, Prof. Rudolf. *DDR.*
Herron, Dr. James Nelson. *USA.*
Hervé, Dr Francisco. *Chile.*
Herzberg, Dr Armin. *BRD.*
Herzberg, Dr. Osnat. *USA.*
Hess, Prof. Dr Heinz. *BRD.*
Hesse, Dr Karl-Friedrich. *BRD.*
Hesse, Dr Rolf S. *Sweden.*
Heuer, Prof. Arthur H. *USA.*
Hewat, Dr A.W. *France.*
Hewat, Dr Elizabeth. *France.*
Heydenreich, Prof Johannes. *DDR.*
Heyding, Dr Robert Donald. *Canada.*
Heymann, Prof Gunter. *DDR.*
Heyns, Prof. Anton Michal. *South Africa.*
Hibbard, Dr. Lyndon Stanley. *USA.*
Hicks, Dr Trevor John. *Australia.*
Hidaka, Mr Tsuneo. *Japan.*
Hiebl, Prof. Dr Kurt. *Austria.*
Higashi, Prof. Akira. *Japan.*
Higatsberger, Prof. Dr Michael Josef. *Austria.*
Higgins, Prof. John Britt. *USA.*
Higgins, Dr. Timothy. *Ireland.*
Highcock. Dr Rona Margaret. *UK.*
Higuchi Dr Yoshiki. *Japan.*
Higuchi, Prof. Taiichi. *Japan.*
Hiismäki, Prof. Pekka Eljas. *Finland.*
Hildebrandt, Prof. Dr Gerhard. *BRD.*

Hilgenfeld, Dr Rolf. *BRD.*
Hill, Mr Christopher Peter. *UK.*
Hill, Dr Roderick Jeffrey. *Australia.*
Hilleard, Dr Ronald James. *UK.*
Hiller, Dr Wolfgang Paul. *BRD.*
Hillman, Dr Harold. *UK.*
Hiltunen, Mr Lassi Ilmari. *Finland.*
Hinawi, Prof. Essam. *Egypt.*
Hinch, Mr. Ralph J. Jr. *USA.*
Hinckley, Prof. Conrad Cutler. *USA.*
Hine, Dr Raymond. *UK.*
Hingerty, Dr. Brian Edward. *USA.*
Hinrichs, Dr Winfried. *BRD.*
Hinrichsen, Prof. Dr Georg. *BRD.*
Hinsch, Dr Thorsten Reinhard. *BRD.*
Hintermann, Dr Hans-Erich. *Switzerland.*
Hinz, Dr Dietrich. *DDR.*
Hinze, Prof. Dr Eckhard. *BRD.*
Hirabayashi, Prof. Makoto. *Japan.*
Hiragi, Dr Yuzuru. *Japan.*
Hirahara, Dr Eiji. *Japan.*
Hirano, Prof. Shin-ichi. *Japan.*
Hirayama, Dr Noriaki. *Japan.*
Hiremath, Mr Chaitanya, N. *India.*
Hirota, Dr Fumio. *Japan.*
Hirotsu, Dr Ken. *Japan.*
Hirotsu, Dr Yoshihiko. *Japan.*
Hirsch, Prof. Sir Peter. *UK.*
Hirshfeld, Prof. Fred. *Israel.*
Hirshfield, Mr. Jordan M. *USA.*
Hirth, Prof. John Price. *USA.*
Hitchcock, Dr P.B. *UK.*
Hite, Prof. Gilbert James. *USA.*
Hjertén, Dr Inger. *Sweden.*
Hjorth, Mr Michael. *Denmark.*
Hlavatá, Mrs Drahomíra. *CSSR.*
Ho, Prof. Chien. *USA.*
Ho, Dr. Douglas M. *USA.*
Hoard, Prof. James Lynn. *USA.*
Hoard, Dr. Laurence Graham. *USA.*
Hobbs, Prof. Linn Walker. *USA.*
Hockly, Dr M. *UK.*
Hocksell, Mr Veli Eerik. *Finland.*
Hodeau, Dr. Jean-Louis. *France.*
Hodgkin, Prof. Dorothy Mary Crowfoot. *UK.*
Hodgkinson, Mr R.A. *UK.*
Hodgson, Prof. Derek John. *USA.*
Hodgson, Prof. Keith Owen. *USA.*
Hodorowicz, Prof. Stanisław. *Poland.*
Hodsdon, Dr. John Marshall. *USA.*
Höbler, Dr Hans-Joachim. *DDR.*
Höche, Dr Hans-Reiner. *DDR.*
Höche, Dr Hellmut. *DDR.*
Höfer, Dr Hans Hermann. *BRD.*
Höfler, Dr Sabine Ida Gerda. *BRD.*
Höhling, Prof. Dr Hans Jürgen. *BRD.*
Höhne, Prof. Ernst. *DDR.*
Höier, Prof. Ragnvald. *Norway.*
Hökelek, Dr Tuncer. *Turkey.*
Hölsä, Dr Jorma Pertti Kalervo. *Finland.*
Hönle, Dr Wolfgang. *BRD.*
Hörl, Prof. Dr Erwin M. *Austria.*
Hötzsch, Dr Günter. *DDR.*
Hoff, Mr. Henry A. *USA.*
Hoff, von, Dr Siegfried. *DDR.*
Hoffmann, Prof. Dr Wolfgang. *BRD.*
Hoffmeister, Dr Wolfgang. *BRD.*
Hofmeister, Dr Herbert. *DDR.*
Hofmeister, Dr Wolfgang. *BRD.*
Hogan, Dr Leonard McNamara. *Australia.*
Hogg, Mr C. *UK.*
Hogg, Dr Joshua Herbert Christopher. *UK.*
Hoggins, Dr. James Thomas. *USA.*
Hogle, Dr. James Martin. *USA.*
Hohlwein, Dr Dietmar. *BRD.*
Hol, Prof. Dr Wim G. J. *Netherlands.*
Holas, Dr hab. Andrzej. *Poland.*
Holben, Mr John. *UK.*
Holbrook, Dr. Stephen Roy. *USA.*
Holden, Dr. Hazel Marguerite. *USA.*

Holden, Dr. James Richard. *USA.*
Holinski, Dr Rüdiger. *BRD.*
Holland, Dr. Hans J. *USA.*
Hollander, Dr. Frederick J. *USA.*
Holldorf, Dr Horst. *DDR.*
Holmes, Mr. Francis Edward. *USA.*
Holmes, Prof. Kenneth Charles. *BRD.*
Holomany, Mr. Mark A. *USA.*
Holt, Dr. Elizabeth Manners. *USA.*
Holt, Dr Ronald Stanley. *UK.*
Holtzberg, Dr. Frederic. *USA.*
Holtzman, Ms. Susan. *USA.*
Holub, Dr Fritz. *Austria.*
Holý, Dr Václav. *CSSR.*
Holzapfel, Prof. Dr Wilfried B. *BRD.*
Hom, Dr. Tommy. *USA.*
Homma, Mr Shigeru. *Japan.*
Homma, Mr Teiichi. *Japan.*
Hon, Dr Ping-Kay. *Hong Kong.*
Hong, Mr Mao-chun. *China.*
Hong, Dr Sam-Hyo. *Sweden.*
Honigmann, Dr Bertold. *BRD.*
Honoré, Dr Tage. *Denmark.*
Honzatko, Prof. Richard E. *USA.*
Hoogsteen, Dr. Karst. *USA.*
Hoonnivathana, Mr Ekachai. *Thailand.*
Hope, Prof. Håkon. *USA.*
Hopkins, Prof. Robert C. *USA.*
Hoppe, Prof. Dr Dr hc Rudolf. *BRD.*
Hoppe, Prof. Günter. *DDR.*
Hordvik, Prof. Asbjörn. *Norway.*
Horioka, Dr Keiji. *Japan.*
Horiuchi, Dr Hiroyuki. *Japan.*
Horiuchi, Dr Shigeo. *Japan.*
Horn, Dr Ernst. *Australia.*
Horn, Dr Jerzy. *Poland.*
Horn, Prof. Manfred Josef. *Peru.*
Horn, Prof. Paul. *France.*
Horne, Mr William. *Saudi Arabia.*
Hornstra, Drs Jan. *Netherlands.*
Hornung, Dr George. *UK.*
Horrigan, Ms. Jane Akerlund. *USA.*
Horrocks, Prof. William DeWitt Jr. *USA.*
Horsey, Dr. Richard Stephen. *USA.*
Horsfield, Mr Edgar Charles. *South Africa.*
Horst, Dr Wolfgang. *BRD.*
Horstmann, Prof. Dr Manfred. *BRD.*
Horton, Mr. John R. *USA.*
Horváth, Dr (Ms) Ilona. *Hungary.*
Horváth, Ing Josef. *CSSR.*
Horvatić, Magistar Davor. *Yugoslavia.*
Horyń, Dr hab. Roman. *Poland.*
Hosaka, Dr Masahiro. *Japan.*
Hoschl, Dr Pavel. *CSSR.*
Hosea, Dr Thomas Jeffrey Cockburn. *Singapore.*
Hosemann, Prof. Dr Dr hc Rolf. *BRD.*
Hoser, Dr Andrzej. *BRD.*
Hoshino, Prof. Sadao. *Japan.*
Hoskins, Dr Bernard Foster. *Australia.*
Hosmane, Prof. Narayan S. *USA.*
Hosoya, Prof. Masahiko. *Japan.*
Hosoya, Prof. Sukeaki. *Japan.*
Hospital, Dr Michel. *France.*
Hossain, Dr. M. Bilayet. *USA.*
Hou, Mr Yong-geng. *China.*
Hough, Dr Edward. *Norway.*
Houng, Prof. Kun-Huang. *Taiwan.*
Hountas, Dr Athanasios. *Greece.*
Houska, Prof. Charles Robert. *USA.*
Housty, Dr Jacques. *France.*
Hovestreydt, Dr Eric Robert. *BRD.*
Hovmöller, Dr Sven. *Sweden.*
Howard, Mr. Andrew Jay. *USA.*
Howard, Dr Christopher John. *Australia.*
Howard, Dr Judith Ann Kathleen. *UK.*
Howard, Dr. Scott A. *USA.*
Howatson, Prof. John. *USA.*
Howe, Ms. Donna-Beth. *USA.*
Howe, Prof. James M. *USA.*
Howell, Dr. Peter Adam. *USA.*

Howes, Ms. Catherine. *USA.*
Howie, Dr Robert Alan. *UK.*
Howie, Prof. Robert Andrew. *UK.*
Howlin, Dr Brendan James. *UK.*
Hoyer, Dr Walter. *DDR.*
Hriljac, Dr Joseph A. *UK.*
Hseu, Prof. Shu-En. *Taiwan.*
Hseu, Prof. Tzong-Hsiung. *Taiwan.*
Hsu, Miss Barbara T. *USA.*
Hsu, Mr. Hsyh-Min. *USA.*
Hsu, Prof. I-Nan. *USA.*
Hsu, Dr. Leh-Yeh Ruth. *USA.*
Htoon, Prof. Sein. *Burma.*
Hu, Mr Guo-zhi. *China.*
Hu, Mr Heng-liang. *China.*
Hu, Dr. Hengliang. *USA.*
Hu, Mr Sheng-zhi. *China.*
Hu, Ms Xiaorui. *BRD.*
Hua, Mr Zi-qian. *China.*
Huang, Bao-quan. *China.*
Huang, Mrs Chi-Yung. *Taiwan.*
Huang, Mr De-bin. *China.*
Huang, Mr. De-Bin. *USA.*
Huang, Prof. Jin-ling. *China.*
Huang, Mr Jin-shun. *China.*
Huang, Mr. Kuei-Shang. *USA.*
Huang, Prof. Liang-ren. *China.*
Huang, Ming-dong. *China.*
Huang, Mr Tai-shan. *China.*
Huang, Dr. Ting Chun. *USA.*
Huang, Mr Tung-Woo. *Taiwan.*
Huang, Mr Zhi-ying. *China.*
Huanosta Tera, Mr Alfonso. *México.*
Hubbard, Dr. Camden Richards. *USA.*
Hubbard, Dr Roderick Eliot. *UK.*
Hubbard, Dr. Stevan R. *USA.*
Hubbell, Mr. John Howard. *USA.*
Hubbert, Mrs Dr Elisabeth. *BRD.*
Huber, Dr Carol P. *Canada.*
Huber, Prof. Robert. *BRD.*
Hubert, Mr Georg. *BRD.*
Huch, Dr Volker. *BRD.*
Hudd, Mr R.C. *UK.*
Huddle, Prof. Benjamin Paul. *USA.*
Hudgens, Mr. Claude R. *USA.*
Hübner, Dr Manfred. *DDR.*
Hübner, Mr Thomas. *BRD.*
Hümmer, Prof. Dr Kurt. *BRD.*
Huffman, Dr. John Curtis. *USA.*
Hughes, Dr Antony Elwyn. *UK.*
Hughes, Dr David Lewis. *UK.*
Hughes, Prof. John M. *USA.*
Hughes, Dr. Robert Edward. *USA.*
Hughes, Mr Thomas Ernest. *UK.*
Hughes, Prof. William Eugene. *USA.*
Hugo, Mr Geoffrey Ronald. *Australia.*
Huiszoon, Dr Cornelis. *Netherlands.*
Hukin, Dr David Ainsworth. *UK.*
Hukins, Dr David William Laurence. *UK.*
Hull, Dr Stephen Edward. *UK.*
Hulliger, Dr Jürg. *Switzerland.*
Hulme, Mr Ralph. *UK.*
Hultzsch, Mr Rainer. *DDR.*
Humble, Dr Sten G. *Sweden.*
Humblet, Dr. Christine. *USA.*
Huml, Dr Karel. *CSSR.*
Hummel, Dr Hans-Ulrich. *BRD.*
Hummel, van, Ing. Gerrit Jan. *Netherlands.*
Hummel, Mr Wolfgang. *Switzerland.*
Humphreys, Prof. Colin John. *UK.*
Hund, Dr Franz Josef. *BRD.*
Hundt, Dr Rudolf. *BRD.*
Hunt, Dr. Lois T. *USA.*
Hunt, Mr. Richard E. *USA.*
Huq, Dr Fazlul. *Australia.*
Hurem, Magistar Zihnija. *Yugoslavia.*
Hurley, Mr Patrick Walter. *UK.*
Hurst, Prof. Vernon James. *USA.*
Hursthouse, Dr Michael Barry. *UK.*
Husain, Dr Abul Hasanat Mohammad. *Bangladesh.*

Jircitano, Dr. Alan John. *USA.*
Joachimiak, Dr Andrzej. *Poland.*
Joel, Dr Nahum. *France.*
Jönsson, Dr Per-Gunnar. *Sweden.*
Jørgensen, Dr Jens-Erik. *Denmark.*
Jørgensen, Mr Ole. *Denmark.*
Joest, Mr Stephan. *BRD.*
Joesten, Prof. Raymond. *USA.*
Johansen, Dr Heinrich. *DDR.*
Johanson, Mr Bo Stefan. *Finland.*
Johansson, Dr Georg. *Sweden.*
Johansson, Mr Karl-Erik. *Sweden.*
Johnsen, Mr Ole. *Denmark.*
Johnson, Prof. Alan Arthur. *USA.*
Johnson, Dr Andrew William Syme. *Australia.*
Johnson, Dr. Carroll K. *USA.*
Johnson, Dr David Julian. *UK.*
Johnson, Dr. Frank Bacchus. *USA.*
Johnson, Prof. Gerald Glenn Jr. *USA.*
Johnson, Prof. John Emil. *USA.*
Johnson, Dr Louise Napier. *UK.*
Johnson, Ms Louise. *South Africa.*
Johnson, Prof. Michael E. *USA.*
Johnson, Dr Michael William. *UK.*
Johnson, Dr N.P. *UK.*
Johnson, Mr Owen. *UK.*
Johnson, Dr. Paul Lorentz. *USA.*
Johnson, Dr Peter Anthony Victor. *UK.*
Johnson, Dr. Quintin C. *USA.*
Johnson, Mrs. Ruth Jeannette Beach. *USA.*
Johnston, Mr Victor James. *Canada.*
Jones, (Milne) Dr Angela Alice. *UK.*
Jones, Mr D. *UK.*
Jones, Prof. Daniel Silas. *USA.*
Jones, Prof. Derry Wynn. *UK.*
Jones, Miss Elizabeth Louise. *South Africa.*
Jones, Dr G.R. *UK.*
Jones, Mr. Glover A. *USA.*
Jones, Dr John Brett. *Australia.*
Jones, Dr. Noel Duane. *USA.*
Jones, Dr Stephen John. *Canada.*
Jones, Prof. T. Alwyn. *Sweden.*
Jones, Dr Tony Cristofer. *New Zealand.*
Jones, Dr William. *UK.*
Jordan, Dr. Steven R. *USA.*
Jordan, Prof. Truman H. *USA.*
Jordanovska, Prof. Dr (Mrs) Vera. *Yugoslavia.*
Jorgensen, Dr. James D. *USA.*
José Yacaman, Dr Miguel. *México.*
Josefsson, Miss Magdalena. *Sweden.*
Joseph, Dr Günter. *Chile.*
Joshi, Prof. Ramesh Vinayak. *India.*
Joshi, Dr Shri Krishna. *India.*
Joshua-Tor, Ms Leemor. *Israel.*
Jost, Dr Karlheinz. *DDR.*
Jostsons, Dr Adam. *Australia.*
Joswig, Dr Werner. *BRD.*
Joubert, Prof. Jean-Claude. *France.*
Jouini, Dr Amor. *Tunisia.*
Jouini, Dr Noureddine. *Tunisia.*
Jouini, Dr Tahar. *Tunisia.*
Jourdan, Dr Claude, René *France.*
Jovanovski, Prof. Dr Gligor. *Yugoslavia.*
Juang, Prof. Tzo-Chuan. *Taiwan.*
Jude, Dr (Mrs) Lidia. *Romania.*
Julian, Prof. Maureen O'Donnell. *USA.*
Julve-Oliva, Dr Miguel. *Spain.*
Jung, Dr Detlef. *BRD.*
Jung, Dr Volkhard. *BRD.*
Jung, Dr Walter. *BRD.*
Jurisch, Dr Manfred. *DDR.*
Jurković, Prof. Dr Ivan. *Yugoslavia.*
Jurnak, Prof. Frances Anne. *USA.*
Juul Jensen, Mrs Dorte. *Denmark.*
Jynge, Mr Knut. *Norway.*
Kaas, Dr Karen. *Denmark.*
Kaat, te, Prof Dr Erich Heinz. *BRD.*
Kabalkina, Prof. Sara Samsonovna. *USSR.*
Kabalov, Dr Yurii Konstantinovich. *USSR.*
Kabelka, Dr Heinz I. *Austria.*

Kabešová, Ing Mária. *CSSR.*
Kabish, Dr Lotfi. *Egypt.*
Kabs, Mr Michael. *BRD.*
Kabsch, Dr Wolfgang. *BRD.*
Kachalov, Dr Oleg Viktorovich. *USSR.*
Kachinsky, Dr Vitol'd Nikolayevch. *USSR.*
Kádár, Dr György. *Hungary.*
Kaduk, Dr. James Albert. *USA.*
Kähkönen, Dr Heikki Antero. *Finland.*
Käll, Mr Per-Olov. *Sweden.*
Kämmel, Dr Thomas. *DDR.*
Kaerlein, Mr Carsten-Peter. *BRD.*
Kaftory, Prof. Menahem. *Israel.*
Kaganer, Dr Vladimir Mikhailovich. *USSR.*
Kagarakis, Prof. Constantine. *Greece.*
Kageyama, Dr Hiroyuki. *Japan.*
Kagotani, Dr Toshio. *Japan.*
Kahlert, Prof. Dr Hartmut. *Austria.*
Kai, Prof. Yasushi. *Japan.*
Kaidalova, Dr Taisiya Alexandrovna. *USSR.*
Kainuma, Prof. Yoshiro. *Japan.*
Kaischew, Prof. Rostislaw. *Bulgaria.*
Kaitner, Dr Branko. *Yugoslavia.*
Kaito, Dr Chihiro. *Japan.*
Kajzar, Dr Franciszek. *Poland.*
Kakudo, Prof. Masao. *Japan.*
Kalanov, Dr Makhmud. *USSR.*
Kalb(Gilboa), Dr Aaron Joseph. *Israel.*
Kalceff, Dr Walter. *Australia.*
Kaldis, Prof. Emanuel. *Switzerland.*
Kaler, Prof. Eric W. *USA.*
Kalicińska-Karut, Mgr Jarosława. *Poland.*
Kalikhman, Dr Vyacheslav Mikhailovich. *USSR.*
Kalinchenko, Dr Anatolii Mikhailovich. *USSR.*
Kalinin, Dr Vadim Rodionovich. *USSR.*
Kalinin, Dr Vladimir Ivanovich. *USSR.*
Kalinkina, Dr Irina Nikolayevna. *USSR.*
Kalinna, Mr Hartmut. *DDR.*
Kalk, Mr Kornelis Harm. *Netherlands.*
Kallal, Dr Ahmed. *Tunisia.*
Kallen, Dr Jörg. *Switzerland.*
Kallio, Mr Pekka Yrjö Juhani. *Finland.*
Kálmán, Prof. Alajos. *Hungary.*
Kálmán, Dr Erika. *Hungary.*
Kalman, Prof. Zwi Heinrich. *Israel.*
Kalnik, Dr. Matthew Walter. *USA.*
Kałuski, Prof. Dr hab. Zygmunt. *Poland.*
Kalweit, Mr Harald. *DDR.*
Kalyanaraman, Dr A. R. *India.*
Kalychak, Dr Yaroslav Mikhailovich. *USSR.*
Kambas, Prof. Kostas. *Greece.*
Kambe, Dr Kyozaburo. *BRD.*
Kamel, Prof. Dr Raafat Wasef. *Egypt.*
Kamenar, Prof. Dr Boris. *Yugoslavia.*
Kamentsev, Dr Igor' Evgenyevich. *USSR.*
Kamer, Mr. Greg. *USA.*
Kamijo, Dr Nagao. *Japan.*
Kaminsky, Dr Alexander Alexandrovich. *USSR.*
Kaminsky, Dr Vladimir Fedorovich. *USSR.*
Kamiya, Dr Kazuhide. *Japan.*
Kamiya, Dr Nobuo. *Japan.*
Kamiya, Prof. Yoshihiro. *Japan.*
Kamminga, Dr H. *UK.*
Kamolchote, Mr Poonsak. *Thailand.*
Kamoun, Dr Slaheddine. *Tunisia.*
Kampf, Mr. Jeff W. *USA.*
Kamphuis, Dr Irenus Gerhardus. *Netherlands.*
Kamprath, Mr Fred-Bodo. *DDR.*
Kamwaya, Dr Mombo E. *Zambia.*
Kanatzidis, Dr. Mercouri G. *USA.*
Kanehisa, Miss Nobuko. *Japan.*
Kanellis, Prof. George. *Greece.*
Kanevskii, Dr Vladimir Mikhailovich. *USSR.*
Kanis, Dr Michael. *DDR.*
Kannan, Dr Kazhiur Kothandapani. *India.*
Kansikas, Dr Jarno Juhani. *Finland.*
Kantardjieff, Prof. Katharine Ann. *USA.*
Kanters, Dr Jan. *Netherlands.*
Kantor, Dr Matvey Matveyevich. *USSR.*
Kapecki, Dr. Jon Alfred. *USA.*

Kaplan, Mr. David B. *USA.*
Kaplunnik, Dr Lidiya Nikolayevna. *USSR.*
Kapon, Dr Moshe. *Israel.*
Kapor-Nahlovski, Dr Agneš. *Yugoslavia.*
Kar(Roy), Dr (Mrs) Tanusree *India.*
Karanović, Dr Ljiljana. *Yugoslavia.*
Karapetyan, Arutyun Arshaluisovich. *USSR.*
Karapinar, Mr Ridvan. *Turkey.*
Karcher, Dr. Barbara Ann. *USA.*
Kardashev, Dr Boris Konstantinovich. *USSR.*
Kardos, Ms Jutta. *Hungary.*
Karipides, Prof. Anastas. *USA.*
Karl, Prof Dr Norbert. *BRD.*
Karle, Dr. Isabella L. *USA.*
Karle, Dr. Jean Marianne. *USA.*
Karle, Prof. Jerome. *USA.*
Karlsson, Dr Bengt E. *Sweden.*
Karlsson, Dr Rolf. *Switzerland.*
Karmazin, Dr Lubomir. *CSSR.*
Karniewicz, Dr Jan. *Poland.*
Karolak-Wojciechowska, Dr Janina. *Poland.*
Karp, Prof. Dr Jan. *Poland.*
Karpinsky, Dr Oleg Georgiyevich. *USSR.*
Karppinen, Dr Markku. *Sweden.*
Kartheuser, Dr Edward Peter. *Belgium.*
Kartusiak, Dr Andrzej Szczepan. *Poland.*
Karup-Møller, Dr Sven. *Denmark.*
Karvinen, Mrs Saila Marjatta. *Finland.*
Karyakina, Dr Tatyana Alexandrovna. *USSR.*
Kas'yanenko, Dr Evgenii Vasil'evich. *USSR.*
Kasai, Prof. Nobutami. *Japan.*
Kashaev, Dr Anvar Akhyarovich. *USSR.*
Kashchiev, Dr Dimcho. *Bulgaria.*
Kashino, Dr Setsuo. *Japan.*
Kashiwase, Dr Yasuji. *Japan.*
Kashyap, Dr Ram Prasad. *India.*
Kasper, Dr. John S. *USA.*
Kassner, Mr Dethard. *BRD.*
Kastner, Dr. Margaret Ellen. *USA.*
Kasturi, Prof Tirumali R. *India.*
Kasuga, Prof. Masanobu. *Japan.*
Kaszkur, Dr Zbigniew. *Poland.*
Katada, Prof. Kinya. *Japan.*
Kataeva, Ol'ga Nikolaevna. *USSR.*
Katagas, Dr Christos. *Greece.*
Katagawa, Dr Takeshi. *Japan.*
Katayama, Dr Chuji. *Japan.*
Katayama, Mr Kenichi. *Japan.*
Katayama, Prof. Mikio. *Japan.*
Katayanagi, Mr Katsuo. *Japan.*
Kato, Dr Akira. *Japan.*
Kato, Mr Ichiro. *Japan.*
Kato, Dr Katsuo. *Japan.*
Kato, Prof. Masanori. *Japan.*
Kato, Prof. Norio. *Japan.*
Kato, Prof. Toshio. *Japan.*
Kato, Mr Yoshihiro. *Japan.*
Katoh, Mr Ichiro. *Japan.*
Kats, Dr Moisey Sukherovich. *USSR.*
Katsanos, Mr Demetrios Evangelos. *Greece.*
Katscher, Dr Hartmut. *BRD.*
Katsnel'son, Prof. Al'bert Anatolyevich. *USSR.*
Katsube, Prof. Yukiteru. *Japan.*
Katsuya, Mr Yoshio. *Japan.*
Katz, Dr Gerald. *Israel.*
Katz, Mr. Henry. *USA.*
Katz, Prof. J. Lawrence. *USA.*
Katz, Prof. Lewis. *USA.*
Katz, Dr. Robert. *USA.*
Katzschmann, Mr Kurt. *DDR.*
Kaub, Mr Jürgen. *BRD.*
Kaučič, Dr Venčeslav. *Yugoslavia.*
Kaufman, Prof. Hershall William. *USA.*
Kaufmann, Dr Thorsten. *DDR.*
Kaus, Dr Gerhard. *BRD.*
Kavounis, Dr Konstantinos. *Greece.*
Kawachi, Dr Yosuke. *New Zealand.*
Kawado, Dr Seiji. *Japan.*
Kawahara, Prof. Akira. *Japan.*
Kawai, Mr Takatoshi. *Japan.*

Kawai, Mr Toshiaki. *Japan.*
Kawaminami, Prof. Masaru. *Japan.*
Kawamori, Miss Asako. *Japan.*
Kawamura, Dr Tsutomu. *Japan.*
Kawano, Prof. Shigeaki. *Japan.*
Kawata, Dr Hiroshi. *Japan.*
Kay, Dr Herbert Frederick. *UK.*
Kay, Dr. Mortimer I. *USA.*
Kayden, Dr Catherine Sheila. *Canada.*
Kayushina, Dr Renata Lvovna. *USSR.*
Kazeeva, Dr Lyudmila Pavlovna. *USSR.*
Kazinets, Dr Maria. *Israel.*
Ke, Mr Heng-ming. *China.*
Keankeo, Miss Watcharaporn. *Thailand.*
Keder, Dr. Nancy Lynn. *USA.*
Keefe, Ms. Lisa J. *USA.*
Keefe, Prof. William Edward. *USA.*
Keem, Dr. John Edward. *USA.*
Keesmann, Prof. Dr Karl-Ingo Ortwin. *BRD.*
Keijser, de, Dr Ir Thomas Henri. *Netherlands.*
Keith, Dr Vepan. *Saudi Arabia.*
Kellenberger, Prof. Eduard. *Switzerland.*
Keller, Dr Egbert. *BRD.*
Keller, Dr Eva Barbara. *Switzerland.*
Keller, Prof. Dr Heimo Jürgen. *BRD.*
Keller, Dr Kurt Wolfgang. *DDR.*
Keller, Dr. Ludwig. *USA.*
Keller, Prof Dr Paul. *BRD.*
Keller, Priv.-Doz. Dr Hans-Lothar. *BRD.*
Keller, Dr hab. Włodzimierz. *Poland.*
Keller, Dr Wolfgang Ludwig. *BRD.*
Kellö, Dr Eleonóra. *CSSR.*
Kelly, Dr Anthony. *UK.*
Kelly, Mrs. Carol J. *USA.*
Kelly, Mr Eric. *UK.*
Kelly, Dr. Judith Ann. *USA.*
Kelly, Dr Patrick Manning. *Australia.*
Kelly, Dr Thomas C. *Ireland.*
Kemme, Dr Andrey Andreyevich. *USSR.*
Kemmler-Sack, Mrs Prof. Dr Sibylle. *BRD.*
Kemp, Dr. Nantelle Smith Pantaleo. *USA.*
Kempa, Mr Paul-Bernd. *BRD.*
Kempster, Dr Charles John Edgar. *UK.*
Kenaan, Dr Feisal. *Saudi Arabia.*
Kendi, Doç. Dr Engin. *Turkey.*
Kennard, Dr Colin Harold Leslie. *Australia.*
Kennard, Dr Olga. *UK.*
Kennedy, Miss D.A. *UK.*
Kennedy, Mr John Matthew. *UK.*
Kennon, Prof Noel Frederick. *Australia.*
Keow-kam-nerd, Dr Kanchana. *Thailand.*
Keppler, Dr Ulrich Hermann. *BRD.*
Keramidas, Mr Konstantinos Georgios. *Greece.*
Kerenović, Magistar Midhat. *Yugoslavia.*
Kern, Prof. Raymond. *France.*
Kerr, Dr Ian Segrave. *UK.*
Kerr, Dr Kathleen Ann. *Canada.*
Kersten, Mr Friedrich. *DDR.*
Kertész, Dr László. *Hungary.*
Keshmiri, Seyyed-Hosein. *Iran.*
Kessenikh, Dr Galina Georgiyevna. *USSR.*
Keszler, Prof. Douglas A. *USA.*
Ketolainen, Prof. Pertti Pekka Juhani. *Finland.*
Kettmann, Ing Viktor. *CSSR.*
Kettunen, Prof. Pentti Olavi. *Finland.*
Keulen, Dr Evert. *Netherlands.*
Khachaturyan, Mr. Armen G. *USA.*
Khadr, Dr Moustafa. *Egypt.*
Khadzhi, Dr Valentin Evstafyevich. *USSR.*
Khaikin, Dr Leonid Solomonovich. *USSR.*
Khakzar, Dr Ahmad. *Iran.*
Khalf, Mr Fuad M. *Libya.*
Khalid, Mr Mohammad. *Pakistan.*
Khalifa, Dr (Mrs) B. Abdel Meguid. *Egypt.*
Khalifa, Prof. (Mrs) Berlant. *Egypt.*
Khalil, Miss Safia. *USA.*
Khan, Mr. A. *USA.*
Khan, Dr Abdul Quadeer. *Pakistan.*
Khan, Dr Ainul Hassan. *Pakistan.*
Khan, Prof. Ali. *Venezuela.*

Khan, Dr Anwarur Rahman. *Bangladesh.*
Khan, Dr. Masood. *USA.*
Khan, Mr Mohammad Afaq. *Pakistan.*
Khandar-Shahabad, Dr Ali-Akbar. *Iran.*
Khantaprab, Dr Chaiyudh. *Thailand.*
Kharchenko, Dr Lyudmila Yulianovna. *USSR.*
Kharchenko, Dr Olga Ivanovna. *USSR.*
Kharitonov, Dr Yury Alexandrovich. *USSR.*
Khasanov, Dr Salavat Salim'yanovich. *USSR.*
Khatanova, Dr Nina Abdulovna. *USSR.*
Khattak, Dr Guldad Khattak. *Saudi Arabia.*
Khawaja, Dr Ehsan Ellahi. *Saudi Arabia.*
Khawaja, Mr Mahmood-ul-Hassan. *Pakistan.*
Kheiker, Prof. Daniel' Moiseyevich. *USSR.*
Kheirov, Dr Mamed Bekovch. *USSR.*
Khidirov, Dr I. *USSR.*
Khidr, Prof. (Mrs) Fatma Abdel Hakim. *Egypt.*
Khimich, Dr Tamara Andranikovna. *USSR.*
Khisina, Dr Nataliya Rafailovna. *USSR.*
Khitrova, Dr Valentina Ivanovna. *USSR.*
Khodashova, Dr Tatyana Semenovna. *USSR.*
Khokhlov, Prof Aleksandr Fedorovich. *USSR.*
Kholeif, Dr Mahmoud. *Egypt.*
Kholov, Dr Alimakhmad. *USSR.*
Khotsyanova, Dr Tatyana Lvovna. *USSR.*
Khundzhua, Dr Andrei Georgievich. *USSR.*
Khurshudyan, Dr Era Khristoforovna. *USSR.*
Khwaja, Dr Farid Akhtar. *Pakistan.*
Kianvash, Dr Abbas. *Iran.*
Kiedrowski, von, Mr Hartmut. *DDR.*
Kiefer, Dr. Charles R. *USA.*
Kiel, Dr Gertrude Lina. *BRD.*
Kiely, Dr Patrick Vincent. *Ireland.*
Kierkegaard, Prof. Peder. *Sweden.*
Kiers, Dr Conradus Theodorus. *Netherlands.*
Kies, Dr Jörg. *DDR.*
Kiessling, Mr Frank-Michael. *DDR.*
Kifune, Dr Kouichi. *Japan.*
Kihara, Dr Kuniaki. *Japan.*
Kihlborg, Prof. Lars. *Sweden.*
Kikuchi, Dr Makoto. *Japan.*
Kikuta, Prof. Seishi. *Japan.*
Killean, Dr Reginald Cameron Gordon. *UK.*
Kim, Ms. Eunice E. *USA.*
Kim, Ms. Heasook. *USA.*
Kim, Prof. Ho Sung. *Korea.*
Kim, Dr Hoon Sup. *Korea.*
Kim, Prof. Jung Ja P. *USA.*
Kim, Dr Key Soo. *Korea.*
Kim, Prof. Kimoon. *Korea.*
Kim, Mr. Kyung Hyun. *USA.*
Kim, Prof. Moon Il. *Korea.*
Kim, Prof. Moon-Jib. *Korea.*
Kim, Dr. Nancy Ellen Kime. *USA.*
Kim, Prof. Peter S. *USA.*
Kim, Dr Sangsoo. *Korea.*
Kim, Prof. Dr Soo Jin. *Korea.*
Kim, Mr. Sukyoung. *USA.*
Kim, Prof. Sung-Hou. *USA.*
Kim, Prof. Dr Yang Bae. *Korea.*
Kim, Prof. Dr Yang. *Korea.*
Kim, Dr Yoonho. *Korea.*
Kimball, Dr. Martha R. *USA.*
Kimmel, Prof. Giora. *Israel.*
Kimura, Prof. Masao. *Japan.*
Kimura, Mr Masao. *Japan.*
Kimyongür, Yrd. Doç. Dr Nurettin. *Turkey.*
King, Prof. Geoffrey Stephen Douglas. *Belgium.*
King, Dr. Hubert Ellis. *USA.*
King, Prof. Hubert Wylam. *Canada.*
King, Dr James Newington. *UK.*
King, Dr. Murray Vernon. *USA.*
Kingma, Ms. Kathleen J. *USA.*
Kingman, Dr. Priscilla Ward. *USA.*
Kini, Mr Ullal Devappa. *340124).*
Kinneging, Dr Albertus Jacobus. *Netherlands.*
Kinzhibalo, Dr Vladimir Vasil'evich. *USSR.*
Kiosse, Dr Georgy Alexandrovich. *USSR.*
Kipling, Miss Susan Jane. *UK.*
Kir'yanova, Dr Elena Viktorovna. *USSR.*

Kirchhoff, Ms. Pamela Moore. *USA.*
Kirchner, Prof. Dr Elisabeth Charlotte. *Austria.*
Kirchner, Prof. Richard Martin. *USA.*
Kirfel, Dr Armin Harald. *BRD.*
Kirichenko, Dr Valentina Vasilyevna. *USSR.*
Kirikov, Dr Vladimir Arkadyevich. *USSR.*
Kirin, Dr (Mrs) Ankica. *Yugoslavia.*
Kiriyama, Prof. Hideko. *Japan.*
Kiriyama, Prof. Ryoiti. *Japan.*
Kirkinsky, Dr Vitaly Alekseyevich. *USSR.*
Kirkman, Dr John Henry. *New Zealand.*
Kirkova, Prof. Elena. *Bulgaria.*
Kirn, Mr. J. F. *USA.*
Kirov, Doc Georgi Nikolov. *Bulgaria.*
Kirov, Mr Georgi Kirilov. *Bulgaria.*
Kirpichnikova, Dr Lyubov' Fedorovna. *USSR.*
Kirschner, Dr. Daniel A. *USA.*
Kirz, Prof. Janos. *USA.*
Kisch, Prof. Hanan Josef. *Israel.*
Kiselev, Dr Nikolay Andreyevich. *USSR.*
Kishi, Dr Kiyoshi. *Japan.*
Kishino, Prof. Seigo. *Japan.*
Kishk, Dr Fawzi Mohamed. *Egypt.*
Kisi, Mr Erich Herold. *Australia.*
Kislovsky, Dr Lev Dmitriyevich. *USSR.*
Kiss, Dr. Klara. *USA.*
Kiss, Mr Sándor. *Hungary.*
Kissinger, Dr. Charles R. *USA.*
Kissinger, Mr. Homer Everett. *USA.*
Kissling, Mr Alexandru. *Romania.*
Kistenmacher, Dr. Thomas John. *USA.*
Kitagawa, Dr Yasuyuki. *Japan.*
Kitahama, Dr Katsuki. *Japan.*
Kitano, Dr Yasuyuki. *Japan.*
Kitano, Mr Yukishige. *Japan.*
Kittl, Mr Pablo. *Chile.*
Kivekäs, Dr Raikko Terjo Ilari. *Finland.*
Kivilahti, Prof. Jorma Kalevi. *Finland.*
Kjær, Dr Kristian. *Denmark.*
Kjekshus, Prof. Arne. *Norway.*
Kjeldgaard, Mr Morten. *Denmark.*
Klanderman, Prof. Kent Arlen. *USA.*
Klapper, Prof. Dr Helmut. *BRD.*
Klaska, Dr Karl-Heinz. *BRD.*
Klassen-Neklyudova, Prof. Marina V. *USSR.*
Klebe, Dr Gerhard. *BRD.*
Klechkovskaya, Dr Vera Vsevolodovna. *USSR.*
Klee, Prof. Dr Wilfrid Edgar. *BRD.*
Klein, Prof. Cheryl Lynn. *USA.*
Kleinert, Dr Peter. *DDR.*
Kleinstück, Prof. Karlheinz. *DDR.*
Kleint, Dr Christian. *DDR.*
Kléman, Dr Maurice. *France.*
Klement, Dr Ulrich. *BRD.*
Klepp, Dr Kurt Otto. *BRD.*
Klepp, Prof. Dr Kurt Otto, *Austria.*
Kleshchinsky, Dr Leonid Innokentyevich. *USSR.*
Klessen, Mr Gerhard. *BRD.*
Klevtsov, Dr Petr Vasilyevich. *USSR.*
Klevtsova, Dr Rimma Fedorovna. *USSR.*
Klewe, Mr Bernt. *Norway.*
Klimakow, Dr. Alexander. *DDR.*
Klimanek, Dr Peter. *DDR.*
Klimm, Dr Detlef. *DDR.*
Klimova, Dr Anna Yuryevna. *USSR.*
Klinga, Dr Martti Evert. *Finland.*
Klinger, Ms. Alexandra. *USA.*
Kliya, Dr Maya Ottovna. *USSR.*
Klöss, Dr. Gerd. *DDR.*
Klooster, Dr. W. *USA.*
Klop, Drs Enno Anton. *Netherlands.*
Klug, Dr Aaron. *UK.*
Klug, Dr (Ms) Annamária. *Hungary.*
Klug, Prof. Harold Philip. *USA.*
Klyavin, Dr Oleg Vladimirovich. *USSR.*
Knab, Dr Galina Grigoryevna. *USSR.*
Kneschke, Dr Götz. *DDR.*
Kniep, Prof Dr Rüdiger. *BRD.*
Knight, Mr Kevin Stephen. *UK.*
Knight, Mr Robert. *UK.*

Knight, Mr Stefan D. *Sweden.*
Knighton, Mr. Daniel R. *USA.*
Knobler, Dr. Carolyn B. *USA.*
Knoch, Dr Falk A. *BRD.*
Knöchel, Dr Claus-Dieter. *BRD.*
Knof, Dr Wolfgang Erich. *BRD.*
Knop, Prof. Osvald. *Canada.*
Knorr, Dr. David B. *USA.*
Knorr, Dr Klaus. *BRD.*
Knott, Mr P.R. *UK.*
Knott, Mr Robert Barry. *Australia.*
Knox, Prof. James Russell. *USA.*
Knox, Dr. Ralph David. *USA.*
Knuuttila, Mrs Hilkka Ritva-Liisa. *Finland.*
Knuuttila, Mr Pekka Juhani. *Finland.*
Kobayashi, Dr Akiko. *Japan.*
Kobayashi, Dr Hayao. *Japan.*
Kobayashi, Prof. Jinzo. *Japan.*
Kobayashi, Mr Masaaki. *Japan.*
Kobayashi, Prof. Nobuyuki. *Japan.*
Kobayashi, Prof. Tadashi. *Japan.*
Kobayashi, Dr Takaaki. *Japan.*
Kobayashi, Dr Takashi. *Japan.*
Kobzareva, Dr Svetlana Alekseyevna. *USSR.*
Koch-Wallraf, Mrs Prof. Dr Maria. *BRD.*
Koch, Dr Beatrix. *Netherlands.*
Koch, Priv.-Doz. Dr Elke. *BRD.*
Koch, Prof. Stephen A. *USA.*
Kocharov, Dr Alexander Georgiyevich. *USSR.*
Kociński, Prof. Dr Jerzy. *Poland.*
Kockel, Dr Andreas. *BRD.*
Kocman, Dr Vladimir. *Canada.*
Koda, Dr Shigetaka. *Japan.*
Kodama, Dr Hideomi. *Canada.*
Kodandapani, Mr R. *India.*
Köhler, Dr Rolf. *DDR.*
Kökçe, Dr Ali. *Turkey.*
Koellner, Mrs Gertraud. *BRD.*
König di Perazzo, Lic Patricia Verónica. *Argentina.*
König, Dr Burkhard. *BRD.*
Köpernik, Mr Horst. *DDR.*
Koeppe, Dr. Roger E. II. *USA.*
Körber, Dr. Fritjof Carl Friedrich. *UK.*
Köszegi, Mr László *Hungary.*
Kötitz, Dr Günther. *DDR.*
Koetzle, Dr. Thomas Frederick. *USA.*
Koh, Dr Lip Lin. *Singapore.*
Kohlbeck, Prof. Dr Franz. *Austria.*
Kohli, Dr Vijay Kumar. *India.*
Kohn, Prof. Jack Arnold. *USA.*
Kohra, Prof. Kazutake. *Japan.*
Kohsari, Dr Amir Hossein. *Iran.*
Koide, Prof. Tsutomu. *Japan.*
Koizumi, Dr Hideo. *Japan.*
Koizumi, Prof. Mitsue. *Japan.*
Kojić-Prodić, Dr Biserka. *Yugoslavia.*
Kojima, Dr Seiji. *Japan.*
Kokkinidis, Dr Michael. *BRD.*
Kokkou, Prof. Socrates Constantinos. *Greece.*
Koknat, Prof. Friedrich Wilhelm. *USA.*
Kokotailo, Dr. George Thomas. *USA.*
Kołakowski, Dr Andrzej. *Poland.*
Kołakowski, Dr Bogdan Józef. *Poland.*
Kolatkar, Mr. Anand Ratnakar. *USA.*
Kolatkar, Mr. Prasanna R. *USA.*
Kolesova, Dr Rimma Vladimirovna. *USSR.*
Kolin, Dr Nikolai Georgievich . *USSR.*
Kolks, Dr. Gary. *USA.*
Koller, Mr Hubert. *BRD.*
Kolobyanina, Dr Tatyana Nikolayevna. *USSR.*
Kolodiyeva, Dr Svetlana Vasilyevna. *USSR.*
Kolontsova, Dr Ekaterina Vasilyevna. *USSR.*
Kolpak, Dr. Francis John. *USA.*
Kolpakov, Dr Andrey Vasilyevich. *USSR.*
Kolster, Prof. Dr Ir Benjamin Harry. *Netherlands.*
Koman, Ing Marián. *CSSR.*
Komarek, Prof. Dr Kurt Ludwig. *Austria.*
Komatsu, Prof. Hiroshi. *Japan.*
Komiya, Dr. Hiromi. *USA.*
Komrska, Dr Jiří. *CSSR.*

Komu, Mr Markku Eino Sakari. *Finland.*
Komura, Prof. Shigehiro. *Japan.*
Komura, Prof. Yukitomo. *Japan.*
Kon, Dr Aviv Yuliseyevich. *USSR.*
Konaka, Prof. Shigehiro. *Japan.*
Kondrashev, Dr Yury Dmitriyevich. *USSR.*
Kondratenko, Dr Lyudmila Konstantinovna. *USSR.*
Kondratyeva, Dr Victoria Victorovna. *USSR.*
Kong, Dr. Eric Siu Wai. *USA.*
Kong, Prof. You-hua. *China.*
Konguetsof, Dr Helen. *Greece.*
Koningsveld, van, Dr Hendrikus. *Netherlands.*
Konitz, Mr Antoni. *Poland.*
Konnert, Dr. John H. *USA.*
Konnert, Mrs. Judith A. *USA.*
Konno, Dr Michiko. *Japan.*
Konstantinov, Dr Ivan. *Bulgaria.*
Konstantinova, Dr Alisa Fedorovna. *USSR.*
Koo, Prof. Dr Chung Hoe. *Korea.*
Kooijman, Drs Huub. *Netherlands.*
Koopmans, Prof. Dr Kasper. *Netherlands.*
Kopelman, Prof. Raoul. *USA.*
Kopf, Dr Jürgen. *BRD.*
Kopka, Mrs. Mary Lou. *USA.*
Koptsik, Prof. Vladimir Alexandrovich. *USSR.*
Koreň, Mr Branislav. *CSSR.*
Koreshkov, Dr Boris Dmitriyevich. *USSR.*
Koritsánszky, Mr Tibor. *Hungary.*
Kormány, Dr Teréz. *Hungary.*
Kornev, Dr Aleksey Nikolayevich. *USSR.*
Korp, Prof. James Douglas. *USA.*
Korsukova, Dr Mariya Mikhailovna. *USSR.*
Korvenranta, Dr Jorma Artturi. *Finland.*
Koryagin, Mr. Vyacheslav Filippovich. *USSR.*
Kosche, Mr Ingeborg. *DDR.*
Kosel, Mr. George Eugene. *USA.*
Kosevich, Prof. Vadim Markovich. *USSR.*
Koshy, Dr Jacob. *Nigeria.*
Kosmachev, Dr Sergey Mikhailovich. *USSR.*
Kosmopoulos, Dr John. *Greece.*
Kosova, Dr Tatyana Borisovna. *USSR.*
Kosovinc, Prof. Dr Ivan. *Yugoslavia.*
Kosten, Mr Klaus. *BRD.*
Kosterin, Dr Evgeny Andreyevich. *USSR.*
Kostiner, Prof. Edward Stephen. *USA.*
Kostorz, Prof. Gernot. *Switzerland.*
Kostov, Prof. Ivan. *Bulgaria.*
Kostov, Dr Ruslan. *Bulgaria.*
Kosturkiewicz, Prof. Dr Zofia. *Poland.*
Koszelak, Mr. Stanley N. *USA.*
Kotani, Prof. Masao. *Japan.*
Kotel'nikova, Dr Elena Nikolayevna. *USSR.*
Koto, Prof. Kichiro. *Japan.*
Kotov, Prof. Nikolay Vladimirovich. *USSR.*
Kotsanidis, Mr Panayotis. *Greece.*
Kotsev, Dr Iosif. *Bulgaria.*
Kotsis, Mr Konstantinos. *Greece.*
Kotur, Dr Bogdan Yaroslavovich. *USSR.*
Koumelis, Dr Christos. *Greece.*
Kountouris, Mr Costas. *Greece.*
Kountz, Dr. Dennis James. *USA.*
Kovachev, Dr Peter. *Bulgaria.*
Kováčová, Dr Katarína. *CSSR.*
Kovács, Prof. Alessandro L. *Italy.*
Koval'chuk, Dr Mikhail Valentinovich. *USSR.*
Kovda, Prof. Leonid Mikhailovich. *USSR.*
Kovyev, Dr Ernest Konstantinovich. *USSR.*
Kowalski, Dr Grzegorz. *Poland.*
Koyama, Prof. Hirozo. *Japan.*
Koyama, Dr Kazutoshi. *Saudi Arabia.*
Koyama, Dr Yasumasa. *Japan.*
Koyano, Mr Kazuo. *Japan.*
Koz'ma, Dr Alexander Alekseyevich. *USSR.*
Koz'min, Dr Petr Alekseyevich. *USSR.*
Kozaki, Prof. Shigeru. *Japan.*
Kozioł, Dr Anna E. *Poland.*
Kožíšek, Ing Jozef. *CSSR.*
Kožíšková, Ing Zlatica. *CSSR.*
Kozlenkov, Dr Alexander Ivanovich. *USSR.*
Kozlova, Dr Olga Gerasimovna. *USSR.*

Kozłowska, Mrs Krystyna. *Poland.*
Kraatz, Dr. Paul. *USA.*
Krabbendam, Drs Hendrik. *Netherlands.*
Krämer, Prof. Dr Volker. *BRD.*
Krajewski, Dr Adriano. *Italy.*
Krajewski, Dr Janusz. *Poland.*
Králík, Dr František. *CSSR.*
Králová, Dr Rudolfa. *CSSR.*
Kramer, Dr Irmtraud. *BRD.*
Krane, Mr Hans-Georg. *BRD.*
Kranjc, Prof. Dr Katarina. *Yugoslavia.*
Kraševec, Dr Viktor. *Yugoslavia.*
Krasner, Prof. Saul. *USA.*
Krasnikov, Dr Vladimir Vladimirovich. *USSR.*
Krasnov, Dr Christoph. *Austria.*
Kratky, Dr Christoph. *Austria.*
Kratky, Prof. em. Dr Dipl.-Ing. Otto. *Austria.*
Kratochvil, Doc.Dr Bohumil. *CSSR.*
Kraus, Doc. Dr Ivo. *CSSR.*
Krause, Mr Christa. *DDR.*
Krause, Dr Christian. *BRD.*
Krause, Dr. Jeannette Alice. *USA.*
Krause, Dr. Stephen. *USA.*
Krausse, Dr Joachim. *DDR.*
Kraut, Prof. Joseph. *USA.*
Krawiec, Mr Mariusz. *Poland.*
Krawitz, Prof. Aaron David. *USA.*
Krebs, Prof. Dr Bernt. *BRD.*
Kremer, Mr Germán. *Chile.*
Krén, Dr Emil. *Hungary.*
Kressner, Dr F.Harry. *DDR.*
Krestev, Mr Venelin. *Bulgaria.*
Kresteva, Dr Manya. *Bulgaria.*
Kretschmer, Dr Rolf-Günther. *DDR.*
Kretsinger, Prof. Robert Harvey. *USA.*
Kreutz, Dr Ernst Wolfgang. *BRD.*
Krever, Dr Maarten. *Netherlands.*
Krimm, Prof. Samuel. *USA.*
Krischner, Prof. Dr Dipl.-Ing. Harald. *Austria.*
Krishna Rao, Prof. K. V. *India.*
Krishna, Prof. Padmanabhan. *India.*
Krishnaiah, Mr Musali. *India.*
Kristensen, Mrs Bente Saustrup. *Denmark.*
Kristmannsdóttir, Cand.Real. Hrefna. *Iceland.*
Kritayakirana, Mrs Rungsri. *Thailand.*
Kritikos, Mr Mikael. *Sweden.*
Krivandina, Dr Elena Alekseyevna. *USSR.*
Krivenko, Dr Vladimir Georgievich. *USSR.*
Krivokoneva, Dr Galina Kirillovna. *USSR.*
Křivý, Dr Ivan. *CSSR.*
Krochuk, Dr Vasilii Maksimovich. *USSR.*
Krogmann, Prof. Dr Klaus. *BRD.*
Krohn, Dr A. *UK.*
Krohn, Prof Martin. *DDR.*
Krol', Dr Inna Mikhailovna. *USSR.*
Kroll, Prof. Dr Herbert. *BRD.*
Kronenburg, Drs Martinus Johannes. *Netherlands.*
Kroon-Batenburg, Dr Louise M. J. *Netherlands.*
Kroon, Dr Jan. *Netherlands.*
Krstanović, Prof. Dr Ilija. *Yugoslavia.*
Krüger, Dr Albrecht. *DDR.*
Krüger, Prof. Dr Carl. *BRD.*
Krueger, Ms. Joanna Katherine. *USA.*
Krug, Prof. Dr Detlef. *BRD.*
Kruger, Dr Gert Jacobus. *South Africa.*
Kruglik, Dr Anatoliy Ivanovich. *USSR.*
Krukowski, Dr Marek. *Poland.*
Krutova, Dr Glafira Ivanova. *USSR.*
Krygowski, prof. Tadeusz Marek. *Poland.*
Krymov, Dr Vladimir Mikhailovich. *USSR.*
Kryshtop, Dr Viktor Mikhailovich. *USSR.*
Kuang, Mrs Bao. *China.*
Kuban, Dr Ralf-Jürgen. *DDR.*
Kuběna, Dr Josef. *CSSR.*
Kubiak, Dr Maria. *Poland.*
Kubiak, Dr hab. Ryszard. *Poland.*
Kubicki, MSc Maciej. *Poland.*
Kubo, Prof. Teruichiro. *Japan.*
Kucab, Dr Marian. *Poland.*
Kuchar, Doz. Dr Friedemar. *Austria.*
Kucharczyk, Dr Damian. *Poland.*

Kucharski, Dr Edward Stanislaw. *Australia.*
Kuchenmeister, Dr. Mark Edward. *USA.*
Kuchitsu, Prof. Kozo. *Japan.*
Kudoh, Dr Yasuhiro. *Japan.*
Kudoh, Dr. Yasuhiro. *USA.*
Kudryavtseva, Dr Galina Petrova. *USSR.*
Kudryavtseva, Dr Rimma Vasil'evna. *USSR.*
Küçükçelebi, Mr Hayrettin. *Turkey.*
Kühlbrandt, Dr Werner. *BRD.*
Kühn, Prof Günther. *DDR.*
Kuehn, Prof. Dr Robert. *BRD.*
Külcü, Doç. Dr Nevzat. *Turkey.*
Küppers, Prof. Dr Horst. *BRD.*
Kürsten, Dr Hans-Dieter. *DDR.*
Kuhs, Dr Werner Friedrich. *BRD.*
Kukharenko, Prof. Alexander Alexandrovich. *USSR.*
Kukina, Dr Galina Alexandrovna. *USSR.*
Kukuy, Dr Anatoly Lvovich. *USSR.*
Kulda, Ing Jiří. *CSSR.*
Kullnig, Dr. Rudolph Karl. *USA.*
Kumagawa, Prof. Masashi. *Japan.*
Kumao, Prof. Akihiro. *Japan.*
Kumar, Dr Rajendra. *India.*
Kumar, Mr Vinay *India.*
Kummer, Drs Ernst Albertus. *Netherlands.*
Kumosinski, Mr. Thomas F. *USA.*
Kumru, Dr Mustafa. *Turkey.*
Kun, Yrd. Doç. Dr Nejat. *Turkey.*
Kundra, Dr Krishan Dev. *India.*
Kundu, Dr Mohanlal. *India.*
Kunrath, Mr José Irineu. *Brasil.*
Kunsch, Dr Dipl.-Ing. Barnabas. *Austria.*
Kunstelj, Dr Dragan. *Yugoslavia.*
Kuntsevich, Dr Tamara Serafimovna. *USSR.*
Kunz, Mr Martin. *Switzerland.*
Kuo, Prof. Ke-hsin. *China.*
Kuo, Prof. Lawrence C. *USA.*
Kuok, Dr Meng Hau. *Singapore.*
Kupčik, Prof. Dr Vladimir. *BRD.*
Kupka, Dr František. *CSSR.*
Kuppuswamy, Dr Nagarajan. *India.*
Kupriyanov, Dr Mikhail Fedotovich. *USSR.*
Kurahashi, Dr Masayasu. *Japan.*
Kuranova, Dr Inna Petrovna. *USSR.*
Kurazhkovskaya, Dr Victoriya Semenovna. *USSR.*
Kurbanov, Dr Khakim Mamadaliyevich. *USSR.*
Kurdyumov, Prof. Georgy Vyacheslavovich. *USSR.*
Kuribayashi, Mr Shunsuke. *Japan.*
Kuriyama, Dr. Masao. *USA.*
Kuriyan, Dr. John. *USA.*
Kurki-Suonio, Prof. Kaarle V. J. *Finland.*
Kurkutova, Prof. Evdokiya Nikitichna. *USSR.*
Kuroda, Prof. Haruo. *Japan.*
Kuroda, Dr Reiko. *UK.*
Kuroya, Dr Hisao. *Japan.*
Kushi, Dr Yoshihiko. *Japan.*
Kusunoki, Mr Masami. *Japan.*
Kusz, Mgr Joachim. *Poland.*
Kutoglu, Dr Ali. *BRD.*
Kutschabsky, Dr Leo. *DDR.*
Kuz'ma, Prof. Yury Bogdanovich. *USSR.*
Kuz'min, Prof. Eduard Alekseyevich. *USSR.*
Kuz'min, Dr Ivan Ivanovich. *USSR.*
Kuz'min, Prof. Runar Nikolayevich. *USSR.*
Kuz'mina, Dr Irina Pavlovna. *USSR.*
Kuz'mina, Dr Lyudmila Georgievna. *USSR.*
Kuz'mina, Dr Mariya Anatol'evna. *USSR.*
Kužel, Dr Radomír. *CSSR.*
Kuzmany, Prof. Dr Hans. *Austria.*
Kuznetsov, Dr Alexander Victorovich. *USSR.*
Kuznetsov, Prof. Fedor Andreyevich. *USSR.*
Kuznetsov, Dr Gennadii Vasil'evich. *USSR.*
Kuznetsov, Prof. Vasily Grigoryevich. *USSR.*
Kuznetsov, Dr Victor Andreyevich. *USSR.*
Kuznetsov, Dr Yurii Georgievich. *USSR.*
Kuzucu, Mr Veysel. *Turkey.*
Kuzyukevich, Dr Anatolii Anatol'evich. *USSR.*
Kvapil, Ing Jiří. *CSSR.*
Kvasnitsa, Dr Victor Nikolayevich. *USSR.*
Kvick, Dr. Ake H. *USA.*

Kwiatkowski, Mgr inż Witold. *Poland.*
Kwietniak, Dr Mark Stefan. *Australia.*
Kwik, Dr. Whei Lu. *Singapore.*
Kwong, Mr. Peter. *USA.*
Kyaw, Dr Htin. *Burma.*
Kyotani, Dr Mutsumasa. *Japan.*
Kyriakos, Prof. Demetrius. *Greece.*
Kyröläinen, Mr Antero Johannes. *Finland.*
Kyutt, Dr Reginal'd Nikolayevich. *USSR.*
La Prade, Dr. Marie Douglas. *USA.*
Labaki, Ms Lucila Chebel. *Brasil.*
Labbé, Dr Philippe. *France.*
Labib, Dr (Mrs) Fawkia. *Egypt.*
Labib, Dr Tarik. *Egypt.*
Labischinski, Dr Harald. *BRD.*
Lacmann, Prof. Dr Ing. Rolf. *BRD.*
Ladd, Dr Marcus Frederick Charles. *UK.*
Ladell, Dr. Joshua. *USA.*
Ladenstein, Dr Rudolf. *BRD.*
Laderman, Dr. Stephen Stromberg. *USA.*
Ladner, Dr. Robert Charles. *USA.*
Lähdeniemi, Dr Matti Juhani Iisakki. *Finland.*
Laggner, Prof. Dr Peter. *Austria.*
Lagiewka, Dr Eugeniusz. *Poland.*
Lagomarsino, Dr Stefano. *Italy.*
Lahiri, Dr Barendra Nath. *India.*
Lahodny-Šarc, Prof. Dr (Mrs) Olga. *Yugoslavia.*
Lahoz, Dr Fernando J. *Spain.*
Lahti, Dr Seppo Ilmari. *Finland.*
Lai, Mr. Hsin-Hsi. *USA.*
Lai, Dr Ting Fong. *Hong Kong.*
Laiho, Dr Reino Toivo Salomo. *Finland.*
Laine, Dr Ensio Sulo Uolevi. *Finland.*
Laing, Dr Mary Elizabeth. *South Africa.*
Laing, Prof. Michael John. *South Africa.*
Lajzerowicz, Prof. Janine. *France.*
Lake, Prof. James A. *USA.*
Laker, Mr Thomas James. *UK.*
Lakshminarayanan, Muthuvijayan. *India.*
Lal, Dr Krishan. *India.*
Lalancette, Prof. Roger A. *USA.*
LaLonde, Ms. Judith. *USA.*
Lam, Dr. Chiu Tin. *USA.*
Lamba, Dr Doriano. *Italy.*
Lambert-Smith, Mr John Ernle Warwick. *Australia.*
Lambert, Mr. Guy. *USA.*
Lambert, Prof. Marianne. *France.*
Lambrianidis, Mr Louie Terry. *Australia.*
Lamm, Mr Viktor Andreas. *BRD.*
Lammert, Mr Barbara. *DDR.*
Lamotte-Brasseur, Dr Josette Marie Louise. *Belgium.*
Lancucki, Mr Christopher Joseph. *Australia.*
Lando, Prof. Jerome B. *USA.*
Landy, Dr. Richard Allen. *USA.*
Lanfranchi, Dr Maurizio. *Italy.*
Lang, Prof. Andrew Richard. *UK.*
Langbein, Prof Dr Werner Dieter. *BRD.*
Lange, Mr Joachim Reinhard. *BRD.*
Langer, Prof. Ebbe Wang. *Denmark.*
Langer, Prof Dr Klaus. *BRD.*
Langer, Dr Vratislav. *Sweden.*
Langford, Dr John Ian. *UK.*
Langlois d'Estaintot, Mme Bétrice. *France.*
Langridge, Prof. Robert. *USA.*
Langs, Dr. David Alan. *USA.*
Lapergauz, Dr Il'ya Samuilovich. *USSR.*
Lappa, Dr Ryszard. *Poland.*
Lappin, Miss Alison. *UK.*
Lara Magaña, Mrs María Eugenia. *México.*
Laredo, Prof. Estrella. *Venezuela.*
Larsen, Mr Finn Krebs. *Denmark.*
Larsen, Dr Ingrid Kjøller. *Denmark.*
Larsen, Mrs Sine. *Denmark.*
Larson, Dr. Allen C. *USA.*
Larson, Dr. Bennett Charles. *USA.*
Larson, Dr. Elizabeth Margaret. *USA.*
Larson, Dr. Steven Bland. *USA.*
Larsson, Miss Ann-Kristin. *Sweden.*
Larsson, Dr Sven. *Sweden.*
Lashewycz-Rubycz, Dr. Romana Alexandra. *USA.*

Lashin, Dr A. Mohamed. *Egypt.*
Lasocha, Dr Wiesław. *Poland.*
Last, Mr Paul Edward. *UK.*
Laswick, Prof. Patty Hall. *USA.*
Lattman, Prof. Eaton Edward. *USA.*
Lauck, Mr Rudolf. *BRD.*
Laudise, Dr. Robert Alfred. *USA.*
Laughlin, Prof. David Eugene. *USA.*
Laugier, Mr Jean. *France.*
Lauher, Dr. Joseph. *USA.*
Lawless, Prof. Kenneth Robert. *USA.*
Lawrence, Dr Michael Colin. *Australia.*
Lawson, Dr. Andrew Cowper. *USA.*
Lawson, Mr David Mark. *UK.*
Lawson, Mr Robert Ian. *UK.*
Lawton, Mr. Stephen Latham. *USA.*
Lawyer, Dr. Carl Henry. *USA.*
Lazar, Mr Constantin. *Romania.*
Lazar, Mr Dušan. *Yugoslavia.*
Lazarenkov, Prof. Vadim Grigor'yevich. *USSR.*
Lazarev, Dr Eduard Mikhailovich. *USSR.*
Lazarev, Dr Valerii Georgievich. *USSR.*
Lazarini, Prof. Dr Franc. *Yugoslavia.*
Le Bars, Dr Michéle. *France.*
Le Cong Quy, Dr. *Vietnam.*
Le Nguyen Soc, Prof. *Vietnam.*
Le Page, Dr Yvon. *Canada.*
Le Roux, Dr Johannes Hendrik. *South Africa.*
Le Roux, Dr Stephanus David. *South Africa.*
Le Trong, Mrs. Isolde. *USA.*
Le Van Huan, Dr. *Vietnam.*
Le, Mr. Liang-Hsien. *USA.*
Lea, Prof. Sydney George. *Canada.*
Leadbetter, Prof. Alan James. *UK.*
Leake, Dr John Anthony. *UK.*
Leban, Prof. Dr Ivan. *Yugoslavia.*
Lebech, Mrs Bente. *Denmark.*
Lebedev, Prof. Vasily Ilyich. *USSR.*
Lebedeva, Dr Marina Vladimirovna. *USSR.*
Lebek, Dr Alexander. *DDR.*
Lebioda, Prof. Lukasz. *USA.*
Leccabue, Dr Fabrizio. *Italy.*
Lechat, Dr Johannes Rudiger. *Brasil.*
Leciejewicz, Prof. Dr Janusz. *Poland.*
Leclaire, Dr André *France.*
Lecong Dzuong, Prof. *Vietnam.*
Ledbetter, Dr. Hassel. *USA.*
Ledesert, Dr Mariannick. *France.*
Lee Moreno, Dr José Luis. *México.*
Lee, Dr. Byungkook. *USA.*
Lee, Dr Chnoong Kheng, *Malaysia.*
Lee, Mrs Florence Lan Fun. *Canada.*
Lee, Prof. James Ching. *USA.*
Lee, Dr John David. *UK.*
Lee, Mrs. Katharine Darby. *USA.*
Lee, Mr. Peter L. *USA.*
Lee, Prof. Wang Chihming. *Taiwan.*
Lee, Dr Xavier. *Canada.*
Lee, Mr. Xuye. *USA.*
Lefaucheux, Dr Françoise. *France.*
Lefebvre, Mr. Kevin R. *USA.*
Lefeld-Sosnowska, Prof. Dr hab. Maria. *Poland.*
Leffers, Dr Torben. *Denmark.*
LeGeros, Dr. Racquel Z. *USA.*
Legros, Prof. Jean-Pierre. *France.*
Lehmann, Mr Christian Wolfgang. *BRD.*
Lehmann, Prof. Dr Gerhard Rudolf. *BRD.*
Lehmann, Dr Gottfried. *DDR.*
Lehmann, Dr Mogens. *France.*
Lehmpfuhl, Dr Gunter. *BRD.*
Lehtinen, Dr Martti Kalevi. *Finland.*
Leipner, Dr Hartmut. *DDR.*
Leipoldt, Prof. Johann Gotlieb. *South Africa.*
Leiro, Dr Jarkko Albert. *Finland.*
Leiserowitz, Prof. Leslie. *Israel.*
Leite, Dr Cirano Rocha. *Brasil.*
Lele, Prof. Shrikant *India.*
Leligny, Mr Henri. *France.*
Lengauer, Mr. Christian Leopold. *Austria.*
Lengauer, Dr Dipl.-Ing. Walter Oskar Franz. *Austria.*

Lenhert, Prof. P. Galen. *USA.*
Lenner, Dr Magnus. *Sweden.*
Lenstra, Prof. Dr Albert Teun Hendrik. *Belgium.*
Lenz, Mrs Andrea. *BRD.*
Leon-Escamilla, Mr. Efigenio Alejandro. *USA.*
Léonard, Dr André Jules Gérard. *Belgium.*
Leonardsen, Mr Erik Sverre. *Denmark.*
Leonhardt, Dr Albrecht. *DDR.*
Leonhardt, Prof Gunter. *DDR.*
Leoni, Prof. Leonardo. *Italy.*
Leonowicz, Dr. Michael Edward. *USA.*
Leonyuk, Dr Lidiya Ivanovna. *USSR.*
Leonyuk, Dr Nikolay Ivanovich. *USSR.*
Leovac, Prof. Dr Vukadin. *Yugoslavia.*
Lepicard, Dr Geneviève. *France.*
Leporati, Prof. Enrico. *Italy.*
Lerf, Dr Anton Eduard. *BRD.*
Leskelä, Dr Markku Antero. *Finland.*
Leslie, Dr Andrew Greig William. *UK.*
Lessinger, Prof. Leslie. *USA.*
Leung, Dr. Peter C. *USA.*
Leute, Prof. Dr Volkmar. *BRD.*
Levan, Mr. Keith R. *USA.*
Levanyuk, Prof. Arkady Petrovich. *USSR.*
Levelut, Dr Anne-Marie. *France.*
Levendis, Dr Demetrius Christos. *South Africa.*
Leventouri, Dr Dora. *Greece.*
Leverett, Dr Peter. *Australia.*
Levi, Dra Laura. *Argentina.*
Levien, Dr. Louise. *USA.*
Levoska, Mr Pentti Juhani. *Finland.*
Lévy, Prof. Francis. *Switzerland.*
Levy, Dr. Henri A. *USA.*
Lewiński, Dr Krzysztof. *Poland.*
Lewis, Dr Eric Leslie Vallance. *UK.*
Lewis, Dr James jr. *BRD.*
Lewis, Mr. Michael D. *USA.*
Lewis, Prof. Michael Harold. *UK.*
Lewit-Bentley Dr Anita. *France.*
Leyerle, Mr. Richard W. *USA.*
Li, Dr. Chi-Tang. *USA.*
Li, Mr Da-ming. *China.*
Li, Prof. De-yu. *China.*
Li, Mr Du. *China.*
Li, Mr. Hui-Ying. *USA.*
Li, Dr Jade. *UK.*
Li, Dr. Naiyin. *USA.*
Li, Prof. Run-shen. *China.*
Li, Mr. Tien-Hsiung. *USA.*
Li, Dr. Yong Ji. *USA.*
Liang, Prof. Dong-cai. *China.*
Liang, Prof. Jing-kui. *China.*
Liang, Prof. Li. *China.*
Liao, Prof. Bo-qang. *China.*
Licci, Dr Francesca. *Italy.*
Licheri, Prof. Giovanni. *Italy.*
Licklider, Mr. Robert A. *USA.*
Liddell, Dr Katharine. *UK.*
Lidin, Mr Sven. *Sweden.*
Liebau, Prof. Dr Friedrich. *BRD.*
Liebertz, Prof. Dr Josef. *BRD.*
Liebman, Dr. Michael N. *USA.*
Liem, Dr D. Hay. *Sweden.*
Lifchitz, Mr Alain. *France.*
Ligenza, Asst. Prof. Sylwester. *Poland.*
Lihl, Prof. em. Dr Franz. *Austria.*
Lii, Dr Kwang-Hwa. *Taiwan.*
Liles, Mr David Charles. *South Africa.*
Lilie, Mr Martin. *DDR.*
Liljas, Prof. Anders. *Sweden.*
Liljas, Dr Lars. *Sweden.*
Lim, Dr. Louis W. *USA.*
Lim, Dr Valery Irovich. *USSR.*
Lima-de-Faria, Dr José. *Portugal.*
Liminga, Prof. Rune. *Sweden.*
Limper, Mr Wolfram. *BRD.*
Lin, Mr Cheng-yi. *China.*
Lin, Prof. Chi-chang. *China.*
Lin, Mr Chuan. *China.*
Lin, Mr Guang-da. *China.*

Lin, Dr Hsi-Che. *Taiwan.*
Lin, Mr. Jar-Shyong. *USA.*
Lin, Ms Jing-ding. *China.*
Lin, Prof. Szu-Bin. *Taiwan.*
Lin, Prof. Tsang-Lang. *Taiwan.*
Lin, Mr Xian-ti. *China.*
Lin, Mr Xing-yuan. *China.*
Lin, Prof. Yong-hua. *China.*
Lin, Mrs Yu-juan. *China.*
Lin, Prof. Zheng-jiong. *China.*
Linares, Dr Jorge. *Peru.*
Lincoln, Dr Francis John. *Australia.*
Lind, Dr. Maurice David. *USA.*
Lindahl, Mr Martin. *Sweden.*
Lindahl, Dr Tommie. *Sweden.*
Lindegaard-Andersen, Prof. Asger. *Denmark.*
Lindeman, Dr Sergei Vital'evich. *USSR.*
Lindemann, Prof. Dr Willi. *BRD.*
Linden, Dr Anthony. *Canada.*
Lindenmeyer, Dr. Paul Henry. *USA.*
Lindgreen, Dr Holger. *Denmark.*
Lindin', Dr Lauma Felixovna. *USSR.*
Lindley, Dr Peter Frank. *UK.*
Lindner, Dr Peter W. *Sweden.*
Lindqvist, Prof. Ingvar. *Sweden.*
Lindqvist, Mr Kristian Vilhelm. *Finland.*
Lindqvist, Prof. Oliver. *Sweden.*
Lindqvist, Dr Ylva Ch. *Sweden.*
Lindroos, Prof. Veikko Kalervo. *Finland.*
Lindström, Dr Rauno. *Finland.*
Lingafelter, Prof. Edward Clay. *USA.*
Linke, Dr Dietmar. *DDR.*
Linke, Dr Walter. *Austria.*
Liopo, Dr Valery Alexandrovich. *USSR.*
Lipka, Mrs Dr Annegret. *BRD.*
Lipkowska, Dr Zofia. *Poland.*
Lipkowski, Asst. Prof. Janusz. *Poland.*
Lippard, Prof. Stephen James. *USA.*
Lippert, Dr. Ernest L. *USA.*
Lippman, Dr. Robert. *USA.*
Lipscomb, Prof. William Nunn. *USA.*
Lipson, Prof. Henry, FRS. *UK.*
Liquori, Prof. Alfonso Maria. *Italy.*
Lis, Dr Tadeusz. *Poland.*
Liso-Rubio, Prof. M. Jesus. *Spain.*
Lisoivan, Dr Vladimir Ivanovich. *USSR.*
Litovchenko, Dr Anatolii Stepanovich. *USSR.*
Little, Mr Thomas W. *Japan.*
Litvin, Dr Alexander Lukich. *USSR.*
Litvin, Dr Boris Nikolayevich. *USSR.*
Litvin, Prof. Daniel Bernard. *USA.*
Litvinov, Dr Igor' Anatol'evich. *USSR.*
Litvinskaya, Mrs Galina Petrovna. *USSR.*
Lityagina, Dr Lyudmila Mitrofanovna. *USSR.*
Liu, Mr Guang-zhao. *China.*
Liu, Prof. Han-qin. *China.*
Liu, Mr. Hansong. *USA.*
Liu, Mr. Hung-Yu. *USA.*
Liu, Prof. Ling-Kang. *Taiwan.*
Liu, Mr Shi-xiong. *China.*
Liu, Mr Tian-liang. *China.*
Liu, Mr Wan. *China.*
Liu, Dr. Xing. *USA.*
Liu, Prof. Xue-lun. *China.*
Liu, Prof. Yong-sheng. *China.*
Liu, Mr Zuo-cai. *China.*
Livnah, Mr Oded. *Israel.*
Ljungström, Dr Evert B. *Sweden.*
Llaguno, Dr Elma Caballes. *Philippines.*
Llanos, Dr Jaime. *Chile.*
Llinás Rivera, Prof. Rubén Darío. *Colombia.*
Lloveras, Dr Joaquim. *Spain.*
Lloyd, Dr Douglas James. *Australia.*
Lo, Mr. Chi-Fung. *USA.*
Lobkovsky, Dr Emil' Borisovich. *USSR.*
Locchi, Prof. Stelio Giovanni. *Italy.*
Lock, Prof. Colin James Lyne. *Canada.*
Loeb, Dr. Arthur L. *USA.*
Loeb, Dr Stephen Joseph. *Canada.*
Löchner, Dr Ulrich. *BRD.*

Löffler, Prof Hans. *DDR.*
Löfgren, Dr Percy. *Sweden.*
Loehlin, Prof. James Herbert. *USA.*
Loeksmanto, Dr Waloejo. *Indonesia.*
Löns, Dr Jürgen. *BRD.*
Lösche, Prof. Artur. *DDR.*
Loghry, Dr. Ray Allen. *USA.*
Loizos, Mr Zafiris. *Greece.*
Lok, Mr. Charles. *USA.*
Lokaj, Ing Ján. *CSSR.*
Lokanatha, Mr S. *India.*
Loll, Mr. Patrick J. *USA.*
Lombard, Dr Anthonie van Altena. *South Africa.*
Lomonov, Dr Vladimir Alekseevich. *USSR.*
Lomov, Dr Andrei Aleksandrovich. *USSR.*
Londos, Dr Charalampos. *Greece.*
Long, Dr. Gabrielle Gibbs. *USA.*
Loopstra, Prof. Dr Bert Onno. *Netherlands.*
Lopes-Vieira, Prof. António. *Portugal.*
Lopez de Lerma, Dr Julian. *Spain.*
Lopez Guerra, Dr.Sc.Silio. *Cuba.*
Lopez-Acevedo, Dr M. Victoria. *Spain.*
Lopez-Castro, Prof. Amparo. *Spain.*
Lopez-Galindo, Dr Alberto. *Spain.*
Lopez-Gonzalez, Prof. Juan de D. *Spain.*
Lopez-Soler, Dr Angel. *Spain.*
Lorentzen, Mr Torben. *Denmark.*
Lorenz, Prof. Dr Wolfgang J. *BRD.*
Loreto, Prof. Lucio. *Italy.*
Lorimer, Dr Gordon Winston. *UK.*
Loshmanov, Dr Arkady Andreyevich. *USSR.*
Lotfy, Dr Mohamed. *Egypt.*
Lottermoser, Dr Werner. *Austria.*
Lou, Mrs Mei-zhen. *China.*
Loub, Dr Josef. *CSSR.*
Louer, Dr Daniel. *France.*
Lough, Dr Alan John. *Canada.*
Loupias, Prof. Geneviève. *France.*
Lovas, Dr György Antal. *Hungary.*
Love, Dr. R. A. *USA.*
Love, Prof. Warner Edwards. *USA.*
Lovell, Dr. Frederick Mauri. *USA.*
Lovell, Mrs S.E. *UK.*
Lovey, Dr Francisco Carlos. *Argentina.*
Low, Prof. Barbara Wharton. *USA.*
Low, Dr John Nicolson. *UK.*
Low, Prof. William Zev. *Israel.*
Lowde, Dr Raymond Douglas. *UK.*
Lowe-Ma, Dr. Charlotte Kathryn. *USA.*
Lowe, Dr Philip Richard. *UK.*
Lowe, Ms Susan Elizabeth. *UK.*
Lozano, Prof. José A. *Colombia.*
Lu, Prof. Jia-xi. *China.*
Lu, Mr Kun-quan. *China.*
Lu, Prof. Qi. *China.*
Lu, Mrs Quang-ying. *China.*
Lu, Mr Shao-fang. *China.*
Lu, Prof. Tian-Huey. *Taiwan.*
Lu, Prof. Yun-jin. *China.*
Lube, Dr Emil' Lvovich. *USSR.*
Lublin, Mr. Paul. *USA.*
Lucas, Dr Brian William. *Australia.*
Lucchetti, Prof. Gabriella. *Italy.*
Ludi, Prof. Andreas. *Switzerland.*
Ludwiczek, Dr Herbert. *Austria.*
Ludwig, Prof. Martha L. *USA.*
Luecke, Mr. Hartmut. *USA.*
Luft, Mr. Joseph. *USA.*
Luger, Prof. Dr Peter. *BRD.*
Lugomer, Dr Stjepan. *Yugoslavia.*
Lugt, van der, Dr Willem. *Netherlands.*
Luić, Magistar Marija. *Yugoslavia.*
Luisi, Dr. Ben Francesco. *USA.*
Łukaszewicz, Prof. Kazimierz. *Poland.*
Lukaszewski, Dr George Michael. *Australia.*
Lumme, Prof. Paavo Olavi. *Finland.*
Luna, Dr Carlos Alfonso. *Colombia.*
Lundberg, Dr Bruno K. S. *Sweden.*
Lundberg, Dr Monica. *Sweden.*

Lundgren, Mr Lennart. *Sweden.*
Lundström, Dr Torsten. *Sweden.*
Lunin, Dr Vladimir Yur'evich. *USSR.*
Luo, Prof. Gu-feng. *China.*
Luo, Mr. Ming. *USA.*
Luo, Mr Yao Guang. *Canada.*
Luss, Mr. Henry Richard. *USA.*
Lustig, Mr. Stanley. *USA.*
Lutz, Mr Dieter. *DDR.*
Lutz, Prof. Dr Heinz Dieter. *BRD.*
Lux, Mr Bernd. *DDR.*
Lvov, Dr Yury Mikhailovich. *USSR.*
Lyakhovitskaya, Dr Vera Aronovna. *USSR.*
Lydon, Dr John Ennis. *UK.*
Lynch, Dr Denis Francis. *Australia.*
Lynch, Dr. V. M. *USA.*
Lyons, Dr Karen. *New Zealand.*
Lyons, Dr Michael Hamilton. *UK.*
Lyons, Mr Paul John. *New Zealand.*
Lyubalin, Dr Mark Dmitriyevich. *USSR.*
Lyubimov, Dr Vasily Nikolayevich. *USSR.*
Lyubitov, Dr Yury Naumovich. *USSR.*
Lyubushkina, Dr, Lyudmila Mikhailovna. *USSR.*
Lyubutin, Dr Igor' Savelyevich. *USSR.*
Lyutin, Dr Vladimir Ivanovich. *USSR.*
Lyutzau, Prof. Vsevolod Grigoryevich. *USSR.*
Lyxell, Mr Dan-Göran. *Sweden.*
Ma, Dr Che-Bao. *Taiwan.*
Ma, Mr. Cheuk Ki. *USA.*
Ma, Prof. Li-dun. *China.*
Ma, Dr Lilian Yan Yan. *Canada.*
Ma, Mr Xing-qi. *China.*
Ma, Prof. Zhe-sheng. *China.*
Maaref, Dr. Saida. *Tunisia.*
Maartmann-Moe, Mr Knut. *Norway.*
Maas, Prof. Dr Gerhard. *BRD.*
Maaskant, Prof. Dr Willem J. A. *Netherlands.*
MacCrone, Prof. Robert Kirsten. *USA.*
Macgillavry, Prof. Dr Carolina H. *Netherlands.*
Machajdík, Ing Daniel. *CSSR.*
Machin, Mr Ken James. *Australia.*
Macía Sanabria, Prof. Carlos A. *Colombia.*
Maciček, Dr Josef. *Bulgaria.*
Maciosowski, Dr Andrzej. *Poland.*
Mack, Mr. John. *USA.*
Mack, Dr. Joseph Philip Grant. *USA.*
Mackay, Prof. Alan Lindsay. *UK.*
Mackay, Dr Maureen Florence. *Australia.*
Mackenzie, Dr James Kenneth. *Australia.*
Mackie, Dr Fiona L. *UK.*
Mackie, Dr. Paul E. *USA.*
Mackinnon, Dr. Ian Donald. *USA.*
MacRae, Dr. Alfred U. *USA.*
Maďar, Doc. Dr Ján. *CSSR.*
Madariaga, Dr Gotzon. *Spain.*
Madden, Prof. John Joseph. *USA.*
Madejski, Ms. Julie. *USA.*
Madhusudan, Mr. *India.*
Madhusudana, Dr N. V. *India.*
Madsen, Mr Ian Charles. *Australia.*
Madureira Filho, Prof. José Barbosa de. *Brasil.*
Maeda, Dr Hironobu. *Japan.*
Mäki, Mr Jouko Kalervo. *Finland.*
Maenhout - Van Der Vorst, Dr W.M.R. *Belgium.*
Mages, Dr Gert Rudolf. *BRD.*
Magini, Dr Mauro. *Italy.*
Magnéli, Prof. Arne. *Sweden.*
Magnus, Dr. Karen A. *USA.*
Magnuson, Arof. Vincent Richard. *USA.*
Magomedova, Dr Nina Samuilovna. *USSR.*
Mahanta, Dr Bhubaneswar. *India.*
Mahata, Dr Akhil *India.*
Mahmood, Mr Khursheed. *Pakistan.*
Mahmoud, Dr Mouyed Mohamed. *Iraq.*
Mahr von Staszewski, Dr Guillermo. *Argentina.*
Mahy, Dr Jan Willem Gaston. *Netherlands.*
Mai, Prof. Zhen-hong. *China.*
Maier, Dr Horst. *BRD.*
Main, Dr Peter. *UK.*
Maiti, Dr Gobinda Chandra. *India.*

Maixner, Dr Jaroslav. *CSSR.*
Maiyer, Prof. Alexander Artemyevich. *USSR.*
Maiza, Dr Pedro José. *Argentina.*
Majchrzak, Dr Stanisław. *Poland.*
Majeste, Prof. Richard J. *USA.*
Majewska, Miss Katarzyna. *Poland.*
Majling, Ing Ján. *CSSR.*
Majumdar, Dr Sunil Kumar. *India.*
Mak, Prof. Thomas Chung-wai. *Hong Kong.*
Makarenko, Dr Igor' Nikolayevich. *USSR.*
Makarov, Prof. Evgeny Sergeyevich. *USSR.*
Makhmudova, Dr Nailiya Kamilovna. *USSR.*
Makinen, Prof. Marvin W. *USA.*
Makovicky, Dr Emil. *Denmark.*
Makowski, Prof. Lee. *USA.*
Maksić, Prof. Dr Zvonimir. *Yugoslavia.*
Maksimova, Dr Nadezhda Vasil'evna. *USSR.*
Mal'tsev, Dr Yurii Fedorovich. *USSR.*
Malagón Castro, Prof. Dimas. *Colombia.*
Malakhova, Dr Lyudmila Fedorovna. *USSR.*
Maleev, Dr Michael. *Bulgaria.*
Malgrange, Prof. Cécile. *France.*
Malicskó, Dr László. *Hungary.*
Malik, Dr Khalifa Mohammad Abdul. *Bangladesh.*
Malin, Dr Anthony Samuel. *Australia.*
Malinenko, Dr Inna Avramovna. *USSR.*
Malinovsky, Dr Stanislav Tadeushevich. *USSR.*
Malinovsky, Prof. Tadeush Iosifovich. *USSR.*
Malinovsky, Dr Yury Alexandrovich. *USSR.*
Malinowski, Dr Mariusz. *Poland.*
Malinowski, Prof. Yordan. *Bulgaria.*
Maliszewski, Dr Edward. *Poland.*
Mallari-Kaballo, Mrs Paz P. *Philippines.*
Malley, Ms. Mary F. *USA.*
Mallinson, Dr Paul Raymond. *UK.*
Mallory, Dr. Chester L. *USA.*
Malm, Dr Jan-Olle. *Sweden.*
Malone, Dr John Francis. *UK.*
Malpezzi, Dr Luciana. *Italy.*
Malssen van, Drs Kees Frederik. *Netherlands.*
Malta, Dr Viscardo. *Italy.*
Maluszyńska, Dr Hanna. *Poland.*
Malý, Mr Karel. *CSSR.*
Malyushitskaya, Dr Zinaida Vladimirovna. *USSR.*
Mammi, Prof. Mario. *Italy.*
Man, Dr Lucia Ivanovna. *USSR.*
Mănăilă, Mrs Rodica. *Romania.*
Manassero, Prof. Mario. *Italy.*
Mancini, Dr Annamaria. *Italy.*
Mandal, Dr Pradip Kumar. *India.*
Mandarino, Dr Joseph Anthony. *Canada.*
Mande, Prof. Chintamani. *India.*
Mande, Mr Sekhar Chintamani. *India.*
Mandel, Prof. Gretchen Sue. *USA.*
Mandel, Prof. Neil Stanley. *USA.*
Mandelkow, Dr Eckard. *BRD.*
Mándy, Mr Tamás. *Hungary.*
Mánek, Ing Břetislav. *CSSR.*
Mangani, Dr Stefano. *Italy.*
Manghi, Lic Estela Margarita. *Argentina.*
Manghnani, Mr. Murli H. *USA.*
Mangia, Prof. Alessandro. *Italy.*
Mani, Mr A. *India.*
Manickkavachgam, Dr Ramanathan. *India.*
Mannami, Dr Michihiko. *Japan.*
Mannan, Prof. Dr Kh. A. I. F. Mafizul. *Bangladesh.*
Manninen, Dr Seppo Olavi. *Finland.*
Manning, Mr David Charles. *UK.*
Manohar, Prof Hattikudur. *India.*
Manojlović-Muir, Dr Ljubica. *UK.*
Manolikas, Prof. Konstantinos. *Greece.*
Manor, Dr. Philip C. *USA.*
Manotti-Lanfredi, Prof. Anna Maria. *Italy.*
Manríquez, Dr Víctor. *Chile.*
Mansikka, Prof. Kauko Antti. *Finland.*
Mansilla - Koblavi, Mrs F. G.M.S. *Côte d'Ivoire.*
Mansour, Mr Saber Moustapha. *Egypt.*
Mantovani, Prof. Giorgio. *Italy.*
Manutchehr-Danai, Dr Mohsen. *Iran.*
Manzoor-I-Khuda, Dr Muhammad. *Bangladesh.*

Maqsood, Dr Asghari. *Pakistan.*
Maquire, Ms. Theresa. *USA.*
Marbec, Lic Ema Rosa. *Argentina.*
Marcano-Fermin, Dr Cenis M. *Spain.*
March, Dr Frank Conroy. *New Zealand.*
Marchessault, Dr Robert Henry. *Canada.*
Marchetti, Dr Fabio. *Italy.*
Marcoen, Dr Jean-Marie. *Belgium.*
Marcos, Dr Celia. *Spain.*
Marega, Dr Carla. *Italy.*
Marezio, Dr Massimo. *France.*
Marfunin, Dr Arnol'd Sergeyevich. *USSR.*
Margarido, Eng. Fernanda. *Portugal.*
Margulis, Prof. Thomas N. *USA.*
Mariano, Dr. Anthony N. *USA.*
Marians, Ms. Carol. *USA.*
Marie, Dr Alain Louis. *BRD.*
Mariezcurrena, Prof Raul Alfredo. *Uruguay.*
Marigo, Prof. Antonio. *Italy.*
Marin-Elena, Dr Jose Manuel. *Spain.*
Marinder, Dr Bengt-Olov. *Sweden.*
Marinković, Prof. Dr Velibor. *Yugoslavia.*
Marinković, Magistar (Mrs) Živka. *Yugoslavia.*
Marinov, Dr Miko. *Bulgaria.*
Mariolacos, Dr Konstantin. *BRD.*
Markgraf, Mr. Steven A. *USA.*
Markley, Prof. John L. *USA.*
Marko, Mr. Eric. *USA.*
Markov, Dr Ivan. *Bulgaria.*
Marković, Magistar Berislav. *Yugoslavia.*
Marković, Magistar Desimir. *Yugoslavia.*
Marković, Dr Vesna. *Yugoslavia.*
Marks, Prof. Laurence Daniel. *USA.*
Marler, Mr Bernd. *BRD.*
Maröy, Dr Kjartan. *Norway.*
Marongiu, Prof. Giaime. *Italy.*
Marquez-Delgado, Prof. Rafael. *Spain.*
Marsau, Prof. Pierre. *France.*
Marsh, Dr. Charles P. *USA.*
Marsh, Mr. Philip. *USA.*
Marsh, Dr. Richard Edward. *USA.*
Marshall, Prof. Garland R. *USA.*
Martelli, Dr Stefano. *Italy.*
Marthinsen, Dr Knut. *Norway.*
Marti-Artoy, Mr Xavier. *Spain.*
Martikainen, Mr Hannu Olavi. *Finland.*
Martin-Ramos, Dr Jose Daniel. *Spain.*
Martin-Vivaldi, Dr Juan Luis. *Spain.*
Martin, Mrs Brigitte. *BRD.*
Martin, Mr. Bruce Alan. *USA.*
Martin, Mr. David C. *USA.*
Martin, Mr. George William. *USA.*
Martin, Dr John Wilson. *UK.*
Martin, Dr. Kimberly Ann. *USA.*
Martin, Miss Lillian Ruth. *Canada.*
Martin, Dr. Philip D. *USA.*
Martin, Dr. Yvonne C. *USA.*
Martinelli, Prof. Giuliano. *Italy.*
Martinez-Carrera, Prof. Sagrario. *Spain.*
Martinez-Ripoll, Prof. Martin. *Spain.*
Martinez, Mr Benjamin. *Spain.*
Martinez, Dr Francisco. *Spain.*
Martinez, Mr Luis Gallego. *Brasil.*
Martinez, Mr. Sergio E. *USA.*
Martorana, Dr Antonino. *Italy.*
Martyshev, Dr Yury Nikolayevich. *USSR.*
Maruha, Prof. Juro. *Japan.*
Marukawa, Prof. Kenzaburo. *Japan.*
Marumo, Prof. Fumiyuki. *Japan.*
Maruse, Prof. Susumu. *Japan.*
Marusin, Dr Evgenii Petrovich. *USSR.*
Maruyama, Prof. Saiyu. *Japan.*
Marvin, Dr Donald Arthur. *UK.*
Marx, Prof. Günter. *DDR.*
Mas, Dr Graciela Raquel. *Argentina.*
Masaki, Dr Norio. *Japan.*
Masakuni, Dr Mayumi. *Japan.*
Mascarenhas, Prof. Sergio. *Brasil.*
Mascarenhas, Prof. Yvonne Primerano. *Brasil.*
Masche, Mr Wolfgang. *DDR.*

Masciocchi, Dr Norberto. *Italy.*
Masi, Mr Dante. *Italy.*
Mašić, Magistar Nikola. *Yugoslavia.*
Maske, Prof. Siegfried. *South Africa.*
Maslen, Dr Edward Norman. *Australia.*
Maslen, Dr Hugh Stafford. *New Zealand.*
Maslennikov, Dr Aleksei Vladimirovich. *USSR.*
Maslov, Dr Andrei Viktorovich. *USSR.*
Mason, Mr. John T. *USA.*
Mason, Dr Kenneth George. *UK.*
Mason, Prof. Paul Robert. *USA.*
Mason, Dr Sax Anton. *France.*
Mason, Prof. Sir Ronald, FRS. *UK.*
Mason, Prof. Stephen Finney. *UK.*
Massa, Prof. Louis. *USA.*
Massa, Prof. Dr Werner. *BRD.*
Massalimov, Dr Ismail Aleksandrovich. *USSR.*
Massalski, Prof. Thaddeus B. *USA.*
Massarotti, Prof. Vincenzo. *Italy.*
Mastacan, Prof. Gheorghe. *Romania.*
Mastropaolo, Dr. Donald. *USA.*
Mastryukov, Dr Vladimir Saidovich. *USSR.*
Masuda, Dr Hideki. *Japan.*
Masuko, Dr Akiyoshi. *Japan.*
Matassa, Dr. Victor Giulio. *USA.*
Mateika, Dr Dieter. *BRD.*
Mateu, Dr. Leonardo. *Venezuela.*
Matherny, Prof. Dr-Ing Mikuláš. *CSSR.*
Mathew, Mr. M. *USA.*
Mathews, Prof. F. Scott. *USA.*
Mathews, Irimpan I. *India.*
Mathieson, Prof Alexander McLeod. *Australia.*
Mathur, Dr Balbir Kumar. *India.*
Matias, Mr. Pedro M. *USA.*
Matias, Dr Pedro. *Portugal.*
Matijašič, Dr (Mrs) Ivanka. *Yugoslavia.*
Matković-Čalogović, Magistar D. *Yugoslavia.*
Matković, Dr Boris. *Yugoslavia.*
Matković, Dr Prosper. *Yugoslavia.*
Matković, Dr Tanja. *Yugoslavia.*
Matkovsky, Dr Orest Ilyarovich. *USSR.*
Matos Beja, Dr Ana Maria. *Portugal.*
Matsubara, Dr Ikuo. *Japan.*
Matsubara, Prof. Takeo. *Japan.*
Matsuda, Prof. Hidehiko. *Japan.*
Matsui, Dr Masanori. *Japan.*
Matsui, Mr Toshiro. *Japan.*
Matsui, Dr Yoshio. *Japan.*
Matsui, Prof. Yoshiro. *Japan.*
Matsumoto, Mr Osamu. *Japan.*
Matsumoto, Prof. Takeo. *Japan.*
Matsuo, Dr Munetsugu. *Japan.*
Matsushima, Dr Norio. *Japan.*
Matsushita, Dr Tadashi. *Japan.*
Matsuura, Dr Yoshiki. *Japan.*
Matsuzaki, Dr Takao. *Japan.*
Mattes, Prof. Dr Rainer. *BRD.*
Matthew, Prof. James Andrew Davidson. *UK.*
Matthews, Prof. Brian W. *USA.*
Matthews, Dr. David Allan. *USA.*
Matthews, Prof. Frederick White. *Canada.*
Matthys, Dr Paul Frederik André Edmond. *Belgium.*
Mattia, Prof. Carlo. *Italy.*
Mattias, Prof. Pierpaolo. *Italy.*
Matveeva, Dr Ol'ga Petrovna. *USSR.*
Matveeva, Mrs Rimma Georgiyevna. *USSR.*
Matyash, Prof Ivan Vasil'evich. *USSR.*
Matyi, Prof. Richard James. *USA.*
Matyja, Dr Elżbieta. *Poland.*
Matyja, Dr Przemysław. *Poland.*
Matz, Prof. Dr Günther. *BRD.*
Matzat, Dr Eckhart. *BRD.*
Maulik, Dr. Prakas R. *USA.*
Maverick, Dr. Andrew William. *USA.*
Maverick, Prof. Emily Fisch. *USA.*
Mavlonov, Dr Sharaf. *USSR.*
Mavridis, Prof. Aristides. *Greece.*
Maximov, Dr Boris Alekseyevich. *USSR.*
Maxwell, Prof. George. *Canada.*
Mayer, Dr Dipl.-Ing. Helmut. *Austria.*

Mayer, Dr Hugo Werner Waldemar. *BRD.*
Mayer, Prof. Itzchak. *Israel.*
Mayo, Mr. William Edward. *USA.*
Mayr, Dr Dipl.-Ing. Michael. *Austria.*
Mazany, Dr. Anthony Michael. *USA.*
Mazey, Dr David John. *UK.*
Mazus, Dr Mark Davidovich. *USSR.*
Mazza, Dr Fernando. *Italy.*
Mazzarella, Prof. Lelio. *Italy.*
Mazzaro, Mr Irineu. *Brasil.*
Mazzi, Prof. Fiorenzo. *Italy.*
Mazzocchi, Ms Vera. *Brasil.*
McAlister, Dr. John Paul. *USA.*
McAllister, Mr Patrick Brian. *UK.*
McArdle, Dr Patrick. *Ireland.*
McAtee, Prof. James L. *USA.*
McCall, Dr Maxine June. *Australia.*
McCallum, Prof. Malcolm Ernest. *USA.*
McCammon, Prof. James Andrew. *USA.*
McCarthy, Prof. Gregory Joseph. *USA.*
McCauley, Dr. James Weymann. *USA.*
McClune, Mr. W. Frank. *USA.*
McClure, Dr. Richard James. *USA.*
McCollor, Mr. Donald P. *USA.*
McConnell, Prof. Duncan. *USA.*
McCrone, Dr. Walter C. *USA.*
McCusker, Dr Lynne Bridget. *Switzerland.*
McDonald, Dr. Robert Charles. *USA.*
McDonald, Dr Walter Stanley. *UK.*
McFarlane III, Dr. Samuel H. *USA.*
McGuire, Dr. Nancy K. *USA.*
McHardy, Dr William James. *UK.*
Mchedlishvili, Dr Boris Victorovich. *USSR.*
McIntyre, Dr Garry, James. *France.*
McKay, Dr. David Bruce. *USA.*
McKee, Dr Vickie. *New Zealand.*
McKenzie, Dr David Robert. *Australia.*
McKenzie, Dr. Thomas Charles. *USA.*
McKeown, Dr. David Alexander. *USA.*
McKie, Dr Christine Hilary. *UK.*
McKie, Dr Duncan. *UK.*
McKinstry, Prof. Herbert Alden. *USA.*
McLaren, Dr Alexander Clark. *Australia.*
McLaren, Prof. Eugene Herbert. *USA.*
McLaughlin, Mr George Millar. *Australia.*
McLean, Dr. W. John. *USA.*
McMahon, Mr Brian. *UK.*
McMillan, Dr. Joyce A. *USA.*
McMullan, Dr. Richard K. *USA.*
McMurdie, Mr. Howard Francis. *USA.*
McNulty, Mr. Thomas. *USA.*
McPartlin, Dr Mary. *UK.*
McPhail, Prof. Andrew Tennent. *USA.*
McPherson, Prof. Alexander. *USA.*
McPherson, Mr. William G. *USA.*
McRee, Dr. Duncan E. *USA.*
McTigue, Ms. Michele A. *USA.*
McWhan, Dr. Denis Bayman. *USA.*
Meagher, Dr Edward Patrick. *USA.*
Mealli, Prof. Carlo. *Italy.*
Medeiros Rodrigues, Dr Maria Mabel. *Brasil.*
Medgyaszay, Dr Márton. *Hungary.*
Médicis, de, Dr Rinaldo M. *Canada.*
Medimorec, Magistar Stanislav. *Yugoslavia.*
Medrud, Dr. Ronald Curtis. *USA.*
Meehan, Prof. Edward Joseph Jr. *USA.*
Meek, Dr Keith Michael Andrew. *UK.*
Meetsma, Drs Auke. *Netherlands.*
Megaw, Dr Helen Dick. *UK.*
Meier, Prof. Dr Hans. *BRD.*
Meier, Prof. Walter M. *Switzerland.*
Meinke, Ms. Gretchen. *USA.*
Meisalo, Prof. Veijo Pauli Juhani. *Finland.*
Meisel, Prof. Armin. *DDR.*
Mejia Cifuentes, Prof. Leonidas. *Colombia.*
Mel, de, Mr. Vidanalage S. J. *USA.*
Mel'nikov, Dr Oleg Konstantinovich. *USSR.*
Mel'nikov, Dr Vitaly Alexandrovich. *USSR.*
Mel'nikov, Dr Vladimir Stepanovich. *USSR.*
Mel'nikova, Dr Alina Mikhailovna. *USSR.*

Melamud, Dr Mordechai. *Israel.*
Melekh, Dr Bernard Abu-Talibovich. *USSR.*
Meleshina, Dr Valentina Alexandrovna. *USSR.*
Melik-Adamyan, Dr Vil'yam Rafailovich. *USSR.*
Melka, Dr Karel. *CSSR.*
Mellini, Prof. Marcello. *Italy.*
Melzer, Mr Rolf. *BRD.*
Menchetti, Prof. Silvio. *Italy.*
Menczel, Dr György. *Hungary.*
Mendelssohn, Dr Monica Jutta. *UK.*
Meng, Prof. Yi-min. *China.*
Menzinger, Prof. Filippo. *Italy.*
Mercer, Dr William Duncan. *UK.*
Mercola, Prof. Daniel Anthony. *USA.*
Mereiter, Prof. Dr Kurt. *Austria.*
Meriani, Dr Sergio. *Italy.*
Mérigoux, Prof. Henri. *France.*
Merino de Matheus, Prof. Lucía Marina. *Colombia.*
Merinov, Dr Boris Vladimirevich. *USSR.*
Merisalo, Dr Matti Juhani. *Finland.*
Merlini, Dr Alfonso Enrico. *Italy.*
Merlino, Prof. Stefano. *Italy.*
Merlo, Prof. Franco. *Italy.*
Merriman, Mr Richard James. *UK.*
Merritt, Dr. Ethan Allen. *USA.*
Merritt, Prof. Lynne Lionel Jr. *USA.*
Mertes, Prof. Kristin Bowman. *USA.*
Mertin, Dr Wilhelm. *BRD.*
Messerschmidt. Dr Albrecht. *BRD.*
Messick, Mr. Julian. *USA.*
Messner, Dr Dieter. *BRD.*
Mestnik Filho, Dr José. *Brasil.*
Mèszáros, Mr Csaba, *Yugoslavia.*
Metcalf, Dr Peter. *BRD.*
Metcalfe, Dr Edward. *UK.*
Metsik, Prof. Mikhail Stepanovich. *USSR.*
Metter, Mr Joachim. *BRD.*
Metz, Dr Bernard. *France.*
Metzger, Prof. Robert Melville. *USA.*
Meunier - Piret, Dr (Mrs) Jacqueline. *Belgium.*
Meurs, van, Dr Ir Frank. *Netherlands.*
Mewis, Prof. Dr Albrecht. *BRD.*
Meyer-Ehmsen, Prof. Dr Gerhard. *BRD.*
Meyer, Prof. Edgar Frederich. *USA.*
Meyer, Prof. Frank Henry. *USA.*
Meyer, Dr Gerd Heinrich. *BRD.*
Meyer, Prof. Dr Hans-Jürgen. *BRD.*
Meyer, Prof. Klaus. *DDR.*
Meyers, Dr. Bernard Lee. *USA.*
Meyers, Prof. Edward Arthur. *USA.*
Mez, Dr Hans-Christian. *Switzerland.*
Mhiri, Dr Tahar. *Tunisia.*
Mian, Dr Mohammad Ashraf. *Pakistan.*
Mian, Mr Muhammad Asghar. *Pakistan.*
Miao, Mr Chun-sheng. *China.*
Miao, Prof. Fang-ming. *China.*
Michaelis de Saenz, Dr Irene Marianne. *Uruguay.*
Michailov, Mr Evgeni. *Bulgaria.*
Michailov, Mr Michail. *Bulgaria.*
Michalec, Ing Rudolf. *CSSR.*
Micheelsen, Prof. Harry. *Denmark.*
Michel, Dr. André Gustave. *Canada.*
Michel, Prof Bernd. *DDR.*
Michel, Dr. David John. *USA.*
Michel, Prof. Karl Heinrich Joseph *Belgium.*
Michel, Prof. Pierre. *France.*
Michell, Mr Ernest William John. *UK.*
Middlemiss, Dr Nora E. *Canada.*
Middleton, Dr Andrew Philip. *UK.*
Mighell, Dr. Alan D. *USA.*
Mihama, Prof. Kazuhiro. *Japan.*
Mihichuk, Prof. Lynn Michael. *Canada.*
Miida, Mr Rokuro. *Japan.*
Mijlhoff, Dr Frans Cornelis. *Netherlands.*
Mikenda, Prof. Dr Werner. *Austria.*
Mikhailov, Dr Al'bert Mikhailovich. *USSR.*
Mikhailov, Dr Igor' Fedorovich. *USSR.*
Mikhailov, Dr Vladimir Ivanovich. *USSR.*
Mikhailov, Dr Yuriy Nikolayevich. *USSR.*
Mikhalenko, Dr Svetlana Ivanovna. *USSR.*

Mikheeva, Dr Irina Victorovna. *USSR.*
Miki, Dr Kunio. *Japan.*
Mikkola, Prof. Donald Emil. *USA.*
Mikler, Dr Helga. *Austria.*
Mikloš, Ing Dušan. *CSSR.*
Mikula, Dr Pavol. *CSSR.*
Milat, Magistar Ognjen. *Yugoslavia.*
Milberg, Dr. Morton Edwin. *USA.*
Milburn, Prof. George Henry William. *UK.*
Milburn, Mr. Michael Vance. *USA.*
Milchev, Dr Alexander. *Bulgaria.*
Milchev, Dr Andrei. *Bulgaria.*
Mildner, Mr. David F. R. *USA.*
Milićev, Prof. Dr Svetozar. *Yugoslavia.*
Milillo, Prof. Frank. *USA.*
Milinski, Prof. Dr Nikola. *Yugoslavia.*
Miliotis, Prof. Demitrios Menelaos. *Greece.*
Milius, Dr Wolfgang. *BRD.*
Millan-Muñoz, Dr Maria. *Spain.*
Millane, Prof. Rick P. *USA.*
Millar, Dr John Joseph. *Australia.*
Milledge, Dr H. Judith. *UK.*
Miller, Prof. Andrew. *UK.*
Miller, Prof. Donald P. *USA.*
Miller, Dr. Kristine Elaine. *USA.*
Miller, Dr. Lance L. *USA.*
Miller, Dr. Maria. *USA.*
Miller, Mr. Mark L. *USA.*
Miller, Dr. Richard Wayne. *USA.*
Miller, Prof. Robert Llewellyn. *USA.*
Miller, Ms Sarah Ann. *Australia.*
Millini, Dr Roberto. *Italy.*
Millionova, Dr Margarita Ivanovna. *USSR.*
Millner, Dr. Ozra Elmo. *USA.*
Mills, Mr Owen S. *UK.*
Miloshev, Prof. Georgi. *Bulgaria.*
Mimault, Prof. Jean. *France.*
Min, Dr Eungi. *Japan.*
Minacheva, Dr Lidiya Khabibovna. *USSR.*
Minagawa, Dr Teruaki. *Japan.*
Minami, Dr. Nobuyuki. *Japan.*
Minami, Mr. Takashi. *Japan.*
Minari, Prof. Fernand. *France.*
Minato, Prof. Hideo. *Japan.*
Mincheva-Stefanova, Prof. Yordanka. *Bulgaria.*
Mineeva, Dr Rimma Mikhailovna. *USSR.*
Minkin, Mrs. Jean Albert. *USA.*
Minkoff, Prof. Isaac. *Israel.*
Minni, Dr Erkki Esa Kalervo. *Finland.*
Minomura, Prof. Shigeru. *Japan.*
Mints, Prof. Rafail Isaakovich. *USSR.*
Mir, Mr Jan Mohammad. *Pakistan.*
Miravitlles, Prof. Carlos. *Spain.*
Mirčeva, Magistar Aneta. *Yugoslavia.*
Mirsky, Dr. Kira. *USA.*
Mirzu-Ghergariu, Mrs Lucretia. *Romania.*
Mishnev, Dr Anatolii Fedorovich. *USSR.*
Misra, Prof Nirmal Kumar. *India.*
Misra, Prof. Somnath. *India.*
Misra, Dr Tripurari. *India.*
Misyul', Dr Sergei Vladimirovich. *USSR.*
Mitchell, Dr Crighton Maurice. *Canada.*
Mitchell, Dr Gary Findlater. *UK.*
Mitchell, Dr Geoffrey Robert. *UK.*
Mitchell, Dr Keith A. R. *Canada.*
Mitra, Prof. Girija Bhushan. *India.*
Mitrprachachon, Dr Pachanee. *Thailand.*
Mitsuda, Dr. Hiromichi. *Japan.*
Mitsuda, Dr. Takeshi. *Japan.*
Mitsui, Prof. Toshio. *Japan.*
Mitsui, Prof. Yukio. *Japan.*
Mitsuishi, Prof. Tomokuni. *Japan.*
Mittemeijer, Prof. Dr Ir Eric Jan. *Netherlands.*
Miura, Dr Yasuhiro. *Japan.*
Miura, Dr Yasunori. *Japan.*
Miuskov, Dr Vasily Fedorovich. *USSR.*
Miyaji, Dr. Hideki. *Japan.*
Miyake, Dr Yasuhiro. *Japan.*
Miyake, Prof. Shizuo. *Japan.*
Miyamae, Dr Hiroshi. *Japan.*

Miyamoto, Dr. Masamichi. *Japan.*
Miyano, Mr. Toshio. *Japan.*
Miyata, Dr. Takeshi. *Japan.*
Miyazawa, Dr. Shintaro. *Japan.*
Miyoshi, Dr Tadahiko. *Japan.*
Mizota, Dr Tadato. *Japan.*
Mizuki, Dr Junichiro. *Japan.*
Mizuno, Dr Hiroshi. *Japan.*
Mizuno, Prof. Joji. *Japan.*
Mlik, Dr Youssef. *Tunisia.*
Mo, Prof. Frode. *Norway.*
Modjtahedi, Dr Mansour. *Iran.*
Modrick, Ms. Michelle Ann. *USA.*
Modrzejewski, Dr hab. Antoni. *Poland.*
Moeckli, Dr Pedro. *Switzerland.*
Möhling, Dr Werner. *DDR.*
Møller, Dr Christian Knakkergaard. *Denmark.*
Möller, Dr Manfred. *BRD.*
Moffat, Prof. John Keith. *USA.*
Mogilevskii, Dr Leonid Yur'evich. *USSR.*
Moguš-Milanković, Magistar Andrea. *Yugoslavia.*
Moh, Prof. Dr Günter Harald. *BRD.*
Mohajeri Moghaddam, Dr Hamid Reza. *Iran.*
Mohamad, Dr Hamzah. *Malaysia.*
Mohanlal, Dr Sembu Krishnaiyer. *India.*
Mohr, Dr Ulrich. *DDR.*
Mohyla, Dr Jury. *Australia.*
Moini, Dr. Ahmad. *USA.*
Mok, Dr Kum-fun. *Singapore.*
Mokeeva, Dr Valentina Ivanovna. *USSR.*
Mokhov, Dr Andrei Vladimirovich. *USSR.*
Mokraya, Dr Ivanna Romanovna. *USSR.*
Mokren, Dr. James David. *USA.*
Molchanov, Dr Vladimir Nikolayevich. *USSR.*
Moldovanova, Prof. Maria. *Bulgaria.*
Molea, Mr. Frank N. *USA.*
Molin, Prof. Gianmario. *Italy.*
Molins, Dr Elies. *Spain.*
Momany, Mr. Cory. *USA.*
Monaco, Dr Hugo Luis. *Italy.*
Monari, Dr Magda. *Italy.*
Moncrief, Dr. J. William. *USA.*
Mondragón, Prof. Alfonso. *USA.*
Monge, Dr M. Angeles. *Spain.*
Mongiorgi, Prof. Romano. *Italy.*
Monroe, Prof. Eugene A. *USA.*
Montenegro de Andrade, Prof. Miguel. *Portugal.*
Montenero, Prof. Angelo. *Italy.*
Montfort, Dr. William R. *USA.*
Montgomery, Prof. (Em.) Henry. *Canada.*
Montgomery, Mr. Thomas S. *USA.*
Moodie, Dr Alexander Forbes. *Australia.*
Moody, Dr Peter Charles Edmund. *UK.*
Moon, Assoc. Prof. Anthony Ronald. *Australia.*
Moor, Dr Robert. *Switzerland.*
Moore, Dr Alan James William. *Australia.*
Moore, Miss Alice Elizabeth. *UK.*
Moore, Dr Anthony Moreton. *UK.*
Moore, Mr. Donald L. *USA.*
Moore, Mrs. Elizabeth J. Weichel. *USA.*
Moore, Dr John Carlton. *UK.*
Moore, Prof. Paul Brian. *USA.*
Moore, Mr Peter Leonard. *UK.*
Mootz, Prof. Dr Dietrich. *BRD.*
Mora de González, Prof. Nery. *Colombia.*
Mora Rodríguez, Prof. Asilloe Jasmina. *Venezuela.*
Moraga, Mr Luís. *Chile.*
Morandi, Prof. Noris. *Italy.*
Moras, Dr Dino. *France.*
Moravcová, Dr Hana. *CSSR.*
Moravec, Ing František. *CSSR.*
Morawiec, Prof. Henryk. *Poland.*
Moreau, Prof. Jean-Michel. *France.*
Moreau, Prof. Jules Francois. *Belgium.*
Moreiras, Dr Damaso. *Spain.*
Moreland, Dr. James Andrew. *USA.*
Moreno-Carretero, Dr Miguel. *Spain.*
Moreno-Echevarria, Dr M. Esperanza. *Spain.*
Moret, Dr Massimo. *Italy.*
Moret, Dr Roger. *France.*

Morffew, Dr Andrew James. *UK.*
Morgan, Dr Colin Harris. *UK.*
Morgan, Dr. Joseph. *USA.*
Morgenroth, Mr. Wolfgang H. *BRD.*
Moriarty, Dr. John Lawrence Jr. *USA.*
Morikawa, Mr Hideki. *Japan.*
Morikawa, Dr. Hiroshi. *Japan.*
Morimoto, Dr. Carl Noboru. *USA.*
Morimoto, Dr. Jun. *Japan.*
Morimoto, Prof. Nobuo. *Japan.*
Morimoto, Dr. Yukio. *Japan.*
Morinaga, Dr Masahiko. *Japan.*
Moring, Dr. Jill. *USA.*
Morino, Prof. Yonezo. *Japan.*
Moritani, Mr Yoshimitsu. *JAPAN.*
Moritz, Prof. Dr Wolfgang Otto. *BRD.*
Mornon, Dr Jean-Paul. *France.*
Moron, Dr M. Carmen. *Spain.*
Morosin, Dr. Bruno. *USA.*
Morosoff, Dr. Nicholas C. *USA.*
Moroz, Dr Ella Mikhailovna. *USSR.*
Morris, Dr Donald Frank Charles. *UK.*
Morris, Mrs. Marlene Cook. *USA.*
Morris, Ms. Nancy L. *USA.*
Morrison, Mr. George. *USA.*
Morrow, Prof. John Charles III. *USA.*
Morrow, Dr. Scott Imlay. *USA.*
Morton, Dr Allan James. *Australia.*
Morvaj, Magistar (Mrs) Jasmina. *Yugoslavia.*
Moseley, Dr Patrick. *UK.*
Moshkin, Dr Sergei Vladimirovich. *USSR.*
Moskvin, Dr Valentin Vasilyevich. *USSR.*
Moss, Dr Barbara Kay. *Australia.*
Moss, Dr David Stanley. *UK.*
Moss, Mr. Simon C. *USA.*
Mosset, Dr Alain. *France.*
Mostad, Mr Arvid. *Norway.*
Motegi, Prof. Hiroshi. *Japan.*
Motherwell, Dr William David Samuel. *UK.*
Mothes, Mr Heinrich. *DDR.*
Motta, Dr Nunzio. *Italy.*
Mottana, Prof. Annibale. *Italy.*
Moudrianakis, Prof. Evangelos N. *USA.*
Moudy, Ms. Lavada Ann. *USA.*
Mourikis, Dr Stamatios. *Greece.*
Moustakali Mavridis, Dr Irene. *Greece.*
Movchan, Dr Nikolai Prokof'evich. *USSR.*
Mowbray, Dr. Sherry Lynn. *USA.*
Moze, Dr Oscar. *Italy.*
Mozzi, Dr. Robert Lewis. *USA.*
Mrafko, Dr Peter. *CSSR.*
Mrayed, Mr Yaseen S. *Libya.*
Mrose, Miss Mary E. *USA.*
Mrvoš-Sermek, Magistar (Mrs) Draginja. *Yugoslavia.*
Mu, Mr Xiang-qi. *China.*
Mucha, Mr Christine. *DDR.*
Muchmore, Ms. Christine R. A. *USA.*
Muchmore, Mr. Steven W. *USA.*
Mucker, Prof. Kenneth. *USA.*
Muddle, Dr Barrington Charles. *Australia.*
Mühlberg, Dr Manfred. *DDR.*
Müller-Buschbaum, Prof. Dr Hanskarl. *BRD.*
Müller-Fahrnow, Mrs. Anke. *BRD.*
Müller-Vogt, Dr German. *BRD.*
Müller, Dr Bernd. *DDR.*
Müller, Dr Brigitte. *DDR.*
Müller, Doz. Dr Mag. Karl Werner. *Austria.*
Müller, Dr Eberhard. *DDR.*
Müller, Prof. Dr Gerd. *BRD.*
Müller, Prof. Dr Gerhard. *BRD.*
Müller, Prof. Horst. *BRD.*
Mueller, Dr. Melvin H. *USA.*
Müller, Dr Paul Hubert. *BRD.*
Müller, Dr Rudolf O. *Switzerland.*
Müller, Prof. Dr Ulrich. *BRD.*
Müller, Prof. Dr Wolfgang Friedrich. *BRD.*
Müllner, Dr Manfred. *BRD.*
Münninghoff, Dr Günter. *BRD.*
Mueser, Mr. Timothy C. *USA.*
Mugnoli, Prof. Angelo. *Italy.*

Muhonen, Dr Heikki Juhani. *Finland.*
Muir, Dr Alastair Kerr. *Canada.*
Muir, Prof. James Alexander. *USA.*
Muir, Dr Kenneth Walter. *UK.*
Muirhead, Dr Hilary. *UK.*
Mujica, Dr Carlos. *Chile.*
Mukai, Prof. Tadasuke. *Japan.*
Mukhamedzhanov, Dr Enver Khamzyaevich. *USSR.*
Mukherjee(mondal), Dr (Mrs) Monika. *India.*
Mukherjee, Dr Alok Kumar. *India.*
Mukherjee, Dr Amal Bikash. *India.*
Mukherjee, Dr Biswanath. *India.*
Mukherjee, Dr Partha Sarathi. *India.*
Mukhopadhyay, Dr (Mrs) Anuradha. *India.*
Mukhopadhyay, Ashis. *India.*
Mukhopadhyay, Mr Bishnu Prasad. *India.*
Mukhopadhyay, Dr Pradip. *India.*
Mukhtarova, Dr Nina Nikolayevna. *USSR.*
Muldawer, Prof. Leonard. *USA.*
Muller, Prof. Jean. *Switzerland.*
Mullica, Dr. Donald Foster. *USA.*
Mundt, Dr Otto. *BRD.*
Munir, Mr Mohammad. *Pakistan.*
Munirathinam, Nethaji. *India.*
Muñoz Picone, Dr Eduardo. *México.*
Munshi, Mr Sanjeev Kumar. *India.*
Munsuz, Prof. Nuri. *Turkey.*
Mura, Dr Pasquale. *Italy.*
Murad, Dr Enver. *BRD.*
Murakami, Dr Takashi. *Japan.*
Murali, Mr. Ramachandran. *USA.*
Muralidharan, Mr K. V. *India.*
Muraoka, Dr Hisashi. *Japan.*
Murasik, Dr Andrzej. *Poland.*
Murata, Prof. Yoshitada. *Japan.*
Murfitt, Mr. Robert R. *USA.*
Murgich, Dr Juan. *Venezuela.*
Murmann, Prof. R. Kent. *USA.*
Murray-Rust, Dr Judith. *UK.*
Murray-Rust, Dr Peter. *UK.*
Murrieta Sánchez, Dr. Héctor, *México.*
Murta, Prof. Clecio. *Brasil.*
Murthy, Dr. Krishna H. M. *USA.*
Murthy, Dr M.R. N. *India.*
Murthy, Dr. N. Sanjeeva. *USA.*
Musaev, Dr Aidyn Alipanakh ogly. *USSR.*
Musatti, Prof. Amos. *Italy.*
Musayev, Dr Faig Nasib ogly. *USSR.*
Muschner, Dr Wolfgang. *DDR.*
Mustafayev, Dr Nariman Mustafa ogly. *USSR.*
Mutikainen, Dr Ilpo Pellervo. *Finland.*
Mutka, Dr Hannu Mika Ilmari. *Finland.*
Mutter, Mrs Graciela. *BRD.*
Mya Mya, Dr Khin. *Burma.*
Myasnikova, Dr Rimma Mikhailovna. *USSR.*
Myer, Prof. George Henry. *USA.*
Mylvaganam, Dr. Sangari Eshwari. *USA.*
Mys'kiv, Dr Mar'yan Grigoryevich. *USSR.*
Myshlyayev, Prof. Mikhail Mikhailovich. *USSR.*
Mzayek, Mr Elias. *Spain.*
Nabarro, Prof. Frank Reginald Nunes. *South Africa.*
Naboka, Dr Marat Nikolayevich. *USSR.*
Nachman, Dr Joseph. *Canada.*
Nadezhina, Dr Tamara Nikolayevna. *USSR.*
Nägele, Dr Walter. *BRD.*
Näsäkkälä, Dr Matti Eerik. *Finland.*
Näsänen, Prof. Reino Olavi. *Finland.*
Nag, Dr Dilip Kumar. *India.*
Nag, Miss Jhumjhumi. *India.*
Naga, Dr Mohamed Abdel Hamid. *Egypt.*
Nagabhushana Rao, Mr Chemboli. *India.*
Nagai, Prof. Ryutaro. *Japan.*
Nagaitsev, Dr Yury Valeryevich. *USSR.*
Nagakura, Prof. Saburo. *Japan.*
Nagakura, Prof. Sigemaro. *Japan.*
Nagano, Dr Kozo. *Japan.*
Nagashima, Dr. Seiichi. *Japan.*
Nagata, Dr Fumio. *Japan.*
Nagl, Dr Antun. *Yugoslavia.*
Nagpal, Dr Kailash Chander. *India.*

Naik, Dr (Mrs) Uma Murlidhar. *India.*
Naiki, Prof. Toshio. *Japan.*
Nakagawa, Dr Atsushi. *Japan.*
Nakahigashi, Dr Kiyotaka. *Japan.*
Nakai, Dr Hisayoshi. *Japan.*
Nakai, Dr. Izumi. *Japan.*
Nakai, Dr. Yasuo. *Japan.*
Nakajima, Dr. Tetuo. *Japan.*
Nakajima, Dr Yoshiharu. *Japan.*
Nakamura, Dr Naotake. *Japan.*
Nakamura, Dr Kazuo. *Japan.*
Nakamura, Dr Osamu. *Japan.*
Nakamura, Prof. Terutaro. *Japan.*
Nakanishi, Dr Hachiro. *Japan.*
Nakanishi, Prof. Norihiko. *Japan.*
Nakano, Prof. Shigeru. *Japan.*
Nakashima, Prof. Shin-ichi. *Japan.*
Nakata, Mr. Kazuaki. *Japan.*
Nakatsu, Prof. Kazumi. *Japan.*
Nakayama, Mr Kan. *Japan.*
Nakazawa, Dr Hiromoto. *Japan.*
Nakazumi, Dr Yoshihide. *Japan.*
Nakhla, Dr Fakhry. *Egypt.*
Nakissa, Dr Manutschehr. *Iran.*
Namba, Dr Yoshiyuki. *Japan.*
Namboodiri, Dr. Krishnan. *USA.*
Namgung, Prof. Dr Hae. *Korea.*
Namikawa, Prof. Kazumichi. *Japan.*
Nančovski, Mr Kosta. *Yugoslavia.*
Nandi, Asok Kumar. *India.*
Nandi, Dr Ranjan Kumar. *India.*
Nanev, Dr Christo. *Bulgaria.*
Nani, Dr Rahim. *Iran.*
Nanni, Mr. Raymond. *USA.*
Napolitano, Prof. Roberto. *Italy.*
Naqvi, Dr Syed Ali Anwar. *Pakistan.*
Narasimhamurthy, Narasappa. *India.*
Narasimhan, Dr P. *India.*
Narayanan, Prof. Palamadi Sundaram. *India.*
Narayanan, Prof. V. Anantha. *USA.*
Nardelli, Prof. Mario. *Italy.*
Nardin, Prof. Giorgio. *Italy.*
Nardov, Dr Andrei Vladimirovich. *USSR.*
Narendra, Dr. Narayana. *USA.*
Narita, Dr. Hajime. *Japan.*
Narita, Mr. Masaki. *Japan.*
Naseif, Dr Abdulah. *Saudi Arabia.*
Nasreen, Miss Shagufta. *Pakistan.*
Nassau, Dr. Kurt. *USA.*
Nassimbeni, Prof. Luigi Renzo. *South Africa.*
Natarajan, Dr Mahadevan. *Canada.*
Natarajan, Dr S. *India.*
Natarajan, Dr Subramanian. *India.*
Natesan, Mr Elango. *India.*
Nath, Mr Kashi. *India.*
Nathan, Dr. Robert. *USA.*
Nathan, Dr Yaacov. *Israel.*
Naud, Dr Jean Marcel. *Belgium.*
Naudon, Dr André. *France.*
Naumova, Dr Inessa Ivanovna. *USSR.*
Navarra, Dr Gabriele. *Italy.*
Navarro, Dr Carmen. *Spain.*
Nave, Dr Colin. *UK.*
Navia, Dr. Manuel Alberto. *USA.*
Nawata, Dr Yoshiharu. *Japan.*
Nawaz, Dr Rab. *UK.*
Naya, Prof. Shigeo. *Japan.*
Nazarboland, Mr Abbas Ali. *Iran.*
Neckel, Prof. Dr Adolf. *Austria.*
Neder, Dr Reinhard Bernhard. *BRD.*
Neels, Prof. Hermann. *DDR.*
Nefedova, Dr Elena Vasilyevna. *USSR.*
Neff, Prof. Dr Hans Josef. *BRD.*
Nehasil, Dr Miroslav. *CSSR.*
Neidhart, Dr. David James. *USA.*
Neidle, Dr Stephen. *UK.*
Neifeind, Dipl-Ing Axel. *BRD.*
Neilson, Dr. George Francis. *USA.*
Nelkowski, Prof. Dr Horst. *BRD.*
Nelmes, Dr Richard J. *UK.*

Nelson, Mr. A. Dwayne. *USA.*
Nemecz, Prof. Ernö. *Hungary.*
Nemiroff, Dr. Michael. *USA.*
Nenner, Miss Ann-Marie. *Sweden.*
Nenonen, Mr Pertti Olavi. *Finland.*
Nenow, Dr Dimiter. *Bulgaria.*
Neronova, Dr Nina Nikolayevna. *USSR.*
Nesper, Dr Reinhard Friedrich. *BRD.*
Nesterov, Dr Vladimir Nikolaevich. *USSR.*
Nesterova, Dr Yaroslava Mikhailovna. *USSR.*
Neubüser, Prof. Joachim Franz F. G.. *BRD.*
Neuman, Mr. Melvin A. *USA.*
Neumann, Prof Hans-Georg. *DDR.*
Neumann, Dr Wolfgang. *DDR.*
Nevskaya, Dr Natalia Aleksandrovna. *USSR.*
Nevskii, Dr Nukolai Nikolaevich. *USSR.*
Newesely, Prof. Dr Heinrich. *BRD.*
Newman, Dr. Robert Alan. *USA.*
Newnham, Prof. Robert Everest. *USA.*
Newsam, Dr. John Michael. *USA.*
Ng, Prof. Ser Choon. *Singapore.*
Ng, Dr Wee Lam. *Malaysia.*
Nguy Tuyet Nhung, Dr. *Vietnam.*
Nguyen An, Prof. *Vietnam.*
Nguyen Hong Quyen, Dr. *Vietnam.*
Nguyen Huu Chu, Dr. *Vietnam.*
Nguyen Khoa Phuc, Dr. *Vietnam.*
Nguyen Van Chi, Dr. *Vietnam.*
Nguyen Van Vuong, Dr. *Vietnam.*
Nguyen, Dr. Khe Thanh. *USA.*
Nguyen, Mr. Thao A. *USA.*
Ni, Prof. Chau-zhou. *China.*
Nicholas, Mr David Michael. *UK.*
Nicholls, Dr Reginald A. *UK.*
Nichols, Mr. Monte Carl. *USA.*
Nicholson, Dr Brian Kenneth. *New Zealand.*
Nicholson, Dr David Graham. *Norway.*
Nicklow, Dr. Robert Merle. *USA.*
Nicolo, Dr Francesco. *Italy.*
Nicolosi, Dr. Joseph Anthony. *USA.*
Nieber, Dr Johannes. *DDR.*
Niebsch, Dr Hans-Hermann. *DDR.*
Niedermayr, Dr Gerhard. *Austria.*
Nieduszynski, Dr Ian Alexander. *UK.*
Nielsen, Dr Anders. *Denmark.*
Nielsen, Mr. Christopher Pine. *USA.*
Nielsen, Mr Kurt. *Denmark.*
Nieminen, Dr Kari Veikko Juhani. *Finland.*
Nieto-Garcia, Dr Fernando. *Spain.*
Nigam, Gur Dayal. *India.*
Nigli, Selina. *India.*
Nigović, Miss Biljana. *Yugoslavia.*
Niimura, Dr. Nobuo. *Japan.*
Niinistö, Dr Lauri. *Finland.*
Nijveldt, Dr Ir Dick. *Netherlands.*
Nikanorov, Dr Stanislav Prokhorovich. *USSR.*
Nikiforov, Prof. Igor' Yakovlevich. *USSR.*
Nikishova, Dr Lidiya Vasilyevna. *USSR.*
Nikitenko, Prof. Valerian Ivanovich. *USSR.*
Nikol'skaya, Dr Larisa Viktorovna. *USSR.*
Nikol'skaya, Dr Natal'ya Kimovna. *USSR.*
Nikolaeva, Mrs Rumiana. *Bulgaria.*
Nikolić, Prof. Dr Pantelija. *Yugoslavia.*
Nikonov, Dr Stanislav Valdimirovich. *USSR.*
Nilakantan, Dr. Ramaswamy. *USA.*
Nilsson, Dr Karin I. *Sweden.*
Nilsson, Dr Rolf O. *Sweden.*
Nimgirawath, Mrs Kloy. *Thailand.*
Nimmo, Dr John Kenneth. *Australia.*
Nishi, Prof. Fumito. *Japan.*
Nishida, Dr Isao. *Japan.*
Nishida, Prof Takashi. *Japan.*
Nishikawa, Dr Masana. *Japan.*
Nishinaga, Prof. Tatau. *Japan.*
Nishino, Dr. Yoichi. *Japan.*
Nishiyama, Dr Tsutomu. *Japan.*
Nishiyama, Prof. Zenji. *Japan.*
Nissen, Prof. Hans-Ude. *Switzerland.*
Nitsche, Prof. Dr Rudolf. *BRD.*
Nitsche, Mr Walter. *DDR.*

Palatnik, Prof. Lev Samoilovich. *USSR.*
Palenik, Prof. Gus Joseph. *USA.*
Palenzona, Prof. Andrea. *Italy.*
Pálinkás, Dr Gábor. *Hungary.*
Palistrant, Prof. Alexander Filippovich. *USSR.*
Paljević, Dr Matija. *Yugoslavia.*
Palkina, Dr Kapitolina Kapitonovna. *USSR.*
Pallai, Dr. Peter V. *USA.*
Palmer, Dr. Kenneth James. *USA.*
Palmer, Dr Rex Alfred. *UK.*
Palomo-Delgado, Dr M. Inmaculada. *Spain.*
Pamplin, Dr Brian Randall. *UK.*
Pan, Prof. Dao-jun. *China.*
Pan, Prof. Ke-zhen. *China.*
Pan, Dr Nitya Ranjan. *India.*
Pan, Prof. Zhao-lu. *China.*
Panagos, Prof. Athanasios. *Greece.*
Panchekha, Dr Petr Alekseyevich. *USSR.*
Pandey, Dr Dhananjai. *India.*
Pandya, Prof. Janardhan Rameshchandra. *India.*
Pandya, Mr Naresh. *Canada.*
Pangarov, Prof. Nikola. *Bulgaria.*
Pangborn, Prof. Robert Northrup. *USA.*
Pangborn, Mr. Walter. *USA.*
Pani, Dr Marcella. *Italy.*
Pannetier, Dr Jean. *France.*
Pannhorst, Dr Wolfgang. *BRD.*
Pant, Dr Arun Kumar. *India.*
Pant, Dr Lalit Mohan. *India.*
Paorici, Prof. Carlo. *Italy.*
Papadakis, Prof. Alexander. *Greece.*
Papadimitraki Chlichlia, Prof. Helena. *Greece.*
Papadopoulos, Mr Demetrius. *Greece.*
Papathanassopoulos, Dr Constantinos. *Greece.*
Papavinasam, Dr E. *India.*
Papiz, Dr Miroslav Zenko. *UK.*
Papp, Mr Gábor. *Hungary.*
Papunen, Prof. Heikki Tapani. *Finland.*
Parasnis, Prof. Arawind Shripad. *India.*
Paretzkin, Mr. Boris. *USA.*
Parge, Dr. Hans Erich. *USA.*
Paris, Dr Eleonora. *Italy.*
Parise, Dr. John B. *USA.*
Parissakis, Prof. George. *Greece.*
Park, Dr Byung Kyu. *Korea.*
Park, Dr. Chang Hoon. *USA.*
Park, Prof. Young Ja. *Korea.*
Párkányi, Dr László. *Hungary.*
Parker, Dr Andrew. *UK.*
Parker, Dr Michael William. *UK.*
Parker, Dr. Robert Louis. *USA.*
Parker, Dr. Sidney Glenn. *USA.*
Parkinson, Dr Gordon Michael. *UK.*
Parks, Ms. Elizabeth Annette Heady. *USA.*
Parks, Dr Terrence Charles. *Australia.*
Parpia, Dr Dawood Yusuf. *UK.*
Parris, Dr. Kevin Delos. *USA.*
Parrish, Dr. William. *USA.*
Parry, Prof. David Anthony Dougall. *New Zealand.*
Parsons, Dr. Donald Frederick. *USA.*
Parsons, Prof. Ian. *UK.*
Parthasarathy, Dr. R. *USA.*
Parthasarathy, Dr S. *India.*
Parthé, Prof. Erwin. *Switzerland.*
Parthier, Mr Lutz. *DDR.*
Partin, Mr. Daniel Edward. *USA.*
Parvez, Dr. Masood. *USA.*
Parvov, Mr Vladimir Fedorovich. *USSR.*
Pascard, Dr Claudine. *France.*
Pascher, Dr Irmin. *Sweden.*
Pasero, Dr Marco. *Italy.*
Pashaev, Dr El'khan Mekhrali-Ogly. *USSR.*
Pashley, Prof. Donald William. *UK.*
Pashov, Mr Nikolai. *Bulgaria.*
Passaglia, Prof. Elio. *Italy.*
Paster, Mr Simeon. *UK.*
Pasti, Mr Fabio. *Italy.*
Patalinghug, Dr Wyona Cruz. *Philippines.*
Patel, Prof. Ambalal Ranchhodbhai. *India.*
Patel, Dr. Jamshed R. *USA.*

Patel, Dr Prabhudas Revandas. *India.*
Patel, Dr Ranjan Prafulbhai. *India.*
Patel, Dr Tankadhar. *Orissa,India.*
Pathak, Prof. Pushkarrai Dalpatram. *India.*
Pathan, Dr Muhammad Taqee. *Pakistan.*
Pathinettam, Dr Padiyan. *India.*
Patmalnieks, Dr Aloisii Alekseevich. *USSR.*
Paton, Mr John Dennis. *UK.*
Patscheider, Mr Jörg. *Switzerland.*
Pattabhi, Dr (Mrs) Vasantha. *India.*
Pattabiraman, Dr. Nagarajan. *USA.*
Pattanayek, Dr. Rekha. *USA.*
Pattridge, Ms. Katherine A. *USA.*
Paturle, Mr. Antoine. *USA.*
Paufler, Prof. Peter. *DDR.*
Paul, Miss Elizabeth A. *USA.*
Paul, Prof. Iain Campbell. *USA.*
Pauling, Prof. Linus. *USA.*
Paulitsch, Prof. Dr Peter. *BRD.*
Paulus, Dr Erich Friedrich. *BRD.*
Pauly, Prof. Hans. *Denmark.*
Paunov, Dr Michael. *Bulgaria.*
Pavalow, Prof. Melvin. *USA.*
Pavel, Dr Nicolae Viorel. *Italy.*
Pavelčík, Ing František. *CSSR.*
Pavie Cardoso, Dr Lisandro. *Brasil.*
Pavkovic, Prof. Stephen Frank. *USA.*
Pavlishin, Dr Vladimir Ivanovich. *USSR.*
Pavlovsky, Dr Alexander Grigoryevich. *USSR.*
Pavlyuk, Dr Anatolii Alekseevich . *USSR.*
Pawlak, Dr Stanisław. *Poland.*
Pawley, Prof. G. Stuart. *UK.*
Payne, Dr Nicholas Charles. *Canada.*
Pazdernik, Prof. LeRoy Joseph. *Canada.*
Peacor, Prof. Donald R. *USA.*
Pearce, Mr I.R. *UK.*
Pebay-Peyroula, Dr Eva. *France.*
Pech, Dr Lucia Yanovna. *USSR.*
Pecharskii, Dr Vitalii Konstantinovich. *USSR.*
Pechstein, Mr Gisela. *DDR.*
Pećina, Miss Planinka. *Yugoslavia.*
Pecoraro, Prof. Vincent L. *USA.*
Pedersen, Dr Berit Fjærtoft. *Norway.*
Pedersen, Prof. Björn. *Norway.*
Pedersen, Dr Jan Skov. *Denmark.*
Pedersen, Prof. Lee G. *USA.*
Pedone, Prof. Carlo. *Italy.*
Peerdeman, Prof. Dr Antonius F. *Netherlands.*
Peeters, Dr Oswald Maurice. *Belgium.*
Peibst, Dr Herbert. *DDR.*
Peiser, Dr. Herbert Steffen. *USA.*
Pelizzi, Prof. Corrado. *Italy.*
Pelizzi, Prof. Giancarlo. *Italy.*
Pellinghelli, Prof. Maria Angela. *Italy.*
Peña, Miss Luzmila. *Chile.*
Penavić, Dr Maja. *Yugoslavia.*
Penco, Prof. Anna Maria. *Italy.*
Peneva, Dr Stefka. *Bulgaria.*
Penfold, Prof. Bruce Russell. *New Zealand.*
Peng, Mr Chang-qi. *China.*
Peng, Ms Ju Lin. *Australia.*
Peng, Mrs Ming-shen. *China.*
Peng, Prof. Shie-ming. *Taiwan.*
Penndorf, Dr Jürgen. *DDR.*
Penner-Hahn, Prof. James E. *USA.*
Pennington, Dr. William T. *USA.*
Pense, Prof. Dr Karl Eduard Jürgen. *BRD.*
Pentinghaus, Dr Horst. *BRD.*
Perales, Dr Aurea. *Spain.*
Percival, Dr Henry Joseph. *New Zealand.*
Perdikatsis, Dr Basilios. *Greece.*
Perdok, Prof. Dr Wiepko Gerhardus. *Netherlands.*
Perego, Dr Giovanni. *Italy.*
Perekalina, Dr Tatyana Mikhailovna. *USSR.*
Perekalina, Dr Zoya Borisovna. *USSR.*
Perelomova, Dr Nataliya Vladislavovna. *USSR.*
Perez-Garcia, Mr Virginia. *Spain.*
Perez-Garrido, Dr Simeon. *Spain.*
Perez-Mato, Dr Juan Manuel. *Spain.*
Pérez, Mrs Carmen. *Chile.*

Perkins, Mr. Herbert O. *USA.*
Perkkiö, Mrs Seija Anneli. *Finland.*
Permér, Miss Lotta. *Sweden.*
Perozzo, Ms. Mary Ann. *USA.*
Perrault, Dr Guy. *Canada.*
Perret, Mr Ramón. *Chile.*
Perrin, Prof. Monique. *France.*
Perry, Dr Alan Leonard. *UK.*
Persdotter, Dr Ingeborg M. *Sweden.*
Pershin, Dr Vitaly Konstantinovich. *USSR.*
Persson, Miss Jeanette. *Sweden.*
Perstnev, Dr Petr Petrovich. *USSR.*
Perthel, Dr Rolf. *DDR.*
Pertlik, Prof. Dr Franz. *Austria.*
Pertsin, Dr Alexander Iosifovich. *USSR.*
Perutz, Dr Max Ferdinand. *UK.*
Pervaiz, Mr Rashed. *Pakistan.*
Perz, Dr Serge. *France.*
Peschar, Dr René. *Netherlands.*
Peshev, Prof Pavel. *Bulgaria.*
Peskin, Dr Vladimir Fedorovich. *USSR.*
Pessa, Prof. Viljo Markus. *Finland.*
Pessen, Dr. Helmut. *USA.*
Pessi Albisu, Malena Silvia. *Uruguay.*
Petcher, Dr Trevor James. *Switzerland.*
Petcov, Mr Alexe. *BRD.*
Péter, Ms Ágnes. *Hungary.*
Peterat, Dr Michael. *BRD.*
Peters, Mr. Charles Richard. *USA.*
Peters, Dr Dieter. *BRD.*
Peters, Dr Karl. *BRD.*
Peterse, Ir Wilhelmus J. A. M. *Netherlands.*
Petersen, Dr. Donald Ralph. *USA.*
Petersen, Prof. Jeffrey L. *USA.*
Petersen, Dr Ole Valdemar. *Denmark.*
Peterson, Prof. John Robert. *USA.*
Peterson, Dr Ronald Charles. *Canada.*
Peterson, Dr. Selmer W. *USA.*
Petiau, Prof. Jacqueline. *France.*
Petipas, Prof. Claude. *France.*
Petraccone, Prof. Vittorio. *Italy.*
Petrás, Mr László. *Hungary.*
Petreus, Mr Ion. *Romania.*
Petříček, Dr Václav. *CSSR.*
Petrik, Ms Anna. *Hungary.*
Petroff, Prof. Jean-François. *France.*
Petropavlov, Dr Nikolay Nikolayevich. *USSR.*
Petrov, Dr Kostadin. *Bulgaria.*
Petrov, Mr Ognyan. *Bulgaria.*
Petrov, Dr Srebri. *Bulgaria.*
Petrov, Dr Thomas Georgiyevich. *USSR.*
Petrov, Dr Vasko. *Bulgaria.*
Petrova, Dr Irina Vladimirovna. *USSR.*
Petrova, Dr Valentina Vasil'evna. *USSR.*
Petrović, Dr Dragoslav. *Yugoslavia.*
Petrovskii, Dr Vitalii Aleksandrovich. *USSR.*
Petrunina, Dr Alla Anastas'yevna. *USSR.*
Petsko, Prof. Gregory Anthony. *USA.*
Pett, Dr. Virginia B. *USA.*
Petter, Prof. Walter. *Switzerland.*
Petukhov, Dr Boris Vladimirovich. *USSR.*
Petz, Mr. John Ignatius. *USA.*
Petzoldt, Dr Jürgen Hugo Hans. *BRD.*
Peyronel, Prof. Giorgio. *Italy.*
Pfefferkorn, Prof. Dr Gerhard Erich. *BRD.*
Pflaum, Mr. Wolfgang Richard. *USA.*
Pfluger, Prof. Clarence Eugene. *USA.*
Pflugrath, Dr. James W. *USA.*
Pflugrath, Dr James William. *BRD.*
Phakey, Dr Prem P. *Australia.*
Phan Ving Phuc, Dr. *Vietnam.*
Phaovibul, Dr Orapin. *Thailand.*
Phavanantha, Dr Phathana. *Thailand.*
Philipov, Mr Alexander. *Bulgaria.*
Philipp, Dr George. *Egypt.*
Philipsborn, von, Prof. Dr Henning. *BRD.*
Phillips, Dr Frederick Lloyd. *Ghana.*
Phillips, Prof. George Neal Jr. *USA.*
Phillips, Prof. James Christopher. *USA.*
Phillips, Mr. Michael W. *USA.*

Phillips, Dr Simon Edward Victor. *UK.*
Phillips, Prof. Sir David Chilton. *UK.*
Phillips, Dr. Susan R. *USA.*
Phillips, Mr. T. J. *USA.*
Phillips, Dr. Theodore II. *USA.*
Phillips, Dr. Walter C. *USA.*
Phizackerley, Dr. Richard Paul. *USA.*
Piazzesi, Prof. AnnaMaria. *Italy.*
Pichert, Mr. Jerome. *USA.*
Pickardt, Prof Dr Joachim. *BRD.*
Picot, Mr. Daniel. *USA.*
Pidzhyan, Prof. Grigory Oganesovich. *USSR.*
Pielaszek, Dr Jerzy. *Poland.*
Pieper, Dr Gerhard. *BRD.*
Pierce-Butler, Dr Melanie Anne. *UK.*
Piermarini, Dr. Gasper John. *USA.*
Pierpont, Prof. Cortlandt G. *USA.*
Pierrot, Dr Marcel. *France.*
Pietraszko, Dr Adam. *Poland.*
Pietraszko, Mrs Donata. *Poland.*
Pietsch, Dr Ullrich. *DDR.*
Pietzsch, Dr Claus. *DDR.*
Pifferi, Dr Augusto. *Italy.*
Pignataro, Mrs. Edith H. *USA.*
Pignedoli, Prof. Anna. *Italy.*
Pigram, Dr William John. *UK.*
Pikin, Dr Sergey Alekseyevich. *USSR.*
Pilati, Dr Tullio. *Italy.*
Pilotti, Dr Anne-Marie. *Sweden.*
Pilz, Prof. Dr Ingrid Edith. *Austria.*
Piña de Noyola, Prof. Maria Cristina. *México.*
Piniella, Dr Juan Francisco. *Spain.*
Pinkerton, Prof. Andrew Alan. *USA.*
Pinsker, Dr Garry Zinov'yevich. *USSR.*
Pinto, Mr Haim. *Israel.*
Pipkin, Dr Noel John. *South Africa.*
Pique, Mr. Michael E. *USA.*
Piret, Prof Paul. *Belgium.*
Pirie, Dr John Douglas. *UK.*
Piro, Dr Oscar Enrique. *Argentina.*
Pirozzi, Prof. Beniamino. *Italy.*
Pisani, Prof. Cesare. *Italy.*
Pisarevsky, Dr Yury Vladimirovich. *USSR.*
Pisutha-Arnond, Dr Visut. *Thailand.*
Pitkänen, Dr Ilkka Pellervo. *Finland.*
Pitkänen, Dr Tuula Esteri. *Finland.*
Pjura, Dr. Phillip Edward. *USA.*
Plakhov, Dr Gennady Fedorovich. *USSR.*
Plana-Llevat, Dr Feliciano. *Spain.*
Plant, Dr John Stewart. *UK.*
Plas, van der, Prof. Dr Leendert. *Netherlands.*
Plastinina, Dr Marina Arkad'evna. *USSR.*
Platikanova, Dr Vesselina. *Bulgaria.*
Platonov, Prof Aleksei Nikolaevich. *USSR.*
Platts, Mr Simon Nicholas. *Australia.*
Plesken, Prof. Wilhelm. *BRD.*
Pleštil, Ing Josef. *CSSR.*
Pletcher, Dr. James F. *USA.*
Pletnev, Dr Vladimir Zakharovich. *USSR.*
Plies, Dr Volker. *BRD.*
Ploc, Dr Robert Allen. *Canada.*
Ploog, Dr Klaus. *BRD.*
Plough-Sørensen, Mrs Gudrun. *Denmark.*
Plust, Dr Heinz-Günther. *BRD.*
Pluth, Dr. Joseph John. *USA.*
Plyasova, Dr Lyudmila Mikhailovna. *USSR.*
Plyusnina, Dr Inga Ivanovna. *USSR.*
Pobedimskaya, Dr Elena Alexandrovna. *USSR.*
Pocev, Prof. Dr Stefan. *Yugoslavia.*
Pochetti, Dr Giorgio. *Italy.*
Pochettino, Dr Alberto Antonio. *Argentina.*
Podberezskaya, Dr Nina Vasilyevna. *USSR.*
Podbrdský, Dr Josef. *CSSR.*
Podder, Dr (Mrs) Aloka. *India.*
Podlahová, Dr Jana. *CSSR.*
Poe, Dr. Martin. *USA.*
Pöllmann, Dr Herbert Josef. *BRD.*
Poharc-Logar, Magistar Vesna. *Yugoslavia.*
Pohl, Prof. Dr Dieter. *BRD.*
Pohl, Prof. Dr Siegfried. *BRD.*

Point, Prof Jean-Jacques. *Belgium.*
Pointer, Dr David John. *UK.*
Polanc, Magistar Ivan. *Yugoslavia.*
Poland, Mr. Virgil Laverne. *USA.*
Polborn, Dr Kurt Volkmar. *BRD.*
Polcarová, Dr Milena. *CSSR.*
Polchovskaya, Dr Tatyana Mikhailovna. *USSR.*
Poleti, Dr Dejan. *Yugoslavia.*
Polidori, Dr Giampiero. *Italy.*
Poljak, Prof. Roberto J. *France.*
Poll, Dr Wolfgang. *BRD.*
Pollack, Dr. Sidney Solomon. *USA.*
Pollard, Dr David Ronald. *UK.*
Pollert, Ing Emil. *CSSR.*
Pollmann, Dipl-Min Siegfried. *BRD.*
Polo, Dr Adriano. *Italy.*
Poltev, Prof. Valerii Ivanovich. *USSR.*
Polvorinos, Dr Angel Jesus. *Spain.*
Polyakova, Dr Irina Nikolaevna. *USSR.*
Polyanskaya, Dr Tamara Mikhailovna. *USSR.*
Polyansky, Dr Evgeny Vasilyevich. *USSR.*
Polychroniades, Prof. Efstathios. *Greece.*
Polynova, Dr Tamara Nikitichna. *USSR.*
Pomés Hernandez, Prof.Dr.Sc. Ramón. *Cuba.*
Pompa, Dr Francesco. *Italy.*
Pon, Mr. George W. *USA.*
Pond, Dr Robert Charles. *UK.*
Pongratz, Dr Dipl.-Ing. Peter. *Austria.*
Pongsapich, Dr Wasant. *Thailand.*
Ponomarev, Dr Vasily Ivanovich. *USSR.*
Pontchour, Miss Cha-on. *Thailand.*
Pontenagel, Dr Wilbert M. G. F. *Netherlands.*
Ponyatovsky, Prof. Evgeny Genrikhovich. *USSR.*
Poojary, Dr M. Damodara *India.*
Poojary, Dr. Maradamoole Damodara. *USA.*
Popelier, Mr Paul Lode Albert. *Belgium.*
Popescu, Conf. Ion. *Romania.*
Popik, Dr Mikhail Vasil'evich. *USSR.*
Popma, Prof. Dr Theo Johan August. *Netherlands.*
Popolitov, Dr Vladislav Ivanovich. *USSR.*
Popov, Dr Aleksandr Nikolaevich. *USSR.*
Popov, Dr Alexander. *Bulgaria.*
Popova, Dr Anastasiya Arsentyevna. *USSR.*
Popović, Prof. Dr Stanko. *Yugoslavia.*
Poppi, Prof. Luciano. *Italy.*
Poppleton, Dr Bruce J. *Australia.*
Porai-Koshits, Prof. Evgeny Alexandrovich. *USSR.*
Porai-Koshits, Prof. Mikhail Alexandrovich. *USSR.*
Porta, Prof. Piero. *Italy.*
Portalone, Dr Gustavo. *Italy.*
Porter, Dr. Leigh Christopher. *USA.*
Portnov, Dr Vadim Nikolayevich. *USSR.*
Portugal, Mr Remberto. *Bolivia.*
Porzio, Dr William. *Italy.*
Posada G., Prof. Enrique. *Colombia.*
Pósfai, Mr Mihály. *Hungary.*
Posner, Prof. Aaron Sidney. *USA.*
Post, Prof. Benjamin. *USA.*
Post, Mr. Jeffrey E. *USA.*
Post, Dr Michael Leonard. *Canada.*
Postma, Dr Johannes Petrus Maria. *BRD.*
Potekhin, Dr Konstantin Al'bertovich. *USSR.*
Potenza, Prof. Joseph Anthony. *USA.*
Potter, Dr Reginald. *UK.*
Potter, Mr. Stephen Anthony. *USA.*
Pouget, Dr. *France.*
Poulieff, Mr Christo. *Bulgaria.*
Poulin, Mrs Suzie. *Canada.*
Poulos, Mr. Thomas L. *USA.*
Pourghazi, Prof Azam. *Iran.*
Povey, Dr David Christopher. *UK.*
Powell, Dr Brian Mathieson. *Canada.*
Powell, Dr. Douglas R. *USA.*
Powell, Prof. Herbert Marcus. *UK.*
Pradhan, Dr Dukhabandhu. *India.*
Prager, Dr Peter Robert. *Australia.*
Pramatus, Miss Supanich. *Thailand.*
Prandl, Prof. Dr Wolfram. *BRD.*
Prasad, Dr Lata. *Canada.*
Prasad, Dr Narayan, *India.*

Prasad, Dr Ravindra. *India.*
Prasad, Dr Satya Murti. *India.*
Prasad, Dr Y. R. Ananth. *India.*
Precigoux, Dr Gilles. *France.*
Predecki, Mr. Paul K. *USA.*
Preisinger, Prof. Dr Anton. *Austria.*
Preiss, Dr Henry, *DDR.*
Preistnall, Mr. Ian. *Netherlands.*
Prelesnik, Dr Bogdan. *Yugoslavia.*
Prendergast, Prof. Franklyn G. *USA.*
Press, Prof. Dr Werner. *BRD.*
Pressprich, Mr. Mark R. *USA.*
Preston, Ms. Kimberly. *USA.*
Pretorius, Dr Jan Andries. *South Africa.*
Preuss, Dr Heinz. *DDR.*
Preut, Dr Hans. *BRD.*
Prevarsky, Dr Anatoly Petrovich. *USSR.*
Prevey, Mr. Paul S. *USA.*
Prewitt, Dr. Charles Thompson. *USA.*
Price, Dr Geoffrey David. *UK.*
Price, Ms. Rebecca Alexis. *USA.*
Prick, Dr Petrus Antonius Johannes. *Netherlands.*
Priestle, Dr John P. *Switzerland.*
Prieto, Dr Manuel. *Spain.*
Priftis, Prof. George. *Greece.*
Prikhod'ko, Dr Leonid Vasilyevich. *USSR.*
Primot, Mr Jacques. *France.*
Prince, Dr. Edward. *USA.*
Pring, Dr Allan. *Australia.*
Pringle, Mr Gordon James. *Canada.*
Pritzkow, Dr Wolfgang. *DDR.*
Privé, Dr. Gilbert Gérard. *USA.*
Prodan, Dr Albert. *Yugoslavia.*
Profeta, Dr. Salvatore Jr. *USA.*
Profi, Mrs Stella. *Greece.*
Protas, Prof. Jean. *France.*
Prout, Dr Charles Keith. *UK.*
Provotorov, Dr Mikhail Viktorovich. *USSR.*
Ptitsyn, Prof. Oleg Borisovich. *USSR.*
Pudovkina, Dr Zoya Vasilyevna. *USSR.*
Puff, Prof. Dr Heinrich. *BRD.*
Puff, Mr Manfred. *DDR.*
Pugachev, Prof. Anatoly Tarasovich. *USSR.*
Puigjaner, Prof. Luis. *Spain.*
Puliti, Dr Raffaella. *Italy.*
Pulliam, Dr. Curtis R. *USA.*
Punge-Witteler, Mrs Dr Barbara. *BRD.*
Punin, Dr Yury Olegovich. *USSR.*
Punkkinen, Dr Matti. *Finland.*
Punte, Dr Graciela. *Argentina.*
Punzi, Dr. John Stephen. *USA.*
Puranik, Dr (Mrs) Vedavati Gururaj. *India.*
Purdy, Mr. Samuel M. *USA.*
Pushcharovsky, Dr Dmitry Yuryevich. *USSR.*
Putnis, Dr Andrew. *UK.*
Puttajakr, Mrs Taswal. *Thailand.*
Puttick, Prof. Keith Ernest. *UK.*
Puxley, Dr David Charles. *UK.*
Pyatenko, Dr Yury Andreyevich. *USSR.*
Pyrros, Dr. Nikos P. *USA.*
Pyykkö, Prof. Veli Pekka. *Finland.*
Qaiser, Mr Mohammad Ali. *Pakistan.*
Qi, Dr Zeng-du. *China.*
Qi, Prof. Zhi-ru. *China.*
Qian, Mrs Min-xie. *China.*
Quader, Prof. Dr Mohammed Abdul. *Bangladesh.*
Quadrado, Prof. Ricardo. *Portugal.*
Quagliata, Prof. Claudio. *Italy.*
Quail, Dr J. Wilson. *Canada.*
Quang Han Khang, Prof. *Vietnam.*
Quaranta Cabral, Dr Ubirajara. *Brasil.*
Quartieri, Dr Simona. *Italy.*
Queiroz do Amaral, Dr Lia. *Brasil.*
Queisser, Prof. Dr Hans Joachim. *BRD.*
Quéré, Prof. Yves. *France.*
Quevedo, Dr Manuel M. *Colombia.*
Quibilan, Mr Edelmiro I., *Philippines.*
Quicksall, Dr. Carl O. *USA.*
Quigley, Prof. Gary Joseph. *USA.*
Quinones Mena, Dr.Sc. Jose. *Cuba.*

Quintana Owen, Mrs Patricia. *México.*
Quintana, Mr. John P. *USA.*
Quiocho, Prof. Florante A. *USA.*
Qurashi, Dr Mazhar Mahmood. *Pakistan.*
Qureshi, Mr Khalid Mahmood. *Pakistan.*
Qureshi, Mr Mohammed Kaleem Akhtar. *Pakistan.*
Raade, Mr Gunnar. *Norway.*
Rabenau, Prof. Dr Albrecht. *BRD.*
Rabenberg, Dr. Llewellyn K. *USA.*
Rabie, Dr (Mrs) Farida Hamed. *Egypt.*
Rabin, Mr Baruch. *Israel.*
Rabinovich, Prof. Dov. *Israel.*
Rabinowich, Prof. D. *UK.*
Rabinowitz, Dr. Israel Nathan. *USA.*
Rachev, Mr Peter. *Bulgaria.*
Rachinger, Prof William Albert. *Australia.*
Radaković, Magistar Aleksandra. *Yugoslavia.*
Radautsan, Prof. Sergey Ivanovich. *USSR.*
Radhakrishnan, Dr. R. *USA.*
Radivojević, Magistar Pavle. *Yugoslavia.*
Radmilović, Magistar Velimir. *Yugoslavia.*
Radnai, Dr Tamás. *Hungary.*
Radonovich, Prof. Lewis J. *USA.*
Radoslovich, Dr Edward William. *Australia.*
Radukić, Prof. Dr Gordana. *Yugoslavia.*
Radulescu, Prof. Dan. *Romania.*
Radwan, Dr Mostafa Mohsen Abdel-Razik. *Egypt.*
Rae, Dr Alan David. *Australia.*
Rae, Dr Alan William James Melville. *UK.*
Rae, Dr Alastair Ian Maxwell. *UK.*
Rafalko, Ms. Patrice White. *USA.*
Raftery, Dr James. *UK.*
Ragab, Dr Abdel Ghani. *Egypt.*
Rager, Dr Helmut. *BRD.*
Raghavacharyulu, Dr Iyyunni Venkata Veera. *India.*
Raghunatha, Chary. *India.*
Raghurama, Mr G. *India.*
Rahman, Dr Asadur. *Bangladesh.*
Rahman, Mr Mohammad Abdul. *Pakistan.*
Rahman, Dr Sheikh Mohammed M. *Bangladesh.*
Raidt, Dr Helmut. *DDR.*
Raines, Prof. Ronald T. *USA.*
Rainov, Dr Nikola. *Bulgaria.*
Raithby, Dr Paul Robert. *UK.*
Rajagopal, Mr Hariharasubramonia Iyer. *India.*
Rajan, Mr R. D. *India.*
Rajan, Dr S. S. *India.*
Rajaram, Dr Ramasamy Karunandam. *India.*
Rajaratnam, Prof. Arthur. *Singapore.*
Rajasekharan, Dr T. *Madras, 1978). Scient.*
Rajeswaran, Dr. Manju. *USA.*
Raju, Dr I. V. K. Bhagavan. *India.*
Raju, Dr K. S. *India.*
Rakin, Dr Vladimir Ivanovich. *USSR.*
Rakova, Dr Elena Vasilyevna. *USSR.*
Ralph, Prof. Brian. *UK.*
Ram Kishore, Dr. *India.*
Ram, Dr Purushottam. *India.*
Rama Rao, Dr B. *India.*
Ramachandran, Prof. Gopalasamudram N. *India.*
Ramakrishnan, Dr Chandrasekhara. *India.*
Ramakumar, Dr Suryanarayana Rao. *India.*
Ramalingam, Dr. Veerappa Pillai. *USA.*
Ramanadham, Dr Muthyala. *India.*
Ramans, Dr Guntis Mirvaldovich. *USSR.*
Ramaswamy, Dr Krishnamachari. *India.*
Ramaswamy, Mr S. *India.*
Ramdas, Dr S. *UK.*
Rammo, Dr Nabil Naim. *Iraq.*
Ramos Bernal, Dr Sergio. *México.*
Ramos Parente, Dr Carlos Benedicto. *Brasil.*
Ramos, Mr Agustin Federico. *Canada.*
Rana, Mr Riaz Ahmad. *Pakistan.*
Randaccio, Prof. Lucio. *Italy.*
Randall, Dr. Clive Alan. *USA.*
Ranganath, Dr G. S. *India.*
Ranganathan, Prof. Srinivasa. *India.*
Range, Prof. Dr Klaus-Jürgen. *BRD.*
Ranger, Dr Georges Joseph. *Canada.*
Ranta, Mr Lasse Kosti. *Finland.*

Rantsordas, Dr Shasikante. *France.*
Rao, Dr. Jejjala Krishna Mohana. *USA.*
Rao, Dr Keshavamurthy Narayana Swamy. *India.*
Rao, Mr Nutakki Nageswara. *Malaysia.*
Rao, Prof. P. Rama. *India.*
Rao, Dr. S. Narasinga. *USA.*
Rao, Dr. Sambhorao Thyagaraja. *USA.*
Rao, Mr. Siram N. V. *USA.*
Rao, Dr. Sudharkara. *USA.*
Rao, Ms. Usha. *USA.*
Rao, Mr Zi He. *Australia.*
Raoux, Dr Denis. *France.*
Rapanut, Dr Teofina Axibal. *Philippines.*
Rappaport, Prof. Harry P. *USA.*
Raptis, Dr. Raphael. *USA.*
Rardin, Mr. R. Lynn. *USA.*
Raselli, Mr Andrea-Raeto. *Switzerland.*
Rashkova, Dr Diana. *Bulgaria.*
Rashkovich, Prof. Leonid Nikolaevich. *USSR.*
Rasines, Prof. Isidoro. *Spain.*
Rasmussen, Dr. Bjarne. *USA.*
Rasmussen, Mrs Hanne. *Denmark.*
Rasmussen, Prof. Svend Erik. *Denmark.*
Rastinejad, Mr. F. *USA.*
Raston, Prof Colin Llewellyn. *Australia.*
Rastsvetaeva, Dr Ramiza Kerarovna. *USSR.*
Ratajczak-Sitarz, Mgr Małgorzata A. *Poland.*
Ratanasthien, Dr Benjavun. *Thailand.*
Rath, Mr. Nigam Prasad. *USA.*
Rath, Prof. Dr Robert. *BRD.*
Rath, Ms. Virginia. *USA.*
Ratna, Miss B. R. *India.*
Ratuszek, Dr Wiktoria Maria. *Poland.*
Ratuszna, Dr Alicja. *Poland.*
Rau, Mr. Robert C. *USA.*
Rau, Dr Tamara Fedorovna. *USSR.*
Rau, Prof. Valery Georgiyevich. *USSR.*
Raudsepp, Dr Mati. *Canada.*
Rausell-Colom, Prof. Jose Antonio. *Spain.*
Rautioaho, Dr Risto Heikki. *Finland.*
Ravaglioli, Dr Antonio. *Italy.*
Raven, Mr Mark Derek. *Australia.*
Ravichandran, Mrs V. *India.*
Ravichandran, Mr. K. G. *USA.*
Ravikumar, Dr. Krishnan. *USA.*
Ravy, Dr Sylvain. *France.*
Rawas, Dr Ahmad. *UK.*
Ray, Prof. Alden Earl. *USA.*
Ray, Dr (Mrs) Gouri. *India.*
Ray, Dr Pankaj Narayan. *India.*
Ray, Dr Pradip Kumar. *India.*
Ray, Prof. Siddhartha. *India.*
Raykhtsaum, Mr. Grigory. *USA.*
Rayment, Prof. Ivan. *USA.*
Raymond, Prof. Kenneth Norman. *USA.*
Read, Dr Randy John. *Canada.*
Rebić, Mr Milenko. *Yugoslavia.*
Rebrov, Dr Aleksandr Nikolaevich. *USSR.*
Rechenberg, Dr Ingrid. *DDR.*
Recker, Prof. Dr Kurt. *BRD.*
Reddy, Dr. B. Swamintha. *USA.*
Redhouse, Dr Alan David. *UK.*
Redler Mr László *Hungary.*
Reeber, Dr. Robert Richard. *USA.*
Reed, Dr. A. Thomas. *USA.*
Reed, Dr. Larry L. *USA.*
Reed, Dr. R. A. *USA.*
Reeder, Prof. Richard James. *USA.*
Reeke, Prof. George Norman Jr. *USA.*
Rees, Prof. Douglas Charles. *USA.*
Regel', Dr Vadim Robertovich. *USSR.*
Regueira Teodósio, Prof. Joel. *Brasil.*
Rehfeldt-Oskierski, Dr Angeline. *BRD.*
Reibenspies, Dr. Joseph. *USA.*
Reiche, Dr Manfred. *DDR.*
Reid, Dr Alan Forrest. *Australia.*
Reid, Dr. Austin Henry Jr. *USA.*
Reid, Dr John Sinclair. *UK.*
Reid, Ms. Susan Sarah. *USA.*
Reidinger, Mr. Franz. *USA.*

Reiher, Dr. Walter E. III. *USA.*
Reimers, Dr Walter. *BRD.*
Reinecke, Mr Kriemhild. *DDR.*
Reinen, Prof. Dr Dirk. *BRD.*
Reinhold, Mr Ingrid. *DDR.*
Reis, Dr. Arthur Henry Jr. *USA.*
Reisner, Dr George. *Israel.*
Reiss, Drs Céleste A. *Netherlands.*
Reissner, Dipl.-Ing. Michael, *Austria.*
Reithmayer, Mr Klaus Thomas. *BRD.*
Rek, Ms. Zofia. *USA.*
Reller, Dr Armin. *Switzerland.*
Remington, Prof. Stephen James. *USA.*
Ren, Prof. Lei-fu. *China.*
Renaud, Dr Anne. *France.*
Rendle, Dr David Forbes. *UK.*
Rendón Diaz Mirón, Dr Luis Emilio. *México.*
Rentsch, Dr Harald. *DDR.*
Rentzeperis, Prof. Panayiotis Ioannis. *Greece.*
Reppart, Dr. William James. *USA.*
Rérat, Dr Claude. *France.*
Restivo, Dr Roderic John. *Canada.*
Restori, Dr. Renzo. *USA.*
Retief, Dr Johannes Jacobus. *South Africa.*
Rettig, Dr Steven John. *Canada.*
Reuber-Kürbs, Mrs Dr Ing Ellen. *BRD.*
Reuter, Mr Dietrich. *DDR.*
Reuter, Dr Hans. *BRD.*
Revkevich, Dr Galina Panteleimonovna. *USSR.*
Rey, Dr. Felix Augusto. *USA.*
Reyes Chumacero, Mr Antonio. *México.*
Reynaers, Prof. Harry Louis. *Belgium.*
Reynhardt, Prof. Eduard Christiaan. *South Africa.*
Reynolds, Dr C.D. *UK.*
Reynolds, Dr Philip Andrew. *Australia.*
Reynolds, Dr. Ross Anthony. *USA.*
Rez, Prof. Iosif Solomonovich. *USSR.*
Rheingold, Dr. Arnold Lange. *USA.*
Rhodes, Prof. Gale. *USA.*
Rhodes, Prof. Rene George. *UK.*
Rhyne, Ms. Kay A. *USA.*
Ribár, Prof. Dr Béla. *Yugoslavia.*
Ribas, Prof. Joan. *Spain.*
Ribbe, Prof. Paul H. *USA.*
Ribbing, Dr Carl-Gustaf. *Sweden.*
Ribeiro Franco, Prof. Rui. *Brasil.*
Ricaldi, Mr Edgar. *Bolivia.*
Ricci, Prof. John S. Jr. *USA.*
Rice, Dr. Catherine Ellen. *USA.*
Rice, Dr David William. *UK.*
Rich, Prof. Alexander. *USA.*
Richard, Prof. Joseph Albert Pierre. *Canada.*
Richards, Dr Brian Peter. *UK.*
Richards, Prof. Frederic Middlebrook. *USA.*
Richards, Dr. Gerald F. *USA.*
Richards, Dr John Philip Gerald. *UK.*
Richardson, Dr. David Claude. *USA.*
Richardson, Dr. James Wyman Jr. *USA.*
Richardson, Mrs. Jane Shelby. *USA.*
Richardson, Dr. John Frederick. *USA.*
Richardson, Dr Mary Frances. *Canada.*
Richardson, Dr Robert Melville. *UK.*
Richman, Prof. Marc Herbert. *USA.*
Richter, Dr Frank. *DDR.*
Richter, Mr Hans. *DDR.*
Richter, Dr Klaus. *DDR.*
Richter, Dr Paul Wilhelm. *South Africa.*
Richter, Dr Rainer. *DDR.*
Richter, Dr Ursula. *BRD.*
Richter, Dr Waltraut. *DDR.*
Rickard, Dr Clifton Edward Frank. *New Zealand.*
Rickards, Mr A.L. *UK.*
Rickert, Prof. Dr Hans. *BRD.*
Ridder, de, Drs Dirk. *Netherlands.*
Rieck, Prof. Dr Gerard Daniel. *Netherlands.*
Riedel, Prof. Dr Erwin. *BRD.*
Rieder, Dr Milan. *CSSR.*
Rieger, Dr Hans Wolfhart. *Switzerland.*
Riegert, Mr. Richard Paul. *USA.*
Riella, Eng. Humberto Gracher. *Brasil.*

Riera, Prof. Victor. *Spain.*
Riesen, Dr Andreas. *Switzerland.*
Riess, Prof. John Karlem. *USA.*
Rietveld, Dr Hugo Marie. *Netherlands.*
Riganti, Prof. Vincenzo. *Italy.*
Rigault de la Longrais, Prof. Germain. *Italy.*
Rigopoulos, Prof. Rigas. *Greece.*
Rigotti, Dr Graciela. *Argentina.*
Rinaldi, Prof. Romano. *Italy.*
Rinaudo, Dr Caterina. *Italy.*
Rincón Saenz, Mr Luis Felipe. *Colombia.*
Rindorf, Mrs Grethe. *Denmark.*
Ringe, Dr. Dagmar. *USA.*
Ringel, Dr Lilli. *DDR.*
Rini, Dr. James. *USA.*
Rios Jara, Dr David. *México.*
Rios Steiner, Mr. Jorge L. *USA.*
Ripamonti, Prof. Alberto. *Italy.*
Ripamonti, Dr Carlo. *Italy.*
Rischák, Mr Géza. *Hungary.*
Risler, Dr Jean-Loup. *France.*
Rissanen, Mr Kari Tapani. *Finland.*
Riste, Prof. Tormod. *Norway.*
Ritschel, Dr Manfred. *DDR.*
Rius, Dr Jordi. *Spain.*
Riva di Sanseverino, Prof. Lodovico. *Italy.*
Riva, Dr Fernando. *Italy.*
Rivera Moras, Mr Vicente. *México.*
Rivera Ocando, Prof. Angela Valentina. *Venezuela.*
Rivera, Prof. Carlos. *Chile.*
Rivero, Dr Blas Eduardo. *Argentina.*
Riveros Rotgé, Dr Héctor Gerardo. *México.*
Rizvi, Prof Adibul Hasan. *Pakistan.*
Rizvi, Dr Syed Sadrul Hassan. *Pakistan.*
Rizzoli, Dr Corrado. *Italy.*
Robbins, Mr. Carl Richard. *USA.*
Robert-Picard, Dr Marie-Claire. *France.*
Roberts, Mr Andrew Clifford. *Canada.*
Roberts, Dr Kevin John. *UK.*
Roberts, Dr. Michael Mark. *USA.*
Roberts, Dr. Sue A. *USA.*
Roberts, Dr. Victoria Anne. *USA.*
Robertson, Prof. Beverly Ellis. *Canada.*
Robertson, Dean B. Ken. *USA.*
Robertson, Dr Glen Bradley. *Australia.*
Robertson, Prof. James David. *USA.*
Robertson, Dr John Harry. *UK.*
Robertson, Prof. John Monteath, FRS. *UK.*
Robertus, Prof. Jon David. *USA.*
Robinson, Dr. Ian Keith. *USA.*
Robinson, Mr. John C. *USA.*
Robinson, Dr Ward Thomas. *New Zealand.*
Robinson, Prof. William Robert. *USA.*
Robl, Dr Christian. *BRD.*
Rochon, Prof. Fernande D. *Canada.*
Rocophyllou Agathonikou, Dr Elsa - Helena. *Greece.*
Roderick, Dr. Steven L. *USA.*
Rodgers, Prof. Allen Lawrence. *South Africa.*
Rodgers, Dr. David William. *USA.*
Rodgers, Dr Kerry Anthony. *New Zealand.*
Rodrigues da Silva, Dr Rilson. *Brasil.*
Rodrigues, Dr Antonio Ricardo Dröher, *Brasil.*
Rodrigues, Dr Edson. *Brasil.*
Rodriguez Castellanos, Dr. Carlos. *Cuba.*
Rodríguez Lara, Prof. Jaime. *Colombia.*
Rodríguez S., Miss Gloria Inés. *Colombia.*
Rodriguez-Carvajal, Dr Juan. *Spain.*
Rodriguez-Clemente, Dr Rafael. *Spain.*
Rodriguez-Gallego, Prof. Manuel. *Spain.*
Rodriguez-Gordillo, Dr Jose. *Spain.*
Rodriguez-Roldan, Dr Ana. *Spain.*
Rodulfo de Gil, Prof. Eldrys Emilia. *Venezuela.*
Roe, Dr. Alfred Lawrence. *USA.*
Roebuck, Dr Peter Hamish Athey. *UK.*
Röller, Mr Klaus. *BRD.*
Römming, Prof. Christian. *Norway.*
Rønsbo, Mr Jørn. *Denmark.*
Rösch, Dr Heinrich. *BRD.*
Röst, Mr Erling. *Norway.*
Roestel, Dr Reinhard. *DDR.*

Roettgers, Mr. Wolbert. *USA.*
Rogacheva, Dr Evelina Danilovna. *USSR.*
Rogers, Prof. Donald. *UK.*
Rogers, Dr K.D. *UK.*
Rogers, Prof. Robin Don. *USA.*
Rogić, Prof. Dr Vinko. *Yugoslavia.*
Roginskaya, Dr Yuliana Eremeyevna. *USSR.*
Rogl, Prof. Dr Peter Franz. *Austria.*
Rognlie, Mr. David G. *USA.*
Rogulić, Prof. Dr Mileva. *Yugoslavia.*
Rohmer, Mr Christian. *BRD.*
Rohrbaugh, Dr. Wayne Joseph. *USA.*
Rohrer, Dr. Douglas C. *USA.*
Roilos, Prof. Minas. *Greece.*
Rojo, Dr Teofilo. *Spain.*
Roldan, Dr. Luis-Gonzalez. *USA.*
Rolim De Camargo, Prof. William Gerson. *Brasil.*
Rolland, Mr Guy. *France.*
Rollett, Dr John Sydney. *UK.*
Romaka (Komarovskaya), Dr Lyubov' P. *USSR.*
Roman, Dr Pascual. *Spain.*
Romanov, Dr Gennady Vasilyevich. *USSR.*
Romero Romo, Dr Mario. *México.*
Romero-Molina, Dr M. Angustias. *Spain.*
Romero, Dr Antonio. *Spain.*
Romero, Dr Miguel. *México.*
Romers, Prof. Dr Cornelis. *Netherlands.*
Roof, Dr. Raymond Bradley. *USA.*
Rories, Dr. Charles C. P. *USA.*
Ros, Dr Josep. *Spain.*
Rosa, Dr Rodolfo. *Italy.*
Rosario, Mr. Nery Rivera. *USA.*
Rosca, Mr Liviu. *Romania.*
Rose-Hansen, Dr John. *Denmark.*
Rose, Dr David Richard. *Canada.*
Rose, Prof. George D. *USA.*
Rose, Mr. Jon Patrick. *USA.*
Rosenberg, Prof. John M. *USA.*
Rosenqvist, Prof. Ivan Thoralf. *Norway.*
Rosenstein, Prof. Robert Daniel. *USA.*
Rosin, Dr Horst. *DDR.*
Ross, Ms. Dawn L. *USA.*
Ross, Dr Donald Keith. *UK.*
Ross, Dr. Frederick Keith. *USA.*
Ross, Dr. Nancy L. *UK.*
Rossell, Dr Henry John. *Australia.*
Rossi, Mr Franco Antonio. *Switzerland.*
Rossi, Dr Marco. *Italy.*
Rossi, Dr. Miriam. *USA.*
Rossmanith, Mrs. Dr Elisabeth. *BRD.*
Rossmann, Prof. Michael G. *USA.*
Rossner, Dr Johannes. *DDR.*
Rossouw, Dr Christopher John. *Australia.*
Rost, Mr Jutta. *DDR.*
Rostami, Dr Farzaneh. *Iran.*
Rotella, Dr. Frank J. *USA.*
Roth, Dr Michel. *France.*
Roth, Dr. Robert Sidney. *USA.*
Roth, Prof. Walter Lester. *USA.*
Rothammel, Mr Walter. *BRD.*
Rothbauer, Dr Richard. *BRD.*
Rott, Dr Volkwin. *BRD.*
Rouffignac, Dr Eric de. *México.*
Roult, Mr Georges. *France.*
Rousseaux, Dr Françoise. *France.*
Rout, Ms Joanne Elizabeth. *UK.*
Roveri, Prof. Norberto. *Italy.*
Rowley, Dr Colin Raymond. *UK.*
Roy, Dr Ajoy Kumer. *Bangladesh.*
Roy, Prof. Rustum. *USA.*
Roychowdhury, Dr Priyobroto. *India.*
Roychowdhury, Dr (Mrs) S. *Lect.*
Royer, Dr. William Edward, Jr. *USA.*
Rozenberg, Dr Yurii Aleksandrovich. *USSR.*
Rozhansky, Dr Vladimir Nikolayevich. *USSR.*
Rozhdestvenskaya, Dr Ira Vasilyevna. *USSR.*
Rozsondai, Dr Béla. *Hungary.*
Rozwarski, Ms. Denise Andrea. *USA.*
Ruban, Prof. Dr Gerhard. *BRD.*
Rubbo, Prof. Marco. *Italy.*

Ruben, Mrs. Helena W. *USA.*
Rubiano Lamouroux, Prof. Manuel. *Colombia.*
Rubin, Dr. Ben Z. *USA.*
Rubin, Dr. Byron. *USA.*
Rubin, Dr. John Ronald. *USA.*
Rubina, Dr Elena Borisovna. *USSR.*
Rubinina, Dr Nataliya Mikhailovna. *USSR.*
Rubio de Cubides, Prof. Julia. *Colombia.*
Ruble, Dr. John Rollo. *USA.*
Rudenko, Mr Sergei Sergeevich. *USSR.*
Ruderman, Dr. I. Warren. *USA.*
Rudert, Mr Rainer. *BRD.*
Rudman, Prof. Reuben M. *USA.*
Rudnick, Dr. Suzanne Ellen. *USA.*
Rudnik, Mr. Paul J. *USA.*
Rudolf, Dr. Philip Reinhold. *USA.*
Rudolph, Prof Peter. *DDR.*
Rüegg, Dr Andreas. *Switzerland.*
Ruiz Mejia, Dr Carlos. *México.*
Ruiz Perez, Mrs Catalina. *BRD.*
Ruiz-Perez, Dr Catalina. *Spain.*
Ruiz-Valero, Dr Caridad. *Spain.*
Ruiz, Mr. Bienvenido. *USA.*
Ruiz, Mr Xavier. *Spain.*
Rukvichal, Mr Surapol. *Thailand.*
Rule, Mr Stephen A. *UK.*
Rumanova, Dr Iskra Mikhailovna. *USSR.*
Rumball, Dr Sylvia Vine. *New Zealand.*
Rundqvist, Prof. Stig O. *Sweden.*
Runje, Dr Vesna. *Yugoslavia.*
Ruppersberg, Prof. Dr Henner. *BRD.*
Rusanovskii, Dr Mikhail Evstaf'evich. *USSR.*
Ruscher, Prof. Christian. *DDR.*
Russell, Dr David Robin. *UK.*
Russell, Mr Nazirullah. *Pakistan.*
Russell, Dr. Thomas Paul. *USA.*
Russev, Dr Krassimir. *Bulgaria.*
Russo, Dr Galina Vladimirovna. *USSR.*
Ruszala, Mr. Ferdinand A. *USA.*
Rutherford, Prof. John Stewart. *South Africa.*
Rutten-Keulemans, Drs E. W. M.. *Netherlands.*
Ruud, Prof. Clayton Olaf. *USA.*
Ruvimov, Dr Segei Sergeevich. *USSR.*
Ružić-Toroš, Dr Živa. *Yugoslavia.*
Ryabchenkov, Dr Vladimir Vasil'evich. *USSR.*
Ryabchikov, Dr Sergei Aleksandrovich. *USSR.*
Ryaboshapka, Dr Karl Petrovich. *USSR.*
Ryadnov, Dr Sergei Nikolaevich. *USSR.*
Ryan, Prof. Clarence A. *USA.*
Ryan, Dr. Robert Reynolds. *USA.*
Ryba, Prof. Earle Richard. *USA.*
Rybakov, Dr Victor Borisovich. *USSR.*
Rychlewska, Dr Urszula. *Poland.*
Rychlý, Ing Rudolf. *CSSR.*
Rydel, Mr. Timothy John. *USA.*
Rypniewski, Dr. Wojciech Robert. *USA.*
Ryvkin, Dr Viktor Adol'phovich. *USSR.*
Ryzhenkov, Dr Alexander Pavlovich. *USSR.*
Rzaigui, Dr Mohamed. *Tunisia.*
Saadinam, Dr Abolfazl. *Iran.*
Saalfeld, Prof. Dr Horst. *BRD.*
Saavedra, Mr Antonio. *Bolivia.*
Sabat, Dr. Michael. *USA.*
Sabat, Dr Michal. *Italy.*
Sabatino, Dr Piera. *Italy.*
Sabbioni, Dr Cristina. *Italy.*
Sabelli, Dr Cesare. *Italy.*
Sabine, Prof Terence Murray. *Australia.*
Sabirov, Dr Vakhobzhon Khusanovich. *USSR.*
Sabrowsky, Prof. Dr Horst. *BRD.*
Sacerdoti, Prof. Michele. *Italy.*
Sack, Dr. John Stuart. *USA.*
Sadanaga, Prof. Ryoichi. *Japan.*
Sadek, Dr Gamil. *Egypt.*
Sadikov, Dr Georgiy Georgievich. *USSR.*
Sadoc, Prof. Jean-çois. *France.*
Sadova, Dr Nina Ivanovna. *USSR.*
Sadowski, Mr. Lucian M. *USA.*
Sadybakasov, Dr Bolot Kemelovich. *USSR.*
Sæthre, Mr Leif Jarle. *Norway.*

Saeed, Mr Syed Mohammad. *Pakistan.*
Sänger. Dr Annette. *BRD.*
Saenger, Prof. Dr Wolfram H. E. *BRD.*
Saf'yanov, Dr Yurii Nikolaevich. *USSR.*
Safa, Dr Mehdi. *Iran.*
Safro, Dr Mark Grigor'evich. *USSR.*
Saghir, Mr Ahmad. *Pakistan.*
Saha, Mr Ajay Prakash. *India.*
Saha, Mr Bishwa Nath. *India.*
Sahalos, Prof. John. *Greece.*
Sahani, Dr Jiwan Lal. *India.*
Sahaymary, Mrs J. James *India.*
Sahl, Prof. Dr Kurt. *BRD.*
Sahle, Dr Wubeshet. *Sweden.*
Sahm, Prof. Dr Ing. Peter Rudolf. *BRD.*
Sahu, Dr Bholanath. *India.*
Sahu, Dr Mahendra. *India.*
Sahu, Dr N. C. *India.*
Sahu, Dr Ram Gopal. *India.*
Sahu, Dr Ramdhani. *India.*
Saif, Dr Saiful-Islam. *Saudi Arabia.*
Saito, Dr Yoshio. *Japan.*
Saito, Mr Yoshiyuki. *Japan.*
Saito, Prof. Norio. *Japan.*
Saito, Prof. Yoshihiko. *Japan.*
Sajó, Mr István. *Hungary.*
Saka, Dr Takashi. *Japan.*
Sakabe, Dr. Noriyoshi. *Japan.*
Sakai, Mr. Katsura. *Japan.*
Sakaki, Prof. Yoneichiro. *Japan.*
Sakamoto, Dr Yosio. *Japan.*
Sakata, Dr Makoto. *Japan.*
Sakellaridis, Prof. Paul. *Greece.*
Sakellariou, Dr Evangelos. *Switzerland.*
Sakharov, Dr Boris Alexandrovich. *USSR.*
Sakkopoulos, Dr Sotirios. *Greece.*
Sakore, Dr. Tukaram D. *USA.*
Sakuma, Dr Takashi. *Japan.*
Sakurai, Dr Junji. *Japan.*
Sakurai, Prof. Tosio. *Japan.*
Sal'dau, Dr El'ga Petrovna. *USSR.*
Salagegheh, Mehdi. *1975).*
Salas-Aparicio, Dr Juan Manuel. *Spain.*
Salas, Dr Guillermo Armando. *México.*
Salazar Orrego, Prof. Ramon. *Peru.*
Saldin, Dr Dilano Kerzaman. *USA.*
Salem, Dr Safia Mahmoud. *Egypt.*
Salemme, Dr. Francis Raymond. *USA.*
Salgado, Dr José. *Portugal.*
Salkind, Prof. Alvin J. *USA.*
Salleh, Dr Mansor bin Haji. *Malaysia.*
Salleh, Dr Mohammad Nawi. *Malaysia.*
Salunke, Dr Dinakar M. *India.*
Salvado Canelhas, Mrs Maria da Graça. *Portugal.*
Salviati, Dr Giancarlo. *Italy.*
Samanta, Chitra. *India.*
Samantaray, Dr Biswas Kumar. *India.*
Samarskaya, Dr Valentina Dmitriyevna. *USSR.*
Samdal, Mr Svein. *Norway.*
Samoilovich, Dr Lidiya Alexandrovna. *USSR.*
Samoilovich, Prof. Mikhail Isaakovich. *USSR.*
Samotin, Dr Nikoloai Dmitrievich. *USSR.*
Sampson, Mr Christopher. *UK.*
Samson, Dr. Sten. *USA.*
Samudzi, Mr. Cleopas Tendai. *USA.*
Samuels, Prof. Robert J. *USA.*
Samuelsen, Prof. Emil J. *Norway.*
Samus', Dr Ivan Dmitriyevich. *USSR.*
Samusina, Mrs. Svetlana Nikolaevna. *USSR.*
Sanadze, Prof. Vladimir Vladimirovich. *USSR.*
Sanchez V., Mr Alfredo. *Colombia.*
Sanchez-Aparicio, Dr Purificacion. *Spain.*
Sanchez-Navas, Mr Antonio. *Spain.*
Sandalaki, Dr Zefi. *Greece.*
Sanderson, Dr Mark Rutherford. *UK.*
Sándor, Dr Endre Elek. *UK.*
Sands, Prof. Donald E. *USA.*
Sandström, Dr Magnus K. E. *Sweden.*
Sangariyavanich, Mrs Archara. *Thailand.*
Sanjeeviraja, Dr C. *India.*

Sanjinés, Mr Orlando. *Bolivia.*
Sankar, Mr B. N. *India.*
Sankaran, Ms Hema. *India.*
Sanni, Dr Bamidele. *Nigeria.*
Sannikov, Dr Daniil Grigoryevich. *USSR.*
Sansoni, Prof. Mirella. *Italy.*
Santarsiero, Dr. Bernard I. Dominic M. *Canada.*
Santini, Dr Antonello. *Italy.*
Santivañez, Mr Reynaldo. *Bolivia.*
Santos, Dr Persio de Souza. *Brasil.*
Santos, Mrs Regina Helena de Almeida. *Brasil.*
Sanz-Aparicio, Prof. Francisco. *Spain.*
Sanz-Aparicio, Dr Juliana. *Spain.*
Saper, Dr. Mark A. *USA.*
Sappenfield, Dr. Eric L. *USA.*
Saravanan, Mr R. *India.*
Sariel, Mr Joseph. *Israel.*
Sarin, Dr Victor Anatol'yevich. *USSR.*
Sarkar, Chitrita. *India.*
Sarkar, Dr Satyabrata. *India.*
Sarkisov, Dr Stepan Ervandovich. *USSR.*
Sarko, Prof. Anatole. *USA.*
Sarma, Prof. Raghupathy. *USA.*
Sarodnik, Dr Reinhard. *DDR.*
Sartori, Prof. Franco. *Italy.*
Sasada, Prof. Yoshio. *Japan.*
Sasaki, Prof. Akio. *Japan.*
Sasaki, Dr Kyoyu. *Japan.*
Sasaki, Dr Satoshi. *Japan.*
Sasaki, Prof. Yukiyoshi. *Japan.*
Sasisekharan, Prof. Viswanathan. *India.*
Sass, Prof. Stephen Louis. *USA.*
Sastry, Dr Bommakanti Sri Rama. *India.*
Sastry, Dr G. V. S. *India.*
Sastry, Dr Medury Dattatreya. *India.*
Sasvári, Dr (Ms) Judit. *Hungary.*
Sasvári, Dr Kálmán. *Hungary.*
Satittada, Miss Gannaga. *Thailand.*
Sato, Mr Hiroki. *Japan.*
Sato, Dr Mamoru. *Japan.*
Sato, Prof. Mitsuo. *Japan.*
Sato, Prof. Shin'ichi. *Japan.*
Sato, Dr Shoichi. *Japan.*
Sato, Mr Tomohiro. *Japan.*
Satow, Prof. Yoshinori. *Japan.*
Satyanarayana Murthy, Dr Keta. *India.*
Satyshur, Dr. Kenneth A. *USA.*
Satz, Mr. Ronald Wayne. *USA.*
Saur, Prof. Dr Ing Eugen. *BRD.*
Sauvage- Simkin, Dr Michèle. *France.*
Savithramma, Miss K. L. *India.*
Sawada, Prof. Akikatsu. *Japan.*
Sawada, Dr Haruo. *Japan.*
Sawada, Dr Toshiyuki. *Japan.*
Sawada, Mr Yasuaki. *Japan.*
Sawaguchi, Prof. Etsuro. *Japan.*
Sawka-Dobrowolska, Dr Wanda. *Poland.*
Sawyer, Dr Jeffrey Frederick. *Canada.*
Sawyer, Dr Lindsay. *UK.*
Sawzik, Ms. Patricia. *USA.*
Sax, Dr. Martin. *USA.*
Sayfarth, Prof Hans-Heinz. *DDR.*
Sayler, Prof. Alice Ann. *USA.*
Sayre, Dr. David. *USA.*
Sazedj-Khosrawan, Dr Feresteh. *Austria.*
Scandale, Prof. Eugenio. *Italy.*
Scapin, Dr Giovanna. *Italy.*
Scaramuzza, Dr Lucio. *Italy.*
Scarbrough, Dr. Frank Edward. *USA.*
Scaringe, Dr. Raymond P. *USA.*
Scatturin, Prof. Vladimiro. *Italy.*
Ščavničar, Prof. Dr Stjepan. *Yugoslavia.*
Schaal, Mr Joachim. *DDR.*
Schäfer, Prof. Dr Herbert Leo. *BRD.*
Schäfer, Prof. Lothar. *USA.*
Schäfer, Dr Peter. *DDR.*
Schaefer, Dr. William Palzer. *USA.*
Schäfer, Dr Wolfgang. *BRD.*
Schagen, Dr Jan-Dirk. *Netherlands.*
Schanda, Dr Friedrich. *BRD.*

Schapink, Dr Frederik Willem. *Netherlands.*
Scharfenberg, Dr Rudolf. *DDR.*
Schattschneider, Prof. Dr Peter. *Austria.*
Scheel, Prof. Hans J. *Switzerland.*
Scheetz, Dr. Barry E. *USA.*
Scheidt, Prof. W. Robert. *USA.*
Schellman, Dr. F. Charlotte. *USA.*
Schenk, Prof. Dr Hendrik. *Netherlands.*
Schenk, Dr Kurt Johann. *Switzerland.*
Schenk, Prof Manfred. *DDR.*
Schevitz, Mr. Richard. *USA.*
Schiavinato, Prof. Giuseppe. *Italy.*
Schieber, Prof. Michael. *Israel.*
Schierbeek, Dr Abraham Johan. *Netherlands.*
Schiffer, Dr. Marianne. *USA.*
Schildberg, Dr Hans Peter. *BRD.*
Schildekamp, Dr. Wilfried. *USA.*
Schilder, Ing Jaroslav. *CSSR.*
Schilling, Mr Hansjoachim. *DDR.*
Schimanski, Dr Uwe Lothar. *BRD.*
Schioler, Dr. Liselotte Jensen. *USA.*
Schippel, Dr Erhard. *DDR.*
Schirber, Dr. James Emmanuel. *USA.*
Schirmer, Dr Ulrich. *BRD.*
Schläfer, Dr Dietrich. *DDR.*
Schläfer, Dr Ursula. *DDR.*
Schlein, Dr Werner. *Chile.*
Schlemper, Prof. Elmer Otto. *USA.*
Schlenker, Prof. Michel. *France.*
Schliephake, Dr Rolf-Werner. *BRD.*
Schloemer, Prof. Dr Hermann J. *BRD.*
Schmalle, Dr Helmut Willi. *Switzerland.*
Schmelczer, Dr Robert. *Switzerland.*
Schmetzer, Dr Karl. *BRD.*
Schmid, Prof. Hans. *Switzerland.*
Schmid, Dr. Michael Francis. *USA.*
Schmidt-Nielsen, Mr Søren. *Denmark.*
Schmidt, Mr Bertram Felix Paul. *BRD.*
Schmidt, Prof. Günter. *DDR.*
Schmidt, Dr. Melvin C. *USA.*
Schmidt, Prof. Paul Woodward. *USA.*
Schmidt, Dr Peter. *DDR.*
Schmidt, Prof. Werner. *DDR.*
Schmidt, Dr. William Charles Jr. *USA.*
Schmirgeld, Dr Lelia. *Argentina.*
Schmitz, Dr Werner. *DDR.*
Schneider, Dr. Diane M. *USA.*
Schneider, Dr. Dieter K. *USA.*
Schneider, Dr Günter. *DDR.*
Schneider, Dr Gunter. *Sweden.*
Schneider, Dr Hartmut. *BRD.*
Schneider, Prof. Herbert. *BRD.*
Schneider, Dr Jochen Richard. *BRD.*
Schneider, Dr Julius. *BRD.*
Schneider, Dr Walter. *BRD.*
Schnering, von, Prof. Dr Dr hc Hans Georg. *BRD.*
Schnick, Dr Wolfgang. *BRD.*
Schobinger-Papamantellos, Dr P. *Switzerland.*
Schoch, Dr Aylva Ernest. *South Africa.*
Schöllhorn, Prof. Dr Robert. *BRD.*
Schoenborn, Dr. Benno Paul. *USA.*
Schönholzer, Mr Peter. *Switzerland.*
Schöning, Prof. Friedrich R. L. *South Africa.*
Scholz, Dr Heinz Werner. *BRD.*
Schomaker, Prof. Verner. *USA.*
Schomburg, Priv.-Doz. Dr Dietmar. *BRD.*
Schoone, Prof. Dr Jean C. *Netherlands.*
Schooneveld, van, Ing. Marinus. *Netherlands.*
Schorin, Dr Hasso. *Venezuela.*
Schott, Prof. Günter. *DDR.*
Schouten, Mr Arie. *Netherlands.*
Schrader, Ms. Patricia Ann. *USA.*
Schrader, Prof. Richard. *DDR.*
Schrag, Dr Joseph D. *Canada.*
Schramm, Dr Volker. *BRD.*
Schranz, Dr Wilfried. *Austria.*
Schrauber, Dr Hannelore. *DDR.*
Schreiter, Dr Peter. *DDR.*
Schreuder, Dr Herman Antony. *Netherlands.*
Schreurs, Drs Antonius Mathias Maria. *Netherlands.*

Schröcke, Prof. Dr Helmut. *BRD.*
Schröder, Dr Friedrich Anton. *BRD.*
Schröder, Mr Jens. *BRD.*
Schroeder, Dr. LeRoy William. *USA.*
Schröder, Dr Winfried. *DDR.*
Schröpfer, Dr Lothar Maximilian. *BRD.*
Schroll, Prof. Dr Erich. *Austria.*
Schryvers, Dr Dominique Maurits. *Belgium.*
Schubert, Dr Gernot. *DDR.*
Schubert, Mr Heike-Kristina. *DDR.*
Schubert, Mr Helmut. *BRD.*
Schubert, Prof. Dr Konrad. *BRD.*
Schülke, Prof Dr Winfried. *BRD.*
Schüller, Prof. Dr Karl-Heinz. *BRD.*
Schürmann, Dr Kay Uwe. *BRD.*
Schulte, Ms. Gayle K. *USA.*
Schultz, Dr. Arthur Jay. *USA.*
Schultz, Dr György. *Hungary.*
Schultz, Prof. Jerold Marvin. *USA.*
Schultze-Rhonhof, Dr Ernst. *BRD.*
Schulz, Dr Georg Eberhardt Bruno. *BRD.*
Schulz, Prof. Dr Heinz Hermann. *BRD.*
Schulz, Dr Manfred. *DDR.*
Schulze, Prof Dietrich. *DDR.*
Schulze, Dr Günter. *DDR.*
Schuman, Mr. Clifford Alan. *USA.*
Schumann, Dr Bernd. *DDR.*
Schunk, Dr Wolfgang. *DDR.*
Schur, Dr Karl. *BRD.*
Schuster, Prof. Dr Hans-Uwe. *BRD.*
Schuster, Prof. Dr Julius Clemens. *Austria.*
Schuster, Dr. Sanford Lee. *USA.*
Schuszter, Dr Ferenc. *Hungary.*
Schutt, Prof. Clarence Ernest. *USA.*
Schutte, Prof. Casper Jan Hendrik. *South Africa.*
Schutte, Drs Willy. *Netherlands.*
Schwabe, Prof. Dr Dietrich Gerhard. *BRD.*
Schwahn, Dr Dietmar. *BRD.*
Schwalbe, Dr Carl Hellmuth Walter. *UK.*
Schwartz, Dr. Lyle Howard. *USA.*
Schwarz, Prof. Dr Karlheinz. *Austria.*
Schwarz, Dr Wolfgang. *BRD.*
Schwarzenbach, Prof. Dieter. *Switzerland.*
Schwarzenberger, Mrs D.R. *UK.*
Schwarzmann, Mrs Dr Sigrid. *BRD.*
Schweda, Dr Eberhard. *BRD.*
Schweinsberg, Dr Heinz Friedrich. *BRD.*
Schweizer, Dr Wolfhard Bernd. *Switzerland.*
Schwomma, Dr Otto. *Austria.*
Sclar, Prof. Charles Bertram. *USA.*
Scordari, Prof. Fernando. *Italy.*
Scott, Prof. Alastair Ian. *USA.*
Scott, Mr. Brian. *USA.*
Scott, Mr. Donald Lee. *USA.*
Scott, Dr Henry Gordon. *Australia.*
Scott, Dr James Douglas. *Canada.*
Scouloudi, Dr Helen. *UK.*
Scrimgeour, Dr Sheelagh Nicoll. *UK.*
Scudder, Dr Marcia Lorraine. *Australia.*
Seabaugh, Mr. Pyrtle W. *USA.*
Seal, Dr Alpana. *India.*
Seal, Prof. Arun Kumar. *India.*
Seal, Dr Michael. *Netherlands.*
Seale, Dr. Steven Keith. *USA.*
Seaton, Prof. Barbara A. *USA.*
Sebastian, Dr Eduardo Manuel. *Spain.*
Sebastian, Dr Mailadil Thomas. *BRD.*
Šebo, Dr Pavel. *CSSR.*
Secco, Prof. Anthony Silvio. *Canada.*
Secco, Dr Luciano. *Italy.*
Seddon, Dr John Michael. *UK.*
Šedivý, Doc. Dr Josef. *CSSR.*
Sedlacek, Dr Paul. *DDR.*
Sedmalis, Prof. Uldis Yanovich. *USSR.*
Sedzik, Dr. Jan. *USA.*
Seeger, Prof. Dr Karlheinz. *Austria.*
Seeholzer, Dr. Steven H. *USA.*
Seely, Prof. Oliver Jr. *USA.*
Seeman, Prof. Nadrian Charles. *USA.*
Seemanu, Dr Hans. *DDR.*

Seetharaman, Mr Venkataramakrishanan. *India.*
Seff, Prof. Karl. *USA.*
Segal, Mr Eugen. *Romania.*
Šegedin, Dr Primož. *Yugoslavia.*
Segmüller, Dr. Armin Paul. *USA.*
Sehnke, Mr. Paul C. *USA.*
Seidel, Dr Peter. *BRD.*
Seidl, Dr Erwin. *Austria.*
Seidl, Ing Vlastimil. *CSSR.*
Seidowski, Dr Eckart. *DDR.*
Seifert, Prof. Dr Hans-Joachim. *BRD.*
Seifert, Dr Karl Josef. *Austria.*
Seifert, Prof. Dr Karl-Friedrich. *BRD.*
Seifert, Dr Wolfgang. *DDR.*
Seiler, Mr Paul. *Switzerland.*
Seitsonen, Dr Sulo Iivari. *Finland.*
Seitz, Dr. Frederick. *USA.*
Sekar, Mr K. *India.*
Sekharudu, Dr. Chandra Y. *USA.*
Seki, Prof. Syuzo. *Japan.*
Sekido, Dr Kiyotane. *Japan.*
Sekizaki, Prof. Masao. *Japan.*
Self, Dr Peter Geoffrey. *Australia.*
Selladurai, S. *India.*
Sellar, Dr Jeffrey Ronald John. *Australia.*
Semenchev', Dr Aleksandr Fedorovich. *USSR.*
Semenova, Dr Tat'yana Fedorovna. *USSR.*
Semitelou, Mrs Julia. *Greece.*
Semmingsen, Dr Dag. *Norway.*
Sen Gupta, Dr Amitava. *India.*
Sen Gupta, Prof. Pradip Kumar. *USA.*
Sen Gupta, Prof. Siba Prasad. *India.*
Sen, Dr Deb Kumar. *India.*
Sen, Miss Mina. *India.*
Sen, Dr Ranjit Kumar. *India.*
Sen, Dr (Mrs) Suchitra. *India.*
Senechal, Prof. Marjorie Lee. *USA.*
Sennett, Ms. Judith Brodkin. *USA.*
Sequeira, Dr Anisbert Stanislaus. *India.*
Serafin, Dr Michael. *BRD.*
Serda, Mr Paweł Poland. *Poland.*
Serdyuk, Dr Igor' Nikolayevich. *USSR.*
Serebryanaya, Dr Nadezhda Ruvimovna. *USSR.*
Sereda, Dr Sergei Vladimirovich. *USSR.*
Sergeev, Dr Yurii Vladimirovich. *USSR.*
Sergienko, Dr Vladimir Semenovich. *USSR.*
Serimaa, Mrs Ritva Elina. *Finland.*
Serra Jones, Mr Alberto. *Cuba.*
Serra, Mr. Michael A. *USA.*
Sersale, Prof. Riccardo. *Italy.*
Servidori, Dr Marco. *Italy.*
Servos, Prof. Kurt. *USA.*
Sevast'yanov, Dr Boris Konstantinovich. *USSR.*
Sewell, Dr Bryan Trevor. *South Africa.*
Sfez, Dr Gérard. *France.*
Sgarabotto, Prof. Paolo. *Italy.*
Sgarlata, Prof. Francesco. *Italy.*
Sgualdino, Dr Giulio. *Italy.*
Shaanan, Dr Boaz. *Israel.*
Shackleton, Miss Judith Mary. *UK.*
Shafer, Ms. Donna. *USA.*
Shaffer, Prof. Lawrence B. *USA.*
Shaffner, Dr. Thomas Jackson. *USA.*
Shaflee, Dr. Fathieh. *USA.*
Shafizade, Prof. Rafik Bekhbud ogly. *USSR.*
Shafranovsky, Prof. Ilarion Ilarionovich. *USSR.*
Shah, Dr Jitendra Shantilal. *UK.*
Shah, Prof Muzaffar Ali. *Pakistan.*
Shahi, Mr Ghulam Nabi. *Pakistan.*
Shaikh, Mr Mohammad Iqbal. *Pakistan.*
Shaikh, Mr Mohammad Sualehin. *Pakistan.*
Shaikh, Mr Qameruddin. *Pakistan.*
Shaked, Prof. Hagai. *Israel.*
Shakked, Prof. Zippora. *Israel.*
Shaldin, Dr Yury Vitalyevich. *USSR.*
Shamburov, Dr Vladimir Alekseyevich. *USSR.*
Shamir, Dr Noah. *Israel.*
Shamray, Dr Vladimir Fedorovich. *USSR.*
Shan, Mr Try-seo. *China.*
Shang, Mr Mao-yu. *China.*

Shannon, Dr. Robert Day. *USA.*
Shao, Prof. Jie-lian. *China.*
Shao, Prof. Mei-cheng. *China.*
Sharipov, Dr Khasan Turabochiv. *USSR.*
Sharma, Dr (Mrs) Aysel. *Nigeria.*
Sharma, Prof. Brahama Datta. *USA.*
Sharma, Mr Braj Bhushan. *India.*
Sharma, Dr Surinder Dutt. *India.*
Sharma, Dr Vinod Chander. *Nigeria.*
Sharp, Prof. David William Arthur. *UK.*
Sharp, Prof. Paul R. *USA.*
Sharpe, Miss Andrea Jane. *UK.*
Sharrah, Prof. Paul Chester. *USA.*
Shashidhar, Dr R. *India.*
Shashkin, Dr Dmitry Petrovich. *USSR.*
Shaw (née Gözen), Dr Leylâ Süheylâ. *UK.*
Shaw, Mr Andrew. *UK.*
Shaw, Dr Martin Philip. *South Africa.*
Shchedrin, Dr Boris Mikhailovich. *USSR.*
Shchepitil'nikov. Dr Boris Vladimirovich. *USSR.*
Shcherbakova, Dr Mira Yakovlevna. *USSR.*
Shea, Dr. Madeline A. *USA.*
Shea, Mr. Michael E. *USA.*
Shebanov, Dr Leonid Anatol'evich. *USSR.*
Shefter, Dr. Eli. *USA.*
Shekhtman, Dr Veniamin Sholomovich. *USSR.*
Sheldon, Dr. Robert Isaly. *USA.*
Sheldrick, Dr Bernard. *UK.*
Sheldrick, Prof. Dr George Michael. *BRD.*
Sheldrick, Prof Dr William Stephen. *BRD.*
Shelley, Dr David. *New Zealand.*
Shen, Mr Cheng. *China.*
Shen, Miss Fu-ling. *China.*
Shen, Prof. Jin-chuan. *China.*
Shen, Dr. Ming-Shing. *USA.*
Shen, Prof. Pooyan. *Taiwan.*
Shen, Dr. Qun. *USA.*
Shenkin, Prof. Peter S. *USA.*
Shepelev, Dr Yury Fedorovich. *USSR.*
Sherfad. Mr Mohamed E. . *Libya.*
Sheridan, Mr. Robert E. *USA.*
Sheriff, Dr. Steven. *USA.*
Sherry, Mrs. Elizabeth Ann Gebert. *USA.*
Sherwood, Prof. John Neil. *UK.*
Shi, Mr Bi-de. *China.*
Shi, Mr Da-shuang. *China.*
Shi, Mr Ni-cheng. *China.*
Shibaeva, Dr Rimma Pavlovna. *USSR.*
Shibata, Prof. Noboru. *Japan.*
Shibata, Prof. Shuzo. *Japan.*
Shibuya, Prof. Iwao. *Japan.*
Shichiri, Dr Takaki. *Japan.*
Shieh, Dr. Huey-Sheng. *USA.*
Shigenari, Prof. Takeshi. *Japan.*
Shihub, Mr Salahedin I. *Libya.*
Shimanouchi, Dr Hirotaka. *Japan.*
Shimaoka, Prof. Kohji. *Japan.*
Shimazu, Dr Masaji. *Japan.*
Shimizu, Prof. Ken'ichi. *Japan.*
Shimoi, Dr Mamoru. *Japan.*
Shimura, Mr Fumio. *Japan.*
Shin, Prof. Hyun So. *Korea.*
Shin, Prof. Dr Whanchul. *Korea.*
Shinn, Dr. Dennis Burton. *USA.*
Shinnaka, Prof. Yasuhiro. *Japan.*
Shintani, Dr Ryuichi. *Japan.*
Shiojiri, Prof. Makoto. *Japan.*
Shiono, Prof. Ryonosuke. *USA.*
Shipley, Prof. G. Graham. *USA.*
Shiro, Dr Motoo. *Japan.*
Shishova, Dr Tatyana Gennadiyevna. *USSR.*
Shiu, Prof. Kom-Bei. *Taiwan.*
Shivaprakash, Dr N. C. *India.*
Shivrin, Dr Oleg Mikolayevich. *USSR.*
Shklover, Dr Valery Efimovich. *USSR.*
Shkol'nikova, Dr Larisa Mikhailovna. *USSR.*
Shlenskii, Dr Aleksei Leonidovich. *USSR.*
Shmueli, Prof. Uri. *Israel.*
Shmyt'ko, Dr Ivan Mikhailovich. *USSR.*
Shnulin, Dr Anatoly Nikolayevich. *USSR.*

Shoda, Prof. Tokugoro. *Japan.*
Shoemaker, Prof. Clara Brink. *USA.*
Shoemaker, Prof. David Powell. *USA.*
Shoham, Dr Gil. *Israel.*
Shoham, Dr Menachem. *Israel.*
Shoham, Mr. Menachem. *USA.*
Shoja, Dr. Massud. *USA.*
Shomaly, Mr. Walid. *USA.*
Shon, Mr. Ki-Joon. *USA.*
Short, Dr. Michael Arthur. *USA.*
Short, Dr. Michael R. *USA.*
Shoukri, Dr Nasri M. *Egypt.*
Shpigler, Mr Bilu. *Israel.*
Shrivastava, Prof. Hari Narayan. *India.*
Shso, Prof. Wei. *China.*
Shternberg, Dr Aleksey Alexandrovich. *USSR.*
Shu, Mr Jin-fu. *China.*
Shuja, Mr Tauqir Ahmad. *Pakistan.*
Shul'pina, Dr Iren Leonidovna. *USSR.*
Shulakov, Dr Evgeniy Vladimirovich. *USSR.*
Shull, Prof. Clifford G. *USA.*
Shumyatskaya, Dr Ninel' Grigoryevna. *USSR.*
Shustov, Dr Alexander Vsevolodovich. *USSR.*
Shuvalov, Dr Aleksandr L'vovich. *USSR.*
Shuvalov, Prof. Lev Alexandrovich. *USSR.*
Shvelashvili, Prof. Arsen Eristovich. *USSR.*
Sianou, Miss Anna. *Greece.*
Siapkas, Prof. Demetrios John. *Greece.*
Sica, Dr Filomena. *Italy.*
Sichevich, Dr Olga Mikhailovna. *USSR.*
Šíchová, Dr Hana. *CSSR.*
Siddiqui, Mr Jawed Ahmad. *Pakistan.*
Siddiqui, Dr Rafiq Ahmad. *Pakistan.*
Sidorenko, Dr Galina Alexandrovna. *USSR.*
Siebels, Mr Hansjörg. *BRD.*
Sieber, Mr Norbert Hermann Wilhelm. *BRD.*
Siegel, Dr. Lester Aaron. *USA.*
Siegel, Dr. Stanley. *USA.*
Sieger, Mr Peter. *BRD.*
Siegrist, Dr. Theo. *USA.*
Sieker, Dr. Larry C. *USA.*
Sielecki, Dr Anita R. *Canada.*
Sieler, Dr Joachim. *DDR.*
Sievers, Dr Rolf. *BRD.*
Sigalovsky, Dr. Julia I. *USA.*
Sigayev, Dr Vladimir Nikolayevich. *USSR.*
Sigler, Prof. Paul Benjamin. *USA.*
Sigvaldason, Dr Gudmundur. *Iceland.*
Siivola, Prof. Jaakko Uolevi. *Finland.*
Sijarić, Dr (Mrs) Galiba. *Yugoslavia.*
Sikirica, Prof. Dr Milan. *Yugoslavia.*
Sikka, Dr Satinder Kumar. *India.*
Sikora, Dr Wiesława. *Poland.*
Sil'vestrova, Dr Iraida Mikhailovna. *USSR.*
Silcox, Prof. John. *USA.*
Silin', Dr Elga Yanovna. *USSR.*
Sills, Mr. Denis. *USA.*
Silong, Dr Sidik bin. *Malaysia.*
Silskulsuk, Mr Buncha. *Thailand.*
Silva, Dr Abelardo M. *Argentina.*
Silva, Dr Elisa. *Chile.*
Silver, Dr Jack. *UK.*
Silverman, Ms. Sheryl. *USA.*
Silverton, Ms. Enid. *USA.*
Silverton, Dr. James V. *USA.*
Sim, Prof. George Andrew. *UK.*
Simard, Mr Michel. *Canada.*
Simard, Dr. Roger Gerard. *USA.*
Sime, Prof. Rodney J. *USA.*
Sime, Dr. Ruth Lewin. *USA.*
Simerská, Dr Marie. *CSSR.*
Simić, Dr Vojislav. *Yugoslavia.*
Simmen, Mr André. *Switzerland.*
Simmins, Dr. John J. *USA.*
Simmons, Dr. Charles J. *USA.*
Simmons, Prof. Ralph O. *USA.*
Simon, Prof. Dr Arndt. *BRD.*
Simon, Dr. David Eugene. *USA.*
Simon, Dr Kálmán. *Hungary.*
Simone, Mr Carlos Alberto de. *Brasil.*

Simonov, Dr Mikhail Alexandrovich. *USSR.*
Simonov, Dr Valentin Ivanovich. *USSR.*
Simonov, Dr Yury Alexandrovich. *USSR.*
Simonsen, Mr Ole. *Denmark.*
Simonsen, Prof. Stanley Harold. *USA.*
Simov, Mr Stefan. *Bulgaria.*
Simpson, Dr. Dale R. *USA.*
Simpson, Dr. Howard D. *USA.*
Sinclair, Dr. Alison. *USA.*
Sinclair, Dr William John. *Australia.*
Singh, Dr Anil Kumar. *India.*
Singh, Dr Bachchan. *India.*
Singh, Dr Bhanu Pratap. *India.*
Singh, Dato' Prof. Dr Chatar. *Malaysia.*
Singh, Dr Govind. *India.*
Singh, Mr K. D. P. *India.*
Singh, Dr. Phirtu. *USA.*
Singh, Dr S.N. *India.*
Singh, Dr. Sarjant. *USA.*
Singh, Mr Surendra Prakash. *India.*
Singh, Prof T.P. *India.*
Singru, Prof. Ramesh Madhao. *India.*
Sinha, Dr. Sunil K. *USA.*
Sinha, Dr Umesh Chandra. *India.*
Sinn, Prof. Ekkehard. *UK.*
Sirdeshmukh, Dr Dinker. *India.*
Siripaisarnpipat, Dr Sutatip. *Thailand.*
Siripitayanon, Dr Jintana. *Thailand.*
Siriratwatanakul, Mr Narin. *Thailand.*
Sironi, Prof. Angelo. *Italy.*
Sirota, Dr Mikhail Isaakovich. *USSR.*
Sirtl, Prof. Dr Erhard. *BRD.*
Sivakumar, Dr K. *India.*
Sivonen, Mr Seppo Juhani. *Finland.*
Sivý, Dr Peter. *CSSR.*
Sizova, Dr Nataliya Leonidovna. *USSR.*
Sjöberg, Prof. Bo. *Sweden.*
Sjöberg, Mr Jörgen. *Sweden.*
Sjölin, Dr H. Lennart G. *Sweden.*
Sjövall, Mr Rune. *Sweden.*
Skakov, Prof. Yurii Aleksandrovich. *USSR.*
Skalicky, Prof. Dr Peter. *Austria.*
Skapski, Dr Andrzej Czeslaw. *UK.*
Skarnulis, Dr Anthony Jerome. *UK.*
Skarstad, Dr Paul Michael. *USA.*
Skellett, Mr C.A. *UK.*
Skelton, Dr Brian Warwick. *Australia.*
Skelton, Dr. Earl Franklin. *USA.*
Skinner, Mr. Doyle P. *USA.*
Skinner, Mr. Matthew M. *USA.*
Skjerpe, Mr Per Martin. *Norway.*
Skolozdra, Dr Roman Vladimirovich. *USSR.*
Skowerenda, Dr Jolanta. *Poland.*
Skowron, Mrs Anecita. *Canada.*
Skrzat, Dr Zofia. *Poland.*
Skrzypczak-Jankun, Dr. Ewa. *USA.*
Slade, Dr Phillip Garland. *Australia.*
Slagle, 3831 Riverside Dr., *USA.*
Slaughter, Prof. Maynard. *USA.*
Ślebarski, Dr Andrzej. *Poland.*
Sleight, Prof. Arthur William. *USA.*
Ślepowroński, Mr Marek. *Poland.*
Sletten, Dr Einar. *Norway.*
Sletten, Prof. Jorunn. *Norway.*
Sliva, Mr. Paul. *USA.*
Śliwiński, Mr Jan. *Poland.*
Šljukić, Prof. Dr Momčilo. *Yugoslavia.*
Sloan, Dr. Gilbert J. *USA.*
Slovenec, Dr Dragutin. *Yugoslavia.*
Slovokhotov, Dr Yurii Leonidovich. *USSR.*
Sluis, van der, Drs Paul. *Netherlands.*
Sly, Prof. William Glenn. *USA.*
Smaalen, van, Dr Sander. *Netherlands.*
Smale, Mr David. *New Zealand.*
Small, Dr Ronald W.H. *UK.*
Smalley, Dr Ian James. *New Zealand.*
Smart, Dr. Bruce E. *USA.*
Smart, Dr Lesley Elizabeth. *UK.*
Smeets, Drs Wilberthus J. J. *Netherlands.*
Smelik, Mr. Eugene. *USA.*

Smetannikova, Dr Olga Gennadiyevna. *USSR.*
Smirnov, Dr Aleksei Evgenievich. *USSR.*
Smirnov, Prof. Yury Mstislavovich. *USSR.*
Smirnova, Dr Nina Lvovna. *USSR.*
Šmit, Dr Ivan. *Yugoslavia.*
Smit, Dr Jan Derk Geert. *Switzerland.*
Smit, Dr Paul H. *Netherlands.*
Smith, Mr. Albert Edward. *USA.*
Smith, Mr. Allan L. *USA.*
Smith, Dr Arnold John. *UK.*
Smith, Dr Bryan Edward. *UK.*
Smith, Dr. Craig Daniel. *USA.*
Smith, Prof. David John. *USA.*
Smith, Prof. Deane Kingsley Jr. *USA.*
Smith, Dr. Dennis H. *USA.*
Smith, Dr. Douglas Lee. *USA.*
Smith, Dr. Francine. *USA.*
Smith, Mr Gallienus William. *UK.*
Smith, Dr. George David. *USA.*
Smith, Dr. Gordon Stuart. *USA.*
Smith, Dr. Graham Monro. *USA.*
Smith, Dr. Harold Glenn. *USA.*
Smith, Prof. Janet Louise. *USA.*
Smith, Prof. John Francis. *USA.*
Smith, Mr John Michael Andrew. *UK.*
Smith, Prof. Joseph Victor. *USA.*
Smith, Dr Katherine Leah. *Australia.*
Smith, Prof. Phillip Ross. *USA.*
Smith, Dr Robert Carr. *UK.*
Smith, Mr. Robert David. *USA.*
Smith, Dr Ross McDowall. *Australia.*
Smith, Dr Sidney Herbert. *UK.*
Smith, Dr. Steven Sidney. *USA.*
Smith, Prof Thomas Frederick. *Australia.*
Smith, Prof. Vedene H., Jr. *Canada.*
Smith, Dr. Ward Whitlock. *USA.*
Smits, Mr Johannes Martinus Maria. *Netherlands.*
Smolander, Dr Kari Juhani. *Finland.*
Smolander, Dr Kimmo Juhani Nils-Eric. *Finland.*
Smolin, Dr Yury Ivanovich. *USSR.*
Smoluchowski, Prof. Roman. *USA.*
Smoorenburg, Mr Henricus C. A. M. *Netherlands.*
Smotrakov, Dr Valery Georgiyevich. *USSR.*
Smout, Ing. *Netherlands.*
Snow, Dr. Mark Edward. *USA.*
Snow, Dr Michael Robert. *Australia.*
Snyder, Prof. Robert L. *USA.*
Soa, Dr Ernst-Adolf. *DDR.*
Soar, Mr Martin. *UK.*
Sobczak, Prof. Dr Rudolf Josef. *Austria.*
Sobolev, Dr Boris Pavlovich. *USSR.*
Sobolev, Dr Chingis Sergeyevich. *USSR.*
Soboleva, Dr Lidiya Victorovna. *USSR.*
Soboleva, Dr Svetlana Vsevolodovna. *USSR.*
Sobry, Dr Roger. *Belgium.*
Soejima, Dr. Yuji. *Japan.*
Sørensen, Dr Alex Mehlsen. *Denmark.*
Sørensen, Mr Ole. *Denmark.*
Søtofte, Mrs Inger. *Denmark.*
Sokol, Dr Anatoly Afanasyevich. *USSR.*
Sokol, Dr Valentina Ivanovna. *USSR.*
Sokolov, Dr Yury Alexandrovich. *USSR.*
Sokolova, Dr Elena Vadimovna. *USSR.*
Sokolova, Dr Nataliya Gavrilovna. *USSR.*
Sola, Dr Joan. *Spain.*
Solans, Prof. Joaquim. *Spain.*
Solans, Prof. Xavier. *Spain.*
Soldánová, Ing Jiřina. *CSSR.*
Soldatos, Prof. Constantinos. *Greece.*
Soldatov, Dr Evgeniy Alexandrovich. *USSR.*
Soledade Jr, Prof. Teomar. *Brasil.*
Soliman, Mr F. Abdel Aal. *Egypt.*
Soliman, Dr Mohamed Soliman. *Egypt.*
Solo'yvev, Dr Sergey Petrovich. *USSR.*
Solorio Munguía, Prof. José Gregorio. *México.*
Solotchina, Dr Emiliya Pavlovna. *USSR.*
Solovyeva, Dr Lidiya Pavlovna. *USSR.*
Soltis, Mr. S. Michael. *USA.*
Soltzberg, Prof. Leonard Jay. *USA.*
Soman, Ms Jayashree. *India.*

Soman, Dr. Kizhake Variyath. *USA.*
Somiya, Prof. Shigeyuki. *Japan.*
Sommer, Dr Joachim. *DDR.*
Sommermann, Mr Günter. *DDR.*
Sommerville, Mrs Polly Baker Melville. *South Africa.*
Sondermann, Dr Ulrich. *BRD.*
Song, Mrs Shi-ying. *China.*
Soni, Dr. Sulil-Datta. *USA.*
Sonin, Prof. Anatoliy Stepanovich. *USSR.*
Sonneveld, Mr Eduard Jan. *Netherlands.*
Sonoike, Prof. Sanemi. *Japan.*
Soós, Mr Miklós. *Hungary.*
Šoptrajanov, Prof. Dr Bojan. *Yugoslavia.*
Šoptrajanova, Prof. Dr Gorica. *Yugoslavia.*
Sorge, Prof Georg. *DDR.*
Soriano Garcia, Dr Manuel. *México.*
Soriano-Calix, Mrs Virginia B. *Philippines.*
Sorokin, Dr Aleksandr Alekseevich. *USSR.*
Sorokin, Dr Lev Mikhailovich. *USSR.*
Sorokina, Dr Natalya Ivanovna. *USSR.*
Sosfenov, Dr Nikita Ilyich. *USSR.*
Sosnowska, Dr hab. Izabela. *Poland.*
Sosnowski, Dr hab. Jerzy. *Poland.*
Soua, Dr Moncef. *Tunisia.*
Soubeyroux, Dr Jean-Louis. *France.*
Šourek, Dr Zbyněk. *CSSR.*
Sousa, de, Mr José Carlos. *Brasil.*
Souza, Mr Carlos, *Chile.*
Souza, Prof. Irineu Marques. *Brasil.*
Sowa, Dr Heidrun. *BRD.*
Soylu, Prof. Dr Hüseyin. *Turkey.*
Spackman, Dr Mark Arthur. *Australia.*
Spadaccini, Dr Nicholas. *Australia.*
Spadon, Prof. Paola. *Italy.*
Spaeth, Prof. Dr Johann-Martin. *BRD.*
Spagna, Dr Riccardo. *Italy.*
Spalding, Dr Dennis Raymond. *South Africa.*
Spangler, Dr. Brenda D. *USA.*
Spanton, Mr. Stephen. *USA.*
Sparks, Dr. Cullie James Jr. *USA.*
Sparks, Dr. Robert Allen. *USA.*
Speier, Mr Peter. *BRD.*
Spek, Dr Anthony Louis. *Netherlands.*
Spence, Prof. John Charles Howorth. *USA.*
Speziali, Mr Nivaldo Lucio. *Switzerland.*
Spielberg, Prof. Nathan. *USA.*
Spindler, Dr Herbert. *DDR.*
Spindler, Mag. Peter. *Austria.*
Spinelli, Dr Silvia Haydeé *Argentina.*
Spink, Mr John Arthur. *Australia.*
Spiridonov, Prof. Victor Pavlovich. *USSR.*
Spirlet, Dr Marie-Rose. *Belgium.*
Spitsyna, Dr Valentina Danilovna. *USSR.*
Spodine, Dr Evgenia, *Chile.*
Spooner, Prof. Stephen. *USA.*
Sprackling, Dr Michael Thomas. *UK.*
Sprang, Dr. Stephen Robert. *USA.*
Spratt, Mr S.B.D. *UK.*
Spreadborough, Dr John. *UK.*
Sprenger, Dr Heinz. *DDR.*
Spriggs, Dr Paul Humphrey. *UK.*
Springer, Dr. James Patrick. *USA.*
Springer, Prof. Dr Tasso. *BRD.*
Sproul, Prof. Gordon Duane. *USA.*
Spruiell, Prof. Joseph Earl. *USA.*
Spurlino, Dr. John C. *USA.*
Spyrellis, Dr Nicolaos. *Greece.*
Spyridelis, Prof. John. *Greece.*
Squattrito, Dr. Philip John. *USA.*
Squire, Dr John Michael. *UK.*
Srdanov, Magistar Gordana. *Yugoslavia.*
Sridhar Prasad, Mr G. *India.*
Srikrishnan, Dr. Thamarapu. *USA.*
Srinivasa, Dr Vishwanathapuram Kalasa. *India.*
Srinivasan, Prof. R. *India.*
Srinivasan, Dr Sampat. *India.*
Srivastava, Dr. Alok Mani. *USA.*
Srivastava, Dr. Ramesh C. *USA.*
Srivastava, Dr Ramesh Chandra. *India.*
Srivastava, Dr Surendra Nath. *India.*

Ståhl, Dr Kenny. *Sweden.*
Stålhandske, Dr Claes. *Sweden.*
Stadermann, Dr Gerd. *BRD.*
Stadler, Dr Hans Peter. *UK.*
Stadler, Mr Maximilian Wolfgang. *BRD.*
Stadnicka, Dr Katarzyna. *Poland.*
Staehlin, Dr Walter. *Switzerland.*
Staikov, Dr Georgy. *Bulgaria.*
Stalick, Dr. Judith Kay. *USA.*
Staliński, Prof. Bohdan. *Poland.*
Stallings, Dr. William C. *USA.*
Stam, Dr Casper Hendrik. *Netherlands.*
Stammers, Dr David Kingsley. *UK.*
Stanfield, Dr. Robyn L. *USA.*
Stanford, Mr Michael John. *UK.*
Stangler, Prof. Dr Ferdinand Karl Ludwig. *Austria.*
Stanisz, Mgr Grzegorz. *Poland.*
Stanko, Prof. Joseph Anthony. *USA.*
Stanković, Dr (Mrs) Slobodanka. *Yugoslavia.*
Stanley, Dr Eric. *Canada.*
Stanley, Prof. George Geoffrey. *USA.*
Stansfield, Dr Robert Frank David. *UK.*
Stanton, Mr. Martin. *USA.*
Starikova, Dr Zoya Alexandrovna. *USSR.*
Starke, Dr Rainer. *DDR.*
Starostina, Dr Lyudmila Sergeevna. *USSR.*
Stasi, Dr Francesca. *Italy.*
Staudenmann, Dr. Jean-Louis. *USA.*
Stauffacher, Dr. Cynthia Vianne. *USA.*
Stearns, Mr. Frederick Stanley. *USA.*
Stec, Dr. Boguslaw. *USA.*
Stecker, Prof. Kurt. *DDR.*
Steeb, Prof. Dr Siegfried. *BRD.*
Steeds, Prof. John Wickham, FRS. *UK.*
Steele, Dr. Ian McKay. *USA.*
Stefanidis, Mr Theodoros. *Sweden.*
Stefanov, Mr Dechko Dimitrov. *Bulgaria.*
Stefanov, Dr Stefan Rashkov. *Bulgaria.*
Stefanović, Magistar Aleksandar. *Yugoslavia.*
Steffen, Dipl. Chem Michael Georg. *BRD.*
Steffen, Dr William Lee. *Australia.*
Steffer-Tun, Mr Wolfgang. *BRD.*
Stegemann, Dr Ing Jürgen. *BRD.*
Stegemeyer, Prof. Dr Horst. *BRD.*
Stegmann, Dr Eleonore. *DDR.*
Steigmann, Dr Gottfried Albert. *UK.*
Steil, Dr Helmut. *DDR.*
Stein, Prof. Richard Stephen. *USA.*
Stein, Mrs Zafra. *Israel.*
Steinberger, Prof. Itzhak T. *Israel.*
Steinbruch, Mr Uta. *DDR.*
Steiner, Prof. Dr Michael. *BRD.*
Steiner, Mr Thomas Johann. *BRD.*
Steiner, Prof. Dr Walter. *Austria.*
Steinfink, Prof. Hugo. *USA.*
Steinhart, Dr Miloš. *CSSR.*
Steinicke, Prof. Ursula. *DDR.*
Steinmann, Dr Gerhard. *BRD.*
Steinrauf, Prof. Larry King. *USA.*
Steinthórsson, Dr Sigurdur. *Iceland.*
Steitz, Prof. Thomas Arthur. *USA.*
Stelzner, Mr Sabine. *DDR.*
Stenberg, Dr Lars. *Sweden.*
Stenkamp, Dr. Ronald Eugene. *USA.*
Stensland, Dr Birgitta. *Sweden.*
Stepanova, Dr Alla Nikolayevna. *USSR.*
Stepanova, Dr Nataliya Stepanovna. *USSR.*
Stepantsov, Dr Evgenii Arkad'evich. *USSR.*
Stephan, Dr D. W. *Canada.*
Stephan, Dr Dieter. *DDR.*
Stephanik, Dr Heinz. *DDR.*
Stephens, Dr. Anthony E. *USA.*
Stephens, Dr Frederick Selwyn. *Australia.*
Stephenson, Dr. Gregory Brian. *USA.*
Stephenson, Prof Neville Charles. *Australia.*
Stępień-Damm, Dr Julia. *Poland.*
Stępień, Dr Andrzej. *Poland.*
Stergiou, Dr Anagnostis Charalambos. *Greece.*
Stergioudis, Dr Georgios Asterios. *Greece.*
Stern, Ms. Charlotte. *USA.*

Sterns, Dr Meta. *Australia.*
Steuhl, Dr Hans Hermann. *BRD.*
Steurer, Dr Walter. *BRD.*
Stevens, Prof. Edwin David. *USA.*
Stevens, Dr. Raymond Charles. *USA.*
Stevenson, Dr Andrew Wesley. *Australia.*
Steward, Prof. Edward George. *UK.*
Stewart, Prof. James McDonald. *USA.*
Stewart, Dr. Martin Van Buren. *USA.*
Stewart, Dr Robert Bruce. *New Zealand.*
Stewart, Prof. Robert Farrell. *USA.*
Stezowski, Prof. Dr John J. *BRD.*
Sthanam, Dr. V. L. Narayana. *USA.*
Stickler, Prof. Dr Roland. *Austria.*
Stillman, Dr Timothy James. *UK.*
Stiopol, Prof. Victoria. *Romania.*
Stipančić, Prof. Dr Mladen. *Yugoslavia.*
Stishov, Dr Sergey Mikhailovich. *USSR.*
Stock, Dr. Ann. *USA.*
Stock, Prof. Stuart R. *USA.*
Stodola, Mr. Robert. *USA.*
Stöckelmann, Dr Diedrich. *BRD.*
Stoeckli-Evans, Prof. (Mrs) H. M. *Switzerland.*
Stölevik, Prof. Reidar. *Norway.*
Stoicovici, Prof. Eugen. *Romania.*
Stoimenos, Prof. John Nikolaos. *Greece.*
Stoinova, Dr Margarita. *Bulgaria.*
Stojadinović, Prof. Dr Slobodan. *Yugoslavia.*
Stojanoff, Dr Vivian. *Brasil.*
Stojanović, Mr Dobrica. *Yugoslavia.*
Stollstorff, Mr. Gregory R. *USA.*
Stomberg, Dr Rolf. *Sweden.*
Storbeck, Prof. Fritz. *DDR.*
Stothart, Dr Philip Hamilton. *UK.*
Stouch, Dr. Terry Richard. *USA.*
Stout, Dr. Charles David. *USA.*
Stout, Dr. George H. *USA.*
Stout, Mr. Thomas J. *USA.*
Stouten, Drs Pieter F. W. *Netherlands.*
Stoyanov, Dr Stoyan. *Bulgaria.*
Stoychev, Mr Nikola. *Bulgaria.*
Strähle, Prof. Dr Joachim. *BRD.*
Strand, Dr Tor Gogstad. *Norway.*
Strandberg, Prof. Bror E. *Sweden.*
Strandberg, Dr Rolf A.G. *Sweden.*
Straver, Drs Leonardus Hendrikus. *Netherlands.*
Streib, Dr. William E. *USA.*
Strell, Mrs Dr Irmtraud. *BRD.*
Strickland, Mr Peter R. *UK.*
Strickler, Dr Peter. *Switzerland.*
Strid, Prof. Karl-Gustav. *Sweden.*
Stridh, Mr Kjell. *Sweden.*
Strocka, Dr Bernhard. *BRD.*
Strogen, Mr Peter. *Ireland.*
Strong, Mr. Roland Kirk. *USA.*
Stroppe, Prof Hilbert. *DDR.*
Stroud, Prof. Robert Michael. *USA.*
Strouse, Prof. Charles Earl. *USA.*
Stróż, Dr Danuta. *Poland.*
Struchkov, Prof. Yury Timofeyevich. *USSR.*
Strübel, Prof. Dr Günter. *BRD.*
Struikmans, Drs Rink. *Netherlands.*
Strukan, Dipl. inž. Neven. *Yugoslavia.*
Strukov, Prof. Boris Anatolyevich. *USSR.*
Strumpel, Dr Marianna Katona. *BRD.*
Strydom, Dr Ockert Andries Wilhelm. *South Africa.*
Stubb, Dr Arne Henrik. *Finland.*
Stubbs, Dr. Gerald James. *USA.*
Stubbs, Dr Milton. *BRD.*
Stubičar, Dr Mirko. *Yugoslavia.*
Stündel, Mr Peter. *DDR.*
Stuhrmann, Prof. Dr Heinrich B. *BRD.*
Stults, Dr. Bailey Ray. *USA.*
Stumpfl, Prof. Dr Eugen Friedrich. *Austria.*
Sturcken, Dr. Edward Francis. *USA.*
Sturkey, Mr. Lorenzo. *USA.*
Sturm, Prof. Dr Dipl.-Ing. Friedwin. *Austria.*
Su, Dr. Shu-Chun. *USA.*
Su, Mr Xiao-dong. *Sweden.*
Su, Mr. Ying-Zhong. *USA.*

Suades, Dr Joan. *Spain.*
Suarez, MSc Nery. *Venezuela.*
Subhadra, Dr K. G. *India.*
Subirana, Prof. Juan A. *Spain.*
Subramanian, Prof. Easwara. *India.*
Subramanian, Dr K. *India.*
Subramanian, Dr. Sethuraman. *USA.*
Subramanya, Mr H. S. *India.*
Subramony, Mr Loganathan. *South Africa.*
Šubrtová, Ing Věra. *CSSR.*
Suck, Dr Dietrich. *BRD.*
Sudarsanam, Mr. Padmanaban. *USA.*
Suddath, Prof. Frederick L. *USA.*
Suddhiprakarn, Dr Anohsloe. *Thailand.*
Sudo, Prof. Toshio. *Japan.*
Sueno, Prof. Shigeho. *Japan.*
Süsse, Prof. Dr Peter. *BRD.*
Suessmann, Dr Hans. *DDR.*
Suffritti, Prof. Giuseppe Baldovino. *Italy.*
Suga, Prof. Hiroshi. *Japan.*
Sugaike, Dr Suezo. *Japan.*
Sugar, Prof. Istvan Peter. *USA.*
Sugawara, Dr Yoko. *Japan.*
Sugihara, Dr Akio. *Japan.*
Sugio, Dr Shigetoshi. *Japan.*
Suguna, Dr K. *India.*
Suh, Prof. Dr Ill-Hwan. *Korea.*
Suh, Prof. Jung Sun. *Korea.*
Suh, Prof. Se Won. *Korea.*
Suhre, Miss Ursula. *BRD.*
Suitch, Dr. Paul Raymond. *USA.*
Suito, Dr Eiji. *Japan.*
Sukapaddhanadhi, Mr Narong. *Thailand.*
Sullenger, Dr. Don Bruce. *USA.*
Sullivan, Dr Richard Arthur. *UK.*
Sultanov, Dr Rafik Mukhadastovich. *USSR.*
Sumin, Dr Vyacheslav Vasil'evich. *USSR.*
Sun, Prof. Dai-sheng. *China.*
Sun, Mr Li-kun. *China.*
Sun, Mr Yi-jian. *China.*
Sunagawa, Prof. Ichiro. *Japan.*
Sundaralingam, Prof. Muttaiya. *USA.*
Sundaramoorthy, Mr M. *India.*
Sundararajan, Dr Pudupadi R. *Canada.*
Sundaresan, Dr Thiagarajan. *UK.*
Sundberg, Dr Margareta. *Sweden.*
Sunder, Dr Sham. *Canada.*
Sundius, Dr Tom Robert. *Finland.*
Sundström, Dr Lorna Jean. *Finland.*
Sunshine, Dr. Steven A. *USA.*
Suoninen, Prof. Eero Juhani. *Finland.*
Suortti, Prof. Pekka. *Finland.*
Supper, Mr. Lee R. *USA.*
Suresh, Mr C. G. *India.*
Suresh, Mr K. A. *India.*
Suri, D. K. *India.*
Surowiec, Dr Marian Ryszard. *Poland.*
Suryanarayana, Dr Challapalli. *India.*
Suryanarayana, Dr Shambhuni V. *India.*
Sussieck-Fornefeld, Dr Cornelia. *BRD.*
Sussman, Prof. Joel Leonard. *Israel.*
Suta, Elizabeth *India.*
Sutherland, Dr Hector Howieson. *UK.*
Sutherland, Dr John Knox. *Canada.*
Sutor, Dr Dorothy June. *UK.*
Sutton, Dr A.L. *UK.*
Sutton, Dr Brian John. *UK.*
Sutton, Mr J.D. *UK.*
Sutton, Mr. Paul A. *USA.*
Suvorov, Dr Ernest Vitalyevich. *USSR.*
Suwalsky, Dr Mario. *Chile.*
Suwińska, Dr Kinga. *Poland.*
Suzuki, Dr Carlos Kenichi. *Brasil.*
Suzuki, Prof. Hideji. *Japan.*
Suzuki, Prof. Ikuo. *Japan.*
Suzuki, Prof. Kazuo. *Japan.*
Suzuki, Dr Michio. *Japan.*
Suzuki, Dr Shigeo. *Japan.*
Suzuki, Prof. Tadasu. *Japan.*
Suzuki, Dr Toshimasa. *Japan.*

Suzuki, Prof. Yoshio. *Japan.*
Sváb, Dr (Ms) Erzsébet. *Hungary.*
Svensson, Dr Christer. *Sweden.*
Svensson, Mr Göran. *Sweden.*
Svensson, Mr Gunnar. *Sweden.*
Svensson, Dr Ing-Britt A. *Sweden.*
Svensson, Dr L. Anders. *Sweden.*
Svergun, Dr Dmitry Ivanovich. *USSR.*
Svinning, Dr Torgeir. *Norway.*
Sviridov, Prof. Dmitry Timofeyevich. *USSR.*
Svisero, Dr Darcy Pedro. *Brasil.*
Swain, Dr. Amy L. *USA.*
Swallow, Dr Arnold Graham. *UK.*
Swaminathan, Dr. P. *USA.*
Swaminathan, Dr. S. *USA.*
Swaminathan, Dr. Vijayalakshmi. *USA.*
Swank, Prof. Duane Douglas. *USA.*
Swanson, Dr. Rosemarie. *USA.*
Swanson, Dr. Stanley M. *USA.*
Swartzendruber, Mr. John K. *USA.*
Sweet, Dr. Robert M. *USA.*
Swenson, Dr. Dale Carl. *USA.*
Swepston, Dr. Paul Nathan. *USA.*
Swerdlow, Mr. Max. *USA.*
Swillens, Mr Eckhard. *DDR.*
Swindells, Dr David Campbell Neil. *UK.*
Swink, Dr. Laurence Nim. *USA.*
Swinnea, Dr. John Steven. *USA.*
Switendick, Dr. Alfred Carl. *USA.*
Syed, Dr A. Sattar. *Bangladesh.*
Syed, Dr. Ashfaquzzaman. *USA.*
Syed, Dr. Rashid. *USA.*
Sygusch, Prof. Jurgen. *Canada.*
Syhre, Dr Hans. *DDR.*
Symerský, Ing Jindřich. *CSSR.*
Syneček, Doc. Dr Vladimír. *CSSR.*
Szarras, Dr Stanisław. *Poland.*
Szebenyi, Dr. Doletha M. E. *USA.*
Szemethy, Miss Andrea. *Hungary.*
Szmid, Dr Zofia. *Poland.*
Sztrókay, Prof. Kálmán. *Hungary.*
Szulzewsky, Dr Klaus. *DDR.*
Szummer, Dr Andrzej. *Poland.*
Szurgot, Dr Marian. *Poland.*
Szymański, Dr Jan Tomasz. *Canada.*
Ta Van That, Prof. *Vietnam.*
Tabata, Dr Hideyo. *Japan.*
Tabira, Mr Yasunori. *Japan.*
Tada, Dr Toshiji. *Japan.*
Tadaki, Dr Tsugio. *Japan.*
Tadini, Prof. Carla. *Italy.*
Tadokoro, Prof. Hiroyuki. *Japan.*
Taeb Dr Abbas. *Iran.*
Tänzer, Dr Dietmar. *DDR.*
Tafeenko, Dr Viktor Aleksandrovich. *USSR.*
Taga, Dr Tooru. *Japan.*
Tagai, Dr Tokuhei. *Japan.*
Tahoun, Prof. Salah. *Saudi Arabia.*
Tai, Dr. Douglas Leung-Tak. *USA.*
Tainer, Dr. John Arthur. *USA.*
Tait, Dr John Mervyn. *UK.*
Tajabor, Dr Nasser. *Iran.*
Takagi, Dr Mieko. *Japan.*
Takagi, Prof. Satio. *Japan.*
Takagi, Dr. Shozo. *USA.*
Takagi, Prof. Yutaka. *Japan.*
Takahashi, Mr Shoichi. *Japan.*
Takahashi, Dr Toshio. *Japan.*
Takahashi, Prof. Yasuhiro. *Japan.*
Takai, Dr Mitsuo. *Japan.*
Takaki, Dr Yoshito. *Japan.*
Takama, Dr Toshihiko. *Japan.*
Takamura, Prof. Jin-ichi. *Japan.*
Takano, Dr Tsunehiro. *Japan.*
Takano, Prof. Yasumasa. *Japan.*
Takano, Dr Yukio. *Japan.*
Takasu, Dr Shin-ichiro. *Japan.*
Takayanagi, Prof. Kunio. *Japan.*
Takazawa, Mr Hiroyuki. *Japan.*
Takeda, Dr Hirofumi. *Japan.*

Takeda, Prof. Hiroshi. *Japan.*
Takeda, Prof. Takayoshi. *Japan.*
Takei, Dr. William J. *USA.*
Takenaka, Dr Akio. *Japan.*
Takenaka, Mr Yasuyuki. *Japan.*
Takeoka, Mr Yoshikatsu. *Japan.*
Takeuchi, Prof. Yoshio. *Japan.*
Takhodzhaev, Dr Bakhodirhodzha. *USSR.*
Takla, Mr Maher Azmi. *Egypt.*
Takusagawa, Dr. Fusao. *USA.*
Talapatra, Dr S. K. *India.*
Talukdar, Dr Amarendra Nath. *India.*
Tamada, Dr Osamu. *Japan.*
Tamasyan, Dr Rafael Arshamovich. *USSR.*
Tamura, Dr Chihiro. *Japan.*
Tan, Prof. Hao-ran. *China.*
Tan, Dr Hock Siew. *Singapore.*
Tanaka, Dr Isao. *Japan.*
Tanaka, Prof. Jiro. *Japan.*
Tanaka, Dr Kiyoaki. *Japan.*
Tanaka, Dr Michiyoshi. *Japan.*
Tanaka, Dr Nobuo. *Japan.*
Tanaka, Dr Nobuo. *Japan.*
Tanemura, Dr Sakae. *Japan.*
Tang, Dr Chia-Pin. *Taiwan.*
Tang, Dr. Sunny C. *USA.*
Tang, Prof. You-qi. *China.*
Tang, Mr Zhi-kai. *China.*
Tani, Dr Katsuhiko. *Japan.*
Tanigaki, Mr Takeshige. *Japan.*
Taniguchi, Mr Tomohiko. *Japan.*
Tanisaki, Prof. Sigetosi. *Japan.*
Tanishiro, Mr Yasumasa. *Japan.*
Tanner, Dr Brian Keith. *UK.*
Tanner, Mr. Lee Elliot. *USA.*
Tanus Alonso, Mrs. Mercedes. *Cuba.*
Tao, Dr. Yuan Kai. *USA.*
Taoka, Prof. Tadami. *Japan.*
Tapfer, Dr Leander. *BRD.*
Tarashchan, Prof Arkadii Nikolaevich. *USSR.*
Tarasov, Dr Yurii Igorevich. *USSR.*
Tardieu, Dr Annette. *France.*
Tardy, Dr Pál. *Hungary.*
Tarimci, Doç. Dr Çelik. *Turkey.*
Tarján, Prof. Imre. *Hungary.*
Tarkhova, Dr Tatyana Nikolayevna. *USSR.*
Tarling, Dr Stephen Edward. *UK.*
Tarna, Mr Toivo Mikael. *Finland.*
Tarnopol'sky, Dr Boris Lvovich. *USSR.*
Taşer, Mr Mehmet. *Turkey.*
Tasker, Mr Michael Peter. *UK.*
Tasker, Dr Peter Anthony. *UK.*
Tasnády, Dr (Ms) Eleonóra. *Hungary.*
Tatarchenko, Dr Vitaly Antonovich. *USSR.*
Tatarinova, Dr Lyudmila Ivanovna. *USSR.*
Tatarsky, Prof. Vitaly Borisovich. *USSR.*
Tate, Dr Cecil. *UK.*
Tate, Dr Isao. *Japan.*
Tatsch, Dr. Clinton Eugene. *USA.*
Tatsuzaki, Prof. Itaru. *Japan.*
Tauber, Dr. Arthur. *USA.*
Tauler, Dr Esperanza. *Spain.*
Tauqir, Dr Anjum. *Pakistan.*
Tavale, Dr Sudam Shankar. *India.*
Távora, Prof. Elysiario. *Brasil.*
Taxer, Dr Karlheinz Jürgen. *BRD.*
Taylor, Prof. Charles Alfred. *UK.*
Taylor, Ms. Denise A. *USA.*
Taylor, Dr Derek. *UK.*
Taylor, Prof. Harry Francis West. *UK.*
Taylor, Dr. Ivan Fate Jr. *USA.*
Taylor, Prof. Jean Ellen. *USA.*
Taylor, Dr John Charles. *Australia.*
Taylor, Dr Max Ronald. *Australia.*
Taylor, Mr Michael William. *South Africa.*
Taylor, Dr Peter. *Canada.*
Taylor, Dr R. *UK.*
Taylor, Prof. Richard T. *USA.*
Taylor, Prof. Robert Cooper. *USA.*
Tazzoli, Prof. Vittorio. *Italy.*

Tchehlarova, Mrs Irina. *Bulgaria.*
Tchuneva, Mrs Vassilka. *Bulgaria.*
Tebbe, Prof. Dr Karl-Friedrich. *BRD.*
Teeter, Dr. Martha Mary. *USA.*
Tegman, Dr Ragnar. *Sweden.*
Teh, Dr Guan Hoe. *Malaysia.*
Teh, Dr Hung Chuan. *Singapore.*
Teh, Dr Ser Kok. *Malaysia.*
Teixeira Mendes, Prof. Antonio Carlos. *Brasil.*
Teixidor, Dr Francesc. *Spain.*
Telegina, Dr Inna Vasilyevna. *USSR.*
Teller, Dr. Raymond. *USA.*
Téllez Ortiz, Mrs Minerva Estela. *México.*
Tellgren, Dr I. G. Roland. *Sweden.*
Tembo, Mr Francis. *Zambia.*
Tempel, Dr Alfred. *DDR.*
Tempelhoff, Dr Klaus. *DDR.*
Tempest, Dr Paul Anthony. *UK.*
Templeton, Prof. David Henry. *USA.*
Templeton, Dr. Lieselotte K. *USA.*
Ten Eyck, Dr. Lynn Forest. *USA.*
Tench, Dr. Alan Howard. *USA.*
Teng, Dr. Siu Tjeng. *USA.*
Tennyson, Mrs Prof. Dr Christel. *BRD.*
Tenon, Mr Abodou Jules. *Côte d'Ivoire.*
Teoh, Mr Lay Hock. *Malaysia.*
Terauchi, Dr Hikaru. *Japan.*
Terent'ev, Dr Evgenii Mikhailovich. *USSR.*
Teresiak, Mr Angelika. *DDR.*
Terwilliger, Dr. Thomas C. *USA.*
Terzis, Dr Aristides. *Greece.*
Teske, Dr Christoph Ludwig. *BRD.*
Teslenko, Dr Valery Fedorovich. *USSR.*
Teuho, Mr Juhani Erkki Tapani. *Finland.*
Tewari, Dr Raghavendra. *India.*
Thabet, Dr Atef. *Egypt.*
Thackeray, Dr Michael Makepeace. *South Africa.*
Thailambal, Dr. Vadakkanthara G. *USA.*
Thanomkul, Dr Srinuan Chaiwasie. *Thailand.*
Thatcher, Mr J.S. *UK.*
Thatcher, Dr. Walter E. *USA.*
Thathachari, Mr. Y. T. *USA.*
Thayer, Dr. Maria Julie McLeod. *USA.*
Théobald, Prof. François. *France.*
Theocharis, Dr C.R. *UK.*
Theodoridou, Dr Irini. *Greece.*
Theodoropoulos, Prof. Dimitrios. *Greece.*
Theodossiou, Prof. Alexandros. *Greece.*
Theophanides, Prof. Theo. *Canada.*
Thewalt, Prof. Dr Ulf. *BRD.*
Thiele, Prof. Dr Gerhard. *BRD.*
Thielke, Mr. Harry G. *USA.*
Thieme, Dr Wolfgang. *DDR.*
Thijsse, Dr Barend Jan. *Netherlands.*
Thinapong, Dr Pongchan Chananont. *Thailand.*
Thirup, Mr Søren. *Denmark.*
Thomas, Dr Charles Richard. *UK.*
Thomas, Mr David Huw. *UK.*
Thomas, Mrs Elizabeth Ann. *UK.*
Thomas, Prof John Meurig, FRS. *UK.*
Thomas, Prof. John O. (Josh). *Sweden.*
Thomas, Dr Pamela Anne. *UK.*
Thomas, Dr. Robert. *USA.*
Thomasson, Mr Ronnie. *Sweden.*
Thompson, Dr Derek Parr. *UK.*
Thompson, Prof. Doris M. *USA.*
Thompson, Dr John Gerard. *Australia.*
Thorkildsen, Dr Gunnar. *Norway.*
Thornton-Pett, Dr M.A. *UK.*
Thorup, Mr Niels. *Denmark.*
Thota, Mr. Narasaiah V. *USA.*
Thozet, Dr Alain Maurice. *France.*
Thudium, Mr. Richard N. *USA.*
Thurn, Dr Herbert. *BRD.*
Tibballs, Dr John Earl. *Norway.*
Tibbitts, Mr. Thomas Theodore. *USA.*
Tibljaš, Mr Darko. *Yugoslavia.*
Tickle, Dr Ian James. *UK.*
Tieghi, Prof. Giuseppe. *Italy.*
Tiekink, Dr Edward Richard Thomas. *Australia.*

Tigges, Mr Hartmut R. *BRD.*
Tiitta, Mr Antero Tapani. *Finland.*
Tikhomirova, Dr Nataliya Alexandrovna. *USSR.*
Tikhonova, Dr Anna Andreyevna. *USSR.*
Tilli, Mr Markku Väinö Kalevi. *Finland.*
Tillinger, Dr. Martin H. *USA.*
Tillmann, Dr Bruno. *BRD.*
Tillmanns, Prof. Dr Ekkehart. *BRD.*
Tilton, Dr. Robert F. Jr. *USA.*
Timchenko, Dr Tamara Iosifovna. *USSR.*
Timmers, Drs Jacob. *Netherlands.*
Timmins, Dr Peter. *France.*
Timofeeva, Dr Valentina Alexandrovna. *USSR.*
Timofeyeva, Dr Tat'yana Vladimirovna. *USSR.*
Tinant, Dr Bernard Guy André François. *Belgium.*
Tintorer Deloado Mr Oscar. *Cuba.*
Tiripicchio-Camellini, Prof. Marisa. *Italy.*
Tiripicchio, Prof. Antonio. *Italy.*
Tischler, Dr. Jonathan Zachary. *USA.*
Tishchenko, Dr Galina Nikolayevna. *USSR.*
Titus, Prof. Donald Dean. *USA.*
Tkachev, Dr Valery Vladimirovich. *USSR.*
Tkalčec, Prof. Dr Emilija. *Yugoslavia.*
Tobelko, Mr Konstantin Ivanovich. *USSR.*
Toby, Dr. Brian H. *USA.*
Todaro, Mr. Louis J. *USA.*
Töpel-Schadt, Dr Jutta. *BRD.*
Törnroos, Mr Karl Wilhelm. *Sweden.*
Törnroos, Assoc. Prof. Ragnar Fredrik. *Finland.*
Tofield, Dr Bruce C. *UK.*
Togawa, Dr Sen-ichi. *Japan.*
Toivonen, Mr Jukka Tapio. *Finland.*
Tokay, Yrd. Doç. Dr Nesrin. *Turkey.*
Tokel, Doç. Dr Selçuk. *Turkey.*
Tokonami, Prof. Masayasu. *Japan.*
Tokushita, Mr Motoyuki. *Japan.*
Tolksdorf, Prof. Dr Wolfgang. *BRD.*
Tollin, Dr Patrick. *UK.*
Tolstikhina, Dr Alla Leonidovna. *USSR.*
Tomas, Dr Milagros. *Spain.*
Tomashpol'sky, Dr Yury Yakovlevich. *USSR.*
Tomasicchio, Dr Michele. *Italy.*
Tomassini, Dr Marco. *Italy.*
Tómasson, Cand.Real. Jens. *Iceland.*
Tomaszewski, Dr Paweł. *Poland.*
Tomchick, Ms. Diana R. *USA.*
Tomeoka, Mr Kazushige. *Japan.*
Tomimitsu, Mr Hiroshi. *Japan.*
Tomita, Dr Katsutoshi. *Japan.*
Tomita, Prof. Ken-ichi. *Japan.*
Tomita, Dr Koichi. *Brasil.*
Tomkeieff, Mr Michael Vamime. *UK.*
Tomlin, Mr. David W. *USA.*
Tomlinson, Dr. Gail Elizabeth. *USA.*
Tomov, Dr Ivan. *Bulgaria.*
Tonejc, Dr (Mrs) Anđelka. *Yugoslavia.*
Tonejc, Dr Anton. *Yugoslavia.*
Tong, Mr Hua. *Australia.*
Tong, Mr. Liang. *USA.*
Tonnard, Dr Victor Edmond. *Belgium.*
Tonomura, Dr Akira. *Japan.*
Tontrakoon, Mr Jeerapong. *Thailand.*
Tooptakong, Dr Uncharee Methong. *Thailand.*
Topalova-Kalitzova, Mrs Maria. *Bulgaria.*
Topić, Dr Mladen. *Yugoslavia.*
Topiol, Dr. Sidney. *USA.*
Topor, Dr Nikolay Dmitriyevich. *USSR.*
Torardi, Dr. Carmine Charlie. *USA.*
Toraya, Dr Hideo. *Japan.*
Toriumi, Dr Koshiro. *Japan.*
Torre, Prof. Louis Peter. *USA.*
Torres Villaseñor, Dr Gabriel. *México.*
Torres-Ruiz, Dr Jose. *Spain.*
Torriani, Dr Iris Linares. *Brasil.*
Toscano, Mr Ruben Alfredo. *México.*
Toshev, Dr Alexander. *Bulgaria.*
Tosi, Prof. Giorgio. *Italy.*
Tosik, Dr Anita. *Poland.*
Tóth, Mr Lajos. *Hungary.*
Tougard, Dr Pierre. *France.*

Toupet, Mr Loic. *France.*
Tourchi, Dr Mohammad. *Iran.*
Toure, Dr Siaka. *Côte d'Ivoire.*
Toussaint, Prof. Jean. *Belgium.*
Tousson, Dr Salama. *Egypt.*
Tovbis, Dr Alexander Borisovich. *USSR.*
Town, Dr Susan Lesley. *Australia.*
Towns, Prof. Robert Lee Roy. *USA.*
Townsend, Dr Stephen Phillip. *UK.*
Toy, Dr Mark. *UK.*
Toyoda, Prof. Koichi. *Japan.*
Trabelsi, Dr Malika. *Tunisia.*
Trætteberg, Prof. Marit. *Norway.*
Traill, Dr Robert James. *Canada.*
Tran Hai Huynh, Dr. *Vietnam.*
Tran Quoc Thang, Dr. *Vietnam.*
Trance, Dr Aurora Serrana. *Philippines.*
Traub, Prof. Wolfie. *Israel.*
Trauth, Mr Jürgen. *BRD.*
Traveria-Cros, Dr Adolfo. *Spain.*
Treasurywala, Dr. Adi M. *USA.*
Trechairusma, Mr Kamchai. *Thailand.*
Trefonas, Prof. Louis Marco. *USA.*
Trefry, Mr Michael George. *Australia.*
Treharne, Dr. Anne-Marie. *USA.*
Treimer, Dr Wolfgang. *BRD.*
Treivus, Dr Evgeny Borisovich. *USSR.*
Tremmel, Mr János. *Hungary.*
Trempler, Mr Jörg. *DDR.*
Tret'yakov, Dr Vyacheslav Nukolaevich. *USSR.*
Trettin, Dr Reinhard. *DDR.*
Treushnikov, Dr Evgeniy Nikolayevich. *USSR.*
Treutmann, Dr Werner. *BRD.*
Trigubo, Lic Alicia Beatriz. *Argentina.*
Trigunayat, Prof. Govind Chandra. *India.*
Trinh Han, Prof. *Vietnam.*
Trinh Le Thu, Dr. *Vietnam.*
Triodina, Dr Nina Sergeyevna. *USSR.*
Trömel, Prof. Dr Martin Gerhard. *BRD.*
Trojko, Magistar Rudolf. *Yugoslavia.*
Trost, Dr Friedrich Karl. *BRD.*
Trosti-Ferroni, Dr Renza. *Italy.*
Trotter, Prof. James. *Canada.*
Troup, Dr. Jan Marshall. *USA.*
Troyanov, Dr Sergei Igorevich. *USSR.*
Trubelja, Prof. Dr Fabijan. *Yugoslavia.*
Trubkin, Dr Nikolai Viktorovich. *USSR.*
Trueblood, Prof. Kenneth N. *USA.*
Trunov, Dr Vadim Konstantinovich. *USSR.*
Trus, Dr. Benes Louis. *USA.*
Truter, Prof. Mary Rosaleen. *UK.*
Tsai, Prof. Chun-che. *USA.*
Tsatis, Mr Demetrius. *Greece.*
Tschulena, Dr Guido. *BRD.*
Tseitlin, Dr Mikhail Nevakhovich. *USSR.*
Tseng, Prof. Poh-Kun. *Taiwan.*
Tsernoglou, Prof. Demetrius. *BRD.*
Tsikhotsky, Dr Evgeny Stanislavovich. *USSR.*
Tsimberis, Mr Nikolaos. *Greece.*
Tsinober, Dr Leonid Iosifovich. *USSR.*
Tsintsadze, Prof. Givi Vasilyevich. *USSR.*
Tsirel'son, Dr Valadimir Grigoryevich. *USSR.*
Tsoli Kataga, Dr (Mrs) Panayota. *Greece.*
Tsolovski, Dr Ilcho. *Bulgaria.*
Tsoukalas, Prof. John. *Greece.*
Tsuda, Mr Noritoshi. *Japan.*
Tsuji, Mr Kazuhiko. *Japan.*
Tsuji, Mr Koji. *Japan.*
Tsukihara, Dr Tomitake. *Japan.*
Tsukimura, Dr Katsuhiro. *Japan.*
Tsunekawa, Dr Shin. *Japan.*
Tsuprun, Dr Vladimir Lvovich. *USSR.*
Tsvankin, Dr Daniel' Yakovlevich. *USSR.*
Tsyganov, Dr Evgeny Matveyevich. *USSR.*
Tuan Sarif, Mr Tuan Besar. *Malaysia.*
Tucker, Mr. Craig Anthony. *USA.*
Tudja, Dr Marijan. *Yugoslavia.*
Tuinstra, Prof. Dr Ir Fokke. *Netherlands.*
Tulinsky, Prof. Alexander. *USA.*
Tulip, Mr William Richard. *Australia.*

Tulloch, Dr Peter Archibald. *Australia.*
Tumanyan, Dr Vladimir Gayevich. *USSR.*
Tun, Mr Saw. *Burma.*
Tun, Dr Zin. *Canada.*
Tunç, Yrd. Doç. Dr Cemil. *Turkey.*
Tunkasiri, Dr Tawee. *Thailand.*
Tuomi, Dr. Donald. *USA.*
Tuomi, Prof. Turkka Olavi. *Finland.*
Turano, Dr. August M. *USA.*
Turley, Dr. June Williams. *USA.*
Turley, Mr. Stewart. *USA.*
Turmezey, Dr Tibor. *Hungary.*
Turner, Dr. Ralph Waldo. *USA.*
Turowska-Tyrk, MSc Ilona. *Poland.*
Turpeinen, Dr Urho Taneli. *Finland.*
Turrillas, Dr Javier. *Spain.*
Turunen, Dr Markus Johannes. *Finland.*
Tuzi, Dr Angela. *Italy.*
Twigg, Mrs. Pamela Dekle. *USA.*
Tyapunina, Dr Nataliya Alexandrovna. *USSR.*
Tykarska, Dr Ewa Maria. *Poland.*
Tyvanchuk, Dr Anna Teodorovna. *USSR.*
Tzeng, Dr. Shih-Ying. *USA.*
Uberbacher, Mr. Edward C. *USA.*
Udalova, Dr Valentina Vasilyevna. *USSR.*
Udubasa, Dr Gheorghe. *Romania.*
Uechi, Mr Tetsuo. *Japan.*
Üçişik, Prof. A. Hikmet. *Turkey.*
Uecker, Mr Reinhard. *DDR.*
Ueda, Prof. Ikuhiko. *Japan.*
Uefuji, Mr Tateki. *Japan.*
Ueki, Dr Tatzuo. *Japan.*
Ülkü, Prof. Dr Dinçer. *Turkey.*
Ünal, Doç. Dr Narin. *Turkey.*
Ueng, Prof. Chuen-Her. *Taiwan.*
Ueno, Mr Tsunehisa. *Japan.*
Uesu, Prof. Yoshiaki. *Japan.*
Uggla, Dr Rolf Åke Magnus. *Finland.*
Ugliengo, Dr Piero. *Italy.*
Ugozzoli, Dr Franco. *Italy.*
Uhl, Dr Eduard. *Austria.*
Ukaji, Prof. Takeshi. *Japan.*
Ukraintsev, Dr Vladimir Alekseevich. *USSR.*
Ulbricht, Prof. Heinz. *DDR.*
Ulický, Doc. Ing Ladislav. *CSSR.*
Ullrich, Prof Hans-Jürgen. *DDR.*
Ulmer, Dr. Kevin M. *USA.*
Ultsch, Mr. Mark. *USA.*
Umansky, Prof. Mark Moiseyevich. *USSR.*
Umegaki, Prof. Yoshiharu. *Japan.*
Umeno, Prof. Masataka. *Japan.*
Unan, Doç. Dr Coşkun. *Turkey.*
Unangst, Prof Dietrich. *DDR.*
Underwood, Dr. Dennis John. *USA.*
Ungar, Dr Goran. *Yugoslavia.*
Ungár, Dr Tamás. *Hungary.*
Ungaretti, Prof. Luciano. *Italy.*
Unge, Dr K. Torsten. *Sweden.*
Unger, Prof. Konrad. *DDR.*
Uno, Prof. Ryosei. *Japan.*
Unoki, Dr Hiromi. *Japan.*
Unonius, Dr Lars-Olof. *Finland.*
Untersteller, Mr Eugen. *BRD.*
Unwalla, Dr. Rayomand. *USA.*
Uragami, Prof. Takuyuki. *Japan.*
Urban, Dr Heinz. *BRD.*
Urban, Prof. Dr Knut Wolf. *BRD.*
Urbańczyk, Prof. Dr hab. Grzegorz. *Poland.*
Urbina de Navarro, Dr Caribay. *Venezuela.*
Urdy, Prof. Charles Eugene. *USA.*
Urland, Prof. Dr Werner. *BRD.*
Urusov, Dr Vadim Sergeyevich. *USSR.*
Urusovskaya, Dr Aida Alexandrovna. *USSR.*
Urzhumtsev, Dr Aleksandr Georgievich. *USSR.*
Usher, Dr Brian Francis. *Australia.*
Usov, Dr Oleg Alekseyevich. *USSR.*
Uszyński, Dr Ignacy. *Poland.*
Uttamasil, Dr Lek. *Thailand.*
Uyeda, Dr Natsu. *Japan.*
Uyeda, Prof. Ryozi. *Japan.*

Uyukin, Dr Evgeny Mikhailovich. *USSR.*
Vaccari, Dr Giuseppe. *Italy.*
Vaciago, Prof. Alessandro. *Italy.*
Vadon, Dr Albert. *France.*
Vähäkangas, Mr Jouko Kaarlo. *Finland.*
Vagg, Dr Robert Sylvester. *Australia.*
Vahrenkamp, Prof. Dr Heinrich. *BRD.*
Vahvaselkä, Dr Aino Margit. *Finland.*
Vahvaselkä, Dr Kaarlo Sakari. *Finland.*
Vainshtein, Prof. Boris Konstantinovich. *USSR.*
Vajda, Dr Erzsébet. *Hungary.*
Val'kovskaya, Dr Margarita Ivanovna. *USSR.*
Val'ter, Prof Anton Antonovich. *USSR.*
Valach, Ing Fedor. *CSSR.*
Valarelli, Dr José Vicente. *Brasil.*
Valassiades, Prof. Odysseus. *Greece.*
Valdrè, Prof. Ugo. *Italy.*
Valencia, Dr Iluminado G. *Philippines.*
Valente, Dr. Edward Joseph. *USA.*
Valenzuela Monjarás, Dr Raúl Alejandro. *México.*
Valera, Dr Anibal Abel. *Peru.*
Valeton, Mrs Prof. Dr Ida Walburga Jakobine. *BRD.*
Valigi, Prof. Mario. *Italy.*
Valin, Dr Mariluz. *Spain.*
Valizadeh, Dr. Mohammad-vali. *Iran.*
Valkonen, Prof. Jussi Uolevi. *Finland.*
Valle, Dr Giovanni. *Italy.*
Vallet-Regi, Dr Maria. *Spain.*
Valvoda, Dr Václav. *CSSR.*
Van Alsenoy, Dr Kris. *Belgium.*
Van den Bosch, Dr Adolf. *Belgium.*
Van den Bossche, Dr Guy Ghislain Remy. *Belgium.*
Van Der Heijden, Dr Simon Petrus Nicolaas. *Canada.*
Van der Helm, Prof. Dick. *USA.*
Van Duyne, Dr. Gregory D. *USA.*
Van Dyk, Dr Martha Sophia. *South Africa.*
Van Engen, Dr. Donna. *USA.*
Van Hove, Dr. Michel Andre. *USA.*
Van Landuyt, Prof. Joseph Florent. *Belgium.*
Van Meerssche, Prof. Maurice. *Belgium.*
Van Meervelt, Dr Luc. *Belgium.*
Van Nordstrand, Mr. Robert Alexander. *USA.*
Van Opdenbosch, Dr. Nicole Marie. *USA.*
Van Roey, Dr. Patrick M. *USA.*
Van Rooyen, Prof. Petrus Hendrik. *South Africa.*
Van Schalkwyk, Prof. Theunis G. D.. *South Africa.*
Van Tendeloo, Prof Gustaaf. *Belgium.*
Vana, Prof. Dr Dipl.-Ing. Norbert Johannes. *Austria.*
Vance, Dr. T. Blake Jr. *USA.*
Vandenberg, Dr. Joanna Maria. *USA.*
Vander Sande, Prof. John Bruce. *USA.*
Vanderah, Ms. Terrell Ann. *USA.*
Vanderhoff, Dr. Peggy A. *USA.*
VanDerveer, Dr. Donald G. *USA.*
Vandlen, Dr. Richard L. *USA.*
Vanghelie, Mr Iulian. *Romania.*
Vanhellemont, Mr Jan Hendrik. *Belgium.*
Vanhouteghem, Mr Frankie Marie. *Belgium.*
Vannerberg, Prof. Nils-Gösta. *Sweden.*
Varela Mora, Prof Juan de Dios. *Colombia.*
Varela, Mr José Arana. *Brasil.*
Varga, Dr László. *Hungary.*
Varga, Mr László. *Hungary.*
Vargas, Mr Edgar. *Bolivia.*
Varghese, Dr Joseph Noozhumurry. *Australia.*
Varnek, Dr Aleksandr Aleksandrovich. *USSR.*
Varschavsky, Mr Ari. *Chile.*
Varughese, Dr. Kottayil Iype. *USA.*
Vasanth, Dr K. L. *India.*
Vasić, Magistar Pavle. *Yugoslavia.*
Vasil'ev, Dr Aleksandr Dmitrievich. *USSR.*
Vasil'ev, Dr Yan Vladimirovich. *USSR.*
Vasil'yev, Dr Alexander Borisovich. *USSR.*
Vasilevsky, Dr. Igor V. *USA.*
Vasilyev, Dr Evgeny Konstantinovich. *USSR.*
Vassilev, Dr Ivan. *Bulgaria.*
Veblen, Prof. David Rodli. *USA.*
Veen, van der, Dr Adriaan Hendrik. *Netherlands.*
Veerapandian, Dr B. *India.*
Vega, Lic Daniel Roberto. *Argentina.*

Vega, Prof. Juan. *Peru.*
Vega, Dr Rosario. *Spain.*
Vegas, Dr Angel. *Spain.*
Veintemillas, Mr Sabino. *Spain.*
Veispals, Dr Aris Arvidovich. *USSR.*
Veith, Prof. Dr Michael. *BRD.*
Velasquez, Prof. Jaime. *Peru.*
Veld, In 't, Ir Gerard Adriaan. *Netherlands.*
Velfe, Mr Hans Dieter. *DDR.*
Velikodnyi, Dr Yury Andreyevich. *USSR.*
Velilla, Dr Nicolas. *Spain.*
Velsko, Dr. Stephan Paul. *USA.*
Vendrell, Dr Marius. *Spain.*
Vene, Mrs Nada. *Yugoslavia.*
Venetopoulos, Dr Cleanthis. *Greece.*
Venevtsev, Prof. Yury Nikolayevich. *USSR.*
Venkatasubramanian, Dr K. *India.*
Venkatesan, Prof. Kailasam. *India.*
Venkobarao, Mr H. N. *India.*
Vennik, Prof. Ir Joost. *Belgium.*
Venturello, Porf. Giovanni. *Italy.*
Venudhar, Dr Y. C. *India.*
Venugopalan, Mr P. *India.*
Veprek, PD Stanislav. *Switzerland.*
Vera Calderón, Mrs Gloria. *México.*
Vera, Mr Rafael. *Chile.*
Verbist, Prof. Jacques Jozef. *Belgium.*
Verdini, Prof. Brunella. *Italy.*
Veremeichik, Dr Tamara Fedorovna. *USSR.*
Verkhovskaya, Dr Kira Alexandrovna. *USSR.*
Verlinde, Dr Christophe Louis-Marie Jos. *Belgium.*
Verma, Dr Ajit Ram *India.*
Verö, Dr Balázs. *Hungary.*
Versaci, Dr Raul Antonio. *Argentina.*
Versic, Dr. Ronald J. *USA.*
Versteeve, Dr Abraham Jan. *Netherlands.*
Verwer, Drs Paul. *Netherlands.*
Very, Dr Jean-Michel. *Switzerland.*
Vesselinov, Mr Iliya. *Bulgaria.*
Vettier, Dr Christian. *France.*
Vezie, Miss Deborah L. *USA.*
Vezzalini, Dr Maria Giovanna. *Italy.*
Vgenopoulos, Prof. Andreas. *Greece.*
Via, Mr. Grayson Hall. *USA.*
Vicente, Prof. Jose. *Spain.*
Vickers, Miss Mary Elizabeth. *UK.*
Vickovič, Dr Ivan. *Yugoslavia.*
Victorio-Gervasio, Mrs Visitacion. *Philippines.*
Viczián, Dr István. *Hungary.*
Vidal, Dr Geneviève. *France.*
Vidal, Lic Haydée Marta. *Argentina.*
Vidal, Dr Jean-Pierre. *France.*
Vielhaber, Dr Edmund Antonius. *BRD.*
Vijay-Kumar, Prof. Senadhi. *USA.*
Vijayakar, Mr Suresh Jaywant. *India.*
Vijayan, Dr (Mrs) Kalyani. *India.*
Vijayan, Prof. Mamannamana. *India.*
Vila, Dr Eladio. *Spain.*
Vilkins, Ms Louise Mary. *Australia.*
Vilkov, Prof Lev Vasilyevich. *USSR.*
Villadsen, Mr Jørgen. *Denmark.*
Villafuerte Castrejón, Mrs María Elena. *México.*
Villalpondo, Mr Abelardo. *Bolivia.*
Villarroel, Prof. Hugo Sergio. *Brasil.*
Villars, Dr. P. *Switzerland.*
Villegas, Dr Mario Oscar. *Bolivia.*
Vincent, Dr Beverly Robert. *Switzerland.*
Vinhas, Dr Laercio Antonio. *Brasil.*
Vinokurov, Prof. Vladimir Mikhailovich. *USSR.*
Visapää, Mr Asko Edvard. *Finland.*
Visser, Drs Jan Willem. *Netherlands.*
Visser, Dr Rudolph Joseph Jacobus. *Netherlands.*
Vistin', Dr Leonard Kazimirovich. *USSR.*
Viswamitra, Prof. Mysore Ananthamurthy. *India.*
Viswanathan, Prof. Dr Krishnamoorthy. *BRD.*
Vitali, Dr Francesca. *Italy.*
Vitali, Prof. Gianfranco. *Italy.*
Vitali, Dr. Jacqueline. *USA.*
Vitanov, Dr Todor. *Bulgaria.*
Viterbo, Prof. Davide Lazzaro Marco. *Italy.*

Viturro, Lic Hector Ruben. *Argentina.*
Vizi, Dr Béla. *Hungary.*
Vlad, Dr Serban-Nicolae. *Romania.*
Vladimirov, Dr Sote. *Yugoslavia.*
Vlak, Dr Wim. *Netherlands.*
Vlasov, Dr Vasily Platonovich. *USSR.*
Vlasse, Dr. Marcus. *USA.*
Vo Vong, Dr. *Vietnam.*
Völlenkle, Prof. Dr Horst. *Austria.*
Voet, Prof. Donald Herman. *USA.*
Vogel, Mrs Sonia. *Chile.*
Vogt, Dr Thomas. *France.*
Voigt, Dr Dieter. *DDR.*
Voigt, Dr Rita. *DDR.*
Voitsekhovsky, Dr Vladimir Nikolayevich. *USSR.*
Vol'kenshtein, Prof. Mikhail Vladimirovich. *USSR.*
Vold, Mr. Carl L. *USA.*
Voliotis, Prof. Stavros. *Greece.*
Volk, Dr Tat'yans Rafailovna. *USSR.*
Volkova, Dr Olga Leonidovna. *USSR.*
Vollstädt, Prof Heiner. *DDR.*
Volodin, Dr Alexander Petrovich. *USSR.*
Volodina, Dr Galina Fedorovna. *USSR.*
Voloshin, Dr Aleksei Eduardovich. *USSR.*
Volz, Mr. Karl W. *USA.*
Von Dreele, Dr. Robert Bruce. *USA.*
Vonk, Dr Christ Gijsbertus. *Netherlands.*
Voort, van der, Drs Elisabeth. *Netherlands.*
Vora, Dr Rasiklal Amulakhbhai. *India.*
Vorbach, Dr Angelika Irene. *BRD.*
Vorderwisch, Dr Peter. *BRD.*
Vorma, Prof., Dr Atso Ilmari. *Finland.*
Voronkova, Dr Valentina Ivanovna, *USSR.*
Voronova, Dr Alexandra Alekseyevna. *USSR.*
Vos-Looyenga, Dr Aafje. *Netherlands.*
Voskresenskaya, Dr Inna Evgenyevna. *USSR.*
Voutsas, Dr George Panayiotis. *Greece.*
Voznyak, Dr Dmitry Konstantinovich. *USSR.*
Vozzhennikov, Dr Valery Mikhailovich. *USSR.*
Vrábel, Ing Viktor. *CSSR.*
Vradis, Mr Alexandros. *Greece.*
Vranka, Mr. Robert G. *USA.*
Vratislav, Ing Stanislav. *CSSR.*
Vrielink, Ms Alice. *UK.*
Vries, de, Drs Johan Louis. *Netherlands.*
Vrublevskaya, Dr Zoya Vasilyevna. *USSR.*
Vucht, van, Dr Johannes H. N. *Netherlands.*
Vukasović, Magistar Momčilo. *Yugoslavia.*
Vyas, Mr K. *India.*
Vyas, Dr. Meenakshi N. *USA.*
Vyas, Dr. Nand Kishore. *USA.*
Waal, van de, Benjamin Willem. *Netherlands.*
Wacker, Dr Friedel Klaus. *BRD.*
Wada, Mr Takeo. *Japan.*
Wadewitz, Dr Heinz. *DDR.*
Wadhawan, Dr Vinod Kumar. *India.*
Waerstad, Mr. Kjell Robert. *USA.*
Wäsch, Dr Elke. *DDR.*
Wagendristel, Prof. Dr Alfred Friedrich. *Austria.*
Wagenfeld, Dr Heinrich Karsten. *Australia.*
Wagner, Dr Anton Johan. *Netherlands.*
Wagner, Prof. Christian Nikolaus Johan. *USA.*
Wagner, Dr Ernst-Heinz. *BRD.*
Wagner, Dr Gerald. *DDR.*
Wagner, Dr Günter. *DDR.*
Wagner, Mr Gunther. *DDR.*
Wahlberg, Dr Anders. *Sweden.*
Wahlberg, Dr Olof. *Sweden.*
Wahlström, Dr Ebba B. *Sweden.*
Wahner, Mr Bettina. *DDR.*
Wajsman, Dr Elżbieta. *Poland.*
Wakabayashi, Dr Katsuzo. *Japan.*
Wakino, Dr Kikuo. *Japan.*
Wakoh, Prof. Shinya. *Japan.*
Wal, van der, Dr Robert Jan. *Netherlands.*
Walcher, Dr Herbert. *BRD.*
Waldmann, Dr Hans. *Switzerland.*
Walitzi, Prof. Dr Eva Maria. *Austria.*
Walker, Dr. Christopher Bland. *USA.*
Walker, Mr. Donald L. *USA.*

Walker, Dr Nigel P. C. *BRD.*
Walker, Dr Peter Jonathan. *UK.*
Walkinshaw, Dr Malcolm Douglas. *Switzerland.*
Wallace, Dr. Bonnie Ann. *USA.*
Wallace, Mr. John C. *USA.*
Wallace, Mr. Peter Lindsay. *USA.*
Wallenberg, Dr L. Reine. *Sweden.*
Waller, Prof. Ivar. *Sweden.*
Wallis, Dr John Douglas. *UK.*
Wallis, Dr Julian Mark. *BRD.*
Wallrafen, Dr Franz. *BRD.*
Wallwork, Dr Stephen Collier. *UK.*
Walsoe de Reca, Dr Elizabeth Noemí. *Argentina.*
Walter, Dr Franz. *Austria.*
Walter, Mr. Mark Harrison. *USA.*
Walter, Mr. Norman Macmillan. *USA.*
Waltersson, Dr Kjell. *Sweden.*
Walther, Mr Christa. *DDR.*
Wan, Dr. Cheng. *USA.*
Wang, Prof. Andrew Hwei-Jiu. *USA.*
Wang, Prof. Bi-Cheng. *USA.*
Wang, Dr Da Neng. *Sweden.*
Wang, Prof. Da-cheng. *China.*
Wang, Dr. Frederick E. *USA.*
Wang, Prof. Gen-yuan. *China.*
Wang, Mr Guan-xin. *China.*
Wang, Mr. Gwo-Ching. *USA.*
Wang, Mr Hong. *Canada.*
Wang, Prof. Jia-huai. *China.*
Wang, Mr. Jinnlung. *Taiwan.*
Wang, Prof. Ju-Chun. *Taiwan.*
Wang, Prof. Kui-ren. *China.*
Wang, Dr Naiding. *BRD.*
Wang, Dr. Po-Wen. *USA.*
Wang, Prof. Pu. *China.*
Wang, Dr. Shou-Dao. *USA.*
Wang, Prof. Shun-jin. *China.*
Wang, Prof. Sue-Lein. *Taiwan.*
Wang, Prof. Wen-kui. *China.*
Wang, Mr Xing-xin. *China.*
Wang, Mrs Yao-ping. *China.*
Wang, Prof. Yu. *Taiwan.*
Wang, Prof. Yu-ming. *China.*
Wang, Mr Zhao-zhou. *China.*
Wang, Prof. Zu-tao. *China.*
Wappler, Dr Gert. *DDR.*
Warchoł, Dr Stanisław. *Poland.*
Warczewski, doc. Jerzy. *Poland.*
Ward, Dr. Donald Leslie. *USA.*
Ward, Mr José. *Chile.*
Ward, Dr. Keith Bolen. *USA.*
Ward, Mr Roger Charles Chavannes. *UK.*
Warhanek, Prof. Dr Hans. *Austria.*
Waring, Dr J.R.S. *UK.*
Warminski, Dr Tadeusz Piotr. *Australia.*
Warner, Ms Joanne Kathleen. *UK.*
Warren, Prof. Bertram Eugene. *USA.*
Warren, Dr. Stephen George. *USA.*
Wartchow, Dr Rudolf. *BRD.*
Waser, Prof. Jurg. *USA.*
Waškowska, Dr Alicja. *Poland.*
Watanabe, Dr Akiteru. *Japan.*
Watanabe, Prof. Denjiro. *Japan.*
Watanabe, Dr Mamoru. *Japan.*
Watanabe, Dr Masaru. *Japan.*
Watanabe, Prof. Takashi. *Japan.*
Watanabe, Mr Yasunari. *Japan.*
Watanabe, Prof. Yasuyoshi. *Japan.*
Watari, Dr Fumio. *Japan.*
Watenpaugh, Dr. Keith Donald. *USA.*
Waters, Dr Joyce Mary. *New Zealand.*
Waters, Dr Thomas Neil Morris. *New Zealand.*
Watkin, Dr David John. *UK.*
Watkins, Prof. Steven F. *USA.*
Watson, Dr David Gilfillan. *UK.*
Watson, Mr. Douglas J. *USA.*
Watson, Dr Herman Charles. *UK.*
Watson, Dr Kenneth John. *Australia.*
Watson, Prof. William Harold. *USA.*
Watt, Mr. William. *USA.*

Watters, Dr William Asher. *New Zealand.*
Watterson, Dr. D. Martin. *USA.*
Watts, Dr Bernard Enrico. *UK.*
Watts, Mrs. Ethel Jean. *USA.*
Watts, Mr John Andrew. *Australia.*
Wawra, Mr Herbert. *DDR.*
Wawrzak, Dr Zdzisław. *Poland.*
Waychunas, Dr. Glenn A. *USA.*
Weakley, Dr Timothy John Ruffer. *UK.*
Weakley, Dr Timothy John Ruffer. *USA.*
Weaver, Dr. Larry H. *USA.*
Weaver, Dr Stephen Donald. *New Zealand.*
Webb, Dr. Lawrence Edward. *USA.*
Webb, Prof. Thomas. *USA.*
Weber, Dr Hans Peter. *Switzerland.*
Weber, Dr Hans-Jürgen. *BRD.*
Weber, Prof. Hans-Peter. *Switzerland.*
Weber, Prof. Dr Harald Wolfgang. *Austria.*
Weber, Dr. Irene Teresa. *USA.*
Weber, Prof. Dr Kurt. *BRD.*
Weber, Dr. Lawrence D. *USA.*
Weber, Dr. Patricia C. *USA.*
Webster, Dr Michael. *UK.*
Wechsler, Dr. Barry Andrew. *USA.*
Weckert, Dr Edgar. *BRD.*
Weeks, Dr. Charles M. *USA.*
Weertman, Dr. Julia Randall. *USA.*
Wegener, Dr Joachim Rolf. *BRD.*
Wegener, Dr Wolter. *Belgium.*
Węglowski, Dr Stanisław. *Poland.*
Wehrenberg, Prof. John Patteson. *USA.*
Wei, Dr. Chin Hsuan. *USA.*
Wei, Mr Ming-xiu. *China.*
Wei, Mr Xin-cheng. *China.*
Weigel, Prof. Dominique. *France.*
Weiner, Dr Karl Ludwig. *BRD.*
Weininger, Prof. Marc Sterling. *USA.*
Weintraub, Dr. Herschel J. R. *USA.*
Weis, Dr Josef. *DDR.*
Weis, Dr. William I. *USA.*
Weise, Dr Günter. *DDR.*
Weiser, Mr. Calvin H. *USA).*
Weiss, Prof. Dr Alarich. *BRD.*
Weiss, Prof. Dr Erwin Ludwig. *BRD.*
Weiss, Mr Hans-Georg. *DDR.*
Weiss, Dr Helmut. *DDR.*
Weiss, Prof. Raymond. *France.*
Weiss, Prof. Richard J. *UK.*
Weiss, Dr Zdeněk. *CSSR.*
Weissman, Dr. Larry. *USA.*
Weissmann, Prof. Sigmund. *USA.*
Weiszburg, Dr Tamás. *Hungary.*
Weitzel, Dr Hans. *BRD.*
Welberry, Dr Thomas Richard. *Australia.*
Welch, Dr Alan Jeffrey. *UK.*
Welch, Dr Dorothy Ann. *UK.*
Welsh, Mr. William James. *USA.*
Wendl, Dr Wolfgang. *BRD.*
Wendland, Mr Bettina. *DDR.*
Weng, Prof. ling-pao. *China.*
Wenig, Prof Dr Werner. *BRD.*
Wenk, Prof. Hans-Rudolf. *USA.*
Werk, Dr Margit L. *Switzerland.*
Werner, Dr Inge. *DDR.*
Werner, Prof. Per-Erik. *Sweden.*
West, Dr Anthony Roy. *UK.*
West, Ms. Elizabeth M. *USA.*
West, Dr N.G. *UK.*
West, Mr. Presbury B. *USA.*
Westbrook, Dr. Edwin M. *USA.*
Westdahl, Miss Marianne. *Sweden.*
Wester, Dr. Dennis Wayne. *USA.*
Westman, Mrs Anna-Karin. *Sweden.*
Westman, Dr Sven. *Sweden.*
Weston, Dr. Norman Ernest. *USA.*
Weston, Mr Simon Allan. *BRD.*
Westphalen, Dr John Arthur. *Australia.*
Weulersse, Prof. Philippe. *France.*
Wheatley, Dr Peter Jaffrey. *UK.*
Wheeler, Dr. George L. *USA.*

Whelan, Dr Michael John. *UK.*
Whillans, Dr Francis David. *Australia.*
Whimp, Dr Peter Olaf. *New Zealand.*
Whiston, Dr Clive David. *UK.*
Whitaker. Miss Claire Rosemary. *Australia.*
Whitaker, Dr Alan. *UK.*
White, Dr Allan Henry. *Australia.*
White, Dr David Nathaniel James. *UK.*
White, Dr Janice Larraine. *UK.*
White, Prof. Joe L. *USA.*
White, Dr. John Greville. *USA.*
White, Prof John William. *Australia.*
White, Dr. Peter Sutherland. *USA.*
White, Dr. Steven P. *USA.*
White, Dr Timothy John. *Australia.*
Whitfield, Dr Harold John. *Australia.*
Whitlow, Dr. Marc David. *USA.*
Whitlow, Dr Simon Hugh. *Canada.*
Whittaker, Dr Eric James William. *UK.*
Whittle, Mr. Robert. *USA.*
Whitworth, Dr Robert William. *UK.*
Whuler, Dr Annick. *France.*
Wickersham, Dr. Charles Edward Jr. *USA.*
Wickham, Dr. Donald Guy. *USA.*
Wickman, Prof. Frans Erik. *Sweden.*
Wicks, Dr Frederick John. *Canada.*
Wiebcke, Dr Michael Helmut. *BRD.*
Wiebenga, Prof. Dr Eelco Herman. *France.*
Wieczorek, Dr Michał *Poland.*
Wieder, Dr Thomas. *BRD.*
Wiegers, Dr Gerrit Adriaan. *Netherlands.*
Wielunski, Dr Leszek Stanisław. *Australia.*
Wierenga, Dr Rikkert Klaas. *BRD.*
Wierzchowski, Dr Wojciech Krzysztof. *UK.*
Wieser, Prof Egbert. *DDR.*
Wiess, Dr. George H. *USA.*
Wieteska, Dr Krzysztof. *Poland.*
Wiewióra, Prof. Dr hab. Andrzej. *Poland.*
Wiff, Dr. Donald Ray. *USA.*
Wignall, Dr. George Denis. *USA.*
Wilchinsky, Dr. Zigmond Walter. *USA.*
Wilde, Dr Wolfgang. *DDR.*
Wildenburg, Mr. Jörg Werner. *BRD.*
Wildman, Dr. Gary C. *USA.*
Wildner, Mr Günter. *DDR.*
Wildner, Mag. Manfred. *Austria.*
Wiley, Prof. Don Craig. *USA.*
Wilford, Dr John Bernard. *UK.*
Wilhelm, Dr Eberhard. *BRD.*
Wilke, Prof. Dr Wolfgang. *BRD.*
Wilken, Dr Gerdt. *BRD.*
Wilkes, Dr. Glenn Richard. *USA.*
Wilkins, Dr Stephen William. *Australia.*
Wilkinson, Dr Anthony Joseph. *UK.*
Wilkinson, Dr Clive. *UK.*
Wilkinson, Dr D. *UK.*
Wilkinson, Dr. Michael Kennerly. *USA.*
Will, Prof. Dr Georg. *BRD.*
Willaime, Prof. Christian. *France.*
Willett, Prof. Roger DuWayne. *USA.*
Williams, Dr. Arthur Ray. *USA.*
Williams, Mr Brian Edward. *Australia.*
Williams, Prof. Donald Elmer. *USA.*
Williams, Dr Geoffrey Allan. *Australia.*
Williams, Dr. Graeme John Bramald. *USA.*
Williams, Dr. Ian Duncan. *USA.*
Williams, Dr. Jack Marvin. *USA.*
Williams, Dr. John A. *USA.*
Williams, Prof. Rickey J. *USA.*
Williams, Dr. Roger L. *USA.*
Williams, Dr. Roger M. *USA.*
Williamson, Dr. Michael Mario. *USA.*
Williard, Prof. Paul Gregory. *USA.*
Willig, Prof. Cesar Dorneles. *Brasil.*
Willis, Prof. Bertram Terence Martin. *UK.*
Willson, Prof. Richard C. *USA.*
Wilsdorf, Prof. Heinz Gerhard Friedrich. *USA.*
Wilson, Dr Alan Richard. *Australia.*
Wilson, Prof. Arthur James Cochran. *UK.*
Wilson, Mr. Charles. *USA.*

Wilson, Mr. David Krause. *USA.*
Wilson, Dr. Frank Charles. *USA.*
Wilson, Prof. Herbert Rees. *UK.*
Wilson, Dr. Ian Andrew. *USA.*
Wilson, Dr Keith Sanderson. *UK.*
Wilson, Dr Michael Jeffrey. *UK.*
Wilson, Dr. Robert L. *USA.*
Wilson, Dr. Roxy B. *USA.*
Wilson, Mr. Scott R. *USA.*
Winborne, Ms. Evon L. *USA.*
Winchell, Prof. Horace. *USA.*
Windle, Dr Alan Hardwick. *UK.*
Windsch, Prof. Wolfgang. *DDR.*
Windsor, Dr Colin George. *UK.*
Wing, Prof. Richard M. *USA.*
Wingert, Dr. Lavinia Meinzer. *USA.*
Wink, Prof. Donald J. *USA.*
Winkler, Dr Fritz Karl. *Switzerland.*
Winkler, Dr Michael. *DDR.*
Winnacker, Prof. Dr Albrecht. *BRD.*
Winslow, Mr. Douglas N. *USA.*
Wintenberger, Dr Micheline. *France.*
Winter, Prof. Dr Dipl.-Ing. Hannspeter. *Austria.*
Winter, Dr Marcus John. *UK.*
Winter, Prof Dr Werner. *BRD.*
Winter, Prof. William Thomas. *USA.*
Wintergrass, Mr. David J. *USA.*
Winzer, Dr Achim. *DDR.*
Wismer, Prof. Robert Kingsley. *USA.*
Wit, de, Drs Martin. *Netherlands.*
With, de, Prof. Dr Gijsbertus. *Netherlands.*
Withers, Dr Raymond Leslie. *Australia.*
Witte, Prof. em. Dr Helmut H. W. *BRD.*
Witters, Prof. Robert Dale. *USA.*
Wittke, Prof. Oscar. *Chile.*
Wlodawer, Dr. Alexander. *USA.*
Wobrauschek, Prof. Dr Dipl.-Ing. Peter. *Austria.*
Wögerbauer, Dr Rupert. *BRD.*
Wölfel, Prof. Dr Erich Richard. *BRD.*
Woensdregt, Drs Cornelis Franciscus. *Netherlands.*
Wokulski, Dr Zygmunt. *Poland.*
Wolbaum, Mr Keith Jonathon. *Canada.*
Wołcyrz, Dr Marek. *Poland.*
Wold, Prof. Aaron. *USA.*
Wolf, Dr Dieter. *BRD.*
Wolf, Prof Dieter. *DDR.*
Wolf, Mr Eberhard. *DDR.*
Wolf, Ms Sharon. *Israel.*
Wolff, de, Dr Pieter Maarten. *Netherlands.*
Wolff, Dr. Gunther Arthur. *USA.*
Wolmershäuser, Dr Gotthelf. *BRD.*
Wolska, Dr Irena. *Poland.*
Woltersdorf, Dr Jörg. *DDR.*
Wonacott, Dr Alan John. *UK.*
Wondratschek, Prof. Dr Hans. *BRD.*
Wondre, Mr Friedrich Rudolf. *UK.*
Wong-Ng, Dr. Winnie Kwai-Wah. *USA.*
Wong, Dr. Ricky Ngok-Shun. *USA.*
Wong, Ms. Rosalind Y. *USA.*
Wong, Dr Yau-Shing. *Hong Kong.*
Wongshaiboon, Dr Sajee. *Thailand.*
Wood, Mr Dermott. *UK.*
Wood, Dr. Elizabeth Armstrong. *USA.*
Wood, Dr Gordon H. *Canada.*
Wood, Dr Graeme John. *Australia.*
Wood, Dr Ian George. *UK.*
Wood, Prof. John Stanley. *USA.*
Wood, Mr. Kenneth J. *USA.*
Wood, Dr Raymond Maurice. *UK.*
Woode, Prof. Moses Kwamena. *USA.*
Woods, Dr Geoffrey Steward. *UK.*
Woolfson, Prof. Michael Mark. *UK.*
Woon, Dr. Tai Chin. *USA.*
Wooster, Mr Antony Martin. *UK.*
Worcester, Prof. David L. *USA.*
Workman, Mr. Samuel Thomas. *USA.*
Worthington, Mr. C. R. *USA.*
Worzala, Prof Horst. *DDR.*
Woźniak, Mr Krzysztof. *Poland.*
Wright, Dr. Christine Gerda Schubert. *USA.*

Wright, Dr. Harlan Tonie. *USA.*
Wright, Dr Helen. *UK.*
Wright, Dr John Albert. *UK.*
Wright, Dr John Dalton. *UK.*
Wright, Dr. William Vaughn. *USA.*
Wroblewski, Dr Thomas. *BRD.*
Wu, Mr Bo-mu. *China.*
Wu, Mr Ding-ming. *China.*
Wu, Dr. Ellen L. *USA.*
Wu, Mrs. Jin. *USA.*
Wu, Mr. Jong-Chang. *USA.*
Wu, Mr. Jong-Ching. *USA.*
Wu, Prof. Nan-Chung. *Taiwan.*
Wu, Prof. Qian-zhang. *China.*
Wu, Mr Shen. *China.*
Wu, Prof. Shou-yu. *China.*
Wu, Prof. Shyi-Kaan. *Taiwan.*
Wu, Prof. Xin-tao. *China.*
Wu, Mr Zi-wu. *China.*
Wüest, Mr Hermann. *Switzerland.*
Wuensch, Prof. Bernhardt John. *USA.*
Wunderlich, Dr Hartmut. *BRD.*
Wunderlich, Dr Jeffrey Alfred. *Australia.*
Wurl, Dr Bernd. *DDR.*
Wyart, Prof. Jean. *France.*
Wyckoff, Prof. Harold Winfield. *USA.*
Wyckoff, Prof. Ralph W. G. *USA.*
Xia, Prof. Zong-xiang. *China.*
Xiao, Prof. Xu-gang. *China.*
Xie, Dr Si-shen. *China.*
Xie, Prof. Xian-de. *China.*
XiMen, Mrs Lu-lu. *China.*
Xu, Prof. Chang-fu. *China.*
Xu, Mr Ji-quan. *China.*
Xu, Prof. Jing-yang. *China.*
Xu, Mr Pei-cang. *China.*
Xu, Prof. Shun-sheng. *China.*
Xu, Prof. Xiao-jie. *China.*
Xu, Dr. Zhang-Bao. *USA.*
Xu, Prof. Zheng-yi. *China.*
Xue, Mrs Ji-yue. *China.*
Xue, Prof. Jun-zhi. *China.*
Xue, Prof. Zhi-lin. *China.*
Xuong, Prof. Nguyen-huu. *USA.*
Yadav, Dr Asheshwar. *India.*
Yadav, Dr Tapaswi. *India.*
Yadav, Dr Vijay Singh. *India.*
Yadava, Dr Bishwanath *India.*
Yağbasan, Dr Rahmi. *Turkey.*
Yağcı, Yrd. Doç. Dr Osman. *Turkey.*
Yagi, Prof. Katsumichi. *Japan.*
Yahalom, Prof. Joseph. *Israel.*
Yahaya, Dr Muhamad. *Malaysia.*
Yakel, Dr. Harry Leonard. *USA.*
Yakhontova, Prof. Liya Konstantinovna. *USSR.*
Yakinthos, Prof. John. *Greece.*
Yakovenko Dr Sergey Sergeyevich. *USSR.*
Yakovlev, Dr Viktor Alekseevich. *USSR.*
Yakubovich, Dr Ol'ga Vselodovna. *USSR.*
Yakushi, Mr Kyuya. *Japan.*
Yakushkin, Dr Evgenii Dmitrievich. *USSR.*
Yalçın, Yrd. Doç. Dr Hüseyin. *Turkey.*
Yalkovsky, Dr. Rafael. *USA.*
Yamada, Dr Yukio. *Japan.*
Yamagata, Dr Yuriko. *Japan.*
Yamaguchi, Mr Hiroshi. *Japan.*
Yamaguchi, Dr Toshio. *Japan.*
Yamaguti, Prof. Tasaburo. *Japan.*
Yamamoto, Dr Akiji. *Japan.*
Yamamoto, Dr Atsushi. *Japan.*
Yamamoto, Dr Naoki. *Japan.*
Yamamoto, Mr Shinichi. *Japan.*
Yamamoto, Prof. Takashi. *Japan.*
Yaman, Doç. Dr Y. Macit. *Turkey.*
Yamanaka, Prof. Takamitsu. *Japan.*
Yamane, Dr Takashi. *Japan.*
Yamashita, Mr Shuji. *Japan.*
Yamazaki, Dr Yohtaro. *Japan.*
Yamnova, Dr Nataliya Arkadyevna. *USSR.*
Yan, Mr. Gao. *USA.*

Yan, Mr Qi-wei. *China.*
Yan, Mr Xiaoqian. *Canada.*
Yan, Mr You-wei. *China.*
Yaneva, Dr Svetlana. *Bulgaria.*
Yang, Mr Chuan-zheng. *China.*
Yang, Prof. Daniel Shun-Chung. *Canada.*
Yang, Mr Guang-di. *China.*
Yang, Ms Guang-ming. *China.*
Yang, Prof. Houng-Yi. *Taiwan.*
Yang, Mr Hua-guang. *China.*
Yang, Mr Hua-hui. *China.*
Yang, Prof. Qi-bin. *China.*
Yang, Prof. Qing-chuan. *China.*
Yang, Prof. Tse Chun. *Taiwan.*
Yang, Miss Wei. *USA.*
Yang, Xian-jue. *China.*
Yang, Prof. Yui-Whei. *Taiwan.*
Yang, Mr Zuo-sheng. *China.*
Yanovskii, Dr Aleksandr Il'ich. *USSR.*
Yanovskii, Dr Vladimir Karlovich. *USSR.*
Yanson, Dr Tamara Ivanovna. *USSR.*
Yanulov, Dr Kirill Paskalyevich. *USSR.*
Yanulova, Dr Lyudmila Alekseevna. *USSR.*
Yanusova, Dr Lyudmila Germanovna. *USSR.*
Yao, Mr Jia-xing. *China.*
Yao, Mr Xin-kan. *China.*
Yarmolyuk, Dr Yaroslav Petrovich. *USSR.*
Yase, Dr Kiyoshi. *Japan.*
Yasinskaya, Dr Angelina Andreyevna. *USSR.*
Yasuda, Prof. Yukio. *Japan.*
Yasuoka, Prof. Noritake. *Japan.*
Yates, Dr. John Harry. *USA.*
Yates, Dr Paul Christopher. *UK.*
Yau, Mr. Jen-Kuan. *USA.*
Yazici, Dr. Rahmi M. *USA.*
Ye, Prof. Heng-qiang. *China.*
Ye, Ms Jinhua. *Japan.*
Yeates, Dr. Todd O. *USA.*
Yelon, Dr. William B. *USA.*
Yeon, Prof. Younghee. *Korea.*
Yewdall, Dr Stephen John. *UK.*
Yin, Prof. Soe. *Burma.*
Ylinen, Dr Eero Elias. *Finland.*
Ymén, Dr B. Ingvar. *Sweden.*
Yoda, Dr Osamu. *Japan.*
Yoel, Mr. David William. *USA.*
Yokomori, Dr Yoshinobu. *Japan.*
Yonath, Prof. Ada. *Israel.*
Yong, Mr Swee Kee. *Malaysia.*
Yoon, Prof. Choon Sup. *Korea.*
Yoon, Dr. Hyo Sub. *USA.*
Yoshiasa, Dr Akira. *Japan.*
Yoshida, Dr Kentaro. *Japan.*
Yoshimatsu, Prof. Mitsuru. *Japan.*
Yoshimura, Dr Junichi. *Japan.*
Yoshimura, Mr Yukio. *Japan.*
Yoshioka, Prof. Hide. *Japan.*
Yoshizawa, Mr Masami. *Japan.*
You, Mr Jun-ming. *China.*
You, Prof. Xiao-zeng. *China.*
Young, Mr Brian Raymond. *UK.*
Young, Dr. Frederick Walter Jr. *USA.*
Young, Prof. R. A. *USA.*
Young, Dr. Victor Gordon Jr. *USA.*
Youngs, Dr. Wiley J. *USA.*
Youssef, Dr I. Mourad. *Egypt.*
Yousufzai, Mr Inayatullah Khan. *Pakistan.*
Ysern, Dr. Xavier. *USA.*
Yu, Mr Kai-bei. *China.*
Yu, Prof. Rui-huang. *China.*
Yu, Prof. Shu-Cheng. *Taiwan.*
Yu, Prof. Wei-hai. *China.*
Yu, Prof. Xiu-fen. *China.*
Yufit, Dr Dmitrii Sergeevich. *USSR.*
Yuhasz, Dr. Stacieann. *USA.*
Yurin, Dr Vladimir Alexandrovich. *USSR.*
Yusaf, Dr Mohammad. *Pakistan.*
Yushin, Dr Yury Yakovlevich. *USSR.*
Yushkin, Prof. Nikolay Pavlovich. *USSR.*
Yvon, Prof. Klaus. *Switzerland.*

Zabel, Dr. Volker. *USA.*
Zábráczki, Mr Jósef. *Hungary.*
Zaccai, Dr Giuseppe. *France.*
Zacharias, Dr. David Edward. *USA.*
Zachau-Christiansen, Dr Birgit. *Denmark.*
Zadeh Kashani, Mr Hassan. *Iran.*
Zadorozhnaya, Dr Lyudmila Alexandrovna. *USSR.*
Zagal'skaya, Dr Yudif' Gertsevna. *USSR.*
Zagari, Prof. Adriana. *Italy.*
Zaghloul, Dr Mohamed Zaki. *Egypt.*
Zahn, Dr Andreas. *DDR.*
Zahn, Dr Gernot. *DDR.*
Zahrt, Dr. John D. *USA.*
Zaitsev, Dr Sergey Mikhailovich. *USSR.*
Zaitseva, Dr Mariya Panteleimonovna. *USSR.*
Zajc, Magistar Andrej. *Yugoslavia.*
Žák, Doc. Dr Lubor. *CSSR.*
Žák, Dr Zdirad. *CSSR.*
Zakharchenko, Dr Irina Nikolayevna. *USSR.*
Zakharov, Dr Lev Nikolaevich. *USSR.*
Zakharov, Dr Nikolay Dmitriyevich. *USSR.*
Zakharova, Prof. Mariya Ivanovna. *USSR.*
Zalba, Dr. Patricia Eugenia. *Argentina.*
Zalessky, Dr Andrey Vladimirovich. *USSR.*
Zalkin, Dr. Allan. *USA.*
Zalutsky, Dr Ivan Ilyich. *USSR.*
Zaluzec, Mr. Eugene J. *USA.*
Zaman, Dr (Mrs) Nazma, *Bangladesh.*
Zamorzayev, Dr Alexander Mikhailovich. *USSR.*
Zanazzi, Prof. Pier Francesco. *Italy.*
Zangrando, Dr Ennio. *Italy.*
Zanjanchi, Dr Mohammad Ali. *Iran.*
Zannetti, Prof. Roberto. *Italy.*
Zanotti, Prof. Giuseppe. *Italy.*
Zanotti, Dr Lucio. *Italy.*
Zappia, Prof. Vincenzo. *Italy.*
Zardas, Mr George. *Greece.*
Zarechnyuk, Dr Oleg Safonovich. *USSR.*
Zarka, Dr Albert. *France.*
Zaslow, Prof. Bertram. *USA.*
Zasorin, Dr Evgeny Zotikovich. *USSR.*
Zatout, Mr Mohamed Abdel Meguid. *Egypt.*
Zauhar, Dr. Randy Jay. *USA.*
Zav'yalova, Anna Arkadyevna. *USSR.*
Zavalii, Dr Petro Yuliyanovich. *USSR.*
Zavodnik, Dr Valerii Efimovich. *USSR.*
Zayakina, Dr Nadezhda Viktorovna. *USSR.*
Zbinden, Mr Peter. *Switzerland.*
Zec, Magistar Slavica. *Yugoslavia.*
Zedler, Dr Achim. *BRD.*
Zeedijk, Ir Hendrik Bastiaan. *Netherlands.*
Zefiro, Dr Livio. *Italy.*
Zegenhagen, Dr. Jorg. *USA.*
Zehnder, PD (Mrs) Margareta. *Switzerland.*
Zeigan, Dr Dieter. *DDR.*
Zeilinger, Prof. Dr Anton Wolfgang. *Austria.*
Zeilinger, Prof. Dr Anton. *BRD.*
Zelada, Mr Gabriel. *Chile.*
Zelaya, Mr José Miguel. *Bolivia.*
Zeldes, Mr Nathan. *Israel.*
Zelingher, Mr Naphtaly. *Israel.*
Zelwer, Dr Charles. *France.*
Zemann, Prof. Dr Josef. *Austria.*
Zemke, Mr Klaus Jürgen. *BRD.*
Zemni, Dr Sadok. *Tunisia.*
Zenginoglou, Mr Charalambos. *Greece.*
Zerbi, Prof. Giuseppe. *Italy.*
Zevin, Prof. Lev. *Israel.*
Zeyfang, Dr Rolf Robert. *BRD.*
Zhang, Mr Bu-sheng. *China.*
Zhang, Mr Ci-he. *China.*
Zhang, Prof. Dao-biau. *China.*
Zhang, Mrs Gen-di. *China.*
Zhang, Prof. Guan-ying. *China.*
Zhang, Mr Guang-rong. *China.*
Zhang, Mr Han-hui. *China.*
Zhang, Mr Han-qing. *China.*
Zhang, Prof. Jiang-hong. *China.*
Zhang, Prof. Le-hui. *China.*
Zhang, Mr Li-xin. *China.*

Zhang, Mrs Rong-ying. *China.*
Zhang, Dr. Rongguang. *USA.*
Zhang, Mr Rui-lin. *China.*
Zhang, Prof. Shao-hui. *China.*
Zhang, Mr Shi-wei. *China.*
Zhang, Ms Shu-de. *China.*
Zhang, Miss Xiaohua. *USA.*
Zhang, Mr Yong-mao. *China.*
Zhang, Prof. Yuan-long. *China.*
Zhang, Mr Yue-ming. *China.*
Zhang, Ms Ze-ying. *China.*
Zhang, Prof. Zong. *China.*
Zhao, Dr. Haiching. *USA.*
Zhao, Mr Jing Tai. *Switzerland.*
Zhao, Prof. Qi-yuan. *China.*
Zhdanov, Prof. German Stepanovich. *USSR.*
Zheludev, Prof. Ivan Stepanovich. *USSR.*
Zheludeva, Dr Svetlana Ivanovna. *USSR.*
Zheng, Prof. Pei-ju. *China.*
Zheng, Prof. Qi-tai. *China.*
Zheng, Prof. Zhe. *China.*
Zhidkov, Dr Nikolay Petrovich. *USSR.*
Zhmurova, Dr Zinaida Ivanovna. *USSR.*
Zhong, Mr Na-tian. *China.*
Zhong, Prof. Wei-zhuo. *China.*
Zhou, Prof. Gong-du. *China.*
Zhou, Mr Gui-en. *China.*
Zhou, Mr Kang-jing. *China.*
Zhou, Mr. Rongsheng. *USA.*
Zhou, Prof. Zhong-yuan. *China.*
Zhou, Mr Zong-hua. *China.*
Zhu, Mr He-bao. *China.*
Zhu, Mr Nai-jue. *China.*
Zhu, Mr Zhong-he. *China.*
Zhuang, Mr Hong-hui. *China.*
Zhuang, Mr Jian. *China.*
Zhukhlistov, Dr Anatoliy Pavlovich. *USSR.*
Zhuze, Prof. Vladimir Panteleimonovich. *USSR.*
Zickert, Mr Kurt. *DDR.*
Zidarova, Mrs Bogdana. *Bulgaria.*
Ziegler, Prof. Dr Manfred Ludwig. *BRD.*
Zielińska-Rohozińska, Dr hab. Elżbieta. *Poland.*
Ziemer, Dr Burkhard. *DDR.*
Zigan, Prof. Dr Franz Martinus. *BRD.*
Zikmund, Dr Zdeněk. *CSSR.*
Ziller, Dr. Joseph W. *USA.*
Zimmerman, Lic Rosa. *Argentina.*
Zimmermann, Dr Helmuth W. *BRD.*
Zimmermann, Dr Ulrich. *Switzerland.*
Zinenko, Dr Victor Ivanovich. *USSR.*
Ziolo, Dr. Ronald F. *USA.*
Ziółowska, Dr Blanka. *Poland.*
Ziołowski, Dr Zbigniew. *Poland.*
Zipper, Prof. Dr Peter. *Austria.*
Zlosilo, Mr Mario. *Chile.*
Żmija, Prof. Józef. *Poland.*
Zobel, Dr Dieter. *BRD.*
Zobetz, Dr Erich. *Austria.*
Zocchi, Prof. Marcello. *Italy.*
Zolensky, Dr. Michael Ewing. *USA.*
Zolliker, Dr. Peter. *USA.*
Zolotoi, Dr Aleksandr Borisovich. *USSR.*
Zoltai, Prof. Tibor Z. *USA.*
Zompa, Dr. Leverett Joseph. *USA.*
Zor, Prof. Dr Muhsin. *Turkey.*
Zorky, Prof. Petr Markovich. *USSR.*
Zorn, Dr Gerhard. *BRD.*
Zosi, Prof. Gianfranco Luigi. *Italy.*
Zotov, Mr Nikolay, *Bulgaria.*
Zou, Ms Xiao-dong, *Sweden.*
Zoutberg, Drs Martinus C. *Netherlands.*
Zschach, Dr Siegfried. *DDR.*
Zschokke-Gränacher, Prof. (Mrs) Iris. *Switzerland.*
Zsoldos, Mrs Éva. *Hungary.*
Zsoldos, Dr Lehel. *Hungary.*
Zubenko, Dr Vasily Vasilyevich. *USSR.*
Zubov, Dr Yuriy Alexandrovich. *USSR.*
Zuccola, Mr. Harmon J. *USA.*
Zuk, Dr. William. *USA.*
Zukerman Schpector, Dr Julio, *Brasil.*

Zulehner, Dr Werner. *BRD.*
Zuñiga, Dr Fco. javier. *Spain.*
Zupko, Ms. Sue A. *USA.*
Zussman, Prof. Jack. *UK.*
Zvezdinskaya, Dr Larisa Vsevolodovna. *USSR.*

Zviedre, Dr Irena Ilyinichna. *USSR.*
Zvinchuk, Dr Rostislav Alekseyevich. *USSR.*
Zvirgzede, Dr Yulia Vil'gel'movna. *USSR.*
Zvirgzede, Dr Yurii Al'fredovich. *USSR.*

Zvyagin, Dr Boris Borisovich. *USSR.*
Zwaan, Dr Pieter Cornelis. *Netherlands.*
Zwell, Mr. Leo. *USA.*
Zwick, Prof. Martin. *USA.*

International Union of Crystallography

GENERAL DESCRIPTION

The formation of the Union was discussed at an international meeting held in London in May 1946; it was accepted by the International Council of Scientific Unions on 7 April 1947. Its objectives are to promote international cooperation in crystallography and to contribute to the advancement of all aspects of crystallography, to promote international publication of crystallographic research, to facilitate standardization of methods, units, nomenclature and symbols used in crystallography, and to form a focus for the relations of crystallography to other sciences.

Since its formation, the Union has remained the focal point for international cooperation in crystallography and, at present, 35 countries belong to the Union, through their National Academy, National Research Council or similar body, or through a scientific society or group of such societies. A list of these members (called Adhering Bodies) is given at the end of this brief description of the Union. Recent triennial Congresses, held in association with the business meetings of the Union (General Assemblies), have been attended by 1,200 - 1,800 scientists. The Union also organizes or sponsors many smaller meetings.

The Union has established 16 Commissions, which are concerned with either a principal publishing activity or a major topic or field of concern to crystallographers. The latter group of Commissions organize many international projects concerned with the establishment of internationally acceptable standards or methods of procedure. In addition, they organize specialist meetings or short courses of instruction for young scientists. Nineteen teaching pamphlets have been published, and more are planned. Details of the work of the Commissions are included in the annual reports of the Executive Committee, published in *Acta Crystallographica*, Section A. The Commissions also submit triennial reports to the General Assembly, which elects their members. In addition to the three publishing Commissions (Journals, *Structure Reports* and *International Tables*) there are Commissions on the following topics:

Apparatus	Neutron Diffraction
Biological Macromolecules	Nomenclature
Charge, Spin and Momentum Densities	Powder Diffraction
Computing	Small Molecules
Crystal Growth and Characterization of Materials	Synchrotron Radiation
Data	Teaching
Electron Diffraction	

The current memberships of the Executive Committee and the Commissions, the names of the Union representatives on other bodies, and the memberships of the National Committees for Crystallography (including the addresses of the Secretaries of these Committees) are given in the appendices to the Reports of the General Assembly and International Congresses of Crystallography, published in *Acta Crystallographica, Section A*. These appendices also give the Statutes and By-Laws of the Union and comprehensive reports on the various activities of the Union. Further information concerning the Union may be obtained from the Executive Secretary, International Union of Crystallography, 5 Abbey Square, Chester CH1 2HU, England.

PUBLISHING ACTIVITIES

Less than a year after its formation the Union was publishing its own scientific journal, *Acta Crystallographica*, in an attempt to reassemble the crystallographic work which was scattered through a wide variety of journals. The journal rapidly became established as the major forum for publication of crystallographic research. Today it runs to 4,000 pages a year and is divided into three sections. The crystallographic community owns and controls the journal and its sister journal, *Journal of Applied Crystallography*, which was created in 1968. Through the Union it appoints the journals' editors and the Union is solely responsible for the finances of the journals.

Other major scientific publishing works which were undertaken right at the start of the life of the Union were *Structure Reports*, which gives critical reports on crystal structure determinations published in all the scientific journals, and *International Tables for X-ray Crystallography*, which contains the theory of crystallographic groups (providing the basic reference work for all crystal structure determinations) and the mathematical, physical and chemical tables required for crystallographic work. Further details of these and other publications of the Union are given at the end of this description of the work of the Union.

ADMINISTRATION

The highest authority of the Union is the General Assembly, which meets triennially and consists of delegates appointed by the Adhering Bodies representing the countries belonging to the Union. The affairs of the Union between General Assemblies are conducted by the Executive Committee, comprising a President, a Vice-President, a General Secretary and a Treasurer (at present these two offices are combined), Immediate Past-President and six ordinary members. The Union secretariat and technical editing office for the journals are directed by an Executive Secretary and are located in Chester, England. The Union is incorporated with its legal domicile in Geneva, Switzerland.

ADHERING BODIES

Country	Category*	Adhering Body
Argentina	I	Consejo Nacional de Investigaciones Cientifícas y Técnicas
Australia	III	Australian Academy of Science
Austria	I	Österreichische Akademie der Wissenschaften
Belgium	II	Académie Royale des Sciences, des Lettres et des Beaux-Arts de Belgique
Brazil	III	Conselho Nacional de Desenvolvimento Cientifico e Tecnologico
Bulgaria	I	Bulgarian Academy of Sciences
Canada	III	National Research Council
Chile	I	Comision Nacional de Investigación Cientifica y Tecnologia
China, People's Rep.	IV	Academia Sinica
Czechoslovakia	I	Československá Akademie Věd
Denmark	I	Royal Danish Academy of Sciences and Letters
Egypt, Arab Rep.	I	Academy of Scientific Research and Technology
Finland	I	Suomen Tiedeakatemiain Valtuuskunta
France	IV	Académie des Sciences (Institut de France)
German Dem. Rep.	I	Vereinigung für Kristallographie in der G.G.W. der D.D.R.
Germany, Fed. Rep.	IV	Arbeitsgemeinschaft Kristallographie
Hungary	I	Magyar Tudományos Akadémia
India	II	Indian National Science Academy
Israel	I	Israel Academy of Sciences and Humanities
Italy	III	Consiglio Nazionale delle Ricerche
Japan	IV	Science Council of Japan
Mexico	I	Consejo Nacional de Ciencia y Tecnologia
Netherlands	II	Stichting voor Fundamenteel Onderzoek der Materie met Röntgen- en Elektronenstralen
New Zealand	I	The Royal Society of New Zealand
Norway	I	Det Norske Videnskaps-Akademi
Poland	I	Polska Akademia Nauk
Portugal	I	Sociedade Portuguesa de Fisica
South Africa	II	South African Council for Scientific and Industrial Research
Spain	III	Consejo Superior de Investigaciones Cientifícas
Sweden	II	Kungliga Vetenskapsakademien
Switzerland	II	Schweizerische Gesellschaft für Kristallographie
UK	V	The Royal Society
USA	V	National Academy of Sciences - National Research Council
USSR	V	Akademija Nauk SSSR
Yugoslavia	I	Jugoslavenska Akademija Znanosti i Umjetnosti

* Adherence to the Union is in one of five Categories, with corresponding voting powers and contributions as set out in Statutes 3.6, 5.5 and 9.4.

FINANCES

The Union receives income to finance its general activities in the form of subscriptions from Adhering Bodies, a subvention from UNESCO through ICSU and yields from investments, which have to be maintained as an essential financial backing for the Union's publishing activities. These activities involve an annual turnover of about 2,500,000 Swiss Francs.

COOPERATION WITH OTHER INTERNATIONAL SCIENTIFIC ORGANIZATIONS

The Union maintains close relations with UNESCO and the International Council of Scientific Unions (ICSU). It is represented on several inter-Union or international bodies which may be concerned with very general or very specific tasks requiring concerted international cooperation. The Union is involved with the work of the ICSU Committees on the Teaching of Science, Science and Technology in Developing Countries, Space Research and Data for Science and Technology, and the International Council of Scientific and Technical Information, as well as committees of other Unions or similar organizations. The International Organization for Crystal Growth and the JCPDS International Centre for Diffraction Data are IUCr Scientific Associates and the European Crystallographic Committee, the Asian Crystallographic Association and the American Crystallographic Association are IUCr Regional Associates.

Publications of the International Union of Crystallography

When the Union was established, it was decided that one of its major tasks should be the promotion of international publication of crystallographic research and of works on crystallography. For this purpose the Union has launched a number of publications which, thanks to the cooperation and efforts of scientists from all over the world, have become leading publications in crystallography. They have become indispensable to all workers in this field, and to those working in solid-state physics and solid-state chemistry, as well as in mineralogy.

ACTA CRYSTALLOGRAPHICA

Publishers: Munksgaard International Publishers Ltd., 35 Nørre Søgade, P.O. Box 2148, DK-1016 Copenhagen K, Denmark

Acta Crystallographica is a scientific journal containing original articles in English, French, German and Russian, dealing with new crystal structures, refinements of known structures, new theoretical and experimental methods of structure determination, the theory of diffraction, computing methods, apparatus, and various other related topics. It is published in three Sections: Section A (foundations of crystallography) and Section B (structural science) are published bi-monthly, and Section C (crystal structure communications) is published monthly. All papers are refereed, and they are accepted for publication only when the material is original and when the contents of the paper are of sufficiently high quality. It may be said that much of the best work in the above-mentioned fields appears in *Acta Crystallographica*.

JOURNAL OF APPLIED CRYSTALLOGRAPHY

Publishers: Munksgaard International Publishers Ltd., 35 Nørre Søgade, P.O. Box 2148, DK-1016 Copenhagen K, Denmark

This scientific journal is published in bi-monthly issues. It is concerned with methods, apparatus, problems, and discoveries in applied crystallography. It deals with the application of existing crystallographic techniques to practical problems and with developments in crystallography that have a prospect of future application. It is intended to meet the needs of those scientists who use crystallographic techniques, especially diffraction methods, to study materials and to control their quality. All full length papers and short communications are refereed. The journal also publishes details of computer programs, crystal data, laboratory notes, computer program abstracts, details of new products and details and lists of forthcoming meetings.

STRUCTURE REPORTS

Publishers: Kluwer Academic Publishers, P.O. Box 17, 3300 AA Dordrecht, The Netherlands

Structure Reports provides reports of virtually all determined crystal structures. The reports are arranged in sections on metals, inorganic compounds and organic compounds. Up to and including 1985, each annual volume was published in two parts: A. Metals and Inorganic Sections, and B. Organic Section (including organometallic compounds). The Organic Section has been terminated as from 1985. The reports generally give: name, formula, papers reported, unit cell and space-group data, details of analysis, atomic positions, and detailed description and discussion of the structure (with bond lengths and angles, and usually with illustrations). The structural data are reported so completely that reference to the original paper is not often necessary. There are extensive indexes in each annual volume, and cumulative ten-year indexes published as separate volumes.

It becomes unnecessary to search hundreds of journals. Each volume gives the essence of one year's worldwide literature on crystal structure determinations of all metal, inorganic, organic, and organometallic materials. The series forms an essential bank of information for all university and research laboratories, and science and reference libraries. In universities, students can learn where and how structural information can be readily obtained. In research laboratories, *Structure Reports* is a ready source of basic data and ideas on materials, which can repay its cost many times over.

INTERNATIONAL TABLES FOR CRYSTALLOGRAPHY

Publishers: Kluwer Academic Publishers, P.O. Box 17, 3300 AA Dordrecht, The Netherlands

Volume A, entitled *Space-Group Symmetry*, was first published in 1983. The latest edition contains xvi + 878 pages. The Commission on International Tables of the International Union of Crystallography had, since 1973, been preparing the material for a totally revised and extended edition of the tables of symmetry groups. The results of these years of collaborative effort led to the production of completely new tables on the 17 plane groups and 230 space groups, comprising about 630 printed pages. This work is complemented by a comprehensive introduction in which symmetry is discussed and the theory and use of the tables is described in detail. This volume replaces Volume I of the previous series of *International Tables for X-ray Crystallography*, but Volumes II, III and IV are still available.

Volumes B and C will be published in 1991 or 1992, replacing Volumes II, III and IV. Volume B is entitled *Reciprocal Space* and Volume C is entitled *Mathematical, Physical and Chemical Tables*. Both volumes will be comparable in size to Volume A.

The *Brief Teaching Edition of Volume A* was first published in 1985. The latest edition contains viii + 120 pages. It consists of 24 selected space-group descriptions and those basic text sections of Volume A which are necessary for the understanding of the space-group examples, for the determination of space groups and for the transformations between the various space-group descriptions, together with a subject index.

FIFTY YEARS OF X-RAY DIFFRACTION

Publishers: Kluwer Academic Publishers, P.O. Box 17, 3300 AA Dordrecht, The Netherlands

This commemorative volume was published in 1962 on the occasion of the 50th anniversary of Max von Laue's discovery of the diffraction of X-rays by crystals, and of the first crystal-structure determinations by W. H. and W. L. Bragg.

The book should be enjoyed not only by the crystallographer but also by the non-specialist, and in particular by all those interested in the development of a well-defined and yet widely branched part of science. Among more than 30 contributors to the book, which is edited by P. P. Ewald, are Sir Lawrence Bragg, N. V. Belov, G. Hägg, Linus Pauling, P. Scherrer, A. Westgren, J. Wyart and R. W. G. Wyckoff.

FIFTY YEARS OF ELECTRON DIFFRACTION

Publishers: Kluwer Academic Publishers, P.O. Box 17, 3300 AA Dordrecht, The Netherlands

This volume, edited by P. Goodman and published in 1981, was compiled in recognition of fifty years of achievement by crystallographers and gas diffractionists in the field of electron diffraction.

The volume is divided into three parts. Part I covers the start of electron diffraction up to 1928. Part II deals with the developments from the formulation of Bethe's theory in 1928 to the present day, in the form of personal memoirs by 36 authors. Part III contains six concise reports on the present art of the subject. The book should be of interest to many scientists outside the field of electron diffraction.

EARLY PAPERS ON DIFFRACTION OF X-RAYS BY CRYSTALS

Publishers: Kluwer Academic Publishers, P.O. Box 17, 3300 AA Dordrecht, The Netherlands

This publication consists of two volumes edited by J. M. Bijvoet, W. G. Burgers and G. Hägg. Volume I (xvi + 372 pages) contains extracts from more than 80 of the most important early papers on X-ray crystallography, arranged in such a way as both to form a history of the science and to serve as a teaching aid. The papers span the period 1912-1934. The five chapters are entitled: The discovery of X-ray diffraction by crystals, interpretations and some of the first structure determinations; The reciprocal lattice; The intensity factors of the kinematical theory; The dynamical theory; The *f*-factor continued, extinction, anomalous scattering. Volume II (xix + 484 pages) covers the development of X-ray crystallography in the 'trial and error' period, the (re)birth of the Fourier method and the discovery of the Patterson synthesis. Both volumes contain a large number of diagrams and together form a fascinating collection of papers and excerpts from papers tracing the development of the science and art of X-ray crystallography. They form an ideal companion to the excellent account of the early period given by Professor Ewald and numerous other contributors in *Fifty Years of X-ray Diffraction.*

SYMMETRY ASPECTS OF M. C. ESCHER'S PERIODIC DRAWINGS

Publishers: Kluwer Academic Publishers, P.O. Box 17, 3300 AA Dordrecht, The Netherlands

Like the intricate mosaics of the Alhambra, Escher's periodic patterns form ideal material for the illustration of the principles of symmetry, and especially of the comparatively new aspects of colour symmetry, in addition to their artistic merit.

The book contains 30 drawings printed in black and white, and 12 four-colour reproductions. Many of these are published in this monograph for the first time. The accompanying text has been written by Professor Caroline H. MacGillavry. A second edition of this book was published in 1976. A Japanese edition has also been published.

IUCr/OXFORD UNIVERSITY PRESS BOOK SERIES

Publishers: Oxford University Press, Walton Street, Oxford OX2 6DP, England.

In 1986 the Union established a book series in crystallography jointly with the Oxford University Press, with three divisions:

IUCr Crystallographic Symposia
IUCr Monographs on Crystallography
IUCr Texts on Crystallography

The first three volumes in the *IUCr Crystallographic Symposia* series are:

1. *Patterson and Pattersons: Fifty Years of the Patterson Function* (1987), edited by Jenny P. Glusker, Betty K. Patterson and Miriam Rossi. Proceedings of a symposium with this title, held in Philadelphia, PA, USA, in November 1984.

2. *Molecular Structure: Chemical Reactivity and Biological Activity* (1988), edited by John J. Stezowski, Jin-Ling Huang and Mei-Cheng Shao. Based on papers presented at an international symposium with this title, held in Beijing, Peoples's Republic of China, in September 1986.

3. *Crystallographic Computing 4: Techniques and New Technologies* (1988), edited by N. W. Isaacs and M. R. Taylor. Proceedings of the International School on Crystallographic Computing, held in Adelaide, Australia, in August 1987.

The first volume in *IUCr Texts on Crystallography* is *The Solid State: from Superconductors to Superalloys,* by André Guinier and Rémi Jullien. This is a translation by W.J.Duffin, of the original French, *La Matière à l'état solide - des supraconducteurs aux superalliages.*

The first volumes in *IUCr Monographs on Crystallography,* expected to be published in 1991, are:

1. *Accurate Molecular Structures,* edited by A. Domenicano and I. Hargittai.

2. *Electron Diffraction Techniques,* by J.M. Cowley.

3. *The Rietveld Method,* edited by R.A. Young. Monograph based on the invited papers presented at an international workshop with this title, held in Petten, The Netherlands, in June 1989.

4. *Paul Peter Ewald - A Memorial Volume,* edited by D.W.J. Cruickshank, N. Kato and H.J. Juretschke. Monograph originating from a commemorative microsymposium held at the IUCr Congress in Perth in August 1987.

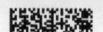